# Lactic Acid Bacteria

Through five editions, and since 1993, *Lactic Acid Bacteria: Microbiological and Functional Aspects* has provided readers with information on how and why fermentation by lactic acid–producing bacteria improves the shelf life, palatability, and nutritive value of perishable foods and also how these microbes have been used as probiotics for decades. Thoroughly updated (with the current lactobacilli taxonomy) and fully revised, with a rearrangement of chapters into four sections, the Sixth Edition covers new findings on health effects, properties, production and stability of LAB as well as regulatory aspects globally. The new edition addresses the technological use of LAB in various fermentations of food, feed, and beverage and their safety considerations. It also includes the rising concept of postbiotics and discusses new targets such as cognitive function, metabolic health, and respiratory health.

## KEY FEATURES

- In 42 chapters, divided into four sections, findings are presented on health effects, properties and stability of LAB as well as production of target-specific LAB
- Covers the revised *'Lactobacillus'* taxonomy
- Addresses novel topics such as postbiotics
- Presents new discoveries related to the mechanisms of actions of lactic acid bacteria
- Covers the benefits of LAB in fermentation of dairy, cereal, meat, vegetable and silage, including non-Western traditional fermented foods from Africa and Asia
- Discusses the less-known role of LAB as food spoilers
- Reports on the health benefits of LAB on humans and animals
- Covers the global regulatory framework related to safety and efficacy

# Lactic Acid Bacteria

## Microbiological and Functional Aspects

## Sixth Edition

Edited by
Gabriel Vinderola, Arthur C. Ouwehand,
Seppo Salminen, and Atte von Wright

CRC Press
Taylor & Francis Group
Boca Raton London New York

CRC Press is an imprint of the
Taylor & Francis Group, an **informa** business

cover: "Community of lactobacilli and yeasts on the surface of sugary kefir grains", credits Melisa Puntillo and CIME (CONICET-UNT).

Sixth edition published 2024
by CRC Press
2385 NW Executive Center Drive, Suite 320, Boca Raton FL 33431

and by CRC Press
4 Park Square, Milton Park, Abingdon, Oxon, OX14 4RN

*CRC Press is an imprint of Taylor & Francis Group, LLC*

© 2024 selection and editorial matter, Gabriel Vinderola, Arthur C. Ouwehand, Seppo Salminen, Atte von Wright; individual chapters, the contributors

Fifth edition published by CRC Press 2019

ISBN: 978-1-032-39938-6 (hbk)
ISBN: 978-1-032-39939-3 (pbk)
ISBN: 978-1-003-35207-5 (ebk)

DOI: 10.1201/9781003352075

Typeset in Times
by Apex CoVantage, LLC

# Contents

## SECTION III   Role of LAB in Health and Disease

# SECTION IV   Safety and Regulation

# Preface

Lactic acid bacteria have throughout history had an important role for human health and well-being as well as food and feed processing but also in food spoilage. There is thus a continued need for a reliable source covering these diverse areas in a multidisciplinary and easy-to-use manner. Our book has been a classic compendium for this broad and changing field.

This is now the 6th edition of *Lactic Acid Bacteria*, which includes the latest developments and updates and expands on the earlier editions of this book. This edition is completely revised and updated to include the most recent developments in lactic acid bacteria research, and it continues in the bold tradition of the earlier versions to provide a comprehensive textbook and also a compendium for both students and teachers as well as other professionals in academia, public service and industry needing information and sources for further development of lactic acid bacteria.

The main driving force for this new revision and update was the recent reclassification of the genus formerly known as *Lactobacillus* and the creation of new genera to accommodate the diverse species. Also, the increased analytical capabilities to study the microbiota have yielded new insights that warranted an update. We have taken advantage of the revision to organize the chapters into four sections: *Taxonomy, Physiology and Molecular Biology*; *Technology and Food Applications*; *Role of LAB in Health and Disease*; and *Safety and Regulation*.

We have progressed considerably compared to previous editions and focus now on the role of lactic acid bacteria and related microbes and their taxonomy, classification and diversity as well as their impact on gut microbiota, animal feeds, food fermentation and many new related areas, including regulatory matters and their less-known role as food spoilers. As the area has grown ever more important and knowledge continues to accumulate, we shall also be prepared to continue updating and expanding the book in the future.

Our book is compiled by four editors with different perspectives on the area varying from technology and microbiology to R&D and human health and regulation. It is intended to supply/support the reader with a compendium of lactic acid bacteria–related science and technology to provide the basis to build science, technology and product development focusing on the area of lactic acid bacteria. The book also provides information on the regulatory frameworks in different geographic regions and areas within which lactic acid bacteria–containing products need to fit. It can be used as a university textbook for multidisciplinary study areas needing knowledge on food microbiology and lactic acid bacteria. Also, scientists in industry will find the book a welcome source for all aspects related to lactic acid bacteria.

# The Editors

**Dr. Gabriel Vinderola** graduated at the Faculty of Chemical Engineering from the National University of Litoral (Santa Fe, Argentina) in 1997 and obtained his Ph.D. in chemistry in 2002 at the same university. He served for a year as a visiting scientist in Canada at the University of Moncton and joined several research teams in Europe (Spain, France, Italy, Germany and Finland). He is presently Principal Researcher of the National Scientific and Technical Research Council (CONICET) and Associate Professor at the Food Technology and Biotechnology Department of his home faculty. He participated in the development of the first commercial cheese-carrying probiotic bacteria from Latin America, released on the market in 1999. In 2011, he was awarded the prize in food technology for young scientists by the National Academy of Natural, Physic and Exact Sciences from Argentina. He has published more than 130 original scientific publications in international refereed journals and book chapters. From 2020, he has served as a member of the board of directors of the International Scientific Association for Probiotics and Prebiotics (ISAPP).

**Dr. Arthur Ouwehand** is Technical Fellow and Digestive Health Platform leader at International Flavors & Fragrances in Kantvik, Finland. He has a research background in both academia and industry. His main interest is in functional foods and dietary supplements, in particular probiotics and prebiotics and their influence on the intestinal microbiota. He is active in the International Life Sciences Institute Europe, the International Dairy Federation, the International Probiotics Association and the International Scientific Association for Probiotics and Prebiotics. Dr. Ouwehand received his M.S. (1992) in cell biology from Wageningen University (the Netherlands) and his Ph.D. (1996) in microbiology from Göteborg University (Sweden). In 1999, he was appointed as an adjunct professor in applied microbiology at the University of Turku (Finland), and he is the author of more than 300 journal articles and book chapters.

**Prof. Seppo Salminen** is professor at the Faculty of Medicine and Director Emeritus of the Functional Foods Forum, University of Turku, Finland. He has been a visiting professor at RMIT University, Melbourne, Australia, and BOKU University, Vienna, Austria. His main research interests are probiotics, prebiotics and intestinal microbiota modulation as well as functional foods and health and regulatory issues in novel foods and health claims. He was a member of the EFSA NDA scientific panel for two terms. He has been active in the International Life Sciences Institute Europe, the International Dairy Federation and the International Scientific Association for Probiotics and Prebiotics (ISAPP, past president and current member of board). He received his M.S. at Washington State University (USA) in 1978, M.Sc. from the University of Helsinki (1979) and Ph.D. from the University of Surrey (United Kingdom) in 1982. He has around 500 journal articles and several textbooks and book chapters, and he has received several international awards, including the IDF-Institute Pasteur Metchnikoff Price, Swiss Price on Modern Nutrition and the Grand Prix du Yoplait.

**Prof. Atte von Wright** graduated from the University of Helsinki (Helsinki, Finland) in 1975 and obtained his Ph.D. in microbiology in 1981 at the University of Sussex, UK. He has a professional background both in industry and academia with research interests ranging from food toxicology to molecular biology and safety aspects of lactic acid bacteria. From 1998 to 2016, Atte von Wright was Professor of Nutritional and Food Biotechnology at the University of Eastern Finland (before 2010 University of Kuopio), Kuopio, Finland, acting today as Professor Emeritus. He has also served in many expert functions of the EU (a member of the Scientific Committee of Animal Nutrition, 1997–2003; a member of the EFSA Scientific Panel on additives and products or substances used in animal feed, 1996–2009; and a member of the EFSA Scientific Panel on Genetically Modified Organisms, 2009–2012).

# Contributors

**Oluwafemi A. Adebo**
Faculty of Science
University of Johannesburg
Gauteng, South Africa

**Leonardo Albarracín**
Laboratory of Immunobiotechnology
Reference Centre for Lactobacilli
    (CERELA-CONICET)
Tucuman, Argentina

**Elisa Ale**
Instituto de Lactología Industrial
    (CONICET-UNL)
Facultad de Ingeniería Química
Universidad Nacional del Litoral
Santa Fe, Argentina

**Silvia Arboleya**
Dairy Research Institute, IPLA-CSIC
Villaviciosa, Spain

**Lars Axelsson**
The Norwegian Food Research Institute Nofima
Ås, Norway

**Kolawole Banwo**
Department of Microbiology
University of Ibadan
Oyo State, Nigeria

**Eveline Bartowsky**
Lallemand Australia The University of Adelaide
School of Agriculture, Food and Wine
Adelaide, Australia

**Shea Beasley**
Sheaps Oy
Ojakkala, Finland

**Marion Bernardeau**
Danisco Animal Nutrition & Health, Innovation
    Division
International Flavors & Fragrances
Leiden, Netherlands
Normandy University
UNICAEN, UNIROUEN
ABTE
Caen, France

**Natalia Biere**
Max Rubner-Institute
Kiel, Germany

**Sylvie Binda**
Rosell Institute for Microbiome and Probiotics
Montreal, Canada

**Ana Binetti**
Instituto de Lactología Industrial
    (CONICET-UNL)
Facultad de Ingeniería Química
Universidad Nacional del Litoral
Santa Fe, Argentina

**Johanna Björkroth**
Department of Food Hygiene and Environmental
    Health
Faculty of Veterinary Medicine
University of Helsinki
Helsinki, Finland

**Juan Borrero**
Sección Departamental de Nutrición y Ciencia de
    los Alimentos
Facultad de Veterinaria
Universidad Complutense de Madrid
Madrid, Spain

**Francesca Bottacini**
Munster Technical University
Biological Sciences
Cork, Ireland

**Patricia Burns**
Instituto de Lactología Industrial
    (CONICET-UNL)
Facultad de Ingeniería Química
Universidad Nacional del Litoral
Santa Fe, Argentina

**Jeremy Burton**
Western University
Lawson Health Research Institute
London (ON), Canada

**Mary I. Butler**
Department of Psychiatry,
Mercy University Hospital
Cork, Ireland

**Patricia Castellano**
Centro de Referencia para Lactobacilos
    (CERELA—CONICET)
San Miguel de Tucumán, Argentina

**Claude P. Champagne**
Research and Development Centre
Agriculture and Agri-Food Canada
Saint-Hyacinthe, QC, Canada

**Ming-Ju Chen**
Department of Animal Science and
    Technology
National Taiwan University
Taipei, Taiwan

**Luis M. Cintas**
Sección Departamental de Nutrición y Ciencia de
    los Alimentos
Facultad de Veterinaria
Universidad Complutense de Madrid
Madrid, Spain

**María Carmen Collado**
Instituto de Agroquímica y Tecnología de los
    Alimentos
Consejo Superior Investigaciones Científicas
    (IATA-CSIC)
Valencia, España

**Fergus W. J. Collins**
APC Microbiome Ireland
Teagasc Food Research Centre
University College Cork
Cork, Co. Cork, Ireland

**Diogo Contente**
Sección Departamental de Nutrición y Ciencia de
    los Alimentos
Facultad de Veterinaria
Universidad Complutense de Madrid
Madrid, Spain

**John F. Cryan**
APC Microbiome Ireland and Department of
    Anatomy and Neuroscience
University College Cork
Cork, Co. Cork, Ireland

**Clara G. de los Reyes-Gavilán**
Dairy Research Institute, IPLA-CSIC
Villaviciosa, Spain

**Paulina Deptula**
University of Copenhagen
Department of Food Science
Frederiksberg, Denmark

**Émilie Desfossés-Foucault**
Biena Inc.
Saint-Hyacinthe, QC, Canada

**Lara Díaz-Formoso**
Sección Departamental de Nutrición y Ciencia de
    los Alimentos
Facultad de Veterinaria
Universidad Complutense de Madrid
Madrid, Spain

**Suzana Dimitrijević**
Department for Biochemical Engineering and
    Biotechnology
Faculty of Technology and Metallurgy
University of Belgrade
Belgrade, Serbia

**Timothy G. Dinan**
APC Microbiome Ireland and Department of
    Psychiatry and Neurobehavioural Science
University College Cork
Cork, Co. Cork, Ireland

**Mariano Elean**
Laboratory of Immunobiotechnology
Reference Centre for Lactobacilli
(CERELA-CONICET)
Tucuman, Argentina.

**Francisco Elegado**
National Institute of Molecular Biology and
Biotechnology
University of the Philippines Los Baños
Laguna, Philippines

**Silvina Fadda**
Centro de Referencia para Lactobacilos
(CERELA—CONICET)
San Miguel de Tucumán, Argentina

**Titilayo D. O. Falade**
International Institute of Tropical Agriculture
Ibadan, Nigeria

**Javier Feito**
Sección Departamental de Nutrición y Ciencia de
los Alimentos
Facultad de Veterinaria
Universidad Complutense de Madrid
Madrid, Spain

**Giovanna E. Felis**
Department of Biotechnology & Verona
University Culture Collection (VUCC-DBT)
University of Verona
Verona, Italy

**Alessandra Fontana**
Università Cattolica del Sacro Cuore
Piacenza, Italy

**Sofia Forssten**
International Flavors & Fragrances
Global Health & Nutrition Sciences
Kantvik, Finland

**Charles M.A.P. Franz**
Max Rubner-Institute
Kiel, Germany

**Hanne Frøkiær**
Institute of Veterinary and Animal Sciences
University of Copenhagen
Frederiksberg C, Denmark

**Michael G. Gänzle**
University of Alberta
Edmonton, AB, Canada

**Nataša Golić**
Institute for Molecular Genetics and Genetic
Engineering
University of Belgrade
Belgrade, Serbia

**Beatriz Gómez-Sala**
Food Biosciences Department
Teagasc Food Research Centre
Moorepark, Fermoy, Co. Cork
APC Microbiome Ireland
University College Cork
Cork, Co. Cork, Ireland

**Corinne Grangette**
Univ. Lille, CNRS UMR9017, Inserm
U1019, CHRU Lille, Institut Pasteur
de Lille, Center for Infection and
Immunity of Lille
Lille, France

**Caroline Gray**
Regulatory Affairs (ANZ)
International Flavors & Fragrances
Auckland, New Zealand

**Miguel Gueimonde**
Instituto de Productos Lácteos de
Asturias
Consejo Superior Investigaciones Científicas
(IPLA-CSIC)
Villaviciosa, Asturias, España

**John Hale**
Blis Technologies
Dunedin, New Zealand

**Xiaomin Han**
Microbiology Lab
China National Center for Food Safety Risk
Assessment (CFSA)
Beijing, China

**Liam Harold**
Blis Technologies
Dunedin, New Zealand

**Pablo E. Hernández**
Sección Departamental de Nutrición y Ciencia de
    los Alimentos
Facultad de Veterinaria
Universidad Complutense de Madrid
Madrid, Spain

**Colin Hill**
APC Microbiome Ireland, Microbiology
    Department
University College Cork
Cork, Co. Cork, Ireland

**Frank Hille**
Max Rubner-Institute
Kiel, Germany

**Noraphat Hwanhlem**
Department of Agricultural Science
Naresuan University
Phitsanulok Thailand

**Fandi Ibrahim**
School of Allied Health Sciences
University of Suffolk
Ipswich, UK

**Geun Eog Ji**
Department of Food & Nutrition
Seoul National University, Seoul, Korea

**Qingru Jiang**
Shenzhen Campus of Sun Yat-sen
    University
Shenzhen, China

**Malee Jirawongsy**
Thailand Food and Drug Administration
Bangkok, Thailand

**Per Johansson**
Department of Food Hygiene and Environmental
    Health
Faculty of Veterinary Medicine
University of Helsinki
Helsinki, Finland

**Haruki Kitazawa**
Laboratory of Animal Food Function
Graduate School of Agricultural Sciences
Tohoku University,
Sendai, Japan.

**Aki Koponen**
Centre for Collaborative Research CCR
Turku School of Economics
University of Turku
Finland

**Andrea Lauková**
Centre of Biosciences of the Slovak Academy of
    Sciences
Institute of animal Physiology
Košiče, Slovakia

**Sarah Lebeer**
University of Antwerp
Department of Bioscience Endineering
Antwerp, Belgium

**Yuan-Kun Lee**
Department of Microbiology & Immunology
National University of Singapore
Singapore

**Irene Lenoir-Wijnkoop**
Department of Pharmaceutical Sciences
Utrecht University
The Netherlands

**Kassem Makki**
University of Gothenburg
Gothenburg, Sweden

**Andrew J. McBain**
Division of Pharmacy and Optometry
School of Health Sciences
Faculty of Biology, Medicine and Health
The University of Manchester, United Kingdom

**Jukka H. Meurman**
Department of Oral and Maxillofacial Diseases
University of Helsinki and Helsinki University
    Hospital
HUS, Finland

**Christian Milani**
Microbial Bioinformatics
Department of Chemistry, Life Sciences and
    Environmental Sustainability
University of Parma
Parma, Italy

**Lorenzo Morelli**
Università Cattolica del Sacro Cuore
Piacenza Italy

**Dagmar Mudroňová**
Department of Microbiology and Immunology
University of Veterinary Medicine and
    Pharmacy
Košice, Slovakia

**Estefanía Muñoz-Atienza**
Sección Departamental de Nutrición y Ciencia de
    los Alimentos
Facultad de Veterinaria
Universidad Complutense de Madrid
Madrid, Spain

**David Obis**
Danone Research & Innovation
Gif sur Yvette, France

**Brian W. O'Mahony**
Department of Psychiatry
Mercy University Hospital
Cork, Ireland

**Catherine A. O'Neill**
Division of Musculoskeletal and Dermatological
    Science,
School of Biological Science
Faculty of Biology, Medicine and health
The University of Manchester, United
    Kingdom

**Myung Soo Park**
Bifido Co. Ltd.
Seoul, Korea

**Jenelle Patterson**
Gowling WLG (Canada) LLP
Ottawa, Canada

**Vasileios Pothakos**
Cargill R&D Centre Europe
Vilvoorde, Belgium

**Jon-Paul Powers**
Gowling WLG (Canada) LLP
Ottawa, Canada

**Scarlett Puebla-Barragan**
Lawson Health Research Institute
The University of Western Ontario
London, ON, Canada

**Melisa Puntillo**
Instituto de Lactología Industrial
    (CONICET-UNL)
Facultad de Ingeniería Química
Universidad Nacional del Litoral
Santa Fe, Argentina

**Mirjana Rajilić-Stojanović**
Department for Biochemical Engineering and
    Biotechnology
Faculty of Technology and Metallurgy
University of Belgrade
Belgrade, Serbia

**Raúl Raya**
Centro de Referencia para Lactobacilos
    (CERELA—CONICET)
San Miguel de Tucumán, Argentina

**Mary C. Rea**
University College Cork
Teagasc Food Research Centre
Cork, Ireland

**Gregor Reid**
Lawson Health Research Institute
The University of Western Ontario
London, ON, Canada

**Minna Rinkinen**
Evidensia
Vaanta, Finland

**R. Paul Ross**
APC Microbiome Ireland
University College Cork
Cork, Co. Cork, Ireland

**Elina Säde**
Department of Food Hygiene and Environmental
    Health
Faculty of Veterinary Medicine
University of Helsinki
Helsinki, Finland

**Ángel Sainz**
Department of Animal Medicine and
    Surgery
Faculty of Veterinary
Complutense University of Madrid
Spain

**Nuria Salazar**
Dairy Research Institute, IPLA-CSIC
Villaviciosa, Spain

**Elisa Salvetti**
Department of Biotechnology & Verona
    University Culture Collection (VUCC-DBT)
University of Verona
Verona, Italy

**Chalat Santivarangkna**
Institute of Nutrition
Mahidol University
Nakhon Pathom, Thailand

**H. Nakibapher Jones Shangpliang**
Department of Microbiology
Sikkim University
Sikkim, India

**Jasvir Singh**
Regulatory & Scientific Affairs (Asia Pacific)
International Flavors & Fragrances
Haryana, India

**Amy B. Smith**
Global Regulatory Affairs, Health
International Flavors & Fragrances
Wilmington, Delaware, USA

**Iva Stamatova**
Department of Oral and Maxillofacial Diseases
University of Helsinki and Helsinki University
    Hospital
HUS, Finland

**Jana Štofilová**
Center of Clinical and Preclinical Research
    MEDIPARK
Faculty of Medicine
Pavol Jozef Šafárik University
Košice, Slovakia

**Ladislav Strojný**
Center of Clinical and Preclinical Research
    MEDIPARK
Faculty of Medicine
Pavol Jozef Šafárik University
Košice, Slovakia

**Ingrid Suryanti Surono**
Food Biotechnology
Bina Nusantara University
Jakarta, Indonesia

**Hania Szajewska**
The Medical University of Warsaw
Department of Paediatrics
Warsaw, Poland

**John Tagg**
Blis Technologies
Dunedin, New Zealand

**Jyoti Prakash Tamang**
School of Life Sciences
Sikkim University
Sikkim, India

**Julie D. Tan**
PhilRootcrops
Visayas State University
Leyte, Philippines

**Yi-Ling Tan**
Program Manager
NYU Langone Health
New York, New York, USA

**Hiroko Tanaka**
Regulatory & Scientific Affairs (Asia Pacific)
International Flavors & Fragrances
Tokyo, Japan

**E-Siong Tee**
TES NutriHealth Strategic Consultancy
Kuala Lumpur, Malaysia

**Lucrecia C. Terán**
Centro de Referencia para Lactobacilos
    (CERELA—CONICET)
San Miguel de Tucumán, Argentina

**Namrata Thapa**
School of Life Sciences
Sikkim University
Sikkim, India

**Daniela Tomei**
Metaregulatoria
São Paulo, Brazil

**Francesca Turroni**
Laboratory of Probiogenomics
Department of Genetics, Biology of
    Microorganisms Anthropology and Evolution
University of Parma
Parma, Italy

**Douwe van Sinderen**
Alimentary Pharmabiotic Centre and Department
    of Biology, BioscienceInstitute
National University of Ireland
Cork, Ireland

**Marco Ventura**
Laboratory of Probiogenomics
Department of Genetics, Biology of Microorganisms
    Anthropology and Evolution
University of Parma
Parma, Italy

**Jean-Paul Vernoux**
Normandy University
UNICAEN, UNIROUEN
ABTE
Caen, France

**Hubert Vidal**
Université Claude Bernard Lyon 1, Charles
    Mérieux Medical School
Oullins, France

**Graciela Vignolo**
Centro de Referencia para Lactobacilos
    (CERELA—CONICET)
San Miguel de Tucumán, Argentina

**Julio Villena**
Laboratory of Immunobiotechnology
Reference Centre for Lactobacilli
    (CERELA-CONICET)
Tucuman, Argentina.

**Le Hoang Vinh**
Regulatory & Scientific Affairs
    (Asia Pacific)
International Flavors & Fragrances
Ho Chi Minh, Vietnam

**Finn K. Vogensen**
University of Copenhagen
Department of Food Science
Frederiksberg, Denmark

**Philip Wescombe**
Lincoln University
Yili Innovation Center Oceania
Lincoln, New Zealand

**Jin Xu**
Microbiology Lab
China National Center for Food Safety Risk
    Assessment (CFSA)
Beijing, China

# Taxonomy, Physiology, and Molecular Biology

# Lactic Acid Bacteria

## *An Introduction to Taxonomy, Physiology, and Molecular Biology*

**1**

Lars Axelsson, Alessandra Fontana, Lorenzo Morelli, and Atte von Wright

## 1.1 BACKGROUND

At the turn of the 20th century, the term "lactic acid bacteria" (LAB) was used to refer to "milk-souring organisms". While similarities between milk-souring organisms and other bacteria producing lactic acid were soon observed, the monograph by Orla-Jensen (1919) formed the basis of the present classification of LAB. The criteria used by Orla-Jensen (cellular morphology, mode of glucose fermentation, temperature ranges of growth, and sugar utilization patterns) are still very important for the classification of LAB, although the advent of more modern taxonomic tools, especially molecular biological methods, has considerably increased the number of LAB genera from the four originally recognized by Orla-Jensen (*Lactobacillus*, *Leuconostoc*, *Pediococcus*, and *Streptococcus*).

LAB have traditionally been associated with food and feed fermentations and are generally considered beneficial microorganisms, some strains even as health-promoting (probiotic) bacteria. However, some genera (*Streptococcus*, *Lactococcus*, *Enterococcus*, *Carnobacterium*) also contain species or strains that are recognized human or animal pathogens. A thorough understanding of the taxonomy, metabolism, and molecular biology of LAB is thus necessary to fully utilize the technological, nutritional, and health-promoting aspects of LAB while avoiding the potential risks.

In the following sections, a brief and concise overview of the present understanding of the taxonomy and physiological, metabolic, and molecular biological characteristics of LAB are presented. The important genera and species are specifically dealt with in the other chapters of this book, and some information will, inevitably, be redundant. However, this general introduction hopefully helps the reader to familiarize with the subject and makes the digestion of the more specific aspects easier.

# 1.2  CURRENT TAXONOMIC POSITION OF LAB

LAB constitutes a group of Gram-positive bacteria united by certain morphological, metabolic, and physiological characteristics. They are nonsporulating, nonrespiring but aerotolerant cocci or rods, which produce lactic acid as one of the main fermentation products of carbohydrates. They lack genuine catalase and are devoid of cytochromes in standard laboratory media. According to the current taxonomic classification, they belong to the phylum *Firmicutes*, class *Bacilli*, and order *Lactobacillales*. Until recently, the different families were considered to include *Aerococcaceae*, *Carnobacteriacea*, *Enterococcaceae*, *Lactobacillaceae*, *Leuconostocaceae*, and *Streptococcaceae* (www.uniprot.org/taxonomy/186826). However, the recent changes of the taxonomy of the Lactobacilli (Zheng et al., 2020), which divide the former genus *Lactobacillus* into 23 different genera, also suggest the fusion of the families *Lactobacillaceae* and *Leuconostocaecae*. In this chapter the former *Lactobacillus* species are referred to with their new genus names. However, the traditional name is also mentioned when considered necessary for the sake of clarity.

The genus *Bifidobacterium* (family *Bifidobacteriaceae*) is historically also considered to belong to the LAB group. However, although *Bifidobacterium* species essentially fit the previous general description, they belong to the phylum *Actinobacteria*, the second major branch of Gram-positive bacteria.

The common LAB genera and their main characteristics are listed in Table 1.1, and more specific taxonomic information is provided in the specific chapters devoted to these LAB groups, including bifidobacteria, in the subsequent sections of this book.

Phylogenetically, LAB can be clustered on the basis of molecular biological criteria, such as rRNA sequencing, and that now exist from whole-genome sequencing, similar and/or more trees have been constructed based on other sets of genes (Zhang et al., 2011; Sun et al., 2015; Zheng et al., 2020). However, the general picture of the phylogenetic position of LAB remains unchanged. The ancestral LAB have apparently been *Bacillus*-like soil organisms, which subsequently have lost several genes and the associated physiological functions while adapting to nutritionally rich ecological niches.

# 1.3  CARBOHYDRATE FERMENTATION PATTERNS

## 1.3.1  Homo- and Heterolactic Fermentation

Because LAB do not possess a functional respiratory system, they have to obtain their energy by substrate-level phosphorylation. With hexoses, there are two basic fermentative pathways. The homofermentative pathway is based on glycolysis (or Embden–Meyerhof–Parnas pathway) and produces virtually only lactic acid (Figure 1.1). Heterofermentative or heterolactic fermentation (also known as pentose phosphoketolase pathway, hexose monophosphate shunt, or 6-phosphogluconate pathway) produces, in addition to lactic acid, significant amounts of $CO_2$ and ethanol or acetate (Figure 1.1, a). As a general rule, pentoses can only be fermented heterofermentatively by entering the pathway as either ribulose-5-phosphate or xylulose-5-phosphate (Kandler, 1983), but then (as is obvious from the fermentation scheme outlined in Figure 1.1b) $CO_2$ is not produced.

Theoretically, homolactic fermentation produces 2 moles of ATP per mole of glucose consumed. In heterolactic fermentation, the corresponding yield is only 1 mole of ATP if the acetyl phosphate formed as an intermediate is reduced to ethanol. However, if acetyl phosphate is converted to acetic acid in the presence of alternative electron acceptors, an extra ATP is formed.

**TABLE 1.1**   Common Genera of LAB and Their Differential Characteristics

| | | | CHARACTERISTICS | | | | | | | |
|---|---|---|---|---|---|---|---|---|---|---|
| FAMILY | GENERA | SHAPE | $CO_2$ FROM GLUCOSE | GROWTH AT 10°C | GROWTH AT 45°C | GROWTH IN 6.5% NaCl | GROWTH IN 18% NaCl | GROWTH AT pH 4.4 | GROWTH AT pH 9.6 | TYPE OF LACTIC ACID |
| Aerococcaceae | *Aerococcus* | Cocci (tetrads) | − | + | − | + | − | − | + | L |
| Carnobacteriaceae | *Carnobacterium* | Rods | Variable[a] | + | − | ND | − | ND | − | L |
| Enterococcaceae | *Enterococcus* | Cocci | − | + | + | + | − | + | + | L |
| | *Tetrageonococcus* | Cocci (tetrads) | − | + | − | + | + | Variable | + | L |
| | *Vagococcus* | Cocci | − | + | − | − | − | + | Variable | ND |
| Lactobacillaceae (including the former Leuconostocaceae) | *Lactobacillus* | Rods | Variable | Variable | Variable | Variable | − | Variable | − | D, L, DL |
| | *Pediococcus* | Cocci (tetrads) | − | Variable | Variable | Variable | − | + | − | L, DL |
| | *Leuconostoc* | Cocci[b] | + | + | − | Variable | − | Variable | − | D |
| | *Fructobacillus* | Rods | + | ± | − | ± | − | Variable | ± | D |
| | *Oenococcus* | Cocci | + | + | − | Variable | − | Variable | − | D |
| | *Weissella Periweissella* | Rods/cocci | + | + | − | Variable | − | Variable | − | D, DL |
| Streptococcaceae | *Lactococcus*[c] | Cocci | − | + | − | − | − | Variable | − | L |
| | *Streptococcus* | Cocci | − | − | Variable | − | − | − | − | L |

*Note:*   ND, not determined.
[a]   When present, $CO_2$ production from glucose by Carnobacteria is weak.
[b]   Some *Weissella* strains and *Periweissella* are rod shaped.
[c]   In older literature, Lactococci are referred to as Group N streptococci.

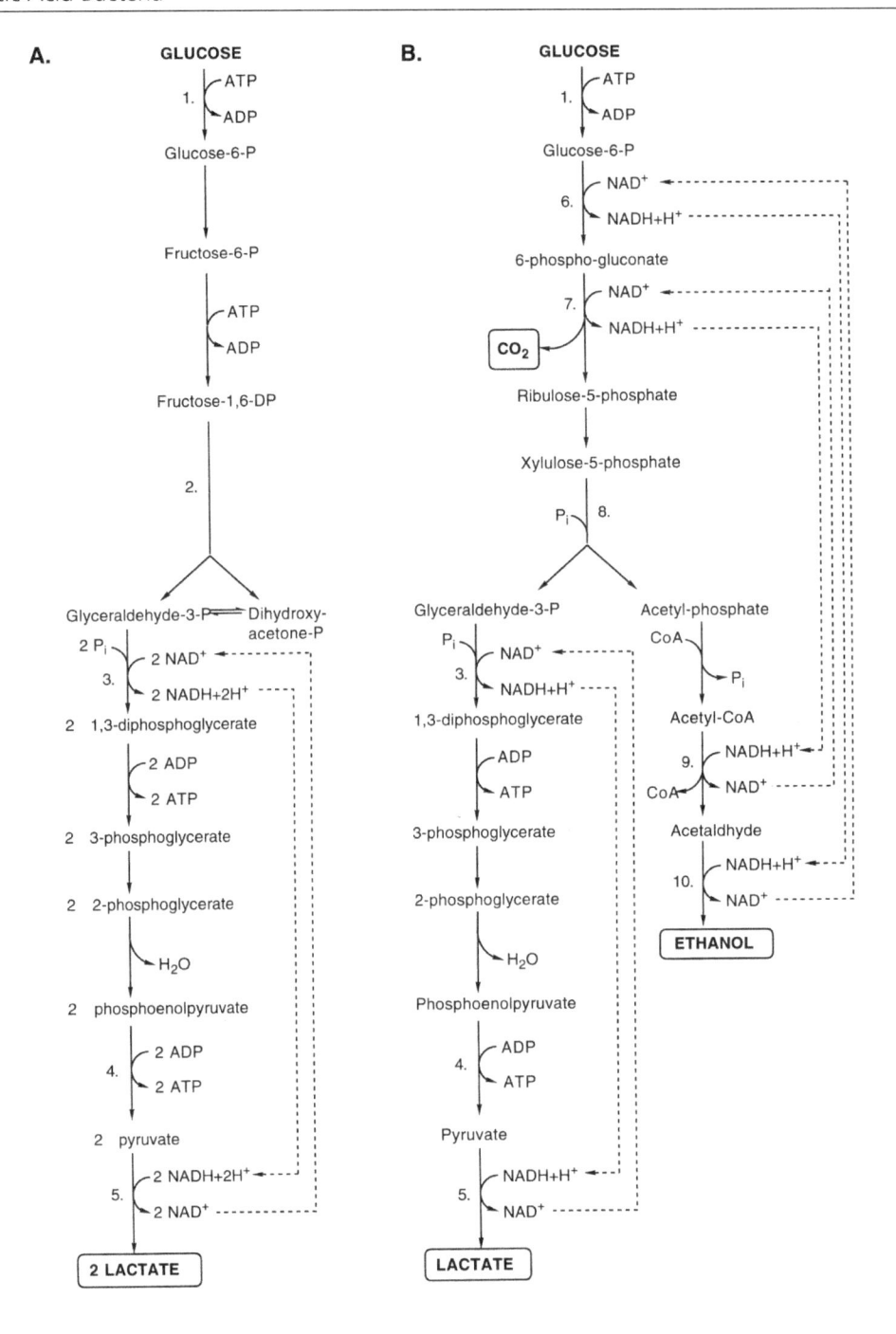

**FIGURE 1.1**   Major fermentation pathways of glucose. (A) Homolactic fermentation (glycolysis, Embden–Meyerhof–Parnas pathway). (B) Heterolactic fermentation (6-phospho-gluconate/phosphoketolase pathway). Selected enzymes are numbered: 1. Glucokinase; 2. Fructose-1,6-diphosphate aldolase; 3. Glyceradehyde-3-phosphate dehydrogenase; 4. Pyruvate kinase; 5. Lactate dehydrogenase; 6. Glucose-6-phosphate dehydrogenase; 7. 6-Phospho-gluconate dehydrogenase; 8. Phosphoketolase; 9. Acetaldehyde dehydrogenase; 10. Alcohol dehydrogenase.

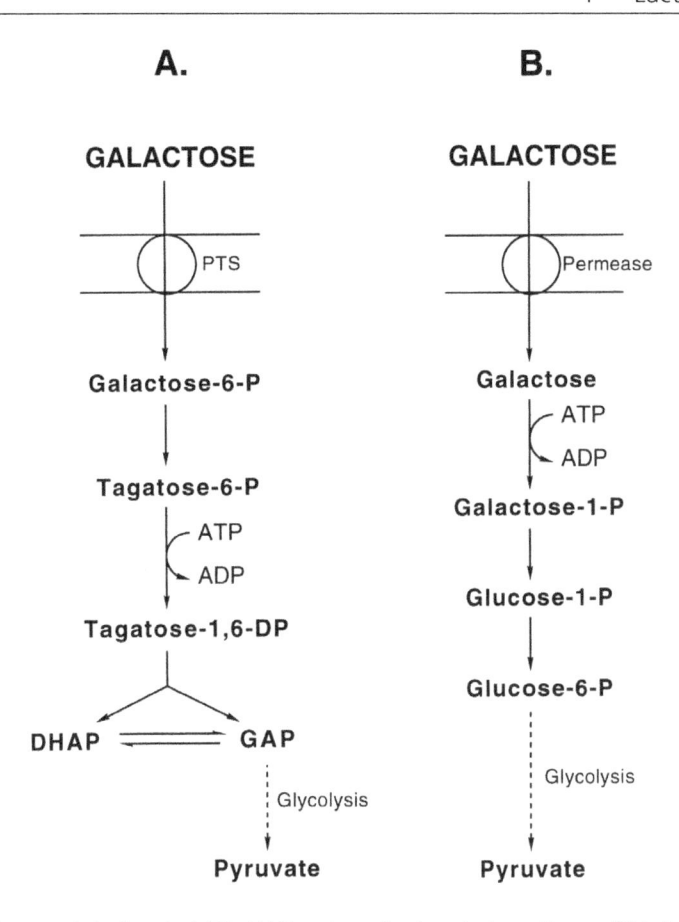

**FIGURE 1.2** Galactose metabolism in LAB. (A) Tagatose-6-phosphate pathway. (B) Leloir pathway.

Hexoses other than glucose (mannose, galactose, fructose) enter the major pathways outlined previously after different isomerization and phosphorylation steps as either glucose-6-phosphate or fructose-6-phosphate. For galactose there are two different pathways, depending on whether it enters the cell as galactose-6-phosphate (via the so-called phospho*enol*pyruvate-dependent phosphotransferase system or PEP:PTS; see Section 1.4.2) as free galactose imported by a specific permease. In the former case the tagatose phosphate pathway is employed (Figure 1.2A) (Bisset and Andersson, 1974) and the so-called Leloir pathway (Figure 1.2B) in the latter (Kandler, 1983).

The fermentation type is an important taxonomic criterion. The genera *Leuconostoc*, *Oenococcus*, *Weissella*, *Periweissella*, and *Fructobacillus* are obligate heterofermentative, as well as the so-called Group III lactobacilli (e.g., *Leviactobacillus brevis*, *Lentilactobacillus buchneri*, *Limosilactobacillus fermentum*, and *Lim. reuteri*). Group I lactobacilli (*Lb. acidophilus*, *Lb. delbrueckii*, *Lb. helveticus*, *Ligilactobacillus salivarius*), on the other hand, are obligate homofermentative (i.e., they cannot metabolize pentoses). Group II or facultatively heterofermentative lactobacilli (*Lacticaseibacillus casei*, *Latilactibacillus curvatus*, *Lat. sakei*, and *Lactiplantibacillus plantarum*), as well as most other LAB, homofermentatively ferment hexoses but may also ferment pentoses through heterolactic fermentation. The division of lactobacilli in three groups (*Thermobacterium*, *Streptobacterium*, and *Betabacterium*) on the basis of their fermentation patterns, as suggested by Orla-Jensen (1919), is still used for pragmatic reasons, although it does not reflect the present phylogeny of the genus.

It should be noted that the outline presented in this chapter represents a generalization, for which there are exceptions, for example, the homolactic fermentation of a pentose (Tanaka et al., 2002) and the homolactic fermentation of fructose by certain obligate heterofermentative lactobacilli (Saier et al., 1996).

## 1.3.2 Fermentation of Disaccharides

Due to the presence of lactose in milk, the metabolism of this disaccharide has been extensively studied, especially in the species used in dairy applications. Lactose can enter the cell either by the means of a specific permease or as lactose phosphate by a lactose-specific PEP:PTS system, and in some cases both systems can coexist (Thompson, 1979). If the transport is permease mediated, lactose is cleaved to glucose and galactose by β-galactosidase, and both of these monosaccharides can subsequently enter the major fermentation pathways. In the case of a PEP:PTS system, another enzyme, phospho-β-d-galactosidase, is needed to split lactose phosphate to glucose and galactose-6-phosphate. Glucose is then processed by the glycolytic pathway, while galactose-6-phosphate enters the tagatose-6-phosphate pathway.

*Lactococcus lactis* typically has a lactose PEP:PTS system, while in many species, such as leuconostocs, *Streptococcus thermophilus*, and thermophilic lactobacilli, the permease system is typical (Hutkins and Morris, 1987; Premi et al., 1972). In *S. thermophilus* and thermophilic lactobacilli, the galactose moiety is not metabolized but excreted into the medium.

Maltose fermentation in LAB has been extensively studied in lactococci, and in this genus the permease system seems to be operational (Sjöberg and Hahn-Hägerdahl, 1989). Another well-known example is *Fructilactobacillus sanfanciscensis*, a lactobacillus found in sourdoughs. This bacterium converts maltose to glucose-1-phosphate and glucose by maltose phosphorylase. Glucose-1-phosphate is used by the bacterium as an energy source, while glucose is excreted into the medium to be used by a yeast, *Candida milleri* (Stolz et al., 1993).

Sucrose fermentation is generally based on the permease system and the action of sucrose hydrolase, which splits the disaccharide to glucose and fructose. In lactococci, a sucrose-specific PEP:PTS system accompanied by sucrose-6-phosphate hydrolase also appears to be functional, producing glucose-6-phosphate and fructose (Thompson and Chassy, 1981). Sucrose may also have a role in exopolysaccharide formation in certain LAB. In *Leuconostoc mesenteroides*, sucrose is cleaved by a cell wall–associated enzyme, dextransucrase, and the glucose moiety is used for dextran synthesis, while fructose is fermented in the usual manner (Cerning, 1990).

## 1.3.3 Alternative Fates of Pyruvate

Pyruvate has a central role in the fermentation pathways, usually acting as an electron acceptor to form lactic acid and thus help to maintain the oxidation-reduction balance in the cell. However, depending on the LAB strain and specific growth conditions, alternative pyruvate utilizing pathways exist. They are summarized in Figure 1.3.

The formation of diacetyl and acetoin/2,3-butanediol occurs specially in certain dairy lactococci in the presence of a pyruvate surplus as a result of the breakdown of the citrate present in the milk to oxaloacetate and acetate. Oxaloacetate is subsequently decarboxylated to pyruvate by oxaloacetate carboxylase. The two alternative pathways leading from pyruvate are outlined in Figure 1.3. The pathways have been reviewed by Hugenholz (1993).

In anaerobic conditions and under substrate limitation, pyruvate can be metabolized to formic acid acetyl-CoA in a reaction catalyzed by pyruvate formate lyase (Thomas et al., 1979; Kandler, 1983). The acetytl-CoA can act as an electron acceptor to yield ethanol or be used for substrate-level phosphorylation and ATP synthesis giving acetate as the end product.

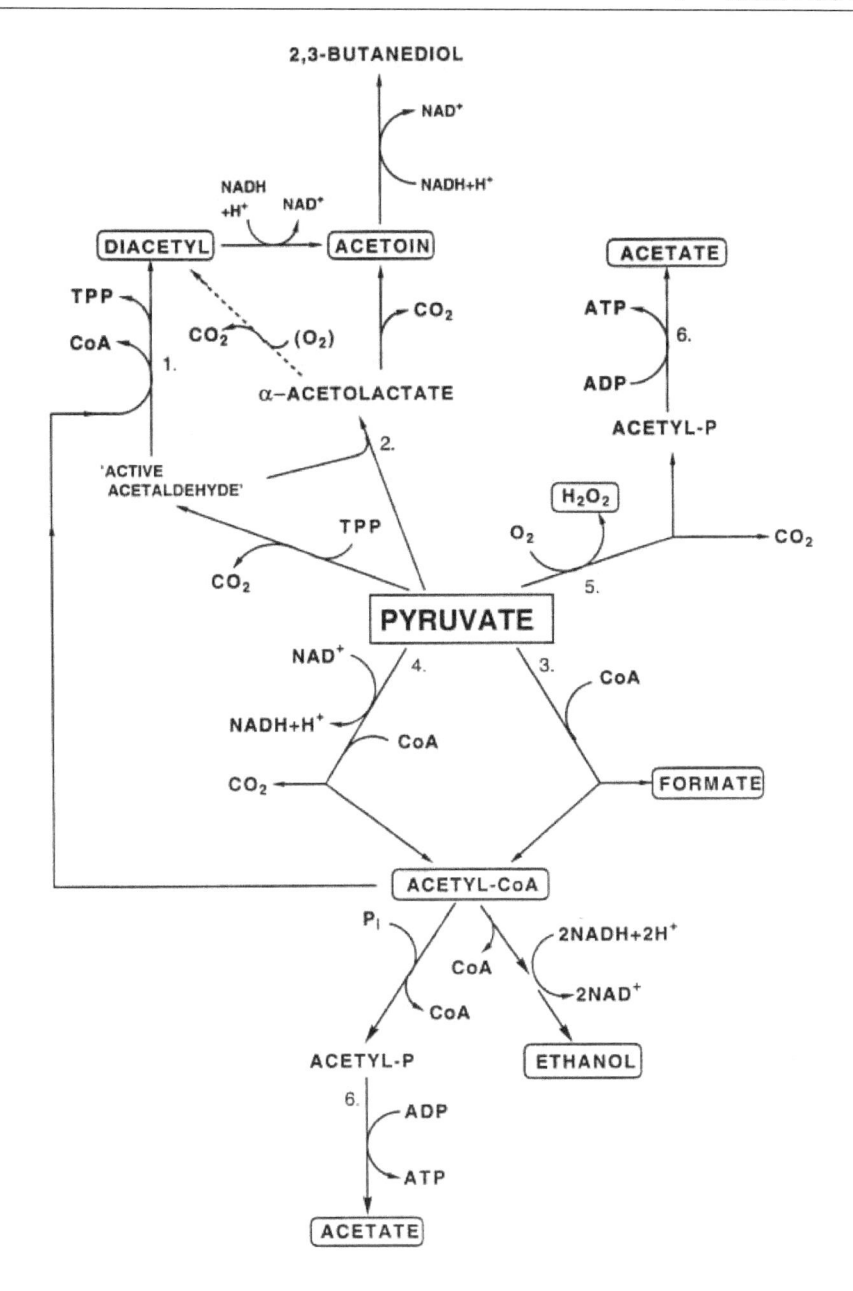

**FIGURE 1.3**  Pathways for the alternative fates of pyruvate. Dashed arrow denotes a nonenzymatic reaction. Important metabolites and end products are framed. Selected enzymatic reactions are numbered: 1. Diacetyl synthase; 2. Acetolactate synthase; 3. Pyruvate–formate lyase; 4. Pyruvate dehydrogenase; 5. Pyruvate oxidase; 6. Acetate kinase.

In the presence of oxygen, pyruvate can be converted to acetate by the action of pyruvate oxidase, $H_2O_2$ being also formed in the reaction. This pathway may lead to a significant aerobic formation of acetate (Sedewitz et al., 1984).

Especially in the lactococci, pyruvate can be oxidized by pyruvate dehydrogenase to acetyl-CoA, while $CO_2$ and NADH are formed (Smart and Thomas, 1987; Cogan et al., 1989). The reaction might provide acetyl-CoA for catabolic functions. The usual end product, however, is acetate.

# 1.3.4 Alternative Electron Acceptors

Besides pyruvate or acetyl-CoA, other electron acceptors such as oxygen or organic compounds can be utilized and lead to profound effects on the energetics and growth rate of LAB.

An example of the stimulatory role of oxygen is the shift from ethanol to acetyl phosphate and finally acetic acid, providing an extra ATP. This reaction, requiring an active NADH oxidase, is apparently very common in leuconostocs and other heterofermentative LAB (Lucey and Condon, 1986; Borch and Molin, 1989). Indeed, some heterofermentatives have a reduced ability to anaerobic growth on glucose due to the lack of acetaldehyde dehydrogenase required for the ethanol branch of the heterofermentative pathway (Eltz and Vandemark, 1960; Stamer and Stoyla, 1967).

In homofermentative LAB, the action of NADH oxidases in aerated cultures may lead to a surplus of pyruvate, which can be shifted to the diacetyl/acetoin pathway (see Section 1.3.3.) (Borch and Molin, 1989).

Polyols are a class of substrates that can often be fermented only in the presence of oxygen. Examples include glycerol fermentation by *Pediococcus pentosaceus* (Dobrogosz and Stone, 1962) and mannitol fermentation of *Lacticaseibacillus casei* (former *Lactobacillus casei*) (Brown and Vandemark, 1968).

Finally, some LAB (enterococci, leuconostocs, and lactococci) are actually able to oxidative phosphorylation when provided with external heme or hemoglobin (Ritchey and Seeley, 1976). The genomic sequence of *L. lactis* revealed the presence of genes required for the synthesis of cytochrome oxidase (Bolotin et al., 2001), leading to a renewed interest in this phenomenon (Gaudu et al., 2002; Koebmann et al., 2008).

Besides oxygen, organic compounds such as oxaloacetate and glycerol can act as electron acceptors in heterofermentative LAB, especially in cofermentation with glucose. Oxaloacetate can be reduced to malate and finally to succinic acid via fumarate (Roos et al., 2000). This pathway is apparently common in LAB isolated from plant material (Kaneuchi et al., 1988). Ramos and Santos (1996) described another type of cofermentation of citrate producing 2,3-butanediol in addition to lactic and acetic acid.

Heterofermentative LAB can also use glycerol as an electron acceptor by dehydrating it first to 3-hydroxypropionaldehyde, which is then reduced to 1,3 propanediol, which is the main fermentation product in addition to lactate, acetate and $CO_2$ (Schütz and Radler, 1984; Talarico and Dobrogosz, 1990). The intermediate, 3-hydroxypropionaldehyde, is a potent antimicrobial substance known as reuterin (Axelsson et al., 1989; Talarico and Dobrogosz, 1989).

## 1.3.4.1 Fructophilic LAB and the Special Case of Fructose Fermentation

Fructophilic LAB are a special group of heterofermentative lactic acid bacteria, which have a special preference for fructose but grow very poorly on glucose (Endo et al., 2009). Fructose fermentation represents an example of the same compound acting both as an electron donor and electron acceptor (Eltz and Vandemark, 1960). The overall balance of the fermentation is:

$$3 \text{ fructose} + 2 \text{ ADP} + 2P_i \longrightarrow 1 \text{ lactate} + 1 \text{ acetate} + 1 \text{ } CO_2 + 2 \text{ mannitol} + 2 \text{ ATP}$$

Taxonomically fructophilic bacteria form two clusters. Four species formerly included in the genus *Leuconostoc* have been renamed as *Fructobacillus* (*F. tropaeoli, F. pseudoficuineus, F. ficuineus, F. fructosus, F. durionis*), while three physiologically fructophilic species of the *Lactobacillus* group (*Apilactobacillus apinorum, A. kunkeei, Fructilactobacillus florum*) cluster near *Lentilactobacillus buchneri* (Endo and Okada, 2008; Endo et al., 2012, 2018). The natural niches of fructophilic LAB include fructose-rich niches (Endo, 2012), particularly honeybees and beehives (Endo and Salminen, 2013).

# 1.4 BIOENERGETICS, SOLUTE TRANSPORT, AND RELATED PHENOMENA

The ATP synthesis in microorganisms is intimately linked with the generation of proton motive force (PMF) across the cell membrane. In respiratory conditions PMF is generated by both the proton gradient–related electrical potential ($\Delta\psi$) and the pH gradient ($\Delta$pH) between the alkaline cytoplasm and the extracellular environment. The influx of protons by PMF can be used to generate ATP by the $H^+$-translocating ATP-synthase.

As fermentative organisms, LAB do not normally possess the electron transport chain, but they have an enzyme system with a reverse activity $H^+$-dependent ATPase. This system maintains the intracellular neutral pH by pumping protons out of the cell in an ATP-consuming reaction. This is essential, since many LAB do not tolerate intracellular pH below 5 (Konings et al., 1989). Since this system consumes ATP, there exists an alternative way to create PMF, so-called energy recycling.

## 1.4.1 Energy Recycling and PMF

Efflux of fermentation end products (lactate, acetate, etc.) can maintain PMF without consuming ATP if the efflux is associated with proton symport (Michels et al., 1979). This has been shown to be the case in *Lc. lactis* at pH > 6.3 and in low external lactate concentrations, resulting in a net charge leaving the cell (ten Brink et al., 1983). In an early stage of growth in a batch culture or in in ecological conditions, where the external lactate is rapidly diluted away, the energetic advantage could be considerable. Another well-described energy-recycling system is the acetate efflux in *Lactiplantibacillus plantarum* (Tseng et al., 1991).

Malolactic fermentation (MLF), or the conversion of malate to lactic acid and $CO_2$ by the action of L-malate: $NAD^+$ decarboxylase (malolactic enzyme), is another example of the generation of PMF (Kunkee, 1991). This system can act as an indirect proton pump by generating a protonated product (lactate) from malate (Poolman et al., 1991; Salema et al., 1996). The PMF generated might be high enough to reverse the function of $H^+$-dependent ATPase and generate ATP, but generally it is considered mainly an energy conservation process.

Citrate metabolism and amino acid decarboxylation may also have similar advantages by the creation of more electroneutral end products from negatively charged compounds (Poolman, 1993).

## 1.4.2 Solute Transport

The three basic systems that are involved in the solute transport in microorganisms are PMF-driven symport, primary transport, and precursor-product antiport. In addition there are group translocation mechanisms, such as the phoshoenolpyruvate-dependent sugar phosphotransferase system.

PMF-driven solute transport is based on specific permeases or carriers, which translocate the solute across the membrane in symport with a proton. This system is particularly relevant for the transport of amino acids in lactococci (Konings et al., 1989; Smid et al., 1989).

The primary transport is associated with the ATP-binding cassette transporters (ABC transporters). The oligo- and dipeptide transport systems of LAB (see Section 1.5) as well as several excretion mechanisms are typical examples of primary transporters (Poolman, 2002).

The arginine deiminase pathway to generate ATP from imported arginine via generation of carbamoyl phosphate and ornithine export is an example of a precursor–product antiport. In lactococci, this exchange reaction has a 1:1 stoichiometry and is mediated by a single membrane-associated arginine/ornithine antiporter (Poolman et al., 1987).

## 1.4.3  Phospho*enol*pyruvate: Sugar Phosphotransferase System and Catabolic Repression

The phosphoenolpyruvate: sugar phosphotransferase system (PEP-PTS) translocates sugars across the membrane and the concentration gradient with a simultaneous phosphorylation. In the process the phosphoryl group of the high-energy compound PEP is transferred by a chain of proteins to a membrane-located enzyme (EIIBC$^S$), which mediates the transport and phosphorylation of the sugar (for a general review of the system, see Postma et al., 1993).

The general features of the system are outlined in Figure 1.4. The proteins EI and Hpr can be shared by several PTS systems, while the rest of the components are sugar specific. In LAB the PEP:PTS is

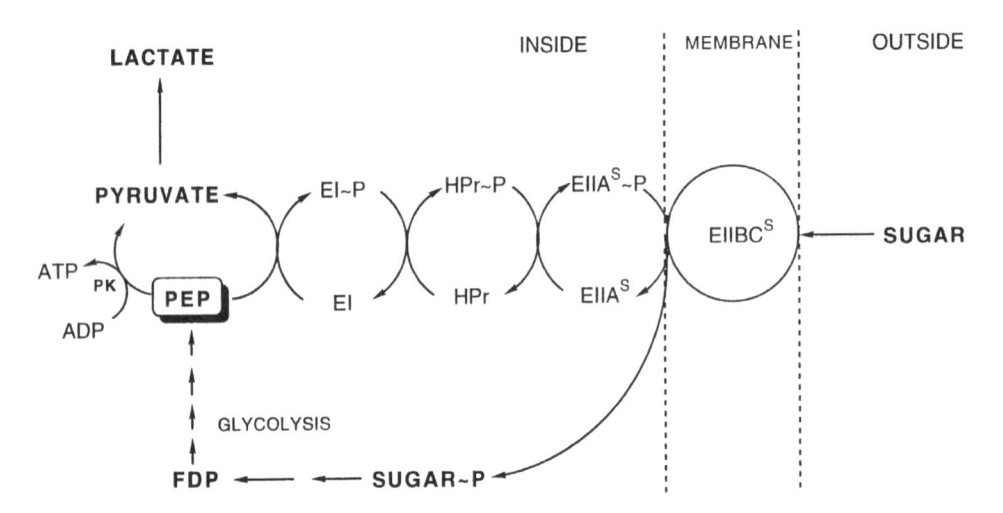

**FIGURE 1.4**    Sugar transport mediated by PEP:PTS system and relation to glycolysis. PK, pyruvate kinase. See text for details.

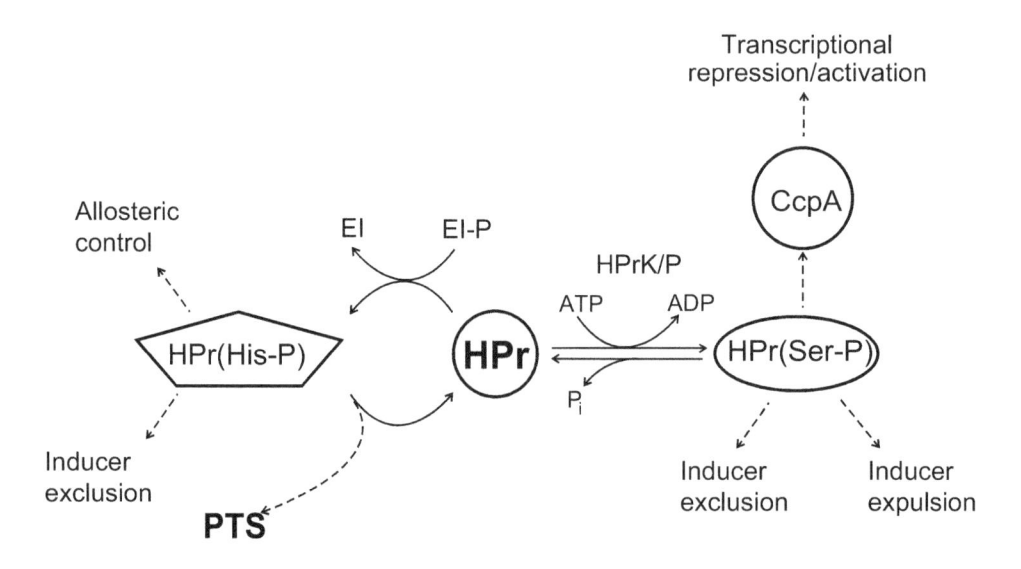

**FIGURE 1.5**    Schematic representation of the central role of HPr in global regulation of carbon transport and metabolism. See text for further details. PTS, phospho*enol*pyruvate:sugar phosphotransferase system.

generally associated with the glycolytic pathway, fructose diphosphate (FDP), and inorganic phosphate acting as a respective activator or repressor of pyruvate kinase (Thompson, 1987). PEP-PTS systems have also been detected in heterofermentative LAB (Saier et al., 1996).

The phosphorylation status of the Hpr protein plays also a central role in catabolic repression (CCR), which regulates the catabolism of other sugars than glucose. The system is based on *trans*-acting catabolite control protein A (CcpA), which acts by binding to a *cis*-acting catabolite responsive element (*cre*) associated with the promoter of the pertinent gene or operon (see the review of Fujita, 2009).

In Gram-positive bacteria the phosphorylation at Ser-46 (instead of the standard His-15) leads to the formation of a complex with CcpA and binding of this complex to the *cre* element repressing the catabolic functions. The phosphorylation status of Hpr in turn is regulated by the level of FDP. In the presence of glucose and glucolysis, the FDP levels are high and stimulate the activity of Hpr-kinase/phosphatase, the enzyme that catalyzes the phosphorylation of Ser-46. An outline of the CCR-system and its activities and the links with the PEP-PTS system is presented in Figure 1.5.

# 1.5 NITROGEN METABOLISM AND PROTEOLYTIC SYSTEM

Many LAB appear to have only a very limited capacity to synthesize amino acids from inorganic nitrogen sources and thus depend on preformed amino acids present in the growth medium. Especially the dairy LAB rely on the proteolytic degradation of external proteins and on the uptake of the resulting peptides and amino acids. The proteolytic machinery has been extensively studied in dairy lactococci, and the system has been thoroughly reviewed by several authors (Kunji et al., 1996; Savijoki et al., 2006; Liu et al., 2010).

Caseinolytic activity is based on the cell wall–associated subtilisin-like serine proteinase (PrtP). The enzyme degrades casein to oligopeptides of variable sizes. Large peptides (4–18 amino acids) are transported by an oligopeptide transport system (Opp), an ABC transporter, while di- and tripeptide transport systems exist for smaller peptides. Two di- and tripeptide transport systems have been characterized, DtpT and Dpp. DtpT is a PMF-driven system, while the Dpp system is an ABC transporter like OPP. Inside the cell, the peptides are further degraded into amino acids by intracellular peptidases. The overall schema of the proteolytic system is shown in Figure 1.6.

The distribution of the different components of the proteolytic system in different lactococcal strains and in other LAB has been reviewed by Liu et al. (2010).

# 1.6 MOLECULAR BIOLOGY OF LAB

## 1.6.1 Core and Pan Genomes of LAB

Lactic acid bacteria are a wide group of prokaryotes characterized by the production of lactic acid as final fermentation catabolite. They are able to reproduce in varied environments, including host-related niches, such as the gastrointestinal tracts (GITs) of humans and animals (Lee et al., 2017; Son et al., 2020; Wang et al., 2020; Zhou et al., 2020) or the urogenital tract (Van Der Veer et al., 2019; Pan et al., 2020), as well as food-related niches, such as dairy (Schmid et al., 2018; Koryszewska-Bagińska et al., 2019; Quilodrán-Vega et al., 2020) or fermented foods (Eisenbach et al., 2018, 2019; Lopez et al., 2022). LAB includes, at its core, five genera: *Lactobacillus*, *Lactococcus*, *Leuconostoc*, *Pediococcus*, and a few

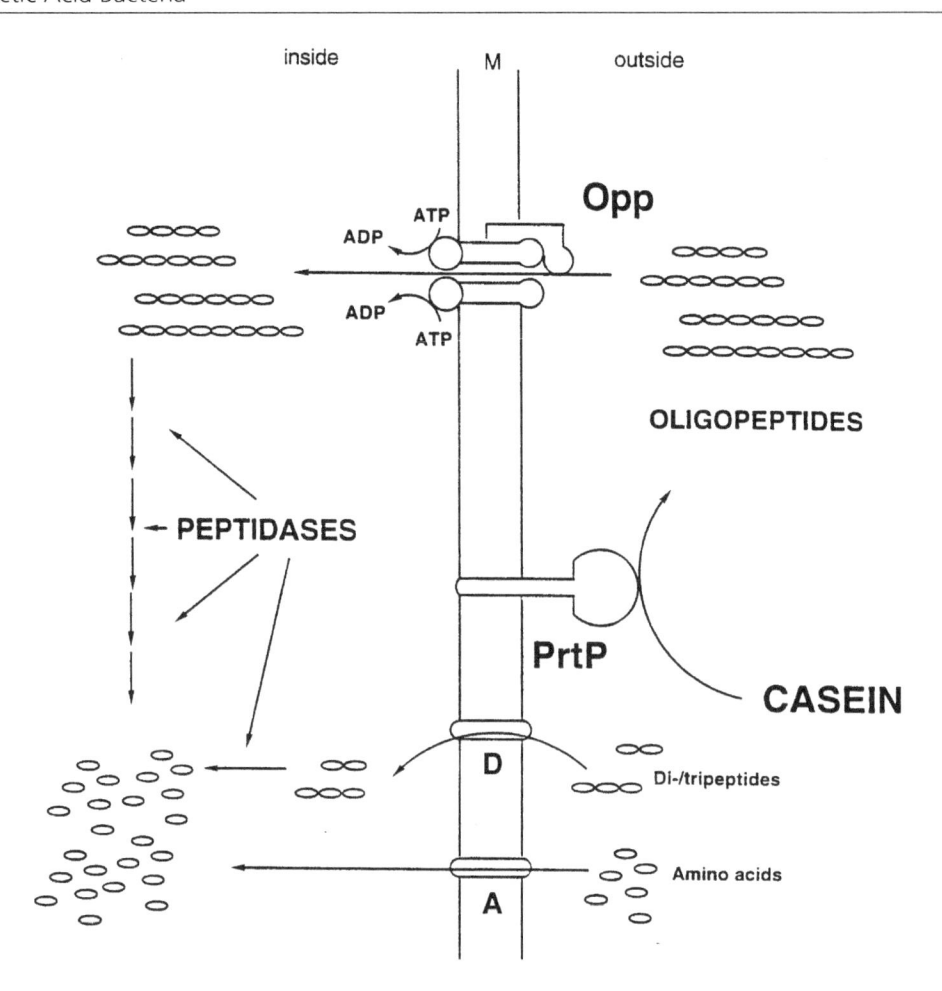

**FIGURE 1.6**  Model of *Lc. lactis* proteolytic pathway. Also included is transport of di- and tripeptides and free amino acids, although their role in growth in milk is limited. PrtP, membrane-anchored proteinase; Opp, oligopeptide transport system; D, di-/tripeptide transport system(s) (the di-/tri-peptide transport system shown is of the PMF-driven type. Note that an Opp-like, ABC transporter-type system for di- and tripeptides also exists); A, amino acid transport system(s); M, cytoplasmic membrane.

species of *Streptococcus* (Goel et al., 2020). *Lactobacillus* represents the major genus of LAB, including species with genome characteristics that vary considerably. Recently, an in-depth comparative genomic analysis has indeed split this genus into 25 genera, based on the large variety of genetic characteristics (Zheng et al., 2020). For instance, the genome size of the different *Lactobacillus* species ranges from 1.23 Mb (*Lactobacillus sanfranciscensis*; currently *Fructilactobacillus sanfranciscensis*) to 4.91 Mb (*Lactobacillus parakefiri*; currently *Lentilactobacillus parakefiri*) (Mendoza et al., 2023). The large difference occurs also with regard to the guanine-cytosine (GC) content, which ranges from 31.93% to 52.07% (Sun et al., 2015). The high difference in genomic characteristics for most *Lactobacillus* species reflects niche adaptation mechanisms that include gene gains or losses, depending on the habitat selective pressure (Martino et al., 2016). Genome size reduction is often related to a high degree of strain specialization to a specific environment or host. This is the case, for instance, of *Lactobacillus delbrueckii* subsp. *bulgaricus* (*L. bulgaricus*) for the dairy niche, *L. reuteri* for the GIT, or *L. iners* for the vaginal niche. Comparative genomics and phylogenomics analyses on *Lactobacillus* at the species level highlighted this specialization trend due to niche adaptation phenomena (Oh et al., 2010; Frese et al., 2011; Martino et al., 2016; Duar et al., 2017). Specifically, comparative genomics studies rely on the identification of different gene categories, defined as the pangenome, including all the genes of a species; the core genome, representing the set of genes shared by all strains (usually the housekeeping genes involved in primary survival) (Zhang et al., 2020; Zhou et al., 2020); the accessory genes, which are those shared by two or more strains within the species; and the unique genes harbored by a specific strain, also identified as the variable or dispensable genome (Tettelin et al., 2008; Zhou et al., 2020). The adaptation of a species to a certain niche mostly relies on the accessory and unique gene content that provides both selective advantages and an increased species diversity (Tettelin et al., 2008; Zheng et al., 2020).

### 1.6.1.1 Lifestyle Adaptation in LAB

Comparative genomics and phylogenomics investigations allowed the split of *Lactobacillus* species into three lifestyle categories: free-living (i.e., plant and environmental isolates), host-adapted (i.e., isolates associated with vertebrates or invertebrates), and nomadic (i.e., species isolated from different habitats) (Duar et al., 2017). As habitat-driven specialization generally involved a genome reduction, the ability to migrate and adapt to different niches showed strains harboring larger genomes with more differentiated functional genes (Martino et al., 2016; Duar et al., 2017).

It is possible to cite different examples of species specialized to food niches and that underwent genome reduction. For instance, *L. bulgaricus* isolated from yogurt showed a genome largely reduced in size, resulting in a 15% higher GC content than the closely related species *Lactobacillus acidophilus* and *Lactobacillus johnsonii* and an increased pseudogene content (Van De Guchte et al., 2006; Mendoza et al., 2023). It is worth noticing that the absence of most enzymes involved in the amino acid biosynthetic pathways within the chromosome of this species relies on an auxotrophic answer to the nutritionally rich milk environment. Another dairy LAB isolate, *Lactobacillus helveticus*, exhibits genomic features of niche adaptation to the dairy environment by presenting gene gains related to proteolytic systems and gene losses for some sugar metabolisms and bile salt tolerance (typical of GIT-adapted species). For instance, a non-functional bile salt hydrolase (*bsh*) gene with a frameshift mutation was found in *Lactobacillus helveticus* DPC4571 (O'Sullivan et al., 2009), and other genomic features retained from a gut-adapted lifestyle (e.g., acidic pH tolerance) have also been shown in dairy *L. helveticus* strains UC1267 and DSM 20075 (Fontana et al., 2019).

Considering fresh meat and fish food niches, *Lactobacillus sakei* (currently *Latilactobacillus sakei*) is a naturally present LAB that has been widely exploited as a starter culture for sausage fermentation (Chaillou et al., 2005; Nyquist et al., 2011; Eisenbach et al., 2019). Genome sequences of strains isolated from the meat environment were characterized by genes mostly related to protein metabolism, redox and oxygen unbalances, high salt levels, resistance to low temperatures during processing (e.g., cold shock proteins), and competition for colonization and dominance (e.g., genes involved in biofilm formation and antimicrobial compounds such as bacteriocins) (Mendoza et al., 2023). Concerning fermented beverages, *Lactobacillus brevis* (currently *Levilactobacillus brevis*) has been identified as one of the common causes of beer spoilage. Strains isolated from beer present larger genomes, including unique genes mostly related to low pH and iso-α-acids from hop compounds (Feyereisen et al., 2019).

With regard to the gastrointestinal tract, in addition to bile salt and acid tolerance, other genes contribute to the survival of *Lactobacillus* species in the gut environment, also depending on the specific host. For instance, *Lactobacillus reuteri* (currently *Limosilactobacillus reuteri*) is a known natural inhabitant of the GITs of humans (Duar et al., 2017) and different animals, such as pigs, rodents, and chickens (Oh et al., 2010; Frese et al., 2011). Comparative genomics revealed that *L. reuteri* strains showed host-specialized genomes differing for specific vitamin biosynthetic gene clusters (e.g., cobalamin), alcohol utilization (e.g., glycerol and propanediol), and antimicrobial compound production (e.g., reuterin) to compete for GIT colonization and resilience (Frese et al., 2011).

In relation to vaginal tract–adapted LAB, *Lactobacillus crispatus*, *Lactobacillus gasseri*, *Lactobacillus jensenii*, and *Lactobacillus iners* are the most known inhabitants (Petrova et al., 2015).

Specifically, the *L. crispatus* genome evolved towards this niche specialization by including genes mostly related to acidic resistance, carbohydrate metabolisms (e.g., pullulanase, sugar binding proteins), and antagonistic compounds to compete for vaginal colonization (Zhang et al., 2020). *L. iners* instead showed strains with the smallest genomes among LAB (e.g., AB-1: 1.3 Mb) due to a large-scale gene loss and acquisition for survival in the vaginal tract through horizontal gene transfer (HGT) events (Macklaim et al., 2011). This genome evolution for *L. iners* revealed the highest degree of specialization to a specific niche among *Lactobacillus* species, suggesting an evolution towards an obligate symbiotic lifestyle (Duar et al., 2017).

Genomic clustering according to the isolation source, however, is not observed in some *Lactobacillus* species, such as *Lactobacillus plantarum* (currently *Lactiplantibacillus plantarum*), *Lactobacillus rhamnosus* (currently *Lacticaseibacillus rhamnosus*), *Lactobacillus casei* (currently *Lacticaseibacillus casei*), and *Lactobacillus paracasei* (currently *Lacticaseibacillus*) (Fontana et al., 2018; Mendoza et al., 2023). *L. plantarum* has been used as a model organism for studying the nomadic lifestyle (Martino et al., 2016; Duar et al., 2017), since this species showed the capability to migrate frequently across different environments.

## 1.6.2 Plasmid Biology of LAB

The observations by L. L. McKay of the spontaneous and acriflavine-induced loss of lactose fermentation ability of lactococcal strains suggested involvement of plasmids in this phenotype (McKay et al., 1972). Subsequent demonstration of extrachromosomal DNA in the lactococci soon led to the identification of several metabolic plasmids and their functions, such as carbohydrate fermentation and proteinase plasmids (Horng et al., 1991; Frere et al., 1993), citrate permease plasmids (Jahns et al., 1991) and phage resistance plasmids (Lucey et al., 1993; Garvey et al., 1995; Gravesen et al., 1995), and plasmids associated with mucoidness or ropiness (van Kranenburg et al., 1997) and bacteriocin production (Davey and Pearce, 1982). Several antibiotic resistance plasmids from different LAB have also been characterized (Clewell et al., 1974; Vescovo et al., 1982; Teuber et al., 1999).

The plasmids of the various LAB genera associated with food production have been reviewed by Cui et al. (2015) and the plasmids specific to the *Lactobacillus* cluster by Davray et al. (2021). In the latter review, 512 plasmids from 282 strains represented by 51 species were analyzed, and the genomic features of plasmids were correlated with the ecological niches in which these species are typically found. The general finding was that the vertebrate-associated lactobacilli have fewer diverse plasmidomes than the so-called free-living and nomadic species. The plasmid-associated phenotypes demonstrated adaptation to the specific environment of the respective species.

### 1.6.2.1 Physical Structure, Replication Mechanisms, Host Range, and Incompatibility

The majority of LAB plasmids belong to the standard type of covalently closed circular, autonomously replicating DNA molecules. The two basic replication mechanisms are the rolling circle and the theta replication.

A small, mostly cryptic, rolling cycle was reviewed by Ruiz-Masó et al. (2015). In this type of replication, the synthesis of the novel strand starts from a specific nick site, the new strand displacing the plus-strand. The result is a circular double-stranded plasmid and a single-stranded intermediate, to which a complementary strand is subsequently synthetized. The final outcome is two double-stranded plasmids. The rolling circle plasmids typically have a broad host range, being able to replicate even in Gram-negative hosts.

The theta-replicating plasmids in LAB are generally of a medium or large size, the previously mentioned metabolic plasmids being illustrative examples. Theta replication is based on a uni- or bidirectional progressive replication fork, and the specific type of theta replication common in LAB plasmids is the so-called theta-A type.

The sequences of many theta A-replicating plasmids share a remarkable degree of homology in their respective replication regions (see the reviews of Krüger et al., 2004, and Kim et al., 2020). The replication protein gene (*repB*) is preceded by an AT-rich origin of replication (*repA*). One of the striking features for the replication region is the presence of several complete or incomplete iterons of typically 22 bp and also a number of inverted repeats and AT-rich direct repeats. The incompatibility determinants of lactococcal theta-replicating plasmids have been tentatively located within this region of iterons and the first inverted repeats (Gravesen et al., 1997).

The lactococcal family of theta-replicating plasmids also seems to be generally compatible with each other. After screening of 12 theta-replicating plasmids, two incompatible pairs—pFV1001 and pFV1201 and pJW565 and pFW094—were found (Gravesen et al., 1997). The incompatibility region could be tentatively located within the previously mentioned region of 22 bp direct repeats and the first inverted repeat.

The presence of megaplasmids (size range 120–490 kbp) has been established in strains of *Ligilactobacillus salivarius* and other lactobacilli (Li et al., 2007). These plasmids can be either circular or linear, the former replicating with the *repA*-based system. The mode of replication of the linear plasmids is not well characterized, and presumably some kind of telomere structures must be present.

## 1.6.3 Gene Transfer in LAB

Gene transfer between bacteria is a fundamental process that drives bacterial genome evolution towards niche adaptation or nomadic lifestyle. Naturally occurring HGT processes indicate the exchange of genetic material between bacteria belonging to the same or different species. They include transformation (natural competence), phage transduction, and conjugation.

### 1.6.3.1 Transformation

Natural competence is a cellular state in which bacterial cells can introduce exogenous DNA through a specific DNA uptake machinery that imports single-stranded DNA. The imported DNA is then actively stabilized and maintained as a plasmid or incorporated into the chromosome (Blokesch, 2016). In the food industry, natural competence was first established in the yogurt-isolated LAB *Streptococcus thermophilus* that expresses the regulator protein of competence ComX after the activation of the quorum-sensing complex ComRS (Fontaine et al., 2010). The concentration increase of ComX induces the expression of the DNA uptake multiprotein complex (ComEA, ComEC, ComFA, and ComFC), along with DNA uptake facilitators (pilus-like proteins ComGA-GG) and DNA stabilization proteins (RecA, SsbA, SsbB, DprA). In other LAB species, such as *Lactobacillus sakei*, natural competence is induced by the overexpression of an alternative sigma factor (Schmid et al., 2012). The phylogenetic conservation of these competence-related genetic traits among *Lactobacillus* genus has been evaluated, revealing that the complete gene set responsible for the cell status of competence within a species is strain dependent. Some strains indeed show mutations in some of these gene sets, inactivating their ability to internalize foreign DNA (Bron et al., 2019).

### 1.6.3.2 Transduction

Bacteriophages are viruses that infect bacterial cells, taking control of host replication, transcription, and translation machineries to force their proliferation. Bacteriophages infecting LAB have been extensively studied, as they are one of the principal causes of fermentation failure in dairy industries (Bron et al., 2019). Phage transduction mediated the transfer of plasmid or chromosomal genes related to sugar fermentation, proteolysis, and antibiotic resistance. An example of plasmid transduction in LAB is the one that occurs in *L. lactis*, regarding an entire plasmid sequence with a proper size being packaged in the bacteriophage head (Wegmann et al., 2012). Bacteriophage infections allow the transfer of genetic material between strains belonging to the same species or even between different species, including the least accessible ones (e.g., *Lactobacillus delbrueckii*) (Ravin et al., 2006; Ammann et al., 2008). The specificity of the bacteriophage to the host relies on its phage-encoded receptor binding proteins (RBPs) that recognize receptors present on the host cell wall (i.e., polysaccharides or proteins) (Mahony et al., 2017). Despite this high specificity of phage–host recognition that could limit the virus range of action, phage-mediated genomic mobilization of plasmids between different species as *L. lactis* and *S. thermophilus* has been shown (Ammann et al., 2008).

### 1.6.3.3 Conjugation

An alternative approach for introducing DNA into poorly transformable or non-transformable LAB is conjugation, or the transfer of DNA from a donor to a recipient by direct cellular contact, which is also a natural route for *in vivo* HGT (Thompson et al., 2001). Conjugation can involve conjugative plasmids or the co-mobilization of a non-conjugative plasmid by the transfer of a conjugative one (Van Kranenburg et al., 2005; Feld et al., 2008). In addition, it is often associated with special sex factors that are responsible for a high frequency of recombination. The sex factors of *Lc. lactis* 712 have been extensively investigated (Gasson et al., 1995; Shearman et al., 1996).

Conjugative plasmids, as well as integrative and conjugative elements (ICEs), are vertically transferred during DNA replication and cell division (Bron et al., 2019). These conjugative mobile genetic elements (MGEs) code for similar type IV secretion mobilization machineries involved in oriT-dependent conjugal transfer to specific recipient cells, along with other genetic functions for chromosomal integration and excision (ICEs) and extra-chromosomal replication (plasmids) (Johnson and Grossman, 2015; Cury et al., 2018; Bron et al., 2019). In addition to these encoded genes, conjugative MGEs also include variable set of accessory genes, named "cargo", that confer additional phenotypic traits to the host cell, often industrially relevant for LAB exploitation (Ainsworth et al., 2014; Johnson and Grossman, 2015). For instance, *L. lactis* includes genes coding for lactose utilization, extracellular proteinase, and polysaccharide production that are located on conjugal plasmids (Ainsworth et al., 2014). However, conjugative MGEs also allow the introduction of undesirable traits like antibiotic resistance. This is particularly evidenced among streptococci, including *S. thermophilus* (Flórez and Mayo, 2017).

## 1.6.4　Mobile Genetic Elements in LAB

Various transposable genetic elements are important mechanisms for the enhancement of genetic mobility and elasticity of bacteria. As typical examples of Mobile Genetic Elements (MGEs) in LAB, we can mention insertion sequences (IS), transposons (Tn), and introns (group II).

### 1.6.4.1 Insertion Sequences and Transposons

Insertion sequences are the simplest form of transposable genetic elements. They have a size range of 750–2000 bp and typically contain only the genes necessary for the transposition, flanked by short, inverted repeats. Numerous IS elements have been characterized, and their properties and occurrence in different bacterial groups, including lactococci, lactobacilli, enterococci, leuconostocs, and pediococci, have been extensively investigated (Mahillon and Chandler, 1998). Differently from IS, transposons contain additional genes, such as antibiotic resistance determinants or genes involved in conjugative gene transfer, flanked by ISs. Conjugative transposons are also called ICEs. An example of genetic traits in LAB encoded on ICEs is the utilization of sucrose and raffinose, as well as the production of the nisin bacteriocin, in *L. lactis* (De Vos et al., 1995; Machielsen et al., 2011). Other examples include *S. thermophilus* ICESt1 and ICESt3 that integrate, in a site-specific manner, into the gene coding for putative fructose-1,6-diphosphate aldolase (Bellanger et al., 2009).

Since conjugative MGEs often code for industrially relevant traits, it is fundamental to deeply investigate their mechanism of transfer and the related host-range limitations. To this extent, it is also important to undercover the role of group II introns (like the one found in the *L. lactis* sex factor) in the modulation of transfer efficiencies of conjugative MGEs (Novikova et al., 2014; Qu et al., 2018).

### 1.6.4.2 Group II Introns

Group II introns are a class of introns that splice with the formation of a closed circular structure ("lariat") similar to the splicing mechanism of eukaryotic mRNAs. The discovery of group II introns in different bacterial genomes is relatively recent (Martínez-Abarca and Toro, 2000). Some group II introns are also MGEs able to home in on analogous alleles that miss the intron (Michel and Ferat, 1995) and interrupt other mobile elements such as IS and transposons. Most of these mobile introns include a multifunctional intron-encoded protein (IEP), which contains reverse transcriptase, DNA endonuclease, and RNA maturase functions for splicing and homing activities (Mohr et al., 1993; Zimmerly et al., 2001). The first bacterial group II intron with proven splicing and genetic mobility abilities was the lactococcal intron Ll.ltrB (Mills et al., 1996, 1997). Like other group II introns, Ll.ltrB includes exon-binding sites (EBSs) that base pair with sites in the upstream exon, namely the intron-binding sites (IBSs) of the homing position (Mills et al., 1996; Matsuura et al., 1997; Mohr et al., 2000). The IEP protein plays a role in positioning Ll.ltrB RNA at the target site and uncoiling the double-stranded DNA to allow RNA–DNA interaction (Mohr et al., 2000). The spliced lariat RNA and IEP allow the formation of a double-stranded cleavage at the analogous intron-less allele (Matsuura et al., 1997; Saldanha et al., 1999). A reverse splicing mechanism then introduces the intron RNA directly into a target DNA site, with the subsequent conversion of integrated RNA to DNA via the activity of the IEP-directed reverse transcriptase (Matsuura et al., 1997; Wank et al., 1999). Since Ll.ltrB mobility is partially nucleotide based, the intron could be redirected into alternative target sites by modifying the EBS recognition sequences to be complementary to new target site IBS sequences (Mills, 2001).

## 1.6.5 Recombinant DNA Techniques and Their Applications

Emerging recombinant DNA techniques are being developed in LAB, such as double-strand DNA (dsDNA) or single-strand DNA (ssDNA) recombineering and CRISPR–Cas systems, which greatly facilitate gene knock-out or knock-in procedures in LAB (Wu et al., 2021).

Recombineering indicates the homologous recombination between exogenous DNA and the bacterial genome to obtain gene insertion, deletion, or replacement (Court et al., 2002; Montiel et al., 2015). DsDNA recombineering is mediated through the λ-Redαβγ and RecET recombinase systems encoded by bacteriophages. Redα/RecE is a 5'-3'dsDNA exonuclease that digests exogenous dsDNA, resulting in 3'-ended ssDNA overhangs. Redβ/RecT is a single-strand annealing protein that binds the ssDNA overhangs, promoting strand exchange. Redγ then inhibits the RecBCD nuclease from attacking linear DNA (Yu et al., 2000). The suitability of λ-Red- and RecET-like systems in LAB was first reported in *L. plantarum* WCFS1, yielding an efficient deletion of the glucosamine-6-phosphate isomerase gene (*gnp*) and replacement of the D-lactate dehydrogenase gene (*ldhD*) (Yang et al., 2015). Selection of positive mutants is marker dependent, and a scar remains at the modified locus after the selection marker is excised. SsDNA recombineering instead does not use selection markers, but it only requires the overexpression of Redβ/RecT to be guided to homologous sequences on the bacterial chromosome (Van Pijkeren and Britton, 2012). The first experiment of ssDNA recombineering in LAB was performed in *L. reuteri* ATCC PTA 6475 and resulted in a good mutation yield (Van Pijkeren and Britton, 2012). However, the absence of antibiotic selection markers leads to a low-efficiency screening of positive mutants.

The CRISPR–Cas system is a genome editing technique based on the constitute adaptive immune systems of prokaryotes that can actively reject the invasion of foreign genetic elements such as phages and plasmids (Horvath and Barrangou, 2010). A Type II CRISPR–Cas9 system from *S. pyogenes* has been exploited as an easy and programmable sequence-based tool for genome editing in some eukaryotes and prokaryotes. The system includes an endonuclease (Cas9), a trans-activating CRISPR RNA (tracrRNA), and a precursor crRNA array containing nuclease guide sequences (spacers) separated by identical direct repeats (Karvelis et al., 2013). The precursor crRNA is processed within the repeat sequences to generate mature crRNA, which then forms a duplex with the tracr-RNA. The duplex interacts with Cas9, targets a specific protospacer adjacent motif (PAM) on the DNA, and binds to proximal chromosomal complementary sequences (protospacers), inducing double-stranded breaks (DSBs) in the chromosome (Cong et al., 2013). The presence of DSBs induces the non-homologous end joining recombination (NHEJ) or homologous recombination (HR) mechanisms for DNA repair, producing the target mutation (Cong et al., 2013). An advantage of the CRISPR–Cas9 technique is that it can be used as a counter selectable marker, as Cas9-induced DSBs in the wild-type allele allow a rapid screening of mutants. Besides the introduction of point mutations, insertions, or deletions in targeted genes, this system can also regulate gene expression through CRISPR interference (CRISPRi) with catalytically inactive variants of Cas9 (dCas9) that lack the endonucleolytic activity but maintain the targeted binding function (Qi et al., 2013). A CRISPRi system was applied in *L. plantarum* for transcriptional regulation of essential cell cycle-related genes (Berlec et al., 2018; Myrbråten et al., 2019).

The described genome editing techniques allow rapid and efficient genetic engineering of LAB. They enable production of probiotics and starter cultures with enhanced traits, as well as LAB with therapeutic functionalities. For instance, modification of exopolysaccharide (EPS) properties to increase gastrointestinal resistance of probiotic LAB, or the introduction of additional metabolic capabilities in starter cultures, such as the ability to ferment galactose by *S. thermophilus*, are possible applications (Rhimi et al., 2011; Hidalgo-Cantabrana et al., 2015). Engineered LAB can be also exploited for delivery of biotherapeutics since they can induce both mucosal and systemic immune responses without the risks of conventional attenuated live pathogens (Tarahomjoo, 2012). Examples include cytokine delivery for the treatment of inflammatory bowel disease (IBD) in colitis-induced mice (Steidler et al., 2000), as well as the production and delivery of allergens by *L. lactis* and *L. plantarum* (e.g., *L. lactis* CHW9 for peanut allergen Ara2, *L. lactis* NZ9800 for birch allergen Bet-v1, and *L. plantarum* NCL21 for cedar pollen allergen Cry j1) to suppress allergen-specific immunoglobulin E response (Daniel et al., 2006; Glenting et al., 2007; Ohkouchi et al., 2012). Moreover, recombinant LAB can be exploited as a potential alternative to current therapies for type I diabetes or anticancer therapeutics, such as the recombinant *L. lactis* NZ9000 expressing fusion protein HSP65–6P277 to improve glucose tolerance or secreting tumor metastasis-inhibiting peptides (Ma et al., 2014; Zhang et al., 2016). Other cancer antigens can be potentially expressed using *L. lactis*, for instance, against human papilloma virus and melanoma (Bermúdez-Humarán et al., 2005; Kalyanasundram et al., 2015).

Besides protein and peptide-based therapeutics, engineered LAB can also produce metabolites with medicinal applications, such as γ-amino butyric acid and hyaluronic acid (Chahuki et al., 2019; Zhong et al., 2019).

# 1.7 CONCLUSION

The evolutionary history of LAB from ancient *Bacillus*-like soil bacteria to a fermentative. metabolically, and physiologically diverse group of families and genera is reflected in their present-day ecological niches and in their molecular biological characteristics. The genetic adaptations detectable both in the chromosomal genomes and the plasmidomes of LAB illustrate this evolutionary process, showing the loss of many ancestral metabolic activities and, on the other hand, demonstrating the compensating changes that make LAB successful competitors in many challenging environments and are also behind their many traditional and modern applications.

# BIBLIOGRAPHY

Ainsworth, S., S. Stockdale, F. Bottacini, J. Mahony and D. van Sinderen. 2014. The *Lactococcus lactis* plasmidome: Much learnt, yet still lots to discover. *FEMS Microbiol Rev* 38. https://doi.org/10.1111/1574-6976.12074.

Ammann, A., H. Neve, A. Geis and K. J. Heller. 2008. Plasmid transfer via transduction from *Streptococcus thermophilus* to *Lactococcus lactis*. *J Bacteriol* 190. https://doi.org/10.1128/JB.01448-07.

Axelsson, L., T. C. Chung, W. J. Dobrogosz and S. E. Lindgren. 1989. Production of a broad spectrum antimicrobial substance by *Lactobacillus reuteri*. *Microb Ecol Health Dis* 2: 131–136.

Bellanger, X., A. P. Roberts, C. Morel, F. Choulet, G. Pavlovic, Mullany, P., et al. 2009. Conjugative transfer of the integrative conjugative elements ICESt1 and ICESt3 from *Streptococcus thermophilus*. *J Bacteriol* 191. https://doi.org/10.1128/JB.01412-08.

Berlec, A., K. Škrlec, J. Kocjan, M. Olenic and B. Štrukelj. 2018. Single plasmid systems for inducible dual protein expression and for CRISPR-Cas9/CRISPRi gene regulation in lactic acid bacterium *Lactococcus lactis*. *Sci Rep* 8. https://doi.org/10.1038/s41598-018-19402-1.

Bermúdez-Humarán, L. G., N. G. Cortes-Perez, F. Lefèvre, V. Guimarães, S. Rabot, J. M. Alcocer-Gonzalez, et al. 2005. A novel nucosal vaccine based on live Lactococci expressing E7 antigen and IL-12 induces systemic and mucosal immune responses and protects mice against human papillomavirus type 16-Induced tumors. *The Journal of Immunology* 175. https://doi.org/10.4049/jimmunol.175.11.7297.

Bisset, D. L. and R. L. Andersson. 1974. Lactose and D-galactose metabolism in group N streptococci: Presence of enzymes for both the D-galactose-1-phosphate and D-tagatose 6-phosphate pathways. *J Bacteriol* 117: 318–320.

Blokesch, M. 2016. Natural competence for transformation. *Curr Biol* 26. https://doi.org/10.1016/j.cub.2016.08.058.

Bolotin, A., P. Wincker, S. Mauger, O. Jaillon, K. Malarme, J. Weissbach, S. D. Ehrlich and A. Sorokin. 2001. The complete genome sequence of the lactic acid bacterium *Lactocococus lactis ssp. lactis* IL1403. *Genome Res* 11: 731–753.

Borch, E. and G. Molin. 1989. The aerobic growth and product formation of *Lactobacillus, Bronchotrix* and *Carnobacterium* batch cultures. *Appl Microbiol Biotechnol* 30: 81–88.

Bron, P. A., B. Marcelli, J. Mulder, S. van der Els, L. P. Morawska, O. P. Kuipers, et al. 2019. Renaissance of traditional DNA transfer strategies for improvement of industrial lactic acid bacteria. *Curr Opin Biotechnol* 56. https://doi.org/10.1016/j.copbio.2018.09.004.

Brown, J. P. and P. J. Vandemark. 1968. Respiration of *Lactobacillus casei*. *Can J Microbiol* 14: 821–825.

Cerning, J. 1990. Exocellular polysaccharides produced by lactic acid bacteria. *FEMS Microbiol Rev* 87: 113–130.

Chahuki, F. F., S. Aminzadeh, V. Jafarian, F. Tabandeh and M. Khodabandeh. 2019. Hyaluronic acid production enhancement via genetically modification and culture medium optimization in *Lactobacillus acidophilus*. *Int J Biol Macromol* 121. https://doi.org/10.1016/j.ijbiomac.2018.10.112.

Chaillou, S., M. C. Champomier-Vergès, M. Cornet, A. M. Crutz-Le Coq, A. M. Dudez, V. Martin, et al. 2005. The complete genome sequence of the meat-borne lactic acid bacterium *Lactobacillus sakei* 23K. *Nat Biotechnol* 23: 1527–1533. https://doi.org/10.1038/nbt1160.

Clewell, D. B., Y. Yagi, G. M. Dunny and S. K. Schultz. 1974. Characterization of three plasmid deoxyribonucleic acid molecules in a strain of *Streptococcus faecalis*: Identification of a plasmid determining erythromycin resistance. *J Bacteriol* 117: 283–289.

Cogan, T. M., D. Walsh and S. Condon. 1989. Impact of aeration on the metabolic end-products formed from glucose and galactose by *Streptococcus lactis*. *J Appl Bacteriol* 66: 74–84.

Cong, L., F. A. Ran, D. Cox, S. Lin, R. Barretto, N. Habib, et al. 2013. Multiplex genome engineering using CRISPR/ Cas system *Supplementary Materials*. *Science* 1979: 339.

Court, D. L., J. A. Sawitzke and L. C. Thomason. 2002. Genetic engineering using homologous recombination. *Annu Rev Genet* 36. https://doi.org/10.1146/annurev.genet.36.061102.093104.

Cui, Y., T. Hu, X. Qu, L. Zhang, Z. Ding and A. Dong. 2015. Plasmids from food lactic acid bacteria: Diversity, similarity and new developments. *Int J Mol Sci*: 13172–13202.

Cury, J., P. H. Oliveira, F. De La Cruz and E. P. C. Rocha. 2018. Host range and genetic plasticity explain the coexistence of integrative and extrachromosomal mobile genetic elements. *Mol Biol Evol* 35. https://doi.org/10.1093/molbev/msy123.

Daniel, C., A. Repa, C. Wild, A. Pollak, B. Pot, H. Breiteneder, et al. 2006. Modulation of allergic immune responses by mucosal application of recombinant lactic acid bacteria producing the major birch pollen allergen Bet v 1. *Allergy: Eur J Allergy Clin Immunol* 61. https://doi.org/10.1111/j.1398-9995.2006.01071.x.

Davey, G. P. and L. E. Pearce. 1982. Production of diplococcin by *Streptococcus cremoris* and its transfer to nonproducing group N streptococci. In *Microbiology-1982*, ed. D. Schlessinger. American Society for Microbiology, Washington, DC, pp. 221–224.

Davray, D., D. Deo and R. Kulkarni. 2021. Plasmids encode niche-specific traits in *Lactobacillaceae*. *Microb Genom* 7(3): mgen000472. https://doi.org/10.1099/mgen.0.000472.

De Vos, W. M., M. M. Beerthuyzen, E. L. Luesink and O. P. Kuipers. 1995. Genetics of the nisin operon and the sucrose-nisin conjugative transposon Tn5276. *Dev Biol Stand* 85.

Dobrogosz, W. J. and R. W. Stone. 1962. Oxidative metabolism of *Pediococcus pentosaceus*. I. Role of oxygen and catalase. *J Bacteriol* 84: 716–723.

Duar, R. M., X. B. Lin, J. Zheng, M. E. Martino, T. Grenier, M. E. Pérez-Muñoz, et al. 2017. Lifestyles in transition: Evolution and natural history of the genus *Lactobacillus*. *FEMS Microbiol Rev* 41. https://doi.org/10.1093/FEMSRE/FUX030.

Eisenbach, L., A. J. Geissler, M. A. Ehrmann and R. F. Vogel. 2019. Comparative genomics of *Lactobacillus sakei* supports the development of starter strain combinations. *Microbiol Res* 221. https://doi.org/10.1016/j.micres.2019.01.001.

Eisenbach, L., D. Janßen, M. A. Ehrmann and R. F. Vogel. 2018. Comparative genomics of *Lactobacillus curvatus* enables prediction of traits relating to adaptation and strategies of assertiveness in sausage fermentation. *Int J Food Microbiol* 286. https://doi.org/10.1016/j.ijfoodmicro.2018.06.025.

Eltz, R. W. and P. J. Vandemark. 1960. Fructose dissimilation by *Lactobacillus brevis*. *J. Bacteriol* 79: 763–776.

Endo, A. 2012. Fructophilic lactic acid bacteria inhabit fructose-rich niches in nature. *Microb Ecol Health Dis*. https://doi.org/10.3402/mehd.v23i0.18563.

Endo, A., Y. Futagawa-Endo and L. M. Dicks. 2009. Isolation and characterization of fructophilic lactic acid bacteria from fructose-rich niches. *Syst Appl Microbiol* 8: 593–600. https://doi.org/10.1016/j.syapm.2009.08.002.

Endo, A., T. Irisawa, Y. Futagawa-Endo, K. Takano, M. du Toit, S. Okada and L. M. T. Dicks. 2012. Characterization and emended description of *Lactobacillus kunkeei* as a fructophilic lactic acid bacterium. *Int J Syst Evol Microbiol* 62: 500–504.

Endo, A., S. Maeno, Y. Tanizawa, W. Kneifel, M. Arita, L. Dicks and S. Salminen. 2018. Fructophilic lactic acid bacteria, a unique group of fructose-fermenting microbes. *Appl Environ Microbiol* 84. https://doi.org/10.1128/AEM.01290-18.

Endo, A. and S. Okada. 2008. Reclassification of the genus *Leuconostoc* and proposals of *Fructobacillus fructosus* gen. nov., comb. nov., *Fructobacillus durionis* comb. nov., *Fructobacillus ficulneus* comb. nov. and *Fructobacillus pseudoficulneus* comb. nov. *Int J Syst Evol Microbiol* 58: 2195–2205.

Endo, A. and S. Salminen. 2013. Honeybees and beehives are rich sources for fructophilic lactic acid bacteria. *Syst Appl Microbiol* 6: 444–448. https://doi.org/10.1016/j.syapm.2013.06.002.

Feld, L., S. Schjørring, K. Hammer, T. R. Licht, M. Danielsen, K. Krogfelt, et al. 2008. Selective pressure affects transfer and establishment of a *Lactobacillus plantarum* resistance plasmid in the gastrointestinal environment. *J Antimicrob Chemother* 61. https://doi.org/10.1093/jac/dkn033.

Feyereisen, M., J. Mahony, P. Kelleher, R. J. Roberts, T. O'Sullivan, J. M. A. Geertman, et al. 2019. Comparative genome analysis of the *Lactobacillus brevis* species. *BMC Genom* 20. https://doi.org/10.1186/s12864-019-5783-1.

Flórez, A. B. and B. Mayo. 2017. Antibiotic resistance-susceptibility profiles of *Streptococcus thermophilus* isolated from raw milk and genome analysis of the genetic basis of acquired resistances. *Front Microbiol* 8. https://doi.org/10.3389/fmicb.2017.02608.

Fontaine, L., C. Boutry, M. H. De Frahan, B. Delplace, C. Fremaux, P. Horvath, et al. 2010. A novel pheromone quorum-sensing system controls the development of natural competence in *Streptococcus thermophilus* and *Streptococcus salivarius*. *J Bacteriol* 192. https://doi.org/10.1128/JB.01251-09.

Fontana, A., I. Falasconi, P. Molinari, L. Treu, A. Basile, A. Vezzi, et al. 2019. Genomic comparison of *Lactobacillus helveticus* strains highlights probiotic potential. *Front Microbiol* 10. https://doi.org/10.3389/fmicb.2019.01380.

Fontana, A., C. Zacconi and L. Morelli. 2018. Genetic signatures of dairy *Lactobacillus casei* Group. *Front Microbiol*. https://doi.org/10.3389/fmicb.2018.02611.

Frère, J., A. Benachour, M. Novel and G. Novel. 1993. Identification of the theta-type minimal replicon of the *Lactococcus lactis* subspecies *lactis* CNRZ270 lactose-protease plasmid pUCL22. *Mol Microbiol* 27: 97–102.

Frese, S. A., A. K. Benson, G. W. Tannock, D. M. Loach, J. Kim, M. Zhang, et al. 2011. The evolution of host specialization in the vertebrate gut symbiont *Lactobacillus reuteri*. *PLoS Genet* 7. https://doi.org/10.1371/journal.pgen.1001314.

Fujita, Y. 2009. Carbon catabolite control of the metabolic network in *Bacillus subtilis*. *Biosci Biotechnol Biochem* 73: 245–259.

Garvey, P. G., G. F. Fitzgerald and C. Hill. 1995. Cloning and DNA sequence analysis of two abortive infection phage resistance determinants from the lactococcal plasmid pNP40. *Appl Environ Microbiol* 61: 4321–4328.

Gasson, M. J., J. J. Godon, C. J. Pillidge, T. J. Eaton, K. Jury and C. A. Shearman. 1995. Characterization and exploitation of conjugation in *Lactococcus lactis*. *Int Dairy J* 5. https://doi.org/10.1016/0958-6946(95)00030-5.

Gaudu, P., K. Vido, B. Cesselin, S. Kulakauskas, J. Tremblay, L. Rezaiki, G. Lamberet, S. Sourice, P. Duvat and A. Gruss. 2002. Respiration capacity and consequences in *Lactococcus lactis*. *Antonie van Leeuwenhoek* 82: 263–269.

Glenting, J., L. K. Poulsen, K. Kato, S. M. Madsen, H. Frøkiær, C. Wendt, et al. 2007. Production of recombinant peanut allergen Ara h 2 using *Lactococcus lactis*. *Microb Cell Fact* 6. https://doi.org/10.1186/1475-2859-6-28.

Goel, A., P. M. Halami and J. P. Tamang. 2020. Genome analysis of *Lactobacillus plantarum* isolated from some Indian fermented foods for bacteriocin production and probiotic marker genes. *Front Microbiol* 11. https://doi.org/10.3389/fmicb.2020.00040.

Gravesen, A., J. Josephsen, A. von Wright and F. K. Vogensen. 1995. Characterization of the replicon of the lactococcal theta-replicating plasmid pJW563. *Plasmid* 34: 105–118.

Gravesen, A., A. von Wright, J. Josephsen and F. K. Vogensen. 1997. Replication regions of two pairs of incompatible lactococcal theta-replicating plasmids. *Plasmid* 38: 115–127.

Hidalgo-Cantabrana, C., B. Sánchez, P. Álvarez-Martín, P. López, N. Martínez-Álvarez, M. Delley, et al. 2015. A single mutation in the gene responsible for the mucoid phenotype of *Bifidobacterium animalis* subsp. *lactis* confers surface and functional characteristics. *Appl Environ Microbiol* 81. https://doi.org/10.1128/AEM.02095-15.

Horng, J. S., K. M. Polzin and L. L. McKay. 1991. Replication and temperature-sensitive maintenance functions of lactose plasmid pSK11L from *Lactococcus lactis* subsp. *cremoris*. *J Bacteriol* 173: 7573–7581.

Horvath, P. and R. Barrangou. 2010. CRISPR/Cas, the immune system of Bacteria and Archaea. *Science* 1979: 327. https://doi.org/10.1126/science.1179555.

Hugenholz, J. 1993. Citrate metabolism in lactic acid bacteria. *FEMS Microbiol Rev* 12: 165–178.

Hutkins, R. W. and H. A. Morris. 1987. Carbohydrate metabolism by *Streptococcus thermophilus*: A review. *J Food Prot* 50: 876–884.

Jahns, A., A. Schafer, A. Geis and M. Teuber. 1991. Identification, cloning and sequencing of the replication region of *Lactococcus lactis* ssp. *lactis* biovar. *diacetylactis* Bu2 citrate plasmid pSL2. *FEMS Microbiol Lett* 64: 253–258.

Johnson, C. M. and A. D. Grossman. 2015. Integrative and Conjugative Elements (ICEs): What they do and how they work. *Annu Rev Genet* 49. https://doi.org/10.1146/annurev-genet-112414-055018.

Kalyanasundram, J., S. L. Chia, A. A. L. Song, A. R. Raha, H. A. Young and K. Yusoff. 2015. Surface display of glycosylated Tyrosinase related protein-2 (TRP-2) tumour antigen on *Lactococcus lactis*. *BMC Biotechnol* 15. https://doi.org/10.1186/s12896-015-0231-z.

Kandler, O. 1983. Carbohydrate metabolism in lactic acid bacteria. *Antonie van Leeuwenhoek* 49: 209–224.

Kaneuchi, C., M. Seki and K. Komagata. 1988. Production of succinic acid from citric acid and related acids by *Lactobacillus* strains. *Appl Environ Micro Biol* 54: 3053–3056.

Karvelis, T., G. Gasiunas, A. Miksys, R. Barrangou, P. Horvath and V. Siksnys. 2013. crRNA and tracrRNA guide Cas9-mediated DNA interference in *Streptococcus thermophilus*. *RNA Biol* 10. https://doi.org/10.4161/rna.24203.

Kim, J. W., V. Bugata, G. Cortés-Cortés, G. Quevedo-Martínez and M. Camps. 2020. Mechanisms of theta plasmid replication in Enterobacteria and implications for adaptation to its host. *EcoSal Plus* 9(1). https://doi.org/10.1128/ecosalplus.ESP-0026-2019.

Koebmann, B., L. M. Blank, C. Solem, D. Petranovic, L. K. Nielsen and P. Ruhrdal Jensen. 2008. Increased biomass yield of *Lactococcus lactis* during energetically limited growth and respiratory conditions. *Biotechnol Appl Biochem* 50: 25–33.

Konings, W. N., B. Poolman and N. Driessen. 1989. Bioenergetics and solute transport in lactococci. *Crit Rev Microbiol* 16: 419–476.

Koryszewska-Bagińska, A., J. Gawor, A. Nowak, M. Grynberg and T. Aleksandrzak-Piekarczyk. 2019. Comparative genomics and functional analysis of a highly adhesive dairy *Lactobacillus paracasei* subsp. *paracasei* IBB3423 strain. *Appl Microbiol Biotechnol* 103. https://doi.org/10.1007/s00253-019-10010-1.

Krüger, R., S. Rakowski and M. Filutovicz. 2004. Participating elements in the replication of iteron-containing plasmids in *Plasmid Biology* Phillips G. and Funnel, B., editors (Washington DC: ASM Press) pp. 25- 45

Kunji, E. R. S., I. Mierau, A. Hagting, B. Poolman and W. N. Konings. 1996. The proteolytic systems of lactic acid bacteria. *Antonie van Leeuwenhoek* 76: 217–246.

Kunkee, R. 1991. Some roles of malic acid and the malolactic fermentation in wine making. *FEMS Microbiol Rev* 88: 52–72.

Lee, J. Y., G. G. Han, E. B. Kim and Y. J. Choi. 2017. Comparative genomics of *Lactobacillus salivarius* strains focusing on their host adaptation. *Microbiol Res* 205. https://doi.org/10.1016/j.micres.2017.08.008.

Li, Y., C. Canchaya, F. Fang, E. Raftis, K. A. Ryan, J. P. van Pijkeren, D. van Sinderen and P. W. O'Toole. 2007. Distribution of megaplasmids in *Lactobacillus salivarius* and other lactobacilli. *J Bacteriol* 189: 6128–6139.

Liu, M., J. R. Bayjanov, B. Renckens, A. Nauta and R. J. Siezen. 2010. The proteolytic system of lactic acid bacteria revisited: A genomic comparison. *BMC Genom* 11: 36. www.biomedcentral.com/1471-2164/11/36.

Lopez, C. M., G. Rocchetti, A. Fontana, L. Lucini and A. Rebecchi. 2022. Metabolomics and gene-metabolite networks reveal the potential of *Leuconostoc* and *Weissella* strains as starter cultures in the manufacturing of bread without baker's yeast. *Food Res Int* 162. https://doi.org/10.1016/j.foodres.2022.112023.

Lucey, C. A. and S. Condon. 1986. Active role of oxygen and NADH oxidase in growth and energy metabolism of *Leuconostoc. J Gen Microbiol* 132: 1789–1796.

Lucey, M., C. Daly and G. Fitzgerald. 1993. Analysis of a region from the bacteriophage resistance plasmid pCI528 involved in conjugative mobilization between *Lactococcus* strains. *J Bacteriol* 175: 6002–6009.

Ma, Y., J. Liu, J. Hou, Y. Dong, Y. Lu, L. Jin, et al. 2014. Oral administration of recombinant *Lactococcus lactis* expressing HSP65 and tandemly repeated P277 reduces the incidence of type I diabetes in non-obese diabetic mice. *PLoS ONE* 9. https://doi.org/10.1371/journal.pone.0105701.

Machielsen, R., R. J. Siezen, S. A. F. T. Van Hijum and J. E. T. Van Hylckama Vlieg. 2011. Molecular description and industrial potential of Tn6098 conjugative transfer conferring alpha-galactoside metabolism in *Lactococcus lactis. Appl Environ Microbiol* 77. https://doi.org/10.1128/AEM.02283-10.

Macklaim, J. M., G. B. Gloor, K. C. Anukam, S. Cribby and G. Reid. 2011. At the crossroads of vaginal health and disease, the genome sequence of *Lactobacillus iners* AB-1. *Proc Natl Acad Sci U S A* 108. https://doi.org/10.1073/pnas.1000086107.

Mahillon, J. and M. Chandler. 1998. Insertion sequences. *Microbiol Mol Biol Rev* 62.

Mahony, J., C. Cambillau and D. Van Sinderen. 2017. Host recognition by lactic acid bacterial phages. *FEMS Microbiol Rev* 41. https://doi.org/10.1093/femsre/fux019.

Martínez-Abarca, F. and N. Toro. 2000. Group II introns in the bacterial world. *Mol Microbiol* 38. https://doi.org/10.1046/j.1365-2958.2000.02197.x.

Martino, M. E., J. R. Bayjanov, B. E. Caffrey, M. Wels, P. Joncour, S. Hughes, et al. 2016. Nomadic lifestyle of *Lactobacillus plantarum* revealed by comparative genomics of 54 strains isolated from different habitats. *Environ Microbiol* 18. https://doi.org/10.1111/1462-2920.13455.

Matsuura, M., R. Saldanha, H. Ma, H. Wank, J. Yang, G. Mohr, et al. 1997. A bacterial group II intron encoding reverse transcriptase, maturase, and DNA endonuclease activities: Biochemical demonstration of maturase activity and insertion of new genetic information within the intron. *Genes Dev* 11. https://doi.org/10.1101/gad.11.21.2910.

McKay, L. L., K. A. Baldwin and E. A. Zottola. 1972. Loss of lactose metabolism in *Streptococcus lactis. Appl Microbiol* 23: 1090–1096.

Mendoza, R. M., S. H. Kim, R. Vasquez, I. C. Hwang, Y. S. Park, H. D. Paik, et al. 2023. Bioinformatics and its role in the study of the evolution and probiotic potential of lactic acid bacteria. *Food Sci Biotechnol* 32. https://doi.org/10.1007/s10068-022-01142-8.

Michel, F. and J. L. Ferat. 1995. Structure and activities of group II introns. *Annu Rev Biochem* 64. https://doi.org/10.1146/annurev.bi.64.070195.002251.

Michels, P. A. M., J. P. J. Michels, J. Boonstra and W. N. Konings. 1979. Generation of an electrochemical proton gradient in bacteria by excretion of metabolic end products. *FEMS Microbiol Lett* 49: 247–257.

Mills, D. A. 2001. Mutagenesis in the post genomics era: Tools for generating insertional mutations in the lactic acid bacteria. *Curr Opin Biotechnol* 12. https://doi.org/10.1016/S0958-1669(00)00254-8.

Mills, D. A., D. A. Manias, L. L. McKay and G. M. Dunny. 1997. Homing of a group II intron from *Lactococcus lactis* subsp. *lactis* ML3. *J Bacteriol* 179. https://doi.org/10.1128/jb.179.19.6107-6111.1997.

Mills, D. A., L. L. Mckay and G. M. Dunny. 1996. Splicing of a group II intron involved in the conjugative transfer of pRS01 in lactococci. *J Bacteriol* 178. https://doi.org/10.1128/jb.178.12.3531-3538.1996.

Mohr, G., P. S. Perlman and A. M. Lambowitz. 1993. Evolutionary relationships among group II intron-encoded proteins and identification of a conserved domain that may be related to maturase function. *Nucleic Acids Res* 21. https://doi.org/10.1093/nar/21.22.4991.

Mohr, G., D. Smith, M. Belfort and A. M. Lambowitz. 2000. Rules for DNA target-site recognition by a lactococcal group II intron enable retargeting of the intron to specific DNA sequences. *Genes Dev* 14. https://doi.org/10.1101/gad.14.5.559.

Montiel, D., H. S. Kang, F. Y. Chang, Z. Charlop-Powers and S. F. Brady. 2015. Yeast homologous recombination-based promoter engineering for the activation of silent natural product biosynthetic gene clusters. *Proc Natl Acad Sci U S A* 112. https://doi.org/10.1073/pnas.1507606112.

Myrbråten, I. S., K. Wiull, Z. Salehian, L. S. Håvarstein, D. Straume, G. Mathiesen, et al. 2019. CRISPR interference for rapid knockdown of essential cell cycle genes in *Lactobacillus plantarum. mSphere* 4. https://doi.org/10.1128/msphere.00007-19.

Novikova, O., D. Smith, I. Hahn, A. Beauregard and M. Belfort. 2014. Interaction between conjugative and retrotransposable elements in Horizontal Gene Transfer. *PLoS Genet* 10. https://doi.org/10.1371/journal.pgen.1004853.

Nyquist, O. L., A. McLeod, D. A. Brede, L. Snipen, Å. Aakra and I. F. Nes. 2011. Comparative genomics of *Lactobacillus sakei* with emphasis on strains from meat. *Mol Genet Genom* 285. https://doi.org/10.1007/s00438-011-0608-1.

Oh, P. L., A. K. Benson, D. A. Peterson, P. B. Patil, E. N. Moriyama, S. Roos, et al. 2010. Diversification of the gut symbiont *Lactobacillus reuteri* as a result of host-driven evolution. *ISME J* 4. https://doi.org/10.1038/ismej.2009.123.

Ohkouchi, K., S. Kawamoto, K. Tatsugawa, N. Yoshikawa, Y. Takaoka, S. Miyauchi, et al. 2012. Prophylactic effect of *Lactobacillus* oral vaccine expressing a Japanese cedar pollen allergen. *J Biosci Bioeng* 113. https://doi.org/10.1016/j.jbiosc.2011.11.025.

Orla-Jensen, S. 1919. *The Lactic Acid Bacteria*. Host and Son, Copenhagen, Denmark.

O'Sullivan, O., J. O'Callaghan, A. Sangrador-Vegas, O. McAuliffe, L. Slattery, P. Kaleta, et al. 2009. Comparative genomics of lactic acid bacteria reveals a niche-specific gene set. *BMC Microbiol* 9. https://doi.org/10.1186/1471-2180-9-50.

Pan, M., C. Hidalgo-Cantabrana, Y. J. Goh, R. Sanozky-Dawes and R. Barrangou. 2020. Comparative analysis of *Lactobacillus gasseri* and *Lactobacillus crispatus* isolated from human urogenital and gastrointestinal tracts. *Front Microbiol* 10. https://doi.org/10.3389/fmicb.2019.03146.

Petrova, M. I., E. Lievens, S. Malik, N. Imholz and S. Lebeer. 2015. *Lactobacillus* species as biomarkers and agents that can promote various aspects of vaginal health. *Front Physiol* 6. https://doi.org/10.3389/fphys.2015.00081.

Poolman, B. 1993. Energy transduction in lactic acid bacteria. *FEMS Microbiol Rev* 12: 125–148.

Poolman, B. 2002. Transporters and their role in LAB cell physiology. *Antonie van Leeuwenhoek* 82: 147–164.

Poolman, B., A. J. M. Driessen and W. N. Konings. 1987. Regulation of arginine-ornithine exchange and the arginine deiminase pathway in *Streptococcus lactis*. *J Bacteriol* 169: 5597–5604.

Poolman, B., D. Molenaar, E. J. Smid, T. Ubbink, T. Abee, P. P. Renault and W. N. Konings. 1991. Malolactic fermentation: Electrogenic malate uptake and malate/lactate antiport generate metabolic energy. *J Bacteriol* 173: 6030–6037.

Postma, P. W., J. W. Lengeler and G. R. Jacobsen. 1993. Phosphoenolpyruvate:carbohydrate phosphotransferase system in lactic acid bacteria. *Microbiol Rev* 57: 543–594.

Premi, L., W. E. Sandine and P. R. Elliker. 1972. Lactose-hydrolyzing enzymes of *Lactobacillus species*. *Appl Microbiol* 24: 51–57.

Qi, L. S., M. H. Larson, L. A. Gilbert, J. A. Doudna, J. S. Weissman, A. P. Arkin, et al. 2013. Repurposing CRISPR as an RNA-guided platform for sequence-specific control of gene expression. *Cell* 152: 1173–1183. https://doi.org/10.1016/j.cell.2013.02.022.

Qu, G., C. L. Piazza, D. Smith and M. Belfort. 2018. Group II intron inhibits conjugative relaxase expression in bacteria by mRNA targeting. *Elife* 7. https://doi.org/10.7554/eLife.34268.

Quilodrán-Vega, S., L. Albarracin, F. Mansilla, L. Arce, B. Zhou, M. A. Islam, et al. 2020. Functional and genomic characterization of *Ligilactobacillus salivarius* TUCO-L2 isolated from lama glama milk: A promising immunobiotic strain to combat infections. *Front Microbiol* 11. https://doi.org/10.3389/fmicb.2020.608752.

Ramos, A. and H. Santos. 1996. Citrate and sugar cofermentation in *Leuconostoc oenos*, a 13C nuclear magnetic resonance study. *Appl Environ Microbiol* 62: 2577–2585.

Ravin, V., T. Sasaki, L. Räisänen, K. A. Riipinen and T. Alatossava. 2006. Effective plasmid pX3 transduction in *Lactobacillus delbrueckii* by bacteriophage LL-H. *Plasmid* 55: 184–193. https://doi.org/10.1016/j.plasmid.2005.12.003.

Rhimi, M., H. Chouayekh, I. Gouillouard, E. Maguin and S. Bejar. 2011. Production of d-tagatose, a low caloric sweetener during milk fermentation using l-arabinose isomerase. *Bioresour Technol* 102. https://doi.org/10.1016/j.biortech.2010.10.078.

Ritchey, T. W. and H. W. J. Seeley. 1976. Distribution of cytochrome-like respiration in streptococci. *J Gen Microbiol* 93: 105–203.

Roos, S., F. Karner, L. Axelsson and H. Jonsson. 2000. *Lactobacillus mucosae* sp. nov., a new species with in vitro mucus binding activity isolated from pig intestine. *Int J Syst Evol Microbiol* 50: 251–258.

Ruiz-Masó, J. A., N. C. Machó, L. Bordanaba-Ruiseco, M. Espinosa, M. Coll and G. Del Solar. 2015. Plasmid rolling-circle replication. *Microbiol Spectr* 3(1): PLAS-0035–2014. https://doi.org/10.1128/microbiolspec.PLAS-0035-2014.

Saier, M. H. J., J.-J. Ye, S. Klinke and E. Nino. 1996. Identification of an anaerobically induced phosphoenolpyruvate-dependent fructose-specific phosphotransferase system and evidence for the Embden-Meyerhof glycolytic pathway in the heterofermentative bacterium *Lactobacillus brevis*. *J Bacteriol* 178: 314–316.

Saldanha, R., B. Chen, H. Wank, M. Matsuura, J. Edwards and A. M. Lambowitz. 1999. RNA and protein catalysis in group II intron splicing and mobility reactions using purified components. *Biochemistry* 38. https://doi.org/10.1021/bi982799l.

Salema, M., J. S. Lolkema, M. V. San Romão and M. C. Lourero Dias. 1996. The proton motive force generated in *Leuconostoc oenos* by L-malate fermentation. *J Bacteriol* 178: 3127–3132.

Savijoki, K., H. Ingmer and P. Varmanen. 2006. Proteolytic systems of lactic acid bacteria. *Appl Microbiol Biotechnol* 71: 394–406.

Schmid, M., J. Muri, D. Melidis, A. R. Varadarajan, V. Somerville, A. Wicki, et al. (2018). Comparative genomics of completely sequenced *Lactobacillus helveticus* genomes provides insights into strain-specific genes and resolves metagenomics data down to the strain level. *Front Microbiol* 9. https://doi.org/10.3389/fmicb.2018.00063.

Schmid, S., C. Bevilacqua and A. M. Crutz-Le Coq. 2012. Alternative sigma factor σ h activates competence gene expression in *Lactobacillus sakei*. *BMC Microbiol* 12. https://doi.org/10.1186/1471-2180-12-32.

Schütz, H. and F. Radler. 1984. Propanediol-1,2-dehydratase and metabolism of glycerol of *Lactobacillus brevis*. *Arch Microbiol* 139: 366–370.

Sedewitz, B., K. H. Schleifer and F. Götz. 1984. Physiological role of pyruvate oxidase in the aerobic metabolism of *Lactobacillus plantarum*. *J Bacteriol* 160: 462–465.

Shearman, C., J. J. Godon and M. Gasson. 1996. Splicing of a group II intron in a functional transfer gene of *Lactococcus lactis*. *Mol Microbiol* 21. https://doi.org/10.1046/j.1365-2958.1996.00610.x.

Sjöberg, A. and B. Hahn-Hägerdahl. 1989. β-Glucose-phosphate, a possible mediator for polysaccharide formation in maltose-assimilating *Lactococcus lactis*. *Appl Environ Microbiol* 55: 1549–1554.

Smart, J. B. and T. D. Thomas. 1987. Effect of oxygen on lactose metabolism in lactic streptococci. *Appl Environ Microbiol* 53: 533–541.

Smid, E. J., A. J. M. Driessen and W. J. Konings. 1989. Mechanism and energetic of dipeptide transport in membrane vesicles of *Lactococcus lactis*. *J Bacteriol* 171: 292–298.

Son, S., J. D. Oh, S. H. Lee, D. Shin and Y. Kim. 2020. Comparative genomics of canine *Lactobacillus reuteri* reveals adaptation to a shared environment with humans. *Genes Genomics* 42. https://doi.org/10.1007/s13258-020-00978-w.

Stamer, J. R. and B. O. Stoyla. 1967. Growth response of *Lactobacillus brevis* to aeration and organic catalysts. *Appl Microbiol* 15: 1025–1030.

Steidler, L., W. Hans, L. Schotte, S. Neirynck, F. Obermeier, W. Falk, et al. 2000. Treatment of murine colitis by *Lactococcus lactis* secreting interleukin-10. *Science* 1979: 289. https://doi.org/10.1126/science.289.5483.1352.

Stolz, P., G. Böcker, R. F. Vogel and W. P. Hammes. 1993. Utilization of maltose and glucose by lactobacilli isolated from sourdough. *FEMS Microbiol Lett* 109: 237–242.

Sun, Z., H. M. Harris, A. McCann, C. Guo, S. Argimon, W. Zhang, X. Yang, et al. 2015. Expanding the biotechnology potential of lactobacilli through comparative genomics of 213 strains and associated genera. *Nat Commun* 6 https://doi.org/10.1038/ncomms9322.

Talarico, T. L. and W. J. Dobrogosz. 1989. Chemical characterization of an antimicrobial substance produced by *Lactobacillus reuteri*. *Antimicrob Agents Chemother* 33: 674–679.

Talarico, T. L. and W. J. Dobrogosz. 1990. Purification and characterization of glycerol dehydratase from *Lactobacillus reuteri*. *Appl Environ Microbiol* 56: 1195–1197.

Tanaka, K., A. Komiyama, K. Sonomoto, A. Ishizaki, S. J. Hall and R. Stanbury. 2002. Two different pathways for D-xylose metabolism and the effect of xylose concentration on the yield coefficient of L-lactate in mixed-acid fermentation by the lactic acid bacterium *Lactococcus lactis* IO-1. *Appl Microbiol Biotechnol* 60: 160–167.

Tarahomjoo, S. 2012. Development of vaccine delivery vehicles based on lactic acid bacteria. *Mol Biotechnol* 51. https://doi.org/10.1007/s12033-011-9450-2.

ten Brink, B., R. Otto, U.-P. Hansen and W. N. Konings. 1983. Energy recycling by lactate efflux in growing and nongrowing cells of *Streptococcus cremoris*. *J Bacteriol* 162: 383–390.

Tettelin, H., D. Riley, C. Cattuto and D. Medini. 2008. Comparative genomics: The bacterial pan-genome. *Curr Opin Microbiol* 11. https://doi.org/10.1016/j.mib.2008.09.006.

Teuber, M., L. Meile and F. Schwarz. 1999. Acquired antibiotic resistance in lactic acid bacteria from food. *Antonie van Leuuwenhoek* 76: 115–137.

Thomas, T. D., D. C. Ellwood and V. M. C. Longyear. 1979. Change from homo- to heterolactic fermentation by *Streptococcus lactis* resulting from glucose limitation in anaerobic chemostat cultures. *J Bacteriol* 138: 109–117.

Thompson, J. 1979. Lactose metabolism in *Streptococcus lactis*: Phosphorylation of galactose and glucose moieties in vivo. *J Bacteriol* 140: 774–785.

Thompson, J. 1987. Regulation of sugar transport and metabolism in lactic acid bacteria. *FEMS Microbiol Rev* 46: 221–231.

Thompson, J. and B. M. Chassy. 1981. Uptake and metabolism of sucrose by *Streptococcus lactis*. *J Bacteriol* 147: 543–551.

Thompson, J. K., K. J. McConville, C. Nicholson and M. A. Collins. 2001. DNA cloning in *Lactobacillus helveticus* by the exconjugation of recombinant mob-containing plasmid constructs from strains of transformable lactic acid bacteria. *Plasmid* 46. https://doi.org/10.1006/plas.2001.1540.

Tseng, C.-P., J.-L. Tsau and T. J. Montville. 1991. Bioenergetic consequences of catabolic shifts by *Lactobacillus plantarum* in response to shifts in environmental oxygen and pH in chemostat cultures. *J Bacteriol* 173: 4411–4416.

Van De Guchte, M., S. Penaud, C. Grimaldi, V. Barbe, K. Bryson, P. Nicolas, et al. 2006. The complete genome sequence of *Lactobacillus bulgaricus* reveals extensive and ongoing reductive evolution. *Proc Natl Acad Sci U S A* 103. https://doi.org/10.1073/pnas.0603024103.

Van Der Veer, C., R. Y. Hertzberger, S. M. Bruisten, H. L. P. Tytgat, J. Swanenburg, A. De Kat Angelino-Bart, et al. 2019. Comparative genomics of human *Lactobacillus crispatus* isolates reveals genes for glycosylation and glycogen degradation: Implications for in vivo dominance of the vaginal microbiota. *Microbiome* 7. https://doi.org/10.1186/s40168-019-0667-9.

Van Kranenburg, R., N. Golic, R. Bongers, R. J. Leer, W. M. De Vos, R. J. Siezen, et al. 2005. Functional analysis of three plasmids from *Lactobacillus plantarum*. *Appl Environ Microbiol* 71. https://doi.org/10.1128/AEM.71.3.1223-1230.2005.

van Kranenburg, R., J. D. Marugg, I. I. van Swam, N. J. Willem and W. M. de Vos. 1997. Molecular characterization of the plasmid-encoded eps gene cluster essential for exopolysaccharide biosynthesis in *Lactococcus lactis*. *Mol Microbiol* 24: 387–397.

Van Pijkeren, J. P. and R. A. Britton. 2012. High efficiency recombineering in lactic acid bacteria. *Nucleic Acids Res* 40. https://doi.org/10.1093/nar/gks147.

Vescovo, M., L. Morelli and V. Bottazzi. 1982. Drug resistance plasmids in *Lactobacillus acidophilus* and *Lactobacillus reuteri*. *Appl Environ Microbiol* 43: 17–20.

Wang, S., B. Yang, R. Paul Ross, C. Stanton, J. Zhao, H. Zhang, et al. 2020. Comparative genomics analysis of *Lactobacillus ruminis* from different niches. *Genes* (Basel) 11. https://doi.org/10.3390/genes11010070.

Wank, H., J. SanFilippo, R. N. Singh, M. Matsuura and A. M. Lambowitz. 1999. A reverse transcriptase/maturase promotes splicing by binding at its own coding segment in a group II intron RNA. *Mol Cell* 4. https://doi.org/10.1016/S1097-2765(00)80371-8.

Wegmann, U., K. Overweg, S. Jeanson, M. Gasson and C. Shearman. 2012. Molecular characterization and structural instability of the industrially important composite metabolic plasmid pLP712. *Microbiology* (United Kingdom) 158. https://doi.org/10.1099/mic.0.062554-0.

Wu, J., Y. Xin, J. Kong and T. Guo. 2021. Genetic tools for the development of recombinant lactic acid bacteria. *Microb Cell Fact* 20. https://doi.org/10.1186/s12934-021-01607-1.

Yang, P., J. Wang and Q. Qi. 2015. Prophage recombinases-mediated genome engineering in *Lactobacillus plantarum*. *Microb Cell Fact* 14. https://doi.org/10.1186/s12934-015-0344-z.

Yu, D., H. M. Ellis, E. C. Lee, N. A. Jenkins, N. G. Copeland and D. L. Court. 2000. An efficient recombination system for chromosome engineering in *Escherichia coli*. *Proc Natl Acad Sci U S A* 97. https://doi.org/10.1073/pnas.100127597.

Zhang, B., A. Li, F. Zuo, R. Yu, Z. Zeng, H. Ma, et al. 2016. Recombinant *Lactococcus lactis* NZ9000 secretes a bioactive kisspeptin that inhibits proliferation and migration of human colon carcinoma HT-29 cells. *Microb Cell Fact* 15. https://doi.org/10.1186/s12934-016-0506-7.

Zhang, Q., L. Zhang, P. Ross, J. Zhao, H. Zhang and W. Chen. 2020. Comparative genomics of *Lactobacillus crispatus* from the gut and vagina reveals genetic diversity and lifestyle adaptation. *Genes* (Basel) 11. https://doi.org/10.3390/genes11040360.

Zhang, Z. G., Z. Q. Ye, L. Yu and P. Shi. 2011. Phylogenomic reconstruction of lactic acid bacteria: An update. *BMC Evol Biol* 11: 1.

Zheng, J., S. Wittouck, E. Salvetti, C. M. A. P. Franz, H. M. B. Harris, P. Mattarelli, et al. 2020. A taxonomic note on the genus *Lactobacillus*: Description of 23 novel genera, emended description of the genus *Lactobacillus beijerinck* 1901, and union of *Lactobacillaceae* and *Leuconostocaceae*. *Int J Syst Evol Microbiol* 70. https://doi.org/10.1099/ijsem.0.004107.

Zhong, Y., S. Wu, F. Chen, M. He and J. Lin. 2019. Isolation of high γ aminobutyric acid producing lactic acid bacteria and fermentation in mulberry leaf powders. *Exp Ther Med*. https://doi.org/10.3892/etm.2019.7557.

Zhou, X., B. Yang, C. Stanton, R. P. Ross, J. Zhao, H. Zhang, et al. 2020. Comparative analysis of *Lactobacillus gasseri* from Chinese subjects reveals a new species-level taxa. *BMC Genom* 21. https://doi.org/10.1186/s12864-020-6527-y.

Zimmerly, S., G. Hausner and X. C. Wu. 2001. Phylogenetic relationships among group II intron ORFs. *Nucleic Acids Res* 29. https://doi.org/10.1093/nar/29.5.1238.

# The Genus *Lactobacillus*—Across the Past and Future

# 2

Fandi Ibrahim, Sarah Lebeer, Elisa Salvetti, and Giovanna E. Felis

## 2.1 THE TAXONOMIC UPDATE

In recent years, there has been an expansion of the number of species belonging to the genus *Lactobacillus* as 261 species identified at the beginning of 2020, presenting scientific and technological challenges. This led to the need for a taxonomic update, which resulted in the subdivision of the genus *Lactobacillus* species into 25 genera. With the new taxonomic note (Zheng et al. 2020), the genus *Lactobacillus* is now restricted to only 44 species identified thus far, which are all phylogenetically closer to the *L. delbrueckii* species than any other species of the new genera. At the family level, all 25 genera arising from the new taxonomy still belong to the phylum *Bacillota* (formerly *Firmicutes*), the class *Bacilli*, the order *Lactobacillales*, and the family *Lactobacillaceae*. Indeed, this update did not come as a surprise since in the previous version of this chapter, it was noted that ideas for more radical subdivision were reported.

### 2.1.1 The Challenges with the Traditional Classification Scheme

The traditional taxonomy of the *Lactobacillus* genus complex encountered two significant challenges. First, it was paraphyletic, meaning that not all descendants of the ancestor belonged to the defined group. In the case of the *Lactobacillus* genus, this paradox extended to include the *Pediococcus* genus and the entire *Leuconostocaceae* family. This incongruence within the phylogenetic tree demanded a re-evaluation. Second, with more than 250 species, the *Lactobacillus* genus complex exhibited an exceptional level of diversity (Sun et al. 2015). Variability was observed in DNA/protein sequences, the number of shared genes, and ecological and metabolic characteristics among its species. This diversity made it challenging to describe the genus complex as a whole. To illustrate this, consider the analogy of classifying animals solely as "mammals" without recognizing smaller groups such as carnivores, marsupials, and primates. Such a simplification would overlook shared evolutionary histories and unique features specific to each subgroup. Therefore, these challenges necessitated the revision of the taxonomic status of all species belonged to the *Lactobacillus* complex genus based on the following criteria: 1) monophyly: new genera

DOI: 10.1201/9781003352075-3

needed to be monophyletic, based on evolutionary history and core genome phylogeny; 2) acceptable diversity: the diversity within new genera should align with that of other genera, ensuring clear separation; and 3) defining characteristics: species within the same genus should share similar ecological niches, metabolic pathways, physiological traits, and signature genes. The details of the methodology followed for the reclassification can be found in the previously published taxonomic note (Zheng et al. 2020), and the conversion from the old names to the new names is facilitated online through lactobacillus.uantwerpen. be and lactobacillus.ualberta.ca.

## 2.1.2 Names of the New Genera and Impact on Commercially Important Species

Following rules of nomenclature but also considering extensive analyses based on stakeholders' acceptance, names were assigned to the new genera following several criteria. The *Lactobacillus* remained to indicate the former *L. delbrueckii* group, with *L. delbrueckii* the type species of the genus. Then, the name *Paralactobacillus*, previously described (Leisner et al. 2000), was revived. For renaming the other remaining 23 genera, three approaches were followed:

- In 19 cases, prefixes were added to *Lactobacillus*, preferably starting with "L" to maintain the same abbreviation. These prefixes reflected shared traits, such as *Limosi-* for slimy characteristics or *Compani-* for a friendly association with other lactobacilli. Other examples are *Latilactobacillus* (widespread), *Liquorilactobacillus* (liquid), *Ligilactobacillis* (ligate), *Bombilactobacillus* (bees), *Levilactobacillus* (leavening).
- In two cases, explicit reference to the former phylogenetic group were made, namely for *L. casei* and *L. plantarum* groups that became *Lacticaseibacillus* and *Lactiplantibacillus*.
- The remaining cases were named to honor influential scientists in the field. Accordingly, *Dellaglioa*, *Holzapfelia* (later modified as *Holzapfeliella*), and *Schleiferilactobacillus* were named in honor of Franco Dellaglio, Wilhelm Holzapfel, and Heinz Schleifer, respectively.

Commercially relevant species of lactobacilli now belong to updated genus *Lactobacillus*, namely *L. acidophilus* and *L. delbrueckii* subsp. *bulgaricus*, or to different genera such as *Lacticaseibacillus* (*L. casei*, *L. paracasei*, and *L. rhamnosus*), *Limosilactobacillus* (*L. fermentum*, *L. reuteri*), *Lactiplantibacillus* (*L. plantarum*), and *Ligilactobacillus* (*L. salivarius*) (Morovic et al. 2016). An updated list is shown in Table 2.1 with the new name and etymology, while complete information can be found in Zheng et al. (2020).

**TABLE 2.1**  List of Old and New Names of the Most Commercially Important Lactobacilli

| *"TRADITIONAL" SPECIES NAME (ALPHABETICAL ORDER)* | *UPDATED SPECIES NAME* | *ETYMOLOGY OF THE GENUS NAME* |
|---|---|---|
| *Lactobacillus acidophilus* | *Lactobacillus acidophilus* | Rod-shaped *Bacillus* from milk |
| *Lactobacillus brevis* | *Levilactobacillus brevis* | *Lactobacillus* with leavening potential |
| *Lactobacillus buchneri* | *Lentilactobacillus buchneri* | Slow (growing) *Lactobacillus* |
| *Lactobacillus casei* | *Lacticaseibacillus casei* | Milk-derived rodlet from the (*Lactobacillus*) *casei* group |
| *Lactobacillus curvatus* | *Latilactobacillus curvatus* | Widespread *Lactobacillus* |
| *Lactobacillus delbrueckii* | *Lactobacillus delbrueckii* | Rod-shaped *Bacillus* from milk |
| *Lactobacillus fermentum* | *Limosilactobacillus fermentum* | Slimy (biofilm-forming) *Lactobacillus* |
| *Lactobacillus gasseri* | *Lactobacillus gasseri* | Rod-shaped *Bacillus* from milk |

*(Continued)*

**TABLE 2.1**  (Continued) List of Old and New Names of the Most Commercially Important Lactobacilli

| "TRADITIONAL" SPECIES NAME (ALPHABETICAL ORDER) | UPDATED SPECIES NAME | ETYMOLOGY OF THE GENUS NAME |
|---|---|---|
| *Lactobacillus helveticus* | *Lactobacillus helveticus* | Rod-shaped *Bacillus* from milk |
| *Lactobacillus jensenii* | *Lactobacillus jensenii* | Rod-shaped *Bacillus* from milk |
| *Lactobacillus paracasei* | *Lacticaseibacillus paracasei* | Milk-derived rodlet from the (*Lactobacillus*) *casei* group |
| *Lactobacillus plantarum* | *Lactiplantibacillus plantarum* | Milk-derived rodlet from the (*Lactobacillus*) *plantarum* group |
| *Lactobacillus pontis* | *Limosilactobacillus pontis* | Slimy (biofilm-forming) *Lactobacillus* |
| *Lactobacillus reuteri* | *Limosilactobacillus reuteri* | Slimy (biofilm-forming) *Lactobacillus* |
| *Lactobacillus rhamnosus* | *Lacticaseibacillus rhamnosus* | Milk-derived rodlet from the (*Lactobacillus*) *casei* group |
| *Lactobacillus sakei* | *Latilactobacillus sakei* | A widespread *Lactobacillus* |
| *Lactobacillus salivarius* | *Ligilactobacillus salivarius* | *Lactobacillus* with a host-associated lifestyle |

## 2.1.3 Benefits of the New Taxonomy

A first important scientific implementation is that the new taxonomy significantly enhances the resolution of sequencing approaches, particularly in metagenomics studies. The previous limitations of 16S rRNA amplicons, which were restricted to the genus level, have been overcome by the division of the *Lactobacillus* genus into phylogenetically and ecologically coherent genera. This higher resolution provides valuable insights into microbial ecosystems, including food fermentations, the human vagina, and the intestines of various animals where lactobacilli play essential roles. Most major 16S rRNA sequence databases have adopted the taxonomic changes, aligning with the new taxonomy's principles. These updates include SILVA, RDP, GTDB, and EzBioCloud, ensuring that the scientific community can work with consistent and up-to-date taxonomy in their research endeavors.

Furthermore, the taxonomy allows the identification of genus-specific traits that were previously obscured by the vast number of species within the genus *Lactobacillus*. These traits include metabolic preferences, such as fermentation pathways and carbohydrate utilization, which are now recognized as genus-level traits. This updated taxonomy thus provides valuable insights into the evolutionary history, habitat adaptation, and potential applications of lactobacilli. For example, the differentiation between homofermentative and heterofermentative organisms is now a genus-level trait, and it has been discovered that heterofermentative lactic acid bacteria form a monophyletic clade.

The taxonomic update also has significant implications for probiotic research. It enables researchers to link phylogeny with metabolism and ecology, providing valuable insights into the mechanisms behind the health benefits of probiotic lactobacilli. Understanding the specific traits and metabolites associated with different genera and species allows for more targeted research into probiotic effector molecules. Researchers can now explore the diversity of lactobacilli more comprehensively and pose intriguing questions, particularly relevant for probiotic research. For example, how do commonly used probiotic genera like *Lactiplantibacillus* and *Lacticaseibacillus* differ in their interactions from host-adapted (vaginal) *Lactobacillus* or *Limosilactobacillus* species? What explains the close relationship between yogurt and vaginal *Lactobacillus* species? When and why did key evolutionary events occur, such as the development of rod-shaped cells, loss of sporulation, metabolic differentiation (e.g., fermentation vs. catalase), mucus interactions, and immune responses?

In conclusion, this taxonomic update facilitates not only scientific discovery but also regulatory approval of *Lactobacillaceae* for use in food and feed, as antibiotic resistance and biogenic amine formation are partly related to the current genus-level taxonomy. The latter properties are crucial criteria for regulatory approval. The new taxonomy, with its well-defined genera and species, simplifies the evaluation of these criteria, ensuring safer applications of *Lactobacillaceae* in the food and feed industries. These taxonomic changes open up exciting avenues for further research, paving the way for a deeper understanding of lactobacilli and their diverse roles in microbiology, health, and industry.

## 2.2 TYPICAL ECOLOGICAL NICHES

Members of the genus *Lactobacillus* complex (family *Lactobacillaceae*) are anaerobic or aerotolerant, rod-shaped, Gram-positive bacteria. Their metabolism was commonly classified as obligately homofermentative or facultatively/obligately heterofermentative, now simplified as homofermentative and heterofermentatives. Lactobacilli are commonly found in a diversity of nutrient-rich environments, including dairy environments, from which they derived their name, and microbial-heavy host habitats such as mucosal surfaces from humans and other mammals, as well as natural ecological niches such as decaying plant material, soil, and insect digestive tracts. They are also widely spread within the members of the animal kingdom from honeybees to humans, forming a part of the natural microbiota of the host animals, and occupy various niches within the host such as the gastrointestinal tract (see Chapters 21 and 23), urogenital tract (see Chapters 21 and 28), oral cavity (see Chapter 27), and skin (see Chapter 21). Lactobacilli are present in dairy environments and are particularly abundant in fermented dairy products (see Chapter 10). In addition, lactobacilli are naturally present in plants and soil. *Lactiplantibacillus plantarum* has been reported to be a dominant naturally occurring bacterial species in vegetables such as cabbage and lettuce (Yang et al. 2010). Lactobacilli are also found in fermented non-dairy foods and have also been isolated from soil samples. Examples include *Lpb. plantarum*, *Lacticaseibacillus paracasei* subsp. *paracasei*, and *Levilactobacillus brevis* (Chen et al. 2005). These lactobacilli do, however, appear to be sensitive to pesticides (Ayansina and Amusan 2013). In humans, lactobacilli are found throughout the gastrointestinal tract, from the oral cavity to fecal material. In the oral cavity, lactobacilli are present in saliva and in dental plaque (Verma et al. 2018). The lactobacilli in the oral cavity may offer protection against harmful microbes, but as acid producers they themselves may also contribute to dental erosion. Probiotic lactobacilli, however, have not been observed to increase dental caries risk (Mahasneh and Mahasneh 2017). Lactobacilli are relatively abundant in the gastrointestinal tract and in fecal samples, from where numerous different *Lactobacillus* species have been isolated. The typical species composition of the intestinal lactobacilli population varies among subjects and geographical regions; examples of typical species include *Lacticaseibacillus rhamnosus*, *L. acidophilus*, *Lpb. plantarum*, *Lcb. paracasei*, and *Lim. reuteri*. While lactobacilli are frequently found in fecal samples, they are present in relatively low levels and normally represent a minor part (0.2%–1.0%) of the fecal microbiota but are present at much higher levels in the upper gastrointestinal tract (Heeney et al. 2018). Lactobacilli are genetically well adapted for the gastrointestinal environment, as reviewed later in this chapter. Intestinal lactobacilli have several interactions with the host and have been linked with numerous health benefits; these will be reviewed in other chapters of this book.

Lactobacilli are important members of a healthy vaginal microbiome (see Chapter 28). Nunn and Forney (2016) characterized the vaginal microbiome of 396 North American women and reported that the communities were clustered into five different groups, four of which were dominated by species of lactobacilli, namely *Lactobacillus iners*, *Lactobacillus crispatus*, *Lactobacillus gasseri*, or *Lactobacillus jensenii*, and a fifth cluster with lower proportion of lactobacilli. The role of *L. iners* in vaginal health is unclear; the species has been detected both in health and microbial dysbiosis (Petrova et al. 2017). Lactobacilli are also naturally and frequently present in human breast milk. Lactobacilli detected in breast milk include *L. acidophilus* and *L. gasseri*, *Lcb. casei* and *Lcb. rhamnosus*, *Lim. fermentum*, and *Lig. salivarius* (Jost et al. 2015). In addition to natural niches of lactobacilli, members of the genus are widely used in food and feed manufacturing and are also used commercially as health-promoting microbes (i.e., probiotics).

# 2.3 LACTOBACILLI IN FOOD AND FEED

Lactobacilli have been part of the normal human diet since the early days of humanity, when the lack of means for preservation meant that much of the stored food became naturally fermented (Ozen and Dinleyici 2015). Today, lactobacilli are present in many different foods and have an excellent record for safety. They are used as starter cultures in food fermentation and as probiotics—health-promoting microbes. Lactobacilli are found in many if not most fermented foods, particularly in dairy products such as yogurt and its regional varieties, cheeses, and fermented milks. Many traditional regional fermented products such as Korean *kimchi* (see Chapter 18) and Caucasian *kefir* (see Chapter 10) are fermented with lactic acid bacteria (LAB), among them strains of *Lactobacillus*. Lactobacilli are important starter cultures in vegetable fermentations such as sauerkraut and pickled vegetables and are used in the making of sourdough bread. Lactobacilli are also found in meats; for example, *Latilactobacillus sakei* is used in fermented meat products (Zagorec and Champomier-Verges 2017). In alcoholic drinks such as beer and wine, lactobacilli may contribute to the flavor of the product but may also act as contaminants. In addition to fermented foods in the human diet, *Lactobacillus* fermentation is also utilized in animal feeds. Species such as *Lpb. plantarum* and *Lentilactobacillus buchneri* are used in the production of silage, a fermented animal feed (Muck et al. 2018).

Lactobacilli are also used commercially as probiotics. Probiotics are used as dietary supplements, commonly administered as capsules or sachets, and in probiotic foods. Probiotic foods may be fermented foods such as yogurt, but they are also used in nonfermented foods and beverages such as probiotic ice cream, probiotic snacks, and probiotic juices. Clinically documented probiotic lactobacilli include, among others, *L. acidophilus* NCFM, *L. acidophilus* La-5, *Lcb. casei* Shirota, *Lcb. casei* DN-114 001, *Lcb. rhamnosus* GG, *Lcb. rhamnosus* HN001, *Lcb. rhamnosus* GR-1, *Lcb. paracasei* F19, *Lpb. plantarum* 299v, and *Lim. reuteri* DSM 17938. Probiotic lactobacilli are linked with numerous beneficial health effects, as reviewed in other chapters of this book. In addition to human probiotics, lactobacilli are also used as probiotic ingredients in feed for farm animals and companion animals. While the field of human probiotics is largely dominated by lactobacilli and bifidobacteria, animal probiotics display greater variability and many commercial animal probiotics come outside these two genera. Nevertheless, lactobacilli have been investigated and utilized as probiotics for animals such as pigs, poultry, and cattle where they are commonly referred to as direct-fed microbials (Uyeno et al. 2015).

## 2.3.1 Carbohydrate Metabolism

### 2.3.1.1 Sugar Transport

Free sugars are transported through either permease systems or an inducible specific phosphotransferase system (PTS). The transport of glucose, fructose, mannose, mannitol, galactose, and lactose can occur through the PTS system to end up as glucose 6-P (Monedero et al. 2007). In a strain of *Lcb. casei*, the PTS seems to be the main mechanism for transporting fructose, mannose, mannitol, sorbose, sorbitol, cellobiose, and lactose (Viana et al. 2000). Free pentoses are taken up by specific permeases and then converted to xylulose 5-phosphate to enter the pentose phosphate pathway for the homofermentative or phosphoketolase pathway for the heterofermentative. Twenty-five complete PTS sugar transport systems were identified in *Lpb. plantarum* WCFS1, reflecting its efficient adaptive capacity (Kleerebezem et al. 2003). PTS$^{Lac}$ has been recently shown to be involved in the transport and metabolism of human milk oligosaccharides, for example, core-2 N-acetylglucosamine by *Lcb. casei* (Bidart et al. 2018) and lacto-N-neotetraose by *L. acidophilus* (Thongaram et al. 2017), while an absence of any PTS was reported in seven *Lim. reuteri* strains (Zhao and Ganzle 2018).

### 2.3.1.2 Sugar Fermentation

Like all LAB, species that make up the genus *Lactobacillus* are chemoheterotrophic, which means they obtain their energy for maintenance and synthesis of macromolecules solely from substrate-level phosphorylation (i.e., fermentation). Glycolysis is the most common pathway, but the phosphoketolase pathway and pentose–phosphate pathway is also utilized in many species. The first step of fermentation pathways is the phosphorylation of free sugars. The phosphorylation of glucose in most species occurs through the ATP-dependent glucokinase reaction, while in some species it occurs via the PTS (Kandler 1983). Homolactic fermentation commonly occurs via glycolysis, while heterolactic fermentation occurs via the phosphoketolase pathway, but homofermentative lactobacilli can also use the pentose phosphate pathway when metabolizing certain substrates (i.e. pentoses). Lactobacilli species are now classed into either homofermentative or heterofermentative, unlike the earlier classification into three groups, which included a facultatively heterofermentative group. The simpler differentiation of homofermenters, which metabolize hexoses via the Embden–Meyerhoff pathway to pyruvate, and heterofermenters, metabolizing hexoses via the phosphoketolase pathway to pyruvate and acetyl-phosphate, is now linked to the presence/absence of the 6-phosphofructokinase gene (Zheng et al. 2020; Salvetti et al. 2018).

## 2.3.2 Alternative Pyruvate Metabolism, Diacetyl Formation

Under anaerobic conditions, pyruvate generated either through glycolysis or the pentose phosphate pathway acts as an electron acceptor to regenerate NAD+. This process leads to the formation of lactic acid regardless of the type of the fermentation pathway. However, there exist several alternative fates for pyruvate under various circumstances. For a different route to take place, there must be either excess pyruvate in the medium or an additional electron acceptor either formed by a different pathway or intentionally added to the growth medium. For instance, adding an exogenous electron acceptor changed the end product's profile for *Lpb. plantarum* generating acetate at the expense of lactate (Ganzle et al. 2007). Examples of pyruvate alternative electron acceptors include $O_2$, citrate, and fructose (De Vuyst et al. 2009). Moreover, some members of the lactobacilli showed an aerobic competency, as genes related to respiratory metabolism have been identified in strains of *Lcb. casei*, *Lpb. plantarum*, and *Lat. sakei*, where oxygen, exogenous heme, and menaquinones have been used as electron acceptors. The respiratory metabolism identified did not lead to a significant increase in growth rate, but it offers physiological advantages that might find application in food industry, as it led to improved tolerance to heat as well as oxidative stress (Ianniello et al. 2015; Zotta et al. 2017). Excess pyruvate can occur naturally in the medium if citrate exists, as in milk and fruits. Fermentation of citrate occurs concomitantly with glycolysis or the pentose phosphate pathway, producing pyruvate, and then spares the pyruvate required as electron acceptor to regenerate $NAD^+$ and leads to changes in its fate to produce formate, acetate, ethanol, and the C4 aroma compounds (diacetyl, acetoin, 2,3-butanediol) in addition to lactic acid (Hugenholtz 1993). These aromatic compounds, in particular the diacetyl aroma, are of significant interest in the dairy industry. Diacetyl is responsible for the characteristic aroma and flavor of butter. It can also be found, undesirably, in other products such as wine and beer (Zapata et al. 2012).

## 2.3.3 Proteolytic System and Peptide Utilization

The proteolytic system is essential for bacterial growth. All lactobacilli require at least some amino acids, depending on the specific species or strain. Some species such as *Lpb. plantarum* require only 3 amino acids, while others such as *L. acidophilus* require 14 amino acids. Consequently, they have a functioning proteolytic system to acquire their amino acids from the growth medium or their natural habitats. Proteolytic systems consist of proteinases, transport systems, and peptidases. The proteinases are secreted extracellularly to hydrolyze proteins into oligopeptides, which are then taken up into the cell by transporters, to be further degraded by intercellular peptidases (Christensen et al. 1999). Lactobacilli generally exhibit low proteolytic activity. Nevertheless, the characteristics of proteolytic activity have been described for several strains belonging to various species, including *L. delbrueckii* subsp. *bulgaricus*, *L. helveticus*, *Lcb. casei*, *Lcb. paracasei*, *Lim. fermentum*, and *L. acidophilus* (El-Gaish et al. 2010), and the proteolytic system has recently been characterized genetically and biochemically for some species (*Lpb. plantarum* and *Fructilactobacillus sanfranciscensis*) (Vermeulen et al. 2005). The level and specificity of proteolytic activity differ between lactobacilli and vary even among the strains of the same species (Oberg et al. 2002). Strains can have different proteolysis rates and generate different peptides; suggesting different proteolytic enzyme specificity (Aguirre et al. 2014). The major factors affecting the proteolytic activity of *Lactobacillus* strains are pH, temperature, metal ions, and the presence of inhibitors (Ganzle 2014).

The effect of the proteolytic activity of 21 strains of *L. helveticus* and *L. delbrueckii* subsp. *bulgaricus* on the functional properties of mozzarella cheese showed different patterns of casein digestion (Oberg et al. 2002). There have not been extensive studies to compare the proteolytic activity among different genera of LAB, but Tobiassen and coworkers found that strains belonging to *Lactobacillus* species exhibited lower activity than those belonging to the genus *Lactococcus* but higher activity than propionibacteria (1997). Proteolytic systems contribute to the biochemical changes occurring during the ripening of various fermented dairy and non-dairy food products, leading to the production of bioactive peptides with immunological and health effects (Shu et al. 2018). They are also responsible for the organoleptic properties of the end products. The functional properties (body, texture, melt, and stretch) of mozzarella cheese depend on the proteolytic system, which has been shown to be strain dependent (Oommen et al. 2002). The proteolytic activity of *Lpb. plantarum* strains may have a role in preventing *Clostridium botulinum* toxin formation in refrigerated foods (Skinner et al. 1999). Recent work suggests that the proteolytic activity of some strains of *L. delberueckii* subsp. *bulgaricus* can help in efforts to develop hypoallergenic dairy products, as they have the capacity to degrade the whey protein β-lactoglobulin (Pescuma et al. 2011).

## 2.3.4 Miscellaneous Metabolic Aspects of *Lactobacillus*

A metabolic route for butyrogenic activity has been identified in two of three strains of *Lpb. plantarum*; production of butyric acid was linked to higher presence and uptake of glutamine in the medium (Botta et al. 2017). The production of bioactive compounds such γ-aminobutyric acid (GABA) and 1,3-propanediol was observed in barley fermented with a strain of *Lim. reuteri*, but that was also accompanied by an increase in the biogenic amine, histamine (Pallin et al. 2016). A *Lvl. brevis* strain has also been optimized to produce higher levels of GABA by modifying the culture parameters and after addition of monosodium glutamate (Wu et al. 2018).

The discovery that some *Lactobacillus* strains can utilize the human milk oligosaccharides as a carbon source (Bidart et al. 2016, 2018) instigated further research on bovine milk oligosaccharides, as human milk is unlikely to be commercially exploitable for its oligosaccharides.

The physiochemical and organoleptic changes occurring during food fermentation by *Lactobacillus* are due to carbohydrate fermentation and/or proteolysis, but changes in non-nutrient contents in the foods occur as well. A reduction of phenolic acids was observed during cherry juice and broccoli puree fermentation by *Lactobacillus* spp., and the resulting phenolic derivatives have presumably higher intestinal bioavailability and hence higher biological activity, as shown *in vivo* and in the SHIME colonic model (Barroso et al. 2014; Filannino et al. 2015). *Lpb. plantarum* efficiently decarboxylated hydroxycinnamic acids (e.g., *o*-coumaric, *m*-coumaric, *p*-coumaric, caffeic, ferulic, and sinapic acids) and reduced them into phenylpropionic acids (Santamaria et al. 2018). The latter have higher antioxidant and biological activity, and the hydroxycinnamate reductase activity was associated with the presence of *hcrAB* genes, which was not widely present among LAB. In an interesting study of malolactic fermentation of red wine, an *Lpb. plantarum* increased the concentration of pyranoanthocyanins, leading to enhanced colour compared to *Oenococcus oeni* (traditional malolactic starter), which decreases the polymerization of anthocyanins (Wang et al. 2018). *L. acidophilus* fermentation of sweet potato quadrupled the levels of caffeic acid, 3,5-dicaffeoylquinic acid, *p*-coumaric acid, and ferulic acid compared to those in raw and boiled sweet potatoes (Shen et al. 2018). An increase in phytosterols as well as *in vitro* inhibitory activity of cancer cells proliferation were also observed with a hydrophilic extract of the fermented products. A *Companilactobacillus crustorum* strain exhibited saponin-degrading capacity through their enzymatic activity of their β-glucuronidase, hence reducing the saponins' hemolytic activity (Qian et al. 2018). The bioconversion of the phenolic compounds might have led to cleavage of the sugar moiety for bacterial growth and externalized their bioactive phytochemicals as a detoxification mechanism (Theilmann et al. 2017). As the bioconversion of glucosinolates into isothiocyanates by intrinsic enzymes or gut microbial myrosinase is essential for their biological activity (e.g., potential anti-cancer properties), the role of *Lactobacillus* fermentation in this bioconversion deserves further investigation. Some lactobacilli showed capacity for phytate hydrolysis, which can have a significant impact of minerals availability from cereals and legume products (Amritha and Venkateswaran 2017). The previously mentioned examples of bioconversion open new opportunities in the fields of functional foods and health-promoting sectors.

In a striking example of the adaptability of the genus *Lactobacillus* members to their niche, the ability of intracellular glycogen synthesis and storage has been reported (Goh and Klaenhammer 2014). The capacity of glycogen synthesis and breakdown is present only in select strains associated with mammalian hosts such as intestinal lactobacilli. Goh and Klaenhammer (2014) attributed the competitive retention of *L. acidophilus* in the mouse intestinal tract to their capacity to synthesize intracellular glycogen. Additionally, some *Lpb. plantarum* strains native to traditionally fermented gruels transiently expressed amylolytic activity (e.g., *L. amylovorus* and *Lacticaseibacillus manihotivorans*) (Humblot et al. 2014). However, amylolytic activity is infrequent in the majority of *Lactobacillus* members, and they are normally assisted by the indigenous enzymes in cereals during fermentation (Hu et al. 2013).

The ability of *Lim. reuteri* strain to chelate selenium (Se) into selenocysteines was proposed for use as vehicle for Se delivery in populations where Se deficiency is common, perhaps in a food-formulated manner (Mangiapane et al. 2014).

# 2.4 EVOLUTIONARY AND GENETIC ASPECTS

Although the new classification is expected to shed light on the evolutionary aspects of the *Lactobacillus* complex, it has been suggested previously that an association exists between the genome size of an organism and the environment it inhabits. Genome reduction among *Lactobacillus* species is related to their adaptation to the gastrointestinal tract of animals and humans. Due to coevolution with their hosts, some species may have acquired a selective association with their hosts, and thus they have lost a considerable part of their genomes through this coevolutionary relationship (Papizadeh et al. 2017). *Lactobacillus* species such as *L. crispatus*, *L. gasseri*, *L. iners*, and *L. jensenii* are commonly detected in the human vagina (Nunn and Forney 2016). Their niche specialization has led to smaller genomes and has reinforced the very narrow ecological distribution of these species. On the other hand, species such as *Lpb. plantarum* and *Lcb. casei* can be detected in a very wide variety of environments (O'Donnell et al. 2013). Such generalist species commonly have larger genomes and have some different evolutionary properties from those of niche specialists (Papizadeh et al. 2017). Because of their economic and health importance, there has been an interest in the genetic modification of lactobacilli. Mutants can be used to help understand the mechanisms of probiotic action of certain strains. There may also be an interest in using LAB as delivery vehicles for therapeutic proteins, as has been done with *Lactococcus lactis* and the local production of interleukin (IL)-10 (Steidler et al. 2009). Other opportunities are changing the immune-modulating activity of probiotics, production or enhanced production of antimicrobials, the expression of specific enzymes, and vaccines (van Pijkeren and Barrangou 2017). The acceptability by the public and regulators of such GMO LAB remains to be determined (Ortiz-Velez and Britton 2017).

# 2.5 CONCLUSIONS

The recent taxonomic changes in the old *Lactobacillus* genus complex represent a significant step toward unraveling the mysteries of these versatile bacteria. By addressing the shortcomings of the old taxonomy and introducing a more precise classification system, researchers are better equipped to explore the evolutionary history, ecological adaptations, and potential applications of lactobacilli. The genera of the *Lactobacillus* complex group are of substantial importance to humans; they are essential members of a healthy human microbiota and serve as technologically, biologically, and functionally important components of a healthy human diet. Important ecological niches of *Lactobacillus* complex include the gastrointestinal tract of animals and the urogenital tract of mammals, as well as breast milk and plants and soil. They are used in fermented foods, and many strains are also consumed as probiotics.

Based on their main sugar fermentation pathways—glycolysis and pentose phosphate pathways—the *Lactobacillus* species can be grouped into homofermentative and heterofermentative species. In addition, under different conditions, lactobacilli feature various alternative pyruvate metabolism pathways involving different electron acceptors. In general, *Lactobacillus* species exhibit low proteolytic activity, but species and strain differences are remarkable, and proteolytic activity of lactobacilli is important, for example, from the perspective of dairy technology. Genomics of lactobacilli reflect the diversity of the environmental niches in which they occur. Recent advancements in molecular biology have greatly benefited the research of lactobacilli and offered new information on the phylogenetic relationships of the genus and the functional diversity of various species.

# BIBLIOGRAPHY

Aguirre, L., Hebert, E. M., Garro, M. S. & Savoy De Giori, G. 2014. Proteolytic activity of Lactobacillus strains on soybean proteins. *LWT—Food Sci Technol*, 59, 780–785.

Amritha, G. K. & Venkateswaran, G. 2017. Use of lactobacilli in cereal-legume fermentation and as potential probiotics towards phytate hydrolysis. *Probiotics Antimicrob Proteins*, 10, 647–653.

Ayansina, A. D. & Amusan, O. A. 2013. Effect of higher concentrations of herbicides on bacterial populations in tropical soil. *Unique Res J Agric Sci*, 1, 1–5.

Barroso, E., Van De Wiele, T., Jimenez-Giron, A., Munoz-Gonzalez, I., Martin-Alvarez, P. J., Moreno-Arribas, M. V., Bartolome, B., Pelaez, C., Martinez-Cuesta, M. C. & Requena, T. 2014. *Lactobacillus plantarum* IFPL935 impacts colonic metabolism in a simulator of the human gut microbiota during feeding with red wine polyphenols. *Appl Microbiol Biotechnol*, 98, 6805–6815.

Bidart, G. N., Rodriguez-Diaz, J., Perez-Martinez, G. & Yebra, M. J. 2018. The lactose operon from *Lactobacillus casei* is involved in the transport and metabolism of the human milk oligosaccharide core-2 N-acetyllactosamine. *Sci Rep*, 8, 7152.

Bidart, G. N., Rodriguez-Diaz, J. & Yebra, M. J. 2016. The extracellular wall-bound beta-*N*-acetylglucosaminidase from *Lactobacillus casei* is involved in the metabolism of the human milk oligosaccharide lacto-*N*-triose. *Appl Environ Microbiol*, 82, 570–577.

Botta, C., Acquadro, A., Greppi, A., Barchi, L., Bertolino, M., Cocolin, L. & Rantsiou, K. 2017. Genomic assessment in *Lactobacillus plantarum* links the butyrogenic pathway with glutamine metabolism. *Sci Rep*, 7, 15975.

Chen, Y. S., Yanagida, F. & Shinohara, T. 2005. Isolation and identification of lactic acid bacteria from soil using an enrichment procedure. *Lett Appl Microbiol*, 40, 195–200.

Christensen, J. E., Dudley, E. G., Pederson, J. A. & Steele, J. L. 1999. Peptidases and amino acid catabolism in lactic acid bacteria. *Antonie Van Leeuwenhoek*, *76*, 217–246.

De Vuyst, L., Vrancken, G., Ravyts, F., Rimaux, T. & Weckx, S. 2009. Biodiversity, ecological determinants, and metabolic exploitation of sourdough microbiota. *Food Microbiol*, *26*, 666–675.

El-Gaish, S., Dalgalarrondo, M., Choiset, Y., Sitohy, M., Ivanova, I. V., Haertle, T. & Chobert, J.-M. 2010. Characterization of a new isolate of *Lactobacillus fermentum* IFO 3956 from Egyptian Ras cheese with proteolytic activity. *Eur Food Res Technol*, *230*, 635–643.

Filannino, P., Bai, Y., Di Cagno, R., Gobbetti, M. & Ganzle, M. G. 2015. Metabolism of phenolic compounds by *Lactobacillus* spp. during fermentation of cherry juice and broccoli puree. *Food Microbiol*, *46*, 272–279.

Ganzle, M. G. 2014. Enzymatic and bacterial conversions during sourdough fermentation. *Food Microbiol*, *37*, 2–10.

Ganzle, M. G. 2015. Lactic metabolism revisited: Metabolism of lactic acid bacteria in food fermentations and food spoilage. *Curr Opin Food Sci*, *2*, 106–117.

Ganzle, M. G., Vermeulen, N. & Vogel, R. F. 2007. Carbohydrate, peptide and lipid metabolism of lactic acid bacteria in sourdough. *Food Microbiol*, *24*, 128–138.

Goh, Y. J. & Klaenhammer, T. R. 2014. Insights into glycogen metabolism in *Lactobacillus acidophilus*: Impact on carbohydrate metabolism, stress tolerance and gut retention. *Microb Cell Fact*, *13*, 94.

Heeney, D. D., Gareau, M. G. & Marco, M. L. 2018. Intestinal Lactobacillus in health and disease, a driver or just along for the ride? *Curr Opin Biotechnol*, *49*, 140–147.

Hu, Y., Ketabi, A., Buchko, A. & Ganzle, M. G. 2013. Metabolism of isomalto-oligosaccharides by *Lactobacillus reuteri* and *bifidobacteria* . *Lett Appl Microbiol*, *57*, 108–114.

Hugenholtz, J. 1993. Citrate metabolism in lactic acid bacteria. *FEMS Microbiol Rev*, *12*, 165–178.

Humblot, C., Turpin, W., Chevalier, F., Picq, C., Rochette, I. & Guyot, J. P. 2014. Determination of expression and activity of genes involved in starch metabolism in *Lactobacillus plantarum* A6 during fermentation of a cereal-based gruel. *Int J Food Microbiol*, *185*, 103–111.

Ianniello, R. G., Zheng, J., Zotta, T., Ricciardi, A. & Ganzle, M. G. 2015. Biochemical analysis of respiratory metabolism in the heterofermentative *Lactobacillus spicheri* and *Lactobacillus reuteri* . *J Appl Microbiol*, *119*, 763–775.

Jost, T., Lacroix, C., Braegger, C. & Chassard, C. 2015. Impact of human milk bacteria and oligosaccharides on neonatal gut microbiota establishment and gut health. *Nutr Rev*, *73*, 426–437.

Kandler, O. 1983. Carbohydrate metabolism in lactic acid bacteria. *Antonie Van Leeuwenhoek*, *49*, 209–224.

Kleerebezem, M., Boekhorst, J., Van Kranenburg, R., Molenaar, D., Kuipers, O. P., Leer, R., Tarchini, R., et al. 2003. Complete genome sequence of *Lactobacillus plantarum* WCFS1. *Proc Natl Acad Sci*, *100*, 1990–1995.

Leisner, J. J., Vancanneyt, M., Goris, J., Christensen, H. & Rusul, G., 2000. Description of *Paralactobacillus selangorensis* gen. nov., sp. nov., a new lactic acid bacterium isolated from chili bo, a Malaysian food ingredient. *Int J Syst Evol Microbiol*, *50*(1), 19–24.

Mahasneh, S. A. & Mahasneh, A. M. 2017. Probiotics: A promising role in dental health. *Dent J (Basel)*, *5*, 26.

Mangiapane, E., Lamberti, C., Pessione, A., Galano, E., Amoresano, A. & Pessione, E. 2014. Selenium effects on the metabolism of a Se-metabolizing *Lactobacillus reuteri*: Analysis of envelope-enriched and extracellular proteomes. *Mol Biosyst*, *10*, 1272–1280.

Monedero, V., Maze, A., Boel, G., Zuniga, M., Beaufils, S., Hartke, A. & Deutscher, J. 2007. The phosphotransferase system of *Lactobacillus casei*: Regulation of carbon metabolism and connection to cold shock response. *J Mol Microbiol Biotechnol*, *12*, 20–32.

Morovic, W., Hibberd, A. A., Zabel, B., Barrangou, R. & Stahl, B. 2016. Genotyping by PCR and high-throughput sequencing of commercial probiotic products reveals composition biases. *Front Microbiol*, *7*, 1747.

Muck, R. E., Nadeau, E. M. G., Mcallister, T. A., Contreras-Govea, F. E., Santos, M. C. & Kung, L., Jr. 2018. Silage review: Recent advances and future uses of silage additives. *J Dairy Sci*, *101*, 3980–4000.

Nunn, K. L. & Forney, L. J. 2016. Unraveling the dynamics of the human vaginal microbiome. *Yale J Biol Med*, *89*, 331–337.

Oberg, C. J., Broadbent, J. R., Strickland, M. & McMahon, D. J. 2002. Diversity in specificity of the extracellular proteinases in *Lactobacillus helveticus* and *Lactobacillus delbrueckii* subsp. *bulgaricus* . *Lett Appl Microbiol*, *34*, 455–460.

O'Donnell, M. M., O'Toole, P. W. & Ross, R. P. 2013. Catabolic flexibility of mammalian-associated lactobacilli. *Microb Cell Fact*, *12*, 48.

Oommen, B. S., McMahon, D. J., Oberg, C. J., Broadbent, J. R. & Strickland, M. 2002. Proteolytic specificity of *Lactobacillus delbrueckii* subsp. *bulgaricus* influences functional properties of mozzarella cheese. *J Dairy Sci*, *85*, 2750–2758.

Ortiz-Velez, L. & Britton, R. 2017. Genetic tools for the enhancement of probiotic properties. *Microbiol Spectr*, *5*.

Ozen, M. & Dinleyici, E. C. 2015. The history of probiotics: The untold story. *Benef Microbes*, *6*, 159–165.

Pallin, A., Agback, P., Jonsson, H. & Roos, S. 2016. Evaluation of growth, metabolism and production of potentially bioactive components during fermentation of barley with *Lactobacillus reuteri* . *Food Microbiol*, *57*, 159–171.

Papizadeh, M., Rohani, M., Nahrevanian, H., Javadi, A. & Pourshafie, M. R. 2017. Probiotic characters of Bifidobacterium and Lactobacillus are a result of the ongoing gene acquisition and genome minimization evolutionary trends. *Microb Pathog*, *111*, 118–131.

Pescuma, M., Hebert, E. M., Rabesona, H., Drouet, M., Choiset, Y., Haertle, T., Mozzi, F., De Valdez, G. F. & Chobert, J. M. 2011. Proteolytic action of *Lactobacillus delbrueckii* subsp. *bulgaricus* CRL 656 reduces antigenic response to bovine beta-lactoglobulin. *Food Chem*, *127*, 487–492.

Petrova, M. I., Reid, G., Vaneechoutte, M. & Lebeer, S. 2017. *Lactobacillus iners*: Friend or Foe? *Trends Microbiol*, *25*, 182–191.

Qian, B., Yin, L., Yao, X., Zhong, Y., Gui, J., Lu, F., Zhang, F. & Zhang, J. 2018. Effects of fermentation on the hemolytic activity and degradation of *Camellia oleifera saponins* by *Lactobacillus crustorum* and *Bacillus subtilis* . *FEMS Microbiol Lett*, 365.

Salvetti, E., Harris, H. M. B., Felis, G. E. & O'Toole, P. W. 2018. Comparative genomics reveals robust phylogroups in the genus *Lactobacillus* as the basis for reclassification. *Appl Environ Microbiol*. https://doi.org/10.1128/AEM.00993-18.

Santamaria, L., Reveron, I., Lopez De Felipe, F., De Las Rivas, B. & Munoz, R. 2018. Ethylphenols formation by *Lactobacillus plantarum*: Identification of the enzyme involved in the reduction of vinylphenols. *Appl Environ Microbiol*. https://doi.org/10.1128/AEM.01064-18.

Shen, Y., Sun, H., Zeng, H., Prinyawiwatkul, W., Xu, W. & Xu, Z. 2018. Increases in phenolic, fatty acid, and phytosterol contents and anticancer activities of sweet potato after fermentation by *Lactobacillus acidophilus* . *J Agric Food Chem*, *66*, 2735–2741.

Shu, G., Huang, J., Chen, L., Lei, N. & Chen, H. 2018. Characterization of goat milk hydrolyzed by cell envelope proteinases from *Lactobacillus plantarum* LP69: Proteolytic system optimization, bioactivity, and storage stability evaluation. *Molecules*, *23*, 1317.

Skinner, G. E., Solomon, H. M. & Fingerhut, G. A. 1999. Prevention of Clostridium botulinum type A, proteolytic B and E toxin formation in refrigerated pea soup by *Lactobacillus plantarum* ATCC 8014. *J Food Sci*, *64*, 724–727.

Steidler, L., Rottiers, P. & Coulie, B. 2009. Actobiotics as a novel method for cytokine delivery. *Ann N Y Acad Sci*, *1182*, 135–145.

Sun, Z., Harris, H. M., McCann, A., Guo, C., Argimón, S., Zhang, W., Yang, X., Jeffcry, I. B., Cooney, J. C., Kagawa, T. F. & Liu, W. 2015. Expanding the biotechnology potential of lactobacilli through comparative genomics of 213 strains and associated genera. *Nat Commun*, *6*(1), 8322.

Theilmann, M. C., Goh, Y. J., Nielsen, K. F., Klaenhammer, T. R., Barrangou, R. & Abou Hachem, M. 2017. *Lactobacillus acidophilus* metabolizes dietary plant glucosides and externalizes their bioactive phytochemicals. *MBio*, *8*, e1421–17.

Thongaram, T., Hoeflinger, J. L., Chow, J. & Miller, M. J. 2017. Human milk oligosaccharide consumption by probiotic and human-associated bifidobacteria and lactobacilli. *J Dairy Sci*, *100*, 7825–7833.

Tobiassen, R. O., Stepaniak, L. & Sorhaug, T. 1997. Screening for differences in the proteolytic systems of *Lactococcus*, *Lactobacillus* and *Propionibacterium* . *Z Lebensm Unters Forsch A*, *2004*, 273–278.

Uyeno, Y., Shigemori, S. & Shimosato, T. 2015. Effect of probiotics/prebiotics on cattle health and productivity. *Microbes Environ*, *30*, 126–132.

Van Pijkeren, J. P. & Barrangou, R. 2017. Genome editing of food-grade Lactobacilli to develop therapeutic probiotics. *Microbiol Spectr*, 5.

Verma, D., Garg, P. K. & Dubey, A. K. 2018. Insights into the human oral microbiome. *Arch Microbiol*, *200*, 525–540.

Vermeulen, N., Pavlovic, M., Ehrmann, M. A., Ganzle, M. G. & Vogel, R. F. 2005. Functional characterization of the proteolytic system of *Lactobacillus sanfranciscensis* DSM 20451T during growth in sourdough. *Appl Environ Microbiol*, *71*, 6260–6266.

Viana, R., Monedero, V., Dossonnet, V., Vadeboncoeur, C., Perez-Martinez, G. & Deutscher, J. 2000. Enzyme I and HPr from *Lactobacillus casei*: Their role in sugar transport, carbon catabolite repression and inducer exclusion. *Mol Microbiol*, *36*, 570–584.

Wang, S., Li, S., Zhao, H., Gu, P., Chen, Y., Zhang, B. & Zhu, B. 2018. Acetaldehyde released by *Lactobacillus plantarum* enhances accumulation of pyranoanthocyanins in wine during malolactic fermentation. *Food Res Int*, *108*, 254–263.

Wu, C. H., Hsueh, Y. H., Kuo, J. M. & Liu, S. J. 2018. Characterization of a potential probiotic *Lactobacillus brevis* RK03 and efficient production of gamma-Aminobutyric acid in batch fermentation. *Int J Mol Sci*, *19*, 143.

Yang, J., Cao, Y., Cai, Y. & Terada, F. 2010. Natural populations of lactic acid bacteria isolated from vegetable residues and silage fermentation. *J Dairy Sci*, *93*, 3136–3145.

Zagorec, M. & Champomier-Verges, M. C. 2017. *Lactobacillus sakei*: A starter for sausage fermentation, a protective culture for meat products. *Microorganisms*, 5, 56.

Zapata, J., Mateo-Vivaracho, L., Lopez, R. & Ferreira, V. 2012. Automated and quantitative headspace in-tube extraction for the accurate determination of highly volatile compounds from wines and beers. *J Chromatogr A*, *1230*, 1–7.

Zhao, X. & Ganzle, M. G. 2018. Genetic and phenotypic analysis of carbohydrate metabolism and transport in *Lactobacillus reuteri* . *Int J Food Microbiol*, *272*, 12–21.

Zheng, J., Ruan, L., Sun, M. & Ganzle, M. 2015. A genomic view of Lactobacilli and pediococci demonstrates that phylogeny matches ecology and physiology. *Appl Environ Microbiol*, *81*, 7233–7243.

Zheng, J., Wittouck, S., Salvetti, E., Franz, C. M., Harris, H. M., Mattarelli, P., O'toole, P. W., Pot, B., Vandamme, P., Walter, J. & Watanabe, K. 2020. A taxonomic note on the genus *Lactobacillus*: Description of 23 novel genera, emended description of the genus *Lactobacillus* Beijerinck 1901, and union of *Lactobacillaceae* and *Leuconostocaceae*. *Int J Syst Evol Microbiol*, *70*(4), 2782–2858.

Zotta, T., Parente, E. & Ricciardi, A. 2017. Aerobic metabolism in the genus *Lactobacillus*: Impact on stress response and potential applications in the food industry. *J Appl Microbiol*, *122*, 857–869.

# Genus *Lactococcus*

# 3

## Atte von Wright

## 3.1 THE TAXONOMIC UNIT DEFINED

The genus *Lactococcus* (formerly "Group N Streptococci") has rapidly expanded, even after the 5th edition of this book, and the taxonomical status of some former species and subspecies has also been revised. The former *Lactococcus lactis* subspecies *cremoris* has been elevated to a distinct species, *L. cremoris*. The former *L. lactis* subspecies *tructae* is now a subspecies of *L cremoris* (Li et al. 2019). Thus the *L. lactis* cluster consists now of *L. lactis* and the subspecies *hordniae* and *L. cremoris* with the subspecies *tructae*.

The other 22 recognized species or subspecies of lactococci are *L. allomyrinae* (Heo et al. 2019), *L. carnosus* and *L. paracarnosus* (Hilgarth et al. 2020), *L. chungangensis* (Cho et al. 2008), *L. formosensis* (with subspecies *bovis*) (Chen et al. 2014; Zhang et al. 2021), *L. fujiensis* (Cai et al. 2010), *L. garvieae* (Collins et al. 1983a, 1983b), *L. hirilactis* (Meucci et al. 2015), *L. hodotermopsidis* (Noda et al. 2020), *L. insecticola* (Noda et al. 2020), *L. kimchi* (Pheng et al. 2020), *L. laudensis* (Meucci et al. 2015), *L. nasutitermitis* (Yang et al. 2016), *L. petauri* (Goodman et al. 2017), *L. piscium* (Williams et al. 1990), *L. plantarum* (Collins 1983), *L. protaetiae* (Heo et al. 2020), *L. raffinolactis* (Schleifer et al. 1985), *L. reticulitermitis* (Yuki et al. 2018), *L. taiwanensis* (Chen et al. 2013), and *L. termiticola* (Noda et al. 2018).

Species identification is based on physiological, chemotaxonomic, and molecular biological criteria. Morphologically, lactococci are Gram-positive cocci of 0.5–1.5 μm in size, forming short chains. They are mesophilic (with the exception of psychrophilic *L. piscium*); ferment hexoses homofermentatively, producing l (+) lactic acid; and have complex growth requirements. Some physiological characteristics differentiating the species are listed in Table 3.1.

According to the molecular criteria (see previous references) *L. allomyriane*, *L. taiwanensis*, *L. protaetiae L. kimchi*, and *L. taiwanensis* form a cluster closely related to the *L. lactis* cluster, while *L. hirilactis L. nasutitermitis*, and *L. fujiensis* appear to form a separate cluster. A third cluster is formed by *L. garviae*, *L. petauri*, and *L. formosensis* and a fourth somewhat heterogenous cluster by *L. reticulotermitis*, *L. raffinolactis*, *L. chungangensis*, *L. laudensis*, *L. piscium*, *L. plantarum*, *L. carnosus*, and *L. paracarnosus*. So far *L. termiticola* appears to be a solitary species outside of clusters. The molecular taxonomic relationships of Lactococci have recently been reviewed by Mahony et al. (2023).

## 3.2 TYPICAL ECOLOGICAL NICHES

Lactococci have been isolated from vegetable materials (Cai et al. 2010; Chen et al. 2014; Pheng et al. 2020); milk (Meucci et al. 2015); meat products stored under modified atmosphere (Hilgarth et al. 2020); and fish or other animal sources, including the human gut (Pérez et al. 2011; Kubota et al. 2010). *L.*

DOI: 10.1201/9781003352075-4

**TABLE 3.1** Differential Phenotypic Characteristics of the Species in Genus *Lactococcus*

| PHENOTYPE / SPECIES | LACTOCOCCUS LACTIS *CLUSTER* | | | | | LACTOCOCCUS FUJIENSIS *CLUSTER* | | | | SPECIES BETWEEN CLUSTERS | LACTOCOCCUS GARVIAE *CLUSTER* | | |
|---|---|---|---|---|---|---|---|---|---|---|---|---|---|
| | *L. LACTIS* | *L. CREMORIS* | *L. ALLOMYRINAE* | *L. TAIWANENSIS* | *L. PROTAETIAE* | *L. KIMCHI* | *L. FUJIENSIS* | *L. NASUTITERMITIS* | *L. HIRILACTIS* | *L. TERMITICOLA* | *L. GARVIAE* | *L. PETAURI* | *L. FORMOSENSIS* |
| Growth at 10 °C | − | + | + | − | + | + | + | + | + | − | − | + | − |
| Growth at 40 °C | −[a] | − | − | − | − | − | − | − | − | ± | + | + | − |
| Growth in 3% NaCl | + | ± | NA | NA | NA | NA | + | − | + | − | + | + | NA |
| Growth in 4% NaCl | + | − | NA | + | NA | − | − | − | − | − | ± | + | + |
| Growth in 6% NaCl | − | − | NA | + | NA | − | − | − | − | − | − | + | + |
| Ammonia from arginine | + | − | + | + | + | NA | NA | NA | + | NA | + | NA | NA |
| Lactose fermentation | +[b] | + | − | + | − | + | + | ± | + | − | − | − | − |
| Mannitol fermentation | − | − | − | + | − | + | + | + | + | + | + | ± | + |
| Raffinose fermentation | −[c] | − | − | − | − | NA | + | − | + | − | − | − | − |
| Starch fermentation | − | − | NA | + | NA | NA | + | ± | + | NA | − | − | − |
| D-ribose fermentation | + | + | − | + | + | NA | + | + | − | − | + | + | + |
| D-xylose fermentation | + | − | − | − | + | + | − | + | − | − | − | − | − |
| Hydrolysis of aesculin | − | + | + | + | + | NA | + | + | + | + | + | NA | + |

(Continued)

**TABLE 3.1** (Continued) Differential Phenotypic Characteristics of the Species in Genus *Lactococcus*

| | | | | LACTOCOCCUS LAUDENSIS *CLUSTER* | | | | | | |
|---|---|---|---|---|---|---|---|---|---|---|
| SPECIES<br>PHENOTYPE | *L.*<br>PISCIUM | *L.*<br>CHUNGANGENSIS | *L.*<br>PLANTARUM | *L.*<br>LAUDENSIS | *L.*<br>RAFFINOLACTIS | *L.*<br>INSECTICOLA | *L.*<br>HODOTERMOPSIDIS | *L.*<br>RETICULITERMITIS | *L.*<br>CARNOSUS | *L.*<br>PARACARNOSUS |
| Growth at 10 °C | − | + | − | + | − | + | − | + | + | + |
| Growth at 40 °C | − | − | − | − | − | − | − | ± | − | − |
| Growth in 3% NaCl | − | − | + | + | − | −d | − | + | ± | ± |
| Growth in 4% NaCl | − | − | NA | + | − | − | − | ± | − | − |
| Growth in 6% NaCl | − | − | − | − | − | − | − | − | − | − |
| Ammonia from arginine | − | − | − | + | − | NA | NA | NA | NA | NA |
| Lactose fermentation | + | − | − | + | − | − | − | − | + | + |
| Mannitol fermentation | − | + | + | + | − | − | − | − | + | + |
| Raffinose fermentation | − | − | − | − | + | − | − | − | + | + |
| Starch fermentation | − | ± | − | − | ± | − | − | + | − | − |
| D-ribose fermentation | − | − | − | − | − | − | − | − | − | − |
| D-xylose fermentation | ± | − | − | + | + | − | + | − | − | − |
| Hydrolysis of aesculin | + | + | + | + | − | NA | NA | NA | NA | NA |

[a] Some *L. lactis* subsp. *lactis* strains may grow at +40 °C
[b] *L. lactis* subsp *hordniae* usually negative
[c] *L. cremoris* subsp. *tructae* usually positive
[d] Grows at 2% NaCl
± = weak or variable reaction, NA = Not available

*chungangensis* was originally isolated from activated sludge foam (Cho et al. 2008) and *L. petauri* from an abscess of a sugar glider (*Petaurus breviceps*). Five recently characterized species, *L. termiticola*, *L. nasutitermitis* and *L. reticulitermitis*, *L. insecticola*, and *L. hodotermopsides* originate from termite gut (Noda et al. 2018, 2020; Yang et al. 2016; Yuki et al. 2018) and *L. protaetiae* from the larvae oriental beetle *Protaetia brevitarsis seulensis* (Heo et al. 2020).

    *L. lactis* and *L. cremoris*, the species most commonly found in raw milk, are thought to arise as a contamination from forage. *L. garvieae* was originally isolated from a mastitis case (Collins et al. 1983) but is better known, together with *L. piscium*, as a fish pathogen. *L. garvieae* typically occurs in warm water species, while *L. piscium* occurs at temperatures below 15 °C (Venderell et al. 2006; Williams et al. 1990). *L. garvieae* can also frequently be isolated from dairy foods (Fortina et al. 2003), and the findings suggest that the fish isolates of *L. garvieae* are lactose negative, while dairy isolates ferment lactose (Fortina et al. 2009). *L. garviae* has been increasingly detected also in human clinical cases (see Section 3.5.2).

# 3.3 DAIRY LACTOCOCCI

*L. lactis* and *cremoris* are the lactococci traditionally used in dairy applications. The phenotypic difference between *cremoris* and *lactis* is the salt tolerance and ability to hydrolyze arginine, both typical for *L. lactis* but absent in *L. cremoris*. The diacetyl-producing variants of *L. lactis* are often referred as biovar. *diacetylactis* (Batt 2000).

    The two subspecies, with so far no dairy applications, are *L. lactis* subsp. *hordniae* (isolated from a grasshopper) and *L. cremoris* subsp. *tructae* (isolated from the intestinal mucus of salmonids). The former has a limited range of fermentable substrates, while the latter differs from the typical *L. cremoris* by possessing arginine dehydrolase activity (Pérez et al. 2011).

## 3.3.1 Starter Use

Dairy lactococci, often together with *Leuconostoc mesenteroides* subsp. *cremoris*, are essential for the mesophilic dairy starters used both for the production of fermented milks and cheeses (Stanley 1998; Courtney 2000). Mesophilic starters are often further classified according to the species composition, the starters containing a single *L. lactis* strain being called 0-starters, while D, L, and DL starters also contain aroma producers, either *L. lactis* subsp. *lactis* biovar. *diacetylactis*, *Lc. mesenteroides*, or both, respectively.

## 3.3.2 Metabolic Characteristics

While dairy lactococci share the basic physiological features of other lactic acid bacteria (LAB), the pathways of their sugar fermentations, proteolytic systems, and genetic basis have been extensively studied owing to their importance to the dairy industry. The availability of genomic and proteomic data and techniques such as *in vivo* nuclear magnetic resonance spectroscopy are also making the metabolomic approaches feasible. In the following paragraphs, some of the key metabolic features of dairy lactococci are briefly summarized.

### 3.3.2.1 Carbohydrate Metabolism

#### 3.3.2.1.1 Sugar Transport
The central hexose transport system in dairy lactococci is the phoshoenolpyruvate phosphotransferase system (PEP-PTS; reviewed by Postma et al. 1993; Deutscher et al. 2006). In PEP-PTS the uptake of

sugars is coupled to their phosphorylation, phosphoenolpyruvate (PEP) acting as the donor of the phosphate group via several steps. PEP first phosphorylates the so-called enzyme I (EI), which subsequently transfers the phosphate group to heat-stable protein (HPr-P). From HPr-P the phosphate is transferred via enzymes EIIA and EIIBC/EIICD to the carbohydrate to be transferred. EI and HPr are nonspecific cytoplasmic proteins, while EII enzymes are sugar specific and linked to the membrane by hydrophobic domains C or D. Glucose transport in dairy lactococci occurs either via the mannose-PTS (the major route) or in some cases by a specific glucose-PTS (Thompson 1987; Thompson and Saier 1981). Regarding the applications, the most important sugar is lactose, which is mainly transported by a specific PTS (de Vos et al. 1990).

In comparison to the extensive information available on hexose transport, the details of pentose transport in lactococci have not received much attention. Apparently they are transported mainly by specific permeases or symporters, as in other bacteria (Erlandson et al. 2000).

### 3.3.2.1.2 Sugar Fermentation

The hexose fermentation of lactococci occurs homofermentatively. While the fermentation of the 6-phosphorylated glucose produced by the PTS system proceeds in a straightforward manner via the glycolytic pathway, the fermentation of phosphorylated lactose requires the action of phospho-β-galactosidase, which splits the molecule to glucose and galactose-6P. Thereafter these two moieties have different fates. Glucose will be phosphorylated and enter glycolysis, while galactose-6P is isomerized to tagatose-6-P. After a further phosphorylation tagatose-1,6P is formed and subsequently split to two triose-3P molecules. These will then be metabolized to pyruvate and finally to lactic acid via the normal glycolytic route. The details of hexose metabolism in *L. lactis* have been reviewed by Neves et al. (2005).

The pentoses have to be processed by different isomerases and epimerases to xylulose-5P, from which the heterolactic phosphoketolase pathway proceeds to yield lactic acid and ethanol/acetaldehyde (Kandler 1983).

## 3.3.2.2 Alternative Pyruvate Metabolism, Diacetyl Formation

While most of the pyruvate formed in sugar metabolism is reduced to lactic acid, in certain circumstances pyruvate is also directed to other metabolic pathways, leading to a variety of end products (diacetyl, acetoin, butanediol, acetate, formate, ethanol, $CO_2$). The detailed pathways are outlined in Chapter 1.

Industrially the most important of these secondary pathways is the formation of diacetyl, the important aroma component of dairy products ("butter flavor"). If hexoses are the only sources of pyruvate, the formation of diacetyl is low. In dairy processes, the citrate, present in milk at 8–9-mM quantities, provides the main source of pyruvate for diacetyl formation. The aroma-producing lactococci (*L. lactis* subsp. *lactis* biovar. *diacetylactis*) possess citrate permease and lyase enzymes. The lyase splits citrate to acetate and oxaloacetate. Oxaloacetate is then decarboxylated to $CO_2$ and pyruvate (Hugenholz 1993). The activity of the citrate transport in lactococci is controlled by external pH (Garciá-Quintáns 1998).

## 3.3.2.3 Proteolytic System and Peptide Utilization

Lactococci are fastidious organisms and strain-dependently auxotrophic for many amino acids, the amount of which in milk is limited. The dairy lactococci have an elaborate proteolytic system coupled with peptide transport, allowing them to obtain the necessary amino acids from the breakdown products of milk proteins. The system consists of a cell-envelope protease (CEP), peptide transport systems, and intracellular peptidases (for reviews, see Savijoki et al. 2006; Doeven et al. 2005). While this machinery or some of its components also operate in other LAB, it is by far best characterized in dairy lactococci.

The CEP of *L. lactis*, PrtP, was the first cloned and characterized LAB CEP (Kok et al. 1988). Subsequently the functional domains of CEPs have been further elucidated (Siezen 1999; see also Chapter 1). The enzyme is a typical serine protease of subtilisin type, anchored to the cell wall by the C-terminus, and cleaves casein to oligopeptides of variable sizes. The generated peptides can be transported inside the cell by several peptide transport systems. The most important of them are the oligopeptide transport systems (Opp and Dpp), which both represent ATP-binding cassette transporters. The Opp system was first identified by cloning a chromosomal fragment enabling a spontaneous *L. lactis* mutant deficient in peptide transport to grow in milk (Tynkkynen et al. 1989). Subsequently, the fragment was shown to contain an operon coding for the oligopeptide binding protein (OppA), two membrane proteins (OppB and OppC), and two ATP-binding proteins (OppD and OppF) (Tynkkynen et al. 1993).

Opp is responsible for the transport of peptides ranging in size up to 18 amino acids. Smaller peptides are handled by di- and tripeptide transporters, such as DtpT and Dpp, Dpp being able to a degree complement the absence of the functional Opp in some lactococcal strains (Hagting et al. 1994; Foucaud et al. 1995).

Once inside, the peptides are broken down by a large number of different peptidases representing metallopeptidases, cysteine peptidases, and serine peptidases. Both endopeptidases, aminopeptidases, di- and tripeptidases, and several proline peptidases have been characterized (Christensen et al. 1999; Liu et al. 2010).

## 3.3.3 Genetics

Lactococci typically contain a large number of plasmids of variable sizes, many of them associated with important functions relevant for their applications in dairy products. Accordingly, the study of lactococcal plasmids dominated the early research, but the rapidly expanding sequence data has shifted the focus of the research on the actual chromosomal genetics of these organisms.

### 3.3.3.1 Sequenced Lactococcal Genomes

The first four published complete genome sequences of dairy lactococci were those of *L. lactis* subsp. *lactis* IL1403, *L. lactis* subsp. *cremoris* SK11, *L. lactis* subsp. *cremoris* MG1363, and *L. lactis* subsp. *lactis* KF147. Characteristic to these genomes is a large number of IS elements and transposons (Bolotin et al. 2001; Makarova et al. 2006), indicating a high degree of plasticity. In the strain KF147, isolated from plant material, several genes coding the degradation of xylan, arabinan, glucans, and fructans and for uptake and conversion of cell wall degradation products were detected, indicating adaptation to the peculiar ecological niche (Siezen et al. 2010).

During the last few years the genomic data of especially dairy lactococci has expanded dramatically, with more than 440 totally or partially sequenced genomes of *L. lactis* or *L. cremoris* in public databases. The new findings have mainly confirmed the conclusions derived from the first genomic analyses. To sum up, the chromosome sizes of dairy lactococci range from 2,250 to 2,589 Mbp (*L. lactis* generally having slightly larger genomes) and typically contain more than 2,300 predicted coding sequences (Kelleher et al. 2017; Yu et al. 2017). The strains tend to genetically cluster according to the niches from which they have been isolated. The history of domestication has resulted in reduced genetic diversity and substrate range in dairy lactococci in comparison to strains derived from the environment or other food sources (see the review of Laroute et al. 2017).

### 3.3.3.2 Lactococcal Plasmid Biology

Plasmids coding for important technological properties have naturally attracted much attention since their discovery in the lactococci (McKay et al. 1972; Cords et al. 1974). In these bacteria, lactose fermentation and proteinase activities are almost invariably associated with relatively large (from 17 kbp to more than 50 kbp) plasmids (McKay 1983). The linkage of citrate permease gene in diacetyl-producing *L. lactis* to small (approx. 8.7 kbp) plasmids was also detected relatively early (Kempler and McKay 1981).

Other plasmid-encoded traits in dairy lactococci include mucoidness or ability to produce extracellular polysaccharides, production of bacteriocins, phage resistance and antibiotic resistance. Many of these aspects are extensively covered in a recent review of Ainsworth et al. (2014)

Mucoidness, or the ability to produce extracellular polysaccharides, is a property of some lactococcal strains that have traditionally been used to give body and texture to certain types of Scandinavian fermented milks (Macura and Townsley 1984). Several lactococcal plasmids associated with mucoid phenotype have been identified (Vedamuthu and Neville 1986; von Wright and Tynkkynen 1987; Neve et al. 1988). Genetic analysis of one of the plasmids, pNZ4000, has indicated the involvement of at least 14 genes in exopolysaccharide production (van Kranenburg et al. 1997; van Kranenburg et al. 2000).

Bacteriocins are bacterial proteins or peptides that inhibit the growth of other bacterial strains or species. Bacteriocins are classified into three groups (Nes et al. 1996): class I containing modified amino acids such as lanthionine and β-methyllanthionine, small heat-stable nonlantibiotics (class II), and large heat-labile bacteriocins (class III). Several plasmid-encoded bacteriocins, such as lacticins (class I) and lactococcins (class II) have been characterized in dairy lactococci (Rincé et al. 1997; Dougherty et al. 1998; Neve et al. 1988; van Belkum et al. 1992).

Three main phage resistance mechanisms with different subdivisions are known to function in lactococci: inhibition of phage adsorption, restriction/modification (R/M) systems, and abortive infection (Abi) or intracellular inhibition of phage development. Tens of plasmid codings for these mechanisms are known (Hill 1993; Daly et al. 1996; Chopin et al. 2005). Even a cursory attempt to describe them in this context is obviously out of the scope of this section. However, it is worth noting that a fourth mechanism, a plasmid encoded CRISPR-Cas type III phage immunity system, has also been characterized (Millen et al. 2012).

A well-characterized 65-kbp conjugative plasmid, pNP40, from *Lc. lactis* ssp. *lactis* biovar *diacetylactis* DRC3 is an illustrative example of lactococcal plasmids associated with multiple functions. pNP40 confers resistance against the bacteriocin nisin and also protects the strain from the attack of the lytic bacteriophage c2 (McKay and Baldwin 1984). Subsequently, the plasmid has been shown to contain genes for at least two Abi systems (AbiE and AbiF) and a restriction–modification system (LlaJI) and for a mechanism-blocking phage DNA injection (Garvey et al. 1995, 1996; O'Driscoll 2004). The fact that this plasmid also codes for cadmium resistance, which can be used as a selective marker, further emphasizes its potential usefulness in engineering dairy starter strains for enhanced phage resistance (Trotter et al. 2001). Among the other genes characterized in this plasmid are a homolog of *recA* (a gene central in DNA recombination and repair) as well as a gene sharing homology with *umuC* (a gene involved in the so-called SOS response to DNA damage) (Garvey et al. 1997). These findings suggest a possible role for genes active in DNA recombination and repair in some of the Abi mechanisms.

Although lactococci are generally sensitive to most antibiotics in clinical use, occasional antibiotic resistances have been observed, based on both chromosomally located and plasmid-linked genes. From the safety point of view, the possibility of the horizontal transfer of these resistances is of particular interest.

A 30-kbp completely sequenced theta-replicating plasmid pK214 coding resistance for streptomycin, tetracycline, and chloramphenicol, with the resistance determinants apparently related to corresponding genes in *Staphylococcus*, *Listeria*, and *Enterococcus*, has been detected in a lactococcal strain isolated from soft cheese. The plasmid also contains five insertion (IS) elements, three of which apparently originate from *Enterococcus faecium* (Teuber et al. 1999).

Plasmids carrying the Tn916 transposon and the associated tetracycline resistance gene *tet(M)* have been detected in lactococci isolated from raw milk (Belén Flórez et al. 2008), and this resistance could be conjugatively transferred to *Lactococcus* and *Enterococcus* recipients.

Another tetracycline resistance gene, *tet(S)*, located on a plasmid of an *L. lactis* strain of fish origin, has been successfully electroporated into *L. garvieae*, which was subsequently used as donor to conjugate the plasmid into *Listeria monocytogenes* (Guglielmetti et al. 2009).

### 3.3.3.3 Genetics of Nisin Biosynthesis

Nisin is a broad-spectrum lantibiotic bacteriocin produced by *L. lactis* and the only bacteriocin that has been accepted as a food additive (E234) to control contaminating Gram-positive bacteria (Courtney 2000).

The genes for nisin biosynthesis and immunity as well as for sucrose utilization are known to reside on a 70-kbp conjugative transposon. This block is flanked by direct repeats of TTTTTG, probably representing a duplication of the target sequence due to transposition. However, no inverted repeats flanking the nisin–sucrose gene block have been identified (Rauch and de Vos 1992). The nisin–sucrose transposons vary slightly from strain to strain, which has resulted in different designations (Tn*5301*, Tn*5276*, Tn*5307*, etc.). The conjugative nature of the nisin–sucrose transposons allow for their introduction even to heterologous hosts, such as dairy enterococci (Broadbent et al. 1995).

### 3.3.3.4 Genetic Modification of Lactococci

Natural gene transfer mechanisms well characterized in lactococci include transduction, observed already in the early 1960s (Sandine et al. 1962; Allen et al. 1963), and conjugation (for a review, see Fitzgerald and Gasson 1988). The so-called sex factors associated with a high frequency of conjugation have also been characterized (Gasson et al. 1995).

Although natural transformation of lactococci have been reported occasionally (Møller-Madsen and Jensen 1962), and complete competence operons (competence meaning the capacity to intake foreign DNA through the cell wall) have been detected in sequenced *L. lactis* genomes (Wydau et al. 2006), natural transformation is not a generally applied gene transfer method in *L. lactis*. The *in vitro* transformation methods include protoplast transformation (Kondo and MacKay 1984; von Wright et al. 1985) and electroporation (Holo and Nes 1987), the latter being the current standard technique.

Since both *in vitro* transformation methods and a range of cloning vectors have been available since the early 1980s, recombinant DNA techniques have been a standard research tool in the molecular biological study of lactococci. The first cloning vectors were based on cryptic lactococcal plasmids with added antibiotic resistance genes as selection markers (de Vos and Simons 1994). Subsequently a variety of expression and integration vectors have been developed (see Mills et al. 2006 for a review). Particularly the nisin-controlled expression system (NICE), based on the ability of the extracellular nisin to act as a transcriptional activation of its own synthesis, has been a useful tool for the expression of heterologous proteins in *L. lactis* (Kleerebeezem et al. 1997). Another controlled expression system, based on stress-induction (SICE) and the promoter of the *GroESL*-operon, has also been described (Benbouziane et al. 2013).

While several food-grade selection systems based on bacteriocin resistance, carbohydrate metabolism, cadmium resistance, and others have also been proposed (Peterbauer et al. 2011; Mills et al. 2006), genetically modified lactococcal strains have not thus far been used in food applications. Instead, their use for the production of therapeutic proteins and oral vaccines could be a promising option (see reviews of Wang et al. 2016 and Wyszyńska et al. 2015). In their review, Bahey-El-Din et al. (2010) list tens of bacterial and viral antigens and immunomodulatory and therapeutic proteins that have been expressed in *L. lactis*.

An *L. lactis* strain expressing human interleukin-10 has actually been successfully used in a human phase I trial in patients with Crohn's disease (Braat et al. 2006). There are several recent reports on the successful use of genetically modified *L. lactis* to ameliorate intestinal inflammation in colitis animal models (Souza et al. 2016; de Cunha et al. 2020; Zurita-Turk et al. 2020), presumably by interfering with the expression of anti-and proinflammatory interleukins.

# 3.4 POTENTIAL USES OF LACTOCOCCI OTHER THAN *L. LACTIS* OR *L. CREMORIS*

Thus far *Lactococcus lactis* and *L. cremoris* are the only lactococci used as dairy starters. The potential of a few strains of *L. lactis* subspecies. *hordniae*, *L. raffinolactis*, *L. garvieae*, *L. piscium*, and *L. plantarum* in dairy applications has been assessed by Holler and Steele (1995). The strains were characterized for phage resistance, lactose fermentation, and growth in milk supplemented with glucose and casein hydrolysate. The most promising were the *L. raffinolactis* strains; however, even they lacked proteinase activity, and the activity was not expressed even after the introduction of the proteinase-associated plasmid from *L. lactis* ssp. *cremoris*. *L. garvieae* has been proposed for starter culture preparations, due to its prevalence in artisanal Italian cheeses and its potential contribution to the typical sensory characteristics of traditional cheeses (Fortina et al. 2007). Even probiotic uses have been proposed for an *L. garvieae* strain isolated from camel milk (Baig et al. 2022). The potential applications should, however, take into consideration the known pathogenicity of certain strains of the species (see Section 3.6.). Probiotic uses have also been proposed to an *L. petauri* originally isolated from the human gut (Li et al. 2021).

A possible field of application for novel lactococci could be vegetable and silage fermentations, considering the common association of lactococci and plant materials (Cai et al. 2010).

# 3.5 SAFETY ASPECTS

## 3.5.1 Dairy Lactococci

*L. lactis* and *L. cremoris* have been consumed in large quantities in dairy products for thousands of years and have a remarkably well-established history of safe use. However, cases of endocarditis (Mansour et al. 2016; Taniguchi et al. 2016; Rostagno et al. 2013; Lin et al. 2010; Halldorsdottir et al. 2002; Mannion and Rothburn 1990; Wood et al. 1985; Pellizzer et al. 1996), septicemia (Gurley et a. 2021; Karaaslan et al. 2011; Uchida et al. 2010; Glikman et al. 2010; Durand et al. 1995), necrotizing pneumonitis (Torre et al. 1990), septic arthritis (Campbell et al. 1993), cerebral abscess (Ahmed et al. 2021; Inoue et al. 2014; Feierabend et al. 2013; Topçu et al. 2011; Akhaddar et al. 2002), infant diarrhea (Karaaslan et al. 2016), peritonitis (Fragkiadakis et al. 2016), and intra-abdominal and liver abscesses (el Hattabi et al. 2021; Lee et al. 2014; Kim et al. 2010; Nakarai et al. 2000) have been reported. The predisposing factors include underlying disease, early age, immunocompromised status, surgical operations, and devices like catheters. However, liver abscess caused by *L. cremoris* in an immunocompetent adult has also been reported (Antolín et al. 2004). Taking into account the extent of human exposure to dairy lactococci, these infections represent extremely rare individual cases and should not be regarded as an indication of human pathogenicity. Moreover, in older cases, the possibility of misidentification of *L. garvieae* as *L. lactis* cannot be ruled out. Consequently, the European Food Safety Authority (EFSA) has included *L. lactis* and *L. lactis* in the QPS list of microorganisms. This means a greatly simplified safety assessment when these bacteria are submitted to EFSA for various applications (EFSA 2018).

In addition to human cases, both *L. lactis* and *L. cremoris* have been occasionally isolated from bovine mastitis (Rodrigues et al. 2016; Plumed-Ferrer et al. 2015a; Werner et al. 2014). Phenotypically mastitis-associated *L. lactis* and *L. cremoris* are able to grow at temperatures ≥ 37 °C. They tolerate lysozyme and gastrointestinal stress factors, have a wide range of fermentable substrates, and adhere better to cultured bovine mammary epithelial cells than dairy starter strains (Plumed-Ferrer et al. 2013). According to the whole-sequence analysis of one of the mastitis strains, the strain has adapted to grow in milk but also has a capsular polysaccharide identical to a known *Streptococcus agalactiae* virulence factor (Plumed-Ferrer et al. 2015b).

## 3.5.2 Other Lactococcal Species

The fish pathogenicity of *L. garvieae* and *L. piscium* was already mentioned in Section 3.2. In addition to pathogenicity, *L. piscium* is increasingly associated with spoilage of meat, seafood, and vegetables (Saraoui et al. 2016).

The pathogenicity and diagnostics of *L. garvieae* have especially been a focus of attention since the species can also be found in clinical specimens of warm-blooded animals and is frequently found as an accidental commensal microorganism in food products. It can, in rare cases, be also associated with severe human infections even in immunocompetent hosts (Gibello et al. 2016; Li et al. 2008 and references therein).

The putative virulence factors of *L. garvieae* have been reviewed by Ture and Altinok (2016) and Gibello et al. (2016). Capsular structures, presence of hemolysin activity, adhesion-associated LPxTG-motif containing surface proteins, siderofores, pilus structures, and possibly enolase ("moonlighting" as a cell surface protein) have all been implicated in fish, veterinary, and human infections, but several other factors may also be involved in *L. garvieae* pathogenicity.

# 3.6 CONCLUSIONS

Lactococci represent a relatively compact LAB genus of 13 species. The dairy species *L. lactis* ssp. *lactis* and ssp. *cremoris* have a long history of safe use in established applications, and both their genetics and physiology have been extensively studied. Their future applications may expand outside traditional food use to biotechnological production of pharmaceuticals. While the other species are currently not used as starters, some of them might have potential uses, provided that their safety can be established and they prove technologically compatible with industrial processes. *L. garvieae* and *L. piscium* are known fish pathogens, the former also occasionally infecting warm-blooded animals. The mechanisms behind their pathogenic potential should be elucidated before their biotechnological applications can be seriously considered.

# BIBLIOGRAPHY

Ahmed, I., K. Aziz, H. Tareen and M.A. Ahmed. 202. Brain abscess caused by *Lactococcus lactis* in a young male. *J Coll Physicians Surg Pak*. 30: 852–854. https://doi.org/10.29271/jcpsp.2021.07.852.

Ainsworth, S., S. Stockdale, F. Bottacini, J. Mahony and D. van Sinderen. 2014. The *Lactococcus lactis* plasmidome: Much learnt, yet still lots to discover. *FEMS Microbiol Rev*. 38(5): 1066–1088.

Akhaddar, A., B. El Mostarchid, M. Gazzaz and M. Boucetta. 2002. Cerebellar abscess due to *Lactococcus lactis*. A new pathogen. *Acta Neurochir*. 144: 305–306.

Allen, L.K., W.E. Sandine and P.R. Elliker. 1963. Transduction in *Streptococcus lactis*. *J Dairy Res*. 30: 351–357.

Antolín, J., R. Cigüenza, I. Salueña, E. Vázquez, J. Hernández and D. Espinós. 2004. Liver abscess caused by *Lactococcus lactis cremoris*: A new pathogen. *Scand J Infect Dis*. 36: 490–491.

Bahey-El-Din, M., C.G.M. Gahan and B.T. Griffin. 2010. *Lactococcus lactis* as a cell factory for delivery of therapeutic proteins. *Curr Gene Ther*. 10: 34–45.

Baig, M.A., M.S. Turner, S.Q. Liu, N.N. Shah and M.M. Ayyash. 2022. Heat, cold, acid, and bile salt induced differential proteomic responses of a novel potential probiotic *Lactococcus garvieae* C47 isolated from camel milk. *Food Chem*. 15(397): 133774. https://doi.org/10.1016/j.foodchem.2022.133774.

Batt, C.A. 2000. *Lactococcus* introduction. In *Encyclopedia of Food Microbiology*, Eds. R.K. Robinson, C.A Batt and P.D. Patel, pp. 1164–1166. Academic Press, San Diego.

Belén Flórez, A., M.S. Ammor and B. Mayo. 2008. Identification of *tet(M)* in two *Lactococcus lactis* strains isolated from a Spanish traditional starter-free cheese made of raw milk and conjugative transfer of tetracycline resistance to lactococci and enterococci. *Int J Food Microbiol*. 121: 189–194.

Benbouziane, B., P. Ribelles, C. Aubry, R. Martin, P. Kharrat, A. Riazi, P. Langella and L.G. Bermudez-Humaran. 2013. Development of a stress-inducible controlled expression (SICE) system in *Lactococcus lactis* for the production and delivery of therapeutic molecules at mucosal surfaces. *J Biotechnol*. 168: 120–129.

Bolotin, A., P. Wincker, S. Mauger, O. Jaillon, K. Malarme, J. Weissenbach, S.D. Ehrlich and A. Sorokin. 2001. The complete genome sequence of the lactic acid bacterium *Lactococcus lactis* ssp. *lactis* IL1403. *Genome Res*. 11: 731–753.

Braat, H., P. Rottiers, D.W. Hommes, N. Huygenbart, E. Reamut, J.P. Remon, S.J. van Deventer, S. Neirynck, M.P. Peppelenbosch and L. Steidler. 2006. A phase I trial with transgenic bacteria expressing interleukin-10 in Crohn's disease. *Clin Gastroenterol Hepatol.* 4: 754–759.

Broadbent, J.R., W.E. Sandine and J.K. Kondo. 1995. Characteristics of Tn5307 exchange and intergeneric transfer of genes associated with nisin production. *Appl Microbiol Biotechnol.* 44: 139–146.

Cai, Y., J. Yang, H. Pang and M. Kitahara. 2010. *Lactococcus fujiensis* sp nov., a lactic acid bacterium isolated from vegetable matter. *Int J Syst Evol Microbiol.* 61: 1590–1594.

Campbell, P., S. Dealler and J.O. Lawton. 1993. Septic arthritis and unpasteurized milk. *J Clin Pathol.* 46: 1057–1058.

Chen, Y., C. Chang, S. Pan, L. Wang, Y. Chang, H. Wu and F. Yanagida. 2013. *Lactococcus taiwanensis* sp. nov., a lactic acid bacterium isolated from fresh cummngcordia. *Int J Syst Evol Microbiol.* 63: 2405–2409.

Chen, Y., M. Otaguro, Y. Lin, S. Pan, S. Ji, C.Y.M. Liou, Y. Chang, H. Wu and F. Yanagida. 2014. *Lactococcus formosensis* sp. nov., a lactic acid bacterium isolated from *yan-tsai-tsin* (fermented broccoli stems). *Int J Syst Evol Microbiol.* 64: 146–151.

Cho, S.-L., S.-W. Nam, J.-H. Yoon, J.-S. Lee, A. Sukhoom and W. Kim. 2008. *Lactococcus chungangensis* sp. nov., a lactic acid bacterium isolated from activated sludge foam. *Int J Syst Evol Microbiol.* 58: 1844–1849.

Chopin, M.-C., A. Chopin and E. Bidnenko. 2005. Phage abortive infection in lactococci: Variations on a theme. *Curr Opin Microbiol.* 8: 473–479.

Christensen, J.E., E.G. Dudley, J.A. Pederson and J.L. Steele. 1999. Peptidases and amino acid catabolism in lactic acid bacteria. *Antonie van Leeuwenhoek.* 76: 217–246.

Collins, M.D., J.A.E. Farrow, B.A. Phillips and O. Kander. 1983a. *Streptococcus garvieae* sp. nov. and *Streptococcus plantarum* spec. nov. *J Gen Microbiol.* 199: 3427–3431.

Collins, M.D., J.A.E. Farrow, B.A. Phillips and O. Kandler. 1983b. *Streptococcus garvieae* sp. nov. *Int J Syst Bacteriol.* 36: 8–12.

Cords, R.B., L.L. McKay and P. Guerry. 1974. Extrachromosomal elements in group N streptococci. *J Bacteriol.* 117: 1149–1152.

Courtney, P.D. 2000. *Lactococcus lactis* Subspecies *lactis* and *cremoris.* In *Encyclopedia of Food Microbiology*, Eds. R.K. Robinson, C.A. Batt and P.D. Patel, pp. 1166–1171. Academic Press, San Diego.

da Cunha, V.P., T.M. Preisser, M.P. Santana, D.C.C. Machado, V.B. Pereira and A. Miyoshi. 2020. Invasive *Lactococcus lactis* producing mycobacterial Hsp65 ameliorates intestinal inflammation in acute TNBS-induced colitis in mice by increasing the levels of the cytokine IL-10 and secretory IgA. *J Appl Microbiol.* 129: 1389–1401. https://doi.org/10.1111/jam.14695.

Daly, C., G.F. Fitzgerald and R. Davis. 1996. Biotechnology of lactic acid bacteria with special reference to bacteriophage resistance. In *Lactic Acid Bacteria: Genetics, Metabolism and Applications*, Eds. G. Venema, J.H.J. Huis in't Veld and J. Hugenholtz, pp. 3–14. Kluyver Academic Publishers, Dordrecht.

Deutscher, J., C. Francke and P.W. Postma. 2006. How phosphotransferase system-related protein phosphorylation regulates carbohydrate metabolism in bacteria. *Microbiol Mol Biol Rev.* 70: 939–1031.

de Vos, W.M., I. Boerrigter, R.J. van Rooyen, B. Reiche and W. Hengstenberg. 1990. Characterization of the lactose-specific enzymes of the phosphotransferase system in *Lactococcus lactis*. *J Biol Chem.* 265: 22554–22560.

de Vos, W.M. and G.F.M. Simons. 1994. Gene cloning and expression in lactococci. In *Genetics and Biotechnology in Lactic Acid Bacteria*, Eds. M.J. Gasson and W.M. de Vos, pp. 52–105. Blackie Academic & Professional, Glasgow.

Doeven, M.K., J. Kok and B. Poolman. 2005. Specificity and selectivity determinants of peptide transport in *Lactococcus lactis* and other microorganisms. *Mol Microbiol.* 57: 640–649.

Dougherty, B.A., C. Hill, J.F. Weidman, D.R. Richardson, J.C. Venter and R.P. Ross. 1998. Sequence and analysis of the 60 kb conjugative, bacteriocin-producing plasmid pMRC01 from *Lactococcus lactis* DPC3147. *Mol Microbiol.* 29: 1029–1038.

Durand, J.M., M.C. Rousseau, J.M. Gandois, G. Kaplanski, M.N. Mallet and J. Soubeyrand. 1995. *Streptococcus lactis* septicemia in a patient with chronic lymphocytic leukemia. *Am J Hematol.* 50: 64–65.

EFSA (European Food Safety Authority). 2018. Update of the list of QPS-recommended biological agents intentionally added to food or feed as notified to EFSA 7: Suitability of taxonomic units notified to EFSA until September 2017. *EFSA J.* 16(1): 5131, 43 pp.

El Hattabi, K., M. Bouali, K. Sylvestre, F.Z. Bensardi, A. El Bakouri, Z. Khalid and A. Fadil. 2021. *Lactococcus lactis ssp lactis* a rare cause of liver abscesses: A case report and literature review. *Int J Surg Case Rep.* 81: 105831. https://doi.org/10.1016/j.ijscr.2021.105831.

Erlandson, K.A., J.-H. Park, W. El Khal, H.-H. Kao, P. Bsaran, S. Brydges and C.A. Batt. 2000. Dissolution of xylose metabolism in *Lactococcus lactis*. *Appl Environ Microbiol.* 66: 3974–3980.

Feierabend, D., R. Reichart, B. Romeike, R. Kalff and J. Walter. 2013. Cerebral abscess due to *Lactococcus lactis cremoris* in a child after sinusitis. *Clin Neurol Neurosurg.* 115: 614–616.

Fitzgerald, G.F. and M.J. Gasson. 1988. In vivo gene transfer systems and transposons. *Biochimie* 70: 489–502.

Fortina, M.G., G. Ricci, A. Acquati, G. Zeppa, A. Gandini and P. Manachini. 2003. Genetic characterization of some lactic acid bacteria occurring in an artisanal protected denomination origin (PDO) Italian cheese, the Toma Piemontese. *Food Microbiol*. 20: 397–404.

Fortina, M.G., G. Ricci and F. Borgo. 2009. A study on the lactose metabolism in *Lactococcus garvieae* reveals a genetic marker for distinguishing between dairy and fish biotypes. *J Food Prot*. 72: 1248–1254.

Fortina, M.G., G. Ricci, F. Foschino, C. Picozzi, P. Dolci, G. Zeppa, L. Cocolin and P.L. Manachini. 2007. Phenotypic typing, technological properties and safety aspects of *Lactococcus garvieae* strains from dairy environments. *J Appl Microbiol*. 103: 445–453.

Foucaud, C., E.R. Kunji, A. Hagting, J. Richard, W.N. Konings, M. Desmazeaud and B. Poolman. 1995. Specificity of peptide transport systems in *Lactococcus lactis:* Evidence for a third system, which transports hydrophobic di- and tripeptides. *J Bacteriol*. 177: 4652–4657.

Fragkiadakis, K., P. Ioannou, E. Barbounakis and G. Samonis. 2016. Intraabdominal abscesses by *Lactococcus lactis ssp. cremoris* in an immunocompetent adult with severe periodontitis and pernicious anemia. *ICD Cases* 11: 27–29.

Garciá-Quintáns, N., C. Magni, D. de Mendoza and P. López. 1998. The citrate transport system of *Lactococcus lactis* subsp *lactis* biovar diacetylactis is induced by acid stress. *Appl Environ Microbiol*. 64: 850–857.

Garvey, P.G., G.F. Fitzgerald and C. Hill. 1995. Cloning and DNA sequence analysis of two abortive infection phage resistance determinants from the lactococcal plasmid pNP40. *Appl Environ Microbiol*. 61: 4321–4328.

Garvey, P.G., C. Hill and G.F. Fitzgerald. 1996. The lactococcal plasmid pNP40 encodes a third bacteriophage resistance mechanism, one which affects phage DNA penetration. *Appl Environ Microbiol*. 62: 676–679.

Garvey, P.G., A. Rince, C. Hill and G.F. Fitzgerald. 1997. Identification of a recA homolog (recALP) on the conjugative lactococcal phage resistance plasmid pNP40: Evidence of a role for chromosomally encoded *recAL* in abortive infection. *Appl Environ Microbiol*. 63: 1244–1251.

Gasson, M.J., J.-J. Godon, C. Pillidge, T.J. Eaton, K. Jury and C.A. Shearman. 1995. Characterization and exploitation of conjugation in *Lactococcus lactis*. *Int Dairy J* 5: 757–762.

Gibello, A., F. Gálan-Sánchez, M. Mar Blanco, M. Rodríguez-Iglesias, L. Dominiguez and J.F. Fernández-Garazábal. 2016. The zoonotic potential of *Lactococcus garvieae*: An overview on microbiology, epidemiology, virulence factors and relationship with its presence in foods. *Res Vet Sci*. 109: 59–70.

Glikman, D., H. Sprecher, A. Chernokozinsky and Z. Weintraub. 2010. *Lactococcus lactis* catheter-related bacteremia in an infant. *Infection* 38: 145–146.

Goodman, L.B., M.R. Lawton, R.J. Franklin-Guild, R.R. Anderson, L. Schaan, A.J. Thacil, M. Wiedmann, C.B. Miller, S.D. Alcaine and J. Kovac. 2017. *Lactococcus petauri* sp. nov. isolated from an abscess of a sugar glider. *Int J Syst Evol Microbiol*. 67: 4397–4404.

Guglielmetti, E., J. Korhonen, J. Heikkinen, L. Morelli and A. von Wright. 2009. Transfer plasmid-mediated resistance to tetracycline in pathogenic bacteria from fish and aquaculture environments. *FEMS Microbiol Lett*. 239: 28–34.

Gurley, A., T. O'Brien, J.M. Garland and A. Finn. 2021. *Lactococcus lactis* bacteraemia in a patient on probiotic supplementation therapy. *BMJ Case Rep*. 14: e243915. https://doi.org/10.1136/bcr-2021-243915.

Hagting, A., E. Kunji, K. Leenhouts, B. Poolman and W. Konings. 1994. The di- and tripeptide transport protein of *Lactococcus lactis*. A new type of bacterial transporter. *J Biol Chem*. 269: 11391–11399.

Halldorsdottir, H.D., V. Haraldsdottir, A. Bodvarsson, G. Porgeirsson and M. Kristjansson. 2002. Endocarditis caused by *Lactococcus cremoris*. *Scand J Infect Dis*. 34: 205–232.

Heo, J., H. Cho, T. Tamura, S. Saitou, K. Park, J.S. Kim, S.B. Hong, S.W. Kwon and S.J. Kim. 2019. *Lactococcus allomyrinae* sp. nov., isolated from gut of larvae of *Allomyrina dichotoma*. *Int J Syst Evol Microbiol*. 69(12): 3682–3688. https://doi.org/10.1099/ijsem.0.003461.

Heo, J., S.J. Kim, M.A. Kim, T. Tamura, S. Saitou, M. Hamada, J.S. Kim, S.B. Hong and S.W. Kwon. 2020. *Lactococcus protaetiae* sp. nov. and *Xylanimonas protaetiae* sp. nov., isolated from gut of larvae of *Protaetia brevitarsis seulensis*. *Antonie Van Leeuwenhoek*. 113: 1009–1021. https://doi.org/10.1007/s10482-020-01413-6.

Hilgarth, M., V. Werum and R.F. Vogel. 2020. *Lactococcus carnosus* sp. nov. and *Lactococcus paracarnosus* sp. nov., two novel species isolated from modified-atmosphere packaged beef steaks. *Int J Syst Evol Microbiol*. 70: 5832–5840. https://doi.org/10.1099/ijsem.0.004481.

Hill, C. 1993. Bacteriophages and bacteriophage resistance in lactic acid bacteria. *FEMS Microbiol Rev*. 12: 87–108.

Holler, B.J. and J.L. Steele. 1995. Characterization of lactococci other than *Lactococcus lactis* for possible use as starter cultures. *Int Dairy J*. 5: 275–279.

Holo, H. and I.F. Nes. 1987. High frequency transformation by electroporation of *Lactococcus lactis* subsp. *cremoris* grown with glysine in osmotically stabilized media. *Appl Environ Microbiol*. 55: 127–132.

Hugenholz, J. 1993. Citrate metabolism in lactic acid bacteria. *FEMS Microbiol Rev*. 12: 165–168.

Inoue, M., A. Saito, H. Kon, H. Uchida, S. Koyama, S. Haryu, T. Sasaki and M. Nishima. 2014. Subdural empyema due to *Lactococcus lactis cremoris*: Case report. *Neurol Med Chir (Tokyo)*. 54: 341–347.

Kandler, O. 1983. Carbohydrate metabolism in lactic acid bacteria. *Antonie van Leeuwenhoek*. 49: 209–224.

Karaaslan, A., A. Soysal, E.K. Kadayifici, N. Yakut, S.O. Demir, G. Akkoc, S. Atici, A. Sarmis, N.U. Toprak and M. Bakir. 2016. *Lactococcus lactis* spp *lactis* infection in infants with chronic diarrhea: Two cases report and literature review in children. *J Infect Dev Ctries* 10: 304–307.

Karaaslan, A., A. Soysal, A. Sarmis, E.K. Kadavifci, K. Cerit, S. Atici, G. Söyletir and M. Bakir. 2011. *Lactococcus lactis* catheter-related bloodstream infection in an infant: Case report. *Jpn J Infect Dis*. 68: 341–342.

Kelleher, P., F. Bottacini, J. Mahony, K.N. Kilcawlwy and D. van Sinderen. 2017. Comparative and functional genomics of *Lactococcus lactis* taxon; insights into evolution and niche adaptation. *BMC Genom*. 18: 267–287.

Kempler, G.M. and L.L. McKay. 1981. Biochemistry and genetics of citrate utilization in *Streptococcus lactis* ssp. *diacetylactis*. *J Dairy Sci*. 64: 1527–1539.

Kim, H.S., D.W. Park, Y.K. Youn, Y.M. Jo, J.Y. Kim, J.Y. Song, J.W. Sohn, H.J. Cheong, W.J. Kim, M.J. Kim and W.S. Choi. 2010. Liver abscess and empyema due to *Lactococcus lactis cremoris*. *J Korean Med Sci*. 25: 1669–1671.

Kleerebeezem, M., M.M. Beerthuyzen, E.E. Vaughan, W.M. de Vos and O.P. Kuipers. 1997. Controlled gene expression systems for lactic acid bacteria: Transferable nisin-inducible expression cassettes for *Lactococcus, Leuconostoc* and *Lactobacillus* spp. *Appl Environ Microbiol*. 63: 4581–4584.

Kok, J., K.J. Leenhouts, A.J. Haandrikman, A.M. Ledeboer and G. Venema. 1988. Nucleotide sequence of the cell wall proteinase gene of *Streptococcus cremoris* WG2. *Appl Environ Microbiol*. 54: 231–238.

Kondo, J.K. and L.L. MacKay. 1984. Plasmid transformation of *Streptococcus lactis* protoplasts: Optimization and use in molecular cloning. *Appl Environ Microbiol*. 48: 252–259.

Kubota, H., H. Tsuji, K. Matsuda, T. Kurakawa, T. Asahara and K. Nojimoto. 2010. Detection of human intestinal catalase-negative Gram-positive cocci by rRNA-targeted reverse transcription—PCR. *Appl Environ Microbiol*. 76: 5440–5451.

Laroute, V., H. Tormo, C. Couderc, M. Mercier-Bonin, P. Le Bourgeois, M. Cocaign-Bousquet and M.-L. Daveran-Mingot. 2017. From Genome to phenotype: An integrative approach to evaluate the biodiversity of *Lactococcus lactis*. *Microorganisms*. https://doi.org/10.3390/microorganisms50200275.

Lee, J.Y., M.Y. Seo, J. Yang, K. Kim, H. Chang, S.C. Kim, S.K. Jo, W. Cho and H.K. Kim. 2014. Polymicrobial peritonitis with *Lactococcus lactis* in a peritoneal dialysis patient. *Chonnam Med J*. 50: 67–69.

Li, O., H. Zhang, W. Wang, Y. Liang, W. Chen, A.U. Din, L. Li and Y. Zhou. 2021. Complete genome sequence and probiotic properties of *Lactococcus petauri* LZys1 isolated from healthy human gut. *J Med Microbiol*. 70. https://doi.org/10.1099/jmm.0.001397.

Li, T.T., W.L. Tian and C.T. Gu. 2019. Elevation of *Lactococcus lactis* subsp. *cremoris* to the species level as *Lactococcus cremoris* sp. nov. and transfer of *Lactococcus lactis* subsp. *tructae* to *Lactococcus cremoris* as *Lactococcus cremoris* subsp. *tructae* comb. nov. *Int J Syst Evol Microbiol*. 71. https://doi.org/10.1099/ijsem.0.004727.

Li, W.-K., Y.-S. Chen, S.-R. Wann, Y.-C. Liu and H.-C. Tsai. 2008. *Lactococcus garvieae* endocarditis with initial presentation of acute cerebral infarction in a healthy immunocompetent man. *Inter Med*. 47: 1143–1146.

Lin, K.H., C.L. Sy, C.S. Chen, C.H. Lee, Y.T. Lin and J.Y. Lin 2010. Infective endocarditis complicated by intracerebral hemorrhage due to *Lactococcus lactis* subsp. cremoris. *Infection* 38: 147–149.

Liu, M., J.R. Bayjanov, B. Renckens, A. Nauta and R.J. Siezen. 2010. The proteolytic system of lactic acid bacteria revisited: A genomic comparison. *BMC Genom*. 11: 36.

Macura, D. and P.M. Townsley. 1984. Scandinavian ropy milk—Identification and characterization of endogenous ropy lactic streptococci and their extracellular excretion. *J Dairy Sci*. 67: 735–744.

Mahony, J., B. Bottacini and D. van Sinderen 2023. Towards he diversification of lactococcal starter and non-starter species in mesophilic dairy culture systems. *Microb Biotechnol*. 16: 1745–1754. https://doi.org/10.1111/1751-7915.14320.

Makarova, K., A. Slesarev, Y. Wolf, A. Sorokin, B. Mirkin, E. Koonin, A. Pavlov, N. Pavlova, V. Karamychev, N. Polouchine, V. Shakhova, I. Grigoriev, Y. Lou, D. Rokshar, S. Lucas, K. Huang, D.M. Goodstein, T. Hawkins, V. Plengvidhya, D. Welker, J. Hughes, Y. Goh, A. Benson, K. Baldwin, J.-H. Lee, I. Díaz-Muñiz, B. Dosti, V. Smeianov, W. Wechter, R. Barabote, G. Lorca, E. Altermann, R. Barrangou, B. Ganesan, Y. Xie, H. Rawsthorne, D. Tamir, C. Parker, F. Breidt, J. Broadbent, R. Hutkins, D. O'Sullivan, J. Steele, G. Unlu, M. Saier, T. Klaenhammer, P. Richardson, S. Kozyavkin, B. Weimer and D. Mills. 2006. Comparative genomics of lactic acid bacteria. *PNAS* 103: 15611–15616.

Mannion, P.T. and M.M. Rothburn. 1990. Diagnosis of bacterial endocarditis caused by *Streptococcus lactis* and assisted by immunoblotting of serum antibodies. *J Infect*. 21: 317–318.

Mansour, B., A. Habib, N. Asli, Y. Geffen, D. Miron and N. Elias. 2016. A case of infective endocarditis and pulmonary septic emboli caused by *Lactococcus lactis*. *Case Rep Pediatr*. 2016: 1024054.

McKay, L.L. 1983. Functional properties of plasmids in lactic streptococci. *Antonie van Leeuwenhoek*. 49: 259–274.

McKay, L.L. and K.A. Baldwin. 1984. Conjugative 40-megadalton plasmid in *Streptococcus lactis* subsp. *diacetylactis* DRC3 is associated with resistance to nisin and bacteriophage. *Appl Environ Microbiol*. 47: 68–74.

McKay, L.L., K.A. Baldwin and E.A. Zottola. 1972. Loss of lactose metabolism in *Streptococcus lactis*. *Appl Microbiol*. 23: 1090–1096.

Meucci, A., M. Zago, L. Rossetti, M.E. Fornasari, B. Bonvini, F. Tidona, M. Povolo, G. Contarini, D. Carminati and G. Giraffa. 2015. *Lactococcus hirilactis* sp. nov., isolated from milk. *Int J Syst Evol Microbiol*. 65: 2091–2096.

Mills, S., O.E. McAuliffe, A. Coffey, G.F. Fitzgerald and R.P. Ross. 2006. Plasmids of lactococci—genetic accessories or genetic necessities. *FEMS Microbiol Rev*. 30: 243–273.

Millen, A., P. Howarth, P. Boyaval and D.A. Romero. 2012 Mibile CRISPR/Cas mediated bacteriophage resistance in *Lactococcus lactis*. *PLoS One* 7(12): e51663.

Møller-Madsen, A. and H. Jensen. 1962. Transformation of *Streptococcus lactis*. In *Proceedings of the XVI International Dairy Congress B*, pp. 255–258.

Nakarai, T., K. Morita, Y. Nojiri, J. Nei and Y. Kawamotori. 2000. Liver abscess due to *Lactococcus lactis cremoris*. *Pediatr Int*. 42: 699–701.

Nes, I.F., D.B. Diep, L.S. Håvarstein, M.B. Brurberg, V. Eijsink and H. Holo. 1996. Biosynthesis of bacteriocins in lactic acid bacteria. In *Lactic Acid Bacteria: Genetics, Metabolism and Applications*, Eds. G. Venema, J.H.J. Huis in't Veld and J. Hugenholtz, pp. 17–32. Kluyver Academic Publishers, Dordrecht.

Neve, H., A. Geis and M. Teuber. 1988. Plasmid-encoded functions of ropy lactic acid streptococcal strains from Scandinavian fermented milk. *Biochimie* 70: 437–442.

Neves, A.R., W.A. Pool, J. Kok, O.P. Kuipers and H. Santos. 2005. Overview on sugar metabolism and its control in *Lactococcus lactis*—The input from in vivo NMR. *FEMS Microbiol Rev*. 29: 531–554.

Noda, S., F. Koyama, C. Aihara, N. Ikeyama, M. Yuki, M. Ohkuma and M. Sakamoto. 2020. *Lactococcus insecticola* sp. nov. and *Lactococcus hodotermopsidis*sp. nov., isolated from the gut of the wood-feeding lower termite *Hodotermopsis sjostedti*. *Int J Syst Evol Microbiol*. 70: 4515–4522. https://doi.org/10.1099/ijsem.0.004309.

Noda, S., M. Sakamoto, C. Aihara, M. Yuki, M. Katsuhara and M. Ohkuma. 2018. *Lactococcus termiticola* sp. nov., isolated from the gut of the wood-feeding higher termite *Nasutitermes takasagoensis*. *Int J Syst Evol Microbiol*. 68: 3832–3836. https://doi.org/10.1099/ijsem.0.003068.

O'Driscoll, J., F. Glynn, O. Cahalane, M. O'Connel-Motherway, G.F. Fitzgerald and D. van Sinderen. 2004. Lactococcal plasmid pNP40 encodes a novel, temperature-sensitive restriction-modification system. *Appl Environ Microbiol*. 70: 5546–5556.

Pellizzer, G., P. Benedetti, F. Biavasco, V. Manfrin, M. Franzett, M. Scagnelli, C. Scarparo and F. de Lalla. 1996. Bacterial endocarditis due to *Lactococcus lactis* subsp. *cremoris*: Case report. *Clin Microbiol Infect*. 2: 230–232.

Pérez, T., J.L. Balcázar, A. Peix, A. Valverde, E. Vélasquez, I. de Blas and I. Ruiz-Zarzuela. 2011. *Lactococcus lactis* subsp. nov. isolated from the intestinal mucus of brown trout (*Salmo trutta*) and rainbow trout (*Oncorhynchus mykiss*). *Int J Syst Evol Microbiol*. 61: 1894–1898.

Peterbauer, C., T. Maischberger and D. Haltrich. 2011. Food-grade gene expression in lactic acid bacteria. *Biotechnol J*. 6: 1147–1161.

Pheng, S., H.L. Han, D.S. Park, C.H. Chung and S.G. Kim. 2020. *Lactococcus kimchii* sp. nov., a new lactic acid bacterium isolated from kimchi. *Int J Syst Evol Microbiol*. 70: 505–510. https://doi.org/10.1099/ijsem.0.003782.

Plumed-Ferrer, C., A. Barberio, R. Franklin-Guild, B. Werner, P. McDonough, J. Bennett, G. Gioia, N. Rota, F. Welcome, D.V. Nydam and P. Moroni. 2015a. Antimicrobial susceptibilities and random amplified polymorphic DNA-PCR fingerprint characterization of *Lactococcus lactis* ssp. *lactis* and *Lactococcus garvieae* isolated from bovine intramammary infections. *J Dairy Sci*. 98: 6216–6225.

Plumed-Ferrer, C., S. Gazzola, C. Fontana, D. Bassi, P.S. Cocconcelli and A. von Wright. 2015b. Genome sequence of *Lactococcus lactis* subsp. *cremoris* Mast36, a strain isolated from bovine mastitis. *Genome Announc*. 3: e00449–15. https://doi.org/10.1128/genomeA.00449-15.

Plumed-Ferrer, C., K. Uusikylä, J. Korhonen and A. von Wright. 2013. Characterization of *Lactococcus lactis* isolates from bovine mastitis. *Vet Microbiol*. 167: 592–599.

Postma, P.W., J.W. Lengeler and G.R. Jacobson. 1993. Phosphoenolpyruvate:carbohydrate phosphotransferase systems of bacteria. *Microbiol Rev*. 57: 543–594.

Rauch, P.J. and W.M. de Vos. 1992. Characterization of the novel nisin-sucrose conjugative transposon Tn5276 and its insertion in *Lactococcus lactis*. *J Bacteriol*. 174: 1280–1287.

Rincé, A., A. Dufour, P. Uguen, J.P. LePennec and D. Haras. 1997. Characterization of the lactacin 481 operon: The *Lactococcus lactis* genes lctF, lctE and lctG encode a putative ABC-transporter involved in bacteriocin immunity. *Appl Environ Microbiol*. 63: 4252–4260.

Rodrigues, M.X., S.F. Lima, C.H. Higgins, S.G. Canniatti-Brazaca and R.C. Bicalho. 2016. The *Lactococcus* genus as a potential emerging mastitis pathogen group: A report on an outbreak investigation. *J Dairy Sci*. 99: 9864–9874.

Rostagno, C., P. Pecile and P.L. Stefano. 2013. Early *Lactococcus lactis* endocarditis after mitral valve repair: A case report and literature review. *Infection* 41: 897–899.

Sandine, W.E., P.R. Elliker, L.K. Allen and W.C. Brown. 1962. Genetic exchange and variability in lactic *Streptococcus* starter organisms. *J Dairy Sci*. 45: 1266–1271.

Saraoui, T., F. Leri, J. Björkroth and M.F. Pilet. 2016. *Lactococcus piscium*: A psychrotrophic lactic acid bacterium with bioprotective or spoilage activity in food—a review. *Appl Microbiol*. 121: 907–918.

Savijoki, K., H. Ingmer and P. Varmanen. 2006. Proteolytic systems of lactic acid bacteria. *Appl Microbiol Biotechnol*. 71: 394–406.

Schleifer, K.H., J. Kraus, C. Dvorac, R. Kilpper-Bälz, M.D. Collins and W. Fisher. 1985. Transfer of *Streptococcus lactis* and related streptococci to the genus *Lactococcus* gen nov. *System Appl Microbiol*. 6: 183–195.

Siezen, R.J. 1999. Multi-domain, cell-envelope proteinases of lactic acid bacteria. *Antonie van Leeuwenhoek*. 76: 139–155.

Siezen, R.J., J. Bayjanov, B. Renckens, M. Wels, S.A.F.T. van Hijum, D. Molenaar and J.E.T. van Hyckama Vlieg. 2010. Complete genome sequence of *Lactococcus lactis* subsp. *lactis* KF147, a plant-associated lactic acid bacterium. *J Bacteriol*. 192: 2649–2650.

Souza, B.M., T.M. Preisser, V.B. Pereira, M. Zurita-Turk, C.P. de Castro, V.P. da Cunha, R.P. de Oliveira, A.C. Gomes-Santos, A.M. de Faria, D.C. Machado, J.M. Chatel, V.A. Azevedo, P. Langella and A. Miyoshi. 2016. *Lactococcus lactis* carrying the pValac eukaryotic expression vector coding for IL-4 reduces chemically-induced intestinal inflammation by increasing the levels of IL-10-producing regulatory cells. *Microb Cell Fact*. 15: 150. https://doi.org/10.1186/s12934-016-0548-x.

Stanley, G. 1998. Cheeses. In *Microbiology of Fermented Foods*, Vol. 1, Ed. B.J.B. Wood, pp. 263–307. Blackie Academic & Professional, London.

Taniguchi, K., M. Nakayama, K. Nakahira, Y. Nakura, N. Kanagawa, I. Yanagihara and S. Miyaishi. 2016. Sudden infant death due to Lactococcal infective endocarditis. *Leg Med (Tokyo)*. 41: 897–899.

Teuber, M., L. Meile and F. Schwarz. 1999. Acquired antibiotic resistance in lactic acid bacteria from food. *Antonie van Leeuwenhoek*. 76: 115–137.

Thompson, J. 1987. Sugar transport in lactic acid bacteria. In *Sugar Transport and Metabolism in Gram-positive Bacteria*, Eds. J. Reizer and A. Peterkofsky, pp. 13–38. Ellis Horwood Limited, Chester.

Thompson, J. and M.H. Saier Jr. 1981. Regulation of methyl-beta-d-thiogalactopyranoside-6-phosophate accumulation in *Streptococcus lactis* by exclusion and expulsion mechanisms. *J Bacteriol*. 146: 885–894.

Topçu, Y., G. Akinci, E. Bayram, S. Hiz and M. Türkmen. 2011. Brain abscess caused by *Lactococcus lactis cremoris* in a child. *Eur J Pediatr*. 170: 1603–1605.

Torre, D., C. Sampietro, G.P. Fiori and F. Luzzaro. 1990. Necrotizing pneumonitis and empyema caused by *Streptococcus cremoris* from milk. *Scand J Infect Dis*. 22: 221–222.

Trotter, M., S. Mills, R.P. Ross, G.F. Fitzgerald and A. Coffey. 2001. The use of cadmium resistance on the phage-resistance plasmid pNP40 facilitates selection for its horizontal transfer to industrial dairy starter lactococci. *Lett Appl Microbiol*. 33: 409–414.

Ture, M. and I. Altinok. 2016. Detection of putative virulence genes of *Lactococcus garvieae*. *Dis Aquat Org*. 119: 59–66.

Tynkkynen, S., G. Buist, E. Kunji, J. Kok, B. Poolman, G. Venema and A. Haandrikman. 1993. Genetic and biochemical characterization of the oligopeptide transport system of *Lactococcus lactis*. *J Bacteriol*. 175: 7523–7532.

Tynkkynen, S., A. von Wright and E.-L. Syväoja. 1989. Peptide utilization encoded by *Lactococcus lactis* SSL135 chromosomal DNA. *Appl Environ Microbiol*. 55: 2690–2695.

Uchida, Y., Morita, H., S. Adachi, T. Asano, T. Taga and N. Kondo. 2010. Bacterial meningitis and septicemia of neonate due to *Lactococcus lactis*. *Pediatr Int*. 53: 119–120.

van Belkum, M.J., J. Kok and G. Venema. 1992. Cloning, sequencing and expression in *Escherichia coli* of *lcnB*, a third bacteriocin determinant from the lactococcal bacteriocin plasmid p9B4–6. *Appl Environ Microbiol*. 58: 572–577.

van Kranenburg, R., M. Kleerebezem and W.M. de Vos. 2000. Nucleotide sequence analysis of the lactococcal EPS plasmid pNZ4000. *Plasmid*. 43(27): 130–136.

van Kranenburg, R., J.D. Marugg, I.I. van Swam, N.J. Willem and V.M. de Vos. 1997. Molecular characterization of the plasmid-encoded eps gene cluster essential for exopolysaccharide biosynthesis in *Lactococcus lactis*. *Mol Microbiol*. 24: 387–397.

Vedamuthu, E.R. and J.M. Neville. 1986. Involvement of a plasmid in production of ropiness (mucoidness) in milk cultures by *Streptococcus cremoris* MS. *Appl Environ Microbiol*. 51: 677–682.

Venderell, D., J.L. Balcázar, I. Ruiz-Zarzuela, I. de Blöas, O. Girones and J.L. Múzquiz. 2006. *Lactococcus garvieae* in fish: A review. *Comp Immun Microbiol Infect Dis*. 29: 177–198.

von Wright, A., A.-M. Taimisto and S. Sivelä. 1985. Effect of Ca²⁺ ions on the plasmid transformation of *Streptococcus lactis* protoplasts. *Appl Environ Microbiol.* 53: 1584–1588.

von Wright, A. and S. Tynkkynen. 1987. Construction of *Streptococcus lactis* subsp. *lactis* strains with a single plasmid associated with mucoid phenotype. *Appl Environ Microbiol.* 53: 1385–1386.

Wang, M., Z. Gao, Y. Zhang and L. Pan. 2016. Lactic acid bacteria as mucosal delivery vehicles: A realistic therapeutic option. *Appl Microbiol Biotechnol.* 100: 5691–5701.

Werner, B., P. Moroni, G. Gioia, L. Lavín-Alconero, A. Yousaf, M.E. Charter, B.M. Carter, J. Bennett, D.V. Nydam, F. Welcome and Y.H. Schucken. 2014. Short communication: Genotypic and phenotypic identification of environmental streptococci and association of *Lactococcus lactis* ssp. *lactis* with intramammary infections among different dairy farms. *J Dairy Sci.* 97: 6964–6969.

Williams, A.M., J.L. Fryer and M.D. Collins. 1990. *Lactococcus piscium* sp. nov. A new *Lactococcus* species from salmonid fish. *FEMS Microbiol Lett.* 68: 109–104.

Wood, H.F., K. Jacobs and M. McCarty. 1985. *Streptococcus lactis* isolated from a patient with subacute bacterial endocarditis. *Am J Med.* 18: 345–347.

Wydau, S., R. Dervyn, J. Anba, S.D. Ehrlich and E. Maguin. 2006. Conservation of key elements of natural competence in *Lactococcus lactis* ssp. *FEMS Microbiol Lett.* 257: 32–42.

Wyszyńska, A., P. Kobierecka, J. Bardowski and E.K. Jagusztyn-Krynicka. 2015. Lactic acid bacteria--20 years exploring their potential as live vectors for mucosal vaccination. *Appl Microbiol Biotechnol.* 99: 2967–2977.

Yang, Y.S., Y. Zheng, Z. Huang, X.-M. Wang and H. Yang. 2016. *Lactococcus nasutitermitis* sp. nov. isolated from a termite gut. *Int J Syst Evol Microbiol.* 64: 146–151.

Yu, J., Y. Song, Y. Ren, Y. Qing, W. Liu and Z. Sun. 2017. Genome level comparisons provide insights into the phylogeny and metabolic diversity of species within the genus *Lactococcus*. *BMC Microbiol.* 17: 213–223.

Yuki, M., M. Sakamoto, Y. Nishimura and M. Ohkuma. 2018. *Lactococcus reticulotermitis* sp. nov. isolated from the gut of the subterranean termite *Reticulitermes speratus*. *Int J Syst Evol Microbiol.* 68: 596–601.

Zhang, H.X., W.L. Tian and C.T. Gu. 2021. Proposal of *Lactococcus garvieae* subsp. *bovis* Varsha and Nampoothiri 2016 as a later heterotypic synonym of *Lactococcus formosensis* Chen *et al.* 2014 and *Lactococcus formosensis* subsp. *bovis* comb. nov. *Int J Syst Evol Microbiol.* 71(8). https://doi.org/10.1099/ijsem.0.004980.

Zurita-Turk, M., B. Mendes Souza, C. Prósperi de Castro, V. Bastos Pereira, V. Pecini da Cunha, T. Melo Preisser, A.M. Caetano de Faria, D. Carmona Cara Machado and A. Miyoshi. 2020. Attenuation of intestinal inflammation in IL-10 deficient mice by a plasmid carrying *Lactococcus lactis* strain. *BMC Biotechnol.* 20(1): 38. https://doi.org/10.1186/s12896-020-00631-0.

# Streptococcus
## *A Brief Update on the Current Taxonomic Status of the Genus*

4

John R. Tagg, Liam K. Harold, John D. F. Hale, Philip A. Wescombe, and Jeremy P. Burton

## 4.1 HISTORICAL PERSPECTIVES

"Streptococcus" derives from the Greek *streptos*—easily twisted like a chain—and *kokkos*—grain or seed—and the term was first used as a descriptor for the chain-forming, coccoid-shaped bacteria commonly detected in wounds and discharges from animal bodies and in about one-half of erysipelas cases (Billroth 1874). Later, Pasteur presented to the French Academy of Medicine on March 11, 1879, an account of his microscopic observations of streptococci isolated from the uterus and blood of women with puerperal sepsis and proposed an etiological association (Alouf and Horaud 1997). However, it was Rosenbach in 1884 who first applied the generic name *Streptococcus* to the chain-forming coccus isolated from suppurative lesions in man (Rosenbach 1884). He designated this bacterium *Streptococcus pyogenes*.

In the ensuing 20 years the association between streptococci and a wide variety of diseases of man and other animals was firmly established, as was the importance of these bacteria in the dairy industry. Early attempts at classification focused on host spectrum and pathogenicity, cultural appearances on agar or gelatin-based media, Gram stain appearance, reactions on blood agar and in milk cultures, and the range of growth temperatures. Hemolysis on blood agar became an important characteristic for identification, with the streptococci from pathological conditions in man almost invariably displaying complete ($\beta$) hemolysis (Schotmuller 1903).

In 1906 Andrewes and Horder examined 1200 streptococci from human, air, and milk sources and on the basis of their sugar fermentation reactions, reduction of neutral red, and growth characteristics in milk distinguished eight groups (Andrewes and Horder 1906): *S. pyogenes*, as previously described by Rosenbach (1884); *Streptococcus equinus*, infrequently disease associated, but commonly present in the intestinal tracts of herbivores and humans and also commonly found in air and dust samples; *Streptococcus mitis*, only occasionally associated with diseases and mainly found in human saliva and intestinal contents; *Streptococcus salivarius*, predominantly isolated from human saliva and intestinal contents; *Streptococcus anginosus*, considered a long-chained pathogenic form of *S. salivarius* associated

DOI: 10.1201/9781003352075-5

with sore throats and also found in the intestine; and *Streptococcus faecalis*, isolated mainly from the human intestine and "pneumococci" distinguished by their capsule formation, but not at that time given a species designation.

Subsequently Orla-Jensen (1919) used fermentation characteristics, tolerance to heat and sodium chloride, temperature limits of growth, and cellular morphology to help define nine groups of streptococci, mainly, however, comprising isolates from dairy sources. It was Sherman (1937), however, who produced the first comprehensive systematic classification of streptococcal isolates from environmental, commensal, and disease-associated sources. He excluded from the genus *Streptococcus* all strictly anaerobic cocci and also the pneumococci because of their extreme sensitivity to bile. Sherman delineated four primary divisions on the basis of their hemolytic activity, group carbohydrate antigens, strong reducing capability, ability to grow at 10 °C and 45 °C, ability to survive heating at 60 °C for 30 min, and growth at pH 9.6 or in the presence of 0.1% methylene blue or in 6.5% sodium chloride. The divisions were termed "pyogenic", "enterococcus", "lactic", and "viridans". The pyogenic division comprised most of the species known then to be pathogenic for man and other animals. These were mostly β-hemolytic and contained a group polysaccharide detectable by the Lancefield precipitin method (Lancefield 1933). Some (but not all) of the viridans cluster produced greening (α-hemolysis) on blood, hence the name "viridans", from the Latin *viridis*, meaning "green". The name has sometimes been misapplied as a general term or even as a species designation, "*Streptococcus viridans*".

The "viridans" streptococci were commonly found as normal inhabitants of the mouth and throat, although some became invasive if host resistance was reduced (as with endocarditis). The designation "lactic" streptococci, although somewhat misleading since all streptococci produce lactic acid, was adopted for the species *Streptococcus lactis* and *Streptococcus cremoris*, since these common milk-souring streptococci had long been referred to as the "lactic acid streptococci". They all expressed the Lancefield group N antigen. The members of the Lancefield group D "enterococcus" division were distinguished by their wide range of growth temperatures and their tolerance to salt and alkali. The physiological/biochemically based Sherman classification scheme, augmented by classical serological characterization, was widely accepted for many years until superseded more recently by schema informed by the application of molecular methodologies.

The introduction of serological grouping following the seminal studies of Lancefield (1941) had a major and enduring influence on the classification of the hemolytic streptococci. Lancefield's recognition of group-specific polysaccharides (the so-called C substances) that could be detected by precipitin reactions provided what at first appeared to be a direct correlation between certain of the serological groups (A to E and N) and particular *Streptococcus* species, as defined on the basis of physiological and biochemical tests (Lancefield 1933). Because of the relative ease with which serological grouping could be conducted using either acid (Lancefield 1933) or formamide (Fuller 1938) extracts of the streptococcal cells, β-hemolytic streptococcal isolates from clinical specimens were often identified as, for example, group A (*S. pyogenes*) or group B (*Streptococcus agalactiae*), and no further testing was done other than serological typing (Griffith 1934) for specialized epidemiological purposes. Because this worked so well, a heavy dependence was placed on serological grouping for the routine identification of hemolytic streptococci. Subsequently, many of the serological groups came to be regarded as homogeneous taxonomic entities (even as distinct species). Exceptions included group C, which was known to contain several species or biotypes; group D, which from the start was reported to comprise *S. faecalis* and *Streptococcus durans*; and group N, which contained both *S. lactis* and *S. cremoris*. With further study it became apparent that streptococcal isolates of quite unrelated species may harbor identical Lancefield antigens, and strains that are genetically related at the species level may have heterogeneous Lancefield antigens. Further insights into the foundation studies of the early streptococcologists can be found in the "History of Streptococcal Research" by Ferretti and Kohler (2022). The outstanding contributions to streptococcal research by Dr. Rebecca Lancefield (affectionately known by her work colleagues as "Mrs. L") over more than 60 years at the Rockefeller University have since 1960 been acknowledged triennially at the Lancefield International Symposia on Streptococci and Streptococcal Diseases (Ferretti and Kohler 2022). The reader is referred to the authoritative compendium *Streptococcus pyogenes: Basic Biology to Clinical Manifestations* (Ferretti et al. 2022) for a specific focus on *S. pyogenes*.

Molecular taxonomic studies have brought major changes to classification of the streptococci; the "lactic" (Lancefield group N) streptococci now constitute the genus *Lactococcus* (Schleifer et al. 1985), and some members of Sherman's (1937) "enterococcus" division have become foundation members of the genus *Enterococcus* (Schleifer and Kilpper-Balz 1984; Sherman 1937). The motile streptococci have been transferred to the genus *Vagococcus* (Collins et al. 1989), and the anaerobic streptococci now belong to the *Peptostreptococcus* genus (Kluyver and van Niel 1936). The "nutritionally variant streptococci" were reassembled within the *Abiotrophia* (Kawamura et al. 1995a), and then some were later displaced into the genus *Granulicatella* (Collins and Lawson 2000).

Although the traditional streptococcal phenotypic criteria (hemolytic reactions and Lancefield serology) are still used in the clinical setting for the indicative classification of streptococcal isolates, these cornerstones of streptococcal classification are now largely being superseded by contemporary molecular taxonomy principles. Multiplex PCR technologies now accurately and efficiently identify a wide variety of clinically relevant streptococcal species (Hatrongjit et al. 2017; Kerdsin et al. 2017). Whole-genome sequencing costs have decreased such that it is now feasible to sequence numerous genomes simultaneously. Since the last iteration of this chapter, dozens of new streptococcal species have been defined or re-named. Most of the newly described species have been isolated from non-human hosts. Reference to the online "List of Prokaryotic Names with Standing in Nomenclature" (Parte 2014) provides an overview that is inclusive of many of the more recently described species.

## 4.2 GENUS CHARACTERISTICS AND SPECIES DISCRIMINATORS

The members of the genus *Streptococcus* are Gram-positive, catalase-negative, cytochrome-negative, facultatively anaerobic, spherical, or ovoid bacteria, less than 2 µm in diameter and with a relatively low G+C content, ranging from 34 to 46%. Taxonomically they belong to the family *Streptococcaceae* within the phylum *Firmicutes*, but more commonly they are regarded as members of the lactic acid bacteria. Division along a single axis causes streptococcal cells to grow as chains, particularly in liquid cultures, and some chains may contain 50 or more cellular units (Ekstedt and Stollerman 1960). Cross-walls form at right angles to the axis of the chain, and after division, an appearance of more tightly conjoined cell pairs within the chain is sometimes evident. Short rod forms can occur, especially during growth on solid media. Streptococci are nonmotile, although some members of the species *Streptococcus sanguinis* exhibit a curious "twitching motility" by using their type IV polar fimbriae in a grappling hook fashion to translocate across surfaces (Henriksen and Henrichsen 1975). Proteinaceous surface fibrils or filaments adorn the surface of many streptococci, sometimes arranged as tufts in some species. These fuzzy coats are known to have important roles in adhesion to surfaces and in protecting the streptococcus against phagocytosis. Most streptococci grow in air, but some, such as *Streptococcus pneumoniae* require additional carbon dioxide for satisfactory growth, a consequence of their adaptation to an oral cavity (carbon dioxide-enriched) habitat. All fail to reduce nitrate. They ferment glucose, predominantly with the formation of lactic acid, but not gas.

## 4.3 GROWTH CHARACTERISTICS *IN VITRO*

Due to their complex nutritional requirements, the growth of most streptococci on basal nutrient media is relatively poor but is aided by enrichment with blood, serum, or fermentable carbohydrate. Their colonies on agar are generally smaller (ca. 1 mm diameter after 24 h at 37 °C) than those of staphylococci and unlike staphylococci; they are almost always nonpigmented, although exceptions under certain culture conditions include most *S. agalactiae* (orange–red) (de la Rosa et al. 1992) and many *Streptococcus mutans* (yellow) (Woltjes et al. 1982). The temperature optimum for most streptococci is around 37 °C, although some species such as *Streptococcus uberis* and *Streptococcus thermophilus* grow at temperatures as low as 10 °C or as high as 45 °C, respectively.

Laboratory identification of some streptococci is facilitated by incorporation of sucrose into the growth medium (e.g., Mitis–Salivarius agar) (Chapman 1944). For example, the size of colonies of *Streptococcus salivarius* and of some of the *bovis/equinus* group of streptococci is augmented by the formation from sucrose of water-soluble extracellular levan (Niven et al. 1941), whereas colonies of the typical dental plaque species *S. sanguinis*, *Streptococcus gordonii*, *Streptococcus oralis*, and most of the mutans group streptococci are hard and cohesive due to their content of insoluble dextrans (MacFadden 1985).

Although the growth of streptococci in liquid cultures is enhanced by the addition of fermentable carbohydrate, this increases lactic acid production, which in turn leads to more rapid cell death in the post-stationary growth phase of the cultures. However, by buffering the media or through regular addition of alkali, heavy growth of most streptococci can be obtained in media such as Todd–Hewitt broth (Todd and Hewitt 1932).

The production by some streptococci of complete zones of hemolysis (β-hemolysis) when growing on the surface of blood agar has long been a key indicator for the presumptive detection of potentially pathogenic streptococci. Other streptococci either produce zones of greenish discoloration (α-hemolysis) or no discernable effect on the red blood cells (the latter perhaps somewhat illogically termed γ-hemolysis). Exceptions occur; some *S. pyogenes* are deficient in hemolysin production (Yoshino et al. 2010), and some members of commensal species such as *S. salivarius* exhibit β-hemolysis on certain blood agar media (Tompkins and Tagg 1987). The β-hemolysis of *S. pyogenes* strains is due to the activity of the well-defined hemolysins, streptolysin O and especially streptolysin S (Nizet 2002), whereas the α-hemolysis is attributed to hydrogen peroxide (formed during growth in the presence of oxygen by some streptococci) inducing the oxidation of the heme iron of hemoglobin with formation of methemoglobin (Barnard and Stinson 1996). This effect is more evident on chocolate agar, since the heating of the blood destroys the native catalase activity of the erythrocytes (Agar et al. 1986).

## 4.4 NUTRITIONAL REQUIREMENTS AND METABOLIC CHARACTERISTICS

The streptococci are in general nutritionally fastidious, indeed more so than their human host in their dependency on preformed amino acids and cofactors (Van der Rijn and Kessler 1980). In the laboratory their growth requirements are generally satisfied by blood agar media comprising a base of peptones, meat extract, and carbohydrate. Genome analyses show that *S. pyogenes*, *S. agalactiae*, and *S. pneumoniae* completely lack any tricarboxylic acid (TCA) cycle capability (Glaser et al. 2002), meaning they are unable to synthesize the precursors of most amino acids. On the other hand, some oral streptococci can grow in the presence of ammonia, sugar, vitamins, and salts. Indeed, the *S. mutans* UA159 genome contains some amino acid biosynthetic pathways and a partial TCA cycle (Ajdic et al. 2002). In nature, the streptococcal requirement for nitrogenous compounds is satisfied by amino acids excreted by companion members of the oral microflora or released by bacterial proteinases acting on tissue proteins. The ability to utilize peptides is dependent on transport mechanisms in the cytoplasmic membrane and intracellular peptidases capable of hydrolyzing the peptides to the constituent amino acids. Oral streptococci have adopted various strategies to cope with environmental acidification, including adaptive acid tolerance responses and production of alkali. Ammonia-generating mechanisms used by some oral streptococci include the hydrolysis of urea by urease, especially by *S. salivarius* (Chen et al. 2000), and degradation of arginine using arginine deiminase by *Streptococcus ratti*, *S. gordonii*, and *S. sanguinis*, a reaction that also yields energy (Griswold et al. 2004).

Streptococci are facultative anaerobes, and most can flourish in the absence of oxygen. *Streptococcus pneumoniae* and many other oral species require elevated carbon dioxide levels for adequate growth, and some other streptococci grow more favorably under anaerobic conditions. Their energy requirements are obtained from the fermentation of carbohydrates, and indeed they are incapable of respiratory metabolism. Among the bacteria capable of growing aerobically, the streptococci are unique in that they are incapable of forming ATP via electron transport systems, and they lack the ability to synthesize porphyrins, cytochromes, or catalase.

# 4.5 CLASSIFICATION

Serial revisions of the genus *Streptococcus* have been implemented since its genesis. The broad subdivision into six major groups and one additional "ill-defined" cluster of species that is outlined in the present chapter is heavily based on 16S rRNA gene sequence data and correlates well with the results of DNA–DNA reassociation experiments and numerical taxonomy studies (Bentley et al. 1991; Facklam 2002; Kawamura et al. 1995b; Kilian 2005; Kohler 2007). The comparative analysis of 138 *Streptococcus* genomes supported an evolutionary clustering of the seven groups into two distinct clades based around (a) the pyogenic, salivarius, bovis/equinus, and mutans groups and (b) the anginosus, mitis, and "ill-defined" groups (Gao et al. 2014). In other studies, the phylogenetic analysis of eight genes in 75 *Streptococcus* species led to the proposal that three additional species groupings be created around *Streptococcus gordonii*, *Streptococcus sobrinus*, and *Streptococcus pluranimalium* (Pontigo et al. 2015). Also, comparative analysis of 46 *Streptococcus* genomes has provided support for the formation of separate taxonomic assemblages based upon (a) *Streptococcus downei* and *Streptococcus criciti* (the Downei group) and (b) *Streptococcus sanguinis* (Richards et al. 2014). It should be noted that many of the more recently described species are as yet represented in the literature only by a very small number of independently isolated strains. At the time of writing, 116 *Streptococcus* species have been approved. Of the more recently proposed species, only *Streptococcus downii* (Martinez-Lamas et al. 2020); *Streptococcus ilei* (Hyun et al. 2021); *Streptococcus rubneri* (Huch et al. 2013); *Streptococcus toyakuensis* (Wajima et al. 2022); and *Streptococcus tigurinus* (Zbinden et al. 2012), later reclassified as *Streptococcus oralis* subsp. *tigurinus* (Jensen et al. 2016), have been primarily isolated from human sources. Although in this chapter we have focused our presentation upon the *Streptococcus* species at present considered principally associates of the human host (Table 4.1), we acknowledge that the domain and influence of this enduring and intriguing cellular entity extends far beyond *Homo sapiens*.

1. Pyogenic group—essentially β-hemolytic species pathogenic for humans and other animals
2. Anginosus group—commensal and occasional opportunistic pathogens found in the oral cavity and the gastrointestinal and genital tracts of humans
3. Mitis group—includes the pathogen *S. pneumoniae* and various oral commensals
4. Salivarius group—comprises dairy streptococci and commensals of the human oral cavity
5. Bovis/equinus group—species principally found in the intestinal tract of several animal species
6. Mutans group—comprises genetically heterogeneous species that are nevertheless phenotypically similar
7. Ill-defined group—species not yet grouped or of uncertain phylogenetic relationships

**TABLE 4.1** Species and Subspecies of the Genus *Streptococcus* Having Associations with Humans

| PHYLOGENETIC GROUP | SPECIES | LANCEFIELD'S GROUP ANTIGEN[a] | HEMOLYSIS[a] | PRINCIPAL HUMAN ASSOCIATIONS | REFERENCE |
|---|---|---|---|---|---|
| Pyogenic group | *S. pyogenes* | A | β | Infections, carriage | Rosenbach (1884) |
| | *S. agalactiae* | B | β (α/—) | Infections, carriage | Nocard and Mollereau (1887) |
| | *S. dysgalactiae* | | | | |
| | subsp. *equisimilis* | C (A, G, L) | β | Infections, carriage | Vandamme et al. (1996) |
| | subsp. *dysgalactiae* | C (L) | α (β/—) | Animal pathogen Pigs, cattle | Vandamme et al. (1996) |
| | *S. equi* | | | | |
| | subsp. *equi* | C | β | Rare zoonosis Animal pathogen Horses, donkeys | Farrow and Collins (1984) |
| | subsp. *zooepidemicus* | C | β | Many animals, including humans | Farrow and Collins (1984) |
| | subsp. *ruminatorum* | C | β | Rare zoonosis Sheep, goats | Fernandez et al. (2004) |
| | *S. uberis* | —, E, C, D, P, U | α/—/β | Cattle pathogen Probiotic for humans | Diernhofer (1932) |
| | *S. parauberis* | | α/— | Cattle, fish pathogen Rare zoonosis | Williams and Collins (1990) |
| | *S. porcinus* | (B), E, P, U, V | β | Pigs Rare zoonosis | Collins et al. (1984) |
| | *S. pseudoporcinus* | — | β | Humans | Bekal et al. (2006) |
| | *S. canis* | G | β | Many animals, humans | Devriese et al. (1986) |
| | *S. iniae* | — | β | Fish Rare zoonosis | Pier and Madin (1976) Gauthier (2015) |
| | *S. urinalis* | — | — | Humans (clinical relevance unknown) | Collins et al. (2000) |
| Anginosus group ("milleri"-group) | *S. anginosus* | —, F, C, A, G | β/α/— | Humans, endocarditis | Whiley and Beighton (1991) |
| | *S. intermedius* | — | α/—/β | Humans, abscess formation | Whiley and Beighton (1991, 1998) |
| | *S. constellatus* | | | Humans | |

(Continued)

**TABLE 4.1** (Continued) Species and Subspecies of the Genus *Streptococcus* Having Associations with Humans

| PHYLOGENETIC GROUP | SPECIES | LANCEFIELD'S GROUP ANTIGEN[a] | HEMOLYSIS[a] | PRINCIPAL HUMAN ASSOCIATIONS | REFERENCE |
|---|---|---|---|---|---|
| | subsp. *pharyngis* | C | β | Humans, oral commensal, abscess formation | Whiley et al. (1999) |
| | subsp. *constellatus* | —, F, C, A, G | β/α/— | Humans, oral commensal, purulent human infections | Whiley et al. (1999) |
| Mitis group | *S. mitis* | —, K, O | α | Humans, oral commensal, endocarditis | Andrewes and Horder (1906); Kilian et al. (1989a) |
| | *S. sanguinis* | —, H, W | α | Humans, opportunist | Kilian et al. (1989a) |
| | *S. oralis* subsp. *oralis* | — | α | Humans, oral commensal, bacterial endocarditis, adult respiratory distress syndrome and streptococcal shock | Jensen et al. (2016) |
| | *S. oralis* subsp. *tigurinus* | | α | Commensal, infective endocarditis Humans | Jensen et al. (2016) |
| | *S. oralis* subsp. *dentisani* | | α | Commensal Humans | Jensen et al. (2016) Lopez-Lopez et al. (2017) |
| | *S. downii* | | α | Humans, oral commensal | Martinez-Lamas et al. (2020) |
| | *S. pneumoniae* | O | α | Humans, pneumonia and invasive disease | Kilpper-Balz et al. (1985) Parks et al. (2015) |
| | *S. gordonii* | —, H | α | Humans, dental plaque, endocarditis | Kilian et al. (1989a) |
| | *S. crista* | | α | Humans, commensal | Handley et al. (1991) |
| | *S. parasanguinis* | —, F, C, G, B | α | Humans, infective endocarditis | Whiley et al. (1990) |
| | *S. australis* | — | | Humans, commensal (children) | Willcox et al. (2001) |
| | *S. ilei* | -, A | α | Humans, oral, opportunist pathogen | Hyun et al. (2021) |
| | *S. rubneri* | | β | Humans, oral commensal | Huch et al. (2013) |
| | *S. toyakuensis* | — | α | Multi-drug–resistant isolate from human blood | Wajima et al. (2022) |
| | *S. lactarius* | | α | Humans, breast milk | Martin et al. (2011) |

| | | | | | |
|---|---|---|---|---|---|
| | S. pseudopnemoniae | | α | Humans, clinical relevance unclear—associated with chronic obstructive pulmonary disease | Arbique et al. (2004) |
| | S. peroris | | α | Humans, oral | Kawamura et al. (1998) |
| | S. infantis | | α | Humans, commensal | Kawamura et al. (1998) |
| | S. sinensis | —, F | α | Humans, endocarditis | Woo et al. (2002) |
| | S. oligofermentans | | α | Humans, oral commensal | Tong et al. (2003) |
| | S. massiliensis | G | — | Humans, skin commensal | Glazunova et al. (2006) |
| | S. timonensis | | α | Human, stomach | Ricaboni et al. (2016) |
| Salivarius group | S. salivarius | —, H, K | —(α/β) | Humans, oral commensal | Andrewes and Horder (1906) |
| | S. vestibularis | — | α | Humans, oral commensal | Whiley and Hardie (1988) |
| | S. thermophilus | — | —/β | Milk, dairy products | Orla-Jensen (1919) |
| Bovis–Equinus group[b] | S. bovis[c] | —, D | α/— | Horses, other ruminants, rare human | Orla-Jensen (1919) |
| | S. equinus[c] | —, D | α | Horses, other ruminants | Andrewes and Horder (1906) |
| | S. gallolyticus | | | | Schlegel et al. (2003b) |
| | (subsp. gallolyticus)[d] | —, D | — | Marsupials, mammals incl. humans, esp. endocarditis and possible colonic carcinoma link | Osawa et al. (1995) Dekker and Lau (2016) Boleij and Tjalsma (2013) |
| | (subsp. macedonicus)[d] (= S. waius) | | | Dairy products | Tsakalidou et al. (1998) Manachini et al. (2002) |
| | (subsp. pasteurianus)[d] | | | Humans, commensal, infective endocarditis (infants) | Poyart et al. (2002) Dekker and Law (2016) |
| | S. infantarius (subsp. infantarius) | —, D | —/α | Humans | Bouvet et al. (1997) |
| | | | | Humans, commensal, infective endocarditis | Schlegel et al. (2003a) |
| | (subsp. colis) | | | Humans, commensals, infective endocarditis | Schlegel et al. (2003a) Parks et al. (2015) |

(Continued)

**TABLE 4.1**  (Continued) Species and Subspecies of the Genus *Streptococcus* Having Associations with Humans

| PHYLOGENETIC GROUP | SPECIES | LANCEFIELD'S GROUP ANTIGEN[a] | HEMOLYSIS[a] | PRINCIPAL HUMAN ASSOCIATIONS | REFERENCE |
|---|---|---|---|---|---|
| | S. alactolyticus | D | —/α | Chickens, pigs, rare human infection | Farrow et al. (1984) |
| | S. lutetiensis | D | α | Humans, bacteremia | Poyart et al. (2002) |
| Mutans group | S. mutans | — | α/—(β) | Humans, dental caries | Clarke (1924) |
| | S. sobrinus | — | —/α | Humans, dental caries | Coykendall (1983) |
| | S. criceti | — | —/α | Hamster, rats, humans | Coykendall (1977) |
| | S. ratti | — | —/α | Rats, humans | Coykendall (1977) |
| | S. downei | — | —/α | Monkeys, caries | Whiley et al. (1988) |
| | S. dentapri | | | Pigs, humans (rare) | Salman et al. (2017) |
| Uncertain grouping or genetic relationship unknown | S. suis | R, S, T | α | Pigs, cattle, humans zoonosis | Kilpper-Balz and Schleifer (1987) |

[a] A "—" represents no reaction with tested group sera or no hemolysis, while a blank indicates no information provided about Lancefield group or hemolysis.

[b] The bovis group is still undergoing significant taxonomic changes, some of which are indicated in this table.

[c] Schlegel et al. (2003b) has indicated that *S. bovis* and *S. equinus* should be combined within the priority species *S. equinus*.

[d] Schlegel et al. (2003b) subsumed *S. gallolyticus*, *S. macedonicus*, and *S. pasteurianus* as subspecies under *S. gallolyticus*, although this has not to date been finally ratified.

# 4.5.1 Pyogenic Group

These streptococci are animal parasites and are typically capable of causing septicemia or respiratory tract infections. Mostly β-hemolytic and displaying cell wall polysaccharide antigens of a variety of Lancefield group specificities, they are also commonly referred to as the hemolytic streptococci. Many of these species seem strongly adapted to particular animal hosts, although some also appear to be capable of causing zoonotic infection of close-contact humans. None is especially resistant to heat or grows at extremes of temperature, pH, or sodium chloride concentration, nor do they exhibit strong reducing capabilities.

By far the best-known member of this group is *S. pyogenes*: versatile, enigmatic, and long considered almost exclusively a human pathogen (Vela et al. 2017). Mice, however, have long been used as an experimental model for *S. pyogenes* pharyngeal infections (Gogos and Federle 2019), and in a concerning recent development, penicillin-macrolide-resistant *S. pyogenes* have been isolated from pet animals (Samir et al. 2020). In humans, *S. pyogenes* is the major cause of puerperal sepsis, and in children it is the most common cause of pharyngitis, scarlet fever, and impetigo. Unfortunately, the clinical signs and symptoms of *S. pyogenes* pharyngitis do not allow its differentiation from pharyngitis due to various other etiologic agents (Shaikh et al. 2012; Thai et al. 2017), and so considerable resources have been directed towards development of effective laboratory technologies to specifically identify *S. pyogenes* and to establish whether the levels detected reflect harmless carriage or active infection (Rich et al. 2021). Historically, the focus upon specifically identifying and treating *S. pyogenes* pharyngitis in school-aged children has been due to the long-established "dogma" that rheumatic fever is a sequel only to *S. pyogenes* pharyngitis (Cunningham 2008). More recently, however, there is increasing evidence that infections of the throat or of the skin by *Streptococcus dysgalactiae* subsp. *equisimilis* may also elicit the disease (Sikder et al. 2017).

*S. pyogenes* can also cause a wide variety of deep and invasive infections—erysiphelas, cellulitis, necrotizing fasciitis, and superantigen-induced streptococcal toxic shock syndrome (Mitchell 2003). Important post-infectious "nonpyogenic" syndromes in addition to rheumatic fever include glomerulonephritis (Cunningham 2008) and also possibly (and more controversially) some pediatric acute-onset neuropsychiatric syndromes (Chiarello et al. 2017; Nielsen et al. 2019). In recent years there have been several important insights into *S. pyogenes* virulence mechanisms, particularly in relationship to adherence, invasion, and immune evasion. While it remains susceptible to most commonly used antibiotics, increasing rates of macrolide and fluoroquinolone resistance have been reported in some regions (Tatara et al. 2020). Several vaccine candidates are currently in development and undergoing evaluation of their efficacy in preventing *S. pyogenes* infections (Harbison-Price et al. 2022). For a comprehensive contemporary overview of the biology and clinical manifestations of *S. pyogenes*, see Ferretti et al. (2022).

In clinical practice the term "group A streptococcus" has been used almost synonymously for *S. pyogenes*. Indeed, with the exception of some uncommon strains that express an A-variant carbohydrate (Elliott et al. 1971), all *S. pyogenes* are group A antigen positive. On the other hand, the converse is not true. Group A carbohydrate antigen reactivity has also been detected in some strains of *S. anginosus* (Facklam 2002), *Streptococcus dysgalactiae* subspecies *equisimilis* (Chochua et al. 2019), *Streptococcus orisratti* (Zhu et al. 2000), and *Streptococcus castoreus* (Mühldorfer et al. 2019).

*S. agalactiae* (commonly referred to as "group B streptococcus"), although long recognized as an important cause of mastitis in cattle, is also commonly resident in the human respiratory, genital, and gastrointestinal tracts and is an important pathogen of humans, especially of infants, where it presents as either sepsis or meningitis (Berardi et al. 2007). Neonatal disease results from colonization of the maternal genital tract (Russell et al. 2017; Paul et al. 2023). Predisposing conditions for infection in nonpregnant adults include diabetes mellitus, cancer, and human immunodeficiency virus infection. As with *S. pyogenes*, the species is well defined and serologically is consistently Lancefield group B, although strains from human sources differ from bovine strains in a number of pathogenic and phenotypic characteristics. Genotyping data have indicated that human and bovine-derived *S. agalactiae* represent mostly distinct populations, but some limited transmission may occur to humans exposed to cattle (Manning et al. 2010; Crestani et al. 2021). *S. agalactiae* has also been identified as an emerging pathogen in aquaculture, associated with considerable morbidity and mortality in farmed fish and shrimp (Kayansamruaj et al. 2023), and moreover is sporadically associated with illness in various other mammalian and nonmammalian host species worldwide (Pereira et al. 2010).

Two subspecies of *S. dysgalactiae* have been proposed as new taxa (Vandamme et al. 1996). *S. dysgalactiae* subspecies *equisimilis* includes isolates from humans and animals that show strong β hemolysis and typically express Lancefield group C or G antigens and occasionally group A or group L. They also express homologs of many *S. pyogenes* virulence genes, including those for many different M proteins, and exhibit similar acute disease manifestations to those of *S. pyogenes* (Oppegaard et al. 2023). On the other hand, *S. dysgalactiae* subspecies *dysgalactiae* is only rarely disease associated in humans (Jordal et al. 2015); shows α, β, or sometimes no hemolysis; and expresses either Lancefield group C or L antigenicity.

Three subspecies of the Lancefield group C *Streptococcus equi* are recognized. *Streptococcus equi* subsp. *equi* causes "strangles" in horses and very rarely causes invasive infections in immunocompromised humans, often following close contact with horses (Torpiano et al. 2020). *Streptococcus equi* subsp. *zooepidemicus* is found in animal and human infections (Facklam 2002). *Streptococcus equi* subsp. *ruminatorum* has been recovered from milk samples from mastitis-affected goats, cows, and sheep (Fernandez et al. 2004). The phenotypically closely related, but genetically distinctive species *S. uberis* and *Streptococcus parauberis* are economically important causative agents of bovine mastitis and in contrast to most other pyogenic streptococci they are α-hemolytic or nonhemolytic. Some strains react with Lancefield group B, E, G, or P antisera (Groschup et al. 1991). Additionally, there have been reports of *S. parauberis* causing infections in non-bovine hosts, such as fish and humans. *S. uberis* strain KJ2$^{sm}$ has had application as an oral probiotic in humans (Zahradnik et al. 2009; Kamble et al. 2022).

The species *Streptococcus porcinus* was proposed in 1984 to accommodate physiologically related β-hemolytic streptococci principally belonging to Lancefield groups E, P, U, and V (Collins et al. 1984). They are associated with pyogenic infections in swine and only rarely from other hosts, including humans. Several isolates from human genitourinary tract specimens were differentiated by 16S rRNA as *Streptococcus pseudoporcinus* (Bekal et al. 2006). Although the prevalence of *S. pseudoporcinus* colonization is low, it has been associated with preterm premature rupture of membranes or spontaneous preterm birth (Grundy et al. 2019).

*Streptococcus canis* are large-colony-forming group G streptococci isolated most commonly from dogs and mastitis-infected cows but also sometimes causing infections in humans (Lam et al. 2007; Pagnossin et al. 2022). Also taxonomically linked to the pyogenic streptococcus group is the emerging zoonotic pathogen *Streptococcus iniae*, isolated from dolphins and fish in aquaculture and communicated to humans by contact with infected aquatic animals (Pier and Madin 1976; Baiano and Barnes 2009), and *Streptococcus urinalis*, a rare cause of urinary tract infections in humans (Collins et al. 2000). Some other species principally from various animal hosts but occasionally isolated from immunocompromised humans include *Streptococcus halichoeri* (marine animals), *Streptococcus castoreus* (beavers), *Streptococcus didelphis* (opossums), and *Streptococcus halichoeri* (grey seals) (Shakir et al. 2021). No reported evidence of human association has yet been reported for *Streptococcus ictaluri* (catfish) (Shewmaker et al. 2007).

## 4.5.2 Anginosus Group

The Anginosus group of streptococci (SAG) are commensals of the oral cavity, gastrointestinal tract, and female urogenital tract that have substantial clinical relevance as agents of pharyngitis and pyogenic infections in various tissues and organs and have a particularly strong association with abscess formation (Fazili et al. 2017). Of late, it has been suggested that SAG may be underrepresented in the clinical cases in the literature because they have been identified as viridans streptococci (Pilarczyk-Zurek et al. 2022). The co-occurrence of SAG in gastrointestinal cancers has been recently described, though its relevance is still to be determined (Liu et al. 2019; Zhou et al. 2022). The members include *Streptococcus anginosus*,

*Streptococcus intermedius*, and *Streptococcus constellatus*—and within the latter, two subspecies are recognized, the subsp. *pharyngis* differing from subsp. *constellatus* in being Lancefield group C and chondroitin sulfatase positive (Grinwis et al. 2010). The taxonomy of this group became confusing due to a lack of international consensus on their nomenclature (Facklam 1984). These streptococci have sometimes been referred to as the *Streptococcus milleri* group, although "*Streptococcus milleri*" has never been an officially approved name (Ruoff 1988).

The Anginosus group of streptococci are small colony formers (<0.5 mm), and although some are β-hemolytic, most are nonhemolytic or occasionally α-hemolytic. Many isolates on blood agar produce a characteristic caramel odor that is due to diacetyl production (Chew and Smith 1992). Their identification is complicated by the diversity of phenotypes and antigenicity. Some strains carry Lancefield group antigens of either A, C, F, or G specificity (Facklam 2002). Recent comparative genomic studies grouped SAG members into *S. intermedius*, *S. constellatus*, and two distinct *S. anginosus* groups. One of the groups contained the *S. anginosus* type strain and other subspecies, whereas group 2 consists of predominantly urinary isolates (Prasad et al. 2023).

## 4.5.3 Mitis Group

Many of the species in the Mitis group have been considered difficult to classify and identify by biochemical methods due to the absence of reliable discriminatory traits. Members of the group are all closely related genetically and share high 16S rRNA gene sequence similarity. Some species are relatively poorly defined, and different species designations have sometimes been applied to the same strain. This has led to studies using new methodologies on older isolates to retrospectively characterize species identification (Sadowy and Hryniewicz 2020; Velsko et al. 2019; Imai et al. 2020; Zheng et al. 2016), including analysis of extracted streptococcal genome reads from 5700-year-old metagenomes (Belman et al. 2022). Core members of the group are *S. mitis*, *S. sanguinis*, *S. oralis*, and *S. pneumoniae*. Most recently MALDI-TOF mass spectrometry has been assessed as a tool to further provide accurate species identification (Marin et al. 2017; Nix et al. 2021; Harju et al. 2017).

*S. sanguinis* initially referred to strains from subacute bacterial endocarditis, and two of its definitive phenotypic characteristics are IgA1 protease activity and extracellular glucan formation from sucrose. Some of the prototypical *S. sanguinis*, such as strain Challis, have now been renamed *S. gordonii*, a species closely similar to *S. sanguinis* but IgA1 protease negative (Kilian et al. 1989b). The Mitis group species *Streptococcus cristatus* (resplendent in its characteristic tufted fibrils) (Handley et al. 1991) and also *Streptococcus parasanguinis* and *Streptococcus australis* differ from *S. sanguinis* in being IgA1 protease negative and glucan negative.

The Mitis group comprises many of the streptococcal species traditionally regarded as "viridans" (greening) streptococci, since they are predominantly α-hemolytic. *S. parasanguinis* is an occasional pathogen of humans (Whiley et al. 1990) and also a cause of asymptomatic mastitis in sheep (Fernandez-Garayzabal et al. 1998). *S. orisratti* is unusual in that it has Lancefield group A antigenicity (Zhu et al. 2000). *Streptococcus lactarius* (Martin et al. 2011) has been detected in breast milk samples from healthy women.

The versatile human pathogen *S. pneumoniae* (optochin sensitive and bile soluble) is placed within the Mitis group on the basis of its phenotypic and genetic similarities to *S. mitis*. *S. pneumoniae*, commonly referred to as 'the pneumococcus", is an important agent of community-acquired pneumonia, sometimes accompanied by bacteremia. Other highly prevalent infections include otitis media, sinusitis, meningitis, and endocarditis (Mitchell and Mitchell 2010). *S. pneumoniae* typically colonizes the upper respiratory tract with no sign of infection (especially in children). Disease results from overactivation or dysregulation of the host inflammatory response (Henriques-Normark and Normark 2010). The genetic flexibility of *S. pneumoniae* as exemplified by expression of a diversity of capsular polysaccharides is driven by its exceptional ability to acquire heterologous genes (by transformation) from cohabitating Mitis, Salivarius, and Anginosus group streptococci in the oral cavity (Kilian et al. 2014). Application of multivalent pneumococcal conjugate vaccines reduces invasive disease due to the vaccine serotypes, although some serotype replacement infection has also ensued (Balsells et al. 2017). In other studies, conjugate vaccines were shown to decrease the metabolic activity but not the nasopharyngeal carriage of *S. pneumoniae* (Andrade et al. 2017). *Streptococcus pseudopneumoniae* was differentiated from *S. pneumoniae* following DNA–DNA hybridization studies and phenotypic tests (it is nonencapsulated and insoluble in bile) (Arbique et al. 2004).

## 4.5.4 Salivarius Group

The Salivarius group is composed of three species: *S. salivarius*, which is an initial colonizer of the human oral mucosa, commonly found on the tongue dorsum, cheeks, and palate, as well as in the digestive tract (Kaci et al. 2014; Mignolet et al. 2016); *Streptococcus vestibularis* (Whiley and Hardie 1988), a mutualistic bacterium present on the vestibulum of human oral mucosa; and *S. thermophilus*, a relatively thermophilic streptococcus used in the production of yogurt and Swiss- or Italian-type cooked cheeses (Schleifer et al. 1991; Delorme et al. 2015).

Phylogenetic analyses suggest a close relationship between *S. vestibularis* and *S. thermophilus*, with *S. salivarius* diverging earlier in the Salivarius group lineage (Pombert et al. 2009; Abdelbary et al. 2021). The question of whether *S. thermophilus* should be classified as a subspecies of *S. salivarius* has been debated, but DNA–DNA re-association experiments indicate that they are separate species (Schleifer et al. 1991). The Salivarius group is closely related to the Bovis-Equinus group, and *S. infantarius* and *S. alactolyticus*, previously in the Salivarius group, are now in the Bovis-Equinus group (Facklam 2002).

Most commonly nonhemolytic on blood agar, *S. salivarius* produces distinctive mucoid colonies on sucrose-containing agar due to the production of water-soluble extracellular polysaccharides (levan). *S. salivarius* strains K12 and M18 are used as probiotics (Tagg et al. 2023), and their excellent competitive capability within the oral ecosystem is attributed to their production of bacteriocin-like inhibitory substances (BLISs), most of which appear to be megaplasmid encoded (Wescombe et al. 2006). *S. vestibularis* does not produce extracellular polysaccharides but, like many *S. salivarius* strains, is urease positive.

## 4.5.5 Bovis/Equinus Group

The Bovis/Equinus group, also known as the *Streptococcus bovis*/*Streptococcus equinus* complex (SBEC), has experienced considerable taxonomic change in recent years, and the situation still appears to be in flux (Jans et al. 2015, 2016; Pompilio et al. 2019). Phenotypic and genetic analyses of the type strains of *S. bovis* and *S. equinus* indicated that the two species should be combined within the priority species *S. equinus* (Schlegel et al. 2003a). Previously, isolates from human infections had been divided into *S. bovis* biotypes I, II/1, and II/2. The biotype I isolates, now reclassified as *Streptococcus gallolyticus* subsp. *gallolyticus* (Schlegel et al. 2003a), were most typically endocarditis associated and are now reported to cause 24% of all streptococcal endocarditis cases, making them very relevant medically (Maciejewska et al. 2022). Interestingly, they have been recognized to have an association with colon tumors since the late 1970s (Klein et al. 1977). Evidence of this being a causative, rather than a correlation, relationship has been gaining in strength, with 65% of patients with infections caused by these bacteria being diagnosed with colorectal cancer in studies (Boleij and Tjalsma 2013). Meanwhile the biotype II/2 isolates were reassigned as *Streptococcus gallolyticus* subsp. *pasteurianus*, and the taxonomically synonymous *Streptococcus macedonicus* and *Streptococcus waius* were combined into *S. gallolyticus* subsp. *macedonicus*. The less-frequently endocarditis-associated biotype II/1 strains formed the new species *S. infantarius* (Schlegel et al. 2000) and *Streptococcus lutetiensis* (Poyart et al. 2002). Another relatively well-separated member of the bovis/equinus group are the porcine and chicken isolates of *S. alactolyticus* (Farrow et al. 1984). Further efforts to improve the taxonomic situation are being made using genome similarities and phylogeny, and the impact of this work on the group will be seen in the next few years. Currently tools such as the bioMerieux VITEK MS (MALDI-TOF MS) system enable medical practitioners to identify most of the pathogenic "viridans streptococci" species, including the Bovis/Equinus Group (Lakshmi and Leela 2022).

## 4.5.6 Mutans Group

The mutans streptococci are a cluster of relatively acidogenic and aciduric streptococcal species having similar characteristics, which include a propensity to bind to the tooth surface and a strong etiological association with the development of dental caries. The prototype species, named *S. mutans* by Clarke in 1924 because of its dimorphic appearance in culture (cocci when growing optimally in liquid media and short rods on agar or in acidic liquid cultures), are mostly α-hemolytic, although some β-hemolytic strains have also been described (Clarke 1924; Crooks et al. 1987). Genome analysis shows *S. mutans* can metabolize a wider range of carbohydrates than any other sequenced Gram-positive bacterium (Ajdic et al. 2002). *S. mutans* can form symbiotic biofilms with *Candida albicans* to drive the development of dental caries through its fermentation of organic acids (Metwalli et al. 2013; Falsetta et al. 2014). Recent work has focused on lessening the impact of *S. mutans* biofilms either using new antimicrobial agents or manipulation of quorum sensing (Cao et al. 2020; Pourhajibagher et al. 2022). A key contributor to progress to caries formation by streptococci is glucosyltransferase (Gtf). The *gtf* genes recently had an updated assessment of their evolutionary history that supported previous findings that the *gtf* genes in streptococci have a common ancestor (Xu et al. 2018). The majority of *gtf* gene sequences in streptococci are found in the mutans group (Xu et al. 2018).

*Streptococcus mutans* and *Streptococcus sobrinus* are the species most commonly isolated from human dental plaque. Although *Streptococcus criceti* and *S. ratti* are occasionally isolated from humans, their primary hosts are hamsters and rats, respectively. Also isolated from rats has been *Streptococcus ferus* (Coykendall 1983), whereas *Streptococcus macacae* and *Streptococcus downei* have been sourced solely from monkeys (Beighton and Hayday 1984)). More recent additions to the mutans group have included *Streptococcus dentirousetti* (Takada and Hirasawa 2008) from the bat oral cavity and *Streptococcus devriesei* from equine teeth (Collins et al. 2004). *Streptococcus orisuis* (Takada and Hirasawa 2007) and *Streptococcus dentapri* (Takada et al. 2010) were both originally isolated from the pig oral cavity, but more recently *S. dentapri* has also been isolated from the dental plaque of a caries-active human (Salman et al. 2017). When the mutans group is phylogenetically assessed on the basis of 136 core genes, it can be broken down into the Downei and Mutans groups (Richards et al. 2014); however, these groups share many phenotypic characteristics, which is why they remain grouped together in this chapter. The Downei group consists of *S. downei* and *S. criceti* (Richards et al. 2014).

## 4.5.7 Ill-Defined Group

The more recently proposed *Streptococcus* species have largely been derived from non-human sources, and most do not fit comfortably within the established six-group classification long applied to the species recovered from *Homo sapiens*. The variety and number of novel species reported is growing and appears likely to be limited in the future only by a depletion in the available resources and exploratory zeal of streptococcologists. Perhaps the most notable of the ill-defined group is *Streptococcus suis*, an important pathogen of pigs worldwide and also a cause of zoonotic infection in the human contacts of infected animals (Kilpper-Balz and Schleifer 1987; Haas and Grenier 2017).

## 4.5.8 Phylogenetic Groupings for the Future

Recent advancements in whole-genome DNA sequencing capabilities have enhanced the ability to achieve the accurate delineation of the relatedness of species over traditional 16S sequencing. The phylogenetic groupings described previously are based on historical analysis of 16S sequencing by Kawamura et al. (1995b) that defined six broad groups and one group of misfits. Gao et al. (2014) concurred with these seven groupings based on the analysis of 278 proteins. Work by Richards et al. (2014) using a core set of 136 genes further broke down these initial seven groups to include the additional groups Sanguinis and Downei, which were split from the Mitis and Mutans groups, respectively. Analysis of eight genes by Pontigo et al. (2015) led to the recommendation of addition of gordonii, pluranimalium, and sobrinus species groups. A more recent and comprehensive analysis by Patel and Gupta (2018) on 70 sequenced members of the streptococcal genus used multiple phylogenetic assessments of different datasets of proteins and simultaneous comparative genomics to identify 134 highly specific converted signature indels to propose a new phylogenetic structure. This proposed structure consists of two major clades of "Mitis-Suis" and "Pyogenes-Equinus-Mutans". Under these major clades lie 14 subclades: Pyogenes, Mutans, Salivarius, Equinus/Bovis, Anginosus, Pneumoniae, Gordonii, Parasanguinis, Suis, Sobrinus, Halotolerans, Porci, Entericus, and Orisratti. Overall, the structure proposed by Patel and Gupta (2018) represents the most thorough and accurate organization of the streptococcal genus to date and should be used as the basis for future work.

It has been proposed that a genomic distance of 95% average nucleotide identity (ANI) is the universal boundary between bacterial and archaeal species (Parks et al. 2020). This, however, causes some issues with the delineation of species within the streptococci. Many species that sit within the Mitis group have ANIs of between 90% and 95% (Jensen et al. 2016), and these were therefore proposed by Parks et al. (2020) to be split into as many as 50 species. New work by Mclean et al. (2022) has, however, challenged this split, showing *S. mitis*, *S. oralis*, and *S. infantis* formed distinct clusters when using ANI, and the respective species shared a common localization phenotype within the mouth. This debate has drawn attention to the need to clearly define if genomic differences alone are sufficient to define new species in spite of their shared phenotypic characteristics.

# 4.6 AGGRESSORS, OPPORTUNISTS, WORKERS, AND PROTECTORS

The technological advances that have made possible the probing of the intricacies of our microbiome continue to open new vistas for exploration and understanding of the roles played by streptococcal bacteria in human existence (Kilian et al. 2016; Marsh 2018). Most of the streptococci are characterized as obligate parasites of mucosal (or sometimes tooth) surfaces of humans and other animals. Indeed, many are long-term, niche-adapted, predominant members of the commensal microflora of the upper respiratory, intestinal, or genital tracts of mammalian species. Exceptions, such as the dairy species *S. thermophilus*, are rare, especially since the reassignment in 1985 of the majority of the dairy streptococci to the genus *Lactococcus* (Schleifer et al. 1985). Most species have a preferred animal host, although some, especially *S. suis*, *S. equi* subsp. *zooepidemicus*, *S. canis*, *S. iniae*, *S. dysgalactiae* subsp. *equisimilis*, and possibly the bovine lineage of *S. agalactiae*, can cause zoonotic infections.

Although some streptococcal commensals have no significant record of disease transgressions, others occasionally function as opportunistic pathogens, either if introduced into competitor-free (normally sterile) tissues or in immunologically compromised hosts. On the other hand, some streptococci are replete with virulence attributes and function as professional pathogens capable of interpersonal spread and of initiating infection in vulnerable (nonimmune protected) individuals.

Members of the disease-associated streptococcal species are known to also commonly occur in non-disease (carriage) relationships in which they are maintained in sub-disease threshold numbers and thereby essentially simulate membership of the indigenous microbiota of the host. This feigned commensal-like behavior of the latent pathogens allows them to avoid the menacing attention of the host's immunological defenses, and so for a time both the host and the streptococcus prosper. In some cases, oral streptococci can employ further subterfuge, entering buccal epithelial cells (LaPenta et al. 1994; Rudney et al. 2005), a state providing them with high-level protection against immune (both innate and adaptive) and antibiotic (both therapeutic and bacteriocin) assault.

## 4.6.1 Aggressors

At one extreme end of the parasitic streptococcal associates of humans are the sometimes aggressively pathogenic species *S. agalactiae*, *S. pneumoniae*, and *S. pyogenes*. The highest occurrence of serious disease associated with these species is found respectively in infants (meningitis and septicemia), the elderly (pneumonia), and young adults (rheumatic carditis). On the other hand, in terms of total disease burden, the most prevalent streptococcal disease of humans is dental caries, and the most highly incriminated etiological agent is *S. mutans*.

The survival of individual streptococcal clones is ultimately dependent on their transmission to new hosts. The initial encounter of streptococcus and host takes place at epithelial surfaces, either the pharynx or skin. For *S. pyogenes* and *S. pneumoniae*, this typically occurs either via infective droplets or by direct contact. For *S. agalactiae* and *S. mutans*, mother-to-child transmission has particular significance—as a source of *S. agalactiae* infection for the neonate or of colonization with *S. mutans* following tooth eruption and especially during a "window of infectivity" period at around 2 years of age (Caufield et al. 1993).

The classic aggressive streptococcus, *S. pyogenes*, harbors a complex virulence repertoire of surface-associated and secreted components (Cunningham 2008; Okumura and Nizet 2014; Mitchell 2003; Ferretti et al. 2022). Key elements of its pathogenicity are the hyaluronic acid capsules that mimic human tissue hyaluronic acid, proteases that specifically destroy the chemotactic signals that attract phagocytic defenses to infecting streptococci (Ly et al. 2017), and surface-bound M proteins that impede the phagocytic defenses and promote invasion of epithelial cells (Brook 2017) and macrophages (Fischetti and Dale 2016). Variations in M protein account for the more than 220 *S. pyogenes* *emm*-types. Vaccine development has been a tortuous process but appears once again to be gaining momentum (Schodel et al. 2017; Sanderson-Smith et al. 2014).

## 4.6.2 Opportunists

For strains of the potentially pathogenic streptococcal species that are temporarily residing in sub-disease threshold proportions within the indigenous microbiota of the human host, there are several changes either to the streptococcal population or to the animal host tissue environment that may effect a change in the relationship with the host to one that is regarded as disease.

(a) Quorum sensing—enhanced upregulation of the expression of streptococcal virulence determinants
(b) Decreased levels of natural competitors within the indigenous microbiota (e.g., following exposure of the host to antibiotics)
(c) Development of defects in the host immune defenses (innate or induced)
(d) Displacement of the potential pathogen to a relatively unprotected niche (e.g., in bacterial endocarditis)
(e) Dietary changes (e.g., influence of sucrose supplementation on proliferation of *S. mutans* and initiation of dental caries)
(f) Zoonotic transmission of a streptococcus from an alternative animal host
(g) Acquisition of virulence determinants from other streptococci by horizontal gene transfer (mediated by transformation, transduction, or transmissible plasmids or transposons)

## 4.6.3 Workers

Streptococci are often maligned for their pathogenic potential in humans and animals. Many streptococcal species, however, are "in-between" companions, in that they can be opportunistic pathogens but are mostly commensal bacteria that play important roles in the maintenance of a healthy microbiota for the host. This is especially the case in the oral cavity and upper respiratory tract and serves as a further gateway for the intestinal and urogenital tracts. These sites within themselves contain unique environmental niches such as the tongue and tooth cavities within which some streptococci have specifically adapted.

Perhaps the most widely used and benign beneficial streptococcal species is *Streptococcus thermophilus*, the primary application of which is to the production of fermented foods such as yogurt and various probiotic applications. While it is not considered to be of human origins, it is often found in the feces due to its consumption in food and dietary supplement products. *S. thermophilus* is a clonal species that emerged relatively recently on the evolutionary timescale, possibly coinciding with the domestication of animals utilized for milk production. Its origin appears to be from a commensal ancestor of the salivarius group (Delorme 2008) with its fame being an adaptation to grow in milk following a series of loss-of-function and likely horizontal gene transfer events (Goh et al. 2011; Hols et al. 2005). Supporting this theory, it has been reported that approximately 10% of the *S. thermophilus* predicted proteome consists of pseudogenes whose original functions were either unnecessary for growth in milk or correspond to surface-exposed proteins involved in host–bacteria interactions (Bolotin et al. 2004; Delorme et al. 2015; Couvigny et al. 2015).

The most well-known role for *S. thermophilus* is its ability to ferment milk to make a range of cheeses and yogurts, and it is widely used as a starter in the dairy industry. Many traditional fermented products including yogurt and various soft (e.g., mozzarella), hard (e.g., cheddar), surface-ripened (e.g., appenzeller), and mold-ripened cheeses (e.g., gorgonzola) also utilize *S. thermophilus* (Beresford et al. 2001; Ganzle 2015), and it is regularly found in spontaneously fermented dairy products such as fermented milks in Mongolia (Yu et al. 2011) and ragusano cheese in Sicily (Licitra et al. 2007). Interestingly, it is also able to ferment plant-based products such as soy juice (Boulay et al. 2020; Harle et al. 2020) and milk kefir grains (Chen et al. 2021; Simova et al. 2002), and it is likely to be more important in the future in meeting the United Nations Sustainable Development Goals of utilizing different agricultural products. Key to its successful role as a giant of industry are traits such as the production of exopolysaccharides, bacteriocins, restriction-modification systems, oxygen tolerance, amino acid metabolism, aromatic compounds, and milk-protein degradation, many of which appear to have been acquired by horizontal gene transfer (Goh et al. 2011; Alexandraki et al. 2019).

While *S. thermophilus* is considered the current workhorse of the genus, other species are also likely to have usefulness for humans. The largest site of streptococcal species diversity resides in the oral cavity, with approximately half of the known human-associated species detected there (Baty et al. 2022). The core genome of *Streptococcus* is considered relatively conserved, with many strains having picked up interesting utilities through horizontal gene transfer or other means and potentially providing useful tools for humans to exploit (Barajas et al. 2019). It is likely that commensal streptococci play important roles in the maintenance of host barrier integrity in places like the oral cavity, as other species do, for example, in the intestinal and urogenital tracts. These include immune interaction, producing antimicrobial molecules such as bacteriocins and enzymes, or simply occupying space with microbes of less detrimental potential to the ones that cause conditions such as tooth decay, periodontal disease, sore throats, and other ailments.

*S. salivarius* and certain other oral streptococci appear to inhibit immune activation by periodontal disease pathogens (MacDonald et al. 2021) and *Candida albicans* (Ishijima et al. 2012; James et al. 2016). Some can also modulate pH through urease production and inhibit *S. mutans* with peroxide (Kreth et al. 2008). In spite of their stereotypic potential disease associations, many strains have been explored as helpful organisms, with examples of functionally useful *Streptococcus salivarius, Streptococcus rattus, Streptococcus uberis*, and *Streptococcus oralis* strains increasingly being tested in human probiotic products for oral cavity and other applications to evaluate their efficacy in the prevention and treatment of various disease conditions (Laleman et al. 2015; Burton et al. 2013).

The specter of escalating antimicrobial resistance has served to catalyze the commercial development of bacteriocins as novel therapeutic agents. The lactic acid bacteria in general and especially the streptococci are recognized as producers of strong candidate bacteriocins and bacteriocin-like inhibitory substances for future development as antimicrobials for the treatment and prevention of recalcitrant infections by current and emerging bacterial and fungal pathogens. A myriad of streptococcal bacteriocins, mostly inhibitory to Gram-positive pathogens, are now being sourced in a purified or cloned formats from either commensal or pathogenic streptococci (Tagg et al. 2023).

Other current industrial activities of streptococci include the production of hyaluronic acid (HA) and collagen-like proteins (for various biomaterial applications) by *S. pyogenes*. HA is an important component of skin, where it helps to retain moisture and maintain elasticity used for skin hydration in cosmetics. Several strains of *Streptococcus*, including *Streptococcus zooepidemicus* (Paiva et al. 2022) and *S. equisimilis* (Jafari et al. 2022), are important sources of HA.

## 4.6.4 Protectors

Streptococci are predominant members of the commensal microbiotas of the mucous membranes of the human oral cavity and to a lesser extent of the nasopharynx and also the intestine. They are also transient members of the skin microbiota. The extent to which commensal streptococci colonize the exposed surfaces of other mammals is as yet unclear since the attention of researchers has to date largely been focused on the streptococcal aggressors. In humans, however, it is known that neonates typically acquire the mother's predominant strain of the commensal species *S. salivarius* within days of birth (Tagg et al. 1983), and this raises the interesting prospect that the inheritance of at least certain components of the indigenous microbiota may predominantly be of a maternal lineage.

It was Pasteur who first introduced the notion of bacteriotherapy: the utilization of "harmless" bacteria to displace pathogenic organisms as a means of treating infection (Pasteur and Joubert 1877). Since commensal streptococci are particularly abundant in the human upper respiratory tract, they have been implicated as microbes having potential to interfere with colonization or infection by potentially pathogenic streptococci. Indeed, children who more frequently acquire *S. pyogenes* have been shown to have relatively fewer oral commensal streptococci exhibiting *in vitro* inhibitory activity against *S. pyogenes* (Crowe et al. 1973; Holm and Grahn 1983). Such findings prompted Roos et al. (1993) to recommend dosing children with a mixture of α-hemolytic streptococci as a supplementary treatment of streptococcal tonsillitis.

Of all the bacterial species known to populate the human oral and nasopharyngeal mucosa in large numbers, *S. salivarius* is perhaps the most innocuous (Burton et al. 2006). It has now been established that certain *S. salivarius* produce multiple bacteriocins, the activity of which is particularly strong against *S. pyogenes*. Children naturally harboring oral populations of *S. salivarius* producing the lantibiotics salivaricin A and/or salivaricin B were significantly less likely to acquire *S. pyogenes* (Dierksen and Tagg 2000). The prototype producer strain of these two bacteriocin activities, *S. salivarius* strain K12, is now widely used as an oral probiotic for the prevention and control of a variety of maladies, including halitosis, pharyngitis, and otitis media (Wescombe et al. 2009; Tagg et al. 2023).

Another approach to the directed implementation of the principles of microbial interference focused on the application of a strongly competitive (bacteriocin-producing) strain of *S. mutans* that had its lactic-acid forming (cariogenic) capability disrupted by genetic modification. The modified strain was used to preemptively colonize and competitively exclude native (acidogenic) *S. mutans* from dental plaque as a strategy for prevention of dental caries (Hillman 2002). The same group has more recently developed a probiotic mixture of *S. rattus*, *S. oralis*, and *S. uberis* for use by humans and companion pets (Zahradnik et al. 2009; Kamble et al. 2022).

Modulation of the microbiota composition by specific introduction of strains that are capable of excluding colonization/infection by target pathogens can be viewed as the controlled manipulation of a process that otherwise occurs only haphazardly in nature. It offers a cost-effective means of achieving protection for the host and it appears that bacterial replacement therapy will have an increasingly prominent role as a strategy for prevention and control of a wide variety of topical bacterial infections of humans and other animals.

# 4.7 CONCLUSION

From relatively humble beginnings, *S. pyogenes*, the twisted chain microbe of Billroth, initially recognized as a common cause of pyogenic infection in humans and other animals, now finds itself positioned as the prototype of a rapidly expanding entourage of companion streptococcal species being isolated from both mammalian and nonmammalian hosts. The availability of increasingly sophisticated molecular methodologies to dissect the cellular content of complex microbiotas appears to have elicited an exponential increase in the discovery of novel streptococci. As researchers continue to probe ever more precisely within the commensal microbial populations of species in addition to *Homo sapiens*, we can anticipate many more twists to the intriguing tale of the *Streptococcus*.

# BIBLIOGRAPHY

Abdelbary, M.M.H., G. Wilms and G. Conrads. 2021. A new species-specific typing method for salivarius group streptococci based on the dephospho-coenzyme A kinase (*coaE*) gene sequencing. *Front. Cell Infect. Microbiol.* 11: 685657. https://doi.org/10.3389/fcimb.2021.685657.

Agar, N.S., S.M. Sadrzadeh, P.E. Hallaway and J.W. Eaton. 1986. Erythrocyte catalase. A somatic oxidant defense? *J. Clin. Invest.* 77: 319–321.

Ajdic, D., W.M. McShan, R.E. McLaughlin, et al. 2002. Genome sequence of *Streptococcus mutans* UA159, a cariogenic dental pathogen. *Proc. Natl. Acad. Sci. U. S. A.* 99: 14434–14439.

Alexandraki, V., M. Kazou, J. Blom, et al. 2019. Comparative genomics of *Streptococcus thermophilus* support important traits concerning the evolution, biology and technological properties of the species. *Front. Microbiol.* 10: 2916.

Alouf, J.E. and T. Horaud. 1997. Streptococcal research at Pasteur Institute from Louis Pasteur's time to date. *Adv. Exp. Med. Biol.* 418: 7–14.

Andrade, D.C., I.C. Borges, M.L. Bouzas, et al. 2017. 10-valent pneumococcal conjugate vaccine (PCV10) decreases metabolic activity but not nasopharyngeal carriage of *Streptococcus pneumoniae* and *Haemophilus influenzae*. *Vaccine* 35: 4105–4111.

Andrewes, F.W. and T.J. Horder. 1906. A study of the streptococci pathogenic for man. *Lancet.* 2: 708–713.

Arbique, J.C., C. Poyart, P. Trieu-Cuot, et al. 2004. Accuracy of phenotypic and genotypic testing for identification of *Streptococcus pneumoniae* and description of *Streptococcus pseudopneumoniae* sp. nov. *J. Clin. Microbiol.* 42: 4686–4696.

Baiano, J.C. and A.C. Barnes. 2009. Towards control of *Streptococcus iniae*. *Emerg. Infect. Dis.* 15: 1891–1896. https://doi.org/10.3201/eid1512.090232.

Balsells, E., L. Guillot, H. Nair and M.H. Kyaw. 2017. Serotype distribution of *Streptococcus pneumoniae* causing invasive disease in children in the post-PCV era: A systematic review and meta-analysis. *PLoS ONE.* 12(5): e0177113. https://doi.org/10.1371/journal.pone.0177113.

Barajas, H.R., M.F. Romero, S. Martínez-Sánchez and L.D. Alcaraz. 2019. Global genomic similarity and core genome sequence diversity of the Streptococcus genus as a toolkit to identify closely related bacterial species in complex environments. *PeerJ* 6: e6233. https://doi.org/10.7717/peerj.6233.

Barnard, J.P. and M.W. Stinson. 1996. The alpha-hemolysin of *Streptococcus gordonii* is hydrogen peroxide. *Infect. Immun.* 64: 3853–3857.

Baty, J.J., S.N. Stoner and J.A. Scoffield. 2022. Oral commensal streptococci: Gatekeepers of the oral cavity. *J. Bact.* 204: e0025722. https://doi.org/10.1128/jb.00257-22.

Beighton, D. and H. Hayday. 1984. The establishment of the bacterium *Streptococcus mutans* in dental plaque and the induction of caries in macaque monkeys (*Macaca fascicularis*) fed a diet containing cooked-wheat flour. *Arch. Oral. Biol.* 29: 369–372.

Bekal, S., C. Gaudreau, R.A. Laurence, E. Simoneau and L. Raynal. 2006. *Streptococcus pseudoporcinus* sp. nov., a novel species isolated from the genitourinary tract of women. *J. Clin. Microbiol.* 44: 2584–2586.

Belman, S., C. Chaguza, N. Kumar, S. Lo and S.D. Bentley. 2022. A new perspective on ancient Mitis group streptococcal genetics. *Microb Genom.* 8(2): 000753. https://doi.org/10.1099/mgen.0.000753.

Bentley, R.W., J.A. Leigh and M.D. Collins. 1991. Intrageneric structure of *Streptococcus* based on comparative analysis of small-subunit rRNA sequences. *Int. J. Syst. Bacteriol.* 41: 487–494.

Berardi, A., L. Lugli, D. Baronciani, et al. 2007. Group B streptococcal infections in a northern region of Italy. *Pediatrics* 120: e487–e493.

Beresford, T.P., N.A. Fitzsimons, N.L. Brennan and T.M. Cogan. 2001. Recent advances in cheese microbiology. *Int. Dairy J.* 11: 259–274.

Billroth, A.W. 1874. *Untersuchungen uber die Vegetationsformen von Coccobacteria septica.* Georg Reimer, Berlin.

Boleij, A. and H. Tjalsma. 2013. The itinerary of *Streptococcus gallolyticus* infection in patients with colonic malignant disease. *Lancet. Infect. Dis.* 13: 719–724. https://doi.org/10.1016/S1473-3099(13)70107-5.

Bolotin, A., B. Quinquis, P. Renault, et al. 2004. Complete sequence and comparative genome analysis of the dairy bacterium *Streptococcus thermophilus. Nat. Biotechnol.* 22: 1554–1558.

Boulay, M., M. Al Haddad and F. Rul. 2020. *Streptococcus thermophilus* growth in soya milk: Sucrose consumption, nitrogen metabolism, soya protein hydrolysis and role of the cell-wall protease PrtS. *Int. J. Food. Microbiol.* 335: 108903.

Bouvet, A., F. Grimont, M.D. Collins, et al. 1997. *Streptococcus infantarius* sp. nov. related to *Streptococcus bovis* and *Streptococcus equinus. Adv. Exp. Med. Biol.* 418: 393–395.

Brook, I. 2017. Treatment challenges of group A beta-hemolytic streptococcal pharyngo-tonsillitis. *Int. Arch. Otorhinolaryngol.* 21: 286–296. https://doi.org/10.1055/s-0036-1584294.

Burton, J.P., B.K. Drummond, C.N. Chilcott, et al. 2013. Influence of the probiotic *Streptococcus salivarius* strain M18 on indices of dental health in children: A randomized double-blind, placebo-controlled trial. *J. Med. Microbiol.* 62: 875–884.

Burton, J.P., P.A. Wescombe, C.J. Moore, et al. 2006. Safety assessment of the oral cavity probiotic *Streptococcus salivarius* K12. *Appl. Environ. Microbiol.* 72: 3050–3053.

Cao, Y., H. Yin, W. Wang, et al. 2020. Killing *Streptococcus mutans* in mature biofilm with a combination of antimicrobial and antibiofilm peptides. *Amino Acids* 52: 1–14. https://doi.org/10.1007/S00726-019-02804-4/METRICS.

Caufield, P.W., G.R. Cutter and A.P. Dasanayake. 1993. Initial acquisition of mutans streptococci by infants: Evidence for a discrete window of infectivity. *J. Dent. Res.* 72: 37–45.

Chapman, G.H. 1944. The isolation of streptococci from mixed cultures. *J. Bacteriol.* 48: 113–114.

Chen, Y.Y., C.A. Weaver and R.A. Burne. 2000. Dual functions of *Streptococcus salivarius* urease. *J. Bacteriol.* 182: 4667–4669.

Chen, Z., T. Liu, T. Ye, X. Yang, et al. 2021. Effect of lactic acid bacteria and yeasts on the structure and fermentation properties of Tibetan kefir grains. *Int. Dairy J.* 114: 104943.

Chew, T.A. and J.M. Smith. 1992. Detection of diacetyl (caramel odor) in presumptive identification of the "*Streptococcus milleri*" group. *J. Clin. Microbiol.* 30: 3028–3029.

Chiarello, F., S. Spitoni, E. Hollander, et al. 2017. An expert opinion on PANDAS/PANS: Highlights and controversies. *Int. J. Psychiatry Clin. Pract.* 21: 91–98.

Chochua, S., J. Rivers, S. Mathis, Z. Li, et al. 2019. Emergent invasive Group A *Streptococcus dysgalactiae* subsp. *equisimilis*, United States, 2015–2018. *Emerg. Infect. Dis.* 25: 1543–1547.

Clarke, J.K. 1924. On the bacterial factor in the aetiology of dental caries. *Br. J. Exp. Pathol.* 5: 141–147.

Collins, M.D., C. Ash, J.A. Farrow, S. Wallbanks and A.M. Williams. 1989. 16S ribosomal ribonucleic acid sequence analyses of lactococci and related taxa. Description of *Vagococcus fluvialis* gen. nov., sp. nov. *J. Appl. Bacteriol.* 67: 453–460.

Collins, M.D., J.A.E. Farrow, V. Katie and O. Kandler. 1984. Taxonomic studies on streptococci of serological groups E, P, U and V: Description of *Streptococcus porcinus* sp. nov. *Syst. Appl. Microbiol.* 5: 402–413.

Collins, M.D., R.A. Hutson, E. Falsen, et al. 2000. An unusual *Streptococcus* from human urine, *Streptococcus urinalis* sp. nov. *Int. J. Syst. Evol. Microbiol.* 50: 1173–1178.

Collins, M.D. and P.A. Lawson. 2000. The genus *Abiotrophia* (Kawamura et al.) is not monophyletic: Proposal of *Granulicatella* gen. nov., *Granulicatella adiacens* comb. nov., *Granulicatella elegans* comb. nov. and *Granulicatella balaenopterae* comb. nov. *Int. J. Syst. Evol. Microbiol.* 50: 365–369.

Collins, M.D., T. Lundstrom, C. Welinder-Olsson, et al. 2004. *Streptococcus devriesei* sp. nov., from equine teeth. *Syst. Appl. Microbiol.* 27: 146–150.

Couvigny, B., C. Therial, C. Gautier, et al. 2015. *Streptococcus thermophilus* biofilm formation: A remnant trait of ancestral commensal life? *PLoS ONE*. 10(6): e0128099.

Coykendall, A.L. 1977. Proposal to elevate the subspecies of *Streptococcus mutans* to species status, based on their molecular composition. *Int. J. Syst. Bacteriol.* 27: 26–30.

Coykendall, A.L. 1983. *Streptococcus sobrinus* nom. rev. and *Streptococcus ferus* nom. rev.: Habitat of these and other mutans streptococci. *Int. J. Syst. Bacteriol.* 33: 883–885.

Crestani, C., T.L. Forde, S.J. Lycett, et al. 2021. The fall and rise of group B *Streptococcus* in dairy cattle: Reintroduction due to human-to-cattle host jumps? *Microb. Genom.* 7(9): 000648. https://doi.org/10.1099/mgen.0.000648.

Crooks, M., S.M. James and J.R. Tagg. 1987. Relationship of bacteriocin-like inhibitor production to the pigmentation and hemolytic activity of mutans streptococci. *Zentralbl. Bakteriol. Mikrobiol. Hyg. A* 263: 541–547.

Crowe, C.C., W.E. Sanders and S. Longley. 1973. Bacterial interference. II. Role of the normal throat flora in prevention of colonization by group A *Streptococcus*. *J. Infect. Dis.* 128: 527–532.

Cunningham, M.W. 2008. Pathogenesis of group A streptococcal infections and their sequelae. *Adv. Exp. Med. Biol.* 609: 29–42. https://doi.org/10.1007/978-0-387-73960-1_3.

Dekker, J.P. and A.F. Lau. 2016. An update on the *Streptococcus bovis* group: Classification, identification, and disease associations. *J. Clin. Microbiol.* 54: 1694–1699. https://doi.org/10.1128/JCM.02977-15.

De la Rosa, M., M. Perez, C. Carazo, et al. 1992. New Granada Medium for detection and identification of group B streptococci. *J. Clin. Microbiol.* 30: 1019–1021.

Delorme, C. 2008. Safety assessment of dairy microorganisms: *Streptococcus thermophilus*. *Int. J. Food. Microbiol.* 126: 274–277.

Delorme, C., A.L. Abraham, P. Renault and E. Guedon. 2015. Genomics of *Streptococcus salivarius*, a major human commensal. *Infect. Genet. Evol.* 33: 381–392.

Devriese, L.A., J. Hommez, R. Kilpper-Balz and K.H. Schleifer. 1986. *Streptococcus canis* sp. nov.: A species of group G streptococci from animals. *Int. J. Syst. Bacteriol.* 36: 422–425.

Dierksen, K.P. and J. Tagg. 2000. The influence of indigenous bacteriocin-producing *Streptococcus salivarius* on the acquisition of *Streptococcus pyogenes* by primary school children in Dunedin, New Zealand. In *Streptococci and Streptococcal Diseases Entering the New Millennium*, Eds. D.R. Martin and J.R. Tagg, pp. 81–85. Securacopy, Auckland.

Diernhofer, K. 1932. Aesculinbouillon as hilfsmittel fur die differenzierung von euter- und milchstreptokokken bei masse untersuchungen. *Milchwirtschaft Forsch.* 13: 368–374.

Ekstedt, R.D. and G.H. Stollerman. 1960. Factors affecting the chain length of group A streptococci. II. Quantitative M-anti-M relationships in the long chain test. *J. Exp. Med.* 112: 687–698.

Elliott, S.D., J. Hayward and T.Y. Liu. 1971. The presence of a group A variant-like antigen in streptococci of other groups with special reference to group N. *J. Exp. Med.* 133: 479–493.

Facklam, R.R. 1984. The major differences in the American and British *Streptococcus* taxonomy schemes with special reference to *Streptococcus milleri*. *Eur. J. Clin. Microbiol.* 3: 91–93.

Facklam, R.R. 2002. What happened to the streptococci: Overview of taxonomic and nomenclature changes. *Clin. Microbiol. Rev.* 15: 613–630.

Falsetta, M.L., M.I. Klein, P.M. Colonne, K. Scott-Anne, et al. 2014. Symbiotic relationship between *Streptococcus mutans* and *Candida albicans* synergizes virulence of plaque biofilms in vivo. *Infect. Immun.* 82: 1968–1981.

Farrow, J.A.E. and M.D. Collins. 1984. Taxonomic studies on streptococci of serological groups C, G and L and possibly related taxa. *Syst. Appl. Microbiol.* 5: 483–493.

Farrow, J.A.E., J. Kruze, B.A. Philips, A.J. Bramley and M.D. Collins. 1984. Taxonomic studies on *Streptococcus bovis* and *Streptococcus equinus*: Description of *Streptococcus alactolyticus* sp. nov. and *Streptococcus saccharolyticus* sp. nov. *Syst. Appl. Microbiol.* 5: 467–482.

Fazili, T., S. Riddell, D. Kiska, et al. 2017. *Streptococcus anginosus* group bacterial infections. *Am. J. Med. Sci.* 354: 257–261.

Fernandez, E., V. Blume, P. Garrido, et al. 2004. *Streptococcus equi* subsp. *ruminatorum* subsp. nov., isolated from mastitis in small ruminants. *Int. J. Syst. Evol. Microbiol.* 54: 2291–2296.

Fernandez-Garayzabal, J.F., E. Fernandez, A. Las Heras, et al. 1998. *Streptococcus parasanguinis*: New pathogen associated with asymptomatic mastitis in sheep. *Emerg. Infect. Dis.* 4: 645–647.

Ferretti, J.J. and W. Köhler. 2022. History of Streptococcal Research. In *Streptococcus Pyogenes: Basic Biology to Clinical Manifestations* [Internet]. Eds. J.J. Ferretti, D.L. Stevens and V.A. Fischetti. Oklahoma City: University of Oklahoma Health Sciences Center.

Ferretti, J.J., D.L. Stevens and V.A. Fischetti (Editors). 2022. *Streptococcus Pyogenes: Basic Biology to Clinical Manifestations [Internet]*. Eds. J.J. Ferretti, D.L. Stevens and V.A. Fischetti. Oklahoma City: University of Oklahoma Health Sciences Center.

Fischetti, V.A. and J.B. Dale. 2016. One more disguise in the stealth behavior of *Streptococcus pyogenes*. *MBio*. 7(3). https://doi.org/10.1128/mBio.00661-16.

Fuller, A.T. 1938. The formamide method for the extraction of polysaccharide from haemolytic streptococci. *Br. J. Exp. Pathol.* 19: 130–139.

Ganzle, M.G. 2015. Lactic metabolism revisited: Metabolism of lactic acid bacteria in food fermentations and food spoilage. *Curr. Opin. Food Sci.* 2: 106–117.

Gao, X.Y., X.Y. Zhi, H.W. Li, H.P. Klenk and W.J. Li. 2014. Comparative genomics of the bacterial genus *Streptococcus* illuminates evolutionary implications of species groups. *PLoS ONE* 9(6): e101229. https://doi.org/10.1371/journal.pone.0101229.

Gauthier, D.T. 2015 Bacterial zoonoses of fishes: A review and appraisal of evidence for linkages between fish and human infections. *Vet. J.* 203: 27–35. https://doi.org/10.1016/j.tvjl.2014.10.028.

Glaser, P., C. Rusniok, C. Buchrieser, et al. 2002. Genome sequence of *Streptococcus agalactiae*, a pathogen causing invasive neonatal disease. *Mol. Microbiol.* 45: 1499–1513.

Glazunova, O.O., D. Raoult and V. Roux. 2006. *Streptococcus massiliensis* sp. nov., isolated from a patient blood culture. *Int. J. Syst. Evol. Microbiol.* 56: 1127–1131.

Gogos, A. and M.J. Federle. 2019. Modelling *Streptococcus pyogenes* pharyngeal colonization in the mouse. *Front. Cell Infect. Microbiol.* 9: 137.

Goh, Y.J., C. Goin, S. O'Flaherty, E. Altermann and R. Hutkins. 2011. Specialized adaptation of a lactic acid bacterium to the milk environment: The comparative genomics of *Streptococcus thermophilus* LMD-9. *Microb. Cell Fact.*10(Suppl 1): S22.

Griffith, F. 1934. The serological classification of *Streptococcus pyogenes*. *J. Hyg.* 34: 542–584.

Grinwis, M.E., C.D. Sibley, M.D. Parkins, et al. 2010. Characterization of *Streptococcus milleri* group isolates from expectorated sputum of adult patients with cystic fibrosis. *J. Clin. Microbiol.* 48: 395–401.

Griswold, A., Y.Y. Chen, J.A. Snyder and R.A. Burne. 2004. Characterization of the arginine deiminase operon of *Streptococcus rattus* FA-1. *Appl. Environ. Microbiol.* 70: 1321–1327.

Groschup, M.H., G. Hahn and J.F. Timoney. 1991. Antigenic and genetic homogeneity of *Streptococcus uberis* strains from the bovine udder. *Epidemiol. Infect.* 107: 297–310.

Grundy, M., N. Suwantarat, M. Rubin, R. Harris, et al. 2019. Differentiating *Streptococcus pseudoporcinus* from GBS: Could this have implications in pregnancy? *Am. J. Obstet. Gynecol.* 220: 490.e1–490.e7.

Haas, B. and D. Grenier. 2017. Understanding the virulence of *Streptococcus suis*: A veterinary, medical, and economic challenge. *Med. Mal. Infect.* https://doi.org/10.1016/j.medmal.2017.10.001.

Handley, P., A. Coykendall, D. Beighton, J.M. Hardie and R.A. Whiley. 1991. *Streptococcus crista* sp. nov., a viridans streptococcus with tufted fibrils, isolated from the human oral cavity and throat. *Int. J. Syst. Bacteriol.* 41: 543–547.

Harbison-Price, N., T. Rivera-Hernandez, J. Osowicki, M.R. Davies, et al. 2022. Current approaches to vaccine development of *Streptococcus pyogenes*. In *Streptococcus Pyogenes: Basic Biology to Clinical Manifestations* [Internet], Eds. J.J. Ferretti, D.L. Stevens and V.A. Fischetti, 2nd ed. University of Oklahoma Health Sciences Center, Oklahoma City, OK. Chapter 31.

Harju, I., C. Lange, M. Kostrzewa, T. Maier, et al. 2017. Improved differentiation of *Streptococcus pneumoniae* and other *S. mitis* group streptococci by MALDI biotyper using an improved MALDI Biotyper database content and a novel result interpretation algorithm. *J. Clin. Microbiol.* 55: 914–922. https://doi.org/10.1128/JCM.01990-16.

Harle, O., H. Falentin, J. Niay, F. Valence, et al. 2020. Diversity of the metabolic profiles of a broad range of lactic acid bacteria in soy juice fermentation. *Food Microbiol.* 89: 103410.

Hatrongjit, R., Y. Akeda, S. Hamada, M. Gottschalk and A. Kerdsin. 2017. Multiplex PCR for identification of six clinically relevant streptococci. *J. Med. Microbiol.* 66: 1590–1595.

Henriksen, S.D. and J. Henrichsen. 1975. Twitching motility and possession of polar fimbriae in spreading *Streptococcus sanguis* isolates from the human throat. *Acta Pathol. Microbiol. Scand. B.* 83: 133–140.

Henriques-Normark, B. and S. Normark. 2010. Commensal pathogens, with a focus on *Streptococcus pneumoniae*, and interactions with the human host. *Exp. Cell Res.* 316: 1408–1414.

Hillman, J.D. 2002. Genetically modified *Streptococcus mutans* for the prevention of dental caries. *Antonie Van Leeuwenhoek*. 82: 361–366.

Holm, S.E. and E. Grahn. 1983. Bacterial interference in streptococcal tonsillitis. *Scand. J. Infect. Dis. Suppl.* 39: 73–78.

Hols, P., F. Hancy, L. Fontaine, et al. 2005. New insights in the molecular biology and physiology of *Streptococcus thermophilus* revealed by comparative genomics. *FEMS Microbiol. Rev.* 29: 435–463.

Huch, M., K. De Bruyne, I. Cleenwerck, A. Bub, et al. 2013. *Streptococcus rubneri* sp. nov., isolated from the human throat. *Int. J. Syst. Evol. Microbiol.* 63: 4026–4032.

Hyun, D.W., J.Y. Lee, M.S. Kim, N.R. Shin, et al. 2021. Pathogenomics of *Streptococcus ilei* sp. nov., a newly identified pathogen ubiquitous in human microbiome. *J. Microbiol.* 59: 792–806.

Imai, K., R. Nemoto, M. Kodana, N. Tarumoto, et al. 2020. Rapid and accurate species identification of Mitis Group Streptococci using the MinION nanopore sequencer. *Front. Cell. Infect. Microbiol.* 10: 11. https://doi.org/10.3389/fcimb.2020.00011.

Ishijima, S.A., K. Hayama, J.P. Burton, et al. 2012. Effect of *Streptococcus salivarius* K12 on the *in vitro* growth of *Candida albicans* and its protective effect in an oral candidiasis model. *Appl. Environ. Microbiol.* 78: 2190–2199.

Jafari, B., M. Keramati, R. Ahangari Cohan, et al. 2022. Development of *Streptococcus equisimilis* Group G mutant strains with ability to produce low polydisperse and low-molecular-weight hyaluronic acid. *Iran Biomed. J.* 26: 454–462.

James, K.M., K.W. MacDonald, R.M. Chanyi, et al. 2016. Inhibition of *Candida albicans* biofilm formation and modulation of gene expression by probiotic cells and supernatant. *J. Med. Microbiol.* 65: 328–336.

Jans, C., T. de Wouters, B. Bonfoh, et al. 2016. Phylogenetic, epidemiological and functional analyses of the *Streptococcus bovis/Streptococcus equinus* complex through an overarching MLST scheme. *BMC Microbiol.* 16: 117. https://doi.org/10.1186/s12866-016-0735-2.

Jans, C., L. Meile, C. Lacroix and M.J. Stevens. 2015. Genomics, evolution, and molecular epidemiology of the *Streptococcus bovis/Streptococcus equinus* complex (SBSEC). *Infect. Genet. Evol.* 33: 419–436.

Jensen, A., C.F.P. Scholz and M. Kilian. 2016. Re-evaluation of the taxonomy of the mitis group of the genus *Streptococcus* based on whole genome phylogenetic analyses, and proposed reclassification of *Streptococcus dentisani* as *Streptococcus oralis* subsp. *dentisani* comb. nov., *Streptococcus tigurinus* as *Streptococcus oralis* subsp. *tigurinus comb.* nov., and *Streptococcus oligofermentans* as a later synonym of *Streptococcus cristatus*. *Int. J. Syst. Evol. Microb.* 66: 4803–4820.

Jordal, S., M. Glambek, O. Oppegaard and B.R. Kittang. 2015. New tricks from an old cow: Infective endocarditis caused by *Streptococcus dysgalactiae* subsp. *dysgalactiae*. *J. Clin. Microbiol.* 53: 731–734.

Kaci, G., D. Goudercourt, V. Dennin, B. Pot, et al. 2014. Anti-inflammatory properties of *Streptococcus salivarius*, a commensal bacterium of the oral cavity and digestive tract. *Appl. Environ. Microbiol.* 80: 928–934. https://doi.org/10.1128/AEM.03133-13.

Kamble, A., Z. Jabin, N. Agarwal and A. Anand. 2022. Effectiveness of oral probiotics in reducing *S. mutans* count in caries-active children: A comparison with chlorhexidine and herbal mouthrinse (Hiora). *Int. J. Clin. Pediatr. Dent.* 15(Suppl 2): S207–S211.

Kawamura, Y., X.G. Hou, F. Sultana, et al. 1995a. Transfer of *Streptococcus adjacens* and *Streptococcus defectivus* to *Abiotrophia* gen. nov. as *Abiotrophia adiacens* comb. nov. and *Abiotrophia defectiva* comb. nov., respectively. *Int. J. Syst. Bacteriol.* 45: 798–803.

Kawamura, Y., X.G. Hou, F. Sultana, H. Miura and T. Ezaki. 1995b. Determination of 16S rRNA sequences of *Streptococcus mitis* and *Streptococcus gordonii* and phylogenetic relationships among members of the genus *Streptococcus*. *Int. J. Syst. Bacteriol.* 45: 406–408.

Kawamura, Y., X.G. Hou, Y. Todome, et al. 1998. *Streptococcus peroris* sp. nov. and *Streptococcus infantis* sp. nov., new members of the *Streptococcus mitis* group, isolated from human clinical specimens. *Int. J. Syst. Bacteriol.* 48: 921–927.

Kayansamruaj, P., N. Dinh-Hung, P. Srisapoome, U. Na-Nakorn and S. Chatchaiphan. 2023. Genomics-driven prophylactic measures to increase streptococcosus resistance in tilapia. *J. Fish Dis.* https://doi.org/10.1111/jfd.13763.

Kerdsin, A., R. Hatrongjit, S. Hamada, Y. Akeda and M. Gottschalk. 2017. Development of a multiplex PCR for identification of beta-hemolytic streptococci relevant to human infections and serotype distribution of invasive *Streptococcus agalactiae* in Thailand. *Mol. Cell Probes.* 36: 10–14.

Kilian, M. 2005. *Streptococcus* and *Lactobacillus*. In *Topley and Wilson's Microbiology and Microbial Infections*, Eds. S.P.P. Boriello, P.R. Murray and G. Funke, pp. 833–881. Hodder Arnold, London.

Kilian, M., I.L. Chapple, M. Hannig, P.D. Marsh, et al. 2016. The oral microbiome—an update for oral healthcare professionals. *Br. Dent. J.* 221: 657–666. https://doi.org/10.1038/sj.bdj.2016.865.

Kilian, M., L. Mikkelsen and J. Henrichsen. 1989a. Taxonomic study of viridans streptococci: Description of *Streptococcus gordonii* sp. nov., and emended description of *Streptococcus sanguis* (White and Niven 1946), *Streptococcus oralis* (Bridge and Sneath 1982), and *Streptococcus mitis* (Andrewes and Horder 1906). *Int. J. Syst. Bacteriol.* 39: 471–484.

Kilian, M., J. Reinholdt, B. Nyvad, E.V. Frandsen and L. Mikkelsen. 1989b. IgA1 proteases of oral streptococci: Ecological aspects. *Immunol. Invest.* 18: 161–170.

Kilian, M., D.R. Riley, A. Jensen, H. Bruggemann and H. Tettelin. 2014. Parallel evolution of *Streptococcus pneumoniae* and *Streptococcus mitis* to pathogenic and mutualistic lifestyles. *MBio.* 5(4): e01490–14. https://doi.org/10.1128/mBio.01490-14.

Kilpper-Balz, R. and K.H. Schleifer. 1987. Description of *Streptococcus suis* nom. rev. *Int. J. Syst. Bacteriol.* 37: 160–162.

Kilpper-Balz, R., P. Wenzig and K.H. Schleifer. 1985. Molecular relationships and classification of some viridans streptococci as *Streptococcus oralis* and emended description of *Streptococcus oralis* (Bridge and Sneath 1982). *Int. J. Syst. Bacteriol.* 35: 482–488.

Klein, R.S., R.A. Recco, M.T. Catalano, et al. 1977. Association of *Streptococcus bovis* with carcinoma of the colon. *N. Engl. J. Med.* 297: 800–802.

Kluyver, A.J. and C.B. van Niel. 1936. Prospects for a natural system of classification of bacteria. *Zentralbl. Bakteriol. Parasitenkd. Infektionskr. Hyg. Abt.* 1: 369–403.

Kohler, W. 2007. The present state of species within the genera *Streptococcus* and *Enterococcus*. *Int. J. Med. Microbiol.* 297: 133–150.

Kreth, J., Y. Zhang and M.C. Herzberg. 2008. Streptococcal antagonism in oral biofilms: *Streptococcus sanguinis* and *Streptococcus gordonii* interference with *Streptococcus mutans*. *J. Bacteriol.* 190: 4632–4640. https://doi.org/10.1128/JB.00276-08.47.

Lakshmi, S.S.J. and K.V. Leela. 2022. A review on updated species list of viridans streptococci causing infective endocarditis. *J. Pure Appl. Microbiol.* 16: 1590–1594.

Laleman, I., E. Yilmaz, O. Ozcelik, C. Haytac, et al. 2015. The effect of a streptococci containing probiotic in periodontal therapy: A randomized controlled trial. *J. Clin. Periodont.* 42: 1032–1041.

Lam, M.M., J.E. Clarridge, E.J. Young and S. Mizuki. 2007. The other group G Streptococcus: Increased detection of *Streptococcus canis* ulcer infections in dog owners. *J. Clin. Microbiol.* 45: 2327–2329.

Lancefield, R.C. 1933. A serological differentiation of human and other groups of hemolytic streptococci. *J. Exp. Med.* 57: 571–595.

Lancefield, R.C. 1941. Specific relationship of cell composition to biological activity of hemolytic streptococci. *Harvey Lectures* 36: 251.

LaPenta, D., C. Rubens, E. Chi and P.P. Cleary. 1994. Group A streptococci efficiently invade human respiratory epithelial cells. *Proc. Natl. Acad. Sci. U. S. A.* 91: 12115–12119.

Licitra, G., J.C. Ogier, S. Parayre, C. Pediliggieri, et al. 2007. Variability of bacterial biofilms of the "Tina" Wood vats used in the Ragusano cheese-making process. *Appl. Environ. Microbiol.* 73: 6980–6987.

Liu, X., L. Shao, X. Liu, et al. 2019. Alterations of gastric mucosal microbiota across different stomach microhabitats in a cohort of 276 patients with gastric cancer. *EBioMedicine*. 40: 336–348. https://doi.org/10.1016/j.ebiom.2018.12.034.

López-López, A., A. Camelo-Castillo, M.D. Ferrer, A. Simon-Soro, et al. 2017. Health-associated niche inhabitants as oral probiotics: The case of *Streptococcus dentisani*. *Front. Microbiol.* 8: 379. https://doi.org/10.3389/fmicb.2017.00379.

Ly, A.T., J.P. Noto, O.L. Walwyn, R.R. Tanz, et al. 2017. Differences in SpeB protease activity among group A streptococci associated with superficial, invasive, and autoimmune disease. *PLoS ONE* 12: e0177784. https://doi.org/10.1371/journal.pone.0177784.

MacDonald, K.W., R.M. Chanyi, J.M. Macklaim, P.A. Cadieux, et al. 2021. *Streptococcus salivarius* inhibits immune activation by periodontal disease pathogens. *BMC Oral. Health*. 21: 245. https://doi.org/10.1186/s12903-021-01606-z.

MacFadden, J.F. 1985. *Media for Isolation—Cultivation—Identification—Maintenance of Medical Bacteria*. Williams & Wilkins, Baltimore.

Maciejewska, A., C. Lugowski and J. Lukasiewicz. 2022. First report on the *Streptococcus gallolyticus* (*S. bovis* Biotype I) DSM 13808 exopolysaccharide structure. *Int. J. Mol. Sci.* 23: 11797.

Manachini, P.L., S.H. Flint, L.J. Ward, et al. 2002. Comparison between *Streptococcus macedonicus* and *Streptococcus waius* strains and reclassification of *Streptococcus waius* (Flint et al. 1999) as *Streptococcus macedonicus* (Tsakalidou et al. 1998). *Int. J. Syst. Evol. Microbiol.* 52: 945–951.

Manning, S.D., A.C. Springman, A.D. Million, et al. 2010. Association of Group B *Streptococcus* colonization and bovine exposure: A prospective cross-sectional cohort study. *PLoS ONE* 5: e8795.

Marin, M., E. Cercenado, C. Sanchez-Carrillo, A. Ruiz, A. Gomez Gonzalez, B. Rodriguez-Sanchez and E. Bouza. 2017. Accurate differentiation of *Streptococcus pneumoniae* from other species within the *Streptococcus mitis* group by peak analysis using MALDI-TOF MS. *Front. Microbiol.* 8: 698. https://doi.org/10.3389/fmicb.2017.00698.

Marsh, P.D. 2018. In sickness and in health—what does the oral microbiome mean to us? An ecological perspective. *Adv. Dent. Res.* 29(1): 60–65.

Martin, V., R. Manes-Lazaro, J.M. Rodriguez and A. Maldonado-Barragan. 2011. *Streptococcus lactarius* sp. nov., isolated from breast milk of healthy women. *Int. J. Syst. Evol. Microbiol.* 61: 1048–1052.

Martinez-Lamas, L., J. Limeres-Posse, P. Diz-Dios and M. Alvarez-Fernandez. 2020. *Streptococcus downii* sp. nov., isolated from the oral cavity of a teenager with Down syndrome. *Int. J. Syst. Evol. Microbiol.* 70: 4098–4104.

McLean, A.R., J. Torres-Morales, F.E. Dewhirst, G.G Borisy, et al. 2022. Site-tropism of streptococci in the oral microbiome. *Mol. Oral. Microbiol.* 37: 229–243. https://doi.org/10.1111/omi.12387.

Metwalli, K.H., S.A. Khan, B.P. Krom and M.A. Jabra-Rizk. 2013. *Streptococcus mutans*, *Candida albicans*, and the human mouth: A sticky situation. *PLoS Pathog.* 9: e1003616. https://doi.org/10.1371/journal.ppat.1003616.

Mignolet, J., L. Fontaine, M. Kleerebezem and P. Hols. 2016. Complete genome sequence of *Streptococcus salivarius* HSISS4, a human commensal bacterium highly prevalent in the digestive tract. *Genome Announc.* 4 (1): e01637–15. https://doi.org/10.1128/genomeA.01637-15.

Mitchell, A.M. and T.J. Mitchell. 2010. *Streptococcus pneumoniae*: Virulence factors and variation. *Clin. Microbiol. Infect.* 16: 411–418.

Mitchell, T.J. 2003. The pathogenesis of streptococcal infections: From tooth decay to meningitis. *Nat. Rev. Microbiol.* 1: 219–230.

Mühldorfer, K., J. Rau, A. Fawzy, C. Heydel, et al. 2019. *Streptococcus castoreus*, an uncommon group A Streptococcus in beavers. *Antonie Van Leeuwenhoek.* 112: 1663–1673.

Nielsen, M.Ø., O. Köhler-Forsberg, C. Hjorthøj, M.E. Benros, et al. 2019. Streptococcal Infections and exacerbations in PANDAS: A systematic review and meta-analysis. *Pediatr. Infect. Dis. J.* 38: 189–194.

Niven, C.F.J., K.L. Smiley and J.M. Sherman. 1941. The production of large amounts of polysaccharide by *Streptococcus salivarius*. *J. Bacteriol.* 168: 1384–1391.

Nix, I.D., E.A. Idelevich, A. Schlattmann, K. Sparbier, et al. 2021. MALDI-TOF mass spectrometry-based optochin susceptibility testing for differentiation of *Streptococcus pneumoniae* from other *Streptococcus mitis* group streptococci. *Microorganisms.* 9(10): 2010. https://doi.org/10.3390/microorganisms9102010.

Nizet, V. 2002. Streptococcal beta-hemolysins: Genetics and role in disease pathogenesis. *Trends Microbiol.* 10: 575–580.

Nocard, M. and R. Mollereau. 1887. Sur une mammite contagieuse des vaches laitieres. *Ann. Inst. Pasteur.* 1: 109–126.

Okumura, C.Y. and V. Nizet. 2014. Subterfuge and sabotage: Evasion of host innate defenses by invasive Gram-positive bacterial pathogens. *Annu. Rev. Microbiol.* 68: 439–458.

Oppegaard, O., M. Glambek, D.H. Skutlaberg, S. Skrede, et al. 2023. *Streptococcus dysgalactiae* bloodstream infections, Norway, 1999–2021. *Emerg. Infect. Dis.* 29: 260–267.

Orla-Jensen, S. 1919. The lactic acid bacteria. *Mem. Acad. R. Soc. Denmark Sect. Sci. Ser.* 8: 181–197.

Osawa, R., T. Fujisawa and L.I. Sly. 1995. *Streptococcus gallolyticus* sp. nov., gallate degrading organism formerly assigned to *Streptococcus bovis*. *Syst. Appl. Microbiol.* 18: 74–78.

Pagnossin, D., A. Smith, K. Oravcová and W. Weir. 2022. *Streptococcus canis*, the underdog of the genus. *Vet. Microbiol.* 273: 109524.

Paiva, W.K.V., W.R.D.B. Medeiros, C.F. Assis, E.S. Dos Santos, et al. 2022. Physicochemical characterization and *in vitro* antioxidant activity of hyaluronic acid produced by *Streptococcus zooepidemicus* CCT 7546. *Prep. Biochem. Biotechnol.* 52: 234–243.

Parks, D.H., M. Chuvochina, P.A. Chaumeil, C. Rinke, et al. 2020. A complete domain-to-species taxonomy for Bacteria and Archaea. *Nat. Biotechnol.* 38: 1079–1086.

Parks, T., L. Barrett and N. Jones. 2015. Invasive streptococcal disease: A review for clinicians. *Br. Med. Bull.* 115: 77–89.

Parte, A.C. 2014. LPSN--list of prokaryotic names with standing in nomenclature. *Nucleic Acids Res.* 42(Database issue): D613–D616.

Pasteur, L. and J.F. Joubert. 1877. Charbon et septicémie. *C. R. Soc. Biol. Paris.* 85: 101–115.

Patel, S. and R.S. Gupta. 2018. Robust demarcation of fourteen different species groups within the genus *Streptococcus* based on genome-based phylogenies and molecular signatures. *Infect. Genet. Evol.* 66: 130–151.

Paul, P., B.P. Gonçalves, K. Le Doare and J.E. Lawn. 2023. 20 million pregnant women with group B streptococcus carriage: Consequences, challenges, and opportunities for prevention. *Curr. Opin. Pediatr.* 35: 223–230.

Pereira, U.P., G.F. Mian, I.C. Oliveira, et al. 2010. Genotyping of *Streptococcus agalactiae* strains isolated from fish, human and cattle and their virulence potential in Nile tilapia. *Vet. Microbiol.* 140: 186–192.

Pier, G.B. and S.H. Madin. 1976. *Streptococcus iniae* sp. nov., a beta-hemolytic streptococcus isolated from an Amazon freshwater dolphin, *Inia geoffrensis*. *Int. J. Syst. Bacteriol.* 26: 445–553.

Pilarczyk-Zurek, M., I. Sitkiewicz and J. Koziel. 2022. The clinical view on *Streptococcus anginosus* Group—opportunistic pathogens coming out of hiding. *Front. Microbiol.* 13: 956677.

Pombert, J.F., V. Sistek, M. Boissinot and M. Frenette. 2009. Evolutionary relationships among salivarius streptococci as inferred from multilocus phylogenies based on 16S rRNA-encoding, *recA*, *secA*, and *secY* gene sequences. *BMC Microbiol.* 9: 232.

Pompilio, A., G. Di Bonaventura and G. Gherardi. 2019. An overview on *Streptococcus bovis/Streptococcus equinus* complex isolates: Identification to the species/subspecies level and antibiotic resistance. *Int. J. Mol. Sci.* 20: 480. https://doi.org/10.3390/ijms20030480.

Pontigo, F., M. Moraga and S.V. Flores. 2015. Molecular phylogeny and a taxonomic proposal for the genus *Streptococcus*. *Genet. Mol. Res.* 14: 10905–10918.

Pourhajibagher, M., M. Alaeddini, S. Etemad-Moghadam, B. Rahimi Esboei, et al. 2022. Quorum quenching of *Streptococcus mutans* via the nano-quercetin-based antimicrobial photodynamic therapy as a potential target for cariogenic biofilm. *BMC Microbiol.* 22(1). https://doi.org/10.1186/s12866-022-02544-8.

Poyart, C., G. Quesne and P. Trieu-Cuot. 2002. Taxonomic dissection of the *Streptococcus bovis* group by analysis of manganese-dependent superoxide dismutase gene (sodA) sequences: Reclassification of 'Streptococcus infantarius subsp. coli' as *Streptococcus lutetiensis* sp. nov. and of *Streptococcus bovis* biotype 11.2 as *Streptococcus pasteurianus* sp. nov. *Int. J. Syst. Evol. Microbiol.* 52: 1247–1255.

Prasad, A., A. Ene, S. Jablonska, J. Du, A.J. Wolfe and C. Putonti. 2023. Comparative genomic study of *Streptococcus anginosus* reveals distinct group of urinary strains. *mSphere.* 8(2): e0068722. https://doi.org/10.1128/msphere.00687-22.

Ricaboni, D., M. Mailhe, J.C. Lagier, C. Michelle, N. Armstrong, F. Bittar, V. Vitton, A. Benezech, D. Raoult and M. Million. 2016. Noncontiguous finished genome sequence and description of *Streptococcus timonensis* sp. nov. isolated from the human stomach. *New Microbes New Infect.* 15: 77–88. https://doi.org/10.1016/j.nmni.2016.11.013. PMID: 28050252; PMCID: PMC5192475.

Rich, S.N., M. Prosperi, E.M. Klann, P.T. Codreanu, et al. 2021. Evaluating the diagnostic paradigm for Group A and non-Group A streptococcal pharyngitis in the college student population. *Open Forum. Infect. Dis.* 8(11): ofab482. https://doi.org/10.1093/ofid/ofab482.

Richards, V.P., S.R. Palmer, P.D. Pavinski Bitar, X. Qin, et al. 2014. Phylogenomics and the dynamic genome evolution of the genus Streptococcus. *Genome Biol. Evol.* 6: 741–753.

Roos, K., E. Grahn, S.E. Holm, H. Johansson and L. Lind. 1993. Interfering alpha-streptococci as a protection against recurrent streptococcal tonsillitis in children. *Int. J. Pediatr. Otorhinolaryngol.* 25: 141–148.

Rosenbach, F.J. 1884. *Mikro-organismen bei den Wund-infections-krankheiten des Menschen.* JF Bergmann, Wiesbaden, Germany.

Rudney, J.D., R. Chen and G. Zhang. 2005. Streptococci dominate the diverse flora within buccal cells. *J. Dent. Res.* 84: 1165–1171.

Ruoff, K.L. 1988. *Streptococcus anginosus* ("*Streptococcus milleri*"): The unrecognized pathogen. *Clin. Microbiol. Rev.* 1: 102–108.

Russell, N.J., A.C. Seale, C. O'Sullivan, K. Le Doare, et al. 2017. Risk of early-onset neonatal Group B streptococcal disease with maternal colonization worldwide: Systematic review and meta-analyses. *Clin. Infect. Dis.* 65(suppl_2): S152–S159.

Sadowy, E. and W. Hryniewicz. 2020. Identification of *Streptococcus pneumoniae* and other Mitis streptococci: Importance of molecular methods. *Eur. J. Clin. Microbiol. Infect. Dis.* 39: 2247–2256. https://doi.org/10.1007/s10096-020-03991-9.

Salman, H.A., R.S. Kumar, N.C. Babu and K. Imran. 2017. First detection and characterization of *Streptococcus dentapri* from caries active subject. *J. Clin. Diagn. Res.* 11(7): DM01–DM03.

Samir, A., K.A. Abdel-Moein and H.M. Zaher. 2020. Emergence of penicillin-macrolide-resistant *Streptococcus pyogenes* among pet animals: An ongoing public health threat. *Comp. Immunol. Microbiol. Infect. Dis.* 68: 101390. https://doi.org/10.1016/j.cimid.2019.101390.

Sanderson-Smith, M., D.M. De Oliveira, J. Guglielmini, D.J. McMillan, et al. 2014. M Protein Study Group. A systematic and functional classification of *Streptococcus pyogenes* that serves as a new tool for molecular typing and vaccine development. *J. Infect. Dis.* 210: 1325–1338.

Schlegel, L., F. Grimont, E. Ageron, P.A. Grimont and A. Bouvet. 2003a Reappraisal of the taxonomy of the *Streptococcus bovis/Streptococcus equinus* complex and related species: Description of *Streptococcus gallolyticus* subsp. *gallolyticus* subsp. nov., *S. gallolyticus* subsp. *macedonicus* subsp. nov. and *S. gallolyticus* subsp. *pasteurianus* subsp. nov. *Int. J. Syst. Evol. Microbiol.* 53: 631–645.

Schlegel, L., F. Grimont, M.D. Collins, et al. 2000. *Streptococcus infantarius* sp. nov., *Streptococcus infantarius* subsp. *infantarius* subsp. nov. and *Streptococcus infantarius* subsp. *coli* subsp. nov., isolated from humans and food. *Int. J. Syst. Evol. Microbiol.* 50: 1425–1434.

Schlegel, L., F. Grimont, P.A. Grimont and A. Bouvet. 2003b. Identification of major streptococcal species by rrn-amplified ribosomal DNA restriction analysis. *J. Clin. Microbiol.* 41: 657–666.

Schleifer, K.H., M. Ehrmann, U. Krusch and H. Neve. 1991. Revival of the species *Streptococcus thermophilus* (ex Orla-Jensen, 1919) nom. rev. *Syst. Appl. Microbiol.* 14: 386–388.

Schleifer, K.H. and R. Kilpper-Balz. 1984. Transfer of *Streptococcus faecalis* and *Streptococcus faecium* to the genus *Enterococcus* norn. rev. as *Enterococcus faecalis* comb. nov. and *Enterococcus faecium* comb. nov. *Int. J. Syst. Bacteriol.* 34: 31–34.

Schleifer, K.H., J. Kraus, C. Dvorak, et al. 1985. Transfer of *Streptococcus lactis* and related streptococci to the genus *Lactococcus. Syst. Appl. Microbiol.* 6: 183–195.

Schodel, F., N.J. Moreland, J.T. Wittes, K. Mulholland, et al. 2017. Clinical development strategy for a candidate group A streptococcal vaccine. *Vaccine.* 35: 2007–2014.

Schotmuller, H. 1903. Die Artunterscheidung der fur den Menschen pathogenen Streptokokken durch Blutagar. *Muench. Med. Wochenschr.* 1: 849–909.

Shaikh, N., N. Swaminathan and E.G. Hooper. 2012. Accuracy and precision of the signs and symptoms of streptococcal pharyngitis in children: A systematic review. *J. Pediatr.* 160: 487–493.

Shakir, S.M., R. Gill, J. Salberg, E. S. Slechta, et al. 2021 Clinical laboratory perspective on *Streptococcus halichoeri*, an unusual nonhemolytic, Lancefield Group B Streptococcus causing human infections. *Emerg. Infect. Dis.* 27: 1309–1316. https://doi.org/10.3201/eid2705.203428.

Sherman, J.M. 1937. The streptococci. *Bacteriol. Rev.* 1: 3–97.

Shewmaker, P.L., A.C. Camus, T. Bailiff, et al. 2007. *Streptococcus ictaluri* sp. nov., isolated from channel catfish *Ictalurus punctatus* broodstock. *Int. J. Syst. Evol. Microbiol.* 57: 1603–1606.

Sikder, S., N.L. Williams, A.E. Sorenson, M.A. Alim, et al. 2017. Group G streptococcus induces an autoimmune IL-17A/IFN-gamma mediated carditis in the Lewis rat model of Rheumatic Heart Disease. *J. Infect. Dis.* https://doi.org/10.1093/infdis/jix637.

Simova, E., D. Beshkova, A. Angelov, T. Hristozova, et al. 2002. Lactic acid bacteria and yeasts in kefir grains and kefir made from them. *J. Ind. Microbiol. Biotechnol.* 28: 1–6.

Tagg, J.R., L.K. Harold, R. Jain and J.D.F. Hale. 2023. Beneficial modulation of human health in the oral cavity and beyond using bacteriocin-like inhibitory substance-producing streptococcal probiotics. *Front. Microbiol.* 14: 1161155.

Tagg, J.R., V. Pybus, L.V. Phillips and T.M. Fiddes. 1983. Application of inhibitor typing in a study of the transmission and retention in the human mouth of the bacterium *Streptococcus salivarius. Arch. Oral. Biol.* 28: 911–915.

Takada, K., K. Hayashi, Y. Sato and M. Hirasawa. 2010. *Streptococcus dentapri* sp. nov., isolated from the wild boar oral cavity. *Int. J. Syst. Evol. Microbiol.* 60: 820–823.

Takada, K. and M. Hirasawa. 2007. *Streptococcus orisuis* sp. nov., isolated from the pig oral cavity. *Int. J. Syst. Evol. Microbiol.* 57: 1272–1275.

Takada, K. and M. Hirasawa. 2008. *Streptococcus dentirousetti* sp. nov., isolated from the oral cavities of bats. *Int. J. Syst. Evol. Microbiol.* 58: 160–163.

Tatara, K., K. Gotoh, K. Okumiya, M. Teramachi, et al. 2020. Molecular epidemiology, antimicrobial susceptibility, and characterization of fluoroquinolone non-susceptible *Streptococcus pyogenes* in Japan. *J. Infect. Chemother.* 26: 280–284.

Thai, T.N., A.P. Dale and M.H. Ebell. 2017. Signs and symptoms of Group A versus Non-Group A strep throat: A meta-analysis. *Fam. Pract.* https://doi.org/10.1093/fampra/cmx072.

Todd, E.W. and L.F. Hewitt. 1932. A new culture medium for the production of antigenic streptococcal hemolysin. *J. Pathol. Bacteriol.* 35: 973–974.

Tompkins, G.R. and J.R. Tagg. 1987. Bacteriocin-like inhibitory activity associated with beta-hemolytic strains of *Streptococcus salivarius. J. Dent. Res.* 66: 1321–1325.

Tong, H., X. Gao and X. Dong. 2003. *Streptococcus oligofermentans* sp. nov., a novel oral isolate from caries-free humans. *Int. J. Syst. Evol. Microbiol.* 53: 1101–1104.

Torpiano, P., N. Nestorova and C. Vella. 2020. *Streptococcus equi* subsp. *equi* meningitis, septicemia and subdural empyema in a child. *IDCases.* 21: e00808. https://doi.org/10.1016/j.idcr.2020.e00808.

Tsakalidou, E., E. Zoidou, B. Pot, et al. 1998. Identification of streptococci from Greek Kasseri cheese and description of *Streptococcus macedonicus* sp. nov. *Int. J. Syst. Bacteriol.* 48: 519–527.

Vandamme, P., B. Pot, E. Falsen, K. Kersters and L.A. Devriese. 1996. Taxonomic study of Lancefield streptococcal groups C, G, and L (*Streptococcus dysgalactiae*) and proposal of S. *dysgalactiae* subsp. *equisimilis* subsp. nov. *Int. J. Syst. Bacteriol.* 46: 774–781.

Van der Rijn, I. and R.E. Kessler. 1980. Growth characteristics of group A streptococci in a new chemically defined medium. *Infect. Immun.* 27: 444–448.

Vela, A.I., P. Villalon, J.A. Saez-Nieto, G. Chacon, et al. 2017. Characterization of *Streptococcus pyogenes* from animal clinical specimens, Spain. *Emerg. Infect. Dis.* 23: 2013–2016.

Velsko, I.M., M.S. Perez and P. Richards. 2019 Resolving phylogenetic relationships for *Streptococcus mitis* and *Streptococcus oralis* through core- and pan-genome analyses. *Genome Biol. Evol.* 11: 1077–1087.

Wajima, T., A. Hagimoto, E. Tanaka, Y. Kawamura, et al. 2022. Identification and characterisation of a novel multidrug-resistant streptococcus, *Streptococcus toyakuensis* sp. nov., from a blood sample. *J. Glob. Antimicrob. Resist.* 29: 316–322.

Wescombe, P.A., J.P. Burton, P.A. Cadieux, et al. 2006. Megaplasmids encode differing combinations of lantibiotics in *Streptococcus salivarius*. *Antonie Van Leeuwenhoek*. 90: 269–280.

Wescombe, P.A., N.C. Heng, J.P. Burton, C.N. Chilcott and J.R. Tagg. 2009. Streptococcal bacteriocins and the case for *Streptococcus salivarius* as model oral probiotics. *Fut. Microbiol.* 4: 819–835.

Whiley, R.A. and D. Beighton. 1991. Emended descriptions and recognition of *Streptococcus constellatus*, *Streptococcus intermedius*, and *Streptococcus anginosus* as distinct species. *Int. J. Syst. Bacteriol.* 41: 1–5.

Whiley, R.A. and D. Beighton. 1998. Current classification of the oral streptococci. *Oral. Microbiol. Immunol.* 13: 195–216.

Whiley, R.A., H.Y. Fraser, C.W. Douglas, et al. 1990. *Streptococcus parasanguis* sp. nov., an atypical viridans *Streptococcus* from human clinical specimens. *FEMS Microbiol. Lett.* 56: 115–121.

Whiley, R.A., L.M. Hall, J.M. Hardie and D. Beighton. 1999. A study of small-colony, beta-haemolytic, Lancefield group C streptococci within the anginosus group: Description of *Streptococcus constellatus* subsp. *pharyngis* subsp. nov., associated with the human throat and pharyngitis. *Int. J. Syst. Bacteriol.* 49: 1443–1449.

Whiley, R.A. and J.M. Hardie. 1988. *Streptococcus vestibularis* sp. nov. from the human oral cavity. *Int. J. Syst. Bacteriol.* 38: 335–339.

Whiley, R.A., R.R.B. Russell, J.M. Hardie and D. Beighton. 1988. *Streptococcus downei* sp. nov. for strains previously described as *Streptococcus mutans* serotype h. *Int. J. Syst. Bacteriol.* 38: 25–29.

Willcox, M.D., H. Zhu and K.W. Knox. 2001. *Streptococcus australis* sp. nov., a novel oral streptococcus. *Int. J. Syst. Evol. Microbiol.* 51: 1277–1281.

Williams, A.M. and M.D. Collins. 1990. Molecular taxonomic studies on *Streptococcus uberis* types I and II. Description of *Streptococcus parauberis* sp. nov. *J. Appl. Bacteriol.* 68: 485–490.

Woltjes, J., H. Legdeur-Velthuis, C.O. Eggink and J. de Graaff. 1982. Beta-haemolysis and pigment production by the oral bacterium *Streptococcus mutans*. *Arch. Oral. Biol.* 27: 279–281.

Woo, P.C., D.M. Tam, K.W. Leung, et al. 2002. *Streptococcus sinensis* sp. nov., a novel species isolated from a patient with infective endocarditis. *J. Clin. Microbiol.* 40: 805–810.

Xu, R.R., W.D. Yang, K.X. Niu, B. Wang, et al. 2018. An update on the evolution of glucosyltransferase (GTF) genes in *Streptococcus*. *Front. Microbiol.* 9(December). https://doi.org/10.3389/fmicb.2018.02979.

Yoshino, M., S.Y. Murayama, K. Sunaoshi, et al. 2010. Nonhemolytic *Streptococcus pyogenes* isolates that lack large regions of the sag operon mediating streptolysin S production. *J. Clin. Microbiol.* 48: 635–638.

Yu, J., W.H. Wang, B.L.G. Menghe, M.T. Jiri, et al. 2011. Diversity of lactic acid bacteria associated with traditional fermented dairy products in Mongolia. *J. Dairy Sci.* 94: 3229–3241.

Zahradnik, R.T., I. Magnusson, C. Walker, et al. 2009. Preliminary assessment of safety and effectiveness in humans of ProBiora3, a probiotic mouthwash. *J. Appl. Microbiol.* 107: 682–690.

Zbinden, A., N.J. Mueller, P.E. Tarr, C. Sproer, et al. 2012. *Streptococcus tigurinus* sp. nov., isolated from blood of patients with endocarditis, meningitis and spondylodiscitis. *Int. J. Syst. Evol. Microbiol.* 62: 2941–2945.

Zheng, W., T.K. Tan, I.C. Paterson, N.V. Mutha, et al. 2016. StreptoBase: An oral *Streptococcus mitis* Group genomic resource and analysis platform. *PLoS ONE*. 11(5): e0151908.

Zhou, C.B., S.Y. Pan, P. Jin, J.W. Deng, et al. 2022. Fecal serignatures of *Streptococcus anginosus* and *Streptococcus constellatus* for noninvasive screening and early warning of gastric cancer. *Gastroenterology*. 162: 1933–1947.

Zhu, H., M.D. Willcox and K.W. Knox. 2000. A new species of oral Streptococcus isolated from Sprague-Dawley rats, *Streptococcus orisratti* sp. nov. *Int. J. Syst. Evol. Microbiol.* 50: 55–61.

# The Genus *Enterococcus*, Bacteria with Beneficial and/or Non-Beneficial Potential?

# 5

Andrea Lauková

Enterococci have been of interest to scientists for centuries. As identification technique development has increased, new enterococcal species strains have been validated. Enterococci, as a lactic acid bacteria group, have been always controversial bacteria; many enterococcal species strains possess beneficial properties (Franz et al. 2011; Soltani et al. 2021), but clinical strains in particular can possess virulence factors such as antibiotic resistance, aggregation factor, gelatinase, and cytolysin which belong among the most common virulence factors (Fisher and Phillips 2009). Therefore, it is necessary to look at this bacterial species from two points of view, beneficial and non-beneficial, to assess each of the two potentials of enterococci for their future use.

## 5.1 TAXONOMY OF THE GENUS *ENTEROCOCCUS*

Currently, the genus *Enterococcus*, as the representative of the family Enterococcacae and of the phylum Firmicutes (Švec and Franz 2014), contains 61 validly described species (www.bacterio.net/enterococcus.html) which were taxonomically allotted on the basis of various types of identification, including genomic analysis (Zhong et al. 2017). But at least four other species have been described: *E. pingfangensis*, *E. dongliensis*, *E. nangangensis* (Li and Gu 2019), which are the species isolated from Chinese traditional pickle juice and the species strain *E. florum* from the cotton flower (Techo et al. 2019). The taxonomy of enterococci, the Gram-positive cocci, has changed considerably, as numerous new species have been described, and they were also re-classified. However, in re-classified groups, according to Sistek et al. (2012) and Franz et al. (2011), not all of the already identified species were involved. They are still outgrouped, such as *E. alcedinis*, *E. caccae*, *E. diestrammenae*, *E. eurekensis*, *E. flavescens*, *E. lactis*, *E. lemanii*, *E. olivae*, *E. plantarum*, *E. porcinus*, *E. rivorum*, *E. rotai*, *E. saccharominimus*, *E. seriolicida*, *E. solitarius*, *E. ureilyticus*, *E. viikkiensis*, and *E. xiangfangensis* (Švec and Franz 2014;

DOI: 10.1201/9781003352075-6

Li et al. 2014; Jin et al. 2017; Zhong et al. 2017). According to Sistek et al. (2012), enterococcal species fall into seven species groups on the basis of 16S rRNA gene similarity. The *Enterococcus avium* group includes the species *E. avium*, *E. devriesei*, *E. gilvus*, *E. malodoratus*, *E. pseudoavium*, *E. raffinosus*; the *Enterococcus cecorum* group includes the species *E. cecorum* and *E. columbae*. The species *E. dispar*, *E. asini*, *E. caninintestini*, *E. hermanniensis*, and *E. pallens* are allotted to the group *Enterococcus dispar*. The *Enterococcus faecalis* group includes eight species, *E. faecalis*, *E. caccae*, *E. haemoperoxidus*, *E. moraviensis*, *E. silesiacus*, *E. termitis*, *E. ureasiticus*, and *E. quebecensis*. The group with the most species is the *Enterococcus faecium* group, involving 11 species: *E. faecium*, *E. canis*, *E. durans*, *E. hirae*, *E. mundtii*, *E. phoeniculicola*, *E. ratti*, *E. villorum*, *E. thailandicus*, *E. gallinarum*, and *E. casseliflavus*. The *Enterococcus gallinarum* group includes the species *E. gallinarum* and *E. casseliflavus*. The species *E. saccharolyticus* has two subspecies, *saccharolyticus* and *taiwanensis* (Chen et al. 2013; Oren and Garrity 2020), and *E. aquimarinus*, *E. camelliae*, *E. italicus*, and *E. sulfureus* belong to the *Enterococcus saccharolyticus* group. In contrast, Holzapfel and Wood (2014) reported enterococcal species grouped into six branches according to the typology of robust phylogenies and previous research. The *Enterococcus faecium* branch contains eight species, *E. faecium*, *E. mundtii*, *E. durans*, *E. hirae*, *E. ratti*, *E. villorum*, *E. thailandicus*, and *E. phoeniculicola*, which were mainly isolated from the bloodstream and the intestinal tract. The *Enterococcus faecalis* branch contains the species *E. faecalis*, *E. termitis*, *E. quebecensis*, *E. moraviensis*, *E. caccae*, *E. haemoperoxidus*, and *E. silesiacus*, mainly isolated from water and the intestinal tract. The *Enterococcus dispar* branch contains *E. dispar*, *E. caninintestini*, and *E. asini*, isolated mainly from the intestinal tract of humans and mammals. The species *E. casseliflavus*, *E. gallinarum*, *E. aquimarinus*, *E. saccharolyticus*, *E. italicus*, *E. sulfureus*, *E. cecorum*, and *E. columbae*, mainly isolated from plant material and intestinal tract of birds, are contained in the *Enterococcus casseliflavus* branch. The *Enterococcus pallens* branch contains the species *E. pallens*, *E. hermanniensis*, *E. devriesei*, *E. gilvus*, *E. malodoratus*, *E. avium*, and *E. raffinosus*, mainly isolated from humans and mammals. The *Enterococcus canis* branch contains only the *E. canis* species, the strain isolated from chronic *otitis externa* in dogs. Zhong et al. (2017) noted that habitat is very important in the evolution of enterococci. Genetic relationships were closer in strains that had similar habitats. This potential relationship between source distribution and genealogy provides researchers with a clue to the evolution of the *Enterococcus* genus. In 2016, more new species were validly identified, such as *E. saigonensis*, *E. bulliens*, and *E. massiliensis*. The *E. saigonensis*–type strain was isolated from retail chicken meat and liver in Saigon (Harada et al. 2016). The species name *E. bulliensis* is associated with bubbling, referring to the gas formation in a catalase test after cultivation on a blood agar medium (Kadri et al. 2015). The species strain *E. massiliensis* was isolated from fresh human stool and named for the city Marseille (France) in Latin (where it was isolated and characterized; Le Page et al. 2016). In 2017, the species *E. crotali* was characterized, isolated from timber rattlesnake *Crotalis horridus* (McLaughlin et al. 2017). The species *E. hulanensis* and *E. songbeiensis* from Chinese traditional pickle juice were validated in 2019 (Li and Gu 2019). *E. hulanensis* was named for the Hulan district in China, the place of the strain isolation, and the name *E. songbeiensis* is after the Songbei district in China. Another new species, *E. mediterraneensis*, was named for the Mediterranean region, where it was isolated from the stool of a 39-years old man from Mbuti (Takakura et al. 2019). *E. burkinafasonensis* (Gouba et al. 2020) is another new species which has been validated recently (in 2020), and it was isolated from human gut microbiota samples originating from Burkina Faso in west Africa. In China (Xinjiang Uyghur Autonomous Region), *E. sinjiangensis* was isolated from yoghurt (Ren et al. 2016). In 2021, the species *E. innesii* (Gooch et al. 2021) was reported, isolated from the wax moth (*Galleria mellonella*). The name of this species pertains to John Innes J.P., the British philanthropist, and the John Innes Centre in Norwich (United Kingdom) where it was characterized.

Enterococci can be associated with a variety of habitats such as soil, plants, water, the gastrointestinal tract (GIT) of humans and animals, rumen, and silage (Lebreton et al. 2014). They are also associated with foods (Aspri et al. 2017); most likely as a result of contamination from plant or animal sources, it has been shown that they have little value as hygiene indicators in the industrial processing of foods (Birollo et al. 2001). Enterococci also seem to play some role in numerous fermented foods. Their wide distribution in nature, when compared with other lactic acid bacteria (LAB), is probably explained by their persistence and their resistance to growth-inhibiting factors such as acidity, salt, drying, heat, and chemical sanitizing agents (Holzapfel et al. 2002). In general, enterococci have an extraordinary genome plasticity and metabolic versatility that enable them to thrive in many diverse environments (Cattoir 2022).

Enterococci belong to the low GC branch of Gram-positive lactic acid bacteria with an optimum growth temperature of 35–37 °C. Relating to the phylogenetic tree with individual strains properties, enterococci can be divided at least into seven groups. Group I members are species that are farthest from other enterococci; they differ phenotypically from the other groups by their unusual negative result in the pyruvate utilization test (Lebreton et al. 2014). The strains of Group II show tropism for mammals (Lebreton et al. 2014). Group III is found ubiquitously along the food chain, but among them have also been found species with tropism to the environment (Lebreton et al. 2014). Species of Groups IV, V, and VI show some tropism for humans and animals, especially mammals, but species in Groups VI and IV are more ubiquitous and can be also found along the food chain and in insects.

## 5.1.1 Beneficial and Non-Beneficial Aspects of Enterococci

When considering enterococci as beneficial, their safety and/or detrimental activities are always debated. Among the best-studied species regarding the potential beneficial/probiotic use are the representatives of the *E. faecium* group, especially the strains of the species *E. faecium* (Franz et al. 2011). However, it is necessary to take each strain into consideration individually, for example, whether it possesses virulence determinants (hemolysin, gelatinase, Esp protein, etc.) as well as beneficial properties attributed to proteolytic, lipolytic, and esterolytic activities, especially in food strains. However, there are differences concerning the virulence factors harbored by enterococci from clinical and other sources. In particular, those isolated from food were found to carry a low incidence of virulence factors (Franz et al. 2001; Fisher and Phillips 2009). The secreted virulence factors of enterococci have a function in pathogenesis. Cytolysin, a so-called hemolysin is, for example, a toxin, the gene for the production of which is located on pheromone-responsive plasmid. Biofilm production is also a fundamental in infection; however, in the case of beneficial strains, it is supposed to be a benefit (Bino 2022). Adhesive abilities are important for sufficient colonization of beneficial strains and for their stability and survival in the ecosystem (Lauková et al. 2004a). To promote safe biotechnology, the German chemical industry has published a list in which microbiota have been classified into specific risk groups (Berufsgenossenschaft der Chemischen Industrie 1995). Based on this classification, Group 1 bacteria are considered not to pose any risk for human or animal health and include the majority of LAB. Group 2 bacteria are considered to have a low potential to cause infection, which is dependent on the immune status of the host, and are not generally regarded as obligatory pathogens. According to this classification, most *Enterococcus* spp. are listed under Group 2; exceptions are indicated for strains *Enterococcus faecium* and *E. durans*, which, on the basis of safe technical experience, may be considered nonrisk strains in the sense of Group 1 organisms. The other interest has to be focused on aminogenic activity of enterococci, especially those from food. Latorre-Moratalla et al. (2010) reported that aminogenic potential of tested starter cultures for fermented meat was strain dependent, although some species had a higher proportion of aminogenic strains than did other species. Enterococci have commonly been reported in tyramine production (Latorre-Moratalla et al. 2010). On the other hand, the lowest hazardous dose of tyramine is 6 mg in classic monoamine oxidase inhibitor treatments; it could easily be reached by consuming 80 g of a sausage containing around 100 mg of tyramine. However, according to the European Union (EU) project Tradisausage QLK1-CT-2002–02240, the usual consumption is 50 g of sausage (Talon 2005). In addition, Lauková et al. (2017a, 2017b) reported that most gelatinase-positive and decarboxylase-positive enterococci isolated from rabbit meat and biogenic amine-producing fecal enterococci from ostriches and pheasants were susceptible to bacteriocins-enterocins. Therefore, bacteriocin production among enterococci (mostly enterocins) has been viewed as a potential beneficial aspect (Franz et al. 2011; Ovchinnikov et al. 2016). According to the selection criteria for promising probiotic (beneficial) strains (Lahtinen et al. 2012), which have been continually updated following new research information, the reason for further investigation is if a potential beneficial strain is also a bacteriocin producer or when bacteriocin-producing strains possess beneficial properties. Bacteriocin-producing beneficial enterococci are used, for example, in cheese manufacturing to minimize the risk of milk and cheese contamination by spoilage bacteria in order to achieve a zero tolerance policy and to prevent late blowing (although chemicals are still used to prevent late gas formation during cheese ripening) (De Vuyst 2004), but some species/strains, for example, *E. durans*, are responsible for typical flavor formation in cheeses. Therefore, the European Food Safety Authority (EFSA 2004) has provided and contributed EU directives to regulate and assess probiotic additive/adjuncts (European Commission 2004). EFSA has also taken responsibility to launch the European initiative toward a qualified presumption of safety (QPS) concept, which, similar to generally recognized as safe (GRAS) in the United States, aims to allow strains with established history and safety status to enter the market without extensive testing requirements. The Nutrition & Health Claims regulation (Reg. 1924/2006) established by the EFSA (2022) was updated by QPS under the EFSA Panel on Biological Hazards (BIOHAZ) during 2020–2022, and the presence of transmissible antibiotic resistance markers in the evaluation of the strains has been established as the most important health criterion. Following these rules, strains that should be claimed as probiotic are supposed to be QPS probiotic and non-QPS probiotic. However, it is necessary to claim each strain in the framework of the species; to claim its antibiotic sensitivity, absence of toxins, and virulence factors; and to claim safety according to toxicological studies such as 3-pack genotoxicity studies and subchronic studies (e.g., 90-day rat study). *E. faecium* strains in the feed additive Bonvital were approved by the EFSA for chicken fattening, and *E. faecium* NCIMB 10415, formerly SF68, has been approved for use as a feed additive for different animals. The British Advisory Committee on Novel Foods and Processes accepted the use of *E. faecium* strain K77D as a starter culture in fermented dairy products in 1996. In contrast to the enterococcal strains, enterococcal bacteriocins (not necessary to evaluate) produced by heterologous hosts or added as cell-free, partially purified preparations have been attractive first for applications in food/feed (Khan et al. 2010). The efficacy of enterocins in cheeses was mentioned by Arvanitoyannis (2009) in connection with the application of the Hazard Analysis and Critical Control Point system and ISO 22000 to foods of animal origin. Ribeiro et al. (2017) reported an anti-listerial effect of enterocins in fresh cheeses. Group II bacteriocins following enterocins are also the most interesting anti-bacterials for the meat industry. The addition of Ent 4231 in Púchov salami led to a reduction in counts of *L. innocua* Li1 by 1.86 log on day 2 compared with the control and the experimental salamis, with the highest reduction on weeks 3 and 4 (2.36 and 2.48 log, respectively) (Lauková and Turek 2004). Hermans et al. (2010) mentioned the approval of Enterocin E-760 by the EFSA for use in poultry meat.

## 5.1.2 Bacteriocins Produced by Enterococci

As mentioned, bacteriocins produced by enterococci are generally referred to as enterocins (Franz et al. 2007; Nes et al. 2014). They are ribosomally synthesized, antimicrobial peptides with activity usually directed against more or less related bacteria. The awareness of bacteriocins produced by enterococci has enormously increased, such that it became necessary to re-classify the enterocin group of bacteriocin itself, however following the general bacteriocin classification. Nes et al. (2014) classified enterocins into the following classes: Class I: Lantibiotic enterocins, among which can be found cytolysin or Enterocin W produced by some *E. faecalis* strains. The enterocin of this class contains lanthionine residues, which dictates that it should be considered a two-component lantibiotic. Class II enterocins are represented by small, non-lantibiotic peptides, subdivided into Class IIa, anti-*Listeria* enterocins of the pediocin family such as Enterocins A, P, durancin GL, Mundticin KS, Hiracin JM79, and the others (Nes et al. 2014); Class IIb, two-peptide (Ent C, X, Q, B etc.) enterocins; Class IIc enterocins synthesized without a leader peptide (Ent 4); Class IId, unmodified, single-peptide enterocins (L50A, L50B); and Class III, high-molecular-weight, heat-labile enterocins such as AS-48 or enterolysins (Wu et al. 2022). Most Class II bacteriocins, which are thermo-stable, small peptides, contain an N-terminal leader peptide which directs the secretion of the bacteriocin and is cleaved off during the secretion process. Their leaders belong to the so-called double-glycine leader type or in some cases a *sec*-dependent leader type, and as a result, the *sec* secretion system is employed for the latter. On the other hand, some bacteriocins are not synthesized with a leader peptide, and their secretion seems to be performed by dedicated ABC transporters with sequence features that are notably different from the ones that externalize the double-glycine leader and the *sec*-leader bacteriocins. Bacteria can produce a single or multiple leaderless bacteriocins. When several bacteriocins are produced by the same strain, the bacteriocin often share strong sequence homology, and their genes are located next to each other. The individual peptides posses antimicrobial activity, but when combined, they exhibit increased potency (Nes et al. 2014; Ovchinnikov et al. 2017). Acedo et al. (2017) presented an *E. canintestini* 49 strain isolated from dog feces with a beneficial potential and production of multiple bacteriocins. In the new century, the interest in natural preservatives has increased enormously, in accordance with consumers' demand for healthy, safe, and fresh products. However, the predominant use of beneficial enterococci and their enterocins is focused on food-producing and social animals.

## 5.1.3 Beneficial Enterococci and Their Enterocins: Application Potential in Animals

Breeders, farmers, and owners have looked for ways to stabilize or improve the health status of farm animals and other animals. Taking into account EU regulations, alternatives to banned antimicrobial growth stimulators are desired. These alternatives are represented by beneficial bacteria with their active metabolites. For years their experimental use has increased with the aim to allow their practical application, which is still under the stringent control of EU institutions such as the EFSA. Although their practical and commercial availability have been gradually improving (Herich et al. 2010; Pogány Simonová et al. 2016; Karaffová et al. 2017; Lauková et al. 2023), there is still not sufficient information regarding the use of their enterocins. In this framework, antimicrobial effects of beneficial enterococci and their enterocins will be controlled, their influence on immunity, biochemical parameters, growth performance, or meat quality which is in association with health of consumers in case of food-producing animals. In other animals (e.g. horse, dogs), the effect of beneficial enterococci and/or their enterocins is related with health of animals or with environmental hygiene. Recently, benefits of probiotic, bacteriocin-producing *E. faecium* AL41/CCM8558 and *E. durans* ED26E/7 strains have been reported in connection with the stimulation of phagocytosis and the respiratory burst of blood polymorphonuclear leucocytes (PMNL) after their application in mice infected with *Trichinella spiralis*. This stimulation can contribute to the destruction of muscle larvae and then a reduced parasite burden in the host (Dvorožňáková et al. 2016). Bacteriocin-producing, probiotic *E. faecium* EF55 and EF2019/CCM7420 stimulated the production of anti-inflammatory cytokine IL-10 in mice (Bucková et al. 2015). Therefore, the efficacy of beneficial, enterocin-producing strains and/or application of their enterocins in different food-producing animals as well as, for example, social animals will be disputed in the next part.

### 5.1.3.1 Beneficial Enterococci and Their Enterocins Applied in Ruminants and Pigs

In agriculture, animal production represents the area with the oldest tradition where beneficial strain implications were established. In ruminants, *E. faecium* strains of non-ruminal origin were applied in 1993 (Kmeť et al. 1993). There are several ways to utilize beneficial strains for their application in ruminants over the developmental periods: in the postnatal period as an adjunct to antibiotic treatments and for stimulation of ruminal digestion in the young. *E. faecium*-containing preparations improved microbial metabolism in the rumen (Kmeť et al. 1993). *E. faecium* M74 (isolated from the stool of a breast-fed infant) was mostly applied to calves to prevent diarrhea cases in the post-weaning period. Moreover, ruminal enterococci-producing bacteriocins were reported (Lauková et al. 1993; Nigutová et al. 2007). Those also active against *Streptococcus bovis* can be used to prevent ruminal acidosis.

Enterococci are part of the endogenous microbiota in pigs (Strompfová et al. 2006). That is why *E. faecium*-based beneficial strains were first applied in pigs with anti-*E. coli* effect to reduce/eliminate diarrhea, specifically *E. faecium* SF68 (NCIMB10415; isolated from an infant's stool). Gnotobiotic pigs were fed the SF68 strain and exposed to *E. coli* strains 0157:K88ac:H19 and 08:K87, K88ab:H19. This induced only mild diarrhea, and none of the pigs died; they continued to eat well and gained weight. Histopathological examinations demonstrated abundant colonization of the intestinal tract by *E. faecium* SF68 (Busing and Zeyner 2015). Although *E. faecium* NCIMB10145 is effective against *E. coli* diarrhea, no protective effect of it was demonstrated on *Salmonella* shedding in weaned pigs (Kreuzer et al. 2012). Tian et al. (2016) reported that another *E. faecium* HDRsEf1 could attenuate enterotoxigenic *Escherichia coli* (ETEC) K88-induced IL-8 secretion and transepithelial electrical resistance (TEER) decrease in IPEC-J2 cells. It is interesting to note that a recent study using jejunal tissue explants in Ussing chambers revealed that increased expression of proinflammatory cytokines was accompanied by an initial TEER increase and reduced permeability towards macromolecules upon ETEC challenge when pigs were fed control but not *E. faecium* strain NCIMB10145-supplemented diets (Lodemann et al. 2017). *E. faecium* CECT 4515 incorporated in the product FECINOR (the company Norel & Nature) showed a high capacity for rapid colonization of the pig intestine during the post-weaning period (Díaz 2006). *E. faecium* is protected by polysaccharide layers, which allow it to pass through the stomach without being affected by the low pH. It also produces enzymes in sufficient amounts, which improves feed digestibility. Feeding of dipeptide Ent A/P-producing beneficial *E. faecium* CCM 7419 ($10^9$ cfu/ml) to piglets 1–14 days of age decreased number of *E. coli* in feces, decreased pH in the duodenum, and increased the concentration of lactic and propionic acids in the colon. The concentrations of total serum protein, calcium, hemoglobin, hematocrit, erythrocytes, and index of phagocytic activity (PA) of leukocytes were significantly higher after CCM 7419 strain application, and cholesterol was significantly lower (Strompfová et al. 2006). It could be disputed that the anti-*E. coli* activity of *E. faecium* beneficial strains could be related to the fact that many of them are available to produce enterocins, which can inhibit coliforms (Lauková et al. 2012); however, *Salmonella* strains are limitedly inhibited by enterocins (Lauková et al. 2004b).

Although bacteriocin production by ruminal enterococci has been already reported (Lauková et al. 1993; Nigutová et al. 2007), use of enterocins to optimize microbiota *in vivo* in ruminants was not reported up to now. The *in vitro* effect of Ent 4231 was demonstrated in rumen fluid (Lauková and Czikková 1998). There staphylococci but also coliforms were inhibited with a log cycle difference from 2.0 or 5.0. A promising application of bacteriocins in ruminants is prevention of mastitis. This beneficial effect of bacteriocins produced by LAB was indicated by Pieterse and Todorov (2010). Lauková et al. (2016) isolated bacteriocin-active enterococci from mastitis milk with inhibition activity, mostly against listeriae. However, substances from *E. faecium* 21 also inhibited *S. aureus* SA5 (from mastitis milk).

More information related to enterocin use was reported concerning pigs (Tian et al. 2016). When a cell-free supernatant (bacteriocin-like substance) of *E. faecium* HDRsEf1 was used, it was noted that it can similarly, as producing strain *E. faecium* HDRsEf1, attenuate ETEC K88-induced IL-8 secretion and TEER decrease in IPEC-J2 cells. ETEC (enterotoxigenic) *E. coli* is the major cause of enteric disease in pigs, responsible for approximately 50% of piglet mortality. Al Atya et al. (2016) showed that combining colistin with enterocin from LAB enhances its *in vitro* antibacterial activity against planktonic and biofilm culture of *E. coli*; colistin disrupted the outer membrane of *E. coli* by acting on lipopolysaccharides, opening the way for the subsequent action of bacteriocins.

### 5.1.3.2 Beneficial Enterococci and Their Enterocins Applied in Poultry

Poultry breeding represents the branch of animal production where beneficial bacterial strains have frequently been utilized to prevent or to reduce the risk of salmonellosis, clostridiae, or campylobacteriosis (Meremäe et al. 2010). Audisio et al. (2000) administered *E. faecium* J96 (chicken isolate, from the crop, with inhibition activity most probably due to lactic acid) orally ($10^9$ cfu/ml) to chickens for both preventive and therapeutic purposes. To test the preventive effect, 30-h-old chickens received *E. faecium* J96 twice a day with an interval of 12 h between each dose for 3 consecutive days. To test the therapeutic effect, *E. faecium* J96 was administered to 4-day-old chickens in the same way and after challenge by *S. enterica* serovar *Pullorum* M97 ($10^5$ cfu/ml). The chickens that were preventively treated with J96 survived the *S. enterica* serovar *Pullorum* challenge. Those that were infected on the first day of the experiment and then inoculated with the J96 strain died 4 days later. Salmonellae were isolated from their livers and spleens. Therefore, J96 can prevent *S. pullorum* infection in chickens but cannot act as a therapeutic agent. This was also confirmed in the experiments using gnotobiotic Japanese quails. Lauková et al. (2003) used *E. faecium* CCM 7419, a dipeptide Ent A/P-producing strain, to check its preventive and therapeutic effect. The birds (3 days old) in the experimental group received *E. faecium* CCM 7419 ($10^9$ cfu/ml), from the start, while the birds in the control group received a placebo. Sixteen hours later, both groups were infected with *S. enterica* serovar *Duesseldorf* SA31 ($10^7$ cfu/ml). A reduction in the count of the SA31 strain, as an effect of CCM 7419, was found in the feces of samples taken at 24 and 48 h from the group with the CCM 7419 strain. Significant reductions were also detected in the count of SA31 cells in the caecum but not in the ileum after the birds were killed (Lauková et al. 2003). When chickens fed *E. faecium* AL41/CCM8558 (non-chicken origin, producing Ent M) were challenged with *S. enterica* serovar *Enteritidis* PT4, a beneficial effect of the AL41 strain on IgA production and secretion in the intestine was demonstrated. Findings also indicated that IgA played important role in decrease of *S. enteritidis* in the intestine, and cytokine TGF-β4 and IL-17 contributed to the increased IgA secretion (Karaffová et al. 2015). Karaffová et al. (2022) also reported that the strain *E. faecium* AL41=CCM 8558 revealed a protective effect and beneficial influence on the local and systemic immune response in *Salmonella enteritidis* PT4-infected chickens. After the AL41 strain, a decrease of staphylococci and enterobacteriae was also noted ($p < 0.001$) in ostriches (Lauková et al. 2015b). When chickens were administered *E. faecium* AL41 strain and were challenged with *Campylobacter jejuni* CCM 6191, the results achieved demonstrated that the AL41 strain can modulate toll-like receptors (TLR) expression and modify the activation of macrophage migration inhibitory factor (MIF), interferon (IFN-β), myeloid differentiation factor 2 (MD-2), and monocyte differentiation antigen (CD14) molecules in the chicken caecum challenged with the CCM 6191 strain. The counts of the AL41 strain were sufficient; as were the counts of enterococci in both caecum and feces (Karaffová et al. 2017). Revajová et al. (2022) also reported that administration of *E. faecium* AL41=CCM 8558 had no substantial effect on the concentrations of acute phase proteins, but a significant increase in β- and ɣ-globulin fractions was noted at the end of the experiment. It may indicate an improvement in the immune status. Also, a significant prolonged stimulatory effect of *E. faecium* AL41=CCM 8558 on the relative expression of molecules (immunoglobulin A, mucin-2) as well as on the dynamic of mucus production in the chicken intestine was observed. Horniaková and Bušta (2006) found a beneficial effect of *E. faecium* 3530 from commercial IMB52 preparation (Biomin GmbH Company, Austria) on egg weight, yolk, and white weight. The early exposure of chickens to *E. faecium* EF55 (isolated from chicken crop, Ent 55-producing) at $10^9$ cfu/ml for 7 consecutive days led to more rapid development of intestinal villi after infection at 8 days of age with *S. enterica* serovar *Enteritidis* compared with the untreated control. Reduced colonization of the intestinal tract by salmonellae in birds treated with *E. faecium* EF55 also preserved the microenvironment of the intestine from harmful effects of the pathogen (Levkut et al. 2009; Herich et al. 2010). Moreover, when broiler chickens were administered the EF55 strain and challenged with *S. Enteritidis* PT 147, the beneficial effect of the EF55 strain was demonstrated on the expression of MUC (mucin gene) and production of IgA+IEL (intraperitoneal lymphocytes) in the caecum after *S. Enteritidis* PT 147 infection (Levkut et al. 2016). The preventive administration of the EF55 strain showed higher mRNA levels of the studied proinflammatory cytokines and LyTac chemokine except of iNOS in the caecum of chicken with EF55 strain and *S. Enteritidis* PT 147, mainly on day 1 after SE 147 infection compared to other groups (Karaffová et al. 2015). There is real evidence of the beneficial effect of bacteriocin-producing enterococci in poultry.

As mentioned earlier, among endogenous infections in poultry, necrotic enteritis caused by clostridia has been reported. Clostridia are present in low counts in healthy poultry; however, after inadequate nutrition, dysfunction of intestinal microbiota can appear, followed by imbalance of host homeostasis. The ruminal strain *E. faecium* CCM 4231, producing Ent 4231 inhibited *in vitro*, the growth of clostridial strains. When dipeptide Ent A/P was administered to Japanese quails following a preventive and therapeutic schedule of experimental application, a significant difference in the SA31 cells was noted in the feces of the group given the preventive treatment compared with control after 8 h and in the group given the therapeutic treatment after 24 h. After 48 h, a lower count of the SA31 strain was noted in both experimental groups. The count of the SA31 strain was also reduced in the caecum and ileum of the group given the therapeutic treatment, but not in the group with the preventive treatment (Lauková et al. 2004b). Application of dipeptide Ent A/P before and after SA31 strain infection showed a protective effect in the duodenal epithelium. Damage to microscopic and submicroscopic structures of enterocytes and goblet cells was less intensive after Ent A/P treatment. However, prominent damage of enterocytes and their necrosis was noted in *Salmonella*-infected quails (Ciganková et al. 2004). Dipeptide Ent A/P showed a stronger therapeutic than prophylactic effect. Concerning the campylobacteriosis, Svetoch and Stern (2010) reported the possible use of bacteriocins to control *Campylobacter* spp. in poultry. Line et al. (2008) showed that Enterocin E-760 isolated from *Enterococcus* spp. NRRL B-30745 reduces colonization of naturally acquired *Campylobacter* spp. in market-age broiler chickens when administered in feeds. Its administration was associated with 8-log reduction in *Campylobacter* counts. Enterocin EM41 is partially purified enterocin produced by *E. faecium* EM41 isolated from ostrich (Lauková et al. 2017c). After its administration in laying hens, reduction of staphylococci, coliforms, and campylobacters was noted and a significant increase in values of PA ($p < 0.05$, $p < 0.01$). García-Vela et al. (2023) reported poultry-origin, bacteriocin-producing strains *E. faecium* X2893 and *E. faecium* X2906 as the most promising candidates for further studies as protective cultures in poultry farming. These strains harbor the genes encoding for Enterocin A and Enterocin B.

### 5.1.3.3 Beneficial Enterococci and Their Enterocins Applied in Broiler Rabbits

In broiler rabbits, the majority of digestive events occur in the caecum, where a wide variety of microbiota resides. Most studies deal with the strictly anaerobic and facultative bacteria (Eshar and Weese 2014); their counts in rabbits depend on the age of the rabbit, and mainly clostridia dominate. Because during the post-weaning period many health problems occur in rabbit husbandry, there is a potential for beneficial, enterocin-producing strains and application of their enterocins. Beneficial enterococci are more useful in broiler rabbit husbandry compared with lactobacilli, for example, because they are present in the rabbit microbiota in substantial numbers (Pogány Simonová et al. 2009a). Successful results have been published on the application of bacteriocin-producing beneficial enterococci and their enterocins to limit unsuitable microbiota in the rabbit digestive tract, to stimulate immune activity, to increase weight gain, and to reduce *Eimeria* spp. oocysts (Pogány Simonová et al. 2009a, 2015, 2020; Lauková et al. 2012, 2017c, 2017e). When *E. faecium* CCM 7420 (isolated from a rabbit) was administered to rabbits via drinking water, an increase in body weight of animals was achieved; in the caecum, a significant reduction of CoPS was noted. Also, the lowest activity of glutathione peroxidase (GPx) was measured, which indicates that administration of this strain did not evoke oxidative stress (Pogány Simonová et al. 2009a, 2020); it is an important result, taking into account that in 2017 EFSA recommended the use of alternatives such as probiotics instead antibiotics or coccidiostats in animals (Murphy et al. 2017). The *in vivo* reduction of *Eimeria* oocysts is in accordance with our *in vitro* studies in which *Eimeria* spp. oocysts were also reduced using beneficial enterococci (Strompfová et al. 2010). When freeze-dried CCM 7420 was administered to rabbits and was compared with a fresh culture, the highest fecal counts of CCM 7420 were measured in samples administered the fresh culture. However, a high energy value of rabbit meat was achieved with both fresh and freeze-dried CCM 7420. Inferring from the higher iron content, it can be concluded that diet supplementation with *E. faecium* CCM 7420 strain may enhance the mineral quality of rabbit meat (Pogány Simonová et al. 2009b, 2016). A decrease of *S. aureus* and *Eimeria* spp. oocysts was noted in both cases, but higher levels of biochemical parameters were found in the serum of the fresh culture group of CCM 7420.

As mentioned, *E. faecium* CCM 8558 is a strain of non-rabbit origin, producing Ent M. Although it slightly colonized the rabbit's GIT, a reduction of CoPS and pseudomonads was noted in the caecum. However, higher PA was measured in rabbits administered *E. faecium* CCM 8558 (AL41) than in control animals (Lauková et al. 2012). Moreover, an increase in villus height to crypt depth ratio in the jejunum indicates that the functionality of enterocytes (intestinal mucosa) was stimulated (Lauková et al. 2017e). It could be also summarized that the AL 41 strain (CCM 8558) could enhance the quality and mineral content of rabbit meat, with a focus on its iron and phosphorus contents (Pogány Simonová et al. 2021). *E. faecium* CCM 4231 is also a strain of non-rabbit origin, and in spite of its slight colonization in rabbits (up to $10^3$ cfu/g), it was found to reduce CoPS and clostridiae (Szabóová et al. 2008). A reduction of *Eimeria* spp. oocysts was also noted, from 117 oocysts per gram (OPG) to 83 OPG. Animals achieved higher weight gain and lower mortality. Moreover, improved amino acid composition in meat and preserved meat quality and nutritional value were observed after CCM 4231 supplementation. The durancin-producing strain *E. durans* ED26E/7 (non-rabbit origin) was found to be sufficiently colonized in the rabbit digestive tract; decrease of coliforms was noted in the caecum and appendix; significant reduction of *Eimeria* spp. oocysts and an increase in PA. Moreover, the values of GPx were lower in the experimental group than in the control, indicating that application of *E. durans* ED26E/7 did not evoke oxidative stress. Biochemical blood parameters and quality of meat were not negatively influenced (Lauková et al. 2017b). It was first use of *E. durans* in rabbit husbandry. Benato (2013) also recommended beneficial enterococci in pet rabbits. Altogether, studies have repeatedly shown the useful effects of beneficial enterococci on rabbits' microbiota and health, independent of the origin of the *E. faecium/E. durans* strain used. The most important effect noted was the reduction of *Eimeria* spp. oocysts, stimulation of phagocytosis, and functionality of intestinal mucosa re-newing. An *E. faecium* EF9a strain isolated from a Pannon White rabbit is a bacteriocinogenic strain with promising application potential in rabbits supporting weight gain (Pogány Simonová et al. 2020).

Up to now (except our Laboratory of Animal Microbiology, Centre of Biosciences of the Slovak Academy of Sciences, Institute of Animal Physiology, Košice, Slovakia, in co-operation with NAFC, Nitra), there is no laboratory testing *in vivo* effect of enterocins in rabbit husbandry. A reduction of fecal coliforms, CoPS involving *S. aureus*, and clostridiae was noted in rabbits with administration of Ent 7420/2019 (produced by rabbit strain *E. faecium* EF2019/CCM7420). After Ent 2019, the lowest activity of GPx was also achieved (Pogány Simonová et al. 2009a, 2013). Moreover, after 21 days of Ent 2019 application, a reduction of *Eimeria* spp. oocysts was recorded (Pogány Simonová et al. 2009a), which is an important result, as was formerly reported in association with banned coccidiostats. Ent 55 is produced by the non-autochthonous strain *E. faecium* EF55. After its application, reduction of staphylococci, clostridiae, coliforms, and pseudomonads was noted; *Eimeria* spp. were also reduced, and PA was stimulated ($p < 0.001$, Lauková et al. 2017d). Ent M applied in rabbits led to a decrease of pseudomonads and staphylococci in feces with a difference of 1.0 cfu/g; in the caecum, coliforms were decreased, and in the appendix, unspecified bacteria were reduced ($p < 0.01$). PA was stimulated, reaching the highest value of 68.8 ± 0.86% in the experimental rabbits compared to control (35.4 ± 0.51%, $p < 0.001$). Moreover, higher average weight gains were detected by Ent M (Lauková et al. 2012). Durancin ED26E/7 reduced coliforms in broiler rabbits, the meat was tasty and juicy with sufficient quality parameters; PA were stimulated significantly ($p < 0.001$; Lauková et al. 2015b; Pogány Simonová et al. 2022a). Using a combination of both Ent M and Durancin in rabbits demonstrated an increase in meat quality parameters. Enterocin A/P even reduced methicillin-resistant staphylococci in broiler rabbits and stimulated an increase in weight gain (Pogány Simonová et al. 2022b). Ent M was also applied to control biofilm-forming *E. hirae* in broiler rabbits (Lauková et al. 2022). It showed a decreased tendency of Kr8+. Using next-generation sequencing, the phyla detected with the highest abundance were Firmicutes, Verrumicrobia, Bacteroidetes, Tenericutes, Proteobacteria, Cyanobacteria, Saccharibacteria, and Actinobacteria. Interaction with some phyla resulted in reduced abundance percentage. Kr8+ did not attack PA, but Ent M supported PA when PA was increased in the Kr8+ group and Ent M with the Kr8+ group as well (Lauková et al. 2022).

## 5.1.3.4 *Beneficial Enterococci and Their Enterocins Applied in Horses*

For horse breeders, a commercially available preparation is Probios, containing a mixture of three probiotic lactobacilli and *E. faecium*, which are not of equine origin. Schoster et al. (2014) introduced a task on the efficacy of beneficial bacteria use in horses. There were mentioned information related to only non-horse origin strains used; they were mostly human-derived. Although Fraga et al. (2008) isolated vaginal LAB from mares and found *Enterococcus* spp. with promising features for use as equine beneficial bacteria, there were not applied *in vivo*. Similarly, Lauková et al. (2008) reported the potential beneficial applications of bacteriocin-producing enterococci isolated from a healthy horse. As formerly reported, *E. faecium* AL41/CCM 8558 is a non-horse origin strain; however, it was applied to 11 horses of different breeds in a pilot experiment for 14 days at a dose of 1 g per animal and day. Horses were fed hay or oats, or they were alternatively grazed (pasture). On day 14, *E. faecium* CCM 8558 strain was detected in average count $2.35 \pm 0.70$ cfu/g in feces of horses. An antimicrobial effect was related to aeromonads; their counts decreased significantly ($p < 0.001$). On day 14, PA value in horses blood showed an increasing tendency compared to days 0–1 ($73.13 \pm 8.55\%$ to $75.11 \pm 8.66\%$). Biochemical parameters in blood were optimized in the physiological range (Lauková et al. 2020). However, Lauková et al. (2023) referred to application of autochthonous, Enterocin 412-producing strain *E. faecium* EF 412 in clinically healthy horses (12) of Slovak warm-blood breed of various ages. After 3 weeks application, the eggs of nematode *Strongylus* spp. as well as *Eimeria* spp. oocysts, were not found in the horses. The phyla Bacteroidetes and Firmicutes dominated, and at the family level, the family Bacteroidales BS11 and S24–7 gut groups and Lentisphaere were detected with the highest abundance. An increasing tendency in PA was also noted after EF 412 application. Biochemical parameters were in physiological range.

Enterocin has never been applied to horses before. At first, Lauková et al. (2018) performed a pilot experimental application of Ent M. The most commonly occurring infections in horses are caused by *Salmonella* spp. or by clostridiae; therefore, to prevent or eliminate their agents using bacteriocins is one of the promising routes. Clinically healthy horses (10) were involved in this pilot experiment. The animals were fed twice a day with hay and oats, or alternatively grazed with access to water *ad libitum*. The experiment lasted 6 weeks. Feces were sampled directly from the rectum and blood from the *vena jugularis*. Each horse itself represented a control animal (compared to its status at the start of the experiment, day 0–1). Sampling was also performed on days 21 and 42 (3 weeks of cessation of Ent M). After initial sampling, the horses were administered 100 µl of Ent M (12 800 AU/ml) in a small feed bolus to ensure it was consumed; Ent M was applied for 3 weeks (21 days). Administration of Ent M led to mathematical reduction of coliforms and campylobacters ($p < 0.05$) and significant reduction of *Clostridium* spp. ($p < 0.001$, $p < 0.001$) on day 21; increase of PA values was noted on days 21 and 42 ($p < 0.05$, $p < 0.0001$); hydrolytic activity or biochemical blood parameters were not influenced. In general, there was a tendency toward beneficial status of the horse. Another Enterocin—Ent 412 produced by the formerly mentioned autochthonous strain *E. faecium* EF 412—was applied in 12 healthy Slovak warm-blood horses. An increasing tendency in PA was noted after 3 weeks of application ($69.0 \pm 3.91$) compared to day 0/1 ($67.89 \pm 3.66$) and to day 42 (when Ent 412 was not applied for 3 weeks, $66.67 \pm 2.74$; Lauková et al. unpublished data). On day 0/1, the dominant phyla were Bacteroidetes, followed by Firmicutes, Proteobacteria, Lentosphaere, Spirochaetae, and Fibrobacteres. The other phyla were represented only in very low percentage abundance (up to 1%). After Ent 412 application, reduction of all formerly mentioned phyla was noted besides the phylum Bacteroidetes.

The Mundticin-like substance EM 41/3 is a bacteriocin/enterocin produced by the autochthonous, fecal strain *E. mundtii* EM 41/3; isolated from the Slovak breed Norik of Muráň (Focková et al. 2022, 2023). This is the first bacteriocin applied in the breeding of the Norik of Muráň breed. Based on bacteriocin characteristics, it is a thermo-stable, small peptide belonging among small bacteriocin peptides, group II, related to enterocin classification. It was applied in 13 clinically healthy mares of the Norik of Muráň breed of different ages. They were administered EM 41/3 substance (100 µl per animal) in the diet/bolus. Horses were also on pasture (grazed) and with obligatory feeding for horses. Each animal itself was control to each sampling. A reduction of coliforms and pseudomonads was noted up to 1.20 log cycle.

PA showed an increasing tendency. GPx activity was lower in horses with EM 41/3 substance comparing to control values (on day 0/1), meaning that EM 41/3 substance application did not evoke oxidative stress (Focková et al. 2023).

### 5.1.3.5 Beneficial Enterococci Applied in Dogs

Up to now, beneficial bacteria involved in probiotic products available on the market are of non-canine origin; for example, Probican paste with *E. faecium* M74 (Medipharm s r.o., Czech Republic), Enteroferm containing *E. faecium* NCIB 10415 (SF68), and FortiFlora (Purina, containing live strains of *E. faecium* SF68 and antioxidant vitamins). Benyacoub et al. (2003) reported that supplementation of dog feed with human-origin *E. faecium* SF68 stimulates the immune functions in young dogs. Marciňáková et al. (2006) administered the bacteriocin-producing beneficial canine-feed–originated strain *E. faecium* EE3 to healthy dogs ($10^9$ cfu/g) for 7 days. Its consumption was not associated with any adverse clinical effects. The average concentration of EE3 strain in dog feces reached a count of $4.85 \pm 2.43 \log_{10}$ cfu/g on day 7, with prolonged survival of up to almost 3 months after cessation of strain feeding. Also, a reduction of staphylococci and pseudomonads was noted. Abnormal cholesterol levels were brought to physiological levels. Moreover, Lauková et al. (unpublished data) tested in dogs the effect of *E. faecium* strains of non-canine origin, both bacteriocin producing and beneficial-*E. faecium* CCM 7419/EK13 and CCM 4231. They were applied for 7 or 14 days (concentration of $10^9$ cfu/g); it was found that they controlled not only the microbiota but also phagocytic activity. *E. faecium* CCM 7419/EK13 is an environmental strain producing dipeptide enterocin A/P; *E. faecium* CCM 4231 is of rumen origin and produces Ent 4231. The average counts of CCM 7419 on day 7 in healthy dogs reached $\log_{10}$ 5.6 cfu/g, persisting up to 3 months at a level of $2.98 \log_{10}$ cfu/g. Counts of $10^5$ cfu/g were detected in healthy dogs after colonization by the CCM 4231 strain; 7 and 14 days after cessation of its administration, $10^3$ cfu/g colonies were still detected. A decrease in the numbers of staphylococci, clostridiae, pseudomonads, and aeromonads was noticed in fecal samples of dogs receiving both strains, CCM 7419 and CCM 4231. Moreover, PA was either increased or not influenced. It seems that the canine GIT is receptive for the colonization of beneficial strains isolated from both canine and non-canine origins. Kubašová et al. (2019) isolated two probiotic strains, *E. faecium* IK25 and D7 from canine feces; *E. faecium* IK25 produced two-peptide Enterocin B/A; Ent B has been already purified (Kubašová et al. 2020). Both strains were applied in healthy dogs. However, after 14 days of application of non-bacteriocin-producing beneficial strain *E. faecium* D7 and Ent B-producing beneficial strain *E. faecium* IK25 in healthy dogs, it was noted that strain D7 was better tolerated without changes of fecal consistency and increased parameters of non-specific cellular immunity (Strompfová et al. 2019). In contrast, after IK25 administration, an increase of Gram-negative bacteria in the post-treatment period was noted, and more liquid fecal consistency was assessed. Following these studies, further studies are needed to postulate if beneficial bacteriocin-producing enterococci are less useful for preventive application in dogs than beneficial strains; however, a non-bacteriocin-producing (Strompfová et al. 2019) *E. faecium* canine strain was deposited in DSM possessing indication DSM 32820, and it was applied to 9 dogs suffering from idiopathic diarrhea for 7 days Kubašová et al. (2022). The 16S rRNA analyses revealed the phylum Firmicutes was predominant, followed by Proteobacteria, Actinobacteria, Fusobacteria, and Bacteroidetes. The abundance of the family Erysipelotrichiacae was higher on day 7 compared to the initial level ($p < 0.05$). Hematological and biochemical parameters were not significantly different, only in individual dogs. Mean fecal dry matter was significantly higher on day 7 after DSM strain application (Kubašová et al. 2022). Enterocins themselves have been not applied in dogs up to now.

### 5.1.3.6 Beneficial Enterococci and Their Enterocins in Association with Human Health

Trichinellosis is a serious food-borne parasitic zoonosis with worldwide distribution (Gozdzik et al. 2017; Dvorožňáková et al. 2022). It is characterized by enteritis (induced by adult worms) and inflammation, with degenerative changes in the skeletal muscles (induced by larvae). As mentioned by Dvorožňáková et al. (2022), *T. spiralis* is highly pathogenic due to the massive production of newborn larvae and elicits a strong host immune response with intestinal inflammatory reactions. Beneficial strains administered to mice were *E. faecium* CCM 8558 and *E. durans* ED26E/7 ($10^9$ cfu/ml), which were infected with *T. spiralis* larvae (400 larvae) on day 7 of treatment. Positive modulation of the gut lymphocyte immunity in *T. spiralis* infection with bacterial strains showed their beneficial effect with the host's anti-parasitic defense. Schofs et al. (2022) mentioned that nematode infections affect a significant percentage of the human population worldwide. And still there are not enough drugs, especially in developing countries, to treat these infections. So, controlling parasitic infections using beneficial strains has emerged as a suitable option. The same authors presented the anti-nematodic effect of *E. faecalis* CECT7121 using *T. spiralis* as a model of nematode infection in mice. This strain in combination with conventional anti-helminthic drugs may be useful for improving clinical and parasitological outcomes.

When beneficial strains *E. faecium* CCM 8558, *E. durans* ED26E/7, and enterocins produced by them (Ent M and Durancin-like ED 26E/7) were applied in Balb/c mice infected with *T. spiralis*, in mice intestine was noted a significant reduction of *T. spiralis* adults but also larval count of *T. spiralis* in muscles was reduced. The strains showed a significant inhibitory effect on female fertility of *T. spiralis*. After individual application of both enterocins produced by formerly mentioned strains, a higher reductive effect on newborn larvae showed Ent M. Reductive activity of muscle larvae after treatment with Ent M was comparable with its producing strain *E. faecium* CCM 8558. Durancin-like ED26E/7 showed lower anti-parasitic activity than its producing strain *E. durans* ED26E/7 (Vargová et al. 2022). Increase of metabolic activity of peritoneal macrophages induced by Ent M and also by both producer strains, *E. faecium* CCM 8558 and *E. durans* ED26E/7, indicated beneficial modulation of host organism immune response during infection with *T. spiralis*. It has indicated also promising use of Ent M in trichinellosis therapy (Dvorožňáková et al. 2021).

# 5.2 CONCLUSIONS

Enterococci are bacteria which can possess beneficial properties, but they can also contain virulence factors genes. Therefore, it is necessary to look at this bacterial species from two points of view, beneficial and non-beneficial, to assess each of these potentials of enterococci for their future use. However, enterococci and their enterocins have been shown to have a beneficial influence on the health of variety of hosts. Combined with their resilience, it makes them excellent probiotic candidates. The issue of antibiotic resistance transfer, however, remains and should be judged on a strain-by-strain basis, as is common for probiotics in any case. Concerning animals, it can be concluded that beneficial enterococci and their enterocins showed mostly antimicrobial effects to control unsuitable microbiota members and stimulate the organism's immune response by increasing phagocytic activity. They did not evoke oxidative stress. They beneficially, or at least not negatively, influenced carcass characteristic or meat quality in the case of food-producing animals. Moreover, some enterococcal strains and their enterocins showed promising results to be used in food–zoonosis–trichinelosis therapy. This is the most encouring possibility for their use.

# BIBLIOGRAPHY

Acedo, J.Z., Ibarra Romero, C., Miyata, S.T., Blaine, A.H., McMullen, L.M., Vederas, J.C. and M.J. van Belkum. 2017. Draft genome sequence of *Enterococcus canintestini* 49, a potential probiotic that produces multiple bacteriocins. *Genome Announc.* 5: e01131–17.

Advisory Committee on Novel Foods and Processes. 1996. Report on *Enterococcus faecium*, strain K77D. In *MAFF Advisory Committee on Novel Foods and Processes*, Report. Ergon House c/o Nobel House, London.

Al Atya, A.K., Abriouel, H., Kempf, I., Jouy, E., Auclair, E., Vacheé, A. and D. Drider. 2016. Effects of colistin and bacteriocins combinations on the in vitro growth of *Escherichia coli* strains from swine origins. *Prob. Antimicrob. Prot.* 8: 183–190.

Arvanitoyannis, I.S. 2009. *HACCP and ISO 22000: Application to Foods of Animal Origin*. Wiley-Blackwell.

Aspri, M., Field, D., Cotter, P.D., Ross, P., Hill, C. and P. Papademas. 2017. Application of bacteriocin-producing *Enterococcus faecium* isolated from donkey milk, in the bio-control of *Listeria monocytogenes* in fresh whey cheese. *Int. Dairy J.* 73: 1–9.

Audisio, M.C., Oliver, G. and M.C. Apella. 2000. Protective effect of *Enterococcus faecium* J96, a potential probiotic strain, on chicks infected with *Salmonella pullorum*. *J. Food Prot.* 63: 1333–1337.

Benato, L. 2013. beneficial bacteria in pet rabbits. *Vet. Times*: 1–8.

Benyacoub, J., Czarnecki-Maulden, G.L., Cavadini, C., Sauthier, T., Anderson, R.E., Schiffrin, E.J. and T. von der Weid. 2003. Supplementation of food with *Enterococcus faecium* SF68 stimulates immune functions in young dogs. *J. Nutr.* 133: 1158–1162.

Berufsgenossenschaft der Chemischen Industrie. 1995. Safety and health promotion aspects of enterococci. FAIR-CT97–3078. In *Enterococci in Food Fermentations. Functional and Safety Aspects*. A Brochure for Practical Users.

Bino, E. 2022. Animal bacteria, a source of bioactive substances in animal husbandry prevention. (in Slovak) Dissertation G385/22, University of Veterinary Medicine and Pharmacy in Košice, Slovakia, CBs SAS IAP, Košice, Slovakia, pp. 1–157.

Birollo, G.A., Reinheimer, J.A. and C.G. Vinderola. 2001. Enterococci vs. nonlactic acid microflora as hygienic indicators in sweetened yogurt. *Food Microbiol.* 18: 597–604.

Bucková, B., Dvorožňáková, M., Revajová, V. and A. Lauková. 2015. Effect of probiotic strains on the T-cell subsets and cytokine production in mice. *Folia Vet.* 59: 18–25.

Busing, K. and A. Zeyner. 2015. Effects of oral *Enterococcus faecium* strain DSM 10663 NCIMB 10415 on diarrhoea patterns and performance of sucking piglets. *Benef. Microbes.* 6: 41–46.

Cattoir, V. 2022. The multifaceted lifestyle of enterococci: Genetic diversity, ecology and risks for public health. *Curr. Opinion Microbiol.* 65: 73–80.

Chen, Y.S., Lin, Y.H., Pan, S.F., Ji, S.H., Chang, Y.C., Yu, C.R., Liou, M.S., Wu, H.C., Otoguro, M., Yanagida, F., Liao, C.C., Chiou, C.M. and B.Q. Huang. 2013. *Enterococcus saccharolyticus* subsp. *taiwanensis* subsp. nov., isolated from brocoli. *Int. J. Syst. Evol. Microbiol.* 63: 4691–4697.

Ciganková, V., Lauková, A., Guba, P. and R. Nemcová. 2004. Effect of enterocin A on the intestinal epithelium of Japanese quails infected by *Salmonella duesseldorf*. *Bull. Vet. Inst. Pulawy* 48: 25–27.

De Vuyst, L. 2004. Technology aspects to the application of functional starter cultures. *Food Technol. Biotechnol.* 38: 105–112.

Díaz, D. 2006. *Enterococcus faecium* CECT 4515: The immediately effective probiotic. *Technical Articles-Pig Industry*, pp. 1–6.

Dvorožňáková, E., Bucková, B., Hurníková, Z., Revajová, V. and A. Lauková. 2016. Effect of probiotic bacteria on phagocytosis and respiratory burst activity of blood polymorphonuclear leucocytes (PMNL) in mice infected with *Trichinella spiralis*. *Vet. Microbiol.* 231: 69–76.

Dvorožňáková, E., Vargová, M., Hurníková, Z., Lauková, A. and V. Revajová. 2022. Modulation of lymphocyte subpopulations in the small intestine of mice treated with probiotic bacterial strains and infected with *Trichinella spiralis*. *J. Appl. Microbiol.* 132: 4430–4439.

Dvorožňáková, E., Vargová, M., Hurníková, Z., Revajová, V. and A. Lauková. 2021. *Slovak Vet. J.* XLVI: 104–106.

EFSA. 2022. Update of the list of QPS-recommended biological agents intentionally added to food or feed as notified to EFSA 15: Suitability of taxonomic units notified to EFSA until September 2021, *EFSA J.* 20: e07045.

Eshar, D. and J.S. Weese. 2014. Molecular analysis of the microbiota in hard faeces from healthy rabbits (Oryctolagus cuniculus) medicated with long term use meloxicam. *BMC Vet. Res.* 10: 623.

European Commission. 2004. List of the authorized additives in feeding stuffs published in application of Article 9t (b) of Council Directive 70/524/EEC concerning additives in feeding stuffs. *Official J. Eur. Union.* C50.

European Food Safety Authority (EFSA). 2004. EFSA scientific colloquium summary report. In *QPS: Qualified Presumption of Safety Microorganisms in Food and Feed*, 13–14 December, Brussels, Belgium.

Fisher, K. and C. Phillips. 2009. The ecology, epidemiology and virulence of *Enterococcus*. *Microbiology* 155: 1749–1757.

Focková, V., Styková, E., Pogány Simonová, M., Maďar, M., Kačírová, J. and A. Lauková. 2022. Horses as a source of bioactive fecal strains *Enterococcus mundtii*. *Vet. Res. Com.* 46: 739–747.

Focková, V., Styková, E., Valocký, I., Bino, E., Pogány Simonová, M., Grešáková, Ľ. and A. Lauková. 2023. Antimikrobiálny potenciál Mundticin EM 41/3 v chove koní (in Slovak), Antimicrobial potential of Mundticin EM 41/3 in horses husbandry. *Slov. Vet. J.* IIIL: 5556. ISSN 1335–0099.

Fraga, M., Perelmuter, K., Delucchi, L., Cidade, E. and P. Zunino. 2008. Vaginal lactic acid bacteria in the mare: Evaluation of the probiotic potential of native *Lactibacillus* spp. and *Enterococcus* spp. *Ant. Leeuwenh.* 93: 71–78.

Franz, C.H.M.A.P., Huch, M., Abriouel, H., Holzapfel, W. and A. Gálvez. 2011. Enterococci as probiotics and their implications in food safety. *Int. J. Food Microbiol.* 151: 125–140.

Franz, C.H.M.A.P., Muscholl-Silberhorn, A.B., Yosif, N.M.K., Vancanneyet, M., Swings, J. and W.H. Holzapfel. 2001. Incidence of virulence factors and antibiotic resistance among enterococci isolated from food. *Appl. Environ. Microbiol.* 67: 4385–4389.

Franz, C.H.M.A.P., van Belkum, M.J., Holzapfel, W.H., Abriouel, H. and A. Gálvez. 2007. Diversity of enterococcal bacteriocins and their grouping in a new classification scheme. *FEMS Microbiol. Rev.* 31: 293–310.

García-Vela, S., Ben Said, L., Soltani, S., Guerbaa, R., Fernández-Fernández, R., Ben Yahia, H., Ben Slama, K., Torres, C. and I. Fliss. 2023. Targeting Enterococci with antimicrobial activity against *Clostridium perfringens* from poultry. *Antibiotics* 12: 231. https://doi.org/10.3390/antibiotics12020231

Gooch, C.C.H., Kiu, R., Rudder, S., Baker, J.D., Hall, J.L. and A. Maxwell. 2021. *Enterococcus innesii* sp. nov., isolated from the wax moth *Galleria mellonella*. *Int. J. Syst. Evol. Microbiol.* 71: 005168.

Gouba, N., Yimagou, E.K., Hassani, Y., Drancourt, M., Fellag, M. and M.D. Mboging Foukou. 2020. *Enterococcus burkinafasoniensis* sp. nov. isolated from human gut microbiota. *New Microbes New Infect.* 36: 100702.

Gozdzik, K., Odoevskaya, I.M., Movsesyan, S.O. and W. Cabaj. 2017. Molecular identification of *Trichinell*a isolates from wildlife animals of the Russian Arctic territories. *Helminthologia* 54: 11–16.

Harada, T., Dang, V.C., Nguyen, D.P., Nguyen, T.A., Sakamoto, M., Ohkuma, M., Motooka, D., Nakamura, S., Uchida, K., Jinnai, M., et al. 2016. *Enterococcus saigonensis* sp. nov., isolated from retail chicken meat and liver. *Int. J. Syst. Evol. Microbiol.* 66: 3779–3785.

Herich, R., Kokinčáková, T., Lauková, A. and M. Levkutová. 2010. Effect of preventive application of *Enterococcus faecium* EF55 on intestinal mucosa during salmonellosis in chickens. *Czech J. Anim. Sci.* 55: 42–47.

Hermans, D., Martel, A., van Deun, K., Verlinden, M., Van Immerseel, F., Garnym, A., Messens, W., Heyndrickx, M., Haesebrouck, F. and I. Pamans. 2010. Intestinal mucus protects *Campylobacter jejuni* in the caeca of colonized broilers chickens against the bactericidal effects of medium-chain fatty acids. *Poult. Sci.* 89: 1444–1155.

Holzapfel, W.H., Guigas, C. and C.H.M.A.P. Franz. 2002. General overview of the enterococci. In *Programme and Book of Abstracts from Conference Enterococci in Foods*, 30–31 May, Berlin, Germany, p. 1.

Holzapfel, W.H. and B.J.B. Wood. 2014. *Lactic Acid Bacteria: Biodiversity and Taxonomy*. Wiley-Blackwell.

Horniaková, E. and L. Bušta. 2006. Production effectivity of feed mixtures with the probiotic *Enterococcus faecium* to the quality of layers' eggs. *Slovak J. Anim. Sci.* 39: 79–83.

Jin, D., Jing, Y., Lu, S., Lai, X.H., Wen Xioung, Y. and J. Xu. 2017. *Enterococcus wangshanyuanii* sp. nov. isolated from faeces of yaks. *Int. J. Syst. Evol. Microbiol.* 67: 5216–5221.

Kadri, Z., Spitaels, F., Cnockaert, M., Praet, J., El Farricha, O., Swings, J. and P. Vandamme. 2015. *Enterococcus bulliens* sp. nov., a novel lactic acid bacterium isolated from camel milk. *Ant. Leeuwenh.* 108: 1257–1265.

Karaffová, V., Marcinková, E., Bobíková, K., Herich, R., Revajová, V., Stašová, D., Kavuľová, A., Levkutová, M., Levkut, M.Jr., Lauková, A., Ševčíková, Z. and M. Levkut, Sr. 2017. TLR4 and TLR21 expression, MIF, IFN-beta, MD-2, CD14 activation, and IgA production in chickens administered with EFAL41 strain challenged with *Campylobacter jejuni*. *Folia Microbiol.* 62: 89–97.

Karaffová, V., Molčanová, M., Herich, R., Revajová, V., Levkutová, M. and M. Levkut. 2015. The detection of relative mRNA expression of cytokine in chickens after *Enterococcus faecium* EF55 administration and *Salmonella enterica* Enteritidis infection. *Slov. Vet. Res.* 52: 173–183.

Karaffová, V., Tóthová, C., Szabóová, R., Revajová, V., Lauková, A., Ševčíková, Z., Herich, R., Levkut, M., Levkut, M., Faixová, Z. and O. Nagy. 2022. The effect of *Enterococcus faecium* AL41 on the acute phase proteins and selected mucosal immune molecules in broiler chickens. *Life* 12: 598. https://doi.org/10.3390/life12040598.

Khan, H., Flint, S. and P.L. Yu. 2010. Enterocins and preservation. *Int. J. Food Microbiol.* 141: 1–10.

Kmeť, V., Flint, H.J. and R.J. Wallace. 1993. Probiotics and manipulation of rumen development and function. *Arch. Anim. Nutr.* 44: 1–10.

Kreuzer, S., Janczyk, P., Assmus, J., Schmidt, M.F., Brockman, G.A. and K. Nockler. 2012. No beneficial effects evident for *Enterococcus faecium* NCIMB 10415 in weaned pigs infected with *Salmonella Enterica* serovar Typhimurium DT104. *Appl. Environ. Microbiol.* 78: 4816–4825.

Kubašová, I., Diep, D.B., Ovchinnikov, K.V., Lauková, A. and V. Strompfová. 2020. Bacteriocin production and distribution of bacteriocin-encoding genes in enterococci from dogs. *Int. J. Antimicrob. Agents.* 55: 105859. https://doi.org/10.1016/j.ijantimicag.2019.11.016.

Kubašová, I., Stempelová, L., Maďari, A., Bujňáková, D., Micenková, L. and V. Strompfová. 2022. Application of canine-derived *Enterococcus faecium* DSM 32820 in dogs with acute idiopathic diarrhoea. *Acta Vet. Beograd.* 72: 167–183.

Kubašová, I., Strompfová, V. and A. Lauková. 2019. Evaluation of enterococci for potential probiotic utilization. *Folia Microbiol.* 64: 177–187.

Lahtinen, S., Ouwehand, A.C., Salminen, S. and A. von Wright. 2012. *Lactic Acid Bacteria. Microbiological and Functional Aspects.* Fourth Ed. CRC Press, Taylor and Group, Boca Raton, London, New York, pp. 1–761.

Latorre-Moratalla, M.L., Bover-Cid, S., Talon, R., Aymerich, T., Garriga, M., Zanardi, E., Ianieri, A., Fraqueza, M.J., Elias, M., Drosinos, E.H., Lauková, A. and M.C. Vidal-Carou. 2010. Distribution of aminogenic activity among potential autochthonous starter cultures for dry fermented sausages. *J. Food Protect.* 73: 524–528.

Lauková, A., Chrastinová, Ľ., Kandričáková, A., Ščerbová, J., Plachá, I., Pogány Simonová, M., Čobanová, K., Formelová, Z., Ondruška, Ľ. and V. Strompfová. 2015b. Bacteriocin substance durancin-like ED26E/7 and their experimental use in broiler rabbits. *Maso* 5: 56–59.

Lauková, A., Chrastinová, Ľ., Micenková, L., Bino, E., Kubašová, I., Kandričáková, A., Gancarčíková, S., Plachá, I., Holodová, M., Grešáková, Ľ., Formelová, Z. and M. Pogány Simonová. 2022. Enterocin M in interaction in broiler rabbits with autochtonous, biofilm-forming *Enterococcus hirae* Kr8 strain. *Prob. Antimicrob. Prot.* 14: 845–853.

Lauková, A., Chrastinová, Ľ., Plachá, I., Szabóová, R., Kandričáková, A., Pogány Simonová, M., Formelová, Z., Ondruška, Ľ., Goldová, M., Chrenková, M. and V. Strompfová. 2017e. Enterocin 55 produced by non rabbit-derived strain *Enterococcus faecium* EF55 in relation with microbiota and selected parameters in broiler rabbits. *Int. J. Environ. Agric. Res. (IJOEAR).* 3: 45–52. ISSN[2454–1850].

Lauková, A., Chrastinová, Ľ., Pogány Simonová, M., Strompfová, V., Plachá, I., Čobanová, K., Formelová, Z., Chrenková, M. and Ľ. Ondruška. 2012. *Enterococcus faecium* AL41: Its Enterocin M and their beneficial use in rabbit husbandry. *Prob. Antimicro. Prot.* 4: 243–249.

Lauková, A. and S. Czikková. 1998. Inhibition effect of enterocin CCM 4231 in the rumen fluid environment. *Lett. Appl. Microbiol.* 26: 215–218.

Lauková, A., Guba, P., Nemcová, R. and M. Mareková. 2004b. Inhibition of *Salmonella enterica* serovar Duesseldorf by enterocin A in gnotobiotic Japanese quails. *Vet. Med. Czech.* 49: 47–51.

Lauková, A., Guba, P., Nemcová, R. and Z. Vasilková. 2003. Reduction of *Salmonella* in gnotobiotic Japanese quails caused by the enterocin A-producing EK13 strain of *Enterococcus faecium. Vet. Res. Commun.* 27: 275–280.

Lauková, A., Kandričáková, A., Buňková, L., Pleva, P. and J. Ščerbová. 2017b. Sensitivity to enterocins of biogenic amine-producing faecal enterococci from ostriches and pheasants. *Prob. Antimicrob. Prot.* 9: 483–491.

Lauková, A., Kandričáková, A. and J. Ščerbová. 2015a. Use of bacteriocin producing, probiotic strain *Enterococcus faecium* AL 41 to control intestinal microbiota in farm ostriches. *Lett. Appl. Microbiol.* 60: 531–535.

Lauková, A., Kandričáková, A., Ščerbová, J., Szaboóvá, R., Plachá, I., Čobanová, K., Pogány Simonová, M. and V. Strompfová. 2017d. *In vivo* model experiment using laying hens treated with *Enterococcus faecium* EM41 from ostrich faeces and its enterocin EM41. *Mac. Vet. Rev.* 40: 157–166.

Lauková, A., Mareková, M. and P. Javorský. 1993. Detection and antimicrobial spectrum of a bacteriocin-like substance produced by *Enterococcus faecium* CCM 4231. *Lett. Appl. Microbiol.* 16: 257–260.

Lauková, A., Micenková, L., Kubašová, I., Bino, E., Kandričáková, A., Plachá, I., Štrkolcová, G., Gálik, B., Kováčik, A., Halo, M. and M. Pogány Simonová. 2023. Microbiota, phagocytic activity, biochemical parameters and parasite control in horses with application of autochthonous, bacteriocin-produicng, probiotic strain *Enterococcus faecium* EF 412. *Prob. Antimicrob. Prot.* 15: 139–148.

Lauková, A., Pogány Simonová, M., Chrastinová, Ľ., Kandričáková, A., Ščerbová, J., Plachá, I., Čobanová, K., Formelová, Z., Ondruška, Ľ., Štrkolcová, G. and V. Strompfová. 2017c. Beneficial effect of bacteriocin-producing strain *Enterococcus durans* ED26E/7 in model experiment using broiler rabbits. *Czech J. Anim. Sci.* 62: 168–177.

Lauková, A., Simonová, M., Strompfová, V., Štyriak, I., Ouwehand, A.C. and M. Várady. 2008. Potential of enterococci isolated from horses. *Anaerobe* 14: 234–236.

Lauková, A., Strompfová, V. and A. Ouwehand. 2004a. Adhesion properties of Enterococci to intestinal mucus of different host. *Vet. Res. Commun.* 28: 647–655. https://doi.org/10.1023/B:VERC.0000045948.04027.a7

Lauková, A., Strompfová, V., Szabóová, R., Slottová, A., Tomáška, M., Kmeť, V. and M. Košta. 2016. Bioactive enterococci isolated from Slovak ewes lump cheese. *Sci. Agricult. Bohemica* 47: 187–193.

Lauková, A., Styková, E., Kubašová, I., Gancarčíková, S., Plachá, I., Mudroňová, D., Kandričáková, A., Miltko, R., Belzecki, G., Valocký, I. and V. Strompfová. 2018. Enterocin M and its beneficial effects in horses-pilot experiment. *Prob. Antimicrob. Prot.* https://doi.org/10.1007/S12602-018-9390-2

Lauková, A., Styková, E., Kubašová, I., Strompfová, V., Gancarčíková, S., Plachá, I., Miltko, R., Belzecki, G., Valocký, I. and M. Pogány Simonová. 2020. Entetrocin-M producing *Enterococcus faecium* CCM 8558 demonstrating probiotic properties in horses. *Prob. Antimicrob. Prot.* 12: 1555–1561. https://doi.org/10.100 7/s/12602-020-09655-6

Lauková, A., Szabóová, R., Pleva, P., Buňková, L. and Ľ. Chrastinová. 2017a. Decarboxylase-positive *Enterococcus faecium* strains isolated from rabbit meat and their sensitivity to enterocins. *Food Sci. Nutr.* 5: 31–37.

Lauková, A. and P. Turek. 2004. Experimental application of bacteriocins during meat products processing (in Slovak). In *Proceedings from the International Conference Agrogood Legislation under Conditions of Globalization, Slovakia.* Štrbské Pleso, November, pp. 164–168.

Lebreton, F., Rob, J., Willems, L. and M.S. Gilmore. 2014. *Enterococcus* diversity, origins in nature, and gut colonization. In Gilmore, M.S., Clewell, D.B., Ike, Y., et al. (eds). *Enterococci: From Commensals to Leading Causes of Drug Resistant Infection* (Internet). Massachusetts Eye and Ear Infirmary. https://www.ncbi.nlm.nih.gov/books/NBK190427/

Le Page, S., Cimmio, T., Togo, A., Million, M., Michelle, C., Khelaifia, S., Lagier, J.C., Raoult, D. and J.M. Rolain. 2016. Noncontiguous finished genome sequence and description of *Enterococcus massiliensis* sp. nov. *New Microbes New Infect.* 12: 90–95.

Levkut, M., Pistl, J., Lauková, A., Revajová, V., Herich, R., Ševčíková, Z., Strompfová, V., Szabóová, R. and T. Kokinčáková. 2009. Antimicrobial activity of *Enterococcus faecium* EF55 against *Salmonella enteritidis* in chicks. *Acta Vet. Hung.* 57: 13–24.

Levkut, M., Revajová, V., Karaffová, V., Lauková, A., Herich, R., Strompfová, V., Ševčíková, Z., Žitňan, R., Levkutová, M. and M. Levkut Sr. 2016. Evaluation of mucin and cytokines expression with intraepithelial lymphocytes determination in the caecum of broilers administered with *Enterococcus faecium* EF55 and challenged with *Salmonella* Enteritidis SE147. *J. Vet. Med. Anim. Health* 8: 2014–2222. ISSN 2141-2529.

Li, C.Y., Tian, F., Zhao, Y.D. and C.T. Gu. 2014. *Enterococcus xiangfangensis* sp. nov., isolated from Chinese pickle. *Int. J. Syst. Evol. Microbiol.* 64: 1012–1017.

Li, Y.Q. and C.T. Gu. 2019. *Enterococcus pingfangensis* sp. nov., *Enterococcus dongliensis* sp. nov., *Enterococcus hulanensis* sp. nov., *Enterococcus nangangensis* sp. nov. and *Enterococcus songbeiensis* sp. nov., isolated from Chinese traditional pickle juice. *Int. J. Syst. Evol. Microbiol.* 69: 3196–3206.

Line, J.E., Svetoch, E.A., Erustanov, B.V., Perelygin, V.V., Mitsevich, I.P., Levchuk, V.P., Svetoch, O.E., Seal, B.S., Siragusa, G.R. and N.J. Stern. 2008. Isolation and purification of enterocin E-760 with a broad antimicrobial activity against Gram-positive and Gram-negative bacteria. *Antimicrob. Agents Chemother.* 52: 1094–1100.

Lodemann, U., Amasheh, S., Radloff, J., Kern, M., Bethe, A., Wieler, L.H., Pieper, R., Zentek, J. and J.R. Aschenbach. 2017. Effects of ex vivo infection with ETEC on jejunal barrier properties and cytokine expression in probiotic-supplemented pigs. *Dig. Dis. Sci.* 62: 922–933.

Marciňáková, M., Simonová, M., Strompfová, V. and A. Lauková. 2006. Oral application of *Enterococcus faecium* strain EE3 in healthy dogs. *Folia Microbiol.* 51: 239–242.

McLaughlin, R.W., Shewmaker, P.L., Whitney, A.M., Humrighouse, B.W., Lauer, A.C., Loparev, V.N., Gulvik, C.A., Cochran, P.A. and S.E. Dowd. 2017. *Enterococcus crotali* sp. nov., isolated from faecal material of a timber rattlesnake. *Int. J. Syst. Evol. Microbiol.* 67: 1984–1989.

Meremäe, K., Elias, P., Tamme, T., Kramarenko, T., Lillenberg, M., Karus, A., Hänninen, M.L. and M. Roasto. 2010. The occurrence of *Campylobacter* spp. in Estonian broiler chicken production in 2002–2007. *Food Control.* 21: 272–275.

Murphy, D., et al. 2017. EMA and EFSA joint Scientific opinion on measures to reduce the need to use antimicrobial agents in animal husbandry in the EU and the resulting on food safety (RONAFA). *EFSA J.* 15IS: e04666.

Nes, I.F., Diep, D.B. and Y. Ike. 2014. Enterococcal bacteriocins and antimicrobial proteins that contribute to niche control. In Gilmore, M.S., Clewel, D.B., Ike, Y., Sankar, N. (eds). *Enterococci: From Commensals to Leading Causes of Drug Resistnt Infection.* Massachusetts Eze and Ear Infirmary.

Nigutová, K., Morovský, M., Pristaš, P., Teather, R.M., Holo, H. and P. Javorský. 2007. Production of enterolysin A by rumen *Enterococcus faecalis* strain and occurrence of enlA homologues among ruminal Gram-positive cocci. *J. Appl. Microbiol.* 102: 563–569.

Oren, A., and G.M. Garrity. 2020. Validation list no. 196. List of new names and new combinations previously effectively, but not validly. *Int. J. Syst. Evol. Microbiol.* 70: 5596–5601.

Ovchinnikov, K.V., Chi, H., Mehmeti, I., Holo, H., Nes, I.F. and D.B. Diep. 2016. Novel group of leaderless multipeptide bacteriocins from Gram-positive bacteria. *Appl. Environ. Microbiol.* 82: 5216–5224.

Ovchinnikov, K.V., Kristiansen, P.E., Straume, D., Jensen, M.S., Aleksandrzak-Piekarczyk, T., Nes, I. F. and D.B. Diep. 2017. The leaderless bacteriocin enterocin K1 is highly potent against *Enterococcus faecium*: A study on structure, target spectrum and receptor. *Front. Microbiol.* 8: 774.

Pieterse, R. and S.P. Todorov. 2010. Bacteriocins—exploring alternatives to antibiotics in mastitis treatment. *Braz. J. Microbiol.* 41(3).

Pogány Simonová, M., Chrastinová, Ľ. and A. Lauková. 2016. Dietary supplementation of a bacteriocinoigennic and probiotic strain of *Enterococcus faecium* CCM7420 and its effect on the mineral content and quality of *Muscullus longissimus dorsi* in rabbits. *Anim. Prod. Sci.* 56: 2140–2145.

Pogány Simonová, M., Chrastinová, Ľ. and A. Lauková. 2020. Effect of beneficial strain *Enterococcus faecium* EF9a isolated from Pannon White rabbits on growth performance and meat quality of rabbits. *Ital. J. Anim. Sci.* 19: 650–655. https://doi.org/10.1080/1828051X.2020.1781553.

Pogány Simonová, M., Chrastinová, Ľ. and A. Lauková. 2021. Effect of *Enterococcus faecium* AL41 (CCM8558) and its Enterocin M on the physicochemical properties and mineral content of rabbit meat. *Agriculture* 11: 1045. https://doi.org/10.3390/agriculture11111045.

Pogány Simonová, M., Chrastinová, Ľ., Ščerbová, J., Focková, V., Plachá, I., Formelová, Z., Chrenková, M. and A. Lauková. 2022b. Preventive potential of dipeptide Enterocin A/P on rabbit health and its effect on growth, microbiota and immune response. *Animals* 12: 1108. https://doi.org/10.3390/ani12091108.

Pogány Simonová, M., Lauková, A., Chrastinová, Ľ., Kandričáková, A., Ščerbová, J., Formelová, Z., Chrenková, M., Žitňan, R., Miltko, R. and G. Belzecki. 2022a. Effect of dite supplementation with *Enterococcus durans* ED26E/7 and its durancin ED 26E/7 on growth performance, caecal enzymatic activity, jejunal morphology and meat properties of broiler rabbits. *Ann. Anim. Sci.* 22: 221–235. https://doi.org/10.2478/aoas-2021-0016.

Pogány Simonová, M., Lauková, A., Chrastinová, Ľ., Strompfová, V., Faix, Š., Vasilková, Z., Ondruška, Ľ., Jurčík, R. and J. Rafay. 2009a. *Enterococcus faecium* CCM 7420, bacteriocin PPB CCM 7420 and their effect in the digestive tract of rabbits. *Czech J. Anim. Sci.* 54: 376–386.

Pogány Simonová, M., Lauková, A., Chrastinová, Ľ., Szabóová, R., Mojto, J., Strompfová, M. and J. Rafay. 2009b. Quality of rabbit meat after application of bacteriocinogenic and probiotic strain *Enterococcus faecium* CCM 4231 in rabbits. *Int. J. Probiotics Prebiotics* 4: 1–6.

Pogány Simonová, M., Lauková, A., Plachá, I., Čobanová, K., Strompfová, V., Szabóová, R. and L. Chrastinová. 2013. Can enterocins affect phagocytosis and glutathione-peroxidase in rabbits? *Centr. Europ. J. Biol.* 8: 730–734. https://doi.org/10.2478/s11535-013-0198-x.

Pogány Simonová, M., Lauková, A., Žitňan, R. and Ľ. Chrastinová. 2015. Effect of rabbit-origin enterocin-produicng strain *Enterococcus faecium* CCM7420 application on growth performance and gut morphometry in rabbits. *Czech J. Anim. Sci.* 60: 509–512.

Ren, X., Li, M. and D. Guo. 2016. *Enterococcus xinjiangensis* sp. nov., Isolated from Yogurt of Xinjiang, China. 2016. *Cur. Microbiol.* 73: 374–378.

Revajová, V., Benková, T., Karaffová, V., Levkut, M., Selecká, E., Dvorožňáková, E., Ševčíková, Z., Herich, R. and M. Levkut. 2022. Influence of immune parameters after *Enterococcus faecium* AL41 administration and *Salmonella* infection in chickens. *Life* 12: 201. https://doi.org/103390/life12020201

Ribeiro, S.C., Ross, R.P., Stanton, C. and G.C.G. Silva. 2017. Characterization and application of antilisterial enterocins on model fresh cheese. *J. Food. Prot.* 80: 1303–1316.

Schofs, L., Sparo, M.D., Guadalupe de yaniz, M., Lissarrague, S., Domínguez, M.P., Álvarez, L.I. and S.F. Sánchez Bruni. 2022. Antinematodic effect of *Enterococcus faecalis* CECT7121 using *Trichinella spiralis* as a model of nematode infection in mice. *Exp. Prasitol.* 241: 108358.

Schoster, A., Weese, J.S. and L. Guardabassi. 2014. Probiotic use in horses-what is the evidence for their clinical efficacy. *J. Vet. Intern. Med.* 28: 1640–1652.

Sistek, V., Maheux, A.F., Boissinot, M., Bernard, K.A., Cantin, P., Cleenwerck, I., De Vos, P. and M.G. Bergeron. 2012. *Enterococcus ureasiticus* sp. nov. and *Enterococcus quebecensis* sp. nov., isolated from water. *Int. J. Syst. Evol. Microbiol.* 62: 1314–1320.

Soltani, S., Hammami, R., Cotter, P.D., Rebuffat, S., Ben Said, L., Gaudreau, H., Bédard, F., Biron, E., Drider, D. and I. Fliss. 2021. Bacteriocins as a new generation of antimicrobials: Toxicity aspects and regulations. *FEMS Microbiol. Rev.* 45: 1–24.

Strompfová, V., Kubašová, I., Ščerbová, J., Maďari, A., Gancarčíková, S., Mudroňová, D., Miltko, R., Belzecki, G. and A. Lauková. 2019. Oral administration of bacteriocin-producing and non-producing strains of *Enterococcus faecium* in dogs. *Appl. Microbiol. Biotechnol.* 103: 4953–4965.

Strompfová, V., Lauková, A., Marciňáková, M. and Z. Vasilková. 2010. Testing of probiotic and bacteriocin-producing lactic acid bacteria towards *Eimeria* sp. *Pol. J. Vet. Sci.* 13: 389–391.

Strompfová, V., Marciňáková, M., Simonová, M., Gancarčíková, S., Jonecová, Z., Sciranková, Ľ., Koščová, J., Buleca, V., Čobanová, K. and A. Lauková. 2006. *Enterococcus faecium* EK13-an enterocin A-producing strain with probiotic character and its effect in piglets. *Anaerobe* 12: 242–248.

Švec, P. and C.M.A.P. Franz. 2014. The family Enterococcacae. In Holzapfel, W.H., Wood, B.J.B. (eds). *Lactic Acid Bacteria: Biodiversity and Taxonomy*. First Ed. John Wiley and Sons, Ltd.

Svetoch, E.A. and N.J. Stern. 2010. Bacteriocins to control *Campylobacter* sp. in poultry—A review. *Poult. Sci.* 89: 1763–1768.

Szabóová, R., Chrastinová, Ľ., Lauková, A., Havairová, M., Simonová, M., Strompfová, V., Faix, Š., Vasilková, Z., Chrenková, M., Plachá, I., Mojto, J. and J. Rafay. 2008. Bacteriocin-producing strain *Enterococcus faecium* CCM 4231 and its use in rabbits. *Int. J. Probiotics Prebiotics* 3: 77–82.

Takakura, T., Francis, R., Anani, H., Bilen, M., Raoult, D. and J. Yaacoub Bou Khalil. 2019. *Enterococcus mediteraneensis* sp. nov., a new bacterium isolated from stool of a 39-year-old Pygmy. *New Microbes New Infect.* 32: 100599.

Talon, R. 2005. *Tradisausage QLK1-CT-2002–02240 Final Report—Assessment and Improvement of Safety of Traditional Dry Sausages from Producers to Consumers*, pp. 1–192. https://cordis.europa.eu/project/iD/QLK1-CT-2002-02240

Techo, S., Shiwa, Y., Tanaka, N., Fujita, N., Miyashita, M., Shibata, C., Booncharaen, A. and S. Tanasupawat. 2019. *Enterococcus florum* sp. nov. isolated from a cotton flower (*Gossynum hirsutum*). *Int. J. Syst. Evol. Microbiol.* 69: 2506–2513. https://doi.org/10.1099/ijsem.0.003524

Tian, Z., Liu, X., Dai, R., Xiao, Y., Wang, X., Bi, D. and D. Shi. 2016. *Enterococcus faecium* HDRsEF1 protects the intestinal epithelium and attenuates ETEC-induced IL-8 secretion in enterocytes. *Mediat. Inflam*: 7474306.

Vargová, M., Dvorožňáková, E., Hurníková, Z., Revajová, V. and A. Lauková. 2022. *Antiparasitic Effect of Enterocin-Producing Bacteria against Trichinella Spiralis Infection.* pre Veda. ISBN 978-80-972360-6-9.

Wu, Y., Pang, X., Wu, Y., Liu, X. and X. Zhang. 2022. Enterocins: Classification, synthesis, antibacterial mechanisms and food applications. *Molecules* 27: 2258.

Zhong, Z., Zhang, W., Song, Y., Liu, W., Xu, H., Xi, X., Menghe, B., Zhang, H. and Z. Sun. 2017. Comparative genomic analysis of the genus *Enterococcus. Microbiol. Res.* 196: 95–105.

# Introduction to the Genera *Pediococcus, Leuconostoc, Weissella, Periweissella* and *Carnobacterium*

# 6

Elina Säde, Per Johansson, and Johanna Björkroth

## 6.1 INTRODUCTION

Although *Enterococcus*, *Lactococcus*, *Lactobacillus*, and *Streptococcus* have been in the core of lactic acid bacteria (LAB) studies, they do not cover the broad taxonomic and functional diversity known to date. This chapter deals with five LAB genera that are less well known because of one or several reasons: (i) they are relatively young genera among LAB, (ii) some species are rarely isolated because they require specific culture conditions, (iii) they are difficult to identify or distinguish from other LAB genera phenotypically, and/or (iv) they reside in underexplored habitats. However, these five genera have relevance to the food and health area. The description of the first genus, *Pediococcus*, is followed by the genus *Leuconostoc*. We continue with *Weissella* and *Periweissella* and close the chapter with a description of *Carnobacterium*, the type genus of the *Carnobacteriaceae*.

## 6.2 GENUS *PEDIOCOCCUS*

### 6.2.1 General Characteristics

Pediococci are LAB occurring in plants and vegetation as well as in many fermented foods and beverages. Pediococci occur as round cells, uniform in size, that often divide in two planes to form tetrads. In contrast to most other coccus-shaped LAB, pediococci usually do not form chains of cells but instead occur in tetrads, in pairs or singly. Some strains may exhibit pseudocatalase activity and decompose peroxides on media with low carbohydrate content by producing non-heme-containing catalase (Dicks and Endo 2014).

DOI: 10.1201/9781003352075-7

*Pediococcus* spp. grow at 30 °C on MRS medium (de Man, Rogosa and Sharpe 1960). Most species grow well at a temperature range of 25 °C –37 °C but some species, namely *Pediococcus acidilactici*, *Pediococcus ethanolidurans*, *Pediococcus pentosaceus*, *Pediococcus siamensis*, and *Pediococcus stilesii*, contain strains that grow at 45 °C or even up to 50 °C. The optimum pH for growth is between 6.0 and 6.5, and most species are reported to initiate growth at pH 4.5 or even below. For most species, the upper pH limit for growth is between 7.5 and 8.0.

Pediococci are facultatively anaerobic or microaerophilic homofermenters, producing mainly lactic acid from glucose. Pediococci produce DL-lactic acid, except for strains of *Pediococcus claussenii*, which convert glucose to L(+)-lactic acid. Fructose and cellobiose are fermented by all species. Most species are ferment also galactose and maltose. Sucrose is also fermented by all species except *Pediococcus inopinatus*, *P. parvulus*, *Pediococcus pentosaceus*, and *P. claussenii*. In contrast, rhamnose, melibiose, melezitose, raffinose, inulin, and α-methyl glucoside-D are not fermented by most pediococci. Nitrate is not reduced, and indole is not formed from tryptophan. Their peptidoglycan type is Lys-D-Asp. Additional physiological and biochemical features of pediococci have been reviewed by Dicks and Endo (2014). During the past decade, the complete genome sequence of several strains of *Pediococcus* has been released, along with studies designed to obtain insights into their metabolic features. These studies have aimed to reveal the molecular background of the metabolic and physiological capabilities of different strains such as the mechanisms allowing strains to adapt to harsh environments (Snauwaert et al. 2015), as well as to understand the phylogeny and evolution of *Pediococcus* spp. (Salvetti et al. 2018).

Like many other LAB, pediococci produce bacteriocins (i.e., pediocins); see Chapter 11. Production of pediocins has been extensively investigated, especially for *P. pentosaceus* and *Pediococcus acidilactici*. A majority of bacteriocins produced by pediococci belong to the bacteriocin group Class IIa, pediocin-like bacteriocins, that have anti-listerial properties and contain a characteristic cationic motif termed "pediocin box" in the N-terminal half. Producer strains of pediocins have mainly been found in the phylogenetically and biochemically related species *Pediococcus acidilactici* and *P. pentosaceus* and, more recently, also in *Pediococcus damnosus* (Papagianni and Anastasiadou 2009).

## 6.2.2 Phylogeny and Taxonomy

The genus *Pediococcus* belongs to the *Lactobacillacea* family within the order *Lactobacillales*. According to the recent genome-based phylogenetic analysis, the genus *Pediococcus* is a sister to the *Lactobacillus buchneri* group and to the members of the former family *Leuconostocaceae* (Salvetti et al. 2018), i.e., *Convivina*, *Fructobacillus*, *Leuconostoc*, *Oenococcus*, *Periweissella* and *Weissella*.

Currently, the genus contains 12 species. For many decades, five species were considered to represent the taxonomic core of the genus *Pediococcus*: *P. damnosus* (the type species of the genus), *P. acidilactici*, *P. pentosaceus*, *Pediococcus parvulus*, and *Pediococcus inopinatus*. Triggered by the use of sequence-based screening approaches for new biodiversity in fermented foods, feeds, and beverages, seven new *Pediococcus* species have been described in the past 20 years: *P. claussenii* (Dobson et al. 2002), *Pediococcus cellicola* (Zhang et al. 2005), *P. stilesii* (Franz et al. 2006), *P. ethanolidurans* (Liu et al. 2006), *P. siamensis* (Tanasupawat et al. 2007), *Pediococcus argentinicus* (De Bruyne et al. 2008a), and *Pediococcus lolii* (Doi et al. 2009). It is noteworthy that many of these new species were described based on taxonomic characterization of a few strains, which limits our current insights into the functional characteristics and ecological role of these organisms.

## 6.2.3 Habitats

The habitats of pediococci are similar to those of other LAB such as *Lactobacillus*, *Leuconostoc*, and *Weissella*, that is, plants and fermented plant material, food and feed, and in the intestinal tract of animals and humans. Of all *Pediococcus* species described to date, *P. acidilactici* and *P. pentosaceus* are the most intensively studied for their ecological significance and biotechnological potential (Dicks and Endo 2014). Both species have been isolated from a large variety of plant materials such as vegetables, fruits, and cereals. Typically, these species occur in relatively small numbers on raw plant materials but often grow to prominent levels during spontaneous fermentation of silage, sauerkraut, beans, cucumbers, olives, and cereals upon which they often establish a stable community together with lactobacilli, leuconostocs, and other LAB. Many types of traditional African foods using fermented cereals such as sorghum as well as various Asian alcoholic beverages are known to contain pediococci (Dicks and Endo 2014). In addition to materials from plant origin, pediococci are also associated with protein-rich animal-derived foods such as raw and fermented sausages, fresh and marinated fish, and cheese.

In addition to their applications in the cheese, wine, and feed industries, pediococci are most commonly exploited as starter cultures in commercial fermentation of meat. More specifically, selected strains of *P. pentosaceus* and *P. acidilactici* are used to produce dry sausages owing to their ability to control the development of undesired and pathogenic bacteria. In contrast to the protective role played by several *Pediococcus* species in food fermentations, other pediococci display undesired properties as spoilers. Owing to their typical resistance to hops, *P. damnosus*, *P. claussenii*, and *P. inopinatus* have been frequently isolated from beer or detected in the brewery environment (Dobson et al. 2002; Sakamoto and Konings 2003; Iijima et al. 2007). Especially *P. damnosus* has been recognized as an important beer spoiler, causing turbidity, acidic off-tastes, and adverse flavors due to diacetyl formation. It has been shown that some strains of *P. damnosus*, commonly referred to as ropy strains, can induce viscosity in beer due to the production of exopolysaccharides (Walling et al. 2005).

Many pediococci are considered commensals of humans and animals, including birds, fish, shrimps, and piglets (Dicks and Endo 2014). Especially in poultry and shrimp industries, strains of *P. acidilactici* have been used as probiotics to promote growth performance and strengthen host defense mechanisms (Castex et al. 2010). *Pediococcus* strains have been detected in saliva and fecal samples of healthy humans but are also recognized as opportunistic pathogens. Representatives of *P. pentosaceus* and *P. acidilactici* have been isolated from a variety of clinical specimens especially those of immune-compromised patients. Because pediococci are vancomycin resistant, they may be confused with enterococci in routine laboratories, which may have resulted in a significant underreporting of these organisms in clinical settings.

## 6.2.4 Identification, Typing, and Detection

At the genus level, phenotypic differentiation of pediococci from the (historically) affiliated genera *Tetragenococcus* and *Aerococcus* is based on the ability of pediococci to grow under acidic conditions (~pH 5) but not under alkaline conditions (~pH 9) and on the production of DL-lactic acid (except for *P. claussenii*) instead of D(+)-lactic acid. The latter characteristic can also be useful to separate pediococci from *Lactobacillus dextrinicus* (a former but atypical member of the genus *Pediococcus*) in conjunction with the ability of this species to produce gas from gluconate (Dicks and Endo 2014).

Although identification of *Pediococcus* species has relied on growth characteristics and sugar fermentation patterns (Dicks and Endo 2014), during the past two decades the molecular methods have taken the lead in the diagnostics and characterization. As with most other bacterial taxa, phenotypic approaches have largely been abandoned in favor of molecular tools for the reliable and accurate identification of pediococci to the species or strain level. Over the years, conventional phenotyping schemes have lost their discriminative value and accuracy due to the steady increase in new LAB species descriptions and a considerable degree of intraspecies variation. For species identification and genotyping, methods such as 16S and 23S gene sequencing, DNA probes, ribotyping, rRNA gene intergenic transcribed spacer (ITS)-PCR analysis, randomly amplified polymorphic DNA (RAPD) polymerase chain reaction (PCR), and pulsed-field gel electrophoresis (PFGE) have been used. As an alternative to 16S rRNA gene sequencing analysis, De Bruyne and coworkers (2008a) proposed a multilocus sequencing scheme using genes encoding the α-subunits of phenylalanyl-tRNA synthase (*pheS*), RNA polymerase (*rpoA*), and ATP synthase (*atpA*) to identify *Pediococcus* spp. Multilocus sequence analysis (MLSA) approaches have also been developed and used to study genomic polymorphisms *Pediococcus* strains (Mora et al. 2000; Calmin et al. 2008).

Next to the identification and typing of *Pediococcus* isolates, DNA-based methods have been used for culture-independent detection of pediococci. As such, a multiplex PCR assay based on the use of 23S rRNA gene sequences was developed for rapid identification of most *Pediococcus* species (Pfannebecker and Fröhlich 2008). *Pediococcus* spp. have also been detected in several studies of employing 16S rRNA gene targeting denaturing gradient gel electrophoresis (DGGE) analysis. In addition, protein-encoding housekeeping genes, such as *rpoB*, have been used to detect and identify pediococci and other LAB from food, feed, and beverages. Furthermore, real-time or quantitative PCR methods have been developed to detect or enumerate specific target genes, such as the *dsp* gene in *P. damnosus* strains that causes ropy spoilage of wine (Delaherche et al. 2004).

# 6.3 GENUS *LEUCONOSTOC*

## 6.3.1 General Characteristics

Leuconostocs are Gram-positive, nonmotile, and asporogenous bacteria, of which cells are ellipsoidal to spherical, often elongated, and usually occur in pairs or chains (Garvie 1986). When grown on a solid medium, cells are elongated and can be mistaken for rods. True cellular capsules are not formed, but many leuconostocs produce extracellular dextran that forms an electron-dense coat on the cell surface. They are facultative anaerobic and catalase negative. Before the genome sequencing project of *Leuconostoc gasicomitatum* LMG 18811[T], it was thought that no functional cytochromes were present. The sequences of fully assembled *Leuconostoc* genomes usually possess genes encoding cytochrome *bd* terminal oxidase and for synthesizing menaquinone. However, different from *L. mesenteroides* DSM 20343[T], *L. gasicomitatum* LMG 18811[T] has a functional electron transport requiring only externally supplied heme for respiration (Johansson et al. 2011). Leuconostocs cannot hydrolyze arginine and do not reduce nitrate. They are nonproteolytic and nonhemolytic. Indole is not formed. Although growth may occur at pH 4.5, leuconostocs prefer an initial medium pH of 6.5. The optimal growth temperature is between 20 °C and 30 °C. The psychrotrophic species *L. carnosum*, *L. gasicomitatum*, *L. gelidum* (both subspecies, i.e., subsp. *gelidum* and *aenigmaticum*), and *L. inhae* grow at refrigerated temperatures; growth of some strains has been detected even at 1 °C.

All leuconostocs need rich media supplemented with growth factors and amino acids. Growth on agar media such as the de Man–Rogosa–Sharpe (MRS) medium is poor without creating an anaerobic atmosphere. Colonies develop usually after 3–5 days, and they are smooth, round, grayish white, and less than 1 mm in diameter. Glucose is fermented to equal amounts of D(−)-lactic acid, $CO_2$, and ethanol or acetate. Major metabolic routes can be mined from the approximately 600 *Leuconostoc* genomes, of which about 60 are completely assembled (www.ncbi.nlm.nih.gov/genome/browse/#!/prokaryotes/Leuconostoc). All of them contain a wide set of genes involved in uptake of sugars, citrate, and amino acids. The genomes include the genes for the phosphoketolase pathway and three alternative pathways for pyruvate utilization by lactate dehydrogenase, pyruvate dehydrogenase, and α-acetolactate synthase.

## 6.3.2 Phylogeny and Taxonomy

The genus *Leuconostoc* belongs to the *Lactobacillaceae* family (https://lpsn.dsmz.de/family/leuconostocaceae) within the order *Lactobacillales*, together with the genera *Convivina, Fructobacillus, Oenococcus, Periweissella*, and *Weissella*. Phylogenetic analyses of the 16S rRNA gene led to the subdivision of *Leuconostoc* into three distinct lineages: the genus *Leuconostoc sensu stricto*, the *Leuconostoc paramesenteroides* group, and *Leuconostoc oenos* (Martinez-Murcia and Collins 1990; Martinez-Murcia et al. 1993). Furthermore, polyphasic taxonomy studies have been leading to several taxonomic revisions within the group *Leuconostoc*. A new genus, *Weissella* (Collins et al. 1993), was described to accommodate members of the so-called *L. paramesenteroides* group (including *L. paramesenteroides* and some atypical, heterofermentative lactobacilli). In addition, *L. oenos* has been reclassified as *Oenococcus oeni* (Dicks et al. 1995). More recently, some atypical leuconostocs of plant origin, including *Leuconostoc durionis*, *Leuconostoc ficulneum*, *Leuconostoc fructosum*, and *Leuconostoc pseudoficulneum*, were assigned to the new genus *Fructobacillus* (Endo and Okada 2008), and the genus *Periweissella* (Bello et al. 2022) was described, with six species considered *Weissella* before. Currently, the genus *Leuconostoc sensu stricto* includes 17 validly published species names (Table 6.1), with *L. mesenteroides* as the type species.

With the exception of *L. fallax*, 16S rRNA gene sequence similarities are high among the type strains of *Leuconostoc* spp., varying from 97.3% to 99.5% (Björkroth and Holzapfel 2006). In addition to the 16S rRNA gene, *atpA*, *dnaK*, *pheS*, *recN*, and *rpoA* loci have been analyzed. The phylogenetic trees of *pheS*, *rpoA*, and *atpA* loci proved to offer discriminatory power for differentiation of species within the genus

**TABLE 6.1**   Leuconostoc Species with Validly Published Names

| SPECIES | FIRST SOURCE OF IDENTIFICATION | REFERENCES |
|---|---|---|
| *L. carnosum* | Vacuum-packaged, cold-stored meat | Shaw and Harding (1989) |
| *L. citreum* | Honey-dew of rye ear | Farrow et al. (1989) |
| *L. falkenbergense* | Fermented string beans and traditional yogurt | Wu and Gu (2021a) |
| *L. fallax* | Sauerkraut | Martinez-Murcia and Collins (1991) |
| *L. gasicomitatum* | Modified atmosphere packaged marinated broiler | Björkroth et al. (2000) Rahkila et al. (2014) comb. nov. Wu and Gu (2021b) Johansson et al. (2022) |
| *L. gelidum* subsp. *gelidum* | Vacuum-packaged, cold-stored meat | Shaw and Harding (1989) Rahkila et al. (2014) comb. nov. |
| *L. gelidum* subsp. *aenigmaticum* | Modified atmosphere packaged pork | Rahkila et al. (2014) |
| *L. holzapfelii* | Ethiopian coffee fermentation | De Bruyne et al. (2007) |
| *L. inhae* | Kimchi | Kim et al. (2003) |
| *L. kimchii* | Kimchi | Kim et al. (2000) |
| *L. lactis*[a] | Dairy product | Garvie (1960) |
| *L. litchii* | Fruit of *Litchi chinensis* | Chen et al. (2020) |
| *L. mesenteroides* subsp. *mesenteroides* | Fermenting olives | Tsenkovskii (1878) Garvie (1983) comb. nov. |
| *L. mesenteriudes* subsp. *cremoris* | Ferment | Garvie (1983) comb. nov. |
| *L. mesenteroides* subsp. *dextranicum* | Not known | Beijernick (1912) Garvie (1983) comb. nov. |
| *L. mesenteroides* subsp. *jonggajibkimchii* | Kimchi product | Jeon et al. (2017) |
| *Leuconostoc miyukkimchii* | Regional kimchi made of brown algae | Lee et al. (2012) |
| *L. palmae* | Palm wine | Ehrmann et al. (2009) |
| *L. pseudomesenteroides* | Cane juice | Farrow et al. (1989) |
| *L. rapi* | Rutabaga | Lyhs et al. (2015) |
| *L. suionicum* | Not known | Gu et al. (2012) Jeon et al. (2017), sp. nov. |

[a] *L. argentinum* has been reclassified as a later synonym of *Leuconostoc lactis* (Vancanneyt et al. 2006).

*Leuconostoc* and were roughly in agreement with 16S rRNA gene-based phylogeny (De Bruyne et al. 2007; Ehrmann et al. 2009). Comparative sequencing of the additional phylogenetic markers *dnaK* and *recA* confirmed the 16S rRNA gene tree topology in the study describing *L. palmae* (Ehrmann et al. 2009). Arahal et al. (2008) tested the discriminatory power of the *recN* locus and concluded that, used either alone or in combination with 16S rRNA encoding gene sequences, *recN* can serve as a phylogenetic marker as well as a tool for species identification. Congruence of evolutionary analyses inside the *Leuconostoc–Oenococcus–Weissella* clade has been assessed by comparative phylogenetic analyses of 16S rRNA, *dnaA*, *gyrB*, *rpoC*, and *dnaK* housekeeping genes (Chelo et al. 2007). Phylogenies obtained with the different genes were in overall good agreement, and a well-supported, almost fully resolved phylogenetic tree was obtained when the combined sequence data were analyzed in a Bayesian approach. Genome sequencing has also resulted in taxonomy changes within leuconostocs. Recently, Johansson et al. (2022) conducted a pangenomic study of the former *L. gelidum* subspecies *aenigmaticum*, *gasicomitatum*, and *gelidum*, that were reclassified by Rahkila et al. (2014), since the genomic taxonomy study of Wu and Gu (2021b) suggested that the species status of *L. gasicomitatum* should be restored. This finding was supported, and only two *L. gelidum* subspecies exist currently. Contrary to the situation of *L. mesenteroides* subspecies, *L. gelidum* subsp. *aenigmaticum* and *gelidum* are phylogenetically distinct (Johansson et al. 2022) and can be distinguished based on either genomic or *pheS*, *rpoA*, and *atpA* sequences.

## 6.3.3 Habitats

Leuconostocs are associated with plants and decaying plant material (Table 6.1). They have been detected in green vegetation and roots (Mundt 1967; Hemme and Foucaud-Scheunemann 2004) and in various fermented vegetable products, such as cucumber, kimchi, cabbage, and olives (Kim et al. 2000, 2003; Mäki 2004; Kim and Chun 2005; Lee et al. 2012). In addition to plant-originated material, leuconostocs are frequent in foods of animal origin, including raw milk and dairy products, meat, poultry, and fish (Kim and Chun 2005; Björkroth and Holzapfel 2006; Vihavainen and Björkroth 2009). However, healthy warm-blooded animals, including humans, are rarely reported to carry *Leuconostoc* in the microbiota of their gut or mucous membranes. It is noteworthy that leuconostocs have been recovered from the intestines of fish (Williams and Collins 1990).

*L. carnosum*, *L. gasicomitatum*, and *L. gelidum* subsp. *gelidum* have often been associated with food spoilage (please see Chapter 19 for more thorough information). Modified atmosphere packaged meat and vegetable-based foods have been prone to *Leuconostoc* spoilage manifested as bulging of the packages, off-odors and smells, and color changes (Susiluoto et al. 2003; Björkroth et al. 1998, 2000; Vihavainen and Björkroth 2007, 2009; Pothakos et al. 2014a, 2014b). Leuconostocs contaminate the foods during manufacture and processing (Hultman et al. 2015; Pothakos et al. 2015a, 2015b), and since the packaging conditions and cold storage do not restrict their growth, contamination may lead to growth and difficult spoilage problems. Processing environment and raw materials are considered the sources of contamination of these bacteria.

Although leuconostocs are not a risk for healthy individuals and are "generally regarded as safe" organisms (Schillinger et al. 2006), some *Leuconostoc* species have been associated with human infections. However, most of the patients involved in these infections had received vancomycin therapy, had an underlying disease, or were premature babies. It is noteworthy that all leuconostocs are intrinsically resistant to vancomycin and other glycopeptide antibiotics (Buu-Hoï et al. 1985; Huygens 1993; Orberg and Sandine 1984; Elisha and Courvalin 1995).

## 6.3.4 Identification, Typing, and Detection

Leuconostocs grow well on MRS and other media designed for LAB (de Man et al. 1960), but there are no selective media for their detection. In some cases, the intrinsic resistance to vancomycin (Johansson et al.

2022) may serve as a method to distinguish *Leuconostoc* from other LAB genera. Anaerobic atmosphere is recommended for obtaining good growth on solid media. An incubation temperature of 25 °C is recommended if the species to be cultivated are not known. The psychrotrophic species *L. carnosum*, *L. gasicomitatum*, *L. gelidum* (both subspecies *gelidum* and *aenigmaticum*), and *L. inhae* do not grow at 37 °C, and some strains may not grow at 28 °C. Identification of leuconostocs to the species level is challenging with biochemical tests. Carbohydrate fermentation profiles vary considerably among *Leuconostoc* species, and yet the profiles are often strain dependent. Therefore, commercial carbohydrate fermentation reaction-based series do not provide reliable identification results (Kulwichit et al. 2007). At the genus level, leuconostocs can be distinguished from homofermentative streptococci and enterococci based on their heterofermentative glucose metabolism using, for example, a tomato-containing medium (Gibson and Abdel-Malek 1945). The other heterofermentative LAB, such as *Lactobacillus*, most *Weissella* spp., and *Periweissella* are rod shaped. The differentiation between leuconostocs and coccoid *Weissella* provides the main challenge since both genera are heterofermentative. This has led to the use of molecular approaches. A genus-specific PCR method (Schillinger et al. 2008) was developed for differentiation between strains of *Leuconostoc* and *Weissella* and a qPCR method was developed for differentiation of *Carnobacterium*, *Lactobacillus*, *Lactococcus* and *Leuconostoc* from inoculated white pudding (Cauchie et al. 2017). In addition, *Leuconostoc* is very well represented in the BRUKER MALDI-Tof database serving as useful tool for the identification of these bacteria.

In addition, numerical analysis of macromolecule patterns (Elliott and Facklam 1993; Björkroth et al. 1998) and gene/genomic sequencing (De Bruyne et al. 2007; Johansson et al. 2022; Nour 1998; Wu and Gu 2021b) have been applied to identify leuconostocs.

A range of DNA-based methods has been applied to characterize leuconostocs at the strain level. Among others, ribotyping based on 16 + 23S rRNA gene restriction fragment length polymorphism (RFLP), PFGE (Björkroth et al. 1998; Vihavainen and Björkroth 2009), and various PCR-based methods such as RAPD (Nieto-Arribas et al. 2010) have been used. Typing methods and 16S rRNA gene amplicon sequencing have been applied to study *Leuconostoc* contamination at food manufacturing facilities or for the identification of specific spoilage organisms (Björkroth 1997; Björkroth and Korkeala 1997; Hultman et al. 2015; Pothakos et al. 2015a). Genome diversity in the genera *Fructobacillus*, *Leuconostoc*, and *Weissella* has also been determined by physical and PFGE mapping (Chelo et al. 2010).

Today, methods based on genomic sequence such as determination of average nucleotide identity (ANI) and digital DNA–DNA hybridization are taking over. This method was used by Jeon et al. (2017) while describing *L. jonggajibkimchii* and reclassifying *L. suionicum*. In addition, the reclassification of *L. gelidum* subspecies (Wu and Gu 2021b; Johansson et al. 2022) was based on genomic taxonomy analyses. A total of 182 genomes of the genus *Leuconostoc* were used for a genomic study by Kumar et al. (2022). They proposed many reclassifications, among them reclassification of *L. gasicomitatum* as *L. inhae*. However, this is apparently due to the poor quality of the genome sequence of the type strain of *L. inhae* they used in the study. It is contaminated with *L. gasicomitatum* DNA, and this has led to misunderstandings. Unfortunately, despite requests to remove this sequence, it still is in the database. However, a good-quality *L. inhae* sequence is available (GCF_019656015) and should be used instead.

# 6.4 GENUS *WEISSELLA*

## 6.4.1 General Characteristics

Weissellas are Gram-positive, nonmotile, and asporogenous short rods with rounded tapered ends, or ovoid (Collins et al. 1993; Björkroth et al. 2009). They occur in pairs or in short chains, and there is tendency toward pleomorphism in some of the species. They have catalase negative phenotype but carry a gene encoding heme-dependent catalase, facultatively anaerobic chemo-organotrophs, and were originally considered not to contain cytochromes. However, like some *Leuconostoc*, *Weissella* species have genes encoding cytochrome *bd*.

Weissellas ferment glucose heterofermentatively. Carbohydrates are fermented via the hexose-monophosphate and phosphoketolase pathways. End products of glucose fermentation are $CO_2$, ethanol, and/or acetate. Depending on the species, the configuration of the lactic acid produced is either DL- or D (–). Weissellas have complex nutritional requirements as amino acids, peptides, fermentable carbohydrates, fatty acids, nucleic acids, and vitamins are generally required for growth. Biotin, nicotine, thiamine, and panthotenic acid or its derivatives are required. Arginine is not hydrolyzed by all species. Growth occurs at 15 °C; some species grow at 42 °C –45 °C. Weissellas have specific peptidoglycan structure based on lysine as a diamino acid, and with the exception of *Weissella kandleri* (Holzapfel and Van Wyk 1982), all contain alanine, or alanine and serine, in the interpeptide bridge. The interpeptide bridge of *W. kandleri* (Lys-L-Ala-Gly-L-Ala$_2$) contains glycine (Holzapfel and Van Wyk 1982). *Weissella* species are usually nonmotile (Collins et al. 1993). A motile species with peritrichous flagella, *Weissella beninensis*, was described (Padonou et al. 2010), but currently it is a member of the recent genus *Periweissella*. *Weissella* species are known to produce various exopolysaccharides (EPSs). Production of dextran, the most well-known EPS formed by heterofermentative LAB, has been recorded for *W. cibaria*, *W. confusa*, *W. kandleri*, and *W. koreensis* (Padonou et al. 2010; Björkroth et al. 2009).

When this chapter was written, there were about 360 genomic *Weissella* sequences available (www.ncbi.nlm.nih.gov/genome/browse/#!/prokaryotes/weissella). Lynch et al. (2015) conducted a comparative genomics study of *W. cibaria* using 10 genomic sequences (4 *W. cibaria*, 1 *W. ceti*, 1 *W. confusa*, 1 *W. halotolerans*, 2 *W. koreensis*, and 1 *W. paramesenteroides*). The genomes had sizes varying from 1.3 to 2.4 Mb. DNA G+C contents ranged from 35 to 45 mol%. The genus pan-proteome was found to comprise 4712 proteins. Analysis of the 4 *W. cibaria* genomes indicated that the core-proteome, consisting of 729 proteins, constituted 69 % of the species pan-proteome. This large core set may explain the divergent niches in which this species has been found. In *W. cibaria*, in addition to a number of phosphotransferase systems conferring the ability to assimilate plant-associated polysaccharides, an extensive proteolytic system was identified.

## 6.4.2 Phylogeny and Taxonomy

The genus *Weissella* was proposed by Collins et al. (1993), and its first members included species previously allocated in the genera *Leuconostoc* or *Lactobacillus*. The species *Leuconostoc paramesenteroides* (Garvie 1967), *Lactobacillus viridescens* (Niven and Evans 1957; Kandler and Abo-Elnaga 1966), *Lactobacillus confusus* (Holzapfel and Van Wyk 1982), *Lactobacillus kandleri* (Holzapfel and Van Wyk 1982), *Lactobacillus minor* (Kandler et al. 1983), and *Lactobacillus halotolerans* (Kandler et al. 1983) kept their specific epithets and were reclassified as *Weissella paramesenteroides*, *Weissella viridescens* (i.e., the type species), *Weissella confusa*, *W. kandleri*, *Weissella minor*, and *Weissella halotolerans*, respectively. These species were followed by inclusion of *Weissella hellenica* (Collins et al. 1993), *Weissella thailandensis* (Tanasupawat et al. 2000), *Weissella cibaria* (Björkroth et al. 2002), *Weissella soli* (Magnusson et al. 2002), and *Weissella koreensis* (Lee et al. 2002). In addition, *Weissella kimchii* was proposed by Choi et al. (2002) but was later recognized as a heterotypic synonym of *Weissella cibaria* (Ennahar and Cai 2004). The most recent species, *W. bombi* (Praet et al. 2015), *W. ceti* (Vela et al. 2011), *W. diastrammennae* (Oh et al. 2013), *W. jogaejeotgali* (Lee et al. 2015), *W. oryzae* (Tohno et al. 2013), and *W. uvarum* (Nisitou et al. 2014), were described as new species of the genus *Weissella* from bumblebees, beaked whales, gut of a camel cricket, rice, and grapes, respectively. Taxonomy, ecology, and biotechnological potential of *Weissella* has been reviewed by Fusco et al. (2015). A set of former *Weissella* i.e., *W. ghanensis* (De Bruyne et al. 2008b), *W. fabaria* (De Bruyne et al. 2010), and *W. fabalis* (Snauwaert et al. 2013b) described during studies dealing with cocoa fermentation and *W. beninensis* (Padonou et al.

2010), first detected from fermented cassava, have recently been separated from *Weissella* as members of *Periweissella* (Bello et al. 2022).

Weissellas share 93.3%–99.6% 16S rRNA gene sequence similarity (Björkroth et al. 2009; Padonou et al. 2010; Oh et al. 2013; Tohno et al. 2013; Nisitou et al. 2014; Praet et al. 2015; Lee et al. 2015). Four main phylogenetic branches exist based on 16S rRNA gene analyses. *W. bombi*, *W. hellenica*, *W. jogaejeotgali*, *W. thailandensis*, and *W. paramesenteroides* are positioned on the same branch, as are *W. cibaria* and *W. confusa*. Another branch is formed by *W. ceti.*, *W. halotolerans*, *W. minor*, *W. uvarum*, and *W. viridescens*. The third branch contains *W. kandleri*, *W. koreensis*, *W. oryzae*, and *W. soli*.

In addition to the 16S rRNA gene phylogeny, analysis with *pheS* (De Bruyne et al. 2010; Snauwaert et al. 2013b; Praet et al. 2015) and *recN* (Arahal et al. 2008) loci has been performed. Congruence of evolutionary relationships inside the *Leuconostoc–Oenococcus–Weissella* clade has been assessed by phylogenetic analyses of 16S rRNA, *dnaA*, *gyrB*, *rpoC*, and *dnaK* (Chelo et al. 2007) housekeeping genes. Phylogenies obtained with the different genes were in overall good agreement, and a well-supported, almost fully resolved phylogenetic tree was obtained when the combined data were analyzed in a Bayesian approach.

## 6.4.3 Habitats

The habitats of *Weissella* species are variable and most commonly involve fermented foods, but the sources of isolation suggest an environmental (e.g., soil, vegetation) origin. The species *W. viridescens*, *W. halotolerans*, and *W. hellenica* have been associated with meat and meat products. More specifically, *W. viridescens* may cause spoilage of cured meat due to green discoloration (Niven and Evans 1957), and it also is a prevailing spoiler of the Spanish blood sausage Morcilla de Burgos (Koort et al. 2006; Diez et al. 2009; Santos et al. 2005). This species is considered moderately heat resistant (Niven et al. 1954), which is not a common property for LAB. Members of *W. cibaria*, *W. confusa*, and *W. koreensis* have been detected in fermented foods of vegetable origin (Björkroth et al. 2002; Lee et al. 2002), whereas *W. confusa* has been associated with Greek salami (Samelis et al. 1994), Mexican pozole (Ampe et al. 1999), and Malaysian chili bo (Leisner et al. 1999). *W. cibaria* and *W. confusa* have also been associated with various types of sourdoughs (Galle et al. 2010; Katina et al. 2009; Scheirlinck et al. 2007; De Vuyst et al. 2002). *W. soli* (Magnusson et al. 2002) is the only species known to originate from soil, although *W. paramesenteroides* has also been detected in soil (Chen et al. 2005). In addition, weissellas have been isolated from sediments of a coastal marsh (Zamudio-Maya et al. 2008) and lake water (Yanagida et al. 2007).

*W. confusa* is considered a member of the normal human intestinal microbiota (Stiles and Holzapfel 1997; Walter et al. 2001). However, *Weissella* species such as *W. cibaria* and *W. confusa* have also been detected in clinical samples of human or animal origin (Björkroth et al. 2002). Strains of *W. confusa* have been associated with rare cases of bacteremia (Olano et al. 2001; Harlan et al. 2010; Salimnia et al. 2011; Lee et al. 2011) and endocarditis (Flaherty et al. 2003) in humans. Identification and significance of *Weissella* infections has been reviewed by Kamboj et al. (2015). *Weissella*-associated infections are mainly due to natural resistance of these species to vancomycin and are usually associated with an underlying disease or immunosuppression of the host. Alterations of the gut microbiota following surgery or chemotherapy are believed to facilitate translocation of *Weissella* spp. due to disruption of the mucosal barrier, predisposing the host to infection with this organism. *Weissella* spp. are inherently resistant to vancomycin. Therefore, early consideration of the pathogenic role of this bacteria and choice of alternate therapy is important to ensure better outcomes. In addition to human cases, *W. confusa* has also been documented as a cause of systemic infection in a healthy primate (*Cercopitheus mona*) (Vela et al. 2003), and unknown *Weissella* strains were isolated from a diseased rainbow trout in China (Liu et al. 2009).

## 6.4.4 Identification, Typing, and Detection

Identification of weissellas both at the genus and species levels is challenging. *Weissella* is routinely cultured using the general growth media for LAB such as MRS medium (De Man et al. 1960), but there is no specific selective medium or enrichment method for the members of this genus. The intrinsic resistance to vancomycin may be useful in certain approaches but does not distinguish between *Weissella* and *Leuconostoc*. It is most difficult to differentiate *Weissella* from *Leuconostoc* and heterofermentative *Lactobacillus* by phenotypic characteristics, for which reason most current identification approaches rely on chemotaxonomic and molecular methods. Numerical analyses of macromolecule patterns have proved useful in the identification of *Weissella* species. *Weissella* and *Leuconostoc* species have been distinguished by comparison of total soluble cell protein patterns (Dicks 1995; Tsakalidou et al. 1997). In addition, 16 + 23S rRNA gene restriction patterns (ribotypes) have been used for differentiating *W. confusa* from *W. cibaria* and grouping the strains into species-specific clusters (Björkroth et al. 2002). Comparison of cellular fatty acid profiles of *Weissella* species (Samelis et al. 1998) correspond well with results recorded in other taxonomic studies and were found valuable in the differentiation of *W. viridescens*, *W. paramesenteroides*, *W. hellenica*, and typical arginine-negative *Weissella* strains isolated from meat.

A genus-specific PCR method was developed for differentiation between the two heterofermentative LAB genera *Leuconostoc* and *Weissella* (Schillinger et al. 2008). For some *Weissella* taxa, species-specific sequences have been located in helix 1007/1022 of the variable region V6 in the 16S rRNA gene (Collins et al. 1993), and comparison of DNA patterns generated after restriction enzyme digests (*Mnl*I, *Mse*I, and *Bce*AI) of a 725 bp 16S rDNA fragment has also been applied (Jang et al. 2002). The *recN* locus may serve as a phylogenetic marker as well as a tool for species identification either alone or in combination with 16S rRNA encoding gene data (Arahal et al. 2008). Phylogenetic analysis of the *pheS* gene has also been used for identification of *Weissella* (de Bruyne et al. 2010; Snauwaert et al. 2013b; Praet et al. 2015).

Typing of weissellas has been based on the use of different macromolecules. Numerical analysis of either *Hind*III or *Eco*RI ribopatterns has been used to characterize *W. viridescens* from Morcilla de Burgos sausages (Koort et al. 2006). Dextran-producing strains of *W. cibaria* and *W. confusa* originating from sourdough were characterized with repetitive element-PCR fingerprinting using (GTG)$_5$-PCR (Bounaix et al. 2010). Genome diversity in the genera *Fructobacillus*, *Leuconostoc*, and *Weissella* was determined by physical and PFGE mapping (Chelo et al. 2010).

MALDI-TOF MS has also been used to distinguish *Weissella* species (Praet et al. 2015), and the members of this genus are very well represented in the BRUCKER database.

# 6.5 GENUS *PERIWEISSELLA*

## 6.5.1 General Characteristics

*Periweissella* resemble *Weissella* in many aspects. They are Gram-stain-positive, rod-shaped or coccobacilli, non-spore-forming, catalase-negative bacteria with heterofermentative metabolism, and ammonia is produced from arginine. *Periweissella*, like *Weissella*, ferment various carbohydrates. The most interesting characteristics distinguishing *Periweissella beninensis* (Padonou et al. 2010) from the other *Periweissella* species is the gliding mobility. Peritrichous flagella were observed in *Periweissella beninensis* using SEM (Padonou et al. 2010). Production of dextran, the most well-known EPS formed by heterofermentative LAB, has been recorded for *P. beninensis*, *P. fabaria* and *P. ghanensis*, (De Bruyne et al. 2010).

## 6.5.2 Phylogeny and Taxonomy

Genus *Periweissella* contains five species at the time of the writing of this chapter: *P. beninensis*, *P. cryptocerci*, *P. fabalis*, *P. fabaria*, and *P. ghanensis*. All five species were formerly considered *Weissella*, but in many phylogenetic studies of *Weissella* species (De Bruyne et al. 2010; De Bruyne et al. 2008b; Heo et al. 2019; Padonou et al. 2010, Snauwaert et al. 2013b), these species branched distinctly from the main clade of *Weissella* species containing the type species (*Weissella viridescens*) of this genus. In a phylogenomic and comparative genomic analyses of *Leuconostocaceae*, Bello et al. (2022) proposed that the species forming "*Weissella* clade 2" should be transferred to *Periweissella* gen. nov. with *P. ghanensis* as the type species. Phylogenetically *P. cryptocerci* is clustered out of the main cluster of the four other species (Heo et al. 2019). The most recent genomic study of *Periweissella* (Qiao et al. 2023) determined the phylogenetic relationship between *Periweissella* and the two closest genera, *Weissella* and *Furfurilactobacillus*, by the phylogenetic analysis and calculation of (core gene) pairwise average amino acid identity. Targeted genomic analysis showed that fructose bisphosphate aldolase was only present in the genome of *P. cryptocerci*. Mannitol dehydrogenase was found in genomes of *P. beninensis*, *P. fabaria*, and *P. fabalis*. Untargeted genomic analysis identified the presence of flagellar genes in *Periweissella* but not in other closely related genera.

## 6.5.3 Habitats

*Periweissella* have mainly been associated with fermented foods. *P. ghanensis*, *P. fabaria*, and *P. fabalis* were detected in association with fermentation of cocoa beans (De Bruyne et al. 2010; De Bruyne et al. 2008b; Snauwaert et al. 2013b). *P. beninensis* (Padonou et al. 2010) originates from submerged fermenting cassava. *P. cryptocerci* (Heo et al. 2019) is an exception to the other *Periweissella* species since it originates from the gut of an insect, *Cryptocercus kyebangensis*.

## 6.5.4 Identification, Typing, and Detection

The recent characterization study leading to distinguishing *Periweissella* from *Weissella* was based on comprehensive phylogenomic and comparative analyses on protein sequences from these species using multiple approaches (Bello et al. 2022). In addition, they performed comparative genomic analyses to identify molecular signatures in the form of conserved signature indels (CSIs) in protein sequences, which are specific for different main clades. Other studies associated with typing and identification than the original species description, Bello et al (2022) and Qiao et al. (2023), were not available at the time this chapter was written.

# 6.6 GENUS *CARNOBACTERIUM*

## 6.6.1 General Characteristics

Carnobacteria are Gram-positive, catalase-negative, and cold-tolerant organisms forming straight, slender, and nonmotile (except for some strains of *Carnobacterium mobile*) rods. They usually occur singly or in pairs, and sometimes in short chains. Some species such as *Carnobacterium funditum* and *Carnobacterium pleistocenium* may form coccobacilli (Hammes and Hertel 2009). One report even describes the infrequent appearance of coccus-shaped cells of *C. mobile*-like clinical isolates when incubated at 37 °C but not at 25 °C (Hoenigl et al. 2010).

At the time of writing, the genus *Carnobacterium* consists of 12 validated species, the majority of which have been isolated from cold environmental niches or cold-stored foods. Most species are mesophiles but psychrotolerant, capable of growing at low temperatures. However, *Carnobacterium iners* is a true psychrophile with optimal growth temperature at 4 °C (Snauwaert et al. 2013a). Carbohydrates are catabolized fermentatively by carnobacteria. Respiration has been shown for *Carnobacterium maltaromaticum* in the presence of heme (Meisel et al. 1994; Hammes and Hertel 2009), and this species consumes substantial amounts of oxygen during exponential growth under aerobic conditions (Borch and Molin 1989). In addition *Carnobacterium divergence* carries genes for respiration and is able to respire in the presence of heme (Iskandar et al.; Kolbeck et al. 2019). Pseudocatalase activity has also been demonstrated for the majority of the *Carnobacterium* species (Ringø et al. 2002). *C. divergens* and *C. maltaromaticum* carry heme-dependent catalase (Yuan et al. 2021). The glycolytic pathway is present in *Carnobacterium divergens*, and production of L-lactic acid has been shown for the majority of carnobacterial species. Acetic acid, formic acid (anaerobically; Borch and Molin 1989), acetoin (aerobically; Borch and Molin 1989), and $CO_2$ may be produced as end products, presumably by decarboxylation/dissimilation of pyruvic acid. Production of lactic acid has not been reported for *C. pleistocenium*, although acetate and ethanol were produced (Pikuta et al. 2005). Also, *C. alterfunditum* and *C. funditum* do not produce lactate from glycerol (Hammes and Hertel 2009). *C. divergens* and *C. maltaromaticum* utilize ribose and gluconic acid as substrates for growth, and acetic acid production can be substantial under aerobic conditions or if glucose is substituted for ribose as a carbohydrate source (Leisner et al. 2007; Hammes and Hertel 2009). This result indicates the possible presence of an inducible phosphoketolase and further suggests that carnobacteria are facultatively or atypical heterofermentative organisms.

Generally, *Carnobacterium* spp. ferment various carbohydrates but with a substantial amount of variation. Glucose, fructose, mannose, ribose, sucrose, and trehalose are utilized by the majority of species. *Carnobacterium jeotgali* is atypical by its narrow spectrum of carbohydrates for fermentation, omitting glucose and ribose (Kim et al. 2009). Two *Carnobacterium* species, *C. divergens* and *C. maltaromaticum*, possess an unusual chitinolytic activity (Leisner et al. 2008), with two putative chitinases showing high amino acid sequence similarities to *Listeria* chitinases (Leisner et al. 2010). Other polysaccharides such as inulin (*C. divergens, Carnobacterium gallinarum*, and some *C. maltaromaticum* strains), starch (*C. pleistocenium*), and glycogen (*C. gallinarum* and *C. mobile*) may also be utilized (Hammes and Hertel 2009).

Production of bacteriocins has been extensively investigated, especially for *C. divergens* and *C. maltaromaticum* and several studies on their potential application have been reported, particularly on their inhibition of *L. monocytogenes* (Leisner et al. 2007; Martin-Visscher et al. 2008).

The species *C. divergens, C. gallinarum, C. maltaromaticum*, and *C. mobile* possess the arginine deiminase pathway, resulting in production of $NH_4^+$ (Collins et al. 1987; Leisner et al. 1994a; Schillinger and Holzapfel 1995), whereas this is not the case for *C. inhibens, C. jeotgali*, or *C. viridans* (Jöborn et al. 1999; Collins et al. 2002; Holley et al. 2002). *C. divergens* and *C. maltaromaticum* are able to decarboxylate tyrosine to tyramine (Leisner et al. 1994b; Laursen et al. 2006) and to generate branched alcohols and aldehydes from valine, leucine, and isoleucine (Laursen et al. 2006; Leisner et al. 2007). Extracellular products arising from metabolism of other amino acids and proteolytic activity have not been reported. Some strains of *C. divergens* and *C. maltaromaticum* metabolize citric acid (Morea et al. 1999; Afzal et al. 2010).

## 6.6.2 Phylogeny and Taxonomy

The genus *Carnobacterium* currently consists of 12 species (Table 6.2). Historically, the two initial species were described as nonaciduric, acetate-sensitive, and atypical heterofermentative *Lactobacillus divergens* and *Lactobacillus piscicola* (synonym *L. carnis*). Collins et al. (1987) proposed that these two species should be reclassified into a new genus, *Carnobacterium*, as *C. divergens* and *C. piscicola* together with two new species, *C. gallinarum* and *C. mobile*. Later, *C. piscicola* was shown to be a synonym of *Lactobacillus maltaromaticus*, resulting in the proposal of the new species name, *C. maltaromaticum*

**TABLE 6.2** Diversity and Biological Distribution of *Carnobacterium* Species

| SPECIES FOODS | ECOLOGICAL GROUP ENVIRONMENT | PHYLOGENETIC GROUP | ISOLATION FREQUENCY | ISOLATED FROM | |
|---|---|---|---|---|---|
| *C. alterfunditum* | II | B | Very low | Shrimp product[a] | Deep sea, live fish, polar lake/sea |
| *C. antarcticum*[b] | II | B | Very low | Not reported | Antarctic sandy soil |
| *C. divergens* | I | A | High | Cheese, fish, meat, and shrimp products | Live fish, polar lake/sea, temperate zone soil, and water |
| *C. funditum*[b] | II | B | Very low | Not reported | Polar lake/sea, live fish, sponge |
| *C. gallinarum*[b] | I | A | Low | Meat products | Not reported |
| *C. iners*[b] | II | B | Low | Not reported | Antarctic pond |
| *C. inhibens*[b] | I | B | Very low | Not reported | Live fish |
| *C. jeotgali*[b] | I | B | Very low | Korean fermented ingredient (seafood) | Not reported |
| *C. maltaromaticum* | I | A | High | Cheese, fish and meat products | Deep sea, insects, live fish, polar lake/sea, temperate zone soil, and water |
| *C. mobile* | I | B | Intermediate | Meat and shrimp products | Live fish |
| *C. pleistocenium*[c] | II | B | Very low | Shrimp product | Permafrost |
| *C. viridans*[b] | I | B | Very low | Meat product | Not reported |

[a] *C. alterfunditum*-like and *C. pleistocenium*-like isolates (Jaffres et al. 2009).

[b] References for species descriptions are *C. alterfunditum* and *C. funditum* (Franzmann et al. 1991); *C. antarcticum* (Zhu et al. 2018); *C. iners* (Snauwaert et al. 2013a); *C. inhibens* (Jöborn et al. 1999); *C. jeotgali* (Kim et al. 2009); *C. viridans* (Holley et al. 2002); remaining species (Collins et al. 1987).

[b] References for species descriptions are *C. alterfunditum* and *C. funditum* (Franzmann et al. 1991); *C. antarcticum* (Zhu et al. 2018); *C. iners* (Snauwaert et al. 2013a); *C. inhibens* (Jöborn et al. 1999); *C. jeotgali* (Kim et al. 2009); *C. viridans* (Holley et al. 2002); remaining species (Collins et al. 1987).

[b] References for species descriptions are *C. alterfunditum* and *C. funditum* (Franzmann et al. 1991); *C. antarcticum* (Zhu et al. 2018); *C. iners* (Snauwaert et al. 2013a); *C. inhibens* (Jöborn et al. 1999); *C. jeotgali* (Kim et al. 2009); *C. viridans* (Holley et al. 2002); remaining species (Collins et al. 1987).

[b] References for species descriptions are *C. alterfunditum* and *C. funditum* (Franzmann et al. 1991); *C. antarcticum* (Zhu et al. 2018); *C. iners* (Snauwaert et al. 2013a); *C. inhibens* (Jöborn et al. 1999); *C. jeotgali* (Kim et al. 2009); *C. viridans* (Holley et al. 2002); remaining species (Collins et al. 1987).

[b] References for species descriptions are *C. alterfunditum* and *C. funditum* (Franzmann et al. 1991); *C. antarcticum* (Zhu et al. 2018); *C. iners* (Snauwaert et al. 2013a); *C. inhibens* (Jöborn et al. 1999); *C. jeotgali* (Kim et al. 2009); *C. viridans* (Holley et al. 2002); remaining species (Collins et al. 1987).

[b] References for species descriptions are *C. alterfunditum* and *C. funditum* (Franzmann et al. 1991); *C. antarcticum* (Zhu et al. 2018); *C. iners* (Snauwaert et al. 2013a); *C. inhibens* (Jöborn et al. 1999); *C. jeotgali* (Kim et al. 2009); *C. viridans* (Holley et al. 2002); remaining species (Collins et al. 1987).

[c] *C. pleistocenium* (Pikuta et al. 2005).

[b] References for species descriptions are *C. alterfunditum* and *C. funditum* (Franzmann et al. 1991); *C. antarcticum* (Zhu et al.

2018); *C. iners* (Snauwaert et al. 2013a); *C. inhibens* (Jöborn et al. 1999); *C. jeotgali* (Kim et al. 2009); *C. viridans* (Holley et al. 2002); remaining species (Collins et al. 1987).

*Source:* Leisner, J.J. et al., *FEMS Microbiol. Rev.*, 31, 592, 2007; Hammes, W.P. and Hertel, C., In *Bergey's Manual of Systematic Bacteriology Vol. 3, The Firmicutes*, Springer, New York, 2009.

(Mora et al. 2003). During the past two decades, novel *Carnobacterium* species have been characterized particularly from samples obtained from permafrost and polar aquatic ecosystems.

*Carnobacterium* spp. can be distinguished in two taxonomic groups, not to be confused with the ecological groups described further in Section 6.5.3, based on phylogenies of 16S rRNA gene sequences (Hammes and Hertel 2009; Kim et al. 2009; Table 6.2). MLSA using three housekeeping genes supported this division (Snauwaert et al. 2013a) as well as genome comparisons (Iskandar et al. 2017). Two of the species belonging to the phylogenetic group A (*C. divergens* and *C. maltaromaticum*) are found in a wide variety of habitats, whereas most of the species belonging to group B have, with the exception of *C. mobile*, been isolated only on a few occasions (Table 6.2).

## 6.6.3 Habitats

The genus can be roughly divided into two ecological groups according to habitats (Table 6.2); however, only two species, *C. divergens* and *C. maltaromaticum*, are frequently encountered in a diverse range of environments, including foods. Group I species were originally isolated from live fish or foods of animal origin. The second group of species, group II, appears to be mainly associated with cold, low-nutrient environments such as Antarctic ice lakes or arctic permafrost (Table 6.2; Leisner et al. 2007; Hammes and Hertel 2009).

Foods that constitute a habitat for *Carnobacterium* spp. include vacuum or modified atmosphere—packed, refrigerated raw, or processed meat products and lightly preserved fish products, milk, and certain types of soft cheese (Leisner et al. 2007). Although *C. divergens* and *C. maltaromaticum* have potential as protective cultures in foods, some strains appear to display undesirable properties as spoilers, where transamination, decarboxylation, and reduction of the amino acids isoleucine, leucine, and valine appear to be particularly important (Leisner et al. 2007). In addition, these two species also produce tyramine, a biogenic amine that constitutes a cause for concern regarding food safety for sensitive individuals, that is, individuals with reduced monoamine oxidase activity (Leisner et al. 2007).

Fish is a common animal reservoir of some carnobacteria, and some isolates of *C. maltaromaticum* can be pathogenic for fish with reduced immune function. On the other hand, strains of this species and of *C. divergens* have also been considered potential probiotics (Toranzo et al. 1993a, 1993b; Ringø et al. 2005; Kim and Austin 2008; Loch et al. 2008; Leisner et al. 2007). Although *Carnobacterium* spp. have rarely been associated with human clinical cases (Leisner et al. 2007), a number of putative virulence factors have been described for *C. maltaromaticum* ATCC 35586, including hemolysins, internalins, a putative capsule synthetic ability, and a gene related to the *Listeria* PrfA virulence regulator (Leisner et al. 2010). β-hemolysis activity has been demonstrated for *C. viridans* and also for a clinical *C. mobile* isolate (Holley et al. 2002; Hoenigl et al. 2010). A number of invertebrates, such as insects and marine sponges, are sources for *C. maltaromaticum* and *C. funditum*, respectively (Shannon et al. 2001; Li and Liu 2006).

## 6.6.4 Identification, Typing, and Detection

Because certain carnobacteria are inhibited by high levels of acetate and low pH, their presence may have been underestimated if acetate-containing media such as MRS or Rogosa agar have been used. Thus, media that do not contain acetate, such as nitrite polymyxin (NP) or *Carnobacterium*-specific agar media, are recommended for culturing or enumeration of carnobacteria (Wasney et al. 2001; Edima et al. 2007).

At the genus level, phenotypic differentiation of carnobacteria from the historically associated genus *Lactobacillus* but also from *Paralactobacillus* and *Weissella* is based on the observation that carnobacteria grow well at pH 9 but not at low pH, for example, on acetate agar (Collins et al. 1987; Hammes and Hertel 2009). They may be differentiated from nonaciduric *Leuconostoc* spp. by the production of L-lactic acid from glucose. *Leuconostoc* and *Weissella* as well as *Isobaculum* and *Desemzia* (both belonging to the *Carnobacteriaceae*) can also be differentiated by not possessing *meso*-diaminopimelic acid in the peptidoglycan (Hammes and Hertel 2009; Holzapfel et al. 2009).

A number of phenotypic methods have been used for species identification and typing of *Carnobacterium*, including biochemical and physiological tests, composition of cellular fatty acids, whole-cell protein profiling, pyrolysis MS, and Fourier transform infrared spectroscopy (Hammes and Hertel 2009; Afzal et al. 2010). Differentiation between the two most commonly encountered species, *C. divergens* and *C. maltaromaticum*, is only possible by using a large number of tests and data evaluations by numerical taxonomy (Laursen et al. 2005; Hammes and Hertel 2009). SDS-PAGE of whole-cell protein extracts is useful as a phenotypic identification method at species and strain level (Laursen et al. 2005).

Today, molecular methods play a key role in identifying carnobacteria from various ecosystems as well as for characterizing the genetic variability of species and strains. MLSA of the housekeeping genes encoding phenylalanyl-tRNA synthase alpha subunit (pheS), RNA polymerase alpha subunit (rpoA), and ATP synthase alpha subunit (atpA) has been shown to give a higher taxonomic resolution, allowing species identification within the genus *Carnobacterium* (16S rRNA gene sequencing has been used as a method per se or in support of results obtained with the other methods). However, this approach fails to differentiate between *C. alterfunditum* and *C. pleistocenium* (Pikuta et al. 2005; Hammes and Hertel 2009). Before the sequencing era, several DNA-based genotyping methods had been evaluated for culture-dependent identification of *Carnobacterium* species, including multiplex PCR, rep-PCR, ribotyping, 16S–23S ISR-based RFLP, AFLP, PFGE, and RAPD (Leisner et al. 2008; Hammes and Hertel 2009; Afzal et al. 2010). RAPD has also been used for typing (Morea et al. 1999), whereas AFLP was not found suitable for this purpose in one study (Laursen et al. 2005). In addition, some of the methods are limited in their applicability because they require specific databases and/or they are difficult to standardize (e.g., 16S–23S ISR-based RFLP; Laursen et al. 2005).

# BIBLIOGRAPHY

Afzal, M.I., T. Jacquet, S. Delaunay, F. Borges, J.-B. Milliere, A.-M. Revol-Junelles and C. Cailliez-Grimal. 2010. *Carnobacterium maltaromaticum*: Identification, isolation tools, ecology and technological aspects in dairy foods. *Food Microbiol. 27*: 573–579.

Alcántara-Hernández, R.J., J.A. Rodríguez-Álvarez, C. Valenzuela-Encinas, F.A. Gutiérrez-Miceli, H. Castañon-González, R. Marsch, T. Ayora-Talavera and L. Dendooven. 2010. The bacterial community in "*taberna*" a traditional beverage of Southern Mexico. *Lett. Appl. Microbiol. 51*: 558–563.

Ampe, F., N. Ben Omar, C. Moizan, C. Wacher and J.-P. Guyot. 1999. Polyphasic study of the spatial distribution of microorganisms in Mexican pozol, a fermented maize dough, demonstrates the need for cultivation-independent methods to investigate traditional fermentations. *Appl. Environ. Microbiol. 65*: 5464–5473.

Arahal, D.R., E. Sanchez, M.C. Macian and E. Garay. 2008. Value of *recN* sequences for species identification and as a phylogenetic marker within the family *Leuconostocaceae* . *Int. Microbiol. 11*: 33–39.

Beijernick, M.W. 1912. Mutation bei Mikroben. *Folia Mikrobiologiya (Delft). 1*: 4–100.

Bello, S., B. Rudra and R.S. Gupta. 2022. Phylogenomic and comparative genomic analyses of *Leuconostocaceae* species: Identification of molecular signatures specific for the genera *Leuconostoc, Fructobacillus* and *Oenococcus* and proposal for a novel genus *Periweissella* gen. nov. *Int. J. System. Evol. Microbiol.* 72: 005284.

Björkroth, J. 1997. DNA-Based Characterisation Methods for Contamination Analysis of Spoilage Lactic Acid Bacteria in Food Processing. PhD Thesis. University of Helsinki, Helsinki, Finland.

Björkroth, J., L.M.T. Dicks and W.H. Holzapfel. 2009. Genus III. *Weissella*. In Bergey's Manual of Systematic Bacteriology Vol. 3, The Firmicutes, Eds. P. De Vos, G.M. Garrity, D. Jones, N.R. Krieg, W. Ludwig, F.A. Rainey, K.-H. Schleifer and W.B. Whitman, pp. 643–654. Springer, New York.

Björkroth, K.J., R. Geisen, U. Schillinger, N. Weiss, P. De Vos, W.H. Holzapfel, H.J. Korkeala and P. Vandamme. 2000. Characterization of *Leuconostoc gasicomitatum* sp. nov., associated with spoiled raw tomato-marinated broiler meat strips packaged under modified-atmosphere conditions. *Appl. Environ. Microbiol.* 66: 3764–3772.

Björkroth, J. and W. Holzapfel. 2006. The genera *Leuconostoc, Oenococcus* and *Weissella*. In The Prokaryotes, Eds. M. Dworkin, S. Falkow, E. Rosenberg, K.-H. Schleifer and E. Stackebrandt. 3rd ed. Vol. 4, pp. 267–319. Springer, New York.

Björkroth, K.J. and H.J. Korkeala. 1997. Use of rRNA gene restriction patterns to evaluate lactic acid bacterium contamination of vacuum-packaged sliced cooked whole-meat product in a meat processing plant. *Appl. Environ. Microbiol.* 63: 448–453.

Björkroth, K.J., U. Schillinger, R. Geisen, N. Weiss, B. Hoste, W.H. Holzapfel, H.J. Korkeala and P. Vandamme. 2002. Taxonomic study of *Weissella confusa* and description of *Weissella cibaria* sp. nov., detected in food and clinical samples. *Int. J. Syst. Evol. Microbiol.* 52: 141–148.

Björkroth, K.J., P. Vandamme and H.J. Korkeala. 1998. Identification and characterization of *Leuconostoc carnosum* associated with production and spoilage of vacuum-packaged sliced cooked ham. *Appl. Environ. Microbiol.* 64: 3313–3319.

Borch, E. and G. Molin. 1989. The aerobic growth and product formation of *Lactobacillus, Leuconostoc, Brochothrix* and *Carnobacterium* in batch cultures. *Appl. Microbiol. Biotechnol.* 30: 81–88.

Bounaix, M.-S., H. Robert, V. Gabriel, S. Morel, M. Remaud-Siméon, B. Gabriel and C. Fontagné-Faucher. 2010. Characterization of dextran-producing *Weissella* strains isolated from sourdoughs and evidence of constitutive dextransucrase expression. *FEMS Microbiol. Lett.* 311: 18–26.

Buu-Hoï, A., C. Branger and J.F. Acar. 1985. Vancomycin-resistant streptococci or *Leuconostoc* sp. *Antimicrob. Agents Chemother.* 28: 458–460.

Cailliez-Grimal, C., R. Miguindou-Mabiala, M. Leseine, A.-M. Revol-Junelles and J.-B. Miliére. 2005. Quantitative polymerase chain reaction used for the rapid detection of *Carnobacterium* species from French soft cheeses. *FEMS Microbiol. Lett.* 250: 163–169.

Calmin, G., F. Lefort and L. Belbahri. 2008. Multi-loci sequence typing (MLST) for two lactic acid bacteria (LAB) species: *Pediococcus parvulus* and *P. damnosus*. *Mol. Biotechnol.* 40: 170–179.

Castex, M., P. Lemaire, N. Wabete and L. Chim. 2010. Effect of probiotic *Pediococcus acidilactici* on antioxidant defences and oxidative stress of *Litopenaeus stylirostris* under *Vibrio nigripulchritudo* challenge. *Fish Shellfish Immunol.* 28: 622–631.

Cauchie, E., M. Gand, G. Kergourlay, B. Taminiau, L. Delhalle, N. Korsak and G. Daube. 2017. The use of 16S rRNA gene metagenetic monitoring of refrigerated food products for understanding the kinetics of microbial subpopulations at different storage temperatures: The example of white pudding. *Int. J. Food Microbiol.* 247: 70–78.

Chelo, I.M., L. Zé-Zé and R. Tenreiro. 2007. Congruence of evolutionary relationships inside the *Leuconostoc—Oenococcus—Weissella* clade assessed by phylogenetic analysis of the 16S rRNA gene, *dnaA, gyrB, rpoC* and *dnaK*. *Int. J. Syst. Evol. Microbiol.* 57: 276–286.

Chelo, I.M., L. Zé-Zé and R. Tenreiro. 2010. Genome diversity in the genera *Fructobacillus, Leuconostoc* and *Weissella* determined by physical and genetic mapping. *Microbiology* 156: 420–430.

Chen, Y.S., L.T. Wang, Y.C. Wu, K. Mori, T. Tamura, C.H. Chang, Y.C. Chang, H.C. Wu, H.H. Yi and P.Y. Wang. 2020. *Leuconostoc litchii* sp. nov., a novel lactic acid bacterium isolated from lychee. *Int. J. System. Evol. Microbiol.* 70: 1585–1590.

Chen, Y.-S., F. Yanagida and T. Shinohara. 2005. Isolation and identification of lactic acid bacteria from soil using an enrichment procedure. *Lett. Appl. Microbiol.* 40: 195–200.

Choi, H.J., C.I. Cheigh, S.B. Kim, J.C. Lee, D.W. Lee, S.W. Choi, J.M. Park and Y.R. Pyun. 2002. *Weissella kimchii* sp. nov., a novel lactic acid bacterium from kimchi. *Int. J. Syst. Evol. Microbiol.* 52: 507–511.

Collins, M.D., J.A.E. Farrow, B.A. Phillips, S. Ferusu and D. Jones. 1987. Classification of *Lactobacillus divergens, Lactobacillus piscicola*, and some catalase-negative, asporogenous, rod-shaped bacteria from poultry in a new genus, *Carnobacterium*. *Int. J. Syst. Bacteriol.* 37: 310–316.

Collins, M.D., R.A. Hytson, G. Foster, E. Falsen and N. Weiss. 2002. *Isobaculum melis* gen. nov., sp nov., a *Carnobacterium*-like organism isolated from the intestine of a badger. *Int. J. Syst. Evol. Microbiol.* 52: 207–210.

Collins, M.D., J. Samelis, J. Metaxopoulos and S. Wallbanks. 1993. Taxonomic studies on some *Leuconostoc*-like organisms from fermented sausages: Description of a new genus *Weissella* for the *Leuconostoc paramesenteroides* group of species. *J. Appl. Bacteriol.* 75: 595–603.

De Bruyne, K., N. Camu, L. De Vuyst and P. Vandamme. 2010. *Weissella fabaria* sp. nov., from a Ghanaian cocoa fermentation. *Int. J. Syst. Evol. Microbiol.* 60: 1999–2005.

De Bruyne, K., N. Camu, K. Lefebvre, L. De Vuyst and P. Vandamme. 2008b. *Weissella ghanensis* sp. nov., isolated from a Ghanaian cocoa fermentation. *Int. J. Syst. Evol. Microbiol.* 58: 2721–2725.

De Bruyne, K., C.M.A.P. Franz, M. Vancanneyt, U. Schillinger, F. Mozzi, G. Font de Valdez, L. De Vuyst and P. Vandamme. 2008a. *Pediococcus argentinicus* sp. nov. from Argentinean fermented wheat flour and identification of *Pediococcus* species by *pheS*, *rpoA* and *atpA* sequence analysis. *Int. J. Syst. Evol. Microbiol.* 58: 2909–2916.

De Bruyne, K., U. Schillinger, L. Caroline, B. Boehringer, I. Cleenwerck, M. Vancanneyt, L. De Vuyst, C.M. Franz and P. Vandamme. 2007. *Leuconostoc holzapfelii* sp. nov., isolated from Ethiopian coffee fermentation and assessment of sequence analysis of housekeeping genes for delineation of *Leuconostoc* species. *Int. J. Syst. Evol. Microbiol.* 57: 2952–2959.

Delaherche, A., O. Claisse and A. Lonvaud-Funel. 2004. Detection and quantification of *Brettanomyces bruxellensis* and 'ropy' *Pediococcus damnosus* strains in wine by real-time polymerase chain reaction. *J. Appl. Microbiol.* 97: 910–915.

de Man, J.C., M. Rogosa and E.M. Sharpe. 1960. A medium for the cultivation of lactobacilli. *J. Appl. Microbiol.* 23: 130–135.

De Vuyst, L., V. Schrijvers, S. Paramithiotis, B. Hoste, M. Vancanneyt, J. Swings, G. Kalantzopoulos, E. Tsakalidou and W. Messens. 2002. The biodiversity of lactic acid bacteria in Greek traditional wheat sourdoughs is reflected in both composition and metabolite formation. *Appl. Environ. Microbiol.* 68: 6059–6069.

Dicks, L.M.T. 1995. Relatedness of *Leuconostoc* species of the *Leuconostoc sensu stricto* line of descent, *Leuconostoc oenos* and *Weissella paramesenteroides* revealed by numerical analysis of total soluble cell protein patterns. *Syst. Appl. Microbiol.* 18: 99–102.

Dicks, L.M.T., F. Dellaglio and M.D. Collins. 1995. Proposal to reclassify *Leuconostoc oenos* as *Oenococcus oeni* [corrig.] gen. nov., comb. nov. *Int. J. Syst. Bacteriol.* 45: 395–397.

Dicks, L.M.T. and A. Endo. 2014. The Family *Lactobacillaceae*: Genera Other than *Lactobacillus*. In The Prokaryotes, Ed. E. Rosenberg, E.F. DeLong, S. Lory, et al., pp. 203–212. Springer, Berlin, Germany.

Diez, A.M., J. Björkroth, I. Jaime and J. Rovira. 2009. Microbial, sensory and volatile changes during the anaerobic cold storage of Morcilla De Burgos previously inoculated with *Weissella viridescens* and *Leuconostoc mesenteroides*. *Int. J. Food Microbiol.* 131: 168–177.

Dobson, C.M., H. Deneer, S. Lee, S. Hemmingsen, S. Glaze and B. Ziola. 2002. Phylogenetic analysis of the genus *Pediococcus*, including *Pediococcus claussenii* sp. nov., a novel lactic acid bacterium isolated from beer. *Int. J. Syst. Evol. Microbiol.* 52: 2003–2010.

Doi, K., Y. Nishizaki, Y. Fujino, T. Ohshima, S. Ohmomo and S. Ogata. 2009. *Pediococcus lolii* sp. nov., isolated from ryegrass silage. *Int. J. Syst. Evol. Microbiol.* 59: 1007–1010.

Edima, H.C., C. Cailliez-Grimal, A.-M. Revol-Junelles, L. Tonti, M. Linder and J.-B. Millière. 2007. A selective enumeration medium for *Carnobacterium maltaromaticum*. *J. Microbiol. Methods* 68: 516–521.

Ehrmann, M.A., S. Freiding and R.F. Vogel. 2009. *Leuconostoc palmae* sp. nov., a novel lactic acid bacterium isolated from palm wine. *Int. J. Syst. Evol. Microbiol.* 59: 943–947.

Elisha, B.G. and P. Courvalin. 1995. Analysis of genes encoding d-alanine: D-Alanine ligase-related enzymes in *Leuconostoc mesenteroides* and *Lactobacillus* spp. *Gene* 152: 79–83.

Elliott, J.A. and R.R. Facklam. 1993. Identification of *Leuconostoc* spp. by analysis of soluble whole-cell protein patterns. *J. Clin. Microbiol.* 31: 1030–1033.

Endo, A. and S. Okada. 2008. Reclassification of the genus *Leuconostoc*, and proposals of *Fructobacillus fructosus*, gen nov., comb. nov., *Fructobacillus durionis*, comb. nov., *Fructobacillus ficulneus*, comb. nov. and *Fructobacillus pseudoficulneus*, comb. nov. *Int. J. Syst. Evol. Microbiol.* 58: 2195–2205.

Ennahar, S. and Y. Cai. 2004. Genetic evidence that *Weissella kimchii* Choi et al. 2002is a later heterotypic synonym of *Weissella cibaria* Björkroth et al. 2002. *Int. J. Syst. Evol. Microbiol.* 54: 463–465.

Farrow, J.A.E., R.R. Facklam and M.D. Collins. 1989. Nucleic acid homologies of some vancomycin-resistant leuconostocs and description of *Leuconostoc citreum* sp. nov. and *Leuconostoc pseudomesenteroides* sp. nov. *Int. J. Syst. Bacteriol.* 39: 279–283.

Flaherty, J.D., P.N. Levett, F.E. Dewhirst, T.E. Troe, J.R. Warren and S. Johnson. 2003. Fatal case of endocarditis due to *Weissella confusa*. *J. Clin. Microbiol.* 41: 2237–2239.

Franz, C.M.A.P., M. Vancanneyt, K. Vandemeulebroecke, M. De Wachter, I. Cleenwerck, B. Hoste, U. Schillinger, W.H. Holzapfel and J. Swings. 2006. *Pediococcus stilesii* sp. nov., isolated from maize grains. *Int. J. Syst. Evol. Microbiol.* 56: 329–333.

Franzmann, P.D., P. Höpfl, N. Weiss and B.J. Tindall. 1991. Psychrotrophic, lactic acid-producing bacteria from anoxic waters in Ace Lake, Antartica; *Carnobacterium funditum* sp. nov. and *Carnobacterium alterfunditum* sp. nov. *Arch. Microbiol.* 156: 255–262.

Fusco, V., G.M. Quero, G.S. Cho, J. Kabisch, D. Meske, H. Neve, W. Bockelmann and C.M. Franz. 2015. The genus *Weissella*. Taxonomy, ecology and biotechnological potential. *Front. Microbiol.* 6: 155.

Galle, S., C. Schwab, E. Arendt and M. Gänzle. 2010. Exopolysaccharide-forming *Weissella* strains as starter cultures for sorghum and wheat sourdoughs. *J. Agric. Food Chem.* 58: 5834–5841.

Garvie, E.I. 1960. The genus *Leuconostoc* and its nomenclature. *J. Dairy Res.* 27: 283–292.

Garvie, E.I. 1967. The growth factor and amino acid requirements of species of the genus *Leuconostoc*, including *Leuconostoc paramesenteroides* (sp. nov.) and *Leuconostoc oenos*. *J. Gen. Microbiol.* 48: 439–447.

Garvie, E.I. 1983. *Leuconostoc mesenteroides* subsp. *cremoris* (Knudsen and Sørensen) comb. nov. and *Leuconostoc mesenteroides* subsp. *dextranicum* (Beijerinck) comb. nov. *Int. J. Syst. Bacteriol.* 33: 118–119.

Garvie, E.I. 1986. Genus *Leuconostoc* van Tieghem 1878, 198[AL] emended Mut. Char. Hucker and Pederson 1930, 66[AL]. In Bergey's Manual of Systematic Bacteriology, Eds. P.H.A. Sneath, N.S. Mair, M.E. Sharpe and J.G. Holt. Vol. 2, pp. 1071–1075. Williams & Wilkins, Baltimore, MD.

Gibson, T. and Y. Abdel-Malek. 1945. The formation of carbon dioxide by lactic acid bacteria and *Bacillus licheniformis* and cultural method of detecting the process. *J. Dairy Res.* 14: 35–44.

Gu, C.T., F. Wang, C.Y. Li., F. Liu and G.C. Huo. 2012. *Leuconostoc mesenteroides* subsp. *suionicum* subsp. nov. *Int. J. Syst. Evol. Microbiol.* 62: 1548–1551.

Hammes, W.P. and C. Hertel. 2009. Genus I. *Carnobacterium*. In Bergey's Manual of Systematic Bacteriology Vol. 3, The Firmicutes, Eds. P. De Vos, G.M. Garrity, D. Jones, N.R. Krieg, W. Ludwig, F.A. Rainey, K.-H. Schleifer and W.B. Whitman, pp. 549–557. Springer, New York.

Harlan, N.P., R.R. Kempker, S.M. Parekh, E.M. Burd and D.T. Kuhar. 2010. *Weissella confusa* bacteremia in a liver transplant patient with hepatic artery thrombosis. *Transpl. Infect. Dis.* https://doi.org/10.1111/j.1399-3062.2010.00579.x.

Hemme, D. and C. Foucaud-Scheunemann. 2004. *Leuconostoc*: Characteristics, use in dairy technology and prospects in functional foods. *Int. Dairy J.* 14: 467–494.

Heo, J., M. Hamada, H. Cho, H.Y. Weon, J.S. Kim, S.B. Hong, S.J. Kim and S.W. Kwon. 2019. *Weissella cryptocerci* sp. nov., isolated from gut of the insect *Cryptocercus kyebangensis*. *Int. J. Syst. Evol. Microbiol.* 69: 2801–2806.

Hoenigl, M., A.J. Grisold, T. Valentin, E. Leitner, G. Zarfel, H. Renner and R. Krause. 2010. Isolation of *Carnobacterium* sp. from a human blood culture. *J. Med. Microbiol.* 59: 493–495.

Holley, R.A., T.Y. Guan, M. Peirson and C.K. Yost. 2002. *Carnobacterium viridans* sp. nov., an alkaliphilic, facultative anaerobe isolated from refrigerated, vacuum-packed bologna sausage. *Int. J. Syst. Evol. Microbiol.* 52: 1881–1885.

Holzapfel, W.H., J. Björkroth and L.M.T. Dicks. 2009. Genus I. *Leuconostoc*. In Bergey's Manual of Systematic Bacteriology Vol. 3, The Firmicutes, Eds. P. De Vos, G.M. Garrity, D. Jones, N.R. Krieg, W. Ludwig, F.A. Rainey, K.-H. Schleifer and W.B. Whitman, pp. 624–635. Springer, New York.

Holzapfel, W.H. and E.P. Van Wyk. 1982. *Lactobacillus kandleri* sp. nov., a new species of the subgenus *Betabacterium* with glycine in the peptidoglycan. *Zentralbl. Bakteriol. Parasitenkd. Infektionskr. Hyg.* 3: 495–502.

Hultman, J., R. Rahkila, J. Ali, J. Rousu and K.J. Björkroth. 2015. Meat processing plant microbiome and contamination patterns of cold-tolerant bacteria causing food safety and spoilage risks in manufacture of vacuum-packaged cooked sausages. *Appl. Environ. Microbiol.* 81: 7088–7097.

Huygens, F. 1993. Vancomycin binding to cell walls of non-streptococcal vancomycin-resistant bacteria. *J. Antimicrob. Chemother.* 32: 551–558.

Iijima, K., K. Suzuki, S. Asano, H. Kuriyama and Y. Kitagawa. 2007. Isolation and identification of potential beer-spoilage *Pediococcus inopinatus* and beer-spoilage *Lactobacillus backi* strains carrying the *horA* and *horC* gene clusters. *J. Inst. Brew.* 113: 96–101.

Iskandar, C.F., F. Borges, B. Taminiau, G. Daube, M. Zagorec, B. Remenant, J.J. Leisner, M.A. Hansen, S.J. Sørensen, C. Mangavel, C. Cailliez-Grimal and A.-M. Revol-Junelles. 2017. Comparative genomic analysis reveals ecological differentiation in the genus *Carnobacterium*. *Front. Microbiol.* 8.

Jaffres, E., D. Sohier, F. Leroi, M.F. Pilet, H. Prevost, J.J. Joffraud and X. Dousset. 2009. Study of the bacterial ecosystem in tropical cooked and peeled shrimps using a polyphasic approach. *Int. J. Food Microbiol.* 131: 20–29.

Jang, J., B. Kim, J. Lee, J. Kim, G. Jeong and H. Han. 2002. Identification of *Weissella* species by the genus-specific amplified ribosomal DNA restriction analysis. *FEMS Microbiol. Lett.* 212: 29–34.

Jeon, H.H., K.H. Kim, B.H. Chun, B.H. Ryu, N.S. Han and C.O. Jeon. 2017. A proposal of *Leuconostoc mesenteroides* subsp. *jonggajibkimchii* subsp. nov. and reclassification of *Leuconostoc mesenteroides* subsp. *suionicum* (Gu *et al.*, 2012) as *Leuconostoc suionicum* sp. nov. based on complete genome sequences. *Int. J. Syst. Evol. Microbiol.* 67: 2225–2230.

Jöborn, A., M. Dorsch, J.C. Olsson, A. Westerdahl and S. Kjelleberg. 1999. *Carnobacterium inhibens* sp. nov., isolated from the intestine of Atlantic salmon (*Salmo salar*). *Int. J. Syst. Evol. Microbiol.* 49: 1891–1898.

Johansson, P., L. Paulin, E. Vihavainen, N. Salovuori, A. Edward, K.J. Björkroth and P. Auvinen. 2011. Genome sequence of a food spoilage lactic acid bacterium *Leuconostoc gasicomitatum* LMG 18811[T] in association with specific spoilage reactions. *Appl. Environ. Microbiol.* 77: 4344–4351.

Johansson, P., E. Säde, J. Hultman, P. Auvinen and J. Björkroth. 2022. Pangenome and genomic taxonomy analyses of *Leuconostoc gelidum* and *Leuconostoc gasicomitatum*. *BMC Genom.* 23(1): 818.

Kamboj, K., A. Vasquez and J.M. Balada-Llasat. 2015. Identification and significance of *Weissella* species infections. *Front. Microbiol. 6*: 1204.

Kandler, O. and I.G. Abo-Elnaga. 1966. Zur Taxonomie der Gattung *Lactobacillus* Beijerinck. IV. *L. corynoides* ein Synonym von *L. viridescens*. *Zentralbl. Bakteriol. Parasitenkd. Infektionskr. Hyg. 120*: 753–754.

Kandler, O., U. Schillinger and N. Weiss. 1983. *Lactobacillus halotolerans* sp. nov., nom. rev. and *Lactobacillus minor* sp. nov., nom. rev. *Syst. Appl. Microbiol. 4*: 280–285.

Katina, K., N.H. Maina, R. Juvonen, L. Flander, L. Johansson, L. Virkki, M. Tenkanen and A. Laitila. 2009. In situ production and analysis of *Weissella confusa* dextran in wheat sourdough. *Food Microbiol. 26*: 734–743.

Kim, B., J. Lee, J. Jang, J. Kim and H. Han. 2003. *Leuconostoc inhae* Sp. Nov., a lactic acid bacterium isolated from kimchi. *Int. J. Syst. Evol. Microbiol. 53*: 1123–1126.

Kim, D.H. and B. Austin. 2008. Characterization of probiotic carnobacteria isolated from rainbow trout (*Oncorhynchus mykiss*) intestine. *Lett. Appl. Microbiol. 47*: 141–147.

Kim, J., J. Chun and H.U. Han. 2000. *Leuconostoc kimchii* sp. nov., a new species from kimchi. *Int. J. Syst. Evol. Microbiol. 50*: 1915–1919.

Kim, M. and J. Chun. 2005. Bacterial community structure in Kimchi, a Korean fermented vegetable food, as revealed by 16S rRNA gene analysis. *Int. J. Food Microbiol. 103*: 91–96.

Kim, M.-S., S.W. Roh, Y.-D. Nam, J.-H. Yoon, and J.-W. Bae. 2009. *Carnobacterium jeotgali*, sp. nov., isolated from a Korean traditional fermented food. *Int. J. Syst. Evol. Microbiol. 59*: 3168–3171.

Kolbeck, S., L. Reetz, M. Hilgarth and R.F. Vogel. 2019. Quantitative oxygen consumption and respiratory activity of meat spoiling bacteria upon high oxygen modified atmosphere. *Front. Microbiol. 10*.

Koort, J., T. Coenye, E.M. Santos, C. Molinero, I. Jaime, J. Rovira, P. Vandamme and J. Björkroth. 2006. Diversity of *Weissella viridescens* strains associated with "Morcilla De Burgos. *Int. J. Food Microbiol. 109*: 164–168.

Kulwichit, W., S. Nilgate, T. Chatsuwan, S. Krajiw, C. Unhasuta and A. Chongthaleong. 2007. Accuracies of *Leuconostoc* phenotypic identification: A comparison of API systems and conventional phenotypic assays. *BMC Infect. Dis. 7*: 69.

Kumar, S., K. Bansal and S.K. Sethi. 2022. Comparative genomics analysis of genus *Leuconostoc* resolves its taxonomy and elucidates its biotechnological importance. *Food Microbiol. 106*.

Laursen, B.G., L. Bay, I. Cleenwerck, M. Vancanneyt, J. Swings, P. Dalgaard and J.J. Leisner. 2005. *Carnobacterium divergens* and *Carnobacterium maltaromaticum* as spoilers or protective cultures in meat and seafood: Phenotypic and genotypic characterization. *Syst. Appl. Microbiol. 28*: 151–164.

Laursen, B.G., J.J. Leisner and P. Dalgaard. 2006. *Carnobacterium* species: Effect of metabolic activity and interaction with *Brochothrix thermosphacta* on sensory characteristics of modified atmosphere packed shrimp. *J. Agric. Food Chem. 54*: 3604–3611.

Lee, J.S., K.C. Lee, J.S. Ahn, T.I. Mheen, Y.R. Pyun and Y.H. Park. 2002. *Weissella koreensis* sp. nov., isolated from kimchi. *Int. J. Syst. Evol. Microbiol. 52*: 1257–1261.

Lee, M.R., Y.T. Huang, C.H. Liao, C.C. Lai, P.I. Lee and P.R. Hsueh. 2011. Bacteraemia caused by *Weissella confusa* at a University Hospital in Taiwan, 1997–2007. *Clin. Microbiol. Infect. 17*: 1226–1231.

Lee, S.H., H.J. Ku, M.J. Ahn, J.S. Hong, S.H. Lee, H. Shin, K.C. Lee, et al. 2015. *Weissella jogaejeotgali* sp. nov., isolated from jogae jeotgal, a traditional Korean fermented seafood. *Int. J. Syst. Evol. Microbiol. 65*: 4674–4681.

Lee, S.H., M.S. Park, J.Y. Juyng and C.O. Jeon. 2012. *Leuconostoc miyukkimchii* sp. nov., isolated from brown algae (*Undaria pinnatifida*) kimchi. *Int. J. Syst. Evol. Microbiol. 62*: 1098–1103.

Leisner, J.J., M.A. Hansen, M.H. Larsen, H. Ingmer, L. Hansen and S.J. Sørensen. 2010. The lactic acid bacterium *Carnobacterium maltaromaticum* possess a diverse range of potential virulence genes some of which are related to similar genes in *Listeria monocytogenes*. *Paper Presented at the 22nd International ICFMH Symposium, Food Micro*, Copenhagen, Denmark.

Leisner, J.J., B.G. Laursen, H. Prévost, D. Drider and P. Dalgaard. 2007. *Carnobacterium*: Positive and negative effects in the environment and in foods. *FEMS Microbiol. Rev. 31*: 592–613.

Leisner, J.J., J.C. Millan, H.H. Huss and L.M. Larsen. 1994b. Production of histamine and tyramine by lactic acid bacteria isolated from vacuum-packed sugar-salted fish. *J. Appl. Bacteriol. 76*: 417–423.

Leisner, J.J., B. Pot, H. Christensen, G. Rusul, J.E. Olsen, B.W. Wee, K. Muhamad and H.M. Ghazali. 1999. Identification of lactic acid bacteria from Chili Bo, a Malaysian food ingredient. *Appl. Environ. Microbiol. 65*: 599–605.

Leisner, J.J., J. Tidemand and L.M. Larsen. 1994a. Catabolism of arginine by *Carnobacterium* spp. isolated from vacuum-packed sugar-salted fish. *Curr. Microbiol. 29*: 95–99.

Leisner, J.J., F.K. Vogensen, J. Kollmann, B. Aideh, P. Vandamme, M. Vancanneyt and H. Ingmer. 2008. α-Chitinase activity among lactic acid bacteria. *Syst. Appl. Microbiol. 31*: 151–156.

Li, Z.-Y. and Y. Liu. 2006. Marine sponge *Craniella austrialiensis*-associated bacterial diversity revelation based on 16S rDNA library and biologically active Actinomycetes screening, phylogenetic analysis. *Lett. Appl. Microbiol. 43*: 410–416.

Liu, J.Y., A.H. Li, C. Ji and W.M. Yang. 2009. First description of a novel *Weissella* species as an opportunistic pathogen for rainbow trout *Oncorhynchus mykiss* (Walbaum) in China. *Vet. Microbiol. 136*: 314–320.

Liu, L., B. Zhang, H. Tong and X. Dong. 2006. *Pediococcus ethanolidurans* sp. nov., isolated from the walls of a distilled-spirit-fermenting cellar. *Int. J. Syst. Evol. Microbiol.* 56: 2405–2408.

Loch, T.P., W. Xu, S.M. Fitzgerald and M. Faisal. 2008. Isolation of a *Carnobacterium maltaromaticum*-like bacterium from systemically infected lake whitefish (*Coregonus clupeaformis*). *FEMS Microbiol. Lett. 288*: 76–84.

Lyhs, U., I. Snauwaert, S. Pihlajaviita, L.D., Vuyst and P. Vandamme. 2015. *Leuconostoc rapi* sp. nov., isolated from sous-vide-cooked rutabaga. *Int. J. Syst. Evol. Microbiol.* 65: 2586–2590.

Lynch, K.M., A. Lucid, E.K. Arendt, R.D. Sleator, B. Lucey and A. Coffey. 2015. Genomics of *Weissella cibaria* with an examination of its metabolic traits. *Microbiol. 161*: 914–930.

Magnusson, J., H. Jonsson, J. Schnurer and S. Roos. 2002. *Weissella soli* sp. nov., a lactic acid bacterium isolated from soil. *Int. J. Syst. Evol. Microbiol.* 52: 831–834.

Mäki, M. 2004. Lactic acid bacteria in vegetable fermentations. In Lactic Acid Bacteria. Microbiological and Functional Aspects, Eds. S. Salminen, A. von Wright and A. Ouwehand. 3rd ed., pp. 419–430. Marcel Dekker, New York.

Martinez-Murcia, A.J. and M.D. Collins. 1990. A phylogenetic analysis of the genus *Leuconostoc* based on reverse transcriptase sequencing of 16S rRNA. *FEMS Microbiol. Lett. 58*: 73–83.

Martinez-Murcia, A.J. and M.D. Collins. 1991. A phylogenetic analysis of an atypical *Leuconostoc*: Description of *Leuconostoc fallax* sp. nov. *FEMS Microbiol. Lett. 66*: 55–59.

Martinez-Murcia, A.J., N.M. Harland and M.D. Collins. 1993. Phylogenetic analysis of some leuconostocs and related organisms as determined from large-subunit rRNA gene sequences: Assessment of congruence of small- and large-subunit rRNA derived trees. *J. Appl. Bacteriol. 74*: 532–541.

Martin-Visscher, L.A., M.J. van Belkum, S. Garneau-Tsodikova, R.M. Whittal, J. Zheng, L.M. McMullen and J.C. Vederas. 2008. Isolation and characterization of carnocyclin A, a novel circular bacteriocin produced by *Carnobacterium maltaromaticum* UAL307. *Appl. Environ. Microbiol. 74*: 4756–4763.

Meisel, J., G. Wolf and W.P. Hammes. 1994. Heme-dependent cytochrome formation in *Lactobacillus maltaromicus* . *Syst. Appl. Microbiol. 17*: 20–23.

Mora, D., M.G. Fortina, C. Parini, D. Daffonchio and P.L. Manachini. 2000. Genomic subpopulations within the species *Pediococcus acidilactici* detected by multilocus typing analysis: Relationships between pediocin AcH/PA-1 producing and non-producing strains. *Microbiology 146*: 2027–2038.

Mora, D., M. Scarpellini, L. Franzetti, S. Colombo and A. Galli. 2003. Reclassification of *Lactobacillus maltaromicus* (Miller et al. 1974) DSM 20342[T] and DSM 20344 and *Carnobacterium piscicola* (Collin et al. 1987) DSM 20730[T] and DSM 20722 as *Carnobacterium maltaromaticum* comb. Nov. *Int. J. Syst. Evol. Microbiol. 53*: 675–678.

Morea, M., F. Baruzzi and P.S. Cocconcelli. 1999. Molecular and physiological characterization of dominant bacterial populations in traditional Mozzarella cheese processing. *J. Appl. Microbiol. 87*: 574–582.

Mundt, J.O. 1967. Spherical lactic acid-producing bacteria of southern-grown raw and processed vegetables. *Appl. Environ. Microbiol. 15*: 1303.

Nieto-Arribas, P., S. Seseña, J.M. Poveda, L. Palop and L. Cabezas. 2010. Genotypic and technological characterization of *Leuconostoc* isolates to be used as adjunct starters in Manchego cheese manufacture. *Food Microbiol. 27*: 85–93.

Nisitou, A., D. Dourou, M.-E. Filippousi, G. Banilas and C. Tassou. 2014. *Weissella uvarum* sp. nov., isolated from wine grapes. *Int. J. Syst. Evol. Microbiol. 64*: 3885–3890.

Niven, C.F. Jr., L.G. Buettner and J.B. Evans. 1954. Thermal tolerance studies on the heterofermentative lactobacilli that cause greening of cured meat products. *Appl. Microbiol. 2*: 26–29.

Niven, C.F. Jr. and J.B. Evans. 1957. *Lactobacillus viridescens* nov. spec., a heterofermentative species that produces a green discoloration of cured meat pigments. *J. Bacteriol. 73*: 758–759.

Nour, M. 1998. Studies on the large subunit rRNA genes and their flanking regions of leuconostocs. *Can. J. Microbiol. 44*: 807–818.

Oh, S.J., N.R. Shin, D.W. Hyun, P.S. Kim, J.Y. Kim, M.S. Kim, J.H. Yun and J.W. Bae. 2013. *Weissella diestrammenae* sp. nov., isolated from the gut of a camel cricket (*Diestrammena coreana*). *Int. J. Syst. Evol. Microbiol. 63*: 2951–2956.

Olano, A., J. Chua, S. Schroeder, A. Minari, M. La Salvia and G. Hall. 2001. *Weissella confusa* (basonym: *Lactobacillus confusus*) bacteremia: A case report. *J. Clin. Microbiol. 39*: 1604–1607.

Orberg, P.K. and W.E. Sandine. 1984. Common occurrence of plasmid DNA and vancomycin resistance in *Leuconostoc* spp. *Appl. Environ. Microbiol. 48*: 1129–1133.

Padonou, S.W., U. Schillinger, D.S. Nielsen, C.M.A.P. Franz, M. Hansen, J.D. Hounhouigan, M.C. Nago and M. Jakobsen. 2010. *Weissella beninensis* sp. nov., a motile lactic acid bacterium from submerged cassava fermentations, and emended description of the genus *Weissella* . *Int. J. Syst. Evol. Microbiol. 60*: 2193–2198.

Papagianni, M. and S. Anastasiadou. 2009. Pediocins: The bacteriocins of pediococci. Sources, production, properties and applications. *Microb. Cell Fact. 8*: 3. https://doi.org/10.1186/1475-2859-8-3.

Pfannebecker, J. and J. Fröhlich. 2008. Use of a species-specific multiplex PCR for the identification of pediococci. *Int. J. Food Microbiol. 128*: 288–296.

Pikuta, E.V., D. Marsic, A. Bej, J. Tang, P. Krader and R.B. Hoover. 2005. *Carnobacterium pleistocenium* sp. nov., a novel psychrotolerant, facultative anaerobe isolated from permafrost of the Fox tunnel in Alaska. *Int. J. Syst. Evol. Microbiol. 55*: 473–478.

Pothakos, V., Y.A. Aulia, I. Van Der Linden, M. Uyttendaele and F. Devlieghere. 2015b. Exploring the strain-specific attachment of *Leuconostoc gelidum* subsp. *gasicomitatum* on food contact surfaces. *Int. J. Food Microbiol. 199*: 41–46.

Pothakos, V., C. Snauwaert, P. De Vos, G. Huys, and F. Devlieghere. 2014b. Psychrotrophic members of *Leuconostoc gasicomitatum*, *Leuconostoc gelidum* and *Lactococcus piscium* dominate at the end of shelf-life in packaged and chilled-stored food products in Belgium. *Food Microbiol. 39*: 61–67.

Pothakos, V., G. Stellato, D. Ercolini and F. Devlieghere. 2015a. Processing environment and ingredients are both sources of *Leuconostoc gelidum*, which emerges as a major spoiler in ready-to-eat meals. *Appl. Environ. Microbiol. 81*: 3529–3541.

Pothakos, V., B. Taminiau, G. Huys, C. Nezer, G. Daube and F. Devlieghere. 2014a. Psychrotrophic lactic acid bacteria associated with production batch recalls and sporadic cases of early spoilage in Belgium between 2010 and 2014. *Int. J. Food Microbiol. 191*: 157–163.

Praet, J., I. Meeus, M. Cnockaert, K. Houf, G. Smagghe and P. Vandamme. 2015. Novel lactic acid bacteria isolated from the bumble bee gut: *Convivina intestini* gen. nov., sp. nov., *Lactobacillus bombicola* sp. nov., and *Weissella bombi* sp. nov. *Antonie van Leeuwenhoek. 107*: 1337–1349.

Qiao, N., J. Bechtner, M. Cnockaert, E. Depoorter, C. Díaz-Muñoz, P. Vandamme, L. De Vuyst and M.G. Gänzle. 2023. Comparative genomic analysis of *Periweissella* and the characterization of novel motile species. *Appl. Environ. Microbiol. 89*.

Rahkila, R., K. De Bruyne, P. Johansson, P. Vandamme and K.J. Björkroth. 2014. Reclassification of *Leuconostoc gasicomitatum* as *Leuconostoc gelidum* subsp. *gasicomitatum* comb. nov., description of *Leuconostoc gelidum* subsp. *aenigmaticum* subsp. nov., designation of *Leuconostoc gelidum* subsp. *gelidum* subsp. nov. and emended description of *Leuconostoc gelidum* . *Int. J. Syst. Evol. Microbiol. 64*: 1290–1295.

Ringø, E., U. Schillinger and W.H. Holzapfel. 2005. Antimicrobial activity of lactic acid bacteria isolated from aquatic animals and the use of lactic acid bacteria in aquaculture. In Microbial Ecology in Growing Animals, Eds. W.H. Holzapfel and P.J. Naughton, pp. 418–453. Elsevier, Edinburgh.

Ringø, E., M. Seppola, A. Berg, R.E. Olsen, U. Schillinger and W. Holzapfel. 2002. Characterization of *Carnobacterium divergens* strain 6251 isolated from intestine of Arctic charr (*Salvelinus alpinus* L.). *Syst. Appl. Microbiol. 25*: 120–129.

Sakamoto, K. and W.N. Konings. 2003. Beer spoilage bacteria and hop resistance. *Int. J. Food Microbiol. 89*: 105–124.

Salimnia, H., G.J. Alangaden, R. Bharadwaj, T.M. Painter, P.H. Chandrasekar and M.R. Fairfax. 2011. *Weissella confusa*: An unexpected cause of vancomycin-resistant Gram-positive bacteremia in immunocompromised hosts. *Transpl. Infect. Dis. 13*: 294–298.

Salvetti, E., H. Harris, G., Felis and P.W. O'Toole. 2018. Comparative genomics reveals robust phylogroups in the genus Lactobacillus as the basis for reclassification. *Appl. Environ. Microbiol.* AEM.00993–18; Accepted manuscript.

Samelis, J., F. Maurogenakis and J. Metaxopoulos. 1994. Characterisation of lactic acid bacteria isolated from naturally fermented Greek dry salami. *Int. J. Food Microbiol. 23*: 179–196.

Samelis, J., J. Rementzis, E. Tsakalidou and J. Metaxopoulos. 1998. Usefulness of rapid GC analysis of cellular fatty acids for distinguishing *Weissella viridescens*, *Weissella paramesenteroides*, *Weissella hellenica* and some non-identifiable, arginine-negative *Weissella* strains of meat origin. *Syst. Appl. Microbiol. 21*: 260–265.

Santos, E.M., A.M. Diez, C. González-Fernández, I. Jaime and J. Rovira. 2005. Microbiological and sensory changes in "Morcilla De Burgos" preserved in air, vacuum and modified atmosphere packaging. *Meat Sci. 71*: 249–255.

Scheirlinck, I., R. Van der Meulen, A. Van Schoor, M. Vancanneyt, L. De Vuyst, P. Vandamme and G. Huys. 2007. Influence of geographical origin and flour type on diversity of lactic acid bacteria in traditional Belgian sourdoughs. *Appl. Environ. Microbiol. 73*: 6262–6269.

Schillinger, U., B. Boehringer, S. Wallbaum, L. Caroline, A. Gonfa, M. Huch Nee Kostinek, W.H. Holzapfel and C.M. Franz. 2008. A genus-specific PCR method for differentiation between *Leuconostoc* and *Weissella* and its application in identification of heterofermentative lactic acid bacteria from coffee fermentation. *FEMS Microbiol. Lett. 286*: 222–226.

Schillinger, U. and W.H. Holzapfel. 1995. The genus *Carnobacterium*. In The Genera of Lactic Acid Bacteria, Eds. B.J.B. Wood and W.H. Holzapfel, pp. 307–326. Blackie Academic & Professional, London.

Schillinger, U., W.H. Holzapfel and K.J. Björkroth. 2006. Lactic acid bacteria. In Food Spoilage Microorganisms, Ed. C.W. Blackburn. 1st ed., pp. 541–578. Woodhead Publishing Limited, Cambridge.

Shannon, A.L., G. Attwood, D.H. Hopcroft and J.T. Christeller. 2001. Characterization of lactic acid bacteria in the larval midgut of the keratinophagous lepidopteran *Hofmannophila pseudospretella* . Lett. Appl. Microbiol. *32*: 36–41.

Shaw, B.G. and C.D. Harding. 1989. *Leuconostoc gelidum* sp. nov. and *Leuconostoc carnosum* sp. nov. from chill-stored meats. *Int. J. Syst. Bacteriol. 39*: 217–223.

Simpson, P.J., G.F. Fitzgerald, C. Stanton and R.P. Ross. 2006. Enumeration and identification of pediococci in powder-based products using selective media and rapid PFGE. *J. Microbiol. Methods 64*: 120–125.

Snauwaert, I., B. Hoste, K. De Bruyne, K. Peeters, L. De Vuyst, A. Willems and P. Vandamme. 2013a. *Carnobacterium iners* sp. nov., a psychrophilic, lactic acid-producing bacterium from the littoral zone of an Antarctic pond. *Int. J. Syst. Evol. Microbiol. 63*: 1370–1375.

Snauwaert, I., Z. Papalexandratou, L. De Vuyst and P. Vandamme. 2013b. Characterization of strains of *Weissella fabalis* sp. nov. and *Fructobacillus tropaeoli* from spontaneous cocoa bean fermentations. *Int. J. Syst. Evol. Microbiol. 63*: 1709–1716.

Snauwaert, I., P. Stragier, L. De Vuyst and P. Vandamme. 2015. Comparative genome analysis of *Pediococcus damnosus* LMG 28219, a strain well-adapted to the beer environment. *BMC Genom. 16*(1): 267.

Stiles, M.E. and W.H. Holzapfel. 1997. Lactic acid bacteria of foods and their current taxonomy. *Int. J. Food Microbiol. 36*: 1–29.

Susiluoto, T., H. Korkeala and J. Björkroth. 2003. *Leuconostoc gasicomitatum* is the dominating lactic acid bacterium in retail modified-atmosphere-packaged marinated broiler meat strip products on sell by day. *Int. J. Food Microbiol. 80:* 89–97.

Tanasupawat, S., A. Pakdeeto, C. Thawai, P. Yukphan and S. Okada. 2007. Identification of lactic acid bacteria from fermented tea leaves (*miang*) in Thailand and proposals of *Lactobacillus thailandensis* sp. nov., *Lactobacillus camelliae* sp. nov., and *Pediococcus siamensis* sp. nov. *J. Gen. Appl. Microbiol. 53*: 7–15.

Tanasupawat, S., O. Shida, S. Okada and K. Komagata. 2000. *Lactobacillus acidipiscis* sp. nov. and *Weissella thailandensis* sp. nov., isolated from fermented fish in Thailand. *Int. J. Syst. Evol. Microbiol. 50*: 1479–1485.

Tohno, M., M. Kitahara, H. Inoque, R. Uegaki, T. Irisawa, M. Ohkuma and K. Tajima. 2013. *Weissella oryzae* sp. nov., isolated from fermented rice grains. *Int. J. Syst. Evol. Microbiol. 63*: 1417–1420.

Toranzo, A.E., B. Novoa, A.M. Baya, F.M. Hetrick, J.L. Barja and A. Figueras. 1993b. Histopathology of rainbow trout, *Oncorhynchus mykiss* (Walbaum), and striped bass, *Morone saxatilis* (Walbaum), experimentally infected with *Carnobacterium piscicola* . *J. Fish Dis. 16*: 261–267.

Toranzo, A.E., J.L. Romalde, S. Nuñez, A. Figueras and J.L. Barja. 1993a. An epizootic in farmed, market-size rainbow trout in Spain caused by a strain of *Carnobacterium piscicola* of unusual virulence. *Dis. Aquat. Org. 17*: 87–99.

Tsakalidou, E., J. Samelis, J. Metaxopoulos and G. Kalantzopoulos. 1997. Atypical *Leuconostoc*-like *Weissella* strains isolated from meat, sharing low phenotypic relatedness with the so far recognized arginine-negative *Weissella* spp. as revealed by SDS-PAGE of whole cell proteins. *Syst. Appl. Microbiol. 20*: 659–664.

Tsenkovskii, L. 1878. Gel formation of sugar beet solutions (trans. title, orig. Russian). *Proc. Soc. Nat. Sci. Imper. Univ. Kharkov. 12*: 137–167.

Vancanneyt, M., M. Zamfir, M. De Wachter, I. Cleenwerck, B. Hoste, F. Rossi, F. Dellaglio, L. De Vuyst and J. Swings. 2006. Reclassification of *Leuconostoc argentinum* as a later synonym of *Leuconostoc lactis* . *Int. J. Syst. Evol. Microbiol. 56*: 213–216.

Vela, A.I., A. Fernández, Y. Bernaldo De Quirós, P. Herráez, L. Domínguez and J.F. Fernández-Garayzábal. 2011. *Weissella ceti* sp. nov., isolated from beaked whales (*Mesoplodon bidens*). *Int. J. Syst. Evol. Microbiol. 61*: 2758–2762.

Vela, A.I., C. Porrero, J. Goyache, A. Nieto, B. Sanchez, V. Briones, M.A. Moreno, L. Dominguez and J.F. Fernandez-Garayzabal. 2003. *Weissella confusa* infection in primate (*Cercopithecus mona*). *Emerg. Infect. Dis. 9*: 1307–1309.

Vihavainen, E.J. and K.J. Björkroth. 2007. Spoilage of value-added, high-oxygen modified-atmosphere packaged raw, beef steaks by *Leuconostoc gasicomitatum* and *Leuconostoc gelidum* . *Int. J. Food Microbiol. 119*: 340–345.

Vihavainen, E.J. and K.J. Björkroth. 2009. Diversity of *Leuconostoc gasicomitatum* associated with meat spoilage. *Int. J. Food Microbiol. 136*: 32–36.

Walling, E., E. Gindreau and A. Lonvaud-Funel. 2005. A putative glucan synthase gene *dps* detected in exopolysaccharide-producing *Pediococcus damnosus* and *Oenococcus oeni* strains isolated from wine and cider. *Int. J. Food Microbiol. 98*: 53–62.

Walter, J., C. Hertel, G.W. Tannock, C.M. Lis, K. Munro and W.P. Hammes. 2001. Detection of *Lactobacillus, Pediococcus, Leuconostoc*, and *Weissella* species in human feces by using group-specific PCR primers and denaturing gradient gel electrophoresis. *Appl. Environ. Microbiol. 67*: 2578–2585.

Wasney, M.A., R.A. Holley and D.S. Jayas. 2001. Cresol red thallium acetate sucrose inulin (CTSI) agar for the selective recovery of *Carnobacterium* spp. *Int. J. Food Microbiol. 64*: 167–174.

Williams, A.M. and M.D. Collins. 1990. Molecular taxonomic studies on *Streptococcus uberis* types I and II. Description of *Streptococcus parauberis* sp. nov. *J. Appl. Bacteriol. 68*: 485–490.

Wu, Y. and C.T. Gu. 2021a. Rejection of the reclassification of *Leuconostoc gasicomitatum* as *Leuconostoc gelidum* subsp. *gasicomitatum* based on whole genome analysis. *Int. J. Syst. Evol. Microbiol. 71:* 005027.

Wu, Y. and C.T. Gu. 2021b. *Leuconostoc falkenbergense* sp. nov., isolated from a lactic culture, fermentating string beans and traditional yogurt. *Int. J. Syst. Evol. Microbiol. 71*: 004602.

Yanagida, F., Y.-S. Chen and M. Yasaki. 2007. Isolation and characterization of lactic acid bacteria from lakes. *J. Basic Microbiol. 47*: 184–190.

Yuan, F., S. Yin, Y. Xu, L. Xiang, H. Wang, Z. Li, K. Fan and G. Pan. 2021. The richness and diversity of catalases in bacteria. *Front. Microbiol. 12*.

Zamudio-Maya, M., J. Narváez-Zapata and R. Rojas-Herrera. 2008. Isolation and identification of lactic acid bacteria from sediments of a coastal marsh using a differential selective medium. *Lett. Appl. Microbiol. 46*: 402–407.

Zhang, B., H. Tong and X. Dong. 2005. *Pediococcus cellicola* sp. nov., a novel lactic acid coccus isolated from a distilled-spirit-fermenting cellar. *Int. J. Syst. Evol. Microbiol. 55*: 2167–2170.

Zhu, S., D. Lin, S. Xiong, X. Wang, Z. Xue, B. Dong, X. Shen, X. Ma, J. Chen, and J. Yang. 2018. *Carnobacterium antarcticum* sp. nov., a psychrotolerant, alkaliphilic bacterium isolated from sandy soil in Antarctica. *Int. J. Syst. Evol. Microbiol. 68*(5): 1672–1677.

# Bifidobacteria
## *General Overview of Ecology, Taxonomy and Genomics*

**7**

Francesca Bottacini, Christian Milani, Francesca Turroni, Marco Ventura, and Douwe van Sinderen

## 7.1 GENERAL FEATURES

Bifidobacteria were first identified from stool samples of breast-fed infants in 1899 by Tissier and termed *Bacillus bifidus* (Tissier 1900). Thereafter, different taxonomic nomenclature designations have been employed, such as *Bacteroides bifidus* and *Lactobacillus bifidus*. Since 1973 bifidobacteria have been classified as a distinct genus (i.e., *Bifidobacterium*), originally including just 11 species (Poupard et al. 1973). Currently, the genus *Bifidobacterium* has expanded significantly (Alessandri et al. 2021) and encompasses over 100 (sub)species (see Table 7.1). Bifidobacteria are characterized as nonmotile, non-sporulating, catalase-negative, anaerobic (or microaerophilic), Gram-positive bacteria, with a relatively high GC content genome (Bergey et al. 2012). The designation of *Bifidobacterium* genus is derived from the fact that their cell morphology can assume a characteristically bifid, branched, or Y-shape. While the

**TABLE 7.1** General Features of *Bifidobacterium* Genomes

| BIFIDOBACTERIUM SPECIES | STATUS | SIZE (MB) | GC% | ORFS | ISOLATION |
|---|---|---|---|---|---|
| B. actinocoloniiforme | Draft | 1.8 | 62.7 | 1484 | Bumblebee digestive tract |
| B. adolescentis | Complete | 2.1 | 59.2 | 1649 | Intestine of adult |
| B. aemilianum | Draft | 2 | 61.1 | 1564 | Carpenter bee digestive tract |
| B. aerophilum | Draft | 3 | 63.6 | 2283 | Feces of tamarin |
| B. aesculapii | Draft | 2.8 | 64.6 | 2172 | Marmoset feces |
| B. amazonense | Draft | 3.3 | 64 | 2735 | Marmoset feces |
| B. angulatum | Draft | 2 | 59.4 | 1523 | Human feces |
| B. animalis subsp. animalis | Complete | 1.9 | 60.5 | 1501 | Rat feces |
| B. animalis subsp. lactis | Complete | 1.9 | 60.5 | 1518 | Fermented milk |
| B. anseris | Draft | 2.2 | 64 | 1757 | Goose feces |
| B. apousia | Draft | 2 | 60.5 | 1632 | Bee digestive tract |

*(Continued)*

**TABLE 7.1** (Continued) General Features of *Bifidobacterium* Genomes

| BIFIDOBACTERIUM SPECIES | STATUS | SIZE (MB) | GC% | ORFS | ISOLATION |
|---|---|---|---|---|---|
| B. apri | Draft | 2.4 | 59.2 | 1871 | Pig feces |
| B. aquikefiri | Draft | 2.4 | 52.3 | 2000 | Household water kefir |
| B. asteroides | Complete | 2.2 | 60 | 1653 | Honeybee hindgut |
| B. avesanii | Draft | 2.7 | 66.3 | 2133 | Tamarin feces |
| B. biavatii | Draft | 3.2 | 63.1 | 2557 | Feces of tamarin |
| B. bifidum | Complete | 2.2 | 62.5 | 1739 | Brest-fed infant feces |
| B. bohemicum | Draft | 2 | 57.4 | 1632 | Bumblebee digestive tract |
| B. bombi | Draft | 1.9 | 56.1 | 1454 | Bumblebee digestive tract |
| B. boum | Draft | 2.1 | 59.3 | 1726 | Bovine rumen |
| B. breve | Complete | 2.5 | 58.9 | 2076 | Infant intestine |
| B. callimiconis | Draft | 3 | 62 | 2330 | Marmoset feces |
| B. callitrichidarum | Draft | 3.1 | 61.8 | 2633 | Tamarin feces |
| B. callitrichos | Draft | 2.9 | 63.5 | 2364 | Feces of common marmoset |
| B. canis | Draft | 2.3 | 57.5 | 1935 | Dog feces |
| B. castoris | Draft | 2.5 | 65.4 | 2020 | Mouse feces |
| B. catenulatum subsp. catenulatum | Draft | 2.1 | 56.1 | 1664 | Adult intestine |
| B. catenulatum subsp. kashiwanohense | Draft | 2.3 | 56.2 | 1948 | Infant feces |
| B. catulorum | Draft | 2.6 | 63.2 | 2101 | Marmoset feces |
| B. cebidarum | Draft | 2.3 | 58.1 | 1936 | Monkey feces |
| B. choerinum | Draft | 2.1 | 65.5 | 1672 | Piglet feces |
| B. choladohabitans | Draft | 2.1 | 61 | 1580 | Bee digestive tract |
| B. choloepi | Draft | 2.2 | 65 | 1584 | Two-toed sloth feces |
| B. colobi | Draft | 2.7 | 58 | 2023 | Colobus feces |
| B. commune | Draft | 1.6 | 53.9 | 1303 | Bumblebee gut |
| B. coryneforme | Complete | 1.7 | 60.5 | 1364 | Honeybee hindgut |
| B. criceti | Draft | 2.1 | 62.5 | 1689 | Cricetus feces |
| B. crudilactis | Draft | 2.4 | 57.7 | 1883 | Raw cow milk |
| B. cuniculi | Draft | 2.5 | 64.9 | 2194 | Rabbit feces |
| B. dentium | Complete | 2.6 | 58.5 | 2129 | Oral cavity |
| B. dolichotidis | Draft | 1.9 | 50.4 | 1417 | Patagonian mara feces |
| B. erythrocebi | – | – | – | – | Primate feces |
| B. eulemuris | Draft | 3 | 62.2 | 2331 | Feces of the black lemur |
| B. felsineum | Draft | 2.4 | 57.1 | 1840 | Tamarin feces |
| B. gallicum | Draft | 2 | 57.6 | 1507 | Adult intestine |
| B. goeldii | Draft | 2.6 | 56.1 | 2014 | Marmoset feces |
| B. hapali | Draft | 2.8 | 54.5 | 2253 | Marmoset feces |
| B. imperatoris | Draft | 2.6 | 56.1 | 2128 | Tamarin feces |
| B. indicum | Complete | 1.7 | 60.5 | 1352 | Insect |
| B. italicum | Draft | 1.8 | 66.3 | 1767 | Rabbit feces |
| B. jacchi | Draft | 2.9 | 62.2 | 2003 | Marmoset feces |
| B. lemurum | Draft | 3 | 62.6 | 2321 | Feces of the ring-tailed lemur |
| B. leontopitheci | Draft | 3 | 64.7 | 2095 | Tamarin feces |
| B. longum subsp. infantis | Complete | 2.8 | 59.9 | 2500 | Intestine of infant |
| B. longum subsp. longum | Complete | 2.4 | 60 | 2028 | Adult intestine |

*(Continued)*

**TABLE 7.1**    (Continued) General Features of *Bifidobacterium* Genomes

| BIFIDOBACTERIUM SPECIES | STATUS | SIZE (MB) | GC% | ORFS | ISOLATION |
|---|---|---|---|---|---|
| B. longum subsp. suis | Draft | 2.3 | 60 | 1955 | Pig feces |
| B. longum subsp. suillum | Draft | 2.3 | 60 | 2015 | Piglet feces |
| B. longum subsp. iuvenis | Draft | 2.5 | 60 | 2056 | Infant feces |
| B. magnum | Draft | 1.8 | 58.7 | 1507 | Rabbit feces |
| B. margollesii | Draft | 2.8 | 61.9 | 2238 | Marmoset feces |
| B. mellis | Draft | 2.1 | 60.7 | 1620 | Bee digestive tract |
| B. merycicum | Draft | 2.3 | 60.3 | 1741 | Bovine rumen |
| B. miconis | Draft | 3.2 | 64.5 | 2383 | Marmoset feces |
| B. miconisargentati | Draft | 3 | 63.2 | 2252 | Marmoset feces |
| B. minimum | Draft | 1.9 | 62.7 | 1590 | Sewage |
| B. mizhiense | Draft | 2.1 | 60.2 | 1599 | Bee digestive tract |
| B. mongoliense | Draft | 2.2 | 62.8 | 1798 | Fermented mare's milk |
| B. moraviense | – | – | – | – | Marmoset feces |
| B. moukalabense | Draft | 2.5 | 59.9 | 2046 | Primate feces |
| B. myosotis | Draft | 2.9 | 62.5 | 2168 | Marmoset feces |
| B. oedipodis | – | – | – | – | Primate feces |
| B. olomucense | – | – | – | – | Primate feces |
| B. panos | – | – | – | – | Primate feces |
| B. parmae | Draft | 2.8 | 65.8 | 2194 | Marmoset feces |
| B. phasiani | Draft | 2.7 | 68.1 | 2021 | Phasianus feces |
| B. platyrrhinorum | Draft | 2.6 | 66.4 | 2049 | Primate feces |
| B. pluvialisilvae | Draft | 2.6 | 63.7 | 1958 | Primate feces |
| B. polysaccharolyticum | Draft | 2.2 | 59.5 | 1675 | Bee digestive tract |
| B. pongonis | Draft | 2.5 | 61.5 | 1863 | Primate feces |
| B. primatium | Draft | 2.7 | 63.2 | 2046 | Tamarin feces |
| B. pseudocatenulatum | Complete | 2.3 | 56.4 | 1791 | Infant feces |
| B. pseudolongum subsp. globosum | Draft | 1.9 | 63.4 | 1574 | Bovine rumen |
| B. pseudolongum subsp. pseudolongum | Draft | 1.9 | 63.1 | 1495 | Swine feces |
| B. psychraerophilum | Draft | 2.6 | 58.7 | 2122 | Pig cecum |
| B. pullorum subsp. pullorum | Draft | 2.1 | 64.3 | 1678 | Feces of chicken |
| B. pullorum subsp. gallinarum | Draft | 2.1 | 64.2 | 1654 | Chicken cecum |
| B. pullorum subsp. saeculare | Draft | 2.3 | 63.7 | 1857 | Rabbit feces |
| B. ramosum | Draft | 3 | 63.5 | 2239 | Tamarin feces |
| B. reuteri | Draft | 2.8 | 60.4 | 2149 | Marmoset feces |
| B. rousetti | Draft | 3 | 64.6 | 2507 | Bat feces |
| B. ruminantium | Draft | 2.2 | 59.2 | 1832 | Bovine rumen |
| B. saguinibicoloris | Draft | 2.7 | 65.9 | 1963 | Primate feces |
| B. saguini | Draft | 2.8 | 56.3 | 2321 | Feces of tamarin |
| B. saimiriisciurei | Draft | 2.8 | 63.6 | 2071 | Feces of tamarin |
| B. samirii | Draft | 2.6 | 66.6 | 1954 | Monkey feces |
| B. santillanense | Draft | 2.8 | 66 | 2148 | Primate feces |
| B. scaligerum | Draft | 2.6 | 58.3 | 2007 | Tamarin feces |
| B. scardovii | Draft | 3.1 | 64.6 | 2480 | Blood |
| B. simiarum | Draft | 2.8 | 63.7 | 2088 | Tamarin feces |

(Continued)

**TABLE 7.1**    (Continued) General Features of *Bifidobacterium* Genomes

| BIFIDOBACTERIUM SPECIES | STATUS | SIZE (MB) | GC% | ORFS | ISOLATION |
|---|---|---|---|---|---|
| *B. simiiventris* | Draft | 2.9 | 63 | 2176 | Primate feces |
| *B. stellenboschense* | Draft | 2.8 | 65.3 | 2202 | Feces of tamarin |
| *B. subtile* | Draft | 2.8 | 60.9 | 2260 | Sewage |
| *B. thermacidophilum* subsp. porcinum | Draft | 2.1 | 60.2 | 1738 | Piglet feces |
| *B. thermacidophilum* subsp. thermacidophilum | Draft | 2.2 | 60.4 | 1823 | Anaerobic digester |
| *B. thermophilum* | Draft | 2.2 | 60.1 | 1756 | Bovine rumen |
| *B. tibiigranuli* | Draft | 2.7 | 60.4 | 2251 | Water kefir |
| *B. tissieri* | Draft | 2.9 | 61 | 2260 | Marmoset feces |
| *B. tsurumiense* | Draft | 2.2 | 52.8 | 1629 | Hamster dental plaque |
| *B. vansinderenii* | Draft | 3.1 | 62.5 | 2522 | Feces of emperor tamarin |
| *B. vespertilionis* | Draft | 3 | 64.2 | 2321 | Bat feces |
| *B. xylocopae* | Draft | 1.8 | 62.8 | 1465 | Bee digestive tract |

majority of the members of this genus is strictly anaerobic, some species such as *Bifidobacterium aster-oides* and *Bifidobacterium animalis* subsp. *lactis* can tolerate oxygen and in fact are able to grow under aerobic conditions (Bottacini et al. 2012; Milani et al. 2013).

# 7.2  ECOLOGY OF BIFIDOBACTERIA

The study of the ecological origin together with genomic data represent important starting points to understand the physiological and metabolic behavior of a bacterium. It is widely accepted that selection pressure represents the driving force for ecological fitness, allowing an organism to successfully survive and multiply in a challenging environment. In terms of ecological origin, bifidobacteria have been isolated prevalently from the mammalian gut (i.e. fecal samples) (Milani et al. 2017b), but some species originate from the intestine of social insects and a small number of bird species (e.g., chicken, peacock and phasian). Notably, three bifidobacterial species have been isolated from environmental sources (*Bifidobacterium minimum*, *Bifidobacterium subtile*, and *Bifidobacterium thermacidophilum* subsp. *thermacidophilum*) (Trovatelli et al. 1974; Dong et al. 2000). Nonetheless, it is likely that such species originated from fecal contamination. Similarly, *Bifidobacterium animalis* subsp. *lactis*, *Bifidobacterium acquikefirii*, *Bifidobacterium mongoliense*, *Bifidobacterium crudilactis* and *B. tibiigranuli* (Meile et al. 1997; Delcenserie et al. 2007; Watanabe et al. 2009; Laureys et al. 2016; Eckel et al. 2020) were isolated from raw milk, fermented milk, or kefir, but it can be speculated that their presence in such environments may be the result of contamination from other sources, possibly also having a fecal origin.

A recent metagenomic profiling study based on amplicon sequencing of an internal transcribed spacer (ITS) (Milani et al. 2014a) in fecal samples collected from 67 different mammalian species revealed that no strict host specialization is observed in this genus and that bifidobacterial species are broadly found across mammals (Milani et al. 2017b). The ITS profiling identified *Bifidobacterium adolescentis*, *Bifidobacterium bifidum*, *Bifidobacterium longum* subsp. *longum*, and *Bifidobacterium pseudolongum* as the most abundant taxa in the gastrointestinal tract of mammalian species (Milani et al. 2017b).

In relation to the human host, it is now clear that bifidobacterial strains can be transmitted from mother to newborn during vaginal delivery (Duranti et al. 2017b; Feehily et al. 2023) and their colonization of and long-term persistence in the gastrointestinal tract is promoted by specialization of certain species according to an individual's diet and life stage (Arboleya et al. 2016). This explains the association of certain species such as *Bifidobacterium breve*, *Bifidobacterium bifidum*, and *Bifidobacterium longum* subsp. *infantis* to an infant-specific bifidobacterial microbiota, while an adult-specific bifidobacterial microbiota mostly encompass members of *Bifidobacterium adolescentis*, *Bifidobacterium catenulatum*, and *Bifidobacterium longum* subsp. *longum* (Turroni et al. 2009; Ventura et al. 2009a; Arboleya et al. 2016). Notably, with aging a general decrease of bifidobacterial diversity and abundance is observed, especially in the elderly population (Yatsunenko et al. 2012).

## 7.2.1 Role of Bifidobacteria in the Human Intestinal Microbiota

The human intestine is a complex ecosystem, where the microbial composition (named gut microbiota) is yet to be fully determined, although it is generally accepted that representatives of four bacterial phyla (Bacteroidota, Bacillota, Actinomycetota, and Pseudomonadota) are the main residents of this ecosystem (Eckburg et al. 2005; Human Microbiome Project 2012; Thursby and Juge 2017). Colonization of the infant gut represents a process that is influenced by several environmental and host factors, such as mode of delivery, type of feeding, and antibiotic usage (Rodriguez et al. 2015; Milani et al. 2017c). Immediately following birth, the neonatal intestine becomes rapidly colonized. In this context, facultative and aero-tolerant microorganisms dominate the intestinal ecosystem, reducing oxygen levels in the intestine, thereby facilitating the subsequent proliferation of a complex community dominated by anaerobic bacteria (Del Chierico et al. 2015). Bifidobacteria are among the first bacteria that colonize the intestine of the neonate (Turroni et al. 2012; Martin et al. 2016; Yassour et al. 2016; Duranti et al. 2017b), and these bacterial gut pioneers are considered key players in modulating mucosal physiology and stimulating the development of the host immune system (Turroni et al. 2014). In adulthood, levels of bifidobacteria significantly decrease (2%–14% relative abundance) yet remain stable (Odamaki et al. 2016). Nevertheless, there does not seem to be a strict infant-versus-adult subdivision of bifidobacterial taxa as bifidobacterial strains have been identified that are shared between mothers and their corresponding children (Milani et al. 2015b; Avershina et al. 2016; Feehily et al. 2023), across multiple generations and between family members (Odamaki et al. 2018). A survey of bifidobacterial communities by metagenomic sequencing was applied to fecal samples retrieved from 25 mother and newborn pairs, revealing the existence of common bifidobacterial strains shared between the mother and the respective child (Duranti et al. 2017b). A recent study based on the analysis of a cohort of 135 mother-infant pairs detected bifidobacterial transfer in almost 50% of cases, using a combination of metagenomics and culturomics, thus supporting the notion that mother-to-child strain transfer occurs relatively frequently (Feehily et al. 2023). Moreover, milk-mediated transmission of bifidobacterial strains may also be supported by their ability to utilize human milk oligosaccharides (HMOs) and/or HMO-derived glycans. It is plausible that the first microbial colonizers, represented by a small number of bifidobacterial species of the human gut include commensals that can metabolize such milk glycans.

# 7.3 GENOMICS AND BIFIDOBACTERIA

## 7.3.1 Biosynthetic Capabilities

Genomic data sets are extremely useful in reconstructing the metabolic capabilities of the organism under investigation. The rapid increase in genomic data available to date for the survey of bifidobacterial genomes has provided a wealth of information for comparative genomic studies. Genome comparisons have allowed large-scale prediction of metabolic pathways across bifidobacterial (sub) species, also supported by the combined use of comparative genomics and functional databases (Hernández-Plaza et al. 2023). The *in silico* analysis of bifidobacterial pan-genomes has indicated that homologs of all enzymes required for the fermentation of glucose and fructose to lactic acid and acetate through the characteristic "fructose-6-phosphate phosphoketolase shunt" (de Vries and Stouthamer 1969), as well as a partial Embden–Meyerhoff pathway required for biosynthetic purposes, are core genome features of this genus (Milani et al. 2014b).

Moreover, genes encoding complete biosynthetic pathways for amino acids, purines, pyrimidines, and vitamins were shown to be variably present among members of the genus *Bifidobacterium*, with generally fewer of such pathways identified in the genomes of bifidobacteria isolated from insects (Milani et al. 2014b).

Interestingly, metabolic pathways for the synthesis of B vitamins (i.e., riboflavin, tetrahydrofolate, thiamine, and pyridoxal 5′-phosphate) have also been identified in the pan-genome of bifidobacteria (Milani et al. 2014b). In the case of riboflavin, a complete biosynthetic operon responsible for the *de novo* production of this vitamin has been characterized in *B. longum* subsp. *infantis*, and homologs of this operon have been identified in 15 other species, mainly originating from primates (Milani et al. 2014b; Solopova et al. 2020). Specifically, riboflavin biosynthesis by *B. longum* subsp. *infantis* ATCC15697 has been shown to be under the transcriptional control of a dedicated FMN riboswitch (Solopova et al. 2020). Notably, tetrahydrofolate is predicted to be synthetized by most of the analyzed species of bifidobacteria isolated from humans (with the sole exception of *Bifidobacterium gallicum*) or other primates (with the sole exception of *B. biavatii*). In contrast, pathways needed for *de novo* biosynthesis of biotin, cobalamin, pantothenate, lipoate, and pyridoxine appear to be partially or completely absent in the bifidobacterial pangenome (Schell et al. 2002; Lee et al. 2008; Sela et al. 2008; Ventura et al. 2009b).

Whether bifidobacteria can synthesize cysteine and how are not fully clear. In this context, genomic exploration for putative genes involved in sulfur-containing amino acid transport did not reveal the presence of the genes required for sulfate transport and reduction to sulfide for *B. longum* subsp. *longum* NCC2705 (Schell et al. 2002) and *B. bifidum* PRL2010 (Ferrario et al. 2015). However, the chromosome sequences of *B. bifidum* PRL2010 encompass the genes related to cysteine synthesis using the homologs of genes encoding cysteine synthase/cysthathione β synthase, O-acetylhomoserine aminocarboxy-propyltransferase, and cystathionine synthase starting from a reduced sulfur source (Ferrario et al. 2015).

These biosynthetic capabilities indicate that *Bifidobacterium* species have adapted to an environment where they cannot rely on an extraneous source of amino acids and certain vitamins. Comparative genome studies complemented by the development of a chemically defined synthetic medium (CDM) for bifidobacteria have confirmed many of these predicted prototrophic and auxotrophic characteristics (Cronin et al. 2012; Ferrario et al. 2015).

## 7.3.2 Metabolic Capabilities

The human body can only metabolize a small fraction of glycans available in the gut (Flint et al. 2012). All other nondigestible carbohydrates, such as those of plant or animal origin, represent critical energy sources for the survival and proliferation of many beneficial microbes of the gut microbiota, including bifidobacteria (El Kaoutari et al. 2013). For this reason, identifying the various molecular strategies employed by gut microbes to harvest and metabolize complex glycans represent a critical step in understanding their adaptation to the challenges encountered in the gastrointestinal tract (GIT) environment. Preliminary *in silico* analyses have revealed that bifidobacterial strains mainly rely on carbohydrates as a source of energy, through the activity of an enzymatic arsenal for saccharides degradation, known as the bifidobacterial glycobiome. Remarkably, it represents one of the largest predicted enzyme resources for carbohydrate metabolism so far identified among known gut commensals (Milani et al. 2015a). The bifidobacterial glycobiome known so far includes 3385 genes encoding predicted carbohydrate-active enzymes, encompassing glycosylhydrolases (GHs), glycosyltransferases (GTs), and carbohydrate esterases

(CEs) that are found across 57 GH, 13 GT, and 7 CE families (Milani et al. 2014b, 2015a, 2015c). In this context, the high abundance of members of the GH13 family, which are known to degrade starch and derived/similar glycans, together with the presence of GH3, GH43, and GH51 (plant-related GH families) in the glycobiomes of certain bifidobacterial taxa (e.g., *B. longum* subsp. *longum*, *B. adolescentis*, and *B. catenulatum*), suggests an adaptation of these bifidobacterial species to a vegetarian or omnivorous diet (Turroni et al. 2017a). Such a high GH content suggests that bifidobacterial strains have adopted a very particular ecological specialization toward the (predominantly) gut environment.

Among the diet-derived, nondigestible carbohydrates utilized by human-associated bifidobacteria are oligosaccharides (raffinose, melezitose, maltotriose, and stachyose), polyols (mannitol and sorbitol), and dietary fibers (resistant starches, maltodextrins, fructans, pectin, cellulose, galactan, xylan, arabinan, arabinogalactan, and arabinoxylan) (Pokusaeva et al. 2011). Bifidobacterial species commonly associated with the infant gastrointestinal tract also utilize various host-derived glycans such as human milk oligosaccharides (HMOs), O-linked glycans (e.g. mucus-derived carbohydrates), or N-linked glycans (e.g. saccharidic components of glycoproteins) (Ojima et al. 2022; Karav et al. 2016). Accordingly, glycobiomes of infant-associated bifidobacterial species are enriched in GH families that are essential for milk and host glycan degradation, such as those representing exo-sialidases, fucosidases, hexosaminidase, and lacto-N-biosidase activities (GH33, GH29, GH95, and GH20) (Milani et al. 2015c; Turroni et al. 2015; James et al. 2019). Glycan utilization in bifidobacteria is achieved through cooperative cross-feeding, where different species contribute to the breakdown of complex carbohydrates in a synergistic manner. In the case of HMOs, strains of *B. longum* subsp. *infantis* utilize a broad range of these glycans through initial uptake and subsequent intracellular degradation, supported by the enzymatic activity of α-fucosidase, α-sialidase, β-galactosidase, and β-N-hexosaminidase (Sela et al. 2008). In contrast, *B. bifidum* primarily conducts extracellular degradation of HMOs and subsequently internalizes the resulting galactose, glucose, and lacto-N-triose units (Turroni et al. 2010b). Others, sometimes called scavenger species, such as *B. breve* and *B. longum* subsp. *longum* can utilize just a limited range of structurally diverse HMOs, thus relying on the foraging activity of *B. bifidum* to access sugar units resulting from the extracellular hydrolysis of milk glycans (Turroni et al. 2018). Notably, utilization of certain fucosylated HMOs (2'-fucosyllactose and 3-fucosyllactose) has been so far observed in *B. longum* subsp. *infantis* and some strains of *B. longum* subsp. *suis*, *B. longum* subsp. *longum*, *B. kashiwanohense*, and *B. pseudocatenulatum* and is associated with the presence of a genomic island containing genes such as α-fucosidase, L-fuconolactone hydrolase, fuconate dehydratase, and L-fucose mutarotase (Garrido et al. 2016; Bunesova et al. 2016; James et al. 2019).

Regarding mucin, functional genomic investigations have outlined that *B. bifidum* species harbors the genetic arsenal responsible for mucin metabolism (sialidases, fucosidases, N-acetylgalactosaminidase, N-acetyl-β-hexosaminidases, and β-galactosidases), thus making the utilisation of this protein decorated by (many) N- and O-linked glycans a unique feature of this taxon (Turroni et al. 2010b; Duranti et al. 2015).

Other examples of bifidobacterial species genetically adapted to access specific glycan sources are those isolated from social insects (e.g., *Bifidobacterium asteroides*, *Bifidobacterium coryneforme*, *Bifidobacterium indicum*, *Bifidobacterium actinocoloniiforme*, *Bifidobacterium bohemicum*, and *Bifidobacterium bombi*) whose glycobiomes are enriched in GH families dedicated to the utilization of simple sugars. Furthermore, *in silico* genome analyses highlighted the presence of a metabolic pathway involved in the utilization of trehalose, which represents a typical glycan storage and blood sugar of many insects (Bottacini et al. 2012).

## 7.3.3 Host Interaction and Colonization

The rapid increase in the number of available genomic sequences of members of the genus *Bifidobacterium* has provided a better understanding of the evolutionary path followed by these microbes and allowed us to reveal the diverse strategies adopted to colonize their host. In particular comparative and functional genomics have highlighted how these microorganisms evolved to utilize a wide range of glycans, which are worthy of continued investigation for their potential roles in enhancing colonization of the human gut. Commensals such as bifidobacteria must coexist with their host and must evade or survive the diversity of responses that the host will generate to combat and eradicate unwanted and pathogenic bacteria.

Bifidobacterial gut colonization relies on the production of extracellular structures to promote adhesion to the intestinal mucosa, evasion of the host immune system, and ability to form biofilms. In relation to the ability of bifidobacteria to attach to the lining of the human intestine, various species of *Bifidobacterium* have been shown to possess genetic loci responsible for the production of sortase-dependent pili, specialized surface appendages, believed to be involved in the binding to cell receptors and mediating cross-talk with the host (Milani et al. 2017a). Among the several bifidobacterial species known to date, *B. dentium* represents an interesting case, as member of this species have been shown to possess the highest number of such pilus biosynthesis-encoding clusters, possibly due to their adaptation to colonize the oral cavity (Milani et al. 2017a). It has recently been shown that the expression of such pili in *Bifidobacterium breve* UCC2003 is modulated by a replication slippage mechanism promoted by the presence of a G-rich tract located upstream of the operon. Further genome-wide comparisons across the *Bifidobacterium* pan-genome have identified such G-rich regions in several bifidobacterial species, thus suggesting that this mechanism is not restricted to *B. breve* (Penno et al. 2022). Sortase-dependent pili are also shown to promote microbe–microbe interactions. In the case of *B. bifidum* PRL2010, sortase-dependent pili are responsible for the binding of fibronectin and collagen but also promote bacterial recruitment and aggregation (Turroni et al. 2014).

A second type of pilus involved in the mechanism of host colonization is the tight adherence (Tad) locus. Functional genomics have demonstrated that this type of pilus is only expressed during *in vivo* colonization and promotes gut persistence in *B. breve* (O'Connell Motherway et al. 2011).

Bifidobacterial gut colonization is also promoted by the presence of an extracellular polysaccharidic layer (EPS) in the cell surface. Some strains of bifidobacteria have been shown to synthesize various types of exopolysaccharides, and further genomic analysis has demonstrated that the genetic locus responsible for EPS biosynthesis is highly variable and strain specific (Ferrario et al. 2016; Fanning et al. 2012; Bottacini et al. 2018b). This extracellular polysaccharide has been shown to play an important role in bacterial colonization, as it seems to enhance the bacterial adhesion to the intestinal epithelium while also functioning as a protective layer suppressing inflammatory host response and preventing clearance from the host immune system (Schiavi et al. 2016; Fanning et al. 2012). In this regard, it has been shown that EPS produced by *Bifidobacterium breve* UCC2003 blocks the maturation of dendritic cells and activation of antigen-specific CD4+ T cells responses to *B. breve*, thus suggesting an immunomodulatory role of this extracellular polysaccharides that is important for immune evasion and colonization of this beneficial bacterium (Hickey et al. 2021).

A recent study conducted in *B. breve* UCC2003 also highlighted how the EPS produced by this strain acts as a nutrient substrate for other members of the infant gut community, thus suggesting its involvement in supporting microbe–microbe interactions in the gut (Püngel et al. 2020). Furthermore, the secretion of an EPS in UCC2003, together with protein interaction, has been shown to promote biofilm formation by facilitating cell accumulation and attachment in response to bile exposure (Kelly et al. 2020).

Altogether, the evidence collected so far suggests that extracellular structures collectively support the ability of *Bifidobacterium* to adhere to the gut lining, resist environmental stress, and compete with potential pathogens, all essential aspects to ensure a successful gut establishment.

## 7.3.4 Mobile Genetic Elements in Bifidobacteria

Genomic analyses have indicated that mobile elements contribute significantly to bifidobacterial genome structure, with the presence of variable numbers of IS elements, transposable elements. (episomal) plasmids, and prophage-like elements (Barrangou et al. 2009; Kim et al. 2009; Lee et al. 2008; Schell et al. 2002; Sela et al. 2008; Ventura et al. 2009b; Ventura et al. 2009c). Eight IS families have been identified in bifidobacterial genomes sequenced thus far: IS3, IS21, IS30, IS110, IS150, IS256, IS607/IS200, and ISL3. IS elements are often associated with genome rearrangement or deletion events (Darling et al. 2008) and are therefore major driving forces in the acquisition or loss of important functions. A recent survey aimed at characterizing the *Bifidobacterium* mobilome identified IS3 family is the most widespread among this genus and identified over 200 mobile genetic hotspots (MGH) associated with horizontal gene transfer (HGT) events and acquisition of antibiotic resistance (AR) genes across bifidobacterial species (Mancino et al. 2019). This study highlighted the presence of DNA integration hotspots and AR in less than 20% of bifidobacterial strains, the majority of which represent isolates from the human gastrointestinal tract, as a result of the widespread occurrence of AR genes in the human gut environment (Mancino et al. 2019).

Extrachromosomal elements such as cryptic plasmids do not occur frequently in *Bifidobacterium*, and they were so far detected only in representatives of *B. longum* subsp. *longum, B. pseudolongum* subsp. *globosum, B. indicum, B. asteroides, B. breve, B. bifidum, B. catenulatum*, and *B. pseudocatenulatum* (Lee and O'Sullivan 2010); however, further sequencing efforts may expand this scenario. Notably, a recent study identified the presence of a >190-Kb extrachromosomal element (megaplasmid pMP7017) in *Bifidobacterium breve* JCM7017 and demonstrated its ability to transfer through conjugation to strains of *Bifidobacterium breve* and *Bifidobacterium longum* subsp. *longum* (Bottacini et al. 2015). Further molecular analysis depicted the replication mechanism of this large plasmid following a theta-type bidirectional replication (Dineen et al. 2021).

Bifidobacteria have been known to be resistant to phage infection, despite the fact that their genomes frequently harbor integrated prophage-like elements (bifidoprophages). A first comparative genomic survey of *Bifidobacterium*-type strains has highlighted the presence of bifidoprophages in many strains (Lugli et al. 2016); however, only a limited number of these elements have been shown to be inducible in *B. breve* and *B. longum* (Mavrich et al. 2018). A further analysis of bifidoprophages extended to 625 bifidobacterial genomes across the whole genus identified 598 putative and likely complete prophage sequences in the majority of strains, where *Bifidobacterium biavatii* DSM 23969, *Bifidobacterium imperatoris* LMG 30297, and *Bifidobacterium cuniculi* LMG 10738 were shown to harbor seven, six, and five prophage-like elements in their respective genomes (Mancino et al. 2019), thus indicating that bifidobacteria are regularly exposed to lysogenic phages in the GIT environment. A recent survey aimed at investigating the diversity of prophage-like elements in 585 human-associated bifidobacteria identified 480 putative bifidoprophages, with an average of almost one element per genome. Further comparisons highlighted a large degree of genomic diversity across these elements; however, the analysis also identified a few candidate bifidophages possibly capable of infecting multiple species (Buckley et al. 2021). Further evidence will need to be collected in order to understand the mechanism of phage-host interaction and the implication in early life colonization of bifidobacterial strains.

Bifidobacteria are known to possess a strong genetic barrier that protect the bacterium from foreign DNA. This barrier consists of the presence of several restriction–modification systems, which employ methylation to protect the cell DNA and cut unmethylated exogenous mobile DNA elements. R-M systems are ubiquitous in bifidobacterial genomes, and the genetic loci responsible for the expression of these multi-subunit enzymes occur at an average of two complete loci per strain (Lugli et al. 2019; Bottacini et al. 2018a). A recent study conducted in *Bifidobacterium breve* compiled the first catalogue of DNA motifs recognized by R-M systems in this species and demonstrated how DNA methylation profiles are highly variable, thus suggesting the existence of strain-specific restriction barriers in this taxon (Bottacini et al. 2018a).

## 7.4 CONCLUSIONS

The genomics era has embraced members of the genus *Bifidobacterium*, providing a significantly better understanding of the evolutionary path followed by this group of bacteria. Furthermore, genomic and comparative genomic analyses of bifidobacteria have highlighted key genes of these microorganisms, such as those involved in carbohydrate metabolism, which are worthy of continued investigation for their potential roles in colonization of the human gut. In this context, ecological surveys based on new culture-independent techniques coupled with the understanding of metabolic capabilities and nutrient requirements of bifidobacteria will provide the identification of novel bifidobacterial species from the mammalian gut. A major achievement is constituted by the recognition of over 100 (sub)species in the *Bifidobacterium* genus, an information resource that has been crucial to obtaining a more robust image of phylogeny and ecology of bifidobacterial species, while it has also generated insights as to how this genus has developed as an independent taxonomic unit within the Eubacteria. From a broader perspective, it can be expected that the availability of a larger number of sequenced genomes will facilitate the development of universal genome sequence analysis schemes, which will allow the adoption of a more natural species concept. Commensals such as bifidobacteria must coexist with their host and must evade or survive the diversity of responses that the host will generate to combat and eradicate unwanted and pathogenic bacteria. Understanding the molecular mechanisms underlying the fascinating ability of the human immune system to differentiate between beneficial and harmful bacteria is a significant scientific challenge for the future.

Despite the advanced insights into microbial composition and genomics, a knowledge gap still remains in the characterization of molecular activity, host–microbe and microbe–microbe interactions for many bifidobacterial species in the intestinal environment. There is also significant limited knowledge on how phage predation affects the establishment of bifidobacterial species across the human lifespan and the underlying phage–host interactions. Further functional genomic applications will allow to expand our knowledge on the implication of genetic features in the colonization of such a complex and interactive environmental niche.

# BIBLIOGRAPHY

Alessandri, G., van Sinderen, D., and Ventura, M. (2021) The genus *Bifidobacterium*: From genomics to functionality of an important component of the mammalian gut microbiota. *Comput Struct Biotechnol J* **19**: 1472–1487.

Arboleya, S., Watkins, C., Stanton, C., and Ross, R.P. (2016) Gut bifidobacteria populations in human health and aging. *Front Microbiol* **7**: 1204.

Avershina, E., Lundgard, K., Sekelja, M., Dotterud, C., Storro, O., Oien, T., et al. (2016) Transition from infant to adult-like gut microbiota. *Environ Microbiol* **18**: 2226–2236.

Barrangou, R., Briczinski, E.P., Traeger, L.L., Loquasto, J.R., Richards, M., Horvath, P., et al. (2009) Comparison of the complete genome sequences of *Bifidobacterium animalis* subsp. *lactis* DSM 10140 and Bl-04. *J Bacteriol* **191**: 4144–4151.

Bergey, D.H., Goodfellow, M., Whitman, W.B., and Parte, A.C. (2012) *The Actinobacteria. Bergey's Manual of Systematic Bacteriology*. Springer, New York.

Bottacini, F., Milani, C., Turroni, F., Sanchez, B., Foroni, E., Duranti, S., et al. (2012) *Bifidobacterium asteroides* PRL2011 genome analysis reveals clues for colonization of the insect gut. *PLoS ONE* **7**: e44229.

Bottacini, F., Morrissey, R., Esteban-Torres, M., James, K., van Breen, J., Dikareva, E., et al. (2018b) Comparative genomics and genotype-phenotype associations in *Bifidobacterium breve*. *Sci Rep* **8**(1): 10633.

Bottacini, F., Morrissey, R., Roberts, R.J., James, K., van Breen, J., Egan, M., et al. (2018a) Comparative genome and methylome analysis reveals restriction/modification system diversity in the gut commensal *Bifidobacterium breve*. *Nucleic Acids Res* **46**(4): 1860–1877.

Bottacini, F., O'Connell Motherway, M., Casey, E., McDonnell, B., Mahony, J., Ventura, M., et al. (2015) Discovery of a conjugative megaplasmid in *Bifidobacterium breve*. *Appl Environ Microbiol* **81**(1): 166–176.

Buckley, D., Odamaki, T., Xiao, J., Mahony, J., van Sinderen, D., and Bottacini, F. (2021) Diversity of human-associated bifidobacterial prophage sequences. *Microorganisms* **9**(12): 2559.

Bunesova, V., Lacroix, C., and Schwab, C. (2016) Fucosyllactose and L-fucose utilization of infant *Bifidobacterium longum* and *Bifidobacterium kashiwanohense*. *BMC Microbiol* **16**(1): 248.

Cronin, M., Zomer, A., Fitzgerald, G.F., and van Sinderen, D. (2012) Identification of iron-regulated genes of *Bifidobacterium breve* UCC2003 as a basis for controlled gene expression. *Bioeng Bugs* **3**: 157–167.

Darling, A.E., Miklos, I., and Ragan, M.A. (2008) Dynamics of genome rearrangement in bacterial populations. *PLoS Genet* **4**: e1000128.

Delcenserie, V., Gavini, F., Beerens, H., Tresse, O., Franssen, C., and Daube, G. (2007) Description of a new species, *Bifidobacterium crudilactis* sp. nov., isolated from raw milk and raw milk cheeses. *Syst Appl Microbiol* **30**: 381–389.

Del Chierico, F., Vernocchi, P., Petrucca, A., Paci, P., Fuentes, S., Pratico, G., et al. (2015) Phylogenetic and metabolic tracking of gut microbiota during perinatal development. *PLoS ONE* **10**: e0137347.

de Vries, W., and Stouthamer, A.H. (1969) Factors determining the degree of anaerobiosis of *Bifidobacterium* strains. *Arch Mikrobiol* **65**: 275–287.

Dineen, R.L., Penno, C., Kelleher, P., Bourin, M.J.B., O'Connell-Motherway, M., and van Sinderen, D. (2021) Molecular analysis of the replication functions of the bifidobacterial conjugative megaplasmid pMP7017. *Microb Biotechnol* **14**(4): 1494–1511.

Dong, X., Xin, Y., Jian, W., Liu, X., and Ling, D. (2000) *Bifidobacterium thermacidophilum* sp. nov., isolated from an anaerobic digester. *Int J Syst Evol Microbiol* **50**(Pt 1): 119–125.

Duranti, S., Lugli, G.A., Mancabelli, L., Armanini, F., Turroni, F., James, K., et al. (2017b) Maternal inheritance of bifidobacterial communities and bifidophages in infants through vertical transmission. *Microbiome* **5**: 66.

Duranti, S., Milani, C., Lugli, G.A., Turroni, F., Mancabelli, L., Sanchez, B., et al. (2015) Insights from genomes of representatives of the human gut commensal *Bifidobacterium bifidum*. *Environ Microbiol* **17**: 2515–2531.

Eckburg, P.B., Bik, E.M., Bernstein, C.N., Purdom, E., Dethlefsen, L., Sargent, M., et al. (2005) Diversity of the human intestinal microbial flora. *Science* **308**: 1635–1638.

Eckel, V.P.L., Ziegler, L.M., Vogel, R.F., and Ehrmann, M. (2020) *Bifidobacterium tibiigranuli sp.* nov. isolated from homemade water kefir. *Int J Syst Evol Microbiol* **70**(3): 1562–1570.

El Kaoutari, A., Armougom, F., Gordon, J.I., Raoult, D., and Henrissat, B. (2013) The abundance and variety of carbohydrate-active enzymes in the human gut microbiota. *Nat Rev Microbiol* **11**: 497–504.

Fanning, S., Hall, L.J., and van Sinderen, D. (2012) *Bifidobacterium breve* UCC2003 surface exopolysaccharide production is a beneficial trait mediating commensal-host interaction through immune modulation and pathogen protection. *Gut Microbes* **3**: 420–425.

Feehily, C., O'Neill, I.J., Walsh, C.J., Moore, R.L., Killeen, S.L., Geraghty, A.A., et al. (2023) Detailed mapping of *Bifidobacterium* strain transmission from mother to infant via a dual culture-based and metagenomic approach. *Nat Commun* **14**(1): 3015.

Ferrario, C., Duranti, S., Milani, C., Mancabelli, L., Lugli, G.A., Turroni, F., et al. (2015) Exploring amino acid auxotrophy in *Bifidobacterium bifidum* PRL2010. *Front Microbiol* **6**: 1331.

Ferrario, C., Milani, C., Mancabelli, L., Lugli, G.A., Duranti, S., Mangifesta, M., et al. (2016) Modulation of the eps-ome transcription of bifidobacteria through simulation of human intestinal environment. *FEMS Microbiol Ecol* **92**: fiw056.

Flint, H.J., Scott, K.P., Duncan, S.H., Louis, P., and Forano, E. (2012) Microbial degradation of complex carbohydrates in the gut. *Gut Microbes* **3**: 289–306.

Garrido, D., Ruiz-Moyano, S., Kirmiz, N., Davis, J.C., Totten, S.M., Lemay, D.G., et al. (2016) A novel gene cluster allows preferential utilization of fucosylated milk oligosaccharides in *Bifidobacterium longum* subsp. *longum* SC596. *Sci Rep* **6**: 35045.

Hernández-Plaza, A., Szklarczyk, D., Botas, J., Cantalapiedra, C.P., Giner-Lamia, J., Mende, D.R., et al. (2023) egg-NOG 6.0: Enabling comparative genomics across 12 535 organisms. *Nucleic Acids Res* **51**(D1): D389–D394.

Hickey, A., Stamou, P., Udayan, S., Ramón-Vázquez, A., Esteban-Torres, M., Bottacini, F., et al. (2021) *Bifidobacterium breve* exopolysaccharide blocks dendritic cell maturation and activation of CD4+ T Cells. *Frontiers Microbiol* **12**: 653587.

Human Microbiome Project, C. (2012) Structure, function and diversity of the healthy human microbiome. *Nature* **486**: 207–214.

James, K., Bottacini, F., Contreras, J.I.S., Vigoureux, M., Egan, M., Motherway, M.O., et al. (2019) Metabolism of the predominant human milk oligosaccharide fucosyllactose by an infant gut commensal. *Sci Rep* **9**(1): 15427.

Karav, S., Le Parc, A., Leite Nobrega de Moura Bell, J.M., Frese, S.A., Kirmiz, N., Block, D.E., Barile, D., et al. (2016) Oligosaccharides released from milk glycoproteins are selective growth substrates for infant-associated bifidobacteria. *Appl Environ Microbiol* **82**(12): 3622–3630.

Kelly, S.M., Lanigan, N., O'Neill, I.J., Bottacini, F., Lugli, G.A., Viappiani, A., et al. (2020) Bifidobacterial biofilm formation is a multifactorial adaptive phenomenon in response to bile exposure. *Sci Rep* **10**(1): 11598.

Kim, J.F., Jeong, H., Yu, D.S., Choi, S.H., Hur, C.G., Park, M.S., et al. (2009) Genome sequence of the probiotic bacterium *Bifidobacterium animalis* subsp. *lactis* AD011. *J Bacteriol* **191**: 678–679.

Laureys, D., Cnockaert, M., De Vuyst, L., and Vandamme, P. (2016) *Bifidobacterium aquikefiri* sp. nov., isolated from water kefir. *Int J Syst Evol Microbiol* **66**: 1281–1286.

Lee, J.H., Karamychev, V.N., Kozyavkin, S.A., Mills, D., Pavlov, A.R., Pavlova, N.V., et al. (2008) Comparative genomic analysis of the gut bacterium *Bifidobacterium longum* reveals loci susceptible to deletion during pure culture growth. *BMC Genom* **9**: 247.

Lee, J.H., and O'Sullivan, D.J. (2010) Genomic insights into bifidobacteria. *MMBR* **74**(3): 378–416.

Lugli, G.A., Duranti, S., Albert, K., Mancabelli, L., Napoli, S., Viappiani, A., et al. (2019) Unveiling genomic diversity among members of the species *Bifidobacterium pseudolongum*, a widely distributed gut commensal of the animal kingdom. *Appl Environ Microbiol* **85**(8): e03065–18.

Lugli, G.A., Milani, C., Turroni, F., Tremblay, D., Ferrario, C., Mancabelli, L., et al. (2016) Prophages of the genus *Bifidobacterium* as modulating agents of the infant gut microbiota. *Environ Microbiol* **18**(7): 2196–2213.

Mancino, W., Lugli, G.A., Sinderen, D.V., Ventura, M., and Turroni, F. (2019) Mobilome and resistome reconstruction from genomes belonging to members of the *Bifidobacterium* genus. *Microorganisms* **7**(12): 638.

Martin, R., Makino, H., Cetinyurek Yavuz, A., Ben-Amor, K., Roelofs, M., Ishikawa, E., et al. (2016) Early-life events, including mode of delivery and type of feeding, siblings and gender, shape the developing gut microbiota. *PLoS ONE* **11**: e0158498.

Mavrich, T.N., Casey, E., Oliveira, J., Bottacini, F., James, K., Franz, C.M.A.P., et al. (2018) Characterization and induction of prophages in human gut-associated *Bifidobacterium* hosts. *Sci Rep* **8**(1): 12772.

Meile, L.W., Reuger, U., Gut, C., Kaufmann, P., Dasen, G., Wenger, S., and Teuber, T. (1997) *Bifidobacterium lactis* sp. nov., a moderately oxygen tolerant species isolated from fermented milk. *Syst Appl Microbiol* **20**: 57–64.

Milani, C., Duranti, S., Bottacini, F., Casey, E., Turroni, F., Mahony, J., et al. (2017c) The first microbial colonizers of the human gut: Composition, activities, and health implications of the infant gut microbiota. *Microbiol Mol Biol Rev* **81**.

Milani, C., Duranti, S., Lugli, G.A., Bottacini, F., Strati, F., Arioli, S., et al. (2013) Comparative genomics of *Bifidobacterium animalis* subsp. *lactis* reveals a strict monophyletic bifidobacterial taxon. *Appl Environ Microbiol* **79**: 4304–4315.

Milani, C., Lugli, G.A., Duranti, S., Turroni, F., Bottacini, F., Mangifesta, M., et al. (2014b) Genomic encyclopedia of type strains of the genus *Bifidobacterium*. *Appl Environ Microbiol* **80**: 6290–6302.

Milani, C., Lugli, G.A., Duranti, S., Turroni, F., Mancabelli, L., Ferrario, C., et al. (2015c) Bifidobacteria exhibit social behavior through carbohydrate resource sharing in the gut. *Sci Rep* **5**: 15782.

Milani, C., Lugli, G.A., Turroni, F., Mancabelli, L., Duranti, S., Viappiani, A., et al. (2014a) Evaluation of bifidobacterial community composition in the human gut by means of a targeted amplicon sequencing (ITS) protocol. *FEMS Microbiol Ecol* **90**: 493–503.

Milani, C., Mancabelli, L., Lugli, G.A., Duranti, S., Turroni, F., Ferrario, C., et al. (2015b) Exploring vertical transmission of bifidobacteria from mother to child. *Appl Environ Microbiol* **81**: 7078–7087.

Milani, C., Mangifesta, M., Mancabelli, L., Lugli, G.A., James, K., Duranti, S., et al. (2017b) Unveiling bifidobacterial biogeography across the mammalian branch of the tree of life. *ISME J* **11**: 2834–2847.

Milani, C., Mangifesta, M., Mancabelli, L., Lugli, G.A., Mancino, W., Viappiani, A., et al. (2017a) The sortase-dependent fimbriome of the Genus *Bifidobacterium*: Extracellular structures with potential to modulate microbe-host dialogue. *Appl Environ Microbiol* **83**.

Milani, C., Turroni, F., Duranti, S., Lugli, G.A., Mancabelli, L., Ferrario, C., et al. (2015a) Genomics of the genus *Bifidobacterium* reveals species-specific adaptation to the glycan-rich gut environment. *Appl Environ Microbiol* **82**: 980–991.

O'Connell Motherway, M., Zomer, A., Leahy, S.C., Reunanen, J., Bottacini, F., Claesson, M.J., et al. (2011) Functional genome analysis of *Bifidobacterium breve* UCC2003 reveals type IVb tight adherence (Tad) pili as an essential and conserved host-colonization factor. *Proc Natl Acad Sci U S A* **108**: 11217–11222.

Odamaki, T., Bottacini, F., Kato, K., Mitsuyama, E., Yoshida, K., Horigome, A., Xiao, J.Z., and van Sinderen, D. (2018) Genomic diversity and distribution of *Bifidobacterium longum* subsp. longum across the human lifespan. *Sci Rep* **8**(1): 85.

Odamaki, T., Kato, K., Sugahara, H., Hashikura, N., Takahashi, S., Xiao, J.Z., et al. (2016) Age-related changes in gut microbiota composition from newborn to centenarian: A cross-sectional study. *BMC Microbiol* **16**: 90.

Ojima, M.N., Jiang, L., Arzamasov, A.A., Yoshida, K., Odamaki, T., Xiao, J., Nakajima, A., et al. (2022) Priority effects shape the structure of infant-type *Bifidobacterium* communities on human milk oligosaccharides. *ISME J* **16**(9): 2265–2279.

Penno, C., Motherway, M.O., Fu, Y., Sharma, V., Crispie, F., Cotter, P.D., et al. (2022) Maximum depth sequencing reveals an ON/OFF replication slippage switch and apparent in vivo selection for bifidobacterial pilus expression. *Sci Rep* **12**(1): 9576.

Pokusaeva, K., Fitzgerald, G.F., and van Sinderen, D. (2011) Carbohydrate metabolism in bifidobacteria. *Genes Nutr* **6**(3): 285–306.

Poupard, J.A., Husain, I., and Norris, R.F. (1973) Biology of the bifidobacteria. *Bacteriol Rev* **37**: 136–165.

Püngel, D., Treveil, A., Dalby, M.J., Caim, S., Colquhoun, I.J., Booth, C., et al. (2020) *Bifidobacterium breve* UCC2003 exopolysaccharide modulates the early life microbiota by acting as a potential dietary substrate. *Nutrients* **12**(4): 948.

Rodriguez, J.M., Murphy, K., Stanton, C., Ross, R.P., Kober, O.I., Juge, N., et al. (2015) The composition of the gut microbiota throughout life, with an emphasis on early life. *Microb Ecol Health Dis* **26**: 26050.

Schell, M.A., Karmirantzou, M., Snel, B., Vilanova, D., Berger, B., Pessi, G., et al. (2002) The genome sequence of *Bifidobacterium longum* reflects its adaptation to the human gastrointestinal tract. *Proc Natl Acad Sci U S A* **99**: 14422–14427.

Schiavi, E., Gleinser, M., Molloy, E., Groeger, D., Frei, R., Ferstl, R., et al. (2016) The surface-associated axopolysaccharide of *Bifidobacterium longum* 35624 plays an essential role in dampening host proinflammatory responses and repressing local TH17 responses. *Appl Environ Microbiol* **82**: 7185–7196.

Sela, D.A., Chapman, J., Adeuya, A., Kim, J.H., Chen, F., Whitehead, T.R., et al. (2008) The genome sequence of *Bifidobacterium longum* subsp. *infantis* reveals adaptations for milk utilization within the infant microbiome. *Proc Natl Acad Sci U S A* **105**: 18964–18969.

Solopova, A., Bottacini, F., Venturi Degli Esposti, E., Amaretti, A., Raimondi, S., Rossi, M., et al. (2020) Riboflavin biosynthesis and overproduction by a derivative of the human gut commensal *Bifidobacterium longum* subsp. *infantis* ATCC 15697. *Front Microbiol* **11**: 573335

Thursby, E., and Juge, N. (2017) Introduction to the human gut microbiota. *Biochem J* **474**: 1823–1836.

Tissier, H. (1900) *Recherchers sur la flora intestinale normale et pathologique du nourisson*. University of Paris, Paris, France.

Trovatelli, L.D., Crociani, F., Pedinotti, M., and Scardovi, V. (1974) *Bifidobacterium pullorum* sp. nov.: A new species isolated from chicken feces and a related group of bifidobacteria isolated from rabbit feces. *Arch Microbiol* **98**: 187–198.

Turroni, F., Bottacini, F., Foroni, E., Mulder, I., Kim, J.H., Zomer, A., et al. (2010b) Genome analysis of *Bifidobacterium bifidum* PRL2010 reveals metabolic pathways for host-derived glycan foraging. *Proc Natl Acad Sci U S A* **107**: 19514–19519.

Turroni, F., Foroni, E., O'Connell Motherway, M., Bottacini, F., Giubellini, V., Zomer, A., et al. (2010a) Characterization of the serpin-encoding gene of *Bifidobacterium breve* 210B. *Appl Environ Microbiol* **76**: 3206–3219.

Turroni, F., Foroni, E., Pizzetti, P., Giubellini, V., Ribbera, A., Merusi, P., et al. (2009) Exploring the diversity of the bifidobacterial population in the human intestinal tract. *Appl Environ Microbiol* 75: 1534–1545.

Turroni, F., Milani, C., Duranti, S., Mahony, J., van Sinderen, D., and Ventura, M. (2017a) Glycan utilization and cross-feeding activities by bifidobacteria. *Trends Microbiol* **26**(4): 339–350.

Turroni, F., Milani, C., Duranti, S., Mahony, J., van Sinderen, D., and Ventura, M. (2018) Glycan utilization and cross-feeding activities by bifidobacteria. *Trends Microbiol* **26**(4): 339–350.

Turroni, F., Ozcan, E., Milani, C., Mancabelli, L., Viappiani, A., van Sinderen, D., et al. (2015) Glycan crossfeeding activities between bifidobacteria under in vitro conditions. *Front Microbiol* **6**: 1030.

Turroni, F., Peano, C., Pass, D.A., Foroni, E., Severgnini, M., Claesson, M.J., et al. (2012) Diversity of bifidobacteria within the infant gut microbiota. *PLoS ONE* 7: e36957.

Turroni, F., Taverniti, V., Ruas-Madiedo, P., Duranti, S., Guglielmetti, S., Lugli, G.A., et al. (2014) *Bifidobacterium bifidum* PRL2010 modulates the host innate immune response. *Appl Environ Microbiol* **80**: 730–740.

Ventura, M., Turroni, F., Canchaya, C., Vaughan, E.E., O'Toole, P.W., and van Sinderen, D. (2009a) Microbial diversity in the human intestine and novel insights from metagenomics. *Front Biosci* (Landmark Ed) **14**: 3214–3221.

Ventura, M., Turroni, F., Lima-Mendez, G., Foroni, E., Zomer, A., Duranti, S., et al. (2009c) Comparative analyses of prophage-like elements present in bifidobacterial genomes. *Appl Environ Microbiol* 75: 6929–6936.

Ventura, M., Turroni, F., Zomer, A., Foroni, E., Giubellini, V., Bottacini, F., et al. (2009b) The *Bifidobacterium dentium* Bd1 genome sequence reflects its genetic adaptation to the human oral cavity. *PLoS Genet* **5**: e1000785.

Watanabe, K., Makino, H., Sasamoto, M., Kudo, Y., Fujimoto, J., and Demberel, S. (2009) *Bifidobacterium mongoliense* sp. nov., from airag, a traditional fermented mare's milk product from Mongolia. *Int J Syst Evol Microbiol* **59**: 1535–1540.

Yassour, M., Vatanen, T., Siljander, H., Hamalainen, A.M., Harkonen, T., Ryhanen, S.J., et al. (2016) Natural history of the infant gut microbiome and impact of antibiotic treatment on bacterial strain diversity and stability. *Sci Transl Med* **8**: 343ra381.

Yatsunenko, T., Rey, F.E., Manary, M.J., Trehan, I., Dominguez-Bello, M.G., Contreras, M., Magris, M., et al. (2012) Human gut microbiome viewed across age and geography. *Nature* **486**(7402): 222–227.

# Bacteriophage and Anti-Phage Mechanisms in Lactic Acid Bacteria

# 8

Frank Hille, Paulina Deptula, Natalia Biere, Charles M. A. P. Franz, and Finn K. Vogensen

## 8.1 INTRODUCTION

Bacteriophages (phages) are viruses that are able to attack bacterial cells. Phages attacking lactic acid bacteria (LAB) may have great influence on the acidification rate in fermented products, as well as influencing product quality and safety. Most LAB phages characterised to date have a narrow host range, attacking only a single or a few strains. In order to avoid phage attack, the LAB, like all other bacteria, have developed phage resistance mechanisms to avoid or reduce phage development. This chapter will review the current status of LAB phages and LAB phage resistance mechanisms. A chapter on bacteriophages and phage resistance mechanisms was reviewed in previous editions of *Lactic Acid Bacteria, Microbiology and Functional Aspects*, Second Ed. (Chapter 14), Third Ed. (Chapter 8), Fourth Ed. (Chapter 9), and Fifth Ed. (Chapter 10).

## 8.2 PHAGES INFECTING LACTIC ACID BACTERIA

Bacteriophages are very diverse and as a result differ in morphology and genome size. Traditionally, taxonomic classification of phages relied on their morphology, notably, Bradley's classification was the first comprehensive attempt at grouping the phages into six distinct morphological types named a–f [1], which was later expanded and diversified into subtypes 1–3 based on the head shape [2]. All LAB phages identified until now are tailed phages with double-stranded DNA genomes (class *Caudoviricetes*) [3]. They may have long contractile tails (myoviruses), long non-contractile tails (siphoviruses) or short non-contractile tails (podoviruses). According to the former Bradley classification, they constitute morphotypes a–c, respectively, and based on their head shape are classified as either subtype 1 (isometric heads) or subtype 2–3 (elongated prolate heads) [4, 5]. (Figure 8.1).

DOI: 10.1201/9781003352075-9

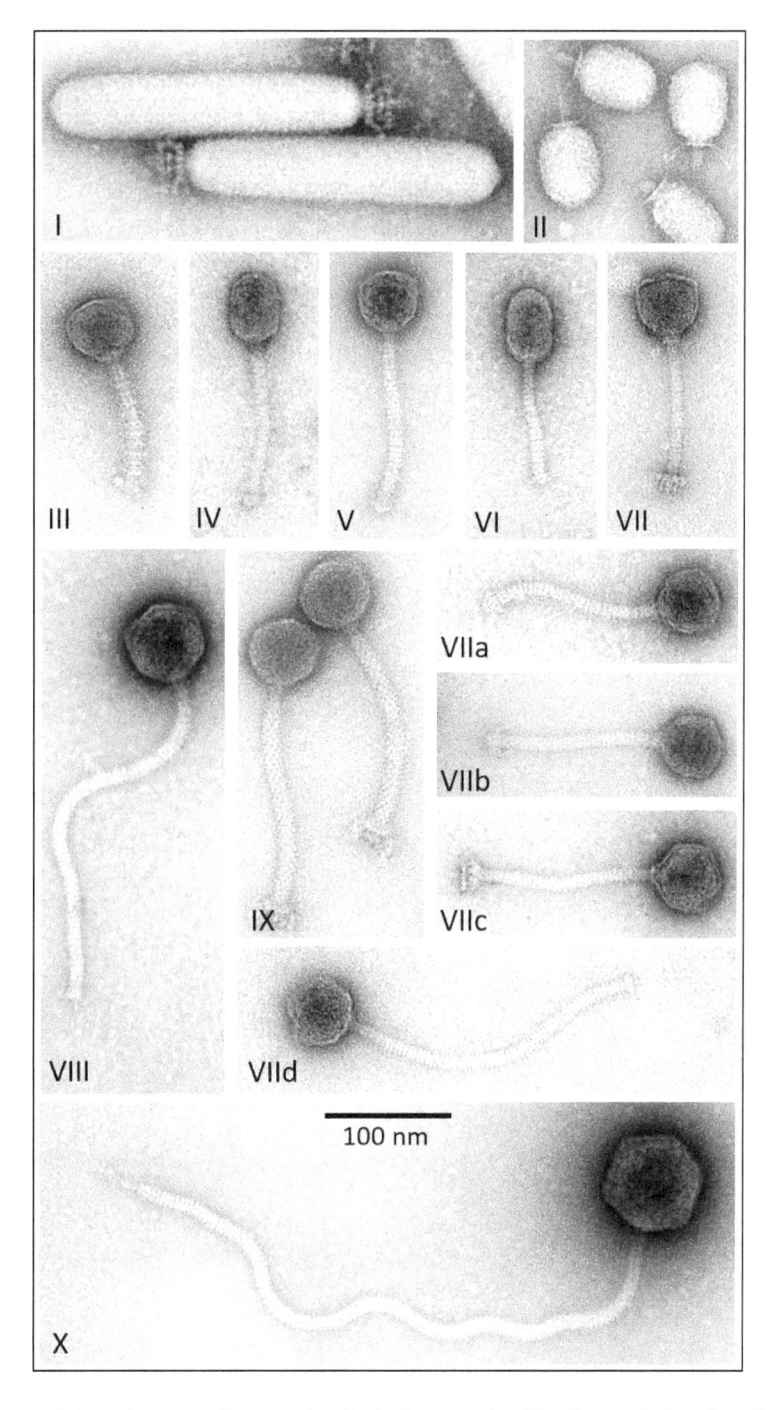

**FIGURE 8.1**   Transmission electron micrographs displaying the classification and diversity of phages infecting *Lactococcus lactis*. *L. lactis* phages are divided into 11 groups (with genus names indicated in parentheses): I—KSY1 (*Chopinvirus*); II—P034; III—1358 (*Whiteheadvirus*); IV—Q54 (*Questintvirus*); V—936 (*Skunavirus*); VI—c2 (*Ceduovirus*); VII—P335 (subgroups: VIIa—Q33; VIIb—r1t; VIIc—TP901-1; VIId—BK5-T (*Sandinevirus*)); VIII—1706 (*Fremauxvirus*); IX—P087 (*Teubervirus*); X—P949 (*Audreyjarvisvirus*); and XI—Nocturne116 (no micrograph available).

The naming conventions of bacteriophages isolated from dairy historically consisted of the name of the bacterial genus on which the phage was isolated, combined with the name of the phage itself. However, phage naming was not very consistent. For example, for phages isolated in English-speaking countries, the canon was to name the phage after the sensitive strain; for example, *Lactococcus* phage c2 was isolated on strain C2 [6], whereas in Germany, phages were indicated by the letter P, followed by the order of isolation, for example, P001 or P335 [7], without indication of isolation strain.

The increase in availability of sequencing data for phage genomes has led to extensive rearrangements in phage taxonomy, which now not only follow a binomial name convention that combines genus and species names, similar to that used in bacteria, but the classification criteria have also been changed. For example, former *Lactococcus* phage c2 is now *Ceduovirus* c2, and *Lactococcus* phage sk1 is now *Skunavirus* sk1. For a phage to be classified as belonging to a certain species, the genome sequence needs to be a minimum of 95% identical over the span of the entire genome to the genome of a representative phage of the species [5].

Bacteriophages differ in their lifestyles, ranging from obligate lytic, through chronic, to temperate [8]. Infection with obligate lytic phages always ends with the cell death and release of accumulated phage progeny until the population of sensitive cells is wiped out. Chronic infection is a result of infection with a filamentous phage, which extrudes individual new phage particles through a pore in the bacterial cell wall without killing the host. There are no known filamentous phages able to infect LAB. Finally, temperate phages are ones that can either enter a lytic lifestyle, where they kill their host at the expense of generating new phage particles, or they can enter a lysogenic cycle, which typically means that the phage genome integrates into the chromosome of the bacterial host as prophage. The integrated phage genome remains dormant and under the control of a phage-encoded repressor that prevents the excision of the phage and it entering the lytic cycle.

## 8.2.1 Phages Infecting *Lactococcus*

Among the five dairy-associated species of *Lactococcus*, *Lactococcus cremoris*, *Lactococcus lactis*, *Lactococcus raffinolactis*, *Lactococcus laudensis* and *Lactococcus hirilactis* [9], phages infecting species used in starter cultures, *L. lactis* and *L. cremoris*, are the best studied. In fact, there are no reports characterising phages infecting the other three species published to date, although prophages predicted bioinformatically in genome sequences have been found [10]. Phages infecting *L. lactis* and *L. cremoris* are considered jointly, as despite the typically narrow host range, many of those phages can cross the species barrier [11]. Phages infecting *Lactococcus* can be divided into 11 groups based on morphology (Figure 8.1) and DNA–DNA hybridisation. Nine of those groups (936, c2, P335, 1358, P087, P034, 949, 1706, and KSY1) consist of multiple related phages [12], while the remaining two, *Qustintvirus* Q54 [12] and nocturne116 [13], each represent a single phage isolate. Among the most prevalent in dairy fermentations are the phages belonging to the taxonomic groups of the genera *Skunavirus* (formerly group 936) and *Ceduovirus* (formerly group c2) and the non-taxonomic group P335. Genera *Ceduovirus* and *Skunavirus* are composed of exclusively lytic phages, currently classified into 34 and 94 species, respectively [14], while group P335 encompasses a very diverse group of both lytic and temperate phages that, despite partial genetic similarity and due to a high level of genetic recombination between the group members, have escaped taxonomic classification so far. Representatives of *Skunavirus* [15], P335 [16], *Audreyjarvisvirus* (949) [17], *Teubervirus* (P087) [18], *Whiteheadvirus* (1358) [19], and *Fremeauxvirus* (1706) [20] use carbohydrate-based receptors to attack their hosts, particularly the so-called cell wall polysaccharide structures (CWPSs) [11], while some of the P335 phages can alternatively use plasmid-encoded exopolysaccharides (EPSs) as receptors [21]. *Ceduoviruses* are exceptional in their use of proteinaceus receptors Pip [22] or structurally similar YjaE [11, 23].

## 8.2.2 Phages Infecting *Streptococcus thermophilus*

Phages attacking *Streptococcus thermophilus*, first discovered in 1922 [24], have been shown to interfere with dairy fermentations, like yoghurt, Swiss-type cheese, and mozzarella. *S. thermophilus* is increasingly used with mesophilic *L. lactis* cultures in mesophilic cheeses like cheddar and cottage cheese. Until approximately 15 years ago, only two groups of *S. thermophilus* phages were recognised [25]: cos-type phages with cohesive ends for packaging DNA in phage heads (now belonging to the *Moineauvirus* genus) and pac-type phages packaging headfuls of DNA in phage heads (now belonging to the *Brussowvirus* genus). Both genera have lytic and temperate members (Figure 8.2). The two genera show DNA cross-hybridising. These two genera are the most commonly encountered *S. thermophilus* phages (65%, and 25%, respectively [26]). They are characterised by having isometric heads and long non-contractile tails with a long spike [26]. Both genera seem to have a fused Tal-RBP gene product as anti-receptor [27]. Within the Dit gene product, a carbohydrate binding domain is identified. The receptors for both phage genera are the rhamnose-glucose polysacharides (RGPs). A new group of *S. thermophilus* phages, the 5093 group (now belonging to the *Vansinderenvirus* genus), was identified in 2011 [28]. Phages in this group have homology to non-dairy streptococci phages. Structurally, they resemble cos-and pac-type phages but do not have a spike but a small baseplate with several small globular proteins [29]. Based on the predicted receptor binding proteins (RBPs), it is suggested that the receptor is of the carbohydrate type [29]. The 987 phage group (now belonging to the *Piorkowskivirus* genus) consists of apparently chimeric phages with structural proteins obtained from the *Lactococcus* pac-type P335 subgroup II phages [30, 31]. These phages have shorter tails and a complex baseplate as seen, for example, in the *Lactococcus* phages Tuc2009 and TP901–1. These phages have probably developed through increased use of *S. thermophilus* in blends with *Lactococcus* strains over

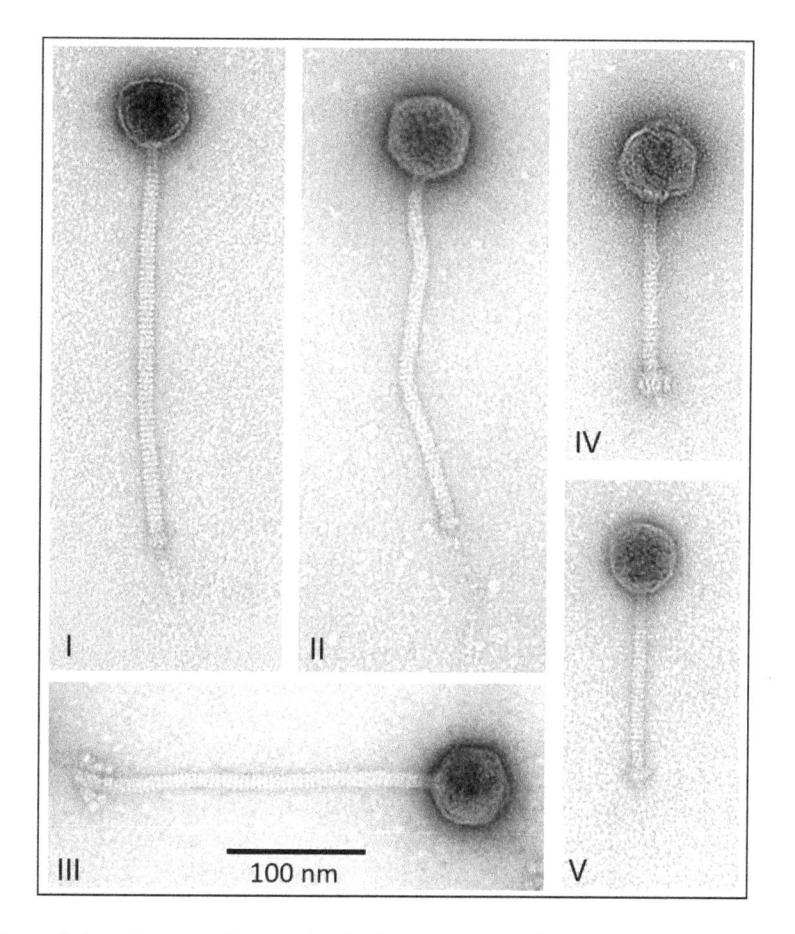

**FIGURE 8.2**  Transmission electron micrographs displaying the classification and diversity of phages infecting *Streptococcus thermophilus*. *S. thermophilus* phages are divided into five groups: I—*Moineauvirus* (former cos-type); II—*Brussowvirus*; (former pac-type) III—*Vansinderenvirus*; IV—*Piorkowskivirus*; and V—P738 phage group.

the last 10–15 years. These phages are increasingly found in dairy samples where *S. thermophilus* and *Lactococcus* are present together [32]. The phage receptor for the *Piorkowskivirus* is exo-polysaccharides [33, 34]. Two different multiplex polymerase chain reaction (PCR) systems have been designed for detection of the four previously mentioned *S. thermophilus* phages [29, 35]. The new P738-group (not assigned to a genus yet) at the moment consists of two phages, P738 and D4446 [36]. These two phages also show homology to non-dairy streptococci. Structurally, they resemble the cos- and pac-type phages with a long spike, but the tail is much shorter (approx. 150 nm) [36].

## 8.2.3 Phages Infecting *Leuconostoc*

Phages against *Leuconostoc* have been found to interfere with dairy fermentations and different plant fermentations, including sauerkraut, kimchi, cucumber and coffee fermentations. A review on phages attacking *Leuconostoc*, *Weissella* and *Oenococcus* was published in 2014 [37]. The first phage against *Leuconostoc* was published in 1946 [38], and since then the majority of phages attacking *Leuconostoc* are of dairy origin. Boizet et al. [39] isolated 19 phages attacking *Leuconostoc mesenteroides* from dairy (3) and coffee fermentations (16) and divided them into six genetic groups based on DNA–DNA homology, morphology and major structural protein analysis. Twelve phages (two of dairy origin) belonged to group I and had genome sizes around 26–27 kb. Three phages of coffee origin belonging to group II had genome sizes of 35–36 kb and cross-hybridised with group I. A single dairy phage isolate (group VI) was unrelated to the other five groups and had a genome of approx. 33 kb. Very interestingly, three unrelated phages isolated from coffee fermentations (Øcc2b from group III, Øcc62 from group IV and Øcc5a from group V) had genomes of 68–70 kb. To our knowledge, no further studies have been conducted on these three phages. In another study, 83 *Leuconostoc* phages of dairy origin (11 dairies and 3 culture collections) were analysed [40]. The majority of these phages (67) belonged to genus *Limdunavirus* and attacked *Leuconostoc pseudomesenteroides*, while 16 belonged to genus *Unaquatrovirus* and attacked *L. mesenteroides*. Four different host ranges were detected, two for *L. mesenteroides* and two for *L. pseudomesenteroides* phages (Figure 8.3) [40]. Pujato et al. [41] isolated nine *L. mesenteroides* phages from blue cheese manufacture with insufficient eye formation and thereby insufficient mould growth. Eight strains from five different starters were used as hosts. Four of the phages appear to be identical based on restriction enzyme digestion and host range. A very broad host range was found, as four of the phages showing different restriction patterns attacked all eight strains. Pujato et al. [42] later found that one of the phages, LDG, was more related to *L. pseudomesenteroides* phages, and the host strain R707 was also later identified as *L. pseudomesenteroides*. Interestingly, they also found cross-species reactions with four of the phages, indicating that low efficiency of plaquing (EOP) cross-species infections are possible. We (Kot & Vogensen, unpublished) have seen similar results, but after careful purifications, these cross-reactions disappeared. It would have been interesting to compare the genome sequence of the cross-species propagated phages with the original phages. In total, eight lytic phages belonging to *Lindunavirus* and seven belonging to *Unaquatrovirus* have been genome sequenced. One temperate *L. pseudomesenteroides* phage, phiMH1, belonging to the genus *Seongbukvirus* was induced and genome sequenced [43].

## 8.2.4 Phages Infecting *Oenococcus oeni*

*Oenococcus oeni* is the main microorganism responsible for malolactic fermentation in wines, thereby reducing acidity and generating aroma compounds. Boizet et al. [39] characterised eight *O. oeni* phages by restriction enzyme analysis, protein profile, and DNA:DNA hybridisations. This revealed two unrelated genetic groups. Seven phages belonged to group I and had two different tail lengths, 185 and 260 nm. The group II phage had a similar structure as group I with a 185-nm tail. Unfortunately, neither of these phages are available. The majority of phages identified and genome sequenced seem to be induced temperate phages [44]. They all contain integrase genes (four different types identified) [45]. These may

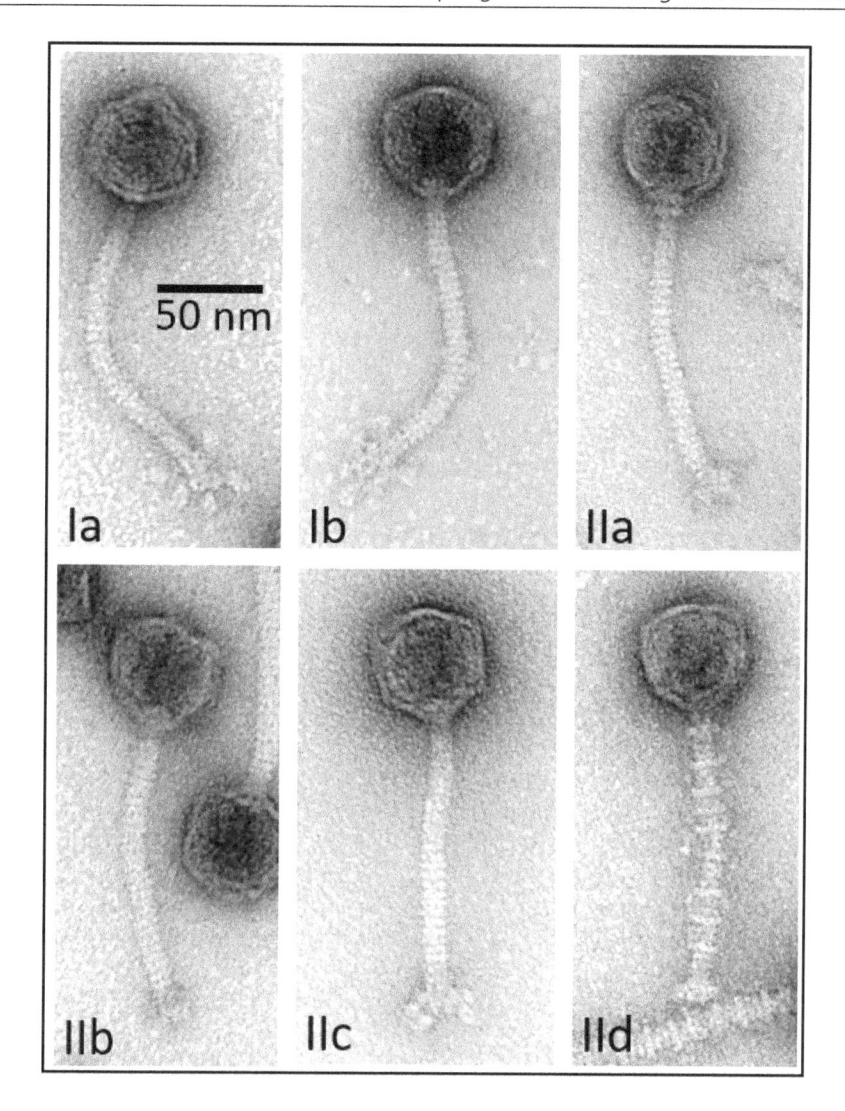

**FIGURE 8.3** Transmission electron micrographs displaying the classification and diversity of phages infecting *Leuconostoc mesenteroides* and *Leuconostoc pseudomesenteroides*. *Ln. mesenteroides* phages (*Unaquatrovirus*) are divided into two morphotypes (Ia and Ib) and *Ln. pseudomesenteroides* phages (*Limdunavirus*) in four morphotypes (IIa–IId).

attack strains of *O. oeni* that are not lysogenic or lysogentic to a different phage [46]. Recently, two lytic phages, OE33PA and Vinitor162, have been identified attacking *O. oeni* [47, 48]. Quite interestingly, the two phages differ in their sensitivity to hydroxycinnamic acids and flavanols in wine [49].

## 8.2.5 Phages Infecting *Lactobacillus*

Temperate phages of *Lactobacillus delbrueckii* could be induced by mitomycin C from approximately ⅓ of strains tested [50]. Ten of these induced phages could be propagated lytically and formed plaques. These can be divided into four groups based on genome sequencing [51, 52]. Group a consists of two phages (LL-H, and phiJB); group b of six phages (Ld3, Ld17, c5, Ld25A, phiLdb, LL-KU and PMBT4, now belonging to the *Cequinquevirus* genus); group c of one phage, the prolated headed JCL1032; and group d of one phage, Jdl1, now belonging to the *Lidleunavirus* genus. The lytic LL-H phage appears to be related to the temperate phage mv4 [53]. A conserved c-terminal of Gp71 from LL-H and of ORF474 from JCL1032 appears to be involved in adsorption of these two phages to their host [54]. Phage LL-H has been shown to use lipoteichoic acids as receptors [55], which depend on D-alanyl and alpha-glucose substitutions of poly(glycerophosphate) [56]. Phage JCL1032 appears to be able to integrate at low frequency [57].

A mitomycin-induced temperate phage, phi0303, was found in *Lactobacillus helveticus* [58]. Prophage remnants were also found in the genome sequence of *L. helveticus* DPC4571 [59]. So far, lytic phages have been detected by pH inhibition or turbidity measurements, but only 6 phages out of 21 have been able to form plaques (phiLH55, phiLH56, phiAQ113, phiAQ114, phiAA18 and phiAB9) [60]. Only phiAQ113 has been genome sequenced [61]. This phage has isometric head and contractive tail (Myovirus).

In the genome sequence of *Lactobacillus johnsonii* NC533, two apparently complete prophages have been identified [62]. These could not be induced by mitomycin. In *Lactobacillus gasseri* NCK102, a temperate phage, phiadh, could be induced by mitomycin [63].

Both temperate and lytic phages have been identified in *Lactiplantibacillus plantarum* strains [51]. In 21 genome-sequenced strains of *L. plantarum* isolated from kimchi, a total of 45 complete prophages were predicted [64]. These could be clustered in three clusters, a Sha1-like cluster, a PM411-like cluster and a third new clutter unrelated to the other two. Six phages that have been propagated or induced have been genome sequenced, and they were arranged into four clusters [51]. The temperate phage Sha1 [65] clusters partly with the lytic phage *Douglaswolinvirus* ATCC8014-B2 [66]. The other lytic phage, *Coetzeevirus* ATCC 8014-B1 [66], clusters together with *Coetzeevirus* phiJL-1 [67], isolated from cucumber fermentations. The third cluster consists of the temperate phage phig1e [68]. The fourth cluster consists of the temperate phage LP65 belonging to the *Salchichonvirus* genus, which has isometric heads and contractile tails [69].

Jarocki et al. [70] analysed more than 400 complete bacterial genome sequences of the *Lacticaseibacillus* genus for prophages. They identified 448 complete prophage genome sequences, of which 129 sequences could be divided into 21 clusters with 45% nucleotide similarities. Older data from sequencing of *Lacticaseibacillus rhamnosus* temperate phages *Behunavirus* BH1 and *Lamunavirus* Lrm1 and the lytic phage *Lacnuvirus* Lc-Nu show they cluster together [71]. The *Lacticaseibacillus casei* temperate phage *Fattrevirus* AT3 also clusters together with *Lamunavirus* Lc-Nu [51] suggesting that lytic and temperate phages of *Lacticaseibacillus* share some homology.

# 8.3 BACTERIAL DEFENCE MECHANISMS (FIGURE 8.4)

## 8.3.1 Prevention of Phages Reaching the Host Cell

One strategy that bacteria have developed to avoid phage infection is to prevent phages from physically reaching the cell. One way to achieve this is to produce outer membrane vesicles (OMVs), which are spherical nanostructures derived from the cells' surface and secreted by the host cells into the surrounding environment that can serve as decoys for phages [72, 73]. OMVs may carry cell materials such as DNA, RNA, proteins and even phages [73–75], and their production is increased due to environmental stresses such as antibiotics and phage infection [73, 76, 75]. These OMVs can also contain phage receptors, and they can thus capture phages which are consequently not able to infect any bacterial cells [73]. For LAB, there has only been limited information on extracellular vesicles, but *Lactococcus lactis* was previously shown to release tailless phage particles within lipid membranes [73, 77].

## 8.3.2 Prevention of Phage Adsorption

The first step in the phage infection process is the adsorption of the viral particle to its host cell's receptor, which requires interaction between the phage's adhesion proteins (referred to as receptor binding proteins,

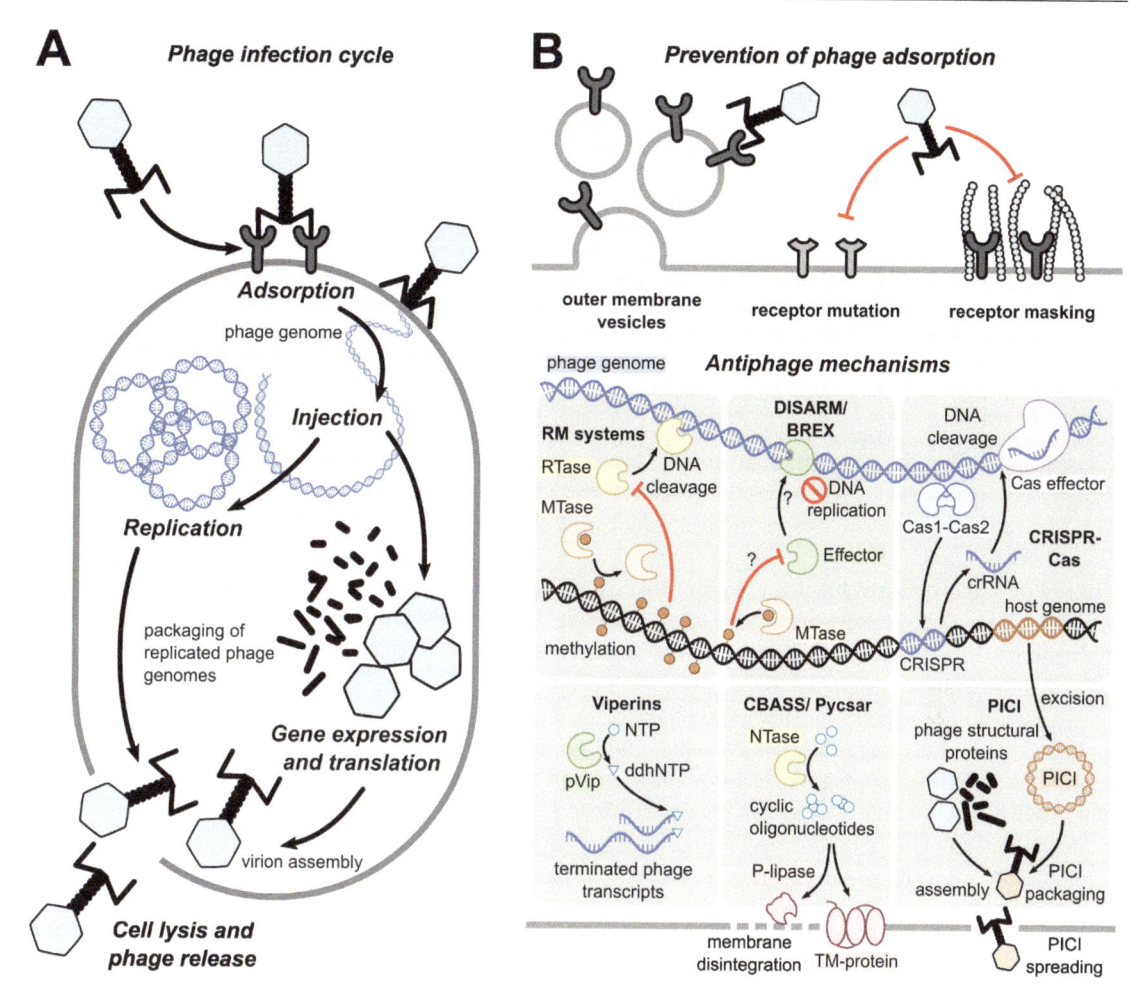

**FIGURE 8.4** Bacterial defence mechanisms. (A) Phage infection cycle. Phage infection commences by the adsorption of the phage to the bacterial cell wall via receptor recognition and binding. The phage subsequently injects its genome into the host where it will replicate. In addition, phage genes will be expressed, altering the host's metabolism and resulting in the production of viral structural proteins. In the late phase, the structural proteins assemble into new phage particles, which are packaged with the replicated phage genomes. Finally, the host cell bursts and new phage particles are release and spread within the bacterial population. (B) Anti-phage mechanisms. The prevention of phage adsorption is often achieved via unspecific mechanisms like the production of outer membrane vesicles (OMVs) and receptor mutations. OMVs can display the phage receptor on their surface and intercept the virus before it reaches the host cell. Mutations in the receptor prevent the recognition of the host cell by the phage and thus prevent adsorption and infection. Moreover, masking the receptor by extracellular polymers physically hinders the phage particle to interact with the bacterial cell wall. Active defence mechanisms that deploy molecular machineries show an incredible diversity within bacteria. Only few are depicted here. Restrictions/modification (RM) systems code for a methyltransferase (MTase) and a nucleolytically active restrictase (RTase). The RTase only recognises unmethylated DNA for degradation and thus interferes with invading phage genomes. The bacterial chromosome is protected due to methylation by the MTase. DISARM (defence island system associated with restriction–modification) and BREX (bacteriophage exclusion) system function similarly, but their exact restriction mechanism remains to be determined. Especially BREX systems seem to block viral replication rather than simply degrading the genome via a nuclease mechanism. While DISARM systems often code for MTases, their activity often has no effect on the antiviral function of the system. CRISPR-Cas systems are adaptive and destroy invaders that have been encountered before. The proteins Cas1 and Cas2 integrate a small portion of the phage genome into the genomic CRISPR locus, which is later transcribed into a CRISPR-RNA (crRNA). Upon a reoccurring attack by the same or a related phage, the crRNA guides the Cas effector to the complementary sequence on the phage genome and cuts the invading

DNA. Viperins are eukaryotic antiviral proteins which have recently been discovered in bacteria as well. The viperin protein (pVip) converts NTP to 3'-deoxy-3',4'-didehydro-NTP (ddhNTP), which are incorporated into the RNA chain during transcription and abort any further elongation of the transcript, abolishing phage gene expression. CBASS and Pycsar systems code for nucleotidyltransferases (NTase) and produce cyclic oligonucleotides upon a phage attack. Those molecules act as second messengers and can activate various proteins like phospholipases (P-lipase) and transmembrane proteins (TM-protein) that often cause membrane instability and result in cell death. This abortive infection mechanism protects the bacterial population. PICIs (phage inducible chromosomal islands) are mobile genetic elements that are integrated in the bacterial chromosome. During a phage infection, PICIs are excised, replicate and modulate the expression of phage proteins, leading to the assembly of altered phage particles which often contain the PICI genome instead of the phage DNA. The PICI then spreads within the bacterial population instead of the phage.

tailspikes, tail fibres or spike proteins) and the bacterial receptors located on the surface of the host. Phage receptors in lactic acid bacteria often include cell wall motifs, exopolysaccharides or polysaccharides [21, 73, 78]. Carbohydrates such as rhamnose, glucose, galactose, glucosamine, galactosamine and ribose present in oligosaccharides have also been identified as receptors in lactococci, streptococci and lactobacilli [79–85]. It is conceivable that the prevention of a productive interaction between the bacterial cell and infective phage will consequently lead to an effective reduction of phage proliferation [86]. Adsorption insensitivity occurs if the phage receptor is absent, altered or masked and is a simple yet effective strategy in establishing phage resistance [87, 81].

In early work on lactic acid bacterial phage resistance, King et al. [88] used ten *Lactococcus cremoris* strains to show that phage-resistant mutants could occur spontaneously at frequencies of ca. $8 \times 10^{-4}$ to $6 \times 10^{-7}$ [88]. While all phage-resistant mutant strains subsequently failed to adsorb phages, the phage-sensitive isolates continued to adsorb phages efficiently. This indicated that the shift from phage-sensitive to phage-resistant condition resulted from a receptor-site mutation. In their natural environment, bacteria are subject to many constant selective pressures that have driven phages and bacteria into an arms race to evolve defence systems and to counter them [72]. This arms race relies on high mutation rates as well as horizontal gene transfer, together leading to a rapid evolution of genetic traits and diversity [72, 89, 90]. The acquisition of point mutations may be the simplest way for bacteria to become resistant to phages. Point mutations may simply arise from frameshift or base substitutions. A PCR-based genome scan of a phage-insensitive lactococcal derivative strain could locate mutations to genes associated with cell wall metabolism, transmembrane- and membrane-associated proteins, prophage components, transcriptional regulators and general metabolism enzymes when compared to the phage-sensitive parent strain [91]. Mutations as a result of transposon integration into a specific phage receptor locus are also known to occur [73].

In addition, the prevention of adhesion of phages has also been attributed to the presence of certain plasmids in, for example, lactococci [92, 93]. The plasmid pME0300 from *Lactococcus lactis* ME2 was the first described to reduce the adsorption of phage 18 to its host [92]. Different plasmids encoding adsorption inhibition were also identified in lactococci, and two of these, plasmids pSK112 [93], and pCI528 [94, 95] appear to prevent phage adsorption by effectively masking phage receptor sites. For plasmid pSK112, a galactose-containing lipoteichoic acid moiety was implicated in the shielding activity [96, 97], while for plasmid pCI528, a cell wall–associated hydrophilic polymer containing rhamnose and galactose was shown to be associated with phage adsorption inhibition [98]. The specific genetic determinants responsible for adsorption inhibition were not characterised, but it was suggested that a large continuous fragment or more than one genetic loci were hypothetically required to achieve the receptor masking effect [87, 98]. Forde et al. [81] also showed a plasmid to encode adsorption inhibition in *L. cremoris* HO2. This plasmid provided protection against phage 712 and partial protection against phage c2 [81]. A plasmid that encoded a 30-kD cell wall protein was identified by Akcelik and Tunail [99] to prevent adsorption of phages to *L. lactis* P25. Another plasmid pKC50 was shown to play a role in adsorption inhibition by directing cell surface antigen synthesis, but the nature of the phage binding prevention still remains unclear [100]. Lastly, phage adsorption may also be impaired by masking receptors as a result of exopolysaccharide production [73, 101].

Bacteriophage insensitive-mutants (BIMs) expressing an adsorption inhibition (Ads) phenotype can be generated rather easily in the laboratory by exposing the bacterial culture to high phage numbers [84, 102]. However, phages can evolve to recognise different cell receptors to evade the Ads mechanism through genome modification. For example, *S. thermophilus* host range phage mutants which exhibited an expanded host specificity were shown to possess point mutations in three different putative tail protein genes, one of which was the phage receptor protein gene [84, 103].

## 8.3.3 Inhibition of DNA Injection or Replication

While blocking phage adsorption to the host cell is one of the initial defences of bacteria against phages, the prevention of phage DNA penetration into the cell is a second simple mechanism adopted by the host cells for such purpose. An early study by Watanabe et al. [104] showed that adsorption of phages may not be sufficient for the phage to translocate DNA into the host cell. In their study, the investigated *Lacticaseibacillus casei* mutants adsorbed phage PL-1 efficiently but did not permit injection of phage DNA into the cell. Subsequently, phage DNA injection was found to be prevented by mutations of the genes encoding the phage infection protein (Pip) [105]. This resulted in host strains being insensitive to phage infection without their growth rate being affected. The latter group then also reported the cloning of host DNA, which appeared to suppress a phage resistance system and thus restored sensitivity to the phage sk1 strain [106, 107]. In *Lactococcus lactis* C2, a reversible adsorption of the phage c2 to the cell wall was followed by an irreversible adsorption to a phage infection protein located in the cell membrane [6]. It could also be shown that a phage-resistant mutant C2 str ain that still adsorbed the phage lacked the phage infection protein and that phage c2 could be inactivated by cell membranes isolated from strain C2 but not by a purified membrane fraction isolated from the Pip-deficient mutant [22, 108].

A further injection-prevention mechanism was reported to be encoded on the *Lactococcus lactis* plasmid pNP40. This mechanism appears to block DNA injection as a result of an alteration at the level of the cell membrane [87] and is especially effective for phage c2 [109, 110]. Despite the availability of the complete nucleotide sequence of plasmid pNP40, the genetic determinants blocking DNA injection could not be identified [111]. The plasmid pNP40 was also described to possess two abortive infection (Abi) phage resistance mechanisms (AbiE and AbiF) [112] as well as a temperature-sensitive restriction/modification system (LlaJ1), which were suggested to lead to a synergistic and enhanced resistance to phage infection [84].

Besides mutation of genes encoding phage infection proteins, prophage-mediated superinfection exclusion (SIE) systems can also directly act on DNA translocation by modifying proteins that are necessary for injection of phage DNA in the host. The first lactococcal DNA injection interference SIE system was ascribed to the P335-type temperate phage Tuc2009 [113]. After integration of the prophage in the chromosome of the *L. lactis* cell, the prophage protein $Sie_{2009}$ is produced and blocks superinfecting phage DNA entry into the cell [113, 114]. McGrath et al. [113] cloned the $sie_{2009}$ gene downstream of a constitutive promotor on the high copy plasmid pNZ44. They could show that *Lactococcus lactis* harbouring pNZ44 $sie_{2009}$ did not protect the bacteria against phages belonging to the c2 or P335 species, whereas complete resistance was obtained against phages of the *Skunavirus*, that is, phages sk1, jj50 and 712 [113]. Such SIE systems were also described for further lactococci [115], as well as for *S. thermophilus* TP-J34, for which the *ltp* gene confers resistance against phages by blocking DNA injection into this specific bacterium (but also in *L. lactis*) [116]. Expression of a *Streptococcus thermophilus* prophage Sfi21 exclusion gene orf203 in trans in high copies indicated that it also conferred resistance to superinfection to a range of heterologous lytic streptococcal phages [117]. While the blocking mechanisms haven't been fully elucidated to date, it has been postulated, for example, that in the case of $Sie_{2009}$, the protein interacts with factors that initiate release of the DNA of the superinfecting phage from the phage head. Alternatively, the protein may interact with cell membrane proteins necessary for DNA translocation [114].

## 8.3.4 Defence Systems

### 8.3.4.1 Restriction-Modification Systems

Once the phage injects its genome into the cell, bacteria evolve multiple mechanisms to abort the infection and abolish phage replication. One mechanism that directly degrades the phage's genome is called restriction modification (RM). RM systems are mainly two-component defence systems that enable the digestion of intracellular DNA, which has been identified as foreign. The discrimination of self and non-self DNA is enabled by a methyltransferase (MTase), which modifies short DNA motifs by the addition of a methyl group, protecting the DNA from digestion by a restriction endonuclease (RTase). While the bacterial genome is methylated, foreign DNA derived from phages is unmodified and lacks protective methylation, which triggers the cleavage of the invading genome by the RTase and thus aborts the infection cycle [73, 114, 118].

To date, four types of RM systems have been described and are classified based on their molecular make-up, cleavage site, target recognition and dependency on co-factors [114, 119]. Type II RM systems are widespread in nature, and most subtypes code for a MTase and a RTase which independently recognise the same short (4- to 8-bp) DNA sequence for methylation and cleavage, respectively. The genes for both enzymes are often encoded on plasmids and therefore can easily spread through horizontal gene transfer [120, 121]. Type I RM systems are more elaborate and code for three subunits that form a complex and exert their function conjointly. The three components are responsible for methylation, DNA cleavage and sequence recognition but are inactive as single proteins [122]. A sub-complex consisting of the methylation and sequence recognition subunit methylates specific DNA sites, whereas target restriction requires all subunits. As opposed to type II systems, DNA digestion depends on the hydrolysis of ATP. Moreover, cleavage by type I RM systems usually occurs far from the recognition site at random DNA sequences, while type II systems cleave within or in the vicinity of the recognition site [122, 114]. Type III RM systems have a similar make-up as type II, as they consist of only two proteins. The MTase is able to methylate specific DNA motifs, whereas target cleavage requires a complex of both the methyltransferase and the restrictase to cleave the target DNA. Similar to type I, the endonucleolytic activity of the complex is energy dependent and relies on ATP hydrolysis [114, 123]. Type IV RM systems are unique, as they lack a methylating component and often consist of only one RTase, which digests independently methylated DNA [114, 121]. At least some type IV RTases require S-adenosyl-methionine (SAM) for their activity, which is unique among RM systems. Usually, SAM is required by the MTase and serves as the donor of the methyl group during DNA methylation in other systems [114].

RM systems are frequently found in the genomes of LAB, with type II RM systems being most abundant. An exception are type IV systems, which have been described only rarely in LABs. However, a comprehensive analysis of type IV McrBC systems revealed an enormous abundance of the systems in various bacteria, indicating that those systems might occur in LAB as well [124]. In Bifidobacteria, type IV RM systems occur as shown in *B. bifidum* and *B. longum*, which code for a type IV GmrSD RM system [125].

Interestingly, there is evidence that RM systems are significantly more prevalent in Lactobacilli species used as cultures in a dairy setting compared to Lactobacilli found in the gut, indicating that the occurrence of those immune systems is niche specific [126]—an observation that might apply to further LAB as well. In *S. thermophilus*, only few phage defence systems are found. With progress in genome sequencing, however, some strains were identified to harbour RM systems belonging to types I–III, which are often encoded on plasmids and integrative conjugative elements [127–129]. Interestingly, if a bacterial cell receives a new RM-system via horizontal gene transfer, its genome is vulnerable to digestion due to the lack of methylation. However, most RM systems circumvent this scenario by being tightly regulated. The MTase is usually over-expressed, and its methylation rate therefore outcompetes the potential detrimental effects of the RTase, as described for the lactococcal LlaDII RM-system. High intracellular MTase concentrations repress its own gene transcription at a later stage, enabling the RTase to locate potential targets and provide protection from potentially invading genomes [130].

## 8.3.4.2 CRISPR-Cas Systems

Clustered regularly interspaced short palindromic repeats (CRISPR)–Cas systems are unique among bacterial anti-phage systems due to their heritable nature and adaptive defence against specific invaders. Specificity is provided by the genomic CRISPR locus, which is composed of identical repeat sequences that are interspaced by so-called spacers [131]. The discovery of those unique genetic structures in bacterial genomes dates back to the 1980s [132], but only two decades later, their role in antiviral defence was proposed due to the realisation that spacer sequences map to transmissible elements like phages and plasmids [133–135]. Experimental validation in *S. thermophilus*—a commonly used starter strain in the dairy industry—finally showed that spacer sequences are derived from invading genomes, and the CRISPR array thus serves as a library of past infections [136]. Upon a reoccurring infection, Cas (CRISPR-associated) nucleases use the sequence information provided by the spacers to cleave matching nucleic acids and thus convey resistance to a particular phage.

The immune system is present in 42% of bacterial and 85% of archaeal genomes [137]. While CRISPR-Cas systems are categorised into two classes, six types and a plethora of sub-types, the general defence mechanism is shared among those systems and encompasses three phases. In the first phase, called adaptation, a small portion of the invading phage's genome is integrated into the bacterial CRISPR locus. This reaction is catalysed by the ubiquitous proteins Cas1 and Cas2, which form a complex that acts as a DNA integrase for site-specific incorporation of the new spacer into the CRISPR array. During the second phase (maturation), the CRISPR array is transcribed into a long precursor CRISPR-RNA (pre-crRNA) and further processed by type-specific Cas proteins or cellular ribonucleases, yielding mature crRNAs that harbour a portion of the repeat and the spacer sequence. In the final stage (interference), the mature crRNA guides Cas nucleases to a spacer-complementary sequence (known as a protospacer) for site-specific cleavage and destruction of the invading genome [131]. In class 1 CRISPR-Cas systems, protospacer cleavage is accomplished by a complex of several Cas proteins, whereas class 2 systems utilise one multi-domain Cas nuclease for interference [137].

Most CRISPR-Cas types evolved an identical mechanism to distinguish self from non-self DNA during interference. This is necessary to avoid targeting of the own genome, as the spacer sequence within the CRISPR array is complementary to the crRNA and therefore a potential target for the crRNA-guided Cas nuclease. The presence of a short motif called a protospacer adjacent motif (PAM) up- or downstream of the protospacer (depending on the type of CRISPR-Cas system) is recognised by the Cas nuclease and crucial for target binding. The PAM is absent in the CRISPR array, thus preventing target recognition and cleavage [131].

The dissemination of CRISPR-Cas systems in LAB varies greatly depending on the bacterial genus and species. CRISPR-Cas types I to III are most widespread in nature and accordingly [137] represent the majority of CRISPR diversity in LAB as well. Most common among the *Lactobacillus* genus are type II CRISPR-Cas systems, with subtype II-A being highly prevalent, as it occurs in approximately 30% of lactobacilli, even exceeding the average occurrence in the domain Bacteria (5%) (Crawley et al. 2018). However, type occurrence differs significantly on the species level; that is, strains of *Lentilactobacillus parabuchneri* mainly harbour type I systems, whereas type III systems are generally underrepresented among Lactobacilli and can be found in only few species, such as *Lactobacillus salivarius* and *Ligilactobacillus ruminis* [138, 139]. The subtype III-A is also underrepresented in *Lactococcus* and can be found only in a few *L. lactis* strains [140].

Type I CRISPR-Cas systems are the most prevalent type in the *Bifidobacterium* genus. Type II systems occur as well [141], while type III systems have not been discovered yet [142]. With regard to further LAB genera, type II systems have been discovered in *Enterococcus* [143] and *Oenococcus* [144].

Intriguingly, multiple CRISPR-Cas systems can occur simultaneously in the genomes of LAB, as exemplified by some *S. thermophilus* strains. The strain *S. thermophilus* DGCC7710 harbours four genomic loci that code for one type I, two type II and one type III CRISPR-Cas systems [145–147], thus illustrating the variety and diversity even within a single genome. The co-occurrence of multiple systems also perfectly exemplifies the high predatory pressure that some LAB species are exposed to.

Besides its significance for antiviral defence, it is noteworthy that the adaptive immune system was rapidly acknowledged by the scientific community due to its potential for practical applications in biotechnology, medicine and agriculture. In particular, the type II-A interference nuclease Cas9 has been extensively studied. The underlying mechanism of target recognition and cleavage enables scientists to easily reprogram the enzyme and utilise it to alter the genomes of numerous different organisms, including human cells [148].

### 8.3.4.3 Abi Systems

Abortive infection systems are not easily classified, as they do not describe a distinct set of molecular machineries or a conserved mode of action. Often, they are defined as a particular phenotype that is triggered by a phage infection and is characterised either by metabolic dormancy or by self-induced cell death. Those two outcomes prevent phage replication and therefore the spread of viral offspring within the bacterial population at the cost of one infected host cell. The underlying pathways and involved proteins, however, can vary significantly [149].

In *Lactococcus*, Abi systems often consist of a single protein and are able to interfere with a variety of different infection stages like phage DNA replication, transcription, translation, phage assembly and premature cell lysis [114]. There have been 23 types of Abi systems described so far in *L. lactis*, and the majority are encoded on mobile elements, particularly on plasmids. Newer studies suggest an even larger diversity within that species [150]. Abi systems are common and can be found in a variety of LAB species. Abiα from *Enterococcus faecalis* and AbiZ from *L. lactis* have both been shown to reduce the burst size of an infecting phage by inducing premature cell lysis, leading to the release of mainly incompletely assembled phage particles [151, 152]. The underlying molecular mechanisms are not always completely understood and vary between different Abi systems. For AbiZ, it has been proposed that the protein prematurely activates the viral holin protein, which regulates cell wall degradation [151, 152].

Interestingly, Abi phenotypes can be caused by other cellular machineries and defence systems: toxin–antitoxin (TA) systems are mainly composed of two components, a toxin and an anti-toxin, which have a variety of functions like stabilisation of mobile genetic elements or stress tolerance. In addition, they can get activated during phage infection, causing the degradation of the anti-toxin and leading to cell death or dormancy induced by the toxin [73, 153, 154]. Several CRISPR-Cas systems have been found to induce dormancy as well. Both classes of CRISPR systems evolved different mechanisms that trigger unspecific degradation of RNA transcripts, either by the effector nuclease [155] or auxiliary proteins [156]. In some CRISPR-Cas systems, however, the underlying mechanism is not yet understood [157].

The discovery of novel phage-defence systems increased due to metagenomic approaches and progress in sequencing techniques. Some of those newly discovered systems have been shown to cause an Abi phenotype, like the cyclic oligonucleotide-based antiphage signalling system (CBASS) [158] and the pyrimidine cyclase system for antiphage resistance (Pycsar). CBASS often consists of four genes and directly interacts with phage components during infection, causing a nucleotidyltransferase to convert ATP and GTP into cyclic di- or trinucleotides, which serve as a second messenger to activate cell-killing effector proteins like phospholipases that interfere with membrane integrity [159]. Similarly, Pycsar systems utilise cyclic second messenger molecules, resulting in an Abi phenotype by altering membrane stability or the depletion of essential cellular metabolites like $NAD^+$. Those systems have been identified in *Leuconostoc*, *Streptococcus*, *Lactococcus* and *Enterococcus* species so far [73, 160].

Similarly, the antiviral properties of retrons—bacterial genetic elements that have been known for decades—was demonstrated only recently [161]. Retrons are a three-component system consisting of a reverse transcriptase, a non-coding RNA and an effector protein. While the exact mechanism remains to be determined, the system arrests the bacterial metabolism when the RecBCD complex—molecular machinery involved in DNA repair and phage defence—gets inhibited by phage proteins [161].

### 8.3.4.4 Novel Defence Systems

Antiviral systems often cluster in certain loci in the bacterial genome [162, 163], making the genomic vicinity of these so-called defence islands a perfect location for the identification of previously unknown phage defence machineries. Researchers have used that approach in the past few years to discover new and diverse systems [164], some of which are mentioned in other sections of this chapter (e.g. CBASS, retrons and phage anti-restriction-induced system [PARIS]). While some systems have been functionally described, their detailed mechanisms, interactions and occurrence—also in LAB species—will be uncovered in the coming years.

The newly described systems bacteriophage exclusion (BREX) and defence island system associated with restriction-modification (DISARM) show similarities to RM systems, as they rely on DNA methylation for defence. However, DISARM systems are more complex and usually code for more genes than RM systems. Even though MTases are encoded by DISARM systems, they do not seem to be crucial for antiviral defence: Methylated phage DNA does not abolish the anti-phage activity of the system, and conversely, the lack of methylation in the bacterial genome does not cause autoimmunity [165–167]. In addition, the exact restriction mechanism of DISARM remains to be determined. Similarly, BREX systems utilise a methylation strategy to distinguish self from non-self DNA. However, their restriction mechanism does not seem to cleave invading DNA but rather block the initiation of phage replication [168–170].

Interestingly, defence mechanisms have been discovered in bacteria and archaea that have been thought to exist only in eukaryotic cells and likely represent the ancestors of viral defence systems in animals and humans. One of those systems utilises proteins called viperins, which produce modified nucleotides during a phage attack that act as RNA polymerisation inhibitors and block viral gene transcription [171]. Similarly, the previously described CBASS system shares ancestry with the eukaryotic cGAS-STING system, which is part of the innate immune response in animals [160]. Moreover, CBASS seems to be widespread in bacteria [160], strongly indicating their occurrence in LAB species as well.

Overall, the number of newly discovered phage defence systems in prokaryotes increased significantly in the past few years, and many of them still await detailed functional characterisation [172]. The same is true for the phylogenetic distribution of those systems among LAB. Some systems, like CRISPR-Cas, RM and Abi, were well studied in members of the LAB genera, and it can be expected that novel systems will be discovered as well. Helpful bioinformatic tools like PADLOC [173, 174] boost the rate of discovery of known defence systems in genomes and plasmids and will allow researchers to identify those systems in LAB species as well.

## 8.4 PHAGE ANTI-DEFENCE MECHANISMS

### 8.4.1 DNA Methylases

Just as bacterial hosts developed varied mechanisms to counteract phage attacks, phages themselves have developed a broad array of opposing strategies. In response to variations in the bacterial cell surface receptors, for example, phages are able to change their tropism by mutating their RPBs. Genes encoding RBP or other recognition proteins related to host recognition were reported to exhibit mutations at high frequency [72]. The mechanisms by which phages can also overcome other phage defence mechanisms vary but include non-specific point mutations in the phage genome and uptake of DNA by co-infecting phages from functional or defective prophages or from various other regions of host DNA, all of which may contribute to phage survival [107].

Phages have developed various ways to prevent inactivation, for example, by restriction modification mechanisms. For example, they may mutate to eliminate restriction sites from their genome and thus avoid recognition by restriction endonucleases [175–177]. Alternatively, they may modify nucleotide bases in sequences recognised by restriction endonucleases (e.g. glucosyl-hydroxymethylcytosine instead of cytosine, as known for phage T4 [121]). DNA modification was also described for *Enterococcus* phages [178], which showed that the phages VPE25 and VFW encoded DNA-modification enzymes that are utilised for glycosylation of their cytosins. Also, phages can change the distance and orientation of restriction sites to avoid restriction by restriction endonucleases that need to recognise two specific sequences spaced at a certain distance and in a specific orientation [179]. Furthermore, they can mask restriction sites with certain proteins that are ejected together with the phage genome [180]. Additionally, they may sequester restriction endonucleases with proteins that mimic the structure of the DNA double helix [181], and lastly, they may acquire genes from the host genome, which encode a methylase that modifies the phage's genome, as was described first by Hill et al. [182] for phage phi50, or may stimulate the activity of the host methylase for the same purpose as described for phage lambda [72, 182, 183]. Other lactic acid bacteria phages were also shown to have diverted methylase genes into their genomes for their protection when infecting strains carrying similar RM systems. These include phages from *Lactobacillus* [64, 184, 185], *Leuconostoc* [186, 187] and *Lactococcus* [188, 189].

## 8.4.2 Anti-Abis

There are many very diverse Abi systems that bacteria employ as an anti-phage mechanism. Because of this great diversity, high numbers of anti-Abi mechanisms have evolved, and these include mutations or deletions of the viral target [190], antitoxin production or antitoxin stabilisation [191–193] and toxin modification [73, 194]. Some of the Abi escaping systems used by lactococcal phages were recently summarised [73] and deal with mutations or deletions in genes for proteins that lead to interference in the phage cycle at the levels of DNA replication or RNA transcription [73].

## 8.4.3 Anti-CRISPR

The escape of CRISPR immunity can be accomplished in various ways by predatory phages. Similar to RM systems, the mutation or deletion of the target sequence (protospacer) in the genome of the phage abolishes CRISPR interference and enables the phage to circumvent destruction [195, 196]. As a response, bacteria often integrate multiple spacers targeting different areas of the phage genome, maintaining immunity. In fact, type I CRISPR-Cas systems evolved a mechanism to reduce predation by phages that escape CRISPR immunity due to mutations. In a process termed primed spacer acquisition, mutations in the protospacer do not result in interference but trigger the uptake of new spacers in the vicinity of the mutated protospacer, resulting in an overall stronger immune response [197–199] due to the presence of more target sites.

In addition to genome alterations, phages evolved more direct mechanisms to circumvent CRISPR immunity by encoding proteins that interrupt the CRISPR immune response. Such anti-CRISPR pathways are enabled by Anti-CRISPR (Acr) proteins which directly interfere with the enzymatic activity of Cas nucleases, preventing degradation of the phage genome [200], and anti-CRISPR-associated (Aca) proteins, which regulate the transcription of Acrs [201, 202]. More than 90 families of Acr proteins have been discovered to date, and they display extreme diversity and vary in their mode of action, sequence conservation, protein size and spectrum of inhibition [203–205]. Identifying new Acrs in the genomes of phages is therefore a daunting task, which often requires practical testing, as bioinformatic predictions are more challenging [73]. For this reason, it can be expected that new Acrs will be identified in various phages infecting LAB species. So far, Acrs of phages infecting *S. thermophilus* have been well characterised, and their activity is mostly directed towards an inhibition of the effector nuclease Cas9 [206–208], preventing CRISPR interference. However, other mechanisms have been proposed, as the protein AcrIIA6 was shown to impede the acquisition of new spacers, preventing *S. thermophilus* from acquiring immunity against its phage [209]. Bioinformatic efforts also led to the discovery of Acrs in further LAB species, that is, Lactobacilli, and further Streptococci and Enterococci, while only a few Acrs were found in members of the Leuconostocaceae family [210].

# 8.5 THE ENEMY OF MY ENEMY IS MY FRIEND: PICIS AND BENEFITS OF CARRYING A PROPHAGE

Despite their perception as mainly dormant parasites, prophages can exert significant physiological effects on their host cell, many of which can be beneficial for the infected bacterium [211]. Increasing fitness and survival by the dissemination of antimicrobial resistance and virulence genes have been well documented [212]. In addition, prophages can be found in LAB species, especially *Lactococcus*, that carry anti-phage systems like Abi (see Section 8.3.4.3) to prevent infections with other phages [213, 214]. In prophages infecting *Oenococcus* and *S. thermophilus*, SIE systems (see the section "Inhibition of DNA Injection or Replication") have been described as well [45, 215].

Another type of recently described mobile genetic elements that are involved in host protection are phage-inducible chromosomal islands (PICI). PICIs are chromosomally integrated satellite phages that lack structural genes and require the presence of a helper phage for proliferation. Upon infection of the helper phage, the PICI genome is excised from the bacterial chromosome and subsequently modulates the synthesis and assembly of new helper phage particles. New phage particles are often smaller and contain the PICI genome instead of the helper phage's genome [120, 216]. Those particles are then released and transduce other bacterial species and sometimes even other bacterial genera [217], thus hindering the dissemination of the helper phage. Some PICIs even have been shown to carry their own anti-phage systems, like PARIS, which produces an Abi phenotype and protects the bacterial population [218]. Interestingly, the system is inactive against the helper phage, allowing proliferation and dissemination of both the PICI element and the helper phage. Only phages that are not able to be hijacked by the PICI element are affected by PARIS [218].

While most PICIs have been mechanistically studied in *Staphylococcus aureus*, related islands have been identified in Streptococci, Lactococci and Enterococci [73, 219–221].

# BIBLIOGRAPHY

1. Bradley DE. Ultrastructure of bacteriophage and bacteriocins. *Bacteriol Rev.* 1967;31(4):230–314.
2. Ackermann HW, Eisenstark A. Present state of phage taxonomy. *Intervirology.* 1974;3(4):201–219; doi:10.1159/000149758.
3. Pujato SA, Quiberoni A, Mercanti DJ. Bacteriophages on dairy foods. *J Appl Microbiol.* 2019;126(1):14–30; doi:10.1111/jam.14062.
4. Mahony J, van Sinderen D. Current taxonomy of phages infecting lactic acid bacteria. *Front Microbiol.* 2014;5:7; doi:10.3389/fmicb.2014.00007.
5. Turner D, Kropinski AM, Adriaenssens EM. A roadmap for genome-based phage taxonomy. *Viruses.* 2021;13(3); doi:10.3390/v13030506.
6. Valyasevi R, Sandine WE, Geller BL. A membrane protein is required for bacteriophage c2 infection of *Lactococcus lactis* subsp. lactis C2. *JBacteriol.* 1991;173(19):6095–6100.
7. Braun Jr V, Hertwig S, Neve H, Geis A, Teuber M. Taxonomic differentiation of bacteriophages of *Lactococcus lactis* by electron microscopy, DNA-DNA hybridization, and protein profiles. *JGenMicrobiol.* 1989;135:2551–2560.
8. Mäntynen S, Laanto E, Oksanen HM, Poranen MM, Díaz-Munoz SL. Black box of phage-bacterium interactions: Exploring alternative phage infection strategies. *Open Biol.* 2021;11(9):210188; doi:10.1098/rsob.210188.

9. White K, Yu J-H, Eraclio G, Dal Bello F, Nauta A, Mahony J, et al. Bacteriophage-host interactions as a platform to establish the role of phages in modulating the microbial composition of fermented foods. *Microbiome Res Rep.* 2022;1(1):3; doi:10.20517/mrr.2021.04.

10. Jung MY, Lee C, Seo MJ, Roh SW, Lee SH. Characterization of a potential probiotic bacterium Lactococcus raffinolactis WiKim0068 isolated from fermented vegetable using genomic and in vitro analyses. *BMC Microbiol.* 2020;20(1):136; doi:10.1186/s12866-020-01820-9.

11. Romero DA, Magill D, Millen A, Horvath P, Fremaux C. Dairy lactococcal and streptococcal phage—host interactions: An industrial perspective in an evolving phage landscape. *FEMS Microbiol Rev.* 2020;44(6):909–932; doi:10.1093/femsre/fuaa048.

12. Deveau H, Labrie SJ, Chopin M-C, Moineau S. Biodiversity and classification of lactococcal phages. *Appl Environ Microbiol.* 2006;72(6):4338–4346.

13. Zrelovs N, Dislers A, Kazaks A. Genome characterization of nocturne116, novel lactococcus lactis-infecting phage isolated from moth. *Microorganisms.* 2021;9(7); doi:10.3390/microorganisms9071540.

14. Zerbini FM, Siddell SG, Lefkowitz EJ, Mushegian AR, Adriaenssens EM, Alfenas-Zerbini P, et al. Changes to virus taxonomy and the ICTV Statutes ratified by the International Committee on Taxonomy of Viruses (2023). *Arch Virol.* 2023;168(7):175; doi:10.1007/s00705-023-05797-4.

15. Dupont K, Janzen T, Vogensen F, Josephsen J, Stuer-Lauridsen B. Identification of *Lactococcus lactis* genes required for bacteriophage adsorption. *Appl Environ Microbiol.* 2004;70(10):5825–5832; doi:10.1128/AEM.70.10.5825-5832.2004.

16. Legrand P, Collins B, Blangy S, Murphy J, Spinelli S, Gutierrez C, et al. The atomic structure of the phage Tuc2009 baseplate tripod suggests that host recognition involves two different carbohydrate binding modules. *mBio.* 2016;7(1):e01781–15; doi:10.1128/mBio.01781-15.

17. Mahony J, Randazzo W, Neve H, Settanni L, van Sinderen D. Lactococcal 949 group phages recognize a carbohydrate receptor on the host cell surface. *Appl Environ Microbiol.* 2015;81(10):3299–3305; doi:10.1128/aem.00143-15.

18. Villion M, Chopin MC, Deveau H, Ehrlich SD, Moineau S, Chopin A. P087, a lactococcal phage with a morphogenesis module similar to an *Enterococcus faecalis* prophage. *Virology.* 2009;388(1):49–56.

19. Farenc C, Spinelli S, Vinogradov E, Tremblay D, Blangy S, Sadovskaya I, et al. Molecular insights on the recognition of a *Lactococcus lactis* cell wall pellicle by phage 1358 receptor binding protein. *J Virol.* 2014; doi:10.1128/jvi.00739-14.

20. Marcelli B, de Jong A, Karsens H, Janzen T, Kok J, Kuipers OP. A specific sugar moiety in the lactococcus lactis cell wall pellicle is required for infection by CHPC971, a member of the rare 1706 phage species. *Appl Environ Microbiol.* 2019;85(19); doi:10.1128/AEM.01224-19.

21. Millen AM, Romero DA, Horvath P, Magill D, Simdon L. Host-encoded, cell surface-associated exopolysaccharide required for adsorption and infection by lactococcal P335 phage subtypes. *Front Microbiol.* 2022;13; doi:10.3389/fmicb.2022.971166.

22. Monteville MR, Ardestani B, Geller BL. Lactococcal bacteriophages require a host cell wall carbohydrate and a plasma membrane protein for adsorption and ejection of DNA. *Appl Environ Microbiol.* 1994;60(9):3204–3211.

23. Millen AM, Romero DA. Genetic determinants of lactococcal C2viruses for host infection and their role in phage evolution. *J Gen Virol.* 2016;97(8):1998–2007; doi:10.1099/jgv.0.000499.

24. Piorkowski G. Beitrag zur Streptokokkenfrage. Anwendung des d'Herelleschen Phänomens auf Streptokokken. *Med Klin.* 1922; XVIII.

25. Le Marrec C, van Sinderen D, Walsh L, Stanley E, Vlegels E, Moineau S, et al. Two groups of bacteriophages infecting Streptococcus thermophilus can be distinguished on the basis of mode of packaging and genetic determinants for major structural proteins. *Appl Environ Microbiol.* 1997;63(8):3246–3253.

26. Hanemaaijer L, Kelleher P, Neve H, Franz CMAP, de Waal PP, van Peij N, et al. Biodiversity of phages infecting the dairy bacterium streptococcus thermophilus. *Microorganisms.* 2021;9(9); doi:10.3390/microorganisms9091822.

27. Lavelle K, Martinez I, Neve H, Lugli GA, Franz C, Ventura M, et al. Biodiversity of streptococcus thermophilus phages in global dairy fermentations. *Viruses.* 2018;10(10); doi:10.3390/v10100577.

28. Mills S, Griffin C, O'Sullivan O, Coffey A, McAuliffe OE, Meijer WC, et al. A new phage on the 'Mozzarella' block: Bacteriophage 5093 shares a low level of homology with other Streptococcus thermophilus phages. *Int Dairy J.* 2011;21(12):963–969; doi:10.1016/j.idairyj.2011.06.003.

29. McDonnell B, Mahony J, Hanemaaijer L, Neve H, Noben JP, Lugli GA, et al. Global survey and genome exploration of bacteriophages infecting the lactic acid bacterium streptococcus thermophilus. *Front Microbiol.* 2017;8:1754; doi:10.3389/fmicb.2017.01754.

30. McDonnell B, Mahony J, Neve H, Hanemaaijer L, Noben JP, Kouwen T, et al. Identification and analysis of a novel group of bacteriophages infecting the lactic acid bacterium streptococcus thermophilus. *Appl Environ Microbiol*. 2016;82(17):5153–5165; doi:10.1128/AEM.00835-16.

31. Szymczak P, Janzen T, Neves AR, Kot W, Hansen LH, Lametsch R, et al. Novel variants of streptococcus thermophilus bacteriophages are indicative of genetic recombination among phages from different bacterial species. *Appl Environ Microbiol*. 2017;83(5); doi:10.1128/AEM.02748-16.

32. Parlindungan E, McDonnell B, Lugli GA, Ventura M, van Sinderen D, Mahony J. Dairy streptococcal cell wall and exopolysaccharide genome diversity. *Microb Genom*. 2022;8(4); doi:10.1099/mgen.0.000803.

33. Szymczak P, Filipe SR, Covas G, Vogensen FK, Neves AR, Janzen T. Cell wall glycans mediate recognition of the dairy bacterium *Streptococcus thermophilus* by bacteriophages. *Appl Environ Microbiol*. 2018;84(23):e01847–18; doi:10.1128/aem.01847-18.

34. Lavelle K, Goulet A, McDonnell B, Spinelli S, van Sinderen D, Mahony J, et al. Revisiting the host adhesion determinants of Streptococcus thermophilus siphophages. *Microbial Biotechnol*. 2020;13(6):1765–1779; doi:10.1111/1751-7915.13593.

35. Szymczak P, Vogensen FK, Janzen T. Novel isolates of bacteriophages from group 5093 identified with an improved multiplex PCR typing method. *Int Dairy J*. 2019;91:18–24; doi:10.1016/j.idairyj.2018.12.001.

36. Philippe C, Levesque S, Dion MB, Tremblay DM, Horvath P, Lüth N, et al. Novel genus of phages infecting streptococcus thermophilus: Genomic and morphological characterization. *Appl Environ Microbiol*. 2020;86(13); doi:10.1128/aem.00227-20.

37. Kot W, Neve H, Heller KJ, Vogensen FK. Bacteriophages of leuconostoc, oenococcus, and weissella. *Front Microbiol*. 2014;5:186; doi:10.3389/fmicb.2014.00186.

38. Mosimann W, Ritter W. Bacteriophages as cause of loss of aroma in butter cultures. (Bakteriophagen als Ursache von Aromaschwund in Rahmsäuerungskulturen). *Schweiz Milchzeitg*. 1946;72:211–212.

39. Boizet B, Mata M, Mignot O, Ritzenthaler P, Sozzi T. Taxonomic characterization of *Leuconostoc mesenteroides* and *Leuconostoc oenos* bacteriophage. *FEMS Microbiol Lett*. 1992;90(3):211–215.

40. Ali Y, Kot W, Atamer Z, Hinrichs J, Vogensen FK, Heller KJ, et al. Classification of lytic bacteriophages attacking dairy leuconostoc starter strains. *Appl Environ Microbiol*. 2013;79(12):3628–3636; doi:10.1128/aem.00076-13.

41. Pujato SA, Guglielmotti DM, Ackermann H-W, Patrignani F, Lanciotti R, Reinheimer JA, et al. Leuconostoc bacteriophages from blue cheese manufacture: Long-term survival, resistance to thermal treatments, high pressure homogenization and chemical biocides of industrial application. *Int J Food Microbiol*. 2014;177:81–88; doi:10.1016/j.ijfoodmicro.2014.02.012.

42. Pujato SA, Guglielmotti DM, Martínez-García M, Quiberoni A, Mojica FJM. Leuconostoc mesenteroides and Leuconostoc pseudomesenteroides bacteriophages: Genomics and cross-species host ranges. *Int J Food Microbiol*. 2017;257:128–137; doi:10.1016/j.ijfoodmicro.2017.06.009.

43. Jang SH, Hwang MH, Chang HI. Complete genome sequence of phiMH1, a Leuconostoc temperate phage. *Arch Virol*. 2010;155(11):1883–1885.

44. Chaib A, Philippe C, Jaomanjaka F, Claisse O, Jourdes M, Lucas P, et al. Lysogeny in the lactic acid bacterium oenococcus oeni is responsible for modified colony morphology on red grape juice agar. *Appl Environ Microbiol*. 2019;85(19); doi:10.1128/AEM.00997-19.

45. Jaomanjaka F, Ballestra P, Dols-lafargue M, Le Marrec C. Expanding the diversity of oenococcal bacteriophages: Insights into a novel group based on the integrase sequence. *Int J Food Microbiol*. 2013;166(2):331–340; doi:10.1016/j.ijfoodmicro.2013.06.032.

46. Poblet-Icart M, Bordons A, Lonvaud-Funel A. Lysogeny of oenococcus oeni (syn. Leuconostoc oenos) and study of their induced bacteriophages. *Curr Microbiol*. 1998;36(6):365–369; doi:10.1007/s002849900324.

47. Jaomanjaka F, Claisse O, Philippe C, Le Marrec C. Complete genome sequence of lytic oenococcus oeni bacteriophage OE33PA. *Microbiol Resour Announc*. 2018;7(6); doi:10.1128/MRA.00818-18.

48. Philippe C, Chaib A, Jaomanjaka F, Claisse O, Lucas PM, Samot J, et al. Characterization of the first virulent phage infecting oenococcus oeni, the queen of the cellars. *Front Microbiol*. 2020;11:596541; doi:10.3389/fmicb.2020.596541.

49. Philippe C, Chaib A, Jaomanjaka F, Cluzet S, Lagarde A, Ballestra P, et al. Wine phenolic compounds differently affect the host-killing activity of two lytic bacteriophages infecting the lactic acid bacterium oenococcus oeni. *Viruses*. 2020;12(11); doi:10.3390/v12111316.

50. Sechaud L, Cluzel P-J, Rousseau M, Baumgartner A, Accolas J-P. Bacteriophages of lactobacilli. *Biochimie*. 1988;70(3):401–410; doi:10.1016/0300-9084(88)90214-3.

51. Casey E, Mahony J, Neve H, Noben JP, Dal Bello F, van Sinderen D. Novel phage group infecting Lactobacillus delbrueckii subsp. lactis, as revealed by genomic and proteomic analysis of bacteriophage Ldl1. *Appl Environ Microbiol*. 2015;81(4):1319–1326; doi:10.1128/AEM.03413-14.

52. Sprotte S, Fagbemigun O, Brinks E, Cho GS, Casey E, Oguntoyinbo FA, et al. Novel phage PMBT4 belonging to the group b L. delbrueckii subsp. bulgaricus phages. *Virus Res.* 2022;308; doi:10.1016/j.virusres.2021.198635.

53. Auad L, Raisanen L, Raya RR, Alatossava T. Physical mapping and partial genetic characterization of the Lactobacillus delbrueckii subsp. bulgaricus bacteriophage lb539. *ArchVirol.* 1999;144(8):1503–1512.

54. Ravin V, Raisanen L, Alatossava T. A conserved C-terminal region in Gp71 of the small isometric-head phage LL-H and ORF474 of the prolate-head phage JCL1032 is implicated in specificity of adsorption of phage to its host, Lactobacillus delbrueckii. *J Bacteriol.* 2002;184(9):2455–2459.

55. Raisanen L, Schubert K, Jaakonsaari T, Alatossava T. Characterization of lipoteichoic acids as lactobacillus delbrueckii phage receptor components. *J Bacteriol.* 2004;186(16):5529–5532.

56. Räisänen L, Draing C, Pfitzenmaier M, Schubert K, Jaakonsaari T, von Aulock S, et al. Molecular interaction between lipoteichoic acids and Lactobacillus delbrueckii phages depends on D-alanyl and alpha-glucose substitution of poly(glycerophosphate) backbones. *J Bacteriol.* 2007;189(11):4135–4140.

57. Riipinen KA, Raisanen L, Alatossava T. Integration of the group c phage JCL1032 of Lactobacillus delbrueckii subsp. lactis and complex phage resistance of the host. *J Appl Microbiol.* 2007;103(6):2465–2475.

58. Deutsch SM, Guezenec S, Piot M, Foster S, Lortal S. Mur-LH, the broad-spectrum endolysin of Lactobacillus helveticus temperate bacteriophage f-0303. *Appl Environ Microbiol.* 2004;70(1):96–103.

59. Callanan M, Kaleta P, O'Callaghan J, O'Sullivan O, Jordan K, McAuliffe O, et al. Genome sequence of Lactobacillus helveticus, an organism distinguished by selective gene loss and insertion sequence element expansion. *J Bacteriol.* 2008;190(2):727–735.

60. Zago M, Bonvini B, Rossetti L, Meucci A, Giraffa G, Carminati D. Biodiversity of Lactobacillus helveticus bacteriophages isolated from cheese whey starters. *J Dairy Res.* 2015;82(2):242–247; doi:10.1017/S0022029915000151.

61. Zago M, Scaltriti E, Rossetti L, Guffanti A, Armiento A, Fornasari ME, et al. Characterization of the genome of the dairy Lactobacillus helveticus bacteriophage PhiAQ113. *Appl Environ Microbiol.* 2013;79(15):4712–4718; doi:10.1128/AEM.00620-13.

62. Ventura M, Canchaya C, Pridmore RD, Brüssow H. The prophages of Lactobacillus johnsonii NCC 533: Comparative genomics and transcription analysis. *Virology.* 2004;320(2):229–242.

63. Altermann E, Klein JR, Henrich B. Primary structure and features of the genome of the Lactobacillus gasseri temperate bacteriophage f adh. *Gene.* 1999;236(2):333–346.

64. Park D-W, Kim S-H, Park J-H. Distribution and characterization of prophages in Lactobacillus plantarum derived from kimchi. *Food Microbiol.* 2022;102:103913; doi:10.1016/j.fm.2021.103913.

65. Yoon BH, Jang SH, Chang HI. Sequence analysis of the Lactobacillus temperate phage Sha1. *Arch Virol.* 2011;156(9):1681–1684; doi:10.1007/s00705-011-1048-2.

66. Briggiler Marco M, Garneau JE, Tremblay D, Quiberoni A, Moineau S. Characterization of two virulent phages of Lactobacillus plantarum. *Appl Environ Microbiol.* 2012;78(24):8719–8734; doi:10.1128/AEM.02565-12.

67. Lu Z, Altermann E, Breidt F, Predki P, Fleming HP, Klaenhammer TR. Sequence analysis of the Lactobacillus plantarum bacteriophage f JL-1. *Gene.* 2005;348:45–54.

68. Kodaira KI, Oki M, Kakikawa M, Watanabe N, Hirakawa M, Yamada K, et al. Genome structure of the Lactobacillus temperate phage f gle: The whole genome sequence and the putative promoter/repressor system. *Gene.* 1997;187(1):45–53.

69. Chibani-Chennoufi S, Dillmann ML, Marvin-Guy L, Rami-Shojaei S, Brüssow H. Lactobacillus plantarum bacteriophage LP65: A new member of the SPO1-like genus of the family Myoviridae *J Bacteriol.* 2004;186(21):7069–7083.

70. Jarocki P, Komoń-Janczara E, Młodzińska A, Sadurski J, Kołodzińska K, Łaczmański Ł, et al. Occurrence and genetic diversity of prophage sequences identified in the genomes of L. casei group bacteria. *Sci Rep.* 2023;13(1):8603; doi:10.1038/s41598-023-35823-z.

71. Jarocki P, Komoń-Janczara E, Podleśny M, Kholiavskyi O, Pytka M, Kordowska-Wiater M. Genomic and proteomic characterization of bacteriophage BH1 spontaneously released from probiotic lactobacillus rhamnosus pen. *Viruses.* 2019;11(12); doi:10.3390/v11121163.

72. Egido JE, Costa AR, Aparicio-Maldonado C, Haas P-J, Brouns SJJ. Mechanisms and clinical importance of bacteriophage resistance. *FEMS Microbiol Rev.* 2022;46(1); doi:10.1093/femsre/fuab048.

73. Philippe C, Cornuault JK, de Melo AG, Morin-Pelchat R, Jolicoeur AP, Moineau S. The never-ending battle between lactic acid bacteria and their phages. *FEMS Microbiol Rev.* 2023;47(4); doi:10.1093/femsre/fuad035.

74. Toyofuku M, Nomura N, Eberl L. Types and origins of bacterial membrane vesicles. *Nat Rev Microbiol.* 2019;17(1):13–24; doi:10.1038/s41579-018-0112-2.

75. Liu Y, Tempelaars MH, Boeren S, Alexeeva S, Smid EJ, Abee T. Extracellular vesicle formation in *Lactococcus lactis* is stimulated by prophage-encoded holin-lysin system. *Microbial Biotechnol.* 2022;15(4):1281–1295; doi:10.1111/1751-7915.13972.

76. Da Silva Barreira D, Lapaquette P, Novion Ducassou J, Couté Y, Guzzo J, Rieu A. Spontaneous prophage induction contributes to the production of membrane vesicles by the Gram-positive bacterium lacticaseibacillus casei BL23. *mBio*. 2022;13(5):e0237522; doi:10.1128/mbio.02375-22.

77. Liu Y, Alexeeva S, Bachmann H, Guerra Martínez JA, Yeremenko N, Abee T, et al. Chronic release of tailless phage particles from lactococcus lactis. *Appl Environ Microbiol*. 2022;88(1):e0148321; doi:10.1128/AEM.01483-21.

78. Ainsworth S, Sadovskaya I, Vinogradov E, Courtin P, Guerardel Y, Mahony J, et al. Differences in lactococcal cell wall polysaccharide structure are major determining factors in bacteriophage sensitivity. *mBio*. 2014;5(3); doi:10.1128/mBio.00880-14.

79. Ishibashi K, Takesue S, Watanabe K, Oishi K. Use of lectins to characterize the receptor sites for bacteriophage PL-1 of lactobacillus casei. *Microbiology (Reading, England)*. 1982;128(10):2251–2259; doi:10.1099/00221287-128-10-2251.

80. Keogh BP, Pettingill G. Adsorption of bacteriophage eb7 on Streptococcus cremoris EB7. *Appl Environ Microbiol*. 1983;45(6):1946–1948.

81. Forde A, Daly C, Fitzgerald GF. Identification of four phage resistance plasmids from *Lactococcus lactis* subsp. cremoris HO2. *Appl Environ Microbiol*. 1999;65(4):1540–1547.

82. Quiberoni A, Stiefel JI, Reinheimer JA. Characterization of phage receptors in Streptococcus thermophilus using purified cell walls obtained by a simple protocol. *J Appl Microbiol*. 2000;89(6):1059–1065.

83. Binetti AG, Quiberoni A, Reinheimer JA. Phage adsorption to Streptococcus thermophilus. Influence of environmental factors and characterization of cell-receptors. *Food Res Int*. 2002;35(1):73–83; doi:10.1016/S0963-9969(01)00121-1.

84. Mills S, Ross RP, Neve H, Coffey A. *Bacteriophage and Anti-Phage Mechanisms in Lactic Acid Bacteria*. 4th edn. Boca Raton: CRC Press; 2012.

85. Bertozzi Silva J, Storms Z, Sauvageau D. Host receptors for bacteriophage adsorption. *FEMS Microbiol Lett*. 2016;363(4); doi:10.1093/femsle/fnw002.

86. Daly C, Fitzgerald GF, Davis R. Biotechnology of lactic acid bacteria with special reference to bacteriophage resistance. *Antonie van Leeuwenhoek*. 1996;70(2–4):99–110.

87. Hill C, Garvey P, Fitzgerald GF. Bacteriophage-host interactions and resistance mechanisms, analysis of the conjugative bacteriophage resistance plasmid pNP40. *Le Lait*. 1996;76(1–2):67–79; doi:10.1051/lait:19961-27.

88. King WR, Collins EB, Barrett EL. Frequencies of bacteriophage-resistant and slow acid-producing variants of streptococcus cremoris. *Appl Environ Microbiol*. 1983;45(5):1481–1485; doi:10.1128/aem.45.5.1481-1485.1983.

89. Takeuchi N, Wolf YI, Makarova KS, Koonin EV. Nature and intensity of selection pressure on CRISPR-associated genes. *J Bacteriol*. 2012;194(5):1216–1225.

90. Hampton HG, Watson BNJ, Fineran PC. The arms race between bacteria and their phage foes. *Nature*. 2020;577(7790):327–336; doi:10.1038/s41586-019-1894-8.

91. Schmidt MT, Olejnik-Schmidt AK, Zaręba A, Pezacki M, Wojewoda I, Grajek W. Induction of Loci Mutation during *Lactococcus lactis* spontaneous conversion to bacteriophage-insensitive phenotype. *Food Biotechnol*. 2010;24(4):332–348; doi:10.1080/08905436.2010.524470.

92. Sanders ME, Klaenhammer TR. Characterization of phage-sensitive mutants from a phage-insensitive strain of streptococcus lactis: Evidence for a plasmid determinant that prevents phage adsorption. *Appl Environ Microbiol*. 1983;46(5):1125–1133; doi:10.1128/aem.46.5.1125-1133.1983.

93. de Vos WM, Underwood HM, Davies F. Plasmid encoded bacteriophage resistance in Streptococcus cremoris SK11. *FEMS Microbiol Lett*. 1984;23(2–3):175–178; doi:10.1111/j.1574-6968.1984.tb01057.x.

94. Costello V. Characterization of bacteriophage-host interaction in Streptococcus cremoris UC503 and related lactic streptococci; Doctoral Thesis. The National University of Ireland, University College, Cork. 1988.

95. Coffey AG, Fitzgerald GF, Daly C. Identification and characterization of a plasmid encoding abortive infection from *Lactococcus lactis* ssp. lactis UC811. *NethMilk Dairy J*. 1989;43:229–244.

96. Sijtsma L, Sterkenburg A, Wouters JT. Properties of the cell walls of lactococcus lactis subsp. cremoris SK110 and their relation to bacteriophage resistance. *Appl Environ Microbiol*. 1988;54(11):2808–2811; doi:10.1128/aem.54.11.2808-2811.1988.

97. Sijtsma L, Jansen N, Hazeleger WC, Wouters JTM, Hellingwerf KJ. Cell surface characteristics of bacteriophage-resistant *Lactococcus lactis* subsp. cremoris SK110 and its bacteriophage-sensitive variant SK112. *Appl Environ Microbiol*. 1990;56(10):3230–3233.

98. Lucey M, Daly C, Fitzgerald GF. Cell-surface characteristics of *Lactococcus lactis* harboring pCI528, a 46 Kb plasmid encoding inhibition of bacteriophage adsorption. *J Gen Microbiol*. 1992;138:2137–2143.

99. Akcelik M, Tunail N. A 30 kd cell wall protein produced by plasmid DNA which encodes inhibition of phage adsorption in *Lactococcus lactis* subsp. lactis P25. *Milchwissenschaft*. 1992;47(4):215–217.

100. Tortorello ML, Trotter KM, Angelos SM, Ledford RA, Dunny GM. Microtiter plate assays for the measurement of phage adsorption and infection in Lactococcus and Enterococcus. *Anal Biochem*. 1991;192(2):362–366.

101. Forde A, Fitzgerald GF. Analysis of exopolysaccharide (EPS) production mediated by the bacteriophage adsorption blocking plasmid, pCI658, isolated from *Lactococcus lactis* ssp. cremoris HO2. *IntDairy J*. 1999;9(7):465–472.

102. Mills S, Coffey A, McAuliffe OE, Meijer WC, Hafkamp B, Ross RP. Efficient method for generation of bacteriophage insensitive mutants of Streptococcus thermophilus yoghurt and mozzarella strains. *J Microbiol Methods*. 2007;70(1):159–164; doi:10.1016/j.mimet.2007.04.006.

103. Duplessis M, Levesque CM, Moineau S. Characterization of streptococcus thermophilus host range phage mutants. *Appl Environ Microbiol*. 2006;72(4):3036–3041.

104. Watanabe K, Hayashida M, Ishibashi K, Nakashima Y. An N-acetylmuramidase induced by PL-1 phage infection of Lactobacillus casei. *J Gen Microbiol*. 1984;130(Pt 2):275–277.

105. Garbutt KC, Kraus J, Geller BL. Bacteriophage resistance in *Lactococcus lactis* engineered by replacement of a gene for a bacteriophage receptor. *J Dairy Sci*. 1997;80(8):1512–1519.

106. Kraus J, Geller BL. Cloning of genomic DNA of *Lactococcus lactis* that restores phage sensitivity to an unusual bacteriophage sk1-resistant mutant. *Appl Environ Microbiol*. 2001;67(2):791–798.

107. Coffey A, Ross RP. Bacteriophage-resistance systems in dairy starter strains: Molecular analysis to application. *Antonie van Leeuwenhoek*. 2002;82(1–4):303–321.

108. Josephsen J, Neve H. Bacteriophages and antiphage mechanisms of lactic acid bacteria. In: Salminen S, von Wright A, Ouwehand A, editors. *Lactic Acid Bacteria Microbiological and Functional Aspects*. New York: Marcel Decker Inc.; 2004, pp. 295–350.

109. Garvey P, Hill C, Fitzgerald GF. The lactococcal plasmid pNP40 encodes a third bacteriophage resistance mechanism, one which affects phage DNA penetration. *Appl Environ Microbiol*. 1996;62(2):676–679.

110. Mills S, McAuliffe OE, Coffey A, Fitzgerald GF, Ross RP. Plasmids of lactococci—genetic accessories or genetic necessities? *FEMS Microbiol Rev*. 2006;30(2):243–273.

111. O'Driscoll J, Glynn F, Fitzgerald GF, van Sinderen D. Sequence analysis of the lactococcal plasmid pNP40: A mobile replicon for coping with environmental hazards. *J Bacteriol*. 2006;188(18):6629–6639.

112. Garvey P, Fitzgerald GF, Hill C. Cloning and DNA sequence analysis of two abortive infection phage resistance determinants from the lactococcal plasmid pNP40. *Appl Environ Microbiol*. 1995;61(12):4321–4328.

113. McGrath S, Fitzgerald GF, van Sinderen D. Identification and characterization of phage-resistance genes in temperate lactococcal bacteriophages. *Mol Microbiol*. 2002;43(2):509–520; doi:10.1046/j.1365-2958.2002.02763.x.

114. Szczepankowska AK, Gorecki RK, Koakowski P, Bardowski JK. Lactic acid bacteria resistance to bacteriophage and prevention techniques to lower phage contamination in dairy fermentation. In: Kongo JM, editor. *Lactic Acid Bacteria—R & D for Food, Health and Livestock Purposes*. London: IntechOpen Limited; 2013; doi:10.5772/51541

115. Mahony J, McGrath S, Fitzgerald GF, van Sinderen D. Identification and characterization of lactococcal-prophage-carried superinfection exclusion genes. *Appl Environ Microbiol*. 2008;74(20):6206–6215.

116. Ali Y, Koberg S, Heßner S, Sun X, Rabe B, Back A, et al. Temperate Streptococcus thermophilus phages expressing superinfection exclusion proteins of the Ltp type. *Front Microbiol*. 2014;5(98); doi:10.3389/fmicb.2014.00098.

117. Bruttin A, Foley S, Brüssow H. DNA-binding activity of the Streptococcus thermophilus phage Sfi21 repressor. *Virology*. 2002;303(1):100–109.

118. Pingoud A, Wilson GG, Wende W. Type II restriction endonucleases--a historical perspective and more. *Nucleic Acids Res*. 2014;42(12):7489–527; doi:10.1093/nar/gku447.

119. Seegers JFML, van Sinderen D, Fitzgerald GF. Molecular characterization of the lactococcal plasmid pCIS3: Natural stacking of specificity subunits of a type I restriction/modification system in a single lactococcal strain. *Microbiology*. 2000;146(Pt 2):435–443.

120. Rostøl JT, Marraffini L. (Ph)ighting phages: How bacteria resist their parasites. *Cell Host Microbe*. 2019;25(2):184–194; doi:10.1016/j.chom.2019.01.009.

121. Labrie SJ, Samson JE, Moineau S. Bacteriophage resistance mechanisms. *Nat Rev Microbiol*. 2010;8(5):317–327; doi:10.1038/nrmicro2315.

122. Murray NE. Type I restriction systems: Sophisticated molecular machines (a legacy of Bertani and Weigle). *Microbiol Mol Biol Rev: MMBR*. 2000;64(2):412–434; doi:10.1128/MMBR.64.2.412-434.2000.

123. Su P, Im H, Hsieh H, Kang'A S, Dunn NW. LlaFI, a type III restriction and modification system in *Lactococcus lactis*. *Appl Environ Microbiol*. 1999;65(2):686–693.

124. Bell RT, Juozapaitis J, Songailiene I, Sahakyan H, Liegute T, Makarova KS, et al. Astonishing diversity and multifaceted biological connections of Type IV restriction-modification systems. *bioRxiv: Preprint Server Biol*. 2023; doi:10.1101/2023.07.31.551357.

125. Machnicka MA, Kaminska KH, Dunin-Horkawicz S, Bujnicki JM. Phylogenomics and sequence-structure-function relationships in the GmrSD family of Type IV restriction enzymes. *BMC Bioinform.* 2015;16:336; doi:10.1186/s12859-015-0773-z.

126. O'Sullivan O, O'Callaghan J, Sangrador-Vegas A, McAuliffe O, Slattery L, Kaleta P, et al. Comparative genomics of lactic acid bacteria reveals a niche-specific gene set. *BMC Microbiol.* 2009;9(1):50.

127. Burrus V, Bontemps C, Decaris B, GuÇdon G. Characterization of a novel type II restriction-modification system, Sth368I, encoded by the integrative element ICESt1 of Streptococcus thermophilus CNRZ368. *Appl Environ Microbiol.* 2001;67(4):1522–1528.

128. Geis A, El Demerdash HAM, Heller KJ. Sequence analysis and characterization of plasmids from Streptococcus thermophilus. *Plasmid.* 2003;50(1):53–69.

129. Solow BT, Somkuti GA. Molecular properties of Streptococcus thermophilus plasmid pER35 encoding a restriction modification system. *CurrMicrobiol.* 2001;42(2):122–128.

130. Christensen LL, Josephsen J. The methyltransferase from the LlaDII restriction-modification system influences the level of expression of its own gene. *J Bacteriol.* 2004;186(2):287–295.

131. Hille F, Richter H, Wong SP, Bratovič M, Ressel S, Charpentier E. The biology of CRISPR-Cas: Backward and forward. *Cell.* 2018;172(6):1239–1259; doi:10.1016/j.cell.2017.11.032.

132. Ishino Y, Shinagawa H, Makino K, Amemura M, Nakata A. Nucleotide sequence of the iap gene, responsible for alkaline phosphatase isozyme conversion in Escherichia coli, and identification of the gene product. *J Bacteriol.* 1987;169(12):5429–5433.

133. Bolotin A, Quinquis B, Sorokin A, Ehrlich SD. Clustered regularly interspaced short palindrome repeats (CRISPRs) have spacers of extrachromosomal origin. *Microbiology.* 2005;151(Pt 8):2551–2561.

134. Mojica FJM, Díez-Villaseñor C, García-Martínez J, Soria E. Intervening sequences of regularly spaced prokaryotic repeats derive from foreign genetic elements. *J Mol Evol.* 2005;60(2):174–182; doi:10.1007/s00239-004-0046-3.

135. Pourcel C, Salvignol G, Vergnaud G. CRISPR elements in Yersinia pestis acquire new repeats by preferential uptake of bacteriophage DNA, and provide additional tools for evolutionary studies. *Microbiology (Reading, England).* 2005;151(Pt 3):653–663; doi:10.1099/mic.0.27437-0.

136. Barrangou R, Fremaux C, Deveau H, Richards M, Boyaval P, Moineau S, et al. CRISPR provides acquired resistance against viruses in prokaryotes. *Science.* 2007;315(5819):1709–1712; doi:10.1126/science.1138140.

137. Makarova KS, Wolf YI, Iranzo J, Shmakov SA, Alkhnbashi OS, Brouns SJJ, et al. Evolutionary classification of CRISPR-Cas systems: A burst of class 2 and derived variants. *Nat Rev Microbiol.* 2020;18(2):67–83; doi:10.1038/s41579-019-0299-x.

138. Roberts A, Barrangou R. Applications of CRISPR-Cas systems in lactic acid bacteria. *FEMS Microbiol Rev.* 2020;44(5):523–537; doi:10.1093/femsre/fuaa016.

139. Kahraman Ilıkkan Ö. Analysis of probiotic bacteria genomes: Comparison of CRISPR/cas systems and spacer acquisition diversity. *Indian J Microbiol.* 2022;62(1):40–46; doi:10.1007/s12088-021-00971-1.

140. Millen AM, Horvath P, Boyaval P, Romero DA. Mobile CRISPR/Cas-mediated bacteriophage resistance in *Lactococcus lactis.* *PLoS ONE.* 2012;7(12):e51663; doi:10.1371/journal.pone.0051663.

141. Pan M, Nethery MA, Hidalgo-Cantabrana C, Barrangou R. Comprehensive mining and characterization of CRISPR-cas systems in bifidobacterium. *Microorganisms.* 2020;8(5); doi:10.3390/microorganisms8050720.

142. Briner AE, Lugli GA, Milani C, Duranti S, Turroni F, Gueimonde M, et al. Occurrence and diversity of CRISPR-Cas systems in the genus bifidobacterium. *PLoS ONE.* 2015;10(7):e0133661; doi:10.1371/journal.pone.0133661.

143. Bonacina J, Suárez N, Hormigo R, Fadda S, Lechner M, Saavedra L. A genomic view of food-related and probiotic Enterococcus strains. *DNA Res: An International Journal for Rapid Publication of Reports on Genes and Genomes.* 2017;24(1):11–24; doi:10.1093/dnares/dsw043.

144. Barchi Y, Philippe C, Chaïb A, Oviedo-Hernandez F, Claisse O, Le Marrec C. Phage encounters recorded in CRISPR arrays in the genus oenococcus. *Viruses.* 2022;15(1); doi:10.3390/v15010015.

145. Horvath P, CoñtÇ-Monvoisin AC, Romero DA, Boyaval P, Fremaux C, Barrangou R. Comparative analysis of CRISPR loci in lactic acid bacteria genomes. *Int J Food Microbiol.* 2009;131(1):62–70.

146. Horvath P, Romero DA, CoñtÇ-Monvoisin AC, Richards M, Deveau H, Moineau S, et al. Diversity, activity, and evolution of CRISPR loci in Streptococcus thermophilus. *J Bacteriol.* 2008;190(4):1401–1412.

147. Horvath P, Barrangou R. CRISPR/Cas, the immune system of bacteria and archaea. *Science (New York, NY).* 2010;327(5962):167–170; doi:10.1126/science.1179555.

148. Wang JY, Doudna JA. CRISPR technology: A decade of genome editing is only the beginning. *Science (New York, NY).* 2023;379(6629):eadd8643; doi:10.1126/science.add8643.

149. Aframian N, Eldar A. Abortive infection antiphage defense systems: Separating mechanism and phenotype. *Trends Microbiol.* 2023;31(10):1003–1012; doi:10.1016/j.tim.2023.05.002.

150. Kelleher P, Mahony J, Bottacini F, Lugli GA, Ventura M, van Sinderen D. The lactococcus lactis pan-plasmidome. *Front Microbiol*. 2019;10(707); doi:10.3389/fmicb.2019.00707.

151. Durmaz E, Klaenhammer TR. Abortive phage resistance mechanism AbiZ speeds the lysis clock to cause premature lysis of phage-infected *Lactococcus lactis*. *J Bacteriol*. 2007;189(4):1417–1425.

152. Lossouarn J, Briet A, Moncaut E, Furlan S, Bouteau A, Son O, et al. *Enterococcus faecalis* countermeasures defeat a virulent picovirinae bacteriophage. *Viruses*. 2019;11(1); doi:10.3390/v11010048.

153. LeRoux M, Laub MT. Toxin-antitoxin systems as phage defense elements. *Ann Rev Microbiol*. 2022;76:21–43; doi:10.1146/annurev-micro-020722-013730.

154. Song S, Wood TK. Post-segregational killing and phage inhibition are not mediated by cell death through toxin/antitoxin systems. *Front Microbiol*. 2018;9:814; doi:10.3389/fmicb.2018.00814.

155. Meeske AJ, Nakandakari-Higa S, Marraffini LA. Cas13-induced cellular dormancy prevents the rise of CRISPR-resistant bacteriophage. *Nature*. 2019;570(7760):241–245; doi:10.1038/s41586-019-1257-5.

156. Niewoehner O, Garcia-Doval C, Rostøl JT, Berk C, Schwede F, Bigler L, et al. Type III CRISPR-Cas systems produce cyclic oligoadenylate second messengers. *Nature*. 2017;548(7669):543–548; doi:10.1038/nature23467.

157. Watson BNJ, Vercoe RB, Salmond GPC, Westra ER, Staals RHJ, Fineran PC. Type I-F CRISPR-Cas resistance against virulent phages results in abortive infection and provides population-level immunity. *Nat Commun*. 2019;10(1):5526; doi:10.1038/s41467-019-13445-2.

158. Duncan-Lowey B, Kranzusch PJ. CBASS phage defense and evolution of antiviral nucleotide signaling. *Curr Opin Immunol*. 2022;74:156–163; doi:10.1016/j.coi.2022.01.002.

159. Cohen D, Melamed S, Millman A, Shulman G, Oppenheimer-Shaanan Y, Kacen A, et al. Cyclic GMP-AMP signalling protects bacteria against viral infection. *Nature*. 2019;574(7780):691–695; doi:10.1038/s41586-019-1605-5.

160. Millman A, Melamed S, Amitai G, Sorek R. Diversity and classification of cyclic-oligonucleotide-based antiphage signalling systems. *Nat Microbiol*. 2020;5(12):1608–1615; doi:10.1038/s41564-020-0777-y.

161. Millman A, Bernheim A, Stokar-Avihail A, Fedorenko T, Voichek M, Leavitt A, et al. Bacterial retrons function in anti-phage defense. *Cell*. 2020;183(6):1551–1561.e12; doi:10.1016/j.cell.2020.09.065.

162. Makarova KS, Wolf YI, Koonin EV. Comparative genomics of defense systems in archaea and bacteria. *Nucleic Acids Res*. 2013;41(8):4360–4377; doi:10.1093/nar/gkt157.

163. Makarova KS, Wolf YI, Snir S, Koonin EV. Defense islands in bacterial and archaeal genomes and prediction of novel defense systems. *J Bacteriol*. 2011;193(21):6039–6056; doi:10.1128/JB.05535-11.

164. Doron S, Melamed S, Ofir G, Leavitt A, Lopatina A, Keren M, et al. Systematic discovery of antiphage defense systems in the microbial pangenome. *Science*. 2018;359(6379); doi:10.1126/science.aar4120.

165. Aparicio-Maldonado C, Ofir G, Salini A, Sorek R, Nobrega FL, Brouns SJJ. Class I DISARM provides anti-phage and anti-conjugation activity by unmethylated DNA recognition. *biorxiv* 2021; https://doi.org/10.1101/2021.12.28.474362.

166. Bravo JPK, Aparicio-Maldonado C, Nobrega FL, Brouns SJJ, Taylor DW. Structural basis for broad anti-phage immunity by DISARM. *Nat Commun*. 2022;13(1):2987; doi:10.1038/s41467-022-30673-1.

167. Ofir G, Melamed S, Sberro H, Mukamel Z, Silverman S, Yaakov G, et al. DISARM is a widespread bacterial defence system with broad anti-phage activities. *Nat Microbiol*. 2018;3(1):90–98; doi:10.1038/s41564-017-0051-0.

168. Shen BW, Doyle LA, Werther R, Westburg AA, Bies DP, Walter SI, et al. Structure, substrate binding and activity of a unique AAA+ protein: The BrxL phage restriction factor. *Nucleic Acids Res*. 2023;51(8):3513–3528; doi:10.1093/nar/gkad083.

169. Goldfarb T, Sberro H, Weinstock E, Cohen O, Doron S, Charpak-Amikam Y, et al. BREX is a novel phage resistance system widespread in microbial genomes. *EMBO J*. 2015;34(2):169–183; doi:10.15252/embj.201489455.

170. Gordeeva J, Morozova N, Sierro N, Isaev A, Sinkunas T, Tsvetkova K, et al. BREX system of Escherichia coli distinguishes self from non-self by methylation of a specific DNA site. *Nucleic Acids Res*. 2019;47(1):253–265; doi:10.1093/nar/gky1125.

171. Bernheim A, Millman A, Ofir G, Meitav G, Avraham C, Shomar H, et al. Prokaryotic viperins produce diverse antiviral molecules. *Nature*. 2021;589(7840):120–124; doi:10.1038/s41586-020-2762-2.

172. Georjon H, Bernheim A. The highly diverse antiphage defence systems of bacteria. *Nat Rev Microbiol*. 2023;21(10):686–700; doi:10.1038/s41579-023-00934-x.

173. Payne LJ, Meaden S, Mestre MR, Palmer C, Toro N, Fineran PC, et al. PADLOC: A web server for the identification of antiviral defence systems in microbial genomes. *Nucleic Acids Res*. 2022;50(W1):W541–W50; doi:10.1093/nar/gkac400.

174. Payne LJ, Todeschini TC, Wu Y, Perry BJ, Ronson CW, Fineran PC, et al. Identification and classification of antiviral defence systems in bacteria and archaea with PADLOC reveals new system types. *Nucleic Acids Res*. 2021;49(19):10868–10878; doi:10.1093/nar/gkab883.

175. Rocha EP, Danchin A, Viari A. Evolutionary role of restriction/modification systems as revealed by comparative genome analysis. *Genome Res.* 2001;11(6):946–958; doi:10.1101/gr.gr-1531rr.

176. Rusinov IS, Ershova AS, Karyagina AS, Spirin SA, Alexeevski AV. Avoidance of recognition sites of restriction-modification systems is a widespread but not universal anti-restriction strategy of prokaryotic viruses. *BMC Genom.* 2018;19(1):885; doi:10.1186/s12864-018-5324-3.

177. Josephsen J, Andersen N, Behrndt H, Brandsborg E, Christiansen G, Hansen MB, et al. An ecological study of lytic bacteriophages of *Lactococcus lactis* subsp. cremoris isolated in a cheese plant over a five year period. *Int Dairy J.* 1994;4(2):123–140.

178. Duerkop BA, Huo W, Bhardwaj P, Palmer KL, Hooper LV. Molecular basis for lytic bacteriophage resistance in enterococci. *mBio.* 2016;7(4); doi:10.1128/mBio.01304-16.

179. Golovenko D, Manakova E, Tamulaitiene G, Grazulis S, Siksnys V. Structural mechanisms for the 5'-CCWGG sequence recognition by the N- and C-terminal domains of EcoRII. *Nucleic Acids Res.* 2009;37(19):6613–6624; doi:10.1093/nar/gkp699.

180. Iida S, Streiff MB, Bickle TA, Arber W. Two DNA antirestriction systems of bacteriophage P1, darA, and darB: Characterization of darA- phages. *Virology.* 1987;157(1):156–166; doi:10.1016/0042-6822(87)90324-2.

181. Zavilgelsky GB, Kotova VY, Rastorguev SM. Antirestriction and antimodification activities of T7 Ocr: Effects of amino acid substitutions in the interface. *Molecular Biol.* 2009;43(1):93–100; doi:10.1134/S0026893309010130.

182. Hill C, Miller LA, Klaenhammer TR. In vivo genetic exchange of a functional domain from a type II A methylase between lactococcal plasmid pTR2030 and a virulent bacteriophage. *J Bacteriol.* 1991;173(14):4363–4370.

183. Loenen WA, Murray NE. Modification enhancement by the restriction alleviation protein (Ral) of bacteriophage lambda. *J Mol Biol.* 1986;190(1):11–22; doi:10.1016/0022-2836(86)90071-9.

184. Kyrkou I, Byth Carstens A, Ellegaard-Jensen L, Kot W, Zervas A, Djurhuus AM, et al. Expanding the diversity of myoviridae phages infecting lactobacillus plantarum-A Novel lineage of lactobacillus phages comprising five new members. *Viruses.* 2019;11(7); doi:10.3390/v11070611.

185. Hui W, Zhang W, Li J, Kwok L-Y, Zhang H, Kong J, et al. Functional analysis of the second methyltransferase in the bacteriophage exclusion system of Lactobacillus casei Zhang. *J Dairy Sci.* 2022;105(3):2049–2057; doi:10.3168/jds.2021-21000.

186. Lu Z, Altermann E, Breidt F, Kozyavkin S. Sequence analysis of Leuconostoc mesenteroides bacteriophage Ø1-A4 isolated from an industrial vegetable fermentation. *Appl Environ Microbiol.* 2010;76(6):1955–1966.

187. Kim S-H, Park J-H. Characterization of prophages in leuconostoc derived from kimchi and genomic analysis of the induced prophage in leuconostoc lactis. *J Microbiol Biotechnol.* 2022;32(3):333–340; doi:10.4014/jmb.2110.10046.

188. Murphy J, Klumpp J, Mahony J, O'Connell-Motherway M, Nauta A, van Sinderen D. Methyltransferases acquired by lactococcal 936-type phage provide protection against restriction endonuclease activity. *BMC Genomics.* 2014;15:831; doi:10.1186/1471-2164-15-831.

189. Murphy J, Bottacini F, Mahony J, Kelleher P, Neve H, Zomer A, et al. Comparative genomics and functional analysis of the 936 group of lactococcal Siphoviridae phages. *Sci Rep.* 2016;6:21345; doi:10.1038/srep21345.

190. Chen B, Akusobi C, Fang X, Salmond GPC. Environmental T4-family bacteriophages evolve to escape abortive infection via multiple routes in a bacterial host employing "altruistic suicide" through type III toxin-antitoxin systems. *Front Microbiol.* 2017;8:1006; doi:10.3389/fmicb.2017.01006.

191. Blower TR, Short FL, Rao F, Mizuguchi K, Pei XY, Fineran PC, et al. Identification and classification of bacterial Type III toxin-antitoxin systems encoded in chromosomal and plasmid genomes. *Nucleic Acids Res.* 2012;40(13):6158–6173; doi:10.1093/nar/gks231.

192. Otsuka Y, Yonesaki T. Dmd of bacteriophage T4 functions as an antitoxin against Escherichia coli LsoA and RnlA toxins. *Mol Microbiol.* 2012;83(4):669–681; doi:10.1111/j.1365-2958.2012.07975.x.

193. Sberro H, Leavitt A, Kiro R, Koh E, Peleg Y, Qimron U, et al. Discovery of functional toxin/antitoxin systems in bacteria by shotgun cloning. *Molecular Cell.* 2013;50(1):136–148; doi:10.1016/j.molcel.2013.02.002.

194. Alawneh AM, Qi D, Yonesaki T, Otsuka Y. An ADP-ribosyltransferase Alt of bacteriophage T4 negatively regulates the Escherichia coli MazF toxin of a toxin-antitoxin module. *Mol Microbiol.* 2016;99(1):188–198; doi:10.1111/mmi.13225.

195. Deveau H, Barrangou R, Garneau JE, Labonte J, Fremaux C, Boyaval P, et al. Phage response to CRISPR-encoded resistance in Streptococcus thermophilus. *J Bacteriol.* 2008;190(4):1390–1400; doi:10.1128/JB.01412-07.

196. Semenova E, Jore MM, Datsenko KA, Semenova A, Westra ER, Wanner B, et al. Interference by clustered regularly interspaced short palindromic repeat (CRISPR) RNA is governed by a seed sequence. *Proc Natl Acad Sci USA.* 2011;108(25):10098–10103; doi:10.1073/pnas.1104144108.

197. Dillard KE, Brown MW, Johnson NV, Xiao Y, Dolan A, Hernandez E, et al. Assembly and translocation of a CRISPR-cas primed acquisition complex. *Cell*. 2018;175(4):934–946.e15; doi:10.1016/j.cell.2018.09.039.

198. Redding S, Sternberg SH, Marshall M, Gibb B, Bhat P, Guegler CK, et al. Surveillance and processing of foreign DNA by the escherichia coli CRISPR-Cas system. *Cell*. 2015;163(4):854–865; doi:10.1016/j.cell.2015.10.003.

199. Babu M, Beloglazova N, Flick R, Graham C, Skarina T, Nocek B, et al. A dual function of the CRISPR-Cas system in bacterial antivirus immunity and DNA repair. *Mol Microbiol*. 2011;79(2):484–502; doi:10.1111/j.1365-2958.2010.07465.x.

200. Bondy-Denomy J, Pawluk A, Maxwell KL, Davidson AR. Bacteriophage genes that inactivate the CRISPR/Cas bacterial immune system. *Nature*. 2013;493(7432):429–432; www.nature.com/nature/journal/v493/n7432/abs/nature11723.html#supplementary-information.

201. Stanley SY, Borges AL, Chen K-H, Swaney DL, Krogan NJ, Bondy-Denomy J, et al. Anti-CRISPR-associated proteins are crucial repressors of anti-crispr transcription. *Cell*. 2019;178(6):1452–1464.e13; doi:10.1016/j.cell.2019.07.046.

202. Shehreen S, Birkholz N, Fineran PC, Brown CM. Widespread repression of anti-CRISPR production by anti-CRISPR-associated proteins. *Nucleic Acids Res*. 2022;50(15):8615–8625; doi:10.1093/nar/gkac674.

203. Bondy-Denomy J, Davidson AR, Doudna JA, Fineran PC, Maxwell KL, Moineau S, et al. A unified resource for tracking anti-CRISPR names. *CRISPR J*. 2018;1:304–305; doi:10.1089/crispr.2018.0043.

204. Li Y, Bondy-Denomy J. Anti-CRISPRs go viral: The infection biology of CRISPR-Cas inhibitors. *Cell Host Microbe*. 2021;29(5):704–714; doi:10.1016/j.chom.2020.12.007.

205. Le H, Yang B, Yi H, Asif A, Wang J, Lithgow T, et al. AcrDB: A database of anti-CRISPR operons in prokaryotes and viruses. *Nucleic Acids Res*. 2021;49(D1):D622–D629; doi:10.1093/nar/gkaa857.

206. Fuchsbauer O, Swuec P, Zimberger C, Amigues B, Levesque S, Agudelo D, et al. Cas9 allosteric inhibition by the anti-CRISPR protein AcrIIA6. *Mol Cell*. 2019;76(6):922–937.e7; doi:10.1016/j.molcel.2019.09.012.

207. Hynes AP, Rousseau GM, Lemay M-L, Horvath P, Romero DA, Fremaux C, et al. An anti-CRISPR from a virulent streptococcal phage inhibits Streptococcus pyogenes Cas9. *Nat Microbiol*. 2017;2(10):1374–1380; doi:10.1038/s41564-017-0004-7.

208. Song G, Zhang F, Tian C, Gao X, Zhu X, Fan D, et al. Discovery of potent and versatile CRISPR-Cas9 inhibitors engineered for chemically controllable genome editing. *Nucleic Acids Res*. 2022;50(5):2836–2853; doi:10.1093/nar/gkac099.

209. Philippe C, Morency C, Plante P-L, Zufferey E, Achigar R, Tremblay DM, et al. A truncated anti-CRISPR protein prevents spacer acquisition but not interference. *Nat Commun*. 2022;13(1):2802; doi:10.1038/s41467-022-30310-x.

210. Yin Y, Yang B, Entwistle S. Bioinformatics identification of anti-CRISPR loci by using homology, guilt-by-association, and CRISPR self-targeting spacer approaches. *mSystems*. 2019;4(5); doi:10.1128/mSystems.00455-19.

211. Howard-Varona C, Hargreaves KR, Solonenko NE, Markillie LM, White RA, Brewer HM, et al. Multiple mechanisms drive phage infection efficiency in nearly identical hosts. *ISME J*. 2018;12(6):1605–1618; doi:10.1038/s41396-018-0099-8.

212. Kondo K, Kawano M, Sugai M. Distribution of antimicrobial resistance and virulence genes within the prophage-associated regions in nosocomial pathogens. *mSphere*. 2021;6(4):e0045221; doi:10.1128/msphere.00452-21.

213. Ruiz-Cruz S, Parlindungan E, Erazo Garzon A, Alqarni M, Lugli GA, Ventura M, et al. Lysogenization of a lactococcal host with three distinct temperate phages provides homologous and heterologous phage resistance. *Microorganisms*. 2020;8(11); doi:10.3390/microorganisms8111685.

214. Eraclio G, Fortina MG, Labrie SJ, Tremblay DM, Moineau S. Characterization of prophages of Lactococcus garvieae. *Sci Rep*. 2017;7(1):1856; doi:10.1038/s41598-017-02038-y.

215. Da Silva Duarte V, Giaretta S, Campanaro S, Treu L, Armani A, Tarrah A, et al. A Cryptic non-inducible prophage confers phage-immunity on the streptococcus thermophilus M17PTZA496. *Viruses*. 2018;11(1); doi:10.3390/v11010007.

216. Penadés JR, Christie GE. The phage-inducible chromosomal Islands: A family of highly evolved molecular parasites. *Ann Rev Virol*. 2015;2(1):181–201; doi:10.1146/annurev-virology-031413-085446.

217. Chen J, Novick RP. Phage-mediated intergeneric transfer of toxin genes. *Science (New York, NY)*. 2009;323(5910):139–141; doi:10.1126/science.1164783.

218. Rousset F, Depardieu F, Miele S, Dowding J, Laval A-L, Lieberman E, et al. Phages and their satellites encode hotspots of antiviral systems. *Cell Host Microbe*. 2022;30(5):740–753.e5; doi:10.1016/j.chom.2022.02.018.

219. Fillol-Salom A, Miguel-Romero L, Marina A, Chen J, Penadés JR. Beyond the CRISPR-Cas safeguard: PICI-encoded innate immune systems protect bacteria from bacteriophage predation. *Curr Opin Microbiol*. 2020;56:52–58; doi:10.1016/j.mib.2020.06.002.

220. Martínez-Rubio R, Quiles-Puchalt N, Martí M, Humphrey S, Ram G, Smyth D, et al. Phage-inducible islands in the Gram-positive cocci. *ISME J.* 2017;11(4):1029–1042; doi:10.1038/ismej.2016.163.

221. Rezaei Javan R, Ramos-Sevillano E, Akter A, Brown J, Brueggemann AB. Prophages and satellite prophages are widespread in Streptococcus and may play a role in pneumococcal pathogenesis. *Nat Commun.* 2019;10(1):4852; doi:10.1038/s41467-019-12825-y.

# SECTION II

# Technology and Food Applications

# Antimicrobials from Lactic Acid Bacteria and Their Potential Applications

# 9

Fergus W. J. Collins, Mary C. Rea, Colin Hill, and R. Paul Ross

## 9.1 INTRODUCTION

The lactic acid bacteria (LAB) are a nonsporulating group of Gram-positive bacteria found in environments ranging from food to the gastrointestinal (GI) tract. LAB have been utilized for millennia due to their food preservative properties and are key to the production of many fermented foods such as yogurt, cheese, sourdough bread, fermented vegetables such as sauerkraut, and fermented meats (see other chapters in this book). The metabolites produced by LAB from the breakdown of the original substrate (i.e., milk, meat or vegetable carbohydrates and proteins) both alter the properties of the product and act to inhibit the growth of competing spoilage microbes (Tamang et al. 2020). A range of antimicrobial metabolites can be produced by LAB (Figure 9.1), primarily in the form of organic acids such as lactic acid, which acidifies the product, thus inhibiting the growth of competing microbes (Fugaban et al. 2022). Many LAB have also been shown to produce antimicrobial peptides known as bacteriocins that target and kill sensitive competing microbes (Alvarez-Sieiro et al. 2016).

Antimicrobials produced by LAB have been shown to inhibit a wide range of food spoilage bacteria, and the addition of bacteriocins such as nisin to food can effectively reduce the levels of pathogens in a variety of products (Mokoena et al. 2021).

A major crisis in modern healthcare is antimicrobial resistance to current therapies, which has been associated with 4.95 million deaths globally in 2019 alone and reported to be directly responsible for 1.27 million deaths (Antimicrobial Resistance Collaborators 2022), figures that are set to double by 2050 unless urgent action is taken (O'Neill 2016). Despite this, the World Health Organization (WHO) has reported an absence of novel leads in the current antimicrobial pipeline (Iskandar et al. 2022). However, the antimicrobials produced by LAB, especially bacteriocins, show potential for applications in the medical sector, with many shown to effectively inhibit multidrug-resistant strains (Fernándes and Jobby 2022). Furthermore, bacteriocins can revive antibiotics rendered ineffective due to antibiotic resistance; indeed, combining nisin Z with the less effective antibiotic methicillin resulted in synergistic killing of methicillin-resistant *Staphylococcus aureus* (MRSA), with the combination being more effective than either

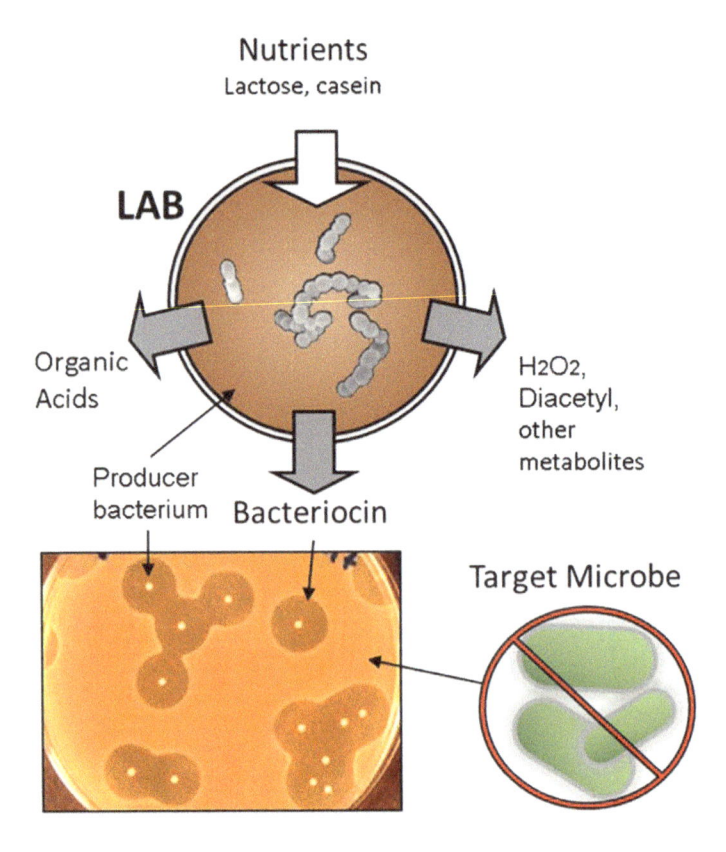

**FIGURE 9.1**  Lactic acid bacteria produce a variety of antimicrobials that can inhibit and kill sensitive microbes. Compounds such as lactic acid are metabolic waste products that can also act as antimicrobials. Bacteriocins are antimicrobial peptides produced by some LAB that target and inhibit competing bacteria.

antimicrobial alone (Ellis et al. 2019). The bacteriocin lacticaseicin 30, produced by *Lacticaseibacillus paracasei*, was proven to be active against several clinical Gram-negative pathogens not commonly reported for LAB bacteriocins, including *Salmonella* species, *Escherichia coli*, *Enterobacter cloacae*, and *Klebsiella* species, which are resistant to antibiotics, and the bacteriocin displayed synergistic activities when combined with colistin against these pathogens (Madi-Moussa et al. 2021).

While much research has focused on the use of these antimicrobials in food, further work needs to be done to realize their potential in medicine (Soltani et al. 2021b). Indeed, some of these antimicrobials, namely bacteriocins and reuterin, are also gaining attention for their anticancer properties. This chapter reviews the range of antimicrobials produced by LAB, outlining their potential applications in industrial, commercial, and medical settings.

## 9.2 BACTERIOCINS

Bacteriocins are heat-stable antimicrobial peptides produced by bacterial cells and, unlike traditional antibiotics, are gene encoded and ribosomally synthesized. The antimicrobial spectrum of these bacteriocins can vary, ranging from a broad spectrum targeting a wide range of bacterial species to a narrow spectrum inhibiting only closely related bacterial species; LAB are known to be prominent bacteriocin producers

(Zacharof and Lovitt 2012). Their mode of action generally involves inhibition of cell wall biosynthesis and disruption of cell membrane integrity. A wide variety of bacteriocins exist, with a common genetic architecture found in many of these gene clusters. The structural gene in bacteriocin operons encodes the active peptide; this usually contains an N-terminal leader sequence. When translated within the cell, the leader sequence reduces or abolishes the antimicrobial activity of the bacteriocin, preventing the cell from being killed by its own antimicrobial. In certain bacteriocins, the leader can also be recognized by modification enzymes, which can then direct the post-translational modification of the active bacteriocin peptide. The leader is also crucial for bacteriocin secretion in that dedicated ATP-binding cassette (ABC) bacteriocin transporters recognize the leader sequence and cleave it at a specific motif (i.e., Gly-Gly) as it is being exported from the cell, thus releasing an active bacteriocin (Oman and van der Donk 2010). Bacteriocins generally have their own secretion system encoded within the operon, which can consist of single or multiple ABC transporters. These transporters also often include protease domains for cleavage of the bacteriocin signal sequence (Havarstein et al. 1995). Some bacteriocins have, however, been shown to be secreted via the Sec translocase system (Herranz and Driessen 2005). Bacteriocin operons also harbor genes encoding immunity proteins that prevent the strains from being killed by their own bacteriocins (Draper et al. 2012). Bacteriocin production can also be under tight regulation in producer cells due to the presence of a two-component regulatory system within the operon which has an important role in quorum sensing (Rohde and Quadri 2006; van der Ploeg 2005). Certain bacteriocins will also have key genes encoding modification and accessory proteins; however, the presence of these depends on the class of bacteriocin.

LAB produce a wide variety of different bacteriocins that can be grouped into classes based on their structure, genetics, and mode of action. Many classification systems for bacteriocins have been proposed, but we will follow the system outlined by Cotter et al. (2013), as this has probably been the most-used classification system in recent years.

## 9.2.1 Class I

Class I bacteriocins undergo enzymatic post-translational modification. Such modifications have important structural and functional roles for these bacteriocins. Class I bacteriocins can be further broken down into subclasses based on how the peptides are modified. Examples of Class I bacteriocins include nisin, a 34 amino acid type I lantibiotic produced by strains of *Lactococcus lactis* that represents one of the more well-known and commercialized bacteriocins (Hooven et al. 1996), and lacticin 3147, a two-peptide lantibiotic encoded on the conjugative plasmid pMRC01 and produced by *Lc. lactis* (Martin et al. 2004).

## 9.2.2 Class II

Class II bacteriocins differ from Class I in that they do not undergo extensive post-translational modification. These bacteriocins are further classified based on their composition and structure. Examples of Class II bacteriocins include pediocin (Papagianni and Anastasiadou 2009), plantaricin JK (Anderssen et al. 1998), enterocin AS-48 (Samyn et al. 1994), and bactofencin (O'Shea et al. 2013).

# 9.3 BACTERIOCIN APPLICATIONS

The antimicrobial activity of bacteriocins coupled with their low toxicity makes them attractive compounds for industry; the fact they are naturally produced by food-grade organisms may also make them more acceptable to modern health-conscious consumers. Bacteriocins are already used in the food industry. Nisin, for example, is an authorized food additive to prevent the outgrowth of food spoilage/ food pathogens (Younes et al. 2017); however, there is the potential to widen the scope of applications, particularly as antimicrobials in the medical field (Cotter et al. 2013).

## 9.3.1  Food Preservation and Safety

The use of bacteriocins in the food industry has primarily focused on food biopreservation and safety. They offer a natural alternative to the addition of chemical preservatives to food and sometimes have actually been shown to improve the flavor of certain fermented foods (O'Sullivan et al. 2002; Younes et al. 2017).

Bacteriocins themselves are used in food manufacture in at least three ways (Figure 9.2), the first of which is the addition of partially purified bacteriocin to the food product (Chikindas et al. 2018). Nisaplin (Danisco), a dried powder containing 1.82% nisin, is one such product (Gough et al. 2017). Nisin was awarded "generally recognized as safe" (GRAS) status by the U.S. Food and Drug Administration in 1988, has been approved by WHO as a food additive, and was also assigned the E number E234 (Younes et al. 2017). Nisin is one of the most commonly used bacteriocin food preservatives, its broad spectrum of activity, heat stability, and history of effectiveness making it an attractive option for the food industry. The addition of Nisaplin to cottage cheese was shown to effectively control the levels of *Listeria mono-cytogenes* (Ferreira and Lund 1996). Nisaplin was also shown to immediately reduce *L. monocytogenes* levels by 3 log CFU/g in queso fresco, a nonfermented cheese (Lourenço et al. 2017).

The second method by which bacteriocins may be added to food is as crude fermentates containing active bacteriocins. One such example is ALTA 2351 (Kerry Bioscience), a pediocin-containing fermentate that has been shown to reduce *L. monocytogenes* numbers in raw sausage batter over 60 days (Knipe and Rust 2009). Another popular example is MicroGARD (Danisco), a product of the fermentation of skimmed milk by LAB. This is extensively used in industry and has been shown to inhibit the spoilage of dairy products such as cottage cheese and yogurt (Makhal et al. 2015; Salih et al. 1990).

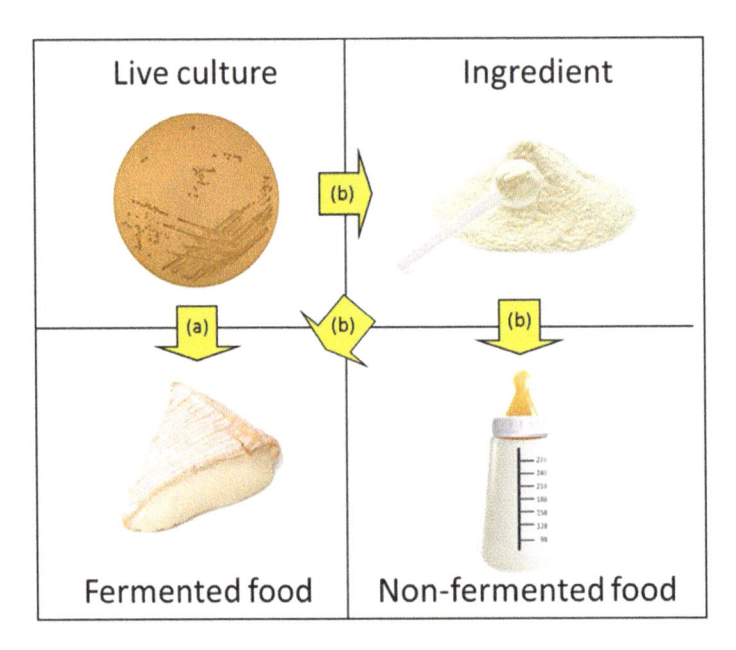

**FIGURE 9.2**  Lactic acid bacteria can be utilized in food biopreservation in several ways. (a) LAB can be used as starter cultures in the production of fermented foods where the *in situ* production of bacteriocins and other antimicrobials can inhibit the growth of spoilage microbes. (b) Bacteriocins from LAB cultures can be concentrated into purified and semi-purified food additives for use as preservatives in the food processing industry.

The third mechanism is to use bacteriocin-producing strains as starter or protective cultures in fermented foods. One advantage of this is that *in situ* production of bacteriocins by starter cultures reduces the need for the addition of external preservatives. The use of bacteriocin-producing cultures has been shown to be effective in the biopreservation of a variety of foods such as fish (Gómez-Sala et al. 2016), meat (Diaz-Ruiz et al. 2012), vegetables (Settanni and Corsetti 2008), and dairy products (Mills et al. 2017). One such example is the use of the nisin producer *Lc. lactis* subsp. *lactis* IFO12007 in the production of miso, a fermented soybean paste. The strain was used as a starter culture and was shown to inhibit the growth of *Bacillus subtilis*, even when this spoilage bacteria is inoculated at a concentration of $10^6$ cells/g (Kato et al. 1999). The combination of a nisin and lacticin 3147–producing *Lc. lactis* strain with a plantaricin-producing *Lactiplantibacillus plantarum* strain in cheese production was shown to cause a reduction of *Listeria* numbers to 0.3 log CFU/g compared to 2.9 log CFU/g in the non-bacteriocin-producing control (Mills et al. 2017).

Bacteriocins are useful as part of a hurdle technology in combination with other methods for food preservation (Hsiao et al. 2016). The use of other compounds and treatments together with the addition of bacteriocins may also lead to synergistic antimicrobial activity. Organic acids, for example, can help increase the net charge of bacteriocins at low pH, thus aiding in bacteriocin translocation through the cell wall; this potential interaction is particularly useful in LAB, which often produce these acids and bacteriocins concurrently (Mills et al. 2011). The role of organic acids as antimicrobials will be discussed later in this chapter. The use of outer membrane permeabilizing agents such as EDTA can also extend the range of activity of these bacteriocins to include Gram-negative bacteria such as *E. coli* O157:H7 in food (Ananou et al. 2005). The combination of lacticin 3147 and the lactoperoxidase system in powdered infant formula, for example, was shown to inhibit the growth of the Gram-negative pathogen *Cronobacter* species up to 12 hours after rehydration of the milk formula (Oshima et al. 2012).

## 9.3.2 Packaging and Nanoparticles

Another potential use for bacteriocins is their incorporation into packaging and other materials (Malhotra et al. 2015). In cases where food comes into contact with the packaging in which it is stored, bacteriocins may simply diffuse from the packaging, thus inhibiting the growth of spoilage microbes on the surface of the food. It serves as an extra hurdle in food processing to prevent food spoilage and extend product shelf life. The incorporation of bacteriocins into packaging has numerous advantages over directly adding them to food through the reduction of nonspecific binding of the bacteriocins to food components rather than the targeted bacterial strains and by reducing the degradation and inactivation of the bacteriocin by food components (Laridi et al. 2003).

There are several ways in which bacteriocins can be incorporated into such materials, for instance, simply coating or absorbing the bacteriocins onto a polymer or directly incorporating the bacteriocins into polymers for packaging (Deshmukh and Thorat 2013). Numerous studies on a range of foods have shown the effectiveness of nisin incorporated into food packaging (Irkin and Esmer 2015). For example, the use of packaging containing immobilized nisin was shown to reduce *Listeria innocua* and *St. aureus* levels as well as those of other microbes in sliced cheese and ham in modified atmosphere packaging, thus improving shelf life (Scannell et al. 2000).

Bacteriocins can also be incorporated into nanoparticles and then used for food safety purposes. Encapsulation of the bacteriocin can help protect it from environmental stresses and ensure controlled release, and nanocarriers used for nisin include nanoliposomes, nanoemulsions, and nanofibers, as examples. (Bahrami et al. 2019). Polydopamine-modified iron oxide nanoparticles covalently immobilized to nisin inactivated *Alicyclobacillus acidoterrestris* cells and spores in apple juice, an acidothermophilic, spore-forming bacterium that survives pasteurization and is known to cause spoilage of beverages (Song et al. 2019). The composites were non-toxic based on *in vitro* assays and did not impact the quality and sensory aspects of the drink. The use of bacteriocin-encapsulated nanoparticles widens the scope for using bacteriocins in liquid foods and food matrices that could degrade the peptides.

### 9.3.3 Bacteriocin-Producing Strains as Probiotics

Probiotics are defined as "live microorganisms which when administered in adequate amounts, confer a health benefit on the host" (FAO/WHO 2001). However, bacteriocin production can also be considered a probiotic trait due to the potential of such strains to inhibit pathogenic bacteria in the GI tract (Dobson et al. 2012). An example of this was demonstrated by Corr et al., who showed that the bacteriocin-producing *Ligilactobacillus salivarius* UCC118 strain displayed the ability to protect mice against infection with *L. monocytogenes*, while a non-bacteriocin-producing mutant failed to show the same effect (2007). Millette et al. also demonstrated that the nisin Z–producing *Lc. lactis* MM19 and pediocin PA-1–producing *Pediococcus acidilactici* MM33 reduced vancomycin-resistant enterococci (VRE) in infected mice (2008). Bacteriocins may also help host strains establish themselves in a complex environment by helping them outcompete the resident microbiota in a particular niche (Dobson et al. 2012). The production of the bacteriocin blpMN by *Streptococcus pneumoniae* were shown to aid the establishment of the strain in the mouse nasopharynx (Dawid et al. 2007). The administration of an *Enterococcus faecalis* strain harboring the bacteriocin 21-encoding plasmid pPD1 was able to colonize and outcompete VRE lacking pPD-1in infected mice, thus indicating the role bacteriocin production may play in the establishment of probiotics and the host microbes in the host (Kommineni et al. 2015). Bacteriocins may also act as useful signaling peptides between the bacteria themselves and also with the host. The production of plantaricins by *Lb. plantarum* WCFS1, for example, was shown to be linked to a change in the levels of the cytokines interleukin 10 and 12 from peripheral blood mononuclear cells, which may offer protection against colitis (Foligne et al. 2007; van Hemert et al. 2010). The potential immunomodulatory properties of nisin have been discussed in a review by Małaczewska and Kaczorek-Łukowska (2021), where studies have reported its capacity to alter cytokine production and innate immune cells.

Due to their specificity, bacteriocins have been touted as microbiome-editing tools, paving the way for precision therapy for microbiome-associated diseases (Heilbronner et al. 2021). For example, the narrow spectrum bacteriocin thuricin CD, produced by *Bacillus thuringiensis*, can kill the gut pathogen *Clostridioides difficile*, responsible for *C. difficile*–associated disease (CDAD), without causing collateral damage to the gut microbiome (Rea et al. 2011).

The benefits of bacteriocin production in probiotic LAB, however, are clearly determined by the producer's ability to actively establish itself and produce bacteriocins in the host, a trait which is not always guaranteed (Dabour et al. 2009).

### 9.3.4 Medical Applications of Bacteriocins

With the increasing prevalence of antibiotic-resistant pathogens, bacteriocins represent a potential novel treatment for the control of these pathogens due to their potency and low toxicity. Much of the work that has been done on this subject has involved the use of animal models; thus more clinical human research must be completed to determine the actual efficacy of bacteriocins in clinical applications. Nonetheless, the results of a small number of clinical trials to date show that bacteriocins have a role to play in medicine. For example, a clinical trial revealed that nisin was effective for the treatment of staphylococcal mastitis in women during lactation (Fernández et al. 2008). The bacteriocins ESL5 produced by *E. faecalis* SL-5 displayed antimicrobial activity against *Cutibacterium acnes*, a key factor in the pathogenesis of acne vulgaris (Kang et al. 2009). The incorporation of the concentrated bacteriocin into a topical lotion was shown to significantly reduce inflammatory lesions in treated patients in comparison to a placebo. A pilot clinical trial with bacteriocin-producing *Lb. salivarius* PS7 with antagonistic activity against otopathogens revealed a significant reduction (84%) in recurring acute otitis media episodes in children (Cárdenas et al. 2019).

Bacteriocins have also exhibited anti-cancer activities (Baindara et al. 2018). A highly concentrated form of the nisin variant, nisin ZP (95%), induced the highest level of apoptosis in head and neck squamous cell carcinoma cells (Kamarajan et al. 2015). In an oral cancer floor-of-mouth mouse model, nisin

ZP reduced tumorigenesis and extended survival (Kamarajan et al. 2015). The anticancer mode of action of nisin ZP include induction of preferential apoptosis, cell cycle arrest, and reduction of cancer cell proliferation, actions it does not exert on primary keratinocytes. Enterocin LNS18 exhibited anticancer activity against HepG2 cells (human liver cancer cells) but did not impact normal fibroblast cells (Al-Madboly et al. 2020). Examples of other anticancer bacteriocins include enterocin LNS18 produced by *Enterococcus thailandicus* (Al-Madboly et al. 2020), pediocin produced by *P. acidilactici* K2a2–3 (Villarante et al. 2011), and microcin E492 produced by *Klebsiella pneumoniae* (Varas et al. 2020). Further research is necessary to fully explore this aspect of bacteriocin functionality and determine how bacteriocins can be used as anticancer therapeutics.

## 9.3.5 Veterinary Applications of Bacteriocins

Bacteriocins also display potential useful benefits for animal care; one such application is the use of bacteriocins to treat mastitis in lactating animals. A study by Cao et al. used an intramammary infusion of nisin to treat mastitis in dairy cows; the results for the nisin treatment had a clinical cure rate similar to that of gentamycin treatment. Nisin, however, is a food-grade product with very few associated issues if it enters the food chain, as opposed to antibiotics such as gentamycin (Cao et al. 2007). A nisin-based teat sealer was shown to reduce *St. aureus* and *E. coli* levels by 3.9 log and 4.22 log, respectively after a 1-minute exposure to the formula, a result comparable to conventional chemical treatments. The nisin formula, however, displayed a lower potential for skin irritation in comparison to the chemical treatments (Sears et al. 1992). A lacticin 3147-containing fermentate was used as a teat dip for mastitis prevention in dairy cows. Here teats were first coated with a pathogen before being treated with the teat dip for 10 minutes. The lacticin 3147-containing fermentate was shown to reduce *Staphylococcus*, *Streptococcus dysgalactiae*, and *Streptococcus uberis* levels by 80%, 97%, and 90%, respectively (Klostermann et al. 2010). In another field trial, a single infusion of a novel formulation containing the lacticin 3147-producing culture to infected teat in mastitic dairy cows showed comparable cure efficacy to the commercial antibiotic Terrexine (Kitching et al. 2019).

The antimicrobial activity of bacteriocins may also be useful for the treatment of GI tract infections in a range of animals. A bacteriocin OR-7 from *Lb. salivarius* was shown to reduce colonization of *Campylobacter jejuni* of chickens. The purified bacteriocin was encapsulated and incorporated into chicken feed and chicks were then challenged with *C. jejuni* strains, where bacteriocin treatment was shown to greatly reduce colonization by the pathogen (Stern et al. 2006). An *in vitro* model of swine intestinal fermentations displayed the potential of pediocin PA-1 to inhibit the growth of pathogenic clostridia in the intestine, again representing an alternative to the use of traditional antibiotics in animal husbandry (Casadei et al. 2009).

Bacteriocins also represent an alternative to the use of subclinical levels of antibiotics to food as a growth promoter. Dietary nisin was shown to increase feed conversion and body weight gain in broiler chickens; this may be a result of the reduced bacterial load in the host, thus increasing the levels of nutrients available. The increased growth may also be due to an improved immune response to pathogenic *Eimeria* parasites (Józefiak et al. 2013). Similarly, the addition of pediocin PA-1 to the feed of chickens challenged with *Clostridium perfringens* improved their growth rate and feed conversion rates (Grilli et al. 2009).

## 9.4 BACTERIOCIN BIOENGINEERING

An appealing aspect of bacteriocins is their gene-encoded nature, which enables the design and development of new peptide structures with improved bioactivity. Most bioengineering studies to date have focused on nisin, where single amino acid changes can result in superior peptides. For example, changing asparagine at position 20 of nisin Z to lysine (resulting in nisin peptide N20K) or changing methionine at position 21 to lysine (peptide M21K) resulted in nisin derivatives capable of targeting the Gram-negative pathogens *Salmonella*, *Pseudomonas*, and *Shigella* (Yuan et al. 2004). Changing serine at position 29 of nisin also resulted in derivatives capable of superior antimicrobial activity against Gram-negative food-associated pathogens including *E. coli*, *Salmonella*, and *Cronobacter sakazakii* (Field et al. 2012). The nisin derivative I4V exhibited improved antimicrobial activity against *Staphylococcus pseudointermedius* and its biofilm compared with native nisin (Field et al. 2015), while nisin derivative M17Q had enhanced activity against *Staphylococcus epidermidis* and its biofilm (Twomey et al. 2020). Bioengineering can also streamline the inhibition spectrum of bacteriocins. For example, nisin derivatives M17Q, T2L, and HTK exhibit increased antimicrobial activity against *Staphylococcus*, an important pathogen in bovine mastitis, but these peptides have reduced anti-*Lactococcus* activity, an important starter culture in downstream dairy fermentations (Field et al. 2021). Bioengineering of nisin has also been utilized to develop derivatives with resistance to the nisin resistance protein NSR by replacing serine 29 with proline (Field et al. 2019). Such derivatives could prove extremely valuable for the treatment of bovine mastitis caused by streptococcal pathogens that are known to carry NSR and are resistant to native nisin (Khosa et al. 2013).

## 9.5 REUTERICYCLIN

Reutericyclin is an antimicrobial *N*-acylated tetrameric acid produced by a number of *Limosilactobacillus reuteri* strains (Gänzle 2004). Reutericyclin is produced by combining a non-ribosomal peptide synthetase (NRPS) and polyketide synthase (PKS) encoded within the *Lb. reuteri* genome. It is unique because it is one of the very few functional NRPS/PKS systems described in LAB (Lin et al. 2015). Reutericyclin acts as a proton-ionophore and dissipates the transmembrane $\Delta$pH of sensitive cells by translocating protons across the cytoplasmic membrane (Gänzle 2004). It is active against a range of Gram-positive bacteria, including pathogens such as *C. difficile* and MRSA (Cherian et al. 2014).

Reutericyclin-producing *Lb. reuteri* strains have a number of potential uses in both food production and as a potential probiotic due to the antimicrobial activity of the compound. Reutericyclin production in sourdough has been shown to help *Lb. reuteri* persist in the environment and remains active against sensitive strains in the dough (Gänzle and Vogel 2003). This suggests that such producing strains may have potential as starter cultures for food preservation (Gänzle and Vogel 2003). As a potential probiotic trait, a reutericyclin-producing strain was found to subtly alter the fecal microbiota of weanling pigs; however, no effect was observed on the presence of clostridial toxins in the hosts after treatment (Yang et al. 2015).

## 9.6 ANTIMICROBIAL METABOLITES

In certain cases, metabolic waste products and intermediates produced by LAB during fermentation can themselves display antimicrobial activity. These antimicrobial metabolites can play an important role in food preservation by limiting the growth of spoilage and pathogenic microbes (Fugaban et al. 2022). The composition and characteristics of these antimicrobials will be further discussed in the following.

### 9.6.1 Organic Acids

Organic acids are the end-product of LAB metabolism; however, in several cases these have been shown to possess antimicrobial activity. The primary acids produced by these cells are lactic acid and acetic acid, but others such as formic acid can also be produced (Fugaban et al. 2022). Their antimicrobial activity may be primarily due to the lowering of the internal pH of sensitive cells. In the uncharged form, these acids are lipid permeable and thus can freely diffuse into the cell's cytoplasm (Hirshfield et al. 2003). Once in the cytoplasm, they can dissociate, causing an accumulation of anions, and can lower the cell's internal pH (pHi) (Salmond et al. 1984). The lowering of the cell's pHi can affect numerous processes in the cells and can lead to internal enzyme denaturation; the increased anion concentration within the cell can also lead to an increase in the transportation of potassium ions into the cell (Roe et al. 2002). This influx of ions increases the turgor pressure within the cell, and in order to balance this, glutamate is then transported out of the cell, which results in the disruption of the cell's osmolarity and thus inhibits cell growth (Warnecke and Gill 2005). A drop in pH can also induce changes in the fatty acid composition of the cell's membrane (Cotter and Hill 2003). The inhibition of cells by weak acids is not, however, solely due to lowering their pHi, as different acids can have specific effects on cells; for instance, the treatment of *E. coli* cells with formate or acetate leads to distinctive transcriptional responses between the two acids. Acetate, for example, was shown to induce the production of proteins found in the RpoS regulon, which is an important controller of the bacterial stress response, while formate caused reduced steady state expression of these genes (Kirkpatrick et al. 2001). Acetic, lactic, and citric acids were also shown to display different levels of antimicrobial activity against *L. monocytogenes*, even when cells had identical pHi values (Young and Foegeding 1993). Such differences may be explained by the distinctive anion pools within the cells after treatments with each acid; the particular lipophilicity of each acid may also affect cells differently as was shown in yeast cells (Capozzi et al. 2009; Hirshfield et al. 2003).

Lactic acid is the primary organic acid produced by LAB, and, as with other organic acids, its activity is not solely due to the lowering of the cell's pHi (Mols and Abee 2011). Upon exposure to lactic acid, *Bacillus cereus* displayed an altered expression of 196 genes, which are not associated with the general response to acidic shock, again indicating a more specific mode of action for this molecule. It was shown to alter the metabolism of these cells by controlling the expression of genes involved in amino acid metabolism (Mols et al. 2010). In addition, lactic acid may also induce oxidative stress within cells (Abbott et al. 2009; Mols et al. 2010; Mols and Abee 2011).

Acetic acid was also found to have a large impact on gene expression in *B. cereus*, altering the regulation of 1430 genes, affecting a variety of pathways involved in oligopeptide and amino acid transport and metabolism, which is similar to the response seen for lactic acid. Acetic acid also altered carbohydrate transport and metabolism in cells, with genes involved in glucose, fructose, lichenan, and trehalose transport and metabolism being downregulated (Mols et al. 2010). As with lactic acid, acetic acid may also inhibit cells by causing oxidative stress (Mols et al. 2010; Mols and Abee 2011).

### 9.6.1.1 Applications

Organic acids can be used in a variety of applications, and due to their broad spectrum of activity, including antifungal activity, and their GRAS status, they are particularly suited to the food industry (Chai et al. 2016; Tomalok et al. 2022; Mani-López et al. 2022).

Solutions containing organic acids are used in the meat processing industry for carcass decontamination in the United States, Canada, Australia, Japan, and the European Union (Loretz et al. 2011; Brashears and Chaves 2017), although in the case of the latter, it is restricted to lactic acid (Brashears and Chaves 2017). They may also be added to juices and beverages, where they can have roles as biopreservatives and as acidity regulators (Quitmann et al. 2014). The neutralization of these organic acids can produce salts that are useful due to their wide spectrum of activity against a range of pathogens and food spoilage microbes such as *E. coli* O157, MRSA, and *Pseudomonas aeruginosa* (Lee et al. 2002; McWilliam Leitch and Stewart 2002). The sodium salts of these organic acids in particular can be used to improve the shelf life of products such as meat, poultry, and fish and to control the growth of pathogens (Maca et al. 1997; Sallam 2007). Levels up to 4.8% by weight of formulation of sodium and potassium lactate are permitted in food to inhibit microbes (Juneja and Thippareddi 2004). These may also be useful if incorporated into food packaging, whereby a controlled release may prolong a product's shelf life by inhibiting spoilage microorganisms (Wang et al. 2015). Often these salts have benefits beyond their antimicrobial activity; sodium lactate, for example, can also act as a humectant and emulsifier (Brewer et al. 1991). Organic acids may also be highly useful in animal feed systems; by reducing the microbial load of feedstuffs, there is an increase in nutrients available to the host. The presence of these acids may reduce the production of ammonia by spoilage microbes, a growth-depressing molecule, and also reduce the pH of the digesta (Dibner and Buttin 2002). Blends of organic acids have proved more beneficial than individual organic acids as performance-enhancing feed additives in broiler chickens (Polycarpo et al. 2017).

The combination of treatment with organic acids and other compounds can greatly enhance the antimicrobial activity of both and can lead to potential novel treatments for food preservation. The combination of lactic acid with the phenolic compound carvacrol has been shown to have synergistic antimicrobial effect against *Shigella sonnei*-infected lettuce leaves (Chai et al. 2016). The combination of organic acids and transition metals was also found to be highly synergistic with up to a thousand-fold increase in antimicrobial activity while greatly improving the effective range of activity against many pathogenic bacteria. Here the organic acids form complexes with these transition metals, thus increasing the permeability of the metals, which leads to an increase in their intracellular concentration. The addition of organic acids to copper sprays currently used in plant and crop treatment could thus greatly increase their antimicrobial potency (Zhitnitsky et al. 2017). There was also found to be synergistic antimicrobial activity between different organic acids and UV-A radiation, even at sublethal levels (de Oliveira et al. 2017).

## 9.6.2 Reuterin

Reuterin is an antimicrobial compound that is an intermediate in the metabolism of glycerol in certain species. The name is derived from its most notable producer, *Lb. reuteri*; however, several other *Lactobacillus* species have also been shown to produce this compound as well as strains from other genera such as certain strains of *Klebsiella* (Martin et al. 2005; Sauvageot et al. 2000; Schütz and Radler 1984; Slininger et al. 1983). Reuterin is composed of a mixture of 3-hydroxypropionaldehyde (3-HPA), its dimer, and its hydrated form (Vollenweider and Lacroix 2004). 3-HPA is an intermediate in the breakdown of glycerol to 1,3-propanediol. Glycerol is first converted to 3-HPA by the adenosylcobalamin-dependent glycerol dehydratase (GDHt); 3-HPA can then be further broken down into 1,3-propanediol by an NAD+-dependent oxidoreductase (Liu and Yu 2015). *Lb. reuteri* is a particularly useful producer of reuterin due to its ability to tolerate larger concentrations of the compound compared to other producing species (Vollenweider and Lacroix 2004). Reuterin inhibits sensitive cells by inducing oxidative stress as the reactive aldehyde in reuterin reacts with thiol groups of small molecules and proteins, which can lead to their inactivation (Schaefer et al. 2010). Reuterin has a broad spectrum of activity, inhibiting a wide range of both Gram-positive and Gram-negative bacteria, along with yeasts, molds, and protozoa (Cleusix et al. 2007).

### 9.6.2.1 Applications

Due to its broad spectrum of activity, reuterin may potentially be a useful antimicrobial. Reuterin-treated mice infected with *Trypanosoma brucei brucei* displayed a 61% reduction in parasitemia levels and had an increased survival rate after a 7-day treatment (Yunmbam and Roberts 1993). Due to its low potential toxicity in the body, along with its inhibitory spectrum against foodborne pathogens and spoilage bacteria, reuterin could also be used in combination with other treatments as a potential food preservative (Fernández-Cruz et al. 2016). It has been reported that at effective antimicrobial concentrations, reuterin is not toxic to eukaryotic cells based on *in vitro* analysis (Soltani et al. 2021a). When added to a Spanish curdled milk product, reuterin alone displayed little antimicrobial activity against *L. monocytogenes* or *St. aureus*. However, when used together with nisin and the lactoperoxidase system, there was synergistic inhibition of these pathogens (Arqués et al. 2008). However, reuterin alone was capable of fungicidal and fungistatic activity in yogurt (Vimont et al. 2019). Reuterin-producing *Lb. reuteri* strains may also be useful potential probiotics. Combining *Lb. reuteri* with glycerol as a symbiotic formulation significantly enhanced the antimicrobial activity of the strain against periodontal pathogens and anaerobic commensals (Van Holm et al. 2022). *Lb. reuteri* itself has been shown to survive gastric transit and has the ability to colonize the intestine (Vollenweider and Lacroix 2004). Models of the colonic epithelium have shown that reuterin production improves the protection offered by *Lb. reuteri* against the adherence, invasion, and intracellular survival of *Salmonella enterica* serovar Typhimurium in a model system (De Weirdt et al. 2012). Glycerol is present in the colon from three sources, diet, microbial production, and release from desquamated epithelial cells (De Weirdt et al. 2010); thus, reuterin can be produced *in situ* in the gut. Reuterin was identified as the most potent inhibitor of colorectal cancer cell growth in a screen of fecal metabolites from wild-type mice (Goyert et al. 2022). However, *Lb. reuteri* and reuterin were found to be downregulated in murine and human colorectal cancer. In the same study, reuterin was shown to inhibit a number of cancer cell lines but not noncancerous human colon cells, and while reuterin-producing *Lb. reuteri* was capable of reducing the growth of implanted tumors in mice, a genetically engineered derivative incapable of reuterin production failed to inhibit tumor growth.

## 9.6.3 Hydrogen Peroxide

Several LAB have also been found to produce hydrogen peroxide ($H_2O_2$) in the presence of oxygen (Hertzberger et al. 2014; Hütt et al. 2016; Schellenberg et al. 2012). The exact mode of action for the antimicrobial activity of $H_2O_2$ is not completely understood; however, it is most likely a combination of DNA damage, protein oxidation, and membrane disruption of the target cell (Imlay et al. 1988; Tamarit et al. 1998). This can be due to the production of reactive hydroxyl radicals formed by Fenton's reaction (Linley et al. 2012). These hydroxyl radicals cause breaks in DNA due to their reaction with the methyl groups of thymine (Di Mascio et al. 1989; Engevik and Versalovic 2017). The small molecular size of the molecule allows it to easily enter the cells where it can react with internal proteins and DNA. The activity of $H_2O_2$ can also be affected by whether the compound is in liquid or gaseous form (Finnegan et al. 2010). $H_2O_2$ is thought to be more effective against Gram-positive than Gram-negative bacteria, while anaerobic strains are thought to be more sensitive to the compound, as they lack the peroxidases and catalase encoded by aerobic bacteria, which allows them to break down $H_2O_2$ (McDonnell and Russell 1999). However, the activity of $H_2O_2$ can also be enhanced synergistically by acting together with lactic acid, which is naturally produced by these bacteria (Atassi and Servin 2010). The membrane damage induced by lactic acid may make cells more susceptible and sensitive to the activity of $H_2O_2$ (Engevik and Versalovic 2017).

### 9.6.3.1 Applications

$H_2O_2$-producing lactobacilli have been associated with protection against the acquisition of bacterial vaginosis, and their absence is associated with a greater risk of acquiring HIV-1 infection (Hawes et al. 1996; Martin Jr et al. 1999). $H_2O_2$ production may also support colonization of the vagina by producing strains (Vallor et al. 2001). While $H_2O_2$ production has been associated with colonization, it has been shown that the levels produced by such strains may not be sufficient to inhibit the growth of vaginal pathogens and that lactic acid may be playing a greater antimicrobial role (Gong et al. 2014; O'Hanlon et al. 2010, 2011; Tachedjian et al. 2018). Rather than acting as an antimicrobial, $H_2O_2$ may have a more important immunomodulatory role in the vagina, lowering the levels of pro-inflammatory cytokines. This may explain the positive correlation between $H_2O_2$-producing lactobacilli and reduced bacterial vaginosis levels (Mitchell et al. 2015). $H_2O_2$ has been reported to increase the sensitivity of opportunistic pathogens to antibiotics such that $H_2O_2$-producing lactobacilli have been suggested as potential "antibiotic assistants" for antibiotic therapy (Sgibnev and Kremleva 2017).

# 9.7 CONCLUSION

Antimicrobials produced by LAB represent an untapped resource in food processing and medical applications. With increasing concerns over food safety and the increase in the prevalence of multidrug-resistant pathogens, there are limited alternatives for novel treatments. Bacteriocins especially may represent a future option to address such issues, as their low toxicity and often high specificity reduce the potential side effects for the host. Many studies have been conducted on nisin, in particular outlining its efficacy against many important pathogens; it has also been approved as a food additive and is extensively used in certain food-processing industries. These antimicrobials can be utilized in many ways by industry. The *in situ* production of bacteriocins and other antimicrobials by starter cultures in fermented foods and probiotics in medicine represent an alternative to the administration of purified or semi-purified bacteriocins, which could prove expensive.

As the growth of antibiotic resistance is an increasing concern, more attention needs to be paid to novel therapeutics such as the use of bacteriocins. A major advantage of bacteriocins is the potential to modify and improve their functionality through bioengineering. As bacteriocins are gene encoded, they can easily be altered through genetic manipulation (Field et al. 2015). This may allow scientists to overcome some of the deficiencies of bacteriocin use, such as improved bioactivity against specific pathogens and reduced risk of resistance development. The narrow spectrum of activity of certain bacteriocins also allows pathogens to be targeted more directly with less collateral damage caused to the rest of the microbiota (Rea et al. 2010). While much of the research into bacteriocins and other antimicrobials from LAB is focused on the food industry, dwindling numbers of novel antibiotics being discovered means that more of a focus must be placed on research into these as therapeutic drugs.

# BIBLIOGRAPHY

Abbott, D. A., Suir, E., Duong, G.-H., De Hulster, E., Pronk, J. T. & Van Maris, A. J. 2009. Catalase overexpression reduces lactic acid-induced oxidative stress in Saccharomyces cerevisiae. *Applied and Environmental Microbiology*, 75, 2320–2325.

Al-Madboly, L. A., El-Deeb, N. M., Kabbash, A., Nael, M. A., Kenawy, A. M. & Ragab, A. E. 2020. Purification, characterization, identification, and anticancer activity of a circular bacteriocin from *Enterococcus thailandicus*. *Frontiers in Bioengineering and Biotechnology*, 8, 450.

Alvarez-Sieiro, P., Montalban-Lopez, M., Mu, D. & Kuipers, O. P. 2016. Bacteriocins of lactic acid bacteria: Extending the family. *Applied Microbiology and Biotechnology*, 100.

Ananou, S., Gálvez, A., Martínez-Bueno, M., Maqueda, M. & Valdivia, E. 2005. Synergistic effect of enterocin AS-48 in combination with outer membrane permeabilizing treatments against *Escherichia coli* O157: H7. *Journal of Applied Microbiology*, 99, 1364–1372.

Anderssen, E. L., Diep, D. B., Nes, I. F., Eijsink, V. G. & Nissen-Meyer, J. 1998. Antagonistic activity of *Lactobacillus plantarum* C11: Two new two-peptide bacteriocins, plantaricins EF and JK, and the induction factor plantaricin A. *Applied and Environmental Microbiology*, 64, 2269–2272.

Antimicrobial Resistance Collaborators. 2022. Global burden of bacterial antimicrobial resistance in 2019: A systematic analysis. *Lancet*, 399, 629–655.

Arqués, J., Rodríguez, E., Nuñez, M. & Medina, M. 2008. Antimicrobial activity of nisin, reuterin, and the lactoperoxidase system on *Listeria monocytogenes* and *Staphylococcus aureus* in Cuajada, a semisolid dairy product manufactured in Spain. *Journal of Dairy Science*, 91, 70–75.

Atassi, F. & Servin, A. L. 2010. Individual and co-operative roles of lactic acid and hydrogen peroxide in the killing activity of enteric strain *Lactobacillus johnsonii* NCC933 and vaginal strain *Lactobacillus gasseri* KS120. 1 against enteric, uropathogenic and vaginosis-associated pathogens. *FEMS Microbiology Letters*, 304, 29–38.

Bahrami, A., Delshadi, R., Jafari, S. M. & Williams, L. 2019. Nanoencapsulated nisin: An engineered natural antimicrobial system for the food industry. *Trends in Food Science and Technology*, 94, 20–31.

Baindara, P., Korpole, S. & Grover, V. 2018. Bacteriocins: Perspective for the development of novel anticancer drugs. *Applied Microbiology and Biotechnology*, *102*, 10393–10408.

Brashears, M. M. & Chaves, B. D. 2017. The diversity of beef safety: A global reason to strengthen our current systems. *Meat Science*, *132*, 59–71.

Brewer, M. S., Mckeith, F., Martin, S. E., Dallmier, A. W. & Meyer, J. 1991. Sodium lactate effects on shelf-life, sensory, and physical characteristics of fresh pork sausage. *Journal of Food Science*, *56*, 1176–1178.

Cao, L., Wu, J., Xie, F., Hu, S. & Mo, Y. 2007. Efficacy of nisin in treatment of clinical mastitis in lactating dairy cows. *Journal of Dairy Science*, *90*, 3980–3985.

Capozzi, V., Fiocco, D., Amodio, M. L., Gallone, A. & Spano, G. 2009. Bacterial stressors in minimally processed food. *International Journal of Molecular Sciences*, *10*, 3076–3105.

Cárdenas, N., Martín, V., Arroyo, R., López, M., Carrera, M., Badiola, C., Jiménez, E. & Rodríguez, J. M. 2019. Prevention of recurrent acute otitis media in children through the use of *Lactobacillus salivarius* PS7, a target-specific probiotic strain. *Nutrients*, *11*, 376.

Casadei, G., Grilli, E. & Piva, A. 2009. Pediocin A modulates intestinal microflora metabolism in swine in vitro intestinal fermentations. *Journal of Animal Science*, *87*, 2020–2028.

Chai, C., Lee, S., Kim, J. & Oh, S. W. 2016. Synergistic antimicrobial effects of organic acids in combination with carvacrol against *Shigella sonnei* . *Journal of Food Safety*, *36*, 360–366.

Cherian, P. T., Wu, X., Maddox, M. M., Singh, A. P., Lee, R. E. & Hurdle, J. G. 2014. Chemical modulation of the biological activity of reutericyclin: A membrane-active antibiotic from *Lactobacillus reuteri* . *Scientific Reports*, *4*, 4721.

Chikindas, M. L., Weeks, R., Drider, D., Chistyakov, V. A. & Dicks, L. M. 2018. Functions and emerging applications of bacteriocins. *Current Opinion in Biotechnology*, *49*, 23–28.

Cleusix, V., Lacroix, C., Vollenweider, S., Duboux, M. & Le Blay, G. 2007. Inhibitory activity spectrum of reuterin produced by *Lactobacillus reuteri* against intestinal bacteria. *BMC Microbiology*, *7*, 101.

Corr, S. C., Li, Y., Riedel, C. U., O'Toole, P. W., Hill, C. & Gahan, C. G. 2007. Bacteriocin production as a mechanism for the antiinfective activity of *Lactobacillus salivarius* UCC118. *Proceedings of the National Academy of Sciences of the United States of America*, *104*, 7617–7621.

Cotter, P. D. & Hill, C. 2003. Surviving the acid test: Responses of Gram-positive bacteria to low pH. *Microbiology and Molecular Biology Reviews*, *67*, 429–453.

Cotter, P. D., Ross, R. P. & Hill, C. 2013. Bacteriocins—a viable alternative to antibiotics? *Nature Reviews Microbiology*, *11*, 95–105.

Dabour, N., Zihler, A., Kheadr, E., Lacroix, C. & Fliss, I. 2009. *In vivo* study on the effectiveness of pediocin PA-1 and *Pediococcus acidilactici* UL5 at inhibiting *Listeria monocytogenes* . *International Journal of Food Microbiology*, *133*, 225–233.

Dawid, S., Roche, A. M. & Weiser, J. N. 2007. The blp bacteriocins of *Streptococcus pneumoniae* mediate intraspecies competition both *in vitro* and *in vivo* . *Infection and Immunity*, *75*, 443–451.

De Oliveira, E. F., Cossu, A., Tikekar, R. V. & Nitin, N. 2017. Enhanced antimicrobial activity based on a synergistic combination of sublethal levels of stresses induced by UV-A light and organic acids. *Applied and Environmental Microbiology*, *83*, e00383–17.

Deshmukh, P. & Thorat, P. 2013. Bacteriocins: A new trend in antimicrobial food packaging. *International Journal of Advanced Research in Engineering and Applied Sciences*, *1*, 1–12.

De Weirdt, R., Crabbe, A., Roos, S., Vollenweider, S., Lacroix, C., Van Pijkeren, J. P., Britton, R. A., Sarker, S., Van De Wiele, T. & Nickerson, C. A. 2012. Glycerol supplementation enhances *L. reuteri*'s protective effect against *S. Typhimurium* colonization in a 3-D model of colonic epithelium. *PLoS ONE*, *7*, e37116.

De Weirdt, R., Possemiers, S., Vermeulen, G., Moerdijk-Poortvliet, T. C., Boschker, H. T., Verstraete, W. & Van de Wiele, T. 2010. Human faecal microbiota display variable patterns of glycerol metabolism. *FEMS Microbiology and Ecology*, *74*, 601–611.

Diaz-Ruiz, G., Omar, N. B., Abriouel, H., Cantilde, M. M. & Galvez, A. 2012. Inhibition of *Listeria monocytogenes* and *Escherichia coli* by bacteriocin-producing *Lactobacillus plantarum* EC52 in a meat sausage model system. *African Journal of Microbiology Research*, *6*, 1103–1108.

Dibner, J. & Buttin, P. 2002. Use of organic acids as a model to study the impact of gut microflora on nutrition and metabolism. *Journal of Applied Poultry Research*, *11*, 453–463.

Di Mascio, P., Wefers, H., Do-Thi, H.-P., Lafleur, M. V. M. & Sies, H. 1989. Singlet molecular oxygen causes loss of biological activity in plasmid and bacteriophage DNA and induces single-strand breaks. *Biochimica et Biophysica Acta (BBA)-Gene Structure and Expression*, *1007*, 151–157.

Dobson, A., Cotter, P. D., Ross, R. P. & Hill, C. 2012. Bacteriocin production: A probiotic trait? *Applied and Environmental Microbiology*, *78*, 1–6.

Draper, L. A., Deegan, L. H., Hill, C., Cotter, P. D. & Ross, R. P. 2012. Insights into lantibiotic immunity provided by bioengineering of LtnI. *Antimicrobial Agents and Chemotherapy*, *56*, 5122–5133.

Ellis, J.-C., Ross, R. P. & Hill, C. 2019. Nizin Z and lacticin 3147 improve efficacy of antibiotics against clinically significant bacteria. *Future Microbiology, 14*, 1573–1587.

Engevik, M. & Versalovic, J. 2017. Biochemical features of beneficial microbes: Foundations for therapeutic microbiology. *Microbiology Spectrum, 5*.

FAO/WHO. 2001. Food and agriculture organization and world health organization expert consultation. *Evaluation of Health and Nutritional Properties of Powder Milk and Live Lactic Acid Bacteria* [Internet]. Available from: www.fao.org/tempref/docrep/fao/meeting/009/y6398e.pdf

Fernándes, A. & Jobby, R. 2022. Bacteriocins from lactic acid bacteria and their potential clinical applications. *Applied Biochemistry and Biotechnology, 194*, 4377–4399.

Fernández, L., Delgado, S., Herrero, H., Maldonado, A. & Rodríguez, J. M. 2008. The bacteriocin nisin, an effective agent for the treatment of staphylococcal mastitis during lactation. *Journal of Human Lactation, 24*, 311–316.

Fernández-Cruz, M. L., Martín-Cabrejas, I., Pérez-Del Palacio, J., Gaya, P., Díaz-Navarro, C., Navas, J. M., Medina, M. & Arqués, J. L. 2016. *In vitro* toxicity of reuterin, a potential food biopreservative. *Food and Chemical Toxicology, 96*, 155–159.

Ferreira, M. & Lund, B. 1996. The effect of nisin on *Listeria monocytogenes* in culture medium and long-life cottage cheese. *Letters in Applied Microbiology, 22*, 433–438.

Field, D., Begley, M., O'Connor, P. M., Daly, K. M., Hugenholtz, F., Cotter, P. D., Hill, C. & Ross, R. P. 2012. Bioengineered nisin A derivatives with enhanced activity against both gram positive and gram negative pathogens. *PLoS ONE 7*, e46884.

Field, D., Blake, T., Mathur, H., O'Connor, P. M., Cotter, P. D., Ross, R. P., & Hill, C. 2019. Bioengineering nisin to overcome the nisin resistance protein. *Molecular Microbiology 111*, 717–731.

Field, D., Considine, K., O'Connor, P. M., Ross, R. P., Hill, C. & Cotter, P. D. 2021. Bio-engineered nisin with increased anti-*Staphylococcus* and selectively reduced anti-*Lactococcus* activity for treatment of bovine mastitis. *International Journal of Molecular Sciences, 22*, 3480.

Field, D., Cotter, P. D., Ross, R. P. & Hill, C. 2015. Bioengineering of the model lantibiotic nisin. *Bioengineered, 6*, 187–192.

Finnegan, M., Linley, E., Denyer, S. P., Mcdonnell, G., Simons, C. & Maillard, J.-Y. 2010. Mode of action of hydrogen peroxide and other oxidizing agents: Differences between liquid and gas forms. *Journal of Antimicrobial Chemotherapy, 65*, 2108–2115.

Foligne, B., Nutten, S., Grangette, C., Dennin, V., Goudercourt, D., Poiret, S., Dewulf, J., Brassart, D., Mercenier, A. & Pot, B. 2007. Correlation between *in vitro* and *in vivo* immunomodulatory properties of lactic acid bacteria. *World Journal of Gastroenterology: WJG, 13*, 236.

Fugaban, J. II, Holzapfel, W. H. & Todorov, S. D. 2022. The overview of natural by-products of beneficial lactic acid bacteria as promising antimicrobial agents. *Applied Food Biotectnology, 9*, 127–143.

Gänzle, M. G. 2004. Reutericyclin: Biological activity, mode of action, and potential applications. *Applied Microbiology and Biotechnology, 64*, 326–332.

Gänzle, M. G. & Vogel, R. F. 2003. Contribution of reutericyclin production to the stable persistence of *Lactobacillus reuteri* in an industrial sourdough fermentation. *International Journal of Food Microbiology, 80*, 31–45.

Gómez-Sala, B., Herranz, C., Díaz-Freitas, B., Hernández, P. E., Sala, A. & Cintas, L. M. 2016. Strategies to increase the hygienic and economic value of fresh fish: Biopreservation using lactic acid bacteria of marine origin. *International Journal of Food Microbiology, 223*, 41–49.

Gong, Z., Luna, Y., Yu, P. & Fan, H. 2014. Lactobacilli inactivate *Chlamydia trachomatis* through lactic acid but not $H_2O_2$. *PLoS ONE, 9*, e107758.

Gough, R., Gómez-Sala, B., O'Connor, P. M., Rea, M. C., Miao, S., Hill, C. & Brodkorb, A. 2017. A simple method for the purification of nisin. *Probiotics and Antimicrobial Proteins, 9*, 363–369.

Goyert, J. W., Bell, H. N. & Shah, Y. M. 2022. PCL22–188: Reuterin in the healthy gut microbiome suppresses colorectal cancer growth through altering redox balance. *Official Journal of the National Comprehensive Cancer Network, 20*, 3–5.

Grilli, E., Messina, M., Catelli, E., Morlacchini, M. & Piva, A. 2009. Pediocin A improves growth performance of broilers challenged with *Clostridium perfringens* . *Poultry Science, 88*, 2152–2158.

Havarstein, L. S., Diep, D. B. & Nes, I. F. 1995. A family of bacteriocin ABC transporters carry out proteolytic processing of their substrates concomitant with export. *Molecular Microbiology, 16*, 229–240.

Hawes, S. E., Hillier, S. L., Benedetti, J., Stevens, C. E., Koutsky, L. A., Wølner-Hanssen, P. & Holmes, K. K. 1996. Hydrogen peroxide—producing lactobacilli and acquisition of vaginal infections. *Journal of Infectious Diseases, 174*, 1058–1063.

Heilbronner, S., Krismer, B., Brötz-Oesterhelt, H. & Peschel, A. 2021. The microbiome-shaping roles of bacteriocins. *Nature Reviews Microbiology, 19*, 726–739.

Herranz, C. & Driessen, A. J. 2005. Sec-mediated secretion of bacteriocin enterocin P by *Lactococcus lactis*. *Applied and Environmental Microbiology*, *71*, 1959–1963.

Hertzberger, R., Arents, J., Dekker, H. L., Pridmore, R. D., Gysler, C., Kleerebezem, M. & De Mattos, M. J. T. 2014. H2O2 production in species of the *Lactobacillus acidophilus* group: A central role for a novel NADH-dependent flavin reductase. *Applied and Environmental Microbiology*, *80*, 2229–2239.

Hirshfield, I. N., Terzulli, S. & O' Byrne, C. 2003. Weak organic acids: A panoply of effects on bacteria. *Science Progress*, *86*, 245–269.

Hooven, H. W., Doeland, C., Kamp, M., Konings, R. N., Hilbers, C. W. & Ven, F. J. 1996. Three-dimensional structure of the lantibiotic nisin in the presence of membrane-mimetic micelles of dodecylphosphocholine and of sodium dodecylsulphate. *The FEBS Journal*, *235*, 382–393.

Hsiao, H. L., Lin, S. B., Chen, L. C. & Chen, H. H. 2016. Hurdle effect of antimicrobial activity achieved by time differential releasing of nisin and chitosan hydrolysates from bacterial cellulose. *Journal of Food Science*, *81*, M1184–M1191.

Hütt, P., Lapp, E., Štšepetova, J., Smidt, I., Taelma, H., Borovkova, N., Oopkaup, H., Ahelik, A., Rööp, T. & Hoidmets, D. 2016. Characterisation of probiotic properties in human vaginal lactobacilli strains. *Microbial Ecology in Health and Disease*, *27*, 30484.

Imlay, J. A., Chin, S. M. & Linn, S. 1988. Toxic DNA damage by hydrogen peroxide through the Fenton reaction *in vivo* and *in vitro*. *Science*, *240*, 640–642.

Irkin, R. & Esmer, O. K. 2015. Novel food packaging systems with natural antimicrobial agents. *Journal of Food Science and Technology*, *52*, 6095–6111.

Iskandar, K., Murugaiyan, J., Hammoudi Halat, D., Hage, S. E., Chibabhai, V., Adukkadukkam, S., Roques, C., Molinier, L., Salameh, P. & Van Dongen, M. 2022. Antibiotic discovery and resistance: The chase and the race. *Antibiotics (Basel)*, *11*, 182.

Józefiak, D., Kierończyk, B., Juśkiewicz, J., Zduńczyk, Z., Rawski, M., Długosz, J., Sip, A. & Højberg, O. 2013. Dietary nisin modulates the gastrointestinal microbial ecology and enhances growth performance of the broiler chickens. *PLoS ONE*, *8*, e85347.

Juneja, V. & Thippareddi, H. 2004. Inhibitory effects of organic acid salts on growth of *Clostridium perfringens* from spore inocula during chilling of marinated ground turkey breast. *International Journal of Food Microbiology*, *93*, 155–163.

Kamarajan, P., Hayami, T., Matte, B., Liu, Y., Danciu, T., Ramamoorthy, A., Worden, F., Kapila, S. & Kapila, Y. 2015. Nisin ZP, a bacteriocin and food preservative, inhibits head and neck cancer tumorigenesis and prolongs survival. *PLoS ONE*, *10*, e0131008.

Kang, B. S., Seo, J.-G., Lee, G.-S., Kim, J.-H., Kim, S. Y., Han, Y. W., Kang, H., Kim, H. O., Rhee, J. H. & Chung, M.-J. 2009. Antimicrobial activity of enterocins from *Enterococcus faecalis* SL-5 against *Propionibacterium acnes*, the causative agent in acne vulgaris, and its therapeutic effect. *The Journal of Microbiology*, *47*, 101–109.

Kato, T., Maeda, K., Kasuya, H. & Matsuda, T. 1999. Complete growth inhibition of *Bacillus subtilis* by nisin-producing lactococci in fermented soybeans. *Bioscience, Biotechnology, and Biochemistry*, *63*, 642–647.

Khosa, S., AlKhatib, Z., & Smits, S. H. J. 2013. NSR from *Streptococcus agalactiae* confers resistance against nisin and is encoded by a conserved *nsr* operon. *Biological Chemistry*, *394*, 1543–1549.

Kirkpatrick, C., Maurer, L. M., Oyelakin, N. E., Yoncheva, Y. N., Maurer, R. & Slonczewski, J. L. 2001. Acetate and formate stress: Opposite responses in the proteome of *Escherichia coli*. *Journal of Bacteriol*, *183*, 6466–6477.

Kitching, M., Mathur, H., Flynn, J., Byrne, N., Dillon, P., Sayers, R., Rea, M. C., Hill, C. & Ross, R. P. 2019. A live bio-therapeutic for mastitis, containing *Lactococcus lactis* DPC3147 with comparable efficacy to antibiotic treatment. *Frontiers in Microbiology*, *10*, 2220.

Klostermann, K., Crispie, F., Flynn, J., Meaney, W. J., Ross, R. P. & Hill, C. 2010. Efficacy of a teat dip containing the bacteriocin lacticin 3147 to eliminate Gram-positive pathogens associated with bovine mastitis. *Journal of Dairy Research*, *77*, 231–238.

Knipe, C. L. & Rust, R. E. 2009. Thermal Processing of Ready-to-Eat Meat Products, John Wiley & Sons, Hoboken, NJ.

Kommineni, S., Bretl, D. J., Lam, V., Chakraborty, R., Hayward, M., Simpson, P., Cao, Y., Bousounis, P., Kristich, C. J. & Salzman, N. H. 2015. Bacteriocin production augments niche competition by enterococci in the mammalian gastrointestinal tract. *Nature*, *526*, 719–722.

Laridi, R., Kheadr, E., Benech, R.-O., Vuillemard, J., Lacroix, C. & Fliss, I. 2003. Liposome encapsulated nisin Z: Optimization, stability and release during milk fermentation. *International Dairy Journal*, *13*, 325–336.

Lee, Y.-L., Cesario, T., Owens, J., Shanbrom, E. & Thrupp, L. D. 2002. Antibacterial activity of citrate and acetate. *Nutrition*, *18*, 665–666.

Lin, X. B., Lohans, C. T., Duar, R., Zheng, J., Vederas, J. C., Walter, J. & Gänzle, M. 2015. Genetic determinants of reutericyclin biosynthesis in *Lactobacillus reuteri*. *Applied and Environmental Microbiology*, *81*, 2032–2041.

Linley, E., Denyer, S. P., Mcdonnell, G., Simons, C. & Maillard, J.-Y. 2012. Use of hydrogen peroxide as a biocide: New consideration of its mechanisms of biocidal action. *Journal of Antimicrobial Chemotherapy*, 67, 1589–1596.

Liu, F. & Yu, B. 2015. Efficient production of reuterin from glycerol by magnetically immobilized *Lactobacillus reuteri* . *Applied Microbiology and Biotechnology*, 99, 4659–4666.

Loretz, M., Stephan, R. & Zweifel, C. 2011. Antibacterial activity of decontamination treatments for cattle hides and beef carcasses. *Food Control*, 22, 347–359.

Lourenço, A., Kamnetz, M. B., Gadotti, C. & Diez-Gonzalez, F. 2017. Antimicrobial treatments to control *Listeria monocytogenes* in queso fresco. *Food Microbiology*, 64, 47–55.

Maca, J., Miller, R. & Acuff, G. 1997. Microbiological, sensory and chemical characteristics of vacuum-packaged ground beef patties treated with salts of organic acids. *Journal of Food Science*, 62, 591–596.

Madi-Moussa, D., Belguesmia, Y., Charlet, A., Drider, D. & Coucheney, F. 2021. Lacticaseicin 30 and colistin as a promising antibiotic formulation against Gram-negative β-lactamase-producing strains and colistin-resistant strains. *Antibiotics (Basel)*, 11, 20.

Makhal, S., Kanawjia, S. & Giri, A. 2015. Effect of MicroGARD on keeping quality of direct acidified Cottage cheese. *Journal of Food Science and Technology*, 52, 936–943.

Małaczewska, J. & Kaczorek-Łukowska, E. 2021. Nisin-A lantibiotic with immunomodulatory properties: A review. *Peptides*, 137, 170479.

Malhotra, B., Keshwani, A. & Kharkwal, H. 2015. Antimicrobial food packaging: Potential and pitfalls. *Frontiers in Microbiology*, 6, 611.

Mani-López, E., Arrioja-Bretón, D. & López-Malo, A. 2022. The impacts of antimicrobial and antifungal activity of cell-free supernatants from lactic acid bacteria *in vitro* and foods. *Comprehensive Reviews in Food Science and Food Safety*, 21, 604–641.

Martin Jr, H. L., Richardson, B. A., Nyange, P. M., Lavreys, L., Hillier, S. L., Chohan, B., Mandaliya, K., Ndinya-Achola, J. O., Bwayo, J. & Kreiss, J. 1999. Vaginal lactobacilli, microbial flora, and risk of human immunodeficiency virus type 1 and sexually transmitted disease acquisition. *Journal of Infectious Diseases*, 180, 1863–1868.

Martin, N. I., Sprules, T., Carpenter, M. R., Cotter, P. D., Hill, C., Ross, R. P. & Vederas, J. C. 2004. Structural characterization of lacticin 3147, a two-peptide lantibiotic with synergistic activity. *Biochemistry*, 43, 3049–3056.

Martin, R., Olivares, M., Marin, M., Xaus, J., Fernández, L. & Rodriguez, J. 2005. Characterization of a reuterin-producing *Lactobacillus coryniformis* strain isolated from a goat's milk cheese. *International Journal of Food Microbiology*, 104, 267–277.

Mcdonnell, G. & Russell, A. D. 1999. Antiseptics and disinfectants: Activity, action, and resistance. *Clinical Microbiology Reviews*, 12, 147–179.

Mcwilliam Leitch, E. & Stewart, C. 2002. Susceptibility of *Escherichia coli* O157 and non-O157 isolates to lactate. *Letters in Applied Microbiology*, 35, 176–180.

Millette, M., Cornut, G., Dupont, C., Shareck, F., Archambault, D. & Lacroix, M. 2008. Capacity of human nisin- and pediocin-producing lactic acid bacteria to reduce intestinal colonization by vancomycin-resistant enterococci. *Applied and Environmental Microbiology*, 74, 1997–2003.

Mills, S., Griffin, C., O'Connor, P., Serrano, L., Meijer, W. C., Hill, C. & Ross, R. P. 2017. A multibacteriocin cheese starter system, comprising nisin and lacticin 3147 in *Lactococcus lactis*, in combination with plantaricin from *Lactobacillus plantarum* . *Applied and Environmental Microbiology*, 83, e00799–17.

Mills, S., Stanton, C., Hill, C. & Ross, R. 2011. New developments and applications of bacteriocins and peptides in foods. *Annual Review of Food Science and Technology*, 2, 299–329.

Mitchell, C., Fredricks, D., Agnew, K. & Hitti, J. 2015. Hydrogen-peroxide producing lactobacilli are associated with lower levels of vaginal IL1B, independent of bacterial vaginosis. *Sexually Transmitted Diseases*, 42, 358.

Mokoena, M. P., Omatola, C. A. & Olaniran, A. O. 2021. Applications of lactic acid bacteria and their bacteriocins against food spoilage microorganisms and foodborne pathogens. *Molecules*, 26, 7055.

Mols, M. & Abee, T. 2011. *Bacillus cereus* responses to acid stress. *Environmental Microbiology*, 13, 2835–2843.

Mols, M., Van Kranenburg, R., Tempelaars, M. H., Van Schaik, W., Moezelaar, R. & Abee, T. 2010. Comparative analysis of transcriptional and physiological responses of *Bacillus cereus* to organic and inorganic acid shocks. *International Journal of Food Microbiology*, 137, 13–21.

O'Hanlon, D. E., Lanier, B. R., Moench, T. R. & Cone, R. A. 2010. Cervicovaginal fluid and semen block the microbicidal activity of hydrogen peroxide produced by vaginal lactobacilli. *BMC Infectious Diseases*, 10, 120.

O'Hanlon, D. E., Moench, T. R. & Cone, R. A. 2011. In vaginal fluid, bacteria associated with bacterial vaginosis can be suppressed with lactic acid but not hydrogen peroxide. *BMC Infectious Diseases*, 11, 200.

Oman, T. J. & Van Der Donk, W. A. 2010. Follow the leader: The use of leader peptides to guide natural product biosynthesis. *Nature Chemical Biology*, 6, 9–18.

O'Neill, J. 2016. Tackling drug-resistant infections globally: Final report and recommendations. *Review on Antimicrobial Resistance*. Available from: https://apo.org.au/node/63983.

O'Shea, E. F., O'Connor, P. M., O'Sullivan, O., Cotter, P. D., Ross, R. P. & Hill, C. 2013. Bactofencin A, a new type of cationic bacteriocin with unusual immunity. *MBio*, *4*, e00498–13.

Oshima, S., Rea, M. C., Lothe, S., Morgan, S., Begley, M., O'Connor, P. M., Fitzsimmons, A., Kamikado, H., Walton, R. & Ross, R. P. 2012. Efficacy of organic acids, bacteriocins, and the lactoperoxidase system in inhibiting the growth of *Cronobacter* spp. in rehydrated infant formula. *Journal of Food Protection*, *75*, 1734–1742.

O'Sullivan, L., Ross, R. P. & Hill, C. 2002. Potential of bacteriocin-producing lactic acid bacteria for improvements in food safety and quality. *Biochimie*, *84*, 593–604.

Papagianni, M. & Anastasiadou, S. 2009. Pediocins: The bacteriocins of pediococci sources, production, properties and applications. *Microbial Cell Factories*, *8*, 3.

Polycarpo, G. V., Andretta, I., Kipper, M., Cruz-Polycarpo, V. C., Dadalt, J. C., Rodrigues, P. H. M. & Albuquerque, R. 2017. Meta-analytic study of organic acids as an alternative performance-enhancing feed additive to antibiotics for broiler chickens. *Poultry Science*, *96*, 3645–3653.

Quitmann, H., Fan, R. & Czermak, P. 2014. Acidic organic compounds in beverage, food, and feed production. *Advances in Biochemical Engineering/Biotechnology*, *143*, 91–141.

Rea, M. C., Dobson, A., O'Sullivan, O., Crispie, F., Fouhy, F., Cotter, P. D., Shanahan, F., Kiely, B., Hill, C. & Ross, R. P. 2011. Effect of broad- and narrow-spectrum antimicrobials on *Clostridium difficile* and microbial diversity in a model of the distal colon. *Proceedings of the National Academy of Sciences of the United States of America*, *108*(Suppl 1), 4639–4644.

Rea, M. C., Sit, C. S., Clayton, E., O' Connor, P. M., Whittal, R. M., Zheng, J., Vederas, J. C., Ross, R. P. & Hill, C. 2010. Thuricin CD, a posttranslationally modified bacteriocin with a narrow spectrum of activity against *Clostridium difficile* . *Proceedings of the National Academy of Sciences of the United States of America*, *107*, 9352.

Roe, A. J., O'Byrne, C., Mclaggan, D. & Booth, I. R. 2002. Inhibition of *Escherichia coli* growth by acetic acid: A problem with methionine biosynthesis and homocysteine toxicity. *Microbiology*, *148*, 2215–2222.

Rohde, B. H. & Quadri, L. E. 2006. Functional characterization of a three-component regulatory system involved in quorum sensing-based regulation of peptide antibiotic production in *Carnobacterium maltaromaticum* . *BMC Microbiology*, *6*, 93.

Salih, M., Sandine, W. & Ayres, J. 1990. Inhibitory effects of Microgard on yogurt and cottage cheese spoilage organisms. *Journal of Dairy Science*, *73*, 887–893.

Sallam, K. I. 2007. Antimicrobial and antioxidant effects of sodium acetate, sodium lactate, and sodium citrate in refrigerated sliced salmon. *Food Control*, *18*, 566–575.

Salmond, C. V., Kroll, R. G. & Booth, I. R. 1984. The effect of food preservatives on pH homeostasis in *Escherichia coli* . *Microbiology*, *130*, 2845–2850.

Samyn, B., Martinez-Bueno, M., Devreese, B., Maqueda, M., Gálvez, A., Valdivia, E., Coyette, J. & Van Beeumen, J. 1994. The cyclic structure of the enterococcal peptide antibiotic AS-48. *FEBS Letters*, *352*, 87–90.

Sauvageot, N., Gouffi, K., Laplace, J.-M. & Auffray, Y. 2000. Glycerol metabolism in *Lactobacillus collinoides*: Production of 3-hydroxypropionaldehyde, a precursor of acrolein. *International Journal of Food Microbiology*, *55*, 167–170.

Scannell, A. G., Hill, C., Ross, R., Marx, S., Hartmeier, W. & Arendt, E. K. 2000. Development of bioactive food packaging materials using immobilised bacteriocins lacticin 3147 and Nisaplin. *International Journal of Food Microbiology*, *60*, 241–249.

Schaefer, L., Auchtung, T. A., Hermans, K. E., Whitehead, D., Borhan, B. & Britton, R. A. 2010. The antimicrobial compound reuterin (3-hydroxypropionaldehyde) induces oxidative stress via interaction with thiol groups. *Microbiology*, *156*, 1589–1599.

Schellenberg, J. J., Dumonceaux, T. J., Hill, J. E., Kimani, J., Jaoko, W., Wachihi, C., Mungai, J. N., Lane, M., Fowke, K. R. & Ball, T. B. 2012. Selection, phenotyping and identification of acid and hydrogen peroxide producing bacteria from vaginal samples of Canadian and East African women. *PLoS ONE*, *7*, e41217.

Schütz, H. & Radler, F. 1984. Anaerobic reduction of glycerol to propanediol-1.3 by *Lactobacillus brevis* and *Lactobacillus buchneri* . *Systematic and Applied Microbiology*, *5*, 169–178.

Sears, P., Smith, B., Stewart, W., Gonzalez, R., Rubino, S., Gusik, S., Kulisek, E., Projan, S. & Blackburn, P. 1992. Evaluation of a nisin-based germicidal formulation on teat skin of live cows. *Journal of Dairy Science*, *75*, 3185–3190.

Settanni, L. & Corsetti, A. 2008. Application of bacteriocins in vegetable food biopreservation. *International Journal of Food Microbiology*, *121*, 123–138.

Sgibnev, A. & Kremleva, E. 2017. Influence of hydrogen peroxide, lactic acid, and surfactants from vaginal lactobacilli on the antibiotic sensitivity of opportunistic bacteria. *Probiotics and Antimicrobial Proteins*, *9*, 131–141.

Slininger, P. J., Bothast, R. J. & Smiley, K. L. 1983. Production of 3-hydroxypropionaldehyde from glycerol. *Applied and Environmental Microbiology*, *46*, 62–67.

Soltani, S., Couture, F., Boutin, Y., Ben Said, L., Cashman-Kadri, S., Subirade, M., Biron, E. & Fliss, I. 2021a. *In vitro* investigation of gastrointestinal stability and toxicity of 3-hyrdoxypropionaldehyde (reuterin) produced by *Lactobacillus reuteri*. *Toxicology Reports*, *8*, 740–746.

Soltani, S., Hammami, R., Cotter, P. D., Rebuffat, S., Said, L. B., Gaudreau, H., Bédard, F., Biron, E., Drider, D. & Fliss, I. 2021b. Bacteriocins as a new generation of antimicrobials: Toxicity aspects and regulations. *FEMS Microbiology Reviews*, *45*, fuaa039.

Song, Z., Wu, H., Niu, C., Wei, J., Zhang, Y. & Yue, T. 2019. Application of iron oxide nanoparticles @ polydopamine-nisin composites to the inactivation of *Alicyclobacillus acidoterrestris* in apple juice. *Food Chemistry*, *287*, 68–75.

Stern, N., Svetoch, E., Eruslanov, B., Perelygin, V., Mitsevich, E., Mitsevich, I., Pokhilenko, V., Levchuk, V., Svetoch, O. & Seal, B. 2006. Isolation of a *Lactobacillus salivarius* strain and purification of its bacteriocin, which is inhibitory to *Campylobacter jejuni* in the chicken gastrointestinal system. *Antimicrobial Agents and Chemotherapy*, *50*, 3111–3116.

Tachedjian, G., O'Hanlon, D. E. & Ravel, J. 2018. The implausible "*in vivo*" role of hydrogen peroxide as an antimicrobial factor produced by vaginal microbiota. *Microbiome*, *6*, 29.

Tamarit, J. P., Cabiscol, E. & Ros, J. 1998. Identification of the major oxidatively damaged proteins in *Escherichia coli* cells exposed to oxidative stress. *Journal of Biological Chemistry*, *273*, 3027–3032.

Tamang, J. P., Cotter, P. D., Endo, A., Han, N. S., Kort, R., Liu, S. Q., Mayo, B., Westerik, N. and Hutkins, R. 2020. Fermented foods in a global age: East meets West. *Comprehensive Reviews in Food Science and Food Safety*, *19*, 184–217.

Tomalok, C. D. G., Wlodarkievicz, M. E., Puton, B. M. S., Colet, R., Zeni, J., Steffens, C., Backes, G. T. & Cansian, R. L. 2022. Organic acids as an alternative to control *Salmonella enterica* serotype *Choleraesuis* and *Listeria monocytogenes* in pork jowl fat. *Journal of Food Safety*, e12999.

Twomey, E., Hill, C., Field, D. and Begley, M. 2020. Bioengineered nisin derivative M17Q has enhanced activity against *Staphylococcus epidermidis*. *Antibiotics*, *9*, 305.

Vallor, A. C., Antonio, M. A., Hawes, S. E. & Hillier, S. L. 2001. Factors associated with acquisition of, or persistent colonization by, vaginal lactobacilli: Role of hydrogen peroxide production. *The Journal of Infectious Diseases*, *184*, 1431–1436.

Van Der Ploeg, J. R. 2005. Regulation of bacteriocin production in *Streptococcus mutans* by the quorum-sensing system required for development of genetic competence. *Journal of Bacteriology*, *187*, 3980–3989.

Van Hemert, S., Meijerink, M., Molenaar, D., Bron, P. A., De Vos, P., Kleerebezem, M., Wells, J. M. & Marco, M. L. 2010. Identification of *Lactobacillus plantarum* genes modulating the cytokine response of human peripheral blood mononuclear cells. *BMC Microbiology*, *10*, 293.

Van Holm, W., Verspecht, T., Carvalho, R., Bernaerts, K., Boon, N., Zayed, N. & Teughels, W. 2022. Glycerol strengthens probiotic effect of *Limosilactobacillus reuteri* in oral biofilms: A synergistic synbiotic approach. *Molecular Oral Microbiology*. Ahead of Print, https://doi.org/10.1111/omi.12386.

Varas, M. A., Muñoz-Montecinos, C., Kallens, V., Simon, V., Allende, M. L., Marcoleta, A. E. & Lagos, R. 2020. Exploiting zebrafish xenografts for testing the *in vivo* antitumorigenic activity of microcin e492 against human colorectal cancer cells. *Frontiers in Microbiology*, *11*, 405.

Villarante, K. I., Elegado, F. B., Iwatani, S., Zendo, T., Sonomoto, K. & de Guzman, E. E. 2011. Purification, characterization and *in vitro* cytotoxicity of the bacteriocin from *Pediococcus acidilactici* K2a2–3 against human colon adenocarcinoma (HT29) and human cervical carcinoma (HeLa) cells. *World Journal of Microbiology and Biotechnology*, *27*, 975–980.

Vimont, A., Fernandez, B., Ahmed, G., Fortin, H. P. & Fliss, I. 2019. Quantitative antifungal activity of reuterin against food isolates of yeasts and moulds and its potential application in yogurt. *International Journal of Food Microbiology*, *289*, 182–188.

Vollenweider, S. & Lacroix, C. 2004. 3-Hydroxypropionaldehyde: Applications and perspectives of biotechnological production. *Applied Microbiology and Biotechnology*, *64*, 16–27.

Wang, H., Zhang, R., Cheng, J., Liu, H., Zhai, L. & Jiang, S. 2015. Functional effectiveness and diffusion behavior of sodium lactate loaded chitosan/poly (L-lactic acid) film with antimicrobial activity. *RSC Advances*, *5*, 98946–98954.

Warnecke, T. & Gill, R. T. 2005. Organic acid toxicity, tolerance, and production in *Escherichia coli* biorefining applications. *Microbial Cell Factories*, *4*, 25.

Yang, Y., Zhao, X., Le, M. H., Zijlstra, R. T. & Gänzle, M. G. 2015. Reutericyclin producing *Lactobacillus reuteri* modulates development of fecal microbiota in weanling pigs. *Frontiers in Microbiology*, *6*, 762.

Younes, M., Aggett, P., Aguilar, F., Crebelli, R., Dusemund, B., Filipič, M., Frutos, M. J., Galtier, P., Gundert-Remy, U. & Kuhnle, G. G. 2017. Safety of nisin (E 234) as a food additive in the light of new toxicological data and the proposed extension of use. *EFSA Journal*, *15*, e05063.

Young, K. M. & Foegeding, P. M. 1993. Acetic, lactic and citric acids and pH inhibition of *Listeria monocytogenes* Scott A and the effect on intracellular pH. *Journal of Applied Microbiology*, *74*, 515–520.

Yuan, J., Zhang, Z. Z., Chen, X. Z., Yang, W. & Huan, L. D. 2004. Site-directed mutagenesis of the hinge region of nisinZ and properties of nisinZ mutants. *Applied Microbiology Biotechnology*, *64*, 806–815.

Yunmbam, M. K. & Roberts, J. F. 1993. *In vivo* evaluation of reuterin and its combinations with suramin, melarsoprol, DL-α-difluoromethylornithine and bleomycin in mice infected with *Trypanosoma brucei brucei* . *Comparative Biochemistry and Physiology Part C: Comparative Pharmacology*, *105*, 521–524.

Zacharof, M. & Lovitt, R. 2012. Bacteriocins produced by lactic acid bacteria a review article. *Apcbee Procedia*, *2*, 50–56.

Zhitnitsky, D., Rose, J. & Lewinson, O. 2017. The highly synergistic, broad spectrum, antibacterial activity of organic acids and transition metals. *Scientific Reports*, *7*, 44554.

# Lactic Acid Bacteria for Fermented Dairy Products

# 10

Sylvie Binda and Arthur C. Ouwehand

## 10.1 INTRODUCTION

The transformation and preservation facilitated by the unique biological process called fermentation has multiplied the possible combinations of lactic acid bacteria (LAB) and their growth variables (such as temperature, salinity, moisture), resulting in plenty of fermented milk product types (e.g., yogurt, dahi, kefir, laban, etc.), making the fermented dairy world full of choice. Since ancient times, humans have used LAB, probably as a fortunate find, as these well-known bacteria can be found widely in nature and can grow and naturally transform milk. The first evidence of milk fermentation dates back approximately 7400 years, when early Europeans are thought to have produced cheese and included fermented dairy foods as major components in their diets (Casanova et al. 2022). As milk is highly perishable, milk fermentation using LAB is a natural way to increase shelf life and at the same time preserve the nutritious components of milk, allowing it to be stored for later use, thereby conveniently dissociating collection from consumption. In addition to obtaining a preserved form of milk, fermentation made milk a more digestible commodity and improved its organoleptic quality and nutritional properties, all good reasons for fermenting dairy by prehistoric farmers. Early dairy fermentations were highly dependent on the spontaneous activity of the indigenous microbes of the milk and of the milking and fermentation environments, while modern ones rely on defined starter cultures with predictable behavior. LAB represent the most extensively studied microorganisms for milk fermentation, and milk itself is known as one of the natural habitats of LAB. The popularity of fermented dairy foods is due to their enhanced shelf life; safety; functionality; and sensory, nutritional, and health-promoting properties (Hill et al., 2017). The latter include the presence of some constituents present in the initial food and substrates of bioactive molecules, vitamins, and minerals and as a source of live organisms (Panahi and Tremblay, 2016). Associated with their popularity, the uniqueness of fermented dairy foods is also due to the simplicity of the artisanal ecological process. By producing organic acids, LAB make preservative agents and generate flavor in products, and they may also produce exopolysaccharides, which are essential for texture. Fermentation has evolved from a natural art, using carefully selected cultures and environments for key scientific understanding while maintaining the simplicity of this art of living food transformation.

Today, more than 5000 papers have been published in peer-reviewed scientific journals on LAB in dairy since the 1970s. It is over the last 10 years that the research has increased tremendously, thanks to the accessibility of new advanced technologies (Walsh et al., 2023) and because of the increasing interest

DOI: 10.1201/9781003352075-12

in dietary solutions to help the general population support and maintain health and wellness. This literature provides information and hypotheses on the bioactivity and mechanisms mediating the effects of fermented dairy foods in health and disease prevention. The health benefits associated with this category and its process could be the result of direct interactions between the host and the ingested live microorganisms or indirect interactions through ingestion of microbial metabolites, fermentation products, or initial matrices themselves (Gille et al., 2018). The beneficial effects of fermented milk products are the result of a variety of bioactive compounds produced by LAB (Hayes et al., 2007). In particular, yogurt has been associated with reducing risk for type 2 diabetes (Aune et al., 2013; Gijsbers et al., 2016, Chen et al., 2017; Gille et al., 2018), reducing symptoms of some food reactions such as lactose maldigestion (Saborido and Leis, 2018), and positively impacting the diversity of microbiota (Zhernakova et al., 2016). Emerging technologies reveal a living community inside the products, and fermented foods have become interesting to scientific communities as models for microbial ecosystems and health properties.

## 10.2 TYPES OF FERMENTED DAIRY FOODS

The technology of milk fermentation is relatively simple and cost effective. Industrial selection and application of LAB as starter cultures are important for standardized fermented milk products manufactured in large-scale production under controlled conditions. There are some important features of LAB starters in fermented milk products that have to be understood in order to be able to select the most appropriate ones. Such starter cultures must be developed with a clear understanding of the ecology of the microbial species associated with the desirable traditional fermentation process and their contribution to a product's safety, quality, and sensory properties. Food transformation is the primary role of starter cultures. The first stage for the selection of strains well adapted to the process specificities is to design the metabolic approach in which acidification rate, biopreservative component production, and exopolysaccharide synthesis will be evaluated. Also, their ability to convert pyruvate and amino acids during various food fermentation processes has to be assessed, as it results in flavor components and hence determines the sensory profiles and variety of fermented foods (Weckx et al., 2009). Usually, optimal fermentation requires different actions of microorganisms in order to obtain desirable changes, and key relations within the bacterial consortia are required. For example, association of LAB can provide other technical properties. This is the case with the urease activity of *Streptococcus thermophilus*. Thanks to this enzyme, *S. thermophilus* can release ammonia from urea. It has been shown that ammonia released by *S. thermophilus* urease activity leads to a more alkaline intracellular pH and a higher lactose conversion, leading to a faster acidification and a stimulation of *L. bulgaricus* (Sieuwerts et al., 2010; Arioli et al., 2017). Novel techniques such as proteomics can be used in order to better understand the interactions between microbes, which is useful to understand the adaptations of LAB in food (Champomier-Vergès et al., 2010).

In recent years, Western consumers have shown interest in fermented dairy products from different geographies, such as skyr, Greek-style yogurt, laban, and kefir. This has expanded the choice of available fermented dairy products, a choice also driven by this new palate of taste and texture.

## 10.3 YOGURT

Fermented milks have been produced for over 10,000 years, initially to extend the shelf life of milk (Tamime and Robinson, 2007). Scientists, from biologists to archaeologists, can tell unique stories about its origins, composition, and fermentation process, with French-Russian Nobel Prize recipient Elie

Metchnikoff as one of the great promotors of yogurt's health benefits. Yogurt has been produced commercially for over a century, first as a pharmaceutical product (Hartley and Denariaz, 1993).

For thousands of years, in many parts of the world, yogurt coexisted with several hundred other fermented milks. According to the Codex Alimentarius, a necessary condition to be called a yogurt is that two specific LAB: *Lactobacillus delbrueckii* subsp. *bulgaricus* (*L. bulgaricus*) and *S. thermophilus*, have to be present in a minimal viable count (sum of microorganisms, min $10^7$ CFU/g). Interestingly, the latter one is the only *Streptococcus* species used in the food industry and is likely to be one of the leading bacteria consumed by humans. A study by Lang et al. (2014) investigating bacterial numbers present in the average adult American diet found that only up to $6 \times 10^6$ bacteria were typically consumed per day. As Americans consume relatively few fermented products but include some yogurt, they are likely to ingest *S. thermophilus*.

The worldwide popularity of yogurt is strongly linked to its bacterial residents, responsible for its pleasant aroma characteristics, specific texture, and some of its health aspects (Marette et al., 2017).

## 10.3.1 Production

Yogurt, with its unique combination of live microorganisms and nutritive components, is an interesting food that can be easily incorporated into a healthy diet. Yogurt is obtained by milk lactic acid fermentation through the action of the characteristic bacterial cultures *L. bulgaricus* and *S. thermophilus* (Figure 10.1).

The importance of *S. thermophilus* in dairy fermentation is well recognized due to its fast acidification capacity as well as the synthesis of other metabolites such as formate, which synergistically enhances the growth of lactobacilli. Also, its ability to produce metabolites such as acetaldehyde or diacetyl contributes to the sensorial properties of fermented dairy products. More specifically, in yogurt starter cultures, interactions between *S. thermophilus* and *L. bulgaricus* are known as a proto-cooperation, a symbiotic relationship between the two species. During lactose hydrolysis, *S. thermophilus* produces $CO_2$ and formic acid, which stimulate the growth of *L. bulgaricus*, while the latter hydrolyses milk proteins, releasing peptides and amino acids that improve *S. thermophilus* growth and activity (Sieuwerts et al., 2008). Globally, the choice of the strains has substantial industrial interest, as this remarkable synergistic relationship improves the yield of the fermentation and the main yogurt characteristics, such as sensory properties and even health benefits (Sawsen, 2012; Thakkar et al., 2018). This well-known bacterial association is increasingly studied at molecular and regulatory levels (Herve-Jimenez et al., 2009) and reveals nutritional exchanges in this symbiotic behavior.

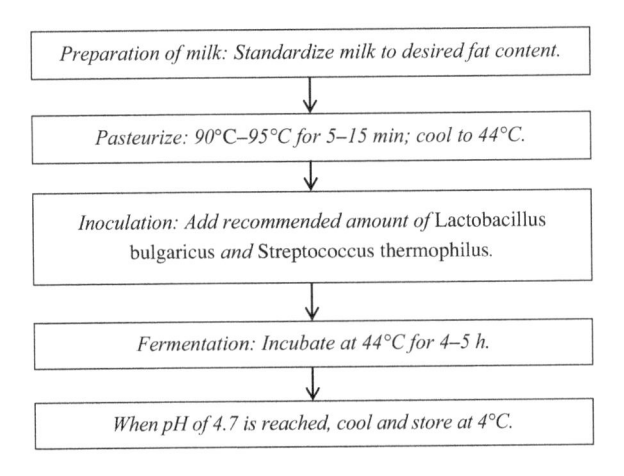

**FIGURE 10.1** Schematic representation of the yogurt production process.

(Modified from Lamoureux, L. et al., *J. Dairy Sci.*, 85, 1058, 2002.)

During the process, lactic acid is produced from the fermentation of lactose; it contributes to the sour taste of yogurt by decreasing pH and allows for the characteristic texture by acting on the milk proteins. When pH drops below pH 5, micelles of caseins (the major milk protein) lose their tertiary structure due to the protonation of their amino acid residues. The denatured protein reassembles by interacting with other hydrophobic molecules, and this intermolecular interaction of caseins creates a structure that allows for the semisolid texture of yogurt (Zourari et al., 1992).

Yogurt can be made from cow, ewe, goat, or buffalo milk and must contain at least 8.25% non-fat solids.

## 10.3.2 Health Benefits

The latest advances in science indicate that LAB in yogurts are of strong interest not just for technological purposes but also because they can exert beneficial health effects (Parvez et al., 2006). Yogurt and individual LAB have shown well-established or promising health effects. The best-known and described effect is on lactose maldigestion, which results from a genetic disposition or acquired deficiency in the enzyme lactase that is required for hydrolyzing lactose to glucose and galactose in the small intestine. If lactose reaches the colon, it is rapidly fermented by the microbiota, leading to gas formation. Lactose intolerance is a condition in which people experience digestive symptoms, such as bloating, diarrhea, gas, and vomiting, after eating or drinking milk or milk products. However, consumption of fermented milk products, especially yogurt, even though they contain lactose, is commonly well tolerated by individuals suffering from lactose maldigestion. This benefit on maldigestion in humans is well established and can be explained by the presence of the lactase-like enzymes β-galactosidase of the yogurt's bacteria *S. thermophilus* and *L. bulgaricus* (Guarner et al., 2005). This bacterial enzyme can compensate for the lack of human lactase, hydrolyzing it in the small intestine and thereby preventing the fermentation of lactose in the large intestine and the corresponding lactose maldigestion symptoms (Marteau et al., 1990). The mechanistic understanding of these yogurt-associated species led to the only current example of live microbes with an European Food Safety Authority (EFSA)-approved health claim. Specifically, the EFSA claim states that "live yogurt cultures in yogurt improve digestion of yogurt lactose in individuals with lactose maldigestion" (2010).

In 2016, Gijsbers et al. published a meta-analysis of the latest relevant data available on yogurt consumption and the risk of type 2 diabetes, based on 12 prospective studies and including 438,140 individuals and 36,125 cases of incident diabetes. In this meta-analysis, a nonlinear inverse association was observed between yogurt consumption and the risk of type 2 diabetes and showed a 14% lower risk of diabetes in those individuals who consumed between 80 and 125 g yogurt/d compared with non-consumers. In line with this, the Food & Drug Administration (FDA 2024) has recently announced a qualified health claim for yogurt and reduced risk of type 2 diabetes. Epidemiological studies have shown that frequent intake and regular consumption of fermented milk and dairy products lower the risk of high blood pressure. This can be explained by the large number and wide variety of peptides that are produced during fermentation with LAB that have a proteolytic activity in milk. Some of these peptides can inhibit the angiotensin I converting enzyme (ACE) that normally raises blood pressure by converting the inactive decapeptide angiotensin I to its active form, angiotensin II, resulting in narrowing of small blood vessels and an increase in blood pressure (Chen et al., 2014; Rai and Sanjukta, 2015). Other epidemiological studies, with various degrees of quality, have reported associations between yogurt consumption and reduced risk for a variety of health conditions (see Table 10.1). As always with epidemiological studies, causality remains to be determined.

Bacterial strains found in yogurt may interact with gut microbiota. Diet can be a major modulator of the gut microbiota (Zhernakova et al., 2016). The abundance of dairy fermentation–related bacteria increased with increasing dairy consumption, indicating the potential of fermented products to act on gut microbiome composition. Other studies describe positive actions for the elimination of pathogenic enteric bacteria as well as the reinforcement of the intestinal barrier (Adolfsson et al., 2004; Rohde et al., 2009).

The health benefits of yogurt consumption and their proposed mechanisms are presented in Figure 10.2.

**TABLE 10.1**   Nonexhaustive List of Health Benefits Associated with Yogurt Consumption in Epidemiological Studies

| HEALTH BENEFIT | SELECTED RECENT STUDIES REFERENCES |
| --- | --- |
| Reduce hypertension | Buendia et al. (2018) |
| Type 2 diabetes | Gijsbers et al. (2016), Ibsen et al. (2017) |
| Cardiovascular disease mortality risk | Farvid et al. (2017), Dehghani et al. (2018) |
| Cataract | Camacho-Barcia et al. (2019) |
| Bone mineral density and hip fracture | Laird et al. (2017), Bian et al. (2018) |
| Physical function in elderly | Laird et al. (2017) |
| Reduced frailty in elderly | Lana et al. (2015) |
| Breast cancer risk | McCann et al. (2017) |
| Atopic dermatitis and food allergy risk in childhood | Shoda et al. (2017) |
| Proteinuric kidney disease risk | Yacoub et al. (2016) |
| Reduced tooth erosion | Salas et al. (2015) |

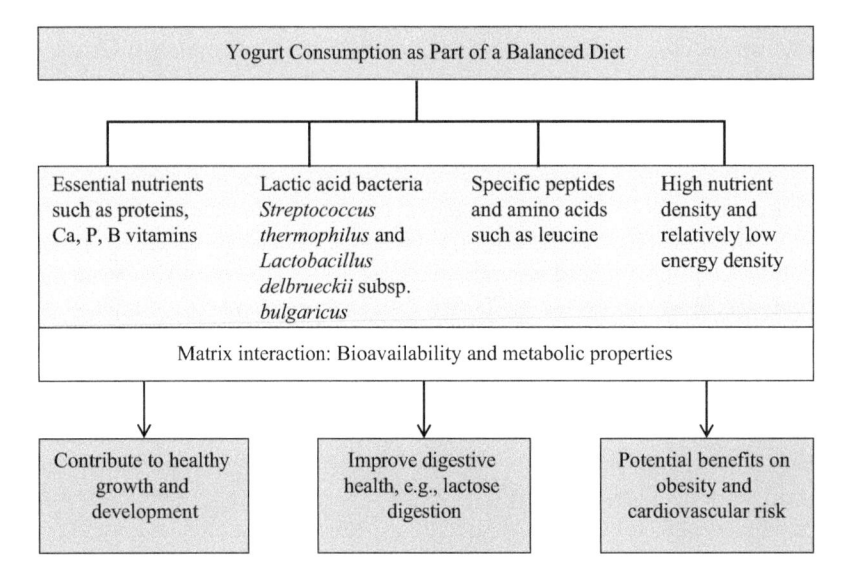

**FIGURE 10.2**   Proposed mechanisms by which yogurt consumption as a part of a balanced diet exerts beneficial health effects.

(Reprinted with permission from Marette, A. et al., *Yogurt: Roles in Nutrition and Impacts on Health*, 2017.)

Last but not least and not discussed here, yogurt is of course an excellent carrier of probiotics and the prime probiotic food on the market (Lourens-Hattingh and Viljoen, 2001).

# 10.4  STRAINED YOGURTS

Strained yogurt is yogurt that has been strained to remove most of its whey, resulting in a thicker and creamier consistency than unstrained yogurt and a change in flavors. Regular yogurt is saltier and sweeter than strained yogurt because the whey contains sodium and lactose. This means that the straining results in some nutritional changes: more protein and calcium due to the concentration and less carbohydrate and sodium because the whey is strained out. Usually, there is a heating step before whey separation that improves the serration yield of strained yogurt. In many different countries around the world these types of yogurt are part of the traditional diet and cooking.

# 10.4.1 Skyr

In the ninth century, skyr ("skeer") was introduced in Iceland by Viking settlers from Norway who considered it a highly valuable food like meat. In order to preserve milk, they turned it into butter or fermented products like skyr. The whey from the skyr production was in turn used to preserve meat. Skyr consumption gradually died out in other northern European areas, eventually being confined principally to Iceland until the 2000s. Since then, it has been gaining popularity in the United States and Europe.

## 10.4.1.1 Production

Skyr is made by fermenting skim or low-fat milk with a small amount of skyr from a previous batch, which provides the bacterial cultures. Once thickened, the whey is slowly drained, leaving a thick, sour product that is comparable to soft cheese (Figure 10.3). Skyr can contain more than 20 grams of protein per cup, with very little fat or added sugar (Gudmundsson and Kristbergsson, 2016). Technically, it is a soft cheese made with skim milk that's been fermented with cultures.

According to Icelandic Regulation 851/2012, skyr is produced from pasteurized skim milk and/or reconstituted skim milk that has been heated to at least 72 °C–78 °C for 15–20 seconds. The skim milk is acidified with skyr culture and rennet is sometimes added to aid coagulation. The whey is separated after fermentation. The curd (*skyr* in Icelandic) can be heated up to 68 °C for 15–20 seconds so most of the skyr culture survive the heating. Skyr should contain a minimum of 16% of milk solids.

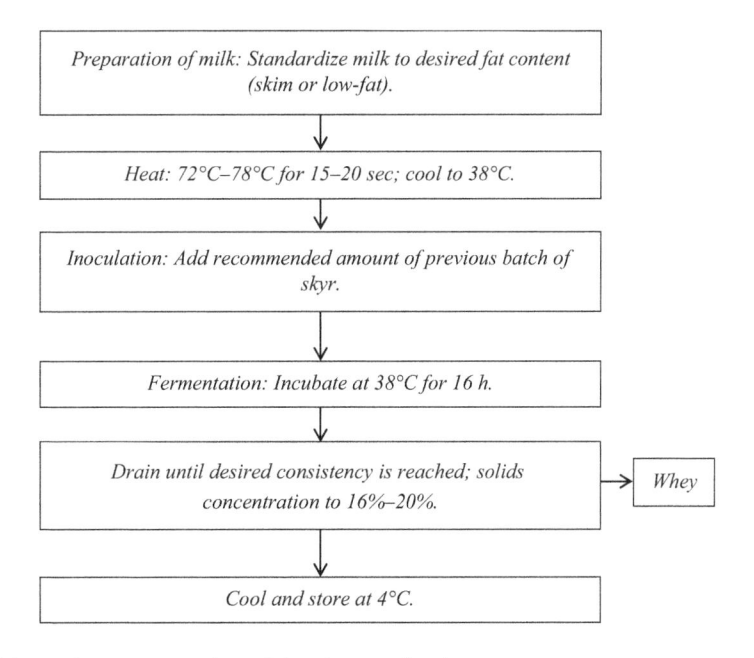

**FIGURE 10.3**  Schematic representation of the skyr production process.

(Modified from Gudmundsson, G., Kristbergsson, K., *Modernization of Traditional Food Processes and Products*, 2016.)

The selection and use of LAB is often based on the product's sensory specificities: complex organoleptic properties with a creamy and cheesy taste, highly acidic, and protein flavor with a shiny and thick final appearance of the product. Some classical skyr culture compositions can contain some *Lactococcus lactis* subsp. *cremoris* and *Lactococcus lactis* subsp. *lactis* especially in order to enhance the creamy and cheese flavors.

### 10.4.1.2 Health Benefits

Although skyr is perceived to be a healthy product, no specific studies on potential health benefits of skyr could be identified. However, the high dairy protein content and low sodium content of skyr suggest that skyr can play an important role as a valuable source of high-quality protein in the diet.

## 10.4.2 Greek-Style Yogurt

Greek-style yogurt is a strained yogurt that has enjoyed substantial popularity for several years.

### 10.4.2.1 Production

Greek-style yogurt is produced using the same process as a regular yogurt, except that the Greek yogurt is strained (Figure 10.4). The straining process is a way to concentrate the protein in the yogurt and make this product ideal for people who want to include more protein in their diet. In fact, to reach the desired solids level, it's important to remove the acid whey. This step is achieved by mechanically separating the whey from the curd using either a centrifugal separator or membrane filtration. The production of large quantities of acid whey presents both economic and environmental challenges. Currently available alternative processes involve fortification of milk with milk protein concentrates to enhance the protein content of the final product (Bong and Moraru, 2014).

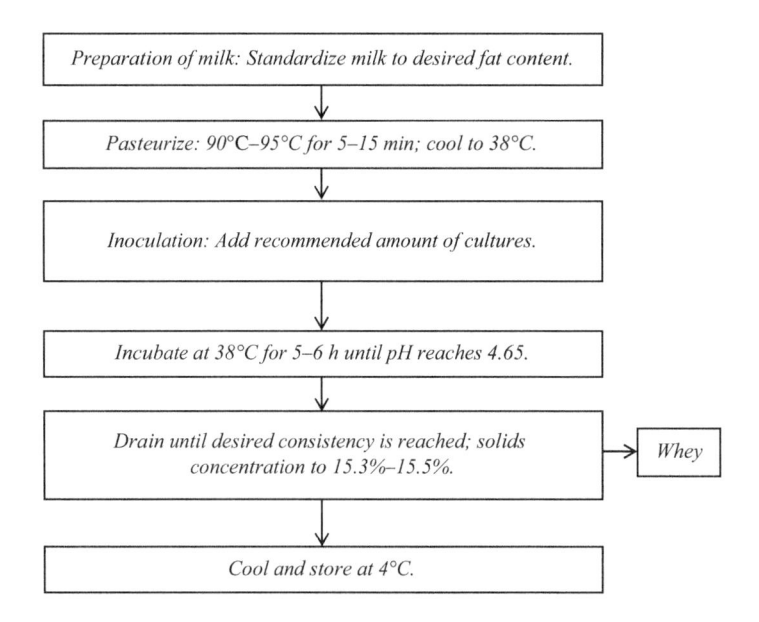

**FIGURE 10.4**  Schematic representation of the Greek-style yogurt production process.

(Modified from Bong, D., Moraru, C., *J. Dairy Sci.*, 97, 1259, 2014.)

### 10.4.2.2 Health Benefits

Recent studies explored the positive link between the consumption of Greek-style yogurt and physical performance and recovery (Bridge et al., 2019). As with skyr, it can be considered an excellent source of high-quality protein.

## 10.4.3 Laban

Laban (also leben) is a fermented milk product produced in Lebanon and some north African and Middle Eastern countries using traditional and industrial manufacturing practices (Surono and Hosono, 2011).

### 10.4.3.1 Production

In some publications, the production of artisanal and industrial laban is explained as a classical lactic fermentation of heat-treated milk using thermophilic starters composed of *S. thermophilus* and *L. bulgaricus* strains (Tamime and Robinson, 2007). Classically, starters used for the artisanal production consist of an unknown number of undefined strains, and industrial starters are characterized by well-defined strains. Artisanal starter cultures are composed of a laban sample taken from the previously produced batch, thus resulting in a variable product, especially in acidity, texture, and flavors. Chammas and coworkers (2006) studied the technological properties of 96 lactic acid bacteria isolated from Lebanese traditional fermented milk "laban". These were classified according to phenotypic and biochemical analyses as *S. thermophilus* and *L. bulgaricus*, thus indicating that laban is likely a fermented milk similar to yogurt. As for yogurt, it can be expected that these two bacteria act in a proto-cooperative way in the laban fermentation process, having positive effects on the growth rate and size of other microbial populations commonly observed in the traditional products.

### 10.4.3.2 Health Benefits

Today, no specific study has investigated the mutualistic relation of these bacteria to the potential health properties of laban beyond its added nutritional value. Laban is considered a source of calcium and protein with low level of fat and carbohydrate.

## 10.4.4 Labneh

Labneh (or labaneh) is a popular fermented milk in the Middle East with a production process slightly more complex than that of laban, as the inoculated cultures come from kefir (Figure 10.5).

### 10.4.4.1 Production

Labneh has a cream or white color, a soft and smooth body, good spreadability with little syneresis, a clean flavor, and slight acidity. It is considered an intermediate product between fermented milks and immature cheeses with high water content, such as quark, Boursin, and petit-Suisse, and has similar physical characteristics due to the syneresis process (Rocha et al., 2014).

### 10.4.4.2 Health Benefits

The nutritional properties of labneh are similar to those of yogurt. Being strained, labneh has 2.5 times higher protein content, 50% more minerals, and a wider microbial diversity than classical yogurt due to

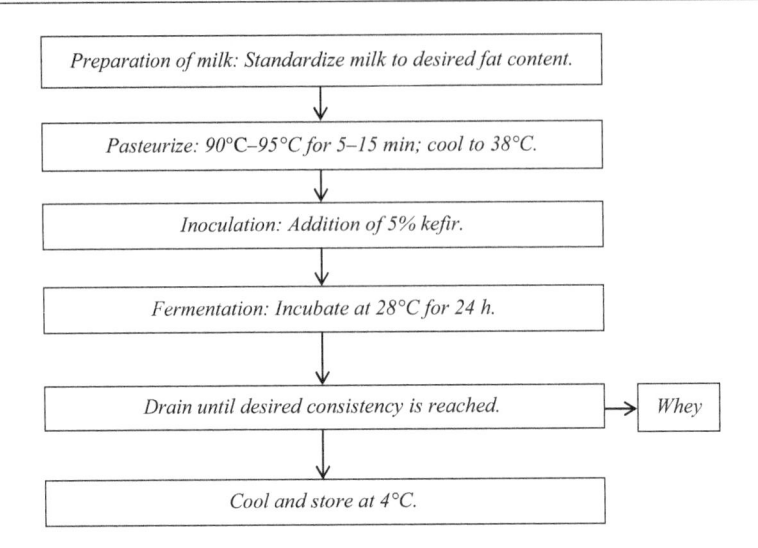

**FIGURE 10.5**   Schematic representation of the labneh production process.
(Modified from Rocha, D.M.U.P., *Food Sci. Technol.*, 34, 694–700, 2014.)

the initial inoculum coming from kefir. The lactose concentration of labneh is also low (approximately 6%) due to its fermentation into lactic acid. No specific studies have been done on labneh's potential health properties. When labneh is produced from kefir, similar health benefits (see the section on kefir) can be expected: antimutagenic and antioxidant effects, hypocholesterolemic properties, β-galactosidase activity, anti-allergenic properties, anti-inflammatory activity, and stimulation of the immune system (Rocha et al., 2014).

Due to its high total solids content, labneh may be considered a suitable matrix for probiotic microorganisms since it offers protection when added, but no significant study was done on such a potential carrier role.

## 10.4.5 Quark

Quark (or quarg) can be classified as a fresh, set, acidic curd cheese. It has culinary applications or is eaten plain or flavored. Quark is common in central and eastern Europe, and different local variations exist. Some scholars suggest that the *lac concretum* Tacitus refers to in his *On the Origin and Situation of the Germans* is actually quark, and its use would thus be ancient (Tacitus, 98).

### 10.4.5.1 Production

In the production of quark (Figure 10.6), the milk is standardized to achieve a low-fat content, then pasteurized and cooled. The traditional production of quark does not involve the use of rennet, and coagulation fully depends on acidification by the mesophilic cultures in the milk. In modern quark production, however, it is common to use rennet in order to achieve a more efficient coagulation.

Common starters used in quark production are mesophilic *Lc. lactis* subsp. *lactis*, *Lc. cremoris*, *Ln. mesenteroides* subsp. *cremoris*, and *Lc lactis* subsp. *lactis* biovar *diacetylactis* (Djuric et al., 2005). When the pH has dropped to 4.6, fermentation is stopped and whey can be strained from the coagulum. Traditionally, whey separation was achieved by straining through a cloth. In large-scale production, whey is separated by centrifugation or ultra-filtration (Puhan et al., 1994).

Quark is usually set-type, but it can be stirred as well to create a thick creamy texture.

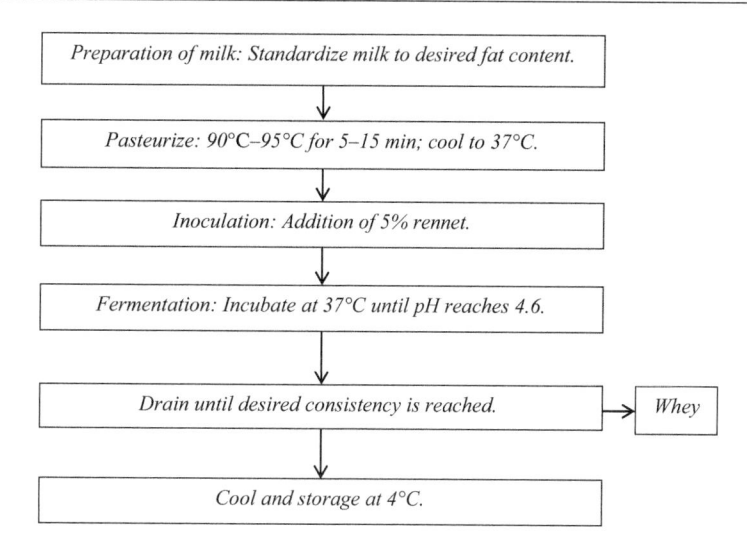

**FIGURE 10.6**  Schematic representation of the quark production process. (Modified from Farkye, N. Y., *Int. J. Dairy Technol.*, 57, 2004.)

### 10.4.5.2 Health Benefits

Although quark has a generally healthy image, few studies have been done on its potential health benefits. Thanks to its thick texture, moderate amounts of quark are well tolerated by lactose-maldigesting individuals. This tolerance does not seem to relate to the presence of microbial β-galactosidase (Shah et al., 1992).

It is likely that quark can be a source of menaquinones (vitamin K2) due to its content of various *Lactococcus* starter cultures, known as vitamin K2–producing bacteria. A recent U.S. study (Fu et al., 2017) investigated phylloquinone (vitamin K1) and the different forms of menaquinones (vitamin K2) in milk, yogurt, Greek yogurt, creams, and cheeses to compare the menaquinone contents of full-fat, reduced-fat, and nonfat dairy products. Dairy products tested contained appreciable amounts of menaquinones, primarily in the forms of MK9, MK10, and MK11 (long-chain menaquinones), and the menaquinone content varied by the fat content of the dairy product.

The diversity of vitamin K forms among dairy products may be related to the microbial species used in the production of fermented dairy products. Most cheese products contain LAB species as starters, which are reported to be the source of various menaquinone forms (Rezaîki et al., 2008), explaining why the vitamin K content seems to be higher in cheese-type process. However, no concrete data could be identified on the actual menaquinone content of quark, as it is highly dependent on the strain and the process used.

Protein is the main dairy solid of quark and 80% of that is casein. Therefore, quark can be considered a good source of dietary protein.

# 10.5 KEFIR

Traditionally, fresh milk (cow, goat, or sheep, though vegetable "milks", such as coconut, rice, or soy, can be used as well) was stored at room temperature in bags made of goat or sheep skin. Grains of a previous kefir culture were added before the bag was suspended in the sun during the day. When the sun went down, the bag was brought inside and was regularly shaken to ensure that the milk and kefir grains remained well mixed as the milk fermented. As kefir was consumed, more milk was added to the bag so that the process could continue uninterrupted (Wszolek et al., 2006).

# 10.5.1 Production

Kefir is a fermented milk obtained by the activity of kefir grains which are composed of LAB as well as acetic acid bacteria (AAB) and yeasts which form a symbiosis. Recent advances in experimental tools have enhanced understanding of how complex multispecies microbial communities are formed with more than 50 different species (Fiorda et al., 2017), how their ecology develops, and how to optimize the conditions that promote the growth of this microbial community of fermented foods. These communities can take many forms, but in the case of kefir, it is a suspended biofilm in liquid. Traditional milk kefir communities (in kefir grains) are composed of yeasts, LAB, AAB, and others bacterial groups within a highly organized biofilm (Marsh et al., 2013). Kefir grains contain the cultures clumped together with casein (milk proteins) and complex sugars. They look like pieces of coral or small clumps of cauliflower and range from the size of a grain of wheat to that of a hazelnut. The grains ferment the milk, incorporating their organisms to create the cultured product. The grains are usually removed with a strainer before consumption of the kefir. Different production processes for kefir are currently used, as shown in Figure 10.7.

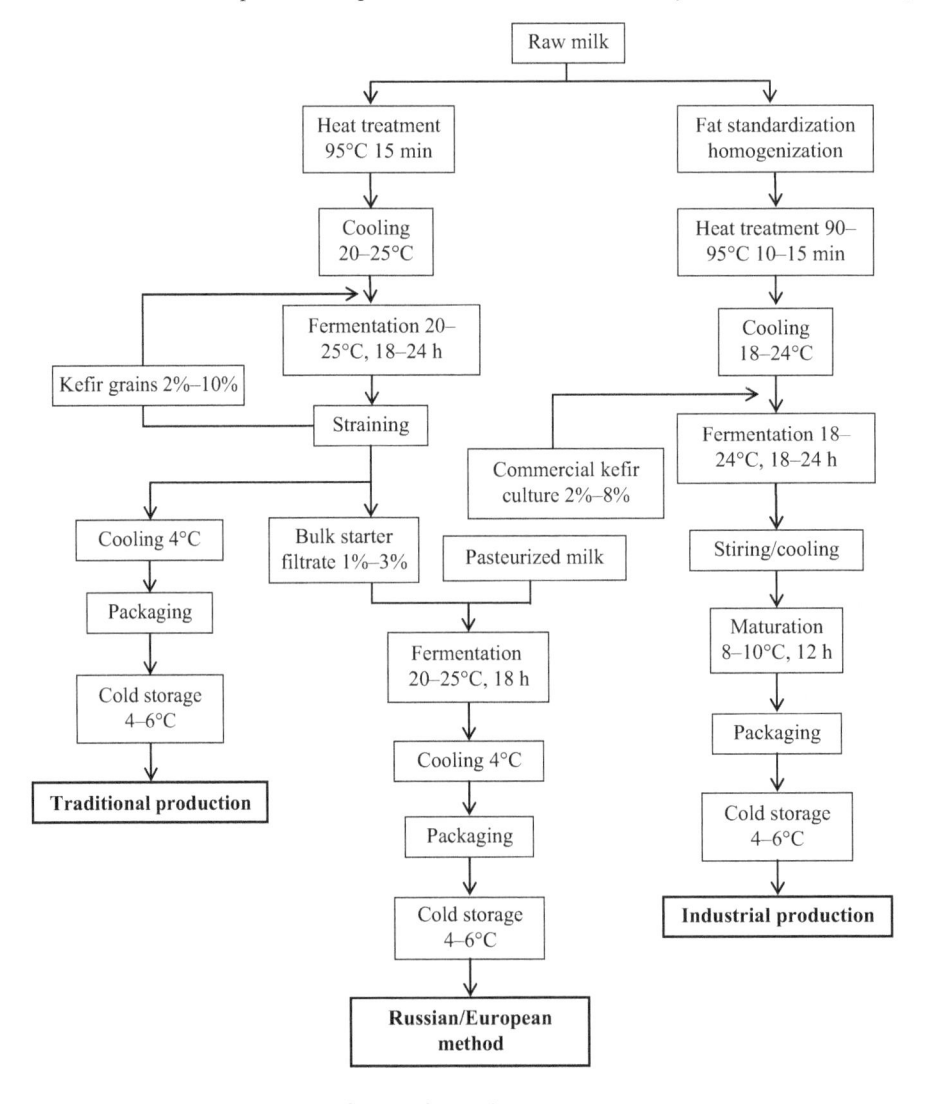

**FIGURE 10.7**  Schematic representation of the kefir production process.

(Modified from Wszolek, M. et al., *Fermented Milks*, Blackwell Publishing, 2006.)

Kefir is defined by the Codex Alimentarius as a fermented milk characterized by specific starter cultures used for fermentation (2003). Kefir cultures are microorganisms prepared from kefir grains, growing in a strong relationship in a food matrix making the biofilm described previously. The Codex Alimentarius Standard defines the starter culture as a mix of the bacteria *Lactobacillus kefiri*; *Lactococcus* spp.; *Leuconostoc* spp.; *Acetobacter* spp.; the lactose-fermenting yeast *Kluyveromyces marxianus*; and the non-lactose-fermenting yeasts *Saccharomyces unisporus*, *S. cerevisiae*, and *S. exiguus*.

## 10.5.2 Health Benefits

Kefir is perceived as a traditional fermented product with intrinsic beneficial properties. A number of studies document kefir's specific health benefits (Rosa et al., 2017). Among the beneficial health effects that have been associated with kefir consumption are modulation of the immune system and inhibition of pathogenic microorganisms. The biological activity of kefir may be attributed to the presence of a complex microbiota as well as the microbial metabolites that are released during fermentation.

Kefir is mentioned in the World Gastroenterology Organisation's guidelines as providing benefits in one very specific case (*Helicobacter pylori* eradication triple-therapy in 1 clinical study) (WGO, 2023).

Specific kefir products have been clinically shown to improve lactose digestion in several studies. For example, Hertzler and Clancy (2003) analyzed the effects of kefir (either plain or fruit flavored) intake in subjects with lactose intolerance; unflavored kefir was similar to plain and flavored yogurt in reducing breath hydrogen, indicative of the ability of kefir to alleviate lactose maldigestion by reducing colic fermentation of lactose. Fruit-flavored kefir, however, performed less well than the tested yogurts and unflavored kefir.

# 10.6 NORTH EUROPEAN FERMENTED DAIRY PRODUCTS

As in other geographic areas, northern Europe has its own specific fermented dairy products. Fondén and coworkers (2006) suggested the classification of traditional Nordic fermented dairy products as presented in Table 10.2. Here, we will describe the manufacturing and suggested health benefits of three characteristic fermented dairy products from northern Europe.

**TABLE 10.2** Classification Scheme of Traditional Nordic Fermented Dairy Products

| MESOPHILIC LACTIC ACID BACTERIA | TEXTURE | TYPE OF MILK | PRODUCT |
|---|---|---|---|
| With exopolysaccharide-producing strains | Viscous | Whole or skimmed | Tätmjölk/Långfil |
| | Gel | Whole | Filbunke/Viili |
| Without exopolysaccharide-producing strains | Viscous | Whole or skimmed | Surmjölk/Filmjölk |
| | Gel | Whole | Filbunke/Viili |
| | Concentrated | Skimmed | Skyr |
| | Fluid | Skimmed | Buttermilk |
| Spontaneous fermentation | Viscous | Whole | Curdled milk |

*Source:* Fondén, R. et al., Nordic/Scandinavian fermented milk products, In: Tamime, A.Y. (ed.), *Fermented Milk*, Blackwell Science, Oxford, UK, 2006.

# 10.6.1 Buttermilk

Buttermilk is a thick liquid dairy product with a rich flavor.

## 10.6.1.1 Production

Traditionally, buttermilk is the aqueous phase that is released during the churning of cream into butter. It contains the same components as the water phase of cream, such as protein, lactose, and minerals, but it also contains the material from the disrupted milk fat globule membrane and therefore contains more phospholipids than whole milk (Sodini et al., 2006). Basically, three processes exist for the production of buttermilk: sweet buttermilk, cultured buttermilk, and whey buttermilk. The traditional production as a by-product from butter manufacturing is described previously. If the milk/cream is not fermented or naturally soured, the product is referred to as sweet buttermilk. If the milk/cream is soured or fermented in a controlled way, the resulting product is referred to as cultured buttermilk. Today, most commercial buttermilk is produced by direct culturing of low-fat milk. Whey buttermilk is produced from whey cream (Fondén et al., 2006; Sodini et al., 2006). The most commonly used organisms in the fermentation of buttermilk are DL-lactate-producing mesophilic starter cultures and may consist of *Lactococcus lactis* subsp. *lactis* (acid production), *Lc. lactis* subsp. *cremoris* (acid and flavor production), *Lc. lactis* subsp. *lactis* biovar *diacetylactis* (acid and flavor production), and *Leuconostoc mesenteroides* subsp. *cremoris* (acid and flavor production) (Mistry, 2001). The production process of cultured buttermilk is presented in Figure 10.8. Citrate may be added to the milk to improve flavor. Citrate, also naturally present in milk, is converted into diacetyl, providing the butter flavor.

## 10.6.1.2 Health Benefits

Buttermilk has traditionally been strongly perceived as healthy. Some contemporary studies seem to lend credence to this.

The main organisms used in the production of cultured buttermilk are *Lc. lactis* subsp. *lactis* and *Lc. lactis* subsp. *cremoris*. These are also the species of lactic acid bacteria that are the highest

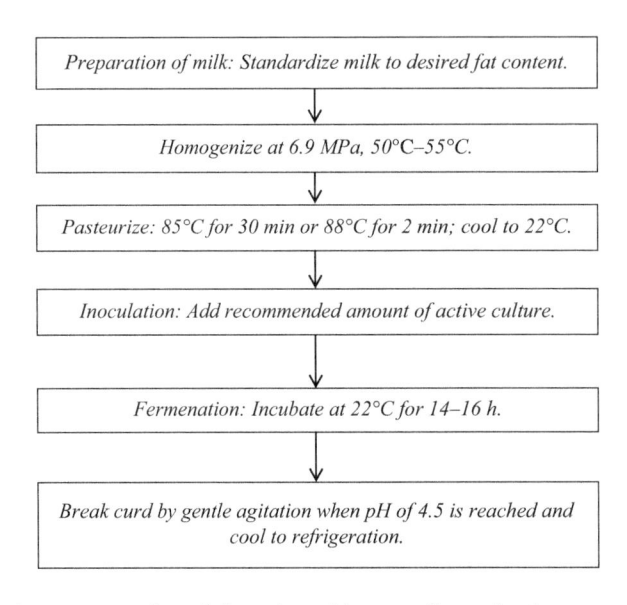

**FIGURE 10.8** Schematic representation of the cultured buttermilk production process.

(Modified from Mistry, V.V., *Applied Dairy Microbiology*, 2nd ed., Marcel Dekker, New York, 2001.)

producers of menaquinones (vitamin K2) (Morishita et al., 1999). Analysis, however, indicated buttermilk contained only 15 ng/g menaquinones (Yasin et al., 2017), which is considerably less then what has been observed in cheese (see subsequently). Nevertheless, the portion size of buttermilk as compared to cheese may make buttermilk a valuable source of menaquinones. Buttermilk and other *Lactococcus*-containing fermented dairy products may thus provide benefits to bone and cardiovascular health (Berenjian et al., 2015).

Consumption of buttermilk has indeed been observed to have beneficial effects on blood pressure (Conway et al., 2014) and total and LDL-cholesterol and triacylglycerol levels (Conway et al., 2013). Whether this is due to the *Lactococcus*-derived menaquinone and/or due to the presence of milk fat globulin membrane remains to determined.

Epidemiological research, however, suggests that consumption of buttermilk was positively correlated with an increased body mass index (Brouwer-Brolsma et al., 2018).

## 10.6.2 Filmjölk

Filmjölk (Sweden) is a traditional Nordic fermented milk marketed under different names in the Nordic countries: Piimä (Finland), Tykmælk (Denmark), and Kulturmelk (Norway).

### 10.6.2.1 Production

Filmjölk is a modern version of traditional Nordic sour milk. Filmjölk is mainly used as a drink. The starter culture consists mainly of *Lc. lactis* ssp. *lactis* as an acid producer and *Lc. lactis* ssp. *lactis* biovar. *diacetylactis* and *Ln. mesenteroides* ssp. *cremoris* as flavor and acid producers (Roginski, 2011). The production process is shown in Figure 10.9.

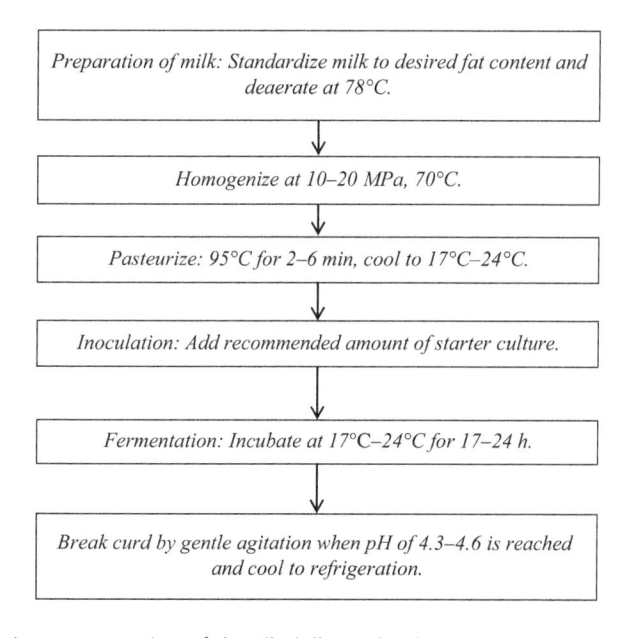

**FIGURE 10.9** Schematic representation of the Filmjölk production process.

(Modified from Fondén, R. et al., *Fermented Milk*, Blackwell Science Ltd, Oxford, UK, 2006; Roginski, H., *Encyclopedia of Dairy Sciences*, Academic Press, New York, 2011.)

## 10.6.2.2 Health Benefits

Due to the use of *Lactococcus* starter cultures, Filmjölk has nutritional benefits similar to buttermilk.

The probiotic versions of Filmjölk that are marketed usually contain various strains of *Lactobacillus acidophilus* and/or *Bifidobacterium animalis* subsp. *lactis*. A probiotic strain of *Lactococcus lactis* L1A has also been used (Carlsson et al., 2009). No research appears to have been done on the potential health benefits of Filmjölk or its correlation to health status in epidemiological studies. Considering Filmjölk is one of the staple dairy products in the Nordic countries, this might be a fruitful field of research.

# 10.6.3  Viili

Viili is a set-type, thick, spoonable, mild acidic, aroma-rich dairy product. The peculiarity of the product is its ropy texture and a thin layer of mold on the top. It is popular in Finland and mainly consumed either plain or flavored.

## 10.6.3.1 Production

The main starter culture is a specific strain of exopolysaccharide- and acid-producing *Lc. lactis* subsp. *cremoris* and *Lc. lactis* subsp. *lactis*. Other LAB strains used in industrially produced viili are *Lc. lactis* subsp. *lactis* biovar *diacetylactis* (contributes to flavor) and *Ln. mesenteroides* subsp. *cremoris* (Kahala et al., 2008). The starter culture also contains the mold species *Geotrichium candidum*. The mold grows only on the surface due to the limited amount of oxygen present in the headspace of the sealed cup. The mold gives the product surface a velvety appearance (Fondén et al., 2006). The production of viili is shown in Figure 10.10. The typical ropy nature is caused by phosphate-containing exopolysaccharide (EPS). The basic structure of viili EPS is mainly composed of D-glucose, D-galactose, L-rhamnose, and phosphate, with an average molecular weight of about 2,000 KDa and repeating unit of $\rightarrow$4-β-Glcp-(1$\rightarrow$4)-β-D-Galp (1$\rightarrow$4)-β-D-Glcp-(1$\rightarrow$ as well as groups of -L-Rhap and -D-Galp-1-p attached to each side of Galp (Nakajima et al., 1992; Wu et al., 2013).

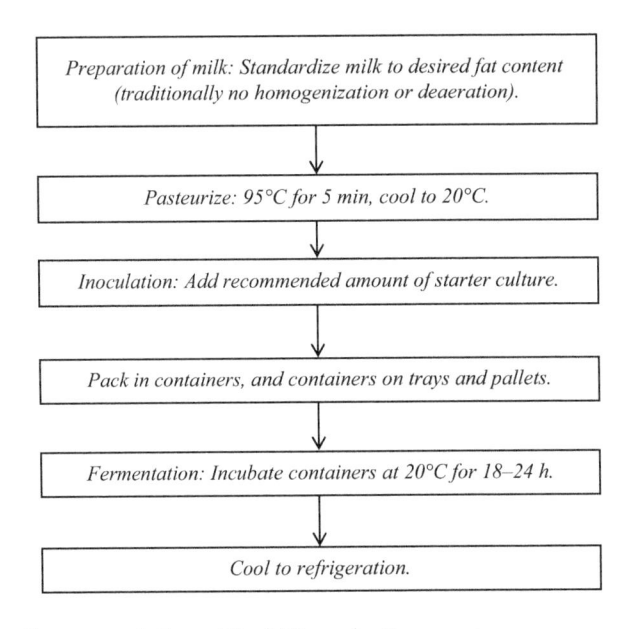

**FIGURE 10.10**  Schematic representation of the Viili production process.

(Modified from Fondén, R. et al., *Fermented Milk*, Blackwell Science Ltd, Oxford, UK, 2006.)

### 10.6.3.2 Health Benefits

No research appears to have been performed on the potential health benefits of viili or its correlation to health status in epidemiological studies. However, *in vitro* studies on the viili exopolysaccharide indicate a health potential. Exopolysaccharides have been observed to be B-cell mitogenic (Kitazawa et al., 1993) and to contribute to macrophage activation (Wu et al., 2013). However, viili exopolysaccharides have also been observed to reduce the adhesion of selected probiotic bacteria *in vitro* but not to affect the *in vitro* adhesion of selected pathogens (Ruas-Madiedo et al., 2006).

Furthermore, cell surface components of the exopolysaccharide-producing *Lc. lactis* subsp. *cremoris* are immunogenic (Kontusaari et al., 1985).

# 10.7 CHEESE

Cheeses come in many variations, and especially in Europe a plethora of local cheeses exist. Despite this diversity, the basic manufacturing process is the same. What differentiates cheeses is the use of different starter cultures, the presence or absence of mold, and whey separation methods (straining, pressing, heating). Differences in ripening methods (time and environmental conditions, use of ripening cultures, smearing of the cheese surface with various preparations, etc.), as well as the use of different types of milk (pasteurized or unpasteurized), also strongly influence the final product. An almost infinite variety of cheeses result from the previously described differences as well as variations in size, shape, added ingredients (especially herbs and spices), and local production peculiarities. Here we will describe cheeses on the basis of their physical appearance: fresh, soft, and hard cheeses. The aim of this section is to provide a brief overview of the different types of cheese available in Europe.

The production of cheese is actually a dehydration process of milk; 6–12 parts of milk yield 1 part of cheese. The basic production process of cheese is as follows: Rennet and starter cultures are added to the milk. The milk coagulates; the coagulum is cut and the whey drained. The remaining curd is pressed into a form and allowed to ripen (see Figure 10.11) (Fox and McSweeney, 2004).

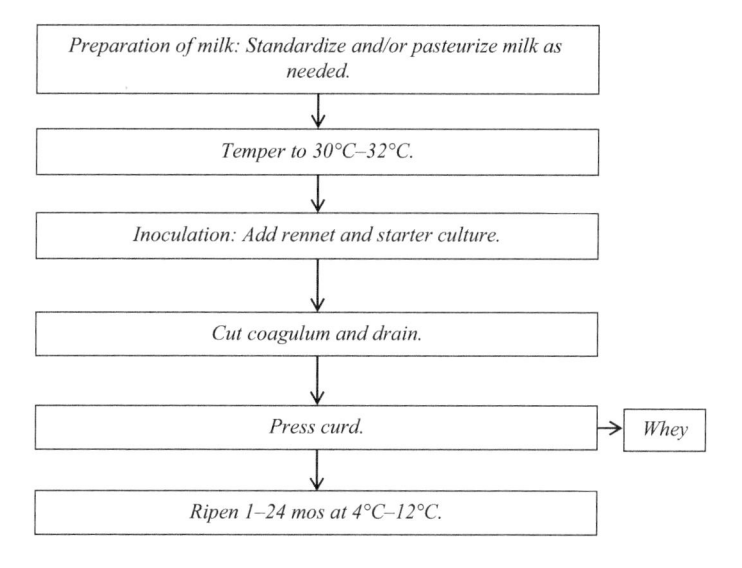

**FIGURE 10.11** Schematic representation of the general cheese production process.

(Modified from Fox, P.F., McSweeney, P.L.H., *Cheese: Chemistry, Physics and Microbiology*, 3rd ed., Elsevier, London, UK, 2004.)

Epidemiological studies indicate that cheese consumption is associated with reduced risk of cardiovascular disease (Chen et al., 2017) and overall cardiovascular mortality (Farvid et al., 2017) and a reduced risk for stroke (Gille et al., 2018).

## 10.7.1 Fresh Cheeses

Examples of fresh cheeses are cottage cheese, mozzarella, and feta(-like) cheese.

### 10.7.1.1 Production

Fresh cheeses are produced similar to the process outlined in Figure 10.11. However, the curd is cut and drained but not pressed. They are either not ripened or ripened only for a short period of time. Mesophilic cultures are commonly used in the production of cottage cheese: *Lc. lactis* subsp. *lactis*, *Lc. lactis* subsp. *cremoris*, and *Lc. lactis* subsp. *lactis* biovar. *diacetylactis* (Ostergaard et al., 2014). In feta cheese produced without starter culture, lactococci, lactobacilli, enterococci, and leuconostoc species are the most commonly used non-starter LAB (Litopoulou-Tzanetaki and Tzanetakis, 2014). In the case of mozzarella, it is mainly thermophilic starters such as *S. thermophilus*, sometimes in combination with *L. bulgaricus* (Pisano et al., 2016).

### 10.7.1.2 Health Benefits

No specific studies appear to have been performed on the impact of fresh cheese consumption on health. Despite the presence of *Lactococcus* ssp., the menaquinone levels in fresh cheeses appear to be rather low (Vermeer et al., 2018). One new area of investigation is the study of EPSs produced by LAB starter cultures in fermented dairy products, which act on the texture of the final product. As after ingestion, these EPSs will ultimately come into contact with the host intestinal microbial population; there's a question around their potential interaction with the microbiome. Tsuda and Miyamoto (2010) compared the prebiotic activity of EPS from a mutant strain of *Lactiplantibacillus plantarum* isolated from traditional homemade cheese with the prebiotic activities of galacto-oligosaccharides (GOSs) and inulin against 37 LAB strains. For the seven strains that were positive for prebiotic utilization, the EPS scored highly for prebiotic activity in comparison to the GOS and inulin prebiotics. It can be concluded that the LAB EPSs from fermented dairy foods indeed have the potential to act as prebiotics for specific microbial strains found in the human gut.

## 10.7.2 Soft Cheeses

Examples of soft cheeses are Camembert and Brie. In addition to LAB, these cheeses also have a surface mold of *Pencillium camemberti* and *G. candidum*. Roquefort is also a soft cheese with blue veins of *Penicillium roqueforti*.

### 10.7.2.1 Production

Production of soft and semi-soft cheeses is similar to the process for general cheese production in Figure 10.11. However, the milk may not always be pasteurized and therefore there may be a contribution of non-starter LAB to the flavor of the cheese. Camembert and Brie use *S. thermophilus* as a starter culture (Wan et al., 1997) and/or *Lc. lactis* subsp. *cremoris*, *Lc. lactis* subsp. *lactis*, *Lc. lactis* subsp. *lactis* biovar *diacetylactis*, and *Ln. mesenteroides* subsp. *cremoris* (Galli et al., 2016).

### 10.7.2.2 Health Effects

Different investigations of health and cheese consumption have been conducted, particularly on folate content and cardiovascular health. For example, Schlienger et al. (2014) sought to evaluate the impact of saturated fatty acids on lipid parameters and blood pressure with regard to Camembert consumption. They observed that over the 5-week daily consumption of two servings of Camembert cheese, blood pressure and serum lipids did not change in moderate hypercholesterolemic subjects. These results suggest that fermented cheese such as Camembert could be consumed daily without affecting serum lipids or blood. Of the soft cheeses, Camembert has also been reported to have the highest menaquinone content, higher than Brie and Roquefort (Vermeer et al., 2018).

The consumption of Camembert has been observed to influence the composition of the intestinal microbiota; *Enterococcus faecalis* levels were found to increase (Firmesse et al., 2007). Furthermore, the levels of organisms found in Camembert (*G. candidum*, *Lc. lactis*, and *Ln. mesenteroides*) were found to be increased in feces (Firmesse et al., 2008).

## 10.7.3 Hard Cheeses

Hard cheeses vary from semi-hard (e.g., Gouda and Edam) to hard (e.g., Parmesan).

### 10.7.3.1 Production

In hard cheeses like Gouda, mesophilic starter cultures are commonly used; *Lc. lactis* subsp. *lactis*, *Lc. cremoris*, *Lc. lactis* subsp. *lactis* biovar *diacetylactis*, and *Leuconostoc* spp. are responsible for diacetyl formation and $CO_2$ (Ayad et al., 2001). Parmesan cheese may also contain *L. helveticus*, *L. delbrueckii* subsp. *lactis*, *L. delbrueckii* subsp. *bulgaricus*, and *Lacticaseibacillus rhamnosus* (Coppola et al., 2000). The production of hard and semi-hard cheeses is similar to the process described in Figure 10.11. The choice of the initial cultures is highly dependent on the ability of the LAB to adapt to technological stresses, such as oxidative stress encountered during stirring in the first stages of the cheese-making process. Different investigations aimed at understanding the response to oxidative stress of different LAB, especially *Lc. lactis*, have been conducted. A recent novel approach, combining metabolic and transcriptomic profiling together with oxygen consumption kinetics (Cretenet et al., 2014), gave new insights into LAB adaptation to oxidative stress, especially during the first stages of the process when oxygen is present. Therefore, LAB well adapted to process constraints should keep technological abilities as acidification and protein reduction should be preserved in the presence of early oxidative stress.

### 10.7.3.2 Health Benefits

Intake of a Gouda-type cheese has been associated with reduced systolic blood pressure (Nilsen et al., 2014) and was observed to reduce serum cholesterol in subjects with metabolic syndrome and elevated serum cholesterol levels (Nilsen et al., 2015). These benefits on cardiovascular health may relate to two different reasons: the presence of antihypertensive peptides present in certain cheeses (Saito et al., 2000) and the presence of menaquinones, which are high in especially Dutch type semi-hard cheeses (Gouda 473–729 ng/g and Edam 674 ng/g) (Vermeer et al., 2018).

Probiotic *Lactobacillus* strains have been successfully included in semi-hard cheeses; this is interesting considering the high water activity of cheese and the extended ripening times. Nevertheless, a clinically relevant dose can be incorporated in a cheese portion (Ouwehand et al., 2011).

# 10.8 CONCLUSION

Fermented dairy products have a long history and a good reputation for high nutritional value (important sources of essential nutrients such as vitamins, minerals, and protein) and many health-promoting effects, such as improvement of lactose metabolism, reduction of risk for type 2 diabetes, and reduction of serum cholesterol. The findings from this bibliographic work also revealed that production processes are elegantly simple, as they generally require very few ingredients and minimal preparation and processing. These fermented foods are a good source of live LAB (unless they are pasteurized after fermentation), including species that reportedly provide human health benefits.

LAB have a crucial role in the production of these fermented dairy products and were initially selected for providing preservation, flavor, and consistency to a rich variety of fermented dairy products. But they do more and can be more than just starters, as explained in this review. Many industrially utilized dairy LAB starter cultures are highly proteolytic. Therefore, the synthesis of bioactive peptides through fermentation can aid development of novel fermented dairy products with specific health benefits. Also, these LAB can be vitamin producers and can promote intestinal barrier integrity. These characteristics have been documented for otherwise recognized probiotic strains. Thus, there may be common mechanisms by which microorganisms in fermented foods and probiotics contribute to human health (Marco et al., 2017). Traditionally, probiotics have been added to fermented dairy products. This is mainly due to the fact that probiotic strains from LAB species are well adapted to dairy-based matrices, in particular yogurt. As a result, these products are good carriers for probiotic bacteria. LAB as probiotics also have unique characteristics to be adapted in various foods applications.

In the future, the development of food is likely to go hand in hand with an increased understanding of the human microbiota. This may lead to a continued and expanding use of LAB in human nutrition.

# BIBLIOGRAPHY

Adolfsson, O., Meydani, S. N. & Russel, R. M. 2004. Yogurt and gut function. *Am J Clin Nutr*, *80* (2), 245–256.

Arioli, S., Della Scala, G., Remagni, M. C., Stuknyte, M., Colombo, S., Guglielmetti, S., De Noni, I., Ragg, E. & Mora, D. 2017. *Streptococcus thermophilus* urease activity boosts *Lactobacillus delbrueckii* subsp. *bulgaricus* homolactic fermentation. *Int J Food Microbiol*, *247*, 55–64.

Aune, D., Norat, T., Romundstad, P. & Vatten, L. J. 2013. Dairy products and the risk of type 2 diabetes: A systematic review and dose-response meta-analysis of cohort studies. *Am J Clin Nutr*, *98* (4), 1066–1083.

Ayad, E. H. E., Verheul, A., Wouters, J. T. M. & Smit, G. 2001. Population dynamics of lactococci from industrial, artisanal and non-dairy origins in defined strain starters for Gouda-type cheese. *Int Dairy J*, *11*, 51–61.

Berenjian, A., Mahanama, R., Kavanagh, J. & Dehghani, F. 2015. Vitamin K series: Current status and future prospects. *Crit Rev Biotechnol*, *35*, 199–208.

Bian, S., Hu, J., Zhang, K., Wang, Y., Yu, M. & Ma, J. 2018. Dairy product consumption and risk of hip fracture: A systematic review and meta-analysis. *BMC Public Health*, *18*, 165.

Bong, D. & Moraru, C. 2014. Use of micellar casein concentrate for Greek-style yogurt manufacturing: Effects on processing and product properties. *J Dairy Sci*, *97* (3), 1259.

Bridge, A., Brown, J., Snider, H., Nasato, M., Ward, W. E., Royj, B. D. & Josse, A. R. 2019. Greek yogurt and 12 weeks of exercise training on strength, muscle thickness and body composition in lean, untrained, university-aged males. *Front Nutr*, *6*, 55.

Brouwer-Brolsma, E. M., Sluik, D., Singh-Povel, C. M. & Feskens, E. J. M. 2018. Dairy shows different associations with abdominal and BMI-defined overweight: Cross-sectional analyses exploring a variety of dairy products. *Nutr Metab Cardiovasc Dis*, *28*, 451–460.

Buendia, J. R., Li, Y., Hu, F. B., Cabral, H. J., Bradlee, M. L., Quatromoni, P. A., Singer, M. R., Curhan, G. C. & Moore, L. L. 2018. Long-term yogurt consumption and risk of incident hypertension in adults. *J Hypertens*, *36*, 1671–1679.

Camacho-Barcia, L., Bullo, M., Garcia-Gavilan, J. F., Martinez-Gonzalez, M. A., Corella, D., Estruch, R., Fito, M., et al. 2019. Dairy products intake and the risk of incident cataracts surgery in an elderly Mediterranean population: Results from the Predimed study. *Eur J Nutr*, *58*, 619–627.

Carlsson, M., Gustafson, Y., Haglin, L. & Eriksson, S. 2009. The feasibility of serving liquid yoghurt supplemented with probiotic bacteria, Lactobacillus rhamnosus LB 21, and *Lactococcus lactis* L1A: A pilot study among old people with dementia in a residential care facility. *J Nutr Health Aging*, *13*, 813–819.

Casanova, E., Knowles, T. D. J., Bayliss, A., Roffet-Salque, M., Heyd, V., Pyzel, J., Classen, E., Domboroczki, L., Ilett, M., Lefranc, P., Jeunesse, C., Marciniak, A., van Wijk, I., Evershed, R. P. 2022. Dating the emergence of dairying by the first farmers of Central Europe using (14)C analysis of fatty acids preserved in pottery vessels. *Proc Natl Acad Sci USA, 119,* e2109325118.

Chammas, G. I., Saliba, R., Corrieu, G. & Beal, C. 2006. Characterisation of lactic acid bacteria isolated from fermented milk "laban". *Int J Food Microbiol, 10* (1), 52–61.

Champomier-Verges, M.-C., Zagorec, M. & Fadda, S. 2010. Proteomics: A tool for understanding lactic acid bacteria adaptation to stressful environments. In: *Biotechnology of Lactic Acid Bacteria: Novel Applications*, Chapter 3, 57–72. Hoboken, NJ: John Wiley & Sons, Ltd. Mozzi, F, Raya, R.R. & Vignolo, G.M.

Chen, G. C., Wang, Y., Tong, X., Szeto, I. M. Y., Smit, G., Li, Z. N. & Qin, L. Q. 2017. Cheese consumption and risk of cardiovascular disease: A meta-analysis of prospective studies. *Eur J Nutr, 56,* 2565–2575.

Chen, Y., Liu, W., Xue, J., Yang, J., Chen, X., Shao, Y., Kwok, L.-Y., Bilige, M., Mang, L. & Zhang, H. 2014. Angiotensin-converting enzyme inhibitory activity of *Lactobacillus helveticus* strains from traditional fermented dairy foods and antihypertensive effect of fermented milk of strain H9. *J Dairy Sci, 97* (11), 6680–6692.

Codex. 2003. *Standard for Fermented Milk* (CODEX STAN 243–2003). Rome: Food and Agriculture Organization of the United Nations and World Health Organization.

Conway, V., Couture, P., Gauthier, S., Pouliot, Y. & Lamarche, B. 2014. Effect of buttermilk consumption on blood pressure in moderately hypercholesterolemic men and women. *Nutrition, 30,* 116–119.

Conway, V., Couture, P., Richard, C., Gauthier, S. F., Pouliot, Y. & Lamarche, B. 2013. Impact of buttermilk consumption on plasma lipids and surrogate markers of cholesterol homeostasis in men and women. *Nutr Metab Cardiovasc Dis, 23,* 1255–1262.

Coppola, R., Nanni, M., Iorizza, M., Sorrentino, A., Sorrentino, E., Chiavari, C. & Grazia, L. 2000. Microbiological characteristics of Parmigiano Reggiano cheese during the cheesemaking and the first months of the ripening. *Lait, 80,* 479–490.

Cretenet, M., Le Gall, G., Wegmann, U., Even, S., Shearman, C., Stentz, R. & Jeanson, S. 2014. Early adaptation to oxygen is key to the industrially important traits of *Lactococcus lactis* ssp. cremoris during milk fermentation. *BMC Genom, 15,* 1054.

Dehghani, M., Mente, A., Rangarajan, S., Sheridan, P., Mohan, V., Iqbal, R., Gupta, R., et al. 2018. Association of dairy intake with cardiovascular disease and mortality in 21 countries from five continents (PURE): A prospective cohort study. *Lancet.* https://doi.org/10.1016/S0140-6736(18)31812-9.

Djuric, M., Panic, M., Milanovic, S., Caric, M., Tekic, M., Krstic, D. & Albijanic, B. 2005. Probiotic starters versus traditional starter in quarg production. *Ann Fac Eng Hunedoara, 3,* 154–163.

EFSA, European Food Safety Authority, 2010. Panel on Dietetic Products NaA: Scientific Opinion on the substantiation of health claims related to live yoghurt cultures and improved lactose digestion (ID 1143, 2976) pursuant to Article 13(1) of Regulation (EC) No 1924/2006. *EFSA J, 8,* 1763–1781.

Farkye, N. Y. 2004. Cheese technology. *Int J Dairy Technol, 57* (2–3).

Farvid, M. S., Malekshah, A. F., Pourshams, A., Poustchi, H., Sepanlou, S. G., Sharafkhah, M., Khoshnia, M., et al. 2017. Dairy food intake and all-cause, cardiovascular disease, and cancer mortality: The golestan cohort study. *Am J Epidemiol, 185,* 697–711.

FDA, Food & Drug Administration, 2024. https://www.fda.gov/food/cfsan-constituent-updates/fda-announces-qualified-health-claim-yogurt-and-reduced-risk-type-2-diabetes

Fiorda, F. A., De Melo Pereira, G. V., Thomaz-Soccol, V., Rakshit, S. K., Pagnoncelli, M. G. B., Vandenberghe, L. P. S. & Soccol, C. R. 2017. Microbiological, biochemical, and functional aspects of sugary kefir fermentation: A review. *Food Microbiol, 66,* 86–95.

Firmesse, O., Alvaro, E., Mogenet, A., Bresson, J. L., Lemee, R., Le Ruyet, P., Bonhomme, C., et al. 2008. Fate and effects of Camembert cheese micro-organisms in the human colonic microbiota of healthy volunteers after regular Camembert consumption. *Int J Food Microbiol, 125,* 176–181.

Firmesse, O., Rabot, S., Bermudez-Humaran, L. G., Corthier, G. & Furet, J. P. 2007. Consumption of Camembert cheese stimulates commensal enterococci in healthy human intestinal microbiota. *FEMS Microbiol Lett, 276,* 189–192.

Fondén, R., Leporanta, K. & Svensson, U. 2006. Nordic/Scandinavian fermented milk products. In: Tamime, A. Y. (ed.) Fermented Milk. Oxford: Blackwell Science Ltd.

Fox, P. F. & McSweeney, P. L. H. 2004. Cheese: An overview. In: Fox, P. F., McSweeney, P. L. H., Cogan, T. M. & Guinee, T. P. (eds.) Cheese: Chemistry, Physics and Microbiology. 3rd ed. London: Elsevier.

Fu, X., Harshman, S. G., Shen, X., Haytowitz, D. B., Karl, J. P., Wolfe, B. E. & Booth, S. L. 2017. Multiple vitamin K forms exist in dairy foods. *Curr Develop Nutr, 1* (6).

Galli, B. D., Martin, J. G. P., Da Silva, P. P. M., Porto, E. & Spoto, M. H. F. 2016. Sensory quality of Camembert-type cheese: Relationship between starter cultures and ripening molds. *Int J Food Microbiol, 234,* 71–75.

Gijsbers, L., Ding, E. L., Malik, V. S., De Goede, J., Geleijnse, J. M. & Soedamah-Muthu, S. S. 2016. Consumption of dairy foods and diabetes incidence: A dose-response meta-analysis of observational studies. *Am J Clin Nutri*, *103* (4), 1111–1124.

Gille, D., Schmid, A., Walther, B. & Vergeres, G. 2018. Fermented food and non-communicable chronic diseases: A review. *Nutrients*, 10.

Guarner, F., Perdignon, G., Corthier, G, Salminen, S., Koletzko, B. & Morelli, L. 2005. Should yoghurt cultures be considered probiotic? *Br J Nutr*, *93* (6), 783–786.

Gudmundsson, G. & Kristbergsson, K. 2016. Modernization of Skyr processing: Icelandic acid-curd soft cheese. In: McElhatton, A. & El Idrissi, M. M. (eds.) Modernization of Traditional Food Processes and Products. London: Springer, 45–53.

Hartley, D. L. & Denariaz, G. 1993. The role of lactic acid bacteria in yogurt fermentation. *Int J Immunother*, *9* (1), 3–17.

Hayes, M., Stanton, C., Fitzgerald, G. F. & Ross, R. P. 2007. Putting microbes to work; Dairy fermentation, cell factories and bioactive peptides. Part II: Bioactive peptide functions. *Biotechnol J*, *2* (4).

Hertzler, S. R. & Clancy, S. M. 2003. Kefir improves lactose digestion and tolerance in adults with lactose maldigestion. *J Am Diet Assoc*, *103* (5), 582–587.

Herve-Jimenez, L., Guillouard, I., Guedon, E., Boudebbouze, S., Hols, P., Monnet, V., Maquin, E. & Rul, F. 2009. Postgenomic analysis of streptococcus thermophilus cocultivated in milk with Lactobacillus delbrueckii subsp. bulgaricus: Involvement of nitrogen, purine, and iron metabolism. *Appl Environ Microbiol*, *75* (7), 2062–2073.

Hill, D., Sugrue, I., Arendt, E., Hill, C., Stanton, C. & Ross, R. P. 2017. Recent advances in microbial fermentation for dairy and health. *F1000Research*, *6*, 751.

Ibsen, D. B., Laursen, A. S. D., Lauritzen, L., Tjonneland, A., Overvad, K. & Jakobsen, M. U. 2017. Substitutions between dairy product subgroups and risk of type 2 diabetes: The Danish Diet, Cancer and Health cohort. *Br J Nutr*, *118*, 989–997.

Kahala, M., Maki, M., Lehtovaara, A., Tapanainen, J. M., Katiska, R., Juuruskorpi, M., Juhola, J. & Joutsjoki, V. 2008. Characterization of starter lactic acid bacteria from the Finnish fermented milk product viili. *J Appl Microbiol*, *105*, 1929–1938.

Kitazawa, H., Yamaguchi, T., Miura, M., Saito, T. & Itoh, T. 1993. B-cell mitogen produced by slime-producing, encapsulates Lactoccus lactis ssp. cremoris isolated from ropy sour milk, viili. *J Dairy Sci*, *76*, 1514–1519.

Kontusaari, S. I., Vuokila, P. T. & Forsen, R. I. 1985. Immunochemical study of triton X-100-soluble surface components of slime-forming, encapsulated Streptococcus cremoris from the fermented milk product viili. *Appl Environ Microbiol*, *50*, 174–176.

Laird, E., Molloy, A. M., McNulty, H., Ward, M., McCarroll, K., Hoey, L., Hughes, C. F., Cunningham, C., Strain, J. J. & Casey, M. C. 2017. Greater yogurt consumption is associated with increased bone mineral density and physical function in older adults. *Osteoporos Int*, *28*, 2409–2419.

Lamoureux, L., Roy, Y. D. & Gauthier, S. 2002. Production of oligosaccharides in Yogurt containing Bifidobacteria and Yogurt cultures. *J Dairy Sci*, *85* (5), 1058.

Lana, A., Rodriguez-Artalejo, F. & Lopez-Garcia, E. 2015. Dairy consumption and risk of frailty in older adults: A prospective cohort study. *J Am Geriatr Soc*, *63*, 1852–1860.

Lang, J. M., Eisen, J. A. & Zivkovic, A. M. 2014. The microbes we eat: Abundance and taxonomy of microbes consumed in a day's worth of meals for three diet types. *Peer J*, *2*, e659.

Litopoulou-Tzanetaki, E. & Tzanetakis, N. 2014. The microfloras of traditional Greek cheeses. *Microbiol Spectr*, *2*, Cm-0009–2012.

Lourens-Hattingh, A. & Viljoen, B. C. 2001. Yogurt as a probiotic carrier food. *Int Dairy J*, *11*, 1–17.

Marco, L. M., Heeney, D., Binda, S., Cifelli, C. J., Cotter, P. D., Foligné, B., Gänzle, M., Kort, R., Pasin, G., Pihlanto, A., Smid, E. J. & Hutkins, R. H. 2017. Health benefits of fermented foods: Microbiota and beyond. *Curr Opinion Biotech*, *44*, 94–102.

Marsh, A. J., O'Sullivan, O., Hill, C., Ross, R. P. & Cotter, P. D. 2013. Sequencing-based analysis of the bacterial and fungal composition of kefir grains and milks from multiple sources. *PLoS ONE*, *8* (7), e69371.

Marette, A., Picard-Deland, E. & Fernandez, M. A. 2017. Yogurt: Roles in Nutrition and Impacts on Health. New York: CRC Press.

Marteau, P., Flourie, B., Pochart, P., Chastang, C., Desjeux, J. F. & Rambaud, J. C. 1990. Effect of the microbial lactase (EC 3.2.1.23) activity in yoghurt on the intestinal absorption of lactose: An *in vivo* study in lactase-deficient humans. *Br J Nutr*, *64*, 71–79.

McCann, S. E., Hays, J., Baumgart, C. W., Weiss, E. H., Yao, S. & Ambrosone, C. B. 2017. Usual consumption of specific dairy foods is associated with breast cancer in the Roswell Park Cancer Institute Data Bank and BioRepository. *Curr Dev Nutr*, *1*, e000422.

Mistry, V. V. 2001. Fermented milks and cream. In: Marth, E. H. & Steele, J. L. (eds.) Applied Dairy Microbiology. 2nd ed. New York: CRC Press.

Morishita, T., Tamura, N., Makino, T. & Kudo, S. 1999. Production of menaquinones by lactic acid bacteria. *J Dairy Sci*, *82*, 1897–1903.

Nakajima, H., Hirota, T., Toba, T., Itoh, T. & Adachi, S. 1992. Structure of the extracellular polysaccharide from slime-forming *Lactococcus lactis* subsp. *cremoris* SBT 0495. *Carbohydr Res*, *224*, 245–253.

Nilsen, R., Hostmark, A. T., Haug, A. & Skeie, S. 2015. Effect of a high intake of cheese on cholesterol and metabolic syndrome: Results of a randomized trial. *Food Nutr Res*, *59*, 27651.

Nilsen, R., Pripp, A. H., Hostmark, A. T., Haug, A. & Skeie, S. 2014. Short communication: Is consumption of a cheese rich in angiotensin-converting enzyme-inhibiting peptides, such as the Norwegian cheese Gamalost, associated with reduced blood pressure? *J Dairy Sci*, *97*, 2662–2668.

Ostergaard, N. B., Eklow, A. & Dalgaard, P. 2014. Modelling the effect of lactic acid bacteria from starter- and aroma culture on growth of Listeria monocytogenes in cottage cheese. *Int J Food Microbiol*, *188*, 15–25.

Ouwehand, A. C., Ibrahim, F., Jorgensen, P., Forssten, S. & Röytiö, H. 2011. Probiotic cheese; new opportunities. *Agro Food Industry Hi Tech*, *22*, 13–16.

Panahi, S. & Tremblay, A. 2016. The potential role of yogurt in weight management and prevention of type 2 diabetes. *J Am Coll Nutr*, *35* (8), 717–731.

Parvez, S., Malik, K. A., Ah Kang, S. & Kim, H. Y. 2006. Probiotics and their fermented food products are beneficial for health. *J Appl Microbiol*, *100* (6).

Pisano, M. B., Scano, P., Murgia, A., Cosentino, S. & Caboni, P. 2016. Metabolomics and microbiological profile of Italian mozzarella cheese produced with buffalo and cow milk. *Food Chem*, *192*, 618–624.

Puhan, Z., Driessen, F. M., Jelen, P. & Tamime, A. Y. 1994. Fresh products—yoghurt, fermented milks, quarg and fresh cheese. *Mljekarstvo*, *44*, 285–298.

Rai, A. K. & Sanjukta, S. 2015. Production of angiotensin I converting enzyme inhibitory (ACE-I) peptides during milk fermentation and their role in reducing hypertension. *Crit Rev Food Sci Nutr*, *57* (13), 2789–2800.

Rezaiki, L., Lamberet, G., Derre, A., Gruss, A. & Gaudu, P. 2008. *Lactococcus lactis* produces short-chain quinones that cross-feed Group B Streptococcus to activate respiration growth. *Mol Microbiol*, *67* (5).

Rocha, D. M. U. P., Martins, J. F. L., Santos, T. S. S. & Moreira, A. V. B. 2014. Labneh with probiotic properties produced from kefir development and sensory evaluation. *Food Sci Technol*, *34* (4), 694–700.

Roginski, H. 2011. Fermented milks/Nordic fermented milks. In: Fuquay, J. W. (ed.) Encyclopedia of Dairy Sciences. New York: Academic Press.

Rohde, C. L., Bartolini, V. & Jones, N. 2009. The use of probiotics in the prevention and treatment of antibiotic-associated diarrhea with special interest in clostridium difficile-associated diarrhea. *Nutr Clin Pract*, *24* (1), 33–40.

Rosa, D. D., Dias, M. M., Grzeskowiak, L. M., Conceicao, L. L. & Peluzio, M. D. C. G. 2017. Milk kefir: Nutritional, microbiological and health benefits. *Nutr Res Rev*, *30*, 82–96.

Ruas-Madiedo, P., Gueimonde, M., De Los Reyes-Gavilan, C. G. & Salminen, S. 2006. Short communication: Effect of exopolysaccharide isolated from "viili" on the adhesion of probiotics and pathogens to intestinal mucus. *J Dairy Sci*, *89*, 2355–2358.

Saborido, R. & Leis, R. 2018. Yogurt and dietary recommendations for lactose intolerance. *Nutr Hospit*, *35*, 45–48.

Saito, T., Nakamura, T., Kitazawa, H., Kawai, Y. & Itoh, T. 2000. Isolation and structural analysis of antihypertensive peptides that exist naturally in Gouda cheese. *J Dairy Sci*, *83*, 1434–1440.

Salas, M. M., Nascimento, G. G., Vargas-Ferreira, F., Tarquinio, S. B., Huysmans, M. C. & Demarco, F. F. 2015. Diet influenced tooth erosion prevalence in children and adolescents: Results of a meta-analysis and meta-regression. *J Dent*, *43*, 865–875.

Sawsen, H. 2012. Evaluation des aptitudes technologiques et probiotiques des bactéries lactiques locales (mémoire). Ouargla: Université Kasdi Merbah-Ouargla.

Schlienger, J. L., Paillard, F., Lecerf, J. M., Romon, M., Bonhomme, C., Schmitt, B., Donazzolo, Y., et al. 2014. Effect on blood lipids of two daily servings of Camembert cheese. An intervention trial in mildly hypercholesterolemic subjects. *Int J Food Sci Nutr*, *65*, 1013–1018.

Shah, N. P., Fedorak, R. N. & Jelen, P. J. 1992. Food consistency effects of quarg in lactose malabsorption. *Int Dairy J*, *2*, 257–269.

Shoda, T., Futamura, M., Yang, L., Narita, M., Saito, H. & Ohya, Y. 2017. Yogurt consumption in infancy is inversely associated with atopic dermatitis and food sensitization at 5 years of age: A hospital-based birth cohort study. *J Dermatol Sci*, *86*, 90–96.

Sieuwerts, S., De Bok, F. A., Hugenholtz, J. & Van Hylckama Vlieg, J. E. 2008. Unraveling microbial interactions in food fermentations: From classical to genomics approaches. *Appl Environ Microbiol*, *74* (16), 4997–5007.

Sieuwerts, S., Molenaar, D., Van Hijum, S. A. F. T., Beerthuyzen, M., Stevens, M. J. A., Janssen, P. W. M., Ingham, C. J., De Bok, F. A. M., De Vos, W. M. & Van Hylckama Vlieg, J. E. T. 2010. Mixed-culture transcriptome analysis reveals the molecular basis of mixed-culture growth in *Streptococcus thermophilus* and *Lactobacillus bulgaricus* . *Appl Environ Microbiol*, *76*, 7775–7784.

Sodini, I., Morin, P., Olabi, A. & Jimenez-Flores, R. 2006. Compositional and functional properties of buttermilk: A comparison between sweet, sour, and whey buttermilk. *J Dairy Sci*, *89*, 525–536.

Surono, I. S. & Hosono, A. 2011. Fermented milks: Types and standards of identity. In: *Encyclopedia of Dairy Sciences*, 470–476. Cambridge MA: Academic Press. Fuquay, J.W., Fox, P.F. & McSweeney, P.L.H.

Tacitus, P. C. 98. De Origine Et Situ Germanorum. Rome.

Tamime, A. Y. & Robinson, R. K. 2007. Science and Technology. 3rd ed. Cambridge: Woodhead Publishing, 808.

Thakkar, P. N., Patel, A. R., Modi, H. A. & Prajapati, J. B. 2018. Evaluation of antioxidative, proteolytic, and ace inhibitory activities of potential probiotic lactic acid bacteria isolated from traditional fermented food products. *Acta Alimentaria*, *47* (1).

Tsuda, H. & Miyamoto, T. 2010. Production of Exopolysaccharide by Lactobacillus plantarum and the prebiotic activity of the Exopolysaccharide. *Food Sci Technol Res*, *16* (1), 87–92.

Vermeer, C., Raes, J., Van 'T Hoofd, C., Knapen, M. H. J. & Xanthoulea, S. 2018. Menaquinone content of cheese. *Nutrients*, 10, 446.

Walsh, A. M., Leech, J., Huttenhower, C., Delhomme-Nguyen, H., Crispie, F., Chervaux, C. & Cotter, P. D. 2023. Integrated molecular approaches for fermented food microbiome research. *FEMS Microbiol Rev*, *47* (2).

Wan, J., Harmark, K., Davidson, B. E., Hillier, A. J., Gordon, J. B., Wilcock, A., Hickey, M. W. & Coventry, M. J. 1997. Inhibition of Listeria monocytogenes by piscicolin 126 in milk and Camembert cheese manufactured with a thermophilic starter. *J Appl Microbiol*, *82*, 273–280.

Weckx, S., Allemeersch, J., Van Der Meulen, R., Vrancken, G., Huts, G., Vandamme, P., Van Hummelen, P. & De Vuyst, L. 2009. Development and validation of a species-independent functional gene microarray that targets lactic acid bacteria. *Appl Environ Microbiol*, 75 (20), 6488–6495.

WGO. 2023. *Guideline on Probiotics and Prebiotics* . https://www.worldgastroenterology.org/UserFiles/file/guidelines/probiotics-and-prebiotics-english-2023.pdf

Wszolek, M., Teahan, B., Skov-Guldager, H. & Tamime, A. Y. 2006. Production of Kefir, Koumiss and other related products. In: Tamime, A. Y. (ed.) Fermented Milks. Oxford: Blackwell Publishing.

Wu, J., Li, M., Liu, L., An, Q., Zhang, J., Zhang, J., Li, M., Duan, W., Liu, D., Li, Z. & Luo, C. 2013. Nitric oxide and interleukins are involved in cell proliferation of RAW264.7 macrophages activated by viili exopolysaccharides. *Inflammation*, *36*, 954–961.

Yacoub, R., Kaji, D., Patel, S. N., Simoes, P. K., Busayavalasa, D., Nadkarni, G. N., He, J. C., Coca, S. G. & Uribarri, J. 2016. Association between probiotic and yogurt consumption and kidney disease: Insights from NHANES. *Nutr J*, *15*, 10.

Yasin, M., Butt, M. S. & Zeb, A. 2017. Vitamin K2 rich food products. In: Gordeladze, J. (ed.) Vitamin *K2*. London: Intechopen.

Zhernakova, A., et al. 2016. Population-based metagenomics analysis reveals markers for gut microbiome composition and diversity. *Science*, *352* (6285), 565–569.

Zourari, A., Accolas, J. P. & Desmazeaud, M. J. 1992. Metabolism and biochemical characteristics of yogurt bacteria. *A review. Le Lait, INRA Editions*, *72* (1), 1–34.

# Lactic Acid Bacteria in the Fermentation of Non-Alcoholic Cereal Products

# 11

Michael G. Gänzle

## 11.1 CEREALS AS A SUBSTRATE

Cereal grains are nutrient dense and can be stored for a long time when dry but can be eaten only after milling or grinding, mixing with water, and fermentation and/or cooking. Taken together, these processing steps make nutrients physically accessible, gelatinize starch, and partially degrade anti-nutritive compounds including bitter-tasting or toxic secondary plant metabolites. Humans are the only primate species with multiple copies of salivary amylase, a copy number variation that was likely selected for in response to the consumption of cooked starchy foods, including cereals (Hardy et al., 2015; Poole et al., 2019). Cereal fermentations to produce (flat)bread and beverages are the oldest food fermentations and likely predate cereal agriculture (Arranz-Otaegui et al., 2018; Hayden et al., 2013; Liu et al., 2018). Globally, fermented non-alcoholic beverages, fermented porridges, and bread are the major groups of fermented cereal foods (Arora et al., 2021; Gänzle, 2022; Jimenez et al., 2022; Pswarayi and Gänzle, 2022; Todorov and Holzapfel, 2015). This chapter aims to provide an overview on cereal-based fermented foods, on community assembly in cereal fermentations, and on lactic acid bacteria that play important roles in cereal fermentations.

Cereals readily support microbial fermentations after grinding and mixing with water. The level of free sugars in matured resting grains is about 1%–3% (Table 11.1). This amount supports the initiation of the fermentation process. In many processes, endogenous cereal enzymes additionally release sugars during fermentation (Gänzle, 2014). Malt or fungal enzymes are often added to break down starch to simple fermentable sugars. Only few strains of lactic acid bacteria (LAB) express extracellular amylases, but amylolytic lactic acid bacteria particularly occur in cereal fermentations of substrates with low amylase activity, for example, maize or tubers (Gänzle and Follador, 2012; Guyot, 2012).

The community assembly in spontaneous cereal fermentations is determined by selection and dispersal limitations (Gänzle and Ripari, 2016; Marco et al., 2021). Cereal grains are associated with a microbiota composed of molds, plant-associated enterobacteria, bacilli, and lactic acid bacteria (Solanki et al., 2019), all of which compete for nutrients after the addition of water to ground cereals. Spontaneous cereal fermentations are generally initiated by *Enterobacteriaceae* and bacilli; lactic acid bacteria become

DOI: 10.1201/9781003352075-13

**TABLE 11.1**  Compositional Data for Whole Dehulled Cereal Grains

| CONSTITUENT | CONTENT (%, DRY MATTER BASIS) |
| --- | --- |
| Polysaccharides (total) | 70–80 |
| Starch | 45–77 |
| Dietary fiber (as nonstarch polysaccharides in lignin) | 9–15 |
| Low molecular weight carbohydrates (total) | 2–5 |
| Fructose | 0.1–0.4 |
| Glucose | 0.1–0.5 |
| Sucrose | 0.5–2 |
| Raffinose | 0.2–0.7 |
| Protein | 8–15 |
| Lipids | 2–6 |
| Ash (minerals) | 1.5–3 |

*Source:* (Salovaara and Simonson, 2004).

dominant at later stages of the fermentation, with *Lactococcus* spp., *Weissella* spp., and *Leuconostoc* spp. dominating in earlier stages of fermentation and acid-tolerant lactobacilli dominating at the end of the fermentation process (Gänzle and Zheng, 2019; Mudoor Sooresh et al., 2023; Pswarayi and Gänzle, 2019). Control of fermentation microbiota by continuous propagation or back-slopping eliminates dispersal limitation, improves fermentation control, and selects for lactic acid bacteria that are highly adapted to the specific fermentation process (Gänzle and Zheng, 2019; Li and Gänzle, 2020).

In addition to the fermentation conditions, the cereal substrate determines the assembly of the communities of microbes in food fermentations. Resting grains of maize and rice have a relatively low amylase activity; fermentation of these cereals often recruits lactobacilli with amylase activity (Guyot, 2012) or includes bacilli as part of the fermentation community (Jimenez et al., 2022; Z. Li et al., 2023). Sorghum and millet have a relatively high content of antimicrobial phenolic compounds, and microbial resistance to secondary plant metabolites impacts community assembly (Pswarayi et al., 2022; Sekwati-Monang et al., 2012). In contrast, wheat and rye have a much lower content of antimicrobial plant compounds but a high amylase activity. The abundant availability of maltose released by β-amylase in wheat and rye sourdoughs support growth of *Fructilactobacillus sanfranciscensis*, which is not known to occur in fermentation of any other cereals (Van Kerrebroeck et al., 2017; Vogel et al., 2011).

## 11.2  LACTIC ACID FERMENTATION IN WET MILLING

Soaking of grains in water before wet milling is customary when corn, sorghum, or millet is ground in traditional food processing. Soaking softens the grain endosperm and greatly reduces the work input required for grinding. Penetration of water into the interior of the kernels takes hours, and simultaneous fermentation occurs. The fermenting microorganisms originate from the surface of the kernels and from the steeping vessel and other equipment (Nout, 2009; Todorov and Holzapfel, 2015; Wacher et al., 1993). The resulting wet starchy material continues to undergo fermentation and carries the sour flavor, which is typical of foods cooked from fermented slurries (Diaz et al., 2019; Vieira-Dalodé et al., 2007).

Lactic acid fermentation is utilized in the preparation of many tropical staple foods (for reviews, see Nout, 2009; Pswarayi and Gänzle, 2022; Todorov and Holzapfel, 2015). In these processes, the lactic acid bacteria not only enhance flavor and texture but also reduce the content of antinutritive compounds (Gänzle, 2020) and inhibit pathogenic and spoilage organisms, mainly by production of organic acids with concomitant reduction of the pH and oxidation/reduction potential (Dinardo et al., 2019; Pswarayi and Gänzle, 2019). Lactic acid bacteria starter cultures have also been studied for controlling undesirable microbial growth in malting (Laitila et al., 2006).

When oats were prepared for food, one technique formerly used was the separation of hulls by slurrying stone-ground oats with water. The procedure enabled the hulls to be strained from the surface, whereas endosperm particles sedimented. The slurry underwent simultaneous sourdough-type fermentation, and this was favored by adding rye sourdough. The sour starchy sediment was used to cook indigenous fermented porridges and gruels, such as Scottish sowens and Karelian kiesa (Fenton, 1974; Salovaara et al., 1990).

# 11.3 LACTIC ACID BACTERIA IN BREAD MAKING

## 11.3.1 Use and Functions of Sourdough

Sourdough is fermented by yeasts and lactic acid bacteria. Historically, dough leavening was the principal technological function of sourdough in baking. The use of sourdough as a leavening agent was replaced only after *Saccharomyces cerevisiae* was industrially produced as a leavening agent in the late 19th century; relatively pure preparations of baker's yeast grown on molasses became commercially available in the middle of the 20th century. The availability of standardized and metabolically active baker's yeast and the substantial reduction of the time required for proofing favored the use of baker's yeast as the sole fermentation organism in wheat baking. Throughout the 20th century, the use of sourdough as a leavening agent in wheat products was limited to specialties, such as Panettone (Italy), specialty baguettes (France), or San Francisco sourdough bread (United States). In these products, sourdough is indispensable to attain the typical appearance and flavor.

The gas-holding capacity of wheat dough is dependent on the properties of polymeric gluten proteins, which are absent in other cereals. The gas-holding capacity of rye dough depends on arabinoxylans, which require solubilization during fermentation for optimum technological functionality. This is one of the reasons the use of sourdough has continued in rye baking. Sourdough is increasingly used also as a baking improver in wheat and rye baking to enhance flavor, texture, and shelf life and to replace additives (Brandt, 2019). Today a substantial proportion of sourdough is not fermented in bakeries but supplied to the baking industry in dried or otherwise stabilized preparations. Approximately 80% of bread produced in Europe and an increasing proportion of bread produced in North America involves the use of sourdough or sourdough products.

Sourdough affects all aspects of bread quality, including bread texture, flavor, and shelf life (Gänzle, 2014). The benefits of lactic fermentation are apparent in all breads but particularly pronounced in rye baking. Excessive sourness in white wheat bread is considered an off flavor by most consumers, whereas in rye bread, sourness is a desirable flavor attribute and is favored over non-acidified rye bread in northern, central, and eastern Europe. Dough for the various flat breads that are made in parts of Asia and Africa as well as for production of steamed bread in East Asia is fermented in a protocol resembling that used for sourdough bread in Europe (Gänzle, 2022). Sourdough is also used for production of injera and kisra flat breads made from sorghum and other local cereals in Ethiopia and Sudan (Nout, 2009).

The functions of sourdoughs in bread making are listed in Table 11.2. Dough leavening by formation of carbon dioxide is achieved by the combined activity of yeasts and heterofermentative lactic acid bacteria (Häggman and Salovaara, 2008a, 2008b; Hammes and Gänzle, 1998). There is also some consumer interest in traditional bread making without added baker's yeast; during the COVID-19 pandemic, the interest in home baking by amateur bakers also increased (Landis et al., 2021).

The technological benefits of sourdough procedures in traditional rye bread making also include the suppression of high endogenous activity of α-amylase and the conversion of insoluble pentosans to soluble polymers to improve the water-holding capacity and gas retention (Gänzle, 2014; Hammes and Gänzle,

**TABLE 11.2**   Functions of Sourdough in Bread Making

Leavening action by yeast and heterofermentative lactic acid bacteria:
Solubilization of arabinoxylans for improved baking properties of doughs, particularly rye doughs
Control of excessive enzymatic activity of rye flour, especially α-amylase
Control and inhibition of microbial spoilage through formation of organic acids and low pH and formation of antifungal compounds
Formation of exopolysaccharides to improve texture and to prevent staling
Formation of taste compounds, including glutamate and γ-glutamyl-peptides
Accumulation of flavor components such as acetic acid and other fermentation products
Accumulation of flavor precursor compounds such as amino acids and reducing carbohydrates
Increase of mineral bioavailability through degradation of phytate
Characterization of the product by a natural image, greater versatility, local and regional products
Modification of starch structure, leading to lower glycemic index values of wheat bread

*Source:* (Gänzle et al., 2023).

1998). Solubilization of arabinoxylans during fermentation also occurs in wheat sourdoughs (Korakli et al., 2001), but in wheat doughs, arabinoxylans are only of minor importance for dough hydration and gas retention in comparison with gluten. Exopolysaccharides produced by lactic acid bacteria during baking also act as water-binding hydrocolloids and improve bread volume and texture (Galle et al., 2012; Galle and Arendt, 2014; Katina et al., 2009). Modification and partial hydrolysis of proteins in wheat and rye sourdoughs occur primarily owing to endogenous enzymes present in the flour (Gänzle et al., 2008; Loponen et al., 2004). Wheat and rye proteinases exhibit optimum activity at the low pH prevailing during sourdough fermentation. Acidification by lactic fermentation also contributes to the technological properties of doughs made from other cereals, which is increasingly exploited in the emerging market for gluten-free bread (Moroni et al., 2009).

Lactic and acetic acids mediate the characteristic taste of sourdough bread (Müller et al., 2021). The conversion of glutamine released from cereal proteins to glutamate also contributes to the taste of sourdough bread (Vermeulen et al., 2007; Zhao et al., 2015). Sourdough fermentation also has a profound influence on the formation of flavor volatiles during dough fermentation and baking. Crumb odor is primarily determined by products of enzymatic and microbial conversions during dough fermentation. Lipid oxidation by cereal enzymes and Ehrlich degradation of amino acids, as well as formation of acetyl esters by yeasts and metabolites from lactic acid bacteria, are major contributors to crumb flavor (Hansen and Schieberle, 2005). Crust odor is dominated by products of thermal reactions during baking. The proteolytic release of amino acid during sourdough fermentation and particularly the conversion of arginine to ornithine by lactobacilli strongly contribute to flavor generation during baking (Hansen and Schieberle, 2005; Thiele et al., 2002).

In addition to the technological benefits and flavor, sourdough bread is characterized by better resistance to microbiological spoilage by molds and rope-forming bacilli (Axel et al., 2017; Pepe et al., 2003; Rosenquist and Hansen, 1998). Molds and rope-forming bacilli are the primary microbial organisms of bread. The major antimicrobial compound in sourdough is acetic acid. Lactic acid reduces pH and increases the proportion of undissociated acetic acid, which is more fungistatic than lactate (Gänzle et al., 1998; Rosenquist and Hansen, 1998). Acidification and acetate levels in sourdough bread suffice to inhibit growth of rope-forming bacilli but do not fully prevent mold growth (Axel et al., 2016; Rosenquist and Hansen, 1998). Several sourdough isolates of *Limosilactobacillus reuteri* produce reutericyclin, a heat-stable tetramic acid derivative which contributes to inhibition of rope-forming bacilli in bread (Gänzle et al., 2000; Z. Li et al., 2020; Lin et al., 2015). Although specific starter cultures were shown to effectively delay fungal spoilage of bread, the contribution of most antifungal metabolites to the prevention of fungal spoilage remains unclear (Ryan et al., 2009b, 2009a). Propionate produced by a mixed culture of *Lentilactobacillus buchneri* and *Lentilactobacillus diolivorans* delayed fungal growth on rye bread

(Zhang et al., 2010). Sourdough fermented with *Levilactobacillus hammesii* delayed mold growth based on the production of antifungal long-chain hydroxy-fatty acids (Black et al., 2013). Phenyllactate has been described as antifungal compound produced by *Lactiplantibacillus plantarum*; however, phenyllactate concentrations in wheat or rye bread remain far below the concentrations required to achieve an antifungal effect (Lavermicocca et al., 2000; Ryan et al., 2009b). Overall, sourdough is effective in preventing ropy spoilage of bread, but inhibition of fungal spoilage is achieved only if sourdough is used with other preservatives or antifungal ingredients.

Sourdough fermentation has been suggested to improve multiple nutritional properties of bread, including improved digestibility of bread, reduced digestibility of starch, improved tolerance for individuals with non-celiac wheat sensitivity or irritable bowel syndrome (IBS), and reduced content of antinutritive compounds (Arora et al., 2021; Poutanen et al., 2009). A review of clinical studies, however, indicated that only few of these claims are supported by randomized clinical trials (Ribet et al., 2023).

Minerals in wheat and rye flour occur as salts of phytate, which have only low bioavailability, as human digestive enzymes do not degrade phytate. Cereal enzymes degrade phytate during sourdough and thus increase the bioavailability of minerals. The reduced phytate content of sourdough bread when compared to straight dough bread relates to extended fermentation time to allow activity of endogenous phytases and to the solubilization of phytate salts at low pH and exceeds phytate degradation in straight dough processes owing to increased phytate solubility at low pH (Hammes and Gänzle, 1998; Poutanen et al., 2009; Tangkongchitr et al., 1982).

Some consumers exhibit react adversely to fermentable oligosaccharides, disaccharides, monosaccharides, and polyols (FODMAPs). Adverse reactions to FODMAPs strongly overlap with IBS and non-celiac wheat intolerance. Comparable to the beneficial effect of prebiotic compounds, adverse health effects of FODMAPs relate to intestinal fermentation. In sensitive individuals, however, rapid intestinal fermentation of FODMAPs and concomitant gas formation cause intestinal discomfort, bloating, and osmotic diarrhea (Schumann et al., 2018; Yan et al., 2018). Sourdough fermentation decreases FODMAP levels through activity of yeast invertase or specific extracellular fructanases produced by yeasts or lactobacilli (Q. Li et al., 2020a; Loponen and Gänzle, 2018; Struyf et al., 2018). Irrespective of the production of specific fructanases by sourdough microbes, yeast invertase activity in combination with the extended fermentation times that are common for sourdough baking processes suffices to achieve a substantial reduction of FODMAPs in wheat and rye bread (Boakye et al., 2022; Menezes et al., 2019; Shewry et al., 2022). Clinical studies indicated that low-FODMAP sourdough bread reduced colonic fermentation of fiber and FODMAPs in IBS patients but did not fully alleviate symptoms of IBS (Laatikainen et al., 2017; Pirkola et al., 2018).

The selection of specific starter cultures and appropriate fermentation condition was also used for enrichment of bioactives, including γ-aminobutyric acid, anti-hypertensive peptides, and kokumio-active γ-glutamyl peptides (Coda et al., 2010; Hu et al., 2011; Stromeck et al., 2011; Tang et al., 2017); however, only few studies quantified these bioactives in sourdough bread, and potential health benefits have not been documented in human trials.

Irrespective of health benefits of metabolites of sourdough microbes, the use of sourdough in baking strongly improves the technological functionality of high-fiber ingredients, such as whole wheat or whole rye flours and bran, largely by improving hydration and solubility of arabinoxylans and other non-starch polysaccharides (Katina et al., 2012, 2006). Sourdough fermentation thus greatly facilitates the use of whole-grain flours as a health-promoting part of the diet (Armet et al., 2022).

## 11.3.2 Sourdough Fermentation Process

Sourdoughs are maintained by continuous propagation or back-slopping, using the previous batch as an inoculum. These sourdoughs are microbiologically and functionally very stable, demonstrating that the processes select for microorganisms that are specifically adapted to the substrate and the fermentation conditions (Böcker et al., 1995; De Vuyst et al., 2014; Gänzle and Zheng, 2019). The long-term stability of the microbiota of specific sourdoughs has been demonstrated at the species level and even at the strain level (Böcker et al., 1995; Gänzle and Vogel, 2003; Spicher and Stephan, 1993). The remarkable stability of sourdoughs that are maintained under carefully controlled conditions also explains why commercial starter cultures consisting of one or more well-defined species or strains of lactic acid bacteria, available as freeze-dried powders or tablets, have not found a substantial market in the baking industry. However, starter sourdoughs containing a stable mixed microbiota are commercially available. The most successful of these starters is probably the Böcker "Reinzucht" sourdough, which contains *Fructilactobacillus sanfranciscensis* (previously *Lactobacillus brevis* var. *lindneri* or *L. sanfranciscensis*) as the dominant species (Böcker et al., 1995; Spicher and Stephan, 1993; Vogel et al., 2011).

Studies on lactic acid bacteria in sourdoughs initially focused on rye sourdoughs produced in central, northern, and eastern Europe (Spicher and Stephan, 1993). The increased use of sourdoughs for use in baking in the past decades resulted in characterization of numerous sourdoughs used for baking or for production of steamed bread globally; currently information on the microbial characterization of more than 1000 sourdoughs used in Europe, North America, or Asia is available (Arora et al., 2021; Gänzle and Zheng, 2019; Landis et al., 2021; Van Kerrebroeck et al., 2017; De Vuyst et al., 2023). Following the growing market for gluten-free cereal foods, reviews have also focused on the microbiological and technological aspects of gluten-free sourdoughs (Moroni et al., 2009). The microbiology of numerous cereal fermentations for production of porridges or non-alcoholic beverages in South Asia, Africa, and South America has also been elucidated (Jimenez et al., 2022; Nout, 2009; Pswarayi and Gänzle, 2022; Tamang et al., 2020; Todorov and Holzapfel, 2015).

Industrial sourdough fermentation processes vary substantially depending on the scale of the operation, degree of automation, and technological aim of the fermentation. Most sourdough processes are batch processes, although continuous propagation systems for large-scale industrial rye bread production have also been developed and are operating in Europe (Meuser, 1995). Sponge dough fermentations are widely used in bread baking and in the production of soda crackers (Gänzle and Zheng, 2019). Sponge doughs are inoculated with baker's yeast, but prolonged fermentation times (>8 h) result in the establishment of a population of lactic acid bacteria to cell counts exceeding $10^8$ colony-forming units (CFU)/g. Traditional sourdoughs used as the sole leavening agent are characterized by frequent back-slopping steps at ambient temperatures (Gänzle and Zheng, 2019; Häggman and Salovaara, 2008a; Vogel et al., 1999). These conditions select for rapidly growing microorganisms. Heterofermentative lactobacilli, particularly *Fl. sanfranciscensis*, typically occur in cell counts of $10^9$ CFU/g, whereas sourdough yeasts occur in cell counts ranging from $10^6$ to $10^8$ CFU/g. Industrial sourdoughs for production of baking improvers are typically fermented for elongated fermentation periods to achieve high levels of acidity (Brandt, 2007). Industrial sourdoughs are dominated by thermophilic and acid-tolerant heterofermentative lactobacilli. In sourdoughs fermented at an elevated temperature (>35 °C), yeasts are essentially absent (Meroth et al., 2003; Vogel et al., 1999). The increased use of sourdough both in industrial applications and for amateur home baking in the last 10 years also resulted in an increased diversity of fermentation processes and, concomitantly, an increased diversity of fermentation microbes that were identified in sourdough (Gänzle and Zheng, 2019; Landis et al., 2021).

Gluten-free sourdoughs are produced by technology and fermentation conditions matching those of wheat and rye sourdoughs but are based on gluten-free cereals maize, rice, sorghum, millet, or teff (Moroni et al., 2009). The use of these cereals in gluten-free sourdough fermentations selects for microbiota that differ from traditional wheat and rye sourdoughs (Meroth et al., 2004; Moroni et al., 2009; Sekwati-Monang et al., 2012). Specifically, *Fl. sanfranciscensis*, a key organism in wheat and rye sourdoughs, has not been identified in sourdoughs prepared with other cereal substrates.

## 11.3.3 Microbial Ecology of Sourdoughs

Whole cereal grains and whole-grain wheat or rye flour contains $10^3$–$10^6$ CFU/g of bacteria and $10^2$–$10^3$ CFU/g of lactic acid bacteria. Spontaneous wheat and rye doughs are dispersal limited (Gänzle and

**TABLE 11.3**  Typical Lactic Acid Bacteria in Sourdoughs

| FERMENTATION TECHNOLOGY | INOCULUM AND FERMENTATION CONDITIONS | KEY ORGANISMS (OTHER FREQUENTLY DETECTED LACTIC ACID BACTERIA) |
|---|---|---|
| Sponge dough fermentation | Baker's yeast, single-batch fermentation for 8–24 hours at ambient temperature | *Lactiplantibacillus plantarum; Levilactobacillus brevis; Latilactobacillus sakei; Pediococcus pentosaceus* |
| Sourdoughs used as sole leavening agent | Five percent to 30% inoculum from previous batch of sourdough and frequent refreshments (two to four times per day) at ambient temperature | *Fructilactobacillus sanfranciscensis (Companilactobacillus (par) alimentarius; Lp. plantarum; Lv. brevis)* |
| Industrial fermentations for use as baking improver | Back-slopping and extended fermentation time, often coupled to high fermentation temperatures | *Limosilactobacillus panis; Lactobacillus amylovorus; Limosilactobacillus pontis; Limosilactobacillus reuteri; Limosilactobacillus fermentum* |

Ripari, 2016); these fermentations are initiated by plant-associated bacilli and *Enterobacteriaceae*, followed by lactic acid bacteria, particularly pediococci, enterococci, and lactococci (Minervini et al., 2015; Van der Meulen et al., 2007). Back-slopping of the sourdough with inoculum from a previous batch results in dominance of lactic acid bacteria. Lactobacilli typically account for more than 99.9% of the bacterial microbiota after 5–10 refreshments (Van der Meulen et al., 2007).

Differences in the fermentation processes and the technological aim that were outlined previously are reflected by the large diversity of lactic acid bacteria and yeasts isolated from sourdoughs. More than 100 species of *Lactobacillaceae* as well as lactococci and *Enterococcus* spp. were isolated in high cell counts from sourdoughs (Arora et al., 2021; Gänzle and Zheng, 2019; Landis et al., 2021; Van Kerrebroeck et al., 2017; De Vuyst et al., 2023). Despite this taxonomic and metabolic diversity of sourdough microbiota, several key organisms were identified that occur frequently in sourdoughs that are propagated with uniform fermentation parameters (Table 11.3).

Sponge doughs are particularly used in wheat baking to improve bread flavor (Hansen and Schieberle, 2005). The use of sponge doughs in the production of soda crackers also aims to degrade the gluten network. Yeast metabolism creates anaerobic conditions and acidifies the dough to a pH of 5.5, which favors growth of lactic acid bacteria over other bacteria that occur in spontaneous sourdoughs without yeast. Commercial baker's yeast preparations are also a source of contamination with lactic acid bacteria. After 24 hours of fermentation, cell counts of lactic acid bacteria typically reach $10^9$ CFU/g, and the dough is acidified to pH 4.0–4.5. Lactobacilli including *Lactiplantibacillus plantarum, Latilactobacillus sakei*, and pediococci were frequently isolated from sponge doughs (Table 11.3).

Sourdoughs used as sole leavening agents typically contain one to three major strains of lactobacilli and a yeast, *Kazachstania humilis* or *Saccharomyces exiguus*. A majority of sourdoughs used as sole leavening agent contain *Fl. sanfranciscensis*. This heterofermentative species was first isolated and described as *L. brevis* spp. *lindneri* from German rye sourdoughs and as *Lactobacillus sanfrancisco* from the San Francisco French bread process (Kline and Sugihara, 1971; Spicher and Stephan, 1993). This species was later identified as the key organism in sourdoughs throughout Europe, North America, and China (Gänzle and Zheng, 2019). *Lp. plantarum* or *Companilactobacillus paralimentarius* are often associated with *Fl. sanfranciscensis* in sourdoughs. *Furfurilactobacillus rossiae, Furfurilactobacillus milii, Levilactobacillus brevis*, and related heterofermentative lactobacilli also occur in these sourdoughs.

The dominance of *Fl. sanfranciscensis* in sourdoughs used as leavening agent is attributable to its rapid growth in the pH range of 4.0–6.0 at ambient temperatures (Gänzle et al., 1998). *Fl. sanfranciscensis* is replaced by thermophilic and acid-tolerant lactobacilli in sourdoughs with extended fermentation times or at increased fermentation temperatures (Meroth et al., 2003). *Fl. sanfranciscensis* utilizes maltose through maltose phosphorylase; some strains do not ferment any other carbohydrate, including glucose (Gänzle, 2015; Neubauer et al., 1994; Stolz et al., 1993; Zheng et al., 2015). *Fl. sanfranciscensis* also makes efficient use of electron acceptors that are available in wheat and rye doughs to increase the energy yield in the phosphoketolase pathway. The availability of hydrogen acceptors leads to a shift in heterofermentative metabolism, and acetate rather than ethanol is produced from acetyl phosphate (Gänzle, 2015). This shift doubles the yield of ATP in heterofermentative metabolism but requires an external hydrogen acceptor to regenerate NADH (Gänzle, 2015). Oxygen, which is reduced to $H_2O$ by NADH peroxidase, and fructose, which is reduced to mannitol by mannitol dehydrogenase, are major hydrogen acceptors in wheat and rye sourdoughs (Stolz et al., 1995). Fructose reduction to mannitol accounts for about 50% and 80% of the acetate produced by *L. sanfranciscensis* in wheat and rye sourdoughs, respectively (Gänzle et al., 2007). Other hydrogen acceptors relevant in sourdough include glutathione, $\alpha$-ketoacids, and aldehydes originating from lipid oxidation (Gänzle, 2015). Understanding the biochemistry of sugar fermentation and acetic acid production in sourdough by heterofermentative lactobacilli has practical uses for improvement of mold-free time, rope prevention, and flavor (see also Table 11.2). In baking applications, the addition of fructose or sucrose has a linear effect on the acetate content of sourdough (Röcken et al., 1992).

The yeast present in a sourdough is acid tolerant and typically forms a stable association with lactobacilli. *Kazachstania humilis* (syn. *Candida humilis* or *C. milleri*) and *S. exiguus* (syn. *S. minor*, *Torulopsis holmii*, *C. holmii*) are the most typical yeast species in sourdoughs. *S. cerevisiae* is also detected. *K. humilis* and *Fl. sanfranciscensis* are symbionts. *K. humilis* is more tolerant to acetic acid than other yeasts; acids produced by lactic acid bacteria thus suppress the growth of competing microbes (Gänzle et al., 1998). Yeast invertase hydrolyses sucrose as well as wheat fructans; hydrolysis of these oligosaccharides supplies fructose for conversion to mannitol by heterofermentative lactobacilli (Gänzle, 2014; Shewry et al., 2022). The stable association of yeasts and lactobacilli in sourdoughs used as leavening agent is also attributable to the lack of competition for nutrients. *Fl. sanfranciscensis* preferentially utilizes maltose, while *K. humilis* is incapable of assimilating maltose, thus eliminating competition for the carbon source (Vogel et al., 1999). Resource partitioning is also observed between homofermentative and heterofermentative lactobacilli, as the former preferentially metabolize glucose, while the latter preferentially metabolize maltose or sucrose (Gänzle, 2015; Tannock et al., 2012; Teixeira et al., 2013). *Fl. sanfranciscensis* preferentially metabolizes peptides, whereas amino acids are the preferred nitrogen source for yeasts (Thiele et al., 2002; Vermeulen et al., 2005). Sourdough yeasts and *Fl. sanfranciscensis* have matching growth requirements with respect to temperature, pH, and ionic strength (Gänzle et al., 1998). Bakers adjust the ratio of yeasts to lactobacilli in sourdough and thus the leavening power and the level of acidity by control of temperature, fermentation time, and dough yield (Gänzle et al., 1998).

The microbiota of fermented industrially at elevated temperatures to achieve high levels of acidity is not as uniform as the microbiota of those sourdoughs that used as leavening agent. Sourdoughs that are fermented industrially for use as acidulants are propagated with a differ substantially with respect to the process conditions. The main process parameters differentiating these sourdoughs used for acidification or leavening are fermentation temperature and the inoculum level or the fermentation time (Gänzle and Zheng, 2019; Meroth et al., 2003). These processes select for acid-tolerant lactobacilli (Table 11.3). Many fermentations are characterized by extended fermentation times of several days to achieve high levels of acidity; these sourdoughs are characterized by dominance of *Limosilactobacillus* spp. and *Lactobacillus* spp. The metabolism of *Limosilactobacillus* spp. is comparable to *Fl. sanfranciscensis* with regard to the use of maltose and hydrogen acceptors (Gänzle, 2015). *Limosilactobacillus reuteri* harbors sucrose phosphorylase for efficient sucrose metabolism (Teixeira et al., 2013). In addition to the growth rate, high levels of acidity select for acid resistant fermentation organisms (Lin and Gänzle, 2014). *Lm. reuteri* uses several mechanisms for acid resistance, which are generally absent in *Fl. sanfranciscensis* (Teixeira et al., 2013; Zheng et al., 2015). The conversion of arginine to ornithine, of glutamine to glutamate and $\gamma$-aminobutyrate, as well as the formation of exopolysaccharides, all contribute to acid resistance of organisms in limosilactobacilli (Gänzle et al., 2007; Q. Li et al., 2020b; Teixeira et al., 2014; Zheng et al., 2015). The continuous supply of amino acids during prolonged fermentation of wheat and rye sourdoughs by cereal substrates particularly supports the effectiveness of acid resistance mechanisms that are based on conversion of amino acids (Gänzle et al., 2008; Thiele et al., 2002). In specific processes, a high fermentation temperature is used to achieve accelerated acidification (Böcker et al., 1995; Meuser, 1995; Vogel et al., 1999).

Gluten-free sourdoughs produced with flour from gluten-free cereals have become available in the past 20 years to cater to the market for gluten-free bread in Europe and North America. The technological function of these sourdoughs—improved texture, flavor, and shelf life of bread—is comparable to the use of sourdough in wheat and rye baking. However, the microbiota of gluten-free sourdoughs are comparable to cereal fermentations in tropical climates that employ the same substrates rather than wheat and rye sourdoughs produced with comparable fermentation conditions (Meroth et al., 2004; Moroni et al., 2009; Pswarayi and Gänzle, 2022). A different carbohydrate supply due to a different level of amylase activities in gluten-free substrates, as well as the presence of antimicrobial polyphenolic compounds in some cereals (e.g., sorghum), contributes to the selection of fermentation microbiota that are specific to the cereal substrate (Sekwati-Monang et al., 2012; Svensson et al., 2010).

## 11.4 LACTIC ACID–FERMENTED PORRIDGES AND NON-ALCOHOLIC BEVERAGES

From a global perspective, a porridges and non-alcoholic beverages are a considerable part of the cereal-based foods made by lactic acid fermentation. Some of these foods are staple foods in parts of Africa, Asia, and South America (Gänzle, 2022; Jimenez et al., 2022; Pswarayi and Gänzle, 2022; Tamang et al., 2020). Maize, rice, sorghum, and millet are used for the preparation of these cereal products; in many cases, malted grains are included that serve both as a source of enzymes and of fermentation microbes (Pswarayi and Gänzle, 2022, 2019). Fermented beverages and porridges are also known in Europe either as drinks, such as kvass and boza made from wheat or rye (Dlusskaya et al., 2008; Todorov and Dicks, 2006), or sowens, flummeries, and other similar oat-based gruel-type products (Fenton, 1974; Salovaara et al., 1990).

Corresponding to the cultural diversity in Africa, a large diversity of fermented cereal foods is produced. Many fermented foods are produced at the household level, which adds to the diversity of cereal fermented products. Information on the composition of fermentation microbes is available for many of the products, including ogi and agidi (Nigeria) (Odunfa and Adeyele, 1985); koko, akassa, and kenkey (Ghana) (Halm et al., 2004; Lei and Jakobsen, 2004); uji, togwa (Mugula et al., 2003); mawè (Houngbédji et al., 2018); and ting and mahewu (southern Africa) (Pswarayi and Gänzle, 2019; Sekwati-Monang and Gänzle, 2011); for review see (Nout, 2009; Pswarayi and Gänzle, 2022; Todorov and Holzapfel, 2015).

In contrast to sourdough fermentations for production of bread or steamed bread, where the communities of fermentation microbes are invariably controlled by back-slopping, the preparation of fermented cereal foods in Africa is generally not controlled by back-slopping (Pswarayi and Gänzle, 2022; Sekwati-Monang and Gänzle, 2011). In fermentation processes that soak grains for 1–3 days prior to milling, lactic acid fermentation takes place during the soaking stage. Soaking softens the grains and makes them easier to crush or wet mill into a slurry, from which hulls, bran particles, and germs can be removed by screening and sieving procedures (Oduro-Yeboah et al., 2016).

Slurrying in water of the material from either wet or dry milling supports fermentation, which is allowed to take place overnight or longer, usually at ambient temperatures. Spontaneous cereal fermentations are initiated by plant-associated microbes and have a characteristic succession of fermentation microbes. *Enterobacteriaceae*, the most abundant facultative anaerobes associated with malt or cereal grains, initiate the fermentation, followed by lactococci, *Leuconostoc* species, and *Weissella* species (Pswarayi and Gänzle, 2019). Increasing levels of acidity consecutively displace *Enterobacteriaceae*, *Leuconostoc* spp., and *Weissella* spp., and more acid-tolerant lactobacilli, predominantly *Limosilactobacillus fermentum* and *Lp. plantarum*, prevail at the end of the fermentation (Dinardo et al., 2019; Pswarayi and Gänzle, 2019).

Source tracking of fermentation microbes with strain-specific quantitative PCR revealed that strains of *Lp. plantarum* were only a very minor component of the microbiota of the raw materials but became dominant after 2 d of fermentation (Pswarayi and Gänzle, 2019). Fermented slurries or porridges are consumed as beverages without further processing or boiled with an appropriate amount of water so that gelatinization of starch occurs and a product of desired consistency is obtained. The final product may be spoonable or stiff and dumpling-like (Nout, 2009; Pswarayi and Gänzle, 2022; Todorov and Holzapfel, 2015).

## 11.5 LACTIC ACID–FERMENTED VINEGARS

In East Asian countries, including China and Japan, cereal vinegars are produced with microbial saccharification cultures, and the fermentation includes lactic acid bacteria as dominant fermentation microbes. In Japan, vinegar fermentation is performed in a single clay vessel that is inoculated with *koji*. The fungal communities are characterized by initial growth of *A. oryzae*, which produces proteolytic and amylolytic enzymes, followed by growth and ethanol formation by *S. cerevisiae*. Bacterial communities are initially composed of *Weissella* spp., *Lactococcus* spp., and *Pediococcus* spp., followed by growth of *Lm. fermentum* and *Lactobacillus acetotolerans* in association with acetic acid bacteria during later stages of the 9-month fermentation (Haruta et al., 2006). In China, cereal vinegars are produced in a two-stage fermentation process. The saccharification culture, *daqu*, is produced by spontaneous fermentation and includes amylolytic and proteolytic fungi and bacilli, bacilli, and *Enterobacteriaceae* but also lactic acid bacteria such as *Weissella* spp., *Lp. plantarum*, and lactococci (Wang et al., 2018; Zheng and Han, 2016). The subsequent mash fermentation, termed *pei*, is started with *daqu* and steamed cereals. The fermentation is initially dominated by lactobacilli, including *Limosilactobacillus* spp., *Lacticaseibacillus* spp., and *Lactobacillus* spp.; acetic acid bacteria increase in abundance in later stages of fermentation (Lu et al., 2016; Wang et al., 2015; Wu et al., 2012).

## 11.6 CEREAL-BASED FOODS CONTAINING LIVE LACTIC ACID BACTERIA

Multiple fermentation processes for the production of beverages or porridges involve fermentation after the gelatinizing heat treatment or fermentation in conjunction with enzymatic starch degradation to obtain products with live lactic acid bacteria. One example of this type of traditional is tarhana, a mixture of soured milk and wheat widely consumed in Turkey and elsewhere. The process involves mixing yogurt with wheat flour derived from boiled, dried, and ground wheat grains. The resulting dough is formed into balls and sun-dried to make the tarhana (Erbaş et al., 2005; Hesseltine, 1979). Other products of same type include mahewu, obushera, munkoyo, kishk, boza, and kvass (Gänzle, 2022; Hesseltine, 1979; Pswarayi and Gänzle, 2022), but comparable numerous comparable products are produced in other parts of Europe, particularly Baltic countries, Africa, and South America.

Live fermentation microbes in spontaneously fermented cereal products typically include high cell counts of *Lm. fermentum* and *Lp. plantarum*, that is, species which include strains with documented probiotic properties. Live dietary microbes from fermented foods are readily detected in fecal samples (Dal Bello et al., 2003; Pasolli et al., 2020) and may constitute up to 10% of intestinal bacterial communities in the terminal ileum (Zaccaria et al., 2023). Human trials documented that a high intake of fermented foods increased the diversity of the intestinal microbiome and decreased inflammation markers (Wastyk et al., 2021). Health benefits of fermented foods were also proposed for fermented cereal products with live fermentation microbes (Marco et al., 2017; Todorov and Holzapfel, 2015). An example of such a cereal food is togwa, a lactic acid–fermented sorghum or maize gruel used in Tanzania as a weaning food or as a beverage which is not heat treated after fermentation (Mugula et al., 2003). The consumption of togwa reduced abundance of enteric pathogens in rectal swabs of children under 5 years old (Kingamkono et al., 1999). The reduction of enteric pathogens has been documented under more controlled conditions in piglets, where wheat fermented with *Lm. reuteri* reduced the intestinal abundance of enterotoxigenic *E. coli* (Chen et al., 2014; Wang et al., 2020). Enteric pathogens, including enterotoxigenic *E. coli* and *Shigella* species, remain major contributors to childhood mortality, particularly in developing countries with low food security (WHO, 2017). Traditional cereal foods with life dietary microbes may thus be a low-cost and widely accepted tool to reduce severity and incidence of childhood diarrhea.

# 11.7 CEREAL-BASED FOODS CONTAINING LIVE PROBIOTIC STRAINS

Like fermented milk-based or vegetable-based foods, lactic-fermented cereal foods also can serve as a vehicle for strains with probiotic properties. Vogel et al. highlighted that the predominant strains present in sourdoughs and in other lactic-fermented cereal foods are closely related to or even identical to species found in the animal and human intestinal tract (1999). This is particularly well illustrated using the example of *Lm. reuteri*, a species that has evolved to colonize the intestinal tract of humans and animals (Frese et al., 2011; F. Li et al., 2023). Sourdough isolates of this species also are of intestinal origin (Su et al., 2012). A comprehensive analysis of the ecology of the genus *Lactobacillus* revealed that several genera of the *Lactobacillaceae* are adapted to vertebrate hosts or insects, while other lactobacilli are adapted to environmental and plant associated habitats or lead a nomadic life that includes transient persistence in environmental and host-associated habitats (Duar et al., 2017; Zheng et al., 2020). Microbiota in back-slopped cereal fermentations predominantly consist of host-adapted microbiota. Examples include *Limosilactobacillus* and *Lactobacillus* species, which are adapted to vertebrate hosts (Duar et al., 2017; Li and Gänzle, 2020), and *Fl. sanfranciscensis*, an insect-associated organism (Boiocchi et al., 2017; Duar et al., 2017). In contrast, spontaneous cereal fermentations are populated predominantly by environmental, plant-associated, and nomadic lactobacilli, including *Lp. plantarum*, *Lv. brevis*, and *Lm. fermentum*.

Despite the increasing recognition of health benefits of live fermentation microbes, probiotic activity of lactobacilli is widely considered a strain-specific trait (Hill et al., 2014; Marco et al., 2021). Corresponding to the intestinal origin of many sourdough lactobacilli, probiotic strains of lactobacilli are generally suitable as starter cultures for cereal foods (Mårtensson et al., 2002; Salovaara and Simonson, 2004; Yang et al., 2015). Many lactobacilli that are currently used as probiotic adjuncts in foods, including *Lm. reuteri*, *Lc. casei*, *Lc. johnsonii*, *Lc. rhamnosus*, *Lp. plantarum*, and *Lm. fermentum* (Hill et al., 2014), are stable members of traditional fermentation microbiota and are highly suitable as starter cultures for probiotic cereal foods and beverages (Dlusskaya et al., 2008; Gänzle and Zheng, 2019; Pswarayi and Gänzle, 2022). For example, one commercial application uses probiotic strains in fermentation of cooked oatmeal and oat bran for the production of yogurt-type alternatives to dairy- and soy-based yogurts (Salovaara and Simonson, 2004). Cereal product alone or in combination with other plant-based ingredients thus provides substantial opportunities to meet the increasing demand for plant-based alternatives for (probiotic) fermented dairy products (Mefleh et al., 2022; Tangyu et al., 2019; Gänzle et al., 2024).

# 11.8 CONCLUSIONS

Lactic acid bacteria are utilized in the production of cereal-based products in many ways. Lactic acid fermentation contributes beneficially both to processing technology and to quality of the end products in terms of nutritional value, flavor, keeping properties, safety, and overall image of the product. Among cereal foods, most scientific research and technological development with respect to lactic acid bacteria has been associated with the sourdough bread-making process.

Traditional fermented cereal foods other than bread (i.e., soured porridges and dumplings) have also received increasing scientific attention. These foods play a major role for millions of people, especially in South America, Africa, and Asia, and deserve research and development input to improve their quality and attractiveness as economical and nutritious staples. Increasing knowledge on the role of lactic acid bacteria in cereal fermentations in tropical climates can contribute to food security and food safety in developing countries and provide templates for development of novel plant-based fermented foods in other areas of the world.

# 11.9 ACKNOWLEDGMENTS

Hannu Salovaara is acknowledged for his contributions to this chapter and in the previous five editions of the book as author and co-author. Despite extensive revisions for the 6th edition, his contributions remain recognizable in the current version.

# BIBLIOGRAPHY

Armet, A.M., Deehan, E.C., O'Sullivan, A.F., Mota, J.F., Field, C.J., Prado, C.M., Lucey, A.J., Walter, J., 2022. Rethinking healthy eating in light of the gut microbiome. *Cell Host Microbe* 30, 764–785. https://doi.org/10.1016/J.CHOM.2022.04.016

Arora, K., Ameur, H., Polo, A., Di Cagno, R., Rizzello, C.G., Gobbetti, M., 2021. Thirty years of knowledge on sourdough fermentation: A systematic review. *Trends Food Sci. Technol.* 108, 71–83. https://doi.org/10.1016/J.TIFS.2020.12.008

Arranz-Otaegui, A., Carretero, L.G., Ramsey, M.N., Fuller, D.Q., Richter, T., 2018. Archaeobotanical evidence reveals the origins of bread 14,400 years ago in northeastern Jordan. *Proc. Natl. Acad. Sci. U. S. A.* 115, 7925–7930. https://doi.org/10.1073/pnas.1801071115

Axel, C., Brosnan, B., Zannini, E., Furey, A., Coffey, A., Arendt, E.K., 2016. Antifungal sourdough lactic acid bacteria as biopreservation tool in quinoa and rice bread. *Int. J. Food Microbiol.* 239, 86–94. https://doi.org/10.1016/j.ijfoodmicro.2016.05.006

Axel, C., Zannini, E., Arendt, E.K., 2017. Mold spoilage of bread and its biopreservation: A review of current strategies for bread shelf life extension. *Crit. Rev. Food Sci. Nutr.* 57, 3528–3542. https://doi.org/10.1080/10408398.2016.1147417

Black, B.A., Zannini, E., Curtis, J.M., Gänzle, M.G., 2013. Antifungal hydroxy fatty acids produced during sourdough fermentation: Microbial and enzymatic pathways, and antifungal activity in bread. *Appl. Environ. Microbiol.* 79, 1866–1873. https://doi.org/10.1128/AEM.03784-12

Boakye, P.G., Kougblenou, I., Murai, T., Okyere, A.Y., Anderson, J., Bajgain, P., Philipp, B., LaPlante, B., Schlecht, S., Vogel, C., Carlson, M., Occhino, L., Stanislawski, H., Ray, S.S., Annor, G.A., 2022. Impact of sourdough fermentation on FODMAPs and amylase-trypsin inhibitor levels in wheat dough. *J. Cereal Sci.* 108, 103574. https://doi.org/10.1016/J.JCS.2022.103574

Böcker, G., Stolz, P., Hammes, W.P., 1995. Neue Erkenntnisse zum Ökosystem Sauerteig und zur Physiologie der sauerteigtypischen Stämme *Lactobacillus sanfrancisco* und *Lactobacillus pontis*. *Getreide Mehl. Brot.* 49, 370–374.

Boiocchi, F., Porcellato, D., Limonta, L., Picozzi, C., Vigentini, I., Locatelli, D.P., Foschino, R., 2017. Insect frass in stored cereal products as a potential source of *Lactobacillus sanfranciscensis* for sourdough ecosystem. *J. Appl. Microbiol.* 123, 944–955. https://doi.org/10.1111/jam.13546

Brandt, M.J., 2007. Sourdough products for convenient use in baking. *Food Microbiol.* 24, 161–164. https://doi.org/10.1016/j.fm.2006.07.010

Brandt, M.J., 2019. Industrial production of sourdoughs for the baking branch—An overview. *Int. J. Food Microbiol.* 302, 3–7. https://doi.org/10.1016/j.ijfoodmicro.2018.09.008

Chen, X.Y., Woodward, A., Zijlstra, R.T., Gänzle, M.G., 2014. Exopolysaccharides synthesized by *Lactobacillus reuteri* protect against enterotoxigenic *Escherichia coli* in piglets. *Appl. Environ. Microbiol.* 80, 5752–5760. https://doi.org/10.1128/AEM.01782-14

Coda, R., Rizzello, C.G., Gobbetti, M., 2010. Use of sourdough fermentation and pseudo-cereals and leguminous flours for the making of a functional bread enriched of γ-aminobutyric acid (GABA). *Int. J. Food Microbiol.* 137, 236–245. https://doi.org/10.1016/J.IJFOODMICRO.2009.12.010

Dal Bello, F., Walter, J., Hammes, W.P., Hertel, C., 2003. Increased complexity of the species composition of lactic acid bacteria in human feces revealed by alternative incubation condition. *Microb. Ecol.* 45, 455–463. https://doi.org/10.1007/s00248-003-2001-z

De Vuyst, L., Gonzalez-Alonso, V., Raditya Wardhana, Y., Pradal, I., 2023. Taxonomy and species diversity of sourdough lactic acid bacteria. In: Gobbetti, M., G¨anzle, B. (Eds.), *Handbook of Sourdough Biotechnology.* Springer, pp. 97–160. https://doi.org/10.1007/978-3-031-23084-4_6.

De Vuyst, L., Van Kerrebroeck, S., Harth, H., Huys, G., Daniel, H.M., Weckx, S., 2014. Microbial ecology of sourdough fermentations: Diverse or uniform? *Food Microbiol.* 37, 11–29. https://doi.org/10.1016/J.FM.2013.06.002

Diaz, M., Kellingray, L., Akinyemi, N., Adefiranye, O.O., Olaonipekun, A.B., Bayili, G.R., Ibezim, J., Plessis, A.S. du, Houngbédji, M., Kamya, D., Mukisa, I., Mulaw, G., Josiah, S.M., Chienjo, W.O., Atter, A., Agbemafle, E., Annan, T., Ackah, N.B., Buys, E.M., Hounhouigan, D.J., Muyanja, C., Nakavuma, J., Odeny, D.A., Sawadogo-Lingani, H., Tefera, A.T., Amoa-Awua, W., Obodai, M., Mayer, M.J., Oguntoyinbo, F.A., Narbad, A., 2019. Comparison of the microbial composition of African fermented foods using amplicon sequencing. *Sci. Rep.* 91(9), 1–8. https://doi.org/10.1038/s41598-019-50190-4

Dinardo, F.R., Minervini, F., De Angelis, M., Gobbetti, M., Gänzle, M.G., 2019. Dynamics of *Enterobacteriaceae* and lactobacilli in model sourdoughs are driven by pH and concentrations of sucrose and ferulic acid. *LWT* 114, 108394. https://doi.org/10.1016/j.lwt.2019.108394

Dlusskaya, E., Jänsch, A., Schwab, C., Gänzle, M.G., 2008. Microbial and chemical analysis of a kvass fermentation. *Eur. Food Res. Technol.* 227, 261–266. https://doi.org/10.1007/s00217-007-0719-4

Duar, R.M., Lin, X.B., Zheng, J., Martino, M.E., Grenier, T., Pérez-Muñoz, M.E., Leulier, F., Gänzle, M., Walter, J., 2017. Lifestyles in transition: Evolution and natural history of the genus *Lactobacillus*. *FEMS Microbiol. Rev.* 41, S27–S48. https://doi.org/10.1093/femsre/fux030

Erbaş, M., Certel, M., Kemal Uslu, M., 2005. Microbiological and chemical properties of Tarhana during fermentation and storage as wet—sensorial properties of Tarhana soup. *LWT—Food Sci. Technol.* 38, 409–416. https://doi.org/10.1016/J.LWT.2004.06.009

Fenton, A., 1974. Sowens in Scotland. *J. Ethnol. Stud.* 12, 41–47. https://doi.org/10.1179/043087774798240938

Frese, S.A., Benson, A.K., Tannock, G.W., Loach, D.M., Kim, J., Zhang, M., Oh, P.L., Heng, N.C.K., Patil, P.B., Juge, N., MacKenzie, D.A., Pearson, B.M., Lapidus, A., Dalin, E., Tice, H., Goltsman, E., Land, M., Hauser, L., Ivanova, N., Kyrpides, N.C., Walter, J., 2011. The evolution of host specialization in the vertebrate gut symbiont *Lactobacillus reuteri*. *PLoS Genet.* 7, e1001314. https://doi.org/10.1371/journal.pgen.1001314

Galle, S., Arendt, E.K., 2014. Exopolysaccharides from sourdough lactic acid bacteria. *Crit. Rev. Food Sci. Nutr.* 54, 891–901. https://doi.org/10.1080/10408398.2011.617474

Galle, S., Schwab, C., Bello, F.D., Coffey, A., Gänzle, M., Arendt, E., 2012. Comparison of the impact of dextran and reuteran on the quality of wheat sourdough bread. *J. Cereal Sci.* 56, 531–537. https://doi.org/10.1016/j.jcs.2012.07.001

Gänzle, M.G., 2014. Enzymatic and bacterial conversions during sourdough fermentation. *Food Microbiol.* 37, 2–10. https://doi.org/10.1016/j.fm.2013.04.007

Gänzle, M.G., 2015. Lactic metabolism revisited: Metabolism of lactic acid bacteria in food fermentations and food spoilage. *Curr. Opin. Food Sci.* 2, 106–117. https://doi.org/10.1016/j.cofs.2015.03.001

Gänzle, M.G., 2020. Food fermentations for improved digestibility of plant foods—an essential ex situ digestion step in agricultural societies? *Curr. Opin. Food Sci.* 32, 124–132. https://doi.org/10.1016/j.cofs.2020.04.002

Gänzle, M.G., 2022. The periodic table of fermented foods: Limitations and opportunities. *Appl. Microbiol. Biotechnol.* 106, 2815–2826. https://doi.org/10.1007/S00253-022-11909-Y

Gänzle, M.G., Ehmann, M., Hammes, W.P., 1998. Modeling of growth of *Lactobacillus sanfranciscensis* and *Candida milleri* in response to process parameters of sourdough fermentation. *Appl. Environ. Microbiol.* 64, 2616–2623. https://doi.org/10.1128/aem.64.7.2616-2623.1998

Gänzle, M.G., Follador, R., 2012. Metabolism of oligosaccharides and starch in lactobacilli: A review. *Front. Microbiol.* 3, 340. https://doi.org/10.3389/fmicb.2012.00340

Gänzle, M.G., Höltzel, A., Walter, J., Jung, G., Hammes, W.P., 2000. Characterization of reutericyclin produced by *Lactobacillus reuteri* LTH2584. *Appl. Environ. Microbiol.* 66, 4325–4333. https://doi.org/10.1128/AEM.66.10.4325-4333.2000

Gänzle, M.G., Loponen, J., Gobbetti, M., 2008. Proteolysis in sourdough fermentations: Mechanisms and potential for improved bread quality. *Trends Food Sci. Technol.* 19, 513–521. https://doi.org/10.1016/j.tifs.2008.04.002

Gänzle, M.G., Monnin, L., Zheng, J., Zhang, L., Coton, M., Sicard, D., Walter, J. Starter culture development and innovation for novel fermented foods. *Ann. Rev. Food Sci. Technol.* 2024. https://doi.org/10.1146/annurev-food-072023-034207

Gänzle, M.G., Qiao, N., Bechtner, J. 2023. The quest for the perfect loaf of sourdough bread continues: Novel developments for selection of sourdough starter cultures. *Int. J. Food Microbiol.* 407, 110421.

Gänzle, M.G., Ripari, V., 2016. Composition and function of sourdough microbiota: From ecological theory to bread quality. *Int. J. Food Microbiol.* 239, 19–25. https://doi.org/10.1016/j.ijfoodmicro.2016.05.004

Gänzle, M.G., Vermeulen, N., Vogel, R.F., 2007. Carbohydrate, peptide and lipid metabolism of lactic acid bacteria in sourdough. *Food Microbiol.* 24, 128–138. https://doi.org/10.1016/j.fm.2006.07.006

Gänzle, M.G., Vogel, R.F., 2003. Contribution of reutericyclin production to the stable persistence of *Lactobacillus reuteri* in an industrial sourdough fermentation. *Int. J. Food Microbiol.* 80, 31–45. https://doi.org/10.1016/S0168-1605(02)00146-0

Gänzle, M.G., Zheng, J., 2019. Lifestyles of sourdough lactobacilli—do they matter for microbial ecology and bread quality? *Int. J. Food Microbiol.* 302, 15–23. https://doi.org/10.1016/j.ijfoodmicro.2018.08.019

Guyot, J.-P., 2012. Cereal-based fermented foods in developing countries: Ancient foods for modern research. *Int. J. Food Sci. Technol.* 47, 1109–1114. https://doi.org/10.1111/J.1365-2621.2012.02969.X

Häggman, M., Salovaara, H., 2008a. Microbial re-inoculation reveals differences in the leavening power of sourdough yeast strains. *LWT—Food Sci. Technol.* 41, 148–154. https://doi.org/10.1016/J.LWT.2007.02.001

Häggman, M., Salovaara, H., 2008b. Effect of fermentation rate on endogenous leavening of *Candida milleri* in sour rye dough. *Food Res. Int.* 41, 266–273. https://doi.org/10.1016/J.FOODRES.2007.12.002

Halm, M., Amoa-Awua, W.K., Jakobsen, M., 2004. Kenkey: An African fermented maize product. In: *Handbook of Food and Beverage Fermentation Technology*. Marcel Dekker, pp. 799–816.

Hammes, W.P., Gänzle, M.G., 1998. Sourdough breads and related products. *Microbiol. Fermented Foods* 199–216. https://doi.org/10.1007/978-1-4613-0309-1_8

Hansen, A., Schieberle, P., 2005. Generation of aroma compounds during sourdough fermentation: Applied and fundamental aspects. *Trends Food Sci. Technol.* 16, 85–94. https://doi.org/10.1016/J.TIFS.2004.03.007

Hardy, K., Brand-Miller, J., Brown, K.D., Thomas, M.G., Copeland, L., 2015. The importance of dietary carbohydrate in human evolution. *Q. Rev. Biol.* 90, 251–268. https://doi.org/10.1086/682587

Haruta, S., Ueno, S., Egawa, I., Hashiguchi, K., Fujii, A., Nagano, M., Ishii, M., Igarashi, Y., 2006. Succession of bacterial and fungal communities during a traditional pot fermentation of rice vinegar assessed by PCR-mediated denaturing gradient gel electrophoresis. *Int. J. Food Microbiol.* 109, 79–87. https://doi.org/10.1016/j.ijfoodmicro.2006.01.015

Hayden, B., Canuel, N., Shanse, J., 2013. What was brewing in the natufian? An archaeological assessment of brewing technology in the Epipaleolithic. *J. Archaeol. Method Theory* 20, 102–150. https://doi.org/10.1007/s10816-011-9127-y

Hesseltine, C.W., 1979. Some important fermented foods of Mid-Asia, the Middle East, and Africa. *J. Am. Oil Chem. Soc.* 56, 367–374. https://doi.org/10.1007/BF02671501

Hill, C., Guarner, F., Reid, G., Gibson, G.R., Merenstein, D.J., Pot, B., Morelli, L., Canani, R.B., Flint, H.J., Salminen, S., Calder, P.C., Sanders, M.E., Net, M.E.S.M., Hill, C., Guarner, F., Reid, G., Gibson, G.R., Merenstein, D.J., Pot, B., Morelli, L., Canani, R.B., Flint, H.J., Salminen, S., Calder, P.C., Sanders, M.E., 2014. The international scientific association for probiotics and prebiotics consensus statement on the scope and appropriate use of the term probiotic. *Nat. Rev. Gastroenterol. Hepatol.* 11, 506–514. https://doi.org/10.1038/nrgastro.2014.66

Houngbédji, M., Johansen, P., Padonou, S.W., Akissoé, N., Arneborg, N., Nielsen, D.S., Hounhouigan, D.J., Jespersen, L., 2018. Occurrence of lactic acid bacteria and yeasts at species and strain level during spontaneous fermentation of mawè, a cereal dough produced in West Africa. *Food Microbiol.* 76, 267–278. https://doi.org/10.1016/j.fm.2018.06.005

Hu, Y., Stromeck, A., Loponen, J., Lopes-Lutz, D., Schieber, A., Gänzle, M.G., 2011. LC-MS/MS quantification of bioactive angiotensin I-converting enzyme inhibitory peptides in rye malt sourdoughs. *J. Agric. Food Chem.* 59, 11983–11989. https://doi.org/10.1021/jf2033329

Jimenez, M.E., O'Donovan, C.M., Ullivarri, M.F. de, Cotter, P.D., 2022. Microorganisms present in artisanal fermented food from South America. *Front. Microbiol.* 13, 941866. https://doi.org/10.3389/FMICB.2022.941866

Katina, K., Juvonen, R., Laitila, A., Flander, L., Nordlund, E., Kariluoto, S., Piironen, V., Poutanen, K., 2012. Fermented wheat bran as a functional ingredient in baking. *Cereal Chem.* 89, 126–134. https://doi.org/10.1094/CCHEM-08-11-0106

Katina, K., Maina, N.H., Juvonen, R., Flander, L., Johansson, L., Virkki, L., Tenkanen, M., Laitila, A., 2009. *In situ* production and analysis of *Weissella confusa* dextran in wheat sourdough. *Food Microbiol.* 26, 734–743. https://doi.org/10.1016/J.FM.2009.07.008

Katina, K., Salmenkallio-Marttila, M., Partanen, R., Forssell, P., Autio, K., 2006. Effects of sourdough and enzymes on staling of high-fibre wheat bread. *LWT—Food Sci. Technol.* 39, 479–491. https://doi.org/10.1016/J. LWT.2005.03.013

Kingamkono, R., Sjögren, E., Svanberg, U., 1999. Enteropathogenic bacteria in faecal swabs of young children fed on lactic acid-fermented cereal gruels. *Epidemiol. Infect.* 122, 23–32. https://doi.org/10.1017/S0950268898001800

Kline, L., Sugihara, T.F., 1971. Isolation and characterization of undescribed bacterial species responsible for the souring activity. *Appl. Microbiol.* 21, 459–465.

Korakli, M., Rossmann, A., Gänzle, M.G., Vogel, R.F., 2001. Sucrose metabolism and exopolysaccharide production in wheat and rye sourdoughs by *Lactobacillus sanfranciscensis*. *J. Agric. Food Chem.* 49, 5194–5200. https://doi.org/10.1021/jf0102517

Laatikainen, R., Koskenpato, J., Hongisto, S.-M., Loponen, J., Poussa, T., Huang, X., Sontag-Strohm, T., Salmenkari, H., Korpela, R., 2017. Pilot study: Comparison of sourdough wheat bread and yeast-fermented wheat bread in individuals with wheat sensitivity and irritable bowel syndrome. *Nutrients* 9, 1215. https://doi.org/10.3390/nu9111215

Laitila, A., Sweins, H., Vilpola, A., Kotaviita, E., Olkku, J., Home, S., Haikara, A., 2006. *Lactobacillus plantarum* and *Pediococcus pentosaceus* starter cultures as a tool for microflora management in malting and for enhancement of malt processability. *J. Agric. Food Chem.* 54, 3840–3851. https://doi.org/10.1021/JF052979J/ASSET/IMAGES/LARGE/JF052979JF00004.JPEG

Landis, E.A., Oliverio, A.M., McKenney, E.A., Nichols, L.M., Kfoury, N., Biango-Daniels, M., Shell, L.K., Madden, A.A., Shapiro, L., Sakunala, S., Drake, K., Robbat, A., Booker, M., Dunn, R.R., Fierer, N., Wolfe, B.E., 2021. The diversity and function of sourdough starter microbiomes. *Elife* 10, 1–24. https://doi.org/10.7554/ELIFE.61644

Lavermicocca, P., Valerio, F., Evidente, A., Lazzaroni, S., Corsetti, A., Gobbetti, M., 2000. Purification and characterization of novel antifungal compounds from the sourdough *Lactobacillus plantarum* strain 21B. *Appl. Environ. Microbiol.* 66, 4084–4090. https://doi.org/10.1128/AEM.66.9.4084-4090.2000/ASSET/A630EB85-567D-4600-BDF3-083AF5846165/ASSETS/GRAPHIC/AM0900386003.JPEG

Lei, V., Jakobsen, M., 2004. Microbiological characterization and probiotic potential of koko and koko sour water, African spontaneously fermented millet porridge and drink. *J. Appl. Microbiol.* 96, 384–397. https://doi.org/10.1046/j.1365-2672.2004.02162.x

Li, F., Li, X., Cheng, C.C., Bujdoš, D., Tollenaar, S., Simpson, D.J., Tasseva, G., Perez-Muñoz, M.E., Frese, S., Gänzle, M.G., Walter, J., Zheng, J., 2023. A phylogenomic analysis of *Limosilactobacillus reuteri* reveals ancient and stable evolutionary relationships with rodents and birds and zoonotic transmission to humans. *BMC Biol.* 21, 1–17. https://doi.org/10.1186/S12915-023-01541-1/FIGURES/7

Li, Q., Gänzle, M., 2020. Host-adapted lactobacilli in food fermentations: Impact of metabolic traits of host adapted lactobacilli on food quality and human health. *Curr. Opin. Food Sci.* 31, 71–80. https://doi.org/10.1016/j.cofs.2020.02.002

Li, Q., Loponen, J., Gänzle, M.G., 2020a. Characterization of the extracellular fructanase FruA in *Lactobacillus crispatus* and its contribution to fructan hydrolysis in breadmaking. *J. Agric. Food Chem.* 68, 8637–8647. https://doi.org/10.1021/acs.jafc.0c02313

Li, Q., Tao, Q., Teixeira, J.S., Su, M.S., Gänzle, M.G., 2020b. Contribution of glutaminases to glutamine metabolism and acid resistance in *Lactobacillus reuteri* and other vertebrate host adapted lactobacilli. *Food Microbiol.* 86, 103343. https://doi.org/10.1016/j.fm.2019.103343

Li, Z., Siepmann, F.B., Rojas Tovar, L.E., Chen, X., Gänzle, M., 2020. Effect of copy number of the $spoVA^{2mob}$ operon, sourdough and reutericyclin on ropy bread spoilage caused by *Bacillus* spp. *Food Microbiol.* 91, 103507. https://doi.org/10.1016/j.fm.2020.103507

Li, Z., Zheng, M., Zheng, J., Gänzle, M.G., 2023. *Bacillus* species in food fermentations: An underappreciated group of organisms for safe use in food fermentations. *Curr. Opin. Food Sci.* 50, 101007. https://doi.org/10.1016/J.COFS.2023.101007

Lin, X.B., Gänzle, M.G., 2014. Effect of lineage-specific metabolic traits of *Lactobacillus reuteri* on sourdough microbial ecology. *Appl. Environ. Microbiol.* 80, 5782–5789. https://doi.org/10.1128/AEM.01783-14

Lin, X.B., Lohans, C.T., Duar, R., Zheng, J., Vederas, J.C., Walter, J., Gänzle, M., 2015. Genetic determinants of reutericyclin biosynthesis in *Lactobacillus reuteri*. *Appl. Environ. Microbiol.* 81, 2032–2041. https://doi.org/10.1128/AEM.03691-14

Liu, L., Wang, J., Rosenberg, D., Zhao, H., Lengyel, G., Nadel, D., 2018. Fermented beverage and food storage in 13,000 y-old stone mortars at Raqefet Cave, Israel: Investigating Natufian ritual feasting. *J. Archaeol. Sci. Rep.* 21, 783–793. https://doi.org/10.1016/j.jasrep.2018.08.008

Loponen, J., Gänzle, M.G., 2018. Use of sourdough in low FODMAP baking. *Foods* 7, 96. https://doi.org/10.3390/foods7070096

Loponen, J., Mikola, M., Katina, K., Sontag-Strohm, T., Salovaara, H., 2004. Degradation of HMW glutenins during wheat sourdough fermentations. *Cereal Chem.* 81, 87–93. https://doi.org/10.1094/CCHEM.2004.81.1.87

Lu, Z.M., Liu, N., Wang, L.J., Wu, L.H., Gong, J.S., Yu, Y.J., Li, G.Q., Shi, J.S., Xu, Z.H., 2016. Elucidating and regulating the acetoin production role of microbial functional groups in multispecies acetic acid fermentation. *Appl. Environ. Microbiol.* 82, 5860–5868. https://doi.org/10.1128/AEM.01331-16

Marco, M.L., Heeney, D., Binda, S., Cifelli, C.J., Cotter, P.D., Foligné, B., Gänzle, M., Kort, R., Pasin, G., Pihlanto, A., Smid, E.J., Hutkins, R., 2017. Health benefits of fermented foods: Microbiota and beyond. *Curr. Opin. Biotechnol.* 44, 94–102. https://doi.org/10.1016/J.COPBIO.2016.11.010

Marco, M.L., Sanders, M.E., Gänzle, M., Arrieta, M.C., Cotter, P.D., De Vuyst, L., Hill, C., Holzapfel, W., Lebeer, S., Merenstein, D., Reid, G., Wolfe, B.E., Hutkins, R., 2021. The international scientific association for probiotics and prebiotics (ISAPP) consensus statement on fermented foods. *Nat. Rev. Gastroenterol. Hepatol.* 18, 196–208. https://doi.org/10.1038/s41575-020-00390-5

Mårtensson, O., Öste, R., Holst, O., 2002. The effect of yoghurt culture on the survival of probiotic bacteria in oat-based, non-dairy products. *Food Res. Int.* 35, 775–784. https://doi.org/10.1016/S0963-9969(02)00074-1

Mefleh, M., Pasqualone, A., Caponio, F., Faccia, M., 2022. Legumes as basic ingredients in the production of dairy-free cheese alternatives: A review. *J. Sci. Food Agric.* 102, 8–18. https://doi.org/10.1002/jsfa.11502

Menezes, L.A.A., Molognoni, L., de Sá Ploêncio, L.A., Costa, F.B.M., Daguer, H., Dea Lindner, J. De, 2019. Use of sourdough fermentation to reducing FODMAPs in breads. *Eur. Food Res. Technol.* 245, 1183–1195. https://doi.org/10.1007/s00217-019-03239-7

Meroth, C.B., Hammes, W.P., Hertel, C., 2004. Characterisation of the microbiota of rice sourdoughs and description of *Lactobacillus spicheri* sp. nov. *Syst. Appl. Microbiol.* 27, 151–159. https://doi.org/10.1078/072320204322881763

Meroth, C.B., Walter, J., Hertel, C., Brandt, M.J., Hammes, W.P., 2003. Monitoring the bacterial population dynamics in sourdough fermentation processes by using PCR-denaturing gradient gel electrophoresis. *Appl. Environ. Microbiol.* 69, 475–482. https://doi.org/10.1128/aem.69.1.475-482.2003

Meuser, F., 1995. Development of fermentation technology in modern bread factories. *Cereal Foods World* 40, 122.

Minervini, F., Celano, G., Lattanzi, A., Tedone, L., de Mastro, G., Gobbetti, M., De Angelis, M., 2015. Lactic acid bacteria in durum wheat flour are endophytic components of the plant during its entire life cycle. *Appl. Environ. Microbiol.* 81, 6736–6748. https://doi.org/10.1128/AEM.01852-15

Moroni, A.V., Dal Bello, F., Arendt, E.K., 2009. Sourdough in gluten-free bread-making: An ancient technology to solve a novel issue? *Food Microbiol.* 26, 676–684. https://doi.org/10.1016/J.FM.2009.07.001

Mudoor Sooresh, M., Willing, B.P., Bourrie, B.C.T., 2023. Opportunities and challenges of understanding community assembly in spontaneous food fermentation. *Foods* 12, 673. https://doi.org/10.3390/FOODS12030673

Mugula, J.K., Nnko, S.A.M., Narvhus, J.A., Sørhaug, T., 2003. Microbiological and fermentation characteristics of togwa, a Tanzanian fermented food. *Int. J. Food Microbiol.* 80, 187–199.

Müller, D.C., Nguyen, H., Li, Q., Schönlechner, R., Miescher Schwenninger, S., Wismer, W., Gänzle, M., 2021. Enzymatic and microbial conversions to achieve sugar reduction in bread. *Food Res. Int.* 143, 110296. https://doi.org/10.1016/J.FOODRES.2021.110296

Neubauer, H., Glaasker, E., Hammes, W.P., Poolman, B., Konings, W.N., 1994. Mechanism of maltose uptake and glucose excretion in *Lactobacillus sanfrancisco*. *J. Bacteriol.* 176, 3007–3012. https://doi.org/10.1128/JB.176.10.3007-3012.1994

Nout, M.J.R., 2009. Rich nutrition from the poorest—Cereal fermentations in Africa and Asia. *Food Microbiol.* 26, 685–692.

Odunfa, S.A., Adeyele, S., 1985. Microbiological changes during the traditional production of ogi-baba, a West African fermented sorghum gruel. *J. Cereal Sci.* 3, 173–180. https://doi.org/10.1016/S0733-5210(85)80027-8

Oduro-Yeboah, C., Mestres, C., Amoa-Awua, W., Fliedel, G., Durand, N., Matignon, B., Michodjehoun, V.L., Saalia, F.K., Sakyi-Dawson, E., Abbey, L., 2016. Steeping time and dough fermentation affect the milling behaviour and quality of white kenkey(nsiho), a sour stiff dumpling prepared from dehulled maize grains. *J. Cereal Sci.* 69, 377–382. https://doi.org/10.1016/J.JCS.2016.05.014

Pasolli, E., De Filippis, F., Mauriello, I.E., Cumbo, F., Walsh, A.M., Leech, J., Cotter, P.D., Segata, N., Ercolini, D., 2020. Large-scale genome-wide analysis links lactic acid bacteria from food with the gut microbiome. *Nat. Commun.* 11, 2610. https://doi.org/10.1038/s41467-020-16438-8

Pepe, O., Blaiotta, G., Moschetti, G., Greco, T., Villani, F., 2003. Rope-producing strains of *Bacillus* spp. from wheat bread and strategy for their control by lactic acid bacteria. *Appl. Environ. Microbiol.* 69, 2321–2329. https://doi.org/10.1128/aem.69.4.2321-2329.2003

Pirkola, L., Laatikainen, R., Loponen, J., Hongisto, S.M., Hillilä, M., Nuora, A., Yang, B., Linderborg, K.M., Freese, R., 2018. Low-FODMAP vs regular rye bread in irritable bowel syndrome: Randomized SmartPill study. *World J. Gastroenterol.* 24, 1259. https://doi.org/10.3748/WJG.V24.I11.1259

Poole, A.C., Goodrich, J.K., Youngblut, N.D., Luque, G.G., Ruaud, A., Sutter, J.L., Waters, J.L., Shi, Q., El-Hadidi, M., Johnson, L.M., Bar, H.Y., Huson, D.H., Booth, J.G., Ley, R.E., 2019. Human salivary amylase gene copy number impacts oral and gut microbiomes. *Cell Host Microbe* 25, 553–564.e7. https://doi.org/10.1016/J.CHOM.2019.03.001

Poutanen, K., Flander, L., Katina, K., 2009. Sourdough and cereal fermentation in a nutritional perspective. *Food Microbiol.* 26, 693–699. https://doi.org/10.1016/j.fm.2009.07.011

Pswarayi, F., Gänzle, M.G., 2019. Composition and origin of the fermentation microbiota of mahewu, a Zimbabwean fermented cereal beverage. *Appl. Environ. Microbiol.* 85, e03130–18. https://doi.org/10.1128/aem.03130-18

Pswarayi, F., Gänzle, M.G., 2022. African cereal fermentations: A review on fermentation processes and microbial composition of non-alcoholic fermented cereal foods and beverages. *Int. J. Food Microbiol.* 378, 109815. https://doi.org/10.1016/J.IJFOODMICRO.2022.109815

Pswarayi, F., Qiao, N., Gaur, G., Gänzle, M., 2022. Antimicrobial plant secondary metabolites, MDR transporters and antimicrobial resistance in cereal-associated lactobacilli: Is there a connection? *Food Microbiol.* 102, 103917. https://doi.org/10.1016/j.fm.2021.103917

Ribet, L., Dessalles, R., Lesens, C., Brusselaers, N., Durand-Dubief, M., 2023. Nutritional benefits of sourdoughs: A systematic review. *Adv. Nutr.* 14, 22–29. https://doi.org/10.1016/J.ADVNUT.2022.10.003

Röcken, W., Rick, M., Reinkemeier, M., 1992. Controlled production of acetic acid in wheat sour doughs. *Z. Lebensm. Unters. Forsch.* 195, 259–263. https://doi.org/10.1007/BF01202806

Rosenquist, H., Hansen, Å., 1998. The antimicrobial effect of organic acids, sour dough and nisin against *Bacillus subtilis* and *B. licheniformis* isolated from wheat bread. *J. Appl. Microbiol.* 85, 621–631. https://doi.org/10.1046/j.1365-2672.1998.853540.x

Ryan, L.A.M., Dal Bello, F., Arendt, E.K., Koehler, P., 2009a. Detection and quantitation of 2,5-diketopiperazines in wheat sourdough and bread. *J. Agric. Food Chem.* 57, 9563–9568. https://doi.org/10.1021/jf902033v

Ryan, L.A.M., Dal Bello, F., Czerny, M., Koehler, P., Arendt, E.K., 2009b. Quantification of phenyllactic acid in wheat sourdough using high resolution gas chromatography-mass spectrometry. *J. Agric. Food Chem.* 57, 1060–1064. https://doi.org/10.1021/JF802578E/ASSET/IMAGES/LARGE/JF-2008-02578E_0004.JPEG

Salovaara, H., Baeckstroem, K., Mantere-Alhonen, S., 1990. "Kiesa", the traditional Carelian fermented oat pudding revisited. In: *Proceedings from the 24th Nordic Cereal Congress.* Stockholm, p. 36. https://doi.org/10.3/JQUERY-UI.JS

Salovaara, H., Simonson, L., 2004. Fermented cereal-based functional foods. In: Hui, Y.H., Meunier-Goddik, L., Josephsen, J., Nip, W.-K., Stanfield, P.S. (Eds.), *Handbook of Food and Beverage Fermentation Technology.* CRC Press, pp. 831–838. https://doi.org/10.1201/9780203913550

Schumann, D., Klose, P., Lauche, R., Dobos, G., Langhorst, J., Cramer, H., 2018. Low fermentable, oligo-, di-, mono-saccharides and polyol diet in the treatment of irritable bowel syndrome: A systematic review and meta-analysis. *Nutrition* 45, 24–31. https://doi.org/10.1016/J.NUT.2017.07.004

Sekwati-Monang, B., Gänzle, M.G., 2011. Microbiological and chemical characterisation of ting, a sorghum-based sourdough product from Botswana. *Int. J. Food Microbiol.* 150, 115–121. https://doi.org/10.1016/j.ijfoodmicro.2011.07.021

Sekwati-Monang, B., Valcheva, R., Gänzle, M.G., 2012. Microbial ecology of sorghum sourdoughs: Effect of substrate supply and phenolic compounds on composition of fermentation microbiota. *Int. J. Food Microbiol.* 159, 240–246. https://doi.org/10.1016/j.ijfoodmicro.2012.09.013

Shewry, P.R., America, A.H.P., Lovegrove, A., Wood, A.J., Plummer, A., Evans, J., van den Broeck, H.C., Gilissen, L., Mumm, R., Ward, J.L., Proos, Z., Kuiper, P., Longin, C.F.H., Andersson, A.A.M., Philip van Straaten, J., Jonkers, D., Brouns, F., 2022. Comparative compositions of metabolites and dietary fibre components in doughs and breads produced from bread wheat, emmer and spelt and using yeast and sourdough processes. *Food Chem.* 374, 131710. https://doi.org/10.1016/J.FOODCHEM.2021.131710

Solanki, M.K., Abdelfattah, A., Britzi, M., Zakin, V., Wisniewski, M., Droby, S., Sionov, E., 2019. Shifts in the composition of the microbiota of stored wheat grains in response to fumigation. *Front. Microbiol.* 10, 1098. https://doi.org/10.3389/FMICB.2019.01098/BIBTEX

Spicher, G., Stephan, H., 1993. *Handbuch Sauerteig*, 4th ed. Behr's Verlag.

Stolz, P., Böcker, G., Hammes, W.P., Vogel, R.F., 1995. Utilization of electron acceptors by lactobacilli isolated from sourdough I. *Lactobacillus sanfrancisco. Zeitschrift für Leb. und Forsch.* 201, 91–96. https://doi.org/10.1007/BF01193208

Stolz, P., Böcker, G., Vogel, R.F., Hammes, W.P., 1993. Utilisation of maltose and glucose by lactobacilli isolated from sourdough. *FEMS Microbiol. Lett.* 109, 237–242. https://doi.org/10.1111/J.1574-6968.1993.TB06174.X

Stromeck, A., Hu, Y., Chen, L., Gäzle, M.G., 2011. Proteolysis and bioconversion of cereal proteins to glutamate and γ-aminobutyrate (GABA) in rye malt sourdoughs. *J. Agric. Food Chem.* 59, 1392–1399. https://doi.org/10.1021/jf103546t

Struyf, N., Vandewiele, H., Herrera-Malaver, B., Verspreet, J., Verstrepen, K.J., Courtin, C.M., 2018. *Kluyveromyces marxianus* yeast enables the production of low FODMAP whole wheat breads. *Food Microbiol.* 76, 135–145. https://doi.org/10.1016/j.fm.2018.04.014

Su, M.S.-W., Oh, P.L., Walter, J., Gänzle, M.G., 2012. Intestinal origin of sourdough *Lactobacillus reuteri* isolates as revealed by phylogenetic, genetic, and physiological analysis. *Appl. Environ. Microbiol.* 78, 6777–6780. https://doi.org/10.1128/AEM.01678-12

Svensson, L., Sekwati-Monang, B., Lutz, D.L., Schieber, R., Gänzle, M.G., 2010. Phenolic acids and flavonoids in nonfermented and fermented red sorghum (*Sorghum bicolor* (L.) Moench). *J. Agric. Food Chem.* 58, 9214–9220. https://doi.org/10.1021/jf101504v

Tamang, J.P., Cotter, P.D., Endo, A., Han, N.S., Kort, R., Liu, S.Q., Mayo, B., Westerik, N., Hutkins, R., 2020. Fermented foods in a global age: East meets West. *Compr. Rev. Food Sci. Food Saf.* 19, 184–217. https://doi.org/10.1111/1541-4337.12520

Tang, K.X., Zhao, C.J., Gänzle, M.G., 2017. Effect of glutathione on the taste and texture of Type I sourdough bread. *J. Agric. Food Chem.* 65, 4321–4328. https://doi.org/10.1021/acs.jafc.7b00897

Tangkongchitr, U., Seib, P.A., Hoseney, R.C., 1982. Phytic Acid. III. Two barriers to the loss of phytate during bread-making. *Cereal Chem.* 59, 216–221.

Tangyu, M., Muller, J., Bolten, C.J., Wittmann, C., 2019. Fermentation of plant-based milk alternatives for improved flavour and nutritional value. *Appl. Microbiol. Biotechnol.* 103, 9263–9275. https://doi.org/10.1007/s00253-019-10175-9

Tannock, G.W., Wilson, C.M., Loach, D., Cook, G.M., Eason, J., O'Toole, P.W., Holtrop, G., Lawley, B., 2012. Resource partitioning in relation to cohabitation of *Lactobacillus* species in the mouse forestomach. *ISME J.* 6, 927–938. https://doi.org/10.1038/ismej.2011.161

Teixeira, J.S., Abdi, R., Su, M.S.W., Schwab, C., Gänzle, M.G., 2013. Functional characterization of sucrose phosphorylase and *scrR*, a regulator of sucrose metabolism in *Lactobacillus reuteri*. *Food Microbiol.* 36, 432–439. https://doi.org/10.1016/j.fm.2013.07.011

Teixeira, J.S., Seeras, A., Sanchez-Maldonado, A.F., Zhang, C., Su, M.S.W., Gänzle, M.G., 2014. Glutamine, glutamate, and arginine-based acid resistance in *Lactobacillus reuteri*. *Food Microbiol.* 42, 172–180. https://doi.org/10.1016/j.fm.2014.03.015

Thiele, C., Gänzle, M.G., Vogel, R.F., 2002. Contribution of sourdough lactobacilli, yeast, and cereal enzymes to the generation of amino acids in dough relevant for bread flavor. *Cereal Chem.* 79, 45–51. https://doi.org/10.1094/CCHEM.2002.79.1.45

Todorov, S.D., Dicks, L.M.T., 2006. Screening for bacteriocin-producing lactic acid bacteria from boza, a traditional cereal beverage from Bulgaria: Comparison of the bacteriocins. *Process Biochem.* 41, 11–19. https://doi.org/10.1016/J.PROCBIO.2005.01.026

Todorov, S.D., Holzapfel, W.H., 2015. Traditional cereal fermented foods as sources of functional microorganisms. In: *Advances in Fermented Foods and Beverages: Improving Quality, Technologies and Health Benefits*. Woodhead Publishing, pp. 123–153. https://doi.org/10.1016/B978-1-78242-015-6.00006-2

Van der Meulen, R., Scheirlinck, I., Van Schoor, A., Huys, G., Vancanneyt, M., Vandamme, P., De Vuyst, L., 2007. Population dynamics and metabolite target analysis of lactic acid bacteria during laboratory fermentations of wheat and spelt sourdoughs. *Appl. Environ. Microbiol.* 73, 4741–4750. https://doi.org/10.1128/AEM.00315-07

Van Kerrebroeck, S., Maes, D., De Vuyst, L., 2017. Sourdoughs as a function of their species diversity and process conditions, a meta-analysis. *Trends Food Sci. Technol.* 68, 152–159. https://doi.org/10.1016/j.tifs.2017.08.016

Vermeulen, N., Gänzle, M.G., Vogel, R.F., 2007. Glutamine deamidation by cereal-associated lactic acid bacteria. *J. Appl. Microbiol.* 103, 1197–1205. https://doi.org/10.1111/j.1365-2672.2007.03333.x

Vermeulen, N., Pavlovic, M., Ehrmann, M.A., Gänzle, M.G., Vogel, R.F., 2005. Functional characterization of the proteolytic system of *Lactobacillus sanfranciscensis* DSM 20451$^T$during growth in sourdough. *Appl. Environ. Microbiol.* 71, 6260–6266. https://doi.org/10.1128/AEM.71.10.6260-6266.2005

Vieira-Dalodé, G., Jespersen, L., Hounhouigan, J., Moller, P.L., Nago, C.M., Jakobsen, M., 2007. Lactic acid bacteria and yeasts associated with gowé production from sorghum in Bénin. *J. Appl. Microbiol.* 103, 342–349. https://doi.org/10.1111/J.1365-2672.2006.03252.X

Vogel, R.F., Knorr, R., Müller, M.R.A., Steudel, U., Gänzle, M.G., Ehrmann, M.A., 1999. Non-dairy lactic fermentations: The cereal world. Antonie van Leeuwenhoek. *Int. J. Gen. Mol. Microbiol.* 76, 403–411. https://doi.org/10.1023/A:1002089515177

Vogel, R.F., Pavlovic, M., Ehrmann, M.A., Wiezer, A., Liesegang, H., Offschanka, S., Voget, S., Angelov, A., Böcker, G., Liebl, W., 2011. Genomic analysis reveals *Lactobacillus sanfranciscensis* as stable element in traditional sourdoughs. *Microb. Cell Fact.* 10, S6. https://doi.org/10.1186/1475-2859-10-S1-S6

Wacher, C., Cañas, A., Cook, P.E., Barzana, E., Owens, J.D., 1993. Sources of microorganisms in pozol, a traditional Mexican fermented maize dough. *World J. Microbiol. Biotechnol.* 9, 269–274. https://doi.org/10.1007/BF00327853

Wang, W., Zijlstra, R.T., Gänzle, M.G., 2020. Feeding *Limosilactobacillus fermentum* K9–2 and *Lacticaseibacillus casei* K9–1, or *Limosilactobacillus reuteri* TMW1.656 reduces pathogen load in weanling pigs. *Front. Microbiol.* 11, 3187. https://doi.org/10.3389/fmicb.2020.608293

Wang, X., Du, H., Zhang, Y., Xu, Y., 2018. Environmental microbiota drives microbial succession and metabolic profiles during Chinese liquor fermentation. *Appl. Environ. Microbiol.* 84. https://doi.org/10.1128/AEM.02369-17

Wang, Z.M., Lu, Z.M., Yu, Y.J., Li, G.Q., Shi, J.S., Xu, Z.H., 2015. Batch-to-batch uniformity of bacterial community succession and flavor formation in the fermentation of Zhenjiang aromatic vinegar. *Food Microbiol.* 50, 64–69. https://doi.org/10.1016/J.FM.2015.03.012

Wastyk, H.C., Fragiadakis, G.K., Perelman, D., Dahan, D., Merrill, B.D., Yu, F.B., Topf, M., Gonzalez, C.G., Van Treuren, W., Han, S., Robinson, J.L., Elias, J.E., Sonnenburg, E.D., Gardner, C.D., Sonnenburg, J.L., 2021. Gut-microbiota-targeted diets modulate human immune status. *Cell* 184, 4137–4153.e14. https://doi.org/10.1016/J.CELL.2021.06.019

WHO, 2017. *Diarrhoeal Disease* [WWW Document]. URL www.who.int/news-room/fact-sheets/detail/diarrhoeal-disease (accessed 9.20.21).

Wu, J.J., Ma, Y.K., Zhang, F.F., Chen, F.S., 2012. Biodiversity of yeasts, lactic acid bacteria and acetic acid bacteria in the fermentation of "Shanxi aged vinegar", a traditional Chinese vinegar. *Food Microbiol.* 30, 289–297. https://doi.org/10.1016/J.FM.2011.08.010

Yan, Y.L., Hu, Y., Gänzle, M.G., 2018. Prebiotics, FODMAPs and dietary fiber—conflicting concepts in development of functional food products? *Curr. Opin. Food Sci.* 20, 30–37. https://doi.org/10.1016/j.cofs.2018.02.009

Yang, Y., Galle, S., Le, M.H.A., Zijlstra, R.T.R.T., Gänzle, M.G., 2015. Feed fermentation with reuteran- and levan-producing *Lactobacillus reuteri* reduces colonization of weanling pigs by enterotoxigenic *Escherichia coli*. *Appl. Env. Microbiol.* 81, 5743–5752. https://doi.org/10.1128/AEM.01525-15

Zaccaria, E., Klaassen, T., Alleleyn, A.M.E., Boekhorst, J., Smokvina, T., Kleerebezem, M., Troost, F.J., 2023. Endogenous small intestinal microbiome determinants of transient colonisation efficiency by bacteria from fermented dairy products: A randomised controlled trial. *Microbiome* 11, 43. https://doi.org/10.1186/S40168-023-01491-4/FIGURES/4

Zhang, C., Brandt, M.J., Schwab, C., Gänzle, M., 2010. Propionic acid production by cofermentation of *Lactobacillus buchneri* and *Lactobacillus diolivorans* in sourdough. *Food Microbiol.* 27, 390–395. https://doi.org/10.1016/j.fm.2009.11.019

Zhao, C.J., Kinner, M., Wismer, W., Gänzle, M.G., 2015. Effect of glutamate accumulation during sourdough fermentation with *Lactobacillus reuteri* on the taste of bread and sodium-reduced bread. *Cereal Chem.* 92, 224–230. https://doi.org/10.1094/CCHEM-07-14-0149-R

Zheng, J., Ruan, L., Sun, M., Gänzle, M., 2015. A genomic view of lactobacilli and pediococci demonstrates that phylogeny matches ecology and physiology. *Appl. Environ. Microbiol.* 81, 7233–7243. https://doi.org/10.1128/AEM.02116-15

Zheng, J., Wittouck, S., Salvetti, E., Franz, C.M.A.P., Harris, H.M.B., Mattarelli, P., O'toole, P.W., Pot, B., Vandamme, P., Walter, J., Watanabe, K., Wuyts, S., Felis, G.E., Gänzle, M.G., Lebeer, S., 2020. A taxonomic note on the genus *Lactobacillus*: Description of 23 novel genera, emended description of the genus *Lactobacillus beijerinck* 1901, and union of *Lactobacillaceae* and *Leuconostocaceae*. *Int. J. Syst. Evol. Microbiol.* 70, 2782–2858. https://doi.org/10.1099/ijsem.0.004107

Zheng, X.W., Han, B.Z., 2016. *Baijiu* (白酒), Chinese liquor: History, classification and manufacture. *J. Ethn. Foods* 3, 19–25. https://doi.org/10.1016/J.JEF.2016.03.001

# Lactic Acid Bacteria in Meat Fermentations

# 12

## Role of Autochthonous Starter Cultures on Quality, Safety, and Health

Silvina Fadda, Patricia Castellano, Lucrecia Terán, Raúl Raya, and Graciela Vignolo

## 12.1 INTRODUCTION

Traditionally, fermentation of meat was considered a method to extend the shelf life of this highly perishable commodity; however, the preservation role of fermentation has become largely outdated since the introduction of the cold chain. Nevertheless, fermented meat products remain very popular and are still produced in large amounts, especially in Europe, where they occur in a vast diversity. This persistence is probably related to their unique and specific sensory properties, their convenience, and their alleged rootedness in cultural heritage (Conter et al., 2008, Aquilanti et al., 2016). Despite the remarkable history of meat fermentation/preservation, profound quality, safety, and health concerns have been triggered, and recently, ecological and ethical awareness related to animal husbandry, climate effects, and global nourishment has surfaced, creating distrust and high market pressure (Leroy et al., 2013). The meat industry should adopt innovation as a key driver for contribute to sustainability and economic growth (Font-i-Furnols and Guerrero, 2014).

The unique sensory and textural characteristics of fermented meats are ascribed to a number of biochemical and physicochemical transformations occurring in the meat during fermentation and ripening (Vignolo et al., 2010b). Flavor development not only depends on the raw materials and processing conditions but also on how these factors affect the composition, dynamic, and metabolism of the sausage's microbiota (Franciosa et al., 2021). Acid production by lactic acid bacteria (LAB) results in a drop of the pH, contributing to the formation of a gel-like texture and acid taste, and has an important role in the inhibition of undesirable microorganisms. Oxidative transformations of fatty acids as well as interactions between the meat myoglobin and nitrogen monoxide originating from the nitrate/nitrite in the curing salt are performed by coagulase-negative staphylococci and the acid-reducing environment (Ravyts et al., 2010). In addition, meat protein degradation is one of the main biochemical events that are catalyzed by muscle and/or

DOI: 10.1201/9781003352075-14

microbial enzymes which contribute to the texture and flavor of fermented sausages. In view of industrially useful innovations, the role of LAB communities existing in fermented meat ecosystems as well as their contribution to biochemical and sensory changes, bio-protection, and health promotion are discussed here. Special attention is paid to two emblematic species, *L. sakei* and *L. curvatus*.

## 12.2 ORIGIN OF LAB IN SPONTANEOUSLY FERMENTED SAUSAGES

Meat ecosystems are a great source of microbial diversity. The ecological strategies adopted by microorganisms to grow in meat are a consequence of the prevailing environmental conditions, in which the intrinsic and extrinsic factors governing microbial growth will determine the type and number of bacteria present in the meat mixture. Among several factors, the physicochemical ones (concentration and availability of nutrients, pH, redox potential, buffering capacity, meat structure, $a_w$, NaCl, nitrate/nitrite, sugars) and those related to storage and processing conditions (temperature and oxygen availability) are the most prevalent. In particular, fermented meat ecosystems favor the growth of highly specific microbial associations that originate in the raw materials used in manufacturing, the environment, and equipment and are highly dependent on the applied technology. Naturally present LAB species are subject to a selection process that takes place during fermentation and eventually defines the final characteristics of fermented sausages (Rantsiou et al., 2005).

It is commonly assumed that meat, as the main input in the production of fermented sausages, can be initially contaminated at the abattoir from the skin, intestines, and carcasses of the animals, and this contamination occurs during the consecutive processing steps of the meat (Sofos, 2008). In addition to carcass cooling that selects for spoilage and pathogenic psychrotrophic bacteria, systems for retail distribution based on vacuum packaging or modified atmosphere packaging of meat cuts using low gas–permeability films are usually applied (Nychas et al., 2008). Among the saprophyte-dominated initial microbiota, the presence of bacteria belonging to *Lactobacillus sensu lato*: *L. sakei, L. curvatus, Dellaglioa algida, Leuconostoc (Ln) (Ln. gelidum, Ln. carnosum, Ln. gasocomitatum)*, and *Carnobacterium (C.) (C. maltaromaticum, C. piscicola, C. divergens)* have been widely reported (Vignolo et al., 2010a; Casaburi et al., 2011). Since dry fermented sausages have been produced for centuries stuffed inside natural casings, these may also be a source of LAB. Indeed, their role as microbiological inoculants for sausage production was recently reported, *L. sakei, Lacticaseibacillus paracasei, L. plantarum*, and *Lactobacillus amylovorus* being the most frequently found (Rebecchi et al., 2015). However, this animal-derived microbiota was less prevalent and less abundant than a core microbiota, psychrotrophic in nature, mainly originating from the environment that constitutes the main reservoir of spoilage bacteria (Chaillou et al., 2014). Microorganisms present in traditional sausages also derived from the manufacturing environment, which are usually called the "house microbiota"; these strains are well adapted to meat and the specific manufacturing process and are therefore able to dominate the microbiota of the products (Talon et al., 2007a). A survey carried out in small European processing units reported the presence of technological microbiota, LAB, and coagulase-negative *Staphylococcus* (CNS) at a level >2 log CFU/cm$^2$ on different surfaces and processing equipment; LAB were found primarily on stuffing/mixing machines, knives, and tables (Talon et al., 2007b).

## 12.3 LAB DIVERSITY FROM TRADITIONAL FERMENTED SAUSAGES

Traditional dry or semidry fermented sausages are manufactured without the addition of starter cultures in small-scale processing units; thus fermentation relies on the indigenous microbiota present in the raw materials and/or the environment (Talon et al., 2007a). Traditional foods should be considered "foods that have been consumed regionally or locally for an extensive time period, reflect cultural inheritance, and are an expression of culture, history, and lifestyle". In this context, as reported by Leroy et al. (2015), fermented meats are traditional archetypal foods, as they have originated from empirical methods for preserving meat in the distant past and have evolved over the centuries into a large number of varieties with strong territorial and socio-cultural connotations. From a technological point of view, the uniqueness of traditional materials, formulations, and production methods referring to the physical, chemical, micro-biological, and organoleptic characteristics of fermented sausages have been specifically highlighted (EuroFIR Consortium, 2017). As such, the microbiological aspects are indeed assumed to be particularly important and often related to the presence of the "house microbiota" (Talon et al., 2007a).

The microbiota in natural environments are complex and often include thousands of genera from a diverse range of species. With the development of high-throughput sequencing (HTS) techniques, investigations have focused on the structure of the core microbiota in a range of traditional fermented products (Cocolin et al., 2013; Ercolini, 2013). In natural environments or under artificially controlled conditions, the core fermentation microbiota uses raw materials to produce a variety of metabolites that will determine the type of final product that shows a significant correlation with the main endogenous factors in the different production stages (Song et al., 2017). In traditional fermented sausages, *L. sakei* and *L. curvatus* are by far the most often isolated species among LAB, but *L. plantarum* and *Pediococcus* spp. have also been isolated. The LAB species isolated from traditional fermented sausages and identified by molecular approaches are presented in Table 12.1.

The prevalence of *L. sakei* and *L. curvatus* during meat fermentation has been widely reported (Fontana et al., 2016; Vignolo et al., 2010a, 2010b, Franciosa et al., 2022; Bassi et al., 2022). Considering the emerged concept of core microbiota (Astudillo-Garcia et al., 2017), these two species may be considered part of the core functional microbiota during meat fermentation. During the last 30 years, numerous studies have focused on *L. sakei* due to its important role in meat technology and its competitiveness for acid production, protein degradation, and bio-preservation potential (Fadda et al., 2010a, 2010b; Zagorec and Champomier-Vergès, 2017, Janßen et al., 2018; Najjari et al., 2020). As a versatile bacterium, *L. plantarum* has been identified as part of fermented sausage microbiota; however, it lacks the meat specialization found in *L. sakei* (Chaillou et al., 2009; Hüfner et al., 2007; Hu et al., 2022). In significantly lower numbers, other *Lactobacillus* species were also identified (Table 12.1).

**TABLE 12.1**  Major Lactic Acid Bacteria Species Isolated from Traditional Fermented Sausages Applying Molecular Approaches

| COUNTRY/SPECIES | IDENTIFICATION APPROACH | SOURCE | REFERENCES |
|---|---|---|---|
| Spain | | | |
| L. sakei/curvatus/plantarum; E. faecium | S-s PCR | Fuet, chorizo | Aymerich et al. (2003) |
| L. sakei/curvatus; Ln. mesenteroides | S-s PCR/RAPD-PCR | Fuet, chorizo | Aymerich et al. (2006) |
| L. curvatus/plantarum/brevis; Ln. mesenteroides; P. pentosaceus/acidilactici | RAPD-PCR | Salchichón, chorizo | Benito et al. (2007) |
| L. sakei/plantarum/paracasei/coryniformis; E. faecium | 16S rDNAs | Dry-fermented sausages | Landeta et al. (2013) |
| Italy | | | |
| L. sakei/curvatus/plantarum/alimentarius/casei/brevis | PCR-TGGE | Fermented sausages | Cocolin et al. (2000) |
| L. sakei/curvatus/plantarum/paracasei | RAPD-PCR | Soppressata | Andrighetto et al. (2001) |
| L. sakei/curvatus/plantarum/brevis/casei/alimentarius | PCR-DGGE | Italian sausages | Cocolin et al. (2001) |

*(Continued)*

**TABLE 12.1** (Continued) Major Lactic Acid Bacteria Species Isolated from Traditional Fermented Sausages Applying Molecular Approaches

| COUNTRY/SPECIES | IDENTIFICATION APPROACH | SOURCE | REFERENCES |
|---|---|---|---|
| L. sakei/curvatus/casei; E. casseliflavus; Lc. lactis; Ln. mesenteroides | PCR-DGGE | Fresh sausages | Cocolin et al. (2004) |
| L. sakei/plantarum/brevis/ paraplantarum; Lc. lactis; E. pseudoavium; Ln. citreum/ mesenteroides; Weissella sp. | RAPD-PCR/ PCR-DGGE | Salami Friulano | Comi et al. (2005) |
| L. sakei/curvatus/paracasei; Lc. garvieae | PCR-DGGE/16S rDNAs | Salami Friulano | Rantsiou et al. (2005) |
| L. sakei/curvatus/plantarum/casei; E. faecium | S-s PCR/PCR-DGGE | Fermented sausages | Urso et al. (2006) |
| L. sakei/curvatus | PCR-DGGE/16S rDNAs | Soppressata | Villani et al. (2007) |
| L. sakei/curvatus/plantarum; Lc. lactis | ARDRA-PCR/ PCR-DGGE | Ciauscolo salami | Aquilanti et al. (2007) |
| L. sakei/plantarum/brevis; P. pentosaceus; Ln. carnosum/ mesenteroides | ARDRA-PCR/ PCR-DGGE | Fermented sausages | Bonomo et al. (2008) |
| L. sakei/curvatus/plantarum/ fermentum/paracasei/salivarius | RAPD-PCR/16S rDNAs | Salame Mantovano | Pisacane et al. (2015) |
| L. sakei/plantarum/coryniformis/ vaginalis/graminis | High-throughput seq | Salame Piacentino | Polka et al. (2015) |
| L. sakei/curvatus; P. pentosaceus | Metagenomics analysis/high-throughput seq | Salame Piemonte | Franciosa et al. (2018, 2021) |
| L. curvatus/sakei | Metataxonomic/ amplification of the V3-V4 region of the 16S rRNA gene | Fermented fish sausages | Belleggia et al. (2022) |
| France | | | |
| L. sakei; Ln. mesenteroides | S-s PCR | Fermented sausages | Ammor et al. (2005) |
| Greece | | | |
| L. sakei/plantarum/curvatus | S-s PCR/16S rDNAs | Fermented sausages | Rantsiou et al. (2006) |
| Argentina | | | |
| L. sakei/curvatus/plantarum/pento sus; E. faecalis/faecium | PCR-DGGE/16S rDNAs | Fermented sausages | Fontana et al. (2005) |
| L. sakei; Ln. inhae/mesenteroides | High-throughput sequencing | Llama meat sausages | Fontana et al. (2016) |
| L. sakei | RAPD-PCR/16S rDNA | Goat-meat fermented sausages | Nediani et al. (2017) |
| Switzerland | | | |
| L. sakei/curvatus; E. faecalis | PCR-RFLP/16S rDNAs | Game sausages | Marty et al. (2012) |
| Portugal | | | |
| L. sakei/plantarum/brevis/paracasei/ paraplantarum/rhamnosus/zeae; E. faecium; Ln. mesenteroides; W. cibaria/viridescens; P. acidilactici/pentosaceous | DGGE | Alheira fermented sausage | Albano et al. (2009) |

(Continued)

**TABLE 12.1**    (Continued) Major Lactic Acid Bacteria Species Isolated from Traditional Fermented Sausages Applying Molecular Approaches

| COUNTRY/SPECIES | IDENTIFICATION APPROACH | SOURCE | REFERENCES |
|---|---|---|---|
| Scandinavia | | | |
| L. sakei/plantarum/alimentarius/ brevis/farciminis | RAPD-PCR/16S rDNA | Fermented sausage | Klingberg et al. (2005) |
| Eastern Europe | | | |
| L. sakei/curvatus; P. pentosaceous; Ln. mesenteroides; E. casseliflavus/ durans | (GTG)5-PCR | Serbian sausages | Danilović et al. (2011) |
| Ln. mesenteroides, L. sakei, E. casseliflavus | Rep-PCR/16S rRNA gene sequencing | Croatian game meat sausages | Maksimovic et al. (2018) |
| Northeast China | | | |
| L. sakei, plantarum, curvatus, W. hellenica, Lc. lactis, | High-throughput sequencing | Traditional dry sausages | Hu et al. (2021) |

L. for former *Lactobacillus*; Lc, Lactococcus; C, Carnobacterium; Ln, Leuconostoc; W, Weissella; P, Pediococcus; E, Enterococcus.

Although pediococci are not a significant part of the microbial community of European fermented sausages, they are used as fast-fermenting starter cultures in North American–style sausages. *P. acidilactici* and *P. pentosaceus* isolated during sausage fermentation were often reported to produce pediocin-like bacteriocins (Bonomo et al., 2008). Moreover, during spontaneous meat fermentation, relatively high numbers of enterococci, particularly *Enterococcus faecium*, were found to contribute to fermentation, along with lactobacilli (Fontana et al., 2009). While the presence of enterococci in food is a matter of controversy, they play an important beneficial role in the production of traditional fermented foods (Table 12.1). In addition, species from the genera *Leuconostoc*, *Carnobacterium*, and *Weissella* were sporadically isolated, underlining possible pitfalls in their identification or simply their low incidence in this ecological niche. Since the main species found in the different traditional fermented sausages reported thus far were *L. sakei* and *L. curvatus*, their robustness during the rigors of meat processing, particularly those present during sausage fermentation, should be highlighted.

# 12.4 *L. SAKEI* AND *L. CURVATUS* AS EMBLEMATIC MEAT LAB

*L. sakei* and *L. curvatus* are two lactic acid bacteria widely used worldwide as starters for meat fermentation. They are phenotypically closely related species and have often been associated as a *sakei/curvatus* group in the past. These species have also been more recently described to be phylogenetically closely related to other *Lactobacillus* species such as *Latilactobacillus fuchuensis* and *Latilactobacillus graminis*. Together they constitute the *Latilactobacillus* genera, in which many strains are psychrotrophic (Zheng et al., 2020). Genomic studies have contributed to a better characterization of each species and elucidated their specificities.

*L. sakei* is a LAB mainly found in fermented or refrigerated meat products, although it has also been isolated from several raw fermented food products of plant origin (Jiménez et al., 2017), spontaneous sourdoughs (Gänzle et al., 2019), the human gastrointestinal tract (GIT) (Chiaramonte et al., 2009), and seafood (Najjari et al., 2008). The ability of *L. sakei* to colonize such different ecological niches is based on its striking differences in physiological and biochemical characteristics compared to other lactobacilli, such as its versatile metabolism and high adaptive capacity (Chaillou et al., 2005; Zagorec and Champomier-Vergès, 2017).

*L. curvatus* is a ubiquitous LAB. Although the type strain DSM20019 was isolated from milk, *L. curvatus* is mainly associated with a diversity of niches, such as fermented vegetables (silage, sauerkraut, sourdough), decaying plant material and manure, beer-fermented meat products, vacuum-packed refrigerated meat, and fish (Terán et al., 2019). *L. curvatus* has also been isolated from cheeses (Antonsson et al., 2003). Some commercial products made with *L. curvatus* FBA2, isolated from a Japanese fermented vegetable, have been patented and marketed in Japan to improve skin health, due to the strain's ability to produce hyaluronic acid and collagen; other kimchi *L. curvatus* isolates have proven properties for the production of interleukin IL-10 in dendritic cells (Terán et al., 2018, 2019). Strain *L. curvatus* CRL 705, isolated from an Argentinean salami, is a model microorganism (Vignolo et al., 1993), mainly to produce the bacteriocins lactocin lac705 and antilisteria AL705. The genome of CRL 705 was the first *L. curvatus* genome sequenced and deposited in GenBank (Hebert et al., 2012). Lactocin lac705 is a type IIb bacteriocin, whose activity depends on the complementary action of two peptides, Lac705-alpha and Lac705-beta, of 33 amino acids each. The AL705 antilisteria activity has not yet been identified. *L. curvatus* CRL 705 can grow and produce bacteriocins in fresh meat vacuum packed and stored cold (2–8 °C), without producing significant changes in meat pH values. Likewise, the CRL 705 strain exerts an effective bioprotective effect on vacuum-packed refrigerated meats without affecting their sensory and structural characteristics (Castellano et al., 2017). On the other hand, the ability of *L. curvatus* CRL705 to hydrolyze meat proteins was evaluated in experimental systems and vacuum-packed fresh meat, as well as the influence of curing additives on bacteriocin production and proteolytic activity (Vignolo et al., 1998; Fadda et al., 2008). The body of knowledge gained on *L. curvatus* CRL705 prompted the development of an autochthonous starter culture and the analysis of its action on various technological and fundamental aspects of meat proteolysis that will be discussed later.

## 12.4.1 Genomic Diversity

The genome of *L. sakei* 23K has revealed a specialized metabolic catalog that reflects the adaptation of the bacterium to meat products, differentiating it from other LAB (Chaillou et al., 2005). Ninety-eight annotated *L. sakei* genomes are available to date, ranging in size from 1.82 to 2.19 Mb and in GC content from to 40.60 to 41.31% GC (www.ncbi.nlm.nih.gov/data-hub/genome/search date: March 11, 2024), suggesting a large inventory of strain-specific genes and therefore a wide potential phenotypic diversity of the species. A microarray-based comparative genomic analysis showed that the divergent regions might result from horizontal gene transfer (Nyquist et al., 2011) coincident with the presence of genomic islands, indicating putative hotspots of gene divergence reported for the *L. sakei* 23K genome (Chaillou et al., 2005). The relationship between isolates was revealed in a collection of strains by the possession of a subset of 60 chromosomal genes belonging to a flexible gene pool, which were used as a PCR tool to classify *L. sakei* strains into 10 clusters (Chaillou et al., 2009). The wide intraspecies diversity of *L. sakei* observed in meat strains is likely a consequence of the broad range of meat ecological niches in which these strains are involved. Possibly, strains of several genotypes could successively dominate the ecological niche during meat storage because of the dynamic variations in microbial competition, fluctuation of nutrient availability, and changing redox conditions (Chaillou et al., 2009). Nevertheless, no link could be clearly established between the genome content of the strains and their ecological origin, whatever the method used (Zagorec and Champomier-Vergès, 2017). Although the first division of *L. sakei* into two subspecies (*L. sakei* subsp. *carnosus* and *L. sakei* subsp. *sakei*) was based on DNA-RAPD and phenotypic characteristics, when a MLST analysis was performed on a larger collection (>200 strains), it was demonstrated that the *L. sakei* population derived from three ancestral lineages, each with a unique population structure; however, a strong pattern corresponding to geographical or food-type origins of strains in each lineage was found. These evolutionary clades strongly indicate that the three lineages may correspond to distinct ecotypes, likely linked to different ecological reservoirs (Chaillou et al., 2009, 2013).

Regarding *L. curvatus* CRL 705, comparative genome analysis with the closely related *L. sakei* 23K strain revealed that they are highly similar and share 1006 CDS (with a cutoff of 80% of identity and 80% of coverage). However, 860 protein-encoding genes are unique to *L. curvatus* CRL 705 and are related to proteins and enzymes involved in the metabolism of carbohydrates, DNA, and fatty acids, as well as in the stress oxidative response and bacteriocin production. However, the genome of *L. curvatus* CRL705 lacks several gene clusters, mainly those related to fatty acid biosynthesis FASII, sucrose utilization, and citrate metabolism, as well as the arginine deiminase pathway, a distinguished characteristic of *L. sakei*. Eighty-eight genomes of *L. curvatus* are available to date in the GenBank database, ranging in size between 1.76 to 2.18 Mb and with a GC content between 41.10 to 42.20% GC (www.ncbi.nlm.nih.gov/data-hub/genome/ search date: March 11, 2024). A pangenome analysis of 13 *L. curvatus* strains evidenced that this species is divided into two ancestral phylogenetic lineages. Traces of the evolutionary pathways are present not only in the allele frequencies of core genes but also in some clusters of conserved metabolic genes, including ribose, maltose and galactitol (Terán et al., 2018). The lifestyle and the ecological niche of the strains has a strong influence on the gene content of the accessory genome, which has led to convergence between strains from both lineages. *L. curvatus* has revealed a wide repertoire of genes for catabolizing plant-derived carbohydrates, and this capacity represents a major difference from the closely related species *L. sakei*.

# 12.5 STARTER CULTURES

In recent years, efforts have been made by the food industry to develop products positioned in a fluctuating space between innovation and tradition to entice consumers without neglecting safety and health benefits. In this context, the need to standardize processing and quality led to the use of starter cultures, thus avoiding reliance on the "in-house" microbiota or "back-sloping" for the meat fermentation process. The great advance in the use of starter cultures in the Unites States was achieved thanks of the work of Deibel and Niven (1957), while in Europe, Niinivaara (1955) introduced micrococci to avoid color and flavor defects. After these first experiences, Nurmi (1966) developed mixed cultures composed of lactobacilli and micrococci. Studies on the ecology of fermented sausages showed that LAB, mainly *Lactobacillus sensu lato* and coagulase negative *Staphylococcus*, represented by *Micrococcaceae*, are the main technologically bacterial groups in sausage fermentation and ripening. Attention should be paid to the functional properties of LAB, which, together with CNS, will perform biochemical, physicochemical, and microbiological reactions during fermentation and ripening of sausages (Figure 12.1). LAB should be added as a starter culture to reach a concentration of $10^7$-$10^8$ CFU/g in the meat batter (Franciosa et al., 2018, 2022).

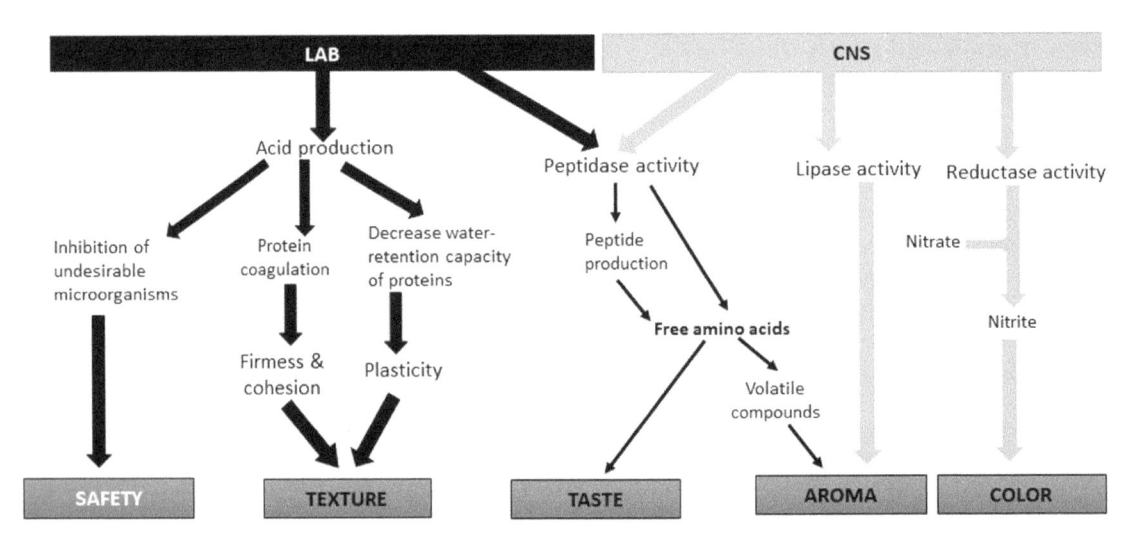

**FIGURE 12.1**  Main activities of lactic acid bacteria and coagulase negative *Staphylococcus* during fermentation and ripening of meat fermented sausages (Fadda et al., 2020).

## 12.5.1 Autochthonous Starter Cultures

Several authors support the hypothesis of the lack of competitiveness of commercial starter cultures produced in northern Europe due to their eventual inability to compete with the native microbiota of other regions of the planet. In addition, the use of commercial starters tends to generate products with great homogeneity, which puts at risk the characteristics of traditional products (Talon et al., 2007a). For this reason, the development of starter cultures formulated with strains selected from microbial communities naturally present in traditional sausages is being promoted. Based on this, the formulation of a starter culture that guarantee hygienic quality and contributes to improving the sensory characteristics of traditional products is a great challenge (Talon et al., 2007a; Franciosa et al., 2022). In this sense, the vast quantity of traditional fermented sausages from different origins constitutes a magnificent biodiversity reserve that can be exploited to design starter cultures with the characteristics of each technology and/or region. Indeed, there are attempts such as the formulation of autochthonous starter cultures composed of *P. acidilactici/Staphylococcus vitulinus* for the production of Iberian fermented sausages (Casquete et al., 2012) or regional sausages based on goat meat from Santiago del Estero (Argentina) consisting of *L. sakei* and *Staphylococcus xylosus* strains (Nediani, 2017).

# 12.6 FUNCTIONALITY OF LAB IN MEAT FERMENTATION

The main functional properties that LAB should present for an optimal dry/semi-dry fermented sausage fermentation are described as follows.

## 12.6.1 Competitiveness

During meat fermentation, competitiveness is strictly related to the ability of LAB to use the nutrients found in this matrix and adapt to the existing environmental conditions, such as temperature variations, osmotic conditions, and oxidative or acidic environments. Due to the high prevalence of *L. sakei* and *L. curvatus* in meat fermentations, their adaptation to this meat ecological niche can be inferred.

Among the few available sugars, *L. sakei* and *L. curvatus* utilize glucose and ribose for growth, sugars that are fermented through different metabolic pathways: glycolytic (homolactic) and phosphoketolase (heterolactic) pathways, respectively (Champomier-Vergés et al., 2002). Despite the basic similarity in *L. sakei* metabolic routes to ferment glucose and ribose, interesting differences were found among strains for ribose utilization. *L. sakei* catabolic machinery for pentose uptake is highly regulated at the transcription level, permitting a fine-tuning of the expression of genes to control the efficient exploitation of available meat carbon sources (McLeod et al., 2011). Moreover, some strains of *L. curvatus* (belonging to lineage 2) have a *rbsU* gene that codifies a transporter for ribose of permease type, characteristic of *L. sakei* 23K (Chaillou et al., 2005), while all the other strains have a *rbsABC* gene that encodes for an ABC-type carrier (Terán et al., 2018). Janßen et al. (2018) found that both ribose operons differ with the ribose transporter system among *L. curvatus* strains. As reported by Chaillou et al. (2005), the gene repertoire of *L. sakei* also suggested the ability of this species to use alternative carbon sources, such as nucleosides or N-acetyl-neuraminic acid, that are present in meat. The catabolism of N-acetylglucosamine and N-acetylmuramic acid is an example of divergence between the two lineages of *L. curvatus* described by Terán et al. (2018), where most strains harbored the genes which encodes the N-acetylglucosamine 6-phosphate deacetylase, while other strains had additional genes which encode the D-lactyl ether N-acetylmuramic 6-phosphate acid esterase for the catabolism of N-acetyl murein.

As a nutritious substrate, meat is a source of amino acids and peptides provided by the degradation of myofibrillar and sarcoplasmic proteins. Transcriptional and proteomic studies were carried out to evaluate

*L. sakei* adaptation to sausage fermentation and to several stressful conditions (low and high NaCl concentration, oxidative shock, etc.) (Hüfner et al., 2007; Champomier-Vergès et al., 2002; Marceau et al., 2004; Belfiore et al., 2013). When *L. sakei* cells were cultured in the presence of myofibrillar and sarcoplasmic meat protein extracts, the upregulation of genes involved in protein translation was found, suggesting that intracellular amino acids resulting from oligopeptide transport and peptidase activity could be used for new protein synthesis (Xu et al., 2015). Indeed, during *L. sakei* 23K growth on meat proteins, proteomic analysis indicated upregulation of peptidases and proteins involved in translation (Fadda et al., 2010c). Accordingly, meat proteins would not represent a stressful environment for *L. sakei*. Furthermore, its psychrotrophic nature and salt tolerance may be due to its ability for the efficient accumulation of osmo- and cryo-protective solutes such as betaine and carnitine, and to its cold stress response; *L. sakei* has more putative cold stress genes than any other lactobacilli (Chaillou et al., 2005). *L. curvatus* CRL 705's ability to adapt to curing additives (CA; NaCl, nitrites, sucrose and ascorbic acid) was studied from a physiologic and proteomic approach. Results indicated that, in response to the more stressful environment produced by CA, this strain slowed down its overall metabolism to adapt and maintain its viability, achieving minor modifications in growth/metabolic parameters and protein expression. These results enlarged the competitive profile of CRL705 as a meat starter and bio-protective culture for fermented sausage production (Terán et al., 2023).

The arginine deiminase (ADI) system present in *L. sakei*, but not in *L. curvatus*, allows cells to produce energy from arginine and contributes to their adaptation to meat fermentation conditions (Champomier-Vergés et al., 1999; Zagorec and Champomier-Vergès, 2017). Finally, the presence of several transporters presumably involved in iron or heme uptake was reported in the *L. sakei* genome (Chaillou et al., 2005). Indeed, this bacterium is one of the rare LAB possessing a heme-dependent catalase; although it is unable to synthesize heme, it could transport heme or heme-carrier molecules. The use of iron sources present in its natural environment was found to ensure *L. sakei* long-term survival (Duhutrel et al., 2010). Several *L. curvatus* strains also harbor an iron-dependent catalase gene (i.e., strains DSM20019 and CRL 705).

## 12.6.2 Meat Protein Degradation

Proteolysis of meat proteins has been widely studied, and the contribution of LAB to peptide and amino acid generation, as well as to the final sensory features of fermented products, is generally accepted. Preliminary approaches to investigating the hydrolytic role of LAB on meat proteins showed the ability of meat-borne LAB to break down beef/pork muscle sarcoplasmic and myofibrillar proteins (Fadda et al., 2010b). The genome of *L. sakei* 23K and *L. curvatus* CRL 705 revealed the absence of genes coding for any extracellular protease (Chaillou et al., 2005; Hebert et al., 2012). However, a major role of bacterial peptidases was attributed to secondary proteolysis; the intracellular peptidases from *L. sakei, L. curvatus*, and *L. plantarum* were responsible for the generation of small peptides and free amino acids (Fadda et al., 2010b). Several studies showed that initial myofibrillar protein hydrolysis was not significantly affected by LAB, the endogenous muscle proteolytic system being responsible for such activity (Spaziani et al., 2009; Fadda et al., 2010b). Nonetheless, meat protein hydrolysis was evident in the presence of lactobacilli, indicating that both microbial and endogenous proteases activated by pH decrease are involved (Fadda et al., 2010c). Analysis of proteins and peptides holds special interest because they play a major role in product quality. Proteomic approaches have been applied to correlate proteolytic profiles to detecting valuable biomarkers as meat quality predictors (Lametsch et al., 2003). Unlike proteomics, peptidomics is suited for comprehensive peptide analysis (Soloviev, 2010). Low molecular weight (LMW) peptides (<3 kDa) in fermented sausages have been regarded as responsible for flavor characteristics (Sentandreu et al., 2003; Sentandreu and Sentandreu, 2011). To define key peptides acting as potential biomarkers, LMW peptides from commercial Argentinean fermented sausages were characterized; while no specific biomarkers relating to commercial brands or quality were recognized, two different types of fermented sausages were distinguished (López et al., 2015a). Furthermore, the autochthonous starter culture (*L. curvatus* CRL705 and *S. vitulinus* GV318) was inoculated in a sausage model to evaluate its role in the proteolytic process. LMW peptides derived from sarcoplasmic and myofibrillar proteins were analyzed by two-dimensional electrophoresis (2-DE) and LC-MS/MS and complemented with amino acid profiles in order to provide a whole map of proteolysis (López et al., 2015b, 2015c) (Figure 12.2). Peptides mainly arose from myoblobin, glyceraldehyde-3-phosphate-dehydrogenase, and fructose-biphosphate-aldolase. Moreover, the hydrolysis of actin, myosin light chain, myosin regulatory light chain 2, and myosin heavy chain was evidenced and the peptide products identified. Overall, protein hydrolysis was enhanced by the starter culture (López et al., 2015b). Peptides, particularly those generated in the inoculated models, could be further proposed as reliable biomarkers specifically of this autochthonous starter culture (López et al., 2015b, 2015c). Moreover, LMW peptides generated by LAB during meat fermentation could be a valuable strategy that can mask or hide off flavors produced by the use of NaCl substituents in order to obtain healthier low-sodium fermented sausages. In fact, Almeida et al. (2018) showed that *L. curvatus* generated the higher number of LMW peptides, while *L. plantarum* and *L. sakei* produced higher free amino acid amounts in a low sodium-fermented sausage model. Each LAB strain generated a unique peptide/amino acid profile.

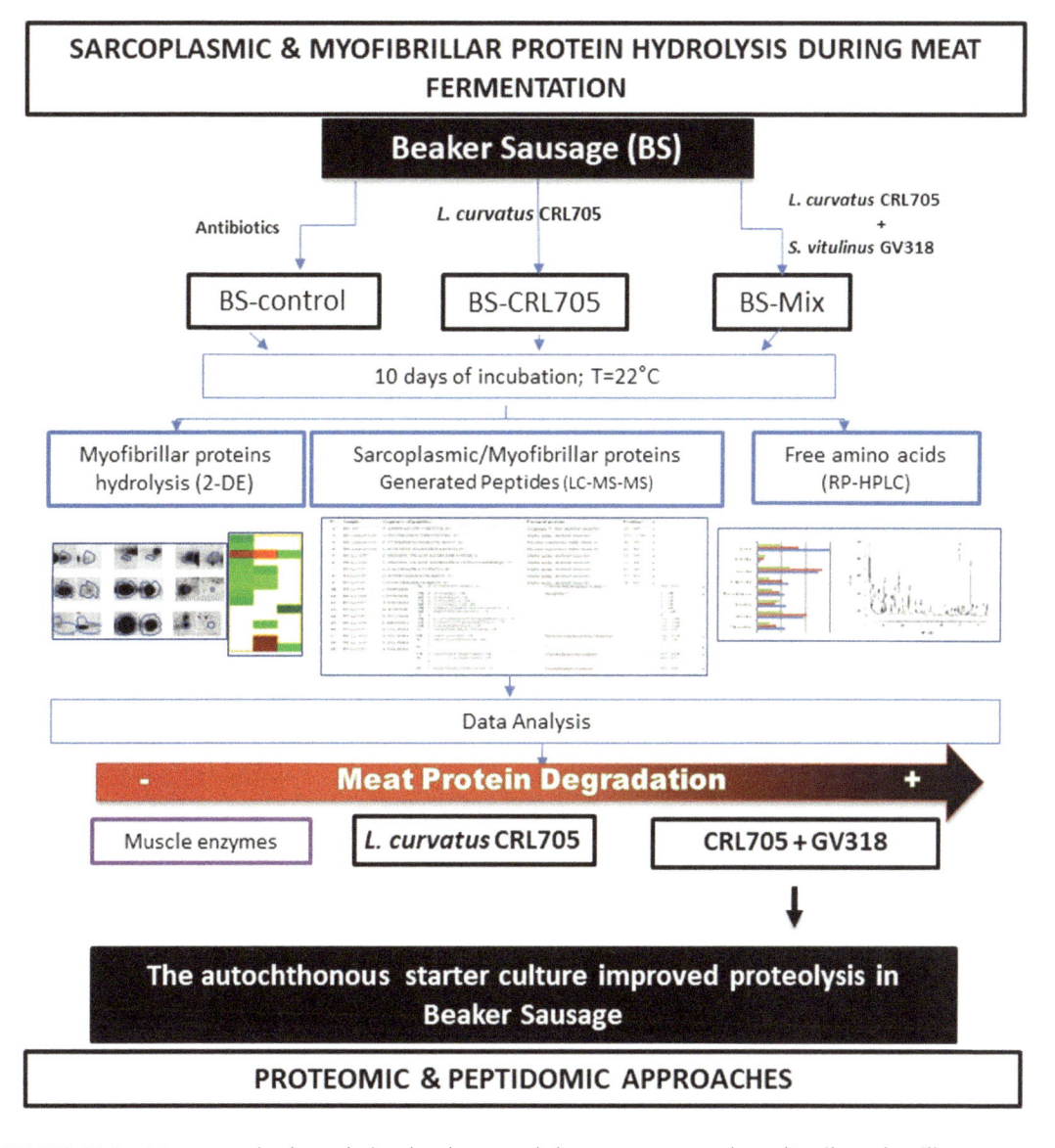

**FIGURE 12.2**   Meat protein degradation by the autochthonous starter culture (*Latilactobacillus curvatus* CRL 705–*Staphylococcus vitulinus* GV 318) evaluated in a beaker sausage model using proteomic and peptidomic approaches (López et al., 2015b, 2015c).

## 12.6.3 Safety

### 12.6.3.1 Bio-Preservation

The globalization of trade and lifestyle ensures that the factors responsible for the emergence of diseases are more present than ever. Despite biotechnology advancements, meat-based foods are still under scrutiny because of the presence of pathogens and unwanted compounds that cause loss of confidence and consequently reduced demand. LAB, as generally recognized as safe (GRAS) and Qualified Presumption of Safety (QPS) organisms, offer an alternative for pathogen-free foods, with minimal processing and fewer additives while maintaining their sensorial characteristics.

The antibacterial activity of organic acids produced by LAB and the ability to decrease pH during fermentation represent the main mechanism for bio-preservation of foods. In fact, LAB are known to produce a wide array of other antimicrobial substances, such as ethanol, diacetyl, hydrogen peroxide, reuterin, reutericyclin, antifungal compounds, and bacteriocins (Woraprayote et al., 2016, 2023). However, not only are these metabolic compounds responsible for the control and/or inhibition of undesirable microorganisms, the presence of LAB interacting with major pathogens such as *Escherichia coli* and *Listeria monocytogenes* is capable of displacing them, as previously demonstrated (Garriga et al., 2015; Orihuel et al., 2018b). In fact, the antagonistic ability of *L. plantarum* CRL681 and *Enterococcus mundtii* CRL35 against *E. coli* O157:H7 and *L. monocytogenes* was reported in meat models (Orihuel et al., 2018a, 2018b). The latest information available on the antimicrobial activity of LAB and their metabolites against meat spoilage microorganisms and foodborne pathogens is systematically revised by Barcenilla et al. (2022).

The use of LAB strains able to produce antimicrobial peptides with an impact on quality and safety during fermentation has been suggested as an extra hurdle and constitutes a promising tool (Castellano et al., 2017; Woraprayote et al., 2023). In fermented sausages, the performance of bacteriocinogenic LAB as starters or co-cultures will depend on the processing technology and the presence of nitrate/nitrite, NaCl, and spices (Leroy et al., 2005; Orihuel et al., 2018a). For the control of pathogens, functional starter cultures that produce bacteriocins and a rapid decrease in pH or a bacteriocinogenic strain in co-culture with an acidogenic LAB have been evaluated (De Souza Barbosa et al., 2015; Casaburi et al., 2016). The genomes of starter cultures of *L. curvatus*, *L. sakei*, and *L. plantarum* that inhibit the growth of *L. monocytogenes* in a variety of fermented sausages are, in fact, rich in bacteriocin genes (Fontana et al., 2015). Both *L. curvatus* and *L. sakei* carried the *sppA*, *sppQ*, and *sapA* structural genes, encoding for the sakacin P, sakacin Q, and curvacin A bacteriocins, respectively; although *L. curvatus* exhibited a higher occurrence of these genes, *L. sakei* strains were found to be more effective at inhibiting different *Listeria* species/strains. In the same study, the structural gene *plantEF* was mostly present in *L. plantarum* strains, and no *pedA* gene was detected in the assayed *P. acidilactici* strains. Although considering *Enterococcus* as safe microorganisms is controversial, a wide distribution of *entA*, *entB*, and *entP* genes in *E. faecium* strains accounted for the high antilisterial activity (Hugas et al., 2003; Fontana et al., 2015). Likewise, the bacteriocin-producing *E. casseliflavus* IM416K1 was used as a natural antagonist to control *L. monocytogenes* in Italian sausages (Sabia et al., 2003). Among the features of technological relevance that bacteriocinogenic strains must have are the good capacity to colonize food and produce bacteriocins in the food matrix, where other preservatives such as nitrites, sodium chloride, organic acids, and pepper are present (Ananou et al., 2005).

## 12.6.4 Security

In addition to the long history of safe use and association with food production and human health, the presence in fermented meat products of biogenic amines (BAs), as well as of transmissible determinants of antibiotic resistance, are risk factors for major concern.

### 12.6.4.1 Biogenic Amine Production

The large amount of proteins present and the proteolytic activity found during the ripening of meat products provide the precursors for biogenic amines, which are produced by amino decarboxylases from LAB during fermentation (Vidal-Carou et al., 2014; Ashaolu et al., 2021). Although amino decarboxylase activity is strain dependent, *L. curvatus* has been reported as a strong BA producer, whereas *L. sakei* and *L. plantarum* were usually reported as weak or non-aminogenic LAB species (Talon and Leroy, 2011; Fontana et al., 2016). Enterococci strains isolated from fermented sausages are well known for their ability to produce BA (Talon and Leroy, 2011). On the other hand, some strains of LAB supported high effectiveness for BA reduction both by the decrease and/or inhibition of endogenous aminogenic microbiota through acid production and/or by metabolizing formed BA through amine oxidase activity (Capozzi et al., 2012; Fadda et al., 2001). The use of decarboxylase-negative starter cultures that are highly competitive and fast acidifiers was reported to prevent the growth of BA producers, leading to nearly BA-free final products, when also high-quality raw materials and appropriate technological conditions are used (Latorre-Moratalla et al., 2012).

### 12.6.4.2 Antibiotic Resistance

The emergence and spread of antimicrobial-resistant bacteria is a growing public health problem. LAB may act as reservoirs of antimicrobial resistance genes that might be transferred to commensal or pathogenic bacteria (Franz et al., 2010). The food chain has been recognized as one of the main routes for the transmission of antibiotic-resistant (AR) bacteria between animal and human populations. The most common antibiotics employed for selection of lactobacilli are tetracycline and erythromycin, followed by chloramphenicol, streptomycin, ampicillin, vancomycin, and clindamycin. Antibiotic resistance in LAB isolated from meat products, such as *L. sakei, L. curvatus*, and *L. plantarum*, has been reported (Zonenschain et al., 2009); the most prevalent genetic determinants for erythromycin (*ermB*) and tetracycline (*tetL, tetM, tetS*) resistance have been identified, suggesting that horizontal gene transfer may have occurred (Zonenschain et al., 2009; Devirgiliis et al., 2013). *L. sakei* strains from spontaneously fermented Argentinean llama sausages were reported as sensitive to erythromycin, tetracycline, and clindamycin, whereas variable resistance to streptomycin and rifampicin was found (Fontana et al., 2016). Therefore, the safety of bacteria to be used as starters or probiotics must be ensured. Isolates can harbor truncated antimicrobial resistance genes without expressing phenotypic resistance; alternatively, strains may show phenotypic resistance to different antimicrobials and not harbor the generally assessed AR genes. Indeed, less dominant or unidentified genes may be involved in the resistance profile. Thus, it is recommended to analyze the main resistance genes, even if the strain does not present phenotypic resistance. If they harbor any resistance gene, the expression of this gene is important to evaluate, as well as the existence of mobile genetic elements that could transfer these determinants to other bacteria (dos Santos Cruxen et al., 2019).

## 12.7 PROBIOTIC LAB IN MEAT FERMENTED PRODUCTS

As mentioned, food technology is deeply involved in conferring positive perception of meat and meat products on consumers. One of its main strategies is the development of meat products with benefits to promote health and reduce the risk of some diseases by replacing specific components or incorporating bioactive components. In this context, the development of probiotic meat products has been discussed in the field of meat science and industry (Arihara, 2006).

Probiotic cultures have been applied mainly in dairy products but also in fruit juices and cereal products. Meat has been proposed as a good vehicle for probiotic bacteria due to its high buffering capacity and, as such, can protect LAB from the low pH environment, as well as against the lethal action of bile (Työppönen et al., 2003). Another positive aspect is that the processing of dry fermented sausage does not involve heating, thus ensuring the viability of the probiotic strains. In addition to the high numbers of LAB harbored at the end of fermentation, the survival of probiotic lactobacilli through the GIT may be promoted due to the encapsulation of microorganisms by the sausage matrix consisting of meat and fat or other materials (Klingberg and Budde, 2006; Vasconcelos et al., 2021). However, the probiotic culture must be well adapted to the sausage conditions to become dominant in the final product, as fermented meat products contain a high natural background microbiota. Certainly, in order to use probiotics as starter cultures, several additional properties are required, including desirable technological performance (i.e. lactic acid production, resistance to NaCl and nitrates), safety (inhibition of foodborne pathogens, absence of antibiotic resistance genes, and lack of aminogenic potential), and an appropriate sensory profile of the meat product, as well as the achievement of a measurable health effect on consumers (Klingberg and Budde, 2006; Ammor and Mayo, 2007). Meat-borne LAB strains have been considered as potential probiotics due to their adaptation to harsh conditions during meat fermentation and their ability to survive during GIT transit (Pennacchia et al., 2006; Ruiz-Moyano et al., 2010). Two probiotic strains, *Lacticaseibacillus rhamnosus* CTC1679 and *L. plantarum* L125, fermented low-sodium and low-fat sausages and reached high bacterial cell numbers at the end of processing without affecting the characteristic sensory properties of the product (Pavli et al., 2020). On the other hand, inoculation of *Limosilactobacillus fermentum* HL57 and *P. acidilactici* SP979 as probiotics in the manufacture of traditional Iberian dry-fermented sausages resulted in a negative impact on color and taste parameters (Ruiz-Moyano et al., 2011). Nonetheless, this matrix was able to support bacterial viability either in the product until the time of consumption or during delivery through the GIT. Thus, the importance of using competitive strains as probiotics has to be highlighted. Although the role of fermented sausages as a vehicle of probiotics was demonstrated, the validation of their effect on human health is still lacking. Indeed, more research at the clinical level is needed to justify the launch of new probiotic strains of meat origin, as well as probiotic fermented sausages.

# 12.8 CONCLUSION

LAB communities are essential constituents of meat fermented products. *L. sakei* and *L. curvatus* are emblematic LAB, considered the functional core in meat fermentations and often used as starter cultures. Their important role in meat technology, competitiveness for growth in meat under curing and drying conditions, acid production, protein degradation, bio-preservation potential, genomic diversity, and proteomic profiles, as well as the feasibility of using fermented sausages as a vehicle for probiotic LAB, were reviewed in this chapter. Fermented sausages are unique products, often represented as elements of food heritage and identity. To deal with a threatening and globalizing trend, contemporary food markets rely on the oxymoronic concept of innovation through tradition. The application of indigenous and carefully selected LAB as starter cultures could help preserve typicality with benefits for both producers and consumers.

# BIBLIOGRAPHY

Albano, H., van Reenen, C.A., Todorov, S.D., et al. (2009) Phenotypic and genetic heterogeneity of lactic acid bacteria isolated from "Alheira," a traditional fermented sausage produced in Portugal. *Meat Sci*, 82:389–398.

Almeida, M.A., Contreras-Castillo, C.J., Saldaña Villa, E., da Silva Pinto, J.S., Palacios, J., Sentandreu, M.A., Fadda, S. (2018) Study of meat protein degradation in low-sodium fermented sausage model using different autochthonous starter cultures. *Food Res Int*, 109:368–379. https://doi.org/10.1016/j.foodres.2018.04.042.

Ammor, M.S., Dufour, E., Zagorec, M., Chaillou, S., Chevallier, I. (2005) Characterization and selection of *Lactobacillus sakei* strains isolated from traditional dry sausage for their potential use as starter cultures. *Food Microbiol*, 22:529–538.

Ammor, M.S., Mayo, B. (2007) Selection criteria for lactic acid bacteria to be used as functional starter cultures in dry sausage production: An update. *Meat Sci*, 76:138–146.

Ananou, S., Garriga, M., Hugas, M., Maqueda, M., Martínez-Bueno, M., Gálvez, A., Valdivia, E. (2005) Control of *Listeria monocytogenes* in model sausages by enterocin AS-48. *Int J Food Microbiol*, 103:179–190.

Andrighetto, C., Zampese, L., Lombardi, A. (2001) RAPD-PCR characterization of lactobacilli isolated from artisanal meat plants and traditional fermented sausages of Veneto region (Italy). *Lett Appl Microbiol*, 33:26–30.

Antonsson, M., Molin, G., Ardö, Y. (2003) *Lactobacillus* strains isolated from Dambo cheese as adjunct cultures in a cheese model system. *Int J Food Microbiol*, 85(1–2):159–169.

Aquilanti, L., Garofalo, C., Osimani, A., Clementi, F. (2016) Ecology of lactic acid bacteria and coagulase negative cocci in fermented dry sausages manufactured in Italy and other Mediterranean countries: An overview. *Int Food Res J*, 23:429–445.

Aquilanti, L., Santarelli, S., Silvestri, G., Osimani, A., Petruzzelli, A., Clementi, F. (2007) The microbial ecology of a typical Italian salami during its natural fermentation. *Int J Food Microbiol*, 120:136–145.

Arihara, K. (2006) Strategies for designing novel functional meat products. *Meat Sci*, 74:219–229.

Ashaolu, T.J., Khalifa, I., Mesak, M.A., Lorenzo, J.M., Farag, M.A. (2021) A comprehensive review of the role of microorganisms on texture change, flavor and biogenic amines formation in fermented meat with their action mechanisms and safety. *Critical Rev Food Sci Nutr*:1–18.

Astudillo-García, C., Bell, J.J., Webster, N.S., Glas, B., Jompa, J., Montoya, J.M., Taylor, M.W. (2017) Evaluating the core microbiota in complex communities: A systematic investigation. *Environ Microbiol*, 19:1450–1462.

Aymerich, T., Martín, B., Garriga, M., Vidal-Carou, M.-C., Bover-Cid, S., Hugas, M. (2006) Safety properties and molecular strain typing of lactic acid bacteria from slightly fermented sausages. *J Appl Microbiol*, 100:40–49.

Barcenilla, C., Ducic, M., López, M., Prieto, M., Álvarez-Ordóñez, A. (2022) Application of lactic acid bacteria for the biopreservation of meat products: A systematic review. *Meat Sci*, 183:108661.

Bassi, D., Milani, G., Belloso Daza, M.V., Barbieri, F., Montanari, C., Lorenzini, S., Montanari, C., Lorenzini, S., Šimat, V., Gardini, F., Tabanelli, G. (2022) Taxonomical identification and safety characterization of Lactobacillaceae from Mediterranean natural fermented sausages. *Foods*, 11(18):2776.

Belfiore, C., Fadda, S., Raya, R., Vignolo, G. (2013) Molecular basis of the adaption of the anchovy isolate *Lactobacillus sakei* CRL1756 to salted environments through a proteomic approach. *Food Res Int*, 54:1334–1341.

Belleggia, L., Ferrocino, I., Corvaglia, M.R., Cesaro, C., Milanović, V., Cardinali, F., Osimani, A. (2022) Profiling of autochthonous microbiota and characterization of the dominant lactic acid bacteria occurring in fermented fish sausages. *Food Res Int*, 154:110990.

Benito, M., Martin, A., Aranda, E., Pérez-Nevado, E., Ruiz-Moyano, S., Córdoba, M. (2007) Characterization and selection of autochthonous lactic acid bacteria isolated from traditional Iberian dry-fermented salchichón and chorizo sausages. *J Food Sci*, 72:M193–M201.

Bonomo, M.G., Ricciardi, A., Zotta, T., Parente, E., Salzano, G. (2008) Molecular and technological characterization of lactic acid bacteria from traditional fermented sausages of Basilicata region (Southern Italy). *Meat Sci*, 80:1238–1248.

Capozzi, V., Russo, P., Ladero, V., Fernández, M., Fiocco, D., Alvarez, M.A., Grieco, F., Spano, G. (2012) Biogenic amines degradation by *Lactobacillus plantarum*: Toward a potential application in wine. *Front Microbiol*, 3:122–127.

Casaburi, A., Di Martino, V., Ferranti, P., Picariello, L., Villani, F. (2016) Technological properties and bacteriocins production by *Lactobacillus curvatus* 54M16 and its use as starter culture for fermented sausage manufacture. *Food Control*, 59:31–45.

Casaburi, A., Nasi, A., Ferrocino, I., Di Monaco, R., Mauriello, G., Villani, F., Ercolini, D. (2011) Spoilage-related activity of *Carnobacterium maltaromaticum* strains in air-stored and vacuum-packed meat. *Appl Environ Microbiol*, 77:7382–7393.

Casquete, R., Benito, M.J., Martín, A., Ruiz-Moyano, S., Pérez-Nevado, F., Córdoba, M.G. (2012) Comparison of the effects of a commercial and an autochthonous *Pediococcus acidilactici* and *Staphylococcus vitulus* starter culture on the sensory and safety properties of a traditional Iberian dry-fermented sausage "salchichón". *Int J Food Sci Technol*, 47:1011–1019.

Castellano, P., Pérez Ibarreche, M., Blanco Massani, M., Fontana, C., Vignolo, G. (2017) Strategies for pathogen biocontrol using lactic acid bacteria and their metabolites: A focus on meat ecosystems and industrial environments. *Microorganisms*, 5:38–62.

Chaillou, S., Champomier-Vergès, M.-C., Cornet, M. (2005) The complete genome sequence of the meat-borne lactic acid bacterium *Lactobacillus sakei* 23K. *Nat Biotechnol*, 23:1527–1533.

Chaillou, S., Chaulot-Talmon, A., Caekebeke, H., Cardinal, M., Christieans, S., Denis, C., Desmonts, M.H., Dousset, X., Feurer, C., Hamon, E., Joffraud, J.J., La Carbona, S., Leroi, F., Leroy, S., Lorre, S., Macé, S., Pilet, M.F., Prévost, H., Rivollier, M., Roux, D., Talon, R., Zagorec, M., Champomier-Vergès, M.C. (2014) Origin and ecological selection of core and food-specific bacterial communities associated with meat and seafood spoilage. *ISME J*, 9:1105–1118.

Chaillou, S., Daty, M., Baraige, F., Dudez, A.M., Anglade, P., Jones, R., Alpert, C.A., Champomier-Vergès, M.C., Zagorec, M. (2009) Intraspecies genomic diversity and natural population structure of the meat-borne lactic acid bacterium *Lactobacillus sakei*. *Appl Environ Microbiol*, 75:970–980.

Chaillou, S., Lucquin, I., Najjari, A., Zagorec, M., Champomier-Vergés, M.C. (2013) Population genetics of *Lactobacillus sakei* reveals three lineages with distinct evolutionary histories. *PLoS ONE*, 8:e73253.

Champomier-Vergès, M.-C., Maguin, E., Mistou, M.Y., Anglade, P., Chich, J.F. (2002) Lactic acid bacteria and proteomics: Current knowledge and perspectives. *J Chromatogr*, 771:329–342.

Champomier-Vergès, M.-C., Zuñiga, M., Morel-Deville, F., Pérez-Martínez, G., Zagorec, M. (1999) Relationship between arginine degradation, pH and survival in *Lactobacillus sakei*. *FEMS Microbiol Lett*, 180:297–304.

Chiaramonte, F., Blugeon, S., Chaillou, S., Langella, P., Zagorec, M. (2009) Behavior of the meat-borne bacterium *Lactobacillus sakei* during its transit through the gastrointestinal tracts of axenic and conventional mice. *Appl Environ Microbiol*, 75:4498–4505.

Cocolin, L, Alessandria, V., Dolci, P., Gorra, R., Rantsiou, K. (2013) Culture independent methods to assess the diversity and dynamics of microbiota during food fermentation. *Int J Food Microbiol*, 167:29–43.

Cocolin, L., Manzano, M., Cantoni, C., Comi, G.H. (2000) Development of a rapid method for the identification of *Lactobacillus* spp. isolated from naturally fermented Italian sausages using a polymerase chain reaction—temperature gradient gel electrophoresis. *Lett Appl Microbiol*, 30:126–129.

Cocolin, L., Manzano, M., Cantoni, C., Comi, G.H. (2001) Denaturing gradient gel electrophoresis analysis of the 16S rRNA gene V1 region to monitor dynamic changes in the bacterial population during fermentation of Italian sausages. *Appl Environ Microbiol*, 67:5113–5121.

Cocolin, L., Rantsiou, K., Iacumin, L., Urso, R., Cantoni, C., Comi, G. (2004) Study of the ecology of fresh sausages and characterization of populations of lactic acid bacteria by molecular methods. *Appl Environ Microbiol*, 70:1883–1894.

Comi, G., Urso, R., Iacumin, L., Rantsiou, K., Cattaneo, P., Cantoni, C., Cocolin, L. (2005) Characterisation of naturally fermented sausages produced in the North East of Italy. *Meat Sci*, 69:381–392.

Conter, M., Zanardi, E., Ghidini, S., Pennisi, L., Vergara, A., Campanini, G., Ianieri, A. (2008) Consumers' behaviour toward typical Italian dry sausages. *Food Control*, 19:609–615.

Danilović, B., Jokovića, N., Petrović, L., Veljović, K., Tolinački, M., Savić, D. (2011) The characterisation of lactic acid bacteria during the fermentation of an artisan Serbian sausage (Petrovská Klobása). *Meat Sci*, 88:668–674.

Deibel, R., Niven, F. (1957) *Pediococcus cerevisiae*: A starter culture for summer sausage. *Bacteriol Proc*, 14:15.

De Souza Barbosa, M., Todorov, S.D., Ivanova, I., Chobert, J-M., Haertl, T., Gombossy de Melo Franco, B.D. (2015) Improving safety of salami by application of bacteriocins produced by an autochthonous *Lactobacillus curvatus* isolate. *Food Microbiol*, 46:254–262.

Devirgiliis, C., Zinno, P., Perozzi, G. (2013) Update on antibiotic resistance in foodborne *Lactobacillus* and *Lactococcus* species. *Front Microbiol*, 4:301–313.

dos Santos Cruxen, C.E., Funck, G.D., Haubert, L., da Silva Dannenberg, G., de Lima Marques, J., Chaves, F.C., da Silva, W.P., Fiorentini, Â.M. (2019) Selection of native bacterial starter culture in the production of fermented meat sausages: Application potential, safety aspects, and emerging technologies. *Food Res Int*, 122:371–382.

Duhutrel, P., Bordat, C., Wu, T.-D., Zagorec, M., Guerquin-Kern, J.-L., Champomier-Vergés, M.-C. (2010) Iron sources used by the nonpathogenic lactic acid bacterium *Lactobacillus sakei* as revealed by electron energy loss spectroscopy and secondary-ion mass spectrometry. *Appl Environ Microbiol*, 76:560–565.

Ercolini, D. (2013) High-throughput sequencing and metagenomics: Moving forward in the culture-independent analysis of food microbial ecology. *Appl Environ Microbiol*, 79:3148–3155.

EuroFIR Consortium (2017) www.eurofir.org/news-2/news/

Fadda, S., Anglade, P., Baraige, F., Zagorec, M., Talon, R., Vignolo, G., Champomier-Vergès, M.C. (2010a) Adaptive response of *Lactobacillus sakei* 23K during growth in the presence of meat extracts: A proteomic approach. *Int J Food Microbiol*, 142:36–43.

Fadda, S., Chambon, C., Champomier-Vergès, M-C., Talon, R., Vignolo, G. (2008) *Lactobacillus* role during conditioning of refrigerated and vacuum-packaged Argentinean meat. *Meat Sci*, 79:603–610.

Fadda, S., López, C., Vignolo, G. (2010b) Role of lactic acid bacteria during meat conditioning and fermentation: Peptides generated as sensorial and hygienic biomarkers. *Meat Sci*, 86:66–79.

Fadda, S., López, C. Vignolo, G. en Alimentos Fermentados, eds. (2020) *Gabriel Vinderola; Ricardo Weill. Embutidos Fermentados Cárnicos: Contribución de Bacterias Lácticas en la calidad global*, Instituto Danone del Cono Sur. Archivo digital. www.danoneinstitute.org/wp-content/uploads/2020/12/Book-Fermented-Food-2020_sp.pdf

Fadda, S., Vignolo, G., Oliver, G. (2001) Tyramine degradation and tyramine/histamine production by lactic acid bacteria and kocuria strains. *Biotech Lett*, 23:2015–2019.

Fadda, S., Vildoza, M.J., Vignolo, G. (2010c) The acidogenic metabolism of *Lactobacillus plantarum* CRL 681 improves sarcoplasmic protein hydrolysis during meat fermentation. *J Muscle Food*, 21:545–556.

Fontana, C., Bassi, D., López, C. Pisacane, V., Otero, M.C., Puglisi, E., Rebecchi, A., Cocconcelli, P.S., Vignolo, G. (2016) Microbial ecology involved in the ripening of naturally fermented llama meat sausages. A focus on lactobacilli diversity. *Int J Food Microbiol*, 236:17–25.

Fontana, C., Cocconcelli, P.S., Vignolo, G. (2005) Monitoring the bacterial population dynamics during fermentation of artisanal Argentinean sausages. *Int J Food Microbiol*, 103:131–142.

Fontana, C., Cocconcelli, P.S., Vignolo, G., Saavedra, L. (2015) Occurrence of antilisterial structural bacteriocins genes in meat borne lactic acid bacteria. *Food Control*, 47:53–59.

Fontana, C., Gazzola, S., Cocconcelli, P.S., Vignolo, G. (2009) Population structure and safety aspects of *Enterococcus* strains isolated from artisanal dry fermented sausages produced in Argentina. *Lett Appl Microbiol*, 49:411–414.

Font-i-Furnols, M., Guerrero, L. (2014) Consumer preference, behavior and perception about meat and meat products: An overview. *Meat Sci*, 98:361–371.

Franciosa, I., Alessandria, V., Dolci, P., Rantsiou, K., Cocolin, L. (2018) Sausage fermentation and starter cultures in the era of molecular biology methods. *Int J Food Microbiol*, 279:26–32.

Franciosa, I., Ferrocino, I., Corvaglia, M.R., Giordano, M., Coton, M., Mounier, J., Cocolin, L. (2022) Autochthonous starter culture selection for Salame Piemonte PGI production. *Food Res Int*, 162:112007.

Franciosa, I., Ferrocino, I., Giordano, M., Mounier, J., Rantsiou, K., Cocolin, L. (2021) Specific metagenomic asset drives the spontaneous fermentation of Italian sausages. *Food Res Int*, 144:110379.

Franz, C.M., Cho, G.-S., Holzapfel, W.H., Gálvez, A. (2010) Safety of lactic acid bacteria. In *Biotechnology of Lactic Acid Bacteria. Novel Applications*, eds. F. Mozzi, R. Raya, G. Vignolo, pp. 341–359. Ames, IA: Wiley-Blackwell.

Gänzle, M.G. Fermented foods. In *Food Microbiology Fundamentals and Frontiers*, eds. M.P. Doyle, F. Diez Gonzalez, C. Hill, 5th ed., pp. 855–900. Washington, DC: ASM Press.

Garriga, M., Rubio, R., Aymerich, T., Ruas-Madiedo, P. (2015) Potentially probiotic and bioprotective lactic acid bacteria starter cultures antagonise the *Listeria monocytogenes* adhesion to HT29 colonocyte-like cells. *Benef Microbes*, 6:337–343. https://doi.org/10.3920/bm2014.0056.

Hebert, E.M., Saavedra, L., Taranto, M.P., Mozzi, F., Magni, C., Nader, M.E.F., Font de Valdez, G., Sesma, F., Vignolo, G., Raya, R.R. (2012) Genome sequence of the bacteriocin-producing *Lactobacillus curvatus* strain CRL705. *J Bacteriol*, 194(2):538–539. pmid:22207745

Hu, Y., Tian, Y., Zhu, J., Wen, R., Chen, Q., Kong, B. (2022) Technological characterization and flavor-producing potential of lactic acid bacteria isolated from traditional dry fermented sausages in northeast China. *Food Microbiol*, 106:104059.

Hu, Y., Wang, H., Kong, B., Wang, Y., Chen, Q. (2021) The succession and correlation of the bacterial community and flavour characteristics of Harbin dry sausage during fermentation. *LWT—Food Sci Technol*, 138:110689. https://doi.org/10.1016/j.lwt.2020.110689

Hüfner, E., Markieton, T., Chaillou, S., Crutz-Le Coq, A.M., Zagorec, M., Hertel, C. (2007) Identification of *Lactobacillus sakei* genes induced during meat fermentation and their role in survival and growth. *Appl Environ Microbiol*, 73:2522–2531.

Hugas, M., Garriga, M., Aymerich, M.T. (2003) Functionality of enterococci in meat products. *Int J Food Microbiol*, 88:223–233.

Hutkins, R.W. (2019) *Microbiology and Technology of Fermented Foods*, 2nd ed. Chigaco, IL: IFT Press.

Janßen, D., Eisenbach, L., Ehrmann, M.A., Vogel, R.F. (2018) Assertiveness of *Lactobacillus sakei* and *Lactobacillus curvatus* in a fermented sausage model. *Int J Food Microbiol*, 285:188–197.

Jiménez, E., Yépez, A., Pérez-Cataluña, A., Ramos Vásquez, E., Zúñiga Dávila, D., Vignolo, G., Aznar, R. (2017) Exploring diversity and biotechnological potential of lactic acid bacteria from Tocosh—traditional Peruvian fermented potatoes—by high-throughput sequencing (HTS) and culturing. *LWT—Food Sci Technol*, 87:567–574.

Klingberg, T.D., Axelsson, L., Naterstad, K., Elsser, D., Budde, B.B. (2005) Identification of potential probiotic starter cultures for Scandinavian-type fermented sausages. *Int J Food Microbiol*, 105:419–431.

Klingberg, T.D., Budde, B.B. (2006) The survival and persistence in the human gastrointestinal tract of five potential probiotic lactobacilli consumed as freeze-dried cultures or as probiotic sausage. *Int J Food Microbiol*, 109:157–159.

Lametsch, R., Karlsson, A., Rosenvold, K., Andersen, H.J., Roepstorff, P., Bendixen, E. (2003) Postmortem proteome changes of porcine muscle related to tenderness. *J Agric Food Chem*, 51:6992–6997.

Landeta, G., Curiel, J.A., Carrascosa, A.V., Muñoz, R., de las Rivas, B. (2013) Technological and safety properties of lactic acid bacteria isolated from Spanish dry-cured sausages. *Meat Sci*, 95:272–280.

Latorre-Moratalla, M.L., Bover-Cid, S., Veciana-Nogués, M.T., Vidal-Carou, M.C. (2012) Control of biogenic amines in fermented sausages: Role of starter cultures. *Front Microbiol*, 3:169–177.

Leroy, F., Geyzen, A., Janssens, M., De Vuyst, L., Scholliers, P. (2013) Meat fermentation at the crossroads of innovation and tradition: A historical outlook. *Trends Food Sci Technol*, 31:130–137.

Leroy, F., Lievens, K., De Vuyst, L. (2005) Modeling bacteriocin resistance and inactivation of *Listeria innocua* LMG 13568 by *Lactobacillus sakei* CTC 494 under sausage fermentation conditions. *Appl Environ Microbiol*, 71:7567–7570.

Leroy, L., Scholliers, P., Amilien, V. (2015) Elements of innovation and tradition in meat fermentation: Conflicts and synergies. *Int J Food Microbiol*, 212:2–8.

López, C.M., Bru, E., Vignolo, G.M., Fadda, S.G. (2015a) Identification of small peptides arising from hydrolysis of meat proteins in dry fermented sausages. *Meat Sci*, 104:20–29.

López, C.M., Sentandreu, M.A., Vignolo, G.M., Fadda, S.G. (2015b) Low molecular weight peptides derived from sarcoplasmic proteins produced by an autochthonous starter culture in a beaker sausage model. *EuPA Open Proteom*, 7:54–63.

López, C.M., Sentandreu, M.A., Vignolo, G.M., Fadda, S.G. (2015c) Proteomic and peptidomic insights on myofibrillar protein hydrolysis in a sausage model during fermentation with autochthonous starter cultures. *Food Res Int*, 78:41–49.

Maksimovic, A.Z., Zunabovic-Pichler, M., Kos, I., Mayrhofer, S., Hulak, N., Domig, K.J., Fuka, M.M. (2018) Microbiological hazards and potential of spontaneously fermented game meat sausages: A focus on lactic acid bacteria diversity. *LWT—Food Sci Technol*, 89:418–426.

Marceau, A., Zagorec, M., Chaillou, S., Méra, T., Champomier-Vergès, M.-C. (2004) Evidence for involvement of at least six proteins in adaptation of *Lactobacillus sakei* to cold temperatures and addition of NaCl. *Appl Environ Microbiol*, 70:7260–7268.

Marty, E., Buchs, J., Eugster-Meier, E., Lacroix, C., Meile, L. (2012) Identification of staphylococci and dominant lactic acid bacteria in spontaneously fermented Swiss meat products using PCR-RFLP. *Food Microbiol*, 29:157–166.

McLeod, A., Snipen, L., Naterstad, K., Axelsson, L. (2011) Global transcriptome response in *Lactobacillus sakei* during growth on ribose. *BMC Microbiol*, 11:145–170.

Najjari, A., Boumaiza, M., Jaballah, S., Boudabous, A., Ouzari, H.I. (2020) Application of isolated *Lactobacillus sakei* and *Staphylococcus xylosus* strains as a probiotic starter culture during the industrial manufacture of Tunisian dry-fermented sausages. *Food Sci Nutr*, 8(8):4172–4184.

Najjari, A., Ouzari, H., Boudabous, A., Zagorec, M. (2008) Method for reliable isolation of *Lactobacillus sakei* strains originating from Tunisian seafood and meat products. *Int J Food Microbiol*, 121:342–351.

National Library of Medicine, National Center for Biotechnology Information (2023) www.ncbi.nlm.nih.gov/data-hub/genome/?taxon=1599&annotated_only=true&refseq_annotation=true&typical_only=true

Nediani, M.T., García, L., Saavedra, L., Martínez, S., Lopez Alzogaray, S., Fadda, S. (2017) Adding value to goat meat: Biochemical and technological characterization of autochthonous lactic acid bacteria to achieve high-quality fermented sausages. *Microorganisms*, 5(2):26.

Niinivaara, F. (1955) Über den einfluss von Bacterienreinkulturen auf die Reifung und Umrötung der Rohwurst. *Acta Agral Fenn*, 85:1–128.

Nurmi, E. (1966) Effect of bacterial inoculation on characteristics and microbial flora of dry sausage. *Acta Agral Fenn*, 108:1–77.

Nychas, G.J., Skandamis, P., Tassou, C., Koutsoumanis, K. (2008) Meat spoilage during distribution. *Meat Sci*, 78:77–89.

Nyquist, O.L., McLeod, A., Brede, D.A., Snipen, L., Aakra, Å., Nes, I.F. (2011) Comparative genomics of *Lactobacillus sakei* with emphasis on strains from meat. *Mol Genet Genomics*, 285:297–311.

Orihuel, A., Bonacina, J., Vildoza, M.J., Vignolo, G., Saavedra, L., Fadda, S. (2018a) Biocontrol of *Listeria monocytogenes* in a meat model using a combination of a bacteriocinogenic strain with curing additives. *Food Res Int*, 107:289–296. https://doi.org/10.1016/j.foodres.2018.02.043.

Orihuel, A., Terán, L., Renaut, J., Vignolo, G.M., De Almeida, A.M., Saavedra, M.L., Fadda, S. (2018b) Differential proteomic analysis of lactic acid bacteria—*Escherichia coli* O157:H7 interaction and its contribution to bioprotection strategies in meat. *Front Microbiol*, 9:1083. https://doi.org/10.3389/fmicb.2018.01083

Pavli, F.G., Argyri, A.A., Chorianopoulos, N.G., Nychas, G.J.E., Tassou, C.C. (2020) Effect of *Lactobacillus plantarum* L125 strain with probiotic potential on physicochemical, microbiological and sensorial characteristics of dry-fermented sausages. *LWT—Food Sci Technol*, 118:108810.

Pennacchia, C., Vaughan, E., Villani, F. (2006) Potential probiotic *Lactobacillus* strains from fermented sausages: Further investigations on their probiotic properties. *Meat Sci*, 73:90–101.

Pisacane, V., Callegari, M.L., Puglisi, E., Dallolio, G., Rebecchi, A. (2015) Microbial analyses of traditional Italian salami reveal microorganisms transfer from the natural casing to the meat matrix. *Int J Food Microbiol*, 207:57–65.

Połka, J., Rebecchi, A., Pisacane, V., Morelli, L., Puglisi, E. (2015) Bacterial diversity in typical Italian salami at different ripening stages as revealed by high-throughput sequencing of 16S rRNA amplicons. *Food Microbiol*, 46:342–356.

Rantsiou, K., Urso, R., Iacumin, L, Cantoni, C., Cattaneo, P., Comi, G., Cocolin, L. (2005) Culture-dependent and -independent methods to investigate the microbial ecology of Italian fermented sausages. *Appl Environ Microbiol*, 71:1977–1986.

Ravyts, F., Steen, L., Goemaere, O., Paelinck, H., De Vuyst, L., Leroy, F. (2010) The application of staphylococci with flavour-generating potential is affected by acidification in fermented dry sausages. *Food Microbiol*, 27:945–954.

Rebecchi, A., Pisacane, V., Miragoli, F., Polka, J., Falasconi, I., Morelli, L., Puglisi, E. (2015) High-throughput assessment of bacterial ecology in hog, cow and ovine casings used in sausages production. *Int J Food Microbiol*, 212:49–59.

Ruiz-Moyano, S., Martín, A., Benito, M.J., Hernández, A., Casquete, R., Córdoba, M.G. (2010) Safety and functional aspects of pre-selected pediococci for probiotic use in Iberian dry-fermented sausages. *Int J Food Sci Technol*, 45:1138–1145.

Ruiz-Moyano, S., Martín, A., Benito, M.J., Hernández, A., Casquete, R., Córdoba, M.G. (2011) Application of *Lactobacillus fermentum* HL57 and *Pediococcus acidilactici* SP979 as potential probiotics in the manufacture of traditional Iberian dry-fermented sausages. *Food Microbiol*, 28:839–847.

Sabia, C., de Niederhäusern, S., Messi, P., Manicardi, G., Bondi, M. (2003) Bacteriocin-producing *Enterococcus casseliflavus* IM 416K1, a natural antagonist for control of *Listeria monocytogenes* in Italian sausages ("cacciatore"). *Int J Food Microbiol*, 87:173–179.

Sentandreu, M.A., Sentandreu, E. (2011) Peptide biomarkers as a way to determine meat authenticity. *Meat Sci*, 89:280–285.

Sentandreu, M.A., Stoeva, S., Aristoy, M.C., Laib, K., Voelter, W., Toldrá, F. (2003) Identification of small peptides generated in Spanish dry cured ham. *J Food Sci*, 68:64–69.

Sofos, J.N. (2008) Challenges to meat safety in the 21st century. *Meat Sci*, 78:3–13.

Soloviev, M. (2010) Peptidomics: *Divide et impera*. In *Peptidomics*, ed. M. Soloviev, pp. 3–9. New York: Humana Press.

Song, Z., Du, H., Zhang, Y., Xu, Y. (2017) Unraveling core functional microbiota in traditional solid-state fermentation by high-throughput amplicons and metatranscriptomics sequencing. *Front Microbiol*, 8:1294–1307.

Spaziani, M., Del Torre, M., Stecchini, M.L. (2009) Changes of physicochemical, microbiological, and textural properties during ripening of Italian low-acid sausages. Proteolysis, sensory and volatile profiles. *Meat Sci*, 81:71–85.

Talon, R., Lebert, I., Lebert, A., Leroy, S., Garriga, M., Aymerich, T., Drosinos, E.H., Zanardi, E., Ianieri, A., Fraqueza, M.J., Patarata, L., Lauková, A. (2007b) Traditional dry fermented sausages produced in small-scale processing units in Mediterranean countries and Slovakia. 1: Microbial ecosystems of processing environments. *Meat Sci*, 77:570–579.

Talon, R., Leroy, S. (2011) Diversity and safety hazards of bacteria involved in meat fermentations. *Meat Sci*, 89:303–309.

Talon, R., Leroy, S., Lebert, I. (2007a) Microbial ecosystems of traditional fermented meat products: The importance of indigenous starters. *Meat Sci*, 77:55–62.

Terán, L.C., Coeuret, G., Raya, R., Zagorec, M., Champomier-Vergès, M.C., Chaillou, S. (2018) Phylogenomic analysis of *Lactobacillus curvatus* reveals two lineages distinguished by genes for fermenting plant-derived carbohydrates. *Genome Biol Evol*, 10(6):1516–1525. https://doi.org/10.1093/gbe/evy106

Terán, L.C., Orihuel, A., Bentencourt, E., Raya, R., Fadda, S. (2023) Role of curing agents in the adaptive response of the bioprotective *Latilactobacillus curvatus* CRL 705 from a physiologic and proteomic perspective. *Bacteria*, 2(4):142–154. https://doi.org/10.3390/bacteria2040011

Terán, L.C., Raya, R.R., Zagorec, M., Champomier-Vergès, M.-C. (2019) Genetics and Genomics of *Lactobacillus sakei* and *Lactobacillus curvatus*. In *Lactobacillus Genomics and Metabolic Engineering*, ed. S.M. Ruzal, pp. 19–30. Caister Academic Press. https://doi.org/10.21775/9781910190890.02

Työppönen, S., Petäjä, E., Mattila-Sandholmet, T. (2003) Bioprotectives and probiotics for dry sausages. *Int J Food Microbiol*, 83:233–244.

Urso, R., Comi, G., Cocolin, L. (2006) Ecology of lactic acid bacteria in Italian fermented sausages: Isolation, identification and molecular characterization. *Syst Appl Microbiol*, 29:671–680.

Vasconcelos, L.I.M., da Silva-Buzanello, R.A., Kalschne, D.L., Scremin, F.R., Bittencourt, P.R.S., Dias, J.T.G., Canan, C., Corso, M.P. (2021) Functional fermented sausages incorporated with microencapsulated *Lactobacillus plantarum* BG 112 in Acrycoat S100. *LWT—Food Sci Technol*, 148:111596.

Vidal-Carou, M.C., Veciana-Nogués, M.T., Latorre-Moratalla, M.L., Bover-Cid, S. (2014) Biogenic amines: Risks and control. In *Handbook of Fermented Meat and Poultry*, eds. F. Toldrá, Y. H. Hui, I. Astiasarán, J.G. Sebranek, R. Talon, 2nd ed., pp. 413–428. West Sussex. Wiley Blackwell.

Vignolo, G., Fadda, S., Kairuz, M., Ruiz Holgado, A., Oliver, G. (1998) Influence of curing additives on the control of *L. monocytogenes* by lactocin 705 in meat slurry. *Food Microbiology*, 15:259–264.

Vignolo, G., Fontana, C., Cocconcelli, P.S. (2010a) New approaches for the study of lactic acid bacteria biodiversity: A focus on meat ecosystems. In *Biotechnology of Lactic Acid Bacteria. Novel Applications*, eds. F. Mozzi, R. Raya, G. Vignolo, pp. 251–271. Ames, IA: Willey-Blackwell.

Vignolo, G., Fontana, C., Fadda, S. (2010b) Semidry and dry fermented sausages. In *Handbook of Meat Processing*, ed. F. Toldrá, pp. 379–398. Ames, IA: Willey-Blackwell.

Vignolo, G., Suriani, F., Holgado, A., Oliver, G. (1993) Antibacterial activity of *Lactobacillus* strains isolated from dry fermented sausages. *J Appl Microbiol*, 75:344–349.

Villani, F., Casaburi, A., Pennacchia, C., Filosa, L., Russo, F., Ercolini, D. (2007) Microbial ecology of the Soppressata of Vallo di Diano, a traditional dry fermented sausage from Southern Italy, and in vitro and in situ selection of autochthonous starter cultures. *Appl Environ Microbiol*, 73:5453–5463.

Woraprayote, W., Janyaphisan, T., Adunphatcharaphon, S., Sonhom, N., Showpanish, K., Rumjuankiat, K., Petchkongkaew, A. (2023) Bacteriocinogenic lactic acid bacteria from Thai fermented foods: Potential food applications. *Food Biosci*, 102385.

Woraprayote, W., Malila, Y., Sorapukdee, S.; Swetwiwathana, A., Benjakul, S., Visessanguan, W. (2016) Bacteriocins from lactic acid bacteria and their applications in meat and meat products. *Meat Sci*, 120:118–132. https://doi.org/10.1016/j.meatsci.2016.04.004.

Xu, H.Q., Gao, L., Jiang, Y.S., Tian, Y., Peng, J., Xa, Q., Chen, Y. (2015) Transcriptome response of *Lactobacillus sakei* to meat protein environment. *J Basic Microbiol*, 55:490–499.

Zagorec, M., Champomier-Vergès, M.-C. (2017) *Lactobacillus sakei*: A starter for sausage fermentation, a protective culture for meat products. *Microorganisms*, 5:56–68.

Zheng, J., Wittouck, S., Salvetti, E., Franz, C.M., Harris, H.M., Mattarelli, P., Lebeer, S. (2020) A taxonomic note on the genus Lactobacillus: Description of 23 novel genera, emended description of the genus *Lactobacillus* Beijerinck 1901, and union of *Lactobacillaceae* and *Leuconostocaceae*. *Int J Syst Evol Microbiol*, 70(4):2782–2858.

Zonenschain, D., Rebecchi, A., Morelli, L. (2009) Erythromycin- and tetracycline-resistant lactobacilli in Italian fermented dry sausages. *J Appl Microbiol*, 107:1559–1568.

# Lactic Acid Bacteria in Vegetable, Fruit, and Seed Fermentations

# 13

## Sofia D. Forssten and Arthur C. Ouwehand

## 13.1 INTRODUCTION

The tradition of fermenting vegetables with salt in low- or no-oxygen conditions is widespread and has helped to preserve the lifespan of foods. It is thought that cucumbers were first fermented around 2000 BCE in the Middle East, early written records of cucumber pickles come from paper fragment remains of a play by the Greek writer Eupolis (429–412 BCE), and pickles are also mentioned several times in the Christian Bible. The Korean-style fermented cabbage kimchi is believed to have originated over 3000 years ago in the primitive pottery age from the natural fermentation of withered vegetables stored in seawater.

Today the primary retail fermented vegetable products produced in Europe and the United States are cucumber pickles, olives, and sauerkraut, while Asia has a broader variety of fermented vegetable products, including pickles and fermented cabbage, notably kimchi in South Korea, although many of these are gaining popularity in the Western countries. Cucumbers, olives, cocoa, capers, and soybeans, as well as coffee, are included in this chapter, as they are consumed at large scale worldwide, although they are considered fruits and seeds, respectively.

Lactic acid bacteria (LAB) are microorganisms representing over 60 genera that can ferment carbohydrates to produce lactic acid, and LAB have been and are still widely used for fermentation of vegetables. LAB are divided into homofermentatives, those that primarily produce lactic acid in glucose fermentation, and heterofermentatives, those that produce ethanol/acetic acid and $CO_2$ in addition to lactic acid. The major LAB found in fermented vegetables are *Lactobacillus acidophilus*, *Lactococcus lactis* subsp. *lactis*, *Lactobacillus delbrueckii* subsp. *lactis*, *Ligilactobacillus. salivarius*, *Lactiplantibacillus plantarum*, *Latilactobacillus sakei* subsp. *sakei*, *Streptococcus thermophilus*, *Pediococcus acidilactici*, *P. damnosus*, *P. pentosaceus*, and *Enterococcus faecalis*, which are classified as homofermentative LAB, and *Leuconostoc mesenteroides*, *Ln. paramesenteroides*, *Ln. dextranicum*, *Levilactobacillus brevis*, *Limosilactobacillus fermentum*, *Weissella confusa*, and *Limosilactobacillus fermentum*, which are classified as heterofermentative LAB (Zaunmüller et al. 2006; Swain et al. 2014; Wang et al. 2021; Sugahara et al. 2022). Some LAB are facultative heterofermentative; that is, they can either produce lactic acid, acetic acid, and carbon dioxide similar to heterofermentative bacteria or only produce lactic acid, depending on the type of sugars available for fermentation (Zaunmüller et al. 2006).

DOI: 10.1201/9781003352075-15

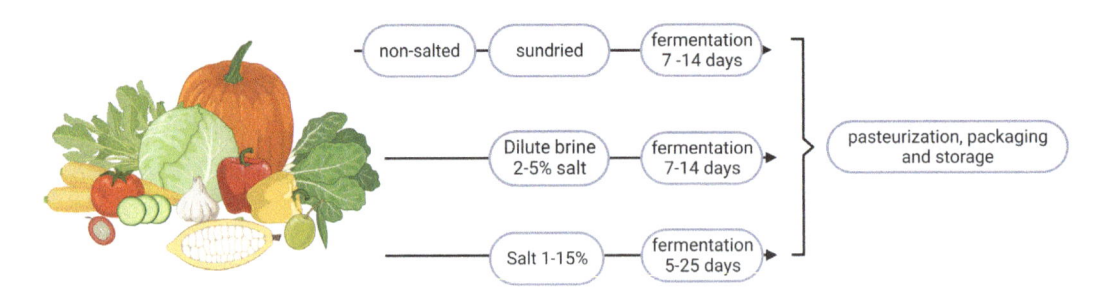

**FIGURE 13.1**  Fermentation processes of vegetables, fruits, and seeds. Figure created with BioRender.com.

Fermentation preserves food; however, it is a process that also improves flavor, texture, and health functionalities, usually without adding preserving agents. Vegetables contain normally Gram-negative aerobic bacteria, yeasts, and Gram-positive LAB. The chemical and physical environment impact the growth of LAB, for example, variations in raw materials, nutrients available, concentration of salt and $O_2$, and pH. The addition of salt suppresses Gram-negative bacteria, yeasts, and molds during early-stage fermentation (Pérez-Díaz et al. 2019). Spoilage and pathogenic microorganisms can be suppressed during vegetable fermentation by the production of lactic or other organic acids and by decreasing the pH (Cheigh and Park 1994). This has been demonstrated through studies of the sequential growth of LAB during sauerkraut fermentation (Harris 1998). Heterofermentative LAB, especially *Ln. mesenteroides*, can initially grow at higher pH. The pH usually starts at about 6.3 in the first stage of the fermentation. When pH is decreased to about 4.5, the growth of *Leuconostoc* species becomes inhibited and more acid-tolerant LAB, homofermentative LAB (e.g., *L. plantarum*), predominate and produce lactic acid, which decreases the pH to 4.0 or less (Harris 1998).

Numerous studies exist on processing methods and fermentation of vegetables, and although previously limited research has been done, a renewed interest in fermented foods has been observed over the last years in Western countries, largely driven by their supposed health benefits (Marco et al. 2017). Figure 13.1 shows the overall fermentation processes for vegetables, fruits, and seeds. The health benefits reported for different fermented vegetables will be presented in Sections 13.3 and 13.4.

# 13.2 PROCESSING METHODS AND MICROORGANISMS IN FERMENTED VEGETABLES

Today vegetable fermentation is done both in households as well as on a large-scale setting in factories across the world. Five components are required independently of what or where the fermentation takes place: the raw material = vegetables, salt, a container, some weight, and the fermentation "cultures". There are a broad variety of raw material being fermented or pickled, such as vegetables, fruits and cassava, but also products that are maybe not immediately recognized as being fermented, such as cacao, coffee, and tea (El Sheikha 2018; Behera et al. 2020).

Salt (without additives) is an important component in the fermentation process, as it not only gives flavor to the final product but also allows water and sugars to be withdrawn from the vegetables that can be used as nutrients for the fermenting organisms, thus favoring the growth of the fermenting bacteria rather than bacteria associated with spoilage. This also allows longer fermentation. Salt hardens the plant pectins while decreasing the activity of pectinase that makes vegetables mushy. The concentration of salt is very important in fermentations and may vary from 1–15%. Salt can be added either directly or by brining, that is, using salt mixed with water (Katz 2020).

As fermentation occurs in salty and acidic conditions, the container material must be chosen accordingly. The container chosen should be easy to clean and free from deep scratches, chips, or pits that can harbor harmful bacteria or that can affect the fermentation. In addition, containers that might leak harmful substances into the fermented vegetables should be avoided. The shape or style of the container does not affect product safety but could affect quality. Some weight needs to be used to keep the vegetables submerged in the container; thus a weight that completely fills the inside of the container and covers the vegetables as completely as possible should be chosen.

Before fermentation, the raw material (vegetables or fresh fruits) contains a variety of microorganisms, such as strains associated with aerobic spoilage like Pseudomonas, Erwinia, and Enterobacter species, as well as yeasts and molds (Nguyen-the and Carlin 1994). These bacterial populations will spoil the vegetables if allowed to grow, and they range from $10^4$ to $10^6$ CFU/g. Hence, brining vegetables for fermentation leads to the production of organic acids and a variety of antimicrobial compounds by LAB (Steinkraus 1992). Initially, LAB are present in lower numbers, $10^2$ to $10^3$ CFU/g, compared with other mesophilic microorganisms, but their growth increases during fermentation due to the diffusion of organic acids into the brine that results in a lower pH (Breidt et al. 2012).

Traditionally fermented pickles are homemade products obtained through spontaneous fermentation; that is, the vegetables are usually fermented by bacteria naturally present in the environment. However, isolated LAB can be used as starter cultures for the development of functional vegetable products with greater consistency and improved functionality. As quality, safety, and mass production issues must be addressed, it requires that raw materials, microbial ecosystems, and fermentation processes be controlled (Lan et al. 2013). If starter cultures are used for the fermentation, the following factors need to be considered: the rate and total production of acids, change in pH, decrease of $NO_3/NO_2$ concentration and production of biogenic amines, type of metabolism, and the ability of the culture to create desirable sensory properties. Bacteriocin-producing starter cultures may be better for producing more controlled and reproducible vegetable fermentations. In addition, the health effect of the fermented vegetables is also an important factor for consumers.

There are two types of fermented pickles, sour or sweet pickles. Sour fermented pickles are made by submerging the raw materials in a dilute brine (2–5% salt). Naturally occurring bacteria grow over 1–2 weeks to produce lactic acid, which then prevents the growth of food poisoning bacteria and other spoilage microorganisms. The amount of salt added controls the type and rate of the fermentation. In sweet fermented pickles, the raw material is preserved by a combination of lactic or acetic acid, sugar, and spices. Fermentation occurs at a salt concentration of 5%, at 20 °C–27 °C for 2–3 weeks. The pH becomes 3.3–3.5 and acidity 1.1% after the fermentation. The fermented pickles are then desalted and processed into various products, potentially including pasteurization at 74 °C for 15 minutes to increase the shelf life of the fermented vegetables (Sarmad Ghazi and Sadiq Jaafir Aziz 2021).

## 13.2.1 Functionality of Plant-Derived LAB

Sugiyama (2009) suggested that plant-derived LAB are superior to animal-derived LAB due to their acid tolerance, immune-stimulating activity, and intestinal regulation. Some strains of plant LAB produced anti-*Helicobacter pylori* substances when cultured in fruit juice. *P. pentosaceus* LP28 (LP28) decreased the amount of subcutaneous fat and hepatic triglyceride and cholesterol levels in a mouse model of diet-induced obesity. Obesity-related genes, as determined by reverse transcription-polymerase chain reaction (PCR), were repressed by oral intake of the LP28 strain. Jin et al. (2010) indicated that plant-derived LAB (*Lb. plantarum* SN13T and SN35N and *Lb. brevis* 925A) were more viable in artificial gastric fluid and bile than animal-derived LAB (*Lc. lactis*, *Lb. bulgaricus*, and *Lb. acidophilus*). It was reported that yogurt prepared with plant-derived *Lb. plantarum* SN13T and SN35N (98:2) exhibited superior probiotic effects for improving constipation, serum lipids, and liver function as compared with the yogurt made with animal-derived LAB *Lc. lactis* A6, *S. thermophilus* 510, and *Lb. bulgaricus* C6 (86.1:13.8:0.1) in human trials (Higashikawa et al. 2010).

## 13.2.2 Sauerkraut

In German, the word "sauerkraut" means sour cabbage, a food very popular in many European countries. Sauerkraut is usually produced by spontaneous fermentation. Northeastern China has a similar food, suancai, pickled Chinese cabbage (napa cabbage) with a similar taste as sauerkraut.

The acid is produced by the natural or starter lactic acid fermentation bacteria in salted-shredded cabbages. Mild-flavored, sweet, white cabbage (*Brassica oleracea*) is used as the main cabbage. The cabbage is thinly sliced in 0.78–16-mm long and finely cut shreds, and salt (2%–3%) is sprinkled over the shredded cabbage. The salted cabbage is packed into a tank and is quickly surrounded by brine and covered with plastic sheeting to ensure air exclusion (Di Cagno et al. 2016). Depending on the temperature, "spontaneous" fermentation will start within a few hours to 1–2 days and will continue between 7 days and several weeks. The ideal temperature and salt (NaCl) concentration are 18 °C and 1.8%–2.25%, respectively. The pH of the final product is from 3.5 to 3.8. Sodium benzoate (0.1%) and potassium metabisulfite can be added as preservatives to unpasteurized products. Sauerkraut can also be pasteurized at 74 °C–82 °C for 3 minutes and then canned or put into glass jars (Montaño et al. 2016).

### 13.2.2.1 Fermentation of Sauerkraut

Sauerkraut fermentation typically relies on a sequential microbial process involving heterofermentative and homofermentative LABs such as *Leuconostoc* spp. and *Weissella* spp. in the early phase and *Lactobacillus* spp., *Lc. lactis*, and *Pediococcus* spp. in the subsequent phases (Di Cagno et al. 2016) The increase of lactic acid concentration inhibits the multiplication of *Ln. mesenteroides*, while it promotes the growth of acid-tolerant species, such as *L. brevis* and in some cases *Lb. sakei* subsp. *carnosus* (formerly *Lactobacillus curvatus*), *Latilactobacillus sakei*, *E. faecalis*, *Lc. lactis*, and *P. pentosaceus*. The end products resulting from both stages of fermentation may include mannitol and acetic acid (~1% each) and lactic acid, which may exceed 2%, depending on how long homolactic fermentation is allowed to continue. *Weissella* and *Leuconostoc citreum* were found in the heterolactic phase of the fermentation; *Lb. plantarum* predominated in the late stage, leading to further acidification and a final pH of ~3.5–3.8. Two LAB species expected to be present (*P. pentosaceus* and *L. brevis*) were apparently minor constituents of the microbiota of commercial sauerkraut fermentations (Harris 1998). Species of LAB found in commercial sauerkraut fermentations by DNA fingerprinting include *Ln. citreum*, *Leuconostoc argentinum*, *Lb. paraplantarum*, *Loigolactobacillus coryniformis*, *Weissella* sp., and *Ln. fallax* (Swain et al. 2014). *Ln. mesenteroides* when grown in cabbage juice had a shorter lag phase and a more rapid generation time than any of the other organisms associated with sauerkraut fermentation (Pederson 1979). Single or mixed starter cultures of *Ln. mesenteroides*, *Lb. brevis*, and *Lb. plantarum* improved the sauerkraut quality over noninoculated natural fermentation. Cabbage fermented with *Ln. mesenteroides* consistently resulted in sauerkraut with a firm texture and reduced off-flavors in 0.5%–2.0% NaCl (Swain et al. 2014). In one fermentation, a mineral salt (3.5 kg) containing 57% NaCl and 28% KCl, 12% Mg sulfate, 2% lysine HCl, and 1% silicon dioxide was prepared. The use of mineral salt with a low level of salt (0.5%) for cabbage (400 kg) fermentation with commercial starter of *Ln. mesenteroides* resulted in mild-tasting sauerkraut juice with good sensory and microbiological quality (Wiander and Ryhanen 2005). When *L. curvatus* 2775 and *L. plantarum* DSM20174 were used as starters, the production of biogenic amines was lowered during sauerkraut fermentation (Halász et al. 1994).

## 13.2.3 Kimchi

Several different types of kimchi exist, due to usage of various raw materials and processing methods. Kimchi is characterized by its palatability, giving sour, sweet, and carbonated tastes. The cabbage is cut and washed before being brined overnight in 10% brine, and then the brined baechu cabbage is rinsed with water before finally draining excess water (Cheigh and Park 1994).

More than 100 types of vegetables can be used to prepare kimchi (Kim and Chun 2005). The content of kimchi depends on family tradition, economics, and seasonal and regional availability of the materials. The most widely used main ingredients are cabbage and radish. Watercress, mustard leaves, pears, apples, pine nuts, chestnuts, gingko nuts, cereals, fish, crabs, meats, and other ingredients can all be incorporated into kimchi (Cheigh and Park 1994). The premixture of mixed or stuffed cabbage is packaged and then fermented at different temperatures, but a low temperature (5 °C) is ideal for preparing a good-tasting product. Baechu (Chinese cabbage; *Brassica rapa* Pekinensis Group) kimchi is the most common among the 161 kinds of kimchi made in Korea. The standardized composition of baechu kimchi ingredients is as follows: brined baechu cabbage (100%) is mixed with 13% sliced radish, 2% green onion, 3.5% red pepper powder, 1.4% garlic, 0.6% ginger, 2.2% fermented anchovy juice, 1.0% sugar, and a final salt level of 2.5% (Cho 1999). The effects of salinity (1.4%, 1.7%, 2.0%, 2.2%, and 2.5%) on kimchi fermentation have also been investigated, and the salinity affected both the abundances of microbial communities and concentrations of metabolites in kimchi (Lee et al. 2021).

## 13.2.3.1 *Microorganisms in Kimchi Fermentation*

Kimchi fermentation is dominated by *Leuconostoc* sp. at 5 °C, but *Lactobacillus* sp. are the major LAB when fermentation occurs at 25 °C or higher temperatures (Jung et al. 2014). Although *Lactobacillus* sp. and *Leuconostoc* sp. were the major LAB participating in the fermentation, the level of *Leuconostoc* sp. was much higher at 5 °C than at 20 °C fermentation (Lee et al. 1992). Cho et al. (2006a) reported that they isolated 970 bacteria during kimchi fermentation in a kimchi refrigerator. The representative genera of 15 species were *Lactobacillus*, *Leuconostoc*, and *Weissella*. *Ln. citrus* and *Ln. gasicomitatum* predominated during the first growth stage; however, *Weissella koreensis* predominated during the second stage. The population dynamics appeared to be markedly influenced by the temperature and seasonal variations in raw materials.

*Lactococcus*, *Lactobacillus*, *Pediococcus*, *Leuconostoc*, and *Weissella* spp. have been isolated from kimchi, with *Leuconostoc*, *Weissella*, and *Lactobacillus* as the key players. Choi et al. (2002) identified the genus *Weissella* from kimchi in the form of the novel species *Weissella kimchii*. The LAB profile during kimchi fermentation varies with pH and acidity. *Ln. mesenteroides* is observed during early fermentation (pH 5.64–4.27 and acidity 0.48%–0.89%), and *Lb. sakei* becomes dominant later in the fermentation (pH 4.15 and acidity 0.98%) (Cho et al. 2009). A distinct subset LAB related to kimchi fermentation is greatly influenced by temperature. *Lb. sakei* predominates in kimjang kimchi, and the strain appears suitable for low fermentation (5 °C–9 °C) and storage (–2 °C) temperatures (Lee et al. 2008a). After fermentation, there is about $10^{8-9}$ CFU LABs per g kimchi. Ahn et al. (2003) isolated 27 bacterial strains from kimchi. They identified *Ln. mesenteroides*, *Ln. carnosum*, *L. curvatus*, *Lb. pentosus*, *W. kimchii*, *W. cibaria*, and *P. pentosaceus*. Most of the clones isolated from five commercially produced baechu kimchi were LAB (*Lactobacillus*, *Leuconostoc*, and *Weissella*).

A metatranscriptomic analysis of LAB gene expression during kimchi fermentation showed that *Lc. mesenteroides* was most active during the early stages of the fermentation, whereas gene expression by *L. sakei* and *W. koreensis* was high during later stages (Jung et al. 2013). However, gene expression by *L. sakei* decreased rapidly after 25 days of fermentation, which was hypothesized to be caused by bacteriophage infection of the species. Many genes related to carbohydrate transport and hydrolysis and lactate fermentation were actively expressed, which indicated typical heterolactic acid fermentation.

Various starters for kimchi fermentation have been studied (Lee et al. 2015, 2020; Lee and Kim 1988; So et al. 1996; Jung et al. 2012; Chang and Chang 2010). By using different LAB starter strains as well as inoculation ratios, the type and concentration of metabolites in kimchi fermentation can be affected, and the manufacturing of kimchi can be standardized. As examples, the use of mixed strains as starters of *Lb. plantarum*, *Lb. brevis*, *Pediococcus cerevisiae*, and *Ln. mesenteroides* isolated from kimchi shortened the fermentation time but also increased the flavor of the kimchi and showed consistent quality (Lee and

Kim 1988). In another study, using psychotropic kimchi LAB as starters resulted in sharp decreases in Gram-negative bacteria and coliforms from the initial stage and shortened fermentation time from 10 to 4 days at 8 °C (So et al. 1996), while by using *Ln. citreum* GJ7, a bacteriocin-producing strain, as a starter, it increased texture, sensory evaluation, and shelf life (Chang and Chang 2010). *Ln. mesenteroides* and *Lb. sakei* were used in one study with the aim to establish a mixed starter culture to standardize the flavor of kimchi. The strains were selected for the culture based on their key roles in kimchi fermentation, and the results showed a distinct difference in kimchi metabolites depending on the LAB starter culture used (Lee et al. 2020).

## 13.2.4 Soybeans

Soybeans are a widely grown crop and one most abundant sources of vegetable proteins, with a composition of about 40% protein, 20% lipids, 35% carbohydrates, 5% minerals, and 10% moisture, in addition to other compounds such as fatty acids, vitamins, flavonoids, isoflavones, phenolic acids, and saponins (Reddy and Duke 2015; Jia et al. 2020). The consumption of fermented soy is widespread in Asia, the main soybean products being natto, miso, tofuyo (Japan); douchi and sufu (China); cheonggukjang, doenjang, kanjang, and meju (Korea); tempeh (Indonesia); thua-nao (Thailand); and kinema, hawaijar, and tungrymbai (India) (do Prado et al. 2022).

### 13.2.4.1 Fermented Soybeans

Fermented soybean products differ from each other due to several parameters, the main one being the microorganisms used. This leads to fermented products with different textures, aromas, and therapeutic and nutraceutical values. For some fermentations, only bacteria are used, while others use only filamentous fungi, although often both microbial groups are used. Miso and tofu are two examples of products where LAB are used for fermentation. The fermented soybean paste miso is prepared by mixing cooled cooked soybeans with koji, a starter culture, often fermented rice, containing mold (*Aspergillus oryzae* or *Aspergillus sojae*), LAB (*Pediococcus halophilus*), and osmophilic yeasts; *Saccharomyces rouxii*, *Candida (Torulopsis) versatilis*, *Candida (Torulopsis) etchellsii* (Sugiyama 1984), and salt water (final concentration is about 10%). This mix is then fermented for several months (Wilson 1995).

Tofu is maybe the best-known food made of soy, and it is produced from water-extracted and salt- or acid-precipitated soybean in the form of a curd. The curd resembles a very firm yogurt or a soft white cheese. Other ways of producing tofu have been evaluated (Li et al. 2017; Wang et al. 2020), such as the use of *Lb. casei* combined with salt coagulants ($MgCl_2$, $MgSO_4$, $CaCl_2$, $CaSO_4$) or fermenting soymilk-specific LAB *Lb. casei and Lb. acidophilus* alone or in combination to improve product stability and quality (Mital and Steinkraus 1979; Serrazanetti et al. 2013). The choice of fermenting organisms is limited to those that can ferment the sugars typical of soy milk, stachyose, raffinose, or sucrose, unless sugars fermented by the desired culture(s) are added to the soy milk. The use of stachyose and raffinose in soy milk should decrease its tendency to produce flatulence in the intestinal tract and therefore improve digestibility and acceptability. Fermented tofu, by means of soymilk fermentation with selected LAB strains and subsequent protein precipitation, has been developed to prevent or delay undesired microbial and chemical spoilage (Serrazanetti et al. 2013).

To produce more stable fermented pickled tofu and "stinky" tofu, a spontaneous fermentative process of the brine surrounding tofu has been used. To produce more stable fermented tofu, dry tofu is superficially slowly fermented with aerobic bacteria added with Chinese wine, vinegar, or red yeast rice (Shurtleff and Aoyagi 2011). "Stinky" tofu is produced by soaking ready-made tofu in naturally fermented stinky brine, where the fermentation metabolites permeate it. From stinky brines, mainly *Lb. fermentum*, *Lb. delbruekii*, *Lb. salivarius*, *Ln. citerum*, *Ln. mesenteroides* and *Ln. pseudomesenteroides,* and *Pediococcus* spp. have been identified (Chao et al. 2008).

## 13.2.5 Fermented Vegetable Juices

Vegetable juices represent an alternative to dairy-fermented products, especially for vegan, strict vegetarian, or allergic consumers. Lactic acid–fermented vegetable juices are produced mainly from cabbage, red beet, horseradish, beet, carrot, celery, and tomato (Karovičová and Kohajdová 2003). Cabbage, pH-adjusted tomato (7.2), carrot, red beet, and spinach are recommended as good vegetables for preparing vegetable juices since they have more fermentable sugars than other vegetables.

Fermented vegetable juices can be prepared by either fermenting the vegetable in the usual way and then pressing the juice or mashing the vegetable or preparing it as a juice (that can be pasteurized) before inoculation with selected starter LAB cultures with $5 \times 10^6$ to $1 \times 10^7$ CFU/g or CFU/mL fermentation. A temperature of 20 °C–30 °C is usually used for the fermentations, and by using lower temperatures, better flavor and taste can be obtained. During fermentation, the pH of juices decreases from 6.0–6.5 to 3.8–4.5.

*Lb. plantarum* C3, *Lb. casei* A4, and *Lb. delbrueckii* D7 were used to produce probiotic cabbage juice (Yoon et al. 2006). The cabbage juice was inoculated ($10^5$ CFU/ml) with 24-h old LAB cultures and incubated at 30 °C for 72 h. After fermentation the samples were stored at 4 °C for 4 weeks. All three strains grew well during fermentation and reached nearly $10 \times 10^8$ CFU/mL after 48 h of fermentation at 30 degrees. While *Lb. plantarum* and *Lb. delbrueckii* were capable of surviving in the fermented cabbage juice during cold storage, *Lb. casei* lost cell viability completely after only 2 weeks of cold storage. In conclusion, *Lb. plantarum* and *Lb. delbrueckii* could be used as probiotic cultures for fermentation of cabbage juice.

As example, different LAB have been used to ferment beet juice. One study using *Lb. acidophilus* Ch-5, *Lb. plantarum* OCK 0858, and *Lb. delbrueckii* subsp. *delbrueckii* OCK 0854 as inoculum for fermentation of sugar beet juice investigated the preservation efficiency. The maximum accumulation of lactic acid (67 to 69%) in the beet juice was obtained after 6 days of cultivation. A preservation efficiency of the fermented juice was shown to exist towards the following: *Escherichia coli* ATCC 25922, *E. coli* OCK 0836, *Enterobacter aerogenes* OCK 0834, *Pseudomonas aeruginosa* ATCC 27853, *Pseudomonas aeruginosa* OCK 0885, *Pseudomonas fluorescens* OCK 0887, *Bacillus subtilis* ATCC 6633, *Bacillus mycoides* OCK 0811, *Fusaruim latenicum* OCK 0508, *Mucor hiemalis* OCK 0519, *Aspergillus niger* OCK 0436, and *Alternaria alternata* OCK 0405 (Klewicka et al. 2004). Another study with red beet juice used *Lb. acidophilus*, *Lb. casei*, *Lb. delbrueckii*, and *Lb. plantarum*. *Lb. acidophilus* and *Lb. plantarum* produced more lactic acid than other cultures and reduced the pH of the beet juice from 6.3 to 4.5 after 2 days at 30 °C. Although all strains gradually lost viability during cold storage (4 weeks at 4 °C), all except *Lb. acidophilus* remained vigorously alive at levels of $10^6$-$10^8$ CFU/ml (Yoon et al. 2005).

One study extracted juice of red beets and then the juice was fermented by two species of probiotic bacteria *Lb. plantarum* and *Lb. paracasei* (both at a dose of $10^6$ CFU/ml) were used for beet juice fermentation, carried at 30 °C for 24 h. Both grew well on the vegetable juice and reached nearly $12.5 \times 10$ (±4.12) and $8.6 \times 10$ (±5.20) CFU/ml, respectively, after fermentation. The pH, acidity (as lactic acid), and viable cell counts were detected weekly for 6 weeks during cold storage at 4 °C. After 6 weeks, the viability of *Lb. plantarum* and *Lb. paracasei* increased (9.03–9.69 log CFU/ml), respectively, while pH was lowered to 2.1 (± 0.020); hence the acidity increased (Jafar and Beyatli 2019).

Beetroot juice contains many minerals such as iron, magnesium, calcium, zinc, copper, and manganese, in addition, beetroot contains a significant amount of folate (90–95 μg/100 g of raw product), carbohydrates, and fiber (1.8 g/100 g of raw product) and is also a source of vitamin C. Thus, a recent study (Janiszewska-Turak et al. 2022) aimed to create powder products using two types of fermentation: spontaneous fermentation and dedicated fermentation with a high amount of LAB, active ingredients, and pigment. As starter culture, *Lb. brevis* KKP 804, *Lb. plantarum* ATCC 4080, and *Lb. fermentum* KKP 811 were used. One percentage (v/v) of NaCl was added to the juice as well as bacterial inoculum of $1 \times 10^7$ CFU/ml, and then kept under anaerobic conditions at 26 °C for 7 days. Daily analyses were performed from the third day of fermentation. The powder was created using spray drying. The results showed that the fermentation process from the third to fifth days kept the bacterial count at the same level, and the juices of those days contained a high content of LAB and had a low pH. However, spontaneous fermentation of the juice did not go well, as it had high pH and low LAB content, with a decrease in bacteria from the third day of fermentation. The powders were stable, although the pigment and LAB results were not satisfactory, and thus further analysis is required.

## 13.2.6 Fermented Cucumbers

Harvested cucumbers (*Cucumis sativus*) are submerged with salt brine and kept under the brine surface with wooden headboards. Fermentation is usually conducted in 5–7% NaCl (Di Cagno et al. 2013). Calcium chloride is typically added to keep the fragile texture and firm the fermented cucumber throughout fermentation and storage (Sarmad Ghazi and Sadiq Jaafir Aziz 2021). After fermentation, the cucumbers are washed to remove excess salt before being packed. Especially *L. plantarum* is used for cucumber fermentations (Breidt et al. 2012; Behera et al. 2018), although other LAB have been identified by PCR in salt-fermented cucumbers such as lactobacilli, *Leuconostoc*, and *Pediococcus* (Di Cagno et al. 2013). Recently, some *Lb. plantarum* strains were identified to be exceptionally salt tolerant, which may be due to the accumulation of osmo- and cryoprotective solutes such as betaine and carnitine (Yao et al. 2020). *Ln. mesenteroides* plays a more important role at low temperatures and low salt concentrations and is the initial major bacteria. Then *P. cerevisiae*, *Lb. brevis*, and *Lb. plantarum* begin to dominate the fermentation. The fermentation is completed by *Lb. brevis* and *Lb. plantarum*. *Ln. mesenteroides*, *P. cerevisiae*, and *Lb. plantarum* are recognized as especially important in cucumber fermentation (Vaughn 1982).

In one study, *E. faecalis*, *Ln. mesenteroides*, *Lb. brevis*, and *P. pentosaceus* were inoculated to cucumber fermentation as starters. *L. plantarum* predominated in the late stage of cucumber fermentation regardless of the species of LAB used for inoculation, due to the greater acid tolerance of this strain (Pederson and Albury 2006).

A mixed-species starter culture consisting of *Lb. plantarum* and *Pediococcus pentosaceus* isolated from spontaneously fermented cucumber was evaluated. The starter culture shortened the fermentation cycle as well as reducing the pathogenic organism population, thus improving the shelf life and quality of fermented cucumber (Ahmed et al. 2021). *Lentilactobacillus buchneri* is beneficial to the silage industry; however, its metabolic capability is detrimental to preservation of cucumbers, as it has been identified as spoiling cucumbers fermented by reduced salt (Suzanne and Roger 2013; Daughtry et al. 2018).

## 13.2.7 Fermented Olive

Olive (*Olea europaea*) is of major agricultural importance in the Mediterranean region. Olive fruits contain a bitter component, phenolic glucoside oleuropein, and have a low concentration of sugar (2.6–6.0%) and a high oil content (12–30%); however, these values may vary due to the degree of maturity and the olive variety. Oleuropein can be removed by alkaline treatment or by brining/salting, fermentation, and acidification (Conte et al. 2020). Olives (> 19 mm in diameter) are placed in vessels and treated with a 1.8%–2.5% NaOH solution for 1 hour, after which 5–8 kg of salt is added. The olives are maintained in this brine for 10–15 days. Finally, a mild washing step is performed to avoid the total elimination of lye (Rejano et al. 2010). There are three kinds of olive processing methods reported (Perpetuini et al. 2020) and these are described as what follows.

### 13.2.7.1 Lye-Treated Black or Green Olives in Brine (California-Style Black/Green Ripe Olives)

California-style black or green ripe olives are obtained through a very similar process. After harvest of the olives at the green–yellow stage, the olives are subjected to three to five applications of

0.5%–2.0% NaOH (lye) under aeration to remove the natural bitterness. Then the olives are washed twice daily for 3–4 days with water to remove most of the residual lye and lower the pH to 7–8. In the case of green olives, ferrous gluconate is not added for darkening. Finally, the washing water is replaced with a mild salt brine, and the olives are canned and heat sterilized at temperatures >110 °C (Fernández et al. 1997) The lactic acid concentration is 0.4%–0.45% after 4–6 weeks. Instead of brine, an acidulant solution containing 0.7% lactic acid, 1.0% acetic acid, 0.3% sodium benzoate, and 0.3% potassium sorbate can be used.

### 13.2.7.2 Lye-Treated Green Olives (Spanish-Style Green Olives)

For Spanish-style green olives, green to straw yellow olives are lye treated to destroy the bitterness, washed to remove the NaOH, and then brined for lactic acid fermentation (Zebin et al. 2018). During washing, food-grade HCl can be added to neutralize the lye. Lye-treated olives are brined in 10%–13% NaCl. Salt is added during the fermentation to maintain NaCl concentration at 5%–6%. This can be increased to more than 7% at the end of fermentation to prevent growth of spoilage microorganisms. The optimum temperature is 24 °C–27 °C, and fermentation proceeds for 3–4 weeks. The fermentation is driven mainly by lactobacilli and to a lesser extent *Leuconostoc* and *Pediococcus* spp. (Corsetti et al. 2012). The pH is 3.8–4.4 with 0.8%–1.2% acidity after fermentation. Olives are packed into glass jars filled with 7% salt brine and sealed. Fermented green olives can be stuffed with strips of red pimento, small onions, and almonds. Pasteurization at 60 °C or hot brine at 80–82 °C can be used.

### 13.2.7.3 Untreated Naturally Ripened Black Olives (Greek-Style Olives)

Greek-style olives (also called kalamata or conservolea) are completely ripened to purple or black and are immersed directly in a brine solution (6%–10%) where they undergo spontaneous fermentation by mixed LAB and yeasts for 8–12 months (Botta and Cocolin 2012). No lye treatment is used, and thus the bitterness remains in the brine. These olives can have a fruity flavor and slightly bitter taste. The mixed microbiota of coliform, yeasts, and lactobacilli are involved in the fermentation. The final acidity of the brine is less than 0.5%, with a pH of 4.3–4.5.

### 13.2.7.4 Microorganisms in Olive Fermentation

Olives are fermented by a similar group of bacteria that ferment sauerkraut and pickles, that is, mainly LAB and yeasts (*Saccharomyces cerevisiae*, *Wickerhamomyces anomalus*, *Candida boidinii*, etc.) (Hurtado et al. 2008; Bonatsou et al. 2017). *Lb. plantarum* and *Lb. pentosus* have been identified as the main LAB isolated from table olives (Lucena-Padrós and Ruiz-Barba 2019; Perpetuini et al. 2020). *Lb. plantarum* has been shown to produce a higher amount of acetic acid during olive fermentation than *Lb. pentosus*, suggesting the lower ability of the latter species to preserve a homofermentative metabolism under stress conditions (Hurtado et al. 2008). LAB are the main bacteria responsible for olive debittering thanks to their enzymatic reservoir (β-glucosidase and esterase). *Lb. pentosus* is characterized by a strong β-glucosidase activity (Franzetti et al. 2011) that catalyzes degradation of oleuropein and the release of glucose and aglycone. Aglycone is converted to non-bitter compounds, such as elenolic acid and hydroxytyrosol (Corsetti et al. 2012). These compounds play an important role in the pH decrease as well as providing microbiological stability to the final product and extended shelf life (Caggia et al. 2004; Corsetti et al. 2012). The lye treatment of olives reduces the initial populations of microorganisms and increases the initial pH to 7.5–8.5. As lye treatments and water washes remove sugars from the olives, nutrients and available sugars are reduced.

As fermentation of olives is still craft based, it is not fully predictable. As other species, such as those belonging to the genera *Aerobacter*, *Escherichia*, and *Bacillus* and yeasts, remain longer in olives than in sauerkraut and pickles, spoilage can easily occur. Thus, starter cultures are highly recommended for fermentation of olives, and either natural starter cultures, that is, microorganisms that spontaneously colonize the raw material [3] or selected starter cultures can be used. As the composition of the natural starter culture is not reproducible, the selected starter cultures are usually represented by a single strain or by a mixture of strains previously selected on the basis of specific features such as high survival capacity in the fermentation environment; high acidifying activity; or ability to hydrolyze phenolic compounds (such as oleuropein), potentially produce volatile molecules, and/or show specific enzymatic activities that contribute positively to the development of the sensory profile of the final product (Corsetti et al. 2012).

## 13.2.8 Caper Berries

The Mediterranean shrub (*Capparis spinosa* L.) is cultivated for its buds and fruits. The Mediterranean countries are the main producers of fermented caper berries, but the final products are exported to central European countries, the United States, and the United Kingdom. Caper berries are collected during the months of June and/or July and immersed in tap water, where the fermentation takes place for approximately 5 to 7 days in a temperature range of 23 to 43 °C, after which the fermented capers are placed in brine and packed (Jiménez-López et al. 2018).

The fermentation of caper fruits is a spontaneous lactic acid fermentation based upon microorganisms present in the raw material and processing environment. One study identified different LAB from different stages during fermentation: *Lb. plantarum, Lb. paraplantarum, Lb. pentosus, Lb. brevis, Lb. fermentum, P. pentosaceus, P. acidilactici*, and *E. faecium. Lb. plantarum* was indicated to be the predominant species (Pérez Pulido et al. 2005)

## 13.2.9 Cocoa

Cocoa beans (*Theobroma cacao* L.) inside cocoa pods are the basis to produce cocoa powder and butter, as well as chocolate. Each cocoa pod contains about 20–50 beans, and over 900 beans are required for 1 kg of chocolate. There are three main nursery subspecies, criollo, forastero, and trinitario, and each subspecies has its own chemical and physical properties. Thus, cocoa, like coffee, is far from being uniform as a raw material. Cocoa pods are harvested by hand, and the preparation process of cocoa starts with the opening of the pods by a wooden club or a cutting tool and then removing the mucilagous pulp surrounding the cocoa beans (Schwan and Wheals 2004; Romel and José 2021), after which a natural microbial fermentation takes place (Cempaka et al. 2014). The fermentation time varies slightly from 2 to 3 days up to between 5 and 7 days, due to the type of cocoa and the technique used (Romel and José 2021). After 3 to 6 days, the fermentation is over, and the next step is drying. After fermentation, some off flavors may be formed by some fungal species. During fermentation, the temperature changes and a temperature between 40 °C and 50 °C can be reached (Schwan and Wheals 2004), favoring both the evolution of microorganisms and a set of enzymes (Guzman et al. 2020; Romel and José 2021). The changes result in the formation of cocoa aroma precursors (amino acids and reducing sugars), and these precursors are enhanced in the drying and roasting stages.

In most cocoa-producing countries, the fermentation process is carried out in an artisanal way, using plastic baskets, wooden drawers, and staggered wooden boxes. These systems are generally covered with banana or heliconia leaves (Figure 13.2). Thus, cocoa bean fermentation can be considered a spontaneous process that is not easily controlled due to natural changes of raw material, different fermentation techniques, and natural variables like weather conditions. Hence, quality fluctuations are often an issue in the chocolate industry. Thus, an alternative non-fermentative processing step, "moist incubation", was published some years ago (Schlüter et al. 2020) that results in a fruitier, more flowery-tasting dark chocolate than the conventional fermentation process.

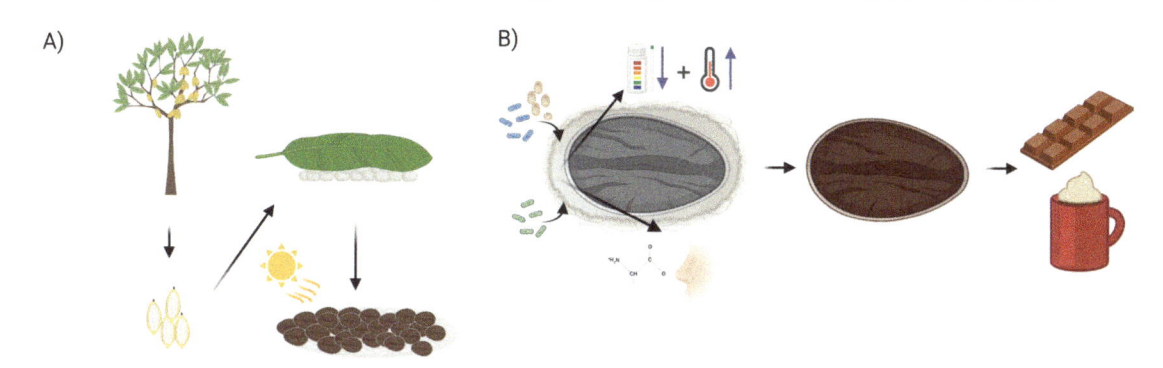

**FIGURE 13.2** The post-harvest stages of cocoa beans. A) After harvesting, the cocoa pod is broken, and for the fermentation stage, the broken pods are placed in a box and covered with banana leaves, and when fermentation is completed, they are dried by the sun. B) The fermentation of the juicy white pulp covering the cocoa beans has two phases, anaerobic and aerobic, after the cocoa pod breaking; the anaerobic phase involves yeast and LAB strains, while the aerobic phase involves acetic acid bacteria (AAB) strains. During fermentation the different microorganisms produce various metabolites, increase the temperature, and decrease the pH of the cotyledon. In addition, flavor precursors are formed. Figure created with BioRender. com and modified from Gutiérrez-Ríos et al. (2022).

## 13.2.9.1 *Microorganisms in Cocoa Fermentation*

The fermentation process is essential to modify the beans; to eliminate the mucilage; and to prepare the grain that requires a battery of enzymes responsible for modifying its color, taste, and smell. During the initial fermentation phase (pulp pH 3.6), yeasts are dominant and increase in number within the first 24 h of fermentation as the cocoa pulp surrounding the cocoa beans is degraded and carbohydrates are metabolized under anaerobic conditions, leading to the formation of ethanol and carbon dioxide and secretion of pectinolytic enzymes. The fermentation process is characterized by a well-known microbial succession. Over 100 microbial species with different metabolic properties have been identified, with two dominant bacterial species, *Lb. fermentum* and *Acetobacter pasteurianus*, and four yeast species, *S. cerevisiae*, *Hanseniaspora thailandica*, *H. opuntiae*, and *Pichia kudriavzevii*, that represent the central bacterium–fungus association in cocoa fermentation in several of the production regions (Meersman et al. 2013; Visintin et al. 2016). The yeasts participate first, then the LAB and, finally, the acetic bacteria (AAB) intervene. Moreover, filamentous fungi and spore-forming Bacillus also participates (Romel and José 2021).

The LAB have the fastest growth rate during the 16–48 h of fermentation and are present in large numbers, although not necessarily in biomass as compared to yeasts for a short period (Nielsen et al. 2005). In studies from Ghana, microaerophilic LAB, citrate-fermenting, acid-tolerant, and ethanol-tolerant *Lb. plantarum* and *Lb. fermentum* strains have been shown to dominate the spontaneous cocoa bean fermentation process (Camu et al. 2007; Nielsen et al. 2007), although the presence of *Lb. fermentum* during fermentation of cocoa pulp-bean remains controversial (Mota-Gutierrez and Cocolin 2022). AAB such as Acetobacter species intervene in the final stage of fermentation. AAB carry out the transformation of the ethanol produced by the yeasts into acetic acid.

As the physiological roles of the predominant micro-organisms are reasonably well understood, as is the pivotal importance of an organized microbial succession, a synthetic microbial cocktail inoculum has been put into use. This cocktail consists of just five species, including members of the three principal groups, to mimic the natural fermentation process and yield good-quality chocolate (Kresnowati et al. 2013). To improve the quality of the processed beans, more research is needed on pectinase production by yeasts, better depulping, fermenter design, and the use of starter cultures (Schwan and Wheals 2004).

## 13.2.10 Coffee

Due to its high popularity worldwide, coffee is one of the most important crops in the world, and the broad variety of coffee beverage tastes results from differences in cultivar, cultivation areas, quality, processing methods, blends, and roasting conditions (Huch and Franz 2015). Fermentation is one of the key processing activities. Like for cocoa, the fermentation process is important for degrading the pulp surrounding the bean as well as positively impacting the coffee's quality attributes. Coffee beans are divided into two main types, Arabica and Robusta. The main difference, besides being different species of the same plant family, comes down to flavor and characteristics of the actual bean. The coffee mucilage and its physical and chemical properties vary according to species and variety (Murthy and Naidu 2011). Mucilage is composed of water (84.2%), protein (8.9%), reducing sugar (4.1%), pectates (0.91%), and ash (0.7%), although its physical and chemical properties vary according to species and variety. Thus, Robusta needs at least 72–100 h for complete dissolution of mucilage, while 12–24 h are enough for Arabica (Velmourougane et al. 2008).

Three different fermentation methods, dry, wet, and semidry, are used to process the coffee fruits immediately after harvesting to allow spontaneous or indigenous fermentation to occur. Depending on the processing type, the time required for fermentation varies, although the main purpose of the fermentation process in all methods is to remove the mucilage layer's rich polysaccharides (pectin), as well as to decrease the water content of the coffee beans. In dry processing, the coffee fruits are cleaned, and floaters are separated immediately after harvesting. Then the entire coffee fruits are dried on the floor or platforms in the sun without prior removal of the pulp (Silva et al. 2008). In wet processing (largely used for Arabica coffee), the ripe coffee fruits go immediately after harvesting through a flotation process to clean debris and remove floaters. The coffee fruit is then pulped, put through a 24- to 48-h underwater tank fermentation process, and dried until the moisture content reaches 10%–12% (Murthy and Madhava Naidu 2012; Haile and Kang 2019). Semidry processing is a combination of both these methods, as the coffee fruits are depulped, with the fermentation process occurring directly on a platform under the sun (Vilela et al. 2010).

### 13.2.10.1 Fermentation of Coffee

The fermentation of coffee relies on the growth and metabolic activities of different groups of microorganisms, including Gram-negative bacteria, bacilli, yeasts and filamentous fungi, acetic acid bacteria, and lactic acid bacteria. Over 50 yeasts and bacterial species that are present during coffee fermentation have been identified (Silva et al. 2013; de Melo Pereira et al. 2014).

To ensure the best quality of coffee and increase the economic benefits for the growers, starter culture in a controlled coffee fermentation can be used (Vinícius de Melo Pereira et al. 2017). However, only limited studies have been reported on the use of a starter culture for coffee fermentation. One study used pectinolytic microbes isolated from spontaneous fermentation (*Lb. brevis* L166, *Erwinia herbicola* C26, *B. subtilis* C12, and *Kluyveromyces fragilis* K211) (Avallone et al. 2002) to study the microbial and physicochemical characteristics of coffee fermentation. However, the authors reported that the inoculation of pectinolytic microbes did not speed up the degradation of the polysaccharide compounds or modify the organoleptic characteristics of the beverage. They suggested that it would be better to use lactic acid bacteria instead to remain as close as possible to natural fermentation.

## 13.2.11 Tea

*Camellia sinensis* is used to produce all different types of tea. The differences between various types come from the different processing steps being used. For fermentation, tea leaves are placed in large, cool, moist, darkened rooms, where they are laid out on flat surfaces with an even layer of about 10 cm. Most often, aluminum or wood is used to avoid a chemical reaction of tea leaf juice and the surface. An ideal combination of a low temperature of about 15 °C and high humidity—about 90%. Fermentation can last from 45 minutes to several hours (Jolvis Pou 2016). Tea fermentation can be classified as non- and light-fermented, semi-fermented, fully fermented, or post-fermented (Ning et al. 2017). Tea fermentation is actually an oxidation process catalyzed by enzymes that are originally present in tea leaves (Fowler et al. 1998), and only some of the teas have undergone microbial fermentation, like Chinese Puer (or Pu-Erh) tea; kombucha; Fuzhuan brick-tea; and Burmese fermented tea, lahpet (Mo et al. 2008) (Megaelectra 2016). Chinese Puer tea is a microbially fermented tea obtained through the action of molds, bacteria, and yeasts on the harvested leaves of the tea plant. Puer tea has very specific aroma, color, and taste depending on the degree of processing. A study investigating the microbial composition of Puer tea found 2.5 million CFU/g leaves, and living organisms, even though in reduced numbers, were detected after the serving procedure, that is, tea leaves brewed at 90 °C for 2 minutes and filtered.

Kombucha, sweetened green, white, or black tea, is fermented by SCOBY, a symbiotic colony of bacteria and yeast. The bacterial component of kombucha cultures has not been extensively studied but is known to comprise several species, including acetic acid bacteria. SCOBY typically includes AAB (*Acetobacter, Gluconobacter*), LAB (lactobacilli, *Lactococcus* spp.), and yeasts (*Saccharomyces* spp., *Zygosaccharomyces* spp.) (Coton et al. 2017). Recently LAB were reported to constitute up to 30% of the bacterial population of kombucha cultures (Marsh et al. 2014; Yang et al. 2022). The interaction of the bacteria and yeast results in a floating cellulose layer on the surface of the fermented tea. The layer becomes thicker the longer it is fermented. For commercial kombucha products, *Bacillus coagulans, Bacillus subtilis, S. cerevisiae* var. *boulardii*, and *Lb. rhamnosus* may be added. As *B. coagulans* is resistant to high temperatures, it is one of the most common (Konuray and Erginkaya 2018; Kim and Adhikari 2020).

# 13.3 COMMON HEALTH EFFECTS OF FERMENTED VEGETABLES AND FRUITS

LAB-fermented foods have been consumed by humans for about 10,000 years for their improved shelf life and palatability characteristics. Today, the consumption of fermented foods is also driven by a health-related perspective, as an array of health-modulating compounds and signal molecules are released in the matrix during fermentation (Castellone et al. 2021). Vegetables and fruits are rich sources of micronutrients (magnesium, calcium, and potassium) and bioactive components consisting of phytochemicals such as polyphenols, dietary fiber, carotenoids, and vitamins, and they contain about 5000 identified bioactive compounds (Wallace et al. 2020). Fermented vegetables might exhibit better functionalities than raw vegetables (Park and Rhee 2005), as the LAB synthesize vitamins and minerals and produce biologically active peptides with enzymes such as proteinase and peptidase, as well as removing some non-nutrients (Melini et al. 2019; Şanlier et al. 2019). Especially cruciferous vegetables are reported to exhibit health benefits (Li et al. 2022). Leafy vegetables contain vitamin C, folate, and carotenoids (Slavin and Lloyd 2012), while cucumbers and Chinese cabbage, for example, contain vitamins C, B1, B2, B3, B6, B11, A, E, and K. Olives contain vitamins B1 (thiamine), B3 (niacin), B6, A, E, and K; black olives also contain vitamin C, while green olives also contain vitamin B11. Green olives contain quantitatively more vitamins than black olives, except for vitamin K, of which both contain 1.4 mg/100 g. Cucumbers, Chinese cabbage, and olives seems to lack vitamins B12 and D. LAB have been reported to produce riboflavin, folate, vitamin B1, and B3, as well as K2 (Paramithiotis et al. 2022).

Fermented vegetables can represent a non-invasive strategy to face different disorders, as during fermentation both biotic and abiotic factors affect the production of vitamins, gamma-aminobutyric acid (GABA), bioactive peptides, and phenolic and organosulfur compounds (Diez-Ozaeta and Astiazaran 2022) and allow amelioration of the intake of nutrients like fiber, vitamins, and minerals, thus improving human health (Saiwal et al. 2019). This is particularly useful since fermented vegetables can be consumed in periods when vegetables are unavailable (Kiczorowski et al. 2022). Vegetables are low in calories and contain vitamins, minerals (including K, Mg, and Se), antioxidants, and phytochemicals as well as dietary fiber, and fermented vegetables have shown preventative and protective roles against various degenerative diseases such as cancer, obesity, diabetes, and cardiovascular diseases, for example, by alteration of the metabolic activities of intestinal microbiota, binding and degrading potential carcinogens, producing antimutagenic and antitumorigenic compounds, enhancing the host's immune response, and other actions (Swain et al. 2014; Zhong et al. 2014). As knowledge about bioactive compounds in fermented vegetables has increased, efforts have been made to improve the general quality of fermented vegetables while also creating a bioactive compound–rich product (Castellone et al. 2021). *Ln. mesenteroides* in combination with *Pediococus dextrinicus* showed good potential to produce bioactive enriched foods. *Lb. sakei* showed a predominance in this feature since its utilization in vegetable fermentations allows the obtaining of foods with three times the concentration of bioactive compounds compared to other studied bacterial strains (Bousquet et al. 2021).

However, although many potential benefits of fermented foods have been identified, recommendations for fermented vegetables are lacking from food guides (Gille et al. 2018; Mukherjee et al. 2022). In the following, some potential health effects of the previously discussed fermented vegetables will be presented.

## 13.3.1 Increased Bioactivity

A recent study investigated the protein, energy, macro-mineral (Na, K, Ca, Mg, P), micro-mineral (Zn, Cu, Fe), heavy metal (Pb, Cd), vitamin C, vitamin A, carotene, and phenolic content in chosen raw and fermented vegetables. During fermentation of the vegetables, the water content was reduced, contributing to a relative increase in the concentration of nutrients, while fat content increased. The fermented vegetables contained lower levels of some basic nutrients, mineral content, vitamin C, and phenols, while vitamin A (23–34%) and carotene (24%) content in the fermented carrots and peppers were increased compared to raw vegetables. All fermented vegetables had lower levels of heavy metals (Kiczorowski et al. 2022).

GABA is a major inhibitory neurotransmitter involved in, for example, stress response. Some LAB strains in vegetable fermentations have been reported to produce GABA *Lb. plantarum, Lb. gasseri*, and *Lb. brevis* (Lim et al. 2017; Moore et al. 2021). Although some GABA producers have been identified in Kimchi, GABA content of naturally fermented kimchi is low. Thus, to increase the GABA content, starter cultures with high GABA production ability should be used (Lee et al. 2021). However, is it still unclear if oral intake results in increased brain GABA concentration.

Although vegetables are not generally rich in protein, some, like cucumbers, Chinese cabbage, and green and black olives, contain 0.65%, 1.5%, 1.03%, and 0.84% protein, respectively (https://fdc.nal.usda.gov/, accessed on 21 August 2021), and thus some bioactive peptides have been reported (Sosalagere et al. 2022). In cucumbers, five peptides with potential anti-hypertensive activity (Fideler et al. 2019) and one peptide from olive seeds exhibiting anti-proliferative capacity on prostate cancer cells and breast cancer cells have been reported (Vásquez-Villanueva et al. 2018).

Phenolic compounds have an important physiological role, as they participate in processes like photosynthesis, respiration, and cell development. In vegetables, the total phenolic content (TPC) is mostly used as a substrate of LAB fermentation and is reported to be rather low, with the amount depending on plant type and variety, cultivation conditions, processing, and storage (Septembre-Malaterre et al. 2018; Babenko et al. 2019). Some *Lb. plantarum* strains have been reported to hydrolyze oleuropein, the main phenolic glucoside of olives (Zago et al. 2013). In addition, other phenolic compounds have been detected in the flesh of fermented olives or the brine (Kaltsa et al. 2015; Durante et al. 2018). Phenolic compounds have also been detected in kimchi, with over-ripened kimchi containing more TPC than short-term fermented (Park et al. 2017). Caffeic acid, naringin, p-coumaric acid, catechin gallate, and epigallocatechin gallate have been found to increase during the first few months of fermentation and then decrease after 3 months. Similarly, sauerkraut has been identified to contain esterified phenolic acids in the first months of fermentation, with a reduction detected after 3–4 months of storage (Kapusta-Duch et al. 2017).

Many LAB, including lactobacilli, lactococci, pediococci, leuconostocs, streptococci, and entero-cocci, display antimicrobial activities in fermented foods, mainly due to production of organic acids but also bacteriocins and antifungal peptides. Most bacteriocin-producing strains have been detected in fermented meat or dairy food, although some bacteriocin producers have been detected in kimchi. The following LAB isolated from kimchi were shown to produce bacteriocins: *Lb. paraplantarum, Lb. plantarum, Lb. sakei, Lc. Lactis, Lb. curvatus, Ln. mesenteroides, Ln. citreum,* and *P. pentosaceus* (Lee et al. 2021).

As standardized fermentation methods may lead to use of a selected and reduced number of strains used as starter culture, it is important to remember that it may lead to products with less bioactivity as well as a lower amount of post- and paraprobiotics in the final product (Castellone et al. 2021), unless starter cultures are specifically chosen for the improvement of bioavailability and absorption of minerals present in vegetables based upon their production of, for example, enzymes like phytase, protease, lipase, and amylase that can hydrolyze lipids, carbohydrates, and proteins into simpler digestible compounds. In addition, microbial enzymes may degrade anti-nutritional factors like protease inhibitors, tannins, and phytates (Brodmann et al. 2017).

# 13.4 HEALTH EFFECTS OF SPECIFIC FERMENTED VEGETABLES, FRUITS, AND SEEDS

## 13.4.1 Sauerkraut

As sauerkraut has been prepared and consumed for a long time, several studies related its health effect have also been performed. A bibliometric study was done some years ago investigating 139 publications over a period of 90 years (1921–2012). Of these publications, 23.7% evaluated the impact of sauerkraut on health, including risk factors or digestive well-being. Sauerkraut consumption has been suggested to contribute to a healthy digestive microbiota. However, the studies could not confirm this, as they only found that sauerkraut induced inflammation locally, and repeated intake may result in diarrhea. According to anecdotal experience, regular intake of small doses (7–10 g daily) of sauerkraut has a very good effect on many patients' gastrointestinal tracts. They report better digestion and less constipation (Raak et al. 2014; Peñas et al. 2017). Consumption of sauerkraut seems to reduce irritable bowel syndrome (IBS) symptoms (Sandberg Nielsen et al. 2018).

## 13.4.2 Kimchi

There are numerous *in vitro, in vivo*, and human studies discussing the health benefits of kimchi, such as antioxidative and antiaging, antimutagenic, anticancer, antiobesity, and other health benefits like immune-stimulatory effects and other probiotic activities (Park et al. 2014; Hongu et al. 2017; Park et al. 2017; Lee et al. 2021).

As kimchi is produced using various raw materials and preparation techniques depending on family or region tradition, there is a broad variation of the end products (Jeong et al. 2013). The health benefits result from the raw materials utilized and their nutritive value as well as the microbiota prevailing in the fermentation (Patra et al. 2016); thus there can be limitations to identify the effects. Vegetables belonging to the Brassicaceae family are used for kimchi, and they have been reported to contain many compounds with health-promoting potential, including dietary fibers, minerals, amino acids, vitamins, carotenoids, glucosinolates, and polyphenols. Kimchi is also rich in phytochemicals, such as indole compounds, b-sitosterol, benzyl isothiocyanate, and thiocyanate. Consumption of kimchi may contribute to enriching the gut microbiota, and various potential probiotic strains have been isolated from kimchi, and their effi-cacy has been reported (Lee et al. 2011; Park et al. 2016; Song et al. 2023). So far, no systematic review is available where all clinical trials of the use of kimchi in the treatment of any condition or symptom have been evaluated. One review is ongoing but not yet published (Kim et al. 2018).

## 13.4.3  Soybean Products

Soy-based food has been rigorously investigated for the past 30 years for its role in chronic disease pre-vention and treatment. The health function of tofu is achieved mainly by high-quality protein, beneficial lipids, vitamins, and minerals, as well as other biologically active compounds such as isoflavones, soy saponins, and others (Messina 2016; Zhang et al. 2018; Messina et al. 2022). Regular consumption of tofu helps reduce the incidence of many diseases such as hypertension, hyperlipidemia, hypercholesterol-emia, arteriosclerosis (Fung et al. 2008; Kesika et al. 2022), coronary heart disease (He and Chen 2013), breast cancer, and others (Riciputi et al. 2016). Moreover, it has been reported that fermented soymilk has immune-modulatory, memory improvement, anti-colitis, and wound-healing activities. However, most of the reports are based on *in vivo* and *in vitro* studies, and only a few studies confirm the health benefits of fermented soy milk for human subjects. The dose and duration of the intervention may vary between the animal model and humans (Kesika et al. 2021). In Asian countries, consumption of fermented soybean products is associated with the reduction of chronic diseases, as the consumption of natural antioxidants is efficient in reducing harmful impacts of reactive oxygen species and in adjusting the body's antioxidant load (Xiao et al. 2015; Do Prado et al. 2021).

## 13.4.4  Vegetable Juices

The biological activity of beetroot juice (Chrobry variety, *Beta vulgaris* L. ssp. *vulgaris*) fermented by *Lb. brevis* 0944 and *Lb. paracasei* 0920 was assessed by investigating the oxidative status of blood serum, kidneys, and liver of rats consuming fermented beetroot juice. The study showed a positive modulation of the gut microbiota and its metabolic activity. The β-glucuronidase activity decreased due to mod-ulation of the gut epithelium microbiota. Some antioxidant capacity of blood serum aqueous fraction was increased, whereas the antioxidant parameters of the blood serum lipid fraction, kidneys, and liver remained unchanged (Klewicka et al. 2015).

Another study investigated the bioactivity of tomato juice fermented by *Lb. plantarum* and *Lb. casei* by measuring the levels of lycopene, total carotenoids, ascorbic acid, total phenolic and volatile com-pounds, 1,1-diphenyl-2-picrylhydrazyl (DPPH), 2,2'-azinobis-3-ethylbenzotiazo-line-6-sulfonic acid (ABTS) radical scavenging capacities, ferric-reducing antioxidant power (FRAP), and *E. coli*, as well as the inhibition of copper-induced human low-density lipoproteins (LDL)-cholesterol oxidation assays. The study showed that the ABTS and DPPH inhibition values, as well as the FRAP and total phenolic content, were significantly increased. Oxidation of LDL-cholesterol was clearly delayed after the addition of the fermented juice. In addition, an *in vitro* inhibitory effect against *E. coli* was substantially increased. Moreover, the results associated with the volatile compounds indicated that fermentation with *Lb. planta-rum* and *Lb. casei* can be a strategy for modifying flavors (Liu et al. 2018).

## 13.4.5  Fermented Olives

As one of the oldest vegetable fermented foods in the Mediterranean area, fermented table olives represent an important healthy food because of their high content of bioactive and health-promoting compounds (Perpetuini et al. 2020). Table olives can be considered an ideal matrix for the survival of probiotics due to the nutrients released by the fruits. In addition, the drupes are coated with a hydrophobic epicuticular wax that promotes microbial adhesion (De Bellis et al. 2010; Valerio et al. 2011; Pérez Montoro et al. 2018). However, further validation by *in vivo* complex animal or human trials should be performed to get a better understanding of their potential health-promoting features for humans (Perpetuini et al. 2020; Anagnostopoulos and Tsaltas 2021).

## 13.4.6 Fermented Capers

Capers are used in many countries as traditional medicine for preventing and curing numerous diseases due to their antioxidant, antimicrobial, anticancer, antirheumatic, tonic, astringent, and diuretic potentials (Rivera et al. 2003). The flower buds of capers have been shown to have high antioxidant activity, with (+)-catechin; 1,2-dihydroxybenzene; apigenin-7-glucoside; and gallic, 3,4-dihydroxybenzoic, and syringic acids as the major bioactive compounds of both fresh and fermented caper buds. The bioactivity seems to be dependent on the size of the caper buds, as small (<8 mm) buds had a higher content as compared to medium (>8 mm and <13 mm) and big (>13 mm) buds (Özcan et al. 2020).

## 13.4.7 Fermented Cocoa

Chocolate has been consumed since at least 400 AD and is rich in polyphenols such as catechins, anthocyanidins, and pro anthocyanidins (Tan et al. 2021). Cacao contains more than 500 different chemical compounds, and some of these have been traditionally used for their antioxidant, anti-carcinogenic, immunomodulatory, vasodilatory, analgesic, and antimicrobial activities. Different effects have been reported, like antioxidant properties, and cocoa offers neuron protection and enhances cognition and positive mood. It lowers immunoglobulin E release in allergic responses. It can affect the immune response and bacterial growth at intestinal levels. It reduces inflammation by inhibiting nuclear factor-κB (Latif 2013). A systematic review detected that chocolate or cocoa product consumption significantly improved triglycerides profiles, while the effects of chocolate on all other outcome parameters were not significantly different. In conclusion, low-to-moderate-quality evidence with a short duration of research (majority 4–6 weeks) showed no significant difference between the effects of chocolate and control groups on parameters related to skin, blood pressure, lipid profile, cognitive function, anthropometry, blood glucose, and quality of life regardless of form, dose, and duration among healthy individuals. It was generally well accepted by study subjects, with gastrointestinal disturbances and unpalatability being the most reported concerns. Cocoa helps in weight loss by improving mitochondrial biogenesis. It increases muscle glucose (Latif 2013).

The term cocobiota was introduced to refer to the microbial association of bacteria and fungi involved in the spontaneous fermentation of cocoa beans, which originate metabolites present in cocoa powder and dark chocolate, which can have beneficial effects on health (Petyaev and Bashmakov 2016; Bastos et al. 2018).

The antimicrobial activity of LAB isolated from fermented cocoa beans has been studied against pathogenic bacteria, where the activity of 66 LAB against *E. coli* NBRC 14237, *S. aureus* NBRC 13276, *B. subtilis* BTCCB, *Salmonella thypii*, and *Listeria monocytogenes* were studied, and of these six isolates showed potential activity against pathogens (Urnemi et al. 2011).

## 13.4.8 Coffee

Coffee intake varies a lot in different countries, and many people consume up to 6 cups of coffee per day, which is about 540 mg of caffeine. The FDA has stated that consumption of 400 mg/day (4–5 cups per day) of caffeine is safe. The health benefits of coffee have been extensively studied in comparison to other fermented foods of plant origin. Coffee may contribute to the prevention of inflammatory and oxidative stress-related diseases, such as obesity, metabolic syndrome, and type 2 diabetes (Lindsay et al. 2012). In addition, consumption of coffee seems to be associated with a lower incidence of several types of cancer and with a reduction in the risk of all-cause mortality (Safe et al. 2023). Meta-analyses have produced controversial results in regard to positive, neutral, or inverse associations with CVD incidence, risk factors, and mortality. The optimal amount of coffee for inverse associations with death, CVD, and CVD-related risk factors was found to be 4 cups of coffee per day (Poole et al. 2017; Tverdal et al. 2020; Simon et al. 2022). The highest consumption levels of coffee tend to be either protective against or not associated with other diet-related chronic disease risks. In particular, the highest consumption level may significantly reduce the risks of type 2 diabetes and cancer (particularly liver, esophagus, oral cavity, colorectum, thyroid, endometrium, colon, pancreas, and breast) by a maximum of 24% and 50%, respectively. However, positive associations have also been observed, as the highest levels of coffee consumption might significantly increase the risks of obesity, other cancers (prostate, urinary tract, bladder), and high blood pressure (Barrea et al. 2021).

## 13.4.9 Tea

Black tea contains polyphenols, including epigallocatechin gallate, theaflavins, thearubigins, the amino acid L-theanine, and several other catechins or flavonoids which provide protection against the onset of several chronic disorders (Dufresne and Farnworth 2000; Rasheed 2019).

Kombucha has been consumed in many countries for a very long time, and several health benefits have been reported from *in vitro* and *in vivo* studies, including antimicrobial benefits, liver and gastrointestinal functions, immune stimulation, detoxification, antioxidant, anti-tumor properties, health prophylactic and recovery effects through immune stimulation; inhibiting the development and progression of cancer, cardiovascular disease, diabetes, and neurodegenerative diseases; and promoting normal central nervous system function (Selvaraj and Gurumurthy 2023). So far only one study has reported results of empirical research (Kapp and Sumner 2019). The biology of SCOBY is complex, the exact microbial composition of kombucha varies (Jayabalan et al. 2014; Kapp and Sumner 2019), and the mechanism of action is not well understood.

# 13.5 POSSIBLE NEGATIVE EFFECTS OF FERMENTED VEGETABLES

Vegetables have historically undergone fermentation, and there are many variables in the fermentation process, including the raw material, microorganisms, and environmental conditions, that lead to many different variations of fermented foods. However, although fermented foods are often produced in unhygienic and contaminated environments, they generally have a very good safety record, even in the developing world.

## 13.5.1 Nitrates, Nitrites, and Nitrosamines

Nitrates and nitrites are widely distributed in nature; they are found in the atmosphere, water, soil, microorganisms, plants, and animals. Nitrosamines are reaction products between nitrogen oxide and secondary amines but can also be generated during fermentation and are classified as carcinogens (Park et al. 2015). Dietary nitrite has been considered harmful to human health, although attention has shifted toward the potential health benefits of dietary nitrate due to *in vivo* formation of nitric oxide (NO) (Hord et al. 2009; Lundberg et al. 2018; Qin and Wang 2022). The nitrate amount in vegetables varies depends on many factors, such as cultivation region and frequency, weather conditions, quality of soil, and productions processes. Vegetables are the most important source of nitrate exposure in the human diet, contributing of more than 80% of nitrate intake (Chou et al. 2003). As example, the maximum nitrate amount to be ingested daily <3.7 mg/kg by body weight (i.e., a person weighing 70 kg should consume less than 255.5 mg of nitrates daily) (EFSA Panel on Food Additives Nutrient Sources added to Food 2017). Vegetables can contain up to 250 mg nitrate/100 g fresh material, where 100–250 mg represent high content and >250 very high content (Hord et al. 2009; Bahadoran et al. 2016).

The nitrate and nitrite contents of acidified and fermented fruits and vegetables were investigated from a variety of commercial sources, and the nitrate concentrations within product types varied largely, likely due to variations in products produced from different batches of raw materials. The nitrate content in pickled products was generally lower than that reported for fresh fruits and vegetables (Ding et al. 2018). Most fermented vegetables are brined during preservation or in finished product processing, which potentially dilutes the nitrate content of the finished products by 30–70% compared to raw material.

In a study focusing on the seven most frequently reported nitrosamines in food, 387 total diet samples (ready-to-eat foods) from the Korean diet were investigated, where the nitrosamine content in the vegetables was higher than reported before (Park et al. 2015). However, as soil microorganisms are known to contribute to nitrosamine formation and the nitrate contents in vegetables can also contribute, fluctuation in nitrosamine contents can be expected (Park et al. 2015).

## 13.5.2 Biogenic Amines

Biogenic amines play essential roles in many physiological processes in the human body, such as cell proliferation, immune response, stomach pH, and gastric acid secretions, and are also active in the nervous system and in the control of blood pressure (Fernández-Reina et al. 2018). They are biologically active natural compounds formed mainly from bacterial decarboxylation of amino acids. Histamine, tyramine, tryptamine, putrescine, and cadaverine are formed from histidine, tyrosine, tryptophan, ornithine, and lysine, respectively. Putrescine is a precursor of the polyamines spermidine and spermine (Halász et al. 1994). Excessive intake of biogenic amines in food causes psychoactive, vasoactive, and tumor growth effects (Kalac et al. 2000). The degree of intoxication depends on the amount and type of biogenic amine ingested and that the detoxification system function is properly working. After food consumption, small quantities of biogenic amines are commonly metabolized in the human gut to physiologically less active forms. Thus, the toxic level of ingested biogenic amines is difficult to establish, as it depends on the individual sensitivity and health status of the consumer. Biogenic amines may also be considered carcinogens because they are able to react with nitrites to form potentially carcinogenic nitrosamines (Doeun et al. 2017; Fong et al. 2021). Hence, it is important to estimate the likelihood of biogenic amines and to prevent their accumulation in food (Spano et al. 2010). High amounts of biogenic amines have been reported for products resulting from the fermentation process, regardless of the food type (Suzzi and Torriani 2015). There is no specific legislation regarding the content of biogenic amines in many fermented products, although it is generally assumed that they should not be allowed to accumulate. The production of biogenic amines by microorganisms varies a lot and is usually strain specific.

The increase of biogenic amine content in fermented foods compared to that of raw materials has been correlated with the LAB driving the fermentation. The level of biogenic amines (except tyramine) in sauerkraut has been reported to be lower than in kimchi (Kalač et al. 2000; Paramithiotis et al. 2022). Generally, reports on the biogenic amine content of fermented cucumbers and olives are lacking in the literature. According to available data, fermented cucumbers seem to contain more biogenic amines than fermented olives but less than kimchi and sauerkraut. Regarding fermented olives, they seem to contain less biogenic amines than kimchi, sauerkraut, and fermented cucumbers (Paramithiotis et al. 2022).

Many studies have reported their presence in nonfermented fruits, vegetables, nuts, and legumes (Sánchez-Pérez et al. 2018), and for sauerkraut, the effect of the raw materials on the production of biogenic amines has been highlighted (Majcherczyk and Surówka 2019; Satora et al. 2021). In addition, a positive correlation between the production of biogenic amines and presence of yeast has been reported, and it is known from alcoholic fermentation that yeasts contribute to the accumulation of biogenic amines (Restuccia et al. 2018).

By using starter cultures, it is possible to have qualitative and quantitative control the production of biogenic amines, that is, to select starter strains that lack biogenic amine-producing activity.

## 13.5.3 Acrylamide

Acrylamide is a toxic compound classified as a probable human carcinogen (International Agency for Research on Cancer 1994. The Agency for Toxic Substances and Disease Registry (ATSDR 2012) and the European Food Safety Authority (EFSA) confirmed that the presence of acrylamide in foods is a public health concern (Benford et al. 2022). It is mainly produced as result of the reaction between the asparagine amino acid and reducing sugars through the Maillard reaction, although different mechanisms appear to be involved in the formation of acrylamide in table olives (Casado et al. 2013; Charoenprasert and Mitchell 2014). Acrylamide concentrations in table olives vary widely, as it seems to be formed only in processing of California-style black and green ripe olives, while it is not detected in other types of table olives made with processing temperatures below 65 °C (Duedahl-Olesen et al. 2022). EFSA considers California-style table olives a potential source of acrylamide since these foods contain similar or even higher levels to those found in other food products such as French fries, cereals, or coffee (Duedahl-Olesen et al. 2022).

In addition to cultivar and temperature, other processing conditions such as storage time (longer than 30 days), reduced lye treatment and washing with water, oxidation, darkening methods, olive harvest, or the presence of additives can affect the acrylamide concentration (Charoenprasert and Mitchell 2014; López-López et al. 2014). This fact explains the higher acrylamide content of California-style black ripe olives when compared with California-style green ripe olives (López-López et al. 2014). The olive variety also has a large influence on the variability of acrylamide levels (Casado ct al. 2013; López-López et al. 2014).

## 13.6 CONCLUSIONS

Fermented vegetables have an important role in the preservation and production of vegetables, as LAB fermentation improves resistance to microbial spoilage. Moreover, LAB-fermented vegetables play important roles in promoting both fermentation and human health by providing a source of live, potentially health-promoting microbes. The type, quality, and variety of the raw materials, climatic conditions, and agricultural practices, as well as the occurrence and conditions of processing and storage, impact the fermentation as well as the end products. As vegetables contain various natural LAB, various fermented end products, flavors, and functional bioactive compounds are produced. Various starter cultures have different characteristics, and appropriate starters can be used for specific purposes when fermenting vegetable products, as microorganisms identified from fermented vegetables or plant-derived LAB provide constant and better nutrient quality as well as functional properties if used as starter cultures for vegetable fermentation and fermented vegetable juices. It is important to remember that the production of bioactive compounds by microorganisms is a strain-dependent characteristic that also depends on the fermentation conditions, such as temperature and time.

Vegetable products fermented with health-promoting LAB starter cultures will contribute to greater health benefits than raw vegetables alone. Regulated fermentation technology might also increase health functionality, improve taste, and increase shelf life of high-added-value products. As interest in fermented food has grown globally, and more products are becoming available for larger public consumption, global regulatory frameworks must also evolve to, for example, prepare guidelines for recommended compositions and intakes and ensure safe production, storage, and transport.

# BIBLIOGRAPHY

Agency for Toxic Substances and Disease Registry (ATSDR) (2012). *Toxicological Profile for Acrylamide*. Atlanta, GA: U.S., Department of Health and Human Services, Public Health Service.

Ahmed, S., F. Ashraf, M. Tariq and A. Zaidi (2021). "Aggrandizement of Fermented Cucumber Through the Action of Autochthonous Probiotic Cum Starter Strains of Lactiplantibacillus Plantarum and Pediococcus Pentosaceus." *Annals of Microbiology* **71**(1): 33.

Ahn, D. K.; Han, T. W.; Shin, H. Y.; Jin, I. N.; Ghim, S. Y. Diversity and antibacterial activity of lactic acid bacteria isolated from kimchi. *Korean J. Microbiol. Biotechnol.* 2003, 31, 191–196.

Anagnostopoulos, D. A. and D. Tsaltas (2021). "Current Status, Recent Advances, and Main Challenges on Table Olive Fermentation: The Present Meets the Future." *Frontiers in Microbiology* **12**: 797295.

Avallone, S., J. Brillouet, B. Guyot, E. Olguin and J. Guiraud (2002). "Involvement of Pectolytic Micro-Organisms in Coffee Fermentation." *International Journal of Food Science & Technology* **37**: 191–198.

Babenko, L. M., O. Smirnov, K. Romanenko, O. Trunova and I. Kosakivska (2019). "Phenolic Compounds in Plants: Biogenesis and Functions." *Ukrainskii Biokhimicheskii Zhurnal* **91**: 5–18.

Bahadoran, Z., P. Mirmiran, S. Jeddi, F. Azizi, A. Ghasemi and F. Hadaegh (2016). "Nitrate and Nitrite Content of Vegetables, Fruits, Grains, Legumes, Dairy Products, Meats and Processed Meats." *Journal of Food Composition and Analysis* **51**: 93–105.

Barrea, L., G. Pugliese, E. Frias-Toral, M. El Ghoch, B. Castellucci, S. P. Chapela, M. L. A. Carignano, D. Laudisio, S. Savastano, A. Colao and G. Muscogiuri (2021). "Coffee Consumption, Health Benefits and Side Effects: A Narrative Review and Update for Dietitians and Nutritionists." *Critical Reviews in Food Science and Nutrition*: 1–24.

Bastos, V. S., M. F. Santos, L. P. Gomes, A. M. Leite, V. M. Flosi Paschoalin and E. M. Del Aguila (2018). "Analysis of the Cocobiota and Metabolites of Moniliophthora Perniciosa-Resistant Theobroma Cacao Beans During Spontaneous Fermentation in Southern Brazil." *Journal of the Science of Food and Agriculture* **98**(13): 4963–4970.

Behera, S. S., A. F. El Sheikha, R. Hammami and A. Kumar (2020). "Traditionally Fermented Pickles: How the Microbial Diversity Associated with Their Nutritional and Health Benefits?" *Journal of Functional Foods* **70**: 103971.

Behera, S. S., R. C. Ray and N. Zdolec (2018). "Lactobacillus Plantarum with Functional Properties: An Approach to Increase Safety and Shelf-Life of Fermented Foods." *BioMed Research International* **2018**: 9361614.

Benford, D., M. Bignami, J. K. Chipman and L. Ramos Bordajandi (2022). "Assessment of the Genotoxicity of Acrylamide." *EFSA Journal* **20**(5): e07293.

Bonatsou, S., C. C. Tassou, E. Z. Panagou and G. E. Nychas (2017). "Table Olive Fermentation Using Starter Cultures with Multifunctional Potential." *Microorganisms* **5**(2).

Botta, C. and L. Cocolin (2012). "Microbial Dynamics and Biodiversity in Table Olive Fermentation: Culture-Dependent and -Independent Approaches." *Frontiers in Microbiology* **3**.

Bousquet, J., J. M. Anto, W. Czarlewski, T. Haahtela, S. C. Fonseca, G. Iaccarino, H. Blain, A. Vidal, A. Sheikh, C. A. Akdis and T. Zuberbier (2021). "Cabbage and Fermented Vegetables: From Death Rate Heterogeneity in Countries to Candidates for Mitigation Strategies of Severe COVID-19." *Allergy* **76**(3): 735–750.

Breidt, F., R. F. McFeeters, I. Perez-Diaz and C.-H. Lee (2012). "Fermented Vegetables." *Food Microbiology*: 841–855.

Brodmann, T., A. Endo, M. Gueimonde, G. Vinderola, W. Kneifel, W. M. de Vos, S. Salminen and C. Gómez-Gallego (2017). "Safety of Novel Microbes for Human Consumption: Practical Examples of Assessment in the European Union." *Frontiers in Microbiology* **8**.

Caggia, C., C. L. Randazzo, M. Di Salvo, F. Romeo and P. Giudici (2004). "Occurrence of Listeria Monocytogenes in Green Table Olives." *Journal of Food Protection* **67**(10): 2189–2194.

Camu, N., T. De Winter, K. Verbrugghe, I. Cleenwerck, P. Vandamme, J. S. Takrama, M. Vancanneyt and L. De Vuyst (2007). "Dynamics and Biodiversity of Populations of Lactic Acid Bacteria and Acetic Acid Bacteria Involved in Spontaneous Heap Fermentation of Cocoa Beans in Ghana." *Applied and Environmental Microbiology* **73**(6): 1809–1824.

Casado, F. J., A. Montaño, D. Spitzner and R. Carle (2013). "Investigations into Acrylamide Precursors in Sterilized Table Olives: Evidence of a Peptic Fraction Being Responsible for Acrylamide Formation." *Food chemistry* **141**(2): 1158–1165.

Castellone, V., E. Bancalari, J. Rubert, M. Gatti, E. Neviani and B. Bottari (2021). "Eating Fermented: Health Benefits of LAB-Fermented Foods." *Foods* **10**(11).

Cempaka, L., L. Aliwarga, S. Purwo and M. T. A. P. Kresnowati (2014). "Dynamics of Cocoa Bean Pulp Degradation During Cocoa Bean Fermentation: Effects of Yeast Starter Culture Addition." *Journal of Mathematical and Fundamental Sciences* **46**(1): 14–25.

Chang, J. Y.; Chang, H. C. Improvements in the quality and shelf life of kimchi by fermentation with the induced bacterion-producing strain, *Leuconostoc citreum* GJ7 as a starter. *J. Food Sci.* 2010, 75, 103–110.

Chang, J. Y. and H. C. Chang (2010). "Improvements in the Quality and Shelf Life of Kimchi by Fermentation with the Induced Bacterion-Producing Strain, *Leuconostoc citreum* GJ7 as a Starter." *Journal of Food Science* 75: 103–110.

Chao, S.-H., Y. Tomii, K. Watanabe and Y.-C. Tsai (2008). "Diversity of Lactic Acid Bacteria in Fermented Brines Used to Make Stinky Tofu." *International Journal of Food Microbiology* 123(1): 134–141.

Charoenprasert, S. and A. Mitchell (2014). "Influence of California-Style Black Ripe Olive Processing on the Formation of Acrylamide." *Journal of Agricultural and Food Chemistry* 62(34): 8716–8721.

Cheigh, H. S. and K. Y. Park (1994). "Biochemical, Microbiological, and Nutritional Aspects of Kimchi (Korean Fermented Vegetable Products)." *Critical Reviews in Food Science and Nutrition* 34(2): 175–203. https://doi.org/10.1080/10408399409527656. PMID: 8011144.

Cho, E. J. (1999). *Standardization and Cancer Chemopreventive Activities of Chinese Cabbage Kimchi*. PhD. Dissertation, Pusan National University, Busan, South Korea.

Cho, J. H.; Lee, D. Y.; Yang, C. N.; Jeon, J. I.; Kim, J. H.; Han, H. U. Microbial population dynamics of kimchi, a fermented cabbage product. *FEMS Microbiol. Lett.* 2006a, 257, 262–267.

Cho, K. M.; Math, R. K.; Islam, S. A.; Lim, W. J.; Hong, S. Y.; Kim, J. M.; Yun, M. G.; Cho J. J.; Yun, H. D. Novel multiplex PCR for the detection of lactic acid bacteria during kimchi fermentation. *Mol. Cell. Probes* 2009, 23, 90–94.

Choi, H. J.; Cheigh, C. I.; Kim, S. B.; Lee, J. C.; Lee, D. W.; Choi, S. W.; Park, J. M.; Pyun, Y. R. *Weissella kimchii* sp. nov., a novel lactic acid bacterium from kimchi. *Int. J. Sys. Evol. Microbiol.* 2002, 52, 507–511.

Chou, S.-S., J.-C. Chung and D.-F. Hwang (2003). "A High Performance Liquid Chromatography Method for Determining Nitrate and Nitrite Levels in Vegetables." *Journal of Food and Drug Analysis Taipei City* 11.

Conte, P., C. Fadda, A. Del Caro, P. P. Urgeghe and A. Piga (2020). "Table Olives: An Overview on Effects of Processing on Nutritional and Sensory Quality." *Foods* 9(4): 514.

Corsetti, A., G. Perpetuini, M. Schirone, R. Tofalo and G. Suzzi (2012). "Application of Starter Cultures to Table Olive Fermentation: An Overview on the Experimental Studies." *Frontiers in Microbiology* 3: 248.

Coton, M., A. Pawtowski, B. Taminiau, G. Burgaud, F. Deniel, L. Coulloumme-Labarthe, A. Fall, G. Daube and E. Coton (2017). "Unraveling Microbial Ecology of Industrial-Scale Kombucha Fermentations by Metabarcoding and Culture-Based Methods." *FEMS Microbiology Ecology* 93(5).

Daughtry, K. V., S. D. Johanningsmeier, R. Sanozky-Dawes, T. R. Klaenhammer and R. Barrangou (2018). "Phenotypic and Genotypic Diversity of Lactobacillus Buchneri Strains Isolated from Spoiled, Fermented Cucumber." *International Journal of Food Microbiology* 280: 46–56.

De Bellis, P., F. Valerio, A. Sisto, S. L. Lonigro and P. Lavermicocca (2010). "Probiotic Table Olives: Microbial Populations Adhering on Olive Surface in Fermentation Sets Inoculated with the Probiotic Strain Lactobacillus Paracasei IMPC2.1 in an Industrial Plant." *International Journal of Food Microbiology* 140(1): 6–13.

de Melo Pereira, G. V., V. T. Soccol, A. Pandey, A. B. Medeiros, J. M. Andrade Lara, A. L. Gollo and C. R. Soccol (2014). "Isolation, Selection and Evaluation of Yeasts for Use in Fermentation of Coffee Beans by the Wet Process." *International Journal of Food Microbiology* 188: 60–66.

Di Cagno, R., R. Coda, M. De Angelis and M. Gobbetti (2013). "Exploitation of Vegetables and Fruits Through Lactic Acid Fermentation." *Food Microbiology* 33(1): 1–10.

Di Cagno, R., P. Filannino and M. Gobbetti (2016). "Fermented Foods: Fermented Vegetables and Other Products." In *Encyclopedia of Food and Health*. Eds. B. Caballero, P. M. Finglas and F. Toldrá. Oxford: Academic Press: 668–674.

Diez-Ozaeta, I. and O. J. Astiazaran (2022). "Fermented Foods: An Update on Evidence-Based Health Benefits and Future Perspectives." *Food Research International* 156: 111133.

Ding, Z., S. D. Johanningsmeier, R. Price, R. Reynolds, V.-D. Truong, S. C. Payton and F. Breidt (2018). "Evaluation of Nitrate and Nitrite Contents in Pickled Fruit and Vegetable Products." *Food Control* 90: 304–311.

Doeun, D., M. Davaatseren and M. S. Chung (2017). "Biogenic Amines in Foods." *Food Science and Biotechnology* 26(6): 1463–1474.

Do Prado, F. G., M. F. Miyaoka, G. V. de Melo Pereira, M. G. B. Pagnoncelli, M. R. M. Prado, S. J. R. Bonatto, M. R. Spier and C. R. Soccol (2021). "Fungal-Mediated Biotransformation of Soybean Supplemented with Different Cereal Grains into a Functional Compound with Antioxidant, Anti-Inflammatory and Antitumoral Activities." *Biointerface Research in Applied Chemistry* 11(1): 8018–8033.

Do Prado, F. G., M. G. B. Pagnoncelli, G. V. de Melo Pereira, S. G. Karp and C. R. Soccol (2022). "Fermented Soy Products and Their Potential Health Benefits: A Review." *Microorganisms* 10(8).

Duedahl-Olesen, L., A. S. Wilde, M. P. Dagnæs-Hansen, A. Mikkelsen, P. T. Olesen and K. Granby (2022). "Acrylamide in Commercial Table Olives and the Effect of Domestic Cooking." *Food Control* 132: 108515.

Dufresne, C. and E. Farnworth (2000). "Tea, Kombucha, and Health: A Review." *Food Research International* 33(6): 409–421.

Durante, M., M. Tufariello, L. Tommasi, M. S. Lenucci, G. Bleve and G. Mita (2018). "Evaluation of Bioactive Compounds in Black Table Olives Fermented with Selected Microbial Starters." *Journal of the Science of Food and Agriculture* 98(1): 96–103.

EFSA Panel on Food Additives Nutrient Sources Added to Food (2017). "Re-evaluation of Potassium Nitrite (E 249) and Sodium Nitrite (E 250) as Food Additives." *EFSA Journal* **15**(6): e04786.

El Sheikha, A. F. (2018). "Revolution in Fermented Foods." *Molecular Techniques in Food Biology*: 239–260.

Fernández, A. G., M. J. F. Díez and M. R. Adams (1997). "Table Olives." In *Production and Processing*. Eds. A. Garrido Fernandez, M. R. Adams, M. J. Fernandez-Diez. New York: Springer Science & Business Media. ISBN0412718103, 9780412718106.

Fernández-Reina, A., J. L. Urdiales and F. Sánchez-Jiménez (2018). "What We Know and What We Need to Know About Aromatic and Cationic Biogenic Amines in the Gastrointestinal Tract." *Foods* **7**(9).

Fideler, J., S. D. Johanningsmeier, M. Ekelöf and D. C. Muddiman (2019). "Discovery and Quantification of Bioactive Peptides in Fermented Cucumber by Direct Analysis IR-MALDESI Mass Spectrometry and LC-QQQ-MS." *Food Chemistry* **271**: 715–723.

Fong, F. L. Y., H. El-Nezami and E. T. P. Sze (2021). "Biogenic Amines—Precursors of Carcinogens in Traditional Chinese Fermented Food." *NFS Journal* **23**: 52–57.

Fowler, M. S., P. Leheup and J. L. Cordier (1998). "Cocoa, Coffee and Tea." In *Microbiology of Fermented Foods*. Ed. B. J. B. Wood. Boston, MA: Springer. https://doi.org/10.1007/978-1-4613-0309-1_5

Franzetti, A., I. Gandolfi, E. Gaspari, R. Ambrosini and G. Bestetti (2011). "Seasonal Variability of Bacteria in Fine and Coarse Urban Air Particulate Matter." *Applied Microbiology and Biotechnology* **90**(2): 745–753.

Fung, W. Y., Y. P. Woo and M. T. Liong (2008). "Optimization of Growth of Lactobacillus Acidophilus FTCC 0291 and Evaluation of Growth Characteristics in Soy Whey Medium: A Response Surface Methodology Approach." *Journal of Agricultural and Food Chemistry* **56**(17): 7910–7918.

Gille, D., A. Schmid, B. Walther and G. Vergères (2018). "Fermented Food and Non-Communicable Chronic Diseases: A Review." *Nutrients* **10**(4).

Gutiérrez-Ríos, H. G., M. L. Suárez-Quiroz, Z. J. Hernández-Estrada, O. P. Castellanos-Onorio, R. Alonso-Villegas, P. Rayas-Duarte, C. Cano-Sarmiento, C. Y. Figueroa-Hernández and O. González-Rios (2022). "Yeasts as Producers of Flavor Precursors During Cocoa Bean Fermentation and Their Relevance as Starter Cultures: A Review." *Fermentation* **8**(7): 331.

Guzman, R., E. Pérez and M. Raymundez (2020). "Structural and Chemical Changes During the Fermentation of Cocoa Beans." In *Theobroma Cacao Production, Cultivation and Uses*. Ed. M. Løvstrøm. Hauppauge, NY: Nova Science Publishers, Inc: 81–104.

Haile, M. and W. Kang (2019). "The Role of Microbes in Coffee Fermentation and Their Impact on Coffee Quality." *Journal of Food Quality* **2019**.

Halász, Á., A. Baráth, L. Simon-Sarkadi and W. Holzapfel (1994). "Biogenic Amines and Their Production by Microorganisms in Food." *Trends in Food Science & Technology* **5**(2): 42–49.

Harris, L. J. (1998). "The Microbiology of Vegetable Fermentations." In *Microbiology of Fermented Foods*. Ed. B. J. B. Wood. Boston, MA: Springer. https://doi.org/10.1007/978-1-4613-0309-1_2

He, F. J. and J. Q. Chen (2013). "Consumption of Soybean, Soy Foods, Soy Isoflavones and Breast Cancer Incidence: Differences Between Chinese Women and Women in Western Countries and Possible Mechanisms." *Food Science and Human Wellness* **2**(3–4): 146–161.

Higashikawa, F., M. Noda, T. Awaya, K. Nomura, H. Oku and M. Sugiyama (2010). "Improvement of Constipation and Liver Function by Plant-Derived Lactic Acid Bacteria: A Double-Blind, Randomized Trial." *Nutrition* **26**(4): 367–374. https://www.sciencedirect.com/science/article/abs/pii/S0899900709002299

Hongu, N., A. S. Kim, A. Suzuki, H. Wilson, K. C. Tsui and S. Park (2017). "Korean Kimchi: Promoting Healthy Meals Through Cultural Tradition." *Journal of Ethnic Foods* **4**(3): 172–180.

Hord, N. G., Y. Tang and N. S. Bryan (2009). "Food Sources of Nitrates and Nitrites: The Physiologic Context for Potential Health Benefits." *The American Journal of Clinical Nutrition* **90**(1): 1–10.

Huch, M. and C. M. A. P. Franz (2015). "21—Coffee: Fermentation and Microbiota." In *Advances in Fermented Foods and Beverages*. Ed. W. Holzapfel. Burlington: Woodhead Publishing: 501–513.

Hurtado, A., C. Reguant, B. Esteve-Zarzoso, A. Bordons and N. Rozès (2008). "Microbial Population Dynamics During the Processing of Arbequina Table Olives." *Food Research International* **41**(7): 738–744.

International Agency for Research on Cancer (1994). "IARC Monographs on the Evaluation for Carcinogenic Risk of Chemicals to Humans." *Lyon, France* **60**: 435–453.

Jafar, N. and N. Beyatli (2019). "Production of Fermented Red Beet Juice Using Probiotic Lactobacilli Bacteria." *Annals of Tropical Medicine and Public Health* **22**: 91–95.

Janiszewska-Turak, E., M. Walczak, K. Rybak, K. Pobiega, M. Gniewosz, Ł. Woźniak and D. Witrowa-Rajchert (2022). "Influence of Fermentation Beetroot Juice Process on the Physico-Chemical Properties of Spray Dried Powder." *Molecules* **27**(3).

Jayabalan, R., R. V. Malbaša, E. S. Lončar, J. S. Vitas and M. Sathishkumar (2014). "A Review on Kombucha Tea-Microbiology, Composition, Fermentation, Beneficial Effects, Toxicity, and Tea Fungus." *Comprehensive Reviews in Food Science and Food Safety* **13**(4): 538–550.

Jeong, S. H., H. J. Lee, J. Y. Jung, S. H. Lee, H. Y. Seo, W. S. Park and C. O. Jeon (2013). "Effects of Red Pepper Powder on Microbial Communities and Metabolites During Kimchi Fermentation." *International Journal of Food Microbiology* **160**(3): 252–259.

Jia, F., S. Peng, J. Green, L. Koh and X. Chen (2020). "Soybean Supply Chain Management and Sustainability: A Systematic Literature Review." *Journal of Cleaner Production* **255**: 120254.

Jiménez-López, J., A. Ruiz-Medina, P. Ortega-Barrales and E. J. Llorent-Martínez (2018). "Phytochemical Profile and Antioxidant Activity of Caper Berries (Capparis Spinosa L.): Evaluation of the Influence of the Fermentation Process." *Food Chemistry* **250**: 54–59.

Jin, H., F. Higashikawa, M. Noda, X. Zhao, Y. Matoba, T. Kumagai and M. Sugiyama (2010). "Establishment of an In Vitro Peyer's Patch Cell Culture System Correlative to In Vivo Study Using Intestine and Screening of Lactic Acid Bacteria Enhancing Intestinal Immunity." *Biological and Pharmaceutical Bulletin* **33**(2): 289–293. https://doi.org/10.1248/bpb.33.289. PMID: 20118555.

Jolvis Pou, K. R. (2016). "Fermentation: The Key Step in the Processing of Black Tea." *Journal of Biosystems Engineering* **41**(2): 85–92.

Jung, J. Y.; Lee, S. H.; Jeon, C. O. Kimchi microflora: History, current status, and perspectives for industrial kimchi production. *Appl. Microbiol. Biotechnol.,* 2014, 98, 2385–2393.

Jung, J. Y.; Lee, S. H.; Jin, H. M.; Hahn, Y.; Madsen, E. L.; Jeon, C. O. Metatranscriptomic analysis of lactic acid bacterial gene expression during kimchi fermentation. *Int J Food Microbiol,* 2013, 163, 171–179.

Jung, J. Y.; Lee, S. H.; Lee, H. J.; Seo, H. Y.; Park, W. S.; JEON, C. O. Effects of Leuconostoc mesenteroides starter cultures on microbial communities and metabolites during kimchi fermentation. *Int. J. Food Microbiol.* 2012, 153, 378–387.

Kalač, P., J. Špička, M. Křížek and T. Pelikánová (2000). "Changes in Biogenic Amine Concentrations During Sauerkraut Storage." *Food Chemistry* **69**(3): 309–314.

Kaltsa, A., D. Papaliaga, E. Papaioannou and P. Kotzekidou (2015). "Characteristics of Oleuropeinolytic Strains of Lactobacillus Plantarum Group and Influence on Phenolic Compounds in Table Olives Elaborated Under Reduced Salt Conditions." *Food Microbiol* **48**: 58–62.

Kapp, J. M. and W. Sumner (2019). "Kombucha: A Systematic Review of the Empirical Evidence of Human Health Benefit." *Annals of Epidemiology* **30**: 66–70.

Kapusta-Duch, J., B. Kusznierewicz, T. Leszczyńska and B. Borczak (2017). "Effect of Package Type on Selected Parameters of Nutritional Quality of Chill-Stored White Sauerkraut." *Polish Journal of Food and Nutrition Sciences* **67**(2): 137–144.

Karovičová, J. and Z. Kohajdová (2003). "Lactic Acid Fermented Vegetable Juices." *Horticultural Science* **30**.

Katz, S. E. (2020). *Vegetable Fermentation.* Raleigh, NC: NC State University Libraries. https://doi.org/10.52750/662109

Kesika, P., B. Sivamaruthi and C. Chaiyasut (2021). "A Review on the Functional Properties of Fermented Soymilk." *Food Science and Technology* **42**.

Kesika, P., B. S. Sivamaruthi and C. Chaiyasut (2022). "A Review on the Functional Properties of Fermented Soymilk." *Food Science and Technology* **42**.

Kiczorowski, P., B. Kiczorowska, W. Samolińska, M. Szmigielski and A. Winiarska-Mieczan (2022). "Effect of Fermentation of Chosen Vegetables on the Nutrient, Mineral, and Biocomponent Profile in Human and Animal Nutrition." *Sci Rep* **12**(1): 13422.

Kim, M. and Chun, J.B. (2005) Bacterial Community Structure in Kimchi, a Korean Fermented Vegetable Food, as Revealed by 16S rRNA Gene Analysis. International Journal of Food Microbiology, 103, 91–96. http://dx.doi.org/10.1016/j.ijfoodmicro.2004.11.030

Kim, J. and K. Adhikari (2020). "Current Trends in Kombucha: Marketing Perspectives and the Need for Improved Sensory Research." *Beverages* **6**(1): 15.

Kim, M. S., H. J. Yang, S. H. Kim, H. W. Lee and M. S. Lee (2018). "Effects of Kimchi on Human Health: A Protocol of Systematic Review of Controlled Clinical Trials." *Medicine (Baltimore)* **97**(13): e0163.

Klewicka, E., I. Motyl and Z. Libudzisz (2004). "Fermentation of Beet Juice by Bacteria of Genus Lactobacillus sp." *European Food Research and Technology* **218**: 178–183.

Klewicka, E., Z. Zduńczyk, J. Juśkiewicz and R. Klewicki (2015). "Effects of Lactofermented Beetroot Juice Alone or with N-nitroso-N-methylurea on Selected Metabolic Parameters, Composition of the Microbiota Adhering to the Gut Epithelium and Antioxidant Status of Rats." *Nutrients* **7**(7): 5905–5915.

Konuray, G. and Z. Erginkaya (2018). "Potential Use of Bacillus Coagulans in the Food Industry." *Foods* **7**(6).

Kresnowati, P., L. Suryani and M. Affifah (2013). "Improvement of Cocoa Beans Fermentation by LAB Starter Addition." *Journal of Medical and Bioengineering* **2**: 274–278.

Lan, C.-H., C.-K. Son, H. P. Ha, H. Florence, L. T. Binh, L.-T. Mai, N. T. H. Tram, T. T. M. Khanh, T. V. Phu, V. Dominique and W. Yves (2013). "Tropical Traditional Fermented Food, a Field Full of Promise. Examples from the Tropical Bioresources and Biotechnology Programme and Other Related French—Vietnamese Programmes on Fermented Food." *International Journal of Food Science & Technology* **48**(6): 1115–1126.

Latif, R. (2013). "Health Benefits of Cocoa." *Current Opinion in Clinical Nutrition & Metabolic Care* **16**(6): 669–674.

Lee, C. W.; Ko, C. Y.; Ha, D. M. Microfloral changes of the lactic acid bacteria during kimchi fermentation and identification of isolates. *Korean J. Appl. Microbiol. Biotechnol.* 1992, 20, 102–109.

Lee, D.; Kim, S.; Cho, J.; Kim, J. Microbial population dynamics and temperature changes during fermentation of kimjang kimchi. *J. Microbiol.* 2008a, 46, 590–593.

Lee, H., H. Yoon, Y. Ji, H. Kim, H. Park, J. Lee, J. Shin and W. Holzapfel (2011). "Functional Properties of Lactobacillus Strains Isolated from Kimchi." *International Journal of Food Microbiology* **145**(1): 155–161.

Lee, J.-J., Y.-J. Choi, M. J. Lee, S. J. Park, S. J. Oh, Y.-R. Yun, S. G. Min, H.-Y. Seo, S.-H. Park and M.-A. Lee (2020). "Effects of Combining Two Lactic Acid Bacteria as a Starter Culture on Model Kimchi Fermentation." *Food Research International* **136**: 109591.

Lee, M.-A., Y.-J. Choi, H. Lee, S. Hwang, H. J. Lee, S. J. Park, Y. B. Chung, Y.-R. Yun, S.-H. Park, S. Min, L.-S. Kwon and H.-Y. Seo (2021). "Influence of Salinity on the Microbial Community Composition and Metabolite Profile in Kimchi." *Fermentation* **7**(4): 308.

Lee, M.-E., J.-Y. Jang, J.-H. Lee, H.-W. Park, H.-J. Choi and T.-W. Kim (2015). "Starter Cultures for Kimchi Fermentation." *Journal of Microbiology and Biotechnology* **25**(5): 559–568.

Lee, S.-J., H.-S. Jeon, J.-Y. Yoo and J.-H. Kim (2021). "Some Important Metabolites Produced by Lactic Acid Bacteria Originated from Kimchi." *Foods* **10**(9): 2148.

Lee, S. H.; Kim S. D. Effect of starter on the fermentation of kimchi. *J. Korean Soc. Food Nutr.* 1988, 17, 342–347.

Lee, S. H. and S. D. Kim (1988). "Effect of Starter on the Fermentation of Kimchi." *Journal of the Korean Society of Food Science and Nutrition* 17: 342–347.

Li, C., X. Rui, Y. Zhang, F. Cai, X. Chen and M. Jiang (2017). "Production of Tofu by Lactic Acid Bacteria Isolated from Naturally Fermented Soy Whey and Evaluation of Its Quality." *LWT—Food Science and Technology* **82**: 227–234.

Li, Y. Z., Z. Y. Yang, T. T. Gong, Y. S. Liu, F. H. Liu, Z. Y. Wen, X. Y. Li, C. Gao, M. Luan, Y. H. Zhao and Q. J. Wu (2022). "Cruciferous Vegetable Consumption and Multiple Health Outcomes: An Umbrella Review of 41 Systematic Reviews and Meta-Analyses of 303 Observational Studies." *Food & Function* **13**(8): 4247–4259.

Lim, H. S., I. T. Cha, S. W. Roh, H. H. Shin and M. J. Seo (2017). "Enhanced Production of Gamma-Aminobutyric Acid by Optimizing Culture Conditions of Lactobacillus Brevis HYE1 Isolated from Kimchi, a Korean Fermented Food." *Journal of Microbiology and Biotechnology* **27**(3): 450–459.

Lindsay, J., P.-H. Carmichael, E. Kröger and D. Laurin (2012). *Coffee: Emerging Health Effects and Disease Prevention.* Ed. Yi-Fang Chu Hoboken, NJ: Wiley-Blackwell: 97–110.

Liu, Y., H. Chen, W. Chen, Q. Zhong, G. Zhang and W. Chen (2018). "Beneficial Effects of Tomato Juice Fermented by Lactobacillus Plantarum and Lactobacillus Casei: Antioxidation, Antimicrobial Effect, and Volatile Profiles." *Molecules* **23**(9).

López-López, A., V. M. Beato, A. H. Sánchez, P. García-García and A. Montaño (2014). "Effects of Selected Amino Acids and Water-Soluble Vitamins on Acrylamide Formation in a Ripe Olive Model System." *Journal of Food Engineering* **120**: 9–16.

Lucena-Padrós, H. and J. L. Ruiz-Barba (2019). "Microbial Biogeography of Spanish-Style Green Olive Fermentations in the Province of Seville, Spain." *Food Microbiol* **82**: 259–268.

Lundberg, J. O., M. Carlström and E. Weitzberg (2018). "Metabolic Effects of Dietary Nitrate in Health and Disease." *Cell Metabolism* **28**(1): 9–22.

Majcherczyk, J. and K. Surówka (2019). "Effects of Onion or Caraway on the Formation of Biogenic Amines During Sauerkraut Fermentation and Refrigerated Storage." *Food Chemistry* **298**: 125083.

Marco, M. L., D. Heeney, S. Binda, C. J. Cifelli, P. D. Cotter, B. Foligné, M. Gänzle, R. Kort, G. Pasin, A. Pihlanto, E. J. Smid and R. Hutkins (2017). "Health Benefits of Fermented Foods: Microbiota and Beyond." *Current Opinion in Biotechnology* **44**: 94–102.

Marsh, A. J., O. O'Sullivan, C. Hill, R. P. Ross and P. D. Cotter (2014). "Sequence-Based Analysis of the Bacterial and Fungal Compositions of Multiple Kombucha (Tea Fungus) Samples." *Food Microbiology* **38**: 171–178.

Meersman, E., J. Steensels, M. Mathawan, P.-J. Wittocx, V. Saels, N. Struyf, H. Bernaert, G. Vrancken and K. J. Verstrepen (2013). "Detailed Analysis of the Microbial Population in Malaysian Spontaneous Cocoa Pulp Fermentations Reveals a Core and Variable Microbiota." *PLoS ONE* **8**(12): e81559.

Megaelectra, A. N. F. (2016). *Investigating Microbial Community of Pu-Erh Teas (Brewed and Non-Brewed) and the Ability of Bacteria Originating from the Tea to Survive in the Gastrointestinal Tract.* Master Master's degree, Lund University.

Melini, F., V. Melini, F. Luziatelli, A. G. Ficca and M. Ruzzi (2019). "Health-Promoting Components in Fermented Foods: An Up-to-Date Systematic Review." *Nutrients* **11**(5).

Messina, M. (2016). "Soy and Health Update: Evaluation of the Clinical and Epidemiologic Literature." *Nutrients* **8**(12).

Messina, M., A. Duncan, V. Messina, H. Lynch, J. Kiel and J. W. Erdman, Jr. (2022). "The Health Effects of Soy: A Reference Guide for Health Professionals." *Frontiers in Nutrition* **9**: 970364.

Mital, B. K. and K. H. Steinkraus (1979). "Fermentation of Soy Milk by Lactic Acid Bacteria. A Review." *Journal of Food Protection* **42**(11): 895–899.

Mo, H., Y. Zhu and Z. Chen (2008). "Microbial Fermented Tea—A Potential Source of Natural Food Preservatives." *Trends in Food Science & Technology* **19**(3): 124–130.

Montaño, A., A. H. Sánchez, V. M. Beato, A. López-López and A. de Castro (2016). "Pickling." In *Encyclopedia of Food and Health*. Eds. B. Caballero, P. M. Finglas and F. Toldrá. Oxford: Academic Press: 369–374.

Moore, J. F., R. DuVivier and S. D. Johanningsmeier (2021). "Formation of γ-Aminobutyric Acid (GABA) During the Natural Lactic Acid Fermentation of Cucumber." *Journal of Food Composition and Analysis* **96**: 103711.

Mota-Gutierrez, J. and L. S. Cocolin (2022). "The Functional and Nutritional Aspects of Cocobiota: Lactobacilli." In *Good Microbes in Medicine, Food Production, Biotechnology, Bioremediation, and Agriculture*. Eds. Frans J. de Bruijn, Hauke Smidt, Luca S. Cocolin, Michael Sauer, David Dowling, Linda Thomashow Hoboken, NJ: Wiley: 199–212.

Mukherjee, A., B. Gómez-Sala, E. M. O'Connor, J. G. Kenny and P. D. Cotter (2022). "Global Regulatory Frameworks for Fermented Foods: A Review." *Frontiers in Nutrition* **9**.

Murthy, P. S. and M. Naidu (2011). "Improvement of Robusta Coffee Fermentation with Microbial Enzymes." *European Journal of Applied Sciences* **3**.

Murthy, P. S. and M. Naidu (2012). "Sustainable Management of Coffee Industry by-Products and Value Addition—A Review." *Resources, Conservation and Recycling* **66**: 45–58.

Nguyen-the, C. and F. Carlin (1994). "The Microbiology of Minimally Processed Fresh Fruits and Vegetables." *Critical Reviews in Food Science and Nutrition* **34**(4): 371–401.

Nielsen, D. S., S. Hønholt, K. Tano-Debrah and L. Jespersen (2005). "Yeast Populations Associated with Ghanaian Cocoa Fermentations Analysed Using Denaturing Gradient Gel Electrophoresis (DGGE)." *Yeast* **22**(4): 271–284.

Nielsen, D. S., O. D. Teniola, L. Ban-Koffi, M. Owusu, T. S. Andersson and W. H. Holzapfel (2007). "The Microbiology of Ghanaian Cocoa Fermentations Analysed Using Culture-Dependent and Culture-Independent Methods." *International Journal of Food Microbiology* **114**(2): 168–186.

Ning, J., J. Sun, S. Li, M. Sheng and Z. Zhang (2017). "Classification of Five Chinese Tea Categories with Different Fermentation Degrees Using Visible and Near-Infrared Hyperspectral Imaging." *International Journal of Food Properties* **20**(sup2): 1515–1522.

Özcan, M. M., I. A. M. Ahmed, F. A. Juhaimi, N. Uslu, M. A. Osman, M. A. Gassem, E. E. Babiker and K. Ghafoor (2020). "The Influence of Fermentation and Bud Sizes on Antioxidant Activity and Bioactive Compounds of Three Different Size Buds of Capparis Ovata Desf. var. Canescens Plant." *Journal of Food Science and Technology* **57**(7): 2705–2712.

Paramithiotis, S., G. Das, H.-S. Shin and J. K. Patra (2022). "Fate of Bioactive Compounds During Lactic Acid Fermentation of Fruits and Vegetables." *Foods* **11**(5): 733.

Park, J. E., J. E. Seo, J. Y. Lee and H. Kwon (2015). "Distribution of Seven N-Nitrosamines in Food." *Toxicol Res* **31**(3): 279–288.

Park, K. and S. Rhee (2005). "Functional Foods from Fermented Vegetable Products: Kimchi (Korean Fermented Vegetables) and Functionality." In *Asian Functional Foods*. Eds. J. Shi, C. Ho and F. Shahidi. Boca Raton, FL: CRC Press: 341–380.

Park, K.-Y., J.-K. Jeong, Y.-E. Lee and J. Daily Iii (2014). "Health Benefits of Kimchi (Korean Fermented Vegetables) as a Probiotic Food." *Journal of Medicinal Food* **17**: 6–20.

Park, K. Y., H. Y. Kim and J. K. Jeong (2017). "Chapter 20—Kimchi and Its Health Benefits." In *Fermented Foods in Health and Disease Prevention*. Eds. J. Frias, C. Martinez-Villaluenga and E. Peñas. Boston: Academic Press: 477–502.

Park, S. Y., H. L. Jang, J. H. Lee, Y. Choi, H. Kim, J. Hwang, D. Seo, S. Kim and J. S. Nam (2017). "Changes in the Phenolic Compounds and Antioxidant Activities of Mustard Leaf (Brassica Juncea) Kimchi Extracts During Different Fermentation Periods." *Food Science and Biotechnology* **26**(1): 105–112.

Park, S. Y., Y. Ji, H. Park, K. Lee, H. Park, B. R. Beck, H. Shin and W. H. Holzapfel (2016). "Evaluation of Functional Properties of Lactobacilli Isolated from Korean White Kimchi." *Food Control* **69**: 5–12.

Patra, J. K., G. Das, S. Paramithiotis and H. S. Shin (2016). "Kimchi and Other Widely Consumed Traditional Fermented Foods of Korea: A Review." *Frontiers in Microbiology* **7**: 1493.

Pederson, C. and M. Albury (2006). "Factors Affecting the Abcterial Flora in Fermenting Vegetables." *Journal of Food Science* **18**: 290–300.

Pederson, C. S. (1979). "Fermented Vegetable Products." In *Microbiology of Food Fermentation*, 2nd ed. Westport, CT: Avi Publishing: 153–209.

Peñas, E., C. Martinez-Villaluenga and J. Frias (2017). "Chapter 24—Sauerkraut: Production, Composition, and Health Benefits." In *Fermented Foods in Health and Disease Prevention*. Eds. J. Frias, C. Martinez-Villaluenga and E. Peñas. Boston: Academic Press: 557–576.

Pérez-Díaz, I. M., J. S. Hayes, E. Medina, A. M. Webber, N. Butz, A. N. Dickey, Z. Lu and M. A. Azcarate-Peril (2019). "Assessment of the Non-Lactic Acid Bacteria Microbiota in Fresh Cucumbers and Commercially Fermented Cucumber Pickles Brined with 6% NaCl." *Food Microbiology* **77**: 10–20.

Pérez Montoro, B., N. Benomar, N. Caballero Gómez, S. Ennahar, P. Horvatovich, C. W. Knapp, E. Alonso, A. Gálvez and H. Abriouel (2018). "Proteomic Analysis of Lactobacillus Pentosus for the Identification of Potential Markers of Adhesion and Other Probiotic Features." *Food Research International* **111**: 58–66.

Pérez Pulido, R., N. Ben Omar, H. Abriouel, R. Lucas López, M. Martínez Cañamero and A. Gálvez (2005). "Microbiological Study of Lactic Acid Fermentation of Caper Berries by Molecular and Culture-Dependent Methods." *Applied and Environmental Microbiology* **71**(12): 7872–7879.

Perpetuini, G., R. Prete, N. Garcia-Gonzalez, M. Khairul Alam and A. Corsetti (2020). "Table Olives More than a Fermented Food." *Foods* **9**(2): 178.

Petyaev, I. M. and Y. K. Bashmakov (2016). "Cocobiota: Implications for Human Health." *Journal of Nutrition and Metabolism* **2016**: 7906927.

Poole, R., O. J. Kennedy, P. Roderick, J. A. Fallowfield, P. C. Hayes and J. Parkes (2017). "Coffee Consumption and Health: Umbrella Review of Meta-Analyses of Multiple Health Outcomes." *BMJ* **359**: j5024.

Qin, L. and S. Wang (2022). "Protective Roles of Inorganic Nitrate in Health and Diseases." *Current Medicine* **1**(1): 4.

Raak, C., T. Ostermann, K. Boehm and F. Molsberger (2014). "Regular Consumption of Sauerkraut and Its Effect on Human Health: A Bibliometric Analysis." *Global Advances Health Medicine* **3**(6): 12–18.

Rasheed, Z. (2019). "Molecular Evidences of Health Benefits of Drinking Black Tea." *International Journal of Health Sciences (Qassim)* **13**(3): 1–3.

Reddy, K. N. and S. O. Duke (2015). "Soybean Mineral Composition and Glyphosate Use." In *Processing and Impact on Active Components in Food*. Ed. Victor Preedy. Cambridge, MA: Academic Press, Elsevier: 369–376.

Rejano, L., A. Montaño, F. J. Casado, A. H. Sánchez and A. de Castro (2010). "Table Olives." In *Olives and Olive Oil in Health and Disease Prevention*. Eds. Victor R. Preedy and Ronald Ross Watson. Cambridge, MA: Academic Press: 5–15.

Restuccia, D., M. R. Loizzo and U. G. Spizzirri (2018). "Accumulation of Biogenic Amines in Wine: Role of Alcoholic and Malolactic Fermentation." *Fermentation* **4**(1): 6.

Riciputi, Y., D. I. Serrazanetti, V. Verardo, L. Vannini, M. F. Caboni and R. Lanciotti (2016). "Effect of Fermentation on the Content of Bioactive Compounds in Tofu-type Products." *Journal of Functional Foods* **27**: 131–139.

Rivera, D., C. Inocencio, M. C. Obón and F. Alcaraz (2003). "Review of Food and Medicinal Uses of *Capparis* L. subgenus *Capparis* (Capparidaceae)." *Economic Botany* **57**: 515–534. https://doi.org/10.1663/0013-0001(2003)057[0515:ROFAMU]2.0.CO;2

Romel, E. G.-A. and G. M.-R. José (2021). "Fermentation of Cocoa Beans." In *Fermentation*. Ed. L. Marta. Rijeka: IntechOpen: Ch. 7.

Safe, S., J. Kothari, A. Hailemariam, S. Upadhyay, L. A. Davidson and R. S. Chapkin (2023). "Health Benefits of Coffee Consumption for Cancer and Other Diseases and Mechanisms of Action." *International Journal of Molecular Sciences* **24**(3).

Saiwal, N., M. Dahiya and H. Dureja (2019). "Nutraceutical Insight into Vegetables and Their Potential for Nutrition Mediated Healthcare." *Current Nutrition & Food Science* **15**(5): 441–453.

Sánchez-Pérez, S., O. Comas-Basté, J. Rabell-González, M. T. Veciana-Nogués, M. L. Latorre-Moratalla and M. C. Vidal-Carou (2018). "Biogenic Amines in Plant-Origin Foods: Are They Frequently Underestimated in Low-Histamine Diets?" *Foods* **7**(12).

Sandberg Nielsen, E., E. Garnås, K. J. Jensen, L. H. Hansen, P. S. Olsen, C. Ritz, L. Krych and D. S. Nielsen (2018). "Lacto-Fermented Sauerkraut Improves Symptoms in IBS Patients Independent of Product Pasteurisation—A Pilot Study." *Food & Function* **9**(10): 5323–5335.

Şanlier, N., B. B. Gökcen and A. C. Sezgin (2019). "Health Benefits of Fermented Foods." *Critical Reviews in Food Science and Nutrition* **59**(3): 506–527.

Sarmad Ghazi, A.-S. and A. Sadiq Jaafir Aziz (2021). "Cucumber Pickles and Fermentations." In *Cucumber Economic Values and Its Cultivation and Breeding*. Ed. W. Haiping. Rijeka: IntechOpen: Ch. 3.

Satora, P., M. Skotniczny, S. Strnad and W. Piechowicz (2021). "Chemical Composition and Sensory Quality of Sauerkraut Produced from Different Cabbage Varieties." *LWT* **136**: 110325.

Schlüter, A., T. Hühn, M. Kneubühl, K. Chatelain, S. Rohn and I. Chetschik (2020). "Novel Time- and Location-Independent Postharvest Treatment of Cocoa Beans: Investigations on the Aroma Formation During 'Moist Incubation' of Unfermented and Dried Cocoa Nibs and Comparison to Traditional Fermentation." *Journal of Agricultural and Food Chemistry* **68**(38): 10336–10344.

Schwan, R. F. and A. E. Wheals (2004). "The Microbiology of Cocoa Fermentation and Its Role in Chocolate Quality." *Critical Reviews in Food Science and Nutrition* **44**(4): 205–221.

Selvaraj, S. and K. Gurumurthy (2023). "An Overview of Probiotic Health Booster-Kombucha Tea." *Chinese Herbal Medicines* **15**(1): 27–32.

Septembre-Malaterre, A., F. Remize and P. Poucheret (2018). "Fruits and Vegetables, as a Source of Nutritional Compounds and Phytochemicals: Changes in Bioactive Compounds During Lactic Fermentation." *Food Research International* **104**: 86–99.

Serrazanetti, D. I., M. Ndagijimana, C. Miserocchi, L. Perillo and M. E. Guerzoni (2013). "Fermented Tofu: Enhancement of Keeping Quality and Sensorial Properties." *Food Control* **34**(2): 336–346.

Shurtleff, W. and A. Aoyagi (2011). *History of Fermented Tofu—A Healthy Nondairy/Vegan Cheese (1610–2011).* Lafayette, CA: Soyinfo Center.

Silva, C. F., L. R. Batista, L. M. Abreu, E. S. Dias and R. F. Schwan (2008). "Succession of Bacterial and Fungal Communities During Natural Coffee (Coffea Arabica) Fermentation." *Food Microbiology* **25**(8): 951–957.

Silva, C. F., D. M. Vilela, C. de Souza Cordeiro, W. F. Duarte, D. R. Dias and R. F. Schwan (2013). "Evaluation of a Potential Starter Culture for Enhance Quality of Coffee Fermentation." *World Journal of Microbiology & Biotechnology* **29**(2): 235–247.

Simon, J., K. Fung, Z. Raisi-Estabragh, N. Aung, M. Y. Khanji, M. Kolossváry, B. Merkely, P. B. Munroe, N. C. Harvey, S. K. Piechnik, S. Neubauer, S. E. Petersen and P. Maurovich-Horvat (2022). "Light to Moderate Coffee Consumption Is Associated with Lower Risk of Death: A UK Biobank Study." *European Journal of Preventive Cardiology* **29**(6): 982–991.

Slavin, J. L. and B. Lloyd (2012). "Health Benefits of Fruits and Vegetables." *Advances in Nutrition* **3**(4): 506–516.

So, M. H., M. Y. Shin and Y. B. Kim (1996). "Effects of Psychrotrophic Lactic Acid Bacterial Starter on Kimchi Fermentation." *Korean Journal of Food Science and Technology* **28**: 806–813.

So, M. H.; Shin, M. Y.; Kim, Y. B. Effects of psychrotrophic lactic acid bacterial starter on kimchi fermentation. *Korean J. Food Sci. Technol.* 1996, 28, 806–813.

Song, E., L. Ang, H.W. Lee, et al. (2023). "Effects of Kimchi on Human Health: A Scoping Review of Randomized Controlled Trials." *Journal of Ethnic Food* **10**. https://doi.org/10.1186/s42779-023-00173-8

Sosalagere, C., B. Adesegun Kehinde and P. Sharma (2022). "Isolation and Functionalities of Bioactive Peptides from Fruits and Vegetables: A Reviews." *Food Chemistry* **366**: 130494.

Spano, G., P. Russo, A. Lonvaud-Funel, P. Lucas, H. Alexandre, C. Grandvalet, E. Coton, M. Coton, L. Barnavon, B. Bach, F. Rattray, A. Bunte, C. Magni, V. Ladero, M. Alvarez, M. Fernández, P. Lopez, P. F. de Palencia, A. Corbi, H. Trip and J. S. Lolkema (2010). "Biogenic Amines in Fermented Foods." *European Journal of Clinical Nutrition* **64**(3): S95–S100.

Steinkraus, K. H. (1992). "Applications of Biotechnology to Fermented Foods: Report of an Ad Hoc Panel of the Board on Science and Technology for International Development." *Lactic Acid Fermentations*. Ed. N. R. C. U. P. O. T. A. O. B. T. T. F. Foods. Washington, DC: National Academies Press: 5.

Sugahara, H., S. Kato, K. Nagayama, K. Sashihara and Y. Nagatomi (2022). "Heterofermentative Lactic Acid Bacteria Such as Limosilactobacillus as a Strong Inhibitor of Aldehyde Compounds in Plant-Based Milk Alternatives." *Frontiers in Sustainable Food Systems* **6**.

Sugiyama, M. (2009). "Characterization of Plant-Derived Lactic Acid Bacteria Producing Anti-pylori Substances and the Strain LP28 Effective to Improve Metabolic Syndrome." Contribution of Lactic Acid Bacteria to Preventive and Pre-symptomatic Medicine. 2009.12.3. Korea–Japan International Symposium, Seoul, Korea.

Sugiyama, S.-I. (1984). "Selection of Micro-Organisms for Use in the Fermentation of Soy Sauce." *Food Microbiology* **1**(4): 339–347.

Suzanne, D. J. and F. M. Roger (2013). "Metabolism of Lactic Acid in Fermented Cucumbers by Lactobacillus Buchneri and Related Species, Potential Spoilage Organisms in Reduced Salt Fermentations." *Food Microbiology* **35**(2): 129–135.

Suzzi, G. and S. Torriani (2015). "Editorial: Biogenic Amines in Foods." *Frontiers in Microbiology* **6**.

Swain, M. R., M. Anandharaj, R. C. Ray and R. Parveen Rani (2014). "Fermented Fruits and Vegetables of Asia: A Potential Source of Probiotics." *Biotechnology Research International* **2014**: 250424.

Tan, T. Y. C., X. Y. Lim, J. H. H. Yeo, S. W. H. Lee and N. M. Lai (2021). "The Health Effects of Chocolate and Cocoa: A Systematic Review." *Nutrients* **13**(9).

Tverdal, A., R. Selmer, J. M. Cohen and D. S. Thelle (2020). "Coffee Consumption and Mortality from Cardiovascular Diseases and Total Mortality: Does the Brewing Method Matter?" *European Journal of Preventive Cardiology* **27**(18): 1986–1993.

Urnemi, S. Syukur, E. Purwati, A. Mustopa, S. Ibrahim and A. Jamsari (2011). *Antimicrobial Activity of Lactic Acid Bacteria Isolated from Cocoa Fermentation in West Sumatra, Indonesia Against Some Pathogenic Bacteria*. Conference paper: the 2nd International Seminar on Chemistry 2011, At: Bandung https://www.researchgate.net/publication/282866939_Antimicrobial_activity_of_lactic_acid_bacteria_isolated_from_cocoa_fermentation_in_West_Sumatra_Indonesia_against_some_pathogenic_bacteria

Valerio, F., S. de Candia, S. L. Lonigro, F. Russo, G. Riezzo, A. Orlando, P. De Bellis, A. Sisto and P. Lavermicocca (2011). "Role of the Probiotic Strain Lactobacillus Paracasei LMGP22043 Carried by Artichokes in Influencing Faecal Bacteria and Biochemical Parameters in Human Subjects." *Journal of Applied Microbiology* **111**(1): 155–164.

Vásquez-Villanueva, R., L. Muñoz-Moreno, M. José Carmena, M. Luisa Marina and M. Concepción García (2018). "In Vitro Antitumor and Hypotensive Activity of Peptides from Olive Seeds." *Journal of Functional Foods* **42**: 177–184.

Vaughn, R. H. (1982). "Lactic Acid Fermentation of Cabbage, Cucumbers, Olives and Other Produce." In *Prescott and Dunn's Industrial Microbiology*, 4th ed. Ed. G. Reed. Westport, CT: Avi Publishing Co: 185–236.

Velmourougane, K., D. R. Shanmukhappa, K. Ventakesh, C. B. Prakasan and J. Jayarama (2008). "Use of Starter Culture in Coffee Fermentation—Effect on Demucilisation and Cup Quality." *Indian Coffee* **72**: 31–34.

Vilela, D. M., G. V. Pereira, C. F. Silva, L. R. Batista and R. F. Schwan (2010). "Molecular Ecology and Polyphasic Characterization of the Microbiota Associated with Semi-Dry Processed Coffee (Coffea Arabica L.)." *Food Microbiology* **27**(8): 1128–1135.

Vinícius de Melo Pereira, G., V. T. Soccol, S. K. Brar, E. Neto and C. R. Soccol (2017). "Microbial Ecology and Starter Culture Technology in Coffee Processing." *Critical Reviews in Food Science and Nutrition* **57**(13): 2775–2788.

Visintin, S., V. Alessandria, A. Valente, P. Dolci and L. Cocolin (2016). "Molecular Identification and Physiological Characterization of Yeasts, Lactic Acid Bacteria and Acetic Acid Bacteria Isolated from Heap and Box Cocoa Bean Fermentations in West Africa." *International Journal of Food Microbiology* **216**: 69–78.

Wallace, T. C., R. L. Bailey, J. B. Blumberg, B. Burton-Freeman, C. y. O. Chen, K. M. Crowe-White, A. Drewnowski, S. Hooshmand, E. Johnson, R. Lewis, R. Murray, S. A. Shapses and D. D. Wang (2020). "Fruits, Vegetables, and Health: A Comprehensive Narrative, Umbrella Review of the Science and Recommendations for Enhanced Public Policy to Improve Intake." *Critical Reviews in Food Science and Nutrition* **60**(13): 2174–2211.

Wang, Y., J. Wu, M. Lv, Z. Shao, M. Hungwe, J. Wang, X. Bai, J. Xie, Y. Wang and W. Geng (2021). "Metabolism Characteristics of Lactic Acid Bacteria and the Expanding Applications in Food Industry." *Frontiers in Bioengineering and Biotechnology* **9**: 612285.

Wang, Y., X. Yang and L. Li (2020). "A New Style of Fermented Tofu by Lactobacillus Casei Combined with Salt Coagulant." *3 Biotech* **10**(2): 81.

Wiander, B. and E. L. Ryhanen (2005). "Laboratory and Large-Scale Fermentation of White Cabbage into Sauerkraut and Sauerkraut Juice by Using Starters in Combination with Mineral Salt with a Low NaCl Content." *European Food Research and Technology* **220**: 191–195.

Wilson, L. A. (1995). "Chapter 22—Soy Foods." In *Practical Handbook of Soybean Processing and Utilization*. Ed. D. R. Erickson. Champaign, IL: AOCS Press: 428–459.

Xiao, Y., L. Wang, X. Rui, W. Li, X. Chen, M. Jiang and M. Dong (2015). "Enhancement of the Antioxidant Capacity of Soy Whey by Fermentation with Lactobacillus Plantarum B1–6." *Journal of Functional Foods* **12**: 33–44.

Yang, J., V. Lagishetty, P. Kurnia, S. M. Henning, A. I. Ahdoot and J. P. Jacobs (2022). "Microbial and Chemical Profiles of Commercial Kombucha Products." *Nutrients* **14**(3).

Yao, W., L. Yang, Z. Shao, L. Xie and L. Chen (2020). "Identification of Salt Tolerance-Related Genes of Lactobacillus Plantarum D31 and T9 Strains by Genomic Analysis." *Annals of Microbiology* **70**(1): 10.

Yoon, K. Y., E. E. Woodams and Y. D. Hang (2005). "Fermentation of Beet Juice by Beneficial Lactic Acid Bacteria." *LWT—Food Science and Technology* **38**(1): 73–75.

Yoon, K. Y., E. E. Woodams and Y. D. Hang (2006). "Production of Probiotic Cabbage Juice by Lactic Acid Bacteria." *Bioresource Technology* **97**: 1427–1430.

Zago, M., B. Lanza, L. Rossetti, I. Muzzalupo, D. Carminati and G. Giraffa (2013). "Selection of Lactobacillus Plantarum Strains to Use as Starters in Fermented Table Olives: Oleuropeinase Activity and Phage Sensitivity." *Food Microbiology* **34**(1): 81–87.

Zaunmüller, T., M. Eichert, H. Richter and G. Unden (2006). "Variations in the Energy Metabolism of Biotechnologically Relevant Heterofermentative Lactic Acid Bacteria During Growth on Sugars and Organic Acids." *Applied Microbiology and Biotechnology* **72**(3): 421–429.

Zebin, G., J. Xiangze, Z. Zhichang, L. Xu, Z. Yafeng, Z. Baodong and X. Jianbo (2018). "Chemical Composition and Nutritional Function of Olive (Olea Europaea L.): A Review." *Phytochemistry Reviews* **17**(5): 1091–1110.

Zhang, Q., C. Wang, B. Li, L. Li, D. Lin, H. Chen, Y. Liu, S. Li, W. Qin, J. Liu, W. Liu and W. Yang (2018). "Research Progress in Tofu Processing: From Raw Materials to Processing Conditions." *Critical Reviews in Food Science and Nutrition* **58**(9): 1448–1467.

Zhong, L., X. Zhang and M. Covasa (2014). "Emerging Roles of Lactic Acid Bacteria in Protection Against Colorectal Cancer." *World Journal of Gastroenterology* **20**(24): 7878–7886.

# Silage Fermentation

# 14

## Arthur C. Ouwehand

## 14.1 INTRODUCTION

Feed constitutes a substantial proportion of the costs of livestock production. Feed availability has traditionally had a seasonal variation. Preservation of forages has therefore always played an important role in livestock production. Silage is a way to preserve and store forage and is used as a major portion of the diet fed to dairy cows and beef cattle in many countries. Silage quality is of prime importance to ensure optimal performance of the livestock (Borreani, Tabacco et al. 2018). Anaerobic fermentation of fresh forage is a preservation method that involves the conversion of plant sugars into organic acids, mainly lactic acid, by epiphytic lactic acid bacteria (LAB), although starters and additives can be used to favor the fermentation. In this sense, silage is not very different from the principle of vegetable fermentations (Chapter 13) and fermented foods (Chapters 10, 11, 12, 13, 17 and 18): "foods made through desired microbial growth and enzymatic conversions of food components" (Marco, Sanders et al. 2021). Minimal nutrient loss is of prime importance to the quality of the fermented food/feed. Production of organic acids lowers the pH of the ensiled forage and inhibits growth of spoilage organisms (Queiroz, Ogunade et al. 2018; Guan, Ran et al. 2021). In addition to oxygen restriction and moisture content of forages, successful preservation of forages as silage is dependent on the activities of the silage microbial population, especially the LAB population (Carvalho, Sales et al. 2021). Silage production needs to be dominated by inoculated bacteria or restricted by additives to optimize fermentation. Improvements in ensiling processes by rapid acidification, maintenance of anaerobic conditions during ensiling, and reducing aerobic spoilage after opening the silo have also helped ensure production of high-quality silage (Fang, Dong et al. 2022; Franco, Tapio et al. 2022).

## 14.2 PHASES OF SILAGE FERMENTATION

Silage fermentation knows, arbitrarily, five phases: 1 aerobic fermentation, 2 start of lactic acid fermentation, 3 completion of lactic acid fermentation, 4 storage, and 5 feed out; see Figure 14.1.

### 14.2.1 Aerobic Phase

This phase is characterized by use of the oxygen trapped within the forage mass. This phase is hence called the aerobic phase. When harvested, the plant material continues to live and will respire for several hours. The plant cells within the cut forage continue to take in oxygen because many cells are still intact and continue to metabolize. Furthermore, plant enzymes that breakdown proteins and carbohydrates

 DOI: 10.1201/9781003352075-16

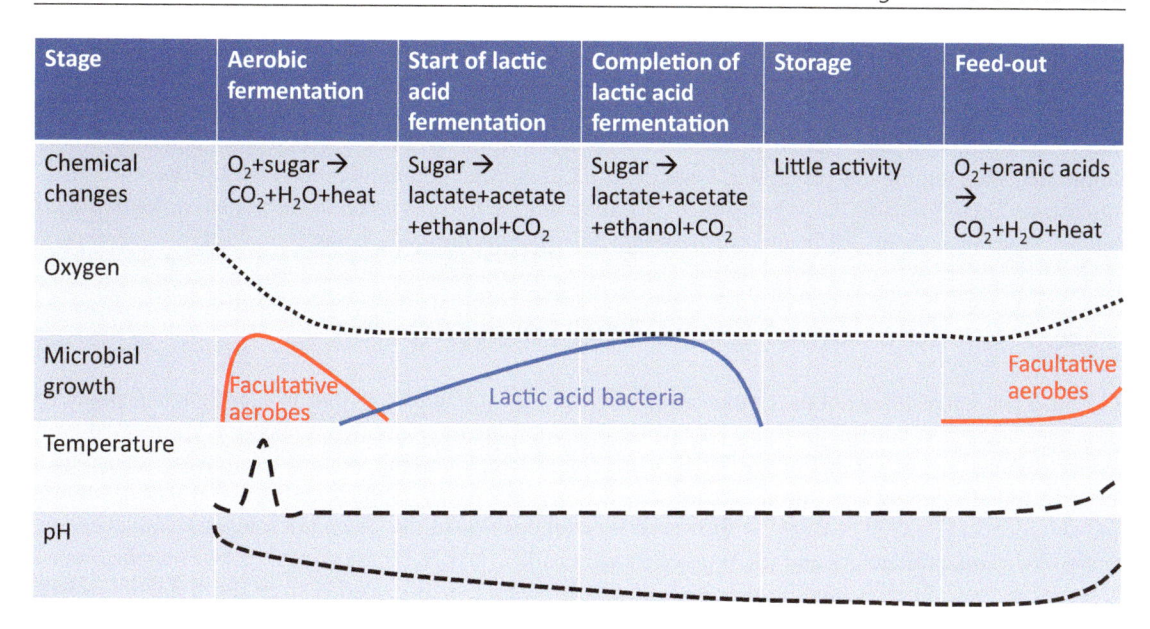

| Stage | Aerobic fermentation | Start of lactic acid fermentation | Completion of lactic acid fermentation | Storage | Feed-out |
|---|---|---|---|---|---|
| Chemical changes | $O_2$+sugar → $CO_2$+$H_2O$+heat | Sugar → lactate+acetate +ethanol+$CO_2$ | Sugar → lactate+acetate +ethanol+$CO_2$ | Little activity | $O_2$+oranic acids → $CO_2$+$H_2O$+heat |
| Oxygen | | | | | |
| Microbial growth | Facultative aerobes | Lactic acid bacteria | | | Facultative aerobes |
| Temperature | | | | | |
| pH | | | | | |

**FIGURE 14.1** Schematic representation of silage fermentation phases (modified after Ishler, Jones et al. 2023).

continue to function and release amino acids and soluble sugars, respectively. As oxygen delays plant cell death and a decrease in pH, the rate of plant cell plasmolysis, which releases the intracellular fluid needed by LAB for synthesis of lactic acid, is also reduced. These processes negatively affect silage quality (Borreani, Tabacco et al. 2018). Also, obligate and facultative aerobic microorganisms such as *Pantoea, Pseudomonas, Sphingomonas*, acetic acid bacteria, and propionic acid bacteria naturally present on the stems and leaves of plants continue to utilize hexose sugars (Ávila and Carvalho 2020; Sun, Bai et al. 2021). The presence of oxygen sustains the growth of molds, yeasts, and undesirable bacteria such as Enterobactereria, *Listeria*, and *Bacteroides* that may later produce toxins or predispose the silage to pathogen growth at feed out (Queiroz, Ogunade et al. 2018; Sun, Bai et al. 2021). Respiration by the plant cells and microbes will lead to the formation of carbon dioxide, water, and heat. During this early stage of the silage process, the temperature can increase to 40°C or higher. Prolonged exposure to higher temperatures my lead to loss of dry matter and denaturation of proteins (Borreani, Tabacco et al. 2018). The respiration phase usually lasts three to five hours, depending on the oxygen supply present. From a management standpoint, the primary goal is to eliminate oxygen as soon as possible and keep it out for the duration of the storage period (Ishler, Jones et al. 2023).

## 14.2.2 Start of Lactic Acid Fermentation

As the supply of oxygen gets depleted, anaerobic bacteria begin to multiply. The primary bacteria during this phase are *Enterobacteria*, which are decreasing, and heterofermentative LAB such as *Lentilactobacillus buchneri, Weissella cibaria, Lactococcus* and *Leuconostoc pseudomesenteroides, Leuconostoc citreum* (Ni, Wang et al. 2015; Sun, Bai et al. 2021). They can tolerate the heat produced during the aerobic phase and are viable in a pH range of 5 to 7 which is found in the fermenting forage at this time. These bacteria produce both acetic and lactic acid but tend to be inefficient at producing these acids relative to nutrients lost in the fermenting crop. This acidifies the forage mass, lowering the pH from about 6.0 in green forage to a pH of about 5.0. The early drop in pH also limits the activity of plant enzymes that break down proteins. This phase of the fermentation process continues for one to two days, no longer than 24 to 72 hours. When the pH drops below 5, homo-fermenters predominate and the next phase of silage fermentation begins (Ishler, Jones et al. 2023). The increased acidity of the forage mass enhances the growth and development of lactic acid–producing bacteria that convert plant carbohydrates to lactic acid, acetic acid, ethanol, mannitol, and carbon dioxide. Homolactic bacteria, or in practice facultative heterofermentative bacteria, are preferred because they can convert plant sugars mainly to lactic acid. Bacterial strains within this group grow in anaerobic conditions, and they require low pH (Ishler, Jones et al. 2023). Species commonly found are *Lactiplantibacillus plantarum*, *Lactiplantibacillus argentoratensis*, *Lactiplantibacillus paraplantarum* and *Lactiplantibacillus pentosus* (Ni, Wang et al. 2015), and *Lacticaseibacillus rhamnosus* (Guo, Wang et al. 2021)

## 14.2.3 Completion of Lactic Acid Fermentation

The third and longest stage of the fermentation process is a continuation of phase 3; lactic acid production continues and peaks during this time. This phase will continue for about two weeks or until the acidity of the forage mass is low enough to restrict all bacterial growth, including the acid-tolerant lactic acid bacteria. Dominating organisms are *L. plantarum*, *L. buchneri*, *Levilactobacillus brevis*, *Weissella minor*, and enterococci (Fu, Sun et al. 2022). However, the composition of the silage microbiota is influenced by many environmental factors and the type of forage ensiled (Okoye, Wang et al. 2023). The silage mass is stable in about 21 days, and fermentation ceases if outside air is excluded from the silage. However, improper ensiling practices will result in an undesirable continuation of the process, as will be discussed in the following.

If silage has undergone proper fermentation, the expected pH will range from 3.5 to 4.5 for corn silage and 4.0 to 5.5 for haylage, depending on forage moisture content. The remainder of phase 3 is the material storage phase. Generally, lack of oxygen prevents the growth of yeast and molds and low pH limits the growth of bacteria during storage (Ishler, Jones et al. 2023).

## 14.2.4 Storage

During the stable storage phase, minimal reactions occur as the silage mass becomes stable, typically within about 21 days. This stability is achieved by maintaining the exclusion of oxygen, which prevents further fermentation and the growth of yeast and molds. In this phase, some reactions, such as the breaking down of complex carbohydrates and proteolysis, may take place, slowly releasing water-soluble carbohydrates to LAB and increasing the digestibility of the starch. Although reactions are minimal, the storage time can affect the fermentative characteristics of the silage, such as microbial population and fermentation products. Proper ensiling practices are crucial during this phase to maintain the desired pH levels and prevent undesirable continuation of the fermentation process (Ávila and Carvalho 2020; Ishler, Jones et al. 2023). During the stable storage phase of silage, the most common microorganisms include lactic acid bacteria, such as *Lactobacillus sensu lato*, *Pediococcus*, and *Enterococcus* species. Especially *L. plantarum* and *L. buchneri* are the most common (Ridwan, Abdelbagi et al. 2023). These bacteria help lower the silage pH by producing lactic acid, resulting in more efficient fermentation and limiting the growth of undesirable organisms (Ishler, Jones et al. 2023). Other microorganisms that may be present include yeasts, butyric acid bacteria, and spore-forming bacteria, such as *Clostridium*, *Bacillus*, and *Paenibacillus*. However, their growth or stabilization depends on various factors, such as pH and oxygen availability (Driehuis, Wilkinson et al. 2018).

## 14.2.5 Storage Deterioration

The general pattern of aerobic deterioration has been known for many decades. When oxygen is introduced to silage, aerobic microorganisms begin to grow, initially respiring soluble substrates and then more complex compounds. Yeasts are generally the initiators of aerobic deterioration, consuming sugars and fermentation acids and raising silage temperature and pH. With increased pH, bacilli and other aerobic bacteria grow, increasing temperature further. Finally, molds complete the silage deterioration (Borreani, Tabacco et al. 2018).

Acetic acid bacteria are aerobic bacteria that are capable of growing at low pH. They grow on ethanol, producing acetic acid. However, once ethanol has been exhausted, they can grow on acetic acid, producing carbon dioxide and water. This will raise pH and permit other aerobic microorganisms to grow. Consequently, acetic acid bacteria can be initiators of aerobic deterioration. At present, acetic acid bacteria have only been reported in corn silage, silage that usually has a low buffering capacity so that lactic and acetic acid concentrations are low and a high sugar content that could permit high ethanol concentrations. They have not been found in C3 grass silages or legume silages. Sugar cane silages and perhaps other similar C4 grass silages may be candidate silages to look for acetic acid bacteria, that is, crops with characteristics similar to corn (Muck 2010).

## 14.2.6 Feed Out

Feed-out begins when the storage phase ends and the silo is opened, which leads to the exposure to oxygen. The presence of oxygen allows for the outgrow of facultative aerobic microorganisms that remained present in the silage or are introduced after opening from the surrounding environment (Ishler, Jones et al. 2023; Okoye, Wang et al. 2023). Organic acids are metabolized by acid-tolerant aerobic organisms, often yeasts and fungi such as *Candida*, *Hansenula*, and *Pichia*. This increases the pH and allows for a next phase where less acid tolerant spoilage organisms to proliferate such as enterobacteria, *Listeria*, and bacilli (Queiroz, Ogunade et al. 2018; Ishler, Jones et al. 2023). Aerobic spoilage is a major cause of nutrient loss in silage and is also often responsible for potential silage pathogenicity and toxicity, which can lead to poor animal performance, diseases, and death (Ogunade, Martinez-Tuppia et al. 2018; Queiroz, Ogunade et al. 2018). Proper feed-out management is therefore essential. Silo design should minimize the size of the silo face, as wider faces facilitate oxygen ingress. The silo face should be smooth to minimize the exposed area. Silages should be fed out at rates that minimize the length of time the face is exposed to the air. Silage stability can be improved by using additives (Queiroz, Ogunade et al. 2018).

# 14.3 SILAGE ADDITIVES

There are many options for additives in silage; formic, propionic, acetic, sorbic, and benzoic acid because they are antifungal and antioxidants such as vitamin E and selenium (Queiroz, Ogunade et al. 2018). Sorbic, benzoic, propionic, and acetic acids improve aerobic stability of the silage at feed out through direct inhibition of yeasts and molds. Sorbic acid also can be inhibitory to some bacteria, such as clostridia, and is a more potent inhibitor of yeasts, molds, and spoilage bacteria than benzoic acid (Muck, Nadeau et al. 2018). To increase the level of propionic acid, the use of *Propionibacterium* additives has been considered (Agarussi, Pereira et al. 2022). However, LAB inoculants also have substantial potential for improving silage quality by enhancing the preservation and fermentation of forage crops. As outlined, these bacteria play a pivotal role in spontaneous silages and as additives may improve the overall efficiency of the fermentation process (Fabiszewska, Zielińska et al. 2019). Furthermore, the exploration of the roles of LAB inoculants in crop silage offers novel insights into their potential impact on animal health, reproduction and the quality of animal products (Okoye, Wang et al. 2023). As *L. plantarum* and *L. buchneri* have been most widely studied as inoculants, the discussion here will be limited to these two species, although it is important to realize that other species/combinations have been studied as well (Okoye, Wang et al. 2023).

Production of high-quality silage is dependent on enhancing fermentation processes during the different phases of forage preservation. Silage additives have been employed for this purpose for many years (Yitbarek and Tamir 2014; Muck, Nadeau et al. 2018). Some of the main classes of silage additives are fermentation stimulants (microbial inoculants, enzymes, and fermentable carbohydrate sources), fermentation inhibitors (acids and formaldehyde), aerobic deterioration inhibitors (LAB, acids, and acid salts), and nutrients (McDonald, Henderson et al. 1991). LAB inoculants have been the subject of considerable research because, relative to chemical additives, they are less expensive, less hazardous, and noncorrosive to farm machinery (Yitbarek and Tamir 2014).

## 14.3.1 *Lentilactobacillus buchneri* as Silage Inoculant

*L. buchneri* is a lactic acid bacterium frequently associated with food bioprocessing and fermentation. It naturally inhabits various ecological niches and has been isolated in fermented cucumber spoilage, grass silage, a bioethanol production plant, the human intestine and oral cavity, cheese, and beer wort (Holzer, Mayrhuber et al. 2003; Heinl and Grabherr 2017; Nethery, Henriksen et al. 2019). The species plays an ambivalent role in many food and feed fermentation processes, where it can act as a useful or detrimental microorganism depending on the application (Heinl and Grabherr 2017). *L. buchneri* is an obligate heterofermentative organism and as such produces lactic acid together with acetic acid and other compounds (Heinl, Spath et al. 2011). In comparison to other lactobacilli, *L. buchneri* is highly resistant to stressors such as ethanol or oxygen (Eikmeyer, Heinl et al. 2015).

The ability of *L. buchneri* to metabolize lactic acid into acetic acid and 1,2-propandiol makes it invaluable to the ensiling process. In the ensiling process, 1,2-propanediol is associated with increased performance and higher silage aerobic stability (Guo, Xu et al. 2023). However, this metabolic activity leads to spoilage in other applications, particularly damaging the cucumber fermentation industry (Nethery, Henriksen et al. 2019). *L. buchneri* has drawn attention due to its extraordinary robustness and unique metabolic pathways, prompting researchers to investigate its genome and potential applications in various industrial settings (Heinl and Grabherr 2017).

The completely annotated genomic sequence of *L. buchneri* CD034, a strain isolated from stable grass silage, providing valuable insights into the species' genomic adaptations such as the identification of genes putatively involved in the breakdown of plant cell wall polymers (Heinl, Wibberg et al. 2012). Further, strains isolated from spoiled fermented cucumber were highly enriched in mobile genetic elements, specifically transposons (Nethery, Henriksen et al. 2019).

In silage the primary function of *L. buchneri* is to increase the stability of silages against deterioration caused by yeasts and molds when exposed to air. This is achieved through the conversion of lactic acid to acetic acid and 1,2-propanediol, which contribute to enhanced stability of silages (Okoye, Wang et al. 2023). *L. buchneri* is a preferred choice as an inoculant, as it contributes to silages with minimal dry matter loss. However, these effects on silage quality are strain specific and dose dependent, as demonstrated in several studies. Inoculating silage with *L. buchneri* has been shown to improve feed performance in animals, such as cows and beef cattle, when compared to uninoculated silage (Holzer, Mayrhuber et al. 2003). Additionally, the use of *L. buchneri* in combination with other lactic acid bacteria, like *L. plantarum* and *Bacillus subtilis*, can further enhance feed and growth performance in animals like calves and bulls (Okoye, Wang et al. 2023).

## 14.3.2 *Lactiplantibacillus plantarum* as Silage Inoculant

*L. plantarum* is a lactic acid bacterium that occupies a diverse range of environmental niches and has an extensive variety in phenotypic properties, metabolic capacity, and industrial applications. It is generally found in environments with high levels of carbohydrates, such as food products (dairy products, fermented meat, sourdoughs) and plant-derived substrates (fermented vegetables and silage). *L. plantarum* occupies different niches in the mammalian body, including the respiratory, gastrointestinal, and urogenital tracts (Siezen and van Hylckama Vlieg 2011). Additionally, in plants it is present at relatively high levels as endophyte (Martínez-Romero, Aguirre-Noyola et al. 2021). *L. plantarum* exhibits a high level of genomic diversity and versatility. One of the main features of its genome appears to be life-style islands consisting of numerous functional gene cassettes, particularly for carbohydrate utilization. These cassettes can be acquired, shuffled, substituted, or deleted in response to niche requirements (Siezen and van Hylckama Vlieg 2011). This provides *L. plantarum* with the ability to ferment a wide range of carbohydrates, which allows it to adapt to various ecological niches.

Because of its natural and ubiquitous presence on and in plant material (Yu, Leveau et al. 2020; Martínez-Romero, Aguirre-Noyola et al. 2021), *L. plantarum* is part of the fermentation microbiota of silage (Okoye, Wang et al. 2023).

For the described reasons, *L. plantarum* is one of the most commonly used silage inoculants. It plays a crucial role in the fermentation process of silage, particularly in whole-plant corn silages. *L. plantarum* grows best at low pH conditions and rapidly acidifies the forage at later stages of fermentation once the pH drops below 5. Although acidification of the silage is one of the main antimicrobial mechanisms, depending on the strain, *L. plantarum* may also produce bacteriocins and exopolysaccharides with anti-fungal activity (Echegaray, Yilmaz et al. 2023). This influences keystone species of LAB that affect the silage fermentation process and inhibits the growth of organisms such as *Enterococcus faecium* and, yeast and molds. The antioxidant capability of *L. plantarum* plays a role in improving the quality of silage. Feruloyl esterase–producing strains, for example, have been reported to show high antioxidant capacity by degrading lignocellulose in alfalfa and releasing free ferulic acid during ensiling (Guo, Xu et al. 2023). This contributes to safer and better-quality silage and subsequent improved animal performance.

# 14.4 CONCLUSION

Silage fermentation is a crucial process in preserving and storing forage for livestock feed. The fermentation process involves the conversion of plant sugars into organic acids by lactic acid bacteria, resulting in a decrease in pH and inhibition of spoilage organisms. The different phases of silage fermentation, including the aerobic phase, start and completion of lactic acid fermentation, storage, and feed out, play distinct roles in the overall preservation and quality of silage. Proper management practices, such as eliminating oxygen, using inoculated bacteria or additives, and preventing aerobic spoilage during feed out, are essential for producing high-quality silage. Silage additives, including organic acids, antioxidants, and microbial inoculants, can further improve fermentation and preserve the forage. The silage process described here is a generalization. Silage is dependent on the forage type, environment, the use of additives, and so on. Overall, understanding and optimizing the silage fermentation process is crucial for ensuring optimal feed quality and animal performance in livestock production.

# BIBLIOGRAPHY

Agarussi, M. C. N., O. G. Pereira, F. E. Pimentel, C. F. Azevedo, V. P. da Silva and F. F. E. Silva (2022). "Microbiome of rehydrated corn and sorghum grain silages treated with microbial inoculants in different fermentation periods." *Sci Rep* **12**(1): 16864.

Ávila, C. L. S. and B. F. Carvalho (2020). "Silage fermentation—updates focusing on the performance of microorganisms." *J Appl Microbiol* **128**(4): 966–984.

Borreani, G., E. Tabacco, R. J. Schmidt, B. J. Holmes and R. E. Muck (2018). "Silage review: Factors affecting dry matter and quality losses in silages." *J Dairy Sci* **101**(5): 3952–3979.

Carvalho, B. F., G. F. C. Sales, R. F. Schwan and C. L. S. Avila (2021). "Criteria for lactic acid bacteria screening to enhance silage quality." *J Appl Microbiol* **130**(2): 341–355.

Driehuis, F., J. M. Wilkinson, Y. Jiang, I. Ogunade and A. T. Adesogan (2018). "Silage review: Animal and human health risks from silage." *J Dairy Sci* **101**(5): 4093–4110.

Echegaray, N., B. Yilmaz, H. Sharma, M. Kumar, M. Pateiro, F. Ozogul and J. M. Lorenzo (2023). "A novel approach to Lactiplantibacillus plantarum: From probiotic properties to the omics insights." *Microbiol Res* **268**: 127289.

Eikmeyer, F. G., S. Heinl, H. Marx, A. Pühler, R. Grabherr and A. Schlüter (2015). "Identification of oxygen-responsive transcripts in the silage inoculant *Lactobacillus buchneri* CD034 by RNA sequencing." *PLoS ONE* **10**(7).

Fabiszewska, A. U., K. J. Zielińska and B. Wróbel (2019). "Trends in designing microbial silage quality by biotechnological methods using lactic acid bacteria inoculants: A minireview." *World J Microbiol Biotechnol* **35**(5): 76.

Fang, D., Z. Dong, D. Wang, B. Li, P. Shi, J. Yan, D. Zhuang, T. Shao, W. Wang and M. Gu (2022). "Evaluating the fermentation quality and bacterial community of high-moisture whole-plant quinoa silage ensiled with different additives." *J Appl Microbiol* **132**(5): 3578–3589.

Franco, M., I. Tapio, J. Pirttiniemi, T. Stefański, T. Jalava, A. Huuskonen and M. Rinne (2022). "Fermentation quality and bacterial ecology of grass silage modulated by additive treatments, extent of compaction and soil contamination." *Fermentation* **8**(4): 156.

Fu, Z., L. Sun, Z. Wang, J. Liu, M. Hou, Q. Lu, J. Hao, Y. Jia and G. Ge (2022). "Effects of growth stage on the fermentation quality, microbial community, and metabolomic properties of Italian ryegrass (Lolium multiflorum Lam.) silage." *Front Microbiol* **13**: 1054612.

Guan, H., Q. Ran, H. Li and X. Zhang (2021). "Succession of microbial communities of corn silage inoculated with heterofermentative lactic acid bacteria from ensiling to aerobic exposure." *Fermentation* **7**(4).

Guo, L., X. Wang, Y. Lin, X. Yang, K. Ni and F. Yang (2021). "Microorganisms that are critical for the fermentation quality of paper mulberry silage." *Food Energy Secur* **10**(4): e304.

Guo, X., D. Xu, F. Li, J. Bai and R. Su (2023). "Current approaches on the roles of lactic acid bacteria in crop silage." *Microb Biotechnol* **16**(1): 67–87.

Heinl, S. and R. Grabherr (2017). "Systems biology of robustness and flexibility: *Lactobacillus buchneri*: A show case." *J Biotechnol* **257**: 61–69.

Heinl, S., K. Spath, E. Egger and R. Grabherr (2011). "Sequence analysis and characterization of two cryptic plasmids derived from *Lactobacillus buchneri* CD034." *Plasmid* **66**(3): 159–168.

Heinl, S., D. Wibberg, F. Eikmeyer, R. Szczepanowski, J. Blom, B. Linke, A. Goesmann, R. Grabherr, H. Schwab, A. Pühler and A. Schlüter (2012). "Insights into the completely annotated genome of *Lactobacillus buchneri* CD034, a strain isolated from stable grass silage." *J Biotechnol* **161**(2): 153–166.

Holzer, M., E. Mayrhuber, H. Danner and R. Braun (2003). "The role of *Lactobacillus buchneri* in forage preservation." *Trends Biotechnol* **21**(6): 282–287.

Ishler, V., C. Jones, A. Heinrichs and G. Roth (2023). "From harvest to feed: Understanding silage management." *Penn State University. College Agric Sci*: 2–11.

Marco, M. L., M. E. Sanders, M. Ganzle, M. C. Arrieta, P. D. Cotter, L. De Vuyst, C. Hill, W. Holzapfel, S. Lebeer, D. Merenstein, G. Reid, B. E. Wolfe and R. Hutkins (2021). "The international scientific association for probiotics and prebiotics (ISAPP) consensus statement on fermented foods." *Nat Rev Gastroenterol Hepatol* **18**(3): 196–208.

Martínez-Romero, E., J. L. Aguirre-Noyola, R. Bustamante-Brito, P. González-Román, D. Hernández-Oaxaca, V. Higareda-Alvear, L. M. Montes-Carreto, J. C. Martínez-Romero, M. Rosenblueth and L. E. Servín-Garcidueñas (2021). "We and herbivores eat endophytes." *Microb Biotechnol* **14**(4): 1282–1299.

McDonald, P., A. R. Henderson and S. J. E. Heron (1991). *The biochemistry of silage*. Chalcombe Publications, Marlow, UK.

Muck, R. E. (2010). "Silage microbiology and its control through additives." *Revista Brasileira de Zootecnia* **39**: 183–191.

Muck, R. E., E. M. G. Nadeau, T. A. McAllister, F. E. Contreras-Govea, M. C. Santos and L. Kung, Jr. (2018). "Silage review: Recent advances and future uses of silage additives." *J Dairy Sci* **101**(5): 3980–4000.

Nethery, M. A., E. D. Henriksen, K. V. Daughtry, S. D. Johanningsmeier and R. Barrangou (2019). "Comparative genomics of eight *Lactobacillus buchneri* strains isolated from food spoilage." *BMC Genom* **20**(1): 902.

Ni, K., Y. Wang, Y. Cai and H. Pang (2015). "Natural lactic acid bacteria population and silage fermentation of whole-crop wheat." *Asian-Australas J Anim Sci* **28**(8): 1123–1132.

Ogunade, I. M., C. Martinez-Tuppia, O. C. M. Queiroz, Y. Jiang, P. Drouin, F. Wu, D. Vyas and A. T. Adesogan (2018). "Silage review: Mycotoxins in silage: Occurrence, effects, prevention, and mitigation." *J Dairy Sci* **101**(5): 4034–4059.

Okoye, C. O., Y. Wang, L. Gao, Y. Wu, X. Li, J. Sun and J. Jiang (2023). "The performance of lactic acid bacteria in silage production: A review of modern biotechnology for silage improvement." *Microbiol Res* **266**: 127212.

Queiroz, O. C. M., I. M. Ogunade, Z. Weinberg and A. T. Adesogan (2018). "Silage review: Foodborne pathogens in silage and their mitigation by silage additives." *J Dairy Sci* **101**(5): 4132–4142.

Ridwan, R., M. Abdelbagi, A. Sofyan, R. Fidriyanto, W. D. Astuti, A. Fitri, M. M. Sholikin, Rohmatussolihat, K. A. Sarwono, A. Jayanegara and Y. Widyastuti (2023). "A meta-analysis to observe silage microbiome differentiated by the use of inoculant and type of raw material." *Front Microbiol* **14**: 1063333.

Siezen, R. J. and J. E. van Hylckama Vlieg (2011). "Genomic diversity and versatility of Lactobacillus plantarum, a natural metabolic engineer." *Microb Cell Fact* **10**(Suppl 1): S3.

Sun, L., C. Bai, H. Xu, N. Na, Y. Jiang, G. Yin, S. Liu and Y. Xue (2021). "Succession of bacterial community during the initial aerobic, intense fermentation, and stable phases of whole-plant corn silages treated with lactic acid bacteria suspensions prepared from other silages." *Front Microbiol* **12**: 655095.

Yitbarek, M. B. and B. Tamir (2014). "Silage additives." *Open J Appl Sci* **4**(5): 44897.

Yu, A. O., J. H. J. Leveau and M. L. Marco (2020). "Abundance, diversity and plant-specific adaptations of plant-associated lactic acid bacteria." *Environ Microbiol Rep* **12**(1): 16–29.

# Lactic Acid Bacteria in Grape Fermentations

*An Example of LAB as Contaminants in Food Processing*

**15**

Eveline Bartowsky

## 15.1 INTRODUCTION

The production of alcoholic beverages dates back over 7000 years. Even though the concept of transforming grape juice into wine does not appear difficult, it can be a complex process to ensure an enjoyable, fault-free, and stable product. The winemaking process used today is not vastly different from that used in the time of the ancient Egyptians and Greeks. However, modern-day winemakers have much more control at the various critical stages from the time of picking grapes through to wine maturation.

Historically, early microbiologists such as Pasteur and Müller-Thurgau had observed the presence of bacteria in wine, and by early 1900, the importance of these bacteria in winemaking was beginning to be understood. They were observed in both sound and spoiled wine. The main function of lactic acid bacteria (LAB) in wine production is to conduct malolactic fermentation (MLF), which is the decarboxylation of L-malic acid to L-lactic acid (Henick-Kling 1993; Möslinger 1901). This increases the wine pH by 0.2–0.5 units, resulting in a softer-tasting wine, and provides microbial stability by the removal of a potential carbon source. Moreover, through bacterial metabolism during MLF, there are various sensory changes that can occur in the wine (Lonvaud-Funel 2000).

Wine LAB can produce a large number of secondary metabolites, with most resulting in favorable sensory outcomes. Several recent reviews cover this aspect of wine LAB (Bartowsky 2005; Bartowsky 2017; Cappello et al. 2017; Virdis et al. 2021). This chapter will discuss the undesirable outcomes of bacterial metabolism during grape fermentations and the various wine spoilage scenarios that can result from LAB contamination in wine.

284

DOI: 10.1201/9781003352075-17

# 15.2 LAB IN GRAPE FERMENTATION

The presence of bacteria and the role that they play in winemaking has been known since the mid- to late 1800s. Grape- and wine-associated bacteria belong to the acetic acid bacteria (AAB) and LAB families. The main role in food fermentation by AAB is the production of acetic acid from ethanol, the basis of vinegar. Under conducive growth conditions, this family of bacteria can cause wine spoilage; they will not be discussed in this chapter, but there are several recent reviews for the interested reader (Bartowsky and Henschke 2008; Cleenwerck and De Vos 2008).

There are four LAB genera associated with wine: *Lactobacillus sensu lato*, *Leuconostoc*, *Oenococcus*, and *Pediococcus*, and they can be readily distinguished morphologically by microscopic examination. Lactobacilli produce short to long slender rod-shaped cells (0.5–1.2 μm × 1.0–10 μm). *Leuconostoc* and *Oenococcus* usually consist of spherical to lenticular cells, most often occurring in pairs or chains (1–2 μm diameter). *Pediococcus* species occur as spherical-shaped cells that are usually found in tetrads (division in two planes; 0.5–0.7 μm diameter, 0.7–1.2 μm length). The species that commonly occur in wine are *Oenococcus oeni*, *Lactiplantibacillus plantarum*, *Levilactobacillus brevis*, *Limosilactobacillus fermentum*, *Lentilactobacillus buchneri*, *Lentilactobacillus hilgardii*, *Fructilactobacillus fructivorans*, *Apilactobacillus kunkeei*, *Liquorilactobacillus oeni*, *Liquorilactobacillus mali*, *Liquorilactobacillus vini*, *Leuconostoc mesenteroides*, *Pediococcus pentosaceus*, *Pediococcus damnosus* (*P. cerevisiae*), *Pediococcus inopinatus*, and *Pediococcus parvulus*. The taxonomic classification of LAB is discussed elsewhere in this book.

The main role of LAB in grape fermentation is conducting MLF, which generally commences as a spontaneous or natural reaction about 1–3 weeks after completion of alcoholic fermentation and lasts 2–12 weeks. LAB originating from the vineyard, on the grapes, and resident in the winery are responsible for the MLF (Bae et al. 2006; Renouf et al. 2005). However, today, winemakers can choose to encourage the onset of MLF by inoculation with commercial cultures of *O. oeni* and *Lb. plantarum*. All wine-associated LAB can conduct MLF, though usually it is conducted by *O. oeni* strains. This species is well adapted to the harsh conditions of wine (low nutrients, high acidity, ethanol concentrations up to or more than 15% v/v). Other LAB genera and species are usually associated with wine spoilage and the production of undesirable aroma and flavor compounds.

*O. oeni* was the only species within the *Oenococcus* genus until the mid-2000s, when *O. kitaharae* was identified in composting distilled shochu residue (Endo and Okada 2006). Over centuries of selective pressure, *O. oeni* has honed and perfected various adaptive strategies that enable it to outcompete with other potential MLF bacteria during the later stages of vinification and thus to dominate in wine (Ribéreau-Gayon et al. 2006). *O. oeni* and *O. kitaharae* differentially encode several carbohydrate utilization and amino acid biosynthesis pathways, which has resulted in adaptation to their individual ecological niches (Borneman et al. 2012). Recently further *Oenococcus* species have been identified, *O. alcoholitolerans*, isolated from Brazilian cachaça (Badotti et al. 2014), and *O. sicerae* from French cider (Cousin et al. 2019). Only *O. kitaharae* is not able to perform MLF due to a point mutation resulting in a premature stop codon in the *mleA* gene encoding the malolactic enzyme (Borneman et al. 2012).

# 15.3 SPOILAGE OF GRAPES AND WINE

All wine-associated LAB species will form secondary metabolites as they grow in wine. The desirability of these compounds is often concentration dependent. Types of wine spoilage are summarized in Figure 15.1 and Table 15.1. Many of the wine spoilage scenarios discussed in the following have been characterized over the last 50 years (Bartowsky 2009; Bartowsky and Pretorius 2008; Sponholz 1993), with few new ones being described. The types of spoilage associated with contaminating LAB in wine range from overtly buttery aromas and vinegary descriptors through to mousy off-flavor, hints of geranium aromas, or viscous wines and compounds that may affect the consumer's health, such as biogenic amines or ethyl carbamate.

**TABLE 15.1**   Wine Spoilage Compounds Due to Bacterial Metabolism during Winemaking

| COMPOUND | ORIGIN OR METABOLISM | SENSORY DESCRIPTOR | AROMA THRESHOLD | BACTERIA (GENUS) |
|---|---|---|---|---|
| Acetic acid | Glucose and fructose Citric acid | Vinegar, sour, pungent | 0.2 g/L | LAB[a] |
| Ethyl acetate | Various, including sugar metabolism | Nail polish remover | 7.5 mg/L | LAB |
| Diacetyl 2,3-butanedione | Citric acid | Buttery, nutty, caramel | 0.1–2 mg/L | Oenococcus Lactobacillus[b] |
| 2-ethoxy-3,5-hexadiene | Sorbic acid | Crushed geranium leaves | 0.1 µg/L | Lactobacillus Pediococcus |
| 2-acetyl-tetrahydropyridine (ACTPY) | Fructose/glucose and lysine | Caged mouse | 4–5 µg/L | Lactobacillus Oenococcus |
| 2-ethyl-tetrahydropyridine (ETPY) | Unclear | Caged mouse | 2–18 µg/L | Lactobacillus Oenococcus |
| 2-acetyl-1-pyrroline (ACPY) | Fructose/glucose and ornithine | Caged mouse | 7–8 µg/L | Lactobacillus Oenococcus |
| Acrolein | Glycerol | Bitterness | 10 mg/L | Lactobacillus Pediococcus |
| β-D-glucan (exopolysaccharide) | Glucose | Ropy, viscous, oily, slimy, thick texture | | Pediococcus |
| Mannitol | Fructose | Viscous, sweet, irritating finish | | Oenococcus |
| Histamine | Decarboxylation of histidine | Physiological reactions | | LAB |
| Biogenic amines | Mainly decarboxylation of amino acid | | | |
| Volatile phenols 4-ethylphenol | p-Coumaric acid | Medicinal, barnyard | 0.14–0.6 mg/L | Lactobacillus |
| Sulfur compounds methanethiol | Methionine | Cooked cabbage, rubber | 0.3 µg/L | Oenococcus (potentially) |
| Dimethyl sulfide 3-(Methylsulfanyl) propan-1-ol 3-(Methylsulfuranyl)-propanoic acid | | Sulfury, floral, fruity, toasted and roasted | | |
| Thiazoles | Cysteine | Popcorn, peanut | 38 µg/L | Oenococcus (potentially) |

[a] LAB includes species from *Lactobacillus*, *Oenococcus*, and *Pediococcus*.
[b] *Lactobacillus sensu lato*

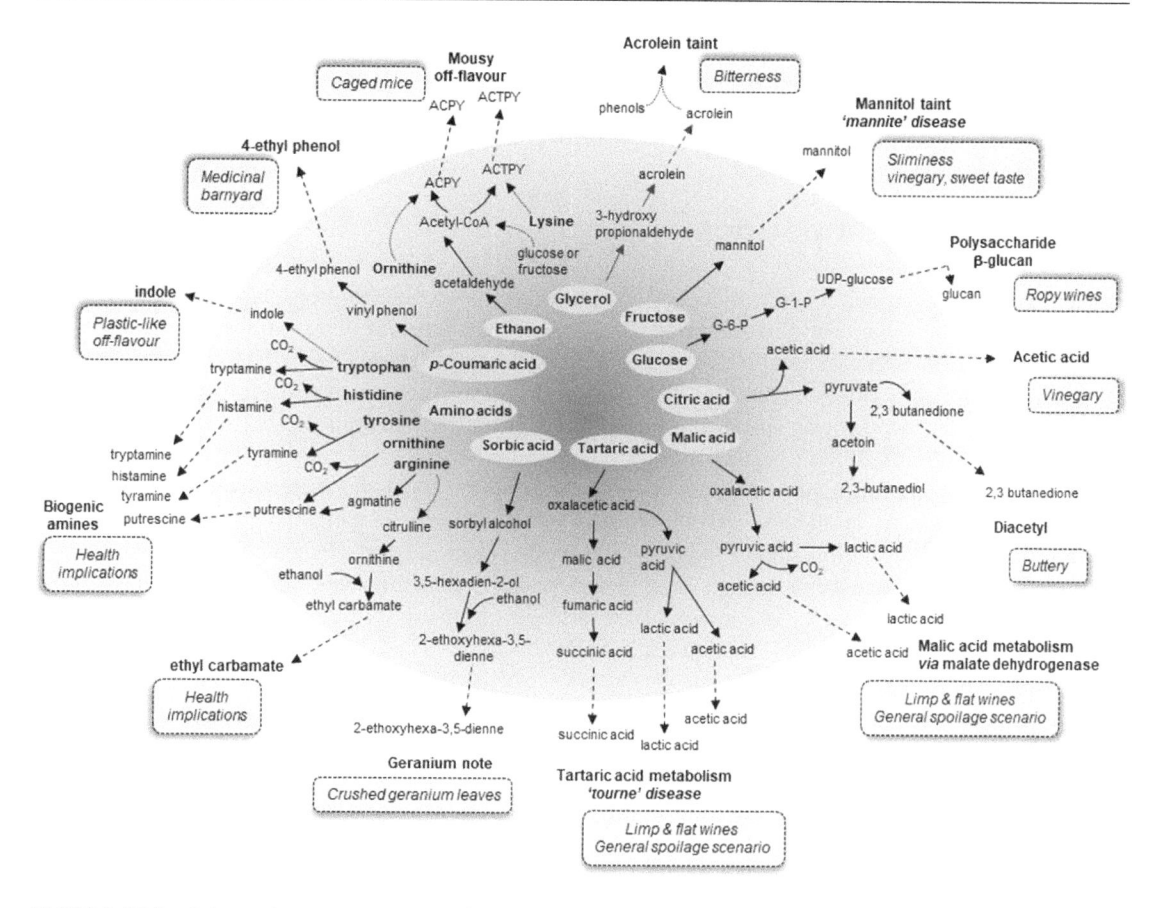

**FIGURE 15.1**    Schematic representation of LAB pathways that can lead to wine spoilage. Typical aroma and flavor descriptors for different metabolites are shown. Not all pathways will necessarily be present in all genera of wine LAB; *Lactobacillus sensu lato, Leuconostoc, Pediococcus,* and *Oenococcus.*

(Compiled from Sponholz, W.-R., *Wine Microbiology and Technology*, Harwood Academic Publishing, Amsterdam, the Netherlands, 1993; Costello, P.J., and Henschke, P.A., *J. Agric. Food Chem.*, 50, 7079, 2002; Wisselink, H.W. et al., *Int. Dairy J.*, 12, 151, 2002; Swiegers, J.H. et al., *Aust. J. Grape Wine Res.*, 11, 139, 2005; Walling, E. et al., *Food Microbiol.*, 22, 71, 2005; Bartowsky, E.J., and Pretorius, I.S., *Biology of Microorganisms on Grapes, in Must and in Wine*, Springer, Heidelberg, Germany, 2008.)

## 15.3.1 Volatile Phenols

Volatile phenols are formed from the hydroxycinnamic acid precursors present in grape must predominantly by yeast (*Dekkera/Brettanomyces*) during fermentation and wine maturation and contribute to off-flavors with descriptors such as bandage, barnyard, medicinal, or stable (Dubois 1983). Several LAB and fungi have the genes encoding phenolic acid decarboxylases (Chatonnet et al. 1995; Du Toit and Pretorius 2000). Strains of *Lb. brevis, Secundilactobacillus. collinoides,* and *Lb. plantarum* have been shown to possess the enzymes to produce 4-ethylphenol from *p*-coumaric acid (Couto et al. 2006). In the same study, pediococci strains were able to metabolize *p*-coumaric acid only through to 4-vinylphenol. Using molecular techniques, this ability of lactobacilli and pediococci to convert *p*-coumaric acid to the 4-vinylphenol was confirmed by identifying the presence of the gene *pdc* (phenol decarboxylase) in numerous strains (De Las Rivas et al. 2009). The same study confirmed the absence of the *pdc* gene in *O. oeni* strains (De Las Rivas et al. 2009). The yeast *Dekkera/Brettanomyces* is able to efficiently conduct the final step to form 4-ethyl phenol in wine, often at very high concentrations, and thus is the main contributor to these spoilage aromas and flavors in wine (Curtin et al. 2007; Heresztyn 1986a).

## 15.3.2 Sulfur Compounds

Sulfur-containing compounds typically occur in wine at very low concentrations, have very low sensory detection thresholds, and generally confer negative sensory attributes to wine, with descriptors such as cabbage, rotten egg, sulfurous, garlic, onion, and rubber (Vermeulen et al. 2005). However, some sulfur-containing compounds can contribute positive aromas to wine, such as strawberry, passion fruit, and grapefruit. The development of sulfur compounds is well understood in yeast but less so in bacteria (Swiegers and Pretorius 2007).

The sulfur-containing amino acids methionine and cysteine can be metabolized by some LAB species. Methionine can be metabolized by *O. oeni* and lactobacilli to methanethiol, dimethyl sulfide, 3-(methylsulfanyl)propan-1-ol, and 3-(methylsulfuranyl)-propanoic acid (Pripis-Nicolau et al. 2004). Cysteine can be the precursor of S-containing heterocycles, such as thiazoles, and there is evidence that *O. oeni* is able to metabolize these compounds, resulting in aroma descriptors, including sulfury, floral, fruity, toasted, and roasted (Pripis-Nicolau et al. 2004).

## 15.3.3 N-Heterocyclic Compounds (Mousy Off-Flavor)

The production of N-heterocyclic compounds in wine results in an unpleasant odor that is reminiscent of caged mice or mouse urine and is often referred to as a mousy off-flavor or mousiness; it was already described as "peculiarly disagreeable flavor in wine" in 1894 (Thudicum 1894). Interestingly, this aroma flavor can only be perceived on the back palate as a persistent aftertaste because of interactions of the wine with the mouth environment; an increase in pH renders the compounds volatile (Tucknott 1977; Tempère et al. 2019). There are three sensory important compounds responsible for the mousy aromas, 2-acetyltetrahydropyridine (ACTPY), 2-acetyl-1-pyrroline (ACPY), and 2-ethyltetrahydropyridine (ETPY) (Heresztyn 1986b).

Mousiness can be caused by *Dekkera/Brettanomyces* yeast, but mostly it is LAB that are the major cause of this wine taint (Grbin and Henschke 2000; Tucknott 1977). The heterofermentative LAB, *O. oeni*, *Lc. mesenteroides*, and some *Lactobacillus* species are capable of synthesizing ACTPY, ACPY, and ETPY (Costello et al. 2001). The homofermentative LAB species, *Pediococcus* and *Lb. plantarum*, produce little or no detectable mousy compounds. The metabolism of lysine and ornithine in the presence of ethanol and fructose by *Lb. hilgardii* has been shown to lead to formation of the three compounds (Costello and Henschke 2002). Recently it has been shown that the development of mousy off-flavor was faster in the presence of metals (iron and copper) in wines (Nikfardjam and Kunz 2021). The risk of mousy off-flavor developing in wine is an ongoing issue, as highlighted in a recent study of Bordeaux red wines produced without added sulfites, where 10% of defects were described in these wines as "Mousy off-flavor" (Pelonmier-Magimel et al. 2020).

## 15.3.4 Malic Acid Metabolism

Malic acid metabolism by LAB is usually via the malolactic enzyme that constitutes MLF; L-malic acid is converted directly to L-lactic acid and $CO_2$ (Möslinger 1901). There are not many microorganisms that can utilize malic acid as a carbon source, and there are few reports on the presence of a malic enzyme or malate dehydrogenase in LAB. *Lacticaseibacillus casei* and *Enterococcus faecalis* can covert malate to pyruvate with a malic enzyme that enables their growth on malate as a carbon source (Landete et al. 2010; London and Meyer 1969; London et al. 1971; Schütz and Radler 1974). *Lb. fermentum* has been demonstrated to produce D- and L-lactate, acetate, succinate, and $CO_2$ from malate (Radler 1986); at low pH (<pH 4), lactate production is favored, while at high pH (>pH 5), mainly succinate and acetate are formed. *Lb. delbrueckii* has been reported to possess a malate dehydrogenase (Caspritz and Radler 1983; Lonvaud-Funel and Strasser de Saad 1982). All these alternative malate metabolisms would be considered spoilage, as most of the secondary metabolites produced are undesirable in wine.

## 15.3.5 Citric Acid Metabolism

Citric acid will be metabolized by many LAB during their growth in red and white wines. Numerous secondary metabolites are produced, of which acetic acid and diacetyl (2,3-butanedione) have the greatest sensory impact. Both compounds will contribute complexity to wine at lower concentrations and at higher concentrations will be considered spoilage. Citric metabolism commences after approximately 60%–75% of malic acid has been metabolized and is usually completed 1 week after MLF is deemed complete (Bartowsky and Henschke 2004; Krieger et al. 2000).

The first step of citric acid metabolism results in the release of acetic acid and through to oxaloacetic acid and then on to pyruvic acid, where under acidic conditions, the metabolism favors the production of diacetyl and the final point of the pathway is 2,3-butanediol (Ramos and Santos 1996). There is always a small increase in volatile acidity (acetic acid) in wine following MLF due to citric acid metabolism; however, this is not the main contributor to undesirable quantities of acetic acid. Elevated concentrations of acetic acid, 0.7–1.1 g/L and higher, impart a vinegar-like character to wine (Corison et al. 1979) and can be due to metabolism of AAB species (Bartowsky et al. 2003; Bartowsky and Henschke 2008); yeast metabolism (Swiegers et al. 2005); or metabolism of sugars by LAB, including *A. kunkeei* (Edwards et al. 1998, 1999).

Diacetyl is an important bacterial secondary metabolite in wine. This compound imparts a buttery, nutty, or butterscotch aroma to wine, and desirability is concentration dependent; up to ~4 mg/L, it contributes to wine complexity, but over 7 mg/L, it is considered objectionable (Martineau et al. 1995; Rankine et al. 1969). The sensory perception of diacetyl is also dependent on the wine matrix, age, style, and origin of the wine (Bartowsky et al. 2002; Martineau and Henick-Kling 1995). The concentration of diacetyl can be relatively easily manipulated during the winemaking process to achieve desired sensory attributes (Bartowsky and Henschke 2004).

## 15.3.6 Tartaric Acid Metabolism

The capacity to metabolize one of the most important organic acids in wine, tartaric acid, is a rare bacterial characteristic. Even though Pasteur described the metabolism of tartaric acid by LAB and named it "tourne" disease, few reports exist on this wine spoilage. Only two *Lactobacillus* species (*Lb. plantarum* and *Lb. brevis*) have been demonstrated to metabolize tartaric acid to acetic acid, succinic acid, and $CO_2$ (Radler and Yannissi 1972; Ribéreau-Gayon et al. 2006; Sponholz 1993). The main enzyme, tartrate dehydratase, is inducible in both species, but with slightly different metabolic pathways. The anaerobic metabolism by homofermentative *Lb. plantarum* from 1 mol of tartrate is 1.5 mol $CO_2$, 0.5 mol acetic acid, and 0.5 mol lactic acid, and from the heterofermentative *Lb. brevis*, 1.33 $CO_2$, 0.67 mol of acetic acid, and 0.3 succinic acid are formed (Radler and Yannissi 1972). Wines demonstrating this spoilage are often described as fizzy, limp, flat, dull, and cloudy. Management of these bacteria is usually by maintaining low wine pH (higher acidity) and a good $SO_2$ regime. Interestingly, there seems to have been a re-emergence of this spoilage in recent years, thought to be a result of climate change and a trend to have wines with lower acidity and not enough sulfites (Bazireau 2022).

## 15.3.7  Lactic Fermentation of Glycerol (Acrolein)

The metabolism of glycerol by LAB, particularly in red wine, can result in bitterness, often referred to as acrolein taint. Again, this is a well-known spoilage disease already described by Pasteur (1873), who associated the presence of rod-shaped bacteria with the loss of glycerol, and Voisenet (1910, 1911), who linked this with bitterness of red wine. Acrolein itself is not bitter, but it reacts with the phenolics groups of anthocyanins to produce a bitter sensation (Rentschler and Tanner 1951); hence it is associated more with red than white wines. Sensory threshold concentrations are not clear; however, acrolein concentrations as low as 10 mg/L have been shown to cause a bitter taint (Margalith 1983).

The key enzyme in anaerobic glycerol metabolism is glycerol dehydratase converting glycerol to 3-hydroxypropionaldehyde, with acrolein spontaneously forming following exposure to heat or long-term storage in acidic solutions (such as wine) (Schütz and Radler 1984; Smiley and Sobolov 1962). The genes for this pathway have been studied in *S. collinoides*, *Lb. hilgardii*, and *Lb. diolivorans* and are organized in an operon of 13 genes, most likely all necessary for the functioning of the three protein subunits of glycerol dehydratase and propane-1,3-diol-dehydrogenase (Gorga et al. 2002). A second pathway for the degradation of glycerol, oxidative branch, uses 3-P-glycerol dehydrogenase resulting in 3-P-dehydroxyacetone, which enters into glycolysis reactions and results in a suite of compounds, including lactic acid, acetic acid, and acetoinic compounds (Bauer et al. 2010a; Sobolov and Smiley 1960).

*Lactobacillus* species appear to be the only LAB linked to glycerol metabolism and acrolein formation (Bauer et al. 2010b; Pasteris and Strasser de Saad 2009; Sauvageot et al. 2000). *P. pentosaceus* can aerobically metabolize glycerol to undesirable flavor compounds, but acrolein is not one of these (Pasteris and Strasser de Saad 2005). In a study of wine-associated LAB, it was demonstrated that the ability to ferment glycerol is limited (1% *O. oeni*, 12% *P. parvulus*, and 31% lactobacilli); however, the assay was conducted under aerobic conditions and does not imply the ability to produce acrolein (Davis et al. 1988). Several other bacteria can transform glycerol to 3-hydroxypropionaldehyde anaerobically: *Bacillus*, *Klebsiella*, *Citrobacter*, *Enterobacter*, and *Clostridium* (Bauer et al. 2010a; Vollenweider and Lacroix 2004).

Management of these LAB in wine to minimize the fermentation of glycerol to produce acrolein mainly involves maintaining low pH and higher $SO_2$ regime to minimize their growth (Ribéreau-Gayon et al. 2006).

## 15.3.8  Sorbic Acid Metabolism (Geranium Off-Flavor)

Sorbic acid (2,4-hexadienoic acid) can be used as a chemical preservative in sweet wines at bottling to prevent yeast fermentation after packaging. At concentrations used in wine (200 mg/L), its antimicrobial activity is ineffective against LAB (Edinger and Splittstoesser 1986a, 1986b). Sorbic acid can be metabolized by LAB species, including *O. oeni* to 2-ethoxyhexa-3,5-diene, which has an odor reminiscent of crushed geranium leaves (*Pelargonium* spp.) (Crowell and Guymon 1975). A small amount of ethanol is required; thus this spoilage is not observed in grape juice, but it can occur if grape juice is added to wine to sweeten it (Sponholz 1993).

## 15.3.9  Mannitol (Fructose Metabolism)

Mannite disease was first described by Pasteur (1873) and is due to fructose reduction by both heterofermentative and homofermentative bacteria resulting in the formation of mannitol, a six-carbon sugar alcohol (polyol), which is perceived as sliminess with a vinegary—estery, slightly sweet—taste in wine (von Weymarn et al. 2002; Wisselink et al. 2002). In general, homofermentative LAB will only produce small amounts of mannitol, whereas some heterofermentative LAB produce substantial amounts of mannitol (Wisselink et al. 2002). The heterofermentative *Lb. brevis* has been shown to produce significant amounts of mannitol from fructose (Martinez et al. 1963). In *O. oeni*, fructose can be metabolized by two different pathways: heterolactic fermentation or mixed heterolactic/mannitol fermentation (Richter et al. 2003). The switch from one fermentation type to the other occurs at the metabolic level and is related to the growth rate.

This type of wine spoilage tends to be complex, as it is also accompanied by high concentrations of acetic acid, D-lactic acid, propanol, 2-butanol, and diacetyl (Sponholz 1993).

## 15.3.10 Exopolysaccharide Metabolism

The production of exopolysaccharides is almost exclusively due to *Pediococcus* growth and metabolism of glucose in wine, usually *P. parvulus* (Dols-Lafargue et al. 2008; Werning et al. 2006). Wines spoiled due to exopolysaccharide production are referred to as having "ropy" or "graisse" disease (as described by Pasteur) and are viscous, slimy, and oily and have a thick texture. Viscosity of wine can add mouthfeel and be a positive feature, although not at the concentrations when produced through the growth of spoilage LAB, particularly *Pediococcus* species. It has been observed that increases in wine viscosity occur when both LAB and AAB are present (Lüthi 1957).

The presence of residual sugar (glucose) in wine, even at low concentrations (<50 mg/L), can with the appropriate growth conditions allow for the formation of exopolysaccharides or β-glucan in wine. Glucan production has been shown to be greater in nutrient-poor media; more glucan is produced in medium containing 0.1 g/L glucose than 2 g/L (Walling et al. 2005a). Where nitrogen is limiting in the same glucose concentration, the production of glucan by *P. parvulus* will be greater. Thus, wines with a high pH, low glucose, and nitrogen concentrations and no agitation are likely to become ropy (Walling et al. 2005a).

In wine, this exopolysaccharide is a high-molecular-weight β-glucan, a glucose homopolymer that consists of a trisaccharide repeating unit with a β-1,3-linked D-glucosyl backbone and branches made up of single β-1,2-linked D-glucopyranosyl residues (Duenas-Chasco et al. 1997, 1998; Llauberes et al. 1990). This type of glucan cannot be removed by enzymatic treatment with currently known enzymes. The pathway for the production of β-D-glucan and polymerization is well characterized (Walling et al. 2005a).

The presence of a plasmid carrying the *dps* gene (glucan synthase responsible for the polymerization of glucan residues) is required for *Pediococcus* production of β-D-glucan and ropy wines (Walling et al. 2005b). Recent studies have shown that some strains of *O. oeni* carry the *dps* gene (Walling et al. 2005b). The glucosyltransferase gene, *gtf*, has been characterized in *P. parvulus* and *O. oeni* (Dols-Lafargue et al. 2008). Recent studies have demonstrated that *O. oeni* strains can produce exopolysaccharides, although usually in much lower concentrations, as found in spoiled ropy wines (Ciezack et al. 2010).

It has been proposed that the biological role of the ropy phenotype in wine LAB could be linked to the ability to tolerate or overcome wine stress conditions (Spano and Massa 2006; Caggianiello et al. 2016).

## 15.3.11 Biogenic Amine Production

Biogenic amines are organic nitrogenous bases of low molecular weight that are formed during the metabolism of living organisms (Smit et al. 2008). They are found in a range of fermented foods and beverages, including wine. When absorbed at too high a concentration, biogenic amines can have undesirable physiological effects, including headaches, as well as gastrointestinal and respiratory distress.

The principal biogenic amines found in wine are histamine, tyramine, putrescine, cadaverine, phenylethylamine, spermidine, spermine, agamatine, and tryptamine. The main microbial source of these compounds in wine is LAB (Coton et al. 1998; Lonvaud-Funel 2001; Lonvaud-Funel and Joyeux 1994; Soufleros et al. 1998). It has been established that red wine generally exhibits higher biogenic amine concentrations than white wine, which is mainly attributed to the greater propensity of red wine to undergo MLF (Bartowsky and Stockley 2011; Vidal-Carou et al. 1990). Biogenic amines in wine have been reviewed by several research groups (Ancin-Azpilicueta et al. 2008; Anli and Bayram 2009; Ferreira and Pinho 2006; Smit et al. 2008).

Histamine is the main biogenic amine in wine and is the most frequently associated with negative health implications. It is formed by the decarboxylation of the amino acid L-histidine, and the International Organization of Vine and Wine has proposed a histamine limit of 10 mg/L in red and white wines (Anli and Bayram 2009). The histidine decarboxylase enzyme has been purified and characterized. The protein is a single polypeptide of 315 amino acids, comprising two subunits, α and β, forming a hexamer (Lonvaud-Funel and Joyeux 1994). The decarboxylation reaction does not directly generate energy; however, the exchange of histidine and histamine at the membrane level creates a proton gradient and a proton motive force, generating ATP (Rollan et al. 1995). In *Lb. hilgardii*, the histidine decarboxylase activity was associated with the presence of an 80-kb plasmid; four genes in a cluster: *hdcP* (histidine/histamine exchanger), *hdcA* (structural protein), *hdcRS* (histidyl-tRNA synthetase), and *hdcB* (unknown product) (Lucas et al. 2005). There is evidence that the same gene cluster is present in other LAB, including *Lactobacillus* 30a and *O. oeni* 9204 (Gevers et al. 2003), and the plasmid-encoded system could be transferred horizontally (Lucas et al. 2005). The ability of wine LAB species to produce histamine varies with both species and strain. Strains of *Lactobacillus*, *Pediococcus*, *Leuconostoc*, and *O. oeni* have been demonstrated to produce histamine (Landete et al. 2005; Marcobal et al. 2006).

Tyramine is formed by the direct decarboxylation of tyrosine with four well-characterized genes (tyrosyl-tRNA synthetase, tyrosine decarboxylase, probable tyrosine permease, and Na+/H$^+$ antiporter) (Lucas and Lonvaud-Funel 2002; Lucas et al. 2003). Tyramine does not appear to be produced by *O. oeni* strains (Moreno-Arribas et al. 2000; Guerrini et al. 2002).

Putrescine is reported to be the most abundant biogenic amine in wine, both qualitatively and quantitatively (Bartowsky and Stockley 2011; Smit et al. 2008). Ornithine is decarboxylated by ornithine decarboxylase (ODC) to putrescine, with the ODC gene first identified in *O. oeni* (Marcobal et al. 2004; Marcobal et al. 2005). Even though there can often be high concentrations of putrescine present in wines following MLF (Bartowsky and Stockley 2011; Gloria et al. 1998), the presence of the ODC gene does not appear to be widespread (Marcobal et al. 2004).

An alternative pathway for the production of putrescine has been proposed from arginine (Mangani et al. 2005). This might explain the high concentrations of putrescine in wine and the minimal presence of the ODC gene in wine LAB. With this pathway, arginine can be catabolized via arginine deiminase (ADI), ornithine transcarbamoylase, and carbamate kinase (Mangani et al. 2005; Smit et al. 2008). This pathway has also been demonstrated in *Lb. hilgardii* (Arena et al. 2001).

## 15.3.12 Indole Production

A high concentration of indole has been linked to "plastic-like" off-aromas, predominantly in wines produced under sluggish fermentation conditions (Capone et al. 2010). Indole has an aroma detection threshold of 23 µg/L in white wine (Capone et al. 2010), and its formation in wine is not understood but is likely to be related to tryptophan metabolism. Tryptophan is an aromatic amino acid, and its metabolism has been linked to other off-flavors such as "untypical aging off-flavor" in wine (Hoenicke et al. 2002). Catabolism of tryptophan by different microorganisms has been observed, including *Lb. casei* and *Lactobacillus helveticus*, which can form aromatic compounds, including indole, that impart putrid, fecal, and unclean flavors to cheese (Gummalla and Broadbent 1999). Strains of *Fructilactobacillus lindneri*, *P. parvulus*, *P. cerevisiae*, and *O. oeni* have been demonstrated to generate indole during MLF, and this ability was dependent on the presence of tryptophan in the medium (Arevalo-Villena et al. 2010).

## 15.3.13 Ethyl Carbamate

Ethyl carbamate, also referred to as urethane, is genotoxic and carcinogenic to animals and is formed through the chemical reaction of ethanol and citrulline, urea, or carbamyl phosphate. Arginine, a quantitatively important amino acid of grape must and wine (Henschke and Jiranek 1993; Ough et al. 1988), is a precursor of citrulline, and LAB vary in their ability to degrade arginine (Granchi et al. 1998).

*O. oeni* and *Lb. buchneri* can metabolize arginine and citrulline (Mira de Orduna et al. 2000). Typical wine parameters (high ethanol, high L-malic acid, low pH) are conducive to ethyl carbamate production; however, concentrations will increase even more at higher pH (Romero et al. 2009). The presence of the *arc* genes, for the arginine–deiminase pathway, has been identified in several genera of wine LAB. A correlation was found between the presence of genes and the ability to degrade arginine; degrading strains included all heterofermentative lactobacilli, *O. oeni*, *P. pentosaceus*, and some strains of *Lc. mesenteroides* and *Lb. plantarum* (Araque et al. 2009).

The role that MLF plays in the potential ethyl carbamate remains unclear; however, experiments conducted in synthetic wine and laboratory vinified wine demonstrated a correlation between arginine degradation, citrulline production, and ethyl carbamate formation. Even though the formation of this compound does not necessarily affect the wine sensory quality, it can potentially have health implications (Weber and Sharypov 2009).

# 15.4 CONCLUSIONS

The best wine is achieved by maximizing the desired aromas and flavors and minimizing the less desirable off-flavors and off-aromas. This chapter has sought to summarize the bacterial and chemical interactions that contribute to undesirable or spoilage wine aromas and flavors. Often it is a fine balance between the concentration of a secondary metabolite being considered desirable or spoilage. Wine-associated LAB cannot always be easily divided into effective or contaminating bacteria. Managing the bacterial microbiota of fermenting grape juice and maturing wine is not always an easy task for the winemaker. Wine acidity (pH) management and timing of sulfur regimes are the most common practices to control contaminant LAB. This chapter did not explore ways to control wine-contaminating LAB; however, there are a few emerging technologies, including UV irradiation, pressure, and electric fields, that have been employed in numerous beverage industries and are beginning to be explored in the wine industry (reviewed by Bartowsky 2009).

# BIBLIOGRAPHY

Ancin-Azpilicueta, C., A. Gonzalez-Marco, and N. Jimenez-Moreno. 2008. Current knowledge about the presence of amines in wine. *Crit Rev Food Sci Nutr 48* (3):257–275.

Anli, R. E. and M. Bayram. 2009. Biogenic amines in wines. *Food Rev Int 25* (1):86–102.

Araque, I., J. Gil, R. Carrete, A. Bordons, and C. Reguant. 2009. Detection of arc genes related with the ethyl carbamate precursors in wine lactic acid bacteria. *J Agric Food Chem 57* (5):1841–1847.

Arena, M. E. and M. C. Manca de Nadra. 2001. Biogenic amine production by *Lactobacillus* . *J Appl Microbiol 90*:158–162.

Arevalo-Villena, M., E. J. Bartowsky, D. Capone, and M. A. Sefton. 2010. Production of indole by wine associated microorganisms under oenological conditions. *Food Microbiol 27* (5):685–690.

Badotti, F., et al. 2014. *Oenococcus alcoholitolerans* sp. nov., a lactic acid bacteria isolated from cachaça and ethanol fermentation processes. *Antonie Van Leeuwenhoek 106*:1259–1267.

Bae, S., G. H. Fleet, and G. M. Heard. 2006. Lactic acid bacteria associated with wine grapes from several Australian vineyards. *J Appl Microbiol 100* (4):712–727.

Bartowsky, E. J. 2005. *Oenococcus oeni* and malolactic fermentation—Moving into the molecular arena. *Aust J Grape Wine Res 11* (2):174–187.

Bartowsky, E. J. 2009. Bacterial spoilage of wine and approaches to minimize it. *Lett Appl Microbiol 48* (2):149–156.

Bartowsky, E. J. 2017. *Oenococcus oeni* and genomic era. *FEMS Microbiol Rev 41*:S84–S94.

Bartowsky, E. J., I. L. Francis, J. R. Bellon, and P. A. Henschke. 2002. Is buttery aroma perception in wines predictable from diacetyl concentration? *Aust J Grape Wine Res 8*:180–185.

Bartowsky, E. J. and P. A. Henschke. 2004. The 'buttery' attribute of wine—diacetyl. Desirability, spoilage and beyond. *Int J Food Microbiol 96*:235–252.

Bartowsky, E. J. and P. A. Henschke. 2008. Acetic acid bacteria spoilage of bottled red wine—A review. *Int J Food Microbiol 125*:60–70.

Bartowsky, E. J. and I. S. Pretorius. 2008. Microbial formation and modification of flavour and flavour compounds in wine. In Biology of Microorganisms on Grapes, in Must and in Wine, edited by H. König, G. Unden and J. Fröhlich. Heidelberg, Germany: Springer.

Bartowsky, E. J. and C. S. Stockley. 2011. Histamine in Australian wines—A survey between 1982 and 2009. *Ann Microbiol 61* (1):167–172.

Bartowsky, E. J., D. Xia, R. L. Gibson, R. L. Fleet, and P. A. Henschke. 2003. Spoilage of bottled red wine by acetic acid bacteria. *Lett Appl Microbiol 36* (5):307–314.

Bauer, R., D. A. Cowan, and A. Crouch. 2010a. Acrolein in wine: Importance of 3-hydroxypropionaldehyde and derivatives in production and detection. *J Agric Food Chem 58* (6):3243–3250.

Bauer, R., M. du Toit, and J. Kossmann. 2010b. Influence of environmental parameters on production of the acrolein precursor 3-hydroxypropionaldehyde by *Lactobacillus reuteri* DSMZ 20016 and its accumulation by wine lactobacilli. *Int J Food Microbiol 137* (1):28–31.

Bazireau, M. 2022. Rotten grape disease on the rise in wines. *Vitisphere*, 11 August, www.vitisphere.com

Borneman, A. R., J. M. McCarthy, P. J. Chambers, and E. J. Bartowsky. 2012. Functional divergence in the genus *Oenococcus* as predicted by genome sequencing of the newly-described species, *Oenococcus kitaharae* . *PLoS ONE 7* (1–10):e29626.

Caggianiello, G., M. Kleerebezem, and G. Spano. 2016. Exopolysaccharides produced by lactic acid bacteria: From health-promoting benefits to stress tolerance mechanisms. *Appl Microbial Biotechnol 100*:3877–3886.

Capone, D. L., K. A. van Leeuwen, and K. H. Pardon, et al. 2010. Identification and analysis of 2-chloro-6-methylph2,6-dichlorophenol and indole—Causes of taints and off-flavours in wines. *Aust J Grape Wine Res 16* (1):210–217.

Cappello, M. S., G. Zapparoli, A. Logrieco, and E. J. Bartowsky. 2017. Linking wine lactic acid bacteria diversity with wine aroma and flavor. *Int J Food Microbiol 243*:16–27.

Caspritz, G. and F. Radler. 1983. Malolactic enzyme of *Lactobacillus plantarum*. Purification, properties, and distribution among bacteria. *J Biol Chem 258*:4907–4910.

Chatonnet, P., D. Dubourdieu, and J. N. Boidron. 1995. The influence of *Brettanomyces/Dekkera* sp. yeasts and lactic acid bacteria on the ethylphenol content of red wines. *Am J Enol Vitic 46* (4):463–468.

Ciezack, G., L. Hazo, G. Chambat, et al. 2010. Evidence for exopolysaccharide production by *Oenococcus oeni* strains isolated from non-ropy wines. *J Appl Microbiol 108* (2):499–509.

Cleenwerck, I. and P. De Vos. 2008. Polyphasic taxonomy of acetic acid bacteria: An overview of the currently applied methodology. *Int J Food Microbiol 125* (1):2–14.

Corison, C. A., C. S. Ough, H. W. Berg, and K. E. Nelson. 1979. Must acetic-acid and ethyl-acetate as mold and rot indicators in grapes. *Am J Enol Vitic 30* (2):130–134.

Costello, P. J. and P. A. Henschke. 2002. Mousy off-flavour of wine: Precursors and biosynthesis of the causative N-heterocycles 2-ethyltetrahydropyridine, 2-acetyltetrahydropyridine, and 2-acetyl-1-pyrroline by *Lactobacillus hilgardii* DSM 20176. *J Agric Food Chem 50* (24):7079–7087.

Costello, P. J., T. H. Lee, and P. A. Henschke. 2001. Ability of lactic acid bacteria to produce N-heterocycles causing mousy off-flavour in wine. *Aus J Grape Wine Res 7*:160–167.

Coton, E., G. Rollan, A. Bertrand, and A. Lonvaud-Funel. 1998. Histamine-producing lactic acid bacteria in wines: Early detection, frequency, and distribution. *Am J Enol Vitic 49* (2):199–204.

Cousin, F. J., R. Le Guellec, C. Chagnot, D. Goux, M. Dalmasso, J.-M. Laplace and M. Cretenet. 2019. *Oenococcus sicerae* sp. nov., isolated from French cider. *Syst Appl Microbiol 42*:302–308.

Couto, J. A., F. M. Campos, A. R. Figueiredo, and T. A. Hogg. 2006. Ability of lactic acid bacteria to produce volatile phenols. *Am J Enol Vitic 57* (2):166–171.

Crowell, E. A. and J. F. Guymon. 1975. Wine constituents arising from sorbic acid addition, and identification of 2-ethoxyhexa-3,5-diene as source of geranium-like off-odor. *Am J Enol Vitic 26* (2):97–102.

Curtin, C. D., J. R. Bellon, P. A. Henschke, P. Godden, and M. de Barros Lopes. 2007. Genetic diversity of *Dekkera bruxellensis* yeasts isolated from Australian wineries. *FEMS Yeast Res 7* (3):471–481.

Davis, C. R., D. Wibowo, G. H. Fleet, and T. H. Lee. 1988. Properties of wine lactic acid bacteria: Their potential enological significance. *Am J Enol Vitic 39* (2):137–142.

De Las Rivas, B., H. Rodriguez, J. A. Curiel, J. M. Landete, and R. Munoz. 2009. Molecular screening of wine lactic acid bacteria degrading hydroxycinnamic acids. *J Agric Food Chem 57* (2):490–494.

Dols-Lafargue, M., H. Y. Lee, C. Le Marrec, et al. 2008. Characterization of *gtf*, a glucosyltransferase gene in the genomes of *Pediococcus parvulus* and *Oenococcus oeni*, two bacterial species commonly found in wine. *Appl Environ Microbiol 74* (13):4079–4090.

Dubois, P. 1983. Volatile phenols in wines. In Flavour of Distilled Beverages, Origins and Developments, edited by J. R. Pifgott. Chichester, UK: Ellis Horwood.

Duenas-Chasco, M. T., M. A. Rodriguez-Carvajal, P. Tejero-Mateo, G. Franco-Rodriguez, J. L. Espartero, A. Irastorza-Iribas, and A. M. Gil-Serrano. 1997. Structural analysis of the exopolysaccharide produced by *Pediococcus damnosus* 2.6. *Carbohydr Res 303* (4):453–458.

Duenas-Chasco, M. T., M. A. Rodriguez-Carvajal, P. Tejero-Mateo, et al. 1998. Structural analysis of the exopolysaccharides produced by *Lactobacillus* spp. G-77. *Carbohydr Res 307* (1–2):125–133.

Du Toit, M. and I. S. Pretorius. 2000. Microbial spoilage and preservation of wine: Using weapons from Nature's own arsenal—A review. *S Afr J Enol Vitic 21* (Special issue):74–96.

Edinger, W. D., and D. F. Splittstoesser. 1986a. Production by lactic-acid bacteria of sorbic alcohol, the precursor of the geranium odor compound. *Am J Enol Vitic 37* (1):34–38.

Edinger, W. D., and D. F. Splittstoesser. 1986b. Sorbate tolerance by lactic-acid bacteria associated with grapes and wine. *J Food Sci 51* (4):1077–1078.

Edwards, C. G., K. M. Haag, M. D. Collins, R. A. Hutson, and Y. C. Huang. 1998. *Lactobacillus kunkeei* sp. nov.: A spoilage organism associated with grape juice fermentations. *J Appl Microbiol 84* (5):698–702.

Edwards, C. G., A. G. Reynolds, A. V. Rodriguez, M. J. Semon, and J. M. Mills. 1999. Implication of acetic acid in the induction of slow/stuck grape juice fermentations and inhibition of yeast by *Lactobacillus* sp. *Am J Enol Vitic 50* (2):204–210.

Endo, A. and S. Okada 2006. *Oenococcus kitaharae* sp. nov., a non-acidophilic and non-malolactic-fermenting *Oenococcus* isolated from a composting distilled shochu residue. *Int J Syst Evol Bacteriol 56*:2345–2348.

Ferreira, I. M. P. L. V. O. and O. Pinho. 2006. Biogenic amines in Portuguese traditional foods and wines. *J Food Prot 69* (9):2293–2303.

Gevers, D., G. Huys, and J. Swings. 2003. In vitro conjugal transfer of tetracycline resistance from *Lactobacillus* isolates to other Gram-positive bacteria. *FEMS Microbiol Lett 225* (1):125–130.

Gloria, M. B. A., B. T. Watson, L. Simon-Sarkadi, and M. A. Daeschel. 1998. A survey of biogenic amines in Oregon Pinot noir and Cabernet Sauvignon wines. *Am J Enol Vitic 49* (3):279–282.

Gorga, A., O. Claisse, and A. Lonvaud-Funel. 2002. Organisation of the genes encoding glycerol dehydratase of *Lactobacillus collinoides, Lactobacillus hilgardii* and *Lactobacillus diolivorans* . *Sci Aliments 22* (1–2):151–160.

Granchi, L., R. Paperi, D. Rosellini, and M. Vincenzini. 1998. Strain variation of arginine catabolism among malolactic *Oenococcus oeni* strains of wine origin. *Ital J Food Sci 10* (4):351–357.

Grbin, P. R. and P. A. Henschke. 2000. Mousy off-flavour production in grape juice and wine by *Dekkera* and *Brettanomyces* yeasts. *Aust J Grape Wine Res 6* (3):255–262.

Guerrini, S., S. Mangani, L. Granchi, and M. Vincenzini. 2002. Biogenic amine production by *Oenococcus oeni* . *Curr Microbiol 44*:374–378.

Gummalla, S. and J. R. Broadbent. 1999. Tryptophan catabolism by *Lactobacillus casei* and *Lactobacillus helveticus* cheese flavor adjuncts. *J Dairy Sci 82* (10):2070–2077.

Henick-Kling, T. 1993. Malolactic fermentation. In Wine Microbiology and Biotechnology, edited by G. H. Fleet. Amsterdam, The Netherlands: Harwood Academic Publisher.

Henschke, P. A. and V. Jiranek. 1993. Yeasts—Metabolism of nitrogen compounds. In Wine Microbiology and Biotechnology, edited by G. H. Fleet. Chur, Switzerland: Harwood Academic Publishers.

Heresztyn, T. 1986a. Metabolism of volatile phenolic-compounds from hydroxycinnamic acids by *Brettanomyces* yeast. *Arch Microbiol 146* (1):96–98.

Heresztyn, T. 1986b. Formation of substituted tetrahydropyridines by species of *Brettanomyces* and *Lactobacillus* isolated from mousy wines. *Am J Enol Vitic 37* (2):127–132.

Hoenicke, K., O. Borchert, K. Gruning, and T. J. Simat. 2002. "Untypical aging off-flavor" in wine: Synthesis of potential degradation compounds of indole-3-acetic acid and kynurenine and their evaluation as precursors of 2-aminoacetophenone. *J Agric Food Chem 50* (15):4303–4309.

Krieger, S. A., E. Lemperle, and M. Ernst. 2000. Management of malolactic fermentation with regard to flavor modification in wine. Paper read at 5th International Symposium on Cool Climate Viticulture and Oenology, 16–20 January, at Melbourne, Australia.

Landete, J. M., S. Ferrer, and I. Pardo. 2005. Which lactic acid bacteria are responsible for histamine production in wine? *J Appl Microbiol 99* (3):580–586.

Landete, J. M., L. Garcia-Haro, A. Blasco, et al. 2010. Requirement of the *Lactobacillus casei* MaeKR two-component system for L-malic acid utilization via a malic enzyme pathway. *Appl Environ Microbiol 76* (1):84–95.

Llauberes, R. M., B. Richard, A. Lonvaud, D. Dubourdieu, and B. Fournet. 1990. Structure of an exocellular beta-d-glucan from *Pediococcus* sp., a wine lactic bacteria. *Carbohydr Res 203* (1):103–107.

London, J. and E. Y. Meyer. 1969. Malate utilization by a group D *Streptococcus*—Physiological properties and purification of an inducible malic enzyme. *J Bacteriol 98* (2):705–711.

London, J., E. Y. Meyer, and S. R. Kulczyk. 1971. Detection of relationships between *Streptococcus faecalis* and *Lactobacillus casei* by immunological studies with 2 forms of malic enzyme. *J Bacteriol 108* (1):196–201.

Lonvaud-Funel, A. 2000. Understanding wine lactic acid bacteria. Progress and prospects in controlling wine quality. Paper Read at ASEV 50th Anniversary Annual Meeting, 19–23 June, at Seattle, Washington, DC.

Lonvaud-Funel, A. 2001. Biogenic amines in wines: Role of lactic acid bacteria. *FEMS Microbiol Lett 199* (1):9–13.

Lonvaud-Funel, A. and A. Joyeux. 1994. Histamine production by wine lactic acid bacteria: Isolation of a histamine-producing strain of *Leuconostoc oenos* . *J Appl Bacteriol 77*:401–407.

Lonvaud-Funel, A. and A. M. Strasser de Saad. 1982. Purification and properties of a malolactic enzyme from a strain of *Leuconostoc mesenteroides* isolated from grapes. *Appl Environ Microbiol 43*:357–361.

Lucas, P. M., J. Landete, M. Coton, E. Coton, and A. Lonvaud-Funel. 2003. The tyrosine decarboxylase operon of *Lactobacillus brevis* IOEB 9809: Characterization and conservation in tyramine-producing bacteria. *FEMS Microbiol Lett 229* (1):65–71.

Lucas, P. M. and A. Lonvaud-Funel. 2002. Purification and partial gene sequence of the tyrosine decarboxylase of *Lactobacillus brevis* IOEB 9809. *FEMS Microbiol Lett 211* (1):85–89.

Lucas, P. M., W. A. M. Wolken, O. Claisse, J. S. Lolkema, and A. Lonvaud-Funel. 2005. Histamine-producing pathway encoded on an unstable plasmid in *Lactobacillus hilgardii* 0006. *Appl Environ Microbiol 71* (3):1417–1424.

Lüthi, H. 1957. Symbiotic problems relating to the bacterial deterioration of wines. *Am J Enol Vitic 8* (4):176–181.

Mangani, S., S. Guerrini, L. Granchi, and M. Vincenzini. 2005. Putrescine accumulation in wine: Role of *Oenococcus oeni* . *Curr Microbiol 51* (1):6–10.

Marcobal, A., B. de las Rivas, M. V. Moreno-Arribas, and R. Munoz. 2004. Identification of the ornithine decarboxylase gene in the putrescine-producer *Oenococcus oeni* BIFI-83. *FEMS Microbiol Lett 239*:213–220.

Marcobal, A., B. de las Rivas, M. V. Moreno-Arribas, and R. Muñoz. 2005. Multiplex PCR method for the simultaneous detection of histamine-, tyramine-, and putrescine-producing lactic acid bacteria in foods. *J Food Prot 68* (4):874–878.

Marcobal, A., P. J. Martin-Alvarez, M. C. Polo, R. Munoz, and M. V. Moreno-Arribas. 2006. Formation of biogenic amines throughout the industrial manufacture of red wine. *J Food Prot 69* (2):397–404.

Margalith, P. Z. 1983. Flavour Microbiology. Edited by P. Z. Margalith. Springfield, IL: Charles C Thomas.

Martineau, B., T. E. Acree, and T. Henick-Kling. 1995. Effect of wine type on the detection threshold for diacetyl. *Food Res Int 28* (2):139–143.

Martineau, B. and T. Henick-Kling. 1995. Performance and diacetyl production of commercial strains of malolactic bacteria in wine. *J Appl Bacteriol 78*:526–536.

Martinez, G., H. A. Barker, and B. L. Horecker. 1963. A specific mannitol dehydrogenase from *Lactobacillus brevis* . *J Biol Chem 238* (5):1598–1603.

Mira de Orduna, R., S.-Q. Liu, M. L. Patchett, and G. J. Pilone. 2000. Kinetics of the arginine metabolism of malolactic wine lactic acid bacteria *Lactobacillus buchneri* CUC-3 and *Oenococcus oeni* Lo111. *J Appl Microbiol 89*:547–552.

Moreno-Arribas, V., S. Torlois, A. Joyeux, A. Bertrand, and A. Lonvaud-Funel. 2000. Isolation, properties and behaviour of tyramine-producing lactic acid bacteria from wine. *J Appl Microbiol 88*:584–593.

Möslinger. 1901. Ueber die Säuren des Weines und den Säurerückgang. *Zeitschr f Untersuchung d Nahr Genussmittel 4*:1120–1130.

Nikfardjam, M. N. and L. Kunz. 2021. Influence of iron and copper on the formation of acetyltetrahydropyridine (ATHP) and 2-acetylpyridine (2-AP) in wine. *Mitt Klosterneuburg 71*:28–36.

Ough, C. S., E. A. Crowell, and B. R. Gutlove. 1988. Carbamyl compound reactions with ethanol. *Am J Enol Vitic 39*:239–242.

Pasteris, S. E. and A. M. Strasser de Saad. 2005. Aerobic glycerol catabolism by *Pediococcus pentosaceus* isolated from wine. *Food Microbiol 22* (5):399–407.

Pasteris, S. E. and A. M. Strasser de Saad. 2009. Sugar-glycerol cofermentations by *Lactobacillus hilgardii* isolated from wine. *J Agric Food Chem 57* (9):3853–3858.

Pasteur, L. 1873. Études sur le Vin. 2nd ed. Paris, France: Savy.

Pelonmier-Magimel, E., P. Mangiorou, P. Darriet, G. de Revel, M. Jourdes, A. Marchal, S. Marchand, A. Pons, L. Riquier, P.-L. Teissedre, C. Thibon, G. Lytra, S. Tempere and J.-C. Barbe. 2020. Sensory characterization of Bordeaux red wines produced without added sulfites. *OENO ONE 54*:687–697.

Pripis-Nicolau, L., G. de Revel, A. Bertrand, and A. Lonvaud-Funel. 2004. Methionine catabolism and production of volatile sulphur compounds by *Oenococcus oeni*. *J Appl Microbiol 96* (5):1176–1184.

Radler, F. 1986. Microbial biochemistry. *Experientia 42*:884–893.

Radler, F. and C. Yannissi. 1972. Weinsäureabbau bei Milchsäurebakterie. *Archiv Mikrobiol 82* (3):219–239.

Ramos, A. and H. Santos. 1996. Citrate and sugar cofermentation in *Leuconostoc oenos*, a [13]C nuclear magnetic resonance study. *Appl Environ Microbiol 62* (7):2577–2585.

Rankine, B. C., J. C. M. Fornachon, and D. A. Bridson. 1969. Diacetyl in Australian dry red wines and its significance in wine quality. *Vitis 8*:129–134.

Renouf, V., O. Claisse, and A. Lonvaud-Funel. 2005. Understanding the microbial ecosystem on the grape berry surface through numeration and identification of yeast and bacteria. *Aust J Grape Wine Res 11* (3):316–327.

Rentschler, H. and H. Tanner. 1951. Das Bitterwerden der Rotweine. *Mitt Lebensmittelunters Hyg 42*:463–475.

Ribéreau-Gayon, J., D. Dubourdieu, B. Donèche, and A. Lonvaud. 2006. Lactic acid bacteria. In Handbook of Enology: The Microbiology of Wine and Vinifications, edited by J. Ribéreau-Gayon, D. Dubourdieu, B. Donèche and A. Lonvaud. Chichester, UK: John Wiley & Sons.

Richter, H., A. A. de Graaf, I. Hamann, and G. Unden. 2003. Significance of phosphoglucose isomerase for the shift between heterolactic and mannitol fermentation of fructose by *Oenococcus oeni*. *Arch Microbiol 180*:465–470.

Rollan, G. C., E. Coton, and A. LonvaudFunel. 1995. Histidine decarboxylase activity of *Leuconostoc oenos* 9204. *Food Microbiol 12* (6):455–461.

Romero, S. V., C. Reguant, A. Bordons, and M. C. Masque. 2009. Potential formation of ethyl carbamate in simulated wine inoculated with *Oenococcus oeni* and *Lactobacillus plantarum*. *Int J Food Sci Technol 44* (6):1206–1213.

Sauvageot, N., K. Gouffi, J. M. Laplace, and Y. Auffray. 2000. Glycerol metabolism in *Lactobacillus collinoides*: Production of 3-hydroxypropionaldehyde, a precursor of acrolein. *Int J Food Microbiol 55* (1–3):167–170.

Schütz, H. and F. Radler. 1984. Anaerobic reduction of glycerol to propanediol-1.3 by *Lactobacillus brevis* and *Lactobacillus buchneri*. *Syst Appl Microbiol 5* (2):169–178.

Schütz, M. and F. Radler. 1974. Das Vorkommen von Malatenzym und Malo-Lactat-Enzym bei verschiedenen Milchsäurebakterien. *Arch Microbiol 96* (4):329–339.

Smiley, K. L. and M. Sobolov. 1962. A cobamide-requiring glycerol dehydrase from an acrolein-forming *lactobacillus*. *Arch Biochem Biophys 97* (3):538–543.

Smit, A. Y., W. J. du Toit, and M. du Toit. 2008. Biogenic amines in wine: Understanding the headache. *S Afr J Enol Vitic 29* (2):109–127.

Sobolov, M. and K. L. Smiley. 1960. Metabolism of glycerol by an acrolein-forming lactobacillus. *J Bacteriol 79* (2):261–266.

Soufleros, E., M.-L. Barrios, and A. Bertrand. 1998. Correlation between the content of biogenic amines and other wine compounds. *Am J Enol Vitic 49* (3):266–278.

Spano, G. and S. Massa. 2006 Environmental stress response in wine lactic acid bacteria: Beyond *Bacillus subtilis*. *Crit Rev Microbiol 32*:77–86.

Sponholz, W.-R. 1993. Wine spoilage by microorganisms. In Wine Microbiology and Technology, edited by G. H. Fleet. Amsterdam, the Netherlands: Harwood Academic Publishing.

Swiegers, J. H., E. J. Bartowsky, P. A. Henschke, and I. S. Pretorius. 2005. Yeast and bacterial modulation of wine aroma and flavour. *Australian Journal of Grape and Wine Research 11*:139–173.

Swiegers, J. H. and I. S. Pretorius. 2007. Modulation of volatile sulfur compounds by wine yeast. *Appl Microbiol Biotechnol 74* (5):954–960.

Tempere, S., B. Chatelet, G. de Revel, M. Dufoir, M. Denat, P.-Y. Ramonet, S. Marchand, M. Sadoudi, N. Richard, P. Lucas, C. Miot-Sertier, O. Claisse, L. Riquier, M.-C. Perello, P. Ballestra. 2019. Comparison between standardized sensory methods used to evaluate the mousy off-flavor in red wine. *OENO ONE 2*:95–105.

Thudicum, J. L. W. 1894. *A Treatise on Wines*. London: George Bell & Sons, p. 378.

Tucknott, O. G. 1977. The Mousy Taint in Fermented Beverages; Its Nature and Origin. Bristol: The University of Bristol.

Vermeulen, C., L. Gijs, and S. Collin. 2005. Sensorial contribution and formation pathways of thiols in foods: A review. *Food Rev Int 21* (1):69–137.

Vidal-Carou, M. C., R. Codony-Salcedo, and A. Marine-Font. 1990. Histamine and tyramine in Spanish wines: Relationships and total sulfur dioxide level, volatile acidity and malo-lactic fermentation intensity. *Food Chem 35*:217–227.

Virdis, C., K. Sumby, E. Bartowsky and V. Jiranek. 2021. Lactic acid bacteria in wine: Technological advances and evaluation of their functional role. *Front Microbiol 11* (612118):1–16.

Voisenet, E. 1910. The formulation of acroleine in the bitter disease in wine. *C R Hebd Seances Acad Sci 150*:1614–1616.

Voisenet, E. 1911. An enzyme of bitterness in wines, an agent of glycerin dehydration. *C R Hebd Seances Acad Sci 153*:363–365.

Vollenweider, S. and C. Lacroix. 2004. 3-Hydroxypropionaldehyde: Applications and perspectives of biotechnological production. *Appl Microbiol Biotechnol 64* (1):16–27.

von Weymarn, N., M. Hujanen, and M. Leisola. 2002. Production of D-mannitol by heterofermentative lactic acid bacteria. *Process Biochem 37*:1207–1213.

Walling, E., M. Dols-Lafargue, and A. Lonvaud-Funel. 2005a. Glucose fermentation kinetics and exopolysaccharide production by ropy *Pediococcus damnosus* IOEB8801. *Food Microbiol 22*:71–78.

Walling, E., E. Gindreau, and A. Lonvaud-Funel. 2005b. A putative glucan synthase gene dps detected in exopolysaccharide-producing *Pediococcus damnosus* and *Oenococcus oeni* strains isolated from wine and cider. *Int J Food Microbiol 98* (1):53–62.

Weber, J. V. and V. I. Sharypov. 2009. Ethyl carbamate in foods and beverages: A review. *Environ Chem Lett 7* (3):233–247.

Werning, M. L., I. Ibarburu, M. T. Duenas, et al. 2006. *Pediococcus parvulus gtf* gene encoding the GTF glycosyltransferase and its application for specific PCR detection of beta-D-glucan-producing bacteria in foods and beverages. *J Food Prot 69* (1):161–169.

Wisselink, H. W., R. A. Weusthuis, G. Eggink, J. Hugenholtz, and G. J. Grobben. 2002. Mannitol production by lactic acid bacteria: A review. *Int Dairy J 12*:151–161.

# Lactic Acid Bacteria in Food Spoilage

**16**

## Vasileios Pothakos and Johanna Björkroth

## 16.1 INTRODUCTION

Lactic acid bacteria (LAB) are ubiquitous to food ecosystems and generally associated with fermented foodstuffs, wherein their activity and acidification deliver improved palatability and preservation (Hutkins, 2006). Throughout their evolution they have been adapted to nutrient-rich environments, such as plant-related habitats, which are considered the archetype source of many taxa (Douillard and de Vos, 2014). This adaptation occurred after the class *Lactobacillales* developed into an individual line of descent from soil *Bacilli* and underwent a considerable loss of ancestral genes associated with pathogenicity, sporulation, motility, and biosynthesis of nutrients (Makarova et al., 2006; Pfeiler and Klaenhammer, 2007). Through these evolutionary events, LAB acquired an auxotrophic character, resulting in a fastidious microbial cohort with significant affinity initially to plants and after the apparition of lactating mammals also to animals (Champomier-Vergès et al., 2002; de Vos, 2011).

Despite the occurrence of LAB as minor populations on intact vegetable or fruit surfaces, animal skin, and grains, as they lack mechanisms to penetrate the outer barriers, they exhibit the remarkable feature to prevail when nutrients become available, as in the case of decaying plants, exuding animal tissue, milk, and flour slurries (Tamang et al., 2016). Their competitiveness in utilizing energy-yielding molecules from food commodities and their ability to acidify the matrix they proliferate in facilitates their rapid outgrowth (Teusink et al., 2011). These traits showcase that their relation and adaptation to food has been orchestrated through evolution (Makarova et al., 2006; Pfeiler and Klaenhammer, 2007). As soon as humans gained control over these amenable microorganisms and started to apply them in dairy, cereal, and meat fermentations, this beneficial group gained technological importance.

However, LAB are also involved in food spoilage. Historically, the earliest reported alterations associated with the outgrowth of LAB species were often linked to intense acidification of alcoholic beverages (i.e., beer, wine, and cider), fermented dairy products (i.e., cheese and yogurt), or condiments such as sauces (Boor et al., 2017; Carr, 1958; Malfeito-Ferreira, 2014; Stiles and Holzapfel, 1997). Moreover, certain dextran-producing species would impede the production of fermented beverages by excessive production of slime (Björkroth and Holzapfel, 2006). Such alterations were typically related to their fermentative nature and were associated with preserving goods at ambient temperature. This type of spoilage microbiota are/were mesophilic, acid-tolerant LAB species from the genera of *Lactiplantibacillus*, *Lactobacillus*, *Latilactobacillus*, *Leuconostoc*, *Levilactobacillus*, and *Pediococcus* (Stiles and Holzapfel, 1997).

Since 1980s, the use of cold chains and polymers for packaging increased the effectiveness of food preservation for retail distribution. Fresh produce, meat, and fishery trade has increasingly been based on refrigeration and the use of suitable barriers in packaging. In these commodities, versatile, mesophilic LAB, such as *Latilactobacillus sakei* and *Leuconostoc mesenteroides*, have frequently been detected as specific spoilage organisms due to their ability to grow at refrigeration temperatures (Chaillou et al., 2009; Dykes et al., 1995). Compared to other food spoilers, such as pseudomonads, enterobacteria, bacilli, yeasts, and molds, LAB are considered the least offensive microorganisms, delivering off-flavors, acidity, blowing due to gas formation, and in some cases biogenic amines, but they are not associated with putrefaction or safety hazards (Borch et al., 1996).

Nowadays, manufacturers apply improved hygiene means, thus controlling to a larger extent microbial contamination cases. However, the expanding production of packaged and cold-stored foodstuffs remains prone to quality deviations. Apart from the aforementioned LAB, novel species started to appear increasingly. Psychrotrophic LAB with lower maximum temperature growth range have been dominating at the end of shelf life in these products (Björkroth and Korkeala, 1997a; Korkeala et al., 1989; Korkeala and Mäkelä, 1989; Pothakos et al., 2015a). They are unable to grow or grow poorly at mesophilic temperatures (i.e., 30°C) routinely used for culturing and usually remain underestimated with the commonly applied microbiological standards (Pothakos et al., 2012; Reuter, 1985). Pioneering work started 35 years ago in Finland, and in the course of time this bacterial subgroup has gained reputation as being responsible for severe quality faults, resulting in significant financial losses for the food industry also in northern Europe, Japan, and Canada.

# 16.2 PACKAGED AND COLD-STORED, HIGHLY PERISHABLE FOODSTUFFS

During the last decades consumer demand for non-invasive preservation methods and the maintenance of highly nutritious/minimally processed foodstuffs triggered the development and combination of novel technologies. Currently, different techniques such as modified-atmosphere packaging (MAP), intelligent packaging, edible coatings, chemical sanitation, and high-pressure processing are implemented to extend the shelf life of perishable foodstuffs (Cutter, 2002; Vanderroost et al., 2014). These advances have facilitated the introduction of convenience goods such as packaged and chilled-stored products. This category encompasses a very broad range of commodities of both animal and plant origin, such as fresh meat, fishery, cooked or marinated meat, dairy, eggs, vegetable or fruit salads, ready-to-eat (RTE) salads, and composite RTE meals. These portioned food products have gained a market share due to the demanding lifestyle of modern societies, and therefore related food industries are advancing rapidly (Kaufman et al., 2000; Ramos et al., 2013).

## 16.2.1 Packaging and Refrigeration

Packaging protects food products from the environment by covering them in an impermeable plastic/polymer wrapping, which acts as a barrier, but at the same time, a selected gas composition (i.e., combination of $O_2$, $CO_2$, and $N_2$) can be used to inhibit microbial growth and/or to control physicochemical parameters (Jacxsens et al., 2003). $N_2$ is inert and thus is used as carrier gas, and $O_2$ serves to sustain the organoleptic acceptability of fresh produce, while $CO_2$ inhibits strictly aerobic and $CO_2$-sensitive microbiota (Egan, 1983). When the appropriate gas composition is coupled with low-temperature storage, the combination is an effective hurdle technology against yeasts, filamentous fungi, aerobic mesophiles (i.e., pseudomonads), and cold-sensitive microaerophilic microbes (i.e., enterobacteriales).

The implementation of high-$O_2$ and high-$CO_2$ packaging enhances the appearance of fresh-cut produce due to the retardation of the physiological deterioration (i.e., respiration rate, loss of textural integrity, discoloration) (Allende et al., 2002; Conesa et al., 2007). In meat technology, high-$O_2$ MAP is generally required to maintain the vivid red color of fresh tissue by keeping the heme pigment in oxymyoglobin form and preventing the development of brown discolorations that render the products uninviting for consumption (Lorenzo and Gómez, 2012; McMillin, 2008). Microbiota can respond differently to this MAP technology coupled to refrigeration, but usually an inhibitory effect on the proliferation of pathogenic or spoiling microbial taxa is achieved.

## 16.2.2 Selection of LAB

The growth of offensive microorganisms belonging to different classes of proteobacteria, bacilli, yeasts, and molds that compose the majority of the diverse microbial assemblages on fresh produce and animal-deriving products is mainly retarded by low temperature and limited $O_2$. However, psychrotrophic LAB that constitute a minor population of the initial food ecosystem are detected more frequently among the most abundant groups at the end of shelf life (Chaillou et al., 2015; Dainty et al., 1979). Apparently, the extrinsic parameters of temperature and modified atmospheres exert a strong selection pressure toward bacteria that possess psychrotrophic character and can sustain a wide range of gas compositions. LAB tend to remain unaffected by superatmospheric $O_2$ packaging because they possess mechanisms to circumvent oxidative stress (e.g., NADH peroxidase system, superoxide dismutase activity, heme-mediated respiration in some LAB), neutralizing the highly reactive oxygen species (Axelsson, 2004). Additionally, they are insensitive to high $CO_2$ due to their acid tolerance, and thanks to their strong psychrotrophic character, they prevail in packaged and refrigerated products. The most reported psychrotrophic LAB belong to the genera *Leuconostoc*, *Lactobacillus*, *Latilactobacillus*, *Carnobacterium*, *Dellaglioa*, *Lactococcus*, and *Paucilactobacillus*, encompassing numerous species novae that have been described the last 30 years.

# 16.3 LAB SPOILAGE OF PACKAGED AND COLD-STORED FOODSTUFFS

## 16.3.1 Fresh and Cooked Meat

Fresh meat, such as lean meat (i.e., beef, pork, lamb) or poultry, constitutes a very susceptible food matrix due to the high $a_w$, neutral pH, and high availability of nutrients such as amino acids, vitamins, carbohydrates, and nucleosides that facilitate microbial proliferation (Nychas et al., 2008). Likewise, cooked meats, such as pasteurized sausage, sliced ham, pâté, foie gras, and cured meat, are prone to spoilage due to the customer demand for more natural, artisanal, or low-in-additive formulation (Vasilopoulos et al., 2013). In general, the origin of the meat tissue and the implementation of technological processes to improve sensory traits, such as marination or moisture enhancement for fresh meat and the reduction of NaCl, nitrate/nitrite salts, or addition of flavorings for cooked meat, in combination with the packaging specifications, select for different LAB to thrive during storage (Björkroth, 2005; Buncic et al., 2014; Nieminen et al., 2012; Vasilopoulos et al., 2010). Microbial contamination of animal carcasses produced under satisfactory hygienic conditions should not exceed an average total 2–4 log CFU/g, depending on the type and processing, after packaging to remain acceptable until the end of shelf life. However, many cases of LAB spoilage occur early during shelf life, reaching very high counts of 8–9 log CFU/g under storage at 2 °C –4 °C. A great diversity of *Leuconostoc gasicomitatum* and *Leuconostoc gelidum*, *Leuconostoc carnosum*, and *Leuconostoc mesenteroides* are some of the most frequently detected spoilers with great abundance in high-$O_2$–containing MAP conditions. These LAB generate organic acids (e.g., acetic acid, lactic acid) and contribute to buttery off-flavors (i.e., diacetyl, acetoin), formation of slime, bulging of packages, and green and yellow discoloration (Diez et al., 2009; Hultman et al., 2020;

Jääskeläinen et al., 2013; Susiluoto et al., 2003; Vihavainen and Björkroth, 2009). *Leuc. gasicomitatum* is capable of heme-mediated respiration, contributing to faster outgrowth in meat under high $O_2$ MAP (Jääskeläinen et al., 2013; Johansson et al., 2011; Johansson et al., 2022). *Latilactobacillus sakei*, *Latilactobacillus curvatus*, *Latilactobacillus fuchuensis*, *Dellaglioa algida*, and *Paucilactobacillus oligofermentans* have been isolated from marinated or minced meat, poultry, pork, and sliced ham stored under vacuum or MAP. The alterations described mainly focus on severe acidification, ropy slime, and off-odors (Audenaert et al., 2010; Doulgeraki et al., 2010; Lyhs and Björkroth, 2008; Sakala et al., 2002; Samelis et al., 2000). *Carnobacterium divergens* and *Carnobacterium maltaromaticum* are mainly associated with vacuum-packed or low-$O_2$–MAP beef, poultry, and pork. Their spoilage manifests mainly through generation of different off-odors (Casaburi et al., 2011; Ercolini et al., 2010; Laursen et al., 2005; Rieder et al., 2012). *Lactococcus piscium* and to a lesser extent *Lactococcus lactis* have been found often in vacuum-packed or MAP meat (Jiang et al., 2010; Rahkila et al., 2012; Saraoui et al., 2016). Similarly, *Enterococcus* spp. have been isolated from meat, albeit not as predominant microbiota (Björkroth et al., 2005; Koort et al., 2004). The aforementioned LAB species are largely encountered in all types of cooked meat, with the addition of members belonging to genus *Weissella*. *Weissella cibaria*, *Weissella confusa*, and *Weissella viridescens* have contributed to blowing of packages and spoilage of delicatessen and minced meat, especially when stored under vacuum (Diez et al., 2009; Nieminen et al., 2011; Zhang et al., 2012).

Recent findings on the temporal transciptome profiles of common meat spoilers (i.e., *Leuc. gelidum*, *Lc. piscium*, and *Pauc. oligofermentans*) grown in co- and tri-cultures *in vitro* provided insights into the interspecies interactions among LAB (Andreevskaya et al., 2018; Duru et al., 2021). The different species employed distinct strategies to overcome the competition. The prevalent *Leuc. gelidum* upregulated carbohydrate catabolic pathways, pyruvate fermentation enzymes, and ribosomal proteins, whereas *Pauc. oligofermentans* overexpressed numerous putative adhesins. Overall, the LAB communities in meat can be diverse and are impacted by the member interactions apart from abiotic factors and meat type.

## 16.3.2 Fishery

This food category encompasses fresh fish, fillets, and seafood that have a very short shelf life due to autolytic enzymatic processes, causing rancidity and proteolysis and great susceptibility to aquatic microbiota and leading to aggressive spoilage that manifests through putrefaction, offensive off-odors, and overall sensorial downgrading (Françoise, 2010). *C. maltaromaticun* and *C. divergens* have been isolated or detected as dominant spoilers from fresh or lightly preserved fishery such as tuna, pangasius, shrimp, sea bream, red drum, rainbow trout, and salmon (Leisner et al., 2007; Parlapani et al., 2015; Silbande et al., 2018). In respect to fishery, these species are very resilient to low temperature, freezing–thawing cycles, and high-pressure processing and are favored by MAP (Laursen et al., 2006; Macé et al., 2012; Noseda et al., 2012). Their spoilage manifestations are associated with off-flavors, such as production of ammonia, alcohols, aldehydes, and ketones. In many cases, spoilage was recorded in processed fishery products or preserves where the use of brine, acetic acid, and marinades facilitated the outgrowth of LAB species. Marinated herring in acetic acid and salt was found spoiled by *Lat. sakei*, *Lat. curvatus*, *Lat. fuchuensis* (Lyhs and Björkroth, 2008), and *Companilactobacillus alimentarius* (Lyhs et al., 2001); broiled lampreys were found spoiled by *Lat. curvatus*, *Leuc. mesenteroides*, and *W. halotolerans* (Merivirta et al., 2005). In the case of an acetic-acid herring preserve, very strong slime was caused by *Leuc. gasicomitatum* and *Leuc. gelidum* (Lyhs et al., 2004). *Lactococcus piscium* and *Vagococcus* spp. have also been increasingly associated with spoilage cases of fishery (Macé et al., 2012; Parlapani et al., 2015); as well, many other LAB species have been isolated but not as dominant spoilers (Françoise, 2010). *Lactococcus piscium* is also considered a bioprotective culture in certain instances when found in fish and does not always contribute to quality downgrading (Saraoui et al., 2016).

## 16.3.3 Eggs and Dairy

For dairy products, the involvement of LAB as starter cultures has been extensively studied for production of fermented products such as yogurt, cheese, buttermilk, and so on (Boor et al., 2017). However, the presence of allochthonous, competitive LAB, despite the occurrence of starter cultures, has been documented in several occasions depicting the fast outgrowth of certain species. Heterofermentative species can result in high viscosity, acetic acid formation, or gas production in yogurt and buttermilk (Lund et al., 2000). In cheesemaking, spoilage due to LAB can occur when excess gas is produced, leading to formation of slits and cracks by $CO_2$ and off-flavors such as gamma-aminobutyric acid by decarboxylation of glutamic acid by lactobacilli and leuconostocs (Boor et al., 2017; Hutkins, 2006). *Paucilactobacillus wasatchensis* has caused both unwanted gas and biogenic amine formation in cheese (Berthoud et al., 2022).

A severe case of contamination was reported in an entire production batch of mozzarella that developed cracks and eyes due to outgrowth of heat-resistant *W. viridescens* (Pothakos et al., 2014a). Also, the use of dairy ingredients (pasteurized milk, milk fat, and cream) for the production of desserts such as custard cream and puddings can lead to a compromised shelf life, as they constitute an appropriate substrate for microbial growth. *Leuc. mesenteroides*, along with a plethora of other microorganisms, was involved in spoilage of custard cream deserts (Arakawa et al., 2008).

Eggs retailed as hard-boiled, pilled, and brined preparations packaged under vacuum developed very strong slime, blowing, and pungent acidity after a few days storage at refrigeration temperature. The causative agent isolated from spoiled samples was identified as *Leuc. gasicomitatum*, responsible for a production batch recall (Pothakos et al., 2014a).

## 16.3.4 Vegetable and Fruit Salads

Post-harvest studies evaluating the microbial diversity of crops bolster the occurrence of mesophilic bacteria, which represent a large portion of the microbiota ranging between 3 and 9 log CFU/g on fresh produce, while Gram-negative taxa are the most dominant (Barth et al., 2009; Di Cagno et al., 2013; Oliveira et al., 2010). Yeasts and filamentous fungi often constitute the majority on raw fruits due to the high sugar content and low pH (Doyle, 2007; Tournas, 2005). Members of LAB, such as genera belonging to family *Lactobacillaceae*, are also commonly found in lower populations around 3 log CFU/g (Abadias et al., 2008; Caponigro et al., 2010; Hutkins, 2006). Generally, dextran-producing strains of *Leuc. mesenteroides* and *Leuc. pseudomesenteroides* have been found in high sucrose niches, such as carrots and sugarcanes, delivering thick, ropy slime (Doyle, 2007). Fresh produce are greatly retailed as RTE salads for convenience purposes and thus undergo washing, mild disinfection, or submersion in acid solutions and slicing before packaging and storing at 4 °C –7 °C. Only a few LAB species have been detected at the end of shelf life of packaged and chill-stored salads. *Leuc. gelidum*, *Leuc. citreum*, and *Lc. piscium* were found at high populations in RTE lettuce in Norway while coexisting with *Pseudomonas* spp. (Rudi et al., 2002). Psychrotrohphic LAB members, namely *Leuc. gasicomitatum* and *Leuc. gelidum*, *Leuc. inhae*, and *Lc. piscium*, were the dominant microbiota at the end of shelf life in chill-stored and packaged mixed salads in a market screening in Belgium, and they also had a strong affinity to sweet bell peppers (Pothakos et al., 2014b). The same species were found as the prevailing microbiota in all the samples originating from one facility handling a wide range of fresh-cut produce and were responsible for severe acidification of the final products (Pothakos et al., 2014c). Apart from RTE salads, composite RTE meals that contain fresh produce were found spoiled by *Leuc. gelidum*, *Lactiplantibacillus* spp. causing acidification in a case study conducted in a processing environment with serious fluctuations in quality (Pothakos et al., 2015b). With regard to fruit, *Leuc. citreum* was isolated from packaged pineapple cubes, while *Leuc. mesenteroides* and *Leuc. gelidum* were found to cause spoilage in honeydew melon (Zhang et al., 2013a, 2013b). Overall, the same psychrotrophic LAB species (unable to grow at 30 °C) are frequently encountered in MAP, RTE salads but in some cases, their contribution to spoilage cannot be inferred (Rudi et al., 2002).

# 16.4 LAB SPOILAGE OF ALCOHOLIC OR ACIDIFIED, AMBIENT-STABLE PRODUCTS

This class of food products have an extended shelf life due to the multiple stressors combined in the hurdle technology implemented for their preservation. The different lines of defense against microbial spoilers encompass adjustment to a low $a_w$, low pH, high ethanol concentration, addition of different acidulants or organic acids (i.e., lactic acid, acetic acid, sorbate and benzoate salts), and occasionally heat treatment in the case of sauces and dressings.

## 16.4.1 Wine and Beer

In large part, the bacterial involvement during winemaking is mainly restricted to malolactic fermentation performed by *Oenococcus oeni* (Luo et al., 2012; Malfeito-Ferreira, 2014). However, other LAB belonging to genera *Lactobacillus* and *Pediococcus* may participate in the process, albeit in few cases, and can contribute to defects. *Levilactobacillus brevis*, *Lentilactobacillus hilgardii*, and *Pediococcus damnosus* are mostly associated with phenomena such as glucan production, glycerol and tartaric acid degradation, or sorbic acid reduction, causing spoilage manifestations like slime, bitterness, and off-flavors, respectively (Bartowsky, 2009).

Beer as an alcoholic beverage exhibits great microbiological stability. It constitutes an unfavorable niche for microorganisms to thrive due to the presence of ethanol (up to 10% m/m) and the addition of hops, which attribute the characteristic bitterness (iso-a-acids) (Sakamoto and Konings, 2003). Moreover, the low pH, the anaerobic environment, and the incidence of low levels of energy-yielding carbohydrates such as maltose and maltotriose deliver an efficient barrier to microbial growth (Geissler et al., 2016). However, genera *Lactobacillus* and *Pediococcus* have been recognized as the most hazardous spoilers for breweries, causing turbidity, ropy slime, off-odors (i.e., diacetyl, acetoin, hydrogen sulfide), and acidity. *Lev. brevis*, *Fructilactobacillus lindneri*, and *P. damnosus* are the most common LAB involved in beer spoilage due to their ability to withstand the antimicrobial effect of hops (Bergsveinson et al., 2016; Kern et al., 2014). It has also been reported that some LAB spoilers enter a viable but non-culturable (VBNC) state as a mechanism for survival under the harsh environmental conditions of beer (Liu et al., 2017a).

## 16.4.2 Fruit Juice and Soft Drinks

LAB species usually associated with spoilage of fruit juices have produced turbidity and excessive acidity, and these implications have involved lactobacilli, leuconostocs, and weissellas (Lund et al., 2000). *Schleiferilactobacillus perolens* was isolated from spoiled soft drinks for the first time as specific spoiling organism (Back et al., 1999).

## 16.4.3 Ketchup, Mayonnaise, and Dressings

Ketchup is a common acidified tomato sauce that, along with salad dressings, spreads, and emulsified dips such as mayonnaise, is widely consumed as a condiment or side dish (Manios et al., 2014). These preparations have reduced $a_w$ by addition of salt, oil, or sugars (e.g., up to 35% total sugars, such as sucrose, high fructose corn syrup, or fructose and glucose) and low pH that renders them stable for months at room temperature (Vermeulen et al., 2007). However, several reports of spoilage involve species of the genus *Lactobacillus*, including *Fructilactobacillus fructivorans*, in ketchup where severe acidification and pungent odor were evaluated (Bjorkroth and Korkeala, 1997a). Moreover, *Lac. plantarum*, *Lev. brevis*, *Lacticaseibacillus casei*, and *Lentilactobacillus buchneri* have occasionally been isolated from spoiled ketchup, vinaigrette, and dressings manifesting with acidification, bulging, and formation of white colonies on the surface of the sauce (Lund et al., 2000).

# 16.5 RECENT ADVANCEMENTS

The great progress in molecular diagnostics and omics technologies has facilitated the close monitoring of food ecosystems as well as the overview of microbial assemblages harbored in food processing environments (Ercolini, 2013; Stellato et al., 2016).

Implementing high-throughput 16S rRNA amplicon sequencing with respect to LAB spoilage allowed for a realistic view of which microbial groups are present without the bias of culture-dependent methods (i.e., media composition and incubation temperature requirements). In the case of packaged and refrigerated products, overlooked members belonging to genera *Leuconostoc*, *Lactobacillus*, *Latilactobacillus*, *Dellaglioa*, *Paucilactobacillus*, and *Lactococcus* unable to grow at 30 °C; *Carnobacterium* species exhibiting poor recovery on acetate-containing media; and *Vagococcus* species having poor recovery on known LAB media were investigated for food quality implications (De Filippis et al., 2018; Jääskeläinen et al., 2016; Nieminen et al., 2012; Parlapani et al., 2015; Pothakos et al., 2014c, 2015b). It has also been possible to dissect microbial communities during the shelf life of foodstuffs (Jääskeläinen et al., 2016) and differentiate sporadic spoilage cases from recurring patterns among production batches, even when the contaminations are very low (Säde et al., 2017). For wine and beer, the limit of VBNC state could be bypassed and all microbial members therefore identified (Liu et al., 2017a, 2017b). The limitation is that low discriminatory power that cannot allow for species identification. Metagenetic studies conducted in meat, fish, fresh produce, and beer have underpinned the key role LAB play in spoilage (Bokulich et al., 2015; Nieminen et al., 2016; Pothakos et al., 2014a).

Generally, the microbiome of a foodstuff is formed by the microbial population associated with the natural habitat of origin, as well as any contamination taking place in the manufacturing premises. Given that technological advancements have accelerated the industrialization of comestibles, food-processing facilities play a substantial role in the establishment of the microbial food consortium (Pothakos et al., 2015c). LAB are ubiquitous to raw material entering food-handling facilities and can be established as house microbiota, contaminating future production batches (Olaimat and Holley, 2012). However, the significant degree of automation in food manufacturing ensures few chances of cross-contamination for thermally treated and ambient-stable products when aseptic filling and pasteurization in the final containers is applied. On the other hand, minimally processed and fresh products remain more susceptible. The performance of source tracking and contamination monitoring studies in food-manufacturing facilities unravels the source of LAB. Artisanal breweries have been studied in order to identify the origin of microorganisms performing the fermentation process and also the points where contamination can occur (Bokulich et al., 2015). In wineries the presence of lactobacilli at the early stages of winemaking correlates with a higher possibility of *Lev. brevis* outgrowth and spoilage manifestations (Hong et al., 2016). A limited number of source tracking studies have evaluated the origin of contamination for spoilage-related microorganisms in RTE salads (Pothakos et al., 2014c, 2015b). The findings of these studies emphasize the adaptation of cold-acclimatized LAB communities to surfaces, equipment, or compartments of the manufacturing facility, wherefrom they contaminate new production batches through direct contact or by air mediation. To date, there have been a number of studies targeting contamination patterns in the case of meat spoilage (Björkroth and Korkeala, 1997b; De Filippis et al., 2013; Hultman et al., 2015; Koo et al., 2013). The premises where foodstuffs are handled provide all the prerequisites for microbial habitation of psychrotrophs since the temperature is consistently low in order to maintain the cold chain, the relative humidity is high, and exudates linger on food contact surfaces and equipment parts. In the case of inadequate sanitation of the production lines, utensils, or hard-to-reach sites, the survival and attachment of psychrotrophic communities can occur (Pothakos et al., 2014c, 2015b, 2015c). A review study emphasizes that among the 80% of the biosphere that has a temperature below 5 °C (encompassing oceans, permafrost regions, alpine areas), human-made habitats like refrigerators and climate-controlled facilities harbor large numbers of ubiquitous psychrotrophic organisms (De Maayer et al., 2014).

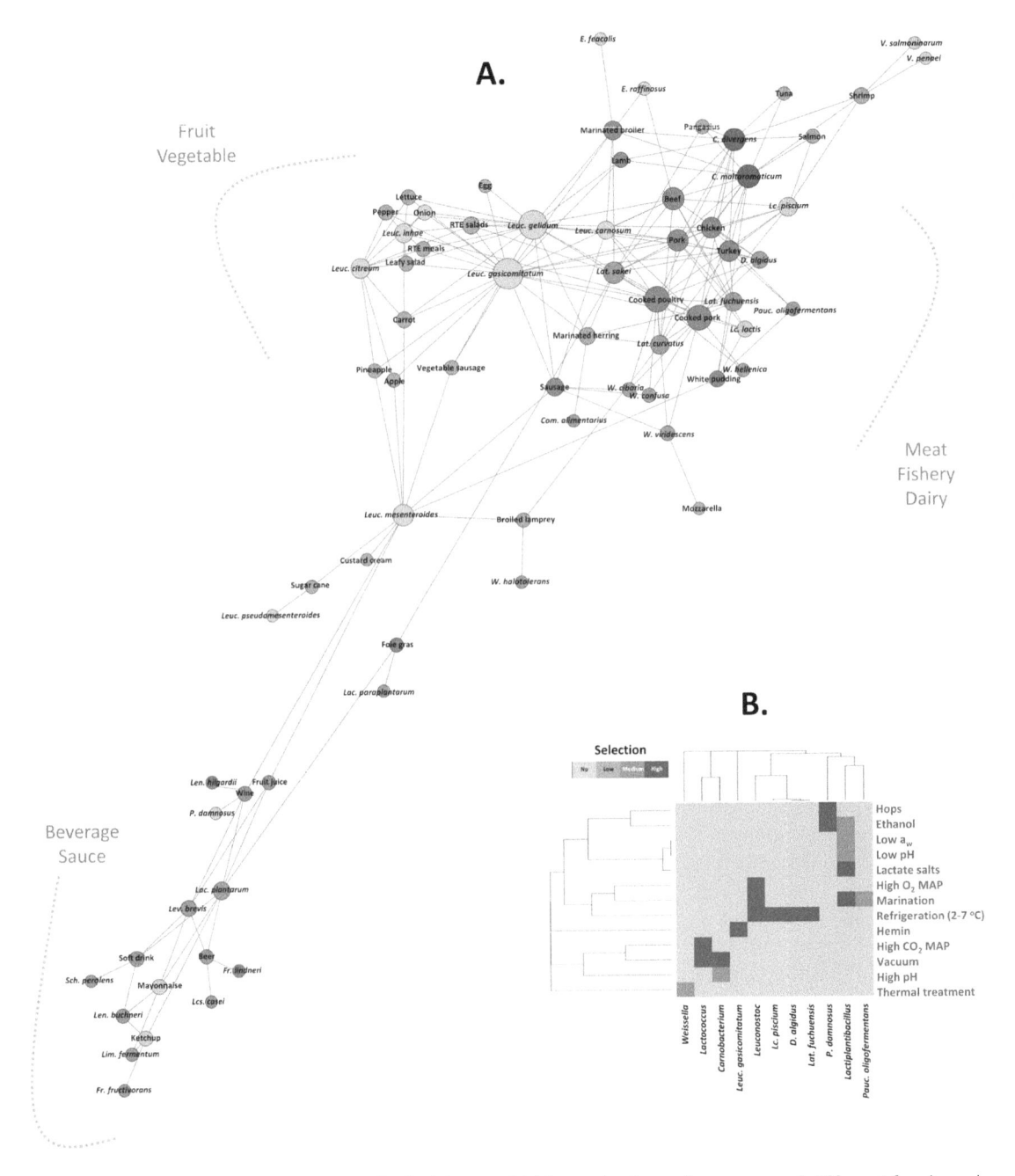

**FIGURE 16.1**   (a) Network presenting the incidence of LAB species in spoilage cases of different food products. All LAB species and foodstuffs are represented as nodes, and the occurrence of the respective species as spoilers in a food matrix is indicated by an edge. The nodes are proportionally sized to the number of edges directed to them, and the edges are color-coded based on the food type (red for meat, fishery, dairy; green for fruit and vegetable; blue for beverage and sauce). (b) Heatmap indicating the selection of certain LAB species or genera based on food constituents and implemented technologies.

# 16.6 CONCLUSION

Overall, LAB have been associated with all types of food products, and in Figure 16.1a all mentions have been summarized and presented in the form of a network. Clearly, there are distinct specificities as to which genera occur in certain types of products, but some species exhibit a remarkably large range of niches. Food matrices and conditions related to the hurdle technology applied can select the LAB to outcompete other microorganisms and become more prevalent (Figure 16.1b).

# BIBLIOGRAPHY

Abadias, M., Usall, J., Anguera, M., Solsona, C., Viñas, I., 2008. Microbiological quality of fresh, minimally-processed fruit and vegetables, and sprouts from retail establishments. *Int. J. Food Microbiol. 123*, 121–129.

Allende, A., Jacxsens, L., Devlieghere, F., Debevere, J., Artés, F., 2002. Effect of superatmospheric oxygen packaging on sensorial quality, spoilage, and *Listeria monocytogenes* and *Aeromonas caviae* growth in fresh processed mixed salads. *J. Food Prot. 65*, 1565–1573.

Andreevskaya, M., Jääskeläinen, E., Johansson, P., Ylinen, A., Paulin, L., Björkroth, J., Auvinen, P., 2018. Food-spoilage-associated *Leuconostoc, Lactococcus*, and *Lactobacillus* species display different survival strategies in response to competition. *Appl. Environ. Microbiol.* https://doi.org/10.1128/AEM.00554-18.

Arakawa, K., Kawai, Y., Iioka, H., Tanioka, M., Nishimura, J., Kitazawa, H., Tsurumi, K., Saito, T., 2008. Microbial community analysis of food-spoilage bacteria in commercial custard creams using culture-dependent and independent methods. *J. Dairy Sci. 91*, 2938–2946.

Audenaert, K., D'Haene, K., Messens, K., Ruyssen, T., Vandamme, P., Huys, G., 2010. Diversity of lactic acid bacteria from modified atmosphere packaged sliced cooked meat products at sell-by date assessed by PCR-denaturing gradient gel electrophoresis. *Food Microbiol. 27*, 12–18.

Axelsson, L., 2004. Lactic acid bacteria: Classification and physiology. In: S. Salminen, A. von Wright, A. Ouwehand (Eds.), Lactic Acid Bacteria. Microbiological and Functional Aspects (3rd ed.). New York: Marcel Dekker.

Back, W., Bohak, I., Ehrmann, M., Ludwig, W., Bruno, P., Kersters, K., Schleifer, K.H., 1999. *Lactobacillus perolens* sp. nov., a soft drink spoilage bacterium. *Syst. Appl. Microbiol. 22*, 354–359.

Barth, M., Hankinson, T.R., Zhuang, H., Breidt, F., 2009. Microbial spoilage of fruits and vegetables. In: W.H. Sperber, M.P. Doyle (Eds.), Compendium of the Microbiological Spoilage of Foods and Beverages. New York: Springer.

Bartowsky, E.J., 2009. Bacterial spoilage of wine and approaches to minimize it. *Lett. Appl. Microbiol. 48*, 149–156.

Bergsveinson, J., Friesen, V., Ziola, B., 2016. Transcriptome analysis of beer-spoiling *Lactobacillus brevis* BSO 464 during growth in degassed and gassed beer. *Int. J. Food Microbiol. 235*, 28–35.

Berthoud, H., Wechsler, D., Irmler, S., 2022. Production of putrescine and cadaverine by *Paucilactobacillus wasatchensis*. *Front. Microbiol. 13*. https://doi.org/10.3389/fmicb.2022.842403

Björkroth, J., 2005. Microbiological ecology of marinated meat products. *Meat Sci. 70*, 477–480.

Björkroth, J.K., Holzapfel, W., 2006. Genera Leuconostoc, Oenococcus and Weissella. In: M. Dworkin, S. Falkow, E. Rosenberg, K.-H. Schleifer, E. Stackebrandt (Eds.), The Prokaryotes. New York: Springer, pp. 267–319.

Björkroth, J.K., Korkeala, H.J., 1997a. *Lactobacillus fructivorans* spoilage of tomato ketchup. *J. Food Prot. 60*, 505–509.

Björkroth, J.K., Korkeala, H.J., 1997b. Use of rRNA gene restriction patterns to evaluate lactic acid bacterium contamination of vacuum-packaged sliced cooked whole-meat product in a meat processing plant. *Appl. Environ. Microbiol. 63*, 448–453.

Björkroth, J.K., Ristiniemi, M., Vandamme, P., Korkeala, H., 2005. *Enterococcus* species dominating in fresh modified-atmosphere-packaged, marinated broiler legs are overgrown by *Carnobacterium* and *Lactobacillus* species during storage at 6°C. *Int. J. Food Microbiol. 97*, 267–276.

Bokulich, N.A., Bergsveinson, J., Ziola, B., Mills, D.A., 2015. Mapping microbial ecosystems and spoilage-gene flow in breweries highlights patterns of contamination and resistance. *Elife 4*, 1–21.

Boor, K.J., Wiedmann, M., Murphy, S., Alcaine, S., 2017. A 100-year review: Microbiology and safety of milk handling. *J. Dairy Sci. 100*, 9933–9951.

Borch, E., Kant-Muermans, M.L., Blixt, Y., 1996. Bacterial spoilage of meat and cured meat products. *Int. J. Food Microbiol. 33*, 103–120.

Buncic, S., Nychas, G.-J., Lee, M.R.F., Koutsoumanis, K., Hébraud, M., Desvaux, M., Chorianopoulos, N., Bolton, D., Blagojevic, B., Antic, D., 2014. Microbial pathogen control in the beef chain: Recent research advances. *Meat Sci. 97*, 288–297.

Caponigro, V., Ventura, M., Chiancone, I., Amato, L., Parente, E., Piro, F., 2010. Variation of microbial load and visual quality of ready-to-eat salads by vegetable type, season, processor and retailer. *Food Microbiol. 27*, 1071–1077.

Carr, J.G., 1958. Lactic acid bacteria as spoilage organisms of fruit juice products. *J. Appl. Bact. 21*, 267–271.

Casaburi, A., Nasi, A., Ferrocino, I., Di Monaco, R., Mauriello, G., Villani, F., Ercolini, D., 2011. Spoilage-related activity of *Carnobacterium maltaromaticum* strains in air-stored and vacuum-packed meat. *Appl. Environ. Microbiol. 77*, 7382–7393.

Chaillou, S., Chaulot-Talmon, A., Caekebeke, H., Cardinal, M., Christieans, S., Denis, C., Hélène Desmonts, M., et al. 2015. Origin and ecological selection of core and food-specific bacterial communities associated with meat and seafood spoilage. *ISME 9*, 1105–1118.

Chaillou, S., Daty, M., Baraige, F., Dudez, A.-M., Anglade, P., Jones, R., Alpert, C.-A., Champomier-Vergès, M.-C., Zagorec, M., 2009. Intraspecies genomic diversity and natural population structure of the meat-borne lactic acid bacterium Lactobacillus sakei. *Appl. Environ. Microbiol. 75*, 970–980.

Champomier-Vergès, M.-C., Maguin, E., Mistou, M.-Y., Anglade, P., Chich, J.-F., 2002. Lactic acid bacteria and proteomics: Current knowledge and perspectives. *J. Chromatogr. B 771*, 329–342.

Conesa, A., Artés-Hernández, F., Geysen, S., Nicolaï, B., Artés, F., 2007. High oxygen combined with high carbon dioxide improves microbial and sensory quality of fresh-cut peppers. *Postharvest Biol. Technol. 43*, 230–237.

Cutter, C.N., 2002. Microbial control by packaging: A review. *Crit. Rev. Food Sci. Nutr. 42*, 151–161.

Dainty, R. H., Shaw, B. G., Harding, C. D. and Michanie, S., 1979. The spoilage of vacuum packaged beef by cold tolerant bacteria. In: A. D. Russell, R. Fuller (Eds.), Cold Tolerant Microbes in Spoilage and the Environment, SAB Technical Series, Vol. 13. London: Academic Press, pp. 83–100.

De Filippis, F., La Storia, A., Villani, F., Ercolini, D., 2013. Exploring the sources of bacterial spoilers in beefsteaks by culture-independent high-throughput sequencing. *PLoS ONE 8*, 1–10.

De Filippis, F., Parente, E., Ercolini, D., 2018. Recent past, present, and future of the food microbiome. *Annu. Rev. Food Sci. Technol. 25*, 589–608.

De Maayer, P., Anderson, D., Cary, C., Cowan, D.A., 2014. Some like it cold: Understanding the survival strategies of psychrophiles. *EMBO Rep. 15*, 508–517.

de Vos, W.M., 2011. Systems solutions by lactic acid bacteria: From paradigms to practice. *Microb. Cell Fact. 10*, 1–13.

Di Cagno, R., Coda, R., De Angelis, M., Gobbetti, M., 2013. Exploitation of vegetables and fruits through lactic acid fermentation. *Food Microbiol. 33*, 1–10.

Diez, A.M., Björkroth, J., Jaime, I., Rovira, J., 2009. Microbial, sensory and volatile changes during the anaerobic cold storage of morcilla de Burgos previously inoculated with *Weissella viridescens* and *Leuconostoc mesenteroides*. *Int. J. Food Microbiol. 131*, 168–177.

Douillard, F.P., de Vos, W.M., 2014. Functional genomics of lactic acid bacteria: From food to health. *Microb. Cell Fact. 13*, S8.

Doulgeraki, A.I., Paramithiotis, S., Kagkli, D.M., Nychas, G.-J.E., 2010. Lactic acid bacteria population dynamics during minced beef storage under aerobic or modified atmosphere packaging conditions. *Food Microbiol. 27*, 1028–1034.

Doyle, M.E., 2007. *Microbial Food Spoilage—Losses and Control Strategies*. Madison, WI: Food Research Institute, University of Wisconsin. https://fri.wisc.edu/files/Briefs_File/2017-07-18_0857_FRI_Brief_Microbial_Food_Spoilage_7_07.pdf

Duru, I.C., Ylinen, A., Belanov, S. Pulido, A.A., Paulin, L., Auvinen, P., 2021. Transcriptomic time-series analysis of cold- and heat-shock response in psychrotrophic lactic acid bacteria. *BMC Genom. 22*, 28.

Dykes, G.A., Cloete, T.E., von Holy, A., 1995. Taxonomy of lactic acid bacteria associated with vacuum-packaged processed meat spoilage by multivariate analysis of cellular fatty acids. *Int. J. Food Microbiol. 28*, 89–100.

Egan, A.F., 1983. Lactic acid bacteria of meat and meat products. *Antonie van Leeuwenhoek. 49*, 3, 327–336.

Ercolini, D., 2013. High-throughput sequencing and metagenomics: Moving forward in the culture-independent analysis of food microbial ecology. *Appl. Environ. Microbiol. 79*, 3148–3155.

Ercolini, D., Ferrocino, I., La Storia, A., Mauriello, G., Gigli, S., Masi, P., Villani, F., 2010. Development of spoilage microbiota in beef stored in nisin activated packaging. *Food Microbiol. 27*, 137–143.

Françoise, L., 2010. Occurrence and role of lactic acid bacteria in seafood products. *Food Microbiol. 27*, 698–709.

Geissler, A.J., Behr, J., von Kamp, K., Vogel, R.F., 2016. Metabolic strategies of beer spoilage lactic acid bacteria in beer. *Int. J. Food Microbiol. 216*, 60–68.

Hong, X., Chen, J., Liu, L., Wu, H., Tan, H., Xie, G., Xu, Q., Zou, H., Yu, W., Wang, L., Qin, N., 2016. Metagenomic sequencing reveals the relationship between microbiota composition and quality of Chinese Rice Wine. *Sci. Rep. 6*, 1–11.

Hultman, J., Johansson, P., Björkroth, J., 2020. Longitudinal metatranscriptomic analysis of a meat spoilage microbiome detects abundant continued fermentation and environmental stress responses during shelf life and beyond. *Appl. Environ. Microbiol.* 86, e01575–20.

Hultman, J., Rahkila, R., Ali, J., Rousu, J., Björkroth, K.J., 2015. Meat processing plant microbiome and contamination patterns of cold-tolerant bacteria causing food safety and spoilage risks in the manufacture of vacuum-packaged cooked sausages. *Appl. Environ. Microbiol.* 81, 7088–7097.

Hutkins, R.W., 2006. Microbiology and Technology of Fermented Foods. Ames, IA: Blackwell.

Jääskeläinen, E., Hultman, J., Parshintsev, J., Riekkola, M.L., Björkroth, J., 2016. Development of spoilage bacterial community and volatile compounds in chilled beef under vacuum or high oxygen atmospheres. *Int. J. Food Microbiol.* 223, 25–32.

Jääskeläinen, E., Johansson, P., Kostiainen, O., Nieminen, T., Schmidt, G., Somervuo, P., Mohsina, M., Vanninen, P., Auvinen, P., Björkroth, J., 2013. Significance of heme-based respiration in meat spoilage caused by *Leuconostoc gasicomitatum* . *Appl. Environ. Microbiol.* 79, 1078–1085.

Jacxsens, L., Devlieghere, F., Ragaert, P., Vanneste, E., Debevere, J., 2003. Relation between microbiological quality, metabolite production and sensory quality of equilibrium modified atmosphere packaged fresh-cut produce. *Int. J. Food Microbiol.* 83, 263–280.

Jiang, Y., Gao, F., Xu, X.L., Su, Y., Ye, K.P., Zhou, G.H., 2010. Changes in the bacterial communities of vacuum-packaged pork during chilled storage analyzed by PCR-DGGE. *Meat Sci.* 86, 889–895.

Johansson, P., Paulin, L., Säde, E., Salovuori, N., Alatalo, E.R., Björkroth, K.J., Auvinen, P., 2011. Genome sequence of a food spoilage lactic acid bacterium *Leuconostoc gasicomitatum* LMG 18811[T] in association with specific spoilage reactions. *Appl. Environ. Microbiol.* 77, 4344–4351.

Johansson, P., Säde, E., Hultman, J., Auvinen, P., Björkroth, K.J., 2022. Pangenome and genomic taxonomy analyses of Leuconostoc gelidum and Leuconostoc gasicomitatum. *BMC Genom.* 23, 818.

Kaufman, P.R., Handy, C.R., Mclaughlin, E.W., Green, G.M., 2000. Understanding the dynamics of produce markets—Consumption and consolidated growth. *USDA Economic Research Service, Agriculture Information Bulletin No. 758* . https://www.ers.usda.gov/publications/pub-details/?pubid=42295

Kern, C.C., Vogel, R.F., Behr, J., 2014. Differentiation of *Lactobacillus brevis* strains using matrix-assisted-laser-desorption-ionization-time-of-flight mass spectrometry with respect to their beer spoilage potential. *Food Microbiol.* 40, 18–24.

Koo, O.K., Mertz, A.W., Akins, E.L., Sirsat, S.A., Neal, J.A., Morawicki, R., Crandall, P.G., Ricke, S.C., 2013. Analysis of microbial diversity on deli slicers using polymerase chain reaction and denaturing gradient gel electrophoresis technologies. *Lett. Appl. Microbiol.* 56, 111–119.

Koort, J., Coenye, T., Vandamme, P., Sukura, A., Björkroth, J., 2004. *Enterococcus hermanniensis* sp. nov., from modified-atmosphere-packaged broiler meat and canine tonsils. *Int. J. Syst. Evol. Microbiol.* 54, 1823–1827.

Korkeala, H., Alanko, T., Mäkelä, P. and Lindroth, S. 1989. Shelf life of vacuum-packed cooked ring sausages at different chill temperatures. *Int. J. Food Microbiol.* 9, 237–247.

Korkeala, H. and Mäkelä, P. 1989. Characterization of lactic acid bacteria isolated from vacuum-packed cooked ring sausages. *Int. J. Food. Microbiol.* 9, 33–43.

Laursen, B.G., Bay, L., Cleenwerck, I., Vancanneyt, M., Swings, J., Dalgaard, P., Leisner, J.J., 2005. *Carnobacterium divergens* and *Carnobacterium maltaromaticum* as spoilers or protective cultures in meat and seafood: Phenotypic and genotypic characterization. *Syst. Appl. Microbiol.* 28, 151–164.

Laursen, B.G., Leisner, J.J., Dalgaard, P., 2006. *Carnobacterium* Species: Effect of metabolic activity and interaction with *Brochothrix thermosphacta* on sensory characteristics of modified atmosphere packed shrimp. *J. Agric. Food Chem.* 54, 3604–3611.

Leisner, J.J., Laursen, B.G., Prévost, H., Drider, D., Dalgaard, P., 2007. *Carnobacterium*: Positive and negative effects in the environment and in foods. *FEMS Microbiol. Rev. 31*, 592–613.

Liu, J., Li, L., Li, B., Peters, B.M., Deng, Y., Xu, Z., Shirtliff, M.E., 2017b. Study on spoilage capability and VBNC state formation and recovery of *Lactobacillus plantarum* . *Microb. Pathog. 110*, 257–261.

Liu, J., Li, L., Li, B., Peters, B.M., Xu, Z., Shirtliff, M.E., 2017a. First study on the formation and resuscitation of viable but nonculturable state and beer spoilage capability of *Lactobacillus lindneri* . *Microb. Pathog. 107*, 219–224.

Lorenzo, J.M., Gómez, M., 2012. Shelf life of fresh foal meat under MAP, overwrap and vacuum packaging conditions. *Meat Sci.* 92, 610–618.

Lund, B.M., Parker, T.C., Gould, G.W., 2000. The Microbiological Safety and Quality of Food. Gaithersburg, MD: Aspen Publishers.

Luo, H., Schmid, F., Grbin, P.R., Jiranek, V., 2012. Viability of common wine spoilage organisms after exposure to high power ultrasonics. *Ultrason. Sonochem. 19*, 415–420.

Lyhs, U., Björkroth, J.K., 2008. *Lactobacillus sakei/curvatus* is the prevailing lactic acid bacterium group in spoiled maatjes herring. *Food Microbiol.* 25, 529–533.

Lyhs, U., Korkeala, H., Vandamme, P., Björkroth, J., 2001. *Lactobacillus alimentarius*: A specific spoilage organism in marinated herring. *Int. J. Food Microbiol. 64*, 355–360.

Lyhs, U., Koort, J., Lundström, H.-S., Björkroth, K.J., 2004. *Leuconostoc gelidum* and *Leuconostoc gasicomitatum* strains dominated the lactic acid bacterium population associated with strong slime formation in an acetic-acid herring preserve. *Int. J. Food Microbiol. 90*, 207–218.

Macé, S., Cornet, J., Chevalier, F., Cardinal, M., Pilet, M.-F., Dousset, X., Joffraud, J.-J., 2012. Characterisation of the spoilage microbiota in raw salmon (*Salmo salar*) steaks stored under vacuum or modified atmosphere packaging combining conventional methods and PCR-TTGE. *Food Microbiol. 30*, 164–172.

Makarova, K., O'Sullivan, O., O'Callaghan, J., Sangrador-Vegas, A., McAuliffe, O., Slattery, L., Kaleta, P., Callanan, M., Fitzgerald, G.F., Ross, R.P., Beresford, T., 2006. Comparative genomics of lactic acid bacteria. *Proc. Natl. Acad. Sci. 103*, 15611–15616.

Malfeito-Ferreira, M., 2014. WINES—Wine Spoilage Yeasts and Bacteria, Encyclopedia of Food Microbiology (2nd ed.). Cambridge, MA: Elsevier.

Manios, S.G., Lambert, R.J.W., Skandamis, P.N., 2014. A generic model for spoilage of acidic emulsified foods: Combining physicochemical data, diversity and levels of specific spoilage organisms. *Int. J. Food Microbiol. 170*, 1–11.

McMillin, K., 2008. Where is MAP going? A review and future potential of modified atmosphere packaging for meat. *Meat Sci. 80*, 43–65.

Merivirta, L.O., Koort, J.M.K., Kivisaari, M., Korkeala, H., Björkroth, K.J., 2005. Developing microbial spoilage population in vacuum-packaged charcoal-broiled European river lamprey (*Lampetra fluviatilis*). *Int. J. Food Microbiol. 101*, 145–152.

Nieminen, T.T., Dalgaard, P., Björkroth, J., 2016. Volatile organic compounds and *Photobacterium phosphoreum* associated with spoilage of modified-atmosphere-packaged raw pork. *Int. J. Food Microbiol. 218*, 86–95.

Nieminen, T.T., Vihavainen, E., Paloranta, A., Lehto, J., Paulin, L., Auvinen, P., Solismaa, M., Björkroth, K.J., 2011. Characterization of psychrotrophic bacterial communities in modified atmosphere-packed meat with terminal restriction fragment length polymorphism. *Int. J. Food Microbiol. 144*, 360–366.

Nieminen, T.T., Välitalo, H., Säde, E., Paloranta, A., Koskinen, K., Björkroth, J., 2012. The effect of marination on lactic acid bacteria communities in raw broiler fillet strips. *Front. Microbiol. 3*, 1–13.

Noseda, B., Islam, M.T., Eriksson, M., Heyndrickx, M., De Reu, K., Van Langenhove, H., Devlieghere, F., 2012. Microbiological spoilage of vacuum and modified atmosphere packaged Vietnamese *Pangasius hypophthalmus* fillets. *Food Microbiol. 30*, 408–419.

Nychas, G., Skandamis, P., Tassou, C., Koutsoumanis, K., 2008. Meat spoilage during distribution. *Meat Sci. 78*, 77–89.

Olaimat, A.N., Holley, R.A., 2012. Factors influencing the microbial safety of fresh produce: A review. *Food Microbiol. 32*, 1–19.

Oliveira, M., Usall, J., Viñas, I., Anguera, M., Gatius, F., Abadias, M., 2010. Microbiological quality of fresh lettuce from organic and conventional production. *Food Microbiol. 27*, 679–684.

Parlapani, F.F., Kormas, K.A., Boziaris, I.S., 2015. Microbiological changes, shelf life and identification of initial and spoilage microbiota of sea bream fillets stored under various conditions using 16S rRNA gene analysis. *J. Sci. Food Agric. 95*, 2386–2394.

Pfeiler, E.A., Klaenhammer, T.R., 2007. The genomics of lactic acid bacteria. *Trends Microbiol. 15*, 546–553.

Pothakos, V., Aulia, Y.A., Van Der Linden, I., Uyttendaele, M., Devlieghere, F., 2015c. Exploring the strain-specific attachment of *Leuconostoc gelidum* subsp. *gasicomitatum* on food contact surfaces. *Int. J. Food Microbiol. 199*, 41–46.

Pothakos, V., Devlieghere, F., Villani, F., Björkroth, J., Ercolini, D., 2015a. Lactic acid bacteria and their controversial role in fresh meat spoilage. *Meat Sci. 109*, 66–74.

Pothakos, V., Samapundo, S., Devlieghere, F., 2012. Total mesophilic counts underestimate in many cases the contamination levels of psychrotrophic lactic acid bacteria (LAB) in chilled-stored food products at the end of their shelf-life. *Food Microbiol. 32*, 437–443.

Pothakos, V., Snauwaert, C., De Vos, P., Huys, G., Devlieghere, F., 2014b. Psychrotrophic members of *Leuconostoc gasicomitatum*, *Leuconostoc gelidum* and *Lactococcus piscium* dominate at the end of shelf-life in packaged and chilled-stored food products in Belgium. *Food Microbiol. 39*, 61–67.

Pothakos, V., Snauwaert, C., De Vos, P., Huys, G., Devlieghere, F., 2014c. Monitoring psychrotrophic lactic acid bacteria contamination in a ready-to-eat vegetable salad production environment. *Int. J. Food Microbiol. 185*, 7–16.

Pothakos, V., Stellato, G., Ercolini, D., Devlieghere, F., 2015b. Processing environment and ingredients are both sources of *Leuconostoc gelidum*, which emerges as a major spoiler in ready-to-eat meals. *Appl. Environ. Microbiol. 81*, 3529–3541.

Pothakos, V., Taminiau, B., Huys, G., Nezer, C., Daube, G., Devlieghere, F., 2014a. Psychrotrophic lactic acid bacteria associated with production batch recalls and sporadic cases of early spoilage in Belgium between 2010 and 2014. *Int. J. Food Microbiol. 191*, 157–163.

Rahkila, R., Nieminen, T., Johansson, P., Säde, E., Björkroth, J., 2012. Characterization and evaluation of the spoilage potential of *Lactococcus piscium* isolates from modified atmosphere packaged meat. *Int. J. Food Microbiol. 156*, 50–59.

Ramos, B., Miller, F. a., Brandão, T.R.S., Teixeira, P., Silva, C.L.M., 2013. Fresh fruits and vegetables—An overview on applied methodologies to improve its quality and safety. *Innov. Food Sci. Emerg. Technol. 20*, 1–15.

Reuter, G., 1985. Elective and selective media for lactic acid bacteria. *Int. J. Food Microbiol. 2*, 55–68.

Rieder, G., Krisch, L., Fischer, H., Kaufmann, M., Maringer, A., Wessler, S., 2012. *Carnobacterium divergens*—A dominating bacterium of pork meat juice. *FEMS Microbiol. Lett. 332*, 122–130.

Rudi, K., Flateland, S.L., Hanssen, J.F., Bengtsson, G., Nissen, H., 2002. Development and evaluation of a 16S ribosomal DNA array-based approach for describing complex microbial communities in ready-to-eat vegetable salads packed in a modified atmosphere. *Appl. Environ. Microbiol. 68*, 1146–1156.

Säde, E., Penttinen, K., Björkroth, J., Hultman, J., 2017. Exploring lot-to-lot variation in spoilage bacterial communities on commercial modified atmosphere packaged beef. *Food Microbiol. 62*, 147–152.

Sakala, R.M., Kato, Y., Hayashidani, H., Murakami, M., Kaneuchi, C., Ogawa, M., 2002. *Lactobacillus fuchuensis* sp. nov., isolated from vacuum-packaged refrigerated beef. *Int. J. Syst. Evol. Microbiol. 52*, 1151–1154.

Sakamoto, K., Konings, W.N., 2003. Beer spoilage bacteria and hop resistance. *Int. J. Food Microbiol. 89*, 105–124.

Samelis, J., Kakouri, A., Rementzis, J., 2000. The spoilage microflora of cured, cooked turkey breasts prepared commercially with or without smoking. *Int. J. Food Microbiol. 56*, 133–143.

Saraoui, T., Leroi, F., Björkroth, J., Pilet, M.F., 2016. *Lactococcus piscium*: A psychrotrophic lactic acid bacterium with bioprotective or spoilage activity in food-a review. *J. Appl. Microbiol. 121*, 907–918.

Silbande, A., Adenet, S., Chopin, C., Cornet, J., Smith-Ravin, J., Rochefort, K., Leroi, F., 2018. Effect of vacuum and modified atmosphere packaging on the microbiological, chemical and sensory properties of tropical red drum (*Sciaenops ocellatus*) fillets stored at 4°C. *Int. J. Food Microbiol. 266*, 31–41.

Stellato, G., La Storia, A., De Filippis, F., Borriello, G., Villani, F., Ercolini, D., 2016. Overlap of spoilage-associated microbiota between meat and the meat processing environment in small-scale and large-scale retail distributions. *Appl. Environ. Microbiol. 82*, 4045–4054.

Stiles, M.E., Holzapfel, W.H., 1997. Lactic acid bacteria of foods and their current taxonomy. *Int. J. Food Microbiol. 36*, 1–29.

Susiluoto, T., Korkeala, H., Björkroth, K.J., 2003. *Leuconostoc gasicomitatum* is the dominating lactic acid bacterium in retail modified-atmosphere-packaged marinated broiler meat strips on sell-by-day. *Int. J. Food Microbiol. 80*, 89–97.

Tamang, J.P., Watanabe, K., Holzapfel, W.H., 2016. Review: Diversity of microorganisms in global fermented foods and beverages. *Front. Microbiol. 7*, 1–28.

Teusink, B., Bachmann, H., Molenaar, D., 2011. Systems biology of lactic acid bacteria: A critical review. *Microb. Cell Fact. 10*, 1–17.

Tournas, V.H., 2005. Moulds and yeasts in fresh and minimally processed vegetables, and sprouts. *Int. J. Food Microbiol. 99*, 71–77.

Vanderroost, M., Ragaert, P., Devlieghere, F., De Meulenaer, B., 2014. Intelligent food packaging: The next generation. *Trends Food Sci. Technol. 39*, 47–62.

Vasilopoulos, C., De Maere, H., De Mey, E., Paelinck, H., De Vuyst, L., Leroy, F., 2010. Technology-induced selection towards the spoilage microbiota of artisan-type cooked ham packed under modified atmosphere. *Food Microbiol. 27*, 77–84.

Vasilopoulos, C., De Vuyst, L., Leroy, F., 2013. Shelf-life reduction as an emerging problem in cooked hams underlines the need for improved preservation techniques. *Crit. Rev. Food Sci. Nutr. 55*, 1425–1443.

Vermeulen, A., Devlieghere, F., Bernaerts, K., Van Impe, J., Debevere, J., 2007. Growth/no growth models describing the influence of pH, lactic and acetic acid on lactic acid bacteria developed to determine the stability of acidified sauces. *Int. J. Food Microbiol. 119*, 258–269.

Vihavainen, E.J., Björkroth, K.J., 2009. Diversity of *Leuconostoc gasicomitatum* associated with meat spoilage. *Int. J. Food Microbiol. 136*, 32–36.

Zhang, B.-Y., Samapundo, S., Pothakos, V., de Baenst, I., Sürengil, G., Noseda, B., Devlieghere, F., 2013a. Effect of atmospheres combining high oxygen and carbon dioxide levels on microbial spoilage and sensory quality of fresh-cut pineapple. *Postharvest Biol. Technol. 86*, 73–84.

Zhang, B.-Y., Samapundo, S., Pothakos, V., Sürengil, G., Devlieghere, F., 2013b. Effect of high oxygen and high carbon dioxide atmosphere packaging on the microbial spoilage and shelf-life of fresh-cut honeydew melon. *Int. J. Food Microbiol. 166*, 378–390.

Zhang, Q.Q., Han, Y.Q., Cao, J.X., Xu, X.L., Zhou, G.H., Zhang, W.Y., 2012. The spoilage of air-packaged broiler meat during storage at normal and fluctuating storage temperatures. *Poult. Sci. 91*, 208–214.

# Examples of Lactic-Fermented Foods of the African Continent

# 17

Kolawole Banwo, Oluwafemi A. Adebo, and Titilayo D. O. Falade

## 17.1 INTRODUCTION

African fermented foods are important for nutrition, health, and economy. The practice of fermenting foods in Africa is deep rooted in culture, dating back centuries in many indigenous communities in the continent, done mostly for preservation that leads to shelf-life extension as well as product development. Other notable benefits of these derived fermented foods are gut immune modulation, sensory enhancement, nutrients improvement and bioavailability, enhanced digestibility, and the delivery of probiotics (Vieira-Dalodé et al., 2007; Anukam and Reid, 2009; Banwo et al., 2020). Lactic acid bacteria (LAB) and yeasts are the predominant groups of microorganisms in African foods. The proliferation of yeasts is encouraged by the acidic natural environment produced by LAB, and the development of bacteria is enhanced by the presence of yeasts, which supply growth factors, such as vitamins and soluble nitrogen compounds. Favorable conditions of humidity and temperature activate fermenting microorganisms, allowing the best adapted ones to dominate the microbiota (Vieira-Dalodé et al., 2007). LAB are non-spore forming, non-motile, Gram-positive, acid-tolerant microorganisms, which are mainly microaerophilic, and their growth is reliant on obtainable sugars as well as other metabolic requirements to make end products such as lactic acid, ethanol, and $CO_2$ (Oliveira et al., 2014).

There are different forms of fermentation, including alkaline, acetic acid, alcoholic, and amino acid/peptide, as well as lactic acid fermentation (Anukam and Reid, 2009). Perhaps due to LAB and yeasts ultimately dominating most African fermented food products that are spontaneously produced, lactic acid fermentation seems to be the most common form of fermentation in Africa. Such fermentation processes are mainly carried out by LAB that belong to the *Lactococcus* (*Lact.*), *Streptococcus* (*S.*), *Leuconostoc* (*Leuc.*), *Lactobacillus* (*Lb.*), *Lactiplantibacillus* (*Lp.*), *Limosilactobacillus* (*Lm.*), *Levilactobacillus* (*Lv.*), *Lacticaseibacillus* (*Lc.*), *Latilactobacillus* (*Lt.*), *Loigolactobacillus* (*Lg.*), *Lentilactobacillus* (*Ln.*), *Ligilactobacillus* (*Lig.*), *Furfurilactobacillus* (*Fur*), *Schleiferilactobacillus* (*Sch.*) (*sensu lato*), and *Pediococcus* (*Ped.*) species, while associated yeast species are *Saccharomyces* (*Sacch.*) and *Candida* (*Can.*) (Gänzle, 2015; Liptakova et al., 2017; Zheng et al., 2020). Since lactic acid is the main by-product of fermentation, the pH of the resultant products is usually in the acidic range. LAB that dominate traditional fermentation systems in Africa also produce antimicrobial properties, hydrogen peroxide, and bacteriocins (small proteinaceous ribosomally generated antimicrobial substances) that inhibit the proliferation of pathogenic microorganisms, thus contributing to food safety.

DOI: 10.1201/9781003352075-19

# 17.1.1  Overview of Fermented Food Products in Northern and Southern Africa

According to Benkerroum (2012), North African countries have an ancient tradition in food technology, where the art of fermentation and preservation of traditional foods has been passed down from one generation to another. Likewise, in the southern part of the African continent, these fermented food products have significant cultural roles, as they are used in ceremonies including weddings and funerals, as well as other traditional rituals. They are also known to have health benefits, including helping to reduce childhood diseases such as diarrhea and malnutrition (Chelule et al., 2015). Fermented food products in these regions exists in numerous forms, mostly cereal based, but some are also produced from dairy and, much less often, roots and tubers in northern and southern Africa (Table 17.1). Notable among these products are *mahewu*, *umqombothi*, and *mabisi* (Figure 17.1).

**TABLE 17.1**  Some Notable Fermented Food Products from Northern and Southern Africa

| PRODUCT | SUBSTRATE(S) USED | PRODUCT FORM | COUNTRY/REGION |
|---|---|---|---|
| **Cereal based** | | | |
| *Bogbe* | Sorghum | Porridge | Botswana |
| *Bouza* | Wheat | Gruel | Egypt |
| *Busa* | Rice or millet | Beverage | Egypt |
| *Chibuku* | Sorghum | Beverage | Zimbabwe |
| *Chikokivana* | Maize or millet | Beverage | Zimbabwe |
| *Hopose* | Wheat (hops) | Beverage | Lesotho |
| *Kachasu* | Maize | Beverage | Zimbabwe |
| *Kishk* | Wheat, oats | Soup | Egypt |
| *Mahewu* | Maize | Gruel | Southern Africa |
| *Mangisi* | Millet | Beverage | Zimbabwe |
| *Masvusvu* | Millet | Beverage | Zimbabwe |
| *Motoho* | Sorghum | Porridge | Lesotho |
| *Munkoyo* | Maize | Beverage | Zambia, Democratic Republic of Congo |
| *Mutwiwa* | Maize | Porridge | Zimbabwe |
| *Oshikundu* | Sorghum or maize | Beverage | Namibia |
| *Sekumukumu* | Sorghum | Beverage | Lesotho |
| *Ting* | Sorghum, maize | Porridge, sourdough | Southern Africa |
| *Umqombothi* | Maize, sorghum | Beverage | Southern Africa |
| **Dairy based** | | | |
| *Amasi* | Cow milk | Gruel | Southern Africa |
| *Kefir* | Cow milk | Beverage | South Africa |
| *Laban rayeb* | | Gruel | Egypt |
| *Leben/lben* | Cow, ewe, goat, camel milk | Sour milk | Algeria, Northern Africa |
| *Mabisi* | Cow milk | Gruel | Zambia |
| *Madila* | Cow milk | Beverage, curd | Botswana |
| *Masse* | Cow milk | Beverage | Mozambique |
| *Zabady* | Buffalo milk (possibly with cow milk) | Beverage | Egypt |

(Modified from Adebo et al., 2023)

**FIGURE 17.1**   Examples of northern and southern African lactic fermented foods and beverages. (A) Baked kisra sheets (Suleiman et al., 2022); (B) umqombothi served in a cup (Hlangwani et al., 2020); (C) mabisi in a cup and plastic bottle (Moonga et al., 2022); (D) emasi (also called amasi) in a serving dish (Simatende et al., 2015).

## 17.1.2  Lactic Acid Bacteria in Fermented Food Products from Northern and Southern Africa

Over the years, several studies have investigated the microbiota of fermented food products from northern and southern Africa with the aim of understanding the microbial population as well as the possibility of isolating, characterizing, and using such microorganisms as starter cultures for subsequent fermentation processes. The prevalence of LAB has thus been reported in fermented products from these regions of Africa (Table 17.2), with the microbiota dependent on a few factors such as climatic and environmental conditions, the source of raw material and its composition, period of fermentation used, and preparation methods.

Microbiota of most African fermented dairy products are thus dominated with LAB (Table 17.2), though often associated with other pathogenic microorganisms (which might have originated from the raw milk), as well as yeasts (Agyei et al., 2020). *Amasi* is a fermented milk usually made from fresh milk commonly consumed in southern African countries such as Zimbabwe, Lesotho, and South Africa. It is usually prepared at the household level and has now gained commercial acceptability because of the nutritional and health benefits it confers to customers (Maleke et al., 2022). Similar to *amasi* are *mabisi* and *madila*, all commonly consumed in southern Africa. From the northern part are dairy products such as *masse* and *zabadi* as well as *leben/lben*. Notable LAB species identified from dairy products in northern and southern Africa include the *Lactobacillaceae* group (*Lb. acidophilus*, *Lv. brevis*, *Lb. delbrueckii*, *Lm. fermentum*, *Lb. helveticus*, *Lb. johnsonii*, *Lb. kefiranofaciens*, *Lc. paracasei*, *Lp. plantarum*, *Lc. rhamnosus*), *Lactococcus* spp. (*Lact. garvieae*, *Lact. lactis*, *Lact. raffinolactis*), *Leuconostoc* spp. (*Leuc. citreum*, *Leuc. lactis*, *Leuc. garlicum*, *Leuc. mesenteroides*, and *Leuc. pseudomesenteroides*) but much less *Pediococcus* spp. (*Ped. pentosaceus*) and *Streptococcus* (*S. equinus*, *S. thermophilus*).

Both southern and northern Africa are home to several notable African fermented cereal products. Notable among these are *bouza* (fermented alcoholic beverage from wheat, commonly consumed in Egypt), *kishk* (mixture of fermented cereal and milk common in Egypt), *mahewu*, *mageu*, *amahewu* (fermented maize porridge of South Africa), and *umqombothi* (an alcoholic beverage made from sorghum in southern Africa). *Lactobacillus* species are much more dominant LAB in cereal-fermented products from northern and southern Africa, with reported LAB including *Lb. amylovorus*, *Lv. brevis*, *Lb. bulgaricus*, *Lc. casei*, *Lb. confuses*, *Lg. coryniformis*, *Lt. curvatus*, *Lb. delbrueckii*, *Lm. fermentum*, *Sch. harbinensis*, *Lm. mucosae*, *Ln. parabuchneri*, *Lp. plantarum*, *Lm. reuteri*, and *Lc. rhamnosus*, though the presence of *Leuconostoc* spp. and *Pediococcus* spp. has been reported in *mahewu*.

**TABLE 17.2**   Lactic Acid Bacteria Identified in Some Fermented Food Products from Northern and Southern Africa

| PRODUCT | LAB IDENTIFIED | REFERENCES |
|---|---|---|
| Amasi | Lb. delbrueckii subsp. lactis, Lp. plantarum, Lact. lactis subsp. lactis, Leuc. mesenteroides subsp. dextranicum, Lactobacillaceae | Maleke et al. (2022) |
| Gariss | Lm. fermentum, Lb. helveticus | Abdelgadir et al. (2008); Jans et al. (2017) |
| Kefir | Lb. delbrueckii subsp. delbrueckii/lactis, Lm. fermentum, Lact. lactis subsp. lactis, Leuc. lactis, Leuc. mesenteroides subsp. mesenteroides/dextranicum | Witthuhn et al. (2004) |
| Leben/lben | Lv. brevis, Lc. paracasei, Lp. plantarum, Lc. rhamnosus, Lc. garvieae, Lact. lactis, Leuc. citreum, Leuc. mesenteroides, Leuc. pseudomesenteroides, Ped. pentosaceus, Streptococcus thermophilus | Ouadghiri et al. (2009) |
| Mabisi | Lv. brevis, Lb. kefiranofaciens, Lp. plantarum, Lact. lactis, Leuc. garlicum, Leuc. pseudomesenteroides, S. equinus, S. thermophilus | Schoustra et al. (2013) |
| Madila | Lb. acidophilus, Lv. brevis, Lb. delbrueckii, Lm. fermentum, Lp. plantarum, Lact. lactis | Ohenhen et al. (2013) |
| Mahewu | Lv. brevis, Lb. bulgaricus, Lb. delbrueckii, Lactobacillus spp., Leuconostoc spp., Pediococcus spp. | Kayitesi et al. (2017), Daji et al. (2022) |
| Masse | Lact. lactis subsp. lactis, Leuc. lactis Leuc. pseudomesenteroides, Leuc. garlicum | Agyei et al. (2020) |
| Merissa | Lactic acid bacteria | Dirar (1992) |
| Ting | Lc. casei, Lg. coryniformis, Lt. curvatus, Lm. fermentum, Sch. harbinensis, Ln. parabuchneri, Lp. plantarum, Lm. reuteri, Lc. rhamnosus | Madoroba et al. (2011); Sekwati-Monang and Gänzle (2011); Adebo et al. (2022) |
| Umqombothi | Lp. plantarum, lactobacilli | Hlangwani et al. (2020) |
| Zabadi | Lb. delbrueckii subsp. bulgaricus, Lb. johnsonii, Lact. garvieae, Lact. raffinolactis, Lact. lactis, S. thermophilus, Leuc. citreum | El-Baradei et al. (2008) |

# 17.1.3  Role of LAB in Fermented Food Products from Northern and Southern Africa

As posited by Gänzle (2015), LAB tend to dominate in most fermented food products through inhibition of other microorganisms, exploitative competition, rapid utilization of available sugars (carbohydrates), and accumulation of organic acids. This explains the significant role of LAB in reducing the pH of fermented food products in northern and southern Africa, subsequently contributing to extended shelf life through its preservative effect. Dirar (1992) reported the preservative effect of LAB during the fermentation of milk in Sudan. In contrast, when LAB fermentation is suppressed (in very hot climatic conditions), growth of pathogenic microorganisms such as *Aerobacter aerogenes* and *Klebsiella pneumoniae* tends to develop (Dirar, 1992). As described by previous studies, the proteolytic ability of LAB enables them to degrade proteins and peptides, subsequently producing metabolites that impact antimicrobial activity and the observed preservative effect (Gänzle, 2015). Accordingly, many strains of LAB have so-called generally recognized as safe (GRAS) status in the US, and many LAB species have qualified presumption of safety (QPS) status in the EU; thus their subsequent use as starter cultures in several food products can be assumed safe.

During the fermentation of these products, LAB fermentation has also been known to improve the nutritional composition of fermented foods by reducing levels of antinutrients known to inhibit nutrient bioavailability. Furthermore, LAB are also known producers of B-group vitamins and vitamin K2 and can serve as a natural and economic way in addition to other known techniques of food fortification. Further, the proteolytic capacities of LAB during fermentation can also release free amino acids and degrade storage protein to increase available amino acid contents and protein levels (Pessione and Cirrincione, 2016; Adebo et al., 2022). LAB have also been implicated in detoxification of foods such as pesticides, mycotoxins, and cyanogenic glucosides in roots and tubers (Adebiyi et al., 2019; Chiocchetti et al., 2019).

## 17.1.4 Overview of Lactic Fermented Foods from Eastern and Western Africa

In eastern and western Africa, traditional fermented foods are produced at the artisanal level for subsistence or informal markets. There are starchy/tuber, cereal, dairy, alcoholic, and non-alcoholic fermented products (Table 17.3). These include cassava-based products such as *gari, fufu, lafun, attiéké, konte*, and *agbelima* (stiff puddings); cereal-based such as *ogi, kunu, koko, agidi, kenkey*, and *uji* (gruels, non-alcoholic drinks, and stiff puddings); dairy-based such as *warakanshi* and *nunu* (cheese and yoghurts); and alcoholic beverages such as palm wine and local beers. These foods and drinks are consumed for nutrition during social events, ceremonies, and celebrations. Examples of notable lactic fermented foods from eastern and western Africa are shown in Figure 17.2.

## 17.1.5 Lactic Acid–Fermented Food Products from Eastern and Western Africa

Several investigations have been conducted to study the microorganisms associated with the fermentations of these foods. The microorganisms involved in the fermentation of traditional foods of eastern and western Africa are shown in Table 17.4.

Root and tuber fermented products are prevalent in the eastern and western African regions. Cassava is common because of its abundance, versatility, and ability to be cultivated in multiple soil types. Fermentation of cassava products enhances the detoxification of antinutritional compounds, thus improving the safety and qualities of the products (Kostinek et al., 2005; Banwo et al., 2020). Lactic acid fermentation in cassava processing is predominantly by LABs, which include homofermentative rods and cocci, facultative heterofermentative rods, heterofermentative rods, and cocci, especially *Lp. plantarum, Lm. fermentum, Lb. delbrueckii, Lv. brevis*, and *Leuconostoc* species (Kostinek et al., 2005; Banwo et al., 2012).

Fermented western African cassava products include *gari, fufu, lafun*, and *agbelima/kokonte*. *Lafun* is a fermented and dried cassava pulp related to *cossettes* in Rwanda and the Democratic Republic of Congo, *mapanga* and *kanyanga* in Malawi, and *makopa* in Tanzania. The dominant LAB in *lafun* and *fufu* have been identified as *Lp. plantarum, Lm. fermentum*, and *Weissella confusa* and a vast spectrum of yeasts (*Pichia scutulata, Hanseniaspora guilliermondii, Saccharomyces cerevisiae, Kluyveromyces marxianus, Pichia rhodanensis, Candida glabrata*). *Gari* and *agbelima* are fermented, toasted grits and the most-consumed food in western and eastern Africa. They have been reported to contain a complex mix of LAB and yeasts such as *Lp. plantarum, Lv. brevis, Leuc. fallax, Weiss. paramesenteroides, Penicillium* spp., *Can. tropicalis, Can. krusei, Zygosaccharomyces* spp., *Streptococcus lactics, Geotrichum candidum*, and *Corynebacterium manihot* (Padonou et al., 2010).

Eastern African products include *kivunde* and *cingwada*. *Kivunde* is a common flour produced from cassava in Tanzania, used for preparing gruels/porridges. It serves as a main source of energy in the eastern region of sub-Saharan Africa. Soft pieces of cassava following the first fermentation are sun-dried after they are washed, kneaded and molded to round shapes/balls. The dried *kivunde* balls can be preserved for up to 2 years, to be later milled into flour powder before boiling into porridge. *Lb. plantarum* is the dominant LAB obtained from *kivunde*. Other cassava products such as *cingwada* and *kokonte* are

**TABLE 17.3**   Some Notable Fermented Food Products from Eastern and Western Africa

| PRODUCT | SUBSTRATE(S) USED | PRODUCT FORM | COUNTRY/REGION |
|---|---|---|---|
| **Cereal-based** | | | |
| *Ogi/akamu* | Maize | Gruel | Nigeria, Benin |
| *Agidi/eko* | Maize | Stiff porridge | Nigeria |
| *Ogi-baba* | Sorghum | Gruel | Nigeria |
| *Koko* | Millet or maize | Gruel | Nigeria, Ghana |
| *Kenkey* | Maize, millet, or sorghum | Stiff dough | Ghana |
| *Banku* | Maize | Stiff dough | Ghana |
| *Aceda* | Sorghum | Porridge | Sudan |
| *Hussuwa* | Sorghum or millet | Porridge | Sudan |
| *Uji* | Maize, millet, or sorghum | Stiff dough | Kenya |
| *Mawè* | Maize | Stiff dough | Benin |
| *Kisra* | Sorghum, pearl millet | Flat bread, sourdough | Sudan |
| *Injera* | Sorghum | Flat bread, sourdough | Ethiopia |
| *Dègè* | Millet | Beverage | Senegal and Côte d'Ivoire |
| *Gowe* | Sorghum or maize | Beverage | Benin |
| *Merissa* | Sorghum and millet | Beverage | Sudan |
| *Burukutu* | Sorghum | Beverage | Nigeria |
| *Kunun-zaki* | Millet | Beverage | Nigeria |
| *Dolo* | Sorghum | Beverage | Burkina Faso |
| *Nasha* | Sorghum | Beverage | Sudan |
| *Obushera* | Millet | Beverage | Uganda |
| *Pito* | Sorghum | Beverage | Benin and Nigeria |
| *Tchapalo* | Millet | Beverage | Côte d'Ivoire |
| *Ikigage* | Sorghum | Beverage | Rwanda |
| *Togwa* | Maize and finger millet | Beverage | East Africa |
| **Dairy-based** | | | |
| *Nunu/nono* | Cow milk | Beverage | Nigeria and Benin |
| *Kule naoto* | Cow milk | Beverage | Kenya |
| *Gariss* | Camel milk | Beverage | Sudan |
| *Gibna bayda/gibna mudaffra* | Cow milk | Cheese | Sudan |
| *Ergo* | Cow milk | Beverage | Ethiopia |
| *Mursik* | Cow milk | Beverage | Kenya |
| *Amabere amaruranu* | Cow milk | Beverage | Kenya |
| *Sussac* | Camel milk | Beverage | Kenya and Somalia |
| *Maziwa lala* | Cow milk | Beverage | Kenya |
| *Rob/roub* | Cow, sheep, and goat milk | Beverage | Sudan |
| *Warakanshi* | Cow milk | Soft cheese | Nigeria and Ghana |
| **Tuber-based** | | | |
| *Gari* | Cassava | Stiff dough | West Africa |
| *Lafun/fufu* | Cassava | Stiff dough | Nigeria |
| *Kivunde* | Cassava | Stiff dough | Tanzania |
| *Cingwada* | Cassava | Stiff dough | East Africa |
| *Agbelima* | Cassava | Stiff dough | Ghana |
| *Kokonte* | Cassava | Stiff dough | Ethiopia |
| *Cossettes* | Cassava | Dough | Rwanda |
| *Mapanga/kanyanga* | Cassava | Dough | Malawi |
| *Makopa* | Cassava | Dough | Tanzania |

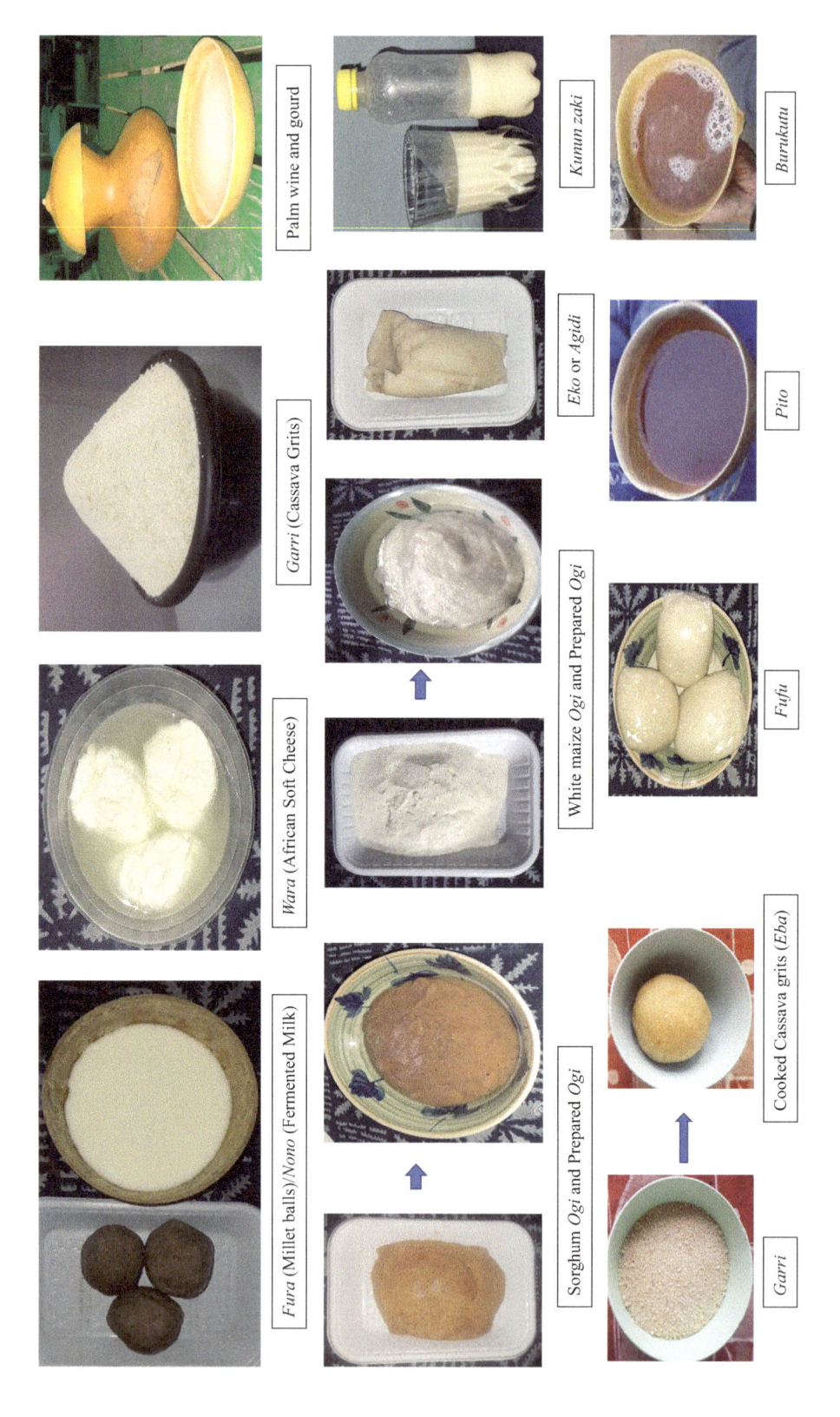

**FIGURE 17.2** Examples of eastern and western African lactic fermented foods and beverages.

**TABLE 17.4** LAB and Yeasts Identified in Some Fermented Food Products from Eastern and Western Africa

| PRODUCT | IDENTIFIED LAB AND YEASTS | REFERENCES |
|---|---|---|
| Ogi/akamu/agidi/ ogi-baba/koko | Lm. fermentum, Lb. helveticus Lb. acidophilus, Lv. brevis, Fur. rossiae, Lact. lactis subsp. cremoris, Weiss. paramesenteroides, Sacch. cerevisiae, Sacch. paradoxus, Rhodotorula graminis, Can. krusei, Can. tropicalis, Clavispora lusitaniae, Geotrichum candidum, and Geo. fermentans | Banwo et al. (2012); Oguntoyinbo and Narbad (2012); Greppi et al. (2013); Sanni et al. (2013); Okeke et al. (2015); Afolayan et al. (2017); Ogunremi et al. (2015) |
| Kenkey | Lp. plantarum, Ped. pentosaceus, Lm. fermentum, Lm. reuteri, Lv. brevis, Sacch. cerevisiae, and Can. krusei | Vieira-Dalodé et al. (2007); Adesulu-Dahunsi et al. (2018a, 2022); Banwo et al. (2020) |
| Hussuwa | Lm. fermentum, Ped. acidilactici, Ped. pentosaceus, and Sacch. cerevisiae | Yousif et al. (2010) |
| Gowe | Lm. fermentum, Lm. mucosae, Ped. acidilactici, Ped. pentosaceus, Weiss. kimchi, Weiss. confusa, Kluyveromyces marxianus, Pichia anomala, Can. krusei, and Can. tropicalis | Vieira-Dalodé et al. (2007) |
| Kunun-zaki | Streptococcus lutetiensis, Lm. fermentum, Lp. plantarum, Lb. delbrueckii subsp. bulgaricus, Ped. pentosaceus, Weiss. confusa, Sacch. cerevisiae, Ped. kudriavzevii, and Pichia kluyveri | Oguntoyinbo and Narbad, (2012); Ogunremi et al. (2015) |
| Kisra | Lb. amylovorus, Lb. brevis, Lb. fermentum, Lb. reuteri, Sacch. cerevisiae, Can. intermedia, and Debaryomyces hansenii | Hamad et al. (1997); Ali and Mustafa (2009) |
| Kivunde | LAB, especially Lb. plantarum | Kimaryo et al. (2000) |
| Gari/lafun/fufu | Lb. plantarum, Lb. brevis, Leuconostoc fallax, Weiss. paramesenteroides, Penicillium spp., Can. tropicalis, Can. krusei, Zygosaccharomyces spp., Pichia scutulata, Hanseniaspora guilliermondii, Sacch. cerevisiae, Kluyveromyces marxianus, Pichia rhodanensis, and Can. glabrata | Kostinek et al. (2005); Banwo et al. (2012); Banwo et al. (2020) |
| Nunu/nono | Lm. fermentum, Lb. helveticus, Lp. plantarum, Leuc. mesenteroides, Lactococcus spp., Ent. faecium, Ent. italicus, Weiss. confusa, Pichia kudriavzevii, Sacch. cerevisiae, Can. parapsilosis, Can. rugosa, Can. tropicalis, and Galactomyces geotrichum | Akabanda et al. (2013); Banwo et al. (2013a, 2013b); Adegboye et al. (2014) |
| Kule naoto | Lp. plantarum, Lm. fermentum, Lc. casei, Lb. acidophilus, Lc. rhamnosus, Ent faecium, Leuc. mesenteroides, Lactococcus spp., and Sacch. species | Mathara et al. (2008); Nduko et al. (2017); Jans et al. (2017) |
| Ergo | Leuc. mesenteroides, Lact. lactis subsp. cremoris, Lact. lactis subsp. lactis, Leuc. cremoris, S. thermophilus, Lb. delbrueckii, Lb. homi, and Micrococcus spp. | Jans et al. (2017) |
| Mursik | Lb. kefiri, Lc. casei, Lc. paracasei, Lc. rhamnosus Bacillus spp., Can. krusei, Can. spaerica, and Can. kefyr, Sacch. fermenti | Nduko et al. (2017); Jans et al. (2017); Banwo et al. (2020) |
| Amabere amaruranu | Lp. plantarum, Leuc. mesenteroides, S. thermophilus, Sacch. cerevisiae, Trichosporon mucoides, Can. famata, Can. albicans, Lb. bulgaricus, and Lb. helveticus | Jans et al. (2017) |

*(Continued)*

**TABLE 17.4**   (Continued) LAB and Yeasts Identified in Some Fermented Food Products from Eastern and Western Africa

| PRODUCT | IDENTIFIED LAB AND YEASTS | REFERENCES |
|---|---|---|
| Sussac | *Ent. faecium, Lb. helveticus, Streptococcus salivarius/thermophilus, Weiss. confusa, Ent. faecalis, Lm. fermentum, Lact. lactis* subsp. *lactis, Leuc. lactis, Leuc. mesenteroides, Lt. curvatus, Lp. plantarum, Lig. salivarius, Lact. raffinolactis, Leuc. mesenteroides* subsp. *mesenteroides, Can. lusitaniae, Cryptococcus laurentii, Rhodotorula mucilaginosa, Sacch. cerevisiae, Trichosporon mucoides,* and *Trichosporon cutaneum* | Jans et al. (2017) |
| Warankashi | *Lp. plantarum, Lm. fermentum, Lp. pentosus, Ped. pentosaceus, Ped. acidilactici, Sacch. cerevisiae,* and *Candida* spp. | Banwo et al. (2012); Adegboye et al. (2014) |
| Palm wine | *Lp. plantarum, Schizosaccharomyces pombe, Kodamaea ohmeri, Hanseniaspora occidentalis, Can. tropicalis, Kloeckera apiculata,* and *Pichia ohmeri* | Amoa-Awua et al. (2007); Banwo et al. (2020) |

fermented by mainly lactic acid bacteria and yeasts (Kimaryo et al., 2000; Kostinek et al., 2005; Banwo et al., 2020).

Various cereals raw materials are used as substrates for fermented beverages or gruels. Some cereals for these include sorghum (*Sorghum bicolor* L. Moench), millets (finger and pearl millets, *Eleusine coracana* and *Pennisetum glaucum*) and maize (*Zea mays*). Sorghum can be fermented into alcoholic beverage (*burukutu*) produced through processes of malting, mashing, boiling, fermentation, and maturation of the grains. *Lactobacillus fermentum* is the dominant LAB during the souring of *burukutu*. Other LAB isolated from fermenting *burukutu* include *Leuc. mesenteroides, Lp. plantarum, Lb. acidophilus, Lactococcus lactis* subsp. *lactis,* and *Lv. brevis.* Yeasts isolated from *burukutu* include *Sacch. cerevisiae* and *Sacch. chavelieria. Pito* is another sorghum-based alcoholic beverage native to different geographical regions of Ghana and Nigeria, while *dolo* is native to Burkina Faso. A consortium of *Lm. fermentum* or *Lb. delbrueckii* and *Sacch. cerevisiae* as a starter culture in the controlled fermentation of *pito* has been reported to yield a better and more preferred taste and aroma. The dominant LAB species of *dolo* are *Lb. fermentum, Lb. delbrueckii* subsp. *delbrueckii, Ped. acidilactici, Lact. lactis,* and *Leuc. lactis* (Atter et al., 2014; Soro-Yao et al., 2014; Djameh et al., 2019).

In Côte d'Ivoire, the traditional beers *chibuku* and *tchapalo* have been fermented by *Lm. fermentum, Lb. acidophilus,* and *Lact. lactis* subsp. *lactis* (Soro-Yao et al., 2014). Starter cultures are from a previous brew from the spontaneous fermentation of the wort. Lyumugabe et al. (2010) reported that *ikigage* beer native to Rwanda is initiated by a traditional starter culture (*umusemburo*), which consists of *Sacch. cerevisiae, Can. inconspicua, Can. magnolia, Can. humilis, Issatchenkia orientalis, Lm. fermentum, Ln. buchneri, Aspergillus* sp., *Staphylococcus aureus,* and *Escherichia coli.* This leads to variation in the beer quality because desirable and undesirable strains are in the starters. However, selected strains of *Sacch. cerevisiae, Lactobacillus* sp., or *Issatchenkia orientalis* gave a more stable beer quality (Lyumugabe et al., 2010). Lactic acid bacteria are responsible for acid production, hence the low pH and sour taste of African beers. Back slopping is mostly practiced in the production of African beers to speed up the fermentation process.

*Ogi* is a cereal gruel produced from the fermentation of maize, sorghum, or millet, widely consumed in Nigeria and many parts of West Africa. It is a breakfast cereal, weaning food for infants, and choice food for the convalescent and elderly. It is also known as *akamu, agidi* (*eko*), and *koko* based on ethnic lingua and product consistency (Oguntoyinbo and Narbad, 2012; Okeke et al., 2015; Banwo et al., 2021b). Lactobacilli and Pediococci are the predominant fermenting microorganisms during the steeping stage, and *Lp. plantarum* predominate the souring stage. Predominant yeast species identified are *Saccharomyces* and *Candida* (Greppi et al., 2013; Ogunremi et al., 2015; Afolayan et al., 2017; Banwo et al., 2021b).

*Kenkey* is a stiff dough produced from the fermentation of maize, similar to *mawè*. This is a popular food product in Ghana. The predominant microorganisms associated with the fermentation of *kenkey* are LAB: *Lp. plantarum, Ped. pentosaceus, Lm. fermentum, Lm. reuteri*, and *Lv. brevis* and yeasts (Banwo et al., 2020).

Palm wine obtained by tapping palm trees such as *Cocos nucifera, Phoenix dactylifera, Raphia hookeri*, and *Elaeis guineensis* is not alcoholic when fresh. When fermented, it contains LAB and yeasts (Amoa-Awua et al., 2007; Banwo et al., 2020).

In East and West Africa, there are several fermented dairy products from cow, goat, and sheep milks. Some include *kule naoto, mursik, amabere amaruranu, sussac, warankashi*, and *nunu*. LAB and yeasts are the major fermenting organisms. There are three different kind of fermented cow's milk native to Kenya: *Kule natao, mursik*, and *Amabere amaruranu. Kule naoto* (native to Kenya), is a 5-day spontaneously fermented cow milk. The LAB associated with *kule naoto* are *Lp. plantarum, Lm. fermentum, Lc. casei, Lb. acidophilus, Lc. rhamnosus, Enterococcus faecium, Leuc. mesenteroides*, and *Lactococcus* spp., while the yeasts are *Saccharomyces* species. *Mursik* has predominant LAB being *Lb. kefiri, Lc. casei, Lc. paracasei*, and *Lc. rhamnosus*; other organisms include *Bacillus* spp., *Can. krusei, Can. spaerica, Can. kefyr*, and *Sacch. fermenti* (Mathara et al., 2008; Jans et al., 2017; Banwo et al., 2020). *Amabere amaruranu* has the major fermenting organisms *Lp. plantarum, Leuc. mesenteroides, S. thermophilus, Sacch. cerevisiae, Trichosporon mucoides, Can. famata*, and *Can. albicans*, while others are *Lb. bulgaricus* and *Lb. helveticus* (Jans et al., 2017).

The predominant LAB and yeasts associated with *warankashi* (soft cheese native to Nigeria and Ghana) are *Lp. plantarum, Lm. fermentum, Sacch. cerevisiae*, and *Candida* spp. (Banwo et al., 2013a, 2013b; Adegboye et al., 2014); those in *nunu/nono* (fermented milk curd often consumed with *fura*, made from fermented millet, together called *fura de nono*) are predominantly *Enterococcus faecium, Ent. italicus, Weiss. confusa, Lb. helveticus*, and *Lactococcus* spp. (Owusu-Kwarteng et al., 2012; Akabanda et al., 2013; Banwo et al., 2013a; 2020).

## 17.1.6 Role of LAB and Yeasts in Fermented Food Products from Eastern and Western Africa

The health-promoting advantages provided by fermented foods are attributable to the bioactive compounds and other metabolites elicited by the microorganisms during the fermentation process. These bioactive properties include exopolysaccharide (EPS) production, which is a natural polymer of sugars that are biologically synthesized by microorganisms during a fermentation process (Adesulu-Dahunsi et al., 2018a, 2018b). *Weissella confusa, Weiss. cibaria*, and *Lactobacillus* strains obtained from *ogi*, a fermented cereal-based non-alcoholic food, possessed EPS and probiotic potentials. The EPSs produced by these microorganisms have the potential to be used in the food industries (Adesulu-Dahunsi et al., 2018a, 2018b). Spontaneous fermentation is common in Africa; however, lactic acid bacteria and yeasts obtained from fermented African foods are good candidates for probiotic development in a controlled fermentation process for improved nutrition and health (Franz and Holzapfel, 2019; Banwo et al., 2020). These organisms have multiple functional properties like resistance to bile salts, gastric acidity, assimilation of cholesterol, adhesion to extracellular matrices, and HT29 MTX cells (Mathara et al., 2008; Yousif et al., 2010; Banwo et al., 2013a; Sanni et al., 2013; Ogunremi et al., 2015; Adesulu-Dahunsi et al., 2018a, 2018b; Motey et al., 2021). Also, LAB and yeasts obtained from *gari* and *ogi* have been reported to ameliorate the effect of lead and cadmium in experimental animals (Ojekunle et al., 2017). Cereal-based foods such as *koko* and *ogi* fermented with starter cultures *Lm. fermentum, Lp. plantarum*, and *Can. tropicalis* possessed higher niacin, riboflavin, magnesium, and potassium contents with high proximate contents (Banwo et al., 2021b).

# 17.2 FOOD SAFETY OF LACTIC-FERMENTED FOODS IN AFRICA

African cuisine includes many fermented foods processed from crops like maize, sorghum, locust beans, melon seeds, cassava, fish, and milk. These fermented foods are beneficial for flavor enhancement, and fermented cereals are used for weaning and post-operative convalescent feeding due to their easier digestibility. This section covers the food safety benefits of lactic fermentation.

## 17.2.1 Reduction of Non-Nutrient Compounds

Some foods are unsafe without the removal of toxic compounds or non-nutrients. To improve the safety of foods like cassava (*Manihot esculenta* Crantz) and sorghum (*S. bicolor* L. Moench), which contain hydrocyanic acid (cyanide) and tannins, respectively (Ingenbleek et al., 2019; Njankouo et al., 2019; Putri et al., 2021), they are fermented. During fermentation, these non-nutrient compounds are reduced (Adeyemo et al., 2016; Adebo et al., 2019). Without their removal, these compounds are unsafe and can be detrimental for proper body functioning and development. Therefore, their detoxification is important for food safety. Reduction in cyanide by 47%, 80%, and 91% has been reported in processing of cassava to chips, *gari*, and *fufu*, respectively (Njankouo et al., 2019). The dominant microbiota that produce linamarase, which hydrolyzes cyanogenic glucosides, are LAB and yeasts, for example, *Lc. lactis*, *Leuc. mesenteroides*, *Lp. plantarum*, and *Lm. fermentum*, while species such as *Lb. coprophilus*, *Lb. delbrueckii*, *Ped. pentosaceus*, and *Pichia exigua* are reported to be able to withstand concentrations of free cyanide from 200 to 800 ppm, making them good candidates for fermentation (Fadahunsi et al., 2020; Banwo et al., 2023).

Tannins are water-soluble, phenolic compounds found in red sorghum varieties. They interact with proteins, forming complexes that reduce digestibility of starch and proteins, inhibit enzymatic reactions, and reduce micronutrient bioavailability (Putri et al., 2021). This makes this food potentially unsafe, especially for vulnerable groups that require high-nutrient diets. To improve their safety, fermentation has been recommended. Up to 77% reduction in tannin has been recorded with fermentation of sorghum by *Lp. plantarum*, which produces tannase and galate decarboxylase that breaks down the tannin to digestible molecules of gallic acid and glucose that are bioavailable. Additional safety benefits are that during fermentation, there is degradation of the disulfide bond in sorghum protein, which makes the proteins more digestible (Putri et al., 2021). Reductions of other antinutritional compounds, including phytate, trypsin inhibitors, and protease inhibitors, have been reported with *Lp. plantarum* and *Lv. brevis* fermentations (Adeyemo et al., 2016; Setiarto and Widhyastuti, 2016).

## 17.2.2 Detoxification of Secondary Metabolites (e.g., Mycotoxins)

Cereals are prone to mycotoxin contamination, especially in the tropics where sub-Saharan Africa is geopositioned. In Africa, there have been multiple reports of mycotoxin contamination in a variety of cereals. These have included aflatoxins (AFs), fumonisins (FBs), deoxynivalenol (DON), zearalenone ZEN), and ochratoxin A (OTA). The removal of mycotoxins from food for safety is critical, as mycotoxins are associated with a myriad of health maladies. AFs, FBs, DON, ZEN, and OTA are causes of hepatocellular carcinoma and are associated with impairments to growth and development of children, immunosuppression, nutrient malabsorption, and kidney disease (Aikore et al., 2019; Ismail et al., 2021; Mollay et al., 2022). It is therefore critical to improve the food safety of staple foods such as cereals that are pre-disposed to mycotoxin contamination.

Detoxification by probiotic organisms during lactic acid fermentation has been surmised to occur via the attachment of mycotoxins to binding sites on the cell membranes of the probiotics. Fermentation with probiotics also has the added benefit of increasing the nutritive properties of fermented cereals. Aflatoxins have been reduced by as much as 70% in cereals for light gruels (Banwo et al., 2021a). However, variations in the extent of mycotoxin reduction are dependent on the species composition of the fermentation systems as well as initial mycotoxin concentration. However, reported lactic acid bacteria for decontamination of mycotoxins include *Lc. rhamnosus*, *Lc. paracasei*, *Lm. reuteri*, *Lp. pentosus*, *Lm. fermentum*, and *Lp. plantarum*. Decontamination has been reported to occur progressively commencing at 24 h (Chlebicz and Śliżewska, 2020).

### 17.2.2.1 Reduction of Spoilage Organisms from pH Reduction (Leading to Increased Shelf Life)

Maize and sorghum gruel (*ogi*) fermentation has led to pH reductions from around 5 to below 4 in 48-h fermentation systems. pH of less than 4 makes the food matrix less favorable for food spoilage organisms that prefer pH of 5 and above. Similar pH reductions to below pH 4 were found with cassava fermentations to *gari* and between pH 4 and 5 during the fermentation process of cassava to *fufu* (Omemu et al., 2018; Awoyale et al., 2021).

## 17.2.3 Gut Conditioning (That Can Lead to Reduction in Diarrhea)

It is suggested that probiotics provide dietary safety to consumers by protecting the natural intestinal microbiota and suppressing the growth of pathogenic organisms. For example, *Clostridium difficile* infection is treated using *Sacch. boulardii* (as a drug) in patients who are seriously ill (Ouwehand et al., 2002; Mc Farland, 2009; Naidu et al., 2012). Gut conditioning from probiotics aids digestion and helps in the development of the microbiome of the gastrointestinal system. However, there have been reports of mild abdominal discomfort in children and bacterial sepsis in individuals with immunocompromised epithelia from probiotics, including Lactobacilli (Naidu et al., 2012). Therefore, administration of probiotics as medication should be done with caution.

## 17.3 CONCLUSIONS AND FUTURE PERSPECTIVES

This chapter has highlighted several traditional lactic acid–fermented foods from all regions of Africa, with the associated microorganisms with great potential. However, the production of African traditional foods is still done by artisans using spontaneous fermentation and back-slopping techniques, which gives inconsistent delivery with every batch. There are still major obstacles in the development of strains with potential for industrial applications for the unique foods of this region (Franz and Holzapfel, 2019; Banwo et al., 2020).

Moreover, further research and inclusivity are essential in the improvement of strains with proven techno-functional characteristics indigenous to Africa in the production of these traditional foods. Cereal-based foods are of paramount significance in the development of probiotic starters with multifunctional properties. With the involvement of government agencies and research institutes, these strains can be distributed at little or no cost. This initiative will go a long way in industrialization in partnership with small- and medium-scale/cottage industries (Franz and Holzapfel, 2019).

# BIBLIOGRAPHY

Abdelgadir, W., Nielsen, D.S., Hamad, S. and Jakobsen, M., 2008. A traditional Sudanese fermented camel's milk product, *gariss*, as a habitat of *Streptococcus infantarius* subsp. *infantarius*. *Int J Food Microbiol*, 127: 215–219.

Adebiyi, J.A., Kayitesi, E., Adebo, O.A., Changwa, R. and Njobeh, P.B. 2019. Food fermentation and mycotoxin detoxification: An African perspective. *Food Control*, 106: 106731.

Adebo, J.A., Njobeh, P.B., Gbashi, S., Oyedeji, A.B., Ogundele, O.M., Oyeyinka, S.A. and Adebo, O.A. 2022. Fermentation of cereals and legumes: Impact on nutritional constituents and nutrient bioavailability. *Fermentation*, 8: 63.

Adebo, O.A., Chinma, C.E., Obadina, A.O., Panda, S., Soares, A. and Gan, R.Y. 2023. An insight into indigenous fermented foods in the tropics. In: *Indigenous Fermented Foods in the Tropics*, Adebo, O.A., Chinma, C.E., Obadina, O., Panda, S., Soares, A. and Gan, R.Y. (Eds.). Elsevier, Amsterdam, Netherlands, pp. 1–12.

Adebo, O.A., Kayitesi, E. and Njobeh, P.B. 2019. Reduction of Mycotoxins during fermentation of whole grain sorghum to whole grain *Ting* (A Southern African food). *Toxins*, 11(3). https://doi.org/10.3390/toxins11030180

Adegboye, B.D., Banwo, K., Ogunremi, O.R. and Sanni, A.I. 2014. Probiotic potentials of yeasts isolated from *Nono* (African fermented milk) and *Wara* (African soft cheese). *Adv Food Sci*, 36(3): 115–124.

Adesulu-Dahunsi, A.T., Dahunsi, S.O. and Ajayeoba, T.A. 2022. Co-occurrence of *Lactobacillus* species during fermentation of African indigenous foods: Impact on food safety and shelf-life extension. *Front Microbiol*, 13: 684730.

Adesulu-Dahunsi, A.T., Jeyaram, K., Sanni, A.I. and Banwo, K. 2018a. Production of exopolysaccharide by strains of *Lactobacillus plantarum* YO175 and OF101 isolated from traditional fermented cereal beverage. *PeerJ*, 6: 1–21. https://doi.org/10.7717/peerj.5326

Adesulu-Dahunsi, A.T., Sanni, A.I., Jeyaram, K., Ojediran, J.O., Ogunsakin, A.O. and Banwo, K. 2018b. Extracellular polysaccharide from *Weissella confusa* OF126: Production, optimization, and characterization. *Int J Biol Macromol*, 111: 514–525.

Adeyemo, S.M., Onilude, A.A. and Olugbogi, D.O. 2016. Reduction of anti-nutritional factors of sorghum by lactic acid bacteria isolated from abacha—an African fermented staple. *Front Sci*, 6(1): 25–30. https://doi.org/10.5923/j.fs.20160601.03

Afolayan, A.O., Ayeni, F.A. and Ruppitsch, W. 2017. Antagonistic and quantitative assessment of indigenous lactic acid bacteria in different varieties of *ogi* against gastrointestinal pathogens. *Pan Afr Med J*, 27: 22–29.

Agyei, D., Owusu-Kwarteng, J., Akabanda, F. and Akomea-Frempong, S. 2020. Indigenous African fermented dairy products: Processing technology, microbiology and health benefits. *Critical Rev Food Sci Nutr*, 60: 991–1006.

Aikore, M.O.S., Ortega-Beltran, A., Eruvbetine, D., Atehnkeng, J., Falade, T.D.O., Cotty, P.J. and Bandyopadhyay, R. 2019. Performance of broilers fed with maize colonized by either toxigenic or atoxigenic strains of *Aspergillus flavus* with and without an aflatoxin-sequestering agent. *Toxins*, 11(10). https://doi.org/10.3390/toxins11100565

Akabanda, F., Owusu-Kwarteng, J., Tano-Debrah, K., Glover, R.L.K., Nielsen, D.S. and Jespersen, L. 2013. Taxonomic and molecular characterization of lactic acid bacteria and yeasts in Nunu, a Ghanaian fermented milk product. *Food Microbiol*, 34(3): 277–283.

Ali, A.A. and Mustafa, M.M. 2009. Isolation, characterization and identification of lactic acid bacteria from fermented sorghum dough used in Sudanese *kisra* preparation. *Pak J Nutr*, 8: 1814–1818.

Amoa-Awua, W.K., Sampson, E. and Tano-Debrah, K. 2007. Growth of yeasts, lactic and acetic acid bacteria in palm wine during tapping and fermentation from felled oil palm (*Elaeis guineensis*) in Ghana. *J Appl Microbiol*, 102(2). https://doi.org/10.1111/j.1365-2672.2006.03074.x

Anukam, K.C. and Reid, G. 2009. African traditional fermented foods and probiotics. *J Med Food*, 12: 1177–1184.

Atter, A., Obiri-Danso, K. and Amoa-Awua, W.K. 2014. Microbiological and chemical processes associated with the production of *burukutu* a traditional beer in Ghana. *Int Food Res J*, 21(5): 1769–1776.

Awoyale, W., Oyedele, H., Adenitan, A.A., Alamu, E.O. and Maziya-Dixon, B. 2021. Comparing the functional and pasting properties of *gari* and the sensory attributes of the *eba* produced using backslopped and spontaneous fermentation methods. *Cogent Food Agric*, 7(1): 1883827. https://doi.org/10.1080/23311932.2021.1883827

Banwo, K., Adesina, T., Aribisala, O. and Falade, T.D.O. 2021a. *Decontamination of Aflatoxins by Selected Probiotics from Cereal Gruel (Ogi) and Changes to Amino Acid Profiles in the Presence of Atoxigenic and Toxigenic Aspergillus Flavus Methods*. International Association of Research Scholars and Fellows (IARSAF) Symposium.

Banwo, K., Asogwa, F.C., Ogunremi, O.R., Adesulu-Dahunsi, A. and Sanni, A.I. 2021b. Nutritional profile and antioxidant capacities of fermented millet and sorghum gruels using lactic acid bacteria and yeasts. *Food Biotechnol*, 35(3): 199–220.

Banwo, K., Ogunremi, O.R. and Sanni, A.I. 2020. Fermentation biotechnology of African traditional foods. In: *Functional Foods and Biotechnology: Biotransformation and Analysis of Functional Foods and Ingredients*, Shetty, K. and Sarkar, D. (Eds.). CRC Taylor and Francis Group, LLC, Boca Raton, FL, pp. 101–124. ISBN: 978-0-367-43522-6.

Banwo, K., Ojetunde, J.T. and Falade, T. 2023. Probiotic and cyanide degrading potentials of *Pediococcus pentosaceus* and *Pichia exigua* isolated from cassava products effluent. *Food Biotechnol*, 37(1): 1–24. https://doi.org/10.1080/08905436.2022.2163252

Banwo, K., Sanni, A.I. and Tan, H.R. 2013a. Technological properties and probiotic potential of *Enterococcus faecium* strains isolated from cow milk. *J Appl Microbiol*, 114(1): 229–241.

Banwo, K., Sanni, A.I. and Tan, H.R. 2013b. Functional properties of *Pediococcus* species isolated from traditional fermented cereal gruel and milk in Nigeria. *Food Biotechnol*, 27(1): 14–38.

Banwo, K., Sanni, A.I., Tan, H.R. and Tian, Y.Q. 2012. Phenotypic and genotypic characterization of lactic acid bacteria isolated from some Nigerian traditional fermented foods. *Food Biotechnol*, 26(2): 124–142.

Benkerroum, N. 2012. Traditional fermented foods of North African countries: Technology and food safety challenges with regard to microbiological risks. *Comp Rev Food Sci Food Safety*, 12: 54–89.

Chelule, P.K., Mokgatle, M.M. and Zungu, L.I. 2015. South African rural community understanding of fermented foods preparation and usage. *J Edu Health Promotion*, 4: 82.

Chiocchetti, G.M., Piedra, C.J., Monedero, V., Zuniga, M., Velez, D. and Devesa, V. 2019. Use of lactic acid bacteria and yeasts to reduce exposure to chemical food contaminants and toxicity. *Crit Rev Food Sci Nutr*, 59: 1534–1545.

Chlebicz, A. and Śliżewska, K. 2020. *In vitro* detoxifcation of aflatoxin B1, Deoxynivalenol, Fumonisins, T-2 toxin and Zearalenone by probiotic bacteria from Genus *Lactobacillus* and *Saccharomyces cerevisiae* yeast. *Probiotics Antimicrob Proteins*, 12: 289–301. https://doi.org/10.1007/s12602-018-9512-x

Daji, G.A., Green, E., Abrahams, A., Oyedeji, A.B., Masenya, K., Kondiah, K. and Adebo, O.A. 2022. Physicochemical properties and bacterial community profiling of optimal *mahewu* (a fermented food product) prepared using white and yellow maize with different inocula. *Foods*, 11: 3171.

Dirar, H.A. 1992. The indigenous fermented foods and beverages of Sudan. In: *Application of Biotechnology to Food Processing*. Expert Group Meeting, Ibadan, Nigeria, UNIDO, pp. 23–40.

Djameh, C., Ellis, W.O., Oduro, I., Saalia, F.K., Haslbeck, K. and Komlaga, G.A. 2019. West African sorghum beer fermented with *Lactobacillus delbrueckii* and *Saccharomyces cerevisiae*: Fermentation by-products. *J Inst Brewing*, 125: 326–332. https://doi.org/10.1002/jib.562

El-Baradei, G., Delacroix-Buchet, A. and Ogier, J.C. 2008. Bacterial biodiversity of traditional *zabady* fermented milk. *Int J Food Microbiol*, 121: 295–301.

Fadahunsi, I.F., Busari, N.K. and Fadahunsi, O.S. 2020. Effect of cultural conditions on the growth and linamarase production by a local species of *Lactobacillus fermentum* isolated from cassava effluent. *Bull Nat Res Centre*, 44(1). https://doi.org/10.1186/s42269-020-00436-3

Franz, C.M.A.P. and Holzapfel, W.H. 2019. Examples of lactic-fermented foods of the African continent. In: *Lactic Acid Bacteria: Microbiological and Functional Aspects*, Vinderola, G., Ouwehand, A.C., Salminen, S. and von Wright, A. (Eds.). CRC Taylor and Francis Group, LLC, Boca Raton, FL, pp. 235–253. ISBN-13: 978-0-8153-6648-5.

Gänzle, M.G. 2015. Lactic metabolism revisited: Metabolism of lactic acid bacteria in food fermentations and food spoilage. *Curr Opin Food Sci*, 2: 106–117.

Greppi, A., Rantsiou, K., Padonou, W., Hounhouigan, J., Jespersen, L., Jakobsen, M. and Cocolin, L. 2013. Determination of yeast diversity in *ogi, mawè, gowé* and *tchoukoutou* by using culture-dependent and—independent methods. *Int J Food Microbiol*, 165: 84–88.

Hamad, S.H., Dieng, M.C., Ehrmann, M.A. and Vogel, R.F. 1997. Characterization of the bacterial flora of Sudanese sorghum, flour and sorghum sourdough. *J Appl Microbiol*, 28: 764–770.

Hlangwani, E., Adebiyi, J.A., Doorsamy, W. and Adebo, O.A. 2020. Processing, characteristics and composition of umqombothi (a South African traditional beer). *Processes*, 8: 1451.

Ingenbleek, L., Sulyok, M., Adegboye, A., Hossou, S.E., Koné, A.Z., Oyedele, A.D., Kisito, C.S.K.J., Dembélé, Y.K., Eyangoh, S., Verger, P., Leblanc, J.C., Le Bizec, B. and Krska, R. 2019. Regional sub-saharan Africa total diet study in Benin, Cameroon, Mali and Nigeria reveals the presence of 164 mycotoxins and other secondary metabolites in foods. *Toxins*, 11(1): 1–23. https://doi.org/10.3390/toxins11010054

Ismail, A., Naeem, I., Gong, Y.Y., Routledge, M.N., Akhtar, S., Riaz, M., Ramalho, L.N.Z., de Oliveira, C.A.F. and Ismail, Z. 2021. Early life exposure to dietary aflatoxins, health impact and control perspectives: A review. *Trends Food Sci Technol*, 112: 212–224. https://doi.org/10.1016/j.tifs.2021.04.002

Jans, C., Meile, L., Kaindi, D.W.M., Kogi-Makau, W., Lamuka, P., Renault, B., Kreikemeyer, B., Lacroix, C., Hattendorf, J., Zinsstag, J., Schelling Fokou, G. and Bonfoh, B. 2017. African fermented dairy products—overview of predominant technologically important microorganisms focusing on African *Streptococcus infantarius* variants and potential future applications for enhanced food safety and security. *Int J Food Microbiol*, 250: 27–36.

Kayitesi, E., Behera, S.K., Panda, S.K., Dlamini, B. and Mulaba-Bafubiandi, A.F. 2017. Amasi and mageu: Expedition from ethnic Southern African foods to cosmopolitan markets. In: *Fermented Food-Part II: Technological Interventions*, Ray, R.C. and Montet, D. (Eds.). CRC Press, Boca Raton, pp. 384–399.

Kimaryo, V.M., Massawe, G.A., Olasupo, N.A. and Holzapfel, W.H. 2000. The use of a starter culture in the fermentation of cassava for the production of *kivunde*, a traditional Tanzanian food product. *Int J Food Microbiol*, 56: 179–190.

Kostinek, M., Specht, I., Edward, V.A., Schillinger, U., Hertel, C., Holzapfel, W.H. and Franz, C.M.A.P. 2005. Diversity and technological properties of predominant lactic acid bacteria from fermented cassava used for the preparation of *Gari*, a traditional African food. *Syst Appl Microbiol*, 28(6): 527–540. https://doi.org/10.1016/j.syapm.2005.03.001

Liptakova, D., Matejcekova, Z. and Valik, L. 2017. Lactic acid bacteria and fermentation of cereals and pseudocereals. In: *Fermentation Processes*, Jozala, A.F. (Ed.). InTech, Croatia, pp. 223–254.

Lyumugabe, F., Kamaliza, G., Bajyana, E. and Thonart, P. 2010. Microbiological and physico-chemical characteristic of Rwandese traditional beer *Ikigage*. *Afr J Biotechnol*, 9(27): 4241–4246.

Madoroba, E., Steenkamp, E.T., Theron, J., Scheirlinck, I., Cloete, T.E. and Huys, G. 2011. Diversity and dynamics of bacterial populations during spontaneous sorghum fermentations used to produce ting, a South African food. *Syst Appl Microbiol*, 34: 227–234.

Maleke, M.S., Doorsamy, W., Abrahams, A.M., Adefisoye, M.A., Masenya, K. and Adebo, O.A. 2022. Influence of fermentation conditions (temperature and time) on the physicochemical properties and bacteria microbiota of *amasi*. *Fermentation*, 8: 57.

Mathara, J.M., Schillinger, U., Guigas, C., Franz, C.A.M.P., Kutima, P.M., Mbugua, S.K., Shin, H.K. and Holzapfel, W.H. 2008. Functional characteristics of *Lactobacillus* spp. from traditional Maasai fermented milk products in Kenya. *Int J Food Microbiol*, 126(1–2): 57–64.

McFarland, L.V. 2009. Evidence-based review of probiotics for antibiotic-associated diarrhea and *Clostridium difficile* infections. *Anaerobe*, 15: 274–280.

Mollay, C., Kimanya, M., Kassim, N. and Stoltzfus, R. 2022. Main complementary food ingredients contributing to aflatoxin exposure to infants and young children in Kongwa, Tanzania. *J Food Control*, 135: 108709. https://doi.org/10.1016/j.foodcont.2021.108709

Moonga, H.B., Schoustra, S.E., Linnemann, A.R., Shindano, J. and Smid, E.J. 2022. Towards valorisation of indigenous traditional fermented milk: *Mabisi* as a model. *Curr Opin Food Sci*, 46: 100835.

Motey, G.A., Owusu-Kwarteng, J., Obiri-Danso, K., Ofori, L.A., Ellis, W.O. and Jespersen, L. 2021. *In vitro* properties of potential probiotic lactic acid bacteria originating from Ghanaian indigenous fermented milk products. *World J Microbiol Biotechnol*, 37(3). https://doi.org/10.1007/s11274-021-03013-6

Naidu, K.S.B., Adam, J.K. and Govender, P. 2012. The use of probiotics and safety concerns: A review. *Afr J Microbiol Res*, 6(41): 6871–6877. https://doi.org/10.5897/ajmr12.1281

Nduko, J.M., Matofari, J.W., Nandi, Z.O. and Sichangi, M.B. 2017. Spontaneously fermented Kenyan milk products: A review of the current state and future perspectives. *Afr J Food Sci*, 11: 1–11.

Njankouo, Y., Mounjouenpou, P., Kansci, G., Josiane, M., Priscile, M., Meguia, F., Natacha, N.S., Eyenga, N., Mikhaïl, M. and Nyegue, A. 2019. Influence of cultivars and processing methods on the cyanide contents of cassava (*Manihot esculenta* Crantz) and its traditional food products. *Sci Afr*, 5: e00119. https://doi.org/10.1016/j.sciaf.2019.e00119

Ogunremi, O.R., Sanni, A.I. and Agrawal, R. 2015. Probiotic potentials of yeasts isolated from some cereal-based Nigerian traditional fermented food products. *J Appl Microbiol*, 119: 797–808.

Oguntoyinbo, F.A. and Narbad, A. 2012. Molecular characterization of lactic acid bacteria and in situ amylase expression during traditional fermentation of cereal foods. *Food Microbiol*, 31: 254–262.

Ohenhen, R.E., Imarenezor, E.P.K. and Kihuha, A.N. 2013. Microbiome of *madila*—a Southern-African fermented milk product. *Int J Basic Appl Sci*, 2: 170–175.

Ojekunle, O., Banwo, K. and Sanni, A.I. 2017. *In vitro* and *in vivo* evaluation of *Weissella cibaria* and *Lactobacillus plantarum* for their protective effect against cadmium and lead toxicities. *Lett Appl Microbiol*, 64: 379–385.

Okeke, C.A., Ezekiel, C.N., Nwangburuka, C.C., Sulyok, M., Ezeamagu, C.O., Adeleke, R.A., Dike, S.K. and Krska, R. 2015. Bacterial diversity and mycotoxin reduction during maize fermentation (steeping) for *Ogi* production. *Front Microbiol*, 6: 1402–1409.

Oliveira, P.M., Zannini, E. and Arendt, E.K. 2014. Cereal fungal infection, mycotoxins, and lactic acid bacteria mediated bioprotection: From crop farming to cereal products. *Food Microbiol*, 37: 78–95.

Omemu, A.M., Okafor, U.I., Obadina, A.O., Bankole, M.O. and Adeyeye, S.A.O. 2018. Microbiological assessment of maize *ogi* co-fermented with pigeon pea. *Food Sci Nutr*, 6(5): 1238–1253. https://doi.org/10.1002/fsn3.651

Ouadghiri, M., Vancanneyt, M., Vandamme, P., Naser, S., Gevers, D., Lefebvre, K., Swings, J. and Amar, M. 2009. Identification of lactic acid bacteria in Moroccan raw milk and traditionally fermented skimmed milk 'lben'. *J Appl Microbiol*, 106: 486–495.

Ouwehand, A.C., Salminen, S. and Isolauri, E. 2002. Probiotics: An overview of beneficial effects. *Anton van Leeuwen*, 82(1/4): 279–289. https://doi.org/10.1023/a:1020620607611

Owusu-Kwarteng, J., Akabanda, F., Nielsen, D.S., Tano-Debrah, K., Glover, R.L.K. and Jespersen, L. 2012. Identifcation of lactic acid bacteria isolated during traditional fura processing in Ghana. *Food Microbiol*, 32: 72–78.

Padonou, S.W., Nielsen, D.S., Akissoe, N.H., Hounhouigan, J.D., Nago, M.C. and Jakobsen, M. 2010. Development of starter culture for improved processing of *Lafun*, an African fermented cassava food product. *J Appl Microbiol*, 109(4): 1402–1410. https://doi.org/10.1111/j.1365-2672.2010.04769.x

Pessione, E. and Cirrincione, S. 2016. Bioactive molecules released in food by lactic acid bacteria: Encrypted peptides and biogenic amines. *Fron Microbiol*, 7: 876.

Putri, S.N.A., Utari, D.P., Martati1, E. and Putri, W.D.R. 2021. Study of sorghum (*Sorghum bicolor* (L.) Moench) grains fermentation with *Lactobacillus plantarum* ATCC 14977 on tannin content. *Int Conf Green Agro-Ind Bioeco*: 1–6. https://doi.org/10.1088/1755-1315/924/1/012037

Sanni, A.I., Franz, C.M.A.P., Schillinger, U., Huch, M., Guigas, C. and Holzapfel, W. 2013. Characterization and technological properties of lactic acid bacteria in the production of *Sorghurt*, a cereal-based product. *Food Biotechnol*, 27(2): 178–198.

Schoustra, S.E., Kasase, C., Toarta, C., Kassen, R. and Poulain, A.J. 2013. Microbial community structure of three traditional Zambian fermented products: *Mabisi, chibwantu* and *munkoyo*. *PLoS ONE*, 8: e63948.

Sekwati-Monang, B. and Gänzle, M.G. 2011. Microbiological and chemical characterization of ting, a sorghum-based sourdough product from Botswana. *Int J Food Microbiol*, 150: 115–121.

Setiarto, R.H.B. and Widhyastuti, N. 2016. Reduction of tannin and phytic acid on sorghum flour by using fermentation of *Rhizopus oligosporus*, *Lactobacillus plantarum* and *Saccharomyces cerevisiae*. *Berita Biologi*, 15: 107–206.

Simatende, P., Gadaga, T.H., Nkambule, S.J. and Siwela, M. 2015. Methods of preparation of Swazi traditional fermented foods. *J Ethnic Foods*, 2: 119–125.

Soro-Yao, A.A., Brou, K., Amani, G., Thonart, P. and Djè, K.M. 2014. The use of lactic acid bacteria starter cultures during the Processing of fermented cereal-based foods in West Africa: A review. *Trop Life Sci Res*, 25(2): 81–100.

Suleiman, A.M.E., Mustafa, W.A. and Osman, O.A. 2022. Selected fermented cereal products of Sudan. In: *African Fermented Food Products—New Trends*, Suleiman, A.M.E., Mariod, A.A. (Eds.). Springer, Cham, Switzerland, pp. 293–312.

Vieira-Dalodé, G., Jespersen, L., Hounhouigan, J., Moller, P.L., Nago, C.M. and Jakobsen, M. 2007. Lactic acid bacteria and yeasts associated with *gowé* production from sorghum in Bénin. *J Appl Microbiol*, 103: 342–349.

Witthuhn, R.C., Schoeman, T. and Britz, T.J. 2004. Isolation and characterization of the microbial population of different South African kefir grains. *Int J Dairy Technol*, 57: 33–37.

Yousif, N.M.K., Huch, M., Schuster, T., Cho, G.S., Dirar, H.A., Holzapfel, W.H. and Franz, C.M.A.P. 2010. Diversity of lactic acid bacteria from *Hussuwa*, a traditional African fermented sorghum food. *Food Microbiol*, 27: 757–768.

Zheng, J., Wittouck, S., Salvetti, E., Franz, C.M.A.P., Harris, H.M.B., Mattarelli, P., O'Toole, P.W., Pot, B., Vandamme, P., Walter, J., Watanabe, K., Wuyts, S., Felis, G.E., Ganzle, M.G. and Lebeer, S. 2020. A taxonomic note on the genus *Lactobacillus*: Description of 23 novel genera, emended description of the genus *Lactobacillus* Beijerinck 1901, and union of *Lactobacillaceae* and *Leuconostocaceae*. *Int J Syst Evol Microbiol*, 70(4): 2782–2858. https://doi.org/10.1099/ijsem.0.004107

# Lactic-Fermented Foods and Alcoholic Beverages in Asia

# 18

H. Nakibapher Jones Shangpliang,
Namrata Thapa, and Jyoti Prakash Tamang

## 18.1 INTRODUCTION

Asia is the largest continent in the world, covering almost 30% of the area, with more than 60% of the total population of the world. It has five major physio-geographical regions with 49 countries: Western Asia, Central Asia, South Asia, Southeast Asia and East Asia (Bear, 2021). Diversity in cuisine and gastronomy is common among the diverse ethnic groups of Asian people, mostly due to antiquity, tradition, food preferences, geographical locations, availability of resources, climatic conditions, religious taboos and socio-political issues. Rice is the most common staple food crop in Asia, mostly in South Asia, Southeast and East Asia, followed by wheat, soybeans, maize, millets and others (Estudillo et al., 2023). Asia has a long history of preparation and consumption of various types of ethnic fermented foods and alcoholic beverages by using 'ethno-microbiological knowledge' of food fermentation of perishable of raw/cooked substrates of plant or animal sources (Tamang, 2022a, 2022b) and also domestication of essential non-toxin and non-pathogenic microorganisms for food production (Tamang et al., 2022). Indigenous Asian fermented foods and alcoholic beverages are hubs of all major groups and include bacteria (lactic acid bacteria (LAB), acetic acid bacteria, non-lactis, bacilli, coagulase-negative staphylococci, etc.), eukaryotes (yeasts, fungi), viruses and archaea (Tamang et al., 2016, 2020; Tamang, 2022a).

## 18.2 ASIAN FERMENTED FOODS AND ALCOHOLIC BEVERAGES

Different types and varieties of ethnic fermented products are consumed in South Asia, Southeast Asia and East Asia based on substrates such as fermented soybeans, fermented vegetables, fermented bamboos, fermented cereals, fermented root-crops, fermented fruits, fermented milk, fermented fish, fermented meats, fermented egg and fermented tea (Rhee et al., 2011; Swain et al., 2014; Tamang, 2016, 2020; Waché et al., 2018; Tamang et al., 2020; Tamang et al., 2021; Muhialdin et al., 2022). Fermented

DOI: 10.1201/9781003352075-20

vegetable products include *kaili* (red sour soup), *paocai, jiangxi yancai, sichuan paocai* and *dongbei suancai* of China; fermented cabbage, *gundruk, inziangsang, khalpi* and *sinki* of India; *kimchi* of Korea; *burong mustasa* of the Philippines; and *paocai* of Taiwan (Chang et al., 2013; Cho et al., 2006; Cho et al., 2013; Dalmacio et al., 2011; Jeong et al., 2013; Kim et al., 2000; Larcia et al., 2011; Li et al., 2021; Liu et al., 2019; Park et al., 2017; Patel et al., 2014; Tamang et al., 2005; Xiao et al., 2020; Yoon et al., 2000). Fermented bamboo products are also very popular in the Asian countries and include *ma* (bamboo shoots) and *suansun* of China and *banstenga, ekung, eup, hirring, mesu, moiya pansung, soibum, soidon* and *soijim* of India (Chen et al., 2022; Das et al., 2023; Dey et al., 2023; Guan et al., 2020; Suwannaphan, 2021; Tamang et al., 2020). Fermented beverages include *kombucha* of China and Turkey; *chyang, lugri* and *toddy* of India; *takju* of Korea; *tapuy* of Philippines; and *shalgam* of Turkey (Akman et al., 2021; Baliyan et al., 2021; Bo et al., 2020; Coelho et al., 2020; Kahraman-Ilıkkan, 2023; Sanico and Medina, 2022, Tamang, 2022a). Fermented cereals include *jiaozi* of China; *dosa, idli, wanti, dhokla, jalebi, raabadi* and *selroti* of India; *tape ketan* of Indonesia; *horreh* of Iran; and *puto* of Philippines (Kavitake et al., 2022; Lavanya et al., 2021; Li et al., 2016; Mandhania et al., 2019, Tamang et al., 2016, Tamang, 2022a). In the Philippines, fruits such as durian (*Durio zibethinus* L.) are commonly fermented to a product locally known as *tempoyak* (Chuah et al., 2016; Yuliana and Dizon, 2011). In Turkey, fruits of *Viburnum opulus* L. (European cranberry bush) are fermented as a juice locally known as *gilaburu* (Akman et al., 2021). In Northeastern India, particularly Nagaland, the leaves of *Colocasia esculenta* L. are fermented to a local product known as *anishi* (Jamir and Deb, 2017). Among all fermented food products, naturally fermented milk (NFM) products harbour a wide variety of fermented products which are also associated with different communities. Fermented milk products are commonly prepared from various animal milks such as buffalo, camel, cow, ewe, goat, mare and yak. These include *borhani, chanar-misti, dahi* and *paneer* of Bangladesh; *dahi* and *datshi* of Bhutan; *ayran, jiaoke, dreg, kurut, koumiss* and *vrum* of China; *meekiri, chhurphe, kalarei, dahi, chhu, chhurpi, mohi, philu, shyow, somar, chhurpi* (hard), *churkam, mar, laal dahi* and *chakka* of India; *dadiah* of Indonesia; *doogh, lighvan* (cheese), *dahi, shiraz, tarkhineh, yoghurt, kashk zard, khameh* and *doogh* of Iran; *aryan* and *shubat* of Kazakhstan; *labneh ambaris* of Lebanon; *aarool, airag, byslag, eejˇgiy, isgelen tarag, qoormog, tarag* and *koumiss* of Mongolia; *laban* of Saudi Arabia; and *meekiri* and *sua chua* of Vietnam (Adikari et al., 2021; Bhagat et al., 2020; Biswal et al., 2022; Das et al., 2020; Davati and Hesami, 2021; Fan et al., 2020; Forouhandeh et al., 2021; Guo et al., 2019; Pakroo et al., 2020; Rashid and Hassanshahian, 2016; Srinivash et al., 2023; Tamang et al., 2016; Uchida et al., 2007; Watanabe et al., 2008). Other fermented products which are very common in the Asian countries are fermented soybean products. These include *sufu, da-jiang* and *douchi* of China; *hawaijar, grep chhurpi, kinema, peha, peron namsing, peruñyaan* and *tungrymbai* of India; *pe poke* of Mynamar; *sieng* of Cambodia, *thua nao* of Thailand, *tempeh* of Indonesia; *miso, shoyu* and *natto* of Japan; and *doenjang* and *cheonggukjang* of Korea (Dey et al., 2023; Huang et al., 2018; Radita et al., 2020; Tamang et al., 2022; Thokchom and Joshi, 2012; Wu et al., 2013; Yulandi et al., 2020; Zhang et al., 2018). In Asia fish is a popular dish and is also traditionally processed/fermented/sun-dried, such as *chouguiyu, suanyu* and *andyucha* of China; *sidra, suka ko maacha, sukuti, namsing, nakham, ngari, sukamas, shidal, hindal, tungtap, hentak, gnuchi, bordia, karati* and *lashim* of India; *inasua* and *masin* of Indonesia; *myeolchiaekjeot* of Korea; *pekasam* of Malaysia; *yegyo ngapi* and *ngachin* of Myanmar; *alamang, burong hipon (tarlac)* and *burong isda* of the Philippines; *plaa-som* and *pla-ra* of Thailand (Jiang et al., 2019; Liu et al., 2021; Sun et al., 2022; Wang et al., 2020; Yang et al., 2020; Zang et al., 2018; Das et al., 2020; Dey et al., 2023; Mahulette et al., 2018; Manguntungi et al., 2020; Narzary et al., 2021; Suwannaphan, 2021). Several meat products are also consumed in Asia which are processed from various animal meats, such as buffalo, cow, goat, pig and yak. These include *arjia*, (beef) *kargyong, chartayshya, faak kargyong, jamma, khyopeh, lang kargyong, lang satchu, pork kargyong, suka ko masu*, (yak) *kargyong* and (yak) *satchu* (Bhutia et al., 2021b; Oki et al., 2011; Rai et al., 2010; Wang et al., 2021).

Wine-making and brewing are not historically part of the food culture of Asian people except the ethnic people of West Asia; instead the Asian people ferment cereals (rice, millets, barley, etc) into mild to moderate alcoholic beverages by using traditionally prepared crude starters (Anal, 2019; Tamang et al., 2016; Tamang, 2020). The crude starter is amylolytic in nature; hence it is also called an amylolytic starter, and it has a round, oval to flat shape and is dry, with variable sizes of 1.0 to 12.0 cm and white/brown/black in colour. Most amylolytic starters are made from rice and include *marcha* of India, Bhutan and Nepal; *phab* of Bhutan; *xaj-pitha, paa, pee* and *phut* of India; *ragi tape* of Indonesia; *nuruk* of Korea; *ragi* of Indonesia; *bubod* of Philippines; *daqu* of China and Taiwan; *loogpang* of Thailand; *benh/men* of Vietnam; and *dombea* or *medombae* of Cambodia (Sujaya et al., 2001; Bal et al., 2014; Das and Goyal, 2014; Bora et al., 2016; Ly et al., 2018; Pradhan and Tamang, 2019; Tamang et al., 2016, 2020). Solid, dry homemade amylolytic starters for alcohol production show the existence of mixed microorganisms represented by different genera and species of mycelial moulds (Anupma and Tamang, 2020) for degrading the starch of fermented cereals to glucose (Thapa and Tamang, 2004), amylase and alcohol-producing yeasts for saccharification (Tamang and Thapa, 2006) and ethanol production (Tsuyoshi et al., 2005) and bacteria for probiotic properties (Pradhan and Tamang, 2021) and antimicrobial activities (Tamang et al., 2007). Mild to moderate alcoholic beverages are prepared from cereals using dry amylolytic starters in South Asia, Southeast Asia and East Asia. Some of these beverages are traditionally distilled to make high-alcohol liquor.

## 18.2.1 Lactic Acid Bacteria in Asian Fermented Foods Identified by Conventional Methods

Conventional methods are those methods where identification of LAB is usually based on the culture-based techniques. These usually include the use of colony and cellular morphology, catalase tests, production of lactic acid and biochemical and physiological methods of characterization (Carr et al., 2002). Identification of LAB in fermented foods has widely been achieved by using analytical profile index (API) systems such as API 50 CHL and API 20 STREP (bioMérieux, France), where LAB strains are characterized based on the carbohydrate utilization pattern and enzymatic reactions. Similar identification system such as BD PhoenixTM100 automated ID/AST systems has also been employed to characterize LAB (Dey et al., 2023). Though these systems can characterize LAB, 16S rRNA gene sequencing is considered the gold standard in bacterial identification (Janda and Abbott, 2007). Furthermore, genotyping methods such as amplified ribosomal DNA restriction analysis (ARDRA), random amplified polymorphic DNA (RAPD), pulse field gel electrophoresis (PFGE), amplified restriction length polymorphism (AFLP), 16S rRNA clone libraries, rep-PCR, species-specific PCR and DNA–DNA hybridization have been widely used to characterize and group isolated strains (Ben Amor et al., 2007; Li et al., 2009). LAB belong to the phylum Bacillota are the most the predominant bacteria in most lactic-fermented foods and beverages. Among LAB species, *Lactiplantibacillus plantarum*, *Levilactobacillus brevis*, *Enterococcus faecium*, *Limosilactobacillus fermentum* and *Leuconostoc mesenteroides* are the most commonly isolated species from different Asian fermented foods and beverages (Table 18.1).

## 18.2.2 LAB in Asian Fermented Foods Identified by PCR-DGGE and Clone Library-16S rRNA Sequencing

Before the intervention of high-throughput sequencing, polymerase chain reaction-denaturing gradient gel electrophoresis (PCR-DGGE) was the most common culture-independent technique used to profile microbial diversity (Tamang et al., 2016). It was widely used in food systems to profile microbial communities through DNA patterns visualized in gel electrophoresis (Ercolini, 2004). This method has been applied to profile both bacterial and yeast communities and is dependent on the target microorganisms and the primers of choice. Extraction of community DNA from the samples followed by gel electrophoresis in the presence of urea and formamide (denaturants) is practiced. Though this process has revolutionized microbial ecology study in fermented foods and beverages, it is very tedious and time consuming. For identification of the DNA fragments, each band is then excised and subjected to the Sanger sequencing

**TABLE 18.1** Lactic Acid Bacteria Reported and Identified from Various Asian Fermented Foods, Starters and Beverages through Culture-Dependent Conventional Methods

| FOOD TYPE | SUBSTRATE | NAME | COUNTRY | METHOD OF IDENTIFICATION | MICROBES* | LITERATURE |
|---|---|---|---|---|---|---|
| Amylolytic starter | Rice | Phab | Bhutan | 16S rRNA gene sequencing | E. durans, E. faecium | Pradhan and Tamang (2019) |
| | | Marcha | | | E. durans, E. faecium, P. pentosaceus | |
| | | Paa | India | | E. faecalis, E. faecium | |
| | | Pee | | | E. durans, E. faecium | |
| | | Phut | | | E. faecium, E. hirae, E. lactis | |
| | | Marcha | | | E. durans, E. faecalis, E. faecium, Ln. mesenteroides, P. acidilactici, P. pentosaceus, W. cibaria | |
| | | | | 16S rRNA and rpoA gene sequencing | L. plantarum ssp. plantarum | Das and Goyal (2014) |
| | | | Nepal | 16S rRNA gene sequencing | L. pentosus, L. plantarum ssp. plantarum | Pradhan and Tamang (2019) |
| | Balinese rice | Ragi tape | Indonesia | 16S rRNA gene sequencing | E. faecium, L. curvatus, P. pentosaceus, W. confusa, W. paramesenteroides | Sujaya et al. (2001) |
| Fermented bamboo | Bamboo | Banstenga | India | BD PhoenixTM100 automated ID/AST system | A. viridans, Ln. mesenteroides, P. acidilactici, P. damnosus, P. pentosaceus | Dey et al. (2023) |
| | | Soidon | | ARDRA, ITS-PCR, ITS-RFLP, RAPD, 16S rRNA gene sequencing | Carnobacterium sp., E. faecium, L. plantarum, L. brevis | Jeyaram et al. (2010) |
| | | Soidon | | ARDRA, 16S rRNA gene sequencing | L. plantarum, Lc. lactis ssp. cremoris, Lc. lactis ssp. lactis, Lc. raffinolactis, Ln. citreum, Ln. lactis, Ln. mesenteroides, L. brevis, P. pentosaceus, V. fluvialis, W. cibaria, W. oryzae | Romi et al. (2015) |
| | | Ekung | | API system | L. casei, L. plantarum, L. brevis, T. halophilus | Tamang and Tamang (2009) |
| | | Eup | | | L. fermentum | |
| | | Hirring | | | Lc. lactis | |

(Continued)

**TABLE 18.1** (Continued) Lactic Acid Bacteria Reported and Identified from Various Asian Fermented Foods, Starters and Beverages through Culture-Dependent Conventional Methods

| FOOD TYPE | SUBSTRATE | NAME | COUNTRY | METHOD OF IDENTIFICATION | MICROBES* | LITERATURE |
|---|---|---|---|---|---|---|
| | | Mesu | | RAPD, rep-PCR, species-specific PCR, DNA–DNA hybridization, 16S rRNA gene sequencing | L. plantarum, L. curvatus, L. brevis, P. pentosaceus | Tamang et al. (2008) |
| | | Soibum | | | E. durans, L. plantarum, Ln. fallax, Ln. lactis, Ln. mesenteroides, L. brevis | |
| | | Soidon | | | Ln. fallax, Ln. lactis, L. brevis | |
| | | Soijim | | | Ln. fallax, Ln. lactis, L. brevis | |
| | | Bamboo shoot | Thailand | 16S rRNA gene sequencing | L. fermentum | Suwannaphan (2021) |
| Fermented beverages | Barley | Chhang | India | 16S rRNA gene sequencing | L. plantarum, L. brevis | Angmo et al. (2016) |
| | Rice/barley/wheat | Lugri | | | L. crustorum, L. paracasei ssp. tolerans, L. paraplantarum, L. pentosus, L. argentoratensis, L. brevis, L. fermentum, L. reuteri, P. acidilactici | Baliyan et al. (2021) |
| | Palm sap | Toddy | | | E. faecalis, L. paracasei, L. plantarum | Das and Tamang (2023) |
| | Rice beer | Chu-as | | API 50 CH kits, 16S rRNA gene sequencing | L. plantarum | Mishra et al. (2017a) |
| | Tea (Camellia sinensis) | Laphet | Myanmar | ARDRA, 16S rRNA gene sequencing | L. pantheris, L. pentosus, L. plantarum, L. suebicus, P. cellicola, P. ethanolidurans, L. collinoides, L. paracollinoides | Bo et al. (2020) |
| | Rice, wheat | Takju | Korea | 16S rRNA gene sequencing | L. paracasei, L. plantarum, L. hilgardii, L. parabuchneri, L. brevis, L. harbinensis, L. paraplantarum, Ln. mesenteroides, Leuconostoc. sp., P. pentosaceus, Pediococcus. sp., W. cibaria, W. confusa, Weissella sp. | Jin et al. (2008); Park and Chung (2014) |
| | Rice | Tapuy | Philippines | 16S rRNA gene sequencing | L. nasuensis, L. paracasei, L. porcinae, L. rhamnosus, L. dextrinicus, L. hilgardii, L. brevis, L. fermentum | Sanico and Medina (2022) |
| | Aqueous black carrot, turnip, salt | Shalgam | Turkey | FTIR, 16S rRNA gene sequencing | L. pentosus, L. plantarum ssp. plantarum, L. fermentum | Akman et al. (2021) |
| Fermented cereal | Rice batter and black lentils | Dosa | India | 16S rRNA gene sequencing | L. plantarum, E. hirae, L. casei, L. fermentum, L. plantarum | Gupta and Tiwari (2014); Kumari et al. (2022); Lavanya et al. (2021) |

| Substrate | Product | Country | Method | Microorganisms | References |
|---|---|---|---|---|---|
| Rice batter and black lentils | | | Morphological, biochemical and physiological keys | *L. plantarum, P. pentosaceus, T. halophilus* | Pal et al. (2005) |
| Rice batter and black lentils | *Idli* | | 16S rRNA gene sequencing | *E. durans, E. faecium, E. faecium, L. plantarum, L. delbrueckii* ssp. *lactis, L. lactis* ssp. *lactis, Lc. lactis* ssp. *cremoris, Ln. garlicum, Ln. pseudomesenteroides, P. acidilactici, P. pentosaceus, P.* sp., *W. cibaria, W. confusa, W. kimchii, W. minor, W. paramesenteroides, P. parvulus, W. cibaria* | Mandhania et al. (2019); Patel et al. (2014) |
| Rice | *Wanti* | | API 50 CH kits, 16S rRNA gene sequencing | *L. plantarum* | Mishra et al. (2017a) |
| | Sour rice | | 16S rRNA gene sequencing | *W. confusa* | Nath et al. (2021) |
| | *Raabadi* | | | *L. plantarum* | Yadav et al. (2016) |
| | *Selroti* | | API system | *E. faecium, L. curvatus, Ln. mesenteroides, P. pentosaceus* | Yonzan and Tamang (2010) |
| Sorghum | Sorghum-Based | | 16S rRNA gene sequencing | *L. pentosus, L. plantarum* | Rao et al. (2015) |
| Bengal gram flour | *Dokhla* | | 16S rRNA gene sequencing, whole genome | *L. plantarum, L. fermentum* | Surve et al. (2022); Patel et al. (2014) |
| Rice | *Puto* | Philippines | Carbohydrate fermentation, PFGE | *Ln. citreum, Ln. fallax, Ln. mesenteroides* ssp. *mesenteroides, Ln. pseudomesenteroides* | Kelly et al. (1995) |
| Dough batter | *Jiaozi* | China | 16S rRNA gene sequencing | *E. durans, E. faecium, L. plantarum, P. pentosaceus* | Li et al. (2016) |
| Wheat mixed with kardeh (*Biarum carduchorum*) | *Horreh* | Iran | 16S rRNA gene sequencing | *E. faecalis, E. faecium, L. plantarum, Ln. citreum, Ln. mesenteroides* ssp. *mesenteroides, L. brevis, L. fermentum, P. pentosaceus, W. cibaria* | Vasiee et al. (2018) |
| Fermented fruit | Durian (*Durio zibethinus* L.) | *Tempoyak* | Philippines | 16S rRNA gene sequencing | *L. fructivorans, F. durionis, L. paracasei* ssp. *paracasei, L. plantarum, Ln. mesenteroides* ssp. *mesenteroides, L. brevis, L. collinoides* | Chuah et al. (2016) |

*(Continued)*

**TABLE 18.1** (Continued) Lactic Acid Bacteria Reported and Identified from Various Asian Fermented Foods, Starters and Beverages through Culture-Dependent Conventional Methods

| FOOD TYPE | SUBSTRATE | NAME | COUNTRY | METHOD OF IDENTIFICATION | MICROBES* | LITERATURE |
|---|---|---|---|---|---|---|
| Fermented juice | Viburnum opulus L. (European cranberry bush) | Gilaburu | Turkey | Biochemical tests | Lactobacillus sp., P. acidilactici, W. paramesenteroides | Yuliana and Dizon (2011) |
|  |  |  |  | Biochemical tests, 16S rRNA gene sequencing | L. plantarum | Akman et al. (2021) |
| Fermented leaves | (Colocasia esculenta L.) | Anishi | India | FTIR, 16S rRNA gene sequencing | L. pentosus, L. plantarum, L. plantarum ssp. plantarum, L. fermentum | Jamir and Deb (2017) |
| Fermented milk | Cow/yak milk | Datshi | Bhutan | 16S rRNA gene sequencing | E. faecium | Shangpliang et al. (2017) |
|  |  | Dahi |  | API 50 CHL | E. durans, E. faecalis, Lc. lactis ssp. cremoris; E. faecium | Dewan and Tamang (2007) |
|  |  | Dahi | India | API system, 16S rRNA gene sequencing | L. alimentarius, Lc. lactis ssp. cremoris, Lc. lactis ssp. lactis, L. bifermentans | Mishra et al. (2017a) |
|  |  |  |  | 16S rRNA gene sequencing | L. fermentum; L. fermentum, L. plantarum, L. fermentum, W. cibaria, E. italicus, Lc. lactis, Lc. lactis ssp. tructae, Ln. mesenteroides, L. delbruecki, L. johnsonii, L. leichmannii, Lc. hircilactis | Das et al. (2020); Patel et al. (2014); Rai and Tamang (2022); Srinivash et al. (2023) |
|  |  | Chhu |  | API system | L. alimentarius, L. farciminis, Lc. lactis ssp. cremoris; L. brevis; L. salivarius; L. bifermentans | Dewan and Tamang (2006) |
|  | Cow/yak/ewe/goat | Chhurphe |  | 16S rRNA gene sequencing | L. plantarum | Angmo et al. (2016) |
|  | Cow/yak milk | Chhurpi |  | API system | L. alimentarius, E. faecium, L. plantarum, L. hilgardii, L. kefiri | Dewan and Tamang (2007) |
|  |  |  |  | 16S rRNA gene sequencing | E. faecalis; E. pseudoavium; Lc. lactis ssp. cremoris; Lc. lactis ssp. hordniae; Ln. mesenteroides; Ln. mesenteroides ssp. jonggajiibkimchii; L. paracasei ssp. tolerans; Lc. lactis ssp. cremoris; Lc. lactis ssp. lactis; L. parabuchneri; Ln. mesenteroides ssp. mesenteroides; L. brevis; L. coryniformis ssp. torquens | Rai and Tamang (2022); Shangpliang and Tamang (2021) |

| | | | | | |
|---|---|---|---|---|---|
| | Chhurpi (hard) | | | Lc. lactis ssp. cremoris; Ln. mesenteroides; Ln. mesenteroides ssp. jonggajibkimchii | Rai and Tamang (2022) |
| | Churkam | | | E. durans; Lc. lactis ssp. cremoris; Lc. lactis ssp. hordniae; Lc. lactis ssp. lactis; Ln. mesenteroides ssp. mesenteroides; L. brevis | Shangpliang and Tamang (2021) |
| | Mar | | | E. durans; L. paracasei ssp. tolerans; Lc. lactis ssp. lactis; Ln. mesenteroides ssp. mesenteroides; L. brevis | |
| | Mohi | | API system | L. alimentarius, Lc. lactis ssp. cremoris, Lc. lactis ssp. lactis | Dewan and Tamang (2007) |
| | | | 16S rRNA gene sequencing | Lc. lactis, Ln. mesenteroides | Rai and Tamang (2022) |
| | Philu | | API system | E. faecium, L. paracasei ssp. paracasei, L. bifermentans | Dewan and Tamang (2007) |
| | Philu | | 16S rRNA gene sequencing | Lc. lactis ssp. cremoris | Rai and Tamang (2022) |
| Yak milk | Shyow | | API 50 CHL | L. casei ssp. pseudoplantarum, L. bifermentans | Dewan and Tamang (2007) |
| Cow/yak milk | Somar | | | L. casei ssp. pseudoplantarum, Lc. lactis ssp. cremoris | |
| Cow milk | Chakka | | 16S rRNA gene sequencing | L. delbrueckii ssp. bulgaricus, L. delbrueckii ssp. indicus, L. delbrueckii ssp. lactis, S. thermophilus | Mahesh et al. (2019) |
| Buffalo milk | Meekiri | | ARDRA, 16S rRNA gene sequencing | L. plantarum, L. acidophilus, L. curvatus, L. fermentum | Adikari et al. (2021) |
| | | Sri Lanka | Biochemical characterization | L. curvatus | Dekumpitiya et al. (2016) |
| | | | Biochemical, species-specific PCR characterization | L. casei ssp. casei, L. plantarum, L. acidophilus, L. delbrueckii, L. delbrueckii ssp. bulgaricus, L. delbrueckii ssp. lactis, L. helveticus, L. sp., Lc. lactis, L. fermentum, S thermophilus | Dekumpitiya et al. (2016) |
| Cow milk | Kalarei | India | 16S rRNA gene sequencing | P. acidilactici, E. faecium, L. plantarum | Bhagat et al. (2020), Bhat and Bajaj (2018), Gupta and Bajaj (2018) |

*(Continued)*

**TABLE 18.1** (Continued) Lactic Acid Bacteria Reported and Identified from Various Asian Fermented Foods, Starters and Beverages through Culture-Dependent Conventional Methods

| FOOD TYPE | SUBSTRATE | NAME | COUNTRY | METHOD OF IDENTIFICATION | MICROBES* | LITERATURE |
|---|---|---|---|---|---|---|
| Ewe milk | | Lighvan cheese | Iran | RAPD, 16S rRNA gene sequencing | L. paracasei, L. rhamnosus, L. fermentum | Forouhandeh et al. (2021) |
| – | | Cheese | | (GTG)5-PCR, ARDRA, 16S rRNA gene sequencing | Lc. lactis ssp. cremoris, Lc. lactis ssp. lactis, Ln. lactis, Ln. mesenteroides ssp. cremoris, Ln. mesenteroides ssp. mesenteroides | Haghshenas et al. (2017) |
| – | | Dahi | | | E. faecalis, Lc. lactis ssp. lactis, Ln. mesenteroides ssp. mesenteroides | |
| – | | Shiraz | | | E. mendtii, L. plantarum, Ln. lactis | |
| – | | Tarkhineh | | | Ln. mesenteroides ssp. mesenteroides | |
| – | | Yogurt | | | E. durans, L. paracasei ssp. paracasei, Lc. lactis ssp. cremoris, Ln. mesenteroides ssp. cremoris | |
| Milk | | Khameh | | 16S rRNA gene sequencing | L. plantarum, Lc. lactis | Rashid and Hassanshahia (2016) |
| Mare milk | | Koumiss | Mongolia | 16S rRNA Clone Library | E. hirae, L. helveticus, L. kefiranofaciens, Lc. lactis, L. kefiri, S. thermophilus | Ringø et al. (2014) |
| | | | | 16S rRNA gene sequencing | L. paracasei, L. pentosus, L. diolivorans, L. hilgardii, L. kefiri, L. parakefiri | |
| Mare milk | | Airag | | API50CHL, 16S rRNA gene sequencing, species-specific primer | L. farciminis, L. paracasei, L. plantarum, L. helveticus, L. kefiri | Uchida et al. (2007) |
| | | | | | E. durans, L. plantarum, Ln. mesenteroides ssp. dextranicum, Ln. mesenteroides ssp. lactis | |
| Cow/ewe/goat milk | | Aarool | | | L. plantarum, Lc. lactis ssp. lactis, Ln. mesenteroides ssp. dextranicum, L. brevis | |
| cow/ewe/goat milk | | Byslag | | | Ln. mesenteroides ssp. dextranicum | |
| Cow/ewe/goat milk | | Eej'giy | | | L. paracasei, L. acetotolerans, L. helveticus, L. kefiri | |
| Cow/ewe/goat milk | | Isgelen tarag | | | L. delbrueckii ssp. bulgaricus, L. helveticus, L. kefiri, L. fermentum | |
| Cow/ewe/goat milk | | Quormog | | | | |

| Cow/ewe/goat milk | Tarag | | | L. delbrueckii ssp. bulgaricus, L. helveticus, L. fermentum | |
|---|---|---|---|---|---|
| Mare milk | Airag | | RAPD, 16S rRNA gene sequencing, species-specific primer | L. farciminis, E. faecium, L. casei, L. plantarum, L. helveticus, L. kefiranofaciens, Lc. lactis ssp. lactis, Lactococcus sp., L. diolivorans, L. hilgardii, L. kefiri, LentiL. parafarraginis, Ln. mesenteroides, Ln. pseudomesenteroides, S. thermophilus | Watanabe et al. (2008) |
| Cow/yak/goat/camel milk | Tarag | | | E. faecium, L. casei, L. paraplantarum, L. plantarum, L. delbrueckii, L. delbrueckii ssp. bulgaricus, L. helveticus, L. kefiranofaciens, L. kefiri, L. fermentum, P. pentosaceus, Streptococcus thermophilus | |
| Yak milk | Yak milk dreg | Tibet (China) | 16S rRNA gene sequencing, DNA-DNA hybridization | E. faecium, L. plantarum, Ln. mesenteroides ssp. dextranicum, Ln. pseudomesenteroides | Duan et al. (2008) |
| Yak milk | Kurut | Tibet (China) | 16S rRNA gene sequencing | E. durans, E. faecalis, E. faecium, L. plantarum, L. delbrueckii ssp. bulgaricus, L. helveticus, Lc. lactis ssp. lactis, Ln. mesenteroides ssp. mesenteroides, L. fermentum, Streptococcus thermophilus | Sun et al. (2010) |
| Cow/mare/goat/ewe milk | Jiaoke | Inner Mongolia (China) | 16S rRNA gene sequencing | E. faecalis, E. faecium, E. gallinarum, E. gilvus, E. italicus, L. casei, L. plantarum, L. helveticus, Lc. garvieae, Lc. lactis, Ln. lactis, L. brevis, S. gallolyticus, S. thermophilus | Fan et al. (2020) |
| Cow milk | Cheese | Sri Lanka | 16S rRNA gene sequencing | L. plantarum | Vanniyasingam et al. (2019) |
| Cow milk | Yoghurt | | | L. plantarum | |
| Buffalo milk | Dadiah | Indonesia | API 50 CHL, 16S rRNA gene sequencing | L. pentosus, L. plantarum ssp. plantarum, Lc. lactis ssp. cremoris, Lc. lactis ssp. lactis, P. pentosaceus | Wirawati et al. (2019) |
| Camel milk | Aryan | Kazakhstan | 16S rRNA gene sequencing | E. faecalis, E. faecium, L. plantarum, Ln. lactis, Ln. mesenteroides | Zhadyra et al. (2021) |

*(Continued)*

**TABLE 18.1**  (Continued) Lactic Acid Bacteria Reported and Identified from Various Asian Fermented Foods, Starters and Beverages through Culture-Dependent Conventional Methods

| FOOD TYPE | SUBSTRATE | NAME | COUNTRY | METHOD OF IDENTIFICATION | MICROBES* | LITERATURE |
|---|---|---|---|---|---|---|
| | Camel milk | Shubat | Kazakhstan | | E. durans, L. plantarum, Lc. lactis, Lc. lactis ssp. hordniae, Ln. citreum, Ln. mesenteroides, Ln. mesenteroides ssp. jonggajibkimchii | Dey et al., 2023 |
| Fermented soyabean | Soybean | Hawaijar | India | BD PhoenixTM100 automated ID/AST system | Lc. plantarum | Kharnaior and Tamang (2023) |
| | | Grep chhurpi | | 16S rRNA gene sequencing | E. faecalis | |
| | | Kinema | | | E. faecium, P. acidilactici | |
| | | Peha | | | P. acidilactici | |
| | | Peron namsing | | | E. faecalis, E. faecium | |
| | | Peruñyaan | | | E. lactis | |
| | | Tungrymbai | | | E. faecium, E. lactis, E. canis, L. plantarum ssp. plantarum, Lc. lactis ssp. lactis, V. lutrae, W. hellenica | Malakar et al. (2017); Thokchom and Joshi (2012) |
| | | | | API sytem, 16S rRNA gene sequencing | L. brevis, L. fermentum | Mishra et al. (2017b) |
| | | Doenjang | Korea | 16S rRNA gene sequencing | E. durans, E. faecium, T. halophilus | Jeong et al. (2014) |
| Fermented vegetables | Cabbage | Fermented cabbage | India | 16S rRNA gene sequencing | L. fermentum, P. parvulus, W. cibaria | Patel et al., 2014 |
| | Rayo-sag' [Brassica rapa L. ssp. campestris (L.) | Gundruk | | Biochemical and physiological keys; RAPD, Species-specific PCR, rep-PCR | P. pentosaceus, L. plantarum | Tamang et al. (2005) |
| | Mustard leaves | Inziangsang | | | P. acidilactici, L. plantarum, L. brevis | |
| | Cucumber | Khalpi | | | L. plantarum, L. brevis, Ln. fallax | |

| | | | | | |
|---|---|---|---|---|---|
| | Fermented radish tap-root | *Sinki* | | | *L. brevis, Ln. fallax* | |
| | Cabbage | *Kimchi* | Korea | API system, 16S rRNA gene sequencing | *L. casei, L. paracasei, L. plantarum* | Cho et al. (2013) |
| | Cabbage | *Kimchi* | Korea | 16S rRNA gene sequencing | *L. pentosus, L. plantarum, L. curvatus, L. sakei, Ln. carnosum, Ln. citreum, Ln. gasicomitatum, Ln. gelidum, Ln. inhae, Ln. lactis, Ln. mesenteroides, W. cibaria, W. confusa, W. koreensis, Ln. kimchii, L. plantarum, L. delbrueckii* ssp. *lactis, L. sakei, Ln. mesenteroides, L. brevis, P. pentosaceus, L. kimchii* | Cho et al. (2006); Kim et al. (2000); Park et al. (2017) Yoon et al. (2000) |
| | Mustard | *Burong mustasa* | Philippine | 16S rRNA gene sequencing | *L. plantarum, Ln. mesenteroides, L. brevis* | Larcia et al. (2011) |
| | Cabbage, carrot, cucumber, radish and bamboo shoot | *Paocai* | Taiwan | API system | *L. casei, L. rhamnosus, L. plantarum* | Chang et al. (2013) |
| Fish products | *Puntius* sp., *Labeo bata* and *Colisa* sp. along with some plant materials like Colocasia | *Namsing* | India | 16S rRNA gene sequencing | *L. plantarum, V. fluvialis* | Chowdhury et al. (2019) |
| | Fish | *Nakham* | | 16S rRNA gene sequencing | *L. rhamnosus, L. helveticus, L. acidipiscis, L. fermentum* | Das et al. (2020) |
| | | *Ngari* | | BD PhoenixTM100 automated ID/AST system | *P. parvulus* | Dey et al. (2023) |
| | | *Sukamas* | | | *P. acidilactici, P. parvulus* | |
| | Fish (*Puntius sophore* and *Setipinna phasa*) | *Shidal* | | 16S rRNA gene sequencing | *E. faecalis, E. faecium, E. hirae, E. lactis, L. plantarum, P. acidilactici, P. lolii, P. pentosaceus* | Gupta et al. (2021) |
| | Fish | *Hindal* | | API 50 CH kits, 16S rRNA gene sequencing | *L. helveticus* | Mishra et al. (2017a) |
| | | *Nakham* | | | *L. rhamnosus, L. fermentum* | |
| | Fish (*Puntius* spp. and *Danio* spp.) | *Tungtap* | | Biochemical tests | *Enterococcus* sp., *Lactobacillus* sp., *Streptococcus* sp. | Rapsang and Joshi (2012) |
| | | | | API system, ARDRA, 16S rRNA gene sequencing | *L. rossiae, L. pentosus, L. plantarum, L. pobuzihii* | Rapsang and Joshi (2015) |

*(Continued)*

**TABLE 18.1** (Continued) Lactic Acid Bacteria Reported and Identified from Various Asian Fermented Foods, Starters and Beverages through Culture-Dependent Conventional Methods

| FOOD TYPE | SUBSTRATE | NAME | COUNTRY | METHOD OF IDENTIFICATION | MICROBES* | LITERATURE |
|---|---|---|---|---|---|---|
| | | | | 16S rRNA gene sequencing | L. pobuzihii | Rapsang et al. (2011) |
| Fish | | Hentak | | API system | L. amylophilus, E. faecium, F. fructosus | Thapa et al. (2004) |
| | | | | Taxonomic keys (Bergey's Manual of Systematic Bacteriology) | L. amylophilus, E. faecium, F. fructosus | |
| Fish (Puntius sophore) | | Ngari | | API system | L. plantarum, Lc. plantarum | |
| | | | | Taxonomic keys (Bergey's Manual of Systematic Bacteriology) | L. plantarum | |
| Fish (Puntius spp. and Danio spp.) | | Tungtap | | API system | F. fructosus, Lc. lactis ssp. cremoris, L. coryniformis ssp. torquens | |
| | | | | Taxonomic keys (Bergey's Manual of Systematic Bacteriology) | F. fructosus, Lc. lactis ssp. cremoris, L. coryniformis ssp. torquens | |
| Fish | | Gnuchi | | API system | E. faecium, P. pentosaceus | Thapa et al. (2006) |
| Fish (Puntius sarana) | | Sidra | | | E. faecalis, Lc. plantarum, W. confusa | |
| Fish (Schizothorax spp.) | | Suka ko maacha | | | E. faecium, Ln. mesenteroides, P. pentosaceus | |
| Fish (Harpodon nehereus) | | Sukuti | | | Lc. lactis ssp. lactis, Lc. plantarum, Ln. mesenteroides | |
| Fish | | Bordia | | API system | L. plantarum, Lc. lactis, Ln. mesenteroides | Thapa et al. (2007) |
| Fish | | Gnuchi | | | E. faecalis, E. faecium, L. plantarum, Lc. lactis, Ln. mesenteroides, P. pentosaceus | |
| Fish | | Karati | | | L. plantarum, Lc. lactis, Ln. mesenteroides | |
| Fish | | Lashim | | | L. plantarum, Lc. lactis, Ln. mesenteroides | |
| Mandarinfish | | Chouguiyu | China | ARDRA, 16S rRNA gene sequencing | E. hermanniensis, Lc. garvieae, Lc. lactis, Lc. raffinolactis, Streptococcus parauberis, Vagococcus sp. | Dai et al. (2013) |

| | | | | | |
|---|---|---|---|---|---|
| | Fish/shrimp/crab | Fermented seafood | China | 16S rRNA gene sequencing | *L. farciminis, L. futsaii, E. faecalis, L. paracasei, L. pentosus, L. plantarum, LentiL. buchneri, L. brevis, L. namurensis, L. acidipiscis, L. fermentum, L. panis, L. pontis, L. reuteri, P. acidilactici, P. pentosaceus, W. cibaria, W. confusa* | Jiang et al. (2019) |
| | Fish | *Suanyu* | China | 16S rRNA gene sequencing | *L. alimentarius, L. farciminis, L. plantarum, L. brevis, L. acidipiscis* | Liu et al. (2021) |
| | Whole fish or fish fillets | *Plaa-som* | Thailand | ARDRA, 16S rRNA gene sequencing | *L. plantarum, Lc. garvieae, L. fermentum, P. pentosaceus, Streptococcus bovis, W. cibaria* | Kopermsub and Yunchalard (2010) |
| | Fish | *Inasua* | Indonesia | 16S rRNA gene sequencing | *L. paracasei, L. rhamnosus, L. plantarum, Ln. mesenteroides* | Mahulette et al. (2018) |
| | Fish | *Yegyo ngapi* | Myanmar | | *T halophilus, T muriaticus* | Kobayashi et al. (2016) |
| | Tinfoil barb (*Barbonymus schwanenfeldii*) | *Ngachin* | Myanmar | PCR-RFLP, 16S rRNA gene sequencing | *CompaniL. farciminis, CompaniL. futsaii, L. paraplantarum, L. reuteri, P. pentosaceus, W. paramesenteroides* | Moe et al. (2015) |
| | Marine fish (*Johnius belangerii*) and freshwater fish (*Thynnichthys thynnoides*) | *Pekasam* | Malaysia | 16S rRNA gene sequencing | *L. pentosus, L. plantarum* | Ida Muryany et al. (2017) |
| | Fish | *Pla-ra* | Thailand | | *E. thailandicus, W. thailandensis* | Suwannaphan (2021) |
| | Freshwater carp or grass carp | *Suan yu* | China | | *L. plantarum, P. pentosaceus* | Zeng et al. (2016) |
| | Grass carp (*Ctenopharyngodon idella*) | *Yucha* | | MALDI-TOF-MS | *L. plantarum, Lc. lactis, L. sakei, L. brevis* | Wang et al. (2020) |
| | | | | 16S rRNA gene sequencing | *L. crustorum, L. farciminis, L. rossiae, L. casei, L. rhamnosus, L. pentosus, L. plantarum, L. buchneri, L. senioris, L. brevis, L. namurensis, L. fermentum, L. coryniformis* | Zhang et al. (2016) |
| Meat products | Goat meat | *Arjia* | India | RAPD, 16S rRNA gene sequencing | *E. faecalis, E. hirae, P. pentosaceus* | Oki et al. (2011) |
| | Goat meat | *Chartayshya* | | | *E. hirae, P. pentosaceus, W. cibaria* | |
| | Goat meat | *Jamma* | | | *E. durans, E. faecium, E. hirae, Ln. citreum, Ln. mesenteroides, P. pentosaceus* | |

*(Continued)*

**TABLE 18.1**    (Continued) Lactic Acid Bacteria Reported and Identified from Various Asian Fermented Foods, Starters and Beverages through Culture-Dependent Conventional Methods

| FOOD TYPE | SUBSTRATE | NAME | COUNTRY | METHOD OF IDENTIFICATION | MICROBES* | LITERATURE |
|---|---|---|---|---|---|---|
| | Pig meat | Faak kargyong | | API 50 CHL test strips | C. maltaromaticum, E. faecium, L. plantarum, Ln. mesenteroides, L. brevis, C. divergens, C. maltaromaticum, E. faecium, L. sanfranciscensis, L. curvatus, L. sakei, Ln. mesenteroides | Rai et al. (2010) |
| | Cow meat | Lang satchu | | | C. maltaromaticum, L. casei, P. pentosaceus | |
| | Buffalo meat | Suka ko masu | | | C. maltaromaticum, E. faecium, L. plantarum | |
| | Yak meat | Yak kargyong | | | C. divergens, C. maltaromaticum, E. faecium, L. sanfranciscensis, L. casei, L. plantarum, L. curvatus, L. sakei, Ln. mesenteroides | |
| | Yak meat | Yak satchu | | | P. pentosaceus | |

*A. = Aerococcus, C. = Carnobacterium, E. = Enterococcus, F. = Fructobacillus, L. = all species previously classified as Lactobacillus (Zheng et al., 2020), Lc. = Lactococcus, Ln. = Leuconostoc, P. = Pediococcus, S. = Streptococcus, T. = Teratogenococcus, V. = Vagococcus, W. = Weissella

method. This method is very powerful, as it allows species identification, but the main limitations include the inability to calculate the relative abundance of the species detected and the multiband signatures (Tamang et al., 2016). Some of the LAB detected and reported in fermented foods and beverages through this technique are given in Table 18.2.

**TABLE 18.2**   Lactic Acid Bacteria Reported and Identified from Various Asian Fermented Foods, Starters and Beverages through the Application of Denaturing Gradient Gel Electrophoresis and Clone Libraries

| TYPES | SUBSTRATE | NAME | COUNTRY | METHOD OF IDENTIFICATION | MICROBES | REFERENCES |
|---|---|---|---|---|---|---|
| Fermented bamboo | Bamboo | Soidon | India | DGGE, 16S rRNA gene sequencing | L. plantarum, Lc. lactis, L. sakei, L. brevis, P. pentosaceus, W. ghanensis, W. oryzae | Romi et al. (2015) |
| Fermented beverages | Tea (Camellia sinensis) | Laphet | Myanmar | | L. camelliae, L. plantarum, L. suebicus, P. ethanolidurans, L. collinoides | Bo et al. (2020) |
| Fermented cereal | Dough batter | Jiaozi | China | | L. alimentarius, L. farciminis, L. plantarum, L. brevis, P. pentosaceus W. sp. | Li et al. (2016) |
| | Rice batter and black lentils | Idli | India | | C. inhibens, E. durans, E. faecium, E. hirae, E. thailandicus, L. acidophilus, L. delbrueckii ssp. bulgaricus, Lc. lactis, L. parabuchneri, L. salivarius, L. antri, L. fermentum, L. oris, L. reuteri, S. infantarius, S. thermophilus, W. cibaria, W. diestrammenae, W. uvarum. | Mandhania et al. (2019) |
| Fermented milk | – | Doogh | Iran | | L. acidophilus, L. helveticus, L. kefiranofaciens | Sayevand et al. (2018) |
| | Mare milk | Koumiss | Mongolia | | Lc. lactis, Lc. lactis ssp. lactis | Ringø et al. (2014) |
| Fermented soyabean | Soyabean | Da-jiang | China | | E. faecium, L. plantarum, Ln. mesenteroides, Ln. gasicomitatum, T. halophilus | Wu et al. (2013) |
| | | Miso | Japan | | Ln. pseudomesenteroides, P. pentosaceus, T. halophilus, W. cibaria | Kim et al. (2010) |
| Fermented vegetables | Mustard | Burongmustasa | Philippines | | L. plantarum, L. plantarum, L. fermentum, L. panis, L. pontis, W. cibaria | Dalmacio et al. (2011); Larcia II et al. (2011) |
| Fish products | Fish | Inasua | Indonesia | | L. apinorum, L. paracasei, L. paracasei, L. curieae, L. hilgardii, L. nagelii, L. sucicola | Mahulette et al. (2018) |

(Continued)

**TABLE 18.2**  (Continued) Lactic Acid Bacteria Reported and Identified from Various Asian Fermented Foods, Starters and Beverages through the Application of Denaturing Gradient Gel Electrophoresis and Clone Libraries

| TYPES | SUBSTRATE | NAME | COUNTRY | METHOD OF IDENTIFICATION | MICROBES | REFERENCES |
|---|---|---|---|---|---|---|
| | Fermented shrimp paste | *Alamang* | Philippines | | *L. fermentum, L. panis* | Dalmacio et al. (2011) |
| | Fermented shrimp and rice | *Burong hipon (Tarlac)* | | | *L. fermentum, L. panis, L. pontis* | Dalmacio et al. (2011) |
| | Fermented fish and rice | *Burongisda* | | | *L. plantarum, L. pontis* | Dalmacio et al. (2011) |
| Fermented milk | Yak milk | *Kurut* | China | 16S rRNA clone library with RFLP screening | *L. delbrueckii ssp. bulgaricus, L. helveticus, Lc. lactis ssp. lactis, S. thermophilus* | Liu et al. (2012) |
| | Cow/mare/ goat/ewe | *Jiaoke* | | SMRT sequencing (16S rRNA) | *L. helveticus, Lc. garvieae, Lc. lactis, Lc. piscium, Lc. raffinolactis, S. agalactiae, S. thermophilus* | Fan et al. (2020) |
| Fermented vegetables | Mustard | *Burongmustasa* | Philippines | Cloned library, 16S rRNA | *L. brevis, S. moniliformis, W. cibaria* | Larcia II et al. (2011) |
| Fermented soyabean | Soyabean | *Tempeh* | Indonesia | gene sequencing | *L. delbrueckii, L. fermentum, L. mucosae* | Radita et al. (2020) |

*Notes:* A. = *Aerococcus*, C. = *Carnobacterium*, E. = *Enterococcus*, F. = *Fructobacillus*, L. = all species previously classified as *Lactobacillus* (Zheng et al., 2020), Lc. = *Lactococcus*, Ln. = *Leuconostoc*, P. = *Pediococcus*, S. = *Streptococcus*, T. = *Teratogenococcus*, V. = *Vagococcus*, W. = *Weissella*.

## 18.2.3  LAB in Asian Fermented Foods Identified by High-Throughput Amplicon Sequencing

High-throughput sequencing such as targeted amplicon sequencing also known as next-generation sequencing (NGS) has revolutionized what we know about microbial diversity (Ercolini, 2013; de Melo Pereira et al., 2022). Microbial ecology study, which profiles the bacterial community, has been widely used to study the different regions of the 16S rRNA gene, in which the common regions are variable regions 1–5 (Table 18.3). In some cases, species-level metataxonomics have been identified; however, most researchers believe that this technique can only give identification up to the genus level. The advantage of amplicon sequencing over DGGE is that the former could resolve the relative abundance of any predominant bacterial community and the detection of even minor members. This method has been widely used in microbial ecology profiling of different LAB associated with fermented food and beverages of Asian countries (Table 18.3). Through high-throughput amplicon sequencing, the predominant genera commonly identified in fermented foods and beverages of Asia belong to the genus *Lactobacillus*, with about 24.4%, followed by *Lactococcus* (16.75%), *Streptococcus* (10.05%), *Leuconostoc* (9.09%), *Weissella* (9.09%), *Enterococcus* (6.7%), *Limosilactobacillus* (4.78%), *Pediococcus* (3.83%), *Lactiplantibacillus* (2.87%), *Lentilactobacillus* (2.39%), *Levilactobacillus* (2.39%), *Ligilactobacillus* (1.44%), *Tetragenococcus* (1.44%), *Lacticaseibacillus* (0.96%), *Latilactobacillus* (0.96%), *Carnobacterium* (0.48%), *Companilactobacillus* (0.48%), *Loigolactobacillus* (0.48%), *Streptobacillus* (0.48%), *Tetragenoccocus* (0.48%) and *Vagococcus* (0.46%) (Shangpliang et al., 2018; Ashaolu and Reale, 2020; Bhutia et al., 2021a, 2021b; Zhang et al., 2021).

**TABLE 18.3** Lactic Acid Bacteria Reported and Identified from Various Asian Fermented Foods, Starters and Beverages through the Application of High-Throughput Sequencing or Amplicon/Targeted Sequencing Methods

| FOOD TYPE | SUBSTRATE | NAME | COUNTRY | METHOD OF IDENTIFICATION | MICROBES | LITERATURE |
|---|---|---|---|---|---|---|
| Fermented bamboo | Bamboo | *Suansun* | China | 16S rRNA (V3-V4) | *L. pentosus, L. acetotolerans, L. zymae, L. fermentum, L. panis* | Guan et al. (2020) |
| | | *Ma* bamboo shoot | | | *Enterococcus* sp., *Lactobacillus* sp., *Lactococcus* sp., *Weissella* sp. | Chen et al. (2022) |
| | | *Soidon* | India | 16S rRNA (V4–V5) | *Lactobacillus* sp., *Lactococcus* sp., *Leuconostoc* sp., *Weissella* sp. | Romi et al. (2015) |
| Fermented beverages | Palm sap | *Toddy* | India | 16S rRNA (V3-V4) | *L. paraplantarum, L.* sp., *Lc. lactis, Lc.* sp., *Ln. fallax, Ln. mesenteroides* | Das and Tamang (2021) |
| | Tea | *Kombucha* | Turkey | | *L. fermentum* | Kahraman-Ilıkkan (2023) |
| | Tea (*Camellia sinensis*) | *Laphet* | Myanmar | | *Lactobacillus* sp., *Pediococcus* sp. | Bo et al. (2020) |
| Fermented cereal | Black gram, rice batter and black lentils | *Idli* | India | 16S rRNA (V3) | *E.* sp., *L.* sp., *Lc.* sp., *W.* sp., *W. beninensis, W. cibaria, W. confusa, W. koreensis, W. oryzae, W. thailandensis, W. viridescens* | Mandhania et al. (2019); Kavitake et al. (2022) |
| Fermented milk | Cow milk | *Borhani* | Bangladesh | 16S rRNA (V4) | *Lactobacillus* sp., *Streptococcus* sp. | Nahidul-Islam et al. (2018) |
| | | *Chanar-misti* | | | *Lactobacillus* sp., *Streptococcus* sp. | |
| | | *Dahi* | | | *Lactobacillus* sp., *Streptococcus* sp. | |
| | | *Paneer* | | | *Lactobacillus* sp., *Lactococcus* sp., *Streptococcus* sp. | |
| | Mare milk | *Koumiss* | China | 16S rRNA (V3-V4) | *Enterococcus* sp., *Lactococcus* sp., *Leuconostoc* sp., *Streptococcus* sp., *Streptococcus* sp. | Xia et al. (2022) |
| | Cow milk | Vrum | | | *L. helveticus, L. kefiranofaciens, L. diolivorans, L. kefiri, Lc. lactis, S. salivarius* | Yamei et al. (2019) |
| | Mare milk | Koumiss | | 16S rRNA (V3-V4) | *Enterococcus* sp., *Lactobacillus* sp., *Lactococcus* sp. | Guo et al. (2019) |
| | Cow | *Dahi* | India | | *Enterococcus* sp., *Lactobacillus* sp., *Lactococcus* sp., *Leuconostoc* sp., *Pediococcus* sp., *S. luteciae, Streptococcus* sp. | (Biswal et al., 2022) |

*(Continued)*

**TABLE 18.3** (Continued) Lactic Acid Bacteria Reported and Identified from Various Asian Fermented Foods, Starters and Beverages through the Application of High-Throughput Sequencing or Amplicon/Targeted Sequencing Methods

| FOOD TYPE | SUBSTRATE | NAME | COUNTRY | METHOD OF IDENTIFICATION | MICROBES | LITERATURE |
|---|---|---|---|---|---|---|
| | Cow/yak milk | Chhurpi | | 16S rRNA (V4-V5) | L. delbrueckii, L. gasseri, L. helveticus, Lc. lactis, Lc. raffinolactis, Ln. lactis, Ln. mesenteroides, Ln. pseudomesenteroides | Shangpliang et al. (2018) |
| | | Churkam | | | L. acidophilus, L. delbrueckii, L. helveticus, L. sakei, Lc. lactis, Lc. raffinolactis, Ln. mesenteroides | |
| | | Dahi | | | L. gasseri, L. helveticus, Lc. lactis, Lc. raffinolactis, Ln. lactis, Ln. mesenteroides, Ln. pseudomesenteroides | |
| | | Mar | | | L. delbrueckii, L. helveticus, Lc. lactis, Lc. piscium, Lc. raffinolactis, Ln. mesenteroides, S. agalactiae | |
| | Cow milk | Kashkzard | Iran | 16S rRNA (V3-V5) | L. zeae, L. plantarum, L. delbrueckii, L. brevis, L. fermentum, L. pontis, L. reuteri, L. vaginalis, P. acidilactici, S. thermophilus | Pakroo et al. (2020) |
| | Camel milk | Aryan | Kazakhstan | 16S rRNA (V3-V4) | L. delbrueckii ssp. bulgaricus, L. helveticus, S. thermophilus, L. delbrueckii ssp. bulgaricus, L. helveticus, Lc. lactis, L. kefiri, Ln. mesenteroides, S. thermophilus | Zhadyra et al. (2021) |
| | Goat milk | LabnehAmbaris | Lebanon | | E. sp., L. rhamnosus, Lactobacillus sp., L. delbrueckii, Lb. helveticus, L. kefiranofaciens, L. sp., L. buchneri, Lentilactobacillus sp., Lc. lactis, P. parvulus, S. parasuis | Khalil et al. (2023) |
| | Camel milk | Laban | Saudi Arabia | | E. lactis, L. acidophilus, Lc. lactis, S. agalactiae, S. gallolyticus, S. infantarius, S. thermophilus, W. cibaria | Yasir et al. (2020) |

| Fermented soyabean | Soybean | *Sufu* | China | 16S rRNA (V3–V4) | *Lactococcus* sp. | Huang et al. (2018) |
|---|---|---|---|---|---|---|
| | | *Da-jiang* | | 16S rRNA (V4) | *Lactobacillus* sp., *Leuconostoc* sp., *Tetragenococcus* sp., *Weissella* sp. | Zhang et al. (2018) |
| | | *Tempeh* | Indonesia | | *Lactococcus* sp., *L. agilis* | Pangastuti et al. (2019) |
| Fermented vegetables | Cabbage | *Paocai* | China | 16S rRNA (V3-V4) | *L. pentosus, L. acetotolerans, L. amylolyticus, L. delbrueckii, L. parabrevis, L. fermentum, L. coryniformis, P. parvulus* | Liu et al. (2019) |
| | | *Jiangxi yancai, Sichuan paocai, Dongbei suancai* | | | *L. acetotolerans, L. sakei* | Xiao et al. (2020) |
| | Chile and tomato | *Kaili* red sour soup | | 16S rRNA (V5-V7) | *Lactobacillus* sp. | Li et al. (2021) |
| | Cabbage | *Kimchi* | Korea | 16S rRNA (V1-V3) | *L. sakei, Ln. gasicomitatum, W. koreensis* | Jeong et al. (2013) |
| Fish products | Fish | *Suanyu* | China | 16S rRNA (V3–V4) | *Lactobacillus* sp., *Tetragenococcus* sp., *Weissella* sp. | (Liu et al., 2021) |
| | Fish/shrimp/crab | Fermented seafood | China | 16S rRNA (V3-V4) | *Enterococcus* sp., *Lactobacillus* sp., *Lactococcus* sp., *Pediococcus* sp., *Weissella* sp. | Jiang et al. (2019) |
| | Freshwater carp (*Cyprinus carpio*) or grass carp (*Ctenopharyngodon idellus*) | *Suanyu* | China | 16S rRNA (V3–V4) | *Enterococcus* sp., *Lactobacillus* sp., *Lactococcus* sp., *Leuconostoc* sp., *Pediococcus* sp., *Weissella* sp. | Zang et al. (2018) |
| | Grass carp (*Ctenopharyngodon idella*) | *Yucha* | China | 16S rRNA (V3–V4) | *Enterococcus* sp., *Lactobacillus* sp., *Lactococcus* sp., *Pediococcus* sp., *Weissella* sp. | Zhang et al. (2016) |
| | Fish (*Harpodon nehereus*) | *Sukuti* | India | 16S rRNA (V3–V4) | *Vagococcus* sp., *Enterococcus* sp. | Bhutia et al. (2021a) |
| | Fish (*Schizothorax* spp.) | *Suka ko maacha* | India | 16S rRNA (V3–V4) | *Enterococcus* sp. | Bhutia et al. (2021a) |

*(Continued)*

**TABLE 18.3**   (Continued) Lactic Acid Bacteria Reported and Identified from Various Asian Fermented Foods, Starters and Beverages through the Application of High-Throughput Sequencing or Amplicon/Targeted Sequencing Methods

| FOOD TYPE | SUBSTRATE | NAME | COUNTRY | METHOD OF IDENTIFICATION | MICROBES | LITERATURE |
|---|---|---|---|---|---|---|
| | Shrimp | *Masin* | Indonesia | 16S rRNA (V3-V4) | *C. crustorum, L. pentosus, L. spicheri, L. acidipiscis, L. salivarius, T. muricatus* | Manguntungi et al. (2020) |
| | Fish | *Myeolchi-aekjeot* | Korea | 16S rRNA (V1-V3) | *Tetragenococcus* sp. | Lee et al. (2015) |
| Meat products | Cow meat | Beef *kargyong* | India | 16S rRNA (V3-V4) | *Lactococcus* sp., *Leuconostoc* sp. | Bhutia et al. (2021b) |
| | Pig meat | Pork *kargyong* | India | 16S rRNA (V3-V4) | *Enterococcus* sp., *Lactobacillus* sp., *Lactococcus* sp., *Leuconostoc* sp., *Weissella* sp. | |
| | Yak meat | *Khyopeh* | India | 16S rRNA (V3-V4) | *Carnobacterium* sp., *Enterococcus* sp., *Lactobacillus* sp. | |

*Notes:* A. = *Aerococcus*, C. = *Carnobacterium*, E. = *Enterococcus*, F. = *Fructobacillus*, L. = all species previously classified as *Lactobacillus* (Zheng et al., 2020), Lc. = *Lactococcus*, Le. = *Leuconostoc*, P. = *Pediococcus*, S. = *Streptococcus*, T. = *Teratogenococcus*, V. = *Vagococcus*, W. = *Weissella*

# 18.2.4 LAB in Asian Fermented Foods Identified by Shotgun Metagenomics

In metataxonomic analysis of fermented foods, the application of shotgun metagenomics is the most reliable and accurate method, as it can discriminate microbial population up to the species level, unlike the other NGS techniques. Shotgun metagenomic analysis is powerful enough to detect all domains of life, including bacteria, eukaryota, archaea and viruses; however, the bacteria domain is the most abundant domain in fermented foods reported through shotgun metagenomics (Tamang et al., 2021; Kharnaior and Tamang, 2022). This method is fairly new, and to date, very few studies have been applied on fermented foods of Asian countries where LAB are associated (Table 18.4). Pseudomonadota, Bacillota and Bacteroidota are the most abundant phyla reported in fermented foods (Yulandi et al., 2020). *Lactiplantibacillus plantarum* and *Lactococcus lactis* are two of the LAB species which are commonly detected through shotgun metagenomics in amylolytic starters, fermented bamboo, fermented cereal and fermented milk. *Levilactobacillus brevis* has been reported on amylolytic starters, fermented bamboo and fermented cereal. *Leuconostoc lactis*, *Leuconostoc pseudomesenteroides*, *Weissella cibaria* and *W. paramesenteroides* have been reported only in amylolytic starters. *Enterococcus hirae*, *Lactiplantibacillus pentosus*, *Lapidilactobacillus dextrinicus*, *Lap. bayanensis* and *Pediococcus stilesii* have been reported only in fermented cereal. A higher number of LAB species are associated with fermented milk products, including *Leuconostoc mesenteroides*, *Enterococcus casseliflavus*, *E. faecalis*, *E. faecium*, *E. italicus*, *Lacticaseibacillus paracasei*, *Lactobacillus delbrueckii*, *L. gallinarum*, *L. helveticus*, *L. kefiranofaciens*, *Lactococcus chungangensis*, *Lc. raffinolactis*, *Lentilactobacillus buchneri*, *Lenti. otakiensis*, *Leuconostoc citreum*, *Limosilactobacillus fermentum*, *Streptococcus macedonicus*, *St. parauberis* and *St. thermophilus* (Table 18.4). In fermented soybean products, *Limosilactobacillus fermentum* and *Enterococcus cecorum* are the only two species which have been reported as abundant species though shotgun metagenomics (Figure 18.1).

**TABLE 18.4** Lactic Acid Bacteria Reported and Identified from Various Asian Fermented Foods, Starters and Beverages through the Application of Shotgun Metagenomics

| FOOD TYPE | SUBSTRATE | NAME | COUNTRY | METHOD OF IDENTIFICATION | ABUNDANT SPECIES | REFERENCES |
|---|---|---|---|---|---|---|
| Amylolytic starter | Rice | *Xaj-pitha* | India | Shotgun metagenomics | *Lactiplantibacillus plantarum, Lactococcus lactis, Leuconostoc lactis, Leu. pseudomesenteroides, Levilactobacillus brevis, W. cibaria, W. paramesenteroides* | Bora et al. (2016) |
| Fermented bamboo | Bamboo | *Moiya pansung, mileye amileye, moiya koshak, midukeye* | | | *Lacticaseibacillus casei, Lactiplantibacillus plantarum, Lactococcus lactis, Leu. gasicomitatum, Leu. kimchii, Leu. mesenteroides, Levilactobacillus brevis, Pediococcus pentosaceus, W. paramesenteroides* | Das et al. (2023) |

*(Continued)*

**TABLE 18.4** (Continued) Lactic Acid Bacteria Reported and Identified from Various Asian Fermented Foods, Starters and Beverages through the Application of Shotgun Metagenomics

| FOOD TYPE | SUBSTRATE | NAME | COUNTRY | METHOD OF IDENTIFICATION | ABUNDANT SPECIES | REFERENCES |
|---|---|---|---|---|---|---|
| Fermented cereal | Wheat flour and black gram | *Jalebi* | | | *Enterococcus hirae*, *E.* sp., *Lactiplantibacillus pentosus*, *Lacti. plantarum*, *Lactiplantibacillus* sp., *Lactococcus lactis*, *Lac.* sp., *Lapidilactobacillus bayanensis*, *Lap. dextrinicus*, *Levilactobacillus brevis*, *Pediococcus stilesii* | Shangpliang and Tamang (2023a) |
| Fermented milk | Cow milk | *Laal dahi* | | | *Enterococcus italicus*, *Lactobacillus delbrueckii*, *Lb. gallinarum*, *Lab. helveticus*, *Lb.* sp., *Lactococcus chungangensis*, *Lac. lactis*, *Lac. raffinolactis*, *Lac.* sp., *Leuconostoc citreum*, *Leu. pseudomesenteroides*, *Leu.* sp., *Streptococcus* sp., *S. thermophilus* | Shangpliang and Tamang (2023b) |
| | Cow/sheep/ goat milk | *Ayran* | China | | *Lactiplantibacillus plantarum*, *Lactobacillus delbrueckii*, *Lab. helveticus*, *Lactococcus lactis*, *Limosilactobacillus fermentum*, *Streptococcus thermophilus* | Kuerman et al. (2023) |
| | Mare milk | *Koumiss* | China | | *Enterococcus casseliflavus*, *E. faecalis*, *E. faecium*, *Lacticaseibacillus paracasei*, *Lactobacillus helveticus*, *Lb. kefiranofaciens*, *Lactococcus lactis*, *Lentilactobacillus buchneri*, *Lentilactobacillus otakiensis*, *Leuconostoc mesenteroides*, *Streptococcus macedonicus*, *S. parauberis* | Yao et al. (2017) |
| Fermented soybean | Soybean | *Tempe* | Indonesia | | *Enterococcus cecorum*, *Limosilactobacillus fermentum* | Yulandi et al. (2020) |

**FIGURE 18.1** Different LAB species as reported through the application of shotgun metagenomics. *Lactiplantibacillus plantarum* and *Lactococcus lactis* are the two commonly detected LAB species in abundance. Additionally, highest number of LAB species are reported from fermented milk products which makes fermented milk products a good source of isolation lactic acid bacteria. In fermented soybean products, only *Limosilactobacillus fermentum* and *Enterococcus cecorum* are detected in abundance.

*Source:* Designed by authors; not copied from any other sources.

# 18.3  ABUNDANT LAB IN DIFFERENT ASIAN FERMENTED FOODS AND BEVERAGES

Lactic acid bacteria consist of several genera, including *Lactobacillus*, *Lactococcus*, *Leuconostoc*, *Pediococcus*, *Streptococcus*, *Aerococcus*, *Alloiococcus*, *Carnobacterium*, *Dolosigranulum*, *Enterococcus*, *Oenococcus*, *Tetragenococcus*, *Vagococcus* and *Weissela*. (Zhang et al., 2023). Recently, *Lactobacillus* genera were further re-classified into 23 new genera (Zheng et al., 2020). About 23 genera have been reported from different fish products, 18 from fermented cereals, 17 from fermented beverages and fermented milk products, 15 from fermented bamboo products, 13 from fermented soybean products, 11 from fermented vegetables and meat products, 10 from fermented fruits, 8 from amylolytic starters, 2 from fermented juice and only 1 from fermented leaves (Tables 18.2–18.4). Genera *Lactiplantibacillus*, *Leuconostoc* and *Levilactobacillus* are the three most commonly found genera found in most fermented products, whereas genera *Amylolactobacillus*, *Apilactobacillus*, *Furfurilactobacillus* and *Liquorilactobacillus* are only reported in fish products. Additionally, *Paucilactobacillus* and *Schleiferilactobacillus* are only reported from fermented beverages, and *Aerococcus* was reported from fermented bamboo products (Figure 18.2). Fermented milk products harbour diverse species of LAB, a total of 146 species, whereas about 114 species have been reported from fish products. About 98 species were reported from fermented cereals, 66 from fermented vegetables, 62 from fermented beverages, 60 from fermented bamboo, 48 from fermented soybean products, 36 from amylolytic starters, 32 from meta products, 18 from fermented fruits, 6 from fermented juice and 2 from fermented leaves. The most common species detected in different fermented foods and beverages are *Lactiplantibacillus pentosus*, *Lactiplantibacillus plantarum*, *Levilactobacillus brevis* and *Leuconostoc mesenteroides*. Other species which are also commonly detected in most products include *Limosilactobacillus fermentum*, *Enterococcus durans*, *Enterococcus faecalis*, *Enterococcus faecium*, *Pediococcus pentosaceus* and *W. cibaria* (Figure 18.3). Overall, from our literature searches, we have gathered information on LAB genera and species associated with different Asian fermented foods and beverages detected through the application of different methods, including culture-dependent methods, denaturing gradient gel electrophoresis, high-throughput amplicon sequencing and shotgun metagenomics. To visualize the lactic acid bacterial distribution in these fermented foods and beverages, a Sankey diagram (https://sankeymatic.com/build/) was constructed showing the distribution of different genera associated with different products (Figure 18.2), and similarly, a heatmap analysis was also constructed to visualize the microbial composition of different LAB species (Figure 18.3).

# 18.4  CONCLUSION

The diversity of a huge number of varieties and types of fermented foods and beverages in Asia is mainly associated with the tradition and culture of diverse communities. Some of these fermented foods include different varieties of fermented milks, fermented cereals, fermented vegetables, fermented bamboo, fermented beverages, fermented soybeans, fermented fruit, fermented juice, fermented leaves, amylolytic starters, fish products and meat products, as well as alcoholic beverages. Asian fermented foods and beverages harbour a wide range of microbial groups, including filamentous moulds, yeasts, bacteria, viruses and archaea; however, LAB belonging to phylum *Bacillota* are of the one of the major microbial groups associated with Asian fermented foods, mainly fermented vegetables, fermented bamboo shoots, fermented cereals, fermented milk products and fermented fish and meat products. The majority of Asian fermented foods and beverages have not been microbiologically studied, except for a few products. Moreover, the microbial communities in the majority of these spontaneously fermented foods of Asia have yet to be profiled by metagenomic shotgun sequencing and multi-omics approaches to understand the biomarkers for health-promoting benefits to consumers.

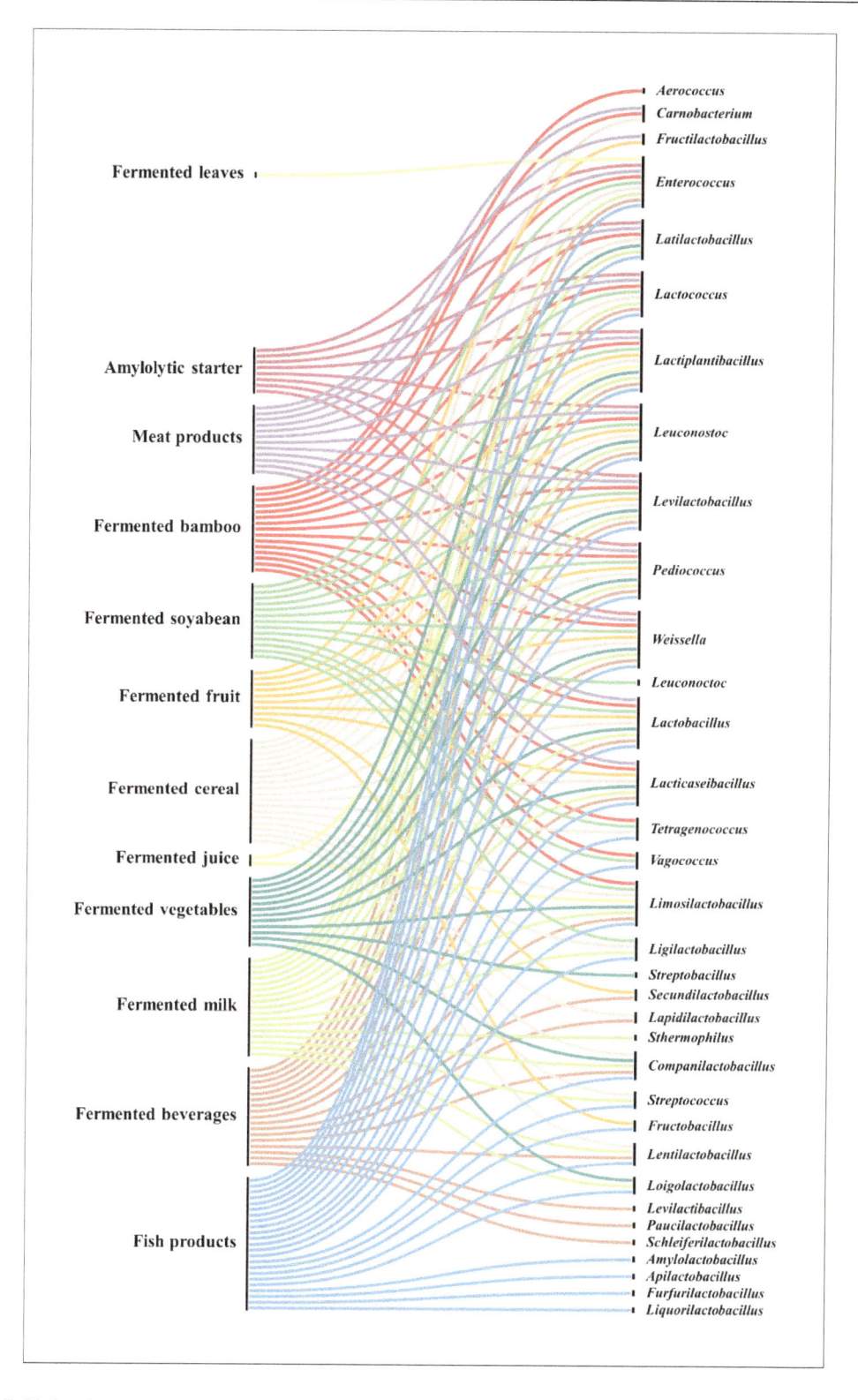

**FIGURE 18.2** Common genera reported from different fermented foods and beverages of Asian countries.

*Source:* Designed by authors; not copied from any other sources.

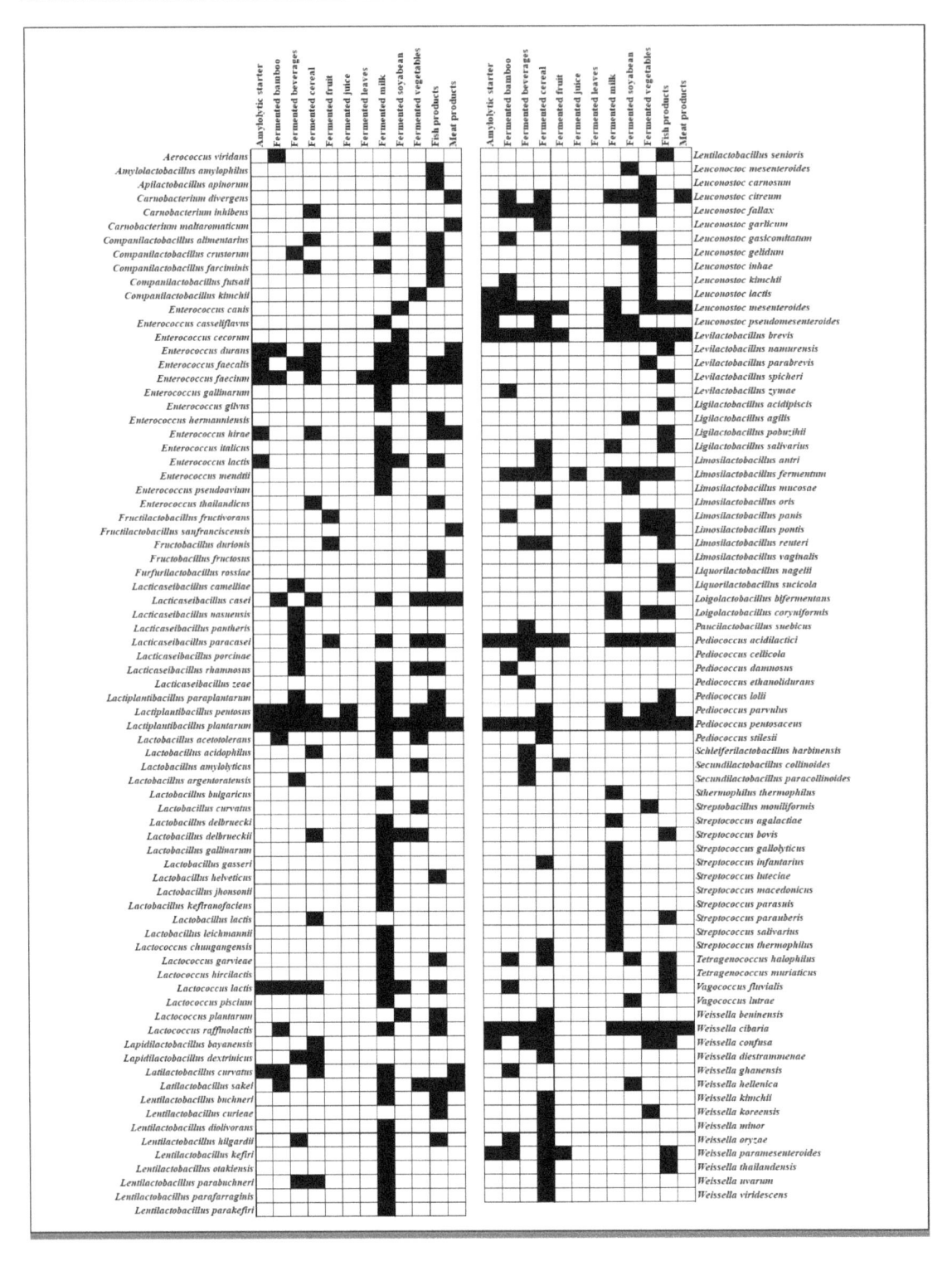

**FIGURE 18.3** Common species reported from different fermented foods and beverages of Asian countries.

*Source:* Designed by authors; not copied from any other sources.

# BIBLIOGRAPHY

Adikari, A., Priyashantha, H., Disanayaka, J., Jayatileka, D., Kodithuwakku, S., Jayatilake, J. and Vidanarachchi, J. (2021). Isolation, identification and characterization of *Lactobacillus* species diversity from Meekiri: Traditional fermented buffalo milk gels in Sri Lanka. *Heliyon* 7(10): e08136. https://doi.org/10.1016/j.heliyon.2021.e08136.

Akman, P.K., Ozulku, G., Tornuk, F. and Yetim, H. (2021). Potential probiotic lactic acid bacteria isolated from fermented gilaburu and shalgam beverages. *LWT* 149: 111705. https://doi.org/10.1016/j.lwt.2021.111705

Anal, A.K. (2019). Quality ingredients and safety concerns for traditional fermented foods and beverages from Asia: A review. *Fermentation* 5(1): 8. https://doi.org/10.3390/fermentation5010008

Angmo, K., Kumari, A. and Bhalla, T.C. (2016). Probiotic characterization of lactic acid bacteria isolated from fermented foods and beverage of Ladakh. *LWT* 66: 428–435.

Anupma, A. and Tamang, J.P. (2020). Diversity of filamentous fungi isolated from some amylase and alcohol-producing starters of India. *Frontiers in Microbiology* 11: 905. http://doi.org/10.3389/fmicb.2020.00905

Ashaolu, T.J. and Reale, A.A. (2020). Holistic review on Euro-Asian lactic acid bacteria fermented cereals and vegetables. *Microorganisms* 8(8): 1176. https://doi.org/10.3390/microorganisms8081176

Bal, J., Yun, S.H., Song, H.Y., et al. (2014). Mycoflora dynamics analysis of Korean traditional wheat-based *nuruk*. *Journal of Microbiology* 52: 1025–1029.

Baliyan, N., Dindhoria, K., Kumar, A., Thakur, A. and Kumar, R. (2021). Comprehensive substrate-based exploration of probiotics from undistilled traditional fermented alcoholic beverage 'Lugri'. *Frontiers in Microbiology* 12: 626964. https://doi.org/10.3389/fmicb.2021.626964

Bear, I. (2021). *Geography of Asia*. Writat Publisher, New Delhi, 260.

Ben Amor, K., Vaughan, E.E. and de Vos, W.M. (2007). Advanced molecular tools for the identification of lactic acid bacteria. *The Journal of Nutrition* 137(3): 741S–747S.

Bhagat, D., Raina, N., Kumar, A., Katoch, M., Khajuria, Y., Slathia, P.S. and Sharma, P. (2020). Probiotic properties of a phytase producing Pediococcus acidilactici strain SMVDUDB2 isolated from traditional fermented cheese product, Kalarei. *Scientific Reports* 10(1): 1–11. https://doi.org/10.1038/s41598-020-58676-2

Bhat, B. and Bajaj, B.K. (2018). Hypocholesterolemic and bioactive potential of exopolysaccharide from a probiotic Enterococcus faecium K1 isolated from kalarei. *Bioresource Technology* 254: 264–267.

Bhutia, M.O., Thapa, N., Shangpliang, H.N.J. and Tamang, J.P. (2021a). High-throughput sequence analysis of bacterial communities and their predictive functionalities in traditionally preserved fish products of Sikkim, India. *Food Research International* 143: 109885. https://doi.org/10.1016/j.foodres.2020.109885

Bhutia, M.O., Thapa, N., Shangpliang, H.N.J. and Tamang, J.P. (2021b). Metataxonomic profiling of bacterial communities and their predictive functional profiles in traditionally preserved meat products of Sikkim state in India. *Food Research International* 140: 110002. https://doi.org/10.1016/j.foodres.2020.110002

Biswal, P., Ghosh, S., Pal, A. and Das, A.P. (2022). Exploration of probiotic microbial biodiversity in acidic environments (curd) and their futuristic pharmaceutical applications. *Geomicrobiology Journal* 39(3–5): 176–185.

Bo, B., Kim, S.-A. and Han, N.S. (2020). Bacterial and fungal diversity in Laphet, traditional fermented tea leaves in Myanmar, analyzed by culturing, DNA amplicon-based sequencing, and PCR-DGGE methods. *International Journal of Food Microbiology* 320: 108508. https://doi.org/10.1016/j.ijfoodmicro.2020.108508

Bora, S.S., Keot, J., Das, S., Sarma, K. and Barooah, M. (2016). Metagenomics analysis of microbial communities associated with a traditional rice wine starter culture (Xaj-pitha) of Assam, India. *3 Biotech* 6: 1–13. https://doi.org/10.1007/s13205-016-0471-1

Carr, F.J., Chill, D. and Maida, N. (2002). The lactic acid bacteria: A literature survey. *Critical Reviews in Microbiology* 28(4): 281–370.

Chang, S.-M., Tsai, C.-L., Wee, W.-C. and Yan, T.-R. (2013). Isolation and functional study of potentially probiotic Lactobacilli from Taiwan traditional paocai. *African Journal of Microbiology Research* 7: 683–691.

Chen, C., Cheng, G., Liu, Y., Yi, Y., Chen, D., Zhang, L., Wang, X. and Cao, J. (2022). Correlation between microorganisms and flavor of Chinese fermented sour bamboo shoot: Roles of Lactococcus and Lactobacillus in flavor formation. *Food Bioscience* 50: 101994. https://doi.org/10.1016/j.fbio.2022.101994

Cho, J., Lee, D., Yang, C., Jeon, J., Kim, J. and Han, H. (2006). Microbial population dynamics of kimchi, a fermented cabbage product. *FEMS Microbiology Letters* 257(2): 262–267.

Cho, Y.-H., Hong, S.-M. and Kim, C.-H. (2013). Isolation and characterization of lactic acid bacteria from kimchi, Korean traditional fermented food to apply into fermented dairy products. *Food Science of Animal Resources* 33(1): 75–82.

Chowdhury, N., Goswami, G., Hazarika, S., Sharma Pathak, S. and Barooah, M. (2019). Microbial dynamics and nutritional status of *namsing*: A traditional fermented fish product of *Mishing* community of Assam. *Proceedings of the National Academy of Sciences, India Section B: Biological Sciences* 89: 1027–1038. https://doi.org/10.1007/s40011-018-1022-9

Chuah, L.-O., Shamila-Syuhada, A.K., Liong, M.T., Rosma, A., Thong, K.L. and Rusul, G. (2016). Physio-chemical, microbiological properties of *tempoyak* and molecular characterisation of lactic acid bacteria isolated from *tempoyak*. *Food Microbiology* 58: 95–104.

Coelho, R.M.D., de Almeida, A.L., do Amaral, R.Q.G., da Mota, R.N. and de Sousa, P.H.M. (2020). *Kombucha*: Review. *International Journal of Gastronomy and Food Science* 22: 100272. https://doi.org/10.1016/j.ijgfs.2020.100272

Dai, Z., Li, Y., Wu, J. and Zhao, Q. (2013). Diversity of lactic acid bacteria during fermentation of a traditional Chinese fish product, Chouguiyu (stinky mandarinfish). *Journal of Food Science* 78(11): M1778–M1783. https://doi.org/10.1111/1750-3841.12289

Dalmacio, L., Angeles, A., Larcia, L., Balolong, M. and Estacio, R. (2011). Assessment of bacterial diversity in selected Philippine fermented food products through PCR-DGGE. *Beneficial Microbes* 2(4): 273–281.

Das, D. and Goyal, A. (2014). Potential probiotic attributes and antagonistic activity of an indigenous isolate *Lactobacillus plantarum* DM5 from an ethnic fermented beverage "Marcha" of North Eastern Himalayas. *International Journal of Food Sciences and Nutrition* 65(3): 335–344.

Das, R., Tamang, B., Najar, I.N., Thakur, N. and Mondal, K. (2023). First report on metagenomics and their predictive functional analysis of fermented bamboo shoot food of Tripura, North East India. *Frontiers in Microbiology* 14: 1158411. https://doi.org/10.3389%2Ffmicb.2023.1158411

Das, S., Mishra, B.K. and Hati, S. (2020). Techno-functional characterization of indigenous *Lactobacillus* isolates from the traditional fermented foods of Meghalaya, India. *Current Research in Food Science* 3: 9–18.

Das, S. and Tamang, J.P. (2021). Changes in microbial communities and their predictive functionalities during fermentation of toddy, an alcoholic beverage of India. *Microbiological Research* 248: 126769. https://doi.org/10.1016/j.micres.2021.126769

Das, S. and Tamang, J.P. (2023). Fermentation dynamics of naturally fermented palm beverages of West Bengal and Jharkhand in India. *Fermentation* 9(3): 301. https://doi.org/10.3390/fermentation9030301

Davati, N. and Hesami, S. (2021). 16S rRNA metagenomic analysis reveals significant changes of microbial compositions during fermentation from Ewe milk to *doogh* with antimicrobial activity. *Food Biotechnology* 35(3): 179–198. https://doi.org/10.1080/08905436.2021.1939045

de Melo Pereira, G.V., de Carvalho Neto, D.P., Maske, B.L., De Dea Lindner, J., Vale, A.S., Favero, G.R., Viesser, J., de Carvalho, J.C., Góes-Neto, A. and Soccol, C.R. (2022). An updated review on bacterial community composition of traditional fermented milk products: What next-generation sequencing has revealed so far? *Critical Reviews in Food Science and Nutrition* 62(7): 1870–1889. https://doi.org/10.1080/10408398.2020.1848787

Dekumpitiya, N., Gamlakshe, D., Abeygunawardena, S. I. and Jayaratne, D. (2016). Identification of the microbial consortium in Sri Lankan buffalo milk curd and their growth in the presence of prebiotics. *Journal of Food Science and Technology Nepal* 9: 20–30.

Dewan, S. and Tamang, J.P. (2006). Microbial and analytical characterization of Chhu-A traditional fermented milk product of the Sikkim Himalayas. *Journal of Scientific & Industrial Research* 65: 747–752.

Dewan, S. and Tamang, J.P. (2007). Dominant lactic acid bacteria and their technological properties isolated from the Himalayan ethnic fermented milk products. *Antonie van Leeuwenhoek* 92(3): 343–352.

Dey, T.K., Lindahl, J.F., Sanjukta, R., Arun Prince Milton, A., Das, S., Kannan, P., Lundkvist, Å., Sen, A. and Ghatak, S. (2023). Characterization of lactic acid bacteria and pathogens isolated from traditionally fermented foods, in relation to food safety and antimicrobial resistance in tribal hill areas of Northeast India. *Journal of Food Quality* 2023. Article ID 6687015. https://doi.org/10.1155/2023/6687015

Duan, Y., Tan, Z., Wang, Y., Li, Z., Li, Z., Qin, G., Huo, Y. and Cai, Y. (2008). Identification and characterization of lactic acid bacteria isolated from Tibetan Qula cheese. *The Journal of General and Applied Microbiology* 54(1): 51–60.

Ercolini, D. (2004). PCR-DGGE fingerprinting: Novel strategies for detection of microbes in food. *Journal of Microbiological Methods* 56(3): 297–314. https://doi.org/10.1016/j.mimet.2003.11.006

Ercolini, D. (2013). High-throughput sequencing and metagenomics: Moving forward in the culture-independent analysis of food microbial ecology. *Applied and Environmental Microbiology* 79(10): 3148–3155. https://doi.org/10.1128/AEM.00256-13

Estudillo, J.P., Kijima, Y. and Sonobe, T. (2023). *Introduction: Agricultural development in Asia and Africa*. Springer, Singapore. http://doi.org/10.1007/978-981-19-5542-6_1

Fan, H., Huo, R., Zhao, J., Zhou, T., Zha, M., Kwok, L.-Y., Zhang, H. and Chen, Y. (2020). Microbial diversity analysis of jiaoke from Xilingol, Inner Mongolia. *Journal of Dairy Science* 103(7): 5893–5905.

Forouhandeh, H., Vahed, S.Z., Ahangari, H., Tarhriz, V. and Hejazi, M.S. (2021). Phenotypic and phylogenetic characterization of *Lactobacillus* species isolated from traditional Lighvan cheese. *Food Production, Processing and Nutrition* 3(1): 1–9. https://doi.org/10.1186/s43014-021-00065-x

Guan, Q., Zheng, W., Mo, J., Huang, T., Xiao, Y., Liu, Z., Peng, Z., Xie, M. and Xiong, T. (2020). Evaluation and comparison of the microbial communities and volatile profiles in homemade suansun from Guangdong and Yunnan provinces in China. *Journal of the Science of Food and Agriculture* 100(14): 5197–5206.

Gupta, A. and Tiwari, S.K. (2014). Probiotic potential of *Lactobacillus plantarum* LD1 isolated from batter of *Dosa*, a South Indian fermented food. *Probiotics and Antimicrobial Proteins* 6: 73–81.

Gupta, M. and Bajaj, B.K. (2018). Functional characterization of potential probiotic lactic acid bacteria isolated from *kalarei* and development of probiotic fermented oat flour. *Probiotics and Antimicrobial Proteins* 10: 654–661. https://doi.org/10.1007/s12602-017-9306-6

Gupta, S., Mohanty, U. and Majumdar, R.K. (2021). Isolation and characterization of lactic acid bacteria from traditional fermented fish product *Shidal* of India with reference to their probiotic potential. *LWT* 146: 111641. https://doi.org/10.1016/j.lwt.2021.111641

Haghshenas, B., Nami, Y., Almasi, A., Abdullah, N., Radiah, D., Rosli, R., Barzegari, A. and Khosroushahi, A.Y. (2017). Isolation and characterization of probiotics from dairies. *Iranian Journal of Microbiology* 9(4): 234.

Huang, X., Yu, S., Han, B. and Chen, J. (2018). Bacterial community succession and metabolite changes during sufu fermentation. *LWT* 97: 537–545.

Ida Muryany, M.Y., Ina Salwany, M.Y., Ghazali, A.R., Hing, H.L. and Nor Fadilah, R. (2017). Identification and characterization of the lactic acid bacteria isolated from Malaysian fermented fish (*Pekasam*). *International Food Research Journal* 24(2): 868–875.

Jamir, B. and Deb, C.R. (2017). Nutritional and microbiological study of anishi: A traditional fermented food product of Nagaland, India. *Journal of Advances in Food Science & Technology* 4(3): 113–121.

Janda, J.M. and Abbott, S.L. (2007). 16S rRNA gene sequencing for bacterial identification in the diagnostic laboratory: Pluses, perils, and pitfalls. *Journal of Clinical Microbiology* 45(9): 2761–2764.

Jeong, D.W., Kim, H.R., Jung, G., Han, S., Kim, C.T. and Lee, J.H. (2014). Bacterial community migration in the ripening of *doenjang*, a traditional Korean fermented soybean food. *Journal of Microbiology and Biotechnology* 24(5): 648–660.

Jeong, S.H., Lee, S.H., Jung, J.Y., Choi, E.J. and Jeon, C.O. (2013). Microbial succession and metabolite changes during long-term storage of kimchi. *Journal of Food Science* 78(5): M763–M769.

Jeyaram, K., Romi, W., Singh, T.A., Devi, A.R. and Devi, S.S. (2010). Bacterial species associated with traditional starter cultures used for fermented bamboo shoot production in Manipur state of India. *International Journal of Food Microbiology* 143(1–2): 1–8. https://doi.org/10.1016/j.ijfoodmicro.2010.07.008

Jiang, S., Ma, C., Peng, Q., Huo, D., Li, W. and Zhang, J. (2019). Microbial profile and genetic polymorphism of predominant species in some traditional fermented Seafoods of the Hainan area in China. *Frontiers in Microbiology* 10: 564. https://doi.org/10.3389/fmicb.2019.00564

Jin, J., Kim, S.-Y., Jin, Q., Eom, H.-J. and Han, N.S. (2008). Diversity analysis of lactic acid bacteria in Takju, Korean rice wine. *Journal of Microbiology and Biotechnology* 18(10): 1678–1682.

Kahraman-Ilıkkan, Ö. (2023). Microbiome composition of kombucha tea from Türkiye using high-throughput sequencing. *Journal of Food Science and Technology* 60: 1826–1833.

Kavitake, D., Suryavanshi, M.V., Kandasamy, S., Devi, P.B., Shouche, Y. and Shetty, P.H. (2022). Bacterial diversity of traditional fermented food, *Idli* by high thorough-put sequencing. *Journal of Food Science and Technology* 59(10): 3918–3927.

Kelly, W., Asmundson, R., Harrison, G. and Huang, C. (1995). Differentiation of dextran-producing Leuconostoc strains from fermented rice cake (*puto*) using pulsed-field gel electrophoresis. *International Journal of Food Microbiology* 26(3): 345–352.

Khalil, A.R., Yvon, S., Couderc, C., Belahcen, L., Jard, G., Sicard, D., Bigey, F., El Rammouz, R., Abi Nakhoul, P. and Eutamène, H. (2023). Microbial communities and main features of labneh Ambaris, a traditional Lebanese fermented goat milk product. *Journal of Dairy Science* 106(2): 868–883.

Kharnaior, P. and Tamang, J.P. (2022). Metagenomic-metabolomic mining of *kinema*, a naturally fermented soybean food of the Eastern Himalayas. *Frontiers in Microbiology* 13: 868383. https://doi.org/10.3389%2Ffmicb.2022.868383

Kharnaior, P. and Tamang, J.P. (2023). Probiotic properties of lactic acid bacteria isolated from the spontaneously fermented soybean foods of the Eastern Himalayas. *Fermentation* 9(5): 461. https://doi.org/10.3390/fermentation9050461

Kim, J., Chun, J. and Han, H.-U. (2000). *Leuconostoc kimchii* sp. nov., a new species from kimchi. *International Journal of Systematic and Evolutionary Microbiology* 50(5): 1915–1919.

Kim, T.W., Lee, J.H., Park, M.H. and Kim, H.Y. (2010). Analysis of bacterial and fungal communities in Japanese-and Chinese-fermented soybean pastes using nested PCR—DGGE. *Current Microbiology* 60: 315–320.

Kobayashi, T., Taguchi, C., Kida, K., Matsuda, H., Terahara, T., Imada, C., Moe, N.K.T. and Thwe, S.M. (2016). Diversity of the bacterial community in Myanmar traditional salted fish *yegyo ngapi*. *World Journal of Microbiology and Biotechnology* 32: 1–9. https://doi.org/10.1007/s11274-016-2127-z

Kopermsub, P. and Yunchalard, S. (2010). Identification of lactic acid bacteria associated with the production of *plaa-som*, a traditional fermented fish product of Thailand. *International Journal of Food Microbiology* 138(3): 200–204. https://doi.org/10.1016/j.ijfoodmicro.2010.01.024

Kuerman, M., Wang, R., Zhou, Y., Tian, X., Cui, Q., Yi, H., Gong, P., Zhang, Z., Lin, K. and Liu, T. (2023). Metagenomic insights into bacterial communities and functional genes associated with texture characteristics of Kazakh artisanal fermented milk Ayran in Xinjiang, China. *Food Research International* 164: 112414. https://doi.org/10.1016/j.foodres.2022.112414

Kumari, V.C., Huligere, S.S., Shbeer, A.M., Ageel, M., MK, J. and Ramu, R. (2022). Probiotic Potential *Lacticaseibacillus casei* and *Limosilactobacillus fermentum* strains isolated from dosa batter inhibit α-glucosidase and α-amylase enzymes. *Microorganisms* 10(6): 1195. https://doi.org/10.3390/microorganisms10061195

Larcia II, L., Estacio, R. and Dalmacio, L. (2011). Bacterial diversity in Philippine fermented mustard (burong mustasa) as revealed by 16S rRNA gene analysis. *Beneficial Microbes* 2(4): 263–271.

Lavanya, B.S., Sreejit, V. and Preetha, R. (2021). *Lactobacillus plantarum* J9, a potential probiotic isolated from cereal/pulses based fermented batter for traditional Indian food and its microencapsulation. *Journal of Food Science and Technology*: 1–10. https://doi.org/10.1007/s13197-021-05258-3

Lee, S.H., Jung, J.Y. and Jeon, C.O. (2015). Bacterial community dynamics and metabolite changes in myeolchi-aekjeot, a Korean traditional fermented fish sauce, during fermentation. *International Journal of Food Microbiology* 203: 15–22.

Li, D., Duan, F., Tian, Q., Zhong, D., Wang, X. and Jia, L. (2021). Physiochemical, microbiological and flavor characteristics of traditional Chinese fermented food Kaili red sour soup. *LWT* 142: 110933. https://doi.org/10.1016/j.lwt.2021.110933

Li, W., Raoult, D. and Fournier, P.-E. (2009). Bacterial strain typing in the genomic era. *FEMS Microbiology Reviews* 33(5): 892–916.

Li, Z., Li, H. and Bian, K. (2016). Microbiological characterization of traditional dough fermentation starter (Jiaozi) for steamed bread making by culture-dependent and culture-independent methods. *International Journal of Food Microbiology* 234: 9–14.

Liu, J., Lin, C., Zhang, W., Yang, Q., Meng, J., He, L., Deng, L. and Zeng, X. (2021). Exploring the bacterial community for starters in traditional high-salt fermented Chinese fish (Suanyu). *Food Chemistry* 358: 129863. https://doi.org/10.1016/j.foodchem.2021.129863

Liu, W., Sun, Z., Zhang, Y., Zhang, C., Yang, M., Sun, T., Bao, Q., Chen, W. and Zhang, H. (2012). A survey of the bacterial composition of kurut from Tibet using a culture-independent approach. *Journal of Dairy Science* 95(3): 1064–1072.

Liu, Z., Peng, Z., Huang, T., Xiao, Y., Li, J., Xie, M. and Xiong, T. (2019). Comparison of bacterial diversity in traditionally homemade paocai and Chinese spicy cabbage. *Food Microbiology* 83: 141–149.

Ly, S., Mith, H., Tarayre, C., Taminiau, B., Daube, G., Fauconnier, M.-L. and Delvigne, F. (2018). Impact of microbial composition of Cambodian Traditional dried starters (dombea) on flavor compounds of rice wine: Combining amplicon sequencing with HP-SPME-GCMS. *Frontiers in Microbiology* 9: 894. https://doi.org/10.3389/fmicb.2018.00894

Mahesh, M., Tushar, J., Pushkar, C. and Sarita, M. (2019). Assessment of lactic acid bacteria isolated from chakka: An Indian fermented dairy product. *Journal of Food and Industrial Microbiology* 5: 132.

Mahulette, F., Mubarik, N.R. and Suwanto, A. (2018). Diversity of lactic acid bacterial in inasua fermentation. *Iranian Journal of Microbiology* 10(5): 314.

Malakar, B., Das, A. and Deka, S. (2017). In vitro probiotic potential of *Enterococcus* species isolated from tungrymbai, a fermented soybean product of Meghalaya, India. *Acta Alimentaria* 46(3): 297–304.

Mandhania, M.H., Paul, D., Suryavanshi, M.V., Sharma, L., Chowdhury, S., Diwanay, S.S., Diwanay, S.S., Shouche, Y.S. and Patole, M.S. (2019). Diversity and succession of microbiota during fermentation of the traditional Indian food idli. *Applied and Environmental Microbiology* 85(13): e00368–00319. https://doi.org/10.1128/AEM.00368-19

Manguntungi, B., Saputri, D., Mustopa, A., Ekawati, N., Afgani, C., Sari, A., Triratna, L., Sukmarini, L., Fatimah, F. and Kusmiran, A. (2020). Metagenomic analysis and biodiversity of Lactic Acid Bacteria (LAB) on masin (fermented sauce) from Sumbawa, West Nusa Tenggara, Indonesia. *Biodiversitas Journal of Biological Diversity* 21(7). https://doi.org/10.13057/biodiv/d210752

Mishra, B.K., Hati, S., Das, S. and Patel, K. (2017a). Bio diversity of *Lactobacillus* cultures associated with the traditional ethnic fermented foods of West Garo Hills, Meghalaya, India. *International Journal of Current Microbiology and Applied Sciences* 6(2): 1090–1102.

Mishra, B.K., Hati, S., Das, S. and Patel, K. (2017b). Identification and characterization of *Lactobacillus* isolates from fermented soya food "Tungrymbai", Meghalaya, India. *International Journal of Current Microbiology and Applied Sciences* 6(2): 1103–1112.

Moe, N.K.T., Thwe, S.M., Shirai, T., Terahara, T., Imada, C. and Kobayashi, T. (2015). Characterization of lactic acid bacteria distributed in small fish fermented with boiled rice in Myanmar. *Fisheries Science* 81: 373–381. https://doi.org/10.1007/s12562-014-0843-6

Muhialdin, B.J., Filimonau, V., Qasem, J.M., et al. (2022). Traditional fermented foods and beverages in Iraq and their potential for large-scale commercialization. *Journal of Ethnic Foods* 9: 18. https://doi.org/10.1186/s42779-022-00133-8

Nahidul-Islam, S.M., Kuda, T., Takahashi, H. and Kimura, B. (2018). Bacterial and fungal microbiota in traditional Bangladeshi fermented milk products analysed by culture-dependent and culture-independent methods. *Food Research International* 111: 431–437.

Narzary, Y., Das, S., Goyal, A.K., et al. (2021). Fermented fish products in South and Southeast Asian cuisine: Indigenous technology processes, nutrient composition, and cultural significance. *Journal of Ethnic Foods* 8: 33. https://doi.org/10.1186/s42779-021-00109-0

Nath, S., Roy, M., Sikidar, J., Deb, B., Sharma, I. and Guha, A. (2021). Characterization and in-vitro screening of probiotic potential of novel *Weissella confusa* strain GCC_19R1 isolated from fermented sour rice. *Current Research in Biotechnology* 3: 99–108.

Oki, K., Rai, A.K., Sato, S., Watanabe, K. and Tamang, J.P. (2011). Lactic acid bacteria isolated from ethnic preserved meat products of the Western Himalayas. *Food Microbiology* 28(7): 1308–1315.

Pakroo, S., Tarrah, A., da Silva Duarte, V., Corich, V. and Giacomini, A. (2020). Microbial diversity and nutritional properties of Persian "Yellow Curd" (Kashk Zard), a promising functional fermented food. *Microorganisms* 8(11): 1658. https://doi.org/10.3390/microorganisms8111658

Pal, V., Jamuna, M. and Jeevaratnam, K. (2005). Isolation and characterization of bacteriocin producing lactic acid bacteria from a South Indian Special *dosa* (Appam) batter. *Journal of Culture Collections* 4(1): 53–60.

Pangastuti, A., Alfisah, R.K., Istiana, N.I., Sari, S.L.A., Setyaningsih, R., Susilowati, A. and Purwoko, T. (2019). Metagenomic analysis of microbial community in over-fermented tempeh. *Biodiversitas Journal of Biological Diversity* 20(4): 1106–1114.

Park, B., Hwang, H., Chang, J.Y., Hong, S.W., Lee, S.H., Jung, M.Y., Sohn, S.-O., Park, H.W. and Lee, J.-H. (2017). Identification of 2-hydroxyisocaproic acid production in lactic acid bacteria and evaluation of microbial dynamics during kimchi ripening. *Scientific Reports* 7(1): 10904. https://doi.org/10.1038/s41598-017-10948-0

Park, J.-H. and Chung, C.-H. (2014). Characteristics of *takju* (a cloudy Korean rice wine) prepared with *nuruk* (a traditional Korean rice wine fermentation starter), and identification of lactic acid bacteria in *nuruk*. *Korean Journal of Food Science and Technology* 46: 153–164.

Patel, A., Prajapati, J., Holst, O. and Ljungh, A. (2014). Determining probiotic potential of exopolysaccharide producing lactic acid bacteria isolated from vegetables and traditional Indian fermented food products. *Food Bioscience* 5: 27–33.

Pradhan, P. and Tamang, J.P. (2019). Phenotypic and genotypic identification of bacteria isolated from traditionally prepared dry starters of the Eastern Himalayas. *Frontiers in Microbiology* 10: 2526. https://doi.org/10.3389/fmicb.2019.02526

Pradhan, P. and Tamang, J.P. (2021). Probiotic properties of lactic acid bacteria isolated from traditionally prepared dry starters of the Eastern Himalayas. *World Journal of Microbiology and Biotechnology* 37: 7. http://doi.org/10.1007/s11274-020-02975-3

Radita, R., Suwanto, A., Kurosawa, N., Wahyudi, A.T. and Rusmana, I (2020). Dynamics of microbial community during tempeh fermentation. *Biotropia* 28(1): 11–20. https://doi.org/10.11598/btb.0.0.0.820

Rai, A.K., Tamang, J.P. and Palni, U. (2010). Microbiological studies of ethnic meat products of the Eastern Himalayas. *Meat Science* 85(3): 560–567. https://doi.org/10.1016/j.meatsci.2010.03.006

Rai, R. and Tamang, J.P. (2022). In vitro and genetic screening of probiotic properties of lactic acid bacteria isolated from naturally fermented cow-milk and yak-milk products of Sikkim, India. *World Journal of Microbiology and Biotechnology* 38(2): 25. https://doi.org/10.1007/s11274-021-03215-y

Rao, P.K., Chennappa, G., Suraj, U., Nagaraja, H., Charith Raj, A. and Sreenivasa, M. (2015). Probiotic potential of *Lactobacillus* strains isolated from sorghum-based traditional fermented food. *Probiotics and Antimicrobial Proteins* 7: 146–156.

Rapsang, G.F. and Joshi, S.R. (2012). Bacterial diversity associated with *tungtap*, an ethnic traditionally fermented fish product of Meghalaya. *Indian Journal of Traditional Knowledge* 11(1): 134–138.

Rapsang, G.F. and Joshi, S.R. (2015). Molecular and probiotic functional characterization of *Lactobacillus* spp. associated with traditionally fermented fish, Tungtap of Meghalaya in northeast India. *Proceedings of the National Academy of Sciences, India Section B: Biological Sciences* 85: 923–933. https://doi.org/10.1007/s40011-013-0234-2

Rapsang, G.F., Kumar, R. and Joshi, S.R. (2011). Identification of *Lactobacillus pobuzihii* from tungtap: A traditionally fermented fish food, and analysis of its bacteriocinogenic potential. *African Journal of Biotechnology* 10(57): 12237–12243.

Rashid, S. and Hassanshahian, M. (2016). Screening, isolation and identification of lactic acid bacteria from a traditional dairy product of Sabzevar, Iran. *International Journal of Enteric Pathogens* 2(4): 3–18393.

Rhee, S.J., Lee, J.E. and Lee, C.H. (2011). Importance of lactic acid bacteria in Asian fermented foods. *Microbial Cell Factory* 10(Suppl 1): S5. https://doi.org/10.1186/1475-2859-10-S1-S5

Ringø, E., Andersen, R., Sperstad, S., Zhou, Z., Ren, P., Breines, E.M., Hareide, E., Yttergård, G.J., Opsal, K., Johansen, H.M., Andreassen, A.K., Kousha, A., Godfroid, J. and Holzapfel, W. (2014). Bacterial community of koumiss from Mongolia investigated by culture and culture-independent methods. *Food Biotechnology* 28(4): 333–353. https://doi.org/10.1080/08905436.2014.964253

Romi, W., Ahmed, G. and Jeyaram, K. (2015). Three-phase succession of autochthonous lactic acid bacteria to reach a stable ecosystem within 7 days of natural bamboo shoot fermentation as revealed by different molecular approaches. *Molecular Ecology* 24(13): 3372–3389.

Sanico, T.C.F. and Medina, P.M.B. (2022). Metagenomic characterization of the culturable bacterial community structure of tapuy, a Philippine Indigenous rice wine, reveals significant presence of potential probiotic bacteria. *Acta Medica Philippina* 56(17). https://doi.org/10.47895/amp.vi0.3232

Sayevand, H.R., Bakhtiary, F., Pointner, A., Remely, M., Hippe, B., Hosseini, H. and Haslberger, A. (2018). Bacterial diversity in traditional *doogh* in comparison to industrial doogh. *Current Microbiology* 75: 386–393.

Shangpliang, H.N.J., Rai, R., Keisam, S., Jeyaram, K. and Tamang, J.P. (2018). Bacterial community in naturally fermented milk products of Arunachal Pradesh and Sikkim of India analysed by high-throughput amplicon sequencing. *Scientific Reports* 8(1): 1–10. https://doi.org/10.1038/s41598-018-19524-6

Shangpliang, H.N.J., Sharma, S., Rai, R. and Tamang, J.P. (2017). Some technological properties of lactic acid bacteria isolated from *Dahi* and *Datshi*, naturally fermented milk products of Bhutan. *Frontiers in Microbiology* 8: 116. https://doi.org/10.3389/fmicb.2017.00116

Shangpliang, H.N.J. and Tamang, J.P. (2021). Phenotypic and genotypic characterizations of lactic acid bacteria isolated from exotic naturally fermented milk (cow and yak) products of Arunachal Pradesh, India. *International Dairy Journal* 118: 105038. http://doi.org/10.1016/j.idairyj.2021.105038

Shangpliang, H.N.J. and Tamang, J.P. (2023a). Metagenomics and metagenome-assembled genomes mining of health benefits in *Jalebi* batter, a naturally fermented cereal-based food of India. *Food Research International* 172: 113130. https://doi.org/10.1016/j.foodres.2023.113130

Shangpliang, H.N.J. and Tamang, J.P. (2023b). Metagenome-assembled genomes for biomarkers of bio-functionalities in *Laal dahi*, an Indian ethnic fermented milk product. *International Journal of Food Microbiology* 402: 113130. https://doi.org/10.1016/j.foodres.2023.113130

Srinivash, M., Krishnamoorthi, R., Mahalingam, P.U., Malaikozhundan, B. and Keerthivasan, M. (2023). Probiotic potential of exopolysaccharide producing lactic acid bacteria isolated from homemade fermented food products. *Journal of Agriculture and Food Research* 11: 100517. https://doi.org/10.1016/j.jafr.2023.100517

Sujaya, I., Amachi, S., Yokota, A., Asano, K. and Tomita, F. (2001). Identification and characterization of lactic acid bacteria in *ragi tape*. *World Journal of Microbiology and Biotechnology* 17: 349–357.

Sun, H., Liu, X., Wang, L., Sang, Y. and Sun, J. (2022). Exploring the fungal community and its correlation with the physicochemical properties of Chinese traditional fermented fish (Suanyu). *Foods* 11(12): 1721. https://doi.org/10.3390/foods11121721

Sun, Z., Liu, W., Gao, W., Yang, M., Zhang, J., Wu, L., Wang, J., Menghe, B., Sun, T. and Zhang, H. (2010). Identification and characterization of the dominant lactic acid bacteria from kurut: The naturally fermented yak milk in Qinghai, China. *The Journal of General and Applied Microbiology* 56(1): 1–10.

Surve, S., Shinde, D.B. and Kulkarni, R. (2022). Isolation, characterization and comparative genomics of potentially probiotic *Lactiplantibacillus plantarum* strains from Indian foods. *Scientific Reports* 12(1): 1940. https://doi.org/10.1038/s41598-022-05850-3

Suwannaphan, S. (2021). Isolation, identification and potential probiotic characterization of lactic acid bacteria from Thai traditional fermented food. *AIMS Microbiology* 7(4): 431. https://doi.org/10.3934%2Fmicrobiol.2021026

Swain, M.R., Anandharaj, M., Ray, R.C. and Rani, R.P. (2014). Fermented fruits and vegetables of Asia: A potential source of probiotics. *Biotechnology Research International* 2014: 250424. https://doi.org/10.1155/2014/250424

Tamang, B. and Tamang, J.P. (2009). Lactic acid bacteria isolated from indigenous fermented bamboo products of Arunachal Pradesh in India and their functionality. *Food Biotechnology* 23(2): 133–147.

Tamang, B., Tamang, J.P., Schillinger, U., Franz, C.M., Gores, M. and Holzapfel, W.H. (2008). Phenotypic and genotypic identification of lactic acid bacteria isolated from ethnic fermented bamboo tender shoots of North East India. *International Journal of Food Microbiology* 121(1): 35–40.

Tamang, J.P. (2016). *Ethnic Fermented Foods and Alcoholic Beverages of Asia*. Springer, New Delhi, 409. ISBN: 978-81-322-2798-4

Tamang, J.P. (2020). *Ethnic Fermented Foods and Alcoholic Beverages of India: Science History and Culture*. Springer Nature, Singapore, 685. ISBN: 978-981-15-1486-9

Tamang, J.P. (2022a). "Ethno-Microbiology" of ethnic Indian fermented foods and alcoholic beverages. *Journal of Applied Microbiology* 133: 145–161. http://doi.org/10.1111/jam.15382

Tamang, J.P. (2022b). Dietary culture and antiquity of the Himalayan fermented foods and alcoholic fermented beverages. *Journal of Ethnic Foods* 9: 30. http://doi.org/10.1186/s42779-022-00146-3

Tamang, J.P., Anupma, A. and Shangpliang, H.N.J. (2022). Ethno-microbiology of *Tempe*, an Indonesian fungal-fermented soybean food and *Koji*, a Japanese fungal starter culture. *Current Opinion in Food Science* 100912. http://doi.org/10.1016/j.cofs.2022.100912

Tamang, J.P., Cotter, P.D., Endo, A., Han, N.S., Kort, R., Liu, S.Q., Mayo, B., Westerik, N. and Hutkins, R. (2020). Fermented foods in a global age: East meets West. *Comprehensive Reviews in Food Science and Food Safety* 19(1): 184–217.

Tamang, J.P., Dewan, S., Tamang, B., Rai, A., Schillinger, U. and Holzapfel, W.H. (2007). Lactic acid bacteria in *Hamei* and *Marcha* of North East India. *Indian Journal of Microbiology* 47(2): 119–125.

Tamang, J.P., Holzapfel, W.H. and Watanabe, K. (2016). Diversity of microorganisms in global fermented foods and beverages. *Frontiers in Microbiology* 7: 377. https://doi.org/10.3389/fmicb.2016.00377

Tamang, J.P., Jeyaram, K., Rai, A.K. and Mukherjee, P.K. (2021). Diversity of beneficial microorganisms and their functionalities in community-specific ethnic fermented foods of the Eastern Himalayas. *Food Research International* 148: 110633. https://doi.org/10.1016/j.foodres.2021.110633

Tamang, J.P., Tamang, B., Schillinger, U., Franz, C.M., Gores, M. and Holzapfel, W.H. (2005). Identification of predominant lactic acid bacteria isolated from traditionally fermented vegetable products of the Eastern Himalayas. *International Journal of Food Microbiology* 105(3): 347–356.

Tamang, J.P. and Thapa, S. (2006). Fermentation dynamics during production of *bhaati jaanr*, a traditional fermented rice beverage of the Eastern Himalayas. *Food Biotechnology* 20(3): 251–261.

Thapa, N., Pal, J. and Tamang, J.P. (2004). Microbial diversity in *ngari*, *hentak* and *tungtap*, fermented fish products of North-East India. *World Journal of Microbiology and Biotechnology* 20: 599–607. https://doi.org/10.1023/B:WIBI.0000043171.91027.7e

Thapa, N., Pal, J. and Tamang, J.P. (2006). Phenotypic identification and technological properties of lactic acid bacteria isolated from traditionally processed fish products of the Eastern Himalayas. *International Journal of Food Microbiology* 107(1): 33–38. https://doi.org/10.1016/j.ijfoodmicro.2005.08.009

Thapa, N., Pal, J. and Tamang, J.P. (2007). Microbiological profile of dried fish products of Assam. *Indian Journal Fisheries* 54(1): 121–125.

Thapa, S. and Tamang, J.P. (2004). Product characterization of *kodo ko jaanr*: Fermented finger millet beverage of the Himalayas. *Food Microbiology* 21: 617–622.

Thokchom, S. and Joshi, S.R. (2012). Antibiotic resistance and probiotic properties of dominant lactic microflora from *Tungrymbai*, an ethnic fermented soybean food of India. *Journal of Microbiology* 50(3): 535–539. https://doi.org/10.1007/s12275-012-1409-x

Tsuyoshi, N., Fudou, R., Yamanaka, S., Kozaki, M., Tamang, N., Thapa, S. and Tamang, J.P. (2005). Identification of yeast strains isolated from *marcha* in Sikkim, a microbial starter for amylolytic fermentation. *International Journal of Food Microbiology* 99(2): 135–146.

Uchida, K., Hirata, M., Motoshima, H., Urashima, T. and Arai, I. (2007). Microbiota of '*airag*', '*tarag*' and other kinds of fermented dairy products from nomad in Mongolia. *Animal Science Journal* 78(6): 650–658.

Vanniyasingam, J., Kapilan, R. and Vasantharuba, S. (2019). Isolation and characterization of potential probiotic lactic acid bacteria isolated from cow milk and milk products. *Agrieast* 13(1): 32–43. http://doi.org/10.4038/agrieast.v13i1.62

Vasiee, A., Alizadeh Behbahani, B., Tabatabaei Yazdi, F., Mortazavi, S.A. and Noorbakhsh, H. (2018). Diversity and probiotic potential of lactic acid bacteria isolated from *horreh*, a traditional Iranian fermented food. *Probiotics and Antimicrobial Proteins* 10: 258–268. https://doi.org/10.1007/s12602-017-9282-x

Waché, Y., Do, T.L., Do, T.B.H., Do, T.Y., Haure, M., Ho, P.H., Anal, K., Le, V.V.M., Li, et al. (2018). Prospects for food fermentation in South-East Asia, topics from the tropical fermentation and biotechnology network at the end of the AsiFood Erasmus+Project. *Frontiers in Microbiology* 9: 2278. https://doi.org/10.3389/fmicb.2018.02278

Wang, S., Han, J., Zhang, J., Lin, X., Liang, H., Li, S., Dong, X. and Ji, C. (2020). Effects of temperature on bacterial biodiversity and qualities of fermented *Yucha* products. *Journal of Aquatic Food Product Technology* 29(1): 43–54. https://doi.org/10.1080/10498850.2019.1693464

Wang, Z., Wang, Z., Ji, L., Zhang, J., Zhao, Z., Zhang, R., Bai, T., Hou, B., Zhang, Y., Liu, D., Wang, W. and Chen, L. (2021). A review: Microbial diversity and function of fermented meat products in China. *Frontiers in Microbiology* 12: 645435. https://doi.org/10.3389/fmicb.2021.645435

Watanabe, K., Fujimoto, J., Sasamoto, M., Dugersuren, J., Tumursuh, T. and Demberel, S. (2008). Diversity of lactic acid bacteria and yeasts in *Airag* and *Tarag*, traditional fermented milk products of Mongolia. *World Journal of Microbiology and Biotechnology* 24: 1313–1325. https://doi.org/10.1007/s11274-007-9604-3

Wirawati, C.U., Sudarwanto, M.B., Lukman, D.W., Wientarsih, I. and Srihanto, E.A. (2019). Diversity of lactic acid bacteria in dadih produced by either back-slopping or spontaneous fermentation from two different regions of West Sumatra, Indonesia. *Veterinary World* 12(6): 823. https://doi.org/10.14202%2Fvetworld.2019.823-829

Wu, J.R., Zhang, J.C., Shi, P., Wu, R., Yue, X.Q. and Zhang, H.P. (2013). Bacterial community involved in traditional fermented soybean paste dajiang made in northeast China. *Annals of Microbiology* 63(4): 1417–1421.

Xia, Y., Oyunsuren, E., Yang, Y. and Shuang, Q. (2022). Comparative metabolomics and microbial communities associated network analysis of black and white horse-sourced koumiss. *Food Chemistry* 370: 130996. https://doi.org/10.1016/j.foodchem.2021.130996

Xiao, M., Huang, T., Huang, C., Hardie, J., Peng, Z., Xie, M. and Xiong, T. (2020). The microbial communities and flavour compounds of Jiangxi yancai, Sichuan paocai and Dongbei suancai: Three major types of traditional Chinese fermented vegetables. *LWT* 121: 108865. https://doi.org/10.1016/j.lwt.2019.108865

Yadav, R., Puniya, A.K. and Shukla, P. (2016). Probiotic properties of *Lactobacillus plantarum* RYPR1 from an indigenous fermented beverage *Raabadi*. *Frontiers in Microbiology* 7: 1683. https://doi.org/10.3389/fmicb.2016.01683

Yamei, Guo, Y.-S., Zhu, J.-J., Xiao, F., Sun, J.-P., Qian, J.-P., Xu, W.-L., Li, C.-D. and Guo, L. (2019). Investigation of physicochemical composition and microbial communities in traditionally fermented *vrum* from Inner Mongolia. *Journal of Dairy Science* 102(10): 8745–8755.

Yang, Z., Liu, S., Lv, J., Sun, Z., Xu, W., Ji, C., Linag, H., Li, S., Yu, C. and Lin, X. (2020). Microbial succession and the changes of flavor and aroma in *Chouguiyu*, a traditional Chinese fermented fish. *Food Bioscience* 37: 100725. https://doi.org/10.1016/j.fbio.2020.100725

Yao, G., Yu, J., Hou, Q., Hui, W., Liu, W., Kwok, L.Y., Menghe, B., Sun, T., Zhang, H. and Zhang, W. (2017). A perspective study of koumiss microbiome by metagenomics analysis based on single-cell amplification technique. *Frontiers in Microbiology* 8: 165. https://doi.org/10.3389/fmicb.2017.00165

Yasir, M., Bibi, F., Hashem, A.M. and Azhar, E.I. (2020). Comparative metagenomics and characterization of antimicrobial resistance genes in pasteurized and homemade fermented Arabian *laban*. *Food Research International* 137: 109639. https://doi.org/10.1016/j.foodres.2020.109639

Yonzan, H. and Tamang, J.P. (2010). Microbiology and nutritional value of selroti, an ethnic fermented cereal food of the Himalayas. *Food Biotechnology* 24(3): 227–247. https://doi.org/10.1080/08905436.2010.507133

Yoon, J.H., Kang, S.S., Mheen, T.I., Ahn, J.S., Lee, H.J., Kim, T.K., Park, C.S., Kho, Y.H., Kang, K.H. and Park, Y.H. (2000). *Lactobacillus kimchii* sp. nov., a new species from kimchi. *International Journal of Systematic and Evolutionary Microbiology* 50(5): 1789–1795.

Yulandi, A., Suwanto, A., Waturangi, D.E. and Wahyudi, A.T. (2020). Shotgun metagenomic analysis reveals new insights into bacterial community profiles in tempeh. *BMC Research Notes* 13(1): 1–7. https://doi.org/10.1186/s13104-020-05406-6

Yuliana, N. and Dizon, E.I. (2011). Phenotypic identification of lactic acid bacteria isolated from *Tempoyak* (fermented durian) made in the Philippines. *International Journal of Biology* 3(2): 145.

Zang, J., Xu, Y., Xia, W., Yu, D., Gao, P., Jiang, Q. and Yang, F. (2018). Dynamics and diversity of microbial community succession during fermentation of *Suan yu*, a Chinese traditional fermented fish, determined by high throughput sequencing. *Food Research International* 111: 565–573.

Zeng, X., Zhang, W. and Zhu, Q. (2016). Effect of starter cultures on the quality of *Suan yu*, a Chinese traditional fermented freshwater fish. *International Journal of Food Science and Technology* 51(8): 1774–1786. https://doi.org/10.1111/ijfs.13140

Zhadyra, S., Han, X., Anapiyayev, B.B., Tao, F. and Xu, P. (2021). Bacterial diversity analysis in Kazakh fermented milks *Shubat* and *Ayran* by combining culture-dependent and culture-independent methods. *LWT* 141: 110877. https://doi.org/10.1016/j.lwt.2021.110877

Zhang, D., Zhang, J., Kalimuthu, S., Liu, J., Song, Z.-M., He, B.-B., Cai, P., Zhong, Z., Feng, C. and Neelakantan, P. (2023). A systematically biosynthetic investigation of lactic acid bacteria reveals diverse antagonistic bacteriocins that potentially shape the human microbiome. *Microbiome* 11(1): 1–20. https://doi.org/10.1186/s40168-023-01540-y

Zhang, J., Song, H.S., Zhang, C., Kim, Y.B., Roh, S.W. and Liu, D. (2021). Culture-independent analysis of the bacterial community in Chinese fermented vegetables and genomic analysis of lactic acid bacteria. *Achieves of Microbiology* 203: 4693–4703.

Zhang, J., Wang, X., Huo, D., Li, W., Hu, Q., Xu, C., Liu, S. and Li, C. (2016). Metagenomic approach reveals microbial diversity and predictive microbial metabolic pathways in Yucha, a traditional Li fermented food. *Scientific Reports* 6(1): 1–9. https://doi.org/10.1038/srep32524

Zhang, P., Wu, R., Zhang, P., Liu, Y., Tao, D., Yue, X., Zhang, Y., Jiang, J. and Wu, J. (2018). Structure and diversity of bacterial communities in the fermentation of da-jiang. *Annals of Microbiology* 68(8): 505–512.

Zheng, J., Wittouck, S., Salvetti, E., Franz, C.M., Harris, H.M., Mattarelli, P., O'Toole, P.W., Pot, B., Vandamme, P. and Walter, J. (2020). A taxonomic note on the genus *Lactobacillus*: Description of 23 novel genera, emended description of the genus *Lactobacillus* Beijerinck 1901, and union of Lactobacillaceae and Leuconostocaceae. *International Journal of Systematic and Evolutionary Microbiology* 70(4): 2782–2858.

# The Production of Lactic Acid Bacteria Starters, Probiotic and Postbiotic Cultures

## *An Industrial Perspective*

19

Gabriel Vinderola, Melisa Puntillo, Émilie Desfossés-Foucault, and Claude P. Champagne

## 19.1 LAB AND BIFIDOBACTERIA IN FOOD AND SUPPLEMENTS

Decades ago, many fermented foods were manufactured by spontaneous fermentation or by cultures that were continuously subcultured at the manufacture plant. Cheese, yogurt, fermented vegetables, dry sausages, and sourdoughs are examples of such food products. Gradually, bacteria that were associated with good products were identified and isolated. As a result, specialized manufacturers developed processes to grow and provide these bacteria for the food manufacturing industry. Not surprisingly, after *Saccharomyces cerevisiae*, lactic acid bacteria (LAB) are the most widely used cultures in food fermentation.

LAB are not only useful to produce fermented foods, but many are also known to offer health benefits when eaten alive in sufficient amounts (Hill et al., 2014) but also not alive, giving raise to the concept of postbiotics. Specific strains may be developed as probiotics and are now manufactured and sold as such. Specific strains of other bacteria, such as bifidobacteria, are also considered probiotics but were not originally isolated from food, as many species currently marketed are most likely to be of human origin. Therefore, for the purpose of this chapter, the term "LAB" will include (1) the cultures that are used for food fermentation and are essential from a technological perspective (e.g., *Lactococcus lactis*) as well as (2) those that are added primarily for their health benefits and considered probiotics. Bifidobacteria will be considered probiotics separate from the LAB.

The manufacturing process of probiotic LAB and bifidobacteria are often similar, if not identical. Thus, many producers of starter cultures also manufacture probiotic bacteria. The goal of this chapter is to describe processes used for the manufacture of LAB and bifidobacteria by specialized suppliers. In some

DOI: 10.1201/9781003352075-21

instances, starter production at the food manufacturers' site is addressed. In this chapter, the term "probiotic bacteria" will mostly refer to cultures based on strains of lactobacilli and bifidobacteria that display probiotic properties, as conditions for the production of probiotics based on genera like *Saccharomyces*, *Propionibacterium*, *Enterococcus*, or *Bacillus* will not be addressed.

A note should also be made that the traditional abbreviations of *L.* used for either for *Lactobacillus* (the former genus reclassified in 2020 in 25 genera (Zheng et al., 2020), *Lactococcus*, or *Leuconostoc* cannot be applied because a given abbreviation cannot represent three different terms; thus the abbreviations *L.*, *Lc.*, and *Ln.*, respectively, will be used, which is not conventional. The species belonging to the former genus *Lactobacillus* cited in this work are:

- *Lactobacillus acidophilus, helveticus, johnsonii, delbrueckii* ssp. *bulgaricus, gasseri*
- *Lacticaseibacillus casei, paracasei, rhamnosus*
- *Lactiplantibacillus plantarum*
- *Limosilactobacillus reuteri*

For purposes of clarity, all these former lactobacilli will be abbreviated with the letter "L".

## 19.2 NUTRITIONAL REQUIREMENTS OF LAB AND BIFIDOBACTERIA AND GROWTH MEDIA

LAB and bifidobacteria have numerous nutritional requirements, and various defined (synthetic) media have been developed to analyze them (Table 19.1). This chapter mainly focuses on the industrial production of starters and probiotic bacteria. In such an industrial setting, defined media are not used because they are too expensive and because requirements vary between strains. Therefore, industry uses ingredients that would contain as many of the essential nutrients as possible (Table 19.1). Obviously, since they are not purified, they are also less expensive.

A goal of this section is not only to provide data on ingredients used for high biomass yields but also to address the links between medium composition and functionality. With respect to functionality, ingredients will influence two main series of attributes:

1. Technological functionality (survival to the various production steps, stability during storage, acidifying activity, aroma production)
2. Health benefits (survival to gastrointestinal conditions, health attributes)

As an example of how modification of the composition of the growth medium can influence technological functionality, Carvalho et al. (2004) showed that freeze-dried cells of *L. bulgaricus* grown in the presence of 2% glucose survived several log orders higher after 10 months of storage than the same culture produced in a medium containing 1% glucose and 1% saccharose. However, since this starter culture it is not grown on lactose, it would presumably not have high levels in β-galactosidase and would be slow in initiating milk fermentation. Other examples are presented in Table 19.1. As for exopolysaccharide (EPS) synthesis, it can either be desirable or undesirable, depending on the application. When producing concentrated freeze-dried cultures, EPS is generally not favored because of increased medium viscosity, which hampers cell recovery by centrifugation or membrane filtration. However, when LAB starters used for dairy fermentation are produced, EPS is favored because it increases the viscosity of fermented milk (Ruas-Madiedo et al., 2002) and could offer a slight protection against bacteriophage attack. Manufacturers need to balance conditions that promote high yields and storage stability with those that generate high bioactivity.

**TABLE 19.1** Growth Factors for LAB/Probiotic Bacteria and How Ingredients Typically Used in Industrial Growth Media (1) Can Provide the Essential Nutrients and (2) Affect the Subsequent Functionality of the Cultures

| INGREDIENT[a] | TYPICAL SOURCE | SOME EFFECTS ON FUNCTIONALITY |
|---|---|---|
| Carbohydrate | Glucose or lactose as purified compounds Dairy-based ingredients (milk, whey) | Greater storage stability of *L. bulgaricus* if grown on these than if grown on a mixture of glucose and sucrose (Carvalho et al., 2004) Faster acidification of dairy products if grown on lactose (Desjardins et al., 1991) More EPS was produced when *B. longum* was grown on lactose rather than on glucose, galactose, or fructose (Audy et al., 2010) |
| Amino acids or peptones | Mainly added through peptones, yeast extracts, or proteins (milk) | Growth on peptones may result in low proteolytic activity and lower acidifying activity when cells are added to milk (St-Gelais et al., 1992) |
| Fatty acids | Tween, acetate | High levels of unsaturated fats in cytoplasmic membranes: Improve survival to freezing Reduce storage stability of dried cultures |
| Nucleosides/ nucleotides | Yeast extracts | None reported; mainly serve as growth factors |
| Minerals: phosphates, magnesium, manganese | Mainly added as salts | Enhance exopolysaccharide production (Macedo et al., 2002) |
| Vitamins | Yeast extracts, meat extracts, and peptones | May enhance fermentation activity but are mainly growth factors |
| Citrate | Purified Dairy-based ingredient | *Lactococcus* starters grown on a citrate-rich medium will subsequently produce higher levels of diacetyl in fermented milks |

[a] Data for ingredients as potential requirements were taken from Cocaign-Bousquet, M. et al., *J. Appl. Bacteriol.*, 79, 108–116, 1995; Morishita, T. et al., *J. Bacteriol.*, 148, 64–71, 1981; and Partanen, L. et al., *Syst. Appl. Microbiol.*, 24, 500–506, 2001.

Since LAB require many amino acids for growth, it would seem logical to enrich the medium with free amino acids, but this is not the case (St-Gelais et al., 1993). Oligopeptides are the main source of nitrogen for *Lactococcus lactis* for growth in milk (Juillard et al., 1995). As a result, peptones are systematically added to growth media for LAB and bifidobacteria. Unfortunately, there is no ideal source of peptone for all species (Potvin et al., 1997). It is obviously impossible to use different peptones for each strain, but special formulations are desirable for high-value strains. If in doubt, blends of peptones are recommendable. Not surprisingly, de Man–Rogosa–Sharpe (MRS) broth, which is one the best media for growth and enumeration of LAB probiotics (Champagne et al., 2011), contains two sources of peptones. In addition to the casein peptones, yeast extracts are found in the medium, as they are not only rich in peptones but also contain most of the other nutrients that are essential for growth: vitamins, nucleosides/ nucleotides, and fatty acids.

Lipid metabolism is little studied, but LAB may have fatty acid requirements. In practice, Tween 80 is added to the medium, and there are different versions of the product depending on chain length (C14 to C18). Presumably, acetate in the medium could serve as building blocks for fatty acid synthesis. Muller et al. (2011) modified the technical properties (heat and acid tolerance) of *L. johnsonii* NCC 533 by supplementing the growth medium with unsaturated fatty acids. Enhanced properties were related to changes in the hydrophobicity and fluidity of the membrane. Also, supraoptimal incubation temperatures result in a decrease in lipid saturation levels of lactobacilli (Schoug et al., 2008). This reduces the fluidity of the membrane, which is undesirable when cultures are to be frozen. However, such changes reduce membrane lipid oxidation during storage, which is beneficial.

Huang et al. (2016) increased the spray-drying resistance of *Propionibacterium freudenreichii* by growing it in hyperconcentrated (from 50 to 300 g/L) sweet whey that triggered a multitolerance response through osmoadaptation. Inclusion of inorganic salts ($KH_2PO_4$, $Na_2HPO_4$, $MnSO_4$, and $MgSO_4$) in the culture medium played a positive role in the thermotolerance of *L. rhamnosus* CRL 1505 to spray-drying (Correa Deza et al., 2017).

EPSs are just one ingredient in numerous bioactives that probiotic bacteria synthesize to exert their health effect. Another example would be short-chain fatty acids (SCFAs), which are now believed to be of the upmost importance in the gastrointestinal system, and probiotics are known to produce acetate, propionate, and butyrate, the last one being the most extensively studied. These SCFAs could have a positive impact on inflammation, cancer, glucose metabolism, lipid metabolism, satiety, and weight loss (Koh et al., 2016). They could also influence our behavior and mood (Sherwin et al., 2016). Furthermore, bifidobacteria are known to increase butyrate production by indigenous bacterial species (Rivière et al., 2016). Other ingredients with potential health benefits include gamma amino butyric acid for hypertension (Inoue et al., 2003), α-galactosidase for gas discomfort (Nobaek et al., 2000), β-galactosidase for lactose discomfort (Kailasapathy et al., 2011), and bile salt hydrolase (BSH) for blood serum cholesterol level (Jones et al., 2012). In many instances, BSH is not synthesized in the absence of bile salts in the medium (Begley et al., 2006). Thus, if a producer of probiotics wishes to have cultures which have high BSH activity, bile salts must be included in the medium.

Two points that are rarely raised when discussing growth media for biomass production of probiotics are (1) the heating treatment that is applied to pasteurize (starters prepared at cheesemaking plants) or sterilize (starter or probiotic manufacturers) the medium and (2) lot variability in ingredients purchased from suppliers. Little is known about the effect of heating temperature on the subsequent growth of cultures, and most data pertain to starters prepared at cheesemaking plants (Feldstein and Westhoff, 1979; Kurultay et al., 2006). Obviously, heating conditions will be detrimental because of vitamin inactivation and precipitation of some minerals. Maillard reactions could reduce the availability of amino acids. On the other hand, Maillard products can reduce the redox level, which is beneficial for strains that are highly sensitive to oxygen. As a result, the effect of medium heating on subsequent growth of cultures varies between strains, and no universally optimum conditions are recommendable. Although it is generally thought that overheating is to be avoided, the important point for reproducible growth of cultures is to avoid variations in the heating and cooling of the media. In industrial conditions, variations will occur due to (1) variable size of the starter tanks and (2) unexpected demands for steam or cooling liquid in other parts of the plant.

Variations between lots (Champagne et al., 1999) and suppliers (Kreft et al., 2001) are another problem. If possible, manufacturers of cultures should pre-test product lots from their ingredient suppliers using automated systems (Potvin et al., 1997).

# 19.3 GROWTH CONDITIONS

Other than the fermentation medium ingredients, four fermentation parameters will affect the growth of probiotics: oxygen, agitation/redox level, temperature, and pH. A few of the optimum conditions for LAB used in fermented foods are presented in Table 19.2. *Streptococcus thermophilus* and *L. delbrueckii* subsp. *bulgaricus* contribute to the reduction of intestinal problems linked to lactose intolerance, surprisingly not in a strain-independent manner but nevertheless species specific (EFSA, 2010).

The temperature of the medium during fermentations is set to the one that is considered optimum for growth (Table 19.2). When the production of EPSs is undesirable, incubating at temperatures slightly above optimum helps to reduce EPS levels (Champagne et al., 2007). Increasing the temperature at the end of the fermentation in order to generate heat stress constitutes an adaptation to temperature profiles designed to improve functionality.

**TABLE 19.2** Some Growth Parameters of Some LAB Used for Food Fermentations

| SPECIES[a] | FOOD PRODUCT | GROWTH PARAMETERS[b] | | | ROLE |
|---|---|---|---|---|---|
| | | pH | °C | $a_w$ | |
| Lc. lactis subsp. lactis | Cheese, sour cream, kefir | 6.0–6.5 | 29–34 | 0.96 | Acidification during fermentation |
| Lc. lactis subsp. cremoris | Cheese, sour cream, kefir | 6.0–6.5 | 28–32 | 0.95 | |
| Lc. lactis subsp. lactis. biovar. diacetylactis | Cheese, sour cream, kefir | 6.0–6.5 | 29–34 | 0.95 | Acidification during fermentation; production of aromas and gas |
| Ln. lactis, Ln. cremoris | Cheese, sour cream, kefir | 5.5–6.0 5.5–6.0 | 20–27 20–27 | 0.96–0.98 0.96–0.98 | Production of dextrans; production of aromas (diacetyl) and gas; slight acidifying activity |
| Ln. mesenteroides | Fermented vegetables | | | | Acidification, particularly with acetic acid |
| S. thermophilus | Yogurt | 6.0–6.5 | 40–42 | 0.94 | Acidification during production; production of exopolysaccharides; production of aroma (acetaldehyde); probiotic (lactose intolerance) |
| L. delbrueckii subsp. bulgaricus | Yogurt | 5.5–6.0 | 42–46 | 0.95 | Acidification during production; production of exopolysaccharides; production of aroma (acetaldehyde); probiotic (lactose intolerance) |
| L. helveticus | Cheese | 5.4–5.9 | 42–47 | | Acidification during production; probiotic (immune system) |
| L. plantarum | Fermented vegetables and meats | | 30–37 | 0.93 | Acidification during production; probiotic |
| L. casei | Cheese | | 30–37 | 0.93 | Cheese ripening; probiotic (diarrhea) |
| Pediococcus acidilactici | Fermented vegetables and meats; probiotic | | 38–42 | | Acidification; probiotic (animal feeds) |

[a] Lc. = Lactococcus; Ln. = Leuconostoc; L. = various new genera derived from Lactobacillus (see Section 20.1); S. = Streptococcus.
[b] Parameters: pH and °C are optimal values, while $a_w$ are minimal levels.

It is recommended to carry out fermentation at the optimal pH. Unfortunately, LAB and most probiotics, excluding *Bacillus* and *Saccharomyces* strains, acidify the medium, which is a challenge for manufacturers. One way to address this problem is to add buffers in the medium. For example, MRS contains four buffers: phosphate, citrate, and, to a lesser extent, peptones and acetate. However, there is a limit to the benefits of adding buffers. For example, cell morphology of *L. bulgaricus* is adversely affected at 2% of phosphate (Wright and Klaenhammer, 1984). As a result, pH control is often practiced for biomass production rather than addition of high levels of buffer salts.

Most growth media are around 0.97–0.98 in $a_w$, which is not inhibitory. However, there are instances in which it can be adjusted to lower values to create a stress that will increase survival of the cultures during drying (Champagne et al., 2012).

The oxygen/redox level is a less-controlled fermentation parameter. The growth of LAB and bifidobacteria is rarely improved by the presence of oxygen. Therefore, as a rule, high agitation of the medium after heating is to be avoided. The reason oxygen can be detrimental is mainly that toxic $H_2O_2$ is produced in the presence of oxygen. With aerobic bacteria, this is not a problem because these cultures synthesize catalase. However, most LAB and bifidobacteria do not produce catalase and rely on NADH oxidase,

NADH peroxidase, and superoxide dismutase to remove toxic metabolites which appear in the presence of oxygen (Talwalkar and Kailasapathy, 2003). Many strains do not produce high levels of these enzymes and do not grow in the presence of oxygen. For such strains, various strategies are available. On a large scale, sparging nitrogen in the medium at the beginning of the fermentation is one method, but simply flushing the headspace with nitrogen might be enough.

# 19.4 FERMENTATION METHODS

## 19.4.1 Type of Process

Batch and fed-batch fermentations are the most common processes used in the industry to produce biomass of LAB and bifidobacteria. The technique has proved useful for the production of probiotic or starter cultures that produce EPS. Because of medium viscosity resulting from strong EPS production, manufacturers of LAB and probiotics have difficulty in concentrating the cultures prior to freezing or drying. EPS synthesis is favored in media with high carbon:nitrogen ratios. Thus, carrying out a fed-batch fermentation, where the carbohydrate is only added gradually to maintain a low carbohydrate level, reduces the viscosity of the medium and enables cell recovery by centrifugation (Champagne et al., 2007).

The production of biomass of *Lactobacillus casei* BPG4 in goat milk whey was compared in batch, fed-batch, and continuous fermentation by Aguirre-Ezkauriatza et al. (2010). Fed-batch cultures at high biomass concentration rendered higher productivity than batch and continuous culture, complete lactose conversion, and a freeze-dried product with higher viable cell counts. It was demonstrated that substrate limitation and product inhibition can be avoided in fed-batch fermentation. Additionally, product inhibition can be eliminated by diluting the product concentration with fresh culture medium (Lee et al., 2007).

## 19.4.2 The Importance of pH Control

Batch and fed-batch fermentations can be carried with or without pH control. LAB, and to a lesser extent bifidobacteria, are relatively acid tolerant, but, without pH control, the accumulation of lactic acid ultimately influences their physiology and microbes cease to grow due to autoacidification rather than depletion of nutrients. As the growth rate slows down when pH drops, controlling pH at values close to neutrality ensures a higher growth rate (Giraud et al., 1991). This can be achieved by automatic addition of ammonium, sodium, or potassium hydroxide or by including in the fermenter $CaCO_3$ or $Ca(OH)_2$ that dissolves as pH drops, neutralizing acidity. The type of neutralizing agent will influence cell yields. With mesophilic lactococci, $NH_4OH$ and $Ca(OH)_2$ generate higher biomass levels than NaOH (Lloyd and Pont, 1973; Peebles et al., 1969). Less information is available with thermophilic starters, and no difference was noted between NaOH and $NH_4OH$ (Champagne et al., 1993).

Up to 10 times more biomass can be obtained when pH is controlled. For example, whereas *L. rhamnosus* 64 achieved cell counts of approximately $1 \times 10^9$ CFU/mL in batch fermentation in dairy-based media without pH control (Lavari et al., 2014), the same strain attained $10^{10}$ CFU/mL after 12 hours in pH 6 controlled batch fermentation in cheese whey permeate properly supplemented, using $Na_2CO_3$/NaOH solution for pH adjustment (Lavari et al., 2015).

Unfortunately, product inhibition occurs even under pH control. Lactate accumulates and becomes inhibitory for growth. With *Lc. lactis*, the growth rate is reduced by 40% when 2% lactate is reached, and no or little growth occurs when 5% lactate is obtained (Bergère and Hermier, 1968). It is important to note that, in appearance, growth still happens under these conditions. Indeed, uncoupling between growth and acidification occurs under various conditions with lactococci: intracellular pH < 5.0 (Nannen and Hutkins, 1991), lactate > 5% (Bergère and Hermier, 1968), incubation temperature > 38 °C (Breheny et al., 1975), or depletion of an essential growth factor. In these instances, acidification can still occur even if CFU increases have stopped, erroneously giving the impression that growth continues as well.

Constant agitation during pH control can introduce oxygen in the medium. As mentioned previously, some LAB and probiotic cultures are sensitive to oxygen. As a result, some pH control fermentations are carried out in a "zone" fashion. In this process, there is no initial agitation of the medium. Acidification is allowed to drop to a given set point (typically at pH = 5.8 for mesophilic cultures), which initiates agitation and then the addition of the alkali. The pH is raised to a second set point (typically pH = 6.0), at which moment alkali addition is stopped, as well as agitation of the medium. The pH is allowed to gradually drop back to the first set point without agitation. Thus agitation, and aeration of the medium, only occurs during the short period whereas pH is raised from set point 1 to set point 2.

The choice of the pH at which fermentation is controlled and the moment of cell harvesting are also aspects to be considered in relation to the technological and the *in vivo* (functional) properties of the culture.

The technological effects include resistance to dehydration, viability during storage, and acidifying activity. Surprisingly, cells grown under pH control above 6.0 show lower survival levels to spray-drying (Silva et al., 2005), air-drying (Champagne et al., 2012), and some freeze-drying assays (Shao et al., 2014) than cultures that are allowed to acidify the medium to pH values below 5.0. However, the detrimental effect of maintaining pH control in the vicinity of pH = 6.0 on subsequent survival to freeze-drying varies between cultures (Champagne and Gardner, 2002). Rault et al. (2010) investigated the effects of freezing and frozen storage (–20 °C) on the survival, cultivability, and acidification activity of *L. bulgaricus* CFL1. Batch fermentation was carried out at pH 5 and 6, or without pH control, and cells were collected at different moments (during and at the end of the log and the stationary phases). It was observed that tolerance of cells enhanced with fermentation time or when the fermentation was carried out at pH 5, as compared with fermentation at pH 6 or without pH control. In some instances, carrying out an acid stress prior to drying enhances the stability of the dried culture during storage (Barbosa et al., 2015), but the contrary was also observed (Saarela et al., 2009). The use of pH control does not only affect biomass yields, it also affects the acidifying activity of the cultures. As a rule, starter cultures that are grown under pH control above pH = 5.5 are subsequently slightly less active in the fermentation of milk than those grown in the traditional fashion when pH is allowed to drop to pH = 4.8 or less (Champagne et al., 1995). Suppliers compensate by recommending inoculation levels with slightly higher CFU levels. From the examples presented, it seems that a rather acidic culture medium (pH controlled close to 5.0) would be useful for enhanced resistance of cultures to technological challenges (freezing or freeze-drying). This will also apply to physiological (gastrointestinal digestion) challenges, which will be discussed subsequently. Since there is variability in the effect of growth pH on subsequent technological properties, assays for each strain are required. However, when preparing biomass under pH control between pH 5.5 and 6.5 for subsequent drying, if in doubt, it seems wise to apply a pH stress at the end of the fermentation.

With respect to probiotics, the selection of a fermentation pH can influence the functionality of the cultures. Examples include resistance to gastrointestinal transit, adherence to the gut, or immune-stimulating activity. Several *in vivo* properties of LAB were reported to be dependent on the pH used for fermentation. Sashihara et al. (2007) produced biomass of the probiotic strain *L. gasseri* OLL2809 with pH controlled, from an initial value of 6.4 to constant values of 6.0, 5.0, or 4.0, using 10% (w/v) KOH. Under all conditions, the same level of biomass was achieved after 10 hours of fermentation, but IL-12 production by splenocytes was greater when they were exposed to bacterial cells produced at lower pH. Freeze-dried cells of *L. rhamnosus* E800 grown at pH 5.0 generally cope better with gastric acidity (pH 3.2) and bile stress than cells grown at pH 5.8 (Saarela et al., 2009). Biomass of the breast milk–derived strain *Bifidobacterium animalis* subsp. *lactis* INL1 was produced in MRS broth at pH 6.5 and 5.0 and submitted to simulated gastric digestion at pH 2 for 90 minutes. Whereas cells produced at pH 6.5 showed a cell decay of 2 log orders after simulated gastric digestion, no cell death was observed for the culture produced at pH 5.0.

## 19.4.3 Application of Other Stress Conditions

Processing parameters can also affect functional properties. The application of single or multiple stresses (osmotic, acidic, oxidative, heat, cold, high pressure) during batch or fed-batch fermentations are strategies to induce a general stress response that may confer cells an enhanced resistance to subsequent processing steps (freezing, freeze- or spray-drying) or to biological challenges (gastrointestinal transit). With the upcoming of the postgenomic era, proteomics has been heavily used for the study of environmental adaptation and stress response (Hussain et al., 2013).

As indicated earlier, mild heat stress (45 °C but not 55 °C) and osmotic stress (0.3 but not 0.7 M NaCl) applied during the growth of *L. rhamnosus* HN001 enhanced survival to drying (Prasad et al., 2003). The application of an acid stress at pH 3.5 and 3.2 did not strongly decrease cell biomass viability of *Oenococcus oeni* SD-2a, but increased freeze-drying survival (Zhao et al., 2009). Páez et al. (2012) demonstrated that the application of a mild heat treatment (52 °C for 15 minutes) before spray-drying enhanced cell survival during storage in a strain-dependent manner. Lavari et al. (2015) reported that incubation in anaerobiosis, harvesting cells at the stationary phase, and the application of a mild heat stress were more effective for reducing cell inactivation during heat challenge, compared to acidic or aerobic stress.

# 19.5 HARVESTING: CENTRIFUGATION VERSUS MEMBRANE FILTRATION

The cell suspensions that are obtained after fermentation are typically between 2 and $5 \times 10^9$ CFU/mL. A concentration step will typically increase the cell density by a factor of 10 to 40. Continuous centrifugation is most used for this purpose.

The process that is used to produce freeze-dried cultures generates cell damages at various levels (Shao et al., 2017). The centrifugation step itself might not significantly affect the viability of cells, as determined by plate counts (Streit et al., 2010). However, damaged cells can subsequently have enhanced sensitivity to the freezing step (Streit et al., 2010). The centrifugation process has three main parameters: temperature, g force, and duration. For *L. bulgaricus*, the centrifugal force and the duration have more influence on cell damages than does the temperature (Streit et al., 2011). In practice, probiotic cultures are centrifuged at g forces approximating 10,000 g and at temperatures below 15 °C. In order to attenuate the cell damage resulting from centrifugation, some processes include a short 30-minute recovery step before freezing (Kurtmann et al., 2009a).

When cultures suffer unacceptable viability losses due to centrifugation, membrane filtration (MF) is applied, although it is a more expensive process. The system can be microfiltration or ultrafiltration but is typically the former. MF can be operated under controlled transmembrane pressure or controlled permeation flux (Boyaval et al., 1996). Under a constant permeation flux, high cell concentrations are eventually reached, and membrane fouling occurs. This is due to a fast, irreversible layer formation and to a reversible cell cake. The microbial deposit characteristics are dependent on the ratio between permeation flux and wall shear stress (Boyaval et al., 1996). In one experiment carried out over 6 hours of filtration, cell viability remained constant up to 3 hours of filtration, but very high viability losses occurred afterward (Boyaval et al., 1996). Viability losses over time in MF systems were also noted by Coutinho and Xavier (2000), and they are due to mechanical stresses applied to cells during pumping. Surprisingly, higher MF flux has been obtained with a 0.45-μm pore size membrane compared to 0.8 and 1.2 μm when applied to lactococci and *Leuconostoc* cultures (Merin et al., 1983). This is presumably linked to the nature of the membrane rather than pore size. This shows that pore size is not the only parameter that will influence concentration rates. The acidifying activity (AA) of a yogurt starter strain (*L. bulgaricus*) is also affected during the MF process, and effects differ from those of centrifugation. Data comparing *L. bulgaricus* cells concentrated by MF or centrifugation suggested the following:

1. At the end of the concentration process itself, AA losses of the cell concentrates were greater during MF
2. AA losses due to freezing itself were generally lower when cells had been concentrated by MF
3. Reduction of AA levels during frozen storage were sometimes lower with MF-concentrated cells, and this occurred when high flow velocities and high transmembrane pressures were applied during MF

An advantage of MF is that a continuous process can be carried out, during a certain period, in order to obtain very high productivity. Thus, a fermentation vat is continuously fed with fresh growth medium. The fermented broth is treated by MF, and the cells are recirculated into the fermentation vat, while the spent medium is eliminated. Such a system can have a 15-fold increase in productivity over batch processes, and populations of up to $10^{11}$ CFU/mL can be reached (Corre et al., 1992). However, cells gradually become damaged due to multiple pumpings, as mentioned, and, although highly productive, to our knowledge, this process is not used extensively in industry.

Unfortunately, no data are available on the effect of temperature during MF or on the consequences of MF on subsequent survival to freezing, drying, or storage.

# 19.6 PROTECTANTS

A multitude of factors related to the culture itself (species, strain, cell size and form, and lipid composition of the membrane), to manufacture (growth medium composition and phase, incubation temperature, pH, osmolarity, aeration, density at freezing, cooling rate) or storage conditions (water activity, moisture, temperature) have been pointed out as able to affect the effectiveness of microorganism preservation. The composition of the medium used to protect cells for freezing or freeze-drying has been pointed out as one of the most important issues to address (Hubálek, 2003). The use of cryoprotectants for freezing or freeze-drying or thermoprotectants for spray-drying is mandatory for cell survival. Cryoprotectants prevent excessive dehydration and the formation of large ice crystals within cells. They have been classified based on their rate of penetration: those that penetrate the cytoplasm quickly (within 30 minutes) (e.g., ethylene glycol), those that penetrate more slowly (glycerol), and those that are nonpenetrating or nonpermeating compounds (oligosaccharides, polymers, proteins) that confer extracellular cryoprotection when present at concentrations of 10%–40% (Hubálek, 2003). The mechanisms of protective action depend on the nature of the ingredients used. The cell membrane is rendered more plastic by permeable cryoprotectants, and they bind intracellular water, avoiding excessive dehydration and preventing the formation of large ice crystals inside the cell. On the other hand, nonpermeable cryoprotectants will adsorb outside the cell, forming a viscous layer, maintaining the structure of ice amorphous in the surroundings of the cell (Saarela et al., 2005). The penetrating compounds are more effective against injuries during slow freezing, whereas the nonpenetrating compounds are more effective against injuries during rapid freezing. In this sense, the almost universal successful use of skim milk as cryoprotectant may be explained based on the presence of small-, medium-, and large-sized molecules that fill in and recover bacterial cells. Milk presents several mechanisms to diminish cellular injury: stabilization of the constituents of the cell membrane, creation of a porous structure that facilitates rehydration, and by providing protective proteins that coat the cells. Yet incorporating additional protective ingredients to milk may enhance its intrinsic protective capacity (Carvalho et al., 2004).

What is called a "glassy state" is a state of high viscosity in which cellular structures are embedded, while, at the same time, molecular mobility and damaging reactions are slowed down (Zhang et al., 2017). The high viscosity of a glassy matrix is due to the low molecular mobility of water. Dehydrated cells are protected during storage if they are stored below the glass transition temperature (Kurtmann et al., 2009b). Increased Tg has been related to storage stability. Trehalose and sucrose share the same molecular weight, but the former has a particular high glass transition temperature of nearly 60 °C higher than the latter (Zhang et al., 2017) and has been pointed to as one of the best ingredients for freeze-drying to ensure long-term (10–20 years) preservation of strains in a culture collection, for example. However, its high cost may render it economically unfeasible for industrial applications, and then lower-cost materials (skim milk, maltodextrin, lactose, mannitol, sucrose, carboxymethylcellulose) are chosen for cultures used as starters or probiotics that will be used within 12–24 months of manufacture. The success of cryopreservation has been related to the choice of ingredients or mixture of ingredients (Table 19.3), with high glass transition temperature or materials that can replace water in the cytoplasm, in the space between the cytoplasmic membrane and the cell wall and outside the cell (Carvalho et al., 2004). The demand for probiotic foods or supplements by vegetarians, vegans, lactose-intolerants, or non-dairy consumers challenges probiotic manufacturers, traditionally used to produce dairy-based cultures, to explore the use of non-dairy ingredients for the incorporation of probiotics into non-dairy foods (e.g., fruit juices or cereal bars). Betaine-trimethylglycine (Saarela et al., 2009), cellobiose (Basholli-Salihu et al., 2014), or a patented mixture of ingredients (dextran, sorbitol, mannitol, sucrose, hydroxyl ethyl starch) that contains no skim milk or any other animal-derived compounds (Corveleyn et al., 2012) have been used in this sense. Maltodextrins are currently popular as protectants. Presumably, this is because this product contains many glucose-based polymers that have different molecular weights. Thus, the benefits of short-chain and long-chain protectants can apply. Also, it is acceptable to various religious groups, does not have allergenic attributes, and is plant based.

**TABLE 19.3**  Diversity of Protectants Used for Freeze- or Spray-Drying Lactic Acid Bacteria and Bifidobacteria

| STRAIN[a] | PROTECTANT | OBSERVATION | REFERENCES |
|---|---|---|---|
| **Freeze-Drying** | | | |
| L. rhamnosus GG L. plantarum NCIMB 8826 | Sucrose, trehalose, and sorbitol, 1%, 5%, and 10% (w/v) | Sucrose offered better protection than trehalose and sorbitol for both strains. | Siaterlis et al. (2009) |
| E. faecium IFA045 L. plantarum IFA278 | Glucose, sucrose, trehalose, maltodextrin (32%) | Nonreducing disaccharides conferred best protection for both strains. | Strasser et al. (2009) |
| L. rhamnosus IMC 501 L. paracasei IMC 502 | Inulin, dextrin, resistant starch, glycerine, sorbitol, mannitol, semi-skimmed milk (10%) | No differences observed among protectants at 4 °C. In storage at room temperature, glycerine conferred highest protection. | Savini et al. (2010) |
| L. bulgaricus CIDCA 333 | 19%–38% galacto-oligosaccharides | Possible dual role of GOS as protectant and prebiotic. | Tymczyszyn et al. (2011) |
| B. longum BIOMA 5920 | Human-like collagen, trehalose, L-cysteine, and glycerol | Response surface methodology was used to optimize the concentration of each ingredient. | Yang et al. (2012) |
| B. lactis INL1 | Sucrose, lactose, skim milk (10%–15%) | Differences in protective capacity were observed only when cells were stressed. | Vinderola et al. (2012) |
| L. paracasei DSM 20258 L. bulgaricus DSM 20081 | Skim milk (up to 6%), trehalose (up to 8%), sodium ascorbate (up to 4%) | The highest survival was observed for maximal concentration assessed of each ingredient. | Jalali et al. (2012) |
| L. acidophilus ATCC 4962 | Sucrose, dextran, sorbitol, monosodium glutamate, glycerol, skim milk | Skim milk was chosen for being cost effective. | Pyar et al. (2014) |

*(Continued)*

**TABLE 19.3**  (Continued) Diversity of Protectants Used for Freeze- or Spray-Drying Lactic Acid Bacteria and Bifidobacteria

| STRAIN[a] | PROTECTANT | OBSERVATION | REFERENCES |
|---|---|---|---|
| **Spray-Drying** | | | |
| Bifidobacteria strains | 30% gelatin, 35% soluble starch, 35% gum arabic, 15% (w/w) skim milk | Greatest reduction for soluble starch and the least with skim milk. | Lian et al. (2002) |
| L. rhamnosus GG | Emulsion-based formulation with whey and starch | No differences in survival compared to freeze-drying. | Ying et al. (2010) |
| L. casei Nad, L. acidophilus A9, L. paracasei A13. | Skim milk (20%), Skim milk (10%) + starch or WPC (10%) | No differences in survival in relation to the ingredients used. | Páez et al. (2012) |
| B. lactis BB12 | Milk with or without inulin, oligofructose-enriched inulin or oligofructose (20%) | Higher stability when prebiotics were included compared to skim milk only. | Fritzen-Freire et al. (2012) |
| L. reuteri DSM20016 | Whey | Whey as substrate and encapsulation matrix within a coupled fermentation and spray-drying process. | Jantzen et al. (2013) |
| L. casei | Maltodextrin | Optimizing the process allowed to achieve 1.1 × $10^{10}$ CFU/g of the strain. | Silva dos Santos et al. (2014) |
| L. plantarum ATCC 8014 | Fresh liquid whey, permeate, and whey retentate | Whey and whey retentate were preferred. | Eckert et al. (2017) |

[a] Lc. = Lactococcus; Ln. = Leuconostoc; L. = Lactobacillus-derived (see section 2.1); S. = Streptococcus.

It should be mentioned that many freeze-drying media also contain magnesium stearate. This is not used to protect the cells. Rather, it is an ingredient that prevents sticking of the powder grains and facilitates the flow of the particles during processing. It is particularly important in tablets.

The effect of the bacterial strain is a frustrating characteristic in the field of freeze-drying LAB or bifidobacteria. Compounds effective in protecting one strain from freezing or drying are ineffective for others (de Valdez et al., 1983, 1985). Thus, individual tests are often needed to identify the correct parameters. Carvalho et al. (2004) released a series of practical recommendations for preparation of freeze-dried LAB. Thus, in the absence of other relevant information on the strain's resistance, (1) skim milk powder should be selected as drying medium as a first approach when studying the survival capacity of the strain to freeze-drying, and (2) ingredient testing should start with carbohydrates. Another ingredient addition that should be done "automatically", if tests cannot be carried out, is ascorbate; this addition is more for storage stability than for surviving freeze-drying.

# 19.7  INDUSTRIAL FORMATS OF STARTER AND PROBIOTIC CULTURES

## 19.7.1  Frozen Cultures

The successive propagation of starter cultures in fermented dairy plants is laborious and risky from the point of view of contamination or bacteriophage attack, and the repeated subcultures make the strain prone to losing some properties due to mutation and auto-selection. Modern dairy plants prefer the use of starter and probiotic cultures in frozen and dried concentrated forms for direct inoculation, called direct vat inoculation (DVI) cultures or direct vat set (DVS) cultures. These cultures contain typically $10^{10}$ to $10^{12}$ CFU/g or mL, depending on whether they are frozen or freeze-dried. As water is removed, higher levels of viable cells can be achieved by freeze-drying. For frozen cultures, the cooling rate is a key factor for cell survival. For low rates, cells are prone to lose water rapidly, and the formation of ice outside cells will harm them virtually, similarly to dehydration. When the process is too fast for cells to lose water rapidly enough, then the osmotic equilibrium is kept by frozen water in the cytoplasm. In this case, viability-threatening intracellular ice formation will occur. The effect of three different freezing temperatures (−20 °C, −80 °C, and −196 °C) on post-freeze-drying survival rates of LAB was studied by Polo et al. (2017), showing that −196 °C ensured the highest post–freeze-drying viability for LAB. In practice, slow freezing is achieved in a freezer or deep freezer (frozen block cultures dispensed in metal cans or plastic containers), whereas a higher cooling rate can be obtained by dropping the cell suspension in a tank with liquid nitrogen or liquid $CO_2$ (pelleted cultures). The droplets congeal almost instantly into pellets, and they are conveyed out of the immersion with a mesh-belt conveyor. From the point of view of the user, laminated paperboard containers may be preferred over plastic pouches, as it is not an uncommon practice to manually break down large clumps of sticked pellets formed when the culture is temporarily or accidentally exposed to inadequate (too high) storage temperatures. Small ice crystals may reorganize into lethally large crystals if fluctuation of the storage temperature (thawing and refreezing) occurs. Commercial frozen concentrated cultures should be stored at a temperature lower than −45 °C to get the shelf life for at least 12 months (Bylund, 2003).

## 19.7.2 Freeze-Drying

Freeze-drying, also called lyophilization, is by far the most conventional process for the industrial production of dried bacterial cultures. Long-term storage is achieved by reducing water activity to values below 0.2. An additional advantage is the lower costs of distribution, as deep-frozen transportation is not needed. Stabilization of bacteria via freeze-drying starts by mixing the bacterial pellet produced with a cryoprotectant solution (usually 10%–20% w/v total solids), followed by a short recovery period, freezing, primary drying (removal of ice by sublimation), and secondary or final drying to remove unfrozen water by desorption (Fonseca et al., 2015). At industrial scale, freeze-drying may require 24–72 hours. Industrial processes are proprietary, but examples of laboratory processes are presented in Table 19.4.

The glassy state is an amorphous metastable state resembling a solid, but with randomly positioned molecules. When a substance in the glassy state is heated above its glass transition temperature, the constituting molecules start getting translational mobility and enter a more liquid-like state. Cells survive better if kept in the glassy state during storage. The role of glassy state on survival of probiotics during each step of production and storage has been reviewed (Santivarangkna et al., 2011). For freeze-dried cultures, water activity, more than moisture, is the key and main factor responsible for cell survival over the shelf life of the culture (Poddar et al., 2014). It is important then to produce cultures with low water activity

**TABLE 19.4**  Examples of Freeze-Drying Processes

| PROCESS | REFERENCES |
|---|---|
| Precooling to −40 °C at atmospheric pressure. Afterward, the pressure is reduced to 0.3 mbar, and the temperature is raised at 0.5 °C/minute until an end temperature of 32 °C is reached. | Kurtmann et al. (2009c) |
| Precooling to −40 °C at atmospheric pressure; 24 hours at 5 °C with 100 mtorr (0.13 mbar) vacuum; 24 hours at 20 °C with 100 mtorr, 12 hours at 20 °C with less than 10 mtorr, and 12 hours at 30 °C with less than 10 mtorr vacuum. | Raymond and Champagne (2015) |

(close to 0.1) to keep cells metabolically inactive but alive and to maintain a low water activity along storage. However, excessive drying may be detrimental too (Ouwehand et al., 2018).

Many factors determine the success of freeze-drying LAB or bifidobacteria. However, this feature is primarily an intrinsic property of the strain. In certain cases, no matter how many optimization efforts are carried out, survival is poor, and the strain will likely not be industrialized. Even for the monomorphic species *B. animalis* subsp. *lactis*, displaying more than 99.9% molecular identity among strains, the production conditions and survival during storage vary among strains (Ouwehand et al., 2018), and optimization must be done on a strain basis. Factors conditioning survival to freeze-drying may appear well before the dehydration process itself. Besides the strain and its shape (rods were reported to be less tolerant to freeze-drying), the composition of the growth medium, the culture conditions, and the moment of harvesting are of importance. The growth medium should have a composition that allows the cytoplasmic accumulation of compatible solutes (NaCl, mannitol, sorbitol, or glutamate, or sources of compatible solutes such as peptones, tryptone, and meat and yeast extracts) that confers protection to osmotic stress and a positive altered fatty acid profile of the membrane (Carvalho et al., 2004), which can be achieved by adding Tween 80 and ascorbic acid to the medium. However, contradictory results can be found. The addition of yeast extract in the propagation medium of a strain of *Lactobacillus delbrueckii* subsp. *bulgaricus* exerted a negative effect on the cryotolerance and decreased survival during lyophilization (Shao et al., 2014). Culture conditions and the application of sublethal treatments (starvation, heat or cold shock, osmotic or oxidative stress) were also shown to affect resistance to freeze-drying due to the synthesis of protective stress proteins during growth. Yet starvation may be tricky: when *L. rhamnosus* GG was produced under limited carbon availability, survival during freeze-drying was significantly decreased (Siaterlis et al., 2009). In this case, a deficient accumulation of cytoplasmic compatible solutes may have been responsible for the poor survival. As mentioned previously, allowing pH to reach a value close to 5 (Rault et al., 2010) or 4.5 (Silva et al., 2005) during biomass production can enhance survival to freeze-drying. Issues related to the protectants used for freeze-drying were already discussed in this chapter. Commercial freeze-dried cultures are expected to have low viability losses (less than 1 log) when stored 12–24 months below 4 °C. In order to know in advance the survival capacity of a specific strain under any given conditions, an accelerated storage test may be used (Tsen et al., 2007). In these tests, cell survival is evaluated after a short period (1–2 months) of storage at an uncommonly high storage temperature (from 37 °C to 70 °C). However, the correlation between the predicted shelf life obtained with these tests (especially at 70 °C) and the real cell survivability remains unclear or, worse, strain specific. As a rule, the higher is the water activity of the powder, the higher is the storage temperature and the higher is the oxygen content in the packaging, then the poorer will be the survival of the bacteria during storage. As well, light could promote oxidation of lipids during storage, causing loss of viability. Thus, packages are always opaque, in addition to being designed to prevent the entrance of oxygen and moisture. Furthermore, bottles having probiotic-containing capsules generally contain sachets that absorb moisture and oxygen, which enter the bottle when consumers open them daily to take a capsule.

## 19.7.3 Spray-Drying

The technique most applied at the industrial level for dehydration of LAB is freeze-drying. In second place of importance and development comes spray-drying, with a cost up to 10 times lower than that of freeze-drying (Schuck et al., 2013). A larger production scale can be attained by spray-drying when compared to freeze-drying. Spray-drying is also a promising tool to produce microencapsulated cultures (Huang et al., 2017a). Spray-drying should not be the first choice for dehydrating a culture, as less survival capacity should be expected, in principle. However, the strain dependency of behaviors may show surprising results, and successful applications of spray-drying to produce starter or probiotic cultures have been reported as well, usually followed by an optimization design due to the many variables (inlet and outlet temperature, feeding rate, air flux) (Behboudi-Jobbehdar et al., 2013). In some instances, an even better survival was reported for spray drying than freeze-drying (Poddar et al., 2014).

One of the difficulties in using spray-drying may be in the scaling up of the process. Much of the laboratory work has been conducted using the gentle spray dryer Buchi B-290 (Huang et al., 2017b), where promising results can be obtained, but this may be difficult to reproduce with a more robust (and less fine-tunable) industrial-scale spray dryer (personal communication). Very comprehensive reviews are periodically published that update aspects of spray-drying such as energy demands and costs, ingredients, and conditions for drying and storage (Broeckx et al., 2016; Huang et al., 2017b; Santivarangkna et al., 2007). Many protectants are the same for freeze- and spray-drying, and much of the knowledge about factors that determines cell survival along storage in freeze-dried cultures may pave the way for developing industrial spray-dried cultures.

It is also worth mentioning that combining spray-drying and air-drying could be explored to further reduce processing costs and widen the scope of strains that can be spray-dried, particularly when production plants are expanding. In spray-drying, high viability losses often occur when the outlet temperature is high, which enables extensive drying (Koc et al., 2010). Thus, a potential approach is to carry out a limited spray-drying in a first step and submit the resulting power to air-drying. Air-drying is two times less expensive than spray-drying (Santivarangkna et al., 2007), and one could even consider combining the air-drying step with fluid-bed microencapsulation.

# 19.8 REACTIVATION OF FROZEN OR DEHYDRATED CULTURES

Significant cell death can occur due to freezing and thawing processes. Cell injuries that result in losses of membrane integrity due to modifications of the phospholipids, in particular a reduction in the fluidity of the cell membrane, may take place (Passot et al., 2015). Resistance to freeze-thawing was related to high membrane fluidity and a homogeneous distribution of fluidity values (Meneghel et al., 2017). In 2011, Champagne et al. (2011) released a series of practical suggestions that should be considered for the viability assessment of probiotics as concentrated cultures or when present in food matrices. In relation to frozen cultures, they recommended rapid thawing at a temperature slightly below those for optimal growth, warning at the same time that CFU counts may sometimes be higher in freshly frozen samples than in the original liquid cell suspension. As chain breakup may occur, homogenization methods should be applied to break chains for improving the precision of the analysis. Even when freeze–thaw steps may break cell chains, it is still advisable to perform a homogenization step on thawed cultures.

A culture that successfully survived freezing, drying, and storage may still partially lose viability if not properly rehydrated. Resuscitation of dried cultures is a key step for having active and effective cultures. The reconstitution of dried cultures may greatly affect not only the survival but also the activity of the bacteria, which is of the most concern for LAB used as starters. Factors affecting survival to rehydration are osmolarity, pH, and composition of the rehydrating solution, as well as the rehydration temperature and rate (Broeckx et al., 2016). Wetting of the particles is often the reconstitution controlling step. The reconstitution media may make the recovery vary up to tenfold (de Valdez et al., 1985). A slow rate of rehydration, carried out at the optimal growth temperature of the culture being rehydrated, is preferred. Champagne et al. (2011) suggested using solutions having between 10% and 20% solids to activate dehydrated cultures. Yet the success of reconstitution is strain dependent, whereas 30-minute reconstitution at pH 8 in the presence of 2% l-arabinose (ratio 1:100 of powder to diluent) was effective for maximum recovery of *B. longum* NCC3001, *L. johnsonii* La1 showed the highest recovery after reconstitution when mixed with maltodextrin at pH 4 (Muller et al., 2010). Special attention should be paid to pH, as very high viability losses were reported when *L. rhamnosus* R0011 was rehydrated in an acid environment such as fruit juice (Reid et al., 2007).

It is also recommended to carry out a 15- to 30-minute incubation period before proceeding to dilutions. Following this rehydration period, a high-shear homogenizing step (ultra-turax blender) is recommended in order to break down any remaining powder particles and even cell chains. With probiotics that are highly sensitive to oxygen, antioxidants (e.g., cysteine) can be added to the dilution and enumeration media (Champagne et al., 2011).

# 19.9 MICROENCAPSULATION

Microencapsulation can be defined in many ways. When applied to ingredients, it can be considered a physio-chemical or mechanical process where a "coating" or "shell" is used for the matrix in which the core material is dispersed (Sarao and Arora, 2017). With respect to probiotics, functionality is introduced, and microencapsulation can be defined as "the process in which cells are retained within an encapsulating membrane to reduce cell injury or cell lost, in a way that result in appropriate microorganism release in the gut" (Martin et al., 2015).

There are four main technologies (extrusion, spray-drying, spray-chilling, spray-coating) and numerous excellent reviews on the intricacies of microencapsulation of probiotics (Martin et al., 2015; Sarao and Arora, 2017). Basically, the coating or the microentrapment matrix protects cells from various stressful conditions from the food matrix or the gastrointestinal environment.

There are two uses of ME with respect to lactic cultures: (1) novel production process for concentrated cells and (2) protection of cultures for application as supplements and foods.

For biomass production, cells are entrapped in alginate beads, which are then added to a fermentation broth. Growth occurs in the beads, and populations up to 20 billion/g can be reached (Figure 19.1). The

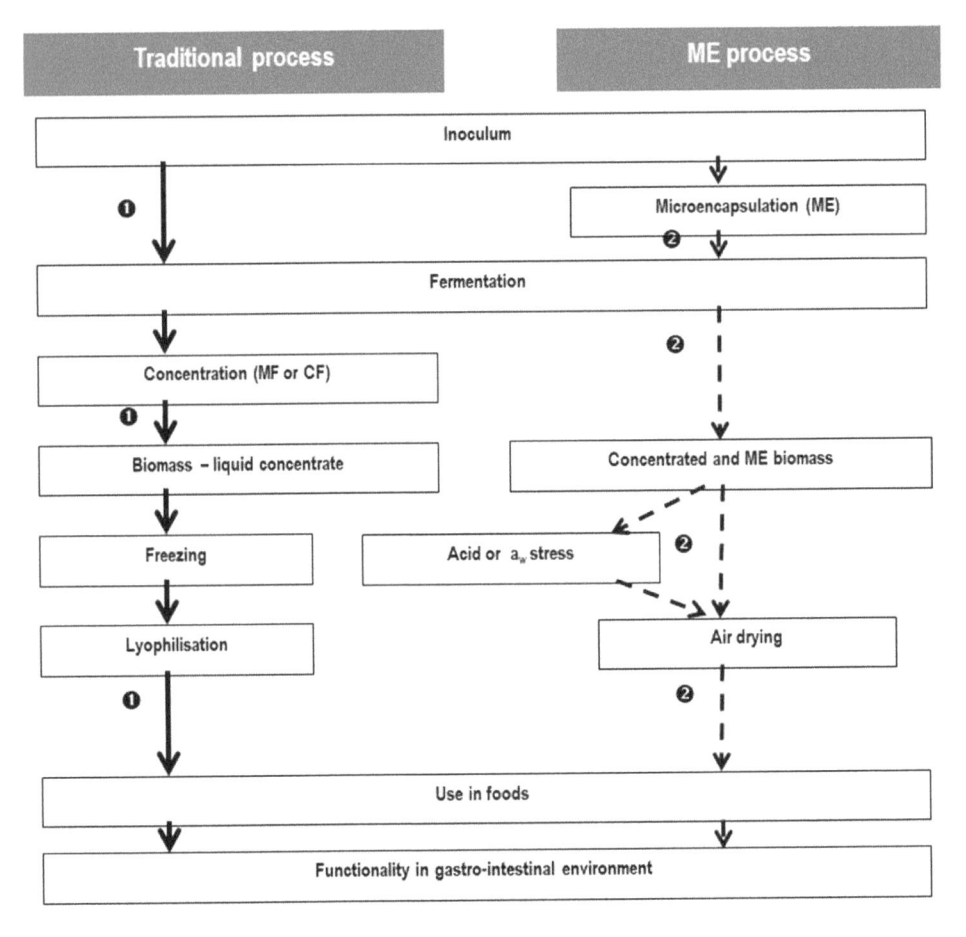

**FIGURE 19.1** Traditional (process ❶) and alginate-bead-microencapsulation technologies (process ❷) for the production of concentrated starters and probiotics.

ME= microencapsulation; CF = centrifugation; MF = membrane filtration.

beads can be recovered, and concentrated biomass particles are obtained without the use of centrifugation or microfiltration (Figure 19.1). Two further advantages are also possible:

1. The beads can be soaked in solutions and easily recovered from them, which can generate acid, temperature or osmotic stresses. The cells are then more resistant to various drying processes (Champagne et al., 2017) or applications in foods or as supplements.
2. They can be air-dried instead of freeze-dried. Air-drying is a much less expensive drying technology, and spray-coating can be carried out on the dried beads for further functionalities.

Spray-coated freeze-dried cultures are available commercially (Champagne et al., 2010). Although microencapsulation has many potential advantages (Champagne, 2006), there are only a limited number of applications in foods (Champagne and Kailasapathy, 2011). This is because they are more expensive and because microencapsulated cultures affect the texture of foods.

It should be stated that a challenge in developing functional foods with microencapsulated probiotics is their enumeration. Classical methodologies underestimate the CFU of microencapsulated cultures. Thus, specific steps must be carried out to release the cells from their capsules before dilution and plating steps are carried out (Champagne et al., 2010).

# 19.10 POSTBIOTICS: WHAT IS NEW?

Cell viability has long been regarded as important for a probiotic to confer a health benefit. However, it has been long recognized that non-viable microbes, their cell components, and their metabolites can also impact health (Vinderola et al., 2022). In 2021, the International Scientific Association for Probiotics and Prebiotics (ISAPP) proposed that the term postbiotics be used to refer to a "preparation of inanimate microorganisms and/or their components that confers a health benefit on the host" (Salminen et al., 2021). Beyond the scope of proposing a definition, the aim of this work was a call to unify into the term postbiotic a group of terms that had been used so far to refer to the health benefits conferred by non-viable microbes or their fragments. These terms included non-viable probiotics, heat-killed probiotics, heat-inactivated probiotics, tyndallized probiotics, ghost probiotics, paraprobiotics, and even postbiotics. Unifying the terms would be of value for communication to stakeholders (industry, consumers, health professionals, regulators) and for future works of systematic reviews and meta-analysis. Probiotics and postbiotics share several features: the microbe/s used should be identified to the strain level, and a well-conducted efficacy study must be available to demonstrate the benefit on the target host. In addition, the finished probiotic or postbiotic product may contain microbial metabolites. The difference is that for a postbiotic, a deliberate technological process for cell viability termination must be applied (heat, radiation, high pressure, lysis, etc.). The microbes constituting a postbiotic may be inanimate, intact cells or may be structural fragments of microbes, such as cell walls. Many preparations of postbiotics also retain microbe-produced substances, such as metabolites, proteins, or peptides, which may contribute to the overall health effect conferred by a postbiotic, but such components are not essential to a postbiotic (Vinderola et al., 2022). Examples of postbiotic products include a partially fermented infant formula produced with *Bifidobacterium breve* C50 and *Streptococcus thermophilus* O65, which are inactivated by spray-drying (Salminen et al., 2020); a combination of two heat-inactivated lactobacilli, including the fermentation products, to treat infant and adult diarrhea (Remes Troche et al., 2020); and a heat-inactivated fermentate of *Saccharomyces cerevisiae* to improve gastrointestinal discomfort (Pinheiro et al., 2017). Postbiotics for animal nutrition were also developed, as in the case of a spray-dried inactivated culture (cells and metabolites) of a specific strain of *Aspergillus oryzae* to attenuate the effects of heat stress (Ríus et al., 2022). Other examples and responses to frequently asked questions about postbiotics have been recently published (Vinderola et al., 2024).

It is important to note that it is not mandatory to start with a probiotic strain to develop a postbiotic. For instance, bacterial lysates made out of inactivated pathogens such *S. aureus, K. pneumoniae,* and *H. influenzae* are used to boost respiratory tract immunity. These products fit the definition of postbiotics (Kaczynska et al., 2022). As mentioned previously (Section 19.2), there are instances where the bacterial cell components are not involved at all, but rather extracellular metabolites. Examples include gamma amino butyric acid and bile salt hydrolase (Jones et al., 2012; Inoue et al., 2003). However, purified metabolites or a mixture of metabolites, in the absence of non-viable microbes or cell components, are not considered postbiotics under the ISAPP perspective.

# 19.11 PERSPECTIVES

It is increasingly understood how experimental conditions will affect clinical results: strain, CFU levels, probiotic format, and moment of consumption of the probiotic. Attempts were made to not only address the parameters that enable the production of high biomass levels of LAB and bifidobacteria but also how production parameters will affect functionality. Hopefully, developing manufacturing processes that enhance the functionality of probiotics will improve clinical results and, further, the development of functional foods and dietary supplements with probiotic bacteria.

# BIBLIOGRAPHY

Aguirre-Ezkauriatza, E.J., Aguilar-Yáñez, J.M., Ramírez-Medrano, A., Alvarez, M.M. 2010. Production of probiotic biomass (*Lactobacillus casei*) in goat milk whey: Comparison of batch, continuous and fed-batch cultures. *Biores Technol 101*:2837–2844.

Audy, J., Labrie, S., Roy, D., LaPointe, G. 2010. Sugar source modulates exopolysaccharide biosynthesis in *Bifidobacterium longum* subsp. *longum* CRC 002. *Microbiol 156*:653–664.

Barbosa, J., Borges, S., Teixeira, P. 2015. Influence of sub-lethal stresses on the survival of lactic acid bacteria after spray-drying in orange juice. *Food Microbiol 52*:77–83.

Basholli-Salihu, M., Mueller, M., Salar-Behzadi, S., Unger, F.M., Viernstein, H. 2014. Effect of lyoprotectants on β-glucosidase activity and viability of *Bifidobacterium infantis* after freeze-drying and storage in milk and low pH juices. *LWT—Food Sci Technol 57*(1):276–282.

Begley, M., Hill, C., Gahan, C.G. 2006. Bile salt hydrolase activity in probiotics. *Appl Environ Microbiol 72*:1729–1738.

Behboudi-Jobbehdar, S., Soukoulis, C., Yonekura, L., Fisk, I. 2013. Optimization of spray-drying process conditions for the production of maximally viable microencapsulated *L. acidophilus* NCIMB 701748. *Drying Technol 31*(11):1247–1283.

Bergère, J.L., Hermier, J. 1968. La production massive de cellules de streptocoques lactiques: II.- Croissance de *Streptococcus lactis* dans un milieu à pH constant. *Le lait 48*:13–30.

Boyaval, P., Lavenant, C., Gésan, G., Daufin, G. 1996. Transient and stationary operating conditions on performance of lactic acid bacteria crossflow microfiltration. *Biotechnol Bioeng 49*:78–86.

Breheny, S., Kanasaki, M., Hillier, A.J., Jago, G.R. 1975. Effect of temperature on the growth and acid production of lactic acid bacteria. The uncoupling of acid production from growth. *Austral J Dairy Technol 30*:145–148.

Broeckx, G., Vandenheuvel, D., Claes, I.J.J., Lebeer, S., Kiekensa, F. 2016. Drying techniques of probiotic bacteria as an important step towards the development of novel pharmabiotics. *Int J Pharm 505*(1–2):303–318.

Bylund, G. 2003. Dairy Processing Handbook. Lund, Sweden: Tetra Pak Processing Systems AB.

Carvalho, A.S., Silva, J., Ho, P., Teixeira, P., Malcata, F.X., Gibbs, P. 2004. Relevant factors for the preparation of freeze-dried lactic acid bacteria. *Int Dairy J 14*:835–847.

Champagne, C.P. 2006. Starter cultures biotechnology: The production of concentrated lactic cultures in alginate beads and their applications in the nutraceutical and food industries. *Chem Indust Chem Engin Quart 12*:11–17.

Champagne, C.P., Gardner, N.J. 2002. Effect of process parameters on the production and drying of *Leuconostoc mesenteroides* cultures. *J Ind Microbiol Biotechnol* 28:291–296.

Champagne, C.P., Gardner, N.J., Lacroix, C. 2007. Fermentation technologies for the production of exopolysaccharide-synthesizing *Lactobacillus rhamnosus* concentrated cultures. *Elect J Biotechnol* 10(2):211–220.

Champagne, C.P., Gaudreau, H., Conway, J., Chartier, N., Fonchy, E. 1999. Evaluation of yeast extracts as growth media supplements for lactococci and lactobacilli by using automated spectrophotometry. *J Gen Appl Microbiol* 45:17–21.

Champagne, C.P., Girard, F., Rodrigue, N. 1993. Production of concentrated suspensions of thermophilic lactic acid bacteria in calcium-alginate beads. *Int Dairy J* 3:257–275.

Champagne, C.P., Kailasapathy, K. 2011. Some current food products with microencapsulated probiotics. *Agro Food Ind Hi-Tech* 22:54–56.

Champagne, C.P., Piette, M., Saint-Gelais, D. 1995. Characteristics of lactococci cultures produced in commercial media. *J Ind Microbiol* 15:472–479.

Champagne, C.P., Raymond, Y., Arcand, Y., 2017. Effects of production methods and protective ingredients on the viability of probiotic *Lactobacillus rhamnosus* R0011 in air-dried alginate beads. *Can J Microbiol* 63:35–45.

Champagne, C.P., Raymond, Y., Simon, J.P. 2012. Effect of water activity and protective solutes on growth and subsequent survival to air-drying of *Lactobacillus* and *Bifidobacterium* cultures. *Appl Microbiol Biotechnol* 95:745–756.

Champagne, C.P., Raymond, Y., Tompkins, T.A. 2010. The determination of viable counts in probiotic cultures microencapsulated by spray coating. *Food Microbiol* 27:1104–1111.

Champagne, C.P., Ross, R.P., Saarela, M., Hansen, K.F., Charalampopoulos, D. 2011. Recommendations for the viability assessment of probiotics as concentrated cultures and in food matrices. *Int J Food Microbiol* 149:185–193.

Cocaign-Bousquet, M., Garrigues, C., Novak, L., Lindley, N.D., Loubiere, P. 1995. Rational development of a simple synthetic medium for the sustained growth of *Lactococcus lactis* . *J Appl Bacteriol* 79:108–116.

Corre, C., Madec, M.N., Boyaval, P. 1992. Production of concentrated *Bifidobacterium bifidum* . *J Chem Technol Biotechnol* 53:189–194.

Correa Deza, M.A., Grillo-Puertas, M., Salva, S., Rapisarda, V.A., Gerez, C.L., Font de Valdez, G. 2017. Inorganic salts and intracellular polyphosphate inclusions play a role in the thermotolerance of the immunobiotic *Lactobacillus rhamnosus* CRL 1505. *PLoS ONE* 12(6): e0179242.

Corveleyn, S., Dhaese, P., Neirynck, S., Steidler, L. 2012. Corveleyn, S., Dhaese, P., Neirynck, S., Steidler, L. 2012 *Cryoprotectants for Freeze Drying of Lactic Acid Bacteria.* The following patents were assigned to Intrexon Actobiotics NV : Europe EP 2424972 A1 and USA US9574169B2.

Coutinho, J.A.P., Xavier, A.M.R.B. 2000. A model for micro/ultrafiltration cell deactivation in cell-recycle reactors. *J Chem Technol Biotechnol* 75:315–319. https://doi.org/10.1002/(SICI)1097-4660(200004)75:4<315::AID-JCTB219>3.0.CO;2-P.

Desjardins, M.-L., Roy, D., Goulet, J., 1991. Beta galactosidase and proteolytic activities of bifidobacteria in milk a preliminary study. *Milchwiss* 46:11–13.

De Valdez, G.F., De Giori, G.S., De Ruiz Holgado, A.P., Oliver, G. 1983. Comparative study of the efficiency of some additives in protecting lactic acid bacteria against freeze-drying. *Cryobiol* 20:560–566.

De Valdez, G.F., De Giori, G.S., De Ruiz Holgado, A.P., Oliver, G. 1985. Effect of drying medium on residual moisture content and viability of freeze-dried lactic acid bacteria. *Appl Environ Microbiol* 49:413–415.

Eckert, C., Garcia Serpa, V., Felipe dos Santos, A.C., da Costa, S.M., Dalpubel, V., Neutzling Lehn, D., Volken de Souza, C.F. 2017. Microencapsulation of *Lactobacillus plantarum* ATCC 8014 through spray drying and using dairy whey as wall materials. *LWT—Food Sci Technol* 82:176–183.

EFSA. 2010. Scientific Opinion on the substantiation of health claims related to live yoghurt cultures and improved lactose digestion (ID 1143, 2976) pursuant to Article 13(1) of Regulation (EC) No 1924/20061 EFSA Panel on Dietetic Products, Nutrition and Allergies (NDA). *EFSA Journal* 8(1763):1761–1718.

Feldstein, F.J., Westhoff, D.C. 1979. The influence of heat treatment of milk on starter activity: What about UHT? *Cult Dairy Prod J* 14:11–15.

Fonseca, F., Cenard, S., Passot, S. 2015. Freeze-drying of lactic acid bacteria. In: Cryopreservation and Freeze-Drying Protocols. Methods in Molecular Biology (Methods and Protocols), Vol. 1257. W. Wolkers and H. Oldenhof (eds). New York: Springer.

Fritzen-Freire, C.B., Prudêncio, E.S., Amboni, R.D.M.C., Pinto, S.S., Negrão-Murakami, A.N., Murakami, F.S. 2012. Microencapsulation of bifidobacteria by spray drying in the presence of prebiotics. *Food Res Int* 45(1):306–312.

Giraud, E., Lelong, B., Raimbault, M. 1991. Influence of pH and initial lactate concentration on the growth of L. *plantarum* . *Appl Microbiol Biotechnol* 36:96–99.

Hill, C., Guarner, F., Reid, G., Gibson, G.R., Merenstein, D.J., Pot, B., Morelli, L., Canani, R.B., Flint, H.J., Salminen, S., Calder, P.C., Sanders, M.E. 2014. Expert consensus document: The international scientific association for probiotics and prebiotics consensus statement on the scope and appropriate use of the term probiotic. *Nat Rev Gastroenterol Hepatol* 11:506–514.

Huang, S., Méjean, S., Rabah, H., Dolivet, A., Le Loir, Y., Chen, Y.D., Jan, G. Jeantet, R., Schuck, P. 2017a. Double use of concentrated sweet whey for growth and spray drying of probiotics: Towards maximal viability in pilot scale spray dryer. *J Food Eng 196*:11–17.

Huang, S., Rabah, H., Jardin, J., Briard-Bion, V., Parayre, S., Maillard, M.-B., Le Loir, Y., Chen, X.D., Schuck, P., Jeantet, R., Jan, G. 2016. Hyperconcentrated sweet whey, a new culture medium that enhances *Propionibacterium freudenreichii* stress tolerance. *Appl Environ Microbiol 82*:4641–4651.

Huang, S., Vignolles, M.L., Chen, Y.D., Le Loir, Y., Jan, G., Schuck, P., Jeantet, R. 2017b. Spray drying of probiotics and other food-grade bacteria. *Trends Food Sci Technol 63*:1–17.

Hubálek, Z. 2003. Protectants used in the cryopreservation of microorganisms. *Cryobiol 46*(3):205–229.

Hussain, M.A., Nezhad, M.H., Sheng, Y., Amoafo, O. 2013. Proteomics and the stressful life of lactobacilli. *FEMS Microbiol Lett 349*:8.

Inoue, K., Shirai, T., Ochiai, H., Kasao, M., Hayakawa, K., Kimura, M., Sansawa, H. 2003. Blood-pressure-lowering effect of a novel fermented milk containing γ-aminobutyric acid (GABA) in mild hypertensives. *Eur J Clin Nut 57*:490–495.

Jalali, M., Abedi, D., Varshosaz, J., Najjarzadeh, M., Mirlohi, M., Tavakoli, N. 2012. Stability evaluation of freeze-dried Lactobacillus paracasei subsp. tolerance and *Lactobacillus delbrueckii* subsp. *bulgaricus* in oral capsules. *Res Pharm Sci 7*(1):31–36.

Jantzen, M., Göpel, A. and Beermann, C. 2013. Direct spray drying and microencapsulation of probiotic *Lactobacillus reuteri* from slurry fermentation with whey. *J Appl Microbiol 115*:1029–1036.

Jones, M.L., Martoni, C.L., Parent, M., Prakash, S. 2012. Cholesterol-lowering efficacy of a microencapsulated bile salt hydrolase-active *Lactobacillus reuteri* NCIMB 30242 yoghurt formulation in hypercholesterolaemic adults. *Brit J Nut 107*:1505–1513.

Juillard, V., Le Bars, D., Kunji, E.R.S., Konings, W.N., Gripon, J.C., Richard, J. 1995. Oligopeptides are the main source of nitrogen for *Lactococcus lactis* during growth in milk. *Appl Environ Microbiol 61*:3024–3030.

Kaczynska, A., Klosinska, M., Janeczek, K., Zarobkiewicz, M., Emeryk, A. 2022. Promising immunomodulatory effects of bacterial lysates in allergic diseases. *Front Immuno 13*:907149.

Kailasapathy, K., Champagne, C., Moore, S. 2011. Synbiotic Yoghurt—A Smart Gut Food: Science, Technology and Applications. New York: Nova Science Publishers, Inc.

Koc, B., Yilmazer, M.S., Balkir, P., Ertekin, F.K. 2010. Spray drying of yogurt: Optimization of process conditions for improving viability and other quality attributes. *Drying Technol 28*:495–507.

Koh, A., De Vadder, F., Kovatcheva-Datchary, P., Backhed, F. 2016. From dietary fiber to host physiology: Short-chain fatty acids as key bacterial metabolites. *Cell 165*:1332–1345.

Kreft, M.E., Champagne, C.P., Jelen, P. 2001. Growth of *Lactobacillus* dairy cultures on two different brands of MRS medium. *Milchwissenschaft 56*:315–317.

Kurtmann, L., Carlsen, C.U., Risbo, J., Skibsted, L.H. 2009a. Storage stability of freeze-dried *Lactobacillus acidophilus* (La-5) in relation to water activity and presence of oxygen and ascorbate. *Cryobiol 58*:175–180.

Kurtmann, L., Carlsen, C.U., Skibsted, L.H., Risbo, J. 2009b. Water activity-temperature state diagrams of freeze-dried *Lactobacillus acidophilus* (La-5): Influence of physical state on bacterial survival during storage. *Biotechnol Prog 25*(1):265–270.

Kurtmann, L., Skibsted, L.H., Carlsen, C.U. 2009c. Browning of freeze-dried probiotic bacteria cultures in relation to loss of viability during storage. *J Agric Food Chem 57*:6736–6741.

Kurultay, Ö., Öksül, Ö., Kaptan, B. 2006. Effects of different heat treatments of milk on some growth characteristics of mixed and single cell cultures of yoghurt bacteria. *Milchwissenschaft 61*:52–55.

Lavari, L., Ianniello, R., Páez, R., Zotta, T., Cuatrin, A., Reinheimer, J., Parente, E., Vinderola, G. 2015. Growth of *Lactobacillus rhamnosus* 64 in whey permeate and study of the effect of mild stresses on survival to spray drying. *LWT Food Sci Technol 63*:322–330.

Lavari, L., Paéz, R., Cuatrin, A., Reinheimer, J., Vinderola, G. 2014. Use of cheese whey for biomass production and spray drying of probiotic lactobacilli. *J Dairy Res 81*:267–274.

Lee, B.B., Tham, H.J., Chan, E.S. 2007. Fed-batch fermentation of lactic acid bacteria to improve biomass production: A theoretical approach. *J Appl Sci 7*(15):2211–2215.

Lian, W.C., Hsiao, H.C., Chou, C.C. 2002. Survival of bifidobacteria after spray-drying. *Int J Food Microbiol 74*(1–2):79–86.

Lloyd, G.T., Pont, E.G. 1973. Some properties of frozen concentrated starters produced by continuous culture. *J Dairy Res 40*:157–167.

Macedo, M.G., Lacroix, C., Gardner, N.J., Champagne, C.P. 2002. Effect of medium supplementation on exopolysaccharide production by *Lactobacillus rhamnosus* RW-9595M in whey permeate. *Int Dairy J 12*:419–426.

Martín, M.J., Lara-Villoslada, F., Ruiz, M.A., Morales, M.E. 2015. Microencapsulation of bacteria: A review of different technologies and their impact on the probiotic effects. *Innov Food Sci Emerg Technol 27*:15–25.

Meneghel, J., Passot, S., Cenard, S., Réfrégiers, M., Jamme, F., Fonseca, F. 2017. Subcellular membrane fluidity of *Lactobacillus delbrueckii* subsp. *bulgaricus* under cold and osmotic stress. *Appl Microbiol Biotechnol* 101(18):6907–6917.

Merin, U., Gordin, S., Tanny, G.B. 1983. Microfiltration of sweetcheese whey. *NZ J Dairy Sci Technol* 18:153–160.

Morishita, T., Deguchi, Y., Yajima, M., Sakurai, T., Yura, T. 1981. Multiple nutritional requirements of lactobacilli: Genetic lesions affecting amino acid biosynthetic pathways. *J Bacteriol* 148:64–71.

Muller, J.A., Ross, R.P., Sybesma, W.F.H., Fitzgerald, G.F., Stanton, C. 2011. Modification of the technical properties of *Lactobacillus johnsonii* NCC 533 by supplementing the growth medium with unsaturated fatty acids. *Appl Environ Microbiol* 77(19):6889–6898.

Muller, J.A., Stanton, C., Sybesma, W., Fitzgerald, G.F., Ross, R.P. 2010. Reconstitution conditions for dried probiotic powders represent a critical step in determining cell viability. *J Appl Microbiol* 108(4):1369–1379.

Nannen, N.L., Hutkins, R.W. 1991. Intracellular pH effects in lactic acid bacteria. *J Dairy Sci* 74:741–746.

Nobaek, S., Johansson, M.L., Molin, G., Ahrné, S., Jeppsson, B. 2000. Alteration of intestinal microflora is associated with reduction in abdominal bloating and pain in patients with irritable bowel syndrome. *Am J Gastroenterol* 95:1231–1238.

Ouwehand, A.C., Sherwin, S., Sindelar, C., Smith, A.B., Stahl, B. 2018. Production of probiotic bifidobacteria. In: The Bifidobacteria and Related Organisms. Biology, Taxonomy, Applications. P. Mattarelli, B. Biavati, W. Holzapfel and B.J.B. Wood (eds). London: Academic Press, pp. 261–269.

Páez, R., Lavari, L., Vinderola, G., Audero, G., Cuatrin, A., Zaritzky, N., Reinheimer, J. 2012. Effect of spray drying on the viability and resistance to simulated gastrointestinal digestion in lactobacilli. *Food Res Int* 48:748–754.

Partanen, L., Marttinen, N., Alatossawa, T. 2001. Fats and fatty acids as growth factors for *Lactobacillus delbrueckii*. *Syst Appl Microbiol* 24:500–506.

Passot, S., Gautier, J., Jamme, F., Cenard, S., Dumas, P., Fonseca, F. 2015. Understanding the cryotolerance of lactic acid bacteria using combined synchrotron infrared and fluorescence microscopies. *Analyst* 140(17):5920–5928.

Peebles, M.M., Gilliland, S.E., Speck, M.L. 1969. Preparation of concentrated lactic *Streptococcus* starters. *Appl Microbiol* 17:805–810.

Pinheiro, I., Robinson, L., Verhelst, A., Marzorati, M., Winkens, B., den Abbeele, P.V., Possemiers, S. 2017. A yeast fermentate improves gastrointestinal discomfort and constipation by modulation of the gut microbiome: Results from a randomized double-blind placebo-controlled pilot trial. *BMC Complement Altern Med* 17(1):441.

Poddar, D., Das, S., Jones, G., Palmer, J., Jameson, G.B., Haverkamp, R.G., Singh, H. 2014. Stability of probiotic *Lactobacillus paracasei* during storage as affected by the drying method. *Int Dairy J* 39(1):1–7.

Polo, L., Mañes-Lázaro, R., Olmeda, I., Cruz-Pio, L.E., Medina, Á., Ferrer, S., Pardo, I.J. 2017. Influence of freezing temperatures prior to freeze-drying on viability of yeasts and lactic acid bacteria isolated from wine. *J Appl Microbiol* 122(6):1603–1614.

Potvin, J., Fonchy, E., Conway, J., Champagne, C.P. 1997. An automatic turbidimetric method to screen yeast extracts as fermentation nutrient ingredients. *J Microbial Methods* 29:153–160.

Prasad, J., McJarrow, P., Gopal, P. 2003. Heat and osmotic stress responses of probiotic *Lactobacillus rhamnosus* HN001 (DR20) in relation to viability after drying. *Appl Environ Microbiol* 69(2):917–925.

Pyar, H., Peh, K.-K. 2014. Cost effectiveness of cryoprotective agents and modified De-man Rogosa Sharpe medium on growth of *Lactobacillus acidophilus*. *J Biol Sci* 17(4):462–471.

Rault, A., Bouix, M., Béal, C. 2010. Cryotolerance of *Lactobacillus delbrueckii* subsp. *bulgaricus* CFL1 is influenced by the physiological state during fermentation. *Int Dairy J* 20(11):792–799.

Raymond, Y., Champagne, C.P. 2015. The use of flow cytometry to accurately ascertain total and viable counts of *Lactobacillus rhamnosus* in chocolate. *Food Microbiol* 46:176–183.

Reid, A.A., Champagne, C.P., Gardner, N., Fustier, P., Vuillemard, J.C. 2007. Survival in food systems of *Lactobacillus rhamnosus* R011 microentrapped in whey protein gel particles. *J Food Sci* 72(1):M031–M037.

Remes Troche, J.M., Coss Adame, E., Valdovinos Díaz, M.A., Gómez Escudero, O., Icaza Chávez, M.E., Chávez-Barrera, J.A., Zárate Mondragón, F., Ruíz Velarde Velasco, J.A., Aceves Tavares, R.G., Lira Pedrín, M.A., Cerda Contreras, E., Carmona Sánchez, R.I., Guerra López, H., Solana Ortiz, R. 2020. *Lactobacillus acidophilus* LB: A useful pharmabiotic for the treatment of digestive disorders. *Therap Adv Gastroenterol* 13:1756284820971201.

Ríus, A.G., Kaufman, J.D., Li, M.M., Hanigan, M.D., Ipharraguerre, I.R. 2022. Physiological responses of Holstein calves to heat stress and dietary supplementation with a postbiotic from Aspergillus oryzae. *Sci Rep* 12(1):1587.

Riviere, A., Selak, M., Lantin, D., Leroy, F., De Vuyst, L. 2016. Bifidobacteria and butyrate-producing colon bacteria: Importance and strategies for their stimulation in the human gut. *Front Microbiol* 7:979.

Ruas-Madiedo, P., Tuinier, R., Kanning, M., Zoon, P. 2002. Role of exopolysaccharides produced by *Lactococcus lactis* subsp. *cremoris* on the viscosity of fermented milks. *Int Dairy J* 12:689–695.

Saarela, M.H., Alakomi, H.L., Puhakka, A., Mättö, J. 2009. Effect of the fermentation pH on the storage stability of *Lactobacillus rhamnosus* preparations and suitability of in vitro analyses of cell physiological functions to predict it. *J Appl Microbiol 106*:1204–1212.

Saarela, M.H., Virkajärvi, I., Alakomi, H.L., Mattila-Sandholm, T., Vaari, A., Suomalainen, T., Mättö, J.J. 2005. Influence of fermentation time, cryoprotectant and neutralization of cell concentrate on freeze-drying survival, storage stability, and acid and bile exposure of *Bifidobacterium animalis* ssp. *lactis* cells produced without milk-based ingredients. *Appl Microbiol 99*(6):1330–1339.

Salminen, S., Collado, M.C., Endo, A., Hill, C., Lebeer, S., Quigley, E.M.M., Sanders, M.E., Shamir, R., Swann, J.R., Szajewska, H., Vinderola, G. 2021. The international scientific association of probiotics and prebiotics (ISAPP) consensus statement on the definition and scope of postbiotics. *Nat Rev Gastroenterol Hepatol 18*(9):649–667.

Salminen, S., Stahl, B., Vinderola, G., Szajewska, H. 2020. Infant formula supplemented with biotics: Current knowledge and future perspectives. *Nutrients 12*(7):1952.

Santivarangkna, C., Aschenbrenner, M., Kulozik, U., Foerst, P. 2011. Role of glassy state on stabilities of freeze-dried probiotics. *J Food Sci 76*(8):R152–R156.

Santivarangkna, C., Kulozik, U., Foerst, P. 2007. Alternative drying processes for the industrial preservation of lactic acid starter cultures. *Biotechnol Prog 23*(2):302–315.

Sarao, L.K., Arora, M. 2017. Probiotics, prebiotics, and microencapsulation: A review. *Crit Rev Food Sci Nut 57*:344–371.

Sashihara, T., Sueki, N., Furuichi, K., Ikegami, S. 2007. Effect of growth conditions of *Lactobacillus gasseri* OLL2809 on the immunostimulatory activity for production of interleukin-12 (p 70) by murine splenocytes. *Int J Food Microbiol 120*(3):274–281.

Savini, M., Cecchini, C., Verdenelli, M.C., Silvi, S., Orpianesi, C., Cresci, A. 2010. Pilot-scale production and viability analysis of freeze-dried probiotic bacteria using different protective agents. *Nutrients 2*(3):330–339.

Schoug, Å., Fischer, J., Heipieper, H., Schnürer, J., Håkansson, S., 2008. Impact of fermentation pH and temperature on freeze-drying survival and membrane lipid composition of Lactobacillus coryniformis Si3. *J Ind Microbiol Biotechnol 35*:175–181.

Schuck, P., Dolivet, A., Méjean, S., Hervé, C., Jeantet, R. 2013. Spray drying of dairy bacteria: New opportunities to improve the viability of bacteria powders. *Int Dairy J 31*(1):12–17.

Shao, Y., Gao, S., Guo, H., Zhang, H.J. 2014. Influence of culture conditions and preconditioning on survival of *Lactobacillus delbrueckii* subspecies *bulgaricus* ND02 during lyophilization. *J Dairy Sci 97*(3):1270–1280.

Shao, Y., Wang, Z., Bao, Q., Zhang, H. 2017. Differential enumeration of subpopulations in concentrated frozen and lyophilized cultures of *Lactobacillus delbrueckii* ssp. *bulgaricus* . *J Dairy Sci 100*:8776–8782.

Sherwin, E., Sandhu, K.V., Dinan, T.G., Cryan, J.F. 2016. May the force be with you: The light and dark sides of the microbiota-gut-brain axis in neuropsychiatry. *CNS Drugs 30*:1019–1041.

Siaterlis, A., Deepika, G., Charalampopoulos, D. 2009. Effect of culture medium and cryoprotectants on the growth and survival of probiotic lactobacilli during freeze drying. *Lett Appl Microbiol 48*(3):295–301.

Silva, J., Carvalho, A.S., Ferreira, R., Vitorino, R., Amado, F., Domingues, P., Teixeira, P., Gibbs, P.A. 2005. Effect of the pH of growth on the survival of *Lactobacillus delbrueckii* subsp. *bulgaricus* to stress conditions during spray-drying. *J Appl Microbiol 98*:775–782.

Silva dos Santos, R.C., Finkler, L., Lamenha Luna Finkler, C. 2014. Microencapsulation of *Lactobacillus casei* by spray drying. *J Microencap 31*(8):759–767.

St-Gelais, D., Roy, D., Hache, S. 1992. Growth and activities of *Lactococcus lactis* in milk enriched with low mineral retentate powders. *J Dairy Sci 75*(9):2344–2352.

St-Gelais, D., Roy, D., Haché, S., Desjardins, M.L., Gauthier, S.F. 1993. Growth of nonproteolytic *Lactococcus lactis* in culture medium supplemented with different casein hydrolyzates. *J Dairy Sci 76*:3327–3337.

Strasser, S., Neureiter, M., Geppl, M., Braun, R., Danner, H. 2009. Influence of lyophilization, fluidized bed drying, addition of protectants, and storage on the viability of lactic acid bacteria. *J Appl Microbiol 107*:167–177.

Streit, F., Athès, V., Bchir, A., Corrieu, G., Béal, C. 2011. Microfiltration conditions modify *Lactobacillus bulgaricus* cryotolerance in response to physiological changes. *Bioprocess Biosyst Eng 34*:197–204.

Streit, F., Corrieu, G., Béal, C. 2010. Effect of centrifugation conditions on the cryotolerance of *Lactobacillus bulgaricus* CFL1. *Food Bioprocess Technol 3*:36–42.

Talwalkar, A., Kailasapathy, K. 2003. Metabolic and biochemical responses of probiotic bacteria to oxygen. *J Dairy Sci 86*:2537–2546.

Tsen, J.-H., Lin, Y.-P., Huang, H.-Y., King, V.A.-E. 2007. Accelerated storage testing of freeze-dried immobilized Lactobacillus acidophilus-fermented banana media. *J Food Proces Preserv 31*:688–701.

Tymczyszyn, E., Gerbino, E., Illanes, A., Gómez-Zavaglia, A. 2011. Galacto-oligosaccharides as protective molecules in the preservation of *Lactobacillus delbrueckii* subsp. bulgaricus. *Cryobiol 62*(2):123–129.

Vinderola, G., Sanders, M. E., Cunningham, M., Hill, C. 2024. Frequently asked questions about the ISAPP postbiotic definition. *Front Microbiol 14*:1324565.

Vinderola, G., Sanders, M.E., Salminen, S. 2022. The concept of postbiotics. *Foods 1*(8):1077.

Vinderola, G., Zacarías, M.F., Bockelmann, W., Neve, H., Reinheimer, J., Heller, K. 2012. Preservation of functionality of *Bifidobacterium animalis* subsp. *lactis* INL1 after incorporation of freeze-dried cells into different food matrices. *Food Microbiol 30*:274–280.

Wright, C.T., Klaenhammer, T.R. 1984. Phosphated milk adversely affects growth, cellular morphology, and fermentative ability of *Lactobacillus bulgaricus* . *J Dairy Sci 67*:44–45.

Yang, C., Zhu, X., Fan, D., Mi, Y., Su, R. 2012. Optimizing the chemical compositions of protective agents for freeze-drying *Bifidobacterium longum* BIOMA 5920. *Chinese J Chem Eng 20*(5):930–936.

Ying, D.Y., Phoon, M.C., Sanguansri, L., Weerakkody, R., Burgar, I., Augustin, M.A. 2010. Microencapsulated *Lactobacillus rhamnosus* GG powders: Relationship of powder physical properties to probiotic survival during storage. *J Food Sci 75*(9):E588–E595.

Zhang, M., Oldenhof, H., Sydykov, B., Bigalk, J., Sieme, H., Wolkers, W.F. 2017. Freeze-drying of mammalian cells using trehalose: Preservation of DNA integrity. *Sci Rep 7*(1):6198.

Zhao, W.Y., Li, H., Wang, H., Li, Z.C., Wang, A.I. 2009. The effect of acid stress treatment on viability and membrane fatty acid composition of *Oenococcus oeni* SD-2a. *Agric Sci China 8*(3):311–316.

Zheng, J., Wittouck, S., Salvetti, E., et al. 2020. A taxonomic note on the genus *Lactobacillus*: Description of 23 novel genera, emended description of the genus *Lactobacillus* Beijerinck 1901, and union of *Lactobacillaceae* and *Leuconostocaceae*. *Int J Syst Evol Microbiol 70*(4):2782–2858.

# Stability of Lactic Acid Bacteria and Bifidobacteria in Foods and Supplements

# 20

Miguel Gueimonde, Clara G. de los Reyes-Gavilán, Noraphat Hwanhlem, Nuria Salazar, and Silvia Arboleya

## 20.1 INTRODUCTION

Lactic acid bacteria (LAB) and bifidobacteria have been part of human nutrition since ancient times, and nowadays strains from these bacterial groups are being extensively used for their health-promoting properties, the so-called *probiotics*. The increasing interest in the inclusion of these microorganisms in foods and supplements has led to the great variety of products currently available. However, in order to achieve the desired health impact, this is necessary that the probiotic microorganisms are present in the products at levels that ensure their activity. Therefore, an accurate enumeration and viability assessment of these microorganisms in the products is of great importance to manufacturers, consumers, and regulatory bodies. The most important quality criteria for probiotic-containing products are the identity of the strains used and the level of viable cells in the product.

In general, LAB used as starters are tolerant to the conditions generated during fermentation and show good stability in the final product. However, the strains used as adjunct cultures for probiotic applications are often sensitive to environmental conditions, showing poor survival and viability. This poor stability constitutes an important challenge since strain stability must be high enough during the whole shelf life of the product. Moreover, the stability of the strains not only in terms of survival but also in terms of metabolic activity is needed to maintain the desired product attributes. Although not yet extensively studied, the introduction of probiotics in foods and supplements also requires stability of the probiotic properties of the strains to ensure that the consumer gets the expected benefit from the product. Moreover, the recent development of the *postbiotics* category, defined as "a preparation of inanimate microorganisms and/or their components that confers a health benefit on the host" (Salminen et al., 2021), brings new challenges to this area and underlines the need for appropriate methods to determine the stability of the functional properties of the products.

DOI: 10.1201/9781003352075-22

Despite the availability of culture-independent molecular tools for stability assessment, most manu-facturers and regulatory bodies still rely on conventional culture techniques. The availability of faster and more accurate methods promises to expand our understanding of the stability of LAB and probiotics, perhaps also providing an approach for assessing postbiotics stability. In this chapter we will describe the factors affecting stability in both foods and supplements and some potential strategies to improve it, as well as the different methods currently available to measure viability of LAB and probiotics.

# 20.2 FACTORS AFFECTING STABILITY

When a microorganism is intended to be used during food manufacture, it has to be able to cope with the stresses present during processing and storage. Among these industrial stresses, the principal factors are strains, food ingredients, low and high temperatures, water activity ($a_w$), oxygen, acidity, humidity (or moisture content), packaging, microencapsulation technique and material, freezing, drying and thawing operations, transportation and storage, and the presence of chemicals or other microorganisms that may inhibit the strain(s) being used. These factors have a profound effect on the survival, stability, and prop-erties of LAB and probiotics. Therefore, a precise understanding of the manufacturing conditions, and how they affect the strains being used, is needed for the development of products containing LAB and/or bifidobacteria (Figure 20.1).

## 20.2.1 Strain and Strain Production Conditions

Resistance to the production processes is of paramount importance for obtaining effective biomass pro-duction. It is well known that some probiotics are fastidious microorganisms that grow poorly in industrial media. Therefore, several parameters have to be taken into account for biomass production, among which optimization of growth and processing conditions and monitoring of viability during manufacturing and storage are of key importance (Tripathi and Giri, 2014).

The effectiveness of large-scale cultivation is dependent on a strain's tolerance to stress, with dif-ferent strains showing large differences in their ability to cope with manufacturing conditions. Several strategies have been developed to improve strain stability; among them, the principal ones are stress adaptation and gene modification (Papadimitriou et al., 2016). Stress-resistant microorganisms usually present a stable phenotype, with changes affecting not only the targeted property but often also presenting

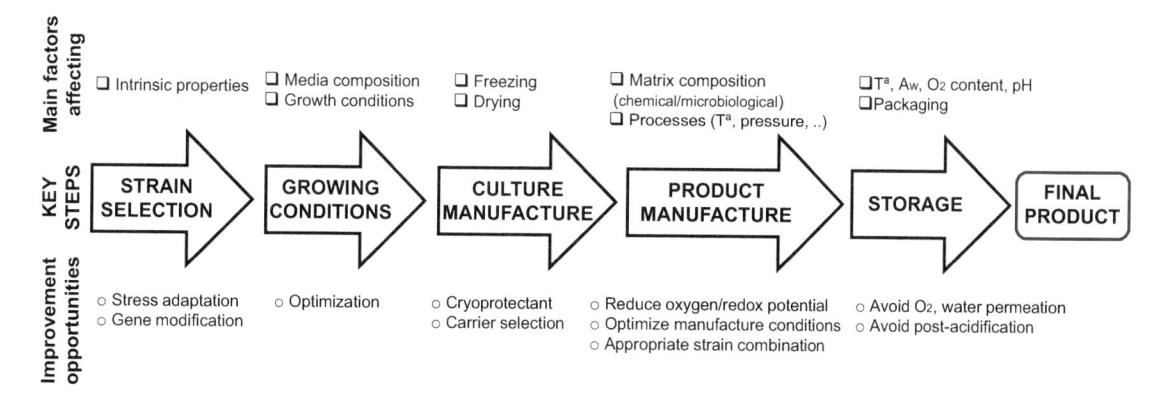

**FIGURE 20.1**   Workflow for the development of products with highly stable LAB or bifidobacteria.

cross-resistance to other stresses. The application of these resistant strains can be useful for improving viability during food processing and gastrointestinal transit (Mathipa and Thantsha, 2015; Nguyen et al., 2016). However, care should be taken, as stress adaptation may alter other physiological and functional properties of the strain (Noriega et al., 2004; Du Toit et al., 2013). An alternative approach is to enhance the stress tolerance of the strains by gene modification. Strains may be genetically modified to accumulate protective compounds (compatible solutes) to improve stress tolerance. Several strategies have been developed in that sense in *Lactobacillus* (*sensu lato*), *Lactococcus*, and *Bifidobacterium* species (Sheehan et al., 2006; Zuo et al., 2020).

In addition to a strain's intrinsic properties, the industrial production processes may themselves have a deep impact on its stability. Among these, freezing and drying conditions, as well as matrix characteristics such as $a_w$ of the product or oxygen concentration, will determine, to a great extent, the strain's stability.

## 20.2.1.1 Freezing

Strains are often supplied as frozen cultures, which presents the advantage of allowing a direct inoculation of vats. Before freezing, bacteria are usually grown in specific culture media until the early stationary phase of growth and are then concentrated by centrifugation. They are resuspended afterward in a suitable medium, including cryoprotectants, to protect bacterial cells from damage due to ice crystal formation. To date, several ways of obtaining frozen concentrates of LAB/probiotics have been developed by using different cryoprotectants (Tripathi and Giri, 2014). The selection of the best-suited cryoprotectant plays a key role for the stability of the strain during the freezing process. However, the use of freezing is being replaced by drying techniques, which are more cost effective and less time consuming (Broeckx et al., 2016).

## 20.2.1.2 Drying

Drying has been extensively used for the production of bacteria included in foods and probiotic products (Broeckx et al., 2016). In general, drying renders high production rates, and dehydrated powders are stable and easy to store and transport.

One of the techniques most commonly used is spray-drying, which comprises the atomization of the bacterial suspension into a drying gas, resulting in fast water evaporation (Huang et al., 2017). This technology is effective but also leads to a loss of viability, as the bacterial cells are subjected to high temperature, mechanical shearing, dehydration, and osmotic pressure. Several studies have reported on the use of different protectants to increase stability and have observed that some strains can be spray-dried without suffering large viability losses (Ávila-Reyes et al., 2014; Chaiongkarn et al., 2019; Arepally et al., 2020).

Freeze-drying is one of the most convenient methods and often considered the best method for obtaining a stable formulation with LAB and bifidobacteria. However, cellular damage due to cytoplasmic membrane injuries has been observed in some cases, which may be partially solved by the addition of cryoprotectants (Tripathi and Giri, 2014; Savedbsoworn et al., 2018; Romyasamit et al., 2022). Choosing an appropriate drying medium is thus of paramount importance. Moreover, spray-drying and freeze-drying can be combined to obtain powders in so-called spray-freeze-drying (SFD). This technique combines the extrusion device from the spray dryers followed by a freeze-drying module, which yields a dry powder with a narrow particle size distribution (Semyonov et al., 2010). However, the overall encapsulation efficiency of SFD was lower than freeze-drying due to the additional stress factors that occurred during atomization and freezing (Kiepś and Dembczyński, 2022).

Industrial exploitation and applications of probiotic strains require efficient preservation technologies to assure viability and metabolic activity. Freezing preservation or freeze-drying operations are widely used for the long-term storage of probiotics but often negatively affect their stability. To prevent or reduce these adverse effects of freezing and thawing stresses, many substances have been used as cryoprotectants, such as skimmed milk, sodium ascorbate, sugars (glucose, sucrose, maltose, trehalose, oligosaccharide, glucose syrup, and lactose), polysaccharides (starch and starch products, alginate, pectin, chitosan, cellulose derivatives, cyclodextrin, maltodextrin, and dextrose), gums (arabic, guar, and carrageenan), proteins (soy proteins, caseins, whey proteins, egg proteins, zein, gelatin, or hydrolysates of these proteins), and their mixtures. However, apart from the use of cryoprotectants, the resistance of the cells to cold damage may effectively preserve cellular activities for a long time (Kwon et al., 2018; Gul and Atalar, 2019).

Finally, it has to be highlighted that due to the high intra- and interspecies variability, a strain's tolerance, the optimum cryoprotectant or carrier molecule, and its appropriate concentration during and following drying have to be chosen case by case (Carvalho et al., 2004). In addition, growth, drying, storage, and rehydration conditions have to be extensively optimized to maximize the recovery and stability of dried cultures.

## 20.2.2 Matrix and Food Manufacturing Processes

Many factors present during food manufacturing processes may have a negative impact on the bacterial viability, thus conditioning the survival of the strains in both food and the gastrointestinal tract (GIT). Among these, the food matrix conditions, such as $a_w$, oxygen, and chemical and microbiological composition, are key factors affecting probiotic survival.

### 20.2.2.1 Water Activity

When applied to food matrices, $a_w$ is defined as the quotient of vapor pressure of water in food and the vapor pressure of pure water. Thus, $a_w$ is a way to measure the free moisture in foods; $a_w$ values range from 0 to 1, with 1 being the value of water activity corresponding to pure water. Probiotics and LAB are usually added to fresh foods with high $a_w$ values, such as milk or juices, which have an expected shelf life of days/weeks. However, probiotics can also be added to dry products with low $a_w$ ($a_w < 0.25$), which in turn have an expected shelf life of months (e.g., infant formula). Viability of probiotics seems to be qualitatively incremented in matrices with low $a_w$ ($\approx 0.11$) when compared with the rapid loss of viability observed at a higher $a_w$ ($\approx 0.43$), indicating its critical effect during the long-term storage (Vesterlund et al., 2012). It has also been reported that low $a_w$ in a food carrier maintains the enzyme activity of bacteria during storage, which may contribute to improved survival (Min et al., 2017). Thus, keeping $a_w$ at low values is a benefit from both the point of view of the strain stability and product shelf life. However, low $a_w$ values are characteristic of dry products, capsules, or supplements rather than food products.

### 20.2.2.2 Oxygen Concentration

With the exception of foods and supplements packed under controlled atmospheres, oxygen is one of the abiotic factors negatively affecting the survival of probiotics (Ladero and Sanchez, 2017). Oxidative damage is induced by the presence of reactive oxygen species (ROS). Bacteria have developed mechanisms, such as glutathione production and antioxidative enzymes (e.g., catalase or superoxide dismutase), for counteracting these oxidative effects (Miller and Britigan, 1997). Probiotics, mainly members of the *Actinomycetota* and *Bacillota* divisions, counteract the toxic effects of ROS with the production of flavoproteins, which in turn results in the production of compounds with high antibacterial activity, such as hydrogen peroxide (Kullisaar et al., 2002).

Since aerobic conditions are present during the manufacturing, transport, and storage of functional foods and dietary supplements, aerotolerance is a desired trait for strains used in industry. However, aerotolerance shows a high strain-to-strain variation and, in some cases, a single species is the most used due just to its good resistance to oxidative and other stresses (for instance the case of *Bifidobacterium animalis* subsp. *lactis*). The problems associated with the lack of aerotolerance are among the main challenges in the probiotic field, especially for the so-called next-generation probiotics (O'Toole et al., 2017).

### *20.2.2.3 Chemical and Microbiological Composition*

Regarding chemical composition, it is obvious that the presence of antimicrobials in the raw material may affect the strain activity. The pH of the product is another limiting factor affecting strain survival; in this regard, preadaptation to low pH has been reported to improve subsequent survival (Saarela et al., 2009). Moreover, the chemical composition of the raw material, such as the fat content, may also have an impact (Vinderola et al., 2000a). It has also been reported that certain food additives, such as flavorings or colorants, may negatively affect the stability of starter and probiotic strains (Vinderola et al., 2002a). In contrast, different carbohydrates, including prebiotics, protein concentrates, gums, or their mixtures, may improve strain stability (Akalin et al., 2007; Kwon et al., 2018; Gul and Atalar, 2019) and the production of health-promoting metabolites (Oliveira et al., 2009).

The interactions among the strains included in bacterial mixes need to be evaluated. Although studies in this area are still scarce, interactions were observed among LAB and probiotic strains (Vinderola et al., 2002b). However, the effect of the presence of other microorganisms is not always detrimental; cofermentation with *Lactococcus lactis* has been found to improve *Bifidobacterium longum* survival in yogurt (Abe et al., 2009a), and the combined use of yeasts and bacteria may also increase survival (Liu and Tsao, 2009). Moreover, interactions among yeast and probiotic bacteria enhance the probiotic properties and metabolism offering augmented protection against pathogens (Zoumpourtikoudi et al., 2018).

It is important to underline that these effects appear to be dependent on the specific compounds and strains being used, and, therefore, a careful case-by-case assessment of the different combinations is always needed.

## 20.2.3 Packaging and Storage

Apart from optimizing strains' production and food manufacture processes, the packaging and storage can also be modified to favor the survival of the microorganisms. Storage temperature, humidity, the use of protectants and oxygen content in the package play a key role in strain survival. In general refrigeration temperatures enhance the survival of the strains, but it is also essential to know the dose in which the probiotic is added to the food and its viability during processing operations and storage (Tripathi and Giri, 2014). A recent meta-analysis studied LAB viability in different refrigerated food matrices and reported that fruit was the type of product that most quickly lost viability (Soto et al., 2023). However, storage of fermented products may cause the so-called post-acidification process, which consists of the production of acid by the starter strains during refrigerated storage; this can affect the stability of probiotics and result in short shelf life of the product, undesirable flavor, and sour taste, making it unacceptable to consumers. Several physical, chemical, and biological methods have also been proposed to avoid this post-acidification process (Zhang et al., 2022).

As stated, the oxygen content may also have an influence. Several methods have been proposed to reduce the oxygen content in packaged probiotic foods, for example, vacuum packaging, addition of antioxidants or oxygen scavengers like ascorbic acid, and genetic manipulation of microorganisms (Terpou et al., 2019). The advances in the field of active packaging and modified atmospheres have also provided means for enhancing strain survival during storage (Ceylan and Atasoy, 2023). Carbon dioxide addition to liquid matrices, such as milk, partially displacing the oxygen, has also been explored and was found to be appropriate for the manufacture of yogurt and other fermented milks (Gueimonde et al., 2002, 2003; Gueimonde and de los Reyes-Gavilán, 2004; Hotchkiss et al., 2006; Sha and Prajapati, 2014). Moreover, this may inhibit the growth of deleterious microorganisms, contributing to the reduction of the risk of contamination of the product (Noriega et al., 2003). The use of probiotics in edible films and coatings as inner packaging is also accepted as an innovative method due to their various advantages such as being consumed with the product and being easily degraded compared to synthetic polymers ( Ceylan and Atasoy, 2023). Lately, a well-designed probiotic encapsulation system was shown to enhance the viability of probiotics in foods (Terpou et al., 2019).

# 20.3 STABILITY IN FOODS AND SUPPLEMENTS

The global probiotics market has been incrementally growing in recent years, as consumers' demand for healthy diets and wellness has continued to increase. Preserving the efficacy of probiotic bacteria exhibits paramount challenges that need to be addressed during the development of functional food products. Several strains of probiotics have been added traditionally to dairy products but also to non-dairy food products like meat, cereals, vegetables, and fruit juices; bakery items; chewing gum; chocolate; infant formula; and non-food carrier types such as capsules, sachets, or tablets are used as well as a source of probiotics. Generally, a progressive loss of stability is observed during storage, which has raised the need for developing formulations that protect bacteria from harsh technological and environmental conditions. To exert their potential health-promoting effects, probiotic bacteria must survive in sufficient numbers for the product to be marketable. In addition, their stability during product storage must be guaranteed, which includes all the properties by which a given bacteria is claimed to be probiotic.

## 20.3.1 Stability in Dairy Products

LAB used as starters normally display appropriate stability in food matrices. When this is not the case, failure in acidification and maturation processes occurs. Probiotic strains often show poorer stability than starter cultures, and to provide health benefits, a minimum daily intake should be ensured (often in the range of $10^8$-$10^9$ CFU/day). Typically, a content of $10^6$ cells per gram or ml is considered the minimum level of probiotics or starters in the product, and methods exist for these determinations (ISO 19344/IDF 232). In addition, in the case of probiotics, not only is the maintenance of viability important, but the functional properties also have to be stable during manufacture and shelf life of the product, which constitutes an area for further research

In dairy products, the acidic conditions and the presence of oxygen represent the main challenges for microorganisms (Meybodi et al., 2020). *B. animalis* subsp. *lactis* is more resistant to oxygen and acid than other *Bifidobacterium* species. For decades, this has made this species the most extensively used bifidobacterial probiotic in dairy products worldwide (Gueimonde et al., 2004; Raeisi et al., 2013; Shori, 2022). The combination of LAB and probiotics could impair the viability of the microorganisms in fermented products, although a positive association has also been observed, depending on the different strains used and the manufacturing conditions (Suharja et al., 2014; Meybodi et al., 2020).

As already indicated, one of the main factors impairing viability and functionality of cultures in dairy products is the post-acidification that may occur in the finished fermented product during refrigerated storage. *Lactobacillus delbrueckii* subsp. *bulgaricus* is mainly responsible for post-acidification in yoghurt and related products. Different strategies have been employed to counteract this deleterious effect. Microbiological strategies range from variation in composition of starter cultures, just avoiding the use of post-acidifying microorganisms (Dan et al., 2015), to the use of flexible ratios of starter cultures (Deshwal et al., 2021). This generally results in longer fermentation times and milder final acidity. Lactase enzyme has been used during production to reduce lactose content, along with modified starter composition (Riis et al., 2019). Some commercial cultures with low post-acidification activity are currently available, including the term "mild" in the brand name (Deshwal et al., 2021). Moreover, some *Streptococcus thermophilus* and *Lc. lactis* strains can act as oxygen scavengers in fermented dairy products (Odamaki et al., 2011).

The use of stress-resistant strains or microorganisms grown under sub-lethal conditions can also be useful for improving stability (Chu-Ky et al., 2014; Ergin et al., 2016; Succi et al., 2017). Some studies indicate that stress-resistant probiotics, although not promoting remarkable changes in acidification profile or the sensory properties of fermented milks, exert a considerable influence on the microbial metabolites produced and hence on the organoleptic quality of yogurt (Sánchez et al., 2010; Settachaimongkon et al., 2015). The selection of next-generation probiotic strains intrinsically more resistant to these factors or obtained by random mutagenesis or adaptation to stress conditions can improve survival of naturally more sensitive microorganisms during processing and storage (Meybodi et al., 2020; Marcos-Fernández et al., 2023).

Apart from the selection and combination of appropriate strains and the control of final pH and post-acidification, other strategies have been used to improve the viability of LAB and probiotics in fermented products. Among them are the addition of protectors and oxygen scavenger compounds, antioxidants, some enzymatic treatments as glucose oxidase, and the use of packaging materials containing oxygen barriers (Chandra et al., 2007; Asli et al., 2017; Dos Santos et al., 2017; Meybodi et al., 2020). Supplementation of fermented milks with bran from different grains or other cereal components has also been used (Hasani et al., 2016; Demirci et al., 2017). Apart from their potential role in human intestinal health, efforts have been made, by using prebiotics, to improve the viability of probiotics in fermented products and afterward during gastric transit (Ross et al., 2005; Akalin et al., 2007; Padilha et al., 2016; Succi et al., 2017). Prebiotics can also be employed as entrapping matrices for probiotics, these combined products being regarded as synbiotics (Ross et al., 2005; Rashidinejad et al., 2022). Cell entrapment techniques by microencapsulation using different methods have been developed by using different polymers for the improvement of bacterial viability (Patrignani et al., 2017; Barajas-Alvarez et al., 2021), alone or in combination with other materials. This is the case of alginate, gellan gum, xanthan gum, starch, cellulose acetate phthalate, gelatin, milk proteins, and κ-carrageenan (Abbas et al., 2022). After drying beads, a surface coating can be applied using gelatin or whey proteins (Reid et al., 2007). The resistance of probiotics and other lactic acid bacteria to environmental stress conditions can be improved in the presence of certain monosaccharides, such as glucose combined with glucose oxidase in yogurts, which positively influences probiotic survival by reducing the oxygen level (Cruz et al., 2012).

Starter LAB survival in curd during cheesemaking is related to inherent strain properties such as temperature and salt sensitivity, presence of prophages, autolytic and permeabilization capacity, envelope proteinases, and metabolic activity; later on during ripening, a decline in starter viability generally occurs (Wilkinson and LaPointe, 2020). Cheese has also been used as a carrier for probiotic bacteria. Its higher pH, more consistent matrix, and higher fat content than fermented milk can act as protectors for microorganisms (Blaiotta et al., 2017). However, $a_w$, cooking of curds, and long ripening times should be mentioned as negative factors affecting viability (Vinderola et al., 2009). It is also important to determine the time and method of adding probiotic bacteria to cheese in order to avoid their loss during whey drainage. There are examples of successful incorporation of probiotics in different types of fresh cheeses (Burns et al., 2008; Jeon et al., 2016; Padilha et al., 2016), ripened cheeses such as Cheddar (Murtaza et al., 2017), Pecorino Siciliano cheese (Pino et al., 2017), Dutch-type cheese (Bielecka and Cichosz, 2017), or Argentinean semi-hard cheese (Bergamini et al., 2005).

Ice cream and frozen dairy products have also been used as carriers for probiotics. Microencapsulation in alginate or co-encapsulation with prebiotics provide protections against stress and good results for viability (Spigno et al., 2015; Shori, 2022). Nevertheless, the formation of ice crystals in the freezing process or the high oxygen content accumulated during whipping can be toxic for anaerobic microorganisms and should be minimized (Park et al., 2015). The $a_w$ and high osmotic pressures due to added sugars and other sweeteners could be an important cause of losing viability (Cruz et al., 2009).

## 20.3.2 Stability in Other Food Matrices

LAB can be also found in non-dairy fermented substrates. Among the meat products, sausages are the most abundant (Casaburi et al., 2016; Quijada et al., 2018). Microorganisms acting as starters commonly are constituted by different species of lactobacilli, enterococci, pediococci, molds, and yeasts, frequently isolated from traditional food sources (Dalla Santa et al., 2014). These microorganisms multiply well in this environment and help to extend shelf life and improve safety and sensory properties of the fermented product.

Fruits and vegetables contain beneficial nutrients which make them a suitable vehicle for LAB and probiotics. LAB belonging to lactobacilli and *Leuconostoc* are the most common bacteria found in natural vegetable lactic acid fermentation (Medina-Pradas et al., 2017). Despite the low pH, probiotics are being increasingly added to fruit beverages due to their good tolerance of acidic conditions in these environments. The viability of microorganisms in fruits and vegetables depends on the strain, final acidity, and the lactic and acetic acid concentrations on the product (Mustafa et al., 2016). Microencapsulation may efficiently contribute to improving viability within fruit juices and vegetables (Gandomi et al., 2016). Prebiotics and other bioactive compounds can also improve the viability of probiotic bacteria in vegetal matrices (Duarte et al., 2017; Pacheco-Ordaz et al., 2018).

Cereal grains, including soya, are important sources of protein, carbohydrates, vitamins, and minerals. In addition, they are a source of nondigestible carbohydrates that can act as prebiotic substrates and/or present other additional beneficial properties. The nutritional quality of whole grains is, however, lower than that of milk, fish, or meat because of the lower protein content, deficiency in essential amino acids (lysine, threonine, and tryptophan), low starch availability, and presence of antinutrients (phytic acid, tannins, and polyphenols). Cereals represent a good substrate for the growth and stability of LAB. Fermentation in traditional and non-traditional fermented cereal products decrease the level of some polysaccharides, improve the protein quality by bacterial synthesis of some amino acids, and increase the availability of group B vitamins. Fermentation also provides optimum pH conditions for enzymatic degradation of phytate and the release of minerals from fiber (Rivera-Espinoza and Gallardo-Navarro, 2010; Shing and Sharma, 2017).

Bakery products with added probiotics have also been developed. *Lacticaseibacillus casei* has been encapsulated to provide resistance to high temperature, avoiding the loss of viability during the baking process without any adverse effects on the flavor or texture (Seyedain-Ardabili et al., 2016). Other products such as savory cereal bars, dark chocolate, and snacks have been also recently developed with encapsulated probiotics (Barajas-Álvarez et al., 2021).

## 20.3.3 Stability in Dry Form

Most LAB cultures used in industrial applications are commercialized in dried form to reduce transportation costs and storage space. The stability and viability of LAB are always questionable when they are exposed to harsh environments during processing and storage. The majority of research into preservation of probiotics focuses on *Lactobacillus* (*sensu lato*) and *Bifidobacterium* species, which typically tolerate different drying techniques, including spray-drying to some extent (Huang et al., 2017). Dried LAB ferments are generally lyophilized to ensure high cell viability, but adverse conditions such as mechanical stress, dehydration, heating, and oxygen exposure can lead to reduced LAB cell viability (Moreira et al., 2021). In the case of dry products, shelf life is directly related to the survival and stability of the probiotic strains included. Generally, viability of the strains decreases during storage (Abe et al., 2009b). Probiotic stability is negatively affected by high $a_w$ (>0.25) and elevated storage temperatures, as well as by the presence of oxygen. Thus, in general, the higher the $a_w$, the lower the stability observed in the product (Abe et al., 2009b; Vesterlund et al., 2012). In addition, as stated, stability is dependent on the drying technique, the carrier type and concentration, or the outlet temperature, among other parameters (Moreira et al., 2021; Ermis, 2022). Low (sugars) and high molecular weight protectants (such as polysaccharides and proteins) have been tested as the most important factors affecting the survival of dried LAB (Moreira et al., 2021).

## 20.4 METHODS FOR DETERMINING STABILITY

Even in the age of genomics, and with the availability of fast and reliable culture-independent molecular tools, manufacturers and regulatory bodies rely on traditional culture techniques to evaluate the stability of the strain(s) in products (Gill, 2017; Jackson et al., 2019). This enumeration of live probiotic bacteria is the gold standard used not only for stability but also for feasibility, formulation, and quality control. Culture-dependent techniques are still crucial to isolate the strains used to determine any physiological or

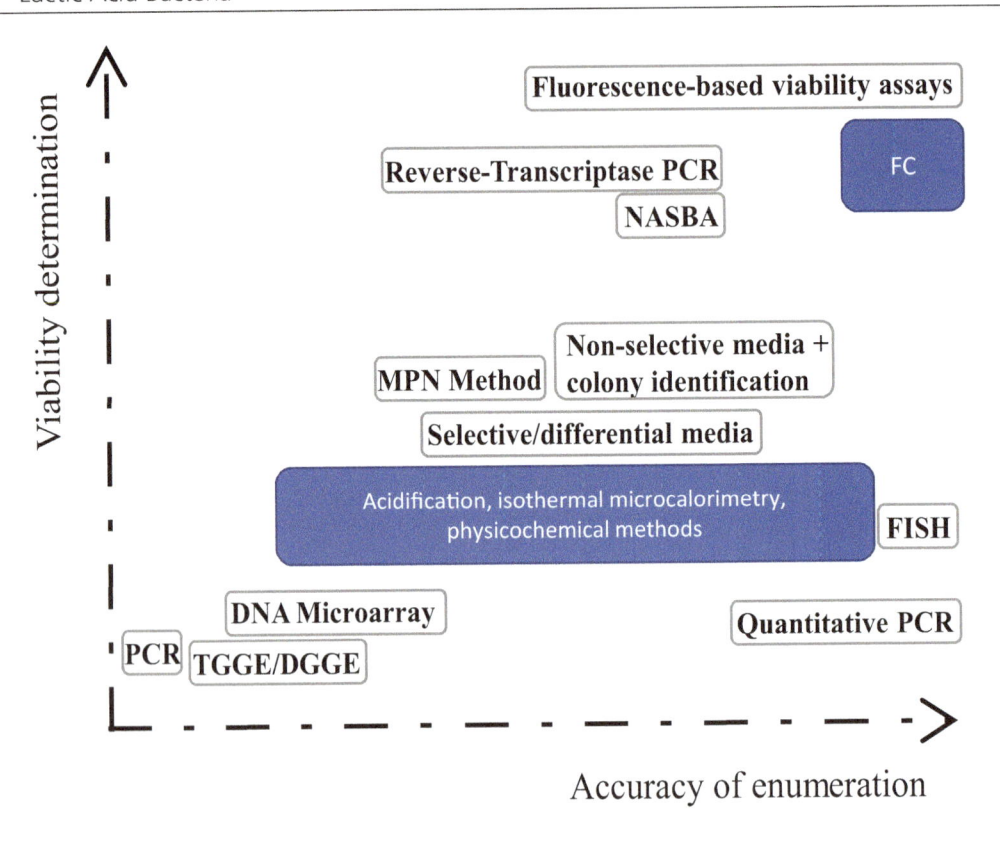

**FIGURE 20.2**  Methods for enumeration and viability determination of bacteria.

biochemical change in the strain during the storage of the product. Isolation and identification of strains is also important for quality control purposes and should be done routinely, as a number of studies carried out over the years have indicated that the identity of the microorganisms does not always correspond to the information stated on the product label (Gueimonde and de los Reyes-Gavilán, 2004; Weese and Martin, 2011; Morovic et al., 2016; Zawistowska-Rojek et al., 2016). Nevertheless, these culture-dependent methods show several limitations, which, together with the advance of molecular biology, have led to the recent development of culture-independent techniques that overcome some of the limitations of the culture-based approaches (Figure 20.2).

## 20.4.1 Culture-Dependent Methods

Differential enumeration of probiotic and starter bacteria is a complex issue. The quantification of probiotics and LAB in food products or supplements has traditionally been carried out by plate counting in selective and differential media (Davis, 2014). Most recognized standards published by the International Organization for Standardization (ISO), the International Dairy Federation (IDF), or the United States Pharmacopeia (USP) for bacterial enumeration use plate count methods (Jackson et al., 2019). International standard methods for sample preparation and enumeration of certain species of LAB and bifidobacteria in milk and dairy products have been developed by the International Dairy Federation (ISO 8261/IDF 122; ISO 20128/IDF 192; ISO 29981/IDF220; ISO 7889/IDF 117; ISO 17792/IDF180; ISO 19344/IDF 232; ISO 6887–1:1999). However, these media and culture conditions may not always be suitable, or applicable, to other food products.

Some challenges are associate with the plate count method. No unique medium is applicable to all the probiotics strains, and very often media are based in a metabolic feature present in some microorganisms of interest but not in others that may be present in the product (Ashraf and Shah, 2011; Süle et al., 2014; Galat et al., 2016). Media based on specific properties of the microorganisms, such as resistance to specific antimicrobial compounds, have also been designed for LAB and bifidobacteria (Karimi et al., 2012; Colombo et al., 2014; Pechar et al., 2014; Bunesova et al., 2015). Aggregates of bacteria and viable but non-culturable cells (VBNCs), frequently observed in probiotic products due the conditions used during manufacture, can make the counting and recovery of the specific species difficult and underestimate the population size (Davis, 2014; Jackson et al., 2019). Moreover, competitive interactions among bacteria in the product may influence population growth and would not be captured in plate counts.

Another technique that has been used for determining bacterial levels in foods is the most probable number (MPN) method. MPN is a qualitative technique for the quantification of bacterial populations based on the determination of the absence/presence of specific attributes of the bacterium (gas production, sugar fermentation, etc.). This technique may be an alternative strategy when establishing accurate bacterial counts is not feasible by plate counting or for estimating very low microbial population sizes (<100 CFU/g). However, it has been shown that MPN has an important bias toward overestimation of microbial levels (Garthright, 1993). Although extensively used for pathogens, the application of this method for probiotics and LAB is rare.

Acidification rate analysis by measuring the change in pH in a liquid medium culture with time is commonly used for assessing the quality of milk starter cultures containing LAB. It was observed to be proportional to cell concentration when comparing resistance to freezing in different LAB, being proposed as a complement or substitute to other viability measurements (Fonseca et al., 2000). In addition, isothermal microcalorimetry technology, based on the idea that heat is produced by a replicating, metabolizing bacterial population and heat production is proportional to the number of bacterial cells in a sample, could be used as method for enumeration of probiotic cultures (Wendel, 2022). Other physical methods based on physicochemical property changes of the medium and the bacterial behavior at solid liquid interface have been proposed for rapid estimation of bacterial concentration in fermented milk (Hosseini et al., 2019).

Notwithstanding the previous, culturing is still considered a unique parameter of microbial viability, but its interpretation entails challenges which can be complemented by other types of methodologies. Moreover, with the emerging trend of postbiotic-containing products, the interest in and application of methods as alternatives to culture will increase.

## 20.4.2 Culture-Independent Methods

Several culture-independent molecular methods such as DNA microarrays, polymerase chain reaction (PCR), temperature gradient gel electrophoresis (TGGE), and denaturing gradient gel electrophoresis (DGGE) analyses can be used for fast and specific detection of bacterial species (Bagheripoor-Fallah et al., 2015). TGGE and DGGE are useful for qualitative analyses of nucleic acids and have been used to study the microbiological composition of food products (Masco et al., 2005; Chen et al., 2014). Nevertheless, these methods do not provide accurate quantitative data.

Molecular enumeration methods have also been developed. The most widely used approaches have been fluorescent *in situ* hybridization (FISH) and real-time quantitative PCR (qPCR). FISH is based on labeling cells with ribosomal RNA (rRNA)-targeting fluorescent probes specific for a certain microbial group. Microbes can then be detected by using, for instance, fluorescence microscopy or flow cytometry (FC). The FISH technique has been used for enumeration of bifidobacteria and LAB in foods (Lahtinen et al., 2006a, 2006b; Babot et al., 2011). FC is a technique often used for counting and characterizing immune cells; however, it is also being used to count bacteria and study their viability and metabolic activity. Although the industry use of FC is still limited, new evidence in FC methodology validations is coming to light (Gorsuch, 2019; Foglia, 2020, Michelutti, 2020). When coupled with fluorescent labeling of the specific microorganisms of interest, FC is a powerful method for the determination of LAB and probiotics (Mudroňová, 2015; Raymond and Champagne, 2015; Pasulka et al., 2021). Label-free biosensors (quartz crystal microbalance and optical waveguide lightmode spectroscopy) have also been applied to the detection and quantification of probiotics in foods and supplements (Szalontai et al., 2014).

Among the wide range of available molecular methods, those based on PCR have been the most broadly used for enumeration of LAB and bifidobacteria in foods (Lahtinen et al., 2006b; Masco et al., 2007). PCR is based on the use of a pair of primers that are complementary to a defined DNA sequence. In qPCR, the DNA being amplified is quantified in real time during the PCR reaction. It is then possible to identify and quantify specific species, or bacterial groups, in a monoculture or in complex microbial communities by qPCR. Several primers have been designed in this way for the identification and quantification of *Bifidobacterium* (Lahtinen et al., 2005; Sheu et al., 2010; Lewis et al., 2016; Arboleya et al., 2020; Kim et al., 2020) and LAB (Sheu et al., 2009; Tsai et al., 2010; Herbel et al., 2013; Shehata et al., 2023) species. A critical step in qPCR is the isolation and purification of DNA, which very often is difficult in complex matrices such as foods.

In the "omics" era, next-generation DNA sequencing can be used as a rapid and advantageous way to identify probiotics in food or dietary supplements. This method can be tailored to the taxonomical level selected, identifying not only the desired bacteria but also possible contaminants or pathogens not reported on the label. Since this technique does not allow determination of the viability of the bacteria, it is necessary to use standard culturing techniques for isolation (Patro et al., 2016).

## 20.4.3 Viability-Oriented Assays

Unfortunately, most molecular methods are not suitable for differentiation between viable and nonviable bacteria. Hence, modifications of the qPCR method have been proposed to enable fast qPCR-based viability measurements (Fittipaldi et al., 2012). These methods require a careful setup and optimization to avoid PCR product amplification from dead cells and to ensure that the treatments used do not compromise the PCR reaction of the DNA isolated from viable cells. They have been used for quantification of bifidobacteria, lactobacilli, and other probiotics (Garcia-Cayuela et al., 2009; Shao et al., 2016; Scariot et al., 2018). In this context, FC could represent an advance in the field, because this technique allows determining VBNC (Foglia et al., 2020; Visciglia et al., 2022). In addition, fluorescence-based methods have been the most widely used for viability assays in the field of probiotics and LAB, and several fluorescent dyes and fluorescent-labeled probes are currently available for quantification of viable cells. Currently used targets of fluorescence-based viability measurements include cytoplasmic membrane integrity, intracellular enzyme activity, intracellular pH, and cell reductase activity (Lahtinen et al., 2006a, 2006b). Nucleic acid intercalating dyes, such as ethidium- or propidium-monoazide, coupled to qPCR have been also proposed for viable cell detection in diverse foods (Scariot et al., 2018; Shi et al., 2022; Wendel, 2022).

Another option for viability assessment is the use of RNA as a target molecule. rRNA has been assessed as an indicator of microbial viability (Lahtinen et al., 2008). However, in normal conditions, the half-life of rRNA is relatively long; therefore, rRNA is not considered a good viability marker. Since mRNA is a highly labile molecule with a short half-life, it has been suggested as a target molecule for viability assays. Viability assays based on the detection of mRNA of the S-layer protein of lactobacilli as a target have also been reported (Saito et al., 2004), and other housekeeping genes' mRNA has been used for viability assessment of *Bifidobacterium* (Reimann et al., 2010). Nevertheless, the short half-life of mRNA also makes the use of this molecule technically challenging. The most commonly used amplification techniques for detecting RNA are reverse transcriptase PCR (RT-PCR), nucleic acid sequence–based amplification (NASBA), and reverse transcriptase-strand displacement amplification (RT-SDA) (Davis, 2014).

Viability assays based on cytoplasmic membrane integrity of probiotics and LAB have been widely used (Lahtinen et al., 2005; Lahtinen et al., 2006a, 2006b; Kramer et al., 2009; Perdana et al., 2012; Gandhi and Shah, 2015). Other targets, such as intracellular enzyme activity (Lahtinen et al., 2006a), ability to maintain intracellular pH (Lahtinen et al., 2006a), membrane potential of bacterial cells (Ben Amor et al., 2002), or production of polyclonal antibodies (Amrouche et al., 2006; Chiron et al., 2018), have also been assessed.

# 20.5 CONCLUSIONS

Developing and/or manufacturing a stable product containing live microorganisms such as LAB or bifidobacteria is always a challenging issue. Several factors, from strain properties and growing conditions to the conditions of storage of the elaborated product, may affect the stability of the microorganisms. It is therefore necessary to have careful case-by-case selection and assessment of the raw material composition, strains and additives used, and manufacturing processes and storage conditions to get a product in which the added microorganisms have the desired stability.

The way stability is defined and how to measure it is another topic of current discussion. This is a matter of increasing interest due to the appearance on the market of the so-called postbiotics, composed by inanimate cells. Traditionally, a microorganism was considered viable if it was able to multiply in an appropriate medium, and stability was much linked to culturability and survival of the strains in the product. For viable cell enumeration, culture-based methods have been traditionally used, but nowadays a large array of culture-independent methods and "omic" techniques are also available for complementing our traditional culture-based knowledge. However, it is important to underline that most food regulations and authorities still rely on the use of culture-based methods. Moreover, postbiotics are, by definition, inanimate; therefore, no culture approach would be appropriate. In addition, probiotic microorganisms can also exist in a not readily cultivable state but still retain metabolic activity and the ability to return to the cultivable state under certain conditions; this is the case of the so-called VBNCs. Therefore, methods measuring multiplication as the sole criterion of viability, although extensively used, hinder the detection of temporally uncultivable cells. For this reason, viability assays apart from those based on multiplication in culture media have also been developed. The availability of these methods is expanding our understanding of bacterial viability, which may be of special importance in this field.

Finally, understanding stability not just as viability or survival but also in terms of changes in strain properties is of special interest. The extensive use of LAB and bifidobacteria as probiotics or postbiotics requires an in-depth knowledge of the stability of the properties that allow strains to interact with hosts and to exert their beneficial effects.

# BIBLIOGRAPHY

Abbas, F. M., F. Saeed, M. Afzaal, et al. 2022. Recent trends in encapsulation of probiotics in dairy and beverage: A review. *J Food Process Presserv* 46:e16689.

Abe, F., H. Miyauchi, A. Uchijima, T. Yaeshima, and K. Iwatsuki. 2009b. Stability of bifidobacteria in powdered formula. *Int J Food Sci Technol* 44:718–724.

Abe, F., S. Tomita, T. Yaeshima, and K. Iwatsuki. 2009a. Effect of production conditions on the stability of a human bifidobacterial species *Bifidobacterium longum* in yogurt. *Lett Appl Microbiol* 49:715–720.

Akalin, A. S., S. Gonc, G. Unal, and S. Fenderya. 2007. Effects of fructooligosaccharide and whey protein concentrate on the viability of starter culture in reduced-fat probiotic yogurt during storage. *J Food Sci* 72:M222–M227.

Amrouche, T., Y. Boutin, O. Moroni, E. Kheadr and I. Fliss. 2006. Production and characterization of anti-bifidobacteria monoclonal antibodies and their application in the development of an immuno-culture detection method. *J Microbiol Methods* 65:159–170.

Arboleya, S., S. Saturio, M. Suárez, et al. 2020. Donated human milk as a determinant factor for the gut bifidobacterial ecology in premature babies. *Microorganisms* 5:760.

Arepally, D., R. S. Reddy, and T. K. Goswami. 2020. Encapsulation of Lactobacillus acidophilus NCDC 016 cells by spray drying: Characterization, survival after in vitro digestion, and storage stability. *Food Funct* 11:8694–8706.

Ashraf, R., and N. P. Shah. 2011. Selective and differential enumerations of *Lactobacillus delbrueckii* subsp. *bulgaricus*, *Streptococcus thermophilus*, *Lactobacillus acidophilus*, *Lactobacillus casei* and *Bifidobacterium* spp. in yoghurt—A review. *Int J Food Microbiol* 149:194–208.

Asli, M. Y., N. Khorshidian, A. M. Mortazavian, and H. Hosseini. 2017. A review on the impact of herbal extracts and essential oils on viability of probiotics in fermented milks. *Curr Nutr Food Sci* 1:6–15.

Ávila-Reyes, S. V., F. J. García-Suarez, T. Jiménez, et al. 2014. Protection of *L. rhamnosus* by spray-drying using two prebiotic colloids to enhance the viability. *Carbohydr Polym* 102:423–430.

Babot, J. D., M. Hidalgo, E. Argañaraz-Martínez, M. C. Apella and A. Perez Chaia. 2011. Fluorescence in situ hybridization for detection of classical propionibacteria with specific 16S rRNA-targeted probes and its application to enumeration in Gruyère cheese. *Int J Food Microbiol* 145:221–228.

Bagheripoor-Fallah, N., A. Mortazavian, H. Hosseini, S. Khoshgozaran-Abras and A. H. Rad. 2015. Comparison of molecular techniques with other methods for identification and enumeration of probiotics in fermented milk products. *Crit Rev Food Sci Nutr* 55:396–413.

Barajas-Álvarez, P., M. González-Ávila, and H. Espinosa-Andrews. 2021. Recent advances in probiotic encapsulation to improve viability under storage and gastrointestinal conditions and their impact on functional food formulation. *Food Rev Int*. http://doi.org/10.1080/87559129.2021.1928691

Ben Amor, K., P. Breeuwer, P. Verbaarschot, et al. 2002. Multiparametric flow cytometry and cell sorting for the assessment of viable, injured, and dead *Bifidobacterium* cells during bile salt stress. *Appl Environ Microbiol* 68:5209–5216.

Bergamini, C. V., E. R. Hynes, A. Quiberoni, V. B. Suarez, and C. A. Zalazar. 2005. Probiotic bacteria as adjunct starters: Influence of the addition methodology on their survival in a semi-hard Argentinean cheese. *Food Res Int* 38:597–604.

Bielecka, M. M., and G. Cichosz. 2017. The influence of an adjunct culture of *Lactobacillus paracasei* LPC-37 on the physicochemical properties of Dutch-type cheese during ripening. *LWT-Food Sci Technol* 83:95–100.

Blaiotta, G., N. Murru, A. Di Cerbo, M. Succi, R. Coppola, and M. Aponte. 2017. Commercially standardized process for probiotic "Italico" cheese production. *LWT-Food Sci Technol* 79:601–608.

Broeckx, G., V. Dieter, I. C. Ingmar, et al. 2016. Drying techniques of probiotic bacteria as an important step towards the development of novel pharmabiotics. *Int J Pharmaceut* 5:303–318.

Bunesova, V., S. Musilova, M. Geigerova, R. Pechar, and V. Rada. 2015. Comparison of mupirocin-based media for selective enumeration of bifidobacteria in probiotic supplements. *J Microbiol Meth* 109:106–109.

Burns, P., F. Patrignani, D. Serrazanetti, et al. 2008. Probiotic Crescenza cheese containing *Lactobacillus casei* and *Lactobacillus acidophilus* manufactured with high-pressure homogenized milk. *J Dairy Sci* 91:500–512.

Carvalho, A. S., J. Silva, P. Ho, P. Teixeira, F. X. Malcata, and P. Gibbs. 2004. Relevant factors for the preparation of freeze-dried lactic acid bacteria. *Int Dairy J* 14:835–847.

Casaburi, A., V. Di Martino, P. Ferranti, L. Picariello, and F. Villani. 2016. Technological properties and bacteriocins production by *Lactobacillus curvatus* 54M16 and its use as starter culture for fermented sausage manufacture. *Food Control* 59:3145.

Ceylan, H. G., and A. F. Atasoy. 2023. New bioactive edible packing systems: Synbiotic edible films/coatings as carries of probiotics and prebiotics. *Food Bioprocess Tech* 16:1413–1426.

Chaiongkarn, A., J. Dathong, W. Phatvej, P. Saman, C. Kuancha, and S. Teetavet. 2019. Stability of spray-dried synbiotics containing *Lactobacillus plantarum* DSM 2648 and exopolysaccharide from Pediococcus acidilactici TISTR 2612 and its vivo effectiveness. *J Food Sci Agr Technol (JFAT)* 5:193–199.

Chandra, R. T. S., S. B. Rachappa, and A. Renu. 2007. Role of oxygen scavengers in improving the stability and viability of *Pediococcus pentosaceus* . *Res J Biotechnol* 2:26–32.

Chen, T., Q. Wu, S. Li, et al. 2014. Microbiological quality and characteristics of probiotic products in China. *J Sci Food Agric* 94(1):131–138.

Chiron, C., T. A. Tompkins, and P. Burguière. 2018. Flow cytometry: A versatile technology for specific quantification and viability assessment of micro-organisms in multistrain probiotic products. *J Appl Microbiol* 124(2):572–584.

Chu-Ky, S., T. K. Bui, T. L. Nguyen, and P. H. Ho. 2014. Acid adaptation to improve viability and X-prolyl dipeptidyl aminopeptidase activity of the probiotic bacterium *Lactobacillus fermentum* HA6 exposed to simulated gastrointestinal tract conditions. *Int J Food Sci Technol* 49:565–570.

Colombo, M., A. E. Zimmermann de Oliveira, A. Fernandes de Carvalho, and L. A. Nero. 2014. Development of an alternative culture medium for the selective enumeration of *Lactobacillus casei* in fermented milk. *Food Microbiol* 39:89–95.

Cruz, A. G., A. E. C. Antunes, A. L. O. P. Sousa, J. A. F. Faria, and S. M. I. Saad. 2009. Ice cream as a probiotic food carrier. *Food Res Int* 42:1233–1239.

Cruz, A. G., W. Castro, J. Faria, et al. 2012. Probiotic yogurts manufactured with increased glucose oxidase levels: Post-acidification, proteolytic patterns, survival of probiotic microorganisms, production of organic acid and aroma compounds. *J Dairy Sci* 95:2261e2269.

Dalla Santa, O. R., R. E. F. de Macedo, H. S. Dalla Santa, C. M. Zanette, R. J. S. de Freitas, and N. N. Terra. 2014. Use of starter cultures isolated from native microbiota of artisanal sausage in the production of Italian sausage. *Food Sci Technol* 34:780–786.

Dan, T., Y. F. Chen, X. Chen, et al. 2015. Isolation and characterization of a *Lactobacillus delbrueckii* subsp. *bulgaricus* mutant with low H+-ATPase activity. *Int J Dairy Technol* 68:527–532.

Davis, C. 2014. Enumeration of probiotic strains: Review of culture-dependent and alternative techniques to quantify viable bacteria. *J Microbiol Meth* 103:9–17.

Demirci, T., K. Aktas, D. Sozeri, H. I. Ozturk, and N. Akin. 2017. Rice bran improve probiotic viability in yoghurt and provide added antioxidative benefits. *J Funct Foods* 36:396–403.

Deshwal, G. K., S. Tiwari, A. Kumar, R. K. Raman, and S. Kadyan. 2021. Review on factors affecting and control of post-acidification in yoghurt and related products. *Trends Food Sci Technol* 109:499–512.

Dos Santos, K. M. O., I. C. de Oliveira, M. A. C. Lopes, A. P. G. Cruz, F. C. A. Buriti, and L. M. Cabral. 2017. Addition of grape pomace extract to probiotic fermented goat milk: The effect on phenolic content, probiotic viability and sensory acceptability. *J Sci Food Agric* 97:1108–1115.

Duarte, F. N. D., J. B. Rodrigues, M. D. Lima, et al. 2017. Potential prebiotic properties of cashew apple (*Anacardium occidentale* L.) agro-industrial byproduct on *Lactobacillus* species. *J Sci Food Agric* 97:3712–3719.

Du Toit, E., S. Vesterlund, M. Gueimonde, et al. 2013. Assessment of the effect of stress-tolerance acquisition on some basic characteristics of specific probiotics. *Int J Food Microbiol* 165:51–56.

Ergin, F., Z. Atamer, A. A. Arslan, et al. 2016. Application of cold-and heat-adapted *Lactobacillus acidophilus* in the manufacture of ice cream. *Int Dairy J* 59:72–79.

Ermis, E. 2022. A review of drying methods for improving the quality of probiotic powders and characterization. *Dry Technol* 40:2199–2126.

Fittipaldi, M., A. Nocker and F. Codony. 2012. Progress in understanding preferential detection of live cells using viability dyes in combination with DNA amplification. *J Microbiol Meth* 91(2):276–289.

Foglia, C., Allesina, S., Amoruso, A., De Prisco, A., and Pane, M. 2020. New insights in enumeration methodologies of probiotic cells in finished products. *J Microbiol Meth* 175:105993.

Fonseca, F., C. Béal, and G. Corrieu. 2000. Method of quantifying the loss of acidification activity of lactic acid starters during freezing and frozen storage. *J Dairy Res* 67(1):83–90.

Galat, A., J. Dufresne, J. Combrisson, et al. 2016. Novel method based on chromogenic media for discrimination and selective enumeration of lactic acid bacteria in fermented milk products. *Food Microbiol* 55:86–94.

Gandhi, A., and N. P. Shah. 2015. Effect of salt on cell viability and membrane integrity of *Lactobacillus acidophilus, Lactobacillus casei* and *Bifidobacterium longum* as observed by flow cytometry. *Food Microbiol* 49:197–202.

Gandomi, H., S. Abbaszadeh, A. Misaghi, S. Bokaie, and N. Noori. 2016. Effect of chitosan-alginate encapsulation with insulin on survival of *Lactobacillus rhamnosus* GG during apple juice storage and under simulated gastrointestinal conditions. *LWT-Food Sci Technol* 69:365–371.

Garcia-Cayuela, T., R. Tabasco, C. Pelaez, and T. Requena. 2009. Simultaneous detection and enumeration of viable lactic acid bacteria and bifidobacteria in fermented milk by using propidium monoazide and real-time PCR. *Int Dairy J* 19:405–409.

Garthright, W. E. 1993. Bias in the logarithm of microbial density estimates from serial dilutions. *Biom J* 35:299–314.

Gill, A. 2017. The importance of bacterial culture to food microbiology in the age of genomics. *Front Microbiol* 8:777.

Gorsuch, J., D. LeSaint, J. VanderKelen, D. Buckman, and C. L. Kitts. 2019. A comparison of methods for enumerating bacteria in direct fed microbials for animal feed. *J Microbiol Methods* 160:124–129.

Gueimonde, M., L. Alonso, T. Delgado, J. C. Bada-Gancedo, and C. G. de los Reyes-Gavilán. 2003. Quality of yogurt made from milk preserved by refrigeration and carbon dioxide addition. *Food Res Int* 36:43–48.

Gueimonde, M., N. Corzo, G. Vinderola, J. Reinheimer, and C.G. de los Reyes-Gavilán. 2002. Evolution of carbohydrate fraction in carbonated fermented milks as affected by beta-galactosidase activity of starter strains. *J Dairy Res* 69:125–137.

Gueimonde, M., S. Delgado, B. Mayo, et al. 2004. Viability and diversity of probiotic *Lactobacillus* and *Bifidobacterium* populations included in commercial fermented milks. *Food Res Int* 37:839–850.

Gueimonde, M., and C. G. de los Reyes-Gavilán. 2004. Reduction of incubation time in carbonated *Streptococcus thermophilus/Lactobacillus acidophilus* fermented milks as affected by the growth and acidification capacity of the starter strains. *Milchwissenschaft* 59:280–283.

Gul, O., amd I. Atalar. 2019. Different stress tolerance of spray and freeze dried *Lactobacillus casei* Shirota microcapsules with different encapsulating agents. *Food Sci Biotechnol 28:*807–816.

Hasani, S., I. Khodadadi, and A. Heshmati. 2016. Viability of *Lactobacillus acidophilus* in rice bran-enriched stirred yoghurt and the physicochemical and sensory characteristics of product during refrigerated storage. *Int J Food Sci Technol 51*:2485–2492.

Herbel, S. R., B. Lauza, M. von Nickisch-Rosenegk, et al. 2013. Species-specific quantification of probiotic lactobacilli in yoghurt by quantitative real-time PCR. *J Appl Microbiol 115*(6):1402–1410.

Hosseini, S., S. Hosseini, and H. Savaloni. 2019. A novel method for rapid estimation of lactic acid bacterial concentration in fermented milk based on superhydrophobic surface wettability. *Int J Food Microbiol 304*:39–48.

Hotchkiss, J. H., B. G. Werner, and E. Y. C. Lee. 2006. Addition of carbon dioxide to dairy products to improve quality: A comprehensive review. *Compr Rev Food Sci Food Saf 5*:158–168.

Huang, S., M. L. Vignolles, X. D. Chen, et al. 2017. Spray drying of probiotics and other food grade bacteria: A review. *Trends Food Sci Technol 63*:1–17.

ISO 6887–1:1999. Microbiology of food and animal feeding stuffs. Preparation of test samples, initial suspension and decimal dilutions for microbiological examination. Part 1: General rules for the preparation of the initial suspension and decimal dilutions.

ISO 7889/IDF 117. Yogurt—Enumeration of characteristic microorganisms—Colony count technique at 37°C.

ISO 8261/IDF 122. Milk and milks products—Preparation of samples and dilutions for microbiological examination.

ISO 17792/IDF 180. Milk, milk products and mesophilic starter cultures—Enumeration of citrate-fermenting lactic acid bacteria—Colony-count technique at 25 degrees C.

ISO 19344/IDF 232. Milk and milk products—Starter cultures, probiotics and fermented products—Quantification of lactic acid bacteria by flow cytometry.

ISO 20128/IDF 192. Milk products—Enumeration of presumptive *Lactobacillus acidophilus*—Colony count technique at 37°C.

ISO 29981/IDF 220. Milk products—Enumeration of presumptive bifidobacteria—Colony count technique at 37°C.

Jackson, S. A., J. L. Schoeni, C. Vegge, et al. 2019. Improving end-user trust in the quality of commercial probiotic products. *Front Microbiol 10*:739.

Jeon, E. B., S. H. Son, R. K. C. Jeewanthi, N. K. Lee, and H. D. Paik. 2016. Characterization of *Lactobacillus plantarum* Lb41, an isolate from kimchi and its application as a probiotic in cottage cheese. *Food Sci Biotechnol 25*:1129–1133.

Karimi, R., A. M. Mortazavian, and A. Amiri-Rigi. 2012. Selective enumeration of probiotic microorganisms in cheese. *Food Microbiol 29*:1–9.

Kiepś, J., and R. Dembczyński. 2022. Current trends in the production of probiotic formulations. *Foods 11*:2330.

Kim, H.-B., E. Kim, S.-M. Yang, S. Lee, M.-J. Kim, and H.-Y. Kim. 2020. Development of real-time PCR assay to specifically detect 22 *Bifidobacterium* species and subspecies using comparative genomics. *Front Microbiol 11*:2087.

Kramer, M., N. Obermajer, M. B. Bogovic, I. Rogelj, and V. Kmetec. 2009. Quantification of live and dead probiotic bacteria in lyophilised product by real-time PCR and by flow cytometry. *Appl Microbiol Biotechnol 84*:1137–1147.

Kullisaar, T., M. Zilmer, M. Mikelsaar, et al. 2002. Two antioxidative lactobacilli strains as promising probiotics. *Int J Food Microbiol 72*:215–224.

Kwon, Y. W., J.-H. Bae, S.-A. Kim, and N. S. Han. 2018. Development of freeze-thaw tolerant *Lactobacillus rhamnosus* GG by adaptive laboratory evolution. *Front Microbiol 9*:2781.

Ladero, V., and B. Sánchez. 2017. Molecular and technological insights into the aerotolerance of anaerobic probiotics: Examples from bifidobacteria. *Curr Opin Food Sci 14*:110–115.

Lahtinen, S. J., H. Ahokoski, J. P. Reinikainen, et al. 2008. Degradation of 16S rRNA and attributes of viability of viable but nonculturable probiotic bacteria. *Lett Appl Microbiol 46*:693–698.

Lahtinen, S. J., M. Gueimonde, A. C. Ouwehand, J. P. Reinikainen, and S. J. Salminen. 2005. Probiotic bacteria may become dormant during storage. *Appl Environ Microbiol 71*:1662–1663.

Lahtinen, S. J., M. Gueimonde, A. C. Ouwehand, J. P. Reinikainen, and S. J. Salminen. 2006b. Comparison of four methods to enumerate probiotic bifidobacteria in a fermented food product. *Food Microbiol 23*:571–577.

Lahtinen, S. J., A. C. Ouwehand, J. P. Reinikainen, J. M. Korpela, J. Sandholm, and S. J. Salminen. 2006a. Intrinsic properties of so-called dormant probiotic bacteria, determined by flow cytometric viability assays. *Appl Environ Microbiol 72*:5132–5134.

Lewis, Z. T., G. Shani, C. Masarweh, et al. 2016. Validating bifidobacterial species and subspecies identity in commercial probiotic products. *Pediatr Res 79*(3):445–452.

Liu, S.-Q., and M. Tsao. 2009. Enhancement of survival of probiotic and non-probiotic lactic acid bacteria by yeasts in fermented milk under non-refrigerated conditions. *Int J Food Microbiol 135*:34–38.

Marcos-Fernández, R., A. Blanco-Míguez, L. Ruíz, A. Margolles, P. Ruas-Madiedo, and B. Sánchez. 2023. Towards the isolation of more robust next generation probiotics: The first aerotolerant *Bifidobacterium bifidum* strain. *Food Res Int 165*:112481.

Masco, L., G. Huys, E. De Brandt, R. Temmerman, and J. Swings. 2005. Culture-dependent and culture-independent qualitative analysis of probiotic products claimed to contain bifidobacteria. *Int J Food Microbiol 102*:221–230.

Masco, L., T. Vanhoutte, R. Temmerman, J. Swings, and G. Huys. 2007. Evaluation of real-time PCR targeting the 16S rRNA and recA genes for the enumeration of bifidobacteria in probiotic products. *Int J Food Microbiol 113*:351–357.

Mathipa, M. G., and M. S. Thantsha. 2015. Cocktail of probiotics pre-adapted to multiple stress factors are more robust under simulated gastrointestinal conditions than their parental counterparts and exhibit enhanced antagonistic capacities against *Escherichia coli* and *Staphylococcus aureus* . *Gut Pathogens 7*:5.

Medina-Pradas, E., I. M. Perez-Diaz, A. Garrido-Fernandez, and F. N. Arroyo-Lopez. 2017. Review of vegetable fermentations with particular emphasis on processing modifications, microbial ecology, and spoilage. In Microbiological Quality of Food: Foodborne Spoilers, ed. A. Bevilacqua, M. R. Corbo, and M. Sinigaglia. Cambridge: Woodhead Publishing Series in Food Science Technology and Nutrition Collections.

Meybodi, N. M., A. M. Mortazavian, M. Arab, and A. Nematollahi. 2020. Probiotic viability in yoghurt: A review of influential factors. *Int Dairy J 109*:104793.

Michelutti, L., M. Bulfoni, and E. Nencioni. 2020. A novel pharmaceutical approach for the analytical validation of probiotic bacterial count by flow cytometry. *J Microbiol Methods 170*:105834.

Miller, R. A., and B. E. Britigan. 1997. Role of oxidants in microbial pathophysiology. *Clin Microbiol Rev 10*:1.

Min, M., C. R. Bunt, S. L. Mason, G. N. Bennett, and M. A. Hussain. 2017. Effect of non-dairy food matrices on the survival of probiotic bacteria during storage. *Microorganisms 5*(3):43.

Moreira, M. T. C., E. Martins, I. T. Perrone, R. de Freitas, L. S. Queiroz, and A. F. de Carvalho. 2021. Challenges associated with spray drying of lactic acid bacteria: Understanding cell viability loss. *Compr Rev Food Sci Food Saf 20*:3267–3283.

Morovic, W., A. A. Hibberd, B. Zabel, R. Barrangou, and B. Stahl. 2016. Genotyping by PCR and high-throughput sequencing of commercial probiotic products reveals composition biases. *Front Microbiol 7*:1747.

Mudroňová, D. 2015. Flow cytometry as an auxiliary tool for the selection of probiotic bacteria. *Benef Microbes 6*(5):727–734.

Murtaza, M. A., N. Huma, M. A. Shabbi, M. S. Murtaza, and M. Anees-Ur-Rehman. 2017. Survival of microorganisms and organic acid profile of probiotic Cheddar cheese from buffalo milk during accelerated ripening. *Int J Dairy Technol 70*:562–571.

Mustafa, S. M., L. S. Chua, H. A. El-Enshasy, F. A. A. Majid, and R. A. Malek. 2016. A review on fruit juice probiotication: Pomegranate. *Curr Nutr Food Sci 12*:4–11.

Nguyen, H. T., D. H. Truong, K. Sonangon, et al. 2016. Biochemical engineering approaches for increasing viability and functionality of Probiotic Bacteria. *Int J Mol Sci 17*:867.

Noriega, L., M. Gueimonde, L. Alonso, and C. G. de los Reyes-Gavilán. 2003. Inhibition of *Bacillus cereus* in carbonated bifidus milk. *Food Microbiol 20*:519–526.

Noriega, L., M. Gueimonde, B. Sanchez, A. Margolles, and C. G. de los Reyes-Gavilán. 2004. Effect of the adaptation to high bile salts concentrations on glycosidic activity, survival at low pH and cross-resistance to bile salts in *Bifidobacterium* . *Int J Food Microbiol 94*:79–86.

Odamaki, T., J. Xiao, S. Yonezawa, T. Yaeshima, and K. Iwatsuki. 2011. Improved viability of bifidobacteria in fermented milk by cocultivation with *Lactococcus lactis* subspecies *lactis*. *J Dairy Sci 94*:1112e1121.

Oliveira, R. O., A. C. Florence, R. C. Silva, et al. 2009. Effect of different prebiotics on the fermentation kinetics, probiotic survival and fatty acids profiles in nonfat symbiotic fermented milks. *Int J Food Microbiol 128*:467–472.

O'Toole, P. W., J. R. Marchesi, and C. Hill. 2017. Next-generation probiotics: The spectrum from probiotics to live biotherapeutics. *Nat Microbiol 2*:17057.

Pacheco-Ordaz, R., A. Wall-Medrano, M. G. Goni, G. Ramos-Clamont-Monfort, J. F. Ayala-Zavala, and G. A. Gonzalez-Aguilar. 2018. Effect of phenolic compounds on the growth of selected probiotic and pathogenic bacteria. *Lett Appl Microbiol 66*:25–31.

Padilha, M., M. L. V. Morales, A. D. S. Vieira, M. G. M. Costa, and S. M. I. Saad. 2016. A prebiotic mixture improved *Lactobacillus acidophilus* and *Bifidobacterium animalis* gastrointestinal in vitro resistance in petit-suisse. *Food Func 7*:2312–2319.

Papadimitriou, K., A. Alegria, P. Bron, et al. 2016. Stress physiology of lactic acid bacteria. *Microbiol Mol Biol Rev 80*:837–890.

Park, S. H., Y. J. Jo, J. Y. Chun, G. P. Hong, M. Davaatseren, and M. J. Choi. 2015. Effect of frozen storage temperature on the quality of premium ice cream. *Korean J Food Sci Anim Resour 35*:793–799.

Pasulka, A. L., A. L. Howes, J. G. Kallet, J. VanderKelen, and C. Villars. 2021. Visualization of probiotics via epifluorescence microscopy and fluorescence in situ hybridization (FISH). *J Microbiol Methods 182*:106151.

Patrignani, F., L. Siroli, D. I. Serrazanetti, et al. 2017. Microencapsulation of functional strains by high pressure homogenization for a potential use in fermented milk. *Food Res Int* 97:250–257.

Patro, J. N., P. Ramachandran, T. Barnaba, M. K. Mammel, J. L. Lewis, and C. A. Elkins. 2016. Culture-independent metagenomic surveillance of commercially available probiotics with high-throughput next-generation sequencing. *mSphere* 1:e00057–16.

Pechar, R., V. Rada, L. Parafati, et al. 2014. Mupirocin-mucin agar for selective enumeration of *Bifidobacterium bifidum* . *Int J Food Microbiol* 191:32–35.

Perdana, J., L. Bereschenko, M. Roghair, et al. 2012. Novel method for enumeration of viable *Lactobacillus plantarum* WCFS1 cells after single-droplet drying. *Appl Environ Microbiol* 78(22):8082–8088.

Pino, A., K. van Hoorde, I. Pitino, et al. 2017. Survival of potential probiotic lactobacilli used as adjunct cultures on Pecorino Siciliano cheese ripening and passage through the gastrointestinal tract of healthy volunteers. *Int J Food Microbiol* 252:42–52.

Quijada, N. M., F. de Filippis, J. J. Sanz, et al. 2018. Different *Lactobacillus* populations dominate in "Chorizo de León" manufacturing performed in different production plants. *Food Microbiol* 70:94–102.

Raeisi, S. N., L. I. I. Ouoba, N. Rarahmand, J. Sutherland, and H. B. Ghoddusi. 2013. Variation, viability and validity of bifidobacteria in fermented milk products. *Food Control* 34:691–697.

Rashidinejad, A., A. Bahrami, A. Rehman, et al. 2022. Co-encapsulation of probiotics with prebiotics and their application in functional/synbiotic dairy products. *Crit Rev Food Sci Nutr* 62:2470–2494.

Raymond, Y., and C. P. Champagne. 2015. The use of flow cytometry to accurately ascertain total and viable counts of *Lactobacillus rhamnosus* in chocolate. *Food Microbiol* 46:176–183.

Reid, A. A., C. P. Champagne, N. Gardner, P. Fustier, and J. C. Vuillemard. 2007. Survival in food systems of *Lactobacillus rhamnosus* R011 microentrapped in whey protein gel particles. *J Food Sci* 72:M031–M037.

Reimann, S., F. Grattepanche, E. Rezzonico, and C. Lacroix. 2010. Development of a real-time RT-PCR method for enumeration of viable *Bifidobacterium longum* cells in different morphologies. *Food Microbiol* 27:236–242.

Riis, S. N., V. Vojinovic, and C. Gilleladen. 2019. *Fermented Milk Product with a Reduced Content of Lactose.* Washington, DC: US Patent and Trademark Office. U.S. Patent Application No. 16/309,892.

Rivera-Espinoza, Y., and Y. Gallardo-Navarro. 2010. Non-dairy probiotic products. *Food Microbiol* 27:1–11.

Romyasamit, C., P. Saengsuwan, P. Boonserm, B. Thamjarongwong, and K. Singkhamanan. 2022. Optimization of cryoprotectants for freeze-dried potential probiotic *Enterococcus faecalis* and evaluation of its storage stability. *Drying Technol* 40:2283–2292.

Ross, R. P., C. Desmond, G. F. Fitzgerald, and C. Stanton. 2005. Overcoming the technological hurdles in the development of probiotic foods. *J Appl Microbiol* 98:1410–1417.

Saarela, M. H., H. L. Alakomi, A. Puhakka, and J. Mättö. 2009. Effect of the fermentation pH on the storage stability of *Lactobacillus rhamnosus* preparations and suitability of in vitro analyses of cell physiological functions to predict it. *J Appl Microbiol* 106:1204–1212.

Saito, Y., M. Sakamoto, S. Takizawa, and Y. Benno. 2004. Monitoring the cell number and viability of *Lactobacillus helveticus* GCL1001 in human feces by PCR methods. *FEMS Microbiol Lett* 231:125–130.

Salminen, S., M. C. Collado, A. Endo, et al. 2021. The international scientific association of probiotics and prebiotics (ISAPP) consensus statement on the definition and scope of postbiotics. *Nat Rev Gastroenterol Hepatol* 18:649–667.

Sánchez, B., M. Fernandez-Garcia, A. Margolles, C. G. de los Reyes-Gavilán, P. Ruas-Madiedo. 2010. Technological and probiotic selection criteria of a bile-adapted *Bifidobacterium animalis* subsp. *lactis* strain. *Int Dairy J* 20:800–805.

Savedboworn, W., K. Teawsomboonkit, S. Surichay, et al. 2018. Impact of protectants on the storage stability of freeze-dried probiotic Lactobacillus plantarum. *Food Sci Biotechnol* 28:795–805.

Scariot, M. C., G. L. Venturelli, E. S. Prudêncio, and A. C. Maisonnave Arisi. 2018. Quantification of *Lactobacillus paracasei* viable cells in probiotic yoghurt by propidium monoazide combined with quantitative PCR. *Int J Food Microbiol* 264:1–7.

Semyonov, D., O. Ramon, Z. Kaplun, L. Levin-Brener, N. Gurevich, and E. Shimoni. 2010. Microencapsulation of *Lactobacillus paracasei* by spray freeze drying. *Food Res Int* 43:193–202.

Settachaimongkon, S., H. J. F. van Valenberg, V. Winata, et al. 2015. Effect of sublethal preculturing on the survival of probiotics and metabolite formation in set-yoghurt. *Food Microbiol* 49:104–115.

Seyedain-Ardabili, M., A. Sharifan, and B. G. Tarzi. 2016. The production of synbiotic bread by microencapsulation. *Food Technol Biotechnol* 54:52–59.

Sha, N., and J. B. Prajapati. 2014. Effect of carbon dioxide on sensory attibutes, physico-chemical parameters and viability of probiotic *L. helveticus* MTCC5463 in fermented milk. *J Food Sci Technol-Mysore* 51:3886–3893.

Shao, Y., Z. Wang, Q. Bao, and H. Zhang. 2016. Application of propidium monoazide quantitative real-time PCR to quantify the viability of *Lactobacillus delbrueckii* ssp. *bulgaricus*. *J Dairy Sci* 99:9570–9580.

Sheehan, V. M., R. D. Sleator, G. F. Fitzgerald, and C. Hill. 2006. Heterologous expression of BetL, a betaine uptake system, enhances the stress tolerance of *Lactobacillus salivarius* UCC118. *Appl Environ Microbiol* 72:2170–2177.

Shehata, H. R., B. Hassane, and S. G. Newmaster. 2023. Real-time polymerase chain reaction methods for strain specific identification and enumeration of strain *Lacticaseibacillus paracasei* 8700:2. *Front Microbiol. 13*:1076631.

Sheu, S. J., W. Z. Hwang, H. C. Chen, Y. C. Chiang, and H. Y. Tsen. 2009. Development and use of tuf gene-based primers for the multiplex PCR detection of *Lactobacillus acidophilus*, *Lactobacillus casei* group, *Lactobacillus delbrueckii*, and *Bifidobacterium longum* in commercial dairy products. *J Food Prot* 72:93–100.

Sheu, S. J., W. Z. Hwang, Y. C. Chiang, W. H. Lin, H. C. Chen, and H. Y. Tsen. 2010. Use of tuf gene-based primers for the PCR detection of probiotic *Bifidobacterium* species and enumeration of bifidobacteria in fermented milk by cultural and quantitative real-time PCR methods. *J Food Sci* 75:M521–M527.

Shi, Z., X. Li, X. Fan, J. Xu, Q. Liu, Z. Wu, and D. Pan. 2022. PMA-qPCR method for the selective quantitation of viable lactic acid bacteria in fermented milk. *Front Microbiol* 13:984506.

Shing, A., and S. Sharma. 2017. Bioactive components and functional properties of biologically activated cereal grains: A bibliographic review. *Crit Rev Food Sci Nutr* 57:3051–3071.

Shori, A. B. 2022. Application of *Bifidobacterium* spp in beverages and dairy food products: An overview of survival during refrigerated storage. *Food Sci Technol* 42:e41520.

Soto, L. P., N. E. Sirini, L. S. Frizzo, et al. 2023. Lactic acid bacteria viability in different refrigerated food matrices: A systematic review and Meta-analysis. *Crit Rev Food Sci Nutr* 18: 63:12178–12206.

Spigno, G., G. Garrido, E. Guidesi, and M. Elli. 2015. Spray-drying encapsulation of probiotics for ice-cream application. *Chem Eng Trans 43*:49–54.

Succi, M., P. Tremonte, G. Pannella, et al. 2017. Pre-cultivation with selected prebiotics enhances the survival and the stress response of *Lactobacillus rhamnosus* strains in simulated gastrointestinal transit. *Front Microbiol* 8:1067.

Suharja, A. A. S., A. Henriksson, and S. Q. Liu. 2014. Impact of *Saccharomyces cerevisiae* on viability of probiotic *Lactobacillus rhamnosus* in fermented milk under ambient conditions. *J Food Process Preserv* 38:326–337.

Süle, J., T. Kõrösi, A. Hucker, and L. Varga. 2014. Evaluation of culture media for selective enumeration of bifidobacteria and lactic acid bacteria. *Braz J Microbiol* 45:1023–1030.

Szalontai, H., N. Adányi, and A. Kiss. 2014. Comparative determination of two probiotics by QCM and OWLS-based immunosensors. *New Biotechnol* 31:395–401.

Terpou, A., A. Papadaki, I. K. Lappa, V. Kachrimanidou, L. A. Bosnea, and N. Kopsahelis. 2019. Probiotics in food systems: Significance and emerging strategies towards improved viability and delivery of enhanced beneficial value. *Nutrients 11*:1591.

Tripathi, M. K., and S. K. Giri. 2014. Probiotic functional foods: Survival of probiotics during processing and storage. *J Funct Foods 9*:225–241.

Tsai, C.-C., C.-H. Lai, B. Yu, and H.-Y. Tsen. 2010. Use of PCR primers and probes based on the 23s rRNA and internal transcription spacer (ITS) gene sequence for the detection and enumeration of *Lactobacillus acidophilus* and *Lactobacillus plantarum* in feed supplements. *Anaerobe 16*:270–277.

Vesterlund, S., K. Salminen, and S. Salminen. 2012. Water activity in dry foods containing live probiotic bacteria should be carefully considered: A case study with *Lactobacillus rhamnosus* GG in flaxseed. *Int J Food Microbiol 157*:319–321.

Vinderola, C. G., N. Bailo, and J. A. Reinheimer. 2000a. Survival of probiotic microflora in Argentinian yogurts during refrigerated storage. *Food Res Int 33*:97–102.

Vinderola, C. G., G. A. Costa, S. Regenhardt, and J. A. Reinheimer. 2002a. Influence of compounds associated with fermented dairy products on the growth of lactic acid starter and probiotic bacteria. *Int Dairy J 12*:579–589.

Vinderola, C. G., C. G. de los Reyes-Gavilán, and J. A. Reinheimer. 2009. Probiotics and prebiotics in fermented dairy products. In Innovation in Food Engineering: New Techniques and Products, ed. M. L. Passos and C. P. Ribeiro. Boca Raton, FL: CRC Press, Taylor & Francis Group.

Vinderola, C. G., P. Mocchiutti, and J. A. Reinheimer. 2002b. Interactions among lactic acid starter and probiotic bacteria used for fermented dairy products. *J Dairy Sci* 85:721–729.

Visciglia, A., S. Allesina, A. Amoruso, et al. 2022. Assessment of shelf-life and metabolic viability of a multi-strain synbiotic using standard and innovative enumeration technologies. *Front Microbiol* 13:989563.

Weese, J. S., and H. Martin. 2011. Assessment of commercial probiotic bacterial contents and label accuracy. *Can Vet J* 52:43–46.

Wendel, U. 2022. Assessing viability and stress tolerance of probiotics—A review. *Front Microbiol* 12:818468.

Wilkinson, M. G., and G. Lapointe. 2020. Invited review: Starter lactic acid bacteria survival in cheese: New perspectives on cheese microbiology. *J Dairy Sci* 103:10963–10985.

Zawistowska-Rojek, A., T. Zareba, A. Mrówka, and S. Tyski. 2016. Assessment of the microbiological status of probiotic products. *Pol J Microbiol* 65:97–104.

Zhang, X., S. Zhang, D. Y. Li, et al. 2022. Niacin inhibits post-acidification of yogurt based on the mining of LDB_RS00370 biomarker gene. *Food Res Int 162*:111929.

Zoumpourtikoudi, V., N. Pyrgelis, M. Chatzigrigoriou, R. N. Tasakis, and M. Touraki. 2018. Interactions among yeast and probiotic bacteria enhance probiotic properties and metabolism offering augmented protection to *Artemia franciscana* against *Vibrio anguillarum*. *Microb Pathog 125*:497–506.

Zuo, F., S. Chen, and H. Marcotte. 2020. Engineer probiotic bifidobacteria for food and biomedical applications—Current status and future prospective. *Biotechnol Adv 45*:107654.

# SECTION III

# Role of LAB in Health and Disease

# Human Microbiota

# 21

## María Carmen Collado and Miguel Gueimonde

In recent years, there has been increasing scientific evidence connecting microbiota and human health, leading to a growing interest in understanding the establishment and factors contributing to microbiota variation from early childhood to old age.

## 21.1  MICROBIOTA AND ITS RELEVANCE TO HUMAN HEALTH

The human microbiota is a complex, dynamic, and diverse community of microorganisms that reside in and on the human body. These microorganisms, including bacteria, viruses, fungi, and archaea, play a fundamental role in health, implicated in important physiological, metabolic and immunological functions and also influencing our disease susceptibility.

The definitions of "microbiota" and "microbiome" are often used interchangeably, but the meaning is different. Microbiota refers to the specific microbial taxa that are associated with the human body, while microbiome refers to the collection of genes from these microbes. The human microbiome refers to the genetic pool of the microbiota.

The distinct human body niches (the gut, vagina, oral cavity, respiratory tract, and skin) harbor a diverse microbial community that co-evolved with humans. Microbes play essential roles in human physiology, including metabolic and immunological activities. Though gut microbiota has been widely studied, however, other microbial niches are understudied (skin, oral, respiratory epithelia, vagina, etc.) and they also play a crucial role in human health. Studies to describe the structure and diversity of the microbiota have been undertaken to understand the influence of microbiota on human health and disease. Of the identified microbes, the phyla Pseudomonadota, Bacillota, Actinomycetota, and Bacteroidota represent approximately 95% of all the species, with 20% of being strictly anaerobes found in mucosal regions, including the oral cavity and gut. The human microbiota is constantly expanding as new studies emerge, increasing our understanding of the microbial repertoire, particularly in the gut.

Although all niches in the human body are populated with bacteria, the gut microbiota is the one with the biggest quantity and diversity of bacteria, and it is also considered a metabolic organ that regulates various digestive and non-digestive processes that have a direct impact on an individual's health. The human gut microbiota is an incredibly complex and constantly evolving microbial community that is unique to each individual and influenced by both environmental and genetic factors.[1,2] It is estimated that each person harbors approximately 0.2 kg of bacteria. The gut community is made up of hundreds of different species that vary in prevalence and abundance throughout the gastrointestinal tract, with the stomach containing a reduced population ($10^3$–$10^4$ CFU/g contents) and the colon containing a more heavily populated one ($10^{11}$–$10^{12}$ CFU/g contents).

The contribution of gut microbiota to human health is significant, not only playing a role in the digestive process but also being critical in the development of both the gut and the immune system.[3] By virtue of their proximity to the mucosa, the bacteria of the gastrointestinal tract are in constant interaction with the mucosa-associated lymphoid tissues, which play a major role in the immune system. This symbiotic relationship explains the important role played by the microbiota in the immunological capacity of the individual, influencing and modulating the behavior of innate and acquired immune cells.

Alterations in microbial composition and diversity are referred to as "dysbiosis", and several studies associate alterations in the microbiota with intestinal diseases and also with an increased risk of developing other diseases, such as allergies, inflammatory bowel problems, diabetes, obesity, and even neurological and cognitive disorders.[4–6]

## 21.2  HOW IS THE HUMAN MICROBIOTA ASSEMBLED? GUT COLONIZATION PROCESS: BIRTH TO THE FIRST YEAR OF LIFE

Massive bacterial exposure occurs at birth when the neonate comes into contact with the maternal and environmental microbiota. Then, the neonate is rapidly colonized by a consortium of pioneer microorganisms, mainly Enterobacteriaceae and lactic acid bacteria, such as those pertaining to the genera formerly known as *Lactobacillus*, which are subsequently displaced by strict anaerobic bacteria such as *Bifidobacterium* and *Bacteroides* spp.[7,8] The neonatal gut microbiota is characterized by low bacterial diversity and a dominance of bacteria belonging to the phyla Pseudomonadota and Actinomycetota. This microbiota develops over the first year, with Bacillota and Bacteroidota becoming the dominant phyla. The initial inoculum is determined by different factors, including gestational age, type of delivery (vaginal delivery versus caesarean section), antibiotic treatment, hygienic environmental conditions, presence of siblings and/or pets, and living environment (rural or urban), among others.[7,9–11] Once the initial microbiota is assembled, the microbial community continues to evolve throughout the lactation period and during the introduction of complementary feeding, and beyond converging towards an adult-type microbiota around 2–5 years of age.[7,12]

## 21.3  FACTORS AFFECTING THE COLONIZATION PROCESS OF THE GUT MICROBIOTA

The establishment of the intestinal microbiota is a step-wise dynamic process, influenced by perinatal and environmental factors from birth to the first years of life. Among the most relevant ones:

**Birth mode** is one of the most impactful moments in the microbiota colonization process, as it has a direct impact on the bacterial diversity that will occupy the different niches in the organism.[13–17] Natural vaginal birth implies direct contact with the maternal vaginal and intestinal microbiota; neonates are colonized by the vaginal and intestinal microbiota,[9,18,19] and newborns are preferentially colonized by *Lactobacillus, Prevotella, Bacteroides, Escherichia/Shigella*, and *Bifidobacterium*. However, this colonization process[20] is altered in neonates born by cesarean section and antibiotic exposure.[14,15,21] In recent decades, there has been an increase in caesarean delivery, in some countries approaching 50% of births.[20] Perhaps because of this, there has been considerable interest in the consequences of this delivery mode for the health of newborns. Cesarean delivery involves the interruption of microbial transfer; the absence of contact with maternal microbiota (vaginal and gut) leads to the proliferation of bacteria from the nosocomial environment and from the mother's skin,[18] which means that the dominant species will differ from those of infants born by natural delivery. Neonates born by caesarean section are colonized by bacteria of the genera *Staphylococcus*, *Streptococcus*, *Corynebacterium*, and *Veillonella* and show an increased presence of pro-inflammatory bacteria such as enterobacteria.[14,15,19] They are also characterized by delayed colonization of *Bacteroidetes* and *Bifidobacterium*. It has been described in epidemiological studies and various meta-analyses that children born vaginally have a lower prevalence of asthma, allergies, respiratory problems, diabetes, and obesity compared to children born by cesarean section, who also show greater morbidity associated with this type of practice,[22–25] and these disorders all have links to microbiota dysbiosis. Thus, potential restoration strategies to alleviate dysbiosis in children born by caesarean section have been reported.[26–31] "*Vaginal seeding*" consists of inoculating neonates with bacteria from the maternal vagina after caesarean section[26,28,32] in order to restore the maternal microbiota in those kids, with consequences in growth and neurodevelopment. In recent years, the "*fecal microbiota transplant*" has been studied as a potential restoration tool from the mother to the cesarean section–delivered infant.[3,27]

**Prematurity** has direct consequences on the status of the neonatal microbiota. Bacterial imbalances have been repeatedly observed in preterm newborns,[34–36] with consequences in health and development. These are characterized by a dominance of bacteria from the Enterobacteriaceae family and an increase in the number of potential pathogens such as *Klebsiella pneumoniae* and *Clostridioles difficile* and a detriment of *Bifidobacterium*.[37,38]

One of the diseases most often associated with microbiota dysbiosis in the preterm neonate is necrotizing enterocolitis (NEC), which affects approximately 13% of premature infants.[39,40] Its pathogenesis is still unknown, although there is evidence that inadequate bacterial colonization and inflammation are factors that play an important role in its development.[41] In this regard, children with necrotizing enterocolitis show high levels of Pseudomonadota,[42] especially Gammaproteobacteria such as *Cronobacter sakazakii*, *Klebsiella* spp., and *Escherichia coli*.[37,40,41]

**Breastfeeding** is the first food choice for newborns, providing them with the energy, macro- and micro-nutrients, and other non-nutritive bioactive compounds they need and favoring the survival and development of newborn infants. Epidemiological studies suggest that continued breastfeeding during the first 6 months of life reduces the risk of developing certain diseases such as allergies and inflammatory bowel disease (IBD) in infants. The immunological benefits of breast milk (BM) include protection against infections and inflammation and also promotion of the proper functioning of the intestinal barrier and maintenance of homeostasis, as well as its contribution to the process of neonatal intestinal microbial colonization.

Exclusively breastfed infants have a different microbiota from those fed with infant formula.[7,43–46] Several studies reported that exclusive breast milk consumption may predict differences in the taxonomic composition of the gut microbiota.[47,48] A recently conducted meta-analysis compared the microbial profiles of exclusively breastfed and non-exclusively breastfed infants.[48] At the order level, higher levels of Bacteroidales and Clostridiales were observed, while at the family level, there was a higher dominance of Bacteroidaceae and Veillonellaceae. Finally, in terms of genera, increases in *Bacteroides*, *Eubacterium*, *Veillonella*, and *Megasphaera* populations were observed. The microbiota of formula-fed infants has a lower proportion of *Bifidobacterium*, *Lactobacillus*, and *Enterococcus* spp., which is associated with an altered maturation of the immune system.[49] Another factor that makes breast milk the ideal food for newborns is its richness in oligosaccharides (HMO), prebiotics essential for the maintenance of bacteria, which play a decisive role in the microbiota[50] and also due immune-related functions and gut–brain axis.[51] In this regard, there is evidence that breastfed infants develop a more uniform and stable microbiota,[45] presenting better intestinal tolerance and a more effective immune response.

Breast milk contains a wide variety of bioactive compounds that play a key role in neonatal development. These bioactive compounds include immunocompetent cells, immunoglobulins, fatty acids, polyamines, oligosaccharides, antimicrobial peptides, lysozyme, lactoferrin and other glycosylated proteins, and, as discussed, bacteria.[52–54] Human milk is a source of commensal, mutualistic, and potentially probiotic bacteria that support the process of intestinal colonization. Despite great interindividual variability, each woman's milk has a unique bacterial composition, including mainly *Streptococcus* and *Staphylococcus*, followed by *Enterobacteriaceae, Bifidobacterium, Enterococcus, Lactococcus*, and *Lactobacillus*, which are among the first colonizers of the infant gut,[54,55] although the presence of some 700 species of bacteria in milk has been described,[56] including the presence of strict anaerobic bacteria present in the intestine such as certain *Clostridium, Faecalibacterium*, and *Akkermansia*. It is estimated that an infant ingesting approximately 500–800 mL of milk per day receives between $10^5$ and $10^7$ bacteria/mL. These microorganisms play a direct role in health, preventing the growth and proliferation of pathogens and intervening in the regulation of immunological, metabolic, and neurodevelopmental processes. It is important to highlight that breast milk microbiota changes throughout the day, lactation time, and between individuals, depending on their geographic location and can be affected by various other factors, including antibiotics.

**Introduction to complementary feeding**: Current nutritional recommendations suggest that introduction to complementary feeding should take place starting at 4–6 months of age.[57,58] In this regard, several studies argue that diet is one of the main factors influencing the gut microbiota, mainly after the introduction of solid foods.[45,59] In a recent work,[60] significant changes in the gut microbiota were observed from 9 months of age onwards due to cessation of breastfeeding and introduction of complementary food. It is known that introducing complementary foods can provide a great opportunity to positively influence a child's microbiota. At this stage, breast milk remains the main food source but begins to be supplemented to meet the infant's nutritional needs. Importantly, breastfeeding beyond 9 months has been associated with an increase in *Lactobacillus* spp. and *Bifidobacterium* spp., particularly *Bifidobacterium longum*.

Weaning modulates the infant gut microbiota structure and functionality towards an adult-like microbiota.[61] The infant microbiota is characterized by microbial species able to use glycans, mucin, and complex carbohydrates and producing SCFA. For example, the gut microbiota during weaning is associated with an increased abundance of *Bacteroides* spp. and members of *Clostridium* groups IV and XIV.[62] In addition, the introduction of solid foods was associated with an increased prevalence of butyrate-producing bacteria such as the *Blautia coccoides* group and the genus *Atopobium* and also, with increases in *Eggerthella, Blautia, Neisseria*, Peptostreptococcaceae, and *Bacteroidetes*, as well as higher relative abundance of *Lactobacillus* and Ruminococcaceae. However, other studies show an increase in *B. longum* subsp. *longum* after the introduction of solid foods.[63] Some previous studies suggest that complementary feeding has a more profound impact on the infant gut microbiota of non-breastfed infants.[64]

**Antibiotic exposure:** It has been shown that the administration of antibiotics is directly related to dysbiosis in children and adults.[65–68] Antibiotic administration seriously affects the balance of the microbiota, especially in the first 36 months, with lasting effects of infant growth.[69] It should also be borne in mind that not all antibiotics have the same impact on the microbiota. An altered microbial colonization profile has been reported for vaginal-born infants with intrapartum antibiotherapy, characterized by lower relative proportions of Actinomycetota and Bacteroidota and an increase in Pseudomonadota and Bacillota phylum. Studies have suggested that the type of antibiotic and the duration of treatment cause variations in adverse effects.[67,70,71] It also appeared that maternal antibiotic consumption during the perinatal period was associated with offspring colonization characterized by higher numbers of *Enterococcus* and *Clostridium* and lower numbers of *Bacteroides* and *Parabacteroides*.

It is necessary to highlight the side effects of antibiotic use, especially in at-risk populations, such as preterm infants, caesarean section infants, or formula-fed infants, whose gut microbiota development has already been disrupted or altered.[13,67] In particular, preterm infants are often treated with antibiotics in order to prevent infection, which also prevents the colonization of beneficial bacteria.[36,37] In addition, it must be taken into account that the environmental factor surrounding infants in intensive care units plays an important role in this process. In some cases, bacteria are not only associated with infections but are also resistant to antibiotics, which can lead to ineffectiveness of some therapies.[72,73]

**Environment**: The family environment (family size and presence of siblings) as well as the presence of animals in the environment are relevant in the process of microbial colonization. Children with siblings have higher proportions of bifidobacteria compared to only children.[74,75] It has been shown that microbial richness is higher in 3-month-old children with pets at home.[76,77] This increase in bacterial richness is associated with a higher population of bacteria belonging to the phylum Bacillota, mainly of the genera *Ruminococcus* and *Oscillospira*, which is associated with a lower risk of developing allergies.

There are also other factors related to the environment that have been associated with the microbiota. It has also been shown that there are differences in the composition of the microbiota in European children depending on geographical location. The microbiota of children from northern Europe had higher levels of *Bifidobacterium* spp., *Clostridium* spp., and *Atopobium* spp. while those from southern Europe had a higher abundance of *Eubacterium* spp., *Lactobacillus* spp., and *Bacteroides* spp.[78] A recent study analyzing 1,020 healthy individuals from 23 populations and six published studies found a correlation between higher latitude and a higher presence of members of the Bacillota phylum and a lower presence of members of the Bacteroidota phylum,[79] which could be explained in part by changes in climate, diet, and lifestyle. In individuals living in rural areas in developing countries, microbial composition and diversity have been observed to differ from those of individuals from urban areas in industrialized countries, such as Europe or the United States.[80] A recent study showed higher levels of *Bifidobacterium* spp., *Bacteroides-Prevotella*, and *Clostridium histolyticum* in Malawian children than in Finnish children aged 6 months,[81] and the intestinal microbiota of children from a rural environment in Burkina Faso was characterized by the presence of *Bacteroides*.[82]

# 21.4 MICROBIOTA FROM CHILDHOOD AND ADOLESCENCE TO ADULTHOOD

Between 3 and 5 years of age, the infant gut microbiota reaches a composition similar to the adult microbiota, which is dominated by three bacterial phyla: Bacillota (Lachnospiraceae and Ruminococcaceae), Bacteroidota (Bacteroidaceae, Prevotellaceae and Rikenellaceae), and Actinomycetota (Bifidobacteriaceae and Coriobacteriaceae). It has been shown[12] that infant gut microbiota undergoes three distinct phases according to the age and microbial evolution: a developmental phase (months 3–14) characterized by changes in alpha diversity and in the most abundant bacterial groups, a transitional phase (months 15–30) in which the abundance of *Bacteroides* and Pseudomonadota increases, and a stable phase (months 31–46) in which phyla and alpha diversity remain unchanged.[12] During this phase, an increased intake of animal protein and plant polysaccharides leads to a decrease in *Bifidobacterium* and an increase in Bacteroidota and Bacillota. However, other studies suggest that the development of the adult microbiota may take longer than 2–5 years of life, suggesting three stages: microbial composition of pre-school children (3 to 6 years), primary school (6 to 12 years), and adolescence (12 to 18 years).[83–85] Reflecting changes in the microbiota, significant differences have also been found between children and adults in genes involved in amino acid degradation, oxidative phosphorylation, and also in vitamin synthesis (B9 and B12) and factors that induce mucosal inflammation.[85]

During adulthood, the microbiota is stable, although our diet determines its composition and diversity. Each person has a unique bacterial footprint that has developed since childhood and adolescence and has adapted to the genetic, physiological, metabolic, and immunological conditions of the host. However, the composition and diversity of the bacterial community depend on the dietary habits maintained from infancy and throughout adulthood. Diet is considered one of the most important modulators of gut microbiota composition, and this has been well documented in several studies.[82,86] A rich and balanced diet is essential to promote microbial diversity and the proper functioning of the gut microbiota. However, our society is exposed to a "*Western lifestyle*", characterized by an excessive intake of foods rich in fats, cholesterol, animal proteins, sugar, and salt and a wide range of processed foods, as well as a lack of physical activity, which favors an inflammatory state, leading to an increased risk of developing various diseases such as obesity, metabolic syndrome, cardiovascular diseases, and colorectal cancer.[87,88] Industrialized diets lead to a predominance of Bacteroidota, while the consumption of fiber, fruit, and vegetables increases the levels of fermentative species such as *Prevotella* spp. and increases levels of SCFA.[89] The beneficial role of the Mediterranean diet on the microbiota and human health is well known. Several studies have shown an increase in beneficial bacteria that favor the biosynthesis of essential nutrients and promote the production of SCFA.[90–92] High adherence to the Mediterranean diet was associated with lower levels of *Escherichia coli* and higher levels of bifidobacteria and acetate.[93] Other studies have shown that the genera *Butyricimonas*, *Desulfovibrio*, and *Oscillospira* are associated with normal-weight adults (BMI < 25), and the genus *Catenibacterium* proliferates with higher adherence to the Mediterranean diet.[92]

Several studies show how variations in location, geography, ethnicity, or lifestyle modulate the microbiota in healthy individuals.[80,94,95] We cannot forget that there are other factors in adulthood that can modify the gut microbiota, such as the use of antibiotics and other drugs,[96] metabolic problems (obesity, diabetes, fatty liver syndrome, etc.),[97] and also pregnancy status, as well as puberty and women's hormonal changes during the lifespan.

## 21.4.1 Maternal Microbiota during Pregnancy and Lactation

Pregnancy and lactation are physiological states in which important biological and metabolic changes take place that are physically demanding for women and may even have a significant impact on their health. In the period leading up to and during pregnancy, it is important to maintain a healthy, balanced, and varied diet to ensure a regular and sufficient supply of nutrients to the fetus. Recent studies have shown that the physiological, immunological, and metabolic changes that occur during pregnancy are paralleled by changes in the composition and diversity of the maternal microbiota.[98,99] In this context, significant changes in bacterial communities appear in women during the first and third trimester, specifically towards increased Pseudomonadota and Actinomycetota.[98] In fact, progesterone, the primary pregnancy hormone, affects the maternal gut microbiota and favors the growth of several *Bifidobacterium* species.[100] During pregnancy, the vaginal microbiota also undergoes changes. A healthy pregnancy is characterized by a vaginal microbiota dominated by *Lactobacillus* spp. and less microbial diversity compared to non-pregnant women.[101,102] Alterations in the vaginal microbiota during pregnancy have been associated with certain problems,[103,104] including a risk for preterm delivery. In addition to all these microbiological changes, pregnancy also affects the subgingival microbiota.[105] Pregnancy is associated with the development and evolution of periodontal disease.[106] According to recent studies, there is a possible relationship between maternal periodontal disease during pregnancy and the risk of preterm delivery and/or low neonatal birth weight. In addition, changes in the maternal microbiota have been associated with adverse pregnancy outcomes, affecting maternal physiological adaptation, placental structure and function, and the intrauterine fetal environment.[107]

Maternal weight before pregnancy, as well as weight gain during pregnancy, has also been related to specific alterations in the maternal and infant intestinal microbiota.[107–109] Obese and overweight pregnant women harbored lower levels of *Bifidobacterium* spp. and *Bacteroides* spp., as well as increased levels of *Staphylococcus* and members of the Enterobacteriaceae family (mainly *Escherichia coli*) compared to normal weight women. These changes in the maternal gut microbiota have also been associated with alterations in biochemical parameters such as folic acid, ferritin, transferrin, and plasma cholesterol levels and are of particular importance during pregnancy in maintaining maternal and infant health status. A positive association has been found between *Bacteroides* spp. and *Bifidobacterium* spp. and folic acid.[110] On the other hand, higher *Bacteroides* counts have been associated with HDL-cholesterol, as well as with lower triglyceride levels. In addition, lower counts of *Bifidobacterium* spp. as well as higher counts of *Enterobacteriaceae* spp., including *E. coli*, have been associated with high plasma ferritin concentration as well as low transferrin.[110]

Diet is capable of modifying the intestinal microbiota of lactating mothers, with a diet richer in nutrients and calories being associated with an increase in Bacillota populations.[111,112] A relationship has been found between animal protein intake and Bacteroides populations, as well as B vitamins, which are associated with an increase in *Prevotella* counts and a decrease in *Bacteroides* levels. Also the intake of minerals such as magnesium, copper, manganese, and molybdenum correlated positively with bacteria of the phylum Bacillota and negatively with the phylum Bacteroidota. However, the impact that these changes may have on infant health is still unknown.

# 21.5 MICROBIOTA IN THE ELDERLY

In the elderly, the microbiota begins to undergo changes, becoming a stage of life in which the microbiota is again unstable and highly susceptible to environmental factors, as it was during early life. The gut microbiota of the elderly is characterized by lower bacterial diversity and also by a decrease in microorganisms considered beneficial and an increase in facultative anaerobic bacteria such as enterobacteria and some opportunistic pathogens such as *Clostridioides difficile*. However, there is a need for longitudinal studies monitoring the microbiota of a cohort of individuals during the transition years from adulthood to senescence as well as cross-sectional studies comparing very old people, such as centenarians, with younger elderly people. Some trends appear repeatedly throughout ageing despite the use of different methodologies in different studies, such as a decrease in the levels of bifidobacteria[113–115] and also reductions observed in the *Clostridium* XIVa group.[116–120] In contrast, for other microbial groups, such as *Bacteroides-Prevotella*, while some studies have reported a reduction, others have not observed this effect.[121–124] Similarly, the evidence for lactobacilli abundance is also unclear, with some researchers observing reduced levels in older people,[120,121,125] while others have reported increases.[119,124] It has also been observed that levels of other important gut microorganisms, such as *Faecalibacterium prausnitzii*, appear to be reduced in older subjects, while other microorganisms, such as *Akkermansia muciniphila*, are increased.[114,118,119,126] Indeed, recent animal studies on naturally aged animals and in a model of progeria have shown that administration of *A. muciniphila* prolongs life, suggesting a protective role for elevated levels of this microorganism.[127,128] Taken together, these observations indicate that the changes in the microbiota of very old subjects are different from those of younger elderly subjects,[118,124,129] suggesting the presence of a specific microbiota profile in these individuals that could be protective by delaying age-related physiological changes.[130] Unfortunately, in the absence of longitudinal studies, it is not possible to determine whether these characteristics of extremely long-lived individuals are age related or if they were already present in the same individuals at a younger age, potentially associated with their lifestyle.[129]

The cascade of effects of microbial shifts associated with ageing can have both beneficial and detrimental consequences. A reduction in microbial diversity, a change in the dominant species, and reduced levels of, for example, butyrate-producing microorganisms or an increase in the number of potentially pathogenic microorganisms are considered undesirable changes. In this regard, the loss of the normal microbiota, defined as the set of microorganisms shared by individuals in a given social group, has been correlated with a decrease in microbial diversity and increased frailty in the elderly.[131,132] In addition to changes in the composition and functionality of the gut microbiota associated with the ageing process itself, with increased oxidative stress and mucosal inflammation,[133] other factors such as infections, drug treatments, dietary changes, and/or institutionalization can introduce new and profound alterations to the gut microbiota.[134,135] It is also important to underline that, in spite of these age-associated microbiota changes, diet is still a primary factor for determining the microbiota composition in aged individuals. in this regard, as is the case for adults, a Mediterranean diet pattern has been associated with increased levels of SCFA and reduced levels of inflammatory markers (IL8) and correlated positively with potentially beneficial microorganisms, such as *F. prausnitzii*, in the elderly.[136] Therefore, also in advanced age, diet is a relevant and modulable factor with impact on the gut microbiota and in human health.

# 21.6 CONCLUSION

Diet, lifestyle, and microbiota are associated with our state of health. There are more and more studies that relate changes in the intestinal microbiota with the risk of suffering certain diseases. One of the most critical periods for the development of an adequate intestinal microbiota are the first years of a person's life, where alterations in the initial microbial colonization can extend into adulthood and can affect our health in the long term. During gestation and lactation, adequate energy intake and a balanced diet are needed to ensure optimal development of the immune system and gut microbiota. There is increasing scientific evidence suggesting that diet is a useful tool in modulating the gut microbiota. Therefore, we should continue to work on the study of the relationship between maternal nutrition and the microbiota during pregnancy, lactation, and infant feeding in early childhood and its role in infant development. This knowledge will drive the development of new and personalized dietary strategies and also, the use of food supplements such as -biotics (probiotics, prebiotics, symbiotics, and postbiotics) targeted to microbial modulation. These strategies are aimed at promoting beneficial microbiological, metabolic, and immunological programming of infant health, thereby improving their long-term quality of life.

# BIBLIOGRAPHY

1. Gilbert J, Blaser MJ, Caporaso JG, Jansson J, Lynch SV, Knight R. Current understanding of the human microbiome. *Nat Med.* 2018;24(4):392–400. doi:10.1038/nm.4517
2. Thursby E, Juge N. Introduction to the human gut microbiota. *Biochem J.* 2017;474(11):1823–1836. doi:10.1042/BCJ20160510
3. Fan Y, Pedersen O. Gut microbiota in human metabolic health and disease. *Nat Rev Microbiol.* 2021;19(1):55–71. doi:10.1038/s41579-020-0433-9
4. Fan Y, Pedersen O. Gut microbiota in human metabolic health and disease. *Nat Rev Microbiol.* 2021;19(1):55–71. doi:10.1038/s41579-020-0433-9
5. Cho I, Blaser MJ. The human microbiome: At the interface of health and disease. *Nat Rev Genet.* 2012;13(4):260–270. doi:10.1038/nrg3182
6. Finlay BB, Humans C, Microbiome. *Are noncommunicable diseases communicable?* Accessed June 27, 2023. www.sciencemagazinedigital.org/sciencemagazine/17_january_2020/MobilePagedArticle.action?articleId=1554024
7. Milani C, Duranti S, Bottacini F, et al. The first microbial colonizers of the human gut: Composition, activities, and health implications of the infant gut microbiota. *Microbiol Mol Biol Rev.* 2017;81(4):e00036–17. doi:10.1128/MMBR.00036-17
8. Koenig JE, Spor A, Scalfone N, et al. Succession of microbial consortia in the developing infant gut microbiome. *Proc Natl Acad Sci USA.* 2011;108(Suppl 1):4578–4585. doi:10.1073/pnas.1000081107
9. Rutayisire E, Huang K, Liu Y, Tao F. The mode of delivery affects the diversity and colonization pattern of the gut microbiota during the first year of infants' life: A systematic review. *BMC Gastroenterol.* 2016;16(1):86. doi:10.1186/s12876-016-0498-0
10. Mesa MD, Loureiro B, Iglesia I, et al. The evolving microbiome from pregnancy to early infancy: A comprehensive review. *Nutrients.* 2020;12(1). doi:10.3390/nu12010133
11. Rodríguez JM, Murphy K, Stanton C, et al. The composition of the gut microbiota throughout life, with an emphasis on early life. *Microb Ecol Health Dis.* 2015;26:26050. doi:10.3402/mehd.v26.26050
12. Stewart CJ, Ajami NJ, O'Brien JL, et al. Temporal development of the gut microbiome in early childhood from the TEDDY study. *Nature.* 2018;562(7728):583–588. doi:10.1038/s41586-018-0617-x
13. Francavilla R, Cristofori F, Tripaldi ME, Indrio F. Intervention for dysbiosis in children born by c-section. *Ann Nutr Metab.* 2018;73(Suppl 3):33–39. doi:10.1159/000490847
14. Salas Garcia MC, Yee AL, Gilbert JA, Dsouza M. Dysbiosis in children born by caesarean section. *Ann Nutr Metab.* 2018;73(Suppl 3):24–32. doi:10.1159/000492168

15. Shin H, Pei Z, Martinez KA, et al. The first microbial environment of infants born by C-section: The operating room microbes. *Microbiome*. 2015;3:59. doi:10.1186/s40168-015-0126-1

16. Shao Y, Forster SC, Tsaliki E, et al. Stunted microbiota and opportunistic pathogen colonization in caesarean-section birth. *Nature*. 2019;574(7776):117–121. doi:10.1038/s41586-019-1560-1

17. Bäckhed F, Roswall J, Peng Y, et al. Dynamics and stabilization of the human gut microbiome during the first year of life. *Cell Host Microbe*. 2015;17(5):690–703. doi:10.1016/j.chom.2015.04.004

18. Kumbhare SV, Patangia DVV, Patil RH, Shouche YS, Patil NP. Factors influencing the gut microbiome in children: From infancy to childhood. *J Biosci*. 2019;44(2):49.

19. Dominguez-Bello MG, Costello EK, Contreras M, et al. Delivery mode shapes the acquisition and structure of the initial microbiota across multiple body habitats in newborns. *Proc Natl Acad Sci USA*. 2010;107(26):11971–11975. doi:10.1073/pnas.1002601107

20. WHO Statement on Caesarean Section Rates. Accessed June 9, 2023. www.who.int/news-room/questions-and-answers/item/who-statement-on-caesarean-section-rates-frequently-asked-questions

21. Montoya-Williams D, Lemas DJ, Spiryda L, et al. The neonatal microbiome and its partial role in mediating the association between birth by cesarean section and adverse pediatric outcomes. *Neonatology*. 2018;114(2):103–111. doi:10.1159/000487102

22. Biasucci G, Benenati B, Morelli L, Bessi E, Boehm G. Cesarean delivery may affect the early biodiversity of intestinal bacteria. *J Nutr*. 2008;138(9):1796S–1800S. doi:10.1093/jn/138.9.1796S

23. Sandall J, Tribe RM, Avery L, et al. Short-term and long-term effects of caesarean section on the health of women and children. *Lancet*. 2018;392(10155):1349–1357. doi:10.1016/S0140-6736(18)31930-5

24. Słabuszewska-Jóźwiak A, Szymański JK, Ciebiera M, Sarecka-Hujar B, Jakiel G. Pediatrics consequences of caesarean section-A systematic review and meta-analysis. *Int J Environ Res Public Health*. 2020;17(21):8031. doi:10.3390/ijerph17218031

25. Tanoey J, Gulati A, Patterson C, Becher H. Risk of type 1 diabetes in the offspring born through elective or non-elective caesarean section in comparison to vaginal delivery: A meta-analysis of observational studies. *Curr Diab Rep*. 2019;19(11):124. doi:10.1007/s11892-019-1253-z

26. Dominguez-Bello MG, De Jesus-Laboy KM, Shen N, et al. Partial restoration of the microbiota of cesarean-born infants via vaginal microbial transfer. *Nat Med*. 2016;22(3):250–253. doi:10.1038/nm.4039

27. Korpela K, Helve O, Kolho KL, et al. Maternal fecal microbiota transplantation in cesarean-born infants rapidly restores normal gut microbial development: A proof-of-concept study. *Cell*. 2020;183(2):324–334 e5. doi:10.1016/j.cell.2020.08.047

28. Song SJ, Wang J, Martino C, et al. Naturalization of the microbiota developmental trajectory of Cesarean-born neonates after vaginal seeding. *Med (N Y)*. 2021;2(8):951–964.e5. doi:10.1016/j.medj.2021.05.003

29. Mueller NT, Differding MK, Sun H, et al. Maternal bacterial engraftment in multiple body sites of cesarean section born neonates after vaginal seeding-a randomized controlled trial. *mBio*. Published online April 19, 2023:e0049123. doi:10.1128/mbio.00491-23

30. Hourigan SK, Dominguez-Bello MG. Microbial seeding in early life. *Cell Host Microbe*. 2023;31(3):331–333. doi:10.1016/j.chom.2023.02.007

31. Fuentes S, de Vos WM. How to manipulate the microbiota: Fecal microbiota transplantation. *Adv Exp Med Biol*. 2016;902:143–153. doi:10.1007/978-3-319-31248-4_10

32. Zhou L, Qiu W, Wang J, et al. Effects of vaginal microbiota transfer on the neurodevelopment and microbiome of cesarean-born infants: A blinded randomized controlled trial. *Cell Host Microbe*. Published online June 7, 2023:S1931–3128(23)00215-9. doi:10.1016/j.chom.2023.05.022

33. Carpén N, Brodin P, de Vos WM, et al. Transplantation of maternal intestinal flora to the newborn after elective cesarean section (SECFLOR): Study protocol for a double blinded randomized controlled trial. *BMC Pediatr*. 2022;22(1):565. doi:10.1186/s12887-022-03609-3

34. Neves LL, Hair AB, Preidis GA. A systematic review of associations between gut microbiota composition and growth failure in preterm neonates. *Gut Microbes*. 2023;15(1):2190301. doi:10.1080/19490976.2023.2190301

35. Aguilar-Lopez M, Dinsmoor AM, Ho TTB, Donovan SM. A systematic review of the factors influencing microbial colonization of the preterm infant gut. *Gut Microbes*. 2021;13(1):1–33. doi:10.1080/19490976.2021.1884514

36. Arboleya S, Binetti A, Salazar N, et al. Establishment and development of intestinal microbiota in preterm neonates. *FEMS Microbiol Ecol*. 2012;79(3):763–772. doi:10.1111/j.1574-6941.2011.01261.x

37. Arboleya S, Sanchez B, Milani C, et al. Intestinal microbiota development in preterm neonates and effect of perinatal antibiotics. *J Pediatr*. 2015;166(3):538–544. doi:10.1016/j.jpeds.2014.09.041

38. Wandro S, Osborne S, Enriquez C, Bixby C, Arrieta A, Whiteson K. The microbiome and metabolome of preterm infant stool are personalized and not driven by health outcomes, including necrotizing enterocolitis and late-onset sepsis. *mSphere*. 2018;3(3):e00104–18. doi:10.1128/mSphere.00104-18

39. Stoll BJ, Hansen NI, Bell EF, et al. Trends in care practices, morbidity, and mortality of extremely preterm neonates, 1993–2012. *JAMA*. 2015;314(10):1039–1051. doi:10.1001/jama.2015.10244

40. Coggins SA, Wynn JL, Weitkamp JH. Infectious causes of necrotizing enterocolitis. *Clin Perinatol*. 2015;42(1):133–154, ix. doi:10.1016/j.clp.2014.10.012

41. Baranowski JR, Claud EC. Necrotizing enterocolitis and the preterm infant microbiome. *Adv Exp Med Biol*. 2019;1125:25–36. doi:10.1007/5584_2018_313

42. Baldassarre ME, Di Mauro A, Capozza M, et al. Dysbiosis and prematurity: Is there a role for probiotics? *Nutrients*. 2019;11(6):1273. doi:10.3390/nu11061273

43. Dai DLY, Petersen C, Hoskinson C, et al. Breastfeeding enrichment of B. longum subsp. infantis mitigates the effect of antibiotics on the microbiota and childhood asthma risk. *Med (N Y)*. Published online December 29, 2022:S2666–6340(22)00518-9. doi:10.1016/j.medj.2022.12.002

44. Gale C, Logan KM, Santhakumaran S, Parkinson JR, Hyde MJ, Modi N. Effect of breastfeeding compared with formula feeding on infant body composition: A systematic review and meta-analysis. *Am J Clin Nutr*. 2012;95(3):656–669. doi:10.3945/ajcn.111.027284

45. Thompson AL, Monteagudo-Mera A, Cadenas MB, Lampl ML, Azcarate-Peril MA. Milk- and solid-feeding practices and daycare attendance are associated with differences in bacterial diversity, predominant communities, and metabolic and immune function of the infant gut microbiome. *Front Cell Infect Microbiol*. 2015;5:3. doi:10.3389/fcimb.2015.00003

46. González S, Selma-Royo M, Arboleya S, et al. Levels of predominant intestinal microorganisms in 1 month-old full-term babies and weight gain during the first year of life. *Nutrients*. 2021;13(7):2412. doi:10.3390/nu13072412

47. Rendina DN, Lubach GR, Phillips GJ, Lyte M, Coe CL. Maternal and breast milk influences on the infant gut microbiome, enteric health and growth outcomes of rhesus monkeys. *J Pediatr Gastroenterol Nutr*. 2019;69(3):363–369. doi:10.1097/MPG.0000000000002394

48. Ho NT, Li F, Lee-Sarwar KA, et al. Meta-analysis of effects of exclusive breastfeeding on infant gut microbiota across populations. *Nat Commun*. 2018;9(1):4169. doi:10.1038/s41467-018-06473-x

49. Mueller NT, Bakacs E, Combellick J, Grigoryan Z, Dominguez-Bello MG. The infant microbiome development: Mom matters. *Trends Mol Med*. 2015;21(2):109–117. doi:10.1016/j.molmed.2014.12.002

50. Borewicz K, Gu F, Saccenti E, et al. The association between breastmilk oligosaccharides and faecal microbiota in healthy breastfed infants at two, six, and twelve weeks of age. *Sci Rep*. 2020;10(1):4270. doi:10.1038/s41598-020-61024-z

51. Bode L. Human milk oligosaccharides: Every baby needs a sugar mama. *Glycobiology*. 2012;22(9):1147–1162. doi:10.1093/glycob/cws074

52. de Weerth C, Aatsinki AK, Azad MB, et al. Human milk: From complex tailored nutrition to bioactive impact on child cognition and behavior. *Crit Rev Food Sci Nutr*. Published online March 30, 2022:1–38. doi:10.1080/10408398.2022.2053058

53. Munblit D, Peroni DG, Boix-Amorós A, et al. Human milk and allergic diseases: An unsolved puzzle. *Nutrients*. 2017;9(8):894. doi:10.3390/nu9080894

54. Selma-Royo M, Calvo Lerma J, Cortes-Macias E, Collado MC. Human milk microbiome: From actual knowledge to future perspective. *Semin Perinatol*. 2021;45(6):151450. doi:10.1016/j.semperi.2021.151450

55. Fernandez L, Langa S, Martin V, et al. The human milk microbiota: Origin and potential roles in health and disease. *Pharmacol Res*. 2013;69(1):1–10. doi:10.1016/j.phrs.2012.09.001

56. Cabrera-Rubio R, Collado MC, Laitinen K, Salminen S, Isolauri E, Mira A. The human milk microbiome changes over lactation and is shaped by maternal weight and mode of delivery. *Am J Clin Nutr*. 2012;96(3):544–551. doi:10.3945/ajcn.112.037382

57. Pearce J, Taylor MA, Langley-Evans SC. Timing of the introduction of complementary feeding and risk of childhood obesity: A systematic review. *Int J Obes (Lond)*. 2013;37(10):1295–1306. doi:10.1038/ijo.2013.99

58. Fewtrell M, Bronsky J, Campoy C, et al. Complementary feeding: A position paper by the European Society for Paediatric Gastroenterology, Hepatology, and Nutrition (ESPGHAN) committee on nutrition. *J Pediatr Gastroenterol Nutr*. 2017;64(1):119–132. doi:10.1097/MPG.0000000000001454

59. Videhult FK, West CE. Nutrition, gut microbiota and child health outcomes. *Curr Opin Clin Nutr Metab Care*. 2016;19(3):208–213. doi:10.1097/MCO.0000000000000266

60. Bergström A, Skov TH, Bahl MI, et al. Establishment of intestinal microbiota during early life: A longitudinal, explorative study of a large cohort of Danish infants. *Appl Environ Microbiol*. 2014;80(9):2889–2900. doi:10.1128/AEM.00342-14

61. Laursen MF, Bahl MI, Michaelsen KF, Licht TR. First foods and gut microbes. *Front Microbiol*. 2017;8:356. doi:10.3389/fmicb.2017.00356

62. Fallani M, Amarri S, Uusijarvi A, et al. Determinants of the human infant intestinal microbiota after the introduction of first complementary foods in infant samples from five European centres. *Microbiology (Reading)*. 2011;157(Pt 5):1385–1392. doi:10.1099/mic.0.042143-0

63. Martin R, Makino H, Cetinyurek Yavuz A, et al. Early-life events, including mode of delivery and type of feeding, siblings and gender, shape the developing gut microbiota. *PLoS ONE*. 2016;11(6):e0158498. doi:10.1371/journal.pone.0158498

64. Davis EC, Wang M, Donovan SM. The role of early life nutrition in the establishment of gastrointestinal microbial composition and function. *Gut Microbes*. 2017;8(2):143–171. doi:10.1080/19490976.2016.1278104

65. Jia J, Xun P, Wang X, et al. Impact of postnatal antibiotics and parenteral nutrition on the gut microbiota in preterm infants during early life. *JPEN J Parenter Enteral Nutr*. 2020;44(4):639–654. doi:10.1002/jpen.1695

66. Candon S, Perez-Arroyo A, Marquet C, et al. Antibiotics in early life alter the gut microbiome and increase disease incidence in a spontaneous mouse model of autoimmune insulin-dependent diabetes. *PLoS ONE*. 2015;10(5):e0125448. doi:10.1371/journal.pone.0125448

67. Neuman H, Forsythe P, Uzan A, Avni O, Koren O. Antibiotics in early life: Dysbiosis and the damage done. *FEMS Microbiol Rev*. 2018;42(4):489–499. doi:10.1093/femsre/fuy018

68. Vangay P, Ward T, Gerber JS, Knights D. Antibiotics, pediatric dysbiosis, and disease. *Cell Host Microbe*. 2015;17(5):553–564. doi:10.1016/j.chom.2015.04.006

69. Uzan-Yulzari A, Turta O, Belogolovski A, et al. Neonatal antibiotic exposure impairs child growth during the first six years of life by perturbing intestinal microbial colonization. *Nat Commun*. 2021;12(1):443. doi:10.1038/s41467-020-20495-4

70. Tamburini S, Shen N, Wu HC, Clemente JC. The microbiome in early life: Implications for health outcomes. *Nat Med*. 2016;22(7):713–722. doi:10.1038/nm.4142

71. Korpela K, Salonen A, Virta LJ, et al. Intestinal microbiome is related to lifetime antibiotic use in Finnish preschool children. *Nat Commun*. 2016;7:10410. doi:10.1038/ncomms10410

72. Zou ZH, Liu D, Li HD, et al. Prenatal and postnatal antibiotic exposure influences the gut microbiota of preterm infants in neonatal intensive care units. *Ann Clin Microbiol Antimicrob*. 2018;17(1):9. doi:10.1186/s12941-018-0264-y

73. Samarra A, Esteban-Torres M, Cabrera-Rubio R, et al. Maternal-infant antibiotic resistance genes transference: What do we know? *Gut Microbes*. 2023;15(1):2194797. doi:10.1080/19490976.2023.2194797

74. Penders J, Thijs C, Vink C, et al. Factors influencing the composition of the intestinal microbiota in early infancy. *Pediatrics*. 2006;118(2):511–521. doi:10.1542/peds.2005-2824

75. Tun HM, Konya T, Takaro TK, et al. Exposure to household furry pets influences the gut microbiota of infant at 3–4 months following various birth scenarios. *Microbiome*. 2017;5(1):40. doi:10.1186/s40168-017-0254-x

76. Azad MB, Konya T, Maughan H, et al. Infant gut microbiota and the hygiene hypothesis of allergic disease: Impact of household pets and siblings on microbiota composition and diversity. *Allergy Asthma Clin Immunol*. 2013;9(1):15. doi:10.1186/1710-1492-9-15

77. Gómez-Gallego C, Forsgren M, Selma-Royo M, et al. The composition and diversity of the gut microbiota in children is modifiable by the household dogs: Impact of a canine-specific probiotic. *Microorganisms*. 2021;9(3):557. doi:10.3390/microorganisms9030557

78. Fallani M, Young D, Scott J, et al. Intestinal microbiota of 6-week-old infants across Europe: Geographic influence beyond delivery mode, breast-feeding, and antibiotics. *J Pediatr Gastroenterol Nutr*. 2010;51(1):77–84. doi:10.1097/MPG.0b013e3181d1b11e

79. Suzuki TA, Worobey M. Geographical variation of human gut microbial composition. *Biol Lett*. 2014;10(2):20131037. doi:10.1098/rsbl.2013.1037

80. Yatsunenko T, Rey FE, Manary MJ, et al. Human gut microbiome viewed across age and geography. *Nature*. Published online 2012. doi:10.1038/nature11053

81. Vatanen T, Kostic AD, d'Hennezel E, et al. Variation in microbiome LPS immunogenicity contributes to autoimmunity in humans. *Cell*. 2016;165(4):842–853. doi:10.1016/j.cell.2016.04.007

82. De Filippo C, Cavalieri D, Di Paola M, et al. Impact of diet in shaping gut microbiota revealed by a comparative study in children from Europe and rural Africa. *Proc Natl Acad Sci USA*. 2010;107(33):14691–14696. doi:10.1073/pnas.1005963107

83. Derrien M, Alvarez AS, de Vos WM. The gut microbiota in the first decade of life. *Trends Microbiol*. 2019;27(12):997–1010. doi:10.1016/j.tim.2019.08.001

84. Ringel-Kulka T, Cheng J, Ringel Y, et al. Intestinal microbiota in healthy U.S. young children and adults—a high throughput microarray analysis. *PLoS ONE*. 2013;8(5):e64315. doi:10.1371/journal.pone.0064315

85. Hollister EB, Riehle K, Luna RA, et al. Structure and function of the healthy pre-adolescent pediatric gut microbiome. *Microbiome*. 2015;3:36. doi:10.1186/s40168-015-0101-x

86. Wastyk HC, Fragiadakis GK, Perelman D, et al. Gut-microbiota-targeted diets modulate human immune status. *Cell.* 2021;184(16):4137–4153 e14. doi:10.1016/j.cell.2021.06.019

87. Sata Y, Marques FZ, Kaye DM. The emerging role of gut dysbiosis in cardio-metabolic risk factors for heart failure. *Curr Hypertens Rep.* 2020;22(5):38. doi:10.1007/s11906-020-01046-0

88. Jardon KM, Canfora EE, Goossens GH, Blaak EE. Dietary macronutrients and the gut microbiome: A precision nutrition approach to improve cardiometabolic health. *Gut.* 2022;71(6):1214–1226. doi:10.1136/gutjnl-2020-323715

89. Cuervo A, Salazar N, Ruas-Madiedo P, Gueimonde M, González S. Fiber from a regular diet is directly associated with fecal short-chain fatty acid concentrations in the elderly. *Nutr Res.* 2013;33(10):811–816. doi:10.1016/j.nutres.2013.05.016

90. Jin Q, Black A, Kales SN, Vattem D, Ruiz-Canela M, Sotos-Prieto M. Metabolomics and microbiomes as potential tools to evaluate the effects of the Mediterranean diet. *Nutrients.* 2019;11(1):207. doi:10.3390/nu11010207

91. De Filippis F, Pellegrini N, Vannini L, et al. High-level adherence to a Mediterranean diet beneficially impacts the gut microbiota and associated metabolome. *Gut.* 2016;65(11):1812–1821. doi:10.1136/gutjnl-2015-309957

92. Garcia-Mantrana I, Selma-Royo M, Alcantara C, Collado MC. Shifts on gut microbiota associated to Mediterranean diet adherence and specific dietary intakes on general adult population. *Front Microbiol.* 2018;9:890. doi:10.3389/fmicb.2018.00890

93. Mitsou EK, Kakali A, Antonopoulou S, et al. Adherence to the Mediterranean diet is associated with the gut microbiota pattern and gastrointestinal characteristics in an adult population. *Br J Nutr.* 2017;117(12):1645–1655. doi:10.1017/S0007114517001593

94. Gupta VK, Paul S, Dutta C. Geography, ethnicity or subsistence-specific variations in human microbiome composition and diversity. *Front Microbiol.* 2017;8:1162. doi:10.3389/fmicb.2017.01162

95. Clemente JC, Pehrsson EC, Blaser MJ, et al. The microbiome of uncontacted Amerindians. *Sci Adv.* 2015;1(3):e1500183. doi:10.1126/sciadv.1500183

96. Vich Vila A, Collij V, Sanna S, et al. Impact of commonly used drugs on the composition and metabolic function of the gut microbiota. *Nat Commun.* 2020;11(1):362. doi:10.1038/s41467-019-14177-z

97. Aron-Wisnewsky J, Vigliotti C, Witjes J, et al. Gut microbiota and human NAFLD: Disentangling microbial signatures from metabolic disorders. *Nat Rev Gastroenterol Hepatol.* 2020;17(5):279–297. doi:10.1038/s41575-020-0269-9

98. Koren O, Goodrich JK, Cullender TC, et al. Host remodeling of the gut microbiome and metabolic changes during pregnancy. *Cell.* 2012;150(3):470–480. doi:10.1016/j.cell.2012.07.008

99. Fuhler GM. The immune system and microbiome in pregnancy. *Best Pract Res Clin Gastroenterol.* 2020;44–45:101671. doi:10.1016/j.bpg.2020.101671

100. Nuriel-Ohayon M, Neuman H, Ziv O, et al. Progesterone increases bifidobacterium relative abundance during late pregnancy. *Cell Rep.* 2019;27(3):730–736 e3. doi:10.1016/j.celrep.2019.03.075

101. Baud A, Hillion KH, Plainvert C, et al. Microbial diversity in the vaginal microbiota and its link to pregnancy outcomes. *Sci Rep.* 2023;13(1):9061. doi:10.1038/s41598-023-36126-z

102. González-Sánchez A, Reyes-Lagos JJ, Peña-Castillo MA, Nirmalkar K, García-Mena J, Pacheco-López G. Vaginal microbiota is stable and mainly dominated by lactobacillus at third trimester of pregnancy and active childbirth: A longitudinal study of ten Mexican women. *Curr Microbiol.* 2022;79(8):230. doi:10.1007/s00284-022-02918-1

103. Keelan JA, Payne MS. Vaginal microbiota during pregnancy: Pathways of risk of preterm delivery in the absence of intrauterine infection? *Proc Natl Acad Sci U S A.* 2015;112(47):E6414. doi:10.1073/pnas.1517346112

104. Gudnadottir U, Debelius JW, Du J, et al. The vaginal microbiome and the risk of preterm birth: A systematic review and network meta-analysis. *Sci Rep.* 2022;12(1):7926. doi:10.1038/s41598-022-12007-9

105. Balan P, Brandt BW, Chong YS, et al. Subgingival microbiota during healthy pregnancy and pregnancy gingivitis. *JDR Clin Trans Res.* 2021;6(3):343–351. doi:10.1177/2380084420948779

106. Ye C, Kapila Y. Oral microbiome shifts during pregnancy and adverse pregnancy outcomes: Hormonal and Immunologic changes at play. *Periodontol 2000.* 2021;87(1):276–281. doi:10.1111/prd.12386

107. Calatayud M, Koren O, Collado MC. Maternal microbiome and metabolic health program microbiome development and health of the offspring. *Trends Endocrinol Metab.* 2019;30(10):735–744. doi:10.1016/j.tem.2019.07.021

108. Collado MC, Isolauri E, Laitinen K, Salminen S. Distinct composition of gut microbiota during pregnancy in overweight and normal-weight women. *Am J Clin Nutr.* 2008;88(4):894–899.

109. Collado MC, Laitinen K, Salminen S, Isolauri E. Maternal weight and excessive weight gain during pregnancy modify the immunomodulatory potential of breast milk. *Pediatr Res.* 2012;72(1):77–85. doi:10.1038/pr.2012.42

110. Santacruz A, Collado MC, Garcia-Valdes L, et al. Gut microbiota composition is associated with body weight, weight gain and biochemical parameters in pregnant women. *Br J Nutr.* 2010;104(1):83–92. doi:10.1017/S0007114510000176

111. Chu DM, Meyer KM, Prince AL, Aagaard KM. Impact of maternal nutrition in pregnancy and lactation on offspring gut microbial composition and function. *Gut Microbes.* 2016;7(6):459–470. doi:10.1080/19490976.2016.1241357

112. Carrothers JM, York MA, Brooker SL, et al. Fecal microbial community structure is stable over time and related to variation in macronutrient and micronutrient intakes in lactating women. *J Nutr.* 2015;145(10):2379–2388. doi:10.3945/jn.115.211110

113. Gavini F, Cayuela C, Antoine JM, et al. Differences in the distribution of bifidobacterial and enterobacterial species in human faecal microflora of three different (CHILDREN, adults, elderly) age groups. *Microb Ecol Health Dis.* 2001;13(1):40–45. doi:10.1080/089106001750071690

114. Hopkins MJ, Sharp R, Macfarlane GT. Age and disease related changes in intestinal bacterial populations assessed by cell culture, 16S rRNA abundance, and community cellular fatty acid profiles. *Gut.* 2001;48(2):198–205. doi:10.1136/gut.48.2.198

115. Biagi E, Candela M, Turroni S, Garagnani P, Franceschi C, Brigidi P. Ageing and gut microbes: Perspectives for health maintenance and longevity. *Pharmacol Res.* 2013;69(1):11–20. doi:10.1016/j.phrs.2012.10.005

116. Hayashi H, Sakamoto M, Kitahara M, Benno Y. Molecular analysis of fecal microbiota in elderly individuals using 16S rDNA library and T-RFLP. *Microbiol Immunol.* 2003;47(8):557–570. doi:10.1111/j.1348-0421.2003.tb03418.x

117. Bartosch S, Fite A, Macfarlane GT, McMurdo MET. Characterization of bacterial communities in feces from healthy elderly volunteers and hospitalized elderly patients by using real-time PCR and effects of antibiotic treatment on the fecal microbiota. *Appl Environ Microbiol.* 2004;70(6):3575–3581. doi:10.1128/AEM.70.6.3575-3581.2004

118. Salazar N, Arboleya S, Fernández-Navarro T, de Los Reyes-Gavilán CG, Gonzalez S, Gueimonde M. Age-associated changes in gut microbiota and dietary components related with the immune system in adulthood and old age: A cross-sectional study. *Nutrients.* 2019;11(8):1765. doi:10.3390/nu11081765

119. Salazar N, López P, Valdés L, et al. Microbial targets for the development of functional foods accordingly with nutritional and immune parameters altered in the elderly. *J Am Coll Nutr.* 2013;32(6):399–406. doi:10.1080/07315724.2013.827047

120. Biagi E, Nylund L, Candela M, et al. Through ageing, and beyond: Gut microbiota and inflammatory status in seniors and centenarians. *PLoS ONE.* 2010;5(5):e10667. doi:10.1371/journal.pone.0010667

121. Claesson MJ, Cusack S, O'Sullivan O, et al. Composition, variability, and temporal stability of the intestinal microbiota of the elderly. *Proc Natl Acad Sci USA.* 2011;108(Suppl 1):4586–4591. doi:10.1073/pnas.1000097107

122. Mäkivuokko H, Tiihonen K, Tynkkynen S, Paulin L, Rautonen N. The effect of age and non-steroidal anti-inflammatory drugs on human intestinal microbiota composition. *Br J Nutr.* 2010;103(2):227–234. doi:10.1017/S0007114509991553

123. Mueller S, Saunier K, Hanisch C, et al. Differences in fecal microbiota in different European study populations in relation to age, gender, and country: A cross-sectional study. *Appl Environ Microbiol.* 2006;72(2):1027–1033. doi:10.1128/AEM.72.2.1027-1033.2006

124. Odamaki T, Kato K, Sugahara H, et al. Age-related changes in gut microbiota composition from newborn to centenarian: A cross-sectional study. *BMC Microbiol.* 2016;16:90. doi:10.1186/s12866-016-0708-5

125. Woodmansey EJ, McMurdo MET, Macfarlane GT, Macfarlane S. Comparison of compositions and metabolic activities of fecal microbiotas in young adults and in antibiotic-treated and non-antibiotic-treated elderly subjects. *Appl Environ Microbiol.* 2004;70(10):6113–6122. doi:10.1128/AEM.70.10.6113-6122.2004

126. Salazar N, Valdés-Varela L, González S, Gueimonde M, de Los Reyes-Gavilán CG. Nutrition and the gut microbiome in the elderly. *Gut Microbes.* 2017;8(2):82–97. doi:10.1080/19490976.2016.1256525

127. Cerro EDD, Lambea M, Félix J, Salazar N, Gueimonde M, De la Fuente M. Daily ingestion of Akkermansia mucciniphila for one month promotes healthy aging and increases lifespan in old female mice. *Biogerontology.* 2022;23(1):35–52. doi:10.1007/s10522-021-09943-w

128. Bárcena C, Valdés-Mas R, Mayoral P, et al. Healthspan and lifespan extension by fecal microbiota transplantation into progeroid mice. *Nat Med.* 2019;25(8):1234–1242. doi:10.1038/s41591-019-0504-5

129. Biagi E, Franceschi C, Rampelli S, et al. Gut microbiota and extreme longevity. *Curr Biol.* 2016;26(11):1480–1485. doi:10.1016/j.cub.2016.04.016

130. Santoro A, Ostan R, Candela M, et al. Gut microbiota changes in the extreme decades of human life: A focus on centenarians. *Cell Mol Life Sci.* 2018;75(1):129–148. doi:10.1007/s00018-017-2674-y

131. Jackson MA, Jeffery IB, Beaumont M, et al. Signatures of early frailty in the gut microbiota. *Genome Med.* 2016;8(1):8. doi:10.1186/s13073-016-0262-7

132. Jeffery IB, Lynch DB, O'Toole PW. Composition and temporal stability of the gut microbiota in older persons. *ISME J.* 2016;10(1):170–182. doi:10.1038/ismej.2015.88

133. Soenen S, Rayner CK, Jones KL, Horowitz M. The ageing gastrointestinal tract. *Curr Opin Clin Nutr Metab Care.* 2016;19(1):12–18. doi:10.1097/MCO.0000000000000238

134. Claesson MJ, Jeffery IB, Conde S, et al. Gut microbiota composition correlates with diet and health in the elderly. *Nature.* 2012;488(7410):178–184. doi:10.1038/nature11319

135. Tarnawski AS, Ahluwalia A, Jones MK. Increased susceptibility of aging gastric mucosa to injury: The mechanisms and clinical implications. *World J Gastroenterol.* 2014;20(16):4467–4482. doi:10.3748/wjg.v20.i16.4467

136. Ruiz-Saavedra S, Salazar N, Suárez A, de Los Reyes-Gavilán CG, Gueimonde M, González S. Comparison of different dietary indices as predictors of inflammation, oxidative stress and intestinal microbiota in middle-aged and elderly subjects. *Nutrients.* 2020;12(12):3828. doi:10.3390/nu12123828

# Lactic Acid Bacteria in the Gut

# 22

Mirjana Rajilić-Stojanović,
Suzana Dimitrijević, and Nataša Golić

## 22.1 INTRODUCTION

The human body is heavily populated by microorganisms. These microorganisms, which are collectively called microbiota, are essential for our well-being, as their metabolism complements ours and contributes to systemic health. The majority of the human-associated microbes reside in the distal gut, where the microbiota is so dense and so active that the colon can be considered a "bioreactor" and gut microbiota a "microbial organ" (Swidsinski et al. 2016). There are still many unknowns regarding gut microbiota, including the timing of its acquisition, the description of its diversity, the definition of tolerable variation within the normal composition, or if its disturbed composition in diseases is a consequence or an etiological factor of the disease. The existence of such open questions in combination with available high-throughput methods has stimulated massive gut microbiota research in recent years. The output of this highly productive research field has confirmed that lactic acid–producing bacteria are a prevalent, abundant, and important part of this complex ecosystem.

In this chapter, we summarise current knowledge about the diversity of lactic acid bacteria in the gut and the factors that contribute to their abundance. We also report diseases that are linked with their disturbed abundance and summarise their hypothetical role in symptom development. Although lactic acid bacteria (LAB) include only members of the Lactobacillales order (the low G+C bacteria), lactic acid can be produced by other gut bacteria. A large proportion of lactic acid is formed by a group of highly prevalent gut microbes—members of the high G+C Bifidobacteriales, which produce acetate and lactate in the molar ratio of 3:2. In addition, there are species belonging to the newly described genera of intestinal lactic acid producing bacteria—such as *Ruthenibacterium lactatiformans* (Shkoporov et al. 2016). This bacterium, similar to several other typical gut bacteria, including *Faecalibacterium prausnitzii*, *Blautia* spp., and *Roseburia* spp., produces lactic acid in addition to the other short-chain fatty acids (SCFAs) (Flint et al. 2015). This indicates that the formation of lactic acid in the gut can be achieved using various pathways and different bacteria. The phenomenon that phylogenetically distant groups perform one function in an ecosystem is known as functional redundancy. Functional redundancy enables sustained symbiotic function despite enormous compositional variations in gut microbiota between individuals. Despite the obvious heterogeneity of the lactic acid-producing bacteria in the gut, this chapter will focus only on the Lactobacillales and the Bifidobacteriales (LAB+B in the text), which are historically recognised as lactic acid–producing bacteria. Other gut lactic acid producers are more relevant for the ecosystem and the host from other perspectives (*e.g.*, *Faecalibacterium* and *Roseburia* because of butyrate production) and will not be addressed in further detail.

# 22.2 DIVERSITY OF LAB+B WITHIN THE GUT MICROBIOTA

## 22.2.1 General Remarks about Gut Microbiota Diversity

The human gut microbiota is an extremely complex and still not fully described ecosystem. Members of all three domains of life—Eukarya, Bacteria, and Archaea—inhabit our digestive tract. Bacteria are by far the most diverse and the most abundant microbiota subpopulation, and bacteria belonging to phyla Bacteroidetes, Firmicutes, Proteobacteria, Actinobacteria, Fusobacteria, and Verrucomicrobia are present in literally all individuals. Although over 1000 cultured species have been detected in gut samples, it is clear that most of the gut inhabitants are still uncultured species (Rajilić-Stojanović and de Vos 2014). The main reason for this is that gut inhabitants are highly adapted to the specific conditions of the gut, which are difficult to mimic in the laboratory. Furthermore, there is an enormous variation in microbiota composition between individuals. The variation is even more evident among individuals from distant ethnic groups (Yatsunenko et al. 2012). Due to this enormous inter-individual variation, the microbiota of tens of thousands more individuals should be profiled before a "normal" gut microbiota can be defined (Falony et al. 2016). The diverse ecosystem in each individual is formed under the influence of various factors, including host–microbial interactions, age, genetics, and gender, but also dietary habits, environment, medication, and stress (Bačić and Rajilić-Stojanović 2022). While the description is ongoing, and novel features are continuously being discovered, there are some generally accepted standpoints about the gut microbiota, as discussed further in detail.

It is clear that the density, diversity, and composition of the microbiota change along the gut. In addition to an undoubted host regulation, this is a direct consequence of a dramatic change in conditions that impact the microbial community. The pH value changes from very low values (pH ~ 1.5) in the stomach (Marieb and Hoehn 2010) to almost neutral values (pH up to 7.5) in the colon (Nugent et al. 2001). Another important environmental factor is oxygen concentration, which gradually decreases along the gut, resulting in strictly anaerobic conditions in the colon. Finally, nutrient flux and nutrient availability differ. In the small intestine, various nutrients are present, but active microbial fermentation is hampered by digestion and nutrient adsorption combined with a relatively short passage time of 2–5 hours (Szarka and Camilleri 2012). The most active place for microbial fermentation is the colon, where indigested food resides for almost 30 hours (Szarka and Camilleri 2012) and is exposed to the activity of the densest microbial ecosystem ever reported (Whitman et al. 1998). In the proximal colon, carbohydrate fermentation occurs, while protein fermentation is dominant in the last, left part of the colon (Cummings and Macfarlane 1991). Since LAB+B are oxygen-tolerant saccharolytic bacteria, they are typically present in the upper parts of the digestive tract and the right colon, with the following distribution: members of the order Lactobacillales typically occupy the small intestine, while bifidobacteria are found in the proximal colon.

Another well-established feature of microbiota is that it changes throughout life. LAB+B are among the first acquired microbes. It is well established that *Bifidobacterium* spp. dominate during the first six months of life in healthy term infants. Bifidobacteria are present in the breast milk of mothers and their establishment in the gut is stimulated by indigestible oligosaccharides of human milk, so-called human milk oligosaccharides (Lawson et al. 2020). Lactobacilli (a term still useful to refer to the old taxon of *Lactobacillus*) are also acquired through the same route, as this group of bacteria constitutes roughly one-third of the total diversity of mothers' milk (Zhang et al. 2020). After the introduction of solid foods, a notable diversification of microbiota occurs, and the prevalence of bifidobacteria decreases dramatically (Bäckhed et al. 2015). For a long time it was believed that microbiota is adult-like already at one year of age (Mitsuoka 1992). However, currently available data show that the maturation of this ecosystem continues until adulthood, although changes are the most pronounced in the first year of life. While some studies could not detect a difference in microbiota between three-year-old children and adults (Yatsunenko et al. 2012), others have reported marked differences between adults and 7–12-year-old children (Hollister et al. 2015) or even between adults and adolescents (Agans et al. 2011). One of the characteristics of younger gut microbiota is a higher abundance of bifidobacteria, as this group of bacteria continuously declines with age, while the other LAB have the lowest abundance in adulthood (Mitsuoka 1992).

## 22.2.2 Diversity of Gut LAB+B

LAB+B are an important subpopulation of gut microbiota. LAB belong to the Bacilli class of the Firmicutes phylum, which is the most abundant and the most diverse group of intestinal bacteria. At this point we would like to point out that phylum Firmicutes was recently renamed to Bacillota (Oren and Garrity 2021), although the majority of publications still use the old term, as will be done in this chapter. B belong to the Actinobacteria, which is normally the third most abundant phylum in the human gut, after Firmicutes and Bacteroidetes (Rajilić-Stojanović and de Vos 2014). The presence of LAB+B has been acknowledged for a very long period, as the first report of lactobacilli presence in the human gut was published in 1900. However, despite a very long period of studying and an undoubted scientific interest in LAB+B, novel species of this group are still being reported, and the most recent report of a member of Lactobacillaceae family spp.—*Limosilactobacillus caccae*—dates from 2021 (Lo et al. 2021).

LAB share the common feature of lactic acid production, and based on this common feature, many distantly related species were initially classified within the *Lactobacillus* genus. Based on the objective diversity of the group, there have been several reclassifications of lactobacilli. In that light, for the first gut LAB, which was isolated in 1900 and denoted as *Bacillus acidophilus* (Moro 1900), it is not clear if it represents *Lactobacillus acidophilus* or one of the other five species derived from the so-called *L. acidophilus* group (Mitsuoka 1992). Parallel with the isolation of novel species, the availability of reliable molecular markers caused the reclassification of LAB. Although reclassifications might appear complex and confusing for part of the scientific community, the reclassifications enable better systematisation of the microbial world, whose complexity is far behind our initial perception.

The introduction of the 16S rRNA gene sequence as a phylogenetic marker enabled the separation of wrongly classified distant relatives of lactobacilli. One such reclassification considers the formation of the *Atopobioum* genus, as in 1992 it was noted that an intestinal LAB *Lactobacillus minutus*, *Lactobacillus rimae*, and *Streptococcus parvulus* cluster into one genus within the phylum of Actinobacteria (Collins and Wallbanks 1992). Currently acknowledged LAB are scattered among five bacterial families within the order Bacilli of the Firmicutes phylum, Lactobacillaceae (which was unified with Leuconostocaceae), Aerococcaceae, Carnobacteriaceae, Enterococcaceae, and Streptococcaceae. The genetic diversity within these families is exceptional, and recently the heterogeneous genus *Lactobacillus* was reclassified into a total of 24 genera (*Lactobacillus sensu stricto* and 23 novel genera; Zheng et al. 2020). Among these, at least ten genera have been detected in the human intestinal samples. Among various human gut lactobacilli, homofermentative *Lactobacillus* and heterofermentative *Limosilactobacillus* are the most diverse (Figure 22.1). In addition to lactobacilli, intestinal samples frequently contain species classified within the genera *Streptococcus* and *Enterococcus*. Based on the currently available data, there is sufficient evidence that members of Lactobacillaceae, Enterococcaceae, and Streptococcaceae are true inhabitants of the human gut. Other LAB genera are scarcely detected; for example, among Carnobacteriacea, *Desemzia incerta* was isolated from intestinal samples, while *Carnobacterium* spp. were detected in several metagenomics studies (e.g., Jeraldo et al. 2016), although not a single human intestinal *Carnobacterium* spp. was isolated and identified at the species level.

Until 2023, more than 100 LAB were reported as inhabitants of the human gut (Figures 22.1 through 22.3). Lactobacilli are one of the predominant groups in the stomach and the small intestine, as shown by culturing and molecular analyses (Reuter 2001; Rajilic-Stojanovic et al. 2020). Due to the availability of cultivation media for the growth of LAB, which were developed already in the 1950s (Rogosa et al. 1951), and oxygen tolerance by these anaerobic bacteria, they are among the most comprehensively studied groups of the human microbiota. Therefore, it is not surprising that there are as many as 49 known cultured species of the Lactobacillaceae family of the human gut (Figure 22.1).

This list will probably be further extended both by the detection of already described species in intestinal samples and by a description of novel species. As stated earlier, lactobacilli are among the

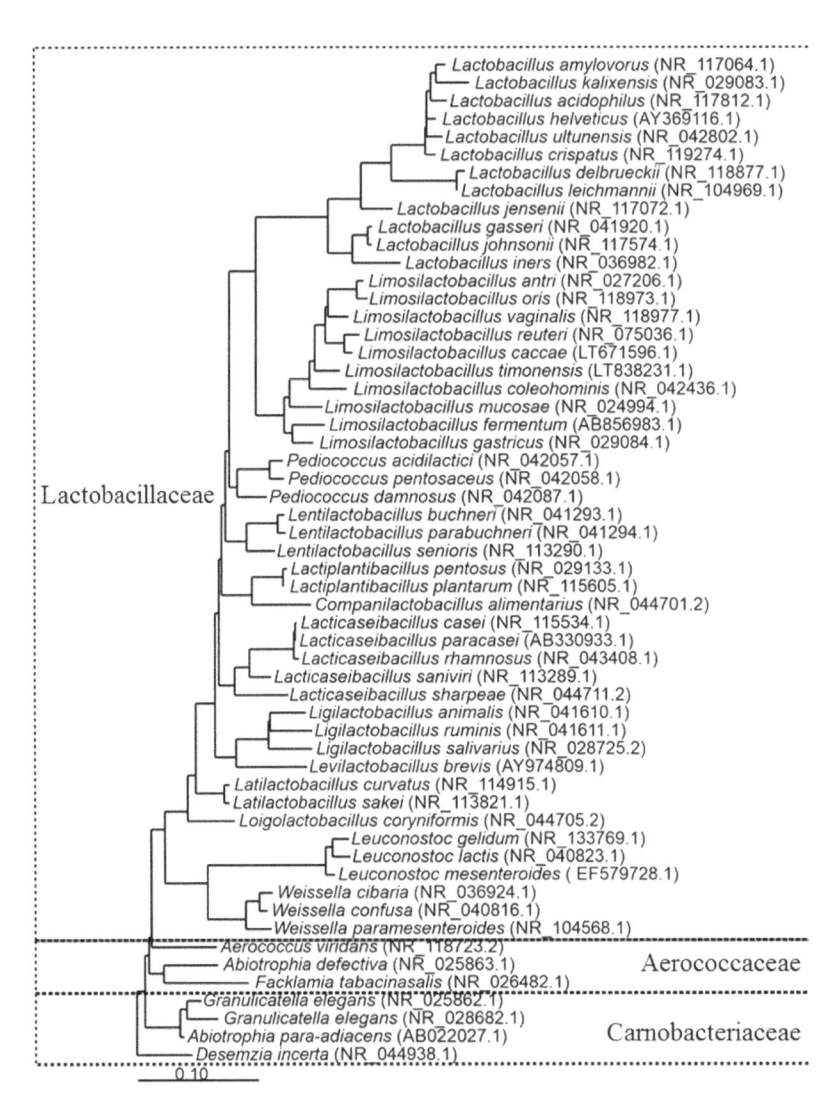

**FIGURE 22.1** Phylogenetic tree showing gut microbiota species that belong to Lactobacillaceae, Carnobacteriaceae, and Aerococcaceae families.

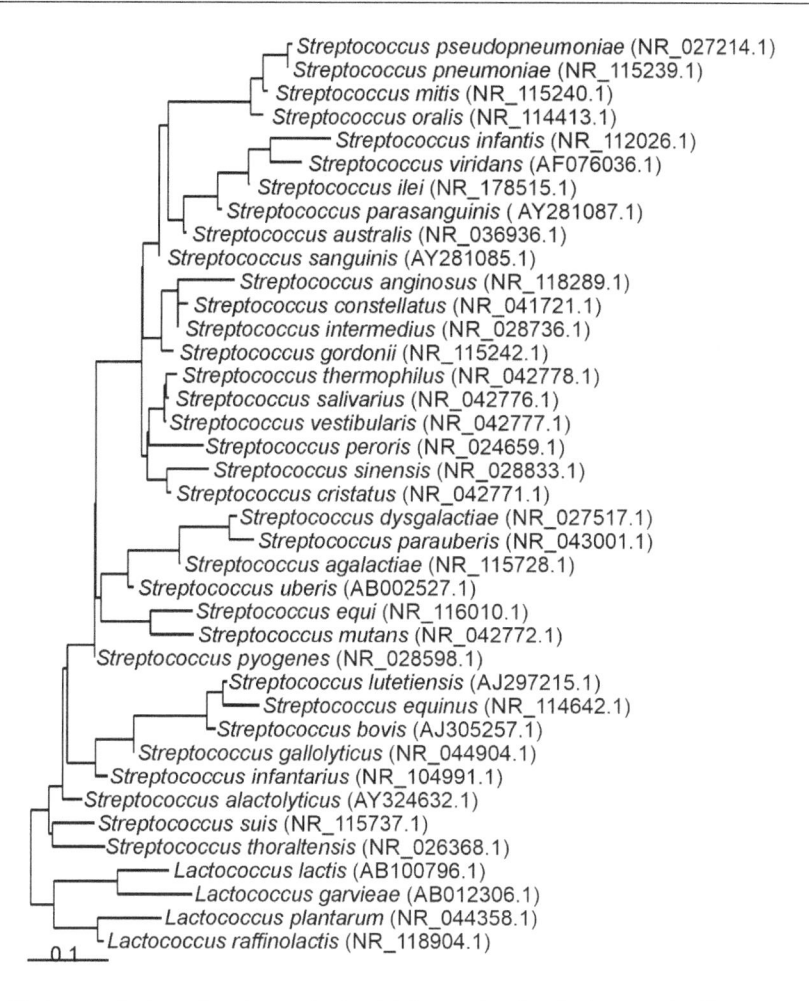

**FIGURE 22.2** Phylogenetic tree showing gut microbiota species that belong to the Streptococcaceae family.

first colonisers of the human intestine. In a developed human, this group of bacteria typically resides in the upper intestinal tract, including the ileum and the stomach. Molecular studies have shown that lactobacilli and streptococci are among the most abundant bacteria in the small bowel (Booijink et al. 2010; Dicksved et al. 2009). These findings have confirmed previous cultivation studies that reported *Streptococcus* and *Enterococcus* species as the dominant bacteria in the small intestine (Reuter 2001; Simon and Gorbach 1986). There is considerable bacterial diversity in both streptococci and enterococci, and similar to lactobacilli, several species were reclassified to different genera within the Streptococcaceae and Enterococcaceae families (Figures 22.2 and 22.3). In total, there are 39 Streptococcaceae and 20 Enterococcaceae gut species. The description of these groups is also not completed, as novel species continue to be reported, with the example of *Streptococcus ilei*, which description was published less than two years ago (Hyun et al. 2021). Both *Streptococcus* and *Enterococcus* spp. are among the first established species in the infant's digestive tract that can be detected already on the first day of life, while *Streptococcus* spp. persist in detectable amounts for a longer period than *Enterococcus* spp. (Gosalbes et al. 2013; Solís et al. 2010).

Although there is considerable evidence that LAB are present in the gut, some authors question if they are true inhabitants of the digestive tract. This scepticism is based on the fact that LAB are not

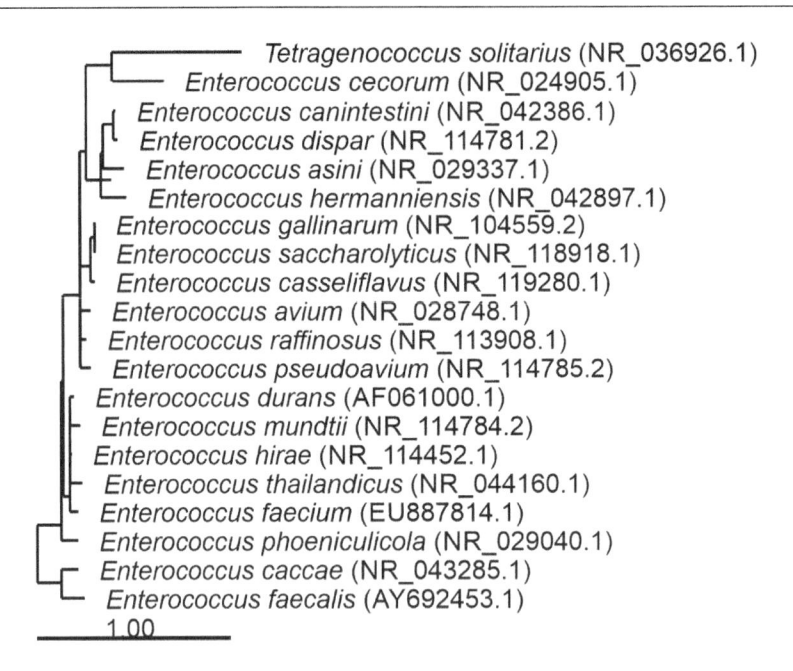

**FIGURE 22.3** Phylogenetic tree showing gut microbiota species that belong to the Enterococcaceae family.

always retrieved from intestinal samples and that they are abundant in the oral cavity. Furthermore, due to their presence in food, it can be argued that some of the detected species were ingested with food (Stolaki et al. 2012). Some authors have suggested that only *Lactobacillus* and *Ligilactobacillus* are native to the human digestive tract of neonates, while in adults, it is possible to find a somewhat larger number of species, including *Lacticaseibacillus rhamnosus* and *Lc. casei/paracasei* (Wall et al. 2007). Undoubtedly, ingested microorganisms can be detected in gut samples, especially when DNA-based techniques are used, but the persistent presence in combination with specific functions that enable survival in the digestive tract should not be ignored. A study that evaluated the influence of diet on the microbiota reported a significant increase in the abundance of foodborne *Lactococcus* and *Pediococcus* species after the switch from a vegan to an omnivore diet (David et al. 2014). It should, however, be noted that these bacteria were found in volunteers even before switching the diets, suggesting that ingested microbes could only reinforce an already resident population. While future research will further elucidate the contribution of food-derived LAB to the gut microbiota, current data already indicate that a diverse LAB community is permanently residing in the gut. Traditionally, *Lactobacillus* spp. (including all genera derived from this genus in recent reclassification) have been recognised as members of the gut ecosystem, while other LAB such as members of genera *Leuconostoc* and *Weissella* would be considered occasional and most likely transient. Contrasting with this view, a study of microbiota in colonic mucosa showed that *Leuconostoc* spp. and *Weissella* spp. are widely distributed, abundant, and even predominant bacteria (Hong et al. 2011). In addition, *Leuconostoc* and *Weissella* spp. were identified as the earliest colonisers of some newborns, as they were detected in meconium samples, while in colonised infants, members of the genus *Leuconostoc* persisted until seven months of age (Gosalbes et al. 2013). Another study detected an even more diverse LAB community in amniotic fluid, placenta membrane, and placenta, and strains belonging to the genera *Lactobacillus, Streptococcus, Lactococcus,* and *Enterococcus* were found either by cultivation-dependent or cultivation-independent techniques (Pelzer et al. 2017). When analysing the microbiota of adults, faecal samples are the most frequently used type of sample. In faeces the density of the microbial ecosystem is extremely high and reaches values of $10^{12}$ cells per g (Whitman et al. 1998). Since LAB predominantly inhabit the upper gut, in faeces they are subdominant—for example, lactobacilli density is between $10^4$ and $10^8$ cells per g of faeces (Rossi et al. 2016). Therefore, even an ultra-deep sequencing analysis of the total microbiota, which typically detects members present at a density of $10^7$ cells per g, would omit the majority of intestinal lactobacilli and other sub-abundant LAB. An alternative approach that enables the detection of species present in very low numbers is to apply procedures that target a bacterial group of interest. As an example, by using group-specific primers, *Latilactobacillus sakei* was detected in faecal samples, although it was present in a very low abundance of $10^3$ cells per gram (Walter et al. 2001). Similarly, a metagenomics study focusing only on the diversity of lactobacilli detected several species for the first time in human gut samples (Rossi et al. 2016).

LAB have several important roles. It has been shown that streptococci take an active role in the fermentation of simple sugars in the small intestine, where the formed lactate may further serve as a substrate to *Veillonella* spp. that convert lactate into propionate (Zoetendal et al. 2012). An in vitro study has shown that streptococci exhibit immunomodulatory activity, although the effect on cytokine production was different if species were studied alone or in co-culture (van den Bogert et al. 2014). The mechanisms for stimulation of the immune system are diverse, but one of the important traits of LAB is the fact that luminal concentration of lactate and pyruvate in the small intestine contributes to the protrusion of the dendritic cells and thereby to the maturation of the immune system (Morita et al. 2019). Some probiotic lactobacilli can produce small molecules that impact metabolic pathways. For example, it has been shown that gut resident lactobacilli produce 5-methoxyindoleacetic acid that activates the Nrf2 antioxidant response pathway in the liver (Saeedi et al. 2020). While the presence of lactobacilli is in general considered beneficial, the role of intestinal streptococci might be dual. Systematic analysis of gastric microbiota indicated that streptococci are dominant bacteria in the stomach of healthy subjects but that their relative abundance increases in gastric cancer patients (Rajilic-Stojanovic et al. 2020). Some common intestinal streptococci have virulence factors and represent opportunistic pathogens that could cause infections, especially in immune-compromised and hospitalised individuals (Hyun et al. 2021; Sinha et al. 2021). Enterococci are gut commensals that have the most controversial status among LAB, and among them, *E. faecalis* and *E. faecium*, in particular, are frequently found in human infections (Popović et al. 2018). Despite the presence of virulent, hospital-adapted strains, there are several probiotic enterococci strains to treat diarrhoea and irritable bowel syndrome, lower cholesterol levels, and improve immunity (Franz et al. 2011).

Bifidobacteriales are a group of lactic acid–producing bacteria that are highly adapted to the digestive tract. These bacteria are only exceptionally detected in other ecological niches or as natural food-borne microbes (e.g., Laureys et al. 2016). Gut lactic acid–producing bacteria from the Bifidobacteriales include 21 *Bifidobacterium* spp. and one member of the *Scardovia* genus (Figure 22.4). The first *Bifidobacterium* spp. was isolated from a breastfed infant's faeces in 1900 by Henri Tissier (1900). These bacteria inhabit the ascending colon and are abundant in faecal samples, where they reach densities between $10^8$ and $10^{10}$ cells/g (Finegold et al. 1974; Moore and Holdeman 1974). Their dominance is typical for breastfed infants due to their ability to degrade human milk oligosaccharides. Genomic sequence analyses of infant-derived bifidobacteria show that these species produce various enzymes, including fucosidases, sialidases, β-hexosaminidase, and β-galactosidases, that are needed for the degradation of milk oligosaccharides. They also contain urease-encoding operon, which enables bifidobacteria to use urea from breast milk as a nitrogen source (Sela 2011). This competitive advantage to utilise the first food better than any other gut bacteria ensures bifidobacterial establishment in the early days of life. Bifidobacteria remain in the human gut throughout the entire life span. This can be explained by their ability to utilise intestinal mucins, which glycan fractions are structurally similar to human milk oligosaccharides, and thereby bifidobacteria can degrade them and use as carbon sources (Belzer 2022). Interestingly, only one isolated group of individuals, the Hazda tribe from Tanzania, harbours no bifidobacteria (Schnorr et al. 2014). *Bifidobacterium* spp. are not affected by the dramatic shifts in diet (David et al. 2014), while almost a decade-long follow-up of microbiota composition of five individuals showed that bifidobacteria are a very stable component of gut microbiota of each analysed person (Rajilić-Stojanović et al. 2013).

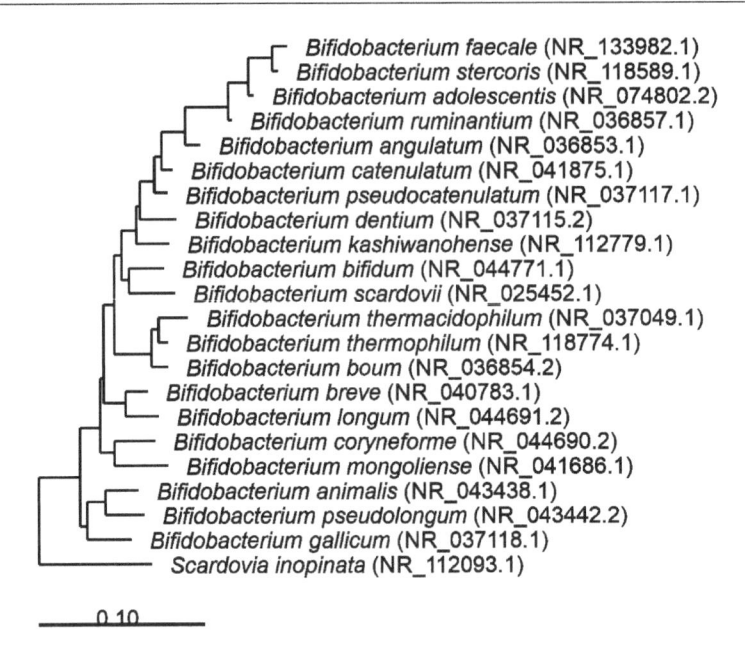

**FIGURE 22.4**  Phylogenetic tree showing gut microbiota species that belong to the Bifidobacteriales order.

# 22.3 ADAPTATION OF LAB+B TO THE GUT AND GUT MICROBIOTA

## 22.3.1 Mechanisms That Enable Survival of LAB+B in the Gut

Only specialised microbes can survive the conditions of the human gut, and gut microbes had to develop strategies to survive various pH conditions, the presence of hydrolytic enzymes, and the presence of bile acids, which are lipid emulsifiers and are particularly aggressive against phospholipids of the cell membrane. Resistance to the acidic environment and the presence of bile salts are very important characteristics for the survival of LAB in the gut (Shehata et al. 2016). Investigation of acid tolerance in some LAB indicated that they possess an adaptive response that can be induced by incubation at a nonlethal acidic pH (van de Guchte et al. 2002). The presumed mechanism involves several transmembrane proteins and histidine decarboxylation that stimulate the formation of ATP and the generation of proton motive force in some lactobacilli (De Angelis and Gobbetti 2004; van de Guchte et al. 2002). The presence of bile salt hydrolase (BSH) genes in some lactobacilli could be important for their ability to survive in the gut (Stolaki et al. 2012), although some researchers have shown that BSH activity and resistance to bile acids were unrelated properties in lactobacilli (De Angelis and Gobbetti 2004). The ability to adhere to the intestinal epithelial cells due to the presence of bacterial cell surface components is an additional factor facilitating the survival of LAB (Haller et al. 2001). *Bifidobacterium* spp. are in principle not equipped with either of these tools, but given that they typically reside in the colon, their adaptation to these harsh conditions of the upper gut is not essential. In the upper part of the digestive tract, another important factor is the presence of oxygen. LAB+B prefer anaerobic conditions, but their ability to produce peroxidase protects them from the toxic effect of oxygen and its by-products (Khalid 2011).

## 22.3.2 Interactions with Other Members of the Ecosystem: The Fate of Lactate

LAB+B are an indispensable part of the gut microbiota, but their major metabolic product—lactic acid/lactate—is typically detected in very low concentrations in the intestinal lumen. The resulting output of carbohydrate metabolism by gut microbiota is composed of gasses (carbon dioxide and hydrogen) and SCFAs, namely acetate, propionate, and butyrate (Cummings 1981). The predominant SCFAs are excreted in a substantial amount (exceeding hundreds of mmol per kg of faeces (Gargari et al. 2016), and the production of lactate contributes to the final output of the major SCFAs. The fate of lactate is dual; part of it is transported through epithelium and further metabolised by the host, while the rest is utilised by so-called lactic-utilising bacteria (Louis et al. 2022). In the human gut, lactate is metabolised by various members of the Firmictues phylum to produce all three SCFAs. The most prevalent and the most abundant human gut utilisers are members of the Lachnospiraceae family *Anaerobutyricum* and *Anaerostipes* species that convert lactate to butyrate. Lactate is transformed to propionate by members of Veillonellaceae family, in particular (Reichardt et al. 2014). Finally, *Eubacterium limosum* converts lactate to butyrate, with net production of acetate (Pham et al. 2019).

In contrast to the widely reported beneficial effects of SCFAs, lactate accumulation is hazardous, as it can cause acidosis and neurological problems (Uribarri et al. 1998). In the healthy colon, lactate concentrations depend on the rate of its production, microbial utilisation, and host absorption. Lactate acidosis occurs rarely as a result of carbohydrate malabsorption, the presence of LAB, and the inability of the host and gut microbiota to metabolise the produced lactate. All these conditions occur in individuals with short bowel syndrome. Those individuals had a substantial small bowel resection, which causes severe nutrient malabsorption. Interestingly, analysis of faecal microbiota of short bowel syndrome patients showed significantly enriched *Lactobacillus* and *Leuconostoc* populations that constituted almost half of the ecosystem, while *Streptococcus* abundance was also increased, although not to such a dramatic extent (Mayeur et al. 2013). Despite the pronounced dysbiosis in all patients, only about half of them accumulated lactate in the gut.

It is interesting to note that in short-bowel syndrome patients, nutrient malabsorption causes a higher concentration of nutrients, including simple sugars in the colon. This nutrient shift is probably the major driver of the microbiota changes, as evident through the increase of LAB abundance. This further suggests that substrate availability is a major determinant that stimulates LAB in the upper gut. LAB are saccharolytic microorganisms, which can use several carbohydrates, while their competitive advantage is their fast growth and the ability to use a great number of substrates for growth and energy (Stolaki et al. 2012; Rajilić-Stojanović 2013).

In addition to the contribution to the cross-feeding between the members of the ecosystem, LAB+B influence the composition of gut microbiota by production of various antimicrobial compounds. These can give them a competitive advantage that prevents their elimination from the ecosystem by other bacteria and provides protective support for hosts against pathogenic invaders. The major antagonistic compounds are lactic and acetic acids, diacetyl, and bacteriocins (Castellano et al. 2017). Some other metabolites of LAB can influence quorum-sensing systems and therefore have an impact on gene regulation, adhesion, and biofilm formation ability of pathogens (Chiu et al. 2017).

## 22.3.3 Association of LAB+B with Health Conditions

LAB+B are known for their health-promoting effects, as they positively influence gut microbiota diversity, modulate immune response and intestinal permeability, and produce bioactive or regulatory metabolites (Hemarajata and Versalovic 2013). However, despite a general view that LAB+B have a beneficial effect on humans, a whole range of different associations of the abundance of these bacteria in the gut microbiota have been detected in various diseases. The list of diseases that are characterised by microbiota dysbiosis is constantly expanding, and some of the associations between intestinal, immunological, but also gut–brain axis disorders and LAB+B abundance was presented earlier (Rajilić-Stojanović et al. 2019). In the following sections, we aim to summarise data on associations between LAB+B and various diseases, decipher the role of gut LAB+B in the development and prevention of disease, and investigate their potential for application as therapeutics.

## 22.3.3.1 LAB+B in Immune-Mediated Diseases

Bifidobacteria are critical early colonisers of the human gut, and their depletion in infants is correlated with autoimmune, bowel-related diseases and allergies (Penders et al. 2007). However, the role of the gut microbiota, which causes or influences systemic immunity in autoimmune diseases, remains elusive. For instance, a lower abundance of *Bifidobacterium* spp. was found in neonates with eczema than in healthy infants, while the opposite trend was observed for *Enterococcus* sp., found more abundant in infants with eczema than in healthy infants (Kalliomäki et al. 2001). Similar results were found in rheumatoid arthritis, a systemic, immune-mediated chronic inflammatory disease (Vaahtovuo et al. 2008). Interestingly, the abundance of *Ligilactobacillus salivarius* was positively correlated with an increase in disease severity (Zhang et al. 2015), pointing to the controversial role, which seems to be strain specific to some LAB members. In a recent study of autoimmune hepatitis, a disease-specific decline of bifidobacteria coupled with a significant increase of streptococci and lactobacilli was detected (Liwinski et al. 2020). Again pointing towards the importance of strain, although lactobacilli are increased in some immune-mediated diseases, probiotic treatments with LAB+B strains, particularly those producing bioactive molecules with immune-modulating properties, exhibited positive effects on patients (Marighela et al. 2019; Mohammed and Elmakhzangy 2017; Sun et al. 2022).

## 22.3.3.2 LAB+B in Intestinal Diseases

A link between changes in microbiota composition and intestinal diseases has been frequently reported. In a recent systematic review, several LAB+B were reported as reproducably detected markers of disbiosis in both inflammatory bowel diseases (IBD)—Crohn's disease or ulcerative colitis (Abdel-Rahman and Morgan 2023). In this systematic meta-analysis, all the abundance of different LAB+B was, surprisingly, increased in active IBD. Although individual studies previously reported the abundance of bifidobacteria, lactobacilli, and streptococci to be decreased in IBD patients (Andoh et al. 2012; Li et al. 2012), the consensus of different studies showed that lactobacilli are increased in both IBDs, while streptococci and bifidobacteria are increased in one of the diseases—Crohn's and ulcerative colitis, respectively. Another LAB—genus *Enterococcus*—was the most consistently reported marker of IBD dysbiosis, which is increased during the active phase of the disease. The relatively high number of these typical small intestinal bacteria could be attributed to their relatively high tolerance for bile salt, which could be present in higher concentration in the large bowel in the active phase of the disease (Abdel-Rahman et al. 2023). Another factor that could boost LAB in IBD could be related to the fact that these bacteria are facultative anaerobes and thereby have better chances to survive treatment with metronidazole, which typically targets anaerobic bacteria. It could be speculated that despite an increased abundance, bifidobacteria and lactobacilli do not contribute to the pathogenesis of IBD since they boost the mucosal defence system (Lukic et al. 2013). Importantly, both bifidobacteria and lactobacilli are able to metabolise tryptophan to produce various arhyl hydrocarbon receptors that induce a beneficial cascade of the immune response (Lamas et al. 2016; Cui et al. 2022).

The number of bifidobacteria and lactobacilli is also decreased in celiac disease (Akobeng et al. 2020). Celiac disease (CD) is a chronic inflammatory condition triggered by the ingestion of gluten (De Palma et al. 2010). Recent studies revealed that CD development could be dependent on breast-feeding and that the presence of *Lacticaseibacillus rhamnosus*, *Lactobacillus gasseri*, *Lactococcus lactis*, *Leuconostoc mesenteroides*, and bifidobacteria in breast milk is crucial to prevent the disease (Parigi et al. 2015; Cenit et al. 2015). The positive effects of bifidobacteria and lactobacilli on CD prognosis could be linked to the ability of LAB to degrade gluten peptides and decrease their immunogenicity and to the regulation of the intestinal permeability of epithelial intestinal cells exposed to gliadin (Caminero et al. 2016).

The role of LAB+B in colorectal cancer (CRC) is not conclusive. In general, *Bifidobacterium* sp. are depleted in CRC, while their application as probiotics has an impact on both the efficacy of therapy and normalisation of the immune response in these patients (Saus et al. 2019). While lactobacilli are also depleted in CRC, *Streptococcus* and *Enterococcus* phylotypes were found to be increased in faecal samples of CRC patients by molecular methods (e.g. Wang et al. 2012). Contrasting these results, analysis of the microbiota in mucosal samples of Chinese CRC patients showed an enriched *Lactococcus* population, whereas *Enterococcus* were depleted (Lu et al. 2016). In contrast to these conflicting results, there is reasonable agreement that *Streptococcus gallolyticus* (previously *S. bovis*) has a deleterious effect on health and contributes to CRC development, and it was for the first time proposed in Klein et al. (1977). While the most prominent microbial marker of CRC is an increase of *Fusobacterium nucleatum*, depletion of both bifidobacteria and lactobacilli could also be relevant for this pathology due to a range of anti-carcinogenic properties including the production of anti-oxidant enzymes and interaction with proteins regulating the cell cycle (Nowak et al. 2019).

### 22.3.3.3 LAB+B in Metabolic Diseases

The growing body of evidence indicates that disturbance of gut microbiota leads to metabolic diseases, such as obesity and diabetes. For instance, in patients with type 2 diabetes (T2D), increased LAB abundance was scored (Larsen et al. 2010; Tilg and Moschen 2014), while a completely opposite trend was observed in type 1 diabetes (T1D) (Murri et al. 2013). Interestingly, the abundance of *Bifidobacterium* and Lactobacillaceae in gut microbiota correlates with serum levels of appetite-regulating hormones, leptin (positive correlation) and ghrelin (negative correlation) (Queipo-Ortuño et al. 2013). Regarding the possible therapeutic effects of LAB, Mihailović et al. (2017) showed that *Lactiplantibacillus paraplantarum* BGCG11 is able to balance the redox homeostasis and reduce the inflammatory response in streptozotocin-induced diabetes in rats, indicating the protective role of *Lactobacilaceae* in this metabolic disorder. On the other hand, another recent study showed that *Lactobacillus fermentum* BGHV110 extends the lifespan of *Caenorhabditis elegans* and improves age-related physiological features, including locomotor function and lipid metabolism (Dinic et al. 2021). This result suggests that autophagy mediated by some LAB strains could be important for the longevity and fitness of the host

## 22.3.4 LAB+B in Gut–Brain–Microbiota Axis

The enteric nervous system (ENS) represents a network of 200–600 million neurons that can be only found around the small and large intestines. The main functions of the ENS are the regulation of intestinal peristalsis, transmucosal fluid flux, local blood flow, the release of intestinal hormones, nutrient absorption, and interactions with the immune system. Intestinal bacteria affect the ENS and modulate intestinal homeostasis by secretion of several molecules that affect neuroendocrine and metabolic pathways communicating with the central nervous system (Carabotti et al. 2015). LAB+B play an important role in this communication, since certain strains of lactobacilli and bifidobacteria produce gamma-aminobutyric acid (GABA) in the intestine (Sokovic et al. 2019). Other bacterial neuroactive molecules include tryptamine, indole-3-propionicacid, serotonin, and SCFAs (Wlodarska et al. 2015).

The gut–brain axis is central for a number of disorders, including irritable bowel syndrome. IBS is a functional intestinal disorder associated with low-grade mucosal inflammation, compromised mucosal barrier function, changed visceral sensitivity, gut–brain axis dysregulation, and microbiota dysbiosis (Enck et al. 2016). Microbiota dysbiosis in IBS is mostly characterised by a significant decrease of bifidobacteria, particularly *B. pseudocatenulatum* (Distrutti et al. 2016; Kassinen et al. 2007; Rajilić-Stojanović et al. 2011). The possible mechanism of the positive effect of bifidobacteria on IBS could be the production of active serine proteinase inhibitors belonging to the SERPIN family that might inactivate IBS-stimulating proteases.

Some studies have also found a reduced abundance of Lactobacilaceae in IBS, although it should be noted that in some cohorts, increased levels of LAB, in particular members of lactobacilli and streptococci, were found in the diarrhoea of predominant IBS patients (Salonen et al. 2010). While this increase of LAB in relation to diarrhoea might only reflect the shift of the community and representation of the small intestinal microbiota, the cases where lactobacilli are decreased are more likely relevant for IBS pathology.

LAB+B could contribute to IBS prevention and treatment by GABA production (Dinan et al. 2015) or through effects on enteric nerves, modulation of vagal afferents, and antinociceptive effect, as shown for *Limosilactobacillus reuteri*. Furthermore, exopolysaccharides produced by LAB also can have antinociceptive and anti-inflammatory activity (Dinic et al. 2018). Interestingly, the direct contact of *Lactobacillus acidophilus* with epithelial cells induces the expression of opioid and cannabinoid receptors in the gut and restores the normal perception of visceral pain (Distrutti et al. 2016). Moreover, lactobacilli improve epithelial barrier function and thereby have systemic effects, and finally, the ability of LAB+B to modulate innate and adaptive immunity also contributes to the prevention of IBS (Bhattarai et al. 2017).

It was found that autism spectrum disorder (ASD), a developmental disorder recognised by social and communication impairments and rigid and repetitive patterns of behaviour and interests, is associated with microbiota dysbiosis (De Angelis et al. 2013; Bojović et al. 2020). Decreased levels of common commensal bacteria enterococci, streptococci, *Lacticaseibacillus rhamnosus*, and bifidobacteria were found in the faecal samples of ASD children compared to healthy controls (De Angelis et al. 2013; Bojović et al. 2020). Interestingly, the levels of SCFAs in ASD children are significantly lower compared to healthy controls, particularly butyrate, which is extremely important for the reinforcement of the mucosal barrier, the regulation of trans-epithelial transport, and modulation of inflammatory and oxidative states of the intestinal mucosa. Bifidobacteria were found to be negatively correlated with propionate and positively with butyrate, while probiotic treatment with *Bifidobacterium* was correlated with decreased anxiety and reduced depressive behaviour (Bercik et al. 2012). There has been an interesting report that treatment with *Lacticaseibacillus rhamnosus* GG in the first six months of life significantly reduced the incidence of attention deficit hyperactivity disorder and Asperger's syndrome when the treated children reached 13 years of age (Pärtty et al. 2015). In multiple sclerosis, the immune-mediated, inflammatory neurodegenerative disease, treatment with several *Bifidobacterium*, Lactobacillaceae, *Lactococcus*, and *Streptococcus* species was able to alleviate disease symptoms (Mielcarz and Kasper 2015). Microbiota analyses in an experimental autoimmune encephalomyelitis (EAE) animal model of multiple sclerosis revealed higher diversity of Lactobacillaceae in genetically resistant rats. Particularly, *Lactobacillus helveticus* and *Ligilactobacillus murinus/animalis* were detected only in EAE-resistant rats (Stanisavljević et al. 2016), pointing to the huge potential of gut microbiota in the context of immune-modulation–based therapies. Recent results even revealed that the efficacy of the myeloid-derived suppressor cells-based therapy of multiple sclerosis (demonstrated in the EAE animal model) is highly dependent on gut microbiota composition (Radojevic et al. 2022).

# 22.4 FACTORS INFLUENCING PRESENCE AND ABUNDANCE OF LAB+B IN GUT MICROBIOTA

LAB+B are considered highly beneficial for humans. They have received considerable scientific attention mainly because of the health claims proposed by Metchnikoff (1908) and their later application as probiotics. Several strategies mainly based on the ingestion of probiotics and prebiotics have been developed in order to elevate the abundance of LAB+B in the gut. In addition to such interventions, several other factors influence the composition of the gut microbiota and the presence and abundance of LAB+B (Figure 22.5).

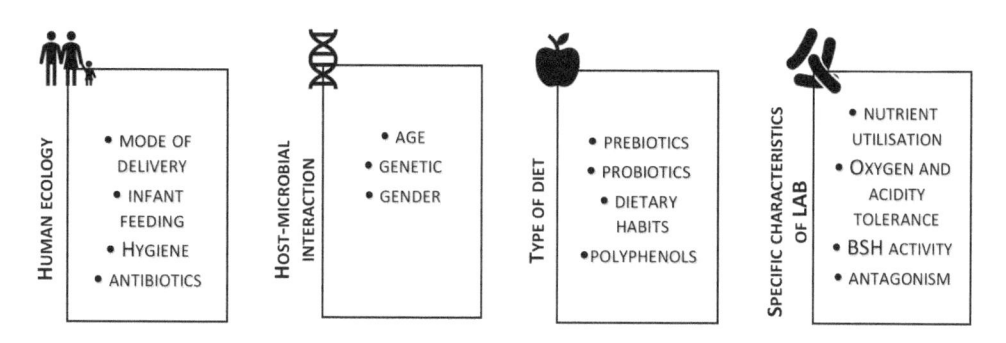

**FIGURE 22.5**   Schematic representation of factors influencing presence and abundance of gut LAB+B.

Since specific characteristics of LAB+B that facilitate their establishment and survival in the gut have already been discussed, other contributing factors will be elucidated in more detail in the next section.

# 22.4.1 Environmental Role in Acquiring and Regulating LAB+B in Gut Microbiota

LAB+B are one of the first human gut colonisers, and recent studies have shown that humans might be exposed to them even in the womb (Pelzer et al. 2017). There are several methods for infant inoculation, and, interestingly, an *E. faecium* strain orally administered to the mothers during pregnancy was detected in meconium samples of mice offspring delivered by caesarean section (Jiménez et al. 2008). This experiment showed that some species can be transmitted in utero from mother to child. Additional inoculation occurs during the delivery, and it plays an important role in shaping the ecosystem in the gut. Even the composition of the vaginal microbiota influences the microbiota of infants. In particular, the presence of group B streptococci in the mother's vagina increases the abundance of enterococci in the faeces of infants, and this effect is evident at least until six months of age (Cassidy-Bushrow et al. 2016). Furthermore, it has been shown that the gut microbiota significantly differs between infants delivered vaginally and via caesarean section. In that respect, the gut microbiota of vaginally delivered newborns is reflective of the vaginal microbiota and has a higher abundance of lactobacilli than the microbiota of newborns delivered by caesarean section, who have more skin-associated microbes (Dominguez-Bello et al. 2010). Epidemiological studies have shown that mode of delivery has a strong impact on an individual's subsequent health (Sevelsted et al. 2015). The increased risks for the development of a range of (auto-)immune diseases in caesarean section–delivered individuals, despite the unknown mechanism behind it, suggest a role of microbiota in the process. Therefore, procedures for inoculation of caesarean section–delivered infants by vaginal microbes have been tested in a pilot setting (Dominguez-Bello et al. 2016). Following the idea that during birth an infant is exposed to the maternal faecal microbes, a study tested the effect of maternal faecal microbiota transplantation on infants born by caesarean section (Korpela et al. 2020). While the long-term impact of such procedures has yet to be determined, it is clear that exposure to vaginal and faecal microbes had an impact on gut microbiota. In the case of inoculation with vaginal microbiota, it is interesting that although LAB predominate in vaginal microbiota, the number of LAB in the infants' gut was not affected.

After birth, the newborn baby becomes exposed to the external conditions that dictate the development of the ecosystem in the gut. Although the predominant first coloniser of newborns' gut is *Escherichia coli*, when an infant is breastfed, the microbiota rapidly becomes dominated by bifidobacteria and remains relatively simple until the introduction of solid foods (Bäckhed et al. 2015). First infant feeding is the most important factor that influences gut microbiota composition and has an impact on health later in life. Mother's milk contains a large proportion of carbohydrates. Lactose, which represents about 54% of the total dry matter, is followed by over 100 indigestible oligosaccharides that contribute to the dry matter of milk by 8% (Kunz et al. 2000). These carbohydrates regulate the colonisation of the infant's gut together with milks' microbiota, which is composed of mainly LAB+B present in a density of $10^5–10^7$ per ml (Martín et al. 2009). Since there are differences in the oligosaccharides composition of mothers' milk, type of mother's milk also influences the gut microbiota (Coppa et al. 2011). This impressive natural regulation of infant gut microbiota indirectly suggests the importance of the ecosystem for systemic health. Interestingly, the imprint of the infant feeding type is lifelong (Ding and Schloss 2014). Preserving the relatively simple and LAB+B-dominant community in the first weeks of life seems to be of tremendous importance for health, and it has been shown that a lower abundance of LAB+B is associated with colic (de Weerth et al. 2013). Furthermore, an early diversification of the microbiota, evident through the presence of various Clostridiales that are typical for adult microbiota, is associated with atopic eczema development (Nylund et al. 2013). In contrast to mothers' milk, cow milk, which serves as the basis for the preparation of the majority of infant feeding formulas, contains a similar percentage of lactose (~48%), while oligosaccharides are present in very low amounts (~0.7%). Due to the lower availability of indigestible carbohydrates, LAB+B are not stimulated as much as in breast-fed infants, and as a consequence, the microbiota is much more diverse and characterised by a lower abundance of LAB+B (Bezirtzoglou et al. 2011). Consequently, the level of lactate in the faeces of formula-fed infants is, significantly, almost fivefold, lower than in breast-fed infants (Ogawa et al. 1992).

In addition to the delivery and the feeding modes, other environmental factors actively shape the gut ecosystem. Among these, hospitalisation and medication, in particular antibiotic use, will have a profound impact on the microbiota. Another important factor is the presence of siblings, and it has been shown that infants with older siblings have significantly more gut bifidobacteria (Penders et al. 2006). The beneficial effect of siblings expands on the so-called hygiene hypothesis. Today's increased use of hygiene practices leads to a reduction in people's contact with microorganisms. Human health and longevity are improved, especially in developed countries, but new allergic and metabolic diseases have arisen without obvious explanation. The "disappearing microbiota" hypothesis postulates that changes in human ecology result in the loss of our ancestral microbiota (Blaser 2006). Many factors had negative impact on gut microbiota diversity, including increased hygiene, immunisation against pathogens, widespread usage of antibiotics, increase in caesarean sections, and reduced breastfeeding (Blaser and Falkow 2009). Finally, smaller family size, one of the new lifestyle trends, lessens the intrafamilial transmission of microbes (Blaser 2006).

## 22.4.2 Host Factors That Influence Gut LAB+B

Considering that every person has a specific genetic makeup, it is very likely that this will influence the gut microbiota (Goodrich et al. 2014). In that line, an increasing degree of gut microbiota similarity was observed with an increase in the level of relatedness (Zoetendal et al. 2001). More studies that followed showed that monozygotic twins have more similar microbiota than any other pair of individuals, indicating the importance of host genetics. In the work of Goodrich et al. (2016) on several UK twins, a correlation between the number of bifidobacteria and the presence of the lactase gene locus in their genome was established. Due to the presence of lactase, individuals hydrolyse lactose. They harbour a lower level of bifidobacteria since lactose is utilised by the host, and thereby there is reduced availability of the substrate for bifidobacteria. Similarly, the lower activity of the sucrase-isomaltase gene influences the absorption of sucrose and impacts gut microbiota (Henström et al. 2016). The influence of host metabolism on the microbial composition is also reflected in the host's ability to produce certain components of the mucus, especially glycoconjugates, which support or prevent the growth of different microbial species in the gut (Nicholson et al. 2012). The host genetics also shape the microbiota through diet preference, which itself is heritable. The influence of gender on the composition of gut microbiota can also be associated with genetic determinants (Santoro et al. 2018).

# 22.4.3 Dietary Components That Influence Gut LAB+B

The primary role of the human gut microbiota is its contribution to digestion. Typically, the microbiota will extract energy from indigestible dietary components. Therefore, diet is one of the major drivers that shape microbiota (Claesson et al. 2012). Since the type of diet is closely related to geographical origin, nationality, cultural habits, and so on, the influence of these factors cannot be ruled out when assessing the impact of food on gut microbiota.

Various intervention diets have a major effect on the abundance of microbiota constituents. For example, a gluten-free diet that is inevitable in CD induces, among other changes, a reduction of lactobacilli and bifidobacteria in gut microbiota (De Palma et al. 2009). This reduction of the LAB+B could be linked with their ability to degrade gluten but also a consequence of a lower polysaccharide intake that is a side effect of a gluten-free diet. Several studies have shown that resistant starch stimulates bifidobacteria (Kovatcheva-Datchary et al. 2009; Martínez et al. 2010), suggesting that the removal of this dietary component would reduce the abundance of a particular bacterial group. Furthermore, one of the intervention diets that can reduce symptoms of IBS and IBD—the diet low in fermentable oligo-, di-, monosaccharides, and polyols (FODMAP)—has an undesired effect on gut microbiota, as it reduces the number of beneficial gut bacteria including *Bifidobacterium* spp. (Halmos et al. 2014). Therefore, it has been suggested that such intervention diets should be applied with caution, as their application might cause microbiota-mediated negative effects on systemic health. Weight loss diets also can have a major impact on the gut microbiota. In animal models, high-fat feeding strongly increases intestinal permeability that is associated with endotoxemia-induced inflammation and dramatic reduction of LAB+B in the intestinal lumen (Cani et al. 2008). Proving the same correlation, reduced calorie intake in humans has been accompanied by structural modulation of the gut microbiota and enrichment of phylotypes that are positively correlated with longevity, including the genus *Lactobacillus* (Zhang et al. 2013). In an obesity treatment program consisting of a calorie-restricted diet and increased physical activity, the microbiota of treated adolescents changed in response to the treatment, which affected the LAB+B population in the gut in a divergent way: *Lactobacillus* spp. increased, while *Bifidobacterium* spp. decreased (Santacruz et al. 2009).

Specific dietary components can also stimulate the abundance of gut LAB+B. Several types of health-promoting dietary polyphenols can boost the abundance of LAB+B in human trials, *in vitro* tests and animal models (Cardona et al. 2013). This is to some extent linked to the ability of these bacteria to deglycosylate complex polyphenols. Furthermore, the deliberate addition of specific dietary components is a strategy to increase the abundance of beneficial gut microbiota members. LAB+B are certainly one of the most often targeted microbial groups that can be stimulated by the use of probiotics and prebiotics, or their combination—synbiotics. They can be introduced through specific probiotic-containing fermented foods such as yogurt (where the food matrix can facilitate their survival through the harsh gut condition) or in the form of supplements (where they are usually protected by a carrier or a coating). The majority of probiotics belong to the LAB+B group, and they are transient or allochthonous gut bacteria that can modulate the gut microenvironment in favour of desirable microbiota shifts. These microbiota-targeted interventions improve health status, especially in elderly populations (Santoro et al. 2018; Valentini et al. 2015) and in cases of antibiotic-caused dysbiosis (Yoon and Yoon 2018). Prebiotics were originally defined as "non-digestible food ingredient(s) that beneficially affect host health by selectively stimulating the growth and/or activity of one or a limited number of bacteria in the colon" (Gibson and Roberfroid 1995). The main established prebiotics are fructans (fructooligosaccharides [FOS] and inulin) and galactans (galactooligosaccharides or GOS) that have a well-documented ability to promote the growth of mainly *Bifidobacterium* spp., and to a lesser extend lactobacilli (Huebner et al. 2007). While LAB+B remain the main target for the vast majority of current prebiotics, recently the definition of prebiotics has been broadened to "a substrate that is selectively utilised by host microorganisms conferring a health benefit" (Gibson et al. 2017). In the case of a synbiotic, the effects of the two components—probiotics and prebiotics—should be synergistic (Swanson et al. 2020). The probiotic may be stimulated to grow in the gut by fermenting the prebiotic, and/or the prebiotic may support a more favourable gut environment in which the probiotic may better compete (Binns 2013).

# 22.5 CONCLUSIONS

In the complex ecosystem of gut microbiota, the LAB+B represent a prevalent, abundant, and highly important subpopulation. They are one of the first colonisers of the human digestive tract that persist in this dynamic ecosystem throughout the entire life span. These bacteria are equipped with various specific functions that enable their survival and indicate an evolutionarily established symbiotic link between humans and the LAB+B. Some mechanisms behind recognised beneficial health effects of these bacteria have already been elucidated, while novel mechanisms are continuously discovered. After more than a century of studying, gut LAB+B remain a not fully described bacterial group that attracts enormous scientific attention. Molecular methods–based insight into gut microbiota diversity showed the presence of its disturbed composition—dysbiosis—in various diseases. While the etiological role of dysbiosis has not yet been determined, it is remarkable that a disturbed abundance of LAB+B is a marker of dysbiosis of several digestive, neural, metabolic, and immune-mediated disorders. This opens new venues for probiotic-based treatments, especially since specific LAB+B strains affect the pathophysiological pathways of these diseases. Although it is clear that the unexplored complexity of the human gut microbiota hides other extremely important microbial companions, based on which new generation probiotics will be developed, the LAB+B persist in being one of the most valuable gut microbial components. Therefore, it is important to understand that several factors, including mode of delivery, infant feeding, family size, hygiene, diet, and medication, can affect the abundance of these bacteria and the host's health. By enabling favourable conditions, complemented with strategies that can stimulate their presence and abundance, humans could ensure a longer and healthier life thanks to their gut LAB+B activity.

# BIBLIOGRAPHY

Abdel-Rahman L. I. H. and X. C. Morgan. 2023. Searching for a consensus among inflammatory bowel disease studies: A systematic meta-analysis. *Inflamm Bowel Dis* 29:125–139.

Agans R., L. Rigsbee, H. Kenche, S. Michail, H. J. Khamis and O. Paliy. 2011. Distal gut microbiota of adolescent children is different from that of adults. *FEMS Microbiol Ecol* 77:404–412.

Akobeng A. K., P. Singh, M. Kumar and S. Al Khodor. 2020. Role of the gut microbiota in the pathogenesis of coeliac disease and potential therapeutic implications. *Eur J Nutr* 59:3369–339.

Andoh A., H. Kuzuoka, T. Tsujikawa, S. Nakamura, F. Hirai, Y. Suzuki, T. Matsui, Y. Fujiyama and T. Matsumoto. 2012. Multicenter analysis of fecal microbiota profiles in Japanese patients with Crohn's disease. *J Gastroenterol* 47:1298–1307.

Bačić A. and M. Rajilić-Stojanović. 2022. Microbiota changes throughout life—An overview. In: *Comprehensive Gut Microbiota*. Ed. Glibetic M. Elsevier, Oxford, pp. 1–12.

Bäckhed F., J. Roswall, Y. Peng, Q. Feng, H. Jia, P. Kovatcheva-Datchary, Y. Li, Y. Xia, H. Xie and H. Zhong. 2015. Dynamics and stabilization of the human gut microbiome during the first year of life. *Cell Host Microbe* 17:690–703.

Belzer C. 2022. Nutritional strategies for mucosal health: The interplay between microbes and mucin glycans. *Trends Microbiol* 30:13–21.

Bercik P., S. Collins and E. Verdu. 2012. Microbes and the gut-brain axis. *Neurogastroenterol Motil* 24:405–413.

Bezirtzoglou E., A. Tsiotsias and G. W. Welling. 2011. Microbiota profile in feces of breast-and formula-fed newborns by using fluorescence in situ hybridization (FISH). *Anaerobe* 17:478–482.

Bhattarai Y., D. A. M. Pedrogo and P. C. Kashyap. 2017. Irritable bowel syndrome: A gut microbiota-related disorder? *Am J Physiol Gastrointest Liver Physiol* 312:G52–G62.

Binns N. 2013. Probiotics, Prebiotics and the Gut Microbiota. ILSI Europe, Brussels, Belgium.

Blaser M. J. 2006. Who are we? Indigenous microbes and the ecology of human diseases. *EMBO Rep* 7:956–960.

Blaser M. J. and S. Falkow. 2009. What are the consequences of the disappearing human microbiota? *Nat Rev Microbiol* 7:887.

Bojović, K., D. Ignjatović, S. Soković Bajić, D. Vojnović Milutinović, M. Tomić, N. Golić and M. Tolinački. 2020. Gut microbiota dysbiosis associated with altered production of short chain fatty acids in children with neurodevelopmental disorders. *Front Cell Infect Microbiol 10*:223.

Booijink C. C., S. El-Aidy, M. Rajilić-Stojanović, H. G. Heilig, F. J. Troost, H. Smidt, M. Kleerebezem, W. M. De Vos and E. G. Zoetendal. 2010. High temporal and inter-individual variation detected in the human ileal microbiota. *Environ Microbiol 12*:3213–3227.

Caminero A., H. J. Galipeau, J. L. McCarville, C. W. Johnston, S. P. Bernier, A. K. Russell, J. Jury, A. R. Herran, J. Casqueiro and J. A. Tye-Din. 2016. Duodenal bacteria from patients with celiac disease and healthy subjects distinctly affect gluten breakdown and immunogenicity. *Gastroenterol 151*:670–683.

Cani P. D., R. Bibiloni, C. Knauf, A. Waget, A. M. Neyrinck, N. M. Delzenne and R. Burcelin. 2008. Changes in gut microbiota control metabolic endotoxemia-induced inflammation in high-fat diet—induced obesity and diabetes in mice. *Diabetes 57*:1470–1481.

Carabotti M., A. Scirocco, M. A. Maselli and C. Severi. 2015. The gut-brain axis: Interactions between enteric microbiota, central and enteric nervous systems. *Ann Gastroenterol 28*:203.

Cardona F., C. Andrés-Lacueva, S. Tulipani, F. J. Tinahones and M. I. Queipo-Ortuño. 2013. Benefits of polyphenols on gut microbiota and implications in human health. *J Nutr Biochem 24*:1415–1422.

Cassidy-Bushrow A. E., A. Sitarik, A. M. Levin, S. V. Lynch, S. Havstad, D. R. Ownby, C. C. Johnson and G. Wegienka. 2016. Maternal group B *Streptococcus* and the infant gut microbiota. *J Dev Origins Health Disease 7*:45–53.

Castellano P., M. Pérez Ibarreche, M. Blanco Massani, C. Fontana and G. M. Vignolo. 2017. Strategies for pathogen biocontrol using lactic acid bacteria and their metabolites: A focus on meat ecosystems and industrial environments. *Microorganisms 5*:38.

Cenit M. C., M. Olivares, P. Codoñer-Franch and Y. Sanz. 2015. Intestinal microbiota and celiac disease: Cause, consequence or co-evolution? *Nutrients 7*:6900–6923.

Chiu L., T. Bazin, M.-E. Truchetet, T. Schaeverbeke, L. Delhaes and T. Pradeu. 2017. Protective microbiota: From localized to long-reaching co-immunity. *Front Immunol 8*:1678.

Claesson M. J., I. B. Jeffery, S. Conde, S. E. Power, E. M. O'Connor, S. Cusack, H. M. Harris, M. Coakley, B. Lakshminarayanan and O. O'Sullivan. 2012. Gut microbiota composition correlates with diet and health in the elderly. *Nature 488*:178.

Collins M. D. and S. Wallbanks. 1992. Comparative sequence analyses of the 16S rRNA genes of *Lactobacillus minutus*, *Lactobacillus rimae* and *Streptococcus parvulus*: Proposal for the creation of a new genus *Atopobium*. *FEMS Microbiol Lett 95*:235–240.

Coppa G. V., O. Gabrielli, L. Zampini, T. Galeazzi, A. Ficcadenti, L. Padella, L. Santoro, S. Soldi, A. Carlucci and E. Bertino. 2011. Oligosaccharides in 4 different milk groups, Bifidobacteria, and *Ruminococcus obeum*. *J Pediatr Gastroenterol Nutr 53*:80–87.

Cui Q.-Y., T. X.-Y. Tian, X. Liang, Z. Zhang, R. Wang, Y. Zhou, H.-X. Yi, P.-M. Gong, K. Lin, T.-J. Liu and L.-W. Zhang. 2022. *Bifidobacterium bifidum* relieved DSS-induced colitis in mice potentially by activating the aryl hydrocarbon receptor. *Food Funct 13*:5115–5123.

Cummings J. H. 1981. Short chain fatty acids in the human colon. *Gut 22*:763.

Cummings J. H. and G. Macfarlane. 1991. The control and consequences of bacterial fermentation in the human colon. *J Appl Microbiol 70*:443–459.

David L. A., C. F. Maurice, R. N. Carmody, D. B. Gootenberg, J. E. Button, B. E. Wolfe, A. V. Ling, A. S. Devlin, Y. Varma and M. A. Fischbach. 2014. Diet rapidly and reproducibly alters the human gut microbiome. *Nature 505*:559.

De Angelis M. and M. Gobbetti. 2004. Environmental stress responses in *Lactobacillus*: A review. *Proteomics 4*:106–122.

De Angelis M., M. Piccolo, L. Vannini, S. Siragusa, A. De Giacomo, D. I. Serrazzanetti, F. Cristofori, M. E. Guerzoni, M. Gobbetti and R. Francavilla. 2013. Fecal microbiota and metabolome of children with autism and pervasive developmental disorder not otherwise specified. *PLoS ONE 8*:e76993.

De Palma G., I. Nadal, M. C. Collado and Y. Sanz. 2009. Effects of a gluten-free diet on gut microbiota and immune function in healthy adult human subjects. *Br J Nutr 102*:1154–1160.

De Palma G., I. Nadal, M. Medina, E. Donat, C. Ribes-Koninckx, M. Calabuig and Y. Sanz. 2010. Intestinal dysbiosis and reduced immunoglobulin-coated bacteria associated with coeliac disease in children. *BMC Microbiol 10*:63.

de Weerth C., S. Fuentes, P. Puylaert and W. M. de Vos. 2013. Intestinal microbiota of infants with colic: Development and specific signatures. *Pediatrics 131*:e550–e558.

Dicksved, J., M. Lindberg, M. Rosenquist, H. Enroth, J. K. Jansson and L. Engstrand. 2009. Molecular characterization of the stomach microbiota in patients with gastric cancer and in controls. *J Med Microbiol 58*:509–516.

Dinan T. G., R. M. Stilling, C. Stanton and J. F. Cryan. 2015. Collective unconscious: How gut microbes shape human behavior. *J Psychiatr Res 63*:1–9.

Ding T. and P. D. Schloss. 2014. Dynamics and associations of microbial community types across the human body. *Nature 509*:357.

Dinić M., M. Herholz, U. Kačarević, D. Radojević, K. Novović, J. Đokić, A. Trifunović and N. Golić. 2021. Host-commensal interaction promotes health and lifespan in *Caenorhabditis elegans* through the activation of HLH-30/TFEB-mediated autophagy. *Aging 13*:8040–8054.

Dinić M., U. Pecikoza, J. Djokic, R. Stepanovic-Petrovic, M. Milenkovic, M. Stevanovic, N. Filipovic, J. Begovic, N. Golic and J. Lukic. 2018. Exopolysaccharide produced by probiotic strain *Lactobacillus paraplantarum* BGCG11 reduces inflammatory hyperalgesia in rats. *Front Pharmacol 9*:1.

Distrutti E., L. Monaldi, P. Ricci and S. Fiorucci. 2016. Gut microbiota role in irritable bowel syndrome: New therapeutic strategies. *World J Gastroenterol 22*:2219.

Dominguez-Bello M. G., E. K. Costello, M. Contreras, M. Magris, G. Hidalgo, N. Fierer and R. Knight 2010. Delivery mode shapes the acquisition and structure of the initial microbiota across multiple body habitats in newborns. *Proc Natl Acad Sci USA 107*:11971–11975.

Dominguez-Bello M. G., K. M. De Jesus-Laboy, N. Shen, L. M. Cox, A. Amir, A. Gonzalez, N. A. Bokulich, S. J. Song, M. Hoashi and J. I. Rivera-Vinas. 2016. Partial restoration of the microbiota of cesarean-born infants via vaginal microbial transfer. *Nat Med 22*:250.

Enck P., Q. Aziz, G. Barbara, A. D. Farmer, S. Fukudo, E. A. Mayer, B. Niesler, E. M. M. Quigley, M. Rajilić-Stojanović, M. Schemann, J. Schwille-Kiuntke, M. Simren, S. Zipfel and R. C. Spiller. 2016. Irritable bowel syndrome. *Nat Rev Dis Primers 2*:16014.

Falony G., M. Joossens, S. Vieira-Silva, J. Wang, Y. Darzi, K. Faust, A. Kurilshikov, M. J. Bonder, M. Valles-Colomer and D. Vandeputte. 2016. Population-level analysis of gut microbiome variation. *Science 352*:560–564.

Finegold S. M., H. R. Attebery and V. L. Sutter. 1974. Effect of diet on human fecal flora: Comparison of Japanese and American diets1'2. *Am J Clin Nutr 27*:1456–1469.

Flint H. J., S. H. Duncan, K. P. Scott and P. Louis. 2015. Links between diet, gut microbiota composition and gut metabolism. *Proc Nutr Soc 74*:13–22.

Franz C. M., M. Huch, H. Abriouel, W. Holzapfel and A. Gálvez. 2011. Enterococci as probiotics and their implications in food safety. *Int J Food Microbiol 151*:125–140.

Gargari, G., V. Taverniti, S. Balzaretti, C. Ferrario, C. Gardana, P. Simonetti and S. Guglielmetti. 2016. Consumption of a *Bifidobacterium bifidum* strain for 4 weeks modulates dominant intestinal bacterial taxa and fecal butyrate in healthy adults. *Appl Environ Microbiol 82*:5850–5859.

Gibson G. R., R. Hutkins, M. E. Sanders, S. L. Prescott, R. A. Reimer, S. J. Salminen, K. Scott, C. Stanton, K. S. Swanson and P. D. Cani. 2017. Expert consensus document: The International Scientific Association for Probiotics and Prebiotics (ISAPP) consensus statement on the definition and scope of prebiotics. *Nat Rev Gastroenterol Hepatol 14*:491.

Gibson G. R. and M. B. Roberfroid. 1995. Dietary modulation of the human colonic microbiota: Introducing the concept of prebiotics. *J Nutr 125*:1401–1412.

Goodrich J. K., E. R. Davenport, M. Beaumont, M. A. Jackson, R. Knight, C. Ober, T. D. Spector, J. T. Bell, A. G. Clark and R. E. Ley. 2016. Genetic determinants of the gut microbiome in UK twins. *Cell Host Microbe 19*:731–743.

Goodrich J. K., J. L. Waters, A. C. Poole, J. L. Sutter, O. Koren, R. Blekhman, M. Beaumont, W. Van Treuren, R. Knight, J. T. Bell, T. D. Spector, A. G. Clark and R. E. Ley. 2014. Human genetics shape the gut microbiome. *Cell 159*:789–799.

Gosalbes M., S. Llop, Y. Valles, A. Moya, F. Ballester and M. Francino. 2013. Meconium microbiota types dominated by lactic acid or enteric bacteria are differentially associated with maternal eczema and respiratory problems in infants. *Clin Exp Allergy 43*:198–211.

Haller D., H. Colbus, M. G. Gänzle, P. Scherenbacher, C. Bode and W. P. Hammes. 2001. Metabolic and functional properties of lactic acid bacteria in the gastro-intestinal ecosystem: A comparative in vitro Studybetween Bacteria of intestinal and fermented food origin. *Syst Appl Microbiol 24*:218–226.

Halmos E. P., C. T. Christophersen, A. R. Bird, S. J. Shepherd, P. R. Gibson and J. G. Muir. 2014. Diets that differ in their FODMAP content alter the colonic luminal microenvironment. *Gut 64*:93–100.

Hemarajata P. and J. Versalovic. 2013. Effects of probiotics on gut microbiota: Mechanisms of intestinal immunomodulation and neuromodulation. *Ther Adv Gastroenterol 6*:39–51.

Henström M., L. Diekmann, F. Bonfiglio, F. Hadizadeh, E.-M. Kuech, M. von Köckritz-Blickwede, L. B. Thingholm, T. Zheng, G. Assadi and C. Dierks. 2016. Functional variants in the sucrase—isomaltase gene associate with increased risk of irritable bowel syndrome. *Gut 67*:263–270.

Hollister E. B., K. Riehle, R. A. Luna, E. M. Weidler, M. Rubio-Gonzales, T.-A. Mistretta, S. Raza, H. V. Doddapaneni, G. A. Metcalf and D. M. Muzny. 2015. Structure and function of the healthy pre-adolescent pediatric gut microbiome. *Microbiome 3*:36.

Hong P.-Y., J. A. Croix, E. Greenberg, H. R. Gaskins and R. I. Mackie. 2011. Pyrosequencing-based analysis of the mucosal microbiota in healthy individuals reveals ubiquitous bacterial groups and micro-heterogeneity. *PLoS ONE* 6:e25042.

Huebner J., R. Wehling and R. Hutkins. 2007. Functional activity of commercial prebiotics. *Int Dairy J* 17:770–775.

Hyun D.-W., J.-Y. Lee, M.-S. Kim, N.-R. Shin, T. W. Whon, K. H. Kim, P. S. Kim, E. J. Tak, M.-J. Jung, J. Y. Lee, H. S. Kim, W. Kang, H. Sung, C. O. Jeon and J. W. Bae. 2021. Pathogenomics of *Streptococcus ilei* sp. nov., a newly identified pathogen ubiquitous in human microbiome. *J Microbiol* 59:792–806.

Jeraldo P., A. Hernandez, H. B. Nielsen, X. Chen, B. A. White, N. Goldenfeld, H. Nelson, D. Alhquist, L. Boardman and N. Chia. 2016. Capturing one of the human gut microbiome's most wanted: Reconstructing the genome of a novel butyrate-producing, clostridial scavenger from metagenomic sequence data. *Front Microbiol* 7:783.

Jiménez E., M. L. Marín, R. Martín, J. M. Odriozola, M. Olivares, J. Xaus, L. Fernández and J. M. Rodríguez. 2008. Is meconium from healthy newborns actually sterile? *Res Microbiol* 159:187–193.

Kalliomäki M., P. Kirjavainen, E. Eerola, P. Kero, S. Salminen and E. Isolauri. 2001. Distinct patterns of neonatal gut microflora in infants in whom atopy was and was not developing. *J Allergy Clin Immunol* 107:129–134.

Kassinen A., L. Krogius-Kurikka, H. Mäkivuokko, T. Rinttilä, L. Paulin, J. Corander, E. Malinen, J. Apajalahti and A. Palva. 2007. The fecal microbiota of irritable bowel syndrome patients differs significantly from that of healthy subjects. *Gastroenterol* 133:24–33.

Khalid K. 2011. An overview of lactic acid bacteria. *Int J Biosc* 1:1–13.

Klein R. S., R. A. Recco, M. T. Catalano, S. C. Edberg, J. I. Casey and N. H. Steigbigel. 1977. Association of *Streptococcus bovis* with carcinoma of the colon. *N Engl J Med* 297:800–802.

Korpela K., O. Helve, K.-L. Kolho, T. Saisto, K. Skogberg, E. Dikareva, V. Stefanovic, A. Salonen, S. Andersson and W. M. de. 2020. Maternal fecal microbiota transplantation in cesarean-born infants rapidly restores normal gut microbial development: A proof-of-concept study. *Cell* 183:324–334.

Kovatcheva-Datchary P., M. Egert, A. Maathuis, M. Rajilić-Stojanović, A. A. De Graaf, H. Smidt, W. M. De Vos and K. Venema. 2009. Linking phylogenetic identities of bacteria to starch fermentation in an in vitro model of the large intestine by RNA-based stable isotope probing. *Environ Microbiol* 11:914–926.

Kunz C., S. Rudloff, W. Baier, N. Klein and S. Strobel. 2000. Oligosaccharides in human milk: Structural, functional, and metabolic aspects. *Ann Rev Nutr* 20:699–722.

Lamas B., M. L. Richard, V. Leducq, H. P. Pham, M. L. Michel, G. Da Costa, C. Bridonneau, S. Jegou, T. W. Hoffmann, J. M. Natividad, L. Brot, S. Taleb, A. Couturier-Maillard, I. Nion-Larmurier, F. Merabtene, P. Seksik, A. Bourrier, J. Cosnes, B. Ryffel, L. Beaugerie, J.-M. Launay, P. Langella, R. J. Xavier and H. Soko. 2016. CARD9 impacts colitis by altering gut microbiota metabolism of tryptophan into aryl hydrocarbon receptor ligands. *Nat Med* 22:598–605.

Larsen N., F. K. Vogensen, F. W. van den Berg, D. S. Nielsen, A. S. Andreasen, B. K. Pedersen, W. A. Al-Soud, S. J. Sørensen, L. H. Hansen and M. Jakobsen. 2010. Gut microbiota in human adults with type 2 diabetes differs from non-diabetic adults. *PLoS ONE* 5:e9085.

Laureys D., M. Cnockaert, L. De Vuyst and P. Vandamme. 2016. *Bifidobacterium aquikefiri* sp. nov., isolated from water kefir. *Int J Syst Evol Microbiol* 66:1281–1286.

Lawson M. A. E., I. J. O'Neill, M. Kujawska, S. G. Javvadi, A. Wijeyesekera, Z. Flegg, L. Chalklen, and L. J. Hall. 2020. Breast milk-derived human milk oligosaccharides promote *Bifidobacterium* interactions within a single ecosystem. *The ISME Journal* 14:635–648.

Li Q., C. Wang, C. Tang, N. Li and J. Li. 2012. Molecular-phylogenetic characterization of the microbiota in ulcerated and non-ulcerated regions in the patients with Crohn's disease. *PLoS ONE* 7:e34939.

Liwinski T., C. Casar, M. C. Ruehlemann, C. Bang, M. Sebode, S. Hohenester, G. Denk, W. Lieb, A. W. Lohse, A. Franke and C. Schramm. 2020. A disease-specific decline of the relative abundance of *Bifidobacterium* in patients with autoimmune hepatitis. *Aliment Pharmacol Ther* 51:1417–1428.

Lo C. I., N. Dione, A. Mbaye, P. Fernández-Mellado G., I. I. Ngom, C. Valles, S. Alibar, J.-C. Lagier, F. Fenollar, P.-E. Fournier, D. Raoult and S. M. Diene. 2021. *Limosilactobacillus caccae* sp. nov., a new bacterial species isolated from the human gut microbiota. *FEMS Microbiol Lett* 368(18):1–9.

Louis P., S. H. Duncan, P. O. Sheridan, A. W. Walker and H. J. Flint. 2022. Microbial lactate utilisation and the stability of the gut microbiome. *Gut Microbiome* 3:e3.

Lu Y., J. Chen, J. Zheng, G. Hu, J. Wang, C. Huang, L. Lou, X. Wang and Y. Zeng. 2016. Mucosal adherent bacterial dysbiosis in patients with colorectal adenomas. *Sci Rep* 6:26337.

Lukic J., I. Strahinic, M. Milenkovic, N. Golic, M. Kojic, L. Topisirovic and J. Begovic. 2013. Interaction of *Lactobacillus fermentum* BGHI14 with rat colonic mucosa: Implications for colitis induction. *Appl Environ Microbiol* 79:5735–5744.

Marieb E. and K. Hoehn. 2010. Human Anatomy and Physiology with Interactive Physiology 10-System Suite. Pearson Publishing, New York.

Marighela T. F., M. I. Arismendi, V. Marvulle, M. K. C. Brunialti, R. Salomão and C. Kayser. 2019. Effect of probiotics on gastrointestinal symptoms and immune parameters in systemic sclerosis: A randomized placebo-controlled trial. *Rheumatology* 58:1985–1990.

Martín R., E. Jiménez, H. Heilig, L. Fernández, M. L. Marín, E. G. Zoetendal and J. M. Rodríguez. 2009. Isolation of bifidobacteria from breast milk and assessment of the bifidobacterial population by PCR-denaturing gradient gel electrophoresis and quantitative real-time PCR. *Appl Environ Microbiol* 75:965–969.

Martínez I., J. Kim, P. R. Duffy, V. L. Schlegel and J. Walter. 2010. Resistant starches types 2 and 4 have differential effects on the composition of the fecal microbiota in human subjects. *PLoS ONE* 5:e15046.

Mayeur C., J.-J. Gratadoux, C. Bridonneau, F. Chegdani, B. Larroque, N. Kapel, O. Corcos, M. Thomas and F. Joly. 2013. Faecal D/L lactate ratio is a metabolic signature of microbiota imbalance in patients with short bowel syndrome. *PLoS ONE* 8:e54335.

Metchnikoff M. E. 1908. Etude sur la flore in- testinale. IV. Le *Bacillus sporogenes* . *Annales de I'Institut Pasteur* 22:942–946.

Mielcarz D. W. and L. H. Kasper. 2015. The gut microbiome in multiple sclerosis. *Curr Treat Options Neurol* 17:18.

Mihailović M., M. Živković, J. A. Jovanović, M. Tolinački, M. Sinadinović, J. Rajić, A. Uskoković, S. Dinić, N. Grdović and N. Golić. 2017. Oral administration of probiotic *Lactobacillus paraplantarum* BGCG11 attenuates diabetes-induced liver and kidney damage in rats. *J Funct Foods* 38:427–437.

Mitsuoka T. 1992. Intestinal flora and aging. *Nutr Rev* 50:438–446.

Mohammed R. H. A. and H. I. I. Elmakhzangy. 2017. The gut microbiota and inflammation in rheumatoid arthritis. In: New Developments in the Pathogenesis of Rheumatoid Arthritis. Ed. Sakkas, L. I. InTech, London, pp. 83–99.

Moore W. and L. V. Holdeman. 1974. Human fecal flora: The normal flora of 20 Japanese-Hawaiians. *Appl Microbiol* 27:961–979.

Morita N., E. Umemoto, S. Fujita, A. Hayashi, J. Kikuta, I. Kimura, T. Haneda, T. Imai, A. Inoue, H. Mimuro, Y. Maeda, H. Kayama, R. Okumura, J. Aoki, N. Okada, T. Kida, M. Ishii, R. Nabeshima and K. Takeda. 2019. GPR31-dependent dendrite protrusion of intestinal CX3CR1+ cells by bacterial metabolites. *Nature* 566:110–114.

Moro E. 1900. Uber den *Bacillus acidophilus* n. sp. *Jahrb. Kinderheilk* 52:38–55.

Murri, M., I. Leiva, J. M. Gomez-Zumaquero, F. J. Tinahones, F. Cardona, F. Soriguer and M. I. Queipo-Ortuño. 2013. Gut microbiota in children with type 1 diabetes differs from that in healthy children: A case-control study. *BMC Med* 11:1–12.

Nicholson J. K., E. Holmes, J. Kinross, R. Burcelin, G. Gibson, W. Jia and S. Pettersson. 2012. Host-gut microbiota metabolic interactions. *Science* 336:1262–1267.

Nowak A., A. Paliwoda and J. Błasiak. 2019. Anti-proliferative, pro-apoptotic and anti-oxidative activity of *Lactobacillus* and *Bifidobacterium* strains: A review of mechanisms and therapeutic perspectives. *Crit Rev Food Sci Nutr* 59:3456–3467.

Nugent S., D. Kumar, D. Rampton and D. Evans. 2001. Intestinal luminal pH in inflammatory bowel disease: Possible determinants and implications for therapy with aminosalicylates and other drugs. *Gut* 48:571–577.

Nylund L., R. Satokari, J. Nikkilä, M. Rajilić-Stojanović, M. Kalliomäki, E. Isolauri, S. Salminen and W. M. De Vos. 2013. Microarray analysis reveals marked intestinal microbiota aberrancy in infants having eczema compared to healthy children in at-risk for atopic disease. *BMC Microbiol* 13:12.

Ogawa K., R. A. Ben, S. Pons and L. F. Bustos. 1992. Volatile fatty acids, lactic acid, and pH in the stools of breast-fed and bottle-fed infants. *J Pediatr Gastroenterol Nutr* 15:248–252.

Oren A. and G. M. Garrity. 2021. Valid publication of the names of forty-two phyla of prokaryotes. *Int J Syst Evol Microbiol* 71:005056.

Parigi S. M., M. Eldh, P. Larssen, S. Gabrielsson and E. J. Villablanca. 2015. Breast milk and solid food shaping intestinal immunity. *Front Immunol* 6:415.

Pärtty A., M. Kalliomäki, P. Wacklin, S. Salminen and E. Isolauri. 2015. A possible link between early probiotic intervention and the risk of neuropsychiatric disorders later in childhood: A randomized trial. *Pediatr Res* 77:823.

Pelzer E., L. F. Gomez-Arango, H. L. Barrett and M. D. Nitert. 2017. Review: Maternal health and the placental microbiome. *Placenta* 54:30–37.

Penders J., E. E. Stobberingh, P. A. van den Brandt and C. Thijs. 2007. The role of the intestinal microbiota in the development of atopic disorders. *Allergy* 62:1223–1236.

Penders J., C. Thijs, C. Vink, F. F. Stelma, B. Snijders, I. Kummeling, P. A. van den Brandt and E. E. Stobberingh. 2006. Factors influencing the composition of the intestinal microbiota in early infancy. *Pediatrics* 118:511–521.

Pham V. T., C. Chassard, E. Riffa, C. Braegger, A. Geirnaert, V. N. Rocha Martin and C. Lacroix. 2019. Lactate metabolism is strongly modulated by fecal inoculum, pH, and retention time in polyferms continuous colonic fermentation models mimicking young infant proximal colon. *mSystems* 4:e00264–e00218.

Popović N., M. Dinic, M. Tolinacki, S. Mihajlovic, A. Terzić-Vidojević, S. Bojic, J. Djokic, N. Golic and K. Veljovic. 2018. New insight into biofilm formation ability, the presence of virulence genes and probiotic potential of *Enterococcus* sp. dairy isolates. *Front Microbiol* 9:78.

Queipo-Ortuño M. I., L. M. Seoane, M. Murri, M. Pardo, J. M. Gomez-Zumaquero, F. Cardona, F. Casanueva and F. J. Tinahones. 2013. Gut microbiota composition in male rat models under different nutritional status and physical activity and its association with serum leptin and ghrelin levels. *PLoS ONE* 8:e65465.

Radojević D., M. Bekić, A. Gruden-Movsesijan, N. Ilić, M. Dinić, A. Bisenić, N. Golić, D. Vučević, J. Đokić and S. Tomić. 2022. Myeloid-derived suppressor cells prevent disruption of the gut barrier, preserve microbiota composition, and potentiate immunoregulatory pathways in a rat model of experimental autoimmune encephalomyelitis. *Gut Microbes* 14(1):2127455.

Rajilić-Stojanović M. 2013. Function of the microbiota. *Best Pract Res Clin Gastroenterol* 27:5–16.

Rajilić-Stojanović M., E. Biagi, H. G. Heilig, K. Kajander, R. A. Kekkonen, S. Tims and W. M. de Vos. 2011. Global and deep molecular analysis of microbiota signatures in fecal samples from patients with irritable bowel syndrome. *Gastroenterol* 141:1792–1801.

Rajilić-Stojanović M. and W. M. de Vos. 2014. The first 1000 cultured species of the human gastrointestinal microbiota. *FEMS Microbiol Rev* 38:996–1047.

Rajilić-Stojanović M., S. Dimitrijević and N. Golić. 2019. Lactic acid bacteria in the gut. In: *Lactic Acid Bacteria*, 5th edn. Eds. G. Vinderola, A. Ouwehand, S. Salminen, A. von Wright. CRC Press, Taylor & Francis Group, Boca Raton, FL, pp. 381–407.

Rajilic-Stojanovic M., C. Figueiredo, A. Smet, R. Hansen, J. Kupcinskas, T. Rokkas, L. Andersen, J. C. Machado, G. Ianiro, A. Gasbarrini, M. Leja, J. P. Gisbert and G. L. Hold. 2020. Systematic review: Gastric microbiota in health and disease. *Aliment Pharmacol & The* 51:582–602.

Rajilić-Stojanović M., H. G. H. J. Heilig, S. Tims, E. G. Zoetendal and W. M. de Vos. 2013. Long-term monitoring of the human intestinal microbiota composition. *Environ Microbiol* 15:1146–1159.

Reichardt N., S. H. Duncan, P. Young, A. Belenguer, C. McWilliam Leitch, K. P. Scott, H. J. Flint and P. Louis. 2014 Phylogenetic distribution of three pathways for propionate production within the human gut microbiota. *ISME J* 8:1323–1335.

Reuter G. 2001. The *Lactobacillus* and *Bifidobacterium* microflora of the human intestine: Composition and succession. *Curr Issues Intest Microbiol* 2:43–53.

Rogosa M., J. A. Mitchell and R. F. Wiseman. 1951. A selective medium for the isolation and enumeration of oral and fecal lactobacilli. *J Bacteriol* 62:132.

Rossi M., D. Martínez-Martínez, A. Amaretti, A. Ulrici, S. Raimondi and A. Moya. 2016. Mining metagenomic whole genome sequences revealed subdominant but constant *Lactobacillus* population in the human gut microbiota. *Environ Microbiol Rep* 8:399–406.

Saeedi B. J., K. H. Liu, J. A. Owens, S. Hunter-Chang, M. C. Camacho, R. U. Eboka, B. Chandrasekharan, N. F. Baker, T. M. Darby, B. S. Robinson, R. M. Jones, D. P. Jones and A. S. Neish. 2020. Gut-resident lactobacilli activate hepatic Nrf2 and protect against oxidative liver injury. *Cell Metab* 31:956–968.e5.

Salonen A., W. M. de Vos and A. Palva. 2010. Gastrointestinal microbiota in irritable bowel syndrome: Present state and perspectives. *Microbiol* 156:3205–3215.

Santacruz A., A. Marcos, J. Wärnberg, A. Martí, M. Martin-Matillas, C. Campoy, L. A. Moreno, O. Veiga, C. Redondo-Figuero and J. M. Garagorri. 2009. Interplay between weight loss and gut microbiota composition in overweight adolescents. *Obesity* 17:1906–1915.

Santoro A., R. Ostan, M. Candela, E. Biagi, P. Brigidi, M. Capri and C. Franceschi. 2018. Gut microbiota changes in the extreme decades of human life: A focus on centenarians. *Cell Mol Life Sci* 75:129–148.

Saus E., S. Iraola-Guzmán, J. R. Willis, A. Brunet-Vega and T. Gabaldón. 2019. Microbiome and colorectal cancer: Roles in carcinogenesis and clinical potential. *Mol Aspects Med* 69:93–106.

Schnorr S. L., M. Candela, S. Rampelli, M. Centanni, C. Consolandi, G. Basaglia, S. Turroni, E. Biagi, C. Peano and M. Severgnini. 2014. Gut microbiome of the Hadza hunter-gatherers. *Nat Commun* 5:3654.

Sela D. A. 2011. Bifidobacterial utilization of human milk oligosaccharides. *Int J Food Microbiol* 149:58–64.

Sevelsted A., J. Stokholm, K. Bønnelykke and H. Bisgaard. 2015. Cesarean section and chronic immune disorders. *Pediatrics* 135:e92–e98.

Shehata M. G., S. A. El Sohaimy, M. A. El-Sahn and M. M. Youssef. 2016. Screening of isolated potential probiotic lactic acid bacteria for cholesterol lowering property and bile salt hydrolase activity. *Ann Agric Sci* 61:65–75.

Shkoporov A. N., A. V. Chaplin, V. A. Shcherbakova, N. E. Suzina, L. I. Kafarskaia, V. K. Bozhenko and B. A. Efimov. 2016. *Ruthenibacterium lactatiformans* gen. nov., sp. nov., an anaerobic, lactate-producing member of the family *Ruminococcaceae* isolated from human faeces. *Int J Syst Evol Miicrobiol* 66:3041–3049.

Simon G. L. and S. L. Gorbach. 1986. The human intestinal microflora. *Dig Dis Sci* 31:147–162.

Sinha D., X. Sun, M. Khare, M. Drancourt, D. Raoult and P.-E. Fournier. 2021. Pangenome analysis and virulence profiling of *Streptococcus intermedius*. *BMC Genom* 22:1–17.

Solís G., C. de Los Reyes-Gavilan, N. Fernández, A. Margolles and M. Gueimonde. 2010. Establishment and development of lactic acid bacteria and bifidobacteria microbiota in breast-milk and the infant gut. *Anaerobe* 16:307–310.

Sokovic Bajic S., J. Djokic, M. Dinic, K. Veljovic, N. Golic, S. Mihajlovic and M. Tolinacki. 2019. GABA-producing natural dairy isolate from artisanal Zlatar cheese attenuates gut inflammation and strengthens gut epithelial barrier *in vitro*. *Front Microbiol* 10:527.

Stanisavljević S., J. Lukić, S. Soković, S. Mihajlovic, M. Mostarica Stojković, D. Miljković and N. Golić. 2016. Correlation of gut microbiota composition with resistance to experimental autoimmune encephalomyelitis in rats. *Front Microbiol* 7:2005.

Stolaki M., W. M. de Vos, M. Kleerebezem and E. G. Zoetendal. 2012. Lactic acid bacteria in the gut. In: *Lactic Acid Bacteria, Microbiological and Functional Aspects*, 4th ed. Eds. S. Lahtinen, S. Salminen and A. von Wright. CRC Press, Taylor & Francis Group, Boca Raton, FL, pp. 385–401.

Sun S., G. Chang and L. Zhang. 2022. The prevention effect of probiotics against eczema in children: An update systematic review and meta-analysis. *J Dermatolog Treat* 33:1844–1854.

Swanson K. S., G. R. Gibson, R. Hutkins, R. A. Reimer, G. Reid, K. Verbeke, K. P. Scott, H. D. Holscher, M. B. Azad, N. M. Delzenne and M. E. Sanders. 2020. The International Scientific Association for Probiotics and Prebiotics (ISAPP) consensus statement on the definition and scope of synbiotics. *Nat Rev Gastroenterol Hepatol* 17(11):687–701.

Swidsinski A., V. Loening-Baucke, S. Schulz, J. Manowsky, H. Verstraelen and S. Swidsinski. 2016. Functional anatomy of the colonic bioreactor: Impact of antibiotics and *Saccharomyces boulardii* on bacterial composition in human fecal cylinders. *Syst Appl Microbiol* 39:67–75.

Szarka L. A. and M. Camilleri. 2012. Methods for the assessment of small-bowel and colonic transit. *Semin Nucl Med* 42(2):113–123.

Tilg H. and A. R. Moschen. 2014. Microbiota and diabetes: An evolving relationship. *Gut* 63:1513–1521.

Tissier H. 1900. Recherches sur la flore intestinale normale et pathologique du nourrisson. University of Paris, Paris, France.

Uribarri J., M. S. Oh and H. J. Carroll. 1998. D-lactic acidosis: A review of clinical presentation, biochemical features, and pathophysiologic mechanisms. *Medicine (Baltimore)* 77:73–82.

Vaahtovuo, J., E. Munukka, M. Korkeamäki, R. Luukkainen and P. Toivanen. 2008. Fecal microbiota in early rheumatoid arthritis. *J Rheumatol* 35:1500–1505.

Valentini L., A. Pinto, I. Bourdel-Marchasson, R. Ostan, P. Brigidi, S. Turroni, S. Hrelia, P. Hrelia, S. Bereswill and A. Fischer. 2015. Impact of personalized diet and probiotic supplementation on inflammation, nutritional parameters and intestinal microbiota—the "RISTOMED project": Randomized controlled trial in healthy older people. *Clin Nutr* 34:593–602.

van de Guchte M., P. Serror, C. Chervaux, T. Smokvina, S. D. Ehrlich and E. Maguin. 2002. Stress responses in lactic acid bacteria. *Antonie Van Leeuwenhoek* 82:187–216.

van den Bogert B., M. Meijerink, E. G. Zoetendal, J. M. Wells and M. Kleerebezem. 2014. Immunomodulatory properties of *Streptococcus* and *Veillonella* isolates from the human small intestine microbiota. *PLoS ONE* 9:e114277.

Wall R., G. Fitzgerald, S. Hussey, T. Ryan, B. Murphy, P. Ross and C. Stanton. 2007. Genomic diversity of cultivable *Lactobacillus* populations residing in the neonatal and adult gastrointestinal tract. *FEMS Microbiol Ecol* 59:127–137.

Walter J., C. Hertel, G. W. Tannock, C. M. Lis, K. Munro and W. P. Hammes. 2001. Detection of *Lactobacillus*, *Pediococcus*, *Leuconostoc*, and *Weissella* species in human feces by using group-specific PCR primers and denaturing gradient gel electrophoresis. *Appl Environ Microbiol* 67:2578–2585.

Wang T., G. Cai, Y. Qiu, N. Fei, M. Zhang, X. Pang, W. Jia, S. Cai and L. Zhao. 2012. Structural segregation of gut microbiota between colorectal cancer patients and healthy volunteers. *ISME J* 6:320.

Whitman W. B., D. C. Coleman and W. J. Wiebe. 1998. Prokaryotes: The unseen majority. *Proc Natl Acad Sci USA* 95:6578–6583.

Wlodarska M., A. D. Kostic and R. J. Xavier. 2015. An integrative view of microbiome-host interactions in inflammatory bowel diseases. *Cell Host Microbe* 17:577–591.

Yatsunenko T., F. E. Rey, M. J. Manary, I. Trehan, M. G. Dominguez-Bello, M. Contreras, M. Magris, G. Hidalgo, R. N. Baldassano and A. P. Anokhin. 2012. Human gut microbiome viewed across age and geography. *Nature* 486:222.

Yoon M. Y. and S. S. Yoon. 2018. Disruption of the gut ecosystem by antibiotics. *Yonsei Med J* 59:4–12.

Zhang C., S. Li, L. Yang, P. Huang, W. Li, S. Wang, G. Zhao, M. Zhang, X. Pang and Z. Yan. 2013. Structural modulation of gut microbiota in life-long calorie-restricted mice. *Nat Commun* 4:2163.

Zhang X., S. Mushajiang, B. Luo, F. Tian, Y. Ni, and W. Yan. 2020. The composition and concordance of *Lactobacillus* populations of infant gut and the corresponding breast-milk and maternal Gut. *Frontier Microbiol 11*.

Zhang, X., D. Zhang, H. Jia, Q. Feng, D. Wang, D. Liang, X. Wu, J. Li, L. Tang, Y. Li, et al. 2015. The oral and gut microbiomes are perturbed in rheumatoid arthritis and partly normalized after treatment. *Nat Med 21*:895–905.

Zheng J., S. Wittouck, E. Salvetti, C. M. A. P. Franz, H. M. B. Harris, P. Mattarelli, P. W. O'Toole, B. Pot, P. Vandamme, J. Walter, K. Watanabe, S. Wuyts, G. E. Felis, M. G. Gänzle and S. Lebeer. 2020. A taxonomic note on the genus *Lactobacillus*: Description of 23 novel genera, emended description of the genus *Lactobacillus* Beijerinck 1901, and union of Lactobacillaceae and Leuconostocaceae. *Int J Syst Evol Microbiol 70*:2782–2858.

Zoetendal E. G., A. D. Akkermans, W. M. Akkermans-van Vliet, J. A. G. de Visser and W. M. de Vos. 2001. The host genotype affects the bacterial community in the human gastronintestinal tract. *Microb Ecol Health Dis 13*:129–134.

Zoetendal E. G., J. Raes, B. Van Den Bogert, M. Arumugam, C. C. Booijink, F. J. Troost, P. Bork, M. Wels, W. M. De Vos and M. Kleerebezem. 2012. The human small intestinal microbiota is driven by rapid uptake and conversion of simple carbohydrates. *ISME J 6*:1415.

# Gastrointestinal Benefits of Probiotics 23

Arthur C. Ouwehand

## 23.1 INTRODUCTION

A wide range of health effects, including enhanced metabolism of dietary compounds, alleviation of disturbed bowel functions, improved resilience of the gastrointestinal (GI) microbiota, resistance against infections within and outside the GI tract, and prevention of allergies, have been linked with the use of specific (combinations of) probiotic microbes. The findings are based on the potential etiological role of the gut microbiota in the pathogenesis of a number of clinical conditions and in the modulation of host immune functions, providing means for modulating the host immune responses both locally and systemically. The proposed and documented GI benefits of probiotics are based on several clinical intervention studies with varying quality and methodology and on meta-analyses of the aforementioned (Table 23.1). This chapter will discuss the outcomes of probiotic intervention studies on GI disorders and diseases. The chapter will focus mainly on *Lactobacillus, sensu lato*, and *Bifidobacterium* probiotics.

**TABLE 23.1** Non-Exhaustive List on the Effects of Probiotics on Gastrointestinal Conditions, with Example References

| CONDITION | PROBIOTIC SELECTION | OUTCOME MEASURES | RESULT | REFERENCES |
|---|---|---|---|---|
| Infectious diarrhea | Unrestricted | Duration of diarrhea | Uncertain | (Collinson et al., 2020) |
| | Unrestricted | Incidence | Effective | (Fagnant et al., 2023) |
| | *Saccharomyces boulardii* | Duration of diarrhea | Effective | (Dinleyici et al., 2012) |
| Rotavirus diarrhea | Unrestricted | Duration of diarrhea | Effective | (Ahmadi et al., 2015) |
| | *Limosilactobacillus reuteri* DSM 17938 | Duration of diarrhea | Effective | (Urbanska et al., 2016) |
| | *Lacicaseibacillus rhamnosus* GG | Duration of diarrhea | Effective | (Ahmadi et al., 2015) |
| | *Limosilactobacillus reuteri* DSM 17938 | Risk of diarrhea | Effective | (Urbanska et al., 2016) |

DOI: 10.1201/9781003352075-26

**TABLE 23.1**   (Continued) Non-Exhaustive List on the Effects of Probiotics on Gastrointestinal Conditions, with Example References

| CONDITION | PROBIOTIC SELECTION | OUTCOME MEASURES | RESULT | REFERENCES |
|---|---|---|---|---|
| Antibiotic-associated diarrhea (AAD) | Unrestricted | Risk of AAD | Effective | (Hempel et al., 2012; Blaabjerg et al., 2017) |
|  | Lactobacilli | Risk of AAD | Effective | (Hempel et al., 2012) |
|  | *Saccharomyces boulardii* | Risk of AAD | Effective | (Blaabjerg et al., 2017) |
| *Clostridiodes difficile–associated diarrhea (CDAD)* | Unrestricted | Risk of CDAD | Effective | (Lau and Chamberlain, 2016; Ma et al., 2020) |
|  |  | Duration of CDAD | Effective | (Ma et al., 2020) |
|  | Lactobacilli | Risk of CDAD | Effective | (Lau and Chamberlain, 2016) |
|  | *Saccharomyces boulardii* | Risk of CDAD | Effective | (Lau and Chamberlain, 2016) |
| Travelers' diarrhea (TD) | Unrestricted | Risk of TD | Limited effective | (McFarland and Goh, 2019) |
| Irritable bowel syndrome (IBS) | Unrestricted | Overall IBS symptoms | Effective | (Niu and Xiao, 2020) |
|  |  | Abdominal pain | Effective | (Niu and Xiao, 2020, Konstantis et al., 2023) |
|  |  | Flatulence | Effective | (Niu and Xiao, 2020) |
|  |  | Bloating | Inconclusive | (Niu and Xiao, 2020, Konstantis et al., 2023) |
|  |  | Adverse events | Increased | (Niu and Xiao, 2020) |
| Constipation | Unrestricted | Defecation frequency | Effective | (van der Schoot et al., 2022) |
|  |  | Reducing colonic transit time | Inconclusive | (Wen et al., 2020; van der Schoot et al., 2022) |
| Ulcerative colitis | Unrestricted | Inducing remission | Effective | (Chen et al., 2021; Kaur et al., 2020) |
|  |  | Maintaining remission | Not effective | (Chen et al., 2021) |
| Crohn's disease | Unrestricted | Inducing remission | Not effective | (Chen et al., 2021) |
|  |  | Maintaining remission | Not effective | (Chen et al., 2021) |
| Pouchitis | Unrestricted | Prevention | Effective | (Xiao et al., 2023) |
|  |  | Maintaining remission | Effective for some formulations | (Dong et al., 2016; Nguyen et al., 2019) |
| Necrotizing enterocolitis (NEC) | Unrestricted | Prevention of NEC | Effective | (Thomas et al., 2017; Sun et al., 2017; Wang et al., 2023a) |
|  |  | Prevention of sepsis | Effective | (Sun et al., 2017; Sawh et al., 2016) |
|  |  | Mortality | Effective | (Thomas et al., 2017; Wang et al., 2023a; Sun et al., 2017) |
| Infant colic | Unrestricted | Treatment | Effective | (Liu et al., 2022) |

*(Continued)*

**TABLE 23.1** (Continued) Non-Exhaustive List on the Effects of Probiotics on Gastrointestinal Conditions, with Example References

| CONDITION | PROBIOTIC SELECTION | OUTCOME MEASURES | RESULT | REFERENCES |
|---|---|---|---|---|
| | *Limosilactobacillus reuteri* DSM 17938 | Decreased crying time | Effective | (Schreck Bird et al., 2017) |
| Postoperative infections in abdominal surgery | Unrestricted | Surgical site infection | Effective | (Araújo et al., 2023; Veziant et al., 2022) |
| | | Sepsis | Effective | (Araújo et al., 2023) |
| *Helicobacter pylori* infection | Unrestricted | Improved eradication rate | Effective | (Zhang et al., 2020; Liang et al., 2023; Wang et al., 2023b) |
| | | Total side effects | Effective | (Zhang et al., 2020; Liang et al., 2023; Wang et al., 2023b) |
| Lactose intolerance | Yogurt | Reduced symptoms | Effective | (Corgneau et al., 2017) |
| | Probiotics | Reduced symptoms | Effective | (Leis et al., 2020) |

# 23.2 DIARRHEA

Numerous clinical studies aiming at the treatment or prevention of an array of diarrheal diseases with probiotics have been published. Studies have been conducted using mixtures of probiotic strains; individual *Lactobacillus* and *Bifidobacterium* strains; or a probiotic yeast, *Saccharomyces boulardii* (Table 23.1). Overall, the duration of diarrhea can be shortened by approximately 1 day with specific probiotics (Allen et al., 2010). While this may seem modest, for synbiotics with a similar effectiveness, savings of 25% for diarrhea-associated healthcare costs have been estimated (Vandenplas and De Hert, 2012). It is reasonable to assume similar savings for probiotics, as the efficacy and cost are similar. In children, shortening of the duration of rotavirus diarrhea by the administration of probiotic bacteria is well established (Vlasova et al., 2016), and probiotics have been accepted and widely used as diarrheal therapy alongside rehydration (Dekate et al., 2013). Moreover, certain probiotics have the potential to play a part in solving the substantial health problem caused by recurrent diarrheal episodes in developing countries (Preidis et al., 2011). However, there is still a need for further community-based intervention trials, especially in developing countries and among underprivileged populations, since most of the current clinical studies have been carried out in hospitals and clinics in developed countries (Guarino et al., 2012).

Antibiotic treatment substantially disrupts the balance of the intestinal microbiota (Ferrer et al., 2017) and may lead to so-called antibiotic-associated diarrhea (AAD) or other digestive disturbances. It also may increase the risk of *Clostridiodes difficile*–associated diarrhea (CDAD) (Napolitano and Edmiston, 2017). Infants may often be asymptomatic carriers of *C. difficile* (Clayton and Toltzis, 2017). However, elderly subjects, subjects with underlying diseases, or otherwise frail individuals are especially prone to CDAD (Asempa and Nicolau, 2017). Because of the seriousness of CDAD and the benefit provided by probiotics (Ma et al., 2020), substantial savings can be obtained with their use (Shen et al., 2017). Due to the high prevalence and potentially detrimental impact of AAD and CDAD on health, the effort to discover efficient probiotics against these diseases has been substantial (McFarland, 2009). According to a meta-analysis on *Lactobacillus* probiotics, lactobacilli can effectively prevent AAD in both children and adults (Blaabjerg et al., 2017) (Table 23.1). While probiotic-containing dairy products, mainly yogurts, appear to be efficacious in reducing risk for AAD (Mantegazza et al., 2018), it is uncertain if non-probiotic-containing dairy products reduce AAD risk. Despite its relative simplicity, very few studies have tested the efficacy of standard yogurt on AAD risk (Patro-Golab et al., 2015). One shortcoming in probiotic intervention studies on AAD is the varying length or complete lack of follow-up after the antibiotic course; since AAD may occur with up to 2 months delay after the antibiotic treatment, inadequate follow-up may result in erroneous interpretations of results (McFarland, 2009). This should, however, not affect the conclusions on the efficacy of probiotics in reducing risk for AAD as it can be assumed that the late occurrence of AAD is equal in both verum and placebo groups.

The use of probiotics against travelers' diarrhea (TD) is popular. However, the clinical evidence for the reducing TD risk remains limited (Table 23.1). Meta-analyses have failed to observe a clear overall benefit for probiotics on TD risk. Only *S. boulardii* was judged to provide a significant reduction of TD risk (McFarland and Goh, 2019). The current evidence is limited by the small number of interventions and variations in the protocols. The wide variety of the potential causes of TD and the difficulties of the volunteers in adhering to the study protocols during traveling lend additional challenges to the probiotic interventions targeting TD.

The suggested potential mechanisms by which probiotics may influence diarrheal disease are many: modulating microbiota composition by production of antimicrobial components, competition for nutrients and binding sites; modulation of the immune system and influencing the intestinal barrier function (Vlasova et al., 2016). While all the mentioned potential mechanisms and many more have been documented *in vitro* and in animal models, it remains to be shown beyond doubt that they are actually involved in the way probiotics work in humans.

There are etiologies of diarrhea and probiotic studies targeting the same diarrhea subtype may use different criteria in subject recruitment and varying outcome measures (Ritchie and Romanuk, 2012), making joint conclusions on probiotics and diarrheal diseases difficult. Nevertheless, probiotics appear to be most effective in the prevention and treatment of rotavirus diarrhea and AAD, while the effect on TD needs further assessment.

## 23.3 IRRITABLE BOWEL SYNDROME

Irregular bowel movements, often accompanied by hard and/or soft stools, are more common than regular bowel movements among Western adults (Heaton et al., 1992). Therefore, the potential market for probiotics that show a benefit on bowel habits among generally healthy adult subjects is substantial. Examples of such benefits include increased stool frequency (Wen et al., 2020) and improved stool consistency (van der Schoot et al., 2022).

Some individuals experience functional bowel symptoms in a recurrent fashion and can have a diagnosis of functional bowel disorders (FBDs), which by definition are devoid of structural and biochemical abnormalities (Drossman, 2017). Consequently, intervention studies conducted with subjects found to have FBDs are applicable for the general population because (i) the time requirements for diagnosis of FBDs are relatively loose (symptoms must have occurred for at least 6 months and have been present for 3 days a month during the preceding 3 months), (ii) no defined etiological factors are known for FBDs, (iii) no inflammatory or structural abnormalities explain FBD symptoms, and (iv) the distinct symptoms encountered in FBDs are also common among the general population lacking GI-related diagnosed abnormalities. In addition, comparison between treatment groups is more efficient when the subjects are likely to experience bowel symptoms during the course of the intervention.

Irritable bowel syndrome (IBS) is an FBD associated with abdominal pain or discomfort, which is relieved by defecation or triggered in association to altered stool form or frequency and often associated with bloating and passage of mucus (Schmulson and Drossman, 2017). Significant differences in levels of selected fecal microbiota members were observed in IBS patients compared with the healthy controls: reductions in lactobacilli, *Bifidobacterium*, and *F. prausnitzii*, whereas the *Bacteroides–Prevotella* group, *Enterococcus*, *Escherichia coli*, *Clostridium coccoides*, and other genera or species displayed no significant differences (Liu et al., 2017a). Analysis of different IBS subtypes indicated that diarrhea-predominant IBS patients had reduced counts of *Lactobacillus* and *Bifidobacterium* compared with the healthy controls, whereas the constipation-predominant IBS patients did not differ (Liu et al., 2017b). Moreover, disturbances of the gut microbiota caused by enteric infections are known to predispose to IBS (Thompson, 2016). Thus, probiotics are an intriguing alternative for alleviation of the symptoms of IBS.

Since IBS is a symptom-based disorder that relies on the subject's own sensation of bowel function, specific subject-filled questionnaires designed for IBS-related symptoms and quality of life are essential for diagnosis. It is, however, uncertain whether these questionnaires are appropriate to detect modest changes potentially induced by probiotics and other treatments. A substantial placebo effect is commonly observed. Studies on the effect of probiotics on IBS should therefore be of sufficient duration and with appropriate numbers of participants. It may also be advisable to focus interventions on a particular subtype of IBS and not a combination to reduce heterogeneity.

A recent meta-analysis of 35 studies concluded that, overall, probiotics provide a significantly better reduction in general symptomology (Niu and Xiao, 2020) than the placebo but no concomitant improvement in quality of life (Konstantis et al., 2023). Niu and Xiao (2020) reported that probiotic use was also associated with an improvement in abdominal pain, bloating, and flatulence compared with placebo. Interestingly, the study also reported a higher incidence of any adverse event in the probiotic group (Niu and Xiao, 2020), although this was not confirmed by Konstantis and co-workers (2023). The beneficial effect was, however, rather heterogeneous between the included studies. Thus, although there does seem to be an overall benefit of probiotics in the management of IBS (Table 23.1), meta-analyses also indicate that many of the studies are not of the quality that would be required to make firm recommendations (Niu and Xiao, 2020).

# 23.4 CONSTIPATION

Constipation has traditionally been defined as less than three defecations per week, but not all people feel constipated with such a defecation frequency, and others may feel constipated even though they have more frequent bowel movements. Thus, the effort needed to defecate and stool consistency may also be taken into account (Drossman, 2017).

It may be concluded that the effects of probiotic bacteria on the bowel habits of healthy subjects are rather modest, which can be seen as a positive finding, as it suggests that the probiotic products are well tolerated. When symptoms are encountered, the distinct symptoms are likely to react to probiotic treatment in a similar manner among healthy subjects with occasional bowel symptoms and subjects found to have FBDs.

A recent meta-analysis of 19 studies investigating the influence of probiotics on stool frequency in constipated subjects indicated a beneficial effect for probiotics (van der Schoot et al., 2022) (Table 23.1). Intestinal transit time has also been reported to be shortened by probiotics (Wen et al., 2020) based on three studies. However, based on six studies, van der Schoot and co-workers (2022) concluded there was no influence on transit time.

Potential mechanisms of action may include changes in the composition and/or activity of the colonic microbiota. The microbiota has been observed to influence colonic contractility. Metabolites, in particular short-chain fatty acids, are thought to play a role in modulating colonic contraction, independent of pH. Also, methane has been found to influence intestinal motility, although no consistent effect has been reported. Bile acid metabolism is also involved in motility, both in diarrhea and constipation (Dimidi et al., 2017).

# 23.5 INFANT COLIC

The most widely used definition for colic is the rule of three: an infant is considered to have colic if the infant fusses or cries for >3 hours, >3 days per week, for >3 weeks (Wessel et al., 1954). These criteria are often modified to span 1 week, instead of 3. Colic seems to peak at 5–6 weeks of life and is rare in infants of 9 weeks and older (Wolke et al., 2017). Although "colic" would suggest an abdominal origin of the condition, it is not certain whether this is the case. The intestinal microbiota may be involved, though; infants who expressed symptoms of colic were found to be colonized with significantly higher levels of Proteobacteria and exhibited lower bacterial diversity compared non-colic infants. Furthermore, colonization levels of *Bifidobacterium* and *Lactobacillus* were inversely related to the amount of crying and fussiness in infants. It is, of course, not clear what is cause and effect here (Dubois and Gregory, 2016).

In particular, one strain, *Limosilactobacillus reuteri* DSM 17938, and its parent strain, ATCC 55730, has been tested for this health target. A meta-analysis of five studies on the strain concluded that it is efficacious in reducing colic (Schreck Bird et al., 2017). Other tested strains are *L. rhamnosus* GG, *B. lactis* BB-12, and a combination of lactobacilli and bifidobacteria, all reported to reduce crying (Liu et al., 2022).

# 23.6 INFLAMMATORY BOWEL DISEASE

Inflammatory bowel disease is a common name given to a group of chronic inflammatory conditions in the GI, including Crohn's disease (CD), ulcerative colitis (UC), and pouchitis. Environmental, genetic, and immunological factors, together with the GI microbiota, are associated with IBD, which presents symptoms and responds to treatment in a considerably subject-specific manner (Rogler and Vavricka, 2015). The role of the GI bacteria in IBD is different from that of conventional pathogens since IBD reacts to anti-inflammatory treatment (Rapozo et al., 2017). Rather, the GI microbiota species composition in IBD appears to be in a dysbiotic state, as specified by decreased numbers of *Faecalibacterium prausnitzii*, *Bacteroides*, *Clostridium coccoides*, *Clostridium leptum*, and *Bifidobacterium* (Prosberg et al., 2016). The role of the intestinal microbiota is further highlighted by the fact that fecal microbial transplantation can be beneficial in the management of ulcerative colitis (Zhou et al., 2021).

Probiotics appear to be effective for inducing remission in active UC (Chen et al., 2021). Further, probiotics do not appear to be significantly different in their effectiveness as compared to mesalazine (Kaur et al., 2020). Probiotics, however, do not appear to be effective in maintaining remission in UC (Chen et al., 2021).

In the case of CD, the results on probiotic intervention studies have thus far not been encouraging. Meta-analyses consistently report that probiotic use neither appears to be beneficial in inducing CD remission nor in maintaining remission (Chen et al., 2021).

Pouchitis is an inflammatory condition that often occurs after ileal pouch anal anastomosis, that is, the construction of an artificial rectum from the distal end of the ileum after surgical removal of the large bowel. Based on eight studies, probiotics have been reported to reduce the risk of pouchitis after anal anastomosis (Xiao et al., 2023). Probiotics have been shown to also hold promise for inducing remission from pouchitis. The "De Simone Formula" performs substantially better than placebo (Nguyen et al., 2019) and similarly to metronidazole but with substantially fewer side effects (Poo et al., 2022).

Taken together, the evidence for a beneficial effect of probiotics on IBD in general seems to be inconclusive, though most promising effects can be expected in pouchitis and the induction of UC remission. However, the number of clinical intervention studies is still limited, and the size of the studies is also often quite small. Interestingly, probiotic supplements that are based on lactobacilli and bifidobacteria or more than one strain are more likely to be beneficial for IBD remission. The dose of $10^{10}$–$10^{12}$ CFU/day may be a reference range for using probiotics to relieve IBD (Zhang et al., 2021).

# 23.7 NECROTIZING ENTEROCOLITIS AND SHORT BOWEL SYNDROME

Necrotizing enterocolitis (NEC) is the most common abdominal emergency of preterm neonates in neonatal intensive care units with a mortality of 15%–30% (Samuels et al., 2017). NEC is thought to be attributed to a disturbed intestinal mucosal barrier and an abnormal intestinal microbiota harbored by most preterm infants, resulting from the reduced exposure to maternal microbiota, the consumption of a sterile diet, and the common use of antibiotics. Many randomized control trials on the prevention of NEC by probiotic therapies have been conducted. Several meta-analyses have been performed and consistently conclude that probiotic administration can reduce the risk for NEC (Thomas et al., 2017; Sun et al., 2017; Wang et al., 2023a). This observation has led to the suggestion that probiotics should be considered standard practice for prevention of NEC in at-risk populations (van den Akker et al., 2020), although this opinion is not shared by all (Poindexter, 2021). Most tested probiotics seem to exert a beneficial effect; they belong mainly to the genera *Bifidobacterium* and *Lactobacillus*. The effectiveness of such a wide selection of probiotic supplements could partially be due to the undeveloped gut microbiota of preterm infants and the fact that infants developing NEC have a higher relative abundance of Proteobacteria and a lower abundance of Fermicutes and Bacteroidetes (Pammi et al., 2017). Interestingly, multi-strain preparations appear to be significantly more efficacious then single-strain products (Chang et al., 2017; Sun et al., 2017).

Another GI impairment encountered by children is short bowel syndrome (SBS), often the result of intentional bowel resection due to NEC or a congenital short bowel. A common complication in SBS is small bowel bacterial overgrowth (SBBO) and D-lactic acidosis (Reddy et al., 2013). Probiotics have been little investigated for SBS and are not recommended (Merras-Salmio and Pakarinen, 2022). Most studies with probiotics for the management of SBBO are case reports, and there is a paucity in reporting. Although potential benefits have been suggested (Avelar Rodriguez et al., 2019), this remains to be thoroughly investigated. Probiotics producing L-lactate have been suggested to be effective in the treatment of D-lactic acidosis in SBS patients (Yilmaz et al., 2018).

# 23.8 ABDOMINAL SURGERY

Abdominal surgery caries a substantial risk for infections; more than 5% of patients develop surgical site infection (SSI). Postoperative infections increase antibiotic use, with the concomitant risk of antibiotic resistance, and prolong hospital stay, with a concomitant increase in medical care expenses (Wu et al., 2016). Probiotics may reduce these infections by modulating the immune system and providing a more benign microbiota, less capable of causing infections.

For SSI, the latest meta-analyses are conclusive, suggesting a significant reduction in SSI (Veziant et al., 2022; Araújo et al., 2023). Also, concerning sepsis risk, there appears to be a clear benefit from the use of probiotics (Araújo et al., 2023). There are also other potential benefits from the use of probiotics in relation to abdominal surgery, including significant reduction of any infection, pneumonia, urinary tract infection; a shortened length of antibiotic therapy, length of ICU stay, and length of hospital stay; and promotion of earlier first defecation and first bowel movement (Tang et al., 2022). Although surgical patients are a particularly vulnerable group, no increase in complications has been reported. Interestingly, subgroup analysis indicated that lower dose ($<10^9$ CFU), longer duration of supplementation (>14 days), and being administrated <5 days before and >10 days after surgery was more effective at reducing the incidence of surgical site infection (Araújo et al., 2023).

# 23.9 *HELICOBACTER PYLORI* ERADICATION

*Helicobacter pylori* infection may result in gastritis, predisposing to peptic ulcers and eventually increasing the risk of gastric cancer. Therefore, the eradication of *H. pylori* is important. Traditionally this has been achieved with a combination of antibiotics and proton-pump inhibitors, often resulting in several adverse effects during treatment.

Probiotics have been found to reduce overall side-effects of *Helicobacter* eradication therapy in children (Liang et al., 2023) and in adults (Wang et al., 2023b). Nausea and diarrhea have been reported to be improved in most studies (Zhang et al., 2020). Reduction of side effects and thereby better compliance with eradication therapy is thought to be one of the reasons for improved success. Most studies and meta-analyses also observe improved eradication rates with the inclusion of probiotics in the standard therapy in children (Liang et al., 2023) and in the general population (Wang et al., 2023b). Interestingly, *Bifidobacterium-Lactobacillus* and *Bifidobacterium-Lactobacillus-Saccharomyces* adjuvant therapy was found to benefit most, with a high eradication rate. Further, administering probiotics before or after triple therapy and longer duration of probiotics can improve therapeutic effect in *H. pylori*–infected individuals (Wang et al., 2023b; Zhang et al., 2020).

# 23.10 LACTOSE MALDIGESTION

With the exception of the populations of northern European origin, lactose maldigestion resulting from lactase deficiency affects the majority of adults worldwide. It is, actually, the normal state for adult mammals and should thus not be seen as a disorder. Lactose maldigesters usually tolerate yogurt better than nonfermented dairy products due to the lactase activity of the yogurt starter cultures (Corgneau et al., 2017). However, it is important to remember that strain-specific differences in lactase activity may vary 100-fold, even between the strains normally used in yogurt manufacture (Sanders et al., 1996). Not all probiotic strains are able to ferment lactose or alleviate lactose intolerance, and the effect is stronger in fermented products than in nonfermented products. Lactase production is strain dependent but also depends strongly on the growth conditions of the bacteria. Lactase production is strong when lactose is the sole source of energy but is reduced significantly in the presence of other energy sources such as glucose (Jiang et al., 1996).

The ability of several lactic acid bacteria starter strains and probiotic strains to improve lactose digestion has been documented (Leis et al., 2020). In addition to bacterial lactase activity, the mechanisms are believed to involve delayed GI transit, improvement of the functions of the commensal microbiota, and reduced sensitivity to symptoms (Savaiano, 2014).

# 23.11 CONCLUSIONS

The well-being of our gut may benefit from supplementation with specific probiotic strains or strain combinations. However, as probiotic effects are strain specific, conclusions based on meta-analyses with several kinds of probiotic supplements included should be drawn with caution. Notwithstanding this, the strongest evidence for probiotic functionality appears to be reducing risk for NEC. Further, in a strain-specific manner, probiotics may reduce the risk for various types of diarrhea, with the notable exception of travelers' diarrhea. Probiotics also appear beneficial in relieving constipation. Although it may seem contradictory that probiotics can be beneficial both for diarrhea and constipation, it is in line with the way probiotics work, supporting normal bodily functions. Probiotics are thus normalizing bowel movements, not just increasing or reducing them. There are also good opportunities for probiotics in IBS; but it is important that specific subgroups of IBS be considered. By reducing treatment side effects, probiotics may improve *Helicobacter* eradication. In particular, *L. reuteri* DSM 17938 appears to improve infant colic. Finally, dairy starter cultures such as *Streptococcus thermophilus* and *Lactobacillus delbrueckii* subsp. *bulgaricus* and selected probiotic strains may support lactose digestion. In most of these cases, the tested probiotics have been lactic acid bacteria and/or bifidobacteria. However, for most probiotic benefits, there remains need for well-designed clinical studies assessing various potential health benefits.

# BIBLIOGRAPHY

Ahmadi, E., Alizadeh-Navaei, R. & Rezai, M. S. 2015. Efficacy of probiotic use in acute rotavirus diarrhea in children: A systematic review and meta-analysis. *Caspian J Intern Med*, 6, 187–195.

Allen, S. J., Martinez, E. G., Gregorio, G. V. & Dans, L. F. 2010. Probiotics for treating acute infectious diarrhoea. *Cochrane Database Syst Rev*, CD003048.

Araújo, M. M., Montalvão-Sousa, T. M., Teixeira, P. D. C., Figueiredo, A. & Botelho, P. B. 2023. The effect of probiotics on postsurgical complications in patients with colorectal cancer: A systematic review and meta-analysis. *Nutr Rev*, 81, 493–510.

Asempa, T. E. & Nicolau, D. P. 2017. Clostridium difficile infection in the elderly: An update on management. *Clin Interv Aging*, 12, 1799–1809.

Avelar Rodriguez, D., Ryan, P. M., Toro Monjaraz, E. M., Ramirez Mayans, J. A. & Quigley, E. M. 2019. Small intestinal bacterial overgrowth in children: A state-of-the-art review. *Front Pediatr*, 7, 363.

Blaabjerg, S., Artzi, D. M. & Aabenhus, R. 2017. Probiotics for the prevention of antibiotic-associated diarrhea in outpatients-a systematic review and meta-analysis. *Antibiotics (Basel)*, 6.

Chang, H. Y., Chen, J. H., Chang, J. H., Lin, H. C., Lin, C. Y. & Peng, C. C. 2017. Multiple strains probiotics appear to be the most effective probiotics in the prevention of necrotizing enterocolitis and mortality: An updated meta-analysis. *PLoS ONE*, 12, e0171579.

Chen, M., Feng, Y. & Liu, W. 2021. Efficacy and safety of probiotics in the induction and maintenance of inflammatory bowel disease remission: A systematic review and meta-analysis. *Ann Palliat Med*, 10, 11821–11829.

Clayton, J. A. & Toltzis, P. 2017. Recent issues in pediatric clostridium difficile infection. *Curr Infect Dis Rep*, 19, 49.

Collinson, S., Deans, A., Padua-Zamora, A., Gregorio, G. V., Li, C., Dans, L. F. & Allen, S. J. 2020. Probiotics for treating acute infectious diarrhoea. *Cochrane Database Syst Rev*, 12, CD003048.

Corgneau, M., Scher, J., Ritie-Pertusa, L., Le, D. T. L., Petit, J., Nikolova, Y., Banon, S. & Gaiani, C. 2017. Recent advances on lactose intolerance: Tolerance thresholds and currently available answers. *Crit Rev Food Sci Nutr*, 57, 3344–3356.

Dekate, P., Jayashree, M. & Singhi, S. C. 2013. Management of acute diarrhea in emergency room. *Indian J Pediatr*, 80, 235–246.

Dimidi, E., Christodoulides, S., Scott, S. M. & Whelan, K. 2017. Mechanisms of action of probiotics and the gastrointestinal microbiota on gut motility and constipation. *Adv Nutr*, 8, 484–494.

Dinleyici, E. C., Eren, M., Ozen, M., Yargic, Z. A. & Vandenplas, Y. 2012. Effectiveness and safety of Saccharomyces boulardii for acute infectious diarrhea. *Expert Opin Biol Ther*, 12, 395–410.

Dong, J., Teng, G., Wei, T., Gao, W. & Wang, H. 2016. Methodological quality assessment of meta-analyses and systematic reviews of probiotics in inflammatory bowel disease and pouchitis. *PLoS ONE*, 11, e0168785.

Drossman, D. A. 2017. Improving the treatment of irritable bowel syndrome with the rome IV multidimensional clinical profile. *Gastroenterol Hepatol (N Y)*, 13, 694–696.

Dubois, N. E. & Gregory, K. E. 2016. Characterizing the intestinal microbiome in infantile colic: Findings based on an integrative review of the literature. *Biol Res Nurs*, 18, 307–315.

Fagnant, H. S., Isidean, S. D., Wilson, L., Bukhari, A. S., Allen, J. T., Agans, R. T., Lee, D. M., Hatch-Mcchesney, A., Whitney, C. C., Sullo, E., Porter, C. K. & Karl, J. P. 2023. Orally ingested probiotic, prebiotic, and synbiotic interventions as countermeasures for gastrointestinal tract infections in nonelderly adults: A systematic review and meta-analysis. *Adv Nutr*, 14(3), 539–554.

Ferrer, M., Mendez-Garcia, C., Rojo, D., Barbas, C. & Moya, A. 2017. Antibiotic use and microbiome function. *Biochem Pharmacol*, 134, 114–126.

Guarino, A., Dupont, C., Gorelov, A. V., Gottrand, F., Lee, J. K., Lin, Z., Lo Vecchio, A., Nguyen, T. D. & Salazar-Lindo, E. 2012. The management of acute diarrhea in children in developed and developing areas: From evidence base to clinical practice. *Expert Opin Pharmacother*, 13, 17–26.

Heaton, K. W., Radvan, J., Cripps, H., Mountford, R. A., Braddon, F. E. & Hughes, A. O. 1992. Defecation frequency and timing, and stool form in the general population: A prospective study. *Gut*, 33, 818–824.

Hempel, S., Newberry, S. J., Maher, A. R., Wang, Z., Miles, J. N., Shanman, R., Johnsen, B. & Shekelle, P. G. 2012. Probiotics for the prevention and treatment of antibiotic-associated diarrhea: A systematic review and meta-analysis. *JAMA*, 307, 1959–1969.

Jiang, T., Mustapha, A. & Savaiano, D. A. 1996. Improvement of lactose digestion in humans by ingestion of unfermented milk containing Bifidobacterium longum. *J Dairy Sci*, 79, 750–757.

Kaur, L., Gordon, M., Baines, P. A., Iheozor-Ejiofor, Z., Sinopoulou, V. & Akobeng, A. K. 2020. Probiotics for induction of remission in ulcerative colitis. *Cochrane Database Syst Rev*, 3, Cd005573.

Konstantis, G., Efstathiou, S., Pourzitaki, C., Kitsikidou, E., Germanidis, G. & Chourdakis, M. 2023. Efficacy and safety of probiotics in the treatment of irritable bowel syndrome: A systematic review and meta-analysis of randomised clinical trials using ROME IV criteria. *Clin Nutr*, 42, 800–809.

Lau, C. S. & Chamberlain, R. S. 2016. Probiotics are effective at preventing Clostridium difficile-associated diarrhea: A systematic review and meta-analysis. *Int J Gen Med*, 9, 27–37.

Leis, R., De Castro, M. J., De Lamas, C., Picans, R. & Couce, M. L. 2020. Effects of prebiotic and probiotic supplementation on lactase deficiency and lactose intolerance: A systematic review of controlled trials. *Nutrients*, 12.

Liang, M., Zhu, C., Zhao, P., Zhu, X., Shi, J. & Yuan, B. 2023. Comparison of multiple treatment regimens in children with *Helicobacter pylori* infection: A network meta-analysis. *Front Cell Infect Microbiol*, 13, 1068809.

Liu, H. N., Wu, H., Chen, Y. Z., Chen, Y. J., Shen, X. Z. & Liu, T. T. 2017a. Altered molecular signature of intestinal microbiota in irritable bowel syndrome patients compared with healthy controls: A systematic review and meta-analysis. *Dig Liver Dis*, 49, 331–337.

Liu, P. C., Yan, Y. K., Ma, Y. J., Wang, X. W., Geng, J., Wang, M. C., Wei, F. X., Zhang, Y. W., Xu, X. D. & Zhang, Y. C. 2017b. Probiotics reduce postoperative infections in patients undergoing colorectal surgery: A systematic review and meta-analysis. *Gastroenterol Res Pract*, 2017, 6029075.

Liu, Y., Ma, D., Wang, X. & Fang, Y. 2022. Probiotics in the treatment of infantile colic: A meta-analysis of randomized controlled trials. *Nutr Hosp*, 39, 1135–1143.

Ma, Y., Yang, J. Y., Peng, X., Xiao, K. Y., Xu, Q. & Wang, C. 2020. Which probiotic has the best effect on preventing Clostridium difficile-associated diarrhea? A systematic review and network meta-analysis. *J Dig Dis*, 21, 69–80.

Mantegazza, C., Molinari, P., D'Auria, E., Sonnino, M., Morelli, L. & Zuccotti, G. V. 2018. Probiotics and antibiotic-associated diarrhea in children: A review and new evidence on *Lactobacillus rhamnosus* GG during and after antibiotic treatment. *Pharmacol Res*, 128, 63–72.

Mcfarland, L. V. 2009. Evidence-based review of probiotics for antibiotic-associated diarrhea and Clostridium difficile infections. *Anaerobe*, 15, 274–280.

Mcfarland, L. V. & Goh, S. 2019. Are probiotics and prebiotics effective in the prevention of travellers' diarrhea: A systematic review and meta-analysis. *Travel Med Infect Dis*, 27, 11–19.

Merras-Salmio, L. & Pakarinen, M. P. 2022. Infection prevention and management in pediatric short bowel syndrome. *Front Pediatr*, 10, 864397.

Napolitano, L. M. & Edmiston, C. E., Jr. 2017. Clostridium difficile disease: Diagnosis, pathogenesis, and treatment update. *Surgery*, 162, 325–348.

Nguyen, N., Zhang, B., Holubar, S. D., Pardi, D. S. & Singh, S. 2019. Treatment and prevention of pouchitis after ileal pouch-anal anastomosis for chronic ulcerative colitis. *Cochrane Database Syst Rev*, 11, Cd001176.

Niu, H. L. & Xiao, J. Y. 2020. The efficacy and safety of probiotics in patients with irritable bowel syndrome: Evidence based on 35 randomized controlled trials. *Int J Surg*, 75, 116–127.

Pammi, M., Cope, J., Tarr, P. I., Warner, B. B., Morrow, A. L., Mai, V., Gregory, K. E., Kroll, J. S., Mcmurtry, V., Ferris, M. J., Engstrand, L., Lilja, H. E., Hollister, E. B., Versalovic, J. & Neu, J. 2017. Intestinal dysbiosis in preterm infants preceding necrotizing enterocolitis: A systematic review and meta-analysis. *Microbiome*, 5, 31.

Patro-Golab, B., Shamir, R. & Szajewska, H. 2015. Yogurt for treating antibiotic-associated diarrhea: Systematic review and meta-analysis. *Nutrition*, 31, 796–800.

Poindexter, B. 2021. Use of probiotics in preterm infants. *Pediatrics*, 147.

Poo, S., Sriranganathan, D. & Segal, J. P. 2022. Network meta-analysis: Efficacy of treatment for acute, chronic, and prevention of pouchitis in ulcerative colitis. *Eur J Gastroenterol Hepatol*, 34, 518–528.

Preidis, G. A., Hill, C., Guerrant, R. L., Ramakrishna, B. S., Tannock, G. W. & Versalovic, J. 2011. Probiotics, enteric and diarrheal diseases, and global health. *Gastroenterology*, 140, 8–14.

Prosberg, M., Bendtsen, F., Vind, I., Petersen, A. M. & Gluud, L. L. 2016. The association between the gut microbiota and the inflammatory bowel disease activity: A systematic review and meta-analysis. *Scand J Gastroenterol*, 51, 1407–1415.

Rapozo, D. C., Bernardazzi, C. & De Souza, H. S. 2017. Diet and microbiota in inflammatory bowel disease: The gut in disharmony. *World J Gastroenterol*, 23, 2124–2140.

Reddy, V. S., Patole, S. K. & Rao, S. 2013. Role of probiotics in short bowel syndrome in infants and children—a systematic review. *Nutrients*, 5, 679–699.

Ritchie, M. L. & Romanuk, T. N. 2012. A meta-analysis of probiotic efficacy for gastrointestinal diseases. *PLoS ONE*, 7, e34938.

Rogler, G. & Vavricka, S. 2015. Exposome in IBD: Recent insights in environmental factors that influence the onset and course of IBD. *Inflamm Bowel Dis*, 21, 400–408.

Samuels, N., Van De Graaf, R. A., De Jonge, R. C. J., Reiss, I. K. M. & Vermeulen, M. J. 2017. Risk factors for necrotizing enterocolitis in neonates: A systematic review of prognostic studies. *BMC Pediatr*, 17, 105.

Sanders, M. E., Walker, D. C., Walker, K. M., Aoyama, K. & Klaenhammer, T. R. 1996. Performance of commercial cultures in fluid milk applications. *J Dairy Sci*, 79, 943–955.

Savaiano, D. A. 2014. Lactose digestion from yogurt: Mechanism and relevance. *Am J Clin Nutr*, 99, 1251S–1255S.

Sawh, S. C., Deshpande, S., Jansen, S., Reynaert, C. J. & Jones, P. M. 2016. Prevention of necrotizing enterocolitis with probiotics: A systematic review and meta-analysis. *PeerJ*, 4, e2429.

Schmulson, M. J. & Drossman, D. A. 2017. What is new in rome IV. *J Neurogastroenterol Motil*, 23, 151–163.

Schreck Bird, A., Gregory, P. J., Jalloh, M. A., Risoldi Cochrane, Z. & Hein, D. J. 2017. Probiotics for the treatment of infantile colic: A systematic review. *J Pharm Pract*, 30(3), 366–374.

Shen, N. T., Leff, J. A., Schneider, Y., Crawford, C. V., Maw, A., Bosworth, B. & Simon, M. S. 2017. Cost-effectiveness analysis of probiotic use to prevent clostridium difficile infection in hospitalized adults receiving antibiotics. *Open Forum Infect Dis*, 4, ofx148.

Sun, J., Marwah, G., Westgarth, M., Buys, N., Ellwood, D. & Gray, P. H. 2017. Effects of probiotics on necrotizing enterocolitis, sepsis, intraventricular hemorrhage, mortality, length of hospital stay, and weight gain in very preterm infants: A meta-analysis. *Adv Nutr*, 8, 749–763.

Tang, G., Huang, W., Tao, J. & Wei, Z. 2022. Prophylactic effects of probiotics or synbiotics on postoperative ileus after gastrointestinal cancer surgery: A meta-analysis of randomized controlled trials. *PLoS ONE*, 17, e0264759.

Thomas, J. P., Raine, T., Reddy, S. & Belteki, G. 2017. Probiotics for the prevention of necrotising enterocolitis in very low-birth-weight infants: A meta-analysis and systematic review. *Acta Paediatr*, 106, 1729–1741.

Thompson, J. R. 2016. Is irritable bowel syndrome an infectious disease? *World J Gastroenterol*, 22, 1331–1334.

Urbanska, M., Gieruszczak-Bialek, D., Szymanski, H. & Szajewska, H. 2016. Effectiveness of lactobacillus reuteri DSM 17938 for the prevention of nosocomial diarrhea in children: A randomized, double-blind, placebo-controlled trial. *Pediatr Infect Dis J*, 35, 142–145.

Van Den Akker, C. H. P., Van Goudoever, J. B., Shamir, R., Domellof, M., Embleton, N. D., Hojsak, I., Lapillonne, A., Mihatsch, W. A., Canani, R. B., Bronsky, J., Campoy, C., Fewtrell, M. S., Mis, N. F., Guarino, A., Hulst, J. M., Indrio, F., Kolacek, S., Orel, R., Vandenplas, Y., Weizman, Z. & Szajewska, H. 2020. Probiotics and preterm infants: A position paper by the ESPGHAN committee on nutrition and the ESPGHAN working group for probiotics and prebiotics. *J Pediatr Gastroenterol Nutr*, 70(5), 664–680.

Vandenplas, Y. & De Hert, S. 2012. Cost/benefit of synbiotics in acute infectious gastroenteritis: Spend to save. *Benef Microbes*, 3, 189–194.

Van Der Schoot, A., Helander, C., Whelan, K. & Dimidi, E. 2022. Probiotics and synbiotics in chronic constipation in adults: A systematic review and meta-analysis of randomized controlled trials. *Clin Nutr*, 41, 2759–2777.

Veziant, J., Bonnet, M., Occean, B. V., Dziri, C., Pereira, B. & Slim, K. 2022. Probiotics/synbiotics to reduce infectious complications after colorectal surgery: A systematic review and meta-analysis of randomised controlled trials. *Nutrients*, 14, 3066.

Vlasova, A. N., Kandasamy, S., Chattha, K. S., Rajashekara, G. & Saif, L. J. 2016. Comparison of probiotic lactobacilli and bifidobacteria effects, immune responses and rotavirus vaccines and infection in different host species. *Vet Immunol Immunopathol*, 172, 72–84.

Wang, H., Meng, X., Xing, S., Guo, B., Chen, Y. & Pan, Y. Q. 2023a. Probiotics to prevent necrotizing enterocolitis and reduce mortality in neonates: A meta-analysis. *Medicine (Baltimore)*, 102, e32932.

Wang, Y., Wang, X., Cao, X. Y., Zhu, H. L. & Miao, L. 2023b. Comparative effectiveness of different probiotics supplements for triple *Helicobacter pylori* eradication: A network meta-analysis. *Front Cell Infect Microbiol*, 13, 1120789.

Wen, Y., Li, J., Long, Q., Yue, C. C., He, B. & Tang, X. G. 2020. The efficacy and safety of probiotics for patients with constipation-predominant irritable bowel syndrome: A systematic review and meta-analysis based on seventeen randomized controlled trials. *Int J Surg*, 79, 111–119.

Wessel, M. A., Cobb, J. C., Jackson, E. B., Harris, G. S., Jr. & Detwiler, A. C. 1954. Paroxysmal fussing in infancy, sometimes called colic. *Pediatrics*, 14, 421–435.

Wolke, D., Bilgin, A. & Samara, M. 2017. Systematic review and meta-analysis: Fussing and crying durations and prevalence of colic in infants. *J Pediatr*, 185, 55–61 e4.

Wu, X. D., Liu, M. M., Liang, X., Hu, N. & Huang, W. 2016. Effects of perioperative supplementation with pro-/synbiotics on clinical outcomes in surgical patients: A meta-analysis with trial sequential analysis of randomized controlled trials. *Clin Nutr*, 37(2), 505–515.

Xiao, W., Zhao, X., Li, C., Huang, Q., He, A. & Liu, G. 2023. The efficacy of probiotics on the prevention of pouchitis for patients after ileal pouch-anal anastomosis: A meta-analysis. *Technol Health Care*, 31, 401–415.

Yilmaz, B., Schibli, S., Macpherson, A. J. & Sokollik, C. 2018. D-lactic acidosis: Successful suppression of d-lactate-producing lactobacillus by probiotics. *Pediatrics*, 142.

Zhang, M., Zhang, C., Zhao, J., Zhang, H., Zhai, Q. & Chen, W. 2020. Meta-analysis of the efficacy of probiotic-supplemented therapy on the eradication of H. pylori and incidence of therapy-associated side effects. *Microb Pathog*, 147, 104403.

Zhang, X. F., Guan, X. X., Tang, Y. J., Sun, J. F., Wang, X. K., Wang, W. D. & Fan, J. M. 2021. Clinical effects and gut microbiota changes of using probiotics, prebiotics or synbiotics in inflammatory bowel disease: A systematic review and meta-analysis. *Eur J Nutr*, 60, 2855–2875.

Zhou, H. Y., Guo, B., Lufumpa, E., Li, X. M., Chen, L. H., Meng, X. & Li, B. Z. 2021. Comparative of the effectiveness and safety of biological agents, tofacitinib, and fecal microbiota transplantation in ulcerative colitis: Systematic review and network meta-analysis. *Immunol Invest*, 50, 323–337.

# Probiotics and the Immune System

# 24

## Hanne Frøkiær

---

## 24.1 INTRODUCTION

---

Like all other bacteria, probiotic bacteria in contact with the body will be recognized by the immune system, which in most cases will exclude the bacteria from entering the body through the surfaces. If unsuccessful in the exclusion, upon entering circulation, the immune system will initiate phagocytosis and killing of the bacteria. In contrast to pathogenic bacteria, probiotic bacteria do not hold virulence factors and hence will not be able to escape immunological recognition, phagocytosis, and killing in healthy individuals. Moreover, probiotics produce metabolites that directly or after transformation by other bacteria can affect cells of the immune system (postbiotics).

Probiotic bacteria are most often administered orally, either through our diet or as capsules. Accordingly, the cells and physiological conditions in the gastrointestinal tract are of relevance for the way by which probiotic bacteria may interact with and influence the immune system. As for other bacteria, entering probiotic bacteria into circulation may stimulate the innate as well as the adaptive immune system, and the response elicited may depend on where in the body bacteria end up. Administering probiotics on the skin for protection and stimulation of the various cells, including immune cells located here, is a new growing application,

This chapter is divided into two sections; the first section gives a brief introduction to the most relevant parts of the immune system and the mechanisms by which the immune system responds to bacteria, with particular focus on probiotic bacteria. The second section discusses the current knowledge and evidence for an effect of probiotics on the immune system and which mechanisms may be involved.

---

## 24.2 THE GUT MUCOSAL BARRIER

---

The epithelial cells form the interface between the body and the vast number of microorganisms present at the surrounding surfaces. In particular, the monolayer of epithelial cells covering the surface of our gastrointestinal tract constitutes a barrier which, due to its role in absorption of nutrients, requires that on one hand the nutrients be able to transverse the surface and on the other that microorganisms be excluded from entering into circulation. Of the entire mucosal barrier, the epithelial barrier of the intestine, especially the small intestine, is by far the largest. To allow such regulated entrance requires a unique architecture and systems of defense (Figure 24.1).

DOI: 10.1201/9781003352075-27

The mucosa of the small intestine is made up of villi that constitute a monolayer of epithelial cells, which are continuously renewed by cells formed from stem cells located in the crypts between the villi. Apart from the monolayer of epithelial cells, an external mucus layer and the underlying *lamina propria* together with the associated lymphoid organs constitute the physical intestinal barrier. The intestinal epithelial cells consist of absorptive enterocytes, mucus-secreting goblet cells, enteroendocrine cells, M cells, and Paneth cells, all of which originate from the stem cells located in the crypts of the villi (Artis, 2008). Of these cells, absorptive enterocytes, Paneth cells, and goblet cells play key roles in the innate defense of the mucosal barrier.

Beneath the epithelial cell layer, immune cells in high numbers are scattered in the *lamina propria*. Here, a rich number of dendritic cells, macrophages, plasma cells, and various other lymphocytes are present. Macrophages and dendritic cells are positioned close to the epithelium, ready to sense and phagocytose antigens and microorganisms that have succeeded in crossing the intestinal epithelium, as well as apoptotic and damaged cells. The plasma cells secrete large amounts of IgA that are shuttled through the epithelial cells to the mucosal surface where IgA binds to the various microbes, thereby inhibiting microbial binding to and penetration through the epithelial layer (Gutzeit et al., 2014).

More ordered assemblies of immune cells are found in Peyer's patches and isolated lymphoid follicles located just below the epithelial cell layer and in the mesenteric lymph nodes, filtering the mesenteric lymph that drains the mucosal sites for activated cells, absorbed lipids, and antigens. These lymphoid structures constitute the inductor sites of the gut immune system, where dendritic cells present food antigens and microbial antigens to naïve T cells, inducing antigen-specific T cells and B cells. Food antigens and bacteria are shuttled through the epithelial cell layer through M cells overlying the Peyer's patches. M cells have a special surface that facilitates uptake of microorganisms. The basolateral side of the M cells are in close contact with antigen-presenting cells. Due to the special conditions here, dendritic cells acquire a specific phenotype, primarily leading to induction of Treg cells and plasma cells producing IgA specific to the food antigens and bacteria. The antigen-specific cells induced here are equipped with homing markers

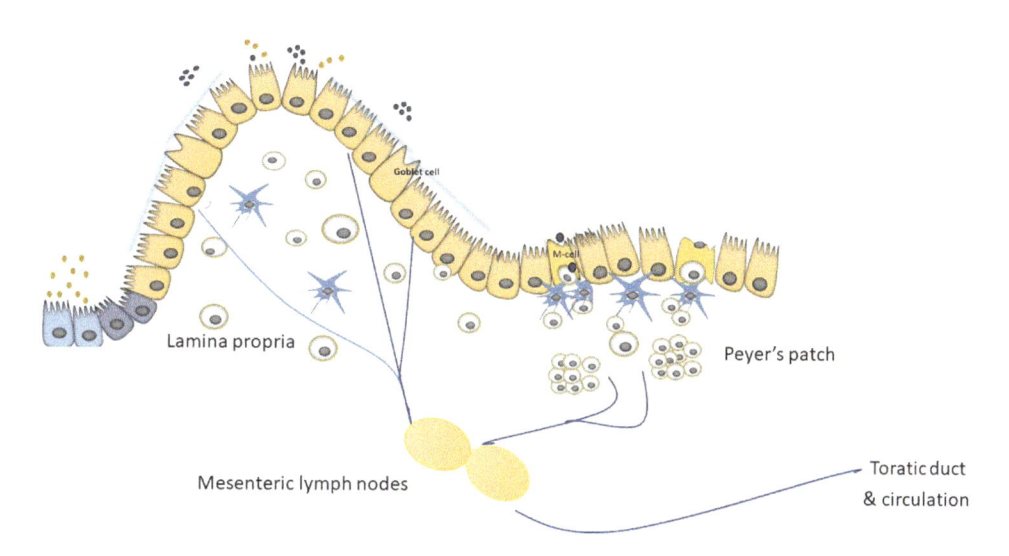

**FIGURE 24.1**   The architecture of the small intestinal mucosa and gut-associated lymph tissue (GALT). A single layer of epithelial cells segregates the cells of the body from the environment. Probiotics and other microorganisms are shuttled through M cells overlaying the intestinal lymph nodes (Peyer's patches), where they initiate an immune response inclusive IgA-producing plasma cells. Microorganisms that enter the *lamina propria* are drained to the mesenteric lymph nodes where an immune response is initiated. The T cells and plasma cells home back to the *lamina propria* (see Figure 24.2 for more details).

directing the mature cells home to the *lamina propria*, the effector site of the gut immune system. Here, antigen-specific cells are present in high numbers together with macrophages and dendritic cells.

Although the intestinal epithelial cells are not classical immune cells, these cells are equipped with many of the receptors found on classical immune cells, and they respond to microbial stimulation through these receptors. Hence, their involvement in the immune response against microorganisms is now well accepted.

The presence of microorganisms at the surface is imperative in order to stimulate the epithelial production of mucins and antimicrobial proteins (AMPs) and recruit immune cells to the mucosal sites. Epithelial cells of the small intestine use recognition of microbial cell components and metabolites to adjust their production of antimicrobial components and to sustain immunological and metabolic homeostasis. In addition, the epithelial cells produce various cytokines that condition the dendritic cells and macrophages located in the *lamina propria* to achieve special phenotypes that are regulatory and less inflammatory.

## 24.3 THE IMMUNOLOGICAL RESPONSE TO PROBIOTIC BACTERIA IN THE GUT

Bacteria present in the gastrointestinal tract may stimulate the intestinal epithelial cells to produce various compounds of importance for a strong barrier as well as molecules (e.g., cytokines) that stimulate and condition immune cells located beneath the epithelial layer (Maldonado-Contreras and McCormick, 2011). The bacteria furthermore produce various metabolites from the degradation of nutrients, which also stimulate or modulate the response of the immune cells.

Despite the efficient intestinal barrier, some bacteria from the gastrointestinal tract eventually succeed in entering into the circulation and may here stimulate a specific response as well as a general homeostasis of the immune system (Figure 24.2, Table 24.1).

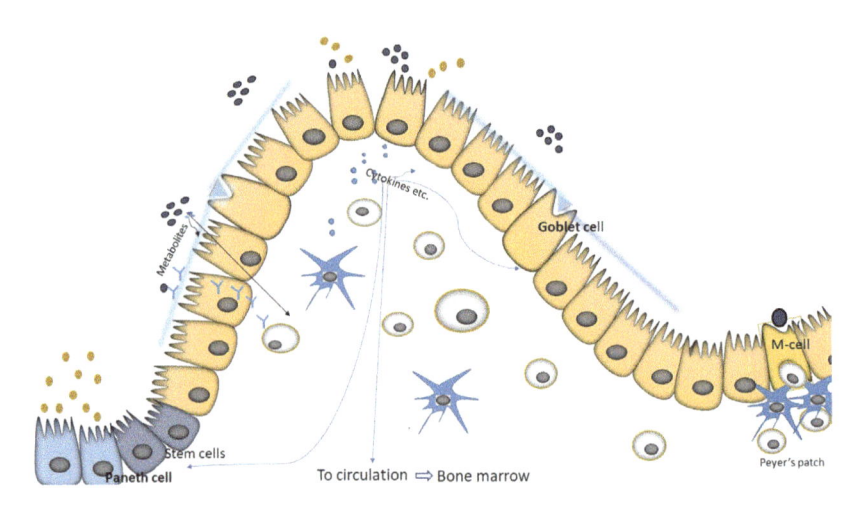

**FIGURE 24.2** Probiotics and other bacteria may stimulate epithelial cells to increase the production and release of mucins from goblet cells, antimicrobial peptides (AMPs) from Paneth cells and absorptive enterocytes. Various cytokines from the epithelial cells condition and stimulate increased mucus and AMP production and stimulate the immune cells scattered in the *lamina propria*, which under healthy conditions lead to a regulatory state. Probiotics may also produce various metabolites, including SCFA, that enhance barrier function and may enhance number and function of regulatory T cells.

**TABLE 24.1**   Mechanisms That May Be Involved in Immune Modulation by Probiotics

| | EFFECT | CONSEQUENCE |
|---|---|---|
| Stimulation of enterocytes | Stimulation of cytokine production | Recruitment of macrophages and dendritic cells<br>Conditioning of cells in *lamina propria* |
| | Stimulation of inflammasomes | Activation of IL18 leading to production of antimicrobial peptides |
| Stimulation of goblet cells | Release of Muc2<br>Generation of more goblet cells | Increased mucus layer<br>Increased mucus production |
| Stimulation of Paneth cells | Production of antimicrobial peptides | Reduction of bacterial load at epithelial surface |
| | Stimulation of stem cell regeneration | Regeneration of epithelium |
| Stimulation of the adaptive immune system in PP/MLN[a] | Production of IgA | Protection of epithelium through IgA binding of bacteria<br>Induction of Treg |
| Stimulation through metabolites produced | Effect of lactate<br>Effect of butyrate | Nutrient for butyrate producers. Direct stimulation of enterocytes and Treg generation<br>Nutrient for enterocytes<br>Treg generation |
| | Effect of propionate | Protects epithelial tissue towards damage during inflammation |
| | Effect of SCFA on histone deacetylases | Increased Treg and Treg function |
| | Effect of indole | Maintenance of epithelial cells and barrier |
| Stimulation of hematopoiesis | Increased concentration of monocytes and neutrophils in blood | Increased phagocytic activity |

[a] *PP/MLN:* Peyer's patches and mesenteric lymph nodes; SCFA: short-chain fatty acid.

# 24.3.1 Stimulation of Intestinal Epithelial Cells

Epithelial cells lining the intestines express various microbial receptors including toll-like receptors (TLRs) and the intracellular nucleotide-binding oligomerization domain-containing (NOD) proteins (Kawai and Akira, 2011; Al Nabhani et al., 2017). Together with TLR1 and TLR6, TLR2 is of most relevance with respect to probiotic bacteria (e.g., lactobacilli). TLR2 recognizes lipoproteins (Hashimoto et al., 2006), found on most bacteria, in conjunction with either TLR1 or TLR6. NOD2 recognizes muramyl dipeptide (MDP), a peptidoglycan motif (Shaw et al., 2008), and NOD1 recognizes d-glutamyl-meso-diaminopimelic acid (DAP), which is primarily found in Gram-negative bacteria. However, a few Gram-positive bacteria, including *Bacillus*, which are also used as probiotics, are also a source of DAP (Chamaillard et al., 2003; Hasegawa et al., 2006). Most probiotics on the market are Gram-positive bacteria. Gram-positive bacteria hold a thick cell wall mainly consisting of peptidoglycan (Malanovic and Lohner, 2016). In addition, dead bacteria constitute a source of peptidoglycan. Intake of probiotic bacteria thus provides a rich source of peptidoglycan to the small intestine.

Microbial stimulation through TLR and NOD plays a key role in maintaining the special segregation between the commensal microbiota and the interior of the host that exists in the gut. It has been demonstrated in knock-out mouse strains that TLR ligand signaling components such as MyD88 and NEMO are indispensable for maintaining this segregation (Rakoff-Nahoum et al., 2004; Nenci et al., 2007). Hence, in the small intestine, where the microbial load is much lower than in the colon, intact or degraded probiotic bacteria may directly stimulate immune defenses of the gut barrier.

## 24.3.2 Goblet Cells and Mucus Production

Goblet cells are the producers of a number of highly glycosylated proteins called mucins of which Mucin 2 (Muc2) constitutes the major part. Muc2 is secreted into the lumen upon microbial stimulation (Birchenough et al., 2016), where it, together with the other mucins, of which the majority are membrane bound on the epithelial cells, forms a viscous layer that keeps most of the microorganisms from the epithelial surface. Thus, the mucus layer is almost impermeable to the majority of microorganisms present in the gut. Hence, a massive contact between microbes and epithelial cells is prevented and, accordingly, exaggerated stimulation of epithelial cells as well as microbial translocation. The number of goblet cells present in the epithelial layer is at least to some extent determined by the microbiota. Recently, it was demonstrated that simultaneous administration of NOD1 and NOD2 ligands to wild-type mice increased the number of Muc2-positive goblet cells (Wang et al., 2015). Probiotic bacteria in the gut are thus expected to stimulate Muc2 production as well as secretion. To this end, several probiotic species have been demonstrated to induce increased expression of Muc2 in the human intestinal cell lines Caco2 and LS 174T (Mattar et al., 2002; Caballero-Franco et al., 2007), and administration of the probiotic mixture VSL#3 to rats led to an increased expression and secretion of Muc2 (Caballero-Franco et al., 2007). Hence, like other microbes, probiotic bacteria hold the capability to stimulate the production and secretion of Muc2 of importance for a thick mucus layer segregating the microbiota from the host.

## 24.3.3 Paneth Cells and Antimicrobial Peptides

Paneth cells present in the crypts of the villi are especially strong producers of a variety of AMPs, including REGIII and α defensins (Bevins and Salzman, 2011). The expression of AMP as well as the release of the AMP from Paneth cells are dependent on microbial stimulation of the cell, but the exact mechanism of stimulation remains to be established.

Similar to stimulating the number of goblet cells, it was recently shown that mice fed probiotic bacteria demonstrated an increase in the number of Paneth cells in the small intestine and increased intestinal fluidic antimicrobial activity (Carzola et al., 2018). Of note, no effect of the probiotics was found in the colon already hosting a vast number of microbes. Hence, the effect may signify that, in particular, mucosal sites with a relatively low microbial burden (the small intestine) will be affected by probiotic intake as this confers an increase of microorganisms present in the small intestine. As the small intestine is the site of infection for many food- and waterborne pathogens such as *Salmonella* and *Listeria*, probiotic intake that increases AMP production in the small intestine might in fact prevent some bacterial gut infections.

In particular, NOD2 is highly expressed in Paneth cells of the small intestine (Blander et al., 2017). Ligation by peptidoglycan to NOD2 induces several cellular events, including secretion of cytokines, epithelial regeneration, and production of AMPs (Nigro et al., 2014; Ramanan et al., 2014). As several probiotic bacteria stimulate through NOD2, probiotics may affect the microbial composition by stimulating the production of AMPs (Wehkamp et al., 2004; Zeuthen et al., 2008a).

Ligation of epithelial NOD1 induces the expression of the chemokine C-C motif chemokine 20 (CCL20), which mediates generation of isolated lymphoid follicles in the intestine and homeostatic bacterial colonization (Bouskra et al., 2008). Hence, probiotic bacteria capable of stimulating through NOD1 (e.g., *Escherichia coli* Nissle and *Bacillus* spp.) might influence bacterial homeostasis through this receptor.

## 24.3.4 Stimulation of Intestinal Epithelial Cells and Recruitment of Immune Cells

The intestinal absorptive cells also respond to receptors of microbial components, such as TLRs and NOD. Upon ligation, signaling pathways are activated leading to transcription of pro-inflammatory cytokines, such as TNF-α, pro-IL1, and pro-IL18, together with chemokines, for example, IL8. TNF-α and IL1β are important cytokines that together with chemokines stimulate the recruitment of macrophages, dendritic cells, and other immune cells to the intestines. Moreover, gut epithelial cells express the protein NLRP6, which in association with ASC (apoptosis-associated speck-like protein containing a CARD) and Caspase-1 constitutes the inflammasome that upon assembly activates pro-IL1β and pro-IL18 by cleavage, resulting in the secretion of IL1β and IL18 (Wlodarska et al., 2014; Levy et al., 2015). Transcription and cleavage of IL18 in enterocytes followed by IL18 binding to the IL18 receptor (IL18R) leads to production of AMPs and induces mucus secretion from goblet cells. Hence, the transcription of inflammasome proteins and the assembly of the inflammasome is required for an appropriate recruitment of immune cells to the *lamina propria*, as well as for production of AMPs and mucus. Many metabolites (e.g. butyrate) produced by the microorganisms also stimulate the production of AMP (Zhao et al., 2018). Hence, provided the presence of butyrate producing bacteria in the gut, lactic acid–producing probiotic bacteria may stimulate epithelial production of AMP.

The macrophages and dendritic cells recruited to the *lamina propria* attain an anti-inflammatory/tolerogenic phenotype due to the conditioning by several cytokines and other mediators secreted by the epithelial cells: thymic stromal lymphopoitin (TSLP), transforming growth factor β (TGF-β), IL-10, and prostaglandin E2 (PGE2), together with various metabolites. The tolerogenic phenotype is not fully dependent on the presence of a microbiota, as TSLP is constitutively expressed by the epithelial cells, but the expression may be upregulated, for example, in response to inflammation and tissue injury or bacterial stimulation. Many probiotic bacterial strains were able to stimulate the upregulation of the production of TSLP and to a lesser extent TGF-β in Caco-2 cells. Gram-negative bacteria including *E. coli* Nissle were in general the most potent stimulators (Zeuthen et al., 2008b).

Activation and polarization of naïve Th cells are crucial for the type of immune response generated. Dendritic cells are the principal stimulators of naïve Th cells. The phenotype of the mucosal dendritic cells varies depending on location. Together with the microbial stimulus, this may determine which type of Th cells are induced (Table 24.2). In the Peyer's patches in the small intestines, dendritic cells expressing CD11b are located in close contact with the epithelial layer and under homeostatic conditions produce IL-10 in response to antigen encounter (Chang et al., 2014). Located in the T-cell area of the Peyer's patches a less abundant type of dendritic cells that does not express CD11b but is capable of producing IL-12 upon stimulation and are involved in antigen presentation. Dendritic cells are also located in the *lamina propria*. They sample antigen coming from the lumen and migrate to the mesenteric lymph nodes to present antigen for T cells. Under homeostatic conditions (in the absence of infection or inflammation), these cells stimulate the generation of regulatory T cells that express gut-homing chemokine receptor CCR9 and integrin $\alpha_4\beta_7$, directing the T cells back to the *lamina propria* (Chang et al., 2014). Even though much research has been done on how various probiotic bacteria stimulate and polarize dendritic cells, it is not documented that probiotic bacteria affect the mucosal environment by directly affecting the polarization of these cells. Most work on how probiotics may stimulate dendritic cells has been done on dendritic cells generated from human peripheral blood or murine bone marrow, which may not be good representatives of the dendritic cells at mucosal sites. Conditioning such dendritic cells with cell supernatant from an intestinal cell line (e.g., Caco-2 cells) results in the induction of more tolerogenic cytokine profiles (Zeuthen et al., 2008b).

Importantly, the entering of probiotic bacteria into the Peyer's patches and mesenteric lymph nodes will lead to the induction of IgA-producing plasma cells equipped with the gut-homing chemokine receptors and integrins, resulting in secretion of probiotic-specific antibodies in the gut mucosa. Recent research points at IgA that is secreted into the lumen and binds to the mucin proteins as a pivotal site of adherence of the microorganisms in the gut along the epithelial surface (Corthesy, 2013). Accordingly, ingestion of probiotic bacteria is likely to induce plasma cells producing IgA that specifically bind to the ingested probiotic bacteria. Thereby, the ingested probiotic bacteria and probably other bacteria cross-reacting with the antibodies will be able to coat/adhere to the epithelium and prevent access of putative pathogenic bacteria to the epithelial surface and accordingly diminish their possibility to cause infection.

Innate lymphoid cells (ILCs) are a recently discovered group of lymphocytes belonging to the innate branch of the immune system. Although the importance of these cells for a well-balanced immune defense

**TABLE 24.2**   T Helper (Th) Cell Subtypes: Cytokine Production and Their Function

|      | CYTOKINES PRODUCED | FUNCTION |
|------|--------------------|----------|
| Th1  | INFγ, TNFα, GM-CSF | Activation of cellular cytotoxicity<br>Macrophage activating |
| Th2  | IL4, IL5, IL13     | Barrier immunity activation<br>Helminth defense |
| Th9  | IL9                | Promotes survival of Treg cells<br>Enhances the suppressive function of Treg cells |
| Th17 | IL17, IL22         | Barrier immunity activation<br>Neutrophil recruitment<br>Suppression of IL10 production |
| Th22 | IL22               | Barrier protection, stimulates expression of antimicrobial peptides<br>Prevents inflammation<br>Provides protection against autoimmune diseases |
| Treg | TGFβ, IL-10        | Suppression of other Th cells<br>Activation of B-cells |

at the mucosal barriers is by now generally accepted, the precise mechanisms through which the ILCs exert their influence on the mucosal barrier remains unclear. It is, however, clear that the influence of the microbiota is important for the maturation and acquisition of the tissue-specific function of the ILCs (Sonnenberg et al., 2012; Hepworth et al., 2013). Apart from the natural killer cells (NK cells), the ILC family consists of the non-cytotoxic innate lymphoid subsets ILC1, ILC2, and ILC3.

NK cells are recruited to the intestinal epithelial site upon microbial stimulation and in particular during viral infection. Many studies involving both humans and animals suggest that ingestion of certain probiotics enhances NK cell activity in blood, indicated by an increased tumoricidal activity and/or increase in the proportion of NK cells (Nagao et al., 2000; Gill et al., 2001).

Upon stimulation by IL1β in the intestine, ILC3s secrete granulocyte-macrophage colony stimulating factor (GM-CSF), which is required for induction of oral tolerance (Mortha et al., 2014). Commensal bacteria may also induce ILC3 cells via mucosal dendritic cells (Sonnenberg and Artis, 2015); thus it is likely that probiotics are good ILC3 stimulators. ILC3 also produces IL22, which induces AMP production in the epithelial cells and expression of the enzyme fucosyltransferase, facilitating fucosylation of surface proteins involved in the barrier defense against enteric pathogens (Goto et al., 2014). Studies have shown that IL22 is expressed at barrier surfaces and that its expression is dysregulated in certain human diseases (e.g., skin inflammation), which suggests a critical role in the maintenance of normal barrier homeostasis (Sonnenberg and Artis, 2015). Although only a little is known about how probiotics may influence ILC regulations, the findings that ILCs play a key role in gut homeostasis and that the gut commensal influence these cells calls for attention and suggests that the modulation of ILCs may represent an important mechanism to be involved in the activity of probiotics.

# 24.4 INDIRECT EFFECTS OF PROBIOTICS THROUGH THE METABOLITES THEY PRODUCE

Probiotic bacteria may also indirectly influence the immune system via the metabolites they produce. Here the production of short-chain fatty acids (SCFA), including acetic acid, propionic acid, and butyric acid, is mostly described. Many probiotic bacteria, in particular bifidobacteria and certain lactobacilli, express enzymes and transporters required to degrade indigestible carbohydrates such as galacto-oligo-saccharides (GOS) and fructo-oligosaccharides (FOS) (Pokusaeva et al., 2011). Indigestible carbohydrate serves as substrate for bacterial fermentation taking place primarily in the colon, but if the carbohydrates are soluble, substantial fermentation may take place in the small intestine as well. As SCFA are absorbed from the intestine at a maximum rate of 0.82 mmol/hr per cm (Schmitt et al., 1977), small intestinal production of SCFA may play a major role in the production of SCFA that reaches circulation. Accordingly, if short-chain indigestible carbohydrates that can be degraded by probiotics are available for the ingested probiotics, this may give rise to increased levels of SCFA in the intestines. In addition to acting as an important nutrient (in particular butyrate) for the intestinal epithelial cells, SCFA hold diverse regulatory functions on the epithelial cells and the immune system (Furusawa et al., 2013; Blander et al., 2017). SCFA are ligands for G protein–coupled receptors (GPRs) and thereby act as signaling molecules influencing the function and expansion of cells expressing GPRs. Both intestinal epithelial cells and immune cells/leucocytes express GPRs (such as GPR43, GPR41, and GPR109A) to which SCFA bind. In response to SCFA, goblet cells enhance the transcription of the mucin genes (Willemsen et al., 2003; Gaudier et al., 2004), and colonization of mice with *Bifidobacterium longum* was shown to induce high levels of acetate, which conferred protection toward the enteropatogenic *E. coli* O157:H7 infection (Fukuda et al., 2011). Binding of SCFA to GPR109A and GPR43 on intestinal epithelial cells *in vitro* activated inflammasome assembly, leading to an increase in secretion of IL18 and AMP (Macia et al., 2015; Zhao et al., 2018). Together, these findings and the fact that bifidobacteria, in particular, are potent producers of SCFA, along with the described mechanisms, may represent key effects of probiotics on the gut epithelium.

Another important effect of SCFA is inhibition of histone deacetylases, which tends to promote a tolerogenic, anti-inflammatory cell phenotype involved in maintaining immune homeostasis. SCFA inhibition of histone deacetylases has been demonstrated to inactivate nuclear factor (NF)-kB in macrophages, dendritic cells, neutrophils, and peripheral blood mononuclear cells (Usami et al., 2008, Singh et al., 2010, Vinolo et al., 2011, Chang et al., 2014). Moreover, SCFA also affect peripheral T cells through histone deacetylase inhibition. In particular, FoxP3-expressing regulatory T cells and their suppressive function is enhanced under homeostatic conditions (Tao et al., 2007). Although a direct effect of a probiotic intervention on inhibition of histone deacetylases in immune cells remains to be demonstrated, it seems plausible that probiotics administered together with dietary fibers or specific prebiotics can induce this effect.

Lactate, the dissociated state of lactic acid, has also been demonstrated to exert effects comparable to SCFA. Lactate may act through the plasma membrane receptor GPR81 as well on the histone acetylases in cells, resulting in effects similar to the effects of SCFA (Manoharan et al., 2021). Of note, lactate activates GPR81 in the concentration range of 1–20 mM, which is 10-fold higher compared to the concentration needed to activate, for example, the butyrate receptor GPR 43. In addition, indole, a metabolite produced by many commensals and probiotic bacteria, is among metabolites that may help maintain a segregation of host and microorganisms. Indole induces several genes involved in the maintenance of epithelial cell structure and function, including genes responsible for tight junction organization, actin cytoskeleton, and mucin production, which together suggest strengthening of epithelial cell barrier properties (Bansal et al., 2010).

# 24.5 PROBIOTIC STIMULATION OF HEMATOPOIESIS

In germ-free mice devoid of a microbiota, myeloid cell development is reduced, and clearance of bacterial infections are accordingly compromised (Khoshravi et al., 2014; Kristensen et al., 2015). Similarly, antibiotic treatment may reduce myelopoiesis (Balmer et al., 2014; Deshmukh et al., 2014; Fuglsang et al., 2018). As the complexity of the microbiota seems to influence the myelopoiesis, broad-spectrum antibiotic treatments are expected to have the strongest influence on the myelopoiesis. The mechanism by which microorganisms stimulate myelopoiesis is not completely clear but may involve the presence of microbial ligands and intact bacteria in the circulation or a microbe-induced upregulation of cytokines involved in the stimulation of myelopoiesis (Semerad et al., 2002; Balmer et al., 2014). Although the complexity of the microbiota seems to matter, administration of a single probiotic strain has been demonstrated to influence the myelopoiesis in mice and may have the same effect in humans. In protein-malnourished mice, *L. rhamnosus* CRL1505 administered orally or nasally restored the neutrophils and monocytes in blood and bone marrow when mice were returned to a normal diet and enhanced the respiratory innate immune response (Herrara et al., 2014). When mice were administered lactobacilli prior to treatment with the chemotherapeutic cyclophosphamide, the myeloid cells in bone marrow and blood increased and had better resistance toward *Candida albicans* infection (Salva et al., 2014). In microbiota-compromised newborn mice, oral administration of lactobacilli likewise restored granulocytes in bone marrow and in the circulation (Fuglsang et al., 2018). Thus, oral administration to immunocompromised mice clearly increases the number of myeloid cells in bone marrow and circulation and the protection against infection. In humans, probiotic administration increased the *ex vivo* production of IL-6 in lipopolysaccharide (LPS)-stimulated blood (Sørensen et al., 2018). The major producers of IL-6 in blood are monocytes (Schirmer et al., 2016), and an increase in IL-6 signifies an increased number of monocytes (Mærkedahl et al., 2018). Interestingly, the microbial-driven alterations in the myeloid cell pool greatly influence susceptibility to a variety of disorders, including infections and allergies. In addition, they have also been shown to regulate the effectiveness of vaccination. Lactate produced by intestinal lactic acid bacteria has been suggested to cause stimulation of hematopoiesis through stimulation of hematopoietic stem cells (Lee et al., 2021). Hence, probiotic- and postbiotic-induced increase in myelopoiesis may represent a key mechanism of action involved in many of the effects related to the innate immune system demonstrated for probiotic bacteria.

## 24.6 EVIDENCE FOR CLINICAL EFFECTS OF PROBIOTICS

A considerable number of human intervention studies have investigated the effect of probiotic administration on a variety of groups of individuals suffering from different disorders or infections. Moreover, some studies assessed preventive effects, while others investigated effects on already established diseases. In addition, numerous studies on effects of probiotics using animal models of different diseases have been published. In particular, diseases in which the microbiota plays a pivotal role may benefit from a positive effect of probiotic administration. Here, only diseases or conditions that are directly linked to the immune system and for which some evidence for an effect of probiotics intervention exists will be mentioned (Table 24.3).

**TABLE 24.3**  Diseases and Conditions in Which Probiotics Have Shown Some Evidence in Clinical Trials

| CONDITION | EFFECT OF PROBIOTIC INTERVENTION |
|---|---|
| Inflammatory bowel disease | Increase in remission |
| Ulcerative colitis | Less clear effects |
| Crohn's disease | |
| Atopic dermatitis | Reduction in severity |
| | Prevention of atopic dermatitis if administered during pregnancy |
| Upper respiratory tract infections (URTIs) | Fewer episodes of URTIs |
| | Shorter duration of episodes |
| | Reduced school absence |
| Necrotizing enterocolitis | Minor reduction in incidence and mortality |
| Pneumonia | Reduction in incidence of ventilator-associated pneumonia |
| Diarrhea[a] | Reduced incidence of diarrhea (children) |
| Antibiotic-associated diarrhea | Reduced incidence of diarrhea |
| *C. difficile*–associated diarrhea | |
| Vaccinations | Effects of antibody titer or seroconversion in both children and adults |

[a]*These effects are most likely not immune related.*

# 24.6.1 Allergy and Atopic Dermatitis

Allergic diseases like hay fever, eczema, asthma, and food allergies have increased dramatically over the last century, coinciding with a more hygienic lifestyle (Bach, 2018). It was accordingly plausible to hypothesize that administration of probiotic bacteria could ameliorate or prevent allergy in affected or predisposed individuals. A high number of *in vitro* studies have shown that many probiotic strains, in particular strains of the genus *Lactobacillus*, exhibit Th1-inducing properties in human or murine dendritic cells (Zeuthen et al., 2006; Weiss et al., 2011). This property has prompted researchers to assume that probiotic bacteria would be able to change a Th2 "atopic status" to a Th1 polarized status and thus prevent or even ameliorate allergy. Enhanced knowledge on mucosal immunology as well as many studies with experimental animals regarding the mechanisms behind the action of bacteria on oral mucosa have revealed a more complex host–microbe interaction. Epithelial cells, metabolites, and the translocation of bacteria/bacterial components also play key roles in maturation and development of a balanced immune system. Furthermore, the only allergic manifestation to which probiotic intake has shown compelling preventive effect is atopic dermatitis.

Atopic dermatitis is the most common chronic inflammatory skin disorder among infants and children, with a current prevalence of 10%–20% (Weidinger and Novak, 2016). Three recent meta-analyses of randomized controlled studies on the effect probiotic administration to individuals <18 years old all revealed that the probiotic administration confers some protection against the occurrence of atopic dermatitis (Pelucchi et al., 2012; Panduru et al., 2015; Huang et al., 2017). However, not all studies report a protective effect. From the studies, it is not possible to identify specific species with strong protective effect nor to establish dose or treatment duration for efficient protection. Of note, one study showed that only prenatal administration followed by postnatal administration, but not postnatal administration alone, protected against the onset of atopic dermatitis, indicating that the perinatal or the very early postnatal period represents a particularly critical period as regards immune maturation and polarization (Panduru et al., 2015). A recent systematic review compared mixtures of different probiotic strains in the prevention of AD found differences between various mixtures and concluded that administration of some mixtures during pregnancy, to the infant, or both demonstrated efficacy in reducing the risk of developing AD (Tan-Lim et al., 2021).

## 24.6.1.1 Suggested Mechanisms Involved

Interestingly, the results from published studies on the effect of probiotics on atopic dermatitis suggest that orally administered probiotics may stimulate systemic immunity in order to influence immune reactions taking place in the skin. In an atopic dermatitis mouse model, where the pathology was already established, administration of an oral cocktail of four probiotic species reduced IgE levels and the cytokines IL-4 and IL-5 while increasing IL-12p40 and IFN-$\gamma$ (Kim et al., 2016). Others have revealed that oral administration of *L. casei* (DN-114 001) reduced the number of CD8$^+$ effector T cells and increased the recruitment of CD4$^+$ effector T cells and Treg cells in the skin of mice with antigen-specific-induced skin inflammation (Hacini-Rachinel et al., 2009). Likewise, oral administration of *L. rhamnosus* Lcr35 increased Treg cells in ovalbumin-induced atopic dermatitis SK-1 hairless mice (Kim et al., 2012). Hence, the increase in regulatory T cells may represent a key effect of oral probiotic administration that may act not only in the gut but also at other barriers, for example, the skin. This is in accordance with the findings that gut and other epithelial cells are crucial in determining the response to allergens and microorganisms because they control the activation of dendritic cells and ILCS, which in turn direct the final T cell response (Hammad and Lambrecht, 2015). If insufficiently stimulated, cells will condition DCs to stimulate the generation of Th2 cells and result in ILC2 generation, while sufficient microbial stimulation results in Treg cells as well as Th1 and Th17 cells. However, in infants of mothers given probiotics during pregnancy, Treg were not increased (Taylor et al., 2007). One study in which a cocktail of probiotic bacteria was administered to pregnant women assessed the T cell population in blood drawn from the 3-month-old children and found that the proportion of Th22 cells was significantly reduced compared to the placebo group (Rø et al., 2017). IL22 produced by the Th22 cells improves the skin epithelial barriers, which may be critical in the first period after birth.

Other studies in animals have reported other mechanisms by which probiotics can attenuate skin inflammation. These mechanisms include stimulating intestinal IgA production and induced tolerance by upregulation of IL-10 levels in Peyer's patches and mesenteric lymph nodes (Inoue et al., 2007; Sawada et al., 2007). Several mechanisms, including these, may support each other in the protection against atopic dermatitis.

## 24.6.2 Inflammatory Bowel Disease

Inflammatory bowel diseases (IBDs) comprise a cluster of diverse chronic relapsing disorders, including Crohn's disease and ulcerative colitis (UC), that affect the gastrointestinal tract (Neurath, 2017). The etiology is multifactorial and includes stenosis, abscesses, and colitis-associated neoplasias and cancer, as is the course of IBD with complex interaction between the immune system, the gut microbiota, and genes leading to the onset of disease. Due to the complexity, the specific courses leading to the onset of IBD are far from understood, but breaches in the gut epithelial barrier, for example, reduction of the mucus layer and loss of epithelial cell integrity as well as increases in abundance of certain bacteria, may be involved (Neurath, 2017). All these factors may be caused by dysbiosis in the microbiota (O'Hara and Shanahan, 2006). As described, dysbiosis may lead to changes in the induction of subtypes of T helper cells, for example, increase in IL-17 T cells at the expense of Treg cells (Omenitti and Pizarro, 2015).

Probiotic bacteria administered, either a single strain or a mixture of probiotic strains, have been shown to ameliorate the symptoms of IBD in humans and mice. A meta-analysis of the use of probiotics to increase the remission of IBD showed that probiotics in patients with UC had significant effect, while the effects of probiotics intervention on patients with Crohn's disease was only marginally significant (Ganji-Arjenaki and Rafieian-Kopaei, 2018).

### 24.6.2.1 Suggested Mechanisms Involved

The exact mechanism by which probiotics exert their beneficial effect is not known, but the common feature is that they can control inflammation through changes in microbial composition, in production of microbial metabolites, and indirect induction of regulatory T-cells (Blander et al., 2017). The probiotic mixture of eight strains, VSL#3, has been shown to induce remission in patients with active UC (Bibiloni et al., 2005), and to maintain patients in remission after operation (Mimura et al., 2004). Many studies employing animal models have shown similar results with VSL#3 or with other bacterial mixtures or individual probiotic strains; several of these studies have been reviewed in Round and Mazmanian (2009). A number of these studies have furthermore demonstrated induction of regulatory T cells (DiGiacinto et al., 2005; Foligne et al., 2007; Amit-Romach et al., 2008; O'Mahony et al., 2008). As an intact epithelial layer well protected by a mucus layer, AMPs, and IgA is essential for induction of regulatory T cell polarization, the exact mechanism leading to induction of a regulatory state remains to be established but may very well be caused be several of the suggested mechanisms.

Moreover, as described, butyrate is a key nutrient for the epithelial cells and furthermore stimulates mucin production by goblet cells and regulatory T cells. Thus, butyrate-producing bacteria and perhaps bacteria producing other metabolites seem to be indispensable for an intact gut barrier.

## 24.6.3 Viral and Bacterial Upper Respiratory Tract Infections

Upper respiratory tract infections include the common cold and inflammation of the trachea and larynx. Most acute URTIs are caused by viral infections and last for 3–7 days (Long and Morris, 2017). It has long been suggested that probiotics through their immune-modulating effect would prevent or reduce the length of episodes of URTIs. This is confirmed by a meta-analyses of a growing number of clinical trial studies (Hao et al., 2015; Coleman et al., 2022). Here, probiotics were found to be better than placebo in reducing the number of participants experiencing episodes of acute URTI by about 47% and the duration of an episode of acute URTI by about 1.89 days. In addition, probiotics intake slightly reduced antibiotic use and cold-related school absence (Hao et al., 2015).

### 24.6.3.1 Suggested Mechanisms Involved

Studies in experimental animals have clearly established that broad-spectrum antibiotic treatment reduces the immune defense against viral and bacterial infections (Ichinohe et al., 2011; Deshmukh et al., 2014). The mechanisms behind the reduced immune defense are only partly understood and involve reduction in the myelopoiesis leading to a reduced number of phagocytic monocytes and neutrophils in circulation. Conversely, oral intake of probiotics increases the number of myelocytes in circulation (Salva et al., 2012). How the number of circulating myelocytes that may prevent viral infection remains to be established, but neutrophils have been shown to pave the way for T cells in infected lung tissue and may thus play an indirect role by streamlining the adaptive immune response toward viruses (Lim et al., 2015). Moreover, probiotics upregulate the expression of receptors on myelocytes involved in phagocytosis of complement and antibody opsonized (complement receptor [CR]1, CR3, Fc-receptor [FcR]1) and microbial degradation (Pelto et al., 1998; Donnet-Hughes et al., 1999; Arunachalam, 2000). The number of NK cells that play a key role in the elimination and confinement of virus-infected cells, especially in the period preceding the establishment of an adaptive immune defense, is also markedly increased by probiotic intake (Drakes et al., 2004), as is the production of antibodies (Link-Amster, 1994; Majamaa et al., 1995). Also, the barrier function in the mouth, gut, and upper respiratory tract may be enhanced by probiotics. This may include the increase in the production of AMPs, mucins, and cytokines recruiting phagocytes and other immune cells, as well as type 1 interferons inducing the expression of viral defense genes in the epithelial cells (Gill, 1998; Meydani and Ha, 2000). Most probably, probiotics may act through several of the suggested mechanisms.

## 24.6.4 Probiotics and Vaccination

Although vaccination does not represent one or a cluster of diseases, it constitutes an extremely important part of public health care. In most Western countries, childhood vaccination programs exist and provide protection against common viruses such as measles, mumps, and rubella, as well as whooping cough. In addition, many people, in particular the elderly, are receiving an annual vaccination against the expected seasonal influenza virus.

### 24.6.4.1 Vaccines for Infants

Vaccines for prevention of diseases common in infancy and childhood include those against diphtheria, tetanus, pertussis, polio, and *Hemophilia influenza* type b to neonates, which are employed routinely. Vaccination against viruses such as rubella, mumps, and measles are likewise administered during early childhood. Accordingly, the effect of probiotic intervention on the antibody level elicited by the vaccination has been assessed in infants and children. Barely half of the studies showed an increasing effect of administration of the probiotics on the antibody level (Zimmermann and Curtis, 2018). The studies show huge variation in strains and the dose administered as well as in the length of administration and age of the infants and children. Such variations may most likely affect the outcome of the studies, as may the type of vaccination.

### 24.6.4.2 Vaccination for Adults

Sixteen intervention studies assessing the effect of a probiotic intervention on the antibody response toward a vaccine in adults have been published (Yeh et al., 2018; Zimmerman and Curtis, 2018). Of these, 12 studies were based on influenza vaccines, and 4 studies were on oral or nasal vaccines. Well over half of the studies report a positive effect of probiotic intervention. Even though getting old is known to weaken the immune system, and stimulation with probiotics thus might have greater influence on the elderly, this cannot be concluded from the published studies. There was no difference in the effect observed in studies enrolling elderly people compared to studies with younger adults. As with the studies of the effect of probiotics on immune response against vaccination in infants, great variation in strains used and in length of intervention exists.

### 24.6.4.3 Suggested Mechanisms Involved

As probiotics are suggested to provide a general stimulation of the immune system, it has been suggested that stimulation of the immune system by probiotic bacteria would lead to an enhanced and prolonged effect of a vaccines, in turn leading to a stronger and/or prolonged protection. The majority of microorganisms, including probiotics, never reach beyond the physical barriers of skin and mucosa and may thus stimulate mainly through stimulation of the epithelial cells, which may respond by production of various growth factors and other cytokines that may exert a systemic effect. Alternatively, it may be that a minor proportion of the ingested probiotics ends up in circulation. In both cases, this may lead to increased number of circulating myelopoietic cells. This remains, however, to be demonstrated.

Alternatively, if vaccines are administered orally, probiotics may confer other kinds of stimulation, for example, by facilitating uptake of the vaccine or by somehow acting as an adjuvant. A review including 45 animal studies of probiotics' effect on viral infection showed that probiotics increased survival rate and reduced virus titer (Wang et al., 2021). In many of the studies, increases in INF-a, INF-g and IL-12, all cytokines stimulating the antiviral response, were revealed. Effects of probiotic administration on vaccination response are most commonly assessed by evaluation of the humoral immune response, the antibody titer elicited by the vaccination, or the seroconversion rate.

## 24.6.5 Effects of Different Probiotic Strains and Doses

There is a great variety among clinical trials in terms of the strains, doses, and length of time with administered probiotics. In addition, the groups of individuals enrolled vary; infants, children, adults, and the elderly, and healthy individuals versus individuals with disorders. So far, there is no clear picture regarding which probiotic strains provide the best effect, and this may depend on the specific condition. One should bear in mind that while only subtle differences may exist within lactobacilli species with regard to their immune modulatory effect, there are large differences in the effect of probiotic strains from different genera (e.g., lactobacilli, bifidobacteria, and *E. coli*), which accordingly may have a greater influence on the outcomes. The number of enrolled individuals in the clinical studies also varies considerably, and the statistical power in many studies may be low, which may cause variable outcomes of the studies.

## 24.7 CONCLUDING REMARKS

The rapidly growing knowledge regarding how commensal microorganisms interact with and affect the immune system has improved our understanding of how probiotics may influence our immune system. An important difference between the action of the gut microbiota and ingested probiotics may be that due to their passage through the entire gastrointestinal tract, probiotics may increase the microbial concentration in the small intestine and can there affect the epithelial cells, either directly or through their metabolites, or increase the number of bacteria and bacterial components reaching circulation. The recognition of the important role of the microbiota and the increased knowledge regarding its interaction with the immune system provide the basis for more targeted studies of the immune-modulating activities of probiotics.

# BIBLIOGRAPHY

Al Nabhani, Z., Dietrich, G., Hugot, J.-P., Barreau, F. (2017) Nod2: The intestinal gate keeper. *PLoS Pathog*, *13*:e1006177.

Amit-Romach, E., Uni, Z., Reifen, R. (2008) Therapeutic potential of two probiotics in inflammatory bowel disease as observed in the trinitrobenzene sulfonic acid model of colitis. *Dis Colon Rectum*, *51*:1828–1836.

Artis, D. (2008) Epithelial-cell recognition of commensal bacteria and maintenance of immune homeostasis in the gut. *Nat Rev Immunol*, 8:411–420.

Arunachalam, K., Gill, H.S., Chandro, R.K. (2000) Enhancement of natural immune function by dietary consumption of *Bifidobacterium lactis* (HN019). *Eur J Clin Nutr*, *54*:263–267.

Bach, J.-F. (2018) The hygiene hypothesis in autoimmunity: The role of pathogens and commensals. *Nat Rev Immunol*, *18*:105–120.

Balmer, M.L., Schürch, C.M., Saito, Y., Geuking, M.B., Li, H., Cuenca, M., Kovtonyuk, L.V., et al. (2014) Microbiota-derived compounds drive steady-state granulopoiesis via MyD88/TICAM signaling. *J Immunol*, *193*:5273–5283.

Bansal, T., Alaniz, R.C., Wood, T.K., Jayaraman, A. (2010) The bacterial signal indole increases epithelial-cell tight-junction resistance and attenuates indicators of inflammation. *Proc Nat Acad Sci*, *107*:228–233.

Bevins, C.L., Salzman, N.H. (2011) Paneth cells, antimicrobials and maintenance of intestinal homeostasis. *Nat Rev Microbiol*, 9:356–368.

Bibiloni, R., Fedorak, R.N., Tannock, G.W., Madsen, K.L., Gionchetti, P., Campieri, M., De Simone, C., Sartor, R.B. (2005) VSL#3 probiotic-mixture induces remission in patients with active ulcerative colitis. *Am J Gastroenterol*, *100*:1539–1546.

Birchenough, G.M., Nyström, E.E., Johansson, M.E., Hansson, G.C. (2016) A sentinel goblet cell guards the colonic crypt by triggering Nlrp6-dependent Muc2 secretion. *Science*, *352*:1535–1542.

Blander, J.M., Longman, R.S., Iliev, I.D., Sonnenberg, G.F., Artis, D. (2017) Regulation of inflammation by microbiota interactions with the host. *Nat Immunol*, *18*:851–860.

Bouskra, D., Brézillon, C., Bérard, M., Werts, C., Varona, R., Boneca, I.G., Eberl, G. (2008) Lymphoid tissue genesis induced by commensals through NOD1 regulates intestinal homeostasis. *Nature*, *456*:507510.

Caballero-Franco, C., Keller, K., De Simone, C., Chadee, K. (2007) The VSL#3 probiotic formula induces mucin gene expression and secretion in colonic epithelial cells. *Am J Physiol Gastrointest Liver Physiol*, *1292*:G315–G32.

Carzola, S.I., Maldonado-Galdeano, C., Weill, R., De Paula, J., Perdigón, G.D.V. (2018) Oral administration of probiotics increases Paneth cells and intestinal antimicrobial activity. *Front Microbiol*, 9:736.

Chamaillard, M., Hashimoto, M., Horie, Y., Masumoto, J., Qiu, S., Saab, L., Ogura, Y., et al. (2003) An essential role for NOD1 in host recognition of bacterial peptidoglycan containing diaminopimelic acid. *Nat Immunol*, 7:702–707.

Chang, P.V., Haob, C.L., Offermann, S., Medzhitov, R. (2014) The microbial metabolite butyrate regulates intestinal macrophage function via histone deacetylase inhibition. *Proc Nat Acad Sci*, *111*:2247–2252.

Chang, S.-Y., Ko, H-J., Kweon, M.-N. (2014) Mucosal dendritic cells shape mucosal immunity. *Exp Mol Med*, *46*:e84.

Coleman, J.L., Hatch-McChesney, A., Small, S.D., Allen, J.T., Sullo, E., Agans, R.T., Fagnant, H.S. (2022) Orally ingested probiotics, prebiotics, and synbiotics as countermeasures for respiratory tract infections in nonelderly adults: A systematic review and meta-analysis. *Adv Nutr*, *13*:2277–2295.

Corthesy, B. (2013) Role of secretory IgA in infection and maintenance of homeostasis. *Autoimmune Rev*, *12*:661–665.

Deshmukh, H.S., Liu, Y., Menkiti, O.R., Mei, J., Dai, N., O'Leary, C.E., Oliver, P.M., Kolls, J.K., Weiser, J.N., Worthen, G.S. (2014) The microbiota regulates neutrophil homeostasis and host resistance to *Escherichia coli* K1 sepsis in neonatal mice. *Nat Med*, *20*:524–530.

DiGiacinto, C., Marinaro, M., Sanchez, M., Strober, W., Boirivant, M. (2005) Probiotics ameliorate recurrent Th1-mediated murine colitis by inducing IL-10 and IL-10-dependent TGF-β-bearing regulatory cells. *J Immunol*, *174*:3237–3246.

Donnet-Hughes, A., Rochat, F., Serrant, P., Aeschlimann, J.M., Schifferin, E.J. (1999) Modulation of nonspecific mechanisms of defense by lactic acid bacteria: Effective dose. *J Dairy Sci*, *82*:863–869.

Drakes, M., Blanchard, T., Czinn, S. (2004) Bacterial probiotic modulation of dendritic cell. *Infect Immun*, *72*:3299–3309. http://doi.org/10.1128/IAI.72.6.3299-3309.2004

Foligne, B., Nutten, S., Grangette, C., Dennin, V., Goudercourt, D., Poiret, S., Dewulf, J., Brassart, D., Mercenier, A., Pot, B. (2007) Correlation between *in vitro* and *in vivo* immunomodulatory properties of lactic acid bacteria. *World J Gastroenterol*, *13*:236–243.

Fuglsang, E., Krych, L., Lundsager, M.T., Nielsen, D.S., Frøkiær, H. (2018) Postnatal administration of *Lactobacillus rhamnosus* HN001 ameliorates perinatal broad-spectrum antibiotic-induced reduction in myelopoiesis and T cell activation in mouse pups. *Mol Nutr Food Res*, in press.

Fukuda, S., Toh, H., Hase, K., Oshima, K., Nakanishi, Y., Yoshimura, K., Tobe, T., et al. (2011) Bifidobacteria can protect from enteropathogenic infection through production of acetate. *Nature*, *469*:543–547.

Furusawa, Y., Obata, Y., Fukuda, S., Endo, T.A., Nakato, G., Takahashi, D., Nakanishi, Y., et al. (2013) Commensal microbe-derived butyrate induces the differentiation of colonic regulatory T cells. *Nature*, *504*:446–450.

Ganji-Arjenaki, M., Rafieian-Kopaei, M. (2018) Probiotics are a good choice in remission of inflammatory bowel diseases: A meta-analysis and systematic review. *J Cell Physiol*, *233*:2091–2103.

Gaudier, E., Jarry, A., Blottière, H.M., de Coppet, P., Buisine, M.P., Aubert, J.P., Laboisse, C., Cherbut, C., Hoebler, C. (2004) Butyrate specifically modulates MUC gene expression in intestinal epithelial goblet cells deprived of glucose. *Am J Physiol Gastrointest Liver Physiol*, *287*:G1168–G1174.

Gill, H.S. (1998) Stimulation of the immune system by lactic cultures. *Int Dairy J*, *8*:535–544.

Gill, H.S., Rutherford, K.J., Cross, M.L. (2001) Dietary probiotic supplementation enhances natural killer activity in the elderly: An investigation of age-related immunological changes. *J Clin Immunol*, *21*:264–271.

Goto, Y., Obata, T., Kunisawa, J., Sato, S., Ivanov, I.I., Lamichhane, A., Takeyama, N., et al. (2014) Innate lymphoid cells regulate intestinal epithelial cell glycosylation. *Science*, *45*:1254009.

Gutzeit, C., Magri, G., Cerutti, A. (2014) Intestinal IgA production and its role in host-microbe interaction. *Immunol Rev*, *260*:76–85.

Hacini-Rachinel, F., Ghei, T.H., Luduec, J.L., Dif, F., Nancey, S., Kaiserlian, D. (2009) Oral probiotic control skin inflammation by acting on both effector and regulatory T cells. *PLoS ONE*, *4*:e4903.

Hammad, H., Lambrecht, B.N. (2015) Barrier epithelial cells and the control of type 2 immunity. *Immunity*, *43*:29–40.

Hao, Q., Dong, B.R., Wu, T. (2015) Probiotics for preventing acute upper respiratory tract infections. *Cochrane Database Syst Rev*, 2:Art. No.:CD006895.

Hasegawa, M., Yang, K., Hashimoto, M., Park, J.H., Kim, Y.G., Fujimoto, Y., Nuñez, G., Fukase, K., Inohara, N. (2006) Differential release and distribution of Nod1 and Nod2 immunostimulatory molecules among bacterial species and environments. *J Biol Chem*, *281*:29054–2906310.

Hashimoto, M., Tawaratsumida, K., Kariya, H., Kiyohara, A., Suda, Y., Krikae, F., et al. (2006) Not lipoteichoic acid but lipoproteins appear to be the dominant immunobiologically active compounds in *Staphylococcus aureus* . *J Immunol*, *177*:3162–3169.

Hepworth, M.R., Monticelli, L.A., Fung, T.C., Ziegler, C.G., Grunberg, S., Sinha, R., Mantegazza, A.R., et al. (2013) Innate lymphoid cells regulate CD4$^+$ T-cell responses to intestinal commensal bacteria. *Nature*, *498*:113–117.

Herrara, M., Salva, S., Villena, J., Barbieri, N., Marranzino, G., Alvarez, S. (2014) Dietary supplementation with Lactobacilli improves emergency granulopoiesis in protein-malnourished mice and enhances respiratory innate immune response. *PLoS ONE*, *9*(4):e90227.

Huang, R., Ning, H., Shen, M., Lpi, J., Zhang, J., Chen, X. (2017) Probiotics for the treatment of atopic dermatitis in children: A systematic review and meta-analysis of randomised controlled trials. *Front Cell Infect Microbiol*, *7*:392.

Ichinohe, T., Pang, I.K., Kumamoto, Y., Peaper, D.R., Ho, J.H., Murray, T.S., Iwasaki, A. (2011) Microbiota regulates immune defense against respiratory tract influenza A virus infection. *Proc Nat Acad Sci*, *108*:5354–5359.

Inoue, R., Utsuka, M., Nishio, A., Ushida, K. (2007) Primary administration of *Lactobacillus johnsonii* NCC533 in weaning period suppresses the elevation of proinflammatory cytokines and CD86 gene expressions in skin lesions in NC/Nga mice. *FEMS Immunol Med Microbiol*, *50*:67–76.

Kawai, T., Akira, S. (2011) Toll-like receptors and their cross talk with other innate receptors in infection and immunity. *Immunity*, *34*:637–650.

Khoshravi, A., Yanez, A., Price, J.G., Chow, A., Merad, M., Goodridge, H.S., Mazmanian, S.K. (2014) Gut microbiota promote hematopoiesis to control bacterial infection. *Cell Host Microbe*, *15*:374–381.

Kim, H.J., Kim, Y.J., Kang, M.J., Seo, J.H., Kim, H.Y., Jeong, S.K., Lee, S.H., Kim, J.M., Hong, S.J. (2012) A novel mouse model of atopic dermatitis with epicutaneous allergen sensitization and the effect of *Lactobacillus rhamnosus* . *Exp Dermatol*, *21*:672–675.

Kim, M.S., Kim, J.E., Yoon, Y.S., Seo, J.G., Chung, M.J., Yum, D.Y. (2016) A probiotic preparation alleviates atopic dermatitis-like skin lesions in murine models. *Toxicol Res*, *32*:149–158.

Kristensen, M.B., Metzdorff, S.B., Damlund, D.S.M., Fink, L.N., Licht, T.R., Frøkiær, H. (2015) Neonatal microbial colonization in mice promotes prolonged dominance of CD11b+Gr-1+ cells and accelerated establishment of the CD4$^+$ T cell population in the spleen. *Imm Infl Dis*, *3*:309–315.

Lee, Y.-S., Kim, T.-Y., Kim, S., Lee, S.-H., Seo, S.-U., Zhou, B.O., Eunju, O., et al. (2021) Microbia-derived lactate promotes hematopoiesis and erythropoiesis by inducing stem cell factor production from leptin receptor+ niche cells. *Exp Mol Med* 53:1319–1331.

Levy, M., Thaiss, C.A., Zeevi, D., Dohnalová, L., Zilberman-Schapira, G., Mahdi, J.A., David, E., et al. (2015) Microbiota-modulated metabolites shape the intestinal microenvironment by regulating NLRP6 inflammasome signaling. *Cell*, *163*:1428–1443.

Lim, K., Hyun, Y.-M., Lambert-Emo, K., Capece, T., Bae, S., Miller, R., Topham, D.J., Kim, M. (2015) Neutrophil trails guide influenza specific CD8+ T cells in the airways. *Science*, *349*:4352.1–10.

Link-Amster, H., Rochat, F., Saudan, K.Y., Mignot, O., Aeschlimann, J.M. (1994) Modulation of a specific humoral immune response and changes in intestinal flora mediated through fermented milk intake. *Imm Med Microbiol*, *10*:55–63.

Long, J.D., Morris, A. (2017) Probiotics in preventing acute upper respiratory tract infections. *Am J Nurs*, *117*:69.

Macia, L., Tan, T., Vieira, A.T., Leach, K., Stanley, D., Luong, S., Maruya, M., et al. (2015) Metabolite-sensing receptors GPR43 and GPR109A facilitate dietary fibre-induced gut homeostasis through regulation of the inflammasome. *Nat Commun*, *6*:6734.

Mærkedahl, R.B., Frøkiær, H., Stenbæk, M.G., Nielsen, C.B., Lind, M.V., Lundtoft, C., Bohr, M.B., et al. (2018) *In vivo* and *ex-vivo* inflammatory markers of common metabolic phenotypes in humans. *Met Syndr Rel Disord*, *16*:29–39.

Majamaa, H., Isolauri, E., Saxelin, M., Vesikari, T. (1995) Lactic acid bacteria in the treatment of acute rotavirus gastroenteritis. *J Ped Gastroenterol Nutr*, *20*:333–338.

Malanovic, N., Lohner, K. (2016) Gram-positive bacteria cell envelopes: The impact on the activity of antimicrobial peptides. *Biochim Biophys Acta*, *1858*:936–946.

Maldonado-Contreras, A.L., McCormick, B.A. (2011) Intestinal epithelial cells and their role in innate mucosal immunity. *Cell Tissue Res*, *343*:5–12.

Manoharan, I., Prasad, P.D., Thangaraju, M., Manicassamy, S. (2021) Lactate-dependent regulation of immune responses by dendritic cells and macrophages. *Front Immunol*, *12*:691134.

Mattar, A.F., Teitelbaum, D.H., Drongowski, R.A., Yongyi, F., Harmon, C.M., Coran, A.G. (2002) Probiotics up-regulate MUC-2 mucin gene expression in a Caco-2 cell-culture model. *Pediatr Surg Int*, *18*:586–590.

Meydani, S.N., Ha, W.K. (2000) Immunologic effects of yogurt. *Am J Clin Nutr*, *71*:861–872.

Mimura, T., Rizzello, F., Helwig, U., Poggioli, G., Schreiber, S., Talbot, I.C., Nicholls, R.J., Gionchetti, P., Campieri, M., Kamm, M.A. (2004) Once daily high dose probiotic therapy (VSL#3) for maintaining remission in recurrent or refractory pouchitis. *Gut*, *53*:108–114.

Mortha, A., Chudnovskiy, A., Hashimoto, D., Bogunovic, M., Spencer, S.P., Belkaid, Y., Merad, M. (2014) Microbiota-dependent cross talk between macrophages and ILC3 promotes intestinal homeostasis. *Science*, *343*:124928.

Nagao, F., Nakayama, M., Muto, T., Okumura, K. (2000) Effects of a fermented milk drink containing *Lactobacillus casei* strain Shirotaon the immune system in healthy human subjects. *Biosci Biotechnol Biochem*, *64*:2706–2708.

Nenci, A., Becker, C., Wullaert, A., Gareus, R., van Loo, G., Danese, S., Huth, M., et al. (2007) Epithelial NEMO links innate immunity to chronic intestinal inflammation. *Nature*, *446*:557–561.

Neurath, M.F. (2017) Current and emerging therapeutic targets for IBD. *Nat Rev Gastroenterol Hepatol*, *14*:269–277.

Nigro, G., Rossi, R., Commere, P.-H., Jay, P., Sansonetti, P.J. (2014) The cytosolic bacterial peptidoglycan sensor Nod2 affords stem cell protection and links microbes to gut epithelial regeneration. *Cell Host Microbe*, *15*:792–798.

O'Hara, A.M., Shanahan, F. (2006) The gut flora as a forgotten organ. *EMBO Rep*, *7*:688–693.

O'Mahony, C., Scully, P., O'Mahony, D., Murphy, S., O'Brien, F., Lyons, A., Sherlock, G., et al. (2008) Commensal-induced regulatory T cells mediate protection against pathogen stimulated NF-κB activation. *PLoS Pathog*, *4*:e1000112.

Omenitti, S., Pizarro, T.T. (2015) The Treg/Th17 axis: A dynamic balance regulated by the gut microbiome. *Front Immunol*, *6*:639.

Panduru, M., Panduru, N.M., Salavestru, C.M., Tiplica, G.-S. (2015) Probiotics and primary prevention of atopic dermatitis: A meta-analysis of randomized controlled studies. *J Eur Acad Dermat Venerol*, *29*:232–242.

Pelto, L., Isolauri, E., Lilius, E.M., Nuutila, J., Salminen, S. (1998) Probiotic bacteria down-regulate the milk-induced inflammatory response in milk-hypersensitive subjects but have an immunostimulatory effect in healthy subjects. *Clin Exp Allergy*, *28*:1474–1474.

Pelucchi, C., Chatenoud, L., Turati, F., Galeone, C., Moja, L., Bach, J.-F., La Vecchia, C. (2012) Probiotics supplementation during pregnancy or infancy for the prevention of atopic dermatitis. *Epidemiol*, *23*:402–414.

Pokusaeva, K., Fitzgerald, G.F., van Sinderen, D. (2011) Carbohydrate metabolism in Bifidobacteria. *Genes Nutr*, *6*:206.

Rakoff-Nahoum, S., Paglino, J., Eslami-Varzaneh, F., Edberg, S., Medzhitov, R. (2004) Recognition of commensal microflora by toll-like receptors is required for intestinal homeostasis. *Cell*, *118*:229–241.

Ramanan, D., Tang, M.S., Bowcutt, R., Loke, P., Cadwell, K. (2014) Bacterial sensor Nod2 prevents small intestinal inflammation by restricting the expansion of the commensal *Bacteroides vulgatus*. *Immunity*, *41*:311–324.

Rø, A.D.B., Simpson, M.R., Rø, T.B., Storrø, O., Johnsen, R., Viden, V., Øien, T. (2017) Reduced Th22 cell proportion and prevention of atopic dermatitis in infants following maternal probiotic supplementation. *Clin Mech Allerg Dis*, *2017*:1–8.

Round, J.L., Mazmanian, S.K. (2009) The gut microbiota shapes intestinal immune responses during health and disease. *Nat Rev Immunol*, *9*:313–323.

Salva, S., Marranzino, G., Villena, J., Agüero, G., Alvarez, S. (2014) Probiotic Lactobacillus strains protect against myelosuppression and immunosuppression in cyclophosphamide-treated mice. *Int Immunopharmacol*, *22*:209–221.

Salva, S., Merino, M.C., Agüero, G., Gruppi, A., Alvarez, S. (2012) Dietary supplementation with probiotics improves hematopoiesis in malnourished mice. *PLoS ONE*, *7*:e31171.

Sawada, J., Morita, H., Tanaka, A., Salminen, S., He, F., Matsuda, H. (2007) Ingestion of heat-treated *Lactobacillus rhamnosus* GG prevents development of atopic dermatitis in NC/Nga mice. *Clin Exp Allergy*, *37*:296–303.

Schirmer, M., Smeekens, S.P., Vlamakis, H., Jaeger, M., Oosting, M., Franzosa, E.A., ter Horst, R., et al. (2016) Linking the human gut microbiome to inflammatory cytokine production capacity. *Cell*, *167*:1125–1136.

Schmitt, M.G., Soergel, K.H., Wood, C.M., Steff, J.J. (1977) Absorption of short-chain fatty acids from the human ileum. *Am J Dig Dis*, *22*:340–347.

Semerad, C.L., Liu, F., Gregory, A.D., Stumpf, K., Link, D.C. (2002) G-CSF is an essential regulator of neutrophil trafficking from the bone marrow to the blood. *Immunity*, *17*:413–423.

Shaw, M.H., Reimer, T., Kim, Y.-G., Nunes, G. (2008) NOD-like Receptors (NLRs): Bona fide intracellular microbial sensors. *Curr Opin Immunol*, *20*:377–382.

Singh, N., Thangaraju, M., Prasad, P.D., Martin, P.M., Lambert, N.A., Boettger, T., Offermanns, S., Ganapathy, V. (2010) Blockade of dendritic cell development by bacterial fermentation products butyrate and propionate through a transporter (Slc5a8)-dependent inhibition of histone deacetylases. *J Biol Chem*, *285*:27601–27608.

Sonnenberg, G.F., Artis, D. (2015) Innate lymphoid cells in the initiation, regulation and resolution of inflammation. *Nat Med*, *21*:698–708.

Sonnenberg, G.F., Monticelli, L.A., Alenghat, T., Fung, T.C., Hutnick, N.A., Kunisawa, J., Shibata, N., et al. (2012) Innate lymphoid cells promote anatomical containment of lymphoid-resident commensal bacteria. *Science*, *336*:1321–1325.

Sørensen, C.A., Fuglsang, E., Jørgensen, C.S., Laursen, R.P., Larnkjær, A., Mølgaard, C., Ritz, C., Michaelsen, K.F., Krogfelt, K.A., Frøkiær, H. (2018) Probiotics and the immunological response to infant vaccinations; a double-blind randomized controlled trial. *Clin Microbiol Infect*, *62*:1800510. http://doi.org/10.1016/j.cmi.2018.07.031.

Tan Lim, C.S.C., Esteban-Ipac, N.A.R., Recto, M.S.T., Castor, M.A.R., Casis-Hao, R.J., Nano, A.L.M. (2021) Comparative effectiveness of probiotic strains on the prevention of pediatric atopic dermatitis: A systematic reviewv and network meta-analysis. *Pediatr Allergy Immunol*, *32*:1255–1270.

Tao, R., de Zoeten, E.F., Ozkaynak, E., Chen, C., Wang, L., Porrett, P.M., Li, B., et al. (2007) Deacetylase inhibition promotes the generation and function of regulatory cells. *Nat Med*, *13*:1299–1307.

Taylor, A.L., Hale, J., Hales, B.J., Dunstan, J.A., Thomas, W.R., Prescott, S.L. (2007) FOXP3 mRNA expression at 6 months of age is higher in infants who develop atopic dermatitis, but is not affected by giving probiotics from birth. *Pediatr Allergy Immunol*, *18*:10–19.

Usami, M., Kishimoto, K., Ohata, A., Miyoshi, M., Aoyama, M., Fueda, Y., Kotani, J. (2008) Butyrate and trichostatin A attenuate nuclear factor kappaB activation and tumor necrosis factor alpha secretion and increase prostaglandin E2 secretion in human peripheral blood mononuclear cells. *Nutr Res*, *28*:321–328.

Vinolo, M.A., Rodrigues, H.G., Hatanaka, E., Sato, F.T., Sampaio, S.C., Curi, R. (2011) Suppressive effect of short-chain fatty acids on production of proinflammatory mediators by neutrophils. *J Nutr Biochem*, *22*:849–855.

Wang, F., Pan, B., Xu, S., Xu, Z., Zhang, T., Zhang, Q., Bao, Y., et al. (2021) A meta-analysis reveals the effectiveness of probiotics and prebiotics against respiratory viral infection. *Bioscience Rep*, *41*: BSR20203638.

Wang, H., Kim, J.J., Denou, E., Gallagher, A., Thornton, D.J., Shajib, M.S., Xia, L., et al. (2015) New role of nod proteins in regulation of intestinal goblet cell response in the context of innate host defense in an enteric parasite infection. *Infect Immun*, *84*:275–285.

Wehkamp, J., Harder, J., Wehkamp, K., Wehkamp, B., Meissner, V., Schlee, M., Enders, C., et al. (2004) NF-κB and AP-1 mediated induction of human β defensin-2 in intestinal epithelial cells by *E. coli* Nissle 1917: A novel effect of a probiotic bacterium. *Infect Immun*, *72*:5750–5758.

Weidinger, S., Novak, N. (2016) Atopic dermatitis. *Lancet*, *387*:1109–1122.

Weiss, G., Christensen, H.R., Zeuthen, L.H., Vogensen, F.K., Jakobsen, M., Frøkiær, H. (2011) Lactobacilli and bifidobacteria induce differential interferon-b profiles in dendritic cells. *Cytokine*, *56*:520–530.

Willemsen, L.E.M., Koetsier, M.A., van Deventer, S.J.H., van Tol, E.A.F. (2003) Short chain fatty acids stimulate epithelial mucin 2 expression through differential effects on prostaglandin E$_1$ and E$_2$ production by intestinal myofibroblasts. *Gut*, *52*:1442–1447.

Wlodarska, M., Thaiss, C.A., Nowarski, R., Henao-Mejia, J., Zhang, J.P., Brown, E.M., Frankel, G., et al. (2014) NLRP6 inflammasome orchestrates the colonic host—microbial interface by regulating goblet cell mucus secretion. *Cell, 156*:1045–1059.

Yeh, T.-L., Shih, P.-C., Liu, S.-J., Lin, C.H., Liu, J.M., Lei, W.T., Lin, C.Y. (2018) The influence of prebiotic or probiotic supplementation on antibody titers after influenza vaccination: A systematic review and meta-analysis of randomized controlled trials. *Drug Des Devel Ther, 12*:217–230.

Zeuthen, L.H., Christensen, H.R., Frøkiær, H. (2006) Lactic acid bacteria inducing a weak Interleukin-12 and tumor necrosis factor-α response in human dendritic cells inhibit strongly stimulating lactic acid bacteria but act synergistically with Gram-negative bacteria. *Clin Vac Immun, 13*:365–375.

Zeuthen, L.H., Fink, L.N., Frøkiær, H. (2008a) Toll-like receptor 2 and nucleotide-binding oligomerization domain-2 play divergent roles in the recognition of gut-derived lactobacilli and bifidobacteria in dendritic cells. *Immunol, 124*:489–502.

Zeuthen, L.H., Fink, L.N., Frøkiær, H. (2008b) Epithelial cells prime the immune response to an array of gut-derived commensals towards a tolerogenic phenotype through distinct actions of thymic stromal lymphopoietin and transforming growth factor-beta. *Immunol, 123*:197–208.

Zimmermann, P., Curtis, N. (2018) The influence of probiotics on vaccine responses—A systematic review. *Vaccine, 36*:207–213.

Zhao, Y., Chen, F., Wu, W., Sun, M., Bilotta, A.J., Yao, S., Xiao, Y., et al. (2018) GPR43 mediates microbiota metabolite SCFA regulation of antimicrobial peptide expression in intestinal epithelial cells via activation of mTOR and STAT3. *Muc Immun, 11*:752–762.

# Targeting the Gut Microbiota in Metabolic Disorders

## Potential Impact of Lactic Acid Bacteria and Next-Generation Probiotics

# 25

Kassem Makki, Hubert Vidal, and Corinne Grangette

## 25.1 INTRODUCTION: LESSONS TO TARGET THE GUT MICROBIOTA IN METABOLIC DISEASES

The human gastrointestinal tract is composed of trillions of microbes, including bacteria, yeast, and viruses (Human Microbiome Project Consortium, 2012). The dominant gut bacteria belong to two main phyla, Bacteroidota (Bacteroidetes) and Bacillota (Firmicutes), and, in lower abundance, Actinomycetota (Actinobacteria), Pseudomonadota (Proteobacteria), and Verrucomicrobiota (Verrucomicrobia) (Eckburg et al., 2005). The diversity of the human microbiome is tremendous, with more than 1000 different species in a collective genome encoding over 3 million genes in comparison to the 23,000 genes of the human genome (J. Li et al., 2014; Malard et al., 2021; Qin et al., 2010). The gut microbiota engraftment and development after birth is a key factor in establishing the host–microbiota interactions, and it is considered an important modulator of early metabolic programming (Cox et al., 2014) and immune development (Gensollen et al., 2016). This symbiotic relationship is influenced by multiple factors, including maternal diet, mode of delivery and feeding, sanitary conditions, use of antibiotics (Vandenplas et al., 2020), and host genetics and epigenetics. Once established, this equilibrium is relatively stable and resilient throughout adult life but can be altered by external events, notably changes in diet and lifestyle, bacterial infections, and antibiotic treatments that could have a critical impact on the composition of the microbial communities (Dogra et al., 2020; Rodríguez et al., 2015). In addition to its digestive role, the

DOI: 10.1201/9781003352075-28

gut microbiota has been shown to impact host physiology and to participate in the regulation of the host immune system, the defense against pathogenic microorganisms, longitudinal growth, tissue development, and many other biological processes important to maintaining a healthy functioning body. Indeed, the gut microbiota regulates the homeostasis of many extra-intestinal organs, notably the liver, adipose tissues, and the lungs, as well as the brain (Gebrayel et al., 2022). Disruption of this symbiotic relationship will lead to so-called gut dysbiosis, which has been shown to increase the incidence of many chronic diseases, such as inflammatory bowel disease (IBD) (Sartor & Muehlbauer, 2007), cardiovascular (Kazemian et al., 2020) and liver diseases (Zheng & Wang, 2021), obesity and diabetes (Ley et al., 2005; Patterson et al., 2016), and neurological disorders (Morais et al., 2020).

Extensive sequencing of the gut microbiota of different human populations representing different societies—westernized, non-industrialized to indigenous ethnic groups—and the use of different experimental models based on nutritional challenges provided direct evidence of changes in the microbial communities in obesity. Even if previous reports highlighted a change in the ratio of the two main bacteria phyla, with a decrease in Bacteroidota and an increase in Bacillota in obese mice (Turnbaugh et al., 2006), more recent meta-analyses showed that the Bacteroidota/Bacillota ratio differed between studies and should not be used as a hallmark of gut dysbiosis in obesity (Sze & Schloss, 2016). It is now well accepted that high microbial diversity, gene richness, and stable microbiome functional cores are features of a healthy gut microbiota, while in both lean individuals and individuals with obesity, low microbial gene richness links with increased adiposity, insulin resistance, inflammation, and dyslipidemia (Human Microbiome Project Consortium, 2012).

The causal link between gut microbes and obesity development was demonstrated when the transfer of the microbiota of obese animals was able to confer the obese phenotype to axenic mice. Interestingly, the bacteria in obese mice extracted more calories from their food, suggesting that differences in the efficiency of caloric extraction may be determined by the microbial composition (Turnbaugh et al., 2006). Similarly, transplanting a faecal microbiota from adult female twin pairs discordant for obesity into germ-free mice provided a unique opportunity to demonstrate the interrelationships between obesity and the gut microbiome (Ridaura et al., 2013). Recently, Thaiss et al. identified that the postdieting weight regain was associated with persistent microbiome alterations in mice which contribute to faster weight regain and metabolic aberrations (Thaiss et al., 2016). This leads to long-term metabolic perturbations.

More specifically, families belonging to Prevotellaceae and Enterobacteriaceae were found enriched in obesity, while Christensenellaceae and Verrucomicrobiaceae were higher in subjects with lower body mass index (BMI) (Tseng & Wu, 2019). Also, *Bifidobacterium* and *Bacteroides* genera were significantly higher in normal-weight individuals (Santacruz et al., 2010). A lower abundance of short-chain fatty acid (SCFA)-producing bacteria, such as *Akkermansia muciniphila* and *Faecalibacterium prausnitzii*, as well as bifidobacteria, was found in subjects with obesity and type 2 diabetes (T2D) (Da Silva et al., 2020a; Dao et al., 2016; Zhang et al., 2013). This was associated with lower bacterial richness and microbiota stability (Le Chatelier et al., 2013). It is now recognized that certain gut symbionts may play either a protective or a pathogenic role in the progression of obesity. Most studies are considering that increased gut-derived SCFA production can be regarded as a biomarker of gut and metabolic health. Studies of twin pairs with discordance for BMI have shown that lower BMI was associated with a more abundant network of primary fiber degraders, while a network of butyrate producers was more prominent in subjects with higher BMI (Tims et al., 2013). Therefore, there is a promising potential for targeting this microbiota dysbiosis in the management of obesity and associated disorders such as type 2 diabetes. However, identifying potential probiotic strains able to efficiently counteract metabolic dysregulation remains a challenge. The following part of this chapter aims to summarize the results that have been obtained through the use of lactic acid bacteria (LAB) and emerging research to identify next-generation probiotics from commensal gut bacteria.

# 25.2  CHALLENGES IN TARGETING THE GUT MICROBIOTA IN OBESITY AND ASSOCIATED METABOLIC DISORDERS USING LACTIC ACID BACTERIA

## 25.2.1  State of the Art

Chronic (low grade) inflammatory diseases (CIDs) such as obesity and T2D are associated with marked alterations in gut microbiota composition and function. The imbalance between gut microbial communities is partly responsible for the disruption of several key biological processes involved in the regulation of host physiology (Makki et al., 2018). Indeed, gut dysbiosis alters intestinal barrier function, which leads to the translocation of bacterial pro-inflammatory compounds such as lipopolysaccharides (LPSs) and to the development of metabolic endotoxemia (Cani et al., 2007). Furthermore, alterations in gut microbiota result in a significant decrease in SCFA production, which are important bacterial metabolites that can signal through G protein–coupled receptors (e.g., GPR41 and GPR43). SCFA regulate gut barrier function, immune responses, thermogenesis, and host glucose metabolism (Koh et al., 2016). Also, gut bacteria play a key role in bile acid metabolism, and alterations in bacterial potential to produce secondary bile acids have been observed in obese subjects and subjects with T2D. Indeed, increased levels of secondary bile acids have been observed in patients with insulin resistance and T2D and have been linked to the severity of the metabolic disturbances as well as to hepatic lipid accumulation and steatosis (Choucair et al., 2020; Haeusler et al., 2013; C. Wang et al., 2022). Bile acids are important signaling molecules targeting several receptors expressed by host cells (Molinaro et al., 2018). Bile acids modulate the activities of farnesoid X receptors (FXRs) and the Takeda G protein–coupled receptor 5 (TGR5), which will affect immune system development, glucose, and lipid metabolisms, thus influencing the development of obesity, hepatic steatosis, and T2D (Molinaro et al., 2018). Thus, targeting alterations within gut microbiota functions to normalize the production of host and microbial bile acids that are considered detrimental at high levels might be a promising strategy to alleviate metabolic disturbances. More recently, subjects with severe obesity and/or T2D have been shown to exhibit lower levels of vitamins such as biotin. Indeed, the bacterial potential of biosynthesis and transport of biotin in humans is altered in a European population with severe obesity and metabolic alteration (Belda et al., 2022). More specifically, a reduction in biotin levels was associated with lower abundance of microbial genes involved in biotin production, which correlated with altered metabolism and inflammation. Furthermore, microbial alterations resulting from a westernized lifestyle have been associated with increased levels of several bacterial metabolites that can be harmful to the host when found in important levels, such as Trimethylamine N-oxide (TMAO) and imidazole propionate (IMP) (Koeth et al., 2013; Molinaro et al., 2020). TMAO is synthesized from trimethylamine (TMA), which is a microbial by-product of the degradation of choline and phosphatidylcholine (Koeth et al., 2013). TMAOs have been strongly associated with increased risk of cardiovascular disease development and contribute to atherosclerosis development through macrophage cholesterol accumulation and inhibition of reverse cholesterol transport (Koeth et al., 2013). Also, the gut microbiota of subjects with T2D is characterized by its ability to produce higher levels of IMP. This newly discovered microbial metabolite is synthesized from urocanate and involves a bacterial enzyme called urocanate reductase. More importantly, higher production of IMP has been shown to be an important contributing factor for the disruption of hepatic insulin signaling and glucose metabolism (Koh et al., 2018). Thus, understanding the complex relationship, called symbiosis, established between host cells and gut bacterial communities will help to design new therapeutic strategies to reduce the production of bacterial harmful metabolites and to limit the development of metabolic disorders. Probiotics, defined as "live microorganisms that when administered in adequate amounts confer a health benefit on the host" (Hill et al., 2014), have been shown to be promising tools to alleviate or prevent the development of obesity and T2D. Most probiotics sold on the market include mainly microorganisms from the genera *Lactobacillus* and *Bifidobacterium* (Zielińska & Kołożyn-Krajewska, 2018). Industrialization of lifestyle, which is an important risk factor for development of CID, is characterized by lower counts of key bacteria that play essential roles in regulating the host immune system and glucose metabolism such as bifidobacteria. Indeed, in a recent elegant study which compared infant microbiomes from industrialized nations vs. non-industrialized populations (infants of Hadza hunter-gatherers of Tanzania), a paucity of *Bifidobacterium infantis* was observed in the microbiome of industrialized infants (Olm et al., 2022). Also, lower counts of *Bifidobacteria* have been found in subjects with overweight or obesity and correlated with higher visceral fat area (Da Silva et al., 2020b; Gong et al., 2022; Kim et al., 2022). All together, these studies suggest that the administration of probiotics from the *Lactobacillus* and *Bifidobacterium* genera could be useful tools in the management of CID such obesity and T2D.

## 25.2.2 Potential Impact of Lactic Acid Bacteria in Humans

Bifidobacteria, together with lactic acid bacteria (*Enterococcus*, *Streptococcus*, and *Lactobacillus*), are among the first colonizers of the infant gut and will influence gut microbiota stability and host physiology (Roswall et al., 2021). Indeed, in preterm infants, supplementation of a mixture of probiotics (FloraBABY, Renew Life, Canada) containing four *Bifidobacterium* strains from species that are common and dominant in the infant gut (*Bifidobacterium breve* HA-129, *Bifidobacterium bifidum* HA-132, *Bifidobacterium longum* subsp. *infantis* HA-116, and *Bifidobacterium longum* subsp. *longum* HA-135) and *Lacticaseibacillus rhamnosus* HA-111 accelerated gut microbiota maturation, which was associated with a favorable metabolic and immune gut milieu (Samara et al., 2022). Also, strains of *Bifidobacterium breve* which possess a highly specialized genetic capability to degrade human milk oligosaccharides were associated with improved intestinal barrier maturation during the first week of life in early preterm infants (B. Ma et al., 2022). Furthermore, lactic acid bacteria and bifidobacteria play an important role in influencing the stability of the microbiome. The stability and resilience of the gut microbiome over time have been linked to better health benefits, while long-term instability of the gut microbiota has been associated with metabolic liver disease and T2D (Frost et al., 2021). A recent study investigating the intra-variability of the microbiome within a disease-free Swedish population over one year found that bifidobacteria abundance correlated positively with a reduced intra-variability of the gut microbes (Olsson et al., 2022), suggesting that bifidobacteria might play a protective role in preventing CID development by stabilizing gut microbial ecology. All together, these studies highlight the importance of bifidobacteria as well as the other lactic acid bacteria in maturing and maintaining a healthy and resilient gut microbiome, which results in a healthy status later in life.

The use of probiotic supplements, yogurts, or fermented foods as strategies to support body weight management and to improve glucose metabolism and reduce fat mass is continuously being evaluated in human intervention studies. A systematic review of randomized clinical trials (RCTs) investigating the use of lactic acid bacteria (*Lactobacillus* and *Bifidobacterium*) highlighted that 23 RCTs reported significant weight loss (Álvarez-Arraño & Martín-Peláez, 2021). Indeed, probiotic or symbiotic intake could lead to significant weight loss mainly after 12 weeks of intervention under different lifestyle conditions, such as low-caloric diets, modified physical activity, or unchanged lifestyle. More specifically, the intake of *Lactobacillus* and *Bifidobacterium* strains were the most efficient at reducing body weight. Indeed, the genus *Lactobacillus* and associated new genera, particularly strains from the species *Lacticaseibacillus rhamnosus*, *Lactobacillus gasseri*, *Lactiplantibacillus plantarum*, *Lacticaseibacillus casei*, *Lactobacillus acidophilus*, *Lactobacillus delbrueckii*, *Limosilactobacillus reuteri*, and *Lactobacillus curvatus*, had positive effects on weight loss and fat mass. For example, two strains of *L. gasseri* (BNR17 and SBT2055) reduced weight and waist and hip circumference (S.-P. Jung et al., 2013; Kadooka et al., 2013) and abdominal adiposity (Kadooka et al., 2010); *L. curvatus* HY7601 and *L. plantarum* KY1032 also reduced body weight and BMI while participants maintained their regular diet and lifestyle (S. Jung et al., 2015); the intake of *L. sakei* CJLS03 under healthy recommendations of lifestyle reduced body fat mass by 0.2

kg and waist circumference by 0.8 cm (Lim et al., 2020). In a RCT performed in overweight adults in Yogyakarta, the consumption of *L. plantarum* Dad-13 shifted the gut microbiota composition, which was associated with reduced BMI and improved intestinal health (Rustanti et al., 2022). Furthermore, daily consumption of beverages containing fragmented *L. amylovorus* improved anthropometric measurements and markers related to lipid and glucose metabolism (Nakamura et al., 2016). Even though some lactobacilli strains might not impact body weight, their use can result in alleviation of some obesity markers. Indeed, supplementation of *L. rhamnosus* HA-114 in a cohort of adults with obesity did not potentiate their body weight loss; however, the bacterial strain did significantly reduce insulin, LDL-cholesterol, and triglyceride levels. More interestingly, *L. rhamnosus* HA-114 supplementation altered the feeding behavior and perception of the subjects and led to a significant reduction in binge-eating tendencies and food cravings (Choi et al., 2022). However, the effects of lactobacilli on host body weight and metabolism might be species and/or strain specific and might result in different health outcomes. Some strains of *L. fermentum*, *L. acidophilus*, and *L. ingluviei* were associated with weight gain in humans, whereas consumption of *L. gasseri* and *L. plantarum* led to weight loss (Million et al., 2012). Furthermore, *Ligilactobacillus salivarius* strain Ls33, which exhibits anti-inflammatory capacities (Foligne et al., 2007), modified microbiota composition in obese adolescents; however, these changes were not associated with improved metabolic parameters and body weight loss (Larsen et al., 2013). Interestingly, we confirmed these results in a murine model of diet-induced obesity (DIO), showing no positive impact of this strain (Alard et al., 2016). Similarly, the administration of *L. paracasei* F19 did not improve metabolic parameters in postmenopausal women or overweight adults (Brahe et al., 2015)

Regarding bifidobacteria, several recent studies have showed promising effects of *Bifidobacterium* strains alone on body weight loss and glucose metabolism in humans. The intake of *B. animalis* ssp. *Lactis* 420 (B420) combined or not with polydextrose had a positive effect on body weight control, which was related to changes in body fat mass (Stenman et al., 2016). Also, *B. animalis* CECT 8145 supplementation combined with healthier dietary recommendations decreased BMI and waist circumference (Pedret et al., 2019). More recently, *B. longum* APC1472 supplementation in healthy overweight/obese adults did not reduce BMI; however, it reduced fasting blood glucose levels, which suggest that *B. longum* APC1472 might be a valuable supplement in reducing specific markers of obesity (Schellekens et al., 2021). A very interesting study performed in adults with obesity and using an autochthonous strain of *Bifidobacterium adolescentis*—strain IVS1—showed improvements in colonic permeability (Krumbeck et al., 2018).

Mixtures of *Lactobacillus* and *Bifidobacterium* strains have been more commonly used in intervention studies and showed more significant effects on reducing body weight in humans. The intake of *L. acidophilus* La5, *B. animalis* subsp. *lactis* BB12 and *L. casei* DN001 in combination with a low-calorie diet reduced BMI, percentage of fat, and leptin levels (Zarrati et al., 2014). In two different cohorts of adults with overweight or obesity during which the subjects maintained their regular lifestyles, the intake of a Lab4P mixture (*L. acidophilus* CUL60 (NCIMB 30157), *L. acidophilus* CUL21 (NCIMB 30156), *L. plantarum* CUL66 (NCIMB 30280), *B. bifidum* CUL20 (NCIMB 30153), and *B. animalis* subsp. *Lactis* CUL34 (NCIMB 30172)) significantly decreased body weight and improved well-being (Michael et al., 2020, 2021). Finally, in a RCT performed in adults with overweight or obesity, supplementation with the VSL#3 mixture composed of eight strains, *B. longum*, *B. infantis*, *B. breve*, *L. acidophilus*, *L. paracasei*, *L. delbrueckii* subsp. *bulgaricus*, *L. plantarum*, and *S. salivarius* subsp. *thermophilus*, affected insulin sensitivity, lipid profile, and atherogenic index favorably and reduced a marker of inflammation (hsCRP) (Rajkumar et al., 2014).

The use of probiotics is not limited to their ability to reduce body weight and fat mass. In fact, probiotic supplementation might have promising effects on liver health, such as non-alcoholic fatty liver disease (NAFLD) and lipid metabolism. In a randomized triple-blind trial conducted among 64 obese children with sonographic NAFLD, the administration of probiotic capsules containing a mixture of *L. acidophilus* ATCC B3208, *B. lactis* DSMZ 32269, *B. bifidum* ATCC SD6576, and *L. rhamnosus* DSMZ 21690 decreased alanine aminotransferase (ALAT), aspartate aminotransferase (ASAT), cholesterol, low-density lipoprotein-C (LDL-C), and triglycerides levels as well as waist circumference without changes in BMI (Famouri et al., 2017). More recently, the supplementation of *L. salivarius* AP-32, *L. rhamnosus* bv-77, and *B. animalis* CP-9 increased high-density lipoprotein (HDL) and adiponectin levels, while total cholesterol, low-density lipoprotein (LDL) and leptin levels were reduced in children with overweight or obesity (A.-C. Chen et al., 2022). Additionally, the use of a multi-strain probiotic composed of *L. paracasei* DSM 24733, *L. plantarum* DSM 24730, *L. acidophilus* DSM 24735 and *L. delbrueckii* subsp. *bulgaricus* DSM 24734, *B. longum* DSM 24736, *B. infantis* DSM 24737, *B. breve* DSM 24732, and *S. thermophilus* DSM 24731 improved significantly the levels of ALAT, cytokines, and endotoxins, as well as liver histology in patients with NAFLD (Duseja et al., 2019).

Although many clinical studies highlighted that probiotics and specifically LAB and bifidobacteria might be a promising therapeutic tool to prevent or alleviate metabolic disease, their efficiency is not clearly demonstrated yet. A better comprehension of bacterial niches in which different probiotics might thrive and how they might interact with the resident gut microbial communities will help to define or choose the right species/strain or mixture to benefit from their effects. Gut microbiota alterations due to westernized conditions differ between and within populations and individuals. This is defined by the Anna Karenina principle, which hypothesizes that all "healthy" microbiomes are similar, and each disease-associated microbiome is "sick" in its own way (Z. S. Ma, 2020). It is important to take this hypothesis into consideration, as human subjects with metabolic diseases might not respond similarly to the same treatment (responder vs. non-responder profile) and might require a personalized therapeutic approach tailored to their own gut microbial alterations. As a result, further studies using experimental models (*in vitro*, bioreactors, and *in vivo* models) are needed to understand the context and the conditions in which LAB and bifidobacteria might be the most effective to exert their beneficial metabolic effects.

## 25.2.3 LAB and Metabolic Diseases: Highlighted Mechanisms from Experimental Studies

Lactobacilli and bifidobacteria supplementations improve host health through different mechanisms. Notably, lactobacilli and bifidobacteria increase the levels of acetate or bile salt hydrolase (*bsh*) activities, which in turn modulate the activity of several host receptors known to regulate the immune system (Foligne et al., 2007) and gut barrier functions (Seth et al., 2008), lower blood cholesterol levels (Bubnov et al., 2017), and stimulate the gut–brain axis (Grasset et al., 2017).

### 25.2.3.1 Impact of LAB on Gut Mucosa Function and Gut Microbiota Composition

The gut mucosa is composed of different cell populations, including enterocytes and secretory cells—Goblet cells, Paneth cells, and enteroendocrine cells—as well as immune cells. In brief, the gut will protect the host from developing infection, inflammation, and metabolic disturbances through the secretion of antimicrobial peptides, the production of mucins to maintain a normal mucus layer, and the expression of tight junction proteins to establish a functional intestinal barrier. All these intestinal functions are impacted by the gut microbiota and have been shown to be altered in the context obesity and T2D. Indeed, a leaky gut barrier, which is characterized by the possible translocation of bacterial components (e.g., LPS) from the gut lumen to the portal blood, leads to low-grade systemic inflammation and insulin resistance. The disruption of the gut barrier function is partly related to an imbalance in gut microbiota composition, and many experimental studies have evaluated the effect of bacteria exhibiting anti-inflammatory abilities and capacities in reinforcing the epithelial barrier. Lactic acid bacteria and bifidobacteria can strengthen or restore a healthy function of the intestinal barrier through different mechanisms. The supplementation of *L. rhamnosus* GG (LGG) or *B. animalis* subsp. *lactis* 420 reduced intestinal permeability, which was associated with lower endotoxemia and improved glucose tolerance in diabetic mice (Honda et al., 2012; Stenman et al., 2014); however, the exact bacterial or molecular mechanisms were not identified.

*Lactobacillus* and *Bifidobacterium* can improve the intestinal gut barrier by the production of SCFA such as acetate. In fact, acetate can either act by itself on host receptors or be further metabolized by other members of the gut bacteria such as *Faecalibacterium prausnitzii* to produce butyrate (Rios-Covian et al., 2015). Acetate and butyrate can signal through GPR41 and GPR43 and regulate the expression of tight junction proteins, which are essential for maintaining a functional intestinal barrier (Pérez-Reytor et al., 2021). Also, probiotics can improve gut barrier function through the transformation of host metabolites such as conjugated bile acids. Indeed, a human-origin probiotic cocktail containing five *Lactobacillus* and five *Enterococcus* strains enriched the amino acid taurine released from the deconjugation of bile acids. Higher taurine levels increased tight junction protein expression and reduced intestinal permeability and inflammation under high-fat diet (HFD) feeding in mice (Ahmadi et al., 2020). In addition, the bacterial cell wall can contain key components that can have an important role in regulating gut permeability. Lipoteichoic acid from the cell wall of a heat-killed *L. paracasei* D3–5 improved gut leakiness under HFD in mice (S. Wang et al., 2019). The enhancement in intestinal barrier function was associated with a significant reduction in inflammation and resulted in an improvement in physical and cognitive functions. More specifically, lipoteichoic acid increased mucin production by modulating the TLR-2/p38-MAPK/NF-kB pathway and improved the gut barrier.

Probiotics do not just target tight junction protein expression and intestinal permeability. LAB play a key role in regulating intestinal stem cell (ISC) regeneration and the number of Paneth and goblet cells. For instance, VSL#3 administration increased crypt height and the number of Lgr5+ ISCs, Paneth cells, and goblet cells in the mouse small intestine. More mechanistically, *Lactobacillus* and *Bifidobacterium* produced lactate and increased ISC proliferation through the activation of the receptor GPR81 (Lee et al., 2018). Paneth and goblet cells can be considered gatekeepers since they play an important key role in keeping gut bacteria away from epithelial cells and fighting against invading pathogens. Thus, finding LAB strains that can maintain a physiological number of these cells and preserving their activities will help to limit metabolic endotoxemia. Indeed, administration of *L. casei* CRL 431 and *L. paracasei* CNCM I-1518 to old mice increased Paneth cell numbers and activity, which in turn reduced epithelial invasion by the pathogens *Staphylococcus aureus* and *Salmonella typhimurium* (Cazorla et al., 2018). Also, in a very elegant study, the supplementation of the probiotic strain *B. longum* NCC 2705 restored colonic mucus layer and mucus growth rate that was impaired by western-style diet feeding in mice (Schroeder et al., 2018). This result suggests that bifidobacteria can contribute to a protection against DIO and glucose intolerance through preserving the thickness of the colonic mucus layer. This mechanism was highlighted in the anti-obesity effect of *Akkermansia muciniphila*, a bacterium that is critical for mucin production by goblet cells and the maintenance of the mucus layer. Indeed, this bacterium restored gut barrier function; normalized metabolic endotoxemia; and reduced weight gain, fat mass development, and glycemia in a murine model of obesity (Everard et al., 2013a).

It is important to highlight that some of the effects of probiotics might be mediated through modifications of the gut microbiota composition and function. Indeed, different bifidobacteria species have been compared in the same experimental model using HFD-fed mice (Aoki et al., 2017). *B. animalis* subsp. *lactis* GLC2505, a strain that is highly proliferative in the gut, reduced visceral fat accumulation and improved glucose tolerance, while *B. longum* subsp. *longum* JCM1217 had no effect. *B. lactis* restructured gut bacteria composition and enriched *Bifidobacterium* and *Lactobacillus* genera, which was associated with an increase in acetate levels. Also, the supplementation of mixed lactobacilli (*L. plantarum* KLDS1.0344 and *L. plantarum* KLDS1.0386) to HFD-fed mice reduced DIO and liver fat accumulation (H. Li et al., 2020). The beneficial effects of these two lactobacilli seem to be mediated partly through a shift in the gut microbiota function and its ability to produce higher levels of SCFA. Indeed, the two strains increased the abundance of certain bacterial genera such as *Ruminococceae*, *Roseburia*, and *Lachnospiraceae*, resulting in a significant increase in acetic acid, propionic acid, and butyric acid. We have also shown that a mixture composed of *L. rhamnosus* LMG S-28148 and the *B. animalis* subsp. *lactis* LMG P-28149 strains protected mice from DIO by restoring the gut microbiota dysbiosis, notably by increasing the abundance of *A. muciniphila* and *Rikenellaceaea* as well as butyrate levels (Alard et al., 2016).

## 25.2.3.2 Impact of LAB on Immune Modulation

Obesity and T2D are associated with a low-grade chronic inflammation provoked a dysregulation of the immune response within the intestine and metabolic organs of the host. As a result, the interest to select strains with anti-inflammatory capacities to dampen the exacerbated inflammatory response in the context of obesity and T2D has gained attention. Indeed, a *B. pseudocatenulatum* strain prevented obesity and low-grade inflammation by restoring the balance between regulatory T cells and B lymphocytes (Moya-Pérez et al., 2015). The use of a multi-strain probiotic derived from traditional fermented dairy product showed stronger anti-inflammatory effects when compared to a single-strain probiotic (LGG) (Roselli et al., 2017), highlighting that using a mixture of different bacteria might be more efficient in affecting inflammation. The multi-strain probiotic could restore the balance between the major subpopulations of immune cells by reducing the number of pro-inflammatory CD8$^+$ T lymphocytes and macrophages and increasing the number of regulatory T lymphocytes under HFD challenge. The three probiotic strains (*L. paracasei* CNCM I-4270, *L. rhamnosus* I-3690, and *B. animalis* subsp. *lactis* I-2494) evaluated by Wang et al. in HFD-fed mice, which attenuated weight gain, glucose intolerance, and hepatic steatosis, were also able to reduce the number of pro-inflammatory macrophages and crown-like structures in the white adipose tissue. However, this was associated with different patterns of gut microbiota alteration and microbial fermentation (J. Wang et al., 2015). Similarly, we have reported that the two bacterial strains, *L. rhamnosus* and *B. lactis*, could reduce weight gain and fat accumulation in adipose tissues and liver, most likely through limiting the recruitment of pro-inflammatory macrophages and restoring the presence of regulatory T cells within the visceral adipose tissue (Alard et al., 2016). Interestingly, when the strains were administrated separately, both bacteria were able to limit adipose tissue inflammation; however, only the *B. lactis* strain prevented weight gain and insulin resistance and restored *A. muciniphila* abundance. These results suggest that some of the anti-inflammatory effects of probiotics can be independent of the impact of the bacteria on body weight gain and fat mass and that probiotics might have direct crosstalk with the immune system to regulate the inflammatory response. Lactobacilli strains can directly modulate immune cells via the production of metabolites or through the components of their cell walls. Indeed, *L. plantarum* OLL2712 possess abilities to modulate the immune system by promoting IL-10 secretion in a TLR2-dependent fashion in murine immune cells *in vitro*. More specifically, *L. plantarum* OLL2712 stimulated TLR2 pathway to release IL-10 (Toshimitsu et al., 2017). Interestingly, this effect was dependent on the growth phase of the bacterium, as *L. plantarum* OLL2712 collected during the exponential growth phase had a stronger effect on TLR2 activation and IL-10 release. This was further confirmed in an *in vivo* model of obese mice where the authors showed that the administration of *L. plantarum* OLL2712 harvested during the exponential growth phase reduced inflammatory markers as well as triglycerides and free-fatty acid levels compared to *L plantarum* collected during the stationary phase (Toshimitsu et al., 2017). Lactobacilli, through the secretion of specific muropeptides, are able to activate the nucleotide-binding oligomerization domain-containing (NOD)2, which resulted in an enhanced anti-inflammatory response (Macho Fernandez et al., 2011). This was further highlighted by Cavallari et al., where the authors showed that muramyl dipeptide (MDP) possesses antidiabetic effects, as its administration to mice on HFD could lower adipose tissue inflammation, endotoxemia, and insulin resistance via NOD2 signaling and IRF4. Importantly, the effects of MDP on host glucose metabolism and inflammatory response were independent of any changes in body weight gain or gut microbiota composition (Cavallari et al., 2017). More recently, the administration of an *L. casei* CRL 431 strain or its cell wall to obese mice could boost thymus function altered during obesity (Balcells et al., 2022). The probiotic supplementation reduced body weight gain and inflammatory cytokine levels and increased IL-10 and IL-7 levels. A recovery of mature T-lymphocyte populations of the thymus has also been observed (Balcells et al., 2022). In addition to the cell wall, lactobacilli strains can secrete metabolites that can act directly on immune cells. A strain of *L. reuteri* isolated from a human sample and capable of producing 3-idoleacetic acid (IAA) was shown to promote the generation of IL-35$^+$ B-cells. Indeed, these IL-35$^+$ B cells reduced adipose tissue accumulation, improved insulin sensitivity and glucose tolerance, and reduced inflammation in obese mice (Su et al., 2022).

All together, these studies emphasize the potential role and capacity of lactobacilli and bifidobacteria to prevent obesity development and its associated metabolic disturbances via the modulation of host immune response.

### 25.2.3.3 Impact on Host Metabolism and Gut–Brain Axis

Obesity- and T2D-associated microbiomes have been often considered important contributors to the disruption of key biological processes involved in glucose metabolism regulation, insulin secretion, and body fat accumulation. LAB supplementations have been shown to be efficient at restoring some of these disrupted metabolic functions.

The incretin system, which relies on the secretion of incretin hormones by enteroendocrine cells harbored in the gut, is among the most important glucose and insulin regulator. GLP-1, a key incretin hormone, is strongly modulated by metabolites and factors produced by gut bacteria. GLP-1 secretion, as well as its receptor (GLP1R), are disrupted in the context of obesity and T2D, and this alteration can be corrected by LAB supplementation. For instance, the administration of *B. lactis* GLC2505 significantly enhanced GLP-1 levels production and secretion (Aoki et al., 2017). The VSL#3 mixture prevented obesity and diabetes development through GLP-1 induction. This was most likely provoked by the increased levels of butyrate that signal through GPR41 and GPR43 (Yadav et al., 2013). Recently, we showed that different LAB strains—*L. lactis* PI23, *B. bifidum* PI22, and *B. longum* PI10—were able to stimulate GLP-1 secretion *in vitro* (Alard et al., 2021). However, the exact molecular mechanisms by which these LAB increased GLP-1 secretion was not explored (Alard et al., 2021). GLP-1 is rapidly degraded by dipeptidyl peptidase-IV (DPP-IV). Interestingly, 13 bifidobacteria strains isolated from fecal samples of breast-fed infants showed varying levels of inhibitory activities against DPP-IV, suggesting that these strains might prolong the half-life of GLP-1 to benefit for its anti-diabetic effects (Zeng et al., 2016). Subjects with obesity and T2D have been characterized by a state of GLP1R resistance, which might be related to a dysbiotic state of the small intestinal microbiota (Grasset et al., 2017). More specifically, a dramatic depletion of *Lactobacillus* genera was observed in mice fed with HFD and correlated with severe GLP1-R resistance in the enteric nervous system (ENS). This led to the disruption of the gut–brain axis and glucose metabolism through alteration in insulin secretion. This important finding suggests that lactobacilli might play an important role in maintaining a functional ENS and host glucose metabolism through the regulation of GLP-1R sensitivity. Also, LGG supplementation to mice modulated enteric nervous system function and gut motility, which is an important factor in regulating host glucose metabolism (Chandrasekharan et al., 2019). Interestingly, adhesion of LGG to intestinal epithelial cells was required to benefit from the effect of the LGG on ENS function and gut motility (Chandrasekharan et al., 2019).

Probiotics can modulate feeding behavior, which can impact body weight gain over time. VSL#3 supplementation decreased the expression of hunger-inducing genes (i.e., agouti-related protein (AgRP) and neuropeptide Y (NpY)) and induced genes involved in satiety expressed in pro-opiomelanocortin neurons within the hypothalamus (Yadav et al., 2013). Our recent data also indicated that daily oral administration of a mixture containing specific bifidobacteria and lactobacilli limited weight gain and reduced obesity-associated metabolic dysfunction and inflammation. This was associated with a hypothalamic upregulation of leptin and leptin receptor gene expression (Alard et al., 2021). Also, *B. longum* APC1472 could reduce ghrelin signaling *in vitro* (Torres-Fuentes et al., 2019), and the supplementation of this strain to obese mice reduced body weight gain, improved glucose metabolism, and reduced insulin and corticosterone levels (Schellekens et al., 2021). More importantly, some of the findings were partially translated to humans, as the supplementation of *B. longum* APC1472 to healthy adults with overweight or obesity had a positive effect on fasting blood glucose levels without impacting body weight (Schellekens et al., 2021). This suggests that probiotics could regulate body weight gain and food intake through the prevention of leptin resistance (Myers et al., 2010).

LAB supplementation can influence energy expenditure and lipid storage in metabolic organs such as adipose tissues and liver. Indeed, the *L. gasseri* BNR17 strain, isolated from human breast milk, significantly increased expression of fatty acid oxidation-related genes and of the main glucose transporter

GLUT4 in adipose tissue, which might have resulted in a reduction in body weight gain and fat accumulation (Kang et al., 2013). The *L. gasseri* SBT2055 strain increased cecal SCFA levels, which were positively correlated with enhanced energy expenditure and lower body weight gain (Shirouchi et al., 2016). *L. reuteri 263* treatment attenuated DIO by decreasing pro-inflammatory factors and enhanced mitochondrial respiration as well as the expression of browning-related genes in WAT (L.-H. Chen et al., 2018). More recently, *B. animalis* subsp. *lactis* GCL2505 was able to prevent body fat accumulation and increase energy expenditure in an acetate-GPR43–dependent fashion (Horiuchi et al., 2020). Bifidobacteria might regulate host metabolism through the secretion of exopolysaccharides (EPSs). An EPS-producing *B. animalis* IPLA R1 strain reduced fasting insulin levels and hepatic triglyceride accumulation in obese mice. In addition, the treatment downregulated the expression of genes involved in the hepatic synthesis of fatty acid and modified gut microbiota composition (Salazar et al., 2019). Finally, CLA produced by LAB, especially the trans-10, cis-12-CLA isomer, has anti-obesity effects by modulating energy expenditure, fatty acid oxidation, and lipogenesis (Lehnen et al., 2015) (Figure 25.1).

Bile acids (BAs) are signaling molecules synthetized in the liver from cholesterol (Molinaro et al., 2018). BAs act through several host receptors such as FXR and TGR5 to modulate host body weight and glucose metabolism (Molinaro et al., 2018). Conjugated BAs, once secreted in the gut lumen, go through bacterial transformations, including deconjugation, epimerization, and dehydroxylation, resulting in the synthesis of secondary BA. Numerous lactobacilli and bifidobacteria can deconjugate BA by expressing bile salt hydrolase enzymes (BSH), thus impacting the activity of FXR and/or TGR5. For example, the administration of VSL#3 probiotic to mice reduced conjugated BA levels, especially tauro-cholic acid (TCA) (Degirolamo et al., 2014). This resulted in an inhibition of the FXR–FGF15 axis and led to increased hepatic BA synthesis, thus affecting both glucose and lipid metabolism (Molinaro et al., 2018). *L. reuteri* strain CGMCC 17942 and *L. plantarum* strain CGMCC 14407 supplementation to mice on a lithogenic diet lowered hyperlipidemia and reduced hepatic steatosis in an intestinal FXR-dependent fashion (Ye et al., 2022). However, it is important to highlight that BSH activities can differ between lactobacilli species and strains, leading to different impacts on host body weight and metabolism. For instance, *L. salivarius* JCM1046 has greater abilities to deconjugate BA than *L. salivarius* UCC118, which correlated with lower body weight gain and lower cholesterol as well as lower triglyceride levels in the liver (Joyce et al., 2014), highlighting the importance of selecting strains properly with specific enzymatic capacities to benefit from their metabolic effects. More recently, supplementation of *B. longum* and *B. bifidum*, two bacterial species markedly reduced in subjects with obesity and high visceral fat mass, protected mice from DIO partly by activating bile acid signaling and oxidative phosphorylation within adipose tissue, leading to reduced body weight and improved hepatic steatosis and glucose tolerance (Kim et al., 2022).

## 25.2.3.4 Conclusion

Probiotics are a promising therapeutic tool that can be used to prevent or alleviate obesity development as well as its associated metabolic markers. However, their efficiency in managing body weight and glucose homeostasis remains controversial. Indeed, while most experimental studies reported beneficial or no impact of LAB on weight gain and metabolic disturbances, some strains have been reported to increase weight gain (Million et al., 2012). These contradictory observations might be explained by the strain specificity effect. Also, confounding factors such as ethnicities, gender, age, medications, origin of the strain, the associated diet, and the dose of the probiotic, as well as the duration and the window time of the administration, should be considered. Furthermore, another important parameter to consider is the microbial ecology in which the probiotics might be able to exert their beneficial effect. The alterations of gut microbiota composition vary markedly between individuals in a diseased context. This strongly suggests that a personalized approach based on the microbiome of each subject is required to choose a suitable single strain or a mixture of probiotics that might better target specific bacterial alterations.

The notion of personalized medicine started to gain more attention over the last decade (Bubnov et al., 2015; Reid et al., 2010) and will help to better determine the niches and/or conditions in which these probiotics can survive and function optimally in order to avoid any negative health outcomes. Indeed, a

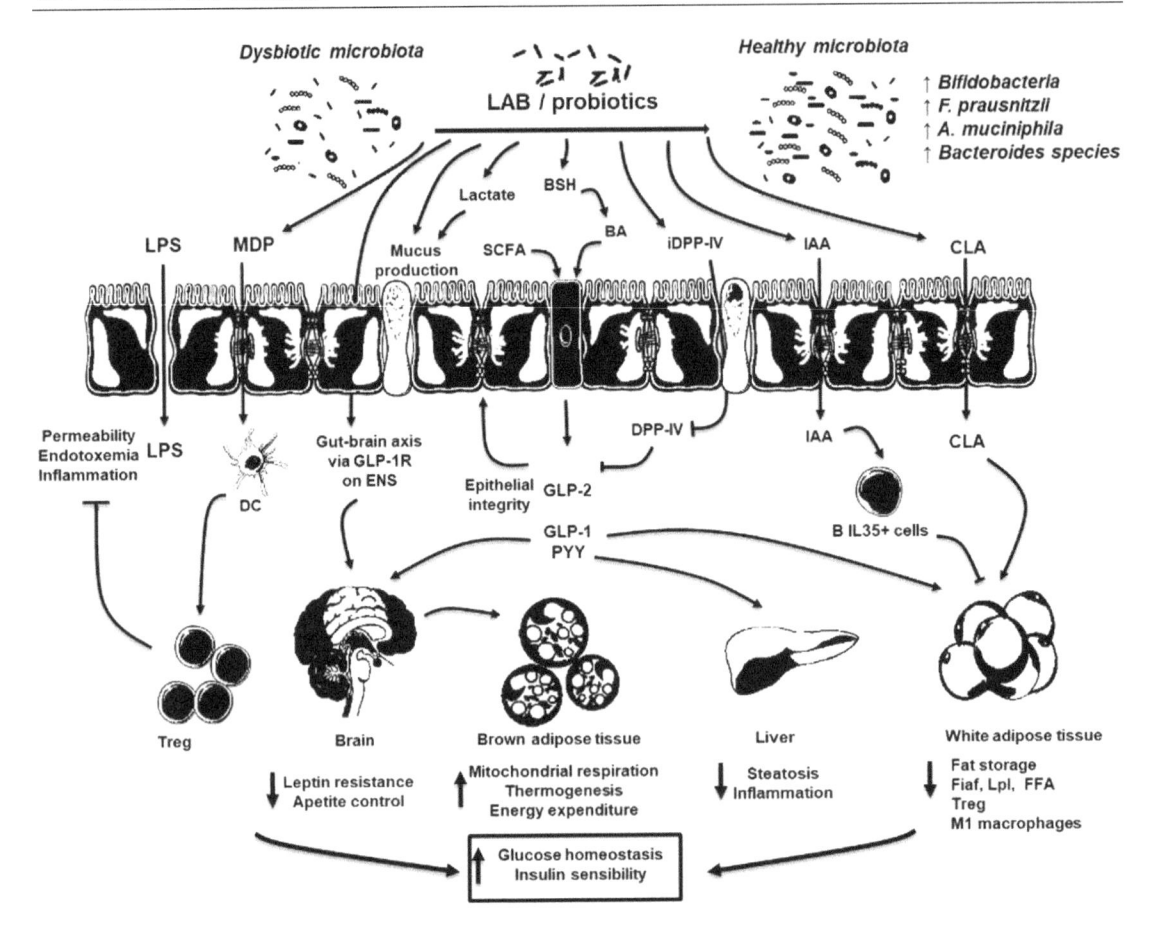

**FIGURE 25.1** Potential mechanisms of action of the anti-obesity effects of LAB/probiotics. LAB may restore the gut microbiota dysbiosis and increase SCFA production, which can activate GPCRs to either induce the secretion of gut peptides (GLP1, GLP-2, PYY) that may allow satiety and glucose homeostasis or maintain the abundance of key immune cells within the gut such as regulatory T cells (Treg). LAB can produce lactate, which will activate its receptor GPR81 expressed on ISC and lead to increased numbers of goblet and Paneth cells. Also, LAB sustain mucin production and secretion by goblet cells and preserve the colonic mucus layer, thus limiting bacterial translocation. LAB with BSH (bile salt hydrolase) activities can modify bile acid (BA) composition, therefore impacting glucose and lipid metabolism. In addition, GLP-1 action is prolonged through LAB-inhibiting activities against DPP-IV or enhanced by increasing the sensitivity of its receptor (GLP-1R) expressed on the ENS in a nitric oxide–dependent fashion. Furthermore, LAB could limit the inflammation and the recruitment of macrophages within the metabolic organs and thus improve insulin sensitivity, notably through the production of MDP (muramyl dipeptide), which binds to NOD2 receptor. Strains that produce IAA (indole-acetic acid) and CLA (conjugated linoleic acid) can favor energy expenditure and limit fat storage, partly through acting on the immune system.

recent study performed in a cohort of subjects with NAFLD showed that the microbiota of these individuals was able to produce ethanol, which contributed to the severity of the disease (Meijnikman et al., 2022). More importantly, the bacterial production of ethanol was strongly linked to higher abundance of lactic acid bacteria such as *Streptococcus* and *Lactobacillus* species (Meijnikman et al., 2022). This finding emphasizes the complexity of the system and the need to develop more comprehensive studies to understand when and how the use of LAB will be efficient to prevent or treat obesity and its associated metabolic disturbances.

# 25.3 NEXT-GENERATION PROBIOTICS IN METABOLIC DISEASES: PERSPECTIVES USING OTHER LIVE BIOTHERAPEUTICS

Microbiome-based therapies are now regarded as new potential strategies for the prevention or treatment of metabolic diseases and other chronic pathologies (Cani & Van Hul, 2015; Chang et al., 2019; O'Toole et al., 2017). Lactobacilli and bifidobacteria species residing or transiting in mammalian gastrointestinal tract benefit from "generally regarded as safe" (GRAS) status associated with "a long history of safe use" as traditional probiotics (Sanders et al., 2019). In parallel, a growing number of specific microorganisms have been reported to be more abundant in healthy populations and to be beneficial to host health, which legitimated the development of new generation of probiotics (NGPs) (Cani & Van Hul, 2015). This section will summarize works concerning intestinal species that do not have GRAS status despite their prevalence in the human gut microbiota. Currently, our knowledge in the field of next-generation probiotics and/or live biotherapeutics (LBP) is mainly based on pre-clinical studies, but in-progress clinical trials will be mentioned.

In agreement with the majority of studies reporting an association between obesity and a decrease of Bacteroidota, the effects of few *Bacteroides* species have been tested in experimental models of obesity. The *B. uniformis* CECT771 strain, the first to be evaluated in DIO in mice, was able to significantly limit weight gain and liver steatosis and improved glucose tolerance. The strain also modulated the composition of the gut microbiota and improved immune dysfunction induced by HFD (Gauffin Cano et al., 2012). A draft genome of the CECT7771 strain and comparative analysis with other *B. uniformis* and *Bacteroides* spp. genomes revealed that *B. uniformis* species exhibit an expanded glycolytic capability when compared with other *Bacteroides* species and is likely to be a mucin degrader that may tightly interact with the host mucosa (Benítez-Páez et al., 2017). The safety assessment of this strain was evaluated in rats after prolonged oral administration, showing no adverse metabolic or tissue alterations (Gómez Del Pulgar et al., 2020). *B. thetaiotaomicron* was found to be decreased in individuals with obesity, and live but not heat-killed *B. thetaiotaomicron* ATCC 29148–5 protected mice against obesity and reduced plasma glutamate concentration. Given the potential association between glutamate and insulin resistance, this may add another mechanism of action of NGPs on human health (Liu et al., 2017). Recently, sphingolipid synthesis by *B. thetaiotaomicron*, notably homoserine dihydroceramide, was determined to ameliorate hepatic lipid accumulation in a mouse model of hepatic steatosis (Le et al., 2022). Yang et al. reported that *B. acidifaciens*, a commensal bacterium previously isolated from mice ceca and human feces, can limit the effect of HFD in mice (2017). The level of these *Bacteroides* species significantly increased in the fecal microbiota of aged mice in which the autophagy-related gene 7 was knocked out in dendritic cells ($Atg7^{\Delta}CD11c$). The mice remained leaner and displayed better metabolic parameters than the $Atg7^{flox/flox(f/f)}$ control old mice (Yang et al., 2017). This highlights that *Bacteroides* species might play a role in regulating age-related obesity and metabolic disturbances. Furthermore, in HFD-fed mice, *B. acidifaciens* JCM10556 administration reduced body weight and fat mass and improved hepatic and peripheral insulin sensitivity. The treatment also resulted in fat oxidation through the activation of the bile acid-TGR5-PPARα axis in adipose tissue, explaining the higher energy expenditure observed in the treated mice. Also, *B. acidifaciens* JCM10556 increased GLP-1 action, thus improving insulin sensitivity and glucose tolerance. Interestingly, the administration of *B. sartorii* JCM17136 (with similar abundance in $Atg7^{\Delta}CD11$ and $Atg7^{f/f}$ mice) did not impact body weight gain under HFD, highlighting the strain-specificity effect of Bacteroides on host body weight and metabolism. Yoshida et al. recently reported that the gut microbiota is an important environmental factor regulating branched-chain amino acid (BCAA) catabolism in brown adipose tissue (BAT). Increased BCAA levels are associated with higher body weight in humans and rodents, and the administration of *B. dorei* and *B. vulgatus* promoted BCAA catabolism in BAT and protected against DIO in mice (Yoshida et al., 2021).

Similarly, *Parabacteroides merdae* protected mice against cardiovascular damage in HFD-fed ApoE-null male mice by enhancing branched-chain amino acid catabolism (Qiao et al., 2022). Since the positive effect of specific polysaccharides in reducing weight gain and metabolic disorders in mice was associated with the growth of *Parabacteroides goldsteinii*, this bacterium was also evaluated in murine models of obesity. Its administration to HFD-fed mice reduced obesity and improved insulin sensitivity, most likely via increased adipose tissue thermogenesis (T.-R. Wu et al., 2019). The metabolic benefit of *P. distasonis* supplementation was also reported. This bacterium reduced weight gain, hyperglycemia, and hepatic steatosis in obese mice (ob/ob and HFD-fed mice) via succinate-activated intestinal gluconeogenesis and secondary bile acid-activated FXR signaling (K. Wang et al., 2019). We also evaluated two strains of *P. distasonis* which exhibited anti-inflammatory activities and abilities to restore gut barrier integrity (Cuffaro et al., 2020, 2021). They were also highly potent in inducing GLP-1 secretion *in vitro*, and oral administration of these strains was highly effective in limiting weight gain and adiposity in obese mice (Cuffaro et al., 2023).

Different species belonging to the Bacillota phylum have been tested in obesity models. Recent studies have reported consistent association of bacteria belonging to Christensenellaceae with leanness and health (X. Li et al., 2020). Indeed, fecal microbiota transplantation of germ-free mice with a human microbiota of a subject with obesity and in which *Christensenella minuta* was added resulted in leaner animals (Goodrich et al., 2014). In a DIO model, the *C. minuta* DSM33407 strain was shown to limit body weight gain, improving metabolic disturbances and gut epithelial integrity restoration. Additionally, in a humanized *in vitro* SHIME model, the strain was able to modulate the intestinal microbiota composition characterized by a decreased Bacillota/Bacteroidota ratio (Mazier et al., 2021). The authors suggested that this strain acts as a keystone bacterium to restore the gut microbiota equilibrium. *C. minuta* DSM33407 supplementation is currently evaluated in a human clinical trial by Ysopia Company and completed the Phase 1 clinical study successfully in 2021, showing excellent safety and tolerability profile (Paquet et al., 2021), as well as the first efficacy results.

The abundance of *F. prausnitzii*, one of the most dominant bacteria of the gut microbiota and a high butyrate producer, was found to be decreased in several human pathologies, such as inflammatory bowel diseases, celiac disease, obesity, and diabetes (Langella et al., 2019; H. Wu et al., 2020). In a murine model of obesity, *F. prausnitzii*-treated mice presented a higher body weight than the HFD-group. However, hepatic fat content, WAT inflammation, and ASAT and ALAT levels were improved, together with increased fatty acid oxidation, adiponectin signaling, and insulin sensitivity (Munukka et al., 2017). Treatment with *F. prausnitzii* also improved muscle mass, possibly through mitochondrial respiration enhancement; modified the gut microbiota composition; and improved intestinal integrity. A human study reported that samples from lean subjects contained the highest count of *F. prausnitzii* genes compared to individuals with obesity and T2D but the lowest count of the *F. prausnitzii*–associated butyryl-CoA:acetate CoA-transferase gene (Hippe et al., 2016). The authors observed different *F. prausnitzii* phylotypes with different abilities to produce butyrate, which may have different capacities to prevent the development of T2D. Indeed, moderate levels seemed to improve insulin sensitivity, while increased concentration may promote inflammation and T2D.

The increase of *Eubacterium hallii* (now designated *Anaerobutyricum soehngenii*) abundance in the small intestine of patients suffering from metabolic syndrome and who received a fecal microbiota transplantation (FMT) of a sample from a lean donor (Vrieze et al., 2012) was associated with improved peripheral insulin sensitivity. The abundance of this bacterium was also inversely associated with BMI in young adults with T1D and overweight or obesity (Igudesman et al., 2022). The effect of live and inactivated *E. hallii*/*A. soehngenii* administrations in obese and diabetic *ob/ob* mice was therefore evaluated (Udayappan et al., 2016). The oral treatment with live bacteria improved insulin sensitivity and energy expenditure. It also slightly modified bile acid composition and increased fecal butyrate concentrations compared to control groups. Building on these encouraging data, a phase I/II pilot study was performed in subjects with metabolic syndrome to determine the safety and efficacy and the dose-response effects on glucose metabolism after a 4-week intervention with the stain *A. soehngenii* L2–7 (Gilijamse et al.,

2020). The treatment was safe, and a significant correlation between the fecal abundance of *A. soehngenii* and improvement in peripheral insulin sensitivity was observed after 4 weeks of treatment. The metabolic response was, however, dependent on the microbiota composition at baseline. Another clinical trial (USA, NCT04529473) has been completed to evaluate the efficacy and safety of a 12-week placebo-controlled study of this bacterium on insulin sensitivity and glycemic control. However, no results have been published yet.

Recently, the heat-shock protein ClpB of *Escherichia coli* was identified as a mimetic of the anorexigenic α-melanocyte stimulating hormone (α-MSH). A food-grade *Hafnia alvei* HA4597 strain, a member of the phylum Pseudomonadata synthetizing the ClpB protein with α-MSH-like motif, was isolated from raw milk as a candidate probiotic and was evaluated in different murine models of obesity. Oral administration of this bacterium reduced food intake, body weight, and fat mass gain in hyperphagic and obese mice (Legrand et al., 2020; Lucas et al., 2019). The clinical efficacy of *H. alvei* HA4597 (HA) was therefore evaluated in a 12-week prospective, double-blind, randomized study including 236 subjects with overweight. In the HA group, an increased feeling of fullness and a greater loss of hip circumference was observed (Déchelotte et al., 2021). This bacterium is now included in the EnteroSatys product provided on the market by TargEDys Company.

Some of the most advanced work concerns *A. muciniphila*, a bacterium of the phylum Verrucomicrobiota, previously identified as a mucin-degrading bacterium living in the gut mucus layer (Derrien et al., 2004). The abundance of this bacterium is decreased in obese and type 2 diabetic mice, and it is inversely correlated with body weight in humans. The administration of this bacterium to mice reversed HFD-induced metabolic disorders and increased endocannabinoid (Acylglycerols) content in the ileum, supporting a direct link between *A. muciniphila* administration, intestinal levels of acylglycerols and glucose, and intestinal homeostasis (Everard et al., 2013b). Interestingly, pasteurization of *A. muciniphila* Muc$^T$ enhanced its effect on fat mass reduction, insulin resistance, and dyslipidemia in HFD-fed mice, and Amuc 100, a purified surface protein of this bacterium acting through TLR2, improved the gut barrier and induced part of the beneficial effects of the strain in mice (Plovier et al., 2017). *A. muciniphila* supplementation to *Apoe$^{-/-}$* mice on HFD reduced endotoxemia and consequently reversed the exacerbation of atherosclerotic lesion formation (J. Li et al., 2016). *A. muciniphila* also reduced immune-mediated liver injury through alleviation of inflammation and hepatocellular death. The effects were associated with a restructuration of the microbiota composition and a restoration of the reduced glutathione/oxidized glutathione (GSH/GSSG) balance (W. Wu et al., 2017; Xia et al., 2022). Also reported was *A. muciniphila* supplementation decreasing islet auto-immunity in NOD mice (Hänninen et al., 2018). A clinical trial regarding the effect of *A. muciniphila* on parameters of metabolic syndrome has been completed in two groups of subjects receiving two different doses of living bacteria and one group supplemented with the heat-killed bacteria (ClinicalTrials.gov ID NCT02637115, Microbes4U). The study showed that treatment with *A. muciniphila*, either alive or pasteurized, is safe and well tolerated and that the pasteurized form limited the worsening of cardiometabolic disorders associated with overweight and obesity (Depommier et al., 2019). Results from multivariate analyses suggested that the beneficial effect of *A. muciniphila* was not linked to an overall modification of the endocannabinoidome, as previously expected (Depommier, Flamand et al., 2021), but the authors suggested that PPARα activation by mono-palmitoyl-glycerols may underlie part of the beneficial metabolic effects (Depommier, Vitale et al., 2021). *A. muciniphila* as a next-generation probiotic received the green light from the European Commission, which authorizes novel foods for use in food supplements and foods for special medical purposes (Overby & Ferguson, 2021). An *Akkermansia*-based nutritional supplement is foreseen to be on the market within 3 years by the spin-off A-Mansia Biotech SA.

Based on the evidence that HFD induced drastic changes in the ileal immune system, Pomié et al. (2016) proposed an original way to prevent HFD-induced insulin resistance and hyperglycemia via a single-dose immunization with diluted microbial extracts from the ileal content of HFD-diabetic mice. This immunization triggered the adaptive immune system to promote CD4 and CD8 T cell proliferation in lymphoid organs and increased cytokines and antibody secretion. The immunization reversed gut microbiota dysbiosis and adoptive transfer of immune cells from immunized mice and improved metabolic features in response to HFD. All together, this pioneering work shows the potential of immunization approaches to use the cross talk between the immune system and the microbiota to improve metabolic control.

In addition, several microbial metabolites could also influence host metabolism by interacting with key proteins involved in metabolic-related signaling pathways, such as G-protein coupled receptor (GPR), AMPK, and PPARγ, and as such could be used as postbiotics (H.-Y. Li et al., 2021; Overby & Ferguson, 2021). Notably, it is now well accepted that increasing SCFA production could be a valuable strategy to prevent gut dysfunction, obesity, and T2D. Notably, SCFA contribute to maintaining the intestinal barrier and could control energy intake through anorectic gut hormone release and gut-brain signaling. They also may increase energy expenditure through induction of thermogenesis in brown adipose tissue as well as browning of the white adipose tissue and fat oxidation. The best-studied mechanisms include the capacity of SCFA to bind to GPR (GPR41 and GPR43) and activate the production of the gut hormone peptide YY (PYY) and glucagon-like peptide 1 (GLP-1), thereby promoting satiety and glucose homeostasis (Blaak et al., 2020). However, the paucity of human studies limits for the moment the use of such molecules in the context of obesity (Vallianou et al., 2020).

Finally, FMT for the treatment of metabolic syndrome was reported in humans (Vrieze et al., 2012), showing an improvement of peripheral insulin sensitivity with a trend to improve hepatic insulin resistance and increased microbiota diversity. Building on this promising first attempt, ten clinical trials testing FMT for treatment of obesity, T2D, and liver steatosis (NAFLD) are planned or in progress according to the ClinicalTrials.gov database.

# 25.4 CONCLUSION

The rapid advances made during the past decade have highlighted the critical role of the gut microbiota in health and diseases, notably its impact on energy extraction and in the regulation of the immune and metabolic homeostasis. Since dysregulations of this microbial ecosystem have been associated with metabolic disorders, targeting this microbiota's dysbiosis is becoming a challenge in the management of obesity. This paves the way for the use of health-promoting beneficial bacteria, so-called probiotics. Indeed, many lactobacilli and bifidobacteria have been shown to positively impact obesity and associated metabolic disorders. Some strains were shown to limit weight gain, adiposity, and insulin resistance and represent interesting tools for weight management. The mechanisms by which these bacteria exhibit their beneficial effects are multifactorial, involving activities both on immune and metabolic functions but also by restructuring the gut microbiota and restoring the gut barrier function and the brain–gut axis. It remains important to better decipher the underlying mechanisms in order to select the best appropriate strains and also to isolate and characterize next-generation probiotic strains. The development of appropriate clinical studies remains of great importance to validate the experimental results and to demonstrate the efficacy of the selected strains to improve metabolic disorders.

Food products and food supplements, containing notably lactobacilli and bifidobacteria, have been at the forefront of the development of microbiome-based therapeutic products. Such products are intended to be used in a healthy population to maintain a healthy state or favor a healthy state in at-risk populations. The increased interest in the clinical application of LBP as medicinal products requires specific attention for their administration in a diseased population. Safety, quality, and efficacy information must be documented to fit with the current regulatory constraints for developers, which in the EU (European Medicines Agency; EMA) are not specifically defined at this moment to allow them to reach the market (Cordaillat-Simmons et al., 2020), while they have been partly laid out by the Food and Drug Administration (FDA) in the US since 2012 (Cordaillat-Simmons et al., 2020). Thus, better understanding of LBP–host interactions and demonstration of their safety, quality, and functionality while considering the global benefit–risk ratio will help to develop efficiently medicinal products to target gut microbiota alterations in obesity and T2D.

# BIBLIOGRAPHY

Ahmadi, S., Wang, S., Nagpal, R., Wang, B., Jain, S., Razazan, A., Mishra, S. P., Zhu, X., Wang, Z., Kavanagh, K., & Yadav, H. (2020, May 7). A human-origin probiotic cocktail ameliorates aging-related leaky gut and inflammation via modulating the microbiota/taurine/tight junction axis. *American Society for Clinical Investigation*. https://doi.org/10.1172/jci.insight.132055

Alard, J., Cudennec, B., Boutillier, D., Peucelle, V., Descat, A., Decoin, R., Kuylle, S., Jablaoui, A., Rhimi, M., Wolowczuk, I., Pot, B., Tailleux, A., Maguin, E., Holowacz, S., & Grangette, C. (2021). Multiple selection criteria for probiotic strains with high potential for obesity management. *Nutrients*, *13*(3), 713. https://doi.org/10.3390/nu13030713

Alard, J., Lehrter, V., Rhimi, M., Mangin, I., Peucelle, V., Abraham, A.-L., Mariadassou, M., Maguin, E., Waligora-Dupriet, A.-J., Pot, B., Wolowczuk, I., & Grangette, C. (2016). Beneficial metabolic effects of selected probiotics on diet-induced obesity and insulin resistance in mice are associated with improvement of dysbiotic gut microbiota. *Environmental Microbiology*, *18*(5), 1484–1497. https://doi.org/10.1111/1462-2920.13181

Álvarez-Arraño, V., & Martín-Peláez, S. (2021). Effects of probiotics and synbiotics on weight loss in subjects with overweight or obesity: A systematic review. *Nutrients*, *13*(10), 10. https://doi.org/10.3390/nu13103627

Aoki, R., Kamikado, K., Suda, W., Takii, H., Mikami, Y., Suganuma, N., Hattori, M., & Koga, Y. (2017). A proliferative probiotic Bifidobacterium strain in the gut ameliorates progression of metabolic disorders via microbiota modulation and acetate elevation. *Scientific Reports*, *7*(1), 1. https://doi.org/10.1038/srep43522

Balcells, F., Martínez Monteros, M. J., Gómez, A. L., Cazorla, S. I., Perdigón, G., & Maldonado-Galdeano, C. (2022). Probiotic consumption boosts thymus in obesity and senescence mouse models. *Nutrients*, *14*(3), 3. https://doi.org/10.3390/nu14030616

Belda, E., Voland, L., Tremaroli, V., Falony, G., Adriouch, S., Assmann, K. E., Prifti, E., Aron-Wisnewsky, J., Debédat, J., Roy, T. L., Nielsen, T., Amouyal, C., André, S., Andreelli, F., Blüher, M., Chakaroun, R., Chilloux, J., Coelho, L. P., Dao, M. C., . . . Clément, K. (2022). Impairment of gut microbial biotin metabolism and host biotin status in severe obesity: Effect of biotin and prebiotic supplementation on improved metabolism. *Gut*, *71*(12), 2463–2480. https://doi.org/10.1136/gutjnl-2021-325753

Benítez-Páez, A., Gómez Del Pulgar, E. M., & Sanz, Y. (2017). The glycolytic versatility of bacteroides uniformis CECT 7771 and its genome response to oligo and polysaccharides. *Frontiers in Cellular and Infection Microbiology*, *7*, 383. https://doi.org/10.3389/fcimb.2017.00383

Blaak, E. E., Canfora, E. E., Theis, S., Frost, G., Groen, A. K., Mithieux, G., Nauta, A., Scott, K., Stahl, B., van Harsselaar, J., van Tol, R., Vaughan, E. E., & Verbeke, K. (2020). Short chain fatty acids in human gut and metabolic health. *Beneficial Microbes*, *11*(5), 411–455. https://doi.org/10.3920/BM2020.0057

Brahe, L. K., Le Chatelier, E., Prifti, E., Pons, N., Kennedy, S., Blædel, T., Håkansson, J., Dalsgaard, T. K., Hansen, T., Pedersen, O., Astrup, A., Ehrlich, S. D., & Larsen, L. H. (2015). Dietary modulation of the gut microbiota—a randomised controlled trial in obese postmenopausal women. *The British Journal of Nutrition*, *114*(3), 406–417. https://doi.org/10.1017/S0007114515001786

Bubnov, R. V., Babenko, L. P., Lazarenko, L. M., Mokrozub, V. V., Demchenko, O. A., Nechypurenko, O. V., & Spivak, M. Y. (2017). Comparative study of probiotic effects of Lactobacillus and Bifidobacteria strains on cholesterol levels, liver morphology and the gut microbiota in obese mice. *The EPMA Journal*, *8*(4), 357–376. https://doi.org/10.1007/s13167-017-0117-3

Bubnov, R. V., Spivak, M. Y., Lazarenko, L. M., Bomba, A., & Boyko, N. V. (2015). Probiotics and immunity: Provisional role for personalized diets and disease prevention. *The EPMA Journal*, *6*(1), 14. https://doi.org/10.1186/s13167-015-0036-0

Cani, P. D., Amar, J., Iglesias, M. A., Poggi, M., Knauf, C., Bastelica, D., Neyrinck, A. M., Fava, F., Tuohy, K. M., Chabo, C., Waget, A., Delmée, E., Cousin, B., Sulpice, T., Chamontin, B., Ferrières, J., Tanti, J.-F., Gibson, G. R., Casteilla, L., . . . Burcelin, R. (2007). Metabolic endotoxemia initiates obesity and insulin resistance. *Diabetes*, *56*(7), 1761–1772. https://doi.org/10.2337/db06-1491

Cani, P. D., & Van Hul, M. (2015). Novel opportunities for next-generation probiotics targeting metabolic syndrome. *Current Opinion in Biotechnology*, *32*, 21–27. https://doi.org/10.1016/j.copbio.2014.10.006

Cavallari, J. F., Fullerton, M. D., Duggan, B. M., Foley, K. P., Denou, E., Smith, B. K., Desjardins, E. M., Henriksbo, B. D., Kim, K. J., Tuinema, B. R., Stearns, J. C., Prescott, D., Rosenstiel, P., Coombes, B. K., Steinberg, G. R., & Schertzer, J. D. (2017). Muramyl dipeptide-based postbiotics mitigate obesity-induced insulin resistance via IRF4. *Cell Metabolism*, 25(5), 1063–1074.e3. https://doi.org/10.1016/j.cmet.2017.03.021

Cazorla, S. I., Maldonado-Galdeano, C., Weill, R., De Paula, J., & Perdigón, G. D. V. (2018). Oral administration of probiotics increases paneth cells and intestinal antimicrobial activity. *Frontiers in Microbiology*, 9, 736. https://doi.org/10.3389/fmicb.2018.00736

Chandrasekharan, B., Saeedi, B. J., Alam, A., Houser, M., Srinivasan, S., Tansey, M., Jones, R., Nusrat, A., & Neish, A. S. (2019). Interactions between commensal bacteria and enteric neurons, via FPR1 induction of ROS, increase gastrointestinal motility in mice. *Gastroenterology*, 157(1), 179–192.e2. https://doi.org/10.1053/j.gastro.2019.03.045

Chang, C.-J., Lin, T.-L., Tsai, Y.-L., Wu, T.-R., Lai, W.-F., Lu, C.-C., & Lai, H.-C. (2019). Next generation probiotics in disease amelioration. *Journal of Food and Drug Analysis*, 27(3), 615–622. https://doi.org/10.1016/j.jfda.2018.12.011

Chen, A.-C., Fang, T.-J., Ho, H.-H., Chen, J.-F., Kuo, Y.-W., Huang, Y.-Y., Tsai, S.-Y., Wu, S.-F., Lin, H.-C., & Yeh, Y.-T. (2022). A multi-strain probiotic blend reshaped obesity-related gut dysbiosis and improved lipid metabolism in obese children. *Frontiers in Nutrition*, 9. www.frontiersin.org/articles/10.3389/fnut.2022.922993

Chen, L.-H., Chen, Y.-H., Cheng, K.-C., Chien, T.-Y., Chan, C.-H., Tsao, S.-P., & Huang, H.-Y. (2018). Antiobesity effect of Lactobacillus reuteri 263 associated with energy metabolism remodeling of white adipose tissue in high-energy-diet-fed rats. *The Journal of Nutritional Biochemistry*, 54, 87–94. https://doi.org/10.1016/j.jnutbio.2017.11.004

Choi, B. S.-Y., Brunelle, L., Pilon, G., Cautela, B. G., Tompkins, T. A., Drapeau, V., Marette, A., & Tremblay, A. (2022). Lacticaseibacillus rhamnosus HA-114 improves eating behaviors and mood-related factors in adults with overweight during weight loss: A randomized controlled trial. *Nutritional Neuroscience*, 1–13. https://doi.org/10.1080/1028415X.2022.2081288

Choucair, I., Nemet, I., Li, L., Cole, M. A., Skye, S. M., Kirsop, J. D., Fischbach, M. A., Gogonea, V., Brown, J. M., Tang, W. H. W., & Hazen, S. L. (2020). Quantification of bile acids: A mass spectrometry platform for studying gut microbe connection to metabolic diseases [S]. *Journal of Lipid Research*, 61(2), 159–177. https://doi.org/10.1194/jlr.RA119000311

Cordaillat-Simmons, M., Rouanet, A., & Pot, B. (2020). Live biotherapeutic products: The importance of a defined regulatory framework. *Experimental & Molecular Medicine*, 52(9), 1397–1406. https://doi.org/10.1038/s12276-020-0437-6

Cox, L. M., Yamanishi, S., Sohn, J., Alekseyenko, A. V., Leung, J. M., Cho, I., Kim, S. G., Li, H., Gao, Z., Mahana, D., Zárate Rodriguez, J. G., Rogers, A. B., Robine, N., Loke, P., & Blaser, M. J. (2014). Altering the intestinal microbiota during a critical developmental window has lasting metabolic consequences. *Cell*, 158(4), 705–721. https://doi.org/10.1016/j.cell.2014.05.052

Cuffaro, B., Assohoun, A. L. W., Boutillier, D., Peucelle, V., Desramaut, J., Boudebbouze, S., Croyal, M., Waligora-Dupriet, A.-J., Rhimi, M., Grangette, C., & Maguin, E. (2021). Identification of new potential biotherapeutics from human gut microbiota-derived bacteria. *Microorganisms*, 9(3), 565. https://doi.org/10.3390/microorganisms9030565

Cuffaro, B., Assohoun, A. L. W., Boutillier, D., Súkeníková, L., Desramaut, J., Boudebbouze, S., Salomé-Desnoulez, S., Hrdý, J., Waligora-Dupriet, A.-J., Maguin, E., & Grangette, C. (2020). In vitro characterization of gut microbiota-derived commensal strains: Selection of *Parabacteroides distasonis* strains alleviating TNBS-induced colitis in mice. *Cells*, 9(9). https://doi.org/10.3390/cells9092104

Cuffaro, B., Boutillier, D., Desramaut, J., Jablaoui, A., Werkmeister, E., Trottein, F., Waligora-Dupriet, A.J., Rhimi, M., Maguin, E., & Grangette, C. (2023). Characterization of two *Parabacteroides distasonis* candidate strains as new live biotherapeutics against obesity. *Cells*, 12(9). 1260. https://doi.org/10.3390/cells12091260

Dao, M. C., Everard, A., Aron-Wisnewsky, J., Sokolovska, N., Prifti, E., Verger, E. O., Kayser, B. D., Levenez, F., Chilloux, J., Hoyles, L., MICRO-Obes Consortium, Dumas, M.-E., Rizkalla, S. W., Doré, J., Cani, P. D., & Clément, K. (2016). Akkermansia muciniphila and improved metabolic health during a dietary intervention in obesity: Relationship with gut microbiome richness and ecology. *Gut*, 65(3), 426–436. https://doi.org/10.1136/gutjnl-2014-308778

Da Silva, C. C., Monteil, M. A., & Davis, E. M. (2020a). Overweight and obesity in children are associated with an abundance of firmicutes and reduction of bifidobacterium in their gastrointestinal microbiota. *Childhood Obesity*, 16(3), 204–210. https://doi.org/10.1089/chi.2019.0280

Da Silva, C. C., Monteil, M. A., & Davis, E. M. (2020b). Overweight and obesity in children are associated with an abundance of firmicutes and reduction of bifidobacterium in their gastrointestinal microbiota. *Childhood Obesity*, 16(3), 204–210. https://doi.org/10.1089/chi.2019.0280

Déchelotte, P., Breton, J., Trotin-Picolo, C., Grube, B., Erlenbeck, C., Bothe, G., Fetissov, S. O., & Lambert, G. (2021). The probiotic strain H. alvei HA4597 improves weight loss in overweight subjects under moderate hypocaloric diet: A proof-of-concept, multicenter randomized, double-blind placebo-controlled study. *Nutrients*, *13*(6), 1902. https://doi.org/10.3390/nu13061902

Degirolamo, C., Rainaldi, S., Bovenga, F., Murzilli, S., & Moschetta, A. (2014). Microbiota modification with probiotics induces hepatic bile acid synthesis via downregulation of the fxr-fgf15 axis in mice. *Cell Reports*, *7*(1), 12–18. https://doi.org/10.1016/j.celrep.2014.02.032

Depommier, C., Everard, A., Druart, C., Plovier, H., Van Hul, M., Vieira-Silva, S., Falony, G., Raes, J., Maiter, D., Delzenne, N. M., de Barsy, M., Loumaye, A., Hermans, M. P., Thissen, J.-P., de Vos, W. M., & Cani, P. D. (2019). Supplementation with Akkermansia muciniphila in overweight and obese human volunteers: A proof-of-concept exploratory study. *Nature Medicine*, *25*(7), 1096–1103. https://doi.org/10.1038/s41591-019-0495-2

Depommier, C., Flamand, N., Pelicaen, R., Maiter, D., Thissen, J.-P., Loumaye, A., Hermans, M. P., Everard, A., Delzenne, N. M., Di Marzo, V., & Cani, P. D. (2021). Linking the endocannabinoidome with specific metabolic parameters in an overweight and insulin-resistant population: From multivariate exploratory analysis to univariate analysis and construction of predictive models. *Cells*, *10*(1), 71. https://doi.org/10.3390/cells10010071

Depommier, C., Vitale, R. M., Iannotti, F. A., Silvestri, C., Flamand, N., Druart, C., Everard, A., Pelicaen, R., Maiter, D., Thissen, J.-P., Loumaye, A., Hermans, M. P., Delzenne, N. M., de Vos, W. M., Di Marzo, V., & Cani, P. D. (2021). Beneficial effects of akkermansia muciniphila are not associated with major changes in the circulating endocannabinoidome but linked to higher mono-palmitoyl-glycerol levels as new PPARα agonists. *Cells*, *10*(1), 185. https://doi.org/10.3390/cells10010185

Derrien, M., Vaughan, E. E., Plugge, C. M., & de Vos, W. M. (2004). Akkermansia muciniphila gen. nov., sp. nov., a human intestinal mucin-degrading bacterium. *International Journal of Systematic and Evolutionary Microbiology*, *54*(Pt 5), 1469–1476. https://doi.org/10.1099/ijs.0.02873-0

Dogra, S. K., Doré, J., & Damak, S. (2020). Gut microbiota resilience: Definition, link to health and strategies for intervention. *Frontiers in Microbiology*, *11*, 572921. https://doi.org/10.3389/fmicb.2020.572921

Duseja, A., Acharya, S. K., Mehta, M., Chhabra, S., Rana, S., Das, A., Dattagupta, S., Dhiman, R. K., & Chawla, Y. K. (2019). High potency multistrain probiotic improves liver histology in non-alcoholic fatty liver disease (NAFLD): A randomised, double-blind, proof of concept study. *BMJ Open Gastroenterology*, *6*(1), e000315. https://doi.org/10.1136/bmjgast-2019-000315

Eckburg, P. B., Bik, E. M., Bernstein, C. N., Purdom, E., Dethlefsen, L., Sargent, M., Gill, S. R., Nelson, K. E., & Relman, D. A. (2005). Diversity of the human intestinal microbial flora. *Science (New York, N.Y.)*, *308*(5728), 1635–1638. https://doi.org/10.1126/science.1110591

Everard, A., Belzer, C., Geurts, L., Ouwerkerk, J. P., Druart, C., Bindels, L. B., Guiot, Y., Derrien, M., Muccioli, G. G., Delzenne, N. M., de Vos, W. M., & Cani, P. D. (2013a). Cross-talk between Akkermansia muciniphila and intestinal epithelium controls diet-induced obesity. *Proceedings of the National Academy of Sciences of the United States of America*, *110*(22), 9066–9071. https://doi.org/10.1073/pnas.1219451110

Everard, A., Belzer, C., Geurts, L., Ouwerkerk, J. P., Druart, C., Bindels, L. B., Guiot, Y., Derrien, M., Muccioli, G. G., Delzenne, N. M., de Vos, W. M., & Cani, P. D. (2013b). Cross-talk between Akkermansia muciniphila and intestinal epithelium controls diet-induced obesity. *Proceedings of the National Academy of Sciences of the United States of America*, *110*(22), 9066–9071. https://doi.org/10.1073/pnas.1219451110

Famouri, F., Shariat, Z., Hashemipour, M., Keikha, M., & Kelishadi, R. (2017). Effects of probiotics on nonalcoholic fatty liver disease in obese children and adolescents. *Journal of Pediatric Gastroenterology and Nutrition*, *64*(3), 413–417. https://doi.org/10.1097/MPG.0000000000001422

Foligne, B., Nutten, S., Grangette, C., Dennin, V., Goudercourt, D., Poiret, S., Dewulf, J., Brassart, D., Mercenier, A., & Pot, B. (2007). Correlation between in vitro and in vivo immunomodulatory properties of lactic acid bacteria. *World Journal of Gastroenterology: WJG*, *13*(2), 236–243. https://doi.org/10.3748/wjg.v13.i2.236

Frost, F., Kacprowski, T., Rühlemann, M., Pietzner, M., Bang, C., Franke, A., Nauck, M., Völker, U., Völzke, H., Dörr, M., Baumbach, J., Sendler, M., Schulz, C., Mayerle, J., Weiss, F. U., Homuth, G., & Lerch, M. M. (2021). Long-term instability of the intestinal microbiome is associated with metabolic liver disease, low microbiota diversity, diabetes mellitus and impaired exocrine pancreatic function. *Gut*, *70*(3), 522–530. https://doi.org/10.1136/gutjnl-2020-322753

Gauffin Cano, P., Santacruz, A., Moya, Á., & Sanz, Y. (2012). Bacteroides uniformis CECT 7771 ameliorates metabolic and immunological dysfunction in mice with high-fat-diet induced obesity. *PLoS ONE*, *7*(7), e41079. https://doi.org/10.1371/journal.pone.0041079

Gebrayel, P., Nicco, C., Al Khodor, S., Bilinski, J., Caselli, E., Comelli, E. M., Egert, M., Giaroni, C., Karpinski, T. M., Loniewski, I., Mulak, A., Reygner, J., Samczuk, P., Serino, M., Sikora, M., Terranegra, A., Ufnal, M., Villeger, R., Pichon, C., . . . Edeas, M. (2022). Microbiota medicine: Towards clinical revolution. *Journal of Translational Medicine*, *20*(1), 111. https://doi.org/10.1186/s12967-022-03296-9

Gensollen, T., Iyer, S. S., Kasper, D. L., & Blumberg, R. S. (2016). How colonization by microbiota in early life shapes the immune system. *Science (New York, N.Y.)*, *352*(6285), 539–544. https://doi.org/10.1126/science.aad9378

Gilijamse, P. W., Hartstra, A. V., Levin, E., Wortelboer, K., Serlie, M. J., Ackermans, M. T., Herrema, H., Nederveen, A. J., Imangaliyev, S., Aalvink, S., Sommer, M., Levels, H., Stroes, E. S. G., Groen, A. K., Kemper, M., de Vos, W. M., Nieuwdorp, M., & Prodan, A. (2020). Treatment with Anaerobutyricum soehngenii: A pilot study of safety and dose—response effects on glucose metabolism in human subjects with metabolic syndrome. *NPJ Biofilms and Microbiomes*, *6*, 16. https://doi.org/10.1038/s41522-020-0127-0

Gómez Del Pulgar, E. M., Benítez-Páez, A., & Sanz, Y. (2020). Safety assessment of bacteroides uniformis CECT 7771, a symbiont of the gut microbiota in infants. *Nutrients*, *12*(2), E551. https://doi.org/10.3390/nu12020551

Gong, H., Gao, H., Ren, Q., & He, J. (2022). The abundance of bifidobacterium in relation to visceral obesity and serum uric acid. *Scientific Reports*, *12*(1), 1. https://doi.org/10.1038/s41598-022-17417-3

Goodrich, J. K., Waters, J. L., Poole, A. C., Sutter, J. L., Koren, O., Blekhman, R., Beaumont, M., Van Treuren, W., Knight, R., Bell, J. T., Spector, T. D., Clark, A. G., & Ley, R. E. (2014). Human genetics shape the gut microbiome. *Cell*, *159*(4), 789–799. https://doi.org/10.1016/j.cell.2014.09.053

Grasset, E., Puel, A., Charpentier, J., Collet, X., Christensen, J. E., Tercé, F., & Burcelin, R. (2017). A specific gut microbiota dysbiosis of type 2 diabetic mice induces GLP-1 resistance through an enteric no-dependent and gut-brain axis mechanism. *Cell Metabolism*, *25*(5), 1075–1090.e5. https://doi.org/10.1016/j.cmet.2017.04.013

Haeusler, R. A., Astiarraga, B., Camastra, S., Accili, D., & Ferrannini, E. (2013). Human insulin resistance is associated with increased plasma levels of 12α-hydroxylated bile acids. *Diabetes*, *62*(12), 4184–4191. https://doi.org/10.2337/db13-0639

Hänninen, A., Toivonen, R., Pöysti, S., Belzer, C., Plovier, H., Ouwerkerk, J. P., Emani, R., Cani, P. D., & De Vos, W. M. (2018). Akkermansia muciniphila induces gut microbiota remodelling and controls islet autoimmunity in NOD mice. *Gut*, *67*(8), 1445–1453. https://doi.org/10.1136/gutjnl-2017-314508

Hill, C., Guarner, F., Reid, G., Gibson, G. R., Merenstein, D. J., Pot, B., Morelli, L., Canani, R. B., Flint, H. J., Salminen, S., Calder, P. C., & Sanders, M. E. (2014). The international scientific association for probiotics and prebiotics consensus statement on the scope and appropriate use of the term probiotic. *Nature Reviews Gastroenterology & Hepatology*, *11*(8), 8. https://doi.org/10.1038/nrgastro.2014.66

Hippe, B., Remely, M., Aumueller, E., Pointner, A., Magnet, U., & Haslberger, A. G. (2016). Faecalibacterium prausnitzii phylotypes in type two diabetic, obese, and lean control subjects. *Beneficial Microbes*, *7*(4), 511–517. https://doi.org/10.3920/BM2015.0075

Honda, K., Moto, M., Uchida, N., He, F., & Hashizume, N. (2012). Anti-diabetic effects of lactic acid bacteria in normal and type 2 diabetic mice. *Journal of Clinical Biochemistry and Nutrition*, *51*(2), 96–101. https://doi.org/10.3164/jcbn.11-07

Horiuchi, H., Kamikado, K., Aoki, R., Suganuma, N., Nishijima, T., Nakatani, A., & Kimura, I. (2020). Bifidobacterium animalis subsp. lactis GCL2505 modulates host energy metabolism via the short-chain fatty acid receptor GPR43. *Scientific Reports*, *10*(1), 4158. https://doi.org/10.1038/s41598-020-60984-6

Human Microbiome Project Consortium. (2012). Structure, function and diversity of the healthy human microbiome. *Nature*, *486*(7402), 207–214. https://doi.org/10.1038/nature11234

Igudesman, D., Crandell, J., Corbin, K. D., Muntis, F., Zaharieva, D. P., Casu, A., Thomas, J. M., Bulik, C. M., Carroll, I. M., Pence, B. W., Pratley, R. E., Kosorok, M. R., Maahs, D. M., & Mayer-Davis, E. J. (2022). The intestinal microbiota and short-chain fatty acids in association with advanced metrics of glycemia and adiposity among young adults with type 1 diabetes and overweight or obesity. *Current Developments in Nutrition*, *6*(10), nzac107. https://doi.org/10.1093/cdn/nzac107

Joyce, S. A., Shanahan, F., Hill, C., & Gahan, C. G. (2014). Bacterial bile salt hydrolase in host metabolism: Potential for influencing gastrointestinal microbe-host crosstalk. *Gut Microbes*, *5*(5), 669–674. https://doi.org/10.4161/19490976.2014.969986

Jung, S., Lee, Y. J., Kim, M., Kim, M., Kwak, J. H., Lee, J.-W., Ahn, Y.-T., Sim, J.-H., & Lee, J. H. (2015). Supplementation with two probiotic strains, Lactobacillus curvatus HY7601 and Lactobacillus plantarum KY1032, reduced body adiposity and Lp-PLA2 activity in overweight subjects. *Journal of Functional Foods*, *19*, 744–752. https://doi.org/10.1016/j.jff.2015.10.006

Jung, S.-P., Lee, K.-M., Kang, J.-H., Yun, S.-I., Park, H.-O., Moon, Y., & Kim, J.-Y. (2013). Effect of *Lactobacillus gasseri* BNR17 on overweight and obese adults: A randomized, double-blind clinical trial. *Korean Journal of Family Medicine*, *34*(2), 80–89. https://doi.org/10.4082/kjfm.2013.34.2.80

Kadooka, Y., Sato, M., Imaizumi, K., Ogawa, A., Ikuyama, K., Akai, Y., Okano, M., Kagoshima, M., & Tsuchida, T. (2010). Regulation of abdominal adiposity by probiotics (Lactobacillus gasseri SBT2055) in adults with obese tendencies in a randomized controlled trial. *European Journal of Clinical Nutrition*, *64*(6), 6. https://doi.org/10.1038/ejcn.2010.19

Kadooka, Y., Sato, M., Ogawa, A., Miyoshi, M., Uenishi, H., Ogawa, H., Ikuyama, K., Kagoshima, M., & Tsuchida, T. (2013). Effect of Lactobacillus gasseri SBT2055 in fermented milk on abdominal adiposity in adults in a randomised controlled trial. *British Journal of Nutrition*, *110*(9), 1696–1703. https://doi.org/10.1017/S0007114513001037

Kang, J.-H., Yun, S.-I., Park, M.-H., Park, J.-H., Jeong, S.-Y., & Park, H.-O. (2013). Anti-obesity effect of Lactobacillus gasseri BNR17 in high-sucrose diet-induced obese mice. *PLoS ONE*, *8*(1), e54617. https://doi.org/10.1371/journal.pone.0054617

Kazemian, N., Mahmoudi, M., Halperin, F., Wu, J. C., & Pakpour, S. (2020). Gut microbiota and cardiovascular disease: Opportunities and challenges. *Microbiome*, *8*(1), 36. https://doi.org/10.1186/s40168-020-00821-0

Kim, G., Yoon, Y., Park, J. H., Park, J. W., Noh, M., Kim, H., Park, C., Kwon, H., Park, J., Kim, Y., Sohn, J., Park, S., Kim, H., Im, S.-K., Kim, Y., Chung, H. Y., Nam, M. H., Kwon, J. Y., Kim, I. Y., . . . Seong, J. K. (2022). Bifidobacterial carbohydrate/nucleoside metabolism enhances oxidative phosphorylation in white adipose tissue to protect against diet-induced obesity. *Microbiome*, *10*(1), 188. https://doi.org/10.1186/s40168-022-01374-0

Koeth, R. A., Wang, Z., Levison, B. S., Buffa, J. A., Org, E., Sheehy, B. T., Britt, E. B., Fu, X., Wu, Y., Li, L., Smith, J. D., DiDonato, J. A., Chen, J., Li, H., Wu, G. D., Lewis, J. D., Warrier, M., Brown, J. M., Krauss, R. M., . . . Hazen, S. L. (2013). Intestinal microbiota metabolism of L-carnitine, a nutrient in red meat, promotes atherosclerosis. *Nature Medicine*, *19*(5), 5. https://doi.org/10.1038/nm.3145

Koh, A., Molinaro, A., Ståhlman, M., Khan, M. T., Schmidt, C., Mannerås-Holm, L., Wu, H., Carreras, A., Jeong, H., Olofsson, L. E., Bergh, P.-O., Gerdes, V., Hartstra, A., Brauw, M. de, Perkins, R., Nieuwdorp, M., Bergström, G., & Bäckhed, F. (2018). Microbially produced imidazole propionate impairs insulin signaling through mTORC1. *Cell*, *175*(4), 947–961.e17. https://doi.org/10.1016/j.cell.2018.09.055

Koh, A., Vadder, F. D., Kovatcheva-Datchary, P., & Bäckhed, F. (2016). From dietary fiber to host physiology: Short-chain fatty acids as key bacterial metabolites. *Cell*, *165*(6), 1332–1345. https://doi.org/10.1016/j.cell.2016.05.041

Krumbeck, J. A., Rasmussen, H. E., Hutkins, R. W., Clarke, J., Shawron, K., Keshavarzian, A., & Walter, J. (2018). Probiotic Bifidobacterium strains and galactooligosaccharides improve intestinal barrier function in obese adults but show no synergism when used together as synbiotics. *Microbiome*, *6*(1), 121. https://doi.org/10.1186/s40168-018-0494-4

Langella, P., Guarner, F., & Martín, R. (2019). Editorial: Next-generation probiotics: From commensal bacteria to novel drugs and food supplements. *Frontiers in Microbiology*, *10*, 1973. https://doi.org/10.3389/fmicb.2019.01973

Larsen, N., Vogensen, F. K., Gøbel, R. J., Michaelsen, K. F., Forssten, S. D., Lahtinen, S. J., & Jakobsen, M. (2013). Effect of Lactobacillus salivarius Ls-33 on fecal microbiota in obese adolescents. *Clinical Nutrition*, *32*(6), 935–940. https://doi.org/10.1016/j.clnu.2013.02.007

Le, H. H., Lee, M.-T., Besler, K. R., & Johnson, E. L. (2022). Host hepatic metabolism is modulated by gut microbiota-derived sphingolipids. *Cell Host & Microbe*, *30*(6). https://doi.org/10.1016/j.chom.2022.05.002

Le Chatelier, E., Nielsen, T., Qin, J., Prifti, E., Hildebrand, F., Falony, G., Almeida, M., Arumugam, M., Batto, J.-M., Kennedy, S., Leonard, P., Li, J., Burgdorf, K., Grarup, N., Jørgensen, T., Brandslund, I., Nielsen, H. B., Juncker, A. S., Bertalan, M., . . . Pedersen, O. (2013). Richness of human gut microbiome correlates with metabolic markers. *Nature*, *500*(7464), 541–546. https://doi.org/10.1038/nature12506

Lee, Y.-S., Kim, T.-Y., Kim, Y., Lee, S.-H., Kim, S., Kang, S. W., Yang, J.-Y., Baek, I.-J., Sung, Y. H., Park, Y.-Y., Hwang, S. W., O, E., Kim, K. S., Liu, S., Kamada, N., Gao, N., & Kweon, M.-N. (2018). Microbiota-derived lactate accelerates intestinal stem-cell-mediated epithelial development. *Cell Host & Microbe*, *24*(6), 833–846. e6. https://doi.org/10.1016/j.chom.2018.11.002

Legrand, R., Lucas, N., Dominique, M., Azhar, S., Deroissart, C., Le Solliec, M.-A., Rondeaux, J., Nobis, S., Guérin, C., Léon, F., do Rego, J.-C., Pons, N., Le Chatelier, E., Ehrlich, S. D., Lambert, G., Déchelotte, P., & Fetissov, S. O. (2020). Commensal Hafnia alvei strain reduces food intake and fat mass in obese mice-a new potential probiotic for appetite and body weight management. *International Journal of Obesity (2005)*, *44*(5), 1041–1051. https://doi.org/10.1038/s41366-019-0515-9

Lehnen, T. E., da Silva, M. R., Camacho, A., Marcadenti, A., & Lehnen, A. M. (2015). A review on effects of conjugated linoleic fatty acid (CLA) upon body composition and energetic metabolism. *Journal of the International Society of Sports Nutrition*, *12*(1), 36. https://doi.org/10.1186/s12970-015-0097-4

Ley, R. E., Bäckhed, F., Turnbaugh, P., Lozupone, C. A., Knight, R. D., & Gordon, J. I. (2005). Obesity alters gut microbial ecology. *Proceedings of the National Academy of Sciences of the United States of America*, *102*(31), 11070–11075. https://doi.org/10.1073/pnas.0504978102

Li, H., Liu, F., Lu, J., Shi, J., Guan, J., Yan, F., Li, B., & Huo, G. (2020). Probiotic mixture of lactobacillus plantarum strains improves lipid metabolism and gut microbiota structure in high fat diet-fed mice. *Frontiers in Microbiology*, *11*, 512. https://doi.org/10.3389/fmicb.2020.00512

Li, H.-Y., Zhou, D.-D., Gan, R.-Y., Huang, S.-Y., Zhao, C.-N., Shang, A., Xu, X.-Y., & Li, H.-B. (2021). Effects and mechanisms of probiotics, prebiotics, synbiotics, and postbiotics on metabolic diseases targeting gut microbiota: A narrative review. *Nutrients*, *13*(9), 3211. https://doi.org/10.3390/nu13093211

Li, J., Jia, H., Cai, X., Zhong, H., Feng, Q., Sunagawa, S., Arumugam, M., Kultima, J. R., Prifti, E., Nielsen, T., Juncker, A. S., Manichanh, C., Chen, B., Zhang, W., Levenez, F., Wang, J., Xu, X., Xiao, L., Liang, S., . . . MetaHIT Consortium. (2014). An integrated catalog of reference genes in the human gut microbiome. *Nature Biotechnology*, *32*(8), 834–841. https://doi.org/10.1038/nbt.2942

Li, J., Lin, S., Vanhoutte, P. M., Woo, C. W., & Xu, A. (2016). Akkermansia muciniphila protects against athero-sclerosis by preventing metabolic endotoxemia-induced inflammation in Apoe-/- Mice. *Circulation*, *133*(24), 2434–2446. https://doi.org/10.1161/CIRCULATIONAHA.115.019645

Li, X., Li, Z., He, Y., Li, P., Zhou, H., & Zeng, N. (2020). Regional distribution of Christensenellaceae and its associa-tions with metabolic syndrome based on a population-level analysis. *PeerJ*, *8*. https://doi.org/10.7717/peerj.9591

Lim, S., Moon, J. H., Shin, C. M., Jeong, D., & Kim, B. (2020). Effect of *Lactobacillus sakei*, a probiotic derived from kimchi, on body fat in Koreans with obesity: A randomized controlled study. *Endocrinology and Metabolism*, *35*(2), 425–434. https://doi.org/10.3803/EnM.2020.35.2.425

Liu, R., Hong, J., Xu, X., Feng, Q., Zhang, D., Gu, Y., Shi, J., Zhao, S., Liu, W., Wang, X., Xia, H., Liu, Z., Cui, B., Liang, P., Xi, L., Jin, J., Ying, X., Wang, X., Zhao, X., . . . Wang, W. (2017). Gut microbiome and serum metabolome alterations in obesity and after weight-loss intervention. *Nature Medicine*, *23*(7), 859–868. https://doi.org/10.1038/nm.4358

Lucas, N., Legrand, R., Deroissart, C., Dominique, M., Azhar, S., Le Solliec, M.-A., Léon, F., do Rego, J.-C., Déchelotte, P., Fetissov, S. O., & Lambert, G. (2019). Hafnia alvei HA4597 strain reduces food intake and body weight gain and improves body composition, glucose, and lipid metabolism in a mouse model of hyperphagic obesity. *Microorganisms*, *8*(1), E35. https://doi.org/10.3390/microorganisms8010035

Ma, B., Sundararajan, S., Nadimpalli, G., France, M., McComb, E., Rutt, L., Lemme-Dumit, J. M., Janofsky, E., Roskes, L. S., Gajer, P., Fu, L., Yang, H., Humphrys, M., Tallon, L. J., Sadzewicz, L., Pasetti, M. F., Ravel, J., & Viscardi, R. M. (2022). Highly specialized carbohydrate metabolism capability in bifidobacterium strains associated with intestinal barrier maturation in early preterm infants. *MBio*, *13*(3), e01299–22. https://doi.org/10.1128/mbio.01299-22

Ma, Z. S. (2020). Testing the Anna Karenina Principle in human microbiome-associated diseases. *IScience*, *23*(4), 101007. https://doi.org/10.1016/j.isci.2020.101007

Macho Fernandez, E., Fernandez, E. M., Valenti, V., Rockel, C., Hermann, C., Pot, B., Boneca, I. G., & Grangette, C. (2011). Anti-inflammatory capacity of selected lactobacilli in experimental colitis is driven by NOD2-mediated recognition of a specific peptidoglycan-derived muropeptide. *Gut*, *60*(8), 1050–1059. https://doi.org/10.1136/gut.2010.232918

Makki, K., Deehan, E. C., Walter, J., & Bäckhed, F. (2018). The impact of dietary fiber on gut microbiota in host health and disease. *Cell Host & Microbe*, *23*(6), 705–715. https://doi.org/10.1016/j.chom.2018.05.012

Malard, F., Dore, J., Gaugler, B., & Mohty, M. (2021). Introduction to host microbiome symbiosis in health and dis-ease. *Mucosal Immunology*, *14*(3), 547–554. https://doi.org/10.1038/s41385-020-00365-4

Mazier, W., Le Corf, K., Martinez, C., Tudela, H., Kissi, D., Kropp, C., Coubard, C., Soto, M., Elustondo, F., Rawadi, G., & Claus, S. P. (2021). A new strain of christensenella minuta as a potential biotherapy for obesity and associ-ated metabolic diseases. *Cells*, *10*(4), 823. https://doi.org/10.3390/cells10040823

Meijnikman, A. S., Davids, M., Herrema, H., Aydin, O., Tremaroli, V., Rios-Morales, M., Levels, H., Bruin, S., de Brauw, M., Verheij, J., Kemper, M., Holleboom, A. G., Tushuizen, M. E., Schwartz, T. W., Nielsen, J., Brandjes, D., Dirinck, E., Weyler, J., Verrijken, A., . . . Nieuwdorp, M. (2022). Microbiome-derived ethanol in nonalco-holic fatty liver disease. *Nature Medicine*, *28*(10), 10. https://doi.org/10.1038/s41591-022-02016-6

Michael, D. R., Davies, T. S., Jack, A. A., Masetti, G., Marchesi, J. R., Wang, D., Mullish, B. H., & Plummer, S. F. (2021). Daily supplementation with the Lab4P probiotic consortium induces significant weight loss in over-weight adults. *Scientific Reports*, *11*(1), 1. https://doi.org/10.1038/s41598-020-78285-3

Michael, D. R., Jack, A. A., Masetti, G., Davies, T. S., Loxley, K. E., Kerry-Smith, J., Plummer, J. F., Marchesi, J. R., Mullish, B. H., McDonald, J. A. K., Hughes, T. R., Wang, D., Garaiova, I., Paduchová, Z., Muchová, J., Good, M. A., & Plummer, S. F. (2020). A randomised controlled study shows supplementation of overweight and obese adults with lactobacilli and bifidobacteria reduces bodyweight and improves well-being. *Scientific Reports*, *10*(1), 1. https://doi.org/10.1038/s41598-020-60991-7

Million, M., Maraninchi, M., Henry, M., Armougom, F., Richet, H., Carrieri, P., Valero, R., Raccah, D., Vialettes, B., & Raoult, D. (2012). Obesity-associated gut microbiota is enriched in Lactobacillus reuteri and depleted in Bifidobacterium animalis and Methanobrevibacter smithii. *International Journal of Obesity (2005)*, *36*(6), 817–825. https://doi.org/10.1038/ijo.2011.153

Molinaro, A., Bel Lassen, P., Henricsson, M., Wu, H., Adriouch, S., Belda, E., Chakaroun, R., Nielsen, T., Bergh, P.-O., Rouault, C., André, S., Marquet, F., Andreelli, F., Salem, J.-E., Assmann, K., Bastard, J.-P., Forslund, S., Le Chatelier, E., Falony, G., . . . Bäckhed, F. (2020). Imidazole propionate is increased in diabetes and associated with dietary patterns and altered microbial ecology. *Nature Communications*, *11*(1), 1. https://doi.org/10.1038/s41467-020-19589-w

Molinaro, A., Wahlström, A., & Marschall, H.-U. (2018). Role of bile acids in metabolic control. *Trends in Endocrinology & Metabolism*, 29(1), 31–41. https://doi.org/10.1016/j.tem.2017.11.002

Morais, L. H., Schreiber, H. L., & Mazmanian, S. K. (2020). The gut microbiota-brain axis in behaviour and brain disorders. *Nature Reviews Microbiology*. https://doi.org/10.1038/s41579-020-00460-0

Moya-Pérez, A., Neef, A., & Sanz, Y. (2015). Bifidobacterium pseudocatenulatum CECT 7765 reduces obesity-associated inflammation by restoring the lymphocyte-macrophage balance and gut microbiota structure in high-fat diet-fed mice. *PLoS ONE*, 10(7), e0126976. https://doi.org/10.1371/journal.pone.0126976

Munukka, E., Rintala, A., Toivonen, R., Nylund, M., Yang, B., Takanen, A., Hänninen, A., Vuopio, J., Huovinen, P., Jalkanen, S., & Pekkala, S. (2017). Faecalibacterium prausnitzii treatment improves hepatic health and reduces adipose tissue inflammation in high-fat fed mice. *The ISME Journal*, 11(7), 1667–1679. https://doi.org/10.1038/ismej.2017.24

Myers, M. G., Leibel, R. L., Seeley, R. J., & Schwartz, M. W. (2010). Obesity and leptin resistance: Distinguishing cause from effect. *Trends in Endocrinology and Metabolism: TEM*, 21(11), 643–651. https://doi.org/10.1016/j.tem.2010.08.002

Nakamura, F., Ishida, Y., Aihara, K., Sawada, D., Ashida, N., Sugawara, T., Aoki, Y., Takehara, I., Takano, K., & Fujiwara, S. (2016). Effect of fragmented Lactobacillus amylovorus CP1563 on lipid metabolism in overweight and mildly obese individuals: A randomized controlled trial. *Microbial Ecology in Health and Disease*, 27. https://doi.org/10.3402/mehd.v27.30312

Olm, M. R., Dahan, D., Carter, M. M., Merrill, B. D., Yu, F. B., Jain, S., Meng, X., Tripathi, S., Wastyk, H., Neff, N., Holmes, S., Sonnenburg, E. D., Jha, A. R., & Sonnenburg, J. L. (2022). Robust variation in infant gut microbiome assembly across a spectrum of lifestyles. *Science*, 376(6598), 1220–1223. https://doi.org/10.1126/science.abj2972

Olsson, L. M., Boulund, F., Nilsson, S., Khan, M. T., Gummesson, A., Fagerberg, L., Engstrand, L., Perkins, R., Uhlén, M., Bergström, G., Tremaroli, V., & Bäckhed, F. (2022). Dynamics of the normal gut microbiota: A longitudinal one-year population study in Sweden. *Cell Host & Microbe*, 30(5), 726–739.e3. https://doi.org/10.1016/j.chom.2022.03.002

O'Toole, P. W., Marchesi, J. R., & Hill, C. (2017). Next-generation probiotics: The spectrum from probiotics to live biotherapeutics. *Nature Microbiology*, 2, 17057. https://doi.org/10.1038/nmicrobiol.2017.57

Overby, H. B., & Ferguson, J. F. (2021). Gut microbiota-derived short-chain fatty acids facilitate microbiota: Host cross talk and modulate obesity and hypertension. *Current Hypertension Reports*, 23(2), 8. https://doi.org/10.1007/s11906-020-01125-2

Paquet, J.-C., Claus, S. P., Cordaillat-Simmons, M., Mazier, W., Rawadi, G., Rinaldi, L., Elustondo, F., & Rouanet, A. (2021). Entering first-in-human clinical study with a single-strain live biotherapeutic product: Input and feedback gained from the EMA and the FDA. *Frontiers in Medicine*, 8. www.frontiersin.org/articles/10.3389/fmed.2021.716266

Patterson, E., Ryan, P. M., Cryan, J. F., Dinan, T. G., Ross, R. P., Fitzgerald, G. F., & Stanton, C. (2016). Gut microbiota, obesity and diabetes. *Postgraduate Medical Journal*, 92(1087), 286–300. https://doi.org/10.1136/postgradmedj-2015-133285

Pedret, A., Valls, R. M., Calderón-Pérez, L., Llauradó, E., Companys, J., Pla-Pagà, L., Moragas, A., Martín-Luján, F., Ortega, Y., Giralt, M., Caimari, A., Chenoll, E., Genovés, S., Martorell, P., Codoñer, F. M., Ramón, D., Arola, L., & Solà, R. (2019). Effects of daily consumption of the probiotic Bifidobacterium animalis subsp. lactis CECT 8145 on anthropometric adiposity biomarkers in abdominally obese subjects: A randomized controlled trial. *International Journal of Obesity (2005)*, 43(9), 1863–1868. https://doi.org/10.1038/s41366-018-0220-0

Pérez-Reytor, D., Puebla, C., Karahanian, E., & García, K. (2021). Use of short-chain fatty acids for the recovery of the intestinal epithelial barrier affected by bacterial toxins. *Frontiers in Physiology*, 12. www.frontiersin.org/articles/10.3389/fphys.2021.650313

Plovier, H., Everard, A., Druart, C., Depommier, C., Van Hul, M., Geurts, L., Chilloux, J., Ottman, N., Duparc, T., Lichtenstein, L., Myridakis, A., Delzenne, N. M., Klievink, J., Bhattacharjee, A., van der Ark, K. C. H., Aalvink, S., Martinez, L. O., Dumas, M.-E., Maiter, D., . . . Cani, P. D. (2017). A purified membrane protein from Akkermansia muciniphila or the pasteurized bacterium improves metabolism in obese and diabetic mice. *Nature Medicine*, 23(1), 107–113. https://doi.org/10.1038/nm.4236

Pomié, C., Blasco-Baque, V., Klopp, P., Nicolas, S., Waget, A., Loubières, P., Azalbert, V., Puel, A., Lopez, F., Dray, C., Valet, P., Lelouvier, B., Servant, F., Courtney, M., Amar, J., Burcelin, R., & Garidou, L. (2016). Triggering the adaptive immune system with commensal gut bacteria protects against insulin resistance and dysglycemia. *Molecular Metabolism*, 5(6), 392–403. https://doi.org/10.1016/j.molmet.2016.03.004

Qiao, S., Liu, C., Sun, L., Wang, T., Dai, H., Wang, K., Bao, L., Li, H., Wang, W., Liu, S.-J., & Liu, H. (2022). Gut Parabacteroides merdae protects against cardiovascular damage by enhancing branched-chain amino acid catabolism. *Nature Metabolism*, 4(10), 1271–1286. https://doi.org/10.1038/s42255-022-00649-y

Qin, J., Li, R., Raes, J., Arumugam, M., Burgdorf, K. S., Manichanh, C., Nielsen, T., Pons, N., Levenez, F., Yamada, T., Mende, D. R., Li, J., Xu, J., Li, S., Li, D., Cao, J., Wang, B., Liang, H., Zheng, H., . . . Wang, J. (2010). A human gut microbial gene catalogue established by metagenomic sequencing. *Nature*, 464(7285), 59–65. https://doi.org/10.1038/nature08821

Rajkumar, H., Mahmood, N., Kumar, M., Varikuti, S. R., Challa, H. R., & Myakala, S. P. (2014). Effect of probiotic (VSL#3) and omega-3 on lipid profile, insulin sensitivity, inflammatory markers, and gut colonization in overweight adults: A randomized, controlled trial. *Mediators of Inflammation*, *2014*, 348959. https://doi.org/10.1155/2014/348959

Reid, G., Gaudier, E., Guarner, F., Huffnagle, G. B., Macklaim, J. M., Munoz, A. M., Martini, M., Ringel-Kulka, T., Sartor, B., Unal, R., Verbeke, K., Walter, J., & International Scientific Association for Probiotics and Prebiotics. (2010). Responders and non-responders to probiotic interventions: How can we improve the odds? *Gut Microbes*, *1*(3), 200–204. https://doi.org/10.4161/gmic.1.3.12013

Ridaura, V. K., Faith, J. J., Rey, F. E., Cheng, J., Duncan, A. E., Kau, A. L., Griffin, N. W., Lombard, V., Henrissat, B., Bain, J. R., Muehlbauer, M. J., Ilkayeva, O., Semenkovich, C. F., Funai, K., Hayashi, D. K., Lyle, B. J., Martini, M. C., Ursell, L. K., Clemente, J. C., . . . Gordon, J. I. (2013). Gut microbiota from twins discordant for obesity modulate metabolism in mice. *Science (New York, N.Y.)*, *341*(6150), 1241214. https://doi.org/10.1126/science.1241214

Rios-Covian, D., Gueimonde, M., Duncan, S. H., Flint, H. J., & de los Reyes-Gavilan, C. G. (2015). Enhanced butyrate formation by cross-feeding between Faecalibacterium prausnitzii and Bifidobacterium adolescentis. *FEMS Microbiology Letters*, *362*(21), fnv176. https://doi.org/10.1093/femsle/fnv176

Rodríguez, J. M., Murphy, K., Stanton, C., Ross, R. P., Kober, O. I., Juge, N., Avershina, E., Rudi, K., Narbad, A., Jenmalm, M. C., Marchesi, J. R., & Collado, M. C. (2015). The composition of the gut microbiota throughout life, with an emphasis on early life. *Microbial Ecology in Health and Disease*, *26*. https://doi.org/10.3402/mehd.v26.26050

Roselli, M., Devirgiliis, C., Zinno, P., Guantario, B., Finamore, A., Rami, R., & Perozzi, G. (2017). Impact of supplementation with a food-derived microbial community on obesity-associated inflammation and gut microbiota composition. *Genes & Nutrition*, *12*, 25. https://doi.org/10.1186/s12263-017-0583-1

Roswall, J., Olsson, L. M., Kovatcheva-Datchary, P., Nilsson, S., Tremaroli, V., Simon, M.-C., Kiilerich, P., Akrami, R., Krämer, M., Uhlén, M., Gummesson, A., Kristiansen, K., Dahlgren, J., & Bäckhed, F. (2021). Developmental trajectory of the healthy human gut microbiota during the first 5 years of life. *Cell Host & Microbe*, *29*(5), 765–776.e3. https://doi.org/10.1016/j.chom.2021.02.021

Rustanti, N., Murdiati, A., Juffrie, M., & Rahayu, E. S. (2022). Effect of probiotic lactobacillus plantarum dad-13 on metabolic profiles and gut microbiota in type 2 diabetic women: A randomized double-blind controlled trial. *Microorganisms*, *10*(9), 9. https://doi.org/10.3390/microorganisms10091806

Salazar, N., Neyrinck, A. M., Bindels, L. B., Druart, C., Ruas-Madiedo, P., Cani, P. D., de Los Reyes-Gavilán, C. G., & Delzenne, N. M. (2019). Functional effects of EPS-producing bifidobacterium administration on energy metabolic alterations of diet-induced obese mice. *Frontiers in Microbiology*, *10*, 1809. https://doi.org/10.3389/fmicb.2019.01809

Samara, J., Moossavi, S., Alshaikh, B., Ortega, V. A., Pettersen, V. K., Ferdous, T., Hoops, S. L., Soraisham, A., Vayalumkal, J., Dersch-Mills, D., Gerber, J. S., Mukhopadhyay, S., Puopolo, K., Tompkins, T. A., Knights, D., Walter, J., Amin, H., & Arrieta, M.-C. (2022). Supplementation with a probiotic mixture accelerates gut microbiome maturation and reduces intestinal inflammation in extremely preterm infants. *Cell Host & Microbe*, *30*(5), 696–711.e5. https://doi.org/10.1016/j.chom.2022.04.005

Sanders, M. E., Merenstein, D. J., Reid, G., Gibson, G. R., & Rastall, R. A. (2019). Probiotics and prebiotics in intestinal health and disease: From biology to the clinic. *Nature Reviews. Gastroenterology & Hepatology*, *16*(10), 605–616. https://doi.org/10.1038/s41575-019-0173-3

Santacruz, A., Collado, M. C., García-Valdés, L., Segura, M. T., Martín-Lagos, J. A., Anjos, T., Martí-Romero, M., Lopez, R. M., Florido, J., Campoy, C., & Sanz, Y. (2010). Gut microbiota composition is associated with body weight, weight gain and biochemical parameters in pregnant women. *The British Journal of Nutrition*, *104*(1), 83–92. https://doi.org/10.1017/S0007114510000176

Sartor, R. B., & Muehlbauer, M. (2007). Microbial host interactions in IBD: Implications for pathogenesis and therapy. *Current Gastroenterology Reports*, *9*(6), 497–507. https://doi.org/10.1007/s11894-007-0066-4

Schellekens, H., Torres-Fuentes, C., van de Wouw, M., Long-Smith, C. M., Mitchell, A., Strain, C., Berding, K., Bastiaanssen, T. F. S., Rea, K., Golubeva, A. V., Arboleya, S., Verpaalen, M., Pusceddu, M. M., Murphy, A., Fouhy, F., Murphy, K., Ross, P., Roy, B. L., Stanton, C., . . . Cryan, J. F. (2021). Bifidobacterium longum counters the effects of obesity: Partial successful translation from rodent to human. *EBioMedicine*, *63*, 103176. https://doi.org/10.1016/j.ebiom.2020.103176

Schroeder, B. O., Birchenough, G. M. H., Ståhlman, M., Arike, L., Johansson, M. E. V., Hansson, G. C., & Bäckhed, F. (2018). Bifidobacteria or fiber protects against diet-induced microbiota-mediated colonic mucus deterioration. *Cell Host & Microbe*, *23*(1), 27–40.e7. https://doi.org/10.1016/j.chom.2017.11.004

Seth, A., Yan, F., Polk, D. B., & Rao, R. K. (2008). Probiotics ameliorate the hydrogen peroxide-induced epithelial barrier disruption by a PKC- and MAP kinase-dependent mechanism. *American Journal of Physiology. Gastrointestinal and Liver Physiology*, *294*(4), G1060–1069. https://doi.org/10.1152/ajpgi.00202.2007

Shirouchi, B., Nagao, K., Umegatani, M., Shiraishi, A., Morita, Y., Kai, S., Yanagita, T., Ogawa, A., Kadooka, Y., & Sato, M. (2016). Probiotic Lactobacillus gasseri SBT2055 improves glucose tolerance and reduces body weight gain in rats by stimulating energy expenditure. *British Journal of Nutrition*, *116*(3), 451–458. https://doi.org/10.1017/S0007114516002245

Stenman, L. K., Lehtinen, M. J., Meland, N., Christensen, J. E., Yeung, N., Saarinen, M. T., Courtney, M., Burcelin, R., Lähdeaho, M.-L., Linros, J., Apter, D., Scheinin, M., Kloster Smerud, H., Rissanen, A., & Lahtinen, S. (2016). Probiotic with or without fiber controls body fat mass, associated with serum zonulin, in overweight and obese adults-randomized controlled trial. *EBioMedicine*, *13*, 190–200. https://doi.org/10.1016/j.ebiom.2016.10.036

Stenman, L. K., Waget, A., Garret, C., Klopp, P., Burcelin, R., & Lahtinen, S. (2014). Potential probiotic Bifidobacterium animalis ssp. lactis 420 prevents weight gain and glucose intolerance in diet-induced obese mice. *Beneficial Microbes*, *5*(4), 437–445. https://doi.org/10.3920/BM2014.0014

Su, X., Zhang, M., Qi, H., Gao, Y., Yang, Y., Yun, H., Zhang, Q., Yang, X., Zhang, Y., He, J., Fan, Y., Wang, Y., Guo, P., Zhang, C., & Yang, R. (2022). Gut microbiota-derived metabolite 3-idoleacetic acid together with LPS induces IL-35+ B cell generation. *Microbiome*, *10*(1), 13. https://doi.org/10.1186/s40168-021-01205-8

Sze, M. A., & Schloss, P. D. (2016). Looking for a signal in the noise: Revisiting obesity and the microbiome. *MBio*, *7*(4), e01018–16. https://doi.org/10.1128/mBio.01018-16

Thaiss, C. A., Itav, S., Rothschild, D., Meijer, M., Levy, M., Moresi, C., Dohnalová, L., Braverman, S., Rozin, S., Malitsky, S., Dori-Bachash, M., Kuperman, Y., Biton, I., Gertler, A., Harmelin, A., Shapiro, H., Halpern, Z., Aharoni, A., Segal, E., & Elinav, E. (2016). Persistent microbiome alterations modulate the rate of post-dieting weight regain. *Nature*. https://doi.org/10.1038/nature20796

Tims, S., Derom, C., Jonkers, D. M., Vlietinck, R., Saris, W. H., Kleerebezem, M., de Vos, W. M., & Zoetendal, E. G. (2013). Microbiota conservation and BMI signatures in adult monozygotic twins. *The ISME Journal*, *7*(4), 707–717. https://doi.org/10.1038/ismej.2012.146

Torres-Fuentes, C., Golubeva, A. V., Zhdanov, A. V., Wallace, S., Arboleya, S., Papkovsky, D. B., Aidy, S. E., Ross, P., Roy, B. L., Stanton, C., Dinan, T. G., Cryan, J. F., & Schellekens, H. (2019). Short-chain fatty acids and microbiota metabolites attenuate ghrelin receptor signaling. *The FASEB Journal*, *33*(12), 13546–13559. https://doi.org/10.1096/fj.201901433R

Toshimitsu, T., Ozaki, S., Mochizuki, J., Furuichi, K., & Asami, Y. (2017). Effects of Lactobacillus plantarum Strain OLL2712 culture conditions on the anti-inflammatory activities for murine immune cells and obese and type 2 diabetic mice. *Applied and Environmental Microbiology*, *83*(7), e03001–16. https://doi.org/10.1128/AEM.03001-16

Tseng, C.-H., & Wu, C.-Y. (2019). The gut microbiome in obesity. *Journal of the Formosan Medical Association = Taiwan Yi Zhi*, *118*(Suppl 1), S3–S9. https://doi.org/10.1016/j.jfma.2018.07.009

Turnbaugh, P. J., Ley, R. E., Mahowald, M. A., Magrini, V., Mardis, E. R., & Gordon, J. I. (2006). An obesity-associated gut microbiome with increased capacity for energy harvest. *Nature*, *444*(7122), 1027–1031. https://doi.org/10.1038/nature05414

Udayappan, S., Manneras-Holm, L., Chaplin-Scott, A., Belzer, C., Herrema, H., Dallinga-Thie, G. M., Duncan, S. H., Stroes, E. S. G., Groen, A. K., Flint, H. J., Backhed, F., de Vos, W. M., & Nieuwdorp, M. (2016). Oral treatment with Eubacterium hallii improves insulin sensitivity in db/db mice. *NPJ Biofilms and Microbiomes*, *2*, 16009. https://doi.org/10.1038/npjbiofilms.2016.9

Vallianou, N., Stratigou, T., Christodoulatos, G. S., Tsigalou, C., & Dalamaga, M. (2020). Probiotics, prebiotics, synbiotics, postbiotics, and obesity: Current evidence, controversies, and perspectives. *Current Obesity Reports*, *9*(3), 179–192. https://doi.org/10.1007/s13679-020-00379-w

Vandenplas, Y., Carnielli, V. P., Ksiazyk, J., Luna, M. S., Migacheva, N., Mosselmans, J. M., Picaud, J. C., Possner, M., Singhal, A., & Wabitsch, M. (2020). Factors affecting early-life intestinal microbiota development. *Nutrition*, *78*, 110812. https://doi.org/10.1016/j.nut.2020.110812

Vrieze, A., Van Nood, E., Holleman, F., Salojärvi, J., Kootte, R. S., Bartelsman, J. F. W. M., Dallinga-Thie, G. M., Ackermans, M. T., Serlie, M. J., Oozeer, R., Derrien, M., Druesne, A., Van Hylckama Vlieg, J. E. T., Bloks, V. W., Groen, A. K., Heilig, H. G. H. J., Zoetendal, E. G., Stroes, E. S., de Vos, W. M., . . . Nieuwdorp, M. (2012). Transfer of intestinal microbiota from lean donors increases insulin sensitivity in individuals with metabolic syndrome. *Gastroenterology*, *143*(4), 913–916.e7. https://doi.org/10.1053/j.gastro.2012.06.031

Wang, C., Wang, Y., Yang, H., Tian, Z., Zhu, M., Sha, X., Ran, J., & Li, L. (2022). Uygur type 2 diabetes patient fecal microbiota transplantation disrupts blood glucose and bile acid levels by changing the ability of the intestinal flora to metabolize bile acids in C57BL/6 mice. *BMC Endocrine Disorders*, *22*(1), 236. https://doi.org/10.1186/s12902-022-01155-8

Wang, J., Tang, H., Zhang, C., Zhao, Y., Derrien, M., Rocher, E., van-Hylckama Vlieg, J. E., Strissel, K., Zhao, L., Obin, M., & Shen, J. (2015). Modulation of gut microbiota during probiotic-mediated attenuation of metabolic syndrome in high fat diet-fed mice. *The ISME Journal*, *9*(1), 1. https://doi.org/10.1038/ismej.2014.99

Wang, K., Liao, M., Zhou, N., Bao, L., Ma, K., Zheng, Z., Wang, Y., Liu, C., Wang, W., Wang, J., Liu, S.-J., & Liu, H. (2019). Parabacteroides distasonis alleviates obesity and metabolic dysfunctions via production of succinate and secondary bile acids. *Cell Reports*, 26(1), 222–235.e5. https://doi.org/10.1016/j.celrep.2018.12.028

Wang, S., Ahmadi, S., Nagpal, R., Jain, S., Mishra, S. P., Kavanagh, K., Zhu, X., Wang, Z., McClain, D. A., Kritchevsky, S. B., Kitzman, D. W., & Yadav, H. (2019). Lipoteichoic acid from the cell wall of a heat killed Lactobacillus paracasei D3–5 ameliorates aging-related leaky gut, inflammation and improves physical and cognitive functions: From C. elegans to mice. *GeroScience*, 42(1), 333–352. https://doi.org/10.1007/s11357-019-00137-4

Wu, H., Tremaroli, V., Schmidt, C., Lundqvist, A., Olsson, L. M., Krämer, M., Gummesson, A., Perkins, R., Bergström, G., & Bäckhed, F. (2020). The gut microbiota in prediabetes and diabetes: A population-based cross-sectional study. *Cell Metabolism*, 32(3), 379–390.e3. https://doi.org/10.1016/j.cmet.2020.06.011

Wu, T.-R., Lin, C.-S., Chang, C.-J., Lin, T.-L., Martel, J., Ko, Y.-F., Ojcius, D. M., Lu, C.-C., Young, J. D., & Lai, H.-C. (2019). Gut commensal Parabacteroides goldsteinii plays a predominant role in the anti-obesity effects of polysaccharides isolated from Hirsutella sinensis. *Gut*, 68(2), 248–262. https://doi.org/10.1136/gutjnl-2017-315458

Wu, W., Lv, L., Shi, D., Ye, J., Fang, D., Guo, F., Li, Y., He, X., & Li, L. (2017). Protective effect of akkermansia muciniphila against immune-mediated liver injury in a mouse model. *Frontiers in Microbiology*, 8, 1804. https://doi.org/10.3389/fmicb.2017.01804

Xia, J., Lv, L., Liu, B., Wang, S., Zhang, S., Wu, Z., Yang, L., Bian, X., Wang, Q., Wang, K., Zhuge, A., Li, S., Yan, R., Jiang, H., Xu, K., & Li, L. (2022). Akkermansia muciniphila ameliorates acetaminophen-induced liver injury by regulating gut microbial composition and metabolism. *Microbiology Spectrum*, 10(1), e0159621. https://doi.org/10.1128/spectrum.01596-21

Yadav, H., Lee, J.-H., Lloyd, J., Walter, P., & Rane, S. G. (2013). Beneficial metabolic effects of a probiotic via butyrate-induced GLP-1 hormone secretion. *The Journal of Biological Chemistry*, 288(35), 25088–25097. https://doi.org/10.1074/jbc.M113.452516

Yang, J.-Y., Lee, Y.-S., Kim, Y., Lee, S.-H., Ryu, S., Fukuda, S., Hase, K., Yang, C.-S., Lim, H. S., Kim, M.-S., Kim, H.-M., Ahn, S.-H., Kwon, B.-E., Ko, H.-J., & Kweon, M.-N. (2017). Gut commensal Bacteroides acidifaciens prevents obesity and improves insulin sensitivity in mice. *Mucosal Immunology*, 10(1), 104–116. https://doi.org/10.1038/mi.2016.42

Ye, X., Huang, D., Dong, Z., Wang, X., Ning, M., Xia, J., Shen, S., Wu, S., Shi, Y., Wang, J., & Wan, X. (2022). FXR signaling-mediated bile acid metabolism is critical for alleviation of cholesterol gallstones by lactobacillus strains. *Microbiology Spectrum*, 10(5), e0051822. https://doi.org/10.1128/spectrum.00518-22

Yoshida, N., Yamashita, T., Osone, T., Hosooka, T., Shinohara, M., Kitahama, S., Sasaki, K., Sasaki, D., Yoneshiro, T., Suzuki, T., Emoto, T., Saito, Y., Ozawa, G., Hirota, Y., Kitaura, Y., Shimomura, Y., Okamatsu-Ogura, Y., Saito, M., Kondo, A., . . . Hirata, K.-I. (2021). Bacteroides spp. promotes branched-chain amino acid catabolism in brown fat and inhibits obesity. *IScience*, 24(11), 103342. https://doi.org/10.1016/j.isci.2021.103342

Zarrati, M., Salehi, E., Nourijelyani, K., Mofid, V., Zadeh, M. J. H., Najafi, F., Ghaflati, Z., Bidad, K., Chamari, M., Karimi, M., & Shidfar, F. (2014). Effects of probiotic yogurt on fat distribution and gene expression of pro-inflammatory factors in peripheral blood mononuclear cells in overweight and obese people with or without weight-loss diet. *Journal of the American College of Nutrition*, 33(6), 417–425. https://doi.org/10.1080/07315724.2013.874937

Zeng, Z., Luo, J. Y., Zuo, F. L., Yu, R., Zhang, Y., Ma, H. Q., & Chen, S. W. (2016). Bifidobacteria possess inhibitory activity against dipeptidyl peptidase-IV. *Letters in Applied Microbiology*, 62(3), 250–255. https://doi.org/10.1111/lam.12510

Zhang, X., Shen, D., Fang, Z., Jie, Z., Qiu, X., Zhang, C., Chen, Y., & Ji, L. (2013). Human gut microbiota changes reveal the progression of glucose intolerance. *PLoS ONE*, 8(8), e71108. https://doi.org/10.1371/journal.pone.0071108

Zheng, Z., & Wang, B. (2021). The gut-liver axis in health and disease: The role of gut microbiota-derived signals in liver injury and regeneration. *Frontiers in Immunology*, 12, 775526. https://doi.org/10.3389/fimmu.2021.775526

Zielińska, D., & Kołożyn-Krajewska, D. (2018). Food-origin lactic acid bacteria may exhibit probiotic properties: Review. *BioMed Research International*, 2018, e5063185. https://doi.org/10.1155/2018/5063185

# Brain–Gut–Microbiota Axis in Mood and Cognition; Impact of Psychobiotics

# 26

Mary I. Butler, Brian W. O'Mahony,
John F. Cryan, and Timothy G. Dinan

## 26.1 INTRODUCTION

The link between our psychological state and gut function has long been appreciated. William James, often considered the father of American psychology, wrote his controversial "Theory of Emotion" in the 1880s. This was one of the first scientific hypotheses proposing a link between an emotion and visceral bodily sensations, including "that peculiar epigastric change felt as precordial anxiety". Today the concept of the gut–brain axis is widely accepted. This bidirectional pathway allows the top-down influence of our central nervous system (CNS) and emotional states on gastrointestinal homeostasis and function as well as a bottom-up modulation of brain function and behavior via neural, endocrine, and immune systems. A more recent addition to the gut–brain axis has been the gut microbiome. The approximately $10^{13-14}$ bacteria that reside in our gut are now believed to be a vital node in the communication pathway with the emergence of an updated term, "the microbiome–gut–brain axis" (MGB axis).[1] While our gut bacteria have historically been regarded as commensal organisms, it is increasingly recognized that their relationship with the host is, in fact, a symbiotic one, impacting the functioning of virtually all organ systems, not least of all the CNS. Could our microbiome hold an etiological, or even therapeutic, key to the mental disorders that are epidemic in today's Western world? Could our mutualistic gut bacteria play a protective role in maintaining psychological health and possibly have the potential to alleviate depressive symptoms or cognitive dysfunction? The implications of our new understanding of the human being as a "superorganism" are only beginning to be realized.

# 26.2 POTENTIAL MECHANISMS BY WHICH THE MICROBIOME AFFECTS CNS FUNCTION

Several approaches have been used preclinically to elucidate the effects of the gut microbiome on the brain and behavior. These include the use of germ-free (GF) animals, animals exposed to probiotics, prebiotics, pathogenic bacterial infections or antibiotics, and fecal microbiome transplantation (FMT). The microbiome appears to play a role in modulating a wide range of systems, including the hypothalamic–pituitary–adrenal (HPA) axis and stress response, the immune system, neurotransmission, tryptophan metabolism, and serotonin synthesis, all processes which are highly relevant to the regulation of mood and cognition.

## 26.2.1 The Gut Microbiome Modulates the Immune System

It has long been known that pathogenic microbes can cause mood disturbances and cognitive dysfunction. Syphilis, a sexually transmitted infection caused by the bacterium *Treponema pallidum*, is a classic example. It can cause a variety of neuropsychological disturbances, including personality change, psychosis, delirium, and dementia.[2] Another example is Lyme disease, a relatively common condition caused by infection with the spirochete *Borrelia burgdorferi*, which is transmitted to humans by infected ticks. Left untreated, this can become multisystem disease that can cause many neuropsychiatric symptoms, including mood disturbance, anxiety, memory problems, and paranoia.[3] Similar neuropsychiatric disturbances can be seen in many nonbacterial infections such as malaria, a plasmodium parasitic infection,[4] and common viral infections such as human immunodeficiency virus, hepatitis C, varicella-zoster, and influenza.[5] That microbes can affect mood and cognition is clear; the method by which they do is less so.

A common denominator between microbial function and neuropsychiatric disturbance is peripheral immune activation and elevated pro-inflammatory biomarkers. Depression has been shown to be associated with a state of low-grade inflammation, characterized by increased levels of pro-inflammatory cytokines such as interleukin-3 (IL-3), interleukin-6 (IL-6), interleukin-12 (IL-12), and tumor necrosis factor alpha (TNF-alpha), as well as raised acute-phase protein CRP.[6] Similarly, schizophrenia has been consistently characterized by higher peripheral levels of IL-6, TNF-$\alpha$, IL-1$\beta$ IL-12, and transforming growth factor-$\beta$ (TGF-$\beta$).[7] Although less well studied, elevations of inflammatory markers are also seen in anxiety and fear-based disorders,[8] as well as in Alzheimer's disease.[9] It is unclear if the pro-inflammatory phenotype observed in neuropsychiatric disturbance is a cause or effect phenomenon, although an etiological role for peripheral cytokines is plausible. The pro-inflammatory cytokine interferon-$\alpha$, widely used in the treatment of cancer and hepatitis C, induces depressive symptoms in 25%–50% of patients,[10] and this can be prevented by antidepressant therapy.[11] Cytokines appear to influence a wide range of neurotransmitter pathways involved in depression, including serotonin, glutamate, and dopamine, through significant effects on synthesis, release, and reuptake. Imaging studies reveal that the brain regions most affected by peripheral inflammation appear to be the basal ganglia and anterior cingulate cortex, areas involved in motor activity, motivation, anxiety, and arousal.[12]

A gut microbiome-driven hypothesis which attributes the source of this peripheral immune activation to a variety of psychopathologies is that of "the leaky gut". The intestinal epithelium is a single cell layer that plays a vital role in maintaining a selectively permeable barrier between the gut lumen and the rest of the body. Stress activates the HPA axis and sympathetic nervous system, and the resulting corticotrophin-releasing hormone (CRH), cortisol, and catecholamines increase intestinal permeability. This leads to an increase in translocation of bacterial components such as lipopolysaccharides (LPS) across the epithelial barrier, which stimulates the production of cytokines by immune cells in the underlying *lamina propria*. Continuous stress-induced impairment of the intestinal barrier may create a feedback loop in which inflammatory cytokines persistently activate the CNS and HPA-axis, resulting in barrier disruption, increased endotoxin translocation, and a pro-inflammatory state.[13] Raised levels of LPS and inflammatory cytokines induce blood–brain barrier dysfunction,[14–16] which is associated with a variety of neuropsychiatric disorders.[17]

The gut microbiome, in close contact with the intestinal epithelial layer, plays a significant role in the homeostasis of the intestinal barrier. A pivotal signaling pathway between bacteria and the host involves toll-like receptors (TLRs). These are transmembrane proteins, expressed predominantly on immune cells, which function as pattern recognition receptors. TLRs recognize characteristic microbial patterns called microbe-associated molecular patterns (MAMPs) and are the gateway to the immune response, signaling the initiation of the cytokine cascade. Without the microbiota, certain TLRs are not fully expressed in the gut, and this can affect the functioning of the immune system.[18]

It is reasonable to surmise that one mechanism by which probiotic bacteria might positively impact brain function may be by attenuation of the pro-inflammatory state seen in depression and stress-related conditions. There is an accumulating body of evidence demonstrating that this is true. In a rat maternal separation model of depression, the probiotic *Bifidobacterium infantis* 35624 was shown to reverse depression-associated behavioral deficits and normalize the immune response.[19] The probiotic *Limosilactobacillus reuteri* and butyrate reduced anxiety-like behaviors through regulation of immune cell homeostasis in mice.[20]

## 26.2.2 The Gut Microbiome Directly Impacts Neurotransmission

A key mechanism by which the commensal gut bacteria communicate with the CNS lies in their ability to influence neurotransmission. They have been shown to do this in a variety of ways, including the exertion of a direct effect on enteric neuronal excitability, modulation of vagal transmission, and production of a range of key neurotransmitters involved in mood and cognition.

The microbiota directly influences the electrophysiological properties of enteric neurons by altering their development and activity. In the first instance, GF mice show abnormal postnatal development of the enteric nervous system with a decrease in nerve density.[21] In addition, the excitability of colonic after-hyperpolarization (AH) neurons is decreased in GF mice,[22] and myenteric neuron excitability was reduced in mice exposed to the probiotics *Bifidobacterium longum* NCC3001[23] and *Limosilactobacillus reuteri*.[24]

The vagus nerve is the major parasympathetic nerve in the body and plays a key role in regulating several organ functions, including heart rate, bronchial constriction, and gut motility. Stimulation of the vagus nerve has been shown to be effective in treating refractory depression[25] and in pain suppression,[26] and various antidepressant and anti-anxiety agents have been shown to modulate vagal transmission.[27,28] Studies using rodents infected with pathogenic bacteria have demonstrated that this nerve is involved in gut–brain signaling, following gastrointestinal infection. In 2002, Wang et al.[29] published a vagotomy (surgical cutting of the vagus nerve) study that demonstrated the role of the vagus nerve in transmitting gut–brain signals in rats infected with *Salmonella enterica*. Two subsequent studies[30,31] infected rodents with *Campylobacter jejuni*, a foodborne pathogen known to cause anxiety-like behaviors. Using c-Fos immunochemistry, they found increased activation in several brain areas that process visceral/autonomic information, including those that are typically activated following vagal stimulation. These studies paved the way for later work into the role of the vagus nerve in mediating the effects of probiotic bacteria. Ingestion of the probiotic *Lactobacillus rhamnosus* (JB-1) positively influenced behaviors relevant to anxiety, depression, and cognition and altered central levels of GABA receptors in mice. Both these neurochemical and behavioral effects were not found in mice who had undergone vagotomy.[32] Another study described mice with a chronic chemical colitis showing associated anxiety-like behaviors that were found to be absent in those mice who had previously been vagotomized.[23] This research team used *Bifidobacterium longum* NCC3001 and found that it had an anxiolytic effect on the mice, but only in those in whom the vagal nerve was intact.

Gut bacteria can also influence neurotransmission through the direct production of key neurotransmitters and neuromodulators. It has been shown that *Lactobacillus* spp. and *Bifidobacterium* spp. produce gamma-aminobutyric acid (GABA); *Escherichia* spp., *Bacillus* spp., and *Saccharomyces* spp. produce noradrenaline; *Bifidobacterium* spp., *Candida* spp., *Streptococcus* spp., *Escherichia* spp., and *Enterococcus* spp. produce serotonin; *Bacillus* spp. produce dopamine; and *Lactobacillus* spp. produce acetylcholine.[33,34] These microbial neurometabolites may represent a common language enabling communication between gut bacteria and host cells. Many of these neuroactive compounds are known play a role in mood disorders, anxiety, and cognitive functioning[35–38] and may provide one explanation for the ability of probiotics to reduce anxiety behaviors and improve memory.

## 26.2.3 The Gut Microbiome Influences the HPA Axis

The HPA axis, along with the sympathetic nervous system, regulates the stress response. A stressor, either physical or psychological, activates the hypothalamus, which, via the release of CRH, stimulates the pituitary gland to produce adrenocorticotropic hormone (ACTH). This in turn acts on the adrenal gland to mediate cortisol release. Cortisol's widespread action regulates a variety of bodily processes, including metabolism, immunity, and brain function. The HPA axis has been shown to be dysregulated in a variety of psychiatric illnesses, in particular depression and anxiety.[39,40]

Landmark studies in 2004 provided evidence that the gut microbiome plays a crucial role in the development of the HPA axis and that there is a critical period in early life when normal development of the HPA axis is contingent on colonization of the gastrointestinal tract (GIT).[41] This research team used GF mice in comparison to specific-pathogen-free (SPF) mice to directly assess the role of the gut microbiome on HPA axis development. They discovered that GF mice exhibited higher ACTH and corticosterone release following a mild restraint stress in comparison to their SPF control counterparts. This exaggerated stress response in GF mice was partially reversed by colonization with feces from SPF mice and completely reversed by mono-association with a specific bacterial strain, *Bifidobacterium infantis*. However, the reversal was time dependent and only occurred if the reconstitution took place at an early stage, thus indicating that there is a narrow window in early life during which gut colonization must occur. Further evidence that the gut microbiome modulates the stress response, including the HPA axis, is that GF and antibiotic-treated mice showed a social deviation which was reversed by inactivation of the HPA axis, and a bacterial species, *Enterococcus faecalis*, promoted social activity and reduced corticosterone levels in mice following social stress.[42]

Further studies exploring the stress response at the neuronal and behavioral levels have yielded inconsistent results. GF animals, with an exaggerated HPA axis response to stress, have exhibited increased anxiety-like behaviors.[43,44] However, several studies[45–47] have reported that GF animals (of different strains and sexes) actually show reduced anxiety-like behaviors in various commonly used animal anxiety measures. Further discrepancies are seen on a neuronal level. Sudo et al.[41] found a decrease in brain-derived-neurotrophic-factor (BDNF), a key neurotrophin involved in neuronal growth and development, along with decreased expression of the NMDA receptor subunit 2A (NR2A) in the hippocampus and cortex of male GF mice. On the other hand, Neufeld et al.[47] reported an increase in hippocampal BDNF mRNA in female mice, while another study reported decreases in hippocampal BDNF mRNA levels and distinct changes in the hippocampal serotoninergic system in male but not female mice.[45] The reasons for these discrepancies are unclear, but it appears that regulation of the MGB axis is sex-dependent.

The microbiome-HPA axis relationship is bidirectional. While the impact of the microbiome on HPA axis development has only recently been demonstrated, it has long been known that the composition of the microbiome is influenced by stress and HPA axis alterations.[48] However, the functional consequences of such changes are only recently being understood. Several studies exploring different animal models of stress consistently report changes in microbiome composition, along with HPA axis alterations. Such models include maternal separation,[49,50] chronic restraint,[51,52] and repeated social stress,[53–55] which have all been associated with changes in microbiome composition.

Such studies offer promise that probiotic treatment has the potential to reduce behavioral and neurochemical stress responses in animals through an influence on the HPA axis, with possible later application for humans. Results from animal studies have been inconsistent. Rats treated with *B. infantis* demonstrated positive behavioral changes but no corresponding reduction in corticosterone.[19] McKernan et al.[56] found that *B. infantis* did reduce corticosterone levels, but the finding did not reach statistical significance, while a *Lactobacillus* probiotic resulted in significantly reduced corticosterone levels in rats.[57] In humans, healthy participants demonstrated a reduction in 24-hour urinary free cortisol following consumption of a combination of *Lactobacillus helveticus* R0052 and *B. longum* R0175,[58] and *Lacticaseibacillus casei* suppressed the stress-related increase of salivary cortisol and the onset of physical symptoms of healthy medical students under academic examination stress.[59]

## 26.2.4 The Gut Microbiome Produces Short-Chain Fatty Acids

Short-chain fatty acids (SCFAs) are produced by gut bacteria, predominantly in the proximal colon, as the result of fermentation of nondigestible carbohydrates and proteins. The main SCFA are acetic acid, propionic acid, and *n*-butyric acid (butyrate), which, following absorption and metabolism, make a small but significant contribution to total body energy production (Macfarlane and Macfarlane, 2012). It has long been appreciated that high-fiber diets are associated with better health outcomes, and the role of SCFA as more than a mere source of nutrients has recently been considered. It is noteworthy that SCFA-producing bacteria are not restricted to a single phylum, suggesting that the ability to produce SCFA must have been a consistent requirement throughout host–microbe co-evolution.[60] The importance of SCFA in depression has recently been uncovered, perhaps caused by low SCFA levels leading to decreased energy and reduced neurotransmitter production.[61]

Following production, SCFAs enter the circulation via the hepatic portal vein. Concentrations of propionate, and to a lesser extent butyrate and acetate, are detectable in peripheral blood at significantly lower levels than in hepatic blood, suggesting that a significant proportion of absorbed SCFAs are metabolized in the liver.[62] SCFAs have been found to act as natural ligands at a variety of transporters and receptors expressed across many different cell types in the body, indicating a role beyond the GIT and liver. These include previously orphan G protein–coupled receptors (GPRs) GPR41 and GPR43, now renamed free fatty acid receptors (FFARs) FFAR2 and FFAR3, respectively,[63] as well as GPR109a, now renamed hydroxycarboxylic acid receptor 2 (HCAR2) (Singh et al., 2014). FFAR2 and FFAR3 are widely expressed on leukocytes, suggesting a role for SCFA in immune regulation.[64] Stimulation of GPR43 by SCFA has been shown to be necessary for the normal resolution of certain inflammatory responses in mouse models of colitis, arthritis, and asthma. GPR43-deficient mice showed unresolving inflammation, which was similar to the dysregulated inflammatory responses seen in GF mice who expressed little or no SCFA.[65] SCFA receptors are also found in adipose tissue and thus thought to play an important role in regulating host adipocyte function and plasma lipid levels.[66]

The direct role of SCFAs within the brain is not yet fully understood. Monocarboxylate transporters (MCTs) are located at the blood–brain barrier (BBB),[67] and these receptors allow SCFAs to gain direct access to the CNS.[68] Additionally, SCFAs have been shown to influence BBB permeability, which may be vital for maintaining CNS homeostasis.[69,70] MCTs are also expressed on glia, astrocytes, and neuronal cells,[67] but the extent to which they transport SCFAs within the CNS is unclear.

At a cellular level, one of the most interesting and well-studied functions of butyrate is that as a histone deacetylation (HDAC) inhibitor. Histone acetylation is an epigenetic modification where acetyl groups are added by histone acetyltransferases (HATs), weakening the attraction between the histone protein and the DNA backbone, thus loosening the chromatin and increasing the propensity for gene transcription. Removal of acetyl groups by histone deacetylases, conversely, results in reduced transcription. This ability of butyrate to inhibit deacetylation has long been realized.[71] Impaired histone acetylation and transcriptional dysfunction are features of some neurodegenerative disorders, and a HDAC inhibitor such as butyrate offers an appealing therapeutic possibility. There have been numerous preclinical studies demonstrating beneficial effects of butyrate administration in facilitating neuronal plasticity and in long-term potentiation, as well as in improving cognition in experimental models of neurodegeneration or cognitive impairment.[60] Butyrate has also been shown to demonstrate antidepressant effects in a mouse model of depression,[72] an action thought to be related to increased expression of the BDNF gene secondary to butyrate's HDAC ability.[73] Multiple studies have shown butyrate to have a possible role in a variety of neuropsychiatric diseases, including Alzheimer's disease,[74] Parkinson's disease,[75] and amyotrophic lateral sclerosis,[76] among others.[77]

## 26.2.5 The Gut Microbiome Influences Tryptophan Metabolism and Serotonin Production

Tryptophan, an essential amino acid which must be ingested in the diet, is the precursor for serotonin (5-hydroxytryptamine, 5-HT), a neurotransmitter most widely known for its role within the CNS in the regulation of mood and other neuropsychological functions. Although it does not appear that a simple deficiency of central serotonin is associated with an increased risk of depression,[78] there is no doubt that the serotonergic system plays an important role, and serotonin receptors are the target of many common and effective antidepressant and anti-anxiety medications. Serotonin has also been shown to play an important role in cognition and memory,[79–81] and its influence extends beyond the CNS. It is an important signaling molecule within the GIT, influencing gut motility, gastric emptying, pancreatic secretion, satiation, and vomiting.[82] Serotonin dysregulation has also been implicated in the pathogenesis of gastrointestinal disorders, including irritable bowel syndrome (IBS)[83] and inflammatory bowel disease (IBD).[84,85] The substantial overlap between the many physiological functions, and indeed psychiatric and gastrointestinal disorders, in which serotonin and the gut microbiome both play a role has led to increased interest in the ability of the microbiome to exert its effects through the regulation of tryptophan metabolism and serotonin synthesis.

More than 90% of the body's 5-HT is synthesized in the gut, predominantly by specialized endocrine cells called enterochromaffin cells (ECs). The metabolic pathway involves the conversion of tryptophan to 5-hydroxytryptophan (5-HTP) by the rate-limiting enzyme tryptophan hydroxylase (TPH). This enzyme is not saturated at normal tryptophan levels, and thus increasing tryptophan concentrations can result in increased metabolite production. 5-HTP is a short-lived intermediate product and quickly converted to 5-HT by an enzyme, aromatic amino acid decarboxylase (AAAD).[86] A key metabolomics study in 2009 revealed that conventionally colonized mice had 2.8 times greater plasma serotonin levels in comparison to GF mice.[87] The mechanisms behind this raised serotonin level were unclear at the time. It was suspected that gut bacteria may mediate direct metabolic transformation in some way, but this was not proven until another important study several years later. Investigators demonstrated that the microbiota directly promotes 5-HT biosynthesis in colonic ECs, which in turn supply 5-HT to the gut mucosa, lumen, and circulating platelets, thus regulating peripheral host serotonin levels. They demonstrated that microbiome-dependent elevation in TPH, the rate-limiting enzyme for 5-HT synthesis in ECs, appeared to account for the increased 5-HT production. Furthermore, this team found that spore-forming microbes from the healthy mouse and human microbiota were sufficient to mediate these microbial effects on serum, colonic, and fecal 5-HT levels.[88]

The brain has limited capacity to store tryptophan, and, as a result, the bioavailability of tryptophan is of vital importance in central serotonin synthesis. Following absorption from the gut, tryptophan enters the circulation, where it exists in both a free and albumin-bound fraction.[89] Although much of the focus historically has been on this serotonin pathway, the vast majority of available tryptophan is metabolized via an alternative route, the kynurenine pathway. This pathway has been known for many decades, but it is only recently being appreciated that dysfunction may have important consequences for

CNS function[90] and indeed play a role in neuropsychiatric disorders such as Alzheimer's disease[91] and depression.[92] The enzymes indoleamine-2,3-dioxygenase (IDO) and tryptophan-2,3-diogenase (TDO) catalyze the initial rate-limiting metabolic step of the kynurenine pathway and lead to the production of kynurenine. This can be metabolized via a number of different routes and ultimately results in the production of neuroactive compounds such as kynurenic acid and quinolinic acid. Kynurenic acid is thought to be a neuroprotective substance and acts as an NMDA antagonist and possibly an α7 nicotinic acetylcholine receptor antagonist;[93] conversely, quinolinic acid is neurotoxic, acting as an NMDA agonist at several receptor subtypes.[94] Preferential metabolism of kynurenine to the potentially neurotoxic quinolinic acid instead of the neuroprotective kynurenic acid has been linked to mood disorders,[92,95] supporting the "neurodegeneration hypothesis of depression".[96,97] TDO is found predominantly in the liver and influenced significantly by glucocorticoid induction.[98] IDO is widely distributed throughout the body. It is subject to control by the immune system, and levels are increased by states such as chronic postinfectious GI inflammation[99] and following the administration of pro-inflammatory cytokines such as interferon-γ.[100]

The gut microbiome appears to directly influence tryptophan breakdown at several points along the metabolic pathway. It impacts circulating concentrations of tryptophan and serotonin, the amount of tryptophan directed toward the tryptophan-kynurenine pathway, and consequently the amount of tryptophan available to the CNS for serotonin synthesis, as well as the preferential direction of metabolism of kynurenine. Although GF mice exhibit lower plasma serotonin levels than conventionally colonized mice, they have 40% greater plasma tryptophan levels.[87] They also show increased 5-HT turnover in the striatum, although not in the hippocampus or frontal cortex.[46] In a recent study,[101] GF mice were used to explore the influence of the gut microbiome on serotonergic neurotransmission in the hippocampus. This study found that male GF mice exhibited a significant elevation in hippocampal concentration of 5-HT compared with conventionally colonized control mice. Concentrations of tryptophan were also increased in the plasma of male GF mice, along with a reduced plasma kynurenine:tryptophan ratio, suggesting that the microbiota may alter central serotonergic neurotransmission via an influence on tryptophan metabolic pathways. The study reported that GF mice exhibited reduced anxiety behaviors in comparison to control mice, a finding which has been replicated consistently by similar GF animal studies across a number of laboratories.[46,47,102,103] Interestingly, although colonization of GF mice post-weaning did restore plasma tryptophan levels as well as normalizing behavioral anxiety measures, it did not reverse the elevated hippocampal 5-HT levels. This is consistent with other microbiome studies in supporting the hypothesis that there is a critical time period for neurodevelopmental programming during which the presence of a normal microbiome plays an important role and that colonization of the gut outside this critical window is ineffective in reversing some central neurochemical abnormalities.[41] Indeed, if not an all-or-nothing critical time window, there is certainly a temporal effect of gut colonization on neuronal development and function.

The probiotic *B. infantis* 35624 is proposed to have antidepressant properties, relating to, among other mechanisms, its influence on tryptophan metabolism.[49] Sprague-Dawley rats treated with this probiotic for 14 days showed a marked increase in plasma tryptophan concentration. A decrease in the kynurenine:tryptophan ratio suggested that this increase in bioavailability of tryptophan was due to a reduction in IDO activity and consequent reduction in the conversion of tryptophan to kynurenine. This corresponded to suppressed pro-inflammatory cytokine production in those rats treated with the probiotic. Furthermore, for the tryptophan that did enter the kynurenine pathway, the probiotic seemed to preferentially direct its metabolism toward the production of the neuroprotective kynurenic acid, as opposed to the more neurotoxic quinolinic acid.[97] Another probiotic, *Lactobacillus johnsonii*, revealed a similar ability to reduce kynurenine concentrations because of reduced IDO activity.[104] Investigators attributed this to its capacity for *in vivo* production of hydrogen peroxide ($H_2O_2$) in the ileum, which has been shown to abolish IDO activity *in vitro*. Given that $H_2O_2$ is produced by many lactic acid bacteria, $H_2O_2$-mediated inhibition of IDO enzyme activity is a plausible mechanism by which the gut microbiome reduces tryptophan metabolism along the kynurenine pathway.

# 26.3 GUT DYSBIOSIS IN NEUROPSYCHIATRIC DISORDERS

The term "dysbiosis" refers to a disturbance in gut microbiota homeostasis. An obvious starting point for human microbiome research is describing the imbalances of gut microbiota seen in various disorders. The microbiome composition has been profiled in several psychiatric illnesses, including depression, schizophrenia, bipolar affective disorder (BPAD), social anxiety disorder (SAD), and posttraumatic stress disorder (PTSD), as well as in neuropsychiatric conditions at both extremes of life, from ASD to Alzheimer's disease and Parkinson's disease.

Depression is one of the most prevalent and disabling conditions affecting the 21st-century Western world. The neurobiological foundations of this disorder, although not completely understood, have been widely studied. Abnormalities of the HPA axis, evidenced by elevated plasma cortisol levels and raised corticotropin-releasing factor (CRF) levels in the cerebrospinal fluid, along with chronic low-grade inflammation with marked increases in the concentration of pro-inflammatory cytokines, are hallmarks of the disease. The recent interest in the gut microbiome is then, perhaps, not surprising when one considers that many of the neurobiological abnormalities seen in depression overlap with those pathways that come under the influence of the gut microbiome. A 2021 systematic review of 17 studies, which limited confounding factors, analyzed the gut microbiota of MDD patients and identified an increase in *Eggerthella*, *Atopobium*, and *Bifidobacterium* (all of the *Actinobacteria* phylum) and a decrease in *Faecalibacterium* in patients with MDD compared with healthy controls.[61] The question as to whether these microbiota differences represent a cause or effect phenomenon must be considered. Several research teams have attempted to address this using FMT from depressed patients to microbiota-depleted or GF rodents.[105,106] They have demonstrated the induction of depressive-like behaviors and accompanying metabolic disturbances in mice harboring "depression microbiota", suggesting that the gut microbiome does indeed have an etiological role in depression.

Several other psychiatric disorders have been studied in relation to gut dysbiosis. Altered gut microbiota composition has been demonstrated in patients with generalized anxiety disorder[107] and social anxiety disorder.[108] A systematic review of 13 studies suggested that low α-diversity and abundance of *Faecalibacterium* and *Bacteroides* may characterize bipolar affective disorder in both a trait- and state-dependent fashion.[109] The relationship between the microbiome and schizophrenia is also becoming increasingly apparent.[110,111] However, studies are complicated by the use of antipsychotic medication, which alters the user's microbiome,[112] and heterogenous populations, with contrasting findings in Chinese[113] and European[114] participants. The microbiome as a biomarker may offer a path to improve schizophrenia diagnosis, with one study using a predictive model with a panel of four bacteria (*Ruminococcus, UCG005, Clostridium sensu stricto 1*, and *Bifidobacterium*) to correctly identify 68% of drug-naïve schizophrenic patients and 93% of risperidone-treated schizophrenia patients.[115] Further support for the importance of the gut microbiome in schizophrenia comes from studies showing gut microbiota transplantation from patients with SCZ provoked schizophrenia-like behaviors in GF recipients.[116,117] Although the causal effect is unknown, it may be secondary to lower glutamate and higher glutamine and GABA in the mice hippocampus, or through manipulation of tryptophan–kynurenine metabolism.

ASD is associated with high rates of gastrointestinal comorbidity,[118] and this was one of the observations that led to interest in the possible role of the gut microbiome. Differences in microbiome composition between subjects with ASD and healthy controls have been identified in several studies. Although findings have varied widely, a general trend which has emerged involves Clostridiaceae, a family within the Firmicutes phylum. Several studies have confirmed increased levels of various *Clostridium* species,[119–122] supporting the findings of an earlier study which demonstrated short-term improvement in ASD symptoms following treatment with vancomycin, an oral antibiotic effective against neurotoxin-producing *Clostridia*.[123] Gut microbiome taxa in ASD children have been found to significantly differ in relative abundance from healthy controls (e.g., *Clostridia, Desulfovibrio, Bifidobacteria, Bacteroides*),[124,125] and these bacteria are strongly associated with GI symptoms, although findings are inconsistent.[126] Thus, while it is difficult to draw any solid conclusions from these human studies, they do appear to support a role for the microbiome in ASD.

Microbiome composition has been profiled in the healthy aging population as well as in patients with age-related neurodegenerative disorders such as Alzheimer's and Parkinson's diseases. A study by Claesson et al.[127] revealed an important relationship between diet, microbiota composition, and the health status of older people. This team found that microbiota composition significantly correlated with measures of frailty, comorbidity, and markers of inflammation in the elderly. Recent studies into the relationship between Alzheimer's disease, and the microbiome suggest a possible association with decreased *Firmicutes* species,[128–131] but conclusions are limited by small sample sizes and presence of several confounding factors, such as heterogeneous diet and medication use.

Parkinson's disease (PD) is a CNS disorder predominantly affecting the motor system. Neuropsychiatric symptoms are extremely common and range from depression and anxiety to psychosis, impulse control disorders, and cognitive disturbance. A role for the gut microbiome in PD is suggested by the observation that gastrointestinal symptoms are very common and often precede the development of motor symptoms. Consistent with this is the finding that pathological alpha-synuclein-aggregates are present in the submucosal and myenteric plexuses of the enteric nervous system prior to their detection in the brain. There have been numerous studies investigating the composition of the microbiome in patients with PD.[132] As with all microbiome studies, there has been considerable variation in results. However, encouragingly clear trends seem to have emerged. *Akkermansia* appears to be overrepresented, as does *Lactobacillus*, while *Prevotella* abundances are reduced. Other findings consistent across studies are the reduced abundance of *Faecalibacterium* and increased levels of *Bifidobacterium*.[133] The vital question as to whether microbiome alterations precede, or are consequential to, the development of PD was alluded to in a recent study. It was demonstrated that 80% of the differential gut microbes in PD versus healthy controls showed similar trends in idiopathic rapid eye movement sleep behavior disorder (RBD), a disorder that is considered a prodrome of PD.[134] This suggests that the microbiome changes do indeed precede the development of PD motor symptoms.

## 26.4 MICROBIOME MANIPULATION IN NEUROPSYCHIATRIC CONDITIONS

The exciting question as to whether microbiome manipulation can serve a therapeutic function is a major focus of research. Much work has been done on attempting to directly modulate the gut microbiota by the introduction of probiotics to the system. However, there are many technical and pragmatic difficulties with this approach. Probiotics need to survive the hostile environment of the acidic stomach to reach the intestine, the site where they can exert a therapeutic effect. In addition, many probiotics merely transit through the gut and are unable to colonize the environment in the long-term. The selection of appropriate strains along with issues in relation to continuity and consistency in batch production make effective probiotic production and use a complex and difficult task. An alternative approach has been to target beneficial host bacteria and selectively enhance their growth within the intestine. "Prebiotics" have this effect and are characterized as nondigestible carbohydrates that are selectively fermented by the bacteria in the large intestine.[135] Prebiotics include substances such as inulin, fructooligosaccharides (FOSs), galactooligosaccharides (GOSs), resistant starch, and other selected dietary fibers. Natural sources of prebiotics include fruits and vegetables such as asparagus, leek, banana, and chicory, as well as grains such as oats and wheat, foodstuffs which have become increasingly lacking in Western-style heavily processed high-fat high-sugar diets.

The term "psychobiotic" was first defined as a probiotic that, ingested in appropriate quantities, had a positive mental health benefit,[136] and the definition has been expanded to include prebiotics and other means of altering the microbiome. Despite the rapid accumulation of preclinical studies in recent years, evidence for the potential of psychobiotics in humans is limited at present. Most studies have been conducted in healthy populations, with only a small number of trials performed in clinical samples. Results are mixed but for the most part tentatively positive, and an evidence base for the use of psychobiotics is slowly emerging.

## 26.4.1 Psychobiotic Effects in Healthy Populations

The majority of studies investigating the effects of probiotics on mood and anxiety symptoms have been performed in healthy subjects and predominantly investigate *Lactobacillus* and *Bifidobacterium* strains. A combination of *Lactobacillus helveticus* and *B. longum*, administered to 66 healthy adults, resulted in subtle improvements in mood and anxiety symptoms.[58] A multispecies probiotic (*B. bifidum*, *Bifidobacterium lactis*, *Lactobacillus acidophilus*, *Levilactobacillus brevis*, *Lacticaseibacillus casei*, *Ligilactobacillus salivarius*, and *Lactococcus lactis*) demonstrated an ability to significantly reduced overall cognitive reactivity to sad mood, thought to represent "vulnerability to depression", in healthy adults,[137] while a triple-blinded RCT of post-menopausal women, using yogurts differing only in their containing *Bifidobacterium lactis* and *Lactobacillus acidophilus*, showed that this combination improved anxiety, stress, and quality of life scores.[138] A large Japanese study involving over 200 healthy participants reported a slight improvement in anxiety levels following 12-week consumption of *Lactobacillus gasseri* and *B. longum*,[139] while a double-blinded RCT of 63 older adults reported that a 12-week course of *Bifidobacterium bifidum* BGN4 and *Bifidobacterium longum* BORI led to improvements in mood, stress, and cognitive flexibility while also increasing levels of BDNF and reducing inflammation-causing gut bacteria[140] It can be difficult to draw conclusions when multiple bacterial strains are used, and, for this reason, trials using single probiotic strains can be more informative. *Lacticaseibacillus casei*, administered to healthy adults, appeared to improve mood, but only in those with a low baseline mood,[141] while a study of 30 young badminton players found that a 6-week course of *Lacticaseibacillus casei* led to reduced stress and anxiety levels, as well as increased aerobic capacity.[142] Another study used a specific Shirota strain of *Lacticaseibacillus casei* (LcS) in university students under exam stress and demonstrated an attenuation of exam-related raised cortisol levels and physical stress symptoms in the probiotic group.[59] Four-week consumption of *B. longum* in healthy male university students resulted in a variety of psychological benefits, including reduced self-reported anxiety and cortisol levels in response to an acute stressor, reduced daily perceived stress, and subtle improvements in visuospatial memory with enhanced frontal midline mobility on EEG.[143] Prebiotics have also been investigated. A research group in Oxford found a significant impact of prebiotics on stress responses in healthy participants.[144] Volunteers received one of two prebiotics (FOS or Bimuno-galactooligosaccharides [B-GOS]) or a placebo (maltodextrin) daily for 3 weeks. Those in the B-GOS group showed significantly reduced waking cortisol responses as well as reduced attention and reactivity to negative emotions.

While the previous studies have revealed subtle beneficial effects of probiotics and prebiotics on mood and anxiety symptoms in humans, others have demonstrated negative results and are important in highlighting the challenges of translating the exciting preclinical advances to human studies. A probiotic blend (*Lacticaseibacillus casei*, *Lactobacillus delbrueckii* subsp. *Bulgaricus*, and *Streptococcus thermophilus*) administered to a student population before and after an examination period did not affect anxiety scores.[145] *Lacticaseibacillus rhamnosus*, a probiotic which had appeared very promising in preclinical work, demonstrated no benefit over placebo in modifying a variety of biochemical and psychological stress-related measures.[146] *Lacticaseibacillus casei*, despite several positive studies, demonstrated no effect on anxiety scores in a study designed to examine its effects on natural killer (NK) cell function in smokers.[147] An interesting imaging study using functional MRI in healthy women revealed that ingestion of a fermented milk product containing *Bifidobacterium animalis* subsp. *lactis*, *S. thermophilus*, *Lactobacillus delbrueckii* subsp. *bulgaricus*, and *Lactobacillus delbrueckii* subsp. *lactis* affected the activity of brain regions controlling central processing of emotion and sensation, although there were no changes in anxiety or depression scores.[148]

Results of human psychobiotic interventional trials in healthy populations are thus mixed, with the hope of positive results tempered by numerous negative trials. This may be partly explained by the fact that mood and anxiety scores in healthy adults are, for the most part, within normal ranges at baseline, thus not allowing much room for improvement. This view has been supported by a meta-analysis of the use of probiotics to alleviate depressive symptoms.[149] Authors included 10 RCTs in the meta-analysis, 7 of which were conducted in healthy subjects. The remaining three studies recruited patients with either self-reported depression symptoms or a formal diagnosis of major depressive disorder. Although the meta-analysis found no significant differences between probiotic and placebo across the 10 trials, a subgroup analysis reported significant benefits of probiotic administration in subjects with mild–moderate depression at baseline. Of course, trials in healthy adults are vital for proof of concept, but studies in people suffering with psychological distress are likely to be more revealing when it comes to assessing true psychobiotic potential. Studies in clinical samples are accumulating and involve non-psychiatric clinical populations such as those with IBS, as well as patients with established psychiatric diagnoses.

## 26.4.2 Psychobiotic Effects in Non-Psychiatric Clinical Populations

IBS has been the subject of many probiotic trials, with evidence available for the beneficial effects of probiotics[150] and fecal transplant[151,152] in alleviating GI symptoms. Because IBS is conceptualized as a stress-related functional GI disorder and often comorbid with depression and anxiety, it has been an ideal prototypic condition for investigating the MGB axis in humans and the potential for probiotics. Given their ability to improve the gastrointestinal symptoms of IBS, probiotics might also offer a treatment approach for its comorbid mood and anxiety symptoms. Several studies using multispecies probiotics, including various combinations of *Lactobacillus*, *Streptococcus*, and *Bifidobacterium* strains, reported no significant impact on anxiety or mood symptoms in patients with IBS.[153–156] However, a double-blind RCT of 44 adults with IBS and comorbid mild-moderate anxiety and/or depression revealed that *B. longum NCC3001* significantly reduced depression scores, although there was no effect on anxiety.[157] In addition, fMRI analysis revealed reduced responses to negative emotional stimuli in multiple brain areas, including the amygdala, an area well known for its role in the regulation of fear and anxiety. The amygdala also has the ability to regulate the HPA axis, a system of vital important in the pathogenesis of depression. A recent exploratory study showed promise in using a combination of *Bifidobacterium longum* strains 1714 and 35642 in treating mild to moderate anxiety and depression in IBS patients.[158] Patients administered this combination showed reduced depression scores and sleep quality, with reduced levels of TNF-α, normalized cortisol awakening response, and improved IBS symptoms.

Inflammatory bowel disease carries significant psychiatric comorbidity, with 66% of patients developing anxiety and up to 34% suffering from co-morbid depression during active disease.[159] The bidirectional relationship of psychological distress and GI symptom flare-up has become increasingly recognized,[160] with probiotics offering a possible treatment option for both. Although there is a lack of human interventional studies of probiotics in IBD patients with psychiatric comorbidities, a 2022 mouse-model study showed that transplantation of Enterobacteriaceae-rich gut microbiota (from human IBD patients) can cause depression and colitis, which was mitigated by oral *Lactiplantibacillus plantarum* subsp. *plantarum* NK151, *Bifidobacterium longum* NK173, and *Bifidobacterium bifidum* NK175 through regulation of pro-inflammatory and anti-inflammatory cytokines.[161]

## 26.4.3 Psychobiotic Effects in Psychiatric Disorders

Given that medication alone is too often ineffective in fully treating patients with psychiatric disorders, the role of probiotics represents a possible augmentation of past approaches. A multispecies probiotic, containing *Lactobacillus acidophilus*, *Lacticaseibacillus casei*, and *B. bifidum*, significantly reduced depressive symptoms in 40 patients with MDD, with beneficial effects in metabolic and inflammatory profiles also noted in the probiotic group.[162] Another RCT, using a probiotic preparation (*Lactobacillus helveticus* and *B. longum*) in patients with at least moderate depressive scores, revealed no benefit over placebo in terms of improving mood or moderating inflammatory or other biomarkers.[163] A 2019 double-blind RCT showed probiotic supplementation (also using *Lactobacillus helveticus and B. longum*) resulted in a significant decrease in depressive symptom severity compared to both a prebiotic and placebo group.[164] Pinto-Sanchez et al.[157] reported significant benefits of *B. longum NCC3001* in alleviating self-reported mild-to-moderate depressive and/or anxiety symptoms. However, the study sample consisted of patients with IBS, an obvious confounder when looking at probiotic potential.

Several probiotic randomized-control trials have been undertaken in patients with schizophrenia. Dickerson et al.[165] hypothesized that given the range of immune system abnormalities associated with schizophrenia, probiotics might, as a result of their immunomodulatory ability, have a beneficial effect. However, consumption of *Lacticaseibacillus rhamnosus* GG and *B. lactis* Bb12 over 14 weeks did not affect schizophrenia symptoms. Despite the absence of any symptomatic response, researchers did find that probiotic treatment altered levels of immunomodulatory proteins, the significance of which is unclear.[166] Two RCTs have investigated use of probiotics alongside other nutritional supplementation in treating patients with schizophrenia. One study combined three bacteria (*Lactobacillus acidophilus, B. bifidum, Limosilactobacillus reuteri,* and *Limosilactobacillus fermentum*) with vitamin D.[167] The combination led to improvement in symptom severity, as well as improvements in fasting plasma glucose levels, cholesterol levels, triglycerides, and insulin concentrations. The second study also found improvement in symptom severity in the group administered a probiotic (*Lactobacillus acidophilus, B. lactis, B. bifidum,* and *B. longum*) and selenium compared to the placebo group.[168] Another avenue for probiotics to improve treatment of schizophrenia is through reduction of the metabolic side-effects of second-generation antipsychotics. One study investigating this in drug-naïve, first-episode patients suggested that dietary fiber is an important adjunct to such probiotic supplementation, as they found that, when used together with olanzapine 15–20mg/day, a probiotic alone (*Bifidobacterium, Lactobacillus,* and *Enterococcus*) did not attenuate weight gain but that the same probiotic alongside 30 g fiber supplementation did.[169]

In bipolar affective disorder, use of an adjunctive probiotic treatment (*Lacticaseibacillus rhamnosus strain GG* and *Bifidobacteriu animalis* subsp. *Lactis* strain Bb12) led to reduced rehospitalization and days of hospitalization in patients with acute mania compared to a control group.[170] Another study was less promising, finding no significant difference in Young Mania Rating Scale (YMRS) and Hamilton's Depression Rating Scale (HDRS) scores between patients given a probiotic (*Bifidobacterium bifidum, Bifidobacterium lactis, Bifidobacterium longum,* and *Lactobacillus acidophilus*) and those given a placebo.[171]

Although not an interventional study, it is worth mentioning an interesting paper by Hilimire et al.,[172] which investigated the relationship between neuroticism, social anxiety, and the consumption of fermented (probiotic-containing) foods. Self-report questionnaires were administered to 710 young adults, and, using an interaction model, the authors showed that exercise, neuroticism, and fermented food consumption significantly and independently predicted social anxiety. Furthermore, for those with high neuroticism scores, a high frequency of fermented food consumption resulted in fewer symptoms of social anxiety. The data suggest that fermented foods containing probiotics may have a protective effect against social anxiety symptoms for those at higher genetic risk, as measured by trait neuroticism.

# 26.5 THE MICROBIOME–GUT–BRAIN AXIS, PSYCHOBIOTICS, AND THE FUTURE—WHERE TO NEXT?

The field of psychiatry desperately needs a therapeutic breakthrough. The frustratingly slow pace of psychiatric drug development is in stark contrast to the rapidly increasing incidence of neuropsychiatric disorders from anxiety to depression, from ASD to Alzheimer's disease, and the ever-growing socioeconomic burden associated with these conditions. There have been virtually no new therapeutic advances in psychiatry over the past 40 years. The 1950s and 1960s were a bountiful period in psychiatric history, with the serendipitous discovery of lithium for the treatment of mania, chlorpromazine with its antipsychotic properties, and imipramine and iproniazid as the first antidepressants. The next two decades saw researchers capitalize on these fortuitous discoveries with the resultant development of serotonin-specific reuptake inhibitors (SSRIs) for depression and a range of typical and atypical antipsychotics for schizophrenia. Fast-forward 20 years, and, despite a swell of advances in genetics, genomics, neuroimaging, cognitive neuroscience, and diagnostic standardization, we are still in a therapeutic desert and using the same medications as our predecessors several decades ago. Could the gut microbiome present a new therapeutic paradigm in the approach to mental illness? Can psychobiotics offer a new class of acceptable and effective treatments for psychiatric patients in the 21st century? The excitement around this new scientific domain is certainly catching. The global probiotic markets were estimated to be USD 34.1 billion in 2020.[173] Many lay publications enthusiastically promote the importance of the gut bacteria in optimizing mental health and well-being. However, despite plausible, encouraging, and rapidly accumulating preclinical evidence, the area is very much in its infancy. In particular, translational studies in both healthy and clinical populations are limited.

The major challenge for microbiome researchers is uncovering which bacteria, of the trillions that reside in our gut, may have psychobiotic potential. The approach to date has been to use a preclinical platform to screen prospective bacterial candidates for an ability to influence various neural, endocrine, or immune pathways known to play a role in depression and anxiety. While this method has yielded some successful results, converting encouraging preclinical data to positive human experience requires a far more extensive understanding of the mechanisms by which probiotics exert their effects. Another major obstacle to human microbiome research is a lack of knowledge as to what constitutes a healthy gut microbiome. The vast quantities of microbes in our gut, along with interindividual variability in microbiome composition, make it difficult to characterize what constitutes "normal". The term "dysbiosis" is used extensively to describe alterations or imbalances in microbiome composition, but the term itself is vague and should be seen more as a concept than a concrete descriptor.[174] As well as the challenge of identifying individual species of bacteria which may be beneficial, knowledge in relation to dosing regimens and optimal duration of intervention is lacking. Even within specific bacterial genera, the probiotic potential of different species can vary considerably, and benefits cannot be presumed to be generalizable between strains. Dosing strategies have varied widely, and there is no consensus as to what a necessary minimum dose should be. Interventional trials have also significantly differed in duration, ranging from 2 weeks to 6 months, with no evidence on which to base decisions regarding the length of time needed for effect.

It is important to remember that probiotics and prebiotics represent only one approach to microbiome manipulation. The most important factor in determining the composition of the gut microbiome of healthy adults is diet. A relationship between diet and psychiatric disorders, in particular anxiety and depression, has long been recognized. Meta-analyses have confirmed that even moderate adherence to a Mediterranean diet is protective against depression and cognitive dysfunction.[175,176] In contrast, Western-style diets that consist of highly processed, high-fat, high-sugar foods, have been shown to be associated with increased risk of depression.[177,178] The novel field of "nutritional psychiatry" is expanding rapidly and promotes the consideration of nutritional factors in mental health.[179] A first-of-its-kind RCT in Australia recently evaluated the efficacy of a dietary improvement program, broadly based on the Mediterranean diet, as an adjunctive treatment in major depressive episodes. They found that the dietary intervention group experienced a significant reduction in depression scores in comparison to a social support control group.[180] Changes in diet can rapidly and reproducibly alter microbiome composition,[181] and it is eminently plausible that the gut microbiome represents the link in the observed association between food and mood. This is an exciting part of the MGB axis puzzle, and it would be unsurprising if dietary change turned out to be the most effective "psychobiotic" of all.

# 26.6 ACKNOWLEDGMENTS

JFC has been an invited speaker at conferences organized by Mead Johnson, Alkermes, Janssen, Ordesa, and Yakult. TGD and JFC have received research funding from GlaxoSmithKline, Mead Johnson, Cremo Nutricia, Pharmavite, Dupont, and 4D Pharma. This support neither influenced nor constrained the contents of this manuscript. There are no other potential conflicts of interest.

# BIBLIOGRAPHY

1. Cryan JF, O'Mahony SM. The microbiome-gut-brain axis: From bowel to behavior. *Neurogastroenterology and Motility: The Official Journal of the European Gastrointestinal Motility Society.* 2011;23(3):187–192. doi:10.1111/j.1365-2982.2010.01664.x

2. Mitsonis CH, Kararizou E, Dimopoulos N, et al. Incidence and clinical presentation of neurosyphilis: A retrospective study of 81 cases. *The International Journal of Neuroscience.* 2008;118(9):1251–1257. doi:10.1080/00207450701239426

3. Fallon BA, Nields JA. Lyme disease: A neuropsychiatric illness. *The American Journal of Psychiatry.* 1994;151(11):1571–1583. doi:10.1176/ajp.151.11.1571

4. Nevin RL, Croft AM. Psychiatric effects of malaria and anti-malarial drugs: Historical and modern perspectives. *Malaria Journal.* 2016;15:332. doi:10.1186/s12936-016-1391-6

5. Coughlin SS. Anxiety and depression: Linkages with viral diseases. *Public Health Reviews.* 2012;34(2):92.

6. Osimo EF, Pillinger T, Rodriguez IM, Khandaker GM, Pariante CM, Howes OD. Inflammatory markers in depression: A meta-analysis of mean differences and variability in 5,166 patients and 5,083 controls. *Brain, Behavior, and Immunity.* 2020;87:901–909.

7. Momtazmanesh S, Zare-Shahabadi A, Rezaei N. Cytokine alterations in schizophrenia: An updated review. *Frontiers in Psychiatry.* 2019;10:892.

8. Michopoulos V, Powers A, Gillespie CF, Ressler KJ, Jovanovic T. Inflammation in fear- and anxiety-based disorders: PTSD, GAD, and beyond. *Neuropsychopharmacology: Official Publication of the American College of Neuropsychopharmacology.* 2017;42(1):254–270. doi:10.1038/npp.2016.146

9. Tan ZS, Beiser AS, Vasan RS, et al. Inflammatory markers and the risk of Alzheimer disease: The framingham study. *Neurology.* 2007;68(22):1902–1908. doi:10.1212/01.wnl.0000263217.36439.da

10. Udina M, Castellvi P, Moreno-Espana J, et al. Interferon-induced depression in chronic hepatitis C: A systematic review and meta-analysis. *The Journal of Clinical Psychiatry.* 2012;73(8):1128–1138. doi:10.4088/JCP. 12r07694

11. McNutt MD, Liu S, Manatunga A, et al. Neurobehavioral effects of interferon-alpha in patients with hepatitis-C: Symptom dimensions and responsiveness to paroxetine. *Neuropsychopharmacology: Official Publication of the American College of Neuropsychopharmacology.* 2012;37(6):1444–1454. doi:10.1038/npp.2011.330

12. Miller AH, Haroon E, Raison CL, Felger JC. Cytokine targets in the brain: Impact on neurotransmitters and neurocircuits. *Depression and Anxiety.* 2013;30(4):297–306. doi:10.1002/da.22084

13. de Punder K, Pruimboom L. Stress induces endotoxemia and low-grade inflammation by increasing barrier permeability. *Frontiers in Immunology.* 2015;6:223. doi:10.3389/fimmu.2015.00223

14. Lv S, Song HL, Zhou Y, et al. Tumour necrosis factor-α affects blood—brain barrier permeability and tight junction-associated occludin in acute liver failure. *Liver International.* 2010;30(8):1198–1210.

15. Rochfort KD, Collins LE, Murphy RP, Cummins PM. Downregulation of blood-brain barrier phenotype by proinflammatory cytokines involves NADPH oxidase-dependent ROS generation: Consequences for interendothelial adherens and tight junctions. *PLoS ONE.* 2014;9(7):e101815.

16. Liu Y, Zhang S, Li X, et al. Peripheral inflammation promotes brain tau transmission via disrupting blood—brain barrier. *Bioscience Reports.* 2020;40(2).

17. Kealy J, Greene C, Campbell M. Blood-brain barrier regulation in psychiatric disorders. *Neuroscience Letters.* 2020;726:133664.

18. O'Hara AM, Shanahan F. The gut flora as a forgotten organ. *EMBO Reports.* 2006;7(7):688–693. doi:10.1038/ sj.embor.7400731

19. Desbonnet L, Garrett L, Clarke G, Kiely B, Cryan JF, Dinan TG. Effects of the probiotic Bifidobacterium infantis in the maternal separation model of depression. *Neuroscience.* 2010;170(4):1179–1188. doi:10.1016/j.neuroscience.2010.08.005

20. Duan C, Huang L, Zhang C, et al. Gut commensal-derived butyrate reverses obesity-induced social deficits and anxiety-like behaviors via regulation of microglial homeostasis. *European Journal of Pharmacology.* 2021;908:174338.

21. Collins J, Borojevic R, Verdu EF, Huizinga JD, Ratcliffe EM. Intestinal microbiota influence the early postnatal development of the enteric nervous system. *Neurogastroenterology and Motility: The Official Journal of the European Gastrointestinal Motility Society.* 2014;26(1):98–107. doi:10.1111/nmo.12236

22. McVey Neufeld KA, Mao YK, Bienenstock J, Foster JA, Kunze WA. The microbiome is essential for normal gut intrinsic primary afferent neuron excitability in the mouse. *Neurogastroenterology and Motility: The Official Journal of the European Gastrointestinal Motility Society.* 2013;25(2):183–e88. doi:10.1111/nmo.12049

23. Bercik P, Park AJ, Sinclair D, et al. The anxiolytic effect of Bifidobacterium longum NCC3001 involves vagal pathways for gut-brain communication. *Neurogastroenterology and Motility: The Official Journal of the European Gastrointestinal Motility Society.* 2011;23(12):1132–1139. doi:10.1111/j.1365-2982.2011.01796.x

24. Ma X, Mao YK, Wang B, Huizinga JD, Bienenstock J, Kunze W. Lactobacillus reuteri ingestion prevents hyperexcitability of colonic DRG neurons induced by noxious stimuli. *American Journal of Physiology Gastrointestinal and Liver Physiology.* 2009;296(4):G868–G875. doi:10.1152/ajpgi.90511.2008

25. Bottomley JM, LeReun C, Diamantopoulos A, Mitchell S, Gaynes BN. Vagus nerve stimulation (VNS) therapy in patients with treatment resistant depression: A systematic review and meta-analysis. *Comprehensive Psychiatry.* 2020;98:152156.

26. Kirchner A, Birklein F, Stefan H, Handwerker HO. Left vagus nerve stimulation suppresses experimentally induced pain. *Neurology.* 2000;55(8):1167–1171.

27. Abdel Salam OM. Fluoxetine and sertraline stimulate gastric acid secretion via a vagal pathway in anaesthetised rats. *Pharmacological Research.* 2004;50(3):309–316. doi:10.1016/j.phrs.2004.01.010

28. Adinoff B, Mefford I, Waxman R, Linnoila M. Vagal tone decreases following intravenous diazepam. *Psychiatry Research.* 1992;41(2):89–97.

29. Wang X, Wang BR, Zhang XJ, Xu Z, Ding YQ, Ju G. Evidences for vagus nerve in maintenance of immune balance and transmission of immune information from gut to brain in STM-infected rats. *World Journal of Gastroenterology.* 2002;8(3):540–545.

30. Gaykema RP, Goehler LE, Lyte M. Brain response to cecal infection with Campylobacter jejuni: Analysis with Fos immunohistochemistry. *Brain, Behavior, and Immunity.* 2004;18(3):238–245. doi:10.1016/j.bbi.2003.08.002

31. Goehler LE, Park SM, Opitz N, Lyte M, Gaykema RP. Campylobacter jejuni infection increases anxiety-like behavior in the holeboard: Possible anatomical substrates for viscerosensory modulation of exploratory behavior. *Brain, Behavior, and Immunity.* 2008;22(3):354–366. doi:10.1016/j.bbi.2007.08.009

32. Bravo JA, Forsythe P, Chew MV, et al. Ingestion of Lactobacillus strain regulates emotional behavior and central GABA receptor expression in a mouse via the vagus nerve. *Proceedings of the National Academy of Sciences of the United States of America.* 2011;108(38):16050–16055. doi:10.1073/pnas.1102999108

33. Roshchina VV. Evolutionary considerations of neurotransmitters in microbial, plant, and animal cells. *Microbial Endocrinology: Interkingdom Signaling in Infectious Disease and Health.* 2010:17–52.

34. Engevik MA, Luck B, Visuthranukul C, et al. Human-derived Bifidobacterium dentium modulates the mammalian serotonergic system and gut–brain axis. *Cellular and Molecular Gastroenterology and Hepatology.* 2021;11(1):221–248.

35. Luscher B, Shen Q, Sahir N. The GABAergic deficit hypothesis of major depressive disorder. *Molecular Psychiatry.* 2011;16(4):383–406.

36. Hashimoto K, Sawa A, Iyo M. Increased levels of glutamate in brains from patients with mood disorders. *Biological Psychiatry.* 2007;62(11):1310–1316.

37. Nutt DJ. Relationship of neurotransmitters to the symptoms of major depressive disorder. *The Journal of Clinical Psychiatry.* 2008;69(Suppl E1):4–7.

38. Geula C, Dunlop SR, Ayala I, et al. Basal forebrain cholinergic system in the dementias: Vulnerability, resilience, and resistance. *Journal of Neurochemistry.* 2021;158(6):1394–1411.

39. Vreeburg SA, Hoogendijk WJ, van Pelt J, et al. Major depressive disorder and hypothalamic-pituitary-adrenal axis activity: Results from a large cohort study. *Archives of General Psychiatry.* 2009;66(6):617–626. doi:10.1001/archgenpsychiatry.2009.50

40. Juruena MF, Eror F, Cleare AJ, Young AH. The role of early life stress in HPA axis and anxiety. *Anxiety Disorders: Rethinking and Understanding Recent Discoveries.* 2020:141–153.

41. Sudo N, Chida Y, Aiba Y, et al. Postnatal microbial colonization programs the hypothalamic-pituitary-adrenal system for stress response in mice. *The Journal of Physiology.* 2004;558(Pt 1):263–275. doi:10.1113/jphysiol.2004.063388

42. Wu Q, Xu Z, Song S, et al. Gut microbiota modulates stress-induced hypertension through the HPA axis. *Brain Research Bulletin.* 2020;162:49–58.

43. Wu W-L, Adame MD, Liou C-W, et al. Microbiota regulate social behaviour via stress response neurons in the brain. *Nature.* 2021;595(7867):409–414.

44. Crumeyrolle-Arias M, Jaglin M, Bruneau A, et al. Absence of the gut microbiota enhances anxiety-like behavior and neuroendocrine response to acute stress in rats. *Psychoneuroendocrinology.* 2014;42:207–217.

45. Clarke G, McKernan DP, Gaszner G, Quigley EM, Cryan JF, Dinan TG. A distinct profile of tryptophan metabolism along the kynurenine pathway downstream of toll-like receptor activation in irritable bowel syndrome. *Frontiers in Pharmacology.* 2012;3:90. doi:10.3389/fphar.2012.00090

46. Diaz Heijtz R, Wang S, Anuar F, et al. Normal gut microbiota modulates brain development and behavior. *Proceedings of the National Academy of Sciences of the United States of America.* 2011;108(7):3047–3052. doi:10.1073/pnas.1010529108

47. Neufeld KM, Kang N, Bienenstock J, Foster JA. Reduced anxiety-like behavior and central neurochemical change in germ-free mice. *Neurogastroenterology and Motility: The Official Journal of the European Gastrointestinal Motility Society.* 2011;23(3):255–264, e119. doi:10.1111/j.1365-2982.2010.01620.x

48. Tannock GW, Savage DC. Influences of dietary and environmental stress on microbial populations in the murine gastrointestinal tract. *Infection and Immunity.* 1974;9(3):591–598.

49. Desbonnet L, Garrett L, Clarke G, Bienenstock J, Dinan TG. The probiotic Bifidobacteria infantis: An assessment of potential antidepressant properties in the rat. *Journal of Psychiatric Research.* 2008;43(2):164–174. doi:10.1016/j.jpsychires.2008.03.009

50. O'Mahony SM, Marchesi JR, Scully P, et al. Early life stress alters behavior, immunity, and microbiota in rats: Implications for irritable bowel syndrome and psychiatric illnesses. *Biological Psychiatry.* 2009;65(3):263–267. doi:10.1016/j.biopsych.2008.06.026

51. Bangsgaard Bendtsen KM, Krych L, Sorensen DB, et al. Gut microbiota composition is correlated to grid floor induced stress and behavior in the BALB/c mouse. *PLoS ONE.* 2012;7(10):e46231. doi:10.1371/journal.pone.0046231

52. Xu M, Wang C, Krolick KN, Shi H, Zhu J. Difference in post-stress recovery of the gut microbiome and its altered metabolism after chronic adolescent stress in rats. *Scientific Reports.* 2020;10(1):1–10.

53. Bailey MT, Dowd SE, Galley JD, Hufnagle AR, Allen RG, Lyte M. Exposure to a social stressor alters the structure of the intestinal microbiota: Implications for stressor-induced immunomodulation. *Brain, Behavior, and Immunity.* 2011;25(3):397–407. doi:10.1016/j.bbi.2010.10.023

54. Bharwani A, Mian MF, Foster JA, Surette MG, Bienenstock J, Forsythe P. Structural & functional consequences of chronic psychosocial stress on the microbiome & host. *Psychoneuroendocrinology.* 2016;63:217–227.

55. Bastiaanssen TF, Gururajan A, van de Wouw M, et al. Volatility as a concept to understand the impact of stress on the microbiome. *Psychoneuroendocrinology.* 2021;124:105047.

56. McKernan DP, Fitzgerald P, Dinan TG, Cryan JF. The probiotic Bifidobacterium infantis 35624 displays visceral antinociceptive effects in the rat. *Neurogastroenterology and Motility: The Official Journal of the European Gastrointestinal Motility Society.* 2010;22(9):1029–1035, e268. doi:10.1111/j.1365-2982.2010.01520.x

57. Gareau MG, Jury J, MacQueen G, Sherman PM, Perdue MH. Probiotic treatment of rat pups normalises corticosterone release and ameliorates colonic dysfunction induced by maternal separation. *Gut.* 2007;56(11):1522–1528. doi:10.1136/gut.2006.117176

58. Messaoudi M, Lalonde R, Violle N, et al. Assessment of psychotropic-like properties of a probiotic formulation (Lactobacillus helveticus R0052 and Bifidobacterium longum R0175) in rats and human subjects. *The British Journal of Nutrition.* 2011;105(5):755–764. doi:10.1017/s0007114510004319

59. Takada M, Nishida K, Kataoka-Kato A, et al. Probiotic Lactobacillus casei strain Shirota relieves stress-associated symptoms by modulating the gut-brain interaction in human and animal models. *Neurogastroenterology and Motility: The Official Journal of the European Gastrointestinal Motility Society.* 2016;28(7):1027–1036. doi:10.1111/nmo.12804

60. Stilling RM, van de Wouw M, Clarke G, Stanton C, Dinan TG, Cryan JF. The neuropharmacology of butyrate: The bread and butter of the microbiota-gut-brain axis? *Neurochemistry International.* 2016;99:110–132. doi:10.1016/j.neuint.2016.06.011

61. Cheung SG, Goldenthal AR, Uhlemann A-C, Mann JJ, Miller JM, Sublette ME. Systematic review of gut microbiota and major depression. *Frontiers in Psychiatry.* 2019;10:34.

62. Cummings JH, Pomare EW, Branch WJ, Naylor CP, Macfarlane GT. Short chain fatty acids in human large intestine, portal, hepatic and venous blood. *Gut.* 1987;28(10):1221–1227.

63. Bolognini D, Tobin AB, Milligan G, Moss CE. The pharmacology and function of receptors for short-chain fatty acids. *Molecular Pharmacology.* 2016;89(3):388–398. doi:10.1124/mol.115.102301

64. Kim CH, Park J, Kim M. Gut microbiota-derived short-chain fatty acids, T cells, and inflammation. *Immune Network.* 2014;14(6):277–288. doi:10.4110/in.2014.14.6.277

65. Maslowski KM, Vieira AT, Ng A, et al. Regulation of inflammatory responses by gut microbiota and chemoat-tractant receptor GPR43. *Nature.* 2009;461(7268):1282–1286. doi:10.1038/nature08530

66. Ge H, Li X, Weiszmann J, et al. Activation of G protein-coupled receptor 43 in adipocytes leads to inhibition of lipolysis and suppression of plasma free fatty acids. *Endocrinology.* 2008;149(9):4519–4526. doi:10.1210/en.2008-0059

67. Pellerin L, Bergersen LH, Halestrap AP, Pierre K. Cellular and subcellular distribution of monocarboxylate transporters in cultured brain cells and in the adult brain. *Journal of Neuroscience Research.* 2005;79(1–2):55–64. doi:10.1002/jnr.20307

68. Vijay N, Morris ME. Role of monocarboxylate transporters in drug delivery to the brain. *Current Pharmaceutical Design.* 2014;20(10):1487–1498.

69. Hoyles L, Snelling T, Umlai U-K, et al. Microbiome—host systems interactions: Protective effects of propionate upon the blood—brain barrier. *Microbiome.* 2018;6(1):1–13.

70. Silva YP, Bernardi A, Frozza RL. The role of short-chain fatty acids from gut microbiota in gut-brain communication. *Frontiers in Endocrinology.* 2020;11:25.

71. Boffa LC, Vidali G, Mann RS, Allfrey VG. Suppression of histone deacetylation in vivo and in vitro by sodium butyrate. *The Journal of Biological Chemistry.* 1978;253(10):3364–3366.

72. Schroeder FA, Lin CL, Crusio WE, Akbarian S. Antidepressant-like effects of the histone deacetylase inhibitor, sodium butyrate, in the mouse. *Biological Psychiatry.* 2007;62(1):55–64. doi:10.1016/j.biopsych.2006.06.036

73. Wei Y, Melas PA, Wegener G, Mathe AA, Lavebratt C. Antidepressant-like effect of sodium butyrate is associated with an increase in TET1 and in 5-hydroxymethylation levels in the Bdnf gene. *The International Journal of Neuropsychopharmacology.* 2014;18(2). doi:10.1093/ijnp/pyu032

74. Ho L, Ono K, Tsuji M, Mazzola P, Singh R, Pasinetti GM. Protective roles of intestinal microbiota derived short chain fatty acids in Alzheimer's disease-type beta-amyloid neuropathological mechanisms. *Expert Review of Neurotherapeutics.* 2018;18(1):83–90.

75. Unger MM, Spiegel J, Dillmann K-U, et al. Short chain fatty acids and gut microbiota differ between patients with Parkinson's disease and age-matched controls. *Parkinsonism & Related Disorders.* 2016;32:66–72.

76. Erber AC, Cetin H, Berry D, Schernhammer ES. The role of gut microbiota, butyrate and proton pump inhibitors in amyotrophic lateral sclerosis: A systematic review. *The International Journal of Neuroscience.* 2020;130(7):727–735. doi:10.1080/00207454.2019.1702549

77. Mirzaei R, Bouzari B, Hosseini-Fard SR, et al. Role of microbiota-derived short-chain fatty acids in nervous system disorders. *Biomedicine & Pharmacotherapy.* 2021;139:111661.

78. Moncrieff J, Cooper RE, Stockmann T, Amendola S, Hengartner MP, Horowitz MA. The serotonin theory of depression: A systematic umbrella review of the evidence. *Molecular Psychiatry.* 2022:1–14.

79. Bostancıklıoğlu M. Optogenetic stimulation of serotonin nuclei retrieve the lost memory in Alzheimer's disease. *Journal of Cellular Physiology.* 2020;235(2):836–847.

80. He J, Hommen F, Lauer N, Balmert S, Scholz H. Serotonin transporter dependent modulation of food-seeking behavior. *PLoS ONE.* 2020;15(1):e0227554.

81. Zhang G, Stackman Jr RW. The role of serotonin 5-HT2A receptors in memory and cognition. *Frontiers in Pharmacology.* 2015;6:225.

82. Mawe GM, Hoffman JM. Serotonin signalling in the gut—functions, dysfunctions and therapeutic targets. *Nature Reviews Gastroenterology & Hepatology.* 2013;10(8):473–486. doi:10.1038/nrgastro.2013.105

83. Luo M, Zhuang X, Tian Z, Xiong L. Alterations in short-chain fatty acids and serotonin in irritable bowel syndrome: A systematic review and meta-analysis. *BMC Gastroenterology.* 2021;21(1):1–13.

84. Costedio MM, Hyman N, Mawe GM. Serotonin and its role in colonic function and in gastrointestinal disorders. *Diseases of the Colon and Rectum.* 2007;50(3):376–388. doi:10.1007/s10350-006-0763-3

85. Pergolizzi S, Alesci A, Centofanti A, et al. Role of serotonin in the maintenance of inflammatory state in crohn's disease. *Biomedicines.* 2022;10(4):765.

86. Berger M, Gray JA, Roth BL. The expanded biology of serotonin. *Annual Review of Medicine.* 2009;60:355–366. doi:10.1146/annurev.med.60.042307.110802

87. Wikoff WR, Anfora AT, Liu J, et al. Metabolomics analysis reveals large effects of gut microflora on mammalian blood metabolites. *Proceedings of the National Academy of Sciences of the United States of America.* 2009;106(10):3698–3703. doi:10.1073/pnas.0812874106

88. Yano JM, Yu K, Donaldson GP, et al. Indigenous bacteria from the gut microbiota regulate host serotonin biosynthesis. *Cell.* 2015;161(2):264–276. doi:10.1016/j.cell.2015.02.047

89. Mc MR, Oncley JL. The specific binding of L-tryptophan to serum albumin. *The Journal of Biological Chemistry.* 1958;233(6):1436–1447.

90. Ruddick JP, Evans AK, Nutt DJ, Lightman SL, Rook GA, Lowry CA. Tryptophan metabolism in the central nervous system: Medical implications. *Expert Reviews in Molecular Medicine.* 2006;8(20):1–27. doi:10.1017/s1462399406000068

91. Almulla AF, Supasitthumrong T, Amrapala A, et al. The tryptophan catabolite or kynurenine pathway in Alzheimer's disease: A systematic review and meta-analysis. *Journal of Alzheimer's Disease.* 2022:1–15.

92. Marx W, McGuinness AJ, Rocks T, et al. The kynurenine pathway in major depressive disorder, bipolar disorder, and schizophrenia: A meta-analysis of 101 studies. *Molecular Psychiatry.* 2021;26(8):4158–4178.

93. Albuquerque EX, Schwarcz R. Kynurenic acid as an antagonist of alpha7 nicotinic acetylcholine receptors in the brain: Facts and challenges. *Biochemical Pharmacology.* 2013;85(8):1027–1032. doi:10.1016/j.bcp.2012.12.014

94. Schwarcz R, Pellicciari R. Manipulation of brain kynurenines: Glial targets, neuronal effects, and clinical opportunities. *The Journal of Pharmacology and Experimental Therapeutics.* 2002;303(1):1–10. doi:10.1124/jpet.102.034439

95. Öztürk M, Yalın Sapmaz Ş, Kandemir H, Taneli F, Aydemir Ö. The role of the kynurenine pathway and quinolinic acid in adolescent major depressive disorder. *International Journal of Clinical Practice.* 2021;75(4):e13739.

96. Myint AM, Kim YK. Cytokine-serotonin interaction through IDO: A neurodegeneration hypothesis of depression. *Medical Hypotheses.* 2003;61(5–6):519–525.

97. Myint A-M, Kim Y-K. Network beyond IDO in psychiatric disorders: Revisiting neurodegeneration hypothesis. *Progress in Neuro-Psychopharmacology and Biological Psychiatry.* 2014;48:304–313.

98. Badawy AAB. Kynurenine pathway of tryptophan metabolism: Regulatory and functional aspects. *International Journal of Tryptophan Research: IJTR.* 2017;10:1178646917691938. doi:10.1177/1178646917691938

99. Bercik P, Verdu EF, Foster JA, et al. Chronic gastrointestinal inflammation induces anxiety-like behavior and alters central nervous system biochemistry in mice. *Gastroenterology.* 2010;139(6):2102–2112.e1. doi:10.1053/j.gastro.2010.06.063

100. Schwarcz R, Bruno JP, Muchowski PJ, Wu HQ. Kynurenines in the mammalian brain: When physiology meets pathology. *Nature Reviews Neuroscience.* 2012;13(7):465–477. doi:10.1038/nrn3257

101. Clarke G, Grenham S, Scully P, et al. The microbiome-gut-brain axis during early life regulates the hippocampal serotonergic system in a sex-dependent manner. *Molecular Psychiatry.* 2013;18(6):666–673. doi:10.1038/mp.2012.77

102. Bercik P, Denou E, Collins J, et al. The intestinal microbiota affect central levels of brain-derived neurotropic factor and behavior in mice. *Gastroenterology.* 2011;141(2):599–609, 609.e1–3. doi:10.1053/j.gastro.2011.04.052

103. Cryan JF, Dinan TG. Mind-altering microorganisms: The impact of the gut microbiota on brain and behaviour. *Nature Reviews Neuroscience.* 2012;13(10):701–712. doi:10.1038/nrn3346

104. Valladares R, Bojilova L, Potts AH, et al. Lactobacillus johnsonii inhibits indoleamine 2,3-dioxygenase and alters tryptophan metabolite levels in BioBreeding rats. *FASEB Journal: Official Publication of the Federation of American Societies for Experimental Biology.* 2013;27(4):1711–1720. doi:10.1096/fj.12-223339

105. Kelly JR, Borre Y, O'Brien C, et al. Transferring the blues: Depression-associated gut microbiota induces neurobehavioural changes in the rat. *Journal of Psychiatric Research.* 2016;82:109–118. doi:10.1016/j.jpsychires.2016.07.019

106. Zheng P, Zeng B, Zhou C, et al. Gut microbiome remodeling induces depressive-like behaviors through a pathway mediated by the host's metabolism. *Molecular Psychiatry.* 2016;21(6):786–796. doi:10.1038/mp.2016.44

107. Jiang HY, Zhang X, Yu ZH, et al. Altered gut microbiota profile in patients with generalized anxiety disorder. *Journal of Psychiatric Research.* 2018;104:130–136. doi:10.1016/j.jpsychires.2018.07.007

108. Butler MI, Bastiaanssen TFS, Long-Smith C, et al. The gut microbiome in social anxiety disorder: Evidence of altered composition and function. *Translational Psychiatry.* 2023;13(1):95. doi:10.1038/s41398-023-02325-5

109. Sublette ME, Cheung S, Lieberman E, et al. Bipolar disorder and the gut microbiome: A systematic review. *Bipolar Disorders.* 2021;23(6):544–564.

110. Nguyen TT, Kosciolek T, Maldonado Y, et al. Differences in gut microbiome composition between persons with chronic schizophrenia and healthy comparison subjects. *Schizophrenia Research.* 2019;204:23–29.

111. Shen Y, Xu J, Li Z, et al. Analysis of gut microbiota diversity and auxiliary diagnosis as a biomarker in patients with schizophrenia: A cross-sectional study. *Schizophrenia Research.* 2018;197:470–477.

112. Yuan X, Zhang P, Wang Y, et al. Changes in metabolism and microbiota after 24-week risperidone treatment in drug naïve, normal weight patients with first episode schizophrenia. *Schizophrenia Research.* 2018;201:299–306.

113. Yan F, Xia L, Xu L, Deng L, Jin G. A comparative study to determine the association of gut microbiome with schizophrenia in Zhejiang, China. *BMC Psychiatry.* 2022;22(1):1–10.

114. Misiak B, Piotrowski P, Cyran A, et al. Gut microbiota alterations in stable outpatients with schizophrenia: Findings from a case—control study. *Acta Neuropsychiatrica*. 2023;35(3):147–155.

115. Gokulakrishnan K, Nikhil J, Viswanath B, et al. Comparison of gut microbiome profile in patients with schizophrenia and healthy controls-A plausible non-invasive biomarker? *Journal of Psychiatric Research*. 2023;162:140–149.

116. Zheng P, Zeng B, Liu M, et al. The gut microbiome from patients with schizophrenia modulates the glutamate-glutamine-GABA cycle and schizophrenia-relevant behaviors in mice. *Science Advances*. 2019;5(2):eaau8317.

117. Zhu F, Guo R, Wang W, et al. Transplantation of microbiota from drug-free patients with schizophrenia causes schizophrenia-like abnormal behaviors and dysregulated kynurenine metabolism in mice. *Molecular Psychiatry*. 2020;25(11):2905–2918.

118. Al-Beltagi M. Autism medical comorbidities. *World Journal of Clinical Pediatrics*. 2021;10(3):15.

119. De Angelis M, Piccolo M, Vannini L, et al. Fecal microbiota and metabolome of children with autism and pervasive developmental disorder not otherwise specified. *PLoS ONE*. 2013;8(10):e76993. doi:10.1371/journal.pone.0076993

120. Finegold SM, Molitoris D, Song Y, et al. Gastrointestinal microflora studies in late-onset autism. *Clinical Infectious Diseases: An Official Publication of the Infectious Diseases Society of America*. 2002;35(Suppl 1):S6–s16. doi:10.1086/341914

121. Parracho HM, Bingham MO, Gibson GR, McCartney AL. Differences between the gut microflora of children with autistic spectrum disorders and that of healthy children. *Journal of Medical Microbiology*. 2005;54(Pt 10):987–991. doi:10.1099/jmm.0.46101-0

122. Song Y, Liu C, Finegold SM. Real-time PCR quantitation of clostridia in feces of autistic children. *Applied and Environmental Microbiology*. 2004;70(11):6459–6465. doi:10.1128/aem.70.11.6459-6465.2004

123. Sandler RH, Finegold SM, Bolte ER, et al. Short-term benefit from oral vancomycin treatment of regressive-onset autism. *Journal of Child Neurology*. 2000;15(7):429–435. doi:10.1177/088307380001500701

124. Dan Z, Mao X, Liu Q, et al. Altered gut microbial profile is associated with abnormal metabolism activity of Autism Spectrum Disorder. *Gut Microbes*. 2020;11(5):1246–1267.

125. Mortera SL, Vernocchi P, Basadonne I, et al. A metaproteomic-based gut microbiota profiling in children affected by autism spectrum disorders. *Journal of Proteomics*. 2022;251:104407.

126. Hughes HK, Rose D, Ashwood P. The gut microbiota and dysbiosis in autism spectrum disorders. *Current Neurology and Neuroscience Reports*. 2018;18:1–15.

127. Claesson MJ, Jeffery IB, Conde S, et al. Gut microbiota composition correlates with diet and health in the elderly. *Nature*. 2012;488(7410):178–184. doi:10.1038/nature11319

128. Cattaneo A, Cattane N, Galluzzi S, et al. Association of brain amyloidosis with pro-inflammatory gut bacterial taxa and peripheral inflammation markers in cognitively impaired elderly. *Neurobiology of Aging*. 2017;49:60–68. doi:10.1016/j.neurobiolaging.2016.08.019

129. Vogt NM, Kerby RL, Dill-McFarland KA, et al. Gut microbiome alterations in Alzheimer's disease. *Scientific Reports*. 2017;7(1):13537. doi:10.1038/s41598-017-13601-y

130. Zhuang Z-Q, Shen L-L, Li W-W, et al. Gut microbiota is altered in patients with Alzheimer's disease. *Journal of Alzheimer's Disease*. 2018;63(4):1337–1346.

131. Liu P, Wu L, Peng G, et al. Altered microbiomes distinguish Alzheimer's disease from amnestic mild cognitive impairment and health in a Chinese cohort. *Brain, Behavior, and Immunity*. 2019;80:633–643.

132. Scheperjans F. The prodromal microbiome. *Movement Disorders: Official Journal of the Movement Disorder Society*. 2018;33(1):5–7. doi:10.1002/mds.27197

133. Romano S, Savva GM, Bedarf JR, Charles IG, Hildebrand F, Narbad A. Meta-analysis of the Parkinson's disease gut microbiome suggests alterations linked to intestinal inflammation. *NPJ Parkinson's Disease*. 2021;7(1):27.

134. Heintz-Buschart A, Pandey U, Wicke T, et al. The nasal and gut microbiome in Parkinson's disease and idiopathic rapid eye movement sleep behavior disorder. *Movement Disorders: Official Journal of the Movement Disorder Society*. 2018;33(1):88–98. doi:10.1002/mds.27105

135. Bindels LB, Delzenne NM, Cani PD, Walter J. Towards a more comprehensive concept for prebiotics. *Nature Reviews Gastroenterology & Hepatology*. 2015;12(5):303–310. doi:10.1038/nrgastro.2015.47

136. Dinan TG, Stanton C, Cryan JF. Psychobiotics: A novel class of psychotropic. *Biological Psychiatry*. 2013;74(10):720–726. doi:10.1016/j.biopsych.2013.05.001

137. Steenbergen L, Sellaro R, van Hemert S, Bosch JA, Colzato LS. A randomized controlled trial to test the effect of multispecies probiotics on cognitive reactivity to sad mood. *Brain, Behavior, and Immunity*. 2015;48:258–264. doi:10.1016/j.bbi.2015.04.003

138. Shafie M, Rad AH, Mohammad-Alizadeh-Charandabi S, Mirghafourvand M. The effect of probiotics on mood and sleep quality in postmenopausal women: A triple-blind randomized controlled trial. *Clinical Nutrition ESPEN.* 2022;50:15–23.

139. Nishihira J, Kagami-Katsuyama H, Tanaka A, Nishimura M, Kobayashi, Y. Elevation of natural killer cell activity and alleviation of mental stress by the consumption of yogurt containing Lactobacillus gasseri SBT2055 and Bifidobacterium longum SBT2928 in a double-blind, placebo-controlled clinical trial. *Journal of Functional Foods.* 2014:261–268.

140. Kim C-S, Cha L, Sim M, et al. Probiotic supplementation improves cognitive function and mood with changes in gut microbiota in community-dwelling older adults: A randomized, double-blind, placebo-controlled, multicenter trial. *The Journals of Gerontology: Series A.* 2021;76(1):32–40.

141. Benton D, Williams C, Brown A. Impact of consuming a milk drink containing a probiotic on mood and cognition. *European Journal of Clinical Nutrition.* 2007;61(3):355–361. doi:10.1038/sj.ejcn.1602546

142. Salleh RM, Kuan G, Aziz MNA, et al. Effects of probiotics on anxiety, stress, mood and fitness of badminton players. *Nutrients.* 2021;13(6):1783.

143. Allen AP, Hutch W, Borre YE, et al. Bifidobacterium longum 1714 as a translational psychobiotic: Modulation of stress, electrophysiology and neurocognition in healthy volunteers. *Translational Psychiatry.* 2016;6(11):e939. doi:10.1038/tp.2016.191

144. Schmidt K, Cowen PJ, Harmer CJ, Tzortzis G, Errington S, Burnet PW. Prebiotic intake reduces the waking cortisol response and alters emotional bias in healthy volunteers. *Psychopharmacology.* 2015;232(10):1793–1801. doi:10.1007/s00213-014-3810-0

145. Marcos A, Warnberg J, Nova E, et al. The effect of milk fermented by yogurt cultures plus Lactobacillus casei DN-114001 on the immune response of subjects under academic examination stress. *European Journal of Nutrition.* 2004;43(6):381–389. doi:10.1007/s00394-004-0517-8

146. Kelly JR, Allen AP, Temko A, et al. Lost in translation? The potential psychobiotic Lactobacillus rhamnosus (JB-1) fails to modulate stress or cognitive performance in healthy male subjects. *Brain, Behavior, and Immunity.* 2017;61:50–59. doi:10.1016/j.bbi.2016.11.018

147. Reale M, Boscolo P, Bellante V, et al. Daily intake of Lactobacillus casei Shirota increases natural killer cell activity in smokers. *The British Journal of Nutrition.* 2012;108(2):308–314. doi:10.1017/s0007114511005630

148. Tillisch K, Labus J, Kilpatrick L, et al. Consumption of fermented milk product with probiotic modulates brain activity. *Gastroenterology.* 2013;144(7):1394–1401, 1401.e1–4. doi:10.1053/j.gastro.2013.02.043

149. Ng QX, Peters C, Ho CYX, Lim DY, Yeo WS. A meta-analysis of the use of probiotics to alleviate depressive symptoms. *Journal of Affective Disorders.* 2018;228:13–19. doi:10.1016/j.jad.2017.11.063

150. Didari T, Mozaffari S, Nikfar S, Abdollahi M. Effectiveness of probiotics in irritable bowel syndrome: Updated systematic review with meta-analysis. *World Journal of Gastroenterology.* 2015;21(10):3072–3084. doi:10.3748/wjg.v21.i10.3072

151. Mizuno S, Masaoka T, Naganuma M, et al. Bifidobacterium-rich fecal donor may be a positive predictor for successful fecal microbiota transplantation in patients with irritable bowel syndrome. *Digestion.* 1961;96(1):29–38.

152. Huang HL, Chen HT, Luo QL, et al. Relief of irritable bowel syndrome by fecal microbiota transplantation is associated with changes in diversity and composition of the gut microbiota. *Journal of Digestive Diseases.* 2019;20(8):401–408.

153. Dapoigny M, Piche T, Ducrotte P, Lunaud B, Cardot J-M, Bernalier-Donadille A. Efficacy and safety profile of LCR35 complete freeze-dried culture in irritable bowel syndrome: A randomized, double-blind study. *World Journal of Gastroenterology: WJG.* 2012;18(17):2067–2075. doi:10.3748/wjg.v18.i17.2067

154. Han K, Wang J, Seo JG, Kim H. Efficacy of double-coated probiotics for irritable bowel syndrome: A randomized double-blind controlled trial. *Journal of Gastroenterology.* 2017;52(4):432–443. doi:10.1007/s00535-016-1224-y

155. Simren M, Ohman L, Olsson J, et al. Clinical trial: The effects of a fermented milk containing three probiotic bacteria in patients with irritable bowel syndrome—a randomized, double-blind, controlled study. *Alimentary Pharmacology & Therapeutics.* 2010;31(2):218–227. doi:10.1111/j.1365-2036.2009.04183.x

156. Whorwell PJ, Altringer L, Morel J, et al. Efficacy of an encapsulated probiotic Bifidobacterium infantis 35624 in women with irritable bowel syndrome. *The American Journal of Gastroenterology.* 2006;101(7):1581–1590. doi:10.1111/j.1572-0241.2006.00734.x

157. Pinto-Sanchez MI, Hall GB, Ghajar K, et al. Probiotic bifidobacterium longum NCC3001 reduces depression scores and alters brain activity: A pilot study in patients with irritable bowel syndrome. *Gastroenterology.* 2017;153(2):448–459.e8. doi:10.1053/j.gastro.2017.05.003

158. Groeger D, Murphy EF, Tan HTT, Larsen IS, O'Neill I, Quigley EM. Interactions between symptoms and psychological status in irritable bowel syndrome: An exploratory study of the impact of a probiotic combination. *Neurogastroenterology & Motility.* 2023;35(1):e14477.

159. Mikocka-Walus A, Knowles SR, Keefer L, Graff L. Controversies revisited: A systematic review of the comorbidity of depression and anxiety with inflammatory bowel diseases. *Inflammatory Bowel Diseases.* 2016;22(3):752–762.

160. Gracie DJ, Guthrie EA, Hamlin PJ, Ford AC. Bi-directionality of brain—gut interactions in patients with inflammatory bowel disease. *Gastroenterology.* 2018;154(6):1635–1646. e3.

161. Yoo J-W, Shin Y-J, Ma X, et al. The alleviation of gut microbiota-induced depression and colitis in mice by anti-inflammatory probiotics Nk151, Nk173, and Nk175. *Nutrients.* 2022;14(10):2080.

162. Akkasheh G, Kashani-Poor Z, Tajabadi-Ebrahimi M, et al. Clinical and metabolic response to probiotic administration in patients with major depressive disorder: A randomized, double-blind, placebo-controlled trial. *Nutrition (Burbank, Los Angeles County, Calif).* 2016;32(3):315–320. doi:10.1016/j.nut.2015.09.003

163. Romijn AR, Rucklidge JJ, Kuijer RG, Frampton C. A double-blind, randomized, placebo-controlled trial of Lactobacillus helveticus and Bifidobacterium longum for the symptoms of depression. *The Australian and New Zealand Journal of Psychiatry.* 2017;51(8):810–821. doi:10.1177/0004867416686694

164. Kazemi A, Noorbala AA, Azam K, Eskandari MH, Djafarian K. Effect of probiotic and prebiotic vs placebo on psychological outcomes in patients with major depressive disorder: A randomized clinical trial. *Clinical Nutrition.* 2019;38(2):522–528.

165. Dickerson FB, Stallings C, Origoni A, et al. Effect of probiotic supplementation on schizophrenia symptoms and association with gastrointestinal functioning: A randomized, placebo-controlled trial. *The Primary Care Companion for CNS Disorders.* 2014;16(1). doi:10.4088/PCC.13m01579

166. Tomasik J, Yolken RH, Bahn S, Dickerson FB. Immunomodulatory effects of probiotic supplementation in schizophrenia patients: A randomized, placebo-controlled trial. *Biomarker Insights.* 2015;10:47–54. doi:10.4137/bmi.s22007

167. Ghaderi A, Banafshe HR, Mirhosseini N, et al. Clinical and metabolic response to vitamin D plus probiotic in schizophrenia patients. *BMC Psychiatry.* 2019;19(1):1–10.

168. Jamilian H, Ghaderi A. The effects of probiotic and selenium co-supplementation on clinical and metabolic scales in chronic schizophrenia: A randomized, double-blind, placebo-controlled trial. *Biological Trace Element Research.* 2021;199:4430–4438.

169. Huang J, Kang D, Zhang F, et al. Probiotics plus dietary fiber supplements attenuate olanzapine-induced weight gain in drug-naïve first-episode schizophrenia patients: Two randomized clinical trials. *Schizophrenia Bulletin.* 2022;48(4):850–859.

170. Dickerson F, Adamos M, Katsafanas E, et al. Adjunctive probiotic microorganisms to prevent rehospitalization in patients with acute mania: A randomized controlled trial. *Bipolar Disorders.* 2018;20(7):614–621.

171. Shahrbabaki ME, Sabouri S, Sabahi A, et al. The efficacy of probiotics for treatment of bipolar disorder-type 1: A randomized, double-blind, placebo controlled trial. *Iranian Journal of Psychiatry.* 2020;15(1):10.

172. Hilimire MR, DeVylder JE, Forestell CA. Fermented foods, neuroticism, and social anxiety: An interaction model. *Psychiatry Research.* 2015;228(2):203–208. doi:10.1016/j.psychres.2015.04.023

173. Kabir SL, Islam SS, Akhter AT. Production, cost analysis, and marketing of probiotics. *Food Microbiology Based Entrepreneurship: Making Money From Microbes.* Springer; 2023:305–326.

174. Hooks KB, O'Malley MA. Dysbiosis and its discontents. *mBio.* 2017;8(5). doi:10.1128/mBio.01492-17

175. Psaltopoulou T, Sergentanis TN, Panagiotakos DB, Sergentanis IN, Kosti R, Scarmeas N. Mediterranean diet, stroke, cognitive impairment, and depression: A meta-analysis. *Annals of Neurology.* 2013;74(4):580–591. doi:10.1002/ana.23944

176. Altun A, Brown H, Szoeke C, Goodwill AM. The Mediterranean dietary pattern and depression risk: A systematic review. *Neurology, Psychiatry and Brain Research.* 2019;33:1–10.

177. Liu C, Xie B, Chou CP, et al. Perceived stress, depression and food consumption frequency in the college students of China Seven Cities. *Physiology & Behavior.* 2007;92(4):748–754. doi:10.1016/j.physbeh.2007.05.068

178. Sanchez-Villegas A, Toledo E, de Irala J, Ruiz-Canela M, Pla-Vidal J, Martinez-Gonzalez MA. Fast-food and commercial baked goods consumption and the risk of depression. *Public Health Nutrition.* 2012;15(3):424–432. doi:10.1017/s1368980011001856

179. Jacka FN. Nutritional psychiatry: Where to next? *EBioMedicine.* 2017;17:24–29. doi:10.1016/j.ebiom.2017.02.020

180. Jacka FN, O'Neil A, Opie R, et al. A randomised controlled trial of dietary improvement for adults with major depression (the 'SMILES' trial). *BMC Medicine.* 2017;15(1):23. doi:10.1186/s12916-017-0791-y

181. David LA, Maurice CF, Carmody RN, et al. Diet rapidly and reproducibly alters the human gut microbiome. *Nature.* 2014;505(7484):559–563. doi:10.1038/nature12820

# Lactic Acid Bacteria and Oral Health

# 27

## Iva Stamatova, Qingru Jiang, and Jukka H. Meurman

Oral health is undeniably an essential component of general health. In its latest definition, it is regarded as multifaceted and includes the ability to speak, smile, smell, taste, touch, chew, swallow, and convey a range of emotions through facial expressions with confidence and without pain, discomfort, and disease of the craniofacial complex (Glick et al. 2016). The oral microbiome establishes early in life; microorganisms colonise the mucosal surfaces within the first hours after birth (Xu et al. 2015). A diversity of over 750 taxa have been characterised in the mouth by cultivation and culture-independent molecular methods (www.home.org). Lactic acid bacteria in the oral cavity are mainly presented by *Lactobacillus* and *Streptococcus* accounting for 29.2% of the oral microbiome (Costalonga and Herzberg 2014).

The most common pathologies in the mouth are microbial in their aetiology. Dental caries, periodontal diseases, endodontic lesions, and alveolar osteitis are directly related to alterations in pathogenicity of oral biofilms. Oral biofilms are from the very nascent stages multispecies communities, and their dynamics is directly dependent on cell–cell interactions. With the advent of high-throughput DNA sequencing, there is more robust evidence gathered that microbiota of diseased sites differs from that of healthy sites (McLaren and Callahan 2018). Advances in gene sequencing and bioinformatics have supported the role of the oral microbiome as primary source for colonisation of the underlying compartments of the gastrointestinal tract (Park et al. 2021), allowing for contemporary interpretation of the focal infection theory.

The role of beneficial healthy bacteria has been related to numerous pathological conditions of the gastrointestinal tract, skin, respiratory system, and mental health. The evidence of probiotics on parameters of oral health has greatly increased over the past two decades (Meurman 2005; Caglar et al. 2005; Seminario-Amez et al. 2017; Twetman and Jørgensen 2021). In this chapter, we discuss the current evidence-based role of oral lactic acid bacteria as probiotic candidates promoting oral health.

## 27.1 DENTAL CARIES

Dental caries remain one of the most prevalent multifactorial oral diseases in children and adults worldwide. They are clinically manifested by localised demineralisation of dental hard tissues by acidic by-products delivered from microbial fermentation of dietary carbohydrates (Selwitz et al. 2007). The oral microbiome plays a key role in the initiation and progression of caries (Dashper et al. 2019). For a long while, *Streptococcus mutans* (MS) has been regarded as the key pathogen responsible for the initiation and disease progression. However, by analysing the 16S rRNA V3–V4 region (Richards et al. 2017) and fungal ITS1 region (O'Connell et al. 2020), researchers have demonstrated that the onset of dental caries

DOI: 10.1201/9781003352075-30

could be linked to multiple species of both bacteria and fungi rather than solely to MS. The ecological plaque hypothesis has been a paradigm shift in understanding of the aetiology of caries and other common oral infectious diseases (Marsh 2004; Marsh et al. 2015). Molecular microbiological methods have also shown that, even with a sugar-rich diet, a much broader spectrum of acidogenic microbes is found in the dental biofilm (Bradshaw and Lynch 2013; Baker et al. 2021). Dental caries in children could be also related to possible interaction of risk factors such as maternal characteristics, environment, child's individual factors, and epigenetics (Fernando et al. 2015). By utilising metatranscriptomics, Benítez-Páez et al. (2014) have observed not only an individual-specific microbiota but also specific dynamics in their activity during biofilm formation. Gene expression of more specialised genes is characteristic for mature biofilms, indicating that the microbial community is adapting as it develops (Solbiati and Frias-Lopez 2018). For targeting prevention of dental caries, a more complex and interdisciplinary approach, including control of biofilm composition, control of dietary sugars in oral biofilm, and prolonged fluoride release, needs to be adopted. Bacteriotherapy emerges as such an alternative preventive measure to prevent dental caries.

To achieve its caries-preventive effect, a probiotic candidate needs to integrate in oral biofilms and lower their cariogenic potential. Most of the probiotic species, however, have limited ability to be permanently retained in the mouth, which necessitates their regular administration.

The first clinical study to address the role of probiotics in caries management has shown that *Lacticaseibacillus rhamnosus* GG (previously known as *Lactobacillus rhamnosus* GG) significantly reduces MS counts in children (Näse et al. 2001). Interest in probiotics to target dental caries has grown through the years, and *L. rhamnosus* has been broadly studied for its caries-reducing effect in high caries–risk children and elderly (Stecksén-Blicks et al. 2009; Glavina et al. 2012; Rodriguez et al. 2016; Petersson et al. 2011). Milk was the main vehicle of administration, and the intervention period correlated directly to the observed outcomes (Lexner et al. 2010).

Evaluating the published data from the interventional studies allows us to conclude that at the saliva level, probiotic administration could lead to lower MS counts (Borrell García et al. 2021; Lai et al. 2021; Manmontri et al. 2020); increase in saliva's buffering capacity, flow rate, and pH (Villavicencio et al. 2018; Ferrer et al. 2020a; Ferrer et al. 2020b); higher secretory IgA level (Pahumunto et al. 2019); and elevated levels of human neutrophil peptides 1–3, salivary ammonia, and calcium (Wattanarat et al. 2021; Ferrer et al. 2020a). Using dental plaque as a test milieu, probiotic species could reduce MS counts (Ratna Sudha et al. 2020); reduce acidogenic potential of dental plaque (Lai et al. 2021); and decelerate biofilm formation, reducing its pathogenic potential (Ferrer et al. 2020a). At a tooth surface level, the observed effects include delayed caries progression (Pahumunto et al. 2018), caries regression (Petersson et al. 2011; Piwat et al. 2020), and generally lower caries prevalence (Stecksén-Blicks et al. 2009; Rodriguez et al. 2016; Stensson et al. 2014). Table 27.1 presents the commonly studied probiotic strains with respect to dental caries. Although the results of the previously mentioned studies could contribute to new strategies for caries prevention, more robust data on duration and vehicle of administration are still needed.

**TABLE 27.1**   Probiotics in Clinical Trials in Relation to Dental Caries

| VEHICLE | SPECIES/STRAINS | SUBJECTS/DURATION | OUTCOMES | REFERENCE |
|---|---|---|---|---|
| Chewable tablet | *Bacillus coagulans* Unique IS2 | Age: 5–15 years, healthy/2 weeks | Reduced MS counts in saliva and plaque | Ratna Sudha et al. (2020) |
| Glycerine | *L. acidophilus, L. rhamnosus, L. casei, L. bulgaricus, L. plantarum, Bifidobacterium longum (2 strains), and S. thermophilus* | Age: 7–12 years, healthy/6 days | Reduced MS in plaque | Patil et al. (2021) |

(Continued)

**TABLE 27.1**   (Continued) Probiotics in Clinical Trials in Relation to Dental Caries

| VEHICLE | SPECIES/STRAINS | SUBJECTS/DURATION | OUTCOMES | REFERENCE |
|---|---|---|---|---|
| Lozenge | L. brevis CD2 | Age: 4–14 years, with type 1 diabetes/60 days | Reduced MS counts in saliva, increased dental plaque pH | Lai et al. (2021) |
| Milk powder in water | L. paracasei SD1 | Age: 1.5–5 years, healthy/3 months | Reduced MS counts in saliva, delayed new caries development | Pahumunto et al. (2018) |
| Milk powder | L. paracasei SD1 | Age: 12–14 years, healthy/6 months | Reduced salivary streptococci and MS, increased salivary IgA (sIgA) | Pahumunto et al. (2019) |
| Milk powder in milk | L. paracasei SD1 | Age: 1–5 years, healthy/6 month | Reduced MS counts in saliva and plaque | Manmontri et al. (2020) |
| Milk powder in water or milk | L. paracasei SD1 | Age: 2–4 years, healthy/6 months | Daily or triweekly consumption both decreased caries risk and increased regressive surfaces | Piwat et al. (2020) |
| Milk powder in milk | L. paracasei SD1 | Age: 1–5 years, healthy/6 months | In severe-ECC (Early Childhood Caries) group, reduced salivary MS levels, decreased caries progression, enhanced salivary human neutrophil peptide 1–3 (HNP1–3) levels | Wattanarat et al. (2021) |
| Drop | L. reuteri ATCC 55730 | Age: 9 years, from families with a history of allergic diseases/13 months | Reduced caries prevalence | Stensson et al. (2014) |
| Tablet | L. reuteri DSM 17938 and ATCC 5289 | Age: 12–18 years, healthy/28 days | Lower rise in MS counts in saliva, did not significantly increase salivary pH | Borrell García et al. (2021) |
| Milk | L. rhamnosus and Bi. longum | Age: 3–4 years, healthy/9 months | Increased salivary buffering capacity | Villavicencio et al. (2018) |
| Milk | L. rhamnosus GG ATCC 53103 | Age: 1–6 years, healthy/7 months | Reduced MS counts in saliva, reduced dental caries, particularly in 3–4-year-old children | Näse et al. (2001) |
| Milk | L. rhamnosus LB21 | Age: 58–84 years, healthy/15 months | Caries remineralisation occurred | Petersson et al. (2011) |
| Milk | L. rhamnosus LB21 | Age: 1–5 years, healthy/21 months | Reduced DMFS | Stecksén-Blicks et al. (2009) |
| Milk | L. rhamnosus SP1 | Age: 2–3 years, healthy with high caries/10 months | Reduced total and severe caries increments | Rodríguez et al. (2016) |

**TABLE 27.1**   (Continued) Probiotics in Clinical Trials in Relation to Dental Caries

| VEHICLE | SPECIES/STRAINS | SUBJECTS/DURATION | OUTCOMES | REFERENCE |
|---|---|---|---|---|
| Bucco-adhesive gel with a dental splint | *S. dentisani* CECT 7746 | Age: 18–65 years, healthy/1 month | Decreased dental plaque amount, beneficial shift in bacteria composition, reduced cariogenic organisms; increased salivary flow rate and salivary ammonia and calcium levels | Ferrer et al. (2020a) |
| Bucco-adhesive gel with a dental splint | *S. dentisani* CECT 7746 | Age: 25–35 years, healthy/a week | Lower MS in plaque in multidose, increased salivary pH | Ferrer et al. (2020b) |

# 27.2 PERIODONTAL DISEASES

The global burden of periodontal diseases remains high, with over 1.1 billion prevalent cases of severe periodontitis worldwide (Chen et al. 2021). The clinical manifestation of periodontal diseases is generally related to gingival swelling, bleeding on probing, and progressive bone loss that originate in the presence of a susceptible host and pathogenicity of the species in the oral biofilms. For successful management of periodontitis, clinicians should have thorough understanding of the pathogenesis, primary aetiology, risk factors, and treatment protocols (Kwon et al. 2021). Because oral biofilm is the cause, periodontitis treatment mainly aims to disrupt, eliminate, and alter the pathogenic potential of supra- and subgingival biofilms. Mechanical plaque removal by professional and home care practices thus remains central in altering the biofilm formation and lowering its pathogenicity. Systemically and locally administered antibiotics, antimicrobial photodynamic therapy (APT), and oral antiseptics have been broadly studied as adjuncts to non-surgical periodontal treatments. These approaches, however, have always raised concerns regarding antibiotic resistance and transfer of resistance genes in biofilm consortia (Elashiry et al. 2021).

The application of probiotics for periodontal diseases appears an attractive preventive modality. Inconsistent findings of clinical and microbiological outcomes of probiotic administration solely, or as an adjunctive method to conventional non-surgical periodontal therapy (NSPT), have been reported in clinical studies. Probiotics given as a sole intervention to healthy subjects and to individuals with experimental gingivitis have shown reduction in gingival and plaque indices and lowered the volume of gingival crevicular fluid (Alanzi et al. 2018; Keller et al. 2018a; Kuru et al. 2017; Lee et al. 2015; Lai et al. 2021). The observed effects could be attributed to probiotic interactions with inflammatory cytokines and lowering TNF-α and IL-8 (Iniesta et al. 2012). A study by Schlagenhauf et al. (2020) in a cohort of navy sailors, allowing for standardisation of environment conditions and diet, showed that the daily administration of *Limosilactobacillus reuteri* (DSM 17938 and PTA 5289) improved and maintained their periodontal health. However, other studies evaluating the effect on gingivitis have not shown positive effects by probiotic administration (Alkaya et al. 2017; Hallström et al. 2016). Clinical and immunological benefits could be achieved when a combined probiotic gel and capsule containing *L. rhamnosus* HN001, *Lacticaseibacillus paracasei* Lpc-37, and *B. animalis* subs. *lactis* HN019 were given after NSPT. Lower levels of IL-1β, IL-6, IL-8, and TNF-α were observed 12 weeks after of a combined probiotic administration (Santana et al. 2022). Table 27.2 presents the common probiotic species used in clinical studies of periodontal diseases.

**TABLE 27.2** Effects of Probiotics as an Adjunct Intervention on Periodontal Diseases

| SPECIES/STRAINS | SUBJECTS | PRE-TREATMENT | VEHICLE AND DURATION | OUTCOMES | REFERENCES |
|---|---|---|---|---|---|
| *Bi. animalis* subsp. *lactis* HN019 | With generalised chronic periodontitis | SRP | Lozenge, 30 days | Decreased PPD, improved clinical attachment gain, reduced counts of periodontal pathogens, and lowered proinflammatory cytokine levels | Invernici et al. (2018) |
| *L. brevis* 7480 CECT and *L. plantarum* 7481 CECT | With periodontitis (stages II and III, grade B) | Subgingival debridement (SD) + light-activated disinfection (LAD) | Gel subgingivally to fill the periodontal pockets after SD and followed by intake of lozenges for 3 months | Combination of SD + LAD + probiotic treatment demonstrated significantly greater reductions in BOP, GIs, and red complex bacteria *P. gingivalis* and *T. forsythia* | Patyna et al. (2021) |
| *L. reuteri* | With chronic periodontitis in smoking patients | SRP | Chewable tablets, 21 days | Reduced BOP, no significant reduction in PD or gain in CAL in moderate and deep pockets compared with control group | Theodoro et al. (2019) |
| *L. reuteri* DSM 17938 and ATCC PTA 5289 | With periodontitis | Whole-mouth scaling, and the residual pockets were subgingivally debrided | Probiotic drops filled all residual pockets after SRP, followed by intake of lozenges for 12 weeks | Drops showed no effects. Lozenges improved the PPD reduction, without an impact on pocket colonisation with periodontopathogens | Laleman et al. (2020b) |
| *L. reuteri* DSM 17938 and ATCC PTA 5289 | With generalized periodontitis stage III and IV, grade C | Session of full mouth guided biofilm therapy (FM-GBT) | Lozenge, 12 weeks | Reduced PPD, higher probing attachment level (PAL) gain, increased BOP reduction | Grusovin et al. (2020) |
| *L. reuteri* DSM 17938 and ATCC PTA 5289 | With periodontitis and molars with deep pockets | S/RSD (scaling and root surface debridement) | Lozenge, 28 days | Improved CAL change at molar sites with ≥ 5 mm deep pockets and conferred a higher probability of shallow residual pocket depth | Pelekos et al. (2020) |
| *L. reuteri* DSM 17938 and ATCC PTA 5289 | With chronic periodontitis | Full-mouth scaling and root surface debridement (S/RSD) | Lozenge, 28 days | No added benefits | Pelekos et al. (2019) |
| *L. reuteri* DSM 17938 and ATCC PTA 5289 | Chronic periodontitis among shamma users | SRP, oral instructions, oral rinse twice a day | Lozenge, 21 days | No added benefits in clinical parameters | Vohra et al. (2020) |
| *L. rhamnosus* SP1 | With stage III periodontitis | SRP/SRP+azithromycin (capsule)/ SRP+probiotic (sachet) | Sachet, 3 months | No adjunctive benefits | Morales et al. (2021) |

| | | | | | |
|---|---|---|---|---|---|
| *L. salivarius* NK02 | Chronic periodontitis, moderate to severe periodontitis | SRP (Scaling and Root Planing) | Mouthwash, 4 weeks | Inhibited bacterial growth on both saliva and sub-gingival crevice and exhibited antibacterial activity against *A. actinomycetemcomitans*, decreased gingival index (GI) and bleeding on probing (BOP), probing pocket depth (PPD) | Sajedinejad et al. (2018) |
| *Streptococcus oralis* KJ3, *S. uberis* KJ2, and *S. rattus* JH145 | With moderate to severe adult periodontitis | SRP | Tablet, 12 weeks | No significant differences in clinical or microbiological parameters but decreased number of sites with plaque and decreased counts of *Prevotella intermedia* | Laleman et al. (2015) |
| *L. reuteri* DSM 17938 and ATCC PTA 5289 | With peri-implant mucositis but without peri-implantitis received supragingival prophylaxis | Supragingival prophylaxis | Tablet, 30 days | Decreased crevicular fluid volume; plaque index; probing depth; gingival index; and concentrations of IL-1b, IL-6, and IL-8 (interleukin 1b, interleukin 6, and interleukin 8) | Flichy-Fernandez et al. (2015) |

The ultimate goal of non-surgical periodontal treatment is the reduction of gingival pocket depth and inflammation, clinically manifested by less bleeding on probing. A number of probiotic preparations have been clinically tested for their effects when given as an adjunct to NSPT (Grusovin et al. 2020; Invernici et al. 2018; Laleman et al. 2020a; Sajedinejad et al. 2018; Theodoro et al. 2019). Furthermore, a combined administration of *Levilactobacillus brevis* and *Lactiplantibacillus plantarum* with ATP could further improve inflammatory parameters in moderate to severe stage II and III periodontitis patients when compared with APT alone (Patyna et al. 2021). Among the most commonly studied probiotics with significant effect on periodontitis is *L. reuteri* (Teughels et al. 2013; Soares et al. 2019; Tekce et al. 2015; Pelekos et al. 2019; Pelekos et al. 2020; Vicario et al. 2013). The plausible mechanisms of probiotic activity are reducing the numbers of periodontal pathogens, such as *Porphyromonas gingivalis*, and the lowering of the concentrations of TNF-α, IL-1β, and IL-17 in gingival crevicular fluid (Teughels et al. 2013). As most of the studied probiotics could only temporarily inhabit the oral cavity, the vehicle of their administration requires particular attention. Common probiotic formulations used are lozenges, tablets, mouth rinses, milk, yoghurt, and ice cream (see Tables 27.1 and 27.2).

Of other methods investigated, hydroxypropylmethylcellulose-based films loaded with *L. brevis* CD2 are attractive vehicles of probiotic release of viable cells that maintain anti-inflammatory enzymatic activity (Abruzzo et al. 2020). A mucoadhesive lipogel incorporating a combination of several probiotics and botanicals was tested for its stability and prolonged release allowing for 5–8 hours of slow release during the night (Giannini et al. 2022). However, more studies are also called for to reach further conclusions.

Sharing similar pathogenesis with periodontal disease, peri-implant mucositis and peri-implantitis are common problems after dental implant placement. *L. reuteri* administered as a lozenge or in a liquid as adjunct to NSPT could improve clinical parameters, including bleeding on probing and probing pocket depth around implants (Galofré et al. 2018; Peña et al. 2019; Laleman et al. 2020b; Alqahtani et al. 2022; Sargolzaei et al. 2022). However, the probiotic intake in those studies has not affected microbial composition in peri-implantitis sites. According to Tada et al. (2018), probiotics indeed prevent inflammation by affecting the host response rather than altering the microta. In general, scientific evidence on probiotics vs. periodontal diseases remains weak. More randomised clinical trials with probiotic supplementation are needed.

## 27.3 ORAL MUCOSAL CONDITIONS

Oral candidiasis remains a highly prevalent mucosal inflammatory condition, especially in the elderly. In the elderly, Hatakka and co-workers observed reduced *Candida* counts in a short-term intervention study where probiotics were administered in cheese (Hatakka et al. 2007). In complete denture wearers, daily consumption of cheese supplemented with either *L. acidophilus* NCFM or *L. rhamnosus* Lr-32 was able to reduce the colonisation of oral *Candida* (Miyazima et al. 2017), and frequent consumption of milk with *L. rhamnosus* SP1 was able to reduce *Candida* counts and the severity of denture stomatitis (Lee et al. 2019).

The observed inhibitory effects are mainly ascribed to probiotic ability directly by competition for adhesion sites or production of secondary metabolites or indirectly by the stimulation of the immune system of the host (Ribeiro et al. 2020). Rossoni and co-workers found that *Lacticaseibacillus paracasei* 28.4 upregulated genes that encode the antifungal peptides galiomicin and gallerymicin (2017). *Candida* grown in the presence of *L. rhamnosus* could significantly lower proteinase and haemolysin activity when compared with the control group (Oliveira et al. 2016). Allonsius et al. (2019) identified a lactobacilli-specific protein with chitinase activity that could break down chitin, the main polymer in the hyphal cell wall of *C. albicans*. This major peptidoglycan hydrolase, Msp1, is found in three closely related taxa: *L. rhamnosus*, *L. casei*, and *L. paracasei*, and could be potentially tested in clinical trials in management of candidiasis.

Oral lichen planus (OLP) is a T-cell mediated chronic inflammatory condition of the skin which often affects oral mucosal membranes. Oral lichen planus may appear as white, lacy patches or red or open sores. The aetiology and pathogenesis of OLP remain unclear. A number of studies suggested that OLP is associated with dysbiosis of the oral microbiome (He et al. 2017; Li et al. 2019; Liu et al. 2021; Wang et al. 2020; Zhong et al. 2020). Li et al. (2019) found that fungal dysbiosis could also be associated with the aggravation of OLP. They have observed higher levels of *Candida* and *Aspergillus* in patients with reticular OLP and higher numbers of *Alternaria* and Sclerotiniaceae in erosive OLP individuals. With the aid of next-generation sequencing technology, Zhong et al. (2020) compared the host cell gene expression profiles and the microbial profiles between OLP patients and matched healthy individuals and identified the activation of the hepatocyte nuclear factor alpha (HNF4A) network in OLP patients. Potential pathogens, including *Corynebacterium matruchotii*, *Fusobacterium periodonticum*, *S. intermedius*, *S. oralis*, and *Prevotella denticola* and furthermore *P. denticola* have been capable of activating the HNF4A gene network (Zhong et al. 2020).

The documented ability of probiotics to modulate the production of inflammatory cytokines and to inhibit MMP-9 expression and mast cell degranulation may make them suitable candidates to favourably affect the pathogenesis in OLP (Han et al. 2017). Individuals with symptomatic OLP who consumed for 30 days a VSL#3 sachet containing eight species of lactic acid strains showed a trend of reduction in salivary IFN-gamma levels but with no other notable changes in pain, disease activity, quality of life, serum/salivary CXCL10, or oral microbiome (Marlina et al. 2022). As an adjunct, probiotic intervention after conventional (i.e., antimycotic or steroid) treatment failed to influence the oral microbiome in patients with OLP (Keller and Kragelund 2018b). More investigations in this field are also needed.

## 27.4 MECHANISMS OF PROBIOTIC EFFECTS IN THE MOUTH

The exact molecular mechanisms of action of probiotics remain to be defined. Four main mechanisms have been proposed to explain probiotic activity. The first is enhancing mucosal barrier function by probiotics, which has mainly been investigated in gut health. Second, probiotics could modulate the immune response. Probiotics reduced the recruitment of T helper 17 cells and downregulated IL-17 cytokine production (Li et al. 2016). Lactobacilli and bifidobacteria reduced the expression of IL-1β, IL-6, IL-7, and TNF-α (Nguyen et al. 2021). Third, probiotics show antagonistic activity against pathogens by metabolites like lactic acid; other organic acids; hydrogen peroxide; or bacteriocins like reuterin from *L. reuteri* (Baca-Castanon et al. 2015), nisin from *Lactococcus lactis* (Heeney et al. 2019), plantaricin from *L. plantarum* (Heeney et al. 2019), and salivaricin from *L. salivarius* and *S. salivarius* (Masdea et al. 2012), respectively. In addition, probiotics could compete with pathogens for adhesion sites. Lactobacilli and bifidobacteria reduced the adhesion of *P. gingivalis*, *F. nucleatum*, and MS to buccal epithelial cells (Invernici et al. 2020), gingival epithelial cells (Albuquerque-Souza et al. 2019), and oral epithelial cells (Mann et al. 2021). Nevertheless, there are still many open questions regarding the mechanisms for observed probiotic effects.

## 27.5 CONCLUSION

Lactic acid bacteria are indigenous members of the oral microbiota, accounting for a third of the entire oral microbiome. The role of beneficial bacteria targeting infectious diseases by modulating biofilm composition is rapidly accumulating. Oral health as an integral part of general health has been shown to be positively affected and promoted by the availability of probiotic lactic acid bacteria in the mouth. Whilst positive results have been published, more multi-centric clinical trials are warranted. Species selection and vehicles of administration tailored for the mouth are of particular importance when addressing probiotics in regard to oral health.

# BIBLIOGRAPHY

Abruzzo A, Vitali B, Lombardi F, Guerrini L, Cinque B, Parolin C, Bigucci F, Cerchiara T, Arbizzani C, Gallucci MC, Luppi B. 2020. Mucoadhesive buccal films for local delivery of *Lactobacillus brevis*. *Pharmaceutics* 12(3):241.

Alanzi A, Honkala S, Honkala E, Varghese A, Tolvanen M, Söderling E. 2018. Effect of *Lactobacillus rhamnosus* and *Bifidobacterium lactis* on gingival health, dental plaque, and periodontopathogens in adolescents: A randomised placebo-controlled clinical trial. *Benef Microbes* 9:593–602.

Albuquerque-Souza E, Balzarini D, Ando-Suguimoto ES, Ishikawa KH, Simionato MRL, Holzhausen M, Mayer MPA. 2019. Probiotics alter the immune response of gingival epithelial cells challenged by *Porphyromonas gingivalis*. *J Periodontal Res* 54:115–127.

Alkaya B, Laleman I, Keceli S, Ozcelik O, Cenk Haytac M, Teughels W. 2017. Clinical effects of probiotics containing *Bacillus* species on gingivitis: A pilot randomized controlled trial. *J Periodont Res* 52:497–504.

Allonsius CN, Vandenheuvel D, Oerlemans EFM, Petrova MI, Donders GGG, Cos P, Delputte P, Lebeer S. 2019. Inhibition of *Candida albicans* morphogenesis by chitinase from *Lactobacillus rhamnosus* GG. *Sci Rep* 9:2900.

Alqahtani F, Alshaikh M, Mehmood A, Alqhtani N, Alkhtani F, Alenazi A. 2022. Efficacy of antibiotic versus probiotics as adjuncts to mechanical debridement for the treatment of peri-implant mucositis. *J Oral Implantol* 48:99–104.

Baca-Castanon ML, De la Garza-Ramos MA, Alcazar-Pizana AG, Grondin Y, Coronado-Mendoza A, Sanchez-Najera RI, Cardenas-Estrada E, Medina-De la Garza CE, Escamilla-Garcia E. 2015. Antimicrobial effect of *Lactobacillus reuteri* on cariogenic bacteria *Streptococcus gordonii*, *Streptococcus mutans*, and periodontal diseases *Actinomyces naeslundii* and *Tannerella forsythia*. *Probiotics Antimicrob Proteins* 7:1–8.

Baker JL, Morton JT, Dinis M, Alvarez R, Tran NC, Knight R. 2021. Deep metagenomics examines the oral microbiome during dental caries, revealing novel taxa and co-occurrences with host molecules. *Genome Res* 31:64–74.

Benítez-Páez A, Belda-Ferre P, Simon-Soro A, Mira A. 2014. Microbiota diversity and gene expression dynamics in human oral biofilms. *BMC Genom* 15:311.

Borrell García C, Ribelles Llop M, García Esparza M, Flichy-Fernández AJ, Marqués Martínez L, Izquierdo Fort R. 2021. The use of *Lactobacillus reuteri* DSM 17938 and ATCC PTA 5289 on oral health indexes in a school population: A pilot randomized clinical trial. *Int J Immunopathol Pharmacol* 35:20587384211031107.

Bradshaw DJ, Lynch RJ. 2013. Diet and the microbial aetiology of dental caries: New paradigms. *Int Dent J* 63(Suppl 2):64–72.

Caglar E, Kargul B, Tanboga I. 2005. Bacteriotherapy and probiotics' role on oral health. *Oral Dis* 11:131–137.

Chen MX, Zhong YJ, Dong QQ, Wong HM, Wen YF. 2021. Global, regional, and national burden of severe periodontitis, 1990–2019: An analysis of the Global Burden of Disease Study 2019. *J Clin Periodontol* 48(9):1165–1188.

Costalonga M, Herzberg M. 2014. The oral microbiome and the immunobiology of periodontal disease and caries. *Immunol Lett* 162:22–38.

Dashper SG, Mitchell HL, Lê Cao KA, Carpenter L, Gussy MG, Calache H, Gladman SL, Bulach DM, Hoffmann B, Catmull DV, Pruilh S, Johnson S, Gibbs L, Amezdroz E, Bhatnagar U, Seemann T, Mnatzaganian G, Manton DJ, Reynolds EC. 2019. Temporal development of the oral microbiome and prediction of early childhood caries. *Sci Rep* 9(1):19732.

Elashiry M, Morandini AC, Cornelius Timothius CJ, Ghaly M, Cutler CW. 2021. Selective antimicrobial therapies for periodontitis: Win the "battle and the war". *Int J Mol Sci* 22(12):6459.

Fernando S, Speicher DJ, Bakr MM, Benton MC, Lea RA, Scuffham PA, Mihala G, Johnson NW. 2015. Protocol for assessing maternal, environmental and epigenetic risk factors for dental caries in children. *BMC Oral Health* 15:167.

Ferrer MD, López-López A, Nicolescu T, Perez-Vilaplana S, Boix-Amorós A, Dzidic M, Garcia S, Artacho A, Llena C, Mira A. 2020a. Topic application of the probiotic *Streptococcus dentisani* improves clinical and microbiological parameters associated with oral health. *Front Cell Infect Microbiol* 10:465.

Ferrer MD, López-López A, Nicolescu T, Salavert A, Méndez I, Cuñé J, Llena C, Mira A. 2020b. A pilot study to assess oral colonization and pH buffering by the probiotic *Streptococcus dentisani* under different dosing regimes. *Odontology* 108:180–187.

Flichy-Fernández AJ, Ata-Ali J, Alegre-Domingo T, Candel-Martí E, Ata-Ali F, Palacio JR, Peñarrocha-Diago M. 2015. The effect of orally administered probiotic *Lactobacillus reuteri*-containing tablets in peri-implant mucositis: A double-blind randomized controlled trial. *J Periodontal Res* 50:775–785.

Galofré M, Palao D, Vicario M, Nart J, Violant D. 2018. Clinical and microbiological evaluation of the effect of *Lactobacillus reuteri* in the treatment of mucositis and peri-implantitis: A triple-blind randomized clinical trial. *J Periodontal Res* 53:378–390.

Giannini G, Ragusa I, Nardone GN, Soldi S, Elli M, Valenti P, Rosa L, Marra E, Stoppoloni D, Merlo Pich E. 2022. Probiotics-containing mucoadhesive gel for targeting the dysbiosis associated with periodontal diseases. *Int J Dent* 2022:5007930.

Glavina D, Gorseta K, Skrinjarić I, Vranić DN, Mehulić K, Kozul K. 2012. Effect of LGG yoghurt on *Streptococcus mutans* and *Lactobacillus* spp. salivary counts in children. *Coll Antropol* 36:129–132.

Glick M, Williams D, Kleinman D, Vujicic M, Watt R, Weyant R. 2016. A new definition for oral health developed by the FDI World Dental Federation opens the door to a universal definition of oral health. *JADA* 147:915–917.

Grusovin MG, Bossini S, Calza S, Cappa V, Garzetti G, Scotti E, Gherlone EF, Mensi M. 2020. Clinical efficacy of *Lactobacillus reuteri*-containing lozenges in the supportive therapy of generalized periodontitis stage III and IV, grade C: 1-year results of a double-blind randomized placebo-controlled pilot study. *Clin Oral Investig* 24:2015–2024.

Hallström H, Lindgren S, Widen C, Renvert S, Twetman S. 2016. Probiotic supplements and debridement of peri-implant mucositis: A randomized controlled trial. *Acta Odontol Scand* 74:60–66.

Han X, Zhang J, Tan Y, Zhou G. 2017. Probiotics: A non-conventional therapy for oral lichen planus. *Arch Oral Biol* 81:90–96.

Hatakka K, Ahola AJ, Yli-Knuuttila H, Richardson M, Poussa T, Meurman JH, Korpela R. 2007. Probiotics reduce the prevalence of oral *Candida* in the elderly—a randomised controlled trial. *J Dent Res* 86:125–130.

He Y, Gong D, Shi C, Shao F, Shi J, Fei J. 2017. Dysbiosis of oral buccal mucosa microbiota in patients with oral lichen planus. *Oral Dis* 23:674–682.

Heeney DD, Zhai Z, Bendiks Z, Barouei J, Martinic A, Slupsky C, Marco ML. 2019. *Lactobacillus plantarum* bacteriocin is associated with intestinal and systemic improvements in diet-induced obese mice and maintains epithelial barrier integrity *in vitro*. *Gut Microbes* 10:382–397.

Iniesta M, Herrera D, Montero E, Zurbriggen M, Matos AR, Marin MJ, Sanchez-Beltran MC, Llama-Palacio A, Sanz M. 2012. Probiotic effects of orally administered *Lactobacillus reuteri*-containing tablets on the subgingival and salivary microbiota in patients with gingivitis. A randomized clinical trial. *J Clin Periodontol* 39:736–744.

Invernici MM, Salvador SL, Silva PHF, Soares MSM, Casarin R, Palioto DB, Souza SLS, Taba M, Jr., Novaes AB, Jr., Furlaneto FAC, Messora MR. 2018. Effects of *Bifidobacterium* probiotic on the treatment of chronic periodontitis: A randomized clinical trial. *J Clin Periodontol* 45:1198–1210.

Invernici MM, Furlaneto FAC, Salvador SL, Ouwehand AC, Salminen S, Mantziari A, Vinderola G, Ervolino E, Santana SI, Silva PHF, Messora MR. 2020. *Bifidobacterium animalis* subsp *lactis* HN019 presents antimicrobial potential against periodontopathogens and modulates the immunological response of oral mucosa in periodontitis patients. *PLoS ONE* 15:e0238425.

Keller MK, Brandsborg E, Holmstrøm K, Twetman S. 2018a. Effect of tablets containing probiotic candidate strains on gingival inflammation and composition of the salivary microbiome: A randomised controlled trial. *Benef Microbes* 9:487–494.

Keller MK, Kragelund C. 2018b. Randomized pilot study on probiotic effects on recurrent candidiasis in oral lichen planus patients. *Oral Dis* 24:1107–1114.

Kuru BE, Laleman I, Yalnızoğlu T, Kuru L, Teughels W. 2017. The influence of a *Bifidobacterium animalis* probiotic on gingival health: A randomized controlled clinical trial. *J Periodontol* 88:1115–1123.

Kwon T, Lamster IB, Levin L. 2021. Current concepts in the management of periodontitis. *Int Dent J* 71:462–476.

Lai S, Lingström P, Cagetti MG, Cocco F, Meloni G, Arrica MA, Campus G. 2021. Effect of *Lactobacillus brevis* CD2 containing lozenges and plaque pH and cariogenic bacteria in diabetic children: A randomised clinical trial. *Clin Oral Investig* 25:115–123.

Laleman I, Pauwels M, Quirynen M, Teughels W. 2020a. A dual-strain *Lactobacilli reuteri* probiotic improves the treatment of residual pockets: A randomized controlled clinical trial. *J Clin Periodontol* 47:43–53.

Laleman I, Pauwels M, Quirynen M, Teughels W. 2020b. The usage of a lactobacilli probiotic in the non-surgical therapy of peri-implantitis: A randomized pilot study. *Clin Oral Implants Res* 31:84–92.

Laleman I, Yilmaz E, Ozcelik O, Haytac C, Pauwels M, Herrero ER, Slomka V, Quirynen M, Alkaya B, Teughels W. 2015. The effect of a streptococci containing probiotic in periodontal therapy: A randomized controlled trial. *J Clin Periodontol* 42(11):1032–1041.

Lee JK, Kim SJ, Ko SH, Ouwehand AC, Ma DS. 2015. Modulation of the host response by probiotic *Lactobacillus brevis* CD2 in experimental gingivitis. *Oral Dis* 21:705–712.

Lee X, Vergara C, Lozano CP. 2019. Severity of *Candida*-associated denture stomatitis is improved in institutionalized elders who consume *Lactobacillus rhamnosus* SP1. *Austr Dent J* 64:229–236.

Lexner MO, Blomqvist S, Dahlen G, Twetman S. 2010. Microbiological profiles in saliva and supragingival plaque from caries-active adolescents before and after a short-term daily intake of milk supplemented with probiotic bacteria—a pilot study. *Oral Health Prev Dent* 8:383–388.

Li Y, Wang K, Zhang B, Tu QC, Yao YF, Cui BM, Ren B, He JZ, Shen X, van Nostrand JD, Zhou JZ, Shi WY, Xiao LY, Lu CQ, Zhou XD. 2019. Salivary mycobiome dysbiosis and its potential impact on bacteriome shifts and host immunity in oral lichen planus. *Int J Oral Sci* 11:13.

Li J, Sung CY, Lee N, Ni Y, Pihlajamaki J, Panagiotou G, El-Nezami H. 2016. Probiotics modulated gut microbiota suppresses hepatocellular carcinoma growth in mice. *Proc Natl Acad Sci USA* 113:e1306–e1315.

Liu H, Chen H, Liao Y, Li H, Shi L, Deng Y, Shen X, Song Z. 2021. Comparative analyses of the subgingival microbiome in chronic periodontitis patients with and without gingival erosive oral lichen planus based on 16S rRNA gene sequencing. *Biomed Res Int* 2021:9995225.

Manmontri C, Nirunsittirat A, Piwat S, Wattanarat O, Pahumunto N, Makeudom A, Sastraruji T, Krisanaprakornkit S, Teanpaisan R. 2020. Reduction of *Streptococcus mutans* by probiotic milk: A multicenter randomized controlled trial. *Clin Oral Investig* 24:2363–2374.

Mann S, Park MS, Johnston TV, Ji GE, Hwang KT, Ku S. 2021. Isolation, characterization and biosafety evaluation of *Lactobacillus fermentum* OK with potential oral probiotic properties. *Probiotics Antimicrob Proteins* 13:1363–1386.

Marlina E, Goodman RN, Mercadante V, Shephard M, McMillan R, Hodgson T, Leeson R, Porter S, Barber JA, Fedele S, Smith AM. 2022. A proof of concept pilot trial of probiotics in symptomatic oral lichen planus (CABRIO). *Oral Dis* 28:2155–2167.

Marsh PD. 2004. Dental plaque as a microbial biofilm. *Caries Res* 38:204–211.

Marsh PD, Head DA, Devine DA. 2015. Ecological approaches to oral biofilms: Control without killing. *Caries Res* 49(Suppl 1):46–54.

Masdea L, Kulik EM, Hauser-Gerspach I, Ramseier AM, Filippi A, Waltimo T. 2012. Antimicrobial activity of *Streptococcus salivarius* K12 on bacteria involved in oral malodour. *Arch Oral Biol* 57:1041–1047.

McLaren MR, Callahan BJ. 2018. In nature, there is only diversity. *mBio* 9:e02149–17.

Meurman JH. 2005. Probiotics: Do they have a role in oral medicine and dentistry? *Eur J Oral Sci* 113:188–196.

Miyazima TY, Ishikawa KH, Mayer M, Saad S, Nakamae A. 2017. Cheese supplemented with probiotics reduced the *Candida* levels in denture wearers-RCT. *Oral Dis* 23:919–925.

Morales A, Contador R, Bravo J, Carvajal P, Silva N, Strauss FJ, Gamonal J. 2021. Clinical effects of probiotic or azithromycin as an adjunct to scaling and root planning in the treatment of stage III periodontitis: A pilot randomized controlled clinical trial. *BMC Oral Health* 21:12.

Näse L, Hatakka K, Savilahti E, Saxelin M, Pönkä A, Poussa T, Korpela R, Meurman JH. 2001. Effect of long-term consumption of a probiotic bacterium, *Lactobacillus rhamnosus* GG, in milk on dental caries and caries risk in children. *Caries Res* 35:412–420.

Nguyen T, Brody H, Radaic A, Kapila Y. 2021. Probiotics for periodontal health-Current molecular findings. *Periodontology 2000* 87:254–267.

O'Connell LM, Santos R, Springer G, Burne RA, Nascimento MM, Richards VP. 2020. Site-specific profiling of the dental mycobiome reveals strong taxonomic shifts during progression of early-childhood caries. *Appl Environ Microbiol* 86:e02825–19.

Oliveira VM, Santos SS, Silva CR, Jorge AO, Leao MV. 2016. *Lactobacillus* is able to alter the virulence and the sensitivity profile of *Candida albicans*. *J Appl Microbiol* 121:1737–1744.

Pahumunto N, Sophatha B, Piwat S, Teanpaisan R. 2019. Increasing salivary IgA and reducing *Streptococcus mutans* by probiotic *Lactobacillus paracasei* SD1: A double-blind, randomized, controlled study. *J Dent Sci* 14:178–184.

Pahumunto N, Piwat S, Chankanka O, Akkarachaneeyakorn N, Rangsitsathian K, Teanpaisan R. 2018. Reducing mutans streptococci and caries development by *Lactobacillus paracasei* SD1 in preschool children: A randomized placebo-controlled trial. *Acta Odontol Scand* 76:331–337.

Park SY, Hwang BO, Lim M, Ok SH, Lee SK, Chun KS, Park KK, Hu Y, Chung WY, Song NY. 2021 Oral-gut microbiome axis in gastrointestinal disease and cancer. *Cancers (Basel)* 13:2124.

Patil RU, Nachan VP, Patil SS, Mhaske RV. 2021. A clinical trial on topical effect of probiotics on oral *Streptococcus mutans* counts in children. *J Indian Soc Pedod Prev Dent* 39:279–283.

Patyna M, Ehlers V, Bahlmann B, Kasaj A. 2021. Effects of adjunctive light-activated disinfection and probiotics on clinical and microbiological parameters in periodontal treatment: A randomized, controlled, clinical pilot study. *Clin Oral Investig* 25:3967–3975.

Pelekos G, Ho SN, Acharya A, Leung WK, McGrath C. 2019. A double-blind, paralleled-arm, placebo-controlled and randomized clinical trial of the effectiveness of probiotics as an adjunct in periodontal care. *J Clin Periodontol* 46(12):1217–1227.

Pelekos G, Acharya A, Eiji N, Hong G, Leung WK, McGrath C. 2020. Effects of adjunctive probiotic *L. reuteri* lozenges on S/RSD outcomes at molar sites with deep pockets. *J Clin Periodontol* 47:1098–1107.

Peña M, Barallat L, Vilarrasa J, Vicario M, Violant D, Nart J. 2019. Evaluation of the effect of probiotics in the treatment of peri-implant mucositis: A triple-blind randomized clinical trial. *Clin Oral Investig* 23(4):1673–1683.

Petersson LG, Magnusson K, Hakestam U, Baiqi A, Twetman S. 2011. Reversal of primary root caries lesions after daily intake of milk supplemented with fluoride and probiotic lactobacilli in older adults. *Acta Odontol Scand* 69:321–327.

Piwat S, Teanpaisan R, Manmontri C, Wattanarat O, Pahumunto N, Makeudom A, Krisanaprakornkit S, Nirunsittirat A. 2020. Efficacy of probiotic milk for caries regression in preschool children: A multicenter randomized controlled trial. *Caries Res* 54:491–501.

Ratna Sudha M, Neelamraju J, Surendra Reddy M, Kumar M. 2020. Evaluation of the effect of probiotic *Bacillus coagulans* Unique IS2 on mutans streptococci and lactobacilli levels in saliva and plaque: A double-blind, randomized, placebo-controlled study in children. *Int J Dentistry* 2020:8891708.

Ribeiro FC, Rossoni RD, de Barros PP, Santos JD, Fugisaki LRO, Leao MPV, Junqueira JC. 2020. Action mechanisms of probiotics on *Candida* spp. and candidiasis prevention: An update. *J Appl Microbiol* 129:175–185.

Richards VP, Alvarez AJ, Luce AR, Bedenbaugh M, Mitchell ML, Burne RA, Nascimento MM. 2017. Microbiomes of site-specific dental plaques from children with different caries status. *Infect Immun* 85(8):e00106–17.

Rodriguez G, Ruiz B, Faleiros S, Vistoso A, Marro ML, Sanchez J, Urzua I, Cabello R. 2016. Probiotic compared with standard milk for high-caries children: A cluster randomized trial. *J Dent Res* 95:402–407.

Rossoni RD, Fuchs BB, de Barros PP, Velloso MD, Jorge AO, Junqueira JC, Mylonakis E. 2017. *Lactobacillus paracasei* modulates the immune system of *Galleria mellonella* and protects against *Candida albicans* infection. *PLoS ONE* 12:e0173332.

Sajedinejad N, Paknejad M, Houshmand B, Sharafi H, Jelodar R, Shahbani Zahiri H, Noghabi KA. 2018. *Lactobacillus salivarius* NK02: A potent probiotic for clinical application in mouthwash. *Probiotics Antimicrob Proteins* 10:485–495.

Santana SI, Silva PHF, Salvador SL, Casarin RCV, Furlaneto FAC, Messora MR. 2022. Adjuvant use of multispecies probiotic in the treatment of peri-implant mucositis: A randomized controlled trial. *J Clin Periodontol* 49(8):828–839.

Sargolzaei N, Arab H, Gerayeli M, Ivani F. 2022. Evaluation of the topical effect of probiotic mouthwash in the treatment of patients with peri-implant mucositis. *J Long Term Eff Med Implants* 32(1):85–91.

Schlagenhauf U, Rehder J, Gelbrich G, Jockel-Schneider Y. 2020. Consumption of *Lactobacillus reuteri*-containing lozenges improves periodontal health in navy sailors at sea: A randomized controlled trial. *J Periodontol* 91:1328–1338.

Selwitz RH. Ismail AI, Pitts NB. 2007. Dental caries. *Lancet* 369:51–59.

Seminario-Amez M, Lopez-Lopez J, Estrugo-Devesa A, Ayuso-Montero R, Jane-Salas E. 2017. Probiotics and oral health: A systematic review. *Med Oral Patol Oral Cir Bucal* 1:e282–e288.

Soares LG, Carvalho EB, Tinoco EMB. 2019. Clinical effect of *Lactobacillus* on the treatment of severe periodontitis and halitosis: A double-blinded, placebo-controlled, randomized clinical trial. *Am J Dent* 32(1):9–13.

Solbiati J, Frias-Lopez J. 2018. Metatranscriptome of the oral microbiome in health and disease. *J Dent Res* 97(5):492–500.

Stecksén-Blicks C, Sjöström I, Twetman S. 2009. Effect of long-term consumption of milk supplemented with probiotic lactobacilli and fluoride on dental caries and general health in preschool children: A cluster-randomized study. *Caries Res* 43:374–381.

Stensson M, Koch G, Coric S, Abrahamsson TR, Jenmalm MC, Birkhed D, Wendt LK. 2014. Oral administration of *Lactobacillus reuteri* during the first year of life reduces caries prevalence in the primary dentition at 9 years of age. *Caries Res* 48:111–117.

Tada H, Masaki C, Tsuka S, Mukaibo T, Kondo Y, Hosokawa R. 2018. The effects of *Lactobacillus reuteri* probiotics combined with azithromycin on peri-implantitis: A randomized placebo-controlled study. *J Prosthodont Res* 62(1):89–96.

Tekce M, Ince G, Gursoy H, Dirikan Ipci S, Cakar G, Kadir T, Yılmaz S. 2015. Clinical and microbiological effects of probiotic lozenges in the treatment of chronic periodontitis: A 1-year follow-up study. *J Clin Periodontol* 42(4):363–372.

Teughels W, Durukan A, Ozcelik O, Pauwels M, Quirynen M, Haytac MC. 2013. Clinical and microbiological effects of *Lactobacillus reuteri* probiotics in the treatment of chronic periodontitis: A randomized placebo-controlled study. *J Clin Periodontol* 40:1025–1035.

Theodoro LH, Cláudio MM, Nuernberg MAA, Miessi DMJ, Batista JA, Duque C, Garcia VG. 2019. Effects of *Lactobacillus reuteri* as an adjunct to the treatment of periodontitis in smokers: Randomised clinical trial. *Benef Microbes* 10:375–384.

Twetman S, Jørgensen MR. 2021. Can probiotic supplements prevent early childhood caries? A systematic review and meta-analysis. *Benef Microbes* 12:231–238.

Vicario M, Santos A, Violant D, Nart J, Giner L. 2013. Clinical changes in periodontal subjects with the probiotic *Lactobacillus reuteri* Prodentis: A preliminary randomized clinical trial. *Acta Odontol Scand* 71:813–819.

Villavicencio J, Villegas LM, Arango MC, Arias S, Triana F. 2018. Effects of a food enriched with probiotics on *Streptococcus mutans* and *Lactobacillus* spp. salivary counts in preschool children: A cluster randomized trial. *J Appl Oral Sci* 26:e20170318.

Vohra F, Bukhari IA, Sheikh SA, Albaijan R, Naseem M, Hussain M. 2020. Effectiveness of scaling and root planing with and without adjunct probiotic therapy in the treatment of chronic periodontitis among shamma users and non-users: A randomized controlled trial. *J Periodontol* 91:1177–1185.

Wang X, Zhao Z, Tang N, Zhao Y, Xu J, Li L, Qian L, Zhang J, Fan Y. 2020. Microbial community analysis of saliva and biopsies in patients with oral lichen planus. *Front Microbiol* 11:629.

Wattanarat O, Nirunsittirat A, Piwat S, Manmontri C, Teanpaisan R, Pahumunto N, Makeudom A, Sastraruji T, Krisanaprakornkit S. 2021. Significant elevation of salivary human neutrophil peptides 1–3 levels by probiotic milk in preschool children with severe early childhood caries: A randomized controlled trial. *Clin Oral Investig* 25:2891–2903.

Xu X, He J, Xue J, Wang Y, Li K, Zhang K, Guo Q, Liu X, Zhou Y, Cheng L, Li M, Li Y, Li Y, Shi W, Zhou X. 2015. Oral cavity contains distinct niches with dynamic microbial communities. *Environ Microbiol* 17:699–710.

Zhong EF, Chang A, Stucky A, Chen X, Mundluru T, Khalifeh M, Sedghizadeh PP. 2020. Genomic analysis of oral lichen planus and related oral microbiome pathogens. *Pathogens* 9:952.

# Human Studies on Probiotics and Endogenous Lactic Acid Bacteria in the Female Urogenital Tract

# 28

Scarlett Puebla-Barragan and Gregor Reid

## 28.1 INTRODUCTION

Lactic acid bacteria (LAB) play a critical role in maintaining urogenital health by reducing the risk of bacterial vaginosis (BV), urinary tract infection (UTI) and vulvovaginal candidiasis (VVC) (Borges et al. 2014; Stapleton 2016). Recent advancements in culturing and molecular techniques have led to a better understanding of the microbial composition of the female urogenital tract. *Lactobacillus acidophilus*, once regarded as the primary species present in the healthy vagina, is in fact not common compared to *Lactobacillus crispatus, Lactobacillus jensenii, Lactobacillus gasseri* and *Lactobacillus iners*, although the latter is also present in dysbiotic states (Tamrakar et al. 2007; Kim et al. 2009; Yamamoto et al. 2009; Forney et al. 2010; Hummelen et al. 2010; Zhou et al. 2010; Zozaya-Hinchliffe et al. 2010; Reid 2016; Petrova et al. 2017). Exactly how *L. iners*, with the smallest genome of LAB reported to date (Macklaim et al. 2010; France et al. 2020), can persist when other lactobacilli have been eliminated remains to be determined, as does the extent of its role in the re-establishment of a healthy state naturally or after antimicrobial use. It has led to the suggestion that there may be clones that somehow confer healthy outcomes and others that may increase the risk of BV (Petrova et al. 2017).

A group of mixed anaerobes, particularly *Gardnerella, Prevotella* and *Atopobium* genera, often in biofilms, are associated with vaginal dysbiosis, particularly BV (Antonio et al. 1999; Dumonceaux et al. 2009; Forney et al. 2010; Srinivasan et al. 2010; Yamamoto et al. 2009; Zhou et al. 2010; McMillan et al. 2015; Lev-Sagie et al. 2022). These can exist without clinical signs or symptoms, raising the question of whether

DOI: 10.1201/9781003352075-31

BV is an infectious or inflammatory state or both. This chapter explores the female urogenital microbiota and the potential use of lactobacilli, particularly as probiotics, for maintaining health and preventing or treating disease.

# 28.2 ENDOGENOUS LAB

## 28.2.1 Microbiota of the Vagina

Culture-independent techniques revolutionized microbiota studies by identifying previously undetected microorganisms in the vagina (Burton and Reid 2002). PCR amplification of 16S rRNA genes followed by denaturing gradient gel electrophoresis detected *L. iners* and *A. vaginae*, which are difficult to identify with culture-based methods (Burton et al. 2003, 2004). Metagenomic community profiling techniques have also been used to deduce bacterial species composition (Yamamoto et al. 2009; Hill et al. 2005; Schellenberg et al. 2009; Srinivasan et al. 2010). However, metagenomic analyses face challenges such as technical variations and the need for careful bioinformatic analyses (Brooks et al. 2015; Gloor and Reid 2016).

Hundreds of bacterial species have been identified in vaginal samples including aerobes, facultative and obligate anaerobes (France et al. 2020). However, the core members of the microbiota are relatively few, depending on whether the environment is healthy or in a state of dysbiosis. Homeostasis is affected by hormones, sexual acts, antimicrobials and a range of environmental and host factors.

Although lactobacilli typically dominate in healthy vagina, a study by Kim et al. (2009) found that one out of eight "healthy" subjects had a microbiota that was not lactobacilli dominated but instead contained significant numbers of *Gardnerella vaginalis*, a bacterium typically associated with BV. This indicates that some BV-like microbiotas do not induce clinical symptoms.

The dysbiotic state associated with the vagina has been termed BV for some time, but this is really an umbrella term that includes a variety of conditions, some symptomatic, some infective and others that produce various metabolites such as gamma hydroxybutyrate and amines (McMillan et al. 2015; Nasioudis et al. 2017; Reid 2017a). Hopefully in due course, BV will be renamed and split into appropriate conditions (Reid 2017a).

## 28.2.2 Microbiota of the Urinary Tract

The discovery of a microbiota in the urinary tract of healthy women has changed the perception of the anatomical region, previously regarded as being sterile (Siddiqui et al. 2011; Wolfe et al. 2012). However, the presence and function of apparently sedentary microorganisms in the bladder and even the kidneys remain unclear (Lewis et al. 2013; Whiteside et al. 2015). Bacteria are known to pass between the vagina and male urethra during sexual intercourse, sometimes requiring male partners to be treated when the female has BV (Nelson et al. 2012; Zozaya et al. 2016).

# 28.3 PROBIOTICS FOR UROGENITAL HEALTH

Probiotics (defined as live microorganisms that, when administered in adequate amounts, confer a health benefit on the host) (FAO/WHO 2001; Hill et al. 2014) have been studied for the prevention of UTI, BV and VVC. At the FDA, they have chosen not yet to adopt the probiotic term except for supplements and foods and currently refer to probiotics for disease prevention and treatment as live biotherapeutics.

## 28.3.1 Origin and Characteristics of Probiotic and Candidate Probiotic Strains

Clinical observations 50 years ago showed that lactobacilli were abundant in the vagina of women who had never experienced a UTI (Bruce et al. 1973), while *E. coli* and uropathogens dominated in women with the disease. This led to the concept of using lactobacilli to displace the uropathogens and restore lactobacilli (Bruce and Reid 1988). From 1985 to 2002, candidate probiotic lactobacilli strains were selected based on their ability to produce antimicrobial substances, adhere to epithelial cells and inhibit urogenital pathogens *in vitro*, as these were deemed attributes suitable for the purpose (Reid et al. 1987; McLean and Rosenstein 2000; Osset et al. 2001; Mastromarino et al. 2002, 2009; Reid 2017b). Hydrogen peroxide, which is produced by lactobacilli, was of particular interest because of its antibacterial properties and isolation in strains from healthy women (Eschenbach et al. 1989; McGroarty et al. 1992). However, studies have since shown that it is not a critical factor in preventing infection or disrupting *G. vaginalis* biofilms (Patterson et al. 2007; Saunders et al. 2007).

Some candidate probiotic lactobacilli strains were tested based on their ability to adhere to epithelial cells, with a view to increasing their persistence when administered as a probiotic. However, only one study has demonstrated a correlation between *in vitro* adherence and increased *in vivo* persistence (Reid et al. 1995). Rather, it appears that probiotic lactobacilli do not persist in the vagina, and it is their beneficial metabolites produced during their stay that are considered more important. As a recent paper indicated, it is quite remarkable that so few of the 200 and more species that belonged to the genus formerly known as *Lactobacillus* are able to so effectively inhabit the vagina (Reid 2023a). How they have adapted to do this is unclear, but they are able to survive low pH, hormonal changes, modulate host responses, outcompete and displace pathogens and survive passage through the gastrointestinal tract if administered orally (Christensen et al. 2002; Burton et al. 2003; Reid et al. 2009). Various mechanisms have been proposed, including disrupting pathogen-containing biofilms (Saunders et al. 2007), interrupting bacteria-bacteria signaling and toxin production (Laughton et al. 2006), inhibiting pathogens through coaggregation (Ekmekci et al. 2009) and producing biosurfactants (Velraeds et al. 1998; Reid et al. 2011).

As an antimicrobial, lactic acid itself, produced by LAB, inhibits some pathogens. However, its spectrum of activity is not sufficiently broad to inhibit or kill pathogens such as enterococci that survive acidic conditions.

Studies have shown that lactobacilli can modulate immunity in favor of responses protective of the host (Kirjavainen et al. 2008; Yeganegi et al. 2010; Rose et al. 2012), including inducing granulocyte colony-stimulating factor (Martins et al. 2009), defensins and IL-8 neutrophil recruitment (Kirjavainen et al. 2008). Immune responses also appear to influence defense against *Candida albicans* (Rizzo et al. 2013), as does increased expression of stress-related genes in fungi exposed to probiotic *Lacticaseibacillus* (formerly *Lactobacillus*) *rhamnosus* GR-1 (Kohler et al. 2012).

## 28.3.2 Application of Probiotics to the Urogenital Tract

*Limosilactobacillus reuteri* RC-14 (previously identified as *Lactobacillus fermentum* and *L. acidophilus*) and *L. rhamnosus* GR-1 are the most studied as probiotics for vaginal health (Chan et al. 1984; Reid et al. 1987; Reid 2017c). Following early clinical studies with *L. rhamnosus* GR-1 on its own and with *L. fermentum* B-54, the *L. reuteri* RC-14 strain replaced B-54 because of its greater *in vitro* activity against Gram-positive pathogens, thereby complementing the activity of GR-1 against Gram-negative pathogens. Clinical studies showed that both strains can be detected in the vagina several weeks after application, making them more effective than gastrointestinal probiotic *L. rhamnosus* GG (Reid et al. 1994; Cadieux et al. 2002; Gardiner et al. 2002; Morelli et al. 2004). The ability of *L. rhamnosus* GR-1 and *L. reuteri* RC-14 to cure BV following daily intravaginal administration (Anukam et al. 2006a, 2006b) was an excellent proof-of-concept study, albeit small and not yet repeated. BV organisms are present in dense biofilms that are recalcitrant to antimicrobial therapy (Swidsinski et al. 2008), and thus while abundance levels may fall post-antibiotic, the pathogens are not fully displaced or killed (Hummelen et al. 2010). Therefore, the intent of the probiotic is to reduce and displace the pathogens and create a favorable environment for indigenous lactobacilli to recover. Oral application can reduce the passage of pathogens from the rectum to the perineum and vagina (Reid et al. 2003). At the very least, the addition of probiotics to antimicrobial therapy should be used to reduce the risk of disease recurrence and side effects of the drugs (Marcone et al. 2008, 2010; Martinez et al. 2009a, 2009b).

Of note, the intent is not for the probiotic to take over the niche. Rather, it is to restore the health status and allow the woman's own protective lactobacilli to return, including after antibiotic treatment (Macklaim et al. 2015).

Other strains have been documented for vaginal application. These include the following products—(i) *L. acidophilus* A-212, *L. rhamnosus* A-119 and *Streptococcus thermophilus* A-336; (ii) *L. rhamnosus* Lcr35; and (iii) *L. rhamnosus* PBO1 and *L. gasseri* EN-153471 (Kern et al. 2012; Larsson et al. 2011; Marcone et al. 2010; Ya et al. 2010). Although not yet clinically available, a product called Lactin V containing *L. crispatus* CTV-05 has shown promise in preventing recurrence of UTI and BV (Hemmerling et al. 2009; Stapleton et al. 2011; Cohen et al. 2020). It is not clear if the plan is to treat BV with antibiotics, then administer CTV-05, which would not be preferable given the poor efficacy of antibiotics. Since there is evidence that the strain can reduce genital inflammation and apparently enhance epithelial integrity (Armstrong et al. 2022), it may be useful prophylactically for women prone to recurrent dysbiosis as well as those at risk of HIV infection.

Many women are afflicted by BV, which may be accompanied by vaginal discomfort and homogeneous malodorous vaginal discharge. Malodor is caused primarily by an increase in biogenic amines, including cadaverine, putrescine, tyramine and trimethylamine, which are mostly produced by pathogens metabolizing amino acids during dysbiosis-related conditions, such as UTI and BV (McMillan et al. 2015; Borgogna et al. 2021; Puebla-Barragan et al. 2020). The potential for a probiotic strain to reduce malodor by breaking down these compounds has been explored. Some vaginal *L. crispatus* isolates were shown to break down amines *in vitro* (Puebla-Barragan et al. 2021). This presents an opportunity for the development of topical products with amine-degrading strains. However, it is important to take inter-strain variability into consideration, since some *L. crispatus* strains can also produce biogenic amines, making them unsuitable for probiotic use (Puebla-Barragan et al. 2021).

The detection of malodor has been a component of the Amsel criteria, a diagnostic method for BV used for many decades. It requires three out of four parameters: milky, homogenous discharge; vaginal pH over 4.5; presence of "clue cells" on microscopic examination of vaginal smear; and/or a positive amine or "Whiff" test (Amsel et al. 1983). The alternative Nugent scoring system only involves performing a Gram stain on a vaginal smear and counting the number of Gram-positive rods versus Gram-negative rods and other bacterial morphotypes (Nugent et al. 1991).

In VVC, infection with *Candida* sp. is not due to absence of lactobacilli and cannot be treated with probiotics alone. Rather, antifungal therapy is needed before probiotics can be considered for preventing recurrences. Having stated that, two studies have demonstrated that *L. rhamnosus* GR-1 and *L. reuteri* RC-14, when used in conjunction with traditional antifungal therapy (fluconazole), can improve cure rate and reduce recurrences (Anukam et al. 2009; Martinez et al. 2009b). In addition, *in vitro* cell culture studies have shown that *L. reuteri* RC-14 alone or in combination with *L. rhamnosus* GR-1 can decrease the number of yeast cells recoverable following treatment, as well as increasing levels of the antimicrobial cytokine IL-8 (Kirjavainen et al. 2008; Martinez et al. 2009c). Another study showed that cocultures of GR-1 and RC-14 with *C. albicans* caused the yeast cells to lose metabolic activity and die (Kohler et al. 2012). Not only that, but as mentioned, the treatment caused an increase in expression of stress-related genes and reduced expression of genes involved in fluconazole resistance, thereby potentially explaining why the probiotic and drug therapy improved eradication of *Candida*. A recent clinical trial of 93 VVC patients treated with *L. crispatus* DSM32720, DSM32718 and DSM32716 orally and intravaginally resulted in reduced discharge and itching/irritation (Mandar et al. 2023). In addition, 89 patients with BV treated orally with *L. crispatus* DSM32717 and DSM32720 had improved Nugent score and lower malodor, discharge and itching/irritation. Despite the mechanisms of these strains not being well investigated, the study illustrates efforts are being made to improve urogenital health.

Another study, albeit in mice, has shown that a probiotic yeast, *Saccharomyces cerevisiae* CNCM I-3856, inserted into the vagina inhibits the expression of *C. albicans* virulent factors, aspartyl proteinases and hyphae-associated proteins Hwp1 and Ece, as well as suppressing the influx of neutrophils (Gabrielli et al. 2018).

With respect to UTI, the prevalence remains high despite the widespread use of antibiotics, affecting millions of women worldwide each year. Probiotics have been investigated as a preventative measure to establish a protective vaginal microbiota that can impede uropathogens from ascending to the bladder. Studies on women with recurrent UTIs have shown that intravaginal implantation of lactobacilli can lead to infection-free periods and fewer recurrences (Bruce and Reid 1988; Reid et al. 1995).

A clinical approach to reducing UTI recurrences, especially in children, is daily antibiotic administration. However, a study comparing daily antibiotics to lactobacilli prophylaxis in children found that the latter therapy was equally effective (Lee et al. 2007). Similarly, oral administration of *L. rhamnosus* GR-1 and *L. reuteri* RC-14 for 1 year reduced the number of symptomatic UTI in elderly women (Beerepoot et al. 2012). These results suggest that restoring lactobacilli in the vagina of women with a history of infection and antibiotic use can be clinically beneficial. However, more clinical trials are needed to prove conclusively that probiotic administration can prevent recurrence of UTI in adults and children (Abdullatif et al. 2021; Emami et al. 2022; Reid 2023b).

## 28.4 FUTURE CHALLENGES AND OPPORTUNITIES

There are several important challenges in order to improve the urogenital health of women. The first is to understand the factors that trigger changes in the vaginal and urinary microbiotas resulting in a healthy state changing to one that causes signs and symptoms of disease. Some of these are known, but the precise mechanisms are not proven, nor is the reason some dysbiosis episodes do not result in clinical infection. Novel technologies might help, such as the "vagina chip" (Mahajan et al. 2022).

Second, an accurate diagnostic test is needed preferably to identify an aberrant vaginal microbiome and host response that can increase the risk of serious infection and preterm labor. This test must differentiate between transient, asymptomatic states and actual disease (Reid 2017a). Artificial intelligence may help in this respect by analyzing as many data points as can be detected (microbiota, metabolites, host factors, symptoms) (Wani et al. 2022).

Third, while more biotic products are becoming available, few have well-documented strains whose properties have been specifically selected for restoration and maintenance of urogenital homeostasis. We are a long way from personalized treatment, but knowing the organisms associated with a woman's healthy state, particularly the *Lactobacillus* species, could narrow the field down to not simply replenishing that strain (as some products appear to be trying to do) but more importantly to counter the main problem. This could be an inflammatory state, overgrowth of pathogenic biofilms, persistent odor, a breakdown in the structure of the indigenous consortium or continuous seeding of a pathogen from the rectum or sexual partner. Options should include intravaginal, oral and locally applied treatments. As these require regulation as drugs under the antiquated regulatory process in some countries, governments should provide incentives and a fast-track route to make products available for safety and efficacy studies followed by consumer access.

In conclusion, the pace of research on the female urogenital tract has been too slow for too long. Hundreds of millions of females around the world annually suffer from these conditions. Diagnosis and treatment are sub-optimal, and recurrences impact quality of life, sexual partnerships and reproduction. It would be good if this chapter and others stimulated change, but the message has already been conveyed for too many decades. For those who continue to undertake research in this field, you are to be applauded. Hopefully, your work will bring remedies to improve the health of women and girls worldwide. In that regard, products need to be affordable to everyone, including those who are economically and socially challenged through no fault of their own in developed and developing countries.

# BIBLIOGRAPHY

Abdullatif, V.A., R.L. Sur, E. Eshaghian, K.A. Gaura, B. Goldman, P.K. Panchatsharam, N.J. Williams, and J.E. Abbott. 2021. Efficacy of probiotics as prophylaxis for urinary tract infections in premenopausal women: A systematic review and meta-analysis. *Cureus. 13*(10): E18843.

Amsel, R., P.A. Totten, C.A. Spiegel, K.C. Chen, D. Eschenbach, and K.K. Holmes. 1983. Nonspecific vaginitis. Diagnostic criteria and microbial and epidemiologic associations. *Am J Med 74*(1): 14–22.

Antonio, M.A.D., S.E. Hawes, and S.L. Hillier. 1999. The identification of vaginal *Lactobacillus* species and the demographic and microbiologic characteristics of women colonized by these species. *J Infect Dis 180*(6): 1950–1956.

Anukam, K.C., M.U. Duru, C.C. Eze, J. Egharevba, A. Aiyebelehin, A.W. Bruce, and G. Reid. 2009. Oral use of probiotics as an adjunctive therapy to fluconazole in the treatment of yeast vaginitis: A study of Nigerian women in an outdoor clinic. *Microb Ecol Health Dis 21*(2): 72–77.

Anukam, K.C., E. Osazuwa, I. Ahonkhai, M. Ngwu, G. Osemene, A.W. Bruce, and G. Reid. 2006a. Augmentation of antimicrobial metronidazole therapy of bacterial vaginosis with oral probiotic *Lactobacillus rhamnosus* GR-1 and *Lactobacillus reuteri* RC-14: Randomized, double-blind, placebo-controlled trial. *Microbes Infect 8*(6): 1450–1454.

Anukam, K.C., E. Osazuwa, G.I. Osemene, F. Ehigiagbe, A.W. Bruce, and G. Reid. 2006b. Clinical study comparing probiotic *Lactobacillus* GR-1 and RC-14 with metronidazole vaginal gel to treat symptomatic bacterial vaginosis. *Microbes Infect 8*(12–13): 2772–2776.

Armstrong, E., A. Hemmerling, S. Miller, K.E. Burke, S.J. Newmann, S.R. Morris, H. Reno, S. Huibner, M. Kulikova, N. Nagelkerke, B. Coburn, C.R. Cohen, and R. Kaul. 2022. Sustained effect of LACTIN-V (*Lactobacillus crispatus* CTV-05) on genital immunology following standard bacterial vaginosis treatment: Results from a randomised, placebo-controlled trial. *Lancet Microbe 3*(6): e435–e442.

Beerepoot, M.A., G. Ter Riet, S. Nys, W.M. van der Wal, C.A. de Borgie, T.M. de Reijke, J.M. Prins, J. Koeijers, A. Verbon, E. Stobberingh, and S.E. Geerlings. 2012. Lactobacilli vs antibiotics to prevent urinary tract infections: A randomized, double-blind, noninferiority trial in postmenopausal women. *Arch Intern Med 172*(9): 704–712.

Borges, S., J. Silva, and P. Teixeira. 2014. The role of lactobacilli and probiotics in maintaining vaginal health. *Arch Gynecol Obstet 289*(3): 479–489.

Borgogna, J.C., M.D. Shardell, S.G. Grace, E.K. Santori, B. Americus, Z. Li, A. Ulanov, L. Forney, T.M. Nelson, R.M. Brotman, J. Ravel, and C.J. Yeoman. 2021. Biogenic amines increase the odds of bacterial vaginosis and affect the growth of and lactic acid production by vaginal *Lactobacillus* spp. *Appl Environmental Microbiol 87*(10): e03068–20.

Brooks, J.P., D.J. Edwards, M.D. Jr. Harwich, M.C. Rivera, J.M. Fettweis, M.G. Serrano, R.A. Reris, N.U. Sheth, B. Huang, P. Girerd, Vaginal Microbiome Consortium, J.F. Strauss, K.K. Jefferson, and G.A. Buck. 2015. The truth about metagenomics: Quantifying and counteracting bias in 16S rRNA studies. *BMC Microbiol 15*: 66.

Bruce, A.W., P. Chadwick, A. Hassan, and G.F. VanCott. 1973. Recurrent urethritis in women. *Can Med Assoc J 108*(8): 973–976.

Bruce, A.W., and G. Reid. 1988. Intravaginal instillation of lactobacilli for prevention of recurrent urinary tract infections. *Can J Microbiol 34*(3): 339–343.

Burton, J.P., P.A. Cadieux, and G. Reid. 2003. Improved understanding of the bacterial vaginal microbiota of women before and after probiotic instillation. *Appl Environ Microbiol 69*(1): 97–101.

Burton, J.P., E. Devillard, P.A. Cadieux, J.-A. Hammond, and G. Reid. 2004. Detection of *Atopobium vaginae* in postmenopausal women by cultivation-independent methods warrants further investigation. *J Clin Microbiol 42*(4): 1829–1831.

Burton, J.P., and G. Reid. 2002. Evaluation of the bacterial vaginal flora of 20 postmenopausal women by direct (Nugent score) and molecular (polymerase chain reaction and denaturing gradient gel electrophoresis) techniques. *J Infect Dis 186*(12): 1770–1780.

Cadieux, P., J. Burton, G. Gardiner, I. Braunstein, A.W. Bruce, C.Y. Kang, and G. Reid. 2002. *Lactobacillus* strains and vaginal ecology. *JAMA 287*(15): 1940–1941.

Chan, R.C., A.W. Bruce, and G. Reid. 1984. Adherence of cervical, vaginal and distal urethral normal microbial flora to human uroepithelial cells and the inhibition of adherence of Gram-negative uropathogens by competitive exclusion. *J Urol 131*(3): 596–601.

Christensen, H.R., H. Frøkiaer, and J.J. Pestka. 2002. Lactobacilli differentially modulate expression of cytokines and maturation surface markers in murine dendritic cells. *J Immunol 168*(1): 171–178.

Cohen, C.R., M.R. Wierzbicki, A.L. French, S. Morris, S. Newmann, H. Reno, L. Green, S. Miller, J. Powell, T. Parks, and A. Hemmerling. 2020. Randomized trial of Lactin-V to prevent recurrence of bacterial vaginosis. *N Engl J Med. 382*(20): 1906–1915.

Dumonceaux, T., J. Schellenberg, V. Goleski, J. Hill, W. Jaoko, J. Kimani, D. Money, T. Ball, F. Plummer, and A. Severini. 2009. Multiplex detection of bacteria associated with normal microbiota and with bacterial vaginosis in vaginal swabs using oligonucleotide-coupled fluorescent microspheres. *J Clin Microbiol 47*(12): 4067–4077.

Ekmekci, H., B. Aslim, and S. Ozturk. 2009. Characterization of vaginal lactobacilli coaggregation ability with *Escherichia coli* . *Microbiol Immunol 53*(2): 59–65.

Emami, E., C. Mt Sherwin, and S. Heidari-Soureshjani. 2022. Effect of probiotics on urinary tract infections in children: A systematic review and meta-analysis. *Curr Rev Clin Exp Pharmacol.* http://doi.org/10.2174/27724328 17666220501114505.

Eschenbach, D.A., P.R. Davick, B.L. Williams, S.J. Klebanoff, K. Young-Smith, C.M. Critchlow, and K.K. Holmes. 1989. Prevalence of hydrogen peroxide-producing *Lactobacillus* species in normal women and women with bacterial vaginosis. *J Clin Microbiol 27*(2): 251–256.

FAO/WHO. 2001. Evaluation of health and nutritional properties of powder milk and live lactic acid bacteria. *Food and Agriculture Organization of the United Nations and World Health Organization Expert Consultation Report* . www.who.int/foodsafety/publications/fs_management/en/probiotics.pdf.

Forney, L.J., P. Gajer, C.J. Williams, G.M. Schneider, S.S.K. Koenig, S.L. McCulle, S. Karlebach, R.M. Brotman, C.C. Davis, K. Ault, and J. Ravel. 2010. Comparison of self-collected and physician-collected vaginal swabs for microbiome analysis. *J Clin Microbiol 48*(5): 1741–1748.

France, M.T., L. Rutt, S. Narina, S. Arbaugh, E. McComb, M.S. Humphrys, B. Ma, M.R Hayward, E.K. Costello, D.A. Relman, D.S. Kwon, and J. Ravel. 2020. Complete genome sequences of six *Lactobacillus iners* strains isolated from the human vagina. *Microbiol Resour Announc 9*(20): e00234–20.

Gabrielli, E., E. Pericolini, N. Ballet, E. Roselletti, S. Sabbatini, P. Mosci, A.C. Decherf, F. Pélerin, P. Perito, P. Jüsten, and A. Vecchiarelli. 2018. *Saccharomyces cerevisiae*-based probiotic as novel anti-fungal and anti-inflammatory agent for therapy of vaginal candidiasis. *Benef Microbes 9*(2): 219–230.

Gardiner, G.E., C. Heinemann, A.W. Bruce, D. Beuerman, and G. Reid. 2002. Persistence of *Lactobacillus fermentum* RC-14 and *Lactobacillus rhamnosus* GR-1 but not *L. rhamnosus* GG in the human vagina as demonstrated by randomly amplified polymorphic DNA. *Clin Diagn Lab Immunol 9*(1): 92–96.

Gloor, G.B., and G. Reid. 2016. Compositional analysis: A valid approach to analyze microbiome high-throughput sequencing data. *Can J Microbiol 62*(8): 692–703.

Hemmerling, A., W. Harrison, A. Schroeder, J. Park, A. Korn, S. Shiboski, and C.R. Cohen. 2009. Phase 1 dose-ranging safety trial of *Lactobacillus crispatus* CTV-05 for the prevention of bacterial vaginosis. *Sex Transm Dis 36*(9): 564–569.

Hill, C., F. Guarner, G. Reid, G.R. Gibson, D.J. Merenstein, B. Pot, L. Morelli, R.B. Canani, H.J. Flint, S. Salminen, P.C. Calder, and M.E Sanders. 2014. Expert consensus document. The international scientific association for probiotics and prebiotics consensus statement on the scope and appropriate use of the term probiotic. *Nat Rev Gastroenterol Hepatol 11*(8): 506–514.

Hill, J.E., S.H. Goh, D.M. Money, M. Doyle, A. Li, W.L. Crosby, M. Links, A. Leung, D. Chan, and S.M. Hemmingsen. 2005. Characterization of vaginal microflora of healthy, nonpregnant women by chaperonin-60 sequence-based methods. *Am J Obstet Gynecol 193*(3 Pt 1): 682–692.

Hummelen, R., J. Macklaim, A. Fernandes, R.J. Dickson, J. Changalucha, G.B. Gloor, and G. Reid. 2010. Deep sequencing of the vaginal microbiome in HIV patients. *PLoS ONE 5*(8): e12078.

Kern, A.M., J.M. Bohbot, and J.M. Cardot. 2012. Preventive treatment of vulvovaginal candidosis with vaginal probiotic Gynophilus (lLr regenerans): Results of the observational study candiflore. *La Lettre du Gynecologue 370*: 33–37.

Kim, T.K., S.M. Thomas, M. Ho, S. Sharma, C.I. Reich, J.A. Frank, K.M. Yeater, D.R. Biggs, N. Nakamura, R. Stumpf, S.R. Leigh, R.I. Tapping, S.R. Blanke, J.M. Slauch, H.R. Gaskins, J.S. Weisbaum, G.J. Olsen, L.L. Hoyer, and B.A. Wilson. 2009. Heterogeneity of vaginal microbial communities within individuals. *J Clin Microbiol 47*(4): 1181–1189.

Kirjavainen, P.V., R.M. Laine, D.E. Carter, J.-A. Hammond, and G. Reid. 2008. Expression of anti-microbial defense factors in vaginal mucosa following exposure to *Lactobacillus rhamnosus* GR-1. *Int J Probiotics 3*(2): 99–106.

Kohler, G., S. Assefa, and G. Reid. 2012. Probiotic interference of *Lactobacillus rhamnosus* GR-1 and *Lactobacillus reuteri* RC-14 with the opportunistic fungal pathogen *Candida albicans*. *Infect Dis Obstetr Gynecol 2012*: 636474.

Larsson, P.G., E. Brandsborg, U. Forsum, S. Pendharkar, K.K. Andersen, S. Nasic, L. Hammarström, and H. Marcotte. 2011. Extended antimicrobial treatment of bacterial vaginosis combined with human lactobacilli to find the best treatment and minimize the risk of relapses. *BMC Infect Dis 11*: 223.

Laughton, J.M., E. Devillard, D.E. Heinrichs, G. Reid, and J.K. McCormick. 2006. Inhibition of expression of a staphylococcal superantigen-like protein by a soluble factor from *Lactobacillus reuteri*. *Microbiology 152*(Pt 4): 1155–1167.

Lee, S.J., Y.H. Shim, S.J. Cho, and J.W. Lee. 2007. Probiotics prophylaxis in children with persistent primary vesico-ureteral reflux. *Pediatr Nephrol 22*(9): 1315–1320.

Lev-Sagie, A., F. De Seta, H. Verstraelen, G. Ventolini, R. Lonnee-Hoffmann, and P. Vieira-Baptista. 2022. The vaginal microbiome: II. Vaginal dysbiotic conditions. *J Low Genit Tract Dis 26*(1): 79–84.

Lewis, D.A., R. Brown, J. Williams, P. White, S.K. Jacobson, J.R. Marchesi, and M.J. Drake. 2013. The human urinary microbiome; bacterial DNA in voided urine of asymptomatic adults. *Front Cell Infect Microbiol 15*(3): 41.

Macklaim, J.M., J.C. Clemente, R. Knight, G.B. Gloor, and G. Reid. 2015. Changes in vaginal microbiota following antimicrobial and probiotic therapy. *Microb Ecol Health Dis 26*: 27799.

Macklaim, J.M., G.B. Gloor, K.C. Anukam, S. Cribby, and G. Reid. 2010. At the crossroads of vaginal health and disease, the genome sequence of *Lactobacillus iners* . *Proc Natl Acad Sci USA 108*(Suppl 1): 4688–4695.

Mahajan, G., E. Doherty, T. To, A. Sutherland, J. Grant, A. Junaid, A. Gulati, N. LoGrande, Z. Izadifar, S.S. Timilsina, V. Horváth, R. Plebani, M. France, I. Hood-Pishchany, S. Rakoff-Nahoum, D.S. Kwon, G. Goyal, R. Prantil-Baun, J. Ravel, and D.E. Ingber. 2022. Vaginal microbiome-host interactions modeled in a human vagina-on-a-chip. *Microbiome 10*(1): 201.

Mändar, R., G. Sõerunurk, J. Štšepetova, I. Smidt, T. Rööp, S. Kõljalg, M. Saare, K. Ausmees, D.D. Lee, M. Jaagura, S. Piiskop, H. Tamm, and A. Salumets. 2023. Impact of *Lactobacillus crispatus*-containing oral and vaginal probiotics on vaginal health: A randomised double-blind placebo controlled clinical trial. *Benef Microbes 1*: 1–10. http://doi.org/10.3920/BM2022.0091

Marcone, V., E. Calzolari, and M. Bertini. 2008. Effectiveness of vaginal administration of *Lactobacillus rhamnosus* following conventional metronidazole therapy: How to lower the rate of bacterial vaginosis recurrences. *New Microbiol 31*(3): 429–433.

Marcone, V., G. Rocca, M. Lichtner, and E. Calzolari. 2010. Long-term vaginal administration of *Lactobacillus rhamnosus* as a complementary approach to management of bacterial vaginosis. *Int J Gynecol Obstetr 110*: 223–226.

Martinez, R.C.R., S.A. Franceschini, M.C. Patta, S.M. Quintana, R.C. Candido, J.C. Ferreira, E.C.P. De Martinis, and G. Reid. 2009b. Improved treatment of vulvovaginal candidiasis with fluconazole plus probiotic *Lactobacillus rhamnosus* GR-1 and *Lactobacillus reuteri* RC-14. *Lett Appl Microbiol 48*(3): 269–274.

Martinez, R.C.R., S.A. Franceschini, M.C. Patta, S.M. Quintana, B. Gomes, E. De Martinis, and G. Reid. 2009a. Improved cure of bacterial vaginosis with single dose of tinidazole (2 g), *Lactobacillus rhamnosus* GR-1, and *Lactobacillus reuteri* RC-14: A randomized, double-blind, placebo-controlled trial. *Can J Microbiol 55*(2): 133–138.

Martinez, R.C.R., S.L. Seney, K.L. Summers, A. Nomizo, E.C.P. De Martinis, and G. Reid. 2009c. Effect of *Lactobacillus rhamnosus* GR-1 and *Lactobacillus reuteri* RC-14 on the ability of *Candida albicans* to infect cells and induce inflammation. *Microbiol Immunol 53*(9): 487–495.

Martins, A.J., P. Colquhoun, G. Reid, and S.O. Kim. 2009. Reduced expression of basal and probiotic-inducible G-CSF in intestinal mononuclear cells is associated with inflammatory bowel disease. *Inflamm Bowel Dis 15*(4): 515–525.

Mastromarino, P., P. Brigidi, S. Macchia, L. Maggi, F. Pirovano, V. Trinchieri, U. Conte, and D. Matteuzzi. 2002. Characterization and selection of vaginal *Lactobacillus* strains for the preparation of vaginal tablets. *J Appl Microbiol 93*(5): 884–893.

Mastromarino, P., S. Macchia, L. Meggiorini, V. Trinchieri, L. Mosca, M. Perluigi, and C. Midulla. 2009. Effectiveness of *Lactobacillus*-containing vaginal tablets in the treatment of symptomatic bacterial vaginosis. *Clin Microbiol Infect 15*(1): 67–74.

McGroarty, J.A., L. Tomeczek, D.G. Pond, G. Reid, and A.W. Bruce. 1992. Hydrogen peroxide production by *Lactobacillus* species: Correlation with susceptibility to the spermicidal compound nonoxynol-9. *J Infect Dis 165*(6): 1142–1144.

McLean, N.W. and I.J. Rosenstein. 2000. Characterisation and selection of a *Lactobacillus* species to re-colonise the vagina of women with recurrent bacterial vaginosis. *J Med Microbiol* 49(6): 543–552.

McMillan, A., S. Rulisa, M. Sumarah, J.M. Macklaim, J. Renaud, J.E. Bisanz, G.B. Gloor, and G. Reid. 2015. A multi-platform metabolomics approach identifies highly specific biomarkers of bacterial diversity in the vagina of pregnant and non-pregnant women. *Sci Reports* 5: 14174.

Morelli, L., D. Zonenenschain, M. Del Piano, and P. Cognein. 2004. Utilization of the intestinal tract as a delivery system for urogenital probiotics. *J Clin Gastroenterol* 38(6 Suppl): S107–S110.

Nasioudis, D., I.M. Linhares, W.J. Ledger, and S.S. Witkin. 2017. Bacterial vaginosis: A critical analysis of current knowledge. *BJOG* 124(1): 61–69.

Nelson, D.E., Q. Dong, B. Van der Pol, E. Toh, B. Fan, B.P. Katz, D. Mi, R. Rong, G.M. Weinstock, E. Sodergren, and J.D. Fortenberry. 2012. Bacterial communities of the coronal sulcus and distal urethra of adolescent males. *PLoS ONE* 7(5): e36298.

Nugent, R.P., M.A. Krohn, and S.L. Hillier. 1991. Reliability of diagnosing bacterial vaginosis is improved by a standardized method of Gram stain interpretation. *J Clin Microbiol* 29(2): 297–301.

Osset, J., R.M. Bartolomé, E. García, and A. Andreu. 2001. Assessment of the capacity of *Lactobacillus* to inhibit the growth of uropathogens and block their adhesion to vaginal epithelial cells. *J Infect Dis* 183(3): 485–491.

Patterson, J.L., P.H. Girerd, N.W. Karjane, and K.K. Jefferson. 2007. Effect of biofilm phenotype on resistance of *Gardnerella vaginalis* to hydrogen peroxide and lactic acid. *Am J Obstet Gynecol* 197(2): 170.e1–7.

Petrova, M.I., G. Reid, M. Vaneechoutte, and S. Lebeer.2017. *Lactobacillus iners*: Friend or foe? *Trends Microbiol* 25(3): 182–191.

Puebla-Barragan, S., P.P. Akouris, K.F. Al, C. Carr, B. Lamb, M. Sumarah, C. van der Veer, R. Kort, J. Burton, and G. Reid. 2021. The two-way interaction between the molecules that cause vaginal malodour and lactobacilli: An opportunity for probiotics. *Int J Mol Sci* 22(22): 12279.

Puebla-Barragan, S., J. Renaud, M. Sumarah, and G. Reid. 2020. Malodorous biogenic amines in *Escherichia coli*-caused urinary tract infections in women-a metabolomics approach. *Sci Rep* 10(1): 9703.

Puebla-Barragan, S., E. Watson, C. van der Veer, J.A. Chmiel, C. Carr, J.P. Burton, M. Sumarah, R. Kort, and G. Reid. 2021. Interstrain variability of human vaginal *Lactobacillus crispatus* for metabolism of biogenic amines and antimicrobial activity against urogenital pathogens. *Molecules* 26(15): 4538.

Reid, G. 2016. Cervicovaginal microbiomes-threats and possibilities. *Trends Endocrinol Metab.* 27(7): 446–454.

Reid, G. 2017a. Is bacterial vaginosis a disease? *Appl Microbiol Biotechnol* 102(2): 553–558.

Reid, G. 2017b. Probiotics use in an infectious disease setting. *Expert Rev Anti Infect Ther* 15(5): 449–455.

Reid, G. 2017c. The development of probiotics for women's health. *Can J Microbiol* 63(4): 269–277.

Reid, G. 2023a. How do lactobacilli search and find the vagina? *Microorganisms* 11(1): 148.

Reid, G. 2023b. Perspective: Microbial interventions for urinary health. *Microbiome Res Reports* 2: 3.

Reid, G., A.W. Bruce, and M. Taylor. 1995. Instillation of *Lactobacillus* and stimulation of indigenous organisms to prevent recurrence of urinary tract infections. *Microecol Ther 23*: 32–45.

Reid, G., D. Charbonneau, J. Erb, B. Kochanowski, D. Beuerman, R. Poehner, and A.W. Bruce. 2003. Oral use of *Lactobacillus rhamnosus* GR-1 and *L. fermentum* RC-14 significantly alters vaginal flora: Randomized, placebo-controlled trial in 64 healthy women. *FEMS Immunol Med Microbiol* 35(2): 131–134.

Reid, G., R.L. Cook, and A.W. Bruce. 1987. Examination of strains of lactobacilli for properties that may influence bacterial interference in the urinary tract. *J Urol* 138(2): 330–335.

Reid, G., J. Dols, and W. Miller. 2009. Targeting the vaginal microbiota with probiotics as a means to counteract infections. *Curr Opin Clin Nutr Metab Care* 12(6): 583–587.

Reid, G., K. Millsap, and A.W. Bruce. 1994. Implantation of *Lactobacillus casei* var *rhamnosus* into vagina. *Lancet 344*(8931): 1229.

Reid, G., J.A. Younes, H.C. Van der Mei, G.B. Gloor, R. Knight, and H.J. Busscher. 2011. Microbiota restoration: Natural and supplemented recovery of human microbial communities. *Nat Rev Microbiol* 9(1): 27–38.

Rizzo, A., A. Losacco, and C.R. Carratelli. 2013. *Lactobacillus crispatus* modulates epithelial cell defense against *Candida albicans* through Toll-like receptors 2 and 4, interleukin 8, and human β-defensins 2 and 3. *Immunol Lett* 156(1–2): 102–109.

Rose, W.A., C.L. McGowin, R.A. Spagnuolo, T.D. Eaves-Pyles, V.L. Popov, and R.B. Pyles. 2012. Commensal bacteria modulate innate immune responses of vaginal epithelial cell multilayer cultures. *PLoS ONE* 7(3): e32728.

Saunders, S., A. Bocking, J. Challis, and G. Reid. 2007. Effect of *Lactobacillus* challenge on *Gardnerella vaginalis* biofilms. *Colloid Surf B* 55(2): 138–142.

Schellenberg, J., M. Links, J. Hill, T. Dumonceaux, G. Peters, S. Tyler, T. Ball, A. Severini, and F. Plummer. 2009. Pyrosequencing of the chaperonin-60 (cpn60) universal target as a tool for determining the composition of microbial communities. *Appl Environ Microbiol* 75(9): 2889–2898.

Siddiqui, H., A.J. Nederbragt, K. Lagesen, S.L. Jeansson, and K.S. Jakobsen. 2011. Assessing diversity of the female urine microbiota by high throughput sequencing of 16S rDNA amplicons. *BMC Microbiol 11*: 244.

Srinivasan, S., C. Liu, C.M. Mitchell, T.L. Fiedler, K.K. Thomas, K.J. Agnew, J.M. Marrazzo, and D.N. Fredricks. 2010. Temporal variability of human vaginal bacteria and relationship with bacterial vaginosis. *PLoS ONE 5*(4): e10197.

Stapleton, A.E. 2016. The vaginal microbiota and urinary tract infection. *Microbiol Spectr. 4*(6). Stapleton, A.E., M. Au-Yeung, T.M. Hooton, D.N. Fredricks, P.L. Roberts, C.A. Czaja, Y. Yarova-Yarovaya, T. Fiedler, M. Cox, and W.E. Stamm. 2011. Randomized, placebo-controlled phase 2 trial of a *Lactobacillus crispatus* probiotic given intravaginally for prevention of recurrent urinary tract infection. *Clin Infect Dis 52*(10): 1212–1217.

Swidsinski, A., W. Mendling, V. Loening-Baucke, S. Swidsinski, Y. Dörffel, J. Scholze, H. Lochs, and H. Verstraelen. 2008. An adherent *Gardnerella vaginalis* biofilm persists on the vaginal epithelium after standard therapy with oral metronidazole. *Am J Obstet Gynecol 198*(1): 97.e1–6.

Tamrakar, R., T. Yamada, I. Furuta, K. Cho, M. Morikawa, H. Yamada, N. Sakuragi, and H. Minakami. 2007. Association between *Lactobacillus* species and bacterial vaginosis-related bacteria, and bacterial vaginosis scores in pregnant Japanese women. *BMC Infect Dis 7*: 128.

Velraeds, M.M., B. van de Belt-Gritter, H.C. van der Mei, G. Reid, and H.J. Busscher. 1998. Interference in initial adhesion of uropathogenic bacteria and yeasts to silicone rubber by a *Lactobacillus acidophilus* biosurfactant. *J Med Microbiol 47*(12): 1081–1085.

Wani, A.K., P. Roy, V. Kumar, and T.U.G. Mir. 2022. Metagenomics and artificial intelligence in the context of human health. *Infect Genet 100*: 105267.

Whiteside, S.A., H. Razvi, S. Dave, G. Reid, and J.P. Burton. 2015. The microbiome of the urinary tract—A role beyond infection. *Nat Rev Urol 12*(2): 81–90.

Wolfe, A.J., E. Toh, N. Shibata, R. Rong, K. Kenton, M. Fitzgerald, E.R. Mueller, P. Schreckenberger, Q. Dong, D.E. Nelson, and L. Brubaker. 2012. Evidence of uncultivated bacteria in the adult female bladder. *J Clin Microbiol 50*(4): 1376–1378.

Ya, W., C. Reifer, and L.E. Miller. 2010. Efficacy of vaginal probiotic capsules for recurrent bacterial vaginosis: A double-blind, randomized, placebo-controlled study. *Am J Obstet Gynecol 203*: 120–e1.

Yamamoto, T., X. Zhou, C.J. Williams, A. Hochwalt, and L.J. Forney. 2009. Bacterial populations in the vaginas of healthy adolescent women. *J Pediatr Adolesc Gynecol 22*(1): 11–18.

Yeganegi, M., C.G. Leung, A. Martins, S.O. Kim, G. Reid, J.R. Challis, and A.D. Bocking. 2010. *Lactobacillus rhamnosus* GR-1-induced IL-10 production in human placental trophoblast cells involves activation of JAK/STAT and MAPK pathways. *Reprod Sci 17*(11): 1043–1051.

Zhou, X., M.A. Hansmann, C.C. Davis, H. Suzuki, C.J. Brown, U. Schütte, J.D. Pierson, and L.J. Forney. 2010. The vaginal bacterial communities of Japanese women resemble those of women in other racial groups. *FEMS Immunol Med Microbiol 58*(2): 169–181.

Zozaya, M., M.J. Ferris, J.D. Siren, R. Lillis, L. Myers, M.J. Nsuami, A.M. Eren, J. Brown, C.M. Taylor, and D.H. Martin. 2016. Bacterial communities in penile skin, male urethra, and vaginas of heterosexual couples with and without bacterial vaginosis. *Microbiome 4*: 16.

Zozaya-Hinchliffe, M., R. Lillis, D.H. Martin, and M.J. Ferris. 2010. Quantitative PCR assessments of bacterial species in women with and without bacterial vaginosis. *J Clin Microbiol 48*(5): 1812–1819.

# Lactic Acid Bacteria and Respiratory Health  **29**

## *Their Beneficial Effects on Viral Infections*

Julio Villena, Leonardo Albarracín, Mariano Elean, and Haruki Kitazawa

## 29.1 INTRODUCTION

Respiratory viruses have been a major problem for humankind for hundreds of years. Pandemics and severe epidemics caused by respiratory viruses have challenged global health systems more than once. Perhaps the most frightening examples of the harmful potential of viral respiratory infections are the 1918–1919 "Spanish flu" pandemic and the COVID-19 pandemic, which are among the deadliest events in recorded human history. The Spanish flu killed an estimated 50 to 100 million persons (Johnson and Mueller 2002), while COVID-19 caused the death of more than 6.8 million persons as of April 4, 2023. Nowadays, acute viral respiratory infections are still a major global public health problem, and they represent a sleeping danger that can severely challenge humanity at the most unexpected times (Pangesti et al. 2018; Kumar et al. 2018)

Despite the progress made by the scientific community with the development of vaccines and antiviral drugs, there are no specific interventions for most respiratory infections of viral origin. These infections continue to cause frequent morbidity and sometimes cause severe outcomes, including death, especially in developing countries (WHO 2013). Respiratory viral infections are implicated in approximately 50% of community-acquired pneumonia in young children and over 90% of bronchiolitis cases in infants and young children seeking medical attention. In addition, they are involved in community-acquired pneumonia of adults and in severe exacerbations of chronic obstructive pulmonary disease (WHO 2013; Pangesti et al. 2018; Kumar et al. 2018). Effective vaccines and appropriate clinical management are currently lacking for most respiratory viral infections. Moreover, current practices for treating these infection diseases include the frequent use of antibiotics that are ineffective and may result in antimicrobial resistance.

Therefore, more research is needed to expand the options for preventing or treating respiratory viral infections. In this regard, research from the last decade has proved that beneficial microbes with the ability to modulate the immune system (immunobiotics) are an interesting alternative to improve the

DOI: 10.1201/9781003352075-32

resistance against respiratory infections. In this chapter, a review of the current knowledge of the positive effects of immunobiotic lactic acid bacteria (LAB) on the modulation of respiratory antiviral immunity and their impact on resistance to viral infections is provided.

# 29.2  RESPIRATORY ANTIVIRAL IMMUNE RESPONSES

In order to understand the effect of beneficial bacteria on viral respiratory infections, it is necessary to remember first some of the immunological mechanisms that are activated in the respiratory tract when a viral attack occurs.

The epithelial cell layer of the respiratory tract represents the first barrier of the host against viral attacks. The mucus layer and the ciliary movement constitute this respiratory barricade. Both mechanisms impair the attachment and internalization of viruses in respiratory epithelial cells (RECs) (Hewitt and Lloyd 2021). When this barrier is overcome and the virus contacts the respiratory mucosa, RECs are able to trigger defense mechanisms through their capacity to recognize the viral molecules. Pattern-recognition receptors (PRRs) expressed in RECs recognize different pathogen-associated molecular patterns (PAMPs) that are exposed during viral replication. These PAMP–PRR interactions activate signaling pathways that initiate the antiviral innate immune response.

Among the PRRs expressed in the respiratory tract that are involved in the recognition of viruses, the most studied are toll-like receptors (TLRs) and the RNA recognition protein RIG-1 (Villena and Kitazawa 2020; Mettelman et al. 2022). TLR3, which is able to recognize viral double-stranded RNA (dsRNA), is expressed at the cell surface or in endosomal compartments of RECs. RIG-I is a cytoplasmic receptor that recognizes single-stranded RNA (ssRNA) and signals through mitochondrial antiviral signaling protein (MAVs). The PAMP–PPR interaction culminates in the activation of nuclear factor κB (NF-κB) and interferon (IFN) regulatory factor 3 (IRF3), leading to the production of type I and III IFNs and inflammatory cytokines (Villena and Kitazawa 2020; Mettelman et al. 2022).

Type I IFNs are released during the earlier stages of viral infection by infected RECs and act in a paracrine or autocrine manner. IFNs activate their receptors (IFNAR) and enhance the expression of hundreds of genes that counteract viral replication. Functional genomic studies identified several of the IFN-induced factors that have important roles in controlling respiratory virus replication (Mettelman et al. 2022) including the 2',5'-oligoadenylate synthetase (OAS)-RNAaseL system (Ronni et al. 1997); MX1 proteins (Cilloniz et al. 2012); and IFN-inducible transmembrane (IFITM) proteins 1, 2, and 3 (Everitt et al. 2012). Therefore, type I IFNs, especially IFN-β, are key to developing an antiviral state in the respiratory tract.

Inflammatory cytokines and chemokines produced as a result of PRR activation are also important for the activation of respiratory antiviral innate immune responses. Infected RECs increase the expression of TNF-α, IL-6, IL-8, CCL2 (MIP-1), CCL5 (RANTES), CCL3 (MIP-1α), and CXCL10 (IP-10) (Goraya et al. 2015). The production of these cytokines is complemented by the activity of inflammasomes that induce the activation of caspase-1 and promote the generation of the active forms of IL-1β and IL-18 (Thomas et al. 2009; Allen et al. 2009). All these inflammatory factors mediate the activation of resident immune cells such as innate lymphoid cells, alveolar macrophages, and dendritic cells (DCs) and induce the recruitment of neutrophils, macrophages, and lymphocytes into the respiratory tract (Goraya et al. 2015; Mettelman et al. 2022). The activation and recruitment of immune cells amplifies the inflammatory response, limits viral spread by the elimination of apoptotic-infected cells, and initiates the adaptive immune responses in the respiratory tract such as the production of specific antibodies or the activation of specific effector T cells.

The respiratory tract contains an elaborate network of DCs that is a focal control point determining the induction of immunity against viruses. These cells perform a unique sentinel function in the respiratory immune response because of their ability to recognize virus through PRRs expression. DCs are activated when PRR–PAMP occurs, and this activation is supported by the inflammatory factors produced by RECs. DCs migrate via the afferent lymphatics to the T cell paracortex of the lung-draining mediastinal lymph nodes (MLNs) to select and activate recirculating naive T cells. T lymphoblasts destined to become virus-specific effector CD4 or CD8 cells at the site of infection leave the lymph node via the efferent lymphatics, reach the circulation, and enter the lung through either the bronchial circulation to populate the conducting airways and pleura or the pulmonary arterial circulation to enter alveolar septa and airspaces (Kaiko et al. 2008).

Effector CD4 Th2 cells support the production of virus-specific antibodies. The type and concentration of antibodies produced in response to viral infection are dependent on the site of exposure. Upper airway exposure results primarily in an IgA response; however, when viruses reach the deep lung after passing through the upper airway, they induce an increased production of specific IgG (Twigg 2005). Secretory IgA binds to critical viral epitopes that have infected the respiratory mucosa and then neutralizes their biological activity, leading to a strong inhibition of viral growth (Sato and Kiyono 2012). In addition, lung-specific IgG is thought to have a direct role in the defense of the lower respiratory tract and together with serum IgG is involved in the reduction of viral spread into extrapulmonary sites (Chiu and Openshaw 2015).

## 29.3 BENEFICIAL EFFECTS OF IMMUNOBIOTIC LACTIC ACID BACTERIA ON INFLUENZA VIRUS INFECTION

Most research regarding the beneficial effects of LAB on respiratory viral infections has focused on the influenza virus (IFV). Studies have demonstrated the ability of immunobiotic LAB to improve respiratory innate antiviral defenses as well as adaptive immunity against IFV. It was reported that orally administered *Lactiplantibacillus plantarum* L-137 improved protection against IFV by increasing type I IFN production (Maeda et al. 2009). The increased production of IFN-β induced by the L-137 strain correlated with the improved survival and the reduction of viral replication in lungs of infected mice. It was also shown that *Lactobacillus gasseri* SBT2055 enhanced resistance of mice to IFV infection (Nakayama et al. 2014). Interestingly, the authors observed that the antiviral factors Mx1 and Oas1a were improved in *Lb. gasseri* SBT2055-treated mice. Moreover, the study demonstrated that the inflammatory response triggered by IFV was differentially regulated by the immunobiotic treatment since *Lb. gasseri* SBT2055-treated mice had reduced lung inflammatory cell infiltration and lower lung damage when compared to controls (Nakayama et al. 2014).

A beneficial regulation of the IFV-triggered inflammatory response by immunobiotics has been also reported by our group (Zelaya et al. 2014; Zelaya et al. 2015; Raya Tonetti et al. 2020). It is well established that lung damage induced by IFV is produced by both virus replication and the uncontrolled inflammatory response (Armstrong et al. 2012). The adequate production of inflammatory factors is necessary to protect against IFV infection together with an appropriate anti-inflammatory regulation in order to prevent the damage of lung tissue. Thus, the proper balance of cytokines is a key factor in determining the outcome of IFV infection. In this regard, we observed that orally (Zelaya et al. 2014) or nasally (Zelaya et al. 2015; Raya Tonetti et al. 2020) administered immunobiotic *Lacticaseibacillus rhamnosus* CRL1505 differentially regulated the levels and kinetics of inflammatory cells and cytokines in mice after IFV challenge. In our experimental model, we observed increased levels of TNF-α, IL-6, neutrophils, and macrophages in the respiratory tract of *Lb. rhamnosus* CRL1505-treated mice early after the challenge with IFV, after which pro-inflammatory cytokines and infiltrated cells started to decrease. In contrast, in untreated control mice, those inflammatory parameters continued to increase during the course of IFV infection. The trend toward lower inflammatory factors and cells registered later during IFV infection in *Lb. rhamnosus* CRL1505-treated mice correlated with a reduced severity of pulmonary damaged when compared to control mice (Zelaya et al. 2014; Zelaya et al. 2015).

Chen et al. (2017) also investigated the ability of orally administered *Enterococcus faecalis* KH2 to beneficially modulate the innate immune response to influenza infection. Researchers observed that KH2 strain protected C57BL/6 mice against IFV as observed by the reduced mortality. IFV enhanced the levels of IL-6, TNF-α, IFN-γ, IL-1β, IL-17, and MCP-1, while treatment with *E. faecalis* significantly diminished the concentrations of pro-inflammatory factors. The work also reported that the protective activity of the KH2 strain was abrogated when recombinant inflammatory chemokines such as MCP-1 were administered concomitantly (Chen et al. 2017). The beneficial modulation of the inflammatory response induced by IFV infection was also reported for the oral administration of *Lb. plantarum* 0111 (Xing et al. 2022).

It was also demonstrated that immunobiotic LAB are able to improve cellular immune response against IFV. Orally administered *Lacticaseibacillus casei* Shirota improved resistance of aged (Hori et al. 2002) and infant mice (Yasui et al. 2004) to IFV infection by enhancing systemic and respiratory NK cell activity and improving the production of IFN-γ and TNF-α by respiratory lymphocytes. Both studies also reported that IFV titers were significantly reduced in mice treated with the *Lb. casei* Shirota strain (Hori et al. 2002; Yasui et al. 2004). Several other studies corroborated these findings by showing similar effects for orally administered lactobacilli (Kawase et al. 2010; Takeda et al. 2011). Immunobiotic strains that belong to the genus previously known as *Lactobacillus* (*Lb. gasseri* TMC0356, *Lb. rhamnosus* GG, or *Lb. plantarum* 06CC2) beneficially modulated NK cells activity and Th1 response against IFV, diminished virus titers, and reduced lung pathological changes (Kawase et al. 2010; Takeda et al. 2011). More recently, Kawahara et al. (2015) described the improvement of respiratory antiviral response by an orally administered bifidobacteria strain. It was shown that *Bifidobacterium longum* MM-2 increased respiratory NK cell activity and IFN-γ production, resulting in improved clinical symptoms, reduced mortality, and decreased virus titers after IFV challenge.

Research work has also demonstrated that nasal administration of immunobiotics is an interesting alternative to improve cellular response against IFV infection (Hori et al. 2001; Izumo et al. 2010; Harata et al. 2010). It was observed that mice nasally treated with non-viable *Lb. casei* Shirota (Hori et al. 2001), *Lactiplantibacillus pentosus* S-PT84 (Izumo et al. 2010), or *Lb. rhamnosus* GG (Harata et al. 2010) had improved levels of Th1 cytokines in the respiratory tract after the challenge with IFV. This improved cellular respiratory immunity correlated with a higher resistance to IFV infection.

The respiratory antiviral humoral immune response can be also favorably modulated by immunobiotic LAB. Yasui et al. (1999) reported that the oral administration of *Bifidobacterium breve* YIT4064 improved the production of anti-IFV IgG antibodies in serum of IFV-infected mice. The YIT4064 strain reduced viral titers, improved the survival rate, and decreased the severity of the symptoms associated with the influenza infection. Similarly, it was shown that orally administered *Lb. pentosus* b240 (Kobayashi et al. 2011), *Levilactobacillus brevis* KB290 (Waki et al. 2014), or *Lb. plantarum* 0111 (Xing et al. 2022) enhanced resistance of mice to IFV challenge by improving the levels of respiratory IgA and IgG specific antibodies. The improved humoral response induced by LAB strains correlated with significant reduction of viral replication and body weight loss.

# 29.4 BENEFICIAL EFFECTS OF IMMUNOBIOTIC LACTIC ACID BACTERIA ON RESPIRATORY SYNCYTIAL VIRUS INFECTION

Respiratory syncytial virus (RSV) is the leading cause of lower respiratory tract illness in infants and young children. Host immune response is implicated in both protective and immunopathological mechanisms during RSV infection. The roles of inflammation in the pathogenesis of RSV have been extensively investigated in mice. Graham et al. (1988) reported the first examples of histopathology of susceptible BALB/c mice challenged with high titers of human RSV, which included perivascular and peribronchial infiltrates, as well as lymphocytes and macrophages in the alveolar spaces. Subsequent studies identified numerous pro-inflammatory mediators produced after RSV challenge, including MCP-1, RANTES, the IFN-γ regulated protein, IP-10, KC, MIP-1α, and IL-17 (Rutigliano et al. 2004; Mukherjee et al. 2011). In addition, Rudd et al. (2005) showed that RSV infection triggers the activation of the TLR3 signaling pathways that regulate the expression of inflammatory factors and further upregulates TLR3 expression in RSV-infected cells. The increased TLR3 expression in the respiratory epithelial cells induced by RSV sensitizes the epithelial cells to subsequent extracellular dsRNA exposure through activation of the inflammation-related transcription factor NF-κB and production of IL-8 (Groskreutz et al. 2006). Therefore, modulating TLR3-triggered inflammatory response is an attractive therapeutic target for reducing RSV-induced lung inflammation.

In this regard, we demonstrated that the oral or nasal administration of *Lb. rhamnosus* CRL1505 is able to differentially modulate respiratory antiviral immunity triggered by TLR3 activation (Villena et al. 2012; Chiba et al. 2013; Tomosada et al. 2013; Albarracín et al. 2020; Garcia-Castillo et al. 2020). To mimic the pro-inflammatory and physiopathological consequences of RNA viral infections in the lung, we used an experimental model of lung inflammation based on the administration of the artificial TLR3 ligand and dsRNA analog poly(I:C). Nasal administration of poly(I:C) to BALB/c mice induced a marked impairment of lung function that was accompanied by the production of pro-inflammatory mediators and inflammatory cell recruitment into the airways (Villena et al. 2012). Exposure to poly(I:C) induced respiratory epithelial cell death and impaired epithelial barrier function as demonstrated by the increased levels lactate-dehydrogenase (LDH) activity and albumin concentration in BAL. Moreover, intranasal administration of poly(I:C) resulted in neutrophils and mononuclear cells influx into the lung and increased levels of inflammatory factors (Villena et al. 2012; Chiba et al. 2013; Garcia-Castillo et al. 2020).

The experimental model used in our works resembled RSV infection since this respiratory virus is able to induce a profile of pro-inflammatory cytokines similar to that observed following in vivo poly(I:C) challenge in mice (Stowell et al. 2009; Groskreutz et al. 2006). In fact, RSV infection in children and experimental RSV inoculation in mice result in prominent local secretion of pro-inflammatory cytokines, such as TNF-α, IL-6, IL-8, MIP-1, RANTES, and MCP-1, and the recruitment and activation of neutrophils and monocytes/macrophages (Bem et al. 2011). In addition, RSV infection induced an increase of IL-10 in the respiratory tract. It was reported that TNF-α contributes to virus clearance during the early stages of RSV infection, which is most likely a result of the NK cell response. However, continued production of TNF-α exacerbates illness and tissue injuries during the late stages of RSV infection (Rutigliano et al. 2004). On the other hand, IL-10 also seems to play a crucial role in controlling disease severity in RSV infection (Sun et al. 2009; Weiss et al. 2011). IL-10 deficiency during RSV challenge did not affect viral load but led to markedly increased disease severity with enhanced weight loss, delayed recovery, and a greater influx of inflammatory cells into the lung and airways and enhanced release of inflammatory mediators (Loebbermann et al. 2012). Then, the capacity of *Lb. rhamnosus* CRL1505 to reduce the production of TNF-α, IL-6, IL-8, and MCP-1 in the respiratory tract after the challenge with poly(I:C) could explain at least partially the reduced lung injuries in the CRL1505-treated group (Villena et al. 2012). Moreover, *Lb. rhamnosus* CRL1505 treatment prior to poly(I:C) challenge induced a significant increase in IL-10 in lung and serum when compared to untreated controls. Consequently, IL-10 would be valuable for attenuating inflammatory damage and pathophysiological alterations in lungs challenged with poly(:IC). *Lb. rhamnosus* CRL1505 treatment would beneficially regulate the balance between pro-inflammatory mediators and IL-10, allowing an effective inflammatory response against infection and avoiding tissue damage.

Oral and nasal treatments with the CRL1505 strain also increased levels of IFN-γ in the respiratory tract after poly(I:C) challenge (Villena et al. 2012; Garcia-Castillo et al. 2020). The higher levels of respiratory IFN-γ in *Lb. rhamnosus* CRL1505-treated mice could be explained by the higher number of CD3$^+$CD4$^+$IFN-γ$^+$ T cells and by an improved activation of these cells by lung DCs. When we analyzed lung DCs in *Lb. rhamnosus* CRL1505-treated mice after the nasal challenge with poly(I:C), we found increased levels of both CD103$^+$ and CD11b$^{high}$ DCs. Moreover, both DC populations showed higher expression of MHC-II when compared with controls. However, IL-12 and IFN-γ were increased only in CD103$^+$ DCs (Villena et al. 2012). Consistent with our results, it has been demonstrated that CD4$^+$CD62L$^{high}$DO11.10 T cells which have been primed with lung CD103$^+$ DCs induced higher frequencies of CD4$^+$ T cells producing IFN-$\gamma$ than IL-4 (Furuhashi et al. 2012).

We also examined whether *Lb. rhamnosus* CRL1505 was able to improve resistance against RSV in infant mice. We demonstrated that oral or nasal administration of *Lb. rhamnosus* CRL1505 to 3-week-old BALB/c mice significantly reduced lung viral loads and tissue injuries after the challenge with RSV (Chiba et al. 2013; Tomosada et al. 2013). The protective effect achieved by the CRL1505 strain was related to its capacity to differentially modulate respiratory antiviral immune response. As observed in poly(I:C) challenge-experiments, CRL1505-treated mice were able to early increase the levels of TNF-$\alpha$ and IL-6 in the respiratory tract after RSV infection. The early increase of these cytokines together with the improved levels of IFN-$\gamma$ should explain the higher capacity of CRL1505-treated mice to reduce viral loads. In addition, *Lb. rhamnosus* CRL1505 significantly increased IL-10 levels that would contribute to protection against inflammatory damage (Chiba et al. 2013; Tomosada et al. 2013). In fact, we demonstrated that both IFN-$\gamma$ and IL-10 are necessary to achieve full protection against RSV in infant mice and that these cytokines are differently involved in the immunoprotective effect of *Lb. rhamnosus* CRL1505. Experiments using blocking anti-IFN-$\gamma$ antibodies significantly abolished the reduction of RSV titers induced by the CRL1505 strain. In addition, the reductions of lung wet:dry weight and BAL albumin concentrations induced by the immunobiotic strain were partially abolished with anti-IFN-$\gamma$ antibodies (Chiba et al. 2013). Treatment of mice with anti-IL-10R antibodies did not induce modification in the reduction of RSV titers. However, blocking IL-10/IL-10R interaction significantly abolished the capacity of the immunobiotic strain to protect against lung tissue damage (Chiba et al. 2013). Furthermore, experiments eliminating alveolar macrophages demonstrated that this respiratory immune cell population is of importance for the protective effects induced by the CRL1505 strain in the context of RSV infection (Garcia-Castillo et al. 2020).

Some recent studies have confirmed the ability of LAB strains, including *Lb. gasseri* SBT2055 (Eguchi et al. 2019), *Lb. rhamnosus* GG (Ji et al. 2021), and *Lb. mucosae* M104R01L3 (Wang et al. 2022), to enhance resistance to RSV infection. In addition, by using the pneumonia virus of mice (PVM), which is also a virus from the *Paramyxoviridae* family, it has been reported that immunobiotic LAB are able to improve protection against pneumoviruses. PVM is a natural rodent pathogen that reproduces many clinical and pathological features of the more severe forms of disease associated with human RSV. Percopo et al. (2015) demonstrated that the nasal administration of *Lb. plantarum* NCIMB 8826 improved survival from acute PVM infection and that this protective effect was associated with suppression of virus replication and diminished expression of virus-induced pro-inflammatory cytokines. Moreover, researchers demonstrated for the first time that immunobiotic lactobacilli are effective at improving protection even after the virus challenge occurred (Percopo et al. 2015; Dyer et al. 2015). Interestingly, it has been found that NOD2 and TLR2 are contributing receptors in the immunomodulator effect of the NCIMB 8826 strain since the protection against lethal virus infection is lost specifically in NOD2$^{-/-}$TLR2$^{-/-}$ mice (Rice et al. 2016).

## 29.5 BENEFICIAL EFFECTS OF IMMUNOBIOTIC LACTIC ACID BACTERIA ON SEVERE ACUTE RESPIRATORY SYNDROME CORONAVIRUS 2 INFECTION

On March 11, 2020, the WHO declared COVID-19 produced by the severe acute respiratory syndrome coronavirus 2 (SARS-CoV-2) a pandemic. The infectious disease caused by the new coronavirus is characterized by an intense inflammatory response and a marked alteration of lung functionality in the most severe cases. As the immunopathological mechanisms of COVID-19 became known, it was evident that it had some similarities with infections caused by IFV and RSV; therefore, the scientific community speculated that immunomodulatory probiotics could have some beneficial roles in the prevention of this disease. Thus, several review articles arose speculating about the potential of immunobiotics in the prevention or reduction of the severity of the infection caused by SARS-CoV-2 (Villena and Kitazawa 2020; Villena et al. 2021; Xavier-Santos et al. 2022). Recently, some concrete scientific evidence was generated in this regard. Earlier studies were performed by our group investigating the effect of the immunomodulatory strains *Lb. plantarum* MPL16 and *Lb. plantarum* CRL1506 on the resistance of human respiratory Calu-3 cells to SARS-CoV-2 infection (Islam et al. 2021a). The pretreatment of Calu-3 cells with the MPL16 or CRL1506 strains modulated the innate antiviral immune response of the RECs triggered by SARS-CoV-2 infection, reducing the virus replication. The early improved production of IFN-β and subsequently enhanced expression of the antiviral factors TLR3, DDX58, Mx1, and OAS1 induced by lactobacilli was associated with the lower SARS-CoV-2 replication. In addition, the MPL16 or CRL1506 strains differentially modulated the production of IL-6, CXCL8, CCL5, and CXCL10 in Calu-3 cells, indicating their capacities to regulate the detrimental inflammatory response (Islam et al. 2021a).

Interestingly, some studies tried to identify potential biomarkers in the respiratory microbiota of SARS-CoV-2-infected patients that could explain differences in COVID-19 severity. The pioneering study of Ventero et al. (2021) described that the genus *Prevotella* was the most abundant in the most critically ill patients. This genus of bacteria was later identified as the most abundant in patients with severe COVID-19 cases by independent studies (Merenstein et al. 2021; Rattanaburi et al. 2022). Furthermore, a negative association between the beneficial LAB genus *Dolosigranulum pigrum* and the increase of COVID-19 severity was observed in these studies (Ventero et al. 2021). Considering these findings, we aimed to investigate the effect of several *D. pigum* strains of human origin on resistance to SARS-CoV-2 infection. Like *Lb. plantarum* MPL16 and *Lb. plantarum* CRL1506, the immunomodulatory *D. pigrum* 040417 reduced viral replication in Calu-3 cells, improved the expression of type I IFNs and antiviral factors, and differentially modulated inflammatory cytokines and chemokines (Islam et al. 2021b). Although further mechanistic and *in vivo* studies are required, it is tempting to speculate that immunomodulatory bacteria, like *Lb. plantarum* MPL16, *Lb. plantarum* CRL1506, or *D. pigrum* 040417, can be an interesting alternative to reduce the severity of COVID-19.

Due to the lack of easily accessible animal models that allow the study of SARS-CoV-2 infection, studies of probiotics in such models are very scarce. It was demonstrated that a probiotic yogurt (containing *B. animalis* Bb-12, *Limosilactobacillus fermentum* PL9988, *Lb. plantarum* SN35N, *L. acidophilus* NCFM, and *B. lactis* Bi-07) administered orally once daily during three weeks before the infection of Syrian golden hamsters with SARS-CoV-2 significantly improved the body weight and lung histopathological alterations, although lung SARS-CoV-2 copy numbers were not reduced (Jeon et al. 2023). On the other hand, some clinical trials demonstrated that early probiotic administration can be a safe and effective adjunctive therapy to reduce the severity of symptoms related to COVID-19 (reviewed in Zhu et al. 2023). Among the most interesting studies are the work of Wang et al. (2021) that evaluated the efficacy of the oropharyngeal probiotic formula containing *St. thermophilus* ENT-K12 on reducing episodes of upper respiratory tract infections in COVID-19 healthcare workers. The study found that probiotic administration was able to diminish the duration of sick days, days absent from work, and days taking antibiotics and anti-viral drugs in frontline medical staff. It was also reported that the administration of a probiotic formula (*Lb. plantarum* KABP022, KABP023, and KAPB033 and *Pediococcus acidilactici* KABP021) to adult symptomatic COVID-19 outpatients reduced nasopharyngeal viral load, lung infiltrates, and duration of both digestive and non-digestive symptoms compared to placebo (Gutiérrez-Castrellón et al. 2022). These beneficial effects were associated with improved levels of anti-SARS-CoV-2 IgM and IgG. These studies clearly show the potential of beneficial LAB to improve resistance to SARS-CoV-2 and reduce the severity of COVID-19.

## 29.6 BENEFICIAL EFFECTS OF IMMUNOBIOTIC LACTIC ACID BACTERIA ON VIRAL–BACTERIAL RESPIRATORY SUPERINFECTIONS

Secondary bacterial pneumonia is a severe complication responsible for high morbidity and mortality associated with viral respiratory infections in high-risk populations (Liu et al. 2012; Muscedere et al. 2013; Bosch et al. 2013). The majority of the clinical observations and experiments in animal models have focused on post-influenza pneumococcal pneumonia (Rynda-Apple et al. 2015). In fact, several studies have evaluated how primary IFV infection enhances susceptibility to secondary pneumococcal disease by increasing bacterial attachment and colonization, disrupting epithelial barriers, and altering the innate immune response in the respiratory tract (Bosch et al. 2013). Other viruses like RVS have been also associated with an increased susceptibility to secondary pneumococcal pneumonia. Clinical and epidemiologic data suggest that RSV enhances the frequency (Weinberger et al. 2013) and severity (Cebey-Lopez et al. 2016) of pneumococcal disease. Mechanisms underlying pneumococcal superinfection include RSV-induced respiratory ciliary dyskinesia that impairs mucociliary clearance in the airways as well as the local destruction of the epithelium (Smith et al. 2014). Elevated pneumococcal adherence to the respiratory tract epithelium is also considered one of the mechanisms facilitating *Streptococcus pneumoniae* infection. RSV virions enhance pneumococcal adherence through the expression of the viral G-protein in epithelial surfaces that serve as an adhesion molecule for pneumococci (Hament et al. 2005; Smith et al. 2014; Avadhanula et al. 2007). Therefore, complex interactions exist between respiratory viruses *Strep. pneumoniae* and the host, which must be fully characterized in order to reduce the severity and mortality of respiratory superinfections caused by these pathogens, especially in young children.

As we mentioned, *Lb. rhamnosus* CRL1505 differentially regulates the antiviral innate immune response in mice after activation of TLR3 or RSV challenge. We have also shown that this protective effect can be obtained with the non-viable bacteria or its peptidoglycan (Villena et al. 2012; Chiba et al. 2013; Tomosada et al. 2013). We hypothesized that the effect of the CRL1505 strain in the respiratory antiviral innate immune response could beneficially influence the resistance to secondary bacterial infections. Therefore, in recent studies, we investigated whether nasal priming with non-viable *Lb. rhamnosus* CRL1505 or its peptidoglycan influences susceptibility to secondary pneumococcal pneumonia in infant mice (Clua et al. 2017; Clua et al. 2020; Raya Tonetti et al. 2022). An important finding of our work was that both the non-viable CRL1505 strain and its peptidoglycan improved respiratory antiviral innate immune response, reduced bacterial transmigration across the lung, and limited pulmonary inflammatory damage caused by *Strep. pneumoniae* after the activation of TLR3 or infection with RSV.

Slight but still significant reductions of pneumococcal cell counts were found in the lungs of treated mice when compared to controls. The effect of CRL1505 peptidoglycan in reducing lung pneumococcal cell counts was modest compared with our own previous studies (Medina et al. 2008; Villena et al. 2009). However, peptidoglycan treatment was able to significantly reduce lung tissue damage and bacterial dissemination into the blood stream. These findings are of importance because studies in clinical trials frequency (Weinberger et al. 2013; Cebey-Lopez et al. 2016) and animal models of RSV-*Strep. pneumoniae* superinfection (Hament et al. 2005; Smith et al. 2014) showed that enhanced lung injuries and elevated levels of bacteremia are critical factors that determine the severity of infection and the rate of mortality. In fact, peptidoglycan-treated infant mice showed a significant improvement of survival after superinfection with RSV and *Strep. pneumoniae* (Clua et al. 2017).

We speculate that the protective effect of CRL1505 peptidoglycan was associated with the reduced replication of RSV allowing lower lung tissue damage and a reduction of *Strep. pneumoniae* adhesion by diminishing the expression of RSV G protein and adhesion molecules in respiratory epithelial cells. In addition, we demonstrated that CRL1505 peptidoglycan induced a differential modulation of the respiratory innate immune response that allowed a reduction of RSV and *Strep. pneumoniae* replication with minimal inflammatory damage of lung tissue. We found that CRL1505 peptidoglycan significantly improved lung $CD3^+CD4^+IFN-\gamma^+$ and $CD3^+CD4^+IL-10^+$ T cells as well as $CD11c^+SiglecF^+IFN-\beta^+$ alveolar macrophages with the consequent increases of IFN-$\gamma$, IL-10, and IFN-$\beta$ in the respiratory tract (Clua et al. 2017). Our results demonstrated that alveolar macrophages play a key role in the beneficial modulation of the respiratory innate immune response and protection against RSV primary infection and secondary pneumococcal pneumonia (Clua et al. 2020). The generation of trained alveolar macrophages and the induction of innate immune memory in the respiratory tract would be essential for the enhanced protection induced by the CRL1505 strain against the respiratory pathogens (Clua et al. 2020; Raya Tonetti et al. 2022).

To the best of our knowledge, there are no other publications in which the effects of immunobiotics on respiratory superinfections have been evaluated. The findings of our work opened the doors for exploring in more detail the microbiological and immunological mechanisms via which immunobiotics interact with immune and non-immune cells of the respiratory tract and beneficially impact the outcome of viral-bacterial respiratory superinfections.

# 29.7 MECHANISMS INVOLVED IN THE BENEFICIAL EFFECTS OF IMMUNOBIOTIC LACTIC ACID BACTERIA ON RESPIRATORY INFECTIONS

Researchers have advanced in the understanding of the immunological mechanisms involved in the beneficial effects of LAB on respiratory health. Three potential mechanisms have been proposed to explain the effect of orally administered LAB on respiratory immunity: a) the activation of systemic and respiratory immune systems by immunomodulatory molecules released in the intestine and transported to distal sites, b) the release of cytokines and the mobilization of immune cells from the gut into the blood and distal mucosal sites, and c) the systemic metabolic reprogramming that induce the production of immunomodulatory metabolites that impact systemic and respiratory immune responses.

a) It was suggested that distal mucosal and peripheral immune cells are directly exposed to bacterial products that activate PRRs in the steady state and help to maintain the normal immune tone. There is evidence that bacterial products from gut commensals such as peptidoglycan can be absorbed and circulate throughout the host and help to modulate the normal development of immune cells (Clarke et al. 2010). In line with this hypothesis, it was speculated that bacterial products from gut commensals trigger PRRs to stimulate immune cells systemically and that factors released by those cells supported steady-state production of pro–IL-1$\beta$, pro–IL-18, and NLR proteins. This idea was sustained by their observation that intestinal injection of TLR ligands restored immune responses to IFV in antibiotic-treated mice (Ichinohe et al. 2011).

b) Kikuchi et al. (2014) showed that *Lb. plantarum* AYA fed to mice impacted in Peyer's patches inducing an activation of antigen-presenting cells and increasing the production of IL-6. Those changes promoted the differentiation of IgA$^+$ B cells into plasma cells and improved the production of mucosal IgA in both the intestine and the respiratory tract. The work also demonstrated an increased resistance to IFV infection related to an improved respiratory humoral response in *Lb. plantarum* AYA-treated mice. In line with these findings, we have demonstrated that the immunobiotic strain *Lb. rhamnosus* CRL1505 not only induce the mobilization of IgA$^+$ B cell from the gut to the respiratory tract (Salva et al. 2010) but also CD4$^+$IFN-$\gamma^+$ T cells (Garcia-Castillo et al. 2020). Considering the capacity of *Lb. rhamnosus* CRL1505 of increasing the number of intestinal CD4$^+$IFN-$\gamma^+$ T cells, we hypothesized that *Lb. rhamnosus* CRL1505 would be able to induce a mobilization of these cells into the respiratory mucosa. We demonstrated that this hypothesis was true since increased numbers of CD4$^+$IFN-$\gamma^+$ T cells were found in the lungs of *Lb. rhamnosus* CRL1505-treated mice (Villena et al. 2012; Chiba et al. 2013). Furthermore, the mobilization of CD4+IFN-γ+ T cells from the intestine to the airways and the higher production of IFN-γ are involved in the improved antiviral state induced by *Lb. rhamnosus* CRL1505. IFN-γ secreted in response to *Lb. rhamnosus* CRL1505 stimulation modulate the pulmonary innate immune microenvironment conducting to the activation of macrophages and DCs, the improvement in the production of type I IFNs, and the generation of a Th1 response with the consequent enhancement of the resistance against subsequent viral challenge (Villena et al. 2012; Chiba et al. 2013; Garcia-Castillo et al. 2020).

c) Fonseca et al. (2017) described significant systemic metabolic reprogramming induced by the oral supplementation with *Lactobacillus johnsonii*, which coincided with protection against RSV-induced immunopathology. The authors described enriched metabolites in *Lb. johnsonii*–supplemented mice during RSV infection that included a range of amino acid, lysolipids, purine, inositol, sterol, and bile metabolites and a profound difference in the concentrations of fatty acids with known immunomodulatory effects. Those metabolic effects were related to changes of the gut microbiome. Interestingly, the immunomodulatory metabolites induced by *L. johnsonii* treatment reduced airway Th2 cytokines, differentially activated DCs, and increased regulatory T cells inducing protection against inflammatory damage during RSV infection.

On the other hand, comparative studies in animal models have demonstrated that the nasal administration of immunobiotics is more efficient than the oral administration to enhance respiratory immunity (Tomosada et al. 2013; Zelaya et al. 2015; Clua et al. 2020). However, it is not clear how immunobiotics initiate cross-talk with the nasal immune system in order to modulate respiratory antiviral immunity. In addition, it is not known whether nasally administered immunobiotics change the composition or the activity of the respiratory microbiota or if those changes affect the respiratory immune responses. More research is necessary to fully understand the cellular and molecular mechanisms involved in the beneficial effect of nasally administered LAB on respiratory health in general and on antiviral immunity in particular.

When considering the mechanisms of actions of immunobiotics in respiratory immunity, the possibility must be considered that the different mechanisms proposed are not mutually exclusive, so that two or more of them may be acting together to beneficially modulate respiratory immunity. In addition, it has also been shown that the immunomodulatory properties of immunobiotics are dependent on the strains. Therefore, studies carried out with certain strains cannot be easily extrapolated to other bacteria, even those of the same genus and species. Consequently, it is still necessary to carry out deeper studies to find the molecular mechanisms by which each individual immunobiotic strain beneficially influences respiratory antiviral immunity.

# 29.8 CONCLUSIONS

As we have reviewed here, research from the last decade has clearly demonstrated that beneficial microorganisms are able to modulate respiratory tract immunity and improve resistance against respiratory viral infections. Studies in animal models have demonstrated that the mucosal priming (oral or nasal stimulation) with immunomodulatory LAB is able to improve the resistance against respiratory viruses including IFV, RSV, PVM, and SARS-CoV-2. Moreover, it was recently reported that the modulation of the antiviral immune response by immunobiotic LAB could also reduce the severity of secondary bacterial respiratory infections.

Studies evaluating the impact of immunomodulatory LAB on respiratory infections in humans have reported controversial results. While some clinical studies have shown that immunobiotic consumption appears to be a feasible way to decrease the incidence of respiratory infections, other works have demonstrated a lack of evidence to support a beneficial effect of immunobiotics on respiratory health (Long and Morris 2017; Amaral et al. 2017; Andrade et al. 2022). Due to the lack of confirmatory studies for some strains evaluated only in animal models and the varied data available, more randomized, double-blind, and placebo-controlled trials in different age populations investigating immunobiotic dose response and different routes of administration, comparing immunobiotic strains/genera, and elucidating the immunological mechanisms are necessary. The realization of appropriate clinical studies would be of fundamental importance to validate the results obtained in animal models and to promote the massive use of immunobiotics as an effective strategy to improve respiratory health worldwide. These clinical studies should also be accompanied by biotechnological developments in parallel, which allow efficient and low-cost administration of immunobiotics such as nutritional supplements or nasal sprays.

# BIBLIOGRAPHY

Albarracin, L., Garcia-Castillo, V., Masumizu, Y., Indo, Y., Islam, M.A., Suda, Y., Garcia-Cancino, A., Aso, H., Takahashi, H., Kitazawa, H., Villena, J. 2020. Efficient selection of new immunobiotic strains with antiviral effects in local and distal mucosal sites by using porcine intestinal epitheliocytes. *Front Immunol*, 11:543.

Allen, I.C., Scull, M.A., Moore, C.B., Holl, E.K., McElvania-TeKippe, E., Taxman, D.J., Guthrie, E.H., Pickles, R.J., Ting, J.P. 2009. The NLRP3 inflammasome mediates *in vivo* innate immunity to influenza A virus through recognition of viral RNA. *Immunity*, 30(4):556–565.

Amaral, M.A., Guedes, G.H.B.F., Epifanio, M., Wagner, M.B., Jones, M.H., Mattiello, R. 2017. Network meta-analysis of probiotics to prevent respiratory infections in children and adolescents. *Pediatr Pulmonol*, 52(6):833–843.

Andrade, B.G.N., Cuadrat, R.R.C., Tonetti, F.R., Kitazawa, H., Villena, J. 2022. The role of respiratory microbiota in the protection against viral diseases: Respiratory commensal bacteria as next-generation probiotics for COVID-19. *Biosci Microbiota Food Health*, 41(3):94–102.

Armstrong, S.M., Wang, C., Tigdi, J., Si, X., Dumpit, C., Charles, S., Gamage, A., Moraes, T.J., Lee, W.L. 2012. Influenza infects lung microvascular endothelium leading to microvascular leak: Role of apoptosis and claudin-5. *PLoS ONE*, 7(10):e47323.

Avadhanula, V., Wang, Y., Portner, A., Adderson, E. 2007. Non-typeable *Haemophilus influenzae* and Streptococcus pneumoniae bind respiratory syncytial virus glycoprotein. *J Med Microbiol*, 56:1133–1137.

Bem, R.A., Domachowske, J.B., Rosenberg, H.F. 2011. Animal models of human respiratory syncytial virus disease. *Am J Physiol Lung Cell Mol Physiol*, 301(2):L148–156.

Bosch, A.A., Biesbroek, G., Trzcinski, K., Sanders, E.A., Bogaert, D. 2013. Viral and bacterial interactions in the upper respiratory tract. *PLoS Pathog*, 9(1):e1003057.

Cebey-López, M., Pardo-Seco, J., Gómez-Carballa, A., Martinón-Torres, N., Martinón-Sánchez, J.M., Justicia-Grande, A. 2016. Bacteremia in children hospitalized with respiratory syncytial virus infection. *PLoS ONE*, 11(2):e0146599.

Chen, M.F., Weng, K.F., Huang, S.Y., Liu, Y.C., Tseng, S.N., Ojcius, D.M., Shih, S.R. 2017. Pretreatment with a heat-killed probiotic modulates monocyte chemoattractant protein-1 and reduces the pathogenicity of influenza and enterovirus infections. *Mucosal Immunol*, 10(1):215–227.

Chiba, E., Tomosada, Y., Vizoso-Pinto, M.G., Salva, S., Takahashi, T., Tsukida, K., Kitazawa, H., Alvarez, S., Villena, J. 2013. Immunobiotic *Lactobacillus rhamnosus* improves resistance of infant mice against respiratory syncytial virus infection. *Int Immunopharmacol*, 17:373–382.

Chiu, C., Openshaw, P.J. 2015. Antiviral B cell and T cell immunity in the lungs. *Nat Immunol*, 16(1):18–26.

Cilloniz, C., Pantin-Jackwood, M.J., Ni, C., Carter, V.S., Korth, M.J., Swayne, D.E., Tumpey, T.M., Katze, M.G. 2012. Molecular signatures associated with Mx1-mediated resistance to highly pathogenic influenza virus infection: Mechanisms of survival. *J Virol*, 86(5):2437–2446.

Clarke, T.B., Davis, K.M., Lysenko, E.S., Zhou, A.Y., Yu, Y., Weiser, J.N. 2010. Recognition of peptidoglycan from the microbiota by Nod1 enhances systemic innate immunity. *Nat Med*, 16(2):228–231.

Clua, P., Kanmani, P., Zelaya, H., Tada, A., Kober, A.K.M.H., Salva, S., Alvarez, S., Kitazawa, H., Villena, J. 2017. Peptidoglycan from immunobiotic *Lactobacillus rhamnosus* improves resistance of infant mice to Respiratory Syncytial Viral infection and secondary pneumococcal pneumonia. *Front Immunol*, 8:948.

Clua, P., Tomokiyo, M., Raya Tonetti, F., Islam, M.A., García Castillo, V., Marcial, G., Salva, S., Alvarez, S., Takahashi, H., Kurata, S., Kitazawa, H., Villena, J. 2020. The role of alveolar macrophages in the improved protection against respiratory syncytial virus and pneumococcal superinfection induced by the peptidoglycan of *Lactobacillus rhamnosus* CRL1505. *Cells*, 9(7):1653.

Dyer, K.D., Drummond, R.A., Rice, T.A., Percopo, C.M., Brenner, T.A., Barisas, D.A., Karpe, K.A., Moore, M.L., Rosenberg, H.F. 2015. Priming of the respiratory tract with immunobiotic *Lactobacillus plantarum* limits infection of alveolar macrophages with recombinant Pneumonia Virus of Mice (rK2-PVM). *J Virol*, 90(2):979–991.

Eguchi, K., Fujitani, N., Nakagawa, H., Miyazaki, T. 2019. Prevention of respiratory syncytial virus infection with probiotic lactic acid bacterium *Lactobacillus gasseri* SBT2055. *Sci Rep*, 9(1):4812.

Everitt, A.R., Clare, S., Pertel, T., John, S.P., Wash, R.S., Smith, S.E., et al. 2012. IFITM3 restricts the morbidity and mortality associated with influenza. *Nature*, 484(7395):519–523.

Fonseca, W., Lucey, K., Jang, S., Fujimura, K.E., Rasky, A., Ting, H.A., Petersen, J., Johnson, C.C., Boushey, H.A., Zoratti, E., Ownby, D.R., Levine, A.M., Bobbit, K.R., Lynch, S.V., Lukacs, N.W. 2017. *Lactobacillus johnsonii* supplementation attenuates respiratory viral infection via metabolic reprogramming and immune cell modulation. *Mucosal Immunol*, 10(6):1569–1580.

Furuhashi, K., Suda, T., Hasegawa, H., Suzuki, Y., Hashimoto, D., Enomoto, N., Fujisawa, T., Nakamura, Y., Inui, N., Shibata, K., Nakamura, H., Chida, K. 2012. Mouse lung CD103+ and CD11bhigh dendritic cells preferentially induce distinct CD4+ T-cell responses. *Am J Respir Cell Mol Biol*, 46:165–172.

Garcia-Castillo, V., Tomokiyo, M., Raya Tonetti, F., Islam, M.A., Takahashi, H., Kitazawa, H., Villena, J. 2020. Alveolar macrophages are key players in the modulation of the respiratory antiviral immunity induced by orally administered *Lacticaseibacillus rhamnosus* CRL1505. *Front Immunol*, 11:568636.

Goraya, M.U., Wang, S., Munir, M., Chen, J.L. 2015. Induction of innate immunity and its perturbation by influenza viruses. *Protein Cell*, 6(10):712–721.

Graham, B.S., Perkins, M.D., Wright, P.F., Karzon, D.T. 1988. Primary respiratory syncytial virus infection in mice. *J Med Virol*, 26:153–162.

Groskreutz, D.J., Monick, M.M., Powers, L.S., Yarovinsky, T.O., Look, D.C. Hunninghake, G.W. 2006. Respiratory syncytial virus induces TLR3 protein and protein kinase R, leading to increased double-stranded RNA responsiveness in airway epithelial cells. *J Immunol*, 176:1733–1740.

Gutiérrez-Castrellón, P., Gandara-Martí, T., Abreu, Y., Abreu, A.T., et al. 2022. Probiotic improves symptomatic and viral clearance in Covid19 outpatients: A randomized, quadruple-blinded, placebo-controlled trial. *Gut Microbes*, 14(1):2018899.

Hament, J.M., Aerts, P.C., Fleer, A., van Dijk, H., Harmsen, T., Kimpen, J.L., Wolfs, T.F. 2005. Direct binding of Respiratory Syncytial Virus to pneumococci: A phenomenon that enhances both pneumococcal adherence to human epithelial cells and pneumococcal invasiveness in a murine model. *Pediatr Res*, 58(6):1198–1203.

Harata, G., He, F., Hiruta, N., Kawase, M., Kubota, A., Hiramatsu, M., Yausi, H. 2010. Intranasal administration of *Lactobacillus rhamnosus* GG protects mice from H1N1 influenza virus infection by regulating respiratory immune responses. *Lett Appl Microbiol*, 50(6):597–602.

Hewitt, R.J., Lloyd, C.M. 2021. Regulation of immune responses by the airway epithelial cell landscape. *Nat Rev Immunol*, 21(6):347–362.

Hori, T., Kiyoshima, J., Shida, K., Yasui, H. 2001. Effect of intranasal administration of *Lactobacillus casei* Shirota on influenza virus infection of upper respiratory tract in mice. *Clin Diagn Lab Immunol*, 8(3):593–597.

Hori, T., Kiyoshima, J., Shida, K., Yasui, H. 2002. Augmentation of cellular immunity and reduction of influenza virus titer in aged mice fed *Lactobacillus casei* strain Shirota. *Clin Diagn Lab Immunol*, 9(1):105–108.

Ichinohe, T., Pang, I.K., Kumamoto, Y., Peaper, D.R., Ho, J.H., Murray, T.S., Iwasaki, A. 2011. Microbiota regulates immune defense against respiratory tract influenza A virus infection. *Proc Natl Acad Sci USA*, 108(13):5354–5359.

Islam, M.A., Albarracin, L., Melnikov, V., Andrade, B.G.N., Cuadrat, R.R.C., Kitazawa, H., Villena, J. 2021b. *Dolosigranulum pigrum* modulates immunity against SARS-CoV-2 in respiratory epithelial cells. *Pathogens*, 10(6):634.

Islam, M.A., Albarracin, L., Tomokiyo, M., Valdez, J.C., Sacur, J., Vizoso-Pinto, M.G., Andrade, B.G.N., Cuadrat, R.R.C., Kitazawa, H., Villena, J. 2021a. Immunobiotic Lactobacilli improve resistance of respiratory epithelial cells to SARS-CoV-2 infection. *Pathogens*, 10(9):1197.

Izumo, T., Maekawa, T., Ida, M., Noguchi, A., Kitagawa, Y., Shibata, H., Yasui, H., Kiso, Y. 2010. Effect of intranasal administration of *Lactobacillus pentosus* S-PT84 on influenza virus infection in mice. *Int Immunopharmacol*, 10(9):1101–1106.

Jeon, H.Y., Kim, K.S., Kim, S. 2023. Effects of yogurt containing probiotics on respiratory virus infections: Influenza H1N1 and SARS-CoV-2. *J Dairy Sci*, 106(3):1549–1561.

Ji, J.J., Sun, Q.M., Nie, D.Y., Wang, Q., Zhang, H., Qin, F.F., Wang, Q.S., Lu, S.F., Pang, G.M., Lu, Z.G. 2021. Probiotics protect against RSV infection by modulating the microbiota-alveolar-macrophage axis. *Acta Pharmacol Sin*, 42(10):1630–1641.

Johnson, N.P.A.S., Mueller, J. 2002. Updating the accounts: Global mortality of the 1918–1920 "Spanish" influenza pandemic. *Bull Hist Med*, 76:105–115.

Kaiko, G.E., Horvat, J.C., Beagley, K.W., Hansbro, P.M. 2008. Immunological decision-making: How does the immune system decide to mount a helper T-cell response? *Immunology*, 123(3):326–338.

Kawahara, T., Takahashi, T., Oishi, K., Tanaka, H., Masuda, M., Takahashi, S., Takano, M., Kawakami, T., Fukushima, K., Kanazawa, H., Suzuki, T. 2015. Consecutive oral administration of *Bifidobacterium longum* MM-2 improves the defense system against influenza virus infection by enhancing natural killer cell activity in a murine model. *Microbiol Immunol*, 59(1):1–12.

Kawase, M., He, F., Kubota, A., Harata, G., Hiramatsu, M. 2010. Oral administration of lactobacilli from human intestinal tract protects mice against influenza virus infection. *Lett Appl Microbiol*, 51(1):6–10.

Kikuchi, Y., Kunitoh-Asari, A., Hayakawa, K., Imai, S., Kasuya, K., Abe, K., Adachi, Y., Fukudome, S., Takahashi, Y., Hachimura, S. 2014. Oral administration of *Lactobacillus plantarum* strain AYA enhances IgA secretion and provides survival protection against influenza virus infection in mice. *PLoS ONE*, 9(1):e86416.

Kobayashi, N., Saito, T., Uematsu, T., Kishi, K., Toba, M., Kohda, N., Suzuki, T. 2011. Oral administration of heat-killed *Lactobacillus pentosus* strain b240 augments protection against influenza virus infection in mice. *Int Immunopharmacol*, 11(2):199–203.

Kumar, B., Asha, K., Khanna, M., Ronsard, L., Meseko, C.A., Sanicas, M. 2018. The emerging influenza virus threat: Status and new prospects for its therapy and control. *Arch Virol*, 163(4):831–844.

Liu, L., Johnson, H.L., Cousens, S., Perin, J., Scott, S., Lawn, J.E., Rudan, I., Campbell, H., Cibulskis, R., Li, M. 2012. Child health epidemiology reference group of WHO and UNICEF. Global, regional, and national causes of child mortality: An updated systematic analysis for 2010 with time trends since 2000. *Lancet*, 379:2151–2161.

Loebbermann, J., Schnoeller, C., Thornton, H., Durant, L., Sweeney, N.P., Schuijs, M., O'Garra, A., Johansson, C., Openshaw, P.J. 2012. IL-10 regulates viral lung immunopathology during acute respiratory syncytial virus infection in mice. *PLoS ONE*, 7:e32371.

Long, J.D., Morris, A. 2017. Probiotics in preventing acute upper respiratory tract infections. *Am J Nurs*, 117(12):69.

Maeda, N., Nakamura, R., Hirose, Y., Murosaki, S., Yamamoto, Y., Kase, T., Yoshikai, Y. 2009. Oral administration of heat-killed *Lactobacillus plantarum* L-137 enhances protection against influenza virus infection by stimulation of type I interferon production in mice. *Int Immunopharmacol*, 9(9):1122–1125.

Medina, M., Villena, J., Salva, S., Vintiñi, E., Langella, P., Alvarez, S. 2008. Nasal administration of *Lactococcus lactis* improves the local and systemic immune responses against *Streptococcus pneumoniae*. *Microbiol Immunol*, 52:399–409.

Merenstein, C., Liang, G., Whiteside, S.A., Cobián-Güemes, A.G., et al. 2021. Signatures of COVID-19 Severity and immune response in the respiratory tract microbiome. *MBio*, 12:e0177721.

Mettelman, R.C., Allen, E.K., Thomas, P.G. 2022. Mucosal immune responses to infection and vaccination in the respiratory tract. *Immunity*, 55(5):749–780.

Mukherjee, S., Lindell, D.M., Berlin, A.A., Morris, S.B., Shanley, T.P., Hershenson, M.B., Lukacs, N.W. 2011. IL-17-induced pulmonary pathogenesis during respiratory viral infection and exacerbation of allergic disease. *Am J Pathol*, 179:248–258.

Muscedere, J., Ofner, M., Kumar, A. 2013. The occurrence and impact of bacterial organisms complicating critical care illness associated with 2009 influenza A (H1N1) infection. *Chest*, 144:39–47.

Nakayama, Y., Moriya, T., Sakai, F., Ikeda, N., Shiozaki, T., Hosoya, T., Nakagawa, H., Miyazaki, T. 2014. Oral administration of *Lactobacillus gasseri* SBT2055 is effective for preventing influenza in mice. *Sci Rep*, 4:4638.

Pangesti, K.N.A., Abd, E. Ghany, M., Walsh, M.G., Kesson, A.M., Hill-Cawthorne, G.A. 2018. Molecular epidemiology of respiratory syncytial virus. *Rev Med Virol*, 28(2).

Percopo, C.M., Rice, T.A., Brenner, T.A., Dyer, K.D., Luo, J.L., Kanakabandi, K., Sturdevant, D.E., Porcella, S.F., Domachowske, J.B., Keicher, J.D., Rosenberg, H.F. 2015. Immunobiotic *Lactobacillus* administered post-exposure averts the lethal sequelae of respiratory virus infection. *Antiviral Res*, 121:109–119.

Rattanaburi, S., Sawaswong, V., Chitcharoen, S., Sivapornnukul, P., Nimsamer, P., Suntronwong, N., Puenpa, J., Poovorawan, Y., Payungporn, S. 2022. Bacterial microbiota in upper respiratory tract of COVID-19 and influenza patients. *Exp Biol Med*, 247:409–415.

Raya Tonetti, F., Clua, P., Fukuyama, K., Marcial, G., Sacur, J., Marranzino, G., Tomokiyo, M., Vizoso-Pinto, G., Garcia-Cancino, A., Kurata, S., Kitazawa, H., Villena, J. 2022. The ability of postimmunobiotics from *L. rhamnosus* CRL1505 to protect against respiratory syncytial virus and pneumococcal super-infection is a strain-dependent characteristic. *Microorganisms*, 10(11):2185.

Raya Tonetti, F., Islam, M.A., Vizoso-Pinto, M.G., Takahashi, H., Kitazawa, H., Villena, J. 2020. Nasal priming with immunobiotic lactobacilli improves the adaptive immune response against influenza virus. *Int Immunopharmacol*, 78:106115.

Rice, T.A., Brenner, T.A., Percopo, C.M., Ma, M., Keicher, J.D., Domachowske, J.B., Rosenberg, H.F., 2016. Signaling via pattern recognition receptors NOD2 and TLR2 contributes to immunomodulatory control of lethal pneumovirus infection. *Antiviral Res*, 132:131–140.

Ronni, T., Matikainen, S., Sareneva, T., Melen, K., Pirhonen, J., Keskinen, P., Julkunen, I. 1997. Regulation of IFN-alpha/beta, MxA, 2',5'-oligoadenylate synthetase, and HLA gene expression in influenza A-infected human lung epithelial cells. *J Immunol*, 158(5):2363–2374.

Rudd, B.D., Burstein, E., Duckett, C.S., Li, X., Lukacs, N.W. 2005. Differential role for TLR3 in respiratory syncytial virus-induced chemokine expression. *J Virol*, 79:3350–3357.

Rutigliano, J.A., Graham, B.S. 2004. Prolonged production of TNF-alpha exacerbates illness during respiratory syncytial virus infection. *J Immunol*, 173:3408–3417.

Rynda-Apple, A., Robinson, K.M., Alcorn, J.F. 2015. Influenza and bacterial superinfection: Illuminating the immunologic mechanisms of disease. *Infect Immun*, 83(10):3764–3770.

Salva, S., Villena, J., Alvarez, S. 2010. Immunomodulatory activity of *Lactobacillus rhamnosus* strains isolated from goat milk: Impact on intestinal and respiratory infections. *Int J Food Microbiol*, 141(1–2):82–89.

Sato, S., Kiyono, H. 2012. The mucosal immune system of the respiratory tract. *Curr Opin Virol*, 2:225–232.

Smith, C.M., Sandrini, S., Datta, S., Freestone, P., Shafeeq, S., Radhakrishnan, P., Williams, G., Glenn, S.M., Kuipers, O.P., Hirst, R.A., Easton, A.J., Andrew, P.W., O'Callaghan, C. 2014. Respiratory Syncytial Virus increases the virulence of *Streptococcus pneumoniae* by binding to penicillin binding protein 1a: A new paradigm in respiratory infection. *Am J Respir Crit Care Med*, 190(2):196–207.

Stowell, N.C., Seideman, J., Raymond, H.A., Smalley, K.A., Lamb, R.J., Egenolf, D.D., Bugelski, P.J., Murray, L.A., Marsters, P.A., Bunting, R.A., Flavell, R.A., Alexopoulou, L., San Mateo, L.R., Griswold, D.E., Sarisky, R.T., Mbow, M.L., Das, A.M. 2009. Long-term activation of TLR3 by poly(I:C) induces inflammation and impairs lung function in mice. *Respir Res*, 10:43.

Sun, J., Madan, R., Karp, C.L., Braciale, T.J. 2009. Effector T cells control lung inflammation during acute influenza virus infection by producing IL-10. *Nat Med*, 15:277–284.

Takeda, S., Takeshita, M., Kikuchi, Y., Dashnyam, B., Kawahara, S., Yoshida, H., Watanabe, W., Muguruma, M., Kurokawa, M. 2011. Efficacy of oral administration of heat-killed probiotics from Mongolian dairy products against influenza infection in mice: Alleviation of influenza infection by its immunomodulatory activity through intestinal immunity. *Int Immunopharmacol*, 11(12):1976–1983.

Thomas, P.G., Dash, P., Aldridge, J.R. Jr., Ellebedy, A.H., Reynolds, C., Funk, A.J., Martin, W.J., Lamkanfi, M., Webby, R.J., Boyd, K.L., Doherty, P.C., Kanneganti, T.D. 2009. The intracellular sensor NLRP3 mediates key innate and healing responses to influenza A virus via the regulation of caspase-1. *Immunity*, 30(4):566–575.

Tomosada, Y., Chiba, E., Zelaya, H., Takahashi, T., Tsukida, K., Kitazawa, H., Alvarez, S. Villena, J. 2013. Nasally administered *Lactobacillus rhamnosus* strains differentially modulate respiratory antiviral immune responses and induce protection against respiratory syncytial virus infection. *BMC Immunol*, 14:40.

Twigg, H.L. 2005. Humoral immune defense (antibodies): Recent advances. *Proc Am Thorac Soc*, 2:417–421.

Ventero, M.P., Cuadrat, R.R.C., Vidal, I., Andrade, B.G.N., Molina-Pardines, C., Haro-Moreno, J.M., Coutinho, F.H., Merino, E., Regitano, L.C.A., Silveira, C.B., Afli, H., López-Pérez, M., Rodríguez, J.C. 2021. Nasopharyngeal microbial communities of patients infected with SARS-CoV-2 that developed COVID-19. *Front Microbiol*, 12:637430.

Villena, J., Barbieri, N.P., Salva, S., Herrera, H.M., Alvarez, S. 2009. Nasal treatment with *Lactobacillus casei* enhances immunty against pneumococcal challenge in malnourished mice. *Microbiol Immunol*, 53:636–646.

Villena, J., Chiba, E., Tomosada, Y., Salva, S., Marranzino, G., Kitazawa, H., Alvarez, S. 2012. Orally administered *Lactobacillus rhamnosus* modulates the respiratory immune response triggered by the viral pathogen-associated molecular pattern poly(I:C). *BMC Immunol*, 13:53.

Villena, J., Kitazawa, H. 2020. The modulation of mucosal antiviral immunity by immunobiotics: Could they offer any benefit in the SARS-CoV-2 pandemic? *Front Physiol*, 11:699.

Villena, J., Li, C., Vizoso-Pinto, M.G., Sacur, J., Ren, L., Kitazawa, H. 2021. *Lactiplantibacillus plantarum* as a potential adjuvant and delivery system for the development of SARS-CoV-2 oral vaccines. *Microorganisms*, 9(4):683.

Waki, N., Yajima, N., Suganuma, H., Buddle, B.M., Luo, D., Heiser, A., Zheng, T. 2014. Oral administration of *Lactobacillus brevis* KB290 to mice alleviates clinical symptoms following influenza virus infection. *Lett Appl Microbiol*, 58(1):87–93.

Wang, Q., Fang, Z., Li, L., Wang, H., Zhu, J., Zhang, P., Lee, Y.K., Zhao, J., Zhang, H., Lu, W., Chen, W. 2022. *Lactobacillus mucosae* exerted different antiviral effects on respiratory syncytial virus infection in mice. *Front Microbiol*, 13:1001313.

Wang, Q., Lin, X., Xiang, X., Liu, W., Fang, Y., Chen, H., Tang, F., Guo, H., Chen, D., Hu, X., Wu, Q., Zhu, B., Xia, J. 2021. Oropharyngeal probiotic ENT-K12 prevents respiratory tract infections among frontline medical staff fighting against COVID-19: A pilot study. *Front Bioeng Biotechnol*, 9:646184.

Weinberger, D.M., Givon-Lavi, N., Shemer-Avni, Y., Bar-Ziv, J., Alonso, W.J., Greenberg, D., Dagan, R. 2013. Influence of pneumococcal vaccines and respiratory syncytial virus on alveolar pneumonia, Israel. *Emerg Infect Dis*, 19:1084–1091.

Weiss, K.A., Christiaansen, A.F., Fulton, R.B., Meyerholz, D.K., Varga, S.M. 2011. Multiple CD4+ T cell subsets produce immunomodulatory IL-10 during respiratory syncytial virus infection. *J Immunol*, 187:3145–3154.

World Health Organization, 2013. *Background Document: Research Needs for the Battle against Respiratory Viruses*. Geneva, Switzerland.

Xavier-Santos, D., Padilha, M., Fabiano, G.A., Vinderola, G., Gomes Cruz, A., Sivieri, K., Costa Antunes, A.E. 2022. Evidence and perspectives of the use of probiotics, prebiotics, synbiotics, and postbiotics as adjuvants for prevention and treatment of COVID-19: A bibliometric analysis and systematic review. *Trends Food Sci Technol*, 120:174–192.

Xing, J.H., Shi, C.W., Sun, M.J., Gu, W., et al. 2022. *Lactiplantibacillus plantarum* 0111 protects against influenza virus by modulating intestinal microbial-mediated immune responses. *Front Microbiol*, 13:820484.

Yasui, H., Kiyoshima, J., Hori, T. 2004. Reduction of influenza virus titer and protection against influenza virus infection in infant mice fed *Lactobacillus casei* Shirota. *Clin Diagn Lab Immunol*, 11(4):675–679.

Yasui, H., Kiyoshima, J., Hori, T., Shida, K. 1999. Protection against influenza virus infection of mice fed *Bifidobacterium breve* YIT4064. *Clin Diagn Lab Immunol*, 6(2):186–192.

Zelaya, H., Tada, A., Vizoso-Pinto, M.G., Salva, S., Kanmani, P., Agüero, G., Alvarez, S., Kitazawa, H., Villena, J. 2015. Nasal priming with immunobiotic *Lactobacillus rhamnosus* modulates inflammation-coagulation interactions and reduces influenza virus-associated pulmonary damage. *Inflamm Res*, 64(8):589–602.

Zelaya, H., Tsukida, K., Chiba, E., Marranzino, G., Alvarez, S., Kitazawa, H., Villena, J. 2014. Immunobiotic lactobacilli reduce viral-associated pulmonary damage through the modulation of inflammation-coagulation interactions. *Int Immunopharmacol*, 19(1):161–173.

Zhu, J., Pitre, T., Ching, C., Zeraatkar, D., Gruchy, S. 2023. Safety and efficacy of probiotic supplements as adjunctive therapies in patients with COVID-19: A systematic review and meta-analysis. *PLoS ONE*, 18(3):e0278356.

# Human Studies on Probiotics for Infants and Children

# 30

Hania Szajewska

## 30.1 INTRODUCTION

Despite many years of extensive research, the role of probiotics in the treatment or prevention of diseases often remains uncertain. In many settings, the use of probiotics continues to rely on health claims made by manufacturers. Until health and nutritional claims regulate the use of probiotics, it is crucial that clinicians understand the various strains and preparations that are commercially available and advise on the use of these products accordingly. In this chapter, the current evidence based on the latest meta-analyses of randomized controlled trials (RCTs) on the efficacy of probiotics in children, mainly for gastrointestinal disorders, is summarized. To identify relevant data, searches of PubMed databases were performed in March 2023, focusing on publications published within the last 5 years. Additionally, PubMed was searched for clinical practice guidelines developed by respected scientific societies or expert groups. As the effects of probiotics are strain specific, emphasis was placed on identifying probiotics with documented efficacy rather than probiotics in general. Caution should be exercised to avoid overinterpreting results when all probiotics are evaluated together. For a summary of the clinical effects of probiotics in children, please see Table 30.1.

**TABLE 30.1** Effects of Probiotics in Children. ESPGHAN and AGA Recommendations

| CONDITION | SOCIETY | RECOMMENDATION |
|---|---|---|
| Treatment of acute gastroenteritis | ESPGHAN 2023 | Conditional (weak) recommendation for<br>• *S. boulardii* (250–750 mg/day, for 5–7 days) (low to very low certainty of evidence)<br>• *L. rhamnosus* GG ($\geq 10^{10}$ CFU/day, typically 5–7 day) (very low certainty of evidence)<br>• *L. reuteri* DSM 17938 ($1 \times 10^8$ to $2 \times 10^8$ to $4 \times 10^8$ CFU/day, for 5 days) (low to very low certainty of evidence)<br>• *L. rhamnosus* 19070–2 and L reuteri DSM 12246 ($2 \times 10^{10}$ CFU of each strain/d, for 5 days) (very low certainty of evidence)<br>Strong recommendation against<br>• *L. helveticus* R0052 and *L. rhamnosus* R0011 (moderate certainty of evidence)<br>Weak recommendation against<br>• *Bacillus clausii* O/C, SIN, N/R, and T (very low certainty of evidence) |

DOI: 10.1201/9781003352075-33

**TABLE 30.1**    (Continued) Effects of Probiotics in Children. ESPGHAN and AGA Recommendations

| CONDITION | SOCIETY | RECOMMENDATION |
|---|---|---|
| | AGA 2020 | Against the use of probiotics in children with acute infectious gastroenteritis in North America (conditional recommendation, moderate quality of evidence) |
| Prevention of AAD | ESPGHAN 2023 | Strong recommendation for:<br>• *L. rhamnosus* GG (moderate quality of evidence)<br>• *S. boulardii* (moderate quality of evidence) |
| | AGA 2020 | Not addressed |
| Prevention of C *difficile* diarrhea | ESPGHAN 2016 | *S. boulardii* (weak recommendation; moderate quality of evidence) |
| | AGA 2020 | • Conditional recommendation, low quality of evidence:<br>• *S. boulardii*;<br>• A two-strain combination of *L. acidophilus* CL 1285 & *L. casei* LBC80R;<br>• A three-strain combination of *L. acidophilus*, *L. delbruekii* subsp. *bulgaricus*, and *B. bifidum*;*<br>• A four-strain combination of *L. acidophilus*, *L. delbruekii* subsp. *bulgaricus*, *B. bifidum*, and *Streptococcus thermophilus** |
| Prevention of nosocomial diarrhea | ESPGHAN 2023 | *L. rhamnosus* GG (at least $10^9$ CFU/day) for the duration of the hospital stay |
| Prevention of necrotizing enterocolitis | ESPGHAN 2023 | Conditional recommendation for<br>• *L. rhamnosus* GG ATCC53103 (at a dose ranging from $1 \times 10^9$ CFU to $6 \times 10^9$ CFU) (low certainty of evidence)<br>• *B. infantis* Bb-02, *B. lactis* Bb-12, and *S. thermophilus* TH-4 at 3.0 to 3.5 × $10^8$ CFU (of each strain) (low certainty of evidence).<br>No recommendation for or against<br>• *L. reuteri* DSM 17938 (very low certainty of evidence).<br>• *B. bifidum* NCDO 1453 and *L. acidophilus* NCDO 1748 (very low certainty of evidence)<br>Conditional recommendation against (low to moderate certainty of evidence)<br>• *B. breve* BBG-001<br>• *S. boulardii* |
| | AGA 2020 | Combination of *Lactobacillus* spp. and *Bifidobacterium* spp.:<br>• *L. rhamnosus* ATCC 53103 and *B. longum* subsp. *infantis*;*<br>• *L. casei* and *B. breve*;*<br>• *L. rhamnosus*, *L. acidophilus*, *L. casei*, *B. longum* subsp. *infantis*, *B bifidum*, and *B. longum* subsp. *longum*;*<br>• *L. acidophilus* and *B. longum* subsp. *infantis*;*<br>• *L. acidophilus* and *B. bifidum*;*<br>• *L. rhamnosus* ATCC 53103 and *B. longum* Reuter ATCC BAA-999;<br>• *L. acidophilus*, *B. bifidum*, *B. animalis* subsp. *lactis*, and *B. longum* subsp. *longum*;*<br>• *B. animalis* subsp *lactis* (including DSM 15954),<br>• *L. reuteri* (DSM 17938 or ATCC 55730),<br>• *L. rhamnosus* (ATCC 53103 or ATCA07FA or LCR 35) |
| *H. pylori* infection | ESPGHAN 2023 | *S. boulardii* (weak recommendation; very low certainty of evidence) |
| Crohn's disease | ESPGHAN 2023 | Not recommended |
| | AGA 2020 | Against the use of probiotics, unless in the context of a clinical trial |

*(Continued)*

**TABLE 30.1** (Continued) Effects of Probiotics in Children. ESPGHAN and AGA Recommendations

| CONDITION | SOCIETY | RECOMMENDATION |
|---|---|---|
| Ulcerative colitis | ESPGHAN & ECCO 2018 | A mixture of eight strains# or *Escherichia coli* Nissle 1917 |
| | ESPGHAN 2023 | Not recommended |
| | AGA 2020 | Against the use of probiotics, unless in the context of a clinical trial |
| Pouchitis | ESPGHAN 2018 | A mixture of eight strains# |
| | AGA 2020 | A mixture of eight strains# |
| Functional abdominal pain disorders, including IBS | ESPGHAN 2023 | *L. reuteri* DSM 17938 (at a dose of $10^8$ CFU to $2 \times 10^8$ CFU/day) for pain intensity reduction (weak recommendation; moderate certainty of evidence: moderate)<br><br>*L. rhamnosus* GG (at a dose of 109 CFU to $3 \times 10^9$ CFU twice daily) for the reduction of pain frequency and intensity in children with IBS (weak recommendation; moderate certainty of evidence: moderate) |
| | AGA 2020 | IBS. Only in the context of a clinical trial |
| Infantile colic | ESPGHAN 2023 | • *L. reuteri* DSM 17938 (at least $10^8$ CFU/day for at least 21 days) in breastfed infants (weak recommendation; moderate certainty of evidence). No recommendation can be made *for* or *against* the use of *L. reuteri* DSM 17938 in formula-fed infants due to insufficient evidence<br>• *B. lactis* BB-12 (at least $10^8$ CFU/day, for 21–28 days) (weak recommendation; moderate certainty of evidence) |
| Functional constipation | ESPGHAN 2023 | Not recommended |

*Abbreviations:* AGA, American Gastroenterology Association; ECCO, European Crohn's and Colitis Organization; ESPGHAN, European Society for Paediatric Gastroenterology, Hepatology and Nutrition; IBS, irritable bowel syndrome; LGG, *Lacticaseibacillus rhamnosus* (formerly known as *Lactobacillus*) *rhamnosus* GG.

* No strain specification was given for any of the strains.
# *L. paracasei* DSM 24733, *L. plantarum* DSM 24730, *L. acidophilus* DSM 24735, *L. delbrueckii subsp. bulgaricus* DSM 24734, *B. longum* DSM 24736, *B. infantis* DSM 24737, *B. breve* DSM 24732, and *S. thermophilus* DSM 247.

# 30.2 TREATMENT OF ACUTE GASTROENTERITIS

Acute gastroenteritis is one of the most common illnesses affecting children. As it is usually a self-limited illness lasting only 5–7 days, treatment focuses on preventing dehydration, metabolic acidosis, and electrolyte imbalances. In most patients with mild to moderate dehydration, oral rehydration solutions are effective in achieving this. However, oral rehydration is often underutilized despite its proven efficacy (Guarino & Albano, 2001). This is primarily because oral rehydration solutions do not reduce the frequency of bowel movements or fluid loss, nor do they shorten the duration of the illness, which reduces their acceptance and generates interest in adjunctive treatments.

Until 2020, many, if not all, professional societies and group of experts advocated use of probiotics with documented efficacy for the management of acute gastroenteritis (Guarino et al., 2014; Gwee et al., 2018; Marchand, 2012; Szajewska et al., 2014). Currently, the recommendations differ, possibly reflecting negative (null) studies questioning the efficacy of some strains with previous positive recommendations (Schnadower et al., 2018; Szymański & Szajewska, 2019).

In 2020, the American Gastroenterology Association (AGA), based on the evaluation of 89 trials, made a conditional recommendation *against* the use of probiotics in children from North America with acute infectious gastroenteritis (moderate quality of evidence)

The rationale for the negative AGA recommendation was that most of the studies were performed outside North America. Moreover, two large, high-quality null trials, performed in Canada and the US, questioned the efficacy of probiotics, or more specifically the probiotics evaluated in these studies, for the management of children with acute gastroenteritis (Freedman et al., 2018; Schnadower et al., 2018).

In 2020 (Szajewska et al.) and then in 2023 (Szajewska et al.), the European Society for Paediatric Gastroenterology, Hepatology and Nutrition (ESPGHAN) Working Group on Probiotics made weak (conditional) recommendations for *S. boulardii, L. rhamnosus* GG, *L. reuteri* DSM 17938, and *L. rhamnosus* 19070–2 and *L. reuteri* DSM 12246 (very low certainty of evidence). Regarding *S. boulardii*, many trials did not specify the strain designation. However, when information was available or assessed retrospectively, the strain most commonly used was recently identified as *S boulardii* CNCM I-745. The Working Group made a strong recommendation against *L. helveticus* R0052 and *L. rhamnosus* R0011 (moderate certainty of evidence) and a weak (conditional) recommendation against *Bacillus clausii* strains O/C, SIN, N/R, and T (very low certainty of evidence).

## 30.3 PREVENTION OF NOSOCOMIAL DIARRHEA

Nosocomial diarrhea is any diarrhea that a patient contracts in a healthcare institution more than 72 hours after admission. Nosocomial diarrhea can prolong hospital stays and increase medical costs (Festini et al., 2010). If probiotics for preventing nosocomial diarrhea in children are considered, the ESPGHAN recommends using *L. rhamnosus* GG at a dose of at least $10^9$ CFU/day for the duration of hospital stay (moderate quality of evidence; strong recommendation) (Hojsak et al., 2018; Szajewska et al., 2023).

## 30.4 PREVENTION OF ANTIBIOTIC-ASSOCIATED DIARRHEA

The frequency of antibiotic-associated diarrhea (AAD) varies with the definitions used. Its incidence ranges from 4.3% to 80%, with a median incidence of 22% (McFarland et al., 2016). Patients with AAD have a mean age ranging from 18 to 48 months (McFarland et al., 2016). The age of the child and the type of antibiotic used are the main risk factors for AAD in children. AAD may occur just a few hours after antibiotic administration or up to 8 weeks later, and it is associated with increased costs and hospital length of stay. The mechanism of AAD remains unclear. A direct effect of the antibiotics on the intestinal mucosa, resulting in alterations in the gut microbiota composition and the overgrowth of pathogens, is one potential mechanism. *Clostridioides difficile* (formerly *Clostridium difficile*) is the most frequent infectious cause of AAD (McFarland et al., 2016).

For preventing AAD, in 2022, the ESPGHAN Working Group on Probiotics recommended using *L. rhamnosus* GG (moderate quality of evidence, strong recommendation) or *S. boulardii* (moderate quality of evidence, strong recommendation). Other strains or combinations of strains have been tested, but sufficient evidence is still lacking (Szajewska et al., 2023). If the use of probiotics for preventing *C difficile*-associated diarrhea is considered, the ESPGHAN suggested using *S. boulardii* (low quality of evidence, conditional recommendation).

In contrast, the AGA (2020) did not formulate any recommendations with regard to the use of probiotics for preventing AAD. However, the AGA conditionally recommended (based on low quality of evidence) certain probiotics for the prevention of *C difficile* infection in children receiving antibiotic treatment. These include *S boulardii*; the two-strain combination of *L. acidophilus* CL 1285 & *L. casei* LBC80R; the three-strain combination of *L. acidophilus, L. delbruekii* subsp. *bulgaricus*, and *B. bifidum*; the four-strain combination of *L. acidophilus, L. delbruekii* subsp. *bulgaricus, B. bifidum*, and *Streptococcus thermophilus*. No strain specification was given for the three-strain and four-strain combinations, which may contribute to confusion for implementation of these recommendations (Su et al., 2020).

# 30.5 NECROTIZING ENTEROCOLITIS

Necrotizing enterocolitis (NEC) is one of the most serious, life-threatening gastrointestinal diseases, which is characterized by various degrees of mucosal or transmural necrosis of the intestine. The highest incidence of NEC occurs in infants with birth weights <1000 g, and the incidence decreases with increasing birth weights. While the exact cause of NEC remains unclear, various factors, in addition to prematurity, play a role in its pathogenesis. These factors include formula feeding rather than breast-feeding, intestinal hypoxia–ischemia, and colonization of the gut with pathogenic microbiota (Neu, 2022). It has been suggested that the enteral administration of probiotics to preterm newborns could prevent infections and NEC as well as reduce antibiotic use.

Several systematic reviews, with or without a meta-analysis, have evaluated the enteral administration of probiotics for the prevention of NEC and mortality in preterm infants (AlFaleh & Anabrees, 2014; Rao et al., 2016; Sharif et al., 2020). These meta-analyses consistently suggest overall efficacy of probiotics when all probiotics have been evaluated together.

However, a 2018 network meta-analysis that aimed to identify the strains with the greatest efficacy (van den Akker et al., 2018), which included 51 RCTs with over 11,000 preterm infants, showed that only a few probiotic treatment combinations significantly reduced mortality rates and NEC incidence. Seven treatments reduced NEC incidence, and there was no clear overlap of strains that were effective on multiple outcomes. This may reflect an inadequate number or size of RCTs or a true lack of effect for certain species (van den Akker et al., 2018).

In 2020, both AGA (Su et al., 2020) and ESPGHAN (van den Akker et al., 2020) published their recommendations on the use of probiotics for preventing NEC. While both were based on pair-wise systematic reviews and network meta-analyses, their conclusions differ. For details and specific recommendations from both societies, see Table 30.1.

# 30.6 *HELICOBACTER PYLORI* INFECTION

Unsatisfactory *H. pylori* eradication rates and therapy-associated side effects are still an issue. Meta-analyses focusing mainly on adults have shown that probiotic supplementation can improve eradication rates and/or reduces side effects of anti-*H. pylori* treatment (Shi et al., 2019; Wang et al., 2017). For pediatric patients, a 2017 systematic review and a network meta-analysis found that probiotics increased eradication rate and reduced total side effects associated with *H. pylori* eradication therapy (Feng et al.). Single-strain meta-analyses found that *S. boulardii* given along with standard triple therapy significantly adverse effects and increased the eradication rate (Zhou et al., 2019). However, data in children for *S. boulardii* were limited (Szajewska et al., 2015). While in both analyses the addition of probiotics to standard triple therapy significantly increased the eradication rate, it was still below the desired level (≥90%) of success.

The 2017 ESPGHAN/NASPGHAN *H. pylori* guidelines (Jones et al., 2017) advised against routinely adding single or combination probiotics to eradication therapy for children. This contrasts with recent adult recommendations that suggest certain probiotics may reduce antibiotic-related side effects and improve *H. pylori* eradication rates (Malfertheiner et al., 2022). The 2023 ESPGHAN guidelines (Szajewska et al., 2023) suggest the use of *S. boulardii* along with *H pylori* therapy to increase eradication rates and reduce gastrointestinal adverse effects, which is in line with the 2022 adults' guidelines.

## 30.7 INFLAMMATORY BOWEL DISEASE

Inflammatory bowel disease (IBD) consists mainly of two distinct disorders: Crohn's disease and ulcerative colitis. A recent systematic review concluded that there is increasing evidence for differences in abundances of some bacteria in patients with IBD compared with controls (Pittayanon et al., 2020), contributing to the initiation and maintenance of inflammation. However, the findings were not consistent, at least partially due to different methods used to analyze the microbiota.

A 2020 Cochrane review found low-certainty evidence suggesting probiotics may induce clinical remission in patients with active ulcerative colitis but not Crohn's disease. The mixture of eight strains of *Escherichia coli* Nissle 1917 may be considered an effective treatment for maintenance in patients with mild ulcerative colitis and *L. reuteri* ATCC 55730 an adjuvant to standard therapy for induction of remission in mild-to-moderate pediatric ulcerative colitis (Kaur et al., 2020; Limketkai et al., 2020).

AGA recommends against the use of probiotics in patients with ulcerative colitis and Crohn's disease except in the context of clinical trials. However, in adults and children with pouchitis, the AGA conditionally recommends the use of the eight-strain combination [*L. paracasei* subsp. *paracasei* DSM 24733, *L. plantarum* DSM 24730, *L. acidophilus* DSM 24735, *L. delbrueckii* subsp. *bulgaricus* DSM 24734, *B. longum* subsp. *longum* DSM 24736, *B. breve* DSM 24732, *B. longum* subsp. *infantis* DSM 24737, and *S. thermophilus* DSM 24731] over no or other probiotics (Su et al., 2020)

ESPGHAN concludes that there is insufficient evidence to make a recommendation for or against the use of probiotics in the management of children with ulcerative colitis or Crohn's disease (Szajewska et al., 2023)

## 30.8 FUNCTIONAL ABDOMINAL PAIN DISORDERS

Functional abdominal pain disorders (FAPDs), particularly irritable bowel syndrome, have been linked to gut microbiota dysbiosis (Pittayanon et al., 2019). However, no clear microbiota signature has been identified, and it is unclear if findings in adults apply to children (Pittayanon et al., 2019).

A 2018 systematic review found insufficient evidence for probiotics in children with FADPs, with only *L. rhamnosus* showing some efficacy (Wegh et al., 2018). A 2021 systematic review and meta-analysis evaluated the effects of strain-specific probiotics on functional abdominal pain in children. Nine randomized controlled trials were included, with *Lacticaseibacillus rhamnosus* GG and *Limosilactobacillus reuteri* DSM 17938 (formerly known as *Lactobacillus* reuteri) being the only two probiotic strains investigated. The results showed that *L. reuteri* DSM 17938 significantly reduced pain intensity and increased the number of pain-free days (Trivic et al., 2021).

The AGA 2020 guidelines acknowledged the abundance of studies on IBS but noted that the significant heterogeneity in study design, outcomes, and probiotics used made it impossible to recommend the use of probiotics for symptomatic children and adults with IBS, except in the context of a clinical trial (Su et al., 2020).

On the other hand, the 2023 ESPGHAN guidelines suggest that *L. reuteri* DSM 17938 (at a dose of $10^8$ CFU to $2 \times 10^8$ CFU/day) may be considered for reducing pain intensity in children with FAPD (certainty of evidence: moderate; grade of recommendation: weak). Additionally, *L. rhamnosus GG* (give at a dose of $10^9$ CFU to $3\times10^9$ CFU twice daily) may be used for reducing pain frequency and intensity in children with IBS (certainty of evidence: moderate; grade of recommendation: weak) (Szajewska et al., 2023).

# 30.9 INFANTILE COLIC

Studies suggest that the gut microbiota in infants with colic differs from that of unaffected infants, and first-pass meconium is associated with subsequent infantile colic (Korpela et al., 2020). In colicky infants, dysbiosis may affect gut motor function and gas production, leading to abdominal pain/colic (Sung & Cabana, 2017).

A 2018 meta-analysis of individual participant data from four RCTs found that the administrating *L. reuteri* DSM 17938 at a dose $1\times10^8$ colony-forming units (CFUs) is likely to reduce crying and/or fussing time in breastfed infants with infantile colic, but its effectiveness in formula-fed infants is unclear (Sung et al., 2014). Other meta-analyses have confirmed these findings (Gutiérrez-Castrellón et al., 2017; Harb et al., 2016). Evidence on other probiotics is limited and inconclusive (Dryl & Szajewska, 2018; Gerasimov et al., 2018; Nocerino et al., 2020; Savino et al., 2020).

A 2019 Cochrane review identified six RCTs (involving 1886 infants) found that *L. rhamnosus* GG and some multi-strain products did not prevent new cases of colic, but reduced crying time compared to placebo (Ong et al., 2019). *L. reuteri* DSM 17938 administered at a dose of $1\times10^8$ CFU to newborns each day for 90 days was effective (Indrio et al., 2014). Other probiotics were also studied; however, evidence is limited (Cabana et al., 2019).

The 2023 ESPGHAN guidelines recommend considering *L. reuteri* DSM 17938 (at least 108 CFU/day for at least 21 days) for managing infant colic in breastfed infants, with moderate certainty of evidence and a weak grade of recommendation. However, there is insufficient evidence to recommend for or against the use of *L. reuteri* DSM 17938 in formula-fed infants. Additionally, *B. lactis* BB-12 (at least $10^8$ CFU/day for 21–28 days) may also be considered for managing infant colic in breastfed infants, with moderate certainty of evidence and a weak grade of recommendation. However, there is currently insufficient evidence to make a recommendation for or against the use of any other probiotics for preventing infant colic.

# 30.10 FUNCTIONAL CONSTIPATION

Functional constipation is a frustrating problem affecting 3% of children globally, and treatment can be challenging and prolonged. Additionally, conventional laxatives may not be well tolerated due to their unpleasant taste. Studies have suggested differences in gut microbiota between patients with and without functional constipation, but inconsistent findings and methods used in the studies make it difficult to draw firm conclusions (Ohkusa et al., 2019).

A 2017 systematic review of seven RCTs involving 515 children with functional constipation found no significant difference in treatment success between probiotics (including *L. rhamnosus casei* Lcr35, *L. rhamnosus* GG; *L. reuteri* DSM 17938; *B lactis* DN-173 010, *B longum*, and a mixture of seven strains and placebo [Wojtyniak & Szajewska, 2017]). These findings were supported by a 2018 systematic review (Wegh et al., 2018). As a result, both ESPGHAN/NASPGHAN in 2014 (Tabbers et al.) and ESPGHAN in 2023 (Szajewska et al.) do not recommend using probiotics for treating functional constipation in children

# 30.11  FOOD ALLERGY PREVENTION AND TREATMENT

Aberrant gut microbiota, possibly caused by factors such as mode of delivery (vaginal vs. caesarean), use of antibiotics during the early neonatal period, and mode of feeding (breast vs. formula feedings), has been linked to the development of allergic diseases (Akagawa & Kaneko, 2022). Probiotics are currently being evaluated as a microbiota-focused intervention for preventing allergic disorders.

In 2015, the World Allergy Organization (WAO) published guidelines on preventing of allergic disease. While the WAO concluded that probiotic supplementation cannot be recommended for reducing the risk of allergy in children, they suggested a potential benefit of using probiotics for preventing eczema in pregnant women at high risk for having an allergic child, in women who breastfeed infants at high risk of developing allergy, and using probiotics in infants at high risk of developing allergy. High risk was defined as the presence of a biologic parent or sibling with asthma, allergic rhinitis, eczema, or food allergy. However, these recommendations were conditional and supported by a very low quality of evidence (Fiocchi et al., 2015).

In 2021, the European Academy of Allergy and Clinical Immunology (EAACI) did not make a recommendation for or against the use of probiotics (as well as prebiotics or synbiotics) during pregnancy, breastfeeding, or infancy for the prevention of food allergies (Halken et al., 2021) Similarly, no recommendation for or against the use of probiotics was formulated for the management of food allergies (Muraro et al., 2022).

# 30.12  RESPIRATORY TRACT INFECTIONS

To benefit patients, families, and society, preventing respiratory tract infections (RTIs), which cause many consultations in young children, is crucial. Probiotics have been investigated for their potential to enhance immune responses and prevent these diseases. A 2018 systematic review evaluated the effects of probiotics on subjects with RTIs and included 15 randomized controlled trials (RCTs) with 5,121 children in day care settings (aged 3 months to 7 years) (Laursen & Hojsak, 2018). *L. rhamnosus* GG reduced the duration of RTIs by about 18 hours compared to placebo based on the pooled results of three RCTs; however, there was no effect on other outcomes. *Bifidobacterium animalis* subsp. *lactis* BB-12 showed no effect on the duration of RTIs or absence from day care based on the results of two studies with 343 participants. Due to limited data and different outcome measures, meta-analyses on other strains or their combination were not possible (Laursen & Hojsak, 2018)

# 30.13  PROBIOTICS IN INFANT FORMULAE

There are several issues related to the addition of probiotic bacteria to infant formula. First, timing—administration is often initiated during early infancy, sometimes at birth, when infant formula may be the only source of feeding for non-breast-fed infants. Second, duration—daily administration of these products is frequently prolonged (several weeks or months). Third, the onset of administration is at a time when the gut microbiota is not yet fully established. Finally, delivery is in the form of a specific matrix (infant formula) that may be the infant's only source of feeding (Braegger et al., 2011). Overall, while some beneficial clinical effects of probiotics are possible, there is currently no robust evidence to recommend their routine use. This conclusion may reflect the small amount of data on specific probiotic strains and outcomes, rather than a genuine lack of effect. The efficacy and safety of each probiotic-supplemented formula should be considered (Skórka et al., 2017)

# 30.14 CONCLUSIONS AND FUTURE DIRECTIONS

Probiotics have the potential to prevent and treat many disorders in the pediatric population. However, guidance is needed on the appropriate use of probiotics to prevent and treat pediatric disorders, including selecting the right microorganism(s) for each condition and determining optimal timing, dosage, and mode of administration. Currently, the best-documented applications are the treatment of acute gastroenteritis and prevention of antibiotic-associated diarrhea. In preterm infants, the most promising application is the prevention of necrotizing enterocolitis, although the routine use of probiotics for this condition remains controversial. For some other indications, more evidence is needed to support their use, and further studies are required to investigate the role of specific probiotics in clinical practice. As not all probiotics are equal in efficacy and safety, each probiotic should be evaluated individually.

# BIBLIOGRAPHY

Akagawa, S., & Kaneko, K. (2022). Gut microbiota and allergic diseases in children. *Allergol Int, 71*(3), 301–309. doi:10.1016/j.alit.2022.02.004

AlFaleh, K., & Anabrees, J. (2014). Probiotics for prevention of necrotizing enterocolitis in preterm infants. *Cochrane Database Syst Rev, 4*, CD005496. doi:10.1002/14651858.CD005496.pub4

Braegger, C., Chmielewska, A., Decsi, T., Kolacek, S., Mihatsch, W., Moreno, L., . . . van Goudoever, J. (2011). Supplementation of infant formula with probiotics and/or prebiotics: A systematic review and comment by the ESPGHAN committee on nutrition. *J Pediatr Gastroenterol Nutr, 52*(2), 238–250. doi:10.1097/MPG.0b013e3181fb9e80

Cabana, M. D., McKean, M., Beck, A. L., & Flaherman, V. (2019). Pilot analysis of early lactobacillus rhamnosus GG for infant colic prevention. *J Pediatr Gastroenterol Nutr, 68*(1), 17–19. doi:10.1097/mpg.0000000000002113

Dryl, R., & Szajewska, H. (2018). Probiotics for management of infantile colic: A systematic review of randomized controlled trials. *Arch Med Sci, 14*(5), 1137–1143. doi:10.5114/aoms.2017.66055

Feng, J. R., Wang, F., Qiu, X., McFarland, L. V., Chen, P. F., Zhou, R., . . . Li, J. (2017). Efficacy and safety of probiotic-supplemented triple therapy for eradication of *Helicobacter pylori* in children: A systematic review and network meta-analysis. *Eur J Clin Pharmacol, 73*(10), 1199–1208. doi:10.1007/s00228-017-2291-6

Festini, F., Cocchi, P., Mambretti, D., Tagliabue, B., Carotti, M., Ciofi, D., . . . de Martino, M. (2010). Nosocomial Rotavirus Gastroenteritis in pediatric patients: A multi-center prospective cohort study. *BMC Infect Dis, 10*, 235. doi:10.1186/1471-2334-10-235

Fiocchi, A., Pawankar, R., Cuello-Garcia, C., Ahn, K., Al-Hammadi, S., Agarwal, A., . . . Schünemann, H. J. (2015). World allergy organization-mcmaster university guidelines for allergic disease prevention (GLAD-P): Probiotics. *World Allergy Organ J, 8*(1), 4. doi:10.1186/s40413-015-0055-2

Freedman, S. B., Williamson-Urquhart, S., Farion, K. J., Gouin, S., Willan, A. R., Poonai, N., . . . Schuh, S. (2018). Multicenter trial of a combination probiotic for children with gastroenteritis. *N Engl J Med, 379*(21), 2015–2026. doi:10.1056/NEJMoa1802597

Gerasimov, S., Gantzel, J., Dementieva, N., Schevchenko, O., Tsitsura, O., Guta, N., . . . Kaprus, V. (2018). Role of lactobacillus rhamnosus (FloraActive) 19070–2 and lactobacillus reuteri (FloraActive) 12246 in infant colic: A randomized dietary study. *Nutrients, 10*(12). doi:10.3390/nu10121975

Guarino, A., & Albano, F. (2001). Guidelines for the approach to outpatient children with acute diarrhoea. *Acta Paediatr, 90*(10), 1087–1095. doi:10.1080/080352501317061413

Guarino, A., Ashkenazi, S., Gendrel, D., Lo Vecchio, A., Shamir, R., Szajewska, H., . . . European Society for Pediatric Infectious, D. (2014). European society for pediatric gastroenterology, hepatology, and nutrition/European society for pediatric infectious diseases evidence-based guidelines for the management of acute gastroenteritis in children in Europe: Update 2014. *J Pediatr Gastroenterol Nutr, 59*(1), 132–152. doi:10.1097/MPG.0000000000000375

Gutiérrez-Castrellón, P., Indrio, F., Bolio-Galvis, A., Jiménez-Gutiérrez, C., Jimenez-Escobar, I., & López-Velázquez, G. (2017). Efficacy of lactobacillus reuteri DSM 17938 for infantile colic: Systematic review with network meta-analysis. *Medicine (Baltimore), 96*(51), e9375. doi:10.1097/md.0000000000009375

Gwee, K. A., Lee, W. W., Ling, K. L., Ooi, C. J., Quak, S. H., Dan, Y. Y., . . . Wong, C. Y. (2018). Consensus and contentious statements on the use of probiotics in clinical practice: A south east Asian gastro-neuro motility association working team report. *J Gastroenterol Hepatol, 33*(10), 1707–1716. doi:10.1111/jgh.14268

Halken, S., Muraro, A., de Silva, D., Khaleva, E., Angier, E., Arasi, S., . . . Roberts, G. (2021). EAACI guideline: Preventing the development of food allergy in infants and young children (2020 update). *Pediatr Allergy Immunol, 32*(5), 843–858. doi:10.1111/pai.13496

Harb, T., Matsuyama, M., David, M., & Hill, R. J. (2016). Infant colic-what works: A systematic review of interventions for breast-fed infants. *J Pediatr Gastroenterol Nutr, 62*(5), 668–686. doi:10.1097/mpg.0000000000001075

Hojsak, I., Szajewska, H., Canani, R. B., Guarino, A., Indrio, F., Kolacek, S., . . . Probiotics/Prebiotics, E. W. G. f. (2018). Probiotics for the prevention of nosocomial diarrhea in children. *J Pediatr Gastroenterol Nutr, 66*(1), 3–9. doi:10.1097/MPG.0000000000001637

Indrio, F., Di Mauro, A., & Riezzo, G. (2014). Prophylactic use of a probiotic in the prevention of colic, regurgitation, and functional constipation—reply. *JAMA Pediatr, 168*(8), 778. doi:10.1001/jamapediatrics.2014.368

Jones, N. L., Koletzko, S., Goodman, K., Bontems, P., Cadranel, S., Casswall, T., . . . Rowland, M. (2017). Joint ESPGHAN/NASPGHAN guidelines for the management of *Helicobacter pylori* in children and adolescents (Update 2016). *J Pediatr Gastroenterol Nutr, 64*(6), 991–1003. doi:10.1097/mpg.0000000000001594

Kaur, L., Gordon, M., Baines, P. A., Iheozor-Ejiofor, Z., Sinopoulou, V., & Akobeng, A. K. (2020). Probiotics for induction of remission in ulcerative colitis. *Cochrane Database Syst Rev, 3*(3), Cd005573. doi:10.1002/14651858.CD005573.pub3

Korpela, K., Renko, M., Paalanne, N., Vänni, P., Salo, J., Tejesvi, M., . . . Tapiainen, T. (2020). Microbiome of the first stool after birth and infantile colic. *Pediatr Res.* doi:10.1038/s41390-020-0804-y

Laursen, R. P., & Hojsak, I. (2018). Probiotics for respiratory tract infections in children attending day care centers-a systematic review. *Eur J Pediatr, 177*(7), 979–994. doi:10.1007/s00431-018-3167-1

Limketkai, B. N., Akobeng, A. K., Gordon, M., & Adepoju, A. A. (2020). Probiotics for induction of remission in Crohn's disease. *Cochrane Database Syst Rev, 7*, Cd006634. doi:10.1002/14651858.CD006634.pub3

Malfertheiner, P., Megraud, F., Rokkas, T., Gisbert, J. P., Liou, J. M., Schulz, C., . . . El-Omar, E. M. (2022). Management of *Helicobacter pylori* infection: The Maastricht VI/Florence consensus report. *Gut.* doi:10.1136/gutjnl-2022-327745

Marchand, V. (2012). Using probiotics in the paediatric population. *Paediatr Child Health, 17*(10), 575–576. doi:10.1093/pch/17.10.575

McFarland, L. V., Ozen, M., Dinleyici, E. C., & Goh, S. (2016). Comparison of pediatric and adult antibiotic-associated diarrhea and Clostridium difficile infections. *World J Gastroenterol, 22*(11), 3078–3104. doi:10.3748/wjg.v22.i11.3078

Muraro, A., de Silva, D., Halken, S., Worm, M., Khaleva, E., Arasi, S., . . . Roberts, G. (2022). Managing food allergy: GA(2)LEN guideline 2022. *World Allergy Organ J, 15*(9), 100687. doi:10.1016/j.waojou.2022.100687

Neu, J. (2022). Prevention of necrotizing enterocolitis. *Clin Perinatol, 49*(1), 195–206. doi:10.1016/j.clp.2021.11.012

Nocerino, R., De Filippis, F., Cecere, G., Marino, A., Micillo, M., Di Scala, C., . . . Berni Canani, R. (2020). The therapeutic efficacy of Bifidobacterium animalis subsp. lactis BB-12() in infant colic: A randomised, double blind, placebo-controlled trial. *Aliment Pharmacol Ther, 51*(1), 110–120. doi:10.1111/apt.15561

Ohkusa, T., Koido, S., Nishikawa, Y., & Sato, N. (2019). Gut microbiota and chronic constipation: A review and update. *Front Med (Lausanne), 6*, 19. doi:10.3389/fmed.2019.00019

Ong, T. G., Gordon, M., Banks, S. S., Thomas, M. R., & Akobeng, A. K. (2019). Probiotics to prevent infantile colic. *Cochrane Database Syst Rev, 3*, Cd012473. doi:10.1002/14651858.CD012473.pub2

Pittayanon, R., Lau, J. T., Leontiadis, G. I., Tse, F., Yuan, Y., Surette, M., & Moayyedi, P. (2020). Differences in gut microbiota in patients with vs without inflammatory bowel diseases: A systematic review. *Gastroenterology, 158*(4), 930–946.e931. doi:10.1053/j.gastro.2019.11.294

Pittayanon, R., Lau, J. T., Yuan, Y., Leontiadis, G. I., Tse, F., Surette, M., & Moayyedi, P. (2019). Gut microbiota in patients with irritable bowel syndrome-a systematic review. *Gastroenterology, 157*(1), 97–108. doi:10.1053/j.gastro.2019.03.049

Rao, S. C., Athalye-Jape, G. K., Deshpande, G. C., Simmer, K. N., & Patole, S. K. (2016). Probiotic supplementation and late-onset sepsis in preterm infants: A meta-analysis. *Pediatrics, 137*(3), e20153684. doi:10.1542/peds.2015-3684

Savino, F., Montanari, P., Galliano, I., Daprà, V., & Bergallo, M. (2020). Lactobacillus rhamnosus GG (ATCC 53103) for the management of infantile colic: A randomized controlled trial. *Nutrients, 12*(6). doi:10.3390/nu12061693

Schnadower, D., Tarr, P. I., Casper, T. C., Gorelick, M. H., Dean, J. M., O'Connell, K. J., . . . Freedman, S. B. (2018). Lactobacillus rhamnosus GG versus placebo for acute gastroenteritis in children. *N Engl J Med, 379*(21), 2002–2014. doi:10.1056/NEJMoa1802598

Sharif, S., Meader, N., Oddie, S. J., Rojas-Reyes, M. X., & McGuire, W. (2020). Probiotics to prevent necrotising enterocolitis in very preterm or very low birth weight infants. *Cochrane Database Syst Rev, 10*, CD005496. doi:10.1002/14651858.CD005496.pub5

Shi, X., Zhang, J., Mo, L., Shi, J., Qin, M., & Huang, X. (2019). Efficacy and safety of probiotics in eradicating *Helicobacter pylori*: A network meta-analysis. *Medicine (Baltimore), 98*(15), e15180. doi:10.1097/md.0000000000015180

Skórka, A., Pieścik-Lech, M., Kołodziej, M., & Szajewska, H. (2017). To add or not to add probiotics to infant formulae? An updated systematic review. *Benef Microbes, 8*(5), 717–725. doi:10.3920/bm2016.0233

Su, G. L., Ko, C. W., Bercik, P., Falck-Ytter, Y., Sultan, S., Weizman, A. V., & Morgan, R. L. (2020). AGA clinical practice guidelines on the role of probiotics in the management of gastrointestinal disorders. *Gastroenterology, 159*(2), 697–705. doi:10.1053/j.gastro.2020.05.059

Sung, V., & Cabana, M. D. (2017). Probiotics for colic-is the gut responsible for infant crying after all? *J Pediatr, 191*, 6–8. doi:10.1016/j.jpeds.2017.09.010

Sung, V., Cabana, M. D., D'Amico, F., Deshpande, G., Dupont, C., Indrio, F., . . . Tancredi, D. (2014). Lactobacillus reuteri DSM 17938 for managing infant colic: Protocol for an individual participant data meta-analysis. *BMJ Open, 4*(12), e006475. doi:10.1136/bmjopen-2014-006475

Szajewska, H., Berni Canani, R., Domellöf, M., Guarino, A., Hojsak, I., Indrio, F., . . . Weizman, Z. (2023). Probiotics for the management of pediatric gastrointestinal disorders: Position paper of the ESPGHAN special interest group on gut microbiota and modifications. *J Pediatr Gastroenterol Nutr, 76*(2), 232–247. doi:10.1097/mpg.0000000000003633

Szajewska, H., Guarino, A., Hojsak, I., Indrio, F., Kolacek, S., Orel, R., . . . Zalewski, B. M. (2020). Use of probiotics for the management of acute gastroenteritis in children: An update. *J Pediatr Gastroenterol Nutr, 71*(2), 261–269. doi:10.1097/mpg.0000000000002751

Szajewska, H., Guarino, A., Hojsak, I., Indrio, F., Kolacek, S., Shamir, R., . . . Weizman, Z. (2014). Use of probiotics for management of acute gastroenteritis: A position paper by the ESPGHAN working group for probiotics and prebiotics. *J Pediatr Gastroenterol Nutr, 58*(4), 531–539. doi:10.1097/mpg.0000000000000320

Szajewska, H., Horvath, A., & Kołodziej, M. (2015). Systematic review with meta-analysis: Saccharomyces boulardii supplementation and eradication of *Helicobacter pylori* infection. *Aliment Pharmacol Ther, 41*(12), 1237–1245. doi:10.1111/apt.13214

Szymański, H., & Szajewska, H. (2019). Lack of efficacy of lactobacillus reuteri DSM 17938 for the treatment of acute gastroenteritis: A randomized controlled trial. *Pediatr Infect Dis J, 38*(10), e237–e242. doi:10.1097/INF.0000000000002355

Tabbers, M. M., DiLorenzo, C., Berger, M. Y., Faure, C., Langendam, M. W., Nurko, S., . . . Benninga, M. A. (2014). Evaluation and treatment of functional constipation in infants and children: Evidence-based recommendations from ESPGHAN and NASPGHAN. *J Pediatr Gastroenterol Nutr, 58*(2), 258–274. doi:10.1097/mpg.0000000000000266

Trivic, I., Niseteo, T., Jadresin, O., & Hojsak, I. (2021). Use of probiotics in the treatment of functional abdominal pain in children-systematic review and meta-analysis. *Eur J Pediatr, 180*(2), 339–351. doi:10.1007/s00431-020-03809-y

van den Akker, C. H. P., van Goudoever, J. B., Shamir, R., Domellöf, M., Embleton, N. D., Hojsak, I., . . . Szajewska, H. (2020). Probiotics and preterm infants: A position paper by the european society for paediatric gastroenterology hepatology and nutrition committee on nutrition and the european society for paediatric gastroenterology hepatology and nutrition working group for probiotics and prebiotics. *J Pediatr Gastroenterol Nutr, 70*(5), 664–680. doi:10.1097/mpg.0000000000002655

van den Akker, C. H. P., van Goudoever, J. B., Szajewska, H., Embleton, N. D., Hojsak, I., Reid, D., . . . Committee on, N. (2018). Probiotics for preterm infants: A strain-specific systematic review and network meta-analysis. *J Pediatr Gastroenterol Nutr, 67*(1), 103–122. doi:10.1097/MPG.0000000000001897

Wang, F., Feng, J., Chen, P., Liu, X., Ma, M., Zhou, R., . . . Zhao, Q. (2017). Probiotics in *Helicobacter pylori* eradication therapy: Systematic review and network meta-analysis. *Clin Res Hepatol Gastroenterol, 41*(4), 466–475. doi:10.1016/j.clinre.2017.04.004

Wegh, C. A. M., Benninga, M. A., & Tabbers, M. M. (2018). Effectiveness of probiotics in children with functional abdominal pain disorders and functional constipation: A systematic review. *J Clin Gastroenterol, 52*(Suppl 1), Proceedings from the 9th Probiotics, Prebiotics and New Foods, Nutraceuticals and Botanicals for Nutrition & Human and Microbiota Health Meeting, held in Rome, Italy from September 10 to 12, 2017, S10–s26. doi:10.1097/mcg.0000000000001054

Wojtyniak, K., & Szajewska, H. (2017). Systematic review: Probiotics for functional constipation in children. *Eur J Pediatr, 176*(9), 1155–1162. doi:10.1007/s00431-017-2972-2

Zhou, B. G., Chen, L. X., Li, B., Wan, L. Y., & Ai, Y. W. (2019). Saccharomyces boulardii as an adjuvant therapy for *Helicobacter pylori* eradication: A systematic review and meta-analysis with trial sequential analysis. *Helicobacter, 24*(5), e12651. doi:10.1111/hel.12651

# Nutrition Economics
## *Adding Value to Fermented Foods and Biotics*

**31**

Irene Lenoir-Wijnkoop, Aki Koponen, and Seppo Salminen

## 31.1 INTRODUCTION

Lactic acid bacteria (LAB) figure among the most studied microorganisms. Recently, a study provided real-world evidence that the consumption of live microbes is linked with various markers of better health: more favorable blood pressure, plasma glucose and insulin levels, and reduction of inflammatory processes, as well as lower waist circumference and body mass index.[1] Beneficial effects may be related to the process of fermentation leading to increased nutritional properties of the concerned food stuff, exerted through direct and indirect interactions between the human body and probiotic strains with a formally proven benefit, in association with prebiotics and/or through inactivated microbial cells or cell components, with or without metabolites, so-called postbiotics.

These "biotics" are an integral part of our daily diet, and scientific evidence has gradually unraveled links between food intake and the human microbiota. Given the mechanisms by which such microbiota-modulating compounds may achieve their desired effects, the current regulatory framework needs definitions that help in understanding the challenges and rewards in determining how microbiota modulating components should be classified.

Over the past 8 years, the ISAPP (www.ISAPPscience.org) has advanced consensus definitions of fermented food, probiotics, prebiotics, synbiotics, and postbiotics, thus improving clarity and understanding among scientists, clinical trial experts, industry, regulators, and consumers and providing a common ground for further development of these fields. The definitions and their background have been published; each definition is based on existing evidence of health-promoting effects (Table 31.1).[2]

Numerous studies have reported the contribution of biotic compounds in better treatment outcomes, reduced risks of disease development, and improved health conditions.[3–5] These reports serve as an evidence base for recommendations and guidance on the use of biotics in spite of existing knowledge gaps.

DOI: 10.1201/9781003352075-34

**TABLE 31.1**  ISAPP Definitions for Fermented Food, Probiotics, Prebiotics, Synbiotics, and Postbiotics

| | | |
|---|---|---|
| Fermented foods | Foods made through desired microbial growth and enzymatic conversions of food components | Marco M et al., *Nat. Rev. Gastroenterol. Hepatol 18*, 196–208 (2021) |
| Probiotics | Live microorganisms that, when administered in adequate amounts, confer a health benefit on the host | Hill C et al., *Nat. Rev. Gastroenterol. Hepatol. 11*, 506–514 (2014) |
| Prebiotics | A substrate that is selectively utilized by host microorganisms, conferring a health benefit | Gibson G et al., *Nat. Rev. Gastroenterol. Hepatol. 14*, 491–502 (2017) |
| Synbiotics | A mixture comprising live microorganisms and substrate(s) selectively utilized by host microorganisms that confers a health benefit on the host | Swanson K et al., *Nat. Rev. Gastroenterol. Hepatol 17*, 687–701 (2020) |
| Postbiotics | Preparation of inanimate microorganisms and/or their components that confers a health benefit on the host | Salminen, S. et al., *Nat Rev Gastroenterol. Hepatol, 18*, 649–667 (2021) |

This chapter provides an overview of the socio-economic value of fermented foods and biotics in health and disease. It discusses the challenges in determining effective prevention strategies for non-diseased or sub-healthy populations and the limitations of existing RCT principles in building evidence on the multiple interactions between food constituents and their targets in the human body. It highlights the potential of computational models in analyzing diverse data and projecting small long-term effects in large populations. The chapter also presents two examples of studies on the socio-economic impact of fermented food and probiotics in different pathological conditions. Furthermore, it discusses the relevance of fermented food and biotics for sustainable health systems and the role of behavioral science in enhancing their use. Finally, it outlines the need for further socioeconomic analyses to optimize the utilization of available healthcare resources and to better promote maintenance of population health and individual well-being.

## 31.2 THE SOCIO-ECONOMIC VALUE OF BIOTICS IN HEALTH AND DISEASE

When conducting trials in the field of food intake and dietary supplements, the existing RCT principles and associated statistical analyses for research provide a reliable evidence base for demonstrating cause–effect relationships in well-identified populations under controlled conditions. This is an important pillar in any kind of health management and decision-making. However, in case of prevention programs in non-diseased or sub-healthy populations the process of determining effective and meaningful strategies is in general more complex than for therapeutic decisions. In both cases efficacy and value outcomes are important considerations, but in addition to the evidence used for clinical treatment protocols, public health instances have to draw on a wide variety of non-randomized evidence data and contextual information sources while taking into account health determinants that are influenced by uncontrolled conditions and lifestyle aspects of the general population.[6,7]

Even the most rigorous setting will not allow to capture the multiple interactions between food constituents and their many different targets in the human body, involving numerous interdependent physiological and metabolic processes, nor the occurrence of very small effects and related health changes only observable over the long term.

Computational models make it possible to analyze diverse and multifaceted data together. Thanks to model-based studies, clinical outcomes obtained within a select study group can be extrapolated to a larger group of similar individuals in the general population. Effects observed during the—by definition limited—study period can thus be projected over a longer period of time. Also, the incremental impact of future changes in a health condition can be incorporated. Furthermore, a diversity of other data can be fed into the model, such as cost data (direct and indirect medical costs, non-medical costs), epidemiological data, measurements of quality of life, population statistics, the consequences of absenteeism and reduced work capacity on productivity loss, and other related socio-economic aspects.

This modelling approach is fully recognized and applied within health economics to calculate the cost-effectiveness of medical interventions as well as to support decision-making in health care delivery and financing.

Public health policies aim to reduce the occurrence of avoidable diseases or to delay the onset of life-course diseases by information campaigns and other interventions conceived to diminish exposure to risk factors. In this setting, biotics have a great potential for limiting impaired health conditions that currently put an unprecedented burden on human well-being and on scarce healthcare resources. This is not only of interest for patients, at-risk citizens, and non-patient populations; it is also crucial for the sustainability of health care resources, as illustrated in the following.

## 31.2.1 Acute Viral Respiratory Tract Infections

Acute viral respiratory tract infections (RTIs) are common worldwide and place a significant burden on national healthcare systems, as well as on individuals, families, and society at large. Treatment mostly relies on symptom control, but even in times of moderate contamination, the cost impact of these very common seasonal respiratory infections is substantial, and, given the lack of satisfactory treatments, prevention is the cornerstone of RTI management.[8] Three meta-analyses confirmed the beneficial impact of probiotics in reducing both the incidence and the duration of symptoms of these flu-like conditions.[9–11] Based on these outcomes, the socio-economic impact associated with the use of probiotics was evaluated according to two scenarios, one considering the duration of illness (YHEC) and the other the incidence (Cochrane). The assessments were performed for France, Canada, and the United States, respectively, in order to take into account country-specific characteristics, related to vaccination coverage, prescription patterns, cost inputs, or differences in sick-leave allowances.

The results for France show that generalized probiotic use would save 2.4 million RTI-days, 291,000 antibiotic courses, and 581,000 sick leave days, based on YHEC data. Applying the Cochrane data, reductions were 6.6 million CRTI days, 473,000 antibiotic courses and 1.5 million sick days. From the NHS perspective, probiotics' economic impact was about €14.6 million saved according to YHEC and €37.7 million according to Cochrane. Higher savings were observed in children, active smokers, and people with more frequent human contacts.[12]

In Canada, probiotic use saved 573,000–2.3 million RTI-days, according to the YHEC—Cochrane scenarios, respectively. These reductions were associated with an avoidance of 52,000–84,000 antibiotic courses and 330,000–500,000 sick-leave days. A projection of corresponding costs reductions amounted to Can$1.3–8.9 million from the healthcare payer perspective and Can$61.2–99.7 million when adding productivity losses.[13]

Finally, for the US population in 2017–2018, generalized probiotic intake would have allowed cost savings for the health care payers of US$4.6 million based on the YHEC scenario and US$373 million for the Cochrane scenario by averting 19 million and 54.5 million RTI sick days, respectively, compared to no probiotics. Antibiotic prescriptions decreased with 1.39–2.16 million courses, whereas absence from work decreased by 3.58–4.2 million days when applying the YHEC and Cochrane data, respectively. When productivity loss is included, total savings for society represented US$784 million or US$1.4 billion for the YHEC and Cochrane scenarios, respectively.[14]

In addition, for all countries, it can be hypothesized that expenses for self-medication and over-the-counter medication as well as costs associated with missed school days and informal care for sick children or the elderly will also decrease.

## 31.2.2 Weight Management and Type 2 Diabetes

A prevalent health problem, preventable through modification of food patterns and lifestyle, is overweight, a condition which bears an increased risk of developing obesity and type 2 diabetes. The global economic costs of overweight and obesity are predicted to rise from a little under US$2 trillion in 2020, to over US$3 trillion by 2030, and more than an astonishing US$18 trillion by 2060 (2019 prices).[15] The rise in costs will especially affect upper middle-income countries as well as higher-income countries and will be seen in every region of the world. Several studies suggest that the consumption of biotics improves metabolic parameters.[16,17]

A large meta-analysis (459,790 individuals) showed that yogurt consumption is associated with a lower risk of developing type 2 diabetes; no such association was seen for dairy products in general. Further to this, a health technology assessment investigated the potential health impact and socio-economic benefits of yogurt intake with regard to type 2 diabetes. The computer model designed for this study took into account the different stages of development of the disease, including the long-term risk of additional morbidity symptoms and loss of quality of life. To account for country-specific data, the analysis focused on the population of the United Kingdom, a country that has many sources of reliable health data. Assuming a causal relationship between yogurt consumption and a lower rate of diabetes cases as reported in scientific publications, the results of this study indicate that increasing average yoghurt consumption by adults over 25 years of age in the UK by 100 g daily could result in nearly 400,000 fewer people developing T2D over 25 years. This could potentially save the UK NHS £2.3bn in direct T2D treatment costs and the costs of treating T2D associated complications. In addition, 267,000 quality-adjusted life years (QALYs) could be generated.

Interestingly, the biggest driver of both cost savings and QALY gains in the model results from the reduction in people with T2D itself rather than a reduction in complications. Savings from direct treatment costs of T2D accounts for 91.5% of the total model savings and approximately 85% of the QALY gains.

Nutrition-economic assessments have mostly been conducted on individual members of the "biotic family" and in particular on the impact of probiotics and prebiotics in healthcare expenditures. Other examples can be found for a variety of indications, such as antibiotic-associated diarrhea,[18–20] necrotizing enterocolitis in neonates,[21] functional gastrointestinal disorders in infants,[22] atopic dermatitis,[23,24] colorectal cancer,[25] and cardiovascular diseases.[26,27]

As the consensus definitions for synbiotics and postbiotics are relatively recent, no cost evaluations are available yet, but the most important research reports and clinical studies have been summarized in the consensus documents.[28] Additional clarifications have been explained as benefits for healthy populations.[29] The scope for further socio-economic analyses with the aim to optimize the spending of available health care resources is very promising.

## 31.3 THE RELEVANCE OF BIOTICS FOR SUSTAINABLE HEALTH SYSTEMS

The economic assessment of benefits in the field of healthcare has developed rapidly, one reason being the rising pressure on healthcare budgets and the growing interest in cost-effective and evidence-based treatments and therapies. This has resulted in an increasing number of cost-effectiveness studies and methodological developments. Additionally, a particular research discipline, called health economics, has evolved, leading to a growing role of health-economic arguments in policy making. However, the fields of functional foods and health economics have only been interconnected over the last decade,[30] and reliable data on the cost implications of nutritional interventions for health management are gradually emerging. The implementation of cost-effective health strategies will contribute to an improvement of socio-economic conditions not only by a lower disease burden but also by a reduction of out-of-pocket costs for individuals and families. It will be of great importance to implement public health strategies that aim to encourage the use of fermented foods and biotics more broadly as part of the daily diet. One can even wonder why the consumption of beneficial live microbes and related compounds is not more frequently recommended as part of a healthy lifestyle or used as an adjuvant of therapeutic approaches. Fermented foods are now promoted in European Union and the United States for health benefits. In the EU, there is a large cost action group working on different aspects of fermented foods and health. One major task for the program is to produce systematic reviews on, for example, fermented foods and bone health and several other health benefits (www.cost.eu/actions/CA20128/).

# 31.4 BEHAVIORAL SCIENCE TO ENHANCE THE USE OF BIOTICS

Healthcare authorities and policymakers are well aware that many of the current disease conditions are determined by our way of life and eating patterns and that prevention and health promotion are not less important than cure and care. This notwithstanding, on average only 3% of the healthcare budget is devoted to prevention and health promotion.[31,32] Most of this extremely low rate of investment concerns the assessment of medical aspects of a technology, followed by economic considerations, organizational impact, and societal consequences. The evaluation of behavioral and lifestyle interventions remains extremely rare (Figure 31.1).[33]

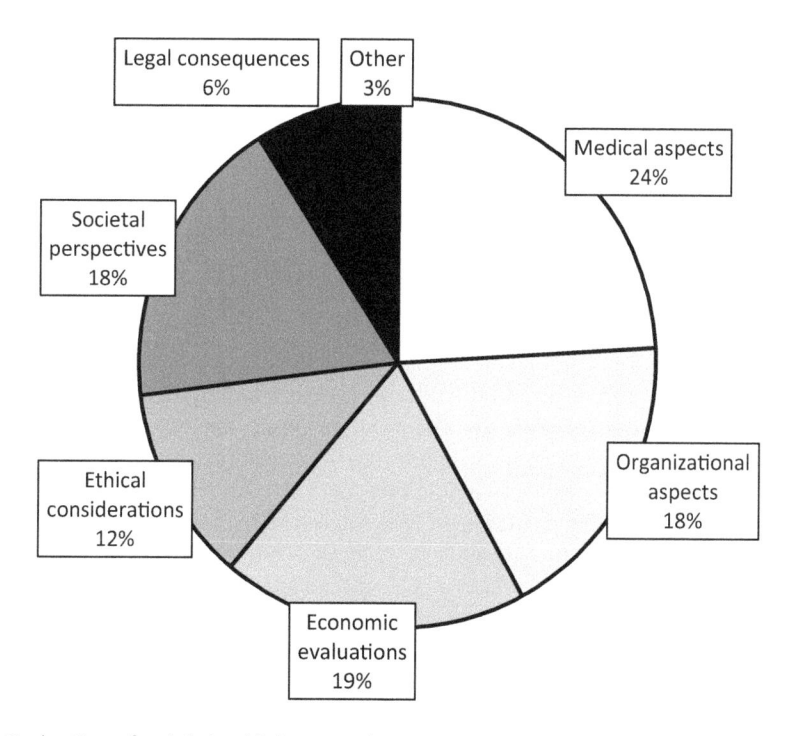

**FIGURE 31.1**   Evaluation of public health interventions.

At present, health-promoting interventions are increasingly devised with the aim to reduce the occurrence of modifiable risk factors and the associated medical and societal burden. However, in the field of nutrition, behavioral determinants seriously compromise effective and consistent adherence to these initiatives because of the difficulty to induce persisting changes in the daily eating habits of non-patient populations. Food choices are driven by lifestyle, taste, genetics, age, psychosocial and socio-economic factors and influenced by personal perception of pleasure and convenience.[34]

In this context the implementation of behavioral science-based methods can effectively guide consumer choice towards biotic-rich products, fostering increased demand and acceptance of pre/probiotics in the market. Behavioral science-based approaches revolve around understanding and leveraging the underlying cognitive processes, biases, and heuristics that influence consumer attitudes. Choice architecture, risk perception, anchoring bias, and the reciprocity principle can be used to influence decisions related to health and well-being.[35] Persuasive strategies such as social norms, scarcity heuristics, and reciprocity can be harnessed to encourage positive consumer behavior.[36] Cognitive biases, including anchoring and risk perception, can impact consumer decision-making processes.[37] Behavioral economics and its effects on decision-making, touching on topics such as anchoring bias, incentive systems, and value-based framing, may be explored,[38] while research on the goal-gradient hypothesis demonstrates how perceived progress toward a goal can accelerate consumer purchases.[39] Lastly, the study on habit formation provides valuable insights into how habits are formed and maintained in the real world, which can be used to encourage long-term adoption of pre/probiotic products.[40]

By applying the insights from psychology, sociology, and economics, policymakers can create interventions that effectively guide individuals toward better choices and outcomes. Methods often focus on "nudging" individuals towards desired behaviors rather than relying on traditional methods like mandates, regulations, or financial incentives. These are typically claimed to be more cost effective and less intrusive and have the potential to induce more sustainable behavior change. Behavioral science thus allows for the design of policies and marketing strategies that are tailored to specific contexts and target audiences, increasing the expected likelihood of achieving desired outcomes. Utilizing social norms through testimonials and endorsements or framing the benefits of pre/probiotics in a manner that aligns with the target audience's values can serve as powerful tools to influence consumer behavior and drive demand for these products. While not specific to pre/probiotics, these principles have been applied successfully across various industries and contexts. Their application to the promotion of fermented and biotic-rich products is a logical extension of their use in other contexts.

# 31.5 CONCLUSION

Fermented foods and biotics can offer a meaningful contribution to health outcomes both in the general population and in patients—a great potential that remains considerably underexploited. Whereas progress has been made with clinical guidelines and applications, translation of biotic research findings into nutritional recommendations and public health policy endorsements has not been achieved in a manner consistent with the amount of the evidence. In addition, the cost-effectiveness of such recommendations remains largely unexplored. Beneficial cultures in fermented foods, probiotics, prebiotics, synbiotics, and postbiotics can contribute to health management in general. A number of challenges must be addressed in order to fully realize biotic benefits, including the need for greater awareness of the accumulated evidence on probiotics and prebiotics among policy makers, strategies to cope with regulatory roadblocks to research, and high-quality human trials that address outstanding research questions in the field. Sustained changes in daily eating habits and other lifestyle aspects are the backbone of preventive public health policies. Behavioral science-based methods can effectively guide consumer choice towards biotic-rich products by leveraging cognitive processes, biases, and heuristics that influence consumer attitudes, and policymakers can create cost-effective interventions tailored to specific contexts and target audiences that increase the likelihood of achieving desired outcomes.

In the longer term, this will not only benefit healthcare resources but also socio-economic conditions and the population at large.

# BIBLIOGRAPHY

1. Hill C, Tancredi DJ, Cifelli CJ, Slavin JL, Gahche J, Marco ML, Hutkins R, Fulgoni VL, Merenstein D, Sanders ME. (2023) Positive Health Outcomes Associated with Live Microbe Intake from Foods, Including Fermented Foods, Assessed using the NHANES Database. *J Nutr.* 22:S0022–3166(23)12622–8. doi:10.1016/j.tjnut.2023.02.019.
2. Salminen S, Collado MC, Endo A, Hill C, Lebeer S, Quigley EMM, Sanders ME, Shamir R, Swann JR, Szajewska H, Vinderola G. (2021) The International Scientific Association of Probiotics and Prebiotics (ISAPP) Consensus Statement on the Definition and Scope of Postbiotics. *Nat Rev Gastroenterol Hepatol.* 18(9):649–667. doi:10.1038/s41575-021-00440-6. Epub 2021 May 4. Erratum in: Nat Rev Gastroenterol Hepatol. 2021 Jun 15: Erratum in: Nat Rev Gastroenterol Hepatol. 2022;19(8):551.
3. www.worldgastroenterology.org/guidelines/probiotics-and-prebiotics/probiotics-and-prebiotics-english
4. Sniffen JC, McFarland LV, Evans CT, Goldstein EJC. (2018) Choosing an Appropriate Probiotic Product for Your Patient: An Evidence-Based Practical Guide. *PLoS ONE.* 13(12):e0209205. doi:10.1371/journal.pone.0209205.
5. Su GL, Ko CW, Bercik P, Falck-Ytter Y, Sultan S, Weizman AV, Morgan RL. (2020) AGA Clinical Practice Guidelines on the Role of Probiotics in the Management of Gastrointestinal Disorders. *Gastroenterology.* 159(2):697–705. doi:10.1053/j.gastro.2020.05.059.
6. Thomson K, Hillier-Brown F, Todd A, et al. (2018) The Effects of Public Health Policies on Health Inequalities in High-Income Countries: An Umbrella Review. *BMC Public Health.* 18:869. https://doi.org/10.1186/s12889-018-5677-1
7. https://joint-research-centre.ec.europa.eu/jrc-news/evidence-informed-policymaking-new-document-foster-discussion-better-use-scientific-knowledge-policy-2022-10-26_en. Accessed 28 Dec 2022.
8. World Health Organization. *Influenza (Seasonal) Fact Sheet N° 211 2014.* www.who.int/news-room/fact-sheets/detail/influenza-(seasonal)
9. Hao Q, Lu Z, Dong BR, Huang CQ, Wu T. (2011) Probiotics for Preventing Acute Upper Respiratory Tract Infections. *Cochrane Database Syst Rev.* 9:CD006895. doi:10.1002/14651858.CD006895.pub2
10. King S, Glanville J, Sanders ME, Fitzgerald A, Varley D. (2014) The Effectiveness of Probiotics on Length of Illness in Healthy Children and Adults Who Develop Common Acute Respiratory Infectious Conditions: A Systematic Review and Meta-Analysis. *Br J Nutr.* 112(1):41–54. doi:10.1017/S0007114514000075.
11. Hao Q, Dong BR, Wu T. (2015) Probiotics for Preventing Acute Upper Respiratory Tract Infections. *Cochrane Database Syst Rev.* 3(2):CD006895. doi:10.1002/14651858.CD006895.pub3
12. Lenoir-Wijnkoop I, Gerlier L, Bresson JL, Le Pen C, Berdeaux G. (2015) Public Health and Budget Impact of Probiotics on Common Respiratory Tract Infections: A Modelling Study. *PLoS ONE.* 10(4):e0122765. doi:10.1371/journal.pone.
13. Lenoir-Wijnkoop I, Gerlier L, Roy D, Reid G. (2016) The Clinical and Economic Impact of Probiotics Consumption on Respiratory Tract Infections: Projections for Canada. *PLoS ONE.* 11(11):e0166232. doi:10.1371/journal.pone.0166232.
14. Lenoir-Wijnkoop I, Merenstein D, Korchagina D, Broholm C, Sanders ME, Tancredi D. (2019) Probiotics Reduce Health Care Cost and Societal Impact of Flu-Like Respiratory Tract Infections in the USA: An Economic Modeling Study. *Front Pharmacol.* 10:980. doi:10.3389/fphar.2019.00980. Erratum in: Front Pharmacol. 2019 Oct 11;10:1182.
15. Okunogbe A, Nugent R, Spencer G, Powis J, Ralston J, Wilding J. (2022) Economic Impacts of Overweight and Obesity: Current and Future Estimates for 161 Countries. *BMJ Global Health.* 7:e009773. doi:10.1136/bmjgh-2022-009773
16. Li HY, Zhou DD, Gan RY, Huang SY, Zhao CN, Shang A, Xu XY, Li HB. (2021) Effects and Mechanisms of Probiotics, Prebiotics, Synbiotics, and Postbiotics on Metabolic Diseases Targeting Gut Microbiota: A Narrative Review. *Nutrients.* 13(9):3211. doi:10.3390/nu13093211.

17. Ferrarese R, Ceresola ER, Preti A, Canducci F. (2018) Probiotics, Prebiotics and Synbiotics for Weight Loss and Metabolic Syndrome in the Microbiome Era. *Eur Rev Med Pharmacol Sci.* 22(21):7588–7605. doi:10.26355/eurrev_201811_16301.

18. Lenoir-Wijnkoop I, Nuijten MJ, Craig J, Butler CC. (2014) Nutrition Economic Evaluation of a Probiotic in the Prevention of Antibiotic-Associated Diarrhea. *Front Pharmacol.* 5:13. doi:10.3389/fphar.2014.00013.

19. Li N, Zheng B, Cai HF, Chen YH, Qiu MQ, Liu MB. (2018) Cost-Effectiveness Analysis of Oral Probiotics for the Prevention of Clostridium Difficile-Associated Diarrhoea in Children and Adolescents. *J Hosp Infect.* 99(4):469–474. doi:10.1016/j.jhin.2018.04.013.

20. Lau VI, Rochwerg B, Xie F, Johnstone J, Basmaji J, Balakumaran J, Iansavichene A, Cook DJ. (2020) Probiotics in Hospitalized Adult Patients: A Systematic Review of Economic Evaluations. *Can J Anaesth.* 67(2):247–261. doi:10.1007/s12630-019-01525-2.

21. Craighead AF, Caughey AB, Chaudhuri A, Yieh L, Hersh AR, Dukhovny D. (2020) Cost-Effectiveness of Probiotics for Necrotizing Enterocolitis Prevention in Very Low Birth Weight Infants. *J Perinatol.* 40(11):1652–1661. doi:10.1038/s41372-020-00790-0.

22. Mahon J, Lifschitz C, Ludwig T, Thapar N, Glanville J, Miqdady M, Saps M, Quak SH, Lenoir Wijnkoop I, Edwards M, Wood H, Szajewska H. (2017) The Costs of Functional Gastrointestinal Disorders and Related Signs and Symptoms in Infants: A Systematic Literature Review and Cost Calculation for England. *BMJ Open.* 7(11):e015594. doi:10.1136/bmjopen-2016-015594.

23. Lenoir-Wijnkoop I, van Aalderen WM, Boehm G, Klaassen D, Sprikkelman AB, Nuijten MJ. (2012) Cost-Effectiveness Model for a Specific Mixture of Prebiotics in The Netherlands. *Eur J Health Econ.* 13(1):101–110. doi:10.1007/s10198-010-0289-4.

24. Sach TH, McManus E, Levell NJ. (2019) Understanding Economic Evidence for the Prevention and Treatment of Atopic Eczema. *Br J Dermatol.* 181(4):707–716. doi:10.1111/bjd.17696.

25. Abdullah MMH, Hughes J, Grafenauer S. (2021) Whole Grain Intakes Are Associated with Healthcare Cost Savings Following Reductions in Risk of Colorectal Cancer and Total Cancer Mortality in Australia: A Cost-of-Illness Model. *Nutrients.* 13(9):2982. doi:10.3390/nu13092982.

26. Abdullah MM, Gyles CL, Marinangeli CP, Carlberg JG, Jones PJ. (2015) Cost-of-Illness Analysis Reveals Potential Healthcare Savings with Reductions in Type 2 Diabetes and Cardiovascular Disease Following Recommended Intakes of Dietary Fiber in Canada. *Front Pharmacol.* 6:167. doi:10.3389/fphar.2015.00167.

27. Murphy MM, Schmier JK. (2020) Cardiovascular Healthcare Cost Savings Associated with Increased Whole Grains Consumption among Adults in the United States. *Nutrients.* 12(8):2323. doi:10.3390/nu12082323.

28. Swanson KS, Gibson GR, Hutkins R, Reimer RA, Reid G, Verbeke K, Scott KP, Holscher HD, Azad MB, Delzenne NM, Sanders ME. (2020) The International Scientific Association for Probiotics and Prebiotics (ISAPP) Consensus Statement on the Definition and Scope of Synbiotics. *Nat Rev Gastroenterol Hepatol.* 17(11):687–701. doi:10.1038/s41575-020-0344-2.

29. Vinderola G, Sanders ME, Salminen S, Szajewska H. (2022) Postbiotics: The Concept and Their Use in Healthy Populations. *Front Nutr.* 9:1002213. doi:10.3389/fnut.2022.1002213.

30. Lenoir-Wijnkoop I. (2022) *The Value of Food—Savoir Vivre = Savoir Manger.* Éditeur: Utrecht University, https://dspace.library.uu.nl/handle/1874/422432. doi:10.33540/1456

31. https://ec.europa.eu/eurostat/fr/web/products-eurostat-news/-/ddn-20210118-1.

32. www.healthsystemtracker.org/chart-collection/what-do-we-know-about-spendingrelated-to-public-health-in-the-u-s-and-comparable-countries/.

33. Stojanovic J, Wübbeler M, Geis S, Reviriego E, Gutiérrez-Ibarluzea I, Lenoir-Wijnkoop I. (2020) Evaluating Public Health Interventions: A Neglected Area in Health Technology Assessment. *Front Public Health.* 8:106. doi:10.3389/fpubh.2020.00106.

34. Munt AE, Partridge SR, Allman-Farinelli M. (2017) The Barriers and Enablers of Healthy Eating Among Young Adults: A Missing Piece of the Obesity Puzzle: A Scoping Review. *Obes Rev.* 18(1):1–17. doi:10.1111/obr.12472

35. Thaler RH, Sunstein CR. (2008) *Nudge: Improving Decisions about Health, Wealth, and Happiness.* New Haven, CT: Yale University Press.

36. Cialdini RB. (2006). *Influence: The Psychology of Persuasion* (Rev. ed.). New York: Harper Collins.

37. Kahneman D. (2011). *Thinking, Fast and Slow.* New York: Farrar, Straus and Giroux.

38. Ariely D. (2008). *Predictably Irrational: The Hidden Forces that Shape Our Decisions.* New York: Harper Collins.

39. Kivetz R, Urminsky O, Zheng Y. (2006) The Goal-Gradient Hypothesis Resurrected: Purchase Acceleration, Illusionary Goal Progress, and Customer Retention. *J Mark Res.* 43(1):39–58.

40. Lally P, Van Jaarsveld CHM, Potts HWW, Wardle J. (2010) How Are Habits Formed: Modelling Habit Formation in the Real World. *Eur J Soc Psychol.* 40(6):998–1009.

# Health-Beneficial Microbes for Companion Animals

# 32

Shea Beasley, Minna Rinkinen, and Ángel Sainz

## 32.1 INTRODUCTION

In companion animal medicine, health-beneficial microbes are increasingly used to treat and prevent various clinical disturbances, mainly of gastrointestinal origin. Disturbed gut microbiota may lead to the development of a multitude of disorders such as diarrhea, allergies, obesity, and stress symptoms (Grzeskowiak et al., 2015), leading to more severe health problems, such as inflammatory bowel disease (IBD) reported both in cats (Marsilio et al., 2019) and in dogs (Xu et al., 2016; Díaz-Regañón et al., 2023). A healthy microbiome has an impact on preventing harmful bacteria occurring as well as transmitting between pets and between pets and humans (Nermes et al., 2013; Song et al., 2013). Studies on pet and human microbial interaction demonstrate that having a dog at home increases the shared skin microbiota between cohabiting adults. Pet parents share more microbiota with their dogs than with other family members (Song et al., 2013). Modifying dog microbiota by canine-specific probiotics can also be reflected in children's microbiota (Gómez Gallego et al., 2021). This leads to the conclusion of the importance of choosing microbes to our companion animals.

The rationale of administering probiotics to companion animals is based on their ability to balance the intestinal microbiota by reducing the colonization of pathogens via competitive exclusion and reinforcing host's immune defense mechanisms, particularly in the intestinal mucosa (Anadon et al., 2006). In light of growing evidence, hosts and microbes create a symbiotic coexistence comprising bidirectional exchange of endocrine, immune, and neural signals with targets in metabolic, immune, humoral, and neural pathways (Moloney et al., 2014; Vuong et al., 2017). Microbial metabolites, including short chain fatty acids (SCFAs), are recognized as important category of therapeutic molecules, as anti-cancer and anti-inflammatory agents as well as modulating the treatment of antibiotic-associated diarrhea, ulcerative colitis, obesity, and cholesterol reduction (Singh et al., 2018). Novel research areas cover the microbial impact on oral health (You et al., 2022), alleviating symptoms of epilepsy (García-Belenguer et al., 2021) and enteropathy in dogs receiving non-steroidal anti-inflammatory drugs (NSAIDs) (Herstad et al., 2022), as well modifying stress-related behavior (Yeh et al., 2022) in dogs.

# 32.2 SPECIES SPECIFICITY

To be able to exert their beneficial effects, such as immunostimulation and metabolome, it is believed that health-promoting microbes and probiotics should be species specific, that is, originate from the same animal species that they are fed to, as bacterial adherence to the gut wall is assumed to be unique to each animal species (Delsuc et al., 2014; Duranti et al., 2021; Fritsch et al., 2023). Diet adaptation is a major driving factor of convergence in gut microbiome composition over evolutionary timescales (Delsuc et al., 2014). As canines are scavengers by nature, the microbial intake of free-ranging domestical dogs is significant (Butler and du Toit, 2002); the change to a modern diet has had an impact on the gut microbiome. This phenomenon tends to hold true in all animal species in contact with humans: The gut microbiomes of coyotes as well as of equines subjected to greater human impact may provide evidence of dysbiosis, indicative of increased physiological stress and reduced health (Ang et al., 2022; Biles et al., 2021).

Although knowledge of the cat microbiome is limited, a recent feline gastrointestinal microbiome study identified healthy domestic cats with 30 fecal bacterial genera, *Prevotella, Bacteroides, Collinsella, Blautia*, and *Megasphaera* being most abundant and *Bacteroides, Blautia, Lachnoclostridium, Sutterella*, and *Ruminococcus gnavus* being most prevalent. The fecal microbiome composition varied with host diet type and living environment (Ganz et al., 2022). Older et al. (2017) presented bacterial communities differing between body sites in the healthy cats and the abundances of the bacterial species between healthy and allergic skin. *Staphylococcus* sp., in addition to other taxa, was more abundant on allergic skin, as is the case in atopic dogs as well (Bradley et al., 2016).

Several canine-obtained bacteria have been characterized for use in dogs, mainly Lactobacilliceae isolated from canine feces (Fujisawa and Mitsuoka, 1996; Beasley et al., 2006; Sharif et al., 2018; Coman et al., 2019; Lin et al., 2020). Fernández and coworkers (2019) reported assessing of *Lacticaseibacillus rhamnosus* MP01 and *Lactiplantibacillus plantarum* MP02 isolated from canine milk for their potential to prevent gastrointestinal infections in weaned puppies. Potential probiotic *Lactobacillus acidophilus* MJCD175 have been isolated from the canine oral cavity, showing a promising effect to support oral health in dogs (You et al., 2022). Dog-derived strains *Bifidobacterium animalis* AHC7 (O'Mahony et al., 2009) and *Limosilactobacillus reuteri* DSM 32203 (NBF-1) (EFSA, 2022) intended for use in dogs, as well as a feline isolate, *L. reuteri* DSM 32264 (NBF-2) (EFSA, 2019), intended for use in cats, have been evaluated but rejected for safety and efficacy for feed additives for cats and dogs. Heritage strains isolated from healthy family-owned cats (*Limosilactobacillus reuteri* F9–3, *Lacticaseibacillus paracasei* F9–4, *Ligilactobacillus animalis* F9–6, *Pediococcus acidilactici* F9–7, *Pediococcus pentosaceus* F9–8, and *Limosilactobacillus reuteri* F9–9) and from healthy family-owned dogs (*Lacticaseibacillus casei* K9–1, *Limosilactobacillus fermentum* K9–2, *Limosilactobacillus reuteri* PCR7, *Enterococcus faecium* PCEF02, and *Pediococcus acidilactici* PCLL) have been approved for use in companion animals in Canada, the US, and China (AAFCO, 2023; Health Canada, 2023). In the US, *Bifidobacterium longum* NCC3001 (BL999) has been shown to reduce anxiety in mice (Bercik et al., 2011); these results were adapted to cats and dogs to maintain calm behavior.

# 32.3 HEALTH-BENEFICIAL MICROBES

Microbial products aimed at pets are gaining popularity in companion animal health care. For admission to the European common market, any microbe intended for animal use is required to comply with the Feed Additive Regulation 1831/2003 requirements, both for safety and efficacy (Becquet, 2003; von Wright, 2005) as well as to fill the WHO requirements (Hill et al., 2014). When a microbe is introduced to feed, the European Food Safety Authority (EFSA) is required to assess the safety of these biological agents when they are proposed for use in regulated products that require market authorization (EFSA, 2018b, 2023).

Although microbes are considered complementary feeds, only a minority of the strains intended for pets are classified as probiotics. A few strains have a technological status, but there are also products available on the market with no feed status. In addition, the scientific evidence to support their beneficial effects is currently quite modest. Only few strains are host specific. A study evaluated 19 commercial pet foods labeled to contain added microbes. No product contained all the microorganisms listed on the package label. Eleven of the 19 feed samples contained additional organisms such as *Pediococcus* spp. (Weese and Arroyo, 2003).

In Europe, *Bacillus velezensis* C-3102 (DSM15544) is approved for use in dogs, *Lactobacillus acidophilus* CECT4529 (D2/CSL) for use in cats and dogs, and *Enterococcus faecium* NCIMB 10415 (SF68, DSM 10663) for use in cats and dogs. These three strains have acceptable regulatory safety assessment and demonstrated a minimum efficacy result for the benefit of the end user.

In addition to the approved strains, other bacteria, such as certain bacilli (*B. cereus, B. licheniformis,* and *B. subtilis*), *Lactobacilliceae* (*L. acidophilus, L. casei, L. delbrueckii, L. fermentum, L. paracasei, L. plantarum,* and *L. rhamnosus*), bifidobacteria (*B. breve, B. longum,* and *B. infantis*), *Pediococcus acidilactitici,* and *Streptococcus* sp., are available in some pet feed brands and complementary feeds without being approved for pet use in Europe. Strains of yeast *Saccharomyces cerevisiae* are on the market as well but approved only for use in equines (Anadon et al., 2006). The variety of microbial products intended for companion animal use is wider in the US and in Canada, including bacterial strains, fecal matter transplantation, and fermented food such as kefir.

Non-viable beneficial microbes and their parts and metabolites have been categorized as feed material in the EU. The definition of a postbiotic has been settled as a "preparation of inanimate microorganisms and/or their components that confers a health benefit on the host" (Salminen et al., 2021). The main reported health benefits of inanimate microbes relate to modulation of the resident microbiota, enhancement of epithelial barrier functions, modulation of local and systemic immune responses, modulation of systemic metabolic responses, and systemic signaling via the nervous system (Salminen et al., 2021; Martorell et al., 2021). Heat-treating probiotics is thought to activate pathways related to host innate immune function (Grzeskowiak et al., 2014; Martorell et al., 2021). Microbial metabolites, including short-chain fatty acids, bile acids, biogenic amines, vitamins, and organic acids, have shown positive effects on pathogenic bacteria control, mineral absorption, glucose metabolism, weight control and obesity, immune response homeostasis, gut barrier improvement, anticancer activity, brain modulation, and regulating inflammation and bioactive lipids that belong to the endocannabinoid system and specific neurotransmitters, such as aminobutyric acid, serotonin, and nitric oxide (Peredo-Lovillo et al., 2020; Rastelli et al., 2019).

Fecal microbiome transplant (FMT) may be beneficial in the treatment of acute intestinal diseases and management of chronic enteropathies (Chaitman and Gaschen, 2021). A single dose of FMT did not have clinical benefit on dogs with acute hemorrhagic diarrhea syndrome, although an increase of SCFA-producing bacterial communities was seen up to 30 days post-FMT (Gal et al., 2021). In a long-term (221 days) case study, the clinical signs of an IBD dog improved remarkably with regard to the changes in the fecal microbiome (Niina et al., 2019).

# 32.4 REGULATION AND SAFETY OF MICROBES INTENDED FOR USE IN COMPANION ANIMALS

Safety and health-beneficial characteristics are strain specific. Hill et al. (2014) concluded that the widespread mechanisms of probiotics can be associated with effects that are observed across taxonomic groups, such as inhibition of potential pathogens or the production of useful metabolites or enzymes. Other effects, including immune effects, are more likely to be strain specific, and claims of benefits can only be made for strains or species in which the mechanistic basis has been demonstrated.

Safety standards on individual strains require at minimum that the probiotic be safe for the intended use and meet European standards of the Qualified Presumption of Safety (QPS) initiative (EFSA, 2023) and EFSA guidance documents on antimicrobial resistance (EFSA, 2018b). North American requirements include the US Food and Drug Administration's (FDA) approaches to be generally recognized as safe (GRAS), with guidance provided by the Association of American Feed Control Officials (AAFCO), Veterinary Health Product (VHP) division of Health Canada, and International Probiotics Association (IPA) *Guidelines to Qualify a Microorganism to be Termed as "Probiotic"* (IPA, 2017). In the US, the National Animal Supplements Council (NASC) provides strict guidelines for product quality assurance to ensure ethical manufacturing and labeling practices are complied with throughout the industry (NASC, 2023).

*B. velezensis* (previously *B. subtilis*) C-3102 strain was found to be not toxigenic and not to show resistance to antibiotics of human and veterinary importance by EFSA. The strain is presumed safe for all target species, including dogs, and for the environment, meeting the criteria for the QPS approach. Four studies investigated the effects of the supplementation to the fecal consistency of dogs, showing a small but significant increase in fecal dry matter content (EFSA, 2017).

*E. faecium* NCIMB 10415 is intended for use in dogs within a dose of $4.5 \times 10^6$–$2 \times 10^9$ CFU/kg feed and in cats at a dose $5 \times 10^6$–$8 \times 10^9$ CFU/kg feed. The strain is susceptible to clinically relevant antibiotics, except for kanamycin. Three studies carried out in dogs demonstrated that the additive has the potential to produce a beneficial effect in dogs when added to feedstuffs at a dose of $2.5 \times 10^9$ CFU kg by increasing the intestinal or serum concentration of IgA and showing inconsistent effects on fecal quality in three studies in which cats were fed the additive at the dose of $7 \times 10^9$ CFU/kg of feed (EFSA, 2013).

*L. acidophilus* CECT 4529 is considered safe for target animals, including cats and dogs, and the environment. *L. acidophilus* CECT 4529 has some potential to reduce the moisture of stools from dogs and cats receiving the additive at $5 \times 10^9$ CFU/kg feed (EFSA, 2018a).

Of the approved probiotics, Lactobacilliceae have a long history in animal and human diet, have a stable genome, and are typically sensitive to antibiotics (Anisimova and Yarullina, 2019; Esteban-Torres et al., 2021). Spore-forming *Bacillus* sp. contains several opportunist pathogen species within the genus but also a stable genome. *B. velezensis* is a plant-originated strain, making it suitable for all animal species. With its spore-forming characteristics, the strain provides stability in shelf life and throughout the intestine (Todorov et al., 2021). Although *E. faecium* NCIMB10415 has been cleared as safe for animal use, the genus Enterococci has a notorious ability to rapidly develop and spread antibiotic resistance (Eaton and Gasson, 2001). Therefore, the safety and suitability of enterococci as probiotics have been challenged (Franz et al., 1999; Liong, 2008; Hanchi et al., 2018; Wang et al., 2021). *E. faecium* may also modify the intestinal microbiota in a potentially harmful direction. In dogs, *E. faecium* increased both the *in vitro* adhesion and fecal shedding of *Campylobacter* sp. (Rinkinen et al., 2003; Vahjen and Männer, 2003). Since multidrug-resistant Enterococci have been isolated from dogs and cats (de Leener et al., 2005), although no correlation to probiotic feeding has been documented (Jackson et al., 2009) and dogs have been noted being a reservoir of ampicillin-resistant *E. faecium* associated with human hospital infections (Damborg et al., 2009), new safe probiotic strains are welcome for companion animals. In addition, each probiotic has unique beneficial characteristics supporting a rich and diverse intestinal microbiome of its host.

Safety of a single-strain microbe is relatively straightforward to determine, whereas assessing the safety of organic material, such as fecal matter, is more complicated. In the consensus statement on the term "probiotic", Hill et al. (2014) raise the issue of donor fecal matter transfer mixtures comprising a number of unknown taxa, including bacteria, yeasts, parasites, and viruses, and it not being known which microbes are responsible for the beneficial effect and which might pose a risk through transfer of antibiotic resistance, production of genotoxic metabolites, or intestinal translocation. Undoubtably, FMT will be difficult to standardize, even with the same donor, due to inter- and intra-individual variability of the gut microbiota, and guidelines and safety procedures are required. A guide reporting standards for preclinical FMT the Guidelines for Reporting Animal Fecal Transplant (GRAFT) was published by Secombe et al. (2021), leading the way to ensure the safety of the patient.

# 32.5 DEVELOPMENT OF INTESTINAL MICROBIOTA

All animals harbor a vast and complex microbiota (Grzeskowiak et al., 2015). Several studies clearly demonstrated microbiota developing in utero and prior to birth in different species (Husso et al., 2021; JaberAlipour et al., 2018; Jiménez et al., 2005). There are yet no published studies on canine or feline microbial colonization prior to delivery. However, similar possibilities can't be ruled out, dogs and cats being mammals and sharing similar physiological and anatomical patterns with humans (Grzeskowiak et al., 2015).

Guard et al. (2017) described the development of the neonatal fecal microbiome in dogs. Feces were collected from puppies and their mothers. Species richness continued to increase significantly from 2 days of age until 42 days of age in puppies. Major phylogenetic changes were noted at all taxonomic levels, with the most profound changes being a shift from primarily Firmicutes in puppies at 2 days of age to a co-dominance of Bacteroidetes, Fusobacteria, and Firmicutes by 21 days of age. Supplementing feed with *E. faecium* NCIMB 10415 and *L. acidophilus* CECT 4529 during pregnancy and lactation improves dog colostrum (Alonge et al., 2020). Manninen and coworkers (2006) noted that adult healthy non-medicated dogs shared *Lactobacillus* strains with their mothers.

The time-limited early-life microbial colonization instructs the immune system and regulates the host gene expression, paving the way for adult host health and disease (Frese et al., 2011; Gensollen et al., 2016; Melandri et al., 2020; Nichols and Davenport, 2021). A diverse microbiota composition plays an important role in the extraction, synthesis, and absorption of many nutrients and metabolites, including bile acids, lipids, amino acids, vitamins, and SCFAs (Rinninella et al., 2019); endocrine and reproductive systems (López-Moreno and Aguilera, 2020); and synthesizing and regulating neurotransmitters, such as serotonin and GABA (Rosado et al., 2010; Saulnier et al., 2013). A diverse microbiome is altered by daily diet, where dietary polyphenols, or duplibiotics, have appeared to distinctly modulate the microbiome (Ephraim et al., 2022; Rodríguez-Daza et al., 2021). Other activities considered a part of a healthy lifestyle, such as exercise and sleep, also build on a diverse microbiome (Allen et al., 2015; Bowers et al., 2020).

Apart from diet, medication, and environmental factors, age of the host is also linked with gut microbial changes. In aging individuals, alterations in levels of microbial metabolites of amino acids, carbohydrates, nucleotides, lipids, and SCFA have been seen, as well as enrichment in opportunistic pro-inflammatory bacteria (Rampelli et al., 2013; Vemuri et al., 2018). Decreases in gut microbiome diversity and lower abundance of lactobacilli have been associated with high age. Also, poor cognitive performance correlates with a lower proportion of Actinobacteria in elderly pet dogs (Kubinyi et al., 2020). Vemuri et al. (2018) suggest *L. acidophilus* DDS-1 has a remodulation effect of gut microbiota, improving the metabolic phenotype in aging mice, and may be worth studying in pets as well.

# 32.6 MODIFICATION OF INTESTINAL MICROBIOTA BY HEALTH-BENEFICIAL MICROBES

Gut microbiota has a crucial immune function against pathogenic bacteria colonization, inhibiting their growth, consuming available nutrients, and/or producing bacteriocins. Gut microbiota also prevent bacteria invasion by maintaining intestinal epithelium integrity and prevent pathogenic colonization by many competition processes: nutrient metabolism, pH modification, antimicrobial peptide secretions, and effects on cell signaling pathways (Reid and Burton, 2002; Rinninella et al., 2019). Some Lactobacilliceae secrete membrane vesicles to deliver bacteriocin peptides to the opportunistic pathogen (Dean et al., 2020). In a recent study, commensal bacteria and their products were discovered to regulate the development, homeostasis, and function of innate and adaptive immune cells (Rinninella et al., 2019).

The entire gastrointestinal tract naturally contains beneficial microbes for the host if undisturbed (Grzeskowiak et al., 2015). Bacterial strains isolated from healthy individuals obtaining specific probiotic characteristics have an impact on host health limited to the strain characteristics. *L. reuteri* strains JCM1081 and TM105 were effective in preventing the *in vitro* binding of *Helicobacter pylori* to host cell receptors (Mukai et al., 2002) due to acid toleration in the stomach. Lactic acid bacteria tolerate the acidic intestinal environment well and have a high intestinal survival rate (Beasley et al., 2006; Conway et al., 1987; Fernández et al., 2019).

The reduction capacity of Enterococci and Lactobacilliceae towards the members of *Clostridium* sp. Group X is well documented. Enterotoxin-producing *Clostridium perfringens* is the most common potential enteric pathogen found in fecal samples of dogs with diarrhea (Bell et al., 2008; Gomez-Gallego et al., 2016; Guard et al., 2017; Suchodolski et al., 2012). Lactobacilliceae reduce the abundance of *C. perfringens* by (1) lowering of pH in the intestinal lumen by significantly increasing the release of short-chain fatty acids, (2) directing inhibition of competing bacteria by the excretion of selectively antimicrobial bacteriocin peptides and other natural compounds known to be produced by the lactobacilli, and (3) competition for adhesion sites on the intestinal wall with other bacteria (Baillon et al., 2004; Biagi et al., 2007; Gomez-Gallego et al., 2016; Marshall-Jones et al., 2006; O'Mahony et al., 2009; Shieh et al., 2011). *E. faecium* significantly decreased the canine *in vitro* mucus adhesion and *in vivo* fecal shedding of *C. perfringens*. On the other hand, in the same studies, *E. faecium* increased both the *in vitro* adhesion and fecal shedding of campylobacters (Rinkinen et al., 2003; Vahjen and Männer, 2003); thus, the seemingly positive outcome on fecal *C. perfringens* numbers may lead to potentially risky spreading of a zoonotic pathogen.

# 32.7  LEAKY GUT

An intact intestinal barrier with a healthy mucosal layer and a diverse microbiome is an integral regulator of health, contributing to gut barrier impairment and its increased permeability (Massier et al., 2021). Leaky gut syndrome is a proposed condition some health practitioners claim is the cause of a wide range of long-term conditions. Stressful disorders such as endurance exercise, nonsteroidal anti-inflammatory drug administration, pregnancy, and surfactants such as bile acids and dietary factors such as emulsifiers increase permeability (Camilleri, 2019). Alterations of tight junctions are important for the gut barrier, since they can lead to an increased influx of bacteria or bacterial components, such as endotoxins, bacterial DNA, and metabolites, into the host circulation (Massier et al., 2021). Permeability alteration is regarded as the basis for the pathogenesis of many diseases, including obesity, increased insulin resistance, infectious enterocolitis, inflammatory bowel disease, irritable bowel syndrome, small intestinal bacterial overgrowth, celiac disease, hepatic fibrosis, food intolerances, and also atopic symptoms (Lopetuso et al., 2015; Lee, 2015; Massier et al., 2021). The modulation of gut barrier function through diet and microbiome modification may represent a potential prevention and treatment target for metabolic diseases (Lopetuso et al., 2015; Massier et al., 2021).

# 32.8  OBESITY

The gut microbiota of obese cats and dogs is different from that of lean individuals (Kim et al., 2023; Tal et al., 2020). The association between feline obesity and the fecal bacterial microbiota was demonstrated in enriched taxa in obese cats compared to lean cats, which may be related to enhanced efficiency of energy harvesting (Tal et al., 2020). In dogs, the gut microbiome composition of obese and normal weight animals is affected by several factors, including diet, age, breed, and disorders (Kim et al., 2023). Obese dogs have been reported to contain Proteobacteria as a predominant phylum, with a high proportion of Pseudomonadales, where in normal weight dogs, the order Clostridiales and genus Lactobacillus were markedly more abundant (Park et al., 2015). Weight loss in obese dogs shifts the fecal microbiota profile to the extent of resembling the microbiota of normal weight dogs (Macedo et al., 2022). An enrichment of Gram-negative bacteria can influence the level of intestinal lipopolysaccharide, and this may be associated with chronic inflammation in obese subjects (Park et al., 2015). Kim and coworkers (2023) also observed that genes related to amino acid biosynthesis and B vitamin biosynthesis were enriched in dogs with a high body mass index. Indigenous bacteria may intervene in the metabolism, increasing energy production from the diet and taking part in the regulation of fatty acid tissue composition (Cani et al., 2012). This was also seen in obese cats, where fecal microbial abundance and biodiversity were only minimally affected during the early phase of a standardized weight loss plan (Tal et al., 2020), suggesting that intake of probiotics may enhance losing weight together with a low energy diet. *Akkermansia muciniphila* has been suggested to suppress high-fat diet–induced obesity and related metabolic disorders in beagle dogs (Lin et al., 2022). Xiao and Kang (2020) emphasized the utilization of bacterial derived short-chain fatty acids by the host as a direct energy source while regulating the appetite of the host through the gut–brain axis, which could support maintaining normal weight. Also, serotonin (5-hydroxytryptamine, 5HT) is produced in the gut (Rastelli et al., 2019) and is involved in hypothalamic regulation of energy consumption. Serotonin levels in the central nervous system are influenced by lower energy conditions in obese beagles compared to normal-weight beagles. Decreased 5HT levels can increase the risk of obesity due to increased appetite (Park et al., 2015).

## 32.9 ACUTE AND CHRONIC ENTEROPATHIES IN DOGS AND CATS—POTENTIAL ROLE OF PROBIOTICS

Digestive signs are very common in small animal practice, reaching around 30% of frequency of the total number of clinical cases (O'Neill et al., 2014). The main objective of the probiotics currently marketed in veterinary medicine for dogs and cats is usually the management of digestive disorders. However, the scientific evidence in this regard is still low, based on the limited number of studies carried out so far. Most studies have evaluated their potential role on acute and chronic enteropathies.

### 32.9.1 Probiotics, Immune System, and Gastrointestinal Health

The relationship among the gastrointestinal mucosa, specifically the gut-associated lymphatic tissue (GALT), diet, and the microbiota, seems to play a significant role in the immune system and in digestive health and disease. The effects of probiotics are not global for all of them but are strain specific. Different effects of probiotics such as immune modulation, improvement of the intestinal barrier, or inhibition of specific entero-pathogens have been described.

Some *in vitro* and *in vivo* studies have evaluated the potential effects of probiotics on systemic immune functions and intestinal health. When administered to healthy puppies, *E. faecium* 10415 SF68 induced significantly higher fecal and serum IgA levels, a higher percentage of circulating B lymphocytes, and distemper virus vaccine-specific circulating IgG and IgA, when compared with placebo, suggesting an enhancement in specific immunological responses (Benyacoub et al., 2003). A similar study with the same probiotic in healthy kittens showed a higher percentage of CD4+ lymphocytes in treated animals but no effect on other specific immune parameters (Veir et al., 2007).

In healthy adult dogs, the administration of *E. faecalis* FK-23 can stimulate nonspecific immune responses, including increased neutrophil phagocytosis and increased lymphocyte blast transformation response (Kanasugi et al., 1997).

Only some studies have evaluated the potential effects of probiotics or synbiotics in healthy dogs and cats using next-generation sequencing methods. When administering a multi-species mixture probiotic (*E. faecium* NCIMB 30183, *Streptococcus salivarus* subsp. *thermophilus* NCIMB 30189, *Bifidobacterium longum* NCIMB 30179, *Lactobacillus acidophilus* NCIMB 30184, *L. casei* subsp. *rhamnosus* NCIMB 30188, *L. plantarum* NCIMB 30187, and *L. delbrueckii* subsp. *bulgaricus* NCIMB 30186) in dogs and cats, no changes were identified in gastrointestinal function or immune markers nor in the major bacterial phyla, except an increase in the abundances of *Enterococcus* and *Streptococcus* spp. (García-Mazcorro et al., 2011). Similarly, a significant rise in Lactobacillaceae was detected in dogs that received another multi-species symbiotic (Gagné et al., 2013).

## 32.9.2 Acute Enteropathies

Most studies evaluating the use of probiotics in dogs with acute gastrointestinal signs show a global better fecal score after weeks of therapy and a shorter duration of the diarrhea.

A probiotic cocktail consisted of *L. acidophilus*, *Pediococcus acidilactici*, *B. subtilis*, *B. licheniformis*, and *L. farciminis* was evaluated in dogs with acute idiopathic gastroenteritis in a randomized, double-blind study. The probiotic administration resulted in a significantly quicker improvement of the diarrhea but not of the vomiting (Herstad et al., 2010).

Similarly, the use of a probiotic based on *B. animalis* AHC7 ($2 \times 10^{10}$ CFU daily) isolated from dogs was evaluated in dogs with acute idiopathic diarrhea, randomly compared to placebo. A rescue treatment with metronidazole was proposed for dogs unresponsive to therapy in both groups. Probiotic administration significantly reduced the time to clinical resolution and the percentage of dogs that were administered metronidazole compared with placebo (Kelley et al., 2009).

A double-blind placebo-controlled study on dogs with acute or intermittent diarrhea evaluated the administration of a sour-milk product containing three canine-derived *Lactobacillus fermentum* VET 9A, *L. rhamnosus* VET 16A, and *L. plantarum* VET 14A ($2 \times 10^9$ CFU/ml) compared to placebo. The treatment was associated with decreased numbers of *C. perfringens* and *E. faecium* in the feces of probiotic-treated dogs and had a normalizing effect on canine stool consistency. Difference among groups in mean fecal score change was achieved on day 7 (Gómez-Gallego et al., 2016).

Regarding shelter dogs, the efficacy of a canine-derived *Bifidobacterium animalis* AHC7 for reducing stress-related gastrointestinal disturbances in dogs relocated to kennels was tested. No difference in diarrhea frequency events could be shown between groups, but significantly better fecal scores at day 21 were obtained in healthy dogs when supplementing with the probiotic prior to and during kennel relocation at doses between $10^7$ and $10^9$ CFU/day (Kelley et al., 2012).

Similarly, a synbiotic supplement (containing *E. faecium* NCIMB 10415 4b1707) also significantly decreased the incidence of diarrhea in shelter dogs (Rose et al., 2017). On the other hand, the percentage of shelter cats with diarrhea ≥2 days treated with *E. faecium* SF68 was significantly lower when compared with placebo (Bybee et al., 2011). Anyway, information available in cats about the potential efficacy of probiotics is currently scarce. A kitten-origin *E. hirae* probiotic preventatively administered to weaned fostered shelter kittens was associated with a decreased incidence of diarrhea (Gookin et al., 2022).

In dogs with acute hemorrhagic diarrhea syndrome without signs of sepsis, a multi-strain probiotic has been evaluated. The mixture contained *L. plantarum*, *S. thermophilus*, *B. breve*, *L. paracasei*, *L. delbrueckii* subsp. *bulgaricus*, *L. acidophilus*, *B. longum*, and *B. infantis*. Probiotic and placebo groups showed a rapid clinical improvement, being significantly quicker in dogs treated with probiotic (Ziese et al., 2018).

Antibiotic administration produces deep and sustained changes in the microbiota. Dogs with antibiotic-induced diarrhea due to lincomycin were treated with a *Saccharomyces boulardii* probiotic, and they had a shorter duration of diarrhea when compared with placebo (Aktas et al., 2007). However, no difference in dysbiosis index was found when comparing placebo and a symbiotic containing *B. bifidum*, *E. faecium*, *E. thermophilus*, *L. acidophilus*, *L. bulgaricus*, *L. casei*, and *L. lantarum* in dogs with antibiotic-induced diarrhea due to clindamycin (Whittemore et al., 2019). Similarly, in cats treated with amoxicillin-clavulanate, no difference in fecal scores was detected when administering *E. faecium* SF68 NCIMB 10415 (Torres-Henderson et al., 2017).

## 32.9.3 Chemotherapy-Induced Gastrointestinal Toxicity

Chemotherapy-induced gastrointestinal toxicity is a serious complication of cancer treatment in human and veterinary medicine. Changes in gastrointestinal microbial populations and secondary structural gastrointestinal changes, altered bacterial metabolic by-products, and increased gastrointestinal inflammatory cytokines are proposed mechanisms (Jugan et al., 2021).

*E. faecalis* FK-23 has been evaluated in dogs receiving cyclophosphamide. Although the probiotic did not inhibit chemotherapy-induced neutropenia, increases in the myeloid/erythroid ratio and neutrophilic lineages were found in the bone marrow of treated dogs (Hasegawa et al., 1996).

In dogs with multicentric lymphoma undergoing multi-agent chemotherapy, a multi-strain probiotic (*S. thermophilus, B. breve, B. longum, B. infantis, L. acidophilus, L. plantarum, L. paracasei*, and *L. delbrueckii* subsp. *bulgaricus*) has recently been evaluated in a pilot study of 10 animals. No dogs receiving probiotics experienced diarrhea compared to four of five receiving placebo. Dogs receiving probiotics had increased fecal *Streptococcus* spp. and *E. coli* (Jugan et al., 2021).

## 32.9.4 Viral and Parasitic Diseases Affecting the Gastrointestinal System

In canine parvoviral enteritis, the administration of a probiotic mixture added to standard treatment decreased the mortality rate compared to the standard treatment alone. The mixture contained *L. acidophilus, P. acidilactici, B. subtilis, B. licheniformis*, and *L. farciminis*. Some clinical biomarkers such as white blood cell and lymphocyte counts also significantly improved in probiotic-treated dogs. However, treatment duration was similar in both groups (Arslan et al., 2012).

Similarly, *Lactobacillus murinus* LbP2 significantly improved different clinical parameters such as fecal consistency, mental status, and appetite compared to placebo in dogs with distemper-associated diarrhea (Delucchi et al., 2017).

Regarding the potential effects of probiotics in gastrointestinal parasitosis, no differences were observed in canine giardiasis between placebo or *E. faecium* SF68 (NCIMB 10415) treatment for giardial cyst shedding or fecal scores in a study where all dogs remained with subclinical parasitosis (Simpson et al., 2009). However, when adding this probiotic to metronidazole, a reduction in *Giardia* cyst elimination and a better clinical response in dogs were described compared with metronidazole alone (Fenimore et al., 2017).

The potential use of probiotics in the control of canine ancylostomiasis has also been evaluated. The administration of a probiotic preparation (*L. acidophilus* ATCC 4536, *L. plantarum* ATCC 8014, and *L. delbrueckii* UFV H2B20) led to a significant temporary reduction of *Ancylostoma* eggs in feces from naturally infected dogs (Coelho et al., 2013).

When adding *E. faecium* DSM 10663 NCIMB 10415 to ronidazole in cats infected by *Tritrichomonas foetus*, the number of relapses was lower than in placebo-treated cats (Lalor et al., 2012). *In vitro* inhibitory effects of enterococci on feline *T. foetus* infection have also been described (Dickson et al., 2019).

## 32.9.5 Chronic Enteropathies

Chronic inflammatory enteropathies are a group of gastrointestinal disorders presenting chronic digestive signs (lasting 3 weeks or longer) that are diagnosed based on the presence of the histopathologic confirmation of intestinal mucosal inflammation after the exclusion of other underlying causes producing similar signs. Chronic enteropathies can further be subdivided retrospectively by response to treatment into food-responsive enteropathy (FRE), antibiotic-responsive enteropathy (ARE), immunosuppressant-responsive enteropathy (IRE), and non-responsive enteropathy (NRE). When the term inflammatory bowel disease is used for dogs, it typically implies that treatment trials with diet and subsequently antibiotics have failed, inflammation has been demonstrated, and an immunosuppressant will be needed (Dandrieux, 2016).

The pathogenesis of this disease is known to be multifactorial, where intestinal barrier dysfunction, immunological dysregulation, and gut microbiota changes play a central role (Díaz-Regañón et al., 2023). Theoretically, supplementation with probiotics is an interesting strategy considering their potential effects on the intestinal epithelial physiology and on the gut and systemic immune responses (Fedorak, 2008). The current scientific evidence suggests that probiotics or synbiotics add little benefit when treating food- or antibiotic-responsive canine chronic enteropathies but could be promising adjunctive treatments in canine inflammatory bowel disease (Schmitz, 2021).

### 32.9.5.1 Food-Responsive Enteropathy

When adding *E. faecium* DSM 10663 on *ex vivo* cultures of duodenal biopsies from dogs with FRE, the effect on different cytokines was very limited. TNF-α gene expression was suppressed, like that found in samples from healthy dogs (Schmitz et al., 2014).

A similar *ex vivo* study was performed with biopsies obtained from dogs with chronic enteropathy, most of them without pathological changes or with mild changes in duodenal biopsies. The addition of canine-derived *L. acidophilus* NCC2628, *L. acidophilus* NCC2766, and *L. johnsonii* NCC2767 significantly increased IL-10 mRNA and protein levels in these samples. These effects on specific regulatory cytokines relative to inflammatory cytokines might contribute to reduction of intestinal inflammation, considering the results obtained in an *ex vivo* culture system of duodenal samples from dogs with chronic enteropathies (Sauter et al., 2005). However, these effects could not be subsequently reproduced *in vivo* (Sauter et al., 2006).

Administration of *L. acidophilus* DSM 13241 improved fecal consistency in a limited number of dogs with non-specific dietary sensitivity (Pascher et al., 2008). However, most randomized clinical trials in dogs with FRE have failed to produce favorable responses in treated dogs. Significant effects on clinical, endoscopic, histopathologic, gene expression of cytokines, and microbial diversity have not been detected in dogs treated with *E. faecium* DSM 10663 (Schmitz et al., 2015; Pilla et al., 2019).

### 32.9.5.2 Antibiotic-Responsive Enteropathy

Antibiotic-responsive enteropathy is recognized as one form of chronic enteropathy that has been associated with intestinal microbiota dysbiosis. In fact, it has been recently proposed as "idiopathic intestinal dysbiosis" (Jergens et al., 2022). Dogs usually improve after an antibiotic trial, based on metronidazole, oxytetracycline, or especially tylosin, and this disease is also called tylosin-responsive diarrhea (Westermarck et al., 2005).

However, the rate of relapsing cases after discontinuing the treatment is high. Furthermore, antibiotic administration causes deep changes in the composition, richness, and diversity of the intestinal microbiota (Stavroulaki et al., 2023), and potential antibiotic resistance is a major global concern.

On the other hand, the role of classically considered enteropathogenic bacteria (e.g., *Salmonella* spp., *Campylobacter jejuni*, specific enteropathogenic *E. coli* strains, *Yersinia* spp., *C. perfringens*, *C. difficile*) is currently controversial. These bacteria have been isolated both in healthy and sick animals (Marks and Kather, 2003). In fact, the usefulness of fecal culture in dogs with chronic diarrhea has been recently evaluated, failing to distinguish between both canine populations (Werner et al., 2021).

Based on these findings, antibiotics should be reserved in dogs with chronic diarrhea, after appropriate dietary trials, for cases with signs of true primary infection (i.e. signs of systemic inflammatory response syndrome or evidence of adherent-invasive bacteria) that justify antibacterial use (Cerquetella et al., 2020). In the remaining cases, alternative treatments to modulate bacterial populations, including the administration of probiotics or synbiotics, have been advocated (Cerquetella et al., 2020). However, the administration of *L. rhamnosus* GG (LGG) instead of tylosin in dogs with "tylosin-responsive diarrhea" failed to prevent diarrhea relapses in all dogs included in a clinical study (Westermarck et al., 2005).

### 32.9.5.3 Immunosuppressant-Responsive Enteropathy

Different clinical trials have shown the potential favorable effects of probiotics in dogs with IRE. A multispecies probiotic (*L. casei, L. plantarum. L. acidophilus, L. delbrueckii* subsp. *bulgaricus, B. longum, B. breve, B. infantis*, and *S. thermophilus*) produces comparable effects in dogs with IRE to a standard therapy based on metronidazole and prednisolone (Rossi et al., 2014). The clinical improvement was slower in probiotic-treated dogs, and reduction of inflammatory cells in duodenal samples was similar in both groups. Dogs treated with probiotic showed an improvement in immune intestinal tolerance, based on the increase of some specific populations such as FoxP3+ T cells. Administration of the probiotic also resulted in significantly increased plasma citrulline concentrations, a marker of global enterocyte mass, suggesting restitution of the mucosal barrier. Furthermore, *Faecalibacterium* spp. abundance increased significantly only in dogs treated with this probiotic (Rossi et al., 2014).

Other clinical trials evaluated the same probiotic, comparing the effects of its administration as adjunctive to the standard therapy (hypoallergenic diet and prednisolone) versus placebo. Both treatments were associated with rapid clinical remission but not improvement in histopathologic inflammation. Probiotic therapy was associated with significantly upregulated tight junction protein expression, suggesting that this probiotic may have beneficial effects on mucosal homeostasis (White et al., 2017).

*S. boulardii* as adjunctive to standard IRE therapy also results in a significant clinical improvement compared to placebo. A significant increase in serum albumin was detected in a limited number of probiotic-treated dogs with protein-losing enteropathy included in the same study (D'Angelo et al., 2018).

A combination of probiotics (*L. acidophilus, L. casei, E. faecium*, and *Bacillus subtilis*), prebiotics, and antibody IgY has recently been evaluated in dogs with chronic enteropathy. Compared to placebo-treated dogs, dogs showed decreased calprotectin, decreased C-reactive protein, increased mucosal *Clostridia* and *Bacteroides*, and decreased *Enterobacteriaceae* in colonic biopsies, suggesting a beneficial effect of this combination on host responses and mucosal microbiota in dogs with chronic enteropathy (Sahoo et al., 2022).

Studies evaluating probiotics in feline chronic enteropathies are very limited. An open-label clinical trial evaluated the efficacy of a probiotic mixture (*E. faecium, Streptococcus (Enterococcus) thermophilus, L. acidophilus, L. bulgaricus, L. casei, Bifidobacterium bifidum*, and *L. plantarum*) in adult cats with chronic diarrhea. The fecal score improved, and owners perceived a clinical improvement of diarrhea after probiotic administration (Hart et al., 2012).

### 32.9.5.4 Feline Chronic Constipation or Idiopathic Megacolon

A mixture probiotic (*S. thermophilus, L. acidophilus, L. plantarum, L. casei, L. helveticus, L. brevis, B. lactis* DSM32246, and *B. lactis* DSM3224) has shown to be beneficial in cats with chronic constipation/megacolon. In a pilot study, 10 cats had significantly reduced clinical severity indexes after treatment. Evaluation of histological parameters suggests a potential anti-inflammatory effect of the probiotic, associated with a reduction of mucosal infiltration, and a restoration of the number of interstitial cells of Cajal (Rossi et al., 2018).

# 32.10 ATOPIC DERMATITIS

Probiotics have been studied extensively in human medicine for the prevention of atopic dermatitis based on the hypothesis that the gut microbiome contributes to the pathophysiology of allergic disease. There have been many trials with mixed results, most in children with a family history of atopy (Arkwright and Koplin, 2023).

Results obtained in different studies in canine atopic dermatitis are not conclusive. A pilot study in an experimental setting including two adult female beagles with severe atopic dermatitis and their 16 puppies checked the potential effect of probiotic supplementation. All puppies were sensitized to a common canine allergen, *Dermatophagoides farinae*. Administration of *L. rhamnosus* GG to puppies appeared to reduce immunologic indicators of atopic dermatitis, although no significant effect in clinical signs was detected (Marsella, 2009). A subsequent study of these animals suggested that early exposure to probiotics had long-term clinical and immunological effects (Marsella et al., 2012).

The administration of the probiotic *L. sakei* probio-65 for 2 months significantly reduced the disease severity index in dogs diagnosed with atopic dermatitis (Kim et al., 2015). However, an oral probiotic supplement with *L. paracasei* K71 did not produce significant effects in atopic dogs (Ohshima-Terada et al., 2015). Similarly, the adjunctive effect of supplementation with *E. faecium* SF68 on oclatinib dose reduction was evaluated in a clinical trial. Clinical improvement was similar in dogs treated with probiotic versus placebo. Supplementation with this probiotic was associated with no difference in oclacitinib dose reduction between groups (Yamazaki et al., 2019).

Finally, the use of a postbiotic spray-based heat-treated lactobacilli resulted in a significant and rapid decrease in the clinical signs associated with canine atopic dermatitis in an open-label, uncontrolled study (Santoro et al., 2021).

# 32.11 CONCLUSIONS

The current scientific evidence suggests that probiotics are a promising strategy that may be useful in canine and feline health, medical therapy, and nutrition. The evaluation of specific strains from different origins for specific clinical situations should be the approach to properly identify the functionality of probiotics. Further well-designed controlled studies are needed to assess the benefits, strengths, and weaknesses of probiotics for dogs and cats. These studies should be focused on their mechanisms and on the clinical effects, without forgetting their safety.

# BIBLIOGRAPHY

AAFCO. AAFCO Official Publication Manual, 2023; Section 36-Fermentation Products, 36.14-Direct-Fed Microorganisms.

Aktas MS, Borku MK, Ozkanlar Y. Efficacy of *Saccharomyces boulardii* as a probiotic in dogs with lincomycin induced diarrhoea. *Bulletin- Vet Inst Pulawy* 2007;51(3):365–369.

Allen JM, Miller ME, Pence BD, Whitlock K, Nehra V, Gaskins HR, White BA, Fryer JD, Woods JA. Voluntary and forced exercise differentially alters the gut microbiome in C57BL/6J mice. *J Appl Physiol* 2015;118:1059–1066.

Alonge S, Aiudi GG, Lacalandra GM, Leoci R, Melandri M. Pre- and probiotics to increase the immune power of colostrum in dogs. *Front Vet Sci* 2020;7:570414. http://doi.org/10.3389/fvets.2020.570414.

Anadon A, Martinez-Larranaga MR, Martinez MA. Probiotics for animal nutrition in the European Union. Regulation and safety assessment. *Regul Toxicol Pharmacol* 2006;45:91–95.

Ang L, Vinderola G, Endo A, Kantanen J, Jingfeng C, Binetti A, Burns P, Qingmiao S, Suying D, Zujiang Y, Rios-Covian D, Mantziari A, Beasley S, Gomez-Gallego C, Gueimonde M, Salminen S. Gut Microbiome Characteristics in feral and domesticated horses from different geographic locations. *Commun Biol* 2022;5(1):172. doi:10.1038/s42003-022-03116-2.

Anisimova EA, Yarullina DR. Antibiotic resistance of lactobacillus strains. *Curr Microbiol.* 2019;76(12):1407–1416. http://doi.org/10.1007/s00284-019-01769-7.

Arkwright PD, Koplin JJ. Impact of a decade of research into atopic dermatitis. *J Allergy Clin Immunol Pract* 2023;11(1):63–71.

Arslan HH, Aksu DS, Terzi G, et al. Therapeutic effects of probiotic bacteria in parvoviral enteritis in dogs. *Rev Vet Med* 2012;163(2):55–59.

Baillon M-LA, Marshall-Jones ZV, Butterwick RF. Effects of probiotic *Lactobacillus acidophilus* strain DSM 13241 in healthy adult dogs. *Am J Vet Res* 2004;65:338–343.

Beasley SS, Manninen TJK, Saris PEJ. Lactic acid bacteria isolated from canine faeces. *J Appl Microbiol* 2006;101:131–138.

Becquet P. EU assessment of enterococci as feed additives. *Int J Food Microbiol* 2003;88(2–3):247–254.

Bell JA, Kopper JJ, Turnbul JA, Barbu NI, Murphy AJ, Mansfield LS. Ecological characterization of the colonic microbiota of normal and diarrheic dogs. *Interdiscip Perspect Inf Dis* 2008. http://doi.org/10.1155/2008/149694.

Benyacoub J, Czarnecki-Maulden GL, Cavadini C, Sauthier T, Anderson RE, Schiffrin EJ, von der Weid T. Supplementation of food with *Enterococcus faecium* (SF68) stimulates immune functions in young dogs. *J Nutr* 2003;133(4):1158–1162.

Bercik P, Park AJ, Sinclair D, Khoshdel A, Lu J, Huang X, Deng Y, Blennerhassett PA, Fahnestock M, Moine D, Berger B, Huizinga JD, Kunze W, Mclean PG, Bergonzelli GE, Collins SM, Verdu EF. The anxiolytic effect of *Bifidobacterium longum* NCC3001 involves vagal pathways for gut–brain communication. *Neurogastroenterol Motil* 2011;23(12):1132–1139. http://doi.org/10.1111/j.1365-2982.2011.01796.x

Biagi G, Cipollini I, Pompei A, Zaghini G, Matteuzzi D. Effect of a *Lactobacillus animalis* strain on composition and metabolism of the intestinal microflora in adult dogs. *Vet Microbiol* 2007;124(1–2):160–165.

Biles T, Beck H, Masters BS. Microbiomes in Canidae. *Ecol Evolut* 2021;11:18531–18539.

Bowers SJ, Vargas F, González A, He S, Jiang P, Dorrestein PC. Repeated sleep disruption in mice leads to persistent shifts in the fecal microbiome and metabolome. *PLoS ONE* 2020;15(2):e0229001. http://doi.org/10.1371/journal.pone.0229001.

Bradley CW, Morris DO, Rankin SC, Cain CL, Misic AM, Houser T, Mauldin EA, Grice EA. Longitudinal evaluation of the skin microbiome and association with microenvironment and treatment in canine atopic dermatitis. *J Invest Dermatol*. 2016;136(6):1182–1190. http://doi.org/10.1016/j.jid.2016.01.023.

Butler and du Toit. Diet of free-ranging domestic dogs (*Canis familiaris*) in rural Zimbabwe: Implications for wild scavengers on the periphery of wildlife reserves. *Anim Conserv* 2002;5:29–37.

Bybee SN, Scorza AV, Lappin MR. Effect of the probiotic *Enterococcus faecium* SF68 on presence of diarrhea in cats and dogs housed in an animal shelter. *J Vet Intern Med* 2011;25:856–860.

Camilleri M. The leaky gut: Mechanisms, measurement and clinical implications in humans. *Gut*. 2019;68(8):1516–1526. http://doi.org/10.1136/gutjnl-2019-318427.

Cani PD, Osto M, Geurts L, Everard A. Involvement of gut microbiota in the development of low-grade inflammation and type 2 diabetes associated with obesity. *Gut Microb* 2012;3(4):279–288.

Cerquetella M, Rossi G, Suchodolski JS, Schmitz SS, Allenspach K, Rodríguez-Franco F, Furlanello T, Gavazza A, Marchegiani A, Unterer S, Burgener IA, Pengo G, Jergens AE. Proposal for rational antibacterial use in the diagnosis and treatment of dogs with chronic diarrhoea. *J Small Anim Pract* 2020;61(4):211–215.

Chaitman J, Gaschen F. Fecal microbiota transplantation in dogs. *Vet Clin North Am Small Anim Pract* 2021;51(1):219–233. http://doi.org/10.1016/j.cvsm.2020.09.012.

Coêlho MD, Coêlho FA, de Mancilha IM. Probiotic therapy: A promising strategy for the control of canine hookworm. *J Parasitol Res* 2013;2013:430413.

Coman MM, Verdenelli MC, Cecchini C, Belà B, Gramenzi A, Orpianesi C, Cresci A, Silvi S. Probiotic characterization of Lactobacillus isolates from canine faeces. *J Appl Microbiol*. 2019;126(4):1245–1256. http://doi.org/10.1111/jam.14197.

Conway PL, Gorbach SL, Goldin BR. Survival of lactic acid bacteria in the human stomach and adhesion to intestinal cells. *J Dairy Sci*. 1987;70:1–12.

Damborg P, Top J, Hendrickx APA, Dawson S, Willems RJL, Guardabassi L. Dogs are a reservoir of ampicillin-resistant *Enterococcus faecium* lineages associated with human infections. *Appl Environ Microbiol*. 2009;75:2360–2365.

Dandrieux JR. Inflammatory bowel disease versus chronic enteropathy in dogs: Are they one and the same? *J Small Anim Pract*. 2016;57(11):589–599.

D'Angelo S, Fracassi F, Bresciani F, Galuppi R, Diana A, Linta N, Bettini G, Morini M, Pietra M. Effect of *Saccharomyces boulardii* in dogs with chronic enteropathies: Double-blinded, placebo-controlled study. *Vet Rec*. 2018;182:258–265.

Dean SN, Rimmer MA, Turner KB, Phillips DA, Caruana JC, Hervey WJ, Leary DH, Walper SA. *Lactobacillus acidophilus* membrane vesicles as a vehicle of bacteriocin delivery. *Front Microbiol*. 2020;11:710. http://doi.org/10.3389/fmicb.2020.00710.

De Leener E, Decostere A, de Graef, Moyaert H, Haesebrouk F. Presence and mechanism of antimicrobial resistance among enterococci from cats and dogs. *Microb Drug Resist*. 2005;11(4).

Delsuc F, Metcalf JL, Wegener Parfrey L, Song SJ, Conzalez A, Knight R. Convergence of gut microbiomes in myr-mecophagous mammals. *Molec Ecol.* 2014;23:1301–1317.

Delucchi L, Fraga M, Zunino P. Effect of the probiotic *Lactobacillus murinus* LbP2 on clinical parameters of dogs with distemper-associated diarrhea. *Can J Vet Res.* 2017;81(2):118–121.

Díaz-Regañón D, García-Sancho M, Villaescusa A, Sainz A, Agulla B, Reyes-Prieto M, Rodríguez-Bertos M, Rodríguez-Franco F. Characterization of the fecal and mucosa—associated microbiota in dogs with chronic inflammatory enteropathy. *Animals.* 2023;13:326.

Dickson R, Vose J, Bemis D, Davis M, Cecere T, Gookin JL, Steiner J, Tolbert MK. The effect of enterococci on feline *Tritrichomonas foetus* infection in vitro. *Vet Parasitol* 2019;273:90–96.

Duranti S, Longhi G, Ventura M, van Sinderen D, Turroni F. Exploring the ecology of bifidobacteria and their genetic adaptation to the mammalian gut. *Microorganisms.* 2021;9:8. http://doi.org/10.3390/microorganisms9010008.

Eaton TJ, Gasson MJ. Molecular screening of enterococcus virulence determinants and potential for genetic exchange between food and medical isolates. *Appl Environ Microbiol.* 2001;67:1628–1635.

EFSA. Scientific opinion on the safety and efficacy of Cylactin (*Enterococcus faecium*) as a feed additive for cats and dogs. *EFSA J.* 2013;11(2):3098.

EFSA. Safety and efficacy of Calsporin (*Bacillus subtilis* DSM 15544) as a feed additive for dogs. *EFSA J.* 2017. http://doi.org/10.2903/j.efsa.2017.4760.

EFSA. Safety and efficacy of *Lactobacillus acidophilus* D2/CSL (*Lactobacillus acidophilus* CECT 4529) as a feed additive for cats and dogs. *EFSA J.* 2018a. http://doi.org/10.2903/j.efsa.2018.5278.

EFSA. Guidance on the characterisation of microorganisms used as feed additives or as production organisms. *EFSA J.* 2018b;16(3):5206. http://doi.org/10.2903/j.efsa.2018.5206.

EFSA. Safety and efficacy of *Lactobacillus reuteri* NBF-2 (DSM 32264) as a feed additive for cats. *EFSA J.* 2019;17(1):5526. http://doi.org/10.2903/j.efsa.2019.5526.

EFSA. Assessment of the efficacy of a feed additive consisting of *Limosilactobacillus reuteri* (formerly *Lactobacillus reuteri*) DSM 32203 for dogs (NBF LANES). *EFSA J.* 2022;20(7):7436. http://doi.org/10.2903/j.efsa.2022.7436.

EFSA. *Qualified Presumption of Safety (QPS);* 2023. www.efsa.europa.eu/en/topics/topic/qualified-presumption-safety-qps.

Ephraim E, Brockman JA, Jewell DE. A diet supplemented with polyphenols, prebiotics and omega-3 fatty acids modulates the intestinal microbiota and improves the profile of metabolites linked with anxiety in dogs. *Biology.* 2022;11:976. http://doi.org/10.3390/biology11070976.

Esteban-Torres M, Ruiz L, Lugli GA, Ventura M, Margolles A, van Sinderen D. Editorial: Role of bifidobacteria in human and animal health and biotechnological applications. *Front Microbiol.* 2021;12:785664. http://doi.org/10.3389/fmicb.2021.785664.

Fedorak RN. Understanding why probiotic therapies can be effective in treating IBD. *J Clin Gastroenterol.* 2008;42:S111–S115.

Fenimore A, Martin L, Lappin MR. Evaluation of metronidazole with and without *Enterococcus Faecium* SF68 in shelter dogs with diarrhea. *Topics in Companion An Med.* 2017;32:100–103.

Fernández L, Martínez R, Pérez M, Arroyo R, Rodríguez JM. Characterization of *Lactobacillus rhamnosus* MP01 and *Lactobacillus plantarum* MP02 and assessment of their potential for the prevention of gastrointestinal infections in an experimental canine model. *Front Microbiol.* 2019;10:1117. http://doi.org/10.3389/fmicb.2019.01117.

Franz CMAP, Holzapfel WH, Stiles ME. Enterococci at the crossroads of food safety. *Int J Food Microbiol.* 1999;47:1–24.

Frese SA, Benson AK, Tannock GW, Loach DM, Kim J, Zhang M, Oh PL, Heng NCK, Patil PB, Juge N, MacKenzie DA, Pearson BM, Lapidus A, Dalin E, Tice H, Goltsman E, Land M, Hauser L, Ivanova N, Kyrpides NC, Walter J. The evolution of host specialization in the vertebrate gut symbiont *Lactobacillus reuteri*. *PLoS Genet.* 2011;7:e1001314. http://doi.org/10.1371/journal.pgen.1001314.

Fritsch DA, Jackson MI, Wernimont SM, Feld GK, Badri DV, Brejda JJ, Cochrane C-Y, Gross KL. Adding a polyphenol-rich fiber bundle to food impacts the gastrointestinal microbiome and metabolome in dogs. *Front Vet Sci.* 2023;9:1039032. http://doi.org/10.3389/fvets.2022.1039032.

Fujisawa T, Mitsuoka T. Homofermentative *Lactobacillus* species predominantly isolated from canine feces. *J Vet Med Sci.* 1996;58:591–593.

Gagné JW, Wakshlag JJ, Simpson KW, Dowd SE, Latchman S, Brown DA, Brown K, Swanson KS, Fahey GC Jr. Effects of a synbiotic on fecal quality, short-chain fatty acid concentrations, and the microbiome of healthy sled dogs. *BMC Vet Res.* 2013;9:246.

Gal A, Barko PC, Biggs PJ, Gedye KR, Midwinter AC, Williams DA, Burchell RK, Pazzi P. One dog's waste is another dog's wealth: A pilot study of fecal microbiota transplantation in dogs with acute hemorrhagic diarrhea syndrome. *PLoS ONE.* 2021;16(4):e0250344. http://doi.org/10.1371/journal.pone.0250344.

Ganz HH, Jospin G, Rojas CA, Martin AL, Dahlhausen K, Kingsbury DD, Osborne CX, Entrolezo Z, Redner S, Ramirez B, Eisen JA, Leahy M, Keaton G, Wong J, Gardy J, Jarett JK. The kitty microbiome project: Defining the healthy fecal "core microbiome" in pet domestic cats. *Vet Sci.* 2022;9:635. http://doi.org/10.3390/vetsci9110635.

García-Belenguer S, Grasa L, Valero O, Palacio J, Luño I, Rosado B. Gut microbiota in canine idiopathic epilepsy: Effects of disease and treatment. *Animals.* 2021;11:3121. http://doi.org/10.3390/ani11113121.

García-Mazcorro JF, Lanerie DJ, Dowd SE, Paddock CG, Grützner N, Steiner JM, Ivanek R, Suchodolski JS. Effect of a multi-species synbiotic formulation on fecal bacterial microbiota of healthy cats and dogs as evaluated by pyrosequencing. *FEMS Microbiol Ecol.* 2011;78(3):542–554.

Gensollen T, Iyer SS, Kasper DL, Blumberg RS. How colonization by microbiota in early life shapes the immune system. *Science.* 2016;352(6285):539–544. http://doi.org/10.1126/science.aad9378.

Gómez-Gallego C, Forsgren M, Selma-Royo M, Nermes M, Collado M, Salminen S, Beasley S, Isolauri E. The composition and diversity of the gut microbiota in children is modifiable by the household dogs: Impact of a canine-specific probiotic. *Microorganisms.* 2021;9:557. http://doi.org/10.3390/microorganisms9030557.

Gómez-Gallego C, Junnila J, Männikkö S, Hämeenoja P, Valtonen E, Salminen S, Beasley S. A canine-specific probiotic product in treating acute or intermittent diarrhea in dogs: A double-blind placebo-controlled efficacy study. *Vet Microbiol.* 2016;197:122–128.

Gookin JL, Strong SJ, Bruno-Bárcena JM, Stauffer SH, Williams S, Wassack E, Azcarate-Peril MA, Estrada M, Seguin A, Balzer J, Davidson G. Randomized placebo-controlled trial of feline-origin *Enterococcus hirae* probiotic effects on preventative health and fecal microbiota composition of fostered shelter kittens. *Front Vet Sci.* 2022;9:923792.

Grzeskowiak L, Collado MC, Beasley S, Salminen S. Pathogen exclusion properties of canine probiotics are influenced by the growth media and physical treatments simulating industrial processes. *J Appl Microbiol.* 2014;116:1308–1314.

Grzeskowiak L, Endo A, Beasley S, Salminen S. Microbiota and probiotics in canine and feline welfare. *Anaerobe.* 2015;34:14–23.

Guard BC, Mila H, Steiner JM, Mariani C, Suchodolski JS, Chastant-Maillard S. Characterization of the fecal microbiome during neonatal and early pediatric development in puppies. *PLOS ONE.* 2017. http://doi.org/10.1371/journal.pone.0175718.

Hanchi H, Mottawea W, Sebei K, Hammami R. The genus enterococcus: Between probiotic potential and safety concerns—An update. *Front Microbiol.* 2018;9:1791. http://doi.org/10.3389/fmicb.2018.01791.

Hart ML, Suchodolski JS, Steiner JM, Webb CB. Open-label trial of a multi-strain synbiotic in cats with chronic diarrhea. *J Feline Med Surg.* 2012;14(4):240–245.

Hasegawa T, Kanasugi H, Hidaka M, Yamamoto T, Abe S, Yamaguchi H. Effect of orally administered heat-killed *Enterococcus faecalis* FK-23 preparation on neutropenia in dogs treated with cyclophosphamide. *Int J Immunopharmacol.* 1996;18(2):103–112.

Health Canada. *List C: Veterinary Health Products 2023.* www.canada.ca/en/public-health/services/antibiotic-antimicrobial-resistance/animals/veterinary-health-products/list-c.html#p1.

Herstad HK, Nesheim BB, Abée-Lund TL, Larse S, Skancke E. Effects of a probiotic intervention in acute canine gastroenteritis—Controlled clinical trial. *J S Anim Pract.* 2010;51:34–38.

Herstad KMV, Vinje H, Skancke E, Næverdal T, Corral F, Llarena A-K, Heilmann RM, Suchodolski JS, Steiner JM, Nyquist NF. Effects of canine-obtained lactic-acid bacteria on the fecal microbiota and inflammatory markers in dogs receiving non-steroidal anti-inflammatory treatment. *Animals.* 2022;12:2519. http://doi.org/10.3390/ani12192519.

Hill C, Guarner F, Reid G, Gibson GR, Merenstein DJ, Pot B, Morelli L, Canani RB, Flint HJ, Salminen S, Calder PC, Sanders ME. The international scientific association for probiotics and prebiotics consensus statement on the scope and appropriate use of the term probiot. *Rev Gastroenterol Hepatol.* 2014;11:506–514.

Husso A, Lietaer L, Pessa-Morikawa T, Grönthal T, Govaere J, Van Soom A, Iivanainen A, Opsomer G, Niku M. The composition of the microbiota in the full-term fetal gut and amniotic fluid: A bovine cesarean section study. *Front Microbiol.* 2021;12:626421. http://doi.org/10.3389/fmicb.2021.626421.

IPA. *IPA Guidelines to Qualify a Microorganism to Be Termed as 'Probiotic';* 2017. IPA-guidelines-to-qualify-a-microorganism-as-probiotic-June-2-2017.pdf

JaberAlipour M, Jalanka J, Pessa-Morikawa T, Kokkonen T, Satokari R, Hynönen U, Iivanainen A, Niku M. The composition of the perinatal intestinal microbiota in cattle. *Sci Rep.* 2018;8:10437. http://doi.org/10.1038/s41598-018-28733-y

Jackson CR, Fedorka-Cray PJ, Davis JA, Barrett JB, Frye JG. Prevalence, species distribution and antimicrobial resistance of enterococci isolated from dogs and cats in the United States. *J Appl Microbiol.* 2009;107:1269–1278.

Jergens AE, Heilmann RM. Canine chronic enteropathy-Current state-of-the-art and emerging concepts. *Front Vet Sci.* 2022;9:923013.

Jiménez E, Fernández L, Marín ML, Martín R, Odriozola JM, Nueno-Palop C, Narbad A, Olivares M, Xaus J, Rodríguez JM. Isolation of commensal bacteria from umbilical cord blood of healthy neonates born by cesarean section. *Curr Microbiol.* 2005;51:270–274.

Jugan MC, Wouda RM, Higginbotham ML. Preliminary evaluation of probiotic effects on gastrointestinal signs in dogs with multicentric lymphoma undergoing multi-agent chemotherapy: A randomised, placebo-controlled study. *Vet Rec Open.* 2021;8(1):e2.

Kanasugi H, Hasegawa T, Goto Y, Ohtsuka H, Makimura S, Yamamoto T. Single administration of enterococcal preparation (FK-23) augments non-specific immune responses in healthy dogs. *Int J Immunopharmacol.* 1997;19:655–659.

Kelley RL, Levy K, Mundell P, et al. Effects of varying doses of a probiotic supplement fed to healthy dogs undergoing kenneling stress. *Int J Appl Res Vet Med.* 2012;10(3):205–216.

Kelley RL, Minikhiem D, Kiely B, O'Mahony L, O'Sullivan D, Boileau T, Park JS. Clinical benefits of probiotic canine-derived *Bifidobacterium animalis* strain AHC7 in dogs with acute idiopathic diarrhea. *Vet Ther.* 2009;10:121–130.

Kim H, Rather IA, Kim H, Kim S, Kim T, Jang J, Seo J, Lim J, Park YH. A double-blind, placebo controlled-trial of a probiotic strain *Lactobacillus sakei* Probio-65 for the prevention of canine atopic dermatitis. *J Microbiol Biotechnol.* 2015;25(11):1966–1969.

Kim H, Seo J, Park T, Seo K, Cho H-W, Chun JL, Kim KH. Obese dogs exhibit diferent fecal microbiome and specific microbial networks compared with normal weight dogs. *Sci Rep.* 2023;13:723. http://doi.org/10.1038/s41598-023-27846-3.

Kubinyi E, Rhali SB, Sándor S, Szabó A, Felföldi T. Gut microbiome composition is associated with age and memory performance in pet dogs. *Animals.* 2020;10:1488. http://doi.org/10.3390/ani10091488.

Lalor S, Gunn-Moore D. Effects of concurrent ronidazole and probiotic therapy in cats with *Tritrichomonas foetus*-associated diarrhoea. *J Feline Med Surg* 2012;14(9):650–658.

Lee SH. Intestinal permeability regulation by tight junction: Implication on inflammatory bowel diseases. *Intest Res.* 2015;13(1):11–18. http://doi.org/10.5217/ir.2015.13.1.11.

Lin C-F, Lin M-Y, Lin C-N, Chiou M-T, Chen J-W, Yang K-C, Wu M-C. Potential probiotic of Lactobacillus strains isolated from the intestinal tracts of pigs and fecel of dogs with antimicrobial activity against multidrug-resistant pathogenic bacteria. *Arch Microbiol.* 2020;202:1849–1860.

Lin X-Q, Chen W, Ma K, Liu Z-Z, Gao Y, Zhang J-G, Wang T, Yang Y-J. *Akkermansia muciniphila* suppresses high-fat diet-induced obesity and related metabolic disorders in beagles. *Molecules.* 2022;27:6074. https://doi.org/10.3390/molecules27186074.

Liong M. Safety of probiotics: Translocation and infection. *Nutr Rev.* 2008;66:192–202.

Lopetuso LR, Scaldaferri F, Bruno G, Petito V, Franceschi F, Gasbarrini A. The therapeutic management of gut barrier leaking: The emerging role for mucosal barrier protectors. *Eur Rev Med Pharmacol Sci.* 2015;19:1068–1076.

López-Moreno A, Aguilera M. Probiotics dietary supplementation for modulating endocrine and fertility microbiota dysbiosis. *Nutrients.* 2020;12:757. http://doi.org/10.3390/nu12030757.

Macedo HT, Fragoso Rentas M, Annibale Vendramini TH, Macegoza MV, Rodrigues Amaral A, Toloi Jeremias J, de Carvalho Balieiro JC, Pfrimer K, Ferriolli E, Pontieri CFF, Brunetto MA. Weight-loss in obese dogs promotes important shifts in fecal microbiota profile to the extent of resembling microbiota of lean dogs. *Anim Microbiome.* 2022;4:1–13.

Manninen TJ, Rinkinen ML, Beasley SS, Saris PE. Alteration of the canine small-intestinal lactic acid bacterium microbiota by feeding of potential probiotics. *Appl Environ Microbiol.* 2006;72:6539–6543.

Marks SL, Kather EJ. Bacterial-associated diarrhea in the dog: A critical appraisal. *Vet Clin North Am Small Anim Pract.* 2003;33:1029–1260.

Marsella R. Evaluation of *Lactobacillus rhamnosus* strain GG for the prevention of atopic dermatitis in dogs. *Am J Vet Res.* 2009;70:735–740.

Marsella R, Santoro D, Ahrens K. Early exposure to probiotics in a canine model of atopic dermatitis has long-term clinical and immunological effects. *Vet Immunol Immunopathol.* 2012;146(2):185–189.

Marshall-Jones ZV, Baillon M-LA, Croft JM, Butterwick RF. Effects of *Lactobacillus acidophilus* DSM13241 as a probiotic in healthy adult cats. *Am J Vet Res.* 2006;57:1005–1011.

Marsilio S, Pilla R, Sarawichitr B, Chow B, Hill SL, Ackermann MR, Estep JS, Lidbury JA, Steiner JM, Suchodolski JS. Characterization of the fecal microbiome in cats with inflammatory bowel disease or alimentary small cell lymphoma. *Sci Rep.* 2019;9:19208. http://doi.org/10.1038/s41598-019-55691.

Martorell P, Alvarez B, Llopis S, Navarro V, Ortiz P, Gonzalez N, Balaguer F, Rojas A, Chenoll E, Ramón D, and Tortajada M. Heat-Treated *Bifidobacterium longum* CECT-7347: A whole-cell postbiotic with antioxidant, anti-inflammatory, and gut-barrier protection properties. *Antioxidants.* 2021;10:536. https://doi.org/10.3390/antiox100405.

Massier L, Blüher M, Kovacs P, Chakaroun RM. Impaired intestinal barrier and tissue bacteria: Pathomechanisms for metabolic diseases. *Front Endocrinol*. 2021;12:616506. http://doi.org/10.3389/fendo.2021.616506.

Melandri M, Aiudi GG, Caira M, Alonge S. A biotic support during pregnancy to strengthen the gastrointestinal performance in puppies. *Front Vet Sci*. 2020;7:417. http://doi.org/10.3389/fvets.2020.00417.

Moloney RD, Desbonnet L, Clarke G, Dinan TG, Cryan JF. The microbiome: Stress, health and disease. *Mamm Genome*. 2014;25:49–74.

Mukai T, Asasaka T, Sato E, Mori K, Matsumoto M, Ohori H. Inhibition of binding of *Helicobacter pylori* to the glycolipid receptors by probiotic *Lactobacillus reuteri*. *FEMS Immunol Med Microbiol*. 2002;32:105–110.

NASC. *National Animal Supplement Council*; 2023. www.nasc.cc

Nermes M, Niinivirta K, Nylund L, Laitinen K, Matomäki J, Salminen S, Isolauri E. Perinatal pet exposure, faecal microbiota, and wheezy bronchitis: Is there a connection? *ISRN Allergy*. 2013;827934.

Nichols RG, Davenport ER. The relationship between the gut microbiome and host gene expression: A review. *Hum Genet*. 2021;140:747–760.

Niina A, Kibe R, Suzuki R, Yuchi Y, Teshima T, Matsumoto H, Kataoka Y, Koyama H. Improvement in clinical symptoms and fecal microbiome after fecal microbiota transplantation in a dog with inflammatory bowel disease. *Vet Med: Res Rep*. 2019;10:197–201.

Ohshima-Terada Y, Higuchi Y, Kumagai T, Hagihara A, Nagata M. Complementary effect of oral administration of *Lactobacillus paracasei* K71 on canine atopic dermatitis. *Vet Dermatol*. 2015;26:350–353.

Older CE, Diesel A, Patterson AP, Meason-Smith C, Johnson TJ, Mansell J, Suchodolski JS, Rodrigues Hoffmann A. The feline skin microbiota: The bacteria inhabiting the skin of healthy and allergic cats. *PLoS ONE*. 2017;12(6):e0178555. https://doi.org/10.1371/journal.pone.0178555.

O'Mahony D, Murphy KB, MacSharry J, Boileau T, Sunvold G, Reinhart G, Kiely B, Shanahan F, O'Mahony L. Portrait of a canine probiotic *Bifidobacterium*—From gut to gut. *Vet Microbiol*. 2009;139:106–112.

O'Neill DG, Church DB, McGreevy PD, Thomson PC, Brodbelt DC. Prevalence of disorders recorded in dogs attending primary-care veterinary practices in England. *PLoS ONE*. 2014;9(3):e90501.

Park HJ, Lee SE, Kim HB, Isaacson RE, Seo KW, Song KH. Association of obesity with serum leptin, adiponectin, and serotonin and gut microflora in beagle dogs. *J Vet Intern Med*. 2015;29(1):43–50. http://doi.org/10.1111/jvim.12455.

Pascher M, Hellweg P, Khol-Parisini A, Zentek J. Effects of a probiotic *Lactobacillus acidophilus* strain on feed tolerance in dogs with non-specific dietary sensitivity. *Arch Anim Nutr*. 2008;62(2):107–116.

Peredo-Lovillo A, Romero-Luna HE, Jiménez-Fernández M. Health promoting microbial metabolites produced by gut microbiota after prebiotics metabolism. *Food Res Int*. 2020;136. http://doi.org/10.1016/j.foodres.2020.109473.

Pilla R, Guard BC, Steiner JM, Gaschen FP, Olson E, Werling D, Allenspach K, Salavati Schmitz S, Suchodolski JS. Administration of a synbiotic containing *Enterococcus faecium* does not significantly alter fecal microbiota richness or diversity in dogs with and without food-responsive chronic enteropathy. *Front Vet Sci*. 2019;6:277.

Rampelli S, Candela M, Turroni S, Biagi E, Collino S, Franceschi C, O'Toole PW, Brigidi P. Functional metagenomic profiling of intestinal microbiome in extreme ageing. *Aging*. 2013;5:12.

Rastelli M, Cani PD, Knau C. The gut microbiome influences host endocrine functions. *Endocr Rev*. 2019;40:1271–1284.

Reid G, Burton J. Use of *Lactobacillus* to prevent infection by hogenic bacteria. *Microbes Infect*. 2002;4:319–324.

Rinkinen M, Jalava K, Westermarck E, Salminen S, Ouwehand AC. Interaction between probiotic lactic acid bacteria and canine enteric pathogens: A risk factor for intestinal *Enterococcus faecium* colonization? *Vet Microbiol*. 2003;92:111–119.

Rinninella E, Raoul P, Cintoni M, Franceschi F, Miggiano GAD, Gasbarrini A, Mele MC. What is the healthy gut microbiota composition? A changing ecosystem across age, environment, diet, and diseases. *Microorganisms*. 2019;7(1):14. http://doi.org/10.3390/microorganisms7010014.

Rodríguez-Daza MC, Pulido-Mateos EC, Lupien-Meilleur J, Guyonnet D, Desjardins Y, Roy D. Polyphenol-mediated gut microbiota modulation: Toward prebiotics and further. *Front Nutr*. 2021;8:689456. http://doi.org/10.3389/fnut.2021.689456.

Rosado B, García-Belenguer S, León M, Chacón G, Villegas A, Palacio J. Blood concentrations of serotonin, cortisol and dehydroepiandrosterone in aggressive dogs. *Appl Anim Beh Sci*. 2010;123:124–130.

Rose L, Rose J, Gosling S, Holmes M. Efficacy of a probiotic-prebiotic supplement on incidence of diarrhea in a dog shelter: A randomized, double-blind, placebo-controlled trial. *J Vet Intern Med*. 2017;31:377–382.

Rossi G, Jergens A, Cerquetella M, Berardi S, Di Cicco E, Bassotti G, Pengo G, Suchodolski JS. Effects of a probiotic (SLAB51) on clinical and histologic variables and microbiota of cats with chronic constipation/megacolon: A pilot study. *Benef Microbes*. 2018;9(1):101–110.

Rossi G, Pengo G, Caldin M, Palumbo Piccionello A, Steiner JM, Cohen ND, Jergens AE, Suchodolski JS. Comparison of microbiological, histological, and immunomodulatory parameters in response to treatment with either combination therapy with prednisone and metronidazole or probiotic VSL#3 strains in dogs with idiopathic inflammatory bowel disease. *PLoS ONE*. 2014;9(4):e94699.

Sahoo DK, Allenspach K, Mochel JP, Parker V, Rudinsky AJ, Winston JA, Bourgois-Mochel A, Ackermann M, Heilmann RM, Köller G, Yuan L, Stewart T, Morgan S, Scheunemann KR, Iennarella-Servantez CA, Gabriel V, Zdyrski C, Pilla R, Suchodolski JS, Jergens AE. Synbiotic-IgY therapy modulates the mucosal microbiome and inflammatory indices in dogs with chronic inflammatory enteropathy: A randomized, double-blind, placebo-controlled study. *Vet Sci.* 2022;10(1):25.

Salminen S, Collado MC, Endo A, Hill C, Lebeer S, Quigley EMM, Sanders ME, Shamir R, Swann JR, Szajewska H, Vinderola G. The international scientific association of probiotics and prebiotics (ISAPP) consensus statement on the definition and scope of postbiotic. *Nat Rev Gastroenterol Hepatol.* 2021;18:649–667. http://doi.org/10.1038/s41575-021-00440-6.

Santoro D, Fagman L, Zhang Y, Fahong Y. Clinical efficacy of spray-based heat-treated lactobacilli in canine atopic dermatitis: A preliminary, open-label, uncontrolled study. *Vet Dermatol.* 2021;32(2):114-e23.

Saulnier DM, Ringel Y, Heyman MB, Foster JA, Bercik P, Shulman RJ, Versalovic J, Verdu EF, Dinan TG, Hecht G, Guarner F. The intestinal microbiome, probiotics and prebiotics in neurogastroenterology. *Gut Microbes.* 2013;1;17–27.

Sauter SN, Allenspach K, Gaschen F, Gröne A, Ontsouka E, Blum JW. Cytokine expression in an ex vivo culture system of duodenal samples from dogs with chronic enteropathies: Modulation by probiotic bacteria. *Domest Anim Endocrinol.* 2005;29:605–622.

Sauter SN, Benyacoup J, Allensbach K, Gaschen F, Ontsouka E, Reuteler G, Cavadini C. Effects of probiotic bacteria in dogs with food responsive diarrhea treated with an elimination diet. *J Anim Physiol Anim Nutr (Berl).* 2006;90:269–277.

Schmitz S. Value of probiotics in canine and feline gastroenterology. *Vet Clin North Am Small Anim Pract.* 2021;51(1):171–217.

Schmitz S, Glanemann B, Garden OA, Brooks H, Chang YM, Werling D, Allenspach K. A prospective, randomized, blinded, placebo-controlled pilot study on the effect of *Enterococcus faecium* on clinical activity and intestinal gene expression in canine food-responsive chronic enteropathy. *J Vet Intern Med.* 2015;29(2):533–543.

Schmitz S, Henrich M, Neiger R, Werling D, Allenspach K. Stimulation of duodenal biopsies and whole blood from dogs with food-responsive chronic enteropathy and healthy dogs with Toll-like receptor ligands and probiotic *Enterococcus faecium. Scand J Immunol.* 2014;80(2):85–94.

Secombe KR, Al-Qadamia GH, Subramaniama CB, Bowena JM, Scotta J, Van Sebillec YZA, Snelson M. Guidelines for reporting on animal fecal transplantation (GRAFT) studies: Recommendations from a systematic review of murine transplantation protocols. *Gut Microbes.* 2021;13(1):e1979878. http://doi.org/10.1080/19490976.2021.1979878.

Sharif NM, Sreedevi B, Chaitanya RK, Sreenivasulu D. Isolation and screening of Lactobacillus species from dogs for probiotic action. *Ind J Anim Res.* 2018;52(12):1739–1744. http://doi.org/10.18805/ijar.B-3429.

Shieh M-J, Shang H-F, Liao F-H, Zhu J-S, Chien Y-W. Lactobacillus fermentum improved intestinal bacteria flora by reducing *Clostridium perfringens. e-SPEN, Eur E J Clin Nutr Metab.* 2011;6:e59ee63.

Simpson KW, Rishniw M, Bellosa M, Liotta J, Lucio A, Baumgart M, Czarnecki-Maulden G, Benyacoub J, Bowman D. Influence of *Enterococcus faecium* SF68 probiotic on giardiasis in dogs. *J Vet Intern Med.* 2009;23(3):476–481.

Singh A, Vishwakarma V, Singhal B. Metabiotics: The functional metabolic signatures of probiotics: Current state-of-art and future research priorities—metabiotics: Probiotics effector molecules. *Adv Biosci Biotechnol.* 2018;9:147–189.

Song SJ, Lauber C, Costello EK, Lozupone CA, Humphrey G, Berg-Lyons D, Caporaso JG, Knight D, Clemente JC, Nakielny S, Gordon JI, Fierer N, Knight R. Cohabiting family members share microbiota with one another and with their dogs. *eLife.* 2013;2:e00458.

Stavroulaki EM, Suchodolski JS, Xenoulis PG. Effects of antimicrobials on the gastrointestinal microbiota of dogs and cats. *Vet J.* 2023;291:105929.

Suchodolski JS, Markel ME, Garcia-Mazcorro JF, Unterer S, Heilmann RM, Dowd SE, Kachroo P, Ivanov I, Minamoto Y, Dillman EM, Steiner JM, Cook AK, Toresson L. The fecal microbiome in dogs with acute diarrhea and idiopathic inflammatory bowel disease. *PLOS ONE.* 2012;7(12):e51907. http://doi.org/10.1371/journal.pone.0051907.

Tal M, Weese JS, Gomez DE, Hesta M, Steiner JM, Verbrugghe A. Bacterial fecal microbiota is only minimally affected by a standardized weight loss plan in obese cats. *BMC Vet Res.* 2020;16:112. http://doi.org/10.1186/s12917-020-02318-2.

Todorov SD, Ivanova IV, Popov I, Richard Weeks R, Chikindas ML. Bacillus spore-forming probiotics: Benefits with concerns? *Crit Rev Microbiol.* 2021. http://doi.org/10.1080/1040841X.2021.1983517.

Torres-Henderson C, Summers S, Suchodolski J, et al. Effect of *Enterococcus faecium* Strain SF68 on gastrointestinal signs and fecal microbiome in cats administered amoxicillin-clavulanate. *Top Companion Anim Med.* 2017;32(3):104–108.

Vahjen W, Männer K. The effect of a probiotic *Enterococcus faecium* product in diets of healthy dogs on bacteriological counts of *Salmonella* spp., *Campylobacter* spp. and *Clostridium* spp. in faeces. *Arch Anim Nutr.* 2003;57:229–233.

Veir JK, Knorr R, Cavadini C, Sherrill SJ, Benyacoub J, Satyaraj E, Lappin MR. Effect of supplementation with *Enterococcus faecium* (SF68) on immune functions in cats. *Vet Ther.* 2007;8(4):229–238.

Vemuri R, Shinde T, Gundamaraju R, Gondalia SV, Karpe AV, Beale DJ, Martoni CJ, Rajaraman E. Lactobacillus acidophilus DDS-1 modulates the gut microbiota and improves metabolic profiles in aging mice. *Nutr.* 2018;10:1255. http://doi.org/10.3390/nu10091255.

von Writgh A. Regulating the safety of probiotics—The European approach. *Curr Pharm Des.* 2005;11:17–23.

Vuong HE, Yano JM, Fung TC, Hsiao EY. The microbiome and host behavior. *Annu Rev Neurosci.* 2017;25(40):21–49. http://doi.org/10.1146/annurev-neuro-072116-031347.

Wang Y, Liang Q, Lu B, Shen H, Liu S, Shi Y, Leptihn S, Li H, Wei J, Liu C, Xiao H, Zheng X, Liu C, Chen H. Whole-genome analysis of probiotic product isolates reveals the presence of genes related to antimicrobial resistance, virulence factors, and toxic metabolites, posing potential health risks. *BMC Genom.* 2021;22(1):210. http://doi.org/10.1186/s12864-021-07539-9.

Weese JS, Arroyo L. Bacteriological evaluation of dog and cat diets that claim to contain probiotics. *Can Vet J.* 2003;44:212–216.

Werner M, Suchodolski JS, Lidbury JA, Steiner JM, Hartmann K, Unterer S. Diagnostic value of fecal cultures in dogs with chronic diarrhea. *J Vet Intern Med.* 2021;35(1):199–208.

Westermarck E, Skrzypczak T, Harmoinen J, Steiner JM, Ruaux CG, Williams DA, Eerola E, Sundbäck P, Rinkinen M. Tylosin-responsive chronic diarrhea in dogs. *J Vet Intern Med.* 2005;19:177–186.

White R, Atherly T, Guard B, Rossi G, Wang C, Mosher C, Webb C, Hill S, Ackermann M, Sciabarra P, Allenspach K, Suchodolski J, Jergens AE. Randomized, controlled trial evaluating the effect of multi-strain probiotic on the mucosal microbiota in canine idiopathic inflammatory bowel disease. *Gut Microbes.* 2017;8(5):451–466.

Whittemore JC, Stokes JE, Price JM, et al. Effects of a synbiotic on the fecal microbiome and metabolomic profiles of healthy research cats administered clindamycin: A randomized, controlled trial. *Gut Microbes.* 2019;10(4):521–539.

Xiao H, Kang S. The role of the gut microbiome in energy balance with a focus on the gut-adipose tissue axis. *Front Genet.* 2020;11:297. http://doi.org/10.3389/fgene.2020.00297.

Xu J, Verbrugghe A, Lourenço M, Janssens GPJ, Liu DJX, Van de Wiele T, Eeckhaut V, Van Immerseel F, Van de Maele I, Niu Y, Bosch G, Junius G, Wuyts B, Hesta M. Does canine inflammatory bowel disease influence gut microbial profile and host metabolism. *BMC Vet Res.* 2016;12:114.

Yamazaki C, Rosenkrantz W, Griffin C. Pilot evaluation of *Enterococcus faecium* SF68 as adjunctive therapy for oclacitinib-responsive adult atopic dermatitis in dogs. *J Small Anim Pract.* 2019;60(8):499–506.

Yeh Y-M, Lye X-Y, Lin H-Y, Wong J-Y, Wu C-C, Huang C-L, Tsai Y-C, Wang L-C. Effects of *Lactiplantibacillus plantarum* PS128 on alleviating canine aggression and separation anxiety. *Appl Anim Beh Sci.* 2022;247.

You I, Mahiddine FY, Park H and Kim MJ. *Lactobacillus acidophilus* novel strain, MJCD175, as a potential probiotic for oral health in dogs. *Front Vet Sci.* 2022;9:94680. http://doi.org/10.3389/fvets.2022.94689.

Ziese AL, Suchodolski JS, Hartmann K, Busch K, Anderson A, Sarwar F, Sindern N, Unterer S. Effect of probiotic treatment on the clinical course, intestinal microbiome, and toxigenic *Clostridium perfringens* in dogs with acute hemorrhagic diarrhea. *PLoS ONE.* 2018;13(9):e0204691. Erratum in: *PLoS ONE* 2023 Jan 12;18(1):e0280539.

# Lactic Acid Bacteria in Aquatic Environments and Their Applications

# 33

Beatriz Gómez-Sala, Diogo Contente, Javier Feito, Lara Díaz-Formoso, Juan Borrero, Estefanía Muñoz-Atienza, Pablo E. Hernández, and Luis M. Cintas

## 33.1 INTRODUCTION

Lactic acid bacteria (LAB) are among the most widely studied microorganisms worldwide. Given the important role that LAB play in different biotechnological processes, it is not surprising that they have received much attention from the scientific community for decades. LAB are widespread in most ecosystems and are commonly isolated from fermented foods, as they are responsible for the fermentation processes, but they are also frequently found in non-fermented foods such as dairy products, meat products, fish and fishery products, fruits, vegetables and cereals (Arbulu et al., 2022).

LAB are used to ensure safety, preserve food quality, develop organoleptic and rheological characteristics and increase the nutritional quality of food. The preservation capability and enhancement of food safety by LAB is due to their ability to inhibit the development of a wide range of spoilage and food-borne pathogenic microorganisms such as *Listeria monocytogenes*, *Clostridium botulinum* and *Staphylococcus aureus*. The main mechanisms of microbial antagonism of LAB are the competition for nutrients from the culture medium or the food substrate in which they develop and the formation of organic acids (mainly lactic and acetic acids), with a concomitant drop in the pH. However, they also produce other antimicrobial metabolites such as ethanol, carbon dioxide, diacetyl, acetaldehyde or hydrogen peroxide and also bacteriocins (Cotter et al., 2005).

Historically, LAB have been associated with spontaneous and industrial food fermentations, and thus they are currently considered safe microorganisms for human and animal consumption; that is, most of them have been assessed by EFSA and included in the list of Qualified Presumption of Safety (QPS) microorganisms for food or feed use or their equivalent generally recognized as safe (GRAS) status awarded by the FDA in the USA (EFSA, 2005a, 2005b, 2007). The latest scientific opinion updating the QPS list has recently been published (EFSA, 2023). In this context, bacteriocinogenic LAB and/or their antimicrobial peptides are getting great attention for their application as: (1) biopreservatives in the food

DOI: 10.1201/9781003352075-36

industry as an additional barrier in the combined strategies of food preservation (hurdle technology) and (2) probiotics in humans and animals in order to reduce the use of antibiotics. Additionally, in the last decade, special attention has been given to the pharmaceutical properties and potential application of bacteriocins in human and veterinary medicine (Todorov et al., 2022; Queiroz et al., 2022). In the context of this chapter, the LAB applications presented here are focused on fish, fishery products and aquaculture.

With a world population expected to reach 9.8 billion by 2050, a sustainable fisheries and aquaculture sector will play a key role in ensuring food security, as the increased demand will challenge fish production over the coming decades. Food loss and waste (FLW) occurs in most, if not all, supply chains. Reducing this loss and waste is becoming increasingly important as demand for fish as food increases (FAO, 2022). Fisheries and aquaculture products are among the most-traded food commodities in the world, a trade that keeps on growing, with the added challenge of being a highly perishable product with a short shelf life not exceeding 1–2 weeks for fresh fish products and 3–4 weeks for lightly preserved ones (Wiernasz et al., 2017; FAO, 2022). Throughout the world, post-harvest fish losses are a major concern and occur in most fish distribution chains, with an estimated 27% of landed fish being lost or wasted due to microbial activity (Gram and Dalgaard, 2002; Leroi, 2015; FAO, 2022). In addition, it has been reported that indigenous bacteria present in aquatic environments or as a result of post contamination during processes are responsible for many cases of food-borne diseases due to fish consumption. It is, therefore, essential to apply adequate preservation technologies to maintain fish safety and quality (Pilet and Leroi, 2011; Wiernasz et al., 2017, 2020).

Fish and fish products are increasingly recognized for their key role in food security and nutrition, not just as a source of protein but also as a unique and extremely diverse provider of essential omega-3 fatty acids and bioavailable micronutrients. We are eating more aquatic foods than ever, and fisheries and aquaculture production reached an all-time record in 2020. However, in order to address the challenges of feeding the world effectively, equitably and sustainably, further changes are needed with their consequent challenges. FAO is committed to Blue Transformation, a visionary strategy that aims to enhance the role of aquatic food systems in feeding the world's growing population. The contribution of aquaculture production now provides now more than half of all fish for human consumption, and it is likely that the future growth of the fisheries sector will come mainly from aquaculture production. To achieve this goal, the sector will have to face significant challenges, including production intensification, disease control and prevention of the environmental deterioration (FAO, 2022).

Aquaculture has been widely recognized as the fastest growing food-producing sector, and, to a large extent, the nourishment of future generations will depend on whether this growth is achieved in a sustainable manner (FAO, 2022). However, this development has not been free of consequences, and the over-intensification of the sector has chronified the incidence of transmissible diseases, so-called pathogens, parasites and pests. In fact, it is estimated that losses suffered by aquaculture globally due to infectious diseases exceed 10%. In view of this situation, the use of antibiotics and other chemical compounds has become widespread not only as therapeutic but also as prophylactic treatments (Chen et al., 2020).

The widespread use of antibiotics as prophylactic and therapeutic agents in fish farming to control bacterial diseases has been associated with the emergence of antibiotic resistances in bacterial pathogens and with the alteration of the aquaculture environment microbiota (Cabello, 2006), which resulted in the ban of antibiotic usage as animal growth promoters in Europe and stringent worldwide regulations on therapeutical antibiotic applications. This scenario has led to an ever-growing interest in the search for and development of alternative strategies for disease control, including adequate hygiene conditions; vaccination programmes; and the use of probiotics, prebiotics and immunostimulants (FAO, 2022; Wiernasz et al., 2017).

# 33.2 MICROBIAL ECOLOGY OF FISH

Fish are the most diverse and widely distributed group of vertebrates, providing an invaluable repertoire of host species to study the nature of vertebrate microbial communities, and yet little is known about the microbial ecology of fishes or the biological and environmental factors that influence fish microbiota (Sylvain et al., 2020; Minich et al., 2022). Fish occupy marine and freshwater habitats across the globe and exhibit a wide variety of physiologies, ecologies and natural histories, with a variety of reproductive strategies and different feeding regimes, all of which suggest that the microbiomes of fish could be highly variable and, consequently, contradictory results in different studies are inevitable within such a diverse group (Egerton et al., 2018; Minich et al., 2022).

In recent years, next-generation sequencing techniques, including amplicon and shot-gun approaches, and associated bioinformatics tools and the exponential growth of aquaculture have boosted the study of fish microbiomes. Simultaneously, DNA database development for reliable classification of taxonomy and functionality has facilitated data interpretation. Due to the integral role in digestion, nutrition and host health and disease control in aquaculture, the fish gut microbiome has been studied much more than those of the skin, gills or mucus (Ghanbari et al., 2015; Leroi, 2015; Egerton et al., 2018). Recently, the fish skin-associated microbiomes and the processes shaping fish skin microbiomes have been the focus of several studies as they play a crucial role in fish holobionts (Berggren et al., 2022).

Teleost microbiome research lags well behind that in mammals, and current knowledge of fish microbiome is still far from being complete. Nevertheless, there is considerable and growing interest in understanding more on this exciting topic from both the standpoint of basic research and biotechnological applications (Luna et al., 2021). In general, teleost's guts, skin mucus and gills support high concentrations of bacterial cells forming highly diverse and structured microbial communities (Sylvain et al., 2020). Like in mammals, most of the studies in the last years have been focused on gaining an understanding of microbes associated with the gut (Perry et al., 2020; Sylvain et al., 2020; Luna et al., 2021; Minich et al., 2022). These studies have identified endogenous (e.g., host ancestry, genotype, or diet) and exogenous factors (e.g., environmental water physicochemical parameters and bacterioplankton composition) that affect the establishment of bacteria in teleost guts (Perry et al., 2020; Sylvain et al., 2020; Luna et al., 2021; Minich et al., 2022).

Even though LAB do not constitute the predominant intestinal microbiota of fish, they are frequently found in the fish intestine and are considered part of their microbiota (Muñoz-Atienza et al., 2013; Leroi, 2015; Ringø et al., 2018; Feito et al., 2022).

There is a wide variation in the composition of fish gut microbiota between species and individuals, but several phyla have been shown to be dominant, including Proteobacteria, Firmicutes, Bacteroidetes, Actinobacteria and Fusobacteria (for a review on this topic, see Ghanbari et al., 2015; Egerton et al., 2018).

# 33.3 OCCURRENCE OF LAB IN AQUATIC ENVIRONMENTS

## 33.3.1 LAB in Living and Fresh Fish

Despite of the high variability in fish microbiota, numerous studies have demonstrated the presence of LAB in fish and their surrounding environments. The first study reporting that fish may contain lactobacilli on the skin, gills and gut was published by Dyer (1947) on Atlantic cod. Later on, Kraus (1961) reported the isolation of lactobacilli from fresh herring, and Kvasnikov et al. (1977) described the presence of LAB in the intestine of various fish species inhabiting ponds at larval, fry and fingerling stages and provided information on the changes in their composition as a function of the season of the year and life-stage of the fish. More recently it has been reported that lactobacilli are part, not dominant, of the native intestinal microbiota of Arctic charr, Atlantic cod, Atlantic salmon and brown trout (Ringø et al., 1997; González et al., 2000; Ringø et al., 2018). In this regard, lactobacilli have been isolated from different locations in different aquatic species such as cod (Lauzon et al., 2010), Persian sturgeon and beluga (Askarian et al., 2009; Ghanbari et al., 2009), anchovies (Belfiore et al., 2013) and various freshwater fish and their environment (González et al., 2000; Banerjee et al., 2013).

In the last years, a considerable diversity of LAB associated with fish and fish environments has been reported, although *Carnobacterium* spp. has been identified as the dominant genus in most studies (González et al., 2000; Pinto et al., 2009; Dallagnol et al., 2021). In this context, *Carnobacterium maltaromaticum* and, to a lesser extent, *Carnobacterium divergens* are considered part of the normal gut microbiota in salmonids (González et al., 2000; Ringø et al., 2018). However, the analysis of salmonids microbiota with culture-independent molecular techniques revealed the presence of other genus belonging to the LAB group different from *Carnobacterium*, such as *Lactobacillus*, *Lactococcus*, *Leuconostoc*, *Streptococcus* and *Weissella*, that can be considered indigenous species. On the other hand, in warm water fish species, *Carnobacterium* has been rarely identified as indigenous gut microbiota, while strains belonging to the genera *Lactobacillus*, *Lactococcus*, *Enterococcus*, *Pediococcus*, *Leuconostoc* and *Weissella* have been frequently identified (Merrifield et al., 2015).

Members of the *Enterococcus* genus have been isolated from fish and fishery products. Cai et al. (1999) reported the isolation of small numbers of enterococci from the intestine of common carp, and González et al. (1999) isolated enterococci from skin, gills and gut of brown trout, wild pike and aquacultured rainbow trout. Moreover, González et al. (2000) isolated 249 LAB strains from freshwater fishes and their surrounding habitats, eight of which were ascribed to the genus *Enterococcus*. On the other hand, Michel et al. (2007) used amplified rDNA gene restriction analysis (ARDRA) to elucidate the nature and prevalence of different fish-associated bacteria belonging to the LAB group, revealing the existence of 12 distinct clusters, two of which included typical enterococci that could not be characterized at the species level by ARDRA. Furthermore, Petersen and Dalsgaard (2003) detected a high prevalence of *Enterococcus faecium* isolates in fish intestinal samples from traditional fish farms, suggesting that enterococci may be members of the normal intestinal microbiota of fish. More recently, Lyons et al. (2017a) revealed that autochthonous *Enterococcus* was present in low abundance in the distal intestine of farmed rainbow trout. In addition to *Enterococcus* spp., *E. faecalis* and *E. faecium* were isolated from the gastro intestinal tract of mrigal (Shahid et al., 2017) and European sea bass (Torrecillas et al., 2018), respectively. Lately, Iseppi et al. (2019) isolated two strains of *Enterococcus mundtii* from red mullet and sardine samples with a strong inhibitory activity against *L. monocytogenes*.

Strains belonging to *Pediococcus* genus have been also isolated from fish and shellfish. Pinto et al. (2009) reported the identification of a bacteriocinogenic *Pediococcus pentosaceus* strain from 78 bacteriocinogenic LAB isolated from non-fermented shellfish. Muñoz-Atienza et al. (2011) described the identification of six *P. pentosaceus* from a total of 74 LAB isolated from aquatic animals, and Gómez-Sala et al. (2015) reported the isolation of *P. pentosaceus* strains from sea bass and cockle. Recently, two strains of *P. pentosaceus* were isolated from rainbow trout, along with six strains of *Pediococcus acidilactici* (Martínez et al., 2017). Similarly, a study carried out by Araújo et al. (2015a, 2016) demonstrated that rainbow trout intestine and rearing environment are potential sources for the isolation of LAB and reported the identification of *P. acidilactici* strains; however, the most common species isolated was *Lactococcus lactis*. In recent years, studies reporting the isolation of *Pediococcus* are focused mainly in fermented fish products around the world. Only one study has revealed *Pediococcus* in the intestine of turbot, finding a significant higher abundance of *Pediococcus* when the fish was fed a diet with 5% stachyose added (Yang et al., 2018).

Several studies have reported the presence of *Weissella* in fish and fish products. *Weisella cibaria* was recently isolated from shrimp gastrointestinal tract (Duc Huy et al., 2020) and *Weissella paramesenteroides* from the intestinal content of pirarucu, one of the largest freshwater fishes in the world (do Vale Pereira et al., 2017). Lyons et al. (2017a) revealed that *Weissella* was present in very low abundance in the distal intestine of farmed rainbow trout. *W. cibaria* was identified by both culture-dependent and -independent methods in the gut of brown trout (Al-Hisnawi et al., 2015) and Atlantic salmon (Hovda et al., 2012), as well as from albacore muscle, cockle and sardine (Muñoz-Atienza et al., 2011). Gómez-Sala et al. (2015) reported the isolation and identification of eight strains of *W. cibaria* from octopus. In a recent study, Araújo et al. (2015a) isolated 18 strains of *Weisella soli* rainbow trout intestine and rearing environment. On the other hand, Sica et al. (2010) studied the LAB microbiota of different fish species, all of them from the same estuary in Argentina, which were fished in two different seasons. Their results showed that, consistent with previous studies, the microbial population of fish varies among seasons. Fish samples collected in the same season had similar LAB composition, and strikingly, the strain *Weissella viridescens* was isolated in fish fished in both seasons; nevertheless, strains belonging to the genus *Leuconostoc* (*Leuconostoc mesenteroides* and *Leuconostoc citreum*) were isolated, but only in one of the seasons (Sica et al., 2010). The genus *Leuconostoc* is associated with fish and fishery products, and specifically *Le. mesenteroides* was isolated from fresh anchovies (Belfiore et al., 2010), stripped weakfish and whitemouth croaker (Sica et al., 2010) and the intestinal tract of snakehead fish (Allameh et al., 2012). In recent years, *Leuconostoc* strains have been isolated from the intestine of rainbow trout, Atlantic salmon and Arctic charr (Lyons et al., 2017b; Rimoldi et al., 2018)

Other LAB genera such as *Streptococcus*, *Aerococcus* and *Vagococcus* have also been isolated from fish and surrounding environments (Merrifield et al., 2015). Streptococci have been isolated from the digestive tract of several fish species, including Artic charr (Ringø et al., 2002), Atlantic salmon (Rimoldi et al., 2018), rainbow trout (Lyons et al., 2017b), beluga and Persian sturgeon (Askarian et al., 2009), grass carp (Wu et al., 2012), European seabass (Torrecillas et al., 2018) and intestine content of pirarucu (do Vale Pereira et al., 2017). Species within the genus *Vagococcus* have frequently been isolated from diseased fish (Merrifield et al., 2015). However, González et al. (2000) reported the isolation of two *Vagococcus fluvialis* strains from the skin, gills and intestines of freshwater fish, and Lyons et al. (2017b) isolated strains belonging to the genera *Vagococcus* from the distal intestine of rainbow trout. To our knowledge, strains within the genera *Oenococcus* and *Tetragenococcus* have not yet been isolated from living or fresh fish.

## 33.3.2 LAB in Lightly Preserved Fish

During storage of fish and fishery products, the microbiota changes due to the ability of microorganisms to tolerate the preservations conditions. Lightly preserved fish products (LPFPs) are defined as uncooked or mildly cooked products with low level of preservatives (<6 % NaCl [w/w] in the aqueous phase) and include products such as cold-smoked fish, marinated fish, gravads, seafood in brine and pickled fish (Leroi, 2010; Pilet and Leroi, 2011; Ghanbari et al., 2013). These processes only reduce the natural microbial population of the raw fish, and for that reason the end products must be stored at chilled temperatures usually under vacuum packaging (VP) or modified atmosphere packaging (MAP) (Leroi, 2010; Ghanbari et al., 2013; Jérôme et al., 2022). Under these conditions, the microbiota typically becomes dominated by LAB after 1–2 weeks of storage at 5 °C, and they can easily reach high counts ($10^7$–$10^8$ CFU $g^{-1}$) (Gram and Dalgaard, 2002; Leroi, 2010; Pilet and Leroi, 2011; Jérôme et al., 2022). VP cold-smoked salmon (CSS) has been the subject of many studies, as it is considered a high risk product by the FAO (2004). Leroi et al. (1998) reported the isolation of LAB at the end of the storage time of CSS, with high numbers of *C. piscicola* and also *Lactobacillus farcimins*, *Latilactobacillus sakei* and *Lactobacillus alimentarius*. Gancel et al. (1997) reported the isolation of 78 *Lactobacillus* strains from fillets of VP smoked and salted herring. LAB have been also isolated from smoked trout (Lyhs et al., 1999) and cooked cold-water shrimp (Dalgaard et al., 2003). Strains belonging to the genus *Enterococcus* have been also isolated from CSS (Tomé et al., 2008; Gómez-Sala et al., 2015). In semi-preserved marinated fish, the nature of the marinating process inhibits the Gram-negative spoilage bacteria but not the LAB, which may spoil the product if not correctly manufactured and stored (Lyhs and Björkroth, 2008). Gómez-Sala et al. (2015) reported a high incidence of LAB isolates in CSS and marinated Atlantic salmon, revealing that these are the dominating bacteria, as previously described (Gram and Dalgaard, 2002). The strains isolated from marinated Atlantic salmon were identified as *Le. mesenteroides* and *L. sakei*, the latter also isolated from CSS (Gómez-Sala et al., 2015). More recently, Maillet et al. (2021) studied the microbial ecology of CSS using culture-dependent and -independent methods during 28-day storage, finding the same dominant genera previously reported in CSS, including *Lactobacillus* and *Carnobacterium*.

Although the cause and role of LAB in LPFP have not yet been extensively studied, it is clear that they are able to adapt to the preservation conditions used in those products (Leroi, 2010; Pilet and Leroi, 2011). The *Carnobacterium* genus is very often found in refrigerated VP and MAP products, as they are resistant to freezing and grow very well at refrigerated temperatures in all packaging conditions and in the presence of many preservatives (Leroi, 2010). Overall, the most frequently dominant species in LPFP are mainly *L. sakei*, *Latilactobacillus curvatus* and *C. maltaromaticum*, although *Leuconostoc* and *Lactococcus* strains are also found (Leroi, 2010; Pilet and Leroi, 2011; Leroi, 2015; Wiernasz et al., 2017; Maillet et al., 2021).

## 33.3.3 LAB in Fermented Fish Products

Fermentation is a traditional food preservation method and is widely used for improving food safety, shelf life and organoleptic and nutritional attributes. Fermented fish products are produced and consumed in different parts of the world and are an integral part of many food cultures. They are especially popular in Asia, Africa and Europe. Many of these countries have their own unique types of fermented products based on differences in raw materials, environmental conditions, microorganisms and dietary traditions (Skåra et al., 2015; Zang et al., 2020). In European cuisines, fermented fish is an old staple food; for instance, the ancient Greeks and Romans made a famous sauce from fermented fish called garum. However, nowadays only a few traditional fermented fish products are still produced in northern Europe, such as hákarl (Iceland), surströmming (Sweden) and rakfisk (Norway) (Skåra et al., 2015). Artisanal fish processing, including fermentation of fish, remains the predominant and most important method of fish preservation in Africa. Fermented fishery products in Africa are usually whole or in cut pieces such as lanhouin (Benin), momone (Ghana) and feseek (Egypt) (El Sheikha et al., 2014; Zang et al., 2020). In Southeast Asia, fermented fish products have a long history, and they are of great nutritional importance and part of the culture of many different ethnic groups. Various different types of fermented fish products including whole fish or fish pieces such as plaa-som (Thailand), sauces such as bakasang (Indonesia) and pastes such as prahok (Cambodia) can be found in Asian markets (Ohshima and Giri, 2014; Zang et al., 2020).

Fermented fish products have traditionally been prepared on the basis of empirical knowledge without any knowledge of the microorganisms involved in the process until the development of modern microbiology and industrial processes. LAB have traditionally been the dominant microorganisms in many fermented fish products, and different LAB starters have been used to produce fermented fish (for a review on this topic, see Zang et al., 2020). The different types of fermented fish products and the LAB involved in the fermentation process can be divided in different groups according to different criteria and are shown in Table 33.1.

**TABLE 33.1** Types and Examples of Traditional Fermented Fish around the World

| TYPES | | PRODUCT/ COUNTRY | RAW MATERIAL | LAB INVOLVED IN THE FERMENTATION PROCESS | REFERENCES |
|---|---|---|---|---|---|
| According to the final product appearance | Fermented whole or in pieces | Hákarl/ Iceland | Greenland shark | *Lactobacillus* sp. | Skåra et al. (2015) |
| | | Plaa-som/ Thailand | Snakehead fish | *Pediococcus pentosaceus, Lactobacillus alimentarius, Weissella* sp., *Lactobacillus planetarium, Lactococcus. garvieae* | Kopermsub and Yunchalard (2010) |
| | Fermented fish pastes | Prahok/ Cambodia | Striped snakehead | *Tetragenococcus* sp. | Chuon et al. (2014) |
| | Fermented fish sauces | Nam-pla/ Thailand | Anchovy/Mrigal carp and salt | No LAB reported | Tanasupawat et al. (2006), Zang et al. (2020) |
| According to the processing method | Spontaneous fermentation | Garum/Italy (Rome) and Greece | Atlantic bluefin tuna/Atlantic mackerel | Not reported | Skåra et al. (2015), Zang et al. (2020) |
| | Fermentation using starter cultures | Som-fug/ Thailand | Bigeye snapper | *Lactiplantibacillus plantarum, Pediococcus acidilactici, P. pentosaceus* | Zang et al. (2020) |
| According to the types of substrates | Products using fish and salt | Lanhouin/ Benin | Cassava fish/Spanish mackerel and salt | No LAB reported | El Sheikha et al. (2014) |
| | Products using fish and salt and carbohydrates | Suan yu/ China | Common carp, roasted rice, salt and sugar | *L. plantarum, P. pentosaceus, Leuconostoc* sp. | Zang et al. (2020) |
| | | Momoni/ Ghana | African jack mackerel/ Snakehead fish, salt, palm syrup, roasted rice | *Lactobacillus* sp., *Pediococcus* sp. | El Sheikha, (2014), Zang et al. (2020) |
| According to the proportion of salt added | High salt products (>20% of the total weight) | Feseekh/ Egypt | Silversides/tigerfish/flathead grey mullet and salt | *Lactobacillus* sp., *Teratogenococcus halophilus* | El Sheikh (2014), Zang et al. (2020) |
| | Low salt products (3–8%) | Rakfish/ Norway | Salmonid, salt | *Lactobacillus* sp. | Skåra et al. (2015) |
| | No salt products | Ngari/ India | Pool barb | *Lactococcus lactis* | Thapa et al. (2004), Zang et al. (2020) |

Adapted from Thapa et al. (2004) and Zang et al. (2020).

# 33.4 BIOPRESERVATION OF FRESH FISH AND FISHERY PRODUCTS WITH LAB

## 33.4.1 Spoilage and Food-Borne Pathogenic Bacteria in Fish and Fishery Products

Soon after the fish dies, endogenous bacteria begin to invade the tissues through the gills, along blood vessels and directly through the skin and the lining of the belly cavity. In addition, the changes brought about by catching, handling and processing are also a source of contamination. In addition, the composition of fish provides sufficient nutrients for the development of spoilage and food-borne pathogenic microorganisms (Gram and Huss, 1996; Leroi, 2010; Ghanbari et al., 2013; Shen and Wang, 2021).

Spoilage and food-borne pathogenic bacteria may be divided into three groups: (1) those naturally present in fish such as *Vibrio* spp. (*V. parahaemolyticus*, *V. cholerae*, *V. vulnificus*), *Cl. botulinum*, *Pleisomonas shigelloides* and *Aeromonas* spp., *Pseudomonas* spp. and *Shewanella* spp.; (2) environmental contaminants such as *L. monocytogenes*, *Cl. botulinum*, *Clostridium perfrigens* and *Bacillus* spp.; and (3) contaminants during handling and processing such as *Salmonella* spp., *Shigella* spp., *Escherichia coli*, *Campylobacter jejuni* and *Staphylococcus aureus* (Huss et al., 2003; Ghanbari et al., 2013; Shen and Wang, 2021).

Spoilage of fresh fish and fishery products is caused by a complex process in which physical, chemical and microbiological factors are involved. Among the microorganisms present in fish, only a fraction of the initial microbiota, known as specific spoilage organisms (SSOs), are responsible for fish spoilage (Gram and Dalgaard, 2002). The high *post mortem* pH (>6), combined with the nutrient availability in fresh fish (low carbohydrate content and the non-protein-nitrogen fraction), allow the quick growth of Gram-negative psychotropic bacteria naturally present in fish like *Pseudomonas* spp. and *Shewanella* spp. At ambient temperature, fish is dominated by mesophilic *Vibrionaceae* and, if the fish is caught in polluted waters, mesophilic *Enterobacteriaceae* (Gram and Huss, 1996; Gram and Dalgaard, 2002; Leroi, 2010; Shen and Wang, 2021). The use of vacuum or MAP packaging selects for the spoilage bacteria *Photobacterium phosphoreum* and also for LAB. In chilled seafood, Gram-positive pathogens are normally of less concern than the Gram-negative spoilage bacteria present, such as *Pseudomonas* spp., *Shewanella* spp. and *Enterobacteriaceae*, which grow more rapidly (Shen and Wang, 2021).

### 33.4.1.1 Spoilage Potential of LAB

In general, LAB are considered to play a minor role in the spoilage of fish and fishery products (Leroi, 2010; Pilet and Leroi, 2011; Tahiluddin et al., 2022). In fresh fish stored at refrigeration temperatures, LAB are not very competitive, and they produce fewer unpleasant odors compared to Gram-negative bacteria such as *S. putrefaciens*, *P. phosphoreum* and *Pseudomonas* spp. (Leroi, 2010). However, during storage of LPFP LAB become predominant, although their role on the spoilage of these products is not very clear. Several authors have found no correlation between total LAB counts and sensory spoilage (Leroi et al., 2001). On the other hand, Joffraud et al. (2001) reported *Lactobacillus* spp. that released large amounts of volatile compounds, those being likely responsible for the off-odors perceived on spoiled vacuum-packaged CSS. Stohr et al. (2001) showed that some *L. sakei* were the main SSO in CSS, while *L. alimentarius* had no effect. The role of *Carnobacterium* species as SSO of LPFP is still unclear (Pilet and Leroi, 2011; Leroi, 2015). Various studies have showed that the inoculation of CSS by strains of *C. maltaromaticum* and *C. divergens* leads to few or no changes in organoleptic quality (Duffes et al., 1999; Nilsson et al., 1999; Brillet et al., 2005). However, Jaffrès et al. (2011) have shown that *C. maltaromaticum* was one of the major SSOs of MAP tropical shrimp. The variability in the results suggests that the bacterial spoiling potential depends on species but is also strain dependent (Pilet and Leroi, 2011; Leroi, 2015).

# 33.4.2 Application of LAB Cultures and Their Bacteriocins in Fish and Fishery Products

Fish is an extremely perishable food commodity which can spoil more rapidly than almost any other food, becoming unfit to eat and possibly dangerous to health due to the microbial growth, chemical change and breakdown of endogenous enzymes (Wiernasz et al., 2020; FAO, 2022). Fish is also a very widely traded commodity, providing abundant opportunities for microbial growth and cross-contamination from different sources, making it more difficult to maintain the hygienic quality required for fish and seafood (Gram and Huss, 1996). Therefore, post-harvest handling, processing, preservation, packaging, storage measures and transportation of fish require particular care in order to maintain the quality and nutritional attributes of fish and avoid waste and losses. Preservation and processing techniques can reduce the rate at which spoilage happens and thus allow fish to be distributed and marketed worldwide. Such techniques include temperature reduction (e.g. chilling and freezing), heat treatment (e.g. smoking), reduction of available water (e.g. drying, salting and smoking) and changing the storage environment (packaging and refrigeration) (FAO, 2022).

In this context, biopreservation technology, which refers to the shelf-life extension and improvement of food safety by using microorganisms and/or their metabolites (Gómez-Sala et al., 2016), rises as an interesting and cost-effective alternative. Biopreservation is an emerging field, having applications along the whole food sector for the processing and preservation of foods. In this respect, LAB may be considered biopreservative agents, as they produce a wide range of antimicrobial metabolites, but mainly because of the production of organic acids and bacteriocins (Cotter et al., 2005; Rathod et al., 2022).

In the last years, bacteriocins and bacteriocin-producing LAB have been the focus of extensive research due to their potential use as natural biopreservatives; however, very few commercial applications have been developed in fish and fishery products. One of the reasons is that the selection of potential protective bacteria for fish and fishery products is still a challenge, as they must be able to: (i) survive steps of food processing; (ii) adapt and develop in an specific media such as fish, with low carbohydrate concentration; and (ii) produce active antagonistic metabolites, but their metabolic activities should not affect the organoleptic characteristics of the product (Leroi, 2010; Pilet and Leroi, 2011; Ghanbari et al., 2013).

Many of the studies regarding seafood biopreservation have been mainly conducted in LPFP because they are considered a risky products of foodstuffs but also because the implementation of LAB strains in fresh fish is more difficult (Ghanbari et al., 2013; Leroi, 2015) (Table 33.2).

## 33.4.2.1 Biopreservation with LAB Cultures

El Bassi et al. (2009) selected two *Lactobacillus* strains (*Lactiplantibacillus plantarum* [formerly *Lactobacillus plantarum*] and *Lactobacillus pentosus*) from 100 isolates from sea products based in their antimicrobial activity and ability to grow at low pH, low temperature and high salt concentration. Strains were inoculated into minced sea bass and samples were vacuum-sealed and stored at 4 °C. The results showed that both strains limited the development of total coliform and slightly reduced the production of trimethylamine (TMA) and total volatile basic nitrogen contents (TVB-N). The biopreservation potential of the multibacteriocinogenic strain *L. curvatus* BCS35 and its bacteriocins was tested in young hake and megrim (Gómez-Sala et al., 2016). The trial carried out at the retail fish market showed that after 14 days of storage at 2 °C, *L. curvatus* BCS35 significantly reduced the microbial counts compared to the control batch in both young hake and megrim. Moreover, according to the fish appraiser from the fish market,

**TABLE 33.2** Bioprotection of Fish and Fish Products Using Lactic Acid Bacteria (LAB) and/or Bacteriocins

| LAB STRAIN | BACTERIOCINOGENIC (Y/N) AND BACTERIOCIN(S) PRODUCEDA | ISOLATION SOURCE | PRODUCT APPLIED | REPORTED EFFECTS | REFERENCE(S) |
|---|---|---|---|---|---|
| Latilactobacillus curvatus BCS35 | Sakacins P and X | Dry-salted cod | Young hake and megrim | Improvement of the organoleptic characteristics, inhibition of Listeria monocytogenes and a higher value in the market | Gómez-Sala et al. (2016) |
| L. curvatus ET06 | Y, ND | CSS | CSS | Inhibition of L. innocua | Tomé et al. (2008) |
| Lacticaseibacillus rhamnosus ATCC 53103 | Y, ND | ND | Sea bass fillets | Inhibition of spoilage bacteria | Kannappan et al. (2017) |
| Latilactobacillus sakei Lb790 | Sakacin P | ND | CSS | Inhibition of L. monocytogenes | Katla et al. (2001) |
| Lactobacillus delbrueckii | ND | Sardine | Minced sardine with salt and sugar | Inhibition of coliforms, Salmonella, Staphylococcus and Clostridium | Ndaw et al. (2008) |
| L. delbrueckii ET32 | Y, ND | CSS | CSS | Inhibition of L. innocua | Tomé et al. (2008) |
| Lactiplantibacillus plantarum 3 | Y, ND | Sea products | Refrigerated and VP minced Dicentrarchus labrax | Suppression of TVB-N and TMA production | El Bassi et al. (2009) |
| Lactobacillus pentosus 7 | N | Sea products | Refrigerated and VP minced D. labrax | Suppression of TVB-N and TMA production | El Bassi et al. (2009) |
| Lacticaseibacillus casei T3 | Y, ND | Dairy | CSS | Inhibition of L. innocua | Vescovo et al. (2006) |
| Enterococcus faecium BNM58 | Enterocin L50 | Albacore | Young hake and megrim | Improvement of the organoleptic characteristics, inhibition of L. monocytogenes and a higher value in the market | Gómez-Sala et al. (2016) |
| E. faecium ET05 | Y, ND | CSS | CSS | Inhibition of L. innocua | Tomé et al. (2008) |
| Enterococcus faecalis L04 | Bacteriocin EFL4 | Sea bass | RTE fresh salmon fillets | Inhibition of spoilage and food-borne pathogenic bacteria | Mei et al. (2020) |
| Lactococcus lactis PSY2 | Bacteriocin PSY2 | Marine perch | Reef cod fillets | Extension of shelf life and better protection against spoilage bacteria | Sarika et al. (2012) |

(Continued)

**TABLE 33.2**   (Continued) Bioprotection of Fish and Fish Products Using Lactic Acid Bacteria (LAB) and/or Bacteriocins

| LAB STRAIN | BACTERIOCINOGENIC (Y/N) AND BACTERIOCIN(S) PRODUCED[a] | ISOLATION SOURCE | PRODUCT APPLIED | REPORTED EFFECTS | REFERENCE(S) |
|---|---|---|---|---|---|
| *Lactococcus piscium* EU2241 | N | Seafood | VP shrimp | Reduction of the number of *L. monocytogenes* and *Staphylococcus aureus* | Matamoros et al. (2009) |
| *Carnobacterium piscicola* A9b | Carnobacteriocin B2 | VP CSS | CSS | Inhibition of *L. monocytogenes* | Nilsson et al. (2004) |
| *C. piscicola* V1 | Piscicocin V1a and V1b | Trout intestine | CSS | Inhibition of *L. monocytogenes* | Brillet et al. (2004) |
| *Carnobacterium divergens* V41 | Divercin V41 | Salmon intestine | CSS | Inhibition of *L. monocytogenes* | Brillet et al. (2004) |
| *C. divergens* M35 | Divergicin M35 | Frozen smoked mussels | CSS | Inactivation of *L. monocytogenes* | Tahiri et al. (2009) |
| *Carnobacterium maltaromaticum* CS526 | Piscicocin CS526 | Surimi | CSS | Decrease the population of *L. monocytogenes* | Yamazaki et al. (2003, 2005) |
| *Leuconostoc gelidum* EU2247 | Y, ND | Seafood | VP shrimp | Reduction of the number of *L. monocytogenes* and *St. aureus* | Matamoros et al. (2009) |
| *Pediococcus acidilactici* ET34 | Y, ND | CSS | CSS | Inhibition of *L. innocua* | Tomé et al. (2008) |

*Abbreviations:* ND, not determined; VP, vacuum packaged; CSS, cold-smoked salmon; RTE, ready-to-eat.

the biopreserved batches were worth a higher price in the market. More recently, Kannappan et al. (2017) inoculated a *Lacticaseibacillus rhamnosus* (formerly *Lactobacillus rhamnosus*) strain into sea bass fillets, and after incubation at 37 °C for 2 days the result showed that this strain was able to regulate the growth of specific and native spoilage bacteria and reduced the production of TMA-N and TVB-N. The strain *L. sakei* Lb790, producing sakacin P (SakP), was compared to a non bacteriocinogenic strain for the inhibition of *L. monocytogenes* in CSS (Katla et al., 2001). The results showed that both strategies applied had a bacteriostatic effect on *L. monocytogenes* at the end of the storage period. Also, a bactericidal effect was observed when the SakP-producing strain was added jointly with SakP. Similarly, Nilsson et al. (2004) carried out a study to assess the contribution of bacteriocins to the inhibition of *L. monocytogenes* by *C. piscicola* strains in a CSS system. In their study, a *Carnobacterium* strain with bacteriocin activity and a non-bacteriocinogenic mutant were examined against *L. monocytogenes*, which revealed that bacteriocins are not indispensable for pathogen reduction since the non-bacteriocinogenic mutant had a marked antilisterial effect. Wiernasz et al. (2020) studied the use of six LAB for salmon dill gravlax biopreservation. Commercial fish samples were inoculated by spraying with the protective cultures to reach an initial concentration of $10^6$ log CFU/g and then microbial ecosystem, biochemical parameters and volatilome was followed for 25 days of storage at 8 °C in vacuum packaging. The results showed that the strains *C. maltaromaticum* SF1944 and *Vagococcus fluvialis* CD264 both have a promising potential as bioprotective cultures to ensure salmon gravlax microbial safety and sensorial quality, respectively (Wiernasz et al., 2020).

## 33.4.2.2 Biopreservation with Bacteriocins

The bacteriocin GP1 with antibacterial activity and produced by *Lacticaseibacillus rhamnosus* (formerly *Lactobacillus rhamnosus*) GP1 was tested for its effect on sensory, chemical and bacteriological quality attributes of fish filets (reef cod) stored at 4 and 0 °C (Sarika et al., 2019). The preservative solutions were sprayed over the surface of the fish samples, and after 28 days of storage, the results showed that the application of bacteriocin GP1 was effective in controlling the growth of coliforms, *Aeromonas* sp., *Lactobacillus* sp. and *Vibrio* sp. in the treated fish samples, while the main chemical parameters remained within the limit of acceptability (Sarika et al., 2019). The biopreservatives potential of bacteriocin PSY2 produced by *Lc. lactis* PSY2 was evaluated using fillets of reef cod (Sarika et al., 2012). The bacteriocin dissolved in sterile distilled water was sprayed into the fish fillets and stored at 4 and −18 °C. The results showed that the biopreservation approach applied at 4 °C extended the shelf life of the fillets and reduced the total count of spoilage bacteria, and the sensory analysis was better in bacteriocin-treated samples than in the control without bacteriocin. Katla et al. (2001) observed a delay in the growth of *L. monocytogenes* when adding SakP or nisin in CSS. The same delay in growth was observed by Duffes et al. (1999) when using semi-purified bacteriocins from *Carnobacterium* spp. on *L. monocytogenes* added to CSS stored at 8 °C and by Nilsson et al. (1997) when using nisin on *L. monocytogenes* added to VP CSS stored at 5 °C. In another study, Behnam et al. (2015) studied the effect of nisin on the biochemical and microbial quality and shelf life of vacuum packaged rainbow trout during 16 days storage at 4 °C. The results showed that treatment with nisin improved the fish quality and also extended the shelf life of the product. The effect of the CFS obtained from *L. curvatus* BCS35 containing sakacin P and sakacin X were tested in young hake and megrim, and the results showed that after 14 days of storage at 2 °C, the microbial counts were significantly reduced compared to the control batch in both fish species (Gómez-Sala et al., 2016). Inhibition studies with pure sakacin P were carried out in CSS, and the results showed that growth of *L. monocytogenes* was completely inhibited for at least 3 weeks (Aasen et al., 2003). Ibrahim and Desouky (2009) carried out a study to test the effect of the antimicrobial metabolites produced by *Lacticaseibacillus casei* (formerly *Lactobacillus casei*) DSM 120011 and *Lactobacillus acidophilus* 1 M in tilapia fillets stored at −18 °C for 90 days. The results showed an improvement of biochemical and microbiological aspects, increasing the overall quality of the fish fillets at the end of the storage period (Ibrahim and Desouky, 2009).

## 33.4.3 Regulatory Issues in Biopreservation of Fish and Fishery Products

Microorganisms used in food production are classified as starter cultures (providing nutritional and organoleptic characteristics) and/or bioprotective (or biopreservative) cultures (providing safety and durability). These properties are inherently linked, such that durability is enhanced by formation of organic acids, which also contribute to the characteristic taste and texture of many fermented foods (Elsser-Gravesen and Elsser-Gravesen, 2014). This means that all starter cultures are bioprotective cultures. However, bioprotective cultures can be used for extended application not only in fermented foods but in foods in general, which means that not all protective cultures are also starter cultures. The distinction between bioprotective and starter culture lies in their intended use. There is no specific legal regulation regarding microbial cultures for foods in Europe. They must fulfil the requirements set out in the General Food Law (EU Regulation No. 178/2002), saying that they must be safe for their intended use, and the sole responsibility for this lies with the distributor.

The safety of the selected strains is a one of the key aspect to consider. In this sense, the QPS approach developed by the EU can be considered a risk assessment tool for any microorganisms which is aimed to enter to the market (EFSA, 2023). Similarly, in the US, the GRAS status exists for food and substances used in food (FDA, 2018). In both cases, regularly updated lists of QPS recommended biological agents or GRAS food substances are provided by the corresponding regulatory bodies (EFSA and FDA, respectively). The first QPS list made in 2007 (EFSA, 2007) consisted of 72 species of microorganisms. The last published update from December 2022 included (EFSA, 2023) *C. divergens*; however, other species commonly considered suitable for seafood biopreservation such as *C. maltaromaticum*, *Lactococcus piscium* or *Leuconostoc gelidum* are not yet included in the list. It should be noted that microorganisms not included on the QPS list are not necessarily considered unsafe. There are more reasons for not being on the QPS list, such as the species not having been evaluated by EFSA. Currently, the QPS list is updated with the new taxonomic units two times a year, and it is reasonably conceivable that new bioprotective strains for seafood products would obtain QPS status if the necessary documentation were sent to EFSA (Leroi, 2010; Laulund et al., 2017)

## 33.5 PROBIOTICS IN AQUACULTURE

The major concerns about the widespread use of antibiotics as prophylactic and therapeutic agents in fish farming to control bacteria include the emergence of antibiotic resistances by fish pathogens, increased number of bacteria acting as reservoirs for antibiotic resistance genes, the alteration of the microbiota of both the fish (dysbiosis) and aquatic environments, suppression of beneficial bacterial in the fish gut and tissue residues harmful for human consumption (Sáenz et al., 2019; Schar et al., 2020; Feito et al., 2022; Ringø et al., 2022; Yilmaz et al., 2022). In this context, a wide range of novel strategies to control bacterial infections in aquaculture have been attracting a growing interest, such as vaccination; the use of prebiotics, probiotics, postbiotics, and bacteriophages; pathogen growth inhibition by short-chain fatty acids and polyhydroxyalkanoates; interference with the regulation of virulence genes (quorum sensing disruption); and the use of antimicrobial peptides (AMP) (Ang et al., 2020; Barroso et al., 2021; Simón et al., 2021; Liu et al., 2022; Mondal & Thomas, 2022; Sudhakaran et al., 2022; Zhang et al., 2022). The use of LAB of aquatic origin as probiotic cultures constitutes an effective, safe, easy to apply, economically efficient and environmentally friendly alternative or complementary strategy to conventional chemotherapy and vaccination for disease control in the aquaculture farming (Ringø et al., 2018; El-Saadony et al., 2021; Simón et al., 2021)

# 33.5.1 Definition of Probiotics

In aquaculture, the interaction between the microbiota and host is not limited to the gastrointestinal tract, as bacteria can be also active on the gills and skin of the host and its surrounding aquatic environment. Therefore, probiotics in aquaculture are considered any microbial cell that either through feed or the rearing environment benefits the host, the farmer or the consumer. This can be achieved by positively modulating the microbial community of the host through several distinct mechanisms of action, including improved resistance to pathogens, immunomodulation, enhanced host nutrition and growth performance, better used of feed, improved carcass and flesh quality, reduced stress, better animal welfare or improving the quality of its rearing environment (Merrifield et al., 2010).

# 33.5.2 Selection of Probiotics

Nowadays, there are no specific national or international guidelines about the evaluation of probiotics intended for use in aquaculture; however, there are international general recommendations for the evaluation of the use of probiotics in food, such as the Guidelines for the Evaluation of Probiotics in Food, made by a joint FAO/WHO working group (FAO/WHO, 2002). Likewise, in the EU, there are several documents published by the Scientific Committee on Animal Nutrition (SCAN) of EFSA about the QPS status, which include the safety assessment of microorganisms used in food and feed, their taxonomical identification, their resistance to antibiotic of clinical importance and their whole genome sequence (EFSA, 2005a, 2005b, 2007, 2021). Additionally, there are several recommendations in recent scientific literature about this issue of great interest and undeniable importance for the public health and environmental protection (Muñoz-Atienza et al., 2013; Pérez-Sánchez et al., 2014; Vieco-Saiz et al., 2019; EFSA, 2021; El-Saadony et al., 2021; Simón et al., 2021; Feito et al., 2022).

# 33.5.3 Action Mechanisms of Probiotics

The main mechanisms of action by which probiotics exert their beneficial effects are the following: (i) inhibition of the pathogenic microorganisms by competitive exclusion, (ii) improvement of the host nutrition due to the nutrient supply and/or enhancement of its digestion, (iii) stimulation of local and systemic immune response by different cell wall components, (iv) improvement of water quality, and finally (v) improved tolerance to stress (Merrifield et al., 2010; Pérez-Sánchez et al., 2014; El-Saadony et al., 2021; Simón et al., 2021; Ringø et al., 2022).

## 33.5.3.1 Competitive Exclusion

Competitive exclusion is a phenomenon where an established microbiota prevents or reduces the colonization of opportunistic and pathogenic microorganisms (Pérez-Sánchez et al., 2014; van Doan et al., 2021; Ringø et al., 2022). In the following, several mechanisms of competitive exclusion are detailed.

### 33.5.3.1.1 Competition by Adhesion Sites

Another mechanism by which probiotics prevent the colonization of pathogens is competition by adhesion sites on the intestinal mucosa and other epithelial surfaces. The ability of probiotic strains to colonize the mucosal surfaces without causing disease in the host is considered a good criterion for the selection of probiotics due to competition by the adhesion sites that could avoid the first stage of infection caused by pathogens. The adhesion of the microorganisms to superficial mucosa can be non-specific (by non-covalent bonds or hydrophobic interactions) or specific (by adhesin molecules, which are expressed on the surface of adherent bacteria and receptors of epithelial cells) (Lazado et al., 2011; Li et al., 2019; Simón et al., 2021; Docando et al., 2022). In this respect, the inhibition of the adhesion of pathogens to fish mucus has been demonstrated *in vitro* by several authors. For instance, Muñoz-Atienza et al. (2014) demonstrated that several LAB, among them *Leuconostoc mesenteroides* subsp. *cremoris* SMM69 and *L. curvatus* BCS35, inhibited the adhesion of several fish pathogens to intestinal mucus of turbot (*Scophthalmus maximus*, L.).

*33.5.3.1.2 Production of Antimicrobial Compounds*
One important mechanism of competition that also has important implications for pathogen control is the production of inhibitory compounds (Simón et al., 2021; van Doan et al., 2021). In general, this antimicrobial activity is due to the production of ethanol, hydrogen peroxide, diacetyl, short-chain fatty acids, $CO_2$, siderophores, lysozymes, proteases, and ribosomal synthesis peptides or proteins (i.e., bacteriocins) (Ringø et al., 2018; El-Saadony et al., 2021; Simón et al., 2021; van Doan et al., 2021; Feito et al., 2022).

## 33.5.3.2 Improvement of Host Nutrition

Probiotics have a direct effect on host growth due to the improvement of its nutrition by providing nutrients (e.g., fatty acids and essential amino acids) or vitamins and/or by improvement of digestibility (e.g., production of lipases and proteases), and, consequently, weight gain (Wang et al., 2020; Simón et al., 2021; Feito et al., 2022). Some *in vivo* studies have reported that the addition of probiotics to fish diet contribute to host nutrition and weight gain and improve the fish metabolome. Nguyen et al. (2018) demonstrated that olive flounder (*Paralichythys olivaceus*) fed with a *Lactococcus lactis* strain for 16 weeks showed improved mean weight gain, length, specific growth rate and feed efficiency. Additionally, the concentration of metabolites such as citrulline, short-chain fatty acids, vitamins and taurine were also higher in the probiotic-fed group. Similarly, Nimalan et al. (2023) observed that the individual or combined use of two probiotic LAB strains (*Lactiplantibacillus plantarum* and *Limosilactobacillus fermentum*) increased the concentration of total fatty acids in the digesta and chyme, improving mucosa health and preventing enteritis in Atlantic salmon (*Salmo salar*).

## 33.5.3.3 Modulation of the Immune System

Among the numerous beneficial effects of probiotics, the modulation of innate and specific immune responses is one of the most commonly reported benefits of probiotics in fish and crustaceans using *in vitro* and *in vivo* assays (Simón et al., 2021). In this respect, a lot of immunomodulatory studies have been performed in several fish using different probiotics and their ability to modulate cytokine production, phagocytic activity, respiratory burst activity, nitric oxide production, peroxidase activity, lysozyme production, complement activity and immunoglobulin production (Muñoz-Atienza et al., 2014; Moroni et al., 2021; Contente et al., 2023).

## 33.5.3.4 Improvement of Water Quality

The water quality in fish and crustacean cultures has been associated with the addition of probiotics, especially Gram-positive bacteria, such as *Bacillus* spp. and *Geobacillus* spp. (Ren et al., 2021; Hassan et al., 2022; Naiel et al., 2022). For instance, Ren et al. (2021) demonstrated that a multi-strain combination of *Bacillus* spp. (two *Bacillus flexus* and one *Bacillus licheniformis*) could effectively improve water quality by removing chemical oxygen demand (COD), ammonia-nitrogen and nitrate in a mariculture system of whiteleg shrimp (*Litopenaeus vannamei*).

## 33.5.3.5 Tolerance to Stress

One of the first studies about the effect of probiotics on the stress tolerance in aquacultured fish was performed by Carnevali et al. (2006) in sea bass, in which they observed a reduction of cortisol levels in fish fed with *Lactobacillus delbrueckii*. More recently, Eissa et al. (2018) reported that the administration of *Bacillus* spp. improved yellow perch (*Perca flavescens*) welfare and recovery after hypoxia and air-exposure by reducing cortisol, heat shock protein Hsp70, glutathione peroxidase and superoxide dismutase.

## 33.5.4 LAB Evaluated as Probiotics in Aquaculture

As previously mentioned, LAB are the group of microorganisms most studied as probiotics for aquaculture, and there are many reviews evaluating their application and results in aquaculture productions (Ringø et al., 2018, 2020, 2022; Simón et al., 2021; Sumon et al., 2022). The following is a compendium of the studies indexed in the ScienceDirect, Scopus, PubMed and Web of Science databases (Table 33.3), carried out with LAB as probiotics for aquaculture (Table 33.4), based on the results published during 2022, in order to avoid overlapping with the aforementioned revisions.

Nowadays, there is a great interest in the use of LAB as probiotics in aquaculture due to several reasons (Ringø et al., 2018, 2020; El-Saadony et al., 2021; Simón et al., 2021). First of all, most LAB are currently considered safe microorganisms for human and animal consumption (GRAS status or QPS status), and nowadays numerous LAB strains are accepted legally as probiotics for use in humans and animal production. In addition, numerous LAB genera are considered part of the intestinal microbiota of farmed fish. Moreover, some LAB of aquatic origin produce antimicrobial compounds that avoid the colonization of pathogenic microorganisms in fish. The main LAB proposed as probiotics in aquaculture have shown numerous beneficial effects, including improvement of the survival rate after the infection with fish pathogens, modulation of the immune system, growth stimulation and improvement of water quality (Araújo et al., 2016; Ringø et al., 2018, 2020; El-Saadony et al., 2021; Simón et al., 2021; Feito et al., 2022; Contente et al., 2023).

The main LAB species that have been evaluated in aquaculture belong to the genera *Lactococcus*, former *Lactobacillus* (which include *Lactobacillus*, *Lactiplantibacillus*, *Latilactobacillus*, *Lacticaseibacillus*, *Ligilactobacillus*, among others), *Leuconostoc*, *Pediococcus*, *Enterococcus* and *Vagococcus*. Muñoz-Atienza et al. (2014) reported that the bath administration of *Le. mesenteroides* subsp. *cremoris* SMM69 and *W. cibaria* P71, isolated from marinated Atlantic salmon and octopus (*Octopus vulgaris*), respectively, to turbot larvae and juveniles, stimulated their immune system, especially the expression of non-specific immunity associated genes (IL-1β, TNF-α, lysozyme, C3, MHC-Iα and MHC-IIα) in mucosal tissues (head-kidney, spleen, liver, intestine and skin). Moreover, Contente et al. (2023) observed that as bacteriocinogenic and non-bacteriocinogenic strains of *Lactococcus cremoris* of aquatic origin exerted immunomodulatory effects on rainbow trout (*Oncorhynchus mykiss*, Walbaum) splenic leukocytes and rainbow trout intestinal epithelial cell line (RTgutGC) *ex vivo* and *in vitro*, respectively. The reported immunomodulatory effects included a strong ability to promote the transcription of several pro- and anti-inflammatory cytokines, AMPs, genes involved with the intestinal barrier integrity and homeostasis, increased respiratory burst activity and stimulated the production of nitric oxide. Therefore, this study demonstrated for the first time the immunomodulatory effects of a bacteriocinogenic LAB over a fish cell line. Araújo et al. (2015b) reported the effectiveness of bacteriocinogenic *Lactococcus cremoris* WA2–67 to protect rainbow trout against infection with the invasive pathogen *Lactococcus garvieae*, demonstrating the relevance of nisin Z (NisZ) production as

**TABLE 33.3** Selected Databases and Search Criteria Used

| DATABASE | SEARCH CRITERIA AND BOOLEAN OPERATORS | WEBSITE |
|---|---|---|
| ScienceDirect | Title, abstract, keywords: lactic acid bacteria and probiotic and aquaculture. Refine by years: 2022 | www.sciencedirect.com/ |
| Scopus | Title-Abs-Key (lactic acid bacteria and probiotic and aquaculture). Refine results: limit to 2022 | www.scopus.com/home.uri |
| PubMed | All fields: (lactic acid bacteria) and (probiotics) and (aquaculture). Results by year: 2022 | https://pubmed.ncbi.nlm.nih.gov/ |
| Web of Science | TS=(lactic acid bacteria and TS=(probiotic)) and TS=(aquaculture). Publication years: 2022 | www.webofscience.com/ |

**TABLE 33.4**  Studies Published during the Year 2022 on the Use of LAB as Probiotics for Aquaculture

| PROBIOTIC | TYPE OF STUDY | DURATION (IN VIVO) | ADMINISTRATION ROUTE | EFFECT | REFERENCE |
|---|---|---|---|---|---|
| **White leg shrimp (Litopenaeus vannamei)** | | | | | |
| Lactiplantibacillus plantarum Ep-M17 | In vitro/in vivo | 4 weeks | $5 \times 10^8$ UFC/g of feed | ↑SGR; ↓FCR; ↑survival rate against *Vibrio parahaemolyticus*; ↑ SOD, CAT, TRY, AKP, LIP, AMS; ↑antimicrobial diversity | Du et al. (2022b) |
| Lactobacillus paracasei | In vivo | 4 weeks/1 week (challenge) | $10^7$ UFC/g of feed | ↑innate immune; ↑genic expression related with immunity; ↑survival rate against *V. parahaemolyticus* | Huang et al. (2022) |
| Lactobacillus acidophilus | In vivo | 60 days/4 days (challenge) | $10^7$ UFC/g of feed | ↑growth; ↑villi height, wall width; ↑intestinal amylase and lipase activity; ↑survival rate against *Vibrio alginolyticus* | Abidin et al. (2022) |
| Pediococcus pentosaceus PP4012, Lactiplantibacillus plantarum LP28, Lacticaseibacillus rhamnosus LRH10, Limosilactobacillus fermentum LF26, Lacticaseibacillus paracasei LPC12 | In vivo | 12 weeks | $10^5$ UFC/g of feed | ↑growth; ↑survival rate against *V. parahaemolyticus*; ↑innate immunity | Lee et al. (2022) |
| Limosilactobacillus reuteri, Pediococcus acidilactici | In vivo | 8 weeks | $10^3$, $10^5$, $10^7$ UFC/g of feed | ↑growth; ↑phenoloxidase activity | Wu et al. (2022) |
| Lactiplantibacillus plantarum 7–40 | In vivo | 8 weeks | $10^8$ UFC/kg of feed | ↑survival rate against *V. alginolyticus*; ↑respiratory burst activity; ↑pexn expression in hemocytes and pen4 in hemocytes and hepatopancreas | Prabawati et al. (2022) |
| Lactococcus lactis S1, Lc. lactis S2, Enterococcus faecalis F3, E. faecalis F7 | In vitro/in vivo | 60 days | $10^6$ UFC/g of feed | Inhibition of ichthyopathogens; ↑growth; ↑survival rate against *Vibrio harveyi* | Cai et al. (2022) |
| Lactiplantibacillus plantarum (NCIMB 30280) | In vitro/in vivo | 4 weeks | $2 \times 10^8$ UFC/g of feed | ↓growth of *Vibrio* spp. (in vitro) | Thompson et al. (2022) |
| L. plantarum HC-2 | In vitro/in vivo | 2 weeks | $5 \times 10^8$ UFC/g | Intestinal epithelial adhesion | Du et al. (2022a) |
| **Red swamp crayfish (Procambarus clarkii)** | | | | | |
| L. fermentum GR-3 | In vitro/in vivo | 30 days | $10^8$ UFC/g of feed | Arsenic removal; lower oxidative stress, dysbiosis and arsenic-related histological lesions | Han et al. (2022) |

| | | | | | |
|---|---|---|---|---|---|
| Common carp (*Cyprinus carpio*) | | | | | |
| *Lactobacillus bulgaricus*, *Streptococcus thermophilus* | *In vivo* | 70 days | Fermented barley flour | ↑growth; ↑use of feed; ↑fillet protein-content; ↑α-amylase, lipase | Qaddoori et al. (2022) |
| Goldfish (*Carassius auratus*) | | | | | |
| *L. rhamnosus* ATCC53103, *Lacticaseibacillus casei* ATCC393, *Lb. plantarum* H8 | *In vitro/in vivo* | 2 weeks | Fermented *Sanguisorba officinalis* | ↑growth; ↓biofilm *Aeromonas hydrophila*; ↑ALP, SOD, CAT; ↑Igs; ↑survival rate against *A. hydrophila* | T. Wang et al. (2022) |
| *Weissella cibaria* C-10 | *In vitro/in vivo* | 5 weeks | $2 \times 10^9$ UFC/kg of feed | ↑inhibition of ichthyopathogens; ↑immune response; ↑survival rate against *Aeromonas veronii* | Zhu et al. (2022) |
| Red drum (*Sciaenops ocellatus*) | | | | | |
| *P. acidilactici* MA18/5M | *In vivo* | 8 weeks | $10^9$ UFC/kg of feed | ↑growth; ↑fillet protein-content | Yamamoto et al. (2022) |
| Siberian sturgeon (*Acipenser baerii*) | | | | | |
| *L. acidophilus* ProLacto | *In vivo* | 8 weeks | $6 \times 10^9$ UFC/kg of feed | ↑growth; ↑survival; ↓eosinophiles | Mocanu et al. (2022) |
| Indian major carp (*Cirrhinus mrigala*) | | | | | |
| *Enterococcus faecium* MC-5 | *In vivo* | 6 weeks | $10^6$–$10^{12}$ UFC/g of feed | ↑growth; ↑AP, SOD; ↑IgM; ↑lysozyme, respiratory burst activity, phagocytic activity; ↑survival rate against *A. hydrophila* | Tilwani et al. (2022) |
| Sea bass (*Dicentrarchus labrax*) | | | | | |
| *P. acidilactici* CNCM I-4622 | *In vivo* | 60 days | $2$–$3 \times 10^{10}$ UFC/ kg of feed | ↑growth; ↑fillet lipidic-content; ↑villi height; ↓water ammoniac | Eissa et al. (2022) |
| Thinlip mullet (*Liza ramada*) | | | | | |
| *L. acidophilus* ATCC4356 | *In vivo* | 8 weeks | $10^8$ UFC/g of feed | ↑growth; ↓stress; ↑CAT, SOD; ↑hepatic health; ↑survival rate against aflatoxin $B_1$ | Khalafalla et al. (2022) |
| Japanese sea cucumber (*Apostichopus japonicus*) | | | | | |
| *L. rhamnosus* M2–4 | *In vivo* | 30 days | $10^9$ UFC/g of feed | ↑beneficial microbiota; ↑lipidic metabolism | Li et al. (2022) |
| Zebrafish (*Danio rerio*) | | | | | |
| *Lc. lactis* KUST48 (LLK48) | *In vitro/in vivo* | 120 hours | $10^4$ UFC intraperitoneal | ↓abundance of *Streptococcus agalactiae*; ↑relative abundance intestinal microbiota; ↑survival rate against *S. agalactiae* | Tan et al. (2022) |

*(Continued)*

| PROBIOTIC | TYPE OF STUDY | DURATION (IN VIVO) | ADMINISTRATION ROUTE | EFFECT | REFERENCE |
|---|---|---|---|---|---|
| *L. rhamnosus* GG | *In vivo* | 2 hours | | ↓ROS; ↑intestinal motility; ↑CAT, SOD, GPx; ↑serotonin | Wang et al. (2022) |
| *Lactobacillus delbrueckii* ATCCBAA2844, *L. acidophilus* ATCC4356 | *In vivo* | 60 days | $10^8$ UFC/g of feed | ↑*villi* height; ↑intra-epithelial lymphocytes; ↑expression of *IL1β, TNFα*; ↑survival rate against *A. hydrophila* | Ehsannia et al. (2022) |
| 19 species of BAL | *In vivo* | 10 days | $10^4$ UFC/mL of water | ↑larval growth; ↑survival; ↑mitochondrial activity; ↑microbiome diversity | Padeniya et al. (2022) |
| *Carnobacterium* sp. T4, *Lc. lactis* TW34, *Lactiplantibacillus pentosus* H16 | *In vitro/in vivo* | 2 weeks | $10^7$ UFC/g of feed | ↑intestinal LAB, ↓*Vibrio* spp. | Sequeiros et al. (2022) |
| *L. plantarum* IBRC-M 10817 | *In vivo* | 10 weeks | $10^5$, $10^6$, and $10^7$ UFC/g de pienso | ↑growth; ↑expression of *IGF-1, GF, LYZ, TNFα*, activin, *FSH-R, A-R* | Safari et al. (2022) |
| Atlantic salmon (*Salmo salar*) | | | | | |
| *Latilactobacillus curvatus* ATCC PTA-127116 and *L. curvatus* ATCC PTA-127117 | *In silico/in vitro/in vivo* | 7 weeks (fresh water) 11 weeks (salt water) | $1{,}0$–$1{,}6 \times 10^8$ UFC/g of feed $6{,}1$–$6{,}3 \times 10^7$ UFC/g of feed | ↑growth | Cathers et al. (2022) |
| Nile tilapia (*Oreochromis niloticus*) | | | | | |
| *P. acidilactici* CNCM I-4622 | *In vivo* | 8 weeks | 0,2% (UFC/kg of feed) | ↑survival rate against *A. hydrophila*; ↑amylase activity; ↑leucocytes; ↑% lymphocytes; serum: ↑total proteins, ↑ALP, ↑Igs, ↑ lysozyme, ↑complement; skin mucus: ↑ lysozyme, ↑complement; ↑SOD, CAT, GPx; ↓malondialdehyde; ↑expression of *IL1β, TNFα* | Mohammadi et al. (2022) |
| *E. faecium* 1, *E. faecium* 2 | *In vivo* | 90 days | $2 \times 10^6$ UFC/g of feed | ↓serum glucose, ↑survival rate against *S. agalactiae* | Dias et al. (2022) |
| *L. rhamnosus* KU985435 | *In vitro/in vivo* | 30 weeks | $4$–$8 \times 10^9$ UFC/kg de pienso | Phytase activity, ↑growth, ↑fillet protein-content | Flefil et al. (2022) |

| Rainbow trout (*Oncorhynchus mykiss*) | | | | | |
|---|---|---|---|---|---|
| *L. acidophilus* VKPM B-3235 | *In vivo* | 45 days | $2 \times 10^8$ UFC/kg of feed | ↑growth; ↓*FCR* | Nikiforov-Nikishin et al. (2022) |
| Combination of *E. faecium, P. acidilactici, L. reuteri* | *In vitro* | – | – | ↓Icthyopathogenic adhesion to epithelial intestinal cells; ↓expression of *IFN, IL8, IL6* in the presence of pathogens; ↑ expression *TNFα, IL10, TLR5M, TLR5* in the absence of pathogens | Pillinger et al. (2022) |
| *Lactobacillus* sp. AM14 | *In vivo* | 8 weeks | UFC/kg of feed | ↑growth; ↑fillet protein-content | Abedi et al. (2022) |
| *P. acidilactici* MA18/5M | *In vivo* | 8 weeks | $10^6$ UFC/g of feed | ↑feed lipid conversion efficiency, ↓abundance of *Mycoplasma* spp.; ↑α diversity; metabolome modulation | Rasmussen et al. (2022) |
| *Latilactobacillus sakei* RDB2 | *In vivo* | 30 days | $10^7$ UFC/g of feed | ↑growth; immunomodulation | Dang et al. (2022) |
| *L. plantarum* S14 | *In vitro/in vivo* | 30 days | $10^8$ UFC/g of feed | Synthesis of selenium nanoparticles; ↑ROS in leucocytes; ↑plasma lysozyme activity; ↑GPx | Yanez-Lemus et al. (2022) |
| Multi-species | | | | | |
| *L. plantarum* r1 | *In vitro* | – | – | Direct and extracellular antimicrobial activity against Gram-negative ichthyopathogens | Zhang et al. (2022) |
| *Lc. lactis* NZ9000, *Lb. rhamnosus* GG, *P. pentosaceus* NCDO990 | *In vitro* | – | – | ↓ *A. hydrophila* growth; ↓biofilm formation | Ibarra-Martínez et al. (2022) |
| *Lc. lactis* subsp. *cremoris* WA2–67 | *In silico* | – | – | Genetic machinery involved in the synthesis of bacteriocins, vitamins, amino acids, stress resistance factors, etc. | Feito et al. (2022) |
| 10 *L. plantarum* strains | *In vitro* | – | – | Production of antioxidant substances, survival in digestive conditions, antagonizes against ichthyopathogens and foodborne pathogens | Iorizzo et al. (2022) |
| *P. pentosaceus* ON495586, *L. plantarum* ON491817, *Lc. lactis* ON479264, *E. faecium* ON478992, *Enterococcus hirae* ON478991, *Enterococcus durans* ON564885 | *In vitro* | – | – | ↓*Vibrio campbelli, V. harveyi, V. parahaemolyticus* growth | Lalitha et al. (2022) |

(*Continued*)

| PROBIOTIC | TYPE OF STUDY | DURATION (IN VIVO) | ADMINISTRATION ROUTE | EFFECT | REFERENCE |
|---|---|---|---|---|---|
| *E. faecium* ST1, *E. faecium* ST9 | *In vitro* | – | – | *In vitro* inhibition of ichthyopathogens | Barros et al. (2022) |
| *Lc. lactis* L1, *Lc. lactis* L2, *E. faecium* 135EF | *In vitro* | – | – | ↓ *S. agalactiae* growth | Pereira et al. (2022) |
| *E. faecium* 11037CHB | *In vitro* | – | – | *In vitro* inhibition of ichthyopathogens | Lopes et al. (2022) |
| *L. acidophilus* F3, *P. pentosaceus* F7, *Lb. plantarum* F12, *P. acidilactici* F15 | *In vitro* | – | – | *In vitro* inhibition of ichthyopathogens | Mazlumi et al. (2022) |
| *Lc. lactis* A12 | *In silico/in vitro* | – | – | *In vitro* inhibition of ichthyopathogens; genetic machinery involved in the synthesis of bacteriocins, amino acids, vitamins, etc. | Melo-Bolívar et al. (2022) |
| *L. fermentum* As4, As11, As20, As21, As41, As48 | *In vitro* | – | – | *In vitro* inhibition of ichthyopathogens; arsenic removal | Bhakta et al. (2022) |

an anti-infective mechanism. This was the first report demonstrating the effective *in vivo* role of bacteriocin (NisZ) production as a mechanism to protect fish against lactococcosis. More recently, Won et al. (2020) suggested that strains of *Lactococcus lactis* and *Pediococcus pentosaceus* were good probiotic candidates for aquaculture, as they demonstrated strong *in vivo* protective effects against *Vibrio parahaemolyticus* infection in whiteleg shrimp.

### 33.5.4.1 Pediococcus acidilactici *CNCM I-4622–MA 18/5M*
### (Bactocell, Lallemand Inc., Cardiff, United Kingdom)

*P. acidilactici* CNCM I-4622–MA 18/5M (Bactocell) is the first and only probiotic microorganism authorized to use in aquaculture in the European Union (Commission Regulation (EC) No 911/2009; Commission Implementing Regulation (EU) No 95/2013). This authorization is based on the recognition of its safety due to its QPS status, as well as its efficacy on the productive improvement in salmonids and shrimps (Commission Regulation (EC) No 911/2009) and all fish in general (Commission Implementing Regulation (EU) No 95/2013). In salmonids, this strain enhances the quality of the final product by means of an increase of fish with a good conformation, preventing vertebral column compression syndrome. In the case of rainbow trout, this syndrome affects more than 20% of the production, causing important economic losses in this sector. Moreover, this probiotic is considered beneficial to the development of all fish, increasing well-formed fish and reducing bone deformation. On the other hand, this strain improves survival and growth in shrimps and increases resistance against infection by *Vibrio* spp.

In Europe, the authorized probiotics used in animal feed are classified as "feed additive" and subjected to assessment of the scientific committee. The use of *P. acidilactici* CNCM I-4622–MA 18/5M (Bactocell) was authorized without a time limit for chickens for fattening by Commission Regulation (EC) No 1200/2005 and for pigs for fattening by Commission Regulation (EC) No 2036/2005; for 10 years for salmonids and shrimps by Commission Regulation (EC) No 911/2009; for weaned piglets by Commission Regulation (EU) No 1120/2010; for laying hens by Commission Regulation (EU) No 212/2011; and for all fish other than salmonids by Commission Implementing Regulation (EU) No 95/2013. Moreover, on 6 May 2013, preparation of *P. acidilactici* CNCM I-4622–MA 18/5M (Bactocell) was authorized as a feed additive for use in water for drinking for weaned piglets, pigs for fattening, laying hens and chickens for fattening by Commission Regulation (EU) No 413/2013. More recently, a new assessment by EFSA Panel on Additives and Products or Substances used in Animal Feed (FEEDAP) suggested the renewal of the authorization for Bactocell was a feed addictive for all fish, including salmonids, and its extension of use for all crustaceans (FEEDAP, 2019).

Regarding its use in aquaculture, the first study was carried out in larval pollack (*Pollachius pollachius*) fed with *Artemia* nauplii enriched with *P. acidilactici* CNCM I-4622–MA 18/5M (Bactocell), which increased their weight (Gatesoupe, 2002). However, studies developed in rainbow trout (Aubin et al., 2005), Nile tilapia (Shelby et al., 2006) and channel catfish (Shelby et al., 2007) showed that this strain did not enhance the growth rate. The first reported beneficial effect of this strain was the prevention of vertebral column compression syndrome in rainbow trout after the administration of this probiotic for five months (Aubin et al., 2005). Moreover, Merrifield et al. (2011) observed that this probiotic colonized the rainbow trout intestine, at least during its feed supplementation, and reduced the condition factor (K) of anatomy conformation, improving the aesthetic quality of fish to consumers. However, in this study, the rainbow trout immune response was not improved, as in other studies developed in Nile tilapia (Shelby et al., 2006) and channel catfish (Shelby et al., 2007). Nevertheless, more recently, Torrecillas et al. (2018) reported the positive effects on growth performance, gut intestinal homeostasis and disease resistance against *V. anguillarum* by supplementing European sea bass (*Dicentrarchus labrax*) juveniles with a synbiotic mixture of prebiotics and Bactocell. Finally, another study carried out in shrimp *Litopenaeus stylirostris* species demonstrated that this probiotic reduced the levels of oxidative stress parameters in fish infected with *Vibrio nigripulchritudo* (Castex et al., 2010).

# 33.6 CONCLUDING REMARKS

The presence of LAB in fish and fishery products is now well established, and both the bioprotective and probiotic potential of many of these strains have been highlighted in the last years. Fish and fishery products are considered of high nutritional value as they are a source of healthy and necessary nutrients. In this context, the growing interest of consumers in nutritional and healthy aspects and change in consumer's habits to fresher and even raw food produced with less or no use of chemical preservatives is influencing consumption decisions. The results obtained in the last years with LAB as bioprotective cultures for improving safety and quality of fish and fishery products are very promising. However, LAB have not been traditionally used in fish and fishery products, and thus biopreservation technology in seafood products is still in its infancy compared to dairy products. In addition, there are still several challenges to overcome such as the ability of the bioprotective culture to develop in a complicate food matrix (pH, carbohydrates content, fat, etc.) and the modification of the sensory characteristics of the product. For these reasons, it is essential to carry out a case-to-case study for the application of biopreservation technology to fish and fishery products. On the other hand, some of the main challenges the aquaculture industry currently faces are the steady supply of safe and quality products and the prevention and control of fish-borne diseases reducing the use of antibiotics. Consequently, there is a need for the development and practical application of alternative or complementary strategies to chemotherapy and vaccination, which are effective, safe, easy to apply, economically efficient and environmentally friendly. In this context, results obtained with the use of LAB as probiotics in aquaculture are very promising and encourage further studies on their safety for fish, humans and the environment, as well as on their probiotic characteristics, not only *in vitro* and *in vivo* but also *in situ* in fish farms. Finally, fish intestinal microbiota has a remarkable impact on fish health, and thus more effort should be put on the elucidation of the effect of the addition of probiotics on the microbiota of fish gastrointestinal tract and their surrounding aquatic environment by using metagenomics, metatranscriptomics and metabolomics combined with next-generation sequencing (NGS) technologies.

# BIBLIOGRAPHY

Aasen, I.M., Markussen, S., Møretrø, T., Katla, T., Axelsson, L., Naterstad, K. (2003) Interactions of the bacteriocins sakacin P and nisin with food constituents. *Int J Food Microbiol*, 87:35–43.

Abedi, S.Z., Yeganeh, S., Moradian, F., Ouraji, H. (2022) The influence of probiotic (isolated based on phytase activity) on growth performance, body composition, and digestibility of rainbow trout, *Oncorhynchus mykiss*. *J World Aquac Soc*, 53(5):1006–1030.

Abidin, Z., Huang, H.-T., Hu, Y.-F., Chang, J.-J., Huang, C.-Y., Wu, Y.-S., Nan, F.-H. (2022) Effect of dietary supplementation with *Moringa oleifera* leaf extract and *Lactobacillus acidophilus* on growth performance, intestinal microbiota, immune response, and disease resistance in whiteleg shrimp (*Penaeus vannamei*). *Fish Shellfish Immunol*, 127:876–890.

Al-Hisnawi, A., Ringo, E., Davies, S.J., Waines, P., Bradley, G., Merrifield, D.L. (2015) First report on the autochthonous gut microbiota of brown trout (*Salmo trutta* Linnaeus). *Aquac Res*, 46:2962–2971.

Allameh, S.K., Daud, H.M., Yusoff, F.M., Saad, C.R., Idris, A. (2012) Isolation, identification and characterization of *Leuconostoc mesenteroides* as a new probiotic isolated from intestine of snakehead fish (*Channa striatus*). *Afr J Biotechnol*, 11:3810–3816.

Amann, R.I., Ludwig, W., Schleifer, K.-H. (1995) Phylogenetic identification and *in situ* detection of individual microbial cells without cultivation. *Microbiol Rev*, 59:143–169.

Ang, C.Y., Sano, M., Dan, S., Leelakriangsak, M., Lal, T.M. (2020) Postbiotics applications as infectious disease control agent in aquaculture. *Biocontrol Sci*, 25(1):1–7.

Araújo, C., Muñoz-Atienza, E., Nahuelquín, Y., Poeta, P., Igrejas, G., Hernández, P.E., Herranz, C., Cintas, L.M. (2015a) Inhibition of fish pathogens by the microbiota from rainbow trout (*Oncorhynchus mykiss*, Walbaum) and rearing environment. *Anaerobe*, 32:7–14.

Araújo, C., Muñoz-Atienza, E., Pérez-Sánchez, T., Poeta, P., Igrejas, G., Hernández, P.E., Herranz, C., Ruiz-Zarzuela, I., Cintas, L.M. (2015b) Nisin Z production by *Lactococcus lactis* subsp. *cremoris* WA2–67 of aquatic origin as a mechanism to protect rainbow trout (*Oncorhynchus mykiss*, Walbaum) against *Lactococcus garvieae*. *Mar Biotechnol*, 17:820–830.

Araújo, C., Muñoz-Atienza, E., Poeta, P., Igrejas, G., Hernández, P.E., Herranz, C., Cintas, L.M. (2016) Characterization of *Pediococcus acidilactici* strains isolated from rainbow trout (*Oncorhynchus mykiss*) feed and larvae: Safety, DNA fingerprinting, and bacteriocinogenicity. *Dis Aquat Org*, 119:129–143.

Arbulu, S., Gómez-Sala, B., Garcia-Gutierrez, E., Cotter, P.D. (2022) Bioprotective cultures and bacteriocins for food. In *Good Microbes in Medicine, Food Production, Biotechnology, Bioremediation and Agriculture*. 1st ed., Eds. Frans J. de Bruijn, Hauke Smidt, Luca S. Cocolin, Michael Sauer, David N. Dowling, Linda Thomashow. John Wiley & Sons Ltd. ISBN: 978-1-119-76254-6.

Askarian, F., Kousha, A., Ringø, E. (2009) Isolation of lactic acid bacteria from the gastrointestinal tract of beluga (*Huso huso*) and Persian sturgeon (*Acipenser persicus*). *J Appl Ichthyol*, 25:91–94.

Aubin, J., Gatesoupe, F.J., Labbé, L., Lebrun, L. (2005) Trial of probiotics to prevent the vertebral column compression syndrome in rainbow trout (*Oncorhynchus mykiss* Walbaum). *Aquac Res*, 36:758–767.

Balcázar, J.L., de Blas, I., Ruiz-Zarzuela, I., Vendrell, D., Calvo, A.C., Márquez, I., Gironés, O., Múzquiz, J.L. (2007a) Changes in intestinal microbiota and humoral immune response following probiotic administration in brown trout (*Salmo trutta*). *Br J Nutr*, 97:522–527.

Balcázar, J.L., de Blas, I., Ruíz-Zarzuela, I., Vendrell, D., Gironés, O., Múzquiz, J.L. (2007b) Enhancement of the immune response and protection induced by probiotic lactic acid bacteria against furunculosis in rainbow trout (*Oncorhynchus mykiss*). *FEMS Immunol Med Microbiol*, 51:185–193.

Balcázar, J.L., Vendrell, D., de Blas, I., Ruiz-Zarzuela, I., Muzquiz, J.L., Girones, O. (2008) Characterization of probiotic properties of lactic acid bacteria isolated from intestinal microbiota of fish. *Aquaculture*, 278:188–191.

Banerjee, S.P, Dora, K.C., Chowdhury, S. (2013) Detection, partial purification and characterization of bacteriocin produced by *Lactobacillus brevis* FPTLB3 isolated from freshwater fish: Bacteriocin from *Lb. brevis* FPTLB3. *J Food Sci Technol*, 50:17–25.

Barros, F.A.L., Silva, L.A., Dias, J.A.R., Abe, H.A., Paixão, P.E.G., Sousa, N.C., Cordeiro, C.A.M., Fujimoto, R.Y. (2022) *In vitro* selection of autochthonous bacterium with probiotic potential for the neotropical fish piauçu *Megaleporinus microcephalus*. *Arq Bras Med Vet Zootec*, 74:327–337.

Barroso, C., Carvalho, P., Nunes, M., Gonçalves, J.F.M., Rodrigues, P.N.S., Neves, J.V. (2021) The Era of antimicrobial peptides: Use of hepcidins to prevent or treat bacterial infections and iron disorders. *Front Immunol*, 12:754437.

Behnam, S., Anvari, M., Rezaei, M., Soltanian, S., Safari, R. (2015) Effect of nisin as a biopreservative agent on quality and shelf life of vacuum packaged rainbow trout (*Oncorhynchus mykiss*) stored at 4 °C. *J Food Sci Technol*, 53:184–192.

Belfiore, C., Björkroth, J., Vihavainen, E., Raya, R., Vignolo, G. (2010) Characterization of *Leuconostoc* strains isolated from fresh anchovy (*Engraulis anchoita*). *J Gen Appl Microbiol*, 56:175–180.

Belfiore, C., Raya, R.R., Vignolo, G.M. (2013) Identification, technological and safety characterization of *Lactobacillus sakei* and *Lactobacillus curvatus* isolated from Argentinean anchovies (*Engraulis anchoita*). *Springerplus*, 2:257.

Berggren, H., Tibblin, P., Yıldırım, Y., Broman, E., Larsson, P., Lundin, D., Forsman, A. (2022) Fish skin microbiomes are highly variable among individuals and populations but not within individuals. *Front Microbiol*, 12:767770.

Bhakta JN, Bhattacharya S, Lahiri S, Panigrahi AK. Probiotic Characterization of Arsenic-resistant Lactic Acid Bacteria for Possible Application as Arsenic Bioremediation Tool in Fish for Safe Fish Food Production. Probiotics Antimicrob Proteins. 2023 Aug;15(4):889-902. doi: 10.1007/s12602-022-09921-9. Epub 2022 Feb 4. PMID: 35119613.

Brillet, A., Pilet, M.F., Prévost, H., Bouttefroy, A., Leroi, F. (2004) Biodiversity of Listeria monocytogenes sensitivity to bacteriocin-producing *Carnobacterium* strains and application in sterile cold-smoked salmon. *J Appl Bacteriol*, 97:1029–1037.

Brillet, A., Pilet, M.F., Prévost, H., Cardinal, M., Leroi, F. (2005) Effect of inoculation of *Carnobacterium divergens* V41, a biopreservative strain against *Listeria monocytogenes* risk, on the microbiological, and sensory quality of cold-smoked salmon. *Int J Food Microbiol*, 104:309–324.

Cabello, F.C. (2006) Heavy use of prophylactic antibiotics in aquaculture: A growing problem for human and animal health and for the environment. *Environ Microbiol*, 8:1137–1144.

Cai, X., Wen, J., Long, H., Ren, W., Zhang, X., Huang, A., Xie, Z. (2022) The probiotic effects, dose, and duration of Lactic Acid Bacteria on disease resistance in *Litopenaeus vannamei*. *Aquac Rep*, 26:101299.

Cai, Y., Suyanandana, P., Saman, P., Benno, Y. (1999) Classification and characterization of lactic acid bacteria isolated from the intestines of common carp and freshwater prawns. *J Gen Appl Microbiol*, 45:177–184.

Carnevali, O., de Vivo, L., Sulpizio, R., Gioacchini, G., Olivotto, I., Silvi, S., Cresci, A. (2006) Growth improvement by probiotic in European sea bass juveniles (*Dicentrarchus labrax*, L.), with particular attention to IGF-1, myostatin and cortisol gene expression. *Aquaculture*, 258:430–438.

Castex, M., Chim, L., Pham, D., Lemaire, P., Wabete, N., Nicolas, J.L., Schmidely, P., Mariojouls, C. (2008) Probiotic *P. acidilactici* application in shrimp *Litopenaeus stylirostris* culture subject to vibriosis in New Caledonia. *Aquaculture*, 275:182–193.

Castex, M., Lemaire, P., Wabete, N., Chim, L. (2009) Effect of dietary probiotic *Pediococcus acidilactici* on antioxidant defences and oxidative stress status of shrimp *Litopenaeus stylirostris*. *Aquaculture*, 294:306–313.

Castex, M., Lemaire, P., Wabete, N., Chim, L. (2010) Effect of probiotic *Pediococcus acidilactici* on antioxidant defenses and oxidative stress of *Litopenaeus stylirostris* under *Vibrio nigripulchritudo* challenge. *Fish Shellfish Immunol*, 28:622–631.

Cathers, H.S., Mane, S.P., Tawari, N.R., Balakuntla, J., Plata, G., Krishnamurthy, M., MacDonald, A., Wolter, M., Baxter, N., Briones, J., Nagireddy, A., Millman, G., Martin, R.E., Kumar, A., Gangaiah, D. (2022) *In silico, in vitro* and *in vivo* characterization of host-associated *Latilactobacillus curvatus* strains for potential probiotic applications in farmed Atlantic salmon (*Salmo salar*). *Sci Rep*, 12:18417.

Chang, C.I., Liu, W.Y. (2002) An evaluation of two probiotic bacterial strains, *Enterococcus faecium* SF68 and *Bacillus toyoi*, for reducing edwardsiellosis in cultured European eel, *Anguilla anguilla* L. *J Fish Dis*, 25:311–315.

Chen, J., Sun, R., Pan, C., Sun, Y., Mai, B., Li, Q.X. (2020) Antibiotics and food safety in aquaculture. *J Agric Food Chem*, 68:11908–11919.

Chiu, C.H., Guu, Y.K., Liu, C.H., Pan, T.M., Cheng, W. (2007) Immune responses and gene expression in white shrimp, *Litopenaeus vannamei*, induced by *Lactobacillus plantarum*. *Fish Shellfish Immunol*, 23:364–377.

Chuon, M.R., Shiomoto, M., Koyanagi, T., Sasaki, T., Michihata, T., Chan, S., Mao, S., Enomoto, T. (2014) Microbial and chemical properties of Cambodian traditional fermented fish products. *J Sci Food Agric*, 94:1124–1131.

Contente, D., Díaz-Rosales, P., Feito, J., Díaz-Formoso, L., Docando, F., Simón, R., Borrero, J., Hernández, P.E., Poeta, P., Muñoz-Atienza, E., Cintas, L.M., Tafalla, C. (2023) Immunomodulatory effects of bacteriocinogenic and non-bacteriocinogenic *Lactococcus cremoris* of aquatic origin on rainbow trout (*Oncorhynchus mykiss*, Walbaum). *Front Immunol*, 14:1178462.

Cosansu, S., Mol, S., Ucok Alakavuk, D., Tosun, S.Y. (2011) Effects of *Pediococcus* spp. on the quality of vacuum-packed horse mackerel during cold storage. *Tar Bil Der*, 17:59–66.

Cotter, P.D., Hill, C., Ross, R.P. (2005) Bacteriocins: Developing innate immunity for food. *Nature Rev Microbiol*, 3:777–788.

Dalgaard, P., Vancanneyt, M., Euras Vilalta, N., Swings, J., Fruekilde, P., Leisner, J.J. (2003) Identification of lactic acid bacteria from spoilage associations of cooked and brined shrimps stored under modified atmosphere between 0 degrees C and 25 degrees C. *J Appl Microbiol*, 94:80–89.

Dallagnol, A.M., Pescuma, M., Espínola, N.G., Vera, M., Vignolo, G.M. 2(021) Hydrolysis of raw fish proteins extracts by *Carnobacterium maltaromaticum* strains isolated from Argentinean freshwater fish. *Biotechnol Rep* (Amst), 29:e00589.

Dang, Y., Sun, Y., Zhou, Y., Men, X., Wang, B., Li, B., Ren, Y. (2022) Effects of probiotics on growth, the toll-like receptor mediated immune response and susceptibility to *Aeromonas salmonicida* infection in rainbow trout *Oncorhynchus mykiss*. *Aquaculture*, 561:738668.

Defoirdt, T., Sorgeloos, P., Bossier, P. (2011) Alternatives to antibiotics for the control of bacterial disease in aquaculture. *Curr Opin Microbiol*, 14:251–258.

Dias, J.A.R., Alves, L.L., Barros, F.A.L., Cordeiro, C.A.M., Meneses, J.O., Santos, T B.R., Santos, C.C.M., Paixão, P.E.G., Filho, R.M.N., Martins, M.L., Pereira, S.A., et al. (2022) Comparative effects of using a single strain probiotic and multi-strain probiotic on the productive performance and disease resistance in *Oreochromis niloticus*. *Aquaculture*, 550:737855.

Docando, F., Nuñez-Ortiz, N., Serra, C.R., Arense, P., Enes, P., Oliva-Teles, A., Díaz-Rosales, P., Tafalla, C. (2022) Mucosal and systemic immune effects of *Bacillus subtilis* in rainbow trout (*Oncorhynchus mykiss*). *Fish Shellfish Immunol*, 124:142–155.

do Vale Pereira, G., da Cunha, D.G., Mourino, J.L.P., Rodiles, A., Jaramillo- Torres, A., Merrifield, D.L. (2017) Characterization of microbiota in Arapaima gigas intestine and isolation of potential probiotic bacteria. *J Appl Microbiol*, 123:1298–1311.

Du, Y., Li, H., Shao, J., Wu, T., Xu, W., Hu, X., Chen, J. (2022b). Adhesion and colonization of the probiotic *Lactobacillus plantarum* HC-2 in the intestine of *Litopenaeus vannamei* are associated with bacterial surface proteins. *Front Microbiol*, 13:878874.

Du, Y., Xu, W., Wu, T., Li, H., Hu, X., Chen, J. (2022a). Enhancement of growth, survival, immunity and disease resistance in *Litopenaeus vannamei*, by the probiotic, *Lactobacillus plantarum* Ep-M17. *Fish Shellfish Immunol*, 129:36–51.

Duffes, F., Corre, C., Leroy, F., Dousset, X., Boyaval, P. (1999) Inhibition of *Listeria monocytogenes* by *in situ* produced and semi-purified bacteriocins of *Carnobacterium* spp. on vacuum-packed, refrigerated cold-smoked salmon. *J Food Prot*, 62:1394–1403.

Dyer, F. E. "Microorganisms from Atlantic Cod". J. Fish. Res. Bd. Cen. 7, (3), 128-136, 1947.

Egerton, S., Culloty, S., Whooley, J., Stanton, C., Ross, R.P. (2018) The gut microbiota of marine fish. *Front Microbiol*, 9:873.

Eichmiller, J.J., Hamilton, M.J., Staley, C., Sadowsky, M.J., Sorensen, P.W. (2016) Environment shapes the fecal microbiome of invasive carp species. *Microbiome*, 4:44.

EFSA. (2005a) QPS-Qualified Presumption of Safety micro-organisms in food and feed. *EFSA Scientific Colloquium, Summary Report*, Octubre 2005.

EFSA. (2005b) Opinion of the Scientific Committee on a request from EFSA related to a generic approach to the safety assessment by EFSA of microorganisms used in food/feed and the production of food/feed additives. *EFSA J*, 226:1–12.

EFSA. (2007) Introduction of a Qualified Presumption of Safety (QPS) approach for assessment of selected microorganisms referred to EFSA. *EFSA J*, 587:1–16.

EFSA. (2021) EFSA statement on the requirements for whole genome sequence analysis of microorganisms intentionally used in the food chain. *EFSA J*, 19:e06506.

EFSA BIOHAZ Panel (EFSA Panel on Biological Hazards). (2023) Scientific opinion on the update of the list of qualified presumption of safety (QPS) recommended microorganisms intentionally added to food or feed as notified to EFSA. *EFSA J*, 21:7747.

Ehsannia, S., Ahari, H., Kakoolaki, S., Anvar, S.A., Yousefi, S. (2022) Effects of probiotics on zebrafish model infected with *Aeromonas hydrophila*: Spatial distribution, antimicrobial, and histopathological investigation. *BMC Microbiol*, 22(1):167.

Eissa, E.-S.H., Baghdady, E.S., Gaafar, A.Y., El-Badawi, A.A., Bazina, W.K., Abd Al-Kareem, O.M., Abd El-Hamed, N.N.B. (2022) Assessing the influence of dietary *Pediococcus acidilactici* probiotic supplementation in the feed of European sea bass (*Dicentrarchus labrax* L.) (Linnaeus, 1758) on farm water quality, growth, feed utilization, survival rate, body composition, blood biochemical parameters, and intestinal histology. *Aquac Nutr*, 2022:1–11.

Eissa, N., Wang, H-P., Yao, H., ElGheit, E.A. (2018) Mixed Bacillus species enhance the innate immune response and stress tolerance in yellow perch subjected to hypoxia and air-exposure stress. *Sci Rep*, 8:6891.

El Bassi, L., Hassouna, M., Shinzato, N., Matsui, T. (2009) Biopreservation of refrigerated and vacuum-packed *Dicentrarchus labrax* by lactic acid bacteria. *J Food Sci*, 74:M335–339.

El-Saadony, M.T., Alagawany, M., Patra, A.K., Kar, I., Tiwari, R., Dawood, M.A.O., Dahma, K., Abdel-Latif, H.M.R. (2021) The functionality of probiotics in aquaculture: An overview. *Fish Shellfish Immunol*, 117:36–52.

El Sheikha, A., Ray, R., Montet, D., Panda, S., Worawattanamateekul, W. (2014) African fermented fish products in scope of risks. *Int Food Res J*, 21(2):425.

Elsser-Gravesen, D., Elsser-Gravesen, A. (2014) Biopreservatives. *Adv Biochem Eng Biotechnol*, 143:29–49.

FAO (2004) Risk assessement of *Listeria monocytogenes* in ready-to-eat foods. *Technical Report*. Microbiological Risk Assessment Series, No 5, Rome, Italy, p. 265. ISBN 92-5-105127-5

FAO (2005) Responsible use of antibiotics in aquaculture. *FAO Fisheries Technical Paper*, 469:1–97.

FAO (2022) *The State of World Fisheries and Aquaculture 2022*. Towards Blue Transformation, Rome, FAO.

FAO/WHO (2002) Guidelines for the evaluation of probiotics in food. *Report of Joint FAO/WHO Working Group*. London, Ontario, Canada.: 11 p.

FDA (2018) Generally Recognized as Safe (GRAS). Available online: www.fda.gov/Food/IngredientsPackagingLabeling/GRAS/ (accessed on 6 June 2018).

FEEDAP (2019) Assessment of the application for renewal of authorization of Bactocell (*Pediococcus acidilactici* CNCM I-4622) as a feed additive for all fish and shrimps and its extension of use for all crustaceans. *EFSA J*, 17(4):e05691.

Feito, J., Araújo, C., Gómez-Sala, B., Contente, D., Campanero, C., Arbulu, S., Saralegui, C., Peña, N., Muñoz-Atienza, E., Borrero, J., del Campo, R., Hernandez, P.E., Cintas, L.M. (2022) Antimicrobial activity, molecular typing and *in vitro* safety assessment of *Lactococcus garvieae* isolates from healthy cultured rainbow trout (*Oncorhynchus mykiss*, Walbaum) and rearing environment. *LWT—Food Sci Technol*, 162:113496.

Feito, J., Contente, D., Ponce-Alonso, M., Díaz-Formoso, L., Araújo, C., Peña, N., Borrero, J., Gómez-Sala, B., del Campo, R., Muñoz-Atienza, E., Hernández, P.E., Cintas, L.M. (2022) Draft genome sequence of *Lactococcus lactis* subsp. *cremoris* WA2–67: A promising nisin-producing probiotic strain isolated from the rearing environment of a Spanish rainbow trout (*Oncorhynchus mykiss*, Walbaum) farm. *Microorganisms*, 10(3):521.

Ferguson, R.M., Merrifield, D.L., Harper, G.M., Rawling, M.D., Mustafa, S., Picchietti, S., Balcázar, J.L., Davies, S.J. (2010) The effect of *Pediococcus acidilactici* on the gut microbiota and immune status of on-growing red tilapia (*Oreochromis niloticus*). *J Appl Microbiol*, 109:851–862.

Flefil, N.S., Ezzat, A., Aboseif, A.M., Negm El-Dein, A. (2022) *Lactobacillus*-fermented wheat bran, as an economic fish feed ingredient, enhanced dephytinization, micronutrients bioavailability, and tilapia performance in a biofloc system. *Biocatal Agric Biotechnol*, 45:102521.

Gancel, F., Dzierszinski, F., Tailliez, R. (1997) Identification and characterization of *Lactobacillus* species isolated from fillets of vacuum-packed smoked and salted herring (*Clupea harengus*). *J Appl Microbiol*, 82:722–728.

Gatesoupe, F.J. (2002) Probiotic and formaldehyde treatments of *Artemia nauplii* as food for larval pollack, *Pollachius pollachius*. *Aquaculture*, 212:347–360.

Ghanbari, M., Jami, M., Domig, K.J., Kneifel, W. (2013) Seafood biopreservation by lactic acid bacteria-A review. *LWT-Food Sci Technol*, 54:315–324.

Ghanbari, M., Kneifel, W., Domig, K.J. (2015) A new view of the fish gut microbiome: Advances from next-generation sequencing. *Aquaculture*, 448:464–475.

Ghanbari, M., Rezaei, M., Jami, M., Nazari, R.M. (2009) Isolation and characterization of *Lactobacillus* species from intestinal contents of beluga (*Huso huso*) and Persian sturgeon (*Acipenser persicus*). *Iran J Vet Res*, 10:152–157.

Giri, S.S., Sukumaran, V., Oviya, M. (2013) Potential probiotic *Lactobacillus plantarum* VSG3 improves the growth, immunity, and disease resistance of tropical freshwater fish, *Labeo rohita*. *Fish Shellfish Immunol*, 34:660–666.

Gómez-Sala, B., Herranz, C., Díaz-Freitas, B., Hernández, P.E., Sala, A., Cintas, L.M. (2016) Strategies to increase the hygienic and economic value of fresh fish: Biopreservation using lactic acid bacteria of marine origin. *Int J Food Microbiol*, 223:41–49.

Gómez-Sala, B., Muñoz-Atienza, E., Sánchez, J., Basanta, A., Herranz, C., Hernández, P.E., Cintas, L.M. (2015) Bacteriocin production of Lactic Acid Bacteria isolated from fish, seafood and fish products. *Eur Food Res Technol*, 241:341–356.

González, C.J., Encinas, J.P., García-López, M.L., Otero, A. (2000) Characterization and identification of lactic acid bacteria from freshwater fishes. *Food Microbiol*, 17:383–391.

González, C.J., López-Díaz, T.M., García-López, M.L., Prieto, M., Otero, A. (1999) Bacterial microflora of wild brown trout (*Salmo trutta*), wild pike (*Esox hucius*) and aquacultured rainbow trout (*Oncorhynchus mykiss*). *J Food Pro*, 62:1270–1277.

Gram, L., Dalgaard, P. (2002) Fish spoilage bacteria—problems and solutions. *Curr Opin Biotechnol*, 13:262–266.

Gram, L., Huss, H.H. (1996) Microbiological spoilage of fish and fish products. *Int J Food Microbiol*, 33:121–137.

Han, R., Khan, A., Ling, Z., Wu, Y., Feng, P., Zhou, T., Salama, E.-S., El-Dalatony, M.M., Tian, X., Liu, P., Li, X. (2022) Feed-additive *Limosilactobacillus fermentum* GR-3 reduces arsenic accumulation in *Procambarus clarkii*. *Ecotoxicol Environ Saf*, 231:113216.

Hassan, M.A., Fathallah, M.A., Elzoghby, M.A., Salem, M.G., Helmy, M.S. (2022) Influence of probiotics on water quality in intensified *Litopenaeus vannamei* ponds under minimum-water exchange. *AMB Expr*, 12:22.

Hovda, M.B., Fontanillas, R., McGurk, C., Obach, A., Rosnes, J.T. (2012) Seasonal variations in the intestinal microbiota of farmed Atlantic salmon (*Salmo salar* L.). *Aquac Res*, 43:154–159.

Huang, H.-T., Hu, Y.-F., Lee, B.-H., Huang, C.-Y., Lin, Y.-R., Huang, S.-N., Chen, Y.-Y., Chang, J.-J., Nan, F.-H. (2022) Dietary of *Lactobacillus paracasei* and *Bifidobacterium longum* improve nonspecific immune responses, growth performance, and resistance against *Vibrio parahaemolyticus* in *Penaeus vannamei*. *Fish Shellfish Immunol*, 128:307–315.

Huss, H.H., Ababouch, L., Gram, L. (2003) Assessment and management of seafood safety and quality. *FAO Fish Tech Pap*, 444.

Huy, N., Ngoc, L., Loc, N., Lan, T., Quang, H., Dung, T., et al. (2020), Isolation of *Weissella cibaria* from Pacific White Shrimp (*Litopenaeus vannamei*) gastrointestinal tract and evaluation of Its pathogenic bacterial inhibition, Indian Journal of Science and Technology, 13(10), 1200–12.

Ibarra-Martínez, D., Muñoz-Ortega, M.H., Quintanar-Stephano, A., Martínez-Hernández, S.L., Ávila-Blanco, M.E., Ventura-Juárez, J. (2022) Antibacterial activity of supernatants of *Lactococcus lactis*, *Lactobacillus rhamnosus*, *Pediococcus pentosaceus* and curcumin against *Aeromonas hydrophila*. *In vitro* study. *Vet Res Commun*, 46(2):459–470.

Ibrahim, S.M., Desouky, S.G. (2009) Effect of antimicrobial metabolites produced by lactic acid bacteria (LAB) on quality aspects of frozen tilapia (*Oreochromis niloticus*) fillets. *WJFMS*, 1:40–45.

Iorizzo, M., Albanese, G., Letizia, F., Testa, B., Tremonte, P., Vergalito, F., Lombardi, S.J., Succi, M., Coppola, R., Sorrentino, E. (2022) Probiotic potentiality from versatile *Lactiplantibacillus plantarum* strains as resource to enhance freshwater fish health. *Microorganisms*, 10(2):463.

Iseppi R, Stefani S, de Niederhausern S, Bondi M, Sabia C, Messi P. Characterization of anti-*Listeria monocytogenes* properties of two bacteriocin-producing *Enterococcus mundtii* isolated from fresh fish and seafood. Curr Microbiol. 2019 Sep;76(9):1010-1019.

Jaffrès, E., Lalanne, V., Macé, S., Cornet, J., Cardinal, M., Sérot, T., Dousset, X., Joffraud, J.J. (2011) Sensory characteristics of spoilage and volatile compounds associated with bacteria isolated from cooked and peeled tropical shrimps using SPME-GC-MS analysis. *Int J Food Microbiol*, 147:195–202.

Jérôme, M., Passerini, D., Chevalier, F., Marchand, L., Leroi, F., Macé, S. (2022) Development of a rapid qPCR method to quantify lactic acid bacteria in cold-smoked salmon. *Int J Food Microbiol*, 363:109504.

Joffraud, J.J., Leroi, F., Roy, C., Berdagué, J.L. (2001) Characterisation of volatile compounds produced by bacteria isolated from the spoilage flora of cold-smoked salmon. *Int J Food Microbiol*, 66:175–184.

Kannappan, S., Sivakumar, K., Sivagnanam, S. (2017) Effect of *Lactobacillus rhamnosus* cells against specific and native fish spoilage bacteria and their spoilage indices on Asian seabass fish chunks. *J Environ Biol*, 38:841–847.

Katla, T., Moretro, T., Aasen, I.M., Holck, A., Axelsson, L., Naterstad, K. (2001) Inhibition of *Listeria monocytogenes* in cold smoked salmon by addition of sakacin P and/or live *Lactobacillus sakei* cultures. *Food Microbiol*, 18:431–439.

Khalafalla, M.M., Zayed, N.F.A., Amer, A.A., Soliman, A.A., Zaineldin, A.I., Gewaily, M.S., Hassan, A.M., Van Doan, H., Tapingkae, W., Dawood, M.A O. (2022) Dietary *Lactobacillus acidophilus* ATCC 4356 relieves the impacts of aflatoxin B1 toxicity on the growth performance, hepatorenal functions, and antioxidative capacity of thinlip grey mullet (*Liza ramada*) (Risso 1826). *Probiotics Antimicrob Proteins*, 14:189–203.

Kim, C.R., Hearnsberger, C.O., Eun, J.B. (1994) Gram-negative bacteria in refrigerated catfish fillets treated with lactic culture and lactic acid. *J Food Prot*, 58:639–643.

Kopermsub, P., Yunchalard, S. (2010) Identification of lactic acid bacteria associated with the production of plaa-som, a traditional fermented fish product of Thailand. *Int J Food Microbiol*, 138:200–204.

Kraus, H. (1961) Mitteilung über das Vorkommen von Lactobazillen auf frischen Heringen. *Arch Lebensmittelhyg*, 12:101–102.

Kvasnikov, E.I., Kovalenko, N.K., Materinskaya, L.G. (1977) Lactic acid bacteria of freshwater fish. *Mikrobiologiya*, 46:619–624.

Lalitha, N., Ronald, B.S.M., Chitra, M.A., Hemalatha, S., Senthilkumar, T.M.A., Muralidhar, M. (2022) Characterization of Lactic Acid Bacteria from the gut of *Penaeus vannamei* as potential probiotic. *Indian J Anim Res*.

Larsen, A.M., Mohammed, H.H., Arias, C.R. (2014) Characterization of the gut microbiota of three commercially valuable warmwater fish species. *J Appl Microbiol*, 116:1396–1404.

Laulund, S., Wind, A., Derkx, F., Zuliani, V. (2017) Regulatory and safety requirements for food cultures. *Microorganisms*, 5:28.

Lauzon, H.L., Gudmundsdottir, S., Petursdottir, S.K., Reynisson, E., Steinarsson, A., Oddgeirsson, M., Bjornsdottir, R., Gudmundsdottir, B.K. (2010) Microbiota of Atlantic cod (*Gadus morhua* L.) rearing systems at pre- and posthatch stages and the effect of different treatments. *J Appl Microbiol*, 109:1775–1789.

Lazado, C.C., Caipang, C.M.A., Brinchmann, M.F., Kiron, V. (2011) *In vitro* adherence of two candidate probiotics from Atlantic cod and their interference with the adhesion of two pathogenic bacteria. *Vet Microbiol*, 148:252–259.

Lee, J.-W., Chiu, S.-T., Wang, S.-T., Liao, Y.-C., Chang, H.-T., Ballantyne, R., Lin, J.-S., Liu, C.-H. (2022) Dietary SYNSEA probiotic improves the growth of white shrimp, *Litopenaeus vannamei* and reduces the risk of *Vibrio* infection via improving immunity and intestinal microbiota of shrimp. *Fish Shellfish Immunol*, 127:482–491.

Leroi, F. (2010) Occurrence and role of lactic acid bacteria in seafood products. *Food Microbiol*, 27:698–709.

Leroi, F. (2015) Role of bacteria in seafood products. In *Seafood Science. Advances in Chemistry, Technology and Applications*, Ed. S.-K. Kim, pp. 458–482. Boca Ratón, FL: CRC Press, Taylor and Francis Group.

Leroi, F., Joffraud, J.J., Chevalier, F., Cardinal, M. (1998) Study of the microbial ecology of cold-smoked salmon during storage at 8°C. *Int J Food Microbiol*, 39:111–121.

Leroi, F., Joffraud, J.J., Chevalier, F., Cardinal, M. (2001) Research of quality indices for cold-smoked salmon using a stepwise multiple regression of microbiological counts and physico-chemical parameters. *J Appl Microbiol*, 90:578–587.

Li, M., Xi, B., Qin, T., Xie, J. (2019) Isolation and characterization of AHL-degrading bacteria from fish and pond sediment. *J Ocean Limnol*, 37:1460–1467.

Li, Y., Zhao, Y., Liu, X., Yuan, L., Liu, X., Wang, L., Sun, H. (2022) Effects of endogenous potential probiotic *Lactobacillus rhamnosus* M2–4 on intestinal microflora and metabolomics in juvenile sea cucumber *Apostichopus japonicus*. *Aquaculture*, 555:738247.

Liu, R., Han, G., Li, Z., Cun, S., Hao, B., Zhang, J., Liu, X. (2022) Bacteriophage therapy in aquaculture: Current status and future challenges. *Folia Microbiol*, 67:573–590.

Lopes, E.M., Silva, A.V., Barros, F.A.L., Santos, A.F.L., Cordeiro, C.A.M., Couto, M.V.S., Paixão, P.E.G., Fujimoto, R.Y., Sousa, N.C. (2022) *In vitro* bacterial probiotic selection from *Nannostomus beckfordi*, an Amazon ornamental fish. *Arq Bras Med Vet Zootec*, 74(1):111–116.

Luna, G.M., Quero, G.M., Kokou, F., Kormas, K. (2021) Time to integrate biotechnological approaches into fish gut microbiome research. *Curr Opin Biotechnol*, 73:121–127.

Lyhs, U., Björkroth, J. (2008) *Lactobacillus sakei/curvatus* is the prevailing lactic acid bacterium group in spoiled maatjes herring. *Food Microbiol*, 25:529–533.

Lyhs, U., Björkroth, J., Korkeala, H. (1999) Characterisation of lactic acid bacteria from spoiled, vacuum-packaged, cold-smoked rainbow trout using ribotyping. *Int J Food Microbiol*, 52:77–84.

Lyons, P.P., Turnbull, J.F., Dawson, K.A., Crumlish, M. (2017a) Exploring the microbial diversity of the distal lumen and mucosa of farmed rainbow trout *Oncorhynchus mykiss* (Walbaum) using next generation sequencing (NGS). *Aquacult Res*, 48:77–91.

Lyons, P.P., Turnbull, J.F., Dawson, K.A., Crumlish, M. (2017b) Effects of low-level dietary microalgae supplementation on the distal intestine microbiome of farmed rainbow trout *Oncorhynchus mykiss* (Walbaum). *Aquacult Res*, 48:2438–2452.

Maillet, A., Denojean, P., Bouju-Albert, A., Scaon, E., Leuillet, S., Xavier, D., Jaffr'es, E., Combrisson, J., Prévost, H. (2021) Characterization of bacterial communities of cold smoked salmon during storage. *Foods*, 10:362.

Martínez, M.P., González Pereyra, M.L., Pena, G.A., Poloni, V., Fernandez Juri, G., Cavaglieri, L.R. (2017) *Pediococcus acidolactici* and *Pediococcus pentosaceus* isolated from a rainbow trout ecosystem have probiotic and ABF1 adsorbing/degrading abilities in vitro. *Food Addit Contam Part A Chem Anal Control Expo Risk Assess*, 34:2118–2130.

Matamoros, S., Leroi, F., Cardinal, M., Gigout, F., Kasbi Chadli, F., Cornet, J., Prévost, H., Pilet, M.F. (2009) Psychrotrophic lactic acid bacteria used to improve the safety and quality of vacuum-packaged cooked and peeled tropical shrimp and cold-smoked salmon. *J Food Prot*, 72:365–374.

Mazlumi, A., Panahi, B., Hejazi, M.A., Nami, Y. (2022) Probiotic potential characterization and clustering using unsupervised algorithms of Lactic Acid Bacteria from saltwater fish samples. *Sci Rep*, 12:11952.

Mei J, Shen Y, Liu W, Lan W, Li N, Xie J. Effectiveness of Sodium Alginate Active Coatings Containing Bacteriocin EFL4 for the Quality Improvement of Ready-to-Eat Fresh Salmon Fillets during Cold Storage. Coatings. 2020; 10(6):506. https://doi.org/10.3390/coatings10060506

Melo-Bolívar, J.F., Ruiz Pardo, R.Y., Junca, H., Sidjabat, H.E., Cano-Lozano, J.A., Villamil Díaz, L.M. (2022) Competitive exclusion bacterial culture derived from the gut microbiome of Nile tilapia (*Oreochromis niloticus*) as a resource to efficiently recover probiotic strains: Taxonomic, genomic, and functional proof of concept. *Microorganisms*, 10:1376.

Merrifield, D.L., Balcázar, J.L., Daniels, C., Zhou, Z., Carnevali, O., Sun, Y.Z., Hoseinifar, S.H., Ringø, E. (2015) Indigenous lactic acid bacteria in fish and crustaceans. In *Aquaculture Nutrition: GutHealth, Probiotics and Prebiotics*, Eds. D.L. Merrifield, E. Ringø, pp. 128–168. Hoboken, NJ: Wiley-Blackwell.

Merrifield, D.L., Bradley, G., Harper, G.M., Baker, R.T.M., Munn, C.B., Davies, S.J. (2011) Assessment of the effects of vegetative and lyophilized *Pediococcus acidilactici* on growth, feed utilization, intestinal colonization and health parameters of rainbow trout (*Oncorhynchus mykiss* Walbaum). *Aquacult Nutr*, 17:73–79.

Merrifield, D.L., Dimitroglou, A., Foey, A., Davies, S.J., Baker, R.T.M., Bøgwald, J., Castex, M., Ringø, E. (2010) The current status and future focus of probiotic and prebiotic applications for salmonids. *Aquaculture*, 302:1–18.

Michel, C., Pelletier, C., Boussaha, M., Douet, D.G., Lautraite, A., Taillez, P. (2007) Diversity of lactic acid bacteria associated with fish and the fish farm environment, established by amplified rRNA gene restriction analysis. *Appl Environ Microbiol*, 73:2947–2955.

Minich, J.J, Härer, A., Vechinski, J., Frable, B.W., Skelton, Z.R., Kunselman, E., Shane, M.A., Perry, D.S., Gonzalez, A., McDonald, D., Knight, R., Michael, T.P., Allen, E.E. (2022) Host biology, ecology and the environment influence microbial biomass and diversity in 101 marine fish species. *Nat Commun*, 13:6978.

Mocanu, E.E., Savin, V., Popa, M.D., Dima, F.M. (2022) The effect of probiotics on growth performance, haematological and biochemical profiles in Siberian sturgeon (*Acipenser baerii* Brandt, 1869). *Fishes*, 7:239.

Mohammadi, G., Hafezieh, M., Karimi, A.A., Azra, M.N., van Doan, H., Tapingkae, W., Abdelrahman, H.A., Dawood, M.A.O. (2022) The synergistic effects of plant polysaccharide and *Pediococcus acidilactici* as a synbiotic additive on growth, antioxidant status, immune response, and resistance of Nile tilapia (*Oreochromis niloticus*) against *Aeromonas hydrophila*. *Fish Shellfish Immunol*, 120:304–313.

Mondal, H., Thomas, J. (2022) A review on the recent advances and application of vaccines against fish pathogens in aquaculture. *Aquac Int*, 30:1971–2000.

Moroni, F., Naya-Catalá, F., Piazzon, M.C., Rimoldi, S., Calduch-Giner, J., Giardini, A., Martínez, I., Brambilla, F., Pérez-Sánchez, J., Terova, G. (2021) The effects of nisin-producing *Lactococcus lactis* strain used as probiotic on gilthead sea bream (*Sparus aurata*) growth, gut microbiota, and transcriptional response. *Front Mar Sci*, 8:659519.

Muñoz-Atienza, E., Araújo, C., Magadán, S., Hernández, P.E., Herranz, C., Santos, Y., Cintas, L.M. (2014) *In vitro* and *in vivo* evaluation of lactic acid bacteria of aquatic origin as probiotics for turbot (*Scophthalmus maximus* L.) farming. *Fish Shellfish Immunol*, 41:570–580.

Muñoz-Atienza, E., Gómez-Sala, B., Araújo, C., Campanero, C., del Campo, R., Hernández, P.E., Herranz, C., Cintas, L.M. (2013) Antimicrobial activity, antibiotic susceptibility and virulence factors of lactic acid bacteria of aquatic origin intended for use as probiotics in aquaculture. *BMC Microbiol*, 13:15.

Muñoz-Atienza, E., Landeta, G., de la Rivas, B., Gómez-Sala, B., Muñoz, R., Hernández, P.E., Cintas, L.M., Herranz, C. (2011) Phenotypic and genetic evaluations of biogenic amine production by lactic-acid bacteria isolated from fish and fish products. *Int J Food Microbiol*, 146:212–216.

Naiel, M.A.E., Abdelghany, M.F., Khames, D.K., El-hameed, S.A.A.A., Mansour, M.G.M., El-Nadi, A.S.M., Shoukry, A.A. (2022) Administration of some probiotic strains in the rearing water enhances the water quality, performance, body chemical analysis, antioxidant and immune responses of Nile tilapia, *Oreochromis niloticus*. *Appl Water Sci*, 12:209.

Ndaw, A., Zinedine, A., Faid, M., Bouseta, A. (2008) Effect of controlled lactic acid bacterial fermentation on the microbiological and chemical qualities of Moroccan sardines (*Sardina pilchardus*). *Acta Microbiol Immunol Hung*, 55:295–310.

Nguyen, T.L., Chun, W.-K., Kim, A., Kim, N., Roh, H.J., Lee, Y., Yi, M., Kim, S., Park, C.-I., Kim, D.-H. (2018) Dietary probiotic effect of *Lactococcus lactis* WFLU12 on low-molecular-weight metabolites and growth of olive flounder (*Paralichthys olivaceus*). *Front Microbiol*, 9:2059.

Nikiforov-Nikishin, A., Nikiforov-Nikishin, D., Kochetkov, N., Smorodinskaya, S., Klimov, V. (2022) The influence of probiotics of different microbiological composition on histology of the gastrointestinal tract of juvenile *Oncorhynchus mykiss*. *Microsc Res Tech*, 85(2):538–547.

Nikoskelainen, S., Ouwehand, A.C., Bylund, G., Salminen, S., Lilius, E.M. (2003) Immune enhancement in rainbow trout (*Oncorhynchus mykiss*) by potential probiotic bacteria (*Lactobacillus rhamnosus*). *Fish Shellfish Immunol*, 15:443–452.

Nikoskelainen, S., Ouwehand, A.C., Salminen, S., Bylund, G. (2001) Protection of rainbow trout (*Oncorhynchus mykiss*) from furunculosis by *Lactobacillus rhamnosus*. *Aquaculture*, 198:229–236.

Nilsson, L., Gram, L., Huss, H.H. (1999) Growth control of *Listeria monocytogenes* on cold-smoked salmon using a competitive lactic acid bacteria flora. *J Food Prot*, 62:336–342.

Nilsson, L., Hus, H.H., Gram, L. (1997) Inhibition of *Listeria monocytogenes* on cold-smoked salmon by nisin and carbon dioxide atmosphere. *Int J Food Microbiol*, 38:217–227.

Nilsson, L., Ng, Y.Y., Christiansen, J.N., Jørgensen, B.L., Grótinun, D., Gram, L. (2004) The contribution of bacteriocin to inhibition of *Listeria monocytogenes* by *Carnobacterium piscicola* strains in cold-smoked salmon systems. *J Appl Microbiol*, 96:133–143.

Nimalan, N., Sørensen, S.L., Fečkaninová, A., Koščová, J., Mudroňová, D., Gancarčíková, S., Vatsos, I.N., Bisa, S., Kiron, V., Sørensen, M. (2023) Supplementation of lactic acid bacteria has positive effects on the mucosal health of Atlantic salmon (*Salmo salar*) fed soybean meal. *Aquac Rep*, 28:101461.

Ohshima, T., Giri, A. (2014) Fermented foods-traditional fish fermentation technology and recent developments. In *Encyclopedia of Food Microbiology*, 2nd ed., Eds. C.A. Batt, M.L. Tortorello, pp. 852–869. Oxford: Academic Press.

Padeniya, U., Larson, E.T., Septriani, S., Pataueg, A., Kafui, A.R., Hasan, E., Mmaduakonam, O.S., Kim, G., Kiddane, A.T., Brown, C.L. (2022) Probiotic treatment enhances pre-feeding larval development and early survival in zebrafish *Danio rerio*. *J Aquat Anim Health*, 34(1):3–11.

Panigrahi, A., Kiron, V., Satoh, S., Hirono, I., Kobayashi, T., Sugita, H., Puangkaew, J., Aoki, T. (2007) Immune modulation and expression of cytokine genes in rainbow trout *Oncorhynchus mykiss* upon probiotic feeding. *Dev Comp Immunol*, 31:372–382.

Pereira, W.A., Piazentin, A.C.M., de Oliveira, R.C., Mendonça, C.M.N., Tabata, Y.A., Mendes, M.A., Fock, R.A., Makiyama, E.N., Corrêa, B., Vallejo, M., Villalobos, E.F., de S. Oliveira, R.P. (2022) Bacteriocinogenic probiotic bacteria isolated from an aquatic environment inhibit the growth of food and fish pathogens. *Sci Rep*, 12(1):5530.

Pérez-Sánchez, T., Balcázar, J.L., Merrifield, D.L., Carnevali, O., Gioacchini, G., de Blas, I., Ruiz-Zarzuela, I. (2011) Expression of immune-related genes in rainbow trout (*Oncorhynchus mykiss*) induced by probiotic bacteria during *Lactococcus garvieae* infection. *Fish Shellfish Immunol*, 31:196–201.

Pérez-Sánchez, T., Ruiz-Zarzuela, I., de Blas, I., Balcázar, J.L. (2014) Probiotics in aquaculture: A current assessment. *Rev Aquaculture*, 6:133–146.

Perry, W.B., Lindsay, E., Payne, C.J., Brodie, C., Kazlauskaite, R. (2020) The role of the gut microbiome in sustainable teleost aquaculture. *Proc Biol Sci*, 287:20200184.

Petersen, A., Dalsgaard, A. (2003) Species composition and antimicrobial resistance genes of *Enterococcus* spp., isolated from integrated and traditional fish farms in Thailand. *Environ Microbiol*, 5:395–402.

Pilet, M.-F., Leroi, F. (2011) Applications of protective cultures, bacteriocins and bacteriophages in fresh seafood and seafood products. In *Protective Cultures, Antimicrobial Metabolites and Bacteriophages for Food and Beverage Biopreservation*, Ed. C. Lacroix, pp. 324–347. Cambridge: Woodhead Publishing.

Pillinger, M., Weber, B., Standen, B., Schmid, M.C., Kesselring, J.C. (2022) Multi-strain probiotics show increased protection of intestinal epithelial cells against pathogens in rainbow trout (*Oncorhynchus mykiss*). *Aquaculture*, 560:738487.

Pinto, A.L., Fernandes, M., Pinto, C., Albano, H., Castilho, F., Teixeira, P., Gibbs, P.A. (2009) Characterization of anti-*Listeria* bacteriocins isolated from shellfish: Potential antimicrobials to control non-fermented seafood. *Int J Food Microbiol*, 129:50–58.

Prabawati, E., Hu, S.-Y., Chiu, S.-T., Balantyne, R., Risjani, Y., Liu, C.-H. (2022) A synbiotic containing prebiotic prepared from a by-product of king oyster mushroom, *Pleurotus eryngii* and probiotic, *Lactobacillus plantarum* incorporated in diet to improve the growth performance and health status of white shrimp, *Litopenaeus vannamei*. *Fish Shellfish Immunol*, 120:155–165.

Qaddoori, M.S., Najim, S.M., Al-Niaeem, K.S. (2022) Effects of some probiotics and synbiotic dietary supplementation on growth performance and digestive enzymes activity of common carp, *Cyprinus carpio*. *J Pharm Negat*, 13(7):175–184.

Queiroz, L.L., Hoffmann, C., Lacorte, G.A., de Melo Franco, B.D.G., Todorov, S.D. (2022) Genomic and functional characterization of bacteriocinogenic lactic acid bacteria isolated from Boza, a traditional cereal-based beverage. *Sci Rep*, 12:1460.

Rasmussen, J.A., Villumsen, K.R., Ernst, M., Hansen, M., Forberg, T., Gopalakrishnan, S., Gilbert, M.T.P., Bojesen, A.M., Kristiansen, K., Limborg, M.T. (2022) A multi-omics approach unravels metagenomic and metabolic alterations of a probiotic and synbiotic additive in rainbow trout (*Oncorhynchus mykiss*). *Microbiome*, 10(1):21.

Rather, I.A., Galope, R., Bajpai, V.K., Lim, J., Paek, W.K., Park, Y.-H. (2017) Diversity of marine bacteria and their bacteriocins: Applications in aquaculture. *Rev Fish Sci Aquac*, 25:257–269.

Rathod, N.B., Nirmal, N.P., Pagarkar, A., Özogul, F., Rocha, J.M. (2022) Antimicrobial impacts of microbial metabolites on the preservation of fish and fishery products: A review with current knowledge. *Microorganisms*, 10:773.

Ren, W., Wu, H., Guo, C., Xue, B., Long, H., Zhang, X., Cai, X., Huang, A., Xie, Z. (2021) Multi-strain tropical *Bacillus* spp. as a potential probiotic biocontrol agent for large-scale enhancement of mariculture water quality. *Front Microbiol*, 12:699378.

Rimoldi, S., Terova, G., Ascione, C., Giannico, R., Brambilla, F. (2018) Next generation sequencing for gut microbiome characterization in rainbow trout (Oncorhynchus mykiss) fed animal by-product meals as an alternative to fish meal protein sources. *PLoS ONE*, 13:e0193652.

Ringø, E., Hoseinifar, S.H., Ghosh, K., Doan, H.V., Beck, B.R., Song, S.K. (2018) Lactic acid bacteria in finfish-an update. *Front Microbiol*, 9:1818.

Ringø, E., Li, X., van Doan, H., Ghosh, K. (2022) Interesting probiotic bacteria other than the more widely used lactic acid bacteria and bacilli in finfish. *Front Mar Sci*, 9:848037.

Ringø, E., Olsen, R.E., Øverli, Ø., Løvik, F. (1997) Effect of dominance hierarchy formation on aerobic microbiota associated with epithelial mucosa of subordinate and dominant individuals of Arctic charr, *Salvelinus alpinus* (L.). *Aquaculture Res*, 28:901–904.

Ringø, E., Seppola, M., Berg, A., Olsen, R.E., Schillinger, U., Holzapfel, W. (2002) Characterization of *Carnobacterium divergens* strain 6251 isolated from intestine of Arctic charr (*Salvelinus alpinus* L.). *Syst Appl Microbiol*, 25:120–129.

Ringø, E., van Doan, H., Lee, S.H., Soltani, M., Hoseinifar, S.H., Harikrishnan, R., Song, S.K. (2020) Probiotics, lactic acid bacteria and bacilli: Interesting supplementation for aquaculture. *J Appl Microbiol*, 129(1):116–136.

Sáenz, J.S., Marques, T.V., Barone, R.S.C., Cyrino, J.E.P., Kublik, S., Nesme, J., Schloter, M., Rath, S., Vestergaard, G. (2019) Oral administration of antibiotics increased the potential mobility of bacterial resistance genes in the gut of the fish *Piaractus mesopotamicus*. *Microbiome*, 7:24.

Safari, R., Imanpour, M.R., Hoseinifar, S.H., Faheem, M., Dadar, M., van Doan, H. (2022) Effects of dietary *Lactobacillus casei* on the immune, growth, antioxidant, and reproductive performances in male zebrafish (*Danio rerio*). *Aquac Rep*, 25:101176.

Sarika, A.R., Lipton, A.P., Aishwarya, M.S. (2019) Biopreservative efficacy of bacteriocin GP1 of *Lactobacillus rhamnosus* GP1 on stored fish filets. *Front Nutr*, 6:29.

Sarika, A.R., Lipton, A.P., Aishwarya, M.S., Dhivya, R.S. (2012) Isolation of a bacteriocin-producing *Lactococcus lactis* and application of its bacteriocin to manage spoilage bacteria in high-value marine fish under different storage temperatures. *Appl Biochem Biotechnol*, 167:1280–1289.

Schar, D., Klein, E.Y., Laxminarayan, R., Gilbert, M., van Boeckel, T.P. (2020) Global trends in antimicrobial use in aquaculture. *Sci Rep*, 10:21878.

Sequeiros, C., Garcés, M.E., Fernández, M., Marcos, M., Castaños, C., Moris, M., Olivera, N.L. (2022) Zebrafish intestinal colonization by three Lactic Acid Bacteria isolated from Patagonian fish provides evidence for their possible application as candidate probiotic in aquaculture. *Aquac Int*, 30(3):1389–1405.

Shahid, M., Hussain, B., Riaz, D., Khurshid, M., Ismail, M., and Tariq, M. (2017). Identification and partial characterization of potential probiotic lactic acid bacteria in freshwater *Labeo rohita and Cirrhinus mrigala*. Aquacult. Res. 48, 1688–1698.

Shelby, R.A., Lim, C., Yildirim-Aksoy, M., Delaney, M.A. (2006) Effects of probiotic diet supplements on disease resistance and immune response of young Nile tilapia, *Oreochromis niloticus*. *J Appl Aquaculture*, 18:23–34.

Shelby, R.A., Lim, C., Yildirim-Aksoy, M., Klesius, P.H. (2007) Effects of probiotic bacteria as dietary supplements on growth and disease resistance in young channel catfish, *Ictalurus punctatus*. *J Appl Aquaculture*, 19:81–91.

Sheng, L., Wang, L. (2021) The microbial safety of fish and fish products: Recent advances in understanding its significance, contamination sources, and control strategies. *Compr Rev Food Sci Food Saf*, 20:738–786.

Sica, M.G., Olivera, N.L., Brugnoni, L.I., Marucci, P.L., López-Cazorla, A.C., Cubitto, M.A. (2010) Isolation, identification and antimicrobial activity of lactic acid bacteria from the Bahía Blanca estuary. *Rev Biol Mar Oceanogr*, 45:389–397.

Simón, R., Docando, F., Nuñez-Ortiz, N., Tafalla, C., Díaz-Rosales, P. (2021) Mechanisms used by probiotics to confer pathogen resistance to teleost fish. *Front Immunol*, 12:653025.

Skåra, T., Axelsson, L., Stefánsson, G., Ekstrand, B., Hagen, H., 2015. Fermented and ripened fish products in the northern European countries. *J Ethn FoodsTanamool*, 2:18–24.

Stohr, V., Joffraud, J.J., Cardinal, M., Leroi, F. (2001) Spoilage potential and sensory profile associated with bacteria isolated from cold-smoked salmon. *Food Res Int*, 34:797–806.

Sudhakaran, G., Guru, A., Haridevamuthu, B., Murugan, R., Arshad, A., Arockiaraj, J. (2022) Molecular properties of postbiotics and their role in controlling aquaculture diseases. *Aquac Res*, 53(9):3257–3273.

Sumon, T.A., Husain, M.A., Sumon, M.A.A., Jang, W.J., Abellan, F.G., Sharifuzzaman, S.M., Brown, C.L., Lee, E-W., Kim, C-H., Hasan, M.T. (2022) Functionality and prophylactic role of probiotics in shellfish aquaculture. *Aquac Rep*, 25:101220.

Sylvain, F.É., Holland, A., Bouslama, S., Audet-Gilbert, É., Lavoie, C., Val, A.L., Derome, N. (2020) Fish skin and gut microbiomes show contrasting signatures of host species and habitat. *Appl Environ Microbiol*, 86:e00789–20.

Tahiluddin, A.B., Maribao, I.P., Amlani, M.Q., Sarri, J.H. (2022) A review on spoilage microorganisms in fresh and processed aquatic food products. *Food Bull*, 1:21–36.

Tahiri, I., Desbiens, M., Benech, R., Kheadr, E., Lacroix, C., Thibault, S., Ouellet, D., Fliss, I. (2004) Purification, characterization and amino acid sequencing of divergicin M35: A novel class IIa bacteriocin produced by *Carnobacterium divergens* M35. *Int J Food Microbiol*, 97:123–136.

Tahiri, I., Desbiens, M., Kheadr, E., Lacroix, C., Fliss, I. (2009) Comparison of different application strategies of divergicin M35 for inactivation of *Listeria monocytogenes* in cold-smoked wild salmon. *Food Microbiol*, 26:783–793.

Tan, C., Li, Q., Yang, X., Chen, J., Zhang, Q., Deng, X. (2022) *Lactococcus lactis'* effect on the intestinal microbiota of *Streptococcus agalactiae*-infected zebrafish (*Danio rerio*). *Microbiol Spectr*, 10(5):e01128–22.

Tanasupawat, S., Pakdeeto, A., Namwong, S., Thawai, C., Kudo, T., Itoh, T. (2006) *Lentibacillus halophilus* sp. nov., from fish sauce in Thailand. *Int J Syst Evol Microbiol*, 56:1859–1863.

Taoka, Y., Maeda, H., Jo, J.-Y., Kim, S.-M., Park, S.-I., Yoshikawa, T., Sakata, T. (2006) Use of live and dead probiotic cells in tilapia *Oreochromis niloticus*. *Fisheries Sci*, 72:755–766.

Thapa, N., Pal, J., Tamang, J.P. (2004) Microbial diversity in Ngari, Hentak and Tungtap, fermented fish products of North-East India. *World J Microbiol Biotechnol*, 20:599.

Thompson, J., Weaver, M.A., Lupatsch, I., Shields, R.J., Plummer, S., Coates, C.J., Rowley, A.F. (2022) Antagonistic activity of Lactic Acid Bacteria against pathogenic vibrios and their potential use as probiotics in shrimp (*Penaeus vannamei*) culture. *Front Mar Sci*, 9:807989.

Tilwani, Y.M., Sivagnanavelmurugan, M., Lakra, A.K., Jha, N., Arul, V. (2022) Enhancement of growth, innate immunity, and disease resistance by probiotic *Enterococcus faecium* MC-5 against *Aeromonas hydrophila* in Indian major carp *Cirrhinus mrigala*. *Vet Immunol Immunopathol*, 253:110503.

Todorov, S.D., Popov, I., Weeks, R., Chikindas, M.L. (2022) Use of bacteriocins and bacteriocinogenic beneficial organisms in food products: Benefits, challenges, concerns. *Foods*, 11:3145.

Tomé, E., Gibbs, P.A., Teixeira, P.C. (2008) Growth control of *Listeria innocua* 2030c on vacuum-packaged cold-smoked salmon by lactic acid bacteria. *Int J Food Microbiol*, 121:285–294.

Torrecillas, S., Rivero-Ramírez, F., Izquierdo, M.S., Caballero, M.J., Makol, A., Suarez-Bregua, P., Fernández-Montero, A., Rotllant, J., Montero, D. (2018) Feeding European sea bass (*Dicentrarchus labrax*) juveniles with a functional synbiotic additive (mannan oligosaccharides and *Pediococcus acidilactici*): An effective tool to reduce low fishmeal and fish oil gut health effects? *Fish Shellfish Immunol*, 81:10–20.

van Doan, H., Soltani, M., Ringø, E. (2021) *In vitro* antagonistic effect and in vivo protective efficacy of Gram-positive probiotics versus Gram-negative bacterial pathogens in finfish and shellfish. *Aquaculture*, 540:736581.

Vendrell, D., Balcázar, J.L., de Blas, I., Ruiz-Zarzuela, I., Gironés, O., Luis Múzquiz, J. (2008) Protection of rainbow trout (*Oncorhynchus mykiss*) from lactococcosis by probiotic bacteria. *Comp Immunol Microbiol Infect Dis*, 31:337–345.

Verschuere, L., Rombaut, G., Sorgeloos, P., Verstraete, W. (2000) Probiotic bacteria as biological control agents in aquaculture. *Microbiol Mol Biol Rev*, 64:655–671.

Vescovo, M., Scolari, G., Zacconi, C. (2006) Inhibition of *Listeria innocua* growth by antimicrobial-producing lactic acid cultures in vacuum-packaged cold-smoked salmon. *Food Microbiol*, 23:689–693.

Vieco-Saiz, N., Belguesmia, Y., Raspoet, R., Auclair, E., Gancel, F., Kempf, I., Drider, D. (2019) Benefits and inputs from lactic acid bacteria and their bacteriocins as alternatives to antibiotic growth promoters during food-animal production. *Front Microbiol*, 10:57.

Wang, C., Chuprom, J., Wang, Y., Fu, L. (2020) Beneficial bacteria for aquaculture: Nutrition, bacteriostasis and immunoregulation. *J Appl Microbiol*, 128(1):28–40.

Wang, J., Wang, L., Shi, S., Cao, Y., Feng, J., Liu, C., Zheng, L. (2022) Probiotic coated with glycol chitosan/alginate relieves oxidative damage and gut dysmotility induced by oxytetracycline in zebrafish larvae. *Food Funct*, 13(20):10476–10490.

Wang, T., Tian, X.-L., Xu, X.-B., Li, H., Tian, Y., Ma, Y.-H., Li, X.-F., Li, N., Zhang, T.-T., Sheng, Y.-D., Tang, Q.-X., et al. (2022) Dietary supplementation of probiotics fermented Chinese herbal medicine *Sanguisorba officinalis* cultures enhanced immune response and disease resistance of crucian carp (*Carassius auratus*) against *Aeromonas hydrophila*. *Fish Shellfish Immunol*, 131:682–696.

Wang, Y.-B., Li, J.-R., Lin, J. (2008a) Probiotics in aquaculture: Challenges and outlook. *Aquaculture*, 281:1–4.

Wang, Y.-B., Tian, Z.-Q., Yao, J.-T., Li, W. (2008b) Effect of probiotics, *Enterococcus faecium*, on tilapia (*Oreochromis niloticus*) growth performance and immune response. *Aquaculture*, 277:203–207.

Weiss, A., Hammes, W.P. (2006) Lactic acid bacteria as protective cultures against Listeria spp. on cold-smoked salmon. *Eur Food Res Technol*, 222:343–346.

Welker, T.L., Lim, C. (2011) Use of probiotics in diets of tilapia. *J Aquac Res Development*, S1:014.

Wiernasz, N., Cornet, J., Cardinal, M., Pilet, M.-F., Passerini, D., Leroi, F. (2017) Lactic acid bacteria selection for biopreservation as a part of a hurdle technology approach applied on seafood. *Front Mar Sci*, 4:119.

Wiernasz, N., Leroi, F., Chevalier, F., Cornet, J., Cardinal, M., Rohloff, J., Passerini, D., Skırnisdottir, S., Pilet, M.F. (2020) Salmon gravlax biopreservation with lactic acid bacteria: A polyphasic approach to assessing the impact on organoleptic properties, microbial ecosystem and volatilome composition. *Front Microbiol*, 10:3103.

Won, S., Hamidoghli, A., Choi, W., Bae, J., Jang, W.J., Lee, S., Bai, S.C. (2020) Evaluation of potential probiotics *Bacillus subtilis* WB60, *Pediococcus pentosaceus*, and *Lactococcus lactis* on growth performance, immune response, gut histology and immune-related genes in whiteleg shrimp, *Litopenaeus vannamei*. *Microorganisms*, 8(2):281.

Wu, S., Wang, G., Angert, E., Wang, W., Li, W., Zou, H. (2012) Composition, diversity, and origin of the bacterial community in Grass carp intestine. *PLoS ONE*, 7:e30440.

Wu, Y.-S., Chu, Y.-T., Chen, Y.-Y., Chang, C.-S., Lee, B.-H., Nan, F.-H. (2022) Effects of dietary *Lactobacillus reuteri* and *Pediococcus acidilactici* on the cultured water qualities, the growth and non-specific immune responses of *Penaeus vannamei*. *Fish Shellfish Immunol*, 127:176–186.

Xing, M., Hou, Z., Yuan, J., Liu, Y., Qu, Y., Liu, B. (2013) Taxonomic and functional metagenomic profiling of gastrointestinal tract microbiome of the farmed adult turbot (*Scophthalmus maximus*). *FEMS Microbiol Ecol*, 86:432–443.

Yamamoto, F.Y., Ellis, M., Bowles, P.R., Suehs, B.A., Carvalho, P.L.P.F., Older, C.E., Hume, M.E., Gatlin, D.M. (2022) Dietary supplementation of a commercial prebiotic, probiotic and their combination affected growth performance and transient intestinal microbiota of red drum (*Sciaenops ocellatus* L.). *Animals*, 12(19):2629.

Yamazaki, K., Suzuki, M., Kawai, Y., Inoue, N., Montville, T.J. (2003) Inhibition of *Listeria monocytogenes* in cold-smoked salmon by *Carnobacterium piscicola* CS526 isolated from frozen surimi. *J Food Prot*, 66:1420–1425.

Yamazaki, K., Suzuki, M., Kawai, Y., Inoue, N., Montville, T.J. (2005) Purification and characterization of a novel class IIa bacteriocin, piscicocin CS526, from surimi-associated *Carnobacterium piscicola* CS526. *Appl Environ Microbiol*, 71:554–557.

Yanez-Lemus, F., Moraga, R., Smith, C.T., Aguayo, P., Sánchez-Alonzo, K., García-Cancino, A., Valenzuela, A., Campos, V.L. (2022) Selenium nanoparticle-enriched and potential probiotic, *Lactiplantibacillus plantarum* S14 strain, a diet supplement beneficial for rainbow trout. *Biology*, 11(10):1523.

Yilmaz, S., Yilmaz, E., Dawood, M.A.O., Ringø, E., Ahmadifar, E., Abdel-Latif, H.M.R. (2022) Probiotics, prebiotics, and synbiotics used to control vibriosis in fish: A review. *Aquaculture*, 547:737514.

Yang, P., Hu, H., Liu, Y., Li, Y., Ai, Q., Xu, W., et al. (2018). Dietary stachyose altered the intestinal microbiota profile and improved the intestinal mucosal barrier function of juvenile turbot, *Scophthalmus maximus* L. Aquaculture 486, 98–106.

Zang, J., Xu, Y., Xia, W., Regenstein, J.M. (2020) Quality, functionality, and microbiology of fermented fish: A review. *Crit Rev Food Sci Nutr*, 60:1228–1242.

Zhang, F., Zhou, K., Xie, F., Zhao, Q. (2022) Screening and identification of Lactic Acid Bacteria with antimicrobial abilities for aquaculture pathogens *in vitro*. *Arch Microbiol*, 204(12):689.

Zhu, X.-K., Yang, B.-T., Hao, Z.-P., Li, H.-Z., Cong, W., Kang, Y.-H. (2022) Dietary supplementation with *Weissella cibaria* C-10 and *Bacillus amyloliquefaciens* T-5 enhance immunity against *Aeromonas veronii* infection in crucian carp (*Carassius auratus*). *Microb Pathog*, 167:105559.

# The Use of Probiotics in Nutrition and Disease Prevention in Farm Animals

# 34

Dagmar Mudroňová, Ladislav Strojný, and Jana Štofilová

## 34.1 INTRODUCTION

Probiotics have already found a stable place in livestock nutrition. A high concentration of animals in large-scale farms is associated with a much easier transfer of pathogens between animals and with a higher level of stress, which significantly increases the importance of the preventive use of probiotic preparations. The results of scientific studies, including testing in real farm conditions, have confirmed the validity of their use both in the prevention of infectious diseases and for increasing the body's resistance during periods of stress. In principle, the role of gut microbiota in animals is similar to those in humans. Colonization of the digestive tract of animals begins immediately after birth/hatching, as healthy neonates are born sterile. During the following weeks, there are significant changes in the composition of the intestinal microbiota; therefore the intestinal microbiome is extremely sensitive during this period. The gut ecosystem of adult animals is stable and changes only due to the effects of external factors of an adequate intensity (e.g. long-lasting change of feed, stress, administration of antibiotics). To ensure optimal growth, production, and health of farm animals, the beneficial microbiota of the gastrointestinal ecosystem can be supported by manipulation of the diet and application of probiotic microorganisms (Bomba et al. 2006). Since the composition of the gastrointestinal microbiota varies among animal species, there are several probiotic preparations available on the market designed for individual farm animal species, and many of them are based on the use of autochthonous strains. As in human medicine, probiotic preparations based on lactic acid bacteria are most commonly used for veterinary purposes.

The use of probiotics for farm animals has specificities. First of all, there is an economic aspect. The increased costs spent on probiotic preparations must be compensated, either by reducing animal morbidity and mortality or by improving production parameters. Furthermore, the preparations must be easily applicable to a large number of animals, that is, either in water or in feed. Another specific feature is the livestock/breeding environment with a high concentration of microorganisms compared to companion animals or humans, which represents a high risk, especially for newborns and young individuals with a developing gut microbiome and immune system. This increases the importance of the use of probiotics, especially for this category of animals. In addition, the preventive application of probiotics to farm animals reduces the need to use synthetic drugs, especially antibiotics, thereby reducing the risk of residues in animal products. This increases the quality and safety of food for human consumption and reduces the burden on the environment.

# 34.2 PROBIOTICS FOR PIGS

The first week of life and weaning can be considered two crucial periods in pig breeding. The period immediately after birth is probably the most critical one in the whole life of a pig. In this period significant growth, morphological changes, and maturation of the gastrointestinal tract take place. The microbial colonization of the porcine intestine begins at birth and follows a rapid succession during the neonatal and weaning period. Following the withdrawal of sow's milk, the piglets are highly susceptible to enteric diseases partly as a result of the altered balance between developing beneficial microbiota and the establishment of intestinal bacterial pathogens. The intestinal immune system of the newborn piglet is poorly developed at birth and undergoes a rapid period of expansion and specialization that is not completed before early weaning (Lallés et al. 2007). It is very important to modulate the gut microbiota of piglets at an early age because in this period, diarrheal diseases with high morbidity and mortality rates present an extraordinarily serious health and economic problem.

The period of weaning is stressful because piglets 3–4 weeks of age are separated from the mother and mixed with piglets from other litters. At weaning, there is a drastic change of diet to dry feed, resulting in remodeling of the gut microbiota structure, dysfunction of the gut barrier, indigestion, and malabsorption, predisposing piglets to enteric infections (Guevarra et al. 2018; Tang et al. 2022). Consequently, a decrease in daily weight gain brings risk for growth retardation. Gastrointestinal disturbances include alterations in small intestine architecture and enzyme activities. The postweaning period is associated with a transient-increased mucosal permeability, disturbed absorptive-secretory electrolyte balance, and altered local inflammatory cytokines (Lallés et al. 2007; Tang et al. 2022).

## 34.2.1 Health Benefits and Application

The use of probiotics for farm animals, including pigs, has been widely reported in the literature. It was shown that probiotics improve growth performance, feed conversion efficiency, nutrient utilization, intestinal microbiota, and gut health and regulate the immune system in pigs (Dowarah et al. 2017a). Probiotics in neonatal piglets can be used to support the development of a stable microbiota, to stimulate the immune system, and to prevent diarrheal diseases. During the weaning and postweaning period, probiotics in pigs are used to prevent postweaning diarrhea (PWD) and stimulate growth (Yang et al. 2015). Bacterial species such as *Limosilactobacillus fermentum, Lactobacillus acidophilus, Limosilactobacillus reuteri, Lactiplantibacillus plantarum, Lactobacillus delbrueckii* subsp. *bulgaricus, Lacticaseibacillus casei, Bifidobacterium lactis, Enterococcus faecium, Enterobacterium faecalis*, and *Bacillus subtilis* are most commonly used for probiotic purposes in pigs (Lim and Tan 2009; Liao and Nyachoti 2017). The use of probiotics in the swine farming industry must have these following characteristics: colonization or metabolic activity in the gut, health promotion, industrial applicability, and safety (Zhang et al. 2023). Probiotics are used in all stages of porcine production namely sow herd, nursery, and growing-finishing pigs (Barba-Vidal et al. 2019).

## 34.2.1.1 Use of Probiotics in Modulation of Gut Microbiota

Microbiological homeostasis of the digestive tract ensures proper nutrient utilization, growth, and development of the host. A dynamic shift is confirmed in the gut microbiota of pigs at different ages and growth phases. In general, *Bacteroides*, *Escherichia*, *Clostridium*, *Lactobacillus*, *Fusobacterium*, and *Prevotella* are dominant in piglets before weaning, then *Prevotella* and *Aneriacter* shift to be the predominant genera, with *Fusobacterium*, *Lactobacillus*, and *Miscellaneous* comparatively minor in postweaned pigs (Yang et al. 2021; Luo et al. 2022). During the weaning period, the gut microbiota undergoes dramatic and partly revocable alterations in the first 7 days, leading to shifts in the intestinal environment (Janczyk et al. 2007). The presence of lactobacilli as a constituent of the normal microbiota of the gastrointestinal tract is considered beneficial to the porcine host (Valeriano et al. 2017; Shin et al. 2019). Various studies have indicated that supplementation of probiotics could help to balance the bacterial community in weaned piglets (Shin et al. 2019; Luise et al. 2019). For example, supplementation of weaned piglets by *L. plantarum* JDFM LP11 increases the diversity and richness in the microbial community, which mainly affected the abundance of *Prevotellaceae*, *Erysipelotrichaceae*, *Sphaerochaetaceae*, *Spirochaetaceae*, and *Christensenellaceae* in comparison with non-treated group (Shin et al. 2019).

Furthermore, the microbiota of the digestive system not only play a role in health maintenance but can also influence the production parameters of pigs, such as quality and quantity of meat (Yang et al. 2018). It was reported that systemic use of probiotics affects the gut microbiota composition and may produce polyunsaturated fatty acid (PUFA)-enriched healthy pork via modulating physiochemical properties (Chang et al. 2018; Rybarczyk et al. 2020). However, the dosage of probiotics is crucial and can significantly affect the intestinal microbiota, which can be related to changes in the quality of the muscles of pigs (Rybarczyk et al. 2020).

## 34.2.1.2 Use of Probiotics in Modulation of Gastrointestinal Ecosystem and Diarrheal Disease Prevention in Neonatal Piglets

Probiotics in neonatal piglets can be effectively used to support the development of a stable microbiota and to prevent diarrheal diseases (Su et al. 2022). The mode of inhibitory action of probiotics against pathogens may be mediated by competition for receptors on the gut mucosa, competition for nutrients, the production of antibacterial substances, and stimulation of the immune system (Zhang et al. 2023).

Gastrointestinal diseases are often the major cause of morbidity and mortality in piglets at an early stage of life. Enterotoxigenic *E. coli* (ETEC) are frequently identified as causative agents of diarrheal diseases of infectious etiology in neonatal and weaned pigs (Luppi 2017; Kim et al. 2022). The ability of ETEC to colonize the gut presents the primary and decisive pathogenic factor since it is known for its virulence. For this reason, effective probiotics should be able to prevent the adhesion of pathogens to intestinal mucosa. *In vitro* studies using porcine epithelial cells IPEC-J2 demonstrated inhibitory activity of several probiotic strains, including *L. plantarum*, *L. rhamnosus*, and *Pediococcus pentosaceus*, on adhesion of ETEC (Zhang et al. 2015; Wang et al. 2018; Yin et al. 2020). Accordingly, the protective effect of probiotics against ETEC-induced diarrhea in piglets was confirmed by several reports (Yang et al. 2015; Li et al. 2018; Wang et al. 2019a; Pupa et al. 2022). For example, oral administration of *L. rhamnosus* GG to newborn piglets decreased incidence of diarrhea associated with alteration of gut microbiota and improvement of jejunal permeability and immunologic barrier (Wang et al. 2019b).

Probiotics are most effective in animals during microbiota development or when microbiota stability is impaired. To obtain the highest effect of probiotic preparations, supplementation of piglets shortly after birth is recommended (Vondrusková et al. 2010). Feeding of probiotics to sows before farrowing and during lactation and to neonatal piglets decreased the numbers of pathogenic microorganisms in sow and piglet feces and resulted in the reduced occurrence of digestive disorders and mortality (Kritas et al. 2015; Konieczka et al. 2023). Supplementation of sows' diets by probiotic product containing *Lactobacillus plantarum* B90 and *Saccharomyces cerevisiae* P beneficially altered colonic microbiome diversity (increased abundance of *Catenibacterium*, *Clostridium*, *Gemmiger*, *Blautia*, and *Roseburia* and reduced *Treponema*) and metabolome profiles of offsprings (Zhu et al. 2022b). Similarly, treatment of sows with probiotic combination of *Lactobacillus helveticus* BGRA43, *L. fermentum* BGHI14, and *Streptococcus thermophilus* BGVLJ1–44 influenced the piglets' gut colonization with beneficial bacteria and reduced the number of *Enterobacteriaceae* in litters and reduced occurrence of diarrhea (Veljovič et al. 2017).

### 34.2.1.3   Use of Probiotics in PWD Prevention and Immune Modulation in Pigs during the Weaning and Postweaning Period

The weaning time is a crucial period in the management of piglets. Thus, it is suggested that husbandry practices must be considered a critical piece in the overall strategy of raising weaned piglets without in-feed antibiotics (Jayaraman and Nyachoti 2017). The process of weaning is one of the most stressful events in the pig's life and can contribute to intestinal and immune system dysfunctions that result in reduced pig health, growth, and feed intake, particularly during the first week after weaning (Campbell et al. 2013). The short period after weaning is characterized by a lower food intake and generally by an energy-deficient state. Functional changes in the small intestine, including morphology, intestinal barrier, mucosal immunity, and gut microbiota are associated with weaning, contributing to low weight gains and predisposition to diarrhea (Zheng et al. 2021).

PWD is one of the most frequent causes of heavy economic losses in pig herds. ETEC *coli* strains are generally considered the main cause of diarrhea at weaning and the period immediately thereafter. Members of the genera *Clostridium*, *Lawsonia*, and *Brachyspira* can be also causes of PWD (Vondrusková et al. 2010). Rotaviruses, coronaviruses, and transmissive gastroenteritis viruses are frequently identified as causative viral agents (Song et al. 2006; Thomsson et al. 2008).

The use of probiotics represents an efficacious strategy in PDW prevention (Su et al. 2022). *L. rhamnosus* GG was effective in ameliorating diarrhea in postweaning piglets induced by *E. coli* K88, possibly via modulation of intestinal microbiota, enhancement of intestinal antibody defense, and regulation of production of systemic inflammatory cytokines (Zhang et al. 2010). Probiotic treatment with *Bacillus subtilis* KN-42 had a positive impact on the incidence of diarrhea, particularly within the first 14 days post-weaning, where the effects of probiotics on PWD were comparable with in-feed antibiotics. *B. subtilis* KN-42 supplementation further improved the bacterial diversity of the intestinal environment, increased the relative number of lactobacilli, and reduced the relative amount of *E. coli* in the feces (Hu et al. 2014). Shu et al. (2001) showed that probiotic treatment using *Bifidobacterium lactis* HN019 reduced weaning diarrhea associated with rotavirus and *E. coli* infection in pigs.

Commensal microorganisms and their metabolites contribute to intestinal mucosal immunity, which is of great importance to the health of the host. There is a two-way communication and mutual influence between the host's immunity and intestinal microorganisms (Peng et al. 2021). Probiotics are well documented for their immunomodulatory activity at both mucosal and systemic level (Roselli et al. 2017). Wang et al. (2009) reported that oral administration of *L. fermentum* I5007 can enhance T-cell differentiation and induce ileum cytokine expression, suggesting that this probiotic strain could modulate immune function in weaned piglets. Administration of probiotic *E. faecium* SF68 significantly reduced levels of cytotoxic T cells (CD8+) in the jejunal epithelium of piglets at an age of 8 weeks. A decline in the frequency of β-hemolytic and O141 serovars of *E. coli* observed in the intestinal contents of probiotic piglets suggesting an explanation for the reduction in cytotoxic T-cell populations (Scharek et al. 2005; Scharek et al. 2007). Azizi et al. (2022) reported that 2 months of probiotic supplementation (*Streptococcus faecalis* T-110, *Clostridium butyricum* TO-A, and *Bacillus mesentericus* TO-A) improved the growth of piglets, and the innate and acquired immunity in the liver of piglets manifested by increased phagocytosis of MHC class II+ cells and the populations of CD4+ cells and IgM+ cells in the liver. Application of *C. butyricum* decreased the levels of IL-1β and IL-18 in the intestines and increased the level of IL-10 in pigs infected with ETEC K88. Furthermore, the same strain of *C. butyricum* improved gut barrier by increasing the expression levels of intestinal tight junction proteins (ZO-1, claudin-3, and occludin) (Li et al. 2018). The addition of *C. butyricum* and *E. faecalis* to the diet of weaned piglets can improve growth performance, which may regulate intestinal morphology by activating the TLR4-mediated MyD88-dependent signaling pathway (Wang et al. 2019a). Several studies confirmed protective effects of *E. faecium* administration in weaned piglets challenged with ETEC or *Salmonella typhimmurium*. *E. faecium* was reported to induce overexpression of tight junction proteins in the ileum and decrease the level of TNF-α in the ileal mucosa and IL-1β in plasma (Peng et al. 2019). Lessard et al. (2009) indicated that *Pediococcus acidilactici* and *Saccharomyces cerevisiae* var. *boulardii* may have the potential to modulate establishment of lymphocyte populations and IgA secretion in the gut and to reduce bacterial translocation to mesenteric lymph node after ETEC infection.

## 34.2.2 Growth-Promoting Effects

Soon after the introduction of the antibiotics for therapy of bacterial infections in production animals, a growth-promoting effect was observed, and antibiotics were used as growth-promoting supplements to feed farm animals. The mode of action of antimicrobial growth promoters is still not exactly known, but at least four modes of action have been proposed to explain the improved antibiotic-mediated animal growth: (1) the inhibition of sub-clinical infections, (2) the reduction of growth-depressing microbial metabolites in the intestines, (3) the increase of nutrient availability via the reduction of microbes sharing the nutrients in the intestines, and (4) the improvement of uptake and use of nutrients through thinner polarized epithelium (Rahman et al. 2022). However, the wide use of antibiotics as growth promoters stimulated the emergence of antibiotic-resistant pathogenic bacteria and contamination of the food chain with residues of antibiotics (Roselli et al. 2005; Vondrusková et al. 2010), resulting in negative effects on human health.

Until 2006, antibiotic growth promoters were included into feed for piglets during the period from birth to weaning to improve the composition of gastrointestinal microbiota and to prevent the occurrence of PWD (Sorensen et al. 2009). The use of growth-promoting antibiotics was banned in the European Union (EU) from 2006 in view of reducing antibiotic resistance phenomena in human therapies. Thus, pig producers today face a prohibition of in-feed antimicrobials and have to find safe and effective alternatives. Various natural substances, including probiotics, have been investigated as efficient alternatives to antibiotic growth promoters (Vondrusková et al. 2010).

The available data from studies and applications in pigs clearly indicate that lactic acid bacteria have great potential as alternatives to in-feed antibiotics (Yang et al. 2015). Many authors have reported the growth-stimulating effect of probiotic lactobacilli to pigs (Dowarah et al. 2017a). Recent meta-analysis (Zhu et al. 2022a) concluded that *Lactobacillus* spp.–based probiotic supplementation improved growth performance of piglets by increasing average daily feed intake, average daily gain, and the gain-to-feed ratio in piglets and in parallel modified the intestinal morphology, especially in the jejunum and ileum. Furthermore, it was shown that lactic acid bacteria could be used also in growing-finishing pigs to improve weight gain and feed conversion and in sows to increase litter weight at birth, weaning litter weight, and the number of piglets at weaning (Yang et al. 2015).

Growth promoting effects were confirmed also for probiotic strains such as *Bifidobacterium animalis* JYBR190 (Pang et al. 2022a), *Bifidobacterium lactis* HN19 (Shu et al. 2001), *Pediococcus acidilactici* FT28 (Dowarah et al. 2017b), or *Bacillus subtilis* (Li et al. 2023), suggesting that probiotics can be regarded as a promising alternative to replace antibiotic growth promoters' usage in pig production. However, there are also studies in which the growth-promoting effects of probiotics have not been confirmed (Kreuzer et al. 2012; Liao and Nyachoti 2017). Optimization of dose and selection of effective probiotic strains, together with consideration of animal age and health status, therefore, remains an important factor in their future application for growth promotion in pigs.

## 34.2.3 Potentiated Probiotics for Pig Industry

If probiotics are to represent a real and effective alternative to antibiotics, it is absolutely necessary to ensure their consistent high effectiveness. The efficacy of probiotics may be potentiated by several methods: the selection of more efficient strains, genetic manipulation, the combination of several strains, and the combination of probiotics and synergistically acting components of a natural origin.

Synbiotics are the combination of both probiotics and prebiotics, which stimulate the growth and/or the activities of both exogenous (probiotic) and endogenous bacteria. Synbiotics seem to be preparations whose potentiated protective and stimulating effects occur in the colon. Potentiated probiotics are defined as biopreparations containing production strains of microorganisms and synergistically acting components of natural origin that potentiate the probiotic effect on both the small intestine and the colon and their beneficial effect on the host by intensifying a mechanism or by extending the range of their probiotic action (Bomba et al. 2002). It seems that to potentiate the effect of probiotics, a number of suitable components may be used such as oligosaccharides, maltodextrin, plants and their extracts, and polyunsaturated fatty acids (PUFAs).

The preventive administration of *L. casei* in combination with maltodextrin during the first week of life decreased the number of *E. coli* colonizing jejunal mucosa in gnotobiotic as well in conventional pigs compared to administration of *L. casei* alone (Bomba et al. 1999). Nemcová et al. (2007) showed that the combination of *L. plantarum*, maltodextrin, and fructooligosaccharides (FOS) proved the most effective one to inhibit the counts of *E. coli* O8:K88 adhering to the intestinal mucosa of the jejunum and colon of conventional piglets at the age of 7 days in comparison with a combination of *L. plantarum* and FOS, a combination of *L. plantarum* and maltodextrin, and *L. plantarum* applied alone. Similarly, prebiotic oligosaccharide lactulose in combination with probiotic strains of *L. plantarum* significantly alleviated postweaning colibacillosis in pigs induced by ETEC K88 oral challenge (Guerra-Ordaz et al. 2014).

Improvement in the colonization of the intestinal mucosa by probiotic bacteria enhances the inhibitory effect of probiotics upon the adhesion of pathogens. It was demonstrated that dietary PUFAs influence the adhesion and growth of probiotic microorganisms. PUFAs modified the adhesion sites on Caco-2 cells, suggesting that dietary PUFAs affect the attachment sites for the gastrointestinal microbiota, possibly by modifying the fatty acid composition of the intestinal wall (Kankaanpaa et al. 2001). The number of *L. paracasei* adhered to jejunal mucosa in the gnotobiotic piglets orally inoculated with seal oil was significantly higher in comparison with the control group (Bomba et al. 2003). The results obtained in another study suggest that the stimulatory effect of PUFAs from flax-seed oil on *L. plantarum* adhesion resulted in enhancement of the inhibitory effect of lactobacilli on *E. coli* K88 in the digestive tract of piglets (Nemcová et al. 2012). It was shown that probiotic *L. plantarum* in combination with flax-seed oil rich in n-3 PUFAs has anti-inflammatory properties, stimulates Th1-mediated cell immunity and phagocytosis, and tends to regulate the inflammatory response induced by ETEC (Chytilová et al. 2013). It was also confirmed that the application of *L. plantarum* and flaxseed oil can modulate mRNA levels of some TLRs, their main adaptor molecule MyD88, and transcription factor NF-κB in various ways and thus regulate pathogen-induced inflammation in the jejunum of gnotobiotic pigs (Chytilová et al. 2014).

Probiotics may represent an attractive alternative to antibiotics and current research should be aimed at improving their efficacy. Regarding probiotics, a combined approach, possibly with other products, will also be needed for use in post-weaning pigs to overcome the complete loss of antimicrobials.

# 34.3 PROBIOTICS FOR RUMINANTS

Ruminants are the major consumer of feed probiotics, with a share of over 40% in the global market. Direct-fed microbials (DFMs) are dietary supplements used primarily in young ruminants to inhibit infection and accelerate establishment of the intestinal microbiota to promote gut health. Further advancement led to more sophisticated probiotic feeds that were targeted at improving fiber digestion and reducing ruminal acidosis in mature cattle, aiming improvements in milk yield, growth, and feed efficiency. Microorganisms that are used in DFM for ruminants are classified as lactic producing bacteria including species of *Lactobacillus*, *Bifidobacterium*, *Enterococcus*, and *Streptococcus* and lactic-utilizing bacteria including species *Megasphaera* and *Propionbacterium*, other bacteria *Prevotella* and *Bacillus*, yeast *Sacharomyces*, and fungus *Aspergillus* (Seo et al. 2010).

The culture of a yeast-fermented product that contains live and dead yeast cells and the spent culture medium containing the metabolites produced by the yeast during fermentation are widely used as feed additives in livestock farming (Pang et al. 2022b). When fed to cattle, yeast cultures have been shown to stimulate cellulolytic bacteria in the rumen, improve fiber digestion, and stabilize rumen pH (Rossi et al. 2006). Yeast also provides vitamins to support the growth of rumen fungi (Hong et al. 2005). Most of these have been shown to be the most active in the lower gut of a ruminant.

Regulatory requirements on application of probiotics in feed of cattle have limited the microbial species that are recognized as safe, such as lactic acid–producing bacteria (e.g., *Lactobacillus*, *Bifidobacteria*, *Enterococcus* spp.), fungi (*Aspergillus oryzae*) or yeast (e.g., *S. cerevisiae*), and *Bacillus* spores (e.g., *B. subtilis* and *B. lichenformis*). One of the other options the use beneficial bacteria is *E. faecium* EF 9296 use as silage inoculations had protect against especially listerial contamination in silages (Marciňáková et al. 2008). The benefits of dietary administration of probiotics and prebiotics on the microbial ecosystem of the gastrointestinal tract of ruminants were summarized by Uyeno et al. (2015).

## 34.3.1 Health Benefits and Applications

The application of probiotics in cattle is recommended in neonatal calves, during the postweaning period, in time of stress, postpartum, and in disease-induced changes in metabolism (subacute and acute acidosis, methane production in rumen).

Probiotics are extensively used in ruminants to stabilize rumen pH and help in various other important functions that are required for their well-being, milk production, and breeding and produce beneficial enzymes and improve nutrient availability and uptake. The use of probiotics in ruminants is more complicated and often depends on whether the target is to combat acidosis, alter the feed-to-weight conversion, reduce the incidence of disease, or decrease methane production (Krehbiel et al. 2003).

Recently, microorganisms that are not part of the natural microbiota in ruminants have been introduced. This includes *Bacillus* spores such as *B. subtilis* and *B. lichenformis*, which are not normal inhabitants of gastrointestinal tract and have been suggested to stimulate the immune system as they are perceived as potentially unharmful (Riddell et al. 2010).

The general beneficial effects of lactic acid bacteria include detoxification of harmful metabolites to improve animal health; displacement of pathogen bacteria from the gut wall through competitive inhibition for attachment, adhesion, or colonization sites in the digestive tract; stimulation of local immune response in the gut; production of B vitamins; production of lactic acid; and creating hostile conditions for pathogenic bacteria.

The discovery of antibiotics after World War II decreased the interest in probiotics, but they were still used to reestablish the intestinal microbiota following aggressive antibiotics treatments (Chiquette 2009).

The use of antibiotics in animal production may contribute to the emergence of antibiotic-resistant bacteria from cattle industry; the establishment of new technologies alternative to antimicrobial agents is strongly needed.

Probiotics are candidates for antibiotic alternatives (Callaway et al. 2004). Antibiotics destroy undesirable but also beneficial microorganisms and expose the rumen to undesirable metabolic changes. Probiotics are employed in cattle production to reduce the use of antibiotics in neonatal and stressed calves, enhance milk production, prevent ruminal acidosis, improve the feed conversion ratio (FCR), enhance the competitive exclusion towards enteropathogens, and rapidly establish a stable microbiota in neonatal calves (Krehbiel et al. 2003).

Chaucheyras-Durand and Fonty (2002) used probiotic yeast, *S. cerevisiae* I-1077, on microbial colonization of the rumen of newborn lambs. This probiotic may be able to accelerate the functionality and/or improve the stability of the rumen ecosystem in young ruminants.

### 34.3.1.1 Use of Probiotics in Neonatal Diarrhea in Calves and Lambs

Inclusion of probiotics in the diet of young calves has been shown to improve performance characteristics, including body weight gain (BWG) and feed conversion as well as average daily gain in the first 2 weeks of life. Primarily in young ruminants, application of probiotics accelerates establishment of the intestinal microbiota involved in feed digestion and promotes gut health.

Gastrointestinal disorders, including diarrhea, are one of the leading causes of mortality and morbidity in neonatal calves, and a reduction in their incidence and duration of diarrhea has been reported in calves consuming probiotics (Kawakami et al. 2010). Neonatal diarrhea is the main cause of calf death and a serious economic problem in the cattle industry. *E. coli*, salmonellae, rotaviruses, and coronaviruses as well as cryptosporidia play an important role in the etiology of diarrheal syndrome in calves. In the young, enterotoxigenic *E. coli* appear to be the most frequent diarrhea-causing agent. The first days following birth and the weaning period are two critical periods where calves have been shown to benefit from probiotic addition to their feed. In the neonate, the microbial population of the gastrointestinal tract is in transition and extremely sensitive (Nousiainen et al. 2004).

Probiotics have been used to decrease diarrhea occurrence in many species. Timmerman et al. (2005) conducted an experiment comparing the difference between multispecies probiotics in milk ruminants and found that they reduced the incidence of diarrhea in veal calves. Magalhaes et al. (2008) found the addition of yeast to calf starter significantly improved fecal scores, along with decreasing mortality rates in calves experiencing high incidence of diarrhea.

Ewaschuk et al. (2004) used LGG isolated from human intestine to maintain viability in the gastrointestinal tract of calves in prevention of diarrhea. This probiotic has been shown to be resistant to acid and bile, have strong adhesive properties to human and rabbit intestinal mucosal cells, suppress bacterial enzyme activity, and produce antimicrobial substances (Lee et al. 2000). Results showed that LGG survives intestinal transit in the young calf, produces no d-lactate and can be administered in an oral rehydration solution. The application of *L. casei* to calves during the first 3 days of age decreased the morbidity and therapy expenses by more than 30% and mortality by more than 50% (Bomba et al. 2006).

Neonate calves are often stressed. The stress leads to scours or diarrhea and weight loss. The other stressors include weaning, transport, vaccination, castration dehorning, high temperatures, and new environment. The probiotic can reduce health problems in young calves. Stressed calves that experience diarrhea have a lower population of lactobacilli in their intestinal tract.

The inhibition effect on diarrhea in calves from 4 to 12 weeks raised in sub-tropical summer was significantly reduced by dietary supplementation of *B. amyloliquefaciens* strain H57 (Le et al. 2016). Stress in animals causing dysbiosis or microbial imbalance in the gastrointestinal tract may be needed for the probiotic to benefit calf health.

### 34.3.1.2 Probiotics and Reduction of E. coli O157:H7

The probiotics that enhance immunoglobulin levels have more positive effect on growth performance, production, and ability to resist disease.

Feedlot cattle have been recognized as a host for *E. coli* O157:H7. This zoonotic pathogen (producing Shiga–toxin) causes hemorrhagic diarrhea and hemolytic uremic syndrome, which can result in acute kidney failure in children (Karmali et al. 2010). Contamination of animal products from infected animals with this pathogen is a serious public health issue. Very few studies have looked at *E. coli* enterotoxemia of food-borne pathogens from feedlot cattle (Tkalcic et al. 2003; Zhao et al. 2003). Brashears et al. (2003a, 2003b) demonstrated a significant reduction in the fecal shedding and carcass contamination of feedlot cattle with *E. coli* O157:H7 using a monostrain probiotic (*L. acidophilus*, NPC 747). Peterson et al. (2007)

reported that *L. acidophilus* strains reduced the shedding of *E. coli* O157:H7 in cattle. In a meta-analysis study, the combination of *L. acidophilus* and *P. freudenreichii* was the most effective probiotic treatment in a dose of $10^9$ CFU/animal/day, reducing the off *E. coli* O157:H57 (Wisener et al. 2014).

### 34.3.1.3 Use of Probiotics in Prevention of Ruminal Acidosis

A probiotic for ruminant animals is a biological preparation that is designed to promote beneficial rumen microbes and to stabilize rumen conditions. The rumen is a complex ecosystem that plays a major role in feed digestion. In adult animals, its volume is about 100 liters, and it harbors bacteria ($10^{11}$ cells/ml), protozoa ($10^5$ cells/ml), fungi ($10^3$ cells/ml), and methanogens ($10^9$ cells/ml) (Chiquette 2009). The rumen microorganisms play a key role in efficient digestion. It is through the rumen that the animal starts deriving energy, protein, and mineral nutrition from pasture and feed. The process is dominated by microorganisms. Ingestion of solid feeds stimulates rumen microbial growth and production of VFA, while calves receiving a liquid diet of milk or milk replacer have minimal development (Heinrichs and Lesmeister 2005).

Ruminal acidosis can be classified clinically as subacute and acute acidosis. In subacute acidosis, elevated levels of short-chain or branched-chain fatty acids, but not lactates, accumulate in the rumen. Acute acidosis consists of high levels of lactic acids in the rumen. Some bacteria with very specific functions in the rumen such as *Butyrivibrio fibrosolvens*, which produce conjugated linoleic acids from linoleic acid, have been proposed as probiotics for ruminants (Fukuda et al. 2006).

Lactic acid bacteria may be useful in reducing ruminal acidosis. A strain of *L. acidophilus* has been shown to reduce total D/L lactate levels and sustain a ruminal pH of 6.0 (Krehbiel et al. 2003). The microbial changes associated with low ruminal pH are an increase in the number of pH-tolerant bacteria (*S. bovis*—lactate producer and *Megasphaera elsdenii*—lactate user). When ruminal pH is below 6.0, the activity of cellulolytic bacteria is seriously decreased, and the number of protozoa declines. Ruminal acidosis can be defined as a low ruminal pH of below 5.6 and high ruminal VFA concentrations (Collins et al. 2009). LAB, including members of genera *Lactobacillus* and *Enterococcus*, are the most extensively studied as probiotics on the pH-stabilizing effect in the rumen (Nocek and Kautz 2006). Production of bacteriocins by some probiotic bacteria (such as *E. faecium*) allows them to control the growth of certain pathogens in the rumen.

Fungi (*A. oryzae*) and yeasts (*S. cerevisiae*) are used as DFM in cattle. It seems to be effective in preventing ruminal acidosis in cattle. Application of live yeasts in feed to cattle can have an effect on the host by removing oxygen in the rumen and boosting the anaerobic environment for anaerobes and subsequently increase the number of lactate-utilizing bacteria that are believed to control acidosis (Collins et al. 2009).

### 34.3.1.4 Use of Probiotics in Control of Methane Production

Methanogens in the rumen convert carbon dioxide into methane by reduction with hydrogen. Methanogens play an important role in the rumen by actively scavenging hydrogen, which is detrimental to rumen digestion (Takahashi et al. 2005).

Different nutritional components and specific rumen microorganisms influence methane production in the rumen. Acetate, carbon dioxide, and hydrogen are the major methane precursors in the rumen, and these metabolites are produced mainly by the breakdown of carbohydrates, in particular cellulose. Therefore, rumen metabolism and subsequent methane production is expressed as the sum of all the different metabolisms depending on the level of carbohydrates provided (Collins et al. 2009). *In vitro* data using twin strains of *S. cerevisiae* demonstrated a small reduction in methane production after 24 h incubation of the yeast in mixed rumen fermentation vessels (Lila et al. 2004).

## 34.3.2 Probiotics and Production of Milk

Probiotics can improve the milk yield in dairy animals. Chiquette et al. (2008) reported increased production of fermentation and milk fat percentage when a newly isolated bacterial strain (*Prevotella bryantii* 25A) was fed to dairy cows from 3 weeks prepartum to 7 weeks postpartum. Stein et al. (2006) reported an 8.5% increase in 4% fat corrected milk in cows receiving $6 \times 10^{10}$ *Propionibacterium*/day from 2 weeks prepartum to 30 weeks postpartum.

Dietary supplementation with a combination of *L. acidophilus* NP51 and *Propionibacterium freudenreichii* NP24 ($4 \times 10^9$) CFU/animal/day) resulted in a 7.6% increase in average daily milk yield in Holstein cows (Boyd et al. 2011).

Maamouri et al. (2014), in feed trials, showed the effect of a probiotic feed supplement containing *Sacharomyces cerevisiae* on milk yield and its composition in Holstein Friesian cows. The supplementation with 2.5 g *S. cerevisiae* ($2.5 \times 10^{10}$ CFU/day) increased milk production by 1.1 kg/cow.

A meta-analysis by Poppy et al. (2012) concluded that commercial probiotics containing *S. cerevisiae* increased milk yield by 1.18 kg/day, fat-corrected milk by 1.61 kg/day, and energy-corrected milk by 1.65 kg/day.

Similarly, dietary supplementation of *S. cerevisiae* increased milk fat yield by 0.06 kg/day and milk protein yield by 0.03 kg/day. Dietary milk intake was increased by 0.62 kg/day during early lactation and 0.78 kg/day during late lactation.

# 34.4 PROBIOTICS FOR POULTRY

The poultry industry during the past two decades has been one of the most dynamic and expanding animal husbandry sectors in the world. Raising healthy animals is very demanding, because this process is affected by many factors. In large-scale rearing facilities, where poultry are exposed to stressful conditions, problems related to enteric diseases result in loss of productivity, increased mortality in flocks, and potential contamination of poultry products, which leads to human food safety concerns. To minimize health problems and enhance poultry production, synthetic hormones and antibiotics have been extensively used in the past. However, as with other farm animal species, the use of hormones and the preventive application of antibiotics was banned in poultry. Therefore, probiotics, as well as a wide range of substances of natural origin, are currently intensively studied as feed additives for better and safer poultry production. However, it is difficult to directly assess different studies using probiotics because the efficacy of a probiotic application depends on many factors: species composition and viability of probiotic, administration level, application method (e.g., spraying, feed, or water), frequency of application (e.g., once, intermittent, or continuous), overall diet, bird age, overall farm hygiene, and environmental stress factors (e.g., temperature, stocking density) (Patterson and Burkholder 2003).

## 34.4.1 Effect on Modulation of Intestinal Microbiota and Health of Poultry

The autochthonous microbiota is considered to act as one of the body's natural defenses and consists of the population of mostly nonpathogenic bacteria normally residing in the GIT. This population is considered to play an important role in the development of "colonization resistance" against potential pathogens. Many studies reported that probiotics which contain lactic acid bacteria, *Bacillus*, and *Saccharomyces* can be used as strategic tools for managing autochthonous microbial populations and inhibition of pathogens. The beneficial effect of probiotics is based on their ability to modify the gut microbiota and suppress pathogenic bacteria through various mechanisms such as competitive exclusion, microbial antagonism, and immune modulation (Rijkers et al. 2010).

Supplementing the diet of chicks and ducks with one dose of *L. fermentum* increased the DNA copies of *Lactobacillus* spp. and *Firmicutes* in the feces, whereas the population of *Bacteroidetes* remained stable or slightly decreased (Angelakis and Raoult 2010). Strompfova et al. (2005) demonstrated that the 4-day application of *L. fermentum* significantly increased the population of LAB in feces and cecum of quail and significantly decreased the counts of *E. coli* in feces. A mixture of *L. agilis* and *L. reuteri or L. salivarius* subsp. *salicinius* significantly increased the presence and diversity of lactic acid bacteria in the chicken jejunum and cecum (Chen et al. 2017; Lan et al. 2004). The probiotic strain *Enterococcus faecium* DSM 7134 significantly increased *Lactobacillus* counts and decreased *E. coli* counts in feces compared with hens that were fed diets without probiotics (Park et al. 2016). Supplementing the diet of broiler chicken with *B. subtilis* C-3102 significantly increased *Lactobacillus* counts in the cecum, ileum, and feces, as well as reducing *E. coli*, *Clostridium perfringens*, and *Salmonella* in the large intestine and feces, compared to a control (Jeong and Kim 2014). Wu et al. (2009) reported that a diet containing a dried *B. subtilis* culture (DBSC) at 250 mg/kg significantly improved the cecal ecosystem of goslings by increasing the *Lactobacillus* and *Bifidobacterium* population and VFA concentration. The application of *S. cerevisiae* at 0.4% or 0.8% into laying hen diets increased lactobacilli counts and reduced bacterial levels of *E. coli*, *Klebsiella* sp., *Staphylococcus* sp., *Micrococcus* sp., *Campylobacter* sp., and *C. perfringens* (Hassanein and Soliman 2010). A probiotic product containing *L. reuteri*, *E. faecium*, *B. animalis*, *P. acidilactici*, and *L. salivarius* resulted in a beneficial modulation of the cecal microbiota, as evidenced by the significant increases in the concentrations of lactic acid bacteria compared with the control and antibiotic treatment (Mountzouris et al. 2007). In addition, the same product was effective at reducing *Salmonella enteritidis* in *Salmonella enteritidis*-challenged broilers (Mountzouris et al. 2009).

The gastrointestinal tract (GIT) of newly hatched chickens is free of microorganisms. The first inoculum originating from the eggshell and environment impacts the further colonization of microbiota and simultaneously the functional development of the intestinal tissue (Apajalahti et al. 2004). Around day 14 of age, a presumably more stable microbiota is not yet established (Torok et al. 2009); however, immunological development of the small intestine has already occurred (Schokker et al. 2009). According to Lee et al. (2010), it takes 2–4 weeks to establish the microbial consortium in the gut of chickens. During the period of microbial colonization of the GIT, the chickens are exposed to the risk of being colonized by pathogens.

The majority of poultry competitive exclusion experiments have focused on the two main zoonoses in which poultry are a major reservoir of *Salmonella* spp. and *Campylobacter* spp. However, recent work has focused on emerging or opportunistic pathogens found in poultry, with, for example, *C. perfringens*, the causative agent of necrotic enteritis in poultry, and *E. coli* O78:K80, the cause of avian colibacillosis and the protozoa *Eimeria* causing coccidiosis.

Since Nurmi and Rantala (1973) first applied the concept of competitive exclusion in poultry to protect chickens against *Salmonella* infection by inoculating them with microbiota from adult birds, numerous studies have demonstrated that the competitive exclusion effects of probiotics can protect hosts against pathogens such as *S. typhimurium*, *S. gallinarum*, *C. jejuni*, *C. perfringens*, *E. coli* O157:H7, and *E. coli* O78:K80 (Timmerman et al. 2004; Casey et al. 2007; Zhang et al. 2007). Many of the competitive exclusion treatments are either defined or undefined cecal contents or multiple bacterial species derived from cecal contents. Some research has focused on early prevention of colonization, which can help to reduce pathogen populations; other studies have demonstrated long-term colonization benefits of direct-fed microbial (DFM) products. Single-strain probiotics, *L. reuteri*, *L. salivarius* CTC2197, *L. johnsonii* FI9785, *L. acidophilus*, *E. faecium*, *B. subtilis*, and *B. longum* PCB 133 were shown to decrease the colonization of chicks and turkeys by *Salmonella*, *E. coli*, *C. jejuni*, and *C. perfringens* (La Ragione et al. 2001, 2004; La Ragione and Woodward 2003; Santini et al. 2010). Vicente et al. (2008) and Higgins et al. (2008) both reported the use of commercial *Lactobacillus* preparation FM-B11 to inhibit *Salmonella enteritidis* in commercial broilers. Willis and Reid (2008) showed that *C. jejuni* was present at a lower level in broiler chickens fed with a standard diet supplemented with a probiotic formulation containing *L. acidophilus*, *L. casei*, *B. thermophilus*, and *E. faecium* with respect to a control. Different *Lactobacillus* strains have been demonstrated to have a protective effect on raw chicken meat against *L. monocytogenes* and *S. enteritidis* (Maragkoudakis et al. 2009). Some studies have showed that feeding broiler chickens with *Pediococcus*, *Enterococcus*, *Bifidobacterium*, *Lactobacillus*, and *Saccharomyces*-based probiotics decreased the population densities of *Eimeria* (Dalloul et al. 2003; Lee et al. 2007; Giannenas et al. 2014; Ritzi et al. 2014).

Different probiotic microorganisms are known to stimulate the immune system of chickens (Brisbin et al. 2008; Seidavi et al. 2017; Fathi et al. 2017; Mohsin et al. 2022). Many authors investigated the underlying immunological mechanisms of the action of probiotics against colonization of the chicken intestine by pathogens. Koenen et al. (2004) reported increases in total IgG and IgM titers in chickens receiving *L. plantarum* and *L. paracasei* and observed increased phagocytic activity of gut-associated immunity cells toward *Salmonella*. Supplementing the diet of chickens with *L. acidophilus, B. bifidum*, and *E. faecalis* is associated with changes in cytokine expression, particularly IFN-γ and IL-12, in the gut-associated lymphoid tissue and this correlates with protection against colonization of *S. typhimurium* (Haghighi et al. 2008). Likewise, Dalloul et al. (2005) suggested a positive impact of the probiotic in stimulation of some of the early immune responses against *Eimeria acervulina*, as characterized by early IFN-γ and IL-2 secretions, resulting in improved local immune defenses against coccidiosis. *L. reuteri, L. salivarius, L. acidophilus*, and cecal microbiota supplied by oral gavage to chicks, challenged or not with *S. enteritidis*, demonstrated through immunohistochemistry the capacity to stimulate the immune system in the form of leukocytic infiltrate by the CD3$^+$, CD4$^+$, and CD8$^+$ lymphocytes in the intestinal epithelium and in the intestinal *lamina propria* of chicks (Noujaim et al. 2008). Similar results were obtained by Vervelde et al. (1998) who identified, also by immunohistochemistry, a great quantity of leukocytic infiltrate constituted by CD3$^+$ lymphocytes, principally of CD4$^+$ and CD8$^+$ cells in the epithelium and in the *lamina propria* of chick intestine, 7 days after the treatment realized with a mixture of recombinant antigen of *Eimeria* and choleric toxin. Likewise, Choi et al. (1999) described the alterations of subpopulations of T cells, among them the CD4$^+$, CD8$^+$, TCR1, and TCR2 lymphocytes, as well as the transcription of IFN-γ and TGF-β4 mRNA in the intestine of chicks, after oral inoculation of *E. acervulina*.

Akbari et al. (2008) demonstrated that infection of young chicks with *S. typhimurium* significantly increased the expression of several of the antimicrobial peptide genes in cecal tonsils. Furthermore, when chickens were treated with probiotics before *Salmonella* infection, the expression of avian β-defensin and cathelicidin genes was reduced to levels comparable with those seen in the negative control group. Probiotic supplementation resulted in increased antibody titers against Newcastle disease virus and infectious bursal disease virus (Haghighi et al. 2005).

## 34.4.2  Effects on Growth Performance

It is important to consider the economic implications of probiotic application. Increasing animal body weight gain (BWG) and improving feed conversion ratio (FCR) are measures that can indicate increased profitability for the producer. The inclusion of probiotics may positively affect these measures in poultry. It was speculated that the beneficial impact of probiotic supplementation on poultry performance would be the outcome of a fine-tuning of the complex gut ecosystem, resulting in improved digestive function, intestinal environment, and broiler health. However, literature data indicate that the effect of probiotic administration on the poultry performance is variable. The differences in the dose and nature of probiotics administered and variation in the physiological state of the birds are likely the reasons (Huyghebaert et al. 2011).

Timmerman et al. (2004, 2005) reported that multispecies preparations have advantages when compared with monostrain or multistrain probiotics. In addition, they provided evidence that species-specific probiotics elicit different health effects than do probiotics derived from another host species. The combination of six *Lactobacillus* strains (isolated from digesta and intestinal tissue samples of healthy chickens) had improved the survival rates of broilers in controlled trials by the addition of probiotics to the drinking water.

Mountzouris et al. (2007) investigated the efficacy of the multibacterial species probiotic product Biomin Poultry5Star (containing *Lactobacillus, Bifidobacterium, Enterococcus*, and *Pediococcus* strains) in broilers' nutrition. Probiotic treatment in feed and water displayed a growth-promoting effect that was comparable to avilamycin treatment. Primalac, a DFM product that contains *L. acidophilus, L. casei, B. thermophilum*, and *E. faecium*, was evaluated in various experiments (Grimes et al. 2008; Russell and

Grimes 2009), resulting in improved turkey and broiler live performance. Kabir et al. (2004) reported the occurrence of significantly ($p < 0.01$) higher live weight gains and carcass yield (weight of leg and breast) in broiler chicks fed with Protexin Boost (containing *L. plantarum*, *L. bulgaricus*, *L. acidophilus*, *L. rhamnosus*, *B. bifidum*, *S. thermophilus*, *E. faecium*, *Aspergillus oryzae*, and *Candida pintolopessi*) on the 2nd, 4th, and 6th week of age both in vaccinated and unvaccinated birds. Supplementation of broiler starter and finisher diets with Protexin 100 g/t in starter and 50 g/t in finisher rations was beneficial in terms of weight gain, feed efficiency, and economic viability. Similarly, Anjum et al. (2005) observed that the dietary supplementation with probiotic preparation Protexin significantly ($p \leq 0.05$) improved body weight gain and FCR in broilers; however, no improvement in feed intake was noted. Jin et al. (1998) noted a dose-dependent response with *Lactobacillus* application. The chickens receiving 0.10% (wt/wt) *Lactobacillus* culture (12 strains of *Lactobacillus* isolated from chicken intestine, which belong to four species—*L. acidophilus*, *L. fermentum*, *L. crispatus*, and *L. brevis*) demonstrated superior FCR. However, higher doses resulted in lower productivity. Many other studies reported significant increases in body weight gain and feed conversion efficiency in broiler chickens fed with multibacterial species probiotics (Mansoub 2010; Salim et al. 2013; Zhang and Kim 2014; Song et al. 2014).

On the other hand, some authors confirmed positive effects of single-strain probiotics on broiler growth performance. Supplementation with Bactocell (*P. acidilactici*) in the levels of 1 and 0.8 g/kg diet significantly increased the body weight and daily weight gain of broiler chicks at late ages (3–6 weeks) (Alkhalf et al. 2010). The increasing dietary probiotic level does not have the best performance. A single dose of *L. fermentum* administered intragastrically improves WG and FCE of broiler chicks (Khan et al. 2007). Angelakis and Raoult (2010), who inoculated one dose of *L. fermentum* (originally isolated from an ostrich) to newborn chicks and ducks, confirmed these results. The animals inoculated with *Lactobacillus* displayed a significant increase not only in their body weight but also in their liver mass. The supplementation of $10^6$ CFU/g of *L. reuteri* Pg4 in the feed of broiler chicks from 0 to 21 days of age increased body weight and ileal villus height (Liu et al. 2007).

Many factors make *Bacillus* a good candidate for probiotic use; it produces organic acids, possesses the capacity to sporulate, secretes enzymes, and is easily cultured in bulk. In addition, in the spore form, it is more resistant to extreme temperatures, which enables its inclusion in the pelleting process used in production of chicken feeds. The administration of *Bacillus*-based direct-fed microbial (*B. coagulans*, *B. subtilis*) via the basal diet had beneficial effects on final weight, daily weight gain, FCR, and survival rate of broiler chickens and goslings (Wu et al. 2008; Zhou et al. 2010; Zhang et al. 2012; Park and Kim 2015; Lei et al. 2015; Hosseindoust et al. 2016). The inclusion of *B. amyloliquefaciens* (Lei et al. 2015) and *Bacillus subtilis* (Jayaraman et al. 2013) in broiler diets led to a better villus height and villus height to crypt depth ratio in the different small intestinal segments associated with better nutrient absorption. In contrast, Jerzsele et al. (2012) and Majidi-Mosleh et al. (2017) reported no effect of *Bacillus*-based direct-fed microbial on the performance of broiler chickens. Mutuş et al. (2006) investigated the effect of diets supplemented with *B. licheniformis* and *B. subtilis* (BioPlus 2B) on morphometric parameters and yield stress of the tibia. They found that thickness of the medial and lateral wall of the tibia, tibiotarsal index, percentage ash, and P content were significantly improved by the probiotic.

Results of studies with *S. cerevisiae* fed to chickens have not been consistent. Many authors (Hooge et al. 2003; Stanley et al. 2004; Shareef et al. 2009) reported that feeding yeast to chicks improves BW gain and feed-to-gain ratio. Other authors reported that active dry yeast effectively increases BW gains without affecting the feed-to-gain ratio in broiler chicks, or, by contrast, supplementation of yeast to broiler diets improves feed-to-gain ratio but not growth rates (Kumprechtova et al. 2000; Karaoglu and Durdag 2005; Zhang et al. 2005).

Some studies showed that probiotics demonstrated greater potential in lower-performing animal-rearing facilities than in those with near-optimal animal performance. Torres-Rodriguez et al. (2007) suggest that administration of the commercial *Lactobacillus*-based probiotic (FM-B11) to turkeys raised under suboptimal conditions increased the average daily gain and market body weight. The observed effects seemed to be due to a better response in subpopulations of flocks with a fair to poor performance history, whereas those with a history of good performance seemed to respond less favorably to the probiotic supplementation. Zulkifli et al. (2000) found that diets containing *Lactobacillus* cultures not only provided enhanced BWG but also improved FCR to chickens reared under stressful environments. High feed conversion efficiency in broilers fed with *Bacillus cereus* were observed even when chickens were raised in unhygienic conditions (Takahashi et al.1997).

## 34.4.3 Effects on Meat Quality and Egg Production

The demand for safe and qualitative meat and egg production on the poultry market has considerably increased nowadays. It seems that probiotics could positively affect these parameters in poultry.

The application of probiotic Protexin in broiler chicks with age ranging from 1 to 35 days showed significant differences in chemical composition, including moisture percentage, crude protein, crude fat, and crude ash between the probiotic-treated groups and the control group (Mahmood et al. 2005). A moderate impact of *E. faecium* M-74 was observed in the carcass yield (80.1% vs. 79.1%), a percentage of carcass trunk in live weight (74.2% vs. 73.6%), and a proportion of breast muscle on bone in live weight in particular (26.5% vs. 24.6%) at 84 days of age in favor of the experimental turkeys compared to the control ones (Chmelnična 2003). Supplementation of probiotics improved the sensory characteristics and microbiological quality of dressed broiler meat during prefreezing and post freezing storage (Kabir et al. 2005). Zhou et al. (2010) evaluated the effect of *B. coagulans* ZjU0616 with different concentrations on meat quality of Guangxi Yellow chicken. The results showed no significant difference in chemical composition of the meat. However, the probiotic had beneficial effects on shear force in raw breast meats of male broilers. The lowest percentage of drip loss was found in the group supplemented with probiotic at $2.0 \times 10^6$ CFU/g. It coincided with the shear force results and showed positive effects of *B. coagulans* ZJU0616 on meat quality. There are trials showing that enrichment of diets with yeast could favorably improve the quality of meat from broilers. For example, edible meats from broiler chicks fed with a diet containing chromium-enriched *S. cerevisiae* or *S. cerevisiae* cell wall and extract exhibited increased tenderness (Zhang et al. 2005) and increased water-holding capacity (Lee et al. 2002). The *S. cerevisiae* cell wall, which contains α-glucan, carboxymethyl glucan, mannans, and some proteinaceous substances, has been reported to display relatively good antioxidative properties (Tsiapali et al. 2001). On the other hand, Loddi et al. (2000) reported that neither probiotic nor antibiotic affected sensory characteristics (intensity of aroma, strange aroma, flavor, strange flavor, tenderness, juiciness, acceptability, characteristic color, and overall aspects) of breast and leg meats.

Dietary supplementation of *B. coagulans* at 100 mg ($6 \times 10^8$ spore) $kg^{-1}$ diet significantly increased eggshell weight, shell thickness, and serum calcium in White Leghorn layer breed (Panda et al. 2008). The addition of commercial probiotic BioPlus 2B (mixture of spray-dried spore-forming *B. subtilis* and *B. licheniformis*) at 250, 500, and 750 $mg.kg^{-1}$ diet increased egg production but decreased the damaged egg ratio ($p < 0.05$), egg yolk cholesterol, and serum cholesterol ($p < 0.001$) levels in Brown-Nick layer hybrids (Kurtoglu et al. 2004). *Bacillus subtilis* (0.10%) and inulin (0.10%), individually or in combination, positively influenced the egg performance, eggshell quality, and calcium retention in aged hens (Abdelqader et al. 2013). *Pediococcus acidilactici* (*PA*) as a probiotic supplement did not significantly affect the egg production of hens but showed potential to improve egg weight and eggshell quality during the early laying period. This effect was dose dependent and was greater at 100 mg *PA* compared with 50 mg *PA* per kg of feed (Mikulski et al. 2012). At a later stage, after 24 wk of feeding PA-supplemented diets, a decrease in yolk cholesterol levels was also noted. ISA brown laying hens fed with the basal diet containing 0.01% *Enterococcus faecium* DSM 7134 had higher egg production than those fed with the basal diet without probiotic from wk 0 to 9 overall ($p < 0.05$). During wk 10 to 18, probiotic supplementation resulted in a significant increase ($p < 0.05$) in egg weight and eggshell thickness (Park et al. 2016). Besides, inclusion of probiotics containing *Lactobacillus sporogenes* (Panda et al. 2008), multi-strain probiotics (Yörük et al. 2004; Khan et al. 2011), and yeast (Park et al. 2002; Yousefi and Karkoodi 2007; Hassanein and Soliman 2010) in laying hens diet enhanced egg production, eggshell thickness, and egg mass and reduced soft or broken eggs of hens. The authors stated that this beneficial effect may be attributable to a favorable environment in the intestinal tract, which may have helped to absorb more calcium. It is known that probiotics might improve the content of calcium, phosphorus, carotenoid, and albumen in serum of layers (Ashmead et al. 1985).

Whereas results of various studies have shown that probiotics could reduce the cholesterol content of egg yolk (Panda et al. 2003; Mahdavi et al. 2005; Ramasamy et al. 2009; Yalçın et al. 2008), studies investigating the effects of probiotics on fatty acid (FA) composition in egg yolk remain scarce. The reduction in yolk cholesterol could be explained by the reduced absorption and/or synthesis of cholesterol in the gastrointestinal tract. It has been reported that probiotics reduce plasma cholesterol and triglycerides, confirming the important roles of gastrointestinal microorganisms in recycling of lipids (Mohan et al. 1995). Yalçın et al. (2010) evaluated the effects of dietary supplementation with yeast *(Saccharomyces cerevisiae) autolysate* on FA composition in egg yolk. They observed significantly ($p < 0.01$) higher levels of total saturated fatty acids and significantly lower levels of total monounsaturated fatty acids ($p < 0.001$) in comparison with unsupplemented control. The ratio of monounsaturated fatty acids to saturated fatty acids was significantly ($p < 0.001$) decreased. In the study reported by Mikulski et al. (2012), supplementation of *Pediococcus acidilactici* at $8.0 \times 10^8$ CFU per kg of diet significantly increased the proportion of PUFAs, including linoleic and linolenic acid, but the PUFAn-6/PUFA n-3 ratio was not modified.

# 34.5 PROBIOTICS FOR OTHER FARM ANIMAL SPECIES

Probiotics can also be effectively used in other farm animals such as honeybees, fur-bearing animals, rabbits, and horses.

## 34.5.1 Honeybees (*Apis mellifera*)

Since 2003, high losses of bee colonies have been reported in Europe and America due to the extensive use of pesticides in agriculture, climatic changes, and inappropriate beekeeping practices, which significantly affect bee health and immunity and consequently contribute to the development of infectious diseases. As in mammals, a healthy gut microbiota has been shown to play a key role in maintaining the health of honey bees by stimulating the immune response and inhibiting pathogens and is therefore critical for protecting bees from disease (Wu et al. 2013). Like in other animal species, also in bees, probiotic bacteria can favorably influence the composition of the intestinal microbiota. However, unlike vertebrates, the diversity of the gut microbiota of bees is very low. The intestinal tract of bees is dominated by 8 to 10 bacterial phylotypes, which make up more than 97% of the microbial community. The main bacterial phylotypes are *Gilliamella, Snodgrassella, Bifidobacterium, Lactobacillus* Firm4 and Firm5, and *Bartonella* (Ellegaard and Engel 2019). Therefore, the balance of the bee microbiota is much more fragile than that of vertebrates. Its disruption results not only in the weakening of immunity but also in other physiological functions. The bee microbiota, above all LAB, plays an important role in many metabolic processes, which include the metabolism of sugars, the synthesis of amino acids, vitamins and short-chain fatty acids, hormonal functions, neurological processes and sensory functions, or pollen fermentation (Vásquez and Olofsson 2009; Zhang et al. 2022). For these reasons, it is very important to be careful when applying probiotic preparations to bee colonies. The results of some studies proved that if probiotics based on non-autochthonous strains were applied to bees, their health even worsened. For example, Andrearczyk et al. (2014) applied to bee colonies a probiotic preparation intended for veterinary use, which contained *L. casei, L. plantarum, S. cerevisiae* and *Rhodopseudomonas palustris*. A higher incidence of *Nosema ceranae*—a widespread microsporidium affecting bee health—and even higher deaths of winter bees were observed in the treated colonies. Similar negative results were obtained by Ptaszyńska and Mułenko (2013) when bees were given a probiotic containing *L. rhamnosus* alone or with the prebiotic inulin during *N. ceranae* infection.

On the contrary, when autochthonous bee strains were used, mostly positive results were achieved. Since LAB are most abundantly represented in the digestive tract of bees and play a key role in the mentioned metabolic processes, they are also most often used as probiotics for honey bees. Potential probiotic LAB are mostly isolated from intestines and honey stomachs of bees, pollen, honey, or hive environment; then they are tested for their inhibitory activity against bee pathogens, technological properties, survival in hive conditions; subsequently characterized; and eventually tested for further required properties (e.g., survival in different application forms such as royal jelly, pollen, sugar, or honey solutions). Several authors have confirmed the strong inhibitory activity of autochthonous LAB strains against *Paenibacillus larvae*—the causative agent of American foulbrood (AFB), which is one of the most serious infectious diseases of bees (Evans and Armstrong 2006; Forsgren et al. 2010; Mudroňová et al. 2011).

Forsgren et al. (2010) isolated 11 LAB phylotypes (members of genera *Lactobacillus* and *Bifidobacterium*) from the honey stomachs of honeybees and tested their inhibition activity against *P. larvae*. The combination of all isolated LAB resulted in total inhibition of all tested *P. larvae* strains; therefore this combination was also used in the infection bioassays. Since young larvae are infected by contaminated food and the bacterial spores germinate and bacteria multiply in the midgut, probiotic LAB were administered to the larval food. The addition of the LAB mixture significantly reduced the number of infected larvae ($p = 0.0007$) and their mortality. Mudroňová et al. (2011) isolated from digestive tracts and honey stomachs of healthy adult honeybees *L. brevis* B50 and *L. plantarum*, which significantly inhibited *P. larvae*. Both lactobacilli showed good technological properties—survival of long-term freezing storage and good growth properties with pH decrease after 24 h to pH 4 for *L. plantarum* and pH 4.7 for *L. brevis*. It has been shown that germination of bacterial spores is inhibited at around pH 4.2, which can be formed by organic acids produced by LAB (De Vuyst and Vandamme 1994). The probiotic strain *L. brevis* B50 showing the best properties was applied to weakened colonies where *P. larvae* were present in the digestive tracts of adult bees, without clinical manifestation of AFB. Within 2 weeks after the first application of the probiotic, this pathogen was completely eliminated from the bees' digestive tract. The composition of the intestinal microbiota was affected, with an increase of LAB counts and, conversely, a decrease of enterobacteria and total aerobes. Treated bee colonies also showed better health status and cleaning activity and higher honey yields, and the occurrence of the *Varroa* mite was even up to 70% lower, which was an unexpected result (Kuzyšinová et al. 2012; Kuzyšinová et al. 2023). Subsequently, it was found that the probiotic preparation significantly increases the expression of genes for antimicrobial peptides (abaecin, defensin-1, hymenoptaecin, apidaecin) but also for toll-like receptors and affects the expression of other immunologically significant molecules (Cactus, Dorsal, PGRP, Relish, Kenny), which are part of immunological activation pathways (Toll, Imd, or JNK signaling pathway) (Maruščáková et al. 2020). A stimulatory effect on gene expression for apidaecin was also noted by Janashia and Alaux (2016), when they applied autochthonous LAB strains to bee larvae. They tested five different strains, and only two of them—*Bifidobacterium asteroides* and *Fructobacillus pseudoficulneus*—showed such a stimulating effect. Also Yoshiyama et al. (2013), after application of a mixture of probiotic LAB strains (*Enterococcus thailandicus*, *Weissella cibaria*, *W. viridescens*, and *L. curvatus*), noted a significant increase in the expression of abaecin, defensin, and hymenoptaecin genes both in adult bees and in larvae.

In recent years, several studies have appeared that confirm the inhibitory effect of lactic acid bacteria against other significant pathogens of bees. Bielik et al. (2021) recorded the elimination of not only *P. larvae* but also *Melisococcus plutonius*, the causative agent of European foulbrood (EFB), from the digestive tract of bees after 2–3 weeks after the first application of the probiotic strain *Apilactobacillus kunkeei*. Similarly, Vásquez et al. (2012) confirmed under *in vitro* as well as *in vivo* conditions antimicrobial activity of *A. kunkeei* against *M. plutonius*. The reduction of the prevalence of EFB after the application of the autochthonous strain *L. plantarum* to bee colonies was also confirmed in field conditions (Pietropaoli et al. 2022). Baffoni et al. (2016) evaluated the effect of the oral administration of LAB on the nosematosis. They noted inhibition of the growth of *N. ceranae* as well as a decrease in the number of the spores. Similar results received Audisio et al. (2015) after administration of *L. johnsonii* and Arredondo et al. (2018) with four *A. kunkeei* strains applied to adult honey bees. Peghaire et al. (2020), who applied the probiotic strain *P. acidilactici* to bees, noted not only a protective effect against nosematosis, but also against applied pesticides (insecticide—thiamethoxam and fungicide boscalid). Both *N. ceranae* and pesticides have been shown to deregulate genes involved in bee development (vitellogenin), immunity (serine protease 40, defensin), and the detoxification system (glutathione-like peroxidase 2, catalase), and these effects were corrected by application of pediococci. The positive effect of probiotics on the development and vitality of the bee colonies, honey yields, and even the life span of bees was also confirmed (Fanciotti et al. 2018; Audisio 2017; Pachla et al. 2021).

These results indicate that beneficial bacteria inhabiting the honeybee play an important role in the resistance to diseases, and therefore probiotic bacteria originating from the honeybee microbiota can be considered for probiotic use in their prevention. Resent research on honey bees showed that the positive influence of beneficial bacteria is mediated not only by direct inhibition of pathogens but also by modulation of important metabolic pathways and immunity of honeybees.

## 34.5.2 Rabbits

Rabbit breeding for meat production has a great economic impact, above all due to the high quality of the meat—low cholesterol and total lipids and high content of protein. Multifactorial intestinal diseases caused by dietary stress in combination with pathogens, above all *E. coli*, *Clostridium* sp., or *Eimeria* sp., are among the most important health problems at rabbit farms (Marlier et al. 2006). Despite the fact that members of the family Lactobacillaceae are not common inhabitants of the digestive tract of rabbits, they are sometimes used for probiotic purposes, but the use of enterococci and yeasts dominates. (Mancini and Paci 2021; Yu and Tsen 1993). Simonová et al. (2005) isolated enterococci from feces of rabbits. Most of isolates showed good survival in the conditions of digestive tract (5% oxgall, pH 3), urease activity, and production of lactic acid. The majority of isolates were characterized as *E. faecium*, others as *E. faecalis*. Moreover, the isolates of *E. faecium* possessed bacteriocinogenic activity (Simonová and Lauková 2004). Bacteriocin-producing strain *E. faecium* with confirmed probiotic properties were administered to growing rabbits (35 days of age) for 3 weeks. The inhibitory effect of *E. faecium* was noted by decreased numbers of *E. coli*, *Clostridium*-like sp., and coagulase-negative staphylococci in cecal contents. The numbers of *Eimeria* sp. oocysts was also reduced. The immunostimulative effect was confirmed by increased phagocytic activity (Szabóová et al. 2008). Various combinations of LAB with bacilli, yeasts, prebiotics, or phytoadditives were tested in several studies. The results of studies conducted under farm conditions are summarized in a review by Mancini and Paci (2021). The presented results showed different effects of probiotic preparations on production parameters, intestinal microbiota, health status, and meat quality, but in most studies at least one of the monitored parameters was positively affected. Several authors reported a positive effect on the gut microbiota composition, which was mostly accompanied by a better feed conversion ratio. Abdel-Samee (1995) found that supplementation of multi-strain probiotics to heat-stressed rabbits at a farm in Egypt significantly increased daily weight gains and improved reproductive parameters (litter size at weaning, weight at birth and at weaning). The incidence of diseases and mortality was also reduced as compared with control rabbits without addition of probiotics. Phuoc and Jamikorn (2017) compared the efficacy of the use of *B. subtilis* and *L. acidophilus* and their combination on production parameters and gut microbiota composition in rabbits. Interestingly, administration of *L. acidophilus* alone or in combination with *B. subtilis* resulted in better results than *B. subtilis*. The authors noted higher numbers of gut beneficial bacteria, nutrient digestibility, feed efficiency, and growth performance in comparison to rabbits supplemented with *B. subtilis* alone. Supuková et al. (2010) fed a combination of *L. fermentum*, *E. faecium*, maltodextrin, and FOS to rabbits. This combination significantly decreased the number of intestinal disorders and increased daily weight gains.

## 34.5.3 Horses

Probiotics in equine practice are used for the prevention of intestinal disorders, for the substitution of beneficial microbiota after or during antibiotic therapy, when horses are stressed, and often also for the prevention or therapy of reproductive problems in mares (Moates 2009).

The horse cecum contains beneficial microorganisms that play an important role in the digestion of food. If its balance is impaired, the horse may have problems with nutrient absorption, which can result in the health disorders such as diarrhea or skin and hoof problems.

Unfortunately, most results received from trial experiments with the per-oral application of probiotic LAB to horses were not very promising (Schoster et al. 2014). Addition of a mixture of lactobacilli, *B. bifidum*, and *E. faecium* into the diet of mature horses had only limited influence on the nutrient digestibility, as only digestibility of fat and certain minerals was increased. Supplementation of this mixture did not demonstrate a reduction of the risk of digestive disorders, such as acidosis, caused by feeding high-starch concentrates to horses (Swyers et al. 2008). *L. pensosus* strain WE7, which showed *in vitro* antagonistic activity against *E. coli*, *Streptococcus zooepidemicus*, *C. difficile*, and *C. perfringens* (Weese et al. 2004), was subsequently administered to neonatal foals. Surprisingly, foals receiving probiotics experienced higher incidence of diarrhea and other clinical disorders (e.g., colic, anorexia, weakness) in comparison with control animals (Weese and Rousseau 2005). Administration of a multi-strain probiotic based on lactobacilli and bifidobacteria to neonatal foals showed only limited potential to modify the gastrointestinal microbiota (Schoster et al. 2016) or to reduce clostridial diarrhea as well as shedding of *Clostridium perfringens* and *Clostridium difficile*, despite *in vitro* activity of these LAB strains (Schoster et al. 2015). One promising pilot study was performed with Enterocin M, a product of probiotic *Enterococcus faecium*. Administration of this bacteriocin to clinically healthy horses led to significant reduction of coliforms, campylobacters, and *Clostridium* spp. and increase of phagocytic activity. Moreover, no negative influence on hydrolytic enzyme profile or biochemical blood parameters was noted (Lauková et al. 2018).

On the other hand, LAB have been used with success in the prevention of reproductive problems in mares because they play an important role in the regulation of the vaginal microbiota. Fraga et al. (2008) isolated lactobacilli and enterococci from the vaginal wall of mares. *L. mucosae*, *L. equi*, and *E. faecalis* have shown the highest antimicrobial activity against reproductive pathogens—*E. coli* and *S. aureus*. Morvayová et al. (2008) also isolated lactobacilli from the vagina of healthy mares and tested their inhibition effect against pathogenic bacteria received from the mare with chronic endometritis—*E. coli*, *S. aureus*, and *S. agalactiae*. Only one of the isolated strains was able to inhibit the growth of *E. coli*. In equine practice, lactobacilli have been successfully used for the therapy of *Candida* infections in mares (Hura 2008, personal communication). Rohrbach et al. (2007) received promising results with the treatment of intravenously administered *Propionibacterium acnes* to mares with persistent endometritis. Results indicated that application of propionibacteria improved successful conception of mares and the number of live delivered foals.

# 34.6 CONCLUSION

Probiotics in farm animals can be effectively used in the modulation of gastrointestinal microbiota, stimulation of the immune system, prevention and treatment of diarrheal diseases in young farm animals, growth stimulation, and optimization of animal production. However, in order to achieve the desired effect, it is necessary to achieve high effectiveness of probiotic preparations. In practice, therefore, only probiotic preparations with a proven and scientifically well-founded effect should be used, preferably based on autochthonous microbial strains. In order to achieve these goals, it is essential to understand the exact modes of action of probiotic microorganisms, to find the correct application scheme and application forms for individual species and categories of farm animals, and to search for possibilities of improving the effectiveness of probiotics by their combination with substances of natural origin.

The ultimate positive effect of probiotics will be reflected not only in the health and productivity of farm animals but also in the quality and safety of food of animal origin and in the lower burden on the environment with synthetic medications.

## 34.7 ACKNOWLEDGMENTS

The work was supported by grant VEGA 1/0454/22 (Scientific Grant Agency of the Ministry of Education, Science, Research and Sport of the Slovak Republic and the Academy of Sciences).

## BIBLIOGRAPHY

Abdelqader, A., A. R. Al-Fataftah, G. Das. 2013. Effects of dietary *Bacillus subtilis* and inulin supplementation on performance, eggshell quality, intestinal morphology and microflora composition of laying hens in the late phase of production. *Anim. Feed Sci. Technol.* 179: 103–111.

Abdel-Samee, A. M. 1995. Using some antibiotics and probiotics for alleviating heat stress on growing and doe rabbits in Egypt. *World Rabbit Sci.* 3(2): 107–111.

Akbari, M. R., H. R. Haghighi, J. R. Chambers, J. Brisbin, L. R. Read, S. Sharif. 2008. Expression of antimicrobial peptides in cecal tonsils of chickens treated with probiotics and infected with *Salmonella enterica* serovar Typhimurium. *Clin. Vaccine Immunol.* 15: 1689–1693.

Alkhalf, A., M. Alhaj, I. Al-homidan. 2010. Influence of probiotic supplementation on blood parameters and growth performance in broiler chickens. *Saudi J. Biol. Sci.* 17: 219–225.

Andrearczyk, S., M. Kadhim, S. Knaga. 2014. Influence of a probiotic on the mortality, sugar syrup ingestion and infection of honeybees with *Nosema* spp. under laboratory assessment. *Med. Weter.* 70(12): 762–765.

Angelakis, E., D. Raoult. 2010. The increase of *Lactobacillus* species in the gut biota of newborn broiler chicks and ducks is associated with weight gain. *PLoS ONE.* 5(5): e10463. http://doi.org/10.1371/journal.pone.0010463.

Anjum, M. I., A. G. Khan, A. Azim, M. Afzal. 2005. Effect of dietary supplementation of multi strain probiotic on broiler growth performance. *Pakistan Vet. J.* 25(1): 25–29.

Apajalahti, J., A. Kettunen, H. Graham. 2004. Characteristics of the gastrointestinal microbial communities, with special reference to the chicken. *Worlds Poult. Sci. J.* 60(2): 223–232.

Arredondo, D., L. Castelli, M. P. Porrini, P. M. Garrido, M. J. Eguaras, P. Zunino, K. Antúnez. 2018. *Lactobacillus kunkeei* strains decreased the infection by honey bee pathogens *Paenibacillus larvae* and *Nosema ceranae*. *Benef. Microbes.* 9(2): 279–290.

Ashmead, H. D., D. J. Graff, H. H. Ashmead. 1985. *Intestinal Absorption of Metal Ions and Chelates*, 171–232. Charles C. Thomas Publ., Springfield, IL.

Audisio, M. C. 2017. Gram-positive bacteria with probiotic potential for the *Apis mellifera* L. honey bee: The experience in the Northwest of Argentina. *Probiotics Antimicrob. Prot.* 9: 22–31.

Audisio, M. C., D. C. Sabaté, M. R. Benítez-Ahrendts. 2015. Effect of Lactobacillus johnsonii CRL1647 on different parameters of honeybee colonies and bacterial populations of the bee gut. *Benef. Microbes.* 25: 1–10.

Azizi, A. F. N., R. Uemura, M. Omori, M. Sueyoshi, M. Yasuda. 2022. Effects of probiotics on growth and immunity of piglets. *Animals (Basel).* 12(14): 1786.

Baffoni, L., F. Gaggìa, D. Alberoni, R. Cabbri, A. Nanetti, B. Biavati, D. Di Gioia. 2016. Effect of dietary supplementation of *Bifidobacterium* and *Lactobacillus* strains in *Apis mellifera* L. against *Nosema ceranae*. *Benef. Microbes.* 7: 45–50.

Barba-Vidal, E., S. M. Martín-Orúe, L. Castillejos. 2019. Practical aspects of the use of probiotics in pig production: A review. *Livest. Sci.* 223: 84–96.

Bielik, B., L. Molnár, V. Vrabec, R. Andrášiová, I. C. Maruščáková, R. Nemcová, J. Toporčák, D. Mudroňová. 2021. Biofilm-forming lactic acid bacteria of honey bee origin intended for potential probiotic use. *Acta Vet. Hung.* 68(4): 345–353.

Bomba, A., Z. Jonecová, S. Gancarčíková, R. Nemcová. 2006. The gastrointestinal microbiota of farm animals. In *Gastrointestinal Microbiology*, eds. A. C. Ouwehand and E. E. Vaughan, 379–397. Taylor & Francis, Taylor & Francis Group, New York.

Bomba, A., R. Nemcová, S. Gancarčíková, R. Herich, R. Kašteľ. 1999. Potentiation of the effectiveness of *Lactobacillus casei* in the prevention of *E. coli* induced diarrhea in conventional and gnotobiotics pigs. In *Mechanisms in the Pathogenesis of Enteric Diseases* 2, eds. P. S. Paul and D. H. Francis, 185–190. Kluwer Academic/Plenum Publishers, New York.

Bomba, A., R. Nemcová, S. Gancarčíková, R. Herich, P. Guba, D. Mudroňová. 2002. Improvement of the probiotic effect of micro-organisms by their combination with maltodextrins, fructo-oligosaccharides and polyunsaturated fatty acids. *Br. J. Nutr.* 88(Suppl. 1): 95–99.

Bomba, A., R. Nemcová, S. Gancarčíková, R. Herich, J. Pistl, V. Révajová, Z. Jonecová, A. Bugarský, M. Levkut, R. Kašteľ, M. Baran, G. Lazar, M. Hluchý, S. Maršálková, J. Pošivák. 2003. The influence of omega-3 polyunsaturated fatty acids (omega-3 pufa) on lactobacilli adhesion to the intestinal mucosa and on immunity in gnotobiotic piglets. *Berl. Munch. Tierärztl. Wochenschr.* 116: 312–316.

Boyd, J., J. West, J. Bernard. 2011. Effects of the addition of direct-fed microbials and glycerol to the diet of lactating dairy cows on milk yield and apparent efficiency of yield. *J. Dairy Sci.* 94: 4616–4622.

Brashears, M. M., M. L. Galyean, G. H. Loneragan, J. E. Mann, K. Killinger-Mann. 2003a. Prevalence of *Escherichia coli* 0157:H7 and performance by beef feedlot cattle given *Lactobacillus* direct-fed microbials. *J. Food Prot.* 66: 748–754.

Brashears, M. M., D. Jaroni, J. Trimble. 2003b. Isolation, selection, and characterization of lactic acid bacteria for a competitive exclusion product to reduce shedding of *Escherichia coli* 0157:H7 in cattle. *J. Food Prot.* 66: 355–363.

Brisbin, J. T., H. Zhou, J. Gong, et al. 2008. Gene expression profiling of chicken lymphoid cells after treatment with *Lactobacillus acidophilus* cellular components. *Dev. Comp. Immunol.* 32: 563–574.

Callaway, T. R., R. C. Anderson, T. S. Edrington, K. J. Genovese, K. M. Bischoff, T. L. Poople, Y. S. Jung, R. B. Harvey, D. J. Nisbet. 2004. What are we doing about *Escherichia coli* 0157:H7 in cattle? *J. Anim. Sci.* 82: 93–99.

Campbell, J. M., J. D. Crenshaw, J. Polo. 2013. The biological stress of early weaned piglets. *J. Anim. Sci. Biotechnol.* 4(1): 19.

Casey, P. G., G. E. Gardiner, G. Casey, et al. 2007. A five-strain probiotic combination reduces pathogen shedding and alleviates disease signs in pigs challenged with *Salmonella enterica* serovar Typhimurium. *Appl. Environ. Microbiol.* 73: 1858–1863.

Chang, S. Y., S. A. Belal, D. R. Kang, Y. Choi, Y. H. Kim, H. S. Choe, J. Y. Heo, K. S. Shim. 2018. Influence of probiotics-friendly pig production on meat quality and physicochemical characteristics. *Korean J. Food Sci. Anim. Resour.* 38(2): 403–416.

Chaucheyras-Durand, F., G. Fonty. 2002. Influence of a probiotic yeast (*Saccharomyces cerevisiae* CNCM I-1077) on microbial colonization and fermentations in the rumen of newborn lambs. *Microb. Ecol. Health Dis.* 14(1): 30–36.

Chen, C. Y., S. W. Chen, H. T. Wang. 2017. Effect of supplementation of yeast with bacteriocin and *Lactobacillus* culture on growth performance, cecal fermentation, microbiota composition and blood characteristics in broiler chickens. *Asian-Aust. J. Anim. Sci.* 30: 211–220.

Chiquette, J. 2009. The role of probiotics in promoting dairy production. *WCDS Adv. Dairy Technol.* 21: 143–157.

Chiquette, J., M. J. Allison, M. A. Rasmussen. 2008. *Prevotella bryantii* 25A used as a probiotic in early lactation dairy cows: Effect on ruminal fermentation characteristics, milk production, and milk composition. *J. Dairy Sci.* 91: 3536–3543.

Chmelnična, L. 2003. Effect of drinking water application of probiotics containing *Enterococcus feacium* M-74 on growth and carcass traits of turkeys. *Acta Fytotech. Zootech.* 6(1): 6–10.

Choi, K. D., H. S. Lillehoj, S. Zalenga. 1999. Changes in local IFN-γ and TGF-β4 mRNA expression and intraepithelial lymphocytes following *Eimeria acervulina* infection. *Vet. Immunol. Immunopathol.* 71: 263–275.

Chytilová, M., D. Mudroňová, R. Nemcová, S. Gancarčíková, V. Buleca, J. Koščová, Ľ. Tkáčiková. 2013. Anti-inflammatory and immunoregulatory effects of flax-seed oil and *Lactobacillus plantarum*—Biocenol LP96 in gnotobiotic pigs challenged with enterotoxigenic *Escherichia coli*. *Res. Vet. Sci.* 95: 103–109.

Chytilová, M., R. Nemcová, S. Gancarčíková, D. Mudroňová, Ľ. Tkáčiková. 2014. Flax-seed oil and *Lactobacillus plantarum* supplementation modulate TLR and NF-κB gene expression in enterotoxigenic *Escherichia coli* challenged gnotobiotic pigs. *Acta Vet. Hungarica* 62: 463–472. http://doi.org/10.1556/AVet.2014.024.

Collins, J. W., R. M. La Ragione, M. J. Woodward, L. E. J. Searle. 2009. Application of prebiotics and probiotics in livestock. In *Prebiotics and Probiotics Science and Technology*, eds. D. Charalampopoulos and R. A. Rastall. Springer, New York, NY. https://doi.org/10.1007/978-0-387-79058-9_30.

Dalloul, R. A., H. S. Lillehoj, T. A. Shellem, J. A. Doerr. 2003. Enhanced mucosal immunity against *Eimeria acervulina* in broilers fed a *Lactobacillus*-based probiotic. *Poult. Sci.* 82(1): 62–66.

Dalloul, R. A., H. S. Lillehoj, N. M. Tamim, T. A. Shellum, J. A. Doerr. 2005. Induction of local protective immunity to *Eimeria acervulina* by a *Lactobacillus*-based probiotic. *Comp. Immunol.* 28: 351–361.

De Vuyst, L., E. J. Vandamme. 1994. *Bacteriocins of Lactic Acid Bacteria*. Blackie Academic & Professional, Oxford. ISBN 0751401749.

Dowarah, R., A. K. Verma, N. Agarwal. 2017a. The use of *Lactobacillus* as an alternative of antibiotic growth promoters in pigs: A review. *Anim. Nutr.* 3: 1–6.

Dowarah, R., A. K. Verma, N. Agarwal, B. H. M. Patel, P. Singh. 2017b. Effect of swine based probiotic on performance, diarrhoea scores, intestinal microbiota and gut health of grower-finisher crossbred pigs. *Livest. Sci.* 195: 74–79.

Ellegaard, K. M., P. Engel. 2019. Genomic diversity landscape of the honey bee gut microbiota. *Nat. Comm.* 10: 446.

Evans, J. D., T. N. Armstrong. 2006. Antagonistic interactions between honey bee bacterial symbionts and implications for disease. *BMC Ecol.* 6: 4. http://doi.org/10.1186/1472-6785-6-4.

Ewaschuk, J. B., J. M. Naylor, G. Chirino-Trejo, G. A. Zello. 2004. *Lactobacillus rhamnosus* strain GG is a potential probiotic for calves. *Can. J. Vet. Res.* 68: 249–253.

Fanciotti, M. N., M. Tejerina, M. R. Benítez-Ahrendts, M. C. Audisio. 2018. Honey yield of different commercial apiaries treated with *Lactobacillus salivarius* A3iob, a new bee-probiotic strain. *Benef. Microbes.* 9: 291–298.

Fathi, M. M., T. A. Ebeid, I. Al-Homidan, N. K. Soliman, O. K. Abou-Emera. 2017. Influence of probiotic supplementation on immune response in broilers raised under hot climate. *Br. Poult. Sci.* 6: 1–5.

Forsgren, E., T. C. Olofsson, A. Vásquez, I. Fries. 2010. Novel lactic acid bacteria inhibiting *Paenibacillus larvae* in honey bee larvae. *Apidologie* 41: 99–108. http://doi.org/10.1051/apido/2009065.

Fraga, M., K. Perelmuter, L. Delucchi, E. Cidade, P. Zunino. 2008. Vaginal lactic acid bacteria in the mare: Evaluation of the probiotic potential of native *Lactobacillus* spp. and *Enterococcus* spp. strains. *Antonie van Leeuwenhoek.* 93: 71–78.

Fukuda, S., Y. Suzuki, M. Murai, N. Asanuma, T. Hino. 2006. Isolation of a novel strain of *Butyrivibrio fibrisolvens* that isomerizes linoleic acid to conjugated linoleic acid without hydrogenation, and its utilization as a probiotic for animals. *J. Appl. Microbiol.* 100: 784–794.

Giannenas, I., E. Tsalie, E. Triantafillou, S. Hessenberger, K. Teichmann, M. Mohnl, D. Tontis. 2014. Assessment of probiotics supplementation via feed or water on the growth performance, intestinal morphology and microflora of chickens after experimental infection with *Eimeria acervulina*, *Eimeria maxima* and *Eimeria tenella*. *Avian Pathol.* 43(3): 209–216.

Grimes, J. L., S. Rahimi, E. Oviedo, B. W. Sheldon, F. B. O. Santos. 2008. Effects of a direct-fed microbial (Primalac) on turkey poults performance and susceptibility to oral *Salmonella* challenge. *Poult. Sci.* 87: 1464–1470.

Guerra-Ordaz, A. A., G. González-Ortiz, R. M. La Ragione, M. J. Woodward, J. W. Collins, J. F. Pérez, S. M. Martín-Orúe. 2014. Lactulose and Lactobacillus plantarum, a potential complementary synbiotic to control postweaning colibacillosis in piglets. *Appl. Environ. Microbiol.* 80(16): 4879–4886.

Guevarra, R. B., S. H. Hong, J. H. Cho, B. R. Kim, J. Shin, J. H. Lee, B. N. Kang, Y. H. Kim, S. Wattanaphansak, R. E. Isaacson, M. Song, H. B. Kim. 2018. The dynamics of the piglet gut microbiome during the weaning transition in association with health and nutrition. *J. Anim. Sci. Biotechnol.* 9: 54.

Haghighi, H. R., M. F. Abdul-Careem, R. A. Dara, J. Chambers, S. Shariff. 2008. Cytokine gene expression in chicken cecal tonsils following treatment with probiotics and *Salmonella* infection. *Vet. Microbiol.* 126: 225–233.

Haghighi, H. R., J. Gong, C. L. Gyles, et al. 2005. Modulation of antibody-mediated immune response by probiotics in chickens. *Clin. Diagn. Lab. Immunol.* 12: 1387–1392.

Hassanein, S. M., N. K. Soliman. 2010. Effect of probiotic (*Saccharomyces cerevisiae*) adding to diets on intestinal microbiota and performance of Hy-Line layers hens. *J. Am. Sci.* 6(11): 159–169.

Heinrichs, A. J., K. E. Lesmeister. 2005. *Rumen Development in the Dairy Calf. Calf and Heifer Rearing*, 53–65. Nottingham University Press, Nottingham.

Higgins, S. E., J. P. Higgins, A. D. Wolfenden, et al. 2008. Evaluation of a *Lactobacillus*-based probiotic culture for the reduction of *Salmonella* Enteritidis in neonatal broiler chicks. *Poult. Sci.* 87: 27–31.

Hong, H. A., L. H. Duc, S. M. Cutting. 2005. The use bacterial spore formers as probiotics. *FEMS Microbiol. Rev.* 29: 813–835.

Hooge, D. M., M. D. Sims, A. E. Sefton, A. Connolly, P. S. Spring. 2003. Effect of dietary mannan oligosaccharide, with or without bacitracin or virginiamycin, on live performance of broiler chickens at relatively high stocking density on new litter. *J. Appl. Poult. Res.* 12: 461–467.

Hosseindoust, A., J. W. Park, I. H. Kim. 2016. Effects of *Bacillus subtilis*, Kefir and beta-glucan supplementation on growth performance, blood characteristics, meat quality and intestine microbiota in broilers. *Korean J. Poult. Sci.* 43: 159–167.

Hu, Y. L., Y. H. Dun, S. A. Li, S. Zhao, N. Peng, Y. X. Liang. 2014. Effects of Bacillus subtilis KN-42 on growth performance, diarrhea and faecal bacterial flora of weaned piglets. *Asian-Australas. J. Anim. Sci.* 27: 1131–1140.

Huyghebaert, G., R. Ducatelle, F. Van Immerseel. 2011. An update on alternatives to antimicrobial growth promoters for broilers. *Vet. J.* 187: 182–188.

Janashia, I., C. Alaux. 2016. Specific immune stimulation by endogenous bacteria in honey bees (Hymenoptera: Apidae). *J. Econom. Entomol.* 109(3): 1474–1477.

Janczyk, P., R. Pieper, H. Smidt, W. B. Souffrant. 2007. Changes in the diversity of pig ileal lactobacilli around weaning determined by means of 16S rRNA gene amplification and denaturing gradient gel electrophoresis. *FEMS Microbiol. Ecol.* 61(1): 132–140.

Jayaraman, B., Ch. M. Nyachoti. 2017. Husbandry practices and gut health outcomes in weaned piglets: A review. *Anim. Nutr.* 3: 205–211.

Jayaraman, S., G. Thangavel, H. Kurian, R. Mani, R. Mukkalil, H. Chirakkal. 2013. *Bacillus subtilis* PB6 improves intestinal health of broiler chickens challenged with *Clostridium perfringens*-induced necrotic enteritis. *Poult. Sci.* 92: 370–374.

Jeong, J. S., I. H. Kim. 2014. Effect of *Bacillus subtilis* C-3102 spores as a probiotic feed supplement on growth performance, noxious gas emission and intestinal microflora in broilers. *Poult. Sci.* 93: 3097–3103.

Jerzsele, A., K. Szeker, R. Csizinszky, E. Gere, C. Jakab, J. J. Mallo, P. Galfi. 2012. Efficacy of protected sodium butyrate, a protected blend of essential oils, their combination and *Bacillus amyloliquefaciens* spore suspension against artificially induced necrotic enteritis in broilers. *Poult. Sci.* 91: 837–843.

Jin, L. Z., Y. W. Ho, N. Abdullah, S. Jalaludin. 1998. Growth performance, intestinal microbial populations and serum cholesterol of broilers fed diets containing *Lactobacillus* cultures. *Poult. Sci.* 77: 1259–1265.

Kabir, S. M. L., M. M. Rahman, M. B. Rahman. 2005. Potentiation of probiotics in promoting microbiological meat quality of broilers. *J. Bangladesh Soc. Agric. Sci. Technol.* 2: 93–96.

Kabir, S. M. L., M. M. Rahman, M. B. Rahman, M. M. Rahman, S. U. Ahmed. 2004. The dynamics of probiotics on growth performance and immune response in broilers. *Int. J. Poult. Sci.* 3: 361–364.

Kankaanpaa, P. E., S. J. Salminen, E. Isolauri, Y. K. Lee. 2001. The influence of polyunsaturated fatty acids on probiotic growth and adhesion. *FEMS Microbiol. Lett.* 194: 149–153.

Karaoglu, M., H. Durdag. 2005. The influence of dietary probiotic (*Saccharomyces cerevisiae*) supplementation and different slaughter age on the performance, slaughter and carcass properties of broilers. *Int. J. Poult. Sci.* 4: 309–316.

Karmali, M. A., V. Gannon, J. M. Sargeant. 2010. Verocytotoxin-producing *Escherichia coli* (VTEC). *Vet. Microbiol.* 140: 360–370.

Kawakami, S. I., Y. Tomoya, N. Naoto, C. Yimin. 2010. Feeding of lactic acid bacteria and yeast on growth and diarrhea of Holstein calves. *J. Anim. Vet. Adv.* 7: 1112–1114.

Khan, M., D. Raoult, H. Richet, H. Lepidi, B. La Scola. 2007. Growth-promoting effects of single-dose intragastrically administered probiotics in chickens. *Br. Poult. Sci.* 48: 732–735.

Khan, S. H., A. Muhammad, M. Nasir, R. Abdul, F. Ghulam. 2011. Effects of supplementation of multi-enzyme and multi-species probiotic on production performance, egg quality, cholesterol level and immune system in laying hens. *J. Applied Anim. Res.* 39: 386–398.

Kim, K., M. Song, Y. Liu, P. Ji. 2022. Enterotoxigenic *Escherichia coli* infection of weaned pigs: Intestinal challenges and nutritional intervention to enhance disease resistance. *Front. Immunol.* 13: 885253.

Koenen, M. E., J. Kramer, R. van der Hulst, L. Heres, S. H. M. Jeurissen, W. J. A. Boersma. 2004. Immunomodulation by probiotic lactobacilli in layer- and meat-type chickens. *Br. Poult. Sci.* 45: 355–366.

Konieczka, P., K. Ferenc, J. N. Jørgensen, L. H. B. Hansen, R. Zabielski, J. Olszewski, Z. Gajewski, M. Mazur-Kuśnirek, D. Szkopek, N. Szyryńska, K. Lipiński. 2023. Feeding Bacillus-based probiotics to gestating and lactating sows is an efficient method for improving immunity, gut functional status and biofilm formation by probiotic bacteria in piglets at weaning. *Anim. Nutr.* 13: 361–372.

Krehbiel, C. R., S. R. Rust, G. Zhang, S. E. Gilliland. 2003. Bacterial direct-fed microbials in ruminant diets: Performance response and mode of action. *J. Anim. Sci.* 81: 120–132.

Kreuzer, S., P. Janczyk, J. Assmus, M. F. Schmidt, G. A. Brockmann, K. Nöckler. 2012. No beneficial effects evident for Enterococcus faecium NCIMB 10415 in weaned pigs infected with Salmonella enterica serovar Typhimurium DT104. *Appl. Environ. Microbiol.* 78(14): 4816–4825.

Kritas, S. K., T. Marubashi, G. Filioussis, E. Petridou, G. Christodoulopoulos, A. R. Burriel, A. Tzivara, A. Theodoridis, M. Pískoriková. 2015. Reproductive performance of sows was improved by administration of a sporing bacillary probiotic (Bacillus subtilis C-3102). *J. Anim. Sci.* 93(1): 405–413.

Kumprechtova, D., P. Zobac, I. Kumprecht. 2000. The effect of *Saccharomyces cerevisiae* Sc47 on chicken broiler performance and nitrogen output. *Czech J. Anim. Sci.* 45: 169–177.

Kurtoglu, V., F. Kurtoglu, E. Seker, B. Coskun, T. Balevi, E. S. Plat. 2004. Effect of probiotic supplementation on laying hen diets on yield performance and serum and egg yolk cholesterol. *Food Addit. Contam.* 21: 817–823.

Kuzyšinová, K., D. Mudroňová, J. Toporčák, L. Molnár. 2012. *In vivo* testing of the efficacy of bee probiotic lactobacilli. *Folia Vet.* 56: 14–16.

Kuzyšinová, K., D. Mudroňová, J. Toporčák, L. Molnár. 2023. Testing of the efficacy of bee probiotic lactobacilli under in vivo conditions. *Folia Vet.* 67: 45–50.

Lallés, J.-P., P. Bosi, H. Smidt, C. R. Stokes. 2007. Nutritional management of gut health in pigs around weaning. *Proc. Nutr. Soc.* 66: 260–268.

Lan, P. T., M. Sakamoto, Y. Benno. 2004. Effects of two probiotic *Lactobacillus* strains on jejunal and cecal microbiota of broiler chicken under acute heat stress condition as revealed by molecular analysis of 16S rRNA genes. *Microbiol. Immunol.* 48: 917–929.

La Ragione, R. M., G. Casula, S. M. Cutting, M. J. Woodward. 2001. *Bacillus subtilis* spores competitively exclude *Escherichia coli* O78:K80 in poultry. *Vet. Microbiol.* 79: 133–142.

La Ragione, R. M., A. Narbad, M. J. Gasson, M. J. Woodward. 2004. *In vivo* characterization of *Lactobacillus johnsonii* FI9785 for use as a defined competitive exclusion agent against bacterial pathogens in poultry. *Lett. Appl. Microbiol.* 38: 197–205.

La Ragione, R. M., M. J. Woodward. 2003. Competitive exclusion by *Bacillus subtilis* spores of *Salmonella enterica* serotype Enteritidis and *Clostridium perfringens* in young chickens. *Vet. Microbiol.* 94: 245–256.

Lauková, A., E. Styková, I. Kubašová, S. Gancarčíková, I. Plachá, D. Mudroňová, A. Kandričáková, R. Miltko, G. Belzecki, I. Valocký, V. Strompfová. 2018. Enterocin M and its beneficial effects in horses—a pilot experiment. *Probiotics Antimicrob. Proteins.* http://doi.org/10.1007/s12602-018-9390-2.

Le, O., P. Dart, K. Harper, D. Zhang, B. Schofield, Callaghan., Lisle, A., Klieve. A & McNeill, D. 2016. Effect of probiotic *Bacillus amyloliquefaciens* strain H57 on productivity and the incidence of diarrhea in dairy calves. *Anim. Prod. Sci.* 57: 912–919.

Lee, D. J., R. A. Drongowski, A. G. Coran, C. M. Harmon. 2000. Evaluation of probiotic treatment in a neonatal animal model. *Pediatr. Surg. Int.* 16: 237–242.

Lee, J. I., Y. D. Kim, D. Y. Kim, et al. 2002. Effects of *Saccharomyces cerevisiae* on growth performance and meat quality of broiler chickens. *Proc. Kor. J. Anim. Sci. Technol.* 34. www.poultryscience.org/ps/paperpdfs/05/p0571015.pdf.

Lee, K., H. S. Lillehoj, G. R. Siragusa. 2010. Direct-fed microbials and their impact on the intestinal microflora and immune system of chickens. *J. Poult. Sci.* 47: 106–114.

Lee, S. H., H. S. Lillehoj, D. W. Park, Y. H. Hong, J. J. Lin. 2007. The effects of *Pediococcus* and *Saccharomyces*-based probiotic (Mitomax) on coccidiosis in broiler chickens. *Comp. Immunol. Microb.* 30: 261–267.

Lei, X., X. Piao, Y. Ru, H. Zhang, A. Peron, H. Zhang. 2015. Effect of *Bacillus amyloliquefaciens*-based direct-fed microbial on performance, nutrient utilization, intestinal morphology and cecal microflora in broiler chickens. *Asian-Australasian J. Anim. Sci.* 28: 239–246.

Lessard, M., M. Dupuis, N. Gagnon, É. Nadeau, J. J. Matte, J. Goulet, J. M. Fairbrother. 2009. Administration of *Pediococcus acidilactici* or *Saccharomyces cerevisiae boulardii* modulates development of porcine mucosal immunity and reduces intestinal bacterial translocation after *Escherichia coli* challenge. *J. Anim. Sci.* 87: 922–934.

Li, H. H., Y. P. Li, Q. Zhu, J. Y. Qiao, W. J. Wang. 2018. Dietary supplementation with *Clostridium butyricum* helps to improve the intestinal barrier function of weaned piglets challenged with enterotoxigenic *Escherichia coli* K88. *J. Appl. Microbiol.* 125: 964–975.

Li, R., J. Liu, Y. Liu, L. Cao, W. Qiu, M. Qin. 2023. Probiotic effects of *Bacillus subtilis* on growth performance and intestinal microecological balance of growing-to-finishing pigs. *J. Food Biochem.* 2023: 7150917.

Liao, S. F., M. Nyachoti. 2017. Using probiotics to improve swine gut health and nutrient utilization. *Anim. Nutr.* 3: 331–343.

Lila, Z. A., N. Mohammed, T. Yasui, Y. Kurokawa, S. Kanda, H. Itabashi H. 2004. Effects of a twin strain of *Saccharomyces cerevisiae* live cells on mixed ruminal microorganism fermentation *in vitro*. *J. Anim. Sci.* 82(6): 1847–1854.

Lim, A., H.-M. Tan. 2009. Effects of farm animals. In *Handbook of Probiotics and Prebiotics*, eds. Y. K. Lee and S. Salminen, 321–375. John Wiley & Sons, Hoboken, NJ.

Liu, J. R, S. F. Lai, B. Yu. 2007. Evaluation of an intestinal *Lactobacillus reuteri* strain expressing rumen fungal xylanase as a probiotic for broiler chickens fed on a wheat-based diet. *Br. Poult. Sci.* 48: 507–514.

Loddi, M. M., E. Gonzalez, T. S. Takita, A. A. Mendes, R. O. Roca, R. Roca. 2000. Effect of the use of probiotic and antibiotic on the performance, yield and carcass quality of broilers. *Rev. Bras. Zootec.* 29: 1124–1131.

Luise, D., M. Bertocchi, V. Motta, C. Salvarani, P. Bosi, A. Luppi, F. Fanelli, M. Mazzoni, I. Archetti, G. Maiorano, B. K. K. Nielsen, P. Trevisi. 2019. *Bacillus* sp. probiotic supplementation diminish the *Escherichia coli* F4ac infection in susceptible weaned pigs by influencing the intestinal immune response, intestinal microbiota and blood metabolomics. *J. Animal Sci. Biotechnol.* 10: 74.

Luo, Y., W. Ren, H. Smidt, A. G. Wright, B. Yu, G. Schyns, U. M. McCormack, A. J. Cowieson, J. Yu, J. He, H. Yan, J. Wu, R. I. Mackie, D. Chen. 2022. Dynamic distribution of gut microbiota in pigs at different growth stages: Composition and contribution. *Microbiol. Spectr.* 10(3): e0068821.

Luppi, A. 2017. Swine enteric colibacillosis: Diagnosis, therapy and antimicrobial resistance. *Porcine. Health Manag.* 8(3): 16.

Maamouri, O., H. Selmi, N. M´hamdi. 2014. Effects of yeast (*Saccharomyces cerevisiae*) feed supplement on milk production and its composition in Tunisian Holstein Friesian cows. Sci. Agric. Bohem. 45: 170–174.

Magalhaes, V. J. A., F. Susca, F. S. Lima, A. F. Branco, I. Yoon, J. E. P. Santos. 2008. Effect of feeding yeast culture on performance, health, and immunocompetence of dairy calves. *J. Dairy Sci.* 91: 1497–1509.

Mahdavi, A. H., H. R. Rahman, J. Pourreza. 2005. Effect of probiotic supplements on egg quality and laying hen's performance. *Int. J. Poult. Sci.* 4(7): 488–492.

Mahmood, T., M. S. Anjum, I. Hussain, R. Perveen. 2005. Effect of probiotic and growth promoters on chemical composition of broiler carcass. *Int. J. Agric. Biol.* 7: 1036–1037.

Majidi-Mosleh, A., A. A. Sadeghi, S. N. Mousavi, M. Chamani, A. Zarei. 2017. Ileal MUC2 gene expression and microbial population, but not growth performance and immune response, are influenced by in ovo injection of probiotics in broiler chickens. *Br. Poult. Sci.* 58: 40–45.

Mancini, S., G. Paci. 2021. Probiotics in rabbit farming: Growth performance, health status, and meat quality. *Animals.* 11(12): 3388.

Mansoub, N. H. 2010. Effect of probiotic bacteria utilization on serum cholesterol and triglycerides contents and performance of broiler chickens. *Global. Vet.* 5: 184–186.

Maragkoudakis, P. A., C. M. Konstantinos, D. Psyrras, et al. 2009. Functional properties of novel protective lactic acid bacteria and application in raw chicken meat against *Listeria monocytogenes* and *Salmonella enteriditis. Int. J. Food Microbiol.* 130: 219–226.

Marciňáková, M., A. Lauková, M. Simonová, V. Strompfová, B. Koréneková, P. Naď. 2008. A new probiotic and bacteriocin-producing strain of *Enterococcus faecium* EF 9296 and its use in grass ensiling. *Czech J. Anim. Sci.* 53: 336–345.

Marlier, D., R. Dewrée, C. Lassence, D. Licois, J. Mainil, P. Coudert, L. Meulemans, R. Ducatelle, H. Vindevogel. 2006. Infectious agents associated with epizootic rabbit enteropathy: Isolation and attempts to reproduce the syndrome. *Vet. J.* 172: 493–500.

Maruščáková, I. C., P. Schusterová, B. Bielik, J. Toporčák, K. Bíliková, D. Mudroňová. 2020. Effect of application of probiotic pollen suspension on immune response and gut microbiota of honey bees (*Apis mellifera*). *Probiotics Antimicrob. Proteins.* 12: 929–936.

Mikulski, D., J. Jankowski, J. Naczmanski, M. Mikulska, V. Demey. 2012. Effects of dietary probiotic (*Pediococcus acidilactici*) supplementation on performance, nutrient digestibility, egg traits, egg yolk cholesterol and fatty acid profile in laying hens. *Poult. Sci.* 91: 2691–2700.

Moates, T. 2009. Probiotic truth. *Am. Q. Horse J.* March: 88–90.

Mohan, B., R. Kadirvel, M. Bhaskaran, A. Natarajan. 1995. An effect of probiotic supplementation on serum/yolk cholesterol and on egg shell thickness in layers. *Br. Poult. Sci.* 36: 799–803.

Mohsin, M., Z. Zhang, G. Yin. 2022. Effect of probiotics on the performance and intestinal health of broiler chickens infected with *Eimeria tenella. Vaccines (Basel).* 10(1): 97.

Morvayová, H., E. Styková, F. Novotný, J. Pošivák, Ľ. Šťavová, I. Valocký. 2008. Inhibition of the pathogenic strains obtained from the mare with chronic endometritis by lactobacilli strains. *Proceedings from International Scientific Conference, Zdrój,* July 22–28, 2008, Polanica, Polland, 133–136.

Mountzouris, K. C., C. Balaskas, I. Xanthakos, A. Tzivinikou, K. Fegeros. 2009. Effects of a multi-species probiotic on biomarkers of competitive exclusion efficacy in broilers challenged with *Salmonella enteritidis. Br. Poult. Sci.* 50(4): 467–478.

Mountzouris, K. C., P. Tsirtsikos, E. Kalamara, S. Nitsch, G. Schatzmayr, K. Fegeros. 2007. Evaluation of the efficacy of probiotic containing *Lactobacillus, Bifidobacterium, Enterococcus,* and *Pediococcus* strains in promoting broiler performance and modulating cecal microbiota composition and metabolic activities. *Poult. Sci.* 86: 309–317.

Mudroňová, D., J. Toporčák, R. Nemcová, S. Gancarčíková, V. Hajdučková, K. Rumanovská. 2011. *Lactobacillus* sp. as a potential probiotic for the prevention of *Paenibacillus larvae* infection in honey bees. *J. Api. Res.* 50(4): 323–324.

Mutuş, R., N. Kocabagli, M. Alp, N. Acar, M. Eren, S. S. Gezen. 2006. The effect of dietary probiotic supplementation on tibial bone characteristics and strength in broilers. *Poult. Sci.* 85: 1621–1625.

Nemcová, R., A. Bomba, S. Gancarčíková, K. Reiffová, P. Guba, J. Koščová, Z. Jonecová, Ľ. Scirankova, A. Bugarský. 2007. Effects of the administration of lactobacilli, maltodextrins and fructooligosaccharides upon the adhesion of *E. coli* O8:K88 to the intestinal mucosa and organic acid levels in the gut contents of piglets. *Vet. Res. Commun.* 31: 791–800.

Nemcová, R., D. Borovská, J. Koščová, S. Gancarčíková, D. Mudroňová, V. Buleca, J. Pistl. 2012. The effect of supplementation of flax-seed oil on interaction of *Lactobacillus plantarum*—Biocenol LP96 and *Escherichia coli* O8:K88ab:H9 in the gut of germ-free piglets. *Res. Vet. Sci.* 93: 39–41.

Nocek, J. E., W. P. Kautz. 2006. Direct-fed microbial supplementation on ruminal digestion, health, and performance of pre- and postpartum dairy cattle. *J. Dairy Sci.* 89: 260–266.

Noujaim, J. C., R. L. Andreatti Filho, E. T. Lima, A. S. Okamoto, R. L. Amorim, R. T. Neto. 2008. Detection of T lymphocytes in intestine of broiler chicks treated with *Lactobacillus* spp. and Challenged with *Salmonella enterica* serovar Enteritidis. *Poult. Sci.* 87: 927–933.

Nousiainen, J., P. Javanainen, J. Setala. 2004. *Lactic Acid Bacteria: Microbiology and Functional Concepts*, 3rd ed., 547–588. Valio Ltd, Helsinki, Finland.

Nurmi, E., M. Rantala. 1973. New aspects of *Salmonella* infection in broiler production. *Nature*. 241: 210–211.

Pachla, A., A. A. Ptaszyńska, M. Wicha, M. Kunat, J. Wydrych, E. Oleńska, W. Małek. 2021. Insight into probiotic properties of lactic acid bacterial endosymbionts of *Apis mellifera* L. derived from the polish apiary. *Saudi J. Biol. Sci.* 28: 1890–1899.

Panda, A. K., S. S. Rama Rao, M. V. L. N. Raju, S. S. Sharma. 2008. Effect of probiotic (*Lactobacillus sporogenes*) feeding on egg production and quality, yolk cholesterol and humoral immune response of White Leghorn layer breeders. *J. Sci. Food Agric.* 88: 43–47.

Panda, A. K., M. R. Reedy, S. V. Rama Rao, N. K. Praharaj.2003. Production performance, serum/yolk cholesterol and immune competence of white leghorn layers as influenced by dietary supplementation with probiotic. *Trop. Anim. Health Pro.* 35: 85–94.

Pang, J., Y. Liu, L. Kang, H. Ye, J. Zang, J. Wang, D. Han. 2022a. Bifidobacterium animalis promotes the growth of weaning piglets by improving intestinal development, enhancing antioxidant capacity, and modulating gut microbiota. *Appl. Environ. Microbiol.* 88(22): e0129622.

Pang, Y., H. Zhang, H. Wen, H. Wan, H. Wu, Y. Chen, S. Li, L. Zhang, X. Sun, B. Li, X. Liu. 2022b. Yeast probiotic and yeast products in enhancing livestock feeds utilization and performance: An overview. *J. Fungi.* 8: 1191.

Park, J. H., I. H. Kim. 2015. The effects of the supplementation of *Bacillus subtilis* RX7 and B2A strains on the performance, blood profiles, intestinal *Salmonella* concentration, noxious gas emission, organ weight and breast meat quality of broiler challenged with *Salmonella typhimurium*. *J. Anim. Physiol. Anim. Nutr.* 99: 326–334.

Park, J. H., G. H. Park, K. S. Ryu. 2002. Effect of feeding organic acid mixture and yeast culture on performance and egg quality of laying hens. *Korea J. Poult. Sci.* 29(2): 109–115.

Park, J. W., J. S. Jeong, S. I. Lee, I. H. Kim. 2016. Effect of dietary supplementation with a probiotic (*Enterococcus faecium*) on production performance, excreta microflora, ammonia emission and nutrient utilization in ISA brown laying hens. *Poult. Sci.* 95: 2829–2835.

Patterson, J. A., K. M. Burkholder. 2003. Application of prebiotics and probiotics in poultry production. *Poult. Sci.* 82: 627–631.

Peghaire, E., A. Moné, F. Delbac, D. Debroas, F. Chaucheyras-Durand, H. El Alaoui. 2020. A *Pediococcus* strain to rescue honeybees by decreasing *Nosema ceranae*—and pesticide-induced adverse effects. *Pest. Biochem. Physiol.* 163: 138–146.

Peng, J., Y. Tang, Y. Huang. 2021. Gut health: The results of microbial and mucosal immune interactions in pigs. *Anim. Nutr.* 7(2): 282–294.

Peng, X., R. Wang, L. Hu, Q. Zhou, Y. Liu, M. Yang, Z. Fang, Y. Lin, S. Xu, B. Feng, J. Li, X. Jiang, Y. Zhuo, H. Li, D. Wu, L. Che. 2019. *Enterococcus faecium* NCIMB 10415 administration improves the intestinal health and immunity in neonatal piglets infected by enterotoxigenic *Escherichia coli* K88. *J. Animal Sci. Biotechnol.* 10: 72.

Peterson, R. E., T. J. Klopfenstein, G. E. Erickson, J. Folmer, S. Hinkley, R. A. Moxley, D. R. Smith. 2007. Effect of *Lactobacillus acidophilus* strain NP51 on *E. coli* 0157:H7 fecal shedding and finishing performance of beef feedlot cattle. *J. Food Prot.* 70: 287–291.

Phuoc, T. L., U. Jamikorn. 2017. Effects of probiotic supplement (*Bacillus subtilis* and *Lactobacillus acidophilus*) on feed efficiency, growth performance, and microbial population of weaning rabbits. *Asian-Australas J. Anim. Sci.* 30(2): 198–205.

Pietropaoli, M., E. Carpana, M. Milito, M. Palazzetti, M. Guarducci, S. Croppi, G. Formato. 2022. Use of *Lactobacillus plantarum* in preventing clinical cases of American and European foulbrood in central Italy. *Appl. Sci.* 12(3): 1388.

Poppy, G., A. Rabiee, I. Lean, W. Sanchez, K. Dorton, P. Morley. 2012. A meta-analysis of the effects of *Saccharomyces cerevisiae* on milk production of lactating dairy cows. *J. Dairy Sci.* 95: 6027–6041.

Ptaszyńska, A. A., W. Mułenko. 2013. Selected aspects of the structure development taxonomy and biology of microsporidian parasites belonging to the genus *Nosema. Med. Weter.* 69(12): 716–726.

Pupa, P., P. Apiwatsiri, W. Sirichokchatchawan, N. Pirarat, T. Nedumpun, D. J. Hampson, N. Muangsin, N. Prapasarakul. 2022. Microencapsulated probiotic *Lactiplantibacillus plantarum* and/or *Pediococcus acidilactici* strains ameliorate diarrhoea in piglets challenged with enterotoxigenic *Escherichia coli. Sci. Rep.* 12: 7210.

Rahman, M. R. T., I. Fliss, E. Biron. 2022. Insights in the development and uses of alternatives to antibiotic growth promoters in poultry and swine production. *Antibiotics (Basel).* 11(6): 766.

Ramasamy, K., N. Abdullah, S. Jalaludin, M. Wong, Y. W. Ho. 2009. Effects of *Lactobacillus* cultures on performance of laying hens and total cholesterol, Lipid and fatty acid composition of egg yolk. *J. Sci. Food Agric.* 89: 482–486.

Riddell, J. B., A. J. Gallegos, D. L. Harmon, K. R. McLeod. 2010. Addition of a *Bacillus* based probiotic to the diet of preruminant calves: Influence on growth, health, and blood parameters. *Intern. J. Appl. Res. Vet. Med.* 8: 78–84.

Rijkers, G. T., S. Bengmark, P. Enck, et al. 2010. Guidance for substantiating the evidence for beneficial effects of probiotics: Current status and recommendations for future research. *J. Nutr.* 140: 671–676.

Ritzi, M. M., W. Abdelrahman, M. Mohnl, R. A. Dalloul. 2014. Effects of probiotics and application methods on performance and response of broiler chickens to an *Eimeria* challenge. *Poult. Sci.* 93: 2772–2778.

Rohrbach, B. W., P. C. Sheerin, C. K. Cantrell, P. M. Matthews, J. V. Steiner, L. E. Dodds. 2007. Effect of adjunctive treatment with intravenously administered *Propionibacterium acnes* on reproductive performance in mares with persistent endometritis. *J. Am. Vet. Med. Assoc.* 231(1): 107–113.

Roselli, M., A. Finamore, M. S. Britti, P. Bosi, I. Oswald, E. Mengheri. 2005. Alternatives to in-feed antibiotics in pigs: Evaluation of probiotics, zinc or organic acids as protective agents for the intestinal mucosa. A comparison of in vitro and in vivo results. *Anim. Sci.* 54: 203–218.

Roselli, M., R. Pieper, C. Rogel-Gaillard, H. de Vries, M. Bailey, H. Smidt, C. Lauridsen. 2017. Immunomodulating effects of probiotics for microbiota modulation, gut health and disease resistance in pigs. *Anim. Feed Sci. Technol.* 233: 104–119.

Rossi, C., A. Sgoifo, V. Dell-Orto, A. L. Bassini, E. Chevaux, G. Savoini. 2006. Effects of live yeast in beef cattle studied. *Feedstuffs*: 11 pp.

Russell, S. M., J. L. Grimes. 2009. The effect of a direct-fed microbial (Primalac) on turkey live performance. *J. Appl. Poult. Res.* 18: 185–192.

Rybarczyk, A., E. Bogusławska-Wąs, A. Łupkowska. 2020. Effect of EM probiotic on gut microbiota, growth performance, carcass and meat quality of pigs. *Livest. Sci.* 241: 104206.

Salim, H. M., H. K. Kang, N. Akter, et al. 2013. Supplementation of Direct-fed microbials as an alternative to antibiotic on growth performance, immune response, cecal microbial population and ileal morphology of broiler chickens. *Poult. Sci.* 92: 2084–2090.

Santini, C., L. Baffoni, F. Gaggia, M. Granata, R. Gasbarri, D. Di Gioia, B. Biavati. 2010. Characterization of probiotic strains: An application as feed additives in poultry against *Campylobacter jejuni. Int. J. Food Microbiol.* 141: 98–108.

Scharek, L., J. Guth, M. Filter, M. F. G. Schmidt. 2007. Influence of the probiotic bacteria *Enterococcus faecium* NCIMB 10415 (SF68) and *Bacillus cereus* var. *toyoi* NCIMB 40112 on the development of serum IgG and fecal IgA of sows and piglets. *Arch. Anim. Nutr.* 61: 223–234.

Scharek, L., J. Guth, K. Reiter, K. D. Weyrauch, D. Taras, P. Schwerk, P. Schierack, M. F. G. Schmidt, L. H. Wieler, K. Tedin. 2005. Influence of a probiotic *Enterococcus faecium* strain on development of the immune system of sows and piglets. *Vet. Immunol. Immunopathol.* 105: 151–161.

Schokker, D., A. J. Hoekman, M. A. Smits, J. M. Rebel. 2009. Gene expression patterns associated with chicken jejunal development. *Dev. Comp. Immunol.* 33(11): 1156–1164.

Schoster, A., L. Guardabassi, H. R. Staempfli, M. Abrahams, M. Jalali, J. S. Weese. 2016. The longitudinal effect of a multi-strain probiotic on the intestinal bacterial microbiota of neonatal foals. *Equine Vet. J.* 48(6): 689–696.

Schoster, A., H. R. Staempfli, M. Abrahams, M. Jalali, J. S. Weese, L. Guardabassi. 2015. Effect of a probiotic on prevention of diarrhea and *Clostridium difficile* and *Clostridium perfringens* shedding in foals. *J. Vet. Intern. Med.* 29(3): 925–931.

Schoster, A., J. S. Weese, L. Guardabassi. 2014. Probiotic use in horses—What is the evidence for their clinical efficacy? *J. Vet. Intern. Med.* 28: 1640–1652.

Seidavi, A., M. Dadashbeiki, M. H. Alimohammadi-Saraei, R. van den Hoven, R. Payan-Carreira, V. Laudadio, V. Tufarelli. 2017. Effects of dietary inclusion level of a mixture of probiotic cultures and enzymes on broiler chickens immunity response. *Environ. Sci. Pollut. Res.* 24: 4637–4644.

Seo, K. J., S.-W. Kim, M. H. Kim, S. D. Upadhaya, D. K. Kam, J. K. Ha. 2010. Direct-fed microbials for ruminant animals. *Asian-Australas. J. Anim. Sci.* 12: 1657–1667.

Shareef, A. M., A. S. A. Al-Dabbagh. 2009. Effect of probiotic (*Saccharomyces cerevisiae*) on performance of broiler chicks. *Iraqi J. Vet. Sci.* 23: 23–29.

Shin, D., S. Y. Chang, P. Bogere, K. Won, J. Y. Choi, Y. J. Choi, H. K. Lee, J. Hur, B. Y. Park, Y. Kim, J. Heo. 2019. Beneficial roles of probiotics on the modulation of gut microbiota and immune response in pigs. *PLoS ONE.* 14(8): e0220843.

Shu, Q., F. Qu, H. S. Gill. 2001. Probiotic treatment using *Bifidobacterium lactis* HN019 reduces weaning diarrhea associated with rotaviruses and *Escherichia coli* infection in piglet model. *J. Pediatr. Gastroenterol. Nutr.* 33: 171–177.

Simonová, M., A. Lauková. 2004. Isolation of faecal *Enterococcus faecium* strains from rabbits and their sensitivity to antibiotics and ability to bacteriocin production. *Bull. Vet. Inst. Pulawy.* 48: 383–386.

Simonová, M., A. Lauková, I. Štyriak, I. 2005. Enterococci from rabbit—Potential feed additive. *Czech J. Anim. Sci.* 50: 416–421.

Song, D. S., B. H. Kang, J. S. Oh, G. W. Ha, J. S. Yang, H. J. Moon, Y. S. Jang, B. K. Park. 2006. Multiplex reverse transcription-PCR for rapid differential detection of porcine epidemic diarrhoea virus, transmissible gastroenteritis virus, and porcine group A rotavirus. *J. Vet. Diagn. Investigat.* 18: 278–281.

Song, J., K. Xiao, Y. L. Ke, et al. 2014. Effect of a probiotic mixture on intestinal microflora, morphology and barrier integrity of broilers subjected to heat stress. *Poult. Sci.* 93: 581–588.

Sorensen, M. T., E. M. Vestergaard, S. K. Jensen, C. Lauridsen, S. Hojsgaard. 2009. Performance and diarrhoea in piglets following weaning at seven weeks of age: Challenge with *E. coli* O 149 and effect of dietary factors. *Livestock Sci.* 123: 314–321.

Stanley, V. G., C. Gray, M. Daley, W. F. Krueger, A. E. Sefton. 2004. An alternative to antibiotic-based drugs in feed for enhancing performance of broilers grown on *Eimeria* spp.- infected litter. *Poult. Sci.* 83: 39–44.

Stein, D. R., D. T. Allen, E. B. Perry, J. C. Bruner, K. W. Gates, T. G. Rehberger, K. Mertz, D. Jones, L. J. Spicer. 2006. Effects of feeding propionibacteria to dairy cows on milk yield, milk components, and reproduction. *J. Dairy Sci.* 89(1): 111–125.

Strompfova, V., M. Marcinakova, S. Gancarcikova, et al. 2005. New probiotic strain *Lactobacillus fermentum* AD1 and its effect in Japanese quail. *Vet. Med.-Czech* 50(9): 415–420.

Su, W., T. Gong, Z. Jiang, Z. Lu, Y. Wang. 2022. The role of probiotics in alleviating postweaning diarrhea in piglets from the perspective of intestinal barriers. *Front. Cell. Infect. Microbiol.* 12: 883107.

Supuková, A., P. Supuka, J. Salaj, K. Oberhauserová, A. Brandeburová. 2010. *Lactobacillus fermentum* and *Enterococcus faecium* in therapy and prevention of digestive disorders in domestic rabbit (*Oryctolagus cuniculus*). In *Conference Proceedings: International Scientific Conference: Probiotics and Prebiotics*, June 15–17, Košice, Slovakia, 181.

Swyers, K. L., A. O. Burk, T. G. Hartsock, E. M. Ungerfeld, J. L. Shelton. 2008. Effects of direct-fed microbial supplementation on digestibility and fermentation end-products in horses fed low- and high-starch concentrates. *J. Anim. Sci.* 86(10): 2596–2608.

Szabóová, R., Ľ. Chrastinová, V. Strompfová, M. Simonová, Z. Vasilková, A. Lauková, I. Plachá, K. Čobanová, M. Chrenková, J. Mojto, R. Jurčík. 2008. Nutrition and digestive physiology. *9th World Rabbit Congress*, June 10–13. Verona, Italy, 821–825.

Takahashi, J., B. Mwenya, B. Santoso, C. Sar, K. Umetsu, T. Kishimoto, K. Nishizaki, K. Kimura, O. Hamamoto. 2005. Mitigation of methane emission and energy recycling in animal agricultural systems. *Asian Aust. J. Anim. Sci.* 18: 1199–1208.

Takahashi, K., Y. Akiba, A. Matsuda. 1997. Effect of probiotics on immune response in broilers chicks under different sanitary conditions or immune activations. *Anim. Sci. Technol.* 68: 537–544.

Tang, X., K. Xiong, R. Fang, M. Li. 2022. Weaning stress and intestinal health of piglets: A review. *Front. Immunol.* 24(13): 1042778.

Thomsson, A., D. Rantzer, J. Botermans, J. Svedsen. 2008. The effect of feeding system at weaning on performance, health and feeding behaviour of pigs of different sizes. *Acta Agr. Scand. Sect. A-Anim. Sci.* 58: 78–83.

Timmerman, H. M., C. J. Koning, L. Mulder, F. M. Rombouts, A. C. Beynen. 2004. Monostrain, multistrain and multispecies probiotics—A comparison of functionality and efficacy. *Int. J. Food Microbiol.* 96: 219–233.

Timmerman, H. M., L. Mulder, H. Everts, et al. 2005. Health and growth of veal calves fed milk replacers with or without probiotics. *J. Dairy Sci.* 88: 2154–2165.

Tkalcic, S., T. Zhao, B. G. Harmon, M. P. Doyle, C. A. Brown, et al. 2003. Fecal shedding of enterohemorrhagic *Escherichia coli* in weaned calves following treatment with probiotic *Escherichia coli*. *J. Food Prot*. 66: 1184–1189.

Torok, V. A., R. J. Hughes, K. Ophel-Keller, M. Ali, R. MacAlpine. 2009. Influence of different litter materials on cecal microbiota colonization in broiler chickens. *Poult. Sci*. 88(12): 2474–2481.

Torres-Rodriguez, A., A. M. Donoghue, D. J. Donoghue, J. T. Barton, G. Tellez, B. M. Hargis. 2007. Performance and condemnation rate analysis of commercial turkey flocks treated with a *Lactobacillus* spp.-based probiotic. *Poult. Sci*. 86: 444–446.

Tsiapali, E., S. Whaley, J. Kalbfleisch, H. Ensley, W. Browder, D. Williams. 2001. Glucans and related natural polymers exhibit weak solution free radical scavenging activity, but stimulate free radical activity in a murine macrophage cell line. *Free Radic. Biol. Med*. 30: 393–402.

Uyeno, Y., S. Shigemori, T. Shimostato. 2015. Minireview. The effect of probiotic/prebiotic on cattle health and productivity. *Microbes Environ*. 30: 126–132.

Valeriano, V. D. V., M. P. Balolong, D. K. Kang. 2017. Probiotic roles of *Lactobacillus* sp. in swine: Insights from gut microbiota. *J. Appl. Microbiol*. 122(3): 554–567.

Vásquez, A., E. Forsgren, I. Fries, R. J. Paxton, E. Flaberg, L. Szekely, T. C. Olofsson. 2012. Symbionts as major modulators of insect health: Lactic acid bacteria and honeybees. *PLoS ONE*. 7(7): http://doi.org/10.1371/annotation/3ac2b867-c013-4504-9e06-bebf3fa039d1.

Vásquez, A., T. C. Olofsson. 2009. The lactic acid bacteria involved in the production of bee pollen and bee bread. *J. Api. Res*. 48: 189–195.

Veljović, K., M. Dinić, J. Lukić, S. Mihajlović, M. Tolinački, M. Živković, J. Begović, I. Mrvaljević, N. Golić, A. Terzić-Vidojević. 2017. Promotion of early gut colonization by probiotic intervention on microbiota diversity in pregnant sows. *Front. Microbiol*. 8: 2028.

Vervelde, L., E. M. Janse, A. N. Vermeulen, S. H. M. Jeurissen. 1998. Induction of local and systemic immune response using cholera toxin as vehicle to deliver antigen in the *lamina propria* of the chicken intestine. *Vet. Immunol. Immunopathol*. 62: 261–272.

Vicente, J. L., A. Torres-Rodriguez, S. E. Higgins, et al. 2008. Effect of a selected *Lactobacillus* spp.-based probiotic on *Salmonella enterica* serovar Enteritidis-infected broiler chicks. *Avian Dis*. 52: 143–146.

Vondruskova, H., R. Slamova, R. Trckova, Z. Zraly, I. Pavlik. 2010. Alternatives to antibiotic growth promoters in prevention of diarrhoea in weaned piglets: A review. *Vet. Med*. 55: 199–224.

Wang, A., H. Yu, X. Gao, X. Li, S. Qiao. 2009. Influence of *Lactobacillus fermentum* I5007 on the intestinal and systemic immune responses of healthy and *E. coli* challenged piglets. *Antonie van Leeuwenhoek, Int. J. Gen. Mol. Microbiol*. 96: 89–98.

Wang, J., Y. Zeng, S. Wang, H. Liu, D. Zhang, W. Zhang, Y. Wang, H. Ji. 2018. Swine-derived probiotic *Lactobacillus plantarum* inhibits growth and adhesion of enterotoxigenic *Escherichia coli* and mediates host defense. *Front. Microbiol*. 9: 1364.

Wang, K., G. Chen, G. Cao, Y. Xu, Y. Wang, C. Yang. 2019a, Effects of Clostridium butyricum and *Enterococcus faecalis* on growth performance, intestinal structure, and inflammation in lipopolysaccharide-challenged weaned piglets. *J. Anim. Sci*. 97: 4140–4151.

Wang, Y., L. Gong, Y. P. Wu, Z. W. Cui, Y. Q. Wang, Y. Huang, X. P. Zhang, W. F. Li. 2019b. Oral administration of Lactobacillus rhamnosus GG to newborn piglets augments gut barrier function in pre-weaning piglets. *J. Zhejiang Univ. Sci. B*. 20(2): 180–192.

Weese, S. J., M. E. Anderson, A. Lowe, R. Penno, T. M. Da Costa, L. Button, K. C. Goth. 2004. Screening of the equine intestinal microbiota for potential probiotic organisms. *Equine Vet. J*. 36(4): 351–355.

Weese, S. J., J. Rousseau. 2005. Evaluation of *Lactobacillus pensosus* WE7 for the prevention of neonatal foal diarrhea. *J. Am. Vet. Med. Assoc*. 226(12): 2031–2034.

Willis, W. L., L. Reid. 2008. Investigating the effects of dietary probiotic feeding regimens on broiler chicken production and *Campylobacter jejuni* presence. *Poult. Sci*. 87: 60.

Wisener, L., J. Sargean, A. O'Connor, M. S. Faires, S. Glass-Kaastra. 2014. The use of direct-fed microbials to reduce shedding of *Escherichia coli* O157 in beef cattle: A systematic review and meta-analysis. *Zoonoses Public Health*. 62: 7589.

Wu, L. Y., Y. J. Fang, R. B. Tan, K. J. Shi. 2009. A comparison of cecal microbiota and volatile fatty acid concentration in goslings fed diets supplemented with or without a dried *Bacillus subtilis* culture. *J. Appl. Anim. Res*. 36: 231–234.

Wu, L. Y., R. B. Tan, K. J. Shi. 2008. Effect of a dried *Bacillus subtilis* culture on gosling growth performance. *Br. Poult. Sci*. 49: 418–422.

Wu, M., Y. Sugimura, D. Taylor, M. Yoshiyama. 2013. Honeybee gastrointestinal bacteria for novel and sustainable disease control strategies. *J. Develop. Sustain. Agricult*. 8: 85–90.

Yalcin, S., B. Ozsoy, H. Erol, S. Yalcin. 2008. Yeast culture supplementation to laying hen diets containing soybean meal or sunflower seed meal and its effect on performance, egg quality traits and blood chemistry. *J. Appl. Poultry Res*.17: 229–236.

Yalcin, S., S. Yalcin, K. Cakin, O. Eltan, L. Dağaşan. 2010. Effects of dietary yeast autolysate (*Saccharomyces cerevisiae*) on performance, egg traits, egg cholesterol content, egg yolk fatty acid composition and humoral immune response of laying hens. *J. Sci. Food Agric.* 90: 1695–1701.

Yang, F., Ch. Hou, X. Zeng, S. Qiao. 2015. The use of lactic acid bacteria as a probiotic in swine diets. *Pathogens* 4: 34–45. http://doi.org/10.3390/pathogens4010034.

Yang, H., Y. Xiang, K. Robinson, J. J. Wang, G. L. Zhang, J. C. Zhao, Y. P. Xiao. 2018. Gut microbiota is a major contributor to adiposity in pigs. *Front. Microbiol.* 9: 3045.

Yang, Y., Y. Liu, J. Liu, H. Wang, Y. Guo, M. Du, C. Cai, Y. Zhao, C. Lu, X. Guo, G. Cao, Z. Duan, B. Li, P. Gao. 2021. Composition of the fecal microbiota of piglets at various growth stages. *Front. Vet. Sci.* 8: 661671.

Yin, H., P. Ye, Q. Lei, Y. Cheng, H. Yu, J. Du, H. Pan, Z. Cao. 2020. In vitro probiotic properties of Pediococcus pentosaceus L1 and its effects on enterotoxigenic Escherichia coli-induced inflammatory responses in porcine intestinal epithelial cells. *Microb Pathog.* 144: 104163.

Yörük, M. A., M. Gül, A. Hayirli, M. Macit. 2004. The effects of supplementation of humate and probiotic on egg production and quality parameters during the late laying period in hens. *Poult. Sci.* 83(1): 84–88.

Yoshiyama, M., M. Wu, Y. Sugimura, N. Takaya, H. Kimoto-Nira, C. Suzuki. 2013. Inhibition of *Paenibacillus larvae* by lactic acid bacteria isolated from fermented materials. *J. Invert. Pathol.* 112: 62–67.

Yousefi, M., K. Karkoodi. 2007. Effect of probiotic *Thepax* and *Saccharomyces cerevisiae* supplementation on performance and egg quality of laying hens. *Int. J. Poult. Sci.* 6(1): 52–54.

Yu, B., H. Y. Tsen. 1993. *Lactobacillus* cells in the rabbit digestive tract and the factors affecting their distribution. *J. Appl. Bacteriol.* 75: 269–275.

Zhang, A. W., B. D. Lee, S. K. Lee, et al. 2005. Effects of yeast (*Saccharomyces cerevisiae*) cell components on growth performance, meat quality, and ileal mucosa development of broiler chicks. *Poult. Sci.* 84: 1015–1021.

Zhang, G., L. Ma, M. P. Doyle. 2007. Potential competitive exclusion bacteria from poultry inhibitory to *Campylobacter jejuni* and *Salmonella. J. Food. Prot.* 70: 867–873.

Zhang, L., Y.-Q. Xu, H.-Y. Liu, T. Lai, J.-L. Ma, J.-F. Wang, Y.-H. Zhu. 2010. Evaluation of *Lactobacillus rhamnosus* GG using an *Escherichia coli* K88 model of piglet diarrhoea: Effects on diarrhoea incidence, faecal microbiota and immune responses. *Vet. Microbiol.* 141: 142–148.

Zhang, W., Y. H. Zhu, J. C. Yang, G. Y. Yang, D. Zhou, J. F. Wang. 2015. A selected lactobacillus rhamnosus strain promotes EGFR-independent akt activation in an enterotoxigenic *Escherichia coli* K88-infected IPEC-J2 cell model. *PLoS ONE.* 10(4): e0125717.

Zhang, Y., Y. Zhang, F. Liu, Y. Mao, Y. Zhang, H. Zeng, S. Ren, L. Guo, Z. Chen, N. Hrabchenko, J. Wu, J. Yu. 2023. Mechanisms and applications of probiotics in prevention and treatment of swine diseases. *Porcine. Health. Manag.* 9(1): 5.

Zhang, Z. F., I. H. Kim. 2014. Effects of multistrain probiotics on growth performance, apparent ileal nutrient digestibility, blood characteristics, cecal microbial shedding, and excreta odor contents in broilers. *Poult. Sci.* 93(2): 364–370.

Zhang, Z. F., X. Mu, Y. Shi, H. Zheng. 2022. Distinct roles of honeybee gut bacteria on host metabolism and neurological processes. *Microbiol. Spectrum.* 10(2): e0243821.

Zhang, Z. F., T. X. Zhou, X. Ao, I. H. Kim. 2012. Effects of beta-glucan and *Bacillus subtilis* on growth performance, blood profiles, relative organ weight and meat quality in broilers fed maize-soybean meal based diets. *Livest. Sci.* 150: 419–424.

Zhao, T., S. Tkalcic, M. P. Doyle, B. G. Harmon, C. A. Brown, et al. 2003. Pathogenicity of enterohemorrhagic *Escherichia coli* in neonatal calves and evaluation of faecal shedding by treatment with probiotic *Escherichia coli. J. Food Prot.* 66: 924–930.

Zheng, L., M. E. Duarte, A. Sevarolli Loftus, S. W. Kim. 2021. Intestinal health of pigs upon weaning: Challenges and nutritional intervention. *Front. Vet. Sci.* 8: 628258.

Zhou, X., Y. Wang, Q. Gu, W. Li. 2010. Effect of dietary probiotic, *Bacillus coagulans*, on growth performance, chemical composition, and meat quality of Guangxi Yellow chicken. *Poult. Sci.* 89: 588–593.

Zhu, C., J. Yao, M. Zhu, C. Zhu, L. Yuan, Z. Li, D. Cai, S. Chen, P. Hu, H. Y. Liu. 2022a. A meta-analysis of *Lactobacillus*-based probiotics for growth performance and intestinal morphology in piglets. *Front. Vet. Sci.* 9: 1045965.

Zhu, Q., M. Song, M. A. K. Azad, Y. Cheng, Y. Liu, Y. Liu, F. Blachier, Y. Yin, X. Kong. 2022b. Probiotics or synbiotics addition to sows' diets alters colonic microbiome composition and metabolome profiles of offspring pigs. *Front. Microbiol.* 13: 934890.

Zulkifli, I., N. Abdullah, N. M. Azrin, Y. W. Ho. 2000. Growth performance and immune response of two commercial broiler strains fed diets containing *Lactobacillus* cultures and oxytetracycline under heat stress conditions. *Br. Poult. Sci.* 41: 59.

# Beneficial Effects of Inactivated Lactic Acid Bacteria

# 35

## Marion Bernardeau and Jean-Paul Vernoux

---

## 35.1 INTRODUCTION

---

Lactic acid bacteria (LAB) constitute a group of Gram-positive bacteria that share certain morphological, metabolic and physiological characteristics (von Wright and Axelsson, 2019). Originally, LAB were classified into six families (*Aerococcaceae, Carnobacteriaceae, Enterococcaceae, Lactobacillaceae, Leuconostocaceae* and *Streptococcaceae*) representing 13 different genera, but they have since been reclassified several times. In 2020, a major taxonomic reorganization of LAB (Zheng et al., 2020) reclassified LAB into a single family, the *Lactobacillaceae*, comprising 31 genera including *Lactobacillus, Paralactobacillus, Pediococcus, Weissella, Fructobacillus, Convivina, Oenococcus* and *Leuconostoc*, as well as 23 new genera containing organisms formerly classified as *Lactobacillus* species.

Lactic acid bacteria have a long and safe history of use in food. For centuries, they have played a crucial role in the production of fermented foods: vegetables, meats and particularly fermented dairy products. Over the past decade, scientific understanding of the biological properties of LAB (in particular their metabolism and functional properties) has expanded considerably. This has enabled the development of more reliable processes of fermentation control during food production and to an expanded range of industrial dairy applications for LAB as starters and adjunct starters/cultures in products that confer health benefits to humans.

Whilst, in theory, bacteria from any genus could exhibit health beneficial properties, bacteria belonging to the LAB group have historically represented the main group of microorganisms whose health-benefiting properties have been studied, recognized and commercialized (Mathur et al., 2020). They meet the probiotic definition established by the FAO and WHO (2002) (Food and Agriculture Organization of the United Nations and the World Health Organization), as being "live microorganisms which when administrated in adequate amounts confer a health benefit on the host". Since that definition, the applications and use of LAB as probiotics have extended from humans (Sauer and Han, 2021) to animals (Deng et al., 2021), as previously noted by Bernardeau and Cretenet (2019).

It is well established that live probiotic bacteria can confer a wide range of health benefits. However, it has become apparent in the past decade that dead microorganisms can also exhibit probiotic (health-benefiting) characteristics, as noted by Adams in 2010 in his discussion of the "probiotic paradox". Dead

DOI: 10.1201/9781003352075-38

(non-viable) microbes offer a potential advantage over live probiotic organisms in being highly stable. They are therefore well suited to use in treating at-risk populations such as young infants (Awad et al., 2010), the elderly (Maruyama et al., 2016), those who are immunocompromised or who have existing inflammatory conditions. This discovery has led to a shift in the focus of research and development away from viable live probiotics to non-viable microbes made from probiotics, as discussed by Bernardeau and Cretenet (2019) in the previous edition of this chapter. As part of this shift, there has been increasing interest in formulating an accepted definition of a dead probiotic, not only from a semantic perspective but also from an application and regulatory perspective. The lack of a clear, common, accepted definition during the past decade has been a significant source of confusion in the scientific literature that was highlighted and discussed in the previous edition of this chapter (Bernardeau and Cretenet, 2019), which advocated for convening a panel of experts to address the issue and formulate a common nomenclature. Progress has been made since that time and, following a panel discussion in 2021, the International Scientific Association of Probiotics and Prebiotics (ISAPP) proposed the term 'postbiotic' to describe a "preparation of inanimate microorganisms and/or their components that confers a health benefit on the host" (Salminen et al., 2021a). In addition to providing a definition and terminology for postbiotics, ISAPP reviewed data on their health benefits, potential modes of action, requirements to meet the stated definition and defined their scope of application (Salminen et al., 2021a). However, the ISAPP definition has not been universally accepted by the scientific community, and there remains debate among groups who previously advocated for alternative terminologies and definitions. Because of this, the current literature remains inconsistent as to the terminologies used to describe the biologically active components of postbiotics, and this is impeding the progress of research and knowledge in this area.

Against this background, this review aims to provide an update on the concept of dead microorganisms that confer health benefits, as applied to LAB. It discusses the latest definitions, areas of consensus and knowledge gaps; the methods applied to produce inactivated microbes and to validate their inactivated status; and their proven health benefits in animal and humans, as well as the current research needs that need to be addressed in order to aid industry in the application of inactivated microorganisms in a regulatory context.

## 35.2 HEALTH BENEFITS OF INACTIVATED MICROORGANISMS—FROM UNDEFINED CONCEPT TO CONTROVERSIAL DEFINITION

Since the formalization of the probiotic concept by the FAO/WHO in 2002, its definition has been continually questioned by parts of the scientific community because of the assertion that probiotic strains must be live microorganisms, thereby excluding dead organisms that may nevertheless also convey a health benefit. Regular publications reinforcing the WHO definition have ensued, each time echoed by other publications questioning the requirement for probiotic microorganisms to be alive (Kataria et al., 2009; Adams, 2010; Lahtinen, 2012). Fueled by such narratives, research has extended beyond probiotic viability to consideration of the functional properties and biologically active components of the microorganism (dead or alive) that may convey a benefit to the host. There is now substantial clinical evidence indicating that inactivated probiotic microorganisms can improve human health. This new knowledge led to the formation of new descriptive terminologies that sit outside of the 2002 FAO/WHO probiotic definition, such as the term 'paraprobiotic' put forward by Taverniti and Guglielmetti (2011) to describe "non-viable microbial cells (intact or broken) or crude cell extracts (i.e. with complex chemical composition), which, when administered (orally or topically) in adequate amounts, confer a benefit on the human or animal consumer". However, 'paraprobiotic' has not been fully adopted by the scientific community, and multiple different terms (including 'inactivated', heat-killed', 'non-viable', 'ghost-probiotics' and 'paraprobiotics') continue to be used without consistency. Exactly ten years since Taverniti and Guglielmetti proposed the term 'paraprobiotic', the ISAPP convened a panel discussion and thereafter proposed a new term, 'post-biotic', to describe a "preparation of inanimate microorganisms and/or their components that confers a health benefit on the host" (Salminen et al., 2021a). In defining the term and conditions for its use, the ISAPP panel considered the totality of scientific and technical knowledge relating to inactivated microorganisms encompassing their health benefits, techniques to achieve their inactivation and methods to verify thereof and the potential of contained biomolecules to exert beneficial effects on the host, and also gave consideration to the environmental drivers of the postbiotic concept, including safety, stability, intellectual property and regulatory considerations. The factors considered by ISAPP in formulating the 'postbiotic' definition are illustrated in Figure 35.1.

Following the initial ISAPP consensus statement and definition of 'postbiotic', adoption of the term has been encouraged via two further publications from authors from the ISAPP panel (Vinderola et al., 2022a, 2022b). These publications have prompted a marked reduction in the use of terms such as 'dead probiotic', 'paraprobiotic' and 'tyndallised probiotic' and a rise in the use of 'postbiotic' in the published literature. However, true consensus in the use of the term is still lacking, particularly among those who previously advocated alternative terminologies and definitions (Aguilar-Toalá et al., 2018). This group of authors argued that the term 'postbiotic' had already been defined before the ISAPP statement, by Tsilingiri and Rescigno in 2013, as "any factor resulting from the metabolic activity of a probiotic or any released molecule capable of conferring beneficial effects to the host in a direct or indirect way", which would encompass metabolic products such as short chain fatty acids, saccharides and bacteriocins. The

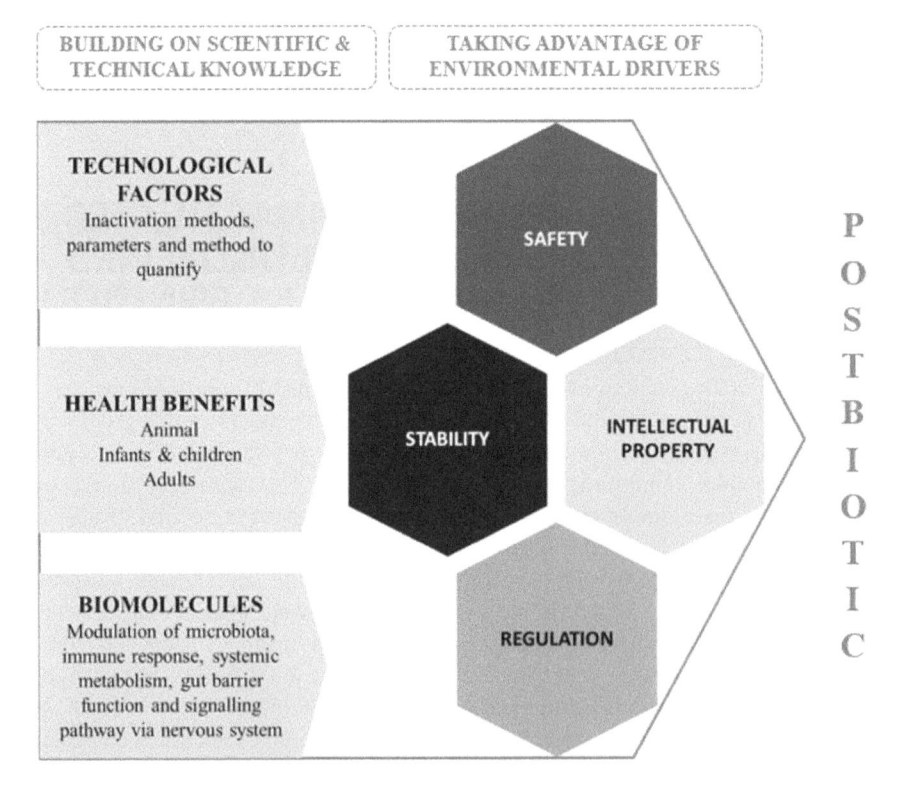

**FIGURE 35.1**   Postbiotics: illustration of the scientific and environmental credentials for defining non-viable microorganisms having health benefits (created based on Salminen et al., 2021a).

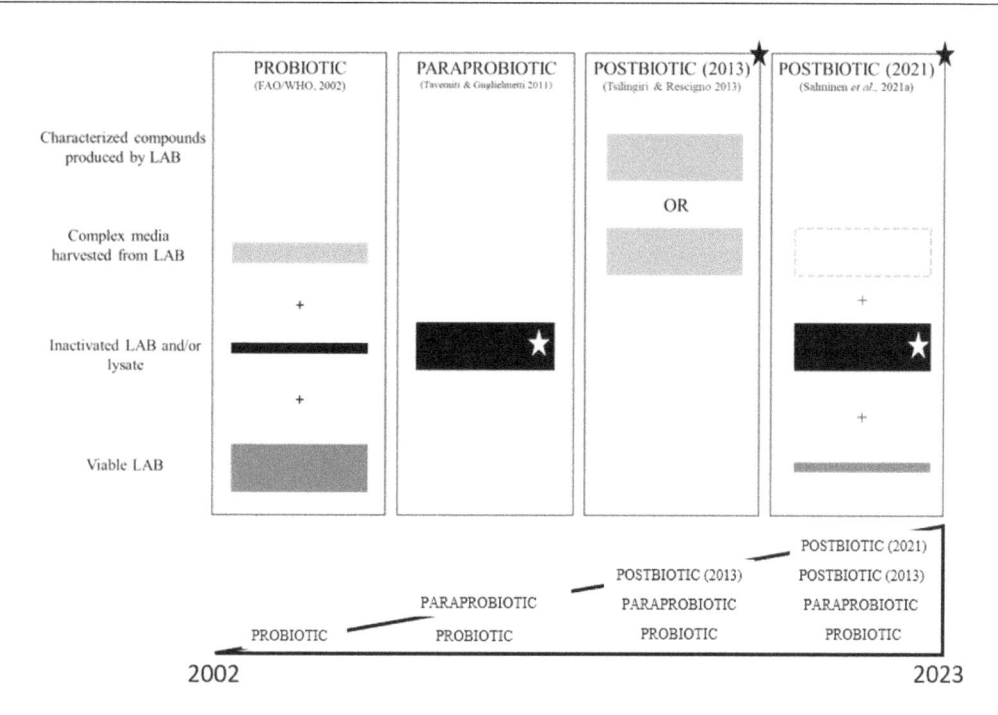

**FIGURE 35.2** Illustration of the accumulation of terms (using the same suffix 'biotic') referring to microorganisms (dead or alive) and their metabolites having health benefits and currently used in the scientific literature in 2023. The size of the shaded boxes indicates the main active 'compounds' expected in the final product, the dotted line refers to active compounds that may be present or not, according to the definition and the stars refer to critical points of confusion (black stars: same terms used to mean two different products; white star: same products referred to by two different terms).

authors also listed a number of limitations to the ISAPP definition, including the uncharacterized nature of the biomolecules, the claimed longer shelf-life stability of inactivated microorganisms, the absence of valid markers to quantify health benefits and the appropriacy of "inanimate" to refer to non-viable or dead microorganisms. Ultimately, Aguilar-Toalá and 37 other scientists co-signed a reply to ISAPP in a correspondence entitled: "Postbiotics—When Simplification Fails to Clarify" (Aguilar-Toala et al., 2021), to which the ISAPP group responded by acknowledging the value of open debate and discussion but maintaining their proposal for use of the term 'postbiotic' as the best terminology to encompass the health benefits of dead and/or inactivated cells that may contain important but non-essential metabolites (Salminen et al., 2021b).

The confusing chronology surrounding the nomenclature of live and dead microorganisms is illustrated in Figure 35.2. It is exemplified by the fact that, since 2021, the term 'postbiotic' is being used equally to describe a complex mixture of dead or inactivated cells or lysate with or without uncharacterized metabolites and a small fraction of viable cells (for which a maximum threshold was not defined by Salminen et al., 2021a) as to describe the acellular fraction harvested from a beneficial microorganism culture or a fully characterized beneficial compound as defined by Tsilingiri and Rescigno in 2013. Terms such as 'postbiotic', 'paraprobiotic' and older terms such as 'heat-killed', 'inactivated' and 'thyndallized' continue to be used in the literature. This makes searching for relevant articles and establishing their relevance to a particular research question very problematic.

Box 35.1 illustrates the three major issues that are raised by the current lack of consensus in the definition of what constitutes an inactivated health-beneficial microorganism.

**BOX 35.1  THE MAJOR ISSUES ARISING FROM THE ABSENCE OF
A COMMON CONSENSUS IN DEFINING INACTIVATED HEALTH-
BENEFICIAL MICROBES IN THE CURRENT LITERATURE**

- **The acellular fraction on its own is not a postbiotic**: The acellular fraction is a complex matrix whose composition is intimately dependent on the growth conditions and genetics of the progenitor. The absence of the progenitor raises the question of safety assessment. Numerous studies have reported beneficial effects of acellular fractions (also referred to in the literature as cell-free culture supernatants) on aspects of human health in the absence of cell biomass or the progenitor.
- **A single characterized molecule extracted from a complex acellular fraction is not a post-biotic**: Even more confusing, and potentially dangerous, is the use of 'postbiotic' to refer to a single extracted, characterized molecule, because this can lead to inference of the product as having the properties of a medicine or drug, which is clearly out of the scope of health beneficial properties gained from ingestion of live or dead microbes or their culture media from food or feed. Examples of this include Park et al. (2022), Fathima et al. (2022) and Abbasi et al. (2021).
- **Health-beneficial compounds produced in situ by undefined microbial populations are not postbiotics**: The in-situ production of health beneficial compounds, either in food/feed or in the gut, is outside the scope of the field of postbiotics since the origin of the microbially produced compounds cannot be fully controlled, characterized or verified; no progenitors are identified; and no intentional inactivation treatment has been applied. Hence, the proposed definition of 'postbiotic' by Moradi et al. (2020) to include bioactive soluble factors (products or metabolic byproducts) produced in complex microbiological cultures, food or the gut that exert a beneficial effect on the food or the consumer is not appropriate. In a similar way, exo-polysaccharides present in kefir are not 'postbiotics', as referred to by Barros et al. (2021a), because they have been generated from an undefined spontaneous fermentation. Health-beneficial compounds released in the gastrointestinal tract from endogenous gut microbiota also do not fit into the postbiotic definition, despite the fact that some of these compounds (e.g. butyrate) have been shown to support gut health. As highlighted by Salminen et al. (2021a), it is mandatory in the context of a 'postbiotic' to have identified, isolated and characterized the progenitor in order to be able to guarantee the safety and quality control of the product.

These examples of misuse of the term 'postbiotic' are fueling debate and discussion among the scientific community. They highlight the growing interest from both scientific and industrial perspectives in the application of inactivated health-beneficial microorganisms and their products to human health. Given the current situation, it will no doubt take a further period of time to refine the various definitions in current use and to align the scientific community as to their meaning. However, this is imperative for the future development of this area of science and for ensuring a transparent, regulatory positioning of inactivated health-beneficial microorganisms that is clearly separated from that of drugs. Questions and points of discussion raised in this chapter have recently been explored by Vinderola et al. (2024).

# 35.3  METHODS FOR INACTIVATING HEALTH-BENEFICIAL MICROORGANISMS

As a preamble, it should be stated that this and the following section (35.4) do not cover methods to extract, characterize and quantify individual soluble factors or biomolecules derived from inactivated microorganisms having a health benefit. This is on the basis that if such factors/biomolecules are treated separately, without any link to the microbial progenitor, they could be considered a pharmaceutical concept and would thus be out of the scope of the topic being reviewed herein.

Whichever terminology is selected ('parabiotic', 'postbiotic' or something else), the inactivation process and associated parameters are central to generating the end product. Bacterial cell inactivation should result from an intentional and deliberate process. The applied technology and associated parameters should be selected for their ability to inactivate cell viability while preserving the integrity of the effector molecules or mechanisms that confer the health benefit(s). The inactivation treatment must alter the cell structure of the microorganism and/or its physiological functioning. As highlighted by de Almada et al. (2016), reference to an inactivation treatment implies that the microorganism has been rendered incapable of growing but has retained the beneficial health effects conferred by its viable form. Inactivation of bacterial cells can be achieved by thermal, non-thermal, physical (mechanical disruption, heat treatment, gamma or UV irradiation, high hydrostatic pressure, freeze-drying, sonication) or chemical (acid deactivation) methods. A great diversity of treatments has been applied across studies published between 1994 and 2018, as previously discussed by Bernardeau and Cretenet (2019). More recently, Zhong et al. (2022) reviewed the different methods applied between 2010 and 2022 to inactivate microorganisms in the context of their applications in metabolic syndrome and reported that 82% of the 22 studies applied heat treatments, highlighting that heat remains the preferred method of inactivation, most likely because it is a reliable, convenient and low-cost method.

To date, insufficient attention has been paid to comparing different inactivation methods or to whether and how the particular method and conditions applied could affect the capacity of the inactivated microorganism to exert its beneficial effect(s) on the host. Among the few studies available, Mehling and Busjahn (2013) compared spray-drying versus freeze-drying of *Limosilactobacillus reuteri* DSMZ 17648 and reported no difference in terms of the resulting beneficial effect of the inactivated microorganism. A study of heat treatment (80 °C for 20 min) versus gamma radiation (8.05 Gy/min) of *L. reuteri* using Cobalt 60 for 20 h by Kamiya et al. (2006) demonstrated that these treatments similarly attenuated cardiac and colonic afferent fiber responses to colorectal distension in Sprague-Dawley rats following oral administration of the live or non-viable (heat-killed or gamma-irradiated) microorganism. Conversely, Wong and Ustunol (2006) showed that heat-inactivated probiotic strains produced higher levels of interleukin 8 in a Caco-2 cell model compared with irradiation-inactivated bacteria. In addition, Ouwehand et al. (2000) observed that gamma irradiation decreased the adhesion to intestinal mucus of most but not all the tested microbial strains (compared to adhesion of the corresponding viable cells), whereas UV irradiation had no effect. More recently, Yokota et al. (2018) studied the anti-inflammatory and protective effects of live and non-viable (heat-killed and UV-treated) *Lactiplantibacillus plantarum* cells (AN1) in a murine inflammatory bowel disease model and observed differences between the effects of heat-killed and ultraviolet irradiation-killed cells on atrophy of colon length, mucosal tissue damage and spleen enlargement. Meanwhile, in a study that assessed the influence of heating temperature and duration on the capacity of 11 non-viable strains of LAB to modulate immune responses in a Caco-2 cell model, Ou et al. (2011) demonstrated that the response is more dependent on the microbial strain than on the heat treatment applied. Similarly, Li et al. (2022) demonstrated that a particular strain of *Lactiplantibacillus plantarum* (L.p H6) administered orally to mice as viable, heat-inactivated or ultrasonically lysed bacterial cells reduced cholesterol level to different degrees and could be a promising postbiotic for regulating cholesterol metabolism. Taken as a whole, the limited body of current literature and apparent high strain-dependency of processing effects makes it difficult to deduce whether a particular treatment condition is more or less likely to preserve the probiotic effect (and associated health benefits) of the inactivated microorganism.

Some new technologies are also emerging for microbial inactivation. These include pulsed electric fields (PEF), power ultrasound (US), high-pressure processing (HPP), ultraviolet (UV), microwave (MW) and ohmic heating methods (OHMs). Although predominantly applied to the preservation of fermented foods, these novel methods have recently been tested to generate inactivated beneficial microorganisms. They offer potential advantages in that they are highly efficient whilst being less aggressive than more traditional methods (Peng et al., 2020).

A list of different methods, including traditional and more modern methods, that have been applied in the field of inactivation of beneficial LAB is provided in Table 35.1, with recent referenced examples.

**TABLE 35.1** Overview of the Diversity of Techniques (Thermal and Non-Thermal) Referenced in the Current Literature to Inactivate Microorganisms. Each listed technique is illustrated by one example of a study investigating the health beneficial effects of the inactivated LAB *in vitro* or *in vivo* in humans or animals, where available (such techniques are highlighted in bold text).

| TECHNIQUE | | SCIENTIFIC PRINCIPLE OF TECHNIQUE | EXAMPLE OF USE IN STUDYING HEALTH-BENEFICIAL EFFECTS OF LAB (IN VITRO OR IN VIVO ) | REFERENCE |
|---|---|---|---|---|
| **Thermal treatments** | **Heat treatment** | – Combination of heat and time to inactivate microorganism.<br>– Alteration of one or more cellular structures or functions (outer and inner membrane, the peptidoglycan cell wall, the nucleoid, the cell's RNA, the ribosomes and diverse enzymes). | *Lacticaseibacillus paracasei* MCC 1849 (without CFS). Pasteurization at 100 °C for 30 min followed by freeze-drying.<br>– *In vivo* (Human) health benefits of heat-killed MCC1849:<br>– immunomodulation;<br>– may prevent aging- or stress-related attenuation of immunity;<br>– may prevent risk of infection in the elderly;<br>– improvement of mood. | Maehata et al. (2021) |
| | **Ohmic heating** | – The resistance of electricity passing agitates atoms and charges particles, causing an increase in temperature.<br>– The treatment affects microbial viability depending on the applied voltages and processing duration.<br>– Preserves bioactive compounds. | *Lactobacillus acidophilus* LA 05 *Lacticaseibacillus casei* subsp. *paracasei* 01, OHM magnitudes 8 V/cm, and 12 V/cm at 60 Hz.<br>– *In vitro*, OHM was highly effective in inactivating cells while causing less damage to the cell membrane integrity.<br>– *In vivo* (in humans), OHM was effective in reducing (↓↑) postprandial glycemia in healthy adults. | Barros et al. (2021b) |
| | Pulsed electric fields | – Short-term electric treatment with a field intensity from 100–300 V/cm to 20–80 kV/cm. | A review of emerging technologies on LAB.<br>The most effective UV light to inactivate LAB is located in the range of UV-C 200–280 nm. | Peng et al. (2020) |
| **Non-thermal treatments** | High-pressure processing | – Utilizes water as a medium to transmit very high pressure to the product.<br>– Inactivation of enzymes and microorganism.<br>– Effectiveness of inactivation depends on matrix, bacterial strain, pressure, temperature and holding time of the HPP processing.<br>– Chemical properties of covalent bond stay intact.<br>– Secondary, tertiary or quaternary structure of proteins, nucleic acids and polysaccharides are affected. | | |

| Method | Description | Details | Reference |
|---|---|---|---|
| Pulsed light combined with UV spectrum | – Application of high-power pulses of electromagnetic radiation at ultraviolet (100–400 nm) wavelengths (highest penetration rates). <br> – Microbial DNA absorbs UV light. <br> – Physicochemical changes in microbial structure. <br> – Treatment hinders replication and gene transcription. | *Lactobacillus brevis*, CCC B1300. <br> 8.9 J/cm² fluence with a clear filter. <br> Applied *in vitro* in beer: <br> ↓↑ spoilage and *Lactobacillus* contamination; <br> ↓↑ *L. brevis* in the beer to undetected level with limited effect on physicochemical properties of the beer. | Pratap-Singh et al. (2023) <br> Barros et al. (2021b) |
| Pulsed magnetic field | – Parameters: magnetic field intensity and pulse number. <br> – High penetrability and low power consumption. <br> – Creation of an ionic current can result in the opening of the cell membrane pores, facilitating electroporation, decrease in energy and carbohydrate metabolism, fragmentation of DNA and enzyme inactivation. | Emerging nonthermal technique currently being explored for its ability to ensure adequate microbial and enzymatic inactivation in food products. Not yet applied in humans/animals. | Basak (2023) |
| Irradiation gamma rays | – Irradiation dose expressed as kilogray (kGy), as function of the radiation energy and exposure time. <br> – Effectiveness at inactivating microorganisms depends on the bacterial strain and its physiological state, applied dose and its ability to repair DNA damage caused by radiation. | *L. acidophilus* LMG 9433, type strain, *Lactiplantibacillus plantarum* subsp. *plantarum* DSM 20205 *Lacticaseibacillus casei* LMG 6904ᵀ *Lacticaseibacillus paracasei* subsp. *paracasei* LMG 12586). <br> Tested doses: 250, 500, 750, 1,000, 1,500, 2,000, 2,500, 3,000 Gy. <br> *In vitro* (in porcine peripheral blood mononuclear cells), stopped the replication of bacteria while preserving their structure and their metabolic activity. <br> Lethal irradiation preserved the membrane integrity and the metabolic activity in LAB. Strain specificity was exhibited. | Porfiri et al. (2022) |
| **Ultrasound (US) and high-intensity ultrasound (HIU)** | – US: characterized by frequencies higher than 100 kHz and intensities below 1 W/cm². <br> – HIU: characterized by frequencies between 20 and 500 kHz, and intensities higher than 1 W/cm². <br> – Cell rupture, microbial and enzymatic inactivation. <br> – Effects dependent on processing parameters (frequency, power, processing time, pulse mode and duration, strain, matrix). | *Lacticaseibacillus casei* H6 (no CFS). <br> The bacterial suspension was treated with an ultrasonic cell pulverizer at 20 kHz for 40 min to prepare the bacterial ultrasonic lysate (uH6). <br> *In vivo* (in male rats): <br> – Prevented total cholesterol and ↓↑ LDL; <br> – Controlled insulin resistance; <br> – Modulated the intestinal microbiome, ↓↑ beneficial bacteria, ↓↑ harmful bacteria; <br> – Attenuated blood pressure. | Brandão et al. (2021) <br> Guimarães et al. (2019) |

*(Continued)*

**TABLE 35.1**  (Continued) Overview of the Diversity of Techniques (Thermal and Non-Thermal) Referenced in the Current Literature to Inactivate Microorganisms. Each listed technique is illustrated by one example of a study investigating the health beneficial effects of the inactivated LAB in vitro or in vivo in humans or animals, where available (such techniques are highlighted in bold text).

| TECHNIQUE | SCIENTIFIC PRINCIPLE OF TECHNIQUE | EXAMPLE OF USE IN STUDYING HEALTH-BENEFICIAL EFFECTS OF LAB (IN VITRO OR IN VIVO ) | REFERENCE |
|---|---|---|---|
| Cold plasma technology | – Consists of free electrons, ions and neutral particles, reactive species in constant interaction, with electrical energy to break covalent bonds and induce numerous chemical reactions.<br>– Causes cell death due to: reactive species etching in cell surfaces formed during plasma generation; volatilization of compounds and intrinsic photodesorption of UV photons; and destruction of genetic material. | *Lentilactobacillus hilgardii* NRRL B-1843. Atmospheric CP dielectric barrier discharge with helium and oxygen as working gases for 5, 10, and 15 min.<br>*In vitro*, 10 min CP exposure resulted in:<br>– Complete inhibition of bacterial growth;<br>– Use of flow cytometer revealed presence of three physiological states: active (14.4%), mid-active (77.5%) and dead (98.8%) of *L. hilgardii* population. Re-cultivation of the sorted subpopulation seemed to confirm that at least some of the mid-active cells were in the viable non-culturable bacteria (VBNC) state. | Niedźwiedź et al. (2020) |
| **Supercritical (SC) carbon dioxide** | – Uses fluids in their SC state.<br>– SC-$CO_2$ inactivates spoilage and pathogenic microorganisms, depending on the processing T°C, pressure and $CO_2$ concentration. SC-$CO_2$ uses moderate T°C (31.1 °C) and pressure (7.38 MPa).<br>– Microorganism inactivation occurs when the $CO_2$ causes oxygen displacement, lowering the pH to inhibit microbial growth. | *Bifidobacterium animalis* subsp. *lactis* Bb-12 (no CFS). Pressure: 10 MPa, 40 °C for 180 min.<br>*In vivo* (in a mouse model) compared to other treatments (irradiation, heat), SC-$CO_2$ inactivated Bb-12:<br>↓ total cholesterol levels in the serum;<br>↓ albumin and creatinine levels, ↓↑ HDL-cholesterol. | Almada et al. (2021) |

CFS, cell-free supernatant; T°C, Temperature.

# 35.4 METHODS FOR MEASURING THE EFFICACY OF INACTIVATION TREATMENTS

Being able to verify the efficacy of an applied inactivation treatment is crucial for several reasons. First, because for some specific bacterial strains, the health benefits are derived directly from the inactivated state of the microorganism. Second, because certain safety parameters rely on the inactivated status of the microorganism. Third, it is critical for defining the product as a postbiotic as distinct from a probiotic, and finally, it is important for quality and regulatory control, including ensuring transparency on food labels, which is expected by the consumer.

Aside from the obvious terms 'dead' and 'live', the literature is littered with other terms that are used interchangeably to describe the viable status of bacterial cells (e.g. 'viable', 'cultivable', 'ghost cells', 'metabolically active cells', 'viable but not cultivable'). As reviewed by Emerson et al. (2017), determining the viability status of a microbial cell can be considered analogous to Erwin Schrödinger's quantum mechanics thought experiment, in which a cell could appear to be simultaneously both alive and dead until a measurement is made. The route from live to dead is a continuum of different metabolic and physiological statuses that have been identified in concert with the improvement of microbiological tools. Depending on the availability of techniques and the way viability is measured, a cell could be assigned live, dead or somewhere in between (Davey, 2011). A viable cell is generally defined as an intact cell that is metabolically active and capable of reproduction (Hammes et al., 2011), whereas cultivability remains the gold standard term used in a variety of cell microbiology applications, such as the measurement of the viability of probiotics or pathogens in the context of food safety.

Traditional methods for assessing cell viability (and absence of viability) include those that rely on a culture-based approach (via colony counts, plate counts or liquid cultures), those that measure cellular and metabolic properties (heat flow, respiration, ATP, Stable-isotope probing) and those based on the measurement of uptake of fluorescent dyes (membrane potential dyes, esterase substrates, membrane integrity dyes) and nucleic acid-based methods (viable PCR, RT-PCR, molecular viability test, nucleic acid sequence-based amplification). These types of methods have been extensively reviewed in recent articles by Emerson et al. (2017) and Kumar and Ghosh (2019) and were discussed in the previous edition of this book chapter. Techniques based on cell membrane integrity and DNA or RNA quantification are also now accepted methods of reference for establishing cell viability, as reviewed by Davis (2014). As the probiotic field continues to grow and because the distinction between probiotics and postbiotics or paraprobiotics is largely based on cell viability, there has become a growing need for more reliable methods to assess the viability status of microorganisms, and it is expected that there will continue to be rapid developments in this area in the coming years.

The current considerations and need for development and refinement of improved methods for quantifying the viability status of cells in inactivated health-beneficial microorganisms, considering the ISAPP definition of postbiotic, are highlighted in Box 35.2.

**BOX 35.2  POINTS OF DISCUSSION CONCERNING THE ISAPP DEFINITION OF 'POSTBIOTIC' (SALMINEN ET AL., 2021A) THAT HIGHLIGHT THE NEED FOR BETTER METHODS TO DETERMINE CELL VIABILITY**

- **Appropriacy of the term 'inanimate':** This term within ISAPP definition consciously allows for the presence of 'substantial amounts of viable bacteria' in the final product but sets no threshold limits, nor any clear definition of what constitutes viability. This creates an area of uncertainty for quality and regulatory control of postbiotics. We suggest replacing the term with 'inanimate' as a more passive term which could encompass a natural change in the viability status of a product after a long period of storage (as already occurs in some probiotic products).

- **Need for accurate method:** The tolerance of persistent cells within the definition of postbiotic requires a definition of viability, per the definition of 'probiotic' (multiplication, metabolic activity . . .), as well as the identification of accurate methods to enumerate cells of differing physiological status in the final product.
- **Intentional active treatment:** Since viable cells can persist in a postbiotic (as defined by ISAPP), the key difference between a probiotic and postbiotic is *the intentional inactivation* treatment. On this basis, we propose a refinement of the ISAPP definition that refers to the "preparation of intentionally inactivated microorganisms".
- **Presence of progenitor is mandatory:** The mandatory presence of the progenitor in the final product is a key characteristic of a postbiotic. We therefore propose an adjustment to the ISAPP definition to achieve greater precision, by replacing "and/or" by "alone or in with their components".
- **New proposed definition:** Taking this together, the new (proposed) definition of 'postbiotic' is as follows: *"A preparation of intentionally inactivated microorganisms, alone or with their components, that confers a health benefit on the host"*.

# 35.5 HEALTH-BENEFICIAL EFFECTS OF INACTIVATED LACTIC ACID BACTERIA

The scientific literature published between 1994 and 2017 relating to the health-beneficial effects of inactivated microorganisms was extensively reviewed in the previous edition of this book chapter (Bernardeau and Cretenet, 2019). Herein, we will update that review with developments relating to inactivated LAB and/or their fermented products from publications within the past 5 years (2019–2023), with a specific focus on their use in animal production. In-depth reviews relating to the human health-beneficial effects of inactivated LAB are already available (for example Mathur et al., 2020; Das et al., 2022).

## 35.5.1 Reported Health Benefits of Inactivated Health-Beneficial LAB in Animal Production Studies Published since 2019

Scientific interest in the potential application of inactivated microorganisms to support the health of farmed animals is not new. However, in the absence of a clear regulatory status for inactivated microorganisms, commercial development of products has been slow. This is in contrast with the situation for viable microorganisms (probiotics) which have a strong legacy in animal production where they are recognized as zootechnical additives and thus benefit from a clear regulatory status (EFSA Regulation, 1831/2003) with published guidance in place to support the review and approval of products for commercial use. These products are commonly named "probiotic" in Europe or "direct-fed microbial" in the USA. Products made of inactivated cells or obtained from fermentation are unregulated and fall into the raw material classification. It is therefore not permitted to claim any kind of benefit from their use, health-related or otherwise. Nevertheless, in light of the relatively recent definitions proposed by the scientific community for postbiotics and paraprobiotics, interest in the development inactivated microorganisms and their fermentation products for use in feed for animal production is growing.

In considering the potential health benefits of inactivated microorganisms to farmed animals, it is important to link the current knowledge of postbiotics, inactivated cells and metabolites produced from fermentation to current animal productions issues and challenges. The short lifespan of animals reared for meat means that chronic diseases such as cancer are rarely encountered. Hence, beneficial effects on these disease processes are of little relevance to the animal production setting. The major current challenge in animal production relates to the need to provide a sufficient, sustainable, supply of high quality, safe meat protein to support a growing global population. The high demand for meat protein means that farming systems must operate at maximal capacity and efficiency. This generates factors of stress that can affect animal health, immunity and growth performance. These stress factors can arise from the production system (density, vaccination, stress, medication), economic and climate change impacts on ingredient availability and supply (low quality raw materials), as well as from environmental (methane reduction) and regulatory constraints (animal welfare, prohibition of antibiotic growth promoters). Since animal health is now legitimately considered a key part of the "one health" concept, in which the inter-relationships between human, animal and environmental health are recognized and supported, any safe and effective 'biotics' (probiotics or postbiotics) have a potential role to play in supporting animal health and ensuring the delivery of safe food to humans.

Zhong et al. (2022) addressed the application of postbiotics (as defined by Salminen et al., 2021a) in animal production for gut health benefits in light of the global reduction of the use of antibiotics as growth promoters. The authors illustrated the current knowledge of how postbiotic components (exopolysaccharides, cellular wall fragments, bacterial DNA, metabolites) and their mode of action (modulation of the immune system, reinforcement of gut barrier function, pathogen inhibition) can lead to improved growth performances and overall gut health status, replacing the need for antibiotic growth promoters.

Inactivated microorganisms and their metabolites from fermentation lend themselves particularly well to animal production systems because their inactivated status is better suited to the constraints of animal feed production. Indeed, non-spore forming probiotics such as LAB are under-represented in the animal production arena because of their sensitivity to the technological processes applied during feed processing. Probiotic *Bacillus*- and *Clostridium*-based feed additives, both of which are spore-forming bacteria, constitute the majority of the current authorizations granted in the EU market, highlighting the importance of product stability in achieving successful commercialization. This is especially true in the poultry feed industry where the harsh environmental conditions encountered during feed processing, including high pressures and temperatures during pelleting and extrusion, may affect the viability of the probiotic. During poultry feed processing, temperatures of 75–85 °C are encountered for 15–20 s with a moisture content of 15 % before pelleting, and the situation is not dissimilar for pig feed production. Scientists have overcome this issue by developing a variety of different approaches and technologies to protect probiotics administered as live cells in food and feed. These include using different food formats (Sanders and Marco, 2010), freeze-drying technology (Fonesca et al., 2015), encapsulating materials (Huq et al., 2013; Riaz and Masud, 2013; Solanki et al., 2013) and immobilization techniques (Mitropoulou et al., 2013). Spore-forming probiotics do not require such protection because they are able to withstand feed production processes intact and can be stored at room temperature over a long period in a desiccated form without any reduction in viability (Bader et al., 2012). Notwithstanding this, there is a continued need for products where the health-beneficial activity of the microorganism is independent of its viability and inactivated health-beneficial microorganisms and/or their metabolites can support this need.

At the time of publication of the previous edition of this book chapter, published studies concerning the potential of health-beneficial inactivated microorganisms for use in animal production were mainly focused on fish and shrimp (8.8% of all identified publications), followed by swine (4.4%) and poultry (4.4%) (Bernardeau and Cretenet, 2019). There was only a single published article on the application of a non-viable bacterium (*Enterococcus* sp.) in ruminants (Tsuruta et al., 2009). Since that time numerous studies have been published on the topic of health and growth performance benefits of inactivated LAB and their metabolites in farming and companion animals. These new studies, published between 2019 and 2023, relate to aquaculture, poultry, swine and ruminants and are summarized in Table 35.2 a–e. For each referenced study, the animal model in which the product was tested, the species of the progenitor, the authors' chosen nomenclature used to describe the tested product, the nature of the product tested, the inactivation process applied and associated processing conditions (where disclosed), the verification method used to confirm inactivation, the overall purpose of the study, the reported beneficial effects and the published reference are given.

The information presented in Table 35.2 exemplifies the aforementioned issues caused by the lack of a clear definition for inactivated microorganisms with health-beneficial properties. There is considerable heterogeneity in the terms used to describe the tested product among the identified studies. Multiple references that refer to the test product as a postbiotic in fact tested the acellular fraction harvested from the centrifugation of a microbial suspension of LAB. Only a few of the studies published since the ISAPP definition of postbiotic (Salminen et al., 2021a) named the test product in accordance with this definition.

**Aquaculture production:** Among the studies relating to aquaculture (Table 35.2 a) that were published before the ISAPP definition (before 2021), two out of four articles referred to heat-killed products and did in fact test the LAB biomass, while the other two tested LAB-fermented food while defining the product as a postbiotic. Among the three studies published since 2021, two misdescribed the tested product as a postbiotic since only the acellular fraction was tested, whilst the other referred to the test product as a paraprobiotic. The reported health benefits from these studies in aquaculture relate to improved growth performance and the prevention of specific bacterial diseases.

**Poultry production:** The literature relating to poultry is more abundant (13 studies published during 2019–2023; Table 35.2 b). This suggests a shift in research effort away from solely focusing on probiotic spore-formers as the microbial entity that can withstand feed processing technologies. Among the listed studies, only one (Abd El-Ghany et al., 2022a) used the term 'postbiotic' in accordance with the ISAPP definition. The majority of the other studies used this term to also refer to the acellular fraction.

**Swine production:** Six studies published during 2019–2023 relating to swine were identified (Table 35.2 c). Five of these were focused on inactivated LAB for use in swine production, whilst one used the pig as a model organism for studying the human health benefits of inactivated LAB. Heat treatment was the inactivation treatment applied in all studies. Only one study (Sato et al., 2019), which tested a commercial product, did not provide any details on the inactivation processing conditions. This may have been due to potential intellectual property concerns. The reported health benefits of inactivated LAB in swine relate to selective modulation of the gut microbiota and its diversity, reduction in the pathogen load, alteration of the expression of immune biomarker genes, improved growth performance and a reduction in diarrhea.

**Ruminant production:** Probiotic LAB have been less commonly applied to ruminant production, and this was reflected in the number of publications relating to the use of inactivated LAB cells. Only four studies were published during 2019–2023, three from the same research group, all concerning the acellular fraction of *Lactiplantibacillus plantarum* RG14. These studies reported beneficial effects of on growth performance, gut bacterial populations including a reduction in methanogens in the rumen of lambs and an increase in fiber degrading bacteria and short chain fatty acid release (Izuddin et al., 2019a; Sasazaki et al., 2020). However, the limited information on effects of inactivated LAB as cells precludes drawing conclusions about the effects of inactivated LAB as intact cells in ruminants.

**Companion animals:** A single published study relating the application of inactivated LAB to companion animals (dogs) was identified (Panasevich et al., 2021), in which the tested product was defined in accordance with the ISAPP definition of postbiotic. The study reported that the product modulated microbial and metabolomic biomarkers associated with gut health, but improved health as such was not reported (Panasevich et al., 2021). Given that there are numerous references online to inactivated LAB products intended for use in companion animals (in non-peer-reviewed publications or other communication materials), the single article identified in the present review should not be taken as reflecting the level of scientific or commercial interest in the use of inactivated LAB for companion animals. More published studies are needed.

Finally, two recent review articles by Abd El-Ghany (2020) and Zamojska et al. (2021) discuss the future of probiotics, paraprobiotics and postbiotics in animal production. However, whilst these articles provide insight into the health properties of metabolites released by beneficial microorganisms, they are limited in their scope and would appear to further contribute to the confusion over nomenclature. Zamojska et al. (2021) provides examples of beneficial effects of plant feed additives in support of the idea that metabolites released by endogenous gut microbiota as a result of consumption of the phytogenics could be regarded as postbiotics.

**TABLE 35.2** Summary Details of Studies Published between 2019 and 2023 on the Health Benefits of Inactivated LAB or Their Metabolites in a) Aquaculture, b) Poultry, c) Pigs, d) Ruminants, and e) Companion Animals.

**a) Aquaculture**

| ANIMAL MODEL | PROGENITOR | DENOMINATION/ EXACT PRODUCT TESTED | INACTIVATION PROCESS | OVERALL PURPOSE OF STUDY | REPORTED BENEFICIAL HEALTH EFFECTS | REFERENCE |
|---|---|---|---|---|---|---|
| *In vitro*— inhibition assay | *Weissella cibaria* isolated from rainbow trout and Nile tilapia | Postbiotic/ harvested CFS | n.d. | Effect on pathogens *Yersinia ruckeri* and *Aeromonas salmonicida* subsp. *salmonicida* | ↓ 4.5 log of *A. salmonicida* subsp. *salmonicida*; Trend towards ↓ growth of *Y. ruckeri*. | Quintanilla-Pineda et al. (2023) |
| *In vitro* and *in vivo*— salmonids | *Lactiplantibacillus plantarum* strain HK L-137, commercial product Feed LP20™ House Wellness Foods Corp., Itami, Japan (20% HK L137 and 80% tapioca dextrin, DM basis) | Paraprobiotic/ inactivated biomass | Heat-killed at 70 °C for 10 min Final concentration in the dry product, 2×10$^{11}$ CFU/g | Effect on epithelial cell line from rainbow trout (*Oncorhynchus mykiss*; RTgutGC) and on pre-smolt Atlantic salmon (*Salmo salar*) | *In vitro* cell line model: ↑ barrier function and production of IL-1b; ↓ production of Anxa1. *In vivo* salmon model: ↓ production of Anxa1 and modulation in the distal intestine of gene expression pathways related to molecular function, biological process and cellular components. | Rocha et al. (2023) |
| *In vivo*— common carp | *Cetobacterium somerae* and *Lactococcus lactis* Stress Worry Free Concentration® SWFC | Postbiotic/ harvested CFS | n.d. | Effect on GP, skin mucus, liver, GH and intestinal microbiota | No effect on GP of *Cyprinus carpio*; ↑ H fish (AOA, IgM, IIM, MD) | Yu et al. (2023) |
| *In vivo*— rainbow trout | Fermented product containing (among others) *Lactobacillus* CECT9882 and non-bitter yeast | Postbiotic/ fermented food | Dry granulated product/no indication on final CFU concentration | Effect on bacterial community composition and structure and against *Lactococcus garvieae* infection | ↑ MD, MR; ↑ protection against *L. garvieae*. | Mora-Sánchez et al. (2020) |
| *In vivo*—Nile tilapia | *Lactiplantibacillus plantarum* (HK L-137) House Wellness Foods Corp., Itami, Japan | Heat-killed | n.d. | Evaluate the synergetic effects of heat-killed *L. plantarum* (HK L-137) and β-glucan (BG) on digestive enzyme activity and intestinal morphology | ↑ final BW and WG, VH, muscle thickness; Upregulation of IGF-1 gene expression; ↑ hematocrit, hemoglobin, triglyceride and glucose levels; ↑ Lipase and protease, upregulation of muscle and liver G6PD gene expression only when HK L-137 was combined with BG. | Dawood et al. (2020) |

(Continued)

**TABLE 35.2** Summary Details of Studies Published between 2019 and 2023 on the Health Benefits of Inactivated LAB or Their Metabolites in a) Aquaculture, b) Poultry, c) Pigs, d) Ruminants, and e) Companion Animals.

| ANIMAL MODEL | PROGENITOR | DENOMINATION/ EXACT PRODUCT TESTED | INACTIVATION PROCESS | OVERALL PURPOSE OF STUDY | REPORTED BENEFICIAL HEALTH EFFECTS | REFERENCE |
|---|---|---|---|---|---|---|
| In vivo—olive flounder (Paralichthys olivaceus) | Heat-killed (HK) Bacillus sp. SJ-10 (B)—KCCM 90078 and HK Lactiplantibacillus plantarum (P) KCCM 1132 alone and in combination | Heat-killed | Autoclaved at 121 °C for 15 min confirmed by the absence of bacterial colonies on agar plate after spreading | Assess the potential of heat-killed probiotic to replace antibiotic growth promotors | The combination product ↑ IIR. The Bacillus HK alone could be advanced in growth-related endocrine, innate immunity, and intestinal morphology, positively affecting a low-fishmeal diet for juvenile olive flounder. | Nguafack et al. (2020) |
| In vivo— Rainbow trout | Fermented food product composed of soy and alfalfa flour, in which two LAB were added in similar concentrations | Postbiotic/ fermented food | The fermented food product was micronized before sending to the manufacturer AQUASOJA | Effect on the intestinal microbiota and development of Lactococcus garvieae infection | ↑ MD, MR; ↑ protection against L. garvieae. | Pérez-Sánchez et al. (2020) |

**b) Poultry**

| ANIMAL MODEL | PROGENITOR | DENOMINATION/ EXACT PRODUCT TESTED | INACTIVATION PROCESS | OVERALL PURPOSE OF STUDY | REPORTED BENEFICIAL HEALTH EFFECTS | REFERENCE |
|---|---|---|---|---|---|---|
| In vivo— colisepticemic broiler chickens | Stabilized non-viable Lactobacilli fermentation product (Culbac®, TransAgra's International Inc., Storm Lake, Iowa, USA) | *Postbiotic/ compound + biomass | n.d. | Effect on H, GP, IR and gut status under E. coli challenge | ↓ disease picture; ↑ GP, IR, bursa of Fabricius/WG ratio; ↓ intestinal coliform count in E. coli-challenged chickens. | Abd El-Ghany et al. (2022a) |
| In vivo— broilers | Lactiplantibacillus plantarum RG11, RI11, and RS5 | Postbiotic/ harvested CFS | n.d. | Assess potential alternative to antibiotic growth promotor | RS5 recorded a significant, highest ($p < 0.05$) final BW, BWG, and significant lowest ($p < 0.05$) FCR; RI11 and RS5 ↑ glutathione peroxidase, SOD, acidic mucin, sulfated mucin and intestinal trefoil factor; RI11 and RS5 upregulated the expression of intestinal mucin 2, occludin, and S-IgA. | Chang et al. (2022) |

| In vivo—broilers | Lactiplantibacillus plantarum RI-11 | Postbiotic and parabrobiotic/harvested CFS | The cell suspension of L. plantarum strains was frozen for 7 days at 30 °C to produce paraprobiotics, no indication on viability rate | Impact on colon mucosa microbiota | RI11 ↓ relative abundance of Proteobacteria and Enterococcus and ↑ abundance of Bacteroides. | Danladi et al. (2022) |
|---|---|---|---|---|---|---|
| In vivo—chickens | Lactobacillus acidophilus species fermentation product, Culbac (TransAgra International Inc., Storm Lake, Iowa) | Postbiotics/dry feed additive and aqueous water version | n.d. | Prevention of necrotic enteritis (NE) in comparison with probiotic and antibiotic | Postbiotic: ↓ severity of NE and mortality; ↑ FCR and European production efficiency factor; ↑ hepatic and IR; Performed better than probiotic and antibiotic. | Abd El-Ghany et al. (2022b) |
| In vitro and on chicken drumsticks | Pediococcus acidilactici (Bactoferm B-LC-20) and Latilactobacillus sakei/Staphylococcus xylosus (1:1 mix) (Bactoferm B-FM) Chr. Hansen Laboratories | Postbiotic/harvested CFS | n.d. | Evaluate antibacterial effects in vitro and on chicken drumsticks | ↓ 5.0 log10 L. monocytogenes in vitro; ↓ S. Typhimurium, L. monocytogenes and TMAB in the chicken drumstick. | İncili et al. (2022) |
| In vivo—broilers | Lactobacillus acidophilus obtained from M/s Unique Biotech Pvt., Ltd, Hyderbad, Telangana, India | Paraprobiotic/biomass | Thermal inactivation: autoclaving the bacterial culture at 121 °C for 15 min confirmed by absence of colony on agar plate | Assess potential as an alternative to antibiotic growth promoters | ↑ GP, BWG, IR, gut H, breast yield; ↓ FI, improves FCR. | Tukaram et al. (2022) |
| In vivo—broilers | Heat-killed Lactiplantibacillus plantarum L-137 | Paraprobiotic | Heat-killed | Effect against subclinical NE caused by Clostridium perfringens. | ↑ VH/crypt depth ratio; ↑ the spleen index value and the cytokine levels, as well as the expression of intestinal β-defensin genes; ↑ expression of cytokines; No cumulative mortality following C. perfringens exposure, compared to other groups. | Pham Thi et al. (2022) |

(Continued)

**TABLE 35.2** (Continued) Summary Details of Studies Published between 2019 and 2023 on the Health Benefits of Inactivated LAB or Their Metabolites in a) Aquaculture, b) Poultry, c) Pigs, d) Ruminants, and e) Companion Animals.

| ANIMAL MODEL | PROGENITOR | DENOMINATION/ EXACT PRODUCT TESTED | INACTIVATION PROCESS | OVERALL PURPOSE OF STUDY | REPORTED BENEFICIAL EFFECTS | REFERENCE |
|---|---|---|---|---|---|---|
| In vivo— broilers | Pediococcus acidilactici (Bactoferm™ B-LC-20, Chr. Hansen Laboratories Copenhagen, Denmark). | Postbiotic/ harvested CFS | n.d. | Evaluate the effects of postbiotics (10% and 50%) alone and in combination with chitosan coating (1%) on the microbial and chemical quality of chicken breast fillets during storage at 4 °C | Only the combination ↓ L. monocytogenes and S. Typhimurium, total viable count, LAB and psychrotrophic bacteria counts and extended the shelf life by up to 12 days. Postbiotic had no effect on pH values or color properties of breast fillets. | İncili et al. (2021) |
| In vivo— Heat-stressed broilers | Lactiplantibacillus plantarum RI11 | Postbiotic/ harvested CFS | n.d. | Assess as a potential alternative to antibiotic growth promoter | RI11 ↑ glutathione peroxidase, CAT, glutathione enzyme activity, upregulated mRNA expression of IL-10, downregulated the IL-8, TNF-alpha, heat shock protein 70, and alpha-1-acid glycoprotein levels; RI11 upregulated ileum zonula occludens-1 and mucin 2 mRNA expression; No effect on ileum claudin 1, ceruloplasmin, IL-6, IL-2, and interferon expression. | Humam et al. (2021) |
| In vivo— broilers | Lactiplantibacillus plantarum strains, RS5, RI11 and UL4 | Postbiotic/ harvested CFS | n.d. | Effect on oxidative stress markers, physiological stress indicators, lipid profile and meat quality in heat-stressed broilers | ↑ T-AOC, CAT, GSH; ↓ α1-AGP, CPN, plasma cholesterol concentration, plasma TG and VLDL; ↓ breast meat pH; ↓ shear force and lightness, drip and cooking loss and yellowness. | Humam et al. (2020) |

| In vivo—yellow-feathered broilers | Heat-inactivated *Bacillus subtilis* and *Lactobacillus acidophilus* BFI (1:1), (Bioforte Biotechnology Co., Ltd. Shenzhen, China). | Heat-inactivated compound probiotics | The heat-inactivated compound probiotics were treated at 80 °C for 30 min. Proved to be inactivated according to bacterium culture test before use. | Effect on growth performance, plasma biochemical indices, and cecal microbiome | ↑ FCR; ↓ plasma cholesterol, creatinine; ↑ β-diversity index of cecal microbiota and pathways related to methane metabolism, transcription machinery, purine metabolism and protein export. | Zhu et al. (2020) |
|---|---|---|---|---|---|---|
| In vivo—broilers | *Lactiplantibacillus plantarum* strains, RS5, RI11 and UL4 | Postbiotic/harvested CFS | n.d. | Effect on growth performance, carcass yield, intestinal morphology, gut microbiota, immune status, and growth hormone receptor and insulin-like growth factor expression in heat stressed broilers | RI11 ↑ final BW, total weight, ADG, FCR; ↑ caecum total bacteria and *Lactobacillus* count and ↓ Salmonella count; No effect on CP; ↑ plasma IgM, hepatic IGF-1 mRNA expression; ↑ VH in the duodenum, jejunum and ileum. ↓ Enterobacteriaceae and *E. coli* counts, caecal pH; hepatic GHR mRNA expression. | Humam et al. (2019) |
| In vivo—broilers | *Pediococcus acidilactici* B-67717), *Limosilactobacillus reuteri* B-67718), *Enterococcus faecium* B-67720), and *Lactobacillus acidophilus* B-67701) Pure Cultures, Inc.; Flock VitalityTM. | Postbiotic | n.d. | Effect on NE, birds challenged with coccidia vaccine and *C. perfringens* | ↑ IgM on jejunal tissue; ↑ IIR; ↓ LS, *C. perfringens* counts and mortality; ↑ WG; ↓ proinflammatory responses in infection model and generated a homeostatic-like response. | Johnson et al. (2019) |

*(Continued)*

**TABLE 35.2** (Continued) Summary Details of Studies Published between 2019 and 2023 on the Health Benefits of Inactivated LAB or Their Metabolites in a) Aquaculture, b) Poultry, c) Pigs, d) Ruminants, and e) Companion Animals.

**c) Pigs**

| ANIMAL MODEL | PROGENITOR | DENOMINATION/ EXACT PRODUCT TESTED | INACTIVATION PROCESS | OVERALL PURPOSE OF STUDY | REPORTED BENEFICIAL EFFECTS | REFERENCE |
|---|---|---|---|---|---|---|
| In vivo— pigs | Lactiplantibacillus plantarum L-137 (HK L-137) LP Pro™ (House Wellness Foods Corp, Itami, Japan) LP ProTM contains 10% of HK L-137 and 90% of whey protein, dextrin, and sunflower lecithin | Paraprobiotic/ cell biomass | Heated at 70 °C for 10 min and lyophilized | Effect of LP Pro™ on growth performance and immunological biomarkers | ↑ WG, ADG, throughout the grower—finisher period; ↑ platelet count in suckling pigs, no effect on other hematological parameters; ↑ relative mRNA expression level of IFN- ß of the suckling and starter pigs; ↑ IR in suckling and starter pigs ↑GP in finishing pigs. | Tartrakoon et al. (2023) |
| In vivo pig as human model | Ligilactobacillus salivarius strain 189 | Postbiotic/ heat-killed lactobacilli | Heat-killed at 90 °C for 15min, absence of colony formation confirmed after for 48h at 37 °C. Concentration of HK LS 189 in the dry product was ca. 1.0× 10^{10} cells/well. | Assess potential anti-obesity effects and the effect on intestinal microorganisms in a pig model | ↓ GP and Prevotella abundance; ↑ Parabacteroides abundance; Showed difference in metabolic pathway and lipid metabolism. | Ryu et al. (2022) |
| In vivo— nursery pigs challenged with F18+Escherichia coli | LBF contained 6 × 10^{10}/g of powder of Lactobacillus (Limosilactobacillus fermentum and L. delbrueckii) (LBiotix, Adare Biome, Houdan, France). | *Postbiotic/ cell biomass + CFS CFS contained peptides, amino acids, carbohydrates and minerals | Heat-inactivated | Effect on intestinal health and prevention of postweaning diarrhea caused by F18+Escherichia coli | ↑ GP and alpha diversity of jejunal mucosa-associated microbiota; Ligilactobacillus salivarius and Propionibacterium acnes ↓ crypt cell proliferation. | Xu et al. (2022) |
| Ex-vivo— jejunal explants from piglets | Lactiplantibacillus plantarum (ATCC 14917) and L. plantarum subsp. plantarum strain G1 | Postbiotics/metabolite extract from acellular fraction. Postbiotic biomass | Heat inactivated by sterilization at 121 °C for 30 Min. The inactivation of bacteria was confirmed by inoculation in MRS plates. | Effect of LP metabolites in reducing deoxynivalenol DON intestinal toxicity | LP metabolites ↑ density of goblet cells in villi and crypts in the DON treated explants; LP metabolites ↓ superoxide anion production and oxidative stress; ↓ toxic effects of DON. | Maidana et al. (2021) |

| Model | LAB strain | Type | Inactivation | Objective | Results | Reference |
|---|---|---|---|---|---|---|
| *In vivo*—weaned pigs | *Lacticaseibacillus rhamnosus* (CJ Cheilledang Biotechnology Research Institute, Seoul, Korea) | Inactivated probiotic | Heat-killed LR was procesed by heating at 80°C for 30 min | Effect on growth performance and immunological biomarkers | ↑ GP, FCR; ↓ post-weaning diarrhea; ↓ 2w concentrations of serum TNF-α, TGF-β1 and cortisol; HK- LR modified IR. | Kang et al. (2021) |
| *In vivo*—post-weaning pigs | *Enterococcus faecium* strain NHRD IHARA (EFNH) alone or in combination with *Clostridium butyricum* strain MIYAIRI 588 (CBM588) (Miyarisan Pharmaceutical Co., Ltd., Tokyo, Japan). | Heat-killed | n.d. | Effect on growth and fecal microbiota composition | HK-EF ↑ GP; No synergistic effect of the combination. | Sato et al. (2019) |

**d) Ruminants**

| Model | LAB strain | Type | Inactivation | Objective | Results | Reference |
|---|---|---|---|---|---|---|
| *In vivo*—post-weaning lambs | *Lactiplantibacillus plantarum* RG14, RG11 and TL1 | Postbiotic/ harvested CFS | n.d. | Assess AOA potential | RG14 ↑ AOA, serum and ruminal fluid AOA; ↓ serum lipid peroxidation and upregulated hepatic AO enzymes and ruminal barrier function. | Izuddin et al. (2020) |
| *In vivo*—Japanese Black calves | *Latilactobacillus sakei* HS-1 | Paraprobiotic | Heat-treated | Effect on health status and blood parameters | ↑ intestinal LAB, etiotropic effects of calf; ↓ number of medical treatments compared with control calves; ↓ number of *E. coli* in feces of control group. | Sasazaki et al. (2020) |
| *In vivo*—post-weaning lambs | *Lactiplantibacillus plantarum* RG14 | Postbiotic/ harvested CFS | n.d. | Effect on GP, ruminal fermentation and microbial profile, blood metabolites | ↑ WG, FI, ND; ↑ butyrate and ruminal ammonia-N; ↑ blood total protein, urea nitrogen and glucose; No effect on blood TG or cholesterol levels; | Izuddin et al. (2019a) |

*(Continued)*

| ANIMAL MODEL | PROGENITOR | DENOMINATION/EXACT PRODUCT TESTED | INACTIVATION PROCESS | OVERALL PURPOSE OF STUDY | REPORTED BENEFICIAL EFFECTS | REFERENCE |
|---|---|---|---|---|---|---|
| | | | | | ↑ population of fiber degrading bacteria; <br> ↓ total protozoa and methanogens in rumen. <br> ↑ mRNA expression of hepatic IGF-1 and ruminal MCT-1. | |
| *In vivo*— post-weaning lambs | *Lactiplantibacillus plantarum* RG14 | Postbiotic/ harvested CFS | n.d. | Effect on gastrointestinal histology, mucosal immunity and barrier function | ↑ ruminal papillae growth, IR and microbial gastrointestinal population. | Izuddin et al. (2019b) |

**e) Companion animals**

| ANIMAL MODEL | PROGENITOR | DENOMINATION/EXACT PRODUCT TESTED | INACTIVATION PROCESS | OVERALL PURPOSE OF STUDY | REPORTED BENEFICIAL EFFECTS | REFERENCE |
|---|---|---|---|---|---|---|
| Dogs | *Lactobacillus acidophilus* (NVL: Culbac; TransAgra, Storm Lake, IA) | Postbiotic/ compound + biomass | n.d. | Effect on health status of dogs | ↓ fecal pH, fecal *E. coli*, *Fusobacterium*; <br> ↑ fecal *Lactobacillus* spp. <br> ↑ fecal acetate and propionate and IgA. | Panasevich et al. (2021) |

*increased/improved:* – reduced/decreased; ADG: average daily gain; AOA: Antioxidant activity; BW: body weight; BWG: body weight gain; CAT: catalase; CFU: colony forming unit; CP: carcass parameters; CPN ceruloplasmin; CFS: cell-free supernatant; FI: feed intake; GP: growth performance; GH: gut health; GSH glutathione; H: health; IgM: immunoglobulin M; IIM: intestinal inflammation markers; IIR: innate immune response; IR: immune response; IM: immune markers; IR: immune response; LS: lesion score; FCR: feed conversion ratio; MD microbiota diversity; MR: microbial richness; n.d. not disclosed or not determined; ND: nutrient digestibility; SOD: superoxide dismutase, T-AOC plasma activity of total-antioxidant capacity; TG: triglycerides; TMAB: total mesophilic aerobic bacteria; VH: villus height; VLDL: very low density lipoprotein; WG: weight gain; α1-AGP alpha-1-acid-glycoprotein. *, used in accordance with the ISAPP definition of 'postbiotic' (Salminen, 2021a).

## 35.5.2  Health Benefits of Inactivated Microorganisms in Humans Reported since 2019

There is substantial clinical evidence that non-viable probiotic bacteria and/or their metabolites derived from fermentation can improve human health, albeit with unknown mechanisms. Some of this evidence refers to beneficial modulation of the immune system as well as to anti-diarrhea effects, anti-blood pressure activity, a reducing effect on blood cholesterol, antioxidant and anticancer activity and the suppression of cell proliferation. Abbasi et al. (2021) recently performed an in-depth review of the bioactivity and health benefits of postbiotics as applied to humans, that included their effects on: growth and inhibition of pathogenic microbes, immunomodulation, modulation of the gut microbiota, recuperation after injury, maintenance of the intestinal barrier function, reduction of bacterial translocation, treatment of diarrhea and colitis, effect on the functioning of the digestive system, reduction in lactose intolerance, improvement of alcohol-induced liver disease, modulation of inflammation, growth inhibition of cancer cells, improvement of food allergy, reduction of blood cholesterol, improvement of colorectal cancer therapies, visceral pain modulation, effect on the functioning of the respiratory system, reduction in the effects of aging, reduction in tooth decay and effect on response to atopic dermatitis treatment. This extensive review can be supplemented with the available published references that cover other applications where inactivation microbes and/or their associated metabolites have shown interest, as discussed subsequently.

**Dental health:** Probiotics have been used in the management of caries. Here, interesting data have demonstrated a superiority of dead- over live-probiotic strains in reducing the cariogenicity of *Streptococcus mutans* (Schwendicke et al., 2014) and the submaxillary salivary concentration in healthy volunteers (Mehling and Busjahn, 2013; Sañudo et al., 2017). It has also been shown that daily intake of heat-killed *L. plantarum* 137 can decrease the depth of periodontal pockets in patients undergoing supportive periodontal therapy (Iwasaki et al., 2016). Chen et al. (2020) compared the antipathogenic activity of viable suspensions versus heat-killed (100 °C for 1 h) *Ligilactobacillus salivarius* (AP-32), *Limosilactobacillus reuteri* (GL-104), *Lacticaseibacillus rhamnosus* (CT-53, F-1), *Lacticaseibacillus paracasei* (ET-66, GL-156), *Lactobacillus helveticus* (RE-78), *L. acidophilus* (TYCA02) and *Bifidobacterium lactis* (BB-115, CP-9) against oral pathogenic *S. mutans, Porphyromonas gingivalis, Fusobacterium nucleatum* and *Aggregatibacter actinomycetemcomitans*. All heat-inactivated cells showed bacteriostatic activities, although with less inhibitory activity compared to the viable cell suspensions.

Periodontitis is a complex chronic inflammatory disease that results from dysregulated crosstalk among the oral microbiota. Multiple clinical studies have shown some improvement with probiotics. However, systematic reviews and meta-analyses have identified little or no benefit. According to Moraes et al. (2022), future studies investigating the value of postbiotics [as defined by Salminen et al. (2021a)] in periodontitis need to take into account the intrinsic complexity and nonlinearity of the disease, in which cause and effect between different individuals are highly variable. Study designs will need to take account of the physiological cues that dictate microbe–microbe and microbe–host interactions within the complex periodontal tissue microenvironment and control for systemic interference in disease progression if they are to have more likelihood of being replicated in *in vivo* trials.

**Women's health:** Interestingly, in the field of women's health, Sawada et al. (2022) demonstrated that daily supplementation of $1 \times 10^{10}$ inactivated cells of *Lactobacillus gasseri* CP2305 by middle-aged women (aged 40–60) suffering from mild menopausal symptoms relieved mild climacteric symptoms and significantly reduced total vasomotor and psychological SMI scores as well total somatic and vasomotor GCS scores after six menstrual cycles, based on questionnaires assessing the Simplified Menopausal Index (SMI) and the Greene Climacteric Scale (GCS). Meanwhile, Komano et al. (2023) reported that the commercial product LC-Plasma (Kyowa Hakko Bio CO., Ltd., Tokyo, Japan) containing a pasteurized, washed culture of *Lactococcus lactis* subsp. *lactis* JCM 5805 that is known to activate plasmacytoid dendritic cells (pDCs) by stimulating toll-like receptor 9 (TLR-9) alleviated fatigue and increased pDC activity caused by a continuous high training load in athletes after 14 days of continuous exercise in a randomized, double-blind, placebo-controlled, parallel-groups study.

**Skin health:** Beyond their use in the treatment of digestive disorders and the improvement of gastro-intestinal functioning, more recent fields of interest for the application of probiotics have included human skin disease. As recently reviewed by Maguire and Maguire (2017), LAB and bifidobacteria are good candidates for use in the prevention of skin diseases and for enhancing skin hydration. The use of non-viable health-beneficial microorganisms in this area has also begun to be considered, although the current evidence is somewhat inconsistent. da Costa Baptista et al. (2013) demonstrated that consumption of the viable (but not the heat-killed) form of *L. rhamnosus* GG improved symptoms of atopic dermatitis in children. On the other hand, Kimoto-Nira (2018) demonstrated that observed beneficial effects of *Lactococcus lactis* H61 consumption on human skin health were independent of the viability of the probiotic cells. The author suggested that the beneficial effect was more likely due to stimulation of the immune response and the nervous system in the case of intake of the heat-killed form of the probiotic and to improvement of the intestinal environment and host-oxidative status in the case of intake of the microorganism as live cells in fermented milk. Tsai et al. (2021) examined the beneficial effect of *Lactiplantibacillus* (formerly *Lactobacillus*) *plantarum* GMNL-6 and *L. paracasei* GMNL-653 in their heat-killed format (washed cells autoclaved at 121 °C for 30 min) on skin wound repair by their promotion of collagen synthesis in Hs68 fibroblast cells. The authors demonstrated that lipoteichoic acid, the major component of *Lactobacillus* cell walls, achieved an anti-fibrogenic effect similar to that of the heat-killed bacteria cells in the TGF-beta stimulated Hs68 fibroblast cell model, indicating potential for its use in injury healing.

**At risk populations:** The safety of the daily use of high doses of probiotic supplementation in at-risk human populations (e.g. immunocompromised, neonatal and hospital patients) has been raised and questioned (Kothari et al., 2019). Inactivated microbes have the potential to offer a safer alternative to probiotics for use in at-risk and vulnerable population groups. For example, it has been shown that LAB inactivated microbes, unlike their probiotic counterparts, are not translocated from the intestine and do not exacerbate local inflammation in pediatric settings (Morniroli et al., 2021; Oglio et al., 2022), post-organ transplant settings (via a xenograft animal model; Kim et al., 2022) or immunosuppressed models (Ali et al., 2022).

**Respiratory health:** Finally, as scientific knowledge has progressed and established the link between the lungs and the gut ('gut–lung axis'), the potential application of probiotics, inactivated LAB or their metabolites to treat respiratory diseases has increased. Using an acute infection mouse model, Chen et al. (2017) demonstrated that oral administration of heat-killed *Enterococcus faecalis* before infection with influenza A virus reduced influenza-mediated morbidity and lung inflammation and proposed a potential mode of action involving the production of monocyte chemoattractant protein-1. In another study involving mice infected with pneumonia virus, Percopo et al. (2015) demonstrated that intranasal inoculation of viable and non-viable (heat-inactivated) forms of *L. plantarum* and *L. reuteri* promoted survival when administered within 24 h of the viral challenge. These effects were reportedly mediated by the suppression of proinflammatory cytokines, limiting of virus recovery and a reduction in neutrophil recruitment to the lung tissue, which occurred irrespective of the viability status of the *Lactobacillus* strains (Percopo et al., 2015). Research using inactivated microorganisms has increased in the light of the SARS-CoV-2 pandemic, reflecting that probiotics in live form may be seen as challenging to an at-risk population. Research suggests that specific strains of inactivated microorganisms and metabolites with antiviral and anti-inflammatory effects could be utilized as promising tools for both adjuvant and inhibition strategies in SARS-CoV-2 patients with no apparent unfavorable adverse effects. Several reviews on this topic have recently become available (Khani et al., 2022; Xavier-Santos et al., 2022) that have identified the two-way strategy by which postbiotics can inhibit the viral S protein and host ACE2 receptor. In an *in vitro* study, Alzahrani et al. (2022) showed that the metabolites derived from *Leuconostoc mesenteroides* GBUT-21 exhibited SARS-CoV-2 inhibitory activity and potent anti-inflammatory activity, which was suggested could help to alleviate the cytokine storm in COVID-19. An antiviral potential of inactivated microorganisms has also been demonstrated against a non-respiratory virus, the herpes virus: Vilhelmova-Ilieva et al. (2022) evaluated the *in vitro* anti-herpes simplex activity (herpes simplex virus type 1, HSV-1) of 11 postbiotic products (lysates or CFS) produced during the fermentation of six candidate-probiotic *Lactobacillus* strains isolated from Bulgarian fermented milk products and identified post-metabolites from *Lactiplantibacillus plantarum* strain L3 as exhibiting the highest anti-herpes virus activity, followed by cell-derived fragments of *Limosilactobacillus fermentum* culture strain S3. This study again highlighted the existence of strain specificity in the degree of effect of the postbiotic product. In a follow-on study, the authors extended their research into another type of herpes virus (Koi KHV) using a different cell line model, the Common carp brain (CCB), and confirmed that the same LAB post-metabolites blocked KHV attachment to the host (CCB) cell, leading to a drop in viral titers, as well as exerting protective effects on CCB cells before they were subjected to viral infection (Vilhelmova-Ilieva et al., 2023).

# 35.6  REGULATORY PERSPECTIVES AND CONCLUSION

During the last two decades, a considerable body of scientific evidence has emerged demonstrating that some specific LAB strains, when ingested in a certain amount, retain and exhibit properties that confer a health benefit to the host (human or animal). These strains are attracting growing research and development interest from the scientific and industrial communities and could form part of an effective global strategy to reduce the overall use of antibiotics in both humans and animals, as well as to support public health. As part of such a strategy, it is important to consider the potential advantage that inactivated health-beneficial microorganisms (the progenitor strain) and associated products (lysates, metabolites, etc.) thereof may have over products based on live strains in certain circumstances. For example, provided that a health-beneficial effect has been proven, products based on inactivated microorganisms could be more suitable for treating at-risk populations or for inclusion in food or feed matrices that are subject to harsh technological processes or to variable storage conditions that might otherwise reduce the viability and efficacy of a live strain product. In this regard, although the health-related outcomes of inactivated microorganisms and their products overlap to some extent with the (well-established) definition of a probiotic, the 'non-viable' status of an inactivated strain that has been intentionally produced, via a well-defined, reproducible process, and that has been verified as to its non-viability, legitimates for the derivation of a new product category for postbiotics within the existing regulatory framework that sets out proper terminology and associated conditions of use.

From a regulatory standpoint, the situation remains complex because, as is the case with the term 'probiotic', the term 'postbiotic' has no legal status within the existing legislation. As stated by ISAPP (Salminen et al., 2021a), progenitor strains of postbiotics that are derived from food-grade microorganisms or species that have been granted a Qualified Presumption of Safety (QPS) status from EFSA might have an easier path to approval compared to those without this status. This reinforces the need (and potential regulatory benefit) of having the progenitor in the final postbiotic product, as is required under the ISAPP definition. In addition, it is critical that any legal definition of postbiotic that is settled upon maintain the positioning of health-benefiting microorganisms as food or feed additives and not as pharmaceutical products, for regulatory purposes. Abd El-Ghany (2020), citing Aguilar-Toalá et al. (2018), describes postbiotics as "fed products such as organic acids and bacteriocins produced by beneficial intestinal microbiota", but this statement does not in fact support the definition of Aguilar-Toalá et al., which states that postbiotics are given as such to the host and have been characterized for the health benefits expected. Furthermore, caution should be applied in including organic acids within the definition of a postbiotic because these products are already defined and covered by feed additive legislation. Similarly, considering a bacteriocin a single specific compound having health benefits would mean that their use would into the legislation for medicines. This would be outside of the intended scope of terms such as probiotic, paraprobiotic and postbiotic that sit firmly within the food and feed space and perhaps explains why the US FDA has recently defined a new "live biotherapeutic products" (LBP) category of drug products, clarifying pharmaceutical expectations of probiotics and other "biotics" products, as recently discussed by Cordaillat-Simmons et al. (2020). Under existing regulatory frameworks, industrial organizations have the choice to either position their product within the feed/food space or to undergo the approval process for a pharmaceutical product, which of course incurs a different level of regulatory approval.

The proposed definition of "postbiotic" that was issued and advocated by ISAPP in 2021 has certainly generated further research interest in the topic and a move towards consensus in the terminology. However, the term and its (ISAPP) definition have not been universally accepted; the topic remains controversial among some groups, and several different terms remain evident in the current literature. In order to avoid further confusion, we propose to refine and thereby strengthen the definition given by Salminen et al. (2021a) to read as follows: "A preparation of intentionally inactivated microorganisms, alone or with their components, that confers a health benefit on the host".

# BIBLIOGRAPHY

Abbasi, A., E. Sheykhsaran, and H. Samadi-Kafil. 2021. Chapter 5. "Bioactivity perspectives and health benefits of postbiotics". In: *Postbiotics: Science, Technology and Applications*. Eds. A. Abbasi, E. Sheykhsaran, and H. Samadi-Kafil, pp. 48–70, Bentham Science Publishers, Sharjah, UAE.

Abd El-Ghany, W. A. 2020. "Paraprobiotics and postbiotics: Contemporary and promising natural antibiotics alternatives and their applications in the poultry field". *Open Vet. J. 10*(3):323–330. http://doi.org/10.4314/ovj.v10i3.11.

Abd El-Ghany, W. A., M. A. Abdel-Latif, F. Hosny, N. M. Alatfeehy, A. E. Noreldin, R. R. Quesnell, R. Chapman, L. Sakai, and A. R. Elbestawy. 2022b. "Comparative efficacy of postbiotic, probiotic, and antibiotic against necrotic enteritis in broiler chickens". *Poult Sci. 101*(8):101988. http://doi.org/10.1016/j.psj.2022.101988.

Abd El-Ghany, W. A., H. Fouad, R. Quesnell, and L. Sakai. 2022a. "The effect of a postbiotic produced by stabilized non-viable Lactobacilli on the health, growth performance, immunity, and gut status of colisepticaemic broiler chickens". *Trop. Anim. Health Prod. 54*(5):286. http://doi.org/10.1007/s11250-022-03300-w.

Adams, C. A. 2010. "The probiotic paradox: Live and dead cells are biological response modifiers". *Nutr. Res. Rev. 23*(1):37–46. http://doi.org/10.1017/S0954422410000090.

Aguilar-Toalá, J. E., S. Arioli, P. Behare, C. Belzer, R. Berni Canani, J. M. Chatel, E. D'Auria, M. Q. de Freitas, E. Elinav, E. A. Esmerino, H. S. García, A. D. da Cruz, A. F. González-Córdova, S. Guglielmetti, J. de Toledo Guimarães, A. Hernández-Mendoza, P. Langella, A. M. Liceaga, M. Magnani, R. Martin, M. T. Mohamad Lal, D. Mora, M. Moradi, L. Morelli, F. Mosca, F. Nazzaro, T. C. Pimentel, C. Ran, C. S. Ranadheera, M. Rescigno, A. Salas, A. S. Sant'Ana, K. Sivieri, H. Sokol, V. Taverniti, B. Vallejo-Cordoba, J. Zelenka, and Z. Zhou. 2021. "Postbiotics- when simplification fails to clarify". *Nat. Rev. Gastroenterol. Hepatol. 18*(11):825–826. http://doi.org/10.1038/s41575-021-00521-6.

Aguilar-Toalá, J. E., R. Garcia-Varela, H. S. Garcia, V. Mata-Haro, A. F. González-Córdova, B. Vallejo-Cordoba, and A. Hernández-Mendoza. 2018. "Postbiotics: An evolving term within the functional foods field". *Trends Food Sci. Technol. 75*:105–114. http://doi.org/10.1016/j.tifs.2018.03.009.

Ali, M. S., E. B. Lee, Y. Quah, B. T. Birhanu, K. Suk, S. K. Lim, and S. C. Park. 2022. "Heat-killed *Limosilactobacillus reuteri* PSC102 ameliorates impaired immunity in cyclophosphamide-induced immunosuppressed mice". *Front Microbiol. 13*:820838. http://doi.org/10.3389/fmicb.2022.820838.

Almada, C. N., C. N. Almada-Érix, A. R. Roquetto, V. A. Santos-Junior, L. Cabral, M. F. Noronha, A. E. S. S. Gonçalves, P. D. Santos, A. D. Santos, J. Martinez, P. C. Lollo, W. K. A. Costa, M. Magnani, and A. S. Sant'Ana. 2021. "Paraprobiotics obtained by six different inactivation processes: Impacts on the biochemical parameters and intestinal microbiota of Wistar male rats". *Int. J. Food Sci. Nutr. 8*:1057–1070. http://doi.org/10.1080/09637486.2021.1906211.

Alzahrani, O. R., Y. M. Hawsawi, A. D. Alanazi, H. E. Alatwi, and I. A. Rather. 2022. "In vitro evaluation of *Leuconostoc mesenteroides* cell-free-supernatant GBUT-21 against SARS-CoV-2". *Vaccines (Basel). 10*(10):1581. http://doi.org/10.3390/vaccines10101581.

Awad, H., H. Mokhtar, S. S. Imam, G. I. Gad, H. Hafez, and N. Aboushady. 2010. "Comparison between killed and living probiotic usage versus placebo for the prevention of necrotizing enterocolitis and sepsis in neonates". *Pak. J. Biol. Sci.: PJBS 13*(6):253–262.

Bader, J., A. Albin, and U. Stahl. 2012. "Spore-forming bacteria and their utilisation as probiotics". *Benef. Microbes 3*(1):67–75. http://doi.org/10.3920/BM2011.0039.

Barros, C. P., L. C. Grom, J. T. Guimarães, C. F. Balthazar, R. S. Rocha, R. Silva, C. N. Almada, T. C. Pimentel, E. L. Venâncio, I. Collopy Junior, P. M. C. Maciel, M. Q. Freitas, E. A. Esmerino, M. C. Silva, M. C. K. H. Duarte, A. S. Sant'Ana, and A. G. Cruz. 2021b. "Paraprobiotic obtained by ohmic heating added in whey-grape juice drink is effective to control postprandial glycemia in healthy adults". *Food Res. Int. 140*:109905. http://doi.org/10.1016/j.foodres.2020.109905.

Barros, S. É. L., C. D. S. Rocha, M. S. B. de Moura, M. P. Barcelos, C. H. T. P. da Silva, and L. I. D. S. Hage-Melim. 2021a. "Potential beneficial effects of kefir and its postbiotic, kefiran, on child food allergy". *Food Funct.* *12*(9):3770–3786. http://doi.org/10.1039/d0fo03182h.

Basak, S. 2023. "The potential of pulsed magnetic field to achieve microbial inactivation and enzymatic stability in foods: A concise critical review". *Future Foods.* *7*:100230. http://doi.org/10.1016/j.fufo.2023.100230.

Bernardeau, M., and M. Cretenet. 2019. Chapter 37 "Probiotic effects of non-viable lactic acid bacteria". In: *Lactic Acid Bacteria, Microbial and Functional Aspects*, 5th edition. Eds. G. Vinderola, A. C. Ouwehand, S. Salminen, and A. von Wright, pp. 609–630, CRC Press, Boca Raton, FL.

Brandão, L. R., J. L. de Brito Alves, W. K. A. da Costa, G. A. H. Ferreira, M. P. de Oliveira, A. Gomes da Cruz, V. A. Braga, J. S. Aquino, H. Vidal, M. F. Noronha, L. Cabral, T. C. Pimentel, and M. Magnani. 2021. "Live and ultrasound-inactivated *Lacticaseibacillus casei* modulate the intestinal microbiota and improve biochemical and cardiovascular parameters in male rats fed a high-fat diet". *Food Funct.* *12*(12):5287–5300. http://doi.org/10.1039/d1fo01064f.

Chang, H. M., T. C. Loh, H. L. Foo, and E. T. C. Lim. 2022. "Lactiplantibacillus plantarum postbiotics: Alternative of antibiotic growth promoter to ameliorate gut health in broiler chickens". *Front. Vet. Sci.* *9*:883324. http://doi.org/10.3389/fvets.2022.883324.

Chen, M.-F., K.-F. Weng, S.-Y. Huang, Y.-C. Liu, S.-N. Tseng, D. M. Ojcius, and S.-R. Shih. 2017. "Pretreatment with a heat-killed probiotic modulates monocyte chemoattractant protein-1 and reduces the pathogenicity of influenza and enterovirus 71 infections". *Mucosal Immunol.* *10*(1):215–227. http://doi.org/10.1038/mi.2016.31.

Chen, Y. T., P. S. Hsieh, H. H. Ho, S. H. Hsieh, Y. W. Kuo, S. F. Yang, and C. W. Lin. 2020. "Antibacterial activity of viable and heat-killed probiotic strains against oral pathogens". *Lett. Appl. Microbiol.* *70*(4):310–317. http://doi.org/10.1111/lam.13275.

Cordaillat-Simmons, M., A. Rouanet, and B. Pot. 2020. "Live biotherapeutic products: The importance of a defined regulatory framework". *Exp. Mol. Med.* *52*(9):1397–1406. http://doi.org/10.1038/s12276-020-0437-6.

da Costa Baptista, I. P., E. Accioly, and P. de Carvalho Padilha. 2013. "Effect of the use of probiotics in the treatment of children with atopic dermatitis; a literature review". *Nutr. Hosp.* *28*(1):16–26. http://doi.org/10.3305/nh.2013.28.1.6207.

Danladi, Y., T. C. Loh, H. L. Foo, H. Akit, N. A. Md Tamrin, and M. Naeem Azizi. 2022. "Effects of postbiotics and paraprobiotics as replacements for antibiotics on growth performance, carcass characteristics, small intestine histomorphology, immune status and hepatic growth gene expression in broiler chickens". *Animals.* *12*:917. http://doi.org/10.3390/ani12070917.

Das, T. K., S. Pradhan, S. Chakrabarti, K. C. Mondal, and K. Ghosh. 2022. "Current status of probiotic and related health benefits". *Appl. Food Res.* *2*:100185. http://doi.org/10.1016/j.afres.2022.100185.

Davey, H. M. 2011. "Life, death, and in-between: Meanings and methods in microbiology". *Appl. Environ. Microbiol.* *77*(16):5571–5576. http://doi.org/10.1128/AEM.00744-11.

Davis, C. 2014. "Enumeration of probiotic strains: Review of culture-dependent and alternative techniques to quantify viable bacteria". *J. Microbiol. Methods.* *103*:9–17. http://doi.org/10.1016/j.mimet.2014.04.012.

Dawood, M. A. O., F. I. Magouz, M. F. I. Salem, Z. I. Elbialy, and H. A. Abdel-Daim. 2020. "Synergetic effects of lactobacillus plantarum and β-glucan on digestive enzyme activity, intestinal morphology, growth, fatty acid, and glucose-related gene expression of genetically improved farmed tilapia". *Probiotics Antimicrob. Proteins.* *12*(2):389–399. http://doi.org/10.1007/s12602-019-09552-7.

de Almada, C. N., C. N. Almada, R. C. R. Martinez, and A. S. Sant'Ana. 2016. "Paraprobiotics: Evidences on their ability to modify biological responses, inactivation methods and perspectives on their application in foods". *Trends Food Sci. Technol.* *58*:96–114. http://doi.org/10.1016/j.tifs.2016.09.011.

Deng, Z., K. Hou, J. Zhao, and H. Wang. 2021. "The probiotic properties of lactic acid bacteria and their applications in animal husbandry". *Curr. Microbiol.* *79*(1):22. http://doi.org/10.1007/s00284-021-02722-3.

EFSA. 2003. "Regulation (EC) No 1831/2003 of the European parliament and of the council of 22 September 2003 on additives for use in animal nutrition (Text with EEA relevance)". *Official J. L.* *268*:29–43.

Emerson, J. B., R. I. Adams, C. M. Betancourt Román, B. Brooks, D. A. Coil, K. Dahlhausen, H. H. Ganz, et al. 2017. "Schrödinger's microbes: Tools for distinguishing the living from the dead in microbial ecosystems". *Microbiome* *5*(1):86. http://doi.org/10.1186/s40168-017-0285-3.

FAO/WHO. 2002. "Guidelines for evaluation of probiotics in food". *Report of a Joint FAO/WHO Working Group on 20. Drafting Guidelines for the Evaluation of Probiotics in Food*. London, Ontario, Canada, April 30 and May 1, 2002.

Fathima, S., R. Shanmugasundaram, D. Adams, and R. K. Selvaraj. 2022. "Gastrointestinal microbiota and their manipulation for improved growth and performance in chickens". *Foods.* *11*(10):1401. http://doi.org/10.3390/foods11101401.

Fonesca, F., S. Cenard, and S. Passot. 2015. "Freeze-drying of lactic acid bacteria". In: *Cryopreservation and Freeze-Drying Protocols. Methods in Molecular Biology*. Eds. W. Wolkers and H. Oldenhof, Vol. 1257, Springer, New York. http://doi.org/10.1007/978-1-4939-2193-5_24.

Guimarães, J. T., C. F. Balthazar, H. Scudino, T. C. Pimentel, E. A. Esmerino, M. Ashokkumar, M. Q. Freitas, and A. G. Cruz. 2019. "High-intensity ultrasound: A novel technology for the development of probiotic and prebiotic dairy products". *Ultrason. Sonochem. 57*:12–21. http://doi.org/10.1016/j.ultsonch.2019.05.004.

Hammes, F., M. Berney, and T. Egli. 2011. "Cultivation-independent assessment of bacterial viability". *Adv. Biochem. Engin./Biotechnol. 124*:123–150. http://doi.org/10.1007/10_2010_95.

Humam, A. M., T. C. Loh, H. L. Foo, W. I. Izuddin, E. A. Awad, Z. Idrus, A. A. Samsudin, and N. M. Mustapha. 2020. "Dietary supplementation of postbiotics mitigates adverse impacts of heat stress on antioxidant enzyme activity, total antioxidant, lipid peroxidation, physiological stress indicators, lipid profile and meat quality in broilers". *Animals (Basel). 10*(6):982. http://doi.org/10.3390/ani10060982.

Humam, A. M., T. C. Loh, H. L. Foo, W. I. Izuddin, I. Zulkifli, A. A. Samsudin, and N. M. Mustapha. 2021. "Supplementation of postbiotic RI11 improves antioxidant enzyme activity, upregulated gut barrier genes, and reduced cytokine, acute phase protein, and heat shock protein 70 gene expression levels in heat-stressed broilers". *Poult. Sci. 100*(3):100908. http://doi.org/10.1016/j.psj.2020.12.011.

Humam, A. M., T. C. Loh, H. L. Foo, A. A. Samsudin, N. M. Mustapha, I. Zulkifli, and W. I. Izuddin. 2019. "Effects of feeding different postbiotics produced by *Lactobacillus plantarum* on growth performance, carcass yield, intestinal morphology, gut microbiota composition, immune status, and growth gene expression in broilers under heat stress". *Animals (Basel) 9*(9):644. http://doi.org/10.3390/ani9090644.

Huq, T., R. A. Khan, B. Riedl, and M. Lacroix. 2013. "Encapsulation of probiotic bacteria in biopolymeric system". *Crit. Rev. Food Sci. Nutr. 53*(9):909–916. http://doi.org/10.1080/10408398.2011.573152.

İncili, G. K., P. Karatepe, M. Akgöl, A. Güngören, A. Koluman, O. I. İlhak, H. Kanmaz, B. Kaya, and A. A. Hayaloğlu. 2022. "Characterization of lactic acid bacteria postbiotics, evaluation in-vitro antibacterial effect, microbial and chemical quality on chicken drumsticks". *Food Microbiol. 104*:104001. http://doi.org/10.1016/j.fm.2022.104001.

İncili, G. K., P. Karatepe, M. Akgöl, B. Kaya, H. Kanmaz, and A. A. Hayaloğlu. 2021. "Characterization of pediococcus acidilactici postbiotic and impact of postbiotic-fortified chitosan coating on the microbial and chemical quality of chicken breast fillets". *Int. J. Biol. Macromol. 184*:429–437. http://doi.org/10.1016/j.ijbiomac.2021.06.106.

Iwasaki, K., K. Maeda, K. Hidaka, K. Nemoto, Y. Hirose, and S. Deguchi. 2016. "Daily intake of heat-killed lactobacillus plantarum L-137 decreases the probing depth in patients undergoing supportive periodontal therapy". *Oral Health Prev. Dent. 14*(3):207–214. http://doi.org/10.3290/j.ohpd.a36099.

Izuddin, W. I., A. M. Human, T. C. Loh, H. L. Foo, and A. A. Samsudin. 2020. "Dietary postbiotic *Lactobacillus plantarum* improves serum and ruminal antioxidant activity and upregulates hepatic antioxidant enzymes and ruminal barrier function in post-weaning lambs". *Antioxidants (Basel). 9*(3):250. http://doi.org/10.3390/antiox9030250.

Izuddin, W. I., T. C. Loh, H. L. Foo, A. A. Samsudin, and A. M. Humam. 2019b. "Postbiotic *L. plantarum* RG14 improves ruminal epithelium growth, immune status and upregulates the intestinal barrier function in post-weaning lambs". *Sci. Rep. 9*(1):9938. http://doi.org/10.1038/s41598-019-46076-0.

Izuddin, W. I., T. C. Loh, A. A. Samsudin, H. L. Foo, A. M. Humam, N. Shazali. 2019a. "Effects of postbiotic supplementation on growth performance, ruminal fermentation and microbial profile, blood metabolite and GHR, IGF-1 and MCT-1 gene expression in post-weaning lambs". *BMC Vet. Res. 15*(1):315. http://doi.org/10.1186/s12917-019-2064-9.

Johnson, C. N., M. H. Kogut, K. Genovese, H. He, S. Kazemi, and R. J. Arsenault. 2019. "Administration of a postbiotic causes immunomodulatory responses in broiler gut and reduces disease pathogenesis following challenge". *Microorganisms. 7*(8):268. http://doi.org/10.3390/microorganisms7080268.

Kamiya, T., L. Wang, P. Forsythe, G. Goettsche, Y. Mao, Y. Wang, G. Tougas, and J. Bienenstock. 2006. "Inhibitory effects of lactobacillus reuteri on visceral pain induced by colorectal distension in sprague-dawley rats". *Gut. 55*(2):191–196. http://doi.org/10.1136/gut.2005.070987.

Kang, J., J. J. Lee, J. H. Cho, J. Choe, H. Kyoung, S. H. Kim, H. B. Kim, and M. Song. 2021. "Effects of dietary inactivated probiotics on growth performance and immune responses of weaned pigs". *J. Anim. Sci. Technol. 63*(3):520–530. http://doi.org/10.5187/jast.2021.e44.

Kataria, J., N. Li, J. L. Wynn, and J. Neu. 2009. "Probiotic microbes: Do they need to be alive to be beneficial?". *Nutr. Rev. 67*(9):546–550. http://doi.org/10.1111/j.1753-4887.2009.00226.x.

Khani, N., R. Abedi Soleimani, G. Noorkhajavi, A. Abedi Soleimani, A. Abbasi, and A. Homayouni Rad. 2022. "Postbiotics as potential promising tools for SARS-CoV-2 disease adjuvant therapy". *J. Appl. Microbiol. 132*(6):4097–4111. http://doi.org/10.1111/jam.15457.

Kim, S., H. H. Lee, W. Choi, C. H. Kang, G. H. Kim, and H. Cho. 2022. "Anti-tumor effect of heat-killed *Bifidobacterium bifidum* on human gastric cancer through Akt-p53-dependent mitochondrial apoptosis in xenograft models". *Int. J. Mol. Sci. 23*(17):9788. http://doi.org/10.3390/ijms23179788.

Kimoto-Nira, H. 2018. "New lactic acid bacteria for skin health via oral intake of heat-killed or live cells". *Anim. Sci. J. = Nihon Chikusan Gakkaiho. 89*(6):835–842. http://doi.org/10.1111/asj.13017.

Komano, Y., K. Fukao, K. Shimada, H. Naito, Y. Ishihara, T. Fujii, T. Kokubo, and H. Daida. 2023. "Effects of ingesting food containing heat-killed *Lactococcus lactis* strain plasma on fatigue and immune-related indices after high training load: A randomized, double-blind, placebo-controlled, and parallel-group study". *Nutrients. 15*(7):1754. http://doi.org/10.3390/nu15071754.

Kothari, D., S. Patel, and S. K. Kim. 2019. "Probiotic supplements might not be universally-effective and safe: A review". *Biomed. Pharmacother. 111*:537–547. http://doi.org/10.1016/j.biopha.2018.12.104.

Kumar, S. S., and A. R. Ghosh. 2019. "Assessment of bacterial viability: A comprehensive review on recent advances and challenges". *Microbiology (Reading). 165*(6):593–610. http://doi.org/10.1099/mic.0.000786.

Lahtinen, S. J. 2012. "Probiotic viability—does it matter?". *Microb. Ecol. Health Dis. 23*. http://doi.org/10.3402/mehd.v23i0.18567.

Li, Y., M. Chen, Y. Ma, Y. Yang, Y. Cheng, H. Ma, D. Ren, and P. Chen. 2022. "Regulation of viable/inactivated/lysed probiotic *Lactobacillus plantarum* H6 on intestinal microbiota and metabolites in hypercholesterolemic mice". *NPJ Sci. Food 6*:50. https://doi.org/10.1038/s41538-022-00167-x.

Maehata, H., S. Arai, N. Iwabuchi, and F. Abe. 2021. "Immuno-modulation by heat-killed *Lacticaseibacillus paracasei* MCC1849 and its application to food products". *Int. J. Immunopathol. Pharmacol. 35*:20587384211008291. http://doi.org/10.1177/20587384211008291.

Maguire, M., and G. Maguire. 2017. "The role of microbiota, and probiotics and prebiotics in skin health." *Arch. Dermatol. Res. 309*(6):411–421. http://doi.org/10.1007/s00403-017-1750-3.

Maidana, L. G., J. Gerez, M. N. S. Hohmann, W. A. Jr Verri, and A. P. F. L. Bracarense. 2021. "*Lactobacillus plantarum* metabolites reduce deoxynivalenol toxicity on jejunal explants of piglets". *Toxicon. 203*:12–21. http://doi.org/10.1016/j.toxicon.2021.09.023.

Maruyama, M., R. Abe, T. Shimono, N. Iwabuchi, F. Abe, and J.-Z. Xiao. 2016. "The effects of non-viable lactobacillus on immune function in the elderly: A randomised, double-blind, placebo-controlled study." *Int. J. Food Sci. Nutr. 67*(1):67–73. http://doi.org/10.3109/09637486.2015.1126564.

Mathur, H., T. P. Beresford, and P. D. Cotter. 2020. "Health benefits of lactic acid bacteria (LAB) fermentates". *Nutrients 12*(6):1679. http://doi.org/10.3390/nu12061679.

Mehling, H., and A. Busjahn. 2013. "Non-viable lactobacillus reuteri DSMZ 17648 (PylopassTM) as a new approach to *Helicobacter Pylori* control in humans". *Nutrients 5*(8):3062–3073. http://doi.org/10.3390/nu5083062.

Mitropoulou, G., V. Nedovic, A. Goyal, and Y. Kourkoutas. 2013. "Immobilization technologies in probiotic food production". *J. Nutr. Met. 2013*:716861. http://doi.org/10.1155/2013/716861.

Moradi, M., S. A. Kousheh, H. Almasi, A. Alizadeh, J. T. Guimarães, N. Yılmaz, and A. Lotfi. 2020. "Postbiotics produced by lactic acid bacteria: The next frontier in food safety". *Compr. Rev. Food Sci. Food Saf. 19*(6):3390–3415. http://doi.org/10.1111/1541-4337.12613.

Moraes, R. M., U. Schlagenhauf, and A. L. Anbinder. 2022. "Outside the limits of bacterial viability: Postbiotics in the management of periodontitis". *Biochem. Pharmacol. 201*:115072. http://doi.org/10.1016/j.bcp.2022.115072.

Mora-Sánchez, B., J. L. Balcázar, and T. Pérez-Sánchez. 2020. "Effect of a novel postbiotic containing lactic acid bacteria on the intestinal microbiota and disease resistance of rainbow trout (Oncorhynchus mykiss)". *Biotechnol. Lett. 42*(10):1957–1962. http://doi.org/10.1007/s10529-020-02919-9.

Morniroli, D., G. Vizzari, A. Consales, F. Mosca, and M. L. Giannì. 2021. "Postbiotic supplementation for children and newborn's health". *Nutrients. 13*(3):781. http://doi.org/10.3390/nu13030781.

Nguafack, T. T., W. J. Jang, M. T. Hasan, Y. H. Choi, S. C. Bai, E. W. Lee, B. J. Lee, S. W. Hur, S. Lee, and I. S. Kong. 2020. Effects of dietary non-viable Bacillus sp. SJ-10, Lactobacillus plantarum, and their combination on growth, humoral and cellular immunity, and streptococcosis resistance in olive flounder (Paralichthys olivaceus). *Res. Vet. Sci. 131*:177–185. http://doi.org/10.1016/j.rvsc.2020.04.026.

Niedźwiedź, I., W. Juzwa, K. Skrzypiec, T. Skrzypek, A. Waśko, M. Kwiatkowski, J. Pawlat, and M. Polak-Berecka. 2020. "Morphological and physiological changes in *Lentilactobacillus hilgardii* cells after cold plasma treatment". *Sci. Rep. 10*:18882. http://doi.org/10.1038/s41598-020-76053-x.

Oglio, F., C. Bruno, S. Coppola, R. De Michele, A. Masino, and L. Carucci. 2022. "Evidence on the preventive effects of the postbiotic derived from cow's milk fermentation with *Lacticaseibacillus paracasei* CBA L74 against pediatric gastrointestinal infections". *Microorganisms 11*(1):10. http://doi.org/10.3390/microorganisms11010010.

Ou, C.-C., S.-L. Lin, J.-J. Tsai, and M.-Y. Lin. 2011. "Heat-killed lactic acid bacteria enhance immunomodulatory potential by skewing the immune response toward Th1 polarization". *J. Food Sci. 76*(5):M260–M267. http://doi.org/10.1111/j.1750-3841.2011.02161.x.

Ouwehand, A. C., S. Tölkkö, J. Kulmala, S. Salminen, and E. Salminen. 2000. "Adhesion of inactivated probiotic strains to intestinal mucus". *Lett. Appl. Microbiol. 31*(1):82–86.

Panasevich, M. R., L. Daristotle, R. Quesnell, G. A. Reinhart, and N. Z. Frantz. 2021. "Altered fecal microbiota, IgA, and fermentative end-products in adult dogs fed prebiotics and a nonviable Lactobacillus acidophilus". *J. Anim. Sci.* 99(12):skab347. http://doi.org/10.1093/jas/skab347.

Park, I., H. Nam, D. H. Goo, S. S. Wickramasuriya, N. Zimmerman, A. H. Smith, T. G. Rehberger, and H. S. Lillehoj. 2022. "Gut microbiota-derived indole-3-carboxylate influences mucosal integrity and immunity through the activation of the aryl hydrocarbon receptors and nutrient transporters in broiler chickens challenged with *Eimeria maxima*". *Front. Immunol.* 13:867754. http://doi.org/10.3389/fimmu.2022.867754.

Peng, K., M. Koubaa, O. Bals, and E. Vorobiev. 2020. "Recent insights in the impact of emerging technologies on lactic acid bacteria: A review". *Food Res. Int.* 137:109544. http://doi.org/10.1016/j.foodres.2020.109544.

Percopo, C. M., T. A. Rice, T. A. Brenner, K. D. Dyer, J. L. Luo, K. Kanakabandi, D. E. Sturdevant, et al. 2015. "Immunobiotic lactobacillus administered post-exposure averts the lethal sequelae of respiratory virus infection". *Antivir. Res.* 121:109–119. http://doi.org/10.1016/j.antiviral.2015.07.001.

Pérez-Sánchez, T., B. Mora-Sánchez, A. Vargas, and J. L. Balcázar. 2020. "Changes in intestinal microbiota and disease resistance following dietary postbiotic supplementation in rainbow trout (Oncorhynchus mykiss)". *Microb. Pathog.* 142:104060. http://doi.org/10.1016/j.micpath.2020.104060.

Pham Thi, H. H., T. V. Phan Thi, N. Pham Huynh, V. Doan, S. Onoda, and T. L. Nguyen. 2022. "Therapeutic effect of heat-killed Lactobacillus plantarum L-137 on the gut health and growth of broilers". *Acta Trop.* 232:106537. http://doi.org/10.1016/j.actatropica.2022.106537.

Porfiri, L., J. Burtscher, R. T. Kangethe, D. Verhovsek, G. Cattoli, K. J. Domig, and V. Wijewardana. 2022. "Irradiated non-replicative lactic acid bacteria preserve metabolic activity while exhibiting diverse immune modulation". *Front. Vet. Sci.* 9:859124. http://doi.org/10.3389/fvets.2022.859124.

Pratap-Singh, A., A. Suwardi, R. Mandal, J. Pico, S. D. Castellarin, D. D. Kitts, and A. Singh. 2023. "Effect of UV filters during the application of pulsed light to reduce lactobacillus brevis contamination and 3-methylbut-2-ene-1-thiol formation while preserving the physicochemical attributes of blonde ale and centennial red ale beers". *Foods.* 12(4):684. http://doi.org/10.3390/foods12040684.

Quintanilla-Pineda, M., C. G. Achou, J. Díaz, A. Gutiérrez-Falcon, M. Bravo, J. I. Herrera-Muñoz, N. Peña-Navarro, C. Alvarado, F. C. Ibañez, and F. Marzo. 2023. "In vitro evaluation of postbiotics produced from bacterial isolates obtained from rainbow trout and nile tilapia against the pathogens *Yersinia ruckeri* and *Aeromonas salmonicida* subsp. *salmonicida.*" *Foods.* 12(4):861. http://doi.org/10.3390/foods12040861.

Riaz, Q. U. A., and T. Masud. 2013. "Recent trends and applications of encapsulating materials for probiotic stability". *Crit. Rev. Food Sci. Nutr.* 53(3):231–244. http://doi.org/10.1080/10408398.2010.524953.

Rocha, S. D. C., P. Lei, B. Morales-Lange, L. T. Mydland, and M. Øverland. 2023. "From a cell model to a fish trial: Immunomodulatory effects of heat-killed Lactiplantibacillus *plantarum* as a functional ingredient in aquafeeds for salmonids". *Front. Immunol.* 14:1125702. http://doi.org/10.3389/fimmu.2023.1125702.

Ryu, S., H. Kyoung, K. I. Park, S. Oh, M. Song, and Y. Kim. 2022. "Postbiotic heat-killed lactobacilli modulates on body weight associated with gut microbiota in a pig model". *AMB Express.* 12(1):83. http://doi.org/10.1186/s13568-022-01424-8.

Salminen, S., M. C. Collado, A. Endo, C. Hill, S. Lebeer, E. M. M. Quigley, M. E. Sanders, R. Shamir, J. R. Swann, H. Szajewska, and G. Vinderola. 2021a. "The international scientific association of probiotics and prebiotics (ISAPP) consensus statement on the definition and scope of postbiotics". *Nat. Rev. Gastroenterol. Hepatol.* 18(9):649–667. http://doi.org/10.1038/s41575-021-00440-6. Erratum in: *Nat. Rev. Gastroenterol. Hepatol.* 2021 Jun 15, Erratum in: *Nat. Rev. Gastroenterol. Hepatol.* 2022 Aug;19(8):551.

Salminen, S., M. C. Collado, A. Endo, C. Hill, S. Lebeer, E. M. M. Quigley, M. E. Sanders, R. Shamir, J. R. Swann, H. Szajewska, and G. Vinderola. 2021b. "Reply to: Postbiotics—when simplification fails to clarify". *Nat. Rev. Gastroenterol. Hepatol.* 18(11):827–828. http://doi.org/10.1038/s41575-021-00522-5. Erratum in: *Nat. Rev. Gastroenterol. Hepatol.* 19(4):275.

Sanders, M. E., and M. L. Marco, 2010. "Food formats for effective delivery of probiotics". *Annu. Rev. Food Sci. Technol.* 1:65–85. http://doi.org/10.1146/annurev.food.080708.100743.

Sañudo, A. I., R. Luque, M. P. Díaz-Ropero, J. Fonollá, and Ó. Bañuelos. 2017. "In vitro and in vivo anti-microbial activity evaluation of inactivated cells of lactobacillus salivarius CECT 5713 against streptococcus mutans". *Arch. Oral Biol.* 84:58–63. http://doi.org/10.1016/j.archoralbio.2017.09.014.

Sasazaki, N., T. Obi, C. Aridome, Y. Fujimoto, M. Furumoto, K. Toda, H. Hasunuma, D. Matsumoto, S. Sato, H. Okawa, O. Yamato, N. Igari, D. Kazami, M. Taniguchi, and M. Takagi. 2020. "Effects of dietary feed supplementation of heat-treated Lactobacillus sakei HS-1 on the health status, blood parameters, and fecal microbes of Japanese Black calves". *J. Vet. Med. Sci.* 82(10):1428–1435. http://doi.org/10.1292/jvms.20-0181.

Sato, Y., Y. Kuroki, K. Oka, M. Takahashi, S. Rao, S. Sukegawa, and T. Fujimura. 2019. "Effects of dietary supplementation with *Enterococcus faecium* and *Clostridium butyricum*, either alone or in combination, on growth and fecal microbiota composition of post-weaning pigs at a commercial farm". *Front. Vet. Sci.* 6:26. http://doi.org/10.3389/fvets.2019.00026.

Sauer, M., and N. S. Han. 2021. "Lactic acid bacteria: Little helpers for many human tasks". *Essays Biochem.* 65(2):163–171. http://doi.org/10.1042/EBC20200133.

Sawada, D., T. Sugawara, T. Hirota, and Y. Nakamura. 2022. "Effects of *Lactobacillus gasseri* CP2305 on mild menopausal symptoms in middle-aged women". *Nutrients.* 14(9):1695. http://doi.org/10.3390/nu14091695.

Schwendicke, F., K. Horb, S. Kneist, C. Dörfer, and S. Paris. 2014. "Effects of heat-inactivated Bifidobacterium BB12 on cariogenicity of Streptococcus mutans in vitro". *Arch. Oral Biol.* 59(12):1384–1390. http://doi.org/10.1016/j.archoralbio.2014.08.012.

Solanki, H., D. D. Pawar, D. A. Shah, V. D. Prajapati, G. K. Jani, A. M. Mulla, and P. M. Thakar. 2013. "Development of microencapsulation delivery system for long-term preservation of probiotics as biotherapeutics agent". *Biomed. Res. Int.* 2013:620719. http://doi.org/10.1155/2013/620719.

Tartrakoon, W., R. Charoensook, T. R. Incharoen, T. S. Numthuam, S. T. Pechrkong, S. T. Onoda, G. S. Shoji, and B. G. Brenig. 2023. "Effects of heat-killed lactobacillus plantarum L-137 supplementation on growth performance, blood profiles, intestinal morphology, and immune gene expression in pigs". *Vet. Sci.* 10:87. http://doi.org/10.3390/vetsci10020087.

Taverniti, V., and S. Guglielmetti. 2011. "The immunomodulatory properties of probiotic microorganisms beyond their viability (ghost probiotics: Proposal of paraprobiotic concept)". *Genes Nutr.* 6(3):261–274. http://doi.org/10.1007/s12263-011-0218-x.

Tsai, W. H., C. H. Chou, T. Y. Huang, H. L. Wang, P. J. Chien, W. W. Chang, and H. T. Lee. 2021. "Heat-killed lactobacilli preparations promote healing in the experimental cutaneous wounds". *Cells.* 10(11):3264. http://doi.org/10.3390/cells10113264.

Tsilingiri, K., and M. Rescigno. 2013. "Postbiotics: What else?". *Benef. Microbes.* 4(1):101–107. http://doi.org/10.3920/BM2012.0046.

Tsuruta, T., R. Inoue, T. Tsukahara, N. Matsubara, M. Hamasaki, and K. Ushida. 2009. "A cell preparation of *Enterococcus faecalis* strain EC-12 stimulates the luminal immunoglobulin a secretion in juvenile calves." *Anim. Sci. J. = Nihon Chikusan Gakkaiho.* 80(2):206–211. http://doi.org/10.1111/j.1740-0929.2008.00621.x.

Tukaram, N. M., A. Biswas, C. Deo, A. J. Laxman, M. Monika, and A. K. Tiwari. 2022. "Effects of paraprobiotic as replacements for antibiotic on performance, immunity, gut health and carcass characteristics in broiler chickens". *Sci. Rep.* 12(1):22619. http://doi.org/10.1038/s41598-022-27181-z.

Vilhelmova-Ilieva, N., G. Atanasov, L. Simeonova, L. Dobreva, K. Mancheva, M. Trepechova, and S. Danova. 2022. "Anti-herpes virus activity of *lactobacillus'* postbiotics". *Biomedicine (Taipei).* 12(1):21–29. http://doi.org/10.37796/2211-8039.1277.

Vilhelmova-Ilieva, N., S. Danova, Z. Petrova, L. Dobreva, G. Atanasov, K. Mancheva, and L. Simeonova. 2023. "Protective and therapeutic capacities of lactic acid bacteria postmetabolites against koi herpesvirus infection in vitro". *Life (Basel).* 13(3):739. http://doi.org/10.3390/life13030739.

Vinderola, G., M. E. Sanders, and S. Salminen. 2022a. "The concept of postbiotics". *Foods.* 11(8):1077. http://doi.org/10.3390/foods11081077.

Vinderola, G., M. E. Sanders, S. Salminen, and H. Szajewska. 2022b. "Postbiotics: The concept and their use in healthy populations". *Front. Nutr.* 9:1002213. http://doi.org/10.3389/fnut.2022.1002213.

Vinderola, G., M. E. Sanders, M. Cunningham, and C. Hill. 2024. "Frequently asked questions about the ISAPP postbiotic definition". *Front. Microbiol.* 14:1324565. http://doi.org/10.3389/fmicb.2023.1324565.

Von Wright, A., and L. Axelsson. 2019. Chapter: 1 "Lactic Acid bacteria: An introduction". In: *Lactic Acid Bacteria, Microbial and Functional Aspects*, 5th edition. Eds. G. Vinderola, A. C. Ouwehand, S. Salminen, and A. von Wright, pp. 1–16, CRC Press, Boca Raton, FL.

Wong, C., and Z. Ustunol. 2006. "Mode of inactivation of probiotic bacteria affects interleukin 6 and interleukin 8 production in human intestinal epithelial-like caco-2 cells". *J. Food Protect.* 69(9):2285–2288.

Xavier-Santos, D., M. Padilha, G. A. Fabiano, G. Vinderola, A. Gomes Cruz, K. Sivieri, and A. E. Costa Antunes. 2022. "Evidences and perspectives of the use of probiotics, prebiotics, synbiotics, and postbiotics as adjuvants for prevention and treatment of COVID-19: A bibliometric analysis and systematic review". *Trends Food Sci. Technol.* 120:174–192. http://doi.org/10.1016/j.tifs.2021.12.033. Erratum in: *Trends Food Sci Technol.* 121:156–160.

Xu, X., M. E. Duarte, and S. W. Kim. 2022. "Postbiotic effects of Lactobacillus fermentate on intestinal health, mucosa-associated microbiota, and growth efficiency of nursery pigs challenged with F18+Escherichia coli". *J. Anim. Sci.* 100(8):skac210. http://doi.org/10.1093/jas/skac210.

Yokota, Y., A. Shikano, T. Kuda, M. Takei, H. Takahashi, and B. Kimura. 2018. "Lactobacillus plantarum AN1 cells increase caecal L. Reuteri in an ICR mouse model of dextran sodium sulphate-induced inflammatory bowel disease". *Int. Immunopharmacol.* 56:119–127. http://doi.org/10.1016/j.intimp.2018.01.020.

Yu, Z., Q. Hao, S. B. Liu, Q. S. Zhang, X. Y. Chen, S. H. Li, C. Ran, Y. L. Yang, T. Teame, Z. Zhang, and Z. G. Zhou. 2023. "The positive effects of postbiotic (SWF concentration) supplemented diet on skin mucus, liver, gut health, the structure and function of gut microbiota of common carp (Cyprinus carpio) fed with high-fat diet". *Fish Shellfish Immunol. 135*:108681. http://doi.org/10.1016/j.fsi.2023.108681.

Zamojska, D., A. Nowak, I. Nowak, and E. Macierzyńska-Piotrowska. 2021. "Probiotics and postbiotics as substitutes of antibiotics in farm animals: A review". *Animals (Basel). 11*(12):3431. http://doi.org/10.3390/ani11123431.

Zheng J., S. Wittouck, E. Salvetti, C. M. A. P. Franz, H. M. B. Harris, P. Mattarelli, P. W. O'Toole, B. Pot, P. Vandamme., J. Walter, K. Watanabe, S. Wuyts, G. E. Felis, M. G. Gänzle, and S. Lebeer. 2020. "A taxonomic note on the genus *Lactobacillus*: Description of 23 novel genera, emended description of the genus *Lactobacillus* Beijerinck 1901, and union of *Lactobacillaceae* and *Leuconostocaceae*". *Int. J. Syst. Evol. Microbiol. 70*:2782–2858. http://doi.org/10.1099/ijsem.0.004107.32293557.

Zhong, Y., T. Wang, R. Luo, J. Liu, R. Jin, and X. Peng. 2022. "Recent advances and potentiality of postbiotics in the food industry: Composition, inactivation methods, current applications in metabolic syndrome, and future trends". *Crit. Rev. Food Sci. Nutr. 20*:1–25. http://doi.org/10.1080/10408398.2022.2158174.

Zhu, C., L. Gong, K. Huang, F. Li, D. Tong, and H. Zhang. 2020. "Effect of heat-inactivated compound probiotics on growth performance, plasma biochemical indices, and cecal microbiome in yellow-feathered broilers". *Front Microbiol. 11*:585623. http://doi.org/10.3389/fmicb.2020.585623.

# Skin Health Benefits of Topical and Enteral Lactic Acid Bacteria

# 36

Catherine A. O'Neill and Andrew J. McBain

## 36.1 LACTIC ACID BACTERIA: KEY PLAYERS IN THE GUT SKIN AXIS

Gastroenterologists and dermatologists alike have long since known that the gut and skin participate in an intimate bi-directional relationship—the gut–skin axis (for review, see [1]). Gut and skin have much in common, perhaps most notably that they are major sites of interface with the environment, are both immune and endocrine organs that are heavily innervated and vascularised, and crucially, have unique microbiota [2–5]. Thus, it is unsurprising that the relationship between gut and skin is often evident in disease: many gastrointestinal diseases have unique skin manifestations and *vice versa* [2, 6]. Whilst the mechanisms underlying this are often unknown and probably involve multiple elements including genetics, the gut microbiota, with its huge metabolic and immune regulatory capacity, is emerging as a central player along the gut-skin axis [7].

As long ago as the 1930s, it was recognised that the gut engages in a relationship with the skin that is mediated, at least in part by the gut microbiota [8]. Two eminent dermatologists, John Stokes and Donald Pillsbury, postulated that acne vulgaris was driven primarily from the gut via a gut–skin–brain axis: They suggested that the emotional state of the patient led to increased intestinal permeability driven by changes to the gut microbiota. This increased permeability contributed to systemic inflammation via leakage of the inflammatory milieu from the gut. Years ahead of their time, Stoke and Pillsbury were also among the first to suggest the use of a lactic acid bacterium (LAB), *Lactobacillus acidophilus*, to alleviate the symptoms of acne [8, 9].

In recent years, there has been an increased understanding of the ability of gut microbes to exert influence at body sites beyond the gut [10–13], and this includes enterally consumed LAB acting along the gut–skin axis, which have been studied for their ability to improve the health of the skin [14, 15], as well as for their ability to treat or prevent skin diseases [16]. Furthermore, the use of topical lactobacilli, either as live organisms or extracts (termed postbiotics) [17], as novel treatments for skin is gaining credence [15]. In this chapter, reports of the efficiency of lactic acid bacteria (LAB) from selected human studies will be reviewed along with new insights from *in vitro* and animal studies that could be translated for human health benefit.

DOI: 10.1201/9781003352075-39

## 36.2  A BRIEF INTRODUCTION TO THE SKIN—A CRITICAL BARRIER

The skin is the major interface of the body with the outside world. This two-way barrier is crucial for preventing excessive water loss from within the body and for preventing the ingress of potentially toxic pathogens and molecules [18]. The importance of skin as a barrier is exemplified in individuals in whom this barrier is severely breached, for example, victims of severe burns and premature babies in which infection and excessive transepidermal water loss (TEWL) are common causes of morbidity and even mortality [19]. Indeed, life as a terrestrial mammal is only possible because of the unique structure of the skin barrier. Skin is composed of three main layers: the epidermis, dermis, and hypodermis [20]—as illustrated in Figure 36.1. The epidermis provides most of the barrier properties of the skin [21]. This squamous epithelium is subdivided into further layers which represent keratinocytes (the epithelial cells of the skin) at various stages of terminal differentiation. The basal layer of the epidermis contains proliferating keratinocytes. Some of these begin to migrate towards the surface, undergoing profound changes in phenotype, associated with the expression of structural proteins and lipids, as they go. Ultimately, in the top layer of the skin, the keratinocytes lose their nuclei and become dead 'corneocytes' embedded in a lipid matrix [21, 22]). This tough, waterproof structure can withstand mechanical stresses as well as prevent water loss.

Ultimately, the corneocytes are shed from the surface of the skin and replaced with new cells in a highly regulated process that in healthy skin takes around 28 days. This delicately balanced process is regulated at least in part by the immune system [23]. Loss of this balance between production and

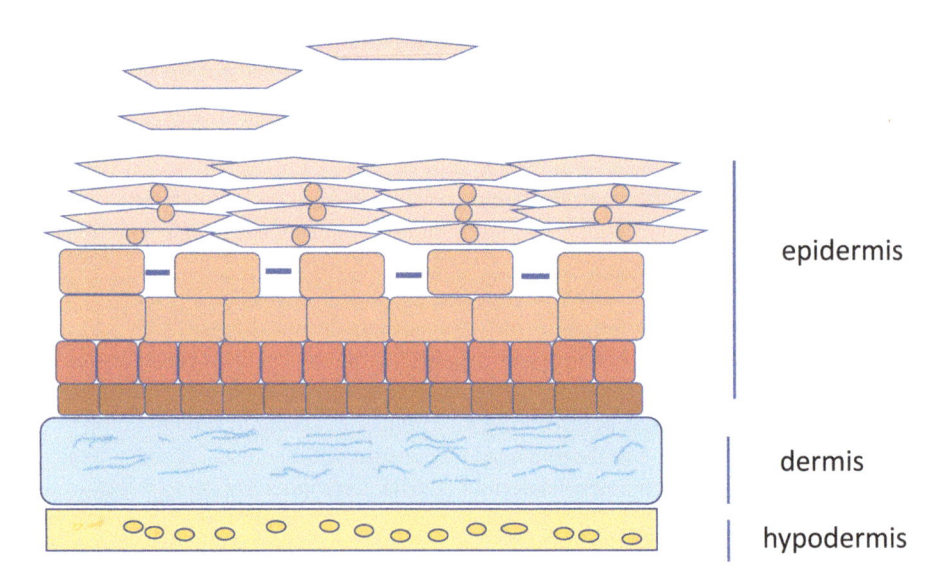

**FIGURE 36.1**   The structure of skin.

*The skin is composed of three main layers:* the epidermis is composed of keratinocytes at different stages of differentiation. Cells in the bottom layer migrate to the top, where they ultimately lose their organelles, become flattened corneocytes, and are ultimately shed from the surface. The dermis is a layer of elastic fibres which provides the mechanical properties of the skin. The hypodermis is an adipose layer which helps with thermoregulation and acts as an energy store.

loss of keratinocytes can be seen in certain skin conditions, such as the auto-immune disease psoriasis which is characterised by 'scaly plaques'. These are due to a build-up of cells at the skin surface because of the overproduction of new keratinocytes in the basal layer and reduced shedding of dead cells at the surface [24]. Underlying the epidermis is the dermis, which is a network of tough elastic fibres, composed mainly of collagens and elastin. These provide skin with its strength and elastic properties of stretch and rebound and allow the skin to engage in the thousands of movements that day that the body engages in [20]. These elastic properties diminish during ageing, and this is mainly associated with damage/loss of these elastic fibres, largely due to sun exposure [25–27]. Many of these elastic fibres are laid down during childhood, but production ceases after puberty. Therefore, once these fibres are lost or damaged, the ability of the skin to rebound diminishes, and the characteristic 'sagging' associated with ageing is observed [28]. Below the dermis is the hypodermis, which is mainly composed of adipose tissue. This layer helps to maintain thermoregulation and can also act as an energy store for the body [20].

## 36.3  CAN LAB ENHANCE THE SKIN BARRIER?

As illustrated in this chapter, probiotics can be applied to the body via the digestive tract or applied directly to the skin (i.e. topical application). We propose that probiotics consumed orally and directed to the enteric tract should be termed 'enteral', as opposed to 'oral', probiotics. This is to distinguish them from probiotics that are directed to the oral cavity to benefit dental health.

Many inflammatory skin conditions and even ageing are associated with reductions in epidermal barrier function and/or its ability to repair following injury. A fully competent barrier also retains water, which improves both the look and feel of the skin. Thus, a major goal in clinical as well as aesthetic dermatology is the improvement or restoration of the epidermal barrier. Since the turnover of the epidermis is largely regulated by the immune system, it is easy to see why LAB, with their ability to modify the immune system, are a promising candidate for the modulation of the skin barrier.

There are now several pivotal studies that suggest improvement of the skin barrier can be achieved via ingestion of oral LAB: A landmark study in mice by the laboratory of Susan Erdman reported that administering *Limosilactobacillus reuteri* to mice in drinking water resulted in several positive changes that can be specified to the integumentary system [29]. The *Limosilactobacillus reuteri* fed mice had thicker, shinier fur compared to their control counterparts. This was associated with increases in sebocyte production and an increase in the acidity of the skin (associated with health). Increased folliculogenesis and dermal thickness were also observed in the LAB-fed mice. These positive changes appear to be mediated by an increase in serum levels of the anti-inflammatory IL-10 (and a concomitant decrease in pro-inflammatory IL-12) because IL-10 deficient mice showed no improvement to their integumentary system when fed *Limosilactobacillus reuteri* [29, 30].

Data in humans also suggest that ingestion of probiotics can improve skin barrier function. A study in healthy humans showed that ingestion of *Lacticaseibacillus paracasei* NCC2461 resulted in reduced 'sensitivity' of the skin to challenge with capsaicin and decreased TEWL in the fed group vs the control group [31]. The *Lacticaseibacillus paracasei*–fed group also had higher levels of circulating TGFβ. Since TGFβ is a known regulator of the epidermal barrier, the improvement in barrier function was attributed to the ability of the LAB to increase circulating levels of the cytokine [31].

The effects of *Lactiplantibacillus plantarum* HY7714 have also been investigated in the context of the skin barrier mostly studied in mouse models, but a double-blind, placebo-controlled human clinical study of 110 human volunteers aged between 41 and 59 years of age has also been carried out. In this study, the oral administration of $1 \times 10^{10}$ CFU/day for 12 weeks is reported to have resulted in a reduction in TEWL at week 12 in the LAB-fed group and increased skin elasticity and 'gloss' [32]. There was also a reported significant reduction in wrinkle depth in the LAB-fed group compared with the placebo-fed group suggestive of effects in the dermis as well as the epidermis. Although the mechanisms underlying these effects were not addressed in humans, previous work in hairless mice showed that the skin-hydrating effects of *L. plantarum* HY7714 were associated with changes to ceramidase and palmotyltransferase expression in the skin [33]. These two proteins are important mediators in the production of lipid components of the barrier. Thus, one hypothesis may be that *L. plantarum* HY7714 induces improvements to the barrier through the modulation of lipid production.

# 36.4  LAB AND BARRIER RESTORATION

A wound represents a breach of the barrier and encompasses several different types of injury, including surgical trauma, burns, and pressure ulcers. Wound healing is a dynamic and highly coordinated process which usually results in rapid restoration of the barrier. However, in some conditions, wound healing is delayed, sometimes for months, and may be associated with infection and/or underlying conditions such as diabetes [34]. A multitude of animal studies have investigated the utility of LAB in wound healing (e.g. [35–38]). Currently, the data on humans is limited. However, a recent systematic review looked specifically at whether oral consumption of 'probiotics' resulted in faster healing of wounds [39]. The review encompassed 348 individuals in seven studies. The authors note that there is no strong evidence to support the role of oral LAB in the treatment of wounds. This is mainly due to the differences in the types of wounds, outcome measures, and even the choice of organisms and the differing dosing regimens employed in the various studies. However, despite this, a more positive outcome in the 'probiotic group' was noted in four of the seven studies, where outcomes such as the reduced need for grafting and fewer wound complications were noted [39]. Of note, none of the studies reviewed reported any adverse effects of oral ingestion of LAB for wounds.

An increasing number of animal and *in vitro* studies suggest a role for topical LAB to improve wound healing. Topical *Lactiplantibacillus plantarum*, for example, has been used in several studies of burns where the presence of the organism is reported to have improved tissue repair [40], reduced mortality [41], and reduced infection [42]. *Limosilactobacillus fermentum* and kefir containing *Lactobacillus lactis* have also been reported to improve the healing of incision wounds in animals [43, 44]. Topical *Lactiplantibacillus plantarum* ATCC 10241 has also been investigated in a small number of human studies: in a burn study comprising 80 individuals, the effectiveness of the organism at healing delayed second and third-degree burns was compared to standard treatment comprising silver sulphadiazine. Topical treatment with the LAB showed promise in decreasing the bacterial load within the wound [45]. It is also claimed that that the organism may have improved the rate of healing if the sample size had been bigger. However, the effectiveness of *Lactiplantibacillus plantarum* in this study depended on whether the burn was second or third degree and the extent of infection before treatment began [45]. Topical *Lactiplantibacillus plantarum* ATCC 10241 has also shown promise for the treatment of venous leg ulcers [46] and diabetic foot ulcers [47], although the sample sizes were small in each case.

Numerous *in vitro* and animal studies point to the possible utility of topical LAB to inhibit wound pathogens and modify tissue physiology to accelerate wound healing (e.g., [48, 49–52]). At present, the data from human studies are difficult to interpret due to the disparities in the studies. Small study sizes, differences in types of wounds investigated, the different organisms used, and the dosing regimens all mean that the offered therapeutic promise of LAB for wound healing is yet to be fulfilled.

**Oral LAB as a treatment for photoageing**: A major environmental influence on the skin is ultraviolet radiation (UVR) from ambient sunlight. UVR is also the main environmental factor inducing premature ageing of the skin [53]. Indeed, 'photoageing' is defined as premature ageing of the skin as a result of repeated exposure to UVR. The most visible signs of photoageing include wrinkle formation, loss of elasticity, epidermal thickening, and hyperpigmentation [54]. UVR induces photoageing via mechanisms including direct damage to proteins and nucleic acids, generation of reactive oxygen species, induction of inflammatory cascades, and immunosuppression. UVR also induces the activity of matrix metalloproteinases, which can directly degrade the elastic fibre network of the dermis [18–20]. LAB and their metabolic products are showing promise in the modulation of some of these effects of UVR. For example, Peguey-Navarro et al. demonstrated in a randomised placebo-controlled human study of 54 volunteers that 6-week oral administration of *Lactobacillus johnsonii* johnsonni La1 could accelerate the recovery of immune homeostasis following UVR-induced immunosuppression in healthy humans [55]. In three clinical studies, Bouilly-Gauthier et al. showed that a 10-week intake of a nutritional supplement also containing La1 was able to reduce the inflammatory infiltrate and increase the minimal dose needed to induce erythema in individuals taking the supplement, compared with volunteers taking a placebo [56]. However, whilst these data are encouraging, it is important to note that the supplement contained carotenoids as well as La1. Carotenoids are known to have many beneficial effects on skin with regard to UVR [57], and so the positive effects seen in these studies could be at least in part due to their presence within the supplement. If LAB are to gain serious traction as interventions for skin, then studies will be required to test their effects alone as well as in potential combinations.

Kim et al. have recently shown that feeding hairless mice a combination of *Bifidobacterium longum* with galactooligosaccharides was protective against UVR-induced photoageing [58]. Either the bacterium or the prebiotic were effective alone, but a greater effect was observed when they were used in combination. The authors noted increased hydration of the skin, increased antioxidant capacity, and abrogation of the pro-inflammatory cytokines associated with exposure to UVR. Of note, none of these effects were observed when the mice were supplemented with collagen (a well-known anti-ageing supplement) instead of the pro/prebiotic combination. Indeed, the pro/prebiotic-induced improvements were associated with increases in the production of short-chain fatty acids (SCFAs), especially acetate. Thus, the authors concluded that the anti-inflammatory effects of SCFAs may be responsible for their observed effects [58].

*Latilactobacillus sakei* Wikim0066 has also been studied for its ability to protect skin from UVR-induced damage [59]. Currently, this organism has only been tested in hairless mice. However, MMP production in response to irradiation was significantly lower in mice fed the LAB for 12 weeks than in the placebo-fed group [59]. Several *in vitro* studies are also suggestive of the idea that LAB, or their derivatives, can modulate ageing and photoageing pathways. For example, tyndallised (heat-inactivated) *Lactobacillus acidophilus* KCCM12625P modulates the expression of MMPs in human skin cells [60], fitting the definition of a postbiotic. *Limosilactobacillus reuteri* DSM 17938, whether used live or as a lysate, can reduce IL-6 and IL-8—both pro-inflammatory—in irradiated human skin equivalent models [61]. Indeed, LAB, either enteral or topical, have been shown to have effects on many of the pathways that are intrinsic to photodamage of the skin [62]. A recent meta-analysis of the literature [63] concluded that LAB are effective in terms of inhibiting MMP pathways and reversing skin barrier damage in rodent models, but the jury is still out in terms of effectiveness for humans!

## 36.5 LAB AS INTERVENTIONS IN DISEASE

LAB, with their antimicrobial and immune-modulatory properties, are obvious candidate therapies for several skin diseases. Here, we discuss the data on four of the most common inflammatory conditions of the skin:

# 36.6 ACNE VULGARIS

Acne vulgaris is a disease of the pilosebaceous unit, which is part of the hair follicle. The disease manifests as inflammatory pustules and papules and noninflammatory comedones (papules) resulting from excessive sebum production along with follicular hyperkeratinisation and the production of inflammatory cytokines [64]. It has long been reported that microbial dysbiosis of both the gut and skin is found in patients with acne vulgaris [65–67]. The disease is more prevalent in developed countries, suggestive of the hypothesis that it is associated with a 'Western diet'. Indeed, a high glycaemic load [68, 69], which is typical of a Western diet, drives the production of insulin-like growth factor (IgF-1). This promotes the overproduction of sebum and keratinocyte proliferation [64, 65]. The excessive production of sebum in turn allows the proliferation of *Cutibacterium acnes*, which for a long time was thought to be the 'pathogen' responsible for acne. Rather, *C. acnes* is a normal part of a healthy skin microbiome, although more recently certain strains have been identified that are more common in acne patients than in the healthy population [70].

The influence of the gut microbiome on acne comes from studies showing that bacteria in the gut can induce IGF-1 [71]. Hence one theory is that the high glycaemic load affects the gut microbiome such that it stimulates the overproduction of IGF-1 [61]. In the study conducted by Stokes and Pilsbury in 1930 [8], a substantial proportion of acne patients are also known to develop hyperchloridia (low acid production). This could allow the migration of bacteria to the distal intestine, thus setting up conditions for an altered microbiome. Today, we also know that small intestinal bacterial overgrowth can increase gut permeability, leading to systemic inflammation [72]. Hence, whilst the role of the gut microbiome in acne is still very much associative, this has prompted investigations into the utility of oral probiotics as a possible therapy.

Acne was one of the first skin diseases in which the efficacy of oral LAB was tested. In 1961, the first clinical trial was conducted. This used a mixture of *Lactobacillus acidophilus* (the original organism suggested by Stokes and Pillsbury, 1930) and *Lactobacillus bulgaricus* and was administered to patients for 8 days. Two weeks of washout was then undertaken, followed by further treatment of 2 weeks. Although varying degrees of clinical improvement were observed, over 80% of patients saw some improvement, although this was better in inflammatory acne [71].

LAB have also been suggested as an adjunct therapy for the treatment of acne. Jung et al. [73] investigated the effects of a probiotic mixture containing *Lactobacillus acidophilus*, *Lactobacillus bulgaricus*, and *Bifidobacterium bifidum* either alone or in combination with conventional minocycline treatment (a commonly used antibiotic for the treatment of acne). The effectiveness of these combinations was tested alongside minocycline alone. Improvement was noted in all three groups, but the greatest clinical improvement was seen in the LAB + minocycline group. A study by Fabbrocini et al. in 2016 [74]) investigated the effects of oral ingestion of *Lacticaseibacillus rhamnosus* SP1 on adults with acne. The study was small ($n = 20$ volunteers with acne) but was double-blind and placebo-controlled. Participants in the treatment group received a dose of $3 \times 10^9$ CFU contained in a liquid daily for 12 weeks, whereas the placebo group only received only the liquid containing no *L. rhamnosus* SP1. At the end of the 12 weeks, biopsies were taken, and the levels of IGF-1 (along with other important markers for acne pathogenesis) were directly measured in the skin. Crucially, this was compared to a similar biopsy taken before the start of treatment. In the LAB-fed group, there was a 32% reduction in IGF1 expression in the skin, and the clinical assessment showed that the acne had improved or markedly improved by physician rating. No such improvements were observed in the placebo group. The study concluded that ingestion of *L. rhamnosus* SP1 normalises the expression of genes associated with acne and improves the appearance of the skin [74].

Several other human studies have also investigated the effects of probiotic mixtures on the clinical outcomes of acne vulgaris. However, in most cases, either several bacteria have been tested in combination or with the addition of other potential active ingredients such as botanical extracts or zinc [75, 76]. Hence, these data are more difficult to interpret. More recently, topical LAB for the treatment/prevention of acne has been tested in a small number of studies: Kang et al. [77] and Sathikulpakdee et al. [78] used lotions derived from the filtered culture media of *Enterococcus faecalis* SL-5 and *Lacticaseibacillus paracasei* MSMC 39–1, respectively, to treat acne lesions. Both studies reported a reduction in the inflammatory lesions, with the *Lacticaseibacillus paracasei*-derived lotion demonstrating equivalent effects to 2.5% benzoyl peroxide, which is a mainstay of topical treatment. More recently an open-label trial of a product containing several live strains of LAB encapsulated to maintain viability provided positive data as to their benefit for the treatment of acne. The product was in the form of a cream containing the encapsulated bacteria, which were released onto the skin when the cream was applied. The cream was applied for 8 weeks and, compared to the placebo, was able to reduce inflammatory lesions. Interestingly, the microbiome of the skin was also transiently altered using the cream, and a reduction in staphylococci and *Cutibacterium acnes* as well as an increase in lactobacilli was observed [79]. Some of the effects of LAB in acne may be directly attributed to the production of lactic acid, since this is known to be beneficial for the skin in acne. Lactic acid is a humectant, and since dehydration of the skin makes acne worse, lactic acid attracts moisture into the skin, which speeds up the rate at which skin cells turn over and produce new healthy skin. Lactic acid also contributes to skin health in acne by unclogging pores [80, 81]. However, given the substantial amounts of *in vitro* data which suggest that LAB can provide functions such as increasing skin barrier properties, speeding up repair, and reducing inflammation, it is likely that the production of lactic acid may not be the sole mechanism whereby LAB are of benefit in the treatment of acne.

# 36.7 ROSACEA

Rosacea is a chronic inflammatory skin disorder affecting around 5% of the adult population. Typically presenting after the third decade of life, it is much more common in fair-skinned women, particularly of northern European ancestry [82]. The condition is characterised by flushing, persistent facial erythema inflammatory papules and pustules and telangiectasias [83]. The disease waxes and wanes and manifests mostly on the cheeks, nose, forehead, and chin.

The pathophysiology of rosacea is yet to be fully elucidated. In some studies, the development of the disease has been associated with colonisation of the skin by the bacterium *Bacillus oleronius* (carried by *Demodex* mites) [84]. Other changes to the skin microbiome have also been observed [85–87]. However, the link between gastrointestinal disturbance and rosacea has been known for many years, with conditions such as malabsorption, indigestion, constipation, and small intestinal bacterial overgrowth being associated with rosacea [88–91]. Perhaps the strongest association is that between *Helicobacter pylori* infection and rosacea [92–94].

Currently, the investigation of LAB for rosacea is in its infancy with a very limited number of studies conducted to date: A case study by Fortuna et al. investigated a rare case of scalp rosacea which was successfully treated with low-dose doxycycline and a probiotic therapy consisting of *Bifodobacterium breve* Br03 and *Ligilactobacillus salivarius* LS01, both at $1 \times 10^9$ twice a day [95]. The patient achieved complete remission of his disease, but of course, the effect of the doxycycline is a major confounder in this study. Buianova et al. [96] compared the 'standard of care', which involved 7 days of oral antibiotics, vitamins, antihistamines, and topical permethrin, to an experimental treatment. This involved the standard of care plus *Bifidobacterium* at $50 \times 10^6$ CFU three times a day and an immune modulator, polyoxidonium, for 21 days. The cohort included 60 rosacea patients, 30 in each group; 56% of patients in the experimental group achieved clinical remission of their disease compared to only 28% in the standard-of-care group. Interestingly, this study also highlighted significantly fewer CFU of lactobacilli and bifidobacteria in the stool of patients before treatment and a significant increase afterwards. These observations perhaps suggest a role for LAB in rosacea management that deserves further exploration, especially in light of evidence from other studies showing the ability of LAB to inhibit *H. pylori*, which is commonly associated with rosacea [97–99].

# 36.8 ATOPIC DERMATITIS

Atopic dermatitis (AD), or eczema, is an inflammatory skin condition affecting up to 20% of children and 3% of adults globally [88]. It is characterised by skin inflammation with dryness and itchiness which may be exacerbated by scratching, leading to infection and potentially scarring. The aetiology is not fully understood but is likely to be multifactorial [89, 90]. The standard of care for (AD) typically involves the application of topical emollients, moisturisers and corticosteroids, along with topical and systemic immunomodulators, allergen avoidance, and behavioural modifications [91]. There is emerging evidence that the use of topical and enteral probiotics, principally LAB, could offer some therapeutic benefits to patients with AD [92]. The microbiome of the skin in individuals with AD is reportedly characterised by a reduced bacterial diversity and a higher abundance of *Staphylococcus aureus* [93], which can exacerbate inflammation and worsen symptoms [94, 95].

As outlined in earlier sections, probiotics can be administered enterally (orally) or applied topically. This section will consider both treatment modalities, focusing on clinical trials.

With respect to enteral probiotics in AD, a study published in 2003 by Rosenfeldt et al. assessed the anti-inflammatory potential of enteral administration of a combination of *L. reuteri* DSM 122460 and *L. rhamnosus* 19070–2 administrated orally to children with AD for 6 weeks [96]. The authors report improvement of eczema symptoms in 56% of the patients, compared to only 15% for those who received the placebo. Decreases in the allergy biomarker eosinophil cationic protein were reported in the treatment group. It is additionally reported that the beneficial effect was more pronounced in patients with elevated IgE concentrations, and also those were tested positive in a skin prick response test. Passerson et al. [97] performed a double-blind prospective randomised study performed on children with AD of 2 years and over. A synbiotic comprising *L. rhamnosus* Lcr35 along with a prebiotic or the prebiotic alone was given orally three times daily over 3 months. The authors report a significant decrease in AD scores in the synbiotic and prebiotic groups, although vehicle control was apparently not included. Fölster-Holst et al. [98] reported on a prospective study with the stated aim of reassessing the efficacy of orally administered *L. rhamnosus* GG (LGG) to treat AD in infants. The study was placebo-controlled, double-blind, and randomised, with 54 infants recruited with moderate to severe AD. The children were administered LGG or placebo orally over 8 weeks. It is reported that LGG was well tolerated, but when the intervention was completed, clinical symptoms did not significantly improve in the test over the placebo group. The authors noted that several factors could explain the lack of positive effects. These include minor differences in the baseline variables, the inclusion of children with more severe AD, and a higher age than studies with reported positive effects.

The three studies outlined above illustrate the general approaches taken in AD clinical trials with the oral administration of high doses (>$10^6$) of *Lactobacillus* cells. However, Cochrane systematic reviews are arguably the best route to assessing the likely efficacy of LAB in AD, at least based on current data in the public domain. Makrgeorgou et al. present the conclusions of such a review [99]. As with other systematic reviews, the conclusions reflect the questions that were evaluated. For example, the Cochrane review compared studies in which the enteral administration of LAB (as live intact bacteria) was compared to a null control, that is, consumption of no LAB. The results of the review appear to identify differences in outcome depending on how the potential effects of the LAB in AD were assessed. Studies in which the investigator rates AD severity were more likely to show positive probiotic effects than those where parents or participants did the assessment. The considerable heterogeneity between the results of individual studies that was noted is difficult to explain and a potential confounder in meta-analysis. Bacterial species, strain, dose, and formulation are all important probiotic-specific factors to consider when developing or evaluating a LAB, as is inter-individual variation.

Several reports have investigated the efficacy of topical LAB in treating AD. For example, Butler et al. [100] compared a topical probiotic based on live *L. reuteri* DSM 17938 to vehicle control. The test formulations were applied twice per day over 8 weeks, in 36 adult volunteers. A clinical improvement that is described as statistically significant is reported based on the SCORing Atopic Dermatitis (SCORAD) tool.

In a study titled 'Effect of the Lactic Acid Bacterium *Streptococcus thermophilus* on Stratum Corneum Ceramide Levels and Signs and Symptoms of Atopic Dermatitis Patients', Di Marzio et al. [101] hypothesised that the reduced abundance of total ceramides could be responsible for the symptoms of atopic dermatitis. They therefore assessed the ability of sonicated *Streptococcus thermophilus* administered topically in a cream to increase skin ceramides. Skin ceramides were reported to increase significantly, which the authors propose may have resulted from the action of bacterial sphingomyelinase. Of even greater note, the direct application of the postbiotic cream to the skin reportedly resulted in symptomatic improvements in all patients. It should be noted, however, that this study included limited comparators.

There is obviously still much to learn about the potential efficacy of enteral and topical LAB used to treat atopic dermatitis. The current literature appears to be quite variable in study design and outcomes. This is reflected in the somewhat pessimistic conclusions of the Cochrane review on the efficacy of enteral probiotics. There are, however, several highly positive and promising clinical trials in the literature on this topic. Intuitively, the application directly to the skin would appear to be a good route of administration for skin conditions. More high-quality clinical studies are required.

## 36.9 PSORIASIS

Psoriasis is a chronic, autoimmune skin condition that results in scaly, inflamed skin. It affects an estimated 2–3% of the global population [88] and is characterised by thick, silvery-white scales on the skin [102]. The aetiology of psoriasis is believed to be related to genetic predisposition, immune system dysfunction, and environmental triggers [102]. The severity of the condition can range from mild to severe, and symptoms can come and go over time. The involvement of both the intestinal and skin microbiomes has been proposed [103, 104]. In terms of the evaluation of the potential probiotics based on LAB for the treatment of psoriasis, Zeng et al. performed a systematic review of randomised controlled trials and pre-clinical trials on the safety and efficacy of enteral probiotic supplements for psoriasis [105]. A reflection of the limited current evidence is the fact that the review only included three RCTs involving a total of 164 participants. The authors state that the data generated in two of these RCTs suggest that the psoriasis area and severity index can be improved by probiotics but that only one of the three RCTs indicates that probiotics can reduce concentrations of biomarkers of inflammation [106].

A placebo-controlled, randomised, double-blind trial was conducted to assess the safety and effectiveness of a probiotic supplement containing *L. rhamnosus* CECT 8, *B. lactis* CECT 8145 and *Bifidobacterium longum* CECT 7347 in treating plaque psoriasis in adults. Ninety participants were randomised into either the probiotic or placebo group. After 12 weeks, the study reported that 66.7% of patients in the probiotic group and 41.9% in the placebo group demonstrated a reduction in psoriasis area and severity index of up to 75% ($p < 0.05$). The authors also noted that analysis of gut microbiota 'confirmed the efficacy of the probiotic in the modulation of the microbiota composition'.

Whilst there is some indication that enterally administered LAB can result in improvements in the severity of psoriasis symptoms, more high-quality studies are required, and there is a clear need for clinical investigations of topic probiotics for psoriasis.

## 36.10 CONCLUSIONS AND FUTURE DIRECTIONS

A growing body of evidence suggests that LAB could be an efficacious intervention for skin health as well as disease (Figure 36.2). However, at present, their potential is limited largely due to a failure of consensus on how to conduct studies. The potential mechanism of both topical and enteral applications includes direct interaction with the local microbiota and direct effects on the host. In general, the rationale for the choice of LAB in many studies is not presented, and the reader is left with a sense that the choice of organisms is a matter of serendipity. Thus, a particular organism is observed to have effects, and the remainder of the study centres around trying to understand the mechanisms underlying these effects. Perhaps the evolution of LAB for skin should turn this on its head: many of the pathophysiologies of skin are starting to be understood mechanistically. One good example of this is psoriasis, which is known to be associated with elevated IL-17 and IL-23 levels. If LAB could be screened for their ability to reduce circulating levels of these cytokines, then these bacteria could be tested as an intervention in the management of psoriasis. A recent study has used this screening approach. The study set out to screen for LAB with the ability to protect skin cells against the effects of UVR. Chen et al. (2022) screened 206 strains of LAB isolated from a variety of sources and found 32 with an ability to protect a skin cell line (HaCaT) from the effect of UVR [103]. Of these isolates, the most efficacious was *Limosilactobacillus fermentum* XJC60. This was attributed to its high production of nicotinamide; a known anti-ageing molecule present in many cosmetics. Thus, the ability of LAB to protect skin from the effects of photoageing may be related not only to effects on the immune system but also to the production of metabolites with anti-ageing effects. Perhaps in the future, this rational approach of identifying key activities required for skin health and then screening for strains with those activities will lead to the development of blends of LAB targeted at specific skin requirements and begin to realise the promise of LAB for the skin.

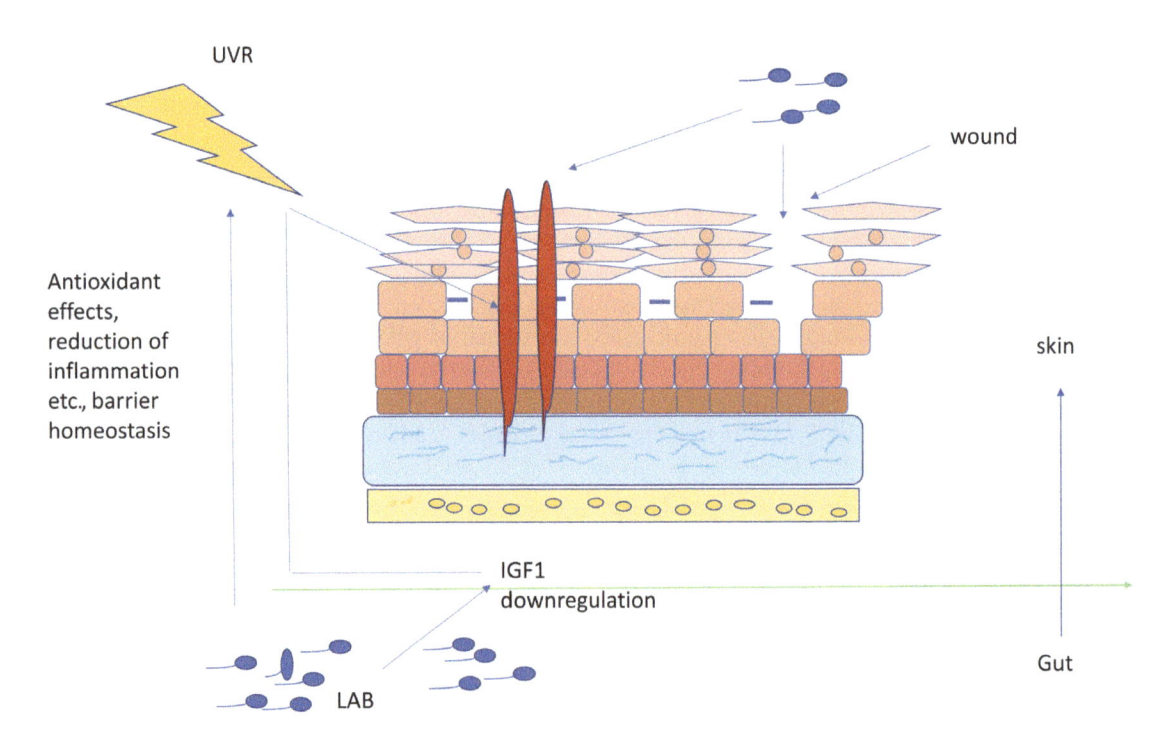

**FIGURE 36.2** Lactic acid bacteria impact skin in many ways.

*Evidence suggests the utility of LAB for several skin conditions:* Inflammatory acne vulgaris (●) has been shown to be reduced with LAB used either topically or orally. LAB also have potential for the treatment of wounds, particularly when applied topically. Photodamaged skin is also improved by ingestion of LAB by a variety of mechanisms.

# BIBLIOGRAPHY

1. O'Neill, C.A., G. Monteleone, J.T. McLaughlin and R. Paus. 2016. The gut-skin axis in health and disease: A paradigm with therapeutic implications. *Bioessays* 38(11): 1167–1176.
2. Fitzpatrick, T.B., L.A. Goldsmith and K. Wolff. 2012. *Fitzpatrick's Dermatology in General Medicine*. McGraw Hill Medical.
3. Cummings, J.H. and G.T. Macfarlane. 1997. Role of intestinal bacteria in nutrient metabolism. *J Parenter Enteral Nutr.* 21(6): 357–365.
4. Bibel, D.J. and J.R. LeBrun. 1975. Effect of experimental dermatophyte infection on cutaneous flora. *J Invest Dermatol.* 64(2): 119–123.
5. Grice, E.A. and J.A. Segre. 2016. The skin microbiome. *Nat Rev Microbiol.* 9(4): 244–253.
6. Goldman, L. and A.I. Schafer. 2016. *Goldman-Cecil Medicine*. Elsevier Saunders.
7. De Pessemier, B., L. Grine, M. Debaere, A. Maes, B. Paetzold and C. Callewaert. 2021. Gut-skin axis: Current knowledge of the interrelationship between microbial dysbiosis and skin conditions. *Microorganisms.* 9(2): 353.
8. Stokes, J.H. and D.M. Pillsbury. 1930. The effect on the skin of emotional and nervous states: Theoretical and practical consideration of a gastrointestinal mechanism. *Arch Derm Syphilol.* 22: 962–993.
9. Bowe, W.P. and A.C. Logan. 2011. Acne vulgaris, probiotics and the gut-brain-skin axis—back to the future? *Gut Pathog.* 3(1): 1.
10. Margolis, K.G., J.F. Cryan and E.A. Mayer. 2021. The microbiota-gut-brain axis: From motility to mood. *Gastroenterology.* 160(5): 1486–1501.
11. Quigley, E.M.M. 2017. Microbiota-brain-gut axis and neurodegenerative diseases. *Curr Neurol Neurosci Rep.* 17(12): 94.
12. Reid, G., T. Abrahamsson, M. Bailey, L.B. Bindels, R. Bubnov, K. Ganguli, C. Martoni, C. O'Neill, H.M. Savignac, C. Stanton, N. Ship, M. Surette, K. Tuohy and S. van Hemert. 2017. How do probiotics and prebiotics function at distant sites? *Benef Microbes.* 8(4): 521–533.
13. Kiousi, D.E., A. Karapetsas, K. Karolidou, M.I. Panayiotidis, A. Pappa and A. Galanis. 2019. Probiotics in extraintestinal diseases: Current trends and new directions. *Nutrients.* 11(4): 788.
14. Elvebakken, H.F., A.B. Bruntse, C. Vedel and S. Kjaerulf. 2023. Topical Lactiplantibacillus plantarum LB244R(R) ointment alleviates skin aging: An exploratory trial. *J Cosmet Dermatol.* 22(6): 1911–1918.
15. Iglesia, S., T. Kononov and A.S. Zahr. 2022. A multi-functional anti-aging moisturizer maintains a diverse and balanced facial skin microbiome. *J Appl Microbiol.* 133(3): 1791–1799.
16. Xie, A., A. Chen, Y. Chen, Z. Luo, S. Jiang, D. Chen and R. Yu. 2023. Lactobacillus for the treatment and prevention of atopic dermatitis: Clinical and experimental evidence. *Front Cell Infect Microbiol.* 13: 1137275.
17. Vinderola, G., M.E. Sanders, S. Salminen and H. Szajewska. 2022. Postbiotics: The concept and their use in healthy populations. *Front Nutr.* 9: 1002213.
18. Fluhr, J. *Bioengineering of the Skin: Water and the Stratum Corneum*. Dermatology; 2005.
19. Kalia, Y.N., L.B. Nonato, C.H. Lund and R.H. Guy. 1998. Development of skin barrier function in premature infants. *J Invest Dermatol.* 111(2): 320–326.
20. Kolarsick, P.A., M.A. Kolarsick and C. Goodwin. 2011. Anatomy and physiology of the skin. *J Dermatol Nurses' Assoc.* 3: 203–213.
21. Madison, K.C. 2003. Barrier function of the skin: "la raison d'etre" of the epidermis. *J Invest Dermatol.* 121(2): 231–241.
22. Jensen, J.M. and E. Proksch. 2009. The skin's barrier. *J Ital Dermatol Venereol.* 144(6): 689–700.
23. Nguyen, A.V. and A.M. Soulika. 2019. The dynamics of the skin's immune system. *Int J Mol Sci.* 20(8): 1811.
24. Griffiths, C.E. and J.N. Barker. 2007. Pathogenesis and clinical features of psoriasis. *Lancet.* 370(9583): 263–271.
25. Watson, R.E., N.K. Gibbs, C.E. Griffiths and M.J. Sherratt. 2014. Damage to skin extracellular matrix induced by UV exposure. *Antioxid Redox Signal.* 21(7): 1063–1077.
26. Wondrak, G.T., M.K. Jacobson and E.L. Jacobson. 2006. Endogenous UVA-photosensitizers: Mediators of skin photodamage and novel targets for skin photoprotection. *Photochem Photobiol Sci.* 5(2): 215–237.
27. Hibbert, S.A., R.E.B. Watson, C.E.M. Griffiths, N.K. Gibbs and M.J. Sherratt. 2019. Selective proteolysis by matrix metalloproteinases of photo-oxidised dermal extracellular matrix proteins. *Cell Signal.* 54: 191–199.
28. Langton, A.K., H.K. Graham, C.E.M. Griffiths and R.E.B. Watson. 2019. Ageing significantly impacts the biomechanical function and structural composition of skin. *Exp Dermatol.* 28(8): 981–984.

29. Levkovich, T., T. Poutahidis, C. Smillie, B.J. Varian, Y.M. Ibrahim, J.R. Lakritz, E.J. Alm and S.E. Erdman. 2013. Probiotic bacteria induce a 'glow of health'. *PLoS ONE.* 8(1): e53867.

30. Erdman, S.E. and T. Poutahidis. 2014. Probiotic 'glow of health': It's more than skin deep. *Benef Microbes.* 5(2): 109–119.

31. Gueniche, A., D. Phillipe, P. Bastien, G. Teuteler, S. Blum, I. Castiel-Higounenc, L. Breton and J. Benyacoub. 2014. Randomised double-blind placebo-controlled study of the effect of Lactobacillus paracasei NCC 2461 on skin reactivity. *Benef Microbes.* 5(2): 137–145.

32. Lee, D.E., C.S. Huh, J. Ra, I.D. Choi, J.W. Jeong, S.H. Kim, J.H. Ryu, Y.K. Seo, J.S. Koh, J.H. Lee, J.H. Sim and Y.T. Ahn. 2015. Clinical evidence of effects of lactobacillus plantarum HY7714 on skin aging: A randomized, double blind, placebo-controlled study. *J Microbiol Biotechnol.* 25(12): 2160–2168.

33. Ra, J., D.E. Lee, S.H. Kim, J.W. Jeong, H.K. Ku, T.Y. Kim, I.D. Dong, W. Jeung, J.H. Sim and Y.T. Ahn. 2014. Effect of oral administration of Lactobacillus plantarum HY7714 on epidermal hydration in ultraviolet B-irradiated hairless mice. *J Microbiol Biotechnol.* 24(12): 1736–1743.

34. Han, G. and R. Ceilley. 2017. Chronic wound healing: A review of current management and treatments. *Adv Ther.* 34(3): 599–610.

35. Moreira, C.F., O. Cassini-Vieira, M.C.C. Canesso, M. Felipetto, H. Ranfley, M.M. Teixeira, J.R. Nicoli, F.S. Martins and L.S. Barcelos. 2021. Lactobacillus rhamnosus CGMCC 1.3724 (LPR) improves skin wound healing and reduces scar formation in mice. *Probiotics Antimicrob Proteins.* 13(3): 709–719.

36. Chen, Z., D. Ceballos-Franciso, F.A. Guardiola and M.A. Esteban. 2020. Dietary administration of the probiotic Shewanella putrefaciens to experimentally wounded gilthead seabream (Sparus aurata L.) facilitates the skin wound healing. *Sci Rep.*10(1): 11029.

37. Tagliari, E., L.F. Campos, T.A.C. Casagrande, T. Fuchs, L. de Noronha and A.C.L. Campos. 2022. Effects of oral probiotics administration on the expression of transforming growth factor beta and the proinflammatory cytokines interleukin 6, interleukin 17, and tumor necrosis factor alpha in skin wounds in rats. *J Parenter Enteral Nutr.* 46(3): 721–729.

38. Tsiouris, C.G., M. Kelesi, G. Vasilopoulos, I. Kalemikerakis and E.G. Papageorgiou. 2017. The efficacy of probiotics as pharmacological treatment of cutaneous wounds: Meta-analysis of animal studies. *Eur J Pharm Sci.* 104: 230–239.

39. Togo, C., A.P. Zidorio, V. Goncalves, P. Botelho, K. de Carvalho and E. Dutra. 2021. Does probiotic consumption enhance wound healing? A systematic review. *Nutrients.* 14(1): 111.

40. Valdez, J.C., M.C. Peral, M. Rachid, M. Santana and G. Perdigon. 2005. Interference of Lactobacillus plantarum with Pseudomonas aeruginosa in vitro and in infected burns: The potential use of probiotics in wound treatment. *Clin Microbiol Infect.* 11(6): 472–479.

41. Argenta, A., L. Satish, P. Gallo, F. Liu and S. Kathju. 2016. Local application of probiotic bacteria prophylaxes against sepsis and death resulting from burn wound infection. *PLoS ONE.* 11(10): e0165294.

42. Satish, L., P.H. Gallo, S. Johnson, C.C. Yates and S. Kathju. 2017. Local probiotic therapy with lactobacillus plantarum mitigates scar formation in rabbits after burn injury and infection. *Surg Infect* (Larchmt). 18(2): 119–127.

43. Rodrigues, K.L., L.R. Caputo, J.C. Carvalho, J. Evangelista and J.M. Schneedorf. 2005. Antimicrobial and healing activity of kefir and kefiran extract. *Int J Antimicrob Agents.* 25(5): 404–408.

44. Jones, M., J.G. Ganopolsky, A. Labbe, M. Gilardino, C. Wahl, C. Martoni and S. Prakash. 2012. Novel nitric oxide producing probiotic wound healing patch: Preparation and in vivo analysis in a New Zealand white rabbit model of ischaemic and infected wounds. *Int Wound J* 9(3): 330–343.

45. Peral, M.C., M.A. Martinez and J.C. Valdez 2009. Bacteriotherapy with Lactobacillus plantarum in burns. *Int Wound J.* 6(1): 73–81.

46. Peral, M.C., M.M. Rachid, N.M. Gobbato, M.A.H. Martinez and J.C. Valdez. 2010. Interleukin-8 production by polymorphonuclear leukocytes from patients with chronic infected leg ulcers treated with Lactobacillus plantarum. *Clin Microbiol Infect.*16(3): 281–286.

47. Arganaraz Aybar, J.N., S.M. Mayor, L. Olea, J.J. Garcia, S. Nisoria, Y. Kolling, C. Melian, M. Rachid, T. Dimani, C. Werenitzky, C. Lorca, S. Salva, N. Gobbato, N. Villena and J.C. Valdez. 2022. Topical administration of lactiplantibacillus plantarum accelerates the healing of chronic diabetic foot ulcers through modifications of infection, angiogenesis, macrophage phenotype and neutrophil response. *Microorganisms* 10(3): 634.

48. Mohtashami, M., M. Mohamadi, M. Azimi-Nezhad, J. Saeidi, F.F. Nia and A. Ghasemi. 2021. Lactobacillus bulgaricus and Lactobacillus plantarum improve diabetic wound healing through modulating inflammatory factors. *Biotechnol Appl Biochem.* 68(6): 1421–1431.

49. Moysidis, M., G. Stavrou, A. Cheva, I.A. Deka, J.K. Tsetis, V. Birba, D. Kapoukranidou, A. Ioannidis and G. Tsaousi. 2022. The 3-D configuration of excisional skin wound healing after topical probiotic application. *Injur.* 53(4): 1385–1393.

50. Dubey, A.K., M. Podia, Priyanka, S. Raut, S. Singh, A.K. Pinnaka and N. Khatri. 2021. Insight into the beneficial role of lactiplantibacillus plantarum supernatant against bacterial infections, oxidative stress, and wound healing in A549 cells and BALB/c mice. *Front. Pharmacol.* 12: 728614.

51. Zaghloul, E.H. and M.I.A. Ibrahim. 2022. Production and characterization of exopolysaccharide from newly isolated marine probiotic lactiplantibacillus plantarum EI6 with in vitro wound healing activity. *Front. Microbiol.* 13: 903363.

52. Nam, Y., J. Baek and W. Kim. 2021. Improvement of cutaneous wound healing via topical application of heat-killed lactococcus chungangensis CAU 1447 on diabetic mice. *Nutrients.* 13(8): 2666.

53. Salminen, A., K. Kaarniranta and A. Kauppinen. 2022. Photoaging: UV radiation-induced inflammation and immunosuppression accelerate the aging process in the skin. *Inflamm. Res.* 71(7–8): 817–831.

54. Singh, M. and C.E. Griffiths. 2006. The use of retinoids in the treatment of photoaging. *Dermatol. Ther.* 19(5): 297–305.

55. Peguet-Navarro, J., C. Dezutter-Dambuyant, T. Buetler, J. Leclaire, H. Smola, S. Blum, P. Bastien, L. Breton and A. Gueniche. 2008. Supplementation with oral probiotic bacteria protects human cutaneous immune homeostasis after UV exposure-double blind, randomized, placebo controlled clinical trial. *Eur J Dermatol.* 18(5): 504–511.

56. Bouilly-Gauthier, D., C. Jeannes, L. Maubert, L. Duteil, C. Queille-Roussel, N. Picardi, C. Montastier, P. Manissier G. Perard and J.P. Ortonne. 2010. Clinical evidence of benefits of a dietary supplement containing probiotic and carotenoids on ultraviolet-induced skin damage. *Br J Dermatol.* 163(3): 536–543.

57. Petruk, G., R. Del Giudice, M.M. Rigano and D.M. MOnti. 2018. Antioxidants from plants protect against skin photoaging. *Oxid Med Cell Longev.* 2018:1454936.

58. Kim, D., K.R. Lee, N.R. Kim, S.J. Park, M. Lee and O.K. Kim. 2021. Combination of bifidobacterium longum and galacto-oligosaccharide protects the skin from photoaging. *J Med Food.* 24(6): 606–616.

59. Park, J.Y., J.Y. Lee, Y. Kim and C.H. Kang. 2023. Latilactobacillus sakei Wikim0066 protects skin through MMP regulation on UVB-irradiated in vitro and in vivo model. *Nutrients.* 15(3): 726.

60. Lim, H.Y., D. Jeong, S.H. Park, K.K. Shin, Y.H. Hong, E. Kim, Y.G. Yu, T.R. Kim, H. Kim, J. Lee and J.Y. Cho. 2020. Antiwrinkle and antimelanogenesis effects of tyndallized lactobacillus acidophilus KCCM12625P. *Int J Mol Sci.* 21(5): 1620.

61. Khmaladze, I., E. Butler, S. Fabre and J.M. Gillbro. 2019. Lactobacillus reuteri DSM 17938-A comparative study on the effect of probiotics and lysates on human skin. *Exp Dermatol.* 28(7): 822–828.

62. Teng, Y., Y. HUang, X. Danfeng, X. Tao and Y. Fan. 2022. The role of probiotics in skin photoaging and related mechanisms: A review. *Clin Cosmet Investig Dermatol.* 15: 2455–2464.

63. Jwo, J.Y., Y.T. Chang and Y.C. Huang. 2023. Effects of probiotics supplementation on skin photoaging and skin barrier function: A systematic review and meta-analysis. *Photodermatol. Photoimmunol. Photomed.* 39(2): 122–131.

64. Hazarika, N. 2021. Acne vulgaris: New evidence in pathogenesis and future modalities of treatment. *J Dermatolog Treat.* 32(3): 277–285.

65. Sanchez-Pellicer, P., L. Navarro-Moratalla, E. Nunez-Delegido, B. Ruzafa-Costas, J. Aguera-Santos and V. Navarro-López. 2022. Acne, microbiome, and probiotics: The gut-skin axis. *Microorganisms.* 10(7): 1303.

66. Deng, Y., H. Wang, J. Zhou, Y. Mou, G. Wang and X. Xiong. 2018. Patients with acne vulgaris have a distinct gut microbiota in comparison with healthy controls. *Acta Derm Venereol.* 98(8): 783–790.

67. Grossi, E., S. Cazzaniga, S. Crotti, L. Naldi, A. Di-Landro, V. Ingordo, F. Cusano, L. Atzori, F.T. Cutri, M.L. Musumeci, E. Pezzarossa, V. Bettoli, M. Caproni and A. Bonci. 2016. The constellation of dietary factors in adolescent acne: A semantic connectivity map approach. *J Eur Acad Dermatol Venereol.* 30(1): 96–100.

68. Cordain, L., S. Lindberg, M. Hurtado, K. Hill, S.B. Eaton and J. Brand-Miller. 2002. Acne vulgaris: A disease of Western civilization. *Arch Dermatol.* 138(12): 1584–1590.

69. Burris, J., W. Rietkerk and K. Woolf. 2014. Relationships of self-reported dietary factors and perceived acne severity in a cohort of New York young adults. *J Acad Nutr Diet.* 114(3): 384–392.

70. Lomholt, H.B., C.F.P. Scholz, H. Bruggemann, H. Tettelin and M. Kilian. 2017. A comparative study of Cutibacterium (Propionibacterium) acnes clones from acne patients and healthy controls. *Anaerobe.* 47: 57–63.

71. Siver, R. 1961. Lactobacillus for the control of acne. *J Med Soc NJ.* 59: 52–53.

72. Fukui, H. and R. Wiest. 2016. Changes of intestinal functions in liver cirrhosis. *Inflamm Intest Dis.* 1(1): 24–40.

73. Jung, G.W., J.E. Tse, I. Guiha and J. Rao. 2013. Prospective, randomized, open-label trial comparing the safety, efficacy, and tolerability of an acne treatment regimen with and without a probiotic supplement and minocycline in subjects with mild to moderate acne. *J Cutan Med Surg.* 17(2): 114–122.

74. Fabbrocini, G., M. Bertona, O. Picazo, H. Pareja-Galeano, G. Monfrecola and E. Emanuele. 2016. Supplementation with Lactobacillus rhamnosus SP1 normalises skin expression of genes implicated in insulin signalling and improves adult acne. *Benef Microbes*. 7(5): 625–630.

75. Rahmayani, T., I.B. Putra and N.K. Jusuf. 2019. The effect of oral probiotic on the interleukin-10 serum levels of acne vulgaris. *Open Access Maced J Med Sci*. 7(19): 3249–3252.

76. Rinaldi, F., L. Marotta, A. Mascolo, A. Amoruso, M. Pane, G, Guiliani and D. Pinto. 2022. Facial acne: A randomized, double-blind, placebo-controlled study on the clinical efficacy of a symbiotic dietary supplement. *Dermatol Ther (Heidelb)*. 12(2): 577–589.

77. Kang, B.S., J.G. Seo, G.S. Lee, J.H. Kim, S.Y. Kim, Y.W. Han, H. Kang H.O. Kim, J.J. Rhee, M.J. Chung and Y.M. Park. 2009. Antimicrobial activity of enterocins from *Enterococcus faecalis* SL-5 against Propionibacterium acnes, the causative agent of acne vulgari and its therapeutic effects. *J Microbiol*.: 101–109.

78. Sathikulpakdee, S., S. Kanokrungsee, P. Vitheejongjaroen, N. Kamanammol, M. Udompataikul and M. Taweechtipatr. 2022. Efficacy of probiotic-derived lotion from lactobacillus paracasei MSMC 39–1 in mild to moderate acne vulgaris, randomized controlled trial. *J Cosmet Dermatol*. 21(10): 5092–5097.

79. Lebeer, S., E.F.M. Oerlemans, I Claes, T. Henkens, L. Delanghe, S. Wuyts, I. Spacova, M.F.L. ven denBroek, I. Tuyaerts, S. Wittouck, I. De Boeck, C.N. Allonsius, F. Kiekens and J. Lambert. 2022. Selective targeting of skin pathobionts and inflammation with topically applied lactobacilli. *Cell Rep Med*. 3(2): 100521.

80. Kontochristopoulos, G. and E. Platsidaki. 2017. Chemical peels in active acne and acne scars. *Clin Dermatol*. 35(2): 179–182.

81. Truchuelo, M., P. Cerda and L.F. Fernandez. 2017. Chemical peeling: A useful tool in the office. *Actas Dermosifiliogr*. 108(4): 315–322.

82. Gether, L., L.K. Overgaards, A. Egeberg and J.P. Thyssen. 2018. Incidence and prevalence of rosacea: A systematic review and meta-analysis. *Br J Dermatol*. 179(2): 282–289.

83. Wilkin, J., M. Dahl, M. Detmar, L. Drake, M.H. Liang, R. Odom and F. Powell. 2004. Standard classification of rosacea: Report of the national rosacea society expert committee on the classification and staging of rosacea. *J Am Acad Dermatol*. 46(4): 584–587.

84. Murillo, N., O. Mediannikov, J. Aubert and D. Raoukt. 2014. Bartonella quintana detection in Demodex from erythematotelangiectatic rosacea patients. *Int J Infect Dis*. 29: 176–177.

85. Buhl, T., M. Sulk, P. Nowak, J. Buddenkotte, I McDonald, J Aubert, I. Carlavan, S. Deret, P. Reiniche, M. Rivier, J.J. Voegel and M. Steinhoff. 2015. Molecular and morphological characterization of inflammatory infiltrate in rosacea reveals activation of Th1/Th17 pathways. *J Invest Dermatol*. 135(9): 2198–2208.

86. Holmes, A.D. 2013. Potential role of microorganisms in the pathogenesis of rosacea. *J Am Acad Dermatol*. 69(6): 1025–1032.

87. Alia, E. and H. Feng. 20220 Rosacea pathogenesis, common triggers, and dietary role: The cause, the trigger, and the positive effects of different foods. *Clin Dermatol*. 40(2): 122–127.

88. Wang, F.Y. and C.C. Chi. 2021. Rosacea, germs, and bowels: A review on gastrointestinal comorbidities and gut-skin axis of Rosacea. *Adv. Ther*. 38(3): 1415–1424.

89. Searle, T., F.R. Ali, S. Carolides and F. Al-Niaimi. 2020. Rosacea and the gastrointestinal system. *Australas J Dermatol*. 61(4): 307–311.

90. Lim, H.G., A. Fischer, M.J Rueda, J. Kendall, S. Kang and A.L. Chien. 2018. Prevalence of gastrointestinal comorbidities in rosacea: Comparison of subantimicrobial modified release doxyclycine versus conventional release doxycycline *J Am Acad Dermatol*. 78(2): 417–419.

91. Egeberg, A., L.N. Weinstock, E.P. Thyssen, G.H. Gislason and J.P. Thyssen. 2017. Rosacea and gastrointestinal disorders:a population-based cohort study. *Br J Dermatol*. 176(1): 100–106.

92. Yang X. 2018. Relationship between *Helicobacter pylori* and Rosacea: Review and discussion. *BMC Infect Dis*. 18(1): 318.

93. Saleh, P., M. Nghavi-Behzad, H. Herizchi, F. Mokhtari, M. Mirza-Aghazadeh-Attari and R. Piri. 2017. Effects of *Helicobacter pylori* treatment of rosacea: A single arm cliical trial study. *J Dermatol*. 44(9): 1033–1037.

94. Wong, F., E. Rayner-Hartley and M.F. Byrne. 2014. Extraintestinal manifestations of *Helicobacter pylori*: A concise review. *World J Gastroenterol*. 20(4): 11950–11961.

95. Leung, D.Y. and T. Bieber. 2003. Atopic dermatitis. *Lancet*. 361(9352): 151–160.

96. Rosenfeldt, V., E. Benfeldt, N.H. Valerius, A. Paerregaard and K.F. Michaelsen. 2004. Effect of probiotics on gastrointestinal symptoms and small intestinal permeability in children with atopic dermatitis. *J Pediatr*. 145(5): 612–616.

97. Passeron, T., J.P. Lacour, E. Fontas and J.P Ortonne. 2006. Prebiotics and synbiotics: Two promising approaches for the treatment of atopic dermatitis in children above 2 years. *Allergy*. 61(4): 431–437.

98. Folster-Holst, R., F. Muller, N. Schnopp, D. Abeck, I. Kreiselmaier, T. Lenz, U. von Ruden, J. Schrezenmeir, E. Christophers and M. Weichenthal. 2006. Prospective, randomized controlled trial on Lactobacillus rhamnosus in infants with moderate to severe atopic dermatitis. *Br J Dermatol*. 155(6): 1256–1261.

99. Makrgeorgou, A., J.L. Bee, F.J. Bath-Hextall, D.F. Murrell, M.L.K. Tang, A. Roberts and R.J. Boyle. 2018. Probiotics for treating eczema. *Cochrane Database Syst Rev*. 11(11): CD006135.

100. Butler, E., C. Lundqvist and J. Axelsson. 2020. Lactobacillus reuteri DSM 17938 as a novel topical cosmetic ingredient: A proof of concept clinical study in adults with atopic dermatitis. *Microorganisms*. 8(7): 1026.

101. Di Marzio, L., C. Centi, B. Cinque, S. Masci, M. Guiliani, A. Arcieri, L. Zicari, C. De Simone and M.G. Cifone. 2003. Effect of the lactic acid bacterium Streptococcus thermophilus on stratum corneum ceramide levels and signs and symptoms of atopic dermatitis patients. *Exp Dermatl*. 12(5): 615–620.

102. Rendon, A. and K. Schakel. 2019. Psoriasis pathogenesis and treatment. *Int J Mol Sci*. 20(6): 1475.

103. Chen, H., Y. Li, X. Xie, M. Chen, L. Xue, J. Wang, Q. Ye, S. Wu, R. Yang, H. Zhao, J. Zhang, Y. Ding and Q. Wu. 2022. Exploration of the molecular mechanisms underlying the anti-photoaging effect of limosilactobacillus fermentum XJC60. *Front Cell Infect Microbiol*. 12: 838060.

104. Thye, A.Y., Y.R. Bah, J.W.F. Law, L.T.H. Tan, Y W. He, S.H. Wong, S. Thuraiiajasingam, K.G. Chan, L.H. Lee and V. Letchumanan. 2022. Gut-skin axis: Unravelling the connection between the gut microbiome and psoriasis. *Biomedicines*. 10(5): 1037.

105. Zeng, L., G. Yu, Y. Wu, W. Hao and H. Chen. 2021. The effectiveness and safety of probiotic supplements for psoriasis: A systematic review and meta-analysis of randomized controlled trials and preclinical trials. *J Immunol Res*. 2021: 7552546.

106. Navarro-Lopez, V., A. Martinez-Andres, A. Ramirez-Bosca, B. Ruzafa-costas, E. Nunez-Delegido, M.A. Carrion-Gutierrez, D. Prieto-Merino, F. Codoner-Cortez, D. Ramon-Vidal, S. Genoves-Martinez, E. Chenoll-Cuadros, J.M. Perez-Orquin, J.A. Pico-Monllor and S. Cgumillas-Lidon. 2019. Efficacy and safety of oral administration of a mixture of probiotic strains in patients with psoriasis: A randomized controlled clinical trial. *Acta Derm Venereol*. 99(12): 1078–1084.

# SECTION IV

# Safety and Regulation

# The Safety of Lactic Acid Bacteria for Use in Foods

# 37

David Obis and Arthur C. Ouwehand

## 37.1 INTRODUCTION

The empiric selection of lactic acid bacteria by humans for food fermentations of animal or plant-based dietary substrates has over the ages contributed, to increase the security, safety, and quality of diets with fermented foods around the world. Today, consumers can enjoy a very wide array of safe, wholesome, and enjoyable fermented food products, taking advantage of raw ingredients diversity, culinary traditions, popular use of home-made fermentation, and the marketing of manufactured fermented food products. As a result, humans have throughout history extensively ingested foods with significant amounts of lactic acid bacteria or foods containing products of their metabolism (Chilton et al. 2015; Bell et al. 2017; Sanlier et al. 2017). This use is at the same time without major or notable side effects reported in the scientific literature. Lactic acid bacteria (LAB) are widely seen as micro-organisms that are inherently safe for human consumption, either directly and/or through their metabolites in fermented foods. The present chapter will focus on safety considerations and safe use of lactic acid bacteria in foods and especially food fermentations by the general, healthy population. We will focus on genera and species documented and used in food fermentations:

*Carnobacterium, Tetragenococcus, Leuconostoc, Enterococcus, Lactococcus, Oenococcus, Pediococcus, Streptococcus, Weissella*, and species from the former genus *Lactobacillus* are now included in at least 18 genera of relevance for use in foods (*Agrilactobacillus, Companilactobacillus, Fructilactobacillus, Lacticaseibacillus, Lactiplantibacillus, Lactobacillus, Lapidilactobacillus, Latilactobacillus, Lentilactobacillus, Levilactobacillus, Ligilactobacillus, Limosilactobacillus, Liquorilactobacillus, Loigolactobacillus, Marinilactibacillus, Paucilactobacillus, Secundilactobacillus, Shleiferilactobacillus*) (Zheng et al. 2020). Animal feed applications, uncommon lactic acid bacteria in food fermentations, yeasts, *Bifidobacterium* spp., and other considerations such as the safety assurance of probiotics are not considered in this chapter.

In preamble, it is worth clarifying the scientific bases and implications of the concepts of Qualified Presumption of Safety (QPS) and generally recognized as safe (GRAS), frequently referred to in the scientific literature, in relation to the topic of this chapter and the selection of safe LAB strains for addition in foods. Other concepts and regulatory frameworks and micro-organism lists exist, as have been reviewed elsewhere (Laulund et al. 2017; IDF 2022).

DOI: 10.1201/9781003352075-41

# 37.1.1 Qualified Presumption of Safety

The QPS concept, used by the European Food Safety Authority (EFSA) since 2007, is often considered a generally accepted recognition of the safety of micro-organisms for use in foods. QPS can be best described as pre-assessment of microorganisms to be added in the food chain, with strains belonging to species having QPS status benefiting from a simplified safety evaluation (Herman et al. 2018). The so-called QPS List included, in its update in 2023, 49 species of lactic acid bacteria out of 113 micro-organisms, with the notable exception of *Enterococcus* species (Table 37.1) (EFSA et al. 2023b). *Ligilactobacillus animalis* (*previously Lactobacillus animalis*), *Lentilactobacillus parafarraginis* (previously *Lactobacillus parafarraginis*), and *Lactiplantibacillus argentoratensis* (previously *Lactobacillus plantarum* subsp. *argentoratensis*) received QPS status since the previous edition of this book (EFSA et al. 2020, 2023b).

The EFSA Panel on Biological Hazards (BIOHAZ) publishes on a regular basis, along with an update of the previously mentioned QPS List, a report on the results of extensive literature surveys to identify and monitor possible safety concerns for human health, applied to a wide array of micro-organisms added

**TABLE 37.1**  The 49 LAB Species Included in the Qualified Presumption of Safety by the EFSA in 2023 (EFSA et al. 2023b)

| GENUS | SPECIES |
|---|---|
| **Carnobacterium** | divergens |
| **Companilactobacillus** | alimentarius |
| | farciminis |
| **Fructilactobacillus** | sanfranciscensis |
| **Lacticaseibacillus** | casei |
| | paracasei |
| | rhamnosus |
| **Lactiplantibacillus** | paraplantarum |
| | pentosus |
| | plantarum |
| **Lactobacillus** | acidophilus |
| | amylolyticus |
| | amylovorus |
| | crispatus |
| | delbrueckii |
| | gasseri |
| | helveticus |
| | johnsonii |
| | kefiranofaciens |
| **Latilactobacillus** | curvatus |
| | sakei |
| **Lentilactobacillus** | buchneri |
| | diolivorans |
| | hilgardii |
| | kefiri |
| | parafarraginis |
| **Leuconostoc** | citreum |
| | lactis |
| | mesenteroides |
| | pseudomesenteroides |

**TABLE 37.1**   (Continued) The 49 LAB Species Included in the Qualified Presumption of Safety by the EFSA in 2023 (EFSA et al. 2023b)

| GENUS | SPECIES |
| --- | --- |
| **Levilactobacillus** | brevis |
| **Ligilactobacillus** | animalis |
| | aviarius |
| | salivarius |
| **Limosilactobacillus** | fermentum |
| | mucosae |
| | panis |
| | pontis |
| | reuteri |
| **Loigolactobacillus** | coryniformis |
| **Lapidilactobacillus** | dextrinicus |
| **Lapidilactobacillus** | dextrinicus |
| **Oenococcus** | oeni |
| **Pediococcus** | acidilactici |
| | parvulus |
| | pentosaceus |
| **Secundilactobacillus** | collinoides |
| **Streptococcus** | thermophilus |
| **Lactococcus** | lactis |

in the food-chain, including a significant number of LAB (Allende et al. 2016; Ricci et al. 2017a, 2018; EFSA et al. 2020, 2023b).

It is important to note that the micro-organisms reviewed by EFSA are those notified to the authority, relevant to the different EFSA mandates in the European Union and through the Safety Assessment it performs. Therefore, this should not be viewed as exhaustive in terms of coverage of micro-organisms added in the food chain. As well, species that are not in the QPS list, such as those of the *Enterococcus* genus, should not be considered unsafe *per se*: this could result from the fact that QPS cannot be ascertained from the existing knowledge of the considered species, yet individual strains may be safe. Also, EFSA may not yet have had the opportunity to assess the safety of any strains of the species through any of its mandates. Fit for the purpose of EFSA tasks and mandates, the approach should not be extrapolated without due consideration of its current scope and usage by the Authority. Interestingly, the update published in 2023 explains in ample details the principle of the QPS approach and the process and methodology applied, with appendices, data sets, and related resources available on an open science platform (EFSA et al. 2023b, 2023c).

QPS status is not strictly equivalent to GRAS and should not be interpreted as unequivocal statement about the safety of use of any strain belonging to the species listed. Rather, the QPS concept provides a pragmatic framework for the safety assessment of the use of lactic acid bacteria in the food chain through the precise identification and characterisation of safety concerns. The qualifications for species used by EFSA panels are also useful to define strain-level safety criteria. At variance with the GRAS concept described subsequently, the QPS status is a pre-assessment applied at the species and subspecies levels, not a final assessment endpoint at the strain level. It can be used as a starting point for safety assessments. Outside its use in EFSA assessments, it is up to the potential user to define final assessment endpoints, and especially to define qualifications necessary for strains and specific use. Indeed, similarly to the GRAS concept, the systematic extensive literature reviews conducted by EFSA are useful to identify potential safety concerns related to LAB. The EFSA QPS list is updated on a regular basis, taking into account new insights in safety. The updates also consider taxonomic changes and, as previously, new LAB species are likely to be added by EFSA in the future as soon as these are included in the Authority's works (Zheng et al. 2020).

## 37.1.2 Generally Recognized as Safe

Lactic acid bacteria, or even foods made with lactic acid bacteria, are frequently designated in the scientific and technical literature as being generally recognized as safe. This American Food and Drug Administration (FDA) designation is, however, often misquoted, as a wide and common acceptance of the intrinsic safety of food substances, including LAB. The designation serves, however, a specific regulatory purpose intended for any substance intentionally added to foods (FDA 2022); furthermore, GRAS status for LAB microorganisms is related to a specific microbial use as food additive and should thus not be interpreted as a generic recognition of the innocuous nature of LAB. GRAS is, however, generic, in essence being based according to the FDA on generally available and accepted scientific data, information, or methods and the application of scientific principles (FDA 2022). Nevertheless, the designation and available data, documentation, and considerations in publicly available GRAS Notifications are useful for those interested in the safety assessment of LAB. There are 39 publicly available GRAS Notifications related to LAB strains as of 2023, summarised in Table 37.2, for which "FDA has no questions" (FDA 2023). While some notices date back from 2002, 20 of the available

**TABLE 37.2** GRAS Notifications per Genus and Species Related to Lactic Acid Bacteria Publicly Available from the FDA Inventory of GRAS Notices (FDA 2023). Species are reported with genus names updated according to changes in *Lactobacillaceae* taxonomy (Zheng et al. 2020).

| GENUS | SPECIES | NUMBER OF GRAS NOTIFICATIONS |
|---|---|---|
| **Carnobacterium** | | **2** |
| | Carnobacterium divergens | 1 |
| | Carnobacterium maltaromaticum | 1 |
| **Lacticaseibacillus** | | **8** |
| | Lacticaseibacillus casei | 2 |
| | Lacticaseibacillus paracasei | 2 |
| | Lacticaseibacillus rhamnosus | 4 |
| **Lactiplantibacillus** | Lactiplantibacillus plantarum | **6** |
| **Lactobacillus** | | **9** |
| | Lactobacillus acidophilus | 6 |
| | Lactobacillus delbrueckii | 1 |
| | Lactobacillus helveticus | 1 |
| | Lactobacillus johnsonii | 1 |
| **Lactococcus** | Lactococcus lactis | **1** |
| **Latilactobacillus** | Latilactobacillus curvatus | **1** |
| **Leuconostoc** | Leuconostoc carnosum | **1** |
| **Limosilactobacillus** | | **7** |
| | Limosilactobacillus fermentum | 3 |
| | Limosilactobacillus reuteri | 4 |

**TABLE 37.2**   (Continued) GRAS Notifications per Genus and Species Related to Lactic Acid Bacteria Publicly Available from the FDA Inventory of GRAS Notices (FDA 2023). Species are reported with genus names updated according to changes in *Lactobacillaceae* taxonomy (Zheng et al. 2020).

| GENUS | SPECIES | NUMBER OF GRAS NOTIFICATIONS |
|---|---|---|
| **Pediococcus** | *Pediococcus acidilactici* | **1** |
| **Streptococcus** | | **4** |
| | *Streptococcus salivarius* | 3 |
| | *Streptococcus thermophilus* | 1 |
| **Weissella** | *Weissella cibaria* | **1** |

notifications were filed between 2018 and 2023, a marked increase since the previous edition of this book. In practice, the GRAS concept is applied at the strain level and for a specific use, as the endpoint of a strain-level safety assessment process. Of note, "Cultured dairy sources, sugars, wheat, malt, and fruit- and vegetable-based sources . . . fermented by "*Streptococcus thermophilus, Lactobacillus acidophilus, Lactobacillus paracasei* subsp. *paracasei, Lactobacillus plantarum, Lactobacillus sakei, Lactobacillus bulgaricus* . . . or mixtures of these strains" were also notified in 2012. Other fermented foods, that is, "that may contain or be derived from microorganisms", are also specified in the US Federal regulations.

# 37.2  HISTORY OF USE OF LACTIC ACID BACTERIA

The wide use of LAB in fermented food is the likely result of their ancient domestication by humans as early as the Neolithic period (Douglas and Klaenhammer 2010). Archaeological evidence for dairy, wine, beer, and bread manufacture in Europe and Asia convincingly account for the prehistoric fermentation of milk, fruit, honey, and grain (Bogucki 1984; Cavalieri et al. 2003; McGovern et al. 2004; Salque et al. 2013). Biomolecular archaeology approaches have been successful in the identification of yeast DNA, and more recently LAB proteins, in archaeological remains of foods and tools, providing a direct association of micro-organisms and LAB with ancient human artefacts (Cavalieri et al. 2003; Yang et al. 2014; Xie et al. 2016). Specific genome characteristics of domesticated microbial species, such as genome decay and loss of metabolic pathways, are also indicative of a recent influence of the evolution of LAB by a domestication process (Campbell-Sills et al. 2015; Gibbons and Rinker 2015; Iskandar et al. 2017). Today's traditional fermented food products are the resulting, living legacy of these early domestication events, accounting for at least a seven-millennia history of use of LAB by humans (Gibbons and Rinker 2015).

Beyond these historical and archaeological studies, since 2002 the European Food and Feed Cultures Association (EFFCA) together with the International Dairy Federation (IDF) has led significant efforts to establish documented history of use for microbial cultures used for food production, especially lactic acid bacteria (Bourdichon et al. 2012; IDF 2022). As of 2022, a total of 127 LAB species (Table 37.3) were included in the so-called "Inventory of Microbial Food Cultures", with the objective to list microbial species with a documented presence in fermented foods, excluding contaminants, species undesirable in foods, and those lacking a clear or relevant role in the fermentation process. This type of initiative is useful for all scientists and technologists interested in LAB and other food microbes and is the largest publicly available data set, documenting 325 species in 617 food usages (IDF 2022). This properly documented multi-millennial history of use provides a high probability of safe use of the LAB species considered.

**TABLE 37.3**   Number of LAB Species with Documented History of Use as Microbial Food Cultures as of 2022. Adapted from (IDF 2022).

| FAMILY | GENUS | NUMBER OF LAB SPECIES |
|---|---|---|
| Carnobacteriaceae | Carnobacterium | 3 |
| | Marinilactibacillus | 2 |
| Enterococcaceae | Enterococcus | 3 |
| | Tetragenococcus | 2 |
| Lactobacillaceae | Agrilactobacillus | 1 |
| | Companilactobacillus | 10 |
| | Fructilactobacillus | 4 |
| | Lacticaseibacillus | 4 |
| | Lactiplantibacillus | 4 |
| | Lactobacillus | 10 |
| | Lapidilactobacillus | 1 |
| | Latilactobacillus | 2 |
| | Lentilactobacillus | 12 |
| | Leuconostoc | 10 |
| | Levilactobacillus | 8 |
| | Ligilactobacillus | 5 |
| | Limosilactobacillus | 7 |
| | Liquorilactobacillus | 7 |
| | Loigolactobacillus | 1 |
| | Oenococcus | 1 |
| | Paucilactobacillus | 2 |
| | Pediococcus | 5 |
| | Shleiferilactobacillus | 2 |
| | Secundilactobacillus | 4 |
| | Weissella | 9 |
| Streptococcacceae | Lactococcus | 6 |
| | Streptococcus | 2 |

# 37.3  POSSIBLE SAFETY CONCERNS ASSOCIATED WITH LAB USE AND HOW TO ADDRESS THEM

Possible theoretical adverse effects associated with the consumption of lactic bacteria from foods may include: (i) infections and other symptoms resulting from the ingestion of bacteria, (ii) transfer of antibiotic resistance determinants contributing to the global resistome, and (iii) production in the food matrix of toxic or deleterious substances. Yet the screening strategy to identify the finest strain often includes, in the characterisation process, criteria that allow discrimination of strains that will present the best safety profile for the intended use and mitigate the previously mentioned safety concerns.

The different safety concerns and possible selection criteria are further discussed in this section.

## 37.3.1  Infections and Other Potential Pathogenicity

One of the theoretical risks associated with ingestion of live bacteria in general, and thus also of LAB, are the possible adverse effects that might be similar to those of well-known pathogenic micro-organisms, such as potential gastro-intestinal symptoms through infection or production of harmful metabolites.

## 37.3.2 Consequences of Exposure to LAB

Different estimates show that humans are typically exposed to daily doses of $10^6$ to $10^9$ colony forming units (CFU) of microorganisms from their diet, while populations that habitually consume fermented foods may ingest as much as $10^{11}$ CFU/day (Lang et al. 2014; Marco et al. 2020). Yet this daily intake seems particularly well tolerated and is not known to generate even mild symptoms, put aside the ingestion of contaminating food-borne pathogens. In addition, the human body is constantly exposed, without noteworthy harm, to significant numbers of bacteria and LAB, including lactobacilli, especially those that reside in the human intestine and are part of the gut microbiota.

The safety of LAB with regard to the potential to cause infections may also be addressed through the examination of the rare published case-studies and epidemiological studies linking clinical cases with LAB, resulting either from food exposure or originating from the human microbiota (Ino et al. 2016; Karaaslan et al. 2016). Those cases may also originate from medical procedures (Aaron et al. 2017). However, as most LAB are not considered pathogenic micro-organisms by clinical microbiologists, the occurrence of LAB in clinical specimens is not systematically examined. Systematic literature searches, conducted by EFSA in the frame of the monitoring of safety concerns for micro-organisms added in the food chain since 2007, confirm that LAB may be involved in only very rare clinical cases of infections (Allende et al. 2015; Allende et al. 2016; Ricci et al. 2017a, 2017b, 2017c, 2018). In those instances, a clear link with the consumption of food with LAB cannot always be made (Reguera-Brito et al. 2016). Most, if not all, clinical cases are linked to severe underlying health conditions. On this subject, the EFSA is systematically monitoring the scientific literature, and the corresponding EFSA opinions provide extensive reviews of the documented cases, including the methodologies used to conduct such bibliographic surveys (Allende et al. 2015, 2016; Ricci et al. 2017, 2018; EFSA et al. 2020, 2023b).

When human intervention studies are performed, there is a basic responsibility and obligation for these studies regardless of the type of bioactive tested and also an excellent opportunity to document potential adverse events. This applies to studies with LAB, in which, documenting the adverse events and determining if they are associated with the 'treatment', a clear and objective assessment can be made on the safety of e.g. LAB, although safety reporting in nutrition trials in general can be improved beyond the statement that the 'treatment was well tolerated', as recently suggested (Bafeta et al. 2018). This is mainly of benefit to specific risk groups, as there is no indication that consumption of LAB as part of food or dietary supplement poses a risk to the general healthy population (Tapiovaara et al. 2016).

Overall, for a significant number of LAB species with a long history of safe use, or that have been the subject to a thorough assessment and monitoring, it can be concluded that there are no safety concerns of possible infections for the general healthy population. Strains that belong to such LAB species may undergo simplified safety assessments, limited only to documented and characterised safety concerns, as long as these are properly identified (see next section). This may help potentially avoiding unnecessary animal testing or screening through less relevant criteria.

# 37.4 THE IMPORTANCE OF PROPER LAB IDENTIFICATION AND CHARACTERIZATION

When selecting or using LAB strains for food fermentation, identification at the lowest taxonomic level possible enables to confirm that the strains or isolates of interest actually belong to species that have a long history of use and/or with the least safety concerns. Classical polyphasic approaches combining genotypic methods such as sequencing adequate markers genes, e.g. encoding 16S RNA, along with carbohydrate fermentation profiles, usually provide an adequate robustness for bacterial species identification. Indeed the use of at least two robust taxonomic markers has been recommended (Salvetti and O'Toole 2017). In specific cases, longer sequences of 16S RNA genes, such as 16S-23S RNA interspacer sequences can be used to further discriminate species with close genetic relationship. Housekeeping genes can also be used as additional taxonomic markers to overcome species ambiguities, such as *pheS*, *rpoA* (Naser et al. 2007) and *recA* (Torriani et al. 2001). These longer sequences or additional markers can be especially useful in the former *Lactobacillus* genus, for instance to differentiate species within the genera *Lactiplantibacilus plantarum paraplantarum/pentosus* or *Lacticaseibacillus casei/paracasei/rhamnosus*.

The limited resolution for LAB, and the labour-intensive aspects of certain identification methods, such as protein profiling, 16S rRNA sequencing, ribotyping, randomly amplified polymorphic DNA (RAPD) fingerprinting, and pulsed-field gel electrophoresis (PFGE), has led to the development of species-level methods for the identification of LAB based on different genetic markers or analytical methods for further species separation. PCR amplification of repetitive bacterial DNA elements (rep-PCR), based on the amplification of specific repetitive bacterial DNA elements, has proven consistently adequate in the case of LAB (Hyytiä-Trees et al. 1999; Sohier et al. 1999; De Urraza et al. 2000; Gevers et al. 2001). Also called GTG-PCR, this approach can be used both for strain identification and typing (Salomskiene et al. 2015).

Clustered regularly interspace short palindromic repeats (CRISPR), another DNA repeat family, and the associated proteins CRISPR-Cas, have as well been proposed and successfully used as tools for bacteria genotyping in complex fermented food matrices (Barrangou and Dudley 2016).

Amplified 16S ribosomal DNA restriction analysis (16S-ARDRA) is also based on the analysis of the gene encoding 16S rRNA. Methods have been developed and used for the identification of strains of lactobacilli, lactococci, and *Leuconostoc* spp. from various food matrices (Ji et al. 2013; Park et al. 2014; Leite et al. 2015). When compared with another bacterial identification technique such as matrix-assisted laser desorption/ionisation time-of-flight mass spectrometry (MALDI-TOF MS), 16S-ARDRA provided comparable results to MALDI-TOF MS. It was only limited, as are all methods based on 16S RNA genes, in the distinction between species with very high 16S rDNA sequence homology, such as *L. casei* and *L. zeae* (Dec et al. 2016). 16S-ARDRA was also used in combination with pulsed-field gel electrophoresis to characterise propionibacteria (Blasco et al. 2015).

Amplification of highly conserved genes, such as the 16S or 23S rRNA genes, can also be analysed with separation techniques such as denaturing gradient gel electrophoresis (DGGE) and classical DNA sequencing. DGGE followed by 16S rDNA sequencing has been used to characterise microbial communities and to identify LAB from diverse foods and non-food environments, such as Chinese liquor and other traditional Asian fermentations (He et al. 2017; Jin et al. 2017), Colombian fermented foods and beverages (Chaves-Lopez et al. 2014), or raw milk (Quigley et al. 2013).

The fast evolution of sequencing technologies has allowed the full sequencing of the genome of a microorganism in a fast, reliable and affordable way. Moreover, an overall genome related index (OGRI) has been proposed to measure the relatedness between strains and type strain of a species by calculating the relatedness between genome sequences, as an alternative to DNA–DNA hybridisation techniques (Richter and Rosselló-Móra 2009; Chun et al. 2018). The microbiologist can also benefit from a full pipeline of tools for genome annotation and analysis (Chun et al. 2018). Whole genome sequencing can therefore enable more accurate species and strain identification. Comparison of genomes among different strains, coupled with classical phenotypic approaches, can be as well applied for further typing and diversity analysis within LAB species (Smokvina et al. 2013; Ceapa et al. 2015, 2016). Although gaps in knowledge may exist and limit the use of curated database for undesirable genetic determinants (Salvetti and O'Toole 2017), the available complete genome of LAB strains should allow their expedited characterisation with regards to safety concerns by assessing the potential presence of genes involved in antimicrobial resistance or in the synthesis of undesirable substances (see subsequently). In 2021, as part of a statement on whole genome sequencing analysis, the EFSA provided guidance on confirming bacterial identity, recommending the use of digital DNA–DNA hybridisation, average nucleotide identity, or phylogenomic methods and also providing guidance if the reference-based read mapping approach is used for identification (EFSA 2021).

For identification purposes, keeping up to date with the current taxonomy is also of paramount importance to avoid any ambiguity in the identification of the strains. The most current taxonomy should be referred to, either through the International Committee on Systematics of Prokaryotes (ICSP—www.the-icsp.org/) or through available publications in the International Journal of Systematic and Evolutionary Microbiology (IJSEM—http://ijs.sgmjournals.org/). Also, the web site "List of Prokaryotic Names with Standing in Nomenclature" founded by Jean P. Euzéby in 1997 is a very useful resource (www.bacterio.net/) (Parte 2018). Reclassification of the *Lactobacillus* genus has been proposed and implemented, as the extraordinary diversity resulted in challenging cloudification issues (Salvetti et al. 2018; Zheng et al. 2020).

# 37.5 ANTIBIOTIC RESISTANCE

The emergence of antibiotic resistance as a public health concern has led to questions, especially in Europe, on the food chain, as a whole, as a possible source of increase of the global resistome, the "reservoir of resistance determinants that can be mobilized into the microbial community" (D'Costa et al. 2006; Hudson et al. 2017). For instance, use of antibiotics in animal production is linked to the propagation of antibiotic resistance as demonstrated by antimicrobial resistance (AMR) in zoonotic species such as *Salmonella* spp., *Campylobacter* spp., *Listeria* spp., *Staphylococcus* spp., and *Escherichia* spp. (Berendonk et al. 2015). Apart from antibiotic use in human or animal medicine, which is believed to be the main driver of the spread antibiotic resistance among human pathogens, the relative contributions of the existing sources of antimicrobial resistance, such as soil or gut microbiota, and other factors, such as chemical pollution, are not fully understood (Perry and Wright 2014; The Lancet Planetary 2018). For instance, a metagenomic study of AMR in the intestinal microbiota concluded that most AMR determinants should be considered intrinsic genes of commensal microbiota with a low risk of transfer to bacterial pathogens (Ruppe et al. 2017).

Regarding the food chain, in a 2008 report, the EFSA concluded that resistance of known pathogenic bacteria to fluoroquinolones as well as to third- and fourth-generation cephalosporins found in a variety of foods and in animals in primary production were priorities in the area of transmission of AMR in food (EFSA 2008).

As described previously and elsewhere, LAB used in foods enjoy a very long, multi-millennial safety record. Nevertheless, the introduction of transferable AMR from intentionally added bacteria such as lactic acid bacteria starter cultures should be avoided. Indeed, horizontally transferred antibiotic genes are the most likely to be transmitted and thus are of special concern.

The EU-funded project ACE-ART concluded that AMR was rare among approximately 1400 isolates of lactic acid bacteria of human, animal, food, and feed origin (EFSA 2008). In the same report, the EFSA concluded AMR could occasionally be isolated from fermented foods, but that antimicrobial resistance in strains used as industrial starter cultures was not widespread (EFSA 2008). Foodborne strains of *Lactobacillus* spp. and *Lactococcus* spp. have been shown to potentially display AMR to tetracycline, erythromycin, and to a lesser extent amino-glycosides (Devirgiliis et al. 2013). Indeed, Teuber et al. (1999) listed a number of antimicrobial resistances detected in lactic acid bacteria isolated from foods. Thus, the selection of LAB for addition in foods with regard to AMR should be done in order to avoid the presence of any transferable AMR genes (Florez et al. 2016; Guo et al. 2017; Anisimova et al. 2018).

The choice of antibiotic substances for a safety screening purpose is of paramount interest both to ensure a pertinent protection of the food chain and to perform appropriate and relevant testing. Since 2005, the EFSA FEEDAP panel, taking over previous work done by the Scientific Committee for Animal Nutrition (SCAN), has developed guidance documents for the assessment of bacterial antimicrobial susceptibility for micro-organisms added in the food chain (EFSA 2005). The work was necessary to provide guidance for LAB applications in the feed area. Nevertheless, the work can be valuable to develop AMR characterisation and screening methods to select LAB for addition in the food chain. The guidance was

**TABLE 37.4** Antibiotics Proposed in EFSA FEEDAP Guidance for Assessment of Bacterial Antimicrobial Susceptibility (EFSA 2012).

| | |
|---|---|
| Ampicillin | Erythromycin |
| Vancomycin | Clindamycin |
| Gentamicin | Tetracycline |
| Kanamycin | Chloramphenicol |
| Streptomycin | |

updated in 2012, providing a list of clinically relevant antibiotics (Table 37.4) to be used for the assessment of bacterial antimicrobial susceptibility (EFSA 2012). Moreover, data on a wider range of antibiotics can be retrieved from literature (Neut et al. 2017).

The availability of representative minimum inhibitory concentration (MIC) distributions in susceptible and resistant isolates is also key to allow correctly distinguishing between susceptible and resistant phenotypes. Isolate identification at least at the genus level and sometimes at the species level is needed for LAB to be able to compare the measured MIC values to the relevant MICs distributions. The EFSA also provided updated guidance on genomic analysis and clarified acquired resistance in a draft guidance published for public consultation in 2023 (EFSA et al. 2023a).

A better understanding of the physiology of LAB has also helped to identify groups of bacteria that are intrinsically non-susceptible to certain antibiotics, thereby limiting the use of testing for these antibiotics or allowing to acknowledge wild type phenotypes not linked with transferable resistance traits. For instance, the intrinsic resistance to most glycopeptides, due to the nature of the cell wall structure of most lactobacilli, was shown to rely on *vanZ* gene. Based on the observed distribution and phylogenetic analysis of the encoded VanZ proteins, congruent with lactobacilli species phylogeny, the best possible explanation of the distribution of the *vanZ* gene in distinct lactobacilli species is likely the result of speciation rather than horizontal transfer (Abriouel et al. 2017).

## 37.5.1 Phenotypic Antibiotic Susceptibility Methods

The use of adapted quantitative antibiotic susceptibility methods is also key. Early studies on the antibiotic susceptibility of LAB have led at least some authors to abandon the use of traditional disc-inhibition methods and clinically relevant test media, to use quantitative methods such as microdilution techniques combined with test media adapted to growth conditions and physiology of LAB (Klare et al. 2005, 2007). This aspect is also important to avoid false positive or negative results, as was documented for different antibiotics, such as trimethoprim. It has been shown, for example, that media composition could interfere with LAB susceptibility results (Danielsen et al. 2004). Existing antibiotic susceptibility methods have therefore been further developed and standardised for this specific purpose (ISO/IDF 2010). In Europe, criteria and guidance documents by the EFSA FEEDAP panel provide a rationalised framework for the evaluation of AMR in LAB, including a reduced the number of antibiotics it recommended for safety assessment in comparison with previous documents (EFSA 2012). However, the 2012 FEEDAP Guidance was identified by EFSA in 2016 as needing further update (EFSA 2016). Further research is needed to continue standardisation of methods and adaptation of assessment criteria (Mayrhofer et al. 2014; EFSA 2016). More data is also needed to measure MIC distributions to antibiotics of interests for LAB from different origins in order to establish more accurate microbiological cut-off values (Florez and Mayo 2017).

In addition, clarification should be provided with interpreting phenotypic antibiotic susceptibility results when growth occurs at the cut-off values (Polka et al. 2016).

## 37.5.2 Genetic Methods for Antibiotic Resistance Determinants

It is also possible to rely on genetic methods for the fast identification of antibiotic resistance determinants in bacterial isolates, such as PCR assays and DNA-DNA hybridisation techniques (Florez et al. 2016; Florez and Mayo 2017; Guo et al. 2017). Whole-genome sequencing of LAB strains may as well allow assessing the presence of genes involved in antimicrobial resistance. These methods need, however, to be targeted with at least prior knowledge of the genetic determinants that must be identified. Unknown antibiotic resistance genes would not likely be detected by these approaches. Furthermore, genes indicated as potential resistance genes may actually serve mainly another essential physiological function, most notably genes encoding for transporter proteins (Morovic et al. 2017).

Metagenomics with functional screening approaches allow also to detect antibiotic resistance genes from DNA extract from the fermented food matrix by screening antibiotic susceptibilities in the obtained cloned metagenomic library (Devirgiliis et al. 2014). These functional screenings could ensure to detect antibiotic resistance encoded by novel, previously undocumented determinants and could be particularly helpful for studying AMR in complex microbial communities. The EFSA provided guidance in 2021 on search for antimicrobial resistance genes regarding the number of databases to use and thresholds of identity at protein or nucleotide levels (EFSA 2021).

Screening LAB strains to avoid AMR is essential to ensure the chosen strains are safe for the intended use, and this question is considered seriously by the industry, as exemplified by the development of standardised analytical methods (ISO/IDF 2010).

# 37.6  RISK FACTORS, DELETERIOUS AND TOXIC METABOLITES

Among the substances produced, biogenic amines and D-lactic acid have been subjected to further characterisation.

## 37.6.1 Production of Biogenic Amines by LAB

Biogenic amine (BA) production in foods has been extensively reviewed (EFSA 2011; Benkerroum 2016; Gardini et al. 2016; Pessione and Cirrincione 2016; Sarkadi 2017). BAs are organic molecules of low molecular weight that may be present in foods as a result of the microbial decarboxylation of amino acids. Histamine and tyramine and to a lesser extent putrescine, cadaverine, tryptamine, 2-phenylethylamine, spermine, spermidine, and agmatine are considered to occur most frequently in foods. Histamine and tyramine are considered of human health importance (EFSA 2011; Gardini et al. 2016). Ingested in high amounts, these compounds may provoke diverse symptoms ranging from moderate to severe reactions (Gardini et al. 2016).

The production of biogenic amines is documented for different species of LAB, although this trait is not systematically present. Histamine production by LAB such as *Tetragenococcus*, lactobacilli, and *Oenococcus* is well documented in fermented products, such as wine, cheese, and fish sauce (Landete et al. 2007). Biogenic amine production or genetic traits encoding this ability can be detected in various LAB genera, including lactobacilli, *Lactococcus*, *Streptococcus*, *Oenococcus*, *Tetragenococcus*, and *Enterococcus* (Ladero et al. 2012; Lorencova et al. 2012; Flasarova et al. 2016; Takebe et al. 2016; Liu et al. 2017; Perin et al. 2017; Poveda et al. 2017).

BA production in fermented food can result, however, from multiple factors in addition to the genetic and enzymatic capabilities of the micro-organisms of interest. Available substrates, temperature, salt concentrations, pH, presence of starter culture, antimicrobial substances, packaging, fermentation process, and additional micro-organisms present in the food can all significantly contribute to enhance or reduce the actual BA content, that is, either in concentration or types of BA present in a food fermented with LAB (Gardini et al. 2016).

The concern for biogenic amine production by LAB in food fermentation processes can be managed by selecting starters lacking known genetic pathways to degrade precursor amino acids into biogenic amines (Landete et al. 2007). Gene clusters encoding decarboxylating enzymes, such as the tyrosine decarboxylase (*tdc*) and histidine decarboxylase (*hdc*) gene clusters, have been identified and characterised in different LAB species. The high similarity either in gene sequence and organisation from different producing LAB of the *tdc* and *hdc* clusters, respectively, have allowed the development of screening methods based on gene detection by PCR in order to identify strains with a genetic potential to produce BA. Optionally, the subsequent sequencing of any eventual amplification products can be performed to avoid false positive results. In addition, this provides a sound basis for identification of BA production gene clusters from whole genome sequencing data. Verification of the actual BA production in the food matrix in the conditions of intended use, through dosage of histamine and tyramine can confirm the results of the genetic analysis.

## 37.6.2 Production of D-Lactic Acid

In the general population, metabolic production of D-lactate in human cells occurs as the result of the methylgloxal pathway. Exogenous sources of D-lactate are the diet (foods containing D-Lactate) and the carbohydrate-fermenting bacteria normally present in the gastrointestinal tract. Endogenous or exogenous D-lactate is then naturally converted into L-lactate by endogenous hepatic enzymes or can be metabolised by the gut microbiota.

However, D-lactic acid is sometimes considered a risk for preterms, neonates, and young infants and for people with short bowel syndrome (SBS). SBS may result in an increase of undigested carbohydrates in the digestive tract. As a result, organic acids produced by intestinal fermentation may exceed the quantity that can be metabolised, resulting in an acidosis. For infants the concern is that the immature physiology is not able to metabolise D-lactate, and this may lead as well to acidosis. Among LAB, *Oenococcus*, *Pediococcus*, *Leuconostoc*, and some lactobacilli (*L. acidophilus*, *L. gasseri*, *L. delbrueckii*, *L. fermentum*, *L. lactis*, *L. brevis*, *L. helveticus*, *L. plantarum*, and *L. reuteri*) can produce D-lactic acid, though the amount of D-lactic acid produced is likely to vary between strains within each species (Pot 2014).

Whether this is a valid concern for healthy infants is questionable. Unless a fermented product is fed to the infant, the administration of D/L-lactate producing LAB does not lead to production of D-lactate in the intestine. This has been shown for D-lactate-producing lactobacilli such as *L. reuteri* and *L. acidophilus* (Connolly et al. 2005). A review of five clinical trials also concluded that the consumption of D-lactate-producing lactobacilli did not cause D-lactic acidosis in healthy children (Lukasik et al. 2018). D-lactate acidosis has been observed in infants with short bowel syndrome but is there related to the consumption of easily fermentable carbohydrates that are not absorbed in this condition. For healthy adults, there is certainly no concern with D-lactate, as there is ample expression of the enzyme D-alpha-hydroxy acid dehydrogenase (Ewaschuk et al. 2005).

## 37.7 CONCLUSIONS AND WAY FORWARD

While the multiple millennial history of use of LAB in foods provides a primary basis for their safe use, potential safety concerns theoretically linked to the ingestion of LAB and fermented foods have been identified and extensively characterised through regulatory agencies, academia, and industry research in the last two decades. The considered safety concerns, including the potential for infections, the transfer of antibiotic resistance determinants, and the production in the food matrix of toxic or deleterious substances, remain either rare or of very limited occurrence. With the current knowledge and available methodologies, it is possible to apply specific criteria and adapted tools to appropriately select LAB strains, thereby avoiding introducing micro-organisms in the food chain that may present a potential safety concern. Furthermore, the ability to fully characterise complex microbial ecosystems in relation to those safety concerns, by determining whole genome sequences of bacterial strains and establishing the functionality of fermented food microbiota, can enable the further assurance of safety for complex and traditional food fermentations. While the available genetic tools will render these assessments more reliable, verification through characterisation of the related specific phenotypic traits and classical analytical methods remains necessary.

# BIBLIOGRAPHY

Aaron, J. G., M. E. Sobieszczyk, S. D. Weiner, S. Whittier and F. D. Lowy (2017). "Lactobacillus rhamnosus endocarditis after upper endoscopy." *Open Forum Infectious Diseases* **4**(2). http://doi.org/10.1093/ofid/ofx085

Abriouel, H., C. W. Knapp, A. Galvez and N. Benomar (2017). *Antibiotic Resistance Profile of Microbes from Traditional Fermented Foods*. Academic Press, Cambridge, MA.

Allende, A., D. Bolton, M. Chemaly, R. Davies, P. S. F. Escamez, R. Girones, L. Herman, K. Koutsoumanis, R. Lindqvist, B. Norrung, A. Ricci, L. Robertson, G. Ru, M. Sanaa, M. Simmons, P. Skandamis, E. Snary, N. Speybroeck, B. Ter Kuile, J. Threlfall, H. Wahlstrom and E. P. B. H. Biohaz (2015). "Statement on the update of the list of QPS-recommended biological agents intentionally added to food or feed as notified to EFSA 3: Suitability of taxonomic units notified to EFSA until September 2015." *EFSA Journal* **13**(12). http://doi.org/10.2903/j.efsa.2015.4331

Allende, A., D. Bolton, M. Chemaly, R. Davies, P. Salvador, F. Escamez, R. Girones, L. Herman, K. Koutsoumanis, R. Lindqvist, B. Norrung, A. Ricci, L. Robertson, G. Ru, M. Sanaa, M. Simmons, P. Skandamis, E. Snary, N. Speybroeck, B. Ter Kuile, J. Threlfall, H. Wahlstrom and E. P. B. H. BIOHAZ (2016). "Update of the list of QPS-recommended biological agents intentionally added to food or feed as notified to EFSA 4: Suitability of taxonomic units notified to EFSA until March 2016." *EFSA Journal* **14**(7). http://doi.org/10.2903/j.efsa.2016.4522

Anisimova, E., N. Bruslik and D. Yarullina (2018). "Antibiotic resistance of dairy and probiotic lactobacilli and its transfer to pathogenic bacteria." *European Journal of Clinical Investigation* **48**: 163–164.

Bafeta, A., M. Koh, C. Riveros and P. Ravaud (2018). "Harms reporting in randomized controlled trials of interventions aimed at modifying microbiota: A systematic review." *Annals of Internal Medicine*. http://doi.org/10.7326/m18-0343

Barrangou, R. and E. G. Dudley (2016). "CRISPR-based typing and next-generation tracking technologies." *Annual Review of Food Science and Technology*, **7**: 395–411.

Bell, V., J. Ferrao and T. Fernandes (2017). "Nutritional guidelines and fermented food frameworks." *Foods* **6**(8). http://doi.org/10.3390/foods6080065

Benkerroum, N. (2016). "Biogenic amines in dairy products: Origin, incidence, and control means." *Comprehensive Reviews in Food Science and Food Safety* **15**(4): 801–826. http://doi.org/10.1111/1541-4337.12212

Berendonk, T. U., C. M. Manaia, C. Merlin, D. Fatta-Kassinos, E. Cytryn, F. Walsh, H. Burgmann, H. Sorum, M. Norstrom, M. N. Pons, N. Kreuzinger, P. Huovinen, S. Stefani, T. Schwartz, V. Kisand, F. Baquero and J. L. Martinez (2015). "Tackling antibiotic resistance: The environmental framework." *Nature Reviews Microbiology* **13**(5): 310–317. http://doi.org/10.1038/nrmicro3439

Blasco, L., M. Kahala, H. Jatila and V. Joutsjoki (2015). "Application of 16S-ARDRA and RFLP-PFGE for improved genotypic characterisation of dairy propionibacteria and combination with characteristic phenotypes." *International Dairy Journal* **50**: 66–71. http://doi.org/10.1016/j.idairyj.2015.06.005

Bogucki, P. I. (1984). "Ceramic sieves of the linear pottery culture and their economic implications." *Oxford Journal of Archaeology* **3**(1): 15–30. http://doi.org/10.1111/j.1468-0092.1984.tb00113.x

Bourdichon, F., S. Casaregola, C. Farrokh, J. C. Frisvad, M. L. Gerds, W. P. Hammes, J. Harnett, G. Huys, S. Laulund, A. Ouwehand, I. B. Powell, J. B. Prajapati, Y. Seto, A. Ter Schure, A. Van Boven, V. Vankerckhoven, A. Zgoda, S. Tuijtelaars and E. B. Hansen (2012). "Food fermentations: Microorganisms with technological beneficial use." *International Journal of Food Microbiology* **154**(3): 87–97. http://doi.org/10.1016/j.ijfoodmicro.2011.12.030

Campbell-Sills, H., M. El Khoury, M. Favier, A. Romano, F. Biasioli, G. Spano, D. J. Sherman, O. Bouchez, E. Coton, M. Coton, S. Okada, N. Tanaka, M. Dols-Lafargue and P. M. Lucas (2015). "Phylogenomic analysis of oenococcus oeni reveals specific domestication of strains to cider and wines." *Genome Biology and Evolution* **7**(6): 1506–1518. http://doi.org/10.1093/gbe/evv084

Cavalieri, D., P. E. McGovern, D. L. Hartl, R. Mortimer and M. Polsinelli (2003). "Evidence for S. cerevisiae fermentation in ancient wine." *Journal of Molecular Evolution* **57**(Suppl 1): S226–S232. http://doi.org/10.1007/s00239-003-0031-2

Ceapa, C., M. Davids, J. Ritari, J. Lambert, M. Wels, F. P. Douillard, T. Smokvina, W. M. de Vos, J. Knol and M. Kleerebezem (2016). "The variable regions of lactobacillus rhamnosus genomes reveal the dynamic evolution of metabolic and host-adaptation repertoires." *Genome Biology and Evolution* **8**(6): 1889–1905. http://doi.org/10.1093/gbe/evw123

Ceapa, C., J. Lambert, K. van Limpt, M. Wels, T. Smokvina, J. Knol and M. Kleerebezem (2015). "Correlation of lactobacillus rhamnosus genotypes and carbohydrate utilization signatures determined by phenotype profiling." *Applied and Environmental Microbiology* **81**(16): 5458–5470. http://doi.org/10.1128/aem.00851-15

Chaves-Lopez, C., A. Serio, C. D. Grande-Tovar, R. Cuervo-Mulet, J. Delgado-Ospina and A. Paparella (2014). "Traditional fermented foods and beverages from a microbiological and nutritional perspective: The colombian heritage." *Comprehensive Reviews in Food Science and Food Safety* **13**(5): 1031–1048. http://doi.org/10.1111/1541-4337.12098

Chilton, S. N., J. P. Burton and G. Reid (2015). "Inclusion of fermented foods in food guides around the world." *Nutrients* **7**(1): 390–404. http://doi.org/10.3390/nu7010390

Chun, J., A. Oren, A. Ventosa, H. Christensen, D. R. Arahal, M. S. da Costa, A. P. Rooney, H. Yi, X. W. Xu, S. De Meyer and M. E. Trujillo (2018). "Proposed minimal standards for the use of genome data for the taxonomy of prokaryotes." *International Journal of Systematic and Evolutionary Microbiology* **68**(1): 461–466. http://doi.org/10.1099/ijsem.0.002516

Connolly, E., T. Abrahamsson and B. Bjorksten (2005). "Safety of D(-)-lactic acid producing bacteria in the human infant." *Journal of Pediatric Gastroenterology and Nutrition* **41**(4): 489–492.

Danielsen, M., H. S. Andersen and A. Wind (2004). "Use of folic acid casei medium reveals trimethoprim susceptibility of Lactobacillus species." *Letters in Applied Microbiology* **38**(3): 206–210. http://doi.org/10.1111/j.1472-765X.2004.01471.x

D'Costa, V. M., K. M. McGrann, D. W. Hughes and G. D. Wright (2006). "Sampling the antibiotic resistome." *Science* **311**(5759): 374–377. http://doi.org/10.1126/science.1120800

Dec, M., A. Puchalski, R. Urban-Chmiel and A. Wernicki (2016). "16S-ARDRA and MALDI-TOF mass spectrometry as tools for identification of Lactobacillus bacteria isolated from poultry." *Bmc Microbiology* **16**. http://doi.org/10.1186/s12866-016-0732-5

De Urraza, P. J., A. Gomez-Zavaglia, M. E. Lozano, V. Romanowski and G. L. De Antoni (2000). "DNA fingerprinting of thermophilic lactic acid bacteria using repetitive sequence-based polymerase chain reaction." *Journal of Dairy Research* **67**(3): 381–392.

Devirgiliis, C., P. Zinno and G. Perozzi (2013). "Update on antibiotic resistance in foodborne Lactobacillus and Lactococcus species." *Frontiers in Microbiology* **4**. http://doi.org/10.3389/fmicb.2013.00301

Devirgiliis, C., P. Zinno, M. Stirpe, S. Barile and G. Perozzi (2014). "Functional screening of antibiotic resistance genes from a representative metagenomic library of food fermenting microbiota." *Biomed Research International*. http://doi.org/10.1155/2014/290967

Douglas, G. L. and T. R. Klaenhammer (2010). "Genomic evolution of domesticated microorganisms." *Annual Review of Food Science and Technology* **1**: 397–414. http://doi.org/10.1146/annurev.food.102308.124134

EFSA (2005). "Opinion of the scientific panel on additives and products or substances used in animal feed (FEEDAP) on the updating of the criteria used in the assessment of bacteria for resistance to antibiotics of human or veterinary importance." *EFSA Journal* **3**(6): 223. http://doi.org/10.2903/j.efsa.2005.223

EFSA (2008). "Foodborne antimicrobial resistance as a biological hazard—scientific opinion of the panel on biological hazards." *EFSA Journal* **6**(8): 765. http://doi.org/10.2903/j.efsa.2008.765

EFSA (2011). "Scientific Opinion on risk based control of biogenic amine formation in fermented foods." *EFSA Journal* **9**(10): 2393. http://doi.org/10.2903/j.efsa.2011.2393

EFSA (2012). "Guidance on the assessment of bacterial susceptibility to antimicrobials of human and veterinary importance." *EFSA Journal* **10**(6): 2740. http://doi.org/10.2903/j.efsa.2012.2740

EFSA (2016). "Analysis of the need for an update of the guidance documents." *EFSA Journal* **14**(5): e04473. http://doi.org/10.2903/j.efsa.2016.4473

EFSA (2021). "EFSA statement on the requirements for whole genome sequence analysis of microorganisms intentionally used in the food chain." *EFSA Journal* **19**(7): e06506. https://doi.org/10.2903/j.efsa.2021.6506

EFSA Panel, O. B. Hazards, K. Koutsoumanis, A. Allende, A. Alvarez-Ordóñez, D. Bolton, S. Bover-Cid, M. Chemaly, A. D. Cesare, F. Hilbert, R. Lindqvist, M. Nauta, R. Nonno, L. Peixe, G. Ru, M. Simmons, P. Skandamis, E. Suffredini, P. S. Cocconcelli, P. S. F. Escámez, M. P. Maradona, A. Querol, L. Sijtsma, J. E. Suarez, I. Sundh, E. N. Fernández, F. Istace, J. Aguillera, R. Brozzi, E. Liébana, B. Guerra, S. Correia and L. Herman (2023a). "Draft statement on how to interpret the QPS qualification on 'Acquired antimicrobial resistance genes'." *EFSA Journal*. http://doi.org/10.2903/j.efsa.202Y.xxxx

EFSA Panel, O. B. Hazards, K. Koutsoumanis, A. Allende, A. Alvarez-Ordóñez, D. Bolton, S. Bover-Cid, M. Chemaly, R. Davies, A. De Cesare, F. Hilbert, R. Lindqvist, M. Nauta, L. Peixe, G. Ru, M. Simmons, P. Skandamis, E. Suffredini, P. S. Cocconcelli, P. S. Fernández Escámez, M. P. Maradona, A. Querol, J. E. Suarez, I. Sundh, J. Vlak, F. Barizzone, S. Correia and L. Herman (2020). "Scientific opinion on the update of the list of QPS-recommended biological agents intentionally added to food or feed as notified to EFSA (2017–2019)." *EFSA Journal* **18**(2): e05966. https://doi.org/10.2903/j.efsa.2020.5966

EFSA Panel, O. B. Hazards, K. Koutsoumanis, A. Allende, A. Álvarez-Ordóñez, D. Bolton, S. Bover-Cid, M. Chemaly, A. de Cesare, F. Hilbert, R. Lindqvist, M. Nauta, L. Peixe, G. Ru, M. Simmons, P. Skandamis, E. Suffredini, P. S. Cocconcelli, P. S. Fernández Escámez, M. P. Maradona, A. Querol, L. Sijtsma, J. E. Suarez, I. Sundh, J. Vlak, F. Barizzone, M. Hempen, S. Correia and L. Herman (2023b). "Update of the list of qualified presumption of safety (QPS) recommended microorganisms intentionally added to food or feed as notified to EFSA." *EFSA Journal* **21**(1): e07747. https://doi.org/10.2903/j.efsa.2023.7747

EFSA Panel, K. Koutsoumanis, A. Allende, A. Alvarez-Ordonez, D. Bolton, S. Bover-Cid, M. Chemaly, A. De Cesare, F. Hilbert, R. Lindqvist, M. Nauta, L. Peixe, G. Ru, M. Simmons, P. Skandamis, E. Suffredini, P. S. Cocconcelli, F. Escámez, P. Salvador, M. M. Prieto and L. Herman (2023c). "Updated list of QPS-recommended microorganisms for safety risk assessments carried out by EFSA [Data set]." *Zenodo*. https://doi.org/10.5281/zenodo.7554079

Ewaschuk, J. B., J. M. Naylor and G. A. Zello (2005). "D-lactate in human and ruminant metabolism." *Journal of Nutrition* **135**(7): 1619–1625. http://doi.org/10.1093/jn/135.7.1619

FDA, U. (2022). "Generally Recognized as Safe (GRAS)." Retrieved 6 June, 2023, from www.fda.gov/food/food-ingredients-packaging/generally-recognized-safe-gras.

FDA, U. (2023). "GRAS Notices." Retrieved 6 June, 2023, from www.cfsanappsexternal.fda.gov/scripts/fdcc/?set=GRASNotices.

Flasarova, R., V. Pachlova, L. Bunkova, A. Mensikova, N. Georgova, V. Drab and F. Bunka (2016). "Biogenic amine production by Lactococcus lactis subsp cremoris strains in the model system of Dutch-type cheese." *Food Chemistry* **194**: 68–75. http://doi.org/10.1016/j.foodchem.2015.07.069

Florez, A. B., I. Campedelli, S. Delgado, A. Alegria, E. Salvetti, G. E. Felis, B. Mayo and S. Torriani (2016). "Antibiotic susceptibility profiles of dairy leuconostoc, analysis of the genetic basis of atypical resistances and transfer of genes in vitro and in a food matrix." *PLoS ONE* **11**(1). http://doi.org/10.1371/journal.pone.0145203

Florez, A. B. and B. Mayo (2017). "Antibiotic resistance-susceptibility profiles of streptococcus thermophilus isolated from raw milk and genome analysis of the genetic basis of acquired resistances." *Frontiers in Microbiology* **8**. http://doi.org/10.3389/fmicb.2017.02608

Gardini, F., Y. Ozogul, G. Suzzi, G. Tabanelli and F. Ozogul (2016). "Technological factors affecting biogenic amine content in foods: A review." *Frontiers in Microbiology* **7**(1218). http://doi.org/10.3389/fmicb.2016.01218

Gevers, D., G. Huys and J. Swings (2001). "Applicability of rep-PCR fingerprinting for identification of Lactobacillus species." *FEMS Microbiology Letters* **205**(1): 31–36. http://doi.org/10.1111/j.1574-6968.2001.tb10921.x

Gibbons, J. G. and D. C. Rinker (2015). "The genomics of microbial domestication in the fermented food environment." *Current Opinion in Genetics & Development* **35**: 1–8. http://doi.org/10.1016/j.gde.2015.07.003

Guo, H. L., L. Pan, L. N. Li, J. Lu, L. Kwok, B. Menghe, H. P. Zhang and W. Y. Zhang (2017). "Characterization of antibiotic resistance genes from lactobacillus isolated from traditional dairy products." *Journal of Food Science* **82**(3): 724–730. http://doi.org/10.1111/1750-3841.13645

He, G. Q., T. J. Liu, F. A. Sadiq, J. S. Gu and G. H. Zhang (2017). "Insights into the microbial diversity and community dynamics of Chinese traditional fermented foods from using high-throughput sequencing approaches." *Journal of Zhejiang University-Science B* **18**(4): 289–302. http://doi.org/10.1631/jzus.B1600148

Herman, L., M. Chemaly, P. S. Cocconcelli, P. Fernandez, G. Klein, L. Peixe, M. Prieto, A. Querol, J. E. Suarez, I. Sundh, J. Vlak and S. Correia (2018). "The qualified presumption of safety assessment and its role in EFSA risk evaluations: 15 years past." *FEMS Microbiology Letters* **366**(1). http://doi.org/10.1093/femsle/fny260

Hudson, J. A., L. J. Frewer, G. Jones, P. A. Brereton, M. J. Whittingham and G. Stewart (2017). "The agri-food chain and antimicrobial resistance: A review." *Trends in Food Science & Technology* **69**: 131–147. http://doi.org/10.1016/j.tifs.2017.09.007

Hyytiä-Trees, E., U. Lyhs, H. Korkeala and J. Björkroth (1999). "Characterisation of ropy slime-producing Lactobacillus sakei using repetitive element sequence-based PCR." *International Journal of Food Microbiology* **50**(3): 215–219. http://doi.org/10.1016/S0168-1605(99)00104-X

IDF (2022). "Inventory of microbial food cultures with safety demonstration in fermented food products." *Bulletin of the IDF* **514**. https://doi.org/10.56169/LJHC7402

Ino, K., K. Nakase, K. Suzuki, A. Nakamura, A. Fujieda and N. Katayama (2016). "Bacteremia due to leuconostoc pseudomesenteroides in a patient with acute lymphoblastic leukemia: Case report and review of the literature." *Case Reports in Hematology*. http://doi.org/10.1155/2016/7648628

Iskandar, C. F., F. Borges, B. Taminiau, G. Daube, M. Zagorec, B. Remenant, J. J. Leisner, M. A. Hansen, S. J. Sorensen, C. Mangavel, C. Cailliez-Grimal and A. M. Revol-Junelles (2017). "Comparative genomic analysis reveals ecological differentiation in the genus carnobacterium." *Frontiers in Microbiology* **8**: 357. http://doi.org/10.3389/fmicb.2017.00357

ISO/IDF "ISO 10932:2010. Milk and milk products—Determination of the minimal inhibitory concentration (MIC) of antibiotics applicable to bifidobacteria and non-enterococcal lactic acid bacteria (LAB)."

Ji, Y., H. Kim, H. Park, J. Lee, H. Lee, H. Shin, B. Kim, C. Franz and W. H. Holzapfel (2013). "Functionality and safety of lactic bacterial strains from Korean kimchi." *Food Control* **31**(2): 467–473. http://doi.org/10.1016/j.foodcont.2012.10.034

Jin, G. Y., Y. Zhu and Y. Xu (2017). "Mystery behind Chinese liquor fermentation." *Trends in Food Science & Technology* **63**: 18–28. http://doi.org/10.1016/j.tifs.2017.02.016

Karaaslan, A., A. Soysal, E. K. Kadayifci, N. Yakut, S. O. Demir, G. Akkoc, S. Atici, A. Sarmis, N. U. Toprak and M. Bakir (2016). "Lactococcus lactis spp lactis infection in infants with chronic diarrhea: Two cases report and literature review in children." *Journal of Infection in Developing Countries* **10**(3): 304–307. http://doi.org/10.3855/jidc.7049

Klare, I., C. Konstabel, S. Muller-Bertling, R. Reissbrodtl, G. Huys, M. Vancanneyt, J. Swings, H. Goossens and W. Witte (2005). "Evaluation of new broth media for microdilution antibiotic susceptibility testing of Lactobacilli, Pediococci, Lactococci, and Bifidobacteria." *Applied and Environmental Microbiology* **71**(12): 8982–8986. http://doi.org/10.1128/aem.71.12.8982-8986.2005

Klare, I., C. Konstabel, G. Werner, G. Huys, V. Vankerckhoven, G. Kahlmeter, B. Hildebrandt, S. Muller-Bertling, W. Witte and H. Goossens (2007). "Antimicrobial susceptibilities of Lactobacillus, Pediococcus and Lactococcus human isolates and cultures intended for probiotic or nutritional use." *Journal of Antimicrobial Chemotherapy* **59**(5): 900–912. http://doi.org/10.1093/jac/dkm035

Ladero, V., M. Fernandez, M. Calles-Enriquez, E. Sanchez-Llana, E. Canedo, M. C. Martin and M. A. Alvarez (2012). "Is the production of the biogenic amines tyramine and putrescine a species-level trait in enterococci?" *Food Microbiology* **30**(1): 132–138. http://doi.org/10.1016/j.fm.2011.12.016

Landete, J. M., B. de las Rivas, A. Marcobal and R. Munoz (2007). "Molecular methods for the detection of biogenic amine-producing bacteria on foods." *International Journal of Food Microbiology* **117**(3): 258–269. http://doi.org/10.1016/j.ijfoodmicro.2007.05.001

The Lancet Planetary, H. (2018). "The natural environment and emergence of antibiotic resistance." *The Lancet Planetary Health* **2**(1): e1. http://doi.org/10.1016/S2542-5196(17)30182-1

Lang, J. M., J. A. Eisen and A. M. Zivkovic (2014). "The microbes we eat: Abundance and taxonomy of microbes consumed in a day's worth of meals for three diet types." *Peerj* **2**. http://doi.org/10.7717/peerj.659

Laulund, S., A. Wind, P. Derkx and V. Zuliani (2017). "Regulatory and safety requirements for food cultures." *Microorganisms* **5**(2): 28.

Leite, A. M. O., M. A. L. Miguel, R. S. Peixoto, P. Ruas-Madiedo, V. M. F. Paschoalin, B. Mayo and S. Delgado (2015). "Probiotic potential of selected lactic acid bacteria strains isolated from Brazilian kefir grains." *Journal of Dairy Science* **98**(6): 3622–3632. http://doi.org/10.3168/jds.2014-9265

Liu, L. B., P. Du, G. F. Zhang, X. Mao, Y. C. Zhao, J. Y. Wang, C. Y. Duan, C. Li and X. D. Li (2017). "Residual nitrite and biogenic amines of traditional northeast sauerkraut in China." *International Journal of Food Properties* **20**(11): 2448–2455. http://doi.org/10.1080/10942912.2016.1239632

Lorencova, E., L. Bunkova, D. Matoulkova, V. Drab, P. Pleva, V. Kuban and F. Bunka (2012). "Production of biogenic amines by lactic acid bacteria and bifidobacteria isolated from dairy products and beer." *International Journal of Food Science and Technology* **47**(10): 2086–2091. http://doi.org/10.1111/j.1365-2621.2012.03074.x

Lukasik, J., S. Salminen and H. Szajewska (2018). "Rapid review shows that probiotics and fermented infant formulas do not cause d-lactic acidosis in healthy children." *Acta Paediatr* **107**(8): 1322–1326. http://doi.org/10.1111/apa.14338

Marco, M. L., C. Hill, R. Hutkins, J. Slavin, D. J. Tancredi, D. Merenstein and M. E. Sanders (2020). "Should there be a recommended daily intake of microbes?" *Journal of Nutrition.* http://doi.org/10.1093/jn/nxaa323

Mayrhofer, S., U. Zitz, F. H. Birru, D. Gollan, A. K. Golos, W. Kneifel and K. J. Domig (2014). "Comparison of the CLSI guideline and ISO/IDF standard for antimicrobial susceptibility testing of lactobacilli." *Microbial Drug Resistance* **20**(6): 591–603. http://doi.org/10.1089/mdr.2013.0189

McGovern, P. E., J. Zhang, J. Tang, Z. Zhang, G. R. Hall, R. A. Moreau, A. Nunez, E. D. Butrym, M. P. Richards, C. S. Wang, G. Cheng, Z. Zhao and C. Wang (2004). "Fermented beverages of pre- and proto-historic China." *Proceedings of the National Academy of Sciences of the United States of America* **101**(51): 17593–17598. http://doi.org/10.1073/pnas.0407921102

Morovic, W., J. M. Roper, A. B. Smith, P. Mukerji, B. Stahl, J. C. Rae and A. C. Ouwehand (2017). "Safety evaluation of HOWARU (R) Restore (Lactobacillus acidophilus NCFM, Lactobacillus paracasei Lpc-37, Bifidobacterium animalis subsp lactis Bl-04 and B. lactis Bi-07) for antibiotic resistance, genomic risk factors, and acute toxicity." *Food and Chemical Toxicology* **110**: 316–324. http://doi.org/10.1016/j.fct.2017.10.037

Naser, S. M., P. Dawyndt, B. Hoste, D. Gevers, K. Vandemeulebroecke, I. Cleenwerck, M. Vancanneyt and J. Swings (2007). "Identification of lactobacilli by pheS and rpoA gene sequence analyses." *International Journal of Systematic and Evolutionary Microbiology* **57**(12): 2777–2789.

Neut, C., S. Mahieux and L. J. Dubreuil (2017). "Antibiotic susceptibility of probiotic strains: Is it reasonable to combine probiotics with antibiotics?" *Médecine et Maladies Infectieuses* **47**(7): 477–483. https://doi.org/10.1016/j.medmal.2017.07.001

Park, J. M., C. Y. Yang, H. Park and J. M. Kim (2014). "Development of a genus-specific PCR combined with ARDRA for the identification of Leuconostoc species in kimchi." *Food Science and Biotechnology* **23**(2): 511–516. http://doi.org/10.1007/s10068-014-0070-z

Parte, A. C. (2018). "LPSN—list of prokaryotic names with standing in nomenclature (bacterio.net), 20 years on." *International Journal of Systematic and Evolutionary Microbiology* **68**(6): 1825–1829. http://doi.org/10.1099/ijsem.0.002786

Perin, L. M., S. Belviso, B. dal Bello, L. A. Nero and L. Cocolin (2017). "Technological properties and biogenic amines production by bacteriocinogenic lactococci and enterococci strains isolated from raw goat's milk." *Journal of Food Protection* **80**(1): 151–157. http://doi.org/10.4315/0362-028x.jfp-16-267

Perry, J. A. and G. D. Wright (2014). "Forces shaping the antibiotic resistome." *Bioessays* **36**(12): 1179–1184. http://doi.org/10.1002/bies.201400128

Pessione, E. and S. Cirrincione (2016). "Bioactive molecules released in food by lactic acid bacteria: Encrypted peptides and biogenic amines." *Frontiers in Microbiology* **7**. http://doi.org/10.3389/fmicb.2016.00876

Polka, J., L. Morelli and V. Patrone (2016). "Microbiological cutoff values: A critical issue in phenotypic antibiotic resistance assessment of lactobacilli and bifidobacteria." *Microbial Drug Resistance* **22**(8): 696–699. http://doi.org/10.1089/mdr.2015.0328

Pot, B. (2014). The genus Lactobacillus. In *Lactic Acid Bacteria: Biodiversity and Taxonomy*. W. H. H. A. B. J. B. Wood. John Wiley & Sons, Ltd.

Poveda, J. M., P. Ruiz, S. Sesena and M. L. Palop (2017). "Occurrence of biogenic amine-forming lactic acid bacteria during a craft brewing process." *Lwt-Food Science and Technology* **85**: 129–136. http://doi.org/10.1016/j.lwt.2017.07.003

Quigley, L., O. O'Sullivan, C. Stanton, T. P. Beresford, R. P. Ross, G. F. Fitzgerald and P. D. Cotter (2013). "The complex microbiota of raw milk." *Fems Microbiology Reviews* **37**(5): 664–698. http://doi.org/10.1111/1574-6976.12030

Reguera-Brito, M., F. Galan-Sanchez, M. M. Blanco, M. Rodriguez-Iglesias, L. Dominguez, J. F. Fernandez-Garayzabal and A. Gibello (2016). "Genetic analysis of human clinical isolates of Lactococcus garvieae: Relatedness with isolates from foods." *Infection Genetics and Evolution* **37**: 185–191. http://doi.org/10.1016/j.meegid.2015.11.017

Ricci, A., A. Allende, D. Bolton, M. Chemaly, R. Davies, R. Girones, L. Herman, K. Koutsoumanis, R. Lindqvist, B. Norrung, L. Robertson, G. Ru, M. Sanaa, M. Simmons, P. Skandamis, E. Snary, N. Speybroeck, B. Ter Kuile, J. Threlfall, H. Wahlstrm, P. S. Cocconcelli, G. K. Deceased, M. P. Maradona, A. Querol, L. Peixe, J. E. Suarez, I. Sundh, J. M. Vlak, M. Aguilera-Gmez, F. Barizzone, R. Brozzi, S. Correia, L. Heng, F. D. Istace, C. Lythgo and P. S. F. Escjmez (2017a). "Scientific opinion on the update of the list of QPS-recommended biological agents intentionally added to food or feed as notified to EFSA." *EFSA Journal* **15**(3). http://doi.org/10.2903/j.efsa.2017.4664

Ricci, A., A. Allende, D. Bolton, M. Chemaly, R. Davies, R. Girones, K. Koutsoumanis, L. Herman, R. Lindqvist, B. Norrung, L. Robertson, G. Ru, M. Sanaa, M. Simmons, P. Skandamis, E. Snary, N. Speybroeck, B. Ter Kuile, J. Threlfall, H. Wahlstrom, P. S. Cocconcelli, G. K. Deceased, L. Peixe, M. P. Maradona, A. Querol, J. E. Suarez, I. Sundh, J. Vlak, S. Correia and P. S. F. Escamez (2017b). "Update of the list of QPS-recommended biological agents intentionally added to food or feed as notified to EFSA 5: Suitability of taxonomic units notified to EFSA until September 2016." *EFSA Journal* **15**(3). http://doi.org/10.2903/j.efsa.2017.4663

Ricci, A., A. Allende, D. Bolton, M. Chemaly, R. Davies, R. Girones, K. Koutsoumanis, R. Lindqvist, B. Norrung, L. Robertson, G. Ru, P. S. F. Escamez, M. Sanaa, M. Simmons, P. Skandamis, E. Snary, N. Speybroeck, B. Ter Kuile, J. Threlfall, H. Wahlstrom, P. S. Cocconcelli, L. Peixe, M. P. Maradona, A. Querol, J. E. Suarez, I. Sundh, J. Vlak, F. Barizzone, S. Correia, L. Herman and E. P. B. Hazards (2018). "Update of the list of QPS-recommended biological agents intentionally added to food or feed as notified to EFSA 7: Suitability of taxonomic units notified to EFSA until September 2017." *EFSA Journal* **16**(1). http://doi.org/10.2903/j.efsa.2018.5131

Ricci, A., A. Allende, D. Bolton, M. Chemaly, R. Davies, R. Girones, K. Koutsoumanis, R. Lindqvist, B. Norrung, L. Robertson, G. Ru, P. S. F. Escamez, M. Sanaa, M. Simmons, P. Skandamis, E. Snary, N. Speybroeck, B. Ter Kuile, J. Threlfall, H. Wahlstrom, P. S. Cocconcelli, L. Peixe, M. P. Maradona, A. Querol, J. E. Suarez, I. Sundh, J. Vlak, S. Correia, L. Herman and E. P. B. Hazards (2017c). "Update of the list of QPS-recommended biological agents intentionally added to food or feed as notified to EFSA 6: Suitability of taxonomic units notified to EFSA until March 2017." *EFSA Journal* **15**(7). http://doi.org/10.2903/j.efsa.2017.4884

Richter, M. and R. Rosselló-Móra (2009). "Shifting the genomic gold standard for the prokaryotic species definition." *Proceedings of the National Academy of Sciences* **106**(45): 19126–19131. http://doi.org/10.1073/pnas.0906412106

Ruppe, E., A. Ghozlane, J. Tap, N. Pons, A.-S. Alvarez, N. Maziers, T. Cuesta, S. Hernando-Amado, J. L. Martinez, T. Coque, F. Baquero, V. F. Lanza, L. Maiz, T. Goulenok, V. de Lastours, N. Amor, B. Fantin, I. Wieder, A. Andremont, W. van Schaik, M. Rogers, X. Zhang, R. J. L. Willems, A. G. De Brevern, J.-M. Batto, H. Blottiere, P. Leonard, V. Lejard, A. Letur, F. Levenez, K. Weiszer, F. Haimet, J. Dore, S. P. Kennedy and S. D. Ehrlich (2017). "Prediction of the intestinal resistome by a novel 3D-based method." *bioRxiv*. http://doi.org/10.1101/196014

Salomskiene, J., A. Abraitiene, D. Jonkuviene, I. Macioniene and J. Repeckiene (2015). "Selection of enhanced anti-microbial activity posing lactic acid bacteria characterised by (GTG)(5)-PCR fingerprinting." *Journal of Food Science and Technology-Mysore* **52**(7): 4124–4134. http://doi.org/10.1007/s13197-014-1512-6

Salque, M., P. I. Bogucki, J. Pyzel, I. Sobkowiak-Tabaka, R. Grygiel, M. Szmyt and R. P. Evershed (2013). "Earliest evidence for cheese making in the sixth millennium BC in northern Europe." *Nature* **493**(7433): 522–525. http://doi.org/10.1038/nature11698

Salvetti, E., H. M. B. Harris, G. E. Felis and P. W. O'Toole (2018). "Comparative genomics reveals robust phylogroups in the genus Lactobacillus as the basis for reclassification." *Applied and Environmental Microbiology*. http://doi.org/10.1128/aem.00993-18

Salvetti, E. and P. W. O'Toole (2017). "When regulation challenges innovation: The case of the genus Lactobacillus." *Trends in Food Science & Technology* **66**: 187–194. http://doi.org/10.1016/j.tifs.2017.05.009

Sanlier, N., B. B. Gokcen and A. C. Sezgin (2017). "Health benefits of fermented foods." *Critical Reviews in Food Science and Nutrition*: 1–22. http://doi.org/10.1080/10408398.2017.1383355

Sarkadi, L. S. (2017). Biogenic amines in fermented foods and health implications. In *Fermented Foods in Health and Disease*. Academic Press, Cambridge, MA.

Smokvina, T., M. Wels, J. Polka, C. Chervaux, S. Brisse, J. Boekhorst, J. E. van Hylckama Vlieg and R. J. Siezen (2013). "Lactobacillus paracasei comparative genomics: Towards species pan-genome definition and exploitation of diversity." *PLoS ONE* **8**(7): e68731. http://doi.org/10.1371/journal.pone.0068731

Sohier, D., J. Coulon and A. Lonvaud-Funel (1999). "Molecular identification of Lactobacillus hilgardii and genetic relatedness with Lactobacillus brevis." *International Journal of Systematic Bacteriology* **49**(Pt 3): 1075–1081. http://doi.org/10.1099/00207713-49-3-1075

Takebe, Y., M. Takizaki, H. Tanaka, H. Ohta, T. Niidome and S. Morimura (2016). "Evaluation of the biogenic amine-production ability of lactic acid bacteria isolated from tofu-misozuke." *Food Science and Technology Research* **22**(5): 673–678. http://doi.org/10.3136/fstr.22.673

Tapiovaara, L., L. Lehtoranta, T. Poussa, H. Makivuokko, R. Korpela and A. Pitkaranta (2016). "Absence of adverse events in healthy individuals using probiotics—analysis of six randomised studies by one study group." *Benef Microbes* **7**(2): 161–169. http://doi.org/10.3920/bm2015.0096

Teuber, M., L. Meile and F. Schwarz (1999). "Acquired antibiotic resistance in lactic acid bacteria from food." *Antonie v Leeuwenhoek* **76**(1–4): 115–137.

Torriani, S., G. E. Felis and F. Dellaglio (2001). "Differentiation of *Lactobacillus plantarum*, L. *pentosus*, and L. *paraplantarum* by recA gene sequence analysis and multiplex PCR assay with recA gene-derived primers." *Applied and Environmental Microbiology* **67**(8): 3450–3454.

Xie, M., A. Shevchenko, B. Wang, A. Shevchenko, C. Wang and Y. Yang (2016). "Identification of a dairy product in the grass woven basket from Gumugou Cemetery (3800 BP, northwestern China)." *Quaternary International* **426**: 158–165. https://doi.org/10.1016/j.quaint.2016.04.015

Yang, Y., A. Shevchenko, A. Knaust, I. Abuduresule, W. Li, X. Hu, C. Wang and A. Shevchenko (2014). "Proteomics evidence for kefir dairy in Early Bronze Age China." *Journal of Archaeological Science* **45**: 178–186. https://doi.org/10.1016/j.jas.2014.02.005

Zheng, J., S. Wittouck, E. Salvetti, C. Franz, H. M. B. Harris, P. Mattarelli, P. W. O'Toole, B. Pot, P. Vandamme, J. Walter, K. Watanabe, S. Wuyts, G. E. Felis, M. G. Gänzle and S. Lebeer (2020). "A taxonomic note on the genus Lactobacillus: Description of 23 novel genera, emended description of the genus Lactobacillus Beijerinck 1901, and union of Lactobacillaceae and Leuconostocaceae." *International Journal of Systematic and Evolutionary Microbiology* **70**(4): 2782–2858. http://doi.org/10.1099/ijsem.0.004107

# Probiotic Regulations in Asian and Australasian Countries

# 38

Yuan-Kun Lee, Jin Xu, Xiaomin Han, Jasvir Singh,
Ingrid Suryanti Surono, Hiroko Tanaka,
Myung Soo Park, Geun Eog Ji, E-Siong Tee,
Julie D. Tan, Francisco Elegado, Yi-Ling Tan,
Ming-Ju Chen, Malee Jirawongsy,
Chalat Santivarangkna, Le Hoang Vinh,
and Caroline Gray

The guidelines for probiotic products in Asia and Australasia are not harmonized; each country has unique product regulations. Some are restrictive, whereas others are elaborate and specific.

## 38.1 CHINA

Probiotics are not new to most Chinese, as they exist in traditional fermented vegetables, soy products, and dietary supplements. China introduced probiotics in fermented dairy products only in the 21st century. The concept of health benefits of probiotics, especially for gut health, was stressed when manufacturers promoted their yogurt (fermented milk) products. Probiotic yogurt has progressed from concept to efficacy by increasing the initial cell count and improving the storage conditions to make sure consumers can receive enough live healthy bacteria to convey the claimed health benefits.

### 38.1.1 Regulations and Standards Related to Probiotics

The production, trading, and distribution of food, food additives, food packaging materials, and other food-related products are strictly regulated in China by specific laws, regulations, and standards.

DOI: 10.1201/9781003352075-42

## 38.1.2 Food Safety Law of the People's Republic of China

To ensure food safety and protect the physical health and safety of the public, the Standing Committee of the National People's Congress passed the Food Safety Law (FSL) at the Seventh Session of the Standing Committee of the Eleventh National People's Congress in 2009 and amended it at the Fourteenth Session of the Standing Committee of the Twelfth National People's Congress in 2015. Subsequently, it was modified for the first time at the Seventh Session of the Standing Committee of the Thirteenth National People's Congress in 2018 and for the second time at the Twentieth Session of the Standing Committee of the Thirteenth National People's Congress in 2021. It is stipulated in this law that people who are engaged in activities such as food production and processing, food sales, and catering services in the territory of the People's Republic of China should abide by this law. The Food Safety Commission of the State Council (FSCSC) was established as a high-level deliberative and coordinating body for food safety of the State Council in 2010. The key responsibilities of the FSCSC were to analyze the food safety situation, research and implement, overall coordinate food safety, put forward important policies and measures for food safety, and urge the implementation of food safety supervision responsibility. During the thirteenth National People's Congress in 2018, the State Administration for Marketing Regulation (SAMR) took over the concrete work of FSCSC. SAMR is responsible for the supervision and administration of food production and operation activities according to FSL and the duties of the State Council. The health administrative department of the State Council (National Health Commission of the People's Republic of China, NHC) shall, in accordance with this law and its functions prescribed by the State

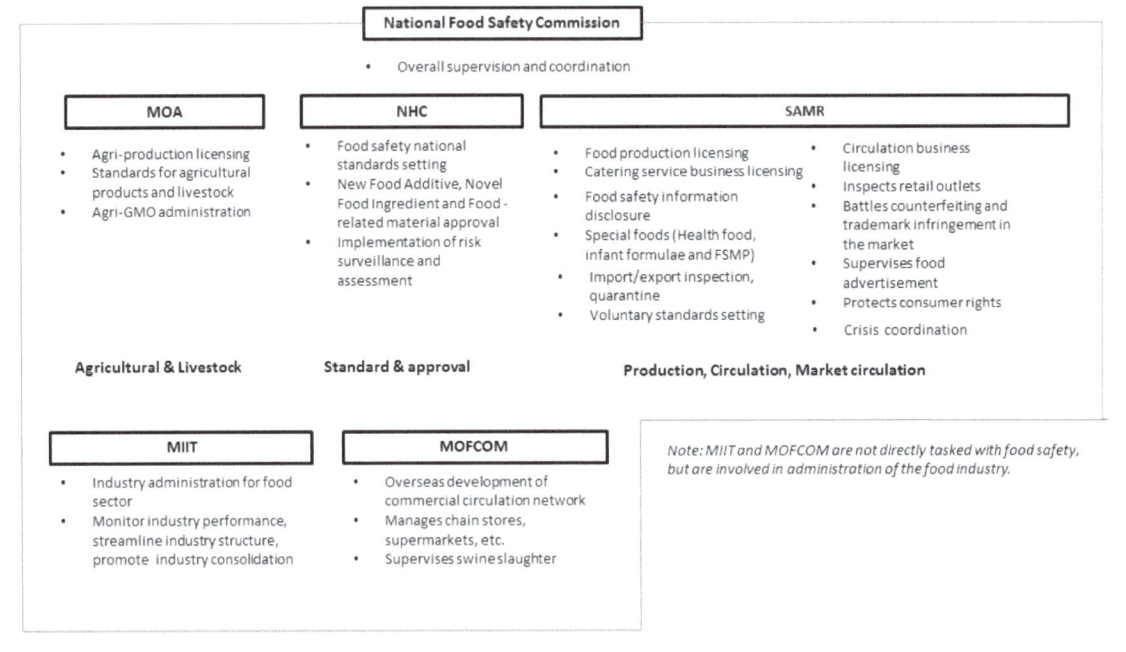

MARA: Ministry of Agriculture and Rural Affairs the People's Republic of China
NHC: National Health Commission of the People's Republic of China
SAMR: State Administration for Marketing Regulation
MIIT: Ministry of Industry and Information Technology of the People's Republic of China
MOFCOM: Ministry of Commerce of the People's Republic of China

**FIGURE 38.1** New setup of government authorities involved in food safety supervision under the new FSL. MARA, Ministry of Agriculture and Rural Affairs the People's Republic of China; NHC, National Health Commission of the People's Republic of China; SAMR, State Administration for Marketing Regulation; MIIT, Ministry of Industry and Information Technology of the People's Republic of China; MOFCOM, Ministry of Commerce of the People's Republic of China.

Council, organize food safety risk monitoring and assessment and develop and publish national food safety standards in conjunction with the food safety supervision and administration department of the State Council.

However, with the new reforms and government agency reshuffles implemented during the National People's Congress in March 2018, the government authorities involved in food safety supervision have been reorganized, as outlined in Figure 38.1.

## 38.1.3 Regulations and Laws for Health Food

Following the provisions of the 2021 revised edition of FSL, the state shall conduct strict supervision and administration of special foods such as health food, formula food for special medical purposes, and formula food for infants. For example, health food that claims to have any health protection functions shall have a scientific basis, and it should not cause any acute, sub-acute, or chronic damage to the human body. The list of raw ingredients for health food and the list of health protection functions which health food may claim shall be developed, adjusted, and published by the food safety supervision and administration department of the State Council in conjunction with the health administrative department of the State Council and the State Administration of Traditional Chinese Medicine. The China Food and Drug Administration (CFDA) issued the Measures for the Registration and Record Filing of Health Food in 2016. Then, food and drug supervision merged into the newly established State Administration for Markert Regulation in 2018, and SAMR revised the measures in 2020. At the same time, SAMR issued the List of Raw Materials for Health Food and the Management Approach for the List of Health Function (2019 edition) and Health Function Interpretation (2022 edition).

According to GB 16740–2014 National Food Safety Standard–Health Food, health food is a food that claims and has specific health functions or aims to supply vitamins and minerals. That is suitable for certain people, able to regulate the body health status, but not for treating a disease or producing any acute, sub-acute, or chronic damage to the human body.

### 38.1.3.1 Approved List of Probiotics for Use in Health Food

The term "probiotic" was legally presented in the Notice on the Evaluation of Health Food with Fungus and Probiotics, issued by the Ministry of Health (MOH) in 2001 (Notice No. 84 [2001]).

Since 2003, the State Food and Drug Administration (SFDA) has taken over the management responsibilities of health food from the MOH. The SFDA released the new regulation "Provisions on the Application and Review of Probiotics Health Food (Trial)", and *Lactobacillus reuteri* was added to the list, but the definition of probiotics was still not given in detail. The Chinese government established the CFDA in 2013 and maintains this list (see Table 38.1).

**TABLE 38.1**   The Approved List of Species of Fungi and Probiotics Used in Health Food

| SPECIES OF FUNGUS USED IN HEALTH FOOD | SPECIES OF PROBIOTICS USED IN HEALTH FOOD |
|---|---|
| *Saccharomyces cerevisiae* | *Bifidobacterium bifidum* |
| *Candida atilis* | *B. infantis* |
| *Kluyveromyces lactis* | *B. longum* |
| *Saccharomyces carlsbergensis* | *B. breve* |
| *Paecilomyces hepiali Chen et Dai,* sp.nov | *B. adolescentis* |
| *Hirsutella hepiali Chen et Shen* | *Lactobacillus. bulgaricus* |
| *Ganoderma lucidum* | *L. acidophilus* |
| *Ganoderma sinensis* | *L. casei* subsp. *caasei* |
| *Ganoderma tsugae* | *Streptococcus thermophilus* |
| *Monacus anka* | |
| *Monacus purpureus* | |

The Chinese government established the SAMR in 2018, and the SAMR was responsible for the probiotic strain approval and approval of health food containing probiotics. In 2019, the SAMR organized the revision of "Guidance for the Registration and Evaluation of Probiotic Health Food" and added and revised the following contents: First, it supplemented the definition of probiotics and amended the definition of probiotic health food simultaneously. Second, the management of probiotic raw materials for health food is raised from the "strain" level to the "culture" level due to the safety and efficacy of different strains being strain specific. Third, the numbers of viable bacteria increased from no less than 106 CFU/mL (g) to no less than 107 CFU/mL (g) to ensure their effects during the shelf life. Fourth, the agency qualification of strain identification was clarified. The government has not approved the recognized probiotics due to the uncertainty of strain-level identification technology and strain-level health functions based on whole-genome sequencing.

### 38.1.3.2 Health Claims for Probiotic Health Food

The SAMR published the rules 'List of Health Functions Claimed by Permitted Health Food Non-Nutrient Supplements' in 2022 (Table 38.2). Most probiotic health foods are approved to have claims to "enhance immunity" and/or "regulate gastrointestinal tract flora" among the 24 permitted health claims.

**TABLE 38.2**   Permitted Health Claims of Health Food in China

| NO. | FUNCTIONS |
|---|---|
| 1 | Enhancing Immune Function |
| 2 | Antioxidative Function |
| 3 | Assisting Memory Improvement Function |
| 4 | Alleviating Eye Fatigue Function |
| 5 | Clear the Throat Function |
| 6 | Sleep Improvement Function |
| 7 | Alleviating Physical Fatigue Function |
| 8 | Enhancing Anoxia Endurance Function |
| 9 | Assisting Control Body Fat |
| 10 | Increasing Bone Density Function |
| 11 | Improving Iron Deficiency Anemia |
| 12 | Assisting Improvement of Acne |
| 13 | Assisting Improvement of Chloasma |
| 14 | Improving Skin Water Content Function |
| 15 | Regulating Gastrointestinal Tract Flora Function |
| 16 | Facilitating Digestion Function |
| 17 | Facilitating Feces Excretion Function |
| 18 | Assisting the Protection of Gastric Mucosa Function |
| 19 | Assisting Maintenance Blood Lipid (Cholesterol/Triglyceride) in Healthy Level |
| 20 | Assisting Maintenance Blood Sugar in Healthy Level |
| 21 | Assisting Maintenance Blood Pressure in Healthy Level |
| 22 | Assisting Protection Against Chemical Injury of Liver Function |
| 23 | Assisting Ionizing Irradiation Hazard Protection Function |
| 24 | Assisting Lead Removal |

## 38.1.4 Laws and Regulations Related to New Food Raw Ingredients

The Food Safety Law of the People's Republic of China specifies that the relevant people in charge should submit safety evaluation materials of products to the health administrative department of the

State Council if people want to use new raw ingredients to produce food. The health administration department of the State Council needs to organize inspection within 60 days from the date of receiving the application. Those materials meeting the food safety requirements will be permitted and published.

NHC is responsible for the management of new food raw ingredients, including the acceptance, administrative review, and licensing for the safety assessment materials. China's National Center for Food Safety Risk Assessment (CFSA) undertook the safety review of new food raw ingredients, such as organizing technical reviews, soliciting social and industrial opinions, and social risk assessment, putting forward comprehensive review conclusions and suggestions. Novel food raw ingredients refer to the following food without traditional eating habits in China (1) animals, plants, and microorganisms; (2) ingredients derived from animals, plants, and microorganisms; (3) food ingredients with original structure changed; (4) other newly researched food raw ingredients.

The departmental rules from the NHC mainly refer to "Administrative Measures for Health Administrative Licensing (2004)" and "Administrative Measures for the Safety Review of New Food Raw Ingredients". The normative documents mainly refer to "Provisions on Application and Acceptance of New Food Raw Ingredients (2013)", "Procedures for Safety Review of New Food Raw Ingredients (2013)", and "Requirements for Application Materials for Safety Assessment Opinions of New Food Raw Ingredients (2013)".

These regulations and rules reflect the transition from the approval of final products to the approval of food raw ingredients, and it is better to avoid repeated approval of similar products and save administrative costs and social resources. It is also necessary to follow safety evaluation principles to ensure safety before market.

## 38.1.5 List of Cultures Used for Food (2022)

The former MOH (the present NHC) issued a "List of Cultures Used for Food" containing 16 cultures in 2010. The MOH implemented the "New Food Raw Material Safety Review Management" in 2013 and approved and announced 16 cultures used in food in succession. To date, the list includes 37 cultures (Table 38.3). The NHC adjusted the classification and nomenclature of the microorganisms in the list based on the latest principles of international microorganism classification and nomenclature in 2022 and set a two-year transition period.

**TABLE 38.3**   List of Cultures Used for Food

| | *UPDATED LIST OF STRAINS* | *ORIGINAL LIST OF STRAINS* |
|---|---|---|
| (1) | *Bifidobacterium* | *Bifidobacterium* |
| 1 | *Bifidobacterium adolescentis* | *Bifidobacterium adolescentis* |
| 2 | *Bifidobacterium animalis* subsp. *animalis* | *Bifidobacterium animals (Bifidobacterium lactis)* |
| 3 | *Bifidobacterium animalis* subsp. *lactis* | |
| 4 | *Bifidobacterium bifidum* | *Bifidobacterium bifidum* |
| 5 | *Bifidobacterium breve* | *Bifidobacterium breve* |
| 6 | *Bifidobacterium longum* subsp. *longum* | *Bifidobacterium longum* |
| 7 | *B. longum* subsp. *infantis* | *Bifidobacterium infantis* |
| (2) | *Lactobacillus* | *Lactobacillus* |
| 1 | *Lactobacillus acidophilus* | *Lactobacillus acidophilus* |
| 2 | *Lactobacillus crispatus* | *Lactobacillus crispatus* |
| 3 | *Lactobacillus brueckii* subsp. *bulgaricus* | *Lactobacillus brueckii* subsp. *bulgaricus* |
| 4 | *Lactobacillus delbrueckii* subsp. *lactis* | *Lactobacillus delbrueckii* subsp. *lactis* |
| 5 | *Lactobacillus gasseri* | *Lactobacillus gasseri* |
| 6 | *Lactobacillus helveticus* | *Lactobacillus helveticus* |

*(Continued)*

**TABLE 38.3**   (Continued) List of Cultures Used for Food

| UPDATED LIST OF STRAINS | | ORIGINAL LIST OF STRAINS |
|---|---|---|
| 7 | Lactobacillus johnsonii | Lactobacillus johnsonii |
| 8 | Lactobacillus kefiranofaciens subsp. kefiranofaciens | Lactobacillus kefiranofaciens subsp. kefiranofaciens |
| (3) | Lacticaseibacillus | Lactobacillus |
| 1 | Lacticaseibacillus casei | Lactobacillus casei |
| 2 | Lacticaseibacillus paracasei | Lactobacillus paracasei |
| 3 | Lacticaseibacillus rhamnosus | Lactobacillus rhamnosus |
| (4) | Limosilactobacillus | Lactobacillus |
| 1 | Limosilactobacillus fermentum | Lactobacillus fermentum |
| 2 | Limosilactobacillus reuteri | Lactobacillus reuteri |
| (5) | Lactiplantibacillus | Lactobacillus |
| 1 | Lactiplantibacillus plantarum | Lactobacillus plantarum |
| (6) | Ligilactobacillus | Lactobacillus |
| 1 | Ligilactobacillus salivarius | Lactobacillus salivarius |
| (7) | Latilactobacillus | Lactobacillus |
| 1 | Latilactobacillus curvatus | Lactobacillus curvatus |
| 2 | Latilactobacillus sakei | Lactobacillus sakei |
| (8) | Streptococcus | Streptococcus |
| 1 | Streptococcus salivarius subsp. thermophilus | Streptococcus thermophilus |
| (9) | Lactococcus | Lactococcus |
| 1 | Lactococcus lactis subsp. lactis | Lactococcus lactis subsp. lactis |
| 2 | Lactococcus lactis subsp. Lactis biovar diacetylactis | Lactococcus lactis subsp. diacetylactis |
| 3 | Lactococcus cremoris | Lactococcus lactis subsp. cremoris |
| (10) | Propionibacterium | Propionibacterium |
| 1 | Propionibacterium freudenreichii subsp. shermanii | Propionibacterium freudenreichii subsp. shermanii |
| (11) | Acidipropionibacterium | Propionibacterium |
| 1 | Acidipropionibacterium acidipropionici | Propionibacterium acidipropionici |
| (12) | Leuconostoc | Leuconostoc |
| 1 | Leuconostoc mesenteroides subsp. mesenteroides | Leuconostoc mesenteroides subsp. mesenteroides |
| (13) | Pediococcus | Pediococcus |
| 1 | Pediococcus acidilactici | Pediococcus acidilactici |
| 2 | Pediococcus pentosaceus | Pediococcus pentosaceus |
| (14) | Weizmannia | Bacillus |
| 1 | Weizmannia coagulans | Bacillus coagulans |
| (15) | Mammaliicoccus | Staphylococcus |
| 1 | Mammaliicoccus vitulinus | Staphylococcus vitulinus |
| (16) | Staphylococcus | Staphylococcus |
| 1 | Staphylococcus xylosus | Staphylococcus xylosus |
| 2 | Staphylococcus carnosus | Staphylococcus carnosus |
| (17) | Kluyveromyces | Kluyveromyces |
| 1 | Kluyveromyces marxianus | Kluyveromyces marxianus |

Note 1. Cultures traditionally used in food processing are still permitted for use; novel strains are regulated under the Novel Food Regulation.

Note 2. The use of the cultures added to the list after 2010 needed to comply with the original announcement.

# 38.1.6 List of Strains Permitted Use in Food for Infants and Young Children

The former MOH (the present NHC) issued a "List of Cultures Used for Infants and Young Children" containing six cultures in four strains in 2011 (Table 38.4). MOH implemented the "New Food Raw Material Safety Review Management" in 2013 and approved and announced eight cultures used in infants and young children in succession. NHC adjusted the classification and nomenclature of the microorganisms in the list based on the latest principles of international microorganism classification and nomenclature in 2022 and set a two-year transition period.

**TABLE 38.4**   List of Strains Permitted for Use in Food for Infants and Young Children

| UPDATED LIST OF STRAINS | ORIGINAL LIST OF STRAINS |
| --- | --- |
| *Lactobacillus acidophilus* NCFM* | *Lactobacillus acidophilus* NCFM* |
| Bb-12 *Bifidobacterium animalis* subsp. *lactis* Bb-12 | *Bifidobacterium animalis* Bb-12 |
| *Bifidobacterium animalis* subsp. *lactis* HN019 | *Bifidobacterium lactis* HN019 |
| *Bifidobacterium animalis* subsp. *lactis* Bi-07 | *Bifidobacterium lactis* Bi-07 |
| *Lacticaseibacillus rhamnosus* GG | *Lactobacillus rhamnosus* LGG |
| *Lacticaseibacillus rhamnosus* HN001 | *Lactobacillus rhamnosus* HN001 |
| *Lacticaseibacillus rhamnosus* MP108 | *Lactobacillus rhamnosus* MP108 |
| *Limosilactobacillus reuteri* DSM 17938 | *Lactobacillus reuteri* DSM 17938 |
| *Limosilactobacillus fermentum* CECT 5716 | *Lactobacillus fermentum* CECT 5716 |
| *Bifidobacterium breve* M-16V | *Bifidobacterium breve* M-16V |
| *Lactobacillus helveticus* R0052 | *Lactobacillus helveticus* R0052 |
| *Bifidobacterium longum* subsp. *infantis* R0033 | *Bifidobacterium infantis* R0033 |
| *Bifidobacterium longum* subsp. *longum* BB536 | *Bifidobacterium longum* subsp. *longum* BB536 |
| *Bifidobacterium bifidum* R0071 | *Bifidobacterium bifidum* R0071 |

*Only limited to use in foods for young children (>1 year)

# 38.1.7 Approved Probiotics

From 1996 to 2022, the health administration departments under the State Council approved 168 health foods containing probiotics. There were 13, 15, 13, and 11 in 1997, 2006, 2009, and 2015, and only 2, 1, and 2 in 2001, 2017, and 2021. No products were approved in 2018 and 2022. For dosage form, 78% of products were in solid form (55 powder products and 42 capsule products), and 22% were in liquid form (20 beverages and 17 oral liquids). The two commonly used strains in health foods are *Lactobacillus* (82.14%) and *Bifidobacterium* (78.0%). *Lactobacillus acidophilus* (61.31%) was the most-used strain, followed by *Bifidobacterium longum* (25%) and *Streptococcus thermophilus* (17.86%). All products were not to the genus level (species level).

# 38.1.8 National Food Safety Standards

According to the new FSL, the national food safety standards are mandatory and exclusive. Besides the food safety standards, no other mandatory standards for food shall be developed. In April 2010, the MOH released 66 dairy standards, including 14 product standards, 2 GMP standards, and 50 analytical standards. These standards became the first food safety standards in China. In 2018, the SAMR was responsible for the approval of probiotics. Due to the controversy over the health function of probiotics, the SAMR did not give statutory permission for the concept of probiotics. Therefore, the SAMR and related departments required the enterprise to revise the relevant dairy products labelled with probiotics gradually since 2018, such as deleting the claim of "probiotics" and changing to "alive strains".

### 38.1.8.1 GB19302–2010 "Fermented Milk"

The standard was revised based on the previous standard GB19302–2003 "Hygienic Standard for Yogurt" and the previous product standard GB2746–1999 Yogurt and "Codex Standard for Fermented Milk". For starter cultures, the standard stipulated that the strains should be *Lactobacillus bulgaricus* (*Lactobacillus delbrueckii* subsp. *bulgaricus*), *Streptococcus thermophilus*, and other cultures permitted by the health administration departments under the State Council. Also, the requirement for the number of *Lactobacillus bulgaricus* in non–heat-treated fermented milk should be ≥1 × 106 CFU/gram (mL) in the standard. Based on the national food safety standard revision plan, this standard has been under revision since 2017, but the draft version is not available to the public.

### 38.1.8.2 GB10765–2021 Infant Formula, GB107672021—Older Infant and Young Children Formula, and GB10766–2021—Older Infant Formula

Both national food safety standard GB 10765–2021—Infant Formula and GB 10767–2021—Older Infants and Young Children Formula had claims of probiotics. Due to the legal interpretation of the Chinese regulatory authorities, the revised standards GB 10765–2021—Infant Formula and GB 10766–2021—Older Infant Formula deleted the claims of "probiotics" and stated only "alive strains". The standard stipulates that total colony count is not applicable to products containing live cultures (aerobic and facultative anaerobic probiotics). The number of alive cultures should be ≥ $1 \times 10^6$ CFU/g (ml).

### 38.1.8.3 GB 4789.34–2016 Microbiological Examination of Food Hygiene—Examination of Bifidobacterium and GB 4789.35–2016 Microbiological Examination of Food Hygiene—Examination of Lactic Acid Bacteria

The standard GB 4789.34–2016 Microbiological Examination of Food Hygiene—Examination of *Bifidobacterium* included six *Bifidobacterium* species, *B. bifidum, B. infantis, B. longum, B. adolescentis, B. animalis,* and *B. breve*. The standard GB 4789.35–2016 Microbiological Examination of Food Hygiene—Examination of Lactic Acid Bacteria stipulated the examination method of *Lactobacillus* in foods containing *Lactobacillus*. It is the official method for identifying and counting the number of live *Lactobacillus* in food products. Three typical media—MRS, modified MRS with Li–Mupirocin, and MC—are used for counting the total number of *Lactobacillus*, *Bifidobacterium*, and *S. thermophilus*, respectively.

### 38.1.8.4 National Food Safety Standards—Food Cultures Used in Food Processing, National Food Safety Standards—Good Manufacturing Practice for Food Cultures Used in Food Processing, and National Food Safety Standards—Safety Evaluation Procedures for Food Cultures Used in Food Production

NHC reviewed the three national food safety standards—Food Cultures Used in Food Processing, Good Manufacturing Practice for Food Cultures Used in Food Processing, and Safety Evaluation Procedures for Food Cultures Used in Food Production in 2022. National Food Safety Standard—Food Cultures Used in Food Processing is applicable to strain preparations used in food processing, including those used in food fermentation or as raw materials added to food, not including the products provided to consumers for direct consumption or traditional production process of wine yeast, red yeast, and so on (GB 7718). Food cultures should conform to the regulations, announcements, and relevant regulations issued by the health administrative department of the State Council. For example, sensory requirements, pollutant limits, microbial limits, and labels should conform to the provisions of GB 7718, marked with their Chinese and Latin name. Infant food strains need to be at the strain level.

National Food Safety Standard—Good Manufacturing Practice for Food Cultures Used in Food Processing is suitable to produce strain preparations used in food processing but not for products in direct consumption and distiller yeast and *Monascus* in the solid-state fermentation process. The standard stipulated the general requirements and management guidelines for places, facilities, and personnel in the raw material purchase, strain use and management, processing, packaging, storage, and transportation in the production of strain preparations used in food processing.

National Food Safety Standard—Safety Evaluation Procedures for Food Cultures Used in Food Production stipulated the strain safety evaluation procedures. They included bacteria, filamentous fungi, yeast, actinomycetes, and single-cell algae. The standard did not apply safety evaluation of genetically modified microorganisms. For the first time, it adopted whole-genome sequencing for strain identification in national standards.

# 38.2 INDIA

India has a strong tradition of consuming ethnic fermented foods as part of daily diets. Many traditional foods, such as idli, dosa, and lassi, among others, are a part of rich culinary heritage that is available almost globally these days. Unfortunately, these foods have not been widely studied using the techniques and principles of modern science. Hence their potential role as probiotics remains driven more by perception than by established peer-reviewed scientific publications. The introduction of modern probiotic products in the Indian market is generally ascribed to Yakult Danone (I) Pvt. Limited, which introduced the Yakult drink to the Indian market in 2007. Yakult is currently manufactured in a single facility in the northern part of India and to date remains a largely urban phenomenon.

It has been seen in many other parts of the world that regulations follow the market developments rather than the other way around, and India is no different in this case. India developed a formal regulatory framework to deal with nutraceutical foods only recently in 2018, which has again undergone significant revision recently.

## 38.2.1 Food Regulatory Framework

Historically, India's food sector was governed by the Prevention of Food Adulteration Act, 1956, for a very long time. It was only in 2006 that the Indian parliament enacted a new comprehensive law, the Food Safety & Standards Act, 2006 (FSSA). Through this act, Food Safety and Standards Authority of India (FSSAI) was established as an independent regulatory body and the highest authority dealing with regulatory framework for food sector. This new framework incorporated a lot of good practices from global frameworks into the FSSA. Unlike in the past, the focus was shifted to food safety and standards using scientific developments as the basis. Risk assessment was conceptualized to be the basis for all risk management measures to be taken by the regulatory authorities. Under this new act, various rules and regulations have been developed and notified. A detailed and updated version can be accessed from www.fssai.nic.in.

## 38.2.2 Nutraceuticals Regulatory Framework

FSSAI started working on creating a framework for regulating the functional foods sector, commonly called nutraceuticals, in 2013. As this is an extremely complex sector, with few globally harmonized approaches available to deal with it, it took a very long time for the authorities to develop a formal framework. Finally, FSSAI notified the Food Safety and Standards (Health Supplements, Nutraceuticals/FSSA-2013, Food for Special Dietary Use, Food for Special Medical Purpose, Functional Food and Novel Food) Regulation (No. 1–4/Nutraceutical/FSSAI-2013, dated December 23, 2016. These regulations came into effect on January 1, 2018.

Issuance and enforcement of this regulation led to realization of challenges associated with implementation of the regulations among all stakeholders. This led to an exercise to review the regulations, which culminated in the form of a revised regulation, Food Safety and Standards (Health Supplements, Nutraceuticals, Food for Special Dietary Use, Food for Special Medical Purpose, and Prebiotic and Probiotic Food) Regulations, 2022. This regulation was operationalized vide letter: F. No. Std/SP-05/T (Nutraceuticals-2022) (E-5184) dated March 29, 2022.

This regulation covers the following categories:

1. Health Supplements (HS)
2. Nutraceuticals (Nutra)
3. Food for Special Dietary Use (FSDU)
4. Food for Special Medical Purpose (FSMP)
5. Prebiotic food and Probiotic food (Pre-Pro)

## 38.2.3 Regulations for Foods Containing Probiotics

As mentioned, foods containing probiotics are regulated through Food Safety and Standards (Health Supplements, Nutraceuticals, Food for Special Dietary Use, Food for Special Medical Purpose, Functional Food and Novel Food) Regulations, 2022. It has also been mandated that all products covered under these regulations intended for children of 2 to 5 years of age shall only be given under medical advice by a recognized medical doctor, dietician, or nutritionist.

This regulation adopted the following definition, which is in line with globally accepted definitions:

"Probiotic Food" means food with live micro-organisms beneficial to human health, which when ingested in adequate numbers as a single strain or as a combination of cultures, confer one or more specified or demonstrated health benefits in human beings.

Further, the regulation has clarified the scope:

Probiotic Food: The foods with added viable microorganisms which when consumed in adequate amount confer health benefits. Provided that the presence of the commonly used starter cultures of lactic acid producing bacteria such as Lactococcus spp., earlier known as *Streptococcus* spp., *Lactobacillus* spp. and other such microorganisms used in the preparation of fermented milk (dahi) and related products shall not be considered as probiotics if the probiotic properties have not been substantiated.

The regulation also specifies that the minimum viable number of added probiotic organisms in probiotic food shall be ≥108 CFU in the recommended serving size per day, provided that a lower viable number may be allowed with proven studies on health benefits, with those numbers subject to the prior approval of the Food Authority.

The guidelines issued by the Indian Council of Medical Research and Department of Biotechnology with respect to probiotics provide additional information on their use and have been an important influence on this regulation. These guidelines act as a reference point when it comes to safety and efficacy evaluation of probiotic strains:

Any new approval of new probiotic strain shall be based on data collected in accordance with guidelines issued by the Indian Council of Medical Research and Department of Biotechnology, Government of India, with respect to probiotics and approval under Food Safety & Standards (Non-Specified Foods & Food Ingredients) Regulations, 2017.

### 38.2.3.1 Approved List of Probiotics

A list of probiotic microorganisms has been provided in the regulation as a schedule. This may be expanded from time to time as and when new organisms are added to it. The current list includes the following:

| S. NO. | NAME OF THE MICROORGANISM |
|---|---|
| 1. | Lactobacillus acidophilus |
| 2. | Lactobacillus plantarum |
| 3. | Lactobacillus reuteri |
| 4. | Lactobacillus rhamnosus |
| 5. | Lactobacillus salivarius |
| 6. | Lactobacillus casei |
| 7. | Lactobacillus brevis |
| 8. | Lactobacillus johnsonii |
| 9. | Lactobacillus delbrueckii subsp. bulgaricus |
| 10. | Bacillus coagulans |
| 11. | Lactobacillus fermentum |
| 12. | Lactobacillus caucasicus |
| 13. | Lactobacillus helveticus |
| 14. | Lactobacillus lactis |
| 15. | Lactobacillus amylovorus |
| 16. | Lactobacillus gallinarum |
| 17. | Lactobacillus delbrueckii (Lactobacillus delbrueckii subsp. Delbrueckii) |
| 18. | Bifidobacterium bifidum |
| 19. | Bifidobacterium lactis (Bifidobacterium animalis subsp. lactis) |
| 20. | Bifidobacterium breve |
| 21. | Bifidobacterium longum (Bifidobacterium longum subsp. longum) |
| 22. | Bifidbacterium animalis (Bifidobacterium animalis subsp. animalis) |
| 23. | Bifidobacterium infantis (Bifidobacterium longum subsp. Infantis) |
| 24. | Streptococcus thermophilus (Streptococcus salivarius subsp. thermophilus) |
| 25. | Saccharomyces boulardii (Saccharomyces cerevisiae subsp. boulardii) |
| 26. | Saccharomyces cerevisiae (Saccharomyces cerevisiae subsp. cerevisiae) |
| 27. | Lactobacillus paracasei |
| 28. | Lactobacillus gasseri |
| 29. | Bacillus clausii |
| 30. | Established probiotic strains of Bacillus subtilis |
| 31. | Bacillus indicus |

Two notes provide additional clarity:

Note 1—These organisms may be used either singly or in combination but shall be declared on the label with full information and must be non-GMO.

Note 2—The Food Authority may add any new strain of microorganism possessing probiotic properties, after proper scientific evaluation, and include it in this schedule.

## 38.2.3.2 *Labelling and Claims*

This regulation mandates following labelling requirements for all products covered under them:

i.   Front of the Pack
   A. The words "HEALTH SUPPLEMENT/NUTRACEUTICAL/FOOD FOR SPECIAL DIETARY USE/FOOD FOR SPECIAL MEDICAL PURPOSE/PREBIOTIC FOOD/ PROBIOTIC FOOD" as applicable to the concerned cate-gory, in capital and bold letters in the immediate proximity of the name or brand name of the product.
   B. A prominent statement indicating the target consumer group and/or age group if the product has been formulated for a specific age group.

ii.  Front or Back of the Pack
   A. The statement "NOT FOR MEDICINAL USE" in capital and bold letters prominently written on label, unless exempted for specific categories under these regulations.
   B. "Recommended usage level".
   C. "Duration of usage", where applicable.
   D. "Not to exceed the recommended daily usage" prominently written.
   E. An advisory warning in cases where a danger may exist with excess consumption.
   F. Warning on any other precautions to be taken while consuming, known side effects if any, contraindications and published product or drug interactions, as applicable.
   G. Statement or warning stating, "product is not to be used as a substitute for a varied diet" except for FSDU and FSMP category.
   H. A warning statement "product is required to be stored out of reach of children".
   I. The quantity of nutrients, expressed in terms of percentage of the relevant recommended daily allowances, unless exempted by any other regulations in force.

iii. Front or Back of the Pack or Accompanied Leaflet.
   A. A declaration on the amount of the nutrients or substances with a nutritional or physiological effect present in the product.
   B. The label, accompanying leaflet or other labelling and advertisement of each type of article of food, referred to in these regulations shall provide sufficient information on the nature and purpose of the article of food and detailed instructions and precautions for its use, and the format of information given shall be appropriate for the intended use of the consumer.

In addition to the previous, specific labeling requirements have been specified for probiotic foods:

i.   Front or Back of Pack
   A. genus and species including strain designation or culture collection number, in brackets where probiotics are mentioned in the list of ingredients; in such cases, internationally accepted short names are allowed.
   B. viable numbers at the end of the shelf life of probiotic strain corresponding to the level at which the efficacy is claimed.
   C. the recommended serving size, which shall deliver the effective viable dose of probiotics related to health claims.
   D. proper storage temperature conditions, and time limit for 'Best Use' after opening the container.

A separate Food Safety & Standards (Advertisement & Claims) regulation, 2018, has been notified which deals with all provisions and processed related to claims. Probiotic foods are required to comply with their provisions. This regulation has provided the following compliance conditions in case a source claim is made on probiotic foods:

Product contains ≥108 CFU in the recommended serving size per day.

All these labelling requirements are in addition to general labelling and display requirements applicable on all foods, as specified in Food Safety & Standards (Labelling & Display) Regulations, 2020.

All the previous regulations are available online on FSSAI website at the following link: www. fssai.gov.in/cms/food-safety-and-standards-regulations.php.

# 38.3 INDONESIA

Under Indonesian regulations, probiotic products are regulated as health supplements by the Directorate of Traditional Medicines, Cosmetics, and Food Supplements Evaluation division, National Agency of Drug and Food Control. Within the Directorate for Standardization of Processed Food, probiotics in processed foods are listed on food labels and advertisements of processed foods as an enclosure/attachment.

In January 2005, the National Agency of Drug and Food Control (NADFC), Republic of Indonesia, established a regulation to control functional food (Badan Pengawas Obat dan Makanan (BPOM), Agency for Drug and Food Control, 2005). However, the terminology of functional foods no longer exists under the regulatory framework. Instead, BPOM adopts the approaches of nutritional claim, nutrient comparative claim, and other specific functional claim. A specific regulation on probiotics was published in May 2016 as an enclosure/attachment of the Regulation for Claim on Label and Food Advertisement of processed foods, which was then revised to the new regulation on probiotics as health supplements in 2021 and an enclosure/attachment of the Regulation for Claim on Label and Food Advertisement of processed foods in 2022.

The Directorate of Traditional Medicines, Cosmetics, and Food Supplements Indonesia issued an approved list of probiotics (Table 38.5), and each registration requires detailed characterization and safety data for the strains being used (NADFC Health Supplement, June 2, 2021, Directorate of Traditional Medicines, Cosmetics, and Food Supplements, concerning safety, benefit/function, and quality).

Companies may apply for registration of probiotic health supplement products upon verification of complete, authentic, valid, and truthful supporting documents related to safety, functionality according to FAO/WHO, and quality such as source of probiotic, product specification, microbiological quality, and other quality requirements referring to Health Supplement Quality regulated in the NADFC to the head of the National Agency of Drugs and Food Control.

Currently, there are 57 probiotic health supplements registered, consisting of 4 local products and 53 imported products, where the only approved functional claim is to maintain gut health.

## 38.3.1 Approved List

An approved list of probiotic strains is provided by the Directorate of Traditional Medicines, Cosmetics, and Food Supplements.

There are 94 approved probiotic strains as of August 2022 (Table 38.5).

**TABLE 38.5**   Approved List of Probiotic Strains by the Directorate of Traditional Medicines, Cosmetics, and Food Supplements. Data as of August 2022

| NO. | PROBIOTIC STRAINS |
| --- | --- |
| 1 | *Lactobacillus reuteri* RC-14 TM |
| 2 | *Lactobacillus reuteri* DSM 17938 |
| 3 | *Lactobacillus reuteri* ATCC PTA 5289 |
| 4 | *Lactobacillus reuteri* RC-14 |
| 5 | *Lactobacillus rhamnosus* GR-1TM |

*(Continued)*

**TABLE 38.5**   (Continued) Approved List of Probiotic Strains by the Directorate of Traditional Medicines, Cosmetics, and Food Supplements. Data as of August 2022

| NO. | PROBIOTIC STRAINS |
| --- | --- |
| 6 | *Lactobacillus rhamnosus* R0011 |
| 7 | *Lactobacillus rhamnosus* GR 1 |
| 8 | *Lactobacillus rhamnosus* LR5 |
| 9 | *Lactobacillus rhamnosus* LMG25626 |
| 10 | *Lactobacillus rhamnosus* EMRO 014 |
| 11 | *Lactobacillus rhamnosus* HA-111 |
| 12 | *Lactobacillus rhamnosus* ATCC 7469 |
| 13 | *Lactobacillus rhamnosus* GG ATCC 53103 |
| 14 | *Lactobacillus acidophilus* ATCC 4356 |
| 15 | *Lactobacillus acidophilus* R0052 |
| 16 | *Lactobacillus acidophilus* W55 |
| 17 | *Lactobacillus acidophilus* CUL-60 |
| 18 | *Lactobacillus acidophilus* CUL-21 |
| 19 | *Lactobacillus acidophilus* BCMC 12130 |
| 20 | *Lactobacillus acidophilus* ATCC1063 |
| 21 | *Lactobacillus acidophilus* LA-G80 |
| 22 | *Lactobacillus acidophilus* LA1 |
| 23 | *Lactobacillus acidophilus* LA-5 |
| 24 | *Lactobacillus acidophilus* LA14 |
| 25 | *Lactobacillus acidophilus* LA85 |
| 26 | *Lactobacillus acidophilus* LA1063 |
| 27 | *Lactobacillus acidophilus* COREE |
| 28 | *Lactobacillus acidophilus* BCRC 14079 |
| 29 | *Lactobacillus acidophilus* W 22 |
| 30 | *Lactobacillus acidophilus* HA-122 |
| 31 | *Lactobacillus helveticus* R0052 |
| 38 | *Lactobacillus casei* ATCC 393 |
| 33 | *Lactobacillus casei* W56 |
| 34 | *Lactobacillus casei* subsp. BCMC 12313 |
| 35 | *Lactobacillus casei* EMRO 002 |
| 36 | *Lactobacillus casei* EMRO 213 |
| 37 | *Lactobacillus casei* HA-108 |
| 38 | *Lactobacillus casei* subsp. *casei* ATCC 393 |
| 39 | *Lactobacillus bulgaricus* ATCC 11842 |
| 40 | *Lactobacillus bulgaricus* EMRO 212 |
| 41 | *Lactobacillus bulgaricus* HA-137 |
| 42 | *Lactobacillus salivarius* W57 |
| 43 | *Lactobacillus lactis* BCMC 12451 |
| 44 | *Lactococcus lactis* W58 |
| 45 | *Lactobacillus plantarum* CBT LP3 |
| 46 | *Lactobacillus plantarum* LP01 AFID 1171 |
| 47 | *Lactobacillus plantarum* EMRO 009 |
| 48 | *Lactiplantibacillus (Lactobacillus) plantarum* Dad-13 |
| 49 | *Lactobacillus fermentum* EMRO 211 |

**TABLE 38.5** (Continued) Approved List of Probiotic Strains by the Directorate of Traditional Medicines, Cosmetics, and Food Supplements. Data as of August 2022

| NO. | PROBIOTIC STRAINS |
| --- | --- |
| 50 | *Lactobacillus fermentum* PCC |
| 51 | *Bifidobacterium breve* ATCC 15700 |
| 52 | *Bifidobacterium breve* HA-129 |
| 53 | *Bifidobacterium longum* ATCC 15707 |
| 54 | *Bifidobacterium longum* NCC 3001 |
| 55 | *Bifidobacterium longum* BCMC 02120 |
| 56 | *Bifidobacterium longum* STCC0986 |
| 57 | *Bifidobacterium longum* BL-G301 |
| 58 | *Bifidobacterium longum* BL-05 |
| 59 | *Bifidobacterium longum* BG7 |
| 60 | *Bifidobacterium longum* OLB6001 |
| 61 | *Bifidobacterium longum* BCRC 14634 |
| 62 | *Bifidobacterium longum* W 108 |
| 63 | *Bifidobacterium longum* BB536 |
| 64 | *Bifidobacterium bifidum* R0071 |
| 65 | *Bifidobacterium bifidum* BCMC 02290 |
| 66 | *Bifidobacterium bifidum* CUL-20 |
| 67 | *Bifidobacterium bifidum* CBT BF3 |
| 68 | *Bifidobacterium bifidum* BF3 |
| 69 | *Bifidobacterium bifidum* BB14 |
| 70 | *Bifidobacterium bifidum* BCRC 11844 |
| 71 | *Bifidobacterium lactis* W52 |
| 72 | *Bifidobacterium lactis* W51 |
| 73 | *Bifidobacterium lactis* BL-07 |
| 74 | *Bifidobacterium lactis* BL-04 |
| 75 | *Bifidobacterium infantis* R0033 |
| 76 | *Bifidobacterium infantis* BCMC 02129 |
| 77 | *Bifidobacterium infantis* HA-116 |
| 78 | *Bifidobacterium animalis* subsp. *lactis* CUL-34 |
| 79 | *Bifidobacterium animalis* subsp. *lactis* HNO19 |
| 80 | *Bifidobacterium animalis* subsp. *lactis* BB12 |
| 81 | *Bifidobacterium animalis* subsp. *lactis* BL-04 |
| 82 | *Streptococcus thermophilus* ATCC 19258 |
| 83 | *Streptococcus thermophilus* CBT ST3 |
| 84 | *Streptococcus thermophilus* ST3 |
| 85 | *Streptococcus thermophilus* HA-110 |
| 86 | *Streptococcus thermophilus* STCC0037 |
| 87 | *Saccharomyces boulardii* Hansen CBS 5926 |
| 88 | *Saccharomyces cerevisiae (boulardii)* CBS 5926 |
| 89 | *Saccharomyces boulardii* CNCM I-745 |
| 90 | *Bacillus coagulans* SANK70258 |
| 91 | *Bacillus coagulans* SNZ 1969 |
| 92 | *Bacillus coagulans* MTCC 5724 |
| 93 | *Bacillus mesentericus* TO-A |
| 94 | *Clostridium butyricum* TO-A |

In the case of a new probiotic strain which is not on the list of probiotic strains approved by the NADFC, companies should apply through written request for assessment to the head of the NADFC through the Directorate of Traditional Medicines, Cosmetics, and Food Supplements.

## 38.3.2 Labelling and Claims for Probiotics in Processed Foods

The new published regulation, Probiotic Food Label and Advertisement, and Control of Claims and Advertising of Processed Food 2022, provides guidelines for probiotic claims, safety, quality, and function of viable microorganisms after written approval from the head of NADFC is published online at the NADFC webpage. The claims should be in accordance with national health and nutritional policy, not for the treatment or prevention of diseases, not misleading, truthfully supported by laboratory analysis of national accredited laboratory, or accredited laboratory from country of origin. Companies are prohibited from claiming processed foods for babies to reduce risk of diseases for young children 1–3 years. With processed food, there should be no claim to prevent, treat, or cure diseases. Regulations on processed foods stipulate that they may contain identified and unidentified starter cultures (in fermented foods), defined microorganisms, or probiotics as food ingredients.

Probiotic(s) in processed foods should have defined types of microbes (bacteria or yeast) classified to the level of genus, species, strain, and types of cells (vegetative or spore), and viable counts (cfu/ml or cfu/g), with a clear method of identification, name of culture collection deposit, commercial trade name, manufacturing process, category, type of processed food, net weight or volume, types of packaging, manufacturing of processed foods, and amount of added probiotics in processed foods. If a product is multi-strain, it should have data on each strain in cfu/g or cfu/ml, minimum number of live probiotic strains at the time of consumption up to the end of the storage period, expiry date, serving size and serving suggestion for processed foods, label design, food composition, and data of similar probiotic-containing processed foods marketed in other countries.

The characteristic of probiotics in *in vitro* and *in vivo* animal experiments, such as gastric and bile tolerance, adhesion properties, antipathogenic bacterial properties, AMR gene transfer, and safety in phase 1 clinical trial, should be provided. If there were multiple strains in the preparation, the safety, compatibility/synergistic relationship, benefits and functions for each of the strain should also be provided at application. The clinical trial should have an outcome such as stool quality, quality and quantity of fecal microorganisms, and short chain fatty acids. Each significant finding should be proved with scientific evidence published in international reputable peer-reviewed journal.

Supporting data for the claim will consist of the proposed claim, clinical trial protocol, ethical clearance, date of research activities, date of accepted publication, list of authors, journal name, and link of publication.

Safety assessment should include, among other information, any side effects (systemic infection, metabolic disorder, overimmune expression on vulnerable groups, AMR gene transfer).

Clinical trial should be on phase 1 in healthy Indonesian subjects and published in a peer-reviewed journal, and strains should have obtained GRAS status.

For multi-strain probiotic preparation, the safety of each strain and synergies as well as antagonisms should be supported by studies. Evaluation of clinical trial for health claim (function claim) and reduce risk of diseases should be based on significant scientific evidence in a double-blind, randomized, placebo-controlled trial or other research design with primary outcome and adequate and valid sample size. Clinical trials conducted outside Indonesia are accepted if they were conducted in countries with food consumption patterns, hygiene, and sanitation practices and public health equivalent to Indonesia, such as Malaysia, Thailand, Vietnam, or other countries with those conditions. If the clinical trial has been conducted outside Indonesia, a confirmation study should be conducted on healthy Indonesian subjects.

Claims for probiotics in processed foods will be assessed case by case. The permitted claim is a probiotic claim for maintaining gut health. Other health claims (functional claims) and claims to reduce the risk of disease will be based on significant findings of clinical trials. Upon approval, the claim can be given on the label.

The label should include the ingredient or composition of processed foods and that it contains probiotics (genus, species, and strain should be in accordance with the nomenclature and viable counts expressed as cfu/serving at the end of the storage period, with valid microbiological assessment). Proper storage instructions for maintaining stable viability should also be mentioned on the label.

## 38.3.3 Summaries of Technical Documents for Probiotics in Processed Foods

   i. Strain identification
  ii. Functional characteristics
 iii. Safety assessment (in vitro/in vivo, WGS, and clinical trial phase 1)
 iv. Efficacy
     • Phase 2 clinical (mandatory in Indonesia with Indonesian population)
     • Phase 3 local (if health claims applied, will consider primary local study as approval of Health Claim)
   v. Product Composition (active ingredients/excipients)
  vi. Directions for use & dosage
 vii. Manufacturing process
viii. Shelf life in accordance to processed food product.

## 38.3.4 Summaries of Technical Documents for Probiotic in Health Supplement

   i. Strain identification
  ii. Functional characteristics
 iii. Safety assessment (in vitro/in vivo & clinical phase 1)
 iv. Efficacy
     • Phase 2 clinical (mandatory in Indonesia with Indonesian population)
   v. Product Composition (active ingredients/excipients)
  vi. Directions for use & dosage
 vii. Manufacturing process
viii. Shelf life

# 38.4 JAPAN

## 38.4.1 Regulation of Food with Health Claims

In 2001, the Ministry of Health, Labor, and Welfare (MHLW) established a regulatory system of food with health claims. This system aims to provide consumers with appropriate information about products claiming to be health foods that satisfy required criteria as food with health claims. The objective of the system is defined in the Food Labelling Standards, Article 9 (Regulatory Systems for Health Claims in Japan 2011).

As of February 2018, foods with health claims are broken down into following three categories:

1. Food for Specified Health Uses (FOSHU): FOSHU refers to foods containing functional ingredients for health that are officially approved by the government to claim physiological effects on the human body. FOSHU is intended to be consumed for the maintenance/promotion of health or special health uses by people who wish to control health conditions, including blood pressure or blood cholesterol. In order to market food as FOSHU, an assessment of the food's safety and effectiveness of its functions for health is required, and the claim must be

approved by Consumer Affairs Agency (CAA), which took over responsibility for food labelling regulations from MHLW in September 2009 (Regulatory Systems of Health Claims in Japan 2011).

2. Food with Nutrient Function Claims (FNFC): FNFC refers to foods containing specified nutrients such as vitamins and/or minerals that can be used to supplement or complement the daily requirement of nutrients which tend to be insufficient in everyday diet. Given that the food product contains certain amounts of nutrients whose function has already been substantiated by scientific evidence, their claims can be applied according to standards without submitting a notification to the government. Consumers can select the product that suits their own purpose to supplement nutrients by referring the claim on the label (Regulatory Systems of Health Claims in Japan 2011).

3. Food with Function Claims (FFC): FFC is a new regulatory system introduced in 2015 by CAA. Applicants (food importers, food manufacturers, food producers, and food retailers) can put health claims on food labelling statements when they submit to the CAA necessary information including scientific evidence on food safety and effectiveness in accordance with the rules prescribed by the law and it is accepted at CAA, before marketing the product. In addition to processed foods, fresh products, such as vegetables, fruits, and fish, also included in the FFC system, and there are some actual examples for such raw agricultural commodities. The FFC system provides opportunities for consumers to make voluntary and reasonable product choices (Japan Consumer Affairs Agency pamphlet 151224–1, 151224–2).

Foods with health claims all make claims about their effectiveness on the maintenance and improvement of health; however, the Japanese regulatory system clearly distinguishes those foods from pharmaceuticals. Foods with health claims stand between pharmaceuticals and general foods; hence healthy adults are the focus of this system. For this reason, health claims on foods must not express medical claims such as "prevent", "cure", "treat", "diagnose", and "for people suffering from diabetes". In addition, claims stating intentional enhancement of health such as "body building" and "skin-whitening" are also prohibited due to concerns about misleading customers (Japan Consumer Affairs Agency pamphlet 151224–1, 151224–2).

In this section, we are going to touch on details about FOSHU and FFC and provide actual examples of probiotic products with health claims under those systems in Japan.

## 38.4.2 Food for Specified Health Uses Definition

FOSHU was established by MHLW in 1991 as a regulatory system to review and approve label statements regarding effects of foods on the human body. As previously mentioned, CAA took over that responsibility after 2009, and currently CAA maintains the guideline.

FOSHU is defined as foods with active constituents that affect the physiological function and biological activities of the human body. This includes foods used in the daily diet that claim to promote a specified health benefit. FOSHU are evaluated by the authority individually according to their substantiation, validity, safety, and quality, and finally approved by CAA.

To facilitate applicants for FOSHU approvals, the following types of FOSHU were introduced in addition to "regular" FOSHU (Regulatory Systems of Health Claims in Japan 2011):

1. Qualified FOSHU: Food with a health function that is not substantiated on scientific evidence which meets the level of FOSHU, or food with certain effectiveness but without an established mechanism of the effective element for the function, will be approved as qualified FOSHU.

2. Standardized FOSHU: Standards and specifications are established for foods with sufficient FOSHU approvals and accumulation of scientific evidence. Standardized FOSHU are approved when they meet the standards and specifications.
3. Reduction of Disease Risk FOSHU: This claim is permitted when reduction of disease risk is clinically and nutritionally established in an ingredient.

## 38.4.3 FOSHU Criteria

Following are the required conditions for FOSHU approval (Notification Shokuanhatsu 0201002, February 1, 2007) (Japan Ministry of Health, Labor and Welfare website):

1. Improvement of dietary habits and contribution to health maintenance and enhancement can be expected by consuming the product.
2. Scientific evidence for the claimed health benefit is available.
3. Clinical and nutritional intake level of the product and/or its functional component is established.
4. The product and/or its functional component is safe for human consumption.
5. The following items regarding functional component are defined:
   a. Physical, chemical, and biological characterization and its methods
   b. Methods of qualitative and quantitative analytical determination
6. The nutrient constituents of the same type of the food are not significantly changed.
7. The food is intended to be consumed daily and not on rare occasions.
8. The product or its functional component is not included in the medical drug list.

## 38.4.4 FOSHU Application

Those who wish to apply for FOSHU labelling on their products are required to submit the following documentation to the CAA:

1. Sample of the entire package with labels and health claims
2. Clinical and nutritional proof of the product and/or functional component for the maintenance of health
3. Clinical and nutritional proof of the amount of the product and/or functional component ingested
4. Safety of the product and functional component, including additional human studies about the eating experience
5. Physical and biological characteristics of the product and functional component
6. Methods of qualitative and quantitative analytical determination of its functional component and analytical assay results of the component in the product
7. Report on the analysis of the designated nutrient constituents and energy content of the product
8. Statement of production process, list of factory equipment, and an explanation of the quality control measures.

Mandated FOSHU documentation can be summarized into three essential requirements for FOSHU approval: (1) effectiveness based on scientific evidence, including clinical studies; (2) safety as assessed from historical consumption pattern data and additional safety studies conducted in humans; and (3) analytical determination of the functional ingredient responsible for the beneficial physiological effect.

Regarding effectiveness, documents should be prepared based on substantiation not only by human intervention studies but also by *in vitro* metabolic and biochemical studies and animal studies. These data should demonstrate statistically significant differences. Basically, a human study should be conducted by using the food in question over a reasonably long-term period (e.g., more than 2 or 3 months). A human study should be approved by a committee on ethics in consideration of the protection of human rights, in accordance with the principles of the Helsinki Declaration. The study should also be well designed, for example, using an appropriate functional marker, appropriate sample size, and enough subjects to prove statistically significant differences. All available literature regarding the related functional ingredients, the related foods, and the related functions should be reviewed.

Any new scientific evidence used to support health-related claims must be published by a suitably qualified journal with expert referees who can review the evidence. Generally, more than two human studies are required for different targeted individuals.

As for safety, both *in vivo* and *in vitro* studies should be carried out to obtain preliminary data confirming the food's safe ingestion by human consumers. Even if an effective component has been consumed as food by a reasonable number of individuals during certain periods, safety data for human consumers should be required for at least three times the minimum effective dosage. The literature regarding related functional ingredients should be reviewed. If the related literature suggests an especially undesirable or adverse health effect, the report should be included as a reference with the scientific explanation or the human study that confirms the product's safety for human consumers.

Concerning analysis, documentation of the methods of analysis of related functional ingredients should submit the suitable and reliable methods of quantitative and qualitative analytical determination. As additional documentation, the stability of related functional ingredients should be confirmed. The effective ingredients and other ingredients with an undesirable or adverse health effect should be confirmed to demonstrate the specified amount using suitable analytical methods. If a product is in the form of tablets or capsules, experiments should be conducted regarding its characteristics of disintegration or dissolution.

It is important within the evaluation process that both the benefits and safety of a given functional food differ from those of a medicine. Functional foods are designed to target healthy people or people in the preliminary stage of a disease or a borderline condition of at-risk groups. Therefore, effectiveness for these people may be reduced as compared with medicine for patients. Generally, foods with functionality have been historically consumed by people and thus can be regarded as safer than innovative pharmaceuticals.

# 38.4.5 FOSHU Labelling

FOSHU comprises foods that would support maintenance and improvement of health, but clearly it is not enough for one's health if a person eats FOSHU only, so a prescribed notice that recommends a balanced diet should be included in the labelling statement. This is like the FFC requirement, which is to be touched on in a later section. The following items must be labeled on a FOSHU package after approval (Japan Ministry of Health, Labor and Welfare website):

1. Foods with health claims (FOSHU)
2. Approved function claims
3. Nutrition facts
4. List of ingredients
5. Total product amount/serving
6. Standard daily intakes
7. Methods of ingestion and warning concerning those methods

8. Percentage of the active substance included in the daily amount of administration to the substance's set DRI
9. Warning concerning cooking methods or preservation.
10. FOSHU logo of approval

As already mentioned, health claims involving FOSHU must not express medical claims such as "prevent", "cure", "treat", and "diagnose". Here are some examples of permitted and prohibited claims for human diseases: (1) Maintain or improve a marker determined by self-diagnosis or a health checkup. An example of a permitted claim is as follows: "This product helps to maintain normal blood pressure, blood sugar, or cholesterol". An example of a prohibited claim is as follows: "This product improves hypertension". (2) Maintain physiological function and organ function of the human body in good condition or improve them. An example of a permitted claim: "This product enhances the absorption of calcium" or "This product helps to improve the movement of the bowel". An example of a prohibited claim: "This product is an effective food for intoxication" or "Enhance fat metabolism".

## 38.4.6 FOSHU Health Claims Classification

As of June 2017, 1099 products have been approved as FOSHU. The existing health claims on FOSHU can be classified into 11 groups (Maeda-Yamamoto 2017):

1. Gastrointestinal condition
2. Blood cholesterol levels
3. Blood glucose levels
4. Blood pressure
5. Dental hygiene
6. Dry skin
7. Mineral absorption
8. Osteogenesis
9. Triacylglycerol
10. Reduction of the risk of specific diseases
11. Prevent neural tube defects (spondyloschisis)

The health benefits of probiotics maybe classified into immunoregulatory activity and effect to regulate the functions of the intestines. For the former benefit, probiotic products are not yet recognized as a FOSHU because the end points are not clear. Their health benefits in reducing disease development risks are therefore limited to calcium intake and osteoporosis and folic acid neural tube defect. For the regulation of intestinal functions, the benefits of alleviating the symptoms of inflammatory bowel disease, severe constipation requiring medical attention, and diarrhea are excluded from consideration, and only limited appeals for improvements in the tendency to cause constipation are allowed. On the other hand, appeals related to improvements in the intestinal environment, for example, an increase in *bifidobacteria* counts, a decrease in harmful bacteria, or an increase in metabolites, seem to be allowed with more flexibility.

Table 38.6 lists the names of probiotic strains contained in foods approved as FOSHU and their health benefits. As can be seen, each product has different claims, but gastrointestinal conditions are the majority. The claims should be supported by the submitted data. To emphasize that bifidobacteria and lactobacilli increase by using a particular product, it is necessary to conduct a study of the relevant product with humans and confirm the actual increase. Similarly, to emphasize that probiotics improve the intestinal environment or maintain good gastrointestinal conditions, it is necessary to present data from a study in humans to support the claim.

**TABLE 38.6**   Probiotics Strains in FOSHU Products[a]

| STRAIN NAME | CLAIM CATEGORY |
| --- | --- |
| *Bifidobacterium breve* BBG-01 | Gastrointestinal conditions |
| *Bifidobacterium lactis* Bb-12 | Gastrointestinal conditions |
| *Bifidobacterium lactis* FK120 | Gastrointestinal conditions |
| *Bifidobacterium lactis* LKM512 | Gastrointestinal conditions |
| *Bifidobacterium longum* BB536 | Gastrointestinal conditions |
| *Bifidobacterium longum* SBT2928 | Gastrointestinal conditions |
| *Lactobacillus casei* NY 1301 | Gastrointestinal conditions |
| *Lactobacillus casei* YIT 9029 | Gastrointestinal conditions |
| *Lactobacillus delbrueckii bulgaricus* 2038 | Gastrointestinal conditions |
| *Lactobacillus gasseri* SBT2055 | Visceral fat |
| *Lactobacillus rhamnosus* GG | Gastrointestinal conditions |
| *Propionibacterium freudenreichii* | Bowel movements |
| *Streptococcus salivarius* subsp. *themophilus* 1131 | Gastrointestinal conditions |

[a] As of December 2017, searched by the author.

## 38.4.7 Food with Function Claim Characteristics

FFC is a regulatory system introduced in 2015 by CAA. Applicants (food importers, food manufacturers, food producers, and food retailers) can put health claims on food labels if they submit the necessary information, including scientific evidence on food safety and effectiveness, to the CAA in accordance with the rules prescribed by the law and it is accepted at CAA, before marketing the product.

In addition to processed foods, fresh products such as vegetables, fruits, and fish fall within the scope of FFC. Receiving an approval for FOSHU product has been challenging to small-to-medium-sized enterprises due to the required costs and long time frames from product development to approval. Hence, they have high expectations of the FFC system, which enables them to enter the foods with health claims market, and have continued their efforts on development of FFC products. In October 2017, the number of FFC products for which notifications were accepted after April 2015 exceeded 1000, which is above the number of approved FOSHU products.

The unique characteristics of the FFC system are summarized as follows (Maeda-Yamamoto 2017):

1. The applicant is responsible for providing scientific evidence of safety and effectiveness of the foods to humans.
2. The functional ingredient has been identified qualitatively and quantitatively using a validated method.
3. The mode of action has been well investigated/characterized using *in vitro* and *in vivo* studies and in human clinical studies.
4. Scientific evidence has been acquired from human clinical trials or research reviews of functional ingredients.
5. Sufficient histories of consumption and safety of the food in Japan.
6. Filed at CAA 60 days prior to the launch and displayed on the packaging.
7. The adequate daily intake is an appropriate volume to eat.

## 38.4.8 FFC Application

Like FOSHU, there are three essential requirements for the FFC dossier: (1) effectiveness based on scientific evidence, (2) safety as assessed from historical consumption pattern data and additional safety studies conducted in humans and (3) analytical method for determination of the functional ingredient responsible for the beneficial physiological effect.

For safety, the applicant substantiates the safety of the product based on the one of the following: (1) history of consumption by humans based on actual intake data, (2) secondary information collected through databases, and (3) completion of a safety study with the finished product or functional ingredient.

Unlike FOSHU, neither an *in vitro/in vivo* animal study nor human clinical study is a part of the mandatory requirement, though of course the applicant is free to submit a study to affirm the safety of their product or the functional ingredient in it.

Considering the increased amount of products in tablet/capsule form in the market recently, applicants should conduct a careful assessment for those types of foods with concentrated functional ingredients, as humans have less consumption history for those foods.

Also, the interaction of the functional ingredient(s) must be evaluated by the applicant. If an interaction is detected, the appropriateness for selling such a product must be explained (Japan Consumer Affair Agency pamphlet 151224–1, 151224–2).

As for effectiveness, there are two ways to provide scientific evidence when evaluating a claimed effect. One is to conduct clinical trial(s) with a finished product. A clinical study is an intervention study that involves healthy human subjects and evaluates the effects of consumption of a certain substance or food on health outcomes. It should be noted that the clinical study must be published in a peer-reviewed scientific journal, and including healthy Japanese subjects is one of the fundamental requirements. Statistical significance from placebo group should be observed by a randomized, double-blind, parallel group comparison study design. In other words, statistical significance by comparison between pre-feed and post-feed is not adequate as scientific evidence to satisfy FFC criteria. Both FFC and FOSHU guidelines should be referred to prior to study protocol development because FFC guidelines recommend following the guidance of FOSHU for clinical study design. This would be a key point for multinational companies to enter the Japanese FFC market. Of course, an ethical study design, informed consent, and a protocol review process in accordance with the Health Declaration are required.

The other way to document is performing a systematic literature review(s). This is approach summarizes the results of available carefully designed clinical studies and/or observation studies (controlled trials) and provides a high level of evidence on the effectiveness of the functional ingredient from a totality-of-evidence point of view. Under the FFC system, bearing a health claim is not permitted in cases where there is a literature that disagrees about the function to human body. In addition, positive results on functionality should be obtained in the clinical trials for supplement type of foods, and in the case of processed foods and fresh foods, positive results should be obtained in clinical and observational studies based on intake of the functional ingredient (Japan Consumer Affairs Agency pamphlet 151224–1, 151224–2).

As for the analytical method, since the applicant is responsible for the product safety and quality under the FFC system, while experts evaluate the raw materials source and manufacturing processes to details in FOSHU, the quality should be confirmed objectively by a third party. CAA entrust a product analysis to local CROs (not disclosed) as a part of Market Basket Survey (an inspection using an actual final product that is picked up from market) to check if the product satisfies the lower limit of the functional ingredient as it was declared in the dossier.

Details of the product recipe are not a requirement for FFC, so the analytical method should be a validated one with performance high enough to quantify the amount of the functional ingredient using a sample which is a complex mixture of various ingredients.

For this reason, it should be noted that an identification method is required for probiotics so that CAA can confirm that the probiotics strain in the product is exactly the notified strain. Rep-PCR, AFLP, RAPD, and/or PFGE are typical examples of identification methods.

## 38.4.9 FFC Labelling

As FFC is a regulatory system, and the government doesn't assess the safety and the effectiveness, it is a required item to clearly state that functionality and safety are not evaluated by the CAA. The following 16 items are mandatory for the labelling statement (Maeda-Yamamoto 2017):

1. Indication of FFC
2. Functional ingredient based on scientific basis and the functionality of the said components or foods containing the said components
3. Adequate daily intakes
4. Nutrition facts
5. Amount of a functional ingredient per serving
6. Acceptance number
7. Contact information of the manufacturer
8. Notice: Functionality and safety are not evaluated by CAA
9. Method of intake
10. Instruction on intake
11. Notice: To promote popularity of the idea of a balanced diet
12. Precaution statement for products which require special attention when cooking or storing
13. Notice: The product does not intend to diagnose, treat, or prevent disease
14. Notice: Not to target one suffering from disease, minors, pregnant women (including women planning to become pregnant), and nursing women
15. Notice: Recommend that one suffering from disease consult with a physician, and that one taking drugs consult a physician or pharmacist
16. Notice: In case of unusual physical change, recommend stopping eating a product immediately and consult a physician

## 38.4.10 FFC Claim Classification

FFC provides food business operators with more flexibility in making health claims than FOSHU, which should be one of the major features of the FFC system. For example, memory and muscle are new categories compared to FOSHU. Again, it should be noted that words such as "prevent", "cure", "treat", and "diagnose" are prohibited.

Table 38.7 lists the names of probiotic strains contained in FFC and their health benefits. So far, health claims received for probiotics are gastrointestinal conditions, bowel movement, visceral fat, and sleep quality. As can be seen, there has been no quality sleep claim case in FOSHU.

**TABLE 38.7** Probiotics Strain Accepted as FFC Products[a]

| STRAIN NAME | |
| --- | --- |
| *Bacillus coagulans* lilac-01 | Improve fecal property and bowel movements |
| *Bifidobacterium breve* B-3 | BMI |
| *Bifidobacterium lactis* BB-12 | Bowel movement |
| *Bifidobacterium lactis* CNCM I-2494 | Gastrointestinal conditions |
| *Bifidobacterium lactis* GCL2505 | Gastrointestinal conditions and bowel movements |
| *Bifidobacterium lactis* HN019 | Bowel movements |
| *Bifidobacterium longum* BB536 | Gastrointestinal conditions |
| *Bifidobacterium longum* JBL01 | Bowel movements |
| *Lactobacillus amylovorus* CP1563 | Function to reduce body fat |
| *Lactobacillus brevis* NTT001 | Gastrointestinal conditions |
| *Lactobacillus gasseri* CP2305 | Gastrointestinal conditions |
| *Lactobacillus gasseri* SBT2055 | Visceral fat |
| *Lactobacillus plantarum* TK61406 | Gastrointestinal conditions |
| *Lactobacillus reuteri* DSM 17938 | Gums health |
| *Saccharomyces cerevisiae* GSP6 | Quality of sleep |

[a] As of December 2017, searched by the author.

Various functionalities of probiotics, including immune and cognitive health, have been reported in academia recently. Inventive ideas are essential for applicants to establish reasonable study protocol, monitoring parameters in conducting clinical trials. They also need to develop a claim based on the clinical data, which should be in the scope of this regulatory system.

## 38.4.11  Future Prospects for FFC

In December 2016, CAA announced the results of an experts' roundtable on nutritional information. The committee recommended that carbohydrates and saccharides, which were not included in the original guideline, should be recognized as functional ingredients. Furthermore, the committee recommended that plant-derived extracts should be added to the scope of functional ingredients. Those were recommendations to enhance FFC. The "extract" that would label the functionality is required to be the equivalent to the "extract" that scientific evidence provided (Maeda-Yamamoto 2017).

In May 2017, the Regulatory Reform Promotion Council recommended that the government establish acceptance criteria for clinical data with 18- and 19-year-old subjects. Simplification of the dossier and shortening of completeness check period at CAA were also recommended, and the guideline was revised on March 28, 2018, to incorporate those recommendations.

The CAA plans to modify the FFC guideline periodically for further improvement.

# 38.5  KOREA

## 38.5.1  Korean Health Functional Food Act

### 38.5.1.1  Article 1 (Purpose)

The purpose of this Act is to contribute to secure the safety of health functional food, improve the quality thereof and promote the sound distribution and sale thereof, thereby contributing to improving the health of nationals and consumer protection.

### 38.5.1.2  Article 18 (Prohibition of False or Exaggerated Labelling and Advertising)

No business operator shall falsely label or exaggeratedly advertise the names, raw materials, manufacturing methods, nutrients, ingredients, usage methods or qualities of health functional foods and the traceability of health functional foods, as follows (amended by Act No. 8941, Mar. 21, 2008):

1. Labeling or advertising any indication that may lead to a misunderstanding that the relevant foods are effective in preventing or treating a disease or the relevant foods are medicines.
2. False labeling or exaggerated advertisements.
3. Labeling or advertising any indication that are likely to deceive, mislead, or confuse consumers.
4. Labeling or advertising any indication that includes names (including the prescriptions of oriental medicines) used only for medicines.
5. Negative labels or advertisements against other companies or their products.

6. Labels or advertisements that haven't deliberated upon under Article 16 (1) or bearing different details from details deliberated upon. (2) The scope of false, exaggerative, or negative labels and advertisements under paragraph (1), and other necessary matters shall be prescribed by Ordinance of the Prime Minister.

### 38.5.1.3 Article 22 (Good Manufacturing Practices, etc.)

The Commissioner of the Food and Drug Administration may determine and publicly notify standards for manufacturing and controlling the quality of good health functional foods (hereinafter referred to as "Good Manufacturing Practice Regulations") to manufacture good health functional foods and control the quality of health functional foods.

### 38.5.1.4 Article 27 (Health Functional Food Deliberation Committee)

1. The Health Functional Food Deliberation Committee shall be established in the Ministry of Health and Welfare to investigate and deliberate on the following matters responding to the questions of the Minister of Health and Welfare or Commissioner of the Korea FDA (amended by Act No. 8852, Feb. 29, 2008; Act No. 9938, Jan. 18, 2010):
   a. Matters concerning policies on health functional foods;
   b. Matters concerning the standards and specifications for health functional foods.
   c. Matters concerning the labels and advertisement of health functional foods; and
   d. Other important matters concerning health functional foods.
2. The Health Functional Food Deliberation Committee may have researchers to examine and study the standards and specifications, and labels and advertisements, etc., of health functional foods.
3. Necessary matters concerning the organization and operation of the Health Functional Food Deliberation Committee under paragraphs (1) and (2) shall be prescribed by the Presidential Decree.

## 38.5.2 Regulations on Probiotics

Probiotics is one of the principal functional ingredients of health functional food approved by the Korea MFDS. In Korea, the acquisition of functional food status for probiotics was supported by human clinical studies based on double-blind, randomized, placebo-controlled designs using human-originated probiotic strains such as *B. bifidum* BGN4. The approved functional claims of the probiotics include "Help to maintain healthy gastrointestinal bacteria population, help to maintain healthy bowl function".

### 38.5.2.1 The Standards and Specifications of the Probiotics

1. Standards for manufacturing
   (1) Raw material: The following microorganism itself or mixed.

| GENUS | SPECIES |
|---|---|
| Lactobacillus | L. acidophilus, L. casei, L. gasseri, L. delbrueckii spp. bulgaricus, L. helveticus, L. fermentum, L. paracasei, L. plantarum, L. reuteri, L. rhamnosus, L. salivarius |
| Lactococcus | Lc. Lactis |
| Enterococcus | E. faecium, E. faecalis |
| Streptococcus | S. thermophilus |
| Bifidobacterium | B. bifidum, B. breve, B. longum, B. animalis spp. Lactis |

    (2) Preparation and/or processing
        (a) It shall be in edible form by culturing and pulverizing the above probiotics. During culturing it shall be meet manufacturing/processing standards and specification (coliforms, *Salmonella* spp., *Listeria monocytogenes*, *Staphylococcus aureus* only) of 'Fermented milk (except fermented milk powder)' in the Food Code.
        (b) *Enterococcus* sp. strain can be used only when there is no antibiotic resistance and toxic gene.
    (3) Content of functional compounds (or marker compounds): Live bacteria shall be contained 100,000,000 CFU/g or more.
2. Specifications
    (1) Appearance: Unique color and flavor, no off-taste and off-flavor
    (2) Probiotics number: No less than labelled amount
    (3) Coliform: Negative
3. Prerequisite for the health functional food
    (1) Health claims: May help to increase the number of beneficial bacteria and control harmful bacteria in the gut maintain healthy bowl function
        MFDS Notice 2020–63, Jul. 10, 2020, Enforcement date: Jul. 11, 2021
    (1) Health claim: May help to increase the number of beneficial bacteria and control harmful bacteria in the gut help to maintain healthy bowel function, maintain gut health
    (2) Daily intake amount: 100,000,000 ~ 10,000,000,000 CFU
    (3) Warning notice for intake
        (a) Consult a health care practitioner prior to intake if you are taking medicines or having disease
        (b) The individual who has an allergy may cause side-effect such as hypersensitivity reaction
        (c) Teach children how to intake daily amount so that they do not over intake
        (d) Consult a health care practitioner and stop intake if you are having adverse event
4. Testing methods
    (1) Appearance: Chapter 4. 2–7 Appearance testing method
    (2) Probiotics number: Chapter 4. 3–57 for lactic acid bacteria, Chapter 4. 3–58 for bacillus- and coccus-form of lactic acid bacteria and bifidobacteria
    (3) Coliform: Referred to Annexed Table 4

## 38.5.2.2 Guidelines for the Safety Evaluation of Probiotics

In September 2022, the Korean MFDS released new guidelines for the safety evaluation of probiotics based on WHO/FAO guidelines for probiotics. In order to evaluate the safety of probiotic strains, this guideline requires the submission of data on identifying strain specificity such as whole genome sequence and its analysis report as well as safety characteristics, including antibiotic resistance, hemolytic activity, toxic substance production, and metabolic characteristics (D-lactic acid production, bile salt deconjugation). The depository number of the strain also should be submitted. This guideline specified microbiological cutoff values of antibiotics against *Enterococcus faecium* and *Enterococcus faecalis*. The pathogenic genes for cytolysin, aggregation substance, hyaluronidase, and gelatinase should be analyzed in the whole genome, and their biological activity should be tested. A single-dose toxicity test from GLP also should be added.

### 38.5.2.3 Scientific Name of Lactobacillus *Will Be Changed*

According to the scientific name of *Lactobacillus* being reclassified, the genus name of the seven *Lactobacillus* species of probiotic raw material will be changed: *Lacticaseibacillus (L. casei, L. paracasei, L. rhamnosus), Limosilactobacillus (L. fermentum, L. reuteri), Lactiplantibacillus (L. plantarum), Ligilactobacillus (L. salivarius)*.

### 38.5.2.4 Health Functional Claims of General Food

Since December 2020, general foods can make health functional claims when there is sufficient scientific evidence supporting the claim. It was intended to promote the vitality of the food industry by inducing the development of functional raw materials in Korea. Currently, 29 functional ingredients, including probiotics, have been approved for use based on evidence that they are effective and safe. The contents of the functional ingredients are required to exceed 30% of the recommended daily intake stipulated in the Health Functional Food Code, and the functional ingredients should be provided by factories in compliance with Health Functional Food Good Manufacturing Practice. Furthermore, these products should be labeled "this is not a Health Functional Food" on the packing to prevent confusion among consumers between general food and health functional food.

# 38.6 MALAYSIA

## 38.6.1 Food Regulatory Framework

The principal food law of Malaysia is under the authority of Food Safety and Quality Division (FSQD) of the Malaysian MOH, which has been developed based on the Food Act 1983 (MOH 1983) and the Food Regulations 1985 (MOH 1985). Numerous amendments have been made to these regulations over the years in response to reviews by FSQD and other government agencies, as well as applications from the food industry. FSQD also enacted various amendments to enable the regulations to be harmonized with standards of the Codex Alimentarius.

All amendments to the food regulations have been processed in a systematic manner via a seven-step process. The whole process also involves consultation with industry and all stakeholders. Relevant expert working groups have been established to focus on and provide expert views on specific subject matter. Expert groups present all proposals for review by the Food Regulations Advisory Committee. All approved amendments are handled by the legal department for gazettement.

The use of probiotic cultures in foods and beverages is regulated under Malaysian Food Regulations 1985.

# 38.6.2 Regulatory Framework for Pharmaceuticals

The Drug Control Authority (DCA), established under the Control of Drugs and Cosmetics Regulations (CDCR) in 1984, is tasked with ensuring the quality, safety, and efficacy of medicinal products through the registration, including quality control, inspection and licensing, and post-registration activities. The National Pharmaceutical Regulatory Division (NPRA) acts as the secretariat to the authority.

Under the CDCR, no person shall manufacture, sell, supply, import, possess, or administer any medicinal product unless it is a registered product. The registration process includes quality control, inspection, and licensing, as well as post-registration of medicinal products.

Medicinal products that are registrable include new drug products, biologics, generics, natural products, and health supplements. The last group includes any product that is used to supplement a diet and to maintain, enhance, and improve the health function of human body. It is presented in small unit dosage forms such as capsules, tablets, powder, or liquids. It may contain one or more or a combination of the following:

1. Vitamins, minerals, amino acids, fatty acids, enzymes, probiotics, and other bioactive substances.
2. Substances derived from natural sources, including animal, mineral, and botanical materials in the forms of extracts, isolates, concentrates, metabolites.
3. Synthetic sources of ingredients mentioned in (1) and (2) if the safety of these has been proven.

# 38.6.3 Regulation on Probiotic Cultures in Foods and Beverages

A regulation to provide a definition for probiotics and permit their addition to foods and specifying the labelling requirements was gazetted in April 2017. Regulation 26A defined "probiotic culture" as live microorganisms which when administered in adequate numbers confer health benefits on the host. The regulation has provided a positive list of bacterial strains that may be recognized as probiotic cultures and permitted to be added to foods. The list currently contains 13 *Bifidobacterium* strains and 19 *Lactobacillus* strains (see Section 38.6.5). Applications can be made to the FSQD for additional strains to be added to the list. An Expert Working Group on Microbiology reviews scientific data that must be submitted with such applications.

When added to food, the probiotic cultures must remain viable with a viable probiotic count of not less than $10^6$ CFU/mL or cfu/g during the shelf life of such food. It is also a requirement that the probiotic cultures used should not contain transmissible antibiotic-resistant genes.

Example of foods that may contain added probiotic cultures are cultured milk or fermented milk, which includes yogurt, cultured cream, or sour cream, provided for in Regulation 113. Cultured milk or fermented milk has been defined as the product prepared by culturing pasteurized milk, sterilized milk, skimmed milk, recombined milk, pasteurized cream, or reduced cream. These foods may contain live probiotic cultures.

Regulation 26A has specified labelling requirements for foods that contain added probiotic cultures. The following must be written on the label of such foods:

1. The words "CONTAINS (state quantity) OF PROBIOTIC CULTURES".
2. The specification of genus, species, and strain of the probiotic cultures; and
3. The direction for storage before and after the package is opened.

In addition, where the media used for propagation and maintenance of the probiotic cultures are derived from animals, the common name of the animal shall be stated on the label of that food by using the words.

The regulation permits a general health claim to be made for foods with added probiotic cultures. The words "Probiotic cultures help in improving intestinal or gut function" or any other words of similar meaning may be used. Companies intending to apply for additional health/other function claims may submit applications to the FSQD, using prescribed forms, accompanied by adequate scientific substantiation.

## 38.6.4 Regulating Probiotic Cultures in Health Supplements

The sale and use of probiotics in pharmaceutical dosage forms are regulated as a health supplement, under the NPRA framework. They are included as one of the registrable medicinal products.

For registration of health supplements, the Drug Registration Guidance Document (DRGD) of the National Pharmaceutical Regulatory Division, Ministry of Health (MOH 2018) serves as the reference guide. The DRGD provides details of the registration process, including quality control, inspection and licensing, and post-registration activities of pharmaceutical and natural products. For registration of products, only web-based online submissions via QUEST at http://npra.moh.gov.my/ are accepted. To conduct transactions via QUEST system, the applicant must first register a membership for QUEST system with NPRA. Full details of the requirements for registration and processing are given in the DRGD.

As stipulated in Regulation 8(8), CDCR 1984, upon registration of a product, the authority notifies the product registration holder and assigns a product registration number (i.e., MAL number) to the registered product via the system. The registration number is specific to the product registered with the name, identity, composition, characteristics, origin (manufacturer), and product registration holder, as specified in the registration documents. It *cannot* be used for any other product. The NPRA website provides a list of the approved registered products (http://npra.moh.gov.my/index.php/recent-updates/new-products-indication/new-products-approved-quest2).

Appendix 4 of the DRGD also provides detailed guidelines for registration of health supplements. Information provided includes definition of health supplement, active ingredients, maximum daily levels of vitamins and minerals, health supplement claims, and specific dossier requirements for registration of health supplements. Also included in the annex is a list of permitted functional claims for several vitamins and minerals as well as probiotics. The permitted claim for probiotics is "helps to improve a beneficial intestinal microflora". It should be noted that the permitted text for a claim is slightly different from that permitted under the Food Regulations, which refers to "improving intestinal or gut function". It is not an exhaustive list, and the authority may consider reviewing applications for other claims with scientific substantiation.

The NPRA website does not provide a list of approved probiotics. However, specific probiotic cultures can be searched from the active ingredients list found on the "product search" tab in the NPRA website (http://npra.moh.gov.my/q3plus/index.php/products-search). For example, keying in "lactobacillus" or "bifidobacterium" will list the various species of these two genera available. The product name, registration number, or notification number as well as the company holding this registration will be displayed.

## 38.6.5 Probiotic Cultures Permitted in Regulation 26A of the Food Regulations Malaysia 1985

1. ***Bifidobacterium* sp.**
   Synonyms: *"Tissieria," "Bifidobacterium"*
   *B. bifidum* Bb-02
   *B. breve* strain Yakult
   *B. breve* M-16V
   *B. animalis* subsp. *lactis* (BB-12)
   *B. lactis* HN019
   *B. lactis* Bl-04
   *B. lactis* Bi-07
   *B. lactis* 420
   *B. lactis* CNCM I-3446
   *B. longum* BB536
   *B. longum* BB-46
   *B. longum* Rosell-175
   *B. longum* ATCC BAA-999

2. ***Lactobacillus* sp**.
   *L. acidophilus* LA-5
   *L. acidophilus* NCFM
   *L. acidophilus* La-14
   *L. acidophilus* Rosell-52
   *L. casei Shirota*
   *L. johnsonii* La 1/Lj 1
   *L. johnsonii* CNCM I-1225
   *L. paracasei* subsp. *paracasei* (L. CASEI 01)
   *L. paracasei* subsp. *paracasei* (L. CASEI 431)
   *L. paracasei* Lpc-37
   *L. paracasei* CNCM I-2116
   *L. plantarum* Lp-115
   *L. rhamnosus* (LGG)
   *L. rhamnosus* Lr-38
   *L. rhamnosus* HN001
   *L. rhamnosus* Rosell-11
   *L. rhamnosus* CGMCC 1.3724
   *L. salivarius* Ls-33
   *L. reuteri* DSM 17938*

**Note***

1. The addition is only allowed in infant formula, follow-up formula, and formulated milk powder for children.
2. A statement "THIS PRODUCT CONTAINS *L. reuteri* DSM 17938 AND NOT RECOMMENDED FOR INFANTS WITH A HISTORY OF GASTROINTESTINAL SURGERY" shall be written in the principal display panel in the label of a package containing infant formula and follow-up formula, in not less than 4-point lettering and in bold.

# 38.7 PHILIPPINES

The guidelines on probiotics in the Philippines are provided in the Circular No. 16 s. 2004 issued by the Food and Drug Administration (FDA). This circular was signed by its former director, Professor Leticia-Barbara B. Gutierrez, on October 26, 2004. The FDA was formerly known as the Bureau of Food and Drugs (BFAD).

The guidelines provide the following definition of probiotics: "Probiotic is a dietary supplement based on living microorganisms which when administered in sufficient quantity, has a beneficial effect on the host organism, improving the equilibrium of the intestinal microflora (Gaarner and Scharfema, 1998)".

According to BFAD, bacterial strains considered probiotics are *Lactobacilli, Bifidobacteria*, non-pathogenic strains of *Streptococcus, Sacchromyces boulardi*, and *Bacillus clausii*. In cases where the bacterial strains do not belong to any of these strains, there is a need to analyze the kind of species, and proof of safe use must be provided. The Guidelines for the Evaluation of Probiotics in Food by WHO-FAO (2002) are used as reference. BFAD emphasizes that the effectivity of a probiotic is due to its survival in the digestive tract, adherence to the mucosal epithelial tracts, and longer viability when used or stored. With probiotics, the host organisms also benefit, and the intestinal ecology are protected and enhanced.

A number of documents are required by BFAD to ensure that the probiotics are safe: (1) the antibiotic resistance patterns should be determined and metabolic activities such as the production of D-lactate and bile salt deconjugation must be assessed; (2) its side effects on humans should also be carefully studied, and there should be an epidemiological surveillance of any adverse incidents among consumers; (3) testing of toxins if the strains are known to be toxin producers should be done following the recommendation of EU Scientific Committee on Animal Nutrition (SCAN, 2000); and (4) hemolytic activity should be determined if the strains possess hemolytic potential.

BFAD also specifies claims for probiotics. Such claims are approved as food supplements and can be reflected on the product labels to advertise and promote the product. These claims generally include improvement of digestion and lactose utilization, and as probiotic strains increase, resistance to infections builds up in the intestine. Probiotics are drugs if they demonstrate a therapeutic effect on humans after sound scientific investigation.

The guidelines on probiotics further describe the evaluation process of probiotics for drug and food use (Figures 38.2 and 38.3). For drug use, the strain is identified phenotypically and genotypically and deposited in the international collection. The functional characterization and safety assessment of identified strains are determined through *in vitro* tests using animals and humans, respectively. To check the efficacy of strain/product, a randomized, double-blind, placebo-controlled (DBPC) human trial is conducted, preferably twice to confirm results. An effectiveness trial follows to compare probiotics with standard treatment of a specific condition. Once proven efficacious, the probiotic drug is labelled indicating the genus/species, minimum numbers of viable bacteria at the end of shelf life, proper storage condition, and corporate contact details for consumer information. When used as food, after identification, functional characterization, and safety assessment, it is also labelled as noted to provide information for the consumer.

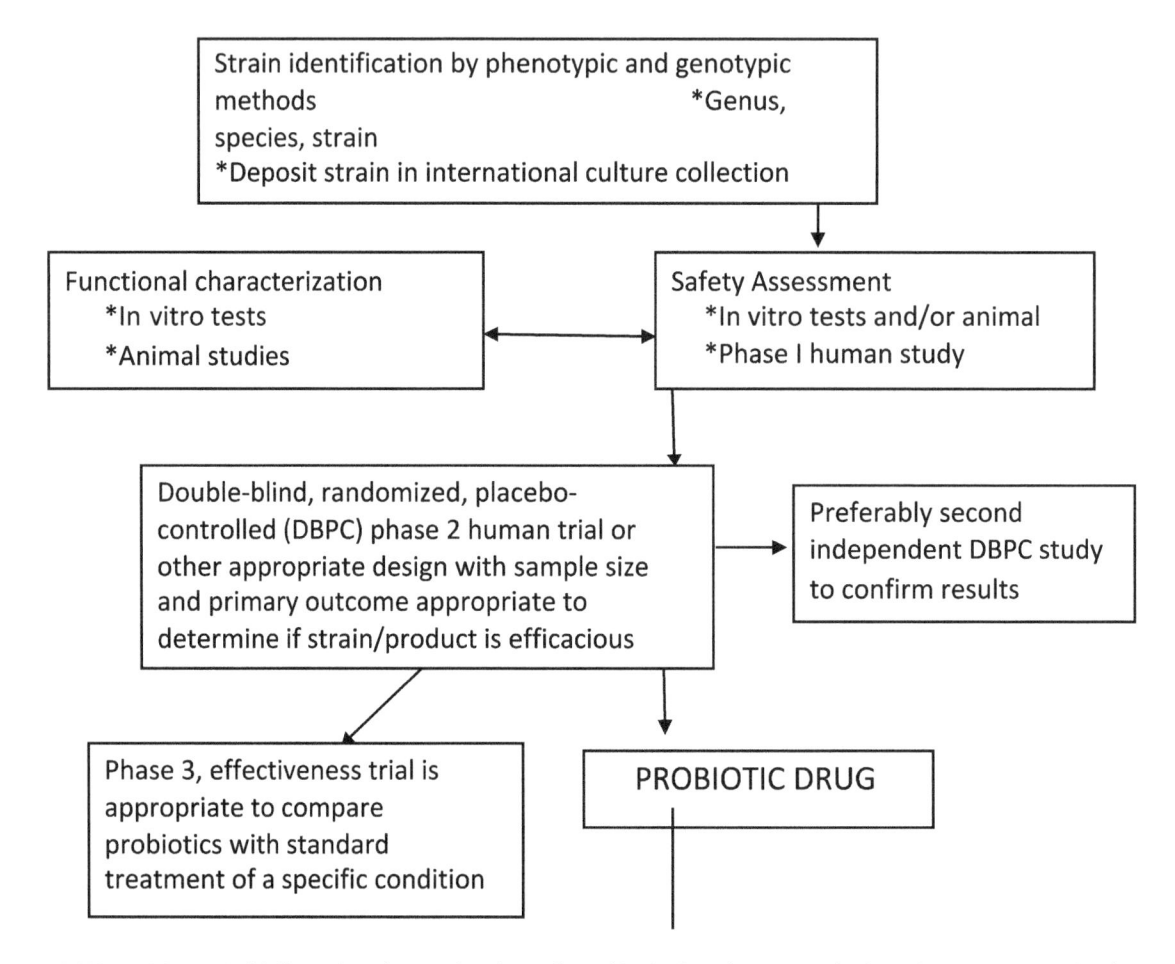

**FIGURE 38.2**  Guidelines for the evaluation of probiotic for drug use. (Taken from Bureau Circular No. 16 s.2004 of BFAD, Philippines.)

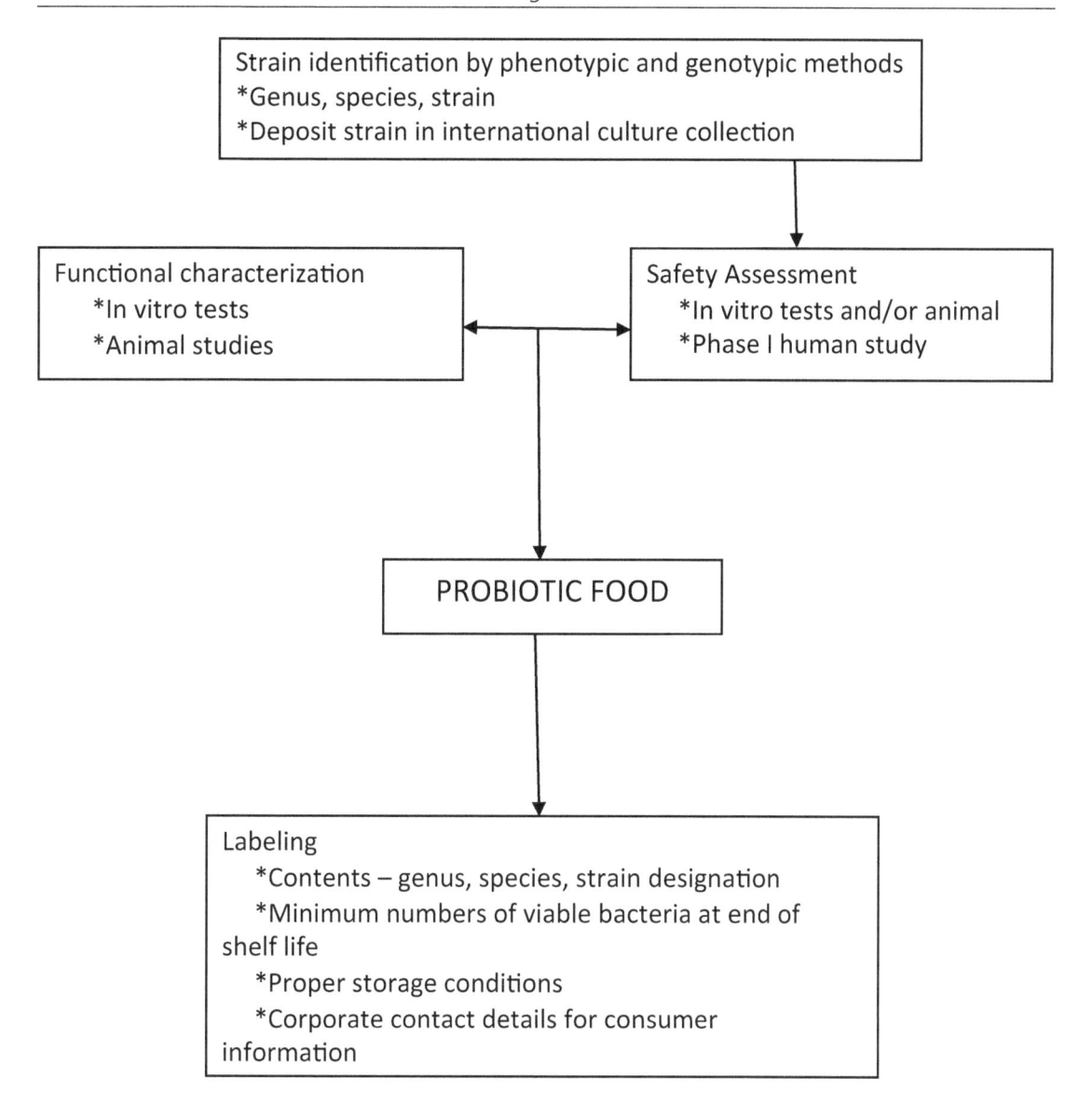

**FIGURE 38.3**   Guidelines for the evaluation of probiotic food use. (Taken from Bureau Circular No. 16 s.2004 of BFAD, Philippines.)

The approved probiotics in the Philippines are as follows:

1. Lactobacilli
2. Bifidobacteria
3. Nonpathogenic strains of *Streptococcus*
4. *Saccharomyces boulardii*
5. *Bacillus clausii*

The list is very general, especially for the genera lactobacilli, bifidobacteria, and streptococci, and has not changed since 2004. There is a need to update this list with the current development of probiotics research wherein health claims should be strain specific.

# 38.8 SINGAPORE

Probiotics may be used in food and health supplements. Health supplements are currently not subject to approval and licensing by the Health Science Authority (HAS) before they can be marketed in Singapore (HSA, 2021). However, dealers of health supplements, including probiotic products, are accountable for ensuring that their products do not contain prohibited ingredients, synthetic drugs, or toxic heavy metals above legally permissible limits (HSA, 2019).

The Singapore Food Agency (SFA) has neither published a list nor a recommended dosing of probiotics-based food products. Dealers are responsible for the safety and quality of the product they produce and sell. In principle, all ingredients and additives used in food must be safe, and probiotics are not excluded.

## 38.8.1 Safety Assessment

When evaluating applications for the use of new food ingredients and additives in Singapore, SFA takes into consideration the safety of the ingredients or additives, as established by the Joint FAO/WHO Expert Committee on Food Additives (JECFA). Determination of maximum permitted levels is done by referencing the recommendations of the Codex Alimentarius Commission (CAC) and the levels permitted in major developed countries, as well as conducting an estimation of the potential intake of the additive by the local population based on the local dietary intake pattern. In situations where ingredients have not been evaluated by JECFA or accepted for use by CAC, SFA takes reference from relevant safety assessments conducted by food safety authorities in major developed countries such as Australia, Canada, the European Union (EU), New Zealand, Japan, and the United States of America (USA). In the case of probiotic microorganisms, SFA also references the guidelines within "FAO Food and Nutrition Paper 85: Probiotics in Food, Health and Nutritional Properties and Guidelines for Evaluation".

Although SFA has not published a list of probiotics allowed for use in food, SFA allowed strains of bifidobacteria and lactobacillus that have a long-proven history of safe use in food (such as *Bifidobacterium bifidum*, *Bifidobacterium lactis*, *Lactobacillus acidophilus*, *Lactobacillus delbrueckii*, *Lactobacillus casei*) to be used in suitable categories of food products. In recent years, SFA has also assessed and allowed the use of *Bacillus coagulans* GBI-30, 6086, and *Bacillus coagulans* SANK-70258 in food.

## 38.8.2 Labelling and Claims

The food regulations do not define the term "probiotics". However, for food products to claim the presence of probiotics, the exact species of the probiotic present in the product must be declared on the product label, and the food manufacturer or importer has to ensure that the viable count of the probiotic present in the product is able to bring about the claimed effect (SFA, 2019, 2021). SFA has allowed the use of the following claims related to probiotics when these criteria are met:

1. Probiotics to help maintain a healthy digestive system.
2. Probiotics to help in digestion.
3. Probiotics to help maintain a desirable balance of beneficial bacterial in the digestive system.

# 38.9  CHINA TAIWAN

At present, the Taiwanese government focuses very intensely on food safety, especially on both chemical and microbiological hazards. The Taiwan Food and Drug Administration (TFDA) is the statutory body responsible for the management of food safety in Taiwan. Food products must comply with the Act Governing Food Safety and Sanitation, together with its enforcement rules and a series of food standards promulgated by the TFDA. The term "foods" as used in this law refers to goods provided to people for eating, drinking, or chewing, and their raw materials (TFDA, Ministry of Health and Welfare, Executive Yuan, Taiwan 2018). In Taiwan, LAB strains used in food can be classified into two categories according to their functions. The first category acts as ingredients or processing aids for food. The second category involves the probiotic functions.

## 38.9.1  Regulating the Use of LAB as Ingredients or Processing Aids for Food

The TFDA in Taiwan lists the raw materials that could be used in food products on the food raw materials integration inquiry platform (https://consumer.fda.gov.tw/Food/Material.aspx?nodeID=160), including microorganisms (Table 38.8). The raw materials in the list require no further toxicity test if they satisfy one of the following two conditions:

1. The raw materials of the product are conventional foodstuffs and are usually consumed as processed food.
2. There is a complete academic literature report on the toxicity safety of the product and a record of human consumption; the raw materials, composition of ingredients, and manufacturing procedure of the product are completely in line with the findings stated in the academic literature report submitted.

**TABLE 38.8**   List of Microorganisms Allowed in Food Products Bulletined by the Department of Health (Executive Yuan, Taiwan, 2009)

| NO. | TYPE | STRAIN |
|---|---|---|
| 1 | Lactic acid bacteria | *Bacillus coagulans* |
| 2 | Saccharolytic bacteria | *Bacillus mesentericus* |
| 3 | | *Bacillus natto, Bacillus subtilis* |
| 4 | Lactic acid bacteria | *Bifidobacterium adolescentis* |
| 5 | Lactic acid bacteria | *Bifidobacterium bifidum* |
| 6 | Lactic acid bacteria | *Bifidobacterium breve* |
| 7 | Lactic acid bacteria | *Bifidobacterium infantis* |
| 8 | Lactic acid bacteria | *Bifidobacterium lactis; Bifidobacterium animalis* subsp. *lactis* |
| 9 | Lactic acid bacteria | *Bifidobacterium longum* |
| 10 | | *Clostridium butyricum* |
| 11 | Lactic acid bacteria | *Lactobacillus acidophilus* |
| 12 | Lactic acid bacteria | *Lactobacillus bifidus* |
| 13 | Lactic acid bacteria | *Lactobacillus brevis* |
| 14 | Lactic acid bacteria | *Lactobacillus bulgaricus* |

*(Continued)*

**TABLE 38.8** (Continued) List of Microorganisms Allowed in Food Products Bulletined by the Department of Health (Executive Yuan, Taiwan, 2009)

| NO. | TYPE | STRAIN |
|---|---|---|
| 15 | Lactic acid bacteria | *Lactobacillus casei,* L *Lactobacillus casei* subsp. *rhamnosus* |
| 16 | Lactic acid bacteria | *Lactobacillus cremoris* |
| 17 | Lactic acid bacteria | *Lactobacillus delbrueckii* |
| 18 | Lactic acid bacteria | *Lactobacillus delbrueckii* subsp. *bulgaricus* |
| 19 | Lactic acid bacteria | *Lactobacillus fermentum* |
| 20 | Lactic acid bacteria | *Lactobacillus gasseri* |
| 21 | Lactic acid bacteria | *Lactobacillus helveticus* |
| 22 | Lactic acid bacteria | *Lactobacillus kefir* |
| 23 | Lactic acid bacteria | *Lactobacillus lactis* |
| 24 | Lactic acid bacteria | *Lactobacillus lactis* subsp. *lactis* |
| 25 | Lactic acid bacteria | *Lactobacillus paracasei* |
| 26 | Lactic acid bacteria | *Lactobacillus plantarum* |
| 27 | Lactic acid bacteria | *Lactobacillus reuteri* |
| 28 | Lactic acid bacteria | *Lactobacillus rhamnosus* |
| 29 | Lactic acid bacteria | *Lactobacillus salivarius* |
| 30 | Lactic acid bacteria | *Lactobacillus sporogenes* |
| 31 | Anka | *Monascus anka, Monascus berkeri, Monascus pilosus, Monascus purpureus, Monascus ruber* |
| 38 | | *Propionibacterium freudenreichii, Popionibacterium shermanii* |
| 33 | Beer yeast, Brewer's yeast | *Saccharomyces cerevisiae* |
| 34 | Zinc yeast | *Saccharomyces cerevisiae* |
| 35 | Selenium yeast | *Saccharomyces cerevisiae* |
| 36 | Chromium yeast | *Saccharomyces cerevisiae* |
| 37 | Iron yeast | *Saccharomyces cerevisiae* |
| 38 | Magnesium yeast | *Saccharomyces cerevisiae* |
| 39 | Potassium yeast | *Saccharomyces cerevisiae* |
| 40 | Molybdenum yeast | *Saccharomyces cerevisiae* |
| 41 | Yeast | *Saccharomyces fragilis, Kluyveromyces fragilis* |
| 42 | Lactic acid bacteria | *Sporolactobacillus inulinus* |
| 43 | Lactic acid bacteria | *Streptococcus lactis* |
| 44 | Lactic acid bacteria | *Streptococcus salivarius* subsp. *thermophilus* |
| 45 | Lactic acid bacteria | *Streptococcus thermophilus* |
| 46 | Yeast | *Saccharomyces cerevisiae* ssp. *chevalieri* |
| 47 | Lactic acid bacteria | *Lactobacillus pentosus* |
| 48 | Lactic acid bacteria | *Streptococcus faecalis* |
| 49 | Yeast | *Candida utilis* |
| 50 | Lactic acid bacteria | *Lactococcus lactis* subsp. *cremoris* |
| 51 | Lactic acid bacteria | *Lactococcus lactis* subsp. *lactis* |
| 52 | Lactic acid bacteria | *Lactococcus lactis* subsp. *lactis* biovar *diacetylactis* |
| 53 | Lactic acid bacteria | *Leuconostoc mesenteroides* subsp. *cremoris* |

The list includes the common LAB genera in food fermentations, such as *Lactobacillus*, *Lactococcus*, *Leuconostoc*, *Bifidobacterium*, and *Streptococcus*. Of these, *Lactobacillus*, *Leuconostoc*, and *Streptococcus* historically represent the major genera in the fermented dairy products, such as yogurt and cheese. Three species of genus *Streptococcus* in the list associated with food fermentations are *S. thermophilus*, *S. lactis*, and *S. faecalis*.

*Bifidobacterium*-fermented products do not possess the typical desirable flavor associated with yogurt or other dairy products due to acetic acid production (Lourens-Hattingh and Viljoen 2001). However, the bifidobacteria group is considered a "probiotic". Bifidobacteria are one of the hundreds of beneficial bacteria that inhabit the body's intestinal tract. *Bifidobacterium* is better to culture along with other lactic acid bacteria to improve the flavor (Modler 1994).

*Bac. coagulans* has also been assessed for safety as a food ingredient by the TFDA. *Bac. coagulans*, initially considered a spore-forming *Lactobacillus*, is a lactic acid–forming bacterial species within the genus *Bacillus*. This strain is often marketed as *L. sporogenes* as probiotics in improving the vaginal flora and improving abdominal pain and bloating in irritable bowel syndrome patients (Sanders et al. 2003).

For importing microorganisms as food ingredients in the TFDA list, besides meeting Taiwan's food import regulations, the following should be documented for TFDA approval:

1. Taxonomy report including scientific name and authority, history of cultures, characteristics of cultures, and pathogenic effects
2. Report of laboratory testing for the purity of the microorganism
3. Photocopy of the import registration license
4. The name and address of the manufacturer

Most information regarding Taiwan's food import regulations is available on the internet at: www.fda.bov. tw. If the LAB strains are not on the list (Table 38.8), the TFDA must first be notified of the new culture before its marketing, and the strain's safety and efficacy must be documented.

In Taiwan, for registration of new raw materials used in food products, including microorganisms, the following documents and materials are required for application of TFDA approval before manufacturing and marketing:

1. An application form including:
   a. The original copy and its duplicate of reports on ingredient list, product specification, and nutrient analysis. The original manufacturer shall issue these reports within 1 year.
   b. The ingredient list report shall specify detailed composition of raw materials and food additives.
   c. A copy of a summarized diagram on the manufacturing process.
   d. Microorganism should include taxonomy (scientific name and authority, history of cultures), characteristics of cultures, and safety aspects.
   e. The ingredient list shall specify the Recommended Daily Dosage of the product.
2. An official certificate attesting to the legitimacy of the original manufacturer.
3. A duplicate copy of the business license of the applicant.

## 38.9.2 Regulating the Use of LAB as Probiotics

1. The Health Food Control Act
   According to the Health Food Control Act, LAB strains claiming health benefits should follow the Health Food Control Act in Taiwan. The Health Food Control Act was first promulgated on February 3, 1999, and amended on May 17, 2006. In the act, health food is defined as "food with specific nutrient or health maintenance effects which is especially labeled or advertised, and do not aim at treating or remedying human diseases". "Health maintenance effects" are effects recognized by the TFDA as those which promote the health of citizens or reduce the risk of serious illnesses. So far, the TFDA has recognized 14 such effects:

a. Regulation of blood lipid
b. Promotion of gastrointestinal functions
c. Maintenance of bone health
d. Maintenance of dental health
e. Regulation of the immune system
f. Regulation of blood sugar level
g. Protection of the liver
h. Postpone aging
i. Anti-fatigue
j. Regulating blood pressure
k. Prevention of accumulating body fat
l. Assistance in iron adsorption
m. Joint health
n. Assistance in adjusting allergic constitution

These effects are not therapeutic, such as treating or remedying human diseases. Health food also includes food in tablet, capsule, powder, or oral liquid forms. According to the TFDA, products with health food claims must prove that they possess the ability to contribute to the health of those consuming them. Any product claiming to be a health food must receive the TFDA's approval before being marketed. Figure 38.4 shows the flowchart of application for a health food permit. There are two tracks of application. If the products satisfy the requirements of the special item standards (so far, only fish oil and anka products), the following reports are no longer needed: safety assessment report, health care effect assessment report, and

(a)

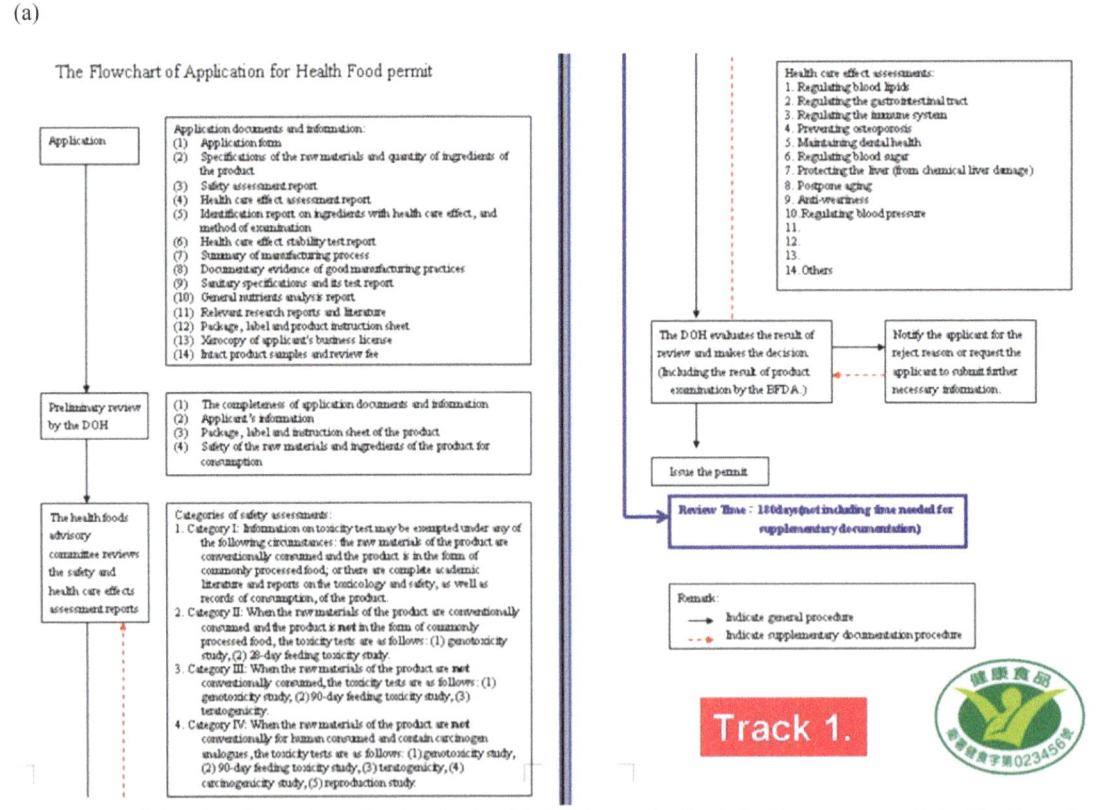

**Figure 38.4(a)** Flowchart of application for health food permit, Track 1. (Department of Health, Executive Yuan, Taiwan, 2009)

(b)

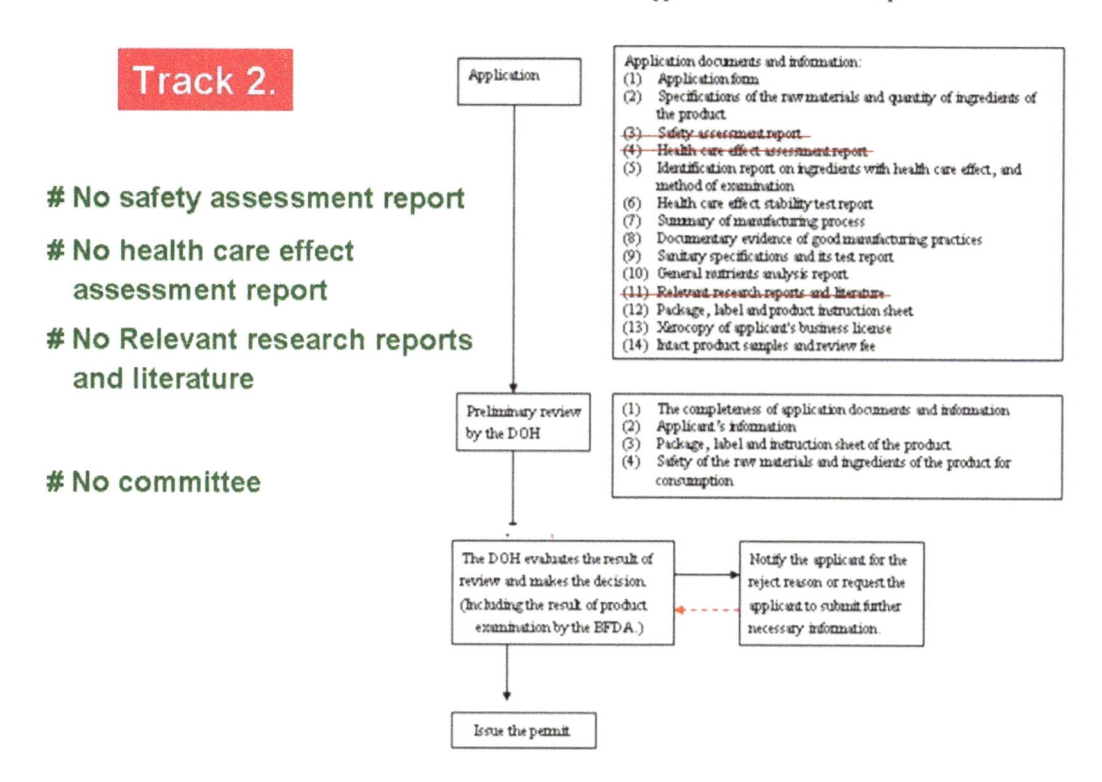

FIGURE 38.4(b)    Flowchart of application for health food permit, Track 2 (Department of Health, Executive Yuan, Taiwan, 2009)

relevant research reports and literatures (Track 2). Otherwise, the product follows the flowchart in Track 1. The information regarding the Health Food Control Act is available at www.fda.gov.tw.

When the products are approved by TFDA as health food, the following information, in Chinese and commonly used symbols, must be conspicuously displayed on the containers, packaging, or written instructions of the health foods:

    a.  Product name
    b.  Name and weight or volume of the contents (if a mixture of two or more components, they must be listed separately)
    c.  Name of food additives
    d.  Expiration date, method, and conditions of preservation
    e.  Name and address of the responsible business operator; the name and address of the importer must be specified if the health food is imported
    f.  The approved health care effects
    g.  Reference number of the permit, the legend of "health food", and standard logo
    h.  Intake amount and other important messages for the consumption of the health food along with other necessary warnings
    i.  Nutrient and its content
    j.  Other material facts designated by the TFDA
    k.  Country of origin

2. Health food related to lactic acid bacteria

Up to now (August 2022), among 428 food products obtaining health food permits authorized by the TFDA (valid and invalid), 43 items are lactic acid bacteria–related products with valid health food certificates (Table 38.2). Dairy-related products, such as drinking yogurts and milk powders, are the major food types. However, LAB products in powder, tablet, and capsule forms are also becoming popular on the Taiwanese market due to their easy consumption and preservation.

Among the 43 LAB-related products, the most common strains associated with health food products are *L. acidophilus*, *L. casei*, *L. delbrueckii*, *L. paracasei*, *B. lactis*, *B. longum*, and *S. thermophilus*. Several other lactic acid bacteria, such as *L. sporogenes*, *L. fermentum*, *L. johnsonii*, *L. salivarius*, *L. gasseri*, and *Bac. coagulans*, are also utilized in health products. The strains used in the 43 LAB-related products all appear on the list of micro-organisms published by TFDA. The safety assessment of these LAB strains is not required. However, the safety assessment of *L. rhamnosus* T cell-1 in TCELL-1LAB powder, including genotoxicity test and 28-day feeding toxicity test, is provided by the manufacturer.

Considering the functional properties linked to lactic acid bacteria, all items fall into the following seven functional categories:

a. Promotion of gastrointestinal functions

This category is the most common functional property associated with LAB. Twenty eight of the 43 LAB-related products are associated with promotion of gastrointestinal functions specified in assistance of increasing population of intestinal probiotic micro-flora, limiting the growth of *Clostridium perfringens*, and keeping the balance of intestinal microflora. The LAB strains involved in promotion of gastrointestinal functions mostly belong to genus *Lactobacillus* (*L. delbrueckii*, *L. acidophilus*, *L. paracasei*, *L. rhamnosus*, and *L. fermentum*) and *Bifidobacterium* (*B. longum*, *B. lactis*, and *B. bifidum*). Other species such as *S. thermophilus* and *Bac. coagulans* are also included in this category.

It is worth mentioning that the AB drinking yogurt manufactured by UniPresident also claims the ability to inhibit *Helicobacter pylori* growth (Table 38.9). LAB are thought to aid in the treatment of *H. pylori* infections (which cause peptic ulcers) in adults when used in combination with standard medical treatments (Eaton et al. 1991; Hamilton-Miller 2003).

b. Regulation of the immune system and assistance in adjusting allergic constitution. Eight of the LAB-related products belong to the assistance in adjusting allergic constitution category. Five of the 43 LAB-related products claim the property of regulation of the immune system, also mainly in anti-allergic effects. Many studies have examined the efficacy of LAB supplementation in immunoregulation (Hong et al. 2009). Although the detailed mechanisms of these effects exerted by LAB have not yet been clarified, cytokines induced by LAB are considered to play key roles in immunoregulation (Shida et al. 2006).

To evaluate the anti-allergic effects of LAB, oral administration of LAB-related products in ovalbumin (OVA)-sensitized mice was conducted. The abilities of enhancing the production of T helper (Th) 1 cytokines [interferon (IFN)-γ, interleukin (IL)-2] and inhibiting the secretion of Th2 cytokines (IL-4, IL-5, IL-10) were determined (Matsuzaki et al. 1998). The effect of suppressing serum IgE and IgG1 secretion in a food allergy model was also investigated for LAB anti-allergic effects (Shida et al. 2002).

The LAB strains involved in the regulation of the immune system and assistance in adjusting allergic constitution, besides one yogurt product with *Streptococcus thermophilus*, all belong to genus *Lactobacillus*, including *L. salivarius* subsp. *salicinius*, *L. johnsonii* EM1, *L. paracasei* 33, *L. plantarum* GMNL-662, *L. casei* Shirota, *L. gasseri* PM-A0005, and *L. salivarius* PM-A006.

c. Regulation of blood lipid, prevention of accumulating body fat, and regulation of blood sugar level.

All three categories are related to metabolic syndrome.

Two (Quaker high-calcium non-fat milk powder and 263 probiotic capsules), one (263 probiotic capsules), and one (ADR-1 probiotic capsules) LAB products received a health food permit issued by the TFDA in regulation of blood lipid, prevention of accumulating body fat, and regulation of blood sugar level, respectively. Regulation of blood lipid is specified in lowering both triglyceride level and low-density lipoprotein (LDL) level as well as increasing high-density lipoprotein (HDL) level. Several studies have demonstrated in animal models that a range of LAB is able to lower serum cholesterol levels. The possible mechanisms involved breaking down bile in the gut, thus inhibiting its reabsorption (which enters the blood as cholesterol). Some human trials also demonstrated that dairy foods fermented with specific LAB can produce modest reductions in total and LDL cholesterol levels (Wollowski et al. 2001).

Prevention of accumulating body fat is specified in reducing fat accumulation under strict nutritional balance and calorie control, as well as proper exercise conditions. The possible mechanism might involve modulation of microbiota and circulating metabolites (such as short chain fatty acids), enhanced susceptibility to reduced body fat accumulation through improvement of energy expenditure, and inhibition of appetite (Chen et al. 2018). For regulation of blood sugar level, previous clinical study (Zhang et al. 2015) suggested that consuming probiotics may improve glucose metabolism by a modest degree, with a potentially greater effect when the duration of intervention is ≥2 months or multiple species of probiotics are consumed.

d. Maintenance of dental health

One item, Dental Lac Troches with *L. paracasei*, is associated with the property of maintaining dental health. *L. paracasei* could limit the growth of *S. mutans*, which is a human odontopathogen. Loesche (1986) indicated that acid uricity appears to be the most consistent attribute of *S. mutans* that is associated with both its colonization at stagnant areas and its cariogenicity. Köhler et al. (1984) suggested that treatment strategies that interfere with the colonization of *S. mutans* may have a profound effect on the incidence of dental decay in human populations.

It is also interesting to note two trends in health food products being marketed in Taiwan. The first one is that one health food item claims more than one functional property. The second is the multiple probiotic strains used in one product. Research is emerging on the potential health benefits of multiple probiotic strains as a health supplement as opposed to a single strain. The human gut has more than 500 types of microbes. It is thought that this diverse environment may benefit from multiple probiotic strains. Different strains may also be associated with different health benefits. More consumers in Taiwan prefer multiple functional products.

Most well-known LAB have a long and safe history of use in food. They are generally considered safe. However, other less-well-known LAB may require more specialized knowledge, as they are isolated and introduced into foods. According to Taiwan food safety regulatory requirements, it is incumbent on the food industry to ensure that these newly discovered microbial food cultures are safe. Additionally, the LAB strains providing health benefits should follow the Health Food Control Act in Taiwan.

**TABLE 38.9**　Lactic Acid–Related Products with Health Food Permits Authorized by TFDA in Taiwan by August 2022

| | PRODUCT BRAND | TYPE | LAB STRAINS | HEALTH BENEFITS |
|---|---|---|---|---|
| 1 | UniPresident AB drinking yogurt | Drinking yogurt | L. acidophilus (La-5)<br>B. lactis (Bb-12) | Promotion of gastrointestinal functions |
| 2 | Bifidobacterium longum BB536 | Powder | B. longum BB536 | Promotion of gastrointestinal functions |
| 3 | Complex probiotics | Powder | B. longum BB536<br>B. lactis Bb12<br>L. acidophilus<br>L. casei | Promotion of gastrointestinal functions: |
| 4 | eN-Lac Intestinal Capsules | Capsules | L. sporogenes<br>L. plantarum<br>L. fermentum | Promotion of gastrointestinal functions |
| 5 | Kuang-Chuan low fat with microencapsulated probiotics | Drinking yogurt | B. longum | Promotion of gastrointestinal functions |
| 6 | King Car lactic acid bacteria | Powder and tablet | B. lactis<br>L. acidophilus<br>L. paracasei | Promotion of gastrointestinal functions |
| 7 | Grape King LGG Probiotics | Powder and tablet | L. casei sp. Rhamnosus GG | Promotion of gastrointestinal functions |
| 8 | Quaker high calcium nonfat milk powder | Milk powder | L. acidophilus<br>L. casei<br>B. lactis | Promotion of gastrointestinal functions<br>Regulation of blood lipid |
| 9 | Yakult | Drinking yogurt | L. casei Shirota | Promotion of gastrointestinal functions |
| 10 | Quaker family low-fat milk powder probiotics formula | Milk powder | L. acidophilus<br>L. casei<br>B. lactis | Promotion of gastrointestinal functions |
| 11 | TCELL-1 LAB powder | Powder | L. rhamnosus T cell-1 | Promotion of gastrointestinal functions |
| 12 | Chang La Probiotics | Tablet | B. longum<br>L. sporogenes | Promotion of gastrointestinal functions |
| 13 | FloraGuard | Beverage | L. acidophilus<br>B. bifidum<br>L. bulgaricus<br>B. longum<br>S. thermophilus | Promotion of gastrointestinal functions |
| 14 | ProBio PCC | Capsule | L. fermentum | Promotion of gastrointestinal functions |
| 15 | Grape King probiotics | tablet | B. longum | Promotion of gastrointestinal functions |

| | | | | |
|---|---|---|---|---|
| 16 | Quaker follow-up milk powder probiotics formula | Milk powder | *L. acidophilus* *L. casei* *B. lactis* | Promotion of gastrointestinal functions |
| 17 | Kuangchuan milk flavor yogurt | Drinking yogurt | *B. lactis* | Promotion of gastrointestinal functions |
| 18 | Yakult Blueberry high calcium yogurt | Drinking yogurt | *L. casei* Shirota *S. thermophilus* | Promotion of gastrointestinal functions |
| 19 | Yakult Strawberry high calcium yogurt | Drinking yogurt | *L. casei* Shirota *S. thermophilus* | Promotion of gastrointestinal functions |
| 20 | Yakult Blueberry high calcium yogurt | Drinking yogurt | *Lactobacillus casei* Shirota *S. thermophilus* | Promotion of gastrointestinal functions |
| 21 | Sint *Lactobacillus salivarius* capsules | Capsule | *L. salivarius* PM-A006 | Regulation of the immune system |
| 22 | Yakult 300 light | Drinking yogurt | *L. casei* Shirota | Promotion of gastrointestinal functions Regulation of the immune system |
| 23 | UniPresident LP33 Functional Drinking Yogurt | Drinking yogurt | *S. thermophilus* *L. paracasei* *L. bulgaricus* | Assistance in adjusting allergic constitution |
| 24 | eN-Lac Plus Capsules | Capsule | *L. paracasei* 33 *L. fermentum* *L. acidophilius* | Assistance in adjusting allergic constitution |
| 25 | Dental Lac Troches | Tablet | *L. paracasei* ADP-1 | Maintenance of dental health |
| 26 | Kuangchuan fresh milk with probiotics | Milk | *B. longum* | Promotion of gastrointestinal functions |
| 27 | TTY BioPharmTen Billion probiotics | Tablet | *L. johnsonii* EM1 | Regulation of the immune system |
| 28 | Brands Xylooligo-Saccharides Plus LAB powder | Powder | *Bacillus coagulans* *L. acidophilus* | Promotion of gastrointestinal functions |
| 29 | LS99 functional drinking yogurt | Drinking yogurt | *L. salivarius* subsp. *salicinius* | Assistance in adjusting allergic constitution |
| 30 | Wei-Jei Capsule | Capsules | *L. casei* sp. *Rhamnosus* *L. gasseri* | Promotion of gastrointestinal functions |
| 31 | Chlorella Plus Bacillus | Tablets | *Bacillus coagulans* | Promotion of gastrointestinal functions |

*(Continued)*

| | PRODUCT BRAND | TYPE | LAB STRAINS | HEALTH BENEFITS |
|---|---|---|---|---|
| 38 | UniPresident AB drinking yogurt (no sugar) | Drinking yogurt | *L. acidophilus* (La-5) *B. lactis* (Bb-12) | Promotion of gastrointestinal functions |
| 33 | Taisugar fructo-oligosaccharide and lactic acid bacteria | Powder | *L. sporogenes* *L. acidophilus* | Promotion of gastrointestinal functions |
| 34 | Yakult 300 | Drinking yogurt | *L. casei Shirota* | Promotion of gastrointestinal functions |
| 35 | 263 lactic acid bacteria capsules | Capsule | *L. reuteri* | Regulation of blood lipid Prevention of accumulating body fat |
| 36 | ADR-1 probiotic capsules | Capsule | *L. reuteri* | Regulation of blood sugar level |
| 37 | Come-Ming Capsule | Capsule | *L. salivarius* | Assistance in adjusting allergic constitution |
| 38 | APF probiotic capsules | Capsule | *L. paracasei* 33 | Regulation of the immune system Assistance in adjusting allergic constitution |
| 39 | Mingshiling Capsules | Capsule | *L. johnsonii EM1* | Assistance in adjusting allergic constitution |
| 40 | IMMUPHYLA LCW23 *Lactobacillus paracasei* Capsules | Capsules | *L. paracasei* | Assistance in adjusting allergic constitution |
| 41 | PROTE 200 Probiotics | Powder | *L. paracasei* | Regulation of the immune system Assistance in adjusting allergic constitution |
| 42 | LG-55A plus | Capsules | *L. gasseri* PM-A0005 | Assistance in adjusting allergic constitution |
| 43 | GM-BMD probiotic capsules | Capsules | *L. plantarum* GMNL-662 | Bone health |

# 38.10 THAILAND

The food control system in Thailand includes setting standards and regulations based on risk analysis, pre-marketing control, and post-marketing control. Entrepreneurs who wish to produce or import food products for sale in the Kingdom of Thailand must apply for premises licenses and product registration/notification, respectively, before the period of sales. Advertising approval is also necessary if it displays any information containing claims about the quality and usefulness of the product.

The main purpose of Thailand's food law is to protect consumers from health hazards resulting from food consumption and to ensure that the information provided on the label of the food they consume is accurate. Therefore, food labels cannot be falsified or misleading, and food advertising must not contain any statement or visual presentation that will likely mislead consumers. This is particular to the label or advertisement related to health benefits.

Microorganisms intended for use in food can be classified depending on how they are consumed and used in food products. For example, they are used as food additives, processing aids, starter cultures, production of food enzymes and additives, and probiotics. Accordingly, essential Notifications of the Ministry of Public Health related to the use of microorganisms in food are:

a. Notification of MoPH no. 281 (2004) Food additives and its amendments, and the Announcement of the Food and Drug Administration (24 June 2005): Food additives shall have the qualities or standards according to Codex Advisory Specification for the Identity and Purity of Food Additives or The Announcement of the Food and Drug Administration. The additives listed include mold starter cultures, yeast and yeast powder, lactic acid bacteria, and enzyme Transglutaminase from *Streptoverticillium mobaraense var.*

b. Notification of MoPH no. 376 (2016), Novel Food: Novel food means: (1) any substance used as food or food ingredients which have been significantly used for human consumption less than fifteen years based on scientific or reliable evidence; (2) any substance used as food or food ingredients to which has been applied a production process not currently used, where that process gives rise to significant changes in the composition or structure of such food which affect their nutritional value, metabolism or level of undesirable substances or; (3) any food product contains either (1) or (2) as an ingredient. However, food additives and food obtained through certain techniques of genetic modification are excluded from this Notification.

c. Notifications and their Amendments announced specifically to the food products. They are, e.g. Notification of MoPH no. 203, 204, 208, 209 (2000) for fish sauces, vinegar, cream, and cheese; Notification of MoPH no. 226, 243 (2001) for clarified butter and Ghee, and meat products; Notification of MoPH no. 289 (2005) for fermented milk; and Notification of MoPH no. 317 (2010) for seasonings and condiments obtained by the hydrolysis of soy proteins.

d. Notification of MoPH no. 339 (2011) and no.346 (2012) The use of probiotic microorganisms in food, and The Announcement of the Food and Drug Administration (8 February 2018): More details are given in the sub-sessions below.

The approval of probiotics for food uses differs among countries and falls into two major groups. The first group consists of countries that do not have an approved list of probiotics, and approval is considered on a case-by-case basis. Each registration must provide detailed characterization and safety data (Indonesia and Singapore), or the consideration is based on other related regulations such as Novel food (Australia and New Zealand), FOSHU and FFC products (Japan). The second group consists of countries that announced positive or approved probiotic lists. These countries are, for example, Malaysia, Philippines, China, South Korea, India, and Canada. Thailand also falls into this group and currently has 24 approved microorganisms.

## 38.10.1 Probiotic Regulation

The use of probiotic microorganisms in foods is regulated by Notification of the Ministry of Public Health no. 339 (2011), as well as the Announcement of the Food and Drug Administration (8 February 2018). The Announcement, aiming to speed up the approval process for probiotics, declared new guidelines for the use of probiotic microorganisms in food. This guideline stipulates three clauses for the use of probiotics in food, and the use of a probiotic should be approved by FDA when it meets the requirements:

a. The probiotic is in the list of the microorganisms annexed to the Notifications of MoPH no. 339 and 346. In addition, the Announcement also added one more approved probiotic (*Lactobacillus plantarum* 299v) to the list.
b. Probable microorganisms remaining in the product should be no less than 106 CFU per 1 g of food throughout the shelf life of the food. The applicant must enclose the analysis results from government agencies, agencies/organizations accredited by the governments, or national or international agencies/organizations accredited by laboratory accreditation agencies according to international standards.
c. Any health claims related to the use of probiotics in food shall be applied for the health claim assessments according to the procedures in the Public Manual of the Food and Drug Administration-Thailand.

## 38.10.2 Definition

Probiotics are live microorganisms which when administered in adequate amounts confer a health benefit on the host, excluding (1) biotherapeutic agents, (2) beneficial microorganisms not used in food, and (3) genetically modified microorganisms (GMM).

## 38.10.3 Guidelines for the Use of Probiotic Microorganisms in Food

The use of probiotic microorganisms in food required FDA approval before markets under the condition set forth in the following notifications:

Condition A. Products contain with probiotics microorganisms listed in the Annex of this Notification (positive list) and viable probiotic microorganisms in such food shall be not less than $10^6$ CFU/g food at the end of its shelf life.
Condition B. Microorganisms other than those specified in the Annex, safety and properties of probiotic microorganism evaluation is required.

## 38.10.4 Guidelines for the Safety Assessment of Probiotic Microorganisms

The safety assessment requires characteristics of such probiotic microorganisms in accordance with the Guidelines for the Evaluation of Probiotics in Food, Joint FAO/WHO 2002 as follows:

1. Testing on identity at genus, species, and strain level, using the most current, valid methodology and nomenclature of microorganisms must conform to the current, scientifically recognized names.
2. The following are tests for probiotic characteristics:
   a. Resistance to gastric acid;
   b. Resistance to bile salt;
   c. Adherence to mucus and/or human epithelial cell and cell line;
   d. Bile salt hydrolase activity; and
   e. Other characteristics (if any) as the case may be.
3. Safety assessment of probiotics *in vitro* or *in vivo* and in human studies to evaluate body reaction to such probiotics are as follows:
   a. Antimicrobial resistance;
   b. Metabolic effect assessments such as D-lactate and bile salt deconjugation;
   c. Side effect in human intervention study;
   d. Epidemiological surveillance of adverse incidents after marketing;
   e. Test for toxin production, if the strain under evaluation belongs to a species that is known toxin producer;
   f. Test for hemolytic activity if the strain under evaluation belongs to a species with known hemolytic potential.

## 38.10.5 Guidelines for Health Claims of Probiotic Microorganisms

Health claims require the support of at least two well-designed human intervention studies for consideration of the efficacy of the probiotic on health. The following criteria should be taken into account in the design of the studies:

a. study groups that are representative of the target group
b. appropriate control group
c. an adequate duration of exposure and follow-up to demonstrate the intended effect
d. characterization of the study groups' background diet and other aspects of relevant of lifestyle
e. an amount of the food or food component consistent with its intended pattern of consumption
f. the influence of the food matrix and dietary context on the functional effect of the component
g. monitoring of subjects' compliance concerning intake of food or food component under test
h. the statistical power to test the hypothesis

## 38.10.6 Guidelines for Labeling of Probiotic Products

Labeling aspects of food products containing probiotic microorganisms should comply with the Notification of Ministry of Public Health entitled "Labeling of Prepackaged Foods, Nutrition Labeling and Additional Statements" as follows:

a. Approval claim statement, or wording such as "Probiotic microorganisms" or "Probiotics";
b. Genus, species, and strain of the probiotic(s);
c. A statement, "This product is not to treat, heal, cure or prevent disease";
d. Quantity and duration recommendation for intake to achieve a health effect;
e. Instructions for use and appropriate conditions for storage;
f. Corporate contact details for consumer information.

## 38.10.7 List of Approved Probiotic Microorganisms

1. *Bacillus coagulans*
2. *Bifidobacterium adolescentis*
3. *Bifidobacterium animalis*
4. *Bifidobacterium bifidum*
5. *Bifidobacterium breve*
6. *Bifidobacterium infantis*
7. *Bifidobacterium lactis*
8. *Bifidobacterium longum*
9. *Bifidobacterium pseudolongum*
10. *Enterococcus durans*
11. *Enterococcus faecium*
12. *Lactobacillus acidophilus*
13. *Lactobacillus crispatus*
14. *Lactobacillus gasseri*
15. *Lactobacillus johnsonii*
16. *Lactobacillus paracasei*
17. *Lactobaacillus plantarum 299V*
18. *Lactobacillus reuteri*
19. *Lactobacillus rhamnosus*
20. *Lactobacillus salivarius*
21. *Lactobacillus zeae*
22. *Propionibacterium arabinosum*
23. *Staphylococcus sciuri*
24. *Saccharomyces cerevisiae* subsp. *boulardii*

# 38.11　VIETNAM

Probiotics and probiotic products are a growing trend in Vietnam as Vietnamese peoples' awareness of health protection is rising together with the growth of the country's gross domestic product (GDP) as the income of the local people has improved regularly. It is also thanks to the state's policy of administrative reform and business environment improvement. As a result, the number of probiotic products in the market is increasing yearly, with contributions by locally produced and imported products.

Probiotic products for humans are under the category of over-the-counter (OTC) drugs and functional foods (supplemented food, health supplements, dietary supplements), in which *Bacillus claussi*, *Bacillus subtilis*, and *Lactobacillus acidophilus*, including combination with vitamins, may be listed as OTC drugs (MOH Vietnam, 2014; Government of Vietnam, 2018). As such, to ensure the safety of patients and the effect of the treatment, OTC drugs must be registered in accordance with the requirements of the ASEAN Common Technical Dossier (ACTD) and the local pharmaceutical law and its regulations.

## 38.11.1 Regulatory Framework of Probiotic Food

### 38.11.1.1 Definition

The specific regulatory framework of probiotic food has not been developed and implemented. Currently, probiotic food is classified as a supplemented food or health supplement as follows:

*Supplemented food* means conventional food supplemented with micronutrients and other elements conducive to health such as vitamins, minerals, amino acids, fatty acids, enzymes, probiotics, prebiotics, and other biologically active substances.

*Dietary supplement* means any product that is used to supplement a diet and to maintain, enhance, improve the healthy function of human body, reduce risk of disease. Dietary supplement contains one or more, or a combination of the following:

1. Vitamins, minerals, amino acids, fatty acids, enzymes, probiotics, and other bioactive substances.
2. Substances derived from natural sources, including animal, mineral, and botanical materials in the forms of extracts, isolates, concentrates, metabolites.
3. Synthetic sources of ingredients mentioned in (1) and (2).

A dietary supplement is presented in processed forms such as capsules, tablets, and preparations of granule, powder, liquids, and other processed forms, and dosage to use in small unit doses.

Decree 15/2018/ND-CP, signed and effective as of February 2, 2018, details and provides guidance on some aspects of the law on food safety in which health supplement/dietary is redefined in line with its definition in ASEAN guidelines. Under this new decree, many provisions of registration, labeling, advertising, quality inspection in customs clearance, and state management are changed, as outlined in the following sections.

## 38.11.2 Registration

Only probiotic foods (imported or local produced) which are classified as dietary supplements must be registered with the Vietnam Food Administration (VFA) as a pre-marketing approval process. The VFA is the highest competent authority at the central level to evaluate and approve the registration of products.

A dossier of application should include:

1. Certificate of free sale (CFS), certificate of exportation, or certificate of health, issued by the competent authority of the manufacturing/export country (applies to imported products).
2. Certificate of analysis, valid within 12 months by a laboratory with certified ISO 17025.
3. Scientific evidence of the functions of the finished product or ingredient. They must all be published in scientific journals or verified by a local scientific research body that is authorized by the competent authority. A version in the Vietnamese language is required.
4. Good Manufacturing Practice Certificate or equivalent, applied by July 1, 2019, for imported products.

## 38.11.3 Labeling

The mandatory contents of the labels of probiotic food should include the following:

| FOOD AND BEVERAGE (SUPPLEMENTED FOOD) | | DIETARY SUPPLEMENT | |
|---|---|---|---|
| 1 | Product name | 1 | Product name |
| 2 | Ingredient name | 2 | Ingredient name |
| 3 | Net content | 3 | Net content |
| 4 | Manufacturing date | 4 | Manufacturing date |
| 5 | Expiry date | 5 | Expiry date |
| 6 | Direction | 6 | Direction |
| 7 | Storage condition | 7 | Storage condition |

*(Continued)*

(Continued)

| FOOD AND BEVERAGE (SUPPLEMENTED FOOD) | DIETARY SUPPLEMENT |
|---|---|
| 8 Name and address of organization, individual responsible for products | 8 Name and address of organization, individual responsible for products |
| 9 Origin country | 9 Origin country |
| 10 Registration number | 10 Registration number |
| 11 Phrase "Supplemented food" | 11 Phrase "Health supplement" or "Dietary supplement" |
| 12 Allergens (if any) | 12 Statement "Attention: This product is not a medicine, and is not a substitute for medicines" |
| 13 Target users | 13 Allergens (if any) |
| 14 Warning | 14 Target users |
| | 15 Warning |

## 38.11.4 Claims

***Content Claim*:**

1. The main ingredients that create the effects of the product must be named and content enumerated. The other ingredients shall be enumerated next and sorted by weight in descending order;
2. A substance shall not be claimed if its content is below 10% RNI;
3. If the content of a substance is from 10% RNI, its content in a serving or 100 g of the product shall be specified.

***Health claim* :** Must reflect the product nature; a claim about the effects of ingredients having a main or combined effects shall be made only when there is scientific evidence; the effects of ingredients must not be claimed as effects of the product.

## 38.11.5 Clinical Study Requirements

Scientific evidence of the functions of the finished product or ingredient should be provided. To claim scientific evidence for the ingredient, the daily dose must be at least 15% of the ingredients described. Requirements for the efficacy test of functions are as follows.

1. The following products must be tested for efficacy on human health:
   a. Products that are claimed to support disease treatment;
   b. Products that have new effects which have not been recognized by other countries in the world;
   c. Products that have new active ingredients which have not been permitted;
   d. Health supplements put on the market for the first time which have different formulae from those of other products with scientific evidence;
   e. Products derived from plants and animals and put on the market for the first time, the composition of which is different from that of traditional medicines published in academic journals.
2. An efficacy test on human health must be conducted in an organization licensed to do scientific research in medicine. Products claimed to support disease treatment must be tested at provincial hospitals (or higher) licensed to do scientific research.
3. If the test for the effects on human health is conducted overseas, it must be performed by agencies accredited by competent authorities of home country, or test results must be published in academic journals.

4. The Food Administration of Vietnam (MOH) shall establish a Scientific Council which consists of experts in appropriate fields to assess the reports on product efficacy tests and announced scientific evidence. The organizational structure and operation of Scientific Council shall comply with the provisions of law.

# 38.12 AUSTRALIA AND NEW ZEALAND

## 38.12.1 Food Regulatory Framework

Australia shares a food regulatory system with New Zealand as a part of a bilateral agreement between these neighboring countries that was forged in the late 1990s. This agreement paved the way for the establishment of Food Standards Australia New Zealand (FSANZ), a statutory authority responsible for setting and maintaining food standards across the two countries (Food Standards Australia New Zealand Act, 1991; Joint Australia New Zealand Food Standards Code, 2000).

The Food Standards Code, the "Code", is the regulatory instrument of this bilateral arrangement. The Code is administered by FSANZ. Amendments to the Code are initiated via an application or proposal-based system by any stakeholder that has a vested interest in the regulation of food in Australia or New Zealand. These applications and proposals are managed in a systematic manner under which FSANZ adheres to statutory requirements relating to time frames, costs, and processing.

While FSANZ is responsible for development and administration of the Code, its enforcement is managed by the authorities of the various states and territories of Australia and New Zealand. These authorities have convened the Implementation Subcommittee for Food Regulation (IFSR) whose objective it is to ensure consistency in enforcement of the Code across the jurisdictions.

## 38.12.2 Regulation on Probiotic Cultures in Food

The Australia New Zealand Food Standards Code applies to all food sold, manufactured, imported, or handled for sale in Australia or New Zealand. Implicit in the Code is the presumption that all substances added to food are food. Thus, cultures and probiotics are considered ingredients, and, as with any other component, must be safe and suitable for use when added to food.

The Code does not define what a probiotic is. Furthermore, FSANZ does not issue a positive list of approved food ingredients other than circumstances where those ingredients are novel foods, nutritive substances, or used as food additives or processing aids or when a food is derived from gene technology. Probiotics are not used as food additives, nor are they used as processing aids, and in instances where they are derived from gene technology, they would be required to undergo a full assessment for approval by the FSANZ application process.

When considering if it is suitable to include a probiotic culture as a food ingredient, the manufacturer must decide if the microorganism would be considered a novel food. The Code defines "novel foods" as follows.

### 38.12.2.1 Definition of Novel Food

In this Code: *novel food* means a non-traditional food* that requires an assessment of the public health and safety considerations with regard to:

a. the potential for adverse effects in humans; or
b. the composition or structure of the food; or

c. the process by which the food has been prepared; or

d. the source from which it is derived; or

e. patterns and levels of consumption of the food; or

f. any other relevant matters.

\* *non-traditional food* means:

a. a food that does not have a history of human consumption in Australia or New Zealand; or

b. a substance derived from a food where that substance does not have a history of human consumption in Australia or New Zealand other than as a component of that food; or

c. any other substance where that substance, or the source from which it is derived, does not have a history of human consumption as a food in Australia or New Zealand.

The food manufacturer has a choice to make a self-assessment against the criteria outlined previously. Alternatively, details on the microorganisms can be submitted to the Advisory Committee on Novel Foods (ACNF) to make a recommendation about whether a food is novel. Note that the ACNF is chaired by FSANZ and representatives from Australian state and territory jurisdictions and the New Zealand Ministry for Primary Industries. The outcome of their assessments is not legally binding, but they are published periodically in a "Record of Views" document on the FSANZ website. If determined to be novel, a probiotic microorganism would require permission in the Code based on history of use and safety considerations.

There are also two specific provisions in the Code pertaining to live microorganisms when added to food. The first exists in Standard 2.9.1 Infant Formula Products, whereby clause 2.9.1–6 permits the addition of L($^+$) lactic acid–producing microorganisms to infant formula products. Live cultures are also mentioned in Schedule 4 to Standard 1.2.7 in relation to the effect of live *Lactobacillus delbrueckii* subsp. *bulgaricus* and *Streptococcus thermophilus* on lactose digestion when present and viable at or above $10^8$ CFU/g.

## 38.12.3 Regulation on Probiotic Health Claims in Food

Health and nutrient content claims in food are regulated under Standard 1.2.7 Nutrition, Health, and Related Claims. This standard follows a three-tiered system under which food producers have a graduated framework of claims to work within. This system includes nutrition content and general-level and high-level health claims.

A nutrition content claim is a claim that:

1. is about:
   a. the presence or absence of any of the following:
      i. a biologically active substance;
      ii. dietary fiber;
      iii. energy;
      iv. minerals;
      v. potassium;
      vi. protein;
      vii. carbohydrate;
      viii. fat;
      ix. the components of any one of protein, carbohydrate, or fat;
      x. salt;
      xi. sodium;
      xii. vitamins; or
   b. glycemic index or glycemic load; and

2. does not refer to the presence or absence of alcohol; and

3. is not a health claim.

In this context, a probiotic would be considered a "biologically active substance". Note that nutrition content claims are permitted in all foods other than infant formula products or kava; therefore, no content claim in relation to probiotic presence can be made in these two product categories.

Under Standard 1.2.7, general-level health claims are defined as "a health claim that is not a high-level health claim", and high-level health claims are those that refer to a serious disease or a biomarker of serious disease. Foods making either a general- or high-level health claim can only do so according to nutrient scoring criteria to ensure that a perceived unhealthy food is not portrayed as healthy.

Currently, general-level health claims can be made either via two means:

1. Pre-approval, under conditions listed in Clause S4–5 of Schedule 4; or

2. If the food producer has met conditions of self-substantiation via systematic review of the literature pertaining to the health effect in food.

High-level health claims can only be made by preapproval via assessment of the evidence by FSANZ. Currently, the only pre-approved health claim concerning probiotics in the Code, as mentioned earlier, is that pertaining to the effect of certain species on lactose digestion.

# 38.13 ACKNOWLEDGMENTS

The authors thank the following for the assistance in the preparation of this chapter: Dr. Nandini Kumar, Retd. Senior Deputy Director General, Indian Council of Medical Research; Professor Wang-June Kim, Department of Food Science and Technology, Dongguk University, Seoul, Korea; Mr. Teng Yong Low, Deputy Director (Regulatory Programmes), Regulatory Administration Group Agri-Food & Veterinary Authority of Singapore; Ms. Lilim Lee and Lynn-Whui Chow, Health Supplements Unit, Complementary Health Products Branch, Health Products Regulation Group, Health Sciences Authority of Singapore.

# BIBLIOGRAPHY

## China

National Food Safety Standards GB 4789.34–2016 Microbiological Examination of Food Hygiene—Examination of Bifidobacterium.
National Food Safety Standards GB 4789.35–2016 Microbiological Examination of Food Hygiene—Examination of Lactic Acid Bacteria.
National Food Safety Standards GB 10765–2021—Infant Formula.
National Food Safety Standards GB 10766–2021—Older Infants Formula.
National Food Safety Standards GB 10767–2021—Older Infants and Young Children Formula.
National Food Safety Standards GB 19302–2010—Fermented Milk.

## India

Food Safety and Standards Authority of India. www.fssai.nic.in

## Indonesia

NADFC. 2021. Regulation No. 17/2021. Pedoman Penilaian Produk Suplemen Kesehatan Mengandung Probiotik (Guidelines on Evaluation of Health Supplement Products Containing Probiotics). National Agency for Drug and Food Control of the Republic of Indonesia, Jakarta. Original Indonesian language document is accessible from: https://jdih.pom.go.id/download/product/1270/17/2021

## Japan

Japan Consumer Affairs Agency Pamphlet: What Are "Food with Function Claims"?—For Consumers www.caa.go.jp/foods/pdf/151224_1.pdf.

Japan Consumer Affairs Agency Pamphlet: The System of "Food with Function Claims" has been launched! Guidance for Industry www.caa.go.jp/foods/pdf/151224_2.pdf.

Japan Ministry of Health, Labor and Welfare website www.mhlw.go.jp/english/topics/foodsafety/fhc/02.html.

Maeda-Yamamoto, M. 2017, Development of functional agricultural products and use of a new health claim system in Japan. *Trends Food Sci. Technol.* 69(2017): 384–338.

Regulatory Systems of Health Claims in Japan, Food Labelling Division, CAA. 2011. https://edisciplinas.usp.br/pluginfile.php/1767953/mod_resource/content/1/Regulatory%20Systems%20of%20Health%20claims%20in%20Japan%20-%20CAA%202011.pdf.

## Korea

As indicated in the text.

## Malaysia

MOH. 1983. *Food Act 1983*. Ministry of Health Malaysia. Accessed 26 March 2018, http://fsq.moh.gov.my/v5/ms/food-act-1983/

MOH. 1985. *Food Regulations Malaysia 1985*. Updated January 2018. Accessed 26 March 2018, http://fsq.moh.gov.my/v5/wp-content/uploads/2018/03/Food-Regulation-1985-update-Jan-20182.pdf

## Philippines

Bureau Circular No. 16 s. 2004. Circular Published by the Bureau of Food and Drugs (October 26, 2004). Filinvest Corporate City, Alabang, Muntinlupa City, Philippines.

## Singapore

HSA. 2019. *List of Health Supplement Claims*. Health Sciences Authority, Singapore. www.hsa.gov.sg/docs/default-source/hprg-tmhs/hsclaimslist.pdf.

HSA. 2021. *In: Regulatory Overview of Health Supplements*. Health Sciences Authority, Singapore. www.hsa.gov.sg/health-supplements/overview.

SFA. 2019. *A Guide to Food Labelling and Advertisements*. Singapore Food Agency, Singapore. www.sfa.gov.sg/docs/default-source/tools-andresources/resources-for businesses/aguidetofoodlabellingandadvertisements.pdf.

SFA. 2021. *Sale of Food Act [Chapter 283, section 56(1)] and Food Regulations*. Singapore Food Agency. www.sfa.gov.sg/docs/defaultsource/legislation/sale-of-food-act/food_regulations.pdf.

## China Taiwan

Chen, T.-S., Yang, N.-S., Lin, Y.-C., Ho, S.-T., Li, K.-Y., Lin, J.-S., Liu, J.-R., and Chen, M.-J. 2018. A combination of *Lactobacillus mali* APS1 and dieting improved the efficacy of obesity treatment via manipulating gut microbiome in mice. *Sci. Rep.* 8: 6153.

Eaton, K.A., Brooks, C.L., Morgan, D., and Krakowka, R.S. 1991. Essential role of urease in pathogenesis of gastritis induced by *Helicobacter pylori* in gnotobiotic piglets. *Infect. Immun.* 59(7): 2470–2475.

Hamilton-Miller, J.M. 2003. The role of probiotics in the treatment and prevention of *Helicobacter pylori* infection. *Int. J. Antimicrob. Agents* 22(4): 360–366.

Hong, W.S., Chen, H.C., Chen, Y.P., and Chen, M.J. 2009. Effects of kefir supernatant and lactic acid bacteria isolated from kefir grain on cytokine production by macrophages. *Int. Dairy J. 19*: 244–251.

Köhler, B., Andreen, I., and Jonsson, B. 1984. The effect of caries-Preventive measures in mothers on dental caries and the oral presence of the bacteria *Streptococcus mutans* and lactobacilli in their children. *Arch. Oral Biol. 29*: 879–883.

Loesche, W.J. 1986. Role of *Streptococcus mutans* in human dental decay. *Microbiol. Rev. 50*: 353–380.

Lourens-Hattingh, A., and Viljoen, B.C. 2001. Yogurt as probiotic carrier food. *Int. Dairy J. 11*: 1–17.

Matsuzaki, T., Yamazaki, R., Hashimoto, S., and Yokokura, T. 1998. The effect of oral feeding of *Lactobacillus casei* strain Shirota on immunoglobulin E production in mice. *J. Dairy Sci. 81*: 48–53.

Ministry of Health and Welfare, Executive Yuan. 2018. *Act Governing Food Safety and Sanitation* . https://law.moj. gov.tw/Eng/LawClass/LawAll.aspx?PCode=L0040001.

Modler, H.W. 1994. Bifidogenic factors sources, metabolism and applications. *Int. Dairy J. 4*(5): 383–407.

Sanders, M.E., Morelli, L., and Tompkins, T.A. 2003. Spore formers as human probiotics: *Bacillus, Sporolactobacillus*, and *Brevibacillus* . *Compr. Rev. Food Sci. Food Saf. 2*: 101.

Shida, K., Kiyoshima-Shibata, J., Nagaoka, M., Watanabe, K., and Nanno, M. 2006. Induction of interleukin-12 by *Lactobacillus* strains having a rigid cell wall resistant to intracellular digestion. *J. Dairy Sci. 89*: 3306–3317.

Shida, K., Takahashi, R., Iwadate, E., Takamizawa, K., Yasui, H., Sato, T., Habu, S., Hachimura, S., and Kaminogawa, S. 2002. *Lactobacillus casei* strain Shirota suppresses serum immunoglobulin E and immunoglobulin G1 responses and systemic anaphylaxis in a food allergy model. *Clin. Exp. Allergy 38*: 563–570.

Taiwan Food and Drug Administration (TFDA). www.doh.gov.tw.

Wollowski, I., Rechkemmer, G., and Pool-Zobel, B.L. 2001. Protective role of probiotics and prebiotics in colon cancer. *Am. J. Clin. Nutr. 73*: 4515–4555.

Zhang, Q.Q., Wu, Y.H., and Xiaoqiang Fei, X.Q. 2015. Effect of probiotics on glucose metabolism in patients with type 2 diabetes mellitus: A meta-analysis of randomized controlled trials. *Medicina. 52*: 28–34.

## Thailand

MoPH. 2011. *Notification of the Ministry of Public Health No. 339 Re: Use of Probiotic Microorganisms in Foods*. Ministry of Public Health, Thailand. http://food.fda.moph.go.th/law/data/announ_moph/V.English/No.%20 339%20Use%20of%20Probiotic%20 Microorganisms%20in%20Foods.pdf.

MoPH. 2012. *Notification of the Ministry of Public Health No. 346 Re: Use of Probiotic Microorganisms in Foods (No.2)*. Ministry of Public Health, Thailand. http://food.fda.moph.go.th/law/data/announ_moph/V.English/ No.%20346%20Use%20of%20Probiotic%20Microorganisms%20in%20Foods%20(No.2).pdf.

Report of a Joint FAO/WHO Working Group on Drafting Guidelines for the Evaluation of Probiotics in Food. London, Ontario, Canada: 2002. April, May. Guidelines for Evaluation of Probiotics in Food. www.who.int/foodsafety/ fs_management/en/probiotic_guidelines.pdf.

## Vietnam

Government of Vietnam. 2018. *Elaboration of Some Articles of the Law of Food Safety*. Decree No. 15/2018/ ND-CP. Socialist Republic of Vietnam, Hanoi. http://van-ban.chinhphu.vn/portal/page/portal/chinhphu/ hethongvanban?class_id=1&page=1&mode=detail&document_id=192829.

MOH Vietnam. 2014. *Regulating the Management of Functional Foods*. Circular No: 43/2014/TT-BYT. Ministry of Health, Socialist Republic of Vietnam, Hanoi. http://vanban.chinhphu.vn/portal/page/portal/chinhphu/ hethongvanban?class_id=1&_page=1&mode=detail&document_id=178185.

## Austrasia

Food Standards Australia New Zealand Act. 1991. www.legislation.gov.au/Details/C2004C00171.

Joint Australia New Zealand Food Standards Code. 2000. www.foodstandards.gov.au/code/Pages/default.aspx.

# Probiotic Regulation in Latin American Countries

# 39

Melisa Puntillo, Ana Binetti, Patricia Burns, Daniela Tomei, Gabriel Vinderola, and Elisa Ale

## 39.1 INTRODUCTION

Functional foods are conventional foods to which specific essential nutrients and/or food components are added for a targeted physiological function, providing a health benefit beyond basic nutrition (Institute of Food Technologist— IFT. Expert Report 2021).

Several food ingredients have been widely used for food enrichment, such as minerals, vitamins, and biotics (prebiotics, probiotics, synbiotics, and postbiotics). Probiotics are defined as "live microorganisms that, when administered in adequate amounts, confer a health benefit on the host" (Hill et al. 2014). Although they have been extensively applied in the food industry, there are important legal gaps in their regulation in most countries of Latin America.

The global market of probiotic supplements was valued at US$6.970.1 million in 2021 and is projected to grow 3.75% on average in the period 2021–2026. Moreover, the global market of probiotic yogurt, plain, flavored, and drinking yogurts, reached US$33.248 million in 2021 and has a projected growth of 2.7% over 2021–2026 (Probiotic Market 2021 and Trend Analysis, International Probiotics Association). In this context, it is important to rely on an appropriate regulatory framework that contributes to this growing market.

Evidence-based communication seems to play a crucial role in consumers' interest in this type of product (Topolska et al. 2021), and the demand for probiotics is closely related to their health and wellness claims (Betz et al. 2015). For this reason, it is crucial to get a clear message across so that consumers can decide whether to choose a particular product based on accurate and reliable information on the label, supported by scientific research.

Regulations governing the incorporation of novel probiotics are sometimes confusing and vary greatly with the geographical region considered. For instance, European countries, Japan, Canada, and the United States have made significant progress in this subject. On the contrary, Latin America presents more diverse and asymmetrical legislation, being Brazil the first country in South America to introduce consistent legislation on functional foods, as well as guidance on health claims for food products.

In Argentina, the institution in charge of dealing with the substantiation of health claims for probiotics in foods is the National Food Institute (Instituto Nacional de Alimentos; INAL), which depends on

DOI: 10.1201/9781003352075-43

the National Administration of Medicines, Food and Medical Technology (Administración Nacional de Medicamentos, Alimentos y Tecnología Médica; ANMAT). Although the regulation of probiotics has been included in the Argentine Food Code (Código Alimentario Argentino; CAA) since 2011, the path to have health claims approved for probiotics is challenging. For example, there are no clear statements about the time each step in the regulation process will take, making uncertain any assessment process in terms of duration of the whole process. Nonetheless, there are a couple of foods (yogurts) and food supplements in Argentina that were already allowed to carry the term probiotic in the label, and some health claims were even approved by the INAL. It would be helpful if the INAL would make publicly available the results of the evaluation process of each submitted dossier, as happens, for instance, in Europe with the European Food Safety Authority (EFSA).

Although each country seeks to create its own legislation, many follow the recommendations of leading countries, such as those belonging to the EU, Australia, and the United States (Myers 2014). These differences in the probiotic regulations among countries mean an additional challenge for manufacturers concerning product innovation, substantiation of claims, labeling, and marketing (Sanders et al. 2011).

As the legislation on functional foods is constantly changing, especially in Latin American countries where regulations for probiotics are generally at an early stage of development, the aim of this chapter is to describe the most recent updates on this matter, focusing solely on the scientific and technical aspects of each particular case.

# 39.2 ARGENTINA

At the national level, there are two organizations responsible for controlling food safety and quality: the National Service of Health and Agri-food Quality (Servicio Nacional de Sanidad y Calidad Agroalimentaria; SENASA), and the INAL, which depends on the ANMAT. The ANMAT possesses economic and financial autonomy and is under the authority of the Ministry of Health. While the SENASA regulates non-processed food, the ANMAT, through the INAL, controls processed and ready-to-eat food by following the guidelines of the Argentine Food Code (Código Alimentario Argentino, CAA), regulations that are continuously revised and updated.

It was not until 2011 that the concept of probiotic was included in article No. 1389 (Chapter XVII) by a joint Resolution of the Secretariat of Policy, Regulation and Institutes (Secretaría de Políticas, Regulación e Institutos, SPReI, Res. N° 261/2011) and the Secretariat of Agriculture, Cattle Farming, and Fishing (Secretaría de Agricultura, Ganadería y Pesca, SAGyP, Res. N° 22/2011). The definition adopted for the term "probiotic" was the one proposed by the Food and Agriculture Organization/World Health Organization (FAO/WHO) in 2002. This article mentions a Protocol for Evaluation of a Probiotic, indicating the minimum requirements a strain should meet to be used as a probiotic food ingredient. Briefly, the protocol includes the following steps: i) identification of the strain at genus, species, and subspecies levels by molecular techniques (16S rDNA sequencing); ii) *in vitro* and *in vivo* characterizations to assess the resistance to the gastric barrier, bile, and lysozyme; iii) *in vivo* and *in vitro* assays to prove the probiotic effects, which must be carried out by national research institutes that are internationally recognized; iv) safety assays to confirm that no bacterial translocation to the liver, spleen, or mesenteric lymph nodes occurs. The article also provides other complementary tests that include the study of antibiotic resistance, hemolytic activity, and toxin release. In addition, the label of the product should report the strain(s) used and the concentration of viable bacteria during its shelf life ($10^6$ to $10^9$ CFU/g). An assessment committee composed of specialists in this topic will evaluate each case and determine if further *in vivo* trials are needed to receive authorization.

In the past 3 years, there were some modifications in the regulations of probiotics in Argentina. For instance, the second and fifth articles from provision No. 2873/2012 (ANMAT) were modified according to provision No. 5893/2021. Currently, the second article mentions an Assessment Committee of Probiotics and Prebiotics which will be coordinated by the INAL and composed of specialized professionals from the competent areas of this institute. When the evaluation of an ingredient requires the opinion of other professionals, this committee might convene experts, organizations, and specialized academic/scientific institutions. In contrast, the previous article stated that the committee would be coordinated by a representative of the INAL and include specialized professionals from the National Health Authority, National Council for Scientific and Technical Research (CONICET), national universities, and other institutions convened for the occasion, together with representatives of the competent authority responsible for the registration.

On the other hand, the fifth article remains similar to the previous one but for the statement about informing, publishing and keeping updated the list of prebiotics and probiotics authorized for their use as food ingredients. Also, the term "*in vivo* assays" was replaced by "assays that demonstrate the physiological effect" of the ingredient.

Regarding the CAA, there have not been any changes in article No. 1389 since 2011, whilst article 1381 about dietary supplements was modified by a joint resolution (3/2020) between the Secretariat of Quality in Health (Secretaría de Calidad en Salud, SCA) and the Secretariat of Food, Bioeconomy and Regional Development (Secretaría de Alimentos, Bioeconomía y Desarrollo Regional, SAByDR). According to this Resolution, article 1381 now includes probiotics among dietary supplements.

# 39.3 BRAZIL

In Brazil, depending on their characteristics, products containing probiotics can be classified as foods, food supplements, or even drugs. Their regulations are set out by two ministries: the Ministry of Health, through a special autarchy, the National Health Surveillance Agency (Agência Nacional de Vigilância Sanitária; ANVISA), and the Ministry of Agriculture. Most beverages and products of animal origin (dairy and meat products) are regulated by the latter, and all other products are under ANVISA jurisdiction.

In order to be legally used in food and food supplements, premarket approval is mandatory. The most important element regarding approval of probiotics in Brazil is the substantiation of identity, safety, and efficacy of the probiotic microorganism. The interested party must submit a petition to ANVISA gathering all the technical documentation required for assessment. This evaluation is an essential requirement for regularization of the products and, without the validation from ANVISA attesting to identity, safety, and efficacy of the strain, no microorganism can be used in any product in Brazil.

ANVISA is a governmental regulatory agency, created in 1999, which operates at the federal level and whose primary goal is to protect and promote public health by exercising health surveillance over products and services, including processes, ingredients, and technologies that may pose any health risks.

Food legislation in Brazil is governed by Law-Decree 986 of October 21, 1969 (DL 986/69) and its normative instructions. It establishes basic food standards in the country, and, to date, it is the principal legal rule governing the production, distribution, storage, and trade of food in this country. The fact that it is a fairly old norm creates some difficulties in the regulatory framework that focuses on the food and food supplements sector, since Decree Law 986/1969 was approved long before ANVISA and subsequent legislation.

## 39.3.1 General Technical Regulation for Safety and Efficacy Demonstration

The authorization for using probiotics in food and food supplements in Brazil requires previous proof of the safety of the probiotic strain and its capacity to promote a beneficial effect on the host. Resolutions ANVISA No. 17 April 30, 1999, and No. 18 April 30, 1999, approve the technical regulation establishing the basic guidelines for the assessment of food safety and risk and for the analysis and verification of functional or health properties alleged in food labeling, respectively.

Resolution 17/99 lays out that the proof of safety must be conducted on the basis of information about the purpose and conditions of use of the food or food ingredient. Moreover, the risk evaluation must be based on scientific evidence, which includes characterization of the product; nutritional, physiological, and/or toxicological assays on experimental animals; epidemiological studies; clinical assays; general evidence from scientific literature, international health organizations, and internationally recognized legislation on the characteristics of the food or ingredient; and evidence of traditional use. Documented information on approval of use of the food or ingredient in other countries, economic blocks, Codex Alimentarius, and other internationally recognized organizations is also considered safety evidence.

Regarding the rules of Resolution 18/99, it is necessary to understand that the legislation applicable to probiotics in Brazil entails the description of a functional or health label claim representing the efficacy of the probiotic strain (ANVISA 2002). These claims must be in accordance with Resolution 18/99, which defines the terms as follows:

*Functional claim*: related to the metabolic or physiological role that the nutrient or non-nutrient plays in the growth, development, maintenance, and other normal functions of the human organism.

*Health claim*: states, suggests, or implies the existence of a connection between the food or ingredient with disease or condition related to health.

It is worth noting that the claim cannot mention that probiotics are able to treat, cure, or prevent any disease, which entails the classification of the product as a drug. Drugs in Brazil are governed by Law 6360 of September 23, 1973, and its normative instructions. Resolution of the Collegiate Board or Resolução da Diretoria Colegiada in Portuguese (RDC) N° 718 of July, 1, 2022, defines probiotic medicines as preparations or products containing defined and viable microorganisms in sufficient quantity to prevent or treat human diseases by interaction with the microbiota, intestinal epithelium, or associated immune cells or by another mechanism of action. The procedures for registration and efficacy demonstration are different from those covered by this chapter.

According to Resolution 18/99, the proof of functional and/or health claims must be conducted based on estimated or recommended intake, purpose, conditions of use, and the following scientific evidence: characterization of the product, biochemical assays, and epidemiological or clinical studies.

These legislations are quite generic and help as a basis for specific technical regulations on probiotics.

## 39.3.2 Specific Rules

Probiotics are regulated by ANVISA through Resolution RDC No. 241, July 26, 2018, concerning the requirements for proving the safety and health benefits of probiotics for use in food, and they were defined as a "live microorganism which, when administered in adequate amounts, confers a health benefit on the individual".

The minimum parameters of evaluation include identification of the strain (phenotypically and genetically); origin and method of production, including whether the microorganism is genetically modified (GMO); production of toxins and bacteriocins; information on the deposit of the strain in an internationally recognized culture bank; evidence of safety by *in vitro*/animal studies, human phase I studies; metabolic activity evaluation; hemolytic activity for species with hemolytic potential; antimicrobial resistance profile and information on the genetic basis of antimicrobial resistance, according to the methodology described by the EFSA; side effects during human studies; and proof of resistance to gastric acidity and bile salts. When probiotics are not isolated from food or human indigenous microbiota, and their safety has not been established at the genus or species level, safety must be proven through the following studies: genotoxicity and mutagenicity, acute toxicity, sub-chronic toxicity, long-term toxicity, and reproductive and developmental toxicity when the strain is intended for children younger than 3 years old and pregnant women (ANVISA 2018).

When probiotics are intended for consumption by pregnant women or children under 3 years of age, randomized, double-blind, placebo-controlled clinical trials conducted with these specific groups should be presented to assess adverse effects on growth and development parameters.

The efficacy must be supported by randomized, double-blind, and placebo-controlled clinical studies, and it is strain specific. Studies' outcomes must demonstrate the relationship between the consumption of the specific probiotic strain object of the petition and the functional effect. The identification and measurement of the effect must be clearly defined. If the effect cannot be measured directly, validated biomarkers that are related to the claimed effect must be identified. Identification of the strain and amounts tested in the studies used as references are essential. The participating population must be the same as the target group of the probiotic under assessment.

## 39.3.3 Claims

The vast majority of food and supplement products containing probiotics in Brazil present claims related to a general benefit of the strain on gastrointestinal health: "(name of the strain) may contribute to gastrointestinal health". This is due to the considerable scientific evidence in this field, which makes approval easier. On the other hand, approval of claims related to the immune system or other effects is more challenging, because general claims are not usually accepted for them, and therefore the level of scientific evidence required is higher.

In order to assist stakeholders on dossier submissions for the approval of probiotics, ANVISA has published guidance with instructions on how to address each topic required by the specific legislation, RDC 241/2018 (Guia orienta sobre instrução processual de probióticos, 2022). In the guidance, the regulatory agency makes it clear that the general claims of probiotics are more easily attributed to the health of the gastrointestinal tract, where these strains act directly. On the other hand, general claims for other body systems often do not communicate adequately the intended benefit and have misleading information (for example: "beneficial to health" or "support immune health").

# 39.4 CHILE

In July 2017, Resolution 860/17 was promulgated by the Ministry of Health, which approved Technical Standard No. 191 on nutritional guidelines to claim health benefits of functional foods. For probiotics (*Lactobacillus* spp., *Bifidobacterium* spp., and other specific bacilli, as indicated in the Technical Standard, article 7), a minimum of $10^7$ CFU/g of each probiotic strain must be present in the food until the end of the shelf life, and the claims may be related to the support of the intestinal microbiota and immune system, and the regulation of intestinal transit. Bacilli should be resistant to gastric acid and other secretions of the digestive tract, and the strain of the microorganism associated with the beneficial effects must be specified in the message. In January 2018, this regulation was modified by Resolution 1671, with only slight modifications concerning some expressions.

In December 2017, Act No. 08/17 was published, presented by an Advisory Committee of Experts on Sanitary Control Regime (Comité de Expertos Asesor en Régimen de Control Sanitario) belonging to the Public Health Institute, Ministry of Health (Instituto de Salud Pública, Ministerio de Salud). In this act, the topic of probiotics was evaluated, recognizing that there are different forms (capsules, tablets, sachets with lyophilized microorganisms) that cannot be controlled by the available guidelines. It also claims that only products containing probiotics in dairy food matrices (yogurt, milk, among others) are regulated. In this sense, the act confusingly relates probiotics to article No. 220 of the Food Sanitary Regulation (Reglamento

Sanitario de Alimentos, RSA), in which the definition of yogurt is described and only the species *Lactobacillus bulgaricus* is mentioned, but no further definition of probiotic is included in this regulation.

Later, Resolution No. 3435 (promulgated in June 2018 by the Institute of Public Health) provided guidelines to evaluate products containing *Lactobacillus* spp., *Bifidobacterium* spp., and other specific bacilli in pharmaceutical oral dosage forms, which were not included in Resolution No. 860/17. This resolution claims that the products containing *Lactobacillus* spp., *Bifidobacterium* spp., and other specific bacilli in pharmaceutical oral forms that can keep balance in the intestinal microbiota and help with the intestinal transit and immune system should be regarded as food, for which Resolution 860/17 must be applied. On the other hand, those products with the bacteria mentioned previously that present effects different from the ones described (or contribute to the balance of the microbiota in other parts than the intestine) must be evaluated as pharmaceutical products.

## 39.5 COLOMBIA

The National Institute of Food and Medicine Surveillance (Instituto Nacional de Vigilancia de Medicamentos y Alimentos; INVIMA) regulates the labeling for probiotics through Resolution No. 333 (2011), which replaced the former Resolution No. 288 (2008). Although this institute depends on the Ministry of Health, it has administrative and financial autonomy and regulates the authorization and commercialization of different products based on the current legislation. According to this new resolution (promulgated by the Ministry of Social Protection), a probiotic must be "a live culture of microorganisms, usually bacteria, that survive the intestinal transit, and particularly resist the stomach acidity, adhering and colonizing the intestine and modifying favorably the microbial balance". Additionally, in Article 22.1, several characteristics need to be fulfilled before a microorganism can be considered a probiotic. These features are briefly described as follows: i) the probiotic must be viable and non-pathogenic, and its natural environment should be the human intestinal tract; ii) it must survive the intestinal tract, having resistance to gastric juice and bile acids; iii) it must be capable of adhering to the intestinal mucus and colonize the intestine; and iv) it must survive during the shelf life of the product to which it is added, presenting levels of at least $1 \times 10^6$ CFU/g. Also, the label should declare that there are other factors apart from the regular consumption of probiotics that might improve the digestive functions, such as diet and physical exercise. For instance, in Article 22.1, there is a model declaration that claims that "an adequate diet and a regular consumption of food with probiotic microorganisms can help to normalize the digestive functions and regenerate the intestinal flora". It should be noted that there are not standardized protocols described in the Colombian regulation to assess any of the requirements stated by Resolution No. 333.

## 39.6 BOLIVIA

Although there is no national regulation of probiotics for human use, the term "probiotic" is mentioned as veterinary additive. The National Service of Agricultural Health and Food Safety (Servicio Nacional de Sanidad Agropecuaria e Inocuidad Alimentaria; SENASAG), created in 2000, included the term "probiotic" as additive in administrative Resolution No. 058/01 of the General Animal Health Regulations (Reglamento General de Sanidad Animal; REGENSA). In 2011, Resolution No. 179/2011 included a list (Appendix IV) of species of microorganisms within the zootechnical category, such as *Lactobacillus acidophilus*, *Lacticaseibacillus rhamnosus*, *Enterococcus faecium*, and *Bacillus subtilis*, among others. In August 2022, the REGENSA was updated with Resolution No. 172/2022, but no changes in terms of probiotics regulation were made.

# 39.7 MEXICO

The Federal Commission for Protection against Health Risks (Comisión Federal para la Protección contra Riesgos Sanitarios; COFEPRIS) is a decentralized organization in charge of the regulation of products and services, as well as their legal authorization. Although there is no regulation for probiotics, the rule NOM-181-SCFI/SAGARPA-2018 about yogurt mentions the use of additional and alternative cultures of the genera formerly called *Lactobacillus*, as well as *Bifidobacterium*. Appendix A of that rule lists some lactobacilli species (*Lactobacillus helveticus*, *Lactobacillus helveticus* spp. *jugurti*, *Lacticaseibacillus casei*, *Lacticaseibacillus paracasei*, *Lactobacillus delbrueckii* subsp. *lactis*, *Lacticaseibacillus rhamnosus*, *Lactiplantibacillus plantarum*, *Lactobacillus johnsonii*, *Lactobacillus acidophilus*, and *Limosilactobacillus reuteri*) and wrongly includes four species of bifidobacteria *(B. longum, B. breve, B. animalis*, and *B. bifidum*) among the most-used lactic acid bacteria. In addition, three commercial probiotic strains are mentioned as well, *Lacticaseibacillus casei* Shirota, *Lacticaseibacillus rhamnosus* GG, and *Lacticaseibacillus casei* Defensis. The additional cultures must reach a level of $10^6$ CFU/g or higher in yoghurt and remain viable, abundant, and active during the shelf life, with the term "active" quite ambiguous. Moreover, the species or a generic name should be always specified, as mentioning that the yogurt contains alternative cultures of lactic acid bacteria would not be enough.

# 39.8 PERÚ

In order to guarantee the physical well-being of people, the Ministry of Agriculture and Irrigation (Ministerio de Desarrollo Agrario y Riego; MINAGRI) established technical and sanitary regulations that bovine dairy products for human consumption must comply with, through the Regulation of Milk and Dairy Products. This regulation, based on the Codex Alimentarius norms, came into force in 2017 in Supreme Decree No. 007–2017-MINAGRI, which was later replaced with Supreme Decree No. 004–2022-MIDAGRI in 2022. The general aim of this regulation is to ensure the manufacture of safe products and prevent practices that may induce error. Although probiotics are not directly covered within the regulation, they are implicitly included, as it is mentioned that those products which are not considered in the regulation must fulfill the requirements of the Codex Alimentarius standards. It is worth noting that infant formulas are excluded from the regulation.

# 39.9 PARAGUAY

In this country, probiotics are classified as biological medicines and defined as "products containing alive, attenuated or killed microorganisms (except vaccines)" by Decree No. 6611/16, which regulates Article 24 of Law 1119/1997 of products for health and others. It is worth clarifying that, although this document was approved in 2016, the definition is not consistent with the consensus definition commonly accepted by the scientific community (Hill et al. 2014). Decree No. 6611 establishes the requirements for the registration and commercialization of biological medicines, whilst the supervision of the fulfillment of the conditions is carried out by The National Directorate of Sanitary Surveillance (Dirección Nacional de Vigilancia Sanitaria; DINAVISA), belonging to the Ministry of Public Health and Social Welfare (Ministerio de Salud Pública y Bienestar Social). General requirements for registration of biological medicines are enumerated in Article 6, and, in case of innovative products, some additional requirements are mentioned in Article 7, such as quality studies of the active ingredient and end product and efficacy, safety, and immunogenicity studies. The criteria applied must be appropriate for each product and follow the WHO or International Council for Harmonisation (ICH) guidelines. In addition, the decree establishes that the therapeutic claims of a registered product may be recognized and extended only if the product is approved by reference regulatory agencies listed in the document after proper evaluation with clinical trials. The reference agencies include high surveillance regulatory agencies from the countries listed in Article 11 of Law No. 3283/2007, European Medicines Agency (EMA), and those recognize by the Pan-American Health Organization (Organización Panamericana de la Salud; OPA). It is crucial to emphasize that probiotics are not manufactured in Paraguay but imported from abroad.

## 39.10 VENEZUELA

In Venezuela, both Sencamer and the Fund for Quality Standardization and Certification (Fondonorma-Fondo para la Normalización y Certificación de la Calidad) work jointly in the regulation of products and services for the food industry, among other sectors. While Sencamer is a governmental institution, Fondonorma is a non-profit civil association with legal autonomy.

In December 2012, Fondonorma established rule NTF 909 for infant milk formula, in which probiotics are defined as "live, non-pathogenic microorganisms of human origin, in sufficient numbers, capable of surviving in the gastrointestinal tract, adhere, colonize and produce an improvement in the intestinal microbial balance, with benefits to the health of the host". Probiotics are mentioned as ingredients, suggesting that the label of the product should indicate genus, species, strain, and the minimum number of viable bacteria at the end of the shelf life. In Appendix A, the most-used species in infant formula are enumerated, among them *Lacticaseibacillus casei, Lacticaseibacillus paracasei, Lactobacillus acidophilus*, and *Bifidobacterium lactis*. It is worth noting that there are no specifications on the minimum level required. A similar rule for infant milk formula was promulgated by Sencamer (formerly called Fodenorca–Development Fund for Standardization, Quality, Certification and Metrology) in 2015 (COVENIN 909 2015), in which other lactobacilli species were mentioned as well (*Limosilactobacillus reuteri*). Nevertheless, the use of other probiotics is allowed as long as they are authorized.

In 2015, Fondonorma promulgated rule NTF 3863 for food supplements, and not only species but also commercial strains are listed. Among the genera, *Bifidobacterium, Lactobacillus, Escherichia, Bacillus, Enterococcus*, and *Lactococcus* were included. More importantly, probiotics should be viable at levels higher than $1 \times 10^4$ (the unit is not specified) during the shelf life of the product. Again, if the strain is not included in Table No. 5 of this rule, new probiotics could be considered if scientific evidence about risks and health benefits is available.

Recently, in April 2022, Fondonorma published another rule (NTF 2393) for the regulation of different types of fermented milk (including yogurt, kefir, and kumis), which mentions the addition of probiotics such as *L. casei, L. acidophilus, Bifidobacterium* spp., and *Lactobacillus* GG, among others that are listed in Table 3D. The genus, species, and strains should be reported on the label, and probiotics should be viable during the entire shelf life of this product at a level of at least $1 \times 10^6$ CFU/g.

## 39.11 COSTA RICA, EL SALVADOR, GUATEMALA, HONDURAS, AND NICARAGUA

The regulation for probiotics in all these countries can be found in Central America Technical Regulation RTCA 67.01.60:10, which regulates the nutrition labeling of pre-packaged food for human consumers older than 3 years old. This document was ratified by the Council of Ministers of Central American Economic Integration (Consejo de Ministros de Integración Económica de Centroamérica; COMIECO) and approved by the Subgroup of Measures of Standardization, composed of the Ministry of Economy of Guatemala (Ministerio de Economía; MINECO); the Salvadorian Technical Regulation Agency (Organismo Salvadoreño de Reglamentación Técnica; OSARTEC); the Ministry of Development, Industry, and Commerce of Nicaragua (Ministerio de Fomento, Industria y Comercio; MIFIC); the Secretariat of Industry and Commerce of Honduras (Secretaría de Industria y Comercio; SIC); and the Ministry of Economy, Industry, and Commerce of Costa Rica (Ministerio de Economía Industria y Comercio; MEIC).

The requirements listed in this document apply to pre-packaged food products that include nutritional information and health or nutritional claims which are commercially available for human consumption in the countries belonging to this subgroup. In the case of Costa Rica and Honduras, fermented and distilled alcoholic beverages are included in this regulation as well.

Health claims must be based on appropriate and strict scientific evidence and contain information about the physiological function of the ingredient or information about a proven diet–health relationship. Additionally, information about the composition of the food product in terms of the physiological function of the ingredient or the recognized diet–health relationship must be provided. Health claims previously approved by another competent entity such as the FDA or UE can be accepted.

Appendix G describes health claims which are allowed and gives examples. The claims that are not included in this appendix must be evaluated by the sanitary authority. The criteria for microorganisms to be regarded as probiotics can be found in Appendix G: being alive; non-pathogenic; isolated from the human digestive tract; and able to survive digestion, adhere to the intestinal mucosa, and colonize the intestine. In addition, the microorganism must be able to survive during the shelf life of the commercial food product at a minimum level of $10^6$ CFU/g. Products containing probiotics must declare the bacterial genus, species, and strain according to recognized international nomenclature; the recommended dose for the probiotic to be effective in relation to the claimed health benefit; and the health-promoting effects it exerts. In addition, the labels must state that consumption of probiotics is not the only factor that can improve the digestive function, as there are others such as diet and physical exercise.

El Salvador is currently working on Technical Regulation RTS 67.06.02:22, published for public consultation in 2022 by the OSARTEC. This document, called "Foods for Special Diets, Nutritional Supplements and Probiotics. Classification, Characteristics, Sanitary Registration Requirements and Labeling", is under revision and pending approval. It establishes the technical specifications, sanitary requirements, and labeling of foods for particular diets, supplements and probiotics manufactured in this country or imported for commercialization. The regulation includes those foods prepared or processed to satisfy nutritional requirements determined by particular physical or physiological conditions or specific diseases or disorders, such as foods for infants and children or for special medical purposes. In this document, probiotics are defined as "live microorganisms which, when administered in adequate quantities, are beneficial to the host's health and are able not only to survive the passage through the digestive tract, but also to proliferate in the intestine". The document includes technical specifications for microbiological quality (Appendix A) and physicochemical analyses (Appendix B) of nutritional supplements and probiotics. Additionally, requirements for modifications of sanitary registration (Appendix C) and amount of sample for analysis of nutritional supplements and probiotics (Appendix D) are described.

# 39.12 ECUADOR

Probiotics have been regulated by the Ecuadorian Institute of Standardization (Instituto Ecuatoriano de Normalización; INEN) since 2011 through the Ecuadorian Technical Norm NTE INEN 1334–3:2011, "Food Products Labelling for Human Consumption. Part 3. Requirements for Nutritional Claims and Health Claims". This technical norm lists the criteria to register a product, specific conditions for acceptable health claims, and some model claims. The requirements for a microorganism to be regarded as a probiotic are the same as those described in RTCA 67.01.60:10 from central American countries.

Additionally, the labels must state that consumption of probiotics is not the only factor that can improve digestive function, as there are other factors such as diet and physical exercise. Health claims must be authorized when all of the following conditions are met: i) health claims must be based on adequate scientific substantiation, the level of evidence must be sufficient to establish the type of effect claimed and its relation to health, and the scientific sustenance must be reviewed in light of new data; ii) any claim about properties must be accepted, or considered acceptable, by the competent authorities of the country where the product is sold; and iii) the claimed benefit must derive from the consumption of a reasonable amount of food or food ingredient in the context of a healthy diet.

The term "probiotic" is also included in the sanitary regulations for the control of food supplements (Resolutions No. ARCSA-DE-028–2016-YMIH and ARCSA-DE-2021–012-AKRG) of the National Agency of Regulation, Control and Sanitary Vigilance (Agencia Nacional de Regulación, Control y Vigilancia Sanitaria; ARCSA). These resolutions provide guidance to regulate food supplements and control the establishments where they are manufactured, stored, distributed, imported, and sold.

In addition, similarly to Bolivia, the Ministry of Agriculture, Cattle Farming, Aquaculture and Fishing regards probiotics as ingredients of veterinary products for aquaculture and fishing. The list of microorganisms that are considered probiotics, and the requirements they must fulfill, can be found in the Manual for the Registration of Companies and Products for Veterinary Use (Manual para el Registro de Empresas y Productos de Uso Veterinario, 2018).

## 39.13 URUGUAY AND PANAMÁ

These countries have no specific regulations concerning probiotics. However, Uruguay recognizes the existence of a direct relationship between the consumption of foods and specific health benefits. Since there is no national regulation on the use of health and/or functional claims, the Sanitary Evaluation Division of the Food Department of the Ministry of Public Health authorized the use of claims approved by international reference organisms such as EFSA (European Union), FDA (USA), ANVISA (Brazil), and Health Canada (Canada) (personal communication). On the other hand, although there is no national regulation of probiotics in Panamá, article 12 of Executive Decree No. 331 (2008) establishes that foods, food additives, and food packaging are subject to the standards and methods approved by the Codex Alimentarius when there is no specific national regulation.

## 39.14 CONCLUSION

Regulations about probiotics, functional foods, and health claims vary widely among Latin American countries. While some of them have been working on creating or improving standards, others do not even have any legislation. In general, countries that lack national norms follow the standards and methods established by the Codex Alimentarius or those approved by international reference organisms such as EFSA, FDA, or ANVISA. Although most countries have updated their legislation regarding probiotics lately, the modifications were slight and poor. On the other hand, Brazil has made considerable progress in regulatory matters, with consistent and science-based legislation, allowing other countries of the region to use it as a basis for their own regulations.

In general, except for Paraguay, definitions of probiotics are consistent with the one proposed by FAO/WHO. In addition, the functionality attributed to a microorganism must be based on scientific evidence on the benefits or effects of the probiotic microorganisms incorporated into the product. In most cases, the essential requirements for a product to have a health claim are the identification of the strain used, the minimum amount of viable cells present until the end of the shelf life, and a clinical study proving the attributed effect.

The continuous and accelerated growth of the functional food and probiotic sector makes it necessary to develop specific and uniform regulations for these products. In this context, it is crucial for Latin American countries to develop internationally harmonized regulations that ensure the application of homogenous criteria in the evaluation of potentially probiotic microorganisms and health claims. More importantly, this approach would also lead to greater transparency for consumers when it comes to choosing functional foods.

## 39.15  ACKNOWLEDGMENTS

We would like to thank Ritva Ann-Mari Repo-Carrasco Valencia (Peru), Jeadran N. Malagon-Rojas (Colombia), María Alejandra Larre (Argentina), Andrea V. Moser (Argentina), Pablo Peña (Paraguay), Marianela Cortés (Costa Rica), Diana Víquez Barrantes (Costa Rica), Dante Salazar (México), Rodrigo Ibáñez (Chile), Luisa D'Angelo (Venezuela), Boris Lemus (El Salvador), and Nidia Rodríguez (El Salvador) for their valuable contributions to this chapter.

## BIBLIOGRAPHY

Administración Nacional de Medicamentos, Alimentos y Tecnología Médica de Argentina (ANMAT). Disposición 2873/2012. www.argentina.gob.ar/normativa/nacional/disposici%C3%B3n-2873-2012-197951/texto

Administración Nacional de Medicamentos, Alimentos y Tecnología Médica de Argentina (ANMAT). Disposición 5893/2021. www.argentina.gob.ar/normativa/nacional/disposici%C3%B3n-5893-2021-353278/texto

Betz, M., Uzueta, A., Rasmussen, H., Gregoire, M., Vanderwall, C., & Witowich, G. 2015. Knowledge, use and perceptions of probiotics and prebiotics in hospitalised patients. *Nutrition & Dietetics*, 72(3): 261–266. https://doi.org/10.1111/1747-0080.12177

Código Alimentario Argentino (CAA), Capítulo XVII: Alimentos de régimen o dietéticos. 2022. www.argentina.gob.ar/sites/default/files/capitulo_xvii_dieteticos_actualiz_2022-12.pdf

Comisión Venezolana de Normas Industriales (COVENIN), Norma COVENIN 909: 2015. www.sencamer.gob.ve/sencamer/normas/909-2015.pdf

Comité de Expertos Asesor en Régimen de Control Sanitario (Chile), Acta 08/17. www.ispch.cl/sites/default/files/documento/2018/04/Acta%20N%C2%B08-17.pdf

Comunicado del Organismo Salvadoreño de Reglamentación Técnica (OSARTEC) Referente al RTS 67.06.02:22: Alimentos para Regímenes Especiales, Suplementos Nutricionales y Probióticos. Clasificación, Características, Requisitos de Registro y Etiquetado. 2022. https://osartec.gob.sv/?p=6093

Congreso de la Nación Paraguaya, Ley 1119/97 de Productos para la Salud y otros. www.mspbs.gov.py/dependencias/dnvs/adjunto/4f2cbc-7.LeyN1119.97DeProductosparalaSaludyOtros.pdf

Congreso de la Nación Paraguaya, Ley 3283 de Protección de la información no divulgada y datos de prueba para los registros farmacéuticos. 2007. www.pj.gov.py/images/contenido/ddpi/leyes/ley-3283-2007-de-proteccion-de-la-informacion-no-divulgada-y-datos-de-prueba.pdf

Decreto-Lei (DL) 986/1969, Brazil. www.gov.br/agricultura/pt-br/assuntos/inspecao/produtos-vegetal/legislacao-1/biblioteca-de-normas-vinhos-e-bebidas/decreto-lei-no-986-de-21-de-outubro-de-1969.pdf/view

Decreto Supremo No. 0004–2022-MIDAGRI, Ministerio de Desarrollo Agrario y Riego, Perú. www.gob.pe/institucion/midagri/normas-legales/2902865-0004-2022-midagri

Food and Agricultural Organization of the United Nations and World Health Organization. 2002. *Joint FAO/WHO Working Group Report on Drafting Guidelines for the Evaluation of Probiotics in Food*. World Health Organization and Food and Agriculture Organization of the United Nations, London Ontario, Canada.

Guia orienta sobre instrução processual de probióticos, Brazil. 2022. www.gov.br/anvisa/pt-br/assuntos/noticias-anvisa/2019/guia-orienta-sobre-instrucao-processual-de-probioticos

Hill, C., Guarner, F., Reid, G., Gibson, G.R., Merenstein, D.J., Pot, B., Morelli, L., Canani, R.B., Flint, H.J., Salminen, S., & Calder, P.C. 2014. The international scientific association for probiotics and prebiotics consensus statement on the scope and appropriate use of the term probiotic. *Nature Reviews Gastroenterology & Hepatology*, 11: 506–514. https://doi.org/10.1038/nrgastro.2014.66

Institute of Food Technologist—IFT. Expert Report. 2021. www.ift.org/career-development/learn-about-food-science/food-facts/food-facts-food-health-and-nutrition/the-411-on-functional-foods

Instituto Ecuatoriano de Normalización (INEN), Norma Técnica Ecuatoriana NTE-INEN 1334–3:2011. www.controlsanitario.gob.ec/wp-content/uploads/downloads/2014/07/ec.nte_.1334.3.2011.pdf

Ministério da Saúde- MS, Agência Nacional de Vigilância Sanitária—ANVISA. Resolução da Diretoria Colegiada (RDC) 17/99. Brazil. www.gov.br/agricultura/pt-br/assuntos/inspecao/produtos-vegetal/legislacao-1/biblioteca-de-normas-vinhos-e-bebidas/resolucao-no-17-de-30-de-abril-de-1999.pdf

Ministério da Saúde- MS, Agência Nacional de Vigilância Sanitária—ANVISA. Resolução da Diretoria Colegiada (RDC) 18/99, Brazil. www.gov.br/agricultura/pt-br/assuntos/inspecao/produtos-vegetal/legislacao-1/biblioteca-de-normas-vinhos-e-bebidas/resolucao-no-18-de-30-de-abril-de-1999.pdf/view

Ministério da Saúde- MS, Agência Nacional de Vigilância Sanitária—ANVISA. Resolução da Diretoria Colegiada (RDC) Nº 2, January 7 2002, Brazil. www.saude.rj.gov.br/comum/code/MostrarArquivo.php?C=MjI1Mw%2C%2C

Ministério da Saúde- MS, Agência Nacional de Vigilância Sanitária—ANVISA. Resolução da Diretoria Colegiada (RDC) No. 241, July 26 2018, Brazil. http://antigo.anvisa.gov.br/documents/10181/3898888/RDC_241_2018_.pdf/941cda52-0657-46dd-af4b-47b4ee4335b7

Ministério da Saúde- MS, Agência Nacional de Vigilância Sanitária—ANVISA. Resolução da Diretoria ColegiadA (RDC) Nº 718, 1st July 2022, Brazil. http://antigo.anvisa.gov.br/documents/10181/2718376/RDC_718_2022_.pdf/caa81352-b1ec-4856-b7bb-dc328dc2dcb9

Ministerio de Agricultura y Ganadería, Agencia de Regulación y Control Fito y Zoosanitario. Manual para el registro de empresas y productos de uso veterinarios. 2018. www.gob.ec/sites/default/files/regulations/2018-10/DOCUMENTO_MANUAL%20PARA%20EL%20REGISTRO%20DE%20EMPRESAS%20Y%20PRODUCTOS%20DE%20USO%20VETERINARIO.pdf

Ministerio de Desarrollo Rural y Tierras, Bolivia. Resolución Administrativa SENASAG No. 172/2022. www.senasag.gob.bo/phocadownload/RESOLUCIONES_ADMINISTRATIVAS/RESOLUCIONES_ADMINISTRATIVAS/SANIDAD_ANIMAL/2022/RA_172_2022.pdf

Ministerio de Protección Social de Colombia. Resolución 333. 2011. https://scj.gov.co/sites/default/files/marco-legal/R_MPS_0333_2011.pdf

Ministerio de Salud de Chile. Resolución 3435. 2018. www.carey.cl/download/filebase/newsalert/res-n-3435-probioticos.pdf

Ministerio de Salud de Chile. Resolución No. 860. 2017. www.bcn.cl/leychile/navegar?idNorma=1105664&idParte=0&idVersion=

Ministerio de Salud de Chile. Resolución No. 1671. 2018. www.bcn.cl/leychile/navegar?idNorma=1113511&idParte=9874214

Ministerio de Salud de Panamá, Decreto 331, Artículo 12. 2008. www.gacetaoficial.gob.pa/pdfTemp/26101/12510.pdf

Ministerio de Salud Pública y Bienestar Social de Paraguay, Decreto No. 6611/16. www.mspbs.gov.py/dependencias/dnvs/adjunto/1a7b18-DecretoN661116.pdf

Myers, S. 2014. South American progress. *Natural Products Insider*, https://www.naturalproductsinsider.com/supplement-regulations/south-american-harmonization.

Normativa Sanitaria para Control de Suplementos Alimenticios, Resolución 28, No. ARCSA-DE-028–2016-YMIH. 2017. www.controlsanitario.gob.ec/wp-content/uploads/downloads/2017/02/Resoluci%C3%B3n_ARCSA-DE-028-2016-YMIH_NTS_SUPLEMENTOS_ALIMENTICIOS.pdf

Organismo Salvadoreño de Reglamentación Técnica (OSARTEC) referente al RTS 67.06.02:22: Alimentos para Regímenes Especiales, Suplementos Nutricionales y Probióticos. Clasificación, Características, Requisitos de Registro y Etiquetado. 2022. http://osartec.gob.sv/wp-content/plugins/download-manager/viewer/viewer.php?dl=http://osartec.gob.sv/wp-content/uploads/download-manager-files/1664505956wpdm_RTS%20APRE-SUPLE-PROBIO_%20VF%2030092022.pdf

Presidência da República, Casa Civil, Subchefia para Assuntos Jurídicos, Brazil. Lei No. 6.360, de 23 de setembro de 1976. www.planalto.gov.br/ccivil_03/leis/l6360.htm

Probiotic Market 2021 and Trend Analysis. *International Probiotics Association (IPA Europe)*. www.ipaeurope.org/legal-framework/market-data/

Reglamento Técnico Centroamericano RTCA 67.01.60:10: Etiquetado Nutricional de productos alimenticios preenvasados para consumo humano para la población a partir de 3 años de edad. 2022. https://extranet.who.int/nutrition/

gina/sites/default/filesstore/COMIECO%202011%20Etiquetado%20Nutricional%20de%20Productos%20 Alimenticios%20Preenvasados%20para%20Consumo%20Humano.pdf

Sanders, M.E., Heimbach, J.T., Pot, B., Tancredi, D.J., Lenoir-Wijnkoop, I., Lahteenmaki-Uutela A., Gueimonde, M., & Banares, S. 2011. Health claims substantiation for probiotic and prebiotic products. *Gut Microbes*, 2(3): 127–133. https://doi.org/10.4161/gmic.2.3.16174

Secretaría de Calidad en Salud, and Secretaría De Alimentos, Bioeconomía y Desarrollo Regional (Argentina). 2020. Joint Resolution 3–2020. www.argentina.gob.ar/normativa/nacional/resoluci%C3%B3n-3-2020-345768/texto

Secretaría de Gobernación de México. Rule NOM-181-SCFI/SAGARPA-2018. https://dof.gob.mx/nota_detalle.php? codigo=5549317&fecha=31/01/2019#gsc.tab=0

Servicio Nacional de Sanidad Agropecuaria e Inocuidad Alimentaria (SENASAG), Bolivia. Resolución Administrativa No. 058/01. www.senasag.gob.bo/phocadownload/RESOLUCIONES_ADMINISTRATIVAS/SANIDAD_ ANIMAL/2001/RA_058_2001.pdf

Servicio Nacional de Sanidad Agropecuaria e Inocuidad Alimentaria (SENASAG), Bolivia. Resolución Administrativa No. 179/2011. www.senasag.gob.bo/phocadownload/RESOLUCIONES_ADMINISTRATIVAS/SANIDAD_ ANIMAL/2011/RA_179_2011.pdf

Topolska, K., Florkiewicz, A., & Filipiak-Florkiewicz, A. 2021. Functional food-consumer motivations and expectations. *International Journal of Environmental Research and Public Health*, 18(10): 5327. https://doi.org/ 10.3390/ijerph18105327

# Regulation of Probiotics in the United States

# 40

Amy B. Smith

## 40.1 BACKGROUND

In the United States, recent trends have demonstrated that probiotics are increasingly added to products geared toward consumer health and wellness (Euromonitor International Limited, Health & Wellness 2023). The recent growth has resulted in consumer education, industry expansion, and increased regulatory scrutiny. This chapter is intended to provide an overview of how probiotics are controlled within the U.S. regulatory landscape. Because probiotics do not neatly fit within a single regulatory category, it is essential to provide a review of the range of possible categories where probiotics are currently included as ingredients today, the breadth of regulatory responsibility of the U.S. Food and Drug Administration (FDA), and how probiotics fit within it.

Probiotics are commercially available within a range of product types regulated by FDA, with primary authority granted by the Federal Food, Drug, and Cosmetics Act (FDCA) (21 USC 9). As outlined by Congress, the jurisdiction of the FDA includes products sold in interstate commerce (Piña and Pines 2012).

Housed within the U.S. Department of Health and Human Services, FDA has jurisdiction over numerous product types sold within the United States as part of interstate commerce and regulates approximately 80% of the U.S. food supply (U.S. FDA 2018b). Under the Office of the Commissioner, there are four directorates that lead the principal functions of FDA: (1) Medical Products and Tobacco, (2) Foods and Veterinary Medicine, (3) Global Regulatory Operations and Policy, and (4) Operations (U.S. FDA 2021a). Within these directorates, FDA is composed of many offices and centers that focus on distinct specialty areas. The classes of products regulated by FDA (U.S. FDA 2022a) can be further categorized as follows:

- Foods (including dietary supplements, bottled water, food additives, infant formula, foods outside of meat and poultry, unbroken eggs, and some egg products)
- Drugs (over-the-counter and prescription)
- Biologics (vaccines, gene therapy, blood, etc.)
- Medical devices (variety of product types, from tongue depressors to surgical implants)
- Electronic products that emit radiation (microwaves, X-ray, etc.)
- Cosmetics (color additives used in cosmetics and personal care products, nail polish, etc.)
- Veterinary products (pet food, animal feed, veterinary drugs, etc.)
- Tobacco products (cigarettes, tobacco, etc.)

DOI: 10.1201/9781003352075-44

This vast array of product types is listed to demonstrate the scale of FDA's jurisdiction, within the mission of protecting public health. Probiotic-specific regulation is lacking; therefore it is the *intended use* of the ingredient that determines how the product might be regulated (Degnan 2008; Sanders *et al.* 2010). As there is no formal acknowledgment or criteria for probiotics as a product type by FDA, one would assume that the FAO/WHO definition (Report 2001) would apply; however, it is not clear if FDA recognizes this definition or criteria within.

# 40.2 PROBIOTICS AS A FOOD INGREDIENT

From an historical perspective, probiotics emerged from the long history of microbial cultures used in food fermentation. The initial benefit to the food could be flavor, preservation, or nutritive value, which results from the biochemical reactions of the microbes in the food (International Dairy Federation 2012). These food cultures, which were initially natural components of traditional foods, were identified as bacteria, fungi, yeast (Stevens and Nabors 2009), and in large portion lactic acid bacteria (LAB) (Johansen 2017). Many of these are the same genera and species used as probiotics today; therefore our initial focus within the U.S. regulation of probiotics will be the *intended use* as a food ingredient.

With the mission of protecting public health, FDA has responsibility for most of the U.S. food supply. This includes all domestic and imported food but does not include meat; poultry; and eggs that are frozen, liquid, or dried (these are the responsibility of the U.S. Department of Agriculture). Within FDA, the Center for Food Safety and Applied Nutrition (CFSAN) has responsibility for food, cosmetics, and dietary supplements (U.S. FDA 2019a) Legal authority is directed through statutes that have evolved since the Food and Drug Act of 1906 and then the more comprehensive FDCA of 1938 (U.S. FDA 2019b; Piña and Pines 2012), which has been amended and modernized numerous times, always maintaining the original structure that allows food law to be enforced.

This framework provides guidance in key areas that are essential for establishing the safety and regulatory status of today's food and food ingredients. Section 201 of the FDCA (21 USC 321) states that a *food* can be defined as

> (1) articles used for food or drink for man or other animals, (2) chewing gum, and (3) articles used for components of any such article.

Section 201 continues, to define the term "food additive" as

> any substance the intended use of which results or may reasonably be expected to result, directly or indirectly, in its becoming a component or otherwise affecting the characteristics of any food (including any substance intended for use in producing, manufacturing, packing, processing, preparing, treating, packaging, transporting, or holding food; and including any source of radiation intended for any such use), if such substance is not generally recognized, among experts qualified by scientific training and experience to evaluate its safety, as having been adequately shown through scientific procedures (or, in the case of a substance used in food prior to January 1, 1958, through either scientific procedures or experience based on common use in food) to be safe under the conditions of its intended use.

Under the Food Additives Amendment (FAA) of 1958, FDA was given authority to require proof of safety before an ingredient, could be legally marketed: a *premarket approval system* (Piña and Pines 2012). This premarket approval pathway is a viable route to market for ingredients that meet the definition of a food additive. However, as noted, the definition provides exemptions if either of two criteria are met, the first being that the ingredient is "generally recognized, among experts . . . to be safe" and the second as "used in food prior to January 1, 1958", each "under the conditions of intended use". These conditions are extremely important, providing exemption from the category of food additive (the premarket approval system) and encompassing the formal "generally recognized as safe" (GRAS) process that is used today to establish the safety of a great number of food ingredients, including probiotic strains, *where a premarket approval is unwarranted.*

When intended to be used as a food ingredient, to be GRAS, a specific set of criteria must be demonstrated for a probiotic strain, where evidence of scientific agreement regarding safety must be provided and may be based on either appropriate testing (scientific procedures) or common use in food prior to 1958. Specific eligibility criteria are stated simply as follows:

> General recognition of safety may be based only on the views of experts qualified by scientific training and experience to evaluate the safety of substances directly or indirectly added to food. The basis of such views may be either (1) scientific procedures or (2) in the case of a substance used in food prior to January 1, 1958, through experience based on common use in food. General recognition of safety requires common knowledge throughout the scientific community knowledgeable about the safety of substances directly or indirectly added to food that there is reasonable certainty that the substance is not harmful under the conditions of its intended use.
>
> (21 CFR 170.30[a])

While provided as one of two options to attain GRAS status, establishing use in food prior to 1958, also known as prior sanctions (21 CFR 170.3[l]), has not been routinely used for probiotic strains.

FDA has released a few lists that indicate which substances are recognized as GRAS, including specific substances in the Code of Federal Regulations, Title 21 Parts 184 and 186 (21 CFR 184; 21 CFR 186), affirmed as generally recognized as safe for direct and indirect food substances, respectively. Additionally, 21 CFR Part 182 (21 CFR 182) lists substances generally recognized as safe, categorized by function or purpose. Among these lists, there are no probiotics listed, although 21 CFR 184 does include several bacterial-derived substances, such as enzymes. However, FDA does *indirectly* recognize the use of microbial cultures in several sections within 21 CFR Part 131, which covers Requirements for Specific Standardized Milk and Cream (21 CFR 131). Specifically, acidified milk may or may not contain "characterizing microbial organisms" (21 CFR 131.111); cultured milk may contain "characterizing microbial organisms" that may be listed with a generic name of the microbes used, including "acidophilus cultured milk" and "cultured buttermilk" when "lactic acid-producing organisms" are used (21 CFR 131.112); "lactic acid-producing bacteria" may or may not be added to acidified sour cream (21 CFR 131.162); yogurt will contain "characterizing bacterial culture" containing "lactic acid-producing bacteria", specifically listed as *Lactobacillus bulgaricus* and *Streptococcus thermophilus* (21 CFR 131.200); the cheddar cheese standard refers to "the action of a lactic acid-producing bacterial culture" (21 CFR 133.113); and "lactic acid-producing bacteria" are listed as an optional ingredient for bread, rolls, and buns (21 CFR 136.110). Outside of these indirect references to "characterizing microbial organisms" and *Lactobacillus bulgaricus* and *Streptococcus thermophilus*, FDA does not formally recognize a specific list of microbial species that are GRAS. Instead, FDA recognizes the use of ingredients that are *derived from* microorganisms, which are listed as either food additives or GRAS (U.S. FDA 2018a). Only a few of these listed species have been recognized to have probiotic potential. Essentially, this leaves us with numerous indirect references to "lactic acid bacteria" or "characterizing microbial organisms" by FDA and does not establish any specific prior sanction that would render individually identified microbial species as GRAS. Instead, GRAS status for probiotics is attained using the alternative GRAS criteria of *scientific procedures.*

As stated in the Code of Federal Regulations,

> Scientific procedures include the application of scientific data (including, as appropriate, data from human, animal, analytical, or other scientific studies), information, and methods, whether published or unpublished, as well as the application of scientific principles, appropriate to establish the safety of a substance under the conditions of its intended use.
>
> (21 CFR 170.3[l])

This standard allows those ingredients that are truly *generally* recognized by experts as safe to circumvent the food additive petition process. Eligibility can be established through history of use in food or by compiling data that demonstrates safety under the proposed conditions of use. In either case, the key aspect of this process is recognition by qualified experts.

FDA's role in the GRAS process has evolved over time (U.S. FDA 2018b), as the original GRAS Affirmation petition process, put in place in 1972, became a GRAS Notification process (62 FR 18938) and became a final rule in 2016 (81 FR 54960). The final rule makes it clear that the procedure for petitioning FDA to affirm GRAS status was being replaced with a voluntary procedure for *notifying* FDA that a substance is GRAS under the conditions of its intended use. Furthermore, the industry received a newly required FDA-defined format for a GRAS dossier with seven distinct parts that are specifically listed and defined in detail within the final rule. These are listed as stated in the GRAS final rule (81 FR 54960):

1. Signed statements and a certification;
2. The identity, method of manufacture, specifications, and physical or technical effect of the notified substance;
3. Dietary exposure;
4. Self-limiting levels of use, in circumstances where the amount of the notified substance that can be added to human food or animal food is limited because the food containing levels of the notified substance above a particular level would become unpalatable or technologically impractical;
5. The history of consumption of the substance for food use by a significant number of consumers (or animals in the case of animal food) prior to January 1, 1958, if a conclusion of GRAS status is based on common use of the substance in food prior to 1958;
6. A narrative that provides the basis for your conclusion of GRAS status, including why the scientific data, information, methods, and principles described in the notice provide a basis for your conclusion that the notified substance is generally recognized, among qualified experts, to be safe under the conditions of its intended use; and
7. A list of the data and information that you discuss in the narrative of your GRAS notice, specifying which of these data and information are generally available, and which of these data and information are not generally available.

A few points should also be highlighted, including the fact that GRAS status is attainable for ingredients intended for use in human food as well as animal feed, and that the notification should be sent to CFSAN's Office of Food Additive Safety (OFAS) for human food and to CVM's Division of Animal Feeds within the Office of Surveillance and Compliance for animal food. Last, it should also be understood that FDA does not distinguish between a conclusion of GRAS status when formally submitted to FDA as a notification versus evidence of self-GRAS status that "remains with its proponent as an independent conclusion (formerly referred to as a 'self-determination') of GRAS status" (81 FR 54960). This indicates that one is not required to formally notify FDA to declare a substance as GRAS, and FDA allows evidence of safety to be housed as an internal dossier, to be disseminated externally as needed by the firm that has prepared the self-GRAS. However, the level of internally housed evidence should be the same *as if* notifying FDA. Many ingredient manufacturers choose the self-GRAS option for a number of ingredient types, including probiotics, despite some within the industry viewing this approach as a negative. It is truly an internal business decision.

There are a number of probiotic strains that have attained GRAS status through notifying FDA. Table 40.1 lists the probiotic species (and in some cases, specific strains) that have been established as safe via submission of GRAS notifications to FDA. FDA publishes all GRAS notifications (withholding confidential information) and the corresponding FDA letter of no objection (U.S. FDA 2024a). Despite the limited number of probiotic species submitted, this GRAS Notice Inventory database provides a great tool for understanding which microbial species have been formally submitted and recognized as safe, according to qualified experts, for inclusion in food and under what specific conditions of use. Information about all

**TABLE 40.1**  GRAS Notifications for Live Microorganisms

| GENUS | SPECIES | PROBIOTIC STRAINS |
|---|---|---|
| Lactobacillus* | casei subsp. paracasei | Lpc-37, F19 |
| | plantarum | Lp-115, 299v, Unique IS2, CECT 7527,7528,7529, DSM33452, ECGC13110402, CBTLP3, MCC0537, NCIMB30562, ATCC202195, DSM34613 |
| | fermentum | CECT5716, LfQi6 |
| | acidophilus | La-14, NCFM, DDS-1 |
| | reuteri | NCIMB 30242, DSM17938 |
| | casei | Shirota, KCTC 12398BP |
| | rhamnosus | HN001, GG, DSM33156, IDCC 3201 |
| | paracasei subsp. paracasei | F-19e |
| | lactis | ATCC PTA-124205 |
| | johnsonii | DSM18775 |
| | curvatus | R0052 |
| | helveticus | |
| Bacillus | coagulans | GBI-30, SANK 70258, SNZ1969, LA-1, DSM17654, SNZ1969 |
| | clausii | 088AE |
| | subtilis | Bss-19, ATCC SD7280, BSMB40, SG188, DE111, ATCC AAN02, ATCC BS50 |
| Propionibacterium | freudenreichii subsp. shermanii | |
| Pediococcus | acidilactici | |
| Streptococcus | salivarius | K12, DB-B5, M18 |
| | thermophilus | Th4 |
| Bifidobacterium | animalis subsp. lactis | HN019, Bi-07, Bl-04, B420, Bf-6, Bb12, AD011, UABla-12, R0421, KCTC 11904BP, IDCC 4301 |
| | breve | M-16V, MCC1274, DSM 33444 |
| | longum | BB536, BORI |
| | longum subsp. infantis | R0033, M-63, SD6720, DSM33361 |
| | bifidum | R0071, BGN4, NITE BP-31 |
| Saccharomyces | cerevisiae | ML01, ECMo01, P1Y0, yBBS002, OYR-185, OYR-243 |
| Staphylococcus | carnosus | DSM25010 |
| Carnobacterium | maltaromatican | CB1 |
| Leuconostoc | carnosum | DSM32756 |
| Anaerobutyricum | soehngenii | CH106 |

* Updated Lactobacillus taxonomy (Zheng et al. 2020) is not reflected in GRAS notification genus listings prior to 2021.

GRAS notices submitted since 1998 are listed within the GRAS Notice Inventory (U.S. FDA 2024a) and has been edited to reflect probiotic strains.

All species and strains listed in Table 40.1 are the substances established as safe by the submitters, for the specified intended use(s) listed, which vary from use in infant formula, general foods. Despite the variety of intended uses among the species/strains, the accompanying FDA letter of no objection confirms a general consensus of safety was determined by qualified experts for all notifications listed. Furthermore, it should be highlighted that while the aspect of safety at the species level is most often evident, FDA has distinctly stated in the GRAS Final Rule (81 FR 54960) that when *biological material* is the substance presented as GRAS, information at the *subspecies* (strain) level should be presented in addition to genus-and species-specific information. This is a change from the earlier proposed rule (62 FR 18938) and indicates FDA's current position on strains as independent ingredients, distinct within a species.

# 40.3  PROBIOTICS IN ANIMAL FOOD/FEED

A second *intended use* for probiotic strains is within animal food and feed. Also covered within the FDCA, animal food ingredients must be recognized as safe, for which either the food additive or the GRAS process is appropriate. Within the FDA organizational structure, the CVM has responsibility for regulating animal feed, which includes pet food. There are some regulations that are common to food for humans and animals (e.g., 21 CFR 170) and some that are specific to animals (e.g., 21 CFR 507, 21 CFR 570). The details will not be presented at length as part of this chapter; however, a brief description is worthwhile. The regulation of animal food in the United States is unique in that while FDA does take a major role in overseeing animal food, partnerships with state and local agencies are essential for effective regulation, especially within the animal feed industry (U.S. FDA 2022b). FDA has partnered with the Association of American Feed Control Officials (AAFCO) (U.S. FDA 2024b), where a memorandum of understanding (U.S. FDA 2019c) defines the roles and responsibilities of each. AAFCO has no regulatory authority but plays a major role in ensuring a safe food supply for animals by providing policies and standards for regulation focusing on manufacturing, labeling, and sale of animal feeds (Association of American Feed Control Officials 2024).

While the benefits of adding probiotics to pet food and animal feed are widely acknowledged (Yirga 2015), FDA does not specifically recognize probiotics as a type of ingredient that is acceptable for use in animal food. AAFCO does address the use of many probiotic strains as "direct-fed microorganisms". The AAFCO Official Publication lists direct-fed microbials (Association of American Feed Control Officials 2018) that have been reviewed by FDA CVM and determined to cause no safety concerns. This list includes several genera and species that have demonstrated probiotic properties.

# 40.4  HUMAN DIETARY SUPPLEMENTS

The next *intended use* is that of a dietary ingredient intended for use *in* or *as* a dietary supplement. As referenced earlier, the FDCA was designed to provide a comprehensive system of regulation that covers a wide variety of products within the food, drugs, and cosmetics marketplace. Understandably, as time goes by, science and innovation can lead to new products that were not anticipated when regulatory structures originated. The discovery of vitamins and minerals, the move to preventive health care by consumers, the increase in life expectancy, and consumer activism all demonstrated a need for a change from the traditional food versus drug legislative doctrine in the early to mid-20th century (Institute of Medicine (US) Food Forum 1999; Bass 2011; Piña and Pines 2012). The enactment of the Nutrition Labeling and Education Act of 1990 (NLEA) by the U.S. Congress (H.R. 3562) revised food labeling requirements and authorized health claims for food products, but a gap remained where regulation was lacking in the space between food and drugs. The impression was that FDA was limiting permissible combinations of vitamin and mineral supplements (Dickinson 2011) and industry and consumer groups influenced Congress to enact the Dietary Supplement Health and Education Act of 1994 (DSHEA) (S.784) (Piña and Pines 2012), which brought about a major change. DSHEA established the statutory definition of dietary supplements which is described in the broad overview condensed within the Passed House Amended Summary (S.784 1994) as

> a product: (1) other than tobacco, intended to supplement the diet that contains a vitamin, mineral, herb, or botanical, dietary substance, or a concentrate, metabolite, constituent, extract, or combination of the above ingredients; (2) that is intended for ingestion, is not represented as food or as a sole item of a meal or diet, and is labeled as a dietary supplement; (3) that includes an article approved as a new drug, certified as an antibiotic, or licensed as a biologic and that was, prior to such approval, certification or licensure, marketed as a dietary supplement or food, unless the conditions of use and dosages are found to be unlawful; and (4) excludes such articles which were not so marketed prior to approval unless found to be lawful. Deems a dietary supplement to be a food. Excludes a dietary supplement from the definition of the term "food additive".

As briefly summarized, DSHEA amended the FDCA and provided tremendous clarity in several areas of previous ambiguity.

What does this mean for probiotics? It is worthwhile to do a deep dive into DSHEA and its application to probiotics, as this regulatory framework allows probiotics to be sold as dietary supplements in the United States today. Keeping in mind that the Passed House Amended Summary highlights key attributes of how dietary supplements are defined by DSHEA, it is valuable to visit the codified terms for an in-depth review.

Section 201 of the FDCA has been amended to include the following (21 USC 321):

(ff) The term "dietary supplement

- (1) means a product (other than tobacco) intended to supplement the diet that bears or contains one or more of the following dietary ingredients:
- (A) a vitamin;
- (B) a mineral;
- (C) an herb or other botanical;
- (D) an amino acid;
- (E) a dietary substance for use by man to supplement the diet by increasing the total dietary intake; or
- (F) a concentrate, metabolite, constituent, extract, or combination of any ingredient described in clause (A), (B), (C), (D), (E);

These criteria are listed out so we may highlight where probiotics most appropriately fit under the definition of a dietary ingredient. While it is not overwhelmingly clear in the text, FDA has stated in public forums that probiotics most likely fall under category E, as a dietary substance for use by humans to supplement the diet by increasing total dietary intake (Welch 2017). To meet this criteria, FDA must agree that the probiotic strain of interest will *supplement* the diet, meaning that it must be currently *in the diet* for the product to qualify as a dietary ingredient.

DSHEA also elaborates in Section 201 of the FDCA (21 USC 321) that a dietary supplement must be *intended to be ingested* and further specifies the acceptable formats of tablet, capsule, powder, softgel, gelcap, or liquid form (21 USC 350). DSHEA indicates that a dietary supplement cannot be represented as the sole item of a meal or diet, must be labeled as a dietary supplement, and is *excluded* from the definition of a food additive. Of utmost importance, dietary supplements are deemed as a food, except when meeting criteria of 201(g), (21 USC 321), which is indicative of a drug. This clear categorization as a food removes the previous ambiguity. In summary, these vital definitions and characteristics are important to be mindful of when marketing a probiotic product in the United States.

Interestingly, DSHEA specifically indicates in Section 4 that the burden of proof for safety of dietary supplements lies with FDA and that the United States must provide proof that a dietary supplement is adulterated (unsafe) (21 USC 342). This is very different from the premarket approval process of the food additive category. However, it is nothing to be taken for granted or ignored. Probiotics manufacturers are highly aware that in addition to the food adulteration standards listed in FDCA (21 USC 342), the DSHEA-specific adulteration standards are extra for dietary supplements, which we will discuss in detail.

DSHEA has also amended FDCA (Section 402) to include indication that a dietary supplement is deemed as adulterated if certain issues are recognized (21 USC 342). First,

> If it is a dietary supplement or contains a dietary ingredient that

> (A)  presents a significant or unreasonable risk of illness or injury under-
>   (i)  conditions of use recommended or suggested in labeling, or
>   (ii)  if no conditions of use are suggested or recommended in the labeling, under ordinary conditions of use;

This first adulteration standard presents *significant* or *unreasonable risk* of illness or injury under the conditions labeled or under ordinary conditions of use. The significant or unreasonable risk is distinct from the *generally* recognized as safe aspect for food ingredients. Again, not to be taken for granted.

DSHEA also amended FDCA (Section 402) to include a second aspect of adulteration (21 USC 342). A dietary supplement is deemed adulterated:

> If it is a dietary supplement or contains a dietary ingredient that
>   B) is a *new dietary ingredient* for which there is inadequate information to provide reasonable assurance that such ingredient does not present a significant or unreasonable risk of illness or injury

Again, the *significant or unreasonable risk* is listed; however, this standard is specific to the expectation that information will provide a "reasonable assurance" of safety for this specific category of a *new dietary ingredient*. New dietary ingredients, or NDIs, will be discussed in detail, but first let's briefly review the topic of safety as presented in Section 4 of DSHEA.

With all the distinctions that FDCA Section 402 brings forth regarding safety and adulteration of dietary supplements, current establishment of the safety of a probiotic strain relies on many of the same aspects that have been included in a GRAS notification of safety for food ingredients. Simply put, DSHEA affirmed the "statutory food safety requirements for dietary supplements" and established a new notification system for NDIs (Dickinson 2011). DSHEA requires food safety standards to be adopted for dietary supplements with the additional requirement, per the adulteration standards, to eliminate a *significant or unreasonable risk of illness or injury* under established conditions. As for applicable food safety standards, there are only two choices: the food additive route or GRAS. DSHEA excludes dietary supplements from the category of food additives; therefore GRAS, the process used for food ingredients, is the likely framework from which to model a process of establishing safety specifically for dietary supplement ingredients. As listed in Table 40.1, demonstrations of safety of probiotic strains have been submitted to FDA as GRAS notifications for use in foods. These probiotic ingredients are GRAS and are commonly added to foods. Therefore, it is important to have exposure to the criteria acceptable for establishing these as ingredients that are *generally recognized as safe*. An assessment of GRAS notifications housed within the GRAS Notice Inventory database (U.S. FDA 2024a) indicates that for probiotic strains, the common themes outlined in Table 40.2 are addressed to establish safety.

While these investigative tools are not absolute, and FDA does not explicitly require these specific attributes to be studied, most probiotic GRAS notifications include a majority of this information to establish the comprehensive safety of a probiotic strain. As stated earlier, FDA has specified in the GRAS final rule (81 FR 54960) that for biological sources, information at the subspecies level should be included. In other words, *strains* are considered distinct ingredients, even those within the same species.

**TABLE 40.2**    Common Topics Included in GRAS Notifications to Establish Safety of Live Microorganisms

| LEVEL OF EVIDENCE | INVESTIGATIVE TOOL | INFORMATION PRESENTED |
| --- | --- | --- |
| Species | Inclusion of species on QPS and/or IDF list(s) | History of use in foods at the species level |
| Species | Extensive literature search of the species | History of demonstration of safety using scientific procedures—evidence of safety available in the public domain |
| Strain | Antimicrobial resistance | Minimal inhibitory concentration (MIC) for clinically relevant antibiotics (FEEDAP 2012) |
| Strain | Whole genome sequence | Unambiguous identification |
| Strain | Mining for virulence factor genes | Evidence of lack of virulence potential |
| Strain | Mining for evidence of toxin production | Evidence of lack of toxin production |
| Strain | Whole genome sequence alignment to type strain or well-characterized strain within the same species | Identifies similarity to strain previously established as safe or studied extensively; identifies differences between the genomes (which must be explained) |
| Strain | Mining for antibiotic resistance genes and proximity to transposable elements | Determination of antibiotic transfer potential |
| Ingredient | Acute toxicity study in rodents | Evidence of lack of acute toxicity |
| Ingredient | Subchronic toxicity study in rodents | Evidence of lack of toxicity with repeated dose delivery (note: this is not routinely included—only on occasion) |
| Ingredient | Animal trial | Evidence of efficacy in animal model (also can be used for safety) |
| Ingredient | Human clinical trial | Evidence of safety and/or efficacy in humans of targeted demographic |

This remains true for dietary supplements, and industry practice is to bridge the safety demonstrated at the species level to that of the subspecies level (a strain) coupled with the inclusion of all relevant strain-specific information.

The level of essential safety testing based on appropriate factors, such as the decision tree created by Pariza et al. (2015) and the consideration of global guidelines (Roe et al. 2022), can be helpful, as these consider all relevant information for establishing safety for probiotic strains. An assessment by Sanders et al. (2010) includes consideration of the health status of target populations in establishing safety.

In summary, the degree of safety discussed previously represents the conventional level accepted, with no objection, by FDA Office of Food Additive Safety (OFAS) in GRAS notifications. The probiotics industry agrees and complies with this evidence level as relevant and customary, as probiotics are most likely generally recognized as safe. This industry practice was deemed acceptable for probiotic strains used in foods and in dietary supplements for many years until FDA's enforcement of NDIs and safety requirements were updated.

To successfully understand the level of diligence required for a probiotic ingredient, it is imperative to understand whether the ingredient may be considered an NDI. Section 8 of DSHEA amended Section 413 of the FDCA to insert information on NDIs (21 USC 350b). The NDI is first mentioned in Section 4 of DSHEA, where safety and adulteration, under FDCA Section 402 (21 U.S.C. 342), are specifically addressed, and *the burden of proof is clearly placed on FDA*. Interestingly, NDIs are also described in detail in Section 8, which immediately references this adulteration standard in FDCA Section 402, with the following description:

A dietary supplement which contains a new dietary ingredient shall be deemed adulterated under section 402(f) unless it meets one of the following requirements:

(1) The dietary supplement contains only dietary ingredients which have been present in the food supply as an article used for food in a form in which the food has not been chemically altered.
(2) There is a history of use or other evidence of safety establishing that the dietary ingredient when used under the conditions recommended or suggested in the labeling of the dietary supplement will reasonably be expected to be safe and, at least 75 days before being introduced or delivered for introduction into interstate commerce, the manufacturer or distributor of the dietary ingredient or dietary supplement provides the Secretary with information, including any citation to published articles, which is the basis on which the manufacturer or distributor has concluded that a dietary supplement containing such dietary ingredient will reasonably be expected to be safe.

As NDIs are mentioned initially in Section 4, it seemed that there was a simple expectation that they should meet the adulteration standard as stated in the Section 4 text. These additional NDI parameters of Section 8 now must be considered, and one of these criteria must be met to be deemed unadulterated.

But first, what is an NDI? As indicated in Section 8 of DSHEA, for purposes of this section, the term "new dietary ingredient" means a dietary ingredient that was not marketed in the United States before October 15, 1994, and does not include any *dietary ingredient* which was marketed in the United States before October 15, 1994.

Now things get interesting. This definition draws a line on the calendar separating dietary ingredients used in or as dietary supplements *before* October 15, 1994 (now known as old dietary ingredients, or ODIs), and *after* this date (NDIs). It is presumed that ingredients commercialized as dietary supplements before DSHEA can be considered safe, however, those commercialized *after* DSHEA must meet the new standard. Considering this, let's review the adulteration standard from Section 4 of DSHEA (21 USC 342), where a dietary supplement is deemed as adulterated, if there is

inadequate information to provide reasonable assurance that such ingredient does not present a significant or unreasonable risk of illness or injury.

Nothing out of the ordinary here. An NDI must be safe, and there must be evidence to prove that it will not pose a risk of illness or injury that is considered significant or unreasonable. However, when we add this standard to the previously described criteria laid out in Section 8, there is potential for confusion as to rationale. Simply put, one of two requirements must be met to avoid adulteration:

1. The first option is a strict requirement of ingredients that have
   a. Been used as an article of food in the food supply; *and*
   b. Have not been chemically altered.
2. The alternate option is simplified as
   a. Requiring the dietary ingredient or supplement to reasonably be anticipated as safe when used under the conditions on the label based on either a history of use or other evidence of safety, *and*
   b. The manufacturer or distributer submits evidence of the set expectation of safety 75 days *before* introduced into interstate commerce, and any citations of published articles used as evidence must be submitted as well.

This leaves two regulatory pathways for NDIs. For probiotics, if the strain has been used as an article in the food supply and has not been chemically altered, then this strain is assumed to be GRAS for use in food. An expectation of safety is satisfied via the GRAS route. This is distinct from the second option listed, as if the probiotic strain has *not* been used as an article of food in the food supply, *then* the expectation of safety has not been met and this evidence must be presented as what is commonly known in the dietary supplement industry today as a New Dietary Ingredient Notification (NDIN). The details required as appropriate evidence of safety as a *reasonable assurance that such ingredient does not present a significant or unreasonable risk of illness or injury.*

To guide the industry in properly gathering the anticipated level of information for NDIN submissions, in 2011, an initial NDI Draft Guidance was released by FDA (U.S. FDA 2011) and in 2016, a second NDI Draft Guidance (U.S. FDA Center for Food Safety and Applied Nutrition 2016a), which established FDA's "current thinking". It is important to note that during the interim between the release of the 2011 and 2016 Draft Guidance documents, FDA's CFSAN formed a new office, the Office of Dietary Supplement Programs. Created in 2015, it predicated an increased focus on dietary supplements within FDA itself and the intent to enforce regulations more effectively (Bill 2017).

The Draft Guidance (U.S. FDA Center for Food Safety and Applied Nutrition 2016a) specifically lists the information that must be submitted in an NDI notification. Information on the identity and composition, the basis for the conclusion of NDI status, conditions of use, history of use or safety evidence, and why the NDI is believed to be safe all must be provided. In line with the GRAS Final Rule, FDA declares in the Draft Guidance that each microbial strain is a separate ingredient, including those within the same species, and requests information on the comparative relationship of the *strain* of interest to others of the same *species*.

FDA includes many key deliberations in the Draft Guidance that are important to be aware of for probiotics. Table 40.3 lists specific positions on NDI requirements presented by FDA and the probiotic industry-specific issues/questions that result.

**TABLE 40.3**  FDA's Interpretation of the NDI Requirement (U.S. FDA Center for Food Safety and Applied Nutrition 2016a) and Potential Issues for the Probiotic Industry

| *FDA'S POSITION* | *FURTHER INFORMATION* | *RESULTING ISSUE* |
|---|---|---|
| A substance must be a dietary ingredient to be an NDI | FDA states in the Draft Guidance that "Bacteria that have never been consumed as food are unlikely to be dietary ingredients". FDA has rejected NDINs submitted for bacterial *strains* that are of species that have a long safe history of use solely because the *strain* itself was not isolated directly from the food supply. | FDA's interpretation represents a lack of understanding of microorganisms. The resulting restriction to only recognize bacterial *strains* that are directly isolated from a food source severely limits safe and efficacious probiotic strains from being introduced into dietary supplements. And this restriction completely contradicts FDA's standard for botanicals, which is not held to the variant level. |
| An NDI is **exempt** from notification if it has been present in the food supply as an article used for food in a form that has not been chemically altered | In the 2016 Guidance document, there are examples of fermentation conditions and media ingredient changes that FDA states as two examples of chemical alteration. | There is scientific evidence to indicate that the identification or safety profile of probiotic microorganisms does not change with a simple change of fermentation media ingredients.* Media changes have been implemented to remove allergens and improve yield as industry practice. Safety and identity is maintained and can be proven. |
| A dietary ingredient marketed outside of the U.S. before Oct. 15, 1994 is an NDI | For consideration of NDI status, marketing must be in the U.S. Marketing in any other country prior to 1994 is irrelevant. | Because of the extremely high burden of safety requirements listed in the NDI guidance, the resources and expense required to submit a successful NDIN for probiotics that have been previously marketed outside of the U.S. is seemingly unnecessary for probiotics with a safe history of use. |

(Continued)

**TABLE 40.3** (Continued) FDA's Interpretation of the NDI Requirement (U.S. FDA Center for Food Safety and Applied Nutrition 2016a) and Potential Issues for the Probiotic Industry

| FDA'S POSITION | FURTHER INFORMATION | RESULTING ISSUE |
| --- | --- | --- |
| An NDI notification can be submitted to cover multiple dietary supplements. | If the dietary ingredients to be combined with your NDI are not the same as the product of the notification, or conditions are different, a new notification may be required. | If an NDI probiotic strain is blended with other safe ingredients, shouldn't evidence of the *lack of interaction* be the primary data set deemed sufficient and submitted in a MF? |
| An NDI Master file (MF) may be submitted and referenced as part of another firm's NDI submission | This would contain ingredients and specifications considered trade secrets and will remain confidential. | There remains no portal to submit a MF in support of NDIN, and no guidance as to content that would be deemed acceptable. |

*(Sanders et al. 2014; Kussell 2013; Quan et al. 2012; Paez-Espino et al. 2015)

It is important to realize that the information presented in Table 40.3 references only a few of the issues that have been puzzling to the probiotics industry regarding FDA's current thinking for NDI guidance. As stated previously, probiotic strains to date are most often those of species with a long history of safe use in foods, and although FDA indirectly references these lactic acid bacteria and the many uses in foods, to date there is no grandfathered list of species that is generally recognized as safe for use in the United States. However, FDA held a public meeting (U.S. FDA 2017b) allowing industry to comment and provide information for FDA consideration in compiling a list of pre-DSHEA (before October 14, 1994) ingredients, also known as ODIs. In line with comments submitted following the release of the Draft Guidance (U.S. FDA Center for Food Safety and Applied Nutrition 2016a), both the Natural Products Association and the International Probiotics Association presented rationales supporting the publication of a list of microorganisms that have a safe history of use in foods *at the species level*. Given the absolute history of safety of many species historically used in foods, the probiotics industry would greatly benefit from a pre-DSHEA list that indicated safety at the species level, as many of the required safety assays listed in the NDI Guidance are expensive, often involve animals, and are labor intensive. Ideally a species-specific list of strains with a safe history of use, even with the inclusion of strain-level qualifications similar to the qualified presumption of safety (QPS) list (EFSA Panel on Biological Hazards 2018) would be agreeable to FDA; there is no list to date, however. Furthermore, it is essentially impossible for the probiotics industry to provide information as proof of marketing at the strain level before 1994. The 21st-century technology of whole-genome sequencing that allows strain distinction was not available at that time. Only genus and species identification was used, and even this was rarely found on a food label, as the generic term "cultures" has been historically used on food label ingredient lists. It is only recently that strain-specific identification has become more widely used. Again, the industry awaits a species-specific list from FDA based on a safe history of use to remove much of the NDI burden for probiotic strains used in dietary supplements.

In summary, the NDI Draft Guidance (U.S. FDA Center for Food Safety and Applied Nutrition 2016a) presents greater regulatory responsibility to the probiotics industry than is indicated in DSHEA (21 U.S.C. 342, 21 U.S.C. 350b), where a *reasonable expectation* was anticipated as evidence of safety. FDA's interpretation of the NDI requirements has been equated to a *premarket approval system* that presents a considerable burden on the dietary supplement industry, where expensive safety trials are now necessary, and FDA is attempting to "recharacterize as many existing dietary ingredients as possible as being NDIs" (Mister and Hathcock 2012). For probiotics specifically, as listed in the NDI Draft Guidance (U.S. FDA Center for Food Safety and Applied Nutrition 2016a), the requirements to establish safety can include 90-day toxicity studies, teratology, and one- or multigenerational rodent studies for each strain. These requirements are received by the probiotics industry as extreme for a strain within a species that has been used in foods throughout history. However, the dietary supplement industry has been active in providing comment to FDA following the release of the 2016 Draft Guidance, and the probiotics industry, including trade associations, probiotic manufacturers, and retail companies, have all provided comments for FDA to consider for probiotic ingredients specifically.

It is highly recommended that if you manufacture or distribute a probiotic product, you remain aware of FDA's response to industry comments from the latest Draft Guidance (U.S. FDA Center for Food Safety and Applied Nutrition 2016a, 2016b) and FDA's proposed compilation of a pre-DSHEA ingredient list. These future FDA actions will determine the legal status of many probiotic supplements. It is also noteworthy to stay abreast of the ongoings of potential regulatory discretion for NDIN submissions for those dietary supplements that contain NDIs for which notifications were not filed, per FDA draft guidance (U.S. FDA 2022c). Furthermore, in March of 2024, CFSAN released the NDI final guidance for procedures and timeframes (U.S. FDA Center for Food Safety and Applied Nutrition 2024), which addresses only a small part of the 2016 draft guidance. While it is useful to have a bit more clarification on who should submit an NDIN and how the process works, there remains confusion regarding details needed to submit a document worthy of FDA acknowledgment.

There are three sections within DSHEA relating to claims and labeling, which we will address briefly in relation to probiotics. Sections 5, 6, and 7 revolutionize labeling standards, specifically including labeling exemptions, where certain scientific evidence is not considered an extension of the label, allowing for health benefit statements outside of standard drug claims (of diagnosis, mitigation, treatment, curing, or prevention of a disease) and addressing exceptional aspects of supplement labeling requirements. Of note, a pertinent reference for probiotic labeling that abides by this set of dietary supplement labeling regulations has been presented in *Best Practices for Probiotics* to yield industry uniformity among probiotic product labels (CRN/IPA 2017).

To begin the review, we will start in Section 5 of DSHEA, Dietary Supplement Claims, which amends 21 U.S.C. 341 by adding Section 403B to the FDCA. It lays out dietary supplement labeling exemptions, specifically addressing that for dietary supplements, it is permissible to present articles, book chapters, and abstracts from peer-reviewed scientific publications concurrent with the sale of a dietary supplement product. Under this section, this would not be considered labeling when prepared by the author/editors, and reprinted in its entirety, if the material:

(1) is not false or misleading;
(2) does not promote a particular manufacturer or brand of a dietary supplement;
(3) is displayed or presented, or is displayed or presented with other such items on the same subject matter, so as to present a balanced view of the available scientific information on a dietary supplement;
(4) if displayed in an establishment, is physically separate from the dietary supplements; and
(5) does not have appended to it any information by sticker or any other method.

Again, it is written that the burden of proof is with the United States to prove that these conditions are not met (21 U.S.C. 341).

This leads us to Section 6 on Statements of Nutritional Support, which amends Section 403 of FDCA to include statements that can be made for dietary supplements (21 USC 343) under the conditions that are listed within. These include the ability to state a benefit provided by a nutrient or dietary ingredient, outside of drug claims. This is commonly known today as a structure/function claim, and the details of legal parameters are as listed in 21 U.S.C. 343:

(A) the statement claims a benefit related to a classical nutrient deficiency disease and discloses the prevalence of such disease in the United States, describes the role of a nutrient or dietary ingredient intended to affect the structure or function in humans, characterizes the documented mechanism by which a nutrient or dietary ingredient acts to maintain such structure or function, or describes general well-being from consumption of a nutrient or dietary ingredient,
(B) the manufacturer of the dietary supplement has substantiation that such statement is truthful and not misleading and
(C) the statement contains, prominently displayed and in boldface type, the following: "This statement has not been evaluated by the Food and Drug Administration. This product is not intended to diagnose, treat, cure, or prevent any disease".

The use of language to indicate the positive effect a probiotic has on the structure/function of the body has become commonplace for probiotic products in the United States, as there is ample scientific evidence to support many health benefits. There is evidence that probiotic efficacy, demonstrated in human clinical trials, is strain specific in most instances (Sanders and Huis in't Veld 1999; Arora and Baldi 2015), although some countries have recognized general digestive health benefits based on species level that meet a set dosage (see Chapter 41 in this volume). The ability to indicate that a product "helps maintain intestinal flora" or "helps maintain regularity" are two specific structure/function phrases that FDA has specifically addressed as examples of acceptable language. In 21 CFR 101, FDA published a final rule (65 FR 1000) which defines the types of structure/function language that are permitted for use on dietary supplement labels, without prior review by the agency. As indicated in Section 403 of the FDCA, manufacturers who include such structure/function language on product labels must, within 30 days after the dietary supplement is first marketed, notify the secretary (of Health and Human Services) of corresponding statements.

Section 7 of DSHEA addresses Dietary Supplement Ingredient Labeling and Nutrition Information Labeling, where Section 403 (21 U.S.C. 343) is further amended. This section includes many features of labeling, which is key because the labeling of a dietary supplement can present unique features that should be used appropriately. A dietary supplement product can be considered misbranded under certain conditions; those relevant to probiotics are presented herein. The product must be labeled as a "dietary supplement" or comparable term, such as "probiotic supplement". The label must contain the name (as described in FDCA 201[ff]) and quantity of each dietary ingredient, and for a proprietary blend, a *total* quantity of all ingredients in that blend must be listed. If the probiotic is covered by specifications of an official compendium, and is listed as such, it must meet those specifications. The probiotic should meet the identity, strength, quality, purity, and set specifications that it is presented as meeting. This section also includes reference to daily intake and details for dietary supplement product types outside of probiotics, such as vitamins, minerals, and herbs or botanicals, so please review if necessary for your type of dietary supplement.

While we have covered those sections of DSHEA most relevant to probiotics, there is value in mentioning Section 9, Good Manufacturing Practices, at least briefly. In Section 402 (21 U.S.C. 342), DSHEA indicates that the Secretary may issue regulations that establish current good manufacturing practices (cGMPs) that are specific for dietary supplements. With advanced notices, outreach activities, a proposed rule and then in 2007, a final rule (U.S. FDA 2017c) (Dickinson 2011), FDA established cGMPs appropriate for dietary supplements (21 CFR 111). These cGMPs are modeled after but distinct from those specific for food and have presented an opportunity for dietary supplement manufacturers to consistently produce dietary supplement products of the highest quality, as anticipated throughout DSHEA.

# 40.5 PROBIOTICS AS MEDICAL FOODS

Next on the list of intended uses for probiotic products is that of medical foods, which are defined in the Orphan Drug Act as,

> a food which is formulated to be consumed or administered enterally under the supervision of a physician and which is intended for the specific dietary management of a disease or condition for which distinctive nutritional requirements, based on recognized scientific principles, are established by medical evaluation.
>
> *(21 U.S.C. 360ee [b] [3])*

The category of medical foods is unique with a few key requirements. The patient must be under the supervision of a physician. The medical food must be formulated to manage a *specific dietary disease condition*, and that condition must have *distinct* nutritional requirements. FDA has released a Guidance for Industry (U.S. FDA Center for Food Safety and Applied Nutrition 2016b) that addresses key exemptions for medical foods, including the Nutrition Labeling of Food (21 CFR 101.9) under certain limiting conditions, the labeling requirements for content claims, and health claims as listed in the Nutrition Labeling and Education Act of 1990 (21 U.S.C. 343[r][5][A]). Otherwise, all applicable food labeling requirements apply. Medical foods are not regulated as drugs, have no preapproval process, and are manufactured under food cGMPs (21 CFR 110). Despite guidance from FDA, some feel that the category of medical foods is not appropriately regulated (DeSimone 2018).

How probiotics fit into the medical food product space is also unique. Probiotics most often meet the safety requirement of GRAS, and they can be manufactured under food cGMPs, but the impasse seems to be whether a specific *dietary management of a disease or condition*, with *distinct nutritional requirements* can be clearly managed by administration of probiotics. It is widely recognized that probiotics do provide a benefit to overall health and promote healthy gastrointestinal microbiota (Hill et al. 2014). There are probiotics that restore disruption of gastrointestinal microbiota, which is *managing* the resulting conditions. It is not clear if FDA would agree, but one could make a case for it. While there are probiotic products that are labeled and sold as medical foods with much success, there remains some concern due to the lack of absolute association of probiotics meeting distinct nutritional requirements as defined for medical foods.

# 40.6 PROBIOTICS AS DRUGS

As we have seen thus far, the uses for probiotics have been under the overall category of foods according to U.S. regulation. The one remaining intended use is within the category of drugs/biological products. For an overview of how probiotics can be categorized as a biological product, Degnan (2008) is recommended, and for an in-depth review of the regulatory landscape for drugs/biologics, please reference Piña and Pines (2012). It is important for probiotic manufacturers and distributors to be aware of regulations that might prohibit maximum flexibility in market potential in the United States, in other words, how to avoid getting "stuck" in the category of drugs and not being able to subsequently sell a probiotic strain as a food or dietary supplement ingredient.

We have discussed FDCA Section 201 (21 U.S.C 321) previously to define a dietary supplement. We have not, however, covered the restrictions of categorization that are also outlined in Section 3 of DSHEA. This statute states that a dietary supplement, as listed in 21 U.S.C.321(ff)(3),

3) does
    (A) include an article that is approved as a new drug under section 505, certified as an antibiotic under section 507, or licensed as a biologic under section 351 of the Public Health Service Act (42 U.S.C. 262) and was, prior to such approval, certification, or license, marketed as a dietary supplement or as a food unless the Secretary has issued a regulation, after notice and comment, finding that the article, when used as or in a dietary supplement under the conditions of use and dosages set forth in the labeling for such dietary supplement, is unlawful under section 402(f); and
    (B) not include
        (i) an article that is approved as a new drug under section 505, certified as an antibiotic under section 507, or licensed as a biologic under section 351 of the Public Health Service Act (42 U.S.C. 262), or
        (ii) an article authorized for investigation as a new drug, antibiotic, or biological for which substantial clinical investigations have been instituted and for which the existence of such investigations has been made public,

which was not before such approval, certification, licensing, or authorization marketed as a dietary supplement or as a food unless the Secretary, in the Secretary's discretion, has issued a regulation, after notice and comment, finding that the article would be lawful under this Act.

Simply put, one should *always* market a probiotic as a food or dietary ingredient prior to marketing as drug/biologic product or even studying and subsequently publishing the results of a probiotic clinical trials with disease endpoints (drug/disease-type studies). As stated, you can always ask the secretary to authorize the probiotic ingredient as a food or dietary ingredient afterwards, through regulation, but this approach is not ideal and has not been done to date.

# 40.7 CONCLUSION

A comprehensive overview of how probiotics are regulated under the category of foods would not be complete without addressing all topics covered in the FDCA in sections 301–399d. However, the focus of this chapter has been limited to the regulatory aspects outside of manufacturing and facility-specific details and requirements. Suffice it to say that in recent years, FDA has invested a great number of resources into modernizing our food safety system. With the advent of the Food Safety Modernization Act (FSMA), which was signed into law in 2011, the food safety system in the United States was essentially overhauled. It is referred to as the most significant revision of food law since the enactment of the FDCA (Piña and Pines 2012). The major themes of the act include prevention, enhanced partnerships, import safety and inspections, compliance, and response (U.S. FDA 2017a) and provide FDA with the regulatory authority to require comprehensive, science-based controls across the food supply that switch the focus from reactive to preventive (U.S. FDA 2018c).

In the United States, the majority of probiotic products fall into the category of dietary supplements, which is a subset of the food category. Therefore, the focus of this chapter has been the current regulatory landscape, as driven by food and dietary supplement regulations. DSHEA absolutely changed the way probiotics, as a dietary supplement, could be marketed in the United States. Hopefully this chapter has left you with a sense of history and a general understanding of the U.S. regulatory landscape for probiotics and brought to light any concerns that you will need to address or research further prior to marketing a probiotic product in the United States.

# 40.8 ACKNOWLEDGMENTS

I would like to thank Elizabeth McCartney and Gary Yingling for their superb editing skills and for painstakingly reviewing this chapter in their spare time. Thank you to Tim Lawlor for market research and citations. I am truly grateful.

# BIBLIOGRAPHY

21 CFR 101. Code of Federal Regulations. Title 21. Part 101. Regulations on Statements Made for Dietary Supplements Concerning the Effect of the Product on the Structure or Function of the Body.
21 CFR 111. Code of Federal Regulations. Title 21. Part 111. Final Rule: Current Good Manufacturing Practice in Manufacturing, Packaging, Labeling, or Holding Operations for Dietary Supplements.
21 CFR 131. Code of Federal Regulations. Title 21. Part 131. Subpart B. Requirements for Specific Standardized Milk and Cream.
21 CFR 131.111. Code of Federal Regulations. Title 21. Part 131. §131.111. Acidified Milk.
21 CFR 131.112. Code of Federal Regulations. Title 21. Part 131. §131.112. Cultured Milk.

21 CFR 131.162. Code of Federal Regulations. Title 21. Part 131. §131.162. Acidified Sour Cream.

21 CFR 131.200. Code of Federal Regulations. Title 21. Part 131. §131.200. Yogurt.

21 CFR 133.113. Code of Federal Regulations. Title 21. Part 133. §133.113. Cheddar Cheese.

21 CFR 136.110. Code of Federal Regulations. Title 21. Part 136. §136.110. Bread, Rolls and Buns.

21 CFR 170.3(l). Code of Federal Regulations. Title 21. Part 170. §170.3(l). Definitions.

21 CFR 170.30(a). Code of Federal Regulations. Title 21. Part 170. §170.30(a). Eligibility for Classification as Generally Recognized as Safe (GRAS).

21 CFR 182. Code of Federal Regulations. Title 21. Part 182. Substances Generally Recognized as Safe.

21 CFR 184. Code of Federal Regulations. Title 21. Part 184. Direct Food Substances Affirmed as Generally Recognized as Safe.

21 CFR 186. Code of Federal Regulations. Title 21. Part186. Indirect Food Substances Affirmed as Generally Recognized as Safe.

21 CFR 507. Code of Federal Regulations. Title 21. Part 507. Current Good Manufacturing Practice, Hazard Analysis, and Risk-Based Preventive Controls for Food for Animals.

21 CFR 570. Code of Federal Regulations. Title 21. Part 570. Food Additives.

21 USC 9. Federal Food, Drug and Cosmetic Act. US Code Title 21, Chapter 9.

21 USC 321. Federal Food, Drug and Cosmetic Act. US Code Title 21, Chapter 9, Section 321. Definitions.

21 USC 342. Federal Food, Drug and Cosmetic Act. US Code Title 21, Chapter 9, Section 342. Adulterated Food.

21 USC 343. Federal Food, Drug and Cosmetic Act. US Code Title 21, Chapter 9, Section 343. Misbranded Food.

21 USC 350. Federal Food, Drug and Cosmetic Act. US Code Title 21, Chapter 9, Section 350. Vitamins and Minerals.

62 FR 18938. Food and Drug Administration. Federal Register. Substances Generally Recognized as Safe. April 17, 1997, pp. 18938–18964.

65 FR 1000. Regulations on Statements Made for Dietary Supplements Concerning the Effect of the Product on the Structure or Function of the Body. January 6, 2000, Vol. 65:4.

81 FR 54960. Food and Drug Administration. Federal Register. Substances Generally Recognized as Safe. Final Rule. August 17, 2016, pp. 54959–55055.

Arora, M., and Baldi, A. 2015. Regulatory Categories of Probiotics Across the Globe: A Review Representing Existing and Recommended Categorization. *Indian Journal of Medical Microbiology 33*(S1):S2–S10.

Association of American Feed Control Officials. 2018. Official Publication. Association of American Feed Control Officials, Inc., Champaign, IL.

Association of American Feed Control Officials. 2024. www.aafco.org/

Bass, S. 2011. Dietary Supplement Regulation: A Comprehensive Guide. Food and Drug Law Institute, Washington, DC.

Bill, F.V. 2017. The State of FDA's Office of Dietary Supplement Programs. *Natural Products Insider*, September 25.

CRN/IPA. 2017. *Best Practices Guidelines for Pro biotics.* www.crnusa.org/sites/default/files/pdfs/CRN-IPA-Best-Practices-Guidelines-for-Probiotics.pdf

Degnan, F.H. 2008. The US Food and Drug Administration and Probiotics: Regulatory Categorization. *Clinical Infectious Diseases 46*:S133–S136.

DeSimone, C. 2018. The Unregulated Probiotic Market. *Clinical Gastroenterology and Hepatology 2018.* http://doi.org/10.1016/j.cgh.2018.01.018

Dickinson, A. 2011. History and Overview of DSHEA. *Fitoterapia 82*:5–10.

EFSA Panel on Biological Hazards (BIOHAZ). 2018. BIOHAZ Statement on QPS: Suitability of Taxonomic Units Notified until September 2017. *EFSA Journal 16*(1):5131.

Euromonitor International Limited, Health & Wellness. 2023. Fortified & Functional Food & Drink by Type. (subscription required)

FEEDAP. 2012. Guidance on the Assessment of Bacterial Susceptibility to Antimicrobials of Human and Veterinary Importance. *EFSA Journal 10*(6):2740.

Hill, C., Guarner, F., Reid, G., et al. (2014). Expert Consensus Document: The International Scientific Association for Probiotics and Prebiotics Consensus Statement on the Scope and Appropriate Use of the Term Probiotic. *Nature Reviews Gastroenterology & Hepatology 11*:506–514.

H.R.3562, Nutrition Labeling and Education Act of 1990. 101st Congress (1989–90). Public Law Number 101–535.

Institute of Medicine (U.S.) Food Forum. 1999. Enhancing the Regulatory Decision-Making Approval Process for Direct Food Ingredient Technologies. Workshop Summary. Appendix A. Legal Aspects of the Food Additive Approval Process. National Academies Press (U.S.), Washington, DC.

International Dairy Federation. 2012. Safety Demonstration of Microbial Food Cultures (MFC) in Fermented Food Products. Bulletin 455/2012.

Johansen, E. 2017. Future Access and Improvement of Industrial Lactic Acid Bacteria Cultures. *Microbial Cell Factories 16*:230.

Kussell, E. 2013. Evolution in Microbes. *Annual Review of Biophysics 42*:493–514.

Mister, S., and Hathcock, J. 2012. Under the Law, FDA Must Grant Different Standards for New Dietary Ingredients and Food Additives. *Regulatory Toxicology and Pharmacology 62*:4456–4458.

Paez-Espino, D., Sharon, I., Morovic, W., Stahl, B., Thomas, B.C., Barrangou, R., and Banfield, J.F. 2015. CRISPR Immunity Drives Rapid Phage Genome Evolution in Streptococcus Thermophilus. *mBio 6*(2):1–9.

Pariza, M., Gillie, K., Kraak-Ripple, S., Leyer, G., and Smith, A.B. 2015. Determining the Safety of Microbial Cultures for Consumption by Humans and Animals. *Regulatory Toxicology and Pharmacology 73*:164–171.

Piña, K.R., and Pines, W.L. 2012. A Practical Guide to FDA's Food and Drug Law Regulation, 4th ed. The Food and Drug Law Institute, Washington, DC.

Quan, S., Ray, J.C.J., Kwota, Z., Duong, T., Balázsi, G., Cooper, T.F., and Monds, R.D. 2012. Adaptive evolution of the lactose utilization network in experimentally evolved populations of Escherichia coli. *PLoS Genetics 8*(1):1–18.

Report of a Joint FAO/WHO Expert Consultation on Evaluation of Health and Nutritional Properties of Probiotics in Food including Powder Milk with Live Lactic Acid Bacteria, Cordoba, Argentina, October 1–4, 2001.

Roe, A.L., Boyte, M.E., Elkins, C.A., Goldman, V.S., Heimbach, J., Madden, E., Oketch-Rabah, H., Sanders, M.E., Sirois, J., and Smith, A. 2022. Considerations for determining safety of probiotics: A USP perspective. *Regulatory Toxicology and Pharmacology 136*:105266. http://doi.org/10.1016/j.yrtph.2022.105266.

S.784—Dietary Supplement Health and Education Act of [19]94. 103rd Congress (1993–4). Public Law Number 103–417.

Sanders, M.E., Akkermans, L.M.A., Haller, D., et al. 2010. Safety Assessment of probiotics for human use. *Gut Microbes 1*(3):164–185.

Sanders, M.E., and Huis in't Veld, J. 1999. Bringing a probiotic-containing functional food to the market: Microbiological, product, regulatory and labeling issues. *Antonie van Leeuwenhoek 76*:293–315.

Sanders, M.E., Klaenhammer, T.R., Ouwehand, A.C., Pot, B., Johansen, E., Heimbach, J.T., and Marco, M.L. 2014. Effects of genetic, processing, or product formulation changes on efficacy and safety of probiotics. *Annals of the New York Academy of Sciences 1309*:1–18.

Stevens, H.C., and Nabors, L.O. 2009. Microbial food cultures: A regulatory update. *Food Technology 63*:36–41.

U.S. FDA. 2017a. Food Safety Modernization Act Overview. www.fda.gov/Food/GuidanceRegulation/FSMA/ucm247546.htm

U.S. FDA. 2017b. Development of a List of Dietary Ingredients that Pre-Date the Dietary Supplement Health and Education Act of 1994 (DSHEA). October 3, 2017, College Park, MD.

U.S. FDA. 2017c. Backgrounder on the Final Rule for Current Good Manufacturing Practices (CGMPs) for Dietary Supplements. www.fda.gov/Food/GuidanceRegulation/CGMP/ucm110863.htm

U.S. FDA. 2018a. Microorganisms & Microbial-Derived Ingredients Used in Food (Partial List). www.fda.gov/food/ingredientspackaginglabeling/gras/microorganismsmicrobialderivedingredients/default.htm (Accessed January 18, 2018).

U.S. FDA. 2018b. How U.S. FDA's GRAS Notification Process Works. www.fda.gov/food/ingredientspackaginglabeling/gras/ucm083022.htm (Accessed January 17, 2023).

U.S. FDA. 2018c. Background on the FDA Food Safety Modernization Act. www.fda.gov/NewsEvents/PublicHealthFocus/ucm239907.htm (Accessed January 19, 2018).

U.S. FDA. 2019a. What We Do At CFSAN. www.fda.gov/about-fda/center-food-safety-and-applied-nutrition-cfsan/what-we-do-cfsan (Accessed March 13, 2024).

U.S. FDA. 2019b. FDA's Legal Authority. www.fda.gov/about-fda/changes-science-law-and-regulatory-authorities/fdas-legal-authority (Accessed March 13, 2024).

U.S. FDA. 2019c. Memorandum of Understanding Between the United States Food and Drug Administration and the Association of American Feed Control Officials. MOU#225–07–7001. https://www.fda.gov/about-fda/domestic-mous/mou-225-07-7001

U.S. FDA. 2021a. FDA Fundamentals. www.fda.gov/about-fda/fda-basics/fda-fundamentals(Accessed March 13, 2024).

U.S. FDA. 2022a. What Does FDA Regulate? www.fda.gov/about-fda/fda-basics/what-does-fda-regulate(Accessed March 13, 2024).

U.S. FDA. 2022b. FDA's Regulation of Pet Food. https://www.fda.gov/animal-veterinary/animal-health-literacy/fdas-regulation-pet-food

U.S. FDA. 2022c. FDA Releases Draft Guidance on NDI Enforcement Discretion. Constituent Update. www.fda.gov/food/cfsan-constituent-updates/fda-releases-draft-guidance-ndi-enforcement-discretion (Accessed March 13, 2024).

U.S. FDA. 2024a. GRAS Notices. www.accessdata.fda.gov/scripts/fdcc/?set=GRASNotices (Accessed March 13, 2024).

U.S. FDA. 2024b. Pet Food. www.fda.gov/animalveterinary/products/animalfoodfeeds/petfood/default.htm

U.S. FDA. Center for Food Safety and Applied Nutrition. 2011. Draft Guidance for Industry: Dietary Supplements: New Dietary Ingredient Notifications and Related Issues. Washington, DC.

U.S. FDA. Center for Food Safety and Applied Nutrition. 2016a. Dietary Supplements: New Dietary Ingredient Notifications and Related Issues: Guidance for Industry. Washington, DC.

U.S. FDA. Center for Food Safety and Applied Nutrition. 2016b. *Frequently Asked Questions About Medical Foods; Second Edition. Guidance for Industry*. Washington, DC.

U.S. FDA. Center for Food Safety and Applied Nutrition. 2024. *Dietary Supplements: New Dietary Ingredient Notification Procedures and Timeframes: Guidance for Industry*. March 2024. Washington, DC.

Welch, C. 2017. Dietary Supplement Regulations: Where We Are and Where We're Going. *IPA Probiotics Workshop*. International Probiotics Association & U.S. Pharmacopeia, October 26.

Yirga, H. 2015. The Use of Probiotics in Animal Nutrition. *Probiotics & Health 3*:2.

Zheng J., Wittouck S., Salvetti E., et al. 2020. A Taxonomic Note on the Genus Lactobacillus: Description of 23 Novel Genera, Emended Description of the Genus Lactobacillus Beijerink 1901, and Union of Lactobacillaceae and Leuconostocaceae. https://doi.org/10.1099/ijsem.0.004107; open access DOI: https://doi.org/10.7939/r3-egnz-m294

# Regulation of Probiotics in Canada

# 41

Jon-Paul Powers and Jenelle Patterson

## 41.1 BACKGROUND

Probiotic microorganisms can be found in a variety of consumer products in the Canadian marketplace, namely as ingredients in foods and natural health products (commonly known as dietary supplements), both of which are regulated by Health Canada. In regulating probiotics in foods, Health Canada has adopted the Food and Agriculture Organization (FAO)/World Health Organization (WHO) definition of probiotics as "live microorganisms which when administered in adequate amounts confer a health benefit on the host" (FAO/WHO, 2006). This definition also forms the basis for the regulation of probiotics in natural health products: the Natural Health Product Regulations specifically define a probiotic as "a monoculture or mixed-culture of live microorganisms that benefit the microbiota indigenous to humans" (Government of Canada, 2008). While probiotics may be used as ingredients in other consumer products, including over-the-counter drugs, prescription drugs, cosmetics, and pet supplements, and as additives and processing agents in a variety of products, a discussion of these particular uses is beyond the scope of this chapter. Instead, we will focus on the regulation of probiotics used as ingredients in foods and natural health products intended for humans.

The classification of a product as either a food or a natural health product depends on a variety of factors including, but not limited to, composition, product format, product representation (including health claims), consumer perception and place of sale, and history of use. For some products, categorization is quite clear. For example, pharmaceutical-type formats, such as tablets and capsules, are consistently viewed as natural health products (i.e., a subset of drugs), whereas formats such as bars and yogurts are consistently viewed as foods. For products with more ambiguous formats (e.g. powdered beverage mixes, ready-to-drink liquid beverages, gummies, etc.), Health Canada has released guidance to assist with determining appropriate categorization (Government of Canada, 2017a). The inclusion of probiotic organisms as ingredients in both foods and natural health products falls under different regulatory frameworks, thus the associated implications of each category will be discussed separately.

## 41.2 FOODS

In Canada, the sale of all foods is governed by the Food and Drugs Act (the "Act") (Government of Canada, 2017b) and the Food and Drug Regulations (the "Regulations") (Government of Canada, 2017c). With the exception of a few specific categories of food (e.g., infant formulas, certain foods for special dietary

DOI: 10.1201/9781003352075-45

purposes, novel foods, etc.), there is no pre-market assessment for food products in Canada, and the marketer is entirely responsible for ensuring compliance with the Regulations. While Health Canada determines policies related to food safety and human health, the Canadian Food Inspection Agency (CFIA) is responsible for enforcement of the Act and Regulations. The Act stipulates the following high-level requirements for foods (noted by specific section of the Act itself) (Government of Canada, 2017b):

- A prohibition of selling or advertising a product to the general public as a treatment, preventative or cure for any of the diseases, disorders or abnormal physical states referred to in Schedule A (namely, a list of serious diseases and health conditions) (Section 3).
- A prohibition of selling any food that (Section 4):
  - Has in or on it any poisonous or harmful substance;
  - Is unfit for human consumption;
  - Consists in whole or in part of any filthy, putrid, disgusting, rotten, decomposed, or diseased animal or vegetable substance;
  - Is adulterated; or
  - Was manufactured, prepared, preserved, packaged, or stored under unsanitary conditions.
- A prohibition of labeling, packaging, treating, processing, selling, or advertising any food in a manner that is false, misleading, or deceptive or is likely to create an erroneous impression regarding its character, value, quantity, composition, merit, or safety (Section 5).

Under the food provisions of the Regulations, live bacterial cultures, including those represented as probiotics, may be considered food ingredients and can generally be added to food products. While there are no specific regulations concerning the use of probiotics in foods, general provisions of the Act and Regulations apply to govern the safety of foods and their ingredients, as well as any associated claims or statements made on food labels and in advertising (discussed in the following section).

As there is no pre-market assessment or notification requirements for the majority of food products in Canada, it is the responsibility of the marketer to ensure that the food or food ingredients are not unsafe. Although there are no prescriptive requirements to substantiate safety of probiotic organisms, safety information may be requested by Health Canada at any time. It is good practice for marketers to ensure they have a complete characterization of all strains of probiotics present in their products, consistent with the requirements summarized for natural health products (outlined in "Strain/Culture Specific Requirements"). Particularly, despite the long history of safe use for many lactic acid bacteria species, Health Canada has expressed concern about the potential for antibiotic resistance and resistance transfer associated with probiotics (Government of Canada, 2009). Thus, marketers are recommended to retain data demonstrating that the strains contained in their products do not pose a significant risk of transferable antimicrobial resistance.

If the probiotic culture or cultures do not have a history of safe use in foods, or the strain is genetically modified, it may fall within the definition of a "novel food" under the Regulations. A novel food is defined in Part B, Division 28 of the Regulations as follows (Government of Canada, 2017c):

a) A substance, including a microorganism, that does not have a history of safe use as a food;
b) A food that has been manufactured, prepared, preserved or packaged by a process that
   i. Has not been previously applied to that food, and
   ii. Causes the food to undergo a major change; and
c) A food that is derived from a plant, animal or microorganism that has been genetically modified such that
   i. The plant, animal or microorganism exhibits characteristics that were not previously observed in that plant, animal or microorganism,
   ii. The plant, animal or microorganism no longer exhibits characteristics that were previously observed in that plant, animal or microorganism, or
   iii. One or more characteristics of the plant, animal, or microorganism no longer fall within the anticipated range for that plant, animal, or microorganism.

If any of these criteria pertain to a particular probiotic strain, the strain would not be compliant with the Regulations and a novel food submission demonstrating the safety of the probiotic strain would be required (Government of Canada, 2006). Health Canada's Food Directorate analyses each submission and, if approved, communicates a decision that no objection is made to the use of the subject product as a food in Canada. Such an approval from Health Canada is required before any novel probiotic can be used in food products sold in Canada.

# 41.3  HEALTH CLAIMS ON FOODS CONTAINING PROBIOTICS

All health claims on foods must be (i) neither false nor misleading, and (ii) supported by adequate and proper substantiation. Indeed, to be considered compliant with Section 5 of the Act, health claims made in respect of probiotics in food should be scientifically validated. The term "probiotic" is considered a health claim in its own right, largely due to the "health benefit" component of the aforementioned definition put forward by the FAO/WHO. Thus, it is recommended that the term "probiotic" and similar terms or representations only be used when accompanied by scientific, validated statements on the benefits or effects of the probiotic microorganism contained in the food. Benefits can either be associated with a species or a particular strain (discussed further subsequently). In general, the following apply to the labelling of foods for which probiotic health claims are made (Government of Canada, 2017d):

a) The term probiotic, and similar terms or representations, should be accompanied by a statement of the demonstrated effect of the probiotic;
b) The claimed effect of the probiotic should be clearly stated in a manner that is not false, misleading, deceptive, or likely to create an erroneous impression with respect to the effect of or benefit from the probiotic as contained in the food;
c) Where a health claim is made, the Latin name (*i.e.*, genus and species) and the strain of the probiotic microorganism that is the subject of the claim should be declared;
d) The level of the probiotic strain expressed in colony forming units (CFU) in a serving of stated size of the food should be declared; and
e) If more than one probiotic strain is added to a food, the above recommendations apply to the mixed culture.

## 41.3.1  Species-Specific Claims

A limited number of species-specific probiotic claims have been accepted for use on foods, provided that the food product contains one or more of the eligible microorganisms at a minimum level of 1 billion colony-forming units (CFUs) per serving of stated size (Government of Canada, 2017d). Health Canada has suggested several of such claims on the basis that certain microorganisms are recognized as naturally forming part of the human gut flora (as summarized in Table 41.1, adapted from (Government of Canada, 2017d)). It is important to note that these claims and associated information are provided by Health Canada as guidance and are not set forth in regulation. With this in mind, marketers may wish to consider strategic opportunities to extrapolate this guidance to species not currently indicated in this table, or to modify label directions and/or CFU quantities. For example, if a marketer has a scientific rationale that attributes any of the species-specific health claims in Table 41.1 to a species not specifically listed or to a dose other than 1 billion CFU/serving, such rationale could be presented to Health Canada's Food Directorate for consideration. While there is no requirement to do so, marketers may request guidance or opinions from Health Canada's Food Directorate prior to implementation.

**TABLE 41.1**   Acceptable Non–Strain-Specific Claims for Probiotics in Food Products

| ELIGIBLE BACTERIAL SPECIES LATIN NAME (ACCEPTABLE NOMENCLATURE) AND SYNONYM WHERE APPLICABLE | ACCEPTABLE NON–STRAIN-SPECIFIC PROBIOTIC CLAIMS |
|---|---|
| *Bifidobacterium adolescentis*<br>*Bifidobacterium animalis* subsp. *Animalis*<br>*Bifidobacterium animalis* subsp. *lactis*<br>  (synonym: *B. lactis*)<br>*Bifidobacterium bifidum*<br>*Bifidobacterium breve*<br>*Bifidobacterium longum* subsp. *infantis*<br>*Bifidobacterium longum* subsp. *longum*<br>*Lactobacillus acidophilus*<br>*Lactobacillus casei*<br>*Lactobacillus fermentum*<br>*Lactobacillus gasseri*<br>*Lactobacillus johnsonii*<br>*Lactobacillus paracasei*<br>*Lactobacillus plantarum*<br>*Lactobacillus rhamnosus*<br>*Lactobacillus salivarius* | Probiotic that naturally forms part of the gut/digestive tract flora.<br>Provides live microorganisms that naturally form part of the gut/digestive tract flora.<br>Probiotic that contributes to healthy gut/digestive tract flora.<br>Provides live microorganisms that contribute to healthy gut/digestive tract flora. |

*Source:* Adapted from Canadian Food Inspection Agency. 2017. Food Labelling for Industry. Health Claims—Probiotic Claims. Table 41.1. Table of acceptable non-strain-specific claims for probiotics. www.inspection.gc.ca/food/labelling/food-labelling-for-industry/health-claims/eng/1392834838383/1392834887794?chap=9 (accessed January 20, 2018).

## 41.3.2 Strain-Specific Claims

While current guidance is helpful in supporting low-level, non–strain-specific probiotic claims on food products, marketers are not limited solely to these statements. Claims regarding the health benefits or effects of specific probiotic strains may also be made. Such claims generally fall into the category of function claims or disease risk reduction and therapeutic claims.

Function claims may be made about physiological effects of probiotic microorganisms in food (e.g., "promotes regularity" and "improves nutrient absorption and aids in digestion"). Acceptable function claims state a specific, scientifically supported physiological effect associated with good health or performance. While there is no requirement to do so, marketers may request guidance or opinions from Health Canada's Food Directorate regarding the acceptability of function claims prior to their use. Importantly, function claims are not made for a food per se; they are made for the nutrients (or probiotics) in the food. For example, a food cannot state "XYZ Brand promotes regularity". An acceptable claim must indicate that "XYZ Brand *contains probiotics* that promote regularity".

Disease risk reduction or therapeutic claims may be permitted for probiotic strains when there is adequate scientific evidence to substantiate the claim. Historically, Health Canada's Food Directorate has reviewed proposed health claims of this nature and determined whether and under what conditions a claim could be made. At this time, Health Canada has not published a health claim assessment related to any probiotic strain. Health claim assessment and acceptance by Health Canada is not specifically required under the Regulations; however, these assessments are relied-on by the Canadian Food Inspection Agency when exercising its enforcement authority over food labelling and advertising to determine whether a food is making a false or misleading claim contrary to the Act.

Disease risk reduction or therapeutic claims related to Schedule A diseases may also be permitted on food products; however, claims of this nature are required to undergo a pre-market assessment by the Food Directorate of Health Canada, as a regulatory amendment is required to allow their use. For such submissions, significant scientific evidence clearly linking the probiotic to the proposed health claim is required. Currently, no probiotic-associated disease risk-reduction claims are permitted for foods.

# 41.4  ADDITIONAL LABELING CONSIDERATIONS FOR FOODS CONTAINING PROBIOTICS

A probiotic claim should be accompanied by the Latin name of the microorganism (i.e., genus and species), along with the identity of the strain, using acceptable nomenclature. It is recommended that the strain be identified using the number assigned by an internationally recognized culture repository. The amount of the probiotic microorganism(s) contained in the product at the end of its shelf life must be declared in CFU per serving of stated size. This statement should appear adjacent to the Nutrition Facts table or the list of ingredients, or in close proximity to the claim.

Foods containing probiotic microorganism(s) must be labeled with a list of ingredients in accordance with sections B.01.008–B.01.010 of the Regulations (Government of Canada, 2017c). The probiotic microorganism(s) must be identified by its (their) common name(s) or by a class name set out in Section B.01.010 of the Regulations. The class name "bacterial culture" may be used to describe all bacterial species added to a food product. When a class name (e.g., bacterial culture) is used in the list of ingredients, the identity (i.e., the genus, species, and strain) of the probiotic bacterial culture(s) should be declared in close proximity to the claim. Marketers should also be aware of Canadian labeling requirements applicable to all foods, including bilingual labeling considerations (English and French).

# 41.5  NATURAL HEALTH PRODUCTS

In Canada, natural health products (frequently abbreviated simply as NHPs) are classified as a subset of drugs and are regulated by Health Canada in accordance with the Natural Health Products Regulations (Government of Canada, 2017c) (the "NHP Regulations") promulgated under the Food and Drugs Act (Government of Canada, 2017b). NHPs are governed by a pre-market approval system, in that all NHPs must be licensed by Health Canada prior to sale. To obtain a product license, referred to as a Natural Product Number or NPN, applicants must submit a product license application to Health Canada's Natural and Non-prescription Health Products Directorate (NNHPD). This application package consists of evidence supporting the safety, efficacy (health claims), and quality of the particular NHP. In contrast to other jurisdictions, all NHPs must have at least one associated health claim. While the evidentiary requirements vary by product type and ingredients, the information presented in the following section will focus on the requirements specifically pertaining to probiotics.

# 41.6  REGULATION OF PROBIOTICS IN NHPS

A natural health product is defined as a substance, or combination of substances, in which all of the medicinal ingredients are set out in Schedule 1 of the NHP Regulations, including homeopathic and traditional medicines, that are manufactured, sold, or represented for use in (Government of Canada, 2008):

a) The diagnosis, treatment, mitigation or prevention of a disease, disorder or abnormal physical state or its symptoms in humans;

b) Restoring or correcting organic functions in humans; or

c) Modifying organic functions in humans, such as modifying those functions in a manner that maintains or promotes health.

Schedule 1, item 8 of the NHP Regulations specifies the generic entry of "a probiotic", and the NHP Regulations themselves define "probiotic" as a monoculture or mixed-culture of live microorganisms that benefit the microbiota indigenous to humans. For completeness, it is important to note that, among other exceptions, Schedule 2 of the NHP Regulations prohibits the use of antibiotics prepared from microorganisms (or synthetic duplicates) as NHPs; these products are classified as over-the-counter or prescription drugs.

# 41.7  LICENSING PROBIOTICS AS NATURAL HEALTH PRODUCTS

As mentioned previously, applicants must submit a product license application supporting the safety, efficacy, and quality of the particular NHP. Health Canada has developed a Probiotics Monograph including several common probiotic species (Government of Canada, 2023), to which applicants may attest to support the safety and efficacy of probiotics provided in their NHPs, if applicable. This monograph contains detailed information on acceptable health claims, doses, dosage forms, subpopulations, microorganisms, and required use and risk information.

In preparing the Probiotics Monograph, the NNHPD divided health claims into two categories: species specific and strain specific. The species-specific health claims relate to generally accepted actions of well-characterized bacterial species and are based on a minimum daily dose of 10 million CFU per day. The strain-specific health claims are higher level in nature, pertaining to disease and disease risk-reduction claims, and are based on a minimum daily dose specific to the strain itself (as supported by the body of cited literature). These species- and strain-specific health claims, and associated usage requirements, are captured in Tables 41.2 and 41.3, respectively.

While the Probiotics Monograph is a useful tool, it is important to note that applicants are not limited to the information specified in the Monograph and product licenses may be sought for different claims, doses, dosage forms, subpopulations, microorganisms, and so on; however, scientific evidence supporting deviations from the Monograph would be required.

While the NNHPD claims to consider the totality of evidence, it is our experience that strain-specific evidence must be clinical in nature and all health claims must be supported by statistically significant results from placebo-controlled studies. Additionally, health claims are assessed based on risk and

**TABLE 41.2**  Acceptable Strain Specific Claims for Bacterial Probiotics in Natural Health Products

| MEDICINAL INGREDIENTS | ACCEPTABLE STRAIN SPECIFIC PROBIOTIC CLAIMS |
|---|---|
| *Lactobacillus johnsonii* strains La1/Lj1/NCC 533 | An adjunct to physician-supervised antibiotic therapy in patients with *Helicobacter pylori* infections ($1.25 \times 10^8$ to $3.6 \times 10^9$ cfu/day) |
| *Lacticaseibacillus rhamnosus* strain GG | Helps to manage acute infectious diarrhea ($6 \times 10^9$ to $1.2 \times 10^{10}$ cfu/day) Helps to manage and/or reduce the risk of antibiotic-associated diarrhea ($1 \times 10^{10}$ to $2 \times 10^{10}$ cfu/day) |

*Source:* Adapted from Health Canada. 2023. Natural Health Product—Probiotics. http://webprod.hc-sc.gc.ca/nhpid-bdipsn/atReq.do?atid=probio&lang=eng (accessed January 20, 2018).

**TABLE 41.3**   Acceptable Non–Strain-Specific Claims for Bacterial Probiotics in Natural Health Products

| MEDICINAL INGREDIENTS | ACCEPTABLE NON–STRAIN-SPECIFIC PROBIOTIC CLAIMS (MINIMUM 107 CFU/DAY) |
|---|---|
| Bifidobacterium adolescentis | Helps support intestinal/gastrointestinal health |
| Bifidobacterium animalis (including subsp. animalis and subsp. lactis) | Could promote a favorable gut flora |
| Bifidobacterium bifidum | Source of probiotics |
| Bifidobacterium breve | |
| Bifidobacterium longum (including subsp. infantis, subsp. longum and subsp. suis) | |
| Lactobacillus acidophilus | |
| Lactobacillus amylolyticus | |
| Lactobacillus amylovorus | |
| Levilactobacillus brevis | |
| Lentilactobacillus buchneri | |
| Lacticaseibacillus casei | |
| Loigolactobacillus coryniformis | |
| Latilactobacillus curvatus | |
| Lactobacillus delbrueckii (including subsp. bulgaricus and subsp. delbrueckii) | |
| Companilactobacillus farciminis | |
| Limosilactobacillus fermentum | |
| Lactobacillus gasseri | |
| Lactobacillus helveticus | |
| Lentilactobacillus hilgardii | |
| Lactobacillus johnsonii | |
| Lactobacillus kefiranofaciens | |
| Lentilactobacillus kefiri | |
| Limosilactobacillus mucosae | |
| Limosilactobacillus panis | |
| Lacticaseibacillus paracasei | |
| Lactiplantibacillus paraplantarum | |
| Lactiplantibacillus plantarum | |
| Limosilactobacillus pontis | |
| Limosilactobacillus reuteri | |
| Lacticaseibacillus rhamnosus | |
| Ligilactobacillus salivarius | |
| Fructilactobacillus sanfranciscensis | |
| Lactococcus lactis | |
| Leuconostoc citreum | |
| Leuconostoc pseudomesenteroides | |
| Leuconostoc lactis | |
| Leuconostoc mesenteroides | |
| Oenococcus oeni | |
| Pediococcus acidilactici | |
| Pediococcus pentosaceus | |
| Propionibacterium freudenreichii (including subsp. shermanii) | |
| Propionibacterium acidipropionici | |
| Lactobacillus crispatus | Source of probiotics |
| Lactobacillus gallinarum | |

*Source:* Adapted from Health Canada. 2023. Natural Health Product—Probiotics. http://webprod.hc-sc.gc.ca/nhpid-bdipsn/atReq.do?atid=probio&lang=eng (accessed February 06, 2023).

evidentiary requirements corresponding to the type and level of the health claim itself. General claims pertaining to health maintenance, support, and promotion (e.g., "supports the immune system") have a low therapeutic impact and thus a relatively low-risk profile. Conversely, treatment claims for major or serious diseases or conditions (e.g., "for the treatment of depressive disorders") are considered to have a high-risk profile and thus must be supported by a more robust level of evidence due to the potential seriousness of any therapeutic failure. For a more exhaustive discussion of health claim categories and associated standards of evidence, we recommend consulting the Health Canada Guidance document "Pathway for Licensing Natural Health Products Making Modern Health Claims" (Government of Canada, 2012).

To fulfil quality requirements, applicants must demonstrate that all products will be manufactured in accordance with good manufacturing practices (GMP) (Government of Canada, 2015a) and will comply with the finished product specifications submitted as part of the product license application. While the specific requirements are beyond the scope of this discussion, evidence supporting GMP requirements for manufacturers, packagers, and labelers may include Health Canada issued site licenses (for domestic parties) or foreign site reference numbers (for foreign parties) (Government of Canada, 2015b). With respect to finished product specifications, additional probiotic requirements are summarized in the following section.

# 41.8 LICENSING REQUIREMENTS SPECIFIC TO PROBIOTICS

Regardless of whether a product is supported by an NNHPD-authored monograph or data submitted by an applicant, Health Canada expects applicants to possess specific data for each probiotic strain contained in an NHP. If applicants attest to matching the NNHPD Probiotics Monograph in support of safety, efficacy, and quality, this data is not required to be submitted as part of the application process, but must be available should it be requested by Health Canada. In instances in which a probiotic organism is not captured in the NNHPD Probiotics Monograph or an applicant chooses, for whatever reason, not to attest to the requirements of the Monograph, this data would be required to be submitted as part of the product license application itself.

The quality data requirements are specified in the NNHPD Probiotics Monograph (Government of Canada, 2023) and can largely be separated into assays applicable to the raw material (i.e., the strain or culture itself) or the finished product, as outlined in the section that follows with wording taken directly from the Government of Canada (2023).

## 41.8.1 Strain/Culture-Specific Requirements

- The species' Latin binomial identification must be up to date and validated.
- Survivability of the microorganisms in the human gut must be demonstrated. *In vitro* gastric acid and bile resistance testing is considered acceptable.
- The microorganism must be identified by phenotype and genotype:
  - Phenotyping must be assessed based on characteristics routinely used to distinguish the species from others. This includes a series of testing for sufficient confirmation of observable traits of the species.
  - Genotyping must be assessed as follows:
    - Species identification by comparison of genome sequence homology in percentage, to both "identical" and "closely related" type strains obtained from an internationally recognized culture collection; and
    - Strain characterization through an up-to-date complete/whole genome sequencing method.

- Absence of virulence of each live microorganism must be established through all of the following:
  - Comparison of antibiotic/antifungal resistance profile to typical species resistance as published by an internationally recognized panel;
  - Explanation of the genetic basis of each atypical antibiotic/antifungal resistance to the species OR demonstration of the absence of all known genetic mechanisms of resistance;
  - Demonstration of lack of horizontal antibiotic/antifungal resistance transfer ability;
  - Demonstration of susceptibility to therapeutic concentrations of at least two commercially available antimicrobial/antifungal agents;
  - Demonstration of the absence of genetic elements responsible for the production of virulence factors characteristic to the genus; and
  - Demonstration of lack of toxigenic activity (i.e., production of toxins) known to the genus.

## 41.8.2 Finished Product Specific Requirements

In addition to meeting the quality requirements for all NHPs (Government of Canada, 2015c), products containing probiotics must also demonstrate the following (Government of Canada, 2023):

- Stability/viability measures, which ensure that a minimum of 80% of the quantity declared on the product label is present at the end of shelf life. This applies to single ingredient counts for live microorganisms cultured separately or total counts for blends where multiple live microorganisms have been cultured together.
- In cases where the live microorganism can interfere with microbial impurity testing, a detailed rationale on how the final product complies is required. Such a rationale should include measures for live microorganism distinguishing at the finished product stage, along with a detailed explanation on how quality assurance measures are put into place to ensure microbial purity.

In addition to these requirements, information on the manufacturing process must be maintained by the applicant (or by the manufacturer on behalf of the applicant) and provided to Health Canada upon request.

As the body of data required to meet NHP quality requirements may be expansive, manufacturers of probiotic strains commonly use Natural Health Product Master Files (NHP-MF) as a way to convey the data to the NNHPD. A NHP-MF is a reference document that contains information about the manufacturing and/or technical specifications of ingredients present in a NHP. Submission of a NHP-MF enables manufacturers to convey proprietary information on their ingredient/product directly to the NNHPD without the need to disclose that information to other product license applicants that seek to incorporate the proprietary ingredient into their NHP formulation(s). By compiling the data into a NHP-MF and providing applicants with a letter authorizing the NNHPD to review the NHP-MF, manufacturers may satisfy regulatory requirements whilst reducing the administrative burden on their customers.

# 41.9 LABELING CONSIDERATIONS FOR PROBIOTICS IN NHPS

All NHPs must be labeled in accordance with Part 5, Section 86, of the NHP Regulations and be consistent with the terms of the market authorization (i.e., the product license) (Government of Canada, 2008). Probiotic medicinal ingredients must be declared by Latin name, including subspecies if applicable, and the associated strain identifier (e.g., *Lactobacillus acidophilus* S123). Quantities of each probiotic medicinal ingredient should be declared in CFU per dosage unit, as applicable (e.g., CFU per capsule, CFU per mL).

# 41.10 FUTURE REGULATORY CONSIDERATIONS

Health Canada is currently proposing revisions to the existing regulatory frameworks that govern cosmetics, NHPs, and non-prescription drugs. The proposal is for the creation of a new regulatory structure, currently dubbed the 'Self-Care Framework', which would effectively group these self-care products into one overriding regulatory system (Government of Canada, 2016). To date, all published information with respect to this proposed framework is presented at a high level and relatively void of details, such that little may be concluded as to the impacts this proposal may have on probiotic ingredients in these products. Given the current stringent evidentiary requirements to support the safety, efficacy, and quality of probiotics, the changes arising as a result of this proposal would likely be minimal. Indeed, setting a higher standard than that currently employed would make little sense given the widespread use of probiotics in the food industry.

Finally, due to the recent nomenclature changes in the list of accepted species described in the NNHPD Probiotics Monograph for NHPs (Government of Canada, 2023), we expect that the Health Canada will soon provide updated guidance with respect to the accepted Latin names for probiotic strains in foods. Notably, these revisions affect several lactic acid bacteria genera listed in Table 41.1 that are approved for food use. It is unclear how any future updates to accepted nomenclature will affect the labeling of current food products.

# BIBLIOGRAPHY

FAO/WHO. (2006). Probiotics in Food: Health and Nutritional Properties and Guidelines for Evaluation. Rome, Italy: Food and Agriculture Organization of the United Nations, World Health Organization. Retrieved March 5, 2018, from www.fao.org/3/a-a0512e.pdf

Government of Canada. (2006). *Guidelines for the Safety Assessment of Novel Foods* . Retrieved March 5, 2018, from www.canada.ca/en/health-canada/services/food-nutrition/legislation-guidelines/guidance-documents/guidelines-safety-assessment-novel-foods-derived-plants-microorganisms/guidelines-safety-assessment-novel-foods-2006.html

Government of Canada. (2008). *Natural Health Products Regulations (SOR/2003–196)* . Retrieved March 5, 2018, from Justice Laws Website http://laws-lois.justice.gc.ca/eng/regulations/SOR-2003-196/

Government of Canada. (2009). *Guidance Document—The Use of Probiotic Microorganisms in Food* . Retrieved March 5, 2018, from www.canada.ca/en/health-canada/services/food-nutrition/legislation-guidelines/guidance-documents/guidance-document-use-probiotic-microorganisms-food-2009.html

Government of Canada. (2012). *Pathway for Licensing Natural Health Products Making Modern Health Claims* . Retrieved March 5, 2018, from www.canada.ca/en/health-canada/services/drugs-health-products/natural-non-prescription/legislation-guidelines/guidance-documents/pathway-licensing-making-modern-health-claims.html

Government of Canada. (2015a). *Good Manufacturing Practices Guidance Document* . Retrieved March 5, 2018, from www.canada.ca/en/health-canada/services/drugs-health-products/natural-non-prescription/legislation-guidelines/guidance-documents/good-manufacturing-practices.html

Government of Canada. (2015b). *Site Licensing Guidance Document* . Retrieved March 5, 2018, from www.canada.ca/en/health-canada/services/drugs-health-products/natural-non-prescription/legislation-guidelines/guidance-documents/site-licensing-guidance-document.html

Government of Canada. (2015c). *Quality of Natural Health Products Guide* . Retrieved March 5, 2018, from www.canada.ca/en/health-canada/services/drugs-health-products/natural-non-prescription/legislation-guidelines/guidance-documents/quality-guide.html

Government of Canada. (2016). *Consulting Canadians on the Regulation of Self-Care Products in Canada* . Retrieved March 5, 2018, from www.canada.ca/en/health-canada/programs/consultation-regulation-self-care-products/consulting-canadians-regulation-self-care-products-canada.html

Government of Canada. (2017a). *Guidance Document: Classification of Products at the Food-Natural Health Product Interface: Products in Food Formats*. Retrieved January 12, 2023, from www.canada.ca/en/health-canada/services/drugs-health-products/natural-non-prescription/legislation-guidelines/guidance-documents/classification-products-at-food-natural-health-product-interface.html

Government of Canada. (2017b). *Food and Drugs Act (R.S.C., 1985, c. F-27)* . Retrieved March 5, 2018, from http://laws-lois.justice.gc.ca/eng/acts/f-27/

Government of Canada. (2017c). *Food and Drug Regulations (C.R.C., c. 870)*. Retrieved March 5, 2018, from http://laws.justice.gc.ca/eng/regulations/c.r.c.,_c._870/index.html

Government of Canada. (2017d). *Health Claims—Probiotic Claims* . Retrieved March 5, 2018, from www.inspection.gc.ca/food/labelling/food-labelling-for-industry/health-claims/eng/1392834838383/1392834887794?chap=9

Government of Canada. (2023). *Natural Health Product—Probiotics*. Retrieved February 6, 2023, from http://webprod.hc-sc.gc.ca/nhpid-bdipsn/atReq.do?atid=probio&lang=eng

# Safety Assessment of Probiotics in the European Union

# 42

Seppo Salminen and Atte von Wright

## 42.1 INTRODUCTION

The European Union (EU) is an association of 27 member states with a total population of approximately 450 million (www.europa.eu). Although the EU is not a federation, the member states have ceded certain parts of their sovereignty to the governing bodies of the EU in order to form a single market area and ensure the free movement of people, capital, goods and services across the Union. Harmonization of the legislation is a central part of this process. Since probiotics are not considered drugs or pharmaceuticals, they fall under the scope of food and feed legislation. Both food and feed are highly regulated, particularly the safety aspects. The European Food Safety Authority (EFSA), established in 2002 and located in Parma, Italy, is the key actor regarding the safety assessment of food, feed and their components, and the guidance documents and opinions of EFSA, although technically considered "soft law", are of utmost importance for anyone wishing to place products on the European market. In the following sections the decision making, the pertinent legislation and the specific tasks of EFSA and their relevance to probiotics are briefly reviewed.

## 42.2 THE DECISION MAKING AND THE LEGAL FRAMEWORK IN THE EU

The central governing body of the EU is the European Commission (Commission), which can be considered the government or the executive branch of the EU, and in which every member state has a commissioner or two. The commissioners do not, however, represent their respective member states but are supposed to impartially focus on the general governance of the EU. Importantly, only the Commission can initiate new legislation in the EU. In the Commission, there are several Directorate-Generals (DGs), which run the executive functions and prepare the legislative initiatives. The central DG regarding food and feed is the DG on Health and Food Safety (SANTE).

The European Parliament is the only elected governing body in the EU. As in the Commission, also in the Parliament, the members do not represent primarily the member states from which they have been elected. Instead they group themselves according to their political and ideological preferences among the different political groupings present in the Parliament.

The Council of the European Union or the Council of Ministers is the final decisive body of the EU. The responsible ministers of each member states are represented in this body, and the laws initiated by the Commission and discussed by the Parliament will be finally approved in the Council of the EU, usually applying the so-called "ordinary legislative procedure", meaning that both the Parliament and the Council agree on the content of the legislation. The decision of the Council requires the backing of 55% of the member states, representing at least 65% of the total population.

## 42.2.1 The Legal Instruments of the EU

There are two main types of union-wide legal documents in the EU, Directives and Regulations. Directives represent measures that the member states have to incorporate into their national legislation within two years after the adoption of the Directive. Regulations are automatically in force in all the EU countries immediately after their promulgation and are included as independent acts, not generally as changes or expansions of national laws, which can be the case with Directives. The general trend in the EU appears to prefer Regulations to Directives, as more and more former Directives are being replaced by Regulations.

# 42.3 THE EUROPEAN FOOD AND FEED LEGISLATION

The foundation of the EU legislation on both food and feed is the Regulation (EC) No 178/2002 (Known as the General Food Law Regulation). This regulation defined most of the general principles of the harmonized legislative measures to be later introduced to ensure the safety of foods and feeds, but most importantly it established the European Food Safety Authority and defined its functions. The mission of EFSA is to provide independent scientific advice and opinions in matters relating to nutrition and food safety. The tasks of EFSA were defined in Article 23 of the Regulation as follows:

- To provide the Community institutions and the Member States with the best possible scientific opinions in all cases provided for by Community legislation and on any question within its mission;
- To promote and coordinate the development of uniform risk assessment methodologies in the fields falling within its mission;
- To provide scientific and technical support to the Commission in the areas within its mission and, when so requested, in the interpretation and consideration of risk assessment

One of the most important tasks of EFSA has been to safety assessment of regulated products, such as food and feed additives, food contact materials, processing aids, genetically modified foods and feeds and novel foods. EFSA also evaluates the food-associated nutritional and functional claims. Thus, EFSA is also the central actor in the assessment of LAB or other microorganisms, or products made thereof, that are to be introduced into the food chain in the EU. This assessment is done by the scientific panels. These panels consist of independent scientists who are not part of the permanent EFSA staff and who are selected on the basis of their recognized expertise.

## 42.3.1 The Authorization Process of a Regulated Product in the EU

In case an applicant wishes to authorize a product, say a food or feed additive, in the EU, the first step is to submit a formal application to the Commission and, at the same time, all the required information related to the product to EFSA. After receiving the application, the Commission will inform the member states and forward also the application to EFSA. In case of a feed additive or a product obtained by genetic modification, three samples of the product should also be sent to the Community reference laboratory. The six Community reference laboratories within the EU coordinate the work of national reference laboratories, develop methods, provide standards and reference materials and training, and two of them are directly involved in regulatory activities, the European Union Reference Laboratory for Genetically Modified Food and Feed and the European Union Reference Laboratory for Feed Additives. The tasks of these two laboratories include the assessment of the analytical method(s) the applicant has proposed for the product.

After the completeness check of all the necessary documents, EFSA is supposed to produce a scientific opinion on the product within a specified time (usually 6–9 months, depending on the case specific legislation). However, because several questions are usually raised during the assessment that require new or complementary data from the applicant, the time can be considerably longer, since the evaluation is discontinued until all the relevant information has been obtained.

The EFSA opinion is submitted to the Commission, which is expected to produce a draft regulation within 3–7 months. The formal approval of the regulation depends of the decision of the relevant section of the Standing Committee on Plants, Animals, Food and Feed (PAFF), which assists the Commission. In the Standing Committee, the member states are represented, and to reach a decision a qualified majority, as in the case of the Council (see Section 42.2.), is needed. If this majority is not obtained, an Appeal Committee (again consisting of member state representatives) can be nominated to discuss the matter further. If even the Appeal Committee cannot reach a conclusion, the Commission may just adopt the original proposal.

## 42.3.2 The Relevant EU Legislation Related to Food Microorganisms or Microbiological Feed Additives

At present, the EU laws require a safety assessment of a food- or feed-associated intentionally added microorganism in following cases:

- The microorganism is genetically modified
- The microorganism is considered a novel food
- The microorganism is used as a feed additive

In the case of human probiotics, the main requirements are not focused on safety but on the verification of any functional or health claim associated with the claim.

### 42.3.1.1 Genetically Modified Microorganisms

The use of genetically modified microorganisms (GMMs) in the EU is covered by i) Directives 2009/41/EC and 2001/18/EC (the former related to the contained use of GMMs and the latter to the deliberate release of GMMs to the environment) and ii) Regulations EC No 1829/2003 on genetically modified food and feed and EC No1830/2003 on the traceability of genetically modified organisms and genetically modified food and feed. The Directives are nowadays not relevant regarding the food or feed use of GMMs, although the principles for their culture at the laboratory or industrial scale would be covered by the national legislation of the member states, into which the requirements of the Directive 2009/41/EC have been incorporated. The deliberate release of GMMs is limited, after the introduction of Regulation EC No 1829/2003, to yet hypothetical cases such as biopesticides or bioremediation agents and would be assessed according to the procedures defined in the Directive. In this process, the national competent authorities will be in the main role, and EFSA would have an advisory function.

Any GMM intended for food or feed use—probiotic or not—would be evaluated according to Regulation EC No 1829/2003, and the actual assessment would be primarily done by the Panel on Genetically Modified Organisms (GMO-Panel). EFSA has published a detailed guideline for the safety assessment (EFSA GMO Panel 2011; Aguilera et al. 2013). Some of the requirements will be discussed in more detail in Section 42.4.2.1. So far, no GMMs have been proposed as human probiotics or micro-biological feed additives, but several enzymes and amino acids biotechnically produced by GMMs and falling under the scope of Regulation EC No 1829/2003 have been assessed by EFSA.

### 42.3.1.2 Microorganisms as Novel Foods

The safety assessment of novel foods in the EU is performed according to the principles outlined in the Regulation (EU) 2015/2283 on Novel Foods. This regulation repeals and replaces two earlier Regulations, (EC) No 258/97 and (EC) No 1852/2001. The main difference in comparison to earlier Regulations is that the safety evaluation is done by EFSA instead of the national competent authorities of the member states, and that the authorizations are now generic and not applicant specific. A novel food, according to the paragraph 2 of Article 3 of the Regulation, means "any food that was not used for human consumption to a significant degree within the Union before 15 May 1997, irrespective of the dates of accession of Member States to the Union". There is also a specific notification procedure for traditional foods from third countries outlined in Section II of the Regulation (articles 14–20).

What comes to microorganisms, "foods consisting of, isolated from or produced from microorganisms, fungi or algae" (Article 3, paragraph 2) are specifically mentioned as a special category of novel foods.

The novel food authorization can start on the initiative of the Commission or following an application submitted in some member state. The Commission will, after receiving the application, inform the member states, and—in most cases—request the opinion of EFSA. In this case the relevant EFSA Panel is the Panel on Dietetic Products, Nutrition and Allergies (NDA-panel).

The final authorization occurs with the comitology procedure outlined in Section 42.3.1. After authorization, the approved food is included in the Union list of novel foods. The few authorized microorganisms are discussed in more detail in Section 42.4.2.2.

### 42.3.1.3 Microorganisms as Feed Additives

Microorganisms intended as feed additives are by far subject to most stringent safety assessment in the EU. The central piece of legislation is Regulation (EC) No 1831/2003 on additives for use in animal nutrition. According to Article 5 of the regulation, a feed additive should not have an adverse effect on animal health, human health or the environment. Human health aspects include both the safety of the user and the consumer. Moreover, a feed additive should either

> favourably affect the characteristics of feed, favourably affect the colour of ornamental fish and birds, satisfy the nutritional needs of animals, favourably affect the environmental consequences of animal production, favourably affect animal performance or welfare, particularly affecting the gastro-intestinal flora or digestibility of feedingstuffs, or have a coccidiostatic or histomonostatic effect.

It is specifically stated that actual antibiotics are not permitted as feed additives in the EU.

In Article 6, feed additives are further divided into five categories: technological additives, sensory additives (enhancing either the organoleptic properties of the feed or the visual characteristics of the food derived from animals), nutritional additives, zootechnical additives and coccidiostats or histomonostats. In Annex I of the Regulation, each category is further divided into different functional groups. For example, in the category "zootechnical additives", the following functional groups are included:

(a) digestibility enhancers: substances which, when fed to animals, increase the digestibility of the diet through action on target feed materials;
(b) gut flora stabilizers: micro-organisms or other chemically defined substances, which, when fed to animals, have a positive effect on the gut flora;
(c) substances which favourably affect the environment;
(d) other zootechnical additives.

Microbiological feed additives are mainly either technological additives (such as silage starters) or zootechnological additives, in practice animal probiotics. Consequently, they are functionally classified either as preservatives (a functional group within the category of technological additives) or gut flora stabilizers.

The authorization process is the one outlined in Section 42.3.1. When finally approved, an additive is included in the European Union Register of Feed Additives (https://ec.europa.eu/food/food-feed-portal/screen/feed-additives/search). Currently there are more than 160 microbial preparations registered as either as silage additives or gut flora stabilizers. The majority of the approximately 95 silage starters are composed either of LAB or propionic acid bacteria, while among the gut flora stabilizers there are, in addition to LAB, several representatives of yeasts (*Saccharomyces cerevisiae*) or Bacilli.

### 42.3.1.4 Human Probiotics and Functional Claims

The EU legislation makes a sharp division between food and medicine and has a strict approach to health claims. Article 7 of Regulation (EU) No 1169/2011 (repealing an earlier directive with essentially the same content) specifically states: "food information shall not attribute to any food the property of preventing, treating or curing a human disease, nor refer to such properties". In the case of functional foods, such as probiotics, and also with many other types of foods with a sharp demarcation between nutritional and health-promoting properties, the interpretation of this rule becomes problematic. In order to address these difficulties, a specific Regulation (EC) 1924/2006 on the nutrition and health claims, was passed.

The acceptable nutrition claims are generic (like "low energy", "low fat", "fat free", etc.), and are listed together with their definitions in the Annex I of the Regulation.

Health claims are divided into two categories, those that are not referring to the reduction of disease risk (Article 13) and those which specifically refer to the reduction of disease risk (Article 14). Note that in both cases the focus on the reduction of disease risk, not on preventing or curing an illness or disease. Regarding the cases covered by Article 13, they can be further divided to general function health claims and new function health claims, the latter based on new scientific evidence or having a proprietary interest. EFSA and its NDA panel have evaluated thousands of health claims submitted by member state competent authorities and provided a series of opinions about their individual acceptability, and further information is available on the EFSA website (www.efsa.europa.eu/en/topics/topic/nutrition-and-health-claims).

Both the new function health claims and the health claims associated with the reduction of disease risk require an application for authorization, which is dealt by the Commission, EFSA and the Standing Committee according to the procedure laid out in Section 42.3.1. Several human probiotics have been assessed as products with new function health claims.

So far, no human probiotic claim has been accepted by EFSA due to deficient characterization of the strains, lack of sound human data showing a cause and effect and poorly defined claims. It should be noted that much of the best evidence of the efficacy of probiotics comes from intervention studies involving individuals with some gastrointestinal disorder, and these types do not qualify to demonstrate the beneficial effects in healthy population.

# 42.4 THE SPECIFIC SAFETY STUDIES REQUIRED BY EFSA FOR MICROORGANISMS

EFSA has two approaches regarding the safety assessment of an individual microorganisms, regardless of its eventual use. There is a generic safety assessment base on the Qualified Perception of Safety (QPS), and then there are specific safety studies required for those microorganisms that do not qualify for the QPS. Both these approaches are described in the following sections.

## 42.4.1  Qualified Presumption of Safety

Qualified Perception of Safety (Leuschner et al. 2010) was introduced in 2007 to simplify the safety assessment procedures and to direct the resources to those microorganisms that have the highest risk potential. To make the QPS list, the evidence of history of safe use at the species level and exposure to humans in food is sufficient to conclude about safety. Evaluation at the strain level is needed to document an antibiotic resistance profile that minimizes a risk of horizontal transfer of antibiotic resistance genes to host microbes. If so, then the regulatory pathway is easier, as no safety concern is usually presented with them. The QPS list and evaluations are always published in the European Food Safety Authority journal, where background for either inclusion in the list is given or reasons for not accepting the microbe to the QPS list are discussed. The list is updated annually by the EFSA Biohazards Panel (BIOHAZ) (2018) (www.efsa.europa.eu/en/topics/topic/qualified-presumption-safety-qps). Its review uses the following criteria:

(a) Taxonomic unit: This is the most crucial part of the assessment procedure. The genus and all known species, including type species, are listed and described according to their main characteristics. The evaluated taxonomic unit must be unambiguously defined. If the assessed bacterial agent cannot be connected to any known species, it will not be recommended for the QPS list. QPS status is also denied if the taxonomic identification is insufficient.
(b) Body of knowledge: This should be related to the history of use for the assessed bacterial agent, the field of application; the existence of sufficient scientific literature about the microorganism; and the possibility to avoid potential adverse effects for humans, livestock and the wider environment.
(c) Possible pathogenicity: The bacterial agent's lack of pathogenic and virulence properties is confirmed, and the findings must be supported by clinical data and scientific literature.

Typically, no specific safety assessments for humans, target animals (in the case of feed additives) or environment are required if a microorganism is considered to have QPS status. Thus the QPS approach really simplifies the safety assessment and has been adopted as a starting point for microbial safety evaluation across all the regulated microbiological products under the mandate of EFSA.

## 42.4.2  The Specific Safety Assessment of Microorganisms in Different Applications

### 42.4.2.1  The Specific Safety Assessment of GMMs

The safety assessment procedures are outlined in the EFSA guidance document (EFSA GMO Panel 2011). The actual assessment procedure is based on the comparative approach, meaning that as much as possible, the safety is established by evaluating the GMM against the comparator, usually a strain that is either considered safe or the risks of which are well characterized, and which has been the parental strain for the GMM.

The specific studies depend very much of the eventual presence of live GMMs or genetically modified DNA in the final product. The products are actually divided into four different categories:

**Category 1:**  Chemically defined purified compounds and their mixtures in which both GMMs and newly introduced genes have been removed (e.g. amino acids, vitamins);
**Category 2:**  Complex products in which both GMMs and newly introduced genes are no longer present (e.g. cell extracts, most enzyme preparations);
**Category 3:**  Products derived from GMMs in which GMMs capable of multiplication or of transferring genes are not present but in which newly introduced genes are still present (e.g. heat-inactivated starter cultures);
**Category 4:**  Products consisting of or containing GMMs capable of multiplication or of transferring genes (e.g. live starter cultures for fermented foods and feed).

A live human or animal probiotic or additive, being in category 4, would need the most stringent evaluation involving assessment of the potential of the modified gene(s) or sequences to spread into other organisms (horizontal gene transfer). Products containing either live GMMs or functional genetically modified DNA are also subject to post-marketing monitoring.

The toxicological studies required depend on the analysis of any compositional changes in the GMM or its metabolites resulting from the genetic modification. The novel proteins should be specifically evaluated using both bioinformatic approaches and, on a case-by-case basis, oral repeated dose animal studies. This could be applied also to other constituents than proteins if significant quantitative or qualitative changes have been observed and no preexisting toxicological information is available. Subchronic (90 day) rodent feeding trials with at least two dose levels are also recommended for the whole GMM (live or dead) if the compositional analysis indicates any concern.

## 42.4.2.2 Safety Requirements for Microorganisms Used as Novel Foods, the Case of Non-Viable Microorganisms

The European safety assessment of microbes in food is focused on the use of live microbes, using the Qualified Presumption of Safety list (see Section 42.4.1) as the starting point. This applies to both microorganisms intended for technological uses (starters, additives, protective cultures, etc.) and to human probiotics. For microorganisms that are not included in the QPS list, whole-genome sequencing and bioinformatic analysis are recommended for the identification of any virulence factors, antibiotic resistances and production of harmful metabolites (such as D-lactate). Both genotoxicity tests and subchronic animal feeding studies and any additional safety studies that an applicant considers relevant can be performed or may be required on case-by-case basis, as outlined in the EFSA Guidance on the preparation and presentation of an application for authorization of a novel food (EFSA NDA Panel 2016).

So far there is only one case of a novel viable bacterium authorized for food use, *Clostridium butyricum* CBM588 authorized as a food supplement (Commission Implementing Decision 2014/907/EU) [the strain has also been authorized as a feed additive (Commission Implementing Regulation (EU) 2021/1411)]. It should be noted that the novel food risk assessment of *C. butyricum* CBM588 was done by the British competent authority instead of EFSA, because the assessment was done according to Regulation (EC) No 257/98, which has been repealed and replaced by the current Regulation (EU) 2015/2283.

The situation is somewhat different with novel, non-viable microorganisms used for probiotic purposes (postbiotics) or for other food uses. In the EU, postbiotics face both novel food regulation and health claim regulation (Salminen et al. 2021). If the microbe, being probiotic or postbiotic, falls into the novel food category, the safety evaluation is more exhaustive and requires toxicological information, among other safety requirements (Gómez-Gallego et al. 2016; EFSA NDA Panel 2021). The EFSA guidance document in the case of safety assessment of live microbes is challenging and lacks clear guidelines. Recent experience demonstrates that safety of inanimate bacteria (potential postbiotics) may be easier to achieve than safety of live bacteria. Safety assessment of inanimate bacteria as novel food is now available for three different preparations (*Bacteroides xylanisolvens*, *Akkermansia muciniphila* and *Mycobacterium setence manresensis*) (EFSA NDA Panel 2015, 2019, 2021, respectively). *B. xylanisolvens* is intended for fermented food products, while *A. muciniphila* and *M. setence manresensis* for food supplements and foods for special medical purposes. These serve as a models for requirements, and all include the means of inactivation of the live microbes. These three microorganisms have not been accepted for QPS status as live preparations. *B. xylanisolvens* and *A. muciniphila* were assessed positively for use as inanimate novel food by EFSA and subsequently authorized by the EU Commission authorization (Commission Implementing Decision (EU) 2015/1291; Commission Implementing Regulation (EU) 2022/168). However, unlike the other two inanimate microbes assessed as safe novel foods, *Mycobacterium setence manresensis* has not yet been authorized by the European Commission. In all three cases, both genotoxicity and repeated dose oral rodent studies were provided to establish the safety of the non-viable bacteria.

Taken together, it appears easier to fulfil regulatory requirements for postbiotics (inanimate microorganisms) compared to probiotics (live microorganisms), even though to our knowledge, no regulators have yet taken a position on the postbiotic definition. The consensus definition discussed herein is therefore especially timely and relevant for postbiotic formulations for food or pharmaceutical applications.

In the European Union, no specific regulation covers postbiotics, but since their consensus definition requires a demonstrated health benefit, current interpretation is that the use of the terms on a food or food supplement would require health claim approval by EFSA and systematic novel food application and approval in Europe before the term can be used in foods or feeds. The recent EU Regulation (EU) 2017/745 for medical devices also has a specific paragraph positioning "living organisms" out of the scope of the Regulation.

## 42.4.2.3 Safety Assessment for Microorganisms Used as Feed Additives

The general requirements for microorganisms listed in the EFSA specific guidance document for microorganisms used as feed additives or production organisms (EFSA FEEDAP Panel 2018) include the proper taxonomic identification and history of the strain, the absence of toxins and virulence factors, lack of production of antibiotic substances and absence of such antimicrobial resistances that could add to the resistance pool of the intestinal bacterial population. The identification should, for bacteria and yeasts, include whole-genome sequencing (WGS), and this data should also be used in the identification of antibiotic resistance genes, virulence factors or other genes of concern. Regarding antibiotic resistances, bacteria should also be phenotypically assessed against a panel of selected antibiotics to detect any resistances that could be considered exceptional. Moreover, if the WGS indicates the presence of genes conveying resistance to antibiotics considered either critically important or highly important by WHO, the minimal inhibitory concentrations (MICs) for these antibiotics should be determined. In general, if there is a phenotypic resistance and corresponding resistance gene(s) can be found, this is considered a hazard.

In case the microorganism belongs to a species having the QPS status, the only additional safety studies required are related to the user safety, and the material to be tested should represent the final formulation of the additive, including carrier materials and other technological components. Typically studies on the skin and eye irritation or delayed type of hypersensitivity and respiratory sensitization are considered, depending on the formulation (EFSA FEEDAP Panel 2012a).

If the microorganism does not have QPS status, its safety should be assessed, in addition to the user, to the target animal, consumer and environment.

The target animal safety can be assessed in three ways: i) one can perform an extensive literature review on the effects of the additive; ii) one can refer to repeated dose toxicity studies performed on laboratory animals—this option, however, is not available for viable microorganisms; or iii) one can perform a tolerance test on the target animal. The minimum requirements for the tolerance study have been defined in a specific EFSA guidance (EFSA 2017a). A minimum of three groups of animals are required, control group, a use-level group and an overdose group. If the overdose is ≥ 100-fold higher than the use level, the parameters followed include visual evidence of clinical signs, performance characteristics and possibly product quality. At lower doses, hematology and blood chemistry should be included, and if the highest tested dose is less than 10 times the use level, gross pathology and, if relevant, histopathology should be performed.

For the consumer safety assessment the EFSA guidance document (EFSA 2017b) states: "For microorganisms and fermentation products, a basic set of toxicity studies should be provided consisting of genotoxicity/mutagenicity tests and a subchronic (90-day) oral toxicity study". The genotoxicity test battery should consist of a bacterial reverse mutation assay and an *in vitro* mammalian cell micronucleus test. There are no specific instructions whether these tests should be performed using the strain or the final product formulation. The main intent is to eliminate the possibility of unknown metabolites that might cause safety concerns. From the toxicological point of view, the proposed procedure is somewhat problematic, and these aspects are further discussed in Section 42.4.3.

Regarding environmental safety, the present EFSA guidance (EFSA FEEDAP Panel 2008) does not specifically address microorganisms. In practice, the safety assessment of microbiological feed additives has been based on the estimation of whether the intended use will significantly increase the specific microorganism in some environmental niche. The guidance is currently under revision, and possibly some more specific instructions on this aspect will be available in the future.

### 42.4.3 Are the Required Safety Studies Appropriate?

Since its introduction, the QPS concept has been a success in reducing the need of unnecessary testing and focusing the resources on the cases where the safety assessment is really needed. However, it can be argued that especially the consumer safety studies required from the non-QPS microorganisms are not always justified. Genotoxicity testing of a microorganism makes little sense, because genotoxic compounds are small molecular weight molecules and would be, if produced, present in the growth medium or culture supernatant. In the case of the authorization of *C. butyricum* (see Section 42.3.1.3), the applicant duly did the genotoxicity studies on a cell lysate (EFSA FEEDAP Panel 2009). This was not considered a suitable method by the FEEDAP panel, but it was concluded that the production process would eliminate eventual secreted genotoxins. Acute, subacute and subchronic tests reported were performed with the actual product. It should, however, be noted that these tests have not been designed to detect bacterial pathogenicity, and the likelihood of eventual unknown toxic metabolites at concentrations sufficient to be detected in animal assays is small. By far the most convincing safety data came from the bioinformatic genomic studies demonstrating the lack of genes associated with any known clostridial toxin.

In the specific case of *Enterococcus faecium*, a non-QPS species but used in many feed additives, the unsuitability of the proposed test battery has been recognized, and specific guidance on the safety assessment of *E. faecium* has been published (EFSA FEEDAP Panel 2012b). The safety evaluation focuses on certain phenotypic and genetic markers that differentiate potentially pathogenic strains from the innocuous ones, such as the minimum inhibitory concentration of ampicillin > 2 mg/L, the presence of IS*16* and two genes for virulence factors (*esp*, $hyl_{Efm}$).

There may well be further examples of tailor-made safety assessment recommendations for specific microorganisms in the future.

## 42.5 CONCLUSIONS

The safety assessment of probiotics or any microorganisms intentionally introduced into the food chain in the EU is limited to the so-called regulated products (GMMs, novel foods, feed additives). For human probiotics, while the products are expected to be safe, the main focus is on the demonstration of efficacy. The safety requirements can be rather extensive and complicated, but the introduction of the QPS concept has greatly simplified the procedures and opened the way for a generic safety assessment. The microorganisms not fulfilling the QPS criteria, such as GMMs and novel types of feed additives representing other microorganisms than LAB, yeasts and bacilli, remain a challenge for consumer safety, because the suitability of the present EFSA guidelines can be questioned.

It is to be assumed that the accumulating experience of bioinformatic analysis of WGS data as well as improved metabolomic understanding of the microorganisms will, in the future, lead to the revision of the present guidance on the consumer safety, at least regarding the testing requirements of non-QPS microorganisms. Because already under the present guidelines, WGS analysis is mandatory for the assessment of bacteria and yeasts, the role of bioinformatics in the safety assessment will, undoubtedly, quite soon become a routine procedure in the safety evaluations of microbiological products.

Certain emerging issues, such as the potential impact of probiotic-induced changes in microbiomes, interactions with drugs and probiotic colonization identified by Merenstein et al. (2023) are not covered by the present regulatory framework, and they might deserve consideration, at least on a case-by-case basis in the future, provided that suitable methodologies will be available.

# BIBLIOGRAPHY

Aguilera, J., A.R. Gomez and I. Olaru. 2013. Principles of the risk assessment of genetically modified microorganisms and their food products in the European Union. *Int. J. Food Microbiol.* 167: 2–7.

Commission Implementing Decision (EU) 2014/907 of 11 December 2014 authorising the placing on the market of *Clostridium butyricum* (CBM 588) as a novel food ingredient under Regulation (EC) No 258/97 of the European Parliament and of the Council. *Official Journal of the European Union* L 359/153–L359/154.

Commission Implementing Decision (EU) 2015/1291 of 23 July 2015 authorising the placing on the market of heat-treated milk products fermented with *Bacteroides xylanisolvens* (DSM 23964) as a novel food under Regulation (EC) No 258/97 of the European Parliament and of the Council. *Official Journal of the European Union* L198/26–L198/27.

Commission Implementing Regulation (EU) 2021/1411 of 27 August 2021concerning the renewal of the authorisation of *Clostridium butyricum* FERM BP-2789 as a feed additive for chickens reared for laying, turkeys for fattening, turkeys reared for breeding, minor avian species (excluding laying birds), weaned piglets and weaned minor porcine species, its authorisation for chickens for fattening, suckling piglets and suckling minor porcine species, and repealing Implementing Regulations (EU) No 373/2011, (EU) No 374/2013 and (EU) No 1108/2014. *Official Journal of the European Union* L304/11–L304/12.

Commission Implementing Regulation (EU) 2022/168 of 8 February 2022 authorising the placing on the market of pasteurised *Akkermansia muciniphila* as a novel food under Regulation (EU) 2015/2283 of the European Parliament and of the Council and amending Commission Implementing Regulation (EU) 2017/2470. *Official Journal of the European Union* L 28/5–L28/7.

EFSA BIOHAZ Panel (EFSA Panel on Biological Hazards). 2018. Statement on the update of the list of QPS-recommended biological agents intentionally added to food or feed as notified to EFSA suitability of taxonomic units notified to EFSA until September 2017. *EFSA J.* 16(1): 5131, 43 pp. https://doi.org/10.2903/j.efsa.2018.5131

EFSA FEEDAP Panel (EFSA Panel on Additives and Products or Substances used in Animal Feed). 2008. Technical Guidance for assessing the safety of feed additives for the environment. *EFSA J.* 842: 1–28.

EFSA FEEDAP Panel (EFSA Panel on Additives and Products or Substances used in Animal Feed). 2009. Scientific Opinion on a request from the European Commission on the safety and efficacy of the product Miya-GoldS (*Clostridium butyricum*) as feed additive for chickens for fattening. *EFSA J.* 1039: 1–16.

EFSA FEEDAP Panel (EFSA Panel on Additives and Products or Substances used in Animal Feed). 2012a. Guidance on studies concerning the safety of use of the additive for users/workers. *EFSA J.* 10(1): 2539. [5 pp.]. http://doi.org/10.2903/j.efsa.2012.2539. Available online: www.efsa.europa.eu/efsajournal

EFSA FEEDAP Panel (EFSA Panel on Additives and Products or Substances used in Animal Feed). 2012b. Guidance on the safety assessment of *Enterococcus faecium* in animal nutrition. *EFSA J.* 10(5): 2682 [10 pp.]. http://doi.org/10.2903/j.efsa.2012.2682. Available online: www.efsa.europa.eu/efsajournal

EFSA FEEDAP Panel (EFSA Panel on Additives and Products or Substances Used in Animal Feed). 2017a. Guidance on the assessment of the safety of feed additives for the target species. *EFSA J.* 15(10): 5021, 19 pp. https://doi.org/10.2903/j.efsa.2017.5021

EFSA FEEDAP Panel (EFSA Panel on Additives and Products or Substances used in Animal Feed). 2017b. Guidance on the assessment of the safety of feed additives for the consumer. *EFSA J.* 15(10): 5022, 17 pp. https://doi.org/10.2903/j.efsa.2017.5022

EFSA FEEDAP Panel (EFSA Panel on Additives and Products or Substances used in Animal Feed). 2018. Guidance on the characterisation of microorganisms used as feed additives or as production organisms. *EFSA J.* 16(3): 5206, 24 pp. https://doi.org/10.2903/j.efsa.2018.5206

EFSA GMO Panel (EFSA Panel on Genetically Modified Organisms (GMO)). 2011. Scientific Opinion on Guidance on the risk assessment of genetically modified microorganisms and their products intended for food and feed use. *EFSA J.* 9(6): 2193 [54 pp.]. http://doi.org/10.2903/j.efsa.2011.2193. Available online: www.efsa.europa.eu/efsajournal.htm

EFSA NDA Panel (EFSA Panel on Dietetic Products, Nutrition and Allergies). 2015. Scientific opinion on the safety of 'heat-treated milk products fermented with *Bacteroides xylanisolvens* DSM 23964' as a novel food. *EFSA J.* 13(1): 3956, 18 pp. http://doi.org/10.2903/j.efsa.2015.3956

EFSA NDA Panel (EFSA Panel on Dietetic Products, Nutrition and Allergies). 2016. Guidance on the preparation and presentation of an application for authorisation of a novel food in the context of Regulation (EU) 2015/2283. *EFSA J.* 14(11): 4594, 24 pp. http://doi.org/10.2903/j.efsa.2016.4594

EFSA NDA Panel (EFSA Panel on Nutrition, Novel Foods and Food Allergens). 2019. Scientific opinion on the safety of heat-killed Mycobacterium setense manresensis as a novel food pursuant to Regulation (EU) 2015/2283. *EFSA J.* 17(11): 5824, 13 pp. https://doi.org/10.2903/j.efsa.2019.5824

EFSA NDA Panel (EFSA Panel on Nutrition, Novel Foods and Food Allergens). 2021. Scientific opinion on the safety of pasteurised *Akkermansia muciniphila* as a novel food pursuant to Regulation (EU) 2015/2283. *EFSA J.* 19(9): 6780, 18pp. https://doi.org/10.2903/j.efsa.2021.6780

EFSA (European Food Safety Authority) Technical Report. 2018. Administrative guidance on the submission of applications for authorisation of a novel food pursuant to Article 10 of Regulation (EU) 2015/2283 EFSA supporting publication 2018:EN-1381. 22 pp. http://doi.org/10.2903/sp.efsa.2018.EN-1381

Gómez-Gallego, C., S. Pohl, S. Salminen, W.M. De Vos and W. Kneifel. 2016. *Akkermansia muciniphila*: A novel functional microbe with probiotic properties. *Benef Microbes*. 7(4): 571–584. http://doi.org/10.3920/BM2016.0009

Leuschner, R., T.P. Robinson, M. Hugas, P.S. Cocconcelli, F. Richard-Forget, G. Klein, T.R. Licht, C. Nguyen-The, A. Querol, et al. 2010. Qualified presumption of safety (QPS): A generic risk assessment approach for biological agents notified to the European Food Safety Authority (EFSA). *Trends Food. Sci. Technol*. 21: 425–435.

Merenstein, D., B. Pot, G. Leyer, A.C. Ouwehand, G.A. Preidis, C.A. Elkins, C. Hill, et al. 2023. Emerging issues in probiotic safety: 2023 perspectives. *Gut Microbes*. 1: 2185034. http://doi.org/10.1080/19490976.2023.2185034

Regulation (EC) No 178/2002 of the European Parliament and of the Council of 28 January 2002 laying down the general principles and requirements of food law, establishing the European Food Safety Authority and laying down procedures in matters of food safety. *Official Journal of the European Communities* L 31/1–L31/24.

Regulation (EU) No 1169/2011 of the European Parliament and of the Council of 25 October 2011 on the provision of food information to consumers, amending Regulations (EC) No 1924/2006 and (EC) No 1925/2006 of the European Parliament and of the Council, and repealing Commission Directive 87/250/EEC, Council Directive 90/496/EEC, Commission Directive 1999/10/EC, Directive 2000/13/EC of the European Parliament and of the Council, Commission Directives 2002/67/EC and 2008/5/EC and Commission Regulation (EC) No 608/2004. *Official Journal of the European Union* L304/18–L304/63.

Regulation (EC) No 1831/2003 of the European Parliament and of the Council of 22 September 2003 on additives for use in animal nutrition. *Official Journal of the European Union* L 268/29–L268/L43.

Regulation (EC) No 1924/2006 of 20 December 2006 on nutrition and health claims made on foods. *Official Journal of the European Union* L404/9–L404/29.

Regulation (EU) No 2283/2015 of the European Parliament and of the Council of 25 November 2015 on novel foods, amending Regulation (EU) No 1169/2011 of the European Parliament and of the Council and repealing Regulation (EC) No 258/97 of the European Parliament and of the Council and Commission Regulation (EC) No 1852/2001. *Official Journal of the European Union* L327/1–L327/22.

Salminen, S., M.C. Collado, A. Endo, C. Hill, S. Lebeer, E.M.M. Quigley, M.E. Sanders, et al. 2021. The international scientific association of probiotics and prebiotics (ISAPP) consensus statement on the definition and scope of postbiotics. *Nat. Rev. Gastroenterol. Hepatol*. 9: 649–667. http://doi.org/10.1038/s41575-021-00440-6

# Index

Note: Page numbers in *italics* indicate a figure and page numbers in **bold** indicate a table on the corresponding page.